HANDBOOK OF EPIGENETICS

HANDBOOK OF EPIGENETICS

The New Molecular and Medical Genetics

THIRD EDITION

Edited by

TRYGVE O. TOLLEFSBOL

Department of Biology, University of Alabama at Birmingham, Birmingham, AL, United States; O'Neal Comprehensive Cancer Center, University of Alabama at Birmingham, Birmingham, AL, United States; Integrative Center for Aging Research, University of Alabama at Birmingham, Birmingham, AL, United States; Nutrition Obesity Research Center, University of Alabama at Birmingham, Birmingham, AL, United States; Comprehensive Diabetes Center, University of Alabama at Birmingham, Birmingham, AL, United States; University Wide Microbiome Center, University of Alabama at Birmingham, Birmingham, AL, United States

Academic Press is an imprint of Elsevier
125 London Wall, London EC2Y 5AS, United Kingdom
525 B Street, Suite 1650, San Diego, CA 92101, United States
50 Hampshire Street, 5th Floor, Cambridge, MA 02139, United States
The Boulevard, Langford Lane, Kidlington, Oxford OX5 1GB, United Kingdom

Copyright © 2023 Elsevier Inc. All rights reserved.

No part of this publication may be reproduced or transmitted in any form or by any means, electronic or mechanical, including photocopying, recording, or any information storage and retrieval system, without permission in writing from the publisher. Details on how to seek permission, further information about the Publisher's permissions policies and our arrangements with organizations such as the Copyright Clearance Center and the Copyright Licensing Agency, can be found at our website: www.elsevier.com/permissions.

This book and the individual contributions contained in it are protected under copyright by the Publisher (other than as may be noted herein).

Notices

Knowledge and best practice in this field are constantly changing. As new research and experience broaden our understanding, changes in research methods, professional practices, or medical treatment may become necessary.

Practitioners and researchers must always rely on their own experience and knowledge in evaluating and using any information, methods, compounds, or experiments described herein. In using such information or methods they should be mindful of their own safety and the safety of others, including parties for whom they have a professional responsibility.

To the fullest extent of the law, neither the Publisher nor the authors, contributors, or editors, assume any liability for any injury and/or damage to persons or property as a matter of products liability, negligence or otherwise, or from any use or operation of any methods, products, instructions, or ideas contained in the material herein.

ISBN: 978-0-323-91909-8

For Information on all Academic Press publications
visit our website at https://www.elsevier.com/books-and-journals

Publisher: Stacy Masucci
Acquisitions Editor: Peter B. Linsley
Editorial Project Manager: Matthew Mapes
Production Project Manager: Omer Mukthar
Cover Designer: Greg Harris

Typeset by MPS Limited, Chennai, India

Contents

List of contributors	xiii

I
Overview

1. Epigenetics Overview   3
TRYGVE O. TOLLEFSBOL

Introduction	3
Molecular Mechanisms of Epigenetics	3
Methods in Epigenetics	4
Model Organisms of Epigenetics	5
Factors Influencing Epigenetic Changes	5
Evolutionary Epigenetics	6
Epigenetic Epidemiology	6
Epigenetics and Human Disease	7
Epigenetic Therapy	7
The Future of Epigenetics	7
Conclusion	8
References	8

II
Molecular Mechanisms of Epigenetics

2. Mechanisms of DNA Methylation and Demethylation During Mammalian Development   11
ZHENGZHOU YING AND TAIPING CHEN

Introduction	11
DNA Methylation	12
DNA Demethylation	17
Conclusions	20
Acknowledgments	21
References	22

3. Mechanisms of Histone Modifications   27
LUDOVICA VANZAN, ATHENA SKLIAS, MARIA BOSKOVIC, ZDENKO HERCEG, RABIH MURR AND DAVID M. SUTER

Introduction	27
Main Histone Modifications	28
Role of Aberrant Histone Modifications in Disease	45
Conclusions	45
References	46

4. The Epigenetics of Noncoding RNA   55
RAVINDRESH CHHABRA

Introduction	55
Conclusions	67
Acknowledgments	67
References	68

5. Prions and prion-like phenomena in epigenetic inheritance   73
PHILIPPE SILAR

Structural Heredity	74
Amyloid Prions of *Saccharomyces cerevisiae*, *Podospora anserina*, and Other Organisms	74
Cis Elements Important for Amyloid Prion Formation	76
Genetic Control of Amyloid Prion Formation and Propagation	76
Amyloid Prion Variants	77
Self-Driven Assembly of Hsp60 Mitochondrial Chaperonin	77
Cytotaxis of Cilia and Other Complex Structures	78
Mixed Heredity: "Nonamyloid Prions" That Propagate by Auto-activation	78
Regulatory Inheritance	79
The Lactose Operon and Its Positive Feedback Loop	79
Crippled Growth, A Self-sustained and Mitotically Inheritable Signaling Pathway in the Filamentous Fungus *Podospora anserina*	80
The White/Opaque Switch of *Candida albicans*, An Epigenetic Switch at the Transcription Level	81
Conclusion	82
References	83

6. Higher-order Chromatin Organization in Diseases, from Chromosomal Position Effect to Phenotype Variegation   89
FRÉDÉRIQUE MAGDINIER AND JÉRÔME D. ROBIN

Introduction	89
Chromosomal Position Effect in Human Pathologies	93
Telomeric Position Effect; Implication in Human Pathologies	98
Conclusions	101
References	102

7. Polycomb-group proteins and epigenetic control of gene activity   111
PRASAD PETHE

Introduction	111

Polycomb Repressive Complexes	112
Mechanism of Gene Regulation by Polycomb Group Proteins	112
PcG-Mediated Control of Gene Expression in Stem Cells to Maintain Homeostasis	114
PcG-mediated Control of Gene Expression in Embryonic Stem Cells	116
PcG-mediated Control of Gene Expression in Cancer Cells	117
Conclusion and Future Perspective	118
Acknowledgment	118
References	118

III

Methods in Epigenetics

8. Methods for Analyzing DNA Cytosine Modifications Genome-wide   123

TIBOR A. RAUCH AND GERD P. PFEIFER

Introduction	123
Methylated DNA Immunoprecipitation	124
MBD Protein-based Affinity Pulldown	124
Methylated-CpG Island Recovery Assay	125
Targeted Bisulfite Sequencing	125
Infinium and EPIC Methylation Bead Chips	127
Whole Genome Bisulfite Sequencing	127
Other Sodium Bisulfite-based Approaches	128
Enzymatic Methyl-sequencing and Pyrimidine Borane-based Methods	129
5-Hydroxymethylcytosine Mapping Methodologies	130
TET-assisted Bisulfite Sequencing	131
Oxidative Bisulfite Sequencing	131
APOBEC-coupled Epigenetic Sequencing	131
SMRT-seq and Nanopore Sequencing	131
Single Cell Whole Genome Methylation Analysis	132
Future Directions and Challenges	133
Acknowledgments	133
References	133

9. Genome-wide Analyses of Histone Modifications in the Mammalian Genome   137

SHULAN TIAN, SUSAN L. SLAGER, ERIC W. KLEE AND HUIHUANG YAN

Introduction	137
Histone Modifications in Regulatory Regions	138
High-throughput Assays for Mapping Histone Modifications	139
Genome-wide Mapping of Histone Modifications	145
Catalog of Histone Modification Profiles	149
Variation of Histone Modifications	149
Conclusions and Perspectives	153
ChIP-seq Data Analysis Workflow	154
Acknowledgments	157
References	158

10. Techniques for Analyzing Genome-wide Expression of Non-coding RNA   163

RENA ONOGUCHI-MIZUTANI, KENZUI TANIUE, KENTARO KAWATA, TOSHIMICHI YAMADA AND NOBUYOSHI AKIMITSU

Introduction	163
cDNA Library Construction	164
Analysis of RNA-seq Data for ncRNAs	169
Estimating the Transcription and Degradation Rates of Non-Coding Transcripts	172
Conclusions	174
Dyrec-seq Protocol	175
Acknowledgments	177
References	178

11. Computational Epigenetics: The Competitive Endogenous RNAs Network Analysis   185

LOO KEAT WEI

Introduction	185
Basic Principles of Competitive Endogenous RNAs	185
The Competitive Endogenous RNAs Network	186
Mathematical Models for Competitive Endogenous RNAs Network Analysis	187
Comparison of Mathematical Models for Competitive Endogenous RNAs Network Analysis	190
Computational Epigenetic Approaches for Competitive Endogenous RNAs Network Analysis	190
Conclusion	195
Acknowledgments	196
Conflict of Interest Statement	196
References	196

IV

Model Organisms of Epigenetics

12. Epigenetic Mechanisms in Bacteria Bridge Physiology, Growth and Host—Pathogen Interactions   201

MARIA MIAH, MIHALY MEZEI AND SHIRAZ MUJTABA

Introduction	201
Protein Phosphorylation in Prokaryotes	201
DNA Methylation in Bacteria	204
Protein Methylation	207
Crosstalks Between Epigenetic Modifications	209
Conclusions and Future Perspectives	209
Summary	210
Protocols for Molecular Modeling	210
References	211
Further Reading	213

13. Drosophila Epigenetics — 215
AKANKSHA BHATNAGAR, ASHLEY M. KARNAY AND FELICE ELEFANT

Introduction: *Drosophila* as a Model Organism in Epigenetic Research	216
Epigenetic Modification of Histone Proteins Regulate Chromatin Packaging and Gene Control in *Drosophila*	217
Position-Effect Variegation	220
The Role of Epigenetics During *Drosophila* Development: Epigenetic Memory	223
Dosage Compensation	229
The Epigenetic Language in Postmitotic Neurons Underlying Cognitive Function	232
Conclusion	238
Protocol	238
References	241

14. Models of Mouse Epigenetic Inheritance: Classification, Mechanisms, and Experimental Strategies — 249
COURTNEY W. HANNA

Introduction	249
Epigenetic Programming in Gametogenesis	250
Epigenetic Re-Programming During Embryogenesis	251
Metastable Epialleles	251
Genomic Imprinting	253
Conclusions	259
Acknowledgments	259
References	259

15. Plant Epigenomics — 263
LEONARDO FURCI, JÉRÉMY BERTHELIER, OSCAR JUEZ, MATIN MIRYEGANEH AND HIDETOSHI SAZE

Introduction	263
Basic Mechanisms of Plant Epigenome Regulation	264
Epigenetic Phenomena in Plants	268
Epigenetic Regulation of Transposable Elements and Interactions With Genes	269
Epigenetic Regulation of Stress Responses in Plants	271
Emerging Technologies for Epigenome Studies in Plants	276
Future Perspectives	278
References	279

V
Factors Influencing Epigenetic Changes

16. Dynamic Changes in Epigenetic Modifications During Mammalian Early Embryo Development — 289
JIE YANG AND WEI JIANG

Introduction	289
DNA methylation during mammalian early embryo development	291
Dynamic changes and function of histone modifications in early embryo development	293
Summary	298
References	299

17. Epigenetic Biomarkers — 303
XIAOTONG HU

Introduction	303
Epigenetic biomarkers offer distinct advantages over genetic biomarkers	304
Minimally invasive tissues are suitable for detecting epigenetic biomarkers	305
Field cancerization and epigenetic biomarkers	306
Potential DNA methylation biomarkers in cancer and other diseases	307
Potential m6A methylation biomarkers in cancer and other diseases	308
Potential histone modification biomarkers in cancer and other diseases	309
Potential non-coding RNA biomarkers in diseases	310
Epigenetic biomarker detection methods in the clinic	313
Challenges and future perspectives	316
References	317

18. Transposable Elements Shaping the Epigenome — 323
KAREN GIMÉNEZ-ORENGA AND ELISA OLTRA

Introduction	323
Classification and structure of transposable elements	324
Transposable elements genomic annotation	328
Epigenetic control of transposable elements	329
Factors influencing transposable elements epigenetics	334
Influence of transposable elements on host epigenetics	335
Concluding remarks	343
References	343

19. Dietary and Metabolic Compounds Affecting Covalent Histone Modifications — 357
GARETH W. DAVISON

Introduction	357
Metabolic and Dietary Control of Histone and Transcriptional Dynamics	359
Regulation of Chromatin Dynamics	360
Histone Demethylation	362
Histone Acetylation	367
Other Histone Modifications	373
Concluding Perspectives	374
Acknowledgments	375
References	375
Further reading	380

20. Epigenetics, Stem Cells, Cellular Differentiation, and Associated Neurological Disorders and Brain Cancer — 381

BHAIRAVI SRINAGESHWAR, GARY L. DUNBAR AND JULIEN ROSSIGNOL

Introduction to Epigenetics	382
Epigenetics and the Human Brain	383
Epigenetics and Glioblastoma	394
Epigenetic Changes in Glioma Stem Cells	394
Genes Involved in Glioblastoma	395
Stem Cell Transplantations in Glioblastoma	396
Neural Stem Cell Transplantation for GB	397
Conclusions	397
References	398

21. Epigenetic Regulation of Skeletal Muscle Regeneration — 403

RODOLFO DANIEL ÁVILA-AVILÉS, CLAUDIA NEGRÓN-LOMAS AND J. MANUEL HERNÁNDEZ-HERNÁNDEZ

Introduction	403
Epigenetic Control in the Maintenance of Quiescence	404
Epigenetic Control of the Activation and Proliferation of SCs	406
Epigenetic Control of SCs Differentiation	408
Small Molecules as a Therapeutic Alternative in the Epigenetic Control of Regeneration	410
Conclusion	412
Acknowledgments	413
References	413

22. Epigenetics of X-chromosome Inactivation — 419

CÍNTIA BARROS SANTOS-REBOUÇAS

Introduction	419
Brief Historical Perspective of XCI	420
X-chromosome Evolution and the Incomplete Nature of XCI	420
Imprinted and Random XCI	422
XCI Regulation and Main Epigenetic Steps	423
XCI Differences Between Mice and Humans	427
X-autosome Dosage Compensation	427
Methods for Exploring XCI Status	429
Physiological and Pathogenic XCI Skewing	429
X-chromosome, Sex Bias and Diseases	432
XCI Plasticity: Opportunities for Epigenetic Therapeutics	434
Concluding Remarks	434
Acknowledgments	434
References	436

23. Epigenetics of Memory Processes — 443

SRAVANI PULYA AND BALARAM GHOSH

Introduction	443
Histone Posttranslational Modifications: Long-term Memory Regulation	444
Histone Acetylation and HATs	445
Histone Deacetylation and Histone Deacetylases	447
Histone Methylation	448
Histone Modifications: Gene Expression	449
Manipulating Histone Modifications	450
DNA Methylation and Long-term Memory	452
Histone Variant Exchange	456
Epitranscriptomics: RNA Epigenetics	457
Summary	458
Acknowledgments	458
References	458

24. Transgenerational Epigenetics — 465

JAMES P. CURLEY, RAHIA MASHOODH AND FRANCES A. CHAMPAGNE

Introduction	465
Epigenetic Consequences of Prenatal Maternal Exposures	466
Postnatal Maternal Regulation of the Epigenome	468
Paternal Influence on Offspring Development	469
Transgenerational Effects of Parental Influence	469
Germline-mediated Transgenerational Inheritance	470
Experience-dependent Epigenetic Inheritance	471
Epigenetics, Plasticity, and Evolving Concepts of Inheritance	472
Summary	473
Acknowledgments	473
References	473

25. DNA Methylation Clocks in Age-related Disease — 479

PETER D. FRANSQUET, JO WRIGGLESWORTH AND JOANNE RYAN

Aging	479
Epigenetic Clocks	480
First Generation Epigenetic Clocks	480
Second Generation Epigenetic Clocks	484
Epigenetic Clocks and Age-related Disease	485
Epigenetic Clocks Without an Age-related Disease Focus	488
Can Epigenetic Age be Modified?	489
Summary	491
References	491

VI

Evolutionary Epigenetics

26. Evolution of Epigenetic Mechanisms in Plants: Insights from H3K4 and H3K27 Methyltransferases — 499

J. ARMANDO CASAS-MOLLANO, ERICKA ZACARIAS AND JULIANA ALMEIDA

Introduction	499
Histone Lysine Methylation	500
Histone Lysine Methyltransferases in Plants	500

Evolution of Plant Histone Lysine Methyltransferases Methylating H3K4	502
Evolution of Plant Histone Lysine Methyltransferases Methylating H3K27	508
Perspectives	514
Acknowledgments	515
References	515

27. Evolution, Functions and Dynamics of Epigenetic Mechanisms in Animals   521
GÜNTER VOGT

Introduction	521
Evolution of Epigenetic Mechanisms in The Animal Kingdom	521
Features of Epigenetic Mechanisms and Role in Gene Regulation	524
Dynamics of Epigenetic Marks During Development	526
Generation of Phenotypic Variation in Populations by Epigenetic Mechanisms	530
Relevance of Epigenetics for Animal Ecology	534
Relevance of Epigenetics for Animal Evolution	538
Conclusions and Perspectives	543
Summary	545
References	545

28. Adaptive evolution and epigenetics   551
ILKKA KRONHOLM

Introduction	551
Epigenetic Variation	551
Modeling Evolution	552
Induced Epigenetic Changes	552
Spontaneous Epigenetic Variation	554
Modeling Spontaneous Epigenetic Variation	555
Limits of Epigenetic Contributions to Adaptation	559
Conclusions and Future Directions	560
Acknowledgments	561
References	561

VII
Epigenetic Epidemiology

29. Epigenetics of Livestock Health, Production, and Breeding   569
EVELINE M. IBEAGHA-AWEMU AND HASAN KHATIB

Introduction	569
Development of Animal Breeding	570
Epigenetic Source of Phenotypic Variation in Livestock Health and Production	571
Revisiting Animal Breeding Planning and Management: The Role of Epigenetic Mechanisms	587
Conclusion	595
Summary	595
Livestock Epigenetics Case Study	595
Acknowledgment	598
References	599

30. Nutritional Epigenetics and Fetal Metabolic Programming   611
HO-SUN LEE

Introduction	611
Metabolic Sensing by Epigenetic Mechanisms	612
Impact of Prenatal Nutrition on Fetal Reprogramming	615
Maternal Diet and Metabolic Epigenome	616
Summary and Perspectives	619
Acknowledgments	620
References	620

31. Epigenetics of Drug Addiction   625
RYAN D. SHEPARD AND FERESHTEH S. NUGENT

Introduction	625
What Is Addiction?	626
Dopamine and Reward Circuits	626
Synaptic Plasticity, Learning and Memory and Addiction	628
Epigenetic Processes in Drug Addiction	630
Utilizing Epigenetic Targets for Diagnosis and Treatments to Combat Addiction and SUDs	632
Conclusion	634
Summary	634
Acknowledgments	634
References	634
Further Reading	637

32. Environmental Influence on Epigenetics   639
MARISOL RESENDIZ, DARRYL S. WATKINS, NAIL CAN ÖZTÜRK AND FENG C. ZHOU

Introduction	639
The Extent (Timeline) of Environmental Influence	641
Exerting Environment	642
Mental or Physiological Environment	643
Hazardous Environmental Pollutants and Chemicals	650
Conclusion and Future Direction	658
Research case study [18]	660
References	663

33. Gut Microbiome Influence on Human Epigenetics, Health, and Disease   669
MARTIN M. WATSON, MARK VAN DER GIEZEN AND KJETIL SØREIDE

Introduction	669
Early Microbiome Exposure and Epigenetic Influence	673
Human Gut Microflora	675
Microbiome, Epigenetics, and Effect on Metabolism	675
Gut microbiota, Inflammation, and Colorectal Carcinogenesis	677

Pathogenic Infections and Cancer	677	Type 1 Diabetes Mellitus	724
Pathogenic Infections and Epigenetic Modifications	680	Systemic Sclerosis	725
Summary	683	Use of Epigenetic Modifications as Potential Biomarkers	726
References	683	Use of Epigenetic Modifiers for Potential Diagnosis and Therapy in Autoimmune Diseases	727
		Summary	728

34. Population Pharmacoepigenomics 687

JACOB PEEDICAYIL

Introduction	687
General Aspects of Population Pharmacoepigenomics	688
Population Variations of Epigenetic Patterns and Population Pharmacoepigenomics	688
Human Epigenome Projects and Population Pharmacoepigenomics	688
Population Pharmacoepigenomics in Relation to Pharmacokinetics	689
Population Pharmacoepigenomics in Relation to Pharmacodynamics	689
3D and 4D Chromatin Structures in the Nuclei of Cells in the Liver and Other Body Organs and Population Pharmacoepigenomics	690
Population Pharmacoepigenomics in Relation to Adverse Drug Reactions and Drug Interactions	691
Conclusions	691
Population Pharmacoepigenomics Case Study	691
References	693

VIII

Epigenetics and Human Disease

35. Cancer Epigenetics 697

MARINA ALEXEEVA, MARCUS ROALSØ AND KJETIL SØREIDE

Introduction	697
Epigenetic Influences Over a Lifetime and Cancer Risk	698
Epigenetics in Cancer: Remodelers, Writers, Readers, and Erasers	699
Genetic and Epigenetic Classification of Cancer	705
Epigenetic Biomarkers in Cancer	708
Epigenetics as Cancer Therapeutic Targets	709
Future Perspectives	710
References	711

36. The Role of Epigenetics in Autoimmune Disorders 715

KERSTIN KLEIN

Epigenetic Mechanisms Influence Autoimmune Processes	715
Mechanisms of Autoimmunity	716
Epigenetics of Immune Cells	719
Epigenetics of Systemic Lupus Erythematosus	720
Epigenetics of Rheumatoid Arthritis	722
Multiple Sclerosis	723

Multiomics and machine learning accurately predict clinical response to adalimumab and etanercept therapy in patients with rheumatoid arthritis [203] [ex: Autoimmunity Epigenetics Case Study] 728
References 730

37. Epigenetics of Brain Disorders 737

ALI JAWAID, ELOÏSE A. KREMER, NANCY V.N. CARULLO AND ISABELLE M. MANSUY

Introduction	738
Important Epigenetic Mechanisms for the Brain	738
Epigenetic Dysregulation in Neurodevelopmental Disorders: The Example of Rett Syndrome	739
Epigenetic Dysregulation in Neurodegenerative Disorders: The Example of Alzheimer's Disease	742
Epigenetic Dysregulation in Psychiatric Disorders: The Example of Depression	745
Epigenetic Dysregulation by Environmental Stress: The Example of Early-life Stress	748
Conclusions and Outlook	749
Summary	750
Research case study	751
Acknowledgments	752
References	752

38. Epigenetics of Metabolic Diseases 761

LINN GILLBERG AND LINE HJORT

Introduction	761
Impact of Age and Lifestyle Factors on the Epigenome	762
Epigenetic Memory, Prenatal Exposure, and Risk of Metabolic Disease	766
Epigenetic Features of Metabolic Diseases	769
Conclusions	771
References	772

39. Imprinting Disorders in Humans 779

THOMAS EGGERMANN

Introduction	779
Clinical Findings in Imprinting Disorders	779
Molecular Findings in Imprinting Disorders	781
Causes of Disturbed Imprinting	782
Cis-acting Factors	783
Trans-acting Factors	784
Maternal Effect Mutations and Multilocus Imprinting Disturbance	784

Translational Use of New Findings in ImpDis and New Methodologies	785
Genetic and Reproductive Counseling	787
Clinical Management	787
Concluding Remarks	787
References	787

IX

Epigenetic Therapy

40. Clinical Applications of Histone Deacetylase Inhibitors   793
ROMAIN PACAUD, JOSE GARCIA, SCOTT THOMAS AND PAMELA N. MUNSTER

Introduction	793
HDACi for the Treatment of Hematological Malignancies	795
HDACi in the Treatment of Solid Tumors	801
Clinical Applications of HDACi for Noncancer Diseases	807
Conclusions and the Future Directions of the Clinical Applications of HDACi	809
References	810

41. Combination Epigenetic Therapy   821
RŪTA NAVAKAUSKIENĖ

Introduction	821
Chromatin-remodeling Agents, Combined Treatment, and Targeted Therapy	822
DNA Methylation/Demethylation and Combined Treatment	823
Histone Modifications and Histone-modifying Enzymes for Epigenetic Treatment	828
Conclusions and Future Perspectives	833
Clinical Trial	833
References	837

X

The Future of Epigenetics

42. New Directions for Epigenetics: Application of Engineered DNA-binding Molecules to Locus-specific Epigenetic Research   843
TOSHITSUGU FUJITA AND HODAKA FUJII

Introduction	843
General Information on Engineered DNA-binding Molecules	844
Locus-specific Epigenome Editing	845
Locus-specific Identification of Epigenetic Molecules that Interact with Target Genomic Regions	858
Conclusions	862
References	862
Index	869

List of contributors

Nobuyoshi Akimitsu Isotope Science Center, The University of Tokyo, Tokyo, Japan

Marina Alexeeva Department of Gastrointestinal Surgery, Stavanger University Hospital, Stavanger, Norway; Gastrointestinal Translational Research Unit, Laboratory for Molecular Biology, Stavanger University Hospital, Stavanger, Norway

Juliana Almeida Centre of Natural Sciences and Humanities, Federal University of ABC, Sao Bernardo do Campo, Brazil

Rodolfo Daniel Ávila-Avilés Laboratory of Epigenetics of Skeletal Muscle Regeneration, Department of Genetics and Molecular Biology, Centre for Research and Advanced Studies-IPN, Mexico City, Mexico

Jérémy Berthelier Plant Epigenetics Unit, Okinawa Institute of Science and Technology Graduate University, Onna-son, Okinawa, Japan

Akanksha Bhatnagar Department of Biology, Drexel University, Philadelphia, PA, United States

Maria Boskovic Laboratory for Cancer Research, University of Split School of Medicine, Split, Croatia

Nancy V.N. Carullo Laboratory of Neuroepigenetics, Brain Research Institute, Medical Faculty of the University of Zurich, Zurich, Switzerland; Zurich Neuroscience Center, ETH and University of Zurich, Zurich, Switzerland

J. Armando Casas-Mollano BioTechnology Institute, University of Minnesota, Twin-Cities, Saint Paul, MN, United States

Frances A. Champagne Department of Psychology, University of Texas, Austin, TX, United States

Taiping Chen The Ministry of Education Key Laboratory of Laboratory Medical Diagnostics, College of Laboratory Medicine, Chongqing Medical University, Chongqing, P.R. China

Ravindresh Chhabra Department of Biochemistry, Central University of Punjab, Ghudda, Punjab, India

James P. Curley Department of Psychology, University of Texas, Austin, TX, United States

Gareth W. Davison Faculty of Life and Health Sciences, Ulster University, Northern Ireland, United Kingdom

Gary L. Dunbar Field Neurosciences Institute, Saginaw, MI, United States

Thomas Eggermann Institut für Humangenetik, RWTH Aachen, Aachen, Germany

Felice Elefant Department of Biology, Drexel University, Philadelphia, PA, United States

Peter D. Fransquet School of Public Health and Preventive Medicine, Monash University, Melbourne, VIC, Australia

Hodaka Fujii Department of Biochemistry and Genome Biology, Hirosaki University Graduate School of Medicine, Aomori, Japan

Toshitsugu Fujita Department of Biochemistry and Genome Biology, Hirosaki University Graduate School of Medicine, Aomori, Japan

Leonardo Furci Plant Epigenetics Unit, Okinawa Institute of Science and Technology Graduate University, Onna-son, Okinawa, Japan

Jose Garcia Division of Hematology and Oncology, University of California, San Francisco, CA, United States

Balaram Ghosh Epigenetic Research Laboratory, Department of Pharmacy, Birla Institute of Technology and Science-Pilani Hyderabad Campus, Shamirpet, Hyderabad, Telangana, India

Linn Gillberg Department of Biomedical Sciences, University of Copenhagen, Copenhagen, Denmark

Karen Giménez-Orenga Centro de Investigación Traslacional San Alberto Magno, Universidad Católica de Valencia San Vicente Mártir, Valencia, Spain

Courtney W. Hanna Centre for Trophoblast Research, University of Cambridge, Cambridge, United Kingdom; Department of Physiology, Development and Neuroscience, University of Cambridge, Cambridge, United Kingdom; Epigenetics Programme, Babraham Institute, Cambridge, United Kingdom

Zdenko Herceg International Agency for Research on Cancer (IARC), Lyon, France

J. Manuel Hernández-Hernández Instituto Politécnico Nacional 2508, San Pedro Zacatenco, Ciudad de México, México

Line Hjort Department of Endocrinology, Copenhagen University Hospital, Copenhagen, Denmark

Xiaotong Hu Sir Run Run Shaw Hospital, Zhejiang University, Hangzhou, Zhejiang, P.R. China

Eveline M. Ibeagha-Awemu Sherbrooke Research and Development Centre, Agriculture and Agri-Food Canada, Sherbrooke, QC, Canada

Ali Jawaid Laboratory of Neuroepigenetics, Brain Research Institute, Medical Faculty of the University of Zurich, Zurich, Switzerland; Department of Health Sciences and

Technology, Institute for Neuroscience, ETH Zurich, Zurich, Switzerland; Zurich Neuroscience Center, ETH and University of Zurich, Zurich, Switzerland

Wei Jiang Frontier Science Center for Immunology and Metabolism, Medical Research Institute, Wuhan University, Wuhan, P.R. China

Oscar Juez Plant Epigenetics Unit, Okinawa Institute of Science and Technology Graduate University, Onna-son, Okinawa, Japan

Ashley M. Karnay Department of Neurobiology & Anatomy, Drexel University College of Medicine, Philadelphia, PA, United States

Kentaro Kawata Isotope Science Center, The University of Tokyo, Tokyo, Japan; Cellular and Molecular Biotechnology Research Institute, National Institute of Advanced Industrial Science and Technology (AIST), Tokyo, Japan

Hasan Khatib Department of Animal and Dairy Sciences, University of Wisconsin, Madison, WI, United States

Eric W. Klee Division of Computational Biology, Department of Quantitative Health Sciences, Mayo Clinic, Rochester, MN, United States

Kerstin Klein Department of BioMedical Research, University of Bern, Bern, Switzerland; Department of Rheumatology and Immunology, Bern University Hospital, Bern, Switzerland

Eloïse A. Kremer Laboratory of Neuroepigenetics, Brain Research Institute, Medical Faculty of the University of Zurich, Zurich, Switzerland; Zurich Neuroscience Center, ETH and University of Zurich, Zurich, Switzerland

Ilkka Kronholm Department of Biological and Environmental Science, University of Jyväskylä, Jyväskylä, Finland

Ho-Sun Lee Interdisciplinary Program in Bioinformatics and Department of Statistics, Seoul National University, Seoul, Republic of Korea; Toxicology Division, National Forensic Service Daegu Institute, Republic of Korea

Frédérique Magdinier Aix Marseille Univ, INSERM, Marseille Medical Genetics, MMG, Marseille, France

Isabelle M. Mansuy Laboratory of Neuroepigenetics, Brain Research Institute, Medical Faculty of the University of Zurich, Zurich, Switzerland; Zurich Neuroscience Center, ETH and University of Zurich, Zurich, Switzerland

Rahia Mashoodh Department of Zoology, University of Cambridge, Cambridge, United Kingdom

Mihaly Mezei Department of Pharmaceutical Sciences, Icahn School of Medicine at Mount Sinai, New York, NY, United States

Maria Miah Department of Biology, Medgar Evers College, City University of New York, Brooklyn, NY, United States

Matin Miryeganeh Plant Epigenetics Unit, Okinawa Institute of Science and Technology Graduate University, Onna-son, Okinawa, Japan

Shiraz Mujtaba Department of Biology, Medgar Evers College, City University of New York, Brooklyn, NY, United States

Pamela N. Munster Division of Hematology and Oncology, University of California, San Francisco, CA, United States

Rabih Murr Faculty of Medicine, Department of Genetic Medicine and Development, University of Geneva, Geneva, Switzerland

Rūta Navakauskienė Life Sciences Center, Vilnius University, Vilnius, Lithuania

Claudia Negrón-Lomas Laboratory of Epigenetics of Skeletal Muscle Regeneration, Department of Genetics and Molecular Biology, Centre for Research and Advanced Studies-IPN, Mexico City, Mexico

Fereshteh S. Nugent Department of Pharmacology & Molecular Therapeutics, F. Edward Hebert School of Medicine, Uniformed Services University of the Health Sciences, Bethesda, MD, United States

Elisa Oltra Centro de Investigación Traslacional San Alberto Magno, Universidad Católica de Valencia San Vicente Mártir, Valencia, Spain; Department of Pathology, School of Medicine and Health Sciences, Universidad Católica de Valencia San Vicente Mártir, Valencia, Spain

Rena Onoguchi-Mizutani Isotope Science Center, The University of Tokyo, Tokyo, Japan

Nail Can Öztürk Department of Anatomy, Mersin University, Mersin, Turkey

Romain Pacaud Division of Hematology and Oncology, University of California, San Francisco, CA, United States

Jacob Peedicayil Department of Pharmacology & Clinical Pharmacology, Christian Medical College, Vellore, Tamil Nadu, India

Prasad Pethe Symbiosis Centre for Stem Cell Research (SCSCR), Symbiosis International University (SIU), Pune, Maharashtra, India

Gerd P. Pfeifer Department of Epigenetics, Van Andel Institute, Grand Rapids, MI, United States

Sravani Pulya Epigenetic Research Laboratory, Department of Pharmacy, Birla Institute of Technology and Science-Pilani Hyderabad Campus, Shamirpet, Hyderabad, Telangana, India

Tibor A. Rauch University of Pécs Medical School, Pécs, Hungary

Marisol Resendiz Stark Neuroscience Research Institute Indiana University School of Medicine, Indianapolis, IN, Uinted States

Marcus Roalsø Department of Gastrointestinal Surgery, Stavanger University Hospital, Stavanger, Norway; Gastrointestinal Translational Research Unit, Laboratory for Molecular Biology, Stavanger University Hospital, Stavanger, Norway; Department of Quality and Health Technology, University of Stavanger, Stavanger Norway

Jérôme D. Robin Aix Marseille Univ, INSERM, Marseille Medical Genetics, MMG, Marseille, France

Julien Rossignol Central Michigan University, Mt. Pleasant, MI, United States

Joanne Ryan School of Public Health and Preventive Medicine, Monash University, Melbourne, VIC, Australia

List of contributors

Cíntia Barros Santos-Rebouças Department of Genetics, Institute of Biology Roberto Alcantara Gomes, State University of Rio de Janeiro, Rio de Janeiro, Rio de Janeiro, Brazil

Hidetoshi Saze Plant Epigenetics Unit, Okinawa Institute of Science and Technology Graduate University, Onnason, Okinawa, Japan

Ryan D. Shepard Department of Pharmacology & Molecular Therapeutics, F. Edward Hebert School of Medicine, Uniformed Services University of the Health Sciences, Bethesda, MD, United States; Synapse and Neural Circuit Research Section, National Institute of Neurological Disorders and Stroke, National Institutes of Health, Bethesda, MD, United States

Philippe Silar Université de Paris, Paris Cité, France

Athena Sklias Center for Integrative Genomics, University of Lausanne, Lausanne, Switzerland

Susan L. Slager Division of Computational Biology, Department of Quantitative Health Sciences, Mayo Clinic, Rochester, MN, United States

Kjetil Søreide Department of Gastrointestinal Surgery, Stavanger University Hospital, Stavanger, Norway; Gastrointestinal Translational Research Unit, Laboratory for Molecular Biology, Stavanger University Hospital, Stavanger, Norway; Department of Clinical Medicine, University of Bergen, Bergen, Norway

Bhairavi Srinageshwar Central Michigan University, Mt. Pleasant, MI, United States

David M. Suter Institute of Bioengineering, School of Life Sciences, Swiss Federal Institute of Technology (EPFL), Lausanne, Switzerland

Kenzui Taniue Isotope Science Center, The University of Tokyo, Tokyo, Japan

Scott Thomas Division of Hematology and Oncology, University of California, San Francisco, CA, United States

Shulan Tian Division of Computational Biology, Department of Quantitative Health Sciences, Mayo Clinic, Rochester, MN, United States

Trygve O. Tollefsbol Department of Biology, University of Alabama at Birmingham, Birmingham, AL, United States; O'Neal Comprehensive Cancer Center, University of Alabama at Birmingham, Birmingham, AL, United States; Integrative Center for Aging Research, University of Alabama at Birmingham, Birmingham, AL, United States; Nutrition Obesity Research Center, University of Alabama at Birmingham, Birmingham, AL, United States; Comprehensive Diabetes Center, University of Alabama at Birmingham, Birmingham, AL, United States; University Wide Microbiome Center, University of Alabama at Birmingham, Birmingham, AL, United States

Mark van der Giezen Department of Chemistry, Bioscience and Environmental Engineering, University of Stavanger, Stavanger, Norway; Biosciences, University of Exeter, Exeter, United Kingdom

Ludovica Vanzan Institute of Bioengineering, School of Life Sciences, Swiss Federal Institute of Technology (EPFL), Lausanne, Switzerland

Günter Vogt Faculty of Biosciences, University of Heidelberg, Heidelberg, Germany

Darryl S. Watkins Stark Neuroscience Research Institute Indiana University School of Medicine, Indianapolis, IN, Uinted States

Martin M. Watson Department of Chemistry, Bioscience and Environmental Engineering, University of Stavanger, Stavanger, Norway

Loo Keat Wei Faculty of Science, Department of Biological Science, Universiti Tunku Abdul Rahman, Kampar, Perak, Malaysia

Jo Wrigglesworth School of Public Health and Preventive Medicine, Monash University, Melbourne, VIC, Australia

Toshimichi Yamada Isotope Science Center, The University of Tokyo, Tokyo, Japan

Huihuang Yan Division of Computational Biology, Department of Quantitative Health Sciences, Mayo Clinic, Rochester, MN, United States

Jie Yang Frontier Science Center for Immunology and Metabolism, Medical Research Institute, Wuhan University, Wuhan, P.R. China

Zhengzhou Ying Department of Epigenetics and Molecular Carcinogenesis, The University of Texas MD Anderson Cancer Center, Houston, TX, United States

Ericka Zacarias BioTechnology Institute, University of Minnesota, Twin-Cities, Saint Paul, MN, United States

Feng C. Zhou Stark Neuroscience Research Institute Indiana University School of Medicine, Indianapolis, IN, Uinted States; Department of Anatomy, Cell Biology and Physiology, IUSM, Indianapolis, IN, United States

SECTION I

Overview

1. Epigenetics overview

CHAPTER

1

Epigenetics Overview

Trygve O. Tollefsbol[1,2,3,4,5,6]

[1]Department of Biology, University of Alabama at Birmingham, Birmingham, AL, United States [2]O'Neal Comprehensive Cancer Center, University of Alabama at Birmingham, Birmingham, AL, United States [3]Integrative Center for Aging Research, University of Alabama at Birmingham, Birmingham, AL, United States [4]Nutrition Obesity Research Center, University of Alabama at Birmingham, Birmingham, AL, United States [5]Comprehensive Diabetes Center, University of Alabama at Birmingham, Birmingham, AL, United States [6]University Wide Microbiome Center, University of Alabama at Birmingham, Birmingham, AL, United States

OUTLINE

Introduction	3	Epigenetic Epidemiology	6
Molecular Mechanisms of Epigenetics	3	Epigenetics and Human Disease	7
Methods in Epigenetics	4	Epigenetic Therapy	7
Model Organisms of Epigenetics	5	The Future of Epigenetics	7
Factors Influencing Epigenetic Changes	5	Conclusion	8
Evolutionary Epigenetics	6	References	8

INTRODUCTION

In 1942, Conrad Waddington first defined epigenetics as the causal interactions between genes and their products that allow for the phenotypic expression [1]. This term has now been somewhat redefined and although there are many variants of the definition today, a consensus definition is that epigenetics is the collective heritable changes in phenotype due to processes that arise independent of changes in the primary DNA sequence. This heritability of epigenetic information was for many years thought to be limited to mitotic cellular divisions. However, it is now apparent that epigenetic processes can be transferred meiotically in organisms from one generation to another [2,3]. This phenomenon was first described in plants [4] and has been expanded to include yeast, *Drosophila*, mouse, and possibly humans [5–7].

MOLECULAR MECHANISMS OF EPIGENETICS

DNA methylation is the most studied of epigenetic processes. It most eukaryotes, DNA methylation consists of transfer of a methyl moiety from S-adenosylmethionine (SAM) to the 5-position of cytosines in certain CpG dinucleotides. This important enzymatic transfer reaction is catalyzed by the DNA methyltransferases (DNMTs). The three major DNMTs are DNMT1, 3A, and 3B. DNMT1 catalyzes what is called maintenance methylation, which occurs during each cellular replication as the DNA duplicates. The other major DNMTs, 3A and 3B, are characterized by their relatively higher de novo methylation activity. This process leads to the introduction of 5-methylcytosines (5mCs) into the genome at sites that were not previously methylated. Notably, the most significant aspect of DNA methylation, which can also influence such processes as X-chromosome inactivation and cellular differentiation, is its effects on gene expression. In general, the more methylated a gene regulatory region, the more likely it is that the gene activity will become downregulated and vice versa, although there are some exceptions to this dogma [8]. Chapter 2 of this book reviews the mechanisms of DNA methylation and demethylation during mammalian development. Major changes in DNA methylation occur during development and this is especially true with respect to early embryonic development. Further, the germ cells also display many changes during gametogenesis. Recent advances have highlighted important roles of the ten-eleven translocation family of dioxygenases. These enzymes convert 5mC to 5-hydroxymethylcytosie (5hmC), 5-formylcytosine (5fC), and 5-carboxylcytosine (5caC) and appear to play important roles in the dramatic DNA demethylation that occurs during early development. The 5mC oxidized derivatives may also serve as epigenetic marks that modulate chromatin regulation.

There are additional important epigenetic changes that occur in the genome. For example, chromatin changes are another central epigenetic process that have an impact not only on gene expression, but also many other biological processes. Posttranslational modifications of histones such as acetylation and methylation occur in a site-specific manner and often influence the binding and activities of other proteins that influence gene regulation. The histone acetyltransferases catalyze histone acetylation and the histone deacetylases (HDACs) result in removal of acetyl groups from key histones that comprise the chromatin. These modifications can occur at numerous sties in the histones and are most common in the amino terminal regions of these proteins as discussed in Chapter 3. In general, increased histone acetylation is associated with greater gene activity and vice versa. Methylation of histones has variable effects on gene activity. Lysine 4 (K4) methylation of histone H3 is often associated with increasing gene activity whereas methylation of lysine 9 (K9) of histone H3 may lead to transcriptional repression. There is also considerable crosstalk between DNA methylation and histone modifications [9] such that cytosine methylation may increase the likelihood of H3K9 methylation and H3K9 methylation may promote cytosine methylation. In addition, histone-modifying epigenetic enzymes have recently been identified that may have key roles that are not dependent on their catalytic activity (Chapter 3).

Among the most exciting advances in epigenetics have been the discoveries of many other processes besides DNA methylation, and histone modifications impacting the epigenetic behavior of cells. RNA is highly versatile and often plays a major role in epigenetic processes. For instance, noncoding RNA (Chapter 4) including both short (<200 nucleotides) and long (>200 nucleotides) forms, often share protein and RNA components with the RNA interference (RNAi) pathway and they may also influence more conventional aspects of epigenetics such as DNA methylation and chromatin marking. These effects appear to be wide-spread and occur in organisms ranging from protists to humans. Notably, the noncoding RNAs may serve as therapeutic targets for a number of human diseases and this area of epigenetics is now attracting considerable attention. Prions are fascinating in that they can influence epigenetic processes independent of DNA and chromatin. In Chapter 5 it is shown that structural heredity also is important in epigenetic expression where alternative states of macromolecular complexes or regulatory networks can have a major effect on phenotypic expression independent of changes in DNA sequences. The prion proteins are able to switch their structure in an autocatalytic manner that can not only influence epigenetic expression, but also lead to human disease although the full potential of prions in epigenetic processes is still on the horizon.

The position of a gene in a given chromosome can also greatly influence its expression (Chapter 6). Upon rearrangement, a gene may be relocated to a heterochromatic region of the genome leading to gene silencing. Moreover, a change in position of a regulatory element may affect the maintenance of chromatin architecture and subsequently cellular functioning. Polycomb mechanisms are another aspect of epigenetics that control all of the major cellular differentiation pathways and are also involved in cell fate. Polycomb protein repression is very dynamic and can be easily reversed by activators. They also raise the threshold of the signals or activators required for transcriptional activation which places these fascinating proteins within the realm of epigenetic processes. Polycomb complexes can generate H2A ubiquitylation and H3K27 methylation that often mediate their repressive functions and H3K27 may serve as an epigenetic memory for Polycomb repression as described in Chapter 7. Therefore, although DNA methylation and histone modifications are mainstays of epigenetics, recent advances have greatly expanded the field of epigenetics to include many other processes such as noncoding RNA, prions, chromosome position effects and Polycomb mechanisms.

METHODS IN EPIGENETICS

Numerous advances in epigenetics that have driven this field for the past two decades can be traced back to the technological breakthroughs that have made the many discoveries possible. We now have a wealth of information about key gene-specific epigenetic changes that occur in a myriad of biological processes. However, among the most exciting advances have been important breakthroughs in analyses of the methylome at high resolution. High-throughput sequencing, which has largely replaced microarray platforms, has made possible new techniques to analyze genome-wide features of epigenetics that are based on uses of methylation-sensitive restriction enzymes, sodium bisulfite conversion and affinity capture with antibodies or proteins that select methylated DNA sequences. Tiling arrays may be employed for analysis of chromosomal segments or the whole genome and whole genome high-throughput sequencing is now a staple for reliable and specific methylomic analyses. Additionally, single-cell analyses of DNA methylation patterns by whole genome epigenetic analyses are emerging as described in Chapter 8. Chromatin immunoprecipitation and sequencing (ChIP-seq) is frequently used in epigenetic analyses and can also be applied to single cells as well as used for mapping two modifications simultaneously as described in Chapter 9. Likewise, genome-wide expression analyses of RNAs have been made possible with RNA-seq that allows detection of a myriad of noncoding RNAs (Chapter 10). Since there has been much information derived from epigenomic approaches, methods to analyze data from ChIP-seq and RNA-seq, for example, are becoming increasingly important and are delineated in Chapter 11. Notably, developments in the tools for assessing epigenetic information have been and will continue to be an important driving force in advancing epigenetics.

MODEL ORGANISMS OF EPIGENETICS

Epigenetic processes are wide-spread and much of our extant knowledge about epigenetics has been derived from model systems, both typical and unique. The ease of manipulation of prokaryotic organisms has facilitated discoveries in the molecular mechanisms of basic epigenetic processes. Although prokaryotic organisms do not contain chromatin, some bacteria are capable of encoding factors which lead to posttranslational modifications of an epigenetic nature as described in Chapter 12. *Drosophila* is a mainstay model in biology in general and the epigenetics field is not an exception in this regard. For example, Chapter 13 discusses how use of *Drosophila* as a model organism has significantly increased knowledge of chromatin organization and may have powerful potential in facilitating understanding of the epigenetics of neurological disorders. Perhaps the most useful model system in epigenetics to date is the mouse model (Chapter 14). Numerous different mouse models have been developed that are important in many different epigenetic processes such as transgenerational epigenetics and imprinting and these models have potential in illuminating a number of human diseases. Analyses of epigenetic processes in development and health of offspring can also be facilitated through the use of mouse models. Plant models (Chapter 15) are of great importance in epigenetics due in part to their plasticity and their ability to silence transposable elements. RNAi silencing in plants has been at the forefront of epigenetics and plant models will likely lead the way in several other epigenetic processes in the future. Use of plant models has also facilitated the development of targeted epigenome editing in plants that may have great significance in enhancing crop performance. Thus, model development, like the advances in techniques, have made many of the most exciting discoveries in epigenetics possible for a number of years.

FACTORS INFLUENCING EPIGENETIC CHANGES

The functions of epigenetics are indeed numerous and it would be next to impossible to do complete justice in one book to this ever-expanding field. However, Chapters 16—25 illustrate a few of the many different functions that epigenetics mediates. For example, Chapter 16 reviews the role of epigenetics in early development and illustrates that both gene-specific and global changes in DNA methylation are central to embryonic development. In addition, perturbations mediated by processes such as stress during development may lead to disease formation later in life. Chapter 17 reviews epigenetic biomarkers that are also very important and serve the key function of informing the staging and classification of disease as well as guiding clinical management. Another factor that can significantly influence epigenetic changes is the activity of transposable elements. These elements insert into coding or noncoding regions of the genome and can also exert effects in *cis* by their insertion or altering of the epigenetic state of the insertional site itself as described in Chapter 18. These mechanisms of transposable elements have not only shaped evolution, but can also contribute to epigenetic-based diseases.

Epigenetics is intricately linked to changes in the metabolism of organisms and these two processes cannot be fully understood separately. SAM is a universal methyl donor and drives many epigenetic processes; the importance of SAM in epigenetic mechanisms is vast. Metabolic functions can also influence the chromatin which is a major mediator of epigenetic processes (Chapter 19). It is now apparent that various environmental influences such as diet and changes in metabolic compounds can regulate the many enzymes that modify histones in mammals. Thus, metabolic processes impacted by dietary changes can greatly influence both DNA methylation and chromatin remodeling, the two major epigenetic mediators, and significantly influence gene regulation in both health and disease.

Stem cells rely in part on signals from the environment and epigenetic mechanisms have central roles in how stem cells respond to environmental influences (Chapter 20) and how therapeutic roles of stem cells may contribute to management of many diseases such as multiple sclerosis, Huntington disease and Parkinson disease. Regenerative medicine is dependent upon stem cells. Regulation of muscle regenerative abilities involves key changes in the epigenome that regulate gene expression and may have potential use to treat skeletal muscle diseases (Chapter 21). Some of the basic tenets of epigenetics were first realized through studies of X-chromosome inactivation which allows for dosage compensation for X-linked gene expression between males and females. As discussed in Chapter 22, advances in the study of the epigenetics of X-chromosome inactivation have revealed that epigenetic modification of the inheritance mode of X-driven phenotypes can impact the plasticity of human conditions and may lead to new strategies for treating dominant X-linked disorders in females. Profound new discoveries have occurred in the area of the epigenetics of memory processes. Recent exciting discoveries have shown that gene regulation through epigenetic mechanisms is necessary for changes in adult brain function and behavior based on life experiences (Chapter 23). Moreover, new drugs that impact epigenetic mechanisms may have future uses in treating or alleviating cognitive dysfunction. Transgenerational inheritance (Chapter 24) is also a form of memory based in part on epigenetics in that early life experiences that impact epigenetic markers can greatly influence adult health and risk for diseases. In addition, the aging process is a form of epigenetic memory and experience in that our genes are epigenetically modified from our parents as well as during our entire life spans that can significantly impact the longevity of humans as well as our risk for the many age-related diseases, many of which are also epigenetically based. As reviewed in Chapter 25, there has been a surge of interest in DNA methylation biomarkers of aging which are referred to as "epigenetic clocks." Environmental factors have been found to influence these epigenetic biomarkers and they have considerable potential for identifying accelerated epigenetic aging, certain epigenetic-based diseases, and may have prognostic capacity for age-related diseases. It is therefore apparent that epigenetics influences numerous different functions and it is highly likely that many additional functions of epigenetics will be discovered in the future.

EVOLUTIONARY EPIGENETICS

Although many think of epigenetic processes as being inherent and static to a specific organism, it is apparent that epigenetics have been a major force behind the evolutionary creation of new species. Chapter 26 reveals that epigenetic mediators such as H3K4 and H3K27 methyltransferases have significantly impacted plant evolution. The expansion of gene families of these enzymes may have allowed for changes in chromatin regulation of new genes and pathways that have modified plant evolution. The epigenetic machinery has also played a major role in the evolution of animals. As described in Chapter 27, DNA methylation may serve as a driver of evolution through its effects on gene regulation. Vertebrates have evolved genome-wide methylation while invertebrates are characterized by mosaic DNA methylation patterns. Epigenotype diversification followed by genetic fixation could be a major mechanism of evolution that is currently understudied and may receive increasing attention in future analyses of epigenetics in evolutionary processes. Adaptive evolution by natural selection requires heritability, and spontaneous epigenetic changes may have been important in adaptive evolution, although there are important differences between genetic variation and epigenetic variation with respect to supply and stability of epigenetic mutations (Chapter 28).

EPIGENETIC EPIDEMIOLOGY

A very new and exciting area within the field of epigenetics is the impact of epigenetic mechanisms on livestock breeding. Chapter 29 indicates that the breeding programs of livestock that are currently in place account for only part of the phenotypic variance in traits and that epigenetic factors of variance need further consideration. In fact, livestock epigenetics may have a major impact on the growth and development of livestock and profoundly affect phenotypic outcomes. Dietary factors are highly variable not only between individuals but also among human populations and various nonhuman species. Many studies have shown that the diet has a profound effect on the epigenetic expression of the genome and therefore on the phenotype. An important example of this phenomenon is the link between the nutritional epigenome and fetal metabolic programming (Chapter 30) in that early epigenetic changes through nutritional means may exacerbate the risk of development of various metabolic conditions later in life. Another factor besides diet that can impact the epigenome is abused use of drugs that induce epigenomic aberrations which are features of drug addiction. Drugs of abuse are also able to recruit epigenetic mechanisms that can impact the reward circuitry which is of central interest in drug abuse as reviewed in Chapter 31. It is possible that modifications of the epigenetic circuitry in cases of chronic drug abuse may lead to therapeutic options of those addicted to certain drugs. Other environmental agents can also greatly impact the epigenome. For example, Chapter 32 reviews the many environmental agents that can lead to alterations in the epigenome thereby inducing toxicity or carcinogenesis. The environment also affects the gut microbiome (Chapter 33) which is known to modulate epigenetic processes. An increased understanding of the dynamic interactions between the gut microbiome and epigenetic modifications may have considerable translational potential. Drugs besides those of addiction can also reshape the epigenome which has opened the new field of pharmacoepigenomics. It is clear that certain populations respond differently to drugs, and much of this variation may be explained by epigenetic factors (Chapter 34). Thus, epidemiological factors have great importance in epigenetics and this is influenced by diet, addictive drugs, environmental agents, the gut microbiome, other drugs besides those of addiction, and likely many other factors as well.

EPIGENETICS AND HUMAN DISEASE

A major interest in epigenetics stems from the role of epigenetic changes in the etiology, progression, and diagnosis of human diseases. Cancer has long been associated with epigenetic alterations and DNA methylation, chromatin modifications, and RNA-dependent regulation have all been shown to affect the incidence and severity of cancer (Chapter 35). Many immune disorders such as systemic lupus erythematosus and rheumatoid arthritis, as well as multiple sclerosis, have been associated with epigenetic aberrations (Chapter 36); epigenetic processes have also been linked to brain disorders (Chapter 37). In the latter case, the Rett syndrome, Alzheimer disease and Huntington disease have been associated in at least some way with epigenetic alterations. Even schizophrenia and depression may have an epigenetic basis in their expression. However, system metabolic disorders may also be related to epigenetic aberrations. For example, obesity, gestational diabetes, and hypertension can influence the fetal chromatin and lead to an increased incidence in adult disease later in life (Chapter 38). Since genomic imprinting is based on epigenetic mechanisms, it may come as no surprise that defects in imprinting can lead to a number of human diseases (Chapter 39). The Prader-Willi syndrome, Angelman syndrome, Silver Russell syndrome, and transient neonatal diabetes mellitus, are all due to imprinting disorders that are based on epigenetic defects. Therefore, the number of diseases impacted by epigenetic processes is large and advances in the treatment of these disorders will likely depend in part on breakthroughs in epigenetic therapy.

EPIGENETIC THERAPY

Although there are many epigenetic therapies that are in use and on the horizon, histone-modifying drugs have probably received the most attention in the clinics. Chief among these are the histone deacetylase (HDAC) inhibitors. Vorinostat (Zolinza), for example, has been approved by the US Food and Drug Administration for use in the treatment of patients with cutaneous T-cell lymphoma (Chapter 40). Many different HDAC inhibitors have been developed, and it is likely that significant improvements will occur for HDAC inhibitors as well as many other drugs that can normalize aberrations in not only histone modifications, but also DNA methylation and perhaps some of the many other epigenetic processes that have been discovered. On the horizon is an increasing interest in combinatorial approaches using epigenetic-modifying agents with other forms of therapy. Moreover, as described in Chapter 41, the combined use of epigenetic-modifying drugs such as DNA methylation and histone methyltransferase, as well as histone acetylase inhibitors, may have significant potential.

THE FUTURE OF EPIGENETICS

For many decades, the field of epigenetics has grown exponentially in part because its many effects on the epigenome has led to a vast number of biological manifestations, involving both natural as well as disease processes. However, a significant impediment to progress in this field has been an outgrowth of the same phenomenon, that is, the vast effects epigenetics has on the genome has made locus-specific studies of epigenetics processes somewhat limited. However, recent developments may overcome this persistent impediment. Exciting new advances in epigenetic editing may contribute significantly to the long sought-after targeting of epigenetic modifications. The development of the clustered regularly interspaced short palindromic repeats (CRISPR) and CRISPR-associated protein (Cas) (CRISPR/Cas) system now enables epigenetics researchers to target loci of interest and to determine the modifiers of interest in a locus-specific manner as described in Chapter 42. This capacity for locus-specific epigenetic editing may spark a new revolution in epigenetic research and finally allow us to treat many diseases that are manifested by epigenetic aberrations with a significantly reduced risk to the patient, due to the more targeted approach it manifests.

CONCLUSION

Increased understanding of the underlying chemistry and enzymatic machinery of epigenetics has served to drive the field of epigenetics well beyond original expectations, and advances continue at a rapid pace. This area of research has also significantly expanded horizontally in that additional epigenetic processes such as noncoding RNA, prion changes, and Polycomb mechanisms have now been established. It is likely that even more epigenetic processes will be discovered in the not too distant future. A major driving force in epigenetics has been the outstanding development of new technology that has not only served to stimulate new discoveries, but has also expanded the field by allowing for novel discoveries possible through the use of these new tools. The development of new model organisms for understanding epigenetic processes has also greatly stimulated the field of epigenetics. We now know that epigenetics is not only intricately associated with metabolism, but also functions in stem cell behavior, tissue regeneration, the transfer of information through generations, neurological memory processes, and even the aging of organisms among may other biological processes. Epigenetics has also played key roles in evolution and has served as a molecular driver of mutations. Moreover, the changing environment is currently reshaping the evolution of many organisms through epigenetic processes. Epidemiological factors such as diet, environmental exposure, the gut microbiome as well as epigenetic-modifying drugs are also influencing our daily lives through changes in epigenetics. Diseases that have been associated with changes in epigenetic processes range from schizophrenia to cancer, and the list of these diseases continues to undergo rapid expansion. Fortunately, the field of epigenetic therapy is also developing rapidly and the hope is that the future will see many novel treatments for the numerous diseases that are derived from heritable as well as environmentally induced epigenetic aberrations. Advances in locus-specific targeting of epigenetics through epigenetic editing and CRISPR/Cas offers but one example, albeit an important, exciting and perhaps revolutionizing example of changes in the future directions of epigenetic research.

References

[1] Waddington CH. The epigenotype. Endeavour 1942;1942(1):18–20.
[2] Li Y, Saldanha SN, Tollefsbol TO. Impact of epigenetic dietary compounds on transgenerational prevention of human diseases. AAPS J 2014;16:27–36.
[3] Tollefsbol TO, editor. Transgenerational epigenetics: evidence and debate. Second Edition Academic Press; 2019.
[4] Brink RA, Styles ED, Axtell JD. Paramutation: directed genetic change. Paramutation occurs in somatic cells and heritably alters the functional state of a locus. Science 1968;159:161–70.
[5] Cavalli G, Paro R. Epigenetic inheritance of active chromatin after removal of the main transactivator. Science 1999;286:955–8.
[6] Grewal SI, Klar AJ. Chromosomal inheritance of epigenetic states in fission yeast during mitosis and meiosis. Cell 1996;86:95–101.
[7] Ghai M, Kader F. A review on epigenetic inheritance of experiences in humans Biochem Genet 2021. Available from: https://doi.org/10.1007/s10528-021-10155-7. Online ahead of print. PMID. Available from: 34792705.
[8] Lai SR, Phipps SM, Liu L, Andrews LG, Tollefsbol TO. Epigenetic control of telomerase and modes of telomere maintenance in aging and abnormal systems. Front Biosci 2005;10:1779–96.
[9] Fuks F, Burgers WA, Brehm A, Hughes-Davies L, Kouzarides T. DNA methyltransferase Dnmt1 associates with histone deacetylase activity. Nat Genet 2000;24:88–91.

SECTION II

Molecular Mechanisms of Epigenetics

2. Mechanisms of DNA methylation and demethylation during mammalian development
3. Molecular mechanisms of histone modifications
4. The epigenetics of non-coding RNA
5. Prions and prion-like phenomena in epigenetic inheritance
6. Higher-order chromatin organization in diseases, from chromosomal position effect to phenotype variegation
7. Polycomb mechanisms and epigenetic control of gene activity

CHAPTER

2

Mechanisms of DNA Methylation and Demethylation During Mammalian Development

Zhengzhou Ying[1,2] and Taiping Chen[1]

[1]Department of Epigenetics and Molecular Carcinogenesis, The University of Texas MD Anderson Cancer Center, Houston, TX, United States [2]The Ministry of Education Key Laboratory of Laboratory Medical Diagnostics, College of Laboratory Medicine, Chongqing Medical University, Chongqing, P. R. China

OUTLINE

Introduction	11	Conclusions	20
DNA Methylation	12	Acknowledgments	21
Maintenance DNA Methylation	12	Glossary	21
De novo DNA Methylation	15	Abbreviations	22
DNA Demethylation	17	References	22
Passive DNA Demethylation	17		
Active DNA Demethylation	18		

INTRODUCTION

DNA methylation, the covalent addition of a methyl group to the fifth position of cytosine (5-methylcytosine, 5mC), is the most common form of DNA modification, which plays important roles in the regulation of chromatin structure and gene expression. DNA methylation is present in various organisms across all kingdoms of eukaryotes. In mammals, DNA methylation mostly occurs in the context of CpG dinucleotides, with 60%–80% of all CpGs in the genome being methylated, although non-CpG (i.e., CpA, CpT, or CpC) methylation is abundant in specific tissues and cell types, including embryonic stem cells (ESCs), oocytes, and brain tissue [1,2]. DNA methylation plays diverse roles in mammalian development, such as gene expression, genomic imprinting, X-chromosome inactivation, and transposon silencing, and in various diseases, including cancer and developmental disorders [3].

5mC is not randomly distributed in the genome. In general, repetitive DNA sequences, including transposable elements and centromeric and pericentric satellite DNA, are heavily methylated. In contrast, CpG islands (CGIs, 1–2 kilobases of CpG-rich regions) present in gene promoters are usually depleted of DNA methylation, with some exceptions. For example, CGIs on the inactive X chromosome in female cells are hypermethylated, and CGIs in imprinting control regions (ICRs) exhibit allele-specific methylation. On the other hand, gene bodies are often highly methylated. Unlike promoter methylation, which correlates with gene silencing, gene body methylation is usually associated with transcriptional activity [3].

DNA methylation patterns and levels are determined by the opposing actions of the methylation and demethylation machineries. The methylation machinery includes DNA methyltransferases (DNMTs), which catalyze the transfer of a methyl group from the methyl donor S-adenosyl-L-methionine (AdoMet or SAM) to the C-5 position of cytosines. There are two families of DNMTs in mammals; DNMT1 is mainly responsible for maintaining DNA methylation patterns through DNA replication, whereas DNMT3A and DNMT3B function primarily as de novo methyltransferases that establish DNA methylation patterns [4]. In rodents, there is an additional de novo methyltransferase, DNMT3C, which is produced by a duplicated gene of *Dnmt3b* and specifically methylates evolutionarily young transposons during spermatogenesis [5]. From a chemical perspective, DNA methylation is considered a stable modification. However, global demethylation occurs in preimplantation embryos and primordial germ cells (PGCs), and locus-specific demethylation takes place during cellular differentiation. DNA demethylation can be achieved by replication-dependent "passive" dilution of 5mC and replication-independent "active" processes [6,7]. Great progress has been made in understanding the mechanisms of demethylation since the discovery that the TET family of proteins—TET1, TET2, and TET3—function as 5mC dioxygenases that convert 5mC, sequentially, to 5-hydroxymethylcytosine (5hmC), 5-formylcytosine (5fC), and 5-carboxylcytosine (5caC) [8–11], which serve as intermediates for DNA demethylation [6,7].

DNA METHYLATION

In 1975, Holliday, Pugh, and Riggs proposed that DNA methylation could be important for cellular memory by serving as a heritable epigenetic mark through cell division [12,13]. Based on the complementarity of CpG/CpG dyads, they reasoned that methylated CpG sites could be replicated semiconservatively during DNA replication. The theory would predict the existence of at least two DNA methylation activities: de novo methylation puts methyl marks at unmodified CpG/CpG dyads to establish DNA methylation patterns, and maintenance methylation converts newly formed hemimethylated CpG sites during DNA replication into fully methylated state to maintain the patterns (Fig. 2.1). The hypothesis was subsequently validated, to a large extent, by the identification of DNA methyltransferases with distinct expression patterns, biochemical properties, and biological functions [4].

Maintenance DNA Methylation

Once DNA methylation patterns are established during embryogenesis, they are maintained in somatic cells in a cell type-specific manner. During each round of DNA replication, DNA becomes hemimethylated, as only CpGs on the parental strand remain methylated while CpGs on the newly replicated daughter strand are unmethylated. To reestablish the symmetry of CpG methylation and keep the specificity, the maintenance DNA methyltransferase activity recognizes hemimethylated CpGs and methylates the corresponding CpGs on the daughter strand. Biochemical, cellular, and genetic evidence suggests that DNMT1 is the major maintenance methyltransferase [4]. In addition, a multidomain protein, UHRF1 (ubiquitin-like with PHD and RING finger domains 1), functions as an essential accessory factor that directs DNMT1 to the right place [14,15].

DNMT1

Mouse *Dnmt1*, the first mammalian DNA methyltransferase gene identified [16], has several transcription start sites and produces two major protein products. Transcripts initiated within a somatic cell-specific exon encode the full-length DNMT1 protein that is expressed in somatic cell types, whereas transcripts initiated within an oocyte-specific exon utilize a downstream AUG as the translation initiation codon, resulting in the DNMT1o isoform that lacks the N-terminal 118 amino acids of full-length DNMT1 (Fig. 2.2A). Both isoforms are equally functional in maintaining DNA methylation, although DNMT1o is more stable than DNMT1 [17].

While all DNMTs contain highly conserved DNA methyltransferase motifs in their C-terminal catalytic domains, the DNMT1 and DNMT3 families show little sequence similarity in their N-terminal regulatory domains (Fig. 2.2A). There are several unique domains in the N-terminal region of DNMT1, which likely contribute to functional specificity. A region at the very N terminus mediates the interaction between DNMT1 and DNA methyltransferase associated protein 1 (DMAP1), a protein implicated in histone acetylation and ATM signaling. The DMAP1-interaction domain is absent in the more stable DNMT1o isoform [17], suggesting that this domain or the interaction between DNMT1 and DMAP1 may be involved in regulating DNMT1 stability. The DMAP1-interaction domain is followed by a proliferating cell nuclear antigen (PCNA) binding domain (PBD), which is required for the interaction with the DNA replication machinery, and a nuclear localization signal (NLS). DNMT1 also contains a motif originally named as the replication foci-targeting sequence (RFTS). Subsequent evidence indicates that RFTS contains a ubiquitin-interacting motif (UIM) that recognizes ubiquitinated lysines in the N-terminal tail of histone H3, which likely plays an important role in targeting DNMT1 to replication foci [18]. Structural data suggest that the RFTS domain also plays an autoinhibitory role in the regulation of DNMT1 activity by

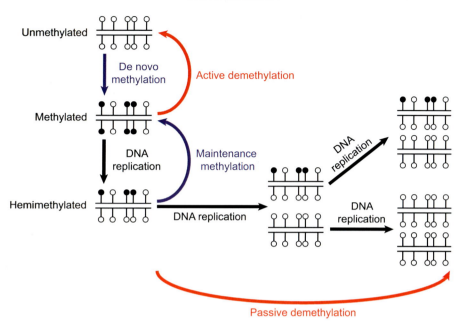

FIGURE 2.1 Overview of mechanisms of DNA methylation and demethylation. During early development, de novo methylation adds methyl groups to specific CpG sites on both DNA strands (resulting in symmetric CpG methylation) and establishes DNA methylation patterns. After each round of DNA replication, the methylated CpG sites become hemimethylated, as the newly replicated daughter DNA strand is unmethylated. Maintenance methylation recognizes hemimethylated CpG sites and "copies" the DNA methylation pattern of the parental strand onto the daughter strand. Failure in maintenance methylation results in replication-dependent loss of DNA methylation, a process known as passive demethylation. DNA demethylation can also be achieved by enzyme-mediated removal of the methyl group or replacement of methylated cytosines (or their derivatives) with unmodified cytosines, a process known as active demethylation, which is independent of DNA replication. Open and filled circles indicate unmethylated and methylated CpG dinucleotides, respectively.

binding to the catalytic domain and blocks the catalytic center [19]. Additionally, DNMT1 contains a CXXC domain, a cysteine-rich motif that binds unmethylated CpGs, and a pair of bromo-adjacent homology (BAH) domains, BAH1 and BAH2. While the precise function of the BAH domains remains unknown, they are required for DNMT1 to be associated with the replication foci during S phase [20].

Biochemical assays reveal that DNMT1 preferentially methylates hemimethylated CpG dinucleotides, although it is capable of methylating unmethylated substrates as well [21]. *Dnmt1* is constitutively expressed in proliferating cells. During S phase, DNMT1 is increased and associates with the replication foci, suggesting that DNMT1 function is coupled with DNA replication [22]. Genetic inactivation of *Dnmt1* in mouse ESCs results in global loss of DNA methylation, but does not affect de novo methylation of integrated provirus DNA [23,24]. These results suggest that DNMT1 functions primarily as a maintenance methyltransferase. Recent studies suggest that DNMT1 is also involved in de novo methylation in specific genomic regions (e.g., retrotransposons) and such activity is probably inhibited in other contexts [25,26].

DNA methylation is dispensable in undifferentiated mouse ESCs, as *Dnmt1* knockout (KO), *Dnmt3a/3b* double KO (DKO), and *Dnmt1/3a/3b* triple KO (TKO) ESCs show no defects in viability and proliferation, but these cells die when induced to differentiate [24,27–30]. Unlike mouse ESCs, human ESCs require *DNMT1*, but not *DNMT3A* and *DNMT3B*, for survival [31]. Mouse and human ESCs represent different pluripotent states, with mouse ESCs being more similar to naïve pluripotent cells and human ESCs being in primed pluripotent state, which may explain the sensitivity of human ESCs to hypomethylation. DNMT1 is also required for the survival of mouse embryonic fibroblasts (MEFs) and the human colorectal cancer cell HCT116 [32,33]. These findings suggest crucial roles for DNMT1 in cellular differentiation and in the viability of differentiated cells. Indeed, complete inactivation of *Dnmt1* results in gastrulation defects and embryonic lethality [24]. In human, missense *DNMT1* mutations in the RFTS of the N-terminal regulatory domain, which likely results in hypomorphic alleles, have been identified in two related neurodegenerative disorders, hereditary sensory, and autonomic neuropathy with dementia and hearing loss type IE (HSAN IE) and autosomal dominant cerebellar ataxia, deafness, and narcolepsy (ADCA-DN) [34].

UHRF1

UHRF1 is a multidomain protein. Genetic studies demonstrate that UHRF1 is essential for maintaining DNA methylation. Similar to the phenotype of *Dnmt1*

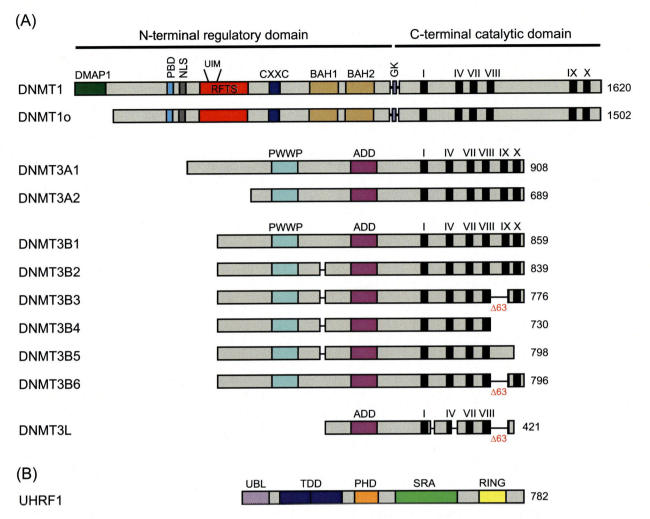

FIGURE 2.2 Schematic diagram of major proteins involved in DNA methylation. (A) Major DNMT isoforms in mouse. The DNMT1 and DNMT3 families of proteins share conserved catalytic motifs (I–X) in their C-terminal catalytic domains. DNMT3L lacks catalytic activity because some essential motifs are missing or mutated. The N-terminal regions of DNMT1 and DNMT3 proteins have little sequence similarity, with distinct domains that contribute to their functional specificities. *DMAP1*, DNA methyltransferase associated protein 1; *PBD*, PCNA-binding domain; *NLS*, nuclear localization signal; *RFTS*, replication foci targeting sequence; *UIM*, ubiquitin interacting motif; *CXXC*, cysteine-rich motif; *BAH*, bromo-adjacent homology; *GK*, glycine/lysine repeats; *I–X*, DNA methyltransferase conserved catalytic motifs; *PWWP*, proline-tryptophan-tryptophan-proline; *ADD*, ATRX-DNMT3-DNMT3L. Note that DNMT3B3 and DNMT3B6 have the exact 63-residue deletion corresponding to that of DNMT3L. (B) Mouse UHRF1. *UBL*, ubiquitin-like domain; *TTD*, tandem tudor domain; *PHD*, plant homeodomain; *SRA*, SET and RING associated; *RING*, really interesting new gene.

deficiency, disruption of *Uhrf1* leads to embryonic lethality and global DNA hypomethylation [14,15,35]. Cellular studies also suggest functional interactions between DNMT1 and UHRF1. Both proteins are enriched at DNA replication foci during the S phase, but DNMT1 fails to localize to these foci in *Uhrf1* KO cells [14,15]. These findings indicate that UHRF1 is a critical regulator of maintenance methylation by directing DNMT1 to hemimethylated CpG sites.

UHRF1 has five conserved domains: a ubiquitin-like (UBL) domain, a tandem tudor domain (TTD), a plant homeodomain (PHD), a Really Interesting New Gene (RING) domain, and a SET and RING associated (SRA) domain (Fig. 2.2B). The SRA domain preferentially binds hemimethylated DNA and likely plays an important role in loading DNMT1 onto newly synthesized DNA substrates [14,15]. The TTD and PHD act in combination to recognize the histone H3 trimethylated at lysine 9 (H3K9me3) mark, and additionally, the PHD binds to histone H3 with unmethylated arginine 2 (H3R2me0) [36,37]. The RING finger domain has E3 ubiquitin ligase activity, and the UBL domain facilitates ubiquitination by stabilizing the E2-E3-chromatin complex [38,39]. UHRF1 has been shown to catalyze monoubiquitination of histone H3 at several lysine residues (K14/K18/K23) in the N-terminal tail, creating DNMT1-binding sites [18,40,41]. Thus, UHRF1 directs DNMT1 to hemimethylated CpG sites through multivalent interactions with chromatin to ensure faithful inheritance of DNA methylation patterns during and/or after DNA replication.

De novo DNA Methylation

After fertilization, both the maternal and paternal genomes undergo global DNA demethylation during preimplantation development. As a result, DNA methylation marks inherited from gametes are largely erased by the blastocyst stage, with the exception of those in ICRs of imprinted loci and some retrotransposons. After implantation, a wave of de novo methylation occurs in the epiblast to establish the initial pattern of DNA methylation. DNA methylation shows further changes during cellular differentiation, and the patterns are then stably maintained in a lineage-specific manner. Similar epigenetic reprograming events also take place during gametogenesis, including global demethylation in PGCs. Subsequently, a new round of de novo DNA methylation takes place during gametogenesis, resulting in different methylation patterns in male and female gametes [3]. De novo methylation is mediated by DNMT3A and DNMT3B (in rodents, DNMT3C also participates in de novo methylation during spermatogenesis). Another member of the DNMT3 family, DNMT3-like (DNMT3L), has no catalytic activity but is an important regulator of de novo methylation (Fig. 2.2A).

DNMT3A and DNMT3B

The *Dnmt3a* and *Dnmt3b* genes were initially identified by searching an expressed sequence tag (EST) database using bacterial type II cytosine-5 methyltransferase sequences as queries. Their protein products have a similar structural organization, including a C-terminal catalytic domain that contains sequence motifs characteristic of all DNA methyltransferases and an N-terminal regulatory domain that is distinct from that of DNMT1 (Fig. 2.2A) [42]. The N-terminal regions of DNMT3A/3B contain two conserved domains implicated in chromatin binding. The proline-tryptophan-tryptophan-proline (PWWP) domain, an ~150-residue domain with a conserved PWWP motif, is necessary for targeting the enzymes to heterochromatin, intergenic and genic regions and mediates binding to H3K36me2 and H3K36me3 marks [43–46]. The ADD (ATRX-DNMT3-DNMT3L) domain interacts with the N-terminal tail of histone H3 with unmodified lysine 4 (H3K4me0) [47,48]. Structural studies revealed that the ADD domain of DNMT3A interacts with its catalytic domain and blocks its DNA-binding affinity, resulting in autoinhibition. Unmodified histone H3 tail (but not H3 tail with H3K4me3) can disrupt the interaction between the ADD and the catalytic domains, leading to DNMT3A activation [49]. DNMT3A/3B also have distinct unstructured regions at the very N termini. Recent work suggests that the unstructured region of DNMT3A is functionally important [50] and interacts with H2AK119ub-modified nucleosomes [51]. These findings indicate that DNMT3A and DNMT3B act as both "writers" and "readers" of epigenetic marks and that their activities and specificities are regulated by specific histone modifications.

The conclusion that DNMT3A and DNMT3B function primarily as de novo methyltransferases is based on several lines of evidence. First, *Dnmt3a* and *Dnmt3b* expression correlates with de novo methylation during development. Specifically, *Dnmt3a* and *Dnmt3b* are highly expressed in early embryos (as well as ESCs) and developing germ cells, and their expression is significantly downregulated in somatic cells and when ESCs are differentiated [42]. In ESCs and germ cells, *Dnmt3a* transcription is mostly driven by an internal promoter, resulting in a shorter isoform known as DNMT3A2, which lacks the N-terminal 219 (mouse) or 223 (human) amino acids of full-length DNMT3A1 (Fig. 2.2A). DNMT3A1 is ubiquitously expressed at low levels in most somatic tissues [52]. *Dnmt3b* produces multiple alternatively spliced isoforms (>30 reported to date), many of which are catalytically inactive (Fig. 2.2A). In ESCs and early embryos, the full-length isoform DNMT3B1, a catalytically active form, is the predominant product, and other isoforms, including the inactive DNMT3B6, are also expressed. In most somatic cells, *Dnmt3b* expression is low, usually with both active (e.g., DNMT3B2) and inactive (e.g., DNMT3B3) isoforms [42,52]. Recent work suggests that the inactive DNMT3B3 and DNMT3B6 isoforms, which, like DNMT3L, lack 63 amino acids in their catalytic domains (Fig. 2.2A), positively regulate de novo methylation by DNMT3A/3B with a preference for DNMT3B [50].

Second, biochemical studies indicate that DNMT3A and DNMT3B behave as de novo methyltransferases. In particular, recombinant DNMT3A/3B proteins show no preference for hemimethylated DNA over unmethylated DNA in vitro, unlike DNMT1, which preferentially methylates hemimethylated DNA [21,42]. Furthermore, in vitro and in vivo evidence indicates that DNMT3A and DNMT3B methylate cytosines at non-CpG sites, such as CpA and CpT, albeit with lower efficiency compared to CpG methylation [53]. Non-CpG methylation, which cannot be maintained during DNA replication, is mediated by de novo methyltransferases.

Third, genetic studies provide definitive evidence for the involvement of DNMT3A and DNMT3B in de novo DNA methylation. Targeted disruption of both *Dnmt3a* and *Dnmt3b* blocks de novo methylation in mouse ESCs and early embryos but has no effect on the maintenance of methylation at imprinted loci [28,54]. While DNMT3A and DNMT3B methylate many genomic loci redundantly, they also have preferred and distinct targets. For example, DNMT3A is more efficient than DNMT3B in methylating major satellite repeats at pericentric heterochromatin, whereas DNMT3B preferentially methylates minor satellite repeats in centromeric regions [29]. Characterization of *Dnmt3a* and *Dnmt3b* KO mice suggests that these enzymes play distinct roles in developmental processes. *Dnmt3a* KO mice develop to term and show no overt defects at birth but die in several weeks of age. In contrast, disruption of *Dnmt3b* leads to embryonic lethality at ~E12.5. *Dnmt3a/3b* DKO embryos exhibit more severe defects and die earlier (before E11.5) than *Dnmt3b* KO embryos [28]. DNA methylation analysis of E9.5 embryos indicate that DNMT3B is largely responsible for methylation of germline-specific genes, pluripotency genes, and many developmental genes and DNMT3A and DNMT3B redundantly methylate some specific genes [55]. Conditional gene KO studies indicate that DNMT3A, but not DNMT3B, is essential for de novo methylation during gametogenesis, including the establishment of DNA methylation imprints [56]. Indeed, DNMT3A, but not DNMT3B, is abundantly expressed in fully grown oocytes [57].

Consistent with their important roles in developmental processes in mice, *DNMT3A* and *DNMT3B* mutations are associated with human diseases [34]. Somatic *DNMT3A* mutations occur frequently in acute myeloid leukemia (AML) and other hematologic malignancies [58]. Many *DNMT3A* mutations have been identified, with the majority (>50%) of cases affecting Arg882 (R882) in the catalytic domain. Although almost all reported *DNMT3A* mutations in leukemia occur in only one allele, there is evidence that DNMT3A R882 mutant proteins have dominant-negative effects by interacting with wild-type DNMT3A to form functionally deficient complexes [59,60]. Germline *DNMT3A* mutations are associated with Tatton-Brown-Rahman syndrome, an autosomal dominant disorder characterized by overgrowth, macrocephaly, dysmorphic facial features and intellectual disability [61]. A recent study identified gain-of-function *DNMT3A* mutations in microcephalic dwarfism [62]. Hypomorphic *DNMT3B* mutations account for 50%−60% of cases with ICF (Immunodeficiency, Centromeric instability, and Facial anomalies) syndrome [28,63,64], a rare recessive autosomal disorder characterized by antibody deficiency, hypomethylation of centromeric DNA and chromosomal abnormalities, and facial dysmorphism [34]. Several other ICF genes—*ZBTB24* (zinc-finger- and BTB domain-containing 24), *CDCA7* (cell division cycle associated 7), and *HELLS* (helicase, lymphoid-specific)—have also been identified [65,66]. HELLS (also known as LSH), a DNA helicase involved in chromatin remodeling, is required for DNA methylation, likely by modulating the accessibility of the DNA methylation machinery to chromatin [67]. Recent studies indicate that ZBTB24 positively regulates *CDCA7* transcription and CDCA7 recruits HELLS to specific genomic regions, including centromeric satellite repeats, to facilitate DNA methylation [68−72].

DNMT3L

Dnmt3L was originally identified by sequence database analysis [73,74]. The protein product contains an ADD domain, but not a PWWP domain, in the N-terminal region. Its C-terminal region shares sequence homology with the catalytic domains of DNMT3A/3B, but lacks some sequence motifs essential for catalytic activity (Fig. 2.2A).

Although DNMT3L has no methyltransferase activity, biochemical and genetic evidence suggests that it is an important regulator of de novo methylation. DNMT3L interacts with DNMT3A and DNMT3B and significantly enhances their catalytic activity [75]. DNMT3L also prevents DNMT3A2 from degradation [76]. Crystallography studies reveal that DNMT3L and DNMT3A interact through their C-terminal domains, with two monomers each forming a tetramer with two active sites [77]. Biochemical and structural data also indicate that the ADD domain of DNMT3L binds the N-terminal tail of histone H3 with H3K4me0, suggesting that DNMT3L plays a role in determining the specificity of de novo methylation [47]. The expression pattern of DNMT3L during development is similar to that of DNMT3A and DNMT3B, with high expression in developing germ cells, early embryos, and ESCs and little expression in somatic tissues [75]. *Dnmt3L* KO mice are viable and grossly normal, suggesting that zygotic DNMT3L is not essential for development. However, both male and female KO mice fail to reproduce [75,78]. *Dnmt3L* KO males are sterile, as they are unable to produce mature sperm (azoospermia). The spermatogenesis defect is due to loss of DNA methylation in germ cells, which results in reactivation of retrotransposons, inducing genomic instability, meiotic catastrophe, and ultimately apoptosis [79]. On the other hand, *Dnmt3L* KO females are able to conceive, but embryos die by midgestation due to failure to establish DNA methylation imprints in oocytes [75,78]. Indeed, the phenotype of *Dnmt3L* KO mice is almost identical to that of mice with conditional *Dnmt3a* deletion in germ cells [56]. Based on genetic evidence and expression patterns, DNMT3L primarily functions as an accessory factor of DNMT3A2 in developing germ cells, and the inactive DNMT3B3 and DNMT3B6 isoforms likely play similar roles as accessory factors of DNMT3B in facilitating de novo methylation during embryonic development and in somatic tissues [50].

DNA DEMETHYLATION

DNA methylation is generally stable in somatic cells. However, two waves of global demethylation take place during development [80]. The first wave occurs in preimplantation embryos, which results in genome-wide erasure of DNA methylation marks inherited from both the paternal and maternal gametes, with the exception of methylation in special regions, such as ICRs and some retrotransposons (e.g., intracisternal-A particles). Reprogramming of the parental genomes in early embryos is important for the establishment of totipotency. The paternal and maternal genomes undergo demethylation through distinct mechanisms. Shortly after fertilization, the male pronucleus rapidly loses its 5mC signal before the onset of DNA replication, suggesting DNA replication-independent "active" demethylation. In contrast, the maternal genome is gradually demethylated during cleavage divisions, mainly through "passive" dilution of 5mC due to insufficient maintenance methylation (Fig. 2.1).

The second wave of global demethylation occurs in PGCs, which is a critical reprogramming event that generates an epigenome for the development of germ cells. DNA methylation imprints at ICRs, which are protected in preimplantation embryos, are erased in PGCs. In mice, PGCs are specified around E7.25 in the epiblast of the developing embryo and then migrate along the embryonic−extraembryonic interface, eventually arriving at the genital ridge. Demethylation takes place during PGC migration. Genome-wide methylation profiling of PGCs at different time points indicates that demethylation occurs in two phases. The first phase, beginning at ∼E8.5, results in genome-wide hypomethylation affecting almost all genomic regions, mainly through passive demethylation. The second phase, occurring from E9.5 to E13.5, affects special loci, including ICRs, germline-specific genes, and CGIs on the X chromosome, and involves both active and passive mechanisms [80].

Passive DNA Demethylation

DNA methylation patterns are maintained during DNA replication by the maintenance methylation machinery that "copies" the methylation pattern from the parental strand onto the daughter strand. Defects in maintenance methylation can therefore lead to replication-dependent dilution of 5mC, known as passive DNA demethylation. Passive demethylation is important for the erasure of DNA methylation marks in both preimplantation embryos and PGCs.

It is generally believed that deficiency in DNMT1 function is the major mechanism for passive demethylation of the maternal genome during early embryogenesis. In preimplantation embryos, the oocyte-specific DNMT1o variant, transmitted from oocytes, is the predominant DNMT1 isoform, although zygotic DNMT1 is also expressed at a low level [57]. In contrast to nuclear localization of DNMT1 in somatic cells, DNMT1o is localized in the cytoplasm of oocytes and preimplantation embryos [81]. The presence or absence of the N-terminal 118 amino acids in different DNMT1 isoforms does not account for the difference in subcellular localization, as DNMT1o has several functional NLSs and, when ectopically expressed in somatic cells, localizes in the nuclei [81]. The mechanism of DNMT1o cytoplasmic retention in oocytes and preimplantation embryos is still unknown. It is worth mentioning that demethylation of the paternal genome, which is initiated by an active process involving TET-mediated 5mC oxidation, also involves passive demethylation, as the oxidative products (i.e., 5hmC, 5fC, and 5caC) become diluted in a replication-dependent manner [82,83].

Despite extensive demethylation during preimplantation development, parental allele-specific DNA methylation in ICRs is faithfully maintained. Genetic studies indicate that DNMT1, but not DNMT3A and DNMT3B, is responsible for maintaining methylation imprints [57,84]. The fact that imprinted loci are protected from this wave of demethylation suggests that DNMT1 is not exclusively retained in the cytoplasm during preimplantation development. While it is poorly understood what confers the specificity of DNMT1, such that methylation is maintained at imprinted genes, but not at other sequences, genetic and epigenetic features may distinguish imprinted loci from other regions. In addition to *Dnmt1*, several genes have been identified that are involved in the maintenance of methylation imprints. For example, the *Krüppel* associated box (KRAB)-containing zinc-finger proteins ZFP57 and ZFP445 play important roles in the maintenance of genomic imprints in mouse and human [85−87]. There is evidence that ZFP57 and ZNF445 specifically binds the methylated allele of ICRs and interacts with KRAB-associated protein 1 (KAP1), a scaffold protein for multiple proteins including DNMTs and UHRF1 [88,89]. Thus, it is likely that nuclear DNMT1 protein in preimplantation embryos, albeit at low levels, is concentrated in special genomic regions, such as ICRs.

In PGCs, passive demethylation is mainly responsible for the first phase of demethylation that erases methylation marks genome-wide. There is evidence that migrating PGCs, which are rapidly cycling, have little de novo or maintenance DNA methylation potential. Specifically, immunofluorescence analysis shows that DNMT3A, DNMT3B, and UHRF1 are repressed in PGCs [90]. The second phase of demethylation in PGCs, which affects special genomic regions including

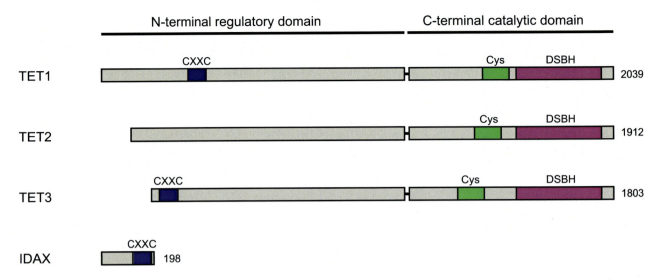

FIGURE 2.3 Schematic diagram of TET family of proteins. Mouse TET proteins (TET1, TET2, and TET3) and IDAX. All three TET proteins share a similar C-terminal catalytic domain with two characteristic features, a cysteine-rich region (Cys) and a double-stranded beta-helix (DSBH) fold. Their N-terminal regions are less conserved, with TET1 and TET3 containing a CXXC zinc finger domain.

ICRs, involves TET-mediated 5mC oxidation, followed by DNA replication-coupled dilution of oxidized derivatives [6,80].

Recent studies suggest that UHRF1 is a major point of regulation that leads to passive demethylation in some biological contexts. For example, in 2C-like totipotent ESCs, UHRF1 is degraded to induce genome-wide hypomethylation so that shortened telomeres are extended, and in cancer cells, overexpression of the protein arginine methyltransferase PRMT6 contributes to global hypomethylation by disrupting the association of UHRF1 with chromatin [91,92].

Active DNA Demethylation

In contrast to DNA replication-dependent passive dilution of 5mC, active DNA demethylation refers to enzyme-mediated removal or modifications of 5mC with the regeneration of unmodified cytosine, which is generally independent of DNA replication. It has been known for a long time that active demethylation takes place during development. However, the mechanisms involved had been elusive until the discovery that TET proteins function as 5mC dioxygenases that convert 5mC into 5hmC, 5fC, and 5caC [8,9]. Studies in recent years have demonstrated that these oxidized derivatives serve as intermediates for demethylation [6,80].

TET Proteins

TET1, the founding member of the TET family, was named for its involvement in the ten-eleven translocation [t(10;11)(q22;q23)] in rare cases of leukemia, which results in the fusion of the *TET1* gene, located on 10q22, with the mixed-lineage leukemia (*MLL*) gene, located on 11q23 [93]. *TET2* and *TET3* were identified on the basis of sequence homology. The TET family belongs to the 2-oxoglutarate (2OG)- and Fe(II)-dependent dioxygenase (2OGFeDO) superfamily. All three TET proteins are capable of oxidizing 5mC into 5hmC, 5fC, and 5caC. They share a similar structural organization, with a C-terminal catalytic domain and an N-terminal regulatory domain (Fig. 2.3). The catalytic domain consists of a cysteine-rich region and a double-stranded β-helix (DSBH) domain characteristic of the 2OGFeDO superfamily. Structural studies reveal that the catalytic core region specifically recognizes CpG dinucleotides, showing preference for 5mC as a substrate and also binding to 5mC derivatives for subsequent oxidation [94,95]. The N-terminal regions of TET proteins are less conserved. TET1 and TET3 contain a CXXC zinc finger domain. The CXXC domain of TET1 binds unmodified, 5mC-modified, and 5hmC-modified CpG-rich DNA, whereas the CXXC domain of TET3 binds unmodified cytosine regardless of whether it is in the CpG context [96,97]. TET2 has lost its CXXC domain during evolution, as a result of a chromosomal inversion event that has converted a portion of the ancestral *TET2* gene into a separate gene, *IDAX*, which encodes a protein containing the original CXXC domain (Fig. 2.3). IDAX binds unmethylated CpG sequences via its CXXC domain, localizes at CGIs, and physically interacts with TET2, suggesting that it may play a role in recruiting TET2 to its genomic targets [98].

TET-dependent Demethylation Pathways

Since direct removal of the methyl group from 5mC by cleaving the C-C bond is a thermodynamically unfavorable reaction, DNA demethylation is most likely achieved by a step-wise process involving intermediates. The finding that TET proteins convert 5mC to 5hmC immediately raised the possibility that 5mC oxidation could be involved in DNA demethylation [8,9]. This notion was strengthened by subsequent work showing that 5hmC can be further oxidized by TET proteins to produce 5fC and 5caC, which can be recognized and excised from DNA by thymine glycosylase (TDG) [10,11]. Two general mechanisms of DNA demethylation involving TET-mediated 5mC oxidation have been proposed (Fig. 2.4). First, as mentioned above, 5hmC, 5fC, and 5caC can be passively diluted due to the lack of maintenance during DNA replication. Indeed, in vitro assays indicate that DNMT1 is much less efficient in utilizing hemihydroxymethylated CpG substrates compared to hemimethylated CpG substrates [99]. Second, excision of 5fC and 5caC by TDG generates abasic sites, which can be repaired by the base excision repair (BER) pathway to restore unmodified cytosines. In support of this mechanism, coexpression of TDG and TET proteins leads to depletion of 5fC and 5caC in HEK293 cells, whereas TDG deficiency results in substantial increases of 5fC and 5caC in mouse ESCs, as well as defects in demethylation and embryonic development in mice [11,100–102]. While there are reports suggesting the possibility of direct removal of 5hmC, 5fC, and 5caC by C-C bond cleavage [103–106], the enzymes (DNA dehydroxymethylases, deformylases, and decarboxylases), if present, remain to be identified. In addition to TET-related pathways, deamination of 5mC to thymine (T) by AID/APOBEC enzymes, followed by replacement of the

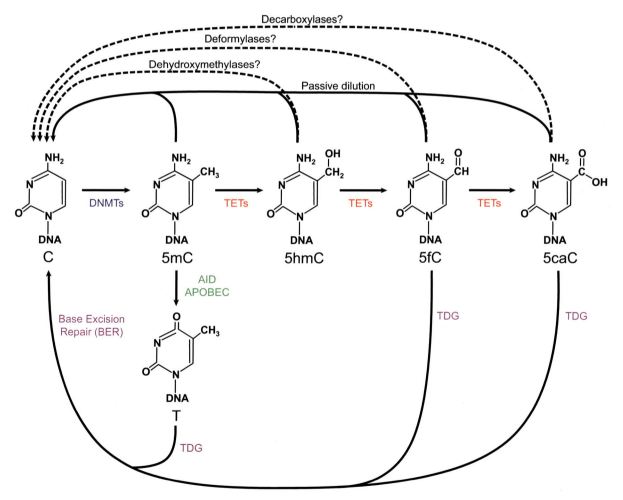

FIGURE 2.4 DNA demethylation pathways involving TET proteins. TET proteins initiate DNA demethylation by oxidizing 5-methylcytosine (5mC) to generate 5-hydroxymethylcytosine (5hmC), which can be further oxidized to 5-formylcytosine (5fC) and 5-carboxylcytosine (5caC). 5mC can also be deaminated by the AID/APOBEC family of deaminases to generate thymine (T). The 5mC derivatives 5fC, 5caC, and T can then be removed by TDG and replaced by unmodified cytosine (C) by the Base Excision Repair (BER) pathway. Because 5hmC, 5fC, and 5caC are poor substrates for maintenance methylation by DNMT1, their removal can also be achieved by DNA replication-dependent passive dilution. While it is possible that 5hmC, 5fC, and 5caC could be directly converted to unmodified cytosine by dehydroxymethylases, deformylases, and decarboxylases, respectively. The existence and identities of these enzymes remain to be determined.

T:G mismatch via BER, has also been implicated in DNA demethylation. There is evidence that this pathway is involved in the establishment of DNA methylation patterns in hematopoietic stem/progenitor cells and germinal center B cells [107,108].

TET Proteins in Development

Although all three TET proteins have similar activity, they show different expression patterns. TET1 and TET2 are expressed in migrating PGCs, whereas TET3 is highly expressed in oocytes and zygotes (fertilized eggs) [109–113]. Consistent with the expression patterns, genetic studies have demonstrated that TET1, TET2, and TET3 have distinct roles in DNA demethylation during development.

In the zygote, the rapid loss of 5mC signal in the male pronucleus is accompanied with dramatic increases in 5hmC, 5fC, and 5caC, suggesting TET-mediated 5mC oxidation [83,109,110]. Methylome analysis reveals that 5mC oxidation also occurs in the female pronucleus, albeit to a lesser extent [114,115]. Since zygotic gene expression does not occur until the 2-cell stage, the TET protein(s) responsible for 5mC oxidation in the zygote must come from the oocyte. Indeed, depletion of maternal TET3, the predominant TET protein in oocytes, blocks 5mC oxidation in both the male and female pronuclei of zygote and inhibits demethylation during preimplantation development [109,111,114]. These results suggest that TET3-mediated 5mC oxidation is a key step that initiates demethylation in preimplantation embryos. Subsequent conversion of 5mC oxidized derivatives to unmodified cytosines likely involves both active mechanisms and passive dilution [81,82,115,116]. While it is not fully understood why the efficiencies of 5mC oxidation differ significantly in the male and female pronuclei, even though they are exposed to the same environment, distinct epigenetic states may be responsible. For example, H3K9me2 is abundant in the female pronucleus but absent in the male pronucleus. EHMT2 (also called G9a) and SETDB1, lysine methyltransferases responsible for H3K9me2 and H3K9me3, respectively, and PGC7 (also known as STELLA and DPPA3), an H3K9me2-binding protein, have been shown to protect the maternal pronucleus from TET3-mediated oxidation of 5mC [116,117]. However, a recent study showed that maternal EHMT2 regulates preimplantation development independent of DNA methylation, calling into question the role for H3K9me2 in protecting DNA methylation in zygotes [118].

TET-mediated 5mC oxidation is also involved in demethylation in PGCs. TET1 and TET2 are expressed between E9.25 and E11.5, whereas TET3 is undetectable [112,113]. The timing of TET1 and TET2 expression, which correlates with increases in 5hmC [113,119], corresponds to the second phase of demethylation that affects special regions, such as ICRs and germline-specific genes. Indeed, genetic ablation of Tet1 or both Tet1 and Tet2 has no effect on global demethylation in PGCs (first phase) but leads to defective demethylation and altered expression of specific genes including imprinted genes and meiosis genes [112,120,121]. A subsequent study also showed that TET1 plays a role in protecting certain regions from remethylation in PGCs [122].

In addition to preimplantation embryos and PGCs, TET proteins are dynamically expressed during development and are present in multiple adult tissues. For example, TET1 and TET2, but not TET3, are highly expressed in mouse ESCs, and differentiation of ESCs leads to downregulation of TET1 and TET2 and upregulation of TET3 [6]. Recent work indicates that TET proteins compete with DNMT3A/3B in ESCs for proper methylation at many CpG sites [123–127]. In addition to ESCs, neuronal cells have high levels of 5hmC [128]. Given that postmitotic neurons do not replicate, it is believed that demethylation in response to stimuli is achieved through active mechanisms. TET enzymes may play important roles in the regulation of DNA methylation and gene expression in neuronal cells. Genetic studies in KO mice suggest that TET proteins have both distinct and redundant functions. Tet1-null mice are viable (although some are slightly smaller at birth) but show reduced fertility and impaired adult neurogenesis [112,129,130]. Tet2-null mice are viable and fertile, however, adult animals develop hematologic phenotypes, characterized by inhibition of hematopoietic stem cell differentiation and myeloid malignancies [131–134]. This phenotype is consistent with the observation that TET2 is frequently mutated in various myeloid malignancies [135]. Zygotic deletion of Tet3 leads to neonatal lethality, and maternal deletion impairs demethylation in preimplantation embryos and high frequencies of embryonic defects, with only ~20% surviving to term [111]. Tet1/2 DKO mice show variable phenotypes: most of them exhibit embryonic defects and die perinatally, and small numbers are viable [121]. Tet1/2/3 TKO mice exhibit gastrulation defects, partly due to increased Lefty–Nodal signaling [136]. Consistent with the developmental phenotypes, Tet1/2/3 TKO ESCs show differentiation defects [123,137]. It is worth mentioning that, most likely, not all developmental phenotypes of TET KO mice are attributable to defects in DNA demethylation. TET proteins, which are large molecules, also have noncatalytic functions [138–140]. Indeed, TET proteins have been shown to regulate histone modifications and form chromatin regulatory complexes [141–143].

CONCLUSIONS

It is generally believed that DNA methylation functions in concert with other epigenetic mechanisms, such as posttranslational modifications of histones, to stably maintain gene silencing and chromatin structure. Genetic manipulations of key components of the DNA methylation and demethylation machineries, including DNMTs and TETs, in mice have greatly facilitated our understanding of the roles of DNA methylation and demethylation in developmental processes in mammals. One area that is still poorly understood is how DNA methylation patterns are specified during development and how DNA methylation is dysregulated in human diseases. Studies in recent years have identified mutations in major components of DNA methylation "writers" (DNMTs), "erasers" (e.g., TETs), and "readers" (e.g., MeCP2) in various diseases [34]. While progress has been made, the mechanisms by which many of these mutations contribute to pathogenesis remain poorly understood. The discovery of TET proteins as 5mC dioxygenases was a breakthrough in the field, and ever since, great progress has been made in understanding the mechanisms involved in active DNA demethylation during development. While the retention of DNMT1 in the cytoplasm is believed to be responsible for passive demethylation of the maternal genome during early embryogenesis, the molecular mechanism involved in keeping DNMT1 in the cytoplasm remains to be elucidated. Moreover, there is evidence that 5mC oxidized derivatives such as 5hmC, in addition to being intermediates of DNA demethylation, act as epigenetic marks for chromatin regulation. Indeed, numerous putative "readers" of these marks have been reported. However, their relevance in binding these marks in vivo and their functional significance in biological processes are largely unknown. In the coming years, we expect to see advances in these areas.

Acknowledgments

Work in our laboratory is supported by grants from the National Institutes of Health (NIH, 1R01DK106418−01 and 1R01AI12140301A1 to T.C.). Z.Y. is supported by a fellowship from the Sam and Freda Davis Fund.

Glossary

Base excision repair (BER) — DNA repair pathway in which a DNA base is removed by a glycosylase enzyme and ultimately replaced by a new base.

Blastocyst — Early stage embryo that has undergone the first cell lineage specification, which results in two primary cell types: the inner cell mass (ICM) and the trophectoderm, which will differentiate into embryonic and extra-embryonic tissues, respectively.

Centromeric chromatin — Type of heterochromatin that is a constituent in the formation of centromeres, which is the region of the chromosome that links sister chromatids to the spindle during cell division. It is flanked by pericentric heterochromatin.

CpG islands (CGIs) — CG-rich genomic regions, ranging from several hundred bp to 2 kb, often located at gene promoters.

CpG sequences — Cytosine followed by a guanine in the 5′−3′ directions on the same strand of DNA linked by phosphate. It is the major substrate for DNA methylation in mammals.

Cytosine methylation — The covalent addition of a methyl group to the fifth position of cytosine (5-methylcytosine, 5mC), mostly occurring within a CpG dinucleotide sequence, is the most common form of DNA modification, which plays important roles in the regulation of chromatin structure and gene expression.

Cytosine hydroxymethylation — DNA modification generated by oxidation of methylcytosine. This modification is mediated by the TET family proteins and is considered both stable and an intermediate in DNA demethylation.

DNA demethylation — Replacement of the methylcytosine with an unmodified cytosine, either through direct or indirect mechanisms.

Dioxygenases — Enzymes that catalyze the addition of the two oxygen atoms from its molecular free form (O_2) to one or two organic substrates.

Embryonic stem cells — Stem cells derived from the undifferentiated inner cell mass from a blastocyst.

Epiblast — Outer layer arising from the inner cell mass in the blastocyst, capable of differentiating into the three primary germ layers (ectoderm, endoderm, and mesoderm) and into the extraembryonic mesoderm of the visceral yolk sac, the allantois, and the amnion.

Genomic imprinting — Epigenetic phenomenon by which certain genes are expressed in a parental origin-specific fashion.

Germline-specific genes — Genes of unique gametogenic function that are tightly regulated outside gametes or the early embryo by DNA methylation. They are specifically demethylated during PGC specification.

Hemimethylated sites — CpGs that are methylated only on one of the two DNA strands.

Heterochromatin — Tightly packed chromatin that is associated with structural functions and gene silencing.

Imprinted locus (imprinted gene) — A gene that shows parental origin-specific monoallelic expression (i.e., Some imprinted genes are expressed only from the paternal allele, and others are expressed only from maternal allele).

Imprinting control regions (ICRs) — Genomic region that acts, in a methylation-sensitive manner, to determine whether imprinted genes are expressed or not according to parent of origin.

Imprints — Allele-specific DNA methylation marks at ICRs.

Inactive X chromosome — In females, one of the two X chromosomes is inactivated to prevent overexpression of X-linked gene products, in order to equalize gene dose between females (with two X chromosomes) and males (with one X chromosome). X chromosome inactivation is achieved through formation of

Major satellite	Tandem repeating DNA sequences of 234 bp that are primarily located in the pericentromeric regions of the mouse genome.
Minor satellite	Tandem repeating AT-rich DNA sequences of 120 bp that are located in the centromeric regions of the mouse genome.
Pericentric heterochromatin	Major repressive heterochromatin domain in the nucleus, which is mainly composed of major satellite repeats in mouse cells.
Pluripotency	The potential ability of a cell to differentiate into all the three embryonic germ layers (endoderm, mesoderm and ectoderm) and to give rise to any embryonic tissue and any adult cell type. For example, ESCs are pluripotent.
Preimplantation embryos	Early embryos from 1-cell to blastocyst stage before implantation.
PGCs	Cells that give rise to both mature germ cells (oocytes and spermatozoa).
Retroelements (retrotransposons)	Endogenous transposable elements that move along the genome via the transcription of an RNA intermediate.
Totipotency	Potential ability of a cell to give rise to all cell types in embryonic and extra-embryonic tissues. For example, zygote is a totipotent cell.
Zygote (fertilized egg)	Cell formed by the fertilization of the oocyte with a sperm cell.

Abbreviations

2OGFeDO	2-oxoglutarate (2OG)- and Fe(II)-dependent dioxygenase
5fC	5-Formylcytosine
5caC	5-Carboxylcytosine
5hmC	5-Hydroxymethylcytosine
5mC	5-Methylcytosine
ADCA-DN	Autosomal dominant cerebellar ataxia, deafness and narcolepsy
ADD	ATRX-DNMT3-DNMT3L
AdoMet (or SAM)	S-adenosyl-L-methionine
AID	Activation-induced cytidine deaminase
AML	Acute myeloid leukemia
APOBEC	Apolipoprotein B mRNA editing enzyme, catalytic polypeptide-like
BAH	Bromo-adjacent homology domain
BER	Base excision repair
CDCA7	Cell division cycle associated 7
CGI	CpG island
CpA	5′-Cytosine nucleotide-phosphate-adenine nucleotide-3′
CpC	5′-Cytosine nucleotide-phosphate-cytosine nucleotide-3′
CpG	5′-Cytosine nucleotide-phosphate-guanine nucleotide-3′
CpT	5′-Cytosine nucleotide-phosphate-thymine nucleotide-3′
CXXC	Cysteine-rich motif
DKO	Double knockout
DMAP1	DNA methyltransferase associated protein 1
DNA	Deoxyribonucleic acid
DNMT	DNA methyltransferase
DSBH	Double-stranded β-helix
ESC	Embryonic stem cell
EST	Expressed sequence tag
H2AK119ub	Monoubiquitination of histone H2A at lysine 119
H3K4me0	Unmodified histone H3 at lysine 4
H3K4me3	Trimethylation of histone H3 at lysine 4
H3K9me2	Dimethylation of histone H3 at lysine 9
H3K9me3	Trimethylation of histone H3 at lysine 9
H3R2me0	Unmodified histone H3 at arginine 2
H3K36me2	Dimethylation of histone H3 at lysine 36
H3K36me3	Trimethylation of histone H3 at lysine 36
HELLS (LSH)	Helicase, lymphoid-specific (lymphocyte-specific helicase)
HSAN IE	Hereditary sensory and autonomic neuropathy with dementia and hearing loss type IE
ICF	Immunodeficiency, centromeric instability, and facial anomalies
ICR	Imprinting control region
KAP1	KRAB-associated protein 1
KO	Knockout
KRAB	*Krüppel* associated box
MeCP2	Methyl-CpG-binding protein 2
MEF	Mouse embryonic fibroblast
MLL	Mixed-lineage leukemia
NLS	Nuclear localization signal
PBD	PCNA-binding domain
PCNA	Proliferating cell nuclear antigen
PGC	Primordial germ cell
PHD	Plant homeodomain
PWWP	Proline-tryptophan-tryptophan-proline
RFTS	Replication foci-targeting sequence
RING	Really Interesting New Gene
SRA	SET and RING associated
TET	Ten-eleven translocation
TDG	Thymine DNA glycosylase
TKO	Triple knockout
TTD	Tandem tudor domain
UBL	Ubiquitin-like domain
UHRF1	Ubiquitin-like with PHD and RING finger domains 1
UIM	Ubiquitin-interacting motif
ZBTB24	Zinc-finger- and BTB domain-containing 24

References

[1] Patil V, Ward RL, Hesson LB. The evidence for functional non-CpG methylation in mammalian cells. Epigenetics. 2014;9(6):823–8.

[2] He Y, Ecker JR. Non-CG methylation in the human genome. Annu Rev Genomics Hum Genet 2015;16:55–77.

[3] Greenberg MVC, Bourc'his D. The diverse roles of DNA methylation in mammalian development and disease. Nat Rev Mol Cell Biol 2019;20(10):590–607.

[4] Chen Z, Zhang Y. Role of mammalian DNA methyltransferases in development. Annu Rev Biochem 2020;89:135–58.

[5] Barau J, Teissandier A, Zamudio N, Roy S, Nalesso V, Herault Y, et al. The DNA methyltransferase DNMT3C protects male germ cells from transposon activity. Science. 2016;354(6314):909–12.

[6] Zhao H, Chen T. Tet family of 5-methylcytosine dioxygenases in mammalian development. J Hum Genet 2013;58(7):421–7.

[7] Wu H, Zhang Y. Reversing DNA methylation: mechanisms, genomics, and biological functions. Cell. 2014;156(1–2):45–68.

REFERENCES

[8] Tahiliani M, Koh KP, Shen Y, Pastor WA, Bandukwala H, Brudno Y, et al. Conversion of 5-methylcytosine to 5-hydroxymethylcytosine in mammalian DNA by MLL partner TET1. Science. 2009;324 (5929):930–5.

[9] Ito S, D'Alessio AC, Taranova OV, Hong K, Sowers LC, Zhang Y. Role of Tet proteins in 5mC to 5hmC conversion, ES-cell self-renewal and inner cell mass specification. Nature. 2010;466 (7310):1129–33.

[10] Ito S, Shen L, Dai Q, Wu SC, Collins LB, Swenberg JA, et al. Tet proteins can convert 5-methylcytosine to 5-formylcytosine and 5-carboxylcytosine. Science. 2011;333(6047):1300–3.

[11] He YF, Li BZ, Li Z, Liu P, Wang Y, Tang Q, et al. Tet-mediated formation of 5-carboxylcytosine and its excision by TDG in mammalian DNA. Science. 2011;333(6047):1303–7.

[12] Holliday R, Pugh JE. DNA modification mechanisms and gene activity during development. Science. 1975;187(4173):226–32.

[13] Riggs AD. X inactivation, differentiation, and DNA methylation. Cytogenet Cell Genet 1975;14(1):9–25.

[14] Bostick M, Kim JK, Esteve PO, Clark A, Pradhan S, Jacobsen SE. UHRF1 plays a role in maintaining DNA methylation in mammalian cells. Science. 2007;317(5845):1760–4.

[15] Sharif J, Muto M, Takebayashi S, Suetake I, Iwamatsu A, Endo TA, et al. The SRA protein Np95 mediates epigenetic inheritance by recruiting Dnmt1 to methylated DNA. Nature. 2007;450(7171):908–12.

[16] Bestor T, Laudano A, Mattaliano R, Ingram V. Cloning and sequencing of a cDNA encoding DNA methyltransferase of mouse cells. The carboxyl-terminal domain of the mammalian enzymes is related to bacterial restriction methyltransferases. J Mol Biol 1988;203(4):971–83.

[17] Ding F, Chaillet JR. In vivo stabilization of the Dnmt1 (cytosine-5)- methyltransferase protein. Proc Natl Acad Sci U S A 2002;99 (23):14861–6.

[18] Qin W, Wolf P, Liu N, Link S, Smets M, La Mastra F, et al. DNA methylation requires a DNMT1 ubiquitin interacting motif (UIM) and histone ubiquitination. Cell Res 2015;25(8):911–29.

[19] Takeshita K, Suetake I, Yamashita E, Suga M, Narita H, Nakagawa A, et al. Structural insight into maintenance methylation by mouse DNA methyltransferase 1 (Dnmt1). Proc Natl Acad Sci U S A 2011;108(22):9055–9.

[20] Yarychkivska O, Shahabuddin Z, Comfort N, Boulard M, Bestor TH. BAH domains and a histone-like motif in DNA methyltransferase 1 (DNMT1) regulate de novo and maintenance methylation in vivo. J Biol Chem 2018;293(50):19466–75.

[21] Pradhan S, Bacolla A, Wells RD, Roberts RJ. Recombinant human DNA (cytosine-5) methyltransferase. I. Expression, purification, and comparison of de novo and maintenance methylation. J Biol Chem 1999;274(46):33002–10.

[22] Leonhardt H, Page AW, Weier HU, Bestor TH. A targeting sequence directs DNA methyltransferase to sites of DNA replication in mammalian nuclei. Cell. 1992;71(5):865–73.

[23] Li E, Bestor TH, Jaenisch R. Targeted mutation of the DNA methyltransferase gene results in embryonic lethality. Cell. 1992;69(6):915–26.

[24] Lei H, Oh SP, Okano M, Juttermann R, Goss KA, Jaenisch R, et al. De novo DNA cytosine methyltransferase activities in mouse embryonic stem cells. Development. 1996;122(10):3195–205.

[25] Li Y, Zhang Z, Chen J, Liu W, Lai W, Liu B, et al. Stella safeguards the oocyte methylome by preventing de novo methylation mediated by DNMT1. Nature. 2018;564(7734):136–40.

[26] Haggerty C, Kretzmer H, Riemenschneider C, Kumar AS, Mattei AL, Bailly N, et al. Dnmt1 has de novo activity targeted to transposable elements. Nat Struct Mol Biol 2021;28(7):594–603.

[27] Tucker KL, Talbot D, Lee MA, Leonhardt H, Jaenisch R. Complementation of methylation deficiency in embryonic stem cells by a DNA methyltransferase minigene. Proc Natl Acad Sci U S A 1996;93(23):12920–5.

[28] Okano M, Bell DW, Haber DA, Li E. DNA methyltransferases Dnmt3a and Dnmt3b are essential for de novo methylation and mammalian development. Cell. 1999;99(3):247–57.

[29] Chen T, Ueda Y, Dodge JE, Wang Z, Li E. Establishment and maintenance of genomic methylation patterns in mouse embryonic stem cells by Dnmt3a and Dnmt3b. Mol Cell Biol 2003;23 (16):5594–605.

[30] Tsumura A, Hayakawa T, Kumaki Y, Takebayashi S, Sakaue M, Matsuoka C, et al. Maintenance of self-renewal ability of mouse embryonic stem cells in the absence of DNA methyltransferases Dnmt1, Dnmt3a and Dnmt3b. Genes Cell 2006;11(7):805–14.

[31] Liao J, Karnik R, Gu H, Ziller MJ, Clement K, Tsankov AM, et al. Targeted disruption of DNMT1, DNMT3A and DNMT3B in human embryonic stem cells. Nat Genet 2015;47(5):469–78.

[32] Jackson-Grusby L, Beard C, Possemato R, Tudor M, Fambrough D, Csankovszki G, et al. Loss of genomic methylation causes p53-dependent apoptosis and epigenetic deregulation. Nat Genet 2001;27(1):31–9.

[33] Chen T, Hevi S, Gay F, Tsujimoto N, He T, Zhang B, et al. Complete inactivation of DNMT1 leads to mitotic catastrophe in human cancer cells. Nat Genet 2007;39(3):391–6.

[34] Hamidi T, Singh AK, Chen T. Genetic alterations of DNA methylation machinery in human diseases. Epigenomics. 2015;7 (2):247–65.

[35] Muto M, Kanari Y, Kubo E, Takabe T, Kurihara T, Fujimori A, et al. Targeted disruption of Np95 gene renders murine embryonic stem cells hypersensitive to DNA damaging agents and DNA replication blocks. J Biol Chem 2002;277(37):34549–55.

[36] Rajakumara E, Wang Z, Ma H, Hu L, Chen H, Lin Y, et al. PHD finger recognition of unmodified histone H3R2 links UHRF1 to regulation of euchromatic gene expression. Mol Cell 2011;43(2):275–84.

[37] Rothbart SB, Dickson BM, Ong MS, Krajewski K, Houliston S, Kireev DB, et al. Multivalent histone engagement by the linked tandem Tudor and PHD domains of UHRF1 is required for the epigenetic inheritance of DNA methylation. Genes Dev 2013;27 (11):1288–98.

[38] Foster BM, Stolz P, Mulholland CB, Montoya A, Kramer H, Bultmann S, et al. Critical role of the UBL domain in stimulating the E3 ubiquitin ligase activity of UHRF1 toward chromatin. Mol Cell 2018;72(4):739–52 e9.

[39] DaRosa PA, Harrison JS, Zelter A, Davis TN, Brzovic P, Kuhlman B, et al. A bifunctional role for the UHRF1 UBL domain in the control of hemi-methylated DNA-dependent histone ubiquitylation. Mol Cell 2018;72(4):753–65 e6.

[40] Nishiyama A, Yamaguchi L, Sharif J, Johmura Y, Kawamura T, Nakanishi K, et al. Uhrf1-dependent H3K23 ubiquitylation couples maintenance DNA methylation and replication. Nature. 2013;502(7470):249–53.

[41] Ishiyama S, Nishiyama A, Saeki Y, Moritsugu K, Morimoto D, Yamaguchi L, et al. Structure of the Dnmt1 reader module complexed with a unique two-mono-ubiquitin mark on histone H3 reveals the basis for DNA methylation maintenance. Mol Cell 2017;68(2):350–60 e7.

[42] Okano M, Xie S, Li E. Cloning and characterization of a family of novel mammalian DNA (cytosine-5) methyltransferases. Nat Genet 1998;19(3):219–20.

[43] Chen T, Tsujimoto N, Li E. The PWWP domain of Dnmt3a and Dnmt3b is required for directing DNA methylation to the major satellite repeats at pericentric heterochromatin. Mol Cell Biol 2004;24(20):9048–58.

[44] Dhayalan A, Rajavelu A, Rathert P, Tamas R, Jurkowska RZ, Ragozin S, et al. The Dnmt3a PWWP domain reads histone

3 lysine 36 trimethylation and guides DNA methylation. J Biol Chem 2010;285(34):26114−20.
[45] Baubec T, Colombo DF, Wirbelauer C, Schmidt J, Burger L, Krebs AR, et al. Genomic profiling of DNA methyltransferases reveals a role for DNMT3B in genic methylation. Nature. 2015;520(7546):243−7.
[46] Weinberg DN, Papillon-Cavanagh S, Chen H, Yue Y, Chen X, Rajagopalan KN, et al. The histone mark H3K36me2 recruits DNMT3A and shapes the intergenic DNA methylation landscape. Nature. 2019;573(7773):281−6.
[47] Ooi SK, Qiu C, Bernstein E, Li K, Jia D, Yang Z, et al. DNMT3L connects unmethylated lysine 4 of histone H3 to de novo methylation of DNA. Nature. 2007;448(7154):714−17.
[48] Otani J, Nankumo T, Arita K, Inamoto S, Ariyoshi M, Shirakawa M. Structural basis for recognition of H3K4 methylation status by the DNA methyltransferase 3A ATRX-DNMT3-DNMT3L domain. EMBO Rep 2009;10(11):1235−41.
[49] Guo X, Wang L, Li J, Ding Z, Xiao J, Yin X, et al. Structural insight into autoinhibition and histone H3-induced activation of DNMT3A. Nature. 2015;517(7536):640−4.
[50] Zeng Y, Ren R, Kaur G, Hardikar S, Ying Z, Babcock L, et al. The inactive Dnmt3b3 isoform preferentially enhances Dnmt3b-mediated DNA methylation. Genes Dev 2020;34(21−22):1546−58.
[51] Weinberg DN, Rosenbaum P, Chen X, Barrows D, Horth C, Marunde MR, et al. Two competing mechanisms of DNMT3A recruitment regulate the dynamics of de novo DNA methylation at PRC1-targeted CpG islands. Nat Genet 2021;53(6):794−800.
[52] Chen T, Ueda Y, Xie S, Li E. A novel Dnmt3a isoform produced from an alternative promoter localizes to euchromatin and its expression correlates with active de novo methylation. J Biol Chem 2002;277(41):38746−54.
[53] Chedin F. The DNMT3 family of mammalian de novo DNA methyltransferases. Prog Mol Biol Transl Sci 2011;101:255−85.
[54] Dahlet T, Argueso Lleida A, Al Adhami H, Dumas M, Bender A, Ngondo RP, et al. Genome-wide analysis in the mouse embryo reveals the importance of DNA methylation for transcription integrity. Nat Commun 2020;11(1):3153.
[55] Borgel J, Guibert S, Li Y, Chiba H, Schubeler D, Sasaki H, et al. Targets and dynamics of promoter DNA methylation during early mouse development. Nat Genet 2010;42(12):1093−100.
[56] Kaneda M, Okano M, Hata K, Sado T, Tsujimoto N, Li E, et al. Essential role for de novo DNA methyltransferase Dnmt3a in paternal and maternal imprinting. Nature. 2004;429(6994):900−3.
[57] Hirasawa R, Chiba H, Kaneda M, Tajima S, Li E, Jaenisch R, et al. Maternal and zygotic Dnmt1 are necessary and sufficient for the maintenance of DNA methylation imprints during preimplantation development. Genes Dev 2008;22(12):1607−16.
[58] Yang L, Rau R, Goodell MA. DNMT3A in haematological malignancies. Nat Rev Cancer 2015;15(3):152−65.
[59] Kim SJ, Zhao H, Hardikar S, Singh AK, Goodell MA, Chen T. A DNMT3A mutation common in AML exhibits dominant-negative effects in murine ES cells. Blood. 2013;122(25):4086−9.
[60] Russler-Germain DA, Spencer DH, Young MA, Lamprecht TL, Miller CA, Fulton R, et al. The R882H DNMT3A mutation associated with AML dominantly inhibits wild-type DNMT3A by blocking its ability to form active tetramers. Cancer Cell 2014;25(4):442−54.
[61] Tatton-Brown K, Seal S, Ruark E, Harmer J, Ramsay E, Del Vecchio Duarte S, et al. Mutations in the DNA methyltransferase gene DNMT3A cause an overgrowth syndrome with intellectual disability. Nat Genet 2014;46(4):385−8.
[62] Heyn P, Logan CV, Fluteau A, Challis RC, Auchynnikava T, Martin CA, et al. Gain-of-function DNMT3A mutations cause microcephalic dwarfism and hypermethylation of Polycomb-regulated regions. Nat Genet 2019;51(1):96−105.
[63] Xu GL, Bestor TH, Bourc'his D, Hsieh CL, Tommerup N, Bugge M, et al. Chromosome instability and immunodeficiency syndrome caused by mutations in a DNA methyltransferase gene. Nature. 1999;402(6758):187−91.
[64] Hansen RS, Wijmenga C, Luo P, Stanek AM, Canfield TK, Weemaes CM, et al. The DNMT3B DNA methyltransferase gene is mutated in the ICF immunodeficiency syndrome. Proc Natl Acad Sci U S A 1999;96(25):14412−17.
[65] de Greef JC, Wang J, Balog J, den Dunnen JT, Frants RR, Straasheijm KR, et al. Mutations in ZBTB24 are associated with immunodeficiency, centromeric instability, and facial anomalies syndrome type 2. Am J Hum Genet 2011;88(6):796−804.
[66] Thijssen PE, Ito Y, Grillo G, Wang J, Velasco G, Nitta H, et al. Mutations in CDCA7 and HELLS cause immunodeficiency-centromeric instability-facial anomalies syndrome. Nat Commun 2015;6:7870.
[67] Zhu H, Geiman TM, Xi S, Jiang Q, Schmidtmann A, Chen T, et al. Lsh is involved in de novo methylation of DNA. EMBO J 2006;25(2):335−45.
[68] Wu H, Thijssen PE, de Klerk E, Vonk KK, Wang J, den Hamer B, et al. Converging disease genes in ICF syndrome: ZBTB24 controls expression of CDCA7 in mammals. Hum Mol Genet 2016;25(18):4041−51.
[69] Jenness C, Giunta S, Muller MM, Kimura H, Muir TW, Funabiki H. HELLS and CDCA7 comprise a bipartite nucleosome remodeling complex defective in ICF syndrome. Proc Natl Acad Sci U S A 2018;115(5):E876−85.
[70] Thompson JJ, Kaur R, Sosa CP, Lee JH, Kashiwagi K, Zhou D, et al. ZBTB24 is a transcriptional regulator that coordinates with DNMT3B to control DNA methylation. Nucleic Acids Res 2018;46(19):10034−51.
[71] Ren R, Hardikar S, Horton JR, Lu Y, Zeng Y, Singh AK, et al. Structural basis of specific DNA binding by the transcription factor ZBTB24. Nucleic Acids Res 2019;47(16):8388−98.
[72] Hardikar S, Ying Z, Zeng Y, Zhao H, Liu B, Veland N, et al. The ZBTB24-CDCA7 axis regulates HELLS enrichment at centromeric satellite repeats to facilitate DNA methylation. Protein Cell 2020;11(3):214−18.
[73] Aapola U, Kawasaki K, Scott HS, Ollila J, Vihinen M, Heino M, et al. Isolation and initial characterization of a novel zinc finger gene, DNMT3L, on 21q22.3, related to the cytosine-5-methyltransferase 3 gene family. Genomics. 2000;65(3):293−8.
[74] Aapola U, Lyle R, Krohn K, Antonarakis SE, Peterson P. Isolation and initial characterization of the mouse Dnmt3l gene. Cytogenet Cell Genet 2001;92(1−2):122−6.
[75] Hata K, Okano M, Lei H, Li E. Dnmt3L cooperates with the Dnmt3 family of de novo DNA methyltransferases to establish maternal imprints in mice. Development. 2002;129(8):1983−93.
[76] Veland N, Lu Y, Hardikar S, Gaddis S, Zeng Y, Liu B, et al. DNMT3L facilitates DNA methylation partly by maintaining DNMT3A stability in mouse embryonic stem cells. Nucleic Acids Res 2019;47(1):152−67.
[77] Jia D, Jurkowska RZ, Zhang X, Jeltsch A, Cheng X. Structure of Dnmt3a bound to Dnmt3L suggests a model for de novo DNA methylation. Nature. 2007;449(7159):248−51.
[78] Bourc'his D, Xu GL, Lin CS, Bollman B, Bestor TH. Dnmt3L and the establishment of maternal genomic imprints. Science. 2001;294(5551):2536−9.
[79] Bourc'his D, Bestor TH. Meiotic catastrophe and retrotransposon reactivation in male germ cells lacking Dnmt3L. Nature. 2004;431(7004):96−9.
[80] Zeng Y, Chen T. DNA methylation reprogramming during mammalian development. Genes (Basel) 2019;10(4):257.

REFERENCES

[81] Cardoso MC, Leonhardt H. DNA methyltransferase is actively retained in the cytoplasm during early development. J Cell Biol 1999;147(1):25–32.

[82] Inoue A, Zhang Y. Replication-dependent loss of 5-hydroxymethylcytosine in mouse preimplantation embryos. Science. 2011;334(6053):194.

[83] Inoue A, Shen L, Dai Q, He C, Zhang Y. Generation and replication-dependent dilution of 5fC and 5caC during mouse preimplantation development. Cell Res 2011;21(12):1670–6.

[84] Howell CY, Bestor TH, Ding F, Latham KE, Mertineit C, Trasler JM, et al. Genomic imprinting disrupted by a maternal effect mutation in the Dnmt1 gene. Cell. 2001;104(6):829–38.

[85] Li X, Ito M, Zhou F, Youngson N, Zuo X, Leder P, et al. A maternal-zygotic effect gene, Zfp57, maintains both maternal and paternal imprints. Dev Cell 2008;15(4):547–57.

[86] Mackay DJ, Callaway JL, Marks SM, White HE, Acerini CL, Boonen SE, et al. Hypomethylation of multiple imprinted loci in individuals with transient neonatal diabetes is associated with mutations in ZFP57. Nat Genet 2008;40(8):949–51.

[87] Takahashi N, Coluccio A, Thorball CW, Planet E, Shi H, Offner S, et al. ZNF445 is a primary regulator of genomic imprinting. Genes Dev 2019;33(1–2):49–54.

[88] Quenneville S, Verde G, Corsinotti A, Kapopoulou A, Jakobsson J, Offner S, et al. In embryonic stem cells, ZFP57/KAP1 recognize a methylated hexanucleotide to affect chromatin and DNA methylation of imprinting control regions. Mol Cell 2011;44(3):361–72.

[89] Zuo X, Sheng J, Lau HT, McDonald CM, Andrade M, Cullen DE, et al. Zinc finger protein ZFP57 requires its co-factor to recruit DNA methyltransferases and maintains DNA methylation imprint in embryonic stem cells via its transcriptional repression domain. J Biol Chem 2012;287(3):2107–18.

[90] Kagiwada S, Kurimoto K, Hirota T, Yamaji M, Saitou M. Replication-coupled passive DNA demethylation for the erasure of genome imprints in mice. EMBO J 2013;32 (3):340–53.

[91] Dan J, Rousseau P, Hardikar S, Veland N, Wong J, Autexier C, et al. Zscan4 inhibits maintenance DNA methylation to facilitate telomere elongation in mouse embryonic stem cells. Cell Rep 2017;20(8):1936–49.

[92] Veland N, Hardikar S, Zhong Y, Gayatri S, Dan J, Strahl BD, et al. The arginine methyltransferase PRMT6 regulates DNA methylation and contributes to global DNA hypomethylation in cancer. Cell Rep 2017;21(12):3390–7.

[93] Lorsbach RB, Moore J, Mathew S, Raimondi SC, Mukatira ST, Downing JR. TET1, a member of a novel protein family, is fused to MLL in acute myeloid leukemia containing the t (10;11)(q22;q23). Leukemia 2003;17(3):637–41.

[94] Hu L, Li Z, Cheng J, Rao Q, Gong W, Liu M, et al. Crystal structure of TET2-DNA complex: insight into TET-mediated 5mC oxidation. Cell. 2013;155(7):1545–55.

[95] Hu L, Lu J, Cheng J, Rao Q, Li Z, Hou H, et al. Structural insight into substrate preference for TET-mediated oxidation. Nature. 2015;527(7576):118–22.

[96] Xu Y, Wu F, Tan L, Kong L, Xiong L, Deng J, et al. Genome-wide regulation of 5hmC, 5mC, and gene expression by Tet1 hydroxylase in mouse embryonic stem cells. Mol Cell 2011;42 (4):451–64.

[97] Xu Y, Xu C, Kato A, Tempel W, Abreu JG, Bian C, et al. Tet3 CXXC domain and dioxygenase activity cooperatively regulate key genes for Xenopus eye and neural development. Cell. 2012;151(6):1200–13.

[98] Ko M, An J, Bandukwala HS, Chavez L, Aijo T, Pastor WA, et al. Modulation of TET2 expression and 5-methylcytosine oxidation by the CXXC domain protein IDAX. Nature. 2013;497(7447):122–6.

[99] Hashimoto H, Liu Y, Upadhyay AK, Chang Y, Howerton SB, Vertino PM, et al. Recognition and potential mechanisms for replication and erasure of cytosine hydroxymethylation. Nucleic Acids Res 2012;40(11):4841–9.

[100] Cortellino S, Xu J, Sannai M, Moore R, Caretti E, Cigliano A, et al. Thymine DNA glycosylase is essential for active DNA demethylation by linked deamination-base excision repair. Cell. 2011;146(1):67–79.

[101] Shen L, Wu H, Diep D, Yamaguchi S, D'Alessio AC, Fung HL, et al. Genome-wide analysis reveals TET- and TDG-dependent 5-methylcytosine oxidation dynamics. Cell. 2013;153 (3):692–706.

[102] Song CX, Szulwach KE, Dai Q, Fu Y, Mao SQ, Lin L, et al. Genome-wide profiling of 5-formylcytosine reveals its roles in epigenetic priming. Cell. 2013;153(3):678–91.

[103] Chen CC, Wang KY, Shen CK. The mammalian de novo DNA methyltransferases DNMT3A and DNMT3B are also DNA 5-hydroxymethylcytosine dehydroxymethylases. J Biol Chem 2012;287(40):33116–21.

[104] Schiesser S, Hackner B, Pfaffeneder T, Muller M, Hagemeier C, Truss M, et al. Mechanism and stem-cell activity of 5-carboxycytosine decarboxylation determined by isotope tracing. Angew Chem Int Ed 2012;51(26):6516–20.

[105] Iwan K, Rahimoff R, Kirchner A, Spada F, Schroder AS, Kosmatchev O, et al. 5-Formylcytosine to cytosine conversion by C-C bond cleavage in vivo. Nat Chem Biol 2018;14(1):72–8.

[106] Schon A, Kaminska E, Schelter F, Ponkkonen E, Korytiakova E, Schiffers S, et al. Analysis of an active deformylation mechanism of 5-formyl-deoxycytidine (fdC) in stem cells. Angew Chem Int Ed 2020;59(14):5591–4.

[107] Kunimoto H, McKenney AS, Meydan C, Shank K, Nazir A, Rapaport F, et al. Aid is a key regulator of myeloid/erythroid differentiation and DNA methylation in hematopoietic stem/progenitor cells. Blood. 2017;129(13):1779–90.

[108] Catala-Moll F, Ferrete-Bonastre AG, Li T, Weichenhan D, Lutsik P, Ciudad L, et al. Activation-induced deaminase is critical for the establishment of DNA methylation patterns prior to the germinal center reaction. Nucleic Acids Res 2021;49(9):5057–73.

[109] Wossidlo M, Nakamura T, Lepikhov K, Marques CJ, Zakhartchenko V, Boiani M, et al. 5-Hydroxymethylcytosine in the mammalian zygote is linked with epigenetic reprogramming. Nat Commun 2011;2:241.

[110] Iqbal K, Jin SG, Pfeifer GP, Szabo PE. Reprogramming of the paternal genome upon fertilization involves genome-wide oxidation of 5-methylcytosine. Proc Natl Acad Sci U S A 2011;108 (9):3642–7.

[111] Gu TP, Guo F, Yang H, Wu HP, Xu GF, Liu W, et al. The role of Tet3 DNA dioxygenase in epigenetic reprogramming by oocytes. Nature. 2011;477(7366):606–10.

[112] Yamaguchi S, Hong K, Liu R, Shen L, Inoue A, Diep D, et al. Tet1 controls meiosis by regulating meiotic gene expression. Nature. 2012;492(7429):443–7.

[113] Hackett JA, Sengupta R, Zylicz JJ, Murakami K, Lee C, Down TA, et al. Germline DNA demethylation dynamics and imprint erasure through 5-hydroxymethylcytosine. Science. 2013;339(6118):448–52.

[114] Guo F, Li X, Liang D, Li T, Zhu P, Guo H, et al. Active and passive demethylation of male and female pronuclear DNA in the mammalian zygote. Cell Stem Cell 2014;15(4):447–58.

[115] Wang L, Zhang J, Duan J, Gao X, Zhu W, Lu X, et al. Programming and inheritance of parental DNA methylomes in mammals. Cell. 2014;157(4):979–91.

[116] Nakamura T, Liu YJ, Nakashima H, Umehara H, Inoue K, Matoba S, et al. PGC7 binds histone H3K9me2 to protect

[117] Zeng TB, Han L, Pierce N, Pfeifer GP, Szabo PE. EHMT2 and SETDB1 protect the maternal pronucleus from 5mC oxidation. Proc Natl Acad Sci U S A 2019;116(22):10834−41.

against conversion of 5mC to 5hmC in early embryos. Nature 2012;486(7403):415−19.

[118] Au Yeung WK, Brind'Amour J, Hatano Y, Yamagata K, Feil R, Lorincz MC, et al. Histone H3K9 methyltransferase G9a in oocytes is essential for preimplantation development but dispensable for CG methylation protection. Cell Rep 2019;27(1):282−93 e4.

[119] Yamaguchi S, Hong K, Liu R, Inoue A, Shen L, Zhang K, et al. Dynamics of 5-methylcytosine and 5-hydroxymethylcytosine during germ cell reprogramming. Cell Res 2013;23(3):329−39.

[120] Yamaguchi S, Shen L, Liu Y, Sendler D, Zhang Y. Role of Tet1 in erasure of genomic imprinting. Nature. 2013;504(7480):460−4.

[121] Dawlaty MM, Breiling A, Le T, Raddatz G, Barrasa MI, Cheng AW, et al. Combined deficiency of Tet1 and Tet2 causes epigenetic abnormalities but is compatible with postnatal development. Dev Cell 2013;24(3):310−23.

[122] Hill PWS, Leitch HG, Requena CE, Sun Z, Amouroux R, Roman-Trufero M, et al. Epigenetic reprogramming enables the transition from primordial germ cell to gonocyte. Nature. 2018;555(7696):392−6.

[123] Verma N, Pan H, Dore LC, Shukla A, Li QV, Pelham-Webb B, et al. TET proteins safeguard bivalent promoters from de novo methylation in human embryonic stem cells. Nat Genet 2018;50(1):83−95.

[124] Gu T, Lin X, Cullen SM, Luo M, Jeong M, Estecio M, et al. DNMT3A and TET1 cooperate to regulate promoter epigenetic landscapes in mouse embryonic stem cells. Genome Biol 2018;19(1):88.

[125] Charlton J, Jung EJ, Mattei AL, Bailly N, Liao J, Martin EJ, et al. TETs compete with DNMT3 activity in pluripotent cells at thousands of methylated somatic enhancers. Nat Genet 2020;52(8):819−27.

[126] Wang Q, Yu G, Ming X, Xia W, Xu X, Zhang Y, et al. Imprecise DNMT1 activity coupled with neighbor-guided correction enables robust yet flexible epigenetic inheritance. Nat Genet 2020;52(8):828−39.

[127] Dixon G, Pan H, Yang D, Rosen BP, Jashari T, Verma N, et al. QSER1 protects DNA methylation valleys from de novo methylation. Science 2021;372(6538):eabd0875.

[128] Kriaucionis S, Heintz N. The nuclear DNA base 5-hydroxymethylcytosine is present in Purkinje neurons and the brain. Science. 2009;324(5929):929−30.

[129] Dawlaty MM, Ganz K, Powell BE, Hu YC, Markoulaki S, Cheng AW, et al. Tet1 is dispensable for maintaining pluripotency and its loss is compatible with embryonic and postnatal development. Cell Stem Cell 2011;9(2):166−75.

[130] Zhang RR, Cui QY, Murai K, Lim YC, Smith ZD, Jin S, et al. Tet1 regulates adult hippocampal neurogenesis and cognition. Cell Stem Cell 2013;13(2):237−45.

[131] Moran-Crusio K, Reavie L, Shih A, Abdel-Wahab O, Ndiaye-Lobry D, Lobry C, et al. Tet2 loss leads to increased hematopoietic stem cell self-renewal and myeloid transformation. Cancer Cell 2011;20(1):11−24.

[132] Ko M, Bandukwala HS, An J, Lamperti ED, Thompson EC, Hastie R, et al. Ten-Eleven-Translocation 2 (TET2) negatively regulates homeostasis and differentiation of hematopoietic stem cells in mice. Proc Natl Acad Sci U S A 2011;108(35):14566−71.

[133] Li Z, Cai X, Cai CL, Wang J, Zhang W, Petersen BE, et al. Deletion of Tet2 in mice leads to dysregulated hematopoietic stem cells and subsequent development of myeloid malignancies. Blood. 2011;118(17):4509−18.

[134] Quivoron C, Couronne L, Della Valle V, Lopez CK, Plo I, Wagner-Ballon O, et al. TET2 inactivation results in pleiotropic hematopoietic abnormalities in mouse and is a recurrent event during human lymphomagenesis. Cancer Cell 2011;20(1):25−38.

[135] Kunimoto H, Nakajima H. TET2: A cornerstone in normal and malignant hematopoiesis. Cancer Sci 2021;112(1):31−40.

[136] Dai HQ, Wang BA, Yang L, Chen JJ, Zhu GC, Sun ML, et al. TET-mediated DNA demethylation controls gastrulation by regulating Lefty-Nodal signalling. Nature. 2016;538(7626):528−32.

[137] Dawlaty MM, Breiling A, Le T, Barrasa MI, Raddatz G, Gao Q, et al. Loss of Tet enzymes compromises proper differentiation of embryonic stem cells. Dev Cell 2014;29(1):102−11.

[138] Wu H, D'Alessio AC, Ito S, Xia K, Wang Z, Cui K, et al. Dual functions of Tet1 in transcriptional regulation in mouse embryonic stem cells. Nature. 2011;473(7347):389−93.

[139] Koh KP, Yabuuchi A, Rao S, Huang Y, Cunniff K, Nardone J, et al. Tet1 and Tet2 regulate 5-hydroxymethylcytosine production and cell lineage specification in mouse embryonic stem cells. Cell Stem Cell 2011;8(2):200−13.

[140] Ito K, Lee J, Chrysanthou S, Zhao Y, Josephs K, Sato H, et al. Non-catalytic roles of Tet2 are essential to regulate hematopoietic stem and progenitor cell homeostasis. Cell Rep 2019;28(10):2480−90 e4.

[141] Chen Q, Chen Y, Bian C, Fujiki R, Yu X. TET2 promotes histone O-GlcNAcylation during gene transcription. Nature. 2013;493(7433):561−4.

[142] Vella P, Scelfo A, Jammula S, Chiacchiera F, Williams K, Cuomo A, et al. Tet proteins connect the O-linked N-acetylglucosamine transferase Ogt to chromatin in embryonic stem cells. Mol Cell 2013;49(4):645−56.

[143] Deplus R, Delatte B, Schwinn MK, Defrance M, Mendez J, Murphy N, et al. TET2 and TET3 regulate GlcNAcylation and H3K4 methylation through OGT and SET1/COMPASS. EMBO J 2013;32(5):645−55.

CHAPTER 3

Mechanisms of Histone Modifications

Ludovica Vanzan[1], Athena Sklias[2], Maria Boskovic[3], Zdenko Herceg[4], Rabih Murr[5] and David M. Suter[1]

[1]Institute of Bioengineering, School of Life Sciences, Swiss Federal Institute of Technology (EPFL), Lausanne, Switzerland [2]Center for Integrative Genomics, University of Lausanne, Lausanne, Switzerland [3]Laboratory for Cancer Research, University of Split School of Medicine, Split, Croatia [4]International Agency for Research on Cancer (IARC), Lyon, France [5]Faculty of Medicine, Department of Genetic Medicine and Development, University of Geneva, Geneva, Switzerland

OUTLINE

Introduction	27	ADP-Ribosylation	44
Nucleosomes and Chromatin Organization	27	Citrullination	44
Epigenetics and Histone Modifications	28	Proline Isomerization	45
Main Histone Modifications	28	Role of Aberrant Histone Modifications in Disease	45
Acetylation	28	Conclusions	45
Methylation	36		
Phosphorylation	39	References	46
Ubiquitination	41		
Sumoylation	43		

INTRODUCTION

Nucleosomes and Chromatin Organization

Eukaryotic genomes are organized in a highly condensed structure called chromatin, that is instrumental for fundamental nuclear processes, such as transcription, replication, and DNA repair. Chromatin exists in at least two distinct functional forms: heterochromatin, a condensed form that generally lacks DNA regulatory activity, and euchromatin, a looser form that provides the environment for DNA regulatory processes. Nucleosomes are the building blocks of chromatin and are formed by two turns of DNA (147 bp) wrapped around an octamer of histones. This octamer is made of two subunits of each of the core histones H2A, H2B, H3, and H4 [1]. An additional histone called H1 (linker histone) binds DNA at the entry and exit from the nucleosomes to increase chromatin compaction [2]. In the genome, canonical histone genes exist in multiple copies clustered together. They are transcribed and translated during S-phase, and freshly synthesized histones are incorporated directly into the newly replicated DNA [3,4].

The canonical core histones may be replaced by several types of histone variants. Their genes are located outside of histone gene clusters, and their expression is not cell cycle-dependent [4]. Deposition of histone variants occurs independently from DNA replication, contributing to the fine-tuning of chromatin function and organization. Because of their sequence and structural differences, histone variants have different effects on chromatin. They may increase nucleosome stability, thus facilitating chromatin compaction and heterochromatin formation, or they may destabilize histone-DNA interactions leading to chromatin opening and transcription factor (TF) binding [4].

Among the most common histone variants are H2A.X, H2A.Z, and H3.3. H2A.X is ubiquitously expressed and is fundamental for DNA repair, as phosphorylation of its Ser 139 (S139; the modified histone is then called γH2A.X) residue signals the site of damage and triggers the DNA repair cascade [5,6]. H2A.Z, however, is localized at gene promoters and particularly at the +1 nucleosome (the first nucleosome after a gene transcription start site [TSS]) [7], and facilitates the association of TFs and nucleosome remodeling complexes with chromatin [8]. Similarly, the H3.3 variant is typical of euchromatic regions, particularly of active regulatory elements [9]. Nucleosomes containing both the H3.3 and H2A.Z variants are less stable than canonical ones [9], and mark active regulatory regions [10].

From a structural point of view, histones are composed by a globular domain at the carboxy (C)-terminal and an intrinsically disordered region at the amino (N)-terminal of the protein. The globular domains of the different histones share the same folding and interact together in forming the histone octamer around which the DNA is wrapped [11,12]. Instead, the N-terminal tails protrude from the nucleosome core and do not form any stable secondary structure [12,13]. Recent in vitro and live-cell imaging studies have shown how the N-terminal disordered histone tails of histones H1 and H2A can drive phase separation [14,15]. This contributes to the general organization of the genome and to the formation of chromatin compartments, which are spatially segregated domains of euchromatin and heterochromatin [16,17].

Epigenetics and Histone Modifications

The chromatin structure fulfills essential functions by condensing DNA, protecting it from damage, and controlling gene expression. However, chromatin must be amenable to conformational changes that regulate accessibility of molecules involved in processes such as transcription, replication, and the detection/repair of DNA breaks. This raises a fundamental question in biology: how is chromatin remodeled in a controlled manner in time and space? Part of the answer resides in the fact that cells have evolved epigenetic mechanisms, notably covalent posttranslational histone modifications, that can alter the structure of chromatin. Histone modifications come in different "flavors" and affect multiple residues, primarily those located at the N-terminal tails of histones that are enriched in positively charged residues [13]. (Figs. 3.1 and 3.2).

The most studied ones are acetylation, methylation, phosphorylation, and ubiquitination. Additional modifications include sumoylation, ADP-ribosylation, citrullination, proline isomerization, and many others (Figs. 3.1 and 3.2). Currently, at least 80 different types of histone modifications have been identified, affecting more than 60 residues.

Traditionally, two main roles were attributed to histone modifications. First, they could affect the nucleosome—nucleosome or DNA—nucleosome interactions through the addition of physical entities or the change in histone charges. Second, they could serve as docking sites for the recruitment or repulsion of specific regulatory proteins. Based on these assumptions, numerous reports raised the possibility that these modifications can be interdependent and act in a combinatorial fashion, thus forming a "histone code" that could further amplify the information potential of the DNA sequence [18,19]. The generality of this hypothesis has been challenged in the past few years by studies showing that the correlation of histone marks with transcriptional states do not necessarily imply a causal relationship [20—23]. Many of these works were related to the role of histone methylation, demonstrating that the presence of a particular mark might be a by-product and not fundamental in the regulation of transcription activation and repression. Instead, noncatalytic activities of the enzymes depositing and removing the histone marks may be the real protagonist of these processes [24]. In the following paragraphs, we will summarize the current knowledge on the potential roles of both histone marks and their modifying enzymes. It has to be noted that, although a new nomenclature for histone-modifying enzymes was suggested [25], we chose the most widely used nomenclature for the sake of clarity. In addition, enzymes responsible for the deposition of a histone mark will be commonly referred to as "writers" and those that remove histone modifications as "erasers."

MAIN HISTONE MODIFICATIONS

Acetylation

Writers and Erasers of Histone Acetylation

Histone acetylation is the process by which an acetyl group from acetyl-CoA is transferred to the ε amino group of lysines on the N-terminal tails of histones. Acetylation can occur on specific lysines in all four histones (Fig. 3.2 and Table 3.1). The positive charge of the lysine is neutralized by the addition of acetyl-CoA, weakening the electrostatic interaction between nucleosomes and the negatively charged DNA. This increases DNA accessibility to TFs and regulatory complexes [26],

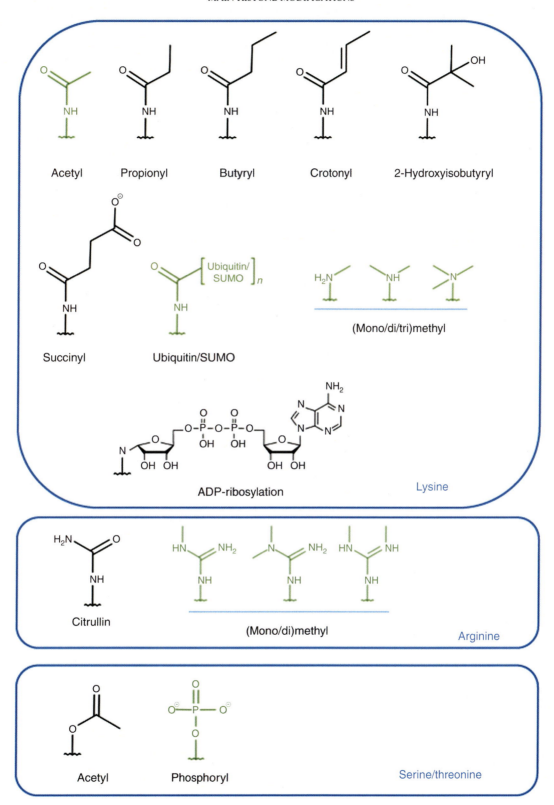

FIGURE 3.1 Nonexhaustive list of posttranslational chemical modifications occurring on histone N-terminal tails in humans and rodents. These modifications are added preferentially to amino acids, such as lysine (K), arginine (R), serine (S), and threonine (T), and impact the DNA compaction and accessibility to transcriptional and DNA repair machineries. The commonly studied histone modifications are highlighted in *green*. SUMO, Small ubiquitin-related modifier protein.

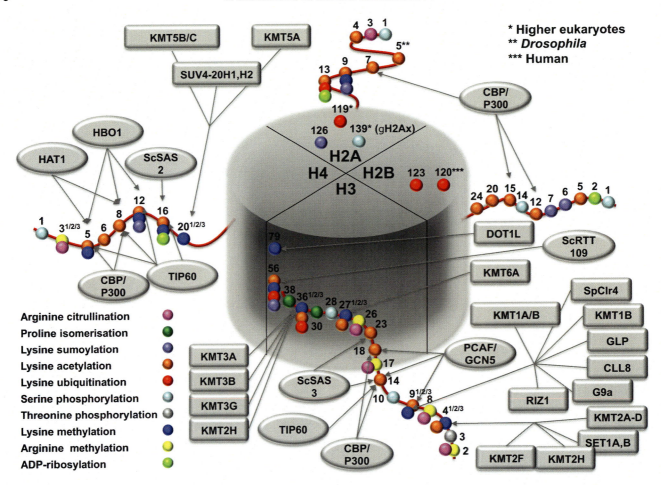

FIGURE 3.2 PTMs of N-terminal tails of core histone H2A, H2B, H3, and H4. Superscripts ($^{1/2/3}$) indicate that specific lysines can be mono-, di-, or tri-methylated. Major histone acetyltransferases (ovals), methyltransferases (rectangles), and their main target residues are indicated. Asterisks indicate if the modification is described exclusively in certain species. PTMs, Posttranslational modifications.

thus making histone acetylation of direct functional relevance for transcription activation. The modification is catalyzed by families of enzymes originally called histone acetyltransferase because they targeted lysine residues on histones. Subsequent studies, however, discovered that their acetylation function extends to nonhistone proteins [27,28]. Their name was therefore changed to lysine acetyltransferases (KATs). Likewise, the erasers of histone acetylation, previously named histone deacetylases (HDACs), are now officially known as KDACs [29]. It should be noted, however, that the old nomenclature it is still commonly used.

Some KATs are nuclear enzymes that catalyze acetylation on already deposited histones in a chromatin context (type-A KATs) [30,31]. Others, instead, acetylate newly synthesized histones in the cytoplasm (B-type KATs) [32,33]. Of the five type-A superfamilies [34], the main ones are the GNAT (Gcn5-related acetyltransferases) family, the MYST (MOZ, YBF2/SAS3, SAS2, and TIP60) family, and the p300/CBP (KAT3A/B) family. Similarly, histone deacetylases are divided into four classes [35], the most important ones being class I and class II KDACs, that interact with the SMRT/N-CoR (silencing mediator for retinoid and thyroid receptors/nuclear receptor corepressor) corepressor complex [36].

The GNAT KAT superfamily is extremely conserved during evolution. The founding member is Gcn5 (general control nonderepressible 5), the first KAT identified in 1995 in Tetrahymena [30]. In mammals, the main representatives are GCN5 (KAT2A) and pCAF (p300/CBP associated factor, KAT2B). The catalytic domain of the GNAT superfamily is composed by four different motifs with various degrees of conservation. Motifs A (AT domain) and B are responsible for acetyl-CoA recognition and binding [37]. Interestingly, motif A is conserved in the MYST superfamily catalytic domain, in addition to a C2HC-type zinc finger [38] and a helix-turn-helix (Esa1) motif [39]. MYST family members sometimes have an additional a chromodomain or a PHD (plant homeodomain) domain [40,41]. GNAT and MYST members often localize to promoters and

TABLE 3.1 Most common histone modifications and their function across different model organisms. Most studied readers, writers and erasers are indicated for the major histone modifications

Histones	Residues	Modifications	Readers	Writers	Erasers	Major Functions	Genomic Contexts
H2A	K5, K9	ac		KAT5/TIP60		DDR	/
	K5	ac		KAT5/TIP60		Activation	Gene bodies
	K126	su				Repression	/
	K119	ub	JARID2/AEBP2		USP16, USP21, BAP1, USP3, USP12, USP46	Repression	/
	S129 (yeast)	ph				DDR	—
H2A.X	K63	ub	RNF168	RNF8/UBC13		DDR	—
	E141	PAR		PARP1–10		DDR	—
	S139	ph	MDC1; RNF8/UBC13	ATM; ATR; DNA-Pk	PP2A, PP4, PP6	DDR	—
	Y142	ph		WSTF	EYA	DDR	—
	N-term	PAR		PARP1–10		DDR	—
H2B	K5	ac				Activation	Gene bodies
	S10	ph				Apoptosis	—
	K34	ub		MSL1/2		Activation	—
	K120	ub	DOT1L	RNF20/40		Activation; elongation	—
	K123 (yeast)	ub	Dot1	Rad6/Bre1		Activation; elongation	—
H3	R2	me2s		PRMT5		Activation	—
	R2	cit		PAD2; PAD4		DDR	Apoptosis
	K4	me1, me2, me3	SGF9; TAF3	KMT2A-D (MLL1–5); KMT2F/G (SETD1A/B, COMPASS complex); SETD7/9; KMT3D (SMYD1); SETMAR; PRDM9 (COMPASS complex); KMT2B/C (MLL3/4) for H3K4me1	KDM1A (LSD1); KDM5C (SMCX); RBP2; KDM5B (PLU1); KDM5D (SMCY/Jarid1ds)	Activation; poised to transcriptional activation	Promoters and enhancers

(Continued)

TABLE 3.1 (Continued)

Histones	Residues	Modifications	Readers	Writers	Erasers	Major Functions	Genomic Contexts
	R8	me2s		PRMT5		Repression	Permissive chromatin
		cit		PAD2; PAD4		Activation	Permissive chromatin
	K9	me1, me2				Activation	Euchromatin
		me3	TIP60; DNMT1; HP1	KMT1A/B (SUV39H1/2), GLP, KMT1E (SETDB1), G9a	KDM1A (LSD1); KDM4B/D (JMJD2B/D); KDM4A (JHDM3A); KDM4C (JMJD2C)	Repression	Pericentric heterochromatin, permanent mark
	S10	ph	14-3-3 family	MSK1/2; AURKB		Activation	Promoters
	R17, R26	me2a		PRMT4		Activation	—
	R26	cit		PAD2; PAD4		Activation	Euchromatin
	K27	me1		KMT6A/B (EZH1/2), G9a/GLP		Activation	Promoters
		me2			KDM6A (UTX)	Activation; enhancer silencing	Promoters and enhancers
		me3				Repression; poised to transcriptional repression	Promoters, development, cell differentiation, X-chromosome silencing
	S28	ph	14-3-3 family			Activation	—
	K36	me1		KMT3B (NSD1) and KMT3G (NSD2) for H3K36me1/2; KMT2H (ASH1) and KMT3A (SETD2) for H3K36me2/3	KDM2A(JHDM1); KDM4A (JHDM3A); KDM4C (JMJD2C); Rbp3 (H3K36me3, yeast)	Activation	Gene bodies
		me2				DDR	—
		me3				Activation; elongation	Exon borders/gene body
	K9, K4, K14, K18, K23, K27, K36, K56	ac	BRG1	p300/CBP (K4, K18, K27); GCN5 (K9); GNAT family (K14, K18, K23); MYST family		Activation; elongation (K9)	Promoters; Enhancers (K18, K27)
	T45	ph				Apoptosis	
	K79	me2, me3	USP10	DOT1L		Activation	Gene bodies, enhancers

H4	S1	ph		DDR	
	R3	me2a	PRMT1	Activation	Promoters
		me2s	PRTM5	Repression	–
	K12	su		Repression	–
	K16	ac	TIP60	DDR	–
	K4, K8, K12	ac	TIP60	Activation	Euchromatin
	K20	me2	L3MBTL1 (H4K20me1/2); DNMT1 (H4K20me3); 53BP1 (H4K20me2)	KMT5A (SET8) for H4K20me; KMT5B/C for H4K20me2/3; Set9 (yeast)	KDM7B (PHF8) Activation; DDR	–
		me3			Repression	–
	T80	ph			Cell cycle progression	–
	K91	ub			DDR	–

Note: Not all histone modifications are present across all model species (*Homo sapiens*, *Saccharomyces cerevisiae*, and *Drosophila melanogaster*). ac, Acetylation; cit, citrullination; DDR, DNA damage response; me, methylation (1, mono; 2, bi; 3, tri; a, asymmetric; s, symmetric); ph, phosphorylation; PAR, poly-ADP-ribosylation; su, sumoylation; ub, ubiquitination; activation, transcriptional activation; repression, transcriptional repression.

gene bodies, where they target histone proteins, positively correlating with gene expression activation [42–44]. Moreover, GCN5 acetylates TFs, functioning as coactivator [45]. As they are deeply involved in most of the cell's regulatory functions, anomalies of KATs can be very damaging for the cell fitness, and indeed members of the MYST family are linked to several types of cancer [46].

P300 and CBP (CREB binding protein) are the two members of a third KAT family, and perhaps some of the most studied transcriptional regulators. While they mostly share structure and function, and are often referred to as a single entity, the two proteins diverge in several properties. First, some studies identified phosphorylation residues specific for each of the two proteins [47]. Second, in response to ionizing radiation, p300, but not CBP, is important for induction of apoptosis [48]. Third, while both proteins are necessary for G1 arrest, induction of the cell cycle inhibitor p21 critically depends on p300, whereas induction of p27 requires CBP [49]. Finally, individual knockouts of the two proteins result in different phenotypes [50–53]. The NCoA-related KAT family members SRC1 (steroid receptor coactivator 1), SRC2 and SRC3 function as coactivators for the recruitment of nuclear steroid receptors [54]. SRC1 preferentially binds to H3 and H4 [55] and recruits CBP [56]. KATs and KDACs usually are part of multi-subunit complexes, interacting with several other cofactors that influence their stability and substrate specificity, and facilitate nucleosome acetylation [57]. Many members of the GNAT family belong to the SAGA (Spt-Ada-Gnc5) complex in yeast or SAGA-like complexes (SLIK, PCAF STAGA, TFTC) in humans [58], and acetylate H3K9/14/18/23 in mammalian cells [57,59]. Other subunits of SAGA complexes are the UBP8 (ubiquitin-specific-processing protease 8, USP8 in mammals), that deubiquitylates H2A and H2B [60,61], and TRRAP (transformation/transcription domain associated protein, Tra1 in yeast) that interacts with TFs and recruits the complex to target chromatin [62,63]. The MYST family KATs usually belong to the KAT7/HBO1 complex, whose primary target is H4 [64] and H3K14 [65], or to the KAT5/TIP60 (Nua4 in yeast) complex, that acetylates H2A and H4 [41]. P300/CBP form complexes with TFs at enhancer and promoter regions [66], where they acetylate H3K27 and H3K18. They may be accompanied by the KAT2B/PCAF (p300/CBP-associated factor) subunit [67].

Role of Histone Acetylation in Transcription Regulation

The first hint of the involvement of histone acetylation in transcription dates back to 1964, when it was observed that chromatin regions of actively transcribed genes tend to have hyperacetylated histones [68]. Several lines of evidence built from this observation, contributing to the current view of histone acetylation being a hallmark for transcription activation. First, the addition of acetyl groups weakens histone-DNA interactions, thereby opening the chromatin structure and facilitating the access of transcription machinery [26]. Work from Craig Peterson's laboratory demonstrated that H4K16ac incorporated into nucleosomal arrays impedes the formation of compacted chromatin fibers and prevents the ATP-mediated chromatin-remodeling factors from mediating nucleosome sliding [69]. Second, general TFs involved in transcription initiation such as TFIID (TF polymerase II D) [70] and TFIIC [71] have histone acetylation abilities or are part of KAT complexes. Indeed, the TAFs (TBP-associated factors) in the SAGA complex recruit TBP (TATA binding protein) to promoters TATA-box [72,73]. Disruption of the SAGA complex strongly reduces transcription [74]. Finally, the TRRAP subunit of SAGA complexes and the p300/CBP KATs acetylate and interact with TFs at active promoter and enhancer regions [62,63,75].

Promoters of transcriptionally active genes are characterized by high levels of histone acetylation, particularly at the H3K9, H2AK9, and H3K56 residues [76,77]. In both yeast and humans, H3 acetylation occurs in nucleosome already decorated by H3K4me3, that is recognized by the SAGA subunit SGF9 (SAGA-complex associated factor 9) [78] and is mediated by GCN5 (H3K9ac [59]) and p300/CBP (H3K4/K18/K27ac [59]). H3R8me2s, instead, blocks H3K9ac deposition [79]. It was recently proposed that H3K9ac may have a function in promoting transcription elongation by facilitating recruitment of the elongation complex and releasing Pol II pause [80]. Interestingly, studies have shown that H3K9ac correlates with transcriptional activation of lineage-specific genes during differentiation [81], and that both H3K9ac and H3K14ac are enriched at bivalent promoters together with H3K4me3 and H3K27me3, although their specific role in this context is yet to be elucidated [82]. Actively transcribed gene bodies need to be accessible by the transcriptional machinery, and as such they too are enriched in acetylation marks, especially H2A/BK5ac, H3K14ac, and H3K23ac [77].

Active enhancer regions are marked by the binding of TFs, general TFs, and RNA Pol II [83,84]. They are also characterized by the RNA Pol II-mediated transcription of eRNAs (enhancer RNAs) [85,86], which are noncoding RNAs that are involved in enhancer-promoter looping and regulation of the transcriptional machinery [87]. Additionally, H3K27ac deposition by p300/CBP [77,88], and that of H3K18ac [59], mark the exit from the H3K4me1-only poised state [88]. Interaction between the CBP catalytic domain and eRNAs stimulates H3K18/K27ac, and it is thought to contribute to the tissue-specificity of enhancer activation during development [89]. However, CBP recruitment to enhancer regions do not depend on eRNAs [75,89]. Enhancer acetylation by p300/CBP also contributes to transcription initiation in a two-step mechanism, first promoting the assembly of the pre-initiation complex (PIC) and the recruitment of RNA Pol II at promoters, and then recruiting BRD4 (bromodomain-containing protein 4), a factor involved in releasing Pol II pause [90]. BRD4, together with RNA Pol II and the Mediator complex, phase-separates to stimulate transcription [91,92]. An interesting recent study demonstrated that, while H3K27ac labels active enhancers, it is dispensable for their activity. Indeed, substitution of K27 with R in the H3.3 variant, while significantly reducing H3K27ac at enhancer regions, did not affect chromatin accessibility, transcription, or embryonic stem cells (ESCs) self-renewal [93]. Acetylation of other lysine residues and noncatalytic activities of p300/CBP may explain this observation.

The presence of KDACs at actively transcribed regions [94] contributes to fine-tune acetylation and subsequently gene expression levels [90]. Deacetylation at transcribed gene bodies is regulated by H3K36me3: in yeast, this methylation mark is recognized by the bromodomain of the Rpd3 (ribonucleic acid polymerase II binding subunit 3) KDAC complex, suppressing spurious transcription initiation [95]. The only acetylation mark reported to correlate with transcriptional repression is H4K20ac. Acetylation of H4K20 only occurs at lowly expressed promoters and associates with the repressive REST (RE1-silencing TF) complex [96]. Its precise role, however, is yet to be elucidated.

For a long time, most studies pointed toward the essential role and nonredundancy of different KATs and their target histone residues, supported by mutation and deletion experiments and the different phenotypes they generated. For example, TIP60 was shown to be essential for early stages of development in mice (embryonic day 3.5, ED 3.5) and for maintaining stem cell pluripotency. In contrast, GCN5 is dispensable for pluripotency, and GCN5-depleted embryos can develop until ED 10.5 [97–99]. However, genome-wide mapping of histone acetylation and KATs points toward redundancy, as various KATs and acetylated histone residues are simultaneously enriched primarily at TSS and promoters of active genes, while others can extend concurrently to the coding sequence [76,94]. This apparent contradiction could be explained by a two-stage model, where a specific acetylation mark is needed for transcription initiation, creating an open chromatin context that allows the recruitment of other KATs, and the deposition of other acetylation marks that allow maintenance of a transcription-friendly environment [100].

Role of Histone Acetylation in DNA Repair

The DNA sequence is highly subjected to different types of damage that if not correctly repaired may greatly endanger cell fitness and viability. Therefore, several mechanisms have evolved to repair damaged DNA quickly and efficiently. These are collectively identified as the DNA damage response (DDR). While mostly known for its role in transcription activation, histone acetylation is also important in signaling DNA damage and in the DDR. Intuitively, an increase in chromatin accessibility appears essential for DDR proteins to access DNA and repair DNA damage. Indeed, upon a double strand break (DSB), TRAPP/TIP60 acetylates several residues on both H4 and H2A and causes nucleosome destabilization [101], possibly by weakening the interaction between the H4 histone tail and the acidic patch of adjacent nucleosomes [102]. Along the same line, deacetylation of H1K85 mediated by KDAC1 further loosens chromatin compaction [103]. The H3K56 residue is also deacetylated, both in response to DSBs [104] and transiently after UV damage [105].

Acetylation of specific residues contributes to selection and fine-tuning of DNA repair pathways. Localized histone H3 and H4 acetylation and deacetylation is triggered by homology directed repair of DSBs, and binding of the NuA4 KAT complex at sites of DNA damage and site-specific histone H4 acetylation occurs concomitantly with H2A phosphorylation after induction of DSBs [106]. Contrarily, transient H4 deacetylation is an early response to DNA damage that facilitates DSB repair by the non homologous end joining (NHEJ) pathway [107,108].

Various studies identified crosstalks between histone acetylation and several other DNA damage-induced histone modifications and chromatin-remodeling events [109]. Following DSB, HP1-β is displaced from the chromatin, making H3K9me3 accessible. TIP60 is then tyrosine phosphorylated, which promotes its binding, via its chromodomain, to H3K9me3 marks [110]. This triggers TIP60-mediated acetylation and activation of the ataxia-telangiectasia mutated (ATM) kinase [111].

In human cells, TIP60 was found to be involved in DNA damage–dependent acetylation of γH2A.X at K5 that promotes its ubiquitination; these sequential events enhance histone dynamics during DDR [112,113]. H4 acetylation by human TIP60 is also required to remodel γH2A.X -containing nucleosome at sites of DNA breaks [114]. Finally, BRG1, the catalytic subunit of SWI/SNF chromatin-remodeling complex, binds to γH2A.X nucleosomes by interacting with acetylated H3 independent from H2A.XS10ph. This suggests that histone H3 acetylation and H2A.X phosphorylation may cooperatively act in a positive feedback loop of DNA repair [115].

Methylation

Writers and Erasers of Histone Methylation

Histone methylation is the covalent addition of a methyl group from the donor S-adenosylmethionine on the side-chain nitrogen atoms of lysine and arginine residues (Table 3.1). Lysines can be mono-, di-, or trimethylated. Arginines are mono- or di-methylated. Additionally, arginine dimethylation exists in both a symmetric and an asymmetric form. Most of histone methylation events occur on H3 and H4. Different residues are regulated by many cellular processes, depending on the type of modification and on the genomic context. Most histone lysine methyltransferases (KMTs) contain an evolutionary conserved domain called SET (Su(var)3-9, Enhancer of zeste, Trithorax) which catalyzes the deposition of the methyl group [116] (Fig. 3.2). The specificity of these enzymes is usually restricted to some residues (Table 3.1) [116–118]. The only KMTs lacking a SET domain are Dot1 (disruptor of telomeric silencing-1) in yeast and its human homolog DOT1L (Dot1-like) which target H3K79 [119]. Protein arginine methyltransferases (PRMTs) are classified type I and II, for asymmetric and symmetric dimethylation respectively. One example is PRMT1 that targets both histone and nonhistone proteins [120]. Although originally classified as a type II enzyme [121], PRMT7 is now generally regarded as the only member of type III, exclusively catalyzing arginine monomethylation [122,123]. The most targeted histone residues are H3R2, H3R17, H3R26, and H4R3 (Fig. 3.2).

The first histone demethylase was identified in 2004 [124]. KDM1A (previously LSD1, lysine specific histone demethylase 1), a nuclear amine oxidase homologous mainly demethylates H3K4, using flavin adenosine dinucleotide (FAD) as a cofactor. KDM1A can also demethylate H3K9, when it is present in a complex with the androgen receptor [124,125]. A number of other related enzymes were subsequently discovered and classified into two families of histone lysine demethylases [126,127]. The JARID1 (Jumonji AT-rich interactive domain 1) family catalyzes a FAD-dependent reaction and the JMD2 (Jumonji C domain–containing proteins) family characterized by a Jumonji domain for the removal of the methyl group via a 2OG-dependent oxidation reaction [126]. Again, members of both families specifically demethylate selected residues [128,129]. In addition, JMJD6, a JmjC-containing protein, was first hypothesized to be identified a direct arginine demethylase [128]. However, other structural and functional studies disputed this proposal, suggesting that JMJD6 is instead a lysine hydroxylase [130,131].

Role of Histone Methylation in Transcription Regulation

Histone Lysine Methylation

Together with acetylation, histone methylation is the most studied epigenetic modification. Contrary to histone acetylation, which is tightly and functionally linked to transcription activation, methylation on histone lysines correlates with different transcriptional states depending on the position of the residue (Table 3.1).

Indeed, H3K4 methylation is associated to transcription initiation and elongation. H3K4me3 is enriched at active gene promoters [132,133], reaching a peak after the first exon [133,134]. H3K4me2 further extends to active gene bodies, together with H3K4me1 that peaks at 5′ end of genes. H3K36, instead, progressively shifts from monomethylation at promoters to trimethylation at 3′ end of genes [133,135,136]. Finally, H3K79me2/3 are highly enriched at TSS of active genes and slowly decline through the first intron. They are also found at several origins of replication in mammals [137,138]. Inactive gene promoters, on the other hand, have high levels of H3K27me3 or H3K9me3 and low levels of H3K4me1/2/3. Active enhancers are marked by H3K27ac, H3K4me1 and H3K79me2/3, while inactive enhancers harbor H3K27me3 [88]. These notions were derived from a large number of ChIP-Seq experiments that generated a detailed picture of the distribution of histone methylation marks. Importantly, low-specificity or cross-reactivity of ChIP-Seq antibodies may generate confusion in the literature concerning the localization and levels of the different forms of methylated marks. This was shown for H3K4me1/2/3 [139], but it is likely occurring in the case of other marks as well.

H3K4 methylation is invariably associated with transcription activation in enhancers and promoters. Early ChIP-Seq experiments in human stem cells showed that 80% of promoters enriched for H3K4me3 correspond to actively transcribed genes [140]. These results, supported by studies demonstrating that H3K4me3 is recognized by the TAF3 subunit of TFIID and favors the recruitment of the whole TFIID complex to gene promoters [141,142], soon led to the hypothesis of H3K4me3-dependent recruitment of transcription regulators. Moreover, recruitment of the H3K4me3-specific Set1 methyltransferase in yeast by RNA Pol II during elongation suggested H3K4me3 could function as the memory of recent transcription activity [143].

The presence of H3K4 methylation is also mechanistically linked to depletion of DNA methylation, a repressive epigenetic mark, at promoters and enhancers [144,145]. Unmethylated CpG islands are recognized by the zinc finger CXXC domain of CFP1 (CpG-binding CXXC domain—containing protein 1), a cofactor of SET1/KMT2F and of the H3K4me3 methyltransferases KMT2A/D [146,147]. Conversely, the presence of H3K4me3 blocks the activity of the H3K4me0-interacting ADD domain contained in DNMT3a and DNMT3L and therefore protects CpG islands from de novo DNA methylation [148].

Interestingly, a small subset of actively transcribed genes present a different H3K4me3 pattern, characterized by lower methylation levels but broader distribution, extending downstream of the promoter and into the gene body [134,149]. These "broad H3K4me3 domains" were identified in correspondence of tumor suppressor genes [149] and in genes essential for the maintenance of cell identity [150]. They display increased transcription elongation compared to "sharp" domains and correlate with a wide range of histone posttranslational modifications (PTMs) [134,149].

However, lack of H3K4me3 does not impede transcription. Both in yeast and in *Drosophila* [20], ablation of H3K4me3 causes defects in cell proliferation but does not impact viability [20,151]. Also, while lack of the COMPASS catalytic subunit SET1/KMT2F is lethal during embryonic development [152], ESCs carrying a catalytically dead SET1 proliferate normally, and they show proliferation defects only after differentiation [153]. Similarly, a very recent paper from the Shilatifard's lab showed that H3K27me3 or DNA methylation-mediated gene silencing following KMT2D knockout can be rescued by inhibition of PRC2 (polycomb repressive complex 2) or DNMT and does not require new H3K4me3 deposition [154]. Taken together, these results suggest that histone modifying enzymes might play fundamental roles in the regulation of gene expression that are independent from their catalytic activity, and therefore depart from the traditional "histone code" hypothesis. Similar observations were also made concerning the role of H3K4me1. Contrary to promoters, enhancers are depleted of H3K4me3 but, when active or poised, are decorated with H3K4me1 by the KMT2B/C (MLL3/4) members of the COMPASS family [155]. The distinction between the active and poised states is given by H3K27ac [88]. The precise function of H3K4me1 is, however, yet to be precisely elucidated. It was shown that the BAF remodeling complex directly binds H3K4me1 to its PHD2 domain, suggesting a role for H3K4me1 in facilitating nucleosome rearrangements [156]. However, cells and flies expressing catalytically dead KMT2B/C have only little defects in gene expression and H3K27ac levels at enhancers, while KMT2B/C knockouts severely damage gene expression and cell differentiation [22,23]. Also, occupancy of Pol II at enhancers and the production of eRNAs seems to be dependent on KMT2B/C but not on H3K4me1 [22], again uncoupling the presence of the methylation mark from the regulatory role of histone modifying enzymes.

Active gene bodies are decorated with H3K36 and H3K79 methylation, which are implicated in both transcription activation and elongation. H3K36me3 is at the center of an intense interaction network of histone and other epigenetic modifications. In yeast, it is exclusively deposited cotranscriptionally by association of Set2 with the phosphorylated Rpb1 subunit of RNA Pol II [157]. In mammals, the histone methyltransferases KMT3B and KMT3G are also involved, and the deposition does not need to be concomitant to transcription elongation [158]. KMT3B (originally called NSD1) mono- and dimethylates H3K36 [159], but dimethylation and trimethylation are catalyzed by KMT2H (ASH1) and KMT3A (SETD2), respectively [160]. H3K36me3 is recognized by the chromodomain of the Eaf3 subunit of the Rpd3S, a histone deacetylase, that deacetylates gene bodies, preventing transcription initiation at cryptic sites [95,161]. Indeed, mutations of both Set2 and Rpd3S lead to hyperacetylation of gene bodies and cryptic transcription initiation [162]. This role is reinforced by high levels of DNA methylation at transcribed gene bodies [163]. The cooccurrence of the two marks depends on the recognition of H3K36me3 by the PWWP domain contained in DNMT3B, that causes de novo DNA methylation in transcribed coding regions [164]. Interestingly, recent studies showed that Set2 requires H2BK120ub for correct positioning on the nucleosome [165] and H3K36me3 deposition. In addition to H3K36me3, Rpd3S interacts with H2BK120ub to prevent transcription initiation from gene bodies [166]. H2BK120ub is also required for methylation of H3K79 by DOT1L (see the section "Role of histone ubiquitination in transcription regulation" for more details) [167,168]. H3K79me has been mostly implicated in the regulation of transcription elongation and of telomere formation. DOT1L interacts with RNA Pol II and the elongation complex [169,170], and H3K79me3 is bound by the Tudor domain of the SMN (survival of motor neuron) protein, that is involved in the formation of the spliceosome [170]. Together, H2Bub and H3K79me also bring the deposition of the H3K4 methylation [171,172]. However, while high H3K79me3 levels strictly correlate with high levels of transcription, DOT1L knockout does not disrupt gene expression [173], while it does cause a reduction of the H3K9me2 and H4K20me heterochromatic marks in telomeric regions and consequent block of cell proliferation [174]. This again hints for a potential separation of the catalytic and not catalytic activities of DOT1L.

Three lysine methylation sites are related to transcriptional repression: H3K9, H3K27, and H4K20me3. H4K20 is monomethylated by KMT5A (also known as SET8) [175] and di- or tri-methylated by KMT5B/C [176]. Interestingly, while KMT5A has roles in DNA repair and cell cycle, several lines of evidence prove that the H4K20me1 mark itself is necessary for a correct regulation of cell cycle progression. Indeed, cells expressing a catalytically dead KMT5A form display low H4K20me1 levels and arrest at the G2/M transition [177]. Moreover, silencing of the H4K20me Jumonji demethylase KDM7B/PHF8 (PHD finger protein 8) in cells delays the G1-S transition [178]. On the other hand, studies on the role of H4K20me1 in transcription generated conflicting results; while some reported a correlation with histone hyperacetylation at transcribed genes and promoters, particularly those with high CpG content [133,179], others have associated H4K20me1 to gene repression. One example is the recruitment of the L3MBTL1 (lethal 3 mbt-like 1) repressor to H4K20me1/2-decorated nucleosomes, inducing chromatin compaction [180]. Reports on H4K20me3, on the other hand, invariably associated it with transcription repression. KMT5C is recruited to mono- and di-methylated residues by its interaction with the heterochromatin protein HP1 [176]. Moreover, recognition by the maintenance DNA methyltransferase DNMT1 of H4K20me3 (via the BAH domain) and H3K9me3 (via the RFTS domain) [181], implicates H4K20me3 in the silencing of the LINE-1 transposable elements [182].

H3K9me2/3 is essential for constitutive heterochromatin formation (telomeres and centromeres), transposons repression and gene silencing [183]. The first H3K9-specific histone methyltransferases identified were the SUV(var)39 family members KMT1A/B [116], followed later by the other SET-domain enzymes KMT1E/SETDB1 and GLP [117,184]. KMT1A/B preferentially methylates histones at centromeric and telomeric regions [185], while KMT1E acts both on telomeres [186] and euchromatic regions [187]. Different mechanisms exist for H3K9me3 deposition: in Schizosaccharomyces pombe small RNAs are bound by two proteins of the RNAi machinery, Argonaute and Dicer, that then interact with the H3K9 HMT complex and target it to the DNA. Alternatively, TFs such as the corepressor KAP1 (KRAB [Kruppel-associated box domain]-associated protein 1) [187], the stress response Atf1/Pcr1 [188] and the telomeric TF Taz1 [189] in yeast, bind specific DNA sequences and recruit HMTs to the DNA. In both cases, H3K9me is then recognized by the chromodomain of HP1, which propagates the heterochromatic state [190]. H3K9me2/3 and HP1, with other related proteins such as SUV39 and the TRIM28 repressor, phase-separate together, likely in order to exclude TFs and activators from heterochromatic regions [191]. In the same spirit, H3K9me3 are physically isolated from the transcribed genome by association with the nuclear lamina [192], a process that in Caenorhabditis elegans is mediated by the G9a methyltransferase [193]. H3K9 and DNA methylation are tightly interconnected: DNMT3a/b are recruited by direct interaction with HP1, and SUV39H knockout in mouse cells leads to alterations in the DNA methylation status [194]. Interestingly, in recent work by the Zaret lab in mouse embryos, they showed that deposition of the H3K9me3 mark is a dynamic process, and that the heterochromatic landscape is reorganized during differentiation to silence genes specific for other lineages [195]. While this discovery awaits deeper investigation, it highlights the importance of H3K9me3-mediated gene silencing.

Methylation of H3K27 is likewise traditionally associated with gene silencing. It is deposited by the KMT6A/EZH2 subunit [118] of the PRC2, that was identified first in Drosophila [196] and subsequently in mammals [197]. As previously mentioned, deposition of H3K27me3 is the result of a tight cross-regulation between PRC1 and H2AK119ub and PRC2 [198–200]. The model proposed based on the current knowledge sees initial H3K27me3 being deposited at specific CG-rich nucleation sites [201]. PRC2 recruitment to nucleosomes is facilitated by its subunits JARID2/AEBP2 (AE binding protein 2) that interact with H2AK119ub [202]. Subsequently, H3K27me3 spreading is adjuvated by an additional allosteric interaction between the PRC2 EED2 subunit with nucleosomes [203]. While EZH1/KMT6B is sufficient to catalyze H3K27me3 deposition at nucleation sites, EZH2/KMT6A is necessary to extend the H3K27me3 domain and prevent histone acetylation [204]. H3K27me3 is excluded from constitutive heterochromatin but contributes to X-chromosome inactivation and plays an active role in the repression of cell-type specific genes [133]. Loss of H3K27me3 due to defective EZH2/KMT6A, for example, causes H3K27 hyperacetylation and activation of tumorigenic genes [205] and impedes ESCs differentiation [204].

Activating H3K4me3 and repressive H3K27me3 marks are often found together at promoters and enhancers of key developmental genes [132]. These bivalent domains are characteristic of "poised" genes, requiring rapid activation or silencing during differentiation. While they were first discovered in ESCs [132], further investigations identified cell type-specific bivalent domains in different types of neurons [206] and adult tissues [207]. Interestingly, while H3K4me3 and H3K27me3 modifications coexists on opposite H3 tails of a single nucleosome [208], H3K27 methylation by PRC2 is hindered by H3K4me3 if on the same histone tail [209]. Whether the H3K4me3-K27me3 bivalency is established during the G1 or M phases of the cell cycle is still debated [210]. DNA methylation and H3K27 methylation, both repressive epigenetic modifications, are often mutually exclusive [211]. Indeed, at bivalent genes promoters, interactions between PRC2 and DNMT3l, a catalytically dead form of DNA methyltransferase, prevents deposition of de novo methylation by DNMT3a/b [212]. Moreover, it was also shown that PRC2 interacts poorly with nucleosomes carrying methylated DNA [213].

Histone Arginine Methylation

The first arginine methyltransferase identified as a transcriptional regulator was PRMT4/CARM1, that deposits the H3R17me2a and the H3R26me2a marks and was described as a coactivator of TFs [214–216]. Monomethylation and dimethylation events by other type I PRMTs such as PRMT1, and PRMT2, also correlate with transcription activation [217,218]. On the other hand, PRMT6 is instead linked to repression [219], while PRMT5 generates both active (H3R2me2s) and repressive (H3R8me2s, H4R3me2s) marks [220,221]. Several studies showed that arginine methylation has a bivalent role in transcription regulation, mediated by the interdependency between histone lysine and arginine methylation. One example is the crosstalk between methylation on H3K4 and H3R2: while symmetric methylation on the arginine residue (H3R2me2s) often colocalizes with H3K4me3 at active genes promoters, H3R2me2a, mediated by PRMT6, prevents methylation of the lysine residue [222]. Again H3K9ac, a mark of transcription activation, is prevented by the presence of H3R8me2s [79], while H3R8me2a prevents HP1 binding to H3K9me3 [223]. Recently, it was also shown that H3R2/8 methylation reduces H3K27me3 deposition, which instead increases in PRMT5 depleted cells. Biochemical studies demonstrated that this particular effect is not due to the lack of PRMT5, but that arginine methylation directly impacts PRC2 activity [224]. H4R3me2a, deposited by PRMT1, is one of the most abundant arginine methylation marks. Both H4R3me2a and PRMT1 are enriched at active gene promoters, and favor the recruitment of p300 and subsequent histone acetylation [225]. Recently, genome-wide exploration of the arginine methylation landscape in chicken showed that H4R3me2a often colocalizes with H3R2me2s at intronic and intergenic regions, and next to nucleosome depleted regions [220]. H4R3me2s, catalyzed by PRMT5, correlates instead with transcriptional repression [1].

Role of Histone Methylation in DNA Repair

As previously mentioned, the first signal of DNA damage is the phosphorylation of γH2AX by ATM [5,6]. Following this first event, the cell needs to decide between homologous recombination (HR) and NHEJ mediated repair. BRCA2 (breast cancer 2) and 53BP play a critical role in the choice between these repair pathways. Binding of H4K20me2 by 53BP1 is central in the activation of the NHEJ pathway. In yeast, H4K20 methylation by Set9 serves as a docking site for the recruitment of Crb2 via its double Tudor domains. Crb2 acts as a DNA damage sensor and checkpoint protein in *S. pombe*, and its recruitment increases cellular survival following genotoxic stress [226]. In mammals, H4K20me2 and H2AK15ub favor 53BP1 recruitment to DNA damage sites [227,228] and the subsequent activation of its downstream effectors [229]. Loss of H4K20me2 writers KMT5B/C causes delays in the formation of 53BP1 foci [230]. On the contrary, H4K16ac prevents 53BP1 binding in order to promote the HR pathway [107].

Other methylated residues are involved in DNA damage repair. For example, H3K9 methylation was linked to activation of KAT complexes (such as TIP60) following DNA damage. Upon DSB, a complex containing Kap1, HP1-β, and SUV39H1/KMT1A is recruited, leading to trimethylation of H3K9. H3K9me3 then recruits and activates TIP60 and enhances the recruitment of the Kap1/HP1-β complex. TIP60 acetylates ATM and H4K16, resulting in subsequent recruitment of BRCA1 and depletion of 53BP1. In a negative feedback loop, ATM phosphorylates Kap1, releasing the Kap1/HP1/ KMT1A complex from the damaged site [231]. Recently it was also found that H2A. XK134me2, methylated by KMT1B/SUV39H2, contributes to the formation of γH2AX, although the exact mechanism is still unclear [232]. The role of arginine methylation in DNA repair is much less understood. Following the repair of a DNA lesion, expression of DNA repair genes needs to be turned off. A recent paper showed that PRMT7 methylates H2AR3 and H4R3 at the promoters of DNA repair genes. Together with the SWI/SNF complex, this downregulates their expression and thereby fine-tunes the DNA damage response. Accordingly, knockdown of PRMT7 in several cell lines increases cell resistance to DNA damaging agents, including chemotherapeutic drugs [233].

Phosphorylation

Phosphorylation, one of the most extensively studied posttranslation modifications, consists in the addition of a phosphate (PO_4) group to the side chain of various amino acids residues. It occurs mostly on (in order of abundance) serine, threonine, or tyrosine, of which it modifies the biochemical characteristics, transforming them from hydrophobic and apolar to hydrophilic and polar residues, with important effects on the global protein folding and on protein-protein interactions. (Table 3.1). It is particularly known for its involvement in the regulation of cell cycle and cell division, but it is also implicated in transcription and DNA damage repair. While a great number of enzymes responsible for the deposition of the phosphate mark (kinases) in response to different kind of stimuli exist, only few erasers of phosphorylation (phosphatases) have been identified [234]. Phosphorylation of histidine residues also exists, but the relative instability of the modified His residues complicated research and only recent studies started to shed light on its functions and on the enzymes involved in its metabolism [235].

Role of Histone Phosphorylation in Transcription Regulation

One of the most multifunctional and studied phosphorylation sites is Ser10 on H3 (H3S10ph). Approximately 30% of the interphasic genome of mouse ESCs is decorated with H3S10ph, with a negative correlation between H3S10ph and H3K9me2 enrichment [236]. H3S10ph is also excluded from lamin associate domains, that are heterochromatin domains adjacent to the nuclear lamina [237] and enriched in H3K9me2/3 [238]. Together, these findings hint at the involvement of H3 phosphorylation in transcription activation [236], and support previously published studies of close interactions between H3S10ph and histone acetylation at promoters of different genes [239,240], sometimes in combination with H3T11ph [240]. Similarly, H3S28ph is correlated with activation of gene expression [241] and is an indication of transcriptional response to cellular stress. Genome-wide studies revealed its presence at more than 50% of stress-induced genes. Deposited by MSK1/2 (mitogen and stress activated kinase 1/2) in macrophages [242], they contribute to the reduction of HDAC occupancy and the promotion of histone acetylation and p300/CBP-dependent transcription [242,243]. Additionally, members of the 14-3-3 protein family, bind to H3S10/28ph and H3K9/14ac [244] or H4K5/8ac [245] and rapidly recruit the SWI/SNF chromatin remodeler to gene promoters in response to different kinds of stimuli, facilitating TF binding and onset of transcription [246]. A study by the Corces's lab in Drosophila, demonstrated the presence of the H3K9acS10ph and H3K27acS28ph dual marks on most enhancers and promoter sequences [247]. Similar results were obtained in mammalian models with imaging studies showing that phosphorylation of H3S10 or H3S28 occurs on different alleles [248].

Another example of the role of phosphorylation in the regulation of gene expression comes from cell division. H3S10ph, deposited by AURKB (Aurora-B kinase) [249], is not erased during mitosis. On the contrary, its levels increase until metaphase, then rapidly decrease in anaphase [236,250]. As such, it is sometimes considered as a hallmark of cell division [251]. Early work on H3 mutants for S10 showed that lack of H3S10ph in mitosis causes defective chromosome condensation and segregation [252], with lasting effects in daughter cells. HP1 normally binds to H3K9me3, a repressive mark, and it is fundamental for chromatin packaging and heterochromatin formation. Phosphorylation on H3S10 disrupts the H3K9me3-HP1 interaction, leading to HP1 dissociation from chromatin in mitosis. When AURKB is depleted, lack of H3S10ph creates a permissive environment for HP1 that stays attached during mitosis [253] and spreads to euchromatic regions [236].

Phosphorylation on other histone residues has similar effects on chromatin structure and accessibility. For example, H3Y41ph, deposited by JAK2 (Janus kinase 2), is also implicated in HP1 exclusion from chromatin [254]. Together with H3K56ac, it increases chromatin accessibility to TFs [255]. H3T118, instead, is located on the dyad, that is, the middle point of the DNA wrapped around the nucleosome [12]. There, it alters the chromatin wrapping by facilitating the creation of nucleosome dimers [256].

Role of Histone Phosphorylation in DNA Damage Response

One of the first and most easily detectable signals of cellular response to DNA damage is phosphorylation of S139 of H2A.X (H2AS129 in yeast, that does not possess the H2A.X variant), mediated by the ATM, ATR (ataxia-telangiectasia Rad3-related) or DNA-Pk serine-threonine kinases [5,6]. This modification, commonly referred to as γH2A.X, spreads over kilobases (in yeast) or 1−2 megabases (in mammalian cells) from DSB sites [257,258] and is required for recruitment and accumulation of DNA repair enzymes [106]. γH2A.X also recruits the INO80 chromatin remodeler complex to repair sites [259].

Interestingly, propagation of γH2A.X is uneven and nonsymmetrical but organized in small clusters [260] and largely confined within TAD borders [261]. The spreading is mediated by the MDC1 (mediator of DNA damage checkpoint 1) complex, which binds to γH2A.X and recruits the MRN (MRE11-RAD50-NBS1) complex to further stimulate ATM activity [262]. Moreover, MDC1 contributes to the activation of CHK1 and CHK2 (checkpoint kinases 1 and 2) in mammals (Rad53 in yeast) to delay cell cycle progression, a fundamental aspect for guaranteeing efficient DNA damage repair [263,264]. γH2A.X is recognized by RNF8/UBC13 complex, that ubiquitinates K63 on the same histone [265], and promotes accumulation of important DDR factors such as 53BP (Rad9 in yeast) or BRCA1 [266,267].

Following efficient DNA repair, S139 phosphorylation on H2A.X needs to be removed to allow the cell cycle to continue. Several different phosphatases (PP2A, PP4, PP6, etc.) remove this phosphorylation mark [268]. In yeast, cell cycle progression is signaled by H4T80ph, that recruits the scaffold protein Rtt107 and triggers DNA damage checkpoint recovery [269]. Another phosphorylation site involved in DNA damage repair is H2A.XY142, deposited by Williams syndrome TF and removed by the EYA (eyes absent) phosphatase at damaged sites [270]. Dephosphorylation at Y142 is slower than phosphorylation at S139, and MDC1 is unable to bind to the double phosphorylated H2A.X. As MCPH1 (microcephalin protein 1), instead, can bind H2A.XS139Y142ph and recruit CHK2, it has been proposed that MCPH1 may be involved in an alternative γH2A.X DNA repair pathway independent from MDC1 [271]. Recent studies showed that MCPH1 may facilitate BRCA2/Rad5-mediated HR at DNA damage sites [263,272]. Alternatively, failed removal of H2A.XY142ph promotes apoptosis over DDR [273]. Other histone phosphorylation sites involved in apoptosis are H2BS10ph (in yeast) and H3T45ph [274]. More studies are necessary to precisely elucidate their roles.

Histone phosphorylation can also occur on H4S1 through the activity of casein kinase II (CK2) in response to DSBs and it facilitates nonhomologous end joining (NHEJ) repair [275]. A study from Côté's laboratory demonstrated that this phosphorylation coincides with a decrease in acetylation, suggesting that it occurs after H4 deacetylation and may regulate chromatin restoration after repair is completed [276].

Ubiquitination

Writers and Erasers of Histone Ubiquitination

While most histone modifications consist in the addition of small chemical groups to histone proteins, ubiquitination refers to the covalent attachment of one (monoubiquitination) or more (polyubiquitination) copies of a 76-amino acid protein to the amino group of a lysine residue. In general, polyubiquitination marks a protein to be degraded via the 26 S proteasome, while monoubiquitination may modify protein function. In the case of histones, ubiquitination has been implicated in the regulation of a wide range of chromatin functions.

Histone H2A was the first histone shown to be ubiquitinated in 1975 [277], followed a few years later by histone H2B (K119, K120, and K143 in mammals) [278]. In the meantime, ubiquitination of H3, H4, and of the linker histone H1 was also reported (Fig. 3.2 and Table 3.1), the monoubiquitinated forms of H2Aub and H2Bub represent the large majority of ubiquitinated histones and have been extensively studied. We will therefore primarily discuss the mechanisms of establishment and the roles of H2A and H2B monoubiquitination (H2Aub and H2Bub) (Table 3.2).

The ubiquitination process requires the sequential involvement of E1-activating, E2-conjugating, and E3-ligase enzymes [279]: E1 collects and transfers one ubiquitin molecule to E2; E2 then binds to E3, which in turn positions the ubiquitin to the target protein. Alternatively, E2-mediated, E3-independent ubiquitination is also possible. Histone E3-ligases mostly belong to the HECT (homologous to E6AP-C terminus) or RING (really interesting new gene) families. H2A ubiquitination is primarily catalyzed by RING domain—containing members of the PRC1 (Polycomb repressive complex 1) complex, which specifically target H2AK118 and H2AK119 in mammals [280]. RING1B, the first identified ubiquitin ligase and member of PRC1, is the main H2AK119ub E3-ligase; its activity is highly dependent on interaction with two other RING domain—containing proteins in the PRC1 complex, RING1A and BMI1 (B lymphoma insertion region 1 homolog) [281,282]. RING-independent proteins, such as the 2A-HUB/hRUL138 and BRCA1, also ubiquitinate H2A, although their actual contribution to the bulk of ubiquitinated histones in vivo is minimal [283]. H2BK123-specific E2 in yeast, Rad6, was the first histone E2 to be identified [284]. Rad6 activity is linked to the RING finger E3-ligase, Bre1. Homologs of Rad6 and Bre1 in humans were also shown to be involved in the formation of H2BK120ub [285,286].

The addition of the ubiquitin moiety to histones is reversible through the activity of deubiquitinating enzymes (DUBs). Several families of DUBs exist in mammals, and about 20 DUBs in yeast. Different mechanisms of deubiquitination were identified: DUBs may recognize and bind either the target protein or the ubiquitin modification itself; moreover, the mark can be removed by cleavage from the end or within a chain [287]. DUBs may act in complexes. For example, in yeast UBP8 is part of the SAGA complex and as such, its activity is required for the transcription of SAGA-regulated genes [60,61]. UBP10 activity, instead, is SAGA independent but Sir dependent [288]. The deubiquitinase activity of its ortholog in human, ubiquitin specific peptidase 36, was shown to be largely conserved [289]. UBP8 and UBP10, likely through their presence in different complexes, are recruited to distinct genomic regions. UBP8 is recruited to regions marked by H3K4me3, such as active promoters, while UBP10 colocalizes with regions marked by H3K79me3, such as telomeres [290]. DUBs specific to H2Aub are USP16, USP21, and BAP1 (BRCA1-associated

TABLE 3.2 Nonexhaustive list of histone modifying enzymes and readers of histone PTMs found in the main chromatin remodeling complexes

Complex	Subunit	Role
COMPASS	KMT2F	H3K4me3 methyltransferase
	PRDM9	H3K4me3 methyltransferase
	KMT2B/C	H3K4me1 methyltransferase
SAGA and SAGA-like	SGF9	Reader of H3K4me3
	GNAT family (GCN5)	H3K9 acetyltransferase
	p300/CBP	H3K4/18/27 acetyltransferase
	UBP8 (yeast)/USP8 (mammals)	H2A/H2B deubiquitination
PRC2	JARID2/AEBP2	Reader of H2AK119ub
	KMT6B/EZH1	H3K27me3 methyltransferase
	KMT6A/EZH2	H3K27me3 methyltransferase
	EED2	Reader of H3K27me3
PRC1	RING1B	Writer of H2AK119 ubiquitination
	CBX proteins	Readers of H3K27me3 and H3K9me3
	L3MBTL2	Reader of H3 and H4 methylation
SWI/SNF	BRG1	Reader of H3 acetylation and H3R26Cit

protein 1). Finally, other DUBs, such as USP3, USP12, and USP46, have a dual specificity toward H2Aub and H2Bub [291].

Role of Histone Ubiquitination in Transcription Regulation

Histone ubiquitination contributes to shaping chromatin architecture, and as such is highly involved in transcription regulation. Current evidence points [60,292,293] towards a preferential association of H2Aub with transcription repression and of H2Bub with gene activation and transcription elongation.

Transcription stimulation by H2Bub occurs through several different mechanisms. In vitro studies demonstrated that H2Bub disrupts chromatin compaction by the means of a small acidic patch in the ubiquitin molecule, preventing tight interactions between neighboring nucleosomes [294,295]. Ubiquitination at H2BK120 is also required to recruit the histone chaperone FACT (facilitates chromatin transaction), that displaces the H2A-H2B dimer from nucleosomes to allow efficient elongation, and restores the octamers after Pol II passing [296–298]. Moreover, writers of H2Bub such as RNF20/40 (RING finger protein 20/40) in mammals (for H2BK120ub), its yeast homologous Bre1, and MSL1 (MSL complex subunit 1)/MSL2 (for H2BK34ub) are recruited to actively transcribed regions by the Pol II complex [299]. Interestingly, it has recently been shown that Bre1 requires Lge1-mediated liquid-liquid phase separation to enrich at gene bodies [300].

The H2Bub landscape also correlates with histone methylation modifications traditionally associated with actively transcribed genes and may facilitate their deposition. A plethora of studies have dissected the relationship between H2BK120ub and the H3K79-specific histone methyltransferase DOT1L (disruptor of telomeric silencing-like), and between their yeast equivalents H2BK123ub and Dot1p [143,167,168,301]. It is now known that DOT1L interacts with both an acidic patch of the H2B protein and the I36 residue of the ubiquitin bound to H2BK120. This interaction reduces DOT1L mobility on the nucleosome and keeps DOT1L in the correct conformation for the deposition of the methylation mark [167,168]. The H4 histone tail was also implicated in this process [301]. Together, H2Bub and H3K79me bring the deposition of the H3K4 methylation, mediated by one of the MLL enzyme family members as a subunit of the COMPASS (complex of protein associated with SET1) complex [171,172]. Contrary to H2Bub, H2Aub plays an important role in many repressive contests, and contributes to the preservation of cell identity. Indeed, many H2A-deubiquitating enzymes are found in subunits of coactivator complexes. Early in vitro experiments showed that H2Aub can inhibit the methylation of H3K4 [292] and that USP10-mediated deubiquitination of the H2A.Z histone variant, enriched at promoters of inducible genes, is required for gene expression activation [293]. Several studies through the years have shown that an elaborate cross-regulation exists between PRC1 and PRC2, that is histone methyltransferase specific for H3K27me3, a marker of repression. Indeed, the H3K27me3 modification favors PRC1 binding, as it is recognized by a CBX (chromobox)-domain family in the PRC1 complex [198]. Conversely, H2AK119ub is recognized by two PRC2 components, JARID2 and AEBP2, enhancing H3K27me3 deposition [199,200]. At the same time, H3K27me3 also attracts more copies of PRC1 [302]. The catalytic activity of PRC is central for the correct functioning of the Polycomb repressing system, both in the formation of Polycomb chromatin domains [303,304] and in the long-range interactions that may be formed by PRC1-repressed loci [302,305]. Loss of H2AK119ub severely destabilizes PRC2 activity, with a consequent loss of H3K27me3 and major activation of previously repressed genes [306]. Interestingly, recent studies showed that *Drosophila* embryos expressing catalytically dead PRC1, lacking therefore H2AK119ub, cannot fully develop, despite the PRC1 target genes still being silenced [307]. These results suggest that PRC1 also has repressive abilities that are independent from H2AK119 ubiquitination, and that H2AK119ub has essential functions that are beyond transcription.

It is also worth mentioning the interesting role played by H3 ubiquitination, deposited by the E3-ligase UHRF1 (ubiquitin like with PHD and RING finger domains 1). During DNA replication, UHRF1 is recruited to the replication fork, where it recognizes the unmodified H3R2 and H3K9me2/3 residues and binds to hemi-CpG methylated DNA [308,309]. The activity of the DNA methyltransferase DNMT1, which copies the pattern of CpG methylation of the mother strand to the daughter strand, requires the interaction of DNMT1 with both UHRF1 and the ubiquitinated residues on H3 [310,311]. H3ub therefore occupies a key position in the quest to maintaining cell identity through generations.

Role of Histone Ubiquitination in DNA Damage Response

Histone ubiquitination is one of the most important PTM in response to DNA damage. Indeed, disruption of ubiquitination leads to defects in the correct engagement of the DSB repair protein complexes [312–314]. Following DSBs, phosphorylation occurs at the K139 position of the histone H2A.X variant. In a chain of events, this leads to the recruitment of the RNF8/UBC13 complex, that ubiquitinates K63 on H2A/γH2A.X [265] and on the linker histone H1 [315]. In turn, H2Aub functions as a docking site for RNF168, another E3-ubiquitin ligase, that ubiquitinates H2AK12/13 to further amplify DDR signaling [227] and the binding of other factors, such as RAP80 (receptor-associated protein 80) and subsequently 53BP1 or BRCA1, two essential proteins for DSB repair [316,317]. The interaction between RNF8 and RNF168 is orchestrated by L3MBTL2 (histone methyl-lysine binding protein 2), a noncanonical PRC1 subunit which binds methylated H3 and H4 and accumulates at the damage site [318]. Other than H2A, H4K91 is modified by the E3-ligase BBAP (B-lymphoma and BAL associated protein) [319]. One other study showed evidence of H3 and H4 ubiquitination as a consequence of DNA damage caused by UV irradiation [320]. Fine-tuning of this process is guaranteed by the activity of DUBs, such as USP26 and USP37 which counteract RNF8 and RNF168 [321], and BRCC36 (BRCA1/2 containing complex 36), targeting H2Aub and RAP80. Loss of BRCC36 causes increased H2AK63ub spreading at the damage site [322].

Sumoylation

Closely related to ubiquitination is sumoylation, another PTM consisting in the covalent attachment of SUMO (small ubiquitin-related modifier), a peptide of about 100 amino acids, to target proteins. It is evolutionary conserved in all eukaryotes, with yeast expressing one SUMO protein and mammals expressing 5 SUMO paralogous (SUMO1–5) [323]. From a biochemical point of view, the enzymatic reaction attaching SUMO to K residues of substrate proteins is similar of that of ubiquitination, although there is a smaller number of enzyme homologs. The first step of the sumoylation cascade is catalyzed by SAE1/SAE2 (SUMO-activating enzyme), a E1-ligase, then SUMO is transferred to the E2 enzyme UBC9 (ubiquitin conjugating enzyme 9) [324], and finally to a E3-enzyme such as RAN binding protein 2 [325] and others [326,327]. Alternatively, UBC9 may directly attach SUMO to the substrate. Loss of UBC9 during embryonic development is lethal [328]. Removal of the SUMO moiety is done by members of the SENP (SUMO/sentrin-specific protease) family [329]. All four core histones, including the H2A.Z variant, can be sumoylated in yeast (Fig. 3.2), specific targets being the H2BK6/7/16/17, H2AK126 and all lysine residues in the N-terminal tail of H4 [330]. In humans, proteomics analysis identified H4K12 as a frequently modified site [331].

The role of histone sumoylation in transcription regulation is still unclear. While some studies reported evidence of acetylation-dependent sumoylation events on H3 and H4 [332], more recent works showed correlative evidence between sumoylation and transcription repression. Indeed, H4K12su is recognized by KMD1A, which mediates the removal of the methyl mark from the H3K4me1/2, normally associated with active chromatin regions [333]. KMD1A then associates with the coREST (corepressor for element 1 silencing TF) repressor and the histone deacetylase HDAC1 to further restrict gene expression [334]. Furthermore, histone desumoylation mediated by the Ulp2 protease promotes transcription elongation in yeast [335]. The implications of histone sumoylation on DDR are yet to be elucidated. However, it is known that H2A.Z is sumoylated following DNA damage, leading to recruitment of the repair protein Rad15 [336].

ADP-Ribosylation

ADP-ribosylation consists in the transfer of one (mono) or more (poly) ADP-ribose moieties from the cofactor NAD+ (nicotinamide adenine dinucleotide) onto a target protein. The two processes are known, respectively as MARylation (mono-ADP-ribosyl-ation) and PARylation (poly-ADP-ribosyl), and the enzymes catalyzing them are called MARTs (mono-ADP-ribosyltransferases) and PARPs (poly ADP-ribose polymerases) respectively. An alternative nomenclature has also been proposed, that classifies MARTs and PARPs families into MARTDs (diphtheria toxin-like) and MARTCs (cholera toxin-like) according to the structure of their catalytic domain [337]. In mammals, most of the histone ribosylation events are catalyzed by PARP1, supported by PARP2/3/7/10 [338]. Early fundamental studies on histone ADP-ribosylation were carried out by the lab of Teni Boulikas [339,340]. ADP-ribosylation was found in all core histones and in the H1 linker histone, often in association with other histone PTM [339,341,342]. However, the high variability of this PTM made it initially difficult to precisely pinpoint the exact histone residues on which it is deposited. The evolution of mass-spectrometry techniques proved to be of great importance in identifying the preferred targets of ribosyl transferases. We now know that Glu, Asp, Lys, Arg, and Ser are all sites of ADP-ribosylation [343–346]. Interestingly, PARP1 often competes with the H1 linker for binding at transcribed promoters [347], where it contributes to the maintenance of an open chromatin conformation by preventing H3K4me3 demethylation [348]. The DSBs generated by topoisomerases at actively transcribed regions may contribute to PARP1 recruitment and activity [349,350]. It also is found at active enhancers bound by nuclear hormone receptors [351,352] and, in *Drosophila*, PARP1 activity promotes enhancer-promoter separation and higher transcription activation [353]. It is to be noted that PARP1 is also further involved in transcription activation and repression by directly PARylating several chromatin modifying enzymes (reviewed in [354]).

The most investigated role of ADP-ribosylation is that in DDR. Residues located on the nucleosomal surface [345], as well as the H1.2 linker variant [103] are the main targets in case of DNA damage [345]. As MAR and PAR add one or more negative charges to the histone proteins, the result is relaxation of the chromatin surrounding the DNA lesion [355] and facilitation of histone removal [356]. PARP1 is primarily involved in this process, and PARP1 inhibitors completely abolish nucleosome eviction at the damage site [356]. Interestingly, recent studies suggest that ADP-ribosylation may be important for the fine-tuning of the cellular response to DNA damage. In the event of oxidative damage, PARP1 targets the H2A.XE141 residue. This leads to the recruitment of the NEIL3 glycosylase, an enzyme involved in the base excision repair (BER) mechanism, but at the same time prevents phosphorylation of S139 on the same histone, which is instead fundamental for activation of the DSB repair pathway [357].

Citrullination

Citrullination is the process converting arginine (R) residues into the nonstandard citrulline amino acid by replacing the imine (NH) group with a ketone (O) group. As at neutral pH arginine residues are positively charged but citrulline ones are not; citrullination causes a reduction in the net positive charge of target proteins. The calcium-dependent enzymes catalyzing the deimination reaction are called peptidylarginine deiminases (PADs). In mammals, five PAD isoforms (PAD1–4 and PAD6) have been identified, but only PAD2 and PAD4 (previously referred to as PAD5) were shown to have a nuclear activity [358,359]. Both citrullinate H3R2, H3R8, H3R17, and H3R26, while H2AR3 and H4R3 have been described to be exclusively citrullinated by PAD4 [358,360,361] (Fig. 3.2). Interestingly, PAD4 exerts a dual enzymatic activity, as it converts arginine and methylated arginine into citrulline via deamination and demethylimination, respectively [128].

PAD4 plays a fundamental role in the formation of NETs (neutrophil extracellular traps), a specialized inflammatory process of neutrophil cell death, part of the innate immune system [128,362]. Recent studies, including time-lapse high resolution microscopy experiments, contributed to dissect the timeline of this process [363,364]. Following a bacterial infection, activated neutrophils migrate to the infection site; here, they release large filamentous structures, the NETs, to trap and kill pathogens. NETs are composed by a scaffold of highly decondensed chromatin decorated with proteases and granular proteins.

Hypercitrullination mediated by PAD4 causes a massive loss of positive charges on histone molecules leading to the large scale chromatin decondensation necessary for NETs formation [363]. This is followed by rupture of the nuclear envelope and DNA release in the cytoplasm first and then, subsequently to the disassembly of the actin cytoskeleton and permeabilization of the plasma membrane, outside of the cell [363]. However, a recent study showed that while citrullination by PAD4 enhances NET formation, this may still occur in a PAD4-deficient context, suggesting that citrullination may not be absolutely required for this process [364].

Citrullination of various histone residues was also linked to pluripotency. H1R54Cit, mediated by PAD4, impairs H1 binding to nucleosomal DNA, contributing to the maintenance of an open chromatin state [365]. More recently, it was shown that BRG1 preferentially binds H3R26Cit histones. Loss of citrullination at this residue correlates with an increase of the H3K9me3 heterochromatin mark, suggesting that the H3R26Cit-BRG1 interaction may suppress heterochromatin formation in pluripotent cells [366].

Proline Isomerization

By definition, two molecules are isomers when they share the same chemical formula but have different chemical structures, which can lead to different physical and chemical properties.

Amino acids exist in two isomeric forms, *cis* and *trans*, according to the spatial position of their functional group on the carbon chain. While most amino acids in a peptide chain usually adopt the *trans* configuration, proline residues oscillate between the *cis* and the *trans* forms, thus affecting protein folding. While the isomerization process can occur spontaneously, enzymes called peptidilprolyl isomerases (PPIases) have evolved to accelerate switching between different conformations [367]. The first evidence of proline isomerization in histones was reported in 2006 by the lab of Tony Kouzarides that identified the budding yeast Frp4 protein as the enzyme catalyzing the isomerization of prolines 30 and 38 on the histone H3 tail (H3P30 and H3P38) (Fig. 3.2) [368]. The *trans*-proline conformation of histone H3P38 was found necessary to induce the Set2-mediated methylation of lysine 36 of histone H3 (H3K36). Accordingly, *cis*-isomerization correlated with lower H3K36 methylation and defective transcription elongation. More recently, a paper from Jane Mellor's group described how acetylation of H3K14 promotes *trans*-isomerization at H3P16. This leads to a reduction of H3K4me3 levels in vivo, suggesting that H3P16 *trans*-isomerization could have a role in transcription repression [369].

ROLE OF ABERRANT HISTONE MODIFICATIONS IN DISEASE

While the progress in identifying different forms of histone modifications and the enzymes responsible for their addition or removal has been remarkably rapid, the way these modifications are disrupted in disease remains incompletely understood. Consistent with the critical role of histone modifications in the proper control of not only gene expression, but also DNA repair, recombination, and genome integrity, disruption of the histone modifying genes has been associated with human diseases, notably cancer [370]. One of the most striking discoveries, resulting from the high-resolution cancer genome sequencing efforts (e.g., The Cancer Genome Atlas), is that recurrent mutations or copy number alterations in the genes encoding histone modifying enzymes are common in human cancers [371,372]. These findings also triggered debate on the role of deregulation of histone modifications in the mechanisms driving tumorigenesis and those underpinning epigenome-wide changes that are widespread in most tumors [371]. Based on the observation that many histone modifying genes are frequently deregulated across human tumors, they have been proposed to act as "drivers" of cancer development and progression, potentially working either as oncogenes or as tumor suppressors [372]. Ongoing and future studies aimed at gaining critical insights into the functional role epigenetic drivers ("epidrivers") should enhance knowledge of the mechanistic basis of epigenome reprogramming in cancer and other diseases. Developing "epigenetic drugs," capable of modulating specific histone modifications, could circumvent the toxicity of general epigenome reprogrammers (such as HDAC inhibitors) and provide a powerful tool for precision treatment.

CONCLUSIONS

Cellular processes such as transcription, DNA replication, and DNA repair are regulated by an intimate crosstalk between histone-modifying complexes and the modifications resulting from their catalytic activities. The potential functional link between histone modifications and transcriptional regulation has for a long time been a central hypothesis in the field, at the expense of DNA repair and DNA replication. Early correlative studies between gene expression and various histone marks culminated in the suggestion that histone modifications dictate the cell transcriptional state. However, with the exception of histone acetylation, the evidence for a causal link between histone modifications and transcriptional activity is limited. Moreover, the bulk of studies has been done on the most common histone modifications, particularly acetylation and methylation, at the cost of less common marks which may perhaps be crucially important for transcriptional regulation.

More recently, new approaches have been developed to focus on the functional relevance of the histone mark deposition. The latest breakthroughs in genome-editing techniques facilitate the generation of catalytically dead histone modifying enzymes and mutations of target histone residues. Progresses in the development of small molecule inhibitors raise similar possibilities. These technological advances, combined with the employment of more quantitative methods, allow us to separate the function of the mark from that of the enzyme, to clarify the exact involvement of histone modifications and histone modifying enzymes in all cellular processes. As highlighted here, exciting new results are showing that histone writers and erasers have in some cases important roles that lie outside of their catalytic one. Similarly, the role played by previously fundamental marks are being re-evaluated.

References

[1] Luger K. Crystal structure of the nucleosome core particle at 2.8 A resolution. Nature 1997;389:251–60.
[2] Fyodorov DV, Zhou B-R, Skoultchi AI, Bai Y. Emerging roles of linker histones in regulating chromatin structure and function. Nat Rev Mol Cell Biol 2018;19:192–206.
[3] Sauer PV, et al. Mechanistic insights into histone deposition and nucleosome assembly by the chromatin assembly factor-1. Nucleic Acids Res 2018;46:9907–17.
[4] Martire S, Banaszynski LA. The roles of histone variants in fine-tuning chromatin organization and function. Nat Rev Mol Cell Biol 2020;21:522–41.
[5] Rogakou EP, Pilch DR, Orr AH, Ivanova VS, Bonner WM. DNA double-stranded breaks induce histone H2AX phosphorylation on serine 139. J Biol Chem 1998;273:5858–68.
[6] Stiff T, et al. ATM and DNA-PK function redundantly to phosphorylate H2AX after exposure to ionizing radiation. Cancer Res 2004;64:2390–6.
[7] Cole L, et al. Multiple roles of H2A.Z in regulating promoter chromatin architecture in human cells. Nat Commun 2021;12:2524.
[8] Hu G, et al. H2A.Z facilitates access of active and repressive complexes to chromatin in embryonic stem cell self-renewal and differentiation. Cell Stem Cell 2013;12:180–92.
[9] Jin C, Felsenfeld G. Nucleosome stability mediated by histone variants H3.3 and H2A.Z. Genes Dev 2007;21:1519–29.
[10] Jin C, et al. H3.3/H2A.Z double variant–containing nucleosomes mark 'nucleosome-free regions' of active promoters and other regulatory regions. Nat Genet 2009;41:941–5.
[11] Mersfelder EL, Parthun MR. The tale beyond the tail: histone core domain modifications and the regulation of chromatin structure. Nucleic Acids Res 2006;34:2653–62.
[12] Cutter AR, Hayes JJ. A brief review of nucleosome structure. FEBS Lett 2015;589:2914–22.
[13] Ghoneim M, Fuchs HA, Musselman CA. Histone tail conformations: a fuzzy affair with DNA. Trends Biochem Sci 2021;46:564–78.
[14] Turner AL, et al. Highly disordered histone H1−DNA model complexes and their condensates. Proc Natl Acad Sci 2018;115:11964–9.
[15] Shakya A, Park S, Rana N, King JT. Liquid-liquid phase separation of histone proteins in cells: role in chromatin organization. Biophys J 2020;118:753–64.
[16] Nuebler J, Fudenberg G, Imakaev M, Abdennur N, Mirny LA. Chromatin organization by an interplay of loop extrusion and compartmental segregation. Proc Natl Acad Sci 2018;115: E6697–706.
[17] Misteli T. The self-organizing genome: principles of genome architecture and function. Cell 2020;183:28–45.
[18] Jenuwein T. Translating the histone code. Science 2001;293:1074–80.
[19] Rothbart SB, Strahl BD. Interpreting the language of histone and DNA modifications. Biochim Biophys Acta BBA - Gene Regul Mech 2014;1839:627–43.
[20] Hödl M, Basler K. Transcription in the absence of histone H3.2 and H3K4 methylation. Curr Biol 2012;22:2253–7.
[21] Pérez-Lluch S, et al. Absence of canonical marks of active chromatin in developmentally regulated genes. Nat Genet 2015;47:1158–67.
[22] Dorighi KM, et al. Mll3 and Mll4 facilitate enhancer RNA synthesis and transcription from promoters independently of H3K4 monomethylation. Mol Cell 2017;66:568–76 e4.
[23] Rickels R, et al. Histone H3K4 monomethylation catalyzed by Trr and mammalian COMPASS-like proteins at enhancers is dispensable for development and viability. Nat Genet 2017; 49:1647–53.
[24] Morgan MAJ, Shilatifard A. Reevaluating the roles of histone-modifying enzymes and their associated chromatin modifications in transcriptional regulation. Nat Genet 2020;52:1271–81.
[25] Allis CD, et al. New nomenclature for chromatin-modifying enzymes. Cell 2007;131:633–6.
[26] Görisch SM, Wachsmuth M, Tóth KF, Lichter P, Rippe K. Histone acetylation increases chromatin accessibility. J Cell Sci 2005;118:5825–34.
[27] Choudhary C, et al. Lysine acetylation targets protein complexes and co-regulates major cellular functions. Science 2009;325:834–40.
[28] Li T, et al. Characterization and prediction of lysine (K)-acetyltransferase specific acetylation sites. Mol Cell Proteomics 2012;11 M111.011080.
[29] Spange S, Wagner T, Heinzel T, Krämer OH. Acetylation of non-histone proteins modulates cellular signalling at multiple levels. Int J Biochem Cell Biol 2009;41:185–98.
[30] Brownell JE, Allis CD. An activity gel assay detects a single, catalytically active histone acetyltransferase subunit in Tetrahymena macronuclei. Proc Natl Acad Sci 1995;92:6364–8.
[31] Blanco-García N, Asensio-Juan E, de la Cruz X, Martínez-Balbás MA. Autoacetylation regulates P/CAF nuclear localization. J Biol Chem 2009;284:1343–52.
[32] Ruiz-Carrillo A, Wangh LJ, Allfrey VG. Selective synthesis and modification of nuclear proteins during maturation of avian erythroid cells. Arch Biochem Biophys 1976;174:273–90.
[33] Allis CD, Chicoine LG, Richman R, Schulman IG. Deposition-related histone acetylation in micronuclei of conjugating Tetrahymena. Proc Natl Acad Sci 1985;82:8048–52.
[34] Roth SY, Denu JM, Allis CD. Histone acetyltransferases. Annu Rev Biochem 2001;81–120.
[35] Park S-Y, Kim J-S. A short guide to histone deacetylases including recent progress on class II enzymes. Exp Mol Med 2020;52:204–12.

REFERENCES

[36] Guenther MG, Barak O, Lazar MA. The SMRT and N-CoR corepressors are activating cofactors for histone deacetylase 3. Mol Cell Biol 2001;21:6091−101.

[37] Albaugh BN, Denu JM. Catalysis by protein acetyltransferase Gcn5. Biochim Biophys Acta BBA - Gene Regul Mech 2021; 1864:194627.

[38] Holbert MA, et al. The human monocytic leukemia zinc finger histone acetyltransferase domain contains DNA-binding activity implicated in chromatin targeting. J Biol Chem 2007;282:36603−13.

[39] Yan Y, Barlev NA, Haley RH, Berger SL, Marmorstein R. Crystal structure of yeast esa1 suggests a unified mechanism for catalysis and substrate binding by histone acetyltransferases. Mol Cell 2000;6:1195−205.

[40] Schiltz RL, et al. Overlapping but distinct patterns of histone acetylation by the human coactivators p300 and PCAF within nucleosomal substrates. J Biol Chem 1999;274:1189−92.

[41] Doyon Y, Selleck W, Lane WS, Tan S, Côté J. Structural and functional conservation of the NuA4 histone acetyltransferase complex from yeast to humans. Mol Cell Biol 2004;24:1884−96.

[42] Govind CK, Zhang F, Qiu H, Hofmeyer K, Hinnebusch AG. Gcn5 promotes acetylation, eviction, and methylation of nucleosomes in transcribed coding regions. Mol Cell 2007;25:31−42.

[43] Ravens S, et al. Mof-associated complexes have overlapping and unique roles in regulating pluripotency in embryonic stem cells and during differentiation. eLife 2014;3:e02104.

[44] Wilde JJ, Siegenthaler JA, Dent SYR, Niswander LA. Diencephalic size is restricted by a novel interplay between GCN5 acetyltransferase activity and retinoic acid signaling. J Neurosci 2017;37:2565−79.

[45] Nagy Z, Tora L. Distinct GCN5/PCAF-containing complexes function as co-activators and are involved in transcription factor and global histone acetylation. Oncogene 2007;26:5341−57.

[46] Avvakumov N, Côté J. The MYST family of histone acetyltransferases and their intimate links to cancer. Oncogene 2007;26: 5395−407.

[47] Kalkhoven E. CBP and p300: HATs for different occasions. Biochem Pharmacol 2004;68:1145−55.

[48] McManus KJ, Hendzel MJ. Quantitative analysis of CBP- and P300-induced histone acetylations in vivo using native chromatin. Mol Cell Biol 2003;23:7611−27.

[49] Kawasaki H, et al. Distinct roles of the co-activators p300 and CBP in retinoic-acid-induced F9-cell differentiation. Nature 1998;393:284−9.

[50] Yao T-P, et al. Gene dosage–dependent embryonic development and proliferation defects in mice lacking the transcriptional integrator p300. Cell 1998;93:361−72.

[51] Kung AL, et al. Gene dose-dependent control of hematopoiesis and hematologic tumor suppression by CBP. Genes Dev 2000; 272−7.

[52] Kasper LH, et al. Conditional knockout mice reveal distinct functions for the global transcriptional coactivators CBP and p300 in T-cell development. Mol Cell Biol 2006;26:789−809.

[53] Fauquier L, et al. CBP and P300 regulate distinct gene networks required for human primary myoblast differentiation and muscle integrity. Sci Rep 2018;8:12629.

[54] Stashi E, York B, O'Malley BW. Steroid receptor coactivators: servants and masters for control of systems metabolism. Trends Endocrinol Metab 2014;25:337−47.

[55] Spencer TE, et al. Steroid receptor coactivator-1 is a histone acetyltransferase. Nature 1997;389:194−8.

[56] Sheppard HM, Harries JC, Hussain S, Bevan C, Heery DM. Analysis of the steroid receptor coactivator 1 (SRC1)-CREB binding protein interaction interface and its importance for the function of SRC1. Mol Cell Biol 2001;21:39−50.

[57] Grant PA, et al. Expanded lysine acetylation specificity of Gcn5 in native complexes. J Biol Chem 1999;274:5895−900.

[58] Grant PA, et al. Yeast Gcn5 functions in two multisubunit complexes to acetylate nucleosomal histones: characterization of an Ada complex and the SAGA (Spt/Ada) complex. Genes Dev 1997;11:1640−50.

[59] Jin Q, et al. Distinct roles of GCN5/PCAF-mediated H3K9ac and CBP/p300-mediated H3K18/27ac in nuclear receptor transactivation: histone acetylation and gene activation. EMBO J 2011;30:249−62.

[60] Henry KW. Transcriptional activation via sequential histone H2B ubiquitylation and deubiquitylation, mediated by SAGA-associated Ubp8. Genes Dev 2003;17:2648−63.

[61] Daniel JA, et al. Deubiquitination of histone H2B by a yeast acetyltransferase complex regulates transcription. J Biol Chem 2004;279:1867−71.

[62] McMahon SB, Wood MA, Cole MD. The essential cofactor TRRAP recruits the histone acetyltransferase hGCN5 to c-Myc. Mol Cell Biol 2000;20:556−62.

[63] Lang SE, McMahon SB, Cole MD, Hearing P. E2F transcriptional activation requires TRRAP and GCN5 cofactors. J Biol Chem 2001;276:32627−34.

[64] Han J, et al. The scaffolding protein JADE1 physically links the acetyltransferase subunit HBO1 with its histone H3−H4 substrate. J Biol Chem 2018;293:4498−509.

[65] Kueh AJ, Dixon MP, Voss AK, Thomas T. HBO1 is required for H3K14 acetylation and normal transcriptional activity during embryonic development. Mol Cell Biol 2011;31:845−60.

[66] Heintzman ND, et al. Distinct and predictive chromatin signatures of transcriptional promoters and enhancers in the human genome. Nat Genet 2007;39:311−18.

[67] Yang X-J, Ogryzko VV, Nishikawa J, Howard BH. A p300/CBP-associated factor that competes with the adenoviral oncoprotein E1A. Nature 1996;382:319−24.

[68] Allfrey VG, Faulkner R, Mirsky AE. Acetylation and methylation of histones and their possible role in the regulation of RNA synthesis. Proc Natl Acad Sci USA 1964;51:786−94.

[69] Shogren-Knaak M. Histone H4-K16 acetylation controls chromatin structure and protein interactions. Science 2006;311: 844−7.

[70] Mizzen CA, et al. The TAFII250 subunit of TFIID has histone acetyltransferase activity. Cell 1996;87:1261−70.

[71] Kundu TK, Wang Z, Roeder RG. Human TFIIIC relieves chromatin-mediated repression of RNA polymerase III transcription and contains an intrinsic histone acetyltransferase activity. Mol Cell Biol 1999;19:1605−15.

[72] Ogryzko VV, et al. Histone-like TAFs within the PCAF histone acetylase complex. Cell 1998;94:35−44.

[73] Papai G, et al. Structure of SAGA and mechanism of TBP deposition on gene promoters. Nature 2020;577:711−16.

[74] Baptista T, et al. SAGA Is a general cofactor for RNA polymerase II transcription. Mol Cell 2017;68:130−43 e5.

[75] Weinert BT, et al. Time-resolved analysis reveals rapid dynamics and broad scope of the CBP/p300 acetylome. Cell 2018;174: 231−44 e12.

[76] Wang Z, et al. Combinatorial patterns of histone acetylations and methylations in the human genome. Nat Genet 2008;40: 897−903.

[77] Rajagopal N, et al. Distinct and predictive histone lysine acetylation patterns at promoters, enhancers, and gene bodies. G3 GenesGenomesGenetics 2014;4:2051−63.

[78] Bian C, et al. Sgf29 binds histone H3K4me2/3 and is required for SAGA complex recruitment and histone H3 acetylation: Sgf29 functions as an H3K4me2/3 binder in SAGA. EMBO J 2011;30:2829−42.

[79] Pal S, Vishwanath SN, Erdjument-Bromage H, Tempst P, Sif S. Human SWI/SNF-associated PRMT5 methylates histone H3 arginine 8 and negatively regulates expression of ST7 and NM23 tumor suppressor genes. Mol Cell Biol 2004;24:9630–45.

[80] Gates LA, et al. Acetylation on histone H3 lysine 9 mediates a switch from transcription initiation to elongation. J Biol Chem 2017;292:14456–72.

[81] Du Y, et al. Nucleosome eviction along with H3K9ac deposition enhances Sox2 binding during human neuroectodermal commitment. Cell Death Differ 2017;24:1121–31.

[82] Karmodiya K, Krebs AR, Oulad-Abdelghani M, Kimura H, Tora L. H3K9 and H3K14 acetylation co-occur at many gene regulatory elements, while H3K14ac marks a subset of inactive inducible promoters in mouse embryonic stem cells. BMC Genomics 2012;13:424.

[83] Koch F, et al. Transcription initiation platforms and GTF recruitment at tissue-specific enhancers and promoters. Nat Struct Mol Biol 2011;18:956–63.

[84] Kouno T, et al. C1 CAGE detects transcription start sites and enhancer activity at single-cell resolution. Nat Commun 2019;10:360.

[85] Kim T-K, et al. Widespread transcription at neuronal activity-regulated enhancers. Nature 2010;465:182–7.

[86] De Santa F, et al. A large fraction of extragenic RNA Pol II transcription sites overlap enhancers. PLoS Biol 2010;8: e1000384.

[87] Arnold PR, Wells AD, Li XC. Diversity and emerging roles of enhancer RNA in regulation of gene expression and cell fate. Front Cell Dev Biol 2020;7:377.

[88] Heintzman ND, et al. Histone modifications at human enhancers reflect global cell-type-specific gene expression. Nature 2009;459:108–12.

[89] Bose DA, et al. RNA binding to CBP stimulates histone acetylation and transcription. Cell 2017;168:135–49 e22.

[90] Narita T, et al. Enhancers are activated by p300/CBP activity-dependent PIC assembly, RNAPII recruitment, and pause release. Mol Cell 2021;81:2166–82 e6.

[91] Sabari BR, et al. Coactivator condensation at super-enhancers links phase separation and gene control. Science 2018;361:eaar3958.

[92] Cho W-K, et al. Mediator and RNA polymerase II clusters associate in transcription-dependent condensates. Science 2018;361:412–15.

[93] Zhang T, Zhang Z, Dong Q, Xiong J, Zhu B. Histone H3K27 acetylation is dispensable for enhancer activity in mouse embryonic stem cells. Genome Biol 2020;21:45.

[94] Wang Z, et al. Genome-wide mapping of HATs and HDACs reveals distinct functions in active and inactive genes. Cell 2009;138:1019–31.

[95] Carrozza MJ, et al. Histone H3 methylation by Set2 directs deacetylation of coding regions by Rpd3S to suppress spurious intragenic transcription. Cell 2005;123:581–92.

[96] Kaimori J-Y, et al. Histone H4 lysine 20 acetylation is associated with gene repression in human cells. Sci Rep 2016;6:24318.

[97] Fazzio TG, Huff JT, Panning B. An RNAi screen of chromatin proteins identifies Tip60-p400 as a regulator of embryonic stem cell identity. Cell 2008;134:162–74.

[98] Ding L, et al. A genome-scale RNAi screen for Oct4 modulators defines a role of the Paf1 complex for embryonic stem cell identity. Cell Stem Cell 2009;4:403–15.

[99] Cai Y, et al. Subunit composition and substrate specificity of a MOF-containing histone acetyltransferase distinct from the male-specific lethal (MSL) complex. J Biol Chem 2010;285:4268–72.

[100] Anamika K, et al. Lessons from genome-wide studies: an integrated definition of the coactivator function of histone acetyl transferases. Epigenetics Chromatin 2010;3:18.

[101] Murr R, et al. Histone acetylation by Trrap–Tip60 modulates loading of repair proteins and repair of DNA double-strand breaks. Nat Cell Biol 2006;8:91–9.

[102] Dhar S, Gursoy-Yuzugullu O, Parasuram R, Price BD. The tale of a tail: histone H4 acetylation and the repair of DNA breaks. Philos Trans R Soc B Biol Sci 2017;372:20160284.

[103] Li Z, et al. Destabilization of linker histone H1.2 is essential for ATM activation and DNA damage repair. Cell Res 2018;28:756–70.

[104] Toiber D, et al. SIRT6 recruits SNF2H to DNA break sites, preventing genomic instability through chromatin remodeling. Mol Cell 2013;51:454–68.

[105] Battu A, Ray A, Wani AA. ASF1A and ATM regulate H3K56-mediated cell-cycle checkpoint recovery in response to UV irradiation. Nucleic Acids Res 2011;39:7931–45.

[106] Downs JA, et al. Binding of chromatin-modifying activities to phosphorylated histone H2A at DNA damage sites. Mol Cell 2004;16:979–90.

[107] Tang J, et al. Acetylation limits 53BP1 association with damaged chromatin to promote homologous recombination. Nat Struct Mol Biol 2013;20:317–25.

[108] Hsiao K-Y, Mizzen CA. Histone H4 deacetylation facilitates 53BP1 DNA damage signaling and double-strand break repair. J Mol Cell Biol 2013;5:157–65.

[109] Dantuma NP, Attikum H. Spatiotemporal regulation of post-translational modifications in the DNA damage response. EMBO J 2016;35:6–23.

[110] Sun Y, et al. Histone H3 methylation links DNA damage detection to activation of the tumour suppressor Tip60. Nat Cell Biol 2009;11:1376–82.

[111] Sun Y, Jiang X, Chen S, Fernandes N, Price BD. A role for the Tip60 histone acetyltransferase in the acetylation and activation of ATM. Proc Natl Acad Sci 2005;102:13182–7.

[112] Ikura T, et al. Involvement of the TIP60 histone acetylase complex in DNA repair and apoptosis. Cell 2000;102:463–73.

[113] Ikura T, et al. DNA damage-dependent acetylation and ubiquitination of H2AX enhances chromatin dynamics. Mol Cell Biol 2007;27:7028–40.

[114] Jha S, Shibata E, Dutta A. Human Rvb1/Tip49 Is required for the histone acetyltransferase activity of Tip60/NuA4 and for the downregulation of phosphorylation on H2AX after DNA damage. Mol Cell Biol 2008;28:2690–700.

[115] Lee H-S, Park J-H, Kim S-J, Kwon S-J, Kwon J. A cooperative activation loop among SWI/SNF, γ-H2AX and H3 acetylation for DNA double-strand break repair. EMBO J 2010;29:1434–45.

[116] Rea S, Eisenhaber F, Ponting CP, Allis CD, Jenuwein T. Regulation of chromatin structure by site-speci®c histone H3 methyltransferases. Nature 2000;406:7.

[117] Dodge JE, Kang Y-K, Beppu H, Lei H, Li E. Histone H3-K9 methyltransferase ESET is essential for early development. Mol Cell Biol 2004;24:2478–86.

[118] Müller J, et al. Histone methyltransferase activity of a drosophila polycomb group repressor complex. Cell 2002;111:197–208.

[119] Feng Q, et al. Methylation of H3-Lysine 79 Is mediated by a new family of HMTases without a SET domain. Curr Biol 2002;12:1052–8.

[120] Tang J, et al. PRMT1 Is the predominant type I protein arginine methyltransferase in mammalian cells. J Biol Chem 2000;275:7723–30.

[121] Lee J-H, et al. PRMT7, a new protein arginine methyltransferase that synthesizes symmetric dimethylarginine. J Biol Chem 2005;280:3656–64.

[122] Miranda TB, Miranda M, Frankel A, Clarke S. PRMT7 Is a member of the protein arginine methyltransferase family with a distinct substrate specificity. J Biol Chem 2004;279:22902–7.

References

[123] Halabelian L, Barsyte-Lovejoy D. Structure and function of protein arginine methyltransferase PRMT7. Life 2021;11:768.

[124] Shi Y, et al. Histone demethylation mediated by the nuclear amine oxidase homolog LSD1. Cell 2004;119:941–53.

[125] Metzger E, et al. LSD1 demethylates repressive histone marks to promote androgen-receptor-dependent transcription. Nature 2005;437:436–9.

[126] Tsukada Y, et al. Histone demethylation by a family of JmjC domain-containing proteins. Nature 2006;439:811–16.

[127] Whetstine JR, et al. Reversal of histone lysine trimethylation by the JMJD2 family of histone demethylases. Cell 2006;125:467–81.

[128] Wang Y, et al. Human PAD4 regulates histone arginine methylation levels via demethylimination. Science 2004;306:279–83.

[129] Cuthbert GL, et al. Histone deimination antagonizes arginine methylation. Cell 2004;118:545–53.

[130] Mantri M, et al. Crystal structure of the 2-oxoglutarate- and Fe(II)-dependent Lysyl hydroxylase JMJD6. J Mol Biol 2010;401:211–22.

[131] Webby CJ, et al. Jmjd6 catalyses Lysyl-hydroxylation of U2AF65, a protein associated with RNA splicing. Science 2009;325:90–3.

[132] Bernstein BE, et al. A bivalent chromatin structure marks key developmental genes in embryonic stem cells. Cell 2006;125:315–26.

[133] Barski A, et al. High-resolution profiling of histone methylations in the human genome. Cell 2007;129:823–37.

[134] Beacon TH, et al. The dynamic broad epigenetic (H3K4me3, H3K27ac) domain as a mark of essential genes. Clin Epigenetics 2021;13:138.

[135] Bannister AJ, et al. Spatial distribution of Di- and Tri-methyl lysine 36 of histone H3 at active genes. J Biol Chem 2005;280:17732–6.

[136] Rao B, Shibata Y, Strahl BD, Lieb JD. Dimethylation of histone H3 at lysine 36 demarcates regulatory and non regulatory chromatin genome-wide. Mol Cell Biol 2005;25:9447–59.

[137] Djebali S, et al. Landscape of transcription in human cells. Nature 2012;489:101–8.

[138] Huff JT, Plocik AM, Guthrie C, Yamamoto KR. Reciprocal intronic and exonic histone modification regions in humans. Nat Struct Mol Biol 2010;17:1495–9.

[139] Shah RN, et al. Examining the roles of H3K4 methylation states with systematically characterized antibodies. Mol Cell 2018;72:162–77 e7.

[140] Zhao XD, et al. Whole-genome mapping of histone H3 Lys4 and 27 trimethylations reveals distinct genomic compartments in human embryonic stem cells. Cell Stem Cell 2007;1:286–98.

[141] Lauberth SM, et al. H3K4me3 interactions with TAF3 regulate preinitiation complex assembly and selective gene activation. Cell 2013;152:1021–36.

[142] Vermeulen M, et al. Selective anchoring of TFIID to nucleosomes by trimethylation of histone H3 lysine 4. Cell 2007;131:58–69.

[143] Ng HH, Robert F, Young RA, Struhl K. Targeted recruitment of Set1 histone methylase by elongating Pol II provides a localized mark and memory of recent transcriptional activity. Mol Cell 2003;11:709–19.

[144] Okitsu CY, Hsieh C-L. DNA methylation dictates histone H3K4 methylation. Mol Cell Biol 2007;27:2746–57.

[145] Stadler MB, et al. DNA-binding factors shape the mouse methylome at distal regulatory regions. Nature 2011;480:490–5.

[146] Lee J-H, Voo KS, Skalnik DG. Identification and characterization of the DNA binding domain of CpG-binding protein. J Biol Chem 2001;276:44669–76.

[147] Xu C, et al. DNA sequence recognition of human CXXC domains and their structural determinants. Structure 2018;26:85–95 e3.

[148] Long HK, King HW, Patient RK, Odom DT, Klose RJ. Protection of CpG islands from DNA methylation is DNA-encoded and evolutionarily conserved. Nucleic Acids Res 2016;44:6693–706.

[149] Chen K, et al. Broad H3K4me3 is associated with increased transcription elongation and enhancer activity at tumor-suppressor genes. Nat Genet 2015;47:1149–57.

[150] Benayoun BA, et al. H3K4me3 breadth is linked to cell identity and transcriptional consistency. Cell 2014;158:673–88.

[151] Briggs SD. Histone H3 lysine 4 methylation is mediated by Set1 and required for cell growth and rDNA silencing in Saccharomyces cerevisiae. Genes Dev 2001;15:3286–95.

[152] Bledau AS, et al. The H3K4 methyltransferase Setd1a is first required at the epiblast stage, whereas Setd1b becomes essential after gastrulation. Development 2014;141:1022–35.

[153] Sze CC, et al. Histone H3K4 methylation-dependent and -independent functions of Set1A/COMPASS in embryonic stem cell self-renewal and differentiation. Genes Dev 2017;31:1732–7.

[154] Douillet D, et al. Uncoupling histone H3K4 trimethylation from developmental gene expression via an equilibrium of COMPASS, Polycomb and DNA methylation. Nat Genet 2020;52:615–25.

[155] Hu D, et al. The MLL3/MLL4 branches of the COMPASS family function as major histone H3K4 monomethylases at enhancers. Mol Cell Biol 2013;33:4745–54.

[156] Local A, et al. Identification of H3K4me1-associated proteins at mammalian enhancers. Nat Genet 2018;50:73–82.

[157] Kizer KO, et al. A novel domain in Set2 mediates RNA polymerase II interaction and couples histone H3 K36 methylation with transcript elongation. Mol Cell Biol 2005;25:3305–16.

[158] Kuo AJ, et al. NSD2 links dimethylation of histone H3 at lysine 36 to oncogenic programming. Mol Cell 2011;44:609–20.

[159] Kudithipudi S, Lungu C, Rathert P, Happel N, Jeltsch A. Substrate specificity analysis and novel substrates of the protein lysine methyltransferase NSD1. Chem Biol 2014;21:226–37.

[160] Dorighi KM, Tamkun JW. The trithorax group proteins Kismet and ASH1 promote H3K36 dimethylation to counteract Polycomb group repression in *Drosophila*. Development 2013;140:4182–92.

[161] Ruan C, Lee C-H, Cui H, Li S, Li B. Nucleosome contact triggers conformational changes of Rpd3S driving high-affinity H3K36me nucleosome engagement. Cell Rep 2015;10:204–15.

[162] Venkatesh S, Li H, Gogol MM, Workman JL. Selective suppression of antisense transcription by Set2-mediated H3K36 methylation. Nat Commun 2016;7:13610.

[163] Lorincz MC, Dickerson DR, Schmitt M, Groudine M. Intragenic DNA methylation alters chromatin structure and elongation efficiency in mammalian cells. Nat Struct Mol Biol 2004;11:1068–75.

[164] Baubec T, et al. Genomic profiling of DNA methyltransferases reveals a role for DNMT3B in genic methylation. Nature 2015;520:243–7.

[165] Bilokapic S, Halic M. Nucleosome and ubiquitin position Set2 to methylate H3K36. Nat Commun 2019;10:3795.

[166] Sansó M, et al. Cdk9 and H2Bub1 signal to Clr6-CII/Rpd3S to suppress aberrant antisense transcription. Nucleic Acids Res 2020;48:7154–68.

[167] Worden EJ, Hoffmann NA, Hicks CW, Wolberger C. Mechanism of cross-talk between H2B ubiquitination and H3 methylation by Dot1L. Cell 2019;176:1490–501 e12.

[168] Valencia-Sánchez MI, et al. Structural basis of Dot1L stimulation by histone H2B lysine 120 ubiquitination. Mol Cell 2019;74:1010–19 e6.

[169] Kim J, Kim H. Recruitment and biological consequences of histone modification of H3K27me3 and H3K9me3. ILAR J 2012;53:232–9.

[170] Chen S, et al. The PZP domain of AF10 senses unmodified H3K27 to regulate DOT1L-mediated methylation of H3K79. Mol Cell 2015;60:319–27.

[171] Hsu PL, et al. Structural basis of H2B ubiquitination-dependent H3K4 methylation by COMPASS. Mol Cell 2019;76:712–23 e4.

[172] Kwon M, et al. H2B ubiquitylation enhances H3K4 methylation activities of human KMT2 family complexes. Nucleic Acids Res 2020;48:5442–56.

[173] Bernt KM, et al. MLL-rearranged leukemia is dependent on aberrant H3K79 methylation by DOT1L. Cancer Cell 2011;20:66–78.

[174] Jones B, et al. The histone H3K79 methyltransferase Dot1L Is essential for mammalian development and heterochromatin structure. PLoS Genet 2008;4:e1000190.

[175] Nishioka K, et al. PR-Set7 is a nucleosome-specific methyltransferase that modifies lysine 20 of histone H4 and is associated with silent chromatin. Mol Cell 2002;9:1201–13.

[176] Schotta G. A silencing pathway to induce H3-K9 and H4-K20 trimethylation at constitutive heterochromatin. Genes Dev 2004;18:1251–62.

[177] Houston SI, et al. Catalytic function of the PR-Set7 histone H4 lysine 20 monomethyltransferase is essential for mitotic entry and genomic stability. J Biol Chem 2008;283:19478–88.

[178] Liu W, et al. PHF8 mediates histone H4 lysine 20 demethylation events involved in cell cycle progression. Nature 2010;466:508–12.

[179] Talasz H, Lindner HH, Sarg B, Helliger W. Histone H4-lysine 20 monomethylation is increased in promoter and coding regions of active genes and correlates with hyperacetylation. J Biol Chem 2005;280:38814–22.

[180] Trojer P, et al. L3MBTL1, a Histone-methylation-dependent chromatin lock. Cell 2007;129:915–28.

[181] Ren W, et al. Direct readout of heterochromatic H3K9me3 regulates DNMT1-mediated maintenance DNA methylation. Proc Natl Acad Sci 2020;117:18439–47.

[182] Ren W, et al. DNMT1 reads heterochromatic H4K20me3 to reinforce LINE-1 DNA methylation. Nat Commun 2021;12:2490.

[183] Richards EJ, Elgin SCR. Epigenetic codes for heterochromatin formation and silencing. Cell 2002;108:489–500.

[184] Tachibana M, Sugimoto K, Fukushima T, Shinkai Y. SET domain-containing protein, G9a, Is a novel lysine-preferring mammalian histone methyltransferase with hyperactivity and specific selectivity to lysines 9 and 27 of histone H3. J Biol Chem 2001;276:25309–17.

[185] Schotta G. Central role of drosophila SU(VAR)3–9 in histone H3-K9 methylation and heterochromatic gene silencing. EMBO J 2002;21:1121–31.

[186] Gauchier M, et al. SETDB1-dependent heterochromatin stimulates alternative lengthening of telomeres. Sci Adv 2019;5:eaav3673.

[187] Schultz DC. SETDB1: a novel KAP-1-associated histone H3, lysine 9-specific methyltransferase that contributes to HP1-mediated silencing of euchromatic genes by KRAB zinc-finger proteins. Genes Dev 2002;16:919–32.

[188] Jia S. RNAi-independent heterochromatin nucleation by the stress-activated ATF/CREB family proteins. Science 2004;304:1971–6.

[189] Kanoh J, Sadaie M, Urano T, Ishikawa F. Telomere binding protein Taz1 establishes Swi6 heterochromatin independently of RNAi at telomeres. Curr Biol 2005;15:1808–19.

[190] Bannister AJ, et al. Selective recognition of methylated lysine 9 on histone H3 by the HP1 chromo domain. Nature 2001;410:120–4.

[191] Wang L, et al. Histone modifications regulate chromatin compartmentalization by contributing to a phase separation mechanism. Mol Cell 2019;76:646–59 e6.

[192] Zhu J. Genome-wide chromatin state transitions associated with developmental and environmental cues. Cell 2013;152:642–54.

[193] Towbin BD, et al. Step-wise methylation of histone H3K9 positions heterochromatin at the nuclear periphery. Cell 2012;150:934–47.

[194] Lehnertz B, et al. Suv39h-mediated histone H3 lysine 9 methylation directs DNA methylation to major satellite repeats at pericentric heterochromatin. Curr Biol 2003;13:1192–200.

[195] Nicetto D, et al. H3K9me3-heterochromatin loss at protein-coding genes enables developmental lineage specification. Science 2019;363:294–7.

[196] Czermin B, et al. Drosophila enhancer of Zeste/ESC complexes have a histone H3 methyltransferase activity that marks chromosomal polycomb sites. Cell 2002;111:185–96.

[197] Kuzmichev A. Histone methyltransferase activity associated with a human multiprotein complex containing the Enhancer of Zeste protein. Genes Dev 2002;16:2893–905.

[198] Gao Z, et al. PCGF homologs, CBX proteins, and RYBP define functionally distinct PRC1 family complexes. Mol Cell 2012;45:344–56.

[199] Kalb R, et al. Histone H2A monoubiquitination promotes histone H3 methylation in Polycomb repression. Nat Struct Mol Biol 2014;21:569–71.

[200] Blackledge NP, et al. Variant PRC1 complex-dependent H2A ubiquitylation drives PRC2 recruitment and polycomb domain formation. Cell 2014;157:1445–59.

[201] Oksuz O, et al. Capturing the onset of PRC2-mediated repressive domain formation. Mol Cell 2018;70:1149–62 e5.

[202] van Mierlo G, Veenstra GJC, Vermeulen M, Marks H. The complexity of PRC2 subcomplexes. Trends Cell Biol 2019;29:660–71.

[203] Poepsel S, Kasinath V, Nogales E. Cryo-EM structures of PRC2 simultaneously engaged with two functionally distinct nucleosomes. Nat Struct Mol Biol 2018;25:154–62.

[204] Lavarone E, Barbieri CM, Pasini D. Dissecting the role of H3K27 acetylation and methylation in PRC2 mediated control of cellular identity. Nat Commun 2019;10:1679.

[205] Piunti A, et al. Therapeutic targeting of polycomb and BET bromodomain proteins in diffuse intrinsic pontine gliomas. Nat Med 2017;23:493–500.

[206] Södersten E, et al. A comprehensive map coupling histone modifications with gene regulation in adult dopaminergic and serotonergic neurons. Nat Commun 2018;9:1226.

[207] Weiner A, et al. Co-ChIP enables genome-wide mapping of histone mark co-occurrence at single-molecule resolution. Nat Biotechnol 2016;34:953–61.

[208] Voigt P, Tee W-W, Reinberg D. A double take on bivalent promoters. Genes Dev 2013;27:1318–38.

[209] Schmitges FW, et al. Histone methylation by PRC2 is inhibited by active chromatin marks. Mol Cell 2011;42:330–41.

[210] Blanco E, González-Ramírez M, Alcaine-Colet A, Aranda S, Di Croce L. The bivalent genome: characterization, structure, and regulation. Trends Genet 2020;36:118–31.

[211] Murphy PJ, et al. Single-molecule analysis of combinatorial epigenomic states in normal and tumor cells. Proc Natl Acad Sci 2013;110:7772–7.

[212] Neri F, et al. Dnmt3L antagonizes DNA methylation at bivalent promoters and favors DNA methylation at gene bodies in ESCs. Cell 2013;155:121–34.

[213] Bartke T, et al. Nucleosome-interacting proteins regulated by DNA and histone methylation. Cell 2010;143:470–84.

[214] Jacques SL, et al. CARM1 preferentially methylates H3R17 over H3R26 through a random kinetic mechanism. Biochemistry 2016;55:1635–44.

[215] Gao W, et al. Arginine methylation of HSP70 regulates retinoid acid-mediated RARβ2 gene activation. Proc Natl Acad Sci USA 2015;112:E3327–36.

[216] Kim JK, et al. PRMT4 is involved in insulin secretion via the methylation of histone H3 in pancreatic β cells. J Mol Endocrinol 2015;54:315–24.

[217] Wu J, Cui N, Wang R, Li J, Wong J. A role for CARM1-mediated histone H3 arginine methylation in protecting histone acetylation by releasing corepressors from chromatin. PLoS One 2012;7:e34692.

[218] An W, Kim J, Roeder RG. Ordered cooperative functions of PRMT1, p300, and CARM1 in transcriptional activation by p53. Cell 2004;117:735–48.

[219] Majumder S, et al. Methylation of histone H3 and H4 by PRMT5 regulates ribosomal RNA gene transcription. J Cell Biochem 2010;553–63.

[220] Beacon TH, Xu W, Davie JR. Genomic landscape of transcriptionally active histone arginine methylation marks, H3R2me2s and H4R3me2a, relative to nucleosome depleted regions. Gene 2020;742:144593.

[221] Girardot M, et al. PRMT5-mediated histone H4 arginine-3 symmetrical dimethylation marks chromatin at G + C-rich regions of the mouse genome. Nucleic Acids Res 2014;42:235–48.

[222] Yuan C-C, et al. Histone H3R2 symmetric dimethylation and histone H3K4 trimethylation are tightly correlated in eukaryotic genomes. Cell Rep 2012;1:83–90.

[223] Rothbart SB, et al. Association of UHRF1 with methylated H3K9 directs the maintenance of DNA methylation. Nat Struct Mol Biol 2012;19:1155–60.

[224] Liu F, et al. PRMT5-mediated histone arginine methylation antagonizes transcriptional repression by polycomb complex PRC2. Nucleic Acids Res 2020;48:2956–68.

[225] Li J, et al. A novel histone H4 arginine 3 methylation-sensitive histone H4 binding activity and transcriptional regulatory function for signal recognition particle subunits SRP68 and SRP72. J Biol Chem 2012;287:40641–51.

[226] Botuyan MV, et al. Structural basis for the methylation state-specific recognition of histone H4-K20 by 53BP1 and Crb2 in DNA repair. Cell 2006;127:1361–73.

[227] Mattiroli F, et al. RNF168 ubiquitinates K13–15 on H2A/H2AX to drive DNA damage signaling. Cell 2012;150:1182–95.

[228] Tuzon CT, et al. Concerted activities of distinct H4K20 methyltransferases at DNA double-strand breaks regulate 53BP1 nucleation and NHEJ-directed repair. Cell Rep 2014;8:430–8.

[229] Zimmerman DL, Boddy CS, Schoenherr CS. Oct4/Sox2 binding sites contribute to maintaining hypomethylation of the maternal Igf2/H19 imprinting control region. PLoS One 2013;8:e81962.

[230] Schotta G, et al. A chromatin-wide transition to H4K20 monomethylation impairs genome integrity and programmed DNA rearrangements in the mouse. Genes Dev 2008;22:2048–61.

[231] Ayrapetov MK, Gursoy-Yuzugullu O, Xu C, Xu Y, Price BD. DNA double-strand breaks promote methylation of histone H3 on lysine 9 and transient formation of repressive chromatin. Proc Natl Acad Sci 2014;111:9169–74.

[232] Sone K, et al. Critical role of lysine 134 methylation on histone H2AX for γ-H2AX production and DNA repair. Nat Commun 2014;5:5691.

[233] Karkhanis V, et al. Protein arginine methyltransferase 7 regulates cellular response to DNA damage by methylating promoter histones H2A and H4 of the polymerase δ catalytic subunit gene, POLD1. J Biol Chem 2012;287:29801–14.

[234] Watson NA, Higgins JMG. Histone kinases and phosphatases. Chromatin Signaling and Diseases. Elsevier; 2016. p. 75–94.

[235] Fuhs SR, Hunter T. pHisphorylation: the emergence of histidine phosphorylation as a reversible regulatory modification. Curr Opin Cell Biol 2017;45:8–16.

[236] Chen CCL, et al. H3S10ph broadly marks early-replicating domains in interphase ESCs and shows reciprocal antagonism with H3K9me2. Genome Res 2018;28:37–51.

[237] Briand N, Collas P. Lamina-associated domains: peripheral matters and internal affairs. Genome Biol 2020;21:85.

[238] Kind J, et al. Single-cell dynamics of genome-nuclear lamina interactions. Cell 2013;153:178–92.

[239] Clayton AL, Rose S, Barratt MJ, Mahadevan LC. Phosphoacetylation of histone H3 on c-*fos*- and c-*jun*-associated nucleosomes upon gene activation. EMBO J 2000;19:3714–26.

[240] Lo W-S, et al. Phosphorylation of serine 10 in histone H3 Is functionally linked in vitro and in vivo to Gcn5-mediated acetylation at lysine 14. Mol Cell 2000;5:917–26.

[241] Li J, et al. Programmable human histone phosphorylation and gene activation using a CRISPR/Cas9-based chromatin kinase. Nat Commun 2021;12:896.

[242] Josefowicz SZ, et al. Chromatin kinases act on transcription factors and histone tails in regulation of inducible transcription. Mol Cell 2016;64:347–61.

[243] Sawicka A, et al. H3S28 phosphorylation is a hallmark of the transcriptional response to cellular stress. Genome Res 2014;24:1808–20.

[244] Walter W, et al. 14-3-3 interaction with histone H3 involves a dual modification pattern of phosphoacetylation. Mol Cell Biol 2008;28:10.

[245] Hu X, et al. Histone cross-talk connects protein phosphatase 1α (PP1α) and histone deacetylase (HDAC) pathways to regulate the functional transition of bromodomain-containing 4 (BRD4) for inducible gene expression. J Biol Chem 2014;289:23154–67.

[246] Drobic B, Pérez-Cadahía B, Yu J, Kung SK-P, Davie JR. Promoter chromatin remodeling of immediate-early genes is mediated through H3 phosphorylation at either serine 28 or 10 by the MSK1 multi-protein complex. Nucleic Acids Res 2010;38:3196–208.

[247] Kellner WA, Ramos E, Van Bortle K, Takenaka N, Corces VG. Genome-wide phosphoacetylation of histone H3 at *Drosophila* enhancers and promoters. Genome Res 2012;22:1081–8.

[248] Khan DH, et al. Mitogen-induced distinct epialleles are phosphorylated at either H3S10 or H3S28, depending on H3K27 acetylation. Mol Biol Cell 2017;28:817–24.

[249] Crosio C, et al. Mitotic phosphorylation of histone H3: spatio-temporal regulation by mammalian aurora kinases. Mol Cell Biol 2002;22:874–85.

[250] Hendzel MJ, et al. Mitosis-specific phosphorylation of histone H3 initiates primarily within pericentromeric heterochromatin during G2 and spreads in an ordered fashion coincident with mitotic chromosome condensation. Chromosoma 1997;106:348–60.

[251] Hsiung CC-S, et al. Genome accessibility is widely preserved and locally modulated during mitosis. Genome Res 2015;25:213–25.

[252] Wei Y, Yu L, Bowen J, Gorovsky MA, Allis CD. Phosphorylation of histone H3 Is required for proper chromosome condensation and segregation. Cell 1999;97:99–109.

[253] Fischle W, et al. Regulation of HP1−chromatin binding by histone H3 methylation and phosphorylation. Nature 2005;438: 1116−22.

[254] Dawson MA, et al. JAK2 phosphorylates histone H3Y41 and excludes HP1α from chromatin. Nature 2009;461:819−22.

[255] Brehove M, et al. Histone core phosphorylation regulates DNA accessibility. J Biol Chem 2015;290:22612−21.

[256] North JA, et al. Histone H3 phosphorylation near the nucleosome dyad alters chromatin structure. Nucleic Acids Res 2014;42:4922−33.

[257] Shroff R, et al. Distribution and dynamics of chromatin modification induced by a defined DNA double-strand break. Curr Biol 2004;14:1703−11.

[258] Clouaire T, et al. Comprehensive mapping of histone modifications at DNA double-strand breaks deciphers repair pathway chromatin signatures. Mol Cell 2018;72:250−62 e6.

[259] Morrison AJ, et al. INO80 and gamma-H2AX interaction links ATP-Dependent Chromatin Remodeling to DNA Damage Repair. Cell 2009;119:767−75.

[260] Natale F, et al. Identification of the elementary structural units of the DNA damage response. Nat Commun 2017;8:15760.

[261] Collins PL, et al. DNA double-strand breaks induce H2Ax phosphorylation domains in a contact-dependent manner. Nat Commun 2020;11:3158.

[262] Stucki M, et al. MDC1 directly binds phosphorylated histone H2AX to regulate cellular responses to DNA double-strand breaks. Cell 2005;123:1213−26.

[263] Wu X, et al. Microcephalin regulates BRCA2 and Rad51-associated DNA double-strand break repair. Cancer Res 2009; 69:5531−6.

[264] Wang J, Gong Z, Chen J. MDC1 collaborates with TopBP1 in DNA replication checkpoint control. J Cell Biol 2011;193: 267−73.

[265] Wang B, Elledge SJ. Ubc13/Rnf8 ubiquitin ligases control foci formation of the Rap80/Abraxas/Brca1/Brcc36 complex in response to DNA damage. Proc Natl Acad Sci 2007;104:20759−63.

[266] Kleiner RE, Verma P, Molloy KR, Chait BT, Kapoor TM. Chemical proteomics reveals a γH2AX-53BP1 interaction in the DNA damage response. Nat Chem Biol 2015;11:807−14.

[267] Liu Y, et al. TOPBP1Dpb11 plays a conserved role in homologous recombination DNA repair through the coordinated recruitment of 53BP1Rad9. J Cell Biol 2017;216:623−39.

[268] Ramos F, Villoria MT, Alonso-Rodríguez E, Clemente-Blanco A. Role of protein phosphatases PP1, PP2A, PP4 and Cdc14 in the DNA damage response. Cell Stress 2019;3:70−85.

[269] Millan-Zambrano G, et al. Phosphorylation of histone H4T80 triggers DNA damage checkpoint recovery. Mol Cell 2018;72: 625−35 e4.

[270] Krishnan N, et al. Dephosphorylation of the C-terminal tyrosyl residue of the DNA damage-related Histone H2A.X Is mediated by the protein phosphatase eyes absent. J Biol Chem 2009;284:16066−70.

[271] Wood JL, Singh N, Mer G, Chen J. MCPH1 functions in an H2AX-dependent but MDC1-independent pathway in response to DNA damage. J Biol Chem 2007;282:35416−23.

[272] Chang H-Y, et al. Microcephaly family protein MCPH1 stabilizes RAD51 filaments. Nucleic Acids Res 2020;48:9135−46.

[273] Cook PJ, et al. Tyrosine dephosphorylation of H2AX modulates apoptosis and survival decisions. Nature 2009;458: 591−6.

[274] Hurd PJ, et al. Phosphorylation of histone H3 Thr-45 Is linked to apoptosis. J Biol Chem 2009;284:16575−83.

[275] Cheung WL, et al. Phosphorylation of histone H4 serine 1 during DNA damage requires casein kinase II in S. cerevisiae. Curr Biol 2005;15:656−60.

[276] Utley RT, Lacoste N, Jobin-Robitaille O, Allard S, Côté J. Regulation of NuA4 histone acetyltransferase activity in transcription and DNA repair by phosphorylation of histone H4. Mol Cell Biol 2005;25:8179−90.

[277] Goldknopf IL, et al. Isolation and characterization of protein A24, a 'histone-like' non-histone chromosomal protein. J Biol Chem 1975;250:7182−7.

[278] West MHP, Bonner WM. Histone 2B can be modified by the attachment of ubiquitin. Nucleic Acids Res 1980;8:4671−80.

[279] Vaughan RM, Kupai A, Rothbart SB. Chromatin regulation through ubiquitin and ubiquitin-like histone modifications. Trends Biochem Sci 2021;46:258−69.

[280] Wang H, et al. Role of histone H2A ubiquitination in Polycomb silencing. Nature 2004;431:873−8.

[281] Cao R, Tsukada Y, Zhang Y. Role of Bmi-1 and Ring1A in H2A ubiquitylation and hox gene silencing. Mol Cell 2005;20: 845−54.

[282] Buchwald G, et al. Structure and E3-ligase activity of the Ring−Ring complex of Polycomb proteins Bmi1 and Ring1b. EMBO J 2006;25:2465−74.

[283] Kalb R, Mallery DL, Larkin C, Huang JTJ, Hiom K. BRCA1 Is a histone-H2A-specific ubiquitin ligase. Cell Rep 2014;8: 999−1005.

[284] Robzyk K, Recht J, Osley MA. Rad6-dependent ubiquitination of histone H2B in yeast. Science 2000;287:501−4.

[285] Koken MH, et al. Structural and functional conservation of two human homologs of the yeast DNA repair gene RAD6. Proc Natl Acad Sci 1991;88:8865−9.

[286] Kim J, Hake SB, Roeder RG. The human homolog of yeast BRE1 functions as a transcriptional coactivator through direct activator interactions. Mol Cell 2005;20:759−70.

[287] Mevissen TET, Komander D. Mechanisms of deubiquitinase specificity and regulation. Annu Rev Biochem 2017;86:159−92.

[288] Emre NCT, et al. Maintenance of low histone ubiquitylation by Ubp10 correlates with telomere-proximal Sir2 association and gene silencing. Mol Cell 2005;17:585−94.

[289] Richardson LA, et al. A conserved deubiquitinating enzyme controls cell growth by regulating RNA polymerase I stability. Cell Rep 2012;2:372−85.

[290] Schulze JM, et al. Splitting the task: Ubp8 and Ubp10 deubiquitinate different cellular pools of H2BK123. Genes Dev 2011;25:2242−7.

[291] Cao J, Yan Q. Histone ubiquitination and deubiquitination in transcription, DNA damage response, and cancer. Front Oncol 2012;2.

[292] Nakagawa T, et al. Deubiquitylation of histone H2A activates transcriptional initiation via trans-histone cross-talk with H3K4 di- and trimethylation. Genes Dev 2008;22: 37−49.

[293] Draker R, Sarcinella E, Cheung P. USP10 deubiquitylates the histone variant H2A.Z and both are required for androgen receptor-mediated gene activation. Nucleic Acids Res 2011; 39:3529−42.

[294] Fierz B, et al. Histone H2B ubiquitylation disrupts local and higher-order chromatin compaction. Nat Chem Biol 2011;7: 113−19.

[295] Debelouchina GT, Gerecht K, Muir TW. Ubiquitin utilizes an acidic surface patch to alter chromatin structure. Nat Chem Biol 2017;13:105−10.

[296] Pavri R, et al. Histone H2B monoubiquitination functions cooperatively with FACT to regulate elongation by RNA polymerase II. Cell 2006;125:703−17.

[297] Fleming AB, Kao C-F, Hillyer C, Pikaart M, Osley MA. H2B ubiquitylation plays a role in nucleosome dynamics during transcription elongation. Mol Cell 2008;31:57−66.

REFERENCES

[298] Chen P, et al. Functions of FACT in breaking the nucleosome and maintaining its integrity at the single-nucleosome level. Mol Cell 2018;71:284–93 e4.

[299] Wu L, Li L, Zhou B, Qin Z, Dou Y. H2B ubiquitylation promotes RNA Pol II processivity via PAF1 and pTEFb. Mol Cell 2014;54:920–31.

[300] Gallego LD, et al. Phase separation directs ubiquitination of gene-body nucleosomes. Nature 2020;579:592–7.

[301] McGinty RK, Kim J, Chatterjee C, Roeder RG, Muir TW. Chemically ubiquitylated histone H2B stimulates hDot1L-mediated intranucleosomal methylation. Nature 2008;453:812–16.

[302] Blackledge NP, et al. PRC1 catalytic activity is central to polycomb system function. Mol Cell 2020;77:857–74 e9.

[303] Brockdorff N. Polycomb complexes in X chromosome inactivation. Philos Trans R Soc B Biol Sci 2017;372:20170021.

[304] Bousard A, et al. The role of *Xist*-mediated polycomb recruitment in the initiation of X-chromosome inactivation. EMBO Rep 2019;20:e48019.

[305] Kundu S, et al. Polycomb repressive complex 1 generates discrete compacted domains that change during differentiation. Mol Cell 2017;65:432–46 e5.

[306] Tamburri S, et al. Histone H2AK119 Mono-ubiquitination is essential for polycomb-mediated transcriptional repression. Mol Cell 2020;77:840–56 e5.

[307] Pengelly AR, Kalb R, Finkl K, Müller J. Transcriptional repression by PRC1 in the absence of H2A monoubiquitylation. Genes Dev 2015;29:1487–92.

[308] Arita K, Ariyoshi M, Tochio H, Nakamura Y, Shirakawa M. Recognition of hemi-methylated DNA by the SRA protein UHRF1 by a base-flipping mechanism. Nature 2008;455:818–21.

[309] Arita K, et al. Recognition of modification status on a histone H3 tail by linked histone reader modules of the epigenetic regulator UHRF1. Proc Natl Acad Sci 2012;109:12950–5.

[310] Qin W, et al. DNA methylation requires a DNMT1 ubiquitin interacting motif (UIM) and histone ubiquitination. Cell Res 2015;25:911–29.

[311] Li T, et al. Structural and mechanistic insights into UHRF1-mediated DNMT1 activation in the maintenance DNA methylation. Nucleic Acids Res 2018;46:3218–31.

[312] Ginjala V, et al. BMI1 Is recruited to DNA breaks and contributes to DNA damage-induced H2A ubiquitination and repair. Mol Cell Biol 2011;31:1972–82.

[313] Moyal L, et al. Requirement of ATM-dependent monoubiquitylation of histone H2B for timely repair of DNA double-strand breaks. Mol Cell 2011;41:529–42.

[314] Pan M-R, Peng G, Hung W-C, Lin S-Y. Monoubiquitination of H2AX protein regulates DNA damage response signaling. J Biol Chem 2011;286:28599–607.

[315] Thorslund T, et al. Histone H1 couples initiation and amplification of ubiquitin signalling after DNA damage. Nature 2016;527:389–93.

[316] Kim H, Chen J, Yu X. Ubiquitin-binding protein RAP80 mediates BRCA1-dependent DNA damage response. Science 2007;316:1202–5.

[317] Sobhian B, et al. RAP80 targets BRCA1 to specific ubiquitin structures at DNA damage sites. Science 2007;316:1198–202.

[318] Nowsheen S, et al. L3MBTL2 orchestrates ubiquitin signalling by dictating the sequential recruitment of RNF8 and RNF168 after DNA damage. Nat Cell Biol 2018;20:455–64.

[319] Yan Q, et al. BBAP monoubiquitylates histone H4 at Lysine 91 and selectively modulates the DNA damage response. Mol Cell 2009;36:110–20.

[320] Wang H, et al. Histone H3 and H4 ubiquitylation by the CUL4-DDB-ROC1 ubiquitin ligase facilitates cellular response to DNA damage. Mol Cell 2006;22:383–94.

[321] Typas D, et al. The de-ubiquitylating enzymes USP26 and USP37 regulate homologous recombination by counteracting RAP80. Nucleic Acids Res 2015;43:6919–33.

[322] Shao G, et al. The Rap80-BRCC36 de-ubiquitinating enzyme complex antagonizes RNF8-Ubc13-dependent ubiquitination events at DNA double strand breaks. Proc Natl Acad Sci 2009;106:3166–71.

[323] Ryu H-Y, Hochstrasser M. Histone sumoylation and chromatin dynamics. Nucleic Acids Res 2021;49:6043–52.

[324] Schwarz SE, Matuschewski K, Liakopoulos D, Scheffner M, Jentsch S. The ubiquitin-like proteins SMT3 and SUMO-1 are conjugated by the UBC9 E2 enzyme. Proc Natl Acad Sci 1998;95:560–4.

[325] Sakin V, Richter SM, Hsiao H-H, Urlaub H, Melchior F. Sumoylation of the GTPase Ran by the RanBP2 SUMO E3 ligase complex. J Biol Chem 2015;290:23589–602.

[326] Johnson ES, Gupta AA. An E3-like factor that promotes SUMO conjugation to the yeast septins. Cell 2001;106:735–44.

[327] Kagey MH, Melhuish TA, Wotton D. The polycomb protein Pc2 Is a SUMO E3. Cell 2003;113:127–37.

[328] Nacerddine K, et al. The SUMO pathway is essential for nuclear integrity and chromosome segregation in mice. Dev Cell 2005;9:769–79.

[329] Kunz K, Piller T, Müller S. SUMO-specific proteases and isopeptidases of the SENP family at a glance. J Cell Sci 2018;131:jcs211904.

[330] Nathan D. Histone sumoylation is a negative regulator in *Saccharomyces cerevisiae* and shows dynamic interplay with positive-acting histone modifications. Genes Dev 2006;20:966–76.

[331] Galisson F, et al. A novel proteomics approach to identify SUMOylated proteins and their modification sites in human cells. Mol Cell Proteomics 2011;10:S1–15.

[332] Hendriks IA, et al. Uncovering global SUMOylation signaling networks in a site-specific manner. Nat Struct Mol Biol 2014;21:927–36.

[333] Dhall A, Weller CE, Chu A, Shelton PMM, Chatterjee C. Chemically sumoylated histone H4 stimulates intranucleosomal demethylation by the LSD1–CoREST complex. ACS Chem Biol 2017;12:2275–80.

[334] Shi Y-J, et al. Regulation of LSD1 histone demethylase activity by its associated factors. Mol Cell 2005;19:857–64.

[335] Ryu H, et al. The Ulp2 SUMO protease promotes transcription elongation through regulation of histone sumoylation. EMBO J 2019;38:e102003.

[336] Kalocsay M, Hiller NJ, Jentsch S. Chromosome-wide Rad51 spreading and SUMO-H2A.Z-dependent chromosome fixation in response to a persistent DNA double-strand break. Mol Cell 2009;33:335–43.

[337] Hottiger MO, Hassa PO, Lüscher B, Schüler H, Koch-Nolte F. Toward a unified nomenclature for mammalian ADP-ribosyltransferases. Trends Biochem Sci 2010;35:208–19.

[338] Liu C, Vyas A, Kassab MA, Singh AK, Yu X. The role of poly ADP-ribosylation in the first wave of DNA damage response. Nucleic Acids Res 2017;45:8129–41.

[339] Boulikas T. At least 60 ADP-ribosylated variant histones are present in nuclei from dimethylsulfate-treated and untreated cells. EMBO J 1988;7:57–67.

[340] Boulikas T. Studies on protein poly(ADP-ribosylation) using high resolution gel electrophoresis. J Biol Chem 1990;265:14627–31.

[341] Ogata N, Ueda K, Kagamiyama H, Hayaishi O. ADP-ribosylation of histone H1. Identification of glutamic acid residues 2, 14, and the COOH-terminal lysine residue as modification sites. J Biol Chem 1980;255:7616−20.

[342] Messner S, et al. PARP1 ADP-ribosylates lysine residues of the core histone tails. Nucleic Acids Res 2010;38:6350−62.

[343] Zhang Y, Wang J, Ding M, Yu Y. Site-specific characterization of the Asp- and Glu-ADP-ribosylated proteome. Nat Methods 2013;10:981−4.

[344] Rosenthal F, Nanni P, Barkow-Oesterreicher S, Hottiger MO. Optimization of LTQ-orbitrap mass spectrometer parameters for the identification of ADP-ribosylation sites. J Proteome Res 2015;14:4072−9.

[345] Karch KR, Langelier M-F, Pascal JM, Garcia BA. The nucleosomal surface is the main target of histone ADP-ribosylation in response to DNA damage. Mol Biosyst 2017;13:2660−71.

[346] Palazzo L, et al. Serine is the major residue for ADP-ribosylation upon DNA damage. eLife 2018;7:e34334.

[347] Krishnakumar R, et al. Reciprocal binding of PARP-1 and histone H1 at promoters specifies transcriptional outcomes. Science 2008;319:819−21.

[348] Krishnakumar R, Kraus WL. PARP-1 Regulates chromatin structure and transcription through a KDM5B-dependent pathway. Mol Cell 2010;39:736−49.

[349] Boulikas T, Bastin B, Boulikas P, Dupuis G. Increase in histone poly(ADP-ribosylation) in mitogen-activated lymphoid cells. Exp Cell Res 1990;187:77−84.

[350] Ju B-G, et al. A topoisomerase IIβ-mediated dsDNA break required for regulated transcription. Science 2006;312:1798−802.

[351] Sawatsubashi S, et al. Ecdysone receptor-dependent gene regulation mediates histone poly(ADP-ribosyl)ation. Biochem Biophys Res Commun 2004;320:268−72.

[352] Schiewer MJ, et al. Dual roles of PARP-1 promote cancer growth and progression. Cancer Discov 2012;2:1134−49.

[353] Benabdallah NS, et al. Decreased enhancer-promoter proximity accompanying enhancer activation. Mol Cell 2019;76:473−84 e7.

[354] Ciccarone F, Zampieri M, Caiafa P. PARP1 orchestrates epigenetic events setting up chromatin domains. Semin Cell Dev Biol 2017;63:123−34.

[355] Poirier GG, de Murcia G, Jongstra-Bilen J, Niedergang C, Mandel P. Poly(ADP-ribosyl)ation of polynucleosomes causes relaxation of chromatin structure. Proc Natl Acad Sci 1982;79:3423−7.

[356] Yang G, et al. Poly(ADP-ribosyl)ation mediates early phase histone eviction at DNA lesions. Nucleic Acids Res 2020;48:3001−13.

[357] Chen Q, et al. ADP-ribosylation of histone variant H2AX promotes base excision repair. EMBO J 2021;40:e104542.

[358] Cherrington BD, Morency E, Struble AM, Coonrod SA, Wakshlag JJ. Potential role for peptidylarginine deiminase 2 (PAD2) in citrullination of canine mammary epithelial cell histones. PLoS One 2010;5:e11768.

[359] Mastronardi FG, et al. Increased citrullination of histone H3 in multiple sclerosis brain and animal models of demyelination: a role for tumor necrosis factor-induced peptidylarginine deiminase 4 translocation. J Neurosci 2006;26:11387−96.

[360] Nakashima K, Hagiwara T, Yamada M. Nuclear localization of peptidylarginine deiminase V and histone deimination in granulocytes. J Biol Chem 2002;277:49562−8.

[361] Zhang X, et al. Peptidylarginine deiminase 2-catalyzed histone H3 arginine 26 citrullination facilitates estrogen receptor target gene activation. Proc Natl Acad Sci 2012;109:13331−6.

[362] Lewis HD, et al. Inhibition of PAD4 activity is sufficient to disrupt mouse and human NET formation. Nat Chem Biol 2015;11:189−91.

[363] Thiam HR, et al. NETosis proceeds by cytoskeleton and endomembrane disassembly and PAD4-mediated chromatin decondensation and nuclear envelope rupture. Proc Natl Acad Sci 2020;117:7326−37.

[364] Tsourouktsoglou T-D, et al. Histones, DNA, and citrullination promote neutrophil extracellular trap inflammation by regulating the localization and activation of TLR4. Cell Rep 2020;31:107602.

[365] Christophorou MA, et al. Citrullination regulates pluripotency and histone H1 binding to chromatin. Nature 2014;507:104−8.

[366] Xiao S, et al. SMARCAD1 contributes to the regulation of naive pluripotency by interacting with histone citrullination. Cell Rep 2017;18:3117−28.

[367] Hanes SD. Prolyl isomerases in gene transcription. Biochim Biophys Acta 2015;1850:2017−34.

[368] Nelson CJ, Santos-Rosa H, Kouzarides T. Proline isomerization of histone H3 regulates lysine methylation and gene expression. Cell 2006;126:905−16.

[369] Howe FS, et al. Lysine acetylation controls local protein conformation by influencing proline isomerization. Mol Cell 2014;55:733−44.

[370] Zhao S, Allis CD, Wang GG. The language of chromatin modification in human cancers. Nat Rev Cancer 2021;21:413−30.

[371] Plass C, et al. Mutations in regulators of the epigenome and their connections to global chromatin patterns in cancer. Nat Rev Genet 2013;14:765−80.

[372] Halaburkova A, et al. Pan-cancer multi-omics analysis and orthogonal experimental assessment of epigenetic driver genes. Genome Res 2020;30:1517−32.

CHAPTER

4

The Epigenetics of Noncoding RNA

Ravindresh Chhabra

Department of Biochemistry, Central University of Punjab, Ghudda, Punjab, India

OUTLINE

Introduction	55	Regulation of Epigenetics by lncRNAs	65
The Noncoding RNA	55	piRNAs and Epigenetics	66
Genetics and Epigenetics	59	CircRNAs and Epigenetics	67
Epigenetic Regulation of miRNA Expression	60	Conclusions	67
Epitranscriptomics of miRNAs	62	Acknowledgments	67
Regulation of Epigenetics by miRNAs	62	Abbreviations	67
Plausible Interdependence between miRNA Targeting and mRNA Methylation	64	References	68
Epigenetic Regulation of lncRNAs	64		
Epitranscriptomics of lncRNA	65		

INTRODUCTION

The Noncoding RNA

The central dogma of life had clearly established the importance of the RNA molecule in the flow of genetic information. The understanding of transcription and translation processes further elucidated three distinct classes of RNA: mRNA, tRNA and rRNA. mRNA carries the information from DNA and gets translated to structural or functional proteins; hence, they are referred to as the coding RNA (RNA which codes for proteins). tRNA and rRNA help in the process of translation among other functions. A major part of the DNA, however, does not code for proteins and was previously referred to as junk DNA. The scientists started realizing the role of the junk DNA in the late 1990s and the ENCODE project, initiated in 2003, proved the significance of junk DNA beyond any doubt [1]. Many RNA types are now known to be transcribed from DNA in the same way as mRNA, but unlike mRNA they do not get translated into any protein; hence, they are collectively referred to as noncoding RNA (ncRNA). The studies have revealed that up to 90% of the eukaryotic genome is transcribed but only 1%–2% of these transcripts code for proteins, the rest all are ncRNAs [2–4]. The ncRNAs less than 200 nucleotides are called small noncoding RNAs and greater than 200 nucleotides are called long noncoding RNAs (lncRNAs) (https://lncipedia.org/). The small ncRNAs can be further classified into miRNAs (http://www.mirbase.org/), siRNAs, piRNAs (http://www.regulatoryrna.org/database/piRNA/) and snoRNAs (http://scottgroup.med.usherbrooke.ca/snoDB/). Circular RNA (circRNA) (http://circatlas.biols.ac.cn/) is another novel class of ncRNA whose size ranges from 100 bases to over 4 kb [5]. The properties of different classes of ncRNAs are briefly discussed in Fig. 4.1A.

The universe of noncoding RNAs continues to expand as the function of many more noncoding RNAs have been elucidated in the recent years. For instance, vault RNA is a distinct class of small ncRNAs with sizes ranging from 88 to 140 nucleotides. Absence of m5c methylation in the vault RNA can further trigger their aberrant

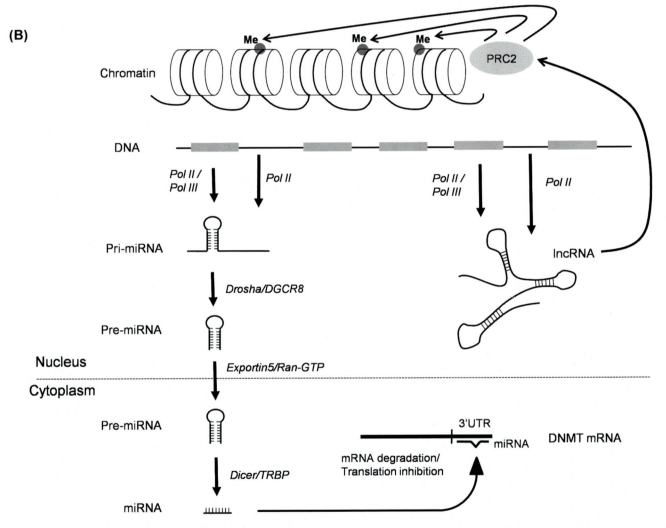

FIGURE 4.1 (A) Different classes of ncRNAs associated with epigenetic phenomenon. The number of RNAs belonging to each class are as per their listing in the referenced databases [6–10]. These numbers are bound to increase in the future. The numbers are not available for endogenous siRNAs. #Certain snoRNAs have been reported to have a length of more than 500 nucleotides [11]. (B) Schematic representation of miRNA and lncRNA biogenesis and their mechanism of action in regulating epigenetic modifications miRNA biogenesis begins in the nucleus with the formation of primary miRNA transcript (Pri-miRNA) which is a few thousand kb in length. The transcription is mediated by either RNA polymerase II (Pol II) or RNA polymerase III (Pol III). The pri-miRNAs are cleaved by Drosha–DGCR8 complex to generate precursor miRNA (pre-miRNA) which is few hundred bases in length. The pre-miRNA is then exported to the cytoplasm by Exportin-5/Ran-GTP and further cleaved by Dicer/TRBP complex to give rise to ~21 nucleotides long mature miRNA. The mature miRNA is capable of mRNA degradation or translational repression of its target mRNAs. Most of the lncRNAs are nucleus bound and do not have such an elaborate process of biogenesis. The intergenic lncRNAs are transcribed by Pol II and the intronic lncRNAs are transcribed by Pol III. To regulate epigenetic mechanisms, the miRNAs target the expression of proteins like DNMTs while the lncRNAs cause chromatin remodeling by recruiting proteins like PRC2 or DNMTs to preferential chromosome locations. The gray blocks represent different genes and the straight lines connecting these blocks represent intergenic regions in DNA. Me denotes methylation.

processing into small-vault RNAs (svRNAs), a few of which have functions similar to miRNAs [12]; one of these svRNAs is necessary for cellular differentiation [13]. The recent advances in high throughput small RNA studies and improved bioinformatics analysis have also highlighted the abundance of small noncoding RNA fragments derived from tRNA and rRNA. In spite of their high abundance, the small RNAs derived from rRNAs and tRNAs were considered as degraded products of no consequence in the RNA sequencing experiment and therefore excluded from further analysis. They are, however, now being studied as relevant ncRNA species with important functions. The small RNA fragments derived from rRNAs and tRNAs, including tRNA-derived fragments (tRFs) and tRNA halves (tiRNAs), are produced from specific shearing of rRNAs and tRNAs [14,15]. tRFs and tiRNAs are responsible for mRNA stability, translation, and epigenetics [15] and rRNAs are involved in mRNA degradation or translational repression [16]. Since very little is known about these small RNAs derived from vault, rRNAs and tRNAs, they are not discussed further in this chapter.

The ncRNAs, both small and long, have been shown to regulate most of the biological processes including apoptosis [17], Wnt signaling [18], epigenetics (Tables 4.1 and 4.2), embryonic development, [19] and tissue morphogenesis [20].

TABLE 4.1　A Brief List of the miRNAs Associated With Epigenetics Phenomenon and the Corresponding Pathological Condition

miRNA	Epigenetic Modification/DNMT Target	Pathological Condition	References
miR-155	Hypomethylated	B-cell lymphoma	[21]
miR-191	Hypomethylated	Hepatocellular carcinoma	[22]
let-7a-3	Hypomethylated	lung adenocarcinoma	[23]
miR-375	Hypomethylated	Er-α positive breast cancer	[24]
miR-200a/b	Hypomethylated	pancreatic cancer	[25]
miR-126	Hypomethylated	Systemic lupus erythematosus	[26]
miR-124−1, -2, -3	Hypermethylated	pancreatic cancer and Acute lymphocytic leukemia	[27,28]
miR-124−1, -2,	Hypermethylated	cervical cancer	[29]
miR-124−3	Hypermethylated	Hepatocellular carcinoma	[30]
miR-34b	Hypermethylated	Acute myeloid leukemia, non-small cell lung cancer	[31,32]
miR-34a	Hypermethylated	prostate cancer	[33,34]
miR-34a, -b, -c	Hypermethylated	colorectal cancer	[35,36]
miR-127	Histone deacetylation and hypermethylation	bladder carcinoma	[37]
miR-29a, -b, -c	DNMT3A	Lung cancer, Acute myeloid leukemia, cerebral ischemia	[38–40]
miR-29a, -b	DNMT3B	Lung cancer, Acute myeloid leukemia	[38,39]
miR-34b[a]	DNMT3B	Prostate cancer	[41]
miR-152	DNMT1	Cholangiocarcinoma, Hepatocellular carcinoma, atherosclerosis, ovarian cancer, endometrial cancer	[42–46]
miR-185	DNMT1	Hepatocellular carcinoma, ovarian cancer	[45,47]
miR-29b[a]	DNMT1	Acute myeloid leukemia	[39]
miR-148	DNMT1	Cholangiocarcinoma	[42]
miR-342	DNMT1	Colorectal cancer	[48]
miR-34b	DNMT1	Prostate cancer	[41]
miR-126, -148, -21[a]	DNMT1	Systemic lupus erythematosus	[26,49]

[a]Indirect inhibitor.

TABLE 4.2 Some of the lncRNAs Involved in Epigenetic Regulation

lncRNA	Function	Biological Role	Cell Line/Tissue	References
HOTTIP	Induces transcription by H3K4me3	Activation of HOXA genes	Foreskin and lung fibroblasts	[50]
ecCEBPA	Associates with DNMT1 and inhibits methylation	Induces transcription of CEBPA	HL-60 and U937 cell line	[51]
DACOR1	Recruits DNMT1 to specific genomic loci and induces methylation	Blocks transcription	HCT116 cell line	[52]
H19	Inhibits SAHH to alter DNMT3b-mediated methylation	Developmental aberrations	Mouse muscle cells	[53]
Evf2	Inhibits methylation of Dlx5/6 enhancer elements and differentially regulates the Dlx5/6 expression by controlling the recruitment of activator and repressor proteins	Differential regulation of genes with shared regulatory elements	Mouse model system	[54]
Kcnq1ot1	Recruits DNMT1 to differentially methylated region	Regulates imprinting of genes within the Kcnq1 domain	Mouse embryonic stem cells	[55]
ANRIL	PRC2 recruitment to and H3K27me3 of p15^{INK4B}	Induces cellular proliferation	WI-38 cell line	[56]
AIR	Targets H3K9 histone methyltransferase G9a to chromatin	Induces transcriptional repression	Mouse models	[57]
HOTAIR	Genome wide induction of H3K27me3 and targeting of PRC2 complex on selective genes	Induces cancer metastasis	MCF7, MDA-MB-231, SK-BR3, breast cancer	[58]

The biogenesis of miRNAs has been extensively reported in literature [59]. The miRNA genes are transcribed by Pol II promoters to produce primary transcripts (pri-transcript) of miRNA which are several thousand nucleotides in length. The pri-transcript is cleaved by Drosha/DGCR8 complex to form about a hundred nucleotide-long precursor miRNA (pre-miRNA). The pre-miRNA is then transported out of the nucleus by Ran-GTP/Exportin-5 complex where it is further cleaved by Dicer/TRBP complex to produce mature miRNA of ∼22 nucleotides in length. The mature miRNA along with Argonaute 2 proteins is loaded onto the RISC complex, binds the target mRNA and causes mRNA degradation or translational repression. In contrast, the lncRNAs do not have an elaborate process of biogenesis and resemble, to a large extent, the transcription of protein-coding genes. The lncRNAs are further classified based on their genomic loci and the direction of their transcription [60,61]. The various types of lncRNAs are long intergenic ncRNAs (lincRNAs, originate from the region between protein-coding genes), intronic lncRNAs (originate from the intronic regions of protein-coding genes), sense lncRNAs (originate from exonic regions of protein-coding genes), antisense lncRNAs (transcribed in the direction opposite to the protein-coding genes), pseudogene lncRNAs (originate from pseudogenes), enhancer lncRNAs (originate from enhancer sequences) and promoter lncRNAs (originate from promoter sequences). Most of the lncRNAs are transcribed by Pol II complex but some of them are also transcribed by Pol III complex [61]. The lncRNAs regulate gene expression in a variety of ways including chromatin reprogramming and as miRNA sponges. The biogenesis of miRNAs and lncRNAs is depicted in Fig. 4.1B.

siRNAs originate from transposon, sense-antisense and stem-loop transcripts. The dsRNA/stem-loop transcripts, formed in the nucleus, are cleaved by Dicer in the cytoplasm to give rise to functional siRNAs. siRNAs, like miRNAs interact with Argonaute 2 protein and cause degradation of their target mRNA [62]. piRNAs are transcribed from intergenic repetitive elements or transposons and are often thought to arise from regions known as piRNA clusters. ssRNA transcripts from these regions act as precursors to piRNA. Unlike miRNAs and siRNAs, they are not dependent on Drosha/Dicer for their biogenesis and interact with PIWI proteins [62]. Circular RNAs were discovered more than 2 decades ago [63] but it is only recently that they have caught the fancy of researchers. Their biogenesis involves back splicing or exon skipping of pre-mRNA transcripts. Although, devoid of 5′ cap and poly-A tail, the circular structure makes them relatively more stable than the linear transcripts. The biogenesis of piRNAs and circRNAs is depicted in Fig. 4.2. Most of the snoRNAs are encoded in the intronic regions with the exception of a few which originate from nonprotein coding genes. The introns excised during mRNA processing are cleaved by exonucleases

INTRODUCTION

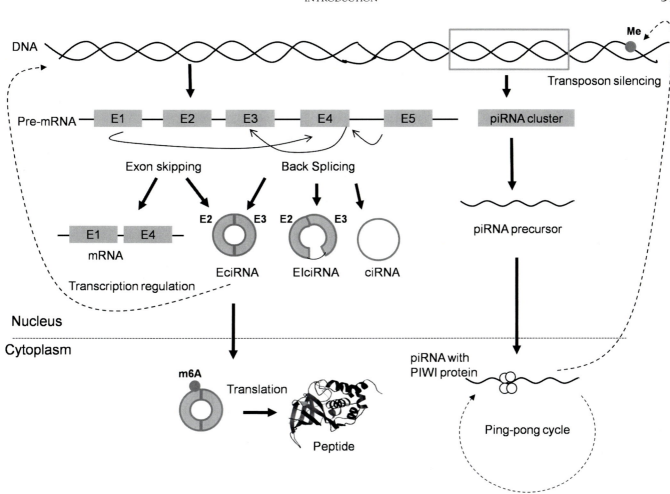

FIGURE 4.2 The biogenesis of piRNA and circRNA. The circRNAs originate through the back splicing or exon skipping of pre-mRNA transcripts. They can further be classified into exonic circRNAs (EciRNA), exon-intron circRNAs (EIciRNA) and circular intronic RNAs (ciRNA). The genomic loci for circRNA seem to have highly active promoter regions with enhanced H3K27Ac. Recent studies have shown that N^6-methyladenosine (m6A) methylations are responsible for the coding potential of circRNAs. The piRNAs are transcribed as piRNA precursors from the specific genomic loci referred to as piRNA clusters. piRNA precursors are exported out of the nucleus and are acted upon by helicases and nucleases in a cyclical process referred to as ping pong cycle to produce mature piRNA. The mature piRNAs have been shown to be responsible for the methylation of transposable elements in the germline. E1, E2, E3, E4, and E5 imply exon 1 to 5, the solid line implies intronic sequence and Me denotes methylation.

to give rise to mature snoRNAs [64]. The snoRNAs are majorly involved in the modification of rRNA but additional functions have also been elucidated including a role similar to mature or precursor miRNAs [65,66].

Genetics and Epigenetics

Genetics is a branch of science which encompasses flow of information from one generation to the next in the form of DNA sequence. It is a science which attributes a particular function or a phenotype of a cell to a specific DNA sequence or sequences. However, there are instances where genotype to phenotype outcomes cannot be explained solely on the basis of DNA sequence. This is because there are additional regulatory factors which determine the genotype to phenotype conversions. Epigenetics is a branch of science which deals with the regulation of the genotype via the modification of nucleotide sequences (DNA and RNA) and/or histone proteins to alter its phenotypic outcome. RD Hotchkiss was the first to report such a modification of DNA sequence [67]. In this seminal work, DNA was found to be methylated at the C5 position of cytosine. Since then, about seven modifications have been reported in DNA and more than 100 such modifications have been elucidated for RNAs [68,69]. Out of these, the most frequently studied and reported modification is methylation both in the case of DNA as well as RNA (including mRNA, tRNA, rRNA, lncRNA, and miRNAs). DNA methylation occurs majorly in promoter regions and to a lesser degree in intragenic regions and is responsible for either inducing or inhibiting transcription [70]. Among the histone modifications, H3K4me3, H3K27me3, H3K9me3, and H3K36me3 are reported to be the key regulators of mRNA expression [71]. RNA methylation is comparatively more complex as it can affect the ribose sugar as well as all the bases [68,72]. Among the RNA species, tRNA and rRNA methylations have been studied in detail because of their abundance, higher stability and availability of techniques to characterize them. The innovations in NGS technology have, however, made it easier to study methylations in mRNA, too. The most common methylations reported in mRNA include m6A (methylation of adenosine at its N^6 position) and m5C (methylation of cytosine at its C^5 position). m6A occurs mostly in 3′-UTRs, internal long exons and around stop codons [73,74] and m5C is enriched in UTR regions and many of the m5C modifications in 3′-UTRs overlap with the binding region of Argonaute proteins [75]. While m6A modification clearly destabilizes the mRNA [74,76], m5C modification may be speculated to stabilize the mRNA [77]. Surprisingly, recent reports have also implicated the role of m6A in enhanced translation, which implies the context driven function of m6A modification [78,79].

Genomic imprinting is a special type of epigenetic phenomenon in which a small fixed set of genes are expressed in a parental-origin specific manner [80]. Genomic imprinting is essential for normal mammalian development and growth. The aberration in genomic imprinting results in a number of diseases including Angelmann syndrome, Prader-Willi syndrome, and autism [80]. The underlying mechanism of DNA and histone modification and the involvement of noncoding RNA are the same for imprinting at developmental stages and the epigenetic regulation at later stages [81].

This chapter will mostly discuss about the epigenetics in context to two major classes of noncoding RNA, miRNAs and lncRNAs. This is because there is sufficient literature available for both these classes in context to epigenetics to arrive at some conclusive inferences. In addition, the recent reports describing the epigenetics of piRNAs and circRNAs are also discussed.

Epigenetic Regulation of miRNA Expression

There are around 1900 miRNA sequences spread across the entire human genome which give rise to ~2600 functional miRNAs [6]. They can be intronic or intergenic miRNAs depending on their genomic loci. The intergenic miRNAs are transcribed independently by their promoters, while the intronic miRNA may either share the promoter with the host gene or may be transcribed independently of the host gene promoter [82]. Epigenetic modification of the miRNA promoter region has been implicated in both physiological and pathological conditions. In fact, an estimated 50% of miRNA genes have been associated with CpG islands, the hotspot for DNA methylation [83].

Hypermethylation/hypomethylation of miRNA promoters has been observed in a number of cancers. Baer et al. in their study on whole miRNome promoter methylation in chronic lymphocytic leukemia (CLL) revealed that 90 miRNA promoters were hypomethylated and 38 promoters were hypermethylated [21]. There are, however, more reports on hypermethylated miRNA promoters in the literature. This may be attributed to the availability of pharmacological demethylating agents which makes it comparatively easier to identify the function of hypermethylated miRNA promoters. Nevertheless, hypomethylated miRNAs have also been implicated in a number of cancers. The increased expression of miR-155 in higher grade B-cell lymphoma [21], miR-191 in hepatocellular carcinoma (HCC) [22], let-7a-3 in lung adenocarcinoma [23], miR-375 in Er-α positive breast cancer [24] and miR-200a/b in pancreatic cancer [25] is because of the hypomethylation of their respective promoters. The hypomethylated miR-375 inhibits the expression of RASD1, an antiproliferative factor in Er-α positive breast cancer [24] and the hypomethylated let-7a-3 causes increase in expression of CDK6, PCNA, PRDX1, and CXCL5 (which induce lung cancer progression) and reduces the expression of PPARG, TGFB2, and SFRP1 (which inhibit lung cancer proliferation) [23]. The direct functional target of hypomethylated miRNAs remains unknown in most of the studies. In addition to hypomethylation, histone modification is also responsible for the elevated levels of miR-375 in Er-α positive breast cancer [24].

The hypermethylation of miR-124 promoter is frequently seen in the tumors of lung, breast, colon, and leukemia and lymphoma [84]. This hypermethylation is however absent in neuroblastoma and sarcoma [84]. miR-124 originates from three different genomic loci and hence are classified as miR-124−1, -124−2 and -124−3, which implies that the sequence of their mature form is the same while the sequence of primary and precursor forms is different. The methylation pattern of all three miR-124 promoters varies in different cancers. While all three types are hypermethylated in pancreatic cancer [27] and acute lymphoblastic leukemia (ALL) [28], miR-124−1 and -2 are hypermethylated in cervical cancer [29] and only miR-124−3 is hypermethylated in HCC [30]. The silencing of miR-124 as a result of hypermethylation induces the expression of vimentin (EMT marker) in HCC [30], Rac1 (oncogene) in pancreatic cancer [27] and CDK6 (cell cycle regulator) in colorectal cancer [84] and ALL [28]. The hypermethylation of miR-124 is correlated with increased mortality and relapse rate in ALL [28] and is responsible for enhanced proliferation and invasion in pancreatic cancer [27].

The promoter hypermethylation is also responsible for silencing of the miR-34 family in acute myeloid leukemia (AML) [31], non–small cell lung cancer (NSCLC) [32], prostate cancer [33,34], and colorectal cancer [35,36]. This aberrant methylation of miR-34 family members induces proliferation in AML and metastasis in NSCLC, colorectal cancer, and prostate cancer. Moreover, the reduced miR-34 induces cancer stem cell phenotype [85], which has been implicated in chemoresistance, metastasis, and relapse of a number of cancers. Since miR-34 silencing caused by hypermethylation could aid cancer progression in different ways, any drug which reverses the methylation pattern of miR-34 would make an excellent therapeutic. Difluorinated-curcumin and 3,3′-diindolylmethane are two such compounds which have been shown to have demethylation effect on miR-34 in the chemoresistant colon cancer cell line [36] and in prostate cancer cells [86], respectively.

In addition to promoter methylation, histone modification can also regulate the expression of miRNAs. At times, both histone modification and DNA methylation coordinate to regulate miRNA expression. For instance, in bladder carcinoma, histone deacetylation and DNA methylation of miRNA promoter was found to be responsible for downregulation of miR-127 and as a result, increase in expression of its target, BCL6 [37]. Similarly, in ALL, DNA methylation and histone modifications (enhanced levels of H3K9me2 and/or low levels of H3K4me3) in CpG islands of around 13 miRNAs downregulated the expression of all 13 miRNAs [87].

An interesting liaison among epigenetics, imprinting and miRNAs were described in mice where miR-127 and miR-136 mediate the function of imprinting to silence the expression of Rtl1, the imprinted gene responsible for placenta formation (Fig. 4.3). The Rtl1 gene is maternally imprinted, as the intergenic differentially methylated region (IG-DMR) downstream of Rtl1 gene is unmethylated while in the paternal chromosome, the methylated IG-DMR prevents the silencing of Rtl1. The unmethylated IG-DMR allows the transcription of antisense-Rtl1 transcript which is further processed to form miR-127 and miR-136. Both these miRNAs bind to the Rtl1 transcript and causes its degradation [88]. It was also observed that it is not just the absence of methylation which induces the expression of miR-127 and miR-136 but it is the epigenetic imprinting which controls their expression [89].

The above discussed examples do not give a complete list of hypermethylated miRNAs and their

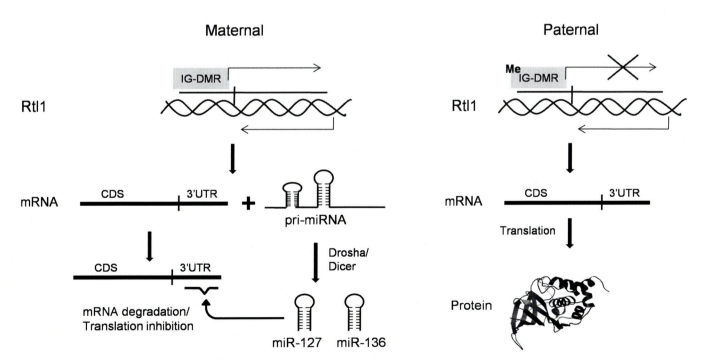

FIGURE 4.3 The imprinted gene, Rtl1 is silenced by miR-127 and miR-136. On the maternal chromosome, IG-DMR downstream of Rtl1 gene is unmethylated which allows the transcription of Rtl1 as well as antisense-Rtl1. The antisense-Rtl1 is further processed to form mature miR-127 and miR-136, which bind to the Rtl1 transcript and silences its expression. On the paternal chromosome, however, IG-DMR is methylated, which precludes the transcription of the antisense-Rtl1 transcript, thus allowing the formation of functional Rtl1 protein. Me denotes methylation. *IG-DMR*, Intergenic differentially methylated region.

associated functional significance, but rather provide a general idea about the significance of epigenetic regulation of miRNAs in the normal functioning of different cell types.

Epitranscriptomics of miRNAs

In contrast to the effect of DNA modifications on the miRNA expression pattern, the direct modifications of miRNA and their associated function are not as extensively reported.

Methylation of miRNAs is a fundamental part of miRNA biogenesis in plants but has rarely been reported in case of animals. In plants, miRNA-mediated degradation of its mRNA-target bound in miRNA-Ago complex is facilitated by uridylation of 3' ends of mRNA by HESO1 (HEN1 suppressor) enzyme. As part of their biogenesis, the plant miRNAs undergo 2'-O-methylation at their 3' ends by the enzyme, HEN1 [90]. The methylated 3' ends of miRNAs prevents their degradation by uridylation and enables their recycling [91]. miRNA-mediated degradation of mRNA in animals also happens by uridylation [92] but herein, very little is known about the recycling of miRNAs. Since many reports in the literature suggest that uridylation reduces the activity as well as stability of miRNAs [93,94], it may be assumed that miRNAs may also get degraded along with its target.

Methylation of miRNAs in humans was first reported in 2014 when it was observed that different forms of miR-125b (primary miRNA, precursor miRNA and mature miRNA) in HeLa cells are methylated at adenosine sites (m6A) by NSUN2 [95]. This is quite surprising as traditionally NSUN2 has been known to add methyl group to C5 of cytosine (m5C). The reason for changed substrate preference of NSUN2 remains unknown. The m6A methylation at any stage of miR-125b biogenesis causes inhibition of the subsequent step in biogenesis. This implies methylation of primary miRNA reduced its processing to precursor form, methylation of precursor miRNA reduced its cleavage to a mature form and methylation of mature miRNA inhibited its function by attenuating the recruitment of RISC by miRNA. Since miR-125b originates from two different sites in the genome, they are of two subtypes, miR-125b-1 and miR-125b-2. Although, all the forms of both miR-125b-1 and -125b-2 were found to be methylated, the functional effect was surprisingly observed only in the case of miR-125b-2. The effect of m6A on miRNAs is not just exclusive to miR-125b but extends to a majority of miRNAs [96]. A rather fundamental role of methylation in miRNA biogenesis in humans was uncovered by Alarcon and co-workers in 2015 [97]. They reported that methyltransferase-like 3 (METTL3) mediates m6A methylation of pri-miRNAs leading to their recognition and processing by DGCR8. They also proposed that RNA-binding protein HNRNPA2B1 could be the one which recognizes the m6A mark on the pri-miRNA and promotes its processing by recruiting DGCR8 [98]. m6A has similar effects (destabilization and protein recruitment) on mRNA and lncRNAs, as well [74,76,99], thus implying a universal function of m6A across various RNA classes.

The additional modifications seen in some miRNAs include DNMT3A/AGO4-mediated m5C [100], METTL1 mediated m7G (7-methylguanosine) methylation [101] and adenosine-to-inosine (A-to-I) editing [102] which regulates the miRNA biogenesis and alters their target specificity.

Regulation of Epigenetics by miRNAs

DNA methylation is one of the most studied epigenetic modifications. In mammals, there are three enzymes known to add methyl group to the specific sites in DNA collectively referred to DNA methyltransferases (DNMT1, DNMT3a, and DNMT3b). DNMT1 is referred to as the maintenance methyltransferase as it maintains the methylation pattern during DNA replication and in contrast, DNMT3a and DNMT3b are called de novo methyltransferases as they carry out de novo methylation during embryonic development. Since methylation patterns are crucial for maintaining the healthy state in individuals, the aberration in expression of the DNMTs results in a number of diseases including systemic lupus erythematosus (SLE) and cancers of liver, prostate, and ovary. In many of these diseases, the aberrant expression of DNMTs is because of miRNAs [38,41–43,49,103]. A few of these miRNA-DNMTs interactions are discussed below.

miR-29b targets DNMT3a and -3b in lung cancer as well as in AML [38,39]. In both these diseases, the expression of DNMT3a and -3b increases as a result of reduced miR-29b. This leads to enhanced methylation and subsequent silencing of tumor suppressor genes, FHIT and WWOX in lung cancer [38] and p15[INK4b] and ESR1 in AML [39]. Apart from directly targeting DNMT3a and -3b, miR-29b also targets DNMT1 indirectly by downregulating its transcriptional activator, Sp1. This makes miR-29b a potent hypomethylating agent which can cause global DNA hypomethylation [39]. The loss of miR-152 causes enhanced expression of its target, DNMT1 in atherosclerosis [44], cholangiocarcinoma [42], HCC [43], and ovarian cancer [45]. The enhanced DNMT1 leads to enhanced methylation and subsequent silencing of an anti-atherosclerotic gene, ERα [44] and tumor-suppressor genes, and CDH1 and

GSTP1 in HCC [43]. The genes targeted for methylation by enhanced DNMT1 in cholangiocarcinoma and ovarian cancer remains unknown. miR-29b and miR-152 therefore act as tumor suppressor genes by keeping the expression of DNMTs in check. Similar functions have also been reported for miR-148 in cholangiocarcinoma [42], miR-185 in ovarian cancer [45], miR-34b in prostate cancer [41], miR-342 in colorectal cancer [48] and miR-152 in endometrial cancer [46]. A rather elaborate mechanism is involved in an indirect activation of de novo methylation by an miR-290 cluster during mouse development. Herein, miR-290 cluster prevents the expression of Rbl2, the inhibitor of DNMT3a and -3b. The absence of miR-290 would thus cause genomic hypomethylation, resulting in elongated telomeres and increased telomere recombination (Fig. 4.4) [104]. Embryonic stem cell maintenance is also dependent on the regulation of de novo methylation by miR-290 cluster. During differentiation of embryonic stem cells, miR-290 mediates indirect activation of DNMTs and represses the expression of Oct4 and Nanog (stem cell marker genes) via promoter methylation [105].

While cancer is associated with hypermethylation by DNMTs, an autoimmune disease, SLE, is associated with hypomethylation caused by reduced expression of DNMT1. Enhanced expression of miR-148 [49] and miR-126 [26] was found to be responsible for this hypomethylation as they both target DNMT1. The hypomethylation of genes CD11a and CD70 and their subsequent activation are speculated to be a cause of SLE [26]. Interestingly, reduced expression of miR-148 causes hypermethylation in cholangiocarcinoma [42], thereby stressing the importance for tight regulation of miRNA expression.

There are a few studies which imply a potential feedback mechanism between the miRNAs and DNMTs. For instance, the hypomethylated promoter region of miR-126 enhances its expression in SLE and miR-126 inhibits the expression of DNMT [26]. Similarly, the reduced expression of miR-34b and miR-152 could be attributed to hypermethylation of their upstream region in prostate cancer [41] and endometrial cancer [46], respectively, which in turn leads to high expression of DNMT1. The mechanism by which the cell decides whether to silence miRNAs or its target DNMTs remains unexplored. Some of the miRNAs associated with epigenetics phenomenon are listed in Table 4.1.

miRNAs can be expected to similarly regulate the m5C methylation of mRNA by targeting the enzymes NSUN2 and TRDMT1, and m6A methylation of mRNA by targeting the enzymes METTL3, METTL14, WTAP, and FTO. In fact, miR-33a and miR-4429 have been reported to inhibit m6A methylation by directly targeting the 3′-UTR of METTL3 in NSCLC and gastric cancer, respectively [106,107]. miR-145, however, regulates m6A in a slightly nonconventional way as it forms a complex with AGO1/FTO/CLIP3 to induce m6A demethylation of CLIP3, a tumor suppressor transcript in glioblastoma stem cells [108]. miRNA-mediated epigenetic regulation has also been implicated in cellular differentiation as well as pluripotency. miRNAs modulate m6A modification of mRNAs by regulating the binding of the methyltransferase, METTL3 to their target mRNAs via a sequencing pairing mechanism [109]. While increased m6A modification promoted the reprogramming of the cells to become pluripotent, reduced m6A had an opposite effect. Wade et al., in 2015, reported that miR-302 promotes the mesendodermal differentiation by inhibiting the expression of BAF170 protein (subunit of the Brg1 chromatin remodeling complex) in human embryonic stem cells [110]. The loss of miR-302 would enhance the expression of BAF170 protein, thereby favoring the differentiation of human embryonic stem cells into ectodermal lineage and hindering the mesendodermal

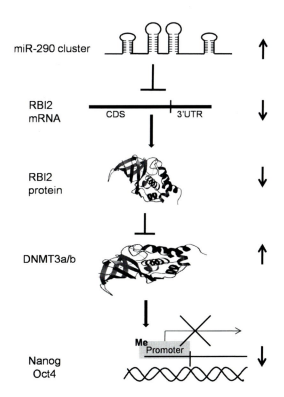

FIGURE 4.4 The indirect regulation of DNMT3a/b by miR-290 cluster affects the cellular differentiation. The de novo methylation of Nanog and Oct4 by DNMT3a/b is essential for the stem cell differentiation during development, but Rbl2 protein blocks DNMT3a/b. During the onset of cellular differentiation, the miR-290 cluster inhibits the expression of Rbl2 and thus allows DNMT3a/b to methylate and silence Nanog and Oct4, the genes responsible for stem cell maintenance. The up arrow indicates enhanced expression and the down arrow indicates reduced expression during cellular differentiation. Me denotes methylation.

differentiation. The literature also shows that miRNA bound RISC complex has a remarkable role in maintaining the stemness of embryonic stem cells. The dysregulation of Dicer, Drosha, or Argonaute proteins is enough to impair the pluripotency of cells. [111]. It is believed that prior to cellular differentiation, the miRNA bound RISC complex allows the translation of key chromatin regulators by releasing its transcriptional control.

Plausible Interdependence between miRNA Targeting and mRNA Methylation

The majority of the 144 types of modifications in RNA [69] were discovered in rRNA or tRNA, owing to their abundance and stability in contrast to the short life span of mRNAs. In the later years, however, the advancement of technology has enabled the study of mRNA modifications. The most commonly studied modifications in mRNA are methylation of adenosine at N^6 position (m6A) [74,112] and methylation of cytosine at C^5 position (m5C) [75,113]. Intriguingly, both m6A and m5C sites are found to be enriched in the 3'-UTR region [73,75,112], the preferred region of miRNAs for binding its target mRNA and subsequently inhibiting its expression. For m5C sites in 3'-UTR, the enrichment was even more in the binding region of Argonaute proteins [75]. Since miRNAs target the 3'-UTR sites by forming a complex with the Argonaute protein, this suggests a potential role of m5C in the miRNA-mediated target inhibition. In the 3'-UTRs where both m6A and TargetScan predicted miRNA-binding sites are present, the m6A precede miRNA-binding sites 62% of the time [73]. miRNAs were speculated to cause this m6A methylation as the majority of the highly expressed miRNAs of brain were shown to have a higher percentage of targets with m6A sites [73].

m6A sites have also been reported to alter the structure of RNA to aid/prevent its binding to specific proteins [98,114]. Specifically, binding of the HUR protein (which blocks miRNA-mRNA interaction) is prevented by m6A methylation [115]. Additionally, while m6A has a negative effect on the stability of mRNA [74,76], m5C may stabilize the mRNA [77]. It is thus worth speculating that the presence of m6A in 3'-UTR regions alters the mRNA structure in a way which makes it amenable for miRNA binding and subsequent degradation in contrast to m5C which somehow prevents the miRNA from inhibiting the expression of its target mRNA. These speculations are however yet to be validated experimentally. This aspect of mRNA methylation in context to miRNA function had been reviewed previously [103].

Epigenetic Regulation of lncRNAs

The mammalian genome encodes thousands of lncRNAs which may be intragenic or intergenic [116]. They may or may not be polyadenylated. lncRNAs have critical roles in various biological processes including embryonic development [19] and tissue morphogenesis [20]. In spite of all these important functions, lncRNAs are poorly conserved in mammalian species [50]. The regulation of lncRNA expression is not yet fully understood but there are many reports which highlight the epigenetic modulation of lncRNAs [71,117,118]. Modification of lysine residues of histone protein H3 at position 4 (H3K4), 9 (H3K9), 27 (H3K27), and 36 (H3K36) play a vital regulatory role in the gene expression. The most common modifications at the aforementioned residues are acetylation and methylation. While H3K4 and H3K36 methylation, as well as H3K9 acetylation, are usually associated with euchromatin implying actively transcribed regions, H3K9 and H3K27 methylation are associated with heterochromatin regions, implying silencing of gene expression in those regions.

Sati et al. in 2012 performed a comprehensive study to locate epigenetic marks across lncRNAs loci in humans [71]. The epigenetic marks discussed in their study included DNA methylation and histone modifications, H3K27me3 and H3K9me3 (associated with reduced expression) and H3K4me3 and H3K36me3 (associated with enhanced expression). They observed that in contrast to the protein-coding genes where *hypo*methylation around or at a transcription start site (TSS) corresponds to higher expression, and *hyper*methylation around or at a TSS corresponds to reduced expression, the methylation density in lncRNAs is high in the downstream region of TSS, irrespective of their expression levels. Thus, suggesting that other factors may be needed to regulate lncRNAs. CpG islands present in lncRNAs were marked by H3K4me3 [71], which generally corresponds to transcriptionally active chromatin [119]. Also, H3K4me3 and H3K36me3 modifications in the TSS and gene body, respectively, corresponded to an increased expression of lncRNAs [71]. Between the repressive marks, while H3K9me3 was present, H3K27me3 was absent in the TSS of the highly expressed lncRNA [71]. In short, many of the epigenetic marks known for the protein-coding genes like DNA methylation and H3K9me3 behave ambiguously or have no effect in cases of lncRNAs but for other epigenetic marks, their effect is similar to that in protein-coding genes. In embryonic stem cells, lncRNAs with high levels of H3K27me3 in their promoter region have reduced expression, and inhibiting this modification by silencing H3K27me3 methyltransferase (Ezh2) induces their expression [117].

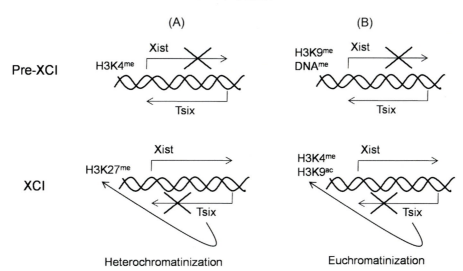

FIGURE 4.5 The proposed mechanisms for X-chromosome inactivation by Xist/Tsix. (A) The research findings of Sun et al. [121] state that prior to X-chromosome inactivation (pre-XCI), Tsix maintains euchromatin state by H3K4 dimethylation and precludes the transcription of Xist. During XCI, Tsix expression is repressed which causes heterochromatinization indicated by the loss of H3K4 dimethylation and gain of H3K27 trimethylation. This leads to transcription of Xist which causes XCI. (B) A contradictory mechanism was proposed by Navarro et al. [120] where during pre-XCI, DNA methylation and H3K9 trimethylation maintains the heterochromatin state of the Xist promoter which precludes the transcription of Xist. During XCI, Tsix expression is repressed which causes euchromatinization of the Xist promoter indicated by gain of H3K4 di/trimethylation and H3K9 acetylation. This allows Xist to express and cause XCI. Me denotes methylation and Ac denotes acetylation.

The expression of XIST, lncRNA responsible for X-chromosome inactivation in females, is regulated by Tsix (noncoding antisense transcript to XIST) via histone modifications [120]. In males, Tsix induces H3K9 trimethylation and DNA methylation [120] and prevents H3K27 trimethylation [120,121] making the embryonic stem cells unable to transcribe XIST, thereby preventing X-inactivation. In the case of females, there is asymmetric regulation of Tsix where Tsix is repressed and XIST expression is induced only on the future inactive X [122]. There are, however, contradictory reports on the mechanism behind Tsix-mediated induction of XIST (Fig. 4.5) [120,121]. On activated X-chromosome, Tsix maintains the repression on XIST expression by directing DNMT3a to methylate the promoter of XIST, thus, allowing the genes of X-chromosome to be expressed.

Epitranscriptomics of lncRNA

Site specific methylation of lncRNAs came to light in 2013 when Amort et al. showed site specific cytosine methylation of XIST and HOTAIR lncRNAs by analyzing a part of the sequences of these two lncRNAs [72]. HOTAIR was always found to be methylated at the cytosine position 1683 in HEK293, NT2, Hs578T, and BT-20 and the HOC7 cell lines. The methylated site C1683 of HOTAIR is in the vicinity of the sequence responsible for interacting with histone demethylase complex LSD1/CoREST/REST, thereby suggesting that this methylation could be important for HOTAIR-LSD1 interaction. While HOTAIR always showed consistent methylation at C1683 position, XIST showed variable methylation pattern at cytosine residues at position 701, 702, 703, 711, and 712 in HEK293 and that, too, only in about one-fifth of the number of tested clones. Moreover, the identical XIST sequence in mice did not show methylation at the corresponding sites. The methylated cytosines might prevent the interaction of XIST with Polycomb-repressive complex 2 (PRC2), which aids in chromatin remodeling via histone modifications.

The m6A and m5C modifications of lncRNA are largely unknown, but there are a few reports that highlight their functional relevance. For instance, the m6A modification of lncRNA, XIST is necessary for XIST-mediated transcriptional silencing of genes on X-chromosome. The silencing of either METTL3, the methyltransferase responsible for m6A modifications, or the RB15, m6A binding protein is sufficient in inhibiting the XIST-mediated gene regulation [123]. The NSUN2 mediated m5C modification of the lncRNA, H19 enhance its stability and promote its binding to G3BP1, an oncoprotein leading to increase in the tumor formation in HCC [124].

Regulation of Epigenetics by lncRNAs

miRNA mediates epigenetic regulation usually by directly or indirectly controlling the expression of

DNMTs. In contrast, the lncRNAs regulate epigenetic phenomenon by chromatin remodeling, which is usually aided by their structure and their ability to recruit proteins like DNMTs and PRC2 to preferential chromosomal locations (Fig. 4.1).

Di Ruscio et al. in 2013 identified a novel lncRNA, ecCEBPA, in the HL-60 and U937 cell line. ecCEBPA is an intragenic lncRNA that is concomitantly expressed along with CEBPA transcript. The ecCEBPA lncRNA is localized in the nucleus and a small region of ecCEBPA is capable of physically associating with DNMT1. The association of ecCEBPA and DNMT1 prevents the methylation of the CEBPA promoter region and thereby induces its transcription [51]. This effect of ecCEBPA may not be limited to CEBPA locus, but extends to other genomic loci too. ecCEBPA is, however, not the exclusive lncRNA that associates with DNMT1. Rather, many nonpolyadenylated lncRNAs have been shown to have a similar function as ecCEBPA. In 2015, a subset of lncRNAs were identified that interact with DNMT1 in a colon cancer cell line, HCT116 [52]. One of these lncRNAs was found to be associated with the 31 genomic loci known to be differentially methylated regions in colon tumors. Additionally, this lncRNA had reduced expression in colon tumors prompting the authors to call it DACOR1 (DNMT1-associated colon cancer repressed lncRNA 1). Just like ecCEBPA, DACOR1 affects DNA methylation levels at multiple sites in the human genome, but unlike ecCEBPA which suppresses DNA methylation, DACOR1 enhances DNA methylation. It seems ecCEBPA competitively binds DNMT1 owing to its higher binding affinity over the genomic loci [51] while DACOR1 may help recruit DNMT1 to specific genomic loci [52]. Also DNMT1-DACOR1 interaction indirectly elevates S-adenosyl methionine (SAM) levels, the preferred methyl donor for DNMT1 activity [52]. The exact mechanism behind this remains unexplored, but the contradictory effect of two lncRNAs points to the complex paradigm of DNMT1-lncRNA interactions. Both these studies indicate that the gene expression pattern dependent on DNA methylation can get significantly altered without any changes in histone complex or DNMT1 expression. lncRNA, RMST was recently shown to induce the expression of DNMT3b by promoting the interaction between HuR and 3'-UTR of DNMT3b [125]. Previously, an indirect regulation of DNMT3b by H19 has also been reported [53]. H19 inhibits the enzyme S-adenosylhomocysteine hydrolase (SAHH), the only mammalian enzyme capable of hydrolyzing S-adenosylhomocysteine (SAH). SAH acts as a feedback inhibitor of S-adenosylmethionine-dependent methyltransferases. Hence, knockdown of H19 increases DNMT3b-mediated methylation of Nctc1 within the Igf2-H19-Nctc1 locus [53]. In a way, H19 lncRNA can affect the methylation by all the DNMTs since SAHH does not bring about this effect by physically interacting with DNMT but rather by inhibiting SAM dependent methyltransferase reaction. Also, the methyltransferases are not equally sensitive to SAHH, hence H19 mediated SAHH knockdown may result in any of the three outcomes-hypermethylation, hypomethylation, or no change in methylation [53,126].

In addition to regulating CpG island methylation, the lncRNAs have also been shown to differentially regulate transcription of adjacent genes via their effect on DNA enhancer elements. One such study was published by Berghoff et al. in 2013 where it was shown that lncRNA Evf2 prevents the site specific methylation of Dlx5/6 ultraconserved enhancer at two sites [54]. Such an effect of lncRNA is referred to as the *trans* effect in contrast to the localized effect of lncRNAs where it is referred to as *cis* effect. Some of the lncRNAs involved in regulation of epigenetics are listed in Table 4.2.

piRNAs and Epigenetics

piRbase, the piRNA database, pegs the number of piRNAs at over 8 million. piRNAs are slightly longer than miRNAs and siRNAs, and they interact with PIWI proteins for inhibiting gene expression in contrast to miRNA/siRNA binding to AGO proteins. The piRNA biogenesis remains elusive in humans and most of our understanding comes from *Drosophila*. It is believed that majority of the piRNAs are transcribed quite like the mRNAs but from the specific genomic loci referred to as piRNA clusters. Some piRNAs are also known to arise from transposons or from the 3'-UTR region of protein-coding genes. piRNA precursors are exported out of the nucleus and are processed to the mature piRNAs by helicases and nucleases [127]. The piRNAs are further methylated at their 3' end by methyltransferase, HEN1, which is supposed to increase their stability. Even though the expression of piRNAs is seen in multiple somatic tissues in mammals, they are predominantly expressed in the testes where they play an important role in the process of spermatogenesis.

The piRNAs in the germline are tasked with the crucial role of de novo methylation of transposable elements. It was recently reported that piRNA-mediated DNA methylation of transposable elements in mouse gonocytes is effected through MIWI2 and a small nuclear protein, SPOCD1 [128]. The dysregulation of this mechanism inhibits spermatogenesis and is responsible for male infertility. Although the piRNAs have been implicated in some aspects of epigenetic regulation, a lot of research is still required to understand their definitive role in epigenetics.

CircRNAs and Epigenetics

There are more than 400,000 known circRNAs in the human genome [7]. The circRNAs are devoid of 5′ cap and 3′ poly-A tail, and their circular shape protects them against exonucleases. They originate through the back splicing or exon skipping of pre-mRNA transcripts. The origin of circRNAs from pre-mRNAs allows us to classify them into four types; exonic circRNAs (contains the sequence from exons only), exon-intron circRNAs (contains sequence from the exon as well as the intron), intergenic circRNAs (formed from the intergenic sequence), and circular intronic RNAs (contains the sequence from introns only) [129]. Approximately 80% of circRNAs belong to the category of exonic circRNAs. Although the detailed mechanism of the biogenesis of circRNAs remains unknown, the genes harboring circRNA seems to have highly active promoter regions with reduced DNA methylation and enhanced H3K27Ac [130]. Moreover, the hypermethylation of promoter regions of circRNAs has been shown to prompt their silencing in cancer specific manner [131].

The functional characterization of circRNA reveals their diversity as they can work as miRNA sponges, interacting partners of RNA-binding proteins and regulators of transcription and translation. Since circRNAs originate from the exons, some of them are also speculated to possess coding potential. It has been reported that circRNAs contain extensive N^6-methyladenosine (m6A) methylations and intriguingly, the introduction of m6A methylation of circRNA makes them amenable to translation [132]. Therefore, majorly circRNA are noncoding in nature but some of them can function as mRNAs and undergo translation in a cap independent manner, owing to the presence of m6A in circRNAs. Moreover, the cytoplasmic export of circRNA has also been attributed to its m6A modification as in the case of circNSUN2 [133].

It, however, remains to be seen whether the circRNAs have any role in epigenetic regulation.

CONCLUSIONS

The investigations into the role of noncoding RNAs in epigenetics have provided a lot of insight into the complexity of gene regulation. In the last decade, the advent of NGS has made it easier to uncover the genome wide histone modifications and methylation patterns and understand the role of noncoding RNAs in the epigenetic phenomenon. The aberrant regulation of epigenetic pathways by miRNAs and lncRNAs has been documented in a number of diseases, but it is still a long way before we could understand the general rules or principles that govern the interactions between noncoding RNA and epigenetics. In addition to miRNAs and lncRNAs, the role of piRNAs in regulating the epigenetics has also come to light but there is comparatively little information and most of it is derived from the work on model organisms including *Caenorhabditis elegans*, *Drosophila melanogaster*, and *Mus musculus*. The research into circRNAs is also at a nascent stage and their regulation of epigenetics remains unknown till date. With continuous innovations in high throughput techniques and improved computational analysis, it can be safely predicted that our knowledge of this area is bound to increase at a very rapid pace. Many of the newer links between noncoding RNA and epigenetics are already on the brink of discovery, including the role of methylations (both m5C and m6A) present in the 3′-UTR of mRNA on the binding capability of miRNAs. It could certainly be assumed that the ncRNA-epigenetic interactions could make for excellent therapeutic targets for pathological conditions resulting from aberrant epigenetic modifications.

Acknowledgments

This work was supported by Research Seed Money Grant (GP-25) awarded to RC by Central University of Punjab, India, UGC-BSR Research Start Up Grant awarded to RC By UGC and DST-FIST grant (SR/FST/LS—I/2018/125(C)) awarded to the Department of Biochemistry, Central University of Punjab, India.

Abbreviations

BCL6	B-cell CLL/lymphoma 6
CDH1	Cadherin 1
CDK6	Cyclin-dependent kinase 6
CXCL5	C-X-C motif chemokine ligand 5
DGCR8	DGCR8, microprocessor complex subunit (DiGeorge syndrome critical region gene 8)
DNMT	DNA methyltransferase
EMT	Epithelial—mesenchymal transition
ERα (official name ESR1)	Estrogen receptor 1
ESR1	Estrogen receptor 1
FHIT	Fragile histidine triad
FTO	FTO, alpha-ketoglutarate dependent dioxygenase (fat mass and obesity associated)
GSTP1	Glutathione S-transferase pi 1
HEN1	double-stranded RNA-binding protein-related
HESO1	Nucleotidyltransferase family protein
HNRNPA2B1	heterogeneous nuclear ribonucleoprotein A2/B1
METTL	methyltransferase-like
NANOG	Nanog homeobox
NSUN2	NOP2/Sun RNA methyltransferase family member 2
OCT4 (official name POU5F1)	POU class 5 homeobox 1

p15^{INK4b} (official name CDKN2B)	Cyclin-dependent kinase inhibitor 2B
PCNA	Proliferating cell nuclear antigen
PPARG	Peroxisome proliferator activated receptor gamma
PRDX1	Peroxiredoxin
Rac1	Ras-related C3 botulinum toxin substrate 1
RASD1	Ras-related dexamethasone induced 1
Rbl2	RB transcriptional corepressor like 2
Rtl1	Retrotransposon-like 1
SFRP1	Secreted frizzled related protein 1
TGFB2	Transforming growth factor beta 2
TRDMT1	tRNA aspartic acid methyltransferase 1
WTAP	Wilms tumor 1 associated protein
WWOX	WW domain containing oxidoreductase

References

[1] Dunham I, Kundaje A, Aldred SF, Collins PJ, Davis CA, Doyle F, et al. An integrated encyclopedia of DNA elements in the human genome. Nature 2012;489:57–74.

[2] Kaikkonen MU, Lam MTY, Glass CK. Non-coding RNAs as regulators of gene expression and epigenetics. Cardiovasc Res 2011;90:430–40.

[3] Frith MC, Pheasant M, Mattick JS. The amazing complexity of the human transcriptome. Eur J Hum Genet 2005;13:894–7.

[4] Djebali S, Davis CA, Merkel A, Dobin A, Lassmann T, Mortazavi A, et al. Landscape of transcription in human cells. Nature 2012;489:101–8.

[5] Lasda E, Parker R. Circular RNAs: diversity of form and function. RNA 2014;20:1829–42.

[6] Kozomara A, Birgaoanu M, Griffiths-Jones S. MiRBase: from microRNA sequences to function. Nucleic Acids Res 2019;47:D155–62.

[7] Wu W, Ji P, Zhao F. CircAtlas: an integrated resource of one million highly accurate circular RNAs from 1070 vertebrate transcriptomes. Genome Biol 2020;21:1–14.

[8] Wang J, Zhang P, Lu Y, Li Y, Zheng Y, Kan Y, et al. PiRBase: a comprehensive database of piRNA sequences. Nucleic Acids Res 2019;47:D175–80.

[9] Bouchard-Bourelle P, Desjardins-Henri C, Mathurin-St-Pierre D, Deschamps-Francoeur G, Fafard-Couture É, Garant JM, et al. SnoDB: an interactive database of human snoRNA sequences, abundance and interactions. Nucleic Acids Res 2020;48:D220–5.

[10] Volders PJ, Anckaert J, Verheggen K, Nuytens J, Martens L, Mestdagh P, et al. Lncipedia 5: towards a reference set of human long non-coding RNAs. Nucleic Acids Res 2019;47:D135–9.

[11] Lestrade L, Weber MJ. snoRNA-LBME-db, a comprehensive database of human H/ACA and C/D box snoRNAs. Nucleic Acids Res 2006;34:D158–62.

[12] S H, AA S, S B, S D, P L, Y S, et al. NSun2-mediated cytosine-5 methylation of vault noncoding RNA determines its processing into regulatory small RNAs. Cell Rep 2013;4:255–61.

[13] Sajini AA, Choudhury NR, Wagner RE, Bornelöv S, Selmi T, Spanos C, et al. Loss of 5-methylcytosine alters the biogenesis of vault-derived small RNAs to coordinate epidermal differentiation. Nat Commun 2019;10:1–13.

[14] Z L, C E, G M, PS M, Y C, B J. Extensive terminal and asymmetric processing of small RNAs from rRNAs, snoRNAs, snRNAs, and tRNAs. Nucleic Acids Res 2012;40:6787–99.

[15] Xie Y, Yao L, Yu X, Ruan Y, Li Z, Guo J. Action mechanisms and research methods of tRNA-derived small RNAs. Signal Transduct Target Ther 2020;5:1–9.

[16] Lambert M, Benmoussa A, Provost P. Small non-coding RNAs derived from eukaryotic ribosomal RNA. Non-Coding RNA 2019;5.

[17] Ambros V. The functions of animal microRNAs. Nature 2004;431:350–5.

[18] Kim NHN-GNH, Kim HS, Kim NHN-GNH, Lee I, Choi H-SH, Li X-YX, et al. p53 and MicroRNA-34 are suppressors of canonical Wnt signaling. Sci Signal 2011;4:ra71.

[19] Sauvageau M, Goff LA, Lodato S, Bonev B, Groff AF, Gerhardinger C, et al. Multiple knockout mouse models reveal lincRNAs are required for life and brain development. Elife 2013;2:e01749.

[20] Grote P, Wittler L, Hendrix D, Koch F, Währisch S, Beisaw A, et al. The tissue-specific lncRNA fendrr is an essential regulator of heart and body wall development in the mouse. Dev Cell 2013;24:206–14.

[21] Baer C, Claus R, Frenzel LP, Zucknick M, Park YJ, Gu L, et al. Extensive promoter DNA hypermethylation and hypomethylation is associated with aberrant microRNA expression in chronic lymphocytic leukemia. Cancer Res 2012;72:3775–85.

[22] He Y, Cui Y, Wang W, Gu J, Guo S, Ma K, et al. Hypomethylation of the hsa-miR-191 locus causes high expression of hsa-mir-191 and promotes the epithelial-to-mesenchymal transition in hepatocellular carcinoma. Neoplasia 2011;13:841–53.

[23] Brueckner B, Stresemann C, Kuner R, Mund C, Musch T, Meister M, et al. The human let-7a-3 locus contains an epigenetically regulated microRNA gene with oncogenic function. Cancer Res 2007;67:1419–23.

[24] de Souza Rocha Simonini P, Breiling A, Gupta N, Malckpour M, Youns M, Omranipour R, et al. Epigenetically deregulated microRNA-375 is involved in a positive feedback loop with estrogen receptor alpha in breast cancer cells. Cancer Res 2010;70:9175–84.

[25] Li A, Omura N, Hong S-M, Vincent A, Walter K, Griffith M, et al. Pancreatic cancers epigenetically silence SIP1 and hypomethylate and overexpress miR-200a/200b in association with elevated circulating miR-200a and miR-200b levels. Cancer Res 2010;70:5226–37.

[26] Zhao S, Wang Y, Liang Y, Zhao M, Long H, Ding S, et al. MicroRNA-126 regulates DNA methylation in CD4 + T cells and contributes to systemic lupus erythematosus by targeting DNA methyltransferase 1. Arthritis Rheum 2011;63:1376–86.

[27] Wang P, Chen L, Zhang J, Chen H, Fan J, Wang K, et al. Methylation-mediated silencing of the miR-124 genes facilitates pancreatic cancer progression and metastasis by targeting Rac1. Oncogene 2014;33:514–24.

[28] Agirre X, Vilas-Zornoza A, Jiménez-Velasco A, Martin-Subero JI, Cordeu L, Gárate L, et al. Epigenetic silencing of the tumor suppressor microRNA Hsa-miR-124a regulates CDK6 expression and confers a poor prognosis in acute lymphoblastic leukemia. Cancer Res 2009;69:4443–53.

[29] Wilting SM, van Boerdonk RAA, Henken FE, Meijer CJLM, Diosdado B, Meijer GA, et al. Methylation-mediated silencing and tumour suppressive function of hsa-miR-124 in cervical cancer. Mol Cancer 2010;9:167.

REFERENCES

[30] Furuta M, Kozaki KI, Tanaka S, Arii S, Imoto I, Inazawa J. miR-124 and miR-203 are epigenetically silenced tumor-suppressive microRNAs in hepatocellular carcinoma. Carcinogenesis 2009;31:766−76.

[31] Pigazzi M, Manara E, Bresolin S, Tregnago C, Beghin A, Baron E, et al. MicroRNA-34b promoter hypermethylation induces CREB overexpression and contributes to myeloid transformation. Haematologica 2013;98:602−10.

[32] Watanabe K, Emoto N, Hamano E, Sunohara M, Kawakami M, Kage H, et al. Genome structure-based screening identified epigenetically silenced microRNA associated with invasiveness in non-small-cell lung cancer. Int J Cancer 2012;130:2580−90.

[33] Liu C, Kelnar K, Liu B, Chen X, Calhoun-Davis T, Li H, et al. The microRNA miR-34a inhibits prostate cancer stem cells and metastasis by directly repressing CD44. Nat Med 2011;17:211−15.

[34] Lodygin D, Tarasov V, Epanchintsev A, Berking C, Knyazeva T, Körner H, et al. Inactivation of miR-34a by aberrant CpG methylation in multiple types of cancer. Cell Cycle 2008;7:2591−600.

[35] Toyota M, Suzuki H, Sasaki Y, Maruyama R, Imai K, Shinomura Y, et al. Epigenetic silencing of microRNA-34b/c and B-cell translocation gene 4 is associated with CpG island methylation in colorectal cancer. Cancer Res 2008;68:4123−32.

[36] Roy S, Levi E, Majumdar AP, Sarkar FH. Expression of miR-34 is lost in colon cancer which can be re-expressed by a novel agent CDF. J Hematol Oncol 2012;5:58.

[37] Saito Y, Liang G, Egger G, Friedman JM, Chuang JC, Coetzee GA, et al. Specific activation of microRNA-127 with downregulation of the proto-oncogene BCL6 by chromatin-modifying drugs in human cancer cells. Cancer Cell 2006;9:435−43.

[38] Fabbri M, Garzon R, Cimmino A, Liu Z, Zanesi N, Callegari E, et al. MicroRNA-29 family reverts aberrant methylation in lung cancer by targeting DNA methyltransferases 3A and 3B. Proc Natl Acad Sci USA 2007;104:15805−10.

[39] Garzon R, Liu S, Fabbri M, Liu Z, Heaphy CEA, Callegari E, et al. MicroRNA-29b induces global DNA hypomethylation and tumor suppressor gene reexpression in acute myeloid leukemia by targeting directly DNMT3A and 3B and indirectly DNMT1. Blood 2009;113:6411−18.

[40] Pandi G, Nakka VP, Dharap A, Roopra A, Vemuganti R. MicroRNA miR-29c down-regulation leading to de-repression of its target DNA methyltransferase 3a promotes ischemic brain damage. PLoS One 2013;8.

[41] Majid S, Dar AA, Saini S, Shahryari V, Arora S, Zaman MS, et al. miRNA-34b inhibits prostate cancer through demethylation, active chromatin modifications, and AKT pathways. Clin Cancer Res 2013;19:73−84.

[42] Braconi C, Huang N, Patel T. Microrna-dependent regulation of DNA methyltransferase-1 and tumor suppressor gene expression by interleukin-6 in human malignant cholangiocytes. Hepatology 2010;51:881−90.

[43] Huang J, Wang Y, Guo Y, Sun S. Down-regulated microRNA-152 induces aberrant DNA methylation in hepatitis B virus-related hepatocellular carcinoma by targeting DNA methyltransferase 1. Hepatology 2010;52:60−70.

[44] Wang YS, Chou WW, Chen KC, Cheng HY, Lin RT, Juo SHH. Microrna-152 mediates dnmt1-regulated DNA methylation in the estrogen receptor?? gene. PLoS One 2012;7.

[45] Xiang Y, Ma N, Wang D, Zhang Y, Zhou J, Wu G, et al. MiR-152 and miR-185 co-contribute to ovarian cancer cells cisplatin sensitivity by targeting DNMT1 directly: a novel epigenetic therapy independent of decitabine. Oncogene 2013;1−9.

[46] Tsuruta T, Kozaki K-i, Uesugi A, Furuta M, Hirasawa A, Imoto I, et al. miR-152 is a tumor suppressor microRNA that is silenced by DNA hypermethylation in endometrial cancer. Cancer Res 2011;71:6450−62.

[47] Qadir XV, Han C, Lu D, Zhang J, Wu T. MicroRNA-185 inhibits hepatocellular carcinoma growth by targeting the DNMT1/PTEN/Akt pathway. Am J Pathol 2014;184:1−11.

[48] Wang H, Wu J, Meng X, Ying X, Zuo Y, Liu R, et al. Microrna-342 inhibits colorectal cancer cell proliferation and invasion by directly targeting DNA methyltransferase 1. Carcinogenesis 2011;32:1033−42.

[49] Pan W, Zhu S, Yuan M, Cui H, Wang L, Luo X, et al. MicroRNA-21 and microRNA-148a contribute to DNA hypomethylation in lupus CD4 + T cells by directly and indirectly targeting DNA methyltransferase 1. J Immunol 2010;184:6773−81.

[50] Wang KC, Yang YW, Liu B, Sanyal A, Corces-Zimmerman R, Chen Y, et al. A long noncoding RNA maintains active chromatin to coordinate homeotic gene expression. Nature 2011;472:120−4.

[51] Di Ruscio A, Ebralidze AK, Benoukraf T, Amabile G, Goff LA, Terragni J, et al. DNMT1-interacting RNAs block gene-specific DNA methylation. Nature 2013;503:371−6.

[52] Merry CR, Forrest ME, Sabers JN, Beard L, Gao XH, Hatzoglou M, et al. DNMT1-associated long non-coding RNAs regulate global gene expression and DNA methylation in colon cancer. Hum Mol Genet 2015;24:6240−53.

[53] Zhou J, Yang L, Zhong T, Mueller M, Men Y, Zhang N, et al. H19 lncRNA alters DNA methylation genome wide by regulating S-adenosylhomocysteine hydrolase. Nat Commun 2015;6:10221.

[54] Berghoff EG, Clark MF, Chen S, Cajigas I, Leib DE, Kohtz JD. Evf2 (Dlx6as) lncRNA regulates ultraconserved enhancer methylation and the differential transcriptional control of adjacent genes. Development 2013;140:4407−16.

[55] Mohammad F, Mondal T, Guseva N, Pandey GK, Kanduri C. Kcnq1ot1 noncoding RNA mediates transcriptional gene silencing by interacting with Dnmt1. Development 2010;137:2493−9.

[56] Kotake Y, Nakagawa K, Kitagawa K, Suzuki S, Liu N, Kitagawa M, et al. Long non-coding RNA ANRIL is required for the PRC2 recruitment to and silencing of p15(INK4B) tumor suppressor gene. Oncogene 2011;30:1956−62.

[57] Nagano T, Mitchell JA, Sanz LA, Pauler FM, Ferguson-Smith AC, Feil R, et al. The air noncoding RNA epigenetically silences transcription by targeting G9a to chromatin. Science 2008;322:1717−20.

[58] Gupta Ra, Shah N, Wang KC, Kim J, Horlings HM, Wong DJ, et al. Long non-coding RNA HOTAIR reprograms chromatin state to promote cancer metastasis. Nature 2010;464:1071−6.

[59] Winter J, Jung S, Keller S, Gregory RI, Diederichs S. Many roads to maturity: microRNA biogenesis pathways and their regulation. Nat Cell Biol 2009;11:228−34.

[60] Kowalczyk MS, Higgs DR, Gingeras TR. Molecular biology: RNA discrimination. Nature 2012;482:310−11.

[61] Khandelwal A, Bacolla A, Vasquez KM, Jain A. Long noncoding RNA: a new paradigm for lung cancer. Mol Carcinog 2015;54:1235−51.

[62] Kim VN, Han J, Siomi MC. Biogenesis of small RNAs in animals. Nat Rev Mol Cell Biol 2009;10:126−39.

[63] Cocquerelle C, Daubersies P, Majerus MA, Kerckaert JP, Bailleul B. Splicing with inverted order of exons occurs proximal to large introns. EMBO J 1992;11:1095−8.

[64] Filipowicz W, Pogači V. Biogenesis of small nucleolar ribonucleoproteins. Curr Opin Cell Biol 2002;14:319−27.

[65] Scott MS, Avolio F, Ono M, Lamond AI, Barton GJ. Human miRNA precursors with box H/ACA snoRNA features. PLoS Comput Biol 2009;5.

[66] Ender C, Krek A, Friedländer MR, Beitzinger M, Weinmann L, Chen W, et al. A Human snoRNA with MicroRNA-Like Functions. Mol Cell 2008;32:519–28.

[67] Hotchkiss RD. Article: the quantitative separation of purines, pyrimidines, and nucleosides by paper. J Biol Chem 1948;175:315–32.

[68] Cantara WA, Crain PF, Rozenski J, McCloskey JA, Harris KA, Zhang X, et al. The RNA modification database, RNAMDB: 2011 update. Nucleic Acids Res 2011;39.

[69] Machnicka MA, Milanowska K, Oglou OO, Purta E, Kurkowska M, Olchowik A, et al. MODOMICS: a database of RNA modification pathways—2013 update. Nucleic Acids Res 2013;41.

[70] Jones PA. Functions of DNA methylation: islands, start sites, gene bodies and beyond. Nat Rev Genet 2012;13:484–92.

[71] Sati S, Ghosh S, Jain V, Scaria V, Sengupta S. Genome-wide analysis reveals distinct patterns of epigenetic features in long non-coding RNA loci. Nucleic Acids Res 2012;40:10018–31.

[72] Amort T, Souliere MF, Wille A, Jia XY, Fiegl H, Worle H, et al. Long non-coding RNAs as targets for cytosine methylation. RNA Biol 2013;10:1003–8.

[73] Meyer KD, Saletore Y, Zumbo P, Elemento O, Mason CE, Jaffrey SR. Comprehensive analysis of mRNA methylation reveals enrichment in 3′UTRs and near stop codons. Cell 2012;149:1635–46.

[74] Fu Y, Dominissini D, Rechavi G, He C. Gene expression regulation mediated through reversible m^{6}A RNA methylation. Nat Rev Genet 2014;15:293–306.

[75] Squires JE, Patel HR, Nousch M, Sibbritt T, Humphreys DT, Parker BJ, et al. Widespread occurrence of 5-methylcytosine in human coding and non-coding RNA. Nucleic Acids Res 2012; 40:5023–33.

[76] Wang Y, Li Y, Toth JI, Petroski MD, Zhang Z, Zhao JC. N6-methyladenosine modification destabilizes developmental regulators in embryonic stem cells. Nat Cell Biol 2014;16: 191–8.

[77] Karikó K, Muramatsu H, Welsh FA, Ludwig J, Kato H, Akira S, et al. Incorporation of pseudouridine into mRNA yields superior nonimmunogenic vector with increased translational capacity and biological stability. Mol Ther 2008;16:1833–40.

[78] Mao Y, Dong L, Liu X-M, Guo J, Ma H, Shen B, et al. m6A in mRNA coding regions promotes translation via the RNA helicase-containing YTHDC2. Nat Commun 2019;10:1–11.

[79] Choe J, Lin S, Zhang W, Liu Q, Wang L, Ramirez-Moya J, et al. mRNA circularization by METTL3–eIF3h enhances translation and promotes oncogenesis. Nature 2018;561:556–60.

[80] Ferguson-Smith AC. Genomic imprinting: the emergence of an epigenetic paradigm. Nat Rev Genet 2011;12:565–75.

[81] Barlow DP. Genomic imprinting: a mammalian epigenetic discovery model. Annu Rev Genet 2011;45:379–403.

[82] Monteys AM, Spengler RM, Wan J, Tecedor L, Lennox KA, Xing Y, et al. Structure and activity of putative intronic miRNA promoters. RNA 2010;16:495–505.

[83] Wang Z, Yao H, Lin S, Zhu X, Shen Z, Lu G, et al. Transcriptional and epigenetic regulation of human microRNAs. Cancer Lett 2013;331:1–10.

[84] Lujambio A, Ropero S, Ballestar E, Fraga MF, Cerrato C, Setién F, et al. Genetic unmasking of an epigenetically silenced microRNA in human cancer cells. Cancer Res 2007;67:1424–9.

[85] Chhabra R, Saini N. microRNAs in cancer stem cells: current status and future directions. Tumour Biol 2014;35:8395–405.

[86] Kong D, Heath E, Chen W, Cher M, Powell I, Heilbrun L, et al. Epigenetic silencing of miR-34a in human prostate cancer cells and tumor tissue specimens can be reversed by BR-DIM treatment. Am J Transl Res 2012;4:14–23.

[87] Roman-Gomez J, Agirre X, Jiménez-Velasco A, Arqueros V, Vilas-Zornoza A, Rodriguez-Otero P, et al. Epigenetic regulation of MicroRNAs in acute lymphoblastic leukemia. J Clin Oncol 2009;27:1316–22.

[88] Seitz H, Youngson N, Lin S-PP, Dalbert S, Paulsen M, Bachellerie J-PP, et al. Imprinted microRNA genes transcribed antisense to a reciprocally imprinted retrotransposon-like gene. Nat Genet 2003;34:261–2.

[89] Cui XS, Zhang DX, Ko YG, Kim NH. Aberrant epigenetic reprogramming of imprinted microRNA-127 and Rtl1 in cloned mouse embryos. Biochem Biophys Res Commun 2009;379:390–4.

[90] Li J, Yang Z, Yu B, Liu J, Chen X. Methylation protects miRNAs and siRNAs from a 3′-end uridylation activity in Arabidopsis. Curr Biol 2005;15:1501–7.

[91] Ren G, Xie M, Zhang S, Vinovskis C, Chen X, Yu B. Methylation protects microRNAs from an AGO1-associated activity that uridylates 5′ RNA fragments generated by AGO1 cleavage. Proc Natl Acad Sci USA 2014;1–6.

[92] Shen B, Goodman HM. Uridine addition after microRNA-directed cleavage. Science 2004;306:997.

[93] Heo I, Joo C, Kim YK, Ha M, Yoon MJ, Cho J, et al. TUT4 in Concert with Lin28 suppresses microRNA biogenesis through pre-microRNA uridylation. Cell 2009;138:696–708.

[94] Heo I, Ha M, Lim J, Yoon MJ, Park JE, Kwon SC, et al. Mono-uridylation of pre-microRNA as a key step in the biogenesis of group II let-7 microRNAs. Cell 2012;151:521–32.

[95] Yuan S, Tang H, Xing J, Fan X, Cai X, Li Q, et al. Methylation by NSun2 represses the levels and function of miR-125b. Mol Cell Biol 2014;34:3630–41.

[96] Berulava T, Rahmann S, Rademacher K, Klein-Hitpass L, Horsthemke B. N6-Adenosine methylation in MiRNAs. PLoS One 2015;10.

[97] Alarcón CR, Lee H, Goodarzi H, Halberg N, Tavazoie SF. N6-methyladenosine marks primary microRNAs for processing. Nature 2015;519:482–5.

[98] Alarcón CR, Goodarzi H, Lee H, Liu X, Tavazoie S, Tavazoie SF. HNRNPA2B1 is a mediator of m(6)A-dependent nuclear RNA processing events. Cell 2015;162:1299–308.

[99] Pan T. N6-methyl-adenosine modification in messenger and long non-coding RNA. Trends Biochem Sci 2013;38:204–9.

[100] Cheray M, Etcheverry A, Jacques C, Pacaud R, Bougras-Cartron G, Aubry M, et al. Cytosine methylation of mature microRNAs inhibits their functions and is associated with poor prognosis in glioblastoma multiforme. Mol Cancer 2020; 19:1–16.

[101] Pandolfini L, Barbieri I, Bannister AJ, Hendrick A, Andrews B, Webster N, et al. METTL1 promotes let-7 microRNA processing via m7G methylation. Mol Cell 2019;74:1278–90 e9.

[102] Y K, B Z, P S, H I, AG H, K N. Redirection of silencing targets by adenosine-to-inosine editing of miRNAs. Science 2007;315: 1137–40.

[103] Chhabra R. miRNA and methylation: a multifaceted liaison. ChemBioChem 2015;16:195–203.

[104] Blasco M, Benetti R, Gonzalo S, Jaco I, Muñoz P, Gonzalez S, et al. A mammalian microRNA cluster controls DNA methylation and telomere recombination via Rbl2-dependent regulation of DNA methyltransferases. Nat Struct Mol Biol 2008;15: 268–79.

[105] Sinkkonen L, Hugenschmidt T, Berninger P, Gaidatzis D, Mohn F, Artus-Revel CG, et al. MicroRNAs control de novo DNA methylation through regulation of transcriptional

repressors in mouse embryonic stem cells. Nat Struct Mol Biol 2008;15:259—67.

[106] Du M, Zhang Y, Mao Y, Mou J, Zhao J, Xue Q, et al. MiR-33a suppresses proliferation of NSCLC cells via targeting METTL3 mRNA. Biochem Biophys Res Commun 2017;482:582—9.

[107] He H, Wu W, Sun Z, Chai L. MiR-4429 prevented gastric cancer progression through targeting METTL3 to inhibit m 6 A-caused stabilization of SEC62. Biochem Biophys Res Commun 2019;517:581—7.

[108] Zepecki JP, Karambizi D, Fajardo JE, Snyder KM, Guetta-Terrier C, Tang OY, et al. miRNA-mediated loss of m6A increases nascent translation in glioblastoma. PLOS Genet 2021;17:e1009086.

[109] Chen T, Hao YJ, Zhang Y, Li MM, Wang M, Han W, et al. m(6)A RNA methylation is regulated by microRNAs and promotes reprogramming to pluripotency. Cell Stem Cell 2015;16:289—301.

[110] Wade SL, Langer LF, Ward JM, Archer TK. MiRNA-mediated regulation of the SWI/SNF chromatin remodeling complex controls pluripotency and endodermal differentiation in human ESCs. Stem Cell 2015;33:2925—35.

[111] Pandolfini L, Luzi E, Bressan D, Ucciferri N, Bertacchi M, Brandi R, et al. RISC-mediated control of selected chromatin regulators stabilizes ground state pluripotency of mouse embryonic stem cells. Genome Biol 2016;17:1—22.

[112] Dominissini D, Moshitch-Moshkovitz S, Schwartz S, Salmon-Divon M, Ungar L, Osenberg S, et al. Topology of the human and mouse m6A RNA methylomes revealed by m6A-seq. Nature 2012;485:201—6.

[113] Hussain S, Aleksic J, Blanco S, Dietmann S, Frye M. Characterizing 5-methylcytosine in the mammalian epitranscriptome. Genome Biol 2013;14:215.

[114] Zhou KI, Parisien M, Dai Q, Liu N, Diatchenko L, Sachleben JR, et al. N6-methyladenosine modification in a long noncoding RNA hairpin predisposes its conformation to protein binding. J Mol Biol 2016;428:822—33.

[115] Warren L, Manos PD, Ahfeldt T, Loh YH, Li H, Lau F, et al. Highly efficient reprogramming to pluripotency and directed differentiation of human cells with synthetic modified mRNA. Cell Stem Cell 2010;7:618—30.

[116] Mercer TR, Dinger ME, Mattick JS. Long non-coding RNAs: insights into functions. Nat Rev Genet 2009;10:155—9.

[117] Wu SC, Kallin EM, Zhang Y. Role of H3K27 methylation in the regulation of lncRNA expression. Cell Res 2010;20:1109—16.

[118] Li Y, Zhang Y, Li S, Lu J, Chen J, Wang Y, et al. Genome-wide DNA methylome analysis reveals epigenetically dysregulated noncoding RNAs in human breast cancer. Sci Rep 2015;5:8790.

[119] Hahn MA, Wu X, Li AX, Hahn T, Pfeifer GP. Relationship between gene body DNA methylation and intragenic H3K9ME3 and H3K36ME3 chromatin marks. PLoS One 2011;6:e18844.

[120] Navarro P, Page DR, Avner P, Rougeulle C. Tsix-mediated epigenetic switch of a CTCF-flanked region of the Xist promoter determines the Xist transcription program. Genes Dev 2006;20:2787—92.

[121] Sun BK, Deaton AM, Lee JT. A transient heterochromatic state in Xist preempts X inactivation choice without RNA stabilization. Mol Cell 2006;21:617—28.

[122] Lee JT, Lu N. Targeted mutagenesis of Tsix leads to nonrandom X inactivation. Cell 1999;99:47—57.

[123] Patil DP, Chen C-K, Pickering BF, Chow A, Jackson C, Guttman M, et al. m6A RNA methylation promotes XIST-mediated transcriptional repression. Nature 2016;537(7620):369—73.

[124] Sun Z, Xue S, Zhang M, Xu H, Hu X, Chen S, et al. Aberrant NSUN2-mediated m5C modification of H19 lncRNA is associated with poor differentiation of hepatocellular carcinoma. Oncogene 2020;39:6906—19.

[125] WX P, P K, W Z, C N, Z W, L Y, et al. lncRNA RMST enhances DNMT3 expression through interaction with HuR. Mol Ther 2020;28:9—18.

[126] Tehlivets O, Malanovic N, Visram M, Pavkov-Keller T, Keller W. S-adenosyl-L-homocysteine hydrolase and methylation disorders: yeast as a model system. Biochim Biophys Acta Mol Basis Dis 2013;1832:204—15.

[127] Wu X, Pan Y, Fang Y, Zhang J, Xie M, Yang F, et al. The biogenesis and functions of piRNAs in human diseases. Mol Ther Nucleic Acids 2020;21:108—20.

[128] Zoch A, Auchynnikava T, Berrens RV, Kabayama Y, Schöpp T, Heep M, et al. SPOCD1 is an essential executor of piRNA-directed de novo DNA methylation. Nature 2020;584:635—9.

[129] Wang F, Nazarali AJ, Ji S. Circular RNAs as potential biomarkers for cancer diagnosis and therapy. Am J Cancer Res 2016;6:1167—76.

[130] Enuka Y, Lauriola M, Feldman ME, Sas-Chen A, Ulitsky I, Yarden Y. Circular RNAs are long-lived and display only minimal early alterations in response to a growth factor. Nucleic Acids Res 2016;44:1370—83.

[131] Ferreira HJ, Davalos V, de Moura MC, Soler M, Perez-Salvia M, Bueno-Costa A, et al. Circular RNA CpG island hypermethylation-associated silencing in human cancer. Oncotarget 2018;9:29208—19.

[132] Yang Y, Fan X, Mao M, Song X, Wu P, Zhang Y, et al. Extensive translation of circular RNAs driven by N 6-methyladenosine. Cell Res 2017;27:626—41.

[133] Chen RX, Chen X, Xia LP, Zhang JX, Pan ZZ, Ma XD, et al. N 6-methyladenosine modification of circNSUN2 facilitates cytoplasmic export and stabilizes HMGA2 to promote colorectal liver metastasis. Nat Commun 2019;10:1—15.

CHAPTER

5

Prions and prion-like phenomena in epigenetic inheritance

Philippe Silar
Université de Paris, Paris Cité, France

OUTLINE

Structural Heredity	74	Mixed Heredity: "Nonamyloid Prions" That Propagate by Auto-activation	78
Amyloid Prions of *Saccharomyces cerevisiae*, *Podospora anserina*, and Other Organisms	74	Regulatory Inheritance	79
Cis Elements Important for Amyloid Prion Formation	76	The Lactose Operon and Its Positive Feedback Loop	79
Genetic Control of Amyloid Prion Formation and Propagation	76	Crippled Growth, A Self-sustained and Mitotically Inheritable Signaling Pathway in the Filamentous Fungus *Podospora anserina*	80
Amyloid Prion Variants	77		
Self-Driven Assembly of Hsp60 Mitochondrial Chaperonin	77	The White/Opaque Switch of *Candida albicans*, An Epigenetic Switch at the Transcription Level	81
Cytotaxis of Cilia and Other Complex Structures	78	Conclusion	82
		References	83

> "This discussion is based on the idea of two types of cellular regulatory systems, both capable of maintaining persistent cellular characteristics but achieving homeostasis by different means. The current concept of a primary genetic material (DNA), replicating by a template mechanism, as opposed to a homeostatic system operating by, perhaps, self-regulating metabolic patterns [1]".
>
> D.L. Nanney 1958

At the beginning of the 20th century, almost as soon as the laws of Mendel were rediscovered, characters that would not follow the rules of classical mendelian segregation were evidenced. Most of these cases of nonmendelian heredity are presently accounted for by mutations in eukaryotic organelle genomes (plastids and mitochondria), by cytoplasmic symbionts, or by viruses and virus-like particles. However, a subset of these phenomena cannot be explained by the presence of nucleic acid-bearing entities in the cytoplasm. Theoretical considerations, first made by Max Delbrück [2], proposed that negatively interacting metabolic networks could generate alternative states, stable enough to be passed on during cell division. Similarly, as early as 1961, a model based on structure inheritance was proposed by Marcou and Rizet to account for a case of nonmendelian inheritance in the fungus *Podospora anserina* [3]. Early experiments with the lactose operon of *Escherichia coli* [4], proved that, indeed, metabolic networks could generate inheritable alternative metabolic states, and studies with paramecia showed that complex sub-cellular structures, that is, cilia, which direct their own assembly in a template-assisted fashion [5], could create alternative states that were inheritable during cell division and even sexual reproduction. This led to the definition of two kinds of epigenetic inheritance [6]: the structural inheritance, based on the transmission of alternative structures of macromolecules and macromolecular complexes, and the regulatory inheritance, based on the alternative states adopted by metabolic or regulatory pathways. As we shall see later, such a clear-cut difference may not be made in some instances, which are clearly a mix of the two types.

It is interesting to note, the acceptance took a long time that DNA, and not protein, is the genetic material. The purification of DNA associated with genetic transformation was the key experiments that permitted final recognition. The discovery in the mid-1960s of mitochondrial and plastid genomic DNAs [7,8] eclipsed for three decades the studies on many "genes" with unorthodox segregation, that could be due to structural or regulatory inheritance. Their analysis was re-ignited by the proposal of R. Wickner in 1994 [9] that two of them, the [PSI+] and [URE3] elements of the yeast Saccharomyces cerevisiae, could be due to inheritable changes in protein structure. At that time, the concern with mad cow disease, which affected a large part of the cattle in the United Kingdom and some other European countries, whose etiologic agent appears to be composed only of proteins, made the scientific community more receptive to unorthodox ideas regarding inheritance. As we shall see, we now can transform cells to alternative "states" with purified prions, a feat that would have postponed the recognition of DNA as the genetic material, and reinforced the hypothesis that proteins were the genetic material, had one of the unorthodox "gene" been chosen in the pioneering transformation experiments!

Here, we will review only a few cases of structural and regulatory inheritance due to space constraint. Indeed, an ever-increasing array of phenomena is now attributed to prions and prion-like elements and interest in these phenomena has greatly increased in recent years, leading to the production of many reviews on the subject [10–18]. Only those prototypic are discussed below.

STRUCTURAL HEREDITY

At the present time, two different kinds of structural heredity have been clearly demonstrated, that of prions, which is based on the structural changes in a single polypeptide, and cytotaxis in which a large macromolecular complex is concerned [19].

AMYLOID PRIONS OF SACCHAROMYCES CEREVISIAE, PODOSPORA ANSERINA, AND OTHER ORGANISMS

The term "prion" for *proteinaceous infectious* particle was first proposed by Prusiner [20] to characterize the etiologic agents of some, at the time, bizarre diseases of mammals called transmissible spongiform encephalopathies (TSEs), including scrapie in sheep, as well as Kuru, Creutzfeldt-Jakob in humans, and later on the mad cow disease. A basic definition of this term is the following: a protein able to adopt two distinct conformations, one of which can convert the other one. Usually, prion proteins may adopt monomeric or oligomeric states, the oligomeric state being amyloid, that is mostly made of fibrils with β-sheets secondary structures that stain easily with Congo Red. The protein can change spontaneously from the monomeric to the oligomeric forms with a low frequency. Importantly, oligomers trigger the switch of the monomers towards the oligomeric and infectious form. This autocatalytic process leads thus to the depletion of the monomers and to the accumulation of the oligomers. TSEs are probably caused by an aberrant folding of the PrP^c protein into the infectious PrP^{Sc} form [21].

Afterwards, this concept was successfully extended to explain the peculiar features of two nonmendelian elements of *S. cerevisiae*, [PSI+] and [URE3] [9]. Indeed, the prion transition alters the function of the protein and consequently the phenotype of the cell. The aggregated state, as well as the associated phenotype, is infectious and stably transmitted from generation to generation by both mitosis and meiosis. Thus, yeast prions act as protein-based genetic elements corresponding to an elegant epigenetic heredity. Several genetic criteria have been retained to suggest a prion behavior for a cellular protein [22]: (1) a prion can be cured, but it can reappear in the cured strain with a constant frequency because the protein able to change to an infectious form is still present, (2) overproduction of a protein capable of becoming a prion increases the frequency of the prion arising *de novo*, and (3) if the prion phenotype is due to the absence of the normal form of the protein, then the null mutant phenotype of the gene for the protein is the same as that of the strain containing the aggregated prion. Also, this gene is required for the prion to propagate. Nowadays, the term prion is thus no longer restricted to TSE agents, but refers to any protein able to adopt an infectious amyloid conformation.

The first two yeast proteins obeying the above criteria and showed to be true amyloid prions in yeast were eRF3 and Ure2p. [PSI+], the prion of the release factor eRF3, also called Sup35p, affects the efficiency of translation termination. This may result in significant morphological or physiological switch when the transition is made [23]. [URE3], the prion of the protein Ure2p alters nitrogen catabolism. Although, the genetic, biochemical and cell biological analysis of these two prions, especially [PSI+], have boosted the comprehension of prion properties, the definitive demonstration that a protein is infectious based on cell transformation was obtained in the ascomycete fungus P. anserina [24]. This organism contains a true amyloid prion that displays the expected properties, the HET-s protein involved in heterokaryon incompatibility [25]. Prion aggregates of HET-s obtained from recombinant protein made in E. coli were introduced by ballistic transformation into prion-free cells of P. anserina and were shown to induce a phenotypic conversion towards the prion containing cells [24]. This demonstration is formally equivalent to transformation experiments conducted by Avery et al. [26] proving that DNA is the support of genetic information. To date, this kind of transformation demonstration was performed for only few additional examples in S. cerevisiae [27–30], but an ever-growing number of S. cerevisiae proteins have been shown to be amyloid prions (Table 5.1). More recently, Schizosaccharomyces

TABLE 5.1 Summary of the Prion and Prion-Like in Different Organisms

Protein	Organism	Prion/Prion-Like	References
Amyloid Prions			
PrP	Mammals	TSE agent	[21]
HET-s	Podospora anserina	HET-s	[24]
Sup35	Saccharomyces cerevisiae	[PSI+]	[27,30,42]
Ure2	S. cerevisiae	[URE3]	[29]
Rnq1	S. cerevisiae	[PIN+]	[28]
New1	S. cerevisiae	[NU+]	[43]
Swi1	S. cerevisiae	[SWI+]	[44,45]
Cyc8[a]	S. cerevisiae	[OCT+]	[46][a]
MOT3	S. cerevisiae	[MOT3+]	[47]
SFP1[b]	S. cerevisiae	[ISP+]	[48]
Mod5	S. cerevisiae	[MOD+]	[49]
Nup100	S. cerevisiae	[NUP100+]	[50]
LSB2[a]	S. cerevisiae	[LSB+]	[51]
Ctr4	Schizosaccharomyces pombe	[CTR+]	[31]
LEF-10	Autographa californica multiple nucleopolyhedrovirus	–	[38]
CPEB	Aplysia californica	–	[32]
LD	Arabidopsis thaliana	–	[35]
Cb-Rho	Clostridium botulinum	–	[36]
cPrD1	Methanosalsum zhilinae	–	[37][c]
Nonamyloid Prion-Like Elements			
PaMpk1 MAPK pathway	P. anserina	C	[52–54]
PrB	S. cerevisiae	[β]	[55]
Pma1	S. cerevisiae	[GAR+]	[56]
Vts1	S. cerevisiae	[SMAUG+]	[57]
Snt1[d]	S. cerevisiae	[ESI+]	[58]

[a]Formal demonstration that these proteins fold into amyloids has not been made, but several data, including sensibility to Hsp104 level, suggest that these proteins are true amyloid prions.
[b]This protein aggregate in nuclei, but the prion is insensitive to hsp104 level. However, nuclear localization of the aggregates may prevent hsp104 to regulate [ISP+].
[c]In this paper, evidences are provided that additional archeal proteins may be prions.
[d]It is not clear whether Snt1 aggregates into amyloid; if so, it will join the true amyloid prions.
Several studies suggest that a dozen additional yeast proteins, not indicated in the table, may also be prion or prion-like traits [43,47,59,60] and that additional animal proteins may be prions [61].

pombe has been shown to support the formation of the [*PSI*⁺] prion and to harbor its own prion [31]. Moreover, protein behaving as prions when expressed in *S. cerevisiae* have been found in animals [32–34], plants [35], bacteria [36], Archaea [37], and even in a virus [38]. The CPEB (Cytoplasmic Polyadenylation Element Binding) proteins of aplysia and mouse have been shown to adopt prion conformation in neurons [34,39] and the aplysia protein to undergo typical structural switch into β-sheet-rich fibers [40]. The cb-rho protein from *Clostridium botulinum* was shown to behave as a prion not only in *S. cerevisiae*, but also in *E. coli* [36] and the viral LEF-10 protein to adopt two different cellular localizations in the native insect host cells [38]. Overall, these data show that true prion proteins exist in all the domains of life. Other prions may include the amyloid-β and tau proteins involved in amyloid fibrils in Alzheimer disease, although this is still a subject of contention [41].

CIS ELEMENTS IMPORTANT FOR AMYLOID PRION FORMATION

Domain analysis showed that the yeast, *P. anserina* and bacterial prions contain a modular prion domain, dispensable for the cellular function but required and sufficient for the prion properties [36,62,63]. Most proteins fused to it behave like a prion. For example, in cells expressing a fusion protein composed of a prion domain and the GFP protein, two populations of cells may be observed, one with a homogenous cytoplasmic fluorescence corresponding to the monomeric form and the other displaying intense punctuated foci due to oligomerization of the fusion protein [63]. Studies have focused on these domains to detect important features. In *S. cerevisiae*, an important feature is richness in glutamines (Q) and asparagines (N), since all known yeast amyloid prion have such a domain. The specific primary sequence is probably not critical, because a prion domain with a sequence randomly shuffled is still able to form prion [64]. A bioinformatic analysis of the yeast genome performed to select gene coding a protein with a Q/N rich domain [65] permitted the identification of the Rnq1 protein (Rich in N and Q) that has prion properties, confirming the importance of this feature. However, it may not be universal, since the prion domains of Het and PrP are not Q/N rich and exhibit no obvious bias in their amino-acid composition. Interestingly, other parts of the protein, not included in the prion domain *sensu stricto*, may be required for the stability of prions *in vivo* by allowing efficient transmission of prion aggregates during cell divisions [66]. Despite their variation in primary sequences, prions appear to adopt a similar conformation into amyloid fibrils (see [67] for a review). The monomer is soluble, rich in α-helix and protease sensitive. The infectious form is oligomeric, rich in β-sheet and partially protease resistant. This fibrous aggregates, identified by a birefringence when stained with Congo Red, are characteristically composed of proteins rich in β-sheet structures. These kind of fibers are detected in prion disease in mammals in the brain of affected individuals and also *in vitro* for recombinant yeast prion proteins. Note that these fibers are not restricted to prion proteins, indeed some native amyloids exist like the curlin protein in *E. coli* and even in the silk of some spiders. Progresses are made in the understanding of the structural basis of infectivity [68,69]. To date only the structure of the Het-s prion has been resolved [70,71] and key residues and inhibitory domains enabling or preventing the folding and polymerization have been identified [72,73].

GENETIC CONTROL OF AMYLOID PRION FORMATION AND PROPAGATION

As expected for a mechanism based on protein conformations, chaperones are implicated in the stability of amyloid prions. Hsp104, Hsp70, and Hsp40 are families of chaperone, but also Hsp70 or Hsp40 activating proteins are involved in this regulation. Hsp104 breaks aggregates into smaller ones, Hsp40 and Hsp70 help to refold the proteins into their native conformation. Hsp104p was the first chaperone identified to modulate prion [74]. Surprisingly both over-expression and depletion of Hsp104 led to the loss of [*PSI*⁺]. Hsp104 is required to break large aggregates of prion producing new oligomers efficient for polymerization or seeds. The normal level of Hsp104 permits the retention of seeds large enough to not be refolded by Hsp40/Hsp70, as well as to have an adequate number of seeds for efficient segregation during cellular division. Among chaperones, Hsp104 appears to be the major player as it is required for all the yeast amyloid prions. Moreover, the first compound destabilizing [*PSI*⁺], guanidine hydrochlorid, is an inhibitor of Hsp104 [75]. However, Hsp104 over-expression affects only [*PSI*⁺], indicating a complex role for Hsp104 with a general effect on yeast prions and another activity specific to [*PSI*⁺]. The effect of over-expression/depletion of members of the families Hsp40 and Hsp70 on the prion stability is more complex than those of Hsp104 (see [76] for a review).

The search for others players modulating prion stability provided one of the most astonishing results in prion studies: the appearance of [*PSI*⁺] is itself controlled by another prion [77]. In absence of this prion, called [*PIN*⁺] for PSI-inducible in yeast cells, PSI is unable to appear. This PIN element is usually due to the prion conformation of Rnq1, but experimentally several other prions including [URE3] permit the formation of [PSI⁺] independently of Rnq1 [43]. The effect of [PIN⁺] is not limited to [PSI⁺], as it also influences the appearance of [URE3] [78]. This kind of interactions seems to be not restricted to positive ones as it is suggested that [PSI⁺] and [URE3] antagonize each other propagation and *de novo* appearance [79]. Destabilizing interactions between [PSI⁺] and [PIN⁺] have also been described [80]. Additionally, the mysterious [NSI⁺] prion-like element was shown to result from the interactions between the [SWI⁺] and [PIN⁺] prions leading to the inhibition of the SUP45 gene expression [81]. The New1 protein responsible for the [NU⁺] prion is able to cleave the [PSI⁺] amyloid fibers, suggesting other levels of complex interactions between prion proteins in both polymerized and nonpolymerized forms and their genes [82]. These data exemplify interactions between prions in a manner similar to interactions between alleles of different genes, emphasizing the similarity between true genes and "prion genes."

Many additional factors participate in the formation/stability of amyloid prions. Some, for example, act by sequestering the prionogenic proteins in a particular location, increasing the probability of the first transitions towards the amyloid form, like the G-Protein Gamma Subunit Ste18 [83]. Moreover, proteins may join the prion amyloid aggregates seemingly without presenting typical prion behavior [84].

AMYLOID PRION VARIANTS

The similarity of prions with true genes is even more pronounced, since allelic variants, called strains, have been discovered for several prions [30,78,85–88] (reviewed in [89]). Indeed, the observation that cells with the prion [PSI⁺] may present different stabilities of [PSI⁺] during cell division and different expressivity of the phenotype led to the proposal that strains of [PSI⁺] exist [85]. The same observation was also made with TSE diseases that would appear to be caused by various strains of infectious agents [87,90,91]. The question is how a protein able to switch to an inactive and infectious state may be connected to several phenotypes? The answer lies in the ability of a unique prion protein to adopt many distinct infectious conformations [27,30,92]. As the first aggregated proteins appear spontaneously, one conformation is adopted among many possibilities. Then, this conformation acts as a template for further aggregation and is accurately transmitted to the successive cycles of conversion. Each conformation gives birth to a strain or variant of prion presenting distinct properties: stability, structure of the amyloid fibers, number of seeds, but also proportion of the prion protein in the aggregated form versus the soluble form [27,30,93,94].

Competition between different forms if present in the same cell usually ends up in one taking over the others [95]. At the level of the organism, these different strains trigger phenotypes with a more or less pronounced severity or effects. Interestingly, some are able to act as a template on proteins with a different primary structure. This enables, for example, some of the conformers from one species, but not all of them, to transmit its conformation to the homologous prion protein from a different species, a process known as species-barrier crossing [96].

Amyloid prions are clearly endowed with the ability to transmit information from one cell generation to the next one and, importantly, the aggregated form can be purified and used in transformation experiments. Some are even responsible for transmitting diseases in mammals, clearly demonstrating their stability outside cells or organisms. They should thus be considered as true "hereditary units" in their ability to carry genetic information [97–99]. At the present time, their actual role in cell physiology is unclear. Mammalian prions are clearly detrimental infectious agents, but in fungi things are not so clear (reviewed in [100]). The *P. anserina* prion may be regarded as beneficial or detrimental [25,101,102], while, in yeast, prions are regarded either as enabling adaptation [23,44,49,103,104] or diseases [105–107] that must be controlled by anti-prion systems [108] (reviewed in [109,110]). Indeed, for most prions, only the monomeric form has a biological activity. Exceptions to this rule includes *P. anserina* HET-s, which appears active only in the oligomeric and infectious conformation [25,111]. Indeed, in this conformation, *P. anserina* can produce a self versus nonself recognition reaction called vegetative or heterokaryon incompatibility. Importantly, other signaling events appear to use similar amyloid formation, especially during self versus nonself recognition, *i.e.*, innate immunity in mammals (reviewed in [112,113]). For [MOD+], polymerization inactivates an isopentenyl transferase, regulating sterol biosynthesis and thereby triggering resistance to antifungal compounds, suggesting that both the prion and nonprion form have some biological activity [48]. The most interesting example of a functional role for a prion protein is that of *aplysia* CPEB [32], a neuronal mRNA translation regulator. Indeed, this protein is likely involved through its prion switch in stabilization of long-term memory [114].

SELF-DRIVEN ASSEMBLY OF HSP60 MITOCHONDRIAL CHAPERONIN

Unlike prions, which can be viewed as abnormal proteins, the *S. cerevisiae* Hsp60 chaperonin provides a clear example of a structure catalyzing its own folding. It demonstrates the necessity for correctly folded preexisting oligomers to ensure the correct folding of further monomers [115]. Some proteins, which are imported from the cytosol into mitochondria, cross the mitochondrial membranes in an unfolded conformation and then are folded in the matrix by Hsp60. Monomers of Hsp60 form a complex, arranged as two stacked 7-mer rings. Once assembled in the matrix of mitochondria, these 14-mer complexes bind unfolded proteins to catalyze their proper folding in an ATP-dependent manner. But Hsp60 proteins are also encoded by a nuclear gene and translated in the cytosol as precursors, which are then translocated into the mitochondrial matrix. So, how could they assemble themselves without preexisting 14-mer complexes to fold them?

To address this question, Cheng et al. took advantage of a temperature-sensitive lethal mutation in the $hsp60$ gene [115]. At 23°C, the $hsp60^{ts}$ mutant cells grew normally, but when the temperature was shifted to 37°C, the mutant cells stopped to grow within one generation, because the impaired Hsp60 complex fails in folding and assembly of imported mitochondrial proteins. An attempt to rescue the growth deficient phenotype of $hsp60^{ts}$ mutant strains was set up with a high copy plasmid, containing the coding sequence of the wild-type Hsp60 precursor, driven by the inducible galactose promoter. Cultures were first shifted from 23°C to 37°C and 2 h later, expression of wild-type $hsp60$ was induced by addition of galactose. In these conditions, the growth deficient phenotype of the mutant strain was not rescued. But strikingly, when expression of wild-type Hsp60 subunits was induced by addition of galactose for 2 h before the temperature shift to 37°C, the mutant cells could grow. This means that wild-type hsp60 complexes can rescue the mutant phenotype at restrictive temperature only when expressed at permissive temperature, indicating that Hsp60 is required for its own assembly. In other words, newly Hsp60 imported subunits can be assembled only if preexisting Hsp60 complexes are present in the matrix of mitochondria. More generally, this study strongly suggests that biogenesis of organelles such as mitochondria are probably not a *de novo* process, but rather relies on preexisting structures, acting as a template. If this template is lost along the path, although the protein subunits are produced, no functional organelles would be made. In mammals, the Hsp60 chaperonin is also essential for viability [116] and some mutations in the human *mHsp60* gene lead to the development of neurodegenerative disorders (see [117] for a review).

CYTOTAXIS OF CILIA AND OTHER COMPLEX STRUCTURES

Prions and Hsp60 are homopolymers of a single protein. However, in the cells most structures are built up from several different polypeptides and additional molecules, such as RNAs, cofactors, etc. While in many cases it has been shown that these complex structures are able to correctly fold themselves spontaneously, often with the help of chaperones, in other cases the preexistence of some structural information is necessary to obtain a correct organization. This was first shown by Beisson and Sonneborn on the orientation of cilia in paramecium [5]. Ciliates, like paramecium, are large cells that display a complex organization. Especially, their cortex is endowed with cilia that are all oriented in the same direction, permitting efficient swimming. Beisson and Sonneborn "grafted" in an inverted orientation rows of cilia in *Paramecium aurelia*, producing "variants" with abnormal swimming behavior [5]. These variants could be maintained over 800 mitotic generations and maintained in sexual crosses. This heritability is due to the fact that preexisting cilia direct the correct insertion and orientation of newly formed cilia (see [118] for a review). This process whereby an old cellular structure orders a new one was called cytotaxis [119]. Cytotaxis of cilia/flagella have also been described in *Tetrahymena* [120], *Chlamydomonas* [121], and *Trypanosoma* [122]. Additional examples for other structures have been described in *Paramecium*, including handedness [123,124] and doublets (see [125] for a review). Even in *S. cevevisiae*, cortical inheritance has been described [126]. The mechanism involved in cilium insertion in *Paramecium* has been analyzed at the ultra-structural level [127], and mutant searches [125] have uncovered *Tetrahymena* and *Paramecium* nuclear mutants with altered cortical elements, which should enable a full understanding at the molecular level how the old structure directs the construction of the new one.

MIXED HEREDITY: "NONAMYLOID PRIONS" THAT PROPAGATE BY AUTO-ACTIVATION

Autocatalytic amyloid formation is not the only way for a protein to produce epigenetic traits. Indeed, whatever autocatalytic modification of a protein may promote an inherited trait if some modified proteins are passed upon daughter cells during division. Especially, many proteins change conformation during regulatory events and are then sequestered into particular compartments. They endow cells with a memory of past events, as seen by the first described one, the whi3 "mnemon" [128,129] (reviewed in [11,13,16,130]). Usually, these compartments with modified proteins are not transmitted to daughter cells and hence are not *bona fide* epigenetic traits. For example, in the case of whi3, it seems that retention of the aggregates by the endoplasmic reticulum of the mother cells prevent this mnemon to become a true prion [131]. Nevertheless, in some cases, the mnemon can be inherited and thus create not only cellular but also epigenetic memory. In recent years, several of these nonamyloid elements now called prion-likes, that stand at the border between structural and regulatory inheritance, have been described.

The first description of such a nonamyloid prion-like element is based on the inheritable autocatalytic cleavage of a protease [55]. The yeast protease B, PrB, is a subtilisin/furin class serine protease derived from a larger, catalytically inactive pro-form encoded by the gene PRB1. The final steps in the maturation of the proenzyme PrB are sequential truncations occurring in the lysosome-like yeast vacuole, catalyzed by protease A, PrA, and finally PrB itself. Mature PrB protease activates other vacuolar hydrolases such as carboxipeptidase Y (CpY), whose activity can be easily assayed.

Deletion of PEP4, encoding the PrA precursor leads to accumulation of the immature form of PrB and therefore loss of its activity, as seen by lack of CpY activity. However, the disappearance of mature PrB after deletion of PEP4 is progressive. CpY activity can be detected in $pep4\Delta$ strains for more than 20 mitotic generations. This hysteresis of PrB activity is referred to as "phenotypic lag" and it is believed to reflect dilution during growth of PEP4 mRNA and PrA protease. The phenotypic lag was initially observed during growth on dextrose medium because dextrose represses PRB1 transcription. Roberts and Wickner [55] tested if this lag might be prolonged after transfer onto glycerol medium which does not repress PRB1 expression. The authors sporulated a diploid heterozygous for PEP4 deletion and germinated the meiotic products on glycerol medium. They found that CpY remained active indefinitely, even in the colonies derived from the $pep4\Delta$ spores. However, when transferred onto dextrose medium, which represses PRB1 transcription, CpY activity of these $pep4\Delta$ cells was progressively lost and not restored by a return to glycerol medium. The "PrB + " state expressing the CpY hydrolase is infectious during cytoduction experiments (*i.e.*, cytoplasmic mixing without caryogamy), even when both donor and recipient $pep4\Delta$ strains are grown on glycerol. This was the demonstration that the PrB + state is triggered by a cytoplasmic and infectious factor called [β]. They further showed that the cells mutated for PrB do not contain [β] and that the over-expression of PRB1 increases the frequency of [β] appearance in $pep4\Delta$ cells that have been previously cured by extended growth on dextrose medium. In this system, PrpA is only needed for the initial conversion of PrB in the absence of [β]. To the authors, the fairly unconventional behavior of the PrB + state is reminiscent to that of structure-based prion, indicating that any enzyme could be similar to a prion, provided that its activity depends on self-modification *in trans* and that there is a mechanism by which it can be transmitted from individual to individual [132]. As we will see below, an additional example of such behavior exhibited by kinases involved in signaling indicates that this is indeed the case. However, because these kinases are involved in regulation, we will discuss them in the next section dealing with regulatory inheritance.

Additional examples of prion-like elements are given in Table 5.1. They include [GAR^+] that affects carbon source utilization in several *Saccharomycotina* yeasts and is linked to modification of the Pma1 proton pump [56,103] and [$SMAUG^+$] that is associated with nonamyloid aggregation of the Vts1 posttranscriptional regulator, and that enable *S. cerevisae* to anticipate return of nutrients after starvation [57,133]. They also include likely [ESI^+] that is linked to aggregation of the Set3C histone deacetylase scaffold Snt1 following its transient phosphorylation [58], although in this latter case it was not shown whether aggregation is amyloid or not.

REGULATORY INHERITANCE

There is now a large body of literature dealing with the behavior of regulatory networks, especially their ability to generate emergent properties enabling cells to finely tune their response to various environmental changes (see [134–136] for reviews). In some cases, these properties result in the generation of bistable states that are inheritable in a more or less faithful fashion [137–142]. Below we will discuss three examples of such regulatory inheritance.

THE LACTOSE OPERON AND ITS POSITIVE FEEDBACK LOOP

In their seminal studies of 1956 [4], Novick and Weiner, and later on Cohn and Horibata [143–145], showed that under defined conditions, it is possible to obtain an epigenetic inheritability of the activation status of the lactose operon in *E. coli*. Indeed, when grown at low concentrations of a gratuitous inducer, *E. coli* are either not induced for their lactose operon or fully induced and are never found in an intermediate state. Noninduced cells can change spontaneously towards the induced state with a constant probability. Thus, when transferred from a medium lacking the inducer towards a medium containing the inducer, the population accumulates more and more cells that have made the transition towards the active state. These cells do not invade the population since the induced cells grow more slowly than the noninduced ones, permitting a dynamic equilibrium in the population. That both the on and off states are inheritable was demonstrated by diluting an equilibrated population in new medium that contained the inducer at an inoculum of one cell per new culture. Two kinds of cultures where obtained, one composed of fully induced cells that originated from a bacterium that was already induced and the other, which accumulated induced cells at the level of the parental culture before dilution. The cause of this behavior is the presence of a positive regulatory loop in the lactose regulation; whereby, the entrance of enough inducer inside the cell activates the operon and especially the production of permease, which in turn allows more inducer to enter the cell. It was demonstrated that the stochastic complete dissociation of the lactose repressor, which binds as a tetramer, triggers the initial burst of production of the permease [146] and that transcription errors may participate in the switch [147]. Another piece of information gained from the study of the lactose operon is that the behavior of the cells is strongly dependent upon the inducer and glucose concentrations [148]. Glucose represses the operon while the inducer activates it. Three main regulatory behaviors can be adopted by the cells, (1) monostable induced at low glucose and high inducer concentrations, (2) monostable uninduced at high glucose and low inducer lever, and (3) the bistable state described above at intermediate concentrations. Importantly, cells that are placed in these intermediate concentrations will behave differently if they originate from the monostable induced or monostable uninduced conditions.

This lactose operon is prototypic of systems with a positive auto-regulatory loop and mathematic models describing their properties are available [148–150]. More complex inheritable units can be envisioned. They are all based on positive auto-regulatory loops or its derivative, the reciprocal repression as the one present in the C1/Cro interaction of the lambda phage. That these transcription factors, which negatively regulate the others, are able to produce an epigenetic inheritable switch has been known for a long time [141,151]. Readers interested in the emergent properties of regulatory networks, including epigenetic inheritance can refer to a number of excellent reviews [134–136]. Most of the research carried out today are performed on man-made regulatory networks. We discuss below two examples of regulatory inheritance encountered in wild organisms: the sectors of filamentous fungi and the yeasts phenotypic switches.

CRIPPLED GROWTH, A SELF-SUSTAINED AND MITOTICALLY INHERITABLE SIGNALING PATHWAY IN THE FILAMENTOUS FUNGUS PODOSPORA ANSERINA

Podospora anserina is a saprophytic filamentous fungus used as a model organism for decades. In the 1990s, Silar and his colleagues noticed that sectors of altered growth could be seen on *P. anserina* growing thalli [52]. This cell degeneration phenomenon, called crippled growth (CG), was easily visible macroscopically, displaying highly pigmented, flat, and female-sterile mycelium as opposed to normal growth. Curiously, the development of these sectors occurred only in special genetic or environmental conditions [52,152]. The switch is controlled in both directions by environmental stimuli [52]. It was rapidly demonstrated that no nucleic acid was involved in the genesis of these sectors and that the presence of C, a cytoplasmic and infectious factor was associated with CG. The mycelium of *P. anserina* can thus exhibit a bistability at the morphological level. Similar phenomena were previously described and reported to be very frequent in filamentous ascomycetes (see [18,153] for reviews). They were generally due to the presence in the cell of cytoplasmic and infectious factors, whose properties appear strikingly similar to prions. Apart from CG, only the "secteur" phenomenon of *Nectria haematococca* has been studied [154], but in this instance no clear model on how the infectious factor is generated is presently available. In the case of CG, a genetic analysis [152] permitted to retrieve numerous genes, which are required to produce C. Some of these *"IDC"* genes (impaired in the development of CG) were cloned and showed to encode a MAP kinase kinase kinase (MAPKKK) [53] and a MAP kinase kinase (MAPKK) [54]. These two proteins are members of a large family of kinases present in all eukaryotes. They act in a sequential manner, *i.e.*, the MAPKKK activates by phosphorylation the MAPKK, which in turn activates by phopshorylation a MAP kinase (or MAPK). These *IDC* mutants unable to produce C were null mutants of either the MAPKKK or MAPKK genes. Further genetic inactivation of the gene coding, the downstream MAP kinase also showed it to be a key element in the genesis of C [54]. Moreover, overexpression of the MAPKKK and MAPK was shown to facilitate the development of CG [53,54]. (1) Presence of a cytoplasmic and infectious factor, (2) necessity of a gene for its propagation, and (3) increased frequency of appearance of the infectious factor when the gene is over-expressed, are properties exhibited by genes coding for prions. Here, the three genes coding—MAPKKK, MAPKK, and MAPK—display these properties. A model related to that of prions but based on an autocatalytic activation loop in the MAPK cascade has thus been proposed to account for the C element [53,54]. Some element(s) downstream of the MAPK would be able to activate directly or indirectly the upstream MAPKKK. In this model, the C element corresponds to components of the cascade in the active state, which are able to activate in *trans* other molecules that are in the inactive state. This results in the complete conversion of the inactive factors to their active form. This strikingly resembles the ability of prions to promote their own aggregation or that of the [β] "prion" in *S. cerevisiae* to promote its own maturation. This model is supported by experiments in *Xenopus* eggs, in which it was demonstrated that the presence of a positive self-regulation in the p42 MAP kinase cascade entails the presence of only two states: one in which no active MAPK is present and the other in which all MAPK molecules are active, the intermediate states being transient [155]. It was also shown that transfer of cytoplasm from an activated egg to an inactivated one results in complete activation of the MAPK cascade in the recipient [155]. This property is conserved over three transfers, in conditions where no cytoplasm originating from the first egg is present. In essence, this is strikingly similar to the cytoplasmic and infectious factors detected in CG and related phenomena.

This regulatory inheritance has many properties in common with prions. However, it displays several differences. First, it relies on many proteins (the whole signaling cascade or at least a subset of the cascade). This implies that the genetic basis is more complex than for prion [152]. In the case of CG, additional factors have been identified and were shown to be necessary for producing the C element, but likely not to be present in the regulatory loop [156–163]. Many genes restricting the spread of C have also been identified, adding another level of complexity [152]. Second, the development and/or spreading of C are highly dependent upon the environmental conditions, a property also exhibited by the [β] factor. Because regulatory networks can adopt complex behavior, depending upon the level of expression of the key "flexibility loci" [164], it is not surprising that the determinism of CG is quite complex and depends upon numerous genetic and environmental factors.

THE WHITE/OPAQUE SWITCH OF CANDIDA ALBICANS, AN EPIGENETIC SWITCH AT THE TRANSCRIPTION LEVEL

Many species of yeast have the ability to switch at various frequencies between different states [23,165–170]. These switches may be caused by classical transcriptional gene silencing (see [171] for a review) or prions [23]. However, regulatory inheritance may also be involved. The most studied of the switches is the white/opaque transition exhibited by *C. albicans* [172]. This transition is present in this diploid fungus when homozygous for the mating type [173,174] or when grown in a condition resembling the environment within its host [175]. The cells may then adopt two morphologies: roundish cells forming white colonies and bigger more elongated cell forming colonies that are more translucent. In fact, the two types of cell differ by an impressive array of differences [176–179] that are not due to genetic differences [180]. *Candida albicans* causes mycosis in humans and the switch likely enables the fungus to adapt to the various niches it will encounter in the human body [181,182].

To understand how the transition is controlled, genes down-regulated by the a1-α2 heterodimers encoded at the *C. albicans* mating type were searched, since the white/opaque switch is present in strains homozygous for their mating-type [173]. Transcriptomic or chromatin-immunoprecipitation approaches identified the WOR1/TOS9 transcription factor as specifically expressed in opaque cells [183–185]. Gene inactivation of WOR1 showed that the cells were locked in the white state. Ectopic expression of WOR1, even as a single pulse or in cells heterozygous for the mating type, was sufficient to convert the whole cell population to the opaque state. Finally, it was shown that WOR1 binds the promoter of its own gene and thereby activates its own expression [185]. Overall, these data permitted the formulation of a model for the white/opaque transition based on the self-activation of the WOR1 transcription factor [183–185]. WOR1 is absent in white cells. Random fluctuation in the transcription of the WOR1 locus permits the expression of a few molecules of WOR1, resulting in further transcription of the WOR1 gene, locking the cell in a state with a high concentration of WOR1. The other targets of WOR1 are then regulated promoting the physiological and morphological changes to the opaque state [186,187]. Although it has not yet been formally demonstrated that the opaque state is infectious towards the white state, the similarity of this bistable system with the ones created by classical amyloid-based prions and by β-type and C-type prion-like factors is evident.

Studies showed that the WOR1 positive regulatory loop is embedded in a complex network of seven transcription factors (WOR1, WOR2, WOR3, WOR4, EFG1, CZF1 and AHR1) with positive and negative feedback loops [188–190]. The multiplicity of the feedbacks appears to ensure a faithful transmission of the white and opaque states through numerous cell generations and accounts for the previously known roles of various transcription factors, corepressors and chromatin remodeling factors in the control of the transition [191–196] (reviewed in [197]). It also enables, in some clinical isolates, the formation of a third state called "gray" that displays a distinct transcriptional profile and physiology [198]. This state does not necessitate WOR1 to appear. However, in these strains, a *wor1 efg1* double mutant appears locked in the gray state, suggesting a connection between this new state and the regulatory network controlling the white/opaque switch. Intriguingly, a recent report shows that *wor1 efg1* double mutants can still undergo the white to opaque switch [199]. The authors postulated that an "alternative opaque pathway" also repressed by EFG1 exist in the investigated strains, and that the white opaque switch is thus more complex than hitherto demonstrated.

Like for the CG of *P. anserina*, the environment controls the switch in both directions. While high temperature triggers the opaque to white switch [172,200], numerous factors are able to trigger the white to opaque transition [201–204]; a study showed that slowing the cell cycle by many means is sufficient to increase the white to opaque switch frequency [205]. The signaling pathways controlling the transition include the Hog1 MAP kinase pathway, which controls stress response in many fungi [206].

CONCLUSION

The various examples presented above show that epigenetic states can be conferred in many ways (Fig. 5.1), provided that an auto-regulatory loop (or a double repression) is present. The presence of this loop ensures that two mutually exclusive states may be exhibited by cells with an identical genome and grown in the same conditions. In the case of the structural inheritance that we have presented, the loops ensure the faithful reproduction of a structure made of proteins (prions, Hsp60, or cilia). There is, however, suggestions that another component of the cell, the membranes, may adopt alternative states [207]. In general, the influence of the environment on this kind of inheritance is moderate. On the contrary, the regulatory inheritance is usually greatly influenced by the environment, since any modification in the concentration of key factors [164] under the influence of external stimuli may drastically alter the behavior of the pathway. Importantly, in the case of regulatory inheritance the phenotype of the cells will depend upon their history, however, not in a directed fashion as in a Lamarckian inheritance. Often, this inheritance results from the emergent properties of complex networks or structural properties of domains that adopt particular structures. It is thus not easy to know whether the inheritable behavior is a by-product of these, or whether it participates in a process essential to the life cycle. If, the white/opaque transition appears to confer a selective advantage to *C. albicans*, what of yeast prions and CG? Certainly in some cases, prions and prion-like elements are involved in differentiation processes [208]. However, human prions clearly cause severe diseases, and it has been proposed that regulatory inheritance may be involved in cancer formation [209]. This regulatory inheritance is presently known in both eukaryotes and prokaryotes, and occurs at all level of gene regulation (transcription, signal transduction, carbon metabolism, etc.). As yet, the full scope of this kind of inheritance based on prion and prion-like phenomena is unknown. We have here presented data obtained with a few model microorganisms, since they are more easily tractable than the multicellular animals and plants. However, we propose that the same epigenetic mechanisms are prevalent in all organisms.

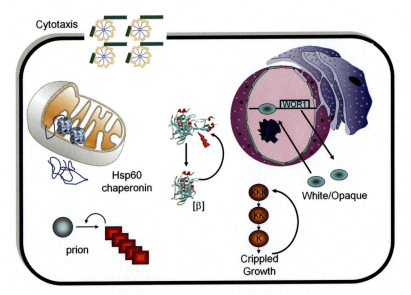

FIGURE 5.1 Schematic diagram of the prions and prion-like elements of eukaryotes discussed in this chapter. See text for detailed explanation on each phenomenon.

References

[1] Nanney DL. Epigenetic control systems. Proc Natl Acad Sci USA 1958;44:712−17.

[2] Delbrück M. Comment of an article by. In: Sonneborn TM, Beale GH, editors. Unités biologiques douées de continuité génétique. Paris France: CNRS; 1949. p. 33−4.

[3] Marcou D. Notion de longévité et nature cytoplasmique du déterminant de la sénescence chez quelques champignons. Ann Sci Natur Bot 1961;2:653−764 Sér. 12.

[4] Novick A, Weiner M. Enzyme induction as an all-or-none phenomenon. Proc Natl Acad Sci USA 1957;43:553−66.

[5] Beisson J, Sonneborn TM. Cytoplasmic inheritance of the organization of the cell cortex in *Paramecium aurelia*. Proc Natl Acad Sci USA 1965;53:275−82.

[6] Beisson J. Non−nucleic acid inheritance and epigenetic phenomena. Cell Biology San Francisco USA. Academic Press Inc; 1977. p. 375−421.

[7] Kirk JTO. Roots: The discovery of chloroplast DNA. Bioessays 1986;4:36−8.

[8] Mounolou JC, Lacroute F. Mitochondrial DNA: an advance in eukaryotic cell biology in the 1960s. Biol Cell 2005;97:743−8.

[9] Wickner RB. [URE3] as an altered URE2 protein: evidence for a prion analog in Saccharomyces cerevisiae. Science (New York, NY) 1994;264:566−9.

[10] Chakravarty AK, Jarosz DF. More than just a phase: prions at the crossroads of epigenetic inheritance and evolutionary change. J Mol Biol 2018;430:4607−18.

[11] Patrick OH, Yasmin L, Fabrice C. Prion-like proteins as epigenetic devices of stress adaptation. Exp Cell Res 2020;112262.

[12] Saad S, Jarosz DF. Protein self-assembly: a new frontier in cell signaling. Curr Opin Cell Biol 2021;69:62−9.

[13] Reichert P, Caudron F. Mnemons and the memorization of past signaling events. Curr Opin Cell Biol 2021;69:127−35.

[14] Manjrekar J, Shah H. Protein-based inheritance. Seminars in cell & developmental biology. Elsevier; 2020. p. 138−55.

[15] Tikhodeyev ON. The mechanisms of epigenetic inheritance: how diverse are they? Biol Rev 2018;93:1987−2005.

[16] Lau Y, Oamen HP, Caudron F. Protein phase separation during stress adaptation and cellular memory. Cells 2020;9:1302.

[17] Tikhodeyev ON, Tarasov OV, Bondarev SA. Allelic variants of hereditary prions: the bimodularity principle. Prion 2017;11:4−24.

[18] Silar P. Phenotypic instability in fungi. Adv Appl Microbiol 2019;107:141−87.

[19] Levkovich SA, Rencus-Lazar S, Gazit E, Bar-Yosef DL. Microbial prions: dawn of a new era. Trends Biochem Sci 2021;46(5):391−405.

[20] Prusiner SB. Novel proteinaceous infectious particles cause scrapie. Science (New York, NY) 1982;216:136−44.

[21] Legname G, Baskakov IV, Nguyen HO, Riesner D, Cohen FE, DeArmond SJ, et al. Synthetic mammalian prions. Science (New York, NY) 2004;305:673−6.

[22] Wickner RB, Shewmaker F, Kryndushkin D, Edskes HK. Protein inheritance (prions) based on parallel in-register beta-sheet amyloid structures. Bioessays 2008;30:955−64.

[23] True HL, Berlin I, Lindquist SL. Epigenetic regulation of translation reveals hidden genetic variation to produce complex traits. Nature 2004;431:184−7.

[24] Maddelein ML, Dos Reis S, Duvezin-Caubet S, Coulary-Salin B, Saupe SJ. Amyloid aggregates of the HET-s prion protein are infectious. Proc Natl Acad Sci USA 2002;99:7402−7.

[25] Coustou V, Deleu C, Saupe S, Begueret J. The protein product of the het-s heterokaryon incompatibility gene of the fungus *Podospora anserina* behaves as a prion analog. Proc Natl Acad Sci USA 1997;94:9773−8.

[26] Avery OT, McLeod CM, McCarty M. Studies on the chemical nature of the substance inducing transformation of pneumococcal types. J Exp Med 1944;79:137−58.

[27] King CY, Diaz-Avalos R. Protein-only transmission of three yeast prion strains. Nature 2004;428:319−23.

[28] Patel BK, Liebman SW. "Prion-proof" for [PIN +]: infection with *in vitro*-made amyloid aggregates of Rnq1p-(132−405) induces [PIN +]. J Mol Biol 2007;365:773−82.

[29] Brachmann A, Baxa U, Wickner RB. Prion generation *in vitro*: amyloid of Ure2p is infectious. EMBO J 2005;24:3082−92.

[30] Tanaka M, Chien P, Naber N, Cooke R, Weissman JS. Conformational variations in an infectious protein determine prion strain differences. Nature 2004;428:323−8.

[31] Sideri T, Yashiroda Y, Ellis DA, Rodríguez-López M, Yoshida M, Tuite MF, et al. The copper transport-associated protein Ctr4 can form prion-like epigenetic determinants in Schizosaccharomyces pombe. Microbial Cell 2017;4:16.

[32] Si K, Lindquist S, Kandel ER. A neuronal isoform of the aplysia CPEB has prion-like properties. Cell 2003;115:879−91.

[33] Heinrich SU, Lindquist S. Protein-only mechanism induces self-perpetuating changes in the activity of neuronal Aplysia cytoplasmic polyadenylation element binding protein (CPEB). Proc Natl Acad Sci 2011;108:2999−3004.

[34] Stephan Joseph S, Fioriti L, Lamba N, Colnaghi L, Karl K, Derkatch Irina L, et al. The CPEB3 protein is a functional prion that interacts with the actin cytoskeleton. Cell Rep 2015;11:1772−85.

[35] Chakrabortee S, Kayatekin C, Newby GA, Mendillo ML, Lancaster A, Lindquist S. Luminidependens (LD) is an *Arabidopsis protein* with prion behavior. Proc Natl Acad Sci USA 2016;113(21):6065−70.

[36] Yuan AH, Hochschild A. A bacterial global regulator forms a prion. Science 2017;355:198.

[37] Zajkowski T, Lee MD, Mondal SS, Carbajal A, Dec R, Brennock PD, et al. The hunt for ancient prions: Archaeal prion-like domains form amyloid-based epigenetic elements. Mol Biol Evol 2021;38:2088−103.

[38] Nan H, Chen H, Tuite MF, Xu X. A viral expression factor behaves as a prion. Nat Commun 2019;10:359.

[39] Si K, Choi Y-B, White-Grindley E, Majumdar A, Kandel ER. Aplysia CPEB can form prion-like multimers in sensory neurons that contribute to long-term facilitation. Cell 2010;140:421−35.

[40] Raveendra BL, Siemer AB, Puthanveettil SV, Hendrickson WA, Kandel ER, McDermott AE. Characterization of prion-like conformational changes of the neuronal isoform of Aplysia CPEB. Nat Struct Mol Biol 2013;20:495−501.

[41] Abbott A. The red-hot debate about transmissible Alzheimer's. Nature 2016;531:294−7.

[42] Sparrer HE, Santoso A, Szoka Jr. FC, Weissman JS. Evidence for the prion hypothesis: induction of the yeast [PSI +] factor by *in vitro*- converted Sup35 protein. Science (New York, NY) 2000;289:595−9.

[43] Derkatch IL, Bradley ME, Hong JY, Liebman SW. Prions affect the appearance of other prions: the story of [PIN(+)]. Cell 2001;106:171−82.

[44] Newby GA, Lindquist S. Pioneer cells established by the [SWI +] prion can promote dispersal and out-crossing in yeast. PLoS Biol 2017;15:e2003476.

[45] Du Z, Park KW, Yu H, Fan Q, Li L. Newly identified prion linked to the chromatin-remodeling factor Swi1 in Saccharomyces cerevisiae. Nat Genet 2008;40:460−5.

[46] Patel BK, Gavin-Smyth J, Liebman SW. The yeast global transcriptional co-repressor protein Cyc8 can propagate as a prion. Nat Cell Biol 2009;11:344−9.

[47] Alberti S, Halfmann R, King O, Kapila A, Lindquist S. A systematic survey identifies prions and illuminates sequence features of prionogenic proteins. Cell 2009;137:146−58.

[48] Rogoza T, Goginashvili A, Rodionova S, Ivanov M, Viktorovskaya O, Rubel A, et al. Non-Mendelian determinant [ISP+] in yeast is a nuclear-residing prion form of the global transcriptional regulator Sfp1. Proc Natl Acad Sci 2010;107:10573−7.

[49] Suzuki G, Shimazu N, Tanaka M. A yeast prion, Mod5, promotes acquired drug resistance and cell survival under environmental stress. Science 2012;336:355−9.

[50] Halfmann R, Wright JR, Alberti S, Lindquist S, Rexach M. Prion formation by a yeast GLFG nucleoporin. Prion 2012;6:391−9.

[51] Chernova TA, Kiktev DA, Romanyuk AV, Shanks JR, Laur O, Ali M, et al. Yeast short-lived actin-associated protein forms a metastable prion in response to thermal stress. Cell Rep 2017;18:751−61.

[52] Silar P, Haedens V, Rossignol M, Lalucque H. Propagation of a novel cytoplasmic, infectious and deleterious determinant is controlled by translational accuracy in Podospora anserina. Genetics 1999;151:87−95.

[53] Kicka S, Silar P. PaASK1, a mitogen-activated protein kinase kinase kinase that controls cell degeneration and cell differentiation in Podospora anserina. Genetics 2004;166:1241−52.

[54] Kicka S, Bonnet C, Sobering AK, Ganesan LP, Silar P. A mitotically inheritable unit containing a MAP kinase module. Proc Natl Acad Sci USA 2006;103:13445−50.

[55] Roberts BT, Wickner RB. Heritable activity: a prion that propagates by covalent autoactivation. Genes Dev 2003;17:2083−7.

[56] Brown JC, Lindquist S. A heritable switch in carbon source utilization driven by an unusual yeast prion. Genes Dev 2009;23:2320−32.

[57] Chakravarty AK, Smejkal T, Itakura AK, Garcia DM, Jarosz DF. A non-amyloid prion particle that activates a heritable gene expression program. Mol Cell 2020;77:251−65 e9.

[58] Harvey ZH, Chakravarty AK, Futia RA, Jarosz DF. A prion epigenetic switch establishes an active chromatin state. Cell 2020;180:928−40 e14.

[59] Chakrabortee S, Byers JS, Jones S, Garcia DM, Bhullar B, Chang A, et al. Intrinsically disordered proteins drive emergence and inheritance of biological traits. Cell 2016;167:369−81 e12.

[60] Byers JS, Jarosz DF. High-throughput screening for protein-based inheritance in S. cerevisiae. J Vis Exp 2017;126:56069.

[61] Galkin A, Velizhanina M, Sopova YV, Shenfeld A, Zadorsky S. Prions and non-infectious amyloids of mammals − similarities and differences. Biochemistry (Moscow). 2018;83:1184−95.

[62] Li L, Lindquist S. Creating a protein-based element of inheritance. Science (New York, NY) 2000;287:661−4.

[63] Patino MM, Liu JJ, Glover JR, Lindquist S. Support for the prion hypothesis for inheritance of a phenotypic trait in yeast. Science (New York, NY) 1996;273:622−6.

[64] Ross ED, Baxa U, Wickner RB. Scrambled prion domains form prions and amyloid. Mol Cell Biol 2004;24:7206−13.

[65] Sondheimer N, Lindquist S. Rnq1: an epigenetic modifier of protein function in yeast. Mol Cell 2000;5:163−72.

[66] Verges KJ, Smith MH, Toyama BH, Weissman JS. Strain conformation, primary structure and the propagation of the yeast prion [PSI+]. Nat Struct Mol Biol 2011;18:493−9.

[67] Baxa U, Cassese T, Kajava AV, Steven AC. Structure, function, and amyloidogenesis of fungal prions: filament polymorphism and prion variants. Adv Protein Chem 2006;73:125−80.

[68] Tessier PM, Lindquist S. Unraveling infectious structures, strain variants and species barriers for the yeast prion [PSI+]. Nat Struct Mol Biol 2009;16:598−605.

[69] Danilov LG, Matveenko AG, Ryzhkova VE, Belousov MV, Poleshchuk OI, Likholetova DV, et al. Design of a new [PSI (+)]-No-more mutation in SUP35 with strong inhibitory effect on the [PSI (+)] prion propagation. Front Mol Neurosci 2019;12:274.

[70] Ritter C, Maddelein ML, Siemer AB, Luhrs T, Ernst M, Meier BH, et al. Correlation of structural elements and infectivity of the HET-s prion. Nature 2005;435:844−8.

[71] Wasmer C, Schutz A, Loquet A, Buhtz C, Greenwald J, Riek R, et al. The molecular organization of the fungal prion HET-s in its amyloid form. J Mol Biol 2009;394:119−27.

[72] Daskalov A, Gantner M, Walti MA, Schmidlin T, Chi CN, Wasmer C, et al. Contribution of specific residues of the beta-solenoid fold to HET-s prion function, amyloid structure and stability. PLoS Pathog 2014;10:e1004158.

[73] Greenwald J, Buhtz C, Ritter C, Kwiatkowski W, Choe S, Maddelein ML, et al. The mechanism of prion inhibition by HET-S. Mol Cell 2010;38:889−99.

[74] Chernoff YO, Lindquist SL, Ono B, Inge-Vechtomov SG, Liebman SW. Role of the chaperone protein Hsp104 in propagation of the yeast prion-like factor [psi+]. Science (New York, NY) 1995;268:880−4.

[75] Jung G, Masison DC. Guanidine hydrochloride inhibits Hsp104 activity in vivo: a possible explanation for its effect in curing yeast prions. Curr Microbiol 2001;43:7−10.

[76] Rikhvanov EG, Romanova NV, Chernoff YO. Chaperone effects on prion and nonprion aggregates. Prion 2007;1:217−22.

[77] Derkatch IL, Bradley ME, Zhou P, Chernoff YO, Liebman SW. Genetic and environmental factors affecting the de novo appearance of the [PSI+] prion in Saccharomyces cerevisiae. Genetics 1997;147:507−19.

[78] Bradley ME, Edskes HK, Hong JY, Wickner RB, Liebman SW. Interactions among prions and prion "strains" in yeast. Proc Natl Acad Sci USA 2002;99(Suppl 4):16392−9.

[79] Schwimmer C, Masison DC. Antagonistic interactions between yeast [PSI(+)] and [URE3] prions and curing of [URE3] by Hsp70 protein chaperone Ssa1p but not by Ssa2p. Mol Cell Biol 2002;22:3590−8.

[80] Bradley ME, Liebman SW. Destabilizing interactions among [PSI(+)] and [PIN(+)] yeast prion variants. Genetics 2003;165:1675−85.

[81] Nizhnikov AA, Ryzhova TA, Volkov KV, Zadorsky SP, Sopova JV, Inge-Vechtomov SG, et al. Interaction of prions causes heritable traits in Saccharomyces cerevisiae. PLoS Genet 2016;12:e1006504.

[82] Inoue Y, Kawai-Noma S, Koike-Takeshita A, Taguchi H, Yoshida M. Yeast prion protein New1 can break Sup35 amyloid fibrils into fragments in an ATP-dependent manner. Genes Cells 2011;16:545−56.

[83] Chernova TA, Yang Z, Karpova TS, Shanks JR, Shcherbik N, Wilkinson KD, et al. Aggregation and prion-inducing properties of the g-protein gamma subunit ste18 are regulated by membrane association. Int J Mol Sci 2020;21:5038.

[84] Sergeeva A, Sopova J, Belashova T, Siniukova V, Chirinskaite A, Galkin A, et al. Amyloid properties of the yeast cell wall protein Toh1 and its interaction with prion proteins Rnq1 and Sup35. Prion 2019;13:21−32.

[85] Derkatch IL, Chernoff YO, Kushnirov VV, Inge-Vechtomov SG, Liebman SW. Genesis and variability of [PSI] prion factors in Saccharomyces cerevisiae. Genetics 1996;144:1375−86.

[86] Schlumpberger M, Wille H, Baldwin MA, Butler DA, Herskowitz I, Prusiner SB. The prion domain of yeast Ure2p induces autocatalytic formation of amyloid fibers by a recombinant fusion protein. Protein Sci 2000;9:440−51.

[87] Morales R, Abid K, Soto C. The prion strain phenomenon: molecular basis and unprecedented features. Biochim Biophys Acta 2007;1772:681−91.

[88] Huang Y-W, King C-Y. A complete catalog of wild-type Sup35 prion variants and their protein-only propagation. Curr Genet 2020;66:97−122.

[89] Wickner RB, Son M, Edskes HK. Prion variants of yeast are numerous, mutable, and segregate on growth, affecting prion pathogenesis, transmission barriers, and sensitivity to anti-prion systems. Viruses 2019;11:238.

[90] Pattison IH, Millson GC. Further experimental observations on scrapie. J Comp Pathol 1961;71:350–9.

[91] Pattison IH, Millson GC. Scrapie produced experimentally in goats with special reference to the clinical syndrome. J Comp Pathol 1961;71:101–9.

[92] Dergalev AA, Alexandrov AI, Ivannikov RI, Ter-Avanesyan MD, Kushnirov VV. Yeast Sup35 prion structure: Two types, four parts, many variants. Int J Mol Sci 2019;20:2633.

[93] Tanaka M, Collins SR, Toyama BH, Weissman JS. The physical basis of how prion conformations determine strain phenotypes. Nature 2006;442:585–9.

[94] Villali J, Dark J, Brechtel TM, Pei F, Sindi SS, Serio TR. Nucleation seed size determines amyloid clearance and establishes a barrier to prion appearance in yeast. Nat Struct Mol Biol 2020;27:540–9.

[95] Yu CI, King CY. Forms and abundance of chaperone proteins influence yeast prion variant competition. Mol Microbiol 2019;111:798–810.

[96] Tanaka M, Chien P, Yonekura K, Weissman JS. Mechanism of cross–species prion transmission: an infectious conformation compatible with two highly divergent yeast prion proteins. Cell 2005;121:49–62.

[97] Chernoff YO. Mutation processes at the protein level: is Lamarck back? Mutat Res/Rev Mutat Res 2001;488:39–64.

[98] Tuite MF. Yeast prions: paramutation at the protein level? Semin Cell Dev Biol 2015;44:51–61.

[99] Wickner RB, Edskes HK, Maddelein M-L, Taylor KL, Moriyama H. Prions of yeast and fungi: proteins as genetic material. J Biol Chem 1999;274:555–8.

[100] Saupe SJ, Jarosz DF, True HL. Amyloid prions in fungi. Microbiol Spectr 2016;4:6.

[101] Debets AJM, Dalstra HJP, Slakhorst M, Koopmanschap B, Hoekstra RF, Saupe SJ. High natural prevalence of a fungal prion. Proc Natl Acad Sci USA 2012;109:10432–7.

[102] Dalstra HJP, Swart K, Debets AJM, Saupe SJ, Hoekstra RF. Sexual transmission of the [Het-s] prion leads to meiotic drive in Podospora anserina. Proc Natl Acad Sci USA 2003;100:6616–21.

[103] Jarosz Daniel F, Lancaster Alex K, Brown Jessica CS, Lindquist S. An evolutionarily conserved prion-like element converts wild fungi from metabolic specialists to generalists. Cell 2014;158:1072–82.

[104] Jarosz Daniel F, Brown Jessica CS, Walker Gordon A, Datta Manoshi S, Ung WL, Lancaster Alex K, et al. Cross-Kingdom chemical communication drives a heritable, mutually beneficial prion-based transformation of metabolism. Cell 2014;158:1083–93.

[105] Nakayashiki T, Kurtzman CP, Edskes HK, Wickner RB. Yeast prions [URE3] and [PSI+] are diseases. Proc Natl Acad Sci USA 2005;102:10575–80.

[106] Wickner RB, Edskes HK, Gorkovskiy A, Bezsonov EE, Stroobant EE. Chapter Four - Yeast and fungal prions: amyloid-handling systems, amyloid structure, and prion biology. In: Theodore Friedmann JCD, Stephen FG, editors. Advances in genetics. Academic Press; 2016. p. 191–236.

[107] Kelly AC, Shewmaker FP, Kryndushkin D, Wickner RB. Sex, prions, and plasmids in yeast. Proc Natl Acad Sci 2012;109:E2683–90.

[108] Son M, Wickner RB. Normal levels of ribosome-associated chaperones cure two groups of [PSI+] prion variants. Proc Natl Acad Sci 2020;117:26298–306.

[109] Wickner RB. Anti-prion systems in yeast. J Biol Chem 2019;294:1729–38.

[110] Wickner RB, Edskes HK, Son M, Wu S, Niznikiewicz M. How do yeast cells contend with prions? Int J Mol Sci 2020;21:4742.

[111] Daskalov A, Habenstein B, Martinez D, Debets AJM, Sabaté R, Loquet A, et al. Signal transduction by a fungal NOD-like receptor based on propagation of a prion amyloid fold. PLoS Biol 2015;13:e1002059.

[112] Daskalov A, Saupe SJ. The expanding scope of amyloid signalling. Prion 2021;15:21–8.

[113] O'Carroll A, Coyle J, Gambin Y. Prions and prion-like assemblies in neurodegeneration and immunity: the emergence of universal mechanisms across health and disease. Semin Cell Dev Biol 2020;99:115–30.

[114] Si K, Kandel ER. The role of functional prion-like proteins in the persistence of memory. Cold Spring Harb Perspect Biol 2016;8:a021774.

[115] Cheng MY, Hartl FU, Horwich AL. The mitochondrial chaperonin hsp60 is required for its own assembly. Nature 1990;348:455–8.

[116] Christensen JH, Nielsen MN, Hansen J, Fuchtbauer A, Fuchtbauer EM, West M, et al. Inactivation of the hereditary spastic paraplegia-associated Hspd1 gene encoding the Hsp60 chaperone results in early embryonic lethality in mice. Cell Stress Chaperones 2010;15:851–63.

[117] Bross P, Magnoni R, Bie AS. Molecular chaperone disorders: defective Hsp60 in neurodegeneration. Curr Top Med Chem 2012;12:2491–503.

[118] Beisson J. Preformed cell structure and cell heredity. Prion 2008;2:1–8.

[119] Sonneborn TM. The determinants and evolution of life. The differentiation of cells. Proc Natl Acad Sci USA 1964;51:915–29.

[120] Ng SF, Frankel J. 180 degrees rotation of ciliary rows and its morphogenetic implications in Tetrahymena pyriformis. Proc Natl Acad Sci USA 1977;74:1115–19.

[121] Feldman JL, Geimer S, Marshall WF. The mother centriole plays an instructive role in defining cell geometry. PLoS Biol 2007;5:e149.

[122] Moreira-Leite FF, Sherwin T, Kohl L, Gull K. A trypanosome structure involved in transmitting cytoplasmic information during cell division. Science (New York, NY) 2001;294:610–12.

[123] Nelsen EM, Frankel J. Maintenance and regulation of cellular handedness in Tetrahymena. Development 1989;105:457–71.

[124] Nelsen EM, Frankel J, Jenkins LM. Non-genic inheritance of cellular handedness. Development 1989;105:447–56.

[125] Frankel J. What do genic mutations tell us about the structural patterning of a complex single-celled organism? Eukaryot Cell 2008;7:1617–39.

[126] Chen T, Hiroko T, Chaudhuri A, Inose F, Lord M, Tanaka S, et al. Multigenerational cortical inheritance of the Rax2 protein in orienting polarity and division in yeast. Science (New York, NY) 2000;290:1975–8.

[127] Iftode F, Fleury-Aubusson A. Structural inheritance in Paramecium: ultrastructural evidence for basal body and associated rootlets polarity transmission through binary fission. Biol Cell 2003;95:39–51.

[128] Caudron F, Barral Y. A super-assembly of Whi3 encodes memory of deceptive encounters by single cells during yeast courtship. Cell 2013;155:1244–57.

[129] Schlissel G, Krzyzanowski MK, Caudron F, Barral Y, Rine J. Aggregation of the Whi3 protein, not loss of heterochromatin, causes sterility in old yeast cells. Science 2017;355:1184–7.

[130] Saarikangas J, Barral Y. Protein aggregation as a mechanism of adaptive cellular responses. Curr Genet 2016;62:711–24.

[131] Lau Y, Parfenova I, Saarikangas J, Nichols RA, Barral Y, Caudron F. A mechanism to prevent transformation of the Whi3 mnemon into a prion. bioRxiv 2020; submitted for publication:2020.03.13.990119.

[132] Roberts BT, Wickner RB. A new kind of prion: a modified protein necessary for its own modification. Cell Cycle (Georgetown, TX) 2004;3:100–3.

[133] Itakura AK, Chakravarty AK, Jakobson CM, Jarosz DF. Widespread prion-based control of growth and differentiation strategies in Saccharomyces cerevisiae. Mol Cell 2020;77: 266–78 e6.

[134] Brandman O, Meyer T. Feedback loops shape cellular signals in space and time. Science (New York, NY) 2008;322:390–5.

[135] Drubin DA, Way JC, Silver PA. Designing biological systems. Genes Dev 2007;21:242–54.

[136] Sprinzak D, Elowitz MB. Reconstruction of genetic circuits. Nature 2005;438:443–8.

[137] Becskei A, Seraphin B, Serrano L. Positive feedback in eukaryotic gene networks: cell differentiation by graded to binary response conversion. EMBO J 2001;20:2528–35.

[138] Gardner TS, Cantor CR, Collins JJ. Construction of a genetic toggle switch in Escherichia coli. Nature 2000;403:339–42.

[139] Kobayashi H, Kaern M, Araki M, Chung K, Gardner TS, Cantor CR, et al. Programmable cells: interfacing natural and engineered gene networks. Proc Natl Acad Sci USA 2004;101:8414–19.

[140] Tchuraev RN, Stupak IV, Tropynina TS, Stupak EE. Epigenes: design and construction of new hereditary units. FEBS Lett 2000;486:200–2.

[141] Toman Z, Dambly C, Radman M. Induction of a stable, heritable epigenetic change by mutagenic carcinogens: a new test system. IARC Sci Publ 1980;243–55.

[142] Ajo-Franklin CM, Drubin DA, Eskin JA, Gee EP, Landgraf D, Phillips I, et al. Rational design of memory in eukaryotic cells. Genes Dev 2007;21:2271–6.

[143] Cohn M, Horibata K. Analysis of the differentiation and of the heterogeneity within a population of Escherichia coli undergoing induced beta-galactosidase synthesis. J Bacteriol 1959;78:613–23.

[144] Cohn M, Horibata K. Inhibition by glucose of the induced synthesis of the beta-galactoside-enzyme system of Escherichia coli. Analysis of maintenance. J Bacteriol 1959;78:601–12.

[145] Cohn M, Horibata K. Physiology of the inhibition by glucose of the induced synthesis of the beta-galactoside enzyme system of Escherichia coli. J Bacteriol 1959;78:624–35.

[146] Choi PJ, Cai L, Frieda K, Xie XS. A stochastic single-molecule event triggers phenotype switching of a bacterial cell. Science (New York, NY) 2008;322:442–6.

[147] Gordon AJ, Satory D, Halliday JA, Herman C. Heritable change caused by transient transcription errors. PLoS Genet 2013;9:e1003595.

[148] Ozbudak EM, Thattai M, Lim HN, Shraiman BI, Van Oudenaarden A. Multistability in the lactose utilization network of Escherichia coli. Nature 2004;427:737–40.

[149] Laurent M, Kellershohn N. Multistability: a major means of differentiation and evolution in biological systems. Trends Biochem Sci 1999;24:418–22.

[150] Santillan M, Mackey MC. Quantitative approaches to the study of bistability in the lac operon of Escherichia coli. J R Soc Interface 2008;5(Suppl 1):S29–39.

[151] Toman Z, Dambly-Chaudiere C, Tenenbaum L, Radman M. A system for detection of genetic and epigenetic alterations in Escherichia coli induced by DNA-damaging agents. J Mol Biol 1985;186:97–105.

[152] Haedens V, Malagnac F, Silar P. Genetic control of an epigenetic cell degeneration syndrome in Podospora anserina. Fungal Genet Biol 2005;42:564–77.

[153] Silar P, Daboussi MJ. Non-conventional infectious elements in filamentous fungi. Trends Genet 1999;15:141–5.

[154] Graziani S, Silar P, Daboussi MJ. Bistability and hysteresis of the 'Secteur' differentiation are controlled by a two-gene locus in Nectria haematococca. BMC Biol 2004;2:18.

[155] Bagowski CP, Ferrell Jr. JE. Bistability in the JNK cascade. Curr Biol 2001;11:1176–82.

[156] Jamet-Vierny C, Debuchy R, Prigent M, Silar P. IDC1, a Pezizomycotina-specific gene that belongs to the PaMpk1 MAP kinase transduction cascade of the filamentous fungus Podospora anserina. Fungal Genet Biol 2007;44:1219–30.

[157] Lacaze I, Lalucque H, Siegmund U, Silar P, Brun S. Identification of NoxD/Pro41 as the homologue of the p22phoxNADPH oxidase subunit in fungi. Mol Microbiol 2015;95:1006–24.

[158] Chan Ho Tong L, Silar P, Lalucque H. Genetic control of anastomosis in Podospora anserina. Fungal Genet Biol 2014;70C:94–103.

[159] Lalucque H, Malagnac F, Brun S, Kicka S, Silar P. A non-mendelian MAPK-generated hereditary unit controlled by a second MAPK pathway in Podospora anserina. Genetics 2012;191:419–33.

[160] Nguyen T-S, Lalucque H, Silar P. Identification and characterization of PDC1, a novel protein involved in the epigenetic cell degeneration crippled growth in Podospora anserina. Mol Microbiol 2018;110:499–512.

[161] Gautier V, Tong L, Nguyen T-S, Debuchy R, Silar P. PaPro1 and IDC4, two genes controlling stationary phase, sexual development and cell degeneration in Podospora anserina. J Fungi 2018;4:85.

[162] Lalucque H, Malagnac F, Green K, Gautier V, Grognet P, Chan Ho Tong L, et al. IDC2 and IDC3, two genes involved in cell non-autonomous signaling of fruiting body development in the model fungus Podospora anserina. Dev Biol 2017;421:126–38.

[163] Timpano H, Chan Ho Tong L, Gautier V, Lalucque H, Silar P. The PaPsr1 and PaWhi2 genes are members of the regulatory network that connect stationary phase to mycelium differentiation and reproduction in Podospora anserina. Fungal Genet Biol 2016;94:1–10.

[164] Bhalla US, Ram PT, Iyengar R. MAP kinase phosphatase as a locus of flexibility in a mitogen-activated protein kinase signaling network. Science (New York, NY) 2002;297:1018–23.

[165] Clemons KV, Hanson LC, Stevens DA. Colony phenotype switching in clinical and non-clinical isolates of Saccharomyces cerevisiae. J Med Vet Mycol 1996;34:259–64.

[166] Lachke SA, Joly S, Daniels K, Soll DR. Phenotypic switching and filamentation in Candida glabrata. Microbiology 2002;148:2661–74.

[167] Lachke SA, Srikantha T, Tsai LK, Daniels K, Soll DR. Phenotypic switching in Candida glabrata involves phase-specific regulation of the metallothionein gene MT-II and the newly discovered hemolysin gene HLP. Infect Immun 2000;68:884–95.

[168] Fries BC, Goldman DL, Casadevall A. Phenotypic switching in Cryptococcus neoformans. Microbes Infect 2002;4:1345–52.

[169] Enger L, Joly S, Pujol C, Simonson P, Pfaller M, Soll DR. Cloning and characterization of a complex DNA fingerprinting probe for Candida parapsilosis. J Clin Microbiol 2001;39:658–69.

[170] Joly S, Pujol C, Schroppel K, Soll DR. Development of two species-specific fingerprinting probes for broad computer-assisted epidemiological studies of Candida tropicalis. J Clin Microbiol 1996;34:3063–71.

[171] Malagnac F, Silar P. Chapter 13 – Epigenetics of eukaryotic microbes. In: Tollefsbol T, editor. Handbook of epigenetics. San Diego, CA: Academic Press; 2011. p. 185–201.

REFERENCES

[172] Slutsky B, Staebell M, Anderson J, Risen L, Pfaller M, Soll DR. "White-opaque transition": a second high-frequency switching system in *Candida albicans*. J Bacteriol 1987;169:189–97.

[173] Miller MG, Johnson AD. White-opaque switching in *Candida albicans* is controlled by mating-type locus homeodomain proteins and allows efficient mating. Cell 2002;110:293–302.

[174] Lockhart SR, Pujol C, Daniels KJ, Miller MG, Johnson AD, Pfaller MA, et al. In *Candida albicans*, white-opaque switchers are homozygous for mating type. Genetics 2002;162:737–45.

[175] Xie J, Tao L, Nobile CJ, Tong Y, Guan G, Sun Y, et al. White-opaque switching in natural MTLa/alpha isolates of *Candida albicans*: evolutionary implications for roles in host adaptation, pathogenesis, and sex. PLoS Biol 2013;11:e1001525.

[176] Soll DR. High-frequency switching in *Candida albicans*. Clin Microbiol Rev 1992;5:183–203.

[177] Soll DR. Gene regulation during high-frequency switching in *Candida albicans*. Microbiology 1997;143(Pt 2):279–88.

[178] Lan CY, Newport G, Murillo LA, Jones T, Scherer S, Davis RW, et al. Metabolic specialization associated with phenotypic switching in *Candida albicans*. Proc Natl Acad Sci USA 2002;99:14907–12.

[179] Scaduto CM, Kabrawala S, Thomson GJ, Scheving W, Ly A, Anderson MZ, et al. Epigenetic control of pheromone MAPK signaling determines sexual fecundity in *Candida albicans*. Proc Natl Acad Sci 2017;114:13780–5.

[180] Beekman CN, Cuomo CA, Bennett RJ, Ene IV. Comparative genomics of white and opaque cell states supports an epigenetic mechanism of phenotypic switching in *Candida albicans*. G3 2021;11:jkab001.

[181] Vargas K, Messer SA, Pfaller M, Lockhart SR, Stapleton JT, Hellstein J, et al. Elevated phenotypic switching and drug resistance of *Candida albicans* from human immunodeficiency virus-positive individuals prior to first thrush episode. J Clin Microbiol 2000;38:3595–607.

[182] Lohse MB, Johnson AD. Differential phagocytosis of white vs opaque *Candida albicans* by Drosophila and mouse phagocytes. PLoS One 2008;3:e1473.

[183] Huang G, Wang H, Chou S, Nie X, Chen J, Liu H. Bistable expression of WOR1, a master regulator of white-opaque switching in *Candida albicans*. Proc Natl Acad Sci USA 2006;103:12813–18.

[184] Srikantha T, Borneman AR, Daniels KJ, Pujol C, Wu W, Seringhaus MR, et al. TOS9 regulates white-opaque switching in *Candida albicans*. Eukaryot Cell 2006;5:1674–87.

[185] Zordan RE, Galgoczy DJ, Johnson AD. Epigenetic properties of white-opaque switching in *Candida albicans* are based on a self-sustaining transcriptional feedback loop. Proc Natl Acad Sci USA 2006;103:12807–12.

[186] Tuch BB, Mitrovich QM, Homann OR, Hernday AD, Monighetti CK, De La Vega FM, et al. The transcriptomes of two heritable cell types illuminate the circuit governing their differentiation. PLoS Genet 2010;6:e1001070.

[187] Si H, Hernday AD, Hirakawa MP, Johnson AD, Bennett RJ. *Candida albicans* white and opaque cells undergo distinct programs of filamentous growth. PLoS Pathog 2013;9:e1003210.

[188] Zordan RE, Miller MG, Galgoczy DJ, Tuch BB, Johnson AD. Interlocking transcriptional feedback loops control white-opaque switching in *Candida albicans*. PLoS Biol 2007;5:e256.

[189] Lohse MB, Johnson AD. Identification and characterization of Wor4, a new transcriptional regulator of white-opaque switching. G3: Genes | Genomes | Genetics 2016;6:721–9.

[190] Hernday AD, Lohse MB, Fordyce PM, Nobile CJ, DeRisi JL, Johnson AD. Structure of the transcriptional network controlling white-opaque switching in *Candida albicans*. Mol Microbiol 2013;90:22–35.

[191] Sonneborn A, Tebarth B, Ernst JF. Control of white-opaque phenotypic switching in *Candida albicans* by the Efg1p morphogenetic regulator. Infect Immun 1999;67:4655–60.

[192] Perez-Martin J, Uria JA, Johnson AD. Phenotypic switching in *Candida albicans* is controlled by a SIR2 gene. EMBO J 1999;18:2580–92.

[193] Klar AJ, Srikantha T, Soll DR. A histone deacetylation inhibitor and mutant promote colony-type switching of the human pathogen *Candida albicans*. Genetics 2001;158:919–24.

[194] Srikantha T, Tsai L, Daniels K, Klar AJ, Soll DR. The histone deacetylase genes HDA1 and RPD3 play distinct roles in regulation of high-frequency phenotypic switching in *Candida albicans*. J Bacteriol 2001;183:4614–25.

[195] Alkafeef SS, Yu C, Huang L, Liu H. Wor1 establishes opaque cell fate through inhibition of the general co-repressor Tup1 in *Candida albicans*. PLoS Genet 2018;14:e1007176.

[196] Frazer C, Staples MI, Kim Y, Hirakawa M, Dowell MA, Johnson NV, et al. Epigenetic cell fate in *Candida albicans* is controlled by transcription factor condensates acting at super-enhancer-like elements. Nat Microbiol 2020;5:1374–89.

[197] Qasim MN, Valle Arevalo A, Nobile CJ, Hernday AD. The roles of chromatin accessibility in regulating the *Candida albicans* white-opaque phenotypic switch. J Fungi 2021;7:37.

[198] Tao L, Du H, Guan G, Dai Y, Nobile CJ, Liang W, et al. Discovery of a white-gray-opaque tristable phenotypic switching system in *Candida albicans*: roles of non-genetic diversity in host adaptation. PLoS Biol 2014;12:e1001830.

[199] Park Y-N, Pujol C, Wessels DJ, Soll DR. *Candida albicans* double mutants lacking both EFG1 and WOR1 can still switch to opaque. mSphere 2020;5:e00918-20.

[200] Rikkerink EH, Magee BB, Magee PT. Opaque-white phenotype transition: a programmed morphological transition in *Candida albicans*. J Bacteriol 1988;170:895–9.

[201] Huang G, Srikantha T, Sahni N, Yi S, Soll DR. CO(2) regulates white-to-opaque switching in *Candida albicans*. Curr Biol 2009;19:330–4.

[202] Ramirez-Zavala B, Reuss O, Park YN, Ohlsen K, Morschhauser J. Environmental induction of white-opaque switching in *Candida albicans*. PLoS Pathog 2008;4:e1000089.

[203] Brenes LR, Lohse MB, Hartooni N, Johnson AD. A set of diverse genes influence the frequency of white-opaque switching in *Candida albicans*. G3: Genes, Genomes, Genetics 2020;10:2593–600.

[204] Dalal CK, Zuleta IA, Lohse MB, Zordan RE, El-Samad H, Johnson AD. A population shift between two heritable cell types of the pathogen *Candida albicans* is based both on switching and selective proliferation. Proc Natl Acad Sci 2019;116:26918–24.

[205] Alby K, Bennett R. Stress-induced phenotypic switching in *Candida albicans*. Mol Biol Cell 2009;20(14):3178–91.

[206] Liang S-H, Cheng J-H, Deng F-S, Tsai P-A, Lin C-H. A novel function for Hog1 stress-activated protein kinase in controlling white-opaque switching and mating in *Candida albicans*. Eukaryot Cell 2014;13:1557–66.

[207] Lockshon D. A heritable structural alteration of the yeast mitochondrion. Genetics 2002;161:1425–35.

[208] Malagnac F, Silar P. Regulation, cell differentiation and protein-based inheritance. Cell Cycle 2006;5:2584–7.

[209] Blagosklonny MV. Molecular theory of cancer. Cancer Biol Ther 2005;4:621–7.

CHAPTER 6

Higher-order Chromatin Organization in Diseases, from Chromosomal Position Effect to Phenotype Variegation

Frédérique Magdinier and Jérôme D. Robin

Aix Marseille Univ, INSERM, Marseille Medical Genetics, MMG, Marseille, France

OUTLINE

Introduction	89
Definition of Chromosomal Position Effects in Model Organisms	90
Genome Topology and Long-distance Interactions	92
Chromosomal Position Effect in Human Pathologies	93
Telomeric Position Effect; Implication in Human Pathologies	98
Conclusions	101
References	102

INTRODUCTION

The genome of eukaryotes is composed of thousands of genes and even more interspersed non-coding sequences. Constraining up to tens of billions of bases within a nucleus of a few microns in diameter requires a high level of DNA compaction that must also exhibit high plasticity in order to allow efficient realization of cellular functions. The structure of the chromatin fiber regulates the accessibility of DNA to the plethora of factors involved in the regulation of gene expression, DNA replication or response to DNA damage.

Each chromatin state can be defined in four dimensions by its level of compaction: the positioning and the spacing of nucleosomes and its histone code predicting how the posttranslational modifications of specific amino acids of the core histones (H2A, H2B, H3, and H4) are translated into distinct information, the presence of histone variants (i.e., H2.AX, H2.AZ, MacroH2A, CENPA, H3.3) and its composition in non-histone binding factors; the covalent modification of the underlying DNA; its topological state the spatial localization within the nucleus and its dynamics during the cell cycle. In particular, mitosis is associated with profound changes in the organization of chromatin, the transient disappearance of the nuclear envelope and most subnuclear organelles, eviction of transcription factors, and gene silencing. Nevertheless, at the end of mitosis, the cell fate-specific transcription program and epigenetic profile is faithfully recapitulated requiring a maintenance mechanism named mitotic bookmarking [1–4].

Generally, open chromatin, where most of the transcription occurs, is referred to as "euchromatin" whereas condensed chromatin, where transcription is generally inhibited, is referred to as "heterochromatin," although various types of chromatin structure are evoked under these denominations. In eukaryotes, heterochromatin is histone hypoacetylated and lacks methylation of H3K4, but is enriched in methylated DNA, histone H3K9 di and tri-methylation, H3K27 and H4K20 methylation, HP1 binding, and can spread over flanking regions thereby inducing transcriptional silencing [5,6]. However, a number of genes escape this strict dichotomy and show a bivalent chromatin signature with both repressive (H3K27me3) and active transcription (H3K4me1−3) marks, allowing poised expression during development and cell lineage differentiation [7−10]. Overall, it has to be kept in mind that chromatin is highly plastic and that transition from one state to another occurs during development or cell differentiation, in particular during the transition from chromatin-relaxed pluripotent cells to cells committed in a specific lineage [7,10].

Genes, regulatory elements and repetitive DNA are interspersed, at the chromosome level, in a mosaic of condensed and open regions. The proximity of different types of chromatin can influence gene expression either positively or negatively [11−13]. However, the identity of chromatin domains is safeguarded by elements such as cis-regulators, chromatin boundary elements, "fuzzy boundaries," or insulators that limit the influence of one region onto the adjacent one [14−16].

The genome can be subjected to structural variations (SVs) corresponding to balanced or unbalanced genomic rearrangements. SVs that include insertions, deletions, duplications, translocations, or inversions of DNA segments are frequent in human diseases. Chromoanagenesis (encompassing chromothripsis, chromoplexy, and chromoanasynthesis) largely explain the occurrence of germline and somatic SVs. Upon rearrangement, a gene relocated in the vicinity of heterochromatin can become silent in a subset of cells, leading to a characteristic variegated pattern of expression as a consequence of this position effect (position effect variegation or PEV). Thus, chromosomal position effect (CPE) or PEV refers to the consequences on gene expression when a gene is positioned in a different chromatin environment. This repositioning can be associated to the stochastic establishment of an epigenetic state modifying the gene's regulatory region, in turn affecting gene expression. Besides, telomere proximity can trigger gene silencing by Telomeric Position Effect (TPE) [17]. Moreover, SVs can interrupt chromatin topology resulting in ectopic interactions of distant regulatory elements.

The goal of this chapter is not to provide a detailed review of all the experimental work that has been published on PEV, CPE, or TPE in different cellular or animal models, but rather to describe their main features and illustrate the evolution of our knowledge on their implication in human pathologies.

Definition of Chromosomal Position Effects in Model Organisms

CPE was originally discovered in flies in the 1920s by Sturtevant, who first described facet changes in the eyes of flies linked to duplication of the Bar locus that influences on expression of the Bar gene, possibly due to a complex mechanism that was named "position effect," involving gene dosage [18]. Later on, Muller showed that an inversion of the X chromosome and relocalization of the white gene close to pericentromeric heterochromatin was associated with a "mottled" phenotype, with each eye having some white (mutant) and red (wild-type) regions with variation from eye to eye, a phenomenon dubbed PEV [11,18−20].

These early observations led to the description of two types of mechanism defined respectively as stable position effect and PEV (Fig. 6.1). Stable position effect is associated with gene silencing upon transgene multimerization (also named repeat-induced silencing) [21,22], while PEV depends on the translocation of euchromatic genes into heterochromatin or insertion of heterochromatin-prone sequences into euchromatin domains and followed by spreading heterochromatin features into euchromatin [23−25]. The extent of this spreading can be variable from cell to cell and leads to the characteristic mosaic appearance of variegated expression at the tissue level. As a consequence, a transgene located in constitutive heterochromatin adopts the compact nucleosomal structure of the insertion site. Moreover, if some transgenes do not variegate when repeated in tandem, certain ones are more prone to silencing, challenging their use in transgenesis.

In flies, the genetic dissection of this process has been performed by means of dominant suppressor (Su(var), suppressor of variegation) and enhancer (E (var), enhancer of variegation) mutations. Around 140 suppressors and 230 enhancers of variegation were identified and approximately 30 are fully characterized [11,26,27]. Most of these modifiers are components of heterochromatin, enzymes that modify histones, non-histone proteins, or nuclear architectural proteins such as the heterochromatin protein 1, HP1, the Su(var)3−9 histone H3K9 methyltransferase (HMTase), the Suv4−20 histone methyltransferase, histone acetyl transferases (HATs), histone deacetylases, the Drosophila homolog of the mammalian LSD1 amine oxidase that demethylates H3K4me2 and H3K4me1, components of the TIP60 complex or Rsf (remodeling and spacing factor) (reviewed in [27]).

Since then, different subgroups of PEV modifiers have been described, depending on their ability to modulate silencing and the chromatin context [28].

Similar screens were also performed in mice [29−31]. Most suppressors or enhancers of PEV

INTRODUCTION

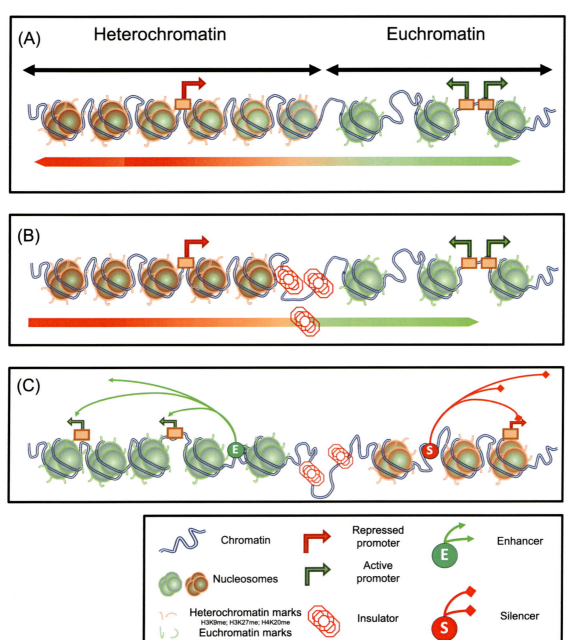

FIGURE 6.1 **Chromatin configuration modulates gene expression and phenotype variegation.** (A) Gene expression can be modulated by the chromatin context. Condensed chromatin (constitutive heterochromatin, repetitive DNA, telomeric DNA) can induce spreading of heterochromatin marks, nucleosomes repositioning or recruitment of chromatin remodeling complexes leading to gene repression. The extend of gene silencing might be different from cell to cell in a given tissues, leading to a variegated expression pattern and mosaic phenotype. This phenomenon is called PEV (position effect variegation) and has been observed in all eukaryotic cells. The presence of "fuzzy boundaries" maintains a dynamic equilibrium between the two types of chromatin and limits heterochromatin spreading. (B) The presence of boundary elements (or insulators) defines strict borders between chromatin regions and separate functional domains by recruiting specific factors, (C) Insulators can also regulate *cis*-regulating elements such as silencer (S) or enhancers (E).

activity encode epigenetic modifiers including the DNA methyltransferases Dnmt1 and Dnmt3b, the Hdac1 histone deacetylase, the Suvar39h1 and Setdb1 histone methyltransferases, the Smarca5, Smarcc1, Pbrm1 or Baz1b chromatin remodeling factors, the Trim28 (Tripartite motif containing 28) transcriptional regulator, and epigenetic modifiers such as Smchd1 or D14Abb1e. Interestingly, screens performed in the mouse led to the identification of additional factors that do not have orthologs in *Drosophila*, especially among genes involved in the regulation of DNA methylation.

So far, screening for modifiers of PEV has been challenging in human cells until the recent advance of genome editing approaches. Non-lethal haploid genetic screen in human KBM7 chronic myeloid leukemia cells carrying a single copy of each chromosome, except chromosome 8 and sex chromosomes [32] uncovered the existence of the HUSH complex (human silencing hub) composed of four factors essential for gene silencing: the SETDB1 histone methyltransferase, MPP8 (M-phase phosphoprotein), periphilin, and FAM208A (TASOR, transgene activation repressor) [33]. The last three factors are associated in the same complex that in turn recruits SETDB1, through interaction with the MPP8 chromodomain. Silenced transgenes probably do not recruit the HUSH complex directly but are rather modified by recruitment of HUSH, SETDB1, and acquiring of H3K9me3 that are present at the heterochromatic sites of insertion, reinforcing thereby silencing in pre-existing heterochromatin domains [34]. HUSH-dependent repression does not require HP1 or TRIM28. Orthologs of the HUSH complex subunits do not exist in flies but are highly conserved from fish to humans [33].

As summarized in this paragraph, position effect was initially defined as the spreading of repressive chromatin marks along a linear genome. However, the definition has recently greatly evolved with our understanding of the way the genome folds and organizes. It is now evident that position effects can also result from more complex interactions as highlighted by the impact of balanced or unbalanced SVs. Our understanding of mechanisms governing position effects will likely evolve again in the near future, in particular, through the implementation of whole genome and long-read next generation sequencing technologies [35].

Genome Topology and Long-distance Interactions

At the nucleus scale, individual chromosomes occupy specific nuclear subdomains defined as chromosome territories. At the chromatin fiber level, active and inactive chromatin domains are often juxtaposed but their respective identity is maintained by specialized elements [36–38] (Fig. 6.1A) such as insulators or boundaries. Boundaries have been initially defined by two non-exclusive properties: enhancer-blocking and barrier activity [37] (Fig. 6.1B,C). Barrier insulators protect genes or regions from the spreading of heterochromatin (Fig. 6.1B). Enhancer-blocking insulators disrupt the communication between a promoter and a cis-regulatory element when placed in-between, without preventing them from interacting with other genetic elements (Fig. 6.1C).

Insulator elements bind specific proteins. In Drosophila, five proteins with enhancer-blocking activity have been identified: Zw5, BEAF-32 [39,40], the GAGA factor [41,42], Su(Hw), and an ortholog of the CCCTC-binding factor CTCF [43,44]. CTCF is the only conserved protein displaying such activity in vertebrates [45]. In human cells, genome-wide studies in different cell types identified several thousands of putative sites for the architectural CTCF protein. Distribution of these sites along chromosomes follow the distribution of genes; CTCF-depleted regions often correspond to clusters of co-expressed genes while CTCF-enriched regions contain genes with multiple alternative promoters [46–48]. These sites participate in long-range interactions and act together with cohesins and the mediator complex [49–51]. So far, little is known on the mechanisms that regulate the binding of CTCF to its target sequences as all CTCF sites are not permanently occupied. At least two protein complexes have been identified as able to compete with CTCF binding: the ChAHP complex (Chd3, Adnp, Hp1) [52] and Smchd1, identified in the mouse as a suppressor of PEV [30,53,54] and more recently as an architectural protein [53,55,56].

Recent experimental procedure developments such as the chromatin conformation capture (3C) [57,58] and derivatives (4C, 5C, HiC), ChIA-PET [59], RNA tagging and recovery of associated proteins (RNA TRAP) or DamID [60] together with next-generation sequencing provided new insights into the spatial organization of chromatin during interphase and the description of nuclear subdomains such as LAD (lamin associated domains) or NAD (nucleolar associated domains) formed by association of chromatin with subnuclear structures, the nuclear lamina and the nucleolus, respectively. We also know that chromatin is organized in active and open domains (A compartments) or inactive and condensed structures (B compartments) where active genes associate with one another while repressed genes cluster together [61]. At the sub-mega base scale, chromatin is organized into topologically associated domains (TADs; median length 800 kb) [61–65]. TADs are required for most enhancer-promoter interactions and are separated by boundaries occupied by CTCF and cohesin or housekeeping genes able to insulate regulatory domains from one another and reduce interactions between adjacent TADs. The identification of TADs has largely contributed to our understanding of enhancers, the identification of tissue-specific enhancers and the possible implication of multiple enhancers in the regulation of a single gene [66], acting as a regulatory hub that orchestrates gene expression regulation [67,68].

The majority of TADs exhibits frequent intrachromatin domain interactions but rarely inter-domain interactions and maintain enhancer-promoter interactions or control spatio-temporal gene expression pattern. However, independent studies in Embryonic Stem Cells (ESC) or ESCs-derived neurons also reported well-defined "insulated neighborhoods" of interactions within TADs [65,69,70]. TADs are conserved between cell types and related species [65,71]. Notably, almost half of the epigenetic marks retained on chromosomes at mitosis correspond to boundaries of topological domains underlining the importance of mitotic bookmarking for the restoration of long-range interaction and TADs after cell division [2]. Within TADs, smaller domains, or sub-TADs (median length 250 kb) are differentially regulated depending on the cell type or differentiation stage.

CHROMOSOMAL POSITION EFFECT IN HUMAN PATHOLOGIES

Chromosomal position effects have been associated with a broad range of phenotypical manifestations and syndromes. A number of evidence suggests a role for CTCF and CTCF binding in pathologies. For instance, microdeletion or microduplications of the CTCF sites at the *IGF2/H19* locus are associated with some cases of nonsyndromic Wilms' tumors [72]. In addition, hypomorphic mutations in factors associated with CTCF chromatin barriers, such as cohesins mutated in the Cornelia de Lange syndrome [73,74], also alter gene expression and regulation through changes in the topology of chromatin. In the ChAHP complex, *ADNP* is mutated in Helsmoortel-Van der Aa syndrome involving intellectual disability and congenital anomalies of different tissues [75]. The gene encoding SMCHD1 is mutated in three distinct rare genetic diseases with distinct phenotypes [76–79]. Thus, in diseases linked either to *ADNP* or *SMCHD1* heterozygous mutation, the diverse phenotype or disease severity might be related to perturbation of the 3D chromatin organization and subsequent pleiotropic effects.

Moreover, all types of chromosomal rearrangements and structural variants are amenable to position effects [80–82] and would in a number of cases likely explain the phenotypical features unexplained by the chromosomal anomaly itself or by disruption of a gene coding sequence. It is estimated that 93% of disease-associated variants are in non-coding DNA [83] with 60% corresponding to known regulatory elements [84–86]. Cytogenetic studies of unselected newborns and control adult males estimate a prevalence of 0.2%–0.5% of balanced cytogenetic abnormalities (BCA) in the general population [87]. Structural variants might be associated with changes in regulatory regions that may interfere with the balance of enhancers, silencers, or insulators, disruption of topological domains, and the aberrant adoption of normally separated *cis*-regulatory elements.

Thus, mechanisms through which CPE can cause human diseases are diverse: separation of the transcription unit from an essential distant regulatory element, juxtaposition of the gene with the enhancer of another gene, competition for the same regulatory element, or classical position effect variegation in which a

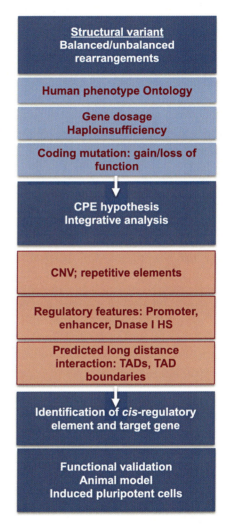

FIGURE 6.2 **Steps associated with the interpretation of structural variations associated with a disease phenotype.** Position effect prediction methods rely on the implementation of databases including the Human Phenotype Ontology or Clinical Genome resource [89] for identification and prioritization of genes driving congenital phenotype after the identification of balanced or unbalanced chromosomal anomalies and structural variants. The position or type of SV will permit the classification of TAD-associated defect (intra TAD SV, TAD fusion, creation of a neo-TAD, TAD shuffling) and possible identification of cis-regulatory elements associated to the phenotype. In most cases, multiple candidate drivers should be considered for explaining the patient's symptoms. Moreover computational prediction tools remain insufficient for in depth functional studies requiring the development of animal or induced pluripotent stem cells models.

gene is moved to a new chromatin environment [88] (Fig. 6.2). CPE-associated pathogenesis has been described in cancers as well as in constitutional pathologies, essentially in the context of chromosomal rearrangements (translocations, deletions, inversions) including BCA i.e., SVs rearrangements of the chromatin structure in the absence of a gain or loss of DNA.

For instance, malignant hemopathies are characterized by acquired chromosomal rearrangements, mainly translocations, which are clonal, non-random, recurrent, and often tumor-type specific. These translocations have two main consequences: the formation of a chimerical gene encoding a new fusion protein or the combination between the coding region of a gene and the promoter/enhancer region of another one leading to inappropriate overexpression of the former gene [90]. A well-known example is the t(8;14)(q24;q32) translocation in Burkitt lymphoma, an aggressive B cell neoplasm that places the c-MYC gene (chr. 8) near the enhancer of the heavy chain of immunoglobulin on chromosome 14 resulting in overexpression of c-MYC in B cells [91]. Other translocations of genes like BCL2 (follicular lymphoma) [92], BCL6 (diffuse large B cell lymphoma) [93], and the TMPRSS2-ETS fusion gene involved in prostate cancer [94] can also be cited. More recently, group 3 and 4 medulloblastomas, a leading cause of cancer-related mortality in children, have been associated with somatic genomic rearrangements associated with GFI1/GFI1B activation. This "enhancer hijacking" induces redistribution of the two genes from regions of transcriptionally silent chromatin to regions enriched in acetylated H3K9 and K27 usually associated with super enhancers [95].

The involvement of CPE in constitutional pathologies is in most cases more difficult to prove even if this mechanism is often proposed to explain disease-associated phenotype in the absence of gene disruption [96–98]. Most associations remained so far hypothetical in the implication of the clinical features and a firm demonstration is rarely found in the literature for several reasons. First, in developmental syndromes, the causative gene often shows a specific spatio-temporal pattern of expression and access to the tissue from the patient is, in most cases, impossible. Second, regulatory elements are often located far from the gene of interest, sometimes included in another neighboring gene and can regulate more than one gene. Finally, if the position effect seems evident when a chromosomal breakpoint occurs in the vicinity of a gene for which the associated phenotype is well known, it is less obvious when the phenotype is not related to the gene or when the gene function is unknown. Examples of pathologies in which the chromosomal position effect have been suspected or proved are summarized in Table 6.1.

Variations that might concur to the variability of phenotypical traits concerns DNA copy number variation (CNV) associated with human genetic diversity and the emergence of the variome concept [161,162]. CNVs and associated variome can be subdivided into two large "subtypes": the sequence variome that takes into account all genome variations and the variome that considers chromosomal heteromorphisms or balanced and unbalanced chromosomal rearrangements. We will focus our review on this second category, keeping in mind that in most instances, the association between CNVs and the disease phenotype remains complex, despite the emergence of novel bioinformatics and system biology tools for integration of large sets of data [161,163,164]. The majority of these CNVs correspond to deletions, insertions, or duplications larger than 1 kb and up to several megabases, likely due to non-allelic homologous recombination [165]. The spectrum of effects of CNVs is very broad with gene dosage impacting specific molecular and cellular pathways to alterations of local chromatin conformation associated with chromatin changes [166–169]. CNVs may alter the expression of genes that are upstream or downstream of the CNV by disrupting regulatory elements [170,171] and has been associated for instance with Crohn disease [172] or skeletal dysplasia and short stature [173]. With regard to CPE, not all gene CNVs result in changes in gene expression levels but a negative correlation between copy number and expression level has been reported for 5%–15% of CNVs [174–177]. Moreover, CNVs can potentially alter chromatin topology by deleting or duplicating chromatin barriers, modifying interactions between cis-regulatory elements [80].

Repetitive elements (RE) which are abundantly distributed in mammalian genome and account for more than half of the human genome [178] also contribute to gene expression variability and chromatin organization. Different subtypes of REs demarcate functional categories of genes based on their expression profile by providing sites for regulatory proteins [179]. Furthermore, recent advances highlight their emerging role in the regulation of the human genome and their implication in disease susceptibility (reviewed in [180]). In particular, tandem repeats are able to functionally impact genome regulation either locally or over long distances [179,181,182], by providing for instance new binding sites for CTCF [183,184]. As a consequence, genome-wide CTCF binding has substantially diverged during evolution, following transposons mediated motif expansion [183–185]. Moreover, transposable elements (TE) carrying their own promoter and regulatory sequences can contribute to the transcriptional regulation of the host genome through positional effect. A number of

TABLE 6.1 Examples of Constitutional Syndromes Involving Position Effect Mechanisms and TAD (Topologically Associated Domain) Disruptions

Gene	Locus	Phenotype	Modification	References
APC	5q22.2	Adenomatous polyposis	Disruption	[99]
ATOH7	10q21.3	nonsyndromic congenital retinal nonattachment	Enhancer deletion	[100]
DLX5/DLX6/DSS 1	7q21.3	Split-hand/foot malformation type I	Disruption	[101] [102–104]
DSC3/2/DSG1/4/2/TTR/DTNA	18q12.1	ASD	Duplication/CNV	[105]
FGFR4/MSXS	5q35	Hunter McAlpine	Duplication	[106]
EPHA4	2q36.1	developmental defect	TAD disruption	[107]
FOXC1	6p25	Primary congenital glaucoma	Disruption	[108]
FOXC2	16q24.3	Lymphedema distichiasis	Disruption	[109]
FOXG1	14q12	Atypical Rett syndrome	Deletion TAD disruption	[110] [111]
FOXL2	3q23	Blepharophimosis Ptosis Epicanthus Inversus syndrome	Disruption/deletion	[112,113]
FOXP1 DPYD TWIST1	3p13	Congenital anomalies, intellectual disability	TAD disruption, loss of enhancer interaction	[114]
FOXP2	7q31	Speech and language disorder	Disruption	[115]
GLI3	7p13	Greig Cephalopolysyndactyly syndrome	Disruption	[116]
HBA HBB	16p13.3 11p15.5	Alpha-Thalassemia Gamma/Beta-Thalassemia	Deletion	[117–119]
HOXB	17q21.3	intellectual disability, hexadactyly	Disruption	[120]
HOXD	2q31	Limb malformations	Disruption	[121,122]
IHH	2q35	polysyndactyly and cranio facial abnormality	TAD disruption	[107]
IKBKG/NEMO	X;2(q23;q33)	Incontinentia Pigmenti-like phenotype	Balanced X.2 translocation	[123]
LCT	2q21	Adult-type hypolactasia	Mutation	[124]
LMNB1	5q23.2	Adult-onset demyelinating Leukodystrophy	Deletion	[125,126]
MEF2C	5q14.3	Intellectual disability Stereotypic movements	Microdeletion TAD disruption	[111,127–129]
MAF	16q23	Cataract, anterior segment dysgenesis, microphthalmia	Disruption	[130]
PAX6	11p13	Aniridia	Disruption	[131] [132]
PITX1	5q31.1	Liebendberg Syndrome	Deletion	[133,134]
PITX2	4q25	Rieger syndrome	Disruption	[135]
PLP1	Xq22	Pelizaeus–Merzbacher syndrome/Spastic Paraplegia, neuropathy	Disruption/Duplication	[136,137]
POU3F4	Xq21.1	X-linked Deafness	Deletion	[138]
REEP3	10q21.3	Autism	Disruption	[139]
RNF135-SUZ12	17q11.2	Neurofibromatosis	microdeletion	[140]
RUNX2	6p21	Cleido Cranial dysplasia	Disruption	[141]

(Continued)

TABLE 6.1 (Continued)

Gene	Locus	Phenotype	Modification	References
SALL1	16q12.1	Townes-Brocks syndrome	Disruption	[142]
SATB2	2q33.1	Glass Syndrome	TAD disruption	[81]
SDC2	8q22	Autism, Multiple Exostoses	Disruption	[143]
SHH	7q36	Holoprosencephaly	Disruption	[144]
SHH	7q36	Pre-axial Polydactyly	Disruption	[145–147]
SHOX	Xp22.3	Leri-Weill Dyschondrosteosis	Deletion/Mutation	
SIX3	2p21	Holoprosencephaly	Disruption	[148]
SOST	17q21	Van Buchem disease	Deletion	[149]
SOX3	Xq27.1	Hypoparathyroidism	Disruption	[150]
SOX4/ID4	6p22.3	Mesomelic dysplasia-hypoplastic tibiae/fistulae	deletion	[151]
SOX9	17q24.3	Campomelic Dysplasia/Pierre Robin sequence	Disruption	[104,154]
		Female-male sex reversal	deletion/mutation	[152,153]
		Cooks syndrome	TAD disruption	[34,154]
SRY	Yp11.3	Sex Reversal	Deletion	[155]
SULF1-SLCO5A1	8q13	Mesomelia synostose syndrome	Deletion	[156]
TFAP2A	6p24.3	Branchio-oculofacial syndrome	Inversion	[157]
			TAD disruption	
TGFB2	1q41	Peters anomaly	Disruption	[158]
TRPS1	8q23.3	Ambras syndrome	Disruption/deletion	[159]
TWIST	7p21.1	Saethre-Chotzen syndrome	Disruption	[160]
		Congenital anomalies, intellectual disability		[114]
WNT6	2q35–36.1	F Syndrome, syndactyly	Inversion	[107]

reports and reviews have documented the role of TE in tissue-specific transcriptional regulation. Among related mechanism, TE can contribute to heterochromatin spreading and gene silencing [186], provide insulator activity [187] or by acting as poised enhancers [188] in diseases such as cancers or auto immune disorders [34,189–191].

Short tandem repeat (STR) expansion diseases are caused by an increase in the number of trinucleotide repeats, which are usually polymorphic in the general population but become unstable beyond a certain number of repeats. More than 25 inherited human diseases are linked to short tandem repeat expansions. These STRs are either found in coding regions resulting in the production of a protein with altered function (i.e., Huntington disease) or in non-coding regions resulting in an altered transcription (Fragile-X syndrome, Friedreich ataxia (FDRA), Steinert myotonic dystrophy (DM)) caused by epigenetic changes and heterochromatin formation [192].

Fragile X syndrome (FXS, OMIM #300624), the most common form of inherited intellectual impairment is caused by expansion of a CGG trinucleotide in the 5′ UTR of the FMR1 gene. In the normal population, the number of repeats varies from 6 to 54. Between 55 to 200 CGG, the repeats become meiotically unstable (premutation) while the full mutation corresponds to more than 200 repeats and is accompanied by CpG hypermethylation of the repeats and neighboring sequences [193], histone H3 and H4 lysine hypoacetylation, H3K9 methylation and FMR1 silencing [194–196]. Another interesting disease linked to triplet expansion is FRDA (OMIM #2293000), an autosomal recessive neurodegenerative disease characterized by difficulties to coordinate movements, dysarthria, loss of reflexes, pes cavus, scoliosis, cardiomyopathy, and diabetes mellitus. FRDA is caused by the expansion of a GAA trinucleotide in the first intron of the FXN gene. Repeats range from 6 to 34 in the general population and over 66 in patients. Expansion is associated with a decrease in FXN transcription level [197] and heterochromatin formation from the expanded GAA triplet-repeat sequence in intron 1 to the *FXN* promoter region [198–200]. As observed in fragile X syndrome, specific CpG sites are hypermethylated in the *FXN* intron 1 compared to control. Hypoacetylation of H3 and H4 histones and hypermethylation of H3K9 modulate *FXN* promoter activity [198–200].

CTCF flanks CTG/CAG trinucleotide repeats at several disease-associated loci, such as the *DM1* locus implicated in myotonic dystrophy (or Steinert disease, OMIM #160900), an autosomal dominant multisystemic disorder characterized by myotonia, muscular dystrophy, cataracts, hypogonadism, cardiac conduction anomaly, and diabetes mellitus [201–203]. The presence of CTCF binding sites upstream and downstream of the expanded CTG tract restricts the extent of antisense transcription and constrains the spreading of heterochromatin from the trinucleotide repeats and DNA methylation [201,202]. Moreover, nearly all disease-associated STRs are located at boundaries of 3D chromatin domains, in particular at boundaries with a markedly higher CpG island density [204]. In FXS, the CGG triplet expansion correlates with the level of FMR1 silencing but also the loss of CTCF occupancy. This leads to the fusion of the FMR1 locus with the upstream TAD that is devoid of an active enhancer suggesting that changes in the 3D conformation of the locus together with the local hypermethylation of the promoter region containing the expanded CGG leads to the pathogenic silencing of the *FMR1* locus.

CPE linked to structural variants has been involved in several developmental syndromes (Table 6.1). Among them, two chromosomal aberrations have been thoroughly characterized allowing the identification of specific subregions causing long-range position effect [152,153,205–208]. The first region is located at the *SOX9* locus on chromosome 17 [209]. Its haploinsufficiency is responsible for campomelic dysplasia (OMIM #114290), a rare autosomal dominant disorder involving shortening and bowing of long bones, skeletal malformations, Pierre Robin sequence (PRS, associating micrognathia, glossoptosis, and cleft palate), hypoplastic lung, and male-to-female sex-reversal. Translocation or inversion breakpoints sparing the *SOX9* sequence allowed the identification of three different clusters associated with either (1) campomelic dysplasia, (2) Pierre-Robin sequence, or (3) an intermediate acampomelic/campomelic phenotype.

The second region is associated with split-hand/foot malformation (SHFM1, OMIM #183600), an autosomal dominant congenital limb defect characterized by a median cleft of the hand or foot (ectrodactyly) resulting in an aspect of "lobster claw" and caused by chromosomal rearrangements of the 7q21q22 region (deletions, translocations, inversions) [101,210–212].

Three distinct subregions within the *SHFM1* locus contain tissue-specific enhancers controlling *DLX5*, *DLX6*, and *DSS1* genes [102,213–215].

An increasing number of reports evidenced the key role of TADs and TAD boundaries in limb formation [107,216], congenital diseases associated with intellectual disabilities [80,82,111,114,217,218], adult-onset demyelinating leukodystrophy [219], or cancer [220,221]. For instance, deletions, inversions, or duplications altering the structure of the TAD-spanning the *WNT6/IHH/EPHA4/PAX3* locus has been implicated in at least three related genetic disorders associated with limb developmental defects and altered digits. The rearrangements disrupt the topology of enhancer-TAD boundaries resulting in inappropriate long-distance interactions and misexpression of protein coding genes without affecting the coding sequences themselves.

In three families affected with a dominantly inherited brachydactyly characterized by short digits, high resolution CGH revealed a heterozygous mutation of 1.75–1.9 Mb on chromosome 2q35–36 including the *EPHA4* gene, a large region of its surrounding TAD and the non-coding part of the *PAX3* TAD removing the *PAX3/EPHA4* boundary [107].

In another family, affected with the F syndrome characterized by severe complex syndactyly of the hand and feet polydactyly, affected individuals carry a 1.1 Mb heterozygous inversion or a 1.4 Mb heterozygous duplication arranged in direct tandem orientation. In both cases, the breakpoint was located in the vicinity of the *WNT6* gene [107].

Finally, in a family affected with a severe polysyndactyly and cranio-facial abnormality, a 900 kb duplication brings the *IHH* gene in proximity to the centromeric potion of the *EPHA4*-containing TAD. The pathogenic effect of the different rearrangements was further confirmed in vivo by the 4 C technique and in engineered mouse models that recapitulate the human phenotype. Wild-type animals show minimal interactions of *Pax3*, *Wnt6*, and *Ihh* with non-coding sequences in the *Epha4* TAD, while all three genes showed increased interaction in strains carrying engineered mutations, mimicking variants found in the patients, with novel interactions within the *Epha4* TAD and a fusion of adjacent TADs. Overall, the different rearrangements described in patients led to deletion of a boundary element that prevents inappropriate cross-TAD chromatin interactions with a cluster of regulatory elements driving limb expression [107].

Genomic rearrangements associated with disease have also been described at the *SOX4* locus in unrelated patients with mesomelic dysplasia where microdeletions involve four genes but also three TADs and two boundaries [151]. Deletion of the TAD boundary was also proposed in the Liebenberg syndrome (OMIM

#186550), an autosomal dominant disease in which the arms acquire morphological features of the leg. In this case, the deletion removes the H2AFY gene in a region 300 kb upstream of the PITX1 gene which determines the hind-limb identity [216]. The H2AFY gene, which encodes the H2A histone family member Y might act as a barrier element that separates the PITX1 TAD from neighboring regulators [133,216]. In a case of autosomal dominant adult-onset demyelinating leukodystrophy, (ADLD; OMIM #169500) causing central nervous system demyelination, a deletion upstream of the LMNB1 gene eliminates the TAD and TAD boundary causing ectopic interactions between nearby TADs, at least three forebrain enhancers and the LMNB1 promoter leading to Lamin B1 overexpression, as observed in other typical ADLD cases [125,126].

Overall, these examples indicate that multiple phenotypes can be associated with disruption of interactions between regulatory sequences located in rearranged regions and distant genes. Besides the different examples described above, TAD disruption associated with SVs has been described for other genes such as FOXG1 [111], SOX9 [34,154] or SATB2 [81], DLX5 and DLX6 [102–104], ATOH7 [100,132], IHH [67,222], EPH4A [107], MEF2C [111,127–129] and estimated to occur in 7% of neurological disorders [111].

The repositioning of TAD boundaries and/or the relocation of enhancers can also cause position effect through a mechanism termed "enhancer adoption" [223] or enhancer hijacking [95]. SVs can also cause deletion of enhancer elements causing a loss of function of cis-regulatory elements [67,95,206] or provoke the fusion of TADs or formation of neo TADs [80,81,111,154]. Still, it remains difficult to provide a conclusive computational prediction for the effect of a given chromosomal rearrangement on long-distance interactions, TAD formation and enhancer-promoter cross-talks without further experimental validation. In addition, CNVs challenge the deciphering of these long-distance interactions and it was estimated that approximately 11.8% of CNVs might be related to the disruption of interactions between regulatory sequences and gene promoters and misregulation of phenotypically relevant genes, underlining the importance of topological domain analysis in deleted regions.

In addition, although still speculative, one might also expect that disruption of chromatin boundaries or chromatin domains might also impact on mitotic bookmarking and precise reactivation of tissue-specific gene expression upon mitosis exit.

Beneficial effects of PEV have also been reported in a few rare cases. An interesting familial case of PLP1-related diseases also highlighted the fact that position effect can modulate disease penetrance and even rescue the phenotype [224]. PLP1-related diseases are disorders of myelin formation in the central nervous system (CNS) and include Pelizaeus–Merzbacher disease (PMD, OMIM #312080) and type 2 spastic paraplegia (OMIM #312920). PMD is an X-linked recessive disease that begins in infancy and manifests by nystagmus, hypotonia, and cognitive impairment progressing to severe spasticity and ataxia caused by duplications, point mutations and deletions of the PLP1 gene. Some cases of PMD, resulting from position effect linked to small rearrangements in the region surrounding the coding sequence, have been described [136,137]. Among which, one particular family composed of a mother, with late-onset PMD, carries a balanced insertion of a small segment of chromosome X including the PLP1 gene into the terminal region of the long arm of chromosome 19 (ins(19;X) (13.4;q22.2q22.2)). One of her sons, with a mild PMD form, inherited the derivative X and carried no PLP1 copy. The second son, who is healthy, inherited the derivative chromosome 19 and two copies of PLP1. However, the PLP1 copy inserted on chromosome 19 might be silenced by the chromatin context and telomere proximity rescuing the phenotype usually associated with abnormal PLP1 dosage. Chromoanagenesis and SVs were also described for its beneficial effect in the remission of Warts, Hypogammaglobulinemia infections and Myelokathexis syndrome (WHIMS, MIM 1936707) where a somatic chromothrypsis caused the deletion of the disabled CXCR4 allele [225].

In conclusion, concomitant association between a clinical phenotype and TAD disruption may underlie a positional effect, potentially associated with changes in gene expression of morbid loci. However, in many instances, break junction could not be captured or fully sequenced due to the presence of repetitive elements. In the near future, the development of whole genome sequencing, novel long-read sequencing technologies, and genome mapping technologies, together with prediction algorithms and the availability of genome-wide maps will likely contribute to our understanding of the regulation of the human genome through identification of gene regulatory elements that constitute the so-called non-coding genome.

TELOMERIC POSITION EFFECT; IMPLICATION IN HUMAN PATHOLOGIES

Besides PEV, silencing also occurs in the vicinity of telomeres from *Saccharomyces cerevisiae* to *Homo sapiens* (Fig. 6.3). In yeast, TPE was first demonstrated by insertion of a construct containing a URA3 auxotrophic marker 1.1 kb from a newly formed telomere. Expression of the URA3 gene allows cell growth in the

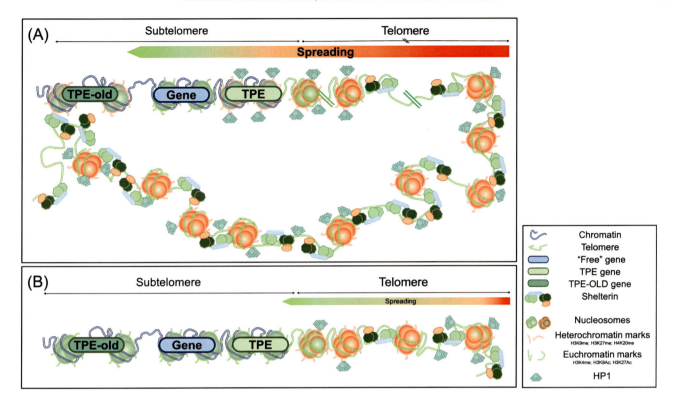

FIGURE 6.3 **Telomere length modulates expression of subtelomeric genes by continuous cis-spreading or long-distance looping.** (A) In eukaryotes, telomeres silence proximal gene (light green) through a mechanism named telomeric position effect (TPE) involving the nucleation of heterochromatin and spreading of chromatin modifications from the telomere to the subtelomeric region. TPE involves a component of the shelterin complex. Composition of subtelomeric regions likely modulate TPE through unknown mechanisms. Some genes are not modulated by TPE (blue) while expression of other genes located up to 10 Mb from the chromosome end (dark green) can be modulated by telomere length suggesting a discontinuous phenomenon. This mechanism involves looping of telomeres toward its target genes and modulates gene expression through telomere position effect over long distance (TPE-OLD). By comparison to PEV, TPE might thus be an alternative and specialized silencing process acting through the interaction between components of the shelterin complex and chromatin remodeling factors. (B) Upon telomere shortening, telomeres and subtelomeric regions are less compacted and loose heterochromatin marks. This euchromatinization alleviates TPE and leads to the induction of subtelomeric gene expression. Upon telomere shortening telomere-associated looping is impaired and expression of TPE-OLD genes increased.

absence of uracil. However, on plates containing 5-fluoro-orotic acid, a drug toxic for cells expressing URA3, 20%–60% of cells were able to grow, suggesting that the subtelomeric URA3 was silenced by telomere proximity [17].

Interestingly, despite structural differences at telomeres formed by transposition of the HeTA and TART retrotransposons to chromosome ends and arrays of telomere associated sequences (TAS) [226,227], *Drosophila melanogaster* also exhibits telomeric silencing [228]. However, genetic modifiers of PEV display little or no effect on TPE suggesting the existence of specialized mechanisms.

In *S. cerevisiae* more than 50 TPE modulators have been identified [17,229,230] but only a few exhibit a specific and complete suppression of telomeric silencing [231]. Among them, the Sir-complex proteins (Sir2p, Sir3p, and Sir4p for silent information regulators) [232], Ku heterodimer components (yKu70p and yKu80p) [233,234], and Rap1p [235] are absolutely required

[12,231]. In other organisms, some of the factors mediating TPE are functional orthologs of *S. cerevisiae* proteins. However, other factors such as tri-methylated H3K9 or HP1 required for TPE in *Drosophila* and fission yeast are absent in *S. cerevisiae* [236–239].

TPE is influenced by telomere length, a variable that is partially inherited, tissue specific, and influenced by environmental driven insults, notwithstanding the length heterogeneity of each chromosome end [12,240].

In humans, the most distal unique regions of chromosomes and telomeres are separated by different types of subtelomeric repeats varying in size from 10 to hundreds of kb. Subtelomeres are highly polymorphic gene rich regions and RNAs produced by these regions include transcripts from multi-copy protein-encoding gene families, single genes, and a large variety of non-coding RNAs (reviewed [226]). Large variations of subtelomeric DNA were among the first examples of CNVs in human cells. The recombination rate at chromosome ends is higher than in the rest of the genome.

Telomeres influence the maintenance and recombination of the adjacent subtelomeric regions and vice versa. Subtelomeres may buffer or facilitate the spreading of silencing that emanates from the telomere as observed in other species. Due to their heterochromatic signature, telomeres and subtelomeres have been considered as transcriptionally silent. Nevertheless, telomeric repeats RNA (TERRA), transcripts, transcribed from subtelomeric promoters, changed this paradigm [241]. Contradictory reports have presented cases where TERRAs are either increased when telomeres are short and down-regulated when telomeres are long [241]; but also opposite observations [242–244].

The first evidence of the existence of TPE in human cells came from the analysis in vivo of replication timing of a human chromosome 22 carrying a chromosomal abnormality [245]. Evidence for transcriptional silencing in the vicinity of human telomeres [246,247] and identification of modulators of TPE [248–250] were finally obtained experimentally by a telomere seeding procedure where natural telomeric regions have been replaced by artificial ones using a construct containing a reporter gene.

Highlighting similarities with other organisms, SIRT6 (Sir2p homolog), HP1, CBX complex and Suv39h, TRF2, H3K4me3, H3K27me3, or DNA methylation were identified as important factors dictating TPE and the chromatin landscape both at telomeres and subtelomeres [251,252]. In humans, the first endogenous TPE gene described was *ISG15* (1p36 locus) [253]. However, unlike *ISG15*, the expression of genes located closer to the 1p telomeres are not influenced by telomere length and TPE factors such as SIRT6 [250] suggesting an alternative mechanism other than based on a continuous spreading (Fig. 6.3A). As in yeast, where a telomere loop represses distal gene while not affecting intervening gene, an equivalent mechanism was discovered in human cells [252]. This extended TPE called TPE-OLD (TPE-over long distances) explains, through formation of a looping structure, the peculiar case of *ISG15* and shows the potential to influence genes located as far as 10 Mb away from telomeres, allowing the identification of new genes regulated by telomeres [252]. Telomere length and architecture along with chromatin remodeling factors orchestrate TPE and TPE-OLD (Fig. 6.3).

The involvement of TPE has been evoked in patients carrying truncated chromosome ends that have been repaired by the process of telomeric healing (de novo addition of telomeric repeats by telomerase at breaks devoid of exact telomeric repeats), telomeric capture (resulting from a break-induced replication event between a truncated chromosome and the distal arm of another chromosome), or formation of ring chromosomes (formed by fusion between the p and q arms within the same chromosome). At present, the molecular mechanisms associated with these inter-chromosomal variations remain poorly investigated. Moreover, differential sensitivity, inherited and insults-driven telomere shortening has been proposed as an explanation for generational differences in some human disorders, including in congenital syndromes associated with telomere deficiency (or telomeropathies) [240,254,255].

Subtelomeric regions are associated with genome evolution and also human disorders. In diseases, subtelomeric imbalances that include deletions, duplications, unbalanced translocations, and complex rearrangements [256] are terminal as well as interstitial and extremely variable in size [257]. Some telomeric polymorphisms and transmitted subtelomeric imbalances are benign and not associated with any phenotypical manifestation [258,259]. However, the importance of subtelomeric rearrangements affecting all chromosomes with the exception of the short arms of acrocentric chromosomes is well established in 5%–10% of idiopathic mental retardation [260–262]. The refinement of diagnostic techniques allowed the identification of a number of genes, but many cases remain poorly characterized [263–265] and rather a few deleterious subtelomeric imbalances are associated with a distinct, recognizable phenotype.

In the cri-du-chat syndrome (5pter deletion) for example, characterized by microcephaly, facial dysmorphism, high-pitched cat-like cry, severe intellectual impairment, and speech delay [266,267] critical regions corresponding to cry (5p15.31), speech delay (5p15.32–15.33) and facial dysmorphism (5p15.31–15.2) have been delineated [160]. The 1p36 monosomy syndrome is the most common subtelomeric microdeletion syndrome with a frequency of 1/5000. It is characterized by intellectual impairment, developmental delay, hearing loss, seizures, growth impairment, hypotonia, heart defect, and a distinctive dysmorphism [268]. In the absence of common breakpoint or common deletion interval, no correlation between the deletion size and the number of clinical features could be clearly made [268] leading to the hypothesis that the 1p36 monosomy syndrome might be due to a positional effect rather than haploinsufficiency of contiguous genes.

Besides subtelomeric rearrangements, hundreds of patients carrying ring chromosomes have been reported with various combinations of malformations, minor abnormalities, growth retardation, and intellectual impairment [269,270]. Ring chromosomes are circularized chromosomes with either a non-supernumerary ring that replaces one of the two alleles, or the presence of a supernumerary ring chromosome. Supernumerary ring chromosomes are usually associated with an abnormal phenotype. Intact rings putatively causing TPE have been reported for the different autosomes

[271–273]. They can be formed by deletion near the end(s) of chromosomes followed by fusion at breakage points or fusion without loss of genetic material. The resulting phenotypes vary greatly depending on the size and nature of the deleted segments. In rings formed by telomere–telomere fusion with little or no loss of chromosomal material the "ring syndrome" might be associated with the silencing of genes in the vicinity of a longer telomere. Moreover, analysis of telomere length in patients with ring chromosomes and their parents suggest that some chromosomes might be prone to ring formation when reaching a critical telomere length [274] suggesting that beside influencing the phenotypic spectrum, telomere length might also influence the familial transmission of the mosaic ring.

To date, the implication of TPE and TPE-OLD in diseases is solely illustrated in Facio-Scapulo-Humeral dystrophy (FSHD) [275–277]. This disease is a skeletal muscle dystrophy with a late age of onset linked, for most cases, to contraction of the *D4Z4* repetitive array on the 4q35 locus [278]. Each *D4Z4* repeat contains the *DUX4* open reading frame encoding a highly conserved double homeobox protein. In FSHD cells, *DUX4* might be transcribed in 1/1000 cell, as opposed to 1/10000 in non-FSHD cells, making *DUX4* a candidate gene for FSHD pathogenesis [279]. It was recently shown that TPE alleviates *DUX4* expression, as telomere length reduction correlates with the increased expression seen in the fraction of DUX4-positive cells expected in FSHD [275]. Besides TPE, TPE-OLD can reshape the chromatin landscape of the 4q35 locus, enhancing in FSHD muscle cells, the expression of the *SORBS2* genes located as far as 5 Mb centromeric to the *D4Z4* array reinforcing the notion of a cooperative effect between telomere length and size of the repetitive region [275–277,280]. TPE and TPE-OLD could thus elegantly explain the late-stage onset of FSHD as well as the wide heterogeneity of its clinical manifestation. Since, TPE-OLD has been involved in the regulation of genes during the physiological response [281,282] or cancer cell differentiation [281,283].

Hence to our knowledge, TPE can extend as far as 80 Kb and TPE-OLD as far as 10 Mb in human cells unraveling potential unthought-of implication of telomeres on gene regulation or pathologies linked to genes located within the "extended" reach of telomeres through position effect mechanisms. According to the work on TPE-OLD, all telomeres are not equal and disparities exists (i.e., tissue specificity), with chromosome ends highly affected by telomere length changes (i.e., numerous genes affected; 3p, 6p, 12p, 20p) or not (no change in gene transcription; 7q, 11q, 13q). Hence, one can argue that telomere dynamics and role in chromatin structure can influence the onset of pathologies and variability in clinical manifestations (due to selective telomere shortening), as seen in FSHD. Altogether, this suggests that a wide spectrum of diseases might be affected by telomere length through long distance regulation of their associated genes. This initial cut-off at 10 Mb but also the tissue-specific diversity of telomere length might underestimate all the possible consequences of telomere length on pathologies, inter-individual variability or age of onset.

Because of the paucity of cases and the variability in the size of the terminal deletion, genotype–phenotype correlations are not established for most syndromes involving either subtelomeric imbalance or ring formation. Epigenetic modifications, chromatin condensation, or loop formation might modulate expression of a number of genes in the vicinity or the rearranged region, or at a distance by modulating the formation or higher-order structures. However, characterization of the rearrangement's effect on gene expression is still needed.

CONCLUSIONS

The extent by which epigenetic changes and position effects contribute to the diversity of human phenotypes is increasingly being recognized, especially when genome-wide association fails to establish a clear genotype/phenotype correlation. Through this review, our aim was to emphasize that besides genetic changes linked to inherited pathologies, alterations of chromatin marks, misorganization of chromatin topology and disruption of the equilibrium maintaining epigenetic information might be associated with disease onset, susceptibility or penetrance. Depending on the size of the genomic region where the genomic rearrangement occurs, several alterations (gene disruption, allelic loss, positional effect) might contribute to the disease phenotype.

Unraveling the complexity of human CPE and TPE in the context of health and diseases is contingent upon our knowledge of epigenetic regulations implicated in pathogenic pathways and decades of extensive research on model organisms should help to validate the hypothetical mechanisms involved. Moreover the emergence of disease models based on the development of induced pluripotent stem cells will become a powerful approach to improve the interpretation of pathogenic complex rearrangements in biologically relevant contexts, through the direct use of cells from patient carrying the genetic variation.

In the future, a better understanding of human position effects, the identification of proteins and pathways involved in its regulation, but also the regional specificities (telomere proximity, euchromatic or heterochromatic regions, subnuclear positioning), might thus be considered for a better understanding of the clinical variability and patho-mechanisms associated with complex diseases. Together with epigenetic regulation in general, deciphering the mechanisms involved in gene variegation is thus a major challenge of the post-genomic era for the understanding and cure of a wide range of human diseases. Altogether, this also underlines the necessity to revisit old findings and observations in light of recent advances in whole genome and large-scale tri-dimensional chromatin analyses since structural variants might perturb TADs structure and result in aberrant long distance interactions. In agreement with this hypothesis, this might also explain how different types of large-scale structural changes might give rise to the same phenotype.

References

[1] Egli D, Birkhoff G, Eggan K. Mediators of reprogramming: transcription factors and transitions through mitosis. Nat Rev Mol Cell Biol 2008;9(7):505–16.

[2] Kadauke S, Blobel GA. Mitotic bookmarking by transcription factors. Epigenetics Chromatin 2013;6(1):6.

[3] Zaidi SK, Grandy RA, Lopez-Camacho C, Montecino M, van Wijnen AJ, Lian JB, et al. Bookmarking target genes in mitosis: a shared epigenetic trait of phenotypic transcription factors and oncogenes? Cancer Res 2014;74(2):420–5.

[4] Caravaca JM, Donahue G, Becker JS, He X, Vinson C, Zaret KS. Bookmarking by specific and nonspecific binding of FoxA1 pioneer factor to mitotic chromosomes. Genes Dev 2013;27(3):251–60.

[5] Peng JC, Karpen GH. Epigenetic regulation of heterochromatic DNA stability. Curr Opin Genet Dev 2008;18(2):204–11.

[6] Trojer P, Reinberg D. Facultative heterochromatin: is there a distinctive molecular signature? Mol Cell 2007;28(1):1–13.

[7] Bernstein BE, Mikkelsen TS, Xie X, Kamal M, Huebert DJ, Cuff J, et al. A bivalent chromatin structure marks key developmental genes in embryonic stem cells. Cell 2006;125(2):315–26.

[8] Mikkelsen TS, Ku M, Jaffe DB, Issac B, Lieberman E, Giannoukos G, et al. Genome-wide maps of chromatin state in pluripotent and lineage-committed cells. Nature 2007;448(7153):553–60.

[9] Margueron R, Trojer P, Reinberg D. The key to development: interpreting the histone code? Curr Opin Genet Dev 2005;15(2):163–76.

[10] Azuara V, Perry P, Sauer S, Spivakov M, Jorgensen HF, John RM, et al. Chromatin signatures of pluripotent cell lines. Nat Cell Biol 2006;8(5):532–8.

[11] Girton JR, Johansen KM. Chromatin structure and the regulation of gene expression: the lessons of PEV in Drosophila. Adv Genet 2008;61:1–43.

[12] Ottaviani A, Gilson E, Magdinier F. Telomeric position effect: from the yeast paradigm to human pathologies? Biochimie 2008;90(1):93–107.

[13] Rabbitts TH, Forster A, Baer R, Hamlyn PH. Transcription enhancer identified near the human C mu immunoglobulin heavy chain gene is unavailable to the translocated c-myc gene in a Burkitt lymphoma. Nature 1983;306(5945):806–9.

[14] Phillips-Cremins JE, Corces VG. Chromatin insulators: linking genome organization to cellular function. Mol Cell 2013;50(4):461–74.

[15] Chetverina D, Aoki T, Erokhin M, Georgiev P, Schedl P. Making connections: insulators organize eukaryotic chromosomes into independent cis-regulatory networks. BioEssays 2014;36(2):163–72.

[16] Wang J, Lawry ST, Cohen AL, Jia S. Chromosome boundary elements and regulation of heterochromatin spreading. Cell Mol Life Sci 2014;71(24):4841–52.

[17] Gottschling DE, Aparicio OM, Billington BL, Zakian VA. Position effect at S. cerevisiae telomeres: reversible repression of Pol II transcription. Cell 1990;63(4):751–62.

[18] Strurtevant A. The effect of unequal crossing over the Bar locus in Drosophila. Genetics 1925;10:117–47.

[19] Muller. Types of visible variations induced by X-rays in Drosophila. J Genet 1930;22:299.

[20] Muller. Further studies on the nature and causes of gene mutations. Proc Sixth Int Congr Genet 1932;1:213–55.

[21] Dorer DR, Henikoff S. Transgene repeat arrays interact with distant heterochromatin and cause silencing in cis and trans. Genetics 1997;147(3):1181–90.

[22] Garrick D, Fiering S, Martin DI, Whitelaw E. Repeat-induced gene silencing in mammals. Nat Genet 1998;18(1):56–9.

[23] Eissenberg JC, James TC, Foster-Hartnett DM, Hartnett T, Ngan V, Elgin SC. Mutation in a heterochromatin-specific chromosomal protein is associated with suppression of position-effect variegation in Drosophila melanogaster. Proc Natl Acad Sci USA 1990;87(24):9923–7.

[24] Festenstein R, Sharghi-Namini S, Fox M, Roderick K, Tolaini M, Norton T, et al. Heterochromatin protein 1 modifies mammalian PEV in a dose- and chromosomal-context-dependent manner. Nat Genet 1999;23(4):457–61.

[25] James TC, Elgin SC. Identification of a nonhistone chromosomal protein associated with heterochromatin in Drosophila melanogaster and its gene. Mol Cell Biol 1986;6(11):3862–72.

[26] Schulze SR, Wallrath LL. Gene regulation by chromatin structure: paradigms established in Drosophila melanogaster. Annu Rev Entomol 2007;52:171–92.

[27] Elgin SC, Reuter G. Position-effect variegation, heterochromatin formation, and gene silencing in Drosophila. Cold Spring Harb Perspect Biol 2013;5(8):a017780.

[28] Phalke S, Nickel O, Walluscheck D, Hortig F, Onorati MC, Reuter G. Retrotransposon silencing and telomere integrity in somatic cells of Drosophila depends on the cytosine-5 methyltransferase DNMT2. Nat Genet 2009;41(6):696–702.

[29] Ashe A, Morgan DK, Whitelaw NC, Bruxner TJ, Vickaryous NK, Cox LL, et al. A genome-wide screen for modifiers of transgene variegation identifies genes with critical roles in development. Genome Biol 2008;9(12):R182.

[30] Blewitt ME, Vickaryous NK, Hemley SJ, Ashe A, Bruxner TJ, Preis JI, et al. An N-ethyl-N-nitrosourea screen for genes involved in variegation in the mouse. Proc Natl Acad Sci USA 2005;102(21):7629–34.

[31] Akhtar W, de Jong J, Pindyurin AV, Pagie L, Meuleman W, de Ridder J, et al. Chromatin position effects assayed by thousands of reporters integrated in parallel. Cell 2013;154(4):914–27.

[32] Kotecki M, Reddy PS, Cochran BH. Isolation and characterization of a near-haploid human cell line. Exp Cell Res 1999;252(2):273–80.

[33] Tchasovnikarova IA, Timms RT, Matheson NJ, Wals K, Antrobus R, Gottgens B, et al. GENE SILENCING. Epigenetic silencing by the HUSH complex mediates position-effect variegation in human cells. Science 2015;348(6242):1481–5.

REFERENCES

[34] Faulkner GJ, Kimura Y, Daub CO, Wani S, Plessy C, Irvine KM, et al. The regulated retrotransposon transcriptome of mammalian cells. Nat Genet 2009;41(5):563–71.

[35] Shaikh TH. Copy number variation disorders. Curr Genet Med Rep 2017;5(4):183–90.

[36] Valenzuela L, Kamakaka RT. Chromatin insulators. Annu Rev Genet 2006;40:107–38.

[37] Gaszner M, Felsenfeld G. Insulators: exploiting transcriptional and epigenetic mechanisms. Nat Rev Genet 2006;7(9):703–13.

[38] Fourel G, Magdinier F, Gilson E. Insulator dynamics and the setting of chromatin domains. Bioessays 2004;26(5):523–32.

[39] Gaszner M, Vazquez J, Schedl P. The Zw5 protein, a component of the scs chromatin domain boundary, is able to block enhancer-promoter interaction. Genes Dev 1999;13(16):2098–107.

[40] Zhao K, Hart CM, Laemmli UK. Visualization of chromosomal domains with boundary element-associated factor BEAF-32. Cell 1995;81(6):879–89.

[41] Ohtsuki S, Levine M. GAGA mediates the enhancer blocking activity of the eve promoter in the Drosophila embryo. Genes Dev 1998;12(21):3325–30.

[42] Belozerov VE, Majumder P, Shen P, Cai HN. A novel boundary element may facilitate independent gene regulation in the Antennapedia complex of Drosophila. EMBO J 2003;22(12):3113–21.

[43] Moon H, Filippova G, Loukinov D, Pugacheva E, Chen Q, Smith ST, et al. CTCF is conserved from Drosophila to humans and confers enhancer blocking of the Fab-8 insulator. EMBO Rep 2005;6(2):165–70.

[44] Gerasimova TI, Lei EP, Bushey AM, Corces VG. Coordinated control of dCTCF and gypsy chromatin insulators in Drosophila. Mol Cell 2007;28(5):761–72.

[45] Bell AC, West AG, Felsenfeld G. The protein CTCF is required for the enhancer blocking activity of vertebrate insulators. Cell 1999;98(3):387–96.

[46] Xie X, Mikkelsen TS, Gnirke A, Lindblad-Toh K, Kellis M, Lander ES. Systematic discovery of regulatory motifs in conserved regions of the human genome, including thousands of CTCF insulator sites. Proc Natl Acad Sci USA 2007;104(17):7145–50.

[47] Kim TH, Abdullaev ZK, Smith AD, Ching KA, Loukinov DI, Green RD, et al. Analysis of the vertebrate insulator protein CTCF-binding sites in the human genome. Cell 2007;128(6):1231–45.

[48] Barski A, Cuddapah S, Cui K, Roh TY, Schones DE, Wang Z, et al. High-resolution profiling of histone methylations in the human genome. Cell 2007;129(4):823–37.

[49] Wendt KS, Yoshida K, Itoh T, Bando M, Koch B, Schirghuber E, et al. Cohesin mediates transcriptional insulation by CCCTC-binding factor. Nature 2008;451(7180):796–801.

[50] Stedman W, Kang H, Lin S, Kissil JL, Bartolomei MS, Lieberman PM. Cohesins localize with CTCF at the KSHV latency control region and at cellular c-myc and H19/Igf2 insulators. EMBO J 2008;27(4):654–66.

[51] Parelho V, Hadjur S, Spivakov M, Leleu M, Sauer S, Gregson HC, et al. Cohesins functionally associate with CTCF on mammalian chromosome arms. Cell 2008;132(3):422–33.

[52] Kaaij LJT, Mohn F, van der Weide RH, de Wit E, Buhler M. The ChAHP complex counteracts chromatin looping at CTCF sites that emerged from SINE expansions in mouse. Cell 2019;178(6):1437–51 e14.

[53] Jansz N, Keniry A, Trussart M, Bildsoe H, Beck T, Tonks ID, et al. Smchd1 regulates long-range chromatin interactions on the inactive X chromosome and at Hox clusters. Nat Struct Mol Biol 2018;25(9):766–77.

[54] Chen K, Hu J, Moore DL, Liu R, Kessans SA, Breslin K, et al. Genome-wide binding and mechanistic analyses of Smchd1-mediated epigenetic regulation. Proc Natl Acad Sci USA 2015;112(27):E3535–44.

[55] Wang CY, Jegu T, Chu HP, Oh HJ, Lee JT. SMCHD1 merges chromosome compartments and assists formation of superstructures on the inactive X. Cell 2018;174(2):406–21 e25.

[56] Gdula MR, Nesterova TB, Pintacuda G, Godwin J, Zhan Y, Ozadam H, et al. The non-canonical SMC protein SmcHD1 antagonises TAD formation and compartmentalisation on the inactive X chromosome. Nat Commun 2019;10(1):30.

[57] Dekker J, Rippe K, Dekker M, Kleckner N. Capturing chromosome conformation. Science 2002;295(5558):1306–11.

[58] Dekker J, Marti-Renom MA, Mirny LA. Exploring the three-dimensional organization of genomes: interpreting chromatin interaction data. Nat Rev Genet 2013;14(6):390–403.

[59] de Wit E, de Laat W. A decade of 3C technologies: insights into nuclear organization. Genes Dev 2012;26(1):11–24.

[60] Guelen L, Pagie L, Brasset E, Meuleman W, Faza MB, Talhout W, et al. Domain organization of human chromosomes revealed by mapping of nuclear lamina interactions. Nature 2008;453(7197):948–51.

[61] Lieberman-Aiden E, van Berkum NL, Williams L, Imakaev M, Ragoczy T, Telling A, et al. Comprehensive mapping of long-range interactions reveals folding principles of the human genome. Science 2009;326(5950):289–93.

[62] Sexton T, Yaffe E, Kenigsberg E, Bantignies F, Leblanc B, Hoichman M, et al. Three-dimensional folding and functional organization principles of the Drosophila genome. Cell 2012;148(3):458–72.

[63] Dixon JR, Selvaraj S, Yue F, Kim A, Li Y, Shen Y, et al. Topological domains in mammalian genomes identified by analysis of chromatin interactions. Nature 2012;485(7398):376–80.

[64] Nora EP, Lajoie BR, Schulz EG, Giorgetti L, Okamoto I, Servant N, et al. Spatial partitioning of the regulatory landscape of the X-inactivation centre. Nature 2012;485(7398):381–5.

[65] Phillips-Cremins JE, Sauria ME, Sanyal A, Gerasimova TI, Lajoie BR, Bell JS, et al. Architectural protein subclasses shape 3D organization of genomes during lineage commitment. Cell 2013;153(6):1281–95.

[66] Moorthy SD, Davidson S, Shchuka VM, Singh G, Malek-Gilani N, Langroudi L, et al. Enhancers and super-enhancers have an equivalent regulatory role in embryonic stem cells through regulation of single or multiple genes. Genome Res 2017;27(2):246–58.

[67] Will AJ, Cova G, Osterwalder M, Chan WL, Wittler L, Brieske N, et al. Composition and dosage of a multipartite enhancer cluster control developmental expression of Ihh (Indian hedgehog). Nat Genet 2017;49(10):1539–45.

[68] Osterwalder M, Barozzi I, Tissieres V, Fukuda-Yuzawa Y, Mannion BJ, Afzal SY, et al. Enhancer redundancy provides phenotypic robustness in mammalian development. Nature 2018;554(7691):239–43.

[69] Dowen JM, Bilodeau S, Orlando DA, Hubner MR, Abraham BJ, Spector DL, et al. Multiple structural maintenance of chromosome complexes at transcriptional regulatory elements. Stem Cell Rep 2013;1(5):371–8.

[70] Ji X, Dadon DB, Powell BE, Fan ZP, Borges-Rivera D, Shachar S, et al. 3D Chromosome regulatory landscape of human pluripotent cells. Cell Stem Cell 2016;18(2):262–75.

[71] Rao SS, Huntley MH, Durand NC, Stamenova EK, Bochkov ID, Robinson JT, et al. A 3D map of the human genome at kilobase resolution reveals principles of chromatin looping. Cell 2014;159(7):1665–80.

[72] Scott RH, Douglas J, Baskcomb L, Huxter N, Barker K, Hanks S, et al. Constitutional 11p15 abnormalities, including heritable

imprinting center mutations, cause nonsyndromic Wilms tumor. Nat Genet 2008;40(11):1329—34.
[73] Watrin E, Kaiser FJ, Wendt KS. Gene regulation and chromatin organization: relevance of cohesin mutations to human disease. Curr Opin Genet Dev 2016;37:59—66.
[74] Cascella M, Muzio MR. Cornelia de lange syndrome. Treasure Island, FL: StatPearls; 2021.
[75] Helsmoortel C, Vulto-van Silfhout AT, Coe BP, Vandeweyer G, Rooms L, van den Ende J, et al. A SWI/SNF-related autism syndrome caused by de novo mutations in ADNP. Nat Genet 2014;46(4):380—4.
[76] Shaw ND, Brand H, Kupchinsky ZA, Bengani H, Plummer L, Jones TI, et al. SMCHD1 mutations associated with a rare muscular dystrophy can also cause isolated arhinia and Bosma arhinia microphthalmia syndrome. Nat Genet 2017;49(2):238—48.
[77] Lemmers RJ, Tawil R, Petek LM, Balog J, Block GJ, Santen GW, et al. Digenic inheritance of an SMCHD1 mutation and an FSHD-permissive D4Z4 allele causes facioscapulohumeral muscular dystrophy type 2. Nat Genet 2012;44(12):1370—4.
[78] Gordon CT, Xue S, Yigit G, Filali H, Chen K, Rosin N, et al. De novo mutations in SMCHD1 cause Bosma arhinia microphthalmia syndrome and abrogate nasal development. Nat Genet 2017;49(2):249—55.
[79] Kinjo K, Nagasaki K, Muroya K, Suzuki E, Ishiwata K, Nakabayashi K, et al. Rare variant of the epigenetic regulator SMCHD1 in a patient with pituitary hormone deficiency. Sci Rep 2020;10(1):10985.
[80] Ibn-Salem J, Kohler S, Love MI, Chung HR, Huang N, Hurles ME, et al. Deletions of chromosomal regulatory boundaries are associated with congenital disease. Genome Biol 2014;15(9):423.
[81] Zepeda-Mendoza CJ, Ibn-Salem J, Kammin T, Harris DJ, Rita D, Gripp KW, et al. Computational prediction of position effects of apparently balanced human chromosomal rearrangements. Am J Hum Genet 2017;101(2):206—17.
[82] Zepeda-Mendoza CJ, Bardon A, Kammin T, Harris DJ, Cox H, Redin C, et al. Phenotypic interpretation of complex chromosomal rearrangements informed by nucleotide-level resolution and structural organization of chromatin. Eur J Hum Genet 2018;26(3):374—81.
[83] Welter D, MacArthur J, Morales J, Burdett T, Hall P, Junkins H, et al. The NHGRI GWAS catalog, a curated resource of SNP-trait associations. Nucleic Acids Res 2014;42(Database issue):D1001—6.
[84] Roadmap Epigenomics C, Kundaje A, Meuleman W, Ernst J, Bilenky M, Yen A, et al. Integrative analysis of 111 reference human epigenomes. Nature 2015;518(7539):317—30.
[85] Parker SC, Stitzel ML, Taylor DL, Orozco JM, Erdos MR, Akiyama JA, et al. Chromatin stretch enhancer states drive cell-specific gene regulation and harbor human disease risk variants. Proc Natl Acad Sci USA 2013;110(44):17921—6.
[86] Farh KK, Marson A, Zhu J, Kleinewietfeld M, Housley WJ, Beik S, et al. Genetic and epigenetic fine mapping of causal autoimmune disease variants. Nature 2015;518(7539):337—43.
[87] McKusick VA, Amberger JS. The morbid anatomy of the human genome: chromosomal location of mutations causing disease. J Med Genet 1993;30(1):1—26.
[88] Kleinjan DJ, van Heyningen V. Position effect in human genetic disease. Hum Mol Genet 1998;7(10):1611—18.
[89] Rehm HL, Berg JS, Brooks LD, Bustamante CD, Evans JP, Landrum MJ, et al. ClinGen—the clinical genome resource. N Engl J Med 2015;372(23):2235—42.
[90] Rabbitts TH. Chromosomal translocations in human cancer. Nature 1994;372(6502):143—9.
[91] Rabbitts TH. Translocations, master genes, and differences between the origins of acute and chronic leukemias. Cell 1991;67(4):641—4.
[92] Korsmeyer SJ. Chromosomal translocations in lymphoid malignancies reveal novel proto-oncogenes. Annu Rev Immunol 1992;10:785—807.
[93] Lossos IS, Akasaka T, Martinez-Climent JA, Siebert R, Levy R. The BCL6 gene in B-cell lymphomas with 3q27 translocations is expressed mainly from the rearranged allele irrespective of the partner gene. Leukemia 2003;17(7):1390—7.
[94] Kumar-Sinha C, Tomlins SA, Chinnaiyan AM. Recurrent gene fusions in prostate cancer. Nat Rev Cancer 2008;8(7):497—511.
[95] Northcott PA, Lee C, Zichner T, Stutz AM, Erkek S, Kawauchi D, et al. Enhancer hijacking activates GFI1 family oncogenes in medulloblastoma. Nature 2014;511(7510):428—34.
[96] State MW, Greally JM, Cuker A, Bowers PN, Henegariu O, Morgan TM, et al. Epigenetic abnormalities associated with a chromosome 18(q21-q22) inversion and a Gilles de la Tourette syndrome phenotype. Proc Natl Acad Sci USA 2003;100(8):4684—9.
[97] Wieacker P, Apeshiotis N, Jakubiczka S, Volleth M, Wieland I. Familial translocation t(1;9) associated with macromastia: molecular cloning of the breakpoints. Sex Dev 2007;1(1):35—41.
[98] Kleinjan DA, Seawright A, Schedl A, Quinlan RA, Danes S, van Heyningen V. Aniridia-associated translocations, DNase hypersensitivity, sequence comparison and transgenic analysis redefine the functional domain of PAX6. Hum Mol Genet 2001;10(19):2049—59.
[99] de Chadarevian JP, Dunn S, Malatack JJ, Ganguly A, Blecker U, Punnett HH. Chromosome rearrangement with no apparent gene mutation in familial adenomatous polyposis and hepatocellular neoplasia. Pediatr Dev Pathol 2002;5(1):69—75.
[100] Ghiasvand NM, Rudolph DD, Mashayekhi M, Brzezinski JAT, Goldman D, Glaser T. Deletion of a remote enhancer near ATOH7 disrupts retinal neurogenesis, causing NCRNA disease. Nat Neurosci 2011;14(5):578—86.
[101] Crackower MA, Scherer SW, Rommens JM, Hui CC, Poorkaj P, Soder S, et al. Characterization of the split hand/split foot malformation locus SHFM1 at 7q21.3-q22.1 and analysis of a candidate gene for its expression during limb development. Hum Mol Genet 1996;5(5):571—9.
[102] Birnbaum RY, Clowney EJ, Agamy O, Kim MJ, Zhao J, Yamanaka T, et al. Coding exons function as tissue-specific enhancers of nearby genes. Genome Res 2012;22(6):1059—68.
[103] Tayebi N, Jamsheer A, Flottmann R, Sowinska-Seidler A, Doelken SC, Oehl-Jaschkowitz B, et al. Deletions of exons with regulatory activity at the DYNC1I1 locus are associated with split-hand/split-foot malformation: array CGH screening of 134 unrelated families. Orphanet J Rare Dis 2014;9:108.
[104] Lango Allen H, Caswell R, Xie W, Xu X, Wragg C, Turnpenny PD, et al. Next generation sequencing of chromosomal rearrangements in patients with split-hand/split-foot malformation provides evidence for DYNC1I1 exonic enhancers of DLX5/6 expression in humans. J Med Genet 2014;51(4):264—7.
[105] Wang P, Carrion P, Qiao Y, Tyson C, Hrynchak M, Calli K, et al. Genotype-phenotype analysis of 18q12.1-q12.2 copy number variation in autism. Eur J Med Genet 2013;56(8):420—5.
[106] Jamsheer A, Sowinska A, Simon D, Jamsheer-Bratkowska M, Trzeciak T, Latos-Bielenska A. Bilateral radial agenesis with absent thumbs, complex heart defect, short stature, and facial dysmorphism in a patient with pure distal microduplication of 5q35.2-5q35.3. BMC Med Genet 2013;14:13.

REFERENCES

[107] Lupianez DG, Kraft K, Heinrich V, Krawitz P, Brancati F, Klopocki E, et al. Disruptions of topological chromatin domains cause pathogenic rewiring of gene-enhancer interactions. Cell 2015;161(5):1012−25.

[108] Akasaka K, Nishimura A, Takata K, Mitsunaga K, Mibuka F, Ueda H, et al. Upstream element of the sea urchin arylsulfatase gene serves as an insulator. Cell Mol Biol (Noisy-le-grand) 1999;45(5):555−65.

[109] Fang J, Dagenais SL, Erickson RP, Arlt MF, Glynn MW, Gorski JL, et al. Mutations in FOXC2 (MFH-1), a forkhead family transcription factor, are responsible for the hereditary lymphedema-distichiasis syndrome. Am J Hum Genet 2000;67(6):1382−8.

[110] Allou L, Lambert L, Amsallem D, Bieth E, Edery P, Destree A, et al. 14q12 and severe Rett-like phenotypes: new clinical insights and physical mapping of FOXG1-regulatory elements. Eur J Hum Genet 2012;20(12):1216−23.

[111] Redin C, Brand H, Collins RL, Kammin T, Mitchell E, Hodge JC, et al. The genomic landscape of balanced cytogenetic abnormalities associated with human congenital anomalies. Nat Genet 2017;49(1):36−45.

[112] Crisponi L, Deiana M, Loi A, Chiappe F, Uda M, Amati P, et al. The putative forkhead transcription factor FOXL2 is mutated in blepharophimosis/ptosis/epicanthus inversus syndrome. Nat Genet 2001;27(2):159−66.

[113] Beysen D, Raes J, Leroy BP, Lucassen A, Yates JR, Clayton-Smith J, et al. Deletions involving long-range conserved nongenic sequences upstream and downstream of FOXL2 as a novel disease-causing mechanism in blepharophimosis syndrome. Am J Hum Genet 2005;77(2):205−18.

[114] Middelkamp S, van Heesch S, Braat AK, de Ligt J, van Iterson M, Simonis M, et al. Molecular dissection of germline chromothripsis in a developmental context using patient-derived iPS cells. Genome Med 2017;9(1):9.

[115] Kosho T, Sakazume S, Kawame H, Wakui K, Wada T, Okoshi Y, et al. De-novo balanced translocation between 7q31 and 10p14 in a girl with central precocious puberty, moderate mental retardation, and severe speech impairment. Clin Dysmorphol 2008;17(1):31−4.

[116] Vortkamp A, Gessler M, Grzeschik KH. GLI3 zinc-finger gene interrupted by translocations in Greig syndrome families. Nature 1991;352(6335):539−40.

[117] Barbour VM, Tufarelli C, Sharpe JA, Smith ZE, Ayyub H, Heinlein CA, et al. Alpha-thalassemia resulting from a negative chromosomal position effect. Blood 2000;96(3):800−7.

[118] Kioussis D, Vanin E, deLange T, Flavell RA, Grosveld FG. Beta-globin gene inactivation by DNA translocation in gamma beta-thalassaemia. Nature 1983;306(5944):662−6.

[119] Driscoll MC, Dobkin CS, Alter BP. Gamma delta beta-thalassemia due to a de novo mutation deleting the 5' beta-globin gene activation-region hypersensitive sites. Proc Natl Acad Sci USA 1989;86(19):7470−4.

[120] Yue Y, Farcas R, Thiel G, Bommer C, Grossmann B, Galetzka D, et al. De novo t(12;17)(p13.3;q21.3) translocation with a breakpoint near the 5' end of the HOXB gene cluster in a patient with developmental delay and skeletal malformations. Eur J Hum Genet 2007;15(5):570−7.

[121] Spitz F, Montavon T, Monso-Hinard C, Morris M, Ventruto ML, Antonarakis S, et al. A t(2;8) balanced translocation with breakpoints near the human HOXD complex causes mesomelic dysplasia and vertebral defects. Genomics 2002;79(4):493−8.

[122] Dlugaszewska B, Silahtaroglu A, Menzel C, Kubart S, Cohen M, Mundlos S, et al. Breakpoints around the HOXD cluster result in various limb malformations. J Med Genet 2006;43(2):111−18.

[123] Genesio R, Melis D, Gatto S, Izzo A, Ronga V, Cappuccio G, et al. Variegated silencing through epigenetic modifications of a large Xq region in a case of balanced X;2 translocation with Incontinentia Pigmenti-like phenotype. Epigenetics 2011;6(10):1242−7.

[124] Enattah NS, Sahi T, Savilahti E, Terwilliger JD, Peltonen L, Jarvela I. Identification of a variant associated with adult-type hypolactasia. Nat Genet 2002;30(2):233−7.

[125] Giorgio E, Robyr D, Spielmann M, Ferrero E, Di Gregorio E, Imperiale D, et al. A large genomic deletion leads to enhancer adoption by the lamin B1 gene: a second path to autosomal dominant adult-onset demyelinating leukodystrophy (ADLD). Hum Mol Genet 2015;24(11):3143−54.

[126] Brussino A, Vaula G, Cagnoli C, Panza E, Seri M, Di Gregorio E, et al. A family with autosomal dominant leukodystrophy linked to 5q23.2-q23.3 without lamin B1 mutations. Eur J Neurol 2010;17(4):541−9.

[127] Saitsu H, Igarashi N, Kato M, Okada I, Kosho T, Shimokawa O, et al. De novo 5q14.3 translocation 121.5-kb upstream of MEF2C in a patient with severe intellectual disability and early-onset epileptic encephalopathy. Am J Med Genet A 2011;155A(11):2879−84.

[128] Le Meur N, Holder-Espinasse M, Jaillard S, Goldenberg A, Joriot S, Amati-Bonneau P, et al. MEF2C haploinsufficiency caused by either microdeletion of the 5q14.3 region or mutation is responsible for severe mental retardation with stereotypic movements, epilepsy and/or cerebral malformations. J Med Genet 2010;47(1):22−9.

[129] Engels H, Wohlleber E, Zink A, Hoyer J, Ludwig KU, Brockschmidt FF, et al. A novel microdeletion syndrome involving 5q14.3-q15: clinical and molecular cytogenetic characterization of three patients. Eur J Hum Genet 2009;17(12):1592−9.

[130] Jamieson RV, Perveen R, Kerr B, Carette M, Yardley J, Heon E, et al. Domain disruption and mutation of the bZIP transcription factor, MAF, associated with cataract, ocular anterior segment dysgenesis and coloboma. Hum Mol Genet 2002;11(1):33−42.

[131] Fantes J, Redeker B, Breen M, Boyle S, Brown J, Fletcher J, et al. Aniridia-associated cytogenetic rearrangements suggest that a position effect may cause the mutant phenotype. Hum Mol Genet 1995;4(3):415−22.

[132] Bhatia S, Bengani H, Fish M, Brown A, Divizia MT, de Marco R, et al. Disruption of autoregulatory feedback by a mutation in a remote, ultraconserved PAX6 enhancer causes aniridia. Am J Hum Genet 2013;93(6):1126−34.

[133] Spielmann M, Brancati F, Krawitz PM, Robinson PN, Ibrahim DM, Franke M, et al. Homeotic arm-to-leg transformation associated with genomic rearrangements at the PITX1 locus. Am J Hum Genet 2012;91(4):629−35.

[134] Clinical observation on Chinese herb Lei-Gong-Teng in treatment of lepra reaction (author's transl). Zhongguo Yi Xue Ke Xue Yuan Xue Bao 1979;1(1):71-74.

[135] Flomen RH, Gorman PA, Vatcheva R, Groet J, Barisic I, Ligutic I, et al. Rieger syndrome locus: a new reciprocal translocation t(4;12)(q25;q15) and a deletion del(4)(q25q27) both break between markers D4S2945 and D4S193. J Med Genet 1997;34(3):191−5.

[136] Lee JA, Madrid RE, Sperle K, Ritterson CM, Hobson GM, Garbern J, et al. Spastic paraplegia type 2 associated with axonal neuropathy and apparent PLP1 position effect. Ann Neurol 2006;59(2):398−403.

[137] Muncke N, Wogatzky BS, Breuning M, Sistermans EA, Endris V, Ross M, et al. Position effect on PLP1 may cause a subset of Pelizaeus-Merzbacher disease symptoms. J Med Genet 2004;41(12):e121.

[138] de Kok YJ, Vossenaar ER, Cremers CW, Dahl N, Laporte J, Hu LJ, et al. Identification of a hot spot for microdeletions in patients with X-linked deafness type 3 (DFN3) 900 kb proximal to the DFN3 gene POU3F4. Hum Mol Genet 1996;5(9):1229−35.

[139] Castermans D, Vermeesch JR, Fryns JP, Steyaert JG, Van de Ven WJ, Creemers JW, et al. Identification and characterization of the TRIP8 and REEP3 genes on chromosome 10q21.3 as novel candidate genes for autism. Eur J Hum Genet 2007;15(4):422−31.

[140] Ferrari L, Scuvera G, Tucci A, Bianchessi D, Rusconi F, Menni F, et al. Identification of an atypical microdeletion generating the RNF135-SUZ12 chimeric gene and causing a position effect in an NF1 patient with overgrowth. Hum Genet 2017;136(10):1329−39.

[141] Fernandez BA, Siegel-Bartelt J, Herbrick JA, Teshima I, Scherer SW. Holoprosencephaly and cleidocranial dysplasia in a patient due to two position-effect mutations: case report and review of the literature. Clin Genet 2005;68(4):349−59.

[142] Marlin S, Blanchard S, Slim R, Lacombe D, Denoyelle F, Alessandri JL, et al. Townes-Brocks syndrome: detection of a SALL1 mutation hot spot and evidence for a position effect in one patient. Hum Mutat 1999;14(5):377−86.

[143] Ishikawa-Brush Y, Powell JF, Bolton P, Miller AP, Francis F, Willard HF, et al. Autism and multiple exostoses associated with an X;8 translocation occurring within the GRPR gene and 3′ to the SDC2 gene. Hum Mol Genet 1997;6(8):1241−50.

[144] Roessler E, Ward DE, Gaudenz K, Belloni E, Scherer SW, Donnai D, et al. Cytogenetic rearrangements involving the loss of the Sonic Hedgehog gene at 7q36 cause holoprosencephaly. Hum Genet 1997;100(2):172−81.

[145] Lettice LA, Heaney SJ, Purdie LA, Li L, de Beer P, Oostra BA, et al. A long-range Shh enhancer regulates expression in the developing limb and fin and is associated with preaxial polydactyly. Hum Mol Genet 2003;12(14):1725−35.

[146] Benito-Sanz S, Thomas NS, Huber C, Gorbenko del Blanco D, Aza-Carmona M, Crolla JA, et al. A novel class of Pseudoautosomal region 1 deletions downstream of SHOX is associated with Leri-Weill dyschondrosteosis. Am J Hum Genet 2005;77(4):533−44.

[147] Sabherwal N, Bangs F, Roth R, Weiss B, Jantz K, Tiecke E, et al. Long-range conserved non-coding SHOX sequences regulate expression in developing chicken limb and are associated with short stature phenotypes in human patients. Hum Mol Genet 2007;16(2):210−22.

[148] Wallis DE, Roessler E, Hehr U, Nanni L, Wiltshire T, Richieri-Costa A, et al. Mutations in the homeodomain of the human SIX3 gene cause holoprosencephaly. Nat Genet 1999;22(2):196−8.

[149] Balemans W, Patel N, Ebeling M, Van Hul E, Wuyts W, Lacza C, et al. Identification of a 52 kb deletion downstream of the SOST gene in patients with van Buchem disease. J Med Genet 2002;39(2):91−7.

[150] Bowl MR, Nesbit MA, Harding B, Levy E, Jefferson A, Volpi E, et al. An interstitial deletion-insertion involving chromosomes 2p25.3 and Xq27.1, near SOX3, causes X-linked recessive hypoparathyroidism. J Clin Invest 2005;115(10):2822−31.

[151] Flottmann R, Wagner J, Kobus K, Curry CJ, Savarirayan R, Nishimura G, et al. Microdeletions on 6p22.3 are associated with mesomelic dysplasia Savarirayan type. J Med Genet 2015;52(7):476−83.

[152] Velagaleti GV, Bien-Willner GA, Northup JK, Lockhart LH, Hawkins JC, Jalal SM, et al. Position effects due to chromosome breakpoints that map approximately 900 Kb upstream and approximately 1.3 Mb downstream of SOX9 in two patients with campomelic dysplasia. Am J Hum Genet 2005;76(4):652−62.

[153] Benko S, Fantes JA, Amiel J, Kleinjan DJ, Thomas S, Ramsay J, et al. Highly conserved non-coding elements on either side of SOX9 associated with Pierre Robin sequence. Nat Genet 2009;41(3):359−64.

[154] Franke M, Ibrahim DM, Andrey G, Schwarzer W, Heinrich V, Schopflin R, et al. Formation of new chromatin domains determines pathogenicity of genomic duplications. Nature 2016;538(7624):265−9.

[155] McElreavey K, Vilain E, Barbaux S, Fuqua JS, Fechner PY, Souleyreau N, et al. Loss of sequences 3′ to the testis-determining gene, SRY, including the Y pseudoautosomal boundary associated with partial testicular determination. Proc Natl Acad Sci USA 1996;93(16):8590−4.

[156] Niejadlik DC, Engelhardt C. An evaluation of the Guest method for determining erythrocyte sedimentation rate. Am J Clin Pathol 1977;68(6):766−8.

[157] Laugsch M, Bartusel M, Rehimi R, Alirzayeva H, Karaolidou A, Crispatzu G, et al. Modeling the pathological long-range regulatory effects of human structural variation with patient-specific hiPSCs. Cell Stem Cell 2019;24(5):736−52 e12.

[158] David D, Cardoso J, Marques B, Marques R, Silva ED, Santos H, et al. Molecular characterization of a familial translocation implicates disruption of HDAC9 and possible position effect on TGFbeta2 in the pathogenesis of Peters' anomaly. Genomics 2003;81(5):489−503.

[159] Fantauzzo KA, Tadin-Strapps M, You Y, Mentzer SE, Baumeister FA, Cianfarani S, et al. A position effect on TRPS1 is associated with Ambras syndrome in humans and the Koala phenotype in mice. Hum Mol Genet 2008;17(22):3539−51.

[160] Zhang X, Snijders A, Segraves R, Niebuhr A, Albertson D, Yang H, et al. High-resolution mapping of genotype-phenotype relationships in cri du chat syndrome using array comparative genomic hybridization. Am J Hum Genet 2005;76(2):312−26.

[161] Iourov IY, Vorsanova SG, Yurov YB. The variome concept: focus on CNVariome. Mol Cytogenet 2019;12:52.

[162] Burn J, Watson M. The human variome project. Hum Mutat 2016;37(6):505−7.

[163] Suwinski P, Ong C, Ling MHT, Poh YM, Khan AM, Ong HS. Advancing personalized medicine through the application of whole exome sequencing and big data analytics. Front Genet 2019;10:49.

[164] Li MJ, Sham PC, Wang J. Genetic variant representation, annotation and prioritization in the post-GWAS era. Cell Res 2012;22(10):1505−8.

[165] Goidts V, Cooper DN, Armengol L, Schempp W, Conroy J, Estivill X, et al. Complex patterns of copy number variation at sites of segmental duplications: an important category of structural variation in the human genome. Hum Genet 2006;120(2):270−84.

[166] Scoles HA, Urraca N, Chadwick SW, Reiter LT, Lasalle JM. Increased copy number for methylated maternal 15q duplications leads to changes in gene and protein expression in human cortical samples. Mol Autism 2011;2(1):19.

[167] Girirajan S, Eichler EE. Phenotypic variability and genetic susceptibility to genomic disorders. Hum Mol Genet 2010;19(R2):R176−87.

[168] Germain ND, Chen PF, Plocik AM, Glatt-Deeley H, Brown J, Fink JJ, et al. Gene expression analysis of human induced pluripotent stem cell-derived neurons carrying copy number variants of chromosome 15q11-q13.1. Mol Autism 2014;5:44.

[169] Andrews T, Honti F, Pfundt R, de Leeuw N, Hehir-Kwa J, Vulto-van Silfhout A, et al. The clustering of functionally

REFERENCES

related genes contributes to CNV-mediated disease. Genome Res 2015;25(6):802–13.
[170] Zhang F, Lupski JR. Non-coding genetic variants in human disease. Hum Mol Genet 2015;24(R1):R102–10.
[171] Chiang C, Scott AJ, Davis JR, Tsang EK, Li X, Kim Y, et al. The impact of structural variation on human gene expression. Nat Genet 2017;49(5):692–9.
[172] McCarroll SA, Huett A, Kuballa P, Chilewski SD, Landry A, Goyette P, et al. Deletion polymorphism upstream of IRGM associated with altered IRGM expression and Crohn's disease. Nat Genet 2008;40(9):1107–12.
[173] Verdin H, Fernandez-Minan A, Benito-Sanz S, Janssens S, Callewaert B, De Waele K, et al. Profiling of conserved non-coding elements upstream of SHOX and functional characterisation of the SHOX cis-regulatory landscape. Sci Rep 2015;5:17667.
[174] Aldred PM, Hollox EJ, Armour JA. Copy number polymorphism and expression level variation of the human alpha-defensin genes DEFA1 and DEFA3. Hum Mol Genet 2005;14(14):2045–52.
[175] Stranger BE, Forrest MS, Dunning M, Ingle CE, Beazley C, Thorne N, et al. Relative impact of nucleotide and copy number variation on gene expression phenotypes. Science 2007;315(5813):848–53.
[176] Botuyan MV, Lee J, Ward IM, Kim JE, Thompson JR, Chen J, et al. Structural basis for the methylation state-specific recognition of histone H4-K20 by 53BP1 and Crb2 in DNA repair. Cell 2006;127(7):1361–73.
[177] McCarroll SA, Hadnott TN, Perry GH, Sabeti PC, Zody MC, Barrett JC, et al. Common deletion polymorphisms in the human genome. Nat Genet 2006;38(1):86–92.
[178] Lander ES, Linton LM, Birren B, Nusbaum C, Zody MC, Baldwin J, et al. Initial sequencing and analysis of the human genome. Nature 2001;409(6822):860–921.
[179] Lu JY, Shao W, Chang L, Yin Y, Li T, Zhang H, et al. Genomic repeats categorize genes with distinct functions for orchestrated regulation. Cell Rep 2020;30(10):3296–311 e5.
[180] Francastel C, Magdinier F. DNA methylation in satellite repeats disorders. Essays Biochem 2019;63(6):757–71.
[181] Gymrek M, Willems T, Reich D, Erlich Y. Interpreting short tandem repeat variations in humans using mutational constraint. Nat Genet 2017;49(10):1495–501.
[182] Gymrek M, Willems T, Guilmatre A, Zeng H, Markus B, Georgiev S, et al. Abundant contribution of short tandem repeats to gene expression variation in humans. Nat Genet 2016;48(1):22–9.
[183] Schmidt D, Schwalie PC, Wilson MD, Ballester B, Goncalves A, Kutter C, et al. Waves of retrotransposon expansion remodel genome organization and CTCF binding in multiple mammalian lineages. Cell 2012;148(1–2):335–48.
[184] Lunyak VV, Prefontaine GG, Nunez E, Cramer T, Ju BG, Ohgi KA, et al. Developmentally regulated activation of a SINE B2 repeat as a domain boundary in organogenesis. Science 2007;317(5835):248–51.
[185] Bourque G, Leong B, Vega VB, Chen X, Lee YL, Srinivasan KG, et al. Evolution of the mammalian transcription factor binding repertoire via transposable elements. Genome Res 2008;18(11):1752–62.
[186] Rowe HM, Kapopoulou A, Corsinotti A, Fasching L, Macfarlan TS, Tarabay Y, et al. TRIM28 repression of retrotransposon-based enhancers is necessary to preserve transcriptional dynamics in embryonic stem cells. Genome Res 2013;23(3):452–61.
[187] Wang J, Vicente-Garcia C, Seruggia D, Molto E, Fernandez-Minan A, Neto A, et al. MIR retrotransposon sequences provide insulators to the human genome. Proc Natl Acad Sci USA 2015;112(32):E4428–37.
[188] Chuong EB, Elde NC, Feschotte C. Regulatory evolution of innate immunity through co-option of endogenous retroviruses. Science 2016;351(6277):1083–7.
[189] Tunbak H, Enriquez-Gasca R, Tie CHC, Gould PA, Mlcochova P, Gupta RK, et al. The HUSH complex is a gatekeeper of type I interferon through epigenetic regulation of LINE-1s. Nat Commun 2020;11(1):5387.
[190] Enriquez-Gasca R, Gould PA, Rowe HM. Host gene regulation by transposable elements: the new, the old and the ugly. Viruses 2020;12(10):1089.
[191] Consortium F, Suzuki H, Forrest AR, van Nimwegen E, Daub CO, Balwierz PJ, et al. The transcriptional network that controls growth arrest and differentiation in a human myeloid leukemia cell line. Nat Genet 2009;41(5):553–62.
[192] Nageshwaran S, Festenstein R. Epigenetics and triplet-repeat neurological diseases. Front Neurol 2015;6:262.
[193] Oberle I, Rousseau F, Heitz D, Kretz C, Devys D, Hanauer A, et al. Instability of a 550-base pair DNA segment and abnormal methylation in fragile X syndrome. Science 1991;252(5010):1097–102.
[194] Coffee B, Zhang F, Ceman S, Warren ST, Reines D. Histone modifications depict an aberrantly heterochromatinized FMR1 gene in fragile x syndrome. Am J Hum Genet 2002;71(4):923–32.
[195] Coffee B, Zhang F, Warren ST, Reines D. Acetylated histones are associated with FMR1 in normal but not fragile X-syndrome cells. Nat Genet 1999;22(1):98–101.
[196] Chutake YK, Lam CC, Costello WN, Anderson MP, Bidichandani SI. Reversal of epigenetic promoter silencing in Friedreich ataxia by a class I histone deacetylase inhibitor. Nucleic Acids Res 2016;.
[197] Sakamoto N, Ohshima K, Montermini L, Pandolfo M, Wells RD. Sticky DNA, a self-associated complex formed at long GAA*TTC repeats in intron 1 of the frataxin gene, inhibits transcription. J Biol Chem 2001;276(29):27171–7.
[198] Al-Mahdawi S, Pinto RM, Ismail O, Varshney D, Lymperi S, Sandi C, et al. The Friedreich ataxia GAA repeat expansion mutation induces comparable epigenetic changes in human and transgenic mouse brain and heart tissues. Hum Mol Genet 2008;17(5):735–46.
[199] Greene E, Mahishi L, Entezam A, Kumari D, Usdin K. Repeat-induced epigenetic changes in intron 1 of the frataxin gene and its consequences in Friedreich ataxia. Nucleic Acids Res 2007;35(10):3383–90.
[200] Saveliev A, Everett C, Sharpe T, Webster Z, Festenstein R. DNA triplet repeats mediate heterochromatin-protein-1-sensitive variegated gene silencing. Nature 2003;422(6934):909–13.
[201] Filippova GN, Thienes CP, Penn BH, Cho DH, Hu YJ, Moore JM, et al. CTCF-binding sites flank CTG/CAG repeats and form a methylation-sensitive insulator at the DM1 locus. Nat Genet 2001;28(4):335–43.
[202] Cho DH, Thienes CP, Mahoney SE, Analau E, Filippova GN, Tapscott SJ. Antisense transcription and heterochromatin at the DM1 CTG repeats are constrained by CTCF. Mol Cell 2005;20(3):483–9.
[203] Lopez Castel A, Nakamori M, Tome S, Chitayat D, Gourdon G, Thornton CA, et al. Expanded CTG repeat demarcates a boundary for abnormal CpG methylation in myotonic dystrophy patient tissues. Hum Mol Genet 2011;20(1):1–15.
[204] Sun JH, Zhou L, Emerson DJ, Phyo SA, Titus KR, Gong W, et al. Disease-associated short tandem repeats co-localize with chromatin domain boundaries. Cell 2018;175(1):224–38 e15.

[205] Foster JW, Dominguez-Steglich MA, Guioli S, Kwok C, Weller PA, Stevanovic M, et al. Campomelic dysplasia and autosomal sex reversal caused by mutations in an SRY-related gene. Nature 1994;372(6506):525–30.

[206] Pfeifer D, Kist R, Dewar K, Devon K, Lander ES, Birren B, et al. Campomelic dysplasia translocation breakpoints are scattered over 1 Mb proximal to SOX9: evidence for an extended control region. Am J Hum Genet 1999;65(1):111–24.

[207] Pop R, Conz C, Lindenberg KS, Blesson S, Schmalenberger B, Briault S, et al. Screening of the 1 Mb SOX9 5' control region by array CGH identifies a large deletion in a case of campomelic dysplasia with XY sex reversal. J Med Genet 2004;41(4):e47.

[208] Wagner T, Wirth J, Meyer J, Zabel B, Held M, Zimmer J, et al. Autosomal sex reversal and campomelic dysplasia are caused by mutations in and around the SRY-related gene SOX9. Cell 1994;79(6):1111–20.

[209] Fonseca AC, Bonaldi A, Bertola DR, Kim CA, Otto PA, Vianna-Morgante AM. The clinical impact of chromosomal rearrangements with breakpoints upstream of the SOX9 gene: two novel de novo balanced translocations associated with acampomelic campomelic dysplasia. BMC Med Genet 2013;14:50.

[210] Scherer SW, Poorkaj P, Allen T, Kim J, Geshuri D, Nunes M, et al. Fine mapping of the autosomal dominant split hand/split foot locus on chromosome 7, band q21.3-q22.1. Am J Hum Genet 1994;55(1):12–20.

[211] Ignatius J, Knuutila S, Scherer SW, Trask B, Kere J. Split hand/split foot malformation, deafness, and mental retardation with a complex cytogenetic rearrangement involving 7q21.3. J Med Genet 1996;33(6):507–10.

[212] van Silfhout AT, van den Akker PC, Dijkhuizen T, Verheij JB, Olderode-Berends MJ, Kok K, et al. Split hand/foot malformation due to chromosome 7q aberrations(SHFM1): additional support for functional haploinsufficiency as the causative mechanism. Eur J Hum Genet 2009;17(11):1432–8.

[213] Scherer SW, Cheung J, MacDonald JR, Osborne LR, Nakabayashi K, Herbrick JA, et al. Human chromosome 7: DNA sequence and biology. Science 2003;300(5620):767–72.

[214] Birnbaum RY, Everman DB, Murphy KK, Gurrieri F, Schwartz CE, Ahituv N. Functional characterization of tissue-specific enhancers in the DLX5/6 locus. Hum Mol Genet 2012;21(22):4930–8.

[215] Kouwenhoven EN, van Heeringen SJ, Tena JJ, Oti M, Dutilh BE, Alonso ME, et al. Genome-wide profiling of p63 DNA-binding sites identifies an element that regulates gene expression during limb development in the 7q21 SHFM1 locus. PLoS Genet 2010;6(8):e1001065.

[216] Kragesteen BK, Spielmann M, Paliou C, Heinrich V, Schopflin R, Esposito A, et al. Dynamic 3D chromatin architecture contributes to enhancer specificity and limb morphogenesis. Nat Genet 2018;50(10):1463–73.

[217] Schluth-Bolard C, Diguet F, Chatron N, Rollat-Farnier PA, Bardel C, Afenjar A, et al. Whole genome paired-end sequencing elucidates functional and phenotypic consequences of balanced chromosomal rearrangement in patients with developmental disorders. J Med Genet 2019;56(8):526–35.

[218] Di Gregorio E, Riberi E, Belligni EF, Biamino E, Spielmann M, Ala U, et al. Copy number variants analysis in a cohort of isolated and syndromic developmental delay/intellectual disability reveals novel genomic disorders, position effects and candidate disease genes. Clin Genet 2017;92(4):415–22.

[219] Giorgio E, Robyr D, Spielmann M, Ferrero E, Di Gregorio E, Imperiale D, et al. A large genomic deletion leads to enhancer adoption by the lamin B1 gene: a second path to autosomal dominant adult-onset demyelinating leukodystrophy (ADLD). Hum Mol Genet 2015;24(11):3143–54.

[220] Hnisz D, Weintraub AS, Day DS, Valton AL, Bak RO, Li CH, et al. Activation of proto-oncogenes by disruption of chromosome neighborhoods. Science 2016;351(6280):1454–8.

[221] Groschel S, Sanders MA, Hoogenboezem R, de Wit E, Bouwman BAM, Erpelinck C, et al. A single oncogenic enhancer rearrangement causes concomitant EVI1 and GATA2 deregulation in leukemia. Cell 2014;157(2):369–81.

[222] Klopocki E, Lohan S, Brancati F, Koll R, Brehm A, Seemann P, et al. Copy-number variations involving the IHH locus are associated with syndactyly and craniosynostosis. Am J Hum Genet 2011;88(1):70–5.

[223] Lettice LA, Daniels S, Sweeney E, Venkataraman S, Devenney PS, Gautier P, et al. Enhancer-adoption as a mechanism of human developmental disease. Hum Mutat 2011;32(12):1492–9.

[224] Inoue K, Osaka H, Thurston VC, Clarke JT, Yoneyama A, Rosenbarker L, et al. Genomic rearrangements resulting in PLP1 deletion occur by nonhomologous end joining and cause different dysmyelinating phenotypes in males and females. Am J Hum Genet 2002;71(4):838–53.

[225] McDermott DH, Gao JL, Liu Q, Siwicki M, Martens C, Jacobs P, et al. Chromothriptic cure of WHIM syndrome. Cell 2015;160(4):686–99.

[226] Mefford HC, Trask BJ. The complex structure and dynamic evolution of human subtelomeres. Nat Rev Genet 2002;3(2):91–102.

[227] Mason JM, Biessmann H. The unusual telomeres of Drosophila. Trends Genet 1995;11(2):58–62.

[228] Gehring WJ, Klemenz R, Weber U, Kloter U. Functional analysis of the white gene of Drosophila by P-factor-mediated transformation. EMBO J 1984;3(9):2077–85.

[229] Craven RJ, Petes TD. Involvement of the checkpoint protein Mec1p in silencing of gene expression at telomeres in Saccharomyces cerevisiae. Mol Cell Biol 2000;20(7):2378–84.

[230] Pryde FE, Louis EJ. Limitations of silencing at native yeast telomeres. EMBO J 1999;18(9):2538–50.

[231] Mondoux MA, Zakian VA. Telomere position effect: silencing near the end. In: De Lange T, Lundblad V, Blackburn EH, editors. Telomeres. 2nd ed. Cold Spring Harbor, NY: Cold Spring Harbor Laboratory Press; 2006. p. 261–316.

[232] Aparicio OM, Billington BL, Gottschling DE. Modifiers of position effect are shared between telomeric and silent mating-type loci in S. cerevisiae. Cell 1991;66(6):1279–87.

[233] Boulton SJ, Jackson SP. Components of the Ku-dependent non-homologous end-joining pathway are involved in telomeric length maintenance and telomeric silencing. EMBO J 1998;17(6):1819–28.

[234] Laroche T, Martin SG, Gotta M, Gorham HC, Pryde FE, Louis EJ, et al. Mutation of yeast Ku genes disrupts the subnuclear organization of telomeres. Curr Biol 1998;8(11):653–6.

[235] Kyrion G, Boakye KA, Lustig AJ. C-terminal truncation of RAP1 results in the deregulation of telomere size, stability, and function in Saccharomyces cerevisiae. Mol Cell Biol 1992;12(11):5159–73.

[236] Cooper JP, Nimmo ER, Allshire RC, Cech TR. Regulation of telomere length and function by a Myb-domain protein in fission yeast. Nature 1997;385(6618):744–7.

[237] Kanoh J, Ishikawa F. spRap1 and spRif1, recruited to telomeres by Taz1, are essential for telomere function in fission yeast. Curr Biol 2001;11(20):1624–30.

[238] Wallrath LL, Elgin SC. Position effect variegation in Drosophila is associated with an altered chromatin structure. Genes Dev 1995;9(10):1263–77.

References

[239] Ekwall K, Javerzat JP, Lorentz A, Schmidt H, Cranston G, Allshire R. The chromodomain protein Swi6: a key component at fission yeast centromeres. Science 1995;269(5229):1429–31.

[240] Laberthonniere C, Magdinier F, Robin JD. Bring it to an end: does telomeres size matter? Cells 2019;8(1):30.

[241] Arnoult N, Van Beneden A, Decottignies A. Telomere length regulates TERRA levels through increased trimethylation of telomeric H3K9 and HP1alpha. Nat Struct Mol Biol 2012;19(9):948–56.

[242] Azzalin CM, Reichenbach P, Khoriauli L, Giulotto E, Lingner J. Telomeric repeat containing RNA and RNA surveillance factors at mammalian chromosome ends. Science 2007;318(5851):798–801.

[243] Smirnova A, Gamba R, Khoriauli L, Vitelli V, Nergadze SG, Giulotto E. TERRA Expression levels do not correlate with telomere length and radiation sensitivity in human cancer cell lines. Front Oncol 2013;3:115.

[244] Schoeftner S, Blasco MA. Developmentally regulated transcription of mammalian telomeres by DNA-dependent RNA polymerase II. Nat Cell Biol 2008;10(2):228–36.

[245] Ofir R, Wong AC, McDermid HE, Skorecki KL, Selig S. Position effect of human telomeric repeats on replication timing. Proc Natl Acad Sci USA 1999;96(20):11434–9.

[246] Baur JA, Zou Y, Shay JW, Wright WE. Telomere position effect in human cells. Science 2001;292(5524):2075–7.

[247] Koering CE, Pollice A, Zibella MP, Bauwens S, Puisieux A, Brunori M, et al. Human telomeric position effect is determined by chromosomal context and telomeric chromatin integrity. EMBO Rep 2002;3(11):1055–61.

[248] Uhlirova R, Horakova AH, Galiova G, Legartova S, Matula P, Fojtova M, et al. SUV39h- and A-type lamin-dependent telomere nuclear rearrangement. J Cell Biochem 2010;109(5):915–26.

[249] Tennen RI, Chua KF. Chromatin regulation and genome maintenance by mammalian SIRT6. Trends Biochem Sci 2011;36(1):39–46.

[250] Tennen RI, Bua DJ, Wright WE, Chua KF. SIRT6 is required for maintenance of telomere position effect in human cells. Nat Commun 2011;2:433.

[251] Mukherjee AK, Sharma S, Sengupta S, Saha D, Kumar P, Hussain T, et al. Telomere length-dependent transcription and epigenetic modifications in promoters remote from telomere ends. PLoS Genet 2018;14(11):e1007782.

[252] Robin JD, Ludlow AT, Batten K, Magdinier F, Stadler G, Wagner KR, et al. Telomere position effect: regulation of gene expression with progressive telomere shortening over long distances. Genes Dev 2014;28(22):2464–76.

[253] Lou Z, Wei J, Riethman H, Baur JA, Voglauer R, Shay JW, et al. Telomere length regulates ISG15 expression in human cells. Aging 2009;1(7):608–21.

[254] Holohan B, Wright WE, Shay JW. Cell biology of disease: telomeropathies: an emerging spectrum disorder. J Cell Biol 2014;205(3):289–99.

[255] Armando RG, Mengual Gomez DL, Maggio J, Sanmartin MC, Gomez DE. Telomeropathies: etiology, diagnosis, treatment and follow-up. Ethical and legal considerations. Clin Genet 2019;96(1):3–16.

[256] Shao L, Shaw CA, Lu XY, Sahoo T, Bacino CA, Lalani SR, et al. Identification of chromosome abnormalities in subtelomeric regions by microarray analysis: a study of 5,380 cases. Am J Med Genet A 2008;146A(17):2242–51.

[257] Ballif BC, Sulpizio SG, Lloyd RM, Minier SL, Theisen A, Bejjani BA, et al. The clinical utility of enhanced subtelomeric coverage in array CGH. Am J Med Genet A 2007;143A(16):1850–7.

[258] Ledbetter DH, Martin CL. Cryptic telomere imbalance: a 15-year update. Am J Med Genet C Semin Med Genet 2007;145C(4):327–34.

[259] Barber JC. Terminal 3p deletions: phenotypic variability, chromosomal non-penetrance, or gene modification? Am J Med Genet A 2008;146A(14):1899–901.

[260] Flint J, Wilkie AO, Buckle VJ, Winter RM, Holland AJ, McDermid HE. The detection of subtelomeric chromosomal rearrangements in idiopathic mental retardation. Nat Genet 1995;9(2):132–40.

[261] Giraudeau F, Aubert D, Young I, Horsley S, Knight S, Kearney L, et al. Molecular-cytogenetic detection of a deletion of 1p36.3. J Med Genet 1997;34(4):314–17.

[262] Horsley SW, Knight SJ, Nixon J, Huson S, Fitchett M, Boone RA, et al. Del(18p) shown to be a cryptic translocation using a multiprobe FISH assay for subtelomeric chromosome rearrangements. J Med Genet 1998;35(9):722–6.

[263] Kleefstra T, Brunner HG, Amiel J, Oudakker AR, Nillesen WM, Magee A, et al. Loss-of-function mutations in euchromatin histone methyl transferase 1 (EHMT1) cause the 9q34 subtelomeric deletion syndrome. Am J Hum Genet 2006;79(2):370–7.

[264] Lamb J, Harris PC, Wilkie AO, Wood WG, Dauwerse JG, Higgs DR. De novo truncation of chromosome 16p and healing with (TTAGGG)n in the alpha-thalassemia/mental retardation syndrome (ATR-16). Am J Hum Genet 1993;52(4):668–76.

[265] Walter S, Sandig K, Hinkel GK, Mitulla B, Ounap K, Sims G, et al. Subtelomere FISH in 50 children with mental retardation and minor anomalies, identified by a checklist, detects 10 rearrangements including a de novo balanced translocation of chromosomes 17p13.3 and 20q13.33. Am J Med Genet A 2004;128(4):364–73.

[266] Bonaglia MC, Giorda R, Mani E, Aceti G, Anderlid BM, Baroncini A, et al. Identification of a recurrent breakpoint within the SHANK3 gene in the 22q13.3 deletion syndrome. J Med Genet 2006;43(10):822–8.

[267] Wilson HL, Wong AC, Shaw SR, Tse WY, Stapleton GA, Phelan MC, et al. Molecular characterisation of the 22q13 deletion syndrome supports the role of haploinsufficiency of SHANK3/PROSAP2 in the major neurological symptoms. J Med Genet 2003;40(8):575–84.

[268] Gajecka M, Mackay KL, Shaffer LG. Monosomy 1p36 deletion syndrome. Am J Med Genet C Semin Med Genet 2007;145C(4):346–56.

[269] Kosztolanyi G. Does "ring syndrome" exist? An analysis of 207 case reports on patients with a ring autosome. Hum Genet 1987;75(2):174–9.

[270] Cote GB, Katsantoni A, Deligeorgis D. The cytogenetic and clinical implications of a ring chromosome 2. Ann Genet 1981;24(4):231–5.

[271] Pezzolo A, Gimelli G, Cohen A, Lavaggetto A, Romano C, Fogu G, et al. Presence of telomeric and subtelomeric sequences at the fusion points of ring chromosomes indicates that the ring syndrome is caused by ring instability. Hum Genet 1993;92(1):23–7.

[272] Sigurdardottir S, Goodman BK, Rutberg J, Thomas GH, Jabs EW, Geraghty MT. Clinical, cytogenetic, and fluorescence in situ hybridization findings in two cases of "complete ring" syndrome. Am J Med Genet 1999;87(5):384–90.

[273] Vermeesch JR, Baten E, Fryns JP, Devriendt K. Ring syndrome caused by ring chromosome 7 without loss of subtelomeric sequences. Clin Genet 2002;62(5):415–17.

[274] Surace C, Berardinelli F, Masotti A, Roberti MC, Da Sacco L, D'Elia G, et al. Telomere shortening and telomere position

effect in mild ring 17 syndrome. Epigenetics Chromatin 2014;7(1):1.
[275] Stadler G, Rahimov F, King OD, Chen JC, Robin JD, Wagner KR, et al. Telomere position effect regulates DUX4 in human facioscapulohumeral muscular dystrophy. Nat Struct Mol Biol 2013;20(6):671–8.
[276] Robin JD, Ludlow AT, Batten K, Gaillard MC, Stadler G, Magdinier F, et al. SORBS2 transcription is activated by telomere position effect-over long distance upon telomere shortening in muscle cells from patients with facioscapulohumeral dystrophy. Genome Res 2015;25(12):1781–90.
[277] Ottaviani A, Schluth-Bolard C, Rival-Gervier S, Boussouar A, Rondier D, Foerster AM, et al. Identification of a perinuclear positioning element in human subtelomeres that requires A-type lamins and CTCF. EMBO J 2009;28(16):2428–36.
[278] Magdinier FUM. Facioscapulohumeral muscular dystrophy: genetics. Encyclopedy of Life Science [Internet]. 2018;.
[279] Schatzl T, Kaiser L, Deigner HP. Facioscapulohumeral muscular dystrophy: genetics, gene activation and downstream signalling with regard to recent therapeutic approaches: an update. Orphanet J Rare Dis 2021;16(1):129.
[280] Arnoult N, Schluth-Bolard C, Letessier A, Drascovic I, Bouarich-Bourimi R, Campisi J, et al. Replication timing of human telomeres is chromosome arm-specific, influenced by subtelomeric structures and connected to nuclear localization. PLoS Genet 2010;6(4):e1000920.
[281] Kim W, Shay JW. Long-range telomere regulation of gene expression: telomere looping and telomere position effect over long distances (TPE-OLD). Differentiation 2018;99:1–9.
[282] Robin JD, Jacome Burbano MS, Peng H, Croce O, Thomas JL, Laberthonniere C, et al. Mitochondrial function in skeletal myofibers is controlled by a TRF2-SIRT3 axis over lifetime. Aging Cell 2020;19(3):e13097.
[283] Kim W, Ludlow AT, Min J, Robin JD, Stadler G, Mender I, et al. Regulation of the human telomerase gene TERT by telomere position effect-over long distances (TPE-OLD): implications for aging and cancer. PLoS Biol 2016;14(12):e2000016.

CHAPTER 7

Polycomb-group proteins and epigenetic control of gene activity

Prasad Pethe

Symbiosis Centre for Stem Cell Research (SCSCR), Symbiosis International University (SIU), Pune, Maharashtra, India

OUTLINE

Introduction	111	PcG-mediated Control of Gene Expression in Embryonic Stem Cells	116
Polycomb Repressive Complexes	112	PcG-mediated Control of Gene Expression in Cancer Cells	117
Mechanism of Gene Regulation by Polycomb Group Proteins	112	Conclusion and Future Perspective	118
PcG-Mediated Control of Gene Expression in Stem Cells to Maintain Homeostasis	114	Acknowledgment	118
		References	118

INTRODUCTION

Cells respond to changes in their environment by expressing genes, which get translated to proteins that help the cells to deal with the changing environment. The mechanism of sensing changes by a cell is a complex phenomenon and regulation of gene activity is an even more complex mechanism. Gene regulatory mechanisms are intricate in mammalian cells with several levels of control such as architectural (heterochromatin), epigenetic, genetic, (enhancers and promoters) and posttranscriptional control by ncRNAs [1]. Epigenetic modifications encompass a wide variety of modifications such as DNA modifications and posttranslational modification of histone proteins [2]. There are dedicated proteins that add the epigenetic modifications ("writers") while others remove these modifications ("erasers"). These proteins have specific domains through which they recognize stretches of DNA ("readers"), which need to be present in protein complexes of both writers and erasers (see Table 7.1). The epigenetic remodelers are in turn regulated by several transcription factors, signaling pathways, relay proteins, and metabolites. This entire machinery enables the cells to fine gene expression to suit their immediate microenvironment and conserve energy.

Polycomb group (PcG) proteins are among the proteins that catalyze posttranslational histone modifications that lead to gene repression. In *Drosophila*, there are specialized appendages called sex combs located on the legs of males, some mutant fruit flies were found to express multiple sex combs and these flies had a mutated Polycomb (Pc) gene [3]. Further work on these sex combs led to the discovery of several genes that led to abnormal regulation of sex combs, which were eventually clubbed together and are called PcG genes, and proteins they encode are called PcG proteins [4–6]. PcG proteins were shown to be

TABLE 7.1 Examples of histone modifying proteins that form part of PRC complexes classified into histone readers, writers and erasers

Readers	Writers	Erasers
CBX2/7	EZH1/2	USP16/22
SWI/SNF	MLL1/2/3/4	BAP
JMJD1/2/3	PRDM9	MYSM1
EED	RIN1A/B	JARID2
BRD2	HAT	LSD1/2
PCL	DOT1L	HDAC 1/2/3

responsible for the expression of homeotic genes such as *Hox* genes. However, later studies, which compared the protein sequences of PcG proteins, found similarities with other proteins that bind to chromatin and carry out histone modifications [7,8]. Reports from *Drosophila* confirmed that PcG proteins bind to specific DNA sequences termed Polycomb response elements (PREs) and bring about histone modifications [9,10].

After their initial discovery in *Drosophila*, PcG protein homologous sequences were reported in mice [11,12]. PcG proteins like BMI1, EZH2, and EED were reported to be oncogenic and shown to be critical for the proliferation and self-renewal of cancer cells [13,14]. Recent studies, which we will discuss in this chapter, have shown that PcG proteins are critical during embryo development, gastrulation, lineage specification, and homeostasis. Research on gene regulation by PcG proteins has greatly enhanced our understanding of gene expression regulation in development and disease, and we are uncovering their intricate role in the mammalian system. The gene regulation by PcG proteins is a vast topic and not every detail can be comprehensively covered in a book chapter. In this chapter, we shall discuss the PcG proteins, the epigenetic modifications they catalyze which leads to controlled gene activity in mammals.

POLYCOMB REPRESSIVE COMPLEXES

Proteins are classified as PcG proteins based on their ability to catalyze histone modifications or assist in proteins that catalyze histone modifications, which ultimately leads to gene repression. PcG proteins assemble into large protein complexes—well documented amongst them are the Polycomb repressive complex 1 and Polycomb repressive complex 2. The Polycomb repressive complex 1 contains RING1B which possesses the catalytic activity to transfer monoubiquitin molecule to lysine 119 of histone H2A; while the Polycomb repressive complex 2 contains EZH2 which catalyzes di- or trimethylation of histone H3 at lysine 27 [15]. Table 7.2 lists the major proteins that constitute the PRC1 and PRC2.

The PcG proteins that catalyze histone modifications are RING1B/RNF2 and EZH2/1 and they form the catalytic core of the PRC1 and PRC2, respectively. Several other proteins do not perform posttranslational modification of chromatin histones but are critical for the activity of the PRCs and are also part of the PRCs. The Polycomb repressive complex 1 includes the core catalytic protein RING1B along with proteins such as PCGF1/2/3/4/5/6, CBX2/4/6/7/8, and RYBP [16,17]. There are six variants of the PRC1 complex depending on the presence of various PCGF proteins into PRC1.1-PRC1.6. The canonical PRC1 (PRC1.4) complex lacks RYBP proteins; whereas, the noncanonical forms possess the RYBP protein [16]. Mice lacking *Ring1b* fail to gastrulate and die around day 7 in the embryo [18], while the mice lacking other PRC1 accessory proteins survive to birth but show several developmental defects. Loss of the catalytic subunit of PRC2, *Ezh2*, leads to embryonic lethality [19]. Noncatalytic components of PRC2 include EED, SUZ12, JARID2, EPOP, and PCL [20,21]. Table 7.2 gives more information about the proteins that constitute the Polycomb repressive complexes (PRCs). The list is certainly not final, as new proteins are being reported, but the proteins listed in Table 7.2 are well studied. The accessory proteins in the PRCs play a critical role in localizing to a specific site in the genome in response to upstream signals.

MECHANISM OF GENE REGULATION BY POLYCOMB GROUP PROTEINS

Gene expression is affected by chromatin accessibility, transcription factors, and several histone modifications.

TABLE 7.2 List of major Polycomb group proteins (PcG) and their functions

PcG Protein	Polycomb Repressive Complex	Function
RNF2/RING1B	Polycomb repressive complex 1	Core catalytic component of PRC1, transfers single ubiquitin molecule to lysine at 119 position of histone H2A. Mice lacking *Ring1b* die at the embryonic stage
RING1A	Polycomb repressive complex 1	Can perform the function of RING1B, but knockout of *Ring1b* is not embryo lethal
PCGF1/2/3/4	Polycomb repressive complex 1	PCGF1—Part of noncanonical PRC1 variants, which performs de novo histone H2A monoubiquitination at lysine 119. Suppresses CDKN1A expression by binding to its promoters PCGF2—Part of noncanonical PRC1, which performs de novo histone H2A monoubiquitination at lysine 119. Mice lacking *Pcgf2* show growth and differentiation defects. Regulates expression genes required for hematopoietic stem cells PCGF3—Part of noncanonical PRC1, which performs de novo histone H2A monoubiquitination at lysine 119 PCGF4—Part of canonical PRC1 complex which requires H3K27me3 mark at promoter sites. It independently controls expression of cell cycle inhibitors and overexpression can leads to several types of cancers. *Bmi1* knockout mice show axial-skeletal malformations PCGF5—Part of noncanonical PRC1 which performs de novo histone H2A monoubiquitination at lysine 119. Can replace PCGF3 in the noncanonical PRC1. PCGF6—part of noncanonical PRC1, which performs de novo histone H2A monoubiquitination at lysine 119
RYBP	Polycomb repressive complex 1	Part of noncanonical PRC1, which performs de novo histone H2A monoubiquitination at lysine 119. It is involved in X chromosome silencing and plays an important role in cancer cell proliferation
CBX1/2/6/7/8	Polycomb repressive complex 1	CBX1—Part of PRC1, which performs histone H2A monoubiquitination at lysine 119 to regulate HOX genes. It can recognize and bind to H3K9me3 and H3K27me3 CBX2—Loss of CBX2 leads to gonadal dysgenesis or sex reversal. It can recognize and bind to H3K9me3 and H3K27me3. It is required for functioning of PRC1 CBX3—Part of PRC1, which performs histone H2A monoubiquitination at lysine 119. Interacts with nonhistone proteins such as Lamin B receptor. It can recognize H3K9me3 CBX4—Part of PRC1, which performs histone H2A monoubiquitination at lysine 119. It can recognize H3K9me3. Regulates expression of TP53 CBX6—Part of PRC1, which performs histone H2A monoubiquitination at lysine 119 CBX—Part of PRC1, which performs histone H2A monoubiquitination at lysine 119. It can recognize H3K9me CBX—Part of PRC1, which performs histone H2A monoubiquitination at lysine 119
JARID1/2	Polycomb repressive complex 2	JARID1—Part of PRC2, and demethylates H3K4me3/2/1. It regulates expression of several genes. Mice lacking Jarid1 show severe neurological developmental disorders JARID2—Part of PRC2, and demethylates H3K4me3/2/1. *Jarid2* knockout mice show several developmental disorders in multiple organs
EZH1/2	Polycomb repressive complex 2	EZH1—part of PRC1 that catalyzes histone H3 methylation. Less efficient than EZH2. More abundant in adult cell and misexpression leads to cancer EZH2—Catalytic component of PRC2, more abundant in embryonic cells. Mice lacking *Ezh2* die in embryonic stage. Transfers methyl group from S-adenosylmethionine (SAM) to lysine residue on histone H3
EED	Polycomb repressive complex 2	Core PRC2 catalytic protein along with EZH2 and SUZ12 that transfers methyl groups to lysine on histone H3 and H4. Helps EZH2 recruit DNMTs at H3K27me3 sites. Mice lacking *Eed* have severely compromised intestinal crypts
SUZ12	Polycomb repressive complex 2	Core PRC2 catalytic protein along with EZH2 and EED that transfers methyl groups to lysine on histone H3 and H4. Mice lacking *Suz12* exhibit embryonic cell death
EPOP	Polycomb repressive complex 2	Allows interaction between PRC2 and other regulatory complexes
PCL	Polycomb repressive complex 2	Binds to activating histone mark H3K36me3 and recruits PRC2. Embryonic stem cells lacking *Pcl* differentiate spontaneously
YY1	Polycomb repressive complex 2	Interacts with EED and EZH2, is involved in hematopoiesis and T cell development, binds to DNA and helps recruit PRCs to HOX genes

The histone modifications such as acetylation and phosphorylation, counteract the positive charge on histone proteins, which allows transcriptional machinery including the "readers" to access and regulate DNA to begin transcription. However, neutral charge molecules such as methylation, sumoylation, and ubiquitinylation provide anchorage points to other proteins that possess specific domains to recognize these histone modifications. These neutral modifications can both positively and negatively regulate gene expression, so it suggests that the genomic context of these molecules determine whether it will lead to activation or repression [22].

Histone modifications are catalyzed by a large number of specialized proteins that can be broadly clubbed as readers, writers, and erasers, and there are proteins clubbed under these same categories which can bind to DNA and bring about DNA modifications. Since PcG proteins introduce posttranslational modifications to histone, here readers, writers, and erasers refer to proteins that recognize and change histone modifications. Table 7.1 shows the list of proteins that comprise epigenetic readers, writers, and erasers. RNA polymerase II is responsible for the transcription of RNA from DNA and PcG proteins regulate events such as Pol II binding to promoters, Pol II elongation, and termination (note the list is not exhaustive). Embryonic stem cells have can differentiate into all three germ lineages, and hence have been extensively used to understand the histone modifications in gene regulation. When embryonic stem cells (ESCs) were compared to NSCs (neural stem cells) and mouse embryonic fibroblasts, it was seen that in ESCs that regions with high CpG content were enriched with H3K4me3 modification; in NPCs, low CpG regions showed no H3K4me3, while ESCs showed more H3K4me3. H3K9me3 and H4K20me3 modifications are associated with centromeres, transposons, and tandem repeats [23].

Histone modifications that occur in the tail region can directly affect the interactions between nucleosomes. Some of the frequent histone marks associated with gene activation are H3K122ac, H3K4me3, H2A.Z and H3K4me1, and H3K27ac and are seen on gene enhancers [24]. Besides these, several other modifications have been reported such as H3K64me3, H3K79me1, H3K79me2, H2B123ub, H4K16ac, H3T118ph, H4T80ph, H3T41ph, H3T45ph, H2AQ104/5me [25]. The histone modifications at gene promoters such as H3K27me3/2/1, H3K9me3, and H2AK119ub1 correlate with gene repression, recent work shows that the H3K27me3 mark even controls imprinted genes at the blastocyst stage [26]. Thus, PcG proteins control gene expression primarily via histone posttranslational modifications at specific sites.

The PcG proteins do not solely regulate gene expression via posttranslational histone modifications. PcG proteins along with ncRNA *XIST* cause X chromosome compaction of one X chromosome in cells containing two X chromosomes [27,28]. Ring1b has been reported to affect gene expression by compacting a specific region and not merely by catalyzing H2AK119ub1 modification in mice ES cells as well as in cancer cells [29–31]. Fig. 7.1 shows a simple representation of one of the mechanisms by which PcG protein regulate gene expression. PRC complexes along with specific transcription factors catalyze specific histone modifications at gene promoters that leads to gene repression. H3K27me3 modification at specific sites can act as silencers, and they silence the gene via dynamic physical interactions such as DNA looping [32]. EZH2, which deposits H3K27me mark, can interact and recruit DNMTs that catalyze DNA methylation and, thus, a DNA stretch is silenced by two epigenetics mechanisms [33,34].

PCG-MEDIATED CONTROL OF GENE EXPRESSION IN STEM CELLS TO MAINTAIN HOMEOSTASIS

In mice pancreas, lineage tracing revealed that Bmi1 is expressed in differentiated acinar cells of the pancreas. Bmi1 allowed the differentiated acinar cells to proliferate in the absence of progenitor cells. Thus, PcG protein might be crucial for maintaining pancreatic homeostasis in mice pancreas. The authors speculate further that Bmi1 overexpression may lead to cancer and terminally differentiated acinar cells might be the cell of origin in pancreatic adenocarcinoma [35]. Inducible conditional *Ring1b* knockout in mice pancreatic progenitor and differentiated cells did not affect the differentiated mice islet cells. However, the removal of *Ring1b* from progenitor cells led to defective islet cells and glucose intolerance. *Ring1b* was not required for repressing nonpancreatic genes in terminally differentiated cells but was required to mark nonpancreatic genes at the progenitor stage [36].

Ying Yang (YY1) protein is one of the few PcG proteins that bind to DNA, unlike many other PcG proteins. Knockout studies using *Yy1* mice showed that loss of Yy1 leads to lower ckit expression. cKit directly regulates the self-renewal and proliferation of hematopoietic stem cells (HSCs). Thus, Yy1 directly regulates the self-renewal and proliferation of HSCs [37]. Recently, it was shown that Yy1 also plays a role in T cells derived from hematopoietic stem cells. Using a conditional knockdown of *Yy1* in mice, it was demonstrated that Yy1 is involved in early T cell development [38].

PcG proteins are involved in X chromosome inactivation in females having two X chromosomes. In mammals, females have two X chromosomes and this double dosage is compensated by the inactivation of

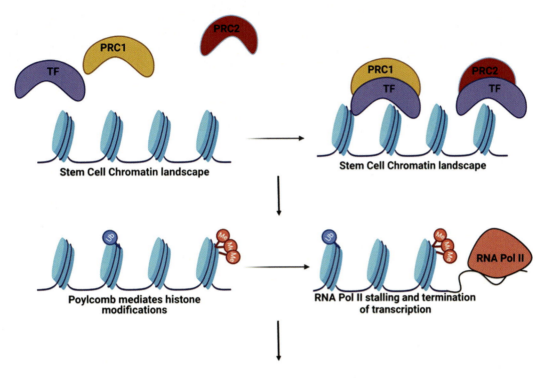

FIGURE 7.1 Polycomb group (PcG) protein control of gene expression: Figure 7.1 shows open chromatin landscape of the pluripotent stem cells in response to a trigger that leads to recruitment of Polycomb repressive complexes (PRC1 and PRC2) along with specific transcription factor (TF) to particular region of DNA. The PRCs catalyze the histone modifications such as H2AK119ub1 and H3K27me3 on promoters of genes which cause the RNA Pol II to stall thereby shutting down gene expression.

one of the X chromosome. This process is initiated by long non-coding RNA *Xist*, which recruits both PRC1 and PRC2. *Pcgf3/5* knockout causes death of female embryos, and *Xist* fails to perform X chromosome inactivation. PRC1 complex containing Pcgf3/5 deposit H2AK119ub1 mark, which recruits PRC2 leading to H3K27me3 deposition. Inhibition of H2AK119ub1 mark deposition dysregulates the *Xist*-mediated X chromosome inactivation [39]. Recently, several mutant *Xist* RNAs were generated and showed the A, B, and C repeats in the *Xist* RNA are responsible for recruiting PRC1 and PRC2 to the X chromosome to be silenced [40]. There are two major types of X chromosome inactivation reported in female mice are— imprinted and random X inactivation. In the imprinted X inactivation process, the paternal X chromosomes are silenced in the extra-embryonic lineages, while the embryonic lineages show random X -inactivation. Using mice oocytes lacking PRC2 protein *Eed* -/- showed that the maternal Eed in oocytes initiates imprinted X inactivation [40]. However, humans do not show imprinted X inactivation and authors indicated human oocytes show a low level of EED.

The heart is an important organ, which is comprised of several different cell types. It is plausible that PcG proteins play their part to generate several cell types from a common pool of progenitor cells via differential gene expression. The anterior heart field in the developing mice heart contributes towards the making of the right ventricle, inter ventricular septum, and outflow tract. Mice with floxed *Ezh2* when crossed with mice with Cre recombinase under *Mef2c* promoters, generated mice that lacked *Ezh2* specifically in particular cardiac cells. This loss of *Ezh2* led to right ventricle enlargement, pulmonary stenosis, and fibrosis. However, when *Ezh2* was removed by placing the Cre recombinase under the *Nkx2.5* promoter which is expressed in differentiated cardiomyocytes, did not show these disruptions. When this dichotomy was further investigated, it was observed that Ezh2 regulates *Six1* and *Eya1*, and these are required for skeletal muscle cells and this disrupts cardiac gene expression. Thus, Ezh2 does not directly regulate cardiac genes via H3K27me3 modifications, but via suppression of noncardiac genes in developing cardiomyocytes [41]. In neonatal mice, it was reported that Ezh2 does not play a part in cardiac regeneration, but suppresses the reactivation of embryonic cardiac genes [42]. Recently, it was shown that inhibition of EZH2 in human fibroblasts helps to directly reprogram them to cardiomyocytes [43].

Thus, it can be inferred that Ezh2 plays an important role in suppressing noncardiac genes in the embryonic stage, but inhibiting cardiac genes in differentiated cardiomyocytes, thus helping in cardiac homeostasis. Hence, targeting Ezh2 could be potentially be used to remove Ezh2 led suppression of cardiac genes and allow the proliferation of resident cardiomyocytes, to treat conditions such as atherosclerosis and fibrosis [44].

Merkel cells are mechanoreceptive cells among various types of cells found in the skin tissue. Mice in which *Ezh2/Ezh1* was conditionally ablated in keratin 14 K14 expressing epidermal stem cells, showed increased proliferation, thus showing that Merkel cells arise from epidermal stem cells. Similar to PcG mechanisms seen in developing cardiomyocytes, Ezh2 controls the expression of transcription factor *Sox2*, which regulates Atoh1; a transcription factor that helps in Merkel cell differentiation from epidermal stem cells [45]. Signaling pathways such as sonic hedgehog (Shh) help differentiation and maturation of Merkel cells. In the epidermis, the Ezh2 containing PRC2 restricts Merkel cell formation, but the area around hair follicles has a high concentration of Shh, this overrides the PRC2 mediated suppression and leads to the formation of Merkel cells [46].

BMI1 regulates the expression of cell cycle inhibitors such as p14ARF and p16INKA in human neural stem cells [47]. *In vitro* study showed that BMI1 regulates proliferation and senescence of human fetal neural stem cells. Neural stem cells differentiate to generate neurons as well as other cells such as astrocytes and oligodendrocytes. Conditional deletion of *Ring1b* in mice neural stem cells led the cells to remain in the neurogenic phase and not proceed to differentiate into astrogenic lineage. In contrast, deletion of *Ring1b* in later stages did not lead to enhanced neuronal differentiation [48]. Deletion of *Ezh2* and *Eed* in neural progenitor cells led to cells remaining in the neurogenic phase and not transition into astrogenic lineage. This PcG regulated mechanism allows the generation of neurons during early development, while later in development when the phase of neurogenesis is slow, PcG proteins initiate differentiation of progenitor cells into astrogenic lineage. When the inhibitor of RING1B PRT4165 was added to human pluripotent stem cells differentiation into neuronal cells, it led to enhanced expression to neuronal markers, thus suggesting that RING1B might favor neuronal formation over other neural cell types [49]. However, the mechanism underlying this observation warrants an investigation.

Cells lining the intestine have rapid turnover and requires frequent cell replacement, a task performed by intestinal stem cells, which reside in the crypts of intestinal villi. Transgenic mice with Cre inducible Bmi1 showed that Bmi1 expressing cells reside in the crypt and it is a marker for intestinal stem cells. Bmi1 allows the intestinal stem cells to proliferate and maintain the intestinal cell lining [50]. The PRC2 protein Eed also regulates intestinal stem cell proliferation. Mice lacking functional *Eed* and *Ezh2* in their intestines were generated using the Cre-Lox system. Compared to control mice, the *Ezh2* knockout mice did not show major change; on the other hand, *Eed* knockout mice had intestinal degeneration as a direct result of the reduced number of proliferating cells. This reduced cell number was due to dysregulation of Wnt pathway genes regulated by Eed [51]. Another intestinal stem cell marker, Lgr5, responds to the Wnt pathway and helps in intestinal regeneration, while Bmi1 has also been identified as intestinal stem cells markers that do not respond to changes in Wnt pathway signaling. Lgr5 intestinal stem cells maintain homeostasis, while Bmi1 intestinal stem cells are normally quiescent but regenerate intestine post-injury [52]. Thus, PcG proteins not only regulate genes required for lineage specification, but also regulate genes in tissue-resident stem cells and help maintain homeostasis.

PCG-MEDIATED CONTROL OF GENE EXPRESSION IN EMBRYONIC STEM CELLS

Embryonic stem cells are derived from the inner cell mass of a blastocyst, under physiological conditions, the inner cell mass (ICM) quickly proceeds to gastrulation [53]. The knockout mice studies for key PcG proteins have shown that the PcG protein are critical for lineage specification. Investigating the early lineage specification *in vivo* is technically challenging, and embryonic stem cells are a good model to understand the role of PcG proteins in early lineage specification. *In vitro* studies using embryonic stem cells can help us understand key aspects of lineage specification by PcG proteins, which cannot be studied in tissue-resident stem cells like neural stem cells (NSCs), hematopoietic stem cells (HSCs), and mesenchymal stem cells (MSCs).

CBX6 and CBX7 are important members of PRC1; these proteins are involved in the maintenance of pluripotency in human embryonic stem cells. When CBX6 in human embryonic stem cells (hESCs) is reduced using shRNA approach, the hESCs spontaneously differentiate [54]. BCL6 corepressor BCOR combines with the PRC1.1 to repress the genes of mesoderm and endoderm, and knockdown of BCOR in hESCs leads to robust differentiation. However, the neuroectodermal genes are suppressed even in absence of BCOR PRC1.1, thus suggesting a different mechanism for suppression of neuroectodermal genes in hESCs [55].

CRISPR/Cas9-mediated removal of *Rybp* from mouse ES cells resulted in impaired differentiation as well as proliferation. Rybp is an important member of the PRC1 that helps to monoubiquitinate histone H2A at lysine 119 [56]. PRT4165 inhibition of RING1B resulted in enhanced neuronal differentiation in human embryonic and human embryonic stem cells, and this RING1B inhibitor could be incorporated into neuronal differentiation protocol [49]. Pcgf6 regulates expression of pluripotency-related genes such as *Oct4, Nanog,* and *Sox2* in mouse embryonic stem cells, knockdown of *Pcgf* leads to downregulation of *Oct4, Nanog,* and *Sox2*, while overexpression of *Pcgf* caused impaired differentiation as cells express pluripotency factors [57].

EZH2 when depleted in hESCs by employing an inducible CRISPR/Cas9 system failed to differentiate efficiently into three germ lineages, although the *EZH2* deficient cells spontaneously differentiated. Curiously as was the case with PRC1.1, lack of *EZH2* led to differentiation towards endoderm and mesoderm but not ectoderm [58]. Later, another group also used inducible CRISPR/Cas9 to knockout both EZH1 and EZH2 in human embryonic stem cells; these cells showed biased differentiation towards endoderm and mesoderm lineages, but not towards ectoderm. Interestingly, when the *EZH1/2* depleted hESCs are converted to naïve state *in vitro* by overexpression of NANOG, the hESCs lacking *EZH1* and *EZH2* did not differentiate, thus indicating that the PRC2 complex has a greater role in primed cells and not in naïve hESCs [59]. Human embryonic stem cells are termed "primed" while mouse embryonic stem cells are termed "naïve" as mouse embryonic stem cells can differentiate into germ cells in addition to the three germ lineages [60]. Knockdown of *EZH1* in human pluripotent stem cells enhanced their T cell differentiation potential, knockdown of *SUZ12* also increased T cell potential but to a lesser extent in comparison to *EZH1* knockdown [61]. The authors isolated developing HSCs in embryonic regions such as aorta-gonad-mesonephros (AGM) and yolk sac (YC) from wild type and *Ezh1* deficient mice. These isolated HSCs when transplanted into NOD SCID mice, produced almost 70% chimerism; on the other hand, transplantation of wild-type HSCs produced less than 5% chimerism in NOD SCID mice. A secondary transplant was done of AGM-derived HSCs from *Ezh1*null mice, which resulted in almost 60% chimerism, while that of wild-type HSCs was less than 4%. The HSCs derived from YC gave similar results to those seen with AGM-derived HSCs. Embryonic stem cells have given insights into the regulation of gene expression not seen in adult stem cells. These studies showed that PcG proteins are crucial determinants of pluripotency and differentiation of embryonic stem cells. PcG-mediated control of gene expression in embryonic stem cells is an active area of basic research that can help us understand how PcG proteins regulate gene expression necessary for lineage specification and more results are awaited before we can definitively describe role of each subunit during development.

PCG-MEDIATED CONTROL OF GENE EXPRESSION IN CANCER CELLS

Cancer cells are characterized by their ability to proliferate uncontrollably, however early detection of cancerous cells is limited due to lack of reliable biomarkers. Human colorectal cancer tissues when compared to normal colorectal tissue showed characteristic expression of EZH2, SUZ12, BMI1, and H3K27me3. Based on this, the tissues were classified into five different categories using the expression of PcG proteins and authors claim they could be used as biomarkers for prognosis on cancer progression [62]. PcG proteins such as EZH2, BMI1, and RING1 were found to be overexpressed in human prostate cancer tissue compared to normal tissue, and expression of EZH2 and BMI1 can be potentially predictive of recurrence [63]. RYBP was downregulated in normal tissues compared to hepatocellular carcinoma (HCC) cells, and reduced RYBP levels predicted poor survival [64].

BMI1 gene amplification and overexpression have been associated with hematological malignancies. BMI1 is expressed in hematopoietic stem cells (HSCs) but not in mature or progenitors cells that arise from HSCs. *BMI1* expression helps in maintaining homeostasis; however, overexpression of *BMI1* negatively correlates with expression of cell cycle inhibitors in lymphomas [65]. In breast cancer cells, *BMI1* overexpression leads to reduced expression of cell cycle inhibitors, while in EZH2 overexpressing cells, the cell cycle inhibitors were found to be expressed. Breast cancer shows a good prognosis when *BMI1* was overexpressed while, with overexpression of *EZH2* there was poor prognosis [66]. PcG protein EZH2 has been previously shown to be correlated with aggressive tumors of the breast, endometrium, prostate, and skin [67]. Epithelial to mesenchymal transition (EMT) is important for cell migration and for cancer cells to metastasize. In colon cancer cell line SW480, TGFβ signaling led to localization of EZH2 to *WNT5A* promoters, thus silencing the *WNT5A*. In the absence of WNT5A, EMT genes were induced, thus TGFβ activates the EMT gene expression program via EZH2-mediated suppression of *WNT5A* [68].

CBX7 in ovarian cancer cells regulates expression of EMT gene *TWIST1*; downregulation of *CBX7* enhanced the expression of *TWIST1* and thus increased tumorigenicity [69]. Recently, in breast cancer cells, transcription factors like SNAIL, SLUG, and TWIST recruit RING1B to *E-CADHERIN* promoter and silence *E-CADHERIN* via histone H2A monoubiquitination. Loss of E-CADHERIN may help cancer cells migrate and metastasize [70]. Further investigations are required to probe whether this recruitment of RING1B to E-CADHERIN is unique to cancer cells or does it occur under normal physiological processes such as wound healing, and what is the process that keeps in check RING1B activity . Pancreatic cancer is one of the most difficult to diagnose and shows high mortality. Bmi1 is expressed in cancerous pancreatic cells, but in several cancers, BMI1 is upregulated. Researchers used transgenic mice that expressed KRas under *PDX1* promoter to induce tumor formation specifically in the pancreas. Moreover, in these mice, *Bmi1* was removed using the Cre-Loxp system, and surprisingly without *Bmi1*, cancer could not be initiated. To confirm if this lack of cancer initiation was not due to the overexpression of the Bmi1 or Bmi1 target Inkα, mice lacking both Bmi1 and Ink4α were generated, but tumor formation was not found. Unlike under normal physiology, in this mice model, Bmi1 did not initiate cancer formation by inhibiting cell cycle inhibitor *Ink4α* expression [71].

The PcG proteins tightly regulate the expression of genes, and misregulation of PcG proteins can result in uncontrolled cell proliferation. Due to the inherent complexity of tumor initiation, it is not always clear if PcG protein misregulation drives tumorigenesis or other mutations drive misexpression of PcG proteins. PcG proteins can, however, be used as biomarkers for monitoring cancer progression and can even be targeted for treating cancers. There are several ongoing clinical trials that involve the use of specific EZH2 inhibitors to limit the growth of synovial sarcoma, non-Hodgkin lymphoma, and mesothelioma among others [72]; http://www.clinicaltrials.gov).

CONCLUSION AND FUTURE PERSPECTIVE

Epigenetic control of gene activity allows differential gene expression and the existence of specialized cell types with the same genetic material. PcG proteins are a group of proteins that alter gene expression via specific posttranslational histone modifications. The PcG proteins are brought together to form large multiprotein complexes termed PRCs, these PRC1 catalyze the addition of histone modifications such as H3K27me2/3 and H2AK119ub1, and these modifications lead to gene repression. The PcG proteins control gene activity in tissue-resident stem cells, thereby allowing them to generate new tissue and replace the tissue lost due to wear or injury. Occasionally, when the PcG proteins are misregulated, it can lead to uncontrolled cell proliferation and tumor formation. Some of the key areas in this field that are being probed include factors or upstream signals that recruit PcGs to specific sites in mammalian cells, the role of noncoding RNA in recruiting PRC to specific sites, formation and disassembly of PRC1 variants, regulation of PcG proteins by posttranslational modifications, interactions between PcG proteins and other histone modifiers, and use of pharmacological inhibitors to regulate PcG proteins to achieve desired control over gene activity. Newer technologies such as two-photon microscopy, CRISPR/Cas9-mediated gene editing, dynamic ChIP-sequencing, and single-cell genomics may reveal the role of PcG proteins in establishing the astonishing heterogeneity within a seemingly homogenous tissue.

Acknowledgment

The author acknowledges the support given by the Symbiosis Centre for Stem Cell Research (SCSCR), Symbiosis International University (SIU).

References

[1] Chen T, Dent SY. Chromatin modifiers and remodellers: regulators of cellular differentiation. Nat Rev Genet 2014;15 (2):93—106.

[2] Goldberg AD, Allis CD, Bernstein E. Epigenetics: a landscape takes shape. Cell 2007;128(4):635—8.

[3] Lewis PH. New mutants report. D. I. S. 1947;21:69.

[4] Ingham PW. A gene that regulates the bithorax complex differentially in larval and adult cells of Drosophila. Cell 1984;37 (3):815—23.

[5] Kassis JA, Kennison JA, Tamkun JW. Polycomb and trithorax group genes in Drosophila. Genetics 2017;206(4):1699—725.

[6] Duncan IM. Polycomblike: a gene that appears to be required for the normal expression of the bithorax and antennapedia gene complexes of Drosophila melanogaster. Genetics 1982;102(1):49—70.

[7] Jones RS, Gelbart WM. Genetic analysis of the enhancer of zeste locus and its role in gene regulation in *Drosophila melanogaster*. Genetics 1990;126(1):185—99.

[8] Eissenberg JC, James TC, Foster-Hartnett DM, Hartnett T, Ngan V, Elgin SC. Mutation in a heterochromatin-specific chromosomal protein is associated with suppression of position-effect variegation in *Drosophila melanogaster*. Proc. Natl Acad Sci USA 1990;87 (24):9923—7.

[9] Chiang A, O'Connor MB, Paro R, Simon J, Bender W. Discrete Polycomb-binding sites in each parasegmental domain of the bithorax complex. Development (Cambridge, England) 1995;121 (6):1681—9.

[10] Fritsch C, Brown JL, Kassis JA, Müller J. The DNA-binding polycomb group protein pleiohomeotic mediates silencing of a Drosophila homeotic gene. Development (Cambridge, England) 1999;126(17):3905—13.

[11] van Lohuizen M, Frasch M, Wientjens E, Berns A. Sequence similarity between the mammalian bmi-1 proto-oncogene and

the Drosophila regulatory genes Psc and Su(z)2. Nature 1991;353(6342):353–5.
[12] Brunk BP, Martin EC, Adler PN. Drosophila genes posterior sex combs and suppressor two of zeste encode proteins with homology to the murine bmi-1 oncogene. Nature 1991;353 (6342):351–3.
[13] van der Lugt NM, Domen J, Linders K, van Roon M, Robanus-Maandag E, te Riele H, et al. Posterior transformation, neurological abnormalities, and severe hematopoietic defects in mice with a targeted deletion of the bmi-1 proto-oncogene. Genes Dev 1994;8(7):757–69.
[14] Alkema MJ, van der Lugt NM, Bobeldijk RC, Berns A, van Lohuizen M. Transformation of axial skeleton due to overexpression of bmi-1 in transgenic mice. Nature 1995;374(6524): 724–7.
[15] Simon JA, Kingston RE. Mechanisms of polycomb gene silencing: knowns and unknowns. Nature reviews. Mol Cell Biol 2009;10(10):697–708.
[16] Chittock EC, Latwiel S, Miller TC, Müller CW. Molecular architecture of polycomb repressive complexes. Biochem Soc Trans 2017;45(1):193–205.
[17] Desai D, Pethe P. Polycomb repressive complex 1: regulators of neurogenesis from embryonic to adult stage. J Cell Physiol 2020;235(5):4031–45.
[18] Voncken JW, Roelen BA, Roefs M, de Vries S, Verhoeven E, Marino S, et al. Rnf2 (Ring1b) deficiency causes gastrulation arrest and cell cycle inhibition. Proc Natl Acad Sci USA 2003;100(5):2468–73.
[19] O'Carroll D, Erhardt S, Pagani M, Barton SC, Surani MA, Jenuwein T. The polycomb-group gene Ezh2 is required for early mouse development. Mol Cell Biol 2001;21(13):4330–6.
[20] Højfeldt JW, Hedehus L, Laugesen A, Tatar T, Wiehle L, Helin K. Non-core subunits of the PRC2 complex are collectively required for its target-site specificity. Mol Cell 2019;76(3):423–36 e3.
[21] Kouznetsova VL, Tchekanov A, Li X, Yan X, Tsigelny IF. Polycomb repressive 2 complex-molecular mechanisms of function. Protein Sci 2019;28(8):1387–99.
[22] Allfrey VG, Faulkner R, Mirsky AE. Acetylation and methylation of histones and their possible role in the regulation of RNA synthesis. Proc Natl Acad Sci USA 1964;51(5):786–94.
[23] Mikkelsen TS, Ku M, Jaffe DB, Issac B, Lieberman E, Giannoukos G, et al. Genome-wide maps of chromatin state in pluripotent and lineage-committed cells. Nature 2007;448 (7153):553–60.
[24] Tropberger P, Pott S, Keller C, Kamieniarz-Gdula K, Caron M, Richter F, et al. Regulation of transcription through acetylation of H3K122 on the lateral surface of the histone octamer. Cell 2013;152(4):859–72.
[25] Lawrence M, Daujat S, Schneider R. Lateral thinking: how histone modifications regulate gene expression. Trends Genet 2016;32(1):42–56.
[26] Santini L, Halbritter F, Titz-Teixeira F, Suzuki T, Asami M, Ma X, et al. Genomic imprinting in mouse blastocysts is predominantly associated with H3K27me3. Nat Commun 2021;12(1): 3804.
[27] Sangiorgi E, Capecchi MR. Bmi1 lineage tracing identifies a self-renewing pancreatic acinar cell subpopulation capable of maintaining pancreatic organ homeostasis. Proc Natl Acad Sci USA 2009;106(17):7101–6.
[28] Margueron R, Reinberg D. The Polycomb complex PRC2 and its mark in life. Nature 2011;469(7330):343–9.
[29] Grau DJ, Chapman BA, Garlick JD, Borowsky M, Francis NJ, Kingston RE. Compaction of chromatin by diverse Polycomb group proteins requires localized regions of high charge. Genes Dev 2011;25(20):2210–21.

[30] van Arensbergen J, García-Hurtado J, Maestro MA, Correa-Tapia M, Rutter GA, Vidal M, et al. Ring1b bookmarks genes in pancreatic embryonic progenitors for repression in adult β cells. Genes Dev 2013;27(1):52–63.
[31] Zhang Y, Liu T, Yuan F, Garcia-Martinez L, Lee KD, Stransky S, et al. The Polycomb protein RING1B enables estrogen-mediated gene expression by promoting enhancer-promoter interaction and R-loop formation. Nucleic Acids Res 2021;49 (17):9768–82.
[32] Bantignies F, Cavalli G. Polycomb group proteins: repression in 3D. Trends Genet 2011;27(11):454–64.
[33] Cai Y, Zhang Y, Loh YP, Tng JQ, Lim MC, Cao Z, et al. H3K27me3-rich genomic regions can function as silencers to repress gene expression via chromatin interactions. Nature Commun 2021;12(1):719.
[34] Jin B, Li Y, Robertson KD. DNA methylation: superior or subordinate in the epigenetic hierarchy? Genes Cancer 2011;2 (6):607–17.
[35] Rose NR, Klose RJ. Understanding the relationship between DNA methylation and histone lysine methylation. Biochim Biophys Acta 2014;1839(12):1362–72.
[36] Lu Z, Hong CC, Kong G, Assumpção A, Ong IM, Bresnick EH, et al. Polycomb group protein YY1 Is an essential regulator of hematopoietic stem cell quiescence. Cell Rep 2018;22(6): 1545–59.
[37] Assumpção A, Fu G, Singh DK, Lu Z, Kuehnl AM, Welch R, et al. A lineage-specific requirement for YY1 Polycomb Group protein function in early T cell development. Development (Cambridge, England) 2021;148(7):dev197319.
[38] Almeida M, Pintacuda G, Masui O, Koseki Y, Gdula M, Cerase A, et al. PCGF3/5-PRC1 initiates Polycomb recruitment in X chromosome inactivation. Science 2017;356(6342):1081–4.
[39] Bousard A, Raposo AC, Żylicz JJ, Picard C, Pires VB, Qi Y, et al. The role of Xist-mediated Polycomb recruitment in the initiation of X-chromosome inactivation. EMBO Rep 2019;20(10): e48019.
[40] Harris C, Cloutier M, Trotter M, Hinten M, Gayen S, Du Z, et al. Conversion of random X-inactivation to imprinted X-inactivation by maternal PRC2. eLife 2019;8:e44258.
[41] Delgado-Olguín P, Huang Y, Li X, Christodoulou D, Seidman CE, Seidman JG, et al. Epigenetic repression of cardiac progenitor gene expression by Ezh2 is required for postnatal cardiac homeostasis. Nat Genet 2012;44(3):343–7.
[42] Ahmed A, Wang T, Delgado-Olguin P. Ezh2 is not required for cardiac regeneration in neonatal mice. PLoS One 2018;13(2): e0192238.
[43] Tang Y, Zhao L, Yu X, Zhang J, Qian L, Jin J, et al. Inhibition of EZH2 primes the cardiac gene activation via removal of epigenetic repression during human direct cardiac reprogramming. Stem Cell Res 2021;53:102365.
[44] Yuan JL, Yin CY, Li YZ, Song S, Fang GJ, Wang QS. EZH2 as an epigenetic regulator of cardiovascular development and diseases. J Cardiovasc Pharmacol 2021;78(2):192–201. Available from: https://doi.org/10.1097/FJC.0000000000001062.
[45] Bardot ES, Valdes VJ, Zhang J, Perdigoto CN, Nicolis S, Hearn SA, et al. Polycomb subunits Ezh1 and Ezh2 regulate the Merkel cell differentiation program in skin stem cells. EMBO J 2013;32(14):1990–2000.
[46] Perdigoto CN, Dauber KL, Bar C, Tsai PC, Valdes VJ, Cohen I, et al. Polycomb-mediated repression and sonic hedgehog signaling interact to regulate merkel cell specification during skin development. PLoS Genet 2016;12(7):e1006151.
[47] Wang Y, Guan Y, Wang F, Huang A, Wang S, Zhang YA. Bmi-1 regulates self-renewal, proliferation and senescence of human fetal neural stem cells in vitro. Neurosci Lett 2010;476(2):74–8.

[48] Hirabayashi Y, Suzki N, Tsuboi M, Endo TA, Toyoda T, Shinga J, et al. Polycomb limits the neurogenic competence of neural precursor cells to promote astrogenic fate transition. Neuron 2009;63(5):600–13.

[49] Desai D, Khanna A, Pethe P. Inhibition of RING1B alters lineage specificity in human embryonic stem cells. Cell Biol Int 2020;44(6):1299–311.

[50] Sangiorgi E, Capecchi MR. Bmi1 is expressed in vivo in intestinal stem cells. Nat Genet 2008;40(7):915–20.

[51] Koppens MA, Bounova G, Gargiulo G, Tanger E, Janssen H, Cornelissen-Steijger P, et al. Deletion of polycomb repressive complex 2 from mouse intestine causes loss of stem cells. Gastroenterology 2016;151(4):684–97 e12.

[52] Yan KS, Chia LA, Li X, Ootani A, Su J, Lee JY, et al. The intestinal stem cell markers Bmi1 and Lgr5 identify two functionally distinct populations. Proc Natl Acad Sci USA 2012;109(2):466–71.

[53] Evans MJ, Kaufman MH. Establishment in culture of pluripotential cells from mouse embryos. Nature 1981;292(5819):154–6.

[54] Santanach A, Blanco E, Jiang H, Molloy KR, Sansó M, LaCava J, et al. The Polycomb group protein CBX6 is an essential regulator of embryonic stem cell identity. Nature communications 2017;8(1):1235.

[55] Wang Z, Gearhart MD, Lee YW, Kumar I, Ramazanov B, Zhang Y, et al. A Non-canonical BCOR-PRC1.1 complex represses differentiation programs in human ESCs. Cell Stem Cell 2018;22(2):235–51 e9.

[56] Zhao W, Liu M, Ji H, Zhu Y, Wang C, Huang Y, et al. The polycomb group protein Yaf2 regulates the pluripotency of embryonic stem cells in a phosphorylation-dependent manner. J Biol Chem 2018;293(33):12793–804.

[57] Yang CS, Chang KY, Dang J, Rana TM. Polycomb group protein Pcgf6 acts as a master regulator to maintain embryonic stem cell identity. Sci Rep 2016;6:26899.

[58] Collinson A, Collier AJ, Morgan NP, Sienerth AR, Chandra T, Andrews S, et al. Deletion of the polycomb-group protein EZH2 leads to compromised self-renewal and differentiation defects in human embryonic stem cells. Cell Rep 2016;17(10):2700–14.

[59] Shan Y, Liang Z, Xing Q, Zhang T, Wang B, Tian S, et al. PRC2 specifies ectoderm lineages and maintains pluripotency in primed but not naïve ESCs. Nat Commun 2017;8(1):672.

[60] Takahashi S, Kobayashi S, Hiratani I. Epigenetic differences between naïve and primed pluripotent stem cells. Cell Mol Life Sci 2018;75(7):1191–203.

[61] Vo LT, Kinney MA, Liu X, Zhang Y, Barragan J, Sousa PM, et al. Regulation of embryonic haematopoietic multipotency by EZH1. Nature 2018;553(7689):506–10.

[62] Benard A, Goossens-Beumer IJ, van Hoesel AQ, Horati H, Putter H, Zeestraten EC, et al. Prognostic value of polycomb proteins EZH2, BMI1 and SUZ12 and histone modification H3K27me3 in colorectal cancer. PLoS One 2014;9(9):e108265.

[63] van Leenders GJ, Dukers D, Hessels D, van den Kieboom SW, Hulsbergen CA, Witjes JA, et al. Polycomb-group oncogenes EZH2, BMI1, and RING1 are overexpressed in prostate cancer with adverse pathologic and clinical features. Eur Urol 2007;52(2):455–63.

[64] Zhu X, Yan M, Luo W, Liu W, Ren Y, Bei C, et al. Expression and clinical significance of PcG-associated protein RYBP in hepatocellular carcinoma. Oncol Lett 2017;13(1):141–50.

[65] Beà S, Tort F, Pinyol M, Puig X, Hernández L, Hernández S, et al. BMI-1 gene amplification and overexpression in hematological malignancies occur mainly in mantle cell lymphomas. Cancer Res 2001;61(6):2409–12.

[66] Pietersen AM, Horlings HM, Hauptmann M, Langerød A, Ajouaou A, Cornelissen-Steijger P, et al. EZH2 and BMI1 inversely correlate with prognosis and TP53 mutation in breast cancer. Breast Cancer Res 2008;10(6) R109.

[67] Bachmann IM, Halvorsen OJ, Collett K, Stefansson IM, Straume O, Haukaas SA, et al. EZH2 expression is associated with high proliferation rate and aggressive tumor subgroups in cutaneous melanoma and cancers of the endometrium, prostate, and breast. J Clin Oncol 2006;24(2):268–73.

[68] Tao J, Shi L, Huang L, Shi H, Chen H, Wang Y, et al. EZH2 is involved in silencing of WNT5A during epithelial-mesenchymal transition of colon cancer cell line. J Cancer Res Clin Oncol 2017;143(11):2211–19.

[69] Li J, Alvero AB, Nuti S, Tedja R, Roberts CM, Pitruzzello M, et al. CBX7 binds the E-box to inhibit TWIST-1 function and inhibit tumorigenicity and metastatic potential. Oncogene 2020;39(20):3965–79.

[70] Wang Y, Sun Y, Shang C, Chen L, Chen H, Wang D, et al. Distinct Ring1b complexes defined by DEAD-box helicases and EMT transcription factors synergistically enhance E-cadherin silencing in breast cancer. Cell Death Dis 2021;12(2):202.

[71] Bednar F, Schofield IIK, Collins MA, Yan W, Zhang Y, Shyam N, et al. Bmi1 is required for the initiation of pancreatic cancer through an Ink4a-independent mechanism. Carcinogenesis 2015;36(7):730–8.

[72] Kim KH, Roberts CW. Targeting EZH2 in cancer. Nat Med 2016;22(2):128–34.

SECTION III

Methods in Epigenetics

8. Methods for assessing DNA cytosine modifications genome-wide
9. Genome-wide analysis of histone modifications in the mammalian genome
10. Techniques for analyzing genome-wide expression of non-coding RNA
11. Computational epigenetics: the CeRNA network analysis

CHAPTER

8

Methods for Analyzing DNA Cytosine Modifications Genome-wide

Tibor A. Rauch[1] and Gerd P. Pfeifer[2]

[1]University of Pécs Medical School, Pécs, Hungary [2]Department of Epigenetics, Van Andel Institute, Grand Rapids, MI, United States

OUTLINE

Introduction	123	5-Hydroxymethylcytosine Mapping Methodologies	130
Methylated DNA Immunoprecipitation	124	TET-assisted Bisulfite Sequencing	131
MBD Protein-based Affinity Pulldown	124	Oxidative Bisulfite Sequencing	131
Methylated-CpG Island Recovery Assay	125	APOBEC-coupled Epigenetic Sequencing	131
Targeted Bisulfite Sequencing	125	SMRT-seq and Nanopore Sequencing	131
Infinium and EPIC Methylation Bead Chips	127	Single Cell Whole Genome Methylation Analysis	132
Whole Genome Bisulfite Sequencing	127	Future Directions and Challenges	133
Other Sodium Bisulfite-based Approaches	128	Acknowledgments	133
Enzymatic Methyl-sequencing and Pyrimidine Borane-based Methods	129	References	133

INTRODUCTION

Historically, DNA methylation has been the first epigenetic modification to be discovered [1,2]. The processes governing enzymatic DNA methylation as well as the biological properties of this modification are reasonably well understood. A functional role of DNA methylation at CpG sequences in epigenetic control, gene regulation, X chromosome inactivation and cell differentiation was first proposed in 1975 [3,4]. Today, the connections between DNA methylation, gene activity, and other epigenetic marks such as histone modifications are being studied in many laboratories. Aberrations in DNA methylation patterns as a cause or consequence of diseases, such as cancer, are analyzed and described in detail.

To explore DNA methylation profiles at the genome scale, a wide range of approaches have been developed. Many of the methods were originally used for detecting methylation changes at the single gene level, but by coupling them with microarray or high throughput sequencing (HTS) platforms, genome-wide analysis tools have become available. On microarray platforms, promoter or CpG island arrays had been used earlier to analyze important gene regulatory regions. Whole genome HTS is now a widely used approach for exploring DNA methylation profiles. Most DNA methylation analysis

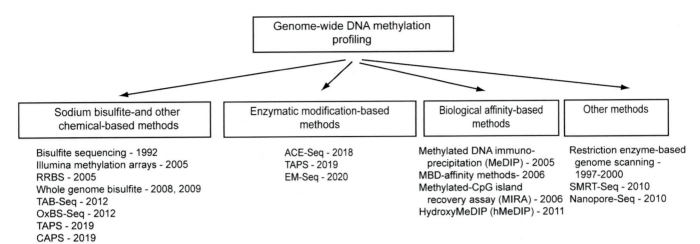

FIGURE 8.1 **Methods for genome-scale analysis of DNA methylation and hydroxymethylation profiles.** The major methods discussed in this chapter are subdivided into sodium bisulfite- and other chemical-based methods, enzymatic conversion-mediated, and affinity-based approaches, and other methods for analysis of DNA methylation. The year(s) when these methods were first introduced are indicated.

methods can be categorized into well-characterized groups based on their principles (Fig. 8.1).

Several techniques are based on antibodies or proteins that bind to methylated DNA and provide a resolution of 100–200 base pairs. Resolution at the single nucleotide level often requires bisulfite-based or other DNA sequencing approaches for which high throughput techniques have been developed. In this review, we will describe several of the more commonly used genome-scale methods for DNA methylation analysis in some detail and will briefly discuss a few of the more recently developed methods, which are still awaiting broader acceptance by the epigenetics community.

The first and second edition of this handbook provided a detailed overview of the development of genome-scale DNA methylation analysis methods, but new technical advances have occurred in this field, and several earlier methods including restriction landmark genomic scanning (RLGS), methylation-sensitive restriction fingerprinting (MSRF), methylation-sensitive representational difference analysis (Ms-RDA) and methylation-specific digital karyotyping (MSDK) are not commonly used anymore. Newer methodologies including Illumina's CpG arrays, reduced representation bisulfite sequencing (RRBS), methods for detection of 5-hydroxymethylcytosine, and enzymatic or chemical base conversion methods have been added.

We begin this chapter with the description of affinity purification methods and then continue on to technologies for base resolution analysis of modified cytosines.

mapping. Methylated DNA fragments can be affinity-purified by using either a 5mC-specific antibody or proteins that specifically bind to methylated DNA sequences [5–8]. Conversely, DNA fragments containing unmethylated CpG sites can be enriched by using specific protein domains, e.g., the CXXC domain that have a selective affinity for unmethylated CpG-containing DNA [9].

A 5-methylcytosine (5mC)-specific antibody has been used for immunoprecipitation of densely methylated DNA sequences [7] (Fig. 8.2). In this method, the fragments generated by sonication of genomic DNA are incubated with anti-5mC antibody and the captured methylated fraction is extracted from the reaction by using protein A/G beads. Isolated DNA fragments are deproteinized, fluorescently labeled and hybridized onto microarray platforms or analyzed by massively parallel sequencing [10]. The main limitation of the Methylated DNA immunoprecipitation (MeDIP) method is the varying quality of 5mC antibodies used in the procedure and the requirement of single-stranded (denatured) DNA for analysis. Methylation profiles can be generated with a resolution of approximately 100–200 base pairs. Like all affinity-based methods, MeDIP cannot measure absolute methylation levels, i.e., the percent methylation at CpG sites (0–100) within a particular region. It can only provide relative levels of methylation as indicated by different peak heights between different genomic compartments.

METHYLATED DNA IMMUNOPRECIPITATION

Affinity purification is the principle of a group of techniques applied to genome-wide DNA methylation

MBD PROTEIN-BASED AFFINITY PULLDOWN

The proteins MBD1, MBD2, MBD3, MBD4, and MeCP2 comprise the core of a small family of nuclear

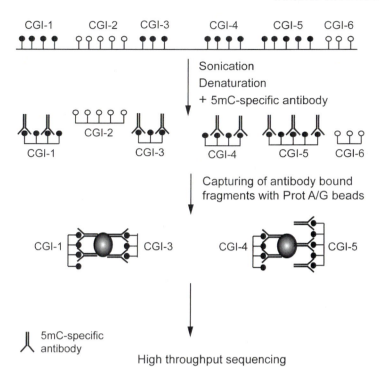

FIGURE 8.2 **Methylated DNA immunoprecipitation (MeDIP-seq).** The MeDIP-seq method is described in the text. Filled and open lollipops mark methylated and unmethylated CpG sites, respectively.

proteins that share a common methyl-CpG binding domain (MBD). Each of these proteins, except perhaps for human MBD3, is capable of binding specifically to CpG-methylated DNA. An affinity column-based method was initially developed that employs the methyl-CpG binding domain of MeCP2 [11,12]. Recombinant MBD fragments can be endowed with an affinity tag such as a His-tag or GST-tag. Incubating fragmented genomic DNA with the tagged MBDs provides a convenient way for enrichment of CpG-methylated DNA fragments. These fragments can then be analyzed, as done earlier using microarrays [8] or currently by using high throughput sequencing (MBD-seq) [13–16].

METHYLATED-CPG ISLAND RECOVERY ASSAY

Among the MBD proteins, the MBD2b isoform has the strongest affinity to methylated DNA and can form heterodimers with other MBD proteins via its C-terminal coiled-coil domain. MBD3L1, a related protein and member of the MBD2/3 sub-family, has no DNA binding domain itself but it is a binding partner of MBD2b via heterodimer formation [17]. The MBD2b/MBD3L1 protein complex has higher affinity to methylated DNA than MBD2b alone [5,6]. In the Methylated-CpG island recovery assay (MIRA) procedure, the fragmented (sonicated) genomic DNA is incubated with the bacterially expressed and purified GST-MBD2b and His-MBD3L1 proteins. The high affinity MBD2b/MBD3L1 complex specifically binds to the methylated genomic DNA fragments and, since MBD2b is GST-tagged, it is easy to purify the complex containing methylated DNA by applying glutathione-coated magnetic beads [5,18–20] (Fig. 8.3). MIRA does not depend on restriction enzyme recognition sites or sodium bisulfite conversion of the DNA, and it works on double-stranded DNA. There is no dependence on DNA sequences other than that it requires a minimum of two methylated CpGs in the captured fragment. MIRA can reliably be performed with a few hundred nanograms of genomic DNA. The MIRA technique has been used to profile DNA methylation patterns at a resolution of 100–200 base pairs in the entire genome of normal human B lymphocytes providing one of the first mammalian "DNA methylome" data sets [21]. MIRA analysis is compatible with different microarray platforms [18,19] and with high-throughput DNA sequencing (MIRA-seq) [22–26].

TARGETED BISULFITE SEQUENCING

The set of techniques described next makes use of the differential sensitivity of cytosine and 5-methylcytosine (5mC) towards chemical modification. This methodology is still challenging when applied to the whole genome, mostly at the level of bioinformatics analysis and is associated with a relatively high cost per sample. The basis of sodium bisulfite sequencing is that cytosine is deaminated to uracil by sodium

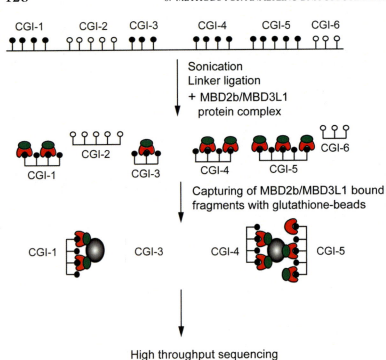

FIGURE 8.3 Methylated-CpG island recovery assay (MIRA-seq). The steps of the MIRA procedure are described in the text. Filled and open lollipops mark methylated and unmethylated CpG sites, respectively.

bisulfite but 5-methylcytosine is resistant to bisulfite-induced deamination [27–29]. Bisulfite sequencing provides single base resolution for analysis of DNA methylation patterns [30]. It was initially based on sequencing of PCR products of bisulfite-treated DNA to profile DNA methylation at specific loci.

Before comprehensive whole genome analysis was possible, researchers have developed the reduced representation bisulfite sequencing (RRBS) method that allows one to analyze methylation profiles at the single nucleotide level in a more limited way, mostly focusing on CpG islands [31] (Fig. 8.4). The original RRBS protocol was combined with Sanger sequencing [32], which was later replaced by HTS. As a first step, genomic DNA is digested with a methylation-insensitive enzyme (e.g., MspI, 5'-CCGG-3') that mainly targets CpG-rich regions, which is followed by adapter ligation and fragment size selection. Fragments in the 150–225 bp size range are isolated and subjected to bisulfite conversion. Bisulfite-treated samples are PCR amplified and sequenced [33]. Although whole genome bisulfite sequencing (WGBS) has a much greater genome-wide coverage of CpGs compared to RRBS, RRBS provides a good alternative for analysis of CpG islands and is substantially more cost-effective [13].

Other more gene-targeted approaches are based on the enrichment of genomic regions using gene-specific hybridization probes. One method captures a subset of genomic targets for single-molecule bisulfite sequencing [34,35]. For example, a set of ~30,000 padlock

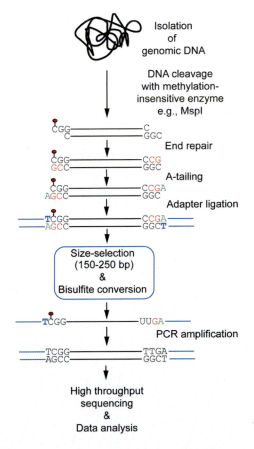

FIGURE 8.4 Reduced representation bisulfite sequencing (RRBS). MspI cleavage enriches genomic fragments from CpG islands and other CpG-rich regions. See text for details.

probes was designed to assess methylation of ~66,000 CpG sites within 2020 CpG islands on human chromosomes 12, 20, and 34 [35]. Other genomic enrichment approaches for targeted bisulfite sequencing have been developed [36]. Locus-specific HTS-based bisulfite sequencing is becoming increasingly useful for detection of methylated alleles in a population of otherwise unmethylated sequences [37–39]. These amplicon sequencing tasks can be accomplished, for example, on the Miseq sequencing platform [40].

INFINIUM AND EPIC METHYLATION BEAD CHIPS

The Illumina Infinium array is a company-developed array platform (Fig. 8.5) [41]. Using the widely used Infinium 450k methylation assays, one can assess the methylation status of ~480,000 cytosines at CpG sequences, which are selected primarily from promoter regions, first exons, gene bodies and 3′UTRs. A newer generation of this platform, the EPIC array, incorporates additional sequences of the genome that are of particular interest, including a large number of putative enhancer sequences to bring the total number of analyzed CpG sites to ~850,000 [42]. Genomic DNA is bisulfite-treated before hybridization onto the microarray chips. The design of oligonucleotides is based on the "all-or-none" concept, surmising that the methylation status is correlated in a 50 bp stretch of DNA, i.e., most neighboring CpGs are thought to be in either a methylated or unmethylated state. The method is based on gene-specific and methylation-dependent single nucleotide primer extension on the bisulfite-converted DNA using primer extension that distinguishes between unmethylated and methylated CpGs after bisulfite conversion. Single-base extension of the probes incorporates a labeled ddNTP, which is subsequently detected by array scanning (Fig. 8.5). The level of methylation for the interrogated locus can be determined by calculating the ratio of the fluorescent signals from the methylated vs the unmethylated state. This technology is used very commonly and there are large numbers of publications in the literature. Even though the genomic CpG coverage is still only a small percentage of all CpGs (28 million on each strand), the design of the probes permits a focused analysis of many genomic regions of interest. However, if methylation changes of interest occur outside of regions covered by the array probes, they will be missed by this technology.

WHOLE GENOME BISULFITE SEQUENCING

The bisulfite approach can be used for whole genome DNA methylation profiling. Several sequencing

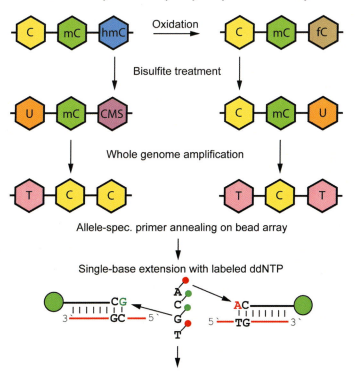

FIGURE 8.5 **Infinium EPIC arrays.** Gene-specific primers are extended to reveal the methylation or hydroxymethylation status of each locus after bisulfite conversion. 5hmC can be converted by chemical oxidation to 5fC to display 5mC only.

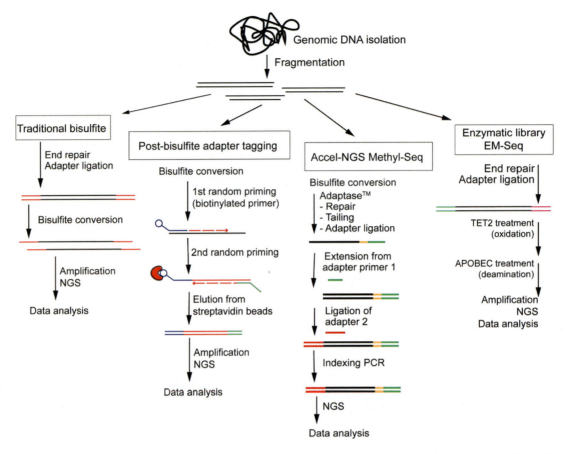

FIGURE 8.6 **Summary of HTS-compatible library preparation methods.** In traditional WGBS approaches (left), methylated adapters (red) are ligated to fragmented DNA before bisulfite conversion and amplification. In the post-bisulfite adapter ligation and Accel-NGS-Methyl-Seq methods (middle), bisulfite treatment both converts and shears genomic DNA and tagging with adapters is performed after bisulfite treatment. EM-seq libraries (right) are generated after fragmentation and adapter ligation to the DNA template.

platforms can be used, such as those from Illumina and ABI/SOLiD. Whole genome bisulfite sequencing (WGBS, also referred to as "Methyl-Seq") has initially been accomplished for the relatively small genome of the plant *Arabidopsis thaliana* [43] and later, for the genome of various mouse and human cell types [44–46]. In the traditional WGBS method, the DNA is fragmented with sonication to 200–300 bp and cytosine-methylated adapters are ligated before bisulfite-conversion. The bisulfite-treated DNA molecules are enriched by PCR initiated from the adapters and then subjected to sequencing. The sequencing reads are then processed and aligned to "computationally bisulfite-converted" genomes. Due to the nature of the initial cytosine deamination step, both strands of each locus are no longer complementary and are sequenced separately. To get quantitative information regarding the percent methylation at individual CpG sites on each strand, a read coverage of at least 20x to 30x is generally recommended. The current cost of sequencing one DNA sample by WGBS is approximately between US $1500 and $2000, still putting larger scale projects with many samples out of reach for most smaller laboratories.

According to the "traditional" protocol, bisulfite treatment is carried out after adapter ligation (Fig. 8.6), and because of extensive degradation of DNA during the bisulfite conversion process, more input material (≥ 1 μg) is usually required when employing this approach. Methylation analyses focusing on rare cell types, early embryos, or on tissue biopsies could be compromised by this treatment. By changing the experimental design, several post-bisulfite adapter ligation methods have been introduced (Fig. 8.6) and smaller amounts (i.e., nanograms) of DNA may be sufficient for analysis [47].

OTHER SODIUM BISULFITE-BASED APPROACHES

After conversion of genomic DNA with sodium bisulfite, medium-throughput approaches have been developed to analyze methylation profiles of several genes simultaneously. In the mass spectrometry-based MALDI-TOF Ms (Sequenom approach), bisulfite-pretreated genomic DNA is used as a template for PCR amplification of specific genes [48]. MALDI-TOF Ms provides direct quantitative methylation data for many genes at the same time [49]. The analysis is usually conducted in multi-well format and is limited by the number of gene targets that are amplified and analyzed in each run.

Another medium throughput approach is multiplexed bisulfite sequencing. To analyze several gene loci simultaneously, barcoded PCR primers can be multiplexed and then the samples are run on a sequencing platform. The advantage of this approach is considerable cost savings relative to WGBS and the ability to obtain quantitative information because tens of thousands of reads can easily be obtained for each locus, and the numbers of methylated vs unmethylated alleles can be counted.

ENZYMATIC METHYL-SEQUENCING AND PYRIMIDINE BORANE-BASED METHODS

Despite the extensive use of the bisulfite sequencing methods, they have drawbacks that should be considered. Bisulfite-mediated conversion of cytosine is only effective under harsh condition (i.e., low pH and high temperature) leading to degradation of the majority (sometimes ≥90%) of the input DNA. This high degree of degradation can be problematic for genome-wide analyses of samples from limited sources (e.g., biopsies or liquid biopsy samples). In addition, sequencing libraries made from bisulfite-treated DNA samples do not properly cover the genome because cytosines/methylated cytosines and the oxidized derivatives are disproportionally distributed [50]. Accordingly, WGBS may produce gaps with no or limited sequence coverage. To solve this issue the sequencing depth can be increased leading to higher sequencing costs.

To overcome bisulfite treatment-related sequencing issues, the enzymatic methyl-sequencing (EM-seq) method was developed (Figs. 8.6 and 8.7), which provides similar information at single nucleotide resolution as WGBS [51–53]. EM-seq relies on enzyme reactions that can be conducted under mild conditions that do not degrade the template DNA. EM-seq specifically detects modified cytosines (i.e., 5mC and 5hmC) by exploiting the catalytic activity of three enzymes ten-eleven translocation 2 (TET2), T4 phage β-glucosyltransferase (T4-βGT) and apolipoprotein mRNA editing enzyme, catalytic polypeptide-like 3 A (APOBEC3A). The enzymes are employed in two consecutive steps leading to protection from deamination of the originally methylated cytosines and deamination of unmethylated cytosines. In the first step, TET2 oxidizes 5mC (and 5hmC) to 5caC, while T4-βGT glycosylates 5hmCs. Both 5caC and glycosylated 5hmC are resistant to subsequent enzymatic deamination. In the second step, APOBEC3A catalyzes deamination of unmodified cytosines to uracil, which is amplified as thymine during PCR. Accordingly, combined application of the three enzymes enables discrimination between unmodified and modified cytosines. Since enzymatic treatment does not destroy the DNA template, EM-seq is ideal for analyzing DNA methylation profiles of samples with limited DNA amounts [53] and may result in better genome mapping quality. The bioinformatics pipelines used to analyze data from WGBS and EM-seq should be equivalent.

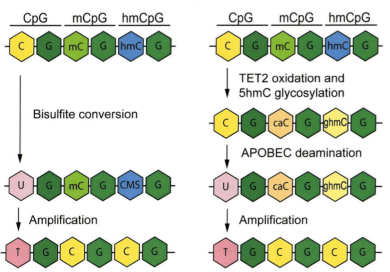

FIGURE 8.7 **Bisulfite treatment-based and enzymatic conversion-mediated DNA methylation profiling.** In bisulfite sequencing (BS-seq), methylated and hydroxymethylated Cs are resistant to bisulfite conversion. EM-seq employs consecutive enzyme treatments of DNA to reveal the methylation status of CpG dinucleotides. *caC*, 5-carboxylcytosine; *CMS*, cytosine-5-methylenesulfonate; *gmC*, glucosylated 5-hydroxymethylcytosine; *hmC*, 5-hydroxymethylcytosine; *mC*, 5-methylcytosine.

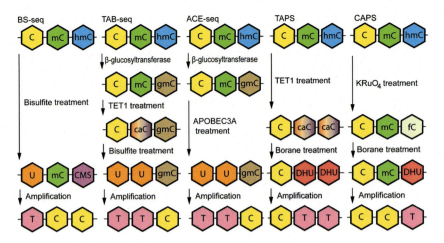

FIGURE 8.8 Comparison of BS-seq, TAB-seq, ACE-seq, TAPS, and CAPS. These methods are used to display 5mC and 5hmC, combined or separately. Whereas regular bisulfite sequencing cannot distinguish between the two modified cytosine bases and provides the sum of 5mC and 5hmC at each position, TAB-seq and ACE-seq provide a direct selective display of 5hmC at base resolution. TAB-seq employs both enzymatic and bisulfite conversion. ACE-Seq exclusively uses enzymes and leaves DNA templates more intact. TAPS and CAPS use a mild chemical conversion method to change 5fC and 5caC into dihydrouracil (DHU). While TAPS converts both 5mC and 5hmC, the perruthenate oxidation step in CAPS will allow selective display of 5hmC. *caC*, 5-carboxylcytosine; *5fC*, 5-formylcytosine; *CMS*, cytosine-5-methylenesulfonate; *gmC*, glucosylated 5-hydroxymethylcytosine; *hmC*, 5-hydroxymethylcytosine; *mC*, 5-methylcytosine. *DHU*, dihydrouracil.

TET-assisted pyrimidine borane sequencing (TAPS) is a new type of chemical and enzymatic conversion-based method [54] (Fig. 8.8). The method is based on TET-mediated oxidation of 5mC and 5hmC to 5-carboxylcytosine (5caC) and pyridine borane reduction of 5caC to dihydrouracil (DHU), which is read as thymine during PCR. TAPS simultaneously scores 5mC and 5hmC as bases converted to thymine. When 5hmC is blocked by glycosylation using beta-glycosyltransferase (TAPS-beta), 5mC can be scored separately [55]. In another modification of this method, chemical-assisted pyridine borane sequencing (CAPS), 5hmC is oxidized first with KRuO$_4$ to 5-formylcytosine (5fC), which is then converted with pyrimidine borane to DHU. CAPS allows selective display of 5hmC [54,55] (Fig. 8.8). Another bisulfite-free method for mapping of 5hmC is hmC-CATCH, which is based on perruthenate oxidation and sequencing of the resulting 5fC after enrichment [56].

The mild chemical reaction conditions used in TAPS and CAPS do not damage DNA and offer a new way of detection: they convert the modified bases directly, unlike bisulfite-seq or EM-seq, which convert the unmodified cytosine base. These methods offer additional advantages such as high sequence quality, higher mapping rates, and simultaneously provide genetic and epigenetic sequencing. In addition, TAPS can be combined with PacBio Single-Molecule Real-Time (SMRT) or nanopore sequencing to improve long-read epigenetic sequencing [57]. Modifications of these methods offer new ways for cost-effective DNA methylation sequencing [58].

5-HYDROXYMETHYLCYTOSINE MAPPING METHODOLOGIES

A recent technological challenge affecting almost all types of DNA methylation analysis has emerged from the discovery of 5-hydroxymethylcytosine (5hmC) in mammalian DNA [59,60]. This base modification is produced by an enzymatic pathway involving the TET family 5mC oxidases. 5hmC appears to be quite abundant in certain mammalian tissues and may have regulatory roles distinct from that of 5mC [61]. It should be noted that 5hmC prevents cleavage by certain (and probably most) methylation-sensitive restriction endonucleases making it indistinguishable from 5-methylcytosine by these approaches [62,63]. Moreover, 5-hydroxymethylcytosine in DNA is resistant to bisulfite-induced deamination similar as is 5-methylcytosine and scores identically in traditional bisulfite sequencing [63,64]. It has been shown that 5hmC reacts with bisulfite and, instead of leading to deamination, this reaction gives rise to cytosine-5-methylenesulfonate as the product [65]. Cytosine-5-methylenesulfonate is only very slowly deaminated by treatment with bisulfite. However, 5hmC is not recognized by both anti-5mC antibodies and MBD family proteins that bind to methylated CpGs [63]. Thus, using existing 5-methylcytosine-related technology, one needs to be careful if the goal is to distinguish between the two modified cytosine bases. This is particularly important in tissues where high levels of 5hmC are present, such as the brain.

Affinity or pulldown approaches have been used to map the distribution of 5hmC in the genome. One approach is based on anti-5hmC antibodies and is called hMeDIP, a technique analogous to MeDIP [66–68]. Another very specific approach for mapping of 5hmC is based on selective chemical labeling of 5hmC. This method employs the T4 bacteriophage beta-glucosyltransferase to transfer a modified glucose containing an azide group onto the hydroxyl group of 5hmC. The azide group is then used for coupling to biotin, affinity enrichment and DNA sequencing [69].

Although bisulfite-treatment cannot be used directly for identifying 5hmC at individual nucleotide positions, by employing some modifications—for example, before the actual bisulfite-treatment—the bisulfite approach can produce base resolution sequence information that unambiguously identifies positions of these modified cytosines in genomic DNA. Currently there are several methods that can be used for identification of 5hmC at single-base resolution. One of these methods, the recently developed CAPS, has been discussed earlier.

TET-ASSISTED BISULFITE SEQUENCING

TET-assisted Bisulfite Sequencing (TAB-seq) relies on the application of two enzymes; it first employs β-glucosyltransferase-mediated glucosylation of 5-hydroxymethylcytosine and then TET-catalyzed oxidation of 5-methylcytosine to 5-carboxylcytosine [70] (Fig. 8.8). This oxidation step requires a highly active TET catalytic domain [71,72]. Genomic DNA samples are treated with β-glucosyltransferase and the glycosylated 5-hydroxymethylcytosines are thus protected from subsequent TET-mediated oxidization. Next, bisulfite conversion of the resulting pretreated DNA samples is conducted, which converts unmodified cytosine and 5-carboxylcytosine residues to uracil, while leaving the glycosylation-protected 5hmC bases intact. Following that step, the BS-treated samples are used for sequencing library generation, and any Cs in the sequencing reads can be interpreted as 5hmC. This method has been used [73,74] but is relatively expensive due to the rarity of the 5hmC base, which requires sequence coverage 5–50 times greater than when 5mC is analyzed (depending on the tissue). A less expensive but more restricted methodology is the combination of TAB-seq and RRBS, which limits the analysis to mostly CpG-rich genomic regions [75].

OXIDATIVE BISULFITE SEQUENCING

In Oxidative bisulfite sequencing (OxBS-Seq), genomic DNA is treated with a strong oxidizing agent (KRuO$_4$) that induces the conversion of 5-hydroxymethylcytosines to 5-formylcytosine but leaves 5-methylcytosine and unmodified cytosine residues intact [76] (Fig. 8.9). Subsequent bisulfite treatment converts 5-formylcytosine and unmodified cytosine residues to uracil. Next, the bisulfite-treated DNA sample is used for sequencing library synthesis and is subjected to HTS. Of note, Ts in the sequencing reads can be either 5-hydroxymethylcytosines or unmodified cytosines; therefore, OxBS-Seq and conventional

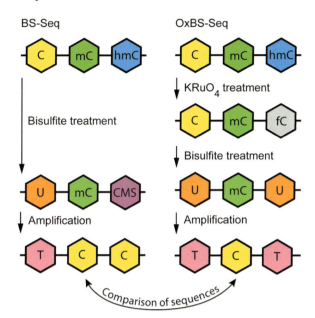

FIGURE 8.9 Oxidative bisulfite sequencing (OxBS-seq). OxBS-seq displays 5mC only and requires a comparison to standard bisulfite sequencing (BS-seq) to infer the sequence positions and abundance of 5hmC and 5mC.

bisulfite-seq data must be compared to each other for revealing the positions of 5hmC residues, which is a disadvantage relative to TAB-seq. The same limitations regarding high sequence coverage and cost also apply here.

APOBEC-COUPLED EPIGENETIC SEQUENCING

This method is based on consecutive enzymatic modifications circumventing many of the bisulfite treatment-associated WGBS challenges (Fig. 8.8) [77]. Unmodified cytosine and 5mC are deaminated by APOBEC3A in an enzymatic reaction during which DNA is not fragmented or degraded. The T4-βGT-glycosylated 5hmC residues, however, are resistant to APOBEC3A conversion and can be distinguished from Cs and 5mCs (Fig. 8.8). Conventional library preparation is employed for APOBEC-coupled epigenetic sequencing (ACE-Seq).

SMRT-SEQ AND NANOPORE SEQUENCING

Although enzyme-based methodologies provide several advantages over the classical bisulfite-based techniques, they are still not user-friendly enough and require specialized highly active enzymes, which may

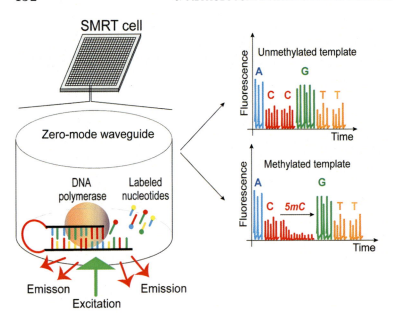

FIGURE 8.10 **Single-molecule, real-time (SMRT) sequencing.** Zero-mode waveguides (wells) on a SMRT cell (platform) are the places of the sequencing reaction. Fluorescently labeled nucleotides are built into single stranded DNA by DNA polymerase and release dye flashes of light. Nucleotide-specific signals are characteristic for C, 5mC and 5hmC and reveal the cytosine modification state at CpG sites.

be difficult to produce or expensive to purchase. In an optimal scenario, there would be no need to employ any chemicals or enzymes to map epigenetic modifications of DNA. The emerging third generation sequencing methods might qualify for the ideal approaches because native DNA is the input material, and no amplification and chemical conversion/enzymatic modifications are required for assessing the epigenetic state of DNA. Currently, there are two platforms, which show good potential towards achieving these goals. Single molecule real-time sequencing (SMRT-seq) and a nanopore-based technology were developed by two companies (i.e., Pacific Bioscience and Oxford Nanopore Technology) [78,79].

In SMRT-seq, a circular DNA molecule is generated by adapter ligation that is combined with a DNA polymerase, which incorporates fluorescently labeled nucleotides (Fig. 8.10). Incorporation of fluorescently labeled nucleotides affects polymerase kinetics allowing discrimination between the various epigenetic modifications including 5mC and 6 mA [78,80]. The SMRT-seq approach might allow mapping of DNA methylation patterns in repetitive sequences because of the long reads. However, because of the low levels of some epigenetic modifications (e.g., 5hmC), a large sequencing coverage is necessary, which significantly increases the cost for larger genomes.

Nanopore sequencing is based on a system in which DNA goes across a microscopic pore and the ionic current between the two separated chambers depends on which nucleotide is squeezing through the pore (Fig. 8.11) [81]. The sequence of the bases is identified by measuring the change in ionic current. Epigenetic modifications have an effect on current changes, which makes it possible to directly detect DNA methylation patterns [82].

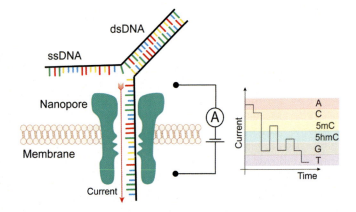

FIGURE 8.11 **Nanopore sequencing.** Nanopores generated from transmembrane proteins in lipid or solid membranes separate electrolytes. In the electric field, single stranded DNA is squeezed through the pores, and nucleotide- and epigenetic modification-specific currents can be measured.

The combination of long-range sequencing with bisulfite-free methods such as EM-seq or TAPS could offer a way to improve accuracy and eliminate the need for high sequencing depth [51,57,83].

SINGLE CELL WHOLE GENOME METHYLATION ANALYSIS

Many questions in biology are beginning to be addressed at the single cell level. DNA methylation analysis of single cells is challenging [84]. With traditional bisulfite sequencing, the degradation of DNA is a major issue, which will lead to dropout of alleles or chromosomal segments. Restriction enzyme cleavage and sequencing based methods can be combined and used for single cell analysis [85]. Also, EPIC arrays and RRBS have been employed for single cell analysis of tumor cell genomes [86,87]. Recent publications have shown that single cell methylation analysis by "whole genome" bisulfite sequencing with reasonably good overall genome coverage is possible for mouse brain cells [88,89]. Further improvements in this area are expected. The high costs for these types of studies are still a barrier that needs to be overcome.

FUTURE DIRECTIONS AND CHALLENGES

The different DNA methylation profiling techniques described in this chapter will continue to be applied in various settings of molecular analysis and probably will at some point move into the clinical arena for disease diagnosis and treatment stratification. The method of choice depends on many parameters, including the scope of analysis that is desired, the exact questions to be pursued, and sample size as well as cost. It is expected that the technological development of high-throughput approaches will continue at a strong pace, allowing cost effective analysis of large sample series with deep sequence coverage. The current challenges and cost surrounding whole genome epigenetic sequencing will probably be solved in the future. However, it may not always be desirable or even necessary to have information on the methylation status of every single one of the ~28 million CpG sites in the haploid human genome. For determining the identity and number of methylated genes, in cancer for example, a lower resolution, affinity-based approach should be fully sufficient, and for disease diagnosis, only a subset of specific genes will most likely be relevant.

Single molecule DNA sequencing includes a set of leading new technologies in which 5-methylcytosine or its oxidation products may be distinguished from the other four standard DNA bases [90]. Until now, these methods have not found widespread application yet; however, further improvements in this area are likely. Another topic of current interest is the cellular heterogeneity inherent to tissues, for example tumors. Single cell analysis of DNA methylation patterns by whole genome epigenetic analysis is challenging but is now beginning to be possible [47,88,89,91,92]. Methods that are based entirely on enzymatic conversion steps should hold the greatest promise in that regard.

Acknowledgments

This work was supported by NIH grants CA234595 and AR079174 to G.P.P.

References

[1] Wyatt GR. Occurrence of 5-methylcytosine in nucleic acids. Nature 1950;166(4214):237–8.
[2] Hotchkiss RD. The quantitative separation of purines, pyrimidines, and nucleosides by paper chromatography. J Biol Chem 1948;175(1):315–32.
[3] Holliday R, Pugh JE. DNA modification mechanisms and gene activity during development. Science 1975;187(4173):226–32.
[4] Riggs AD. X inactivation, differentiation, and DNA methylation. Cytogenet Cell Genet 1975;14(1):9–25.
[5] Rauch T, Li H, Wu X, Pfeifer GP. MIRA-assisted microarray analysis, a new technology for the determination of DNA methylation patterns, identifies frequent methylation of homeodomain-containing genes in lung cancer cells. Cancer Res 2006;66(16):7939–47.
[6] Rauch T, Pfeifer GP. Methylated-CpG island recovery assay: a new technique for the rapid detection of methylated-CpG islands in cancer. Lab Invest 2005;85(9):1172–80.
[7] Weber M, Davies JJ, Wittig D, Oakeley EJ, Haase M, Lam WL, et al. Chromosome-wide and promoter-specific analyses identify sites of differential DNA methylation in normal and transformed human cells. Nat Genet 2005;37(8):853–62.
[8] Gebhard C, Schwarzfischer L, Pham TH, Schilling E, Klug M, Andreesen R, et al. Genome-wide profiling of CpG methylation identifies novel targets of aberrant hypermethylation in myeloid leukemia. Cancer Res 2006;66(12):6118–28.
[9] Illingworth R, Kerr A, Desousa D, Jorgensen H, Ellis P, Stalker J, et al. A novel CpG island set identifies tissue-specific methylation at developmental gene loci. PLoS Biol 2008;6(1):e22.
[10] Li D, Zhang B, Xing X, Wang T. Combining MeDIP-seq and MRE-seq to investigate genome-wide CpG methylation. Methods 2015;72:29–40.
[11] Cross SH, Charlton JA, Nan X, Bird AP. Purification of CpG islands using a methylated DNA binding column. Nat Genet 1994;6(3):236–44.
[12] Zhang X, Yazaki J, Sundaresan A, Cokus S, Chan SW, Chen H, et al. Genome-wide high-resolution mapping and functional analysis of DNA methylation in arabidopsis. Cell 2006;126(6):1189–201.
[13] Harris RA, Wang T, Coarfa C, Nagarajan RP, Hong C, Downey SL, et al. Comparison of sequencing-based methods to profile DNA methylation and identification of monoallelic epigenetic modifications. Nat Biotechnol 2010;28(10):1097–105.
[14] Serre D, Lee BH, Ting AH. MBD-isolated Genome Sequencing provides a high-throughput and comprehensive survey of DNA methylation in the human genome. Nucleic Acids Res 2010;38(2):391–9.
[15] Brinkman AB, Simmer F, Ma K, Kaan A, Zhu J, Stunnenberg HG, et al. Whole-genome DNA methylation profiling using MethylCap-seq. Methods 2010;52(3):232–6.
[16] Subhash S, Andersson PO, Kosalai ST, Kanduri C, Kanduri M. Global DNA methylation profiling reveals new insights into epigenetically deregulated protein coding and long noncoding RNAs in CLL. Clin Epigenetics 2016;8:106.
[17] Jiang CL, Jin SG, Pfeifer GP. MBD3L1 is a transcriptional repressor that interacts with methyl-CpG-binding protein 2 (MBD2) and components of the NuRD complex. J Biol Chem 2004;279(50):52456–64.
[18] Rauch T, Wang Z, Zhang X, Zhong X, Wu X, Lau SK, et al. Homeobox gene methylation in lung cancer studied by genome-wide analysis with a microarray-based methylated CpG island recovery assay. Proc Natl Acad Sci U S A 2007;104(13):5527–32.

[19] Rauch TA, Zhong X, Wu X, Wang M, Kernstine KH, Wang Z, et al. High-resolution mapping of DNA hypermethylation and hypomethylation in lung cancer. Proc Natl Acad Sci U S A 2008;105(1):252−7.

[20] Rauch TA, Pfeifer GP. The MIRA method for DNA methylation analysis. Methods Mol Biol 2009;507:65−75.

[21] Rauch TA, Wu X, Zhong X, Riggs AD, Pfeifer GP. A human B cell methylome at 100-base pair resolution. Proc Natl Acad Sci U S A 2009;106(3):671−8.

[22] Jung M, Kadam S, Xiong W, Rauch TA, Jin SG, Pfeifer GP. MIRA-seq for DNA methylation analysis of CpG islands. Epigenomics 2015;7(5):695−706.

[23] Jin SG, Xiong W, Wu X, Yang L, Pfeifer GP. The DNA methylation landscape of human melanoma. Genomics 2015;106(6):322−30.

[24] Almamun M, Levinson BT, Gater ST, Schnabel RD, Arthur GL, Davis JW, et al. Genome-wide DNA methylation analysis in precursor B-cells. Epigenetics 2014;9(12):1588−95.

[25] Lee EJ, Luo J, Wilson JM, Shi H. Analyzing the cancer methylome through targeted bisulfite sequencing. Cancer Lett 2013;340(2):171−8.

[26] Benjamin AL, Green BB, Crooker BA, McKay SD, Kerr DE. Differential responsiveness of Holstein and Angus dermal fibroblasts to LPS challenge occurs without major differences in the methylome. BMC Genomics 2016;17:258.

[27] Hayatsu H. Discovery of bisulfite-mediated cytosine conversion to uracil, the key reaction for DNA methylation analysis—a personal account. Proc Jpn Acad Ser B Phys Biol Sci 2008;84(8):321−30.

[28] Hayatsu H, Wataya Y, Kazushige K. The addition of sodium bisulfite to uracil and to cytosine. J Am Chem Soc 1970;92(3):724−6.

[29] Clark SJ, Harrison J, Paul CL, Frommer M. High sensitivity mapping of methylated cytosines. Nucleic Acids Res 1994;22(15):2990−7.

[30] Frommer M, McDonald LE, Millar DS, Collis CM, Watt F, Grigg GW, et al. A genomic sequencing protocol that yields a positive display of 5-methylcytosine residues in individual DNA strands. Proc Natl Acad Sci U S A 1992;89(5):1827−31.

[31] Meissner A, Mikkelsen TS, Gu H, Wernig M, Hanna J, Sivachenko A, et al. Genome-scale DNA methylation maps of pluripotent and differentiated cells. Nature 2008;454(7205):766−70.

[32] Meissner A, Gnirke A, Bell GW, Ramsahoye B, Lander ES, Jaenisch R. Reduced representation bisulfite sequencing for comparative high-resolution DNA methylation analysis. Nucleic Acids Res 2005;33(18):5868−77.

[33] Gu H, Smith ZD, Bock C, Boyle P, Gnirke A, Meissner A. Preparation of reduced representation bisulfite sequencing libraries for genome-scale DNA methylation profiling. Nat Protoc 2011;6(4):468−81.

[34] Ball MP, Li JB, Gao Y, Lee JH, LeProust EM, Park IH, et al. Targeted and genome-scale strategies reveal gene-body methylation signatures in human cells. Nat Biotechnol 2009;27(4):361−8.

[35] Deng J, Shoemaker R, Xie B, Gore A, LeProust EM, Antosiewicz-Bourget J, et al. Targeted bisulfite sequencing reveals changes in DNA methylation associated with nuclear reprogramming. Nat Biotechnol 2009;27(4):353−60.

[36] Lister R, Ecker JR. Finding the fifth base: genome-wide sequencing of cytosine methylation. Genome Res 2009;19(6):959−66.

[37] Du Y, Li M, Chen J, Duan Y, Wang X, Qiu Y, et al. Promoter targeted bisulfite sequencing reveals DNA methylation profiles associated with low sperm motility in asthenozoospermia. Hum Reprod 2016;31(1):24−33.

[38] Pabinger S, Ernst K, Pulverer W, Kallmeyer R, Valdes AM, Metrustry S, et al. Analysis and visualization tool for targeted amplicon bisulfite sequencing on Ion Torrent sequencers. PLoS One 2016;11(7):e0160227.

[39] Hahn MA, Jin SG, Li AX, Liu J, Huang Z, Wu X, et al. Reprogramming of DNA methylation at NEUROD2-bound sequences during cortical neuron differentiation. Sci Adv 2019;5 eaax0080.

[40] Masser DR, Berg AS, Freeman WM. Focused, high accuracy 5-methylcytosine quantitation with base resolution by benchtop next-generation sequencing. Epigenetics Chromatin 2013;6(1):33.

[41] Dedeurwaerder S, Defrance M, Calonne E, Denis H, Sotiriou C, Fuks F. Evaluation of the infinium methylation 450K technology. Epigenomics 2011;3(6):771−84.

[42] Moran S, Arribas C, Esteller M. Validation of a DNA methylation microarray for 850,000 CpG sites of the human genome enriched in enhancer sequences. Epigenomics 2016;8(3):389−99.

[43] Cokus SJ, Feng S, Zhang X, Chen Z, Merriman B, Haudenschild CD, et al. Shotgun bisulphite sequencing of the Arabidopsis genome reveals DNA methylation patterning. Nature 2008;452(7184):215−19.

[44] Lister R, Pelizzola M, Dowen RH, Hawkins RD, Hon G, Tonti-Filippini J, et al. Human DNA methylomes at base resolution show widespread epigenomic differences. Nature 2009;462:315−22.

[45] Mo A, Luo C, Davis FP, Mukamel EA, Henry GL, Nery JR, et al. Epigenomic landscapes of retinal rods and cones. eLife 2016;5:e11613.

[46] Schultz MD, He Y, Whitaker JW, Hariharan M, Mukamel EA, Leung D, et al. Human body epigenome maps reveal noncanonical DNA methylation variation. Nature 2015;523(7559):212−16.

[47] Smallwood SA, Lee HJ, Angermueller C, Krueger F, Saadeh H, Peat J, et al. Single-cell genome-wide bisulfite sequencing for assessing epigenetic heterogeneity. Nat Methods 2014;11(8):817−20.

[48] Ehrich M, Nelson MR, Stanssens P, Zabeau M, Liloglou T, Xinarianos G, et al. Quantitative high-throughput analysis of DNA methylation patterns by base-specific cleavage and mass spectrometry. Proc Natl Acad Sci U S A 2005;102(44):15785−90.

[49] Olk-Batz C, Poetsch AR, Nollke P, Claus R, Zucknick M, Sandrock I, et al. Aberrant DNA methylation characterizes juvenile myelomonocytic leukemia with poor outcome. Blood 2011;117(18):4871−80.

[50] Olova N, Krueger F, Andrews S, Oxley D, Berrens RV, Branco MR, et al. Comparison of whole-genome bisulfite sequencing library preparation strategies identifies sources of biases affecting DNA methylation data. Genome Biol 2018;19(1):33.

[51] Sun Z, Vaisvila R, Hussong LM, Yan B, Baum C, Saleh L, et al. Nondestructive enzymatic deamination enables single-molecule long-read amplicon sequencing for the determination of 5-methylcytosine and 5-hydroxymethylcytosine at single-base resolution. Genome Res 2021;31(2):291−300.

[52] Feng S, Zhong Z, Wang M, Jacobsen SE. Efficient and accurate determination of genome-wide DNA methylation patterns in Arabidopsis thaliana with enzymatic methyl sequencing. Epigenetics Chromatin 2020;13(1):42.

[53] Vaisvila R, Ponnaluri VKC, Sun Z, Langhorst BW, Saleh L, Guan S, et al. Enzymatic methyl sequencing detects DNA methylation at single-base resolution from picograms of DNA. Genome Res 2021;31(7):1280−9.

[54] Liu Y, Siejka-Zielinska P, Velikova G, Bi Y, Yuan F, Tomkova M, et al. Bisulfite-free direct detection of 5-methylcytosine and 5-hydroxymethylcytosine at base resolution. Nat Biotechnol 2019;37(4):424−9.

REFERENCES

[55] Liu Y, Hu Z, Cheng J, Siejka-Zielinska P, Chen J, Inoue M, et al. Subtraction-free and bisulfite-free specific sequencing of 5-methylcytosine and its oxidized derivatives at base resolution. Nat Commun 2021;12(1):618.

[56] Zeng H, He B, Xia B, Bai D, Lu X, Cai J, et al. Bisulfite-Free, nanoscale analysis of 5-hydroxymethylcytosine at single base resolution. J Am Chem Soc 2018;140(41):13190–4.

[57] Liu Y, Cheng J, Siejka-Zielinska P, Weldon C, Roberts H, Lopopolo M, et al. Accurate targeted long-read DNA methylation and hydroxymethylation sequencing with TAPS. Genome Biol 2020;21(1):54.

[58] Cheng J, Siejka-Zielinska P, Liu Y, Chandran A, Kriaucionis S, Song CX. Endonuclease enrichment TAPS for cost-effective genome-wide base-resolution DNA methylation detection. Nucleic Acids Res 2021;49(13):e76.

[59] Kriaucionis S, Heintz N. The nuclear DNA base 5-hydroxymethylcytosine is present in Purkinje neurons and the brain. Science 2009;324(5929):929–30.

[60] Tahiliani M, Koh KP, Shen Y, Pastor WA, Bandukwala H, Brudno Y, et al. Conversion of 5-methylcytosine to 5-hydroxymethylcytosine in mammalian DNA by MLL partner TET1. Science 2009;324(5929):930–5.

[61] Hahn MA, Szabo PE, Pfeifer GP. 5-Hydroxymethylcytosine: a stable or transient DNA modification? Genomics 2014;104:314–23.

[62] Tardy-Planechaud S, Fujimoto J, Lin SS, Sowers LC. Solid phase synthesis and restriction endonuclease cleavage of oligodeoxynucleotides containing 5-(hydroxymethyl)-cytosine. Nucleic Acids Res 1997;25(3):553–9.

[63] Jin SG, Kadam S, Pfeifer GP. Examination of the specificity of DNA methylation profiling techniques towards 5-methylcytosine and 5-hydroxymethylcytosine. Nucleic Acids Res 2010;38(11):e125.

[64] Huang Y, Pastor WA, Shen Y, Tahiliani M, Liu DR, Rao A. The behaviour of 5-hydroxymethylcytosine in bisulfite sequencing. PLoS One 2010;5(1):e8888.

[65] Hayatsu H, Shiragami M. Reaction of bisulfite with the 5-hydroxymethyl group in pyrimidines and in phage DNAs. Biochemistry 1979;18(4):632–7.

[66] Jin SG, Wu X, Li AX, Pfeifer GP. Genomic mapping of 5-hydroxymethylcytosine in the human brain. Nucleic Acids Res 2011;39:5015–24.

[67] Stroud H, Feng S, Morey Kinney S, Pradhan S, Jacobsen SE. 5-Hydroxymethylcytosine is associated with enhancers and gene bodies in human embryonic stem cells. Genome Biol 2011;12(6):R54.

[68] Williams K, Christensen J, Pedersen MT, Johansen JV, Cloos PA, Rappsilber J, et al. TET1 and hydroxymethylcytosine in transcription and DNA methylation fidelity. Nature 2011;473(7347):343–8.

[69] Song CX, Szulwach KE, Fu Y, Dai Q, Yi C, Li X, et al. Selective chemical labeling reveals the genome-wide distribution of 5-hydroxymethylcytosine. Nat Biotechnol 2011;29(1):68–72.

[70] Yu M, Hon GC, Szulwach KE, Song CX, Jin P, Ren B, et al. Tet-assisted bisulfite sequencing of 5-hydroxymethylcytosine. Nat Protoc 2012;7(12):2159–70.

[71] Huang Z, Meng Y, Szabo PE, Kohli RM, Pfeifer GP. High-resolution analysis of 5-hydroxymethylcytosine by TET-assisted bisulfite sequencing. Methods Mol Biol 2021;2198:321–31.

[72] Huang Z, Yu J, Johnson J, Jin SG, Pfeifer GP. Purification of TET proteins. Methods Mol Biol 2021;2272:225–37.

[73] Chen K, Zhang J, Guo Z, Ma Q, Xu Z, Zhou Y, et al. Loss of 5-hydroxymethylcytosine is linked to gene body hypermethylation in kidney cancer. Cell Res 2016;26(1):103–18.

[74] Hahn MA, Qiu R, Wu X, Li AX, Zhang H, Wang J, et al. Dynamics of 5-hydroxymethylcytosine and chromatin marks in Mammalian neurogenesis. Cell Reports 2013;3(2):291–300.

[75] Hahn MA, Li AX, Wu X, Pfeifer GP. Single base resolution analysis of 5-methylcytosine and 5-hydroxymethylcytosine by RRBS and TAB-RRBS. Methods Mol Biol 2015;1238:273–87.

[76] Booth MJ, Ost TW, Beraldi D, Bell NM, Branco MR, Reik W, et al. Oxidative bisulfite sequencing of 5-methylcytosine and 5-hydroxymethylcytosine. Nat Protoc 2013;8(10):1841–51.

[77] Schutsky EK, DeNizio JE, Hu P, Liu MY, Nabel CS, Fabyanic EB, et al. Nondestructive, base-resolution sequencing of 5-hydroxymethylcytosine using a DNA deaminase. Nat Biotechnol 2018;36:1083–90.

[78] Flusberg BA, Webster DR, Lee JH, Travers KJ, Olivares EC, Clark TA, et al. Direct detection of DNA methylation during single-molecule, real-time sequencing. Nat Methods 2010;7(6):461–5.

[79] Laszlo AH, Derrington IM, Brinkerhoff H, Langford KW, Nova IC, Samson JM, et al. Detection and mapping of 5-methylcytosine and 5-hydroxymethylcytosine with nanopore MspA. Proc Natl Acad Sci U S A 2013;110(47):18904–9.

[80] Tse OYO, Jiang P, Cheng SH, Peng W, Shang H, Wong J, et al. Genome-wide detection of cytosine methylation by single molecule real-time sequencing. Proc Natl Acad Sci U S A 2021;118(5) e2019768118.

[81] Stoddart D, Heron AJ, Mikhailova E, Maglia G, Bayley H. Single-nucleotide discrimination in immobilized DNA oligonucleotides with a biological nanopore. Proc Natl Acad Sci U S A 2009;106(19):7702–7.

[82] Schatz MC. Nanopore sequencing meets epigenetics. Nat Methods 2017;14(4):347–8.

[83] Sakamoto Y, Zaha S, Nagasawa S, Miyake S, Kojima Y, Suzuki A, et al. Long-read whole-genome methylation patterning using enzymatic base conversion and nanopore sequencing. Nucleic Acids Res 2021;49(14):e81.

[84] Karemaker ID, Vermeulen M, Single-Cell DNA. Methylation profiling: technologies and biological applications. Trends Biotechnol 2018;36(9):952–65.

[85] Niemoller C, Wehrle J, Riba J, Claus R, Renz N, Rhein J, et al. Bisulfite-free epigenomics and genomics of single cells through methylation-sensitive restriction. Commun Biol 2021;4(1):153.

[86] Chaligne R, Gaiti F, Silverbush D, Schiffman JS, Weisman HR, Kluegel L, et al. Epigenetic encoding, heritability and plasticity of glioma transcriptional cell states. Nat Genet 2021;53(10):1469–79.

[87] Johnson KC, Anderson KJ, Courtois ET, Gujar AD, Barthel FP, Varn FS, et al. Single-cell multimodal glioma analyses identify epigenetic regulators of cellular plasticity and environmental stress response. Nat Genet 2021;53(10):1456–68.

[88] Liu H, Zhou J, Tian W, Luo C, Bartlett A, Aldridge A, et al. DNA methylation atlas of the mouse brain at single-cell resolution. Nature 2021;598(7879):120–8.

[89] Zhang Z, Zhou J, Tan P, Pang Y, Rivkin AC, Kirchgessner MA, et al. Epigenomic diversity of cortical projection neurons in the mouse brain. Nature 2021;598(7879):167–73.

[90] Korlach J, Turner SW. Going beyond five bases in DNA sequencing. Curr Opin Struct Biol 2012;22(3):251–61.

[91] Farlik M, Sheffield NC, Nuzzo A, Datlinger P, Schonegger A, Klughammer J, et al. Single-cell DNA methylome sequencing and bioinformatic inference of epigenomic cell-state dynamics. Cell Rep 2015;10(8):1386–97.

[92] Gravina S, Dong X, Yu B, Vijg J. Single-cell genome-wide bisulfite sequencing uncovers extensive heterogeneity in the mouse liver methylome. Genome Biol 2016;17(1):150.

CHAPTER

9

Genome-wide Analyses of Histone Modifications in the Mammalian Genome

Shulan Tian, Susan L. Slager, Eric W. Klee and Huihuang Yan

Division of Computational Biology, Department of Quantitative Health Sciences, Mayo Clinic, Rochester, MN, United States

OUTLINE

Introduction	137
Histone Modifications in Regulatory Regions	138
Promoter and Enhancer	138
Silencer	139
High-throughput Assays for Mapping Histone Modifications	139
Histone Modification Profiling with Low Input and Single Cell	140
Joint Profiling of Histone Modification and Transcriptome in Single Cell	144
Genome-wide Mapping of Histone Modifications	145
Acylation	145
Homocysteinylation	148
Monoaminylation	148
O-GlcNAcylation	148
Catalog of Histone Modification Profiles	149
Variation of Histone Modifications	149
Genetic Variants alter Histone Modifications	149
Alteration of Histone Modifications in Disease	153
Conclusions and Perspectives	153
ChIP-seq Data Analysis Workflow	154
Equipment	154
Test Data	154
Procedure	154
Timing	157
Expected Outcomes	157
Troubleshooting	157
Acknowledgments	157
Abbreviations	157
References	158

INTRODUCTION

The core histones are subject to a diverse array of posttranslational modifications, mainly on the N-terminal tails, which are responsible for the establishment and maintenance of distinct chromatin states [1,2]. Beyond histone methylation, acetylation, phosphorylation, and ubiquitination, numerous novel types of modifications were recently discovered [3].

Chromatin organization is disrupted during DNA replication. Two repressive marks, H3K27me3 and H3K9me3, but none of the active marks, are found epigenetically inherited; after DNA replication, euchromatin is re-established by H4K16ac [4]. However, chromatin occupancy after replication and sequencing (ChOR-seq) data support that, for both active and repressive marks, positional information of the modified parental histones in the genome is faithfully inherited to daughter DNA strands during DNA replication [2]. After DNA replication, the restoration of modification level via de novo histone methylation occurs in regions pre-demarcated by modified parental histones.

Histone modifications play central roles in controlling gene expression [1,5,6]. They are lightly associated with other cellular processes as well, including chromatin architecture [7], cell cycle [8], cell memory [2], DNA damage repair [8–10], and alternative splicing [11–16]. This chapter will focus on the recent development of epigenomic assays for histone modification profiling from low input and single cells, novel histone marks and their genome-wide distribution, epigenetic features of cis-regulatory elements, as well as changes of histone marks during development and disease progression. At the end, areas for future research are also described.

HISTONE MODIFICATIONS IN REGULATORY REGIONS

There are two major classes of cis-regulatory elements: promoters and enhancers that promote gene expression, in contrast to silencers that repress gene expression [17,18] and insulators that block promoters from interactions with other regulatory elements [19]. While promoters are capable of initiating gene expression locally, enhancers can activate gene expression remotely in a cell type specific manner [6,20].

Promoter and Enhancer

Active promoter regions show enrichment of H3K4me3 and acetylation of histone H3 or H4 (like H3K27ac and H3K9ac), and are bound by chromatin regulatory complexes [20]. H3K4me1 is known to play a role in delineating the boundaries of active promoters [20]. Active enhancers are located in open chromatin regions and typically enriched with H3K4me1, H3K27ac and histone variants H2A.Z [21,22]. In several cell types, such as MLL-AF4 leukemia cells, human embryonic stem cells, and GM12878 lymphoblastoid cells, a new type of enhancers are identified that are enriched with H3K79me2/3 instead [23,24].

Mechanistically, MLL3/4 binds to enhancers, which facilitates recruitment of P300 to acetylate H3K27 [25]. H3K4me1 is a hallmark of the initial enhancer activation [22]. H3K4me1 supports H3K27ac, as reflected by partial reduction of H3K27ac level in H3K4me1-lack, MLL3/4 catalytically deficient mouse embryonic stem cells [25]. Two additional roles were established for H3K4me1 at enhancers. First, H3K4me1 increases the binding of the BAF complex, a SWI/SNF chromatin-remodeling complex, to enhancers, where H3K4me1-modified nucleosomes are more efficiently remodeled by BAF in vitro [20]. Second, in mouse embryonic stem cells, H3K4me1 plays an active role in establishing long-range chromatin interactions by facilitating the recruitment of the cohesin complex at enhancers [22]. Enhancer regulates target gene expression by the recruitment of transcription factors (TFs) and cofactors [22,26].

Several databases have been developed to host human tissue-specific enhancers [27], experimentally validated human and mouse enhancers [28], cell type-specific transcribed enhancers in human [29], as well as super-enhancers identified from over 240 human tissues and cell types [30]. These public databases provide valuable resources to study cell type-specific gene regulation and functionally annotate disease risk loci.

Super-enhancer

Super-enhancers are large clusters of active enhancers with unusually high activity, binding of TFs, cofactors and chromatin regulators, as well as higher frequencies of chromatin interactions compared with typical enhancers [30–32]. Super-enhancers are preferentially associated with genes encoding cell type-specific TFs with lineage-specific roles [30,32]. It has been well established that tumor cells often acquire super-enhancers to drive oncogenic gene expression [32].

Many super-enhancers are hierarchically organized, containing small functional sub-domains and large tracts of dispensable elements that lack enhancer activity [31]. The functional sub-domains bound by cohesin and TFs are responsible for super-enhancer function in activating gene expression [31,33].

Shadow Enhancer

While a single enhancer can target multiple genes, a group of enhancers, called shadow enhancers, can work in concert to regulate expression of the same gene [34]. Enhancer redundancy is commonly seen in the human and mouse genome. By correlating expression level of enhancer RNAs (eRNAs) with that of transcription start sites (TSSs) of 2206 genes over hundreds of human cell lines and tissues, on average each TSS was found to be associated with five enhancers within 500 kb [35]. A separate study of 1058 genes uncovered that five or more functionally redundant enhancers were linked to the same gene during mouse development [36].

Given the high functional redundancy, many loss-of-function mutations in shadow enhancers will cause only subtle phenotypes, if any, in humans [34,36]. Thus, enhancer redundancy may represent a protective mechanism, by which noncoding mutations in individual enhancers will not affect target gene expression [34].

To better understand the extent of functional redundancy, CRISPR-Cas9 genome editing was used to create 23 deletion lines and inter-crosses in the mouse, including deletions of a single or two enhancers simultaneously from seven regulatory regions required for limb development [36]. For the ten events, each involving a single enhancer deletion, none of them caused noticeable changes in limb morphology and H3K27ac level was overall unaffected for the untargeted enhancers. However, deleting pairs of limb enhancers near the same gene caused discernible phenotypes. A similar phenomenon was observed for the regulation of HoxA gene expression in mouse embryonic stem cells [37]. While the deletion of two retinoic acid responsive enhancers individually had no obvious impact on the expression of HoxA genes, double-knockout of both enhancers significantly reduced the expression of HoxA genes upon retinoic acid induction. These findings support the redundant roles for shadow enhancers in regulating gene expression.

Dual Activity of Promoter and Enhancer

Enhancers and promoters are surprisingly similar in chromatin state and sequence architectures; both are located in nucleosome-free regions, bound by RNA polymerase II (RNAPII) and TFs, and are capable of initiating bidirectional transcription [18].

At active promoters, aside from the transcription of protein-coding genes, short noncoding RNAs, known as upstream antisense RNAs (uaRNAs), are also transcribed from divergently oriented TSSs by a separate RNAPII complex that is positioned at the upstream edge of nucleosome-free region in the proximal promoter [26]. Similarly at active enhancers, enhancer RNAs (eRNAs) are transcribed bidirectionally from TF binding sites at the edges of nucleosome-free region [18,26,38]. Of the ~65,000 enhancers annotated by the FANTOM project [35], ~16,000 had detectable levels of eRNAs based on RNA-seq data from 8928 cancer samples in The Cancer Genome Atlas (TCGA), suggesting widespread transcription at active enhancers [39].

There is compelling evidence that promoters can have enhancer activity and vice versa [6,18]. For example, active enhancers can work as promoters and initiate transcription locally at their boundaries [18]. For the promoters with enhancer activity, their function as enhancer vs as core-promoter often does not co-occur [26]; their role as enhancers to activate transcription from core promoters also involves long-range chromatin interactions [6,26,40]. Using CapSTARR-seq, 2.4% and 3% of core promoters from all coding genes showed enhancer activity in human HeLa and K562 cells, respectively [41]. Whole-genome STARR-seq, which was designed on top of the Self-Transcribing Active Regulatory Regions sequencing (STARR-seq) previously developed in fly and CapSTARR-seq, identified about 1% of the active enhancers that overlap TSSs in LNCaP human prostate cancer cell line [6,42].

These lines of evidence indicate that enhancers and promoters may be functionally interchangeable rather than totally exclusive. It is proposed that cis-regulatory elements can potentially have varying capability of being a promoter and an enhancer [18].

Silencer

Silencers are regulatory DNA elements in the genome that can repress gene expression [43]. Emerging evidence supports a role for silencers in determining lineage specificity and cell fate [44]. Unlike enhancers that have been the focus of extensive investigations in terms of their genomic locations, cell type specificity, functionality, regulatory mechanisms, and experimental validation, there is a knowledge gap in our understanding of silencers, for which the identification and validation remain challenging [17,44–46].

H3K27me3 is a well-known mark for silencers [17,46]. Repressive marks H4K20me1 and H3K9me3 are also associated with silencers [46]. In addition, in mouse embryonic stem cells, systematic analysis of Polycomb repressive complex 2 (PRC2)-bound silencers revealed co-enrichment of H3K27me3 and H3K4me1, along with enrichment of H3K4me3 and assay for transposase-accessible chromatin sequencing (ATAC-seq) signal [45]. Besides histone marks, transcriptional repressors are also found to bind to silencers [17,44].

The identity of a regulatory region functioning as an enhancer or a silencer may be context-dependent, relying on the binding of coactivators or corepressors [45,47]. It is estimated 16% of candidate silencers in mouse and 25% of those in human from one cell type can act as enhancers in other cell type(s), showing enrichment of enhancer marks (DNase I hypersensitive sites [DHS], H3K4me1 and H3K27ac) [44]. Resembling the mechanisms of enhancer action, distant silencers are thought to direct transcriptional repression via long-range chromatin interactions [43,45]. Within the topologically associating domain (TAD), candidate silencers can each target multiple genes and multiple silencers can also interact with a single gene [44,46].

HIGH-THROUGHPUT ASSAYS FOR MAPPING HISTONE MODIFICATIONS

Conventional ChIP-seq typically requires over 10,000 cells due to sample loss over multiple steps and low immunoprecipitation efficiency [48,49]. It also suffers from low signal and high noise, thus requiring high sequencing depth to identify sufficient peaks above backgrounds [50,51]. In particular, the epigenome, such as DNA methylation and histone modifications, is highly cell-type specific, which cannot be resolved with ChIP-seq in bulk tissues [52]. Thus, developing alternative methods for high-throughput mapping of histone modifications from low input, and even at single-cell resolution, will reveal cell-type specific regulatory mechanisms during development and disease progression [50,53]. While single-cell ATAC-seq (scATAC-seq) has accelerated the discovery of a cell-type specific gene regulation program, single-cell technologies for mapping histone modifications have just started to emerge [50,54], including those for joint analysis with transcriptome [52,55].

Histone Modification Profiling with Low Input and Single Cell

Numerous ChIP-seq alternatives have been developed to map histone modifications with low input. Most methods use the enzyme/antibody-tethering approaches to achieve chromatin cleavage or tagmentation in situ under native conditions (Table 9.1). These aforementioned methods, as described below, have the advantage of high signal-to-noise ratios and requiring low sequencing depths. They use protein A-MNase (pA-MN) fusion protein [51,56–58], protein A-Tn5 transposase (pA-Tn5) fusion protein [5,49,50,53,55,57,59], Tn5 transposase tagging [48,60], or antibody-MNase/protein A-MNase conjugate [54]. On the other hand, several microfluidics-based protocols have also been developed (Table 9.1). For example, the scChIP-seq protocol, which adopts droplet microfluidics coupled with single-cell DNA barcoding, was used to map H3K27me3 in breast cancer patient-derived xenograft samples [61].

ChIL-seq and scChIL-seq

Harada et al. [48] developed chromatin integration labeling (ChIL), a immunoprecipitation-free technique combining immunostaining, Tn5 transposase tagging and RNA-mediated linear amplification. ChIL-seq can map H3K4me3, H3K27ac, and H3K27me3 using only 100 cells, and the single-cell protocol (scChIL-seq) reduces the number of cells to 5. The main drawbacks are the reduced sensitivity to the detection of heterochromatin marks and long time (3–4 days) due to several steps.

CUT&RUN and CUT&Tag

Cleavage under targets and release using nuclease (CUT&RUN) combines chromatin immunocleavage (ChIC) with sequencing to profile histone modifications with only 100 cells and TF binding with 1000 cells [56,67]. CUT&RUN maps regions of chromatin protein occupancy with antibody binding by tethering the pA-MNase fusion protein in permeabilized cells. After activation of MNase cleavage with Ca^{++} addition, DNA fragments from targeted sites are extracted from the supernatant, and subjected to library preparation and sequencing [49,56]. A hybrid protein A-protein G-MNase (pAG-MNase) fusion protein, was later developed with enhanced binding to most commercial antibodies [51]. Because cleavages are confined only to regions around binding sites and undigested chromatin from vast majority of the genome is largely insoluble, CUT&RUN has a much lower background and requires only about 10% of the sequencing depth of ChIP-seq. This method is effective in mapping protein-DNA interactions and repressive marks (such as H3K27me3) in compacted chromatin regions [56]. Importantly, the carry-over *Escherichia coli* DNA from purification of pA-MNase or pAG-MNase can be used to normalize the data, eliminating the need of spike-in control. uliCUT&RUN, short for ultra-low-input CUT&RUN, is a variant of CUT&RUN, with key modifications to minimize background signal and reduce number of cells in starting material [57]. uliCUT&RUN can map CTCF and H3K4me3 from only 10 mouse embryonic stem cells [58].

Cut&RUN requires DNA end polishing and adapter ligation in constructing sequencing libraries and is not easily transformed into single-cell application [49]. cleavage under targets and tagmentation (CUT&Tag) uses a Tn5 transposase-protein A (pA-Tn5) fusion protein, which achieves tagmentation of chromatin at antibody-bound sites following activation of transposase by addition of Mg^{++} [49]. CUT&Tag can profile histone modifications (like H3K4me1, H3K4me2, and H3K27me3), RNAPII and CTCF binding from as few as 60 cells. With the exceptionally high efficiency, CUT&Tag needs only one-fourth of the sequencing depth of CUT&RUN. Single-cell CUT&Tag (scCUT&Tag) works for both H3K27me3 and H3K4me2 [49]. Another scCUT&Tag protocol combines bulk CUT&Tag with droplet-based 10x Genomics scATAC-seq library preparation [50]. It was used to map both active (H3K4me3, H3K27ac and H3K36me3) and repressive histone marks (H3K27me3), as well as TF (OLIG2) and the cohesin complex subunit RAD21. Analysis of single-cell histone mark data identified all major cell types in the central nervous system from the mouse brain, and further revealed H3K4me3 heterogeneity of subpopulations by integrating with scRNA-seq data [50]. scCUT&Tag profiles help deconvolute key regulatory principles such as H3K4me3-H3K27me3 bivalency at promoters, lineage-restricted spreading of H3K4me3, and promoter-enhancer connections [50].

TABLE 9.1 Methods for Mapping Histone Modifications and DNA Binding Proteins with Low Input

Method	Mark	Cell type	Cell number	Key features	Ref.
CUT&RUN	H3K4me2, H3K27ac, and H3K27me3	Human K562 cell line	100–8000	Protein A fused to MNase (pA-MN fusion protein) or a hybrid protein A-protein G-MNase construct (pAG-MNase)	[51,56]
uliCUT&RUN	H3K4me3	Mouse embryonic stem cells	10–500,000	pA-MN fusion protein	[57,58]
CUT&Tag	H3K4me1, H3K4me2, and H3K27me3	Human K562 cell line	60–100,000	Protein A fused to Tn5 transposase (pA-Tn5 fusion protein)	[49]
scCUT&Tag	H3K4me2 and H3K27me3	Human K562 cell line	Single cell	pA-Tn5 fusion protein	[49]
scCUT&Tag	H3K4me3, H3K27ac, H3K27me3, and H3K36me3	Mouse brain	Single cell	Bulk CUT&Tag combined with droplet-based 10x Genomics single-cell ATAC-seq	[50]
CUTAC	H3K4me2 and H3K4me3	Human K562 cell line	60–60,000	Low-salt CUT&Tag method with antibody against H3K4me2 or H3K4me3, which enables antibody-tethered tagmentation of accessible DNA regions	[59]
ChIL-seq/ scChIL-seq	H3K4me3, H3K27ac, H3K27me3, and H3K9me3	Mouse C2C12 myoblast cells	100–1000 (down to 5 cells by scChIL-seq)	Oligonucleotide-conjugated antibody; combines immunostaining, Tn5 transposase tagging and linear amplification	[48]
ACT-seq/iACT-seq	H3K4me1, H3K4me3, H3K4me3, H2K27ac, and H2A.Z	Human HEK293T cells	1000–50,000 (single cell by iACT-seq)	pA-Tn5 fusion protein is first bound to an antibody to allow antibody-guided chromatin fragmentation	[5]
scChIC-seq	H3K4me3 and H3K27me3	Human white blood cells	Single cell	Targeted chromatin cleavage by MNase via antibody-MNase or protein A-MNase conjugate	[54]
CoBATCH	H3K27ac, H3K36me3, and RNAPII	Mouse embryo (for H3K27ac) and heart tissue (for H3K36me3 and RNAPII)	Single cell	pA-Tn5 fusion protein-tethered chromatin fragmentation with or without crosslinking	[53]
itChIP-seq/sc-itChIP-seq	H3K4me3, H3K27me3, H3K27ac, and P300	Mouse embryonic stem cells; mouse heart and embryo tissue (for H3K27ac)	100–10,000 (for H3K4me3, H3K27me3, and P300); single cell (for H3K27ac)	Tn5 transposase-based chromatin tagmentation coupled with primers for barcoding and PCR amplification	[60]
MOWChIP-seq	H3K4me3, H3K27ac, H3K27me3, H3K9me3, H3K36me3, and H3K79me2	Human lymphoblastoid cell line (H3K4me3 and H3K27ac); mouse fetal liver (H3K27ac); mouse prefrontal cortex (all marks)	100–10,000 cells; 1000 cells (mouse prefrontal cortex)	A microfluidics-based protocol for low-input ChIP-seq; improve the quantity and quality of ChIP-enriched DNA through the use of a packed bed of antibody-coated beads for high-efficiency ChIP, followed by effective oscillatory washing to remove nonspecific adsorption; support up to eight reactions in parallel with the eight-unit device	[62,63]
SurfaceChIP-seq	H3K4me3, H3K27ac, and H3K27me3	Human lymphoblastoid cell line (H3K4me3 and H3K27me3); mouse brain (all 3 marks)	30–5000 cells (H3K4me3) and 100–10,000 cells (H3K27me3); 1000 cells (mouse brain)	A microfluidic platform developed for low-input ChIP-seq; allow up to 8 reactions in parallel using polydimethyl siloxane (PDMS)/ glass channel; chromatin is fragmented with MNase treatment	[64]

(Continued)

TABLE 9.1 (Continued)

Method	Mark	Cell type	Cell number	Key features	Ref.
PnP-ChIP-seq	H3K4me1, H3K4me3, H3K27ac, H3K36me3, H3K27me3, and H3K9me3	Mouse embryonic stem cells	100–3000 cells (optimal results with 3000 cells)	A fully automated microfluidic platform developed on Fluidigm C1 controller for low-input ChIP-seq; support up to 24 ChIP experiments in parallel using disposable PDMS-based plates; the whole process takes <1d with only 30 min hands-on time	[65]
FloChIP	H3K4me1, H3K4me3, H3K27ac, and H3K9me3	Human lymphoblastoid cell lines	500–100,000 (500 cells only tested with H3K27ac)	A microfluidic implementation of ChIP protocol enabling highly automatic profiling of histone marks and TFs, both for individual marks and sequentially with two marks; it applies Tn5 transposase-based chromatin tagmentation and direct on-chip indexing of chromatin-bound DNA	[66]
CoTECH	H3K4me3 and H3K27me3	Mouse embryonic stem cells	Single cell	pA-Tn5 fusion protein with combinatorial indexing strategy for adding T5 and T7 barcodes; a co-assay of single-cell chromatin occupancy and transcriptome	[55]
Paired-Tag	H3K4me1, H3K4me3, H3K27ac, H3K27me3, and H3K9me3	Mouse frontal cortex and hippocampus	Single cell	Joint profiling of chromatin modification and transcriptome in single cells; permeabilized nuclei are incubated with a given antibody to target pA-Tn5 binding to chromatin, allowing subsequent tagmentation and reverse transcription	[52]

Similar to ATAC-seq, CUT&Tag also uses Tn5 transposase to cut the genome, but the latter allows antibody-tethered tagmentation of accessible DNA sites. As H3K4me2 and H3K4me3 are present at promoters and enhancers, Cleavage under targeted accessible chromatin (CUTAC), which is based on low-salt H3K4me2 or H3K4me3 CUT&Tag tagmentation, is able to map transcription-coupled chromatin accessible sites that are highly comparable to standard ATAC-seq [59]. A large proportion of CUTAC sites match nucleosome-free regions that represent transcription-coupled regulatory elements.

ACT-seq and scChIC-seq

The antibody-guided chromatin tagmentation sequencing (ACT-seq) method also utilizes a pA-Tn5 fusion protein that is targeted to chromatin sites by a specific antibody, thus enabling chromatin fragmentation at genomic sites with antibody binding [5]. ACT-seq was used to map histone modifications (H3K4me1, H3K4me2, H3K4me3, and H3K27ac), histone variant (H2A.Z), as well as chromatin-binding protein (BRD4) from as few as 1000 human HEK293T cells. By incorporating an index multiplexing strategy into ACT-seq, iACT-seq can map thousands of cells at single-cell resolution.

The same group also developed scChIC-seq, a CUT&RUN-based single-cell chromatin immunocleavage sequencing method, for single-cell epigenetic profiling with or without crosslinking [54]. In scChIC-seq, MNase is recruited to chromatin regions by a specific antibody via antibody-MNase or protein A-MNase conjugate, resulting in chromatin cleavage at sites with a given histone modification or TF binding. To minimize DNA loss, both target and off-target DNA fragments are recovered for adapter ligation; the former are preferentially amplified and sequenced due to their relatively smaller sizes. Using human white blood cells, scChIC-seq generated about 100,000 unique reads per cell for H3K4me3; on average, 50% of the reads were mapped to the peak regions identified from bulk cells. Clustering analysis of informative H3K4me3 peaks from 242 single-cell libraries identified seven clusters, of which four were assigned to T cells, natural killer cells, B cells, and monocytes on the basis of cell type-specific peaks identified from bulk ChIP-seq data. A major challenge for scChIC-seq is the low mapping rate [53].

CoBATCH and itChIP-seq

Wang et al. [53] developed an in situ ChIP strategy for epigenomic profiling from low input and CoBATCH (combinatorial barcoding and targeted chromatin release) for single-cell profiling under native conditions or with crosslinking. In situ ChIP was tested for H3K4me3, H3K27ac, and H3K27me3 with as few as 100 mouse embryonic stem cells. It can be implemented directly on intact tissues with scarce cells, such as mouse embryos, with antibodies against H3K4me3 and H3K27ac. On the other hand, single-cell CoBATCH can generate about 10,000 unique reads per cell with little background [53]. To understand epigenetic heterogeneity of cell subpopulations in complex tissue types, CoBATCH was used to profile H3K27ac in Cdh5-traced endothelial cell lineages from 10 organs of mouse embryos. Cluster analysis of H3K27ac peaks for 2758 single cells defined three major clades according to four enhancer-regulated modules.

The itChIP-seq (simultaneous indexing and tagmentation-based ChIP-seq) and sc-itChIP-seq methods, both developed by the above group, allow genome-wide mapping of histone and non-histone marks from low-input and single-cell samples, respectively [60]. itChIP-seq adopts combinatorial dual-indexing strategy, such that each sequencing read contains index from transposase plate and a second index from the PCR plate to help discriminate individual cells [68]. itChIP-seq demonstrated robust performance for H3K4me3, P300 and H3K27me3 with 100–10,000 cells, and for RNAPII and EZH2 with 500 cells. By single-cell H3K27ac profiling of cardiac lineage in mice from the formation of the cardiac crescent to the onset of tube formation, sc-itChIP-seq revealed cell type-specific usage of enhancers following the differentiation of bipotent cardiac progenitors into endothelial and cardiomyocyte cells [60].

Microfluidics-based ChIP-seq Platforms

Several microfluidics-based protocols have been developed to address the challenges in standardization, parallelization, and automation of low-input ChIP-seq experiments, including MOWChIP-seq [62,63], SurfaceChIP-seq [64], and more recently, FloChIP [66] and PnP-ChIP-seq [65]. Of these, the FloChIP protocol can work for TFs [66].

MOWChIP-seq is a semi-automated microfluidics-based ChIP-seq protocol working with as few as 100 cells [62]. It improves ChIP efficiency by using a packed bed of antibody-coated beads, coupled with oscillatory washing to remove nonspecific adsorption. A parallelized protocol can support up to eight reactions in an eight-unit device [63]. MOWChIP-seq has been tested with multiple histone marks in both human B-cell cell line and mouse tissues [62,63]. Data from ChIP experiment starting with 100 or 600 cells are largely comparable (r = 0.84 on average) with data using the same number of cells from stock chromatin of 10,000 cells [62]. MOWChIP-seq with H3K27ac revealed that over half of the active enhancers in hematopoietic stem and progenitor cells from mouse fetal liver are unique compared to those identified by Lara-Astiaso et al. [69] from samples collected from over 16 hematopoietic differentiation stages [62]. A major limitation is that MOWChIP-seq depends on custom-fabricated microfluidic device [63].

The same group further developed SurfaceChIP-seq [64], which utilizes an antibody-coated polydimethyl siloxane (PDMS)/glass channel surface rather than immunomagnetic beads used by MOWChIP-seq. It allows up to eight reactions in parallel. For H3K4me3, even with only 30 cells from lymphoblastoid cell line GM12878, SurfaceChIP-seq can generated about 3 million unique de-duplicated reads, with a Pearson correlation of 0.9 between replicates. High levels of enrichment were obtained for both H3K4me3 and H3K27me3 with 500–1000 cells. Using 1000 cells in SurfaceChIP-seq, Ma et al. [64] profiled H3K4me3, H3K27me3, and H3K27ac across NeuN+ (neuronal) and NeuN- (glial) fractions of both prefrontal cortex (PFC) and cerebellum in mouse brain. They identified marked differences in the epigenetic landscape between these two functionally distinct regions. Between PFC and cerebellum NeuN+, 46% of the differentially expressed genes are associated with differential histone modifications in their promoters. In particular, between 73% and 82% of the super-enhancers are cell-type specific; NeuN+ specific super-enhancers from the two regions are enriched with binding motifs of the TFs with critical roles in neuronal development and differentiation.

FloChIP uses a microfluidic system to streamline the preparation of ChIP library by Tn5 transposase-based indexing of chromatin-bound DNA directly on-chip [66]. It can be used for profiling single marks, including TFs, and a combination of two marks sequentially, up to eight reactions with the eight IP lanes. In the human lymphoblastoid cell line GM12878, FloChIP can generate reproducible results with 500 cells for H3K27ac ($r = 0.78$ compared to bulk ENCODE data). With 100,000 cells, FloChIP data is highly comparable ($r \geq 0.85$) with ENCODE data for three other histone marks: H3K4me1, H3K4me3 and H3K9me3. Importantly, FloChIP allows straightforward sequential ChIP (re-ChIP) for measuring genome-wide co-occupancy of two marks, like H3K27me3/H3K4me3 in bivalent promoters in mouse embryonic stem cells. Finally, by profiling TF MEF2A binding across 32 B-cell lymphoblastoid cell lines, Dainese et al. [66] revealed its regulatory role in lymphoblastoid cell proliferation that is driven by interactions with other main factors including RUNX3.

Finally, the plug and play (PnP)-ChIP-seq is a fully automated microfluidic platform, built on a commercially available Fluidigm C1 controller rather than sophisticated custom-made equipment [65]. It can complete 24 ChIP reactions simultaneously using disposable polydimethyl siloxane (PDMS)-based plates within 1d, with only 30 min hands-on time needed for pipetting and harvesting DNA samples. The MNase-based native PnP-ChIP-seq protocol is robust with 3000 cells when tested on all six histone marks required by the International Human Epigenome Consortium (IHEC). For H3K4me3, PnP-ChIP-seq with 100 cells can capture over 50% of the peaks generated by bulk ChIP-seq. It can potentially be used to profile histone modifications in rare cell subpopulations, as demonstrated for the rare *2-cell–like* cells present in the heterogeneous mouse embryonic stem cells. The feasibility of applying PnP-ChIP-seq on TFs and other non-histone proteins remains to be tested [65].

Joint Profiling of Histone Modification and Transcriptome in Single Cell

Technology advances in multi-omics platforms have enabled single-cell profiling of multiple layers of epigenome separately, such as scChIP-seq, scATAC-seq, and single-cell Hi-C [52,55]. New methods have also been developed to profile transcriptome and chromatin accessibility in the same cells. Nevertheless, regions of repressive chromatin regions, which can be targeted using ChIP-seq with repressive histone marks or chromatin binding proteins, are less accessible to ATAC-seq mapping [49,61]. Complementarily, computational algorithms have been developed to integrate unpaired single-cell multi-omics datasets, which is complicated by the uncertainty in cell type assignment across different modalities [55]. Most recently, two methods enable joint profiling of both transcriptome and histone modification in the same cells, which should provide a more accurate view of cell type-resolved epigenetic regulation of gene expression in complex tissues [52,55].

CoTECH

Combined assay of transcriptome and enriched chromatin binding (CoTECH) uses a combinatorial barcoding method for co-assay of chromatin state and transcription in single cells [55]. It is designed on top of CoBATCH, a single-cell assay used for profiling histone modifications and DNA binding proteins [53], with modifications to also allow simultaneous measurement of mRNA expression. CoTECH was used to investigate transcription together with H3K4me3 or H3K27me3 from naive to primed embryonic stem cells in mouse. About 70% and 80% of the quality-filtered single cells showed concordant cluster assignment between the two modalities for mRNA-H3K4me3 and mRNA-H3K27me3, respectively. Based on the dynamics of expression pattern, genes were clustered into three groups; of which, two showed a trend of expression from low to high and one showed a reverse trend. On the other hand, the associated H3K4me3 showed a trend of high to low occupancy for all the three groups of genes, while H3K27me3 general showed a trend of low to high or the opposite, indicating that both marks play context-dependent roles in regulating gene expression program during cell-fate dynamics.

Paired-Tag

Paired-Tag, or parallel analysis of individual cells for RNA expression and DNA from targeted tagmentation by sequencing, can be used for joint profiling of single-cell histone modifications and transcriptome [52]. Paired-Tag represents an extension of Paired-seq, a technique developed for co-assay of open chromatin and gene expression in single cells [70], by combining with the CUT&Tag strategy used for profiling histone modifications. Paired-Tag is aimed to streamline integration of gene expression with multiple histone marks. It was used to jointly profile transcriptome with each of five histone modifications, H3K4me1, H3K4me3, H3K27ac, H3K27me3, and H3K9me3, in the mouse frontal cortex and hippocampus. Using the paired transcriptomic profiles to annotate cell clusters of each histone mark, Paired-Tag generated cell type-specific maps of epigenome and transcriptome for 22 mouse brain cell types. Such joint analyses have provided insights into regulatory programs of different gene sets in these distinct cell types.

GENOME-WIDE MAPPING OF HISTONE MODIFICATIONS

The second edition of this epigenetics handbook describes genome-wide profiles for major types of histone modifications including methylation, acetylation, ubiquitination and phosphorylation. Several additional types of histone modifications have been discovered, such as acylation, homocysteinylation, monoaminylation, O-GlcNAcylation, lipidation, and glycation [71,72]. The first four were mapped using ChIP-seq (Table 9.2), as described below.

Acylation

High-sensitivity mass spectrometry (Ms)-based proteomic analyses have identified at least 10 histone lysine (K) acylations, which include 2-hydroxyisobutyrylation (Khib) [82,87,88], β-hydroxybutyrylation (Kbhb) [75,88], benzoylation (Kbz) [71,73,75], butyrylation (Kbu) [72,75,87,88], crotonylation (Kcr) [72,75,87,88], glutarylation (Kglu) [71,75,87,88], lactylation (Kla) [71,80,87], malonylation (Kmal) [75,87,88], propionylation (Kpr) [75,87,88], and succinylation (Ksuc) [72,75,87,88]. Histone acylations are similar to the widely studied histone acetylation in the enzymatic regulation [3], potentially linking cell metabolic state to chromatin architecture [3,73,74,78].

Considered as acetylation-competing marks, histone acylations preferentially occupy genomic regions where histone acetylations are highly enriched [76]. Emerging evidence suggests that they seemingly play similar roles in regulating gene expression [88]. However, the dynamics and distinct molecular functions of histone acylations are just beginning to be revealed, showing that the competitive histone acetylation and butyrylation/crotonylation impact the dynamic binding of bromodomain and extraterminal (BET) proteins to chromatin [76,77]. Goudarzi et al. [77] studied acetylation and butyrylation on H4 K5 and K8 during sperm cell differentiation in meiotic cells (spermatocytes) and post-meiotic cells (round spermatids) from mouse testis. ChIP-seq profiles indicated that TSS regions of highly expressed genes are occupied by all the four marks, in contrast to much reduced expression of the genes whose TSS regions lack any of the marks. The differential expression between spermatocytes and round spermatids is coupled with differential occupancy of these marks in gene TSS regions. Also, vast majority of the TSS regions bound by BRDT, a bromodomain-containing protein, were also co-occupied by the four marks. In vitro binding assay demonstrated that, in line with co-occurrence of these marks based on ChIP-seq data, the replacement of H4K8ac by H4K8bu had only marginal effect on BRDT binding; surprisingly, substitution of H4K5ac with H4K5bu abolished the BRDT binding to histone H4. Thus, the alternating of these two competing marks, H4K5ac and H4K5bu, possibly through their rapid turnover, modulates the dynamics of BRDT binding to and dissociation from the respectively modified chromatin.

Further, Gao et al. [76] discovered that the H4K5 acyl/acetyl ratio regulates BRD4 binding to chromatin. They used a precursor B-cell acute lymphoblastic leukemia cell line REH, in which *FASTKD1* expression downregulates mitochondrial activity responsible for fatty acid synthesis/β-oxidation, ultimately leading to reduction of histone butyrylation and crotonylation. *FASTKD1* knockout (via CRISPR-Cas9) in REH cells resulted in a clear increase of H4K5bu-cr, but not H4K5ac. By comparing BRD4-chromatin binding in the wild-type vs FASTKD1 knockout REH cells with immunoblotting, they found that the higher H4K5bu-cr/H4K5ac ratios observed in the latter were associated with higher amounts of BRD4 in supernatants and lower amounts in pellets, suggesting weaker BRD4 binding to chromatin.

TABLE 9.2 ChIP-seq Mapping of Novel Histone Modifications

Modification	Mark	Function	Sample	Distribution	Ref.
Benzoylation	Pan Kbz	Transcriptional activation	Human HepG2 cell line	Preferentially associated with promoters	[73]
Butyrylation	H3K14bu	Transcriptional activation	Mouse liver	Preferentially associated with promoters	[74]
Butyrylation	H3K23bu	Transcriptional activation	Mouse myoblast cell line	Preferentially associated with promoters	[75]
Butyrylation	H4K5bu	Transcriptional activation	B-cell acute lymphoblastic leukemia cell line	Preferentially associated with promoters	[76]
Butyrylation	H4K5bu	Transcriptional activation	Spermatogenic cells from mouse testis	Preferentially associated with promoters	[77]
Butyrylation	H4K8bu	Transcriptional activation	Spermatogenic cells from mouse testis	Preferentially associated with promoters	[77]
Crotonylation	H3K18cr	Transcriptional activation	Mouse colon epithelium	Preferentially associated with promoters	[78]
Crotonylation	H3K23cr	Transcriptional activation	Mouse myoblast cell line	Preferentially associated with promoters	[75]
Crotonylation	H4K5cr	Transcriptional activation	B-cell acute lymphoblastic leukemia cell line	Preferentially associated with promoters	[76]
Glutarylation	H4K91glu	Transcriptional activation	Human HepG2 cell line	Mainly enriched in promoters of highly expressed genes; also enriched within gene-body	[79]
Lactylation	H3K18la	Transcriptional activation	Mouse bone marrow-derived macrophage	Enriched in promoters	[80]
Propionylation	H3K14pr	Transcriptional activation	Mouse liver	Preferentially associated with promoters	[74]
Propionylation	H3K23pr	Transcriptional activation	Mouse myoblast cell line	Preferentially associated with promoters	[75]
β-hydroxybutyrylation	H3K9bhb	Transcriptional activation	Mouse liver	Preferentially associated with promoters	[81]
2-hydroxyisobutyrylation	H4K8hib	Transcriptional activation	Mouse meiotic and post-meiotic cells	Preferentially associated with promoters	[82]
Homocysteinylation	H3K79Hcy	Transcriptional activation	Mouse neural stem cells	Over half of the peaks located in intergenic regions and about 30% in introns	[83]
Serotonylation	H3K4me3Q5ser	Permissive gene expression	Mouse brain; human induced pluripotent stem cells and derived neurons	Enriched in promoters	[84]
O-GlcNAcylation	H2BS112GlcNAc	Transcriptional activation	Human HeLa cell line	Over half of the peaks in intergenic regions and >40% in introns	[85]
O-GlcNAcylation	H2AS40GlcNAc	Transcriptional activation	Mouse (C57BL/6 N strain)	Over half of the peaks in gene-body; 21%–40% in intergenic; 9%–28% in promoters	[86]

ChIP-seq data further showed that, within TSS ± 2 kb regions, the H4K5bu/H4K5ac and H4K5cr/H4K5ac ratios accelerate with increasing H4K5ac occupancy, a trend more conspicuous in FASTKD1 knockout cell lines. BRD4 ChIP-seq signals were significantly more enriched around TSSs of the most highly expressed genes, thus supporting a model that an overall reduction of genome-wide BRD4 binding, driven by increased H4K5bu-cr/H4K5ac ratio, is accompanied by redistribution of BRD4 binding to sites elsewhere, including most active gene promoters. This may reflect a common mechanism underlying the regulation of genomic distribution for bromodomain-containing proteins.

Butyrylation, Crotonylation, and Propionylation

Focusing on H3 lysine residue 23, Jo et al. [75] identified H3K23bu, H3K23cr, and H3K23pr in mouse myoblast cell line C2C12. ChIP-seq showed that the three marks were preferentially associated with gene promoters, suggesting their roles in promoter regulation [75]. In addition, H3K14pr and H3K14bu were also enriched in promoters of active genes, highly overlapping with H3K9ac and H3K4me3 peaks [74]. Besides K23, crotonylation was discovered on additional H3 lysine residues including K9, K14, K18, K27, and K56 [78]. Butyrylation was also identified on H3 K9, K18, K27, K36, K37, K79, and K122, on H4 K5, K8, K12, and K16, as well as on H2B K5 and K20 [77]. Of those, H4K5bu and H4K8bu, like their competing marks H4K5ac and H4K8ac, were enriched within gene promoter regions as well [77]. ChIP-seq in mouse colon epithelium revealed that H3K18cr was enriched in the promoter regions, with higher levels of gene expression associated with increased H3K18cr occupancy over TSSs [78]. H3K18cr is highly dynamic during the cell cycle, showing increased signals in S and G2-M phase compared with G1 cells [78]. Unlike butyrylation and propionylation, however, histone crotonylation also represses gene transcription. While H2BK12cr, H3K4cr, H3K18cr, and H3K23cr promote transcription [75,89], H3K9cr [90] and H3K27cr inhibit transcription [89].

By utilizing the yeast metabolic cycle (YMC) that shows periodic gene expression between a low and a high oxygen consumption phase, Gowans et al. [90] found that H3K9cr and H3K9ac are temporally segregated over the YMC. While H3K9ac signal peaks in a high oxygen consumption phase when acetyl-CoA is generated, H3K9cr signal peaks concomitantly with the decrease of H3K9ac following the transition to low oxygen consumption when energy becomes limited and crotonyl-CoA is generated. ChIP-seq showed that both marks are enriched around TSSs and transcriptional termination sites of metabolic genes. Unexpectedly, as cells transit into the low oxygen consumption phase, H3K9cr represses the expression of energy-consuming genes timely by interacting with TAF14 (AF9 in human), a major reader of H3K9cr whose YEATS domain shows a notable preference for H3K9cr over H3K9ac [91]. Mutation in the TAF14 YEATS domain abolished its binding, resulting in reactivation of these genes. This study highlights a key role for TAF14 and histone crotonylation in linking metabolic state dynamics to the periodicity of gene expression program.

Benzoylation

Histone lysine benzoylation can be stimulated by sodium benzoate, a molecule widely used as food preservative, through generation of benzoyl-CoA [3]. Proteomic analysis in sodium benzoate- stimulated human HepG2 and mouse RAW cell lines together identified 22 Kbz sites from core histones and linker histone H1, mainly on the N-terminal tails [73]. ChIP-seq with a pan anti-Kbz antibody identified >20,000 peaks in sodium benzoate-treated HepG2 cells, with over half located in the TSS ± 3 kb regions. Kbz enrichment is positively correlated with gene expression from RNA-seq data, suggesting a role in promoting transcription.

Glutarylation

Histone lysine glutarylation (Kglu) was identified at 27 residues on the core histones in human HeLa cells [79]. ChIP-seq showed that H4K91glu is largely enriched over promoter regions of highly expressed genes, and within gene body as well. Downregulation of H4K91glu is tightly linked to chromatin condensation during mitosis and DNA damage response [79].

Lactylation

Lactate under glycolysis regulation serves as a precursor to stimulate histone lysine lactylation. A total of 28 Kla sites were identified on core histones in human HeLa and mouse bone marrow-derived macrophage cells [80]. To understand H3K18la dynamics, H3K18la and H3K18ac ChIP-seq, as well as RNA-seq, were performed on mouse bone marrow-derived macrophage cells treated with lipopolysaccharide and interferon-γ. The analysis revealed that both histone marks were enriched in gene promoters. Of the genes marked by increased H3K18la, two-thirds lacked a significant increase of H3K18ac signal, indicating different temporal dynamics of lactylation from acetylation. Together with RNA-seq data, genes with increased H3K18la occupancy were activated following the treatment, indicating its direct roles in stimulating gene transcription [80].

2-hydroxyisobutyrylation and β-hydroxybutyrylation

Histone lysine 2-hydroxyisobutyrylation (Khib) was identified at 63 residues on core and linker histone H1 from human and mouse [82]. ChIP-seq showed that over half of H4K8ac peaks overlapped with H4K8hib peaks, with over 90% of the genes that were marked by H4K8ac were also co-marked by H4K8hib; in contrast, over 70% of H4K8hib peaks are unique without H4K8ac occupancy, suggesting possibly distinct role of H4K8hib [82]. H4K8hib is positively associated with gene transcription, with signals enriched in promoters in mouse meiotic and post-meiotic cells.

Histone lysine β-hydroxybutyrylation (Kbhb) was identified at 44 sites, including 38 in β-hydroxybutyrate-treated human HEK293 cells and 26 in mouse liver cells [81]. In mice livers, Kbhb levels are significantly induced in response to elevated levels of β-hydroxybutyrate under physiologic states of starvation or streptozotocin-induced diabetic ketoacidosis. To understand genome-wide distribution of histone Kbhb marks, ChIP-seq was done for H3K4bhb, H3K9bhb, and H4K8bhb in mouse liver cells. For H3K4bhb and H4K8bhb, about 10% peaks are in the promoter regions, with the vast majority in the introns and intergenic regions. In contrast, over 40% of the H3K9bhb peaks are located in the gene promoters, highly correlated with H3K4me3 and H3K9ac, two active promoter marks [81]. In particular, H3K9bhb are associated with a unique set of upregulated genes during starvation that are not marked by H3K9ac and H3K4me3 and are enriched in starvation-responsive metabolic pathways, suggesting non-overlapping roles of histone Kbhb vs acetylation and methylation.

Homocysteinylation

Homocysteine (Hcy) could act as a substrate in histone modification. A total of 39 lysine sites are identified to show homocysteinylation in human embryonic brain samples [83]. Most of them occur at sites that also show other types of modifications. Excessive maternal homocysteine during pregnancy increases the risk of neural tube defects, which are serious congenital malformations. To understand the roles of KHcy in neural tube defects, H3K79Hcy ChIP-seq and RNA-seq were performed in mouse neural stem cells [83]. About 30% of the peaks were found in the introns and >50% in intergenic regions, versus <5% in gene promoters. Elevated gene expression was generally coupled with an increase of H3K79Hcy within gene-body. However, increased H3K79Hcy level down-regulated a subset of genes related to neural tube defects, highlighting its role in offspring neural tube defects.

Monoaminylation

Histone monoaminylation, which is recently discovered to play roles in regulating neural functions and behaviors, occurs at glutamine residues by reacting with monoamine neural transmitters serotonin and dopamine [3]. In organisms that produce serotonin (also called 5-hydroxytryptamine), serotonin can be covalently attached to histone H3 at glutamine residue 5 by transglutaminase 2. H3Q5ser frequently occurs on H3K4me3-modified histones to form dual modification H3K4me3Q5ser, even though transglutaminase 2-mediated serotonylation does not require prior H3K4me3 [3,84,92]. As a reader of H3Q5ser, WDR5 preferentially binds H3Q5ser-modified over H3K4me3-modified and unmodified H3 [93]. H3K4me3Q5ser signal was identified broadly across nerve, gut, heart and peripheral blood cells, more enriched in the first two [84,94].

ChIP-seq revealed that H3K4me3Q5ser peaks were predominately located in promoters in embryonic mouse brain [84]. Neuronal genes that are activated during mouse brain development showed marked gains of H3K4me3 and H3K4me3Q5ser. In cells ectopically expressing a serotonylation-lack H3 mutant, there is a significant alteration of expression for genes targeted by H3K4me3Q5ser, which causes deficits in differentiation [84]. In human neuroblastoma cell line, WDR5 was found to colocalize with H3Q5ser in the promoters of cancer-promoting genes and enhance their expression to induce cell proliferation [93]. This result suggests a novel role for WDR5-H3Q5ser-mediated epigenetic mechanism in promoting tumorigenesis.

In dopaminergic neurons, transglutaminase 2-mediated dopaminylation attaches dopamine to H3 at glutamine 5 to form H3Q5dop [95]. Transglutaminase 2 equally dopaminylates H3K4me3-carrying histone, resulting in dual modification H3K4me3Q5dop. H3Q5dop was found to play a central role in cocaine addiction-induced transcriptional plasticity in the midbrain. Unlike H3K4me3Q5dop, H3Q5dop expression was significantly decreased in the midbrain ventral tegmental area of cocaine users. Reversely, withdrawal from cocaine was accompanied by an accumulation of H3Q5dop in the ventral tegmental area in rats. Decreasing H3Q5dop during withdrawal reduced cocaine-seeking behavior of rats, indicating a causal role for H3Q5dop in cocaine addiction behaviors [95].

O-GlcNAcylation

H2A, H2B, and H2A variants (H2A.X and H2A.Z) can be GlcNAcylated [85]. When glucose levels drop below normal, H2B GlcNAcylation at serine 112 (H2BS112GlcNAc) responds to the decrease of glucose by promoting H2B lysine monoubiquitination (H2BK120ub1), which is associated with active transcription. ChIP-seq identified 47,375 H2BS112GlcNAc peaks in HeLa cells, with >40% located in introns and >50% in intergenic regions [85]. About 8% of the H2BS112GlcNAc sites overlapped with H2BK120ub1 peaks. An aggregate plot of H2BS112GlcNAc signals over TSS ± 3 kb regions revealed signal peaks around TSSs, suggesting that H2BS112GlcNAc is correlated with active transcription.

H2AS40 in the L1 loop of the H2A globular domain can also undergo O-GlcNAcylation [86]. ChIP-seq in mouse trophoblast stem cells and the derived trophoblast subtypes showed that >50% of the H2AS40GlcNAc peaks were located within gene-body, of which about 35% and 55% were in exons in these two cell types, respectively; another 40% and 21% were found in intergenic regions, leaving the remaining 9% and 28% in promoter regions. Integration of ChIP-seq with expression array data revealed that H2AS40GlcNAc-marked genes showed relatively higher levels of expression compared with all genes in trophoblast stem cells, suggesting that H2AS40GlcNAc within gene-body positively regulates gene expression. A similar trend of correlation was also seen for H2AS40GlcNAc at TSS and gene-body in trophoblast subtypes [86].

CATALOG OF HISTONE MODIFICATION PROFILES

ChIP-seq has been widely used to map genome-wide occupancy of histone modifications, histone variants and DNA-binding proteins. The ChIP-Atlas database, for example, uniformly processed public ChIP-seq data from over 144,000 experiments, including 100 different histone modifications and variants in human and 97 in mouse [96]. The ENCODE and Roadmap epigenomics projects have used ChIP-seq to map histone modifications over a large collection of cell and tissue types (Table 9.3).

The ENCODE phase III generated 4834 and 1158 new experiments in human and mouse tissues or cells, respectively, representing a significant expansion of previous release [97]. Specifically, 11 histone modifications and two histone variants were mapped by ChIP-seq across 79 human cell and tissue types, plus eight histone modifications across 6–12 tissues at each of the eight developmental stages in mouse [97]. Together, 926,535 and 339,815 candidate cis-regulatory elements were defined in human and mouse [97]. In complementary, Hi-C chromatin conformation maps were generated for 33 human tissue and cell types and ATAC-seq in 48 primary human tissues [97]. As part of the ENCODE project, Grubert et al. [98] mapped H3K27ac in 22 human cell lines, collectively identifying 288,711 genomic regions with enhancer activity. About 23% of the cohesin-mediated looping interactions, identified by ChIA-PET, are anchored by the identified enhancers.

Many epigenetic maps have limitations of being under-represented for many disease-relevant tissues, as well as a large variation of data quality due to factors such as experimental noise, different lab protocols, and data analysis pipelines. Boix et al. [99] presented EpiMap, which is a compendium of epigenomic maps across >800 samples, with defined chromatin states, enhancer locations, and target genes. They uniformly processed epigenetic data from 859 biosamples, generated mostly by ENCODE and Roadmap projects, covering 471 tissues, 183 cell lines, 170 primary cells, and 35 in vitro differentiated cells. After filtering, the retained 833 high-quality reference epigenomes are from 33 tissue categories. In summary, the analyses identified 2.1 million active enhancers, 3.6 million open chromatin regions, and 3.3 million tissue-specific enhancer-gene connections with a median distance of ~42 kb [99]. This invaluable epigenomics resource was used to annotate 30,000 genetic loci associated with 540 traits, revealing putative causal variants within tissue-specific enhancers and their candidate target genes.

To understand the chromatin dynamics during mouse fetal development, Gorkin et al. [100] performed 1128 ChIP-seq on 8 histone modifications (H3K4me1, H3K4me2, H3K4me3, H3K9ac, H3K9me3, H3K27ac, H3K27me3, and H3K36me3) and 132 ATAC-seq across 72 distinct combinations of stage (8) and tissue type (8–12) (Table 9.3). They identified a confident set of candidate enhancers, from which a subset of dynamic enhancers was identified that showed H3K27ac activity changes between sequential stages. Most (median 84%) dynamic enhancers overlapped open chromatin regions, but only 5%–35% (median 14%) overlapped differential open chromatin regions, which likely reflects temporal differences between H3K27ac and chromatin accessibility during fetal development. Overall, ~1.3% of the genome differed in chromatin state between adjacent stages within the same tissue and ~1.2% between tissues at the same stage. This study provided the most systematic analysis of chromatin dynamics during mouse fetal development [100].

VARIATION OF HISTONE MODIFICATIONS

Genetic Variants alter Histone Modifications

Over 90% of genome-wide association study (GWAS)-identified disease associated risk variants are located in noncoding regions of the human genome [99]. Their roles in disease are likely regulatory, but understanding the exact molecular mechanisms underlying the associations remains a major challenge [102,105,113]. In particular, the causal variants often co-exist in strong linkage disequilibrium with many nonfunctional ones [103].

TABLE 9.3 Examples of Large-Scale Mapping of Histone Modifications and hQTL (Histone Modification QTL) Detection

Feature	Mark	Sample[a]	Outcome	Ref.
Genome-wide map	H3K4me1, H3K4me2, H3K4me3, H3K9ac, H3K9me2, H3K9me3, H3K27ac, H3K27me3, H3K36me3, H3K79me2, H4K20me1, H2A.Z, and H3.3	Human (79 tissue and cell types; from ENCODE)	Defined 926,535 candidate cis-regulatory elements	[97]
Genome-wide map	H3K4me1, H3K4me2, H3K4me3, H3K9ac, H3K9me3, H3K27ac, H3K27me3, and H3K36me3	Mouse (6–12 tissues at each of 8 developmental stages; from ENCODE)	Defined 339,815 candidate cis-regulatory elements	[97]
Genome-wide map	H3K27ac	Human (22 cell lines; from ENCODE)	Defined 288,711 enhancers present in ≥2 cell types; ~23% of cohesin-mediated loops were anchored by enhancers	[98]
Genome-wide map	H2AK5ac, H2AK9ac, H2BK5ac, H2BK12ac, H2BK15ac, H2BK20ac, H2BK120ac, H3K4ac, H3K4me1, H3K4me2, H3K4me3, H3K9ac, H3K9me1, H3K9me2, H3K9me3, H3K14ac, H3K18ac, H3K23ac, H3K23me2, H3K27ac, H3K27me3, H3K36me3, H3K56ac, H3K79me1, H3K79me2, H3T11ph, H4K5ac, H4K8ac, H4K12ac, H4K20me1, H4K91ac, H2A.Z, and H3.3	Human (833 biosamples in 33 tissue categories; mostly from ENCODE/Roadmap)	Identified 2.1 million active enhancers and 3.6 million open chromatin regions	[99]
Genome-wide map	H3K4me1, H3K4me2, H3K4me3, H3K9ac, H3K9me3, H3K27ac, H3K27me3, and H3K36me3	Mouse fetal tissue (8–12 tissues at each of eight developmental stages; from ENCODE)	~1.2% of the genome differed in chromatin state between tissues at the same stage and ~1.3% between adjacent stages within the same tissue; 21,142 enhancer-promoter interactions were predicted	[100]
Allelic imbalance	H3K4me3, H3K4me1, H3K27me3, H3K9me3, H3K27ac, and H3K36me3	Human (35 tissue and cell types)	0.24%–1.32% peaks showed allelic imbalance for H3K4me1, H3K4me3, H3K27ac and H3K36me3, versus 9.76% and 24.65% for H3K27me3 and H3K9me3; H3K27ac and H3K4me3 were more enriched on the allele with low levels of DNA methylation	[101]
hQTL	H3K27ac	Human LCL (n = 57)	8,764 hQTLs were identified in peaks with G-SCI method developed by the study; they were highly linked to autoimmune disease-associated SNPs	[102]
hQTL	H3K4me1 and H3K27ac	Human LCL with SLE (n = 25)	4,858, 817 and 586 hQTLs in H3K27ac, H3K4me1 and both peaks; 386 hQTLs were mapped to 44 autoimmune disease risk haplotypes	[103]
hQTL	H3K4me1, H3K4me3, and H3K27ac	Human LCL (n = 271) and fibroblast (n = 78)	Generated genome-wide map linking regulatory elements to genes; cis-regulatory domains form trans-regulatory hubs to regulate gene expression	[104]
hQTL	H3K4me1 and H3K27ac	Human CD14+ monocytes, CD16+ neutrophils, and CD4+ naive T cells (n = 104–174)	189 H3K27ac and 190 H3K4me1 hQTLs overlapped with autoimmune disease risk loci	[105]
hQTL	H3K4me3 and H3K27me3	Human neutrophils (n = 106 and 107)	26,154 hQTLs were identified for H3K4me3 and 124,541 for H3K27me3 (within 1 Mb); 621 and 367 of these SNPs also impacted PU.1 binding	[106]

(Continued)

TABLE 9.3 (Continued)

Feature	Mark	Sample[a]	Outcome	Ref.
hQTL	H3K27ac	Human heart LV tissue from nonfailing ontrol (n = 34) and end-stage heart failure (n = 36)	1680 cardiac hQTLs were identified, including 62 in colocalization with heart-related GWAS SNPs	[107]
hQTL	H3K4me3 and H3K27ac	Human liver wedge biopsies (n = 9 and 18)	51 hQTLs were identified for H3K4me3 and 921 for H3K27ac (within 10 kb); 116 gene-peak pairs are likely regulated by the hQTLs that are in strong linkage with eQTLs	[108]
hQTL	H3K27ac	Human brain PFC, TC, and CB from ADS (n = 45) and control (n = 49) individuals	1912–2255 hQTLs identified in peaks for the three brain regions; four of them are potential causal variants for psychiatric diseases	[109]
hQTL	H3K9ac	Human brain PFC (n = 433)	1681 H3K9ac peaks were impacted by hQTL SNPs within 1 Mb; hQTLs were enriched in Alzheimer's disease, bipolar, and schizophrenia risk loci	[110]
hQTL	H3K27ac	Human brain PFC from control (n = 50) and ADS (n = 31) samples; PFC NeuN + and NeuN − fraction from control (n = 117) and schizophrenia (n = 109) samples	8464 peaks were impacted by 7976 hQTL SNPs within 1 Mb; 2477 hQTLs overlapped eQTLs	[111]
hQTL	H3K4me3 and H3K27ac	Human brain PFC and ACC from healthy individuals (n = 19); both NeuN + and NeuN − fraction	Between 6695 and 8042 hQTLs were identified for H3K27ac and 1565 and 3517 for H3K4me3 (within 10 kb of peaks); neuronal hQTLs showed strong enrichment for schizophrenia risk loci	[112]

[a] Number in parentheses indicates number of sampels used in ChIP-seq, unless specified otherwise.
ACC, anterior cingulate cortex; ADS, autism spectrum disorder; CB, cerebellum; LCL, lymphoblastoid cell line; PFC, prefrontal cortex; SLE, systemic lupus erythematosus; TC, temporal cortex.

Analysis of eQTLs, i.e., the genetic variants significantly correlated with variation in gene expression level, has been widely used to identify potential regulatory variants; however, eQTL SNPs are largely located outside of regulatory regions [102]. In addition, genetic variants can impact histone modification levels in an allele-dependent manner, referred to as histone modification QTLs (hQTLs) [109]. While some hQTL analyses restricted the association tests solely to the SNPs located within regions enriched with a given histone modification (i.e., peaks) [102,103], others included those within a certain distance of a peak, from 10 kb [108] up to 1 Mb [106,110,111]. Mechanistically, hQTL SNPs can alter TF binding affinity and looping interaction between regulatory elements [103]. hQTLs strategy has been used to prioritize functional variants [102,107–109], especially those from disease-relevant tissues and cell types [103,112], as illustrated in Fig. 9.1.

Most early hQTL work used human cell lines or blood cell types for obvious reason that they are relatively easy to collect [107,108] (Table 9.3). Three studies identified hQTLs in lymphoblastoid cell lines for H3K27ac [102–104], H3K4me1 [103,104], and H3K4me3 [104]. Up to three-fourths of the eQTL SNPs also affect cis-regulatory domain activity [104]. Intersecting hQTLs with eQTLs and dsQTLs (DNase I sensitivity QTLs) revealed that, of the eQTLs and dsQTLs SNPs that are located within H3K27ac peaks, 42% and 29% were highly linked to hQTLs ($R^2 \geq 0.8$), respectively [102]. This suggests the prevalence for SNPs to alter both H3K27ac and chromatin accessibility, together leading to the change of gene expression. Pelikan et al. [103] found that 83% of the H3K27ac hQTLs were located in the H3K27ac peaks that mediated looping interactions. Further, in two of the above studies, the identified hQTLs were used to annotate GWAS disease risk loci, strongly linking hQTLs to autoimmune disease-associated SNPs [102,103].

Two studies applied hQTL analysis to three blood cell types in healthy individuals [105,106]. Mapping H3K4me3 and H3K27me3 in human neutrophils uncovered hQTL SNPs that also impacted the binding of PU.1, a lineage-specific TF [106]. These variants tended to affect PU.1 binding and H3K4me3 in the

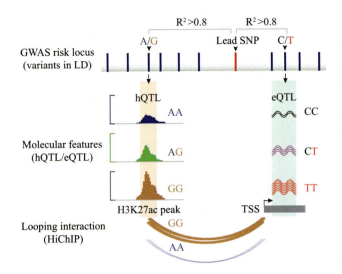

FIGURE 9.1 Regulatory variants within GWAS-identified disease risk locus. Regulatory variants can alter histone modification levels (hQTL) or be associated with the variation of gene expression (eQTL). hQTL and eQTL both may colocalize with the GWAS lead SNP, or highly link ($R^2 > 0.8$) with each other and with GWAS lead SNP. hQTL SNP may also be associated with differential looping interactions between the SNP-carrying enhancer and the target gene promoter. *eQTL*, gene expression QTL; *GWAS*, genome-wide association study; *HiChIP*, in situ Hi-C combined with chromatin immunoprecipitation; *hQTL*, histone modification QTL; *LD*, linkage disequilibrium.

same direction, but PU.1 and H3K27me3 in the opposite direction, suggesting coordinated effects of genetic variants on multiple molecular traits [106]. Studies in monocytes, neutrophils, and naive T cells revealed colocalization of H3K4me1 and H3K27ac QTLs with 62% and 54% autoimmune disease risk loci, respectively [105]. As demonstrated in lymphoblastoid cell lines [102], over 40% of eQTLs matched or were tightly ($R^2 \geq 0.8$) associated with H3K4me1 or H3K27ac hQTLs [105].

hQTL analysis was also expanded to bulk brain tissues [109,110], sorted brain cell types [111,112], and to other less accessible tissue types such as liver [108] and heart [107]. Numerous hQTL studies focused on bulk tissues from different brain areas implicated in psychiatric and neurodegenerative disease, and sorted neuronal and neuron-depleted cell fraction [109–112]. In these cases, the identified hQTLs reflect brain region- or cell type-specific variation in the regulatory regions. These studies supported the findings in cell lines and blood cells that SNP effect on gene expression variation is partly mediated by hQTLs [110,111]. Brain hQTLs from bulk tissue homogenate are enriched in GWAS risk loci for Alzheimer's disease [110] and psychiatric disease (such as bipolar and schizophrenia) [109,110]. However, for psychiatric disease, there are striking enrichments of risk variants in neuronal chromatin but not in non-neuronal cells, confirming cell type as a key source of epigenome variation [112].

Using ChIP-seq on liver wedge biopsies taken prior to transplantation surgery, 51 H3K4me3 and 921 H3K27ac hQTLs were identified [108]. Comparing to ENCODE ChIP-seq data in HepG2 cells revealed that these liver hQTL-associated peaks were highly enriched for binding sites of both hepatocyte nuclear factors (HNF4A, HNF4G, and FOXA2) and TFs known to be involved in hepatocellular remodeling (JUN and JUND). One hundred and sixteen gene-peak pairs are likely co-regulated by the hQTLs and eQTLs that are in strong linkage with each other ($R^2 > 0.8$).

H3K27ac ChIP-seq was performed on left ventricle free-wall tissue collected at the time of surgery from 36 end-stage heart failure subjects and 34 nonfailing controls, which identified 1680 hQTLs within the peaks [107]. Of these, 167 hQTLs were strongly linked ($R^2 > 0.8$) to human left ventricle eQTLs identified by the GTEx project, and 62 were colocalized with SNPs identified through heart-related GWAS, suggesting possible roles for hQTLs in mediating heart related disease associations.

The effect of common (>5% minor allele frequency) noncoding variants on chromatin variation can be tested using association-based methods, such as those used for hQTL discovery. In addition, heterozygous SNPs within peaks from individual samples can be used to identify allelic imbalance, i.e., reads containing one allele significantly outnumber other reads containing the alternative allele [101,102]. Allelic imbalance analysis is robust to the presence of strong confounding factors in a population, and more powerful in detecting rare and private variants with regulatory potential [105]. Allelic imbalance maps were generated for six histone marks (H3K4me1, H3K4me3, H3K9me3, H3K27ac, H3K27me3, and H3K36me3) and DNA methylation across 36 distinct cell and tissue types from 13 donors (Table 9.3) [101]. The average percentage of peaks with allelic imbalance ranged from only 0.24% for H3K4me1 and 0.6% for H3K27ac, up to 9.8% for H3K27me3 and 24.7% for H3K9me3. The large differences between the active and repressive marks may, in part, be due to an overall much larger peak size for both H3K27me3 and H3K9me3. At promoters, H3K27ac and H3K4me3 were more abundant on the allele that is less DNA methylated, and a reverse trend was found for H3K9me3. At enhancers, H3K27ac tended to occur more frequently on the allele with less DNA methylation as well [101].

Alteration of Histone Modifications in Disease

Alteration of epigenome is associated with gene expression reprogramming in disease [114,115]. To understand the role of histone modifications in Alzheimer disease, H3K27ac ChIP-seq was performed on bulk entorhinal cortex samples from disease cases (n = 24) and age-matched controls (n = 23) [116]. Differential peaks were largely clustered into disease and control groups, suggesting widespread alterations of H3K27ac in the entorhinal cortex of patients, including regulatory variation near several known risk genes. Rather than relying on bulk tissue, Nott et al. [113] generated H3K27ac and H3K4me3 ChIP-seq data from four major cell types sorted from cortical brain tissue: microglia, neurons, oligodendrocyte, and astrocyte. As expected, enhancer landscape is highly cell-type specific. While over three-fourths of the active promoters are shared among all four cell types, active enhancers are largely non-overlapping between cell types [113]. Importantly, for psychiatric disorders, GWAS risk variants are strongly enriched within enhancers and promoters from neurons, as opposed to Alzheimer disease variants that are largely located within microglia enhancers. This study highlights the importance of selecting the most relevant cell type(s) for functional dissection of disease risk loci.

Aberrant regulatory activity, especially that of enhancers, is a hallmark of many cancers. Epigenetic alterations can occur locally, such as locus-restricted gain of enhancers that target oncogenes, or globally through TF-mediated re-distribution of regulatory elements [21,115]. It is well established that epigenetic alterations can be caused by mutations and misregulation of chromatin regulators [117], insertions/deletions or copy number variation [21,118], among others. In the glioblastoma model, for example, analysis of The Cancer Genome Atlas (TCGA) data identified a focal amplification covering the 130-kb noncoding upstream region and EGFR itself. ChIP-seq with H3K27ac confirmed that this noncoding region harbors a gained enhancer that drives oncogenic EGFR expression [115]. In addition, a systematic analysis of 102 tumor cell genomes revealed small insertions whose presence in enhancer DNA sequences increases H3K27ac signal, leading to the formation of enhancers near known oncogenes [118]. SMARCB1 is a core subunit of the SWI/SNF chromatin-remodeling complex. SMARCB1 is highly mutated in pediatric rhabdoid tumors, which impairs the binding of SWI/SNF to typical enhancers, but not to super-enhancers. These SWI/SNF bound super-enhancers target genes with essential roles in rhabdoid tumor survival [117].

There is growing evidence that epigenetic mechanisms contribute to disease progression and metastasis [119]. Pancreatic ductal adenocarcinoma contains subpopulations with distinct epigenome profiles. Unlike patients with regional (peritoneal) metastatic disease, tumor sections from patients diagnosed with widespread distant metastases (liver and lung) underwent progressive loss of H3K9me2 and H3K9me3 from heterochromatin domains during subclonal evolution [120]. A similar pattern of H3K9me2 and H3K9me3 loss was observed in patient-derived cell lines, in which increased signals were detected accordingly in distant metastases for active marks H3K9ac, H3K27ac, and H4K16ac. These findings support a global reprogramming of epigenetic landscape during the evolution of distant metastasis. Using an organoid culture system, Roe et al. [21] mapped enhancers during pancreatic ductal adenocarcinoma progression. H3K27ac profiling revealed no change of global enrichment between primary tumor and paired metastasis-derived organoids, but there are 857 regions with increased and another 1709 regions with decreased H3K27ac in metastasis. These gained regions are enriched with the binding of FOXA1, a pioneer TF that is confirmed to drive the massive alterations in enhancers during metastatic transition [21].

Neuroblastoma is a very rare type of pediatric neoplasm characterized by a paucity of gene mutations and intratumoral divergence of gene expression. ChIP-seq was performed for H3K27ac and H3K4me3 on cell lines derived from the same patients: the CD133- adrenergic cells, which correspond to classic neuroblastoma cells, and CD133 + mesenchymal cells, which are more chemoresistant and enriched in relapse tumors. Analysis of enhancer landscape identified a few TFs with known roles in adrenergic differentiation that were marked with super-enhancers in mesenchymal cells, including PRRX1, a homeobox TF. Importantly, inducible expression of PRRX1 could drive gene expression profile and super-enhancer landscape in adrenergic cells toward a mesenchymal state, suggesting an epigenetic mechanism underlying intratumoral heterogeneity in neuroblastoma [121].

CONCLUSIONS AND PERSPECTIVES

High-sensitivity mass spectrometry-based proteomic analysis has identified multiple novel types of histone modifications involved in transcriptional regulation [3,71]. In addition, by mapping major active and repressive histone marks, reference epigenomes have been generated for large collections of cell lines and primary tissue and cell types [97,99]. These invaluable resources are broadly utilized to functionally annotate the risk loci associated with complex disease [99].

In particular, many studies have illustrated the dynamics of histone modifications in development, disease, and treatment response [100,122], as well as the interactions between genetic and epigenetic factors. Further effort will be needed to expand the sample sources and increase the mapping resolution, to better understand the biological functions and combinatorial roles of histone modifications, and to address computational challenges in data integration.

For many of the novel histone modifications, their biological functions remain largely unknown [71,123]. The chromatin regulators that deposit, bind, and erase these marks are not fully characterized yet. In this regard, deployment of CRISPR-based technologies, such as the deactivated CRISPR-Cas9 (dCas9), for precise epigenome editing, should allow an in-depth functional characterization of novel histone marks [124–126]. By tethering a dCas9-fused chromatin regulator complex to specific genomic locations, such tools have been successfully used to measure the levels of locus-specific histone modifications and their downstream effects on the perturbation of gene expression [125,127,128].

In addition, many tissue types, developmental stages and pertinent environmental conditions are still under-represented or missed in the current catalog of mapped histone modifications [99]. Also, due to the technical challenges, the vast majority of ChIP-seq data are generated from heterogeneous bulk tissues or cell lines, limiting our understanding of treatment responsive cell subpopulation [61], cell type-specific regulatory mechanisms [100] and target genes [97,113]. Fortunately, several alternative ChIP-seq protocols have been developed to handle low input. Among them, CUT&RUN and CUT&Tag are increasingly used to profile histone modifications, TF binding, and occupancy of chromatin regulators from limited cells at a higher resolution and with a much lower sequencing depth. Continuous improvement in the sensitivity and reproducibility should accelerate the transformation of single-cell protocols into routine applications [60,97].

Finally, more powerful computational algorithms are needed to model histone modification profiles for knowledge discovery and to integrate with other omics data. Machine learning algorithms, such as random forest, support vector machine, and XGBoost classifier, have been used to model histone modifications in relation to gene expression and RNA splicing. For example, they were used to reveal histone modification patterns associated with alternative splicing events [14–16], to identify human silencers by modeling epigenetic profiles correlated with gene expression [17], and to predict enhancer target genes based on correlation between gene expression levels and the activity of histone marks within nearby enhancers [99].

CHIP-SEQ DATA ANALYSIS WORKFLOW

Equipment

Software and Resources	URLs
BEDTools	https://github.com/arq5x/bedtools2/releases/ (v2.27.1)
BWA	https://sourceforge.net/projects/bio-bwa/files/ (v0.7.10)
ChIPseeker	https://www.bioconductor.org/packages/release/bioc/src/contrib/ChIPseeker_1.26.0.tar.gz (v1.26.0)
deepTools	https://github.com/deeptools/deepTools/ (v3.3.2)
IGV	https://data.broadinstitute.org/igv/projects/downloads/ (v2.3.93)
MACS	https://pypi.org/project/MACS2/ (v2.0.10)
Picard	https://broadinstitute.github.io/picard/ (v1.67)
Samtools	https://sourceforge.net/projects/samtools/files/samtools/ (v1.9)
Blacklist	https://github.com/Boyle-Lab/Blacklist/tree/master/lists/,hgxx-blacklist.v2.bed.gz (xx is 19 or 38)
Human reference	https://hgdownload-test.gi.ucsc.edu/goldenPath/hgxx/bigZips/hgxx.fa.gz (xx is 19 or 38)

Test Data

In this protocol, we use public H3K27ac ChIP-seq data from diffuse large B-cell lymphoma cell line OCI-LY1. The IP library was sequenced in paired-end mode (GSM1703914_H3K27ac.R1.fastq.gz and GSM1703914_H3K27ac.R2.fastq.gz, 20.3 million pairs of reads). The input library was sequenced in single-end mode (GSM1703915_input.R1.fastq.gz, 21.1 million reads). The paired-end and single-end reads require different parameter settings in mapping and post-alignment processing.

Procedure

Map Reads to the Reference Genome

1. Build index of the reference
   $ bwa index -a bwtsw hg19.fa
2. Align reads to the genome
   First specify name prefix output files:
   $ IP_prefix = GSM1703914_H3K27ac
   $ INPUT_prefix = GSM1703915_input
   1. Map paired-end reads from IP:
      $ bwa aln -l 32 -k 2 /path_to/hg19.fa ${IP_prefix}.R1.fastq.gz > ${IP_prefix}.R1.sai

```
$ bwa aln -l 32 -k 2 /path_to/hg19.fa
${IP_prefix}.R2.fastq.gz > ${IP_prefix}.R2.sai
   $ bwa sampe -s /path_to/hg19.fa ${IP_prefix}.
R1.sai ${IP_prefix}.R2.sai ${IP_prefix}.R1.fastq.gz
${IP_prefix}.R2.fastq.gz | samtools view -Sbh - >
${IP_prefix}.bam
```
In the above commands:
-l 32 specifies the first 32 base pairs as the seed
-k 2 allows <=2 bp mismatches in the seed
/path_to/ points to the directory where the pre-built reference index files are saved

2. Map single-end reads from input:
```
$ bwa aln -l 32 -k 2 /path_to/hg19.fa
${INPUT_prefix}.R1.fastq.gz > ${INPUT_prefix}.R1.sai
   $ bwa samse -f ${INPUT_prefix}.sam
/path_to/hg19.fa ${INPUT_prefix}.R1.sai
${INPUT_prefix}.R1.fastq.gz
   $ samtools view -Sbh ${INPUT_prefix}.sam >
${INPUT_prefix}.bam
```
(Optionally) delete temporary files:
```
$ rm ${IP_prefix}.R[1,2].sai ${INPUT_prefix}.R1.sai ${INPUT_prefix}.sam
```

3. Check mapping status
```
$ samtools flagstat ${IP_prefix}.bam
$ samtools flagstat ${INPUT_prefix}.bam
```
? Troubleshooting

Post-alignment Processing

4. Select uniquely mapped reads
   1. To extract properly paired unique reads with high mapping quality (>=30) for IP. Proper-paired reads have SAM flags of 83/163 and 99/147, and uniquely-mapped reads are marked with the tag of "XT:A:U." Read pairs are first labeled as "1" if meeting these criteria or "0" otherwise in the ${IP_prefix}.id.txt file below. Read pairs with label of "1" are then extracted for further analysis.
      ```
      $ samtools view -H ${IP_prefix}.bam > ${IP_prefix}.sam
         $ samtools view ${IP_prefix}.bam | awk 'BEGIN {FS = OFS = "\t"} {print $2,$5,$12}' | awk 'NR%2 {printf $0"\t";next;}1' | awk 'BEGIN {FS = OFS = "\t"} {if ((($1 = = 83) || ($1 = = 99)) && ($2 >= 30) && ($5 >= 30) && ($3 ~ /XT:A:U/) && ($6 ~ /XT:A:U/)) print "1\n1"; else print "0\n0"}' > ${IP_prefix}.id.txt
         $ samtools view ${IP_prefix}.bam | paste ${IP_prefix}.id.txt - | awk 'BEGIN {FS = OFS = "\t"} {if ($1 = = 1) print substr($0, index($0, $2))}' >> ${IP_prefix}.sam
         $ samtools view -Sbh ${IP_prefix}.sam > ${IP_prefix}.filtered.bam
      ```
   2. To extract uniquely mapped reads with mapping quality >=30 for input
      ```
      $ samtools view -h -q 30 -F 0x04 ${INPUT_prefix}.bam | samtools view -Sbh - > ${INPUT_prefix}.filtered.bam
      ```
      (Optionally) delete temporary files:
      ```
      $ rm ${IP_prefix}.id.txt ${IP_prefix}.sam
      ```

5. Sort by coordinates and remove duplicates with Picard
   ```
   $ java -jar /path_to/SortSam.jar INPUT = ${IP_prefix}.filtered.bam OUTPUT = ${IP_prefix}.sorted.bam SO = coordinate TMP_DIR = ./ VALIDATION_STRINGENCY = SILENT
      $ java -jar /path_to/MarkDuplicates.jar INPUT = ${IP_prefix}.sorted.bam OUTPUT = ${IP_prefix}.dedup.bam METRICS_FILE = ${IP_prefix}.dedup_metrics.txt REMOVE_DUPLICATES = TRUE ASSUME_SORTED = true TMP_DIR = ./ VALIDATION_STRINGENCY = SILENT
   ```
   Sort and remove duplicates from input following the same commands.

6. Get duplicate rate
   ```
   $ cat ${IP_prefix}.dedup_metrics.txt | grep "Unknown Library" | awk 'BEGIN {FS = OFS = "\t"} {gsub(/[ ] + /,"\t",$0); printf("%0.4 f\n",$9)}'
   ```
   Use the same command for the input by simply replace ${IP_prefix} with ${INPUT_prefix}
   ? Troubleshooting

Peak calling and Peak Distribution

7. Peak calling with MACS
   Considering that IP reads are from paired-end sequencing and input reads are from single-end sequencing, user can extract the first end from IP for peak calling.
   ```
   $ samtools view -h ${IP_prefix}.dedup.bam | awk 'BEGIN {FS = OFS = "\t"} {if (($1 ~ /\@/) || (($2 = = 83) || ($2 = = 99)) print $0}' | samtools view -Sbh - > ${IP_prefix}.dedup.end1.bam
   ```
   a) Call peaks at FDR = 0.01 as follows:
      ```
      $ macs2 callpeak -t ${IP_prefix}.dedup.end1.bam -c ${INPUT_prefix}.dedup.bam -n ${IP_prefix}.macs2 -f BAM -g hs —keep-dup all -q 0.01 —nomodel —shiftsize = 100
      ```
      Critical for histone marks with broad occupancy, such as H3K27me3, H3K9me3, and H3K36me3, add the option "—broad" to the above command.
      If no input, simply omits the "-c" parameter.
   b) Filter out peaks that overlap blacklisted regions
      ```
      $ cat ${IP_prefix}.macs2_peaks.encodePeak | intersectBed -a stdin -b $BLACKLIST —v > ${IP_prefix}.macs2_peaks.filtered.encodePeak
      ```

8. Estimate FRiP (fraction of reads in peaks)
    $ samtools view -c ${IP_prefix}.dedup.bam
    $ cat ${IP_prefix}.macs2_peaks.filtered.encodePeak | cut -f 1–3 | intersectBed –abam ${IP_prefix}.dedup.bam -b stdin | samtools view -c -

    The first command estimates the total usable reads and the second one estimates the number of reads falling into peaks. Their ratio represents FRiP, which is a key quality metric.
    ? Troubleshooting
9. (Optional) Overlap with public ChIP-seq data from relevant cell types
    This will give the user an idea about the overall data quality. For major histone marks, a good source of public data is provided by the Roadmap project, which is available at https://egg2.wustl.edu/roadmap/data/byFileType/peaks/consolidated/.
10. Estimate peak overlap over different genomic features
    Use the R package "ChIPseeker" to plot peak distribution in a genome (Fig. 9.2), as follows:
    > library(ChIPseeker)
    > library(TxDb.Hsapiens.UCSC.hg19.knownGene)
    > txdb <- TxDb.Hsapiens.UCSC.hg19.knownGene
    > library(org.Hs.e.g.,db)
    >peakAnno <- annotatePeak ("GSM1703914_H3K27ac.macs2_peaks.filtered.encodePeak", tssRegion = c(-2000, 2000), TxDb = txdb, annoDb = "org.Hs.e.g.,db")
    >plotAnnoPie(peakAnno, cex = 0.7)

Signal Visualization

11. Generate bigWig file with deepTools
    1) Index BAM file
        $ samtools index ${IP_prefix}.dedup.bam
    2) Estimate scaling factor for normalization based on total usable reads
        $ SF = $(samtools view -c ${IP_prefix}.dedup.bam |awk -v s = 1000000 '{print s/$1}')
    3) Generate signal track
        $ bamCoverage -b ${IP_prefix}.dedup.bam -o ${IP_prefix}.bw -of bigwig -bs 50 —scaleFactor $SF —smoothLength 150 —ignoreDuplicates

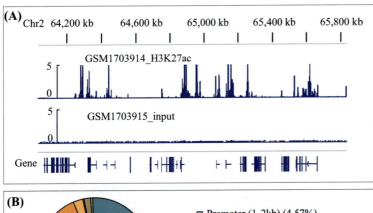
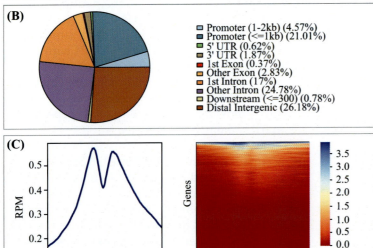

FIGURE 9.2 Visualization of H3K27ac enrichment in the public data GSM1703914. (A) An IGV snapshot of H3K27ac signal and input over a 1.7-Mb region on chr2. (B) Distribution of peaks across different genomic features, generated using the ChIPseeker package. (C) Average profile and heatmap of signal over TSS ± 2 kb, generated by the deepTools package.

Here, a bin size of 50 bp is used. By specifying a smoothLength of 150 bp, in calculating signal for a given 50-bp bin, reads falling within both left and right bins are also considered.

Use the above commands to generate bigWig file for the input by adding the parameter "−extendReads 200" in the above bamCoverage command, as input was sequenced in single-end mode. The bigWig files can be visualized in IGV (Fig. 9.2).

Critical user can also estimate the scaling factor based on the number of reads in peaks, or based on the number of usable reads from the spike-in control if an exogenous spike-in is used in ChIP.

12. Plot signal enrichment around TSSs

Use deepTools commands to generate aggregate signal plot and heatmap around TSS ± 2000 bp (Fig. 9.2). A 6-column tab-delimited bed file (chr, start, end, gene Symbol, gene ID, and strand) containing gene model is needed, e.g., gencode.hg19.bed shown below, along with the bigWig (.bw) file.
$ computeMatrix reference-point −referencePoint TSS -b 2000 -a 2000 -R gencode.hg19.bed -S ${IP_prefix}.bw −skipZeros -o ${IP_prefix}.TSS.gz −outFileSortedRegions ${IP_prefix}_gene.bed
$ plotHeatmap -m ${IP_prefix}.TSS.gz -o ${IP_prefix}.TSS.pdf −sortUsing sum

Timing

Step 1: Building index, ~1 h
Steps 2−3: Read mapping, ~3 h
Steps 4−6: Post-alignment processing, ~20 min
Steps 7−10: Peak calling and peak distribution, ~15 min
Steps 11−12: Signal visualization, <1.5 h

Expected Outcomes

For a ChIP-seq data, mapping rate typically should be ~80% or higher. Duplicate rate tends to be high for deeply sequenced samples and for factors with high levels of enrichment (i.e., with a high FRiP). Based on the ENCODE guideline, the duplicate rate should be <20% for 10 million uniquely mapped reads.

Overall data quality can be assessed visually in IGV, as shown in Fig. 9.2. Samples with few peaks have poor quality. For active marks like H3K27ac, samples with <10,000 peaks is less optimal and better to be excluded from differential analysis. Peak distribution in the genome is mark dependent. For H3K4me3, a significant portion of the peaks would be in the promoters. For H3K4me1 and H3K27ac, a significant portion of the peaks should be in introns and intergenic regions.

Troubleshooting

Step	Problem	Possible Reason	Solution
3	Low mapping rate	Sample contamination; adapter sequence	Trim adapter sequence; check possible contamination
6	High duplicate rate	Low input; over amplification	Use more starting material; reduce number of PCR cycle
8	No/weak signal or high background	The antibody does not work or lacks specificity	Evaluate antibody quality

Acknowledgments

This work is supported by the Mayo Clinic Center for Individualized Medicine.

Abbreviations

ACT-seq	Antibody-guided chromatin tagmentation sequencing
ATAC-seq	Assay for transposase-accessible chromatin sequencing
CapSTARR-seq	Capturing self-transcribing active regulatory regions sequencing
CUTAC	cleavage under targeted accessible chromatin
CUT&RUN	Cleavage under targets and release using nuclease
CUT&Tag	Cleavage under targets and tagmentation
ChIP-seq	Chromatin immunoprecipitation and sequencing
ChIA-PET	Chromatin interaction analysis with paired-end tag sequencing
ChIL-seq	Chromatin integration labeling and sequencing
ChOR-seq	Chromatin occupancy after replication and sequencing
CRISPR-Cas9	Clustered regularly interspaced short palindromic repeats/CRISPR associated protein 9
CoBATCH	Combinatorial barcoding and targeted chromatin release
CoTECH	Combined assay of transcriptome and enriched chromatin binding
DHS	DNase I hypersensitive site
dsQTL	DNase I sensitivity quantitative trait locus
ENCODE	Encyclopedia of DNA elements
eRNA	Enhancer RNA
eQTL	Expression quantitative trait locus
FAIRE-seq	Formaldehyde-assisted isolation of regulatory elements followed by sequencing
GWAS	Genome-wide association study
hQTL	Histone modification quantitative trait locus
iACT-seq	Indexing and antibody-guided chromatin tagmentation sequencing
itChIP-seq	Indexing and tagmentation-based chromatin immunoprecipitation and sequencing

MOWChIP-seq	Microfluidic oscillatory washing-based ChIP-seq
Paired-seq	Parallel analysis of individual cells for RNA expression and DNA accessibility by sequencing
Paired-Tag	Parallel analysis of individual cells for RNA expression and DNA from targeted tagmentation by sequencing
PnP-ChIP-seq	Plug and play ChIP-seq
QTL	Quantitative trait locus
RNAPII	RNA polymerase II
STARR-seq	Self-transcribing active regulatory regions sequencing
scATAC-seq	Single-cell assay for transposase-accessible chromatin sequencing
scChIC-seq	Single-cell chromatin immunocleavage sequencing
scChIP-seq	Single-cell chromatin immunoprecipitation and sequencing
scChIL-seq	Single-cell chromatin integration labeling and sequencing
scCUT&Tag	Single-cell cleavage under targets and tagmentation
sc-itChIP-seq	Single-cell Indexing and tagmentation-based chromatin immunoprecipitation and sequencing
scRNA-seq	Single-cell RNA sequencing
SNP	Single nucleotide polymorphism
SurfaceChIP-seq	ChIP on antibody-coated channel surface followed by sequencing
TCGA	The Cancer Genome Atlas
TF	Transcription factor
TSS	Transcription start site
uliCUT&RUN	Ultra-low-input cleavage under targets and release using nuclease
uaRNA	Upstream antisense RNA

References

[1] Talbert PB, Henikoff S. The Yin and Yang of histone marks in transcription. Annu Rev Genomics Hum Genet 2021; 22:147-170;.

[2] Reveron-Gomez N, Gonzalez-Aguilera C, Stewart-Morgan KR, Petryk N, Flury V, Graziano S, et al. Accurate recycling of parental histones reproduces the histone modification landscape during DNA replication. Mol Cell 2018;72(2):239–49 e5.

[3] Dai Z, Ramesh V, Locasale JW. The evolving metabolic landscape of chromatin biology and epigenetics. Nat Rev Genet 2020;21(12):737–53.

[4] Reinberg D, Vales LD. Chromatin domains rich in inheritance. Science 2018;361(6397):33–4.

[5] Carter B, Ku WL, Kang JY, Hu G, Perrie J, Tang Q, et al. Mapping histone modifications in low cell number and single cells using antibody-guided chromatin tagmentation (ACT-seq). Nat Commun 2019;10(1):3747.

[6] Medina-Rivera A, Santiago-Algarra D, Puthier D, Spicuglia S. Widespread enhancer activity from core promoters. Trends Biochem Sci 2018;43(6):452–68.

[7] Zylicz JJ, Bousard A, Zumer K, Dossin F, Mohammad E, da Rocha ST, et al. The implication of early chromatin changes in X chromosome inactivation. Cell 2018;182–97 e23.

[8] Schmitz ML, Higgins JMG, Seibert M. Priming chromatin for segregation: functional roles of mitotic histone modifications. Cell Cycle 2020;19(6):625–41.

[9] Clouaire T, Rocher V, Lashgari A, Arnould C, Aguirrebengoa M, Biernacka A, et al. Comprehensive mapping of histone modifications at DNA double-strand breaks deciphers repair pathway chromatin signatures. Mol Cell 2018;72(2):250–62 e6.

[10] Kim J, Sturgill D, Sebastian R, Khurana S, Tran AD, Edwards GB, et al. Replication stress shapes a protective chromatin environment across fragile genomic regions. Mol Cell 2018;69(1):36–47 e7.

[11] Li T, Liu Q, Garza N, Kornblau S, Jin VX. Integrative analysis reveals functional and regulatory roles of H3K79me2 in mediating alternative splicing. Genome Med 2018;10(1):30.

[12] Xu Y, Wang Y, Luo J, Zhao W, Zhou X. Deep learning of the splicing (epi)genetic code reveals a novel candidate mechanism linking histone modifications to ESC fate decision. Nucleic Acids Res 2017;45(21):12100–12.

[13] Xu Y, Zhao W, Olson SD, Prabhakara KS, Zhou X. Alternative splicing links histone modifications to stem cell fate decision. Genome Biol 2018;19(1):133.

[14] Hu Q, Kim EJ, Feng J, Grant GR, Heller EA. Histone posttranslational modifications predict specific alternative exon subtypes in mammalian brain. PLoS Comput Biol 2017;13(6):e1005602.

[15] Hu Q, Greene CS, Heller EA. Specific histone modifications associate with alternative exon selection during mammalian development. Nucleic Acids Res 2020;48(9):4709–24.

[16] Agirre E, Oldfield AJ, Bellora N, Segelle A, Luco RF. Splicing-associated chromatin signatures: a combinatorial and position-dependent role for histone marks in splicing definition. Nat Commun 2021;12(1):682.

[17] Huang D, Petrykowska HM, Miller BF, Elnitski L, Ovcharenko I. Identification of human silencers by correlating cross-tissue epigenetic profiles and gene expression. Genome Res 2019;29(4):657–67.

[18] Andersson R, Sandelin A. Determinants of enhancer and promoter activities of regulatory elements. Nat Rev Genet 2020;21(2):71–87.

[19] Özdemir I, Gambetta MC. The role of insulation in patterning gene expression. Genes (Basel) 2019;10(10):767.

[20] Local A, Huang H, Albuquerque CP, Singh N, Lee AY, Wang W, et al. Identification of H3K4me1-associated proteins at mammalian enhancers. Nat Genet 2018;50(1):73–82.

[21] Roe JS, Hwang CI, Somerville TDD, Milazzo JP, Lee EJ, Da Silva B, et al. Enhancer reprogramming promotes pancreatic cancer metastasis. Cell 2017;170(5):875–88.

[22] Yan J, Chen SA, Local A, Liu T, Qiu Y, Dorighi KM, et al. Histone H3 lysine 4 monomethylation modulates long-range chromatin interactions at enhancers. Cell Res 2018;28(3):387.

[23] Godfrey L, Crump NT, Thorne R, Lau IJ, Repapi E, Dimou D, et al. DOT1L inhibition reveals a distinct subset of enhancers dependent on H3K79 methylation. Nat Commun 2019;10(1):2803.

[24] Godfrey L, Crump NT, O'Byrne S, Lau IJ, Rice S, Harman JR, et al. H3K79me2/3 controls enhancer-promoter interactions and activation of the pan-cancer stem cell marker PROM1/CD133 in MLL-AF4 leukemia cells. Leukemia 2021;35(1):90–106.

[25] Dorighi KM, Swigut T, Henriques T, Bhanu NV, Scruggs BS, Nady N, et al. Mll3 and Mll4 facilitate enhancer RNA synthesis and transcription from promoters independently of H3K4 monomethylation. Mol Cell 2017;66(4):568–76 e4.

[26] Haberle V, Stark A. Eukaryotic core promoters and the functional basis of transcription initiation. Nat Rev Mol Cell Biol 2018;19(10):621–37.

[27] Visel A, Minovitsky S, Dubchak I, Pennacchio LA. VISTA enhancer browser—a database of tissue-specific human enhancers. Nucleic Acids Res 2007;35(Database issue):D88–92.

[28] Bai X, Shi S, Ai B, Jiang Y, Liu Y, Han X, et al. ENdb: a manually curated database of experimentally supported enhancers for human and mouse. Nucleic Acids Res 2020;48(D1) D51-d7.

[29] Wang J, Dai X, Berry LD, Cogan JD, Liu Q, Shyr Y. HACER: an atlas of human active enhancers to interpret regulatory variants. Nucleic Acids Res 2018;47(D1):D106–12.

[30] Jiang Y, Qian F, Bai X, Liu Y, Wang Q, Ai B, et al. SEdb: a comprehensive human super-enhancer database. Nucleic Acids Res 2018;47(D1):D235–43.

[31] Huang J, Li K, Cai W, Liu X, Zhang Y, Orkin SH, et al. Dissecting super-enhancer hierarchy based on chromatin interactions. Nat Commun 2018;9(1):943.

[32] Hnisz D, Abraham BJ, Lee TI, Lau A, Saint-Andre V, Sigova AA, et al. Super-enhancers in the control of cell identity and disease. Cell 2013;155(4):934–47.

[33] Barakat TS, Halbritter F, Zhang M, Rendeiro AF, Perenthaler E, Bock C, et al. Functional dissection of the enhancer repertoire in human embryonic stem cells. Cell Stem Cell 2018;23(2):276–88 e8.

[34] Kvon EZ, Waymack R, Gad M, Wunderlich Z. Enhancer redundancy in development and disease. Nat Rev Genet 2021;22(5):324–36.

[35] Andersson R, Gebhard C, Miguel-Escalada I, Hoof I, Bornholdt J, Boyd M, et al. An atlas of active enhancers across human cell types and tissues. Nature 2014;507(7493):455–61.

[36] Osterwalder M, Barozzi I, Tissieres V, Fukuda-Yuzawa Y, Mannion BJ, Afzal SY, et al. Enhancer redundancy provides phenotypic robustness in mammalian development. Nature 2018;554(7691):239–43.

[37] Cao K, Collings CK, Marshall SA, Morgan MA, Rendleman EJ, Wang L, et al. SET1A/COMPASS and shadow enhancers in the regulation of homeotic gene expression. Genes Dev 2017;31(8):787–801.

[38] Azofeifa JG, Allen MA, Hendrix JR, Read T, Rubin JD, Dowell RD. Enhancer RNA profiling predicts transcription factor activity. Genome Res 2018;28(3):334–44.

[39] Chen H, Li C, Peng X, Zhou Z, Weinstein JN, Liang H. A pan-cancer analysis of enhancer expression in nearly 9000 patient samples. Cell 2018;173(2):386–99 e12.

[40] Mumbach MR, Satpathy AT, Boyle EA, Dai C, Gowen BG, Cho SW, et al. Enhancer connectome in primary human cells identifies target genes of disease-associated DNA elements. Nat Genet 2017;49(11):1602–12.

[41] Dao LTM, Galindo-Albarrán AO, Castro-Mondragon JA, Andrieu-Soler C, Medina-Rivera A, Souaid C, et al. Genome-wide characterization of mammalian promoters with distal enhancer functions. Nat Genet 2017;49(7):1073–81.

[42] Liu Y, Yu S, Dhiman VK, Brunetti T, Eckart H, White KP. Functional assessment of human enhancer activities using whole-genome STARR-sequencing. Genome Biol 2017;18(1):219.

[43] Cai Y, Zhang Y, Loh YP, Tng JQ, Lim MC, Cao Z, et al. H3K27me3-rich genomic regions can function as silencers to repress gene expression via chromatin interactions. Nat Commun 2021;12(1):719.

[44] Doni Jayavelu N, Jajodia A, Mishra A, Hawkins RD. Candidate silencer elements for the human and mouse genomes. Nat Commun 2020;11(1):1061.

[45] Ngan CY, Wong CH, Tjong H, Wang W, Goldfeder RL, Choi C, et al. Chromatin interaction analyses elucidate the roles of PRC2-bound silencers in mouse development. Nat Genet 2020;52(3):264–72.

[46] Pang B, Snyder MP. Systematic identification of silencers in human cells. Nat Genet 2020;52(3):254–63.

[47] Panigrahi A, O'Malley BW. Mechanisms of enhancer action: the known and the unknown. Genome Biol 2021;22(1):108.

[48] Harada A, Maehara K, Handa T, Arimura Y, Nogami J, Hayashi-Takanaka Y, et al. A chromatin integration labelling method enables epigenomic profiling with lower input. Nat Cell Biol 2019;21(2):287–96.

[49] Kaya-Okur HS, Wu SJ, Codomo CA, Pledger ES, Bryson TD, Henikoff JG, et al. CUT&Tag for efficient epigenomic profiling of small samples and single cells. Nat Commun 2019;10(1):1930.

[50] Bartosovic M, Kabbe M, Castelo-Branco G. Single-cell CUT&Tag profiles histone modifications and transcription factors in complex tissues. Nat Biotechnol 2021;39(7):825–35.

[51] Meers MP, Bryson TD, Henikoff JG, Henikoff S. Improved CUT&RUN chromatin profiling tools. Elife 2019;8:e46314.

[52] Zhu C, Zhang Y, Li YE, Lucero J, Behrens MM, Ren B. Joint profiling of histone modifications and transcriptome in single cells from mouse brain. Nat Methods 2021;18(3):283–92.

[53] Wang Q, Xiong H, Ai S, Yu X, Liu Y, Zhang J, et al. CoBATCH for high-throughput single-cell epigenomic profiling. Mol Cell 2019;76(1):206–16 e7.

[54] Ku WL, Nakamura K, Gao W, Cui K, Hu G, Tang Q, et al. Single-cell chromatin immunocleavage sequencing (scChIC-seq) to profile histone modification. Nat Methods 2019;16(4):323–5.

[55] Xiong H, Luo Y, Wang Q, Yu X, He A. Single-cell joint detection of chromatin occupancy and transcriptome enables higher-dimensional epigenomic reconstructions. Nat Methods 2021;18(6):652–60.

[56] Skene PJ, Henikoff S. An efficient targeted nuclease strategy for high-resolution mapping of DNA binding sites. Elife 2017;6:e21856.

[57] Patty BJ, Hainer SJ. Transcription factor chromatin profiling genome-wide using uliCUT&RUN in single cells and individual blastocysts. Nat Protoc 2021;16(5):2633–66.

[58] Hainer SJ, Bošković A, McCannell KN, Rando OJ, Fazzio TG. Profiling of pluripotency factors in single cells and early embryos. Cell 2019;177(5):1319–29 e11.

[59] Henikoff S, Henikoff JG, Kaya-Okur HS, Ahmad K. Efficient chromatin accessibility mapping in situ by nucleosome-tethered tagmentation. Elife 2020;9:e63274.

[60] Ai S, Xiong H, Li CC, Luo Y, Shi Q, Liu Y, et al. Profiling chromatin states using single-cell itChIP-seq. Nat Cell Biol 2019;21(9):1164–72.

[61] Grosselin K, Durand A, Marsolier J, Poitou A, Marangoni E, Nemati F, et al. High-throughput single-cell ChIP-seq identifies heterogeneity of chromatin states in breast cancer. Nat Genet 2019;51(6):1060–6.

[62] Cao Z, Chen C, He B, Tan K, Lu C. A microfluidic device for epigenomic profiling using 100 cells. Nat Methods 2015;12(10):959–62.

[63] Zhu B, Hsieh YP, Murphy TW, Zhang Q, Naler LB, Lu C. MOWChIP-seq for low-input and multiplexed profiling of genome-wide histone modifications. Nat Protoc 2019;14(12):3366–94.

[64] Ma S, Hsieh YP, Ma J, Lu C. Low-input and multiplexed microfluidic assay reveals epigenomic variation across cerebellum and prefrontal cortex. Sci Adv 2018;4(4):eaar8187.

[65] Dirks RAM, Thomas PC, Wu H, Jones RC, Stunnenberg HG, Marks H. A plug and play microfluidic platform for standardized sensitive low-input chromatin immunoprecipitation. Genome Res 2021;31(5):919–33.

[66] Dainese R, Gardeux V, Llimos G, Alpern D, Jiang JY, Meireles-Filho ACA, et al. A parallelized, automated platform enabling individual or sequential ChIP of histone marks and transcription factors. Proc Natl Acad Sci U S A 2020;117(24):13828–38.

[67] Skene PJ, Henikoff JG, Henikoff S. Targeted in situ genome-wide profiling with high efficiency for low cell numbers. Nat Protoc 2018;13(5):1006–19.

[68] Vitak SA, Torkenczy KA, Rosenkrantz JL, Fields AJ, Christiansen L, Wong MH, et al. Sequencing thousands of single-cell genomes with combinatorial indexing. Nat Methods 2017;14(3):302–8.

[69] Lara-Astiaso D, Weiner A, Lorenzo-Vivas E, Zaretsky I, Jaitin DA, David E, et al. Immunogenetics. Chromatin state dynamics during blood formation. Science 2014;345(6199):943–9.

[70] Zhu C, Yu M, Huang H, Juric I, Abnousi A, Hu R, et al. An ultra high-throughput method for single-cell joint analysis of open chromatin and transcriptome. Nat Struct Mol Biol 2019;26(11):1063—70.

[71] Chan JC, Maze I. Nothing is yet set in (Hi)stone: novel post-translational modifications regulating chromatin function. Trends Biochem Sci 2020;45(10):829—44.

[72] Maksimovic I, David Y. Non-enzymatic covalent modifications as a new chapter in the histone code. Trends Biochem Sci 2021; 46(9):718—730;.

[73] Huang H, Zhang D, Wang Y, Perez-Neut M, Han Z, Zheng YG, et al. Lysine benzoylation is a histone mark regulated by SIRT2. Nat Commun 2018;9(1):3374.

[74] Kebede AF, Nieborak A, Shahidian LZ, Le Gras S, Richter F, Gómez DA, et al. Histone propionylation is a mark of active chromatin. Nat Struct Mol Biol 2017;24(12):1048—56.

[75] Jo C, Park S, Oh S, Choi J, Kim EK, Youn HD, et al. Histone acylation marks respond to metabolic perturbations and enable cellular adaptation. Exp Mol Med 2020;52(12):2005—19.

[76] Gao M, Wang J, Rousseaux S, Tan M, Pan L, Peng L, et al. Metabolically controlled histone H4K5 acylation/acetylation ratio drives BRD4 genomic distribution. Cell Rep 2021;36(4):109460.

[77] Goudarzi A, Zhang D, Huang H, Barral S, Kwon OK, Qi S, et al. Dynamic competing histone H4 K5K8 acetylation and butyrylation are hallmarks of highly active gene promoters. Mol Cell 2016;62(2):169—80.

[78] Fellows R, Denizot J, Stellato C, Cuomo A, Jain P, Stoyanova E, et al. Microbiota derived short chain fatty acids promote histone crotonylation in the colon through histone deacetylases. Nat Commun 2018;9(1):105.

[79] Bao X, Liu Z, Zhang W, Gladysz K, Fung YME, Tian G, et al. Glutarylation of histone H4 Lysine 91 regulates chromatin dynamics. Mol Cell 2019;76(4):660—75 e9.

[80] Zhang D, Tang Z, Huang H, Zhou G, Cui C, Weng Y, et al. Metabolic regulation of gene expression by histone lactylation. Nature 2019;574(7779):575—80.

[81] Xie Z, Zhang D, Chung D, Tang Z, Huang H, Dai L, et al. Metabolic regulation of gene expression by histone lysine beta-hydroxybutyrylation. Mol Cell 2016;62(2):194—206.

[82] Dai L, Peng C, Montellier E, Lu Z, Chen Y, Ishii H, et al. Lysine 2-hydroxyisobutyrylation is a widely distributed active histone mark. Nat Chem Biol 2014;10(5):365—70.

[83] Zhang Q, Bai B, Mei X, Wan C, Cao H, Dan L, et al. Elevated H3K79 homocysteinylation causes abnormal gene expression during neural development and subsequent neural tube defects. Nat Commun 2018;9(1):3436.

[84] Farrelly LA, Thompson RE, Zhao S, Lepack AE, Lyu Y, Bhanu NV, et al. Histone serotonylation is a permissive modification that enhances TFIID binding to H3K4me3. Nature 2019;567(7749):535—9.

[85] Fujiki R, Hashiba W, Sekine H, Yokoyama A, Chikanishi T, Ito S, et al. GlcNAcylation of histone H2B facilitates its monoubiquitination. Nature 2011;480(7378):557—60.

[86] Hirosawa M, Hayakawa K, Yoneda C, Arai D, Shiota H, Suzuki T, et al. Novel O-GlcNAcylation on Ser(40) of canonical H2A isoforms specific to viviparity. Sci Rep 2016;6:31785.

[87] Rousseaux S, Khochbin S. Histone acylation beyond acetylation: Terra Incognita in chromatin biology. Cell J 2015;17(1):1—6.

[88] Sabari BR, Zhang D, Allis CD, Zhao Y. Metabolic regulation of gene expression through histone acylations. Nat Rev Mol Cell Biol 2017;18(2):90—101.

[89] Li K, Wang Z. Histone crotonylation-centric gene regulation. Epigenetics Chromatin 2021;14(1):10.

[90] Gowans GJ, Bridgers JB, Zhang J, Dronamraju R, Burnetti A, King DA, et al. Recognition of histone crotonylation by Taf14 links metabolic state to gene expression. Mol Cell 2019;76(6):909—21 e3.

[91] Andrews FH, Shinsky SA, Shanle EK, Bridgers JB, Gest A, Tsun IK, et al. The Taf14 YEATS domain is a reader of histone crotonylation. Nat Chem Biol 2016;12(6):396—8.

[92] Anastas JN, Shi Y. Histone serotonylation: can the brain have "happy" chromatin? Mol Cell 2019;74(3):418—20.

[93] Zhao J, Chen W, Pan Y, Zhang Y, Sun H, Wang H, et al. Structural insights into the recognition of histone H3Q5 serotonylation by WDR5. Sci Adv 2021;7(25):eabf4291.

[94] Fu L, Zhang L. Serotonylation: a novel histone H3 marker. Signal Transduct Target Ther 2019;4:15.

[95] Lepack AE, Werner CT, Stewart AF, Fulton SL, Zhong P, Farrelly LA, et al. Dopaminylation of histone H3 in ventral tegmental area regulates cocaine seeking. Science 2020;368(6487):197—201.

[96] Oki S, Ohta T, Shioi G, Hatanaka H, Ogasawara O, Okuda Y, et al. ChIP-Atlas: a data-mining suite powered by full integration of public ChIP-seq data. EMBO Rep 2018;19(12):e46255.

[97] Moore JE, Purcaro MJ, Pratt HE, Epstein CB, Shoresh N, Adrian J, et al. Expanded encyclopaedias of DNA elements in the human and mouse genomes. Nature 2020;583(7818):699—710.

[98] Grubert F, Srivas R, Spacek DV, Kasowski M, Ruiz-Velasco M, Sinnott-Armstrong N, et al. Landscape of cohesin-mediated chromatin loops in the human genome. Nature 2020;583(7818):737—43.

[99] Boix CA, James BT, Park YP, Meuleman W, Kellis M. Regulatory genomic circuitry of human disease loci by integrative epigenomics. Nature 2021;590(7845):300—7.

[100] Gorkin DU, Barozzi I, Zhao Y, Zhang Y, Huang H, Lee AY, et al. An atlas of dynamic chromatin landscapes in mouse fetal development. Nature 2020;583(7818):744—51.

[101] Onuchic V, Lurie E, Carrero I, Pawliczek P, Patel RY, Rozowsky J, et al. Allele-specific epigenome maps reveal sequence-dependent stochastic switching at regulatory loci. Science 2018;361(6409):eaar3146.

[102] del Rosario RC, Poschmann J, Rouam SL, Png E, Khor CC, Hibberd ML, et al. Sensitive detection of chromatin-altering polymorphisms reveals autoimmune disease mechanisms. Nat Methods 2015;12(5):458—64.

[103] Pelikan RC, Kelly JA, Fu Y, Lareau CA, Tessneer KL, Wiley GB, et al. Enhancer histone-QTLs are enriched on autoimmune risk haplotypes and influence gene expression within chromatin networks. Nat Commun 2018;9(1):2905.

[104] Delaneau O, Zazhytska M, Borel C, Giannuzzi G, Rey G, Howald C, et al. Chromatin three-dimensional interactions mediate genetic effects on gene expression. Science 2019;364(6439).

[105] Chen L, Ge B, Casale FP, Vasquez L, Kwan T, Garrido-Martin D, et al. Genetic drivers of epigenetic and transcriptional variation in human immune cells. Cell 2016;167(5):1398—414 e24.

[106] Watt S, Vasquez L, Walter K, Mann AL, Kundu K, Chen L, et al. Genetic perturbation of PU.1 binding and chromatin looping at neutrophil enhancers associates with autoimmune disease. Nat Commun 2021;12(1):2298.

[107] Tan WLW, Anene-Nzelu CG, Wong E, Lee CJM, Tan HS, Tang SJ, et al. Epigenomes of human hearts reveal new genetic variants relevant for cardiac disease and phenotype. Circ Res 2020;127(6):761—77.

[108] Caliskan M, Manduchi E, Rao HS, Segert JA, Beltrame MH, Trizzino M, et al. Genetic and epigenetic fine mapping of complex trait associated loci in the human liver. Am J Hum Genet 2019;105(1):89—107.

[109] Sun W, Poschmann J, Cruz-Herrera Del Rosario R, Parikshak NN, Hajan HS, Kumar V, et al. Histone acetylome-wide

association study of autism spectrum disorder. Cell 2016;167 (5):1385−97 e11.
[110] Ng B, White CC, Klein HU, Sieberts SK, McCabe C, Patrick E, et al. An xQTL map integrates the genetic architecture of the human brain's transcriptome and epigenome. Nat Neurosci 2017;20(10):1418−26.
[111] Wang D, Liu S, Warrell J, Won H, Shi X, Navarro FCP, et al. Comprehensive functional genomic resource and integrative model for the human brain. Science 2018;362(6420):eaat8464.
[112] Girdhar K, Hoffman GE, Jiang Y, Brown L, Kundakovic M, Hauberg ME, et al. Cell-specific histone modification maps in the human frontal lobe link schizophrenia risk to the neuronal epigenome. Nat Neurosci 2018;21(8):1126−36.
[113] Nott A, Holtman IR, Coufal NG, Schlachetzki JCM, Yu M, Hu R, et al. Brain cell type-specific enhancer-promoter interactome maps and disease-risk association. Science 2019;366(6469):1134−9.
[114] Ai R, Laragione T, Hammaker D, Boyle DL, Wildberg A, Maeshima K, et al. Comprehensive epigenetic landscape of rheumatoid arthritis fibroblast-like synoviocytes. Nat Commun 2018;9(1):1921.
[115] Morton AR, Dogan-Artun N, Faber ZJ, MacLeod G, Bartels CF, Piazza MS, et al. Functional enhancers shape extrachromosomal oncogene amplifications. Cell 2019;179(6):1330−41 e13.
[116] Marzi SJ, Leung SK, Ribarska T, Hannon E, Smith AR, Pishva E, et al. A histone acetylome-wide association study of Alzheimer's disease identifies disease-associated H3K27ac differences in the entorhinal cortex. Nat Neurosci 2018;21(11):1618−27.
[117] Wang X, Lee RS, Alver BH, Haswell JR, Wang S, Mieczkowski J, et al. SMARCB1-mediated SWI/SNF complex function is essential for enhancer regulation. Nat Genet 2017;49(2):289−95.
[118] Abraham BJ, Hnisz D, Weintraub AS, Kwiatkowski N, Li CH, Li Z, et al. Small genomic insertions form enhancers that misregulate oncogenes. Nat Commun 2017;8:14385.
[119] Chatterjee A, Rodger EJ, Eccles MR. Epigenetic drivers of tumourigenesis and cancer metastasis. Semin Cancer Biol 2018;51:149−59.
[120] McDonald OG, Li X, Saunders T, Tryggvadottir R, Mentch SJ, Warmoes MO, et al. Epigenomic reprogramming during pancreatic cancer progression links anabolic glucose metabolism to distant metastasis. Nat Genet 2017;49(3):367−76.
[121] van Groningen T, Koster J, Valentijn LJ, Zwijnenburg DA, Akogul N, Hasselt NE, et al. Neuroblastoma is composed of two super-enhancer-associated differentiation states. Nat Genet 2017;49(8):1261−6.
[122] Beekman R, Chapaprieta V, Russinol N, Vilarrasa-Blasi R, Verdaguer-Dot N, Martens JHA, et al. The reference epigenome and regulatory chromatin landscape of chronic lymphocytic leukemia. Nat Med 2018;24(6):868−80.
[123] Xu H, Wu M, Ma X, Huang W, Xu Y. Function and mechanism of novel histone posttranslational modifications in health and disease. Biomed Res Int 2021;2021:6635225.
[124] Nakamura M, Gao Y, Dominguez AA, Qi LS. CRISPR technologies for precise epigenome editing. Nat Cell Biol 2021;23(1):11−22.
[125] Adli M. The CRISPR tool kit for genome editing and beyond. Nat Commun 2018;9(1):1911.
[126] Rees HA, Liu DR. Base editing: precision chemistry on the genome and transcriptome of living cells. Nat Rev Genet 2018;19(12):770−88.
[127] Li J, Mahata B, Escobar M, Goell J, Wang K, Khemka P, et al. Programmable human histone phosphorylation and gene activation using a CRISPR/Cas9-based chromatin kinase. Nat Commun 2021;12(1):896.
[128] Li K, Liu Y, Cao H, Zhang Y, Gu Z, Liu X, et al. Interrogation of enhancer function by enhancer-targeting CRISPR epigenetic editing. Nat Commun 2020;11(1):485.

CHAPTER

10

Techniques for Analyzing Genome-wide Expression of Non-coding RNA

Rena Onoguchi-Mizutani[1], Kenzui Taniue[1], Kentaro Kawata[1,2], Toshimichi Yamada[1] and Nobuyoshi Akimitsu[1]

[1]Isotope Science Center, The University of Tokyo, Tokyo, Japan [2]Cellular and Molecular Biotechnology Research Institute, National Institute of Advanced Industrial Science and Technology (AIST), Tokyo, Japan

OUTLINE

Introduction	163
cDNA Library Construction	164
Depletion of Ribosomal RNA from cDNA Libraries	165
Library Preparation for Strand-specific RNA-seq	166
Detection of Low-abundance RNA Species	167
Analysis of RNA-seq Data for ncRNAs	169
Algorithm for lncRNA Detection	169
Algorithm for Small ncRNA Detection	169
Algorithm for Circular RNAs Detection	170
Algorithm for Fusion RNA Detection	171
Estimating the Transcription and Degradation Rates of Non-Coding Transcripts	172
A Genome-wide Approach to Determine RNA Stability	172
Detecting Unstable ncRNAs by Inhibiting their Decay	173
Estimation of Transcription Rate by the Detection of Nascent Transcripts	173
Kinetic Determination of RNA Production and Degradation	174
Conclusions	174
Dyrec-seq Protocol	175
Before you Begin	175
Key Resources Table	175
Step-by-Step Method Details	175
Quantification and Statistical Analysis	176
Expected Outcomes	177
Advantages	177
Limitations	177
Optimization and Troubleshooting	177
Acknowledgments	177
Glossary	177
References	178

INTRODUCTION

Genome-wide transcriptome analyses provide an unprecedented opportunity to study gene regulation. Much of the genome is transcribed and transcripts include not only coding, but also non-coding RNAs (ncRNAs).

Classification of ncRNAs is generally based on their length; small ncRNAs (<200 nucleotides (nt)) and long ncRNAs (lncRNAs; >200 nt) [1,2]. Small ncRNAs include microRNAs (miRNAs), transfer RNAs (tRNAs) and piwi-interacting RNAs (piRNA). Based on their associations with genomic features, lncRNAs are further classified as follows; long intergenic RNAs (lincRNAs) from intergenic regions, enhancer RNAs (eRNA) from enhancer regions, sense ncRNAs, which overlap and share some sequence with the same strand of a protein-coding mRNA, but do not have coding potential, antisense RNAs from the antisense strand of protein-coding genes, and circular RNAs (circRNAs) [2]. It has become clear that these ncRNAs are involved in gene regulation in several ways; for example, control of the epigenetic state (reviewed in [3,4]), formation of phase-separated domains [5], or regulation of cellular stress responses [6–10]. To reveal the gene regulation system globally, improvements in transcriptome analysis are critically important.

Early transcriptome studies mainly relied on microarray technology. This strategy has some limitations, a major shortcoming being that it cannot detect unknown genes or variant patterns. This made it difficult to comprehensively annotate transcripts and to quantify transcriptomes, including unknown ones. To overcome this limitation, tiling microarrays, containing oligonucleotide probes encompassing an entire defined genomic region, were employed. These enabled novel transcripts to be identified without prior knowledge of the genes within the defined genomic region [11]. This methodology was applied to survey the lncRNAs expressed from the human *HOX* loci, resulting in the discovery of the HOX antisense intergenic RNA (*HOTAIR*), which is an antisense lncRNA that is embedded in the *HOXC* locus and represses *in trans* transcription of the *HOXD* locus [12]. However, the use of tiling microarrays for genome-wide analysis has declined because they are difficult to manufacture and are expensive.

The emergence of massively parallel sequencing technologies to sequence cDNA derived from RNA (i.e., RNA-seq) has revolutionized transcriptome analysis. RNA-seq technology can directly uncover a sequence's identity and can, therefore, detect novel transcripts. RNA-seq is now the method of choice to study gene expression comprehensively. However, some limitations of transcriptome analysis remain, especially for ncRNAs. These limitations are attributed to characteristics of ncRNAs, which are different from mRNAs. NcRNAs tend to be expressed at lower levels, have no or shortened polyadenylated (polyA) tails, and are unstable transcripts [13–15]. Improvements in library construction protocols and RNA-seq data analysis for ncRNAs are ongoing.

RNA-seq technology can be used to study many different aspects of RNA biology, including single-cell gene expression, translation, nascent RNAs, RNA-protein interaction, RNA structure and transcriptome architecture [16–22]. RNA-seq technology can be divided into three platforms: short-read cDNA sequence (Illumina), long-read cDNA sequence (Pacific Biosciences, Oxford Nanopore) and long-read direct RNA sequence (Oxford Nanopore) [23]. Short-read technology uses mRNA fragmentation and size selection of fragments during bead-based library purification to yield sequences of 200–300 nt in length [24]. This method has become the *de facto* method to detect and quantify transcriptome-wide gene expression, both because it is cheaper and easier to perform than microarrays or long-read sequencing methods, and because it produces high-quality transcriptome expression data [23]. However, long-read approaches overcome some of the problems associated with short-read approaches, such as more complete identification and quantification of isoform diversity because ambiguity in the mapping of sequence reads is reduced [23]. Also, long-read sequencing technology reduces false-positive splice-junction detection by short-read RNA-seq computational tools and detects them more accurately. In addition, direct long-read sequencing technology can read the RNA sequence directly without modification, cDNA synthesis and/or PCR amplification during library preparation [25]. Long-read sequencing technologies do, however, have some limitations. For example, long-read technologies feature much lower throughput and much higher error rates than short-read sequencing methods, which can affect the sensitivity and specificity of sequencing results. Moreover, long-read RNA-seq technology is currently more expensive than short-read RNA-seq. In this review, we focus on short-read sequencing because Illumina short-read RNA-seq currently dominates the field. We review the short-read RNA-seq technologies that are optimized to search for ncRNAs. First, we summarize key points of constructing cDNA libraries that focus on ncRNAs. Second, we introduce annotation methods developed to identify ncRNAs. Finally, we present RNA-seq-based methods to monitor rates of transcription and degradation.

cDNA LIBRARY CONSTRUCTION

Traditional RNA-seq libraries were designed to efficiently enrich polyadenylated mRNAs. Recent studies have revealed the characteristic features of ncRNAs that are different from mRNAs, including short or no poly (A) tail, strand specificity, and relatively low expression level. Therefore, traditional RNA-seq library construction has a tendency to underestimate ncRNA expression. For instance, phenol/chloroform extraction, which is a common method of purifying RNA from cells or tissues, results in underestimation of the expression of certain scaffold RNAs, such as nuclear paraspeckle assembly transcript (*NEAT1*) lncRNA [26]. *NEAT1* promotes the formation of paraspeckle nuclear bodies by binding multiple RNA-binding proteins (RBPs) [27]. The multivalent interactions between *NEAT1* and RBPs result in the trapping of most *NEAT1* in the protein phase during conventional phenol/chloroform RNA extraction [26]. *NEAT1* extraction was increased 20-fold after extensive needle shearing or heat treatment of the cell lysate [26]. Other lncRNAs in phase-separated ribonucleoprotein (RNP) structures share this property of *NEAT1* (reviewed in [28]), indicating that RNA extraction methods using needle shearing or heating should be extended to the analyses of scaffold lncRNA expression levels.

In this section, we introduce strategies of cDNA library construction to improve the quality of ncRNA analysis, including ribosomal RNA depletion, strand-specific library construction, and methods to detect low abundance transcripts.

Depletion of Ribosomal RNA from cDNA Libraries

To maximize the information available from RNA-seq analysis, removal of ribosomal RNA (rRNA) is very important because rRNA is the most abundant RNA species in a total RNA sample, and its presence prevents the acquisition of sufficient sequencing-depth for other RNAs. For mRNA detection, a standard solution is to enrich for poly(A) transcripts using oligo d(T)-based affinity substrates. However, this strategy is not optimal for the detection of ncRNAs because it excludes transcripts that lack or have shortened poly(A) tails, such as small RNAs generated by RNA polymerase III and other kinds of lncRNA [13]. To overcome these problems, several rRNA depletion strategies have been developed. We introduce four procedures for preparing cDNA libraries without rRNAs, namely RNase H treatment, Ribo-zero treatment, duplex-specific nuclease (DSN) treatment, and the not-so-random primers (NSR) method (Fig. 10.1).

To selectively degrade rRNAs using RNase H endoribonuclease, fragmented total RNA is hybridized to tiling antisense DNA oligonucleotides against rRNAs. The RNA strand of DNA-RNA hybrids is specifically digested by RNase H. The RNase H treatment followed by DNase treatment degrades both rRNAs and residual antisense DNA oligos. The rRNA-depleted RNAs can then be used to synthesize cDNA libraries [29].

Ribo-zero treatment removes rRNA from RNA-seq libraries by the specific capturing of rRNAs. In this method, fragmented total RNA is hybridized to rRNA sequence-specific biotinylated probes that are then captured with streptavidin beads. The supernatant represents rRNA-depleted RNAs [30]. A Ribo-zero kit is commercially available (Epicentre Biotechnologies) for the following species: human, mouse, yeast, bacteria, and plant leaf. Analysis of other species may require a custom designed kit.

DSN treatment depletes highly abundant transcripts, such as rRNAs, to enrich for rare RNAs in a cDNA library. DSN selectively degrades double-stranded DNAs during heat denaturation of the cDNA with subsequent re-annealing. The hybridization rate for each cDNA is proportional to the square of its concentration; therefore, in the re-annealing step, cDNAs that are generated from abundant transcripts from double-stranded DNA more

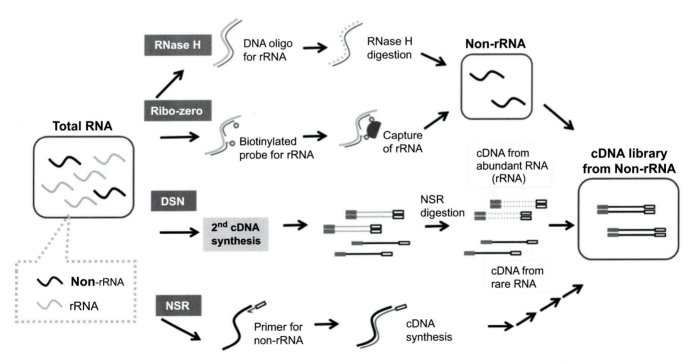

FIGURE 10.1 Construction of cDNA libraries without rRNA. Relevant steps of the rRNA-depletion protocols are shown. Black wavy lines indicate non-rRNAs and gray wavy lines indicate rRNAs. RNase H represents Ribo-zero methods that remove rRNA from a total RNA sample. The DSN method depletes cDNAs generated from abundant RNAs, including rRNA. The NSR approach generates cDNA libraries directly from total RNA, using a not-so-random primer.

efficiently than cDNAs from rare transcripts, and are therefore more effectively degraded by DSN. Remaining single-stranded cDNAs from lowly expressed transcripts are subjected to double-strand cDNA library synthesis [31,32]. This strategy does not require any pre-designed oligonucleotides; therefore, it is suitable for analyzing whole transcriptomes of non-model organisms.

The last method uses not-so-random (NSR) primers that bind to non-rRNA transcripts during the reverse transcription step. These NSR primers were selected computationally from a random hexamer library. The NSR approach also allows the generation of strand-specific cDNA libraries (for details see the section "Library preparation for strand-specific RNA-seq"). The NSR method generates rRNA-depleted cDNA libraries from small amounts of RNA, and a commercial kit, named Ovation RNA-seq system (NuGEN), is available [33].

Each depletion strategy has strengths and weaknesses. For example, three of the methods (RNase H treatment, Ribo-zero treatment, and the NSR method) require similar amounts of time to prepare cDNA libraries; however, the DSN method takes an additional day. Adiconis et al. compared these methods of eliminating rRNAs. RNase H treatment and Ribo-zero treatment are good for low quality RNA samples, and the Ovation RNA-seq system has the advantage of working with low quantity RNA samples [34] (Table 10.1). Application of these rRNA-depletion approaches to RNA-seq enabled the detection of larger numbers of non-poly(A) transcripts, including ncRNAs, compared with poly(A) selection [35]. Moreover, these methods make it possible to use low quality RNAs, such as degraded RNA from fixed clinical samples, or small quantities of input RNA [34].

Library Preparation for Strand-specific RNA-seq

Several studies have identified antisense transcripts involved in epigenetic regulation; for example, *Kcnq1ot1*, *HOTAIR*, *AIR*, and *ANRIL* (reviewed in [36]). Genome-wide approaches have also identified many antisense transcripts with potential regulatory roles [16].

TABLE 10.1 Summary of Advantages and Aisadvantages of cDNA Library Construction Without rRNAs

Library	Advantage	Disadvantage
RNase H	Applicable for low quality sample	
Ribo-Zero	Applicable for low quality sample commercial available kit	Per-sample cost is relatively high
NSR	Applicable for low quantity sample commercial available kit	Per-sample cost is relatively high
DSN	Per-sample cost is relatively low	

Interestingly, the epigenetic state is regulated by switching the strand of ncRNA transcription. The polycomb group (PcG) and trithorax group (TrxG) proteins, which are well known chromatin regulators, provide competitive regulation through a nearby enhancer known as the PcG response element/TrxG response element (PRE/TRE). The PRE/TRE consists of DNA elements that reversibly switch between silencing and activating epigenetic states [37]. This PRE/TRE switching is regulated by lncRNAs, which are transcribed from the opposite strands of the PRE/TRE, at the vestigial locus in *Drosophila*. A genome-wide analysis identified many PcG-regulated sites that might switch their function between repression and activation by switching the strand of ncRNA transcription [38]. Consequently, strand information has grown in importance in RNA-seq analysis. However, for standard cDNA libraries for RNA-seq, the synthesis of second strand-cDNAs by random priming followed by adaptor attachment results in loosing strand information. In this section, we introduce methods to prepare strand-specific cDNA libraries. There are two main strategies; one is based on ligating different adaptors at 5′ and 3′ ends of each RNA fragment (as adopted by the different adaptor ligation method, the NSR approach, and the SMART method) and the other relies on marking one strand of the double-stranded cDNA with a chemical modification or specifically degrading one strand of a double-stranded cDNA (adopted by the ASSAGE [Asymmetric Strand-Specific Analysis of Gene Expression] and deoxy-UTP second strand approaches).

The different adaptor ligation method relies on the ligation of two distinct adaptors, one to the 5′ end and one to the 3′ end of the RNA. Initially, the RNA is fragmented by metal hydrolysis, subjected to gel electrophoresis and the small RNA fraction is excised from the gel. The 3′ and 5′ RNA adaptors are then ligated to RNA fragments, respectively. Reverse transcription PCR is then conducted to obtain the cDNA libraries [39].

As noted in the "rRNA depletion" section, the NSR approach generates strand-specific cDNA libraries directly from total RNA, using NSR primers, which bind to non-rRNA transcripts. Total RNAs are first reverse transcribed using NSR antisense primers containing an adaptor at the 5′ end. Subsequently, RNA templates are digested by RNase H treatment. Double-strand cDNAs are then synthesized using NSR sense primers with another adaptor at the 5′ end. As a result, different adapter sequences are added to the sense and antisense orientation of each cDNA during first- and second-strand cDNA synthesis. This method needs neither an rRNA removal step nor a ligation step [33].

The SMART method uses two intrinsic properties of Moloney murine leukemia virus (MMLV) reverse transcriptase. One is its terminal transferase activity, and the other is template switching [40]. This method commences with RNA fragmentation, followed by single-strand cDNA synthesis, which is primed using a tagged random hexamer. Via the terminal transferase activity of MMLV reverse transcriptase, a short cytosine (C) tail is attached to the 3′ end of newly synthesized single-stranded cDNAs. A tagged primer containing a short guanine (G) tail is then hybridized to the C tail, leading to double-strand cDNA generation (known as a template switching reaction). The cDNA library is amplified using primers complementary to 5′ and 3′ tag sequences [41].

The key feature of the ASSAGE approach is to treat RNAs with bisulfite, which converts all cytidine residues to uridine residues. Bisulfite-treated RNAs are reverse transcribed into cDNAs; therefore, these cDNA sequences can be matched to only the converted strand but not to the other strand [42].

In the deoxy-UTP (dUTP) second strand approach, single-strand cDNA is synthesized using normal dNTPs; however, double-strand cDNA is generated with dUTP as a substitute for dTTP. The double-strand cDNA incorporating dUTP is fragmented and a Y-shaped adaptor is attached to both the 5′ and 3′ end of the cDNA. The dUTP-containing strand of the double-strand cDNA is selectively destroyed by uracil-N-glycosylase, which hydrolyzes first- or second-strand DNA containing uracil. As a result, the first cDNA strand remains intact. All remaining cDNA molecules are sequenced in the same direction because the Y-shape adaptor is used for library preparation, and each arm of the Y-shaped adaptor has a distinct index sequence [43].

Levin et al. compared these cDNA library construction methods for strand-specific RNA-seq and summarized their relative advantages and disadvantages [44]. The different adaptor ligation method has high strand specificity, but this method requires large amounts of RNA for multiple size selection. The NSR and ASSAGE approaches are relatively simple; however, quality is relatively low. Levin et al. mentioned that the dUTP second strand approach is a well-balanced strategy across their criteria because library construction is similar to standard sample preparation and it produces high overall quality [44].

Detection of Low-abundance RNA Species

The expression levels of ncRNAs tend to be lower than those of protein coding genes [45]. Low abundance transcripts are disadvantaged in RNA-seq analysis because of the extended dynamic range; i.e., highly expressed transcripts account for most of the mapped reads, and lowly expressed RNAs account for a small percentage of sequencing reads [46]. The limitation of finite sequence read coverage precludes precise transcript assembly or quantification. To resolve this problem, several approaches have been developed. In this section, we introduce methods to enrich RNAs of interest (RNA CaptureSeq, chromatin associated-RNA-seq, RIP-seq) or to label target transcripts (Digital RNA-seq), with a focus on ncRNA analysis.

RNA CaptureSeq overcomes the insufficient depth of coverage of lowly expressed transcripts (Fig. 10.2A) [47]. This approach focuses on genes of interest, enriches sequence read coverage, and allows identification or quantification of rare transcripts. RNA CaptureSeq is similar to the exosome sequencing approach [48]. The first step is the design of oligonucleotide probes against targeted transcripts. In the second step, oligonucleotide probes are hybridized to prepared cDNA libraries followed by washing out non-targeted cDNAs. This enriches sequencing reads for target transcripts [49]. CaptureSeq improves sequencing coverage by several 10-fold compared with standard RNA-seq and has the advantages of analyzing lncRNA expression and identifying novel lncRNA isoforms [43,50,51].

Chromatin-associated RNA-seq (Chrom-seq) was developed based on the idea that ncRNAs are involved in epigenetic regulation or that other nascent RNAs are mainly localized in the chromatin fraction. The Chrom-seq method fractionates whole cells biochemically into cytoplasmic, nucleoplasmic and chromatin associated fractions. RNA isolated from each fraction is depleted of rRNAs, random primed, and used to generate strand-specific cDNA libraries (Fig. 10.2B). Some ncRNAs, such as *XIST* or spliceosomal small nuclear RNAs (snRNAs), were confirmed to be enriched in the chromatin-associated fraction [13,52]. Chrom-seq enables the detection of lowly expressed RNAs localized in the chromatin-associated fraction; therefore, this approach was used to detect eRNAs and has contributed to revealing the mechanism of eRNA biogenesis [53]. In another study, thousands of chromatin-enriched RNAs (cheRNAs) were discovered [54]. The number of cheRNAs was larger than the number of previously recognized eRNAs, and there was positive correlation between gene expression and proximity to a cheRNA. This cheRNA expression data might be useful to annotate de novo enhancer regions [54].

A large fraction of RNAs interact with RBPs and are involved in gene regulation; therefore, focusing on RBPs that interact with ncRNAs is useful for the detection of ncRNAs. RNA immunoprecipitation (RIP) following deep sequencing (RIP-seq) permits the enrichment of RNAs of interest by immunoprecipitation with an RBP-specific antibody and analysis of

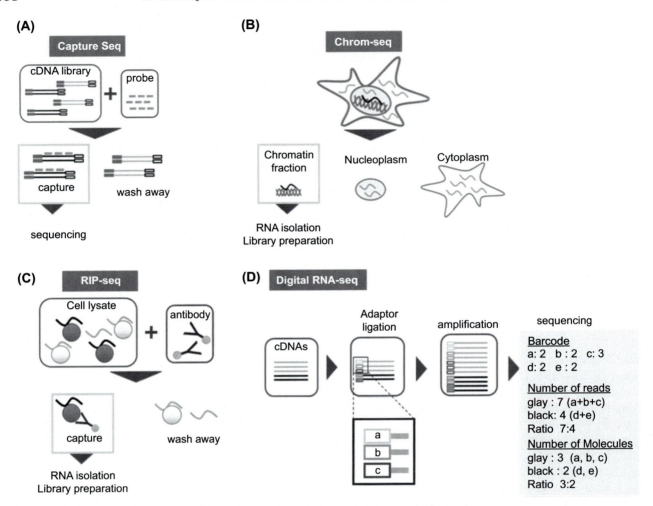

FIGURE 10.2 RNA-seq methods to detect lowly expressed transcripts, such as ncRNAs. Relevant RNA-seq steps for the detection of lowly expressed RNAs. Black wavy lines indicate target RNAs and gray wavy lines indicate non-target RNAs. (A) CaptureSeq is based on capture and enrichment of genes of interest by hybridization of DNA oligonucleotide probes to cDNA libraries. (B) Chrom-seq focuses on chromatin-associated RNAs that are biochemically fractionated. (C) RIP-seq involves the immunoprecipitation of RNA-binding proteins of interest and enriching RNAs bound to them. (D) In digital RNA-seq, cDNA libraries are labeled with distinct barcodes before amplification. The ratios of RNA molecules are calculated from unique barcode numbers.

coprecipitated RNA by RNA-seq (Fig. 10.2C). RIP-seq has been used to identify ncRNAs involved in epigenetic regulation. RIP of native EZH2, which is a component of polycomb repressive complex 2 (PRC2), revealed that EZH2 cofactors contain some lncRNAs, such as *Xist*, *Tsix*, and *RepA* [55]. Consequently, RIP-seq against EZH2 was conducted with strand-specific RNA-sequencing, which identified many additional lncRNAs that bind to EZH2. It was also suggested that *Gtl2*, a representative EZH2 cofactor RNA, regulates imprinting of the *Dlk1* locus via PRC2-mediated histone H3 lysine 27 trimethylation (H3K27me3) [56].

Digital RNA-seq reduces sequence-dependent PCR amplification bias by labeling almost all cDNA molecules with distinct barcodes before amplification. These barcodes are optimized to minimize bias during the ligation and amplification phase. rRNA-depleted RNA is reverse transcribed to cDNA, and then random adaptors with unique barcodes are ligated to each 3′ and 5′ end. The excess adaptors are removed, and cDNAs containing the barcodes are amplified by PCR. These PCR amplicons are sequenced, and the barcodes are used to determine the ratio of RNA molecules. Digital RNA-seq has several advantages, such as low amplification bias during PCR, which provides information about RNA abundance and enables detection of lowly expressed RNAs (Fig. 10.2D). This method does have a disadvantage in that it requires high sequence read coverage [57].

To select transcripts of interest from cDNA libraries, a combinatorial approach of molecular indexing with RNA-targeted sequencing was proposed. After labeling cDNA molecules with barcodes as in digital RNA-seq, cDNAs are hybridized to biotinylated capture probes, trapped by streptavidin beads, eluted and analyzed by deep sequencing [58].

ANALYSIS OF RNA-seq DATA FOR ncRNAs

An advantage of RNA-seq for identifying ncRNAs is that little or no a priori information of an RNA's sequence is required and, therefore, all required information can be obtained directly from the data. This advantage, however, is heavily dependent on the ability to reconstruct a transcriptome from millions of short reads. Reconstruction of all full-length transcripts from short reads with considerable sequencing error rates gives rise to computational challenges; quantifying the diverse expression of transcripts, correcting sequencing biases and errors and managing alternative splicing. There are two alternative strategies for transcriptome reconstruction. Mapping-first approaches (e.g., Augustus [59], iReckon [60], Scripture [61], SLIDE [62] and StringTie [63]) first align all reads to the reference genome, and then merge the sequences using overlapping alignments and spanning splice junctions with reads and paired-ends. Alternatively, assembly-first methods (e.g., Oases [64], Trans-ABySS [65], Trinity [66] and Velvet [67]) assemble transcripts directly from the reads, which can be mapped subsequently to a reference genome. The algorithmic details of short read mapping and companions of annotating methods are beyond the scope of this review (these are reviewed in [68]). Here we briefly introduce algorithms that detect specific ncRNA classes, including lncRNAs, small RNAs, circRNAs, and fusion RNAs.

Algorithm for lncRNA Detection

To analyze lncRNA expression, there are several pipelines that unify the read aligners, such as TopHat [69], GSNAP [70], STAR [71,72], and HISAT [73], in addition to transcriptome reconstruction algorithms, including Scripture [61], Cufflinks [74], and StringTie [75]. A challenging task in lncRNA investigation is to evaluate protein coding potential after annotation. Similar to mRNAs, most lncRNAs are RNA polymerase II (RNAPII) products, and can be capped and polyadenylated. The pathways that produce lncRNAs are the same or similar to those that produce mRNAs and are affected by histone-modification profiles, splicing signals, and exon/intron lengths. A strategy to distinguish lncRNAs from mRNAs is to calculate the codon substitution frequency (CSF) score [61,76], especially using PhyloCSF [77]. A disadvantage of using CSF is that it requires an extremely long computation time. An alternative approach is classification based on support vector machine (SVM), which is a typical machine learning method, to identify lncRNAs accurately and quickly. For example, coding or non-coding [78], coding potential calculator [79], iSeeRNA [80] and lncRScan-SVM [81] have been developed using this strategy.

In recent years, large-scale transcriptome sequencing projects have provided several kinds of database for lncRNA annotation, such as Ref-seq [82] and GENCODE [83,84]. lncRNA annotation is continuously updated. For example, 17,858 novel human lncRNA transcripts were added to GENCODE v31 (released in June 2019), an approximate 60% increase compared with the previous release [83,84]. These findings indicate the existence of unannotated lncRNAs; therefore, there have been no easy-to-use public RNA-seq pipelines for lncRNA detection.

Algorithm for Small ncRNA Detection

Although RNA-seq can survey small RNAs, specifying the type of small ncRNA from the pool of other sequenced small RNAs or degradation products is a central problem in annotating small RNAs. miRDeep is believed to be the first stand-alone tool to address this problem. It can detect both known and novel miRNAs in *Caenorhabditis elegans*, humans, and dogs [85]. miRDeep uses a probabilistic model of miRNA biogenesis to score compatibility of the position and frequency of a sequenced small RNA with the structural information of the miRNA precursor. Based on miRDeep, several algorithms have been developed (e.g., miRDeep2 [86] and miRDeep* [87]) to improve the sensitivity and accuracy of the prediction. In addition to miRDeep, there are numerous other miRNA search methods, including miRanalyzer [88], miRspring [89] and miRTRAP [90].

piRNA is the largest class of small RNA molecules expressed in mammalian cells, especially in germ cells, and associate with PIWI proteins. piRNAs have different features to miRNAs. piRNAs have no clear secondary structure, are usually 25—32 nt long, and lack primary sequence conservation. Moreover, the presence of a 5′ uridine is common in piRNAs. Betel et al. developed the first de novo algorithm to identify piRNAs by applying the position-specific use of 10 bases upstream and 10 bases downstream of the 5′ U [91]. Using this algorithm, they identified mouse piRNAs with the precision of about 65%. Zhang et al. developed a *k*-mer based algorithm, named piRNApredictor, to improve precision and to explore piRNAs in organisms for which genome information is lacking [92]. They characterized piRNA sequences in five model species for the training set, and consequently reached a prediction precision of over 90% and sensitivity of over 60%. When piRNApredictor was applied to the deep-sequenced small RNA data of the migratory locust (*Locusta migratoria*), it identified 87,536 piRNAs with the same precision and sensitivity as in five model species. Wang et al. developed the piRNA annotation program (Piano) using piRNA-transposon interaction information [93]. A support vector machine was used to classify real piRNAs and pseudo piRNAs with over 90% prediction sensitivity, specificity, and accuracy.

tRNAs are classical ncRNAs that transfer amino acids to the ribosome-mRNA complex for translation [94]. Recent studies have discovered tRNA-derived small RNAs, which are produced through tRNA cleavage by a specific nuclease [95]. Major species of tRNA-derived small RNAs are tRNA halves, which are stress-induced fragments derived from the anticodon loop of tRNAs, and tRNA-derived fragments, which act in similar way to miRNAs [95]. Difficulties exist in detecting and quantifying tRNA-derived small RNAs because of the features of tRNAs, such as repetitive sequences, multiple copies in the genome, and heavily modification. Based on these problems, optimized tools for tRNA-derived small RNAs have been developed, including tDRmapper [96], MINTmap [97], and SPORTS1.0 [98].

There are other classes of ncRNAs, such as small nucleolar RNAs (snoRNAs) and small nuclear RNAs (snRNAs). These ncRNAs can be analyzed by preexisting tools for ncRNA analysis [99]. For example, iSRAP [100], miARma-seq [101], and sRNAnalyzer [102] can be used for the identification and quantification of snoRNAs [99]. Integrated tools to analyze several classes of small ncRNA, including miRNAs, are DARIO [103] and ncPRO-seq [104]. These methods interrogate and perform detailed profiling of small RNAs derived from annotated non-coding regions in several ncRNA public databases.

Algorithm for Circular RNAs Detection

circRNAs are covalently closed, endogenous biomolecules [105–107] that are generated by a non-canonical splicing event called back-splicing, which is regulated by specific cis-acting elements and trans-acting factors [105–107]. circRNAs are generally expressed at low levels [108–111], are highly stable [109,112,113], and are often expressed in cell-type and tissue-specific patterns [114–116]. Unlike most linear RNAs, circRNAs lack 3'-poly(A) tails; therefore, they are rarely observed in poly(A)+ RNA-seq profiles [105,107]. Unlike standard poly(A)+ RNA-seq, RNA-seq of rRNA-depleted total RNA (e.g., using Ribo-Zero technology) can be used to discover novel circRNAs [109,110,117,118]. In addition, circRNA candidates are often validated using RNase R, an exonuclease that specifically degrades linear RNAs [109]. rRNA-depleted and RNase R-digested RNA were used to prepare libraries enriched for circRNAs and to differentiate circRNAs from mRNAs [119].

A number of bioinformatic pipelines have been developed to identify and quantify circRNA expression from RNA-seq datasets. Most genome-wide circRNA detection tools generally identify the unique backsplice junction region [105,107]. However, the development of a circRNA detection tool is difficult because (1) a large proportion of circRNAs are at relatively low abundance; (2) read lengths vary in different sequencing data sets; and (3) the complexities of eukaryotic transcription may generate other non-canonical transcripts, such as fusion genes. There are numerous prediction algorithms for circRNAs [105] (Table 10.2): find_circ [109], MapSplice [120], CIRCexplorer [108], circRNA_finder [117], CIRI (CircRNA Identifier) [121], KNIFE [122], segmehl [123], CIRI2 (CircRNA Identifier2) [124], ACSF [125], CIRCexplorer2 [126], DCC [127], and Uroborus [128]. Hansen et al. compared the output from five different algorithms (find_circ, MapSplice, CIRCexplorer, circRNA_finder, and CIRI) [129]. CIRCexplorer and MapSplice produce the most reliable lists of circRNAs. However, these two algorithms require a gene annotation list and 2–3 days to complete the prediction. They suggested combining circRNA_finder and find_circ for a faster and almost equally reliable output. To predict circRNAs in poorly annotated organisms, find_circ, circRNA_finder, and CIRI are useful because these algorithms work *de novo* without knowledge of gene annotations and exon-intron structures. Importantly, Hansen et al. pointed out that the five algorithms yield highly divergent results and a high level of false positives, which indicates the need for predictions to be validated. Hansen subsequently extended this analysis to 11 algorithms [130], and concluded that the algorithms mostly agreed on highly expressed circRNAs with algorithm-specific false positives. Hansen recommended that the circRNA prediction output be combined and merged with other algorithms to identify circRNAs. Reconstructing the sequence of circRNAs from short read RNA-seq data has proved challenging given the similarity of circRNAs and their corresponding linear mRNAs. Zhang et al. presented an experimental and computational method (CIRI-long) for the extensive profiling of full-length circRNAs using nanopore sequencing technology [131], which enabled the identification of a new type of intronic self-ligated circRNA that exhibits particular splicing and expression patterns.

Considering the algorithm-specific false-positives generated by the available computational tools [130], the validation of computational prediction results by experimental approaches is required to select high-confidence circRNAs for further study [105,107]. RT and quantitative PCR (RT-qPCR) assays using divergent primers flanking the backsplice junction are often performed to confirm these sites. However, PCR-based validation does not guarantee the presence of circRNAs

TABLE 10.2 Available Bioinformatic Tools to Detect circRNAs

Tool	Alignment	URL
find_circ	Bowtie2	https://github.com/marvin-jens/find_circ
MapSplice	Bowtie	http://www.netlab.uky.edu/p/bioinfo/MapSplice
CIRCexplorer	TopHat2, Bowtie, STAR	https://github.com/YangLab/CIRCexplorer
circRNA_finder	STAR	https://github.com/bioxfu/circRNAFinder
CIRI	BWA	https://sourceforge.net/projects/ciri/files/
KNIFE	Bowtie, Bowtie2	https://github.com/lindaszabo/KNIFE
CIRI2	BWA	https://sourceforge.net/projects/ciri/files/CIRI2/
ACSF	BWA	https://github.com/arthuryxt/acfs
CIRCexplorer2	Multiple	https://github.com/YangLab/CIRCexplorer2
DCC	STAR	https://github.com/dieterich-lab/DCC
Uroborus	TopHat2, Bowtie	https://github.com/WGLab/UROBORUS

because linear RNAs with the same sequence as the entire backsplice junction site can be amplified. A more direct and accurate method to confirm circRNA existence is northern blot analysis because it does not include RT or amplification steps. In both the northern blotting and PCR methods, total RNA can be pretreated with RNase R to further verify the presence of circRNA [105,107]. In addition, in situ hybridization can be used to visualize circRNAs in fixed cells with probes that span the backsplice junction. In summary, circRNA studies can be validated by the techniques used to study linear RNAs; however, it is necessary to perform experiments to distinguish in vitro-circularized RNAs from linear RNAs with the same sequence.

Algorithm for Fusion RNA Detection

Fusion RNAs result from either chromosomal rearrangements or splicing mechanisms involving nonchromosomal rearrangements [132]. In 2009, Maher et al. reported two studies identifying novel fusion transcripts using high-throughput sequencing [133,134]. In 2010, Sboner et al. released the first fusion detection tool, FusionSeq, and since then, more than 40 tools have been developed for detecting fusion transcripts using RNA-seq data and/or whole-genome sequence data [135,136]. RNA-seq only detects fusion transcripts that occur at the RNA level and multiple alternative splice variants resulting from fusions. However, using both RNA-seq reads and whole-genome sequence reads enables the differentiation of fusion events that occur at the DNA or RNA level only. Most fusion detection tools require both genome and transcriptome sequences as references, although some consider either only the genome (i.e., MapSlice [120]) or transcriptome (i.e., EricScript [137] and JAFFA [138]) sequences. Using only transcriptome sequences as a reference limits the detection of fusion transcripts with novel exons.

The initial step of fusion detection has two approaches [135]. Most tools align reads to the reference sequence before finding fusion transcript breakpoints from resulting alignment patterns. However, some tools first assemble sequenced reads into de novo contigs, and then align the contigs to reference sequences. This latter strategy was used to discover several de novo fusions transcripts relevant to acute myeloid leukemia [139].

Fusion transcript detection tools require lots of time and computational memory, and vary in their hardware requirements, installation complexity, and prediction accuracy. Moreover, they may detect false-positive fusions, with tool performance being dependent on the datasets. This performance has been previously investigated. In 2016, Li et al. reported the comparative evaluation of 15 fusion transcript detection tools using synthetic datasets of different coverage, read length, insert size, and background noise, in addition to three real datasets in which all true fusion transcripts were not fully known [140] (Table 10.3). They concluded that the performance of SOAPfuse [141] was best for both synthetic and real datasets, followed by FusionCatcher [142] and JAFFA. Their results also indicated that EricScript and ChimeraScan [143] performed well on synthetic data, but poorly on the three real datasets. Moreover, they recommended applying several different pipelines and combining the results from applications to identify fusion transcript candidates. At almost the same time, Kumar et al. compared the performance of 12 fusion detection software packages [144], and demonstrated that EricScript was balanced in terms of sensitivity, specificity, and computational resources required. A recent

TABLE 10.3 Available Bioinformatic Tools to Detect Fusion RNAs

Tool	Alignment	URL
FusionSeq	ELAND	http://archive.gersteinlab.org/proj/rnaseq/fusionseq/
MapSlice	Bowtie	http://www.netlab.uky.edu/p/bioinfo/MapSplice
EricScript	BWA/BLAT	https://sourceforge.net/projects/ericscript/
JAFFA	Bowtie/BLAT	https://github.com/Oshlack/JAFFA/wiki
SOAPfuse	Soap2/BWA/BLAT	https://github.com/Nobel-Justin/SOAPfuse
FusionCatcher	Bowtie/Bowtie2/STAR/BLAT	https://github.com/ndaniel/fusioncatcher
ChimeraScan	Bowtie/BWA	https://code.google.com/archive/p/chimerascan/
STAR-Fusion	STAR	https://github.com/STAR-Fusion/STAR-Fusion
Arriba	STAR	https://github.com/suhrig/arriba
STAR-SEQR	STAR	https://github.com/ExpressionAnalysis/STAR-SEQR
TrinityFusion	STAR using the Trinity assembler	https://github.com/trinityrnaseq/TrinityFusion

comparative study by Haas et al. found that STAR-Fusion [71,72,145], Arriba [146], and STAR-SEQR [147] were the most accurate and rapid for fusion detection from cancer transcriptome data compared with 19 other methods [148]. They also demonstrated that although de novo assembly-based fusion detection methods, such as TrinityFusion [66,149], achieved a lower level of sensitivity compared with read mapping-based approaches, they could reconstruct fusion isoforms and tumor viruses, which are both important in cancer research. Thus, de novo assembly-based fusion detection methods can offer more complete fusion isoform sequence evidence and reconstruct foreign transcripts, such as tumor viruses. Finally, fusions detected by sequencing technology should be carefully validated because transcripts predicted by software tools may be artifacts [132,136].

ESTIMATING THE TRANSCRIPTION AND DEGRADATION RATES OF NON-CODING TRANSCRIPTS

Recent studies have revealed that ncRNAs have unique features in terms of their biogenesis and fate that distinguish them from mRNAs [150]. Most ncRNAs are short-lived, unstable transcripts that are present at a very low steady-state level [14], making them difficult to detect by standard RNA-seq. RNA degradation not only limits the number of transcripts produced from the genome, but also regulates the function of ncRNAs [151,152]. In this section, we introduce a method to estimate the degradation kinetics of ncRNAs (BRIC-seq). We then describe several strategies to detect ncRNAs in RNA-seq, such as knock out of degradation factors and detection of nascent transcripts, which can also estimate the transcription rate.

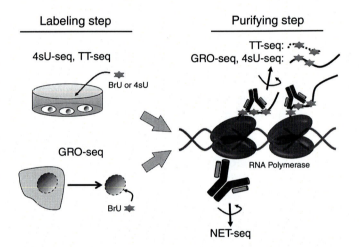

FIGURE 10.3 Workflows for 4sU-seq, TT-seq, and GRO-seq. In GRO-seq, nuclei are isolated for run-on in vitro transcription in the presence of BrU, whereas other methods directly label nascent RNA in vivo. Before RNA isolation, TT-seq fragments the labeled RNA. The labeled RNA is then captured by immunoprecipitation using BrU or 4sU antibodies. In NET-seq, nascent RNA is co-purified via immunoprecipitation of the RNAPII elongation complex.

These methods include GRO-seq, PRO-seq, NET-seq, and CAGE-seq. Finally, we introduce recent methods that calculate both transcription and degradation rates, including TT-seq, SLAM-seq, TimeLapse-seq, TUC-seq, TUC-seq dual, and Dyrec-seq (Fig. 10.3).

A Genome-wide Approach to Determine RNA Stability

RNA stabilities are measured by the decrease in RNA over time after inhibiting RNA transcription. However, these methods contain intrinsic noise originating from transcription inhibition, which obscures the true half-lives of transcripts. Previously, we developed an inhibitor-free method, 5'-bromo-uridine (BrU) immunoprecipitation chase-deep sequencing analysis (BRIC-seq) [14,153]. BRIC-seq labels endogenous transcripts by culturing cells with BrU. After removal of BrU from the culture medium, total RNAs containing BrU-labeled RNAs (BrU-RNA) are then isolated from cells at sequential time points. BrU-RNAs are then recovered by immunoprecipitation. The half-life of each transcript is calculated from the decreasing amount of BrU-RNA measured by RNA-seq. BRIC-seq was used to determine the half-lives of 11,052 mRNAs and 1,418 ncRNAs in HeLa cells [14].

Detecting Unstable ncRNAs by Inhibiting their Decay

Deletion of genes encoding nuclear degradation machinery (Rrp6 and Trf4) or the cytoplasmic 5'-3' Xrn1 exonuclease enables the identification of unstable transcripts in yeast. RNA-seq of a $rrp6\Delta$ $depl-trf4$ strain identified novel cryptic unstable transcripts (CUTs) [154]. Interestingly, most of the identified CUTs corresponded to transcripts that were divergent from the promoter regions of genes. RNA-seq for the wild type and a $xrn1\Delta$ strain identified 1658 Xrn1-sensitive unstable transcripts (XUTs), of which 66% were antisense to the open reading frame [155]. The XUTs included 183 CUTs, 543 stable uncharacterized transcripts, and 932 novel ncRNAs. Notably, chromatin immunoprecipitation sequencing (ChIP-seq) analysis of $xrn1\Delta$ strains revealed a significant decrease of RNAPII occupancy over 273 genes with antisense XUTs, indicating antisense ncRNA-mediated regulation of gene expression. In mammalian cells, PROMPTs, which are unstable polyadenylated transcripts transcribed at 0.5–2.5 kb upstream of active transcription start sites in an antisense direction, were identified by the depletion of exosome components [156].

Estimation of Transcription Rate by the Detection of Nascent Transcripts

To provide a snapshot of transcriptional activity, global run-on sequencing (GRO-seq) was developed. GRO-seq maps the genome-wide positions, amount and orientation of transcriptionally engaged RNA polymerases [16]. In this method, transcription is halted *in vivo* and then reinitiated in isolated nuclei under conditions that allow labeling of nascent transcripts with BrU (nuclear run-on RNA; NRO-RNA), thereby enabling them to be distinguished from bulk RNA. BrU-labeled NRO-RNA is then triple-selected via immunoprecipitation. This procedure results in a 10,000-fold enrichment of the NRO-RNA; therefore, assaying nascent RNAs dramatically increases the sensitivity of unstable ncRNA detection. In the first report of GRO-seq, Core et al. found that most promoters have an engaged polymerase that is upstream and in the opposite orientation to the gene [157]. This divergent polymerase was associated with active genes, but did not elongate beyond the promoter, which was similar to the results for antisense CUTs identified in the $rrp6\Delta$ $depl-trf4$ strain [154]. Importantly, GRO-seq can detect signal-dependent transcriptional responses of eRNAs. By combining temporal measurement of transcription factor binding, histone acetylation and methylation, with nascent RNA transcription, Kakkonen et al. observed a sequence of events at *de novo* enhancers proceeding from (1) unmarked chromatin, to (2) transcription factor binding, to (3) histone H4K5/8 acetylation coupled to eRNA transcription, followed by (4) progressive mono- and demethylation of H3K4 [158]. Although these findings indicated that transcription-coupled H3K4 methylation of *de novo* enhancers did not require the eRNA products, other GRO-seq studies demonstrate that eRNA itself induces specific enhancer−promoter looping and gene activation following cell stimulation [159,160].

Several modifications of GRO-seq have been made to improve its resolution. Precision nuclear run-on sequencing (PRO-seq) maps the genome-wide distribution of transcriptionally engaged RNAPIIs at base pair resolution [161]. PRO-seq uses biotin-labeled NTPs (biotin-NTPs) for nuclear run-on analysis. Supplying only one of the four biotin-NTPs restricts RNAPII to incorporate a single or a few identical bases, resulting in sequence reads that have the same 3' end base within each library. Another approach is termed GRO-cap, which detects a nascent RNA with a 5' cap [157,162]. By precisely defining eRNA start sites using GRO-cap, Lam et al. showed that transfer of the full enhancer activity to a target promoter requires both the sequences mediating transcription-factor binding and the specific sequence encoding the eRNA [163]. By combining GRO-seq data with cap analysis gene expression (CAGE) data, Core et al. classified transcription start site pairs on the basis of the transcripts' stability (unstable transcripts are detected only by GRO-cap, whereas stable transcripts are detected by both GRO-cap and CAGE). They analyzed these transcription start sites together with DNA sequence and functional genomic data to categorize the precise nature of the structure and chromatin content at each initiation site [157]. These analyses identified a common architecture of initiation, histone modifications, highly positioned flanking nucleosomes, and two modes of transcription factor binding. These results support a unified model of transcription initiation at promoters and enhancers.

In contrast to GRO-seq, native elongating transcript sequencing (NET-seq) exploits the highly stable DNA-RNA-RNAP complex to monitor transcription at nucleotide resolution [164–166]. To facilitate purification, the NET-seq protocol involves preparation of *Saccharomyces cerevisiae* expressing 3× Flag tagged RNAPII [164]. After cell lysis, RNA associated with RNAPII is efficiently immunoprecipitated for RNA-seq. NET-seq is an ideal tool to investigate divergent transcription (antisense CUTs) and has uncovered several fundamental properties of this phenomenon, such as, most promoters show strong directionality, favoring sense transcription, and suppression of antisense transcripts is enforced by (1) Rpd3S-mediated histone H4 deacetylation that prevents the initiation of antisense transcription, and (2) the Nrd1-Nab3-Sen1 complex, which terminates antisense transcripts and shuttles antisense CUTs to the exosome for degradation.

Kinetic Determination of RNA Production and Degradation

Cellular RNA levels are determined by the integration of RNA production, processing and degradation. Rabani et al. combined metabolic labeling of RNA with 4sU at high temporal resolution with RNA quantification and computational modeling to estimate RNA transcription and degradation rates [167]. Total RNA levels reflect the effects of RNA transcription and degradation, whereas newly transcribed RNA (RNA-4sU) contains only RNA that was actively transcribed during the labeling pulse. When the labeling time is sufficiently short, the RNA-4sU is still in the nucleus, thus reflecting the average transcription rate. A computational model resolves the RNA levels into separate contributions of RNA production and degradation and estimates changes in degradation rate. Recently, this computational framework was improved to quantify the level, editing sites and transcription, processing, and degradation rates of each transcript at splice junction resolution [168].

Transient transcriptome sequencing (TT-seq) is an approach that maps entire transcriptionally active regions uniformly and estimates the rate of RNA synthesis and degradation [169]. TT-seq is also based on 4sU-seq. A disadvantage of 4sU-seq is that it fails to map transcripts uniformly, because only a short 3′ region of nascent transcripts are labeled with 4sU, and long preexisting 5′ regions dominate the RNA-seq data. To remove this 5′ bias, TT-seq fragments the 4sU-RNA before isolation. This fragmentation permits the immunoprecipitation of only newly transcribed 4sU-RNA fragments, which is a unique feature compared with other approaches. TT-seq maps complete transcribed regions, complementing GRO-cap, which detects RNA 5′ ends. TT-seq also monitors RNA synthesis, whereas GRO-seq, PRO-seq, and NET-seq detect RNAs attached to a polymerase. In addition, TT-seq can estimate the transcription rate and stability of RNAs by calculating the rate of phosphodiester bond formation or breakage at each transcribed position. TT-seq results show that mRNAs and lincRNAs have the highest synthesis rates and longest half-lives (50 min), whereas other ncRNAs, such as eRNAs, have low synthesis rates and short half-lives (less than 10 min) [169]. TT-seq analysis also indicated that the short half-lives of eRNAs are likely to result from fewer U1 small nuclear ribonucleoprotein-binding sites, which inhibit early termination of RNA synthesis and lead to RNA stabilization, compared with mRNAs [169].

In recent years, three similar methods have been proposed to measure RNA dynamics using biochemical nucleoside conversion approaches: thiol(SH)-linked alkylation for the metabolic sequencing of RNA (SLAM-seq) [170], TimeLapse-seq [171], and thiouridine-to-cytidine sequencing (TUC-seq) [172]. In these three methods, newly transcribed RNAs are labeled with 4sU, then point mutations in RNA-seq data introduced by reverse transcription-dependent 4sU-to-C conversion are identified. These methods have been used to detect and measure the kinetics of ncRNAs [172–174]. For example, SLAM-seq analysis determined the kinetics of miRNAs in *Drosophila* S2 cells [174]. Early steps of miRNA biogenesis, the processing of primary and precursor miRNAs, occurs in a matter of minutes, faster than the generation of mRNAs. However, late steps of the miRNA biogenesis pathway, the formation of RNA-induced silencing complex with Argonaute (Ago) proteins, start approximately one hour after the beginning of miRNA production, indicating that Ago-loading is a key kinetic bottleneck to ensure faithful miRNA loading [174].

TUC-seq dual [175] and Dyrec-seq [176] are two novel methods of improving these 4sU-to-C conversion methods to enable quantification of RNA synthesis and degradation. TUC-seq dual is a modified version of TUC-seq, which uses two modified nucleosides, 4sU and 6-thioguanosine (6sG), in subsequent pulses. By simultaneously detecting 4sU-to-C and 6sG-to-A conversion, TUC-seq measures both RNA synthesis and decay. Dyrec-seq uses the combined methods of SLAM-seq and BRIC-seq to simultaneously quantify RNA synthesis and degradation rates [176]. Each method has advantages and disadvantages (reviewed in [177]), but continuing improvements in analyzing RNA dynamics are overcoming their limitations and enhancing ncRNA measurements.

CONCLUSIONS

Recent progress in the development of experimental and computational RNA-seq methods has revealed unexpected and profound insights. These include the identification of novel ncRNAs, estimates of the kinetics of transcription and degradation, and identification of expression patterns that are indicators of cellular states. Importantly, combined with ChIP-seq data, RNA-seq provides opportunities to understand epigenetic mechanisms and their intricate relationships with gene regulation. For example, GRO-cap and ChIP-seq demonstrated how eRNA transcription and epigenetic modification coordinate in a signal-dependent manner, leading to a better understanding of gene expression in eukaryotic cells [158,159]. Current RNA-seq technologies have both advantages and disadvantages, as illustrated in this review. Thus, caution is necessary in choosing an appropriate RNA-seq and/or a reliable analytical approach for the specific biological question of interest. This review might help researchers to choose the optimal RNA-seq technology for genome-wide studies of ncRNAs.

ncRNAs were recently reported to be key molecules for the assembly of phase-separated membraneless organelles, such as paraspeckles [27], nuclear stress bodies [178], telomere foci [179], and chromatin foci in X-chromosome inactivation [180]. Because lncRNAs are capable of multivalent interactions with the protein components of phase-separated RNP complexes, they serve as molecular scaffolds in these complexes [26,27]. RNA secondary structure also contributes to their function as molecular scaffolds in phase-separated condensates [28,181]. Moreover, mutations and gene fusions cause aberrant assembly of membraneless organelles in cancer [182]. Additionally, some previously annotated lncRNAs have coding potential (reviewed in [164]). We also found that some transcripts annotated as ncRNAs encode peptides, and that unannotated peptides may perform important roles in cells [183]. Future research will need to focus on the translational activity of transcripts with short open reading frames to define true ncRNAs.

In this review, we have discussed analysis of ncRNA expression, but RNA-seq can also be used for functional studies of ncRNAs [184], such as identifying RNA secondary structure, mapping RNA-protein interactions, or the global profiling of RNA–RNA interactions [185,186]. Furthermore, a novel spatial transcriptome approach, known as ascorbate peroxidase (APEX)-seq, was recently proposed to define RNA localization in distinct subcellular compartments [187]. In this technique, Fazal et al. fused the peroxidase enzyme, APEX2, to a marker protein or peptide in nine distinct subcellular locations, and biotinylated the RNAs near APEX2-fused proteins. They then enriched RNAs by streptavidin-biotin pulldown and performed RNA sequencing analysis. This method contributes to the comprehensive analysis of ncRNA function. Finally, we emphasize the importance of single-cell RNA-seq [188–190], which has already had an impact on our understanding of diverse biological processes, with broad implications for both basic and clinical research. Examples include investigation of mono-allelic gene expression, lineage tracing during cellular differentiation and developmental processes, and bi-stable gene expression. Similar to RNA-seq from collections of cells, various types of single-cell RNA-seq have been developed and each technology has its own strengths and limitations. In conclusion, technologies for transcriptomic studies are still a work in progress, and exciting novel methods will be developed in the future.

Dyrec-seq PROTOCOL

Before you Begin

Depending on the cell type and conditions, nucleotide analogs to label intracellular RNA may not be incorporated. It is necessary to first measure incorporation efficiency with multiple cell types using high performance liquid chromatography (HPLC) and select the most appropriate one for the experiment.

Key Resources Table

Reagent or Resource	Source	Identifier
Antibodies		
Anti-BrdU antibody	MBL	Clone 2B1
Chemicals, peptides, and recombinant proteins		
5′-bromouridine (BrU)	Wako	320–34741
4′-thiouridine (4sU)	Sigma	T4509
Critical commercial assays		
SLAM seq kinetics kit	Lexogen	Cat. No. 061
Software and algorithms		
SlamDunk v0.3.3	[170]	

Step-by-Step Method Details

Metabolic Labeling of Intracellular RNAs

Timing: 48–60 hours

1. Culture cells in appropriate medium for 1 day.
2. Replace medium with medium containing 150 μM BrU. Culture cells for 12 hours to label intracellular RNAs.
3. Replace medium with medium containing 200 μM 4sU.

4. Wash cells twice with phosphate-buffered saline (PBS), and isolate the intracellular RNAs in a time series (e.g., 0, 15, 30, 45, 60, 120, 240, 480, and 720 min) using RNAiso Plus (TAKARA).

Alkylation of 4sU Labeled RNAs
Timing: 3–6 hours

5. Alkylate the RNA (5.0 μg each; step 4) using the SLAM seq kinetics kit (Lexogen).

Preparation of BrU Labeled Luciferase RNA
Timing: 2–3 hours

6. Transcribe BrU labeled luciferase RNA using the T7 RiboMax Express Large Scale RNA Production System (Promega) with 0.3 mM BrUTP, 7.5 mM rNTPs, and 1.6 μg of linearized p-GEM Luc2.
7. Purify the BrU-labeled luciferase RNA using the QIAGEN RNeasy MinElute Cleanup Kit (QIAGEN).

Preparation of Protein G Sepharose Conjugated with Anti-BrdU Antibody
Timing: 2–3 hours

8. Wash Protein G Sepharose 4 Fast Flow (GE Healthcare) slurry (20 μL per sample) three times with 1.0 mL 0.1% bovine serum albumin (BSA) in PBS.
9. Resuspend the pellet in 300 μL 0.1% BSA in PBS. Add 1.0 μg of the anti-BrdU antibody (Medical & Biological Laboratories Co., Ltd [MBL]) to each sample.
10. Incubate the mixture with rotation for 1 hour at room temperature.
11. Wash the beads three times with 1 mL 0.1% BSA in PBS.
12. Resuspend the pellet in 200 μL 0.1% BSA in PBS containing 100 U RNasin plus RNase inhibitor (Promega).

Immunoprecipitation of BrU Labeled RNAs
Timing: 2–3 hours

13. Dissolve the RNA sample (12.0 μg each; step 4) in buffer A (25 mM Tris–HCl at pH 7.4, 6.25 mM EDTA).
14. As a spike-in, add 1.0 ng BrU-labeled luciferase RNA (step 7) to the RNA.
15. Denature the RNA mixture by heating for 1 min at 80°C and then add the anti-BrdU antibody-conjugated beads (step 12).
16. Incubate the mixture for 1 h at room temperature with rotation.
17. Wash the beads bound with BrU-labeled RNA three times with 1 mL 0.1% BSA in PBS.
18. Resuspend the final pellet in 100 μL buffer B (10 mM Tris–HCl at pH 7.4, 50 mM NaCl).
19. Add 300 μL ISOGEN LS (Nippon Gene) to the resuspended pellet, and isolate the BrU-labeled RNA in accordance with the manufacturer's instructions.

Sequencing the Alkylated and the Immunoprecipitated RNAs
Timing: 24–48 hours

20. Evaluate the quality of the alkylated RNA (step 5) and the immunoprecipitated RNA (step 19) using an Agilent RNA Nano 6000 kit (Agilent) on the Agilent Bioanalyzer 2100 (Agilent).
21. Reverse transcribe each of the labeled RNA samples (5.0 ng each), and subsequently remove the template RNA.
22. Initiate second-strand synthesis using a random primer containing Illumina-compatible linker sequences and appropriate in-line barcodes, then perform magnetic bead-based purification.
23. Amplify the library using PCR (12 cycles for alkylated RNAs, 18 cycles for immunoprecipitated RNAs), and then purify using AMPure XP.
24. Generate libraries for 100-bp end reads using standard Illumina protocols, and sequence on the HiSeq 3000 platform (Illumina).

Quantification and Statistical Analysis

Quantification of 4sU and BrU-labeled RNA
Timing: 6–8 hours

25. Obtain human genome sequence and annotation data (GRCh38) from the Ensembl database (release 92).
26. Generate genome-wide 3′-UTR sequences as *bed* files from the annotation data.
27. Quantify the newly synthesized and preexisting polyadenylated RNAs based on the sequence data using SlamDunk 0.3.3v with the following default parameters:

    $ slamdunk all -r [*reference fasta file*] -b [*annotation bed file*] -o [*output directory*] -t 4 [*fastq file*]

28. Estimate the amount of BrU-labeled RNA as counts per million of each gene, based on immunoprecipitated RNA.
29. Estimate the amount of 4sU-labeled RNA as counts per million of each gene, based on alkylated RNA and their T-to-C conversion ratio.

Estimation of RNA Synthesis and Degradation Rates
Timing: 2–3 hours

30. Based on time series of BrU-labeled RNA (x_{BrU}) and 4sU-labeled RNA (x_{4sU}), estimate RNA

synthesis rate (k_s) and degradation rate (k_d) by fitting to the following equations:

$$x_{4sU} = \frac{k_s}{k_d}\left(1 - e^{-k_d t}\right)$$

$$x_{BrU} = \alpha e^{-k_d t}$$

where t indicates time of sampling, and α is estimated as a scaling factor.

Expected Outcomes

This method, named Dyrec-seq, enables users to estimate synthesis and degradation rates of RNA derived from multiple individual genes simultaneously and comprehensively.

Advantages

Some methods used to estimate either RNA synthesis or degradation rates are not completely comparable because of varying conditions. Dyrec-seq uses multiple nucleotide analogs to estimate these rates simultaneously, which enables accurate comparison.

Limitations

Depending on the cell type or conditions, the nucleotide analogs may not be incorporated in cells. As described in the "Before You Begin" section, it is necessary to measure nucleotide analog incorporation efficiency using HPLC or other methods.

Optimization and Troubleshooting

The Cell Does Not Incorporate Nucleotide Analogs

Some types of cell do not incorporate nucleotide analogs to label intracellular RNA. If this occurs, a different cell type should be used.

The cell Dies During RNA Labeling

Some nucleotide analogs, including 4sU, can be cytotoxic. Cell viability can be enhanced by decreasing nucleotide analog concentration.

Incorporation Efficiency of Nucleotide Analogs is Too Low

If the amount of RNA is sufficient for sequencing, severe problems calculating the degradation rate based on BrU-labeled RNA do not occur; however, calculation of synthesis rate based on 4sU-labeled RNA may be inaccurate. Calculate the incorporation efficiency (μ) from synthesis rates ($\hat{k}_s$) and degradation rates ($\hat{k}_d$) of total RNA estimated from the total number of reads including T-to-C conversion (X) as follows, and correct the synthesis rate.

$$X = \frac{\hat{k}_s}{\hat{k}_d}\left(1 - e^{-\hat{k}_d t}\right)$$

$$\mu = \frac{\hat{k}_d}{10^6 \times \hat{k}_s}$$

Acknowledgments

This work was supported by grants from The Ministry of Education, Culture, Sports, Science and Technology (MEXT) KAKENHI (20K15988, 21H04792, 21H00243, 21H02758). We thank Edanz Group (https://en-author-services.edanz.com/ac) for editing a draft for this article.

Glossary

4sU	4-thiouridine
AIR	antisense of IGF2R non-protein coding RNA
ANRIL	CDKN2B antisense RNA 1
APEX	ascorbate peroxidase
asRNA	antisense RNA
ASSAGE	asymmetric strand-specific analysis of gene expression
BRIC-seq	5′-bromo-uridine immunoprecipitation chase-deep sequencing analysis
BrU	5′-bromo-uridine
CAGE	cap analysis gene expression
cheRNA	chromatin-enriched RNA
ChIP-seq	chromatin immunoprecipitation sequencing
Chrom-seq	chromatin-associated RNA-seq
circRNA	circular RNA
conRNA	convergent RNA
CSF	codon substitution frequency
CUT	cryptic unstable transcript
Digital RNA-seq	RNA-seq based approach with barcoding of RNA molecules before amplification
Dlk1	delta like non-canonical Notch ligand 1
DSN	duplex-specific nuclease
ENCODE	Encyclopedia of DNA Elements Project
eRNA	enhancer RNA
EZH2	enhancer of zeste 2 polycomb repressive complex 2 subunit
GRO-cap	RNA-seq based approach to detect a nascent RNA with a 5′ cap
GRO-seq	global run-on sequencing
Gtl2	paternally-imprinted transcript, which regulates expression of nearby Dlk1
HOTAIR	HOX transcript antisense RNA
HOX	homeotic genes
HOXD	homeobox D cluster
Kcnq1ot1	KCNQ1 opposite strand/antisense transcript 1
lincRNA	large intergenic ncRNA
lncRNA	long ncRNA
miRNA	microRNA
ncRNA	non-coding RNA
NET-seq	native elongating transcript sequencing
NRO-RNA	nuclear run-on RNA
NSR	"not-so-random" primers that target non-rRNA transcripts

PcG	polycomb group
piRNA	Piwi-interacting RNA
PRC2	polycomb repressive complex 2
PRE/TRE	PcG response element/TrxG response element
PROMPT	promoter upstream transcript
PRO-seq	precision nuclear run-on sequencing
RBP	RNA-binding protein
RepA	A small internal non-coding transcript from the *Xist* locus
RIP-seq	RNA immunoprecipitation followed by deep sequencing
RNA CaptureSeq	RNA-seq based approach to capture and enrich genes of interest
RNA-seq	high-throughput RNA sequencing
Rrp6	RNA exosome component
sincRNA	short intergenic ncRNA
siRNA	short interfering RNA
SMART	switching mechanism at the 5′ end of the RNA transcript
snoRNA	small nucleolar RNA
snRNA	spliceosomal small nuclear RNAs
SVM	support vector machine
TARDIS	targeted RNA directional sequencing method for rare RNA discovery
Trf4	poly(A) polymerase
TrxG	trithorax group
Tsix	inactive X chromosome-specific antisense transcript across the 3′ end of the *Xist* locus
TT-seq	transient transcriptome sequencing
usRNA	upstream antisense RNA
XIST	inactive X specific transcripts
Xrn1	5′-3′ exoribonuclease 1
XUT	Xrn1-sensitive unstable transcript

References

[1] Taniue K, Akimitsu N. The functions and unique features of lncrnas in cancer development and tumorigenesis. Int J Mol Sci 2021;22:1–20. Available from: https://doi.org/10.3390/ijms22020632.

[2] St. Laurent G, Wahlestedt C, Kapranov P. The Landscape of long noncoding RNA classification. Trends Genet 2015;31:239–51. Available from: https://doi.org/10.1016/j.tig.2015.03.007.

[3] Cech TR, Steitz JA. The noncoding RNA revolution – Trashing old rules to forge new ones. Cell 2014;157:77–94. Available from: https://doi.org/10.1016/j.cell.2014.03.008.

[4] Holoch D, Moazed D. RNA-mediated epigenetic regulation of gene expression. Nat Rev Genet 2015;16:71–84. Available from: https://doi.org/10.1038/nrg3863.

[5] Yamazaki T, Nakagawa S, Hirose T. Architectural RNAs for membraneless nuclear body formation. Cold Spring Harb Symp Quant Biol 2019;84:227–37. Available from: https://doi.org/10.1101/sqb.2019.84.039404.

[6] Shirahama S, Onoguchi-Mizutani R, Kawata K, Taniue K, Miki A, Kato A, et al. Long noncoding RNA U90926 is crucial for herpes simplex virus type 1 proliferation in murine retinal photoreceptor cells. Sci Rep 2020;10:19406. Available from: https://doi.org/10.1038/s41598-020-76450-2.

[7] Onoguchi-Mizutani R, Kishi Y, Ogura Y, Nishimura Y, Imamachi N, Suzuki Y, et al. Identification of novel heat shock-induced long non-coding RNA in human cells. J Biochem 2021;169:497–505. Available from: https://doi.org/10.1093/jb/mvaa126.

[8] Onoguchi-Mizutani R, Kirikae Y, Ogura Y, Gutschner T, Diederichs S, Akimitsu N. Identification of a heat-inducible novel nuclear body containing the long noncoding RNA MALAT1. J Cell Sci 2021;134:jcs253559. Available from: https://doi.org/10.1242/jcs.253559.

[9] Imamura K, Imamachi N, Akizuki G, Kumakura M, Kawaguchi A, Nagata K, et al. Long noncoding RNA NEAT1-dependent SFPQ relocation from promoter region to paraspeckle mediates IL8 expression upon immune stimuli. Mol Cell 2014;53:393–406. Available from: https://doi.org/10.1016/j.molcel.2014.01.009.

[10] Shirahama S, Taniue K, Mitsutomi S, Tanaka R, Kaburaki T, Sato T, et al. Human U90926 orthologous long non-coding RNA as a novel biomarker for visual prognosis in herpes simplex virus type-1 induced acute retinal necrosis. Sci Rep 2021;11:12164. Available from: https://doi.org/10.1038/s41598-021-91340-x.

[11] Bertone P, Stolc V, Royce TE, Rozowsky JS, Urban AE, Zhu X, et al. Global identification of human transcribed sequences with genome tiling arrays. Science 2004;306:2242–6. Available from: https://doi.org/10.1126/science.1103388.

[12] Rinn JL, Kertesz M, Wang JK, Squazzo SL, Xu X, Brugmann SA, et al. Functional demarcation of active and silent chromatin domains in human HOX loci by noncoding RNAs. Cell 2007;129:1311–23. Available from: https://doi.org/10.1016/j.cell.2007.05.022.

[13] Derrien T, Johnson R, Bussotti G, Tanzer A, Djebali S, Tilgner H, et al. The GENCODE v7 catalog of human long noncoding RNAs: analysis of their gene structure, evolution, and expression. Genome Res 2012;22:1775–89. Available from: https://doi.org/10.1101/gr.132159.111.

[14] Tani H, Mizutani R, Salam KA, Tano K, Ijiri K, Wakamatsu A, et al. Genome-wide determination of RNA stability reveals hundreds of short-lived noncoding transcripts in mammals. Genome Res 2012;22:947–56. Available from: https://doi.org/10.1101/gr.130559.111.

[15] Tani H, Akimitsu N. Genome-wide technology for determining RNA stability in mammalian cells: historical perspective and recent advantages based on modified nucleotide labeling. RNA Biol 2012;9:1233–8. Available from: https://doi.org/10.4161/rna.22036.

[16] Core LJ, Waterfall JJ, Lis JT. Nascent RNA sequencing reveals widespread pausing and divergent initiation at human promoters. Sci (80-) 2008;322:1845–8. Available from: https://doi.org/10.1126/science.1162228.

[17] Underwood JG, Uzilov AV, Katzman S, Onodera CS, Mainzer JE, Mathews DH, et al. FragSeq: transcriptome-wide RNA structure probing using high-throughput sequencing. Nat Methods 2010;7:995–1001. Available from: https://doi.org/10.1038/nmeth.1529.

[18] Licatalosi DD, Mele A, Fak JJ, Ule J, Kayikci M, Chi SW, et al. HITS-CLIP yields genome-wide insights into brain alternative RNA processing. Nature 2008;456:464–9. Available from: https://doi.org/10.1038/nature07488.

[19] Ståhl PL, Salmén F, Vickovic S, Lundmark A, Navarro JF, Magnusson J, et al. Visualization and analysis of gene expression in tissue sections by spatial transcriptomics. Science 2016;353:78–82. Available from: https://doi.org/10.1126/science.aaf2403.

[20] Ingolia NT, Ghaemmaghami S, Newman JRS, Weissman JS. Genome-wide analysis in vivo of translation with nucleotide resolution using ribosome profiling. Science 2009;324:218–23. Available from: https://doi.org/10.1126/science.1168978.

[21] Kertesz M, Wan Y, Mazor E, Rinn JL, Nutter RC, Chang HY, et al. Genome-wide measurement of RNA secondary structure in yeast. Nature 2010;467:103–7. Available from: https://doi.org/10.1038/nature09322.

[22] Tang F, Barbacioru C, Wang Y, Nordman E, Lee C, Xu N, et al. mRNA-Seq whole-transcriptome analysis of a single cell. Nat Methods 2009;6:377–82. Available from: https://doi.org/10.1038/nmeth.1315.

[23] Stark R, Grzelak M, Hadfield J. RNA sequencing: the teenage years. Nat Rev Genet 2019;20:631–56. Available from: https://doi.org/10.1038/s41576-019-0150-2.

[24] Wang Z, Gerstein M, Snyder M. RNA-Seq: a revolutionary tool for transcriptomics. Nat Rev Genet 2009;10:57–63. Available from: https://doi.org/10.1038/nrg2484.

[25] Garalde DR, Snell EA, Jachimowicz D, Sipos B, Lloyd JH, Bruce M, et al. Highly parallel direct RNA sequencing on an array of nanopores. Nat Methods 2018;15:201–6. Available from: https://doi.org/10.1038/nmeth.4577.

[26] Chujo T, Yamazaki T, Kawaguchi T, Kurosaka S, Takumi T, Nakagawa S, et al. Unusual semi-extractability as a hallmark of nuclear body-associated architectural noncoding RNAs. EMBO J 2017;36:1447–62. Available from: https://doi.org/10.15252/embj.201695848.

[27] Yamazaki T, Souquere S, Chujo T, Kobelke S, Chong YS, Fox AH, et al. Functional domains of NEAT1 architectural lncRNA induce paraspeckle assembly through phase separation. Mol Cell 2018;70:1038–53. Available from: https://doi.org/10.1016/j.molcel.2018.05.019 e7.

[28] Fay MM, Anderson PJ. The role of RNA in biological phase separations. J Mol Biol 2018;430:4685–701. Available from: https://doi.org/10.1016/j.jmb.2018.05.003.

[29] Morlan JD, Qu K, Sinicropi DV. Selective depletion of rRNA enables whole transcriptome profiling of archival fixed tissue. PLoS One 2012;7:1–8. Available from: https://doi.org/10.1371/journal.pone.0042882.

[30] Huang R, Jaritz M, Guenzl P, Vlatkovic I, Sommer A, Tamir IM, et al. An RNA-seq strategy to detect the complete coding and non-coding transcriptome including full-length imprinted macro ncrnas. PLoS One 2011;6:e0027288. Available from: https://doi.org/10.1371/journal.pone.0027288.

[31] Zhulidov PA, Bogdanova EA, Shcheglov AS, Vagner LL, Khaspekov GL, Kozhemyako VB, et al. Simple cDNA normalization using kamchatka crab duplex-specific nuclease. Nucleic Acids Res 2004;32. Available from: https://doi.org/10.1093/nar/gnh031.

[32] Yi H, Cho YJ, Won S, Lee JE, Jin Yu H, Kim S, et al. Duplex-specific nuclease efficiently removes rRNA for prokaryotic RNA-seq. Nucleic Acids Res 2011;39. Available from: https://doi.org/10.1093/nar/gkr617.

[33] Armour CD, Castle JC, Chen R, Babak T, Loerch P, Jackson S, et al. Digital transcriptome profiling using selective hexamer priming for cDNA synthesis. Nat Methods 2009;6:647–9. Available from: https://doi.org/10.1038/nmeth.1360.

[34] Adiconis X, Borges-Rivera D, Satija R, Deluca DS, Busby MA, Berlin AM, et al. Comparative analysis of RNA sequencing methods for degraded or low-input samples. Nat Methods 2013;10:623–9. Available from: https://doi.org/10.1038/nmeth.2483.

[35] Kapranov P, St Laurent G, Raz T, Ozsolak F, Reynolds CP, Sorensen PHB, et al. The majority of total nuclear-encoded non-ribosomal RNA in a human cell is "dark matter" un-annotated RNA. BMC Biol 2010;8:149. Available from: https://doi.org/10.1186/1741-7007-8-149.

[36] Wang KC, Chang HY. Molecular mechanisms of long noncoding RNAs. Mol Cell 2011;43:904–14. Available from: https://doi.org/10.1016/j.molcel.2011.08.018.

[37] Hekimoglu B, Ringrose L. Non-coding RNAs in polycomb/trithorax regulation. RNA Biol 2009;6:129–37. Available from: https://doi.org/10.4161/rna.6.2.8178.

[38] Herzog VA, Lempradl A, Trupke J, Okulski H, Altmutter C, Ruge F, et al. A strand-specific switch in noncoding transcription switches the function of a Polycomb/Trithorax response element. Nat Genet 2014;46:973–81. Available from: https://doi.org/10.1038/ng.3058.

[39] Lister R, O'Malley RC, Tonti-Filippini J, Gregory BD, Berry CC, Millar AH, et al. Highly integrated single-base resolution maps of the epigenome in arabidopsis. Cell 2008;133:523–36. Available from: https://doi.org/10.1016/j.cell.2008.03.029.

[40] Zhu YY, Machleder EM, Chenchik A, Li R, Siebert PD. Reverse transcriptase template switching: a SMART approach for full-length cDNA library construction. Biotechniques 2001;30:892–7. Available from: https://doi.org/10.2144/01304pf02.

[41] Cloonan N, Forrest ARR, Kolle G, Gardiner BBA, Faulkner GJ, Brown MK, et al. Stem cell transcriptome profiling via massive-scale mRNA sequencing. Nat Methods 2008;5:613–19. Available from: https://doi.org/10.1038/nmeth.1223.

[42] He Y, Vogelstein B, Velculescu VE, Papadopoulos N, Kinzler KW. The antisense transcriptomes of human cells. Science 2008;322:1855–7. Available from: https://doi.org/10.1126/science.1163853.

[43] Parkhomchuk D, Borodina T, Amstislavskiy V, Banaru M, Hallen L, Krobitsch S, et al. Transcriptome analysis by strand-specific sequencing of complementary DNA. Nucleic Acids Res 2009;37. Available from: https://doi.org/10.1093/nar/gkp596.

[44] Levin JZ, Yassour M, Adiconis X, Nusbaum C, Thompson DA, Friedman N, et al. Comprehensive comparative analysis of strand-specific RNA sequencing methods. Nat Methods 2010;7:709–15. Available from: https://doi.org/10.1038/nmeth.1491.

[45] Djebali S, Davis CA, Merkel A, Dobin A, Lassmann T, Mortazavi A, et al. Landscape of transcription in human cells. Nature 2012;489:101–8. Available from: https://doi.org/10.1038/nature11233.

[46] Jiang L, Schlesinger F, Davis CA, Zhang Y, Li R, Salit M, et al. Synthetic spike-in standards for RNA-seq experiments. Genome Res 2011;21:1543–51. Available from: https://doi.org/10.1101/gr.121095.111.Freely.

[47] Clark MB, Mercer TR, Bussotti G, Leonardi T, Haynes KR, Crawford J, et al. Quantitative gene profiling of long noncoding RNAs with targeted RNA sequencing. Nat Methods 2015;12:339–42. Available from: https://doi.org/10.1038/nmeth.3321.

[48] Teer JK, Bonnycastle LL, Chines PS, Hansen NF, Aoyama N, Swift AJ, et al. Systematic comparison of three genomic enrichment methods for massively parallel DNA sequencing. Genome Res 2010;20:1420–31. Available from: https://doi.org/10.1101/gr.106716.110.

[49] Mercer TR, Gerhardt DJ, Dinger ME, Crawford J, Trapnell C, Jeddeloh JA, et al. Targeted RNA sequencing reveals the deep complexity of the human transcriptome. Nat Biotechnol 2012;30:99–104. Available from: https://doi.org/10.1038/nbt.2024.

[50] Vitorino R, Guedes S, Amado F, Santos M, Akimitsu N. The role of micropeptides in biology. Cell Mol Life Sci 2021;78:3285–98. Available from: https://doi.org/10.1007/s00018-020-03740-3.

[51] Bussotti G, Leonardi T, Clark MB, Mercer TR, Crawford J, Malquori L, et al. Improved definition of the mouse transcriptome via targeted RNA sequencing. Genome Res 2016;26:705–16. Available from: https://doi.org/10.1101/gr.199760.115.

[52] Bhatt DM, Pandya-Jones A, Tong AJ, Barozzi I, Lissner MM, Natoli G, et al. Transcript dynamics of proinflammatory genes revealed by sequence analysis of subcellular RNA fractions. Cell 2012;150:279–90. Available from: https://doi.org/10.1016/j.cell.2012.05.043.

[53] Lai F, Gardini A, Zhang A, Shiekhattar R. Integrator mediates the biogenesis of enhancer RNAs. Nature 2015;525:399–403. Available from: https://doi.org/10.1038/nature14906.

[54] Werner MS, Ruthenburg AJ. Nuclear fractionation reveals thousands of chromatin-tethered noncoding RNAs adjacent to active genes. Cell Rep 2015;12:1089—98. Available from: https://doi.org/10.1016/j.celrep.2015.07.033.

[55] Zhao J, Sun BK, Erwin JA, Song J-J, Lee JT. Polycomb proteins targeted by a short repeat RNA to the mouse X chromosome. Science 2008;322:750—6. Available from: https://doi.org/10.1126/science.1163045.

[56] Zhao J, Ohsumi TK, Kung JT, Ogawa Y, Grau DJ, Sarma K, et al. Genome-wide identification of polycomb-associated RNAs by RIP-seq. Mol Cell 2010;40:939—53. Available from: https://doi.org/10.1016/j.molcel.2010.12.011.

[57] Shiroguchi K, Jia TZ, Sims PA, Xie XS. Digital RNA sequencing minimizes sequence-dependent bias and amplification noise with optimized single-molecule barcodes. Proc Natl Acad Sci U S A 2012;109:1347—52. Available from: https://doi.org/10.1073/pnas.1118018109.

[58] Fu GK, Xu W, Wilhelmy J, Mindrinos MN, Davis RW, Xiao W, et al. Molecular indexing enables quantitative targeted RNA sequencing and reveals poor efficiencies in standard library preparations. Proc Natl Acad Sci U S A 2014;111:1891—6. Available from: https://doi.org/10.1073/pnas.1323732111.

[59] Stanke M, Keller O, Gunduz I, Hayes A, Waack S, Morgenstern B. AUGUSTUS: ab initio prediction of alternative transcripts. Nucleic Acids Res 2006;34:435—9. Available from: https://doi.org/10.1093/nar/gkl200.

[60] Mezlini AM, Smith EJM, Fiume M, Buske O, Savich GL, Shah S, et al. IReckon: simultaneous isoform discovery and abundance estimation from RNA-seq data. Genome Res 2013;23:519—29. Available from: https://doi.org/10.1101/gr.142232.112.

[61] Guttman M, Garber M, Levin JZ, Donaghey J, Robinson J, Adiconis X, et al. Ab initio reconstruction of cell type-specific transcriptomes in mouse reveals the conserved multi-exonic structure of lincRNAs. Nat Biotechnol 2010;28:503—10. Available from: https://doi.org/10.1038/nbt.1633.

[62] Li JJ, Jiang CR, Brown JB, Huang H, Bickel PJ. Sparse linear modeling of next-generation mRNA sequencing (RNA-Seq) data for isoform discovery and abundance estimation. Proc Natl Acad Sci U S A 2011;108:19867—72. Available from: https://doi.org/10.1073/pnas.1113972108.

[63] Pertea M, Kim D, Pertea GM, Leek JT, Salzberg SL. Transcript-level expression analysis of RNA-seq experiments with HISAT, StringTie and Ballgown. Nat Protoc 2016;11:1650—67. Available from: https://doi.org/10.1038/nprot.2016.095.

[64] Schulz MH, Zerbino DR, Vingron M, Birney E. Oases: robust de novo RNA-seq assembly across the dynamic range of expression levels. Bioinformatics 2012;28:1086—92. Available from: https://doi.org/10.1093/bioinformatics/bts094.

[65] Robertson G, Schein J, Chiu R, Corbett R, Field M, Jackman SD, et al. De novo assembly and analysis of RNA-seq data. Nat Methods 2010;7:909—12. Available from: https://doi.org/10.1038/nmeth.1517.

[66] Grabherr MG, Haas BJ, Yassour M, Levin JZ, Thompson DA, Amit I, et al. Full-length transcriptome assembly from RNA-Seq data without a reference genome. Nat Biotechnol 2011;29:644—52. Available from: https://doi.org/10.1038/nbt.1883.

[67] Zerbino DR, Birney E. Velvet: algorithms for de novo short read assembly using de Bruijn graphs. Genome Res 2008;18:821—9. Available from: https://doi.org/10.1101/gr.074492.107.

[68] Steijger T, Abril JF, Engström PG, Kokocinski F, Akerman M, Alioto T, et al. Assessment of transcript reconstruction methods for RNA-seq. Nat Methods 2013;10:1177—84. Available from: https://doi.org/10.1038/nmeth.2714.

[69] Trapnell C, Pachter L, Salzberg SL. TopHat: discovering splice junctions with RNA-Seq. Bioinformatics 2009;25:1105—11. Available from: https://doi.org/10.1093/bioinformatics/btp120.

[70] Wu TD, Nacu S. Fast and SNP-tolerant detection of complex variants and splicing in short reads. Bioinformatics 2010;26:873—81. Available from: https://doi.org/10.1093/bioinformatics/btq057.

[71] Dobin A, Davis CA, Schlesinger F, Drenkow J, Zaleski C, Jha S, et al. STAR: ultrafast universal RNA-seq aligner. Bioinformatics 2013;29:15—21. Available from: https://doi.org/10.1093/bioinformatics/bts635.

[72] Dobin A, Gingeras TR. Optimizing RNA-Seq mapping with STAR. Methods Mol Biol 2016;1415:245—62. Available from: https://doi.org/10.1007/978-1-4939-3572-7_13.

[73] Kim D, Langmead B, Salzberg SL. HISAT: a fast spliced aligner with low memory requirements. Nat Methods 2015;12:357—60. Available from: https://doi.org/10.1038/nmeth.3317.

[74] Trapnell C, Williams BA, Pertea G, Mortazavi A, Kwan G, Van Baren MJ, et al. Transcript assembly and quantification by RNA-Seq reveals unannotated transcripts and isoform switching during cell differentiation. Nat Biotechnol 2010;28:511—15. Available from: https://doi.org/10.1038/nbt.1621.

[75] Pertea M, Pertea GM, Antonescu CM, Chang T-C, Mendell JT, Salzberg SL. StringTie enables improved reconstruction of a transcriptome from RNA-seq reads. Nat Biotechnol 2015;33:290—5. Available from: https://doi.org/10.1038/nbt.3122.

[76] Guttman M, Amit I, Garber M, French C, Lin MF, Feldser D, et al. Chromatin signature reveals over a thousand highly conserved large non-coding RNAs in mammals. Nature 2009;458:223—7. Available from: https://doi.org/10.1038/nature07672.

[77] Lin MF, Jungreis I, Kellis M. PhyloCSF: a comparative genomics method to distinguish protein coding and non-coding regions. Bioinformatics 2011;27:275—82. Available from: https://doi.org/10.1093/bioinformatics/btr209.

[78] Jia H, Osak M, Bogu GK, Stanton LW, Johnson R, Lipovich L. Genome-wide computational identification and manual annotation of human long noncoding RNA genes. Rna 2010;16:1478—87. Available from: https://doi.org/10.1261/rna.1951310.

[79] Kong L, Zhang Y, Ye ZQ, Liu XQ, Zhao SQ, Wei L, et al. CPC: assess the protein-coding potential of transcripts using sequence features and support vector machine. Nucleic Acids Res 2007;35:345—9. Available from: https://doi.org/10.1093/nar/gkm391.

[80] Sun K, Chen X, Jiang P, Song X, Wang H, Sun H. iSeeRNA: identification of long intergenic non-coding RNA transcripts from transcriptome sequencing data. BMC Genom 2013;14. Available from: https://doi.org/10.1186/1471-2164-14-S2-S7.

[81] Sun L, Liu H, Zhang L, Meng J. lncRScan-SVM: a tool for predicting long non-coding RNAs using support vector machine. PLoS One 2015;10:1—16. Available from: https://doi.org/10.1371/journal.pone.0139654.

[82] O'Leary NA, Wright MW, Brister JR, Ciufo S, Haddad D, McVeigh R, et al. Reference sequence (RefSeq) database at NCBI: current status, taxonomic expansion, and functional annotation. Nucleic Acids Res 2016;44:D733—45. Available from: https://doi.org/10.1093/nar/gkv1189.

[83] Frankish A, Diekhans M, Jungreis I, Lagarde J, Loveland JE, Mudge JM, et al. GENCODE 2021. Nucleic Acids Res 2021;49: D916—23. Available from: https://doi.org/10.1093/nar/gkaa1087.

[84] Frankish A, Diekhans M, Ferreira A-M, Johnson R, Jungreis I, Loveland J, et al. GENCODE reference annotation for the human and mouse genomes. Nucleic Acids Res 2019;47:

D766–73. Available from: https://doi.org/10.1093/nar/gky955.

[85] Friedländer MR, Chen W, Adamidi C, Maaskola J, Einspanier R, Knespel S, et al. Discovering microRNAs from deep sequencing data using miRDeep. Nat Biotechnol 2008;26:407–15. Available from: https://doi.org/10.1038/nbt1394.

[86] Friedländer MR, MacKowiak SD, Li N, Chen W, Rajewsky N. MiRDeep2 accurately identifies known and hundreds of novel microRNA genes in seven animal clades. Nucleic Acids Res 2012;40:37–52. Available from: https://doi.org/10.1093/nar/gkr688.

[87] An J, Lai J, Lehman ML, Nelson CC. MiRDeep*: an integrated application tool for miRNA identification from RNA sequencing data. Nucleic Acids Res 2013;41:727–37. Available from: https://doi.org/10.1093/nar/gks1187.

[88] Hackenberg M, Sturm M, Langenberger D, Falcón-Pérez JM, Aransay AM. miRanalyzer: a microRNA detection and analysis tool for next-generation sequencing experiments. Nucleic Acids Res 2009;37:68–76. Available from: https://doi.org/10.1093/nar/gkp347.

[89] Humphreys DT, Suter CM. MiRspring: a compact standalone research tool for analyzing miRNA-seq data. Nucleic Acids Res 2013;41. Available from: https://doi.org/10.1093/nar/gkt485.

[90] Hendrix D, Levine M, Shi W. MiRTRAP, a computational method for the systematic identification of miRNAs from high throughput sequencing data. Genome Biol 2010;11:R39. Available from: https://doi.org/10.1186/gb-2010-11-4-r39.

[91] Betel D, Sheridan R, Marks DS, Sander C. Computational analysis of mouse piRNA sequence and biogenesis. PLoS Comput Biol 2007;3:2219–27. Available from: https://doi.org/10.1371/journal.pcbi.0030222.

[92] Zhang Y, Wang X, Kang L. A k-mer scheme to predict piRNAs and characterize locust piRNAs. Bioinformatics 2011;27:771–6. Available from: https://doi.org/10.1093/bioinformatics/btr016.

[93] Wang K, Liang C, Liu J, Xiao H, Huang S, Xu J, et al. Prediction of piRNAs using transposon interaction and a support vector machine. BMC Bioinform 2014;15:1–8. Available from: https://doi.org/10.1186/s12859-014-0419-6.

[94] Anderson P, Ivanov P. tRNA fragments in human health and disease. FEBS Lett 2014;588:4297–304. Available from: https://doi.org/10.1016/j.febslet.2014.09.001.

[95] Rashad S, Niizuma K, Tominaga T. tRNA cleavage: a new insight. Neural Regen Res 2020;15:47–52. Available from: https://doi.org/10.4103/1673-5374.264447.

[96] Selitsky SR, Sethupathy P. tDRmapper: challenges and solutions to mapping, naming, and quantifying tRNA-derived RNAs from human small RNA-sequencing data. BMC Bioinform 2015;16:354. Available from: https://doi.org/10.1186/s12859-015-0800-0.

[97] Loher P, Telonis AG, Rigoutsos I. MINTmap: fast and exhaustive profiling of nuclear and mitochondrial tRNA fragments from short RNA-seq data. Sci Rep 2017;7:41184. Available from: https://doi.org/10.1038/srep41184.

[98] Shi J, Ko E-A, Sanders KM, Chen Q, Zhou T. SPORTS1.0: a tool for annotating and profiling non-coding RNAs optimized for rRNA- and tRNA-derived small RNAs. Genom Proteom Bioinform 2018;16:144–51. Available from: https://doi.org/10.1016/j.gpb.2018.04.004.

[99] Di Bella S, La Ferlita A, Carapezza G, Alaimo S, Isacchi A, Ferro A, et al. A benchmarking of pipelines for detecting ncRNAs from RNA-Seq data. Brief Bioinform 2020;21:1987–98. Available from: https://doi.org/10.1093/bib/bbz110.

[100] Quek C, Jung C-H, Bellingham SA, Lonie A, Hill AF. iSRAP – a one-touch research tool for rapid profiling of small RNA-seq data. J Extracell Vesicles 2015;4:29454. Available from: https://doi.org/10.3402/jev.v4.29454.

[101] Andrés-León E, Núñez-Torres R, Rojas AM. miARma-Seq: a comprehensive tool for miRNA, mRNA and circRNA analysis. Sci Rep 2016;6:25749. Available from: https://doi.org/10.1038/srep25749.

[102] Wu X, Kim T-K, Baxter D, Scherler K, Gordon A, Fong O, et al. sRNAnalyzer-a flexible and customizable small RNA sequencing data analysis pipeline. Nucleic Acids Res 2017;45:12140–51. Available from: https://doi.org/10.1093/nar/gkx999.

[103] Fasold M, Langenberger D, Binder H, Stadler PF, Hoffmann S. DARIO: a ncRNA detection and analysis tool for next-generation sequencing experiments. Nucleic Acids Res 2011;39:112–17. Available from: https://doi.org/10.1093/nar/gkr357.

[104] Chen CJ, Servant N, Toedling J, Sarazin A, Marchais A, Duvernois-Berthet E, et al. NcPRO-seq: a tool for annotation and profiling of ncRNAs in sRNA-seq data. Bioinformatics 2012;28:3147–9. Available from: https://doi.org/10.1093/bioinformatics/bts587.

[105] Kristensen LS, Andersen MS, Stagsted LVW, Ebbesen KK, Hansen TB, Kjems J. The biogenesis, biology and characterization of circular RNAs. Nat Rev Genet 2019;20:675–91. Available from: https://doi.org/10.1038/s41576-019-0158-7.

[106] Patop IL, Wüst S, Kadener S. Past, present, and future of circRNAs. EMBO J 2019;38:e100836. Available from: https://doi.org/10.15252/embj.2018100836.

[107] Li X, Yang L, Chen L-L. The biogenesis, functions, and challenges of circular RNAs. Mol Cell 2018;71:428–42. Available from: https://doi.org/10.1016/j.molcel.2018.06.034.

[108] Zhang X-O, Wang H-B, Zhang Y, Lu X, Chen L-L, Yang L. Complementary sequence-mediated exon circularization. Cell 2014;159:134–47. Available from: https://doi.org/10.1016/j.cell.2014.09.001.

[109] Memczak S, Jens M, Elefsinioti A, Torti F, Krueger J, Rybak A, et al. Circular RNAs are a large class of animal RNAs with regulatory potency. Nature 2013;495:333–8. Available from: https://doi.org/10.1038/nature11928.

[110] Guo JU, Agarwal V, Guo H, Bartel DP. Expanded identification and characterization of mammalian circular RNAs. Genome Biol 2014;15:409. Available from: https://doi.org/10.1186/s13059-014-0409-z.

[111] Jeck WR, Sharpless NE. Detecting and characterizing circular RNAs. Nat Biotechnol 2014;32:453–61. Available from: https://doi.org/10.1038/nbt.2890.

[112] Enuka Y, Lauriola M, Feldman ME, Sas-Chen A, Ulitsky I, Yarden Y. Circular RNAs are long-lived and display only minimal early alterations in response to a growth factor. Nucleic Acids Res 2016;44:1370–83. Available from: https://doi.org/10.1093/nar/gkv1367.

[113] Jeck WR, Sorrentino JA, Wang K, Slevin MK, Burd CE, Liu J, et al. Circular RNAs are abundant, conserved, and associated with ALU repeats. RNA 2013;19:141–57. Available from: https://doi.org/10.1261/rna.035667.112.

[114] Xia S, Feng J, Lei L, Hu J, Xia L, Wang J, et al. Comprehensive characterization of tissue-specific circular RNAs in the human and mouse genomes. Brief Bioinform 2017;18:984–92. Available from: https://doi.org/10.1093/bib/bbw081.

[115] Maass PG, Glažar P, Memczak S, Dittmar G, Hollfinger I, Schreyer L, et al. A map of human circular RNAs in clinically relevant tissues. J Mol Med (Berl) 2017;95:1179–89. Available from: https://doi.org/10.1007/s00109-017-1582-9.

[116] Salzman J, Chen RE, Olsen MN, Wang PL, Brown PO. Cell-type specific features of circular RNA expression. PLoS Genet 2013;9:e1003777. Available from: https://doi.org/10.1371/journal.pgen.1003777.

[117] Westholm JO, Miura P, Olson S, Shenker S, Joseph B, Sanfilippo P, et al. Genome-wide analysis of drosophila circular RNAs reveals their structural and sequence properties and age-dependent neural accumulation. Cell Rep 2014;9:1966–80. Available from: https://doi.org/10.1016/j.celrep.2014.10.062.

[118] Yang L, Duff MO, Graveley BR, Carmichael GG, Chen L-L. Genomewide characterization of non-polyadenylated RNAs. Genome Biol 2011;12:R16. Available from: https://doi.org/10.1186/gb-2011-12-2-r16.

[119] Zhang Y, Zhang X-O, Chen T, Xiang J-F, Yin Q-F, Xing Y-H, et al. Circular intronic long noncoding RNAs. Mol Cell 2013;51:792–806. Available from: https://doi.org/10.1016/j.molcel.2013.08.017.

[120] Wang K, Singh D, Zeng Z, Coleman SJ, Huang Y, Savich GL, et al. MapSplice: accurate mapping of RNA-seq reads for splice junction discovery. Nucleic Acids Res 2010;38:e178. Available from: https://doi.org/10.1093/nar/gkq622.

[121] Gao Y, Wang J, Zhao F. CIRI: an efficient and unbiased algorithm for de novo circular RNA identification. Genome Biol 2015;16:1–16. Available from: https://doi.org/10.1186/s13059-014-0571-3.

[122] Szabo L, Morey R, Palpant NJ, Wang PL, Afari N, Jiang C, et al. Statistically based splicing detection reveals neural enrichment and tissue-specific induction of circular RNA during human fetal development. Genome Biol 2015;16:1–26. Available from: https://doi.org/10.1186/s13059-015-0690-5.

[123] Hoffmann S, Otto C, Doose G, Tanzer A, Langenberger D, Christ S, et al. A multi-split mapping algorithm for circular RNA, splicing, trans-splicing and fusion detection. Genome Biol 2014;15. Available from: https://doi.org/10.1186/gb-2014-15-2-r34.

[124] Gao Y, Zhang J, Zhao F. Circular RNA identification based on multiple seed matching. Brief Bioinform 2018;19:803–10. Available from: https://doi.org/10.1093/bib/bbx014.

[125] You X, Vlatkovic I, Babic A, Will T, Epstein I, Tushev G, et al. Neural circular RNAs are derived from synaptic genes and regulated by development and plasticity. Nat Neurosci 2015;18:603–10. Available from: https://doi.org/10.1038/nn.3975.

[126] Zhang XO, Dong R, Zhang Y, Zhang JL, Luo Z, Zhang J, et al. Diverse alternative back-splicing and alternative splicing landscape of circular RNAs. Genome Res 2016;26:1277–87. Available from: https://doi.org/10.1101/gr.202895.115.

[127] Cheng J, Metge F, Dieterich C. Specific identification and quantification of circular RNAs from sequencing data. Bioinformatics 2016;32:1094–6. Available from: https://doi.org/10.1093/bioinformatics/btv656.

[128] Song X, Zhang N, Han P, Moon B-S, Lai RK, Wang K, et al. Circular RNA profile in gliomas revealed by identification tool UROBORUS. Nucleic Acids Res 2016;44:e87. Available from: https://doi.org/10.1093/nar/gkw075.

[129] Hansen TB, Venø MT, Damgaard CK, Kjems J. Comparison of circular RNA prediction tools. Nucleic Acids Res 2015;44:e58. Available from: https://doi.org/10.1093/nar/gkv1458.

[130] Hansen TB. Improved circRNA identification by combining prediction algorithms. Front Cell Dev Biol 2018;6:20. Available from: https://doi.org/10.3389/fcell.2018.00020.

[131] Zhang J, Hou L, Zuo Z, Ji P, Zhang X, Xue Y, et al. Comprehensive profiling of circular RNAs with nanopore sequencing and CIRI-long. Nat Biotechnol 2021;39:836–45. Available from: https://doi.org/10.1038/s41587-021-00842-6.

[132] Taniue K, Akimitsu N. Fusion genes and RNAs in cancer development. Non-Coding RNA 2021;7:1–14. Available from: https://doi.org/10.3390/ncrna7010010.

[133] Maher CA, Kumar-Sinha C, Cao X, Kalyana-Sundaram S, Han B, Jing X, et al. Transcriptome sequencing to detect gene fusions in cancer. Nature 2009;458:97–101. Available from: https://doi.org/10.1038/nature07638.

[134] Maher CA, Palanisamy N, Brenner JC, Cao X, Kalyana-Sundaram S, Luo S, et al. Chimeric transcript discovery by paired-end transcriptome sequencing. Proc Natl Acad Sci U S A 2009;106:12353–8. Available from: https://doi.org/10.1073/pnas.0904720106.

[135] Kumar S, Razzaq SK, Vo AD, Gautam M, Li H. Identifying fusion transcripts using next generation sequencing. Wiley Interdiscip Rev RNA 2016;7:811–23. Available from: https://doi.org/10.1002/wrna.1382.

[136] Shi X, Singh S, Lin E, Li H. Chimeric RNAs in cancer. Adv Clin Chem 2021;100:1–35. Available from: https://doi.org/10.1016/bs.acc.2020.04.001.

[137] Benelli M, Pescucci C, Marseglia G, Severgnini M, Torricelli F, Magi A. Discovering chimeric transcripts in paired-end RNA-seq data by using EricScript. Bioinformatics 2012;28:3232–9. Available from: https://doi.org/10.1093/bioinformatics/bts617.

[138] Davidson NM, Majewski IJ, Oshlack A. JAFFA: high sensitivity transcriptome-focused fusion gene detection. Genome Med 2015;7:43. Available from: https://doi.org/10.1186/s13073-015-0167-x.

[139] Ley TJ, Miller C, Ding L, Raphael BJ, Mungall AJ, Robertson AG, et al. Genomic and epigenomic landscapes of adult de novo acute myeloid leukemia. N Engl J Med 2013;368:2059–74. Available from: https://doi.org/10.1056/NEJMoa1301689.

[140] Liu S, Tsai WH, Ding Y, Chen R, Fang Z, Huo Z, et al. Comprehensive evaluation of fusion transcript detection algorithms and a meta-caller to combine top performing methods in paired-end RNA-seq data. Nucleic Acids Res 2015;44:e47. Available from: https://doi.org/10.1093/nar/gkv1234.

[141] Jia W, Qiu K, He M, Song P, Zhou Q, Zhou F, et al. SOAPfuse: an algorithm for identifying fusion transcripts from paired-end RNA-Seq data. Genome Biol 2013;14:R12. Available from: https://doi.org/10.1186/gb-2013-14-2-r12.

[142] Edgren H, Murumagi A, Kangaspeska S, Nicorici D, Hongisto V, Kleivi K, et al. Identification of fusion genes in breast cancer by paired-end RNA-sequencing. Genome Biol 2011;12:R6. Available from: https://doi.org/10.1186/gb-2011-12-1-r6.

[143] Iyer MK, Chinnaiyan AM, Maher CA. ChimeraScan: a tool for identifying chimeric transcription in sequencing data. Bioinformatics 2011;27:2903–4. Available from: https://doi.org/10.1093/bioinformatics/btr467.

[144] Kumar S, Vo AD, Qin F, Li H. Comparative assessment of methods for the fusion transcripts detection from RNA-Seq data. Sci Rep 2016;6:21597. Available from: https://doi.org/10.1038/srep21597.

[145] Dobin A, Gingeras TR. Mapping RNA-seq reads with STAR. Curr Protoc Bioinformatics 2015;51:11.14.1–11.14.19. Available from: https://doi.org/10.1002/0471250953.bi1114s51.

[146] Uhrig S, Ellermann J, Walther T, Burkhardt P, Fröhlich M, Hutter B, et al. Accurate and efficient detection of gene fusions from RNA sequencing data. Genome Res 2021;31:448–60. Available from: https://doi.org/10.1101/gr.257246.119.

[147] GitHub. *ExpressionAnalysis/STAR-SEQR: RNA Fusion Detection and Quantification*. n.d. https://github.com/ExpressionAnalysis/STAR-SEQR.

[148] Haas BJ, Dobin A, Li B, Stransky N, Pochet N, Regev A. Accuracy assessment of fusion transcript detection via read-

mapping and de novo fusion transcript assembly-based methods. Genome Biol 2019;20:213. Available from: https://doi.org/10.1186/s13059-019-1842-9.
[149] Haas BJ, Papanicolaou A, Yassour M, Grabherr M, Blood PD, Bowden J, et al. De novo transcript sequence reconstruction from RNA-seq using the Trinity platform for reference generation and analysis. Nat Protoc 2013;8:1494–512. Available from: https://doi.org/10.1038/nprot.2013.084.
[150] Quinn JJ, Chang HY. Unique features of long non-coding RNA biogenesis and function. Nat Rev Genet 2016;17:47–62. Available from: https://doi.org/10.1038/nrg.2015.10.
[151] Chatterjee S, Großhans H. Active turnover modulates mature microRNA activity in Caenorhabditis elegans. Nature 2009;461:546–9. Available from: https://doi.org/10.1038/nature08349.
[152] Pefanis E, Wang J, Rothschild G, Lim J, Kazadi D, Sun J, et al. RNA exosome-regulated long non-coding RNA transcription controls super-enhancer activity. Cell 2015;161:774–89. Available from: https://doi.org/10.1016/j.cell.2015.04.034.
[153] Imamachi N, Tani H, Mizutani R, Imamura K, Irie T, Suzuki Y, et al. BRIC-seq: a genome-wide approach for determining RNA stability in mammalian cells. Methods 2014;67:55–63. Available from: https://doi.org/10.1016/j.ymeth.2013.07.014.
[154] Neil H, Malabat C, D'Aubenton-Carafa Y, Xu Z, Steinmetz LM, Jacquier A. Widespread bidirectional promoters are the major source of cryptic transcripts in yeast. Nature 2009;457:1038–42. Available from: https://doi.org/10.1038/nature07747.
[155] Van Dijk EL, Chen CL, Daubenton-Carafa Y, Gourvennec S, Kwapisz M, Roche V, et al. XUTs are a class of Xrn1-sensitive antisense regulatory non-coding RNA in yeast. Nature 2011;475:114–19. Available from: https://doi.org/10.1038/nature10118.
[156] Preker P, Nielsen J, Kammler S, Lykke-Andersen S, Christensen MS, Mapendano CK, et al. RNA exosome depletion reveals transcription upstream of active human promoters. Science 2008;322:1851–4. Available from: https://doi.org/10.1126/science.1164096.
[157] Core LJ, Martins AL, Danko CG, Waters CT, Siepel A, Lis JT. Analysis of nascent RNA identifies a unified architecture of initiation regions at mammalian promoters and enhancers. Nat Genet 2014;46:1311–20. Available from: https://doi.org/10.1038/ng.3142.
[158] Kaikkonen MU, Spann NJ, Heinz S, Romanoski CE, Allison KA, Stender JD, et al. Remodeling of the enhancer landscape during macrophage activation is coupled to enhancer transcription. Mol Cell 2013;51:310–25. Available from: https://doi.org/10.1016/j.molcel.2013.07.010.
[159] Wang D, Garcia-Bassets I, Benner C, Li W, Su X, Zhou Y, et al. Reprogramming transcription by distinct classes of enhancers functionally defined by eRNA. Nature 2011;474:390–7. Available from: https://doi.org/10.1038/nature10006.
[160] Li W, Notani D, Ma Q, Tanasa B, Nunez E, Chen AY, et al. Functional roles of enhancer RNAs for oestrogen-dependent transcriptional activation. Nature 2013;498:516–20. Available from: https://doi.org/10.1038/nature12210.
[161] Kwak H, Fuda NJ, Core LJ, Lis JT. Precise maps of RNA polymerase reveal how promoters direct initiation and pausing. Science 2013;339:950–3. Available from: https://doi.org/10.1126/science.1229386.
[162] Kruesi WS, Core LJ, Waters CT, Lis JT, Meyer BJ. Condensin controls recruitment of RNA polymerase ii to achieve nematode X-chromosome dosage compensation. Elife 2013;2013:e00808. Available from: https://doi.org/10.7554/eLife.00808.
[163] Lam MTY, Cho H, Lesch HP, Gosselin D, Heinz S, Tanaka-Oishi Y, et al. Rev-Erbs repress macrophage gene expression by inhibiting enhancer-directed transcription. Nature 2013;498:511–15. Available from: https://doi.org/10.1038/nature12209.
[164] Churchman LS, Weissman JS. Nascent transcript sequencing visualizes transcription at nucleotide resolution. Nature 2011;469:368–73. Available from: https://doi.org/10.1038/nature09652.
[165] Mayer A, Di Iulio J, Maleri S, Eser U, Vierstra J, Reynolds A, et al. Native elongating transcript sequencing reveals human transcriptional activity at nucleotide resolution. Cell 2015;161:541–54. Available from: https://doi.org/10.1016/j.cell.2015.03.010.
[166] Nojima T, Gomes T, Grosso ARF, Kimura H, Dye MJ, Dhir S, et al. Mammalian NET-seq reveals genome-wide nascent transcription coupled to RNA processing. Cell 2015;161:526–40. Available from: https://doi.org/10.1016/j.cell.2015.03.027.
[167] Rabani M, Levin JZ, Fan L, Adiconis X, Raychowdhury R, Garber M, et al. Metabolic labeling of RNA uncovers principles of RNA production and degradation dynamics in mammalian cells. Nat Biotechnol 2011;29:436–42. Available from: https://doi.org/10.1038/nbt.1861.
[168] Rabani M, Raychowdhury R, Jovanovic M, Rooney M, Stumpo DJ, Pauli A, et al. High-resolution sequencing and modeling identifies distinct dynamic RNA regulatory strategies. Cell 2014;159:1698–710. Available from: https://doi.org/10.1016/j.cell.2014.11.015.
[169] Schwalb B, Michel M, Zacher B, Frühauf K, Demel C, Tresch A, et al. TT-seq maps the human transient transcriptome. Science 2016;352:1225–8. Available from: https://doi.org/10.1126/science.aad9841.
[170] Herzog VA, Reichholf B, Neumann T, Rescheneder P, Bhat P, Burkard TR, et al. Thiol-linked alkylation of RNA to assess expression dynamics. Nat Methods 2017;14:1198–204. Available from: https://doi.org/10.1038/nmeth.4435.
[171] Schofield JA, Duffy EE, Kiefer L, Sullivan MC, Simon MD. TimeLapse-seq: adding a temporal dimension to RNA sequencing through nucleoside recoding. Nat Methods 2018;15:221–5. Available from: https://doi.org/10.1038/nmeth.4582.
[172] Riml C, Amort T, Rieder D, Gasser C, Lusser A, Micura R. Osmium-mediated transformation of 4-thiouridine to cytidine as key to study RNA dynamics by sequencing. Angew Chem Int Ed 2017;56:13479–83. Available from: https://doi.org/10.1002/anie.201707465.
[173] Yin Y, Lu JY, Zhang X, Shao W, Xu Y, Li P, et al. U1 snRNP regulates chromatin retention of noncoding RNAs. Nature 2020;580:147–50. Available from: https://doi.org/10.1038/s41586-020-2105-3.
[174] Reichholf B, Herzog VA, Fasching N, Manzenreither RA, Sowemimo I, Ameres SL. Time-resolved small RNA sequencing unravels the molecular principles of microRNA homeostasis. Mol Cell 2019;75:756–68. Available from: https://doi.org/10.1016/j.molcel.2019.06.018 e7.
[175] Gasser C, Delazer I, Neuner E, Pascher K, Brillet K, Klotz S, et al. Thioguanosine conversion enables mRNA-lifetime evaluation by RNA sequencing using double metabolic labeling (TUC-seq DUAL). Angew Chem Int Ed 2020;59:6881–6. Available from: https://doi.org/10.1002/anie.201916272.
[176] Kawata K, Wakida H, Yamada T, Taniue K, Han H, Seki M, et al. Metabolic labeling of RNA using multiple ribonucleoside analogs enables the simultaneous evaluation of RNA synthesis and degradation rates. Genome Res 2020;30:1481–91. Available from: https://doi.org/10.1101/gr.264408.120.
[177] Yamada T, Akimitsu N. Contributions of regulated transcription and mRNA decay to the dynamics of gene expression. Wiley Interdiscip Rev RNA 2019;10:e1508. Available from: https://doi.org/10.1002/wrna.1508.
[178] Ninomiya K, Adachi S, Natsume T, Iwakiri J, Terai G, Asai K, et al. Lnc RNA-dependent nuclear stress bodies promote

intron retention through SR protein phosphorylation. EMBO J 2020;39:1–20. Available from: https://doi.org/10.15252/embj.2019102729.

[179] Min J, Wright WE, Shay JW. Clustered telomeres in phase-separated nuclear condensates engage mitotic DNA synthesis through BLM and RAD52. Genes Dev 2019;33:814–27. Available from: https://doi.org/10.1101/gad.324905.119.

[180] Cerase A, Armaos A, Neumayer C, Avner P, Guttman M, Tartaglia GG. Phase separation drives X-chromosome inactivation: a hypothesis. Nat Struct Mol Biol 2019;26:331–4. Available from: https://doi.org/10.1038/s41594-019-0223-0.

[181] Jain A, Vale RD. RNA phase transitions in repeat expansion disorders. Nature 2017;546:243–7. Available from: https://doi.org/10.1038/nature22386.

[182] Taniue K, Akimitsu N. Aberrant phase separation and cancer. FEBS J 2021;. Available from: https://doi.org/10.1111/febs.15765.

[183] Yeasmin F, Imamachi N, Tanu T, Taniue K, Kawamura T, Yada T, et al. Identification and analysis of short open reading frames (sORFs) in the initially annotated noncoding RNA LINC00493 from human cells. J Biochem 2021;169:421–34. Available from: https://doi.org/10.1093/jb/mvaa143.

[184] Mortimer SA, Kidwell MA, Doudna JA. Insights into RNA structure and function from genome-wide studies. Nat Rev Genet 2014;15:469–79. Available from: https://doi.org/10.1038/nrg3681.

[185] Engreitz JM, Sirokman K, McDonel P, Shishkin AA, Surka C, Russell P, et al. RNA-RNA interactions enable specific targeting of noncoding RNAs to nascent pre-mRNAs and chromatin sites. Cell 2014;159:188–99. Available from: https://doi.org/10.1016/j.cell.2014.08.018.

[186] Cai Z, Cao C, Ji L, Ye R, Wang D, Xia C, et al. RIC-seq for global in situ profiling of RNA-RNA spatial interactions. Nature 2020;582:432–7. Available from: https://doi.org/10.1038/s41586-020-2249-1.

[187] Fazal FM, Han S, Parker KR, Kaewsapsak P, Xu J, Boettiger AN, et al. Atlas of subcellular RNA localization revealed by APEX-seq. Cell 2019;178:473–90. Available from: https://doi.org/10.1016/j.cell.2019.05.027 e26.

[188] Stegle O, Teichmann SA, Marioni JC. Computational and analytical challenges in single-cell transcriptomics. Nat Rev Genet 2015;16:133–45. Available from: https://doi.org/10.1038/nrg3833.

[189] Saliba A-E, Westermann AJ, Gorski SA, Vogel J. Single-cell RNA-seq: advances and future challenges. Nucleic Acids Res 2014;42:8845–60. Available from: https://doi.org/10.1093/nar/gku555.

[190] Hrdlickova R, Toloue M, Tian B. RNA-seq methods for transcriptome analysis. Wiley Interdiscip Rev RNA 2017;8:1364. Available from: https://doi.org/10.1002/wrna.1364.

CHAPTER

11

Computational Epigenetics: The Competitive Endogenous RNAs Network Analysis

Loo Keat Wei

Faculty of Science, Department of Biological Science, Universiti Tunku Abdul Rahman, Kampar, Perak, Malaysia

OUTLINE

Introduction	185	Computational Epigenetic Approaches for Competitive Endogenous RNAs Network Analysis	190
Basic Principles of Competitive Endogenous RNAs	185		
The Competitive Endogenous RNAs Network	186	Conclusion	195
Mathematical Models for Competitive Endogenous RNAs Network Analysis	187	Acknowledgments	196
		References	196
Comparison of Mathematical Models for Competitive Endogenous RNAs Network Analysis	190		

INTRODUCTION

MicroRNA (miRNA) is a noncoding, an evolutionary conserved, 21–24 bases long component with a hairpin structure, which modulates the activities of messenger RNA (mRNA), noncoding RNA (ncRNA) or long noncoding RNA (lncRNA) [1]. Recently, it has been reported that an altered microRNA expression in COVID-19 patients allows the researchers to determine the presence of SARS-CoV-2 infection [2]. The miRNA contributes to less than 5% of the genome of phylum Annelida, phylum Arthropod, and phylum Chordate, but can target at least 60% of the protein-coding genes. By regulating the protein translation and mRNA decay rates, miRNA is able to control posttranscriptional activities endogenously [3]. For example, the 3′ untranslated region of mRNA, which is comprised of AU-rich sequences, can be regulated by several miRNAs via the miRNA response elements (MREs) [4]. The mRNA-miRNA interaction that mediates gene transcription activity is influenced by the presence of competitive endogenous RNAs (ceRNAs) [3]. Hence, we highlight the importance of ceRNA in the transcriptional regulation in this updated chapter of computational epigenetics. In addition, the strategies to choose an appropriate mathematical model for ceRNA network analysis as well as the context-based and genome-wide based approaches for computational epigenetics analyses are discussed, summarized, and compared.

BASIC PRINCIPLES OF COMPETITIVE ENDOGENOUS RNAS

CeRNA is the endogenous RNA transcript originating from a cluster of ncRNAs, which is comprised of transcribed pseudogenic RNA, lncRNA, and circular RNA [3]. The experiment findings involving *in vivo* and *in vitro* models that support the ceRNA regulatory system are beyond the scope of this chapter. However, the important players of the ceRNA network and their working mechanisms in regulating an epigenetic event will be discussed in the following paragraphs.

Transcribed pseudogenic RNAs are the main elements that mimic the identified genes, and usually malfunctional within the genome [5]. This is because the translational activities of pseudogenic RNAs often intermittent with frameshift mutations, deletions and insertions, which in turn unable to encode the functional proteins. Thus far, more than 15,000 pseudogenic elements have been transcribed, following the natural selection of mammalian development. Some of which are processed into pseudogenic mRNA elements by sharing similar conserved sequences as miRNAs [5]. This phenomenon suggests that pseudogenic mRNA elements can utilize distal transcriptional factors, repressors, or enhancers for different transcriptional activities [5].

Another player within the ceRNA network, namely lncRNA, is comprised of more than 300 bases that serve as the enhancer for various types of RNAs. This element scrubbers between miRNA and other RNA species for a variety of cell functions [6]. It actively participates in most of the transcriptional and posttranscriptional activities as well as the chromatin remodeling events. Despite its existence for more than 300 million years, the sequences are not well conversed within the phylum Chordate. However, this fast-evolving lncRNA element still resembles the features of a processed mRNA [6]. For example, lncRNA has the capability of retaining posttranscriptional activities such as 5′ capping, consisting of at least two exons, and regulating alternative splicing over the intronic regions. It has been reported at least 60% of lncRNAs consist of polyA tails [6]. A significant amount of lncRNA genes are existed in an antisense manner and positioned within 10 kb of the protein-coding genes. Due to its active participation in the physiological processes, it can be easily detected in different types of tissues but predominated in nuclease and cytoplasm. The lncRNA was once thought to be relatively fragile and unstable. However, Clark et al. [7] have proven that most of the lncRNAs are enormously stable, where their half-lives are more than 2 hours. In addition, a small portion of these lncRNAs are having half-lives more than 12 hours. Regardless of their half-lives, lncRNAs tend to mediate transcriptional activities by sequestering various regulatory factors through chromatin remodeling events. The regulatory factors include not limited to miRNA, distal transcription factors, and other ceRNA complex [7].

Circular RNA encompasses 1–5 exons with various lengths of intronic regions. In the presence of multiple Alu repeats, the length of the intronic regions can equip up to three times more bases compared to the respective linear RNA. Such a feature allows circular RNA to splice the intronic regions easily and to accommodate its circularization into a closely looped structure [8]. There are more than 20,000 types of circular RNAs identified so far. Circular RNA plays various roles in transcriptional regulations. For example, circular RNA regulates the expression of parental genes by sponging miRNAs and mRNAs, which contributes to protein complex scaffolding as well as interfering with the interactions between mRNAs and proteins [8].

THE COMPETITIVE ENDOGENOUS RNAS NETWORK

The ceRNA network exhibits vast applications as diagnostic biomarkers for diseases such as Parkinson [9], cardiovascular [10], multiple sclerosis [11], depression [12], etc. For example, ceRNA network revealed that *FBXL7* and *PTBP2* as well as *lncRNA NEAT1*—which are involved in autophagy, DNA repair mechanism, and vesicle transport—are the new targets for Parkinson disease pathogenesis [9]. In addition, it has also been suggested in a ceRNA network analysis that a malfunctional *TNF-TGFBR2* contributes to cardiac dysfunction by enhancing TGF-β expression, which in turns leads to cardiovascular disease development [10]. Nonetheless, the ceRNA network analysis revealed that *XIST-miR-326-HNRNPA1* are the therapeutic target for multiple sclerosis. These ceRNA components are active for spliceosome and RNA transport activities as well as the mTOR signaling pathway [11]. Moreover, network analysis suggested that ceRNA protocadherin α subfamily C2 and ceRNA Cyclin dependent kinase 6 are related to the occurrence and development of depression [12].

Within the ceRNA regulatory network, the ceRNAs tend to bind with a couple of communal miRNA and regulate their function as well as cross-regulate each other at the posttranscriptional level. The ceRNA network rival for miR binding sites, where multiple ceRNAs in a network can modulate one or more miRNA molecules. All these miRNAs are freely available in a cell and sharing identical MREs recognition sites. Alternatively, ceRNA also interacts with other ceRNAs in the network and regulates different sets of MREs indirectly [13]. The binding between miRNA-MREs is thought to be the repressor for the targeted gene. Each miRNA shows the ability to anchorage multiple MREs when repressing hundreds of these transcripts within the transcriptome. Thus far, about 0.2 million protein-coding genes have been identified throughout the human genome. Of which, more than a quarter of them are tightly shielded with MREs [14].

MREs serve as the ceRNA language that ensures the cross-talk between ceRNA and their respective components in the network are successful. The effectiveness of the cross-talk between ceRNA-ceRNA, ceRNA-mRNA, and ceRNA-miRNA in the network is highly dependent on the relative content of ceRNAs and their respective miRNAs. This includes (1) the uniqueness of miRNA sequences, (2) the amount of miRNAs and ceRNA, (3) the magnitude of seed region complementarity with the target, (4) the subcellular distribution of different types of RNA species in each cells, and (5) how miRNA reduces the availability of mRNAs [13,15,16].

It has been suggested that a huge fluctuation of ceRNA expression is warranted to enhance or suppress the miRNA expression [14]. In addition, the complex interactions within the ceRNA network allow the ceRNA to act as the sponge within the cell. This sponging ceRNA competitively binds and enriches the miRNAs, while ensuring sufficient concentration of miRNAs that are available to repress the miR-targeted mRNA. Of which, the targeted mRNA is shielded from degradation upon miR binding [13]. Nonetheless, sometimes the amount of MREs is not equivalent to the concentration of ceRNAs. Even though two MREs may bind to the similar miR, however, the nucleotide composition of these two MREs may not be the same, thus, affecting the effectiveness of MREs to bind with the miR [13]. Likewise, a single miRNA that binds to at least ten RNAs may not repress the gene expression at the similar momentum. Hence, it is important to establish a computational framework and employ appropriate mathematical models when analyzing the ceRNA network [13].

MATHEMATICAL MODELS FOR COMPETITIVE ENDOGENOUS RNAS NETWORK ANALYSIS

Several mathematical models for ceRNA have been developed based on the assumption that there is a stoichiometry reaction among the players in the posttranscriptional regulation, including miRNAs, ceRNAs and miRNA-targeted transcripts. When miRNAs and ceRNAs are in equimolar, the ceRNA crosstalk tends to exhibit the greatest effect on posttranscriptional activity [15,16]. However, in the phenomenon where the concentration of ceRNA is more than the available amount of miRNAs, the ceRNA crosstalk is operated at the maximal range [17,18]. These conflicting statements are predominantly implicated by the mathematical models used (Table 11.1), with the assumption that the concentration of miRNAs must exceed the number of transcript target sites (Table 11.1). Thus, different variables such as the concentrations of miRNAs, ceRNA, miRNA targets, which can implicate the outcomes of a mathematical model, are discussed in the following paragraphs. To ease the comparison and selection for the recommended mathematical models, the assumption made, as well as the highlight and limitations of each mathematical model, are summarized in Table 11.1.

The rate-equation based miRNA-ceRNA interaction model (Table 11.1) is unable to provide a long-term interference on the posttranscriptional rates [16]. It is pertinent to recognize the fact that there is transcriptional noise within the ceRNA networks that can contribute to different degree of gene expression for a cell. The application of Gaussian mathematical model for steady state fluctuations within the ceRNA network that disregards molecular noise entirely may contribute to a bias estimation (*Model 1*). Hence, the development of a mathematical model that purely depends on the steady state fluctuation of miRNA-mRNA or miRNA-ceRNA ratio for testing the efficiency of ceRNA network is meaningless.

Model 1 [16]

$$N = 2 \ ceRNA \ species;$$

$$M = 1 \ miR \ species$$

$P(r) = normal \ distribution \ of \ transcription \ rates, \ r = \{b_1, b_2, \beta\}$

$$\Sigma = correlation \ matrix \ of \ inputs$$

$$P(r) \propto \exp\left[\frac{1}{2}(r-\bar{r})^T \Sigma^{-1}(r-\bar{r})\right]$$

To overcome the influence of steady state fluctuation within the ceRNA network, a more sophisticated mathematical models which rely on the Pearson's correlation coefficient to detect the miRNA modulated crosstalk may be a better choice [16].

Model 2 [16]

$$\sigma_k^2 = variance \ of \ r_k$$

$$\rho_{12} = \frac{\Sigma_k \chi_{1k} \chi_{2k} \sigma_k^2}{\sqrt{(\Sigma_k \chi_{1k}^2 \sigma_k^2)(\Sigma_k \chi_{2k}^2 \sigma_k^2)}}$$

The mathematical model that emphasizes the Pearson's correlation coefficient is developed from the gene expression data. However, the outcome generated from this type of mathematical model is highly dependent on the autonomous gene transcription rates (*Model 2*). Ala et al. [15] applied the Pearson's correlation coefficient to determine the most potential

TABLE 11.1 The Most Important Mathematical Model for CeRNA (Competitive Endogenous RNAs) Network

Author	Year	Mathematical Model	Assumption/Hypothesis	Highlight	Limitation	References
Figliuzzi	2013	A minimal, steady state, rate-equation based model	A condition where at least partial degradation of miR-ceRNA needs to work in a stoichiometrically manner when applying the Gaussian distribution model	Competitive miR-ceRNA interaction through asymmetric and symmetric couplings, where ceRNAs acts an intermediate and miRNA serves as the responsive component, makes ceRNA-ceRNA crosstalk rather sparse, even with the present of abundances miRNA. Nonetheless, miR-ceRNA network topology may connect and boost ceRNA-ceRNA crosstalk	A stationary miR-ceRNA interaction contributes to a transient interference on the posttranscriptional rates	[16]
Ala	2013	A mass-action model	A flawless ceRNA-mediated cross regulation of miR take places at a near-equimolar equilibrium within the network	A catalytic interaction between miR and targeted ceRNAs in a ratio of 1:2, suggested that the ceRNA network is cross-regulated through a simple titration mechanism in a threshold-like manner	This mass-action model only allows for ceRNA documentation and qualitative prediction, while, disallows for quantitative prediction	[15]
Bosia	2013	A stochastic model	Vectors are distributed based on a multivariate Gaussian distribution	Stochasticity model that combines the multivariate Gaussian approximation and the moment generating function, to determine out-of-equilibrium and equilibrium states of a ceRNA network	Stochasticity only becoming of increasing importance when modest amounts of molecules are involved in the ceRNA network	[13]

crosstalk that near an equimolar state of a ceRNA network. Their findings showed that an altered expression of a single ceRNA can affect the expression levels of other ceRNAs within the network. However, the true representation of miRNA modulated crosstalk within the ceRNA network is unable to be determined by using the models developed by Figliuzzi et al. [16].

It is important to noted that silencing a highly expressed ceRNA may impose a greater impact on the ceRNA network than silencing the ceRNA where its gene is expressed at negligible amount. The kinetic model developed by Ala et al. [15] is based on the dynamic properties of two ceRNA species and one miRNA species of the molecular titration mechanism (Model 3, Table 11.1). However, regardless of which types of ceRNAs (i.e. lncRNA, pseudogenic RNAs or circular RNA) chosen, silencing a pair of ceRNAs at the equimolar conditions may pose a great impact on the ceRNA network ([15], Table 11.1).

Model 3 [15]

$$S = free\ miRs;$$

$$R_1 = ceRNA\ species,\ target\#1;$$

$$R_2 = ceRNA\ species,\ target\#2$$

$$S_{SS} \to \frac{k_s - \alpha\ (k_{R_1} + k_{R_2})}{g_S},$$

$$if\ \alpha\ (k_{R_1} + k_{R_2}) < k_S\ (0\ otherwise),$$

$$R_{1SS} \to \frac{k_{R_1}}{g_R}\left(1 - \frac{k_s}{\alpha\ (k_{R_1} + k_{R_2})}\right),$$

$$if\ \alpha\ (k_{R_1} + k_{R_2}) \geq k_S\ (0\ otherwise),$$

$$R_{2SS} \to \frac{k_{R_2}}{g_R}\left(1 - \frac{k_s}{\alpha\ (k_{R_1} + k_{R_2})}\right),$$

$$if\ \alpha\ (k_{R_1} + k_{R_2}) \geq k_S\ (0\ otherwise),$$

Likewise, there are also arguments on different numbers of MREs that may affect the efficiency of ceRNA-miRNA cross-talk within the ceRNA network [15,16]. In fact, the dosage of ceRNA and miRNA within a network may suggest the efficiency of cross-regulation of all components within the ceRNA network. For instance, the effectiveness of ceRNA in modulating the cross-regulation is highly dependent on the miRNA oscillation within the mass-action model. The oscillation of miRNAs suggested that the efficiency of ceRNA cross-regulation activities decelerates with the amount of shared miRNAs. On the contrary, in the case of abundant miRNAs that are not being shared by the ceRNAs, the efficiency of cross-regulation has found to be decreased significantly [15]. When validating the proposed mass-action model using phosphatase and tensin homolog and ceRNA vesicle-associated membrane protein-associated protein A, it is confirmed that the efficiency of ceRNA cross-talk is enhanced with the increasing number of shared miRNAs. On the other hand, the cross-regulation of ceRNA is reduced when a pair of ceRNA is targeted by the nonshared miRNAs that are available abundantly within the network [15].

The ceRNA network is not a standalone tool that regulates the gene expression profiles within the computational epigenetic context. These sponging components also work closely with various types of transcriptional factors to regulate the transcriptional framework within the epigenome. An extensive analysis of ceRNA network and gene ontology revealed a significant overexpression of transcriptional genes within the mass-action model (*Model 3*) developed by Ala et al. [15]. Higher values of the Pearson's correlation coefficient for the expression of transcriptional factors indicate that the subtle changes in the miRNA concentrations can have impactful consequences on the efficiency ceRNA network crosstalk [15]. However, the activities of transcription factors on regulating their targets are relatively steady, which is independent from the expression levels of miRNAs. The importance of regulatory effect of transcription factor may be implicated for diseases related to the cellular and organismal physiology [15]. For example, it is observed that ceRNA network may firmly engage with transcription factors in regulating the pathological and physiological environment of the cancerous cells [15].

To a lesser extent, this mass-action model is preferably meant for ceRNA documentation and qualitative prediction, not for users to quantify their prediction [15]. Typically, most of the gene expression data are either presented as a skewed curve or exhibited as nonlinearity. The mathematical models suggested by Figliuzzi et al. [16] are contented within the mean-field limit. Therefore, to have a broader view on how ceRNA network is operated, the stochastic model that analyzes out-of-equilibrium and equilibrium states of miRNA, and the targeted mRNA within the ceRNA network has been proposed [13].

The mature miRNA are the main limiting factor for a ceRNA network. Specifically, a threshold-like design between two interacting components plays equally important roles in the ceRNA network. For example, a titration mechanism where Gene X that shares a common miRNA with Gene Y is preferable to muddle with the mRNA of Gene X. This contributes to a decelerated repression of Gene Y. The relative concentration of these two components must fall within the thresholds limit to ensure the validity of the model. Such a titrative fashion modulation also suggests the plausibility that all molecules are hypertensive within the ceRNA network [13].

A threshold effect proposed by Bosia et al. [13] demonstrates that stochasticity model (*Model 4*), i.e., combination of two approaches can determine the intrinsic nonlinearity and complication evolved from the ceRNA network. The two approaches, which beyond the mean-field approximation, are the moment generating function and multivariate Gaussian approximation (*Model 4*, Table 11.1). By combining these approaches, we can determine the expression level and correlation of molecular components that are not contented to the transcriptional noise generated, but also the intrinsic Pearson's correlation coefficients (*Model 4*).

Model 4 [13]

Model 4.1: Multivariate Gaussian approximation—mean and covariances

$$N = 2 \; ceRNA \; species;$$

$$M = 1 \; miR \; species$$

$$\vec{X} = (X_1, \ldots, X_{N+M}): (R_1, \ldots, R_N, S_1, \ldots, S_N)$$

$$\mu_1 = E(X_i)$$

$$c_{ij} = E(X_i X_j) - E(X_i)E(X_j)$$

$$E(X_i X_j X_k) = c_{ij}\mu_k + c_{ik}\mu_j + c_{jk}\mu_i$$

$$E(X_i X_j X_k X_l) = c_{ij}c_{lk} + c_{ik}c_{jl} + c_{il}c_{jk}$$

Model 4.2: Variances, Fano factor, coefficient of variation

$$\sigma_{<X_i>} = \sqrt{<X_i^2> <X_i>^2}$$

$$f_{xi} = \frac{\sigma^2_{<X_i>}}{<X_i>}$$

$$CV_{xi} = \frac{\sigma_{<X_i>}}{<X_i>}$$

Model 4.3: Pearson's correlation coefficients

$$\rho_{x_i,x_j} = \frac{<X_iX_j> - <X_i><X_j>}{\sigma_{<X_i>}\sigma_{<X_j>}}$$

Model 4.1 outperforms the common linear noise approximation, as it captures the threshold of a closed system. The expressions of means are regarded as the control parameters of a threshold, which may affect the transcriptional noise and Pearson's correlation coefficients [13]. Moreover, *Model 4.2* enables us to determine the transcriptional noise within the ceRNA network by measuring the variance, Fano factor, and coefficient of variation for each molecular component. It has been observed that the transcriptional rate of a single ceRNA can accelerate other ceRNAs within the network, while decelerating the transcriptional noise profile of miRNA (*Model 4.2*). *Model 4.3* enables us to determine whether both ceRNA and miRNA are positively or negatively correlated with each other. *Model 4* suggests that when there is a small number of interacting species exists within the ceRNA network, the great values of transcriptional noise profile and Pearson's correlation coefficient are observed. Despite that stochasticity *Model 4* is applicable for both out-of-equilibrium and equilibrium state analysis, the stochasticity only becomes increasingly important in computational epigenetics when modest amounts of players are involved in the ceRNA network [13].

COMPARISON OF MATHEMATICAL MODELS FOR COMPETITIVE ENDOGENOUS RNAS NETWORK ANALYSIS

Every mathematical model for ceRNA network analysis comes with its strengths and limitations. By exploring the strengths of each model, enable researchers to perform ceRNA analysis to the fullest and limiting the chances of obtaining false positive/false negative results. For example, a Gaussian distribution model proposed by Figliuzzi et al. [16] always exhibits the characteristics of a normal distribution. By which, a Gaussian distribution model always follow a bell-shaped curve, presumed the measured mean values shadows a normal distribution. Despite a minimum sample size of three is needed when applying this model, the Gaussian distribution model is classified as the most extensively used among all types of distributions. This is supported by the fact that the Gaussian distribution model can divide a normal distribution curve into their respective quartiles, and the cumulative effects of ceRNA activities can also be observed easily with this model. However, one has to take note that the reliability of Gaussian distribution model is highly dependent on the calculation that applied on a ceRNA network. The multivariate Gaussian distribution outperformed the Gaussian distribution model. Unlike the Gaussian distribution, the multivariate Gaussian distribution can cater for the multifactorial factorial/covariates that is involved in the ceRNA network. For example, large numbers of random variables that present in the ceRNA network can be tested according to a particular distribution that focused on Central Limit theorem. Moreover, the forward- and backward-recursions when performing the stepwise logistic regression that associated with the multivariate Gaussian distribution analysis, enable researchers to determine the true players of the ceRNA network.

COMPUTATIONAL EPIGENETIC APPROACHES FOR COMPETITIVE ENDOGENOUS RNAS NETWORK ANALYSIS

Several computational approaches can be used for ceRNA network analysis: (1) mathematical modeling that have been discussed in the previous section, (2) pair-wise correlation, and (3) partial association. Regardless of which approach is chosen to establish the ceRNA network, researchers tend to initiate the analysis by identifying potential miRNA targets. There are numerous methods that can be used to determine miRNA targets within the epigenome. More information on how miRNA affects the gene expression, and which methods to be used for a miRNA analysis, can refer to Computational Epigenetics and Diseases [3]. There are several useful software/tools for miRNA determination, not limited to miRanda [19] and TargetScan [20]. No matter which methods have been chosen for miRNA analysis, one needs to tackle the false positive problems that arise from the analysis. Likewise, various databases or tools for ceRNA network analysis are presented in Table 11.2.

In addition, we would like to discuss and compare the global-based and context-based ceRNA prediction approaches in this section. The global-based ceRNA prediction approaches can be sub-divided into: (1) proportion-based and (2) hypergeometric-text-based [61]. Furthermore, the context-based ceRNA prediction approaches can be sub-divided into: (1) hypergeometric test-coexpression, (2) sensitivity-based correlation, and (3) conditional-based mutual information [61].

When viewing the ceRNA network at a global view, it is important to note that not all ceRNA molecules are functioning at a similar propensity. The proportion-based approach classifies the potential genes in the epigenome

TABLE 11.2 Databases and Tools for ceRNA (Competitive Endogenous RNAs) Network Analysis

Databases/Tools	Key Point	Accessibility/URL	References
AnnoLnc	Web portal for systematically annotating novel human lncRNAs for genomic location, expression patterns, secondary structure, miRNA interaction, transcriptional regulation, protein interaction as well as genetic association and evolution based on at least 700 data sources	http://annolnc.cbi.pku.edu.cn	[21]
Cancerin	Cancerin is an R-software that integrates several algorithms such as miRanda, DIANA-microT when combining the information of copy number alteration, DNA methylation, gene and miR expression datasets. These algorithms aids in determining the cancer-associated ceRNA interactions	https://github.com/bozdaglab/Cancerin	[22]
Cerina	Competing Endogenous RNA for INtegrative Annotations (Cerina) is an integrative framework that anticipates circRNAs biological functions through Pareto Frontier Analysis when constructing the ceRNA model. This tool was developed based on the combinatory effects of R packages such as circlize, dendextend, shinyjs, htmltools, shinydashboard, shinyBS, reshape, DT, shinycssloaders, heatmaply, tidyverse, visNetwork, Matrix, fastcluster, and igraph, ggolot, and plotrix. Of which, commercial software Lucidchart (http://www.lucidchart.com) gather all final version of figures	https://githu.b.com/jcard.enas14/CERINA	[23]
Circ2Traits	A database that links circRNAs-miR with diseases by calculating the likelihood of association	http://gyanxet-beta.com/circdb/	[24]
CircAtlas2.0	It incorporates more than one million of circRNAs as well as their expression and functional profiles of human, rat, mouse, mcaca, chicken and pig. This interactive and open-access webpage calculates multiple conservation score for individuals, tissues, and species prior annotating the circRNAs through their co-expression and regulatory networks. It also combines several pipelines such as circad, circRNADisease, circR2Disease to correlate circRNAs with diseases	http://159.226.67.237:8080/new/index.php	[25]
Circbank	A web-accessible database which provides novel naming system for circRNAs according to their host genes. It also identified the miRNA binding site, conservation information, m6A modification, mutational information as well as protein-coding potential of circRNAs. The output consists of predictable miRNA-circRNA pairs are analyzed by Targetscan and Miranda algorithms	http://www.circbank.cn/	[26]
CircFunBase	A web-accessible database of 7000 manually curated functional circRNA entries of *Homo sapiens, Mus musculus* and others, that enable researchers to visualize the circRNA-miRNA interaction networks	http://bis.zju.edu.cn/CircFunBase/	[27]
CircIMPACT	An R package that provides an integrative circRNA-gene expression analysis and determines the genes expression which are based on changes in circRNA expression	https://github.com/AFBuratin/circIMPACT	[28]
CircInteractome	circRNA interactome is a web tool that searches through public databases for circRNA, miRNA, and RBP. It aids researchers to determine potential circRNAs which can serve as RBP sponges and determine potential internal ribosomal entry sites. In addition, this tool also allows users to design junction-spanning primers as well as designing siRNAs for circRNA silencing	https://circinteractome.nia.nih.gov/	[29]
CircNet	This database allows users to determine circRNAs and its isoform level expression profiles as well as its genomic annotations and sequences. This public database also allows users to integrate miRNA-target networks, examine circRNA-miRNA-gene regulatory networks and tissue-specific circRNA expression profiles, and illustrates an integrated regulatory network of circRNAs-miRNAs-genes	http://circnet.mbc.nctu.edu.tw/	[30]
CircR2Disease	Manually curated database on experimentally validated circRNA-disease associations which published before 31 March 2018. This database consists of 725 experimentally supported associations, 661 circRNAs and one hundred diseases	http://bioinfo.snnu.edu.cn/CircR2Disease/	[31]

(Continued)

TABLE 11.2 (Continued)

Databases/Tools	Key Point	Accessibility/URL	References
circRNA disease	Manually curated, open-access database for searching disease-circRNAs association	http://cgga.org.cn:9091/circRNADisease/	[32]
CoInDiViNE	It is the extension of DiViNE framework starting from version 2.5, which used a novel state space generation algorithm that applies component-interaction automata (CI) and a property specification CI-LTL logic. This tool facilitates the partial order reduction when it is run in a parallel setting	https://divine.fi.muni.cz/whatsnew.html	[33]
Co-LncRNA	Web portal to determine GO annotations and KEGG pathways which may be exaggerated by co-expressed protein-coding genes of a single or multiple lncRNAs, based on human RNA-Seq data	https://ngdc.cncb.ac.cn/databasecommons/database/id/6923	[34]
Crinet	Crinet contemplates all miRNAs, lncRNAs, and pseudogenes as potential ceRNAs. This tool eliminates spurious ceRNA pairs through a network deconvolution algorithm, when constructing the ceRNA network	https://github.com/ziynet/Crinet	[35]
CSCD	The cancer-specific circular RNAs (CSCD) is the first database that predicts MREs, RNA binding protein sites and open reading frames as well as splicing events in linear transcripts for each circRNA	http://gb.whu.edu.cn/CSCD/	[36]
Cupid	Cupid enables users to predict both interactions of microRNA–target and their mediated ceRNA simultaneously	http://cupidtool.sourceforge.net/	[37]
DEsubs	An R package for pathway network creation and processing, subpathway mining, visualization and enrichment analysis through a case-specific approach	http://bioconductor.org/packages/DEsubs/	[38]
DIANA-LncBase v2	An updated version of DIANA-LncBase v1 which comprises of massive miRNA:lncRNA interactions. The DIANA-LncBase v1 provides MREs on lncRNAs through the experimentally supported and in silico predicted data via DIANA-microT algorithm. It also enables users to explore cell type specific miRNA:lncRNA regulation for mouse and human	http://www.microrna.gr/LncBase	[39]
DiViNE 4.3	The distributed verification environment project (DiViNE) furnishes a big framework in assisting the distributed verification on clusters via (1) a DiViNE library, (2) a DiViNE Tool, and (3) a ready-to use cluster	https://divine.fi.muni.cz/whatsnew.html	[40]
lnCeDB	lnCeDB contains human lncRNAs (based on GENCODE 19 version) which can potentially act as ceRNAs. This database deploys data from TargetScan, StarBase and miRCode. By which, an lncRNA-mRNA pair is calculated through the ceRNA score and hypergeometric test	http://gyanxet-beta.com/lncedb/	[41]
LncACTdb2.0	The LncACTdb2.0 consists of manually curating ceRNA interactions which include circular RNAs and pseudogenes, that fit for 23 species and 213 phenotypes. In addition, there are 33 types of cancer curated from TCGA data which could illustrate the ceRNA networks that are associated with cancers. Of which, the LncACTdb2.0 has incorporated various online tools such as LncACT-Get, LncACTFunction, LncACT-Survival, LncACT-Network and LncACTBrowse	http://www.bio-bigdata.net/LncACTdb/	[42]
LnCeCell	A database of predicted lncRNA-associated ceRNA networks at single-cell resolution, which allows users to visualize the global map of ceRNA subcellular location and its gene regulatory networks within a single cell	http://www.bio-bigdata.net/LnCeCell/ http://bio-bigdata.hrbmu.edu.cn/LnCeCell/	[43]
LAceModule	LAceModule deploys a dynamic correlation measure that integrates the conservative Pearson's correlation coefficient and dynamic correlation liquid association when estimating the sensitivity of the correlation of ceRNAs to microRNAs	https://github.com/GaoLabXDU/LAceModule	[44]
Linc2GO	The first database that employs ceRNA hypothesis when predicting and annotating the lincRNA functions	https://rdrr.io/github/tpq/miSciTools/man/linc2go.html	[45]
LnCeVar	It is the database showing the genomic variations that disturb lncRNA-associated ceRNA network regulation. In addition the information is curated from the published literature and high-throughput data sets	http://www.bio-bigdata.net/LnCeVar/	[46]

(Continued)

TABLE 11.2 (Continued)

Databases/Tools	Key Point	Accessibility/URL	References
lnc-GFP	Long noncoding RNA global function predictor ("lnc-GFP") is a bi-colored network, global probable functions predictor for lncRNA at massive scale, through the integration of gene expression data and protein interaction data	–	[47]
LncRNA2Function	The first ontology-driven user-friendly web portal for lncRNAs. It deploys hypergeometric test when annotating the functions of a single lncRNA or a set of lncRNAs	http://mlg.hit.edu.cn/lncrna2function	[48]
LncRNAs2Pathways	It is an R-based tool that categorizes the functional pathways impacted by the combinatorial effects of a set of lncRNAs of interest, based on Kolmogorov–Smirnov-like statistical test, through a global network propagation algorithm	https://cran.r-project.org/web/packages/LncPath/index.html	[49]
LncRNAdb v2.0	An all-inclusive, manually curated reference eukaryotic lncRNAs database which integrates with nucleotide sequence information, Illumina Body Atlas expression profiles, and a BLAST search tool. This database complies with International Nucleotide Sequence Database Collaboration. Outputs can be exported as a REST API	http://lncrnadb.org/	[50]
lncRNAtor	Database for human, mouse, zebrafish, fruit fly, worm, and yeast lncRNAs. This web interface consists of annotation, gene expression, sequence analysis, phylogenetic conservation and protein binding of lncRNAs	http://lncrnator.ewha.ac.kr/	[51]
LncSEA	A human lncRNA sets database that integrates transcription factor ChIP-seq, ATAC-seq, DNase-seq and H3K27ac ChIP-seq data. It facilitates lncRNA annotations and enrichment analyses for at least 40,000 lncRNA reference sets over 18 categories as well as 66 subcategories	http://bio.liclab.net/LncSEA/index.php	[52]
MUST	An online calculation tool for Minimal Unsatisfiable Subsets Enumeration (MUST). This tool deploys three domain agnostic algorithms, including MARCO, TOME and ReMUS, which allows users to perform the calculation in SAT domain, SMT domain, and LTL domain	https://www.fi.muni.cz/~xbendik/research/must	[53]
ncFANs v2.0	Integrative platform for functional annotation of ncRNAs, which consists of the updated version of ncFANs-NET and ncFANs-eLnc as well as ncFANs-CHIP	http://bioinfo.org/ncfans http://ncfans.gene.ac	[54]
NONCODEV6	Updated database of NONCODE for long ncRNA annotation in animals and plants. This version exhibits human lncRNA-disease relationships including cancers, lncRNA-tissue expression annotation profiles for five plants, conservation annotation of lncRNAs at transcript level for 23 plant species	http://www.noncode.org/	[55]
Noncoder	Database for exon array-based detection of lncRNAs	http://noncoder.mpi-bn.mpg.de	[56]
riboCIRC	Comprehensive database of translatable circRNAs	http://www.ribocirc.com/	[57]
RPmirDI	miRNA target prediction is enhanced by a cascaded, semisupervised machine learning reciprocal perspective method	cu-bic.ca/RPmirDIP	[58]
SPONGEdb	A pan-cancer database that visualize ceRNA interactions through R and Python packages	https://github.com/spnge/sponge-db	[59]
starBase v2.0	An interactive web portal that decrypted miRNA-ceRNA, protein-RNA and miRNA-ncRNA interaction networks from large-scale CLIP-Seq data	https://starbase.sysu.edu.cn/starbase2/	[60]

by calculating the proportion of common miRNAs within a ceRNA network (*Formula 1*) [62]. This approach may gather all the potential genes within the ceRNA network for further analysis without considering the parasitic components in the network. However, overpresentation of certain genes within the ceRNA network may contribute to skewed results. For instance, a fair view of potential gene sets showed that the highest contribution to the ceRNA network is unable to rule out. The ratio-based approach is calculated with *Formula 1*.

Formula 1: Proportion-based approach

$$G = gene\ sets;$$

$$x = gene\ x\ from\ gene\ sets\ G;$$

$$miR_y = miR\ sets\ that\ regulate\ gene\ Y;$$

$$miR_x = miR\ sets\ that\ regulate\ gene\ X$$

$$R(x) = \frac{miR_y \cap miR_x}{miR_x},\ x \varepsilon G$$

Next, the hypergeometric text-based approach can determine the presence of several pairs of genes that are co-mediated by similar miRNAs in the ceRNA network (*Formula 2*). This approach is preferred over the proportion-based approach, as it ranked the RNAs through a probability model [63]. The true players for the ceRNA network are determined based on RNA ranking, which also generated a false discovery rate concurrently. RNA components with a false discovery rate of more than 0.05 ($q > 0.05$) shall not be considered for the subsequent analyses to avoid the chances of obtaining false positive outcomes for the ceRNA network [64].

Formula 2: Hypergeometric text-based approach

$$N = all\ miRs;$$

$$N_X = total\ number\ of\ miRs\ whch\ regulate\ RNA\ X$$

$$N_Y = total\ number\ of\ miRs\ whch\ regulate\ RNA\ Y$$

$$N_{XY} = number\ of\ shared\ miRs$$

$$P = 1 - F(N_{XY} - 1 | N, N_X, N_Y)$$

$$P = 1 - \sum_{z=0}^{N_{XY}-1} \frac{\binom{N_X}{z}\binom{N-N_X}{N_Y-z}}{\binom{N}{N_Y}}$$

Despite that both the proportion-based and hypergeometric text-based approaches can determine multiple sets of important components for a ceRNA network, the prediction results are highly dependent on the total number of genes/miRNAs obtained from the primary searching. The final results of the ratio-based approach is a proportion of useful genes from a pool of ceRNA components within the network, while those of the hypergeometric text-based approach are based on the p-value and q-value of a probabilistic model. Noticeably, the chances of obtaining the false positive prediction are still unavoidable if the total number of genes/miRNAs used in both approaches are relatively small (e.g., $n < 30$) [64].

Similarly, the context-based ceRNA prediction approach using the Pearson's correlation coefficient may be a better solution for the ceRNA network prediction [65]. Context-based hypergeometric test-coexpression approach that integrates *Formula 2* and *Formula 3* can gather multiple pairs of active ceRNA-ceRNA components in the network. This method scored and graded all the potential genes/miRs in the ceRNA network through false discovery rate obtained from *Formula 2*; prior to the integration of the specific genes/miRNAs expression data using the Pearson's correlation coefficient (*Formula 3*). When assessing and grading the potential genes/miRs and their expression data independently through *Formula 2* and *Formula 3*, the chances of obtaining bias prediction may be reduced. More importantly, *Formula 3* enables the researchers to calculate the mean rank for all the important components in the ceRNA network with minimal bias (*Formula 3*).

Formula 3: Hypergeometric test-coexpression (Pearson's correlation coefficient) approach

$$x_i = RNA\ X\ or\ Gene\ X\ expression\ level\ in\ sample\ i;$$

$$y_i = RNA\ Y\ or\ Gene\ Y\ expression\ level\ in\ sample\ i;$$

$$\bar{x} = mean\ RNA\ X\ or\ Gene\ X\ expression\ level;$$

$$\bar{y} = mean\ RNA\ Y\ or\ Gene\ Y\ expression\ level;$$

$$R = \frac{\sum_{i=1}^{n}(x_i - \bar{x})(y_i - \bar{y})}{\sqrt{\sum_{i=1}^{n}(x_i - \bar{x})^2}\sqrt{\sum_{i=1}^{n}(y_i - \bar{y})^2}}$$

Sensitivity-based correlation approach integrates the enormous genome-wide data of gene-miRNA expression profiles for constructing a context-based ceRNA network [66]. This approach categorizes multiple gene pairs of highly correlated RNAs that share similar miRNAs (*Formula 4*). It generates sets of sensitivity-based correlation S scores from the random matching between different pairs of RNAs and their corresponding miRNAs before suggesting the significant correlation of RNAs-miRNAs within the ceRNA network (*Formula 4*).

Formula 4: Sensitivity-based correlation approach

$$R_X = RNA\ X$$

$$R_Y = RNA\ Y$$

$$M = miR\ M\ that\ co-regulates\ RNA\ X\ and\ RNA\ Y$$

$$R_{AB} = Pearson's\ correlation\ coefficient\ for\ RNA\ X\ and\ RNA\ Y$$

$$R_{AM} = Pearson's\ correlation\ coefficient\ for\ RNA\ X\ and\ miR\ M$$

$$R_{MB} = Pearson's\ correlation\ coefficient\ for\ miR\ M\ and\ RNA\ Y$$

$$R_{XY|M} = \frac{R_{XY} - R_{XM}R_{MY}}{\sqrt{1-R_{XY}^2}\sqrt{1-R_{MY}^2}}$$

$$S = R_{XY} - R_{XY|M}$$

Although Pearson's correlation coefficient approaches (*Formula 3* and *Formula 4*) may be useful in determining the important components within the ceRNA network, the validity of such a prediction is limited by sample sizes and outliers. The results of Pearson's correlation coefficient may only valid with minimum/no outliers captured in the analysis. In fact, the value of Pearson's correlation coefficient is highly affected by the presence of outliers as well as the sample sizes involved in the analysis. Smaller sample size that contains appropriate number of outliers can decrease the value of Pearson's correlation coefficient significantly.

On the other hands, the conditional-based mutual information allows the researcher to identify the direct dependencies of RNAs within the ceRNA network [67]. It is the most widely used approach for a context-based analysis, because, different gene pairs of RNAs within the ceRNA network are identified through the p-values generated from their respective miRs (*Formula 5*). This approach suggested that the regulatory activities of ceRNA components (X) on their targets (Y) are measured by the miRNA expression (M) (*Formula 5*). According to *Formula 5*, both Fisher's exact test (*Formula 5.1*) and Fisher's method (*Formula 5.3*) are the main calculations involved in the ceRNA network prediction. Specifically, the dimensions of miRNAs (Y_1; Y_2) that are in common for RNA-RNA pairs (X_1 and X_2) must attain statistical significance in the Fisher's exact test (*Formula 5.1*), before it can be further analyzed with the Fisher's method (*Formula 5.3*). However, being the intermediate step of *Formula 5.1* and *Formula 5.3*, *Formula 5.2* grades the statistical significance of each miRNA in the ceRNA network. *Formula 2* helps to eliminate the poorly expressed RNA components within the network by performing 1000 times shuffling. Subsequently, all the p-values obtained from each combinations of $ceRNAX_1 - miR - ceRNAX_2$ are ranked before one can obtain the χ^2 value through *Formula 5.3*.

Formula 5: Conditional-based mutual information approach

Formula 5.1 Fisher's exact test

$$X_1 = RNA\ X_1$$

$$X_2 = RNA\ X_2$$

$$Y_1 = the\ target\ for\ X_1$$

$$Y_2 = the\ target\ for\ X_2$$

$$\pi_{miR}\ (Y_1;Y_2) = \pi_{miR}\ (Y_1) \cap \pi_{miR}\ (Y_2)$$

Formula 5.2

$$YM = the\ expression\ between\ target\ for\ X\ and\ miR$$

$$Y = target\ for\ X$$

$$I[miR_iYM] > I[miR_iY]$$

Formula 5.3: Fisher's method

$$i = miR\ of\ i$$

$$N = total\ number\ of\ miRs$$

$$\chi^2 = -2 \sum_{i-1}^{N} \ln(p_i)$$

CONCLUSION

CeRNAs, the sponging materials within the epigenome, have shown their importance in regulating genes transcriptional activities. The ceRNAs tend to squeeze with miRNAs through the recognition sites of MREs. Different combinations of ceRNA-miRNA-ceRNA have shown different efficiency in regulating the gene transcriptional activities. The efficiency of each component in the ceRNA network can be determined by computational epigenetic approach and mathematical modeling approach. We proposed stochasticity *Model 4* as the best mathematical model for overcoming transcriptional noise generated along with the ceRNA network analysis, which allows researchers to perform out-of-equilibrium and equilibrium state analysis concurrently. Pearson's correlation coefficient has been widely used to determine the ceRNA components at both genome-wide and context-based levels. However, the conditional-based mutual information may outperform the Pearson's correlation coefficient approach because it allows researchers to evaluate the direct dependencies of RNAs within the ceRNA network. Regardless of which mathematical model or computational epigenetic approach is chosen for ceRNA network analysis, it is important to avoid the false positive or false negative findings. It is possible to combine different mathematical modeling and computational epigenetic approach when developing the best pipeline for ceRNA network analysis.

Acknowledgments

This research was supported by Ministry of Higher Education, Malaysia through Fundamental Research Grant Scheme FRGS/1/2021/SKK0/UTAR/02/4 and Universiti Tunku Abdul Rahman Research Fund (UTARRF) IPSR/RMC/UTARRF/2021-C1/L07.

Conflict of Interest Statement

None declared.

References

[1] Siddika T, Heinemann IU. Bringing microRNAs to light: methods for microRNA quantification and visualization in live cells. Front Bioeng Biotechnol 2021;8:1534.

[2] Farr RJ, Rootes CL, Rowntree LC, Nguyen THO, Hensen L, et al. Altered microRNA expression in COVID-19 patients enables identification of SARS-CoV-2 infection. PLoS Pathog 2021;17(7):e1009759.

[3] Wei LK. Computational epigenetics and disease. Computational epigenetics and diseases. Academic Press; 2019. p. 1–9.

[4] Wei LK, Au A. Computational epigenetics. Handbook of epigenetics. Elsevier; 2017. p. 167–90.

[5] Pink RC, Wicks K, Caley DP, Punch EK, Jacobs L, Carter DR. Pseudogenes: pseudo-functional or key regulators in health and disease? RNA 2011;17(5):792–8.

[6] Ulitsky I, Bartel DP. lincRNAs: genomics, evolution, and mechanisms. Cell. 2013;154(1):26–46.

[7] Clark MB, Johnston RL, Inostroza-Ponta M, Fox AH, Fortini E, Moscato P, et al. Genome-wide analysis of long noncoding RNA stability. Genome Res 2012;22(5):885–98.

[8] Lu M. Circular RNA: functions, applications and prospects. ExRNA 2020;2(1):1–7.

[9] Zhang X, Feng S, Fan Y, Luo Y, Jin L, Li S. Identifying a comprehensive ceRNA network to reveal novel targets for the pathogenesis of Parkinson's disease. Front Neurol 2020;11:810.

[10] Song C, Zhang J, Qi H, Feng C, Chen Y, Cao Y, et al. The global view of mRNA-related ceRNA cross-talks across cardiovascular diseases. Sci Rep 2017;7(1):1–2.

[11] Ding Y, Li T, Yan X, Cui M, Wang C, Wang S, et al. Identification of hub lncRNA ceRNAs in multiple sclerosis based on ceRNA mechanisms. Mol Genet Genom 2021;296(2):423–35.

[12] Lang Y, Zhang J, Yuan Z. Construction and dissection of the ceRNA-ceRNA network reveals critical modules in depression. Mol Med Rep 2019;19(5):3411–20.

[13] Bosia C, Pagnani A, Zecchina R. Modelling competing endogenous RNA networks. PLoS One 2013;8(6):e66609.

[14] Cai Y, Yu X, Hu S, Yu J. A brief review on the mechanisms of miRNA regulation. Genom Proteom Bioinform 2009;7(4):147–54.

[15] Ala U, Karreth FA, Bosia C, Pagnani A, Taulli R, Léopold V, et al. Integrated transcriptional and competitive endogenous RNA networks are cross-regulated in permissive molecular environments. Proc Natl Acad Sci 2013;110(18):7154–9.

[16] Figliuzzi M, Marinari E, De Martino A. MicroRNAs as a selective channel of communication between competing RNAs: a steady-state theory. Biophys J 2013;104(5):1203–13.

[17] Hausser J, Zavolan M. Identification and consequences of miRNA–target interactions—beyond repression of gene expression. Nat Rev Genet 2014;15(9):599–612.

[18] Jens M, Rajewsky N. Competition between target sites of regulators shapes post-transcriptional gene regulation. Nat Rev Genet 2015;16(2):113–26.

[19] Turner D. An overview of Miranda. ACM Sigplan Not 1986;21(12):158–66.

[20] Oliveira AC, Bovolenta LA, Nachtigall PG, Herkenhoff ME, Lemke N, Pinhal D. Combining results from distinct MicroRNA target prediction tools enhances the performance of analyses. Front Genet 2017;8:59.

[21] Hou M, Tang X, Tian F, Shi F, Liu F, Gao G. AnnoLnc: a web server for systematically annotating novel human lncRNAs. BMC Genom 2016;17(1):931.

[22] Do D, Bozdag S. Cancerin: a computational pipeline to infer cancer-associated ceRNA interaction networks. PLoS Comput Biol 2018;14(7):e1006318.

[23] Cardenas J, Balaji U, Gu J. Cerina: systematic circRNA functional annotation based on integrative analysis of ceRNA interactions. Sci Rep 2020;10(1):1–4.

[24] Ghosal S, Das S, Sen R, Basak P, Chakrabarti J. Circ2Traits: a comprehensive database for circular RNA potentially associated with disease and traits. Front Genet 2013;4:283.

[25] Wu W, Ji P, Zhao F. CircAtlas: an integrated resource of one million highly accurate circular RNAs from 1070 vertebrate transcriptomes. Genome Biol 2020;21(1):1–4.

[26] Liu M, Wang Q, Shen J, Yang BB, Ding X. Circbank: a comprehensive database for circRNA with standard nomenclature. RNA Biol 2019;16(7):899–905.

[27] Meng X, Hu D, Zhang P, Chen Q, Chen M. CircFunBase: a database for functional circular RNAs. Database 2019;2019.

[28] Buratin A, Gaffo E, Dal Molin A, Bortoluzzi S. CircIMPACT: an R package to explore circular RNA impact on gene expression and pathways. Genes. 2021;12(7):1044.

[29] Dudekula DB, Panda AC, Grammatikakis I, De S, Abdelmohsen K, Gorospe M. CircInteractome: a web tool for exploring circular RNAs and their interacting proteins and microRNAs. RNA Biol 2016;13(1):34–42.

[30] Liu YC, Li JR, Sun CH, Andrews E, Chao RF, Lin FM, et al. CircNet: a database of circular RNAs derived from transcriptome sequencing data. Nucleic Acids Res 2016;44(D1):D209–15.

[31] Fan C, Lei X, Fang Z, Jiang Q, Wu FX. CircR2Disease: a manually curated database for experimentally supported circular RNAs associated with various diseases. Database 2018;2018.

[32] Zhao Z, Wang K, Wu F, Wang W, Zhang K, Hu H, et al. circRNA disease: a manually curated database of experimentally supported circRNA-disease associations. Cell Death Dis 2018;9(5):1–2.

[33] Beneš N, Černá I, Křivánek M. Coindivine: parallel distributed model checker for component-based systems. arXiv preprint arXiv 2011;1111:0373.

[34] Zhao Z, Bai J, Wu A, Wang Y, Zhang J, Wang Z, et al. Co-LncRNA: investigating the lncRNA combinatorial effects in GO annotations and KEGG pathways based on human RNA-Seq data. Database 2015;2015.

[35] Kesimoglu ZN, Bozdag S. Crinet: a computational tool to infer genome-wide competing endogenous RNA (ceRNA) interactions. BioRxiv 2020;.

[36] Xia S, Feng J, Chen K, Ma Y, Gong J, Cai F, et al. CSCD: a database for cancer-specific circular RNAs. Nucleic Acids Res 2018;46(D1):D925–9.

[37] Chiu HS, Llobet-Navas D, Yang X, Chung WJ, Ambesi-Impiombato A, Iyer A, et al. Cupid: simultaneous reconstruction of microRNA-target and ceRNA networks. Genome Res 2015;25(2):257–67.

[38] Vrahatis AG, Balomenos P, Tsakalidis AK, Bezerianos A. DEsubs: an R package for flexible identification of differentially expressed subpathways using RNA-seq experiments. Bioinformatics 2016;32(24):3844–6.

[39] Paraskevopoulou MD, Vlachos IS, Karagkouni D, Georgakilas G, Kanellos I, Vergoulis T, et al. DIANA-LncBase v2: indexing microRNA targets on non-coding transcripts. Nucleic Acids Res 2016;44(D1):D231–8.

[40] Barnat J, Brim L, Cerná I, Moravec P, Ročkai P, Šimeček P. DiVinE – a tool for distributed verification. International Conference on Computer Aided Verification. Berlin, Heidelberg: Springer; 2006. p. 278–81.

[41] Das S, Ghosal S, Sen R, Chakrabarti J. In Ce DB: database of human long noncoding RNA acting as competing endogenous RNA. PLoS One 2014;9(6):e98965.

[42] Wang P, Li X, Gao Y, Guo Q, Wang Y, Fang Y, et al. LncACTdb 2.0: an updated database of experimentally supported ceRNA interactions curated from low-and high-throughput experiments. Nucleic Acids Res 2019;47(D1):D121–7.

[43] Wang P, Guo Q, Hao Y, Liu Q, Gao Y, Zhi H, et al. LnCeCell: a comprehensive database of predicted lncRNA-associated ceRNA networks at single-cell resolution. Nucleic Acids Res 2021;49(D1):D125–33.

[44] Wen X, Gao L, Hu Y. LAceModule: identification of competing endogenous RNA modules by integrating dynamic correlation. Front Genet 2020;11:235.

[45] Liu K, Yan Z, Li Y, Sun Z. Linc2GO: a human LincRNA function annotation resource based on ceRNA hypothesis. Bioinformatics 2013;29(17):2221–2.

[46] Wang P, Li X, Gao Y, Guo Q, Ning S, Zhang Y, et al. LnCeVar: a comprehensive database of genomic variations that disturb ceRNA network regulation. Nucleic Acids Res 2020;48(D1):D111–17.

[47] Guo X, Gao L, Liao Q, Xiao H, Ma X, Yang X, et al. Long non-coding RNAs function annotation: a global prediction method based on bi-colored networks. Nucleic Acids Res 2013;41(2):e35.

[48] Jiang Q, Ma R, Wang J, Wu X, Jin S, Peng J, et al. LncRNA2Function: a comprehensive resource for functional investigation of human lncRNAs based on RNA-seq data. BMC Genom 2015;16(3):1–11.

[49] Han J, Liu S, Sun Z, Zhang Y, Zhang F, Zhang C, et al. LncRNAs2Pathways: identifying the pathways influenced by a set of lncRNAs of interest based on a global network propagation method. Sci Rep 2017;7(1):1–4.

[50] Quek XC, Thomson DW, Maag JL, Bartonicek N, Signal B, Clark MB, et al. lncRNAdb v2. 0: expanding the reference database for functional long noncoding RNAs. Nucleic Acids Res 2015;43(D1):D168–73.

[51] Park C, Yu N, Choi I, Kim W, Lee S. lncRNAtor: a comprehensive resource for functional investigation of long non-coding RNAs. Bioinformatics 2014;30(17):2480–5.

[52] Chen J, Zhang J, Gao Y, Li Y, Feng C, Song C, et al. LncSEA: a platform for long non-coding RNA related sets and enrichment analysis. Nucleic Acids Res 2021;49(D1):D969–80.

[53] Bendík J, Černá I. MUST: minimal unsatisfiable subsets enumeration tool. In: International conference on tools and algorithms for the construction and analysis of systems, Cham: Springer; 2020. p. 135–52.

[54] Zhang Y, Bu D, Huo P, Wang Z, Rong H, Li Y, et al. ncFANs v2. 0: an integrative platform for functional annotation of noncoding RNAs. Nucleic Acids Res 2021;49(W1):W459–68.

[55] Zhao L, Wang J, Li Y, Song T, Wu Y, Fang S, et al. NONCODEV6: an updated database dedicated to long noncoding RNA annotation in both animals and plants. Nucleic Acids Res 2021;49(D1):D165–71.

[56] Gellert P, Ponomareva Y, Braun T, Uchida S. Noncoder: a web interface for exon array-based detection of long non-coding RNAs. Nucleic Acids Res 2013;41(1):e20.

[57] Li H, Xie M, Wang Y, Yang L, Xie Z, Wang H. riboCIRC: a comprehensive database of translatable circRNAs. Genome Biol 2021;22(1):1.

[58] Kyrollos DG, Reid B, Dick K, Green JR. RPmirDIP: reciprocal perspective improves miRNA targeting prediction. Sci Rep 2020;10(1):1–3.

[59] Hoffmann M, Pachl E, Hartung M, Stiegler V, Baumbach J, Schulz MH, et al. SPONGEdb: a pan-cancer resource for competing endogenous RNA interactions. NAR Cancer. 2021;3(1):zcaa042.

[60] Li JH, Liu S, Zhou H, Qu LH, Yang JH. starBase v2. 0: decoding miRNA-ceRNA, miRNA-ncRNA and protein–RNA interaction networks from large-scale CLIP-Seq data. Nucleic Acids Res 2014;42(D1):D92–7.

[61] Li Y, Jin X, Wang Z, Li L, Chen H, Lin X, et al. Systematic review of computational methods for identifying miRNA-mediated RNA-RNA crosstalk. Brief Bioinform 2019;20(4):1193–204.

[62] Xu J, Li Y, Lu J, Pan T, Ding N, Wang Z, et al. The mRNA related ceRNA–ceRNA landscape and significance across 20 major cancer types. Nucleic Acids Res 2015;43(17):8169–82.

[63] Leong HS, Kipling D. Text-based over-representation analysis of microarray gene lists with annotation bias. Nucleic Acids Res 2009;37(11):e79.

[64] Korthauer K, Kimes PK, Duvallet C, Reyes A, Subramanian A, Teng M, et al. A practical guide to methods controlling false discoveries in computational biology. Genome Biol 2019;20(1):1–21.

[65] Martirosyan A, Marsili M, De Martino A. Translating ceRNA susceptibilities into correlation functions. Biophys J 2017;113(1):206–13.

[66] Ray R, Pandey P. Surveying computational algorithms for identification of miRNA–mRNA regulatory modules. Nucleus 2017;60(2):165–74.

[67] Hornakova A, List M, Vreeken J, Schulz MH. JAMI: fast computation of conditional mutual information for ceRNA network analysis. Bioinformatics. 2018;34(17):3050–1.

SECTION IV

Model Organisms of Epigenetics

12. Epigenetic mechanisms in bacteria bridge physiology, growth, and host-pathogen interactions
13. *Drosophila* epigenetics
14. Mouse models of epigenetic inheritance: classification, mechanisms, and experimental strategies
15. Plant epigenomics

CHAPTER 12

Epigenetic Mechanisms in Bacteria Bridge Physiology, Growth and Host—Pathogen Interactions

Maria Miah[1], Mihaly Mezei[2] and Shiraz Mujtaba[1]

[1]Department of Biology, Medgar Evers College, City University of New York, Brooklyn, NY, United States
[2]Department of Pharmaceutical Sciences, Icahn School of Medicine at Mount Sinai, New York, NY, United States

OUTLINE

Introduction	201	Conclusions and Future Perspectives	209
Protein Phosphorylation in Prokaryotes	201	Summary	210
Protein Acetylation in Bacteria	*203*	Protocols for Molecular Modeling	210
DNA Methylation in Bacteria	204	References	211
Protein Methylation	207	Further Reading	213
Crosstalks Between Epigenetic Modifications	209		

INTRODUCTION

Epigenetic mechanisms regulate the cellular processes by mediating DNA methylation and protein modifications. Subsequently, these genomic and proteomic changes modulate gene activities without changing the target DNA sequence [1,2]. Given that the complexity of the human genome is less understood, it has been reported to contain 20,000—25,000 protein-encoding genes only. Comparatively, the human proteome can mediate myriad of functions during development, infection, cell division, and metabolism. The chemical alterations facilitated by posttranslational modifications (PTMs) and protein—protein interactions provide a multi-tiered regulatory capacity to the biological processes. PTMs allow covalent attachments of various functional groups which serve to regulate many cellular processes. Further, mutually exclusive modifications on a single amino acid residue, reversible in nature, modulate the cellular response to external and internal changes. Together, these dynamic modifications regulate vital cellular processes, including growth, development, cell signaling, and even modulate responses to stress and external stimuli. Some exhaustively studied PTMs include methylation, acetylation, and phosphorylation of proteins. While epigenetic mechanisms are well understood in eukaryotes, the advent of new genomic and proteomic platforms has unraveled these mechanisms in prokaryotes, which are the focus of this chapter.

PROTEIN PHOSPHORYLATION IN PROKARYOTES

The reversible protein phosphorylation plays a pivotal role in modulating cellular functions, including metabolism, cell cycle, DNA repair, and apoptosis [3,4]. Despite its ubiquitous nature, mainly three amino acids in eukaryotes are phosphorylation targets, namely, serine, threonine, and tyrosine. The kinases form one large family of enzymes that catalyzes phosphorylation. The kinase activity is reversed by phosphatases, which belong to three distinct families: phosphoprotein phosphatase, metal-dependent protein phosphatase, and protein-tyrosine phosphatase [5–8]. The process of phosphorylation and dephosphorylation acts as a molecular switch regulating several cellular processes during stress and development [9–13]. Due to the regulatory effects of protein phosphorylation on cell cycle progression, the active site of kinases is targeted to treat tumors [14–16].

Numerous bacterial processes are dependent on protein phosphorylation, including but not limited to the two-component system (TCS), cell-wall biosynthesis, and biofilm formation [17,18]. In bacteria, a wide range of amino acid residues can undergo phosphorylation, including histidine, arginine, lysine, aspartate, and cysteine (Table 12.1). These protein kinases have been classified into five different groups. These groups are sensor histidine kinases (SHK), Tyr kinases (BYK), ARG kinases, Hanks-type Ser/Thr kinases (STKs), and atypical Ser kinases [17,18]. The TCS has been identified as one of the major processes that prokaryotes can communicate with their environment [19]. The TCS serve as important mediators of signal transduction that allow bacteria and lower eukaryotes to detect physicochemical changes. Following detection, the signal is transmitted through the cytoplasm to the bacterial nucleoid, which ultimately modulates the expression of downstream target genes. Typically, TCS consists of a transmembrane-bound receptor SHK and a DNA-binding response regulator (RR) [19]. The SHKs are homodimer transmembrane proteins that consist of a phosphotransferase domain and an ATP-binding domain [20,21]. A change in the extracellular environment induces phosphorylation of SHK, which subsequently triggers the transfer of phosphate moiety from the ATP-binding domain to a specific histidine residue (Fig. 12.1) [20,21]. Later, the RR facilitates the transfer of the phosphoryl group to an aspartate residue. After the RR is phosphorylated, a conformational change leads to cellular response. Typically, this is mediated by either stimulating or repressing the expression of specific genes [20,21].

One of the well-studied histidine kinases is chemotaxis protein A (CheA) which plays a crucial role in regulating chemotaxis [22–24]. After phosphorylation, CheA donates its phosphoryl group from histidine to the aspartate residue on the RR. Following this, the phosphoryl group is quickly transferred to chemotaxis protein Y (CheY), which is directly involved in flagellar movement [25–27]. In *Escherichia coli*, the phosphorylated CheY binds to the motor protein FliM at the cytoplasmic face, which induces a clockwise rotation of the flagella. After dephosphorylation, decreased phosphorylation levels of CheY results in the counterclockwise rotation of the flagella [25–27]. In the case of *E. coli*, the TCS facilitates chemotaxis, but in the case of *Streptococcus suis*, the TCS enhances pathogenicity and the ability to respond to varying levels of stress. Over 15 different groups of TCSs have been predicted in *S. suis*, which are speculated to enhance the virulence of the bacterium. The vancomycin-resistance-associated sensor/regulator system has been identified for its role in *S. suis*

TABLE 12.1 Protein Phosphorylation Residues: Eukaryotes and Prokaryotes Are Both Capable of Undergoing Posttranslational Modification of Protein Phosphorylation, the Covalent Addition of a Phosphate Group

Target Residues for Protein Phosphorylation	Eukaryotes	Prokaryotes
Arginine		✓
Aspartate		✓
Cysteine		✓
Histidine		✓
Serine	✓	✓
Thereonine	✓	✓
Tyrosine	✓	✓

This table highlights the residues which are targets of protein phosphorylation. Prokaryotes carry a wider range of residues subject to the posttranslational modification of protein phosphorylation.

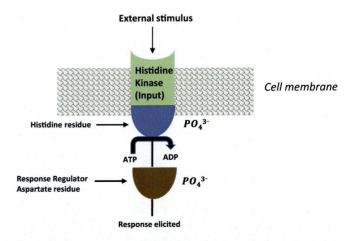

FIGURE 12.1 Histidine kinase contains two specific domains involved in facilitating the two-component system. When an external stimulus is detected by the input domain, the histidine residue is then autophosphorylated. Subsequently, the phosphoryl group is then transferred onto the response regulator, aspartate, which elicit a response.

developing multidrug resistance to hydrogen peroxide, survival in human blood and much more.

Mycobacterium tuberculosis possesses 11 different STKs [28–30]. For instance, *pknA* and *pknB* regulate the synthesis of the cell wall and shape [28–30]. In addition to cell wall and shape, protein kinase B (PknB) has been identified for its functional role in the formation of cellular envelope. Previous studies suggest that the overexpression of these kinases shrinks the cellular morphology [28–30]. Conversely, reduced expression of these kinases induces elongated shape. Besides, phosphorylation of FtsZ in *M. tuberculosis* comprises the septum formation [28–30]. Although the specific mechanism is still elusive, it is speculated that the GTP-binding site is affected upon phosphorylation [28–30].

GlmU synthesizes an integral precursor to peptidoglycan, uridine diphosphate (UDP)-N-acetyl-glucosamine [28–30]. PknB possesses partial control over the acetyltransferase activity of GlmU through the phosphorylation of the threonine residues at the C-terminal domain. CwlM, an additional PknB substrate, has been characterized for its peptidoglycan synthesis through nutrient dependence. The UDP-N-acetyl-glucosamine 1-carboxyvinyltransferase MurA, is the first enzyme in the peptidoglycan synthesis pathway. Furthermore, it is only activated once CwlM is phosphorylated. Bacteria lacking adequate nutrients fail to possess phosphorylated CwlM, resulting in the diminished activity of MurA. A consequence of this is decreased peptidoglycan synthesis and a newly developed tolerance to several antibiotics [28–30]. Taken together, these studies underscore the role of phosphorylation in bacterial physiology and cellular response.

Protein Acetylation in Bacteria

Protein acetylation results from the enzyme-mediated covalent linkage of the acetyl moiety that is transferred from the acetyl-CoA to the ε-sidechain of the amino acid lysine by the KATs (lysine acetyltransferases) [31]. Subsequently, the acetylated lysine moieties serve as a hub for the recruitment of chromatin-associated factors that trigger chromatin remodeling and molecular interactions, which serve as cues for distinct biological outcomes and cell fate decisions [32,33]. Genome-wide ChIP-on-CHIP data revealed that most KATs function as a coactivator, which act in tandem with transcription factors and RNA polymerase II for the activation of downstream target genes [34]. Further, site-specific lysine acetylation on the N-terminal tail of histone proteins neutralize the positive charge of amino acid lysine, which induces chromatin remodeling to facilitate the access of DNA by transcriptional machinery [31,35–37]. Besides chromatin, KAT-mediated acetylation upregulates the functions of transcription factors by enhancing its stability and triggering protein–protein interactions thereby stimulating the activities of target genes. For quite some time, the mechanisms, and physiological enhancements of PTMs were largely isolated to eukaryotes. However, recent progress in the genomic and proteomic technologies has revealed a variety of PTMs also existing in prokaryotic systems. Furthermore, although the specific roles are still elusive, certain physiological adaptations have been identified including altered protein activity and increased stability, metabolic regulation, signal transduction and virulence. Protein acetylation mechanistically is a nucleophilic acyl substitution that involves the transfer of an acetyl group from acetyl coenzyme A (AcCoA) onto the amino terminus of a protein (N_α) or the ε-amino-terminal of a lysine residue (N_ε) [38–40]. Contrary to the initial notion of acetylation being a eukaryotic phenomenon, recent studies suggest that N_ε acetylation extends to the bacterial world. This reaction in bacteria, is also facilitated by KATs, which catalyze the transfer of the acetyl group from AcCoA to the target lysine residue (Fig. 12.2) [38–40]. In bacteria, the two common acetyltransferases are PAT/YfiQ. In the nonenzymatic process, acetyl-phosphate (AcP) is used as the acetyl donor [41].

In *Salmonella*, a gram-negative facultative pathogen, acetylation results in decreased protein activity and DNA binding [42]. PhoP/PhoQ is a well-known TCS that undergoes acetylation and subsequent deacetylation. It has recently been discovered that when *Salmonella* is exposed to extreme stress in the external environment, there is a profound decrease in the frequency of site-specific acetylation at lysine 201 of

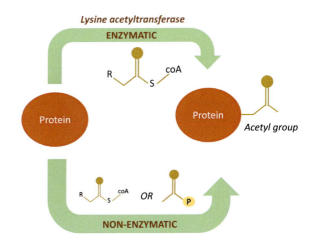

FIGURE 12.2 Enzymatic protein acetylation uses the enzyme lysine acetyltransferase and acetyl-CoA as the acetyl donor. Alternatively, nonenzymatic protein acetylation can proceed with either acetyl-CoA or acetyl-phosphate.

PhoP. Furthermore, if the PAT gene was deleted, then PhoP binding to the DNA was enhanced. Interestingly, when lysine was substituted with glutamine, there was a significant decline in pathogenicity. These findings suggest that lysine acetylation of PhoP, inversely affects the pathogenicity of Salmonella.

In yet another example, lysine acetylation negatively affects the function of topoisomerase 1 in E. coli [43]. Topoisomerase 1, encoded by topA, regulates DNA supercoiling [43]. While acetylation decreases the ability of topoisomerase 1 for binding to DNA, the deacetylase CobB reverses both enzymatic and nonenzymatic lysine acetylation [43]. To relieve supercoiling, topoisomerase 1 must be able to bind to DNA but when acetylation occurs, this activity is inhibited. In E. coli the acetylation of the CheY protein has been associated with facilitating chemotaxis speed [44–52]. It was observed that mutants lacking the enzyme acetyl-CoA synthetase (Acs), required significantly more time to respond to external stimuli. As mentioned previously, chemotaxis is also facilitated by phosphorylation. Phosphorylation mediates initial binding to the motor switch whereas acetylation further enhances the bindings. Accelerated adaptation has been found to be related to CheY acetylation [44–52]. When a mutant lacking the acetylating enzyme Acs was compared to a wild-type parent, which did contain the enzyme, it was found that the mutant required significantly longer time to adapt to an attractant or repellent. Furthermore, overexpression of Acs significantly shortened adaptation time, making it even faster than the wild-type parent. This suggests that protein acetylation of CheY affects overall adaptation time in E. coli chemotaxis. When Acs is lacking in the system, the inhibitory effects can be avoided by fostering conditioning that favor the production of acetyl-CoA which allows for Acs-independent CheY autoacetylation [44–52].

Acetylation impacts bacterial acid tolerance as noted in many pathogens that require optimal conditions of low pH to continue invading their hosts. As such, the ability to sense and adequately respond to an acidic environment is critical for pathogenesis and Salmonella is an excellent example. CyaA and crp code for adenylate cyclase and cyclic AMP receptor protein, respectively on PAT and function as positive regulator [53]. During incidents of acid stress, PAT is downregulated. However, the PAT deletion mutant displays a significantly higher survival rate under acid stress and maintains a stable intracellular pH. The implication is that acetylation contributes to the bacterial response during times of acid stress and may also contribute to acid resistance.

HilD has for long been identified as a critical transcriptional regulator for SP-1 in Salmonella [54]. It is noted that without the acetylation of HilD, SP-1 cannot be activated [55]. Acetylation of K297 stabilizes HilD but decreases its affinity to bind to DNA. When mouse models were compared, the acetylated K297 experienced impaired virulence. This suggests that deacetylation of K297 is integral to facilitate Salmonella virulence [42,56].

Peptidoglycan plays a crucial role in providing durability to bacterial cells. Evidence suggests that the peptidoglycan of both gram-positive and -negative bacteria are subjected to O-acetylation [57,58]. Lysozymes are naturally occurring antimicrobials that cleave peptidoglycan, resulting in bacterial death. The bacteria, Staphylococcus aureus, O-acetylates its peptidoglycan and thus are resistant to lysozymes [57,58]. Conversely, bacteria that do not exhibit O-acetylation of their peptidoglycan, are sensitive to lysozymes such as E. coli and Bacillus anthracis. Collectively, acetylation in bacteria plays multiple roles, like in eukaryotes. However, availability of tools such as highly specific antibodies for genome-wide CHIP assays will directly establish acetylation to regulating gene expression in bacteria.

DNA METHYLATION IN BACTERIA

DNA methylation serves as a pre-transcriptional epigenetic mechanism for switching on and off the expression of downstream genes [59–61]. CpG islands are the dense concentrations of cytosine and guanine base pairs [60]. In these pairs, cytosine is modified by addition of methyl group (-CH3) (Fig. 12.3) [62,63]. While the methylated patches regulate gene silencing, the unmethylated areas are poised for gene expression. DNA methylation is facilitated by a family of enzymes known as DNA methyltransferases (DNMTs), which need S-adenosyl-L-methionine (SAM) as the methyl donor for transferring the methyl group to either an adenine or cytosine. DNMTs are substrate specific as an adenine methylation generates N-6-methyladenine (6 mA), also known to possess greater activity in prokaryotes. C-5-methylcytosine (5mC), is significantly more common in eukaryotes [64]. In addition to 6 mA, N-4-methylcytosine (4mC) methylation has also been identified in bacterial DNA (Table 12.2) [64]. In bacteria, DNA methylation has mostly been associated with the sequence-specific restriction and modification system (RM). While initially it was presumed that almost all DNMTs are associated with the RM system, increasing evidence suggests that orphan methyltransferases are also very common throughout bacteria. DNA adenine methyltransferase (DAM) falls within this category and is one of the most widely studied enzymes contributing to epigenetic modifications within bacteria [65–73].

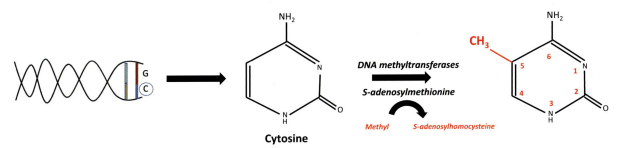

FIGURE 12.3 A DNA methyltransferase catalyzes the transfer of a methyl group onto the fifth carbon of the cytosine nucleotide. S-adenosylmethionine (SAM) is used as the methyl donor during the reaction and is converted to S-andenosyl-L-homocysteine upon loss of the methyl group via a DNA methyltransferase.

TABLE 12.2 DNA Methylation in Eukaryotes and Prokaryotes

Mtases	Eukaryotes	Prokaryotes
6mA	✓	✓*
5mC	✓*	✓
4mC		✓

Three different forms of DNA methyltransferases exist in bacterial systems, 6mA, 5mC, and 4mC. Of the three, only 6mA and 5mC exist in eukaryotes. Note: "*" Indicates this form of methylation is most prevalent in that specific system.

The K12 E. coli strain has approximately 130 DAM enzymes. Mechanistically, DAMs facilitate the transfer of a methyl group from SAM to the N6 position of adenine in a guanine adenine thymine cytosine (GATC) sequence. Further analysis reveals that DAM contains two binding sites for SAM, one is the primary catalytic site and the other is the allosteric site. In a hemi-methylated DNA one of the strands has the methylated moiety, but the site on the complementary strand, does not. Once DAM is bound to the template, roughly 55 GATCs positions are methylated simply from one binding event [68,74].

DAM has frequently been associated with correcting errors during DNA replication through methyl-directed mismatch repair (MMR) [70,74–78]. The MMR complex in E. coli requires key components for functionality and of them, three are critical to the initiation of MMR process, specifically, MutS (MutatorS), MutL (MutatorL) and MutH (MutatorH). MutH, belongs to the family of type-II restriction endonucleases and recognizes hemi-methylated GATC sequences [76,79]. MutS has been coined as the "mismatch recognition protein," for its ability to recognize base-base mismatches, alongside insertion and deletion mispairs [70,74–78]. MutL serves as an enhancement protein and physically binds with MutS, which then allows for the recruitment and activation of MutH. During replication, errors are frequent and typically, the two proteins MutH and SeqA will arrive before DAM and attempt repairs and adjust supercoiling, respectively. SeqA protein preferentially attaches to hemi-methylated DNA [70,74–78]. DNA methylation serves a vital role in assisting the cells to determine which strand is the template strand. When a DNA strand is methylated, following replication, only one strand will be methylated, hence, hemi-methylated. The daughter strand however will be unmethylated. In E. coli, the hemi-methylated region of DNA presents itself immediately behind the replication fork, where the parent strand is methylated at the N6 position of adenine in GATC sequences, but the complementary daughter strand remains unmethylated. This variation is how the parent strand is discriminated from the newly synthesized daughter strand. A cascade of events follows to successfully complete mismatch repair. In the presence of ATP, the protein MutS will bind to a mismatch on DNA which allows for the recruitment of MutL. Protein MutH endonuclease will then bind to the hemi-methylated GATC sites and once activated, it proceeds to cleave the unmethylated daughter strand of hemi-methylated dGATC [70,74–78].

Increased rates of mutation are observed when MutH and SeqA are prevented from carrying out their respective functions. This is frequently seen when Dam numbers are increased and as a result, there is a decrease in hemi-methylated DNA. With little hemi-methylated DNA, the two proteins are unable to carry out their roles. Interestingly, both, an underproduction and an overproduction of DAM contribute to the same consequence, elevated levels of mutation frequency. When there is too much of DAM then premature DNA becomes methylated. The alternative, too little DAM results in the inability to determine which strand is the parent strand and which is the template strand. In this event, there's a likelihood that the incorrect strand be used for mismatch repair [75].

Deinococcus radiodurans serves as a model organism for the examination of DNA repair and genomic stability because of its abundant 4mC levels. For a long time, *D. radiodurans* was understood to lack DNA methylation. However, it was discovered that *D. radiodurans* displays 6 mA abilities [67,80–82]. Notably, despite exposure to high doses of gamma radiation, *D. radiodurans* maintains the integrity of its outer layer as well as rescue its genome simply within a few hours. A novel 4mC modification was recently identified in *D. radiodurans* and its presence was owed entirely to the *M.DraR1* gene. *M.DraR1* was understood to methylate and prevent the target region from being excised by the restriction endonuclease. Further, while examining bacteria lacking *M.DraR1*, it was established that they exhibit differential genomic expression associated with a lack of MTase. This included an inability to repair DNA along with transcription and nucleotide processing. These findings suggest that the 4mC is critical to *D. radiodurans* sustainability. In order to properly establish whether *M.DraR1* functioned with 4mC or 6 mA, a knockout strain was created with a deleted *M.DraR1* gene. When compared to their wild-type counterparts, it was found that the removal of *M.DraR1* gene reduced the 4mC ratio but led to little change on the levels of 6 mA. In order to discover the motif that *M.DraR1* specifically methylates, purification techniques were employed and various oligonucleotides were produced through annealing methods. It was revealed that *M.DraR1* is bound to "CCCGCGG" and used the traditional SAM as the methyl donor [67,80–82].

DNA methylation is an epigenetic modification critical to the progression of the cell cycle in *Caulobacter*, along with other alphaproteobacteria. CcrM is a DNA adenine orphan type II DNA methyltransferases, which mediates cell cycle-regulated DNA methyltransferase. CcrM functions similarly to DAM, in that, it transfers a methyl group from SAM to the N6 of an adenine on a 5'-GANTC-3' recognition sites of hemi-methylated DNA. CcrM along with the proteins, DnaA, GcrA, and CtrA, are involved in the expression of more than 200 genes [66,83].

With every division, *Caulobacter* will replicate its circular chromosome once and the product will be two new cells with differences in functionality and shape. Unique to Alphaproteobacteria is their highly specific and asymmetrical cellular development. Ultimately, the two-progeny generated are, namely: stalked and swarmer cells. Further, both possess differences in their phenotypes along with their gene regulation. The stalked cell can replicate its chromosome. The progeny swarmer cell (G1 phase) is unable to initiate chromosome replication until it differentiates into a stalked cell. However, the progeny stalked cell immediately initiates chromosome replication and enters S phase [83,84]. Towards the late S phase, a spike in the CcrM protein converts the hemi-methylated chromosomes into methylated chromosomes. Thus, the methylation states of these GATC motifs largely contribute to the successful transcription of the genes critical to the cell cycle. As previously mentioned, these genes include DnaA, GcrA and CtrA. DnaA is regarded as the initiator of chromosomal replication and facilitates the transcription of almost 50 other genes that are also regulated by the cell-cycle. More specifically, DnaA promotes the unwinding of DNA at origin of replication. Essentially, without the presence of ample amounts of DnaA protein, the initiation phase of DNA replication cannot proceed. For *dnaA* to be transcribed, the GANTC site strictly within the promoter region is required to be fully methylated. Following this, when replication is initiated, as the process makes its way through the replication fork, *dnaA* promoter is gradually converted from its methylated state to a Hemi-methylated state. As expected, this decreases the transcription of *dnaA* [66,83].

Furthermore, a second critical gene, *ctrA*, also changes from its methylated state to a hemi-methylated state that leads to the promoter activation. Two specific promoters are involved in the transcription of *ctrA* and one of them, ctrAP1, is critically regulated by DNA methylation. The ctrAP promoter is only activated hemi-methylation that happens once the replication fork slides through it. During the presence of methylated targets, GcrA is then able to recruit and stabilize RNA polymerase and serves to form a transcription bubble near the ctrAP1 [71,84,85]. Functionally, DnaA and CtrA possess antagonistic roles where the former controls the initiation of DNA replication allows for the transcription of the *dnaA* only when the promoter is fully methylated. Conversely, CtrA inhibits initiation of DNA replication and transcription is activated only when the promoter is hemi-methylated. CcrM has been found to be present only during a very specific time and for a short duration during the progression of the cell cycle. The duration is meant to overlap with their transcription and translation times. Overexpression of CcrM leads to misregulation of a portion of cell cycle-controlled genes. Therefore, limiting the concentration of CcrM to a specific and short window, is essential for controlling cell cycle progression and fundamental to avoiding errors [71,85].

Helicobacter pylori is a bacterium that is characterized for its ability to cause gastric cancer in humans [86–88]. *H pylori* strain 26695 possesses a 4mC methyltransferase, M2.HpyAII that recognizes the 5'TCTTC 3' sequence and methylates the very first cytosine [69,87,89]. To analyze the role of the 4mC modification in this strain, an *M2.HpyAII* deletion strain was generated. This deletion strain revealed reduced binding to the host cells that decreased inflammation and apoptosis. Further, *H. pylori* typically induced apoptosis by inhibiting the G1 to S phase in the host cells. However, this was not seen in high frequency in the presence of the deletion strain. With the 4mC modification lost, roughly 102 genes coding for virulence and ribosome assembly in the strain experienced differential expression. Furthermore, the deletion strain also exhibited reduced transformation capacity and struggled to uptake both point-mutated DNA and larger fragments These findings suggest that 4mC modification is a key regulator in many pathogenic processes of *H. pylori* [69,87,89].

A phasevarion, also called phase-variable regulon, facilitates expression of several genes and proteins through the phase variation of a single DNMT [90–96]. This allows for various methylation sites throughout the bacterial genome and in turn, epigenetic mechanisms establish a diverse variety of genomic expression. This diversity allows survival in a certain environment, thereby increasing its longevity. Phase variation is crucial to the bacterial genome because it allows for changes in pathogenesis, and a means to rapidly adapt to changes in the environment. This rapid response is attributed to the ON-OFF switching of multiple allelic variants. Through global changes in methylation, phase-variable methyltransferases can contribute to pathogenicity while also making room for other epigenetic phenomena [90–96].

One of the earlier bacterial pathogens studied for existing phasevarions include, *Haemophilus influenzae* [91,92,97–100]. NTH1 strain of *H. influenzae* lacks a polysaccharide capsule. It is known for being one of the main causes of chronic and recurrent otitis media in children worldwide [100]. NTH1 has been identified for having only one phase variable methyltransferase, ModA, which has 21 specific alleles [91,92,97–100]. An experiment was done to analyze the different strains for NTHI from a group of healthy children and children with otitis media. Furthermore, it was revealed that five of the phase-variable alleles, respectively: modA2, modA4, modA5, ModA9, and modA10, were components of at least two thirds of the isolates. Previously it had been shown that the modA2 ON status may occur within the middle ear during experimental disease in a chinchilla model of otitis media. In that study, inoculated NTH1 that was ON, remained in the ON position. Conversely, when inoculated OFF, there was a constant switch from ON to OFF inside the middle ear. This suggests that in the middle ear, the ON position is more likely to survive than the OFF-phase variable [91,92,97–100].

PROTEIN METHYLATION

Protein methylation is catalyzed by a group of enzymes known as protein methyltransferases, which methylate histone and nonhistone proteins [101–103]. Enzymatically, a methyl group is transferred from the S-5′-adnosyl-L-methionine (SAM) to the lysine or arginine residues of a histone or a nonhistone substrate [101–103]. The two classes of methyltransferases are lysine methyltransferases (KMTs) and protein arginine methyltransferases (RMTs) [104–106]. The lysine residue can undergo mono-, di-, and trimethylation, whereas the arginine residue can be mono- or dimethylated (Fig. 12.4). SET domains within the methyltransferases that are comprised of 130–140 amino acids catalyze methylation [107].

Recent evidence suggests that protein methylation may have a role in modulating bacterial virulence. In bacteria, lysine, aspartate, glutamate, and glutamine are the targets of methylation [108]. The outer membrane proteins (Omps) facilitate cell-cell interaction and cell signaling [109]. The Omps are subjected to PTMs, including protein methylation [109]. The outer membrane protein precursor (OmpB) from the virulent strains of *R. prowazekii* were compared to bacteria lacking virulence, which revealed that the virulent strains possess significantly more methylated-lysine residues [109]. Additionally, OmpB plays an important role in organizing the intracellular movement within the cell. To further support the notion that methylated-lysine residues contribute to *Rickettsia prowazekii* virulence, the gene which codes for lysine methyltransferase, *Rp028/027*, was thoroughly examined [110,111]. When

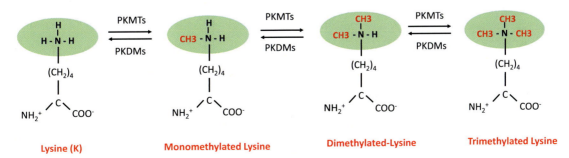

FIGURE 12.4 This diagram reflects histone modification on a lysine residue. Protein lysine methyltransferases can facilitate the addition of one, two, or three methyl groups onto the accepting residue. Protein lysine methyltransferases catalyze the forward reaction using S-adenosyl-l-methionine (SAM) as the methyl donor and the reverse reaction is catalyzed by protein lysine demethyltransferases.

compared to their avirulent counterparts, it appeared that the avirulent organisms, *R. prowazekii* Madrid E strain, had a frameshift mutation in *Rp028/027* gene. Interestingly, the virulent strain BreinL did not reveal similar type of methylation. Bechah et al. examined four separate *R. prowazekii*, all four of which differed in their levels of virulence [112]. The findings revealed that the spectrum of PTMs on surface proteins contribute to bacterial virulence. Similarly, specifically N-methylation in the *Leptospira* species influence virulence. The LipL32 has been identified as an Omp and when the urine of infected rats was analyzed, it was revealed that LipL32 was trimethylated at many of the lysine residues [113]. Together, these data support the crucial role of methylation in regulating the bacterial virulence.

In the pathogenesis of *Bacillus anthracis* toxins play a pivotal role by overwhelming the host's defenses. Nevertheless, in addition to toxins, *B. anthracis* expresses small modular proteins, whose functions in pathogenesis are unclear. One of such factors is suppressor-of-variegation, enhancer-of-zeste, trithorax protein in *B. anthracis* (*Ba*SET) methylates human histone H1, resulting in repression of NF-κB functions. Notably, *Ba*SET is secreted and undergoes nuclear translocation to enhance H1 methylation in *B. anthracis*-infected macrophages. Collectively, *Ba*SET is required for repression of host transcription as well as proper *B. anthracis* growth, making it a potentially unique virulence determinant [114].

We examined the structure of several methyl transferases from various organisms, stretching from viral to human: *E. coli* (3tka), *Methanosarcana antivorans* (6mro), *Methanosarcana mazei* (3mgg), *Paramecium bursaria chlorella* virus X-ray and NMR (3kmj, 2g46, resp.), *Micromonospora carbonacea* (5t39), human (4mi5). While they show significant structural variability, three of them (2g46, 3kmj, 4mi5) are similar enough that structural alignment could delineate secondary structure elements that are conserved. Fig. 12.5 shows the structural alignment (done with the program Pymol [115,116]) of the three similar structures and the residue TYR105 (of 2g46) that is suggested to initiate the transfer process [117,118].

The parts that are conserved, a set of beta sheets and a loop, are forming a cage. As TYR105, a residue found critical for activity on the 2g46 structure, is also part of the cage, this indicates the location of the active site. Examination of the other structures discussed failed to find enough structural similarity to produce a meaningful structural alignment. However, all of them contained a beta sheet. Furthermore, in structures where the cysteine-containing cofactor is present, the relative position of the cofactor and the beta sheet suggests that the common structural element of the active sites of methyl transferases is the presence of a beta sheet.

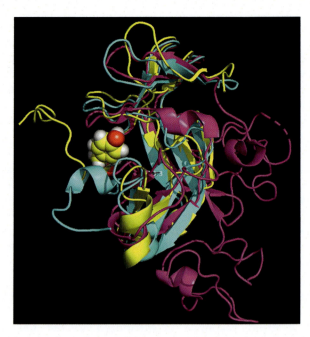

FIGURE 12.5 Structural alignment of the cartoon representation of human and viral methyltransferases. Viral (3kmj): light blue; human, X-ray (4mi5): pink; human, NMR (2g46): yellow. TYR105 of 2g46 is shown as VdW spheres.

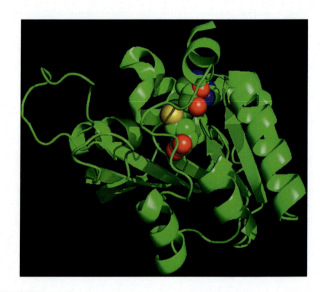

FIGURE 12.6 Cartoon representation of *Methanosarcana antivorans* (6mro) with the cofactor SAH shown as VdW spheres.

Fig. 12.6 shows the structure of *Methanosarcana antivorans* (6mro) with the cofactor SAH as an example.

In addition, we also generated homology models for the SET domain of *B. anthracis* (Genbank AJI40906.1) and *Bacillus cereus* (Genbank ALZ64087.1) based on the structure of the pea methyltransferase (PDB ID 2h23) using the program Modeller [119,120]. Fig. 12.7 shows the three structures as aligned by Modeller. The binding site residues (based on the PDB annotation) are

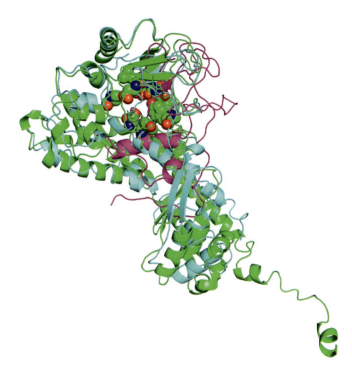

FIGURE 12.7 Cartoon representation of the target and model SET domain structures. Green: pea (PDB ID 2h23), red: *Bacillus anthracis* (Genbank AJI40906.1), light blue: *Bacillus cereus* (Genbank ALZ64087.1). The binding site is shown as VdW spheres.

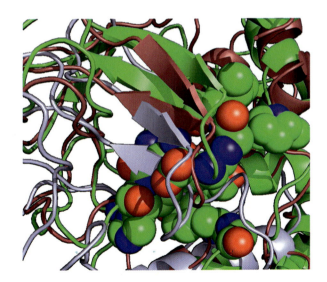

FIGURE 12.8 Close-up view of the aligned beta sheets (rotated by c. 180° from the view on Figure +3). Color scheme is the same as that of Figure +3.

shown as spheres. While *B. anthracis* has several secondary structure elements aligned to the template, *B. cereus* (a much smaller protein than either of the other two) shows little obvious similarity. All three, however, contain beta sheets aligned in the vicinity of the highlighted binding site residues, shown on Fig. 12.8, providing further support of our observation above of the putative importance of these beta sheets.

CROSSTALKS BETWEEN EPIGENETIC MODIFICATIONS

As epigenetic mechanisms are continued to be investigated in the bacterial kingdom, new epigenetic modification sites are being uncovered [17,18,121,122]. A few of the modifications common to bacteria and eukaryotes include protein phosphorylation, DNA methylation, protein acetylation, and protein methylation. Bacteria possess a wide range of residues as targets of various epigenetic modifications, allowing the functional adaptations to be endless. Although a single modification can provide plasticity to bacterial physiology and cellular response, there is evidence that the molecular interplay between two distinct modifications can cooperatively intertwine to achieve a desired functionality. As previously mentioned, the TCS is widely used by bacterial organisms to recognize and respond to stimulants and repellants. The CheY protein in *E. coli* has been identified as a bacterial system that undergoes both phosphorylation and acetylation to respond to environmental stimuli [45,123]. When CheY is phosphorylated by CheA, there is a reduction in their binding abilities. Conversely, CheY ability to bind FliM and phosphatase CheZ is increased [45,123]. Consequently, there is a clockwise rotation. Like phosphorylation, acetylation can also lead to an activated CheY, leading to a clockwise rotation. When CheY is phosphorylated, autoacetylation is inhibited [26,27,45,123]. However, in the event of dephosphorylation of CheY, autophosphorylation is promoted. Based on this, it appears that CheY is dually regulated by both acetylation and phosphorylation [26,27,45,123]. In a similar manner, there is evidence that interplay exists between the chemosensory systems that govern chemotaxis and biofilm formation in the organism, *Comamonas testosteroni* [124]. It was established that as many as 17 different chemoreceptors from *C. testosteroni* were also identified to have a role in the chemotaxis pathways [124]. Of the 17, three are expected to directly affect chemotaxis [124]. The suggestion is that Che pathways regulate chemotaxis and the Flm pathway regulates biofilm formation, but interestingly they can also influence each other [124].

CONCLUSIONS AND FUTURE PERSPECTIVES

A better understanding of the epigenetic mechanisms within the world of lower organisms could lead to novel therapeutic modalities. Delineating the mechanistic determinants of pathways emerging from various epigenetic modifications will present possible targets to inhibit or promote specific responses. As more and more pathogens are discovered, the world of emerging infectious diseases grows progressively larger. Yet, many pathways in the bacterial system are still poorly understood due to a lack of investigations. The means to further analyze the intricate pathways in their biology can be facilitated with the development of novel proteomic devices.

Phosphorylation is widespread in regulating specific bacterial responses, and as such, it is speculated that phosphorylation controls bacterial pathogenesis. Despite this understanding, there is inadequate information supporting the mechanism. Kinases can often facilitate the growth and survival of pathogens, and if there is the ability to inhibit kinases, then there is a potential to reduce the survival rate. Further analysis of phosphorylation is necessary to develop novel antibacterial medications. Similarly, acetylation has been shown to reduce pathogenesis in multiple bacteria, whereas deacetylation has been shown to enhance virulence and binding to the host target. Targeting the enzymes that facilitate the addition and removal of acetyl groups, KATs and KDACs may also lead to potentially novel drug treatments.

The O-methyltransferases are characterized for their extreme substrate specificity. Despite their specificity, they also function as versatile biomolecules that can be used as intermediates in various biosynthetic processes [125]. Among these processes, some include antibiotic production, human catecholamine neurotransmitters and lastly, plant phenylpropanoids. Although there are many O-methyltransferases present in the database, few have been characterized with their functionality. However, knowing their respective functionalities paves the way to use them as biocatalysts in future research. Further research in this area will allow the development of novel therapeutics tools for mechanistic interventions to modulate disease-related functions.

SUMMARY

In the past two decades, our understanding of epigenetically mediated gene regulation has advanced considerably due to the development of genomic and proteomic tools. Accumulating studies have already established the role of epigenetic mechanisms in bacteria. This chapter comprehensively showcased the advancement in the knowledge of epigenetic mechanisms in controlling the bacterial life cycle. We have highlighted the roles of DNA methylation and acetylation, methylation, and phosphorylation of proteins. Our overall goal was to provide the reader with a growing understanding of the epigenetic mechanisms in bacteria.

PROTOCOLS FOR MOLECULAR MODELING

1. Analyze experimental structures.
   1. Download the structure from the Protein Data Bank (PDB) at https://www.rcsb.org.
   2. Display the structure with a molecular graphics program, such as VMD (free from https://www.ks.uiuc.edu) or Pymol (free education license or paid from https://pymol.org). These programs allow the user to measure distances, angles and use a variety of display formats.
   3. The program Simulaid, run interactively from the command line (free for academic use from https://mezeim01.u.hpc.mssm.edu/simulaid) can be used to perform a number of analyses such as list contacts between regions/domains, hydrogen bonds, salt bridges and separate external and internal waters.
2. Generate homology models.
   We used the program Modeller (free for academic use from https://salilab.org/modeller) to generate the homology models discussed in this paper. It is run via a Python script; below is an example of the script used in our work.

Python script running the creation of a homology model.

```
from modeller import *
# First read two sequences and align them
# Target sequence in the file bacA.pir in PIR format
# Template sequence extracted from the PDB file with PDBid 2h23
log.verbose()
env = environ()
env.io.atom_files_directory = "./"
aln = alignment(env, file = 'bacA.pir')
mdl = model(env)
code = '2h23'
mdl.read(file = code,   model_segment = ('FIRST:@', 'END:'))
aln.append_model(mdl, align_codes = code, atom_files = code)
aln.write(file = '2h23.ali')
aln.malign(gap_penalties_1d = (-600, -400))
aln.write(file = '2h23_tr.ali')
# Now generate the model(s) using the alignment just created and
# the structure of the template in the PDB file
from modeller.automodel import *
```

```
a = automodel(env, alnfile = '2h23_tr.ali',
knowns = '2h23', sequence = 'bacA')
a.starting_model = 1
a.ending_model = 5
a.make()
```

References

[1] Becker PB, Workman JL. Nucleosome remodeling and epigenetics. Cold Spring Harb Perspect Biol 2013;5(9):a017905.

[2] Goldberg AD, Allis CD, Bernstein E. Epigenetics: a landscape takes shape. Cell 2007;128(4):635–8.

[3] Bonasio R, Tu S, Reinberg D. Molecular signals of epigenetic states. Science 2010;330(6004):612–16.

[4] Patel DJ, Wang Z. Readout of epigenetic modifications. Annu Rev Biochem 2013;82:81–118.

[5] Cordeiro MH, Smith RJ, Saurin AT. A fine balancing act: a delicate kinase-phosphatase equilibrium that protects against chromosomal instability and cancer. Int J Biochem Cell Biol 2018;96:148–56.

[6] Shaltiel IA, et al. Distinct phosphatases antagonize the p53 response in different phases of the cell cycle. Proc Natl Acad Sci USA 2014;111(20):7313–18.

[7] Vallardi G, Cordeiro MH, Saurin AT. A kinase-phosphatase network that regulates kinetochore-microtubule attachments and the SAC. Prog Mol Subcell Biol 2017;56:457–84.

[8] Vallardi G, Saurin AT. Mitotic kinases and phosphatases cooperate to shape the right response. Cell Cycle 2015;14(6):795–6.

[9] Dhanushkodi NR, Mohankumar V, Raju R. Sindbis virus induced phosphorylation of IRF3 in human embryonic kidney cells is not dependent on mTOR. Innate Immun 2012;18(2):325–32.

[10] Dong Z, Bode AM. The role of histone H3 phosphorylation (Ser10 and Ser28) in cell growth and cell transformation. Mol Carcinog 2006;45(6):416–21.

[11] Dong Z, Qi X, Fidler IJ. Tyrosine phosphorylation of mitogen-activated protein kinases is necessary for activation of murine macrophages by natural and synthetic bacterial products. J Exp Med 1993;177(4):1071–7.

[12] Geraghty KM, et al. Regulation of multisite phosphorylation and 14-3-3 binding of AS160 in response to IGF-1, EGF, PMA and AICAR. Biochem J 2007;407(2):231–41.

[13] Gioeli D, et al. Androgen receptor phosphorylation. Regulation and identification of the phosphorylation sites. J Biol Chem 2002;277(32):29304–14.

[14] Akoulitchev S, Reinberg D. The molecular mechanism of mitotic inhibition of TFIIH is mediated by phosphorylation of CDK7. Genes Dev 1998;12(22):3541–50.

[15] Chen XG, et al. Rapamycin regulates Akt and ERK phosphorylation through mTORC1 and mTORC2 signaling pathways. Mol Carcinog 2010;49(6):603–10.

[16] Fenton AK, et al. Phosphorylation-dependent activation of the cell wall synthase PBP2a in Streptococcus pneumoniae by MacP. Proc Natl Acad Sci USA 2018;115(11):2812–17.

[17] Macek B, et al. Phosphoproteome analysis of E. coli reveals evolutionary conservation of bacterial Ser/Thr/Tyr phosphorylation. Mol Cell Proteomics 2008;7(2):299–307.

[18] Macek B, et al. The serine/threonine/tyrosine phosphoproteome of the model bacterium Bacillus subtilis. Mol Cell Proteomics 2007;6(4):697–707.

[19] Pruss BM. Involvement of two-component signaling on bacterial motility and biofilm development. J Bacteriol 2017;199:18.

[20] Muok AR, Briegel A, Crane BR. Regulation of the chemotaxis histidine kinase CheA: a structural perspective. Biochim Biophys Acta Biomembr 2020;1862(1):183030.

[21] Muok AR, et al. Atypical chemoreceptor arrays accommodate high membrane curvature. Nat Commun 2020;11(1):5763.

[22] Wang X, et al. The linker between the dimerization and catalytic domains of the CheA histidine kinase propagates changes in structure and dynamics that are important for enzymatic activity. Biochemistry 2014;53(5):855–61.

[23] Wang X, et al. CheA-receptor interaction sites in bacterial chemotaxis. J Mol Biol 2012;422(2):282–90.

[24] Wang X, et al. Computational and experimental analyses reveal the essential roles of interdomain linkers in the biological function of chemotaxis histidine kinase CheA. J Am Chem Soc 2012;134(39):16107–10.

[25] Lombard DB, et al. Mammalian Sir2 homolog SIRT3 regulates global mitochondrial lysine acetylation. Mol Cell Biol 2007;27(24):8807–14.

[26] Sarkar MK, Paul K, Blair D. Chemotaxis signaling protein CheY binds to the rotor protein FliN to control the direction of flagellar rotation in Escherichia coli. Proc Natl Acad Sci USA 2010;107(20):9370–5.

[27] Wheatley P, et al. Allosteric priming of E. coli CheY by the flagellar motor protein FliM. Biophys J 2020;119(6):1108–22.

[28] Anandan T, et al. Phosphorylation regulates mycobacterial proteasome. J Microbiol 2014;52(9):743–54.

[29] Mir M, et al. Mycobacterial gene cuvA is required for optimal nutrient utilization and virulence. Infect Immun 2014;82(10):4104–17.

[30] Prisic S, Husson RN. Mycobacterium tuberculosis Serine/Threonine Protein kinases. Microbiol Spectr 2014;2(5).

[31] Roth SY, Allis CD. Histone acetylation and chromatin assembly: a single escort, multiple dances? Cell 1996;87(1):5–8.

[32] Marmorstein R. Structural and chemical basis of histone acetylation. Novartis Found Symp 2004;259:78–98 discussion 98–101, 163-9.

[33] Marmorstein R, Zhou MM. Writers and readers of histone acetylation: structure, mechanism, and inhibition. Cold Spring Harb Perspect Biol 2014;6(7):a018762.

[34] Sandmann T, Jakobsen JS, Furlong EE. ChIP-on-chip protocol for genome-wide analysis of transcription factor binding in Drosophila melanogaster embryos. Nat Protoc 2006;1(6):2839–55.

[35] Edmondson DG, et al. In vivo functions of histone acetylation/deacetylation in Tup1p repression and Gcn5p activation. Cold Spring Harb Symp Quant Biol 1998;63:459–68.

[36] Munoz-Galvan S, et al. Histone H3K56 acetylation, Rad52, and non-DNA repair factors control double-strand break repair choice with the sister chromatid. PLoS Genet 2013;9(1):e1003237.

[37] Pelletier G, et al. Competitive recruitment of CBP and Rb-HDAC regulates UBF acetylation and ribosomal transcription. Mol Cell 2000;6(5):1059–66.

[38] Sheikh BN, Akhtar A. The many lives of KATs — detectors, integrators and modulators of the cellular environment. Nat Rev Genet 2019;20(1):7–23.

[39] O'Garro C, et al. The biological significance of targeting acetylation-mediated gene regulation for designing new mechanistic tools and potential therapeutics. Biomolecules 2021;11(3):455.

[40] Patel J, Pathak RR, Mujtaba S. The biology of lysine acetylation integrates transcriptional programming and metabolism. Nutr Metab (Lond) 2011;8:12.

[41] de Diego Puente T, et al. The protein acetyltransferase PatZ from Escherichia coli is regulated by autoacetylation-induced oligomerization. J Biol Chem 2015;290(38):23077–93.

[42] Ren J, et al. Acetylation of lysine 201 inhibits the DNA-binding ability of PhoP to regulate salmonella virulence. PLoS Pathog 2016;12(3):e1005458.

[43] Zhou Q, et al. Biochemical basis of E. coli topoisomerase I relaxation activity reduction by nonenzymatic lysine acetylation. Int J Mol Sci 2018;19(5):1439.

[44] AbouElfetouh A, et al. The E. coli sirtuin CobB shows no preference for enzymatic and nonenzymatic lysine acetylation substrate sites. Microbiologyopen 2015;4(1):66−83.

[45] Barak R, et al. Acetylation of the chemotaxis response regulator CheY by acetyl-CoA synthetase purified from Escherichia coli. J Mol Biol 2004;342(2):383−401.

[46] Christensen DG, et al. Post-translational protein acetylation: an elegant mechanism for bacteria to dynamically regulate metabolic functions. Front Microbiol 2019;10:1604.

[47] Hu LI, et al. Acetylation of the response regulator RcsB controls transcription from a small RNA promoter. J Bacteriol 2013;195(18):4174−86.

[48] Hu LI, Lima BP, Wolfe AJ. Bacterial protein acetylation: the dawning of a new age. Mol Microbiol 2010;77(1):15−21.

[49] Lima BP, et al. Involvement of protein acetylation in glucose-induced transcription of a stress-responsive promoter. Mol Microbiol 2011;81(5):1190−204.

[50] Schilling B, et al. Global lysine acetylation in Escherichia coli results from growth conditions that favor acetate fermentation. J Bacteriol 2019;201(9):e00768-18.

[51] Schilling B, et al. Protein acetylation dynamics in response to carbon overflow in Escherichia coli. Mol Microbiol 2015;98(5):847−63.

[52] Wolfe AJ. Bacterial protein acetylation: new discoveries unanswered questions. Curr Genet 2016;62(2):335−41.

[53] Koo H, et al. Regulation of gene expression by protein lysine acetylation in Salmonella. J Microbiol 2020;58(12):979−87.

[54] Jones BD. Salmonella invasion gene regulation: a story of environmental awareness. J Microbiol 2005;43:110−17 **Spec No**.

[55] Hung CC, Eade CR, Altier C. The protein acyltransferase Pat post-transcriptionally controls HilD to repress Salmonella invasion. Mol Microbiol 2016;102(1):121−36.

[56] Sang Y, et al. Acetylation regulating protein stability and DNA-binding ability of HilD, thus modulating salmonella typhimurium virulence. J Infect Dis 2017;216(8):1018−26.

[57] Clarke AJ. Peptidoglycan: another brick in the wall. Nat Chem Biol 2017;13(7):695−6.

[58] Sychantha D, et al. Mechanistic pathways for peptidoglycan O-Acetylation and De-O-acetylation. Front Microbiol 2018;9:2332.

[59] Ahuja N, et al. Aging and DNA methylation in colorectal mucosa and cancer. Cancer Res 1998;58(23):5489−94.

[60] Ayton PM, Chen EH, Cleary ML. Binding to nonmethylated CpG DNA is essential for target recognition, transactivation, and myeloid transformation by an MLL oncoprotein. Mol Cell Biol 2004;24(23):10470−8.

[61] Baylin SB. DNA methylation and gene silencing in cancer. Nat Clin Pract Oncol 2005;2(Suppl 1):S4−11.

[62] Rountree MR, et al. DNA methylation, chromatin inheritance, and cancer. Oncogene 2001;20(24):3156−65.

[63] Urnov FD. Methylation and the genome: the power of a small amendment. J Nutr 2002;132(8 Suppl):2450S−6S.

[64] Willbanks A, et al. The evolution of epigenetics: from prokaryotes to humans and its biological consequences. Genet Epigenet 2016;8:25−36.

[65] Botting CH, et al. Extensive lysine methylation in hyperthermophilic crenarchaea: potential implications for protein stability and recombinant enzymes. Archaea 2010;2010:106341.

[66] Gonzalez D, et al. The functions of DNA methylation by CcrM in Caulobacter crescentus: a global approach. Nucleic Acids Res 2014;42(6):3720−35.

[67] Li S, et al. N (4)-Cytosine DNA methylation is involved in the maintenance of genomic stability in deinococcus radiodurans. Front Microbiol 2019;10:1905.

[68] Lobner-Olesen A, Marinus MG, Hansen FG. Role of SeqA and Dam in Escherichia coli gene expression: a global/microarray analysis. Proc Natl Acad Sci USA 2003;100(8):4672−7.

[69] Ma B, et al. Biochemical and structural characterization of a DNA N6-adenine methyltransferase from Helicobacter pylori. Oncotarget 2016;7(27):40965−77.

[70] Marinus MG. DNA methylation and mutator genes in Escherichia coli K-12. Mutat Res 2010;705(2):71−6.

[71] Reisenauer A, Shapiro L. DNA methylation affects the cell cycle transcription of the CtrA global regulator in Caulobacter. EMBO J 2002;21(18):4969−77.

[72] Sanchez-Romero MA, Cota I, Casadesus J. DNA methylation in bacteria: from the methyl group to the methylome. Curr Opin Microbiol 2015;25:9−16.

[73] Wang S, et al. A deficiency in S-adenosylmethionine synthetase interrupts assembly of the septal ring in Escherichia coli K-12. Mol Microbiol 2005;58(3):791−9.

[74] Marinus MG, Casadesus J. Roles of DNA adenine methylation in host-pathogen interactions: mismatch repair, transcriptional regulation, and more. FEMS Microbiol Rev 2009;33(3):488−503.

[75] Li GM. Mechanisms and functions of DNA mismatch repair. Cell Res 2008;18(1):85−98.

[76] Putnam CD. Evolution of the methyl directed mismatch repair system in Escherichia coli. DNA Repair (Amst) 2016;38:32−41.

[77] Marinus MG, Lobner-Olesen A. DNA Methylation. EcoSal Plus 2009;3(2).

[78] Marinus MG, Lobner-Olesen A. DNA Methylation. EcoSal Plus 2014;6(1).

[79] Mendillo ML, Putnam CD, Kolodner RD. Escherichia coli MutS tetramerization domain structure reveals that stable dimers but not tetramers are essential for DNA mismatch repair in vivo. J Biol Chem 2007;282(22):16345−54.

[80] Patil NA, et al. Putative DNA modification methylase DR_C0020 of Deinococcus radiodurans is an atypical SAM dependent C-5 cytosine DNA methylase. Biochim Biophys Acta Gen Subj 2017;1861(3):593−602.

[81] Schmier BJ, et al. The structure and enzymatic properties of a novel RNase II family enzyme from Deinococcus radiodurans. J Mol Biol 2012;415(3):547−59.

[82] Shaiwale NS, et al. DNA adenine hypomethylation leads to metabolic rewiring in Deinococcus radiodurans. J Proteomics 2015;126:131−9.

[83] Collier J. Cell division control in Caulobacter crescentus. Biochim Biophys Acta Gene Regul Mech 2019;1862(7):685−90.

[84] Horton JR, et al. The cell cycle-regulated DNA adenine methyltransferase CcrM opens a bubble at its DNA recognition site. Nat Commun 2019;10(1):4600.

[85] Reisenauer A, Quon K, Shapiro L. The CtrA response regulator mediates temporal control of gene expression during the Caulobacter cell cycle. J Bacteriol 1999;181(8):2430−9.

[86] Blaser MJ, Chen Y, Reibman J. Does Helicobacter pylori protect against asthma and allergy? Gut 2008;57(5):561−7.

[87] Kumar S, et al. N4-cytosine DNA methylation regulates transcription and pathogenesis in Helicobacter pylori. Nucleic Acids Res 2018;46(7):3815.

[88] Kumar S, et al. Prevalence of Helicobactor pylori in patients with perforated duodenal ulcer. Trop Gastroenterol 2004;25(3):121−4.

[89] Kumar R, Mukhopadhyay AK, Rao DN. Characterization of an N6 adenine methyltransferase from Helicobacter pylori strain 26695 which methylates adjacent adenines on the same strand. FEBS J 2010;277(7):1666−83.

[90] Blakeway LV, et al. The Moraxella catarrhalis phase-variable DNA methyltransferase ModM3 is an epigenetic regulator that affects bacterial survival in an in vivo model of otitis media. BMC Microbiol 2019;19(1):276.

[91] Brockman KL, et al. The ModA2 Phasevarion of nontypeable *Haemophilus influenzae* regulates resistance to oxidative stress and killing by Human Neutrophils. Sci Rep 2017;7(1):3161.

[92] Brockman KL, et al. ModA2 phasevarion switching in non-typeable *Haemophilus influenzae* increases the severity of experimental otitis media. J Infect Dis 2016;214(5):817–24.

[93] Phillips ZN, et al. Phasevarions of bacterial pathogens – phase-variable epigenetic regulators evolving from restriction-modification systems. Microbiology (Reading) 2019;165(9):917–28.

[94] Phillips ZN, et al. Phase-variable bacterial loci: how bacteria gamble to maximise fitness in changing environments. Biochem Soc Trans 2019;47(4):1131–41.

[95] Seib KL, et al. Epigenetic regulation of virulence and immunoevasion by phase-variable restriction-modification systems in bacterial pathogens. Annu Rev Microbiol 2020;74:655–71.

[96] VanWagoner TM, et al. The modA10 phasevarion of nontypeable Haemophilus influenzae R2866 regulates multiple virulence-associated traits. Microb Pathog 2016;92:60–7.

[97] Atack JM, et al. The nontypeable Haemophilus influenzae major adhesin Hia Is a dual-function lectin that binds to human-specific respiratory tract sialic acid glycan receptors. mBio 2020;11(6):e02714-20.

[98] Atack JM, et al. A biphasic epigenetic switch controls immunoevasion, virulence and niche adaptation in non-typeable Haemophilus influenzae. Nat Commun 2015;6:7828.

[99] Robledo-Avila FH, et al. A bacterial epigenetic switch in non-typeable Haemophilus influenzae modifies host immune response during otitis media. Front Cell Infect Microbiol 2020;10:512743.

[100] Thornton RB, et al. Panel 7 – Pathogenesis of otitis media – a review of the literature between 2015 and 2019. Int J Pediatr Otorhinolaryngol 2020;130(Suppl 1):109838.

[101] Bannister AJ, Schneider R, Kouzarides T. Histone methylation: dynamic or static? Cell 2002;109(7):801–6.

[102] Butler JS, et al. Histone-modifying enzymes: regulators of developmental decisions and drivers of human disease. Epigenomics 2012;4(2):163–77.

[103] Greer EL, Shi Y. Histone methylation: a dynamic mark in health, disease and inheritance. Nat Rev Genet 2012;13(5):343–57.

[104] Bedford MT, Clarke SG. Protein arginine methylation in mammals: who, what, and why. Mol Cell 2009;33(1):1–13.

[105] Dacwag CS, et al. Distinct protein arginine methyltransferases promote ATP-dependent chromatin remodeling function at different stages of skeletal muscle differentiation. Mol Cell Biol 2009;29(7):1909–21.

[106] Herrmann F, et al. Human protein arginine methyltransferases in vivo–distinct properties of eight canonical members of the PRMT family. J Cell Sci 2009;122(Pt 5):667–77.

[107] Aravind L, Iyer LM. Provenance of SET-domain histone methyltransferases through duplication of a simple structural unit. Cell Cycle 2003;2(4):369–76.

[108] Zhang M, et al. Systematic proteomic analysis of protein methylation in prokaryotes and eukaryotes revealed distinct substrate specificity. Proteomics 2018;18(1).

[109] Yang DCH, et al. Outer membrane protein OmpB methylation may mediate bacterial virulence. Trends Biochem Sci 2017;42(12):936–45.

[110] Chao CC, et al. Proteome analysis of madrid E strain of *Rickettsia prowazekii*. Proteomics 2004;4(5):1280–92.

[111] Zhang JZ, et al. A mutation inactivating the methyltransferase gene in avirulent Madrid E strain of *Rickettsia prowazekii* reverted to wild type in the virulent revertant strain Evir. Vaccine 2006;24(13):2317–23.

[112] Bechah Y, et al. Genomic, proteomic, and transcriptomic analysis of virulent and avirulent *Rickettsia prowazekii* reveals its adaptive mutation capabilities. Genome Res 2010;20(5):655–63.

[113] Witchell TD, et al. Post-translational modification of LipL32 during Leptospira interrogans infection. PLoS Negl Trop Dis 2014;8(10):e3280.

[114] Mujtaba S, et al. Anthrax SET protein: a potential virulence determinant that epigenetically represses NF-kappaB activation in infected macrophages. J Biol Chem 2013;.

[115] Seeliger D, de Groot BL. Ligand docking and binding site analysis with PyMOL and Autodock/Vina. J Comput Aided Mol Des 2010;24(5):417–22.

[116] Simmons AD, et al. Using a PyMOL activity to reinforce the connection between genotype and phenotype in an undergraduate genetics laboratory. PLoS One 2014;9(12):e114257.

[117] Qian C, et al. Structural insights of the specificity and catalysis of a viral histone H3 lysine 27 methyltransferase. J Mol Biol 2006;359(1):86–96.

[118] Qian C, Zhou MM. SET domain protein lysine methyltransferases: structure, specificity and catalysis. Cell Mol Life Sci 2006;63(23):2755–63.

[119] Eswar N, et al. Protein structure modeling with MODELLER. Methods Mol Biol 2008;426:145–59.

[120] Pieper U, et al. MODBASE, a database of annotated comparative protein structure models and associated resources. Nucleic Acids Res 2009;37(Database issue):D347–54.

[121] Aivaliotis M, et al. Ser/Thr/Tyr protein phosphorylation in the archaeon Halobacterium salinarum – a representative of the third domain of life. PLoS One 2009;4(3):e4777.

[122] Soufi B, et al. Proteomics reveals evidence of cross-talk between protein modifications in bacteria: focus on acetylation and phosphorylation. Curr Opin Microbiol 2012;15(3):357–63.

[123] Barak R, Eisenbach M. Co-regulation of acetylation and phosphorylation of CheY, a response regulator in chemotaxis of *Escherichia coli*. J Mol Biol 2004;342(2):375–81.

[124] Huang Z, et al. Cross talk between chemosensory pathways that modulate chemotaxis and biofilm formation. mBio 2019;10(1):e02876-18.

[125] Haslinger K, Hackl T, Prather KLJ. Rapid in vitro prototyping of O-methyltransferases for pathway applications in *Escherichia coli*. Cell Chem Biol 2021;28(6):876–86 e4.

Further Reading

Choudhary C, et al. The growing landscape of lysine acetylation links metabolism and cell signalling. Nat Rev Mol Cell Biol 2014;15(8):536–50.

Crawford GE, et al. DNase-chip: a high-resolution method to identify DNase I hypersensitive sites using tiled microarrays. Nat Methods 2006;3(7):503–9.

CHAPTER

13

Drosophila Epigenetics

Akanksha Bhatnagar[1], Ashley M. Karnay[2] and Felice Elefant[1]

[1]Department of Biology, Drexel University, Philadelphia, PA, United States [2]Department of Neurobiology & Anatomy, Drexel University College of Medicine, Philadelphia, PA, United States

OUTLINE

Introduction: *Drosophila* as a Model Organism in Epigenetic Research	216
Epigenetic Modification of Histone Proteins Regulate Chromatin Packaging and Gene Control in *Drosophila*	217
Histone Acetylation	218
Histone Methylation	219
Noncoding RNAs	219
DNA Methylation	219
Position-Effect Variegation	220
Heterochromatin Within the Drosophila Genome	220
Protein Regulators of Position-Effect Variegation	221
Epigenetic Histone Modifications Regulate Position-Effect Variegation in Drosophila	221
The Role of Epigenetics During *Drosophila* Development: Epigenetic Memory	223
Epigenetic Modifications Maintain Patterns of Hox Gene Expression Throughout Drosophila Development	223
Mechanisms of Silencing and Activation: Bivalent Chromatin	223
PcG Proteins and Gene Repression	223
TrxG Proteins and Gene Activation	226
PREs and PcG-TrxG-Mediated Gene Expression	226
Transgenerational Epigenetic Inheritance: "Epigenetic Memory"	227
Epigenetics During S-Phase	227
Epigenetics During M-Phase	228
Dosage Compensation	229
The Male-Specific Lethal Complex/Dosage Compensation Complex	229
Regulation of Dosage Compensation Complex Targeting to the X Chromosome	230
Dosage Compensation Complex Spreading and Target Recognition	231
Transcriptional Activation of Active X-Linked Genes by the Dosage Compensation Complex	231
The Epigenetic Language in Postmitotic Neurons Underlying Cognitive Function	232
Activity-Dependent Histone Acetylation and Cognitive Function	232
Histone Acetyltransferase: Histone Deacetylase Interplay in Cognitive Function	233
Use of Drosophila Disease Models to Study Dysregulation of HAT Activity in the Etiology of Human Neurodegenerative Diseases	234
Histone Acetylation Targeted Therapeutics for Neurodegenerative Disorders	235
Environmental Enrichment Improves Learning and Memory in Drosophila	237
Conclusion	238
Protocol	238
Before You Begin	238
Key Resources Table	239
Materials and Equipment	239
Step-by-Step Method Details	239
Expected Outcomes	240
Advantages	240
Limitations	240
Optimization and Troubleshooting	240
Glossary	241
Abbreviations	241
References	241

INTRODUCTION: DROSOPHILA AS A MODEL ORGANISM IN EPIGENETIC RESEARCH

Epigenetics is the mechanism underlying stable maintenance of gene expression that involves physically "marking" DNA or its associated proteins, which allows genotypically identical cells to be phenotypically distinct, without alterations in the nucleotide sequences [1]. *Drosophila melanogaster* has been utilized for over a century as a powerful model organism to help elucidate much of the fundamental concepts of human epigenetics. Conrad H. Waddington conducted studies in 1942, showing that altering the developmental environment of *Drosophila* embryos resulted in mature flies with varying thorax and wing structures [2]. Herman J. Muller followed with his research using X-rays to induce genetic mutations after which he observed a mosaic-like expression of the mutant and wild-type phenotype in the offspring, a phenomenon he termed "position effect variegation" [3,4]. Findings from subsequent studies in *Drosophila* by Jack Shultz and later by Ed Lewis, led to the characterization of the family of genes whose protein products are responsible not only for the regulation of this epigenetic phenomena but also for regulation of normal development and the transmission of epigenetic programs across generations [5].

There are many advantages that make *Drosophila* an ideal model system for the study of epigenetic gene control in a variety of biological processes. First, *Drosophila* contains fewer genes than humans, indicating less overall genetic redundancy [6]. Additionally, more than 60% of the protein-coding genes within the fly genome have human homologs [7], and similarly, 75% of human disease-linked genes have homologs in *Drosophila* [8], making it an excellent model for translational or comparative genetic and epigenetic studies. Further, the polytene chromosome located within the larval salivary glands of *Drosophila*, provide scientists the unique ability to directly visualize structural changes that result from the epigenetic modifications of chromatin [9,10]. These chromosomes develop from normal chromosomes and are a result of specialized cells undergoing multiple rounds of replication without dividing. They have a characteristic banding pattern that arises as a result of regional chromatin condensation, with areas of densely packaged chromatin forming dark bands that contrast with areas of less compacted light bands. Epigenetic modifications to the chromatin result in the localized unwinding of the chromatin, which can be directly visualized as chromosome "puffs" identifying regions of active transcription [11]. Recent findings indicate a correlation between the positioning of cis-regulatory elements, patterns of epigenetic modifications and association with regulatory proteins between the polytene chromosome and diploid interphase chromosomes making it an ideal model for the study of chromatin dynamics and regulation across cell types [12,13]. Another advantage of using *Drosophila* is its short life cycle, which translates to short generation times, allowing researchers the ability to rapidly assess the transgenerational effects of epigenetic modifications. Lastly, *Drosophila's* distinct morphological developmental stages allow researchers to study the genetic and epigenetic changes at specific embryonic, larval, pupal transition or adult fly stages. Initial studies on *Drosophila* embryos, such as the discovery that discrete genes regulate different aspects of development [14], have been instrumental in guiding the present-day developmental fields. Additionally, several evolutionary conserved developmental mechanisms including the PcG and TrxG proteins mediated transcriptional regulation of developmental homeotic (*Hox*) genes were first characterized in *Drosophila* [15].

Drosophila has also proven to be a valuable tool for the study of epigenetic mechanisms underlying neurobiology and human neurodegenerative diseases. The fly nervous system is complex and highly evolved, and shares important regulatory genes and molecular pathways with the human nervous system [16]. The fly brain contains approximately 200,000 neurons forming a complex synaptic network that mediates its ability to perform measurable higher-order functions similar to those in humans, such as learning and memory, feeding, courtship, aggression and circadian rhythm, to name a few [17,18]. Further, when compared to the 86 million neurons in the human brain, the compact fly brain enables researchers to dissect individual neuronal circuits many of which carry out distinct and human-conserved higher order function. Recently, the adult *Drosophila* central brain has been successfully mapped out into a detailed connectome comprising of 25,000 neurons with approximately 20 million chemical synapses between them [19]. Being the most comprehensive wiring map of the fruit fly brain, the *Drosophila* connectome is a powerful tool kit that allows researchers to identify individual cell types and neuronal circuits, develop neuro-circuitry models and track upstream or downstream neuronal partners [19].

Many genes found in the *Drosophila* genome have structural and/or functional homologs in humans and as a result, specific anatomical structures within the fly have become attractive model systems for studying epigenetic gene control mechanisms underlying the underpinnings of biological processes mirrored in humans. The *Drosophila* neuromuscular junction is an excellent example of this. Genetic manipulation of synaptic components coupled with sensitive electrophysiological techniques, have enabled its detailed characterization, making it a valuable model system for studying synapse function, regulation, and plasticity [20]. Further, fly synapses have a predictable structure, are relatively large in size, easy to manipulate, and morphologically and functionally comparable to those of humans, making them pertinent for the study of numerous aspects of the mammalian central nervous system, including synaptogenesis, the events underlying synaptic transmission, and pathogenesis of neurodegenerative diseases [21]. The *Drosophila* visual system is also a powerful neural circuit used to provide insights into the overt toxic effects of individual human disease-causing genes or neurodegenerative agents [7]. This is because eye degeneration defects are not lethal and can be easily quantitatively measured based on direct visualization of cell loss within the organized pattern of cells. This system has widely been used to measure the neurodegenerative effects of expanded polyglutamine repeats characteristic of Huntington's disease and spinocerebellar ataxia and amyloid β aggregates in Alzheimer disease [22,23]. Importantly, this system has proved instrumental in enabling the identification of epigenetic modulators that can rescue such neurodegenerative defects.

Epigenetic control mechanisms involved in the integration of higher-order sensory information and learning and memory have also been successfully studied in *Drosophila* using the mushroom body as a well characterized cognitive model. The mushroom body (MB) is a bilateral brain structure composed of a complex network of unmyelinated axons, glial processes, dendrites and specialized neurons called Kenyon cells [24,25]. The *Drosophila* MB is known to regulate a range of behavioral and physiological functions from olfactory learning and memory, courtship conditioning to decision-making under uncertain conditions [26–30]. Coupled with the multitude of experimental paradigms aimed at analyzing the behavioral consequences of genetic interventions, the MB provides scientists with an easy-to-use cognition-linked circuit model of the brain [31]. Importantly, *Drosophila* can perform a multitude of tasks utilized in a variety of behavioral paradigms that can be studied and quantified. These behaviors range from the simple, such as aggression [32] to more complex functions like locomotor or circadian sleep/wake cycles [33]. In particular, courtship conditioning in *Drosophila* is a complex behavioral learning paradigm that requires multimodal sensory input, involving chemosensory, mechanosensory, visual and olfactory pathways and is thus well suited to study experience dependent synaptic plasticity [34–37]. Complex behaviors such as associative learning and memory can be quantitatively studied using a variety of different assays that quantitatively measure courtship conditioning and the ability of *Drosophila* to learn to avoid a specific object or odor after experiencing it in conjunction with a painful stimulus [38].

In summary, as a result of the short life cycle, ease of environmental manipulation, the high degree of genetic similarity between fly and human, the facility of genetic manipulation, and the growing number of experimental paradigms available to study the consequences of such manipulations, *Drosophila melanogaster* continues to be an ideal model organism in the study of epigenetics in humans.

EPIGENETIC MODIFICATION OF HISTONE PROTEINS REGULATE CHROMATIN PACKAGING AND GENE CONTROL IN *DROSOPHILA*

Epigenetic alterations in gene expression profiles, which can be stable or transient, are the result of chromatin remodeling factors and posttranslational modifications (PTM) to chromatin components that result in changes to the degree of chromatin compaction which in turn impact the accessibility of the transcriptional machinery to the DNA template [39,40]. Such epigenetic gene regulatory mechanisms have long been studied effectively using *Drosophila* as a model system. In eukaryotic cells, the DNA is wrapped around octamers of four histone proteins (H2A, H2B, H3, H4) to form the basic unit of chromatin called nucleosomes [16,41]. This organization presents a major physical hurdle for the transcriptional machinery and must be overcome to enable gene transcription in vivo. One of the main mechanisms utilized by the cell to overcome the repressive nucleosomal environment is through the covalent, posttranslational modification of the core histone proteins. These modifications include acetylation, methylation, phosphorylation, ubiquitination, sumoylation, and ADP-ribosylation [42]. Histones are often concurrently modified on several residues and there is also a dynamic interplay between histone modifications and DNA modifications (such as DNA methylation), thus creating staggering combinatorial possibilities for gene regulation [43]. These modifications not only work to modulate the structure of the chromatin itself but can also serve as distinct docking sites for other protein complexes that participate in gene transcription or repression [39]. Moreover, the recruitment of additional chromatin-interacting proteins often depends not solely on the presence of a particular PTM but rather also relies on a combination of local marks and the overall context within which those marks are presented [44]. In this way, epigenetic patterns serve as a "code" that is generated by enzymes that function as "writers" to add specific PTMs to histones, recognized by "readers" that are comprised of regulatory proteins that control chromatin packaging, and removed by enzymes that serve as "erasers." These chromatin regulatory proteins work in concert to determine the transcriptional programs employed under transient cellular conditions, timing-specific developmental stages, or alternatively stably maintained across generations [45]. Two of the most well characterized epigenetic histone PTMs that are fully modeled in *Drosophila* will be discussed in the following sections.

Histone Acetylation

One of the most extensively studied PTMs involved in gene regulation is histone acetylation. This epigenetic modification is generated by members of the family of histone acetyltransferase (HAT) enzymes. These enzymes catalyze the transfer of an acyl moiety from acyl coenzyme A to lysine residues on the n-terminal tails of the histones [40]. Acetylation alters the chromatin state in favor of a more relaxed, open conformation that is conducive to transcriptional activation and achieved through a number of mechanisms (Fig. 13.1) [45,46]. Based on the sequence similarity, substrate preferences and functions, HATs can be divided into two major superfamilies: GCN5/PCAF family and MYST family. The GCN5/PCAF family, such as Gcn5 and PCAF HATs, contains bromodomain, targets transcriptional activators and preferentially acetylates H3 [47]. In contrast, the MYST family, such as MOF, Tip60, and HBO HATs, contain chromodomain and preferentially targets H4 [47]. Acetylation is believed to neutralize the positive charge on the lysine residue, thus reducing the affinity between histones and DNA, allowing for a more open or relaxed chromatin state [16,48]. More recent evidence points to acetylated lysine residues serving as molecular docking sites for proteins that act to further relax the chromatin structure such as bromodomain-containing ATP-dependent chromatin remodeling complexes and various members of the transcriptional machinery [46,49]. Indeed, Tip60 (Tat-interactive protein-60 kDa) a member of the MYST HAT super family that is recruited to chromatin, is a member of the Tip60 protein complex shown to possess not only HAT activity but also ATPase, DNA helicase, as well as DNA binding activities [49]. The relaxed hyperacetylated chromatin conformation [48] is marked by an increase in the sensitivity of the chromatin to DNase digestion, as well as to an increase in the flexibility of the nucleosomal DNA, due to the reduction in conformation hindrance provided by nucleosomal compaction [50]. This chromatin unwinding allows access of the transcriptional machinery to the regulatory elements of the genes, facilitating their expression. It has been shown that TAF_{II} (TBP-associated factor), which is part of the transcription initiation factor TFIID complex, has intrinsic HAT activity, thus enabling the unwinding of

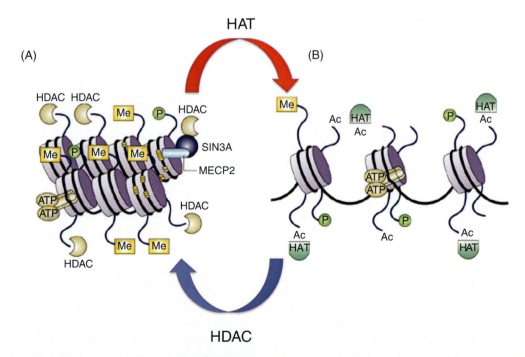

FIGURE 13.1 **Chromatin structure regulates transcriptional activity.** Histone acetylation is regulated by the antagonistic activities of histone acetyltransferases (HATs) that acetylate histones and histone deacetylases (HDACs) that deacetylate histones. (A) Histone deacetylation induces a closed chromatin configuration with concomitant transcriptional repression. (B) Histone acetylation opens chromatin packaging to allow for transcriptional activation [56]. *Source: Adapted from Johnstone RW. Hisonte-deacetylase inhibitors: novel drugs for the treatment of cancer. Nat Rev 2002;1:287–99.*

regions around specific genes to allow TFIID access to previously repressed genes [51]. Others have also shown increased access to gene promoter regions via histone acetylation, which facilitates the ability of RNA Pol II to initiate transcription at specific genetic loci [52]. Histone acetylation is a reversible process that is conversely regulated by histone deacetylase (HDACs) enzymes. HDACs catalyze the removal of the acyl moiety from the histone tail lysine residues, leading to the formation of a more compact and transcriptionally repressed chromatin state [53]. Drosophila has 5 HDAC genes: Rpd3 and HDAC3, HDAC4, Sir2 and HDAC6, belonging to mammalian HDAC classes I, II, III, and IV, with each HDAC presenting distinct temporal expression pattern and transcriptional regulation of a unique set of genes [54,55].

Histone Methylation

Posttranslational methylation of histones is a well-studied modification known to influence an array of biological processes, especially in the context of development. Lysine resides on histones 3 and 4 are primarily methylated on all endogenous basic residues with mono-, di-, or tri-methylation being observed [57–59]. Methyl groups tend to exhibit a more stable association with histones relative to acetyl groups; therefore, it is no surprise that they are most often implicated in transgenerational epigenetic inheritance, a process studied very effectively in Drosophila. SET (Su(var)3-9, Enhancer-of-zeste and Trithorax) domain containing lysine methyltransferases (KMTs) are capable of modifying both histone and non-histone proteins, catalyzing the transfer of methyl moieties from S-adenosyl methionine to the ε-amino groups of lysine residues [57]. In Drosophila, dSet1 is the main H3K4 di- and tri-methyltransferase throughout development [60]. Proteins with PHD finger and chromodomains are most commonly associated with the binding of methyl residues, however, a variety of other methyl-binding domains exist [61]. Like acetylation, histone methylation is a reversible process, regulated by a variety of lysine demethylases, which are specific for each degree of methylation [44]. Presently, Drosophila is predicted to contain 13 predicted histone methylases that are classified as amine oxidase demethylases or JmjC-domain containing demethylases [62,63]. Histone methylation in Drosophila is further discussed in upcoming sections.

The stepwise and variable pattern of histone methylation provides a heterogeneous platform for dynamically regulated transcription. Because methylation can occur on multiple sites within the same histone and different lysine residues are capable of varying degrees of methylation, a combination of methyl moieties are possible and translate to differential gene expression states. For example, H3K4me3 has been associated with regions of active transcription while H3K27me3 marks are indicative of transcriptional silencing [64]. Leading to further complexity, methylation marks can be associated with opposing activities; H4K20me is highly enriched within the promoter and coding regions of actively transcribed genes [65], while the same mark has also been shown to be a negative regulator of transcription, demonstrated in knock down assays [66]. Methylation can also indirectly lead to the reorganization of the 3D structure of chromatin, by designating regions of DNA for recognition by chromatin remodeling complexes, resulting in the induction of higher-order chromatin conformations that suppress transcription [67–69].

Noncoding RNAs

There are a variety of types of noncoding RNAs (ncRNAs) identified in Drosophila that mediate epigenetic gene expression are conserved in mammalian systems. For example, long ncRNAs (lncRNAs) are a class of ncRNAs that exceed 200 nucleotides. Elegant work has demonstrated that lncRNAs mediate a wide variety of biological processes and have emerged as critical regulators in epigenetic gene control. LncRNAs are mainly localized in the nucleus and are predominant in many organisms including Drosophila, mice, and humans. Recent advances in technology have revealed that lncRNA regulatory mechanisms in Drosophila are similar to those in humans and other mammalian systems making the fly an excellent model to explore lncRNAs function in development and disease. lncRNAs have been shown to be in involved in epigenetic gene control by recruiting or sequestering chromatin regulatory enzymes that include DNA methyltransferases, histone deacetylases and chromatin remodeling complexes to control chromatin packaging. Additional information regarding the roles of lncRNA in the maintenance of gene expression patterns can be found in a number of recent reviews [70–72]. Further, there are additional types of small ncRNAs that exist in Drosophila that are processed from double stranded RNA that include microRNAs (miRNAs), small interfering RNAs (siRNAs) and PIWI interacting RNAs (piRNAs), all of which have been implicated in epigenetic gene control via similar chromatin regulatory protein recruitment mechanisms as for lnRNAs.

DNA Methylation

DNA methylation is also a critical player in controlling chromatin structure to regulate gene expression.

Typically, high levels of DNA methylation lead to condensed chromatin and gene repression. Whether DNA methylation exists in *Drosophila* has been historically controversial based on conflicting data [73]. However, recent studies demonstrate that *Drosophila* contains an ortholog for DNA methyltransferase 2 (Dnmt2) and the dTet gene that encodes a DNA hydroxymethylase enzyme [74,75]. Further, high resolution LC-Ms/MS and HPLC technology confirm the presence of 5mC in adult *Drosophila* [76]. A more recent study assessed genome-wide DNA methylation levels in flies using high throughput sequencing techniques [77]. These studies revealed differentially methylated regions localized on genes involved in mitosis and chromosomal segregation, suggesting a role for DNA methylation in epigenetic control of genes that mediate cell cycle processes. Future research should clarify further the biological function of DNA methylation in *Drosophila*.

In summary, acetylation and methylation, as well as phosphorylation, ubiquitination, sumoylation, and ADP-ribosylation of histones, comprise a complex array of possible epigenetic modification that confers flexibility and the ability for rapid modulation to the temporal and spatial requirements of gene transcription during development and environmental change.

POSITION-EFFECT VARIEGATION

Covalent histone modifications have the ability to regulate complex gene expression profiles by modulating the degree of compaction of the chromatin fiber. Heterochromatin is characterized by a more compact chromatin state generally correlating with the repression of transcription. It is established and maintained by combinations of different marks dictated by the interaction between histone modifying enzymes and chromosomal structural proteins, leading to demarcation of structurally and functionally distinct chromosomal domains. In contrast, euchromatin is less compacted and reflects areas of active transcription. Although traditionally chromatin was divided into heterochromatin and euchromatin, more recent classification has five principal chromatin types in *Drosophila*, with three repressive and two active states [78]. The repressive states have either PcG binding with H3K27 methylation, Heterochromatin Protein 1 (HP1) binding with H3K9 methylation or absence of classic heterochromatin markers. In contrast, one active state shows binding with several proteins including the nucleosome-remodeling ATPase *brahma* and chromatin structure regulator *Su(var)2−10*, while the other state shows predominant binding of MRG15 and H3K36 trimethylation [78]. Further studied have analyzed genome-wide chromatin landscape based on 18 histone modifications in *Drosophila* [79] and found conservation of chromatin organization with the human genome [80].

Once heterochromatin formation is initiated, it is capable of spreading to and silencing nearby euchromatic genes, resulting in the formation of facultative heterochromatin, an epigenetics-mediated mechanism for permanent gene inactivation [4,81]. This process is well modeled in *Drosophila*, when genes that are normally expressed are aberrantly moved, either by rearrangement or transposition, to a site near the euchromatin-heterochromatin (Eu-Het) transition zone and as a result, are abnormally packaged into a heterochromatic form and silenced [4,82]. This phenomenon termed "position-effect variegation" (PEV), was first observed in 1925 by Alfred Sturtevant while studying mutations in the *Bar* eye gene responsible for establishing the number of facets within the *Drosophila* eye during development. He observed that gene duplication mutations resulted in the unexpected increase in the number of copies of the *Bar* genes but intriguingly, the resultant phenotype was that of a reduction in the eye size. This suggested that the phenotype was not due to the number of copies of the gene but rather to the positioning or arrangement of the genes within the chromosome, an influence he called "position effect" [83]. Following up on this work, Herman Muller used x-rays to induce mutations within the *white* gene, a mutation that results in the expression of a white eye instead of the wild-type red eye. After x-ray exposure, he observed a unique phenotype in which each eye expressed a combination of the red and white pigmentation, i.e., a "mottled" or "variegated" phenotype. He concluded that the gene itself was not mutated, evidenced by the successful expression in a subset of photoreceptor cells, but that aberrant repositioning of the gene likely led to abnormal silencing in some cells, resulting in the variegated phenotype [4,83]. Genetic rearrangements resulting in the repositioning of euchromatic genes near a Eu-Het transition zone are now known to lead to the repressing chromatin packaging and silencing of normally expressed genes, a phenomenon called PEV. The regulation and maintenance of these dynamic regions of the genome clearly have an important influence on gene expression and *Drosophila* has proven to be an ideal model system for the analysis of the epigenetic processes controlling heterochromatin dynamics and PEV.

Heterochromatin Within the *Drosophila* Genome

Heterochromatin, originally identified *via* cytological studies using the polytene chromosome, is defined as chromatin regions that maintain a dense staining and condensation pattern throughout the cell cycle [81,83]. These domains are composed mostly of tandem repeat motifs called "satellite" DNA sequences and transposable elements, such as DNA transposons and retroviruses. In contrast, euchromatin has a variable condensation and staining pattern, reflecting regions of active gene transcription [4,82]. In *Drosophila*, these chromatin domains can be characterized by their expression of unique combinations of histone marks. In an extensive analysis performed by Yin et al. (2011) utilizing chromatin immunoprecipitation in combination with high-throughput sequencing (ChIP-Seq) in the adult fly, a high resolution whole-genome map of the key histone modifications corresponding to regions of euchromatin and heterochromatin regions was generated [79]. In *Drosophila*, euchromatic regions are generally enriched in histone lysine methylation at H3K4, H3K79 and H3K36 [84], histone acetylation of H3K9, H4K16, and H3K14 [82,84] and phosphorylation of H3S10 [4,85]. In contrast, within the *Drosophila* genome there are distinct regions of heterochromatin, each displaying a unique combination of H3K9, H3K27, H4K20 histone methylation marks [64]. For example, heterochromatin found at the chromosomal telomeres contain H3K9 mono-, di-, and tri-methylated histone states [81]. Regions of heterochromatin within the chromosome's centromeres are termed centric, or chromocenter heterochromatin and are specifically enriched in the marks H3K9me3 and H3K27me [64,81]. Pericentric heterochromatin, regions flanking centric heterochromatin, contains H4K20me3 [81,84] and H3K9me3 [86,87]. In contrast to euchromatin, histones within heterochromatic regions are generally hypoacetylated, with only H4K12ac found within chromocenter heterochromatin [88].

Protein Regulators of Position-Effect Variegation

Loss-of-function (LOF) mutational analysis has allowed the identification of a number of genes whose protein products are associated with the establishment and maintenance of heterochromatin and PEV. These genes are classified as either enhancers of PEV [E(var)], whose LOF results in an increase in gene silencing or suppressors of PEV [Su(var)] whose LOF results in a loss of gene silencing [4]. The action of these proteins, which are components of heterochromatin as well as enzymes involved in the modification of chromatin and histone proteins, correlates with changes in the chromatin state seen in PEV [4,81].

E(var) genes encode proteins responsible for chromatin decondensation and increased gene expression; therefore mutations in these genes lead to the enhancement of heterochromatic gene silencing. One such gene unique to *Drosophila* is *E(var)3-9*, encoding a zinc finger protein, which was originally identified following its LOF by the appearance of the variegated eye pigmentation phenotype, indicative of the induction of heterochromatin spreading into and silencing of the adjacent *white (wt)* gene [89]. E(var) proteins have been shown to counteract the suppressing effects of heterochromatin in a variety of ways. The main mechanism of action involves epigenetic-mediated chromatin remodeling, leading to increased accessibility of gene promoter regions to the transcriptional machinery [90]. Many of the identified *E(var)* genes, such as *jumeaux* and *E(var)3-93E*, encode general transcription factors, enhancing gene expression in a way independent of chromatin modification [91,92]. E(var) proteins have also been proposed to work by inhibiting the formation of heterochromatin-mediated silencing complexes, thereby maintaining a euchromatic state conducive to gene expression [93,94].

Conversely, LOF of *Su(var)* genes leads to the suppression of heterochromatic gene silencing while additional copies of the genes enhance heterochromatin formation and spreading in *Drosophila* [81]. These proteins are either structural components of heterochromatin itself or histone-modifying enzymes responsible for the epigenetic-regulation of heterochromatin [95]. Three important *Su(var)* genes play key roles in PEV-mediated gene silencing in *Drosophila*: *Su(var)3-9*, *Su(var)2-5*, *Su(var)4-20*. *Su(var)3-9* encodes an H3K9 histone methyltransferase containing two evolutionarily conserved domains: (1) a C-terminal SET domain with methyltransferase activity, which shares homology with *Su(var)3-9*, *Enhancer of zestes*, and *Trithorax* proteins, and (2) a methyl-binding chromodomain. Both domains are essential for the association of *Su(var)3-9* with heterochromatin [96]. *Su(var)2-5* encodes the structural protein, HP1, which possesses multiple protein and methyl-binding domains. For example, its chromodomain allows it to bind to heterochromatin-specific H3K9me2/3 marks [4,81], while its C-terminal chromo shadow domain provides with the ability to dimerize with itself and to interact with *Su(var)3-9* [95]. The *Su(var)4-20* protein is a H4K20-specific histone methyltransferase that controls the H4K20me3 characteristic of pericentric heterochromatin [4]. Additional genes implicated in the spread of pericentric heterochromatin observed in PEV include *Su(var)3-1* encoding the protein kinase JIL-1, *Su(var)3-3* encoding a H3K4-specific demethylase homologous to the human LSD1 protein [97], and *Su(var)3-26* whose protein product is the histone deacetylase HDAC1 [98,99].

Epigenetic Histone Modifications Regulate Position-Effect Variegation in *Drosophila*

Because different sets of specific histone modifications are found within euchromatin and heterochromatin, it is not surprising that the formation of pericentric heterochromatin and its encroachment into adjacent regions of euchromatin require a sequence of events involving the removal of one set and the replacement of another set of histone modifications (Fig. 13.2). To initiate the transition from euchromatin to heterochromatin, the active chromatin marks H3K9ac and H3K4me2 are removed by the histone deacetylase *Su(var)3−26* (HDAC1) and the demethylase *Su(var)3−3* (LSD1), respectively. Removal of these marks are prerequisites for the di- and tri-methylation of H3K9, the histone modification hallmark of repressed chromatin regions, by *Su(var)3−9* [100]. H3K9me2/3 then serves as a binding platform for HP1, which, through its multiple diverse domains, is capable of recruiting a variety of proteins to participate in the *cis*-spreading of the newly initiated heterochromatin region into neighboring domains. Through a similar mechanism, transposable elements (TEs) embedded within *Drosophila* euchromatin are epigenetically silenced by a class of intervening RNAs called piwi-interacting RNAs (piRNAs) which mediate the deposition of H3K9me marks and spreading of heterochromatin to adjacent gene regions, selectively suppressing TE's deleterious effects on the *Drosophila* genome [4,101,102]. HP1 has been shown to recruit H4K20-specific HMTase *Su(var)4−20*, which catalyzes the trimethylation of H4K20, an evolutionarily conserved mark of pericentric heterochromatin [84]. *Su(var)3−7*, a zinc-finger protein, binds directly with HP1 within multiple heterochromatic regions, as evidenced by co-immunoprecipitation assays and direct visualization on polytene and embryonic chromosomes [103]. Based on the observation that HP1 and *Su(var)3−7* remain closely associated with pericentric heterochromatin regions long after its induction, it has been proposed that these proteins serve as stable structural elements, maintaining the compacted chromatin state characteristic of these transcriptionally repressed regions, however, the exact role in PEV remains unknown [103,104]. It should be noted that there are additional proteins that can also initiate chromatin repression at certain genomic locations that include the well characterized Polycomb repressive complexes 1 and 2 (PRC1 and PRC2) that are discussed in more detail in section 'PcG Proteins and Gene Repression'. Briefly, PRC1 recognizes and binds the repressive H3K27me3 epigenetic mark deposited by PRC2 to mediate chromatin compaction and heritable gene repression.

It has been proposed that the genetic rearrangements resulting from aberrant gene translocations results in the removal of a specific barrier zone or boundary element demarcating adjacent euchromatin and heterochromatin regions, thus allowing the migration of chromatin compaction not normally seen in nuclei with structurally normal chromosomes [4,94]. Although the molecular nature of these buffering elements remains uncharacterized, it is speculated that their disruption or removal leads to the initiation of heterochromatin, which can subsequently propagate as far as 175 kb into nearby genes, resulting in the shifting of euchromatin-heterochromatin junctions relative to proximal genes and PEV gene silencing [105]. In *Drosophila*, this process is mediated both by HP1 itself and by the coordinated cross talk between multiple

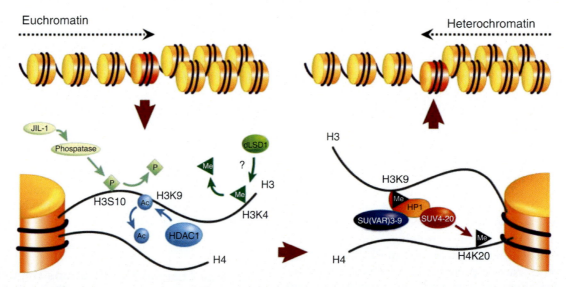

FIGURE 13.2 **Transition from euchromatin to heterochromatin.** To initiate the transition from euchromatin to heterochromatin, the active chromatin marks H3K9ac and H3K4me2 are removed by the histone deacetylase *Su(var)3−26* (HDAC1) and the demethylase *Su(var)3−3* (LSD1), respectively. Removal of these marks are prerequisites for the di- and tri-methylation of H3K9, the histone modification hallmark of repressed chromatin regions, by *Su(var)3−9* [100]. H3K9me2/3 then serves as a binding platform for HP1, which, through its multiple diverse domains, is capable of recruiting a variety of proteins to participate in the *cis*-spreading of the newly initiated heterochromatin region into neighboring domains. *Source: Adapted from Grewal SIS, Elgin SCR. Transcription and RNAi in the formation of heterochromatin. Nat Rev 2007;447:399−406.*

histone modifications. The interdependence of various posttranslational modifications of histone proteins provides a means by which chromatin dynamics can be regulated throughout the Drosophila genome; PEV-mediated gene silencing is an important example.

In summary, the large body of work in Drosophila on PEV has substantially elevated our understanding of heterochromatin and euchromatin dynamics, and the regulatory influence of epigenetics.

THE ROLE OF EPIGENETICS DURING DROSOPHILA DEVELOPMENT: EPIGENETIC MEMORY

Epigenetic Modifications Maintain Patterns of Hox Gene Expression Throughout Drosophila Development

During the early stages of development, the identity and function of each cell within an organism is determined by a unique pattern of gene transcription determined *via* the integration of endogenous and exogenous signals. Initiated by transcription factors, these complex changes in the cellular transcriptome are epigenetically maintained and inherited as part of cellular differentiation and development in a process referred to as "epigenetic memory" [106,107]. The repressed and activated expression states of developmental genes are maintained by the antagonistic and highly conserved function of the Polycomb group (PcG) and Trithorax group (TrxG) proteins [67,108,109]. While PcG and TrxG systems regulate a plethora of cellular processes that include genomic imprinting, stem cell biology, and X chromosome inactivation, here we will focus on their well-characterized role in Hox developmental gene regulation as an example of how the two systems mechanistically control complex gene expression profiles *via* both activation and repression. First characterized in Drosophila, PcG and TrxG proteins are required for the persistent regulation of appropriate expression patterns of the developmental homeotic (Hox) genes. Hox genes encode transcription factors whose regulated expression is responsible for the specification of segmental identity in the developing fly embryo [110]. These genes are clustered into two separate gene complexes: the Antennapedia and Bithorax complexes, with the former regulating development of the head and regions of the thorax and the latter regulating the abdominal sections and remaining regions of the thorax (Fig. 13.3) [110–112]. During the initial stages of development, after a period of transcriptional repression, maternally inherited transcription factors establish a spatial and temporal pattern of Hox gene expression that will eventually drive specific tissue differentiation and body segment development [113]. Multimeric PcG and TrxG protein complexes epigenetically mediate the long-term maintenance of these gene expression programs throughout the life of the fly [67,108–110].

Mechanisms of Silencing and Activation: Bivalent Chromatin

The temporal and spatial regulation of Hox gene expression correlates with specific histone modifications and is established and maintained by TrxG activator and PcG repressor proteins. Polycomb repressive complexes 1 and 2 (PRC1 and PRC2) are the most well characterized of the PcG complexes and mediate gene silencing by PRC2 catalyzing deposition of the repressive H3K27me3 histone mark on transcriptionally inactive PcG target genes [115]. Subsequently, PRC1 recognizes and binds this H3K27me3 mark, leading to chromatin compaction and heritable gene repression [115]. Conversely, TrxG complexes activate PcG target genes by establishing activating H3K4me3 and H3K27ac marks [116]. The epigenetic marks that PcG and TrxG generate control gene expression profiles critical for cell fate decisions that include neoplastic growth, tissue regeneration, and stem cell biology. Recent compelling evidence demonstrates that the PcG and TrxG systems act synergistically to generate bivalent (poised or paused) chromatin domains that comprise both activating H3K4m4 and repressing H3K27me3 histone modifications at the same location [117]. The combination of these epigenetic marks at regulatory regions within specific genomic loci that are termed bivalent domains, are considered to poise developmental gene expression for rapid activation while maintaining repression in the absences of differentiation signals [118]. While bivalency remains not fully understood functionally and mechanistically, it is thought to play a critical role in guiding developmental cell fates and mediating neuronal diversity in mammalian systems. Intriguingly, recent elegant work in Drosophila utilizing a PcG reporter fly line revealed the presence of bivalent domains at endogenous PcG target genes that included Hox genes in the fly embryo [117]. These studies demonstrate the existence of bivalency at PcG target genes in Drosophila, support a role for bivalency in regulation of Hox genes, and establish Drosophila as a potential powerful model to further investigate bivalency mechanistic function during in vivo development.

PcG Proteins and Gene Repression

The Polycomb system is made up of three protein complexes: PhoRC (Pho repressive complex), PRC2 (Polycomb repressive complex 2), and PRC1 (Polycomb repressive complex 1) [109]. Each complex is composed of a heterogenous group of factors that work

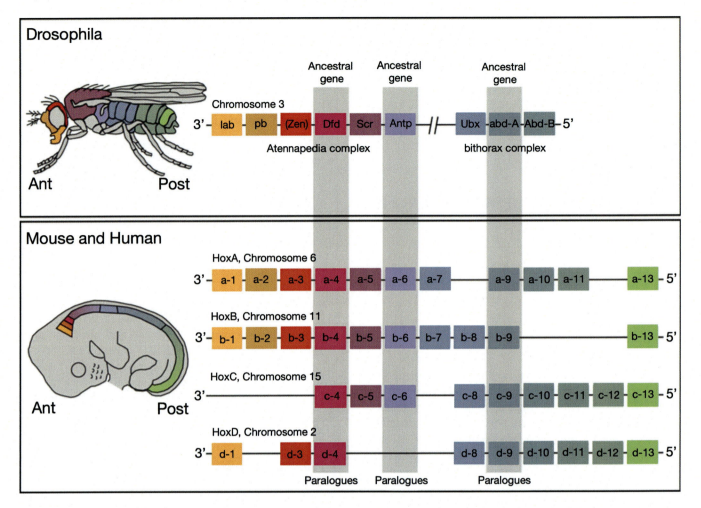

FIGURE 13.3 **Organization of *Hox* clusters in *Drosophila* and mammals.** The *Drosophila Hox* genes (top) are grouped into two genomic clusters: the Antennapedia complex and the Bithorax complex. Expression domains of the individual *Hox* genes within the Antennapedia and Bithorax complexes along the anteroposterior axis of the *Drosophila* embryo match the array of the genes along the chromosome, displaying a property termed collinearity. Similar to *Drosophila*, mice also possess a set of *Hox* genes (bottom) that show conserved temporal and structural collinearity with the embryonic axis such that the lower numbered paralogues are expressed earlier and more anterior on the embryonic axis than the higher numbered paralogues. (See color match between genes and their expression domains on the embryonic axis.) An important difference between these two systems is that the mouse has four clusters (*Hoxa*, *Hoxb*, *Hoxc*, and *Hoxd*) instead of one, as a result of two rounds of gene duplication. The *Hox* complement of the mouse also reveals that some individual *Hox* genes have been duplicated whereas others have been lost in each cluster. Three sets of paralogues and their corresponding ancestral genes are designated by the gray bars [114]. *Source: Adapted from Pang D, Thompson DNP. Embryology and bony malformations of the craniovertebral junction. Childs Nerv Syst 2011;27 (4):523–64.*

cooperatively in a stepwise manner to modify and remodel the chromatin of repressed target *Hox* genes (Fig. 13.4A). PhoRC is the only PcG complex capable of directly binding to target *Hox* DNA, a unique function mediated by one of its members, Pho. PRC2 is recruited to the DNA bound PhoRC complex, and its subunit EZ (enhancer of zeste), a SET domain lysine methyltransferase, catalyzes the tri-methylation of H3K27 on proximal histones [67,108,109]. Additional PRC2 complexes are capable of binding to this modification through the Esc (extra sex combs) subunit, proposed to result in the propagation of H3K27 tri-methylation and the coating of the remainder of the gene body, as is commonly observed in repressed *Hox* chromatin [67]. The PRC1 complex recognizes and binds H3K27me3 through the chromodomain of its component Pc (Polycomb) [107,108]. The main catalytic subunit of PRC1 is the E3 ubiquitin ligase Sce (sex combs extra), and is responsible for catalyzing the ubiquitylation of H2AK118 (H2AK119ub1 in mammals), a histone modification associated with transcriptional repression [107,109,119]. It should be noted that additional proteins other than Pho may also be sufficient for PRC1/2 recruitment and tethering to target *Hox* genes [120]. Furthermore, recent studies showed that PRC1 can also be involved in

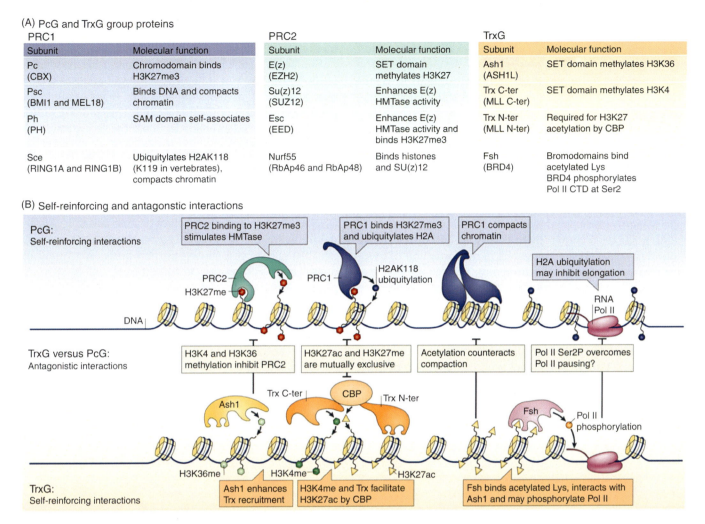

FIGURE 13.4 **Mechanisms of silencing and activation by PcG and TrxG protein complexes.** (A) Subunits of Polycomb repressive complex 1 (PRC1) and PRC2, a selected Trithorax group (TrxG) of proteins are shown. The *Drosophila* protein names are shown with selected vertebrate homologs listed in brackets. The Polycomb group (PcG) and TrxG proteins are recruited to Pc and Trx response elements (PREs) through a platform of sequence-specific DNA-binding proteins. (B) Molecular mechanisms. PREs are thought to be depleted of nucleosomes, and chromatin modifications occur on the flanking nucleosomes. Self-reinforcing (top and bottom) and antagonistic (middle) interactions of PcG and TrxG with chromatin are depicted. Top: Polycomb-mediated silencing. Bottom: TrxG mediated activation. *Source: Adapted from Steffen PA, Leonie R. What are memories made of? How Polycomb and Trithorax proteins mediate epigenetic memory. Nat Rev Mole Cell Biol 2014;15:340–56.*

the recruitment of PRC2, suggesting a bidirectional order of chromosomal tethering that challenges the classical unidirectional model [121].

Several mechanisms underlying PcG-mediated negative regulation of transcription have been proposed based on the purification of each protein complex and the characterization of the interactions between each component protein and their target proteins or genes (Fig. 13.4B) [122]. First, the histone modifications catalyzed by PRC1 and 2 promote chromatin compaction and this transcriptionally repressive structural alteration is reinforced by the interaction of other members of the PcG system with these histone modifications. Second, PcG-mediated transcriptional repression may also be a consequence of its physical interference with various steps required for transcription. For example, PRC1-mediated H2A ubiquitylation has been shown to have a direct role in preventing activated RNA Pol II release from promoter regions, limiting transcription elongation [123]. Similarly, H3K27me3 deposition throughout *Hox*o gene transcriptional units may also impede elongation and Pol II processivity [67]. Third, PcG complexes may also serve to silence gene expression by removing activating histone modifications and/or blocking the recruitment of chromatin remodeling factors responsible for decondensing chromatin. For example, a variant of the PRC1 complex, CHRASCH, contains the histone deacetylase HDAC1, responsible for the removal of activating histone acetyl marks and subsequent chromatin compaction [108]. Lastly, PRC1, specifically Psc, directly disrupts ATP-dependent nucleosome remodeling by the SWI/SNF complex, a homolog of the *Drosophila* Brm complex, preventing chromatin unwinding and thus maintaining a transcriptionally silent chromatin structure [124,125].

Within the past few years, a new model for the simultaneous and coordinated PcG-mediated silencing of multiple distant *Hox* genes has emerged. Electron microscopy (EM) has revealed that that the PRC1 complex, as well as multiple of its core components individually, is capable of converting the open "beads-on-a-string" conformation into a highly compacted, complex 3D chromatin structure [126]. It has been shown that H3K27me3 laid down by EZ bound to target sequences results in the formation of looping structures within the chromatin, restricting access to transcriptional machinery and silencing any promoters within [67–69]. These chromatin loops may facilitate the colocalization of multiple *Hox* genes normally located hundreds of kilobase pairs apart, allowing their interaction and coordinated repression within distinct regions of the nucleus [110]. In agreement with this model, repressed genes from both the Antennapedia and Bithorax clusters, separated within the genome by approximately 10 Mb, have been found within the same PcG repressive structure in the nucleus of the *Drosophila* embryo [127,128]. Recent studies also support that chromatin has an intrinsic capacity to undergo phase separation into functionally distinct, but physically adjacent domains via liquid-liquid phase separation [129]. In support of this, the Chromobox2 (CBX2) subunit of PRC1 complex has been proposed to mediate the phase separation of PcG-repressed chromatin that is necessary for chromatin compaction and proper development [130].

TrxG Proteins and Gene Activation

The Trithorax (TrxG) system is composed of a family of proteins involved in maintaining the activation of *Hox* gene expression and opposing the gene silencing activity of the Polycomb group (Fig. 13.4a,b). TrxG proteins maintain *Hox* gene expression by mediating the histone modification-based open chromatin formation and active chromatin remodeling [131,132]. Ash1 (absent, small, or homeotic disks 1) is a SET domain lysine methyltransferase responsible for the trimethylation of H3K36 [108] and it along with the TrxG protein Trx (Trithorax) are capable of trimethylating H3K4 [67,107]. These marks are not only associated with decondensed chromosomal states but can also directly inhibit the PRC2-mediated deposition of repressive H3K27 tri-methylation marks in certain circumstances [133,134]. Trx is also capable of directly binding the H3K27-specific HAT CBP (CREB binding protein), whose deposition results in both chromatin unwinding and inhibition of PRC2-mediated histone methylation [107,135].

Similar to regulation by PcG proteins, the maintenance of the transcriptional state of TrxG target genes is a consequence of gene accessibility and Pol II processivity. The posttranslational modifications catalyzed by TrxG proteins or TrxG-associated proteins facilitate the reduction in electrostatic interactions between the histone proteins and DNA, thus increasing the accessibility of the DNA template to members of the transcriptional machinery. It has been proposed that TrxG proteins work to modify nucleosomes downstream of the transcription start site thereby facilitating the unencumbered travel of RNA Pol II. Additionally, the vertebrate bromodomain containing protein BRD4 has been shown to bind acetylated lysine residues and promote transcriptional elongation by phosphorylating the C-terminal domain of Pol II [136]. It is speculated that BRD4's *Drosophila* homolog Fsh (female sterile homeotic) may function in a similar manner [107]. Additional TrxG genes have been identified by the ability of their encoded proteins to counteract PcG-mediated silencing and upon further analysis, their specific regulatory functions were shown to involve the modulation of nucleosome positioning in the context of chromatin. One such TrxG gene *brm* (*brahma*) encodes an ATP-dependent chromatin remodeling protein, SWI/SNF, responsible for altering the chromatin structure of positively regulated *Hox* genes, facilitating gene expression [137].

PREs and PcG-TrxG-Mediated Gene Expression

The PcG and TrxG proteins maintain their repressive or activating influence on their target genes by binding to Polycomb group response elements (PREs) or trithorax group response elements (TREs), respectively [108,138]. Only characterized within *Drosophila*, PREs/TREs are required for the maintenance of *Hox* gene expression patterns through embryogenesis [107]. These *cis*-regulatory sequences have been shown to range in size from between a few hundred to a thousand base pairs [108] and are located dozens of kilobases away from *Hox* gene transcription start sites [122]. Shown to be devoid of nucleosomes, their regulatory function is based upon the recruitment and persistent binding of TrxG and PcG proteins resulting in the stable modification and modulation of nearby histones and the establishment and reinforcement of expression patterns.

These domains contain binding sites for DNA-binding proteins involved in the recruitment of PcG and TrxG protein complexes [108]. A few of these targeting proteins characterized in *Drosophila* include PHO, PHOL, PSQ, GAF, Zeste, GRH, and DPS1 [108,122]. It has been proposed that in the repressed state, DNA-binding proteins recruit PcG complexes to PREs, while in the active state DNA-binding proteins recruit TrxG proteins to TREs [132].

Taken together, recruitment of PcG or TrxG proteins to the *cis*-regulatory elements appears to be dependent upon the interaction between the DNA-binding proteins, as well as other local features of the chromatin itself, such as the existing chromatin landscape or the presence of nearby enhancer sequences [107,108]. Importantly, recent evidence suggests that the association of these recruiting proteins with PREs or TREs is not mutually exclusive, and that both groups can be differentially and constitutively bound to both PREs and TREs. Additionally, chromatin-associated lncRNA has been implicated as a factor influencing the bidirectional modulation between TrxG-mediated gene expression and PcG-mediated repression through their interaction with various TrxG/PcG factors and regulatory elements. Thus, it has been proposed that PcG and TrxG factors provide a regulatory platform that can be rapidly switched between expression states based on the developmental needs of the organism at a particular developmental time point [67,107].

While PREs and TREs have not yet been characterized in mammals, the idea that CpG islands, or 1–2 kb DNA regions enriched in C+G content, may serve as the mammalian analog is becoming increasingly popular. Evidence supporting this notion is reviewed in [139,140].

In summary, the TrxG and PcG are well-characterized families of enzymes responsible for the long-term epigenetic control of *Hox* gene transcription. Studies in *Drosophila* elucidating their individual and interdependent functions in modulating gene expression have served as an important foundation for further investigation into their roles in vertebrate and mammalian developmental gene expression.

Transgenerational Epigenetic Inheritance: "Epigenetic Memory"

Once the epigenetic programs regulated by TrxG and PcG are established, they must be faithfully maintained and replicated during division in every cell cycle in a process referred to as transgenerational epigenetic inheritance. The existence of this process is exemplified by the observation that an epigenetically silenced transgene can maintain its repressed state in cells propagated in vitro for more than a year [141]. The primary impediments against epigenetic inheritance during DNA replication and mitosis are the potential disruption of DNA-protein interactions during the passing of the replication machinery along the genome and the possible replacement of parental histones with newly synthesized, unmarked histone proteins. *Drosophila* provides an excellent model to study strategies that have evolved to ensure that epigenetic landscapes are successfully inherited.

Epigenetics During S-Phase

During S-phase, the parental histone octamers dissociate into H3-H4 tetramers and H2A-H2B dimers and are then, along with newly synthesized histones, randomly reassembled on the daughter DNA duplexes [142]. In order to maintain previously established gene expression patterns, the specific combinations of histone modifications must be maintained throughout or reinstalled after replication. It was originally thought that all regulatory protein complexes, including histones and their modifications, were completely removed from parental DNA to allow the passage of the replication machinery and thus required de novo re-establishment after replication. These findings begged the question that if all of the marks are removed, how is the parental pattern of histone modifications reinstated in the absence of any previously established guiding landmarks? Insight into this question was provided by chromatin immunoprecipitation assays in S2 *Drosophila* cell cultures that revealed a dramatic increase in PRC2 binding to PREs in early S-phase with a subsequent drop in binding in late S phase (Fig. 13.5A). These experiments suggested that faithful transmission of parental marks relies upon the excessive deposition of chromatin marks prior to S phase completion rather than de novo deposition mediated by the efforts of histone modifying enzymes after replication [143]. However, a more recent alternative model suggests that histone modifying enzymes may actually remain closely associated with the DNA during replication and thus are able to immediately direct the re-establishment of epigenetic marks onto the newly synthesized and assembled histone proteins [144]. In support of this model, Trx, Pc, and EZ are found within close proximity to the replication fork, and all are bound to their respective PREs/TREs on the nascent DNA near the replication fork (Fig. 13.5B) [145]. A similar mechanism has been proposed for chromatin remodeling proteins that are capable of directly binding to members of the replication machinery [146].

While *Drosophila* must ensure the proper transmission of proteins and histone modifications that activate euchromatic gene expression, it is equally important to direct the proper formation of heterochromatic locations within the daughter chromatin for sustained gene repression. HP1 directs the stable transmission of both itself and repressive H3K9 methylation marks throughout the cell cycle, thus facilitating the inheritance of the parental pattern of epigenetic marks and of structural proteins associated with heterochromatin. During nucleosome reorganization after replication fork passage, the *Drosophila* chaperone protein CAF-1 guides newly

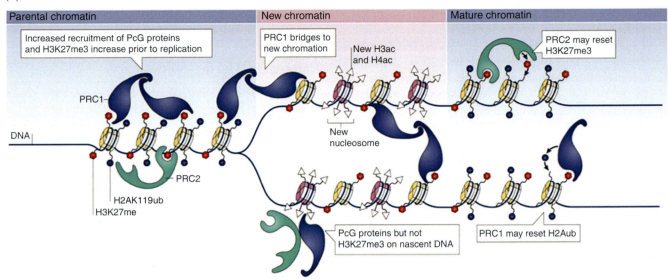

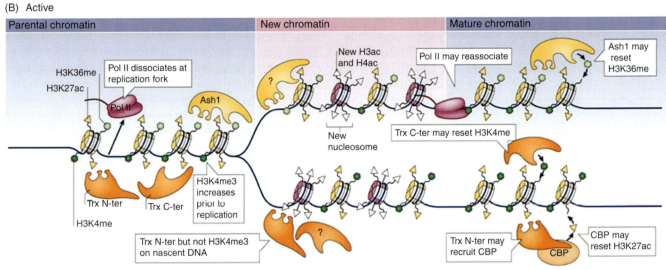

FIGURE 13.5 Replication and maintenance of silent (A) and active (B) states in chromatin. "Generic" active and silent chromatin is depicted, although specific modifications are differently distributed at Polycomb/Trithorax response elements (PREs), promoters and gene bodies. PREs are thought to be depleted of nucleosomes, and chromatin modifications occur on the nucleosomes flanking them. Parental chromatin, newly replicated chromatin and mature chromatin are shown. Source: *Adapted from Steffen PA, Leonie R. What are memories made of? How Polycomb and Trithorax proteins mediate epigenetic memory. Nat Rev Mole Cell Biol 2014;15:340–56.*

synthesized histone proteins to the replicated DNA, where they join inherited parental histones to form the daughter nucleosomes. HP1 forms a complex with CAF-1, functioning to guide HP1 to the site of replication where it binds inherited parental histones displaying the repressive H3K9me3 marks via its chromodomain [131]. In a process similar to heterochromatin spreading during PEV, HP1 binding then mediates the recruitment of the histone methyltransferase *Su(var)3–9* catalyzing the deposition of additional H3K9me3 marks on to neighboring histones and facilitating the spreading of H3K9me3-rich heterochromatic domains onto the appropriate regions of daughter DNA [131,147].

Epigenetics During M-Phase

After replication, mitosis requires the coordinated condensation of chromosomes that, once again, disrupts protein-chromatin interactions and higher-order chromatin structures [131]. Therefore, any proteins directly mediating the formation of compact or open daughter chromatin states, such as HP1 within heterochromatic regions or PRC1 dimers, must re-associate following cytokinesis. For HP1, its dynamic interaction with chromatin during m-phase is dictated by an epigenetic switch referred to as the "methyl/phos switch" [148]. In *Drosophila*, a conserved family of protein kinases called the Aurora-like kinases directs proper chromosome segregation [149]. At the onset of mitosis, Aurora B catalyzes the phosphorylation of H3S10, the residue which neighbors H3K9. Intriguingly, H3S10ph deposition leads to the displacement of HP1 from chromatin, while conversely, depletion of Aurora B results in the aberrant association of HP1 with chromatin during mitosis [150]. Completion of cytokinesis results in the removal of H3S10ph by the protein phosphatase PP1, allowing HP1 to rebind to the H3K9 methyl marks [16,131]. Despite the removal of HP1, the H3K9 methylation mark previously bound by HP1 remains intact throughout mitosis, not only maintaining the epigenetic landscape but providing a scaffold for HP1 re-association after cytokinesis; this allows it to resume its role as a heterochromatic marker in the newly divided cell. Similarly, H3S10ph mediates the removal of HP1 to allow proper m-phase progression while simultaneously serving as an epigenetic placeholder for HP1. Conversely, unlike the repressor HP1, some transcriptional activator proteins like ASH1 and Trx have been shown to remain bound to chromatin during mitosis to faithfully transmit the pattern of gene expression in newly divided cells by both maintaining the landscape of activating H3K4me3 marks and by antagonizing the deposition of PcG-mediated silencing marks within the next generation (Fig. 13.5b). Recently, HP1 in *Drosophila* has also been shown to undergo liquid phase separation that is suggested to facilitate heterochromatin formation in early embryos [151].

In summary, *Drosophila* has served as an important model system to elucidate how epigenetic gene control mechanisms are faithfully inherited so that cell fate and more dynamic regulatory processes such as signal transduction can be effectively maintained after cell division.

DOSAGE COMPENSATION

In many organisms, evolution has resulted in the unequal distribution of the X and Y sex chromosomes with females having two X chromosomes and males having one X and one Y chromosome. This genetic aneuploidy requires a means of compensating for the resulting difference in X-linked gene representation between the sexes. Different organisms have developed different mechanisms to adjust for this imbalance in processes termed dosage compensation. For example, many mammals including *Mus musculus* and humans randomly inactivate one of the two female X chromosomes [152]. In *Drosophila*, dosage compensation is achieved through the coordinated, multi-step efforts of protein complexes and ncRNAs culminating in the epigenetically-mediated enhanced transcription of active genes within the entire X-chromosome in males (reviewed in [153,154]).

The Male-Specific Lethal Complex/Dosage Compensation Complex

The phenomenon of dosage compensation in *Drosophila* was first uncovered in the seminal finding by Mukherjee and Berman. They discovered that the level of [^{3}H]uridine incorporation, indicative of RNA synthesis, by the male X chromosome was equivalent to that of the two female X chromosomes [153]. This was followed by the finding that LOF of four specific genes resulted in the equalization of [^{3}H]uridine incorporation between the sexes and in a dramatic reduction in male viability without altering female phenotypes 32. These genes, *msl* (male-specific lethal)*1*, *msl2*, *msl3*, *mle* (maleless) have since been shown to be a part of the protein/RNA multisubunit complex responsible for dosage compensation in *Drosophila* called the MSL complex, also known as the DCC (dosage compensation complex) [5,154,155].

The DCC is made up of a multi-protein core including MSL1, MSL2, MSL3, MLE, and MOF (males absent of first), along with one of two male-specific ncRNAs *roX1*, *roX2* (RNA of the X) (Fig. 13.6) [154]. MSL2 has been implicated as the director of DCC assembly, initiating formation by binding to and stabilizing MSL1, possibly *via* the interaction between a RING finger domain of MSL2

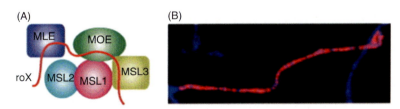

FIGURE 13.6 **A hierarchy of binding sites for the dosage compensation complex (DCC).** Image depicts polytene chromosomes stained for MSL1 in red and DNA in blue. In the wild-type male, the DCC binds to hundreds of sites on the X chromosome. The DCC is made up of a multiprotein core including MSL1, MSL2, MSL3, MLE, and MOF (males absent of first), along with one of two male-specific noncoding RNAs *roX1*, *roX2* (RNA of the X) [154]. MSL2 has been implicated as the director of DCC assembly, initiating formation by binding to and stabilizing MSL1, possibly *via* the interaction between a RING finger domain of MSL2 and a coiled-coil domain within MSL1 [167]. Source: *Adapted from Gilfillan GD, Dahlsveen IK, Becker PB. Lifting a chromosome: dosage compensation in* Drosophila melanogaster. *FEBS Lett 2004;567:8–14.*

and a coiled-coil domain within MSL1 [156,157]. This MSL2/MSL1 core regulates the chromatin-binding activity of the DCC and provides a platform for complex nucleation [157,158]. MOF encodes a member of the MYST family of HATs shown to direct the acetylation of H4K16 and plays an important role in DCC-mediated transcriptional enhancement [153]. The MLE protein has inherent ATPase helicase and single-stranded RNA and DNA-binding abilities [159] and has been proposed to have a dual role within the DCC and in dosage compensation. Its helicase activity is important in DCC-mediated enhancement of transcription and in incorporation of the X-chromosome-encoded roX RNAs into the protein complex [160]. These two ncRNAs are functionally redundant within the DCC, with single mutant males showing no phenotype [161]. However, their incorporation is critical for proper DCC targeting as male flies containing mutant roX RNAs display a decrease in the localization of DCC proteins along the X chromosome [161]. Although the two ncRNAs are different in size and in the majority of their sequence, they do share a sequence called the roX box, comprised of highly structured tandem stem loops [162], which enhance MOF-mediated histone acetylation in vivo [163]. Incorporation of roX RNA into the DCC complex is a multi-step process. First, the RNA helicase activity of MLE remodels critical stem-loop structures in the roX RNA to release alternative secondary structures [164,165]. This RNA remodeling further unmasks binding sites for MSL2 allowing for their integration [164,165]. In addition to MSL2, MSL3 and MOF also have direct RNA binding abilities through their N-terminal chromodomain, and binding directly with roX RNA is essential for their integration into the DCC [154,165,166]. Therefore, roX1/2 RNA serve as a platform for assembly of a functional DCC on the male X-chromosome.

Regulation of Dosage Compensation Complex Targeting to the X Chromosome

Proper upregulation of active male X-linked genes is dependent upon appropriate assembly of the DCC along the male X chromosome, while avoiding the female X chromosomes or the autosomes. The sex-specific chromosomal targeting of the DCC is crucial for viability [153] and thus, Drosophila has evolved a mechanism to ensure dosage compensation is present in males but absent in females. SXL (sex-lethal) is an RNA−binding protein whose fully functional form is only expressed in females. SXL recognizes and binds to polypyrimidine tracts within the 5′ and 3′ UTR of the *msl2* mRNA [168], where it recruits UNR (upstream N-Ras). With both bound to the *msl2* transcript, they then act cooperatively to inhibit MSL2 translation in females [153,169,170]. Conversely, in Drosophila males, alternative splicing of the *sxl* transcript results in the expression of a non-functional SXL protein [171], thereby enabling successful MSL2 protein synthesis and DCC formation in males. Additionally, the assembly of the DCC complex exclusively on the X chromosome is a result of the recognition of specific sites within the X-encoded *roX* genes by members of the DCC along with the co-transcriptional incorporation of *roX* RNA into the DCC complex. Studies show that MSL2 is responsible for *roX1* and *roX2* expression [172,173], suggesting that this interaction may serve as a nucleation site for X-linked DCC complex formation. Similarly, is has been demonstrated that the incorporation of *roX* RNAs into partial DCC complexes occurs at the site of *roX* RNA synthesis [160]. These findings support a model by which the role of X-encoded *roX* genes is to serve as the nucleation site for DCC assembly, mediating complex formation specifically on the X chromosome.

Intriguingly, approximately 35 chromatin sites along the X chromosome have been shown to bind the MSL1/MSL2 core with only two of those sites corresponding to the *roX1* and *roX2* genes [174], indicating additional targeting sequences exist. Chromatin immunoprecipitation performed in conjunction with high-throughput sequencing (ChIP-seq) identified 150 DCC targeting sequences dubbed chromatin entry sites (CES) or high-affinity sites (HAS) (Fig. 13.7), with further analysis characterizing a GA-repeat motif present in 91% of identified HASs and enriched on the X chromosome compared to autosomes [175,176]. This latter

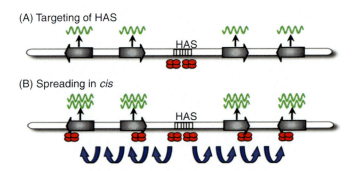

FIGURE 13.7 Model of dosage compensation. The two-step model of MSL targeting to the X chromosome. (A) The MSL complex (*red circles*) targets specific HAS on the X chromosome in a sequence-dependent manner. HAS represent a subset of the MSL-bound regions in wild-type Drosophila that also recruit the MSL complex under more stringent conditions (such as when inserted into an autosome, or when integral subunits of the MSL complex are missing). (B) After initial targeting, the MSL complex spreads along in *cis* from the entry sites (shown by *blue arrows*), and predominantly binds to the 3′ end of actively transcribed genes. MSL binding causes acetylation at histone H4 and results in a global change of the chromatin structure, facilitating a twofold transcriptional upregulation of X-linked genes in males (*green lines*). Source: *Adapted from Bachtrog D, Toda NRT, Lockton S. Dosage compensation and demasculinization of X chromosomes in Drosophila. Curr Biol 2010;16:1–16.*

motif, named the MSL recognition element (MRE), was found to be, in addition to the roX DHS sequences, sufficient for DCC recognition and reporter gene expression [175]. Compelling evidence points to the zinc finger protein CLAMP (chromatin-linked adaptor for MSL proteins) as the main regulator of the chromosome-specific targeting of the DCC. The CLAMP protein is capable of both binding MREs and recruiting the DCC to these recognition elements located on the X chromosome, providing a direct link between DCC and the X chromosome [177,178]. Therefore, taken together, these findings suggest that sequences within *roX* genes, in concert with a defined distribution of HASs recognized by the CLAMP protein, likely play key roles as regulators of the recognition and localization of DCCs along the length of the X chromosome.

While dosage compensation via assembly of the DCC on the X chromosome in males is critical, it is equally important that the autosomes are untouched by this mechanism. Insights into how this chromosome-specific regulation occurs came with the discovery that the *mof* gene resides on the X chromosome and that the recruitment of the DCC complex proteins to the X chromosome is dependent upon its transcriptionally active state [180]. In addition, an alternative mechanism by which X-linked MOF expression aids in the targeting of DCC away from autosomes has been proposed. In this model, termed the inverse dosage model, expression and subsequent incorporation of MOF into the DCC, along with its association with the X chromosome, results in the sequestration of the MOF protein away from autosomes [180,181]. This alternative model suggests that the purpose of the DCC is not only to increase the expression of active X-linked genes, but also to prevent any H4K16ac-mediated transcriptional enhancement of autosomal genes by DCC-independent MOF activity [182].

Dosage Compensation Complex Spreading and Target Recognition

While the *roX* gene and HAS sequences are critical for DCC complex formation and localization on the X chromosome, they are not universally found among the target active genes, implicating the existence of another set of targeting sequences that mediate DCC recognition of gene targets. It has been proposed that the DCC, after recognizing and binding to specific HASs on the X chromosome, initiates the epigenetic-based remodeling of the local chromatin, which would allow the diffusion or spreading of DCCs along the X to lower affinity binding sites that correspond to the bodies of active genes (Fig. 13.7) [179]. Whether DCC spreading is a linear process involving the continuous DCC-chromatin association as it scans along the chromatin or one involving a bind-and-release mechanism of chromatin sampling remains unclear. However, this process requires the action of MLE's ATP-dependent helicase activity and/or MOF's histone acetyltransferase activity, indicating that local chromatin remodeling is likely the underlying mechanism for DCC spreading [160,183–186]. It has been proposed that DNA duplex unwinding by MLE and chromatin conformation relaxation mediated by MOF-directed histone acetylation potentially results in both the increased ability of the DCC to travel unfettered along the chromatin and may also expose nearby low affinity sites for DCC recognition and binding. Alternatively, it has been suggested that DCCs spreading away from HASs is a direct result of elevated DCC concentrations around the entry sites [182]. In this model, the high-affinity sites would serve as DCC "hubs," constantly recruiting complexes to their location after the previous complex leaves, thereby driving the increase in the proximal concentrations of DCC.

Transcriptional Activation of Active X-Linked Genes by the Dosage Compensation Complex

Although enhancement of transcription is the basis for dosage compensation, the exact mechanism by which this is achieved remains unclear. Two main theories, both involving epigenetic modifications, have been proposed and it is not unreasonable to believe that both can play a role in the targeted upregulation of transcription during dosage compensation. In the first model, epigenetic-based chromatin remodeling results in the unwinding of chromatin providing the transcription machinery with a more open conformation, facilitating elongation and RNA Pol II progression through the target gene region. This hypothesis is based on the findings of high levels of H4K16ac marks, which are generally transcriptionally active, found mostly within the body and especially within the 3′ end of the active transcriptional units [187,188]. In fact, global run-on sequencing (GRO-seq) demonstrated more efficient RNA Pol II movement through X-linked gene transcribed units than through autosomal genes, and a dependence of efficiency on the presence of H4K16ac [189,190]. These findings support a model by which H4K16ac-dependent chromatin relaxation leads to transcriptional upregulation *via* facilitation of elongation by possibly presenting Pol II with fewer obstacles during elongation providing a more favorable platform for gene transcription progression. In the second model, histone acetylation may actively counteract factors responsible for condensation of chromatin, which would normally disrupt DCC function. H4K16ac has been shown in vivo to antagonize the

ISWI family of ATP-dependent chromatin remodeling agents involved in the maintenance of higher-order chromatin structure [191] by establishing a chromatin structure inaccessible to ISWI proteins at gene promotors, thereby regulating its influence on the degree of chromatin compaction and gene expression during dosage compensation.

In summary, insights into the epigenetic mechanisms that underlie multiple facets of dosage compensation in flies, including DCC formation, targeting, binding, and spreading, culminating in X-chromosome-specific gene transcriptional activation and survival in male flies has been elucidated. Although mechanisms underlying dosage compensation vary in mammals with X-inactivation in females, early genetic and biochemical experiments in *Drosophila* have guided our understanding of epigenetic processes compensating for expression of genes on the sex chromosomes.

THE EPIGENETIC LANGUAGE IN POSTMITOTIC NEURONS UNDERLYING COGNITIVE FUNCTION

Postmitotic neurons within the brain have the remarkable ability to shape and refine their synaptic connections in response to external stimuli. This adaptability relies on the precise regulation of specific gene programs and is the foundation of transient and sustained synaptic plasticity. Epigenetic histone modifications have emerged as the primary mechanism by which neuroadaptation is transcriptionally achieved in response to external cues. Because the addition and removal of histone acetyl marks can be an extremely rapid process, it is well suited as the dynamic epigenetic link between the constantly changing environment and the molecular responses necessary for experience-driven behavioral changes. Furthermore, the balance between HAT and HDAC functions is important for establishing and maintaining the epigenetic landscape required for directing important neuronal functions such as learning and memory. The disruption of this delicate balance results in the pathogenesis of neurodegenerative disorders. *Drosophila* has proven to be a powerful model system for elucidating epigenetic gene control mechanisms underlying both learning and memory as well as cognitive deficits associated neurological disorders [192–194].

Activity-Dependent Histone Acetylation and Cognitive Function

Neural networks are able to constantly adapt to the ever changing environment by converting a transient external stimuli to long-lasting alterations in neuronal structure and function. These synaptic plasticity changes serve as the basis for long-term memory (LTM) formation [195], and are driven by epigenetic mechanisms that regulate dynamic gene transcription in response to neuronal activation [196,197]. The initiation of these changes in synaptic strength occurs at the level of the individual synapse, where influences from the extracellular environment trigger downstream signaling pathways that eventually terminate in epigenetic-mediated expression alterations. Beginning this process, neurotransmitters, growth factors, or cytokines are released from surrounding neurons and facilitate the influx of calcium into the posy-synaptic neuron through binding to their cognate receptors. Intracellular calcium then serves as a powerful second messenger, activating a multitude of downstream signaling cascades, some of which have been shown to modulate the activity of specific HATs and HDACs, which can influence learning and memory via their specific chromatin mediating enzymatic activities. Nuclear shuttling of HATs in response to external cues was first identified in Tip60 HAT [198,199]. Upon extracellular stimulation, Tip60 was subcellularly redistributed inside the hippocampal neurons, resulting in a significant increase in the nuclear accumulation of Tip60 [198,199]. This increase in Tip60 nuclear localization corresponded with increased chromatin binding at synaptic gene loci resulting in activation of synaptic gene expression [199] (Fig. 13.8). Further, DNA fluorescent in-situ hybridization (DNA-FISH) experiments revealed that hippocampal stimulation induces colocalization of these synaptic plasticity genes and Tip60 with RNA-Pol II-rich transcription factories, which is suggested to play an integral step in activity-dependent regulation of gene transcription [199]. The role of Tip60 in memory formation is further supported by the finding that loss of Tip60 in the *Drosophila* brain results in histone hypoacetylation and gene dysregulation with concomitant defects in axon formation and immediate-recall memory [194,200]. Similar to this mechanism, the transcription factor CREB (cAMP response element binding protein) is activated by calcium-mediated phosphorylation which then recruits the HAT CBP (CREB-binding protein) to chromatin mediating specific genes expression changes [84,201,202]. Moreover, targeted knockdown of CBP HAT activity within the *Drosophila* mushroom body results in the attenuation of LTM formation [203]. Activity-dependent transcriptional activation has also been suggested to be a consequence of HDAC export out of the nucleus, permitting HAT and transcription factor-mediated gene activation [204]. Specifically, ClassIIa HDACs (HDAC 4,5,7,9) have been shown to undergo signal-dependent phosphorylation which brings about nuclear export of the enzymes and de-repression of their targets [205,206]. Taken together, these findings indicate that the activation of intracellular signaling pathways leads to enzymatic alterations in the epigenetic landscape, which mediate the transcriptional

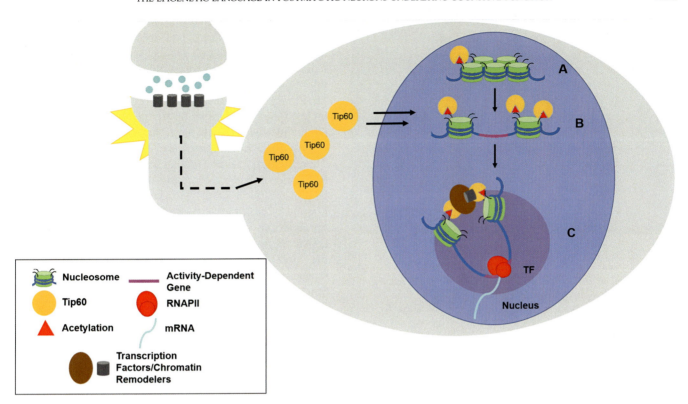

FIGURE 13.8 **Extracellular stimulation induces nuclear shuttling of Tip60 HAT for activity-dependent gene expression.** (A) Under basal conditions, Tip60 is found both in the cytoplasm and the nucleus. Intranuclear Tip60 may be associated with certain synaptic plasticity genes prior to stimulation, representing a feature of rapidly inducible genes. (B) Upon KCl-mediated extracellular stimulation, Tip60 shuttles inside the nucleus and is recruited to synaptic plasticity genes. Tip60's histone acetyltransferase activity then facilitates chromatin relaxation, potentially exposing regulatory sequences and recruiting additional transcription factors and/or chromatin remodelers, mediating higher-order chromatin conformational changes. (C) The resulting higher-order chromatin conformational changes facilitate the recruitment of synaptic plasticity genes to RNA Polymerase II (RNAPII)-rich transcription factories, thereby inducing their rapid, activity-dependent co-transcription. Source: *Adapted from Karnay A, Karisetty BC, Beaver M, Elefant F. Hippocampal stimulation promotes intracellular Tip60 dynamics with concomitant genome reorganization and synaptic gene activation. Mol Cell Neurosci 2019;101:103412.*

changes necessary for neuronal functional flexibility and cognitive function.

Histone Acetyltransferase: Histone Deacetylase Interplay in Cognitive Function

Drosophila is an attractive model for studies focused on epigenetic dissection of components of memory formation due to the existence of genetic tools that enable restricting gene expression manipulation to specific subregions of the brain and the availability of reproducible memory assays [207–209]. In particular, the *Drosophila* mushroom body (MB) is known to regulate a variety of behavioral and physiological functions such as from learning and memory, courtship conditioning, and decision making under uncertain conditions [26–30]. While the cellular and molecular mechanisms involved in memory formation and maintenance have long been a topic of research, new evidence indicates that epigenetic regulation plays a key role in these fundamental processes. Specifically, the regulation of global histone acetylation by HATs and HDACs has been shown to be of critical importance in the epigenetics-based regulation of cognitive function. In a series of experiments within the MB of the fly, appropriate levels of Tip60 HAT activity was shown to be crucial for immediate recall memory [210] (Fig. 13.9). Conversely, the reduction in acetylation levels induced by the targeted overexpression of the HDAC Rpd3 and HDAC4 within the adult *Drosophila* mushroom body resulted in severely impaired long-term courtship memory [138,211]. These studies support the premise that under normal conditions, maintaining the balance between HAT and HDAC levels and activity is crucial for establishing appropriate histone modification patterns that serve to control both stable and rapidly changing gene expression profiles critical for both neuronal homeostasis, and appropriate neurophysiological response outputs such as long-term potentiation, and learning and memory, respectively [212].

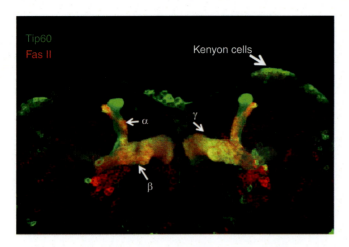

FIGURE 13.9 *Drosophila* mushroom body (MB) neurons. Shown, are MB neurons co-immunostained with antibodies that label Tip60 shown in green and Fasciclin II (Fas II) shown in red. Fas II is a cell adhesion molecule that is expressed strongly in the MB α/β lobes and weakly in the γ lobe in the MB of the adult fly brain. Appropriate levels of Tip60 are required for axon outgrowth in the adult *Drosophila* MB.

Appropriate HAT levels within the central nervous system have been shown to be required for the proper development and function of neural morphology and functional processes implicated in cognition. Experiments in *Drosophila* have shown that reducing Tip60 HAT levels within the mushroom body causes significantly shorter mushroom body lobes in the adult fly brain, indicative of defects in axonal outgrowth [194]. Further, loss of Tip60 HAT activity in motor neurons causes axonal transport stalling [210], and loss of Tip60 in the sLNv neurons that regulate sleep causes repressed axonal outgrowth [213]. Functional consequences of these Tip60-mediated defects include impairment in short-term memory, sleep/wake cycles and locomotor activity [194,213,214]. Additionally, Tip60 HAT activity has been shown to regulate proper synaptic bouton expansion at the *Drosophila* neuromuscular junction through modulation of the synaptic microtubule cytoskeleton [92]. Intriguingly, the activity-dependent remodeling events that shape the formation of neural circuits during development, mirror the events underlying changes in synaptic connectivity required for learning and memory formation. Experience-dependent neuroadaptation requires increases in synaptic bouton densities and dendrite extensions that are similar to those essential during development [215]. Therefore, it has been hypothesized that the signaling cascades that result in experience-driven changes in cognitive function lead to the modulation of epigenetic-mediated transcriptional control of genes that promote synaptic growth and neuronal homeostasis. Although the identity of the genes responsible for the morphological and molecular changes underlying learning and memory has not been completely characterized, advances in high-throughput sequencing technology has begun to tease apart the epigenetically-regulated genes involved in synaptic plasticity and neuronal development. For example, microarray analysis of Tip60 HAT-mutant flies revealed that Tip60 regulates the expression of genes whose functions include neuronal development and synaptic function [200]. Further studies have since revealed that several synaptic plasticity genes are *bona fide* direct targets of Tip60, where loss of Tip60 activity in mutant flies results in reduced Tip60 chromatin enrichment and synaptic gene repression [216]. Moreover, ChIP-Seq analysis in *Drosophila* reveals that Tip60 target genes are enriched for functions in cognitive processes and accordingly, key genes representing these pathways are misregulated in the Tip60 HAT mutant fly brain [194]. Together, these studies using the *Drosophila* model highlight the importance of specific HATs such as Tip60 in the epigenetic regulation of cognitive function.

Use of *Drosophila* Disease Models to Study Dysregulation of HAT Activity in the Etiology of Human Neurodegenerative Diseases

Neurodegenerative disorders, such as Alzheimer disease (AD), Parkinson disease (PD), Huntington disease (HD) and Amyotrophic Lateral Sclerosis (ALS), are marked by progressive loss of neurons within the central nervous system. Due to the complex nature of these disorders and our limited understanding of the underlying pathogenic mechanisms, current treatments for most neurodegenerative disorders only target the associated symptoms without curing the disease itself [217]. Although involvement of genetic factors, such as DNA mutations, have long been associated with several of these disorders, recent studies support a complex interplay between genetic and environmental factors as the major causative factor [218–220]. More importantly, histone acetylation dysregulation has emerged as a common epigenetic mechanism underlying multiple kinds of neurodegenerative diseases [212,221–224], suggesting histone acetylation-mediated gene regulation is critical for proper neuronal function and its disruption triggers neuronal susceptibility to degeneration.

Alzheimer disease (AD) is one of the major age-related neurodegenerative disorders that accounts for 60%–80% of all dementia cases [225,226]. The selective atrophy of the hippocampus and cerebral cortex and the progressive cognitive deterioration characteristic of AD are associated with aberrant aggregations of intraneuronal Aβ (amyloid beta) plaques and interneuronal tau neurofibrillary tangles. Several powerful models of human AD have been developed using *Drosophila* as a model system, providing scientists with the ability to study the epigenetic mechanisms behind this neurodegenerative disease from early presymptomatic to late terminal stages [7]. These models utilize the overexpression of β-amyloid, amyloid precursor protein (APP), presinilin, and human tau proteins resulting in phenotypes ranging from physical abnormalities such as eye degeneration, locomotor and cognitive deficits, to subcellular defects such as accumulations of amyloid plaques, neurofibrillary tangles, and neuromuscular junction (NMJ) and axonal degeneration [7]. Accumulating evidence garnered from such AD Drosophila models suggests that the clinical symptoms associated with AD are the result of early disease impairment of epigenetic mechanisms of gene control that cause synaptic degeneration [220,227]. For example, AD model flies which display deficits in both learning and short-term memory harbor a significant reduction in histone acetylation levels, loss of cognition-associated gene expression, and an increase in pro-apoptotic gene expression early during disease progression that then induces neurodegeneration [192,214,228]. Notably, tau-induced neurodegeneration in a Drosophila AD model has shown to alter chromatin structure by loss of heterochromatin, which in turn activates expression of genes that were otherwise silenced in controls [229].

Importantly, AD research specifically studying Tip60 has provided a wealth of intriguing and exciting information regarding the molecular mechanisms underlying AD pathogenesis as well as potential therapeutic strategies for the future. These studies show that there is an imbalance between Tip60 HAT and HDAC2 in the fly brain under APP induced neurodegenerative conditions, such that loss of Tip60 and gain of HDAC2 results in reduced histone acetylation at several synaptic gene loci that in turn represses synaptic gene expression in the APP fly brain [216]. Remarkably, increasing Tip60 HAT levels in the Drosophila nervous system under APP induced neurodegenerative conditions is sufficient to rescue AD-associated neuronal impairments such as apoptotic neurodegeneration in the central nervous system (CNS) [228], axonal outgrowth [213,230], synaptic vesicle transport in motor neurons [214], and mushroom body morphology [216]. Increasing Tip60 also restored associated disrupted complex functional abilities impaired in AD that include sleep cycles [213,230], locomotor function [214], learning and memory defects [194,216], and rescue of some genes critical for the function of these neural processes [214,216,228]. Similarly, disruption of Tip60/HDAC2 balance was also observed in the Drosophila Aβ (amyloid beta) induced neurodegenerative conditions, where increasing Tip60 HAT levels protected against several AD-associated phenotypes including Aβ plaque formation, short-term memory, locomotion, and longevity [231]. Additionally, similar to the Drosophila APP and Aβ induced neurodegenerative conditions, alterations in the Tip60/HDAC2 balance has been observed to be an early event common to several other Drosophila neurodegenerative models, such as the mutant β-synuclein (PD), the polyglutamine repeats in the Huntingtin gene (HD) and the overexpression of VAMP-associated protein (ALS) induced neurodegenerative conditions in the Drosophila nervous system [232]. These alterations in Tip60 and HDAC2 expression levels in the neurodegenerative fly brain directly affect their chromatin binding at target gene loci, therefore resulting in reduced histone acetylation and repression of synaptic genes with concomitant defects in synapse formation at the NMJ and associated behavioral phenotypes, such as locomotion and short-term memory [232]. Moreover, rescue of Tip60 HAT levels in the Drosophila mushroom body protected against locomotor and short-term memory deficits in multiple neurodegenerative conditions [232], suggesting Tip60 provides neuroprotection against a broad range of neurodegenerative phenotypes and conditions. Together, these findings demonstrate that Tip60-mediated restoration of neural histone acetylation homeostasis is generally neuroprotective and exemplify Drosophila as an effective tool to dissect epigenetic mechanisms underlying neurodegenerative disorders (Fig. 13.10).

Histone Acetylation Targeted Therapeutics for Neurodegenerative Disorders

Based on epigenetics studies in Drosophila and rodent models, restoration of histone acetylation based therapy is currently being tested for treatment of multiple neurodegenerative conditions like AD [233,234], PD [235], HD [236] and ALS [224]. HDAC inhibition-mediated histone acetylation restoration has been shown to be promising where feeding HDAC inhibitors to flies rescued altered gene expression [237], neurotoxicity [237], neuronal degeneration [237–239], locomotion, [240] and longevity [238] under several Drosophila neurodegenerative conditions. However, applicability of HDAC inhibitors as safe-cognition promoting therapeutics is limited by their lack of specificity as they act to increase global acetylation [234,241,242]. Alternatively, since many HATs have nonredundant functions and increasing specific HAT activity has been shown to rescue several neurodegenerative phenotypes [212,216,243], HAT activators have emerged as novel potential alternatives to HDAC inhibitors for neuroprotection. In Drosophila Aβ42-induced neurodegeneration, increasing CBP HAT levels is capable of rescuing neurodegenerative phenotypes modeled in the Drosophila eye, neuronal apoptosis, and axonal targeting defects [244]. Similarly, Tip60 HAT protects against neurological impairments in different neurodegenerative diseases via similar modes of action [216,231,232]. A recent genome-

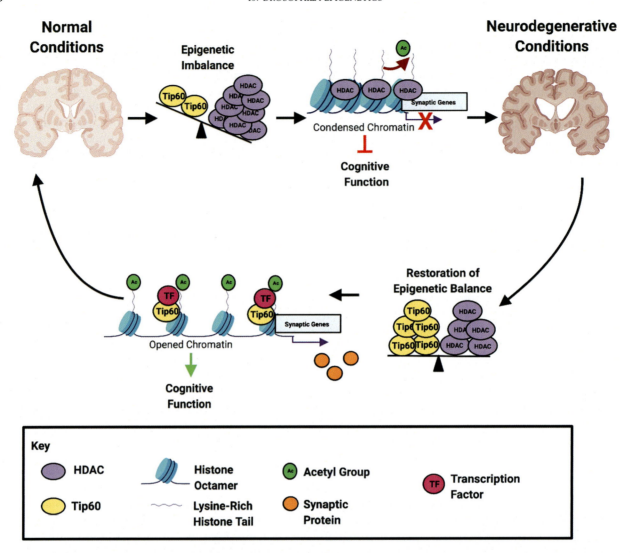

FIGURE 13.10 **Tip60 mediated restoration of histone acetylation homeostasis is neuroprotective in multiple neurodegenerative conditions.** Neurodegenerative diseases are characterized by early impaired acetylation homeostasis that consequently lead to altered neuronal transcription profiles and cognitive decline. Conversely, modulation of cellular levels of specific HATs, such as Tip60, protects against altered acetylation homeostasis in the brain to maintain appropriate neuroplasticity gene expression profiles and neural health. Thus, targeting specific HATs for therapeutic intervention may offer more promising alternatives for neurodegenerative diseases than currently available HDAC inhibitors. *Source: Adapted from Beaver M, Bhatnagar A, Panikker P, Zhang H, Snook R, Parmar V, et al. Disruption of Tip60 HAT mediated neural histone acetylation homeostasis is an early common event in neurodegenerative diseases. Sci Rep 2020;10(1):18265.*

wide chromatin immunoprecipitation and sequencing study in the *Drosophila* APP neurodegenerative brain has uncovered mechanisms underlying Tip60/HDAC2 chromatin binding and its implications for Alzheimer disease [245]. Tip60 and HDAC2 were found to be co-recruited to similar and in some cases identical genomic loci within the neural genome *via* common transcription factors. Accordingly, Tip60 and HDAC2 co-regulate expression of a similar set of genes that function in cognition linked neural processes disrupted early in AD progression [245]. Further, increased HDAC2 in the APP brain was shown to displace Tip60 chromatin recruitment that may be initiated at co-Tip60/HDAC2 docking sites, causing harmful changes in gene expression that persist and worsen during disease progression (Fig. 13.11). Increasing Tip60 levels in the APP brain was found to exert its neuroprotective role by either reducing HDAC2 levels [216] and/or by displacing inappropriately enhanced HDAC2 chromatin binding to restore Tip60 mediated gene regulation [245]. Moreover, since HDAC2 binding was detected at far more genes that were altered in the APP brain, HDAC2 is suggested to have reduced target gene specificity as compared to Tip60 [245]. Lastly, since Tip60 chromatin binding is altered in the APP larval brain well before amyloid plaque accumulation and lethality [216,245], loss of Tip60-mediated chromatin modeling is an initial event in neurodegenerative progression that can be targeted early in disease treatment. Together, these

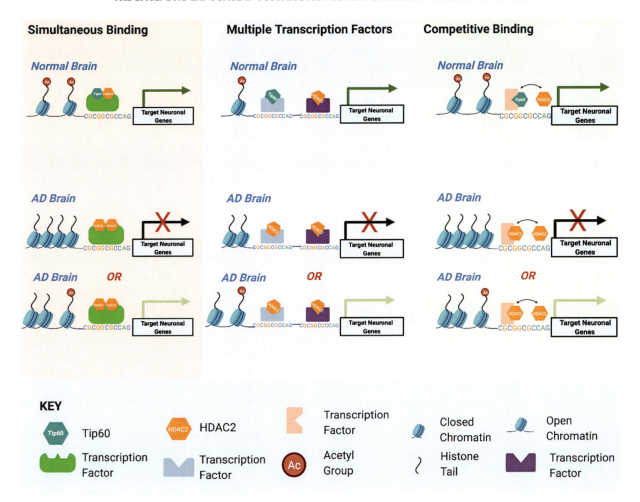

FIGURE 13.11 Mechanisms underlying Tip60/HDAC2 co-mediated neuronal gene regulation in normal brain vs Alzheimer brain. Transcription factors (TF) serve as docking sites for recruitment of Tip60 and HDAC2 to similar set of neuronal genes either simultaneously to the same TF, separately to multiple TFs within close proximity to one another or competitively to a given TF. Tip60/HDAC2 imbalance in Alzheimer disease causes enhanced HDAC2 recruitment with concomitant gene disruption where gene expression is either increased or decreased, resulting in impaired neuronal processes that contribute towards neurodegeneration. Increasing Tip60 levels displaces inappropriately enhanced HDAC2 binding to re-gain control over neuronal gene expression, therefore providing neuroprotection early in the neurodegenerative process. Source: *Adapted from Beaver M, Karisetty BC, Zhang H, Bhatnagar A, Armour E, Parmar V, et al. Chromatin and transcriptomic profiling uncover dysregulation of the Tip60 HAT/HDAC2 epigenomic landscape in the neurodegenerative brain. bioRxiv 2021:2021.02.27.433179.*

findings underscore use of HAT activation as a potential therapeutic alternative to HDAC inhibition for treatment of neurodegenerative conditions like Alzheimer disease.

Environmental Enrichment Improves Learning and Memory in *Drosophila*

The ability of the neuronal circuitry to morphologically adapt to changing environmental conditions is the basis for experience-driven higher-order behavioral functions such as learning and memory. Enhanced sensory and motor stimulation through interactions with an "enriched" environment that is comprised of positive social reinforcements has been shown to lead to alterations in gene transcription profiles, modulation of the morphology of synaptic connection, and even facilitation of learning and memory, in a variety of organisms (reviewed in [246]). Environmental enrichment (EE) conditions have also been shown to have neuroprotective benefits under neuropathological conditions such as AD [247–250]. These observations have allowed scientists to gain insights into the interactions between genes and the environment and the impact of environmental enrichment on neuroadaptation. While experimental EE conditions may vary between studies exploring EE neuroadaptive benefits, one critical and non-variable EE component widely conserved amongst species is social environmental enrichment

[251,252]. Well established studies using *Drosophila* show that similar to mammals, social EE promotes significant beneficial structural changes in regions throughout the fly brain that include the mushroom body (MB) [26–28,30,253]. Social EE promotes enhanced MB axon and dendrite formation, synaptic plasticity, and neuronal MB Kenyon cell growth [251,254]. Recent studies demonstrate that EE benefits require epigenetic gene regulation involving induction of specific histone acetylation profiles [248,255–257]. For example, CBP has been shown to extrinsically regulate environment-induced adult neurogenesis, suggesting CBP-dependent transcriptional neuroadaptation is an important mediator of EE-induced benefits [255]. Nevertheless, how specific HATs mediate cognitive gene expression programs in response to changing environmental cues and the select HATs involved in this process remain largely unknown.

Recent studies have exploited the power of *Drosophila* genetics and the behavioral and physiological conservation between flies and mammals in terms of their positive neuroadaptive response to investigate whether Tip60 HAT action is required for an EE induced beneficial neuroadaptive response. These studies revealed that flies raised under EE conditions displayed enhanced MB axonal outgrowth, synaptic marker protein production, histone acetylation induction and transcriptional activation of cognition linked genes when compared to their genotypically identical siblings raised under isolated conditions [198]. Further, these beneficial changes were impaired in both Tip60 HAT mutant flies and APP neurodegenerative flies. While EE conditions provided some beneficial neuroadaptive changes in the APP neurodegenerative fly MB, such positive changes were significantly enhanced by increasing MB Tip60 HAT levels [198]. Together, these results implicated Tip60 as a critical mediator of EE-induced benefits, and provide broad insights into synergistic behavioral and epigenetic-based therapeutic approaches for treatment of cognitive impairment.

CONCLUSION

In conclusion, *Drosophila melanogaster* has proven to be a powerful model system for elucidating the fundamental role epigenetic control mechanisms play in various conserved biological processes. In this chapter, we have summarized studies utilizing *Drosophila* that have proven invaluable for shedding light into fundamental concepts regarding the function of the generation and maintenance of specific epigenetic patterns in chromatin that govern processes unique to the fly such as X-linked dosage compensation, as well as in processes conserved in humans such as epigenetic mechanisms underlying gene control, PEV, developmental programming, epigenetic inheritance, and neuronal plasticity guiding learning and memory. Examples are also given in how *Drosophila* represents a uniquely powerful tool for analyzing the detrimental consequences of aberrations in epigenetic patterns in neurological disorders and thus represents a valuable resource for the development of potential epigenetic-based therapeutic interventions for cognitive deficits that are a hallmark of neurodegenerative disorders.

With regards to future directions in the field of *Drosophila* epigenetics, while it is clear that histone acetylation is a key mechanism through which the constantly changing environment can influence cognitive functions, exciting new research in *Drosophila* is aimed at tearing apart precisely how histone acetylation mediates environmental enrichment-induced neuronal survival, and the mechanisms through which these changes can enhance learning and memory [198]. Conversely, understanding how disruption of the normally tightly regulated mechanisms of acetylation can result in abnormal changes in neuronal morphology and circuitry and lead to learning and memory impairment is key to understanding neurodegenerative diseases [194,198,210]. Recognition of the pivotal importance of histone acetylation has led to the hypothesis that HAT-based epigenetic therapeutic strategies can provide neuroprotective benefits in AD, a concept currently being developed in *Drosophila* and being ascertained in ongoing research for their applicability in mammalian systems.

PROTOCOL

Immunohistochemistry in adult *Drosophila* brains

Before You Begin

Timing: 60–120 minutes

1. Prepare the following solutions:
   a. 1X PBS, pH 7.2.
   b. 0.2% Triton-X in 1X PBS (PBT).
   c. 4% paraformaldehyde in PBS (fixing solution).
   d. 5% normal goat serum in PBS (block solution).
2. Prepare Sylgard dissection dishes.

Note: For best results, prepare fresh 4% paraformaldehyde solution every 1–2 weeks if kept at room temperature or up to 4 weeks if kept at 4°C.

Key Resources Table

Reagent or Resource	Source	Identifier
Antibodies		
Horseradish peroxidase (HRP, neuronal membrane)	Abcam	ab34885
Fasciclin II (Fas II, mushroom body)	DSHB	1D4
Bruchpilot (BRP, synaptic active zones)	DSHB	nc82
Prospero (neural stem cells)	DSHB	Mr1A
Tip60/KAT5 HAT	Abcam	ab23886
GFP	Abcam	ab290
Alexa Fluor 488	Abcam	ab150077
Alexa Fluor 568	Abcam	ab175473
Alexa Fluor 647	Abcam	ab150115
Chemicals, Peptides, and Recombinant Proteins		
1X PBS, pH 7.2	Gibco	20012043
Triton-X100	Sigma-Aldrich	X100
Paraformaldehyde	Sigma-Aldrich	252549
Normal goat serum (NGS)	Abcam	ab7481
Sylgard 182 Silicone Elastomer Kit	Dow Corning	4019601
Vectashield Mounting Media	Vector Laboratories	H-1000–10
Experimental Models: Organisms/Strains		
Drosophila adult flies	Bloomington Drosophila Stock Center	–
Other		
Carbon dioxide flow bed for anaesthetization	Genesee Scientific	59–172
Dissecting forceps - #5	Dumont	11,251–10
Dissecting forceps - #55	Dumont	11,295–51

Materials and Equipment

Materials

- Sharp forceps.
- Dissection dishes.
- 24-well immunostaining plate.
- Pipette and tips.
- Dry wipes.
- Microscope slides.
- Cover slips.
- Nail polish for sealing.

Equipment

- Standard fly culturing equipment.
- Carbon dioxide fly pad.
- Dissecting stereomicroscope.
- Orbital shaker.
- Confocal microscope.

Reagents

- PBS.
- Triton-X.
- Paraformaldehyde.
- Normal goat serum.
- Primary antibody of interest.
- Secondary antibody.
- Vectashield mounting media.

Step-by-Step Method Details

Adult Brain Dissection

Timing: 10–20 minutes

1. Anesthetize adult flies on carbon dioxide fly pad.
2. Transfer to a dissection dish containing ice cold PBS under the dissection microscope.
3. Using the non-dominant forceps, grasp the abdomen to keep hold of the fly. Then, with the help of the dominant forceps, grab and pull the proboscis to separate head from body.
4. Cut through the connective tissue between proboscis and eye to expose brain. Once brain is visible, pull trachea and cuticle to isolate the brain.
5. Clean the brain with gentle manipulations until all attachments are removed.

Note: Dissections must be performed quickly to avoid tissue degradation. Dissections within 5 minutes per brain is a good starting point. Immediately transfer the dissected brains to the fixing solution.

Fixation and Blocking

Timing: 90 minutes

6. Carefully transfer dissected brains to a 24-well immunostaining plate containing fixing solution.
7. Incubate on shaker for 30 minutes at room temperature.
8. Wash with PBT 3 times, 10 minutes each while shaking at room temperature.
9. Replace PBT with blocking solution and keep on shaker for 30 minutes at room temperature.

Antibody Staining
Timing: 4 hours to overnight

10. Dilute primary antibodies in blocking solution to obtain desirable concentration.
11. Incubate with primary antibody solution on a shaker for 2–4 hours at room temperature or overnight at 4°C (preferred).
12. Wash with PBT 3 times, 10 minutes each while shaking at room temperature.
13. Incubate with appropriate secondary antibodies at suggested dilution for 2 hours at room temperature in dark.
14. Wash with PBS 3 times, 10 minutes each while shaking at room temperature.

Note: Refer to manufacturer's instructions and previous literature to determine the appropriate antibody concentration for use in immunohistochemistry. Dilutions usually range between 1:200 to 1:1000.

Note: After secondary antibody incubation, all subsequent incubations should be done in dark.

Mounting
Timing: 30 minutes

15. To avoid squishing of brains on the slide, place a spacer or coverslip on each side of the slide.
16. Add a drop of PBS in the center of the slide and carefully place the adult brains in the desired orientation.
17. Remove the PBS around the brains using dry wipes and gently add 30 μL Vectashield mounting media.
18. Place a coverslip on the brains by slowly lowering the coverslip from one side. Remove any excess mounting media with dry wipes.
19. Using nail polish, seal the edges of the coverslip and the slide is ready for imaging.

Note: For best results, slides should be stored at 4°C in dark and imaged within one week of preparation.

Imaging
Timing: 60–180 minutes

20. Use a compound fluorescent microscope or a confocal microscope (preferred) for imaging.
21. View the slide under 20X objective lens to image the whole brain and then use 40X objective lens to image smaller regions of the brain.

Expected Outcomes

This protocol allows for staining and visualization of neural proteins, circuits, and cells in a whole mount *Drosophila* adult brain. Using this protocol, we have visualized mushroom body (MB) neurons in a 7-day old adult brain by co-staining Tip60 protein (green) and Fasciclin II (red) (Fig. 13.9). MB are prominent bilateral structures in the fly brain that are critical for learning and memory behavior. Fas II, a cell adhesion molecule that is commonly used as a mushroom body marker, is expressed strongly in the MB α/β lobes and weakly in the γ lobe in the MB of the adult fly brain. By co-staining with Fas II, Tip60 histone acetyltransferase enzyme was observed to be enriched in the adult MB, consistent with its function of regulating learning and memory in *Drosophila*.

Advantages

Immunohistochemistry in *Drosophila* adult brains has several advantages. First, immunostaining can be performed on whole-mount brains, therefore obviating the need for cryo-sectioning. Second, due to its small size, complete *Drosophila* brain can be visualized in a single frame. Third, the adult *Drosophila* brain has been successfully mapped out into a detailed connectome compromising of 25,000 neurons, therefore allowing users to identify specific neural circuitry in whole brains.

Limitations

Since a specific antibody is required for targeting each desired protein, immunohistochemical approach can be limited by the antibody availability for *Drosophila* species.

Optimization and Troubleshooting
High Background Staining

The high level of background staining obscures tissue imaging.

Probable Causes and Potential Solution

The most probable causes are listed below along with their solutions:

- Insufficient blocking: Increase blocking concentration and/or incubation time.
- High primary antibody concentration: Dilute the antibody further to its optimal range.
- Non-specific binding of secondary antibody: Run a control without any primary antibody.
- Insufficient washing: Increase washing time to remove any unbound antibodies.
- Incubation temperature is too high: Incubate samples at 4°C overnight.

Glossary

Dosage Compensation — A term describing the mechanism of X-linked gene expression equalization between the sexes.

Epigenetics — The study of phenotypic variations arising from the heritable changes in gene expression that result from alterations in the genome that do not involve changes in the nucleotide sequence.

Epigenetic memory — The maintenance of epigenetically-mediated gene transcriptional states across cell generations.

Histone acetylation — The transfer of an acyl moiety from acyl coenzyme A to lysine residues on the n-terminal tails of the histones by histone acetyltransferase (HAT) enzymes

Histone methylation — The transfer of methyl moieties from S-adenosyl methionine to the ε-amino groups of lysine residues of histone proteins by lysine methyltransferase (KMT) enzymes.

Position-effect variegation — The process by which genes that are normally expressed are aberrantly moved, either by rearrangement or transposition, to a site near the euchromatin-heterochromatin (Eu-Het) transition zone and as a result, are abnormally packaged into a heterochromatic form and silenced.

Transgenerational epigenetic inheritance — The faithful maintenance and replication of epigenetic patterns through the cell cycle.

Abbreviations

Aβ	Amyloid beta
AD	Alzheimer disease
ALS	Amyotrophic Lateral Sclerosis
APP	Amyloid precursor protein
Ash1	Absent, small, or homeotic disks 1
Brm	Brahma
CBP	CREB-binding protein
CES	Chromatin entry site
CLAMP	Chromatin-linked adaptor for MSL protein
CNS	Central nervous system
CREB	cAMP response element–binding protein
DCC	Dosage compensation complex
E(var)	Enhancer of variegation
EE	Environmental enrichment
EM	Electron microscopy
Esc	Extra sex combs
Eu–Het	Euchromatin–heterochromatin
EZ	Enhancer of zeste
Fsh	Female sterile homeotic
HAS	High-affinity site
HAT	Histone acetyltransferase
HD	Huntington's disease
HDAC	Histone deacetylase
HOX genes	Homeotic genes
HP1	Heterochromatin protein 1
KMT	Lysine methyltransferase
lncRNA	Long noncoding RNA
LOF	Loss of function
LTM	Long-term memory
MB	Mushroom body
MLE	Maleless
MOF	Males absent of first
MRE	MSL recognition element
MSL	Male-specific lethal
Pc	Polycomb
PcG	Polycomb group
PD	Parkinson disease
PEV	Position-effect variegation
PhoRC	Pho repressive complex
piRNA	Piwi-interacting RNA
PRC1	Polycomb repressive complex 1
PRC2	Polycomb repressive complex 2
PRE	Polycomb group response element
PTM	Posttranslational modification
Sce	Sex combs extra
SET	Su(var)3-9, Enhancer-of-zeste and Trithorax.
Su(var)	Suppressor of variegation
SXL	Sex lethal
TAF$_{II}$	TBP-associated factor
TE	Transposable element
Tip60	Tat-interactive protein 60 kDa
TRE	Trithorax group response element
Trx	Trithorax
TrxG	Trithorax group
UNR	Upstream N-Ras

References

[1] Levenson JM, Sweatt JD. Epigenetic mechanisms in memory formation. Nat Rev Neurosci 2005;6(2):108–18.

[2] Noble D. Conrad Waddington and the origin of epigenetics. J Exp Biol 2015;218:816–18.

[3] Rusche LN, Lynch PJ. Assembling heterochromatin in the appropriate places: a boost is needed. J Cell Physiol 2015;219(3):525–8.

[4] Elgin SC, Reuter G. Position-effect variegation, heterochromatin formation, and gene silencing in Drosophila. Cold Spring Harb Perspect Biol 2013;5(8):a017780.

[5] Rubin GM, Yandell MD, Wortman JR, Gabor Miklos GL, Nelson CR, Hariharan IK, et al. Comparative genomics of the eukaryotes. Science (New York, NY) 2000;287(5461):2204–15.

[6] Stapleton M, Liao G, Brokstein P, Hong L, Carninci P, Shiraki T, et al. The Drosophila gene collection: identification of putative full-length cDNAs for 70% of D. melanogaster genes. Genome Res 2002;12(8):1294–300.

[7] Pandey UB, Nichols CD. Human disease models in *Drosophila melanogaster* and the role of the fly in therapeutic drug discovery. Pharmacol Rev 2011;63(2):411–36.

[8] Bier E. Drosophila, the golden bug, emerges as a tool for human genetics. Nat Rev Genet 2005;6(1):9–23.

[9] Min K. Drosophila as a model to study human–brain degenerative diseases. Parkinsonism Relat Disord 2001;7(3):165–9.

[10] Zhimulev IF, Boldyreva LV, Demakova OV, Poholkova GV, Khoroshko VA, Zykova TY, et al. Drosophila polytene chromosome bands formed by gene introns. Dokl Biochem Biophys 2016;466(1):57–60.

[11] Zhimulev IF, Belyaeva ES, Vatolina TY, Demakov SA. Banding patterns in *Drosophila melanogaster* polytene chromosomes correlate with DNA-binding protein occupancy. BioEssays 2012;34(6):498–508.

[12] Zhimulev IF, Zykova TY, Goncharov FP, Khoroshko VA, Demakova OV, Semeshin VF, et al. Genetic organization of interphase chromosome bands and interbands in *Drosophila melanogaster*. PLoS One 2014;9(7):e101631.

[13] Eagen KP, Hartl TA, Kornberg RD. Stable chromosome condensation revealed by chromosome conformation capture. Cell 2015;163(4):934–46.

[14] Nüsslein-Volhard C, Wieschaus E. Mutations affecting segment number and polarity in Drosophila. Nature 1980;287(5785):795–801.

[15] Jürgens G. A group of genes controlling the spatial expression of the bithorax complex in Drosophila. Nature 1985;316(6024):153–5.

[16] Boros IM. Histone modification in Drosophila. Brief Funct Genomics 2012;11(4):319–31.

[17] Mukherjee K, Twyman RM, Vilcinskas A. Insects as models to study the epigenetic basis of disease. Prog Biophys Mol Biol 2015;118(1–2):69–78.

[18] Raji JI, Potter CJ. The number of neurons in Drosophila and mosquito brains. PLoS One 2021;16(5):e0250381.

[19] Scheffer LK, Xu CS, Januszewski M, Lu Z, Takemura S-Y, Hayworth KJ, et al. A connectome and analysis of the adult Drosophila central brain. Elife 2020;9:e57443.

[20] Budnik V. Synapse maturation and structural plasticity at Drosophila neuromuscular junctions. Curr Opin Neurobiol 1996;6(6):858–67.

[21] Menon KP, Andrews S, Murthy M, Gavis ER, Zinn K. The translational repressors Nanos and Pumilio have divergent effects on presynaptic terminal growth and postsynaptic glutamate receptor subunit composition. J Neurosci 2009;29(17):5558–72.

[22] Sarkar A, Irwin M, Singh A, Riccetti M, Singh A. Alzheimer's disease: the silver tsunami of the 21(st) century. Neural Regen Res 2016;11(5):693–7.

[23] Burr AA, Tsou WL, Ristic G, Todi SV. Using membrane-targeted green fluorescent protein to monitor neurotoxic protein-dependent degeneration of Drosophila eyes. J Neurosci Res 2014;92(9):1100–9.

[24] Ito K, Awano W, Suzuki K, Hiromi Y, Yamamoto D. The Drosophila mushroom body is a quadruple structure of clonal units each of which contains a virtually identical set of neurones and glial cells. Development 1997;124(4):761–71.

[25] Barnstedt O, Owald D, Felsenberg J, Brain R, Moszynski JP, Talbot CB, et al. Memory-relevant mushroom body output synapses are cholinergic. Neuron 2016;89(6):1237–47.

[26] Heisenberg M. Mushroom body memoir: from maps to models. Nat Rev Neurosci 2003;4(4):266–75.

[27] Guven-Azkan T, Davis RL. Functional neuroanatomy of Drosophila olfactory memory formation. Learn Mem 2014;21(519–526):19.

[28] Farris SM. Evolution of complex higher brain centers and behaviors: behavioral correlates of mushroom body elaboration in insects. Brain Behav Evol 2013;82(1):9–18.

[29] Akalai DB, Wilson CF, Zong L, Tanaka NK, Ito K, Davis RL. Roles for Drosophila mushroom body neurons in olfactory learning and memory. Learn Mem 2006;13(5):659–68.

[30] Margulies C, Tully T, Dubnau J. Deconstructing memory in Drosophila. Curr Biol 2005;15(17):R700–13.

[31] Heisenberg M. What do the mushroom bodies do for the insect brain? An introduction. Learn Mem 1998;5(1):1–10.

[32] Lee G, Hall JC. A newly uncovered phenotype associated with the fruitless gene of Drosophila melanogaster: aggression-like head interactions between mutant males. Behav Genet 2000;30(4):263–75.

[33] Zordan MA, Benna C, Mazzotta G. Monitoring and analyzing Drosophila circadian locomotor activity. Methods Mole Biol 2007;362:67–81.

[34] Mehren JE, Ejima A, Griffith LC. Unconventional sex: fresh approaches to courtship learning. Curr Opin Neurobiol 2004;14(6):745–50.

[35] Keene AC, Waddell S. Drosophila olfactory memory: single genes to complex neural circuits. Nat Rev Neurosci 2007;8:341–54.

[36] Busto GU, Cervantes-sandoval I, Davis RL. Olfactory learning in Drosophila. Physiology 2010;26(6):338–46.

[37] Dubnau J, Tully T. Functional anatomy: from molecule to memory. Curr Biol 2001;11(6):R240–3.

[38] Quinn WG, Harris WA, Benzer S. Conditioned behavior in Drosophila melanogaster. Proc Natl Acad Sci USA 1974;71(3):708–12.

[39] Kuo MH, Allis CD. Roles of histone acetyltransferases and deacetylases in gene regulation. BioEssays 1998;20(8):615–26.

[40] Grunstein M. Histone acetylation in chromatin structure and transcription. Nature 1997;389(6649):349–52.

[41] Peterson CL, Laniel MA. Histones and histone modifications. Curr Biol 2004;14(14):R546–51.

[42] Schneider A, Chatterjee S, Bousiges O, Selvi BR, Swaminathan A, Cassel R, et al. Acetyltransferases (HATs) as targets for neurological therapeutics. Neurotherapeutics 2013;10(4):568–88.

[43] Wood MA, Hawk JD, Abel T. Combinatorial chromatin modifications and memory storage: a code for memory? Learn Mem 2006;13(3):241–4.

[44] Scharf AN, Imhof A. Every methyl counts—epigenetic calculus. FEBS Lett 2011;585(13):2001–7.

[45] Day JJ, Sweatt JD. Epigenetic treatments for cognitive impairments. Neuropsychopharmacology 2012;37(1):247–60.

[46] Peixoto L, Abel T. The role of histone acetylation in memory formation and cognitive impairments. Neuropsychopharmacology 2013;38(1):62–76.

[47] Marmorstein R, Roth SY. Histone acetyltransferases: function, structure, and catalysis. Curr Opin Genet Dev 2001;11(2):155–61.

[48] Krajewski WA, Becker PB. Reconstitution of hyperacetylated, DNase I-sensitive chromatin characterized by high conformational flexibility of nucleosomal DNA. Proc Natl Acad Sci USA 1998;95(4):1540–5.

[49] Ikura T, Ogryzko VV, Grigoriev M, Groisman R, Wang J, Horikoshi M, et al. Involvement of the TIP60 histone acetylase complex in DNA repair and apoptosis. Cell 2000;102(4):463–73.

[50] Mizzen CA, Yang XJ, Kokubo T, Brownell JE, Bannister AJ, Owen-Hughes T, et al. The TAF(II)250 subunit of TFIID has histone acetyltransferase activity. Cell 1996;87(7):1261–70.

[51] Nightingale KP, Wellinger RE, Sogo JM, Becker PB. Histone acetylation facilitates RNA polymerase II transcription of the Drosophila hsp26 gene in chromatin. EMBO J 1998;17(10):2865–76.

[52] Liu D. From histone modification to gene repression: epigenetic regulation by holocarboxylase synthetase-containing repression complex and FAD-dependent lysine specific demethylase [CHES thesis]. University of Nebraska-Lincoln; 2014.

[53] Das C, Lucia MS, Hansen KC, Tyler JK. CBP/p300-mediated acetylation of histone H3 on lysine 56. Nature 2009;459(7243):113–17.

[54] Milazzo G, Mercatelli D, Di Muzio G, Triboli L, De Rosa P, Perini G, et al. Histone deacetylases (HDACs): evolution, specificity, role in transcriptional complexes, and pharmacological actionability. Genes 2020;11(5):556.

[55] Cho Y, Griswold A, Campbell C, Min K-T. Individual histone deacetylases in Drosophila modulate transcription of distinct genes. Genomics 2005;86(5):606–17.

[56] Johnstone RW. Hisonte-deacetylase inhibitors: novel drugs for the treatment of cancer. Nat Rev 2002;1:287–99.

[57] Murray K. The occurrence of epsilon-N-methyl Lysine in histones. Biochemistry 1964;3:10–15.

[58] Hempel K, Lange HW, Birkofer L. Epsilon-N-trimethyllysine, a new amino acid in histones. Naturwissenschaften 1968;55(1):37.

REFERENCES

[59] Paik WK, Kim S. ε-N-dimethyllysine in histones. Biochem Biophys Res Commun 1967;27(4):479—83.

[60] Hallson G, Hollebakken RE, Li T, Syrzycka M, Kim I, Cotsworth S, et al. dSet1 is the main H3K4 di-and tri-methyltransferase throughout Drosophila development. Genetics 2012;190(1):91—100.

[61] Greer EL, Shi Y. Histone methylation: a dynamic mark in health, disease and inheritance. Nat Rev Genet 2012;13(5):343—57.

[62] Holowatyj A, Yang Z-Q, Pile LA. Histone lysine demethylases in Drosophila melanogaster. Fly (Austin Tex) 2015;9(1):36—44.

[63] Hyun K, Jeon J, Park K, Kim J. Writing, erasing and reading histone lysine methylations. Exp Mol Med 2017;49(4):e324.

[64] Ebert A, Schotta G, Lein S, Kubicek S, Krauss V, Jenuwein T, et al. Su(var) genes regulate the balance between euchromatin and heterochromatin in Drosophila. Genes Dev 2004;18(23):2973—83.

[65] Talasz H, Lindner HH, Sarg B, Helliger W. Histone H4-lysine 20 monomethylation is increased in promoter and coding regions of active genes and correlates with hyperacetylation. J Biol Chem 2005;280(46):38814—22.

[66] Congdon LM, Houston SI, Veerappan CS, Spektor TM, Rice JC. PR-Set7-mediated monomethylation of histone H4 lysine 20 at specific genomic regions induces transcriptional repression. J Cell Biochem 2010;110(3):609—19.

[67] Schuettengruber B, Chourrout D, Vervoort M, Leblanc B, Cavalli G. Genome regulation by polycomb and trithorax proteins. Cell 2007;128(4):735—45.

[68] Cleard F, Moshkin Y, Karch F, Maeda RK. Probing long-distance regulatory interactions in the Drosophila melanogaster bithorax complex using Dam identification. Nat Genet 2006;38(8):931—5.

[69] Comet I, Savitskaya E, Schuettengruber B, Negre N, Lavrov S, Parshikov A, et al. PRE-mediated bypass of two Su(Hw) insulators targets PcG proteins to a downstream promoter. Dev Cell 2006;11(1):117—24.

[70] Li D, Ge Y, Zhao Z, Zhu R, Wang X, Bi X. Distinct and coordinated regulation of small non-coding RNAs by E2f1 and p53 during Drosophila development and in response to DNA damage. Front Cell Dev Biol 2021;9.

[71] Li K, Tian Y, Yuan Y, Fan X, Yang M, He Z, et al. Insights into the functions of LncRNAs in Drosophila. Int J Mol Sci 2019;20(18):4646.

[72] Soleimani S, Valizadeh Arshad Z, Moradi S, Ahmadi A, Davarpanah SJ, Azimzadeh Jamalkandi S. Small regulatory noncoding RNAs in Drosophila melanogaster: biogenesis and biological functions. Brief Funct Genomics 2020;19(4):309—23.

[73] Dunwell TL, Pfeifer GP. Drosophila genomic methylation: new evidence and new questions. Epigenomics 2014;6(5):459—61.

[74] Wang F, Minakhina S, Tran H, Changela N, Kramer J, Steward R. Tet protein function during Drosophila development. PLoS One 2018;13(1):e0190367.

[75] Schaefer M, Steringer JP, Lyko F. The Drosophila cytosine-5 methyltransferase Dnmt2 is associated with the nuclear matrix and can access DNA during mitosis. PLoS One 2008;3(1):e1414.

[76] Capuano F, Mülleder M, Kok R, Blom HJ, Ralser M. Cytosine DNA methylation is found in Drosophila melanogaster but absent in *Saccharomyces cerevisiae, Schizosaccharomyces pombe*, and other yeast species. Anal Chem 2014;86(8):3697—702.

[77] Deshmukh S, Abhyankar V, Vellarikkal SK, Sivasubbu S, Scaria V, Deobagkar D. Genome-wide comparison of DNA methylation between life cycle stages of Drosophila melanogaster using high-throughput sequencing techniques. bioRxiv 2020;.

[78] Filion GJ, van Bemmel JG, Braunschweig U, Talhout W, Kind J, Ward LD, et al. Systematic protein location mapping reveals five principal chromatin types in Drosophila cells. Cell 2010;143(2):212—24.

[79] Kharchenko PV, Alekseyenko AA, Schwartz YB, Minoda A, Riddle NC, Ernst J, et al. Comprehensive analysis of the chromatin landscape in *Drosophila melanogaster*. Nature 2011;471(7339):480—5.

[80] Ho JW, Jung YL, Liu T, Alver BH, Lee S, Ikegami K, et al. Comparative analysis of metazoan chromatin organization. Nature 2014;512(7515):449—52.

[81] Ebert A, Lein S, Schotta G, Reuter G. Histone modification and the control of heterochromatic gene silencing in Drosophila. Chromosome Res 2006;14(4):377—92.

[82] Yasuhara JC, Wakimoto BT. Molecular landscape of modified histones in Drosophila heterochromatic genes and euchromatin-heterochromatin transition zones. PLoS Genet 2008;4(1):e16.

[83] Girton JR, Johansen KM. Chromatin structure and the regulation of gene expression: the lessons of PEV in Drosophila. Adv Genet 2008;61:1—43.

[84] Schotta G, Lachner M, Sarma K, Ebert A, Sengupta R, Reuter G, et al. A silencing pathway to induce H3-K9 and H4-K20 trimethylation at constitutive heterochromatin. Genes Dev 2004;18(11):1251—62.

[85] Zhang W, Deng H, Bao X, Lerach S, Girton J, Johansen J, et al. The JIL-1 histone H3S10 kinase regulates dimethyl H3K9 modifications and heterochromatic spreading in Drosophila. Development 2006;133(2):229—35.

[86] Berger SL. The complex language of chromatin regulation during transcription. Nature 2007;447(7143):407—12.

[87] Riddle NC, Minoda A, Kharchenko PV, Alekseyenko AA, Schwartz YB, Tolstorukov MY, et al. Plasticity in patterns of histone modifications and chromosomal proteins in Drosophila heterochromatin. Genome Res 2011;21(2):147—63.

[88] Turner BM, Birley AJ, Lavender J. Histone H4 isoforms acetylated at specific lysine residues define individual chromosomes and chromatin domains in Drosophila polytene nuclei. Cell 1992;69(2):375—84.

[89] Weiler KS. E(var)3—9 of Drosophila melanogaster encodes a zinc finger protein. Genetics 2007;177(1):167—78.

[90] Farkas G, Gausz J, Galloni M, Reuter G, Gyurkovics H, Karch F. The Trithorax-like gene encodes the Drosophila GAGA factor. Nature 1994;371(6500):806—8.

[91] Seum C, Spierer A, Pauli D, Szidonya J, Reuter G, Spierer P. Position-effect variegation in Drosophila depends on dose of the gene encoding the E2F transcriptional activator and cell cycle regulator. Development 1996;122(6):1949—56.

[92] Hofmann A, Brunner M, Korge G. The winged-helix transcription factor JUMU is a haplo-suppressor/triplo-enhancer of PEV in various tissues but exhibits reverse PEV effects in the brain of Drosophila melanogaster. Chromosome Res 2009;17(3):347—58.

[93] Sass GL, Henikoff S. Comparative analysis of position-effect variegation mutations in *Drosophila melanogaster* delineates the targets of modifiers. Genetics 1998;148(2):733—41.

[94] Locke J, Kotarski MA, Tartof KD. Dosage-dependent modifiers of position effect variegation in Drosophila and a mass action model that explains their effect. Genetics 1988;120(1):181—98.

[95] Grewal SIS, Elgin SCR. Transcription and RNAi in the formation of heterochromatin. Nat Rev 2007;447:399—406.

[96] Schotta G, Ebert A, Krauss V, Fischer A, Hoffmann J, Rea S, et al. Central role of Drosophila SU(VAR)3—9 in histone H3-K9 methylation and heterochromatic gene silencing. EMBO J 2002;21(5):1121—31.

[97] Rudolph T, Yonezawa M, Lein S, Heidrich K, Kubicek S, Schafer C, et al. Heterochromatin formation in Drosophila is initiated through active removal of H3K4 methylation by the LSD1 homolog SU(VAR)3-3. Mol Cell 2007;26(1):103—15.

[98] Czermin B, Schotta G, Hulsmann BB, Brehm A, Becker PB, Reuter G, et al. Physical and functional association of SU(VAR) 3−9 and HDAC1 in Drosophila. EMBO Rep 2001;2(10):915−19.

[99] Mottus R, Sobel RE, Grigliatti TA. Mutational analysis of a histone deacetylase in *Drosophila melanogaster*: missense mutations suppress gene silencing associated with position effect variegation. Genetics 2000;154(2):657−68.

[100] Vermaak D, Malik HS. Multiple roles for heterochromatin protein 1 genes in Drosophila. Annu Rev Genet 2009;43:467−92.

[101] Lee YC. The Role of piRNA-mediated epigenetic silencing in the population dynamics of transposable elements in *Drosophila melanogaster*. PLoS Genet 2015;11(6):e1005269.

[102] de Vanssay A, Bouge AL, Boivin A, Hermant C, Teysset L, Delmarre V, et al. piRNAs and epigenetic conversion in Drosophila. Fly (Austin Tex) 2013;7(4):237−41.

[103] Cleard F, Delattre M, Spierer P. SU(VAR)3−7, a Drosophila heterochromatin-associated protein and companion of HP1 in the genomic silencing of position-effect variegation. EMBO J 1997;16(17):5280−8.

[104] James TC, Eissenberg JC, Craig C, Dietrich V, Hobson A, Elgin SC. Distribution patterns of HP1, a heterochromatin-associated nonhistone chromosomal protein of Drosophila. Eur J Cell Biol 1989;50(1):170−80.

[105] Vogel MJ, Pagie L, Talhout W, Nieuwland M, Kerkhoven RM, van Steensel B. High-resolution mapping of heterochromatin redistribution in a Drosophila position-effect variegation model. Epigenetics Chromatin 2009;2(1):1.

[106] Cao R, Tsukada Y, Zhang Y. Role of Bmi-1 and Ring1A in H2A ubiquitylation and Hox gene silencing. Mol Cell 2005;20(6):845−54.

[107] Steffen PA, Leonie R. What are memories made of? How Polycomb and Trithorax proteins mediate epigenetic memory. Nat Rev Mole Cell Biol 2014;15:340−56.

[108] Grimaud C, Negre N, Cavalli G. From genetics to epigenetics: the tale of Polycomb group and trithorax group genes. Chromosome Res 2006;14(4):363−75.

[109] Schwartz YB, Pirrotta V. Polycomb silencing mechanisms and the management of genomic programmes. Nat Rev Genet 2007;8(1):9−22.

[110] Mallo M, Alonso CR. The regulation of Hox gene expression during animal development. Development 2013;140(19):3951−63.

[111] Lewis RA, Wakimoto BT, Denell RE, Kaufman TC. Genetic analysis of the antennapedia gene complex (Ant-C) and adjacent chromosomal regions of *Drosophila melanogaster*. II. Polytene chromosome segments 84A-84B1,2. Genetics 1980;95(2):383−97.

[112] Lewis EB. A gene complex controlling segmentation in Drosophila. Nature 1978;276(5688):565−70.

[113] Kmita M, Duboule D. Organizing axes in time and space; 25 years of colinear tinkering. Science (New York, NY) 2003;301(5631):331−3.

[114] Pang D, Thompson DNP. Embryology and bony malformations of the craniovertebral junction. Childs Nerv Syst 2011;27(4):523−64.

[115] Cao R, Wang L, Wang H, Xia L, Erdjument-Bromage H, Tempst P, et al. Role of histone H3 lysine 27 methylation in Polycomb-group silencing. Science (New York, NY) 2002;298(5595):1039−43.

[116] Schuettengruber B, Martinez A-M, Iovino N, Cavalli G. Trithorax group proteins: switching genes on and keeping them active. Nat Rev Mol Cell Biol 2011;12(12):799−814.

[117] Akmammedov A, Geigges M, Paro R. Bivalency in Drosophila embryos is associated with strong inducibility of Polycomb target genes. Fly (Austin Tex) 2019;13(1−4):42−50.

[118] Voigt P, Tee W-W, Reinberg D. A double take on bivalent promoters. Genes Dev 2013;27(12):1318−38.

[119] Lavigne M, Francis NJ, King IF, Kingston RE. Propagation of silencing; recruitment and repression of naive chromatin in trans by polycomb repressed chromatin. Mol Cell 2004;13(3):415−25.

[120] Brown JL, Sun M-A, Kassis JA. Global changes of H3K27me3 domains and Polycomb group protein distribution in the absence of recruiters Spps or Pho. Proc Natl Acad Sci 2018;115(8):E1839−48.

[121] van Kruijsbergen I, Hontelez S, Veenstra GJC. Recruiting polycomb to chromatin. Int J Biochem Cell Biol 2015;67:177−87.

[122] Muller J, Kassis JA. Polycomb response elements and targeting of Polycomb group proteins in Drosophila. Curr Opin Genet Dev 2006;16(5):476−84.

[123] Stock JK, Giadrossi S, Casanova M, Brookes E, Vidal M, Koseki H, et al. Ring1-mediated ubiquitination of H2A restrains poised RNA polymerase II at bivalent genes in mouse ES cells. Nat Cell Biol 2007;9(12):1428−35.

[124] Shao Z, Raible F, Mollaaghababa R, Guyon JR, Wu CT, Bender W, et al. Stabilization of chromatin structure by PRC1, a Polycomb complex. Cell 1999;98(1):37−46.

[125] Francis NJ, Saurin AJ, Shao Z, Kingston RE. Reconstitution of a functional core polycomb repressive complex. Mol Cell 2001;8(3):545−56.

[126] Francis NJ, Kingston RE, Woodcock CL. Chromatin compaction by a polycomb group protein complex. Science (New York, NY) 2004;306(5701):1574−7.

[127] Bantignies F, Roure V, Comet I, Leblanc B, Schuettengruber B, Bonnet J, et al. Polycomb-dependent regulatory contacts between distant Hox loci in Drosophila. Cell 2011;144(2):214−26.

[128] Tolhuis B, Blom M, Kerkhoven RM, Pagie L, Teunissen H, Nieuwland M, et al. Interactions among Polycomb domains are guided by chromosome architecture. PLoS Genet 2011;7(3):e1001343.

[129] Llorens-Giralt P, Camilleri-Robles C, Corominas M, Climent-Cantó P. Chromatin organization and function in Drosophila. Cells 2021;10(9):2362.

[130] Plys AJ, Davis CP, Kim J, Rizki G, Keenen MM, Marr SK, et al. Phase separation of Polycomb-repressive complex 1 is governed by a charged disordered region of CBX2. Genes Dev 2019;33(13−14):799−813.

[131] Blomen VA, Boonstra J. Stable transmission of reversible modifications: maintenance of epigenetic information through the cell cycle. Cell Mol Life Sci 2011;68(1):27−44.

[132] Kiefer JC. Epigenetics in development. Dev Dyn 2007;236(4):1144−56.

[133] Yuan W, Xu M, Huang C, Liu N, Chen S, Zhu B. H3K36 methylation antagonizes PRC2-mediated H3K27 methylation. J Biol Chem 2011;286(10):7983−9.

[134] Schmitges FW, Prusty AB, Faty M, Stutzer A, Lingaraju GM, Aiwazian J, et al. Histone methylation by PRC2 is inhibited by active chromatin marks. Mol Cell 2011;42(3):330−41.

[135] Tie F, Banerjee R, Stratton CA, Prasad-Sinha J, Stepanik V, Zlobin A, et al. CBP-mediated acetylation of histone H3 lysine 27 antagonizes Drosophila Polycomb silencing. Development 2009;136(18):3131−41.

[136] Devaiah BN, Lewis BA, Cherman N, Hewitt MC, Albrecht BK, Robey PG, et al. BRD4 is an atypical kinase that phosphorylates serine2 of the RNA polymerase II carboxy-terminal domain. Proc Natl Acad Sci USA 2012;109(18):6927−32.

[137] Armstrong JA, Papoulas O, Daubresse G, Sperling AS, Lis JT, Scott MP, et al. The Drosophila BRM complex facilitates global transcription by RNA polymerase II. EMBO J 2002;21(19):5245−54.

[138] Fitzsimons HL, Schwartz S, Given FM, Scott MJ. The histone deacetylase HDAC4 regulates long-term memory in Drosophila. PLoS One 2013;8(12):e83903.

[139] Bauer M, Trupke J, Ringrose L. The quest for mammalian Polycomb response elements: are we there yet? Chromosoma 2016;125(3):471–96.

[140] Aranda S, Mas G, Di Croce L. Regulation of gene transcription by Polycomb proteins. Sci Adv 2015;1(11):e1500737.

[141] Feng YQ, Desprat R, Fu H, Olivier E, Lin CM, Lobell A, et al. DNA methylation supports intrinsic epigenetic memory in mammalian cells. PLoS Genet 2006;2(4):e65.

[142] Ehrenhofer-Murray AE. Chromatin dynamics at DNA replication, transcription and repair. Eur J Biochem 2004;271(12):2335–49.

[143] Lanzuolo C, Lo Sardo F, Diamantini A, Orlando V. PcG complexes set the stage for epigenetic inheritance of gene silencing in early S phase before replication. PLoS Genet 2011;7(11):e1002370.

[144] Petruk S, Black KL, Kovermann SK, Brock HW, Mazo A. Stepwise histone modifications are mediated by multiple enzymes that rapidly associate with nascent DNA during replication. Nat Commun 2013;4:2841.

[145] Petruk S, Sedkov Y, Johnston DM, Hodgson JW, Black KL, Kovermann SK, et al. TrxG and PcG proteins but not methylated histones remain associated with DNA through replication. Cell 2012;150(5):922–33.

[146] Escobar TM, Oksuz O, Saldaña-Meyer R, Descostes N, Bonasio R, Reinberg D. Active and repressed chromatin domains exhibit distinct nucleosome segregation during DNA replication. Cell 2019;179(4):953–63 e11.

[147] Huang H, Yu Z, Zhang S, Liang X, Chen J, Li C, et al. Drosophila CAF-1 regulates HP1-mediated epigenetic silencing and pericentric heterochromatin stability. J Cell Sci 2010;123(Pt 16):2853–61.

[148] Fischle W, Wang Y, Allis CD. Binary switches and modification cassettes in histone biology and beyond. Nature 2003;425 (6957):475–9.

[149] Giet R, Glover DM. Drosophila aurora B kinase is required for histone H3 phosphorylation and condensin recruitment during chromosome condensation and to organize the central spindle during cytokinesis. J Cell Biol 2001;152 (4):669–82.

[150] Hirota T, Lipp JJ, Toh BH, Peters JM. Histone H3 serine 10 phosphorylation by Aurora B causes HP1 dissociation from heterochromatin. Nature 2005;438(7071):1176–80.

[151] Strom AR, Emelyanov AV, Mir M, Fyodorov DV, Darzacq X, Karpen GH. Phase separation drives heterochromatin domain formation. Nature 2017;547(7662):241–5.

[152] Heard E, Disteche CM. Dosage compensation in mammals: fine-tuning the expression of the X chromosome. Genes Dev 2006;20(14):1848–67.

[153] Lucchesi JC, Kuroda MI. Dosage compensation in Drosophila. Cold Spring Harb Perspect Biol 2015;7(5).

[154] Conrad T, Akhtar A. Dosage compensation in *Drosophila melanogaster*: epigenetic fine-tuning of chromosome-wide transcription. Nat Rev Genet 2011;13(2):123–34.

[155] Lucchesi JC. Dosage compensation in Drosophila and the "complex' world of transcriptional regulation. BioEssays 1996;18(7):541–7.

[156] Scott MJ, Pan LL, Cleland SB, Knox AL, Heinrich J. MSL1 plays a central role in assembly of the MSL complex, essential for dosage compensation in Drosophila. EMBO J 2000;19(1):144–55.

[157] Copps K, Richman R, Lyman LM, Chang KA, Rampersad-Ammons J, Kuroda MI. Complex formation by the Drosophila MSL proteins: role of the MSL2 RING finger in protein complex assembly. EMBO J 1998;17(18):5409–17.

[158] Hallacli E, Lipp M, Georgiev P, Spielman C, Cusack S, Akhtar A, et al. Msl1-mediated dimerization of the dosage compensation complex is essential for male X-chromosome regulation in Drosophila. Mol Cell 2012;48(4):587–600.

[159] Lee CG, Chang KA, Kuroda MI, Hurwitz J. The NTPase/helicase activities of Drosophila maleless, an essential factor in dosage compensation. EMBO J 1997;16(10):2671–81.

[160] Meller VH, Gordadze PR, Park Y, Chu X, Stuckenholz C, Kelley RL, et al. Ordered assembly of roX RNAs into MSL complexes on the dosage-compensated X chromosome in Drosophila. Curr Biol 2000;10(3):136–43.

[161] Meller VH, Rattner BP. The roX genes encode redundant male-specific lethal transcripts required for targeting of the MSL complex. EMBO J 2002;21(5):1084–91.

[162] Ilik IA, Quinn JJ, Georgiev P, Tavares-Cadete F, Maticzka D, Toscano S, et al. Tandem stem-loops in roX RNAs act together to mediate X chromosome dosage compensation in Drosophila. Mol Cell 2013;51(2):156–73.

[163] Park SW, Kang Y, Sypula JG, Choi J, Oh H, Park Y. An evolutionarily conserved domain of roX2 RNA is sufficient for induction of H4-Lys16 acetylation on the Drosophila X chromosome. Genetics 2007;177(3):1429–37.

[164] Wutz A. Noncoding roX RNA remodeling triggers fly dosage compensation complex assembly. Mol Cell 2013;51(2):131–2.

[165] Keller CI, Akhtar A. The MSL complex: juggling RNA–protein interactions for dosage compensation and beyond. Curr Opin Genet Dev 2015;31:1–11.

[166] Akhtar A, Zink D, Becker PB. Chromodomains are protein-RNA interaction modules. Nature 2000;407(6802):405–9.

[167] Gilfillan GD, Dahlsveen IK, Becker PB. Lifting a chromosome: dosage compensation in *Drosophila melanogaster*. FEBS Lett 2004;567:8–14.

[168] Bashaw GJ, Baker BS. The msl-2 dosage compensation gene of Drosophila encodes a putative DNA-binding protein whose expression is sex specifically regulated by Sex-lethal. Development 1995;121(10):3245–58.

[169] Abaza I, Coll O, Patalano S, Gebauer F. Drosophila UNR is required for translational repression of male-specific lethal 2 mRNA during regulation of X-chromosome dosage compensation. Genes Dev 2006;20(3):380–9.

[170] Beckmann K, Grskovic M, Gebauer F, Hentze MW. A dual inhibitory mechanism restricts msl-2 mRNA translation for dosage compensation in Drosophila. Cell 2005;122(4):529–40.

[171] Salz HK. Sex determination in insects: a binary decision based on alternative splicing. Curr Opin Genet Dev 2011;21 (4):395–400.

[172] Bai X, Alekseyenko AA, Kuroda MI. Sequence-specific targeting of MSL complex regulates transcription of the roX RNA genes. EMBO J 2004;23(14):2853–61.

[173] Rattner BP, Meller VH. Drosophila male-specific lethal 2 protein controls sex-specific expression of the roX genes. Genetics 2004;166(4):1825–32.

[174] Kelley RL, Meller VH, Gordadze PR, Roman G, Davis RL, Kuroda MI. Epigenetic spreading of the Drosophila dosage compensation complex from roX RNA genes into flanking chromatin. Cell 1999;98(4):513–22.

[175] Alekseyenko AA, Peng S, Larschan E, Gorchakov AA, Lee OK, Kharchenko P, et al. A sequence motif within chromatin entry sites directs MSL establishment on the Drosophila X chromosome. Cell 2008;134(4):599–609.

[176] Straub T, Grimaud C, Gilfillan GD, Mitterweger A, Becker PB. The chromosomal high-affinity binding sites for the Drosophila dosage compensation complex. PLoS Genet 2008;4 (12):e1000302.

[177] Kuzu G, Kaye EG, Chery J, Siggers T, Yang L, Dobson JR, et al. Expansion of GA dinucleotide repeats increases the density of CLAMP binding sites on the X-chromosome to promote drosophila dosage compensation. PLoS Genet 2016;12(7): e1006120.

[178] Soruco MM, Chery J, Bishop EP, Siggers T, Tolstorukov MY, Leydon AR, et al. The CLAMP protein links the MSL complex to the X chromosome during Drosophila dosage compensation. Genes Dev 2013;27(14):1551–6.

[179] Bachtrog D, Toda NRT, Lockton S. Dosage compensation and demasculinization of X chromosomes in Drosophila. Curr Biol 2010;16:1–16.

[180] Kind J, Akhtar A. Cotranscriptional recruitment of the dosage compensation complex to X-linked target genes. Genes Dev 2007;21(16):2030–40.

[181] Bhadra U, Pal-Bhadra M, Birchler JA. Role of the male specific lethal (msl) genes in modifying the effects of sex chromosomal dosage in Drosophila. Genetics 1999;152(1):249–68.

[182] Tie F, Banerjee R, Conrad PA, Scacheri PC, Harte PJ. Histone demethylase UTX and chromatin remodeler BRM bind directly to CBP and modulate acetylation of histone H3 lysine 27. Mol Cell Biol 2012;32(12):2323–34.

[183] Morra R, Smith ER, Yokoyama R, Lucchesi JC. The MLE subunit of the Drosophila MSL complex uses its ATPase activity for dosage compensation and its helicase activity for targeting. Mol Cell Biol 2008;28(3):958–66.

[184] Gu W, Wei X, Pannuti A, Lucchesi JC. Targeting the chromatin-remodeling MSL complex of Drosophila to its sites of action on the X chromosome requires both acetyl transferase and ATPase activities. EMBO J 2000;19(19):5202–11.

[185] Gu W, Szauter P, Lucchesi JC. Targeting of MOF, a putative histone acetyl transferase, to the X chromosome of Drosophila melanogaster. Dev Genet 1998;22(1):56–64.

[186] Gelbart ME, Kuroda MI. Drosophila dosage compensation: a complex voyage to the X chromosome. Development 2009;136(9):1399–410.

[187] Smith ER, Allis CD, Lucchesi JC. Linking global histone acetylation to the transcription enhancement of X-chromosomal genes in Drosophila males. J Biol Chem 2001;276(34):31483–6.

[188] Bell O, Conrad T, Kind J, Wirbelauer C, Akhtar A, Schubeler D. Transcription-coupled methylation of histone H3 at lysine 36 regulates dosage compensation by enhancing recruitment of the MSL complex in Drosophila melanogaster. Mol Cell Biol 2008;28(10):3401–9.

[189] Larschan E, Bishop EP, Kharchenko PV, Core LJ, Lis JT, Park PJ, et al. X chromosome dosage compensation via enhanced transcriptional elongation in Drosophila. Nature 2011;471(7336):115–18.

[190] Gilfillan GD, Straub T, de Wit E, Greil F, Lamm R, van Steensel B, et al. Chromosome-wide gene-specific targeting of the Drosophila dosage compensation complex. Genes Dev 2006;20(7):858–70.

[191] Corona DF, Clapier CR, Becker PB, Tamkun JW. Modulation of ISWI function by site-specific histone acetylation. EMBO Rep 2002;3(3):242–7.

[192] Pirooznia K, Elefant F. Modulating epigenetic HAT activity: A promising theraputic option for neurological disease? J Mole Cloning Genet Recomb 2012;1(1):1–3.

[193] Pirooznia K, Elefant F. Targeting specific HATs for neurodegenerative disease treatment: translating basic biology to therapeutic possibilities. Front Cell Neurosci 2013;7(30):1–18.

[194] Xu S, Wilf R, Menon T, Panikker P, Sarthi J, Elefant F. Epigenetic control of learning and memory in Drosophila by Tip60 HAT action. Genetics 2014;198(4):1571–86.

[195] Kandel ER, Pittenger C. The past, the future and the biology of memory storage. Philos Trans R Soc Lond B Biol Sci 1999;354(1392):2027–52.

[196] Zovkic IB, Guzman-Karlsson MC, Sweatt JD. Epigenetic regulation of memory formation and maintenance. Learn Mem 2013;20(2):61–74.

[197] Lubin FD, Gupta S, Parrish RR, Grissom NM, Davis RL. Epigenetic mechanisms: critical contributors to long-term memory formation. Neuroscientist 2011;17(6):616–32.

[198] Xu S, Panikker P, Iqbal S, Elefant F. Tip60 HAT action mediates environmental enrichment induced cognitive restoration. PLoS One 2016;11(7):e0159623.

[199] Karnay A, Karisetty BC, Beaver M, Elefant F. Hippocampal stimulation promotes intracellular Tip60 dynamics with concomitant genome reorganization and synaptic gene activation. Mol Cell Neurosci 2019;101:103412.

[200] Lorbeck M, Pirooznia K, Sarthi J, Zhu X, Elefant F. Microarray analysis uncovers a role for Tip60 in nervous system function and general metabolism. PLoS One 2011;6(4):e18412.

[201] Lonze BE, Ginty DD. Function and regulation of CREB family transcription factors in the nervous system. Neuron 2002;35(4):605–23.

[202] Watts RJ, Hoopfer ED, Luo L. Axon pruning during Drosophila metamorphosis: evidence for local degeneration and requirement of the ubiquitin-proteasome system. Neuron 2003;38(6):871–85.

[203] Hirano Y, Masuda T, Naganos S, Matsuno M, Ueno K, Miyashita T, et al. Fasting launches CRTC to facilitate long-term memory formation in Drosophila. Science (New York, NY) 2013;339(6118):443–6.

[204] Chawla S, Vanhoutte P, Arnold FJ, Huang CL, Bading H. Neuronal activity-dependent nucleocytoplasmic shuttling of HDAC4 and HDAC5. J Neurochem 2003;85(1):151–9.

[205] Parra M. Class II a HDAC s—new insights into their functions in physiology and pathology. FEBS J 2015;282(9):1736–44.

[206] Schlumm F, Mauceri D, Freitag HE, Bading H. Nuclear calcium signaling regulates nuclear export of a subset of class IIa histone deacetylases following synaptic activity. J Biol Chem 2013;288(12):8074–84.

[207] Kahsai L, Zars T. Learning and memory in Drosophila: behaviour, genetics and neural systems. Int Rev Neurobiol 2011;99:139–67.

[208] Fiala A. Olfaction and olfactory learning in Drosophila: recent progress. Curr Opin Neurobiol 2007;17(6):720–6.

[209] Siwicki KK, Ladewski L. Associative learning and memory in Drosophila: beyond olfactory conditioning. Behav Processes 2003;64(2):225–38.

[210] Xu S, Elefant F. Tip off the HAT- Epigenetic control of learning and memory by Drosophila Tip60. Fly (Austin Tex) 2015;9(1):22–8.

[211] Fitzsimons HL, Scott MJ. Genetic modulation of Rpd3 expression impairs long-term courtship memory in Drosophila. PLoS One 2011;6(12):e29171.

[212] Saha RN, Pahan K. HATs and HDACs in neurodegeneration: a tale of disconcerted acetylation homeostasis. Cell Death Differ 2006;13(4):539–50.

[213] Pirooznia SK, Chiu K, Chan MT, Zimmerman JE, Elefant F. Epigenetic regulation of axonal growth of Drosophila pacemaker cells by histone acetyltransferase tip60 controls sleep. Genetics 2012;192(4):1327–45.

[214] Johnson AA, Sarthi J, Pirooznia SK, Reube W, Elefant F. Increasing Tip60 HAT Levels rescues axonal transport defects and associated behavioral phenotypes in a Drosophila Alzheimer's disease model. J Neurosci 2013;33(17):7535–47.

[215] Mayford M, Siegelbaum SA, Kandel ER. Synapses and memory storage. Cold Spring Harb Perspect Biol 2012;4(6).

[216] Panikker P, Xu S-J, Zhang H, Sarthi J, Beaver M, Sheth A, et al. Restoring Tip60 HAT/HDAC2 balance in the neurodegenerative brain relieves epigenetic transcriptional repression and reinstates cognition. J Neurosci 2018;38(19):4569–83.

[217] Chen X, Pan W. The treatment strategies for neurodegenerative diseases by integrative medicine. Integr Med Int 2014;1(4):223–5.

[218] Lovrečić L, Maver A, Zadel M, Peterlin B. The role of epigenetics in neurodegenerative diseases. Neurodegener Dis 2013;345.

[219] Hwang JY, Aromolaran KA, Zukin RS. The emerging field of epigenetics in neurodegeneration and neuroprotection. Nat Rev Neurosci 2017;18(6):347–61.

[220] Berson A, Nativio R, Berger SL, Bonini NM. Epigenetic regulation in neurodegenerative diseases. Trends Neurosci 2018;41(9):587–98.

[221] Konsoula Z, Barile FA. Epigenetic histone acetylation and deacetylation mechanisms in experimental models of neurodegenerative disorders. J Pharmacol Toxicol Methods 2012;66(3):215–20.

[222] Feng Y, Jankovic J, Wu Y-C. Epigenetic mechanisms in Parkinson's disease. J Neurol Sci 2015;349(1–2):3–9.

[223] Karisetty BC, Bhatnagar A, Armour EM, Beaver M, Zhang H, Elefant F. Amyloid-β peptide impact on synaptic function and neuroepigenetic gene control reveal new therapeutic strategies for Alzheimer's disease. Front Mol Neurosci 2020;13:577622.

[224] Rossaert E, Pollari E, Jaspers T, Van Helleputte L, Jarpe M, Van Damme P, et al. Restoration of histone acetylation ameliorates disease and metabolic abnormalities in a FUS mouse model. Acta Neuropathol Commun 2019;7(1):107.

[225] No authors listed. 2020 Alzheimer's disease facts and figures. Alzheimer's Dementia 2020;16(3):391–460.

[226] M Niedowicz D, T Nelson P, Paul Murphy M. Alzheimer's disease: pathological mechanisms and recent insights. Curr Neuropharmacol 2011;9(4):674–84.

[227] Kramer JM, Kochinke K, Oortveld MAW, Marks H, Kramer D, de Jong EK, et al. Epigenetic regulation of learning and memory by Drosophila EHMT/G9a. PLoS Biol 2011;9(1):e1000569.

[228] Pirooznia SK, Sarthi J, Johnson AA, Toth MS, Chiu K, Koduri S, et al. Tip60 HAT activity mediates APP induced lethality and apoptotic cell death in the CNS of a Drosophila Alzheimer's disease model. PLoS One 2012;7(7):e41776.

[229] Frost B, Hemberg M, Lewis J, Feany MB. Tau promotes neurodegeneration through global chromatin relaxation. Nat Neurosci 2014;17(3):357–66.

[230] Pirooznia SK, Elefant F. A HAT for sleep?: epigenetic regulation of sleep by Tip60 in Drosophila. Fly (Austin Tex) 2013;7(2):99–104.

[231] Zhang H, Karisetty BC, Bhatnagar A, Armour EM, Beaver M, Roach TV, et al. Tip60 protects against amyloid-β-induced transcriptomic alterations via different modes of action in early vs late stages of neurodegeneration. Mol Cell Neurosci 2020;109:103570.

[232] Beaver M, Bhatnagar A, Panikker P, Zhang H, Snook R, Parmar V, et al. Disruption of Tip60 HAT mediated neural histone acetylation homeostasis is an early common event in neurodegenerative diseases. Sci Rep 2020;10(1):18265.

[233] Xu K, Dai X-L, Huang H-C, Jiang Z-F. Targeting HDACs: a promising therapy for Alzheimer's disease. Oxid Med Cell Longev 2011;2011:143269.

[234] Yang S-S, Zhang R, Wang G, Zhang Y-F. The development prospection of HDAC inhibitors as a potential therapeutic direction in Alzheimer's disease. Transl Neurodegener 2017;6(1):19.

[235] Sharma S, Taliyan R. Targeting histone deacetylases: a novel approach in Parkinson's disease. Parkinson's Disease 2015;2015:303294.

[236] Shukla S, Tekwani BL. Histone deacetylases inhibitors in neurodegenerative diseases, neuroprotection and neuronal differentiation. Front Pharmacol 2020;11(537).

[237] Outeiro TF, Kontopoulos E, Altmann SM, Kufareva I, Strathearn KE, Amore AM, et al. Sirtuin 2 inhibitors rescue alpha-synuclein-mediated toxicity in models of Parkinson's disease. Science (New York, NY) 2007;317(5837):516–19.

[238] Steffan JS, Bodai L, Pallos J, Poelman M, McCampbell A, Apostol BL, et al. Histone deacetylase inhibitors arrest polyglutamine-dependent neurodegeneration in Drosophila. Nature 2001;413(6857):739–43.

[239] Pallos J, Bodai L, Lukacsovich T, Purcell JM, Steffan JS, Thompson LM, et al. Inhibition of specific HDACs and sirtuins suppresses pathogenesis in a Drosophila model of Huntington's disease. Hum Mol Genet 2008;17(23):3767–75.

[240] St Laurent R, O'Brien LM, Ahmad ST. Sodium butyrate improves locomotor impairment and early mortality in a rotenone-induced Drosophila model of Parkinson's disease. Neuroscience 2013;246:382–90.

[241] Didonna A, Opal P. The promise and perils of HDAC inhibitors in neurodegeneration. Ann Clin Transl Neurol 2015;2(1):79–101.

[242] Chuang D-M, Leng Y, Marinova Z, Kim H-J, Chiu C-T. Multiple roles of HDAC inhibition in neurodegenerative conditions. Trends Neurosci 2009;32(11):591–601.

[243] Selvi BR, Cassel JC, Kundu TK, Boutillier AL. Tuning acetylation levels with HAT activators: therapeutic strategy in neurodegenerative diseases. Biochim Biophys Acta 2010;1799(10–12):840–53.

[244] Cutler T, Sarkar A, Moran M, Steffensmeier A, Puli OR, Mancini G, et al. Drosophila eye model to study neuroprotective role of CREB Binding Protein (CBP) in Alzheimer's disease. PLoS One 2015;10(9):e0137691.

[245] Beaver M, Karisetty BC, Zhang H, Bhatnagar A, Armour E, Parmar V, et al. Chromatin and transcriptomic profiling uncover dysregulation of the Tip60 HAT/HDAC2 epigenomic landscape in the neurodegenerative brain. bioRxiv 2021; 2021.02.27.433179.

[246] Nithiananthanrajah J, Hannan AJ. Enriched environments, experience-dependent plasticity and disorders of the nervous system. Nat Rev Neurosci 2006;7(9):697–709.

[247] Carulli D, Foscarin S, Rossi F. Activity-dependent plasticity and gene expression modifications in the adult CNS. Front Mol Neurosci 2011;4(50):1–11.

[248] Sweatt JD. Experience-dependent epigenetic modifications in the central nervous system. Biol Psychiatry 2009;65(3):191–7.

[249] Fischer A, Sananbenesi F, Wang X, Dobbin M, Tsai LH. Recovery of learning and memory is associated with chromatin remodelling. Nature 2007;447(7141):178–82.

[250] Praag HV, Kempermann G, Gage FH. Neural consequences of environmental enrichment. Nat Neurosci 2000;1:191–8.

[251] Donlea JM, Shaw PJ. Sleeping together: using social interactions to understand the role of sleep in plasticity. Adv Genet 2009;68:57–81.

[252] Heisenberg M, Heusipp M, Wanke C. Structural plasticity in the Drosophila brain. J Neurosci 1995;15(3 Pt 1):1951–60.

[253] Aso Y, Grubel K, Busch S, Friedrich AB, Siwanowicz I, Tanimoto H. The mushroom body of adult Drosophila characterized by GAL4 drivers. J Neurogenet 2009;23(1–2):156–72.

[254] Technau GM. Fiber number in the mushroom bodies of adult Drosophila melanogaster depends on age, sex and experience. J Neurogenet 2007;21:183–96.

[255] Lopez-Atalaya JP, Ciccarelli A, Viosca J, Valor LM, Jimenez-Minchan M, Canals S, et al. CBP is required for environmental enrichment-induced neurogenesis and cognitive enhancement. EMBO J 2011;30(20):4287–98.

[256] Sweatt JD. Creating stable memories. Science (New York, NY) 2011;331:869–71.

[257] Penner MR, Roth TL, Barnes CA, Sweatt JD. An epigenetic hypothesis of aging-related cognitive dysfunction. Front Aging Neurosci 2010;2:9.

CHAPTER

14

Models of Mouse Epigenetic Inheritance: Classification, Mechanisms, and Experimental Strategies

Courtney W. Hanna[1,2,3]

[1]Centre for Trophoblast Research, University of Cambridge, Cambridge, United Kingdom [2]Department of Physiology, Development and Neuroscience, University of Cambridge, Cambridge, United Kingdom [3]Epigenetics Programme, Babraham Institute, Cambridge, United Kingdom

OUTLINE

Introduction	249	Mechanisms of Imprinting: Germline Differentially Methylated Domains	255
Epigenetic Programming in Gametogenesis	250	Cataloging Imprinted Domains	255
Epigenetic Re-Programming During Embryogenesis	251	Tissue and Cell-Specific Imprinting	256
Metastable Epialleles	251	Function of Imprinted Genes	256
Discovery of Metastable Epialleles	251	Noncanonical Imprinting	257
Mechanisms of Metastability	252	Transient Imprinting	258
Environmental Modulation of Epigenetic Inheritance	252	Conclusions	259
Genomic Imprinting	253	Acknowledgments	259
Discovery of Genomic Imprinting	253	References	259

INTRODUCTION

The mammalian embryonic genome is comprised of two sets of chromosomes, one maternally inherited and one paternally inherited through the fertilization of an oocyte by a sperm. Beyond the genetic information provided by these germ cells, the inherited genetic material also carries numerous epigenetic modifications. These include chemical modifications to the DNA itself, but also to its associated histone proteins. Epigenetic inheritance represents the transmission of epigenetic information from the parent to offspring via the germline. A long-standing interest in the field has been: to what extent does epigenetic inheritance occur, and how does it influence the health and viability of offspring?

The mouse model has been pivotal in building our understanding of epigenetic inheritance due to the variety of available isogenic inbred strains, embryo manipulation techniques, and genetic engineering tools. While the mouse model offers the considerable advantages of being a tractable system to study epigenetic inheritance, it is important to note that epigenetic mechanisms are not always fully recapitulated in human development. Therefore, the knowledge gained from studies in mouse should be extrapolated with caution.

Due to the extensive epigenetic re-programming that occurs in early embryogenesis, there are relatively few regions of the genome that inherit an epigenetic state. Generally, these domains fall into two broad categories: metastable epialleles and genomic imprints. While genomic imprints are established during gametogenesis, metastable epialleles are regions that are relatively resistant to epigenetic re-programming events. The transmission of epigenetic states from parent to offspring at these loci contributes to non-Mendelian inheritance of traits and disease, highlighting the importance of understanding their underlying mechanisms.

In this chapter, I will discuss the insights we have gained from studies in mouse, including (1) the distinct epigenetic landscapes established in sperm and oocytes, (2) epigenetic re-programming in the pre- and postimplantation embryo, and (3) how metastable epialleles and genomic imprints are established and able to propagate their epigenetic state through phases of re-programming. I will also discuss recent advances in our understanding of epigenetic mechanisms of genomic imprinting in mouse, comparing classic imprinting with the more recently discovered modes of transient and noncanonical imprinting. This chapter will highlight how tools and techniques, in combination with mouse transgenic models, have been foundational in our understanding of epigenetic inheritance.

EPIGENETIC PROGRAMMING IN GAMETOGENESIS

In early mouse embryonic development, by embryonic day (E) 7.25, a small cluster of epiblast cells are specified as primordial germ cells (PGCs) [1–3]. These cells begin to proliferate and migrate towards the forming genital ridge. Following PGC migration, there is widespread erasure of repressive epigenetic marks, including DNA methylation and H3K9 di- and trimethylation [4–8] (Fig. 14.1). This re-programming is predominantly due to passive loss of these marks across cell divisions through the nuclear exclusion of the maintenance DNA methyltransferase 1 (DNMT1) and histone methyltransferases and/or their co-factors [8,9]. The erasure of these repressive marks is not only essential to remove inherited genomic imprints, but to enable activation of genes essential for the initiation of meiosis and gametogenesis [10]. The specificity of this activation program, termed germline-re-programming responsive genes, appears to be mediated by binding of TET1, which may act as an activator through its recruitment of chromatin modifiers or demethylation of enhancers [10].

Epigenetic marks are then re-established during different developmental windows, depending on the sex of the embryo. Epigenetic programming occurs prenatally in male spermatogonia and postnatally during oocyte maturation in females [12,13] (Fig. 14.1). Beyond this difference in timing, sperm and oocytes

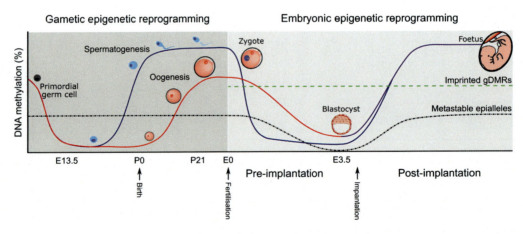

FIGURE 14.1 Epigenetic re-programming in mouse development. DNA methylation, among other epigenetic modifications, undergoes two waves of re-programming during development, the first occurs during gametogenesis and the second during embryogenesis. DNA methylation undergoes widespread erasure in primordial germ cells, which is completed by embryonic day (E)13.5. DNA methylation is then re-established during gametogenesis at differing times, depending on the sex of the embryo. In males, DNA methylation is reset before birth in spermatogonial stem cells, while in females, during oocyte growth between postnatal (P) days 5–21. Upon fertilization, the paternally inherited DNA undergoes rapid and widespread removal of DNA methylation (blue line), while maternally inherited DNA methylation is passively lost through cleavage-stage cell divisions (red line). After implantation of the blastocyst, DNA methylation is re-established prior to lineage specification. Imprinted germline differentially methylated regions (gDMRs) are protected from the pre- and postimplantation re-programming, maintaining allelic DNA methylation throughout development into adulthood (green dashed line). Conversely, DNA methylation at metastable epialleles appears to be re-programmed during embryogenesis, re-acquiring similar levels to that of the parental genome (black dashed line) [11]. *Adapted from Hanna CW, Demond H, Kelsey G. Epigenetic regulation in development: is the mouse a good model for the human?. Hum Reprod Update 2018;24:556–76. Creative Commons License http://creativecommons.org/licenses/by/4.0/.*

acquire very distinct patterns of epigenetic marks and chromatin packaging.

Spermatogonia acquire very high levels of DNA methylation, predominantly through the action of de novo DNMT3A, rodent-specific DNMT3C, and catalytically inactive co-factor DNMT3L [14–16]. The targeting of de novo DNMTs in spermatogonia is dependent on two complementary mechanisms. The piRNA pathway recruits DNMTs to a subset of repetitive elements and the paternally imprinted domain *Rasgrf1* [17], while the interaction of DNMT3A and DNMT3L with euchromatic H3K36me2 permits widespread accumulation of DNA methylation throughout the majority of the genome [18]. During spermatogenesis, the vast majority of histones are replaced with protamines, enabling the DNA to be densely packaged into the compact mature sperm head. As such, very few histones are present on paternally inherited DNA (~1%), including a subset of CpG-rich promoters and gene-poor regions that retain bivalently marked histones (H3K4me3 and H3K27me3) [19–21].

In females, epigenetic programming occurs in meiotically arrested oocytes during oocyte growth and for the majority of epigenetic marks that have been assessed to date, shows distinctive genomic distributions. Fully grown oocytes have a bimodal pattern of DNA methylation, with DNA methylation almost exclusively restricted to transcribed gene bodies and inter-genic regions remaining unmethylated [22,23]. The interaction of DNMT3A and DNMT3L with SETD2-mediated H3K36me3 is key to this patterning [14,18,24,25]. The unique interaction of DNMT3A with chromatin via H3K36 methylation likely explains the divergent patterning of DNA methylation between oocytes and sperm [18]. Histone modifications associated with both active (H3K4me3) and repressive (H3K27me3, H2AK119ub and H3K9me2) chromatin all show a mutual exclusivity with DNA methylation in the oocyte and hence predominantly occupy broad inter-genic domains, patterns not typically observed across other cell types [26–31]. The oocyte, unlike sperm, retains its modified histones after fertilization and thus maternal DNA is capable of transmitting far more epigenetic information to the embryo.

EPIGENETIC RE-PROGRAMMING DURING EMBRYOGENESIS

After fertilization, the maternal and paternal DNA remain separated as pronuclei until just prior to the first cleavage-stage cell division. This separation allows the paternal chromatin to be repackaged, through the replacement of protamines with maternal histones [32]. Paternal DNA also undergoes rapid and widespread erasure of DNA methylation [33] (Fig. 14.1). The mechanisms underlying this erasure have remained elusive [34,35], but it is thought to involve DNA repair mechanisms [36,37]. Consequently, very few epigenetic marks persist on paternal DNA and this is reflected in fact that the vast majority of epigenetic inheritance identified-to-date is maternal in origin.

Conversely, maternal chromatin undergoes asynchronous phases of re-programming upon fertilization. The broad domains of H3K4me3 and H2AK119ub are both quite rapidly replaced by more canonical distributions at the 2-cell stage after zygotic genome activation, with H3K4me3 localizing at transcribed gene promoters and H2AK119ub at silent developmental gene promoters [26,28,31]. In contrast, the broad domains of H3K27me3 and the majority of DNA methylation are passively lost during the preimplantation cleavage divisions, and are not reset until postimplantation development [27] (Fig. 14.1).

Despite the dramatic re-programming of most inherited epigenetic marks in the embryo, there are two types of loci that have been shown to be protected, at least to some extent, from these mechanisms: metastable epialleles and imprinted domains. In the following sections, I will discuss how these regions were discovered and what we currently know about the mechanisms that underlie their establishment and persistence.

METASTABLE EPIALLELES

Metastable epialleles are regions of the genome that show variable epigenetic states between genetically identical individuals and evidence of epigenetic inheritance between generations. Consequently, true metastable epialleles likely fail to undergo complete epigenetic re-programming during gametogenesis and embryogenesis, and hence represent regions where a parental epigenetic state can be either completely or partially transmitted to offspring. Isogenic inbred mice represent the ideal system for studying epigenetic variability at metastable epialleles because all offspring are genetically identical, and inheritance patterns can be evaluated across a single litter, hence controlling for in utero environment.

Discovery of Metastable Epialleles

One of the classic examples of a metastable epiallele is the *Agouti viable yellow* (A^{vy}) allele, which is the consequence of a spontaneous insertion of an intracisternal A-type particle (IAP) endogenous retroviral element approximately 100-kb upstream of the *Agouti* coat color gene [38,39]. The IAP insertion drives ectopic expression of *Agouti*, resulting in yellow, obese mice. There is a variably penetrant phenotype associated with the A^{vy}

allele, which is a result of variable silencing of the IAP insertion by DNA methylation [40]. A differing range of phenotypes have been observed among offspring from mothers with a methylated or unmethylated A^{vy} allele, revealing an extent of epigenetic inheritance. This phenotypic shift was not observed with paternal transmission, suggesting that the epigenetic state of this metastable epiallele may be more stably maintained on the maternal allele through epigenetic re-programming events. Furthermore, the epigenetic inheritance at this locus is susceptible to maternal environmental exposures (discussed further in section Environmental Modulation of Epigenetic Inheritance.

There have been continued efforts to identify additional metastable epialleles in the mouse genome. The *Axin fused* (*Axin^{Fu}*) allele was identified through the striking variability in its kinked tail phenotype observed among isogenic mice, similar to the A^{vy} allele, and showed evidence of epigenetic inheritance through both maternal and paternal transmission [41,42]. The *Axin^{Fu}* allele is a consequence of an IAP insertion into an intron of the *Axin* gene, which results in expression of an aberrantly truncated protein. Subsequent genome-wide screens have sought to identify candidate metastable epialleles in an unbiased manner, leading to the identification of several more IAP-derived transcripts with variable DNA methylation [43–46]. However, few examples beyond the A^{vy} and *Axin^{Fu}* alleles show evidence of epigenetic inheritance. In a genome-wide screen of DNA methylation and gene expression in C57BL/6J, Kazachenka and co-authors were able to identify a single novel IAP among many variably methylated IAPs, at the *Gm13849* locus, where the maternal DNA methylation status impacted the range of DNA methylation values observed in offspring [45]. Importantly, this DNA methylation was inversely correlated with gene expression of the IAP-fused *Gm13849* transcript. Overall, their data supports that the vast majority of IAPs undergo efficient epigenetic re-programming during gametogenesis and/or embryogenesis, with few metastable epialleles exhibiting epigenetic inheritance.

Mechanisms of Metastability

Metastable epialleles have been shown to have similar DNA methylation levels across tissues within a single individual [45,47,48], including the germline [49,50]. However, unlike in somatic cell types, a consistent DNA methylation state in the germline requires protection against DNA methylation erasure in PGCs and re-methylation during gametogenesis. To date, the dynamics and mechanisms underpinning the stability of DNA methylation at metastable IAPs through gametic epigenetic re-programming remains unexplored. However, it has been reported that subsets of IAPs show resistance to demethylation in PGCs [4,6,7,51], so it is possible that metastable epialleles are part of this more general mechanism to maintain silencing of specific repeat classes in the germline (Fig. 14.1).

After fertilization, DNA methylation levels at metastable epialleles will also need to be protected from or demarcated in some way during embryonic epigenetic re-programming to facilitate epigenetic inheritance. Surprisingly, characterization of DNA methylation at the A^{vy} allele across preimplantation development revealed that both maternal and paternal alleles lose inherited DNA methylation [49] (Fig. 14.1). This would suggest that, at least for this IAP, an alternative epigenetic mark may act as the "memory" for the DNA methylation state in the parent. However, the range of DNA methylation values seen across offspring for metastable epialleles supports that there is considerable instability in the faithful maintenance and/or re-establishment of the parental epigenetic state in offspring.

Key molecular players acting in the epigenetic inheritance of metastable epialleles have been identified in an extensive genetic screen, termed "modifiers of murine metastable epialleles" (*Mommes*) [52–54]. Using a mouse line carrying a GFP reporter transgene that variably expresses GFP in red blood cells (RBC) and N-ethyl-N-nitrosourea mutagenesis, mutations were identified that negatively or positively impacted the proportion of RBCs expressing the GFP reporter in isogenic mice [53]. Many of the candidate genes identified are epigenetic modifiers, including a number of DNA and histone methyltransferases associated with maintenance of imprinted germline differentially methylated regions (gDMRs), including *Dnmt1*, *Setdb1*, and *Trim28* [52]. Importantly, when candidates were then tested for modulation of the *Agouti* phenotype, mutations were shown to result in similar de-repression of the A^{vy} allele in offspring (i.e., a shift towards the yellow phenotype) as was observed for the GFP transgene [52]. While the mechanisms underpinning epigenetic inheritance of metastable epialleles still remain elusive, undoubtedly, continued characterization of the candidate *Mommes* genes will be essential in uncovering further critical regulators.

Environmental Modulation of Epigenetic Inheritance

The variable nature and direct link between epigenetic state and phenotype of metastable epialleles A^{vy} and *Axin^{Fu}* have made these mouse models a preferred system to study the impact of environment on epigenetic inheritance. Maternal and in utero environmental exposures have revealed that these metastable epialleles are highly susceptible to these perturbations. Supplementing maternal diet with methyl donors during pregnancy

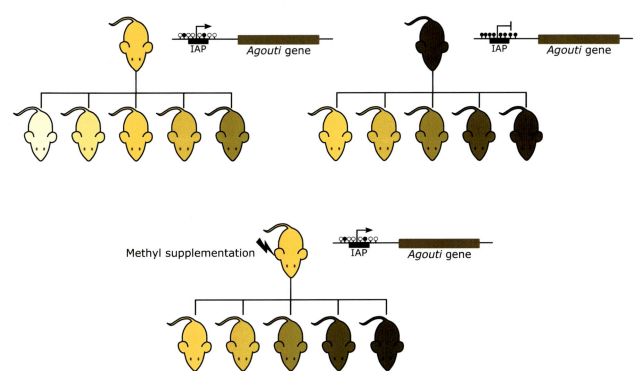

FIGURE 14.2 **Epigenetic inheritance of the A^{vy} metastable epiallele.** The *Agouti* (A^{vy}) mouse model carries an intracisternal A-particle (IAP) insertion upstream of the *Agouti* coat color gene. When this A^{vy} metastable epiallele is unmethylated, it drives ubiquitous expression of the *Agouti* gene, resulting in yellow, obese mice; whereas, when it is methylated, it remains silent and mice remain brown and healthy. A^{vy} is epigenetically unstable, with a spectrum of methylation and, consequently, coat colors observed among offspring. A^{vy} shows evidence of epigenetic inheritance, as the range of methylation values and phenotypes seen in the offspring differ depending on the methylation status of the mother (top panel). Maternal environmental exposures, such as dietary supplementation of methyl donors, can alter the epigenetic inheritance pattern seen in offspring, shifting the A^{vy} to a more methylated state (bottom panel).

increased DNA methylation of the A^{vy} and $Axin^{Fu}$ alleles and consequently altered the phenotypic outcomes of the offspring [47,48] (Fig. 14.2). Similarly, in utero exposure to phytoestrogen genistein or phalates increased DNA methylation of the A^{vy} allele [55,56], while *in utero* exposure to bisphenol A, a synthetic estrogen, showed the opposite effect [57]. Importantly, recent data would suggest that the epigenetic susceptibility of these metastable epialleles may be the exception, not the rule. Bertozzi and co-authors evaluated the epigenetic effects of maternal environmental exposures on the repertoire of variably methylated IAPs identified in mouse [45,58]. They observed no significant effects on DNA methylation at these IAPs in response to maternal exposure to bisphenol A, obesogenic diet or methyl supplementation.

GENOMIC IMPRINTING

Early evidence that distinct epigenetic marks may be transmitted from each of the parental gametes to the embryo was demonstrated through embryo manipulation experiments in mouse models (Fig. 14.3). Diploid embryos generated with only maternal (gynogenetic or parthenogenetic) or paternal (androgenetic) DNA were shown to be embryonic lethal [59,60]. These experiments definitively demonstrated that the maternal and paternal genomic contributions to the embryo were non-equivalent. It was hypothesized that maternal and paternal DNA carried "imprints," allowing certain genes to only be expressed from one inherited copy based on parent-of-origin. In the decades since, there have been extensive efforts to identify all imprinted genes in the mouse genome and understand the mechanisms regulating genomic imprinting.

Discovery of Genomic Imprinting

The discovery of the first putative imprinted regions in the mouse genome began through the use of genetic and embryonic manipulation techniques in mouse models. There were examples of large deletions associated with distinctive visible phenotypes, such as the "hairpin-tail" and related "T-associated maternal effect (*Tme*)" locus on chromosome 17, that showed non-Mendelian inheritance [61,62]. Mice heterozygous for the hairpin-tail or *Tme* deletions showed differing phenotypes, depending on the parent-of-origin, supporting that these loci contained

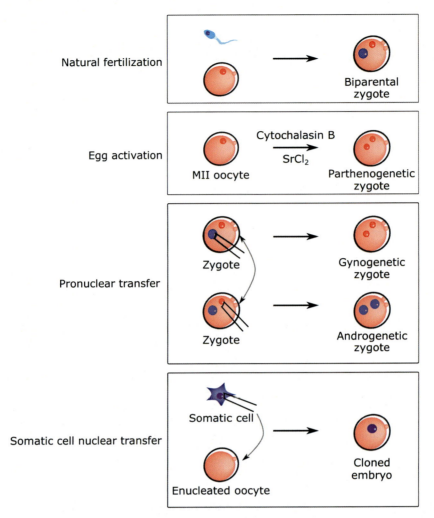

FIGURE 14.3 **Embryo manipulation techniques in mouse.** Embryo manipulation techniques have been instrumental in building our understanding of the importance of parental genetic and epigenetic contributions to the embryo. Natural fertilization occurs when a sperm enters an oocyte to create a biparental zygote, which is able to support healthy embryonic development. These are two approaches to creating uniparental zygotes: (1) A meiotic stage II (MII) oocyte can be artificially activated by cytochalasin B and SrCl$_2$, causing it to retain its second polar body and begin cleavage stage divisions. This approach generates a bi-maternal embryo, termed parthenogenetic. (2) The maternal or paternal pronuclei can be removed from a naturally fertilized zygote and replaced to generate bi-paternal (androgenetic) or bi-maternal (gynogenetic) embryos. A bi-parental nuclei can be removed from a somatic cell and transferred to an enucleated oocyte, termed somatic cell nuclear transfer. The oocyte will then re-program the somatic nuclei to a totipotent state, which can then go on to support the development of a cloned embryo.

imprinted genes. Uniparental disomies (UPDs) are chromosomal abnormalities, where the genomic content remains diploid, but both copies of an entire chromosome or a portion of a chromosome are inherited from one parent. Through the identification of UPDs associated with embryonic lethality, a number of chromosomal regions containing putative imprinted genes were identified [63]. Years later, the generation of diploid mouse embryos with only maternal or paternal DNA provided direct evidence that the DNA inherited from oocyte and sperm were not equivalent. While both failed to survive embryonic development, they showed distinctive developmental phenotypes; androgenetic embryos had an overgrowth of placental trophoblast with almost no embryonic development, whereas gynogenetic embryos failed to develop placental tissues [59,60]. Together these findings supported that genomic imprinting was essential for normal mouse development.

Building on these early models and approaches, the first imprinted genes were identified in 1991, including maternally expressed *H19* [64] and *Igf2r* [65], and paternally expressed *Igf2* [66,67]. These and the majority of the imprinted genes identified since are located within clusters and their imprinted expression is dependent on a central imprinting control region (ICR), which often directly regulates the allelic expression of a non-coding RNA. Genomic imprinting was subsequently shown to be an epigenetic phenomenon, with imprinted DNA methylation being the critical regulator of ICRs and often imprinted gene promoters [68].

Mechanisms of Imprinting: Germline Differentially Methylated Domains

Evidence that DNA methylation was an epigenetic regulator of imprinted gene expression first came from imprinted transgene constructs in mouse. It was shown that DNA methylation at a transgene was erased and re-established during gametogenesis, and that DNA methylation was maintained distinctly based on parent-of-origin in the embryo [69]. The use of Filial (F)1 hybrid embryos generated from breeding distantly related inbred mouse strains subsequently allowed the assessment of allelic DNA methylation at imprinted regions, using strain-specific single nucleotide polymorphisms (SNPs). These investigations identified parent-of-origin allelic DNA methylation patterns at imprinted ICRs [70]. The essential role for DNA methylation in imprinting was demonstrated through disruption of germline DNA methylation patterning. Conditional ablation of the de novo DNA methyltransferase 3 A (DNMT3A) or deletion of co-factor DNMT3L in oocytes and sperm in transgenic mouse models revealed that both were critical for the establishment of germline DNA methylation [14,24,71,72]. Disrupted DNA methylation in spermatogenesis results in meiotic catastrophe and azoospermia, while oogenesis is seemingly unaffected [14,24,73]. Fertilization of *Dnmt3a* or *Dnmt3l* knockout oocytes resulted in embryos with a loss of imprinting at all maternally methylated ICRs (also termed maternal gDMRs) and was shown to be embryonic lethal at around mid-gestation [14,24].

The number of regions that are differentially methylated in oocytes and sperm are far more extensive than the 25 known imprinted gDMRs [74,75], as a consequence of the differing genome-wide patterns of DNA methylation (discussed in section Epigenetic Programming During Gametogenesis). While the majority of the allelic asymmetries in DNA methylation are erased or over-written in the wave of embryonic epigenetic re-programming (described in section Epigenetic Reprogramming During Embryogenesis), imprinted gDMRs are uniquely protected from these dynamics. Methylated CpG-containing motifs within gDMRs are bound by krab zinc-finger proteins ZFP57 and/or ZNF445, enabling the recruitment of TRIM28 (also known as KAP1), H3K9 methyltransferase SETDB1 and maintenance DNMT1 to the methylated allele [76–79] (Fig. 14.4). Ablation of any of these components compromises the inherited allelic methylation state at these gDMRs, shown *in vivo* using transgenic mouse models and/or in embryonic stem cells [76,77,80–83].

Cataloging Imprinted Domains

There have been extensive efforts using high throughput sequencing approaches to comprehensively identify imprinted gene expression and DNA methylation patterns in the genome. In recent years, systematic screens have been done using two main strategies in mouse models. Sequencing data from reciprocal hybrid F1 mice, generated from crossing two inbred strains, allows for the allelic mapping of data using strain-specific SNPs. Optimally, distantly related inbred strains are used, such as C57BL/6 and CAST/Ei, which contain a strain-specific SNP every 120 bps on average, allowing the majority of regions in the genome to be assessed [85]. Nevertheless, this approach is limited by the distribution and density of strain-specific SNPs in the hybrids used and hence is not informative for all genomic loci. Thus, as an alternative, studies have used androgenetic and gynogenetic (or parthenogenetic) embryos. The use of these manipulated embryos offers the advantage that data is derived from only paternal or maternal material and is therefore informative across the genome. The disadvantage with this approach is that these embryos do not successfully progress through embryogenesis and hence, gene expression or DNA methylation differences may reflect the abnormal developmental trajectories of these uniparental embryos.

The evaluation of gene expression to identify mono-allelically expressed genes with a parent-of-origin bias has been done with RNA-sequencing or array-based technologies. The culmination of numerous studies over the past three decades have led to the identification of more than 150 imprinted genes in the mouse genome (http://www.geneimprint.org/), summarized in [86]. Obtaining a complete list of imprinted genes has proven challenging however, primarily because the evaluation of allelic gene expression is limited to those genes that are expressed in the developmental window and/or tissue evaluated. Hence, the imprinting status of a number of genes remains contentious or provisional http://www.geneimprint.org/.

Conversely, imprinted gDMRs are inherited through early epigenetic re-programming in the epiblast and are hence present in most derivative tissues, irrespective of whether associated imprinted genes are expressed. Thus, the evaluation of allelic differences in DNA methylation has been an alternative approach to identify imprinted domains [87]. The advent of ultra-low input bisulfite sequencing allowed the comparative DNA methylation analyses between oocyte, sperm and early cleavage stage embryos, which has been pivotal in identifying candidate imprinted gDMRs [23,75]. To date, 3 paternal and 22 maternal imprinted gDMRs have been identified in the mouse genome [74], although a number of gDMRs appear to be transiently maintained in the preimplantation embryo (discussed further in section Transient Imprinting).

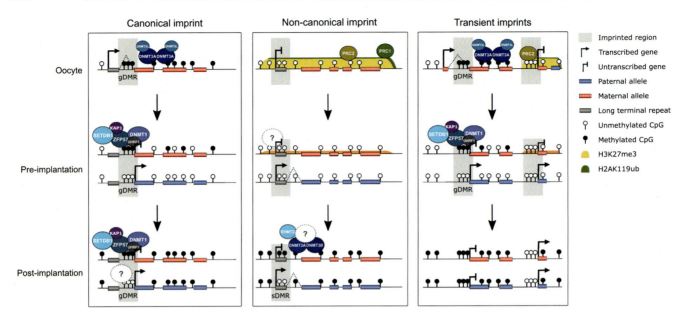

FIGURE 14.4 **Molecular mechanisms of maternal genomic imprinting.** Canonical imprints are established during oogenesis, as de novo DNA methylation is targeted to transcribed gene bodies through the activity of DNMT3A guided by DNMT3L. Alternative upstream promoters, often corresponding to long terminal repeat (LTR) insertions, are widespread in the oocyte, resulting in DNA methylation of gDMR located at canonical promoters. In pre- and postimplantation embryo development, a complex of ZFP57 (or ZNF445), KAP1 and SETDB1 binds to methylated gDMRs to recruit DNMT1 and UHRF1 to maintain the DNA methylation status through cell divisions. In the postimplantation embryo, it is postulated that unknown factors may protect the unmethylated allele at gDMRs from becoming aberrantly methylated. Noncanonical imprints are established by PRC2 targeting of H3K27me3 during oogenesis, which is at least partially dependent on ubiquitination of H2AK119 by PRC1. Both H3K27me3 and H2AK119ub1 are lowly enriched throughout untranscribed regions of the oocyte genome. In the preimplantation embryo, maternally-inherited H3K27me3 is progressively lost. It remains unknown whether factors or an alternative epigenetic mark are necessary to demark non-canonically imprinted maternal alleles. In postimplantation development, noncanonical imprints become restricted to the extra-embryonic tissues, through the tissue-specific expression of underlying LTR insertions. Noncanonical imprints transition from maternal H3K27me3 to DNA methylation through the activity of de novo DNMT3A and 3B and EHMT2, creating secondary DMRs (sDMRs). In the embryonic lineages, noncanonical imprinted domains become biallelically methylated and hence lose imprinting status (not shown). Transient imprints can be mediated by DNA methylation or H3K27me3 in the oocyte and their establishment is identical to canonical and noncanonical imprints, respectively. Furthermore, their persistence through preimplantation development is likely facilitated by the same mechanisms by which DNA methylation- and H3K27me3-dependent imprints are maintained. Transient imprints are unique in that they are not protected from postimplantation epigenetic programming, resulting in an epigenetic synchronization of the two parental alleles, either through gain or loss of epigenetic marks [84]. *Adapted from Hanna CW, Kelsey G. Features and mechanisms of canonical and noncanonical genomic imprinting. Genes Dev 2021;35:821—34. Creative Commons License http://creativecommons.org/licenses/by/4.0/.*

Tissue and Cell-Specific Imprinting

The genome-wide screens for genes with imprinted allelic expression has revealed that there is heterogeneity between tissues [88,89]. This is, in part, attributable to imprinted genes being expressed in only a subset of tissues, but there are clear examples of genes showing imprinted expression in some tissues and bi-allelic expression in others [88,90]. Furthermore, recent data from studies using reporter constructs have demonstrated that there can be heterogeneity even at the cellular level within a tissue. These studies have inserted distinctive fluorescent or luciferase reporters allelically within specific imprinted loci, allowing a direct readout at the single-cell level of whether allelic repression is maintained at imprinted loci [91,92].

Utilizing these luciferase reporter systems, it has been demonstrated that environmental exposures *in utero* can impact allelic gene expression and DNA methylation. For example, the *Cdkn1c* locus showed persistent changes in imprinted allelic expression in the offspring exposed to an altered maternal diet [91]. Recently, using a similar reporter system in the *Dlk1* locus, maternal diet-induced obesity was shown to result in de-repression of the imprinted maternal allele not only in offspring exposed *in utero*, but in the F2 generation as well [93]. These reporter systems have provided an incredibly sensitive way to detect heterogeneity in the fidelity of imprinting at a given locus between cells and developmental windows.

Function of Imprinted Genes

Genomic imprinting is essential for embryonic development, as demonstrated by the failure of androgenetic and parthenogenetic embryos to survive [59,60]. Recently, a study used gene targeting to "balance out" the allelic ratios of a subset of imprinted genes in androgenetic and parthenogenetic embryos, enabling a rescue of these embryonic lethal phenotypes. The authors demonstrated that if all three paternally imprinted gDMRs were deleted in a parthenogenetic ESC, when injected into an MII oocyte, could result in a normally growing bi-maternal embryo [94]. Conversely, deletion of seven maternal gDMRs in androgenetic ESCs resulted in viable bi-paternal offspring, when co-injected with sperm into an empty MII oocyte. This study is reminiscent of earlier work that demonstrated a deletion of the H19 locus in non-growing oocytes, which have acquired no maternal imprints, can be used to fertilize a normal fully-grown oocyte to generate parthenogenetic live animals [95]. Together these studies have revealed a critical set of imprinted domains necessary for development, but studies interrogating the importance of single imprinted clusters or genes have revealed that many imprinted domains play a role in regulating fetal growth, placentation, and modulating maternal physiology.

Many mouse models have been generated carrying either deletion of imprinted genes themselves or the regulatory ICRs. Characterization of these knockout models has demonstrated that there is an essential and reproducible role for imprinting in regulating fetal growth, development, and metabolism (reviewed in [96]). This accumulating body of evidence supports the "parental conflict hypothesis," which proposes that genomic imprinting arose in placental mammals due to the competing interests of the maternally and paternally inherited genomes to regulate maternal resource allocation during pregnancy [97]. The parental conflict hypothesis posits that paternally expressed genes promote fetal growth, while maternally expressed genes restrict it to preserve maternal resources for future pregnancies.

One of key ways in which imprinted genes may modulate fetal growth is through regulation of placental development and function. This has been demonstrated through detailed characterization of placental structure and function in mice with deletion and/or overexpression of imprinted genes such as *Igf2*, *Ascl2*, *Phlda2* and *Peg3* [98–101]. Notably, imprinting in the placenta may be particularly susceptible to environmental exposures. Female mice exposed to bisphenol A early in pregnancy had embryos with abnormal imprinted DNA methylation and gene expression in the placenta, which consequently led to placental abnormalities [102]. Intriguingly, it has been shown that correct imprinting of *Phlda2* in the placenta not only impacts fetal development, but is critical for modulating maternal physiology and behavior. Embryo transfer experiments require fertilized embryos to be generated in vitro or isolated from a pregnant female to be transferred into a pseudo-pregnant female of a different genotype; this approach allows one to disentangle the effects of a genetic change in the mother versus the offspring. Creeth and co-authors were able to show that wildtype females carrying pups with loss of *Phlda2* expression had distinct changes in their hypothalamus and hippocampus during pregnancy, and showed increased nurturing of pups after birth [103]. The placenta, which forms the maternal-fetal interface during pregnancy, is uniquely positioned to be able to influence both fetal and maternal physiology during pregnancy, and hence may be a critical site for the effects of imprinted genes.

Noncanonical Imprinting

The mouse placenta has been shown to express many imprinted genes, yet reports of placental-specific imprinting have been contentious. Allelic gene expression in the placenta is uniquely challenging to evaluate due to contaminating maternal cells, which can skew the allelic ratios in some datasets. However, rigorously validated data has demonstrated that not only do some imprinted gene clusters contain placental-specific imprinted genes (e.g., *Airn/Igf2r* and *Kcnq1ot1/Kcnq1* clusters), but that there are a number of isolated placental-specific imprinted genes in the mouse genome [88,104,105]. The vast majority are expressed from the paternal allele and hence are maternally silenced. However, when maternally inherited DNA methylation has been ablated, a number of these placental-specific imprinted genes showed persistent allelic expression [105]. Thus, it has been suggested that inherited epigenetic marks, beyond DNA methylation, could mediate genomic imprinting, particularly in the placenta. Only recently was it definitively demonstrated that maternally inherited H3K27me3 can confer genomic imprinting, termed noncanonical imprinting.

Using a DNase hypersensitivity assay to identify regions of open chromatin, genome-wide profiles were compared between androgenetic and parthenogenetic embryos [106]. A number of regions showed an open chromatin signature on the paternal allele but were not silenced by maternal DNA methylation, rather they co-localized with maternal H3K27me3. The authors tested the requirement for maternal H3K27me3 by overexpressing an H3K27 demethylase in the early embryo, which resulted in a loss of maternal repression. Later, this was demonstrated using a conditional deletion of Polycomb repressor complex 2 (PRC2) component *Eed* in oocytes, resulting in a loss of maternally-inherited H3K27me3 [107].

Intriguingly, the majority of these non-canonically imprinted domains are only transiently present in the preimplantation embryo [106,108] (discussed further in section Transient Imprinting). Postimplantation, it appears that no non-canonically imprinted regions are retained in embryonic lineages, while a handful persist in extra-embryonic lineages and several correspond to the previously identified placental-specific imprints [105]. The lineage specificity of noncanonical imprinting in the postimplantation extra-embryonic tissues has been linked to the tissue-specific expression of underlying retrotransposon-derived long terminal repeats, which act as imprinted promoters for protein-coding and noncoding RNA transcripts in the yolk sac and placenta [109].

The mechanisms of establishment and maintenance of non-canonically imprinted regions are still being elucidated, but recent research has shed some light on these processes (Fig. 14.4). Deposition of H3K27me3 in the oocyte is dependent on PRC2 activity, which is, at least in part, directed by PRC1-mediated ubiquitinoylation of H2AK119 [28,31]. As discussed in section Epigenetic Programming During Gametogenesis, H3K27me3 and H2AK119ub are found mutually exclusively with DNA methylation in the oocyte, with low, broad enrichment across the transcriptionally silent fraction of the genome. Inheritance of maternal H3K27me3 from the oocyte is necessary for setting up allelic asymmetry in the preimplantation embryo [106], resulting in paternal-specific expression of upwards of 200 domains [108]. However, maternal enrichment for H3K27me3 is progressively diminished during cleavage-stage divisions [27]. At a subset of noncanonical imprints, maternal H3K27me3 is replaced with maternal DNA methylation, resulting in acquisition of somatic (or secondary) imprinted DMRs (sDMRs) exclusively in the postimplantation extra-embryonic lineages [109,110]. The mechanisms underlying this transition are yet to be determined, but may involve repressive H3K9me2, as acquisition of allelic methylation at noncanonical imprints is dependent on H3K9 methyltransferase EHMT2 (also known as G9A) [111,112].

Imprinted silencing of the paternal X chromosome is a phenomenon that occurs in mouse extra-embryonic tissues. It was shown to be independent of inherited DNA methylation many years ago [113], but only recently was it revealed that mechanisms underpinning noncanonical imprinting also mediate imprinted XCI. The *Xist* promoter is coated in H3K27me3 in the oocyte, resulting in imprinted paternal-specific expression of *Xist* in the preimplantation embryo and postimplantation extra-embryonic tissues [114]. Ablation of maternally inherited H3K27me3 results in loss of imprinted XCI [107]. Recently, it was also demonstrated that there is a continued requirement for PRC2, and to a lesser extent PRC1, for imprinted XCI in early embryogenesis [112].

Somatic cell nuclear transfer (SCNT) techniques have been used to demonstrate the importance H3K27me3-mediated noncanonical imprinting in embryo development. SCNT involves injecting a nucleus from a somatic cell into an enucleated oocyte [115,116] (Fig. 14.3). In the context of the oocyte, the somatic nucleus is re-programmed to a totipotent state and can result in a live "cloned" offspring. The efficiency of the SCNT process is exceptionally poor, with only 2% of mouse SCNT embryos surviving to birth [117]. The efficiency of SCNT has been improved to almost 15% with variations to the technique, in particular employing methods to overcome incomplete re-programming of the epigenetic landscape of the somatic nuclei [118]. However, even when these barriers can be partially overcome through genetic and epigenetic manipulations, a failure to establish noncanonical imprinting impairs postimplantation development and placentation [119].

Transient Imprinting

As techniques have advanced to be able to assay epigenetic marks in fewer and fewer cells, it has become possible to characterize inherited allelic epigenetic marks in mouse preimplantation embryos. Initially through the evaluation of DNA methylation and gene expression, but more recently also by profiling histone modifications, it has been revealed that many genomic loci show transient imprinting. This phenomenon appears to be almost exclusively maternally driven by epigenetic marks inherited from the oocyte. Their transient nature is a result of both alleles either becoming silenced or activated during postimplantation epigenetic programming, resulting in a loss of imprinting. Importantly, transient allelic enrichment for repressive marks can regulate allelic gene expression in the preimplantation embryo. While we still do not have a full understanding of the biological importance of transient imprinting, there are clear examples where it can have lasting impacts on genomic regulation and embryonic development.

Transiently imprinted maternal gDMRs were first identified through a screen to identify regions that showed a loss of DNA methylation in embryos derived from *Dnmt3l* null females [120]. Ablation of DNMT3L in females results in a failure to establish oocyte DNA methylation patterns, and consequently when fertilized, embryos lack maternally-inherited gDMRs. Proudhon and co-authors identified several novel maternal gDMRs that were only transiently present in the preimplantation embryo, becoming biallelically methylated postimplantation (Fig. 14.4). A subsequent study utilized detailed DNA methylation analyses across development to identify a further 200 putative transient maternal gDMRs that resolved to a biallelically unmethylated state in the postimplantation embryo [121].

Upon the discovery of noncanonical imprinting, it was noted that the majority of H3K27me3-mediated imprints were transient in nature [107], with only a few persisting past the blastocyst stage. This phenomenon was recently further characterized by RNA sequencing of F1 hybrid blastocyst embryos [108]. The authors identified 71 previously unreported genes that showed transiently imprinted expression in preimplantation development. They demonstrated that the vast majority of these transient imprints were associated with and dependent upon maternal H3K27me3. Notably, similar to canonical imprinting, many of these novel noncanonical transiently imprinted genes were located in clusters, either distal to known imprinted gene clusters or within novel domains.

While the biological importance of transient imprints remains almost entirely unexplored, but the characterization of the *Zdbf2/Gpr1* locus has revealed that even fleeting imprinted expression can have lasting impacts. A transient maternal gDMR was identified at the *Zdbf2/Gpr1* locus, which regulates the allelic expression of transcript *Liz* in the preimplantation embryo [122]. Allelic *Liz* expression results in the establishment of a downstream imprinted sDMR on the paternal allele near the canonical *Zdbf2* promoter, which persists throughout development. Loss of transient imprinting of *Liz* leads to the failure to establish the imprinted sDMR, which is essential for establishing the correct gene expression pattern in the postnatal brain and leads to a postnatal growth deficiency [123].

CONCLUSIONS

The study of epigenetic inheritance and re-programming in mouse models have provided unparalleled insights into the molecular mechanisms underpinning these processes. Inheritance of epigenetic marks from parental gametes, in the form of genomic imprints or metastable epialleles, are critical for development and health of the offspring. These domains appear to be particularly sensitive to environmental perturbations during epigenetic re-programming events in early development, which can lead to lasting epigenetic, and consequently phenotypic, changes in offspring.

Acknowledgments

Thank you to Dr. Kathleen Stewart-Morgan at the Biotech Research & Innovation Centre, Copenhagen, Denmark, for providing valuable comments and feedback on the manuscript.

Courtney Hanna is funded by a Next Generation Fellowship from the Centre for Trophoblast Research, Cambridge, UK, and Sir Henry Dale Fellowship jointly funded by the Wellcome Trust and the Royal Society (Grant Number 222582/Z/21/Z).

References

[1] Chiquoine AD. The identification, origin, and migration of the primordial germ cells in the mouse embryo. Anat Rec 1954;118:135–46.

[2] Ginsburg M, Snow MH, McLaren A. Primordial germ cells in the mouse embryo during gastrulation. Development 1990;110:521–8.

[3] Saitou M, Barton SC, Surani MA. A molecular programme for the specification of germ cell fate in mice. Nature 2002;418:293–300.

[4] Guibert S, Forne T, Weber M. Global profiling of DNA methylation erasure in mouse primordial germ cells. Genome Res 2012;22:633–41.

[5] Hajkova P, Ancelin K, Waldmann T, Lacoste N, Lange UC, Cesari F, et al. Chromatin dynamics during epigenetic reprogramming in the mouse germ line. Nature 2008;452:877–81.

[6] Hajkova P, Erhardt S, Lane N, Haaf T, El-Maarri O, Reik W, et al. Epigenetic reprogramming in mouse primordial germ cells. Mech Dev 2002;117:15–23.

[7] Seisenberger S, Andrews S, Krueger F, Arand J, Walter J, Santos F, et al. The dynamics of genome-wide DNA methylation reprogramming in mouse primordial germ cells. Mol Cell 2012;48:849–62.

[8] Seki Y, Yamaji M, Yabuta Y, Sano M, Shigeta M, Matsui Y, et al. Cellular dynamics associated with the genome-wide epigenetic reprogramming in migrating primordial germ cells in mice. Development 2007;134:2627–38.

[9] Kagiwada S, Kurimoto K, Hirota T, Yamaji M, Saitou M. Replication-coupled passive DNA demethylation for the erasure of genome imprints in mice. EMBO J 2013;32:340–53.

[10] Hill PWS, Leitch HG, Requena CE, Sun Z, Amouroux R, Roman-Trufero M, et al. Epigenetic reprogramming enables the transition from primordial germ cell to gonocyte. Nature 2018;555:392–6.

[11] Hanna CW, Demond H, Kelsey G. Epigenetic regulation in development: is the mouse a good model for the human? Hum Reprod Update 2018;24:556–76.

[12] Ueda T, Abe K, Miura A, Yuzuriha M, Zubair M, Noguchi M, et al. The paternal methylation imprint of the mouse H19 locus is acquired in the gonocyte stage during foetal testis development. Genes Cells 2000;5:649–59.

[13] Hiura H, Obata Y, Komiyama J, Shirai M, Kono T. Oocyte growth-dependent progression of maternal imprinting in mice. Genes Cells 2006;11:353–61.

[14] Kaneda M, Okano M, Hata K, Sado T, Tsujimoto N, Li E, et al. Essential role for de novo DNA methyltransferase Dnmt3a in paternal and maternal imprinting. Nature 2004;429:900–3.

[15] Kato Y, Kaneda M, Hata K, Kumaki K, Hisano M, Kohara Y, et al. Role of the Dnmt3 family in de novo methylation of imprinted and repetitive sequences during male germ cell development in the mouse. Hum Mol Genet 2007;16:2272–80.

[16] Barau J, Teissandier A, Zamudio N, Roy S, Nalesso V, Herault Y, et al. The DNA methyltransferase DNMT3C protects male germ cells from transposon activity. Science 2016;354:909–12.

[17] Watanabe T, Tomizawa S, Mitsuya K, Totoki Y, Yamamoto Y, Kuramochi-Miyagawa S, et al. Role for piRNAs and noncoding RNA in de novo DNA methylation of the imprinted mouse Rasgrf1 locus. Science 2011;332:848–52.

[18] Shirane K, Miura F, Ito T, Lorincz MC. NSD1-deposited H3K36me2 directs de novo methylation in the mouse male germline and counteracts Polycomb-associated silencing. Nat Genet 2020;52:1088–98.

[19] Brykczynska U, Hisano M, Erkek S, Ramos L, Oakeley EJ, Roloff TC, et al. Repressive and active histone methylation

mark distinct promoters in human and mouse spermatozoa. Nat Struct Mol Biol 2010;17:679—87.
[20] Carone BR, Hung JH, Hainer SJ, Chou MT, Carone DM, Weng Z, et al. High-resolution mapping of chromatin packaging in mouse embryonic stem cells and sperm. Dev Cell 2014;30:11—22.
[21] Erkek S, Hisano M, Liang CY, Gill M, Murr R, Dieker J, et al. Molecular determinants of nucleosome retention at CpG-rich sequences in mouse spermatozoa. Nat Struct Mol Biol 2013;20:868—75.
[22] Veselovska L, Smallwood SA, Saadeh H, Stewart KR, Krueger F, Maupetit-Mehouas S, et al. Deep sequencing and de novo assembly of the mouse oocyte transcriptome define the contribution of transcription to the DNA methylation landscape. Genome Biol 2015;16 209-015-0769-z.
[23] Kobayashi H, Sakurai T, Imai M, Takahashi N, Fukuda A, Yayoi O, et al. Contribution of intragenic DNA methylation in mouse gametic DNA methylomes to establish oocyte-specific heritable marks. PLoS Genet 2012;8:e1002440.
[24] Bourc'his D, Xu GL, Lin CS, Bollman B, Bestor TH. Dnmt3L and the establishment of maternal genomic imprints. Science 2001;294:2536—9.
[25] Shirane K, Toh H, Kobayashi H, Miura F, Chiba H, Ito T, et al. Mouse oocyte methylomes at base resolution reveal genome-wide accumulation of non-CpG methylation and role of DNA methyltransferases. PLoS Genet 2013;9:e1003439.
[26] Zhang B, Zheng H, Huang B, Li W, Xiang Y, Peng X, et al. Allelic reprogramming of the histone modification H3K4me3 in early mammalian development. Nature 2016;537:553—7.
[27] Zheng H, Huang B, Zhang B, Xiang Y, Du Z, Xu Q, et al. Resetting epigenetic memory by reprogramming of histone modifications in mammals. Mol Cell 2016;63:1066—79.
[28] Chen Z, Djekidel MN, Zhang Y. Distinct dynamics and functions of H2AK119ub1 and H3K27me3 in mouse preimplantation embryos. Nat Genet 2021;53:551—63.
[29] Dahl JA, Jung I, Aanes H, Greggains GD, Manaf A, Lerdrup M, et al. Broad histone H3K4me3 domains in mouse oocytes modulate maternal-to-zygotic transition. Nature 2016;537:548—52.
[30] Au Yeung WK, Brind'Amour J, Hatano Y, Yamagata K, Feil R, Lorincz MC, et al. Histone H3K9 methyltransferase G9a in oocytes is essential for preimplantation development but dispensable for CG methylation protection. Cell Rep 2019;27:282—93 e4.
[31] Mei H, Kozuka C, Hayashi R, Kumon M, Koseki H, Inoue A. H2AK119ub1 guides maternal inheritance and zygotic deposition of H3K27me3 in mouse embryos. Nat Genet 2021;53:539—50.
[32] Torres-Padilla ME, Bannister AJ, Hurd PJ, Kouzarides T, Zernicka-Goetz M. Dynamic distribution of the replacement histone variant H3.3 in the mouse oocyte and preimplantation embryos. Int J Dev Biol 2006;50:455—61.
[33] Santos F, Hendrich B, Reik W, Dean W. Dynamic reprogramming of DNA methylation in the early mouse embryo. Dev Biol 2002;241:172—82.
[34] Amouroux R, Nashun B, Shirane K, Nakagawa S, Hill PW, D'Souza Z, et al. De novo DNA methylation drives 5hmC accumulation in mouse zygotes. Nat Cell Biol 2016;18:225—33.
[35] Hill PW, Amouroux R, Hajkova P. DNA demethylation, Tet proteins and 5-hydroxymethylcytosine in epigenetic reprogramming: an emerging complex story. Genomics 2014;104:324—33.
[36] Wossidlo M, Arand J, Sebastiano V, Lepikhov K, Boiani M, Reinhardt R, et al. Dynamic link of DNA demethylation, DNA strand breaks and repair in mouse zygotes. EMBO J 2010;29:1877—88.
[37] Hajkova P, Jeffries SJ, Lee C, Miller N, Jackson SP, Surani MA. Genome-wide reprogramming in the mouse germ line entails the base excision repair pathway. Science 2010;329:78—82.

[38] Dickies MM. A new viable yellow mutation in the house mouse. J Hered 1962;53:84—6.
[39] Duhl DM, Vrieling H, Miller KA, Wolff GL, Barsh GS. Neomorphic agouti mutations in obese yellow mice. Nat Genet 1994;8:59—65.
[40] Morgan HD, Sutherland HG, Martin DI, Whitelaw E. Epigenetic inheritance at the agouti locus in the mouse. Nat Genet 1999;23:314—18.
[41] Vasicek TJ, Zeng L, Guan XJ, Zhang T, Costantini F, Tilghman SM. Two dominant mutations in the mouse fused gene are the result of transposon insertions. Genetics 1997;147:777—86.
[42] Reed SC. The inheritance and expression of fused, a new mutation in the house mouse. Genetics 1937;22:1—13.
[43] Weinhouse C, Anderson OS, Jones TR, Kim J, Liberman SA, Nahar MS, et al. An expression microarray approach for the identification of metastable epialleles in the mouse genome. Epigenetics 2011;6:1105—13.
[44] Druker R, Bruxner TJ, Lehrbach NJ, Whitelaw E. Complex patterns of transcription at the insertion site of a retrotransposon in the mouse. Nucleic Acids Res 2004;32:5800—8.
[45] Kazachenka A, Bertozzi TM, Sjoberg-Herrera MK, Walker N, Gardner J, Gunning R, et al. Identification, characterization, and heritability of murine metastable epialleles: implications for non-genetic inheritance. Cell 2018;175:1259—71 e13.
[46] Oey H, Isbel L, Hickey P, Ebaid B, Whitelaw E. Genetic and epigenetic variation among inbred mouse littermates: identification of inter-individual differentially methylated regions. Epigenetics Chromatin 2015;8:54 54-015-0047-z. eCollection 2015.
[47] Waterland RA, Dolinoy DC, Lin JR, Smith CA, Shi X, Tahiliani KG. Maternal methyl supplements increase offspring DNA methylation at Axin Fused. Genesis 2006;44:401—6.
[48] Waterland RA, Jirtle RL. Transposable elements: targets for early nutritional effects on epigenetic gene regulation. Mol Cell Biol 2003;23:5293—300.
[49] Blewitt ME, Vickaryous NK, Paldi A, Koseki H, Whitelaw E. Dynamic reprogramming of DNA methylation at an epigenetically sensitive allele in mice. PLoS Genet 2006;2:e49.
[50] Rakyan VK, Chong S, Champ ME, Cuthbert PC, Morgan HD, Luu KV, et al. Transgenerational inheritance of epigenetic states at the murine Axin(Fu) allele occurs after maternal and paternal transmission. Proc Natl Acad Sci U S A 2003;100:2538—43.
[51] Lane N, Dean W, Erhardt S, Hajkova P, Surani A, Walter J, et al. Resistance of IAPs to methylation reprogramming may provide a mechanism for epigenetic inheritance in the mouse. Genesis 2003;35:88—93.
[52] Blewitt M, Whitelaw E. The use of mouse models to study epigenetics. Cold Spring Harb Perspect Biol 2013;5:a017939.
[53] Daxinger L, Harten SK, Oey H, Epp T, Isbel L, Huang E, et al. An ENU mutagenesis screen identifies novel and known genes involved in epigenetic processes in the mouse. Genome Biol 2013;14:R96 R96—2013-14-9-r96.
[54] Ashe A, Morgan DK, Whitelaw NC, Bruxner TJ, Vickaryous NK, Cox LL, et al. A genome-wide screen for modifiers of transgene variegation identifies genes with critical roles in development. Genome Biol 2008;9:R182 2008-9—12-r182.
[55] Dolinoy DC, Weidman JR, Waterland RA, Jirtle RL. Maternal genistein alters coat color and protects Avy mouse offspring from obesity by modifying the fetal epigenome. Environ Health Perspect 2006;114:567—72.
[56] Neier K, Cheatham D, Bedrosian LD, Dolinoy DC. Perinatal exposures to phthalates and phthalate mixtures result in sex-specific effects on body weight, organ weights and intracisternal A-particle (IAP) DNA methylation in weanling mice. J Dev Orig Health Dis 2019;10:176—87.

REFERENCES

[57] Dolinoy DC, Huang D, Jirtle RL. Maternal nutrient supplementation counteracts bisphenol A-induced DNA hypomethylation in early development. Proc Natl Acad Sci U S A 2007;104:13056−61.

[58] Bertozzi TM, Becker JL, Blake GET, Bansal A, Nguyen DK, Fernandez-Twinn DS, et al. Variably methylated retrotransposons are refractory to a range of environmental perturbations. Nat Genet 53: 2021;1233−42.

[59] Surani MA, Barton SC, Norris ML. Development of reconstituted mouse eggs suggests imprinting of the genome during gametogenesis. Nature 1984;308:548−50.

[60] McGrath J, Solter D. Completion of mouse embryogenesis requires both the maternal and paternal genomes. Cell 1984;37:179−83.

[61] Winking H, Silver LM. Characterization of a recombinant mouse T haplotype that expresses a dominant lethal maternal effect. Genetics 1984;108:1013−20.

[62] Johnson DR. Hairpin-tail: a case of post-reductional gene action in the mouse egg. Genetics 1974;76:795−805.

[63] Searle AG, Beechey CV. Complementation studies with mouse translocations. Cytogenet Cell Genet 1978;20:282−303.

[64] Bartolomei MS, Zemel S, Tilghman SM. Parental imprinting of the mouse H19 gene. Nature 1991;351:153−5.

[65] Barlow DP, Stoger R, Herrmann BG, Saito K, Schweifer N. The mouse insulin-like growth factor type-2 receptor is imprinted and closely linked to the Tme locus. Nature 1991;349:84−7.

[66] DeChiara TM, Robertson EJ, Efstratiadis A. Parental imprinting of the mouse insulin-like growth factor II gene. Cell 1991;64:849−59.

[67] Ferguson-Smith AC, Cattanach BM, Barton SC, Beechey CV, Surani MA. Embryological and molecular investigations of parental imprinting on mouse chromosome 7. Nature 1991;351:667−70.

[68] Ferguson-Smith AC. Genomic imprinting: the emergence of an epigenetic paradigm. Nat Rev Genet 2011;12:565−75.

[69] Chaillet JR, Vogt TF, Beier DR, Leder P. Parental-specific methylation of an imprinted transgene is established during gametogenesis and progressively changes during embryogenesis. Cell 1991;66:77−83.

[70] Bartolomei MS, Webber AL, Brunkow ME, Tilghman SM. Epigenetic mechanisms underlying the imprinting of the mouse H19 gene. Genes Dev 1993;7:1663−73.

[71] Hata K, Kusumi M, Yokomine T, Li E, Sasaki H. Meiotic and epigenetic aberrations in Dnmt3L-deficient male germ cells. Mol Reprod Dev 2006;73:116−22.

[72] Hata K, Okano M, Lei H, Li E. Dnmt3L cooperates with the Dnmt3 family of de novo DNA methyltransferases to establish maternal imprints in mice. Development 2002;129:1983−93.

[73] Bourc'his D, Bestor TH. Meiotic catastrophe and retrotransposon reactivation in male germ cells lacking Dnmt3L. Nature 2004;431:96−9.

[74] Ferguson-Smith AC, Bourc'his D. The discovery and importance of genomic imprinting. Elife 2018;7:e42368. Available from: https://doi.org/10.7554/eLife.42368.

[75] Smallwood SA, Tomizawa S, Krueger F, Ruf N, Carli N, Segonds-Pichon A, et al. Dynamic CpG island methylation landscape in oocytes and preimplantation embryos. Nat Genet 2011;43:811−14.

[76] Takahashi N, Coluccio A, Thorball CW, Planet E, Shi H, Offner S, et al. ZNF445 is a primary regulator of genomic imprinting. Genes Dev 2019;33:49−54.

[77] Li X, Ito M, Zhou F, Youngson N, Zuo X, Leder P, et al. A maternal-zygotic effect gene, Zfp57, maintains both maternal and paternal imprints. Dev Cell 2008;15:547−57.

[78] Mackay DJ, Callaway JL, Marks SM, White HE, Acerini CL, Boonen SE, et al. Hypomethylation of multiple imprinted loci in individuals with transient neonatal diabetes is associated with mutations in ZFP57. Nat Genet 2008;40:949−51.

[79] Quenneville S, Verde G, Corsinotti A, Kapopoulou A, Jakobsson J, Offner S, et al. In embryonic stem cells, ZFP57/KAP1 recognize a methylated hexanucleotide to affect chromatin and DNA methylation of imprinting control regions. Mol Cell 2011;44:361−72.

[80] Strogantsev R, Krueger F, Yamazawa K, Shi H, Gould P, Goldman-Roberts M, et al. Allele-specific binding of ZFP57 in the epigenetic regulation of imprinted and non-imprinted monoallelic expression. Genome Biol 2015;16:112 112-015-0672-7.

[81] Leung D, Du T, Wagner U, Xie W, Lee AY, Goyal P, et al. Regulation of DNA methylation turnover at LTR retrotransposons and imprinted loci by the histone methyltransferase Setdb1. Proc Natl Acad Sci U S A 2014;111:6690−5.

[82] Li E, Beard C, Jaenisch R. Role for DNA methylation in genomic imprinting. Nature 1993;366:362−5.

[83] Messerschmidt DM, de Vries W, Ito M, Solter D, Ferguson-Smith A, Knowles BB. Trim28 is required for epigenetic stability during mouse oocyte to embryo transition. Science 2012;335:1499−502.

[84] Hanna CW, Kelsey G. Features and mechanisms of canonical and noncanonical genomic imprinting. Genes Dev 2021;35:821−34.

[85] Keane TM, Goodstadt L, Danecek P, White MA, Wong K, Yalcin B, et al. Mouse genomic variation and its effect on phenotypes and gene regulation. Nature 2011;477:289−94.

[86] Li Y, Li J. Technical advances contribute to the study of genomic imprinting. PLoS Genet 2019;15:e1008151.

[87] Xie W, Barr CL, Kim A, Yue F, Lee AY, Eubanks J, et al. Base-resolution analyses of sequence and parent-of-origin dependent DNA methylation in the mouse genome. Cell 2012;148:816−31.

[88] Andergassen D, Dotter CP, Wenzel D, Sigl V, Bammer PC, Muckenhuber M, et al. Mapping the mouse Allelome reveals tissue-specific regulation of allelic expression. Elife 2017;6:e25125. Available from: https://doi.org/10.7554/eLife.25125.

[89] Babak T, DeVeale B, Tsang EK, Zhou Y, Li X, Smith KS, et al. Genetic conflict reflected in tissue-specific maps of genomic imprinting in human and mouse. Nat Genet 2015;47:544−9.

[90] Ferrón SR, Charalambous M, Radford E, McEwen K, Wildner H, Hind E, et al. Postnatal loss of Dlk1 imprinting in stem cells and niche astrocytes regulates neurogenesis. Nature 2011;475:381−5.

[91] Van de Pette M, Abbas A, Feytout A, McNamara G, Bruno L, To WK, et al. Visualizing changes in Cdkn1c expression links early-life adversity to imprint mis-regulation in adults. Cell Rep 2017;18:1090−9.

[92] Stelzer Y, Wu H, Song Y, Shivalila CS, Markoulaki S, Jaenisch R. Parent-of-origin DNA methylation dynamics during mouse development. Cell Rep 2016;16:3167−80.

[93] Van de Pette M, Galvao A, Millership SJ, To W, Dimond A, Prodani C, et al. Epigenetic change induced by *in utero* dietary challenge provokes phenotypic variability across multiple generations of mice. bioRxiv 2020; 2020.08.07.241034.

[94] Li ZK, Wang LY, Wang LB, Feng GH, Yuan XW, Liu C, et al. Generation of bimaternal and bipaternal mice from hypomethylated haploid ESCs with imprinting region deletions. Cell Stem Cell 2018;23:665−76 e4.

[95] Kono T, Obata Y, Wu Q, Niwa K, Ono Y, Yamamoto Y, et al. Birth of parthenogenetic mice that can develop to adulthood. Nature 2004;428:860−4.

[96] Charalambous M, da Rocha ST, Ferguson-Smith AC. Genomic imprinting, growth control and the allocation of nutritional resources: consequences for postnatal life. Curr Opin Endocrinol Diabetes Obes 2007;14:3−12.

[97] Moore T, Haig D. Genomic imprinting in mammalian development: a parental tug-of-war. Trends Genet 1991;7: 45–9.

[98] Tunster SJ, Boqué-Sastre R, McNamara GI, Hunter SM, Creeth HDJ, John RM. Peg3 deficiency results in sexually dimorphic losses and gains in the normal repertoire of placental hormones. Front Cell Dev Biol 2018;6:123.

[99] Tunster SJ, Creeth HDJ, John RM. The imprinted Phlda2 gene modulates a major endocrine compartment of the placenta to regulate placental demands for maternal resources. Dev Biol 2016;409:251–60.

[100] Tunster SJ, McNamara GI, Creeth HDJ, John RM. Increased dosage of the imprinted Ascl2 gene restrains two key endocrine lineages of the mouse Placenta. Dev Biol 2016;418: 55–65.

[101] Aykroyd BRL, Tunster SJ, Sferruzzi-Perri AN. Igf2 deletion alters mouse placenta endocrine capacity in a sexually dimorphic manner. J Endocrinol 2020;246:93–108.

[102] Susiarjo M, Sasson I, Mesaros C, Bartolomei MS. Bisphenol a exposure disrupts genomic imprinting in the mouse. PLoS Genet 2013;9:e1003401.

[103] Creeth HDJ, McNamara GI, Tunster SJ, Boque-Sastre R, Allen B, Sumption L, et al. Maternal care boosted by paternal imprinting in mammals. PLoS Biol 2018;16:e2006599.

[104] Calabrese JM, Starmer J, Schertzer MD, Yee D, Magnuson T. A survey of imprinted gene expression in mouse trophoblast stem cells. G3 (Bethesda) 2015;5:751–9.

[105] Okae H, Hiura H, Nishida Y, Funayama R, Tanaka S, Chiba H, et al. Re-investigation and RNA sequencing-based identification of genes with placenta-specific imprinted expression. Hum Mol Genet 2012;21:548–58.

[106] Inoue A, Jiang L, Lu F, Suzuki T, Zhang Y. Maternal H3K27me3 controls DNA methylation-independent imprinting. Nature 2017;547:419–24.

[107] Inoue A, Chen Z, Yin Q, Zhang Y. Maternal Eed knockout causes loss of H3K27me3 imprinting and random X inactivation in the extraembryonic cells. Genes Dev 2018;32: 1525–36.

[108] Santini L, Halbritter F, Titz-Teixeira F, Suzuki T, Asami M, Ma X, et al. Genomic imprinting in mouse blastocysts is predominantly associated with H3K27me3. Nat Commun 2021;12: 3804–021-23510-4.

[109] Hanna CW, Pérez-Palacios R, Gahurova L, Schubert M, Krueger F, Biggins L, et al. Endogenous retroviral insertions drive non-canonical imprinting in extra-embryonic tissues. Genome Biol 2019;20:225 225-019-1833-x.

[110] Chen Z, Yin Q, Inoue A, Zhang C, Zhang Y. Allelic H3K27me3 to allelic DNA methylation switch maintains noncanonical imprinting in extraembryonic cells. Sci Adv 2019;5:eaay7246.

[111] Zeng T, Pierce N, Szabó PE. H3K9 methyltransferase EHMT2/G9a controls ERVK-driven non-canonical imprinted genes. bioRxiv 2021; 2021.03.29.437617.

[112] Andergassen D, Smith ZD, Rinn JL, Meissner A. Diverse mechanisms for epigenetic imprinting in mammals. bioRxiv 2021; 2021.04.30.442087.

[113] Chiba H, Hirasawa R, Kaneda M, Amakawa Y, Li E, Sado T, et al. De novo DNA methylation independent establishment of maternal imprint on X chromosome in mouse oocytes. Genesis 2008;46:768–74.

[114] Inoue A, Jiang L, Lu F, Zhang Y. Genomic imprinting of Xist by maternal H3K27me3. Genes Dev 2017;31:1927–32.

[115] Campbell KH, McWhir J, Ritchie WA, Wilmut I. Sheep cloned by nuclear transfer from a cultured cell line. Nature 1996;380:64–6.

[116] Gurdon JB. The developmental capacity of nuclei taken from intestinal epithelium cells of feeding tadpoles. J Embryol Exp Morphol 1962;10:622–40.

[117] Wakayama T, Perry AC, Zuccotti M, Johnson KR, Yanagimachi R. Full-term development of mice from enucleated oocytes injected with cumulus cell nuclei. Nature 1998;394:369–74.

[118] Miyamoto K, Tajima Y, Yoshida K, Oikawa M, Azuma R, Allen GE, et al. Reprogramming towards totipotency is greatly facilitated by synergistic effects of small molecules. Biol Open 2017;6:415–24.

[119] Matoba S, Wang H, Jiang L, Lu F, Iwabuchi KA, Wu X, et al. Loss of H3K27me3 imprinting in somatic cell nuclear transfer embryos disrupts post-implantation development. Cell Stem Cell 2018;23:343–54 e5.

[120] Proudhon C, Duffie R, Ajjan S, Cowley M, Iranzo J, Carbajosa G, et al. Protection against de novo methylation is instrumental in maintaining parent-of-origin methylation inherited from the gametes. Mol Cell 2012;47:909–20.

[121] Smith ZD, Chan MM, Humm KC, Karnik R, Mekhoubad S, Regev A, et al. DNA methylation dynamics of the human preimplantation embryo. Nature 2014;511:611–15.

[122] Duffie R, Ajjan S, Greenberg MV, Zamudio N, Escamilla del Arenal M, Iranzo J, et al. The Gpr1/Zdbf2 locus provides new paradigms for transient and dynamic genomic imprinting in mammals. Genes Dev 2014;28:463–78.

[123] Greenberg MV, Glaser J, Borsos M, Marjou FE, Walter M, Teissandier A, et al. Transient transcription in the early embryo sets an epigenetic state that programs postnatal growth. Nat Genet 2017;49:110–18.

CHAPTER 15

Plant Epigenomics

Leonardo Furci, Jérémy Berthelier, Oscar Juez, Matin Miryeganeh and Hidetoshi Saze

Plant Epigenetics Unit, Okinawa Institute of Science and Technology Graduate University, Onna-son, Okinawa, Japan

OUTLINE

Introduction	263
Basic Mechanisms of Plant Epigenome Regulation	264
DNA Methylation	264
Histone Modifications and Chromatin Structure	266
Epigenetic Phenomena in Plants	268
Vernalization	268
Paramutation	268
Genome Imprinting	268
Mobile sRNAs	268
Noncoding RNAs	269
Epigenetic Regulation of Transposable Elements and Interactions With Genes	269
Epigenetics Regulation of Transposable Elements in Genic Regions	269
Epigenetics Regulation of Transposable Elements Located Within Genes	270
Epigenetic Regulation of Stress Responses in Plants	271
Plant Epigenetics and Abiotic Stress	271
Plant Epigenetics and Biotic Stress	273
Emerging Technologies for Epigenome Studies in Plants	276
From Bulk to Single-Cell: The Emergence of Single-Cell Epigenomic Informatics	276
Advances in Epigenome Editing	278
Future Perspectives	278
References	279

INTRODUCTION

Plants conserve various epigenome regulation mechanisms common to other eukaryotic species, including fungi and animals. These include DNA methylation at cytosine residues, histone modifications, and RNA interference-based chromatin modifications. However, plants have also evolved specific epigenetic mechanisms related to their life cycles and survival strategies in natural environments. Owing to their sessile nature, plants need to cope with fluctuating surrounding environmental stresses, including heat, cold, UV light, soil nutrition, drought, salinity, flooding, and pathogen challenges. Epigenetic regulation is exploited to sense and respond to changes in the surrounding environment, in which the chromatin states may be maintained as epigenetic memory in nuclei and transmitted through mitosis to daughter cells, or often to progenies through meiosis. In addition, plants use epigenetic regulation to suppress parasitic DNA elements in the genome and transcriptional gene regulation during development and reproduction.

Recent developments in high-throughput sequencing-based technologies, including chromatin analysis using Hi-C and related techniques, RNA-seq analysis in single-cell resolutions, and detection of chromatin modifications in a small number of cells, allow researchers to understand chromatin dynamics of plant tissues and cells at unprecedented depths and resolutions.

This chapter summarizes the basic epigenome regulation mechanisms in plants and recent findings in related fields of plant epigenomics research. As the research of plant epigenomics has been broadened in the past decades, we cannot fully cover all aspects of topics and findings in the field in this chapter. Therefore, in this chapter, we focus mainly on recent findings of epigenome regulation in plants, emphasizing interactions between plants and environments in relatively well-studied model plant species.

BASIC MECHANISMS OF PLANT EPIGENOME REGULATION

DNA Methylation

DNA Methylation and Its Regulation

DNA cytosine methylation is a widely conserved epigenetic modification in bacteria, fungi, animals, and plants [1]. The primary function of DNA methylation in plants, as in other organisms, is to transcriptionally suppress parasitic genome elements called transposable elements (TEs) [2,3]. In plant genomes, TEs and repeats are heavily methylated at symmetric Cytosine-Guanine dinucleotide (CG) and CHG sites and asymmetric CHH sites (H can be A, T, C) (Fig. 15.1A). Loss of DNA methylation leads to the reactivation and transposition of TEs in the genome, suggesting the importance of DNA methylation in TE silencing. On the other hand, ~30% of protein-coding genes are associated with methylation in their bodies (gene-body methylation; GbM), which is conserved among plant species [4,5]. The role of GbM is still under intensive study. Recent reports suggests that GbM could affect alternative splicing [6,7], and is also required to suppress spurious antisense transcription initiation from gene body [8]. On the other hand, a low level of N6-adenine methylation (6 mA) has been detected recently in the genomes of plants as in animals plants [1]. In plant genomes, 6 mA is often associated with actively transcribed genes [9,10], but the function of m6A requires further elucidation.

DNA cytosine methylation is controlled by other chromatin modifications, including histone H3 Lysine9 methylation (H3K9me) and small interfering RNA (siRNA). Complex interactions of those pathways and molecules involved in these pathways are well described in other excellent reviews [11–14]; here, we briefly summarize the regulation of DNA methylation in plant genomes (Fig. 15.1). In plants, the establishment of DNA methylation in unmethylated DNA (de novo methylation) is triggered by an RNA interference (RNAi)-based mechanism called RNA-directed DNA methylation (RdDM) (Fig. 15.1B). Plants contain a specific RNA polymerase complex named RNA polymerase IV (Pol IV), which produces noncoding RNA precursors from repetitive genome regions. RNAi factors, including RNA DEPENDENT RNA POLYMERASES (RDRs) and DICER-LIKE proteins, process the precursors and generate 21–24 nucleotide (nt) siRNAs, which are incorporated into ARGONAUTE (AGO) protein complex. The AGO-siRNA complex targets scaffold RNAs produced by another plant-specific RNA polymerase V (Pol V), which recruits the de novo methylase DOMAIN REARRANGED METHYLTRANSFERASES to introduce DNA methylation of cytosines in all target sequence contexts. In addition, the non-canonical RdDM pathway can be activated via aberrant RNA synthesis by RNA polymerase II (Pol II) and RDR6 [14] (Fig. 15.1B, Inset). In contrast, H3K9me recruits plant-specific methylases named CHROMOMETHYLASES (CMTs), containing chromodomain and mediating CHG/CHH methylations in TEs. Once established, CG methylation is maintained by METHYLTRANSFERASE1 (MET1), the ortholog of DNA METHYLTRANSFERASE1 (DNMT1) in animals, through DNA replication (Fig. 15.1A). The maintenance of CG methylation through DNA replication requires conserved proteins VARIATION IN MATHYLATIONS (orthologous to Ubiquitin Like With PHD And Ring Finger Domains 1 (UHRF1) in mammals) that target hemi-methylated DNA strands [11] and recruits MET1 to methylate the newly synthesized DNA strand. The maintenance of DNA methylation also requires the chromatin remodeling factor DECREASE IN DNA METHYLATION 1 (DDM1), recruiting DNA methylases to TEs to maintain repressed chromatin states called heterochromatin [15]. Notably, a recent report described the animal de novo methylase DNMT3 homologs in basal plants (green algae, moss, fern, and gymnosperms) but not in angiosperms (flowering plants); these homologs are involved in CG and CHH de novo methylation of heterochromatin [16].

DNA demethylation activities by DNA glycosylases/lyases can counteract DNA cytosine methylation (Fig. 15.1A). In contrast to DNA demethylation by ten-eleven translocation enzymes that mediate oxidation of the methyl group of cytosine in animals, DNA glycosylases remove methylated cytosine bases to generate abasic sites, which are eventually removed by base excision DNA repair pathway and replaced with unmethylated cytosine [17].

DNA Methylation Dynamics During the Plant Life Cycle

DNA methylation levels are dynamically controlled during the plant life cycles (Fig. 15.2). During embryogenesis in *Arabidopsis thaliana*, CHH methylation levels, especially in TEs, are increased, which is likely correlated with cell cycle activity in highly dividing cells [18–22].

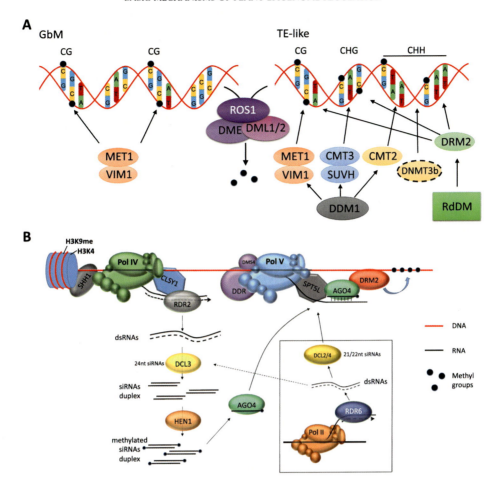

FIGURE 15.1 Mechanisms of DNA methylation establishment, maintenance, and removal in plants. (A) In plants, cytosine methylation can be found at CG contexts in gene bodies (gene body methylation, GbM), or at a combination of CG, CHG, and asymmetric CHH contexts (where H can be A, T, C) in transposable elements (TEs) and repeats (TE-like). The panel displays DNA methyltransferases responsible for DNA methylation maintenance in each context (see text for more details). (B) DNA methylation establishment, and maintenance in asymmetric CHH context, is carried out by small interfering RNAs (siRNA)-directed DNA methylation (RdDM) pathway. Plant-specific RNA polymerase Pol IV transcribes genomic regions to be methylated into noncoding RNA precursors, which are then converted into double-stranded RNAs by DEPENDENT RNA POLYMERASES 2 (RDR2) and processed into 24nt siRNAs by DICER-LIKE 3 (DCL3). The 24nt siRNAs are incorporated into ARGONAUTE 4 (AGO4), and the AGO-siRNA complex targets scaffold RNAs produced by the second plant-specific RNA polymerase Pol V. The scaffold recruits de novo methyltransferase DOMAINS REARRANGED METHYLTRANSFERASE 2 (DRM2) to introduce DNA methylation in all target sequence contexts. Inset: non-canonical RdDM pathway. Aberrant RNA synthesis by RNA polymerase Pol II triggers conversion into double-stranded RNA molecules by RDR6, which are subsequently processed into 24nt siRNAs (by DCL3) or 21–22nt siRNAs (by DCL2/4).

In *Arabidopsis*, dividing stem cells in the shoot apical meristem (SAM) show DNA hypermethylation because of high activities of RdDM, MET1, and CMTs [23–25]. In the root apical meristem, columella cells show high CHH methylation levels because of the abundance of 24nt siRNAs and RdDM factors [26]. Similarly, in rice, SAM shows higher CHH methylation levels in TEs because of increased RdDM activities compared to leaf tissues, suggesting a robust silencing of harmful TEs in highly dividing stem cell lineages [27].

In addition, plant reproductive tissues show drastic changes in DNA methylation (Fig. 15.2). During meiosis process in males, CHH sites are globally hypomethylated in meiocytes whereas CG and CHG methylation are stably maintained [28]. After meiosis, a single microspore undergoes two rounds of mitosis (male gametogenesis) to produce two sperm nuclei and one vegetative nucleus, forming mature pollen [29,30] (Fig. 15.2). In the vegetative nucleus, active DNA demethylation of CG sites occurs, which is mediated by the DNA demethylases DEMETER (DME) and REPRESSOR OF SILENCING 1 (ROS1) (Fig. 15.1A) [31,32]. In contrast, CHH sites in the vegetative nucleus are hypermethylated, especially in TEs, by RdDM and CMT; this hypermethylation is associated with the production of siRNAs from TEs [32]. The vegetative nucleus does not genetically contribute to the next generation; whereas, siRNAs generated from TE sequences are thought to be transmitted and reinforce DNA methylation in sperm nuclei to suppress potentially active TEs in the subsequent generation [29,33].

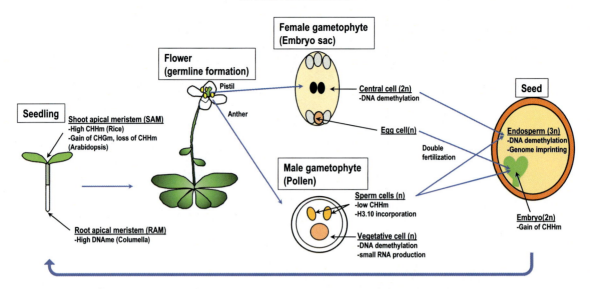

FIGURE 15.2 Plant life cycle and epigenome dynamics (*Arabidopsis* as an example). During embryogenesis in *Arabidopsis thaliana*, CHH (H = C, T, A) methylation levels, especially in Transposable Elements (TEs), are increased. In *Arabidopsis*, dividing stem cells in the shoot apical meristem (SAM) show DNA hypermethylation because of high activities of DNA methylases. In the root apical meristem (RAM), columella cells show high CHH methylation levels because of the abundance of 24nt siRNAs and RNA-directed DNA methylation (RdDM) factors. Similarly, in rice, SAM shows higher CHH methylation levels in TEs because of increased RdDM activities. During meiosis process in males, CHH sites are globally hypomethylated in meiocytes whereas CG and CHG methylation are stably maintained. After meiosis, a single microspore undergoes two rounds of mitosis (male gametogenesis) to produce two sperm nuclei and one vegetative nucleus, forming mature pollen. In the vegetative nucleus, active DNA demethylation of CG sites occurs, which is mediated by the DNA demethylases DEMETER (DME) and REPRESSOR OF SILENCING 1 (ROS1) (Fig. 15.1A). CHH sites in the vegetative nucleus are hypermethylated, especially in TEs, by RdDM and CHROMOMETHYLASES (CMTs); this hypermethylation is associated with the production of siRNAs from TEs. In contrast, megaspores generated through female meiosis undergo three rounds of mitosis (female gametogenesis) to form female gametophytes consisting of seven cell types, including an egg cell (n) and central cells (2n). Fertilization between an egg and a sperm produces an embryo (2n), whereas that between a central cell and a sperm produces an endosperm (3n), known as a double fertilization. Central cells and endosperms undergo genome-wide DNA demethylation by DME activity. In the endosperm, maternally and paternally inherited chromosomes show asymmetric DNA methylation patterns in TEs in promoter regions, allowing the parent-of-origin expression of TE-associated genes, called genome imprinting.

In contrast, megaspores generated through female meiosis undergo three rounds of mitosis (female gametogenesis) to form female gametophytes consisting of seven cell types, including an egg cell (n) and central cells (2n) (Fig. 15.2). Fertilization between an egg and a sperm produces an embryo (2n), whereas that between a central cell and a sperm produces an endosperm (3n); the phenomenon is known as a double fertilization. As vegetative cells in pollen, central cells, and endosperms undergo genome-wide DNA demethylation by DME activity [34]. In the endosperm, maternally and paternally inherited chromosomes show asymmetric DNA methylation patterns in TEs in promoter regions, allowing the parent-of-origin expression of TE-associated genes, called genome imprinting (See section "Genome Imprinting").

Histone Modifications and Chromatin Structure

Histone Modifications and Modifying Enzymes

As in other eukaryotes, plants conserve various histone modifications, including acetylation, methylation, phosphorylation, and ubiquitination [35,36]. These histone modifications are regulated by specific "writer" and "eraser" enzymes and lead to recruitments of "reader" proteins to regulate chromatin states [37]. The distribution patterns of histone modifications in the genome of plants are similar to those of other organisms. Active genes are associated with active histone marks including histone H3/H4 acetylation, H3K4 mono-, di-, and tri-methylation (H3K4me1/2/3), and H3K36 tri-methylation (H3K36me3), whereas TEs and silenced genes are associated with repressive histone marks such as H3K9 mono- and di-methylation (H3K9me1/2) and H3K27 tri-methylation (H3K27me3). In plants, heterochromatin is also associated with H3K27 monomethylation (H3K27me1) [38]. In terms of histone acetylation, there are genes encoding 13 histone acetyltransferases (HATs; writer)—including those belonging to GNAT, MYST, and CBP classes—and 18 histone deacetylases (HDACs; eraser)—including those belonging to RPD3 and Sirtuin classes—in the *Arabidopsis* genome [39,40]. Recent studies showed that RPD3-class HDACs are involved in the regulation of flowering time, and also of the circadian clock that is a conserved molecular mechanism required for diurnal transcriptional control of endogenous genes for adaptation to light/dark cycles. It has been shown that *Arabidopsis* RPD3-class HISTONE DEACETYLASE 9 regulates acetylation in the promoter of timing of CAB EXPRESSION 1 (TOC1) gene that is required for generation of circadian rhythms [41]. In addition, another RPD3-class HDA6 represses expression of *TOC1* as well as other core circadian clock component genes *LATE ELONGATED HYPOCOTYL (LHY)* and *CIRCADIAN CLOCK ASSOCIATED 1 (CCA1)* [42]. On the other hand, HDA19 is required for regulation of photoperiod-dependent flowering in *Arabidopsis* [43].

Histone methylases (writers) and demethylases (erasers) are also well conserved in plants. There are ASHH-class (ASH1 HOMOLOG), ATX-class (*Arabidopsis trithorax*), SUVH-class (SU(VAR)3−9 HOMOLOG), and EZ-class (enhancer of zeste) SET-domain histone methyltransferases conserved in plants (in *Arabidopsis*, also named SET-domain group proteins; SDGs) [44]. Generally, ASHH-class SDGs regulate H3K36 methylation; whereas, ATX-class SDGs regulate H3K4 methylation at genic regions. However, in *Arabidopsis*, ATX-class SDGs ATXR5 and ATXR6 mediate H3K27me1 in heterochromatin, preventing ectopic DNA replication initiation from centromeric regions [45]. Heterochromatic H3K9me is mediated by SUVH-class histone methylases, which recruit CMTs to direct non-CG methylation at TE sequences. SUVH-class histone methylases contain SRA domains that bind to methylated DNA, allowing methylases to reinforce H3K9 methylation to heterochromatic loci associated with DNA methylation [11]. H3K27me3 is mediated by Polycomb repressive complex 2 (PRC2), containing EZ-class histone methyltransferases essential for the repression of developmental genes in plants [46]. On the other hand, plants also conserve histone-demethylating enzymes, including mammalian lysine-specific demethylase (LSD) homologs, and Jumonji domain-containing proteins (JMJs) [47−49]. LSD-class demethylases specifically act on H3K4 demethylation, whereas JMJ-class demethylases can remove H3K4 methylation, H3K9 methylation, and H3K27 methylation. Histone demethylases are essential for epigenome resetting and reprogramming during the reproductive phase and early embryogenesis. In *Arabidopsis*, JMJ-class H3K27 demethylases EARLY FLOWERING 6 (ELF6) and RELATIVE OF ELF6 (REF6), and JMJ13 are involved in H3K27 demethylation of developmental genes, including reactivation of the flowering repressor gene *Flowering Locus C (FLC)* [50−52]. A recent report suggested that these demethylases are essential for removing H3K27 methylation during male gametogenesis, facilitating the activation of genes involved in spermatogenesis and embryogenesis in the next generation [50].

Various "reader" proteins can recognize histone methylation within specific domains. Although heterochromatin protein (HP1) in yeast and animals can bind to H3K9me to repress transcription of TEs, the plant homolog of HP1 (LIKE-HETEROCHROMATIN1 (LHP1)) instead binds to H3K27me3 to repress developmental genes [53]. Recent reports have shown that the Agenet domain-containing protein ADCP1 can bind to H3K9me and maintain TE silencing and DNA methylation in *Arabidopsis* [54,55]. H3K9me is also recognized by the Bromo-Adjacent Homology (BAH) domain and chromodomain present in CMTs to reinforce non-CG methylation in heterochromatic regions [56]. In plants, "bivalent" modifications of H3K27me3 and H3K4me3 are bound by EARLY BOLTING IN SHORT DAY and SHORT LIFE proteins via its BAH and PHD (Plant homeodomain) domains that regulate floral transition [57−59].

Histone Variants

In addition to canonical core histones H2A, H2B, H3, and H4, plants also contain histone variants, including H2A.X, H2A.Z, and H3 variant CENP-A [60]. Regarding the H2A variants, H2A.Z is associated with genes responsive to environmental stresses [61] and has a dual role for activation or repression, depending on their localization and monoubiquitination state [62]. In addition, the plant-specific H2A variant H2A.W is incorporated into heterochromatin, required for proper chromatin organization and compaction [63,64]. For histone H3, similar to other eukaryotes, plants have replicative H3.1 incorporated into nucleosomes during DNA replication; conversely, H3.3 is deposited independently of replication. There are four amino acid differences between H3.1 and H3.3. Histone methylases ATRX5/6 specifically recognize H3.1, introducing H3K27me1 at heterochromatin [65]. Recent reports have shown that plant histone H3 variants H3.10 and H3.15 are resistant to H3K27 methylation, playing a role in epigenetic reprogramming during plant callus formation and spermatogenesis (Fig. 15.2) [50,66]. In addition to core histones, linker histone H1 is required for DNA methylation and heterochromatin condensation [67−69].

Regulation of Higher-Order Chromatin Structures in Plants

Hi-C analysis of *Arabidopsis* revealed structural chromatin domains such as A and B compartments in other eukaryotic nuclei [70,71]. These studies further identified inter- and intra-chromosomal interactive heterochromatic domains in the genome, named KNOT [70], which is formed by KNOT-engaged elements (~10 loci in the *Arabidopsis* genome) enriched in TE sequences [72]. KNOT is similar to the flamenco locus in *Drosophila*, which acts as a TE trap to produce piRNAs for TE silencing. The inter- and intra-chromosomal interactions of KNOT and other chromosome features, including pericentromeres, are altered by heat stress [73], suggesting a direct effect of environmental stress on chromatin reorganization in plant nuclei. Plant species with larger and more complex genomes, including rice, cotton, and tomato, show topologically associated domain (TAD)-like structures [74–76]. In contrast to animals, TAD boundaries in plant chromosomes tend to be associated with actively transcribed genes [74].

An assay for transposase-accessible chromatin combined with high-throughput sequencing (ATAC-seq) analysis of 13 plant species identified conserved *cis*-regulatory elements (CREs), which are often far away from genes and can interact with genes by forming chromatin loops [77,78]. In animals, these distal CREs often act as enhancers of gene transcription marked with H3K4me1 and H3K27ac, whereas plant distal CREs are associated with H3K56ac, H3K27me3, or unmodified states [77]. These results suggest that, as in animals, the long-range interactions of chromatin and gene regulation are prevalent in plant species, especially with large genomes.

EPIGENETIC PHENOMENA IN PLANTS

Vernalization

Some plant species require a continuous cold period to induce flowering, that is, the transition from the vegetative to the reproductive phase. The process is called vernalization, in which the experience of cold period is maintained as the "epigenetic memory" of winter that promotes flowering in spring [79,80]. Although the molecular mechanisms of vernalization are well studied in wheat and *Arabidopsis*, the molecular mechanisms and factors involved in the process are not similar [81]. In winter-annual *Arabidopsis* strains, the floral repressor gene *FLC*, a MADS-box transcription factor, is upregulated and represses induction flowering before vernalization. Vernalization triggers the accumulation of H3K27me3 by PRC2 and recruitment of LHP1, leading to transcriptional silencing of the *FLC* locus, with coordinated actions of noncoding RNAs and chromatin modifiers [79]. The switching of *FLC* expression from ON to OFF occurs cell-autonomously, which induces a quantitative response to vernalization [82]. The H3K27me3 and repressed state of *FLC* are stably maintained during mitosis, although re-programmed by the action of H3K27 demethylase ELF6, and *FLC* becomes active in the next generation embryo [52]. The plant requires vernalization again for flowering.

Paramutation

Paramutation is a classic epigenetic phenomenon first observed in plants [83], which shows a non-Mendelian inheritance of particular traits. Similar phenomena have been observed in other eukaryotes, including mammals [84,85]. Paramutation induces changes in gene activities to allelic loci after genome hybridization, which can be inherited by progenies for multiple generations without genetic changes. Studies in maize have revealed that the establishment of paramutations requires RdDM factors and siRNAs [85].

Genome Imprinting

The differential expression of genes depending on the parental origin, called genome imprinting, has been intensively studied in mammals and flowering plants. Genome imprinting in mammals is observed in embryonic lineages and placenta; however, genome imprinting in plants is mainly observed in the endosperm, the tissue with triploid cells containing two sets of maternally inherited chromosomes from the central cell and one set of paternally inherited chromosomes from a sperm cell (Fig. 15.2) [86]. DME actively demethylates maternal alleles associated with methylated TEs, inducing the maternal-specific expression of genes in plants, including *Arabidopsis*, rice, and maize [34,87]. In addition, deposition of H3K27me3 by the PRC2 complex containing the H3K27 methylase MEDEA (MEA) promotes asymmetric gene expression between maternal and paternal alleles in the endosperm [86]. Loss of imprinting in DME and MEA mutants causes seed abortion, suggesting the importance of parent-of-origin gene expression for proper endosperm and seed development.

Mobile sRNAs

Cell-to-cell movement of siRNAs has been observed in plants, which can direct epigenome changes in the genome DNA of recipient cells. In plant tissues, cells are connected by the channel structure on cell membrane called plasmodesmata that allows cell-to-cell transmission of sRNAs [88]. Furthermore, the living tissue called phloem allows a long-distance siRNA movement from shoot to root, together with translocation of organic compounds, hormones, proteins, and nucleic acids from source to sink (e.g., sucrose produced in leaf tissue to rootstock). Elegant grafting experiments using *Arabidopsis dcl2/3/4* triple mutant deficient in the production of siRNAs showed that 22–24nt siRNAs can move from shoot to root through graft union, which direct DNA methylation, especially on TE sequences in root cells [89,90]. A recent study demonstrated that grafting of wild-type shoot to the mutant of *MutS HOMOLOG1* (*MSH1*) rootstock in *Arabidopsis* and tomato showed RdDM-dependent heritable growth vigor in the progeny [91], suggesting that mobile sRNAs can induce meiotically heritable epigenetic variations in plants.

Noncoding RNAs

Plants utilize noncoding RNAs, especially long noncoding RNAs (lncRNAs, typically >200 nt in size), for regulation of gene activities in response to environmental and developmental cues. One example of such regulatory lncRNAs is *COOLAIR* in *Arabidopsis*, which is produced in the antisense orientation from the *FLC* locus during vernalization (See section "Vernalization") [79]. *COOLAIR* promotes repression of *FLC* locus via demethylation of histone H3K4 and deposition of H3K27me3 by PRC2. Similarly, the cold-induced noncoding RNA *SVALKA*, produced from the antisense strand of the C-repeat/dehydration-responsive element binding factor1 (CBF1) gene locus, inhibits *CBF1* transcripts and affects freezing tolerance in *Arabidopsis* [92]. These natural antisense transcripts (NATs) are identified in 15%−20% of Arabidopsis loci [93−95], and about 60% of loci in the rice genome [96], while most of the NATs are yet to be investigated for their function. On the other hand, the lncRNA *APOLO* expression can be induced by the plant hormone auxin, and can modulate chromatin loop formation in *Arabidopsis* [97]. A recent study further showed that *APOLO* can act not only *in cis* but also *in trans* by forming DNA-RNA hybrid called R-loop in promoter regions to regulate transcription of auxin-responsive loci [98].

EPIGENETIC REGULATION OF TRANSPOSABLE ELEMENTS AND INTERACTIONS WITH GENES

TEs are important DNA components of eukaryotic organisms and have been intensively studied in plants. They represent a genomic content of approximately 20% in *A. thaliana*, 35% in rice *Oryza sativa*, and more than 80% in wheat *Triticum aestivum* [99−101]. TEs can induce diverse impacts on gene regulation and are important drivers of plant evolution, adaptation, and domestication [3,102]. Although TEs are generally thought to generate de novo insertions in random genomic locations, some transposons prefer insertions in heterochromatic regions, whereas others target gene-rich euchromatic regions [103,104].

Epigenetics Regulation of Transposable Elements in Genic Regions

In addition to TE silencing mechanisms, plants have evolved mechanisms to prevent the spread of DNA and H3K9 methylation from the TE sequence to flanking genic regions. For example, DNA demethylase ROS1 removes DNA methylation of TEs near gene promoters to activate downstream genes [105−107]. In addition, H3K9 demethylase Increase IN BONSAI METHYLATION 1 (IBM1) removes H3K9 methylation from gene bodies [108]. In contrast, the H3K4 demethylase LSD1 homolog FLOWERING LOCUS D (FLD), LSD1-LIKE1 and LSD1-LIKE2 (LDL1, LDL2), and JMJ14 carry out the demethylation of H3K4 methylation marks from TEs [109]. Thus, these enzymes can consolidate the heterochromatin-euchromatin boundary to silence TEs without disturbing the activity of surrounding genes [110]. However, despite these mechanisms, de novo TE insertions can deposit new epigenetic marks that affect surrounding chromatin states and potentially perturb the expression of neighboring genes. Such events are called TE-associated epiallele formation, where the epigenetic state of TE and associated genes are often stably maintained for multiple generations [111].

Historically, the first TE-associated epiallele was discovered in a maize line characterized by a colorless kernel [112]. Changes in DNA methylation of a TE located upstream affect the expression of a gene involved in pigment synthesis, which causes the colorless kernel phenotype. Thus, TE-associated epialleles can be formed by spreading DNA and histone methylation marks of TEs toward regulatory elements of genes, thereby affecting gene expression and consequently plant phenotypes and/or fitness. In the cultivar Flying Saucers of *Ipomoea tricolor*, the spreading of DNA methylation from a Mutator TE located upstream of the DFR-B gene involved in anthocyanin synthesis triggers the hypermethylation of the promoter and disturbs gene expression [113]. The change in the DNA methylation state of the TE gives rise to variegated flowers. Recently, a similar phenomenon was observed in the Radish *Raphanus sativus* L. cultivar, which has a red-flesh phenotype that frequently turns to a white-fleshed phenotype in the next generations [114]. This change is also triggered by spreading of DNA methylation from a hypermethylated CACTA TE in the promoter region of *RsMYB1*, involved in anthocyanin synthesis. The change in the DNA methylation state of the promoter is heritable, causes downregulation of *RsMYB1*, and inhibits anthocyanin synthesis in the white-fleshed radish phenotype [114].

In addition to color changes, TE-associated epialleles affecting fitness, vitamin synthesis, or controlling sexual differentiation have also been described in plants. For example, in tomato *Solanum lycopersicum*, the change in DNA methylation of a short interspersed nuclear element (SINE) located in the promoter of the gene *VTE3 (1)* affects its expression and causes an accumulation of vitamin E in the fruit [115]. In rice, a spontaneous

TE-associated epiallele affects leaf angle. The hypomethylation of a miniature inverted-repeat transposable element located in the promoter of the gene *RAV6*, involved in brassinosteroid homeostasis regulation, causes ectopic expression of the gene and changes in leaf angle [116]. Recently, another natural TE-associated epiallele, which correlates with local climate adaptation, has been revealed in ecotypes of *A. thaliana*. Various ecotypes of *A. thaliana* bear a retrotransposon called *NMR19* in the promoter of the gene *PHEOPHYTIN PHEOPHORBIDE HYDROLASE* (*PPH*), and DNA methylation of this TE has been suggested to control leaf senescence by regulating the expression of *PPH* [117].

TE-associated epialleles also affect the sex determination and flowering timing of plants. In melon, the change in methylation of an hAT TE located near the transcription factor gene *CmWIP1* induces DNA hypermethylation and promotes gene silencing. The activation of *CmWIP1* leads to the development of male flowers, whereas its silencing promotes the formation of female flowers [118]. The gene *FLOWERING WAGENINGEN* (*FWA*) in *A. thaliana* is expressed in the wild-type endosperm but is transcriptionally silenced in other developmental stages, because of the DNA hypermethylation of SINE repeats near the transcription starting site (TSS). However, in the *ddm1* mutant of *A. thaliana*, the SINE repeats are hypomethylated, and ectopic expression of *FWA* occurs in other developmental tissues, resulting in a late-flowering phenotype [119]. Thus, by inserting closely to regulator elements, TEs can indirectly affect gene expression, which has a significant impact on the phenotypes of host plants.

Epigenetics Regulation of Transposable Elements Located Within Genes

TE insertions have also been found within the intragenic region, including the intronic, exonic and 5′/3′-UTR regions of plant genomes [120–124]. Intragenic TEs are mainly located in the intronic region [121,124], and most of them are associated with DNA methylation, H3K9 methylation, and targeted by siRNAs; Indeed, many intronic TEs are associated with heterochromatic epigenetic marks within plant genomes, including *A. thaliana*, rice, *Glycine max*, and maize [121,123–125]. This suggests that plants require mechanisms to allow RNA Pol II to transcribe heterochromatic TEs in introns to ensure proper transcription of associated genes. The transcription regulator ENHANCED DOWNY MILDEW 2 (EDM2) in *Arabidopsis* was identified as an essential factor for the transcription of the gene *RECOGNITION OF PERONOSPORA PARASITICA 7* (*RPP7*), which contains heterochromatic TEs in the introns [126]. In addition, other factors involved in this pathway, such as INCREASE IN BONSAI METHYLATION 2 (IBM2) and ASI1-IMMUNOPRECIPITATED PROTEIN 1 (AIPP1), have been identified [127–129]. In mutants deficient in these proteins, transcription of genes containing intronic heterochromatin is prematurely terminated by cryptic polyadenylation sites provided by intronic TE sequences [127]. Thus, the epigenetic masking mechanism of intronic TEs allows the accumulation of TEs in genes and the expansion of intronic sequences in plant genomes [130].

The *FLC* in *A. thaliana* provides an interesting example of how epigenetic mechanisms targeting intronic TEs can affect host gene transcription (Fig. 15.3A) [131]. *FLC* is involved in flowering time regulation, and *A. thaliana* ecotypes have natural variations in TE insertions at this locus. For instance, the ecotype Ler bears a Mutator TE in the first intronic region of *FLC*; however, this TE is not present in other ecotypes such as Col-0. Although *FLC* is properly transcribed in Col-0, the intronic TE in Ler is targeted by the siRNA that mediates gene silencing [132]. This TE-associated epiallele causes an early flowering phenotype (Fig. 15.3A) [133]. More recently, the RdDM pathway has been found to repress the expression of a cryptic transcript initiated from the intronic Mutator TE by promoting heterochromatinization of the TE [134]. In *Arabidopsis hallerii*, *FLC* is also associated with DNA methylation because of the presence of intronic SINE. Notably, the TE displays a seasonal change in CHH methylation and may control *FLC* expression and flowering [135]. *FLC* in *Capsella rubella* shows high TE insertion frequency as *A. thaliana*; moreover, a rare TE insertion at the 3′-UTR of the *FLC* locus seems to control the flowering time by affecting the mRNA stability [136]. In contrast, in the winter wheat cultivar *T. aestivum*, vernalization treatment has been found to induce the DNA hypermethylation of TEs located in the first intronic region of the gene *VRN-A1* and is associated with changes in the expression of the gene [137]. Changes in the methylation state of TE inserted in the 3′-UTR region may also lead to TE-associated epiallele formation. In *A. thaliana*, the *ddm1* mutant after repeated self-pollination shows phenotypic abnormalities, including a dwarf phenotype called *bonsai* [138]. The dwarf phenotype is caused by the silencing of *BONSAI* (*BNS*) genes because of a spreading of DNA methylation from LINE located in the 3′-UTR of *BNS*. Changes in DNA and H3K9 methylation of intronic TEs can also affect gene function by affecting full-length transcript production. In *A. thaliana*, the mutant of H3K9 methylase KRYPTONITE (KYP) and its homologs cause loss of H3K9me and an accumulation of short alternative transcripts because of cryptic polyadenylation sites in intronic TEs. This leads to a reduction in the full-length transcript of *RPP7* and a defect in resistance to pathogens [139]. Similar transcriptional defects occur in other

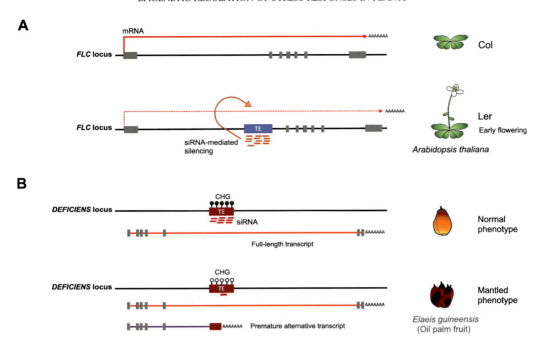

FIGURE 15.3 Versatile roles of TE-associated epialleles in plants. (A) In *Arabidopsis thaliana* Ler ecotype, *FLC* gene has an insertion of TE in an intron targeted by siRNA, which mediates gene silencing and leads to an early flowering phenotype. (B) In the oil palm *Elaeis guineensis*, the hypomethylation of an intronic TE in the gene *DEFICIENS* triggers an alternative isoform of the mRNA, resulting in an abnormal "Mantled" fruit phenotype.

loci in *A. thaliana ddm1* or *cmt3* mutants [121]. Recently, mutants of the RNA-binding protein FPA, suggested as an antagonist of IBM2 [140], have also been shown to promote alternative polyadenylation for genes containing intragenic TEs such as *RPP4* [141]. In addition to *A. thaliana*, alternative splicing caused by an intronic TE has been described for the oil palm *Elaeis guineensis*. The hypomethylation of a LINE named Karma, located within an intron of the gene *DEFICIENS*, triggers alternative splicing and premature termination of the gene, resulting in severe abnormal fruit development (Fig. 15.3B) [142].

Overall, plants have evolved epigenetic mechanisms to tolerate TEs inserted near or within genes by balancing TE silencing and gene transcription. Notably, some TE-associated epialleles cause abnormal plant phenotypes (e.g., Karma TE and *DEFICIENS* in oil palm), but others are linked to modulation of phenotypes for the environmental adaptation (e.g., Mutator TE and *FLC* in *A. thaliana*), emphasizing a versatile role of TE-associated epialleles in plant genomes.

EPIGENETIC REGULATION OF STRESS RESPONSES IN PLANTS

Plant Epigenetics and Abiotic Stress

Owing to unpredictable climate change, plants are often exposed to abiotic stresses, including salt stress, heat stress, and drought stress, reducing their productivity. Epigenetic mechanisms such as histone modification and DNA methylation are known to modulate stress-responsive gene expression at the transcriptional and posttranscriptional levels by altering the chromatin status of the genes [80]. Transcription factors and RdDM play critical roles in the regulation of gene expression under abiotic stress in plants. In addition, some epigenetic mechanisms, such as histone modification, play essential roles in stress memory, which may be inherited by offspring [143]. Abiotic stresses mainly include high/low temperature, drought, high salinity, malnutrition, and heavy metal toxicity. It has been shown that when genetically identical plants are exposed to various stresses, they display significant variations in DNA methylation. For example, Verhoeven et al. [144] exposed apomictic dandelion plants to various stresses, and individual plants in all groups displayed significant variations in DNA methylation. Their results indicated that DNA methylation variation caused by stress is common and mostly heritable, highlighting the independent role of epigenetic inheritance in adaptation and evolution along with genetic inheritance. Another study performed a comparative analysis of the methylome and gene expression in 60 annual clones of the stress-tolerant poplar *Populus simonii*, and investigated the effect of four abiotic stress treatments (salinity, osmotic, heat, and cold) from 3 to 24 h [145]. The DNA methylation level was increased after 3 h of all stress treatments, with the highest level after heat stress treatment. This section reviews some of these representative studies and provides a brief overview of the contribution of epigenetic mechanisms to plant responses to some important abiotic stresses.

Heat Stress

Epigenetic modifications such as DNA methylation, histone modifications, histone variants, ATP-dependent chromatin remodeling, histone chaperones, small RNAs, long noncoding RNAs, and other undefined epigenetic mechanisms can regulate the expression of heat-responsive genes and function to prevent heat-related damage [146]. DNA methylation processes have been widely studied in response to heat stress, especially through the activity of methyltransferases. Shen et al. [147] identified an association between CMT2 and temperature seasonality in *A. thaliana*, where *cmt2* mutants showed low sensitivity to heat stress, implying that genetic regulation of epigenetic modifications might be the underlying reason for natural adaptation to heat.

Heat shock proteins in plants play an essential role in heat resistance. Heat stress causes the accumulation of H3K9Ac and H3K4me3 on heat shock protein genes such as *HSP18*, *HSP22.0*, *APX2*, and *HSP70*. In addition, histone modification and DNA methylation via the RdDM pathway play essential roles in plant responses to heat stress [148,149]. *Arabidopsis thaliana* imprinted gene *SDC* (*SUPPRESSOR OF DRM1 DRM2 CMT3*), which is a target of the RdDM pathway and is silenced by DNA methylation during the vegetative phase, is activated by heat stress to regulate the expression of genes that help recover from heat stress [150]. Therefore, heat stress may disable the silencing effect of RdDM via the regulation of transcription of heat tolerance genes [151]. It has also been reported that the upregulation of epigenetic modulators, such as DNA methyltransferase, may be responsible for increased genome methylation in *Arabidopsis* under heat stress [152]. The thermal sensitivity of seed size in rice is also caused by changes in DNA and histone methylation (H3K9me2) during endosperm development [153]. Histone acetylation has also been reported under heat stress [154]. The length of the heat stress period has different effects on epigenetic mechanisms, suggesting the complexity of the epigenetic regulation mechanism behind heat stress [146].

Cold Stress

Cold stress affects metabolism and gene expression in plants. Plants have the ability of "cold acclimation," in which low but non-freezing temperatures increase their tolerance to subsequent freezing temperatures [155]. The C REPEAT BINDING FACTOR (CBF)-COLD-RESPONSIVE (COR) pathway is a well-characterized mechanism in the cold stress response of plants [156]. Cold stress increases the levels of CBF transcription factors, which then upregulate COR genes [156]. A recent study in *Arabidopsis* has shown that a chromatin remodeler gene named *PICKLE* (*PKL*) participates in the CBF-dependent cold stress response and is required for proper cold stress tolerance in plants. As *PKL* is involved in the RdDM [157] and H3K27me3 deposition [158] pathways, this suggests that both H3K27me3 and DNA methylation may affect cold tolerance stress memory in plants.

WD40 repeat-containing protein, HOS15, is involved in histone deacetylation and cold tolerance in *Arabidopsis* and functions as a targeting protein in the ubiquitination-proteasome degradation pathway, and HISTONE DEACETYLASE 2C (HD2C) is one of its interacting partners. Park et al. [159] reported that HOS15 interacts with HD2C and that both proteins are associated with the promoters of COR genes. Their results demonstrated that cold stress induces HOS15-mediated chromatin modifications by degrading HD2C.

The cold stress treatment of maize seedlings has also shown genome-wide DNA methylation [160]. Studies in crops have shown the role of epigenetic changes in low temperature-induced dormancy [161]. Histone acetylation also helps plants respond to cold stress, and it is enriched in the gene-body of COR genes [159]. This is regulated by HATs and HDACs. Histone acetylation has also been reported to be induced upon cold treatment in the promoter regions of some COR genes [162].

Salt Stress

High salinity has a massive negative effect on plant life events, mainly because of sodium ion toxicity, osmotic stress, and secondary stress, such as oxidative damage [155]. It has been shown that under salinity stress, DNA methylation has a significant impact on regulating gene expression. Ferreira et al. [163] reported DNA hypomethylation in response to high salinity. They showed a difference in DNA methylation between the salt-tolerant and salt-sensitive genotypes of rice, also associated with expression of DRM2 that was upregulated under salt stress in salt-sensitive but not in salt-tolerant genotypes. Studies on crop plants have shown that stress-induced DNA methylation may be independent of genetic variations [164]. HIGH-AFFINITY K + CHANNEL1 (HKT1), mediating Na + influx in plants, is a transporter that coordinates with the SALT OVERLY SENSITIVE pathway and results in salt resistance [165]. In wild-type *Arabidopsis*, a putative small RNA target region was identified approximately 2.6 kb upstream of the ATG start codon of *HKT1*; its promoter region was shown to be heavily methylated in all sequence contexts [166]. The *HKT1* gene body has been shown to have a high level of

H3K27me3. Additionally, high salinity stress upregulates *HKT1*, possibly because of the removal of H3K27me3 [167]. Studies on wheat have shown similar results, where high methylation levels resulted from high salt stress associated with the downregulation of HKT genes [168]. *MYB74*, regulated mainly by RdDM, is a salt-induced transcription factor. Similar to HKT1, its promoter is highly methylated by the RdDM pathway under normal conditions; therefore, *MYB74* shows a low expression level. However, under high salinity stress, DNA methylation and 24-nt siRNA levels have been shown to be extremely low at *MYB74*, associated with the higher expression of *MYB74* [169].

Histone modification dynamics also play essential roles in plant responses to salinity stress. Salt stress usually causes the deposition of active histone marks, such as H3K9K14Ac and H3K4me3, and reduces the deposition of repressive histone marks, such as H3K9me2 and H3K27me3, on salt tolerance genes [170]. Studies on *Arabidopsis* mutants have shown that the transcriptional adaptor ADA2b (a modulator of histone acetyltransferase activity) plays an essential role in salt stress susceptibility. ADA2b upregulates salt-responsive genes through locus-specific acetylation of histones H4 and H3 [171].

H3 Ser-10 phosphorylation (H3S10p), usually related to chromatin density, is another histone modification that can be induced by salinity treatment. Using tobacco and *Arabidopsis* cell lines, Sokol et al. [172] studied the nucleosomal response of plant cells to high salinity and reported that cells from both plants showed rapid upregulation of H3S10p, followed by upregulation of H3ac and histone H4ac. The three types of nucleosomal responses observed were strictly correlated with the expression of stress-type specific genes. HDA6 is important for the accumulation of H3K4me3 on salt-responsive genes; similar to *ada2b-1*, mutations in *HDA6* result in increased salt sensitivity and decreased expression level of abiotic stress tolerance genes in *Arabidopsis* [173]. Therefore, the crosstalk between genome-wide DNA methylation and histone modifications induced by high salinity is known to synchronize their roles against salt stress [174].

Drought Stress

Drought stress causes abscisic acid (ABA) production, resulting in increased resistance to water deficits [175]. ABA controls multiple developmental events, such as seed germination, stomatal closure, and root growth. It also plays a crucial role in growth arrest in seedlings when ABA-dependent transcription factors alter the transcriptional expression patterns. The epigenetic mechanisms underlying ABA stress responses have been widely studied (e.g., [176,177]). It has also been shown that the expression of genes related to drought stress is closely related to changes in histone dynamics [154]. Ding et al. [178] reported that drought stress increases deposition of H3K4me3 within the *NINE CISEPOXYCAROTENOID DIOXYGENASE 3* gene body region, a key enzyme involved in ABA synthesis. Another study reported that under drought stress, H3K4me3 and H3K9ac levels in the promoter regions of drought-responsive genes increase and cause higher expression of these genes [179]. When drought stress occurs, and after plants recover from dehydration, H3K4me3 and H3K9ac are removed from drought genes [180]. Ramirez-Prado et al. [181] also reported that the reduction of H3K27me3 deposition within the gene body of drought response transcription factors is important for drought tolerance in *Arabidopsis*.

In addition to histone modifications, DNA methylation also contributes to drought tolerance in plants. Studying *Populus trichocarpa*, Liang et al. [182] reported that drought stress changes the levels of DNA methylation and thereby the expression patterns of many drought stress response genes. Therefore, changes in DNA methylation could regulate the expression of drought response genes on a genome-wide scale. Wang et al. [183] detected genotype-specific patterns of drought-induced DNA methylation sites in rice. Another study in rice detected hypermethylation related to drought sensitivity and hypomethylation related to drought tolerance [184]. These studies suggest that chromatin changes caused by histone modifications and DNA methylation play important roles in drought tolerance in plants [185].

Plant Epigenetics and Biotic Stress

Recent studies have highlighted the importance of the epigenetic mechanisms of plants in modulating the immune response against pathogens. Chromatin status and DNA methylation can fine-tune the transcriptional regulation of defense genes and preserve information through cell division and meiosis, providing within- and across-generation short-term adaptation, respectively. Understanding epigenetic regulation of the plant immune system has provided a novel mechanistic explanation for plant-specific phenomena identified before the era of molecular biology.

Plant Immune System

Plants possess a sophisticated immune system that allows them to resist attacks by different microbial pathogens and herbivores. Unlike animals, which have specialized cells and tissues devoted to immunity, each plant cell can mount a localized immune response upon pathogen perception and send systemic signals that "prime" distal unaffected areas for enhanced immune response.

Pattern recognition receptors (PRRs) initiate an immune response by recognizing pathogen-associated molecular patterns or damage-associated molecular patterns released into the extracellular space in response to damage caused by pathogens [186]. Pathogen recognition triggers kinase- and hormone-mediated signaling, which results in the activation of specific downstream defenses, such as the production of reactive oxygen species, deposition of the β-1,3-glucan callose, and production of antimicrobial and/or antiherbivoral proteins [187]. The phytohormones salicylic acid (SA), jasmonic acid, ethylene, and ABA play key roles in pathogen defense signaling [188]. In addition, multiple defense hormones can be produced at once and can act in an antagonistic or synergistic manner (hormonal signal signature), ensuring activation of the appropriate defenses against the correct attacker while minimizing fitness costs [188]. This combination of defense responses constitutes the pathogen-triggered immunity (PTI) branch of the plant immune system. PTI relies on a large number of highly polymorphic loci and yields broad-spectrum resistance against diverse pathogens with similar lifestyles and pathogenic strategies (quantitative resistance) [189].

To increase virulence, plant pathogens have evolved secreted protein effectors that counteract PTI by masking detection by plant immune receptors, blocking downstream signaling, or manipulating plant signaling to activate inappropriate suites of defenses [190]. In response, plants have evolved a highly specialized intracellular nucleotide-binding domain and leucine-rich repeat receptors (NLRs) called resistance (R) genes, which recognize these effectors and trigger a programmed cell death called hypersensitive response (HR) [191]. Effector-triggered immunity (ETI) relies on the specific interaction between a pathogen strain possessing a specific effector and host plants expressing the matching R-gene (compatible interaction), resulting in the full immunity of the plant host (qualitative resistance) [191]. Due to the precise nature of compatible interactions, plant genomes carry multiple R-genes poised to rapidly evolve in response to the selective pressure from pathogen virulence, resulting in an evolutionary arms race between the pathogen and host [192].

Recent evidence has highlighted the association rather than the separation between PTI and ETI immune signaling [187], and epigenetic mechanisms have been shown to affect both branches of the plant immune system.

Epigenetic Memory and Plant Immune Priming

Plants retain a "memory" of environmental stresses perceived in their lifetime, allowing them to adapt and respond more effectively to recurring stressors [193].

Early plant pathology studies identified a spatial and temporal form of induced immunity called systemic acquired resistance (SAR), well before the era of molecular biology [194]. Following HR induction from pathogen attack, SAR is characterized by enhanced resistance to a subsequent attack by a pathogen with a similar lifestyle in distal tissues from the initial attack [194]. Subsequent molecular characterization identified several chemical agents involved in distal SAR signaling, such as SA, pipecolic acid, and azelaic acid, as well as the identification of faster and stronger transcription of defense genes at the distal attack site as the mechanism for broad-spectrum enhanced resistance [194]. The enhanced responsiveness state of the plant immune system characteristics of SAR is referred to as plant immune priming or induced resistance (IR) and appears in response to diverse stimuli such as exogenous defense hormone spray, specific priming chemicals, or signaling from beneficial rhizobacteria [195]. Notably, in *Arabidopsis* plants, IR from the priming agent β-aminobutyric acid was found to last for up to 4 weeks following the initial stimulus [196]. Multiple studies have shown that IR can be maintained trans-generationally from the parental plant to isogenic progenies across a stress-free generation (transgenerational IR, t-IR) [197,198]. These findings prompted further studies into the role of epigenetic mechanisms such as chromatin modifications and DNA methylation in the regulation of plant immune "memory" and IR. Moreover, unlike animals, some plants have been shown to only partially reset DNA methylation patterns during meiosis [18], resulting in *meta*-stable (yet reversible) transgenerational epialleles that differ in their DNA methylation patterns [199].

Chromatin Modification and Plant Immune Priming

Jaskiewicz et al. were the first to show that the SAR-related priming of WRKY gene promoters in *Arabidopsis* is associated with increased permissive marks H3K4me3 and H3K9ac [200]. Additionally, the SA analog benzothiadiazole (BTH)-primed wild-type *Arabidopsis* plants and constitutively primed mutants indicated accumulation of H3K4me3 and H3K9ac at the promoter site of the SA marker gene *PR1* [201]. Similarly, treatment with novel defense priming agent sulforaphane (SFN) provokes the covalent modification (K4me3, K9ac) of histone H3 in the promoter and promoter-proximal region of defense genes *WRKY6* and *PDF1.2* but not *PR1*, in *Arabidopsis* plants [202]. SFN-triggered H3K4me3 and H3K9ac coincide with chromatin unpacking in the *WRKY6* and *PDF1.2* regulatory regions, primed *WRKY6* expression, and reduced susceptibility to the biotrophic pathogen

Hyaloperonospora arabidopsidis, an agent of downy mildew disease [202].

Using formaldehyde-assisted isolation of regulatory elements (FAIRE)-seq, a recent study generated a genome-wide map of regulatory DNA elements characteristic of defense priming and SAR in *Arabidopsis* and associated with open chromatin status. In total, 10,129 priming/SAR-associated sites of open chromatin in the 5′ leader region of 3025 genes were identified [203].

These results indicate that, under unstressed conditions, defense genes are under reversible epigenetic repression, most likely to avoid the unnecessary production of resource-costly antimicrobial proteins and metabolites. Deposition of permissive chromatin marks at the promoter of defense regulators or downstream genes provides a potential mechanistic explanation for the enhanced transcriptional responsiveness characteristic of IR, as these marks allow poised binding of Pol II without transcription initiation, but faster transcription initiation once a pathogen attack is perceived [204].

DNA Methylation and Transgenerationally-Induced Resistance

Chromatin modifications at specific genomic regions are highly correlated with the underlying DNA methylation levels. Although dynamic regulation of chromatin marks provides a potential mechanistic explanation for within-generation IR, these marks are reset across the germline, with no evidence of transgenerational inheritance of chromatin status yet found in plants [50,205]. Plants do not fully reset DNA methylation patterns across generations, leading to multiple studies to identify naturally and artificially induced epialleles that remain stable over multiple plant generations and affect their development and flowering regulation [206–208]. Owing to its role in determining chromatin status and transmitting transgenerational information, DNA methylation affects within-generation IR and t-IR. Indeed, DNA hypomethylated *Arabidopsis* mutants are primed to activate SA-dependent defense genes, resulting in resistance against biotrophic pathogens [209]. In rice and *Arabidopsis*, DNA methylation is dynamically modulated in response to biotic stressors, such as the biotrophic pathogen *Pseudomonas syringae*; viral infection; hyperosmotic salt stress; heat stress; and water deficit stress [199]. In *Arabidopsis*, repeated exposure to these stresses triggers t-IR, increasing resistance to the stress to which the parental plants have been exposed [192]. The t-IR response is associated with global changes in DNA methylation proportional to the stress intensity levels experienced in previous generations [210] and can provide an ecological advantage in matched environments [211]. Although the mechanism by which hypomethylated epialleles are transmitted is not yet fully understood, it has been proposed that a lack of DNA methylation maintenance in TE-rich DNA regions is responsible [199].

Artificial hypomethylated epialleles have been introgressed into wild-type *Arabidopsis* plants by crossing with DNA hypomethylated mutants *met1* and *ddm1* of identical genetic backgrounds. The genetic mutation was removed through backcrossing, selection, and selfing across eight generations, and the recombinant epimutations stabilized in a genetic wild-type background [212,213]. In the *ddm1*-derived epigenetic recombinant inbred lines (epiRILs), hypomethylated regions are meta-stable for up to 16 generations, despite the fully functional DNA methylation establishment and maintenance machinery in these lines [214].

Using *ddm1*-based epiRILs, two studies have found that specific hypomethylated epialleles in *Arabidopsis* control quantitative resistance against the biotrophic downy mildew pathogen *H. arabidopsidis* [215], necrotrophic fungus *Plectosphaerella cucumerina* [215], and protist *Plasmodiophora brassicae* agent of clubroot disease [216].

Epigenetic Regulation of R-Genes

R-genes are often found in gene clusters and show unusually high variations in copy numbers within the same species. This enhanced duplication rate can be attributed to the high incidence of retrotransposons within R-gene clusters, as retrotransposons can increase gene duplication through nonallelic homologous recombination [217]. Duplicated genes are theorized to accelerate the evolution rate, as additional copies are subjected to less selective pressure and can freely accumulate genetic mutations, allowing plants to maintain the evolutionary arms race with microbial pathogens with a fast life cycle [192]. Owing to this close association, TEs are often found in intronic or UTR regions of R-genes, forming spurious TSSs or transcription termination sites, resulting in truncated or incorrect mRNA isoforms [130] (See section "Epigenetics Regulation of Transposable Elements Located Within Genes"). To maintain proper transcription of R-genes, plants epigenetically silence TE regions within or near R-genes through heterochromatic mark deposition, even though these marks allow Pol II passage and are maintained upon transcription [111,130]. Through an unknown mechanism, transcriptional regulators IBM2 and EDM2 must ignore the spurious signals within heterochromatic regions of R-genes such as *RPP7* and *RPP4* [127,139]. Recently, these two proteins have been found to target multiple NLR genes in the *Arabidopsis* genome [218].

A recent study identified mild upregulation of specific subsets of PRR/NLR genes in different mutants lacking DNA methylation or repressive H3K9me2 mark, indicating that these genes are also epigenetically suppressed at the transcriptional level [219].

EMERGING TECHNOLOGIES FOR EPIGENOME STUDIES IN PLANTS

Technological advances that allow DNA or histone modification analysis at the single-cell level, coupled with new approaches for altering the epigenome, have the potential to unveil unknown epigenetic mechanisms in plants. Recent technological improvements provide unprecedented opportunities to expand our understanding of plant epigenomics by enhancing our ability to monitor gene expression and genome structure at a level of detail that was unimaginable just a few years ago [220–223]. Here, we provide an overview of the improvements in methodologies underlying current single-cell and epigenome editing techniques, their potential significance, and future perspectives in plant epigenomics.

From Bulk to Single-Cell: The Emergence of Single-Cell Epigenomic Informatics

In multicellular organisms, different cell types tend to be organized into separate groups according to their functions. Plants have evolved highly organized arrangements of cells and tissues into organs that perform specialized functions. Therefore, cellular diversity within a tissue is more complex than that estimated by bulk analysis of the cell population, which can only return averaged results and is insufficient for studying heterogeneous systems. Recent developments in bulk-based omic studies have allowed smaller and smaller samples, eventually enabling single-cell examinations that provide powerful methods to dissect the biological processes of plants. Remarkably, improvements in sensitivity and throughput in single-cell epigenomics are giving us a unique opportunity for insights into epigenomic variation and the stochastic nature of gene expression.

Advances in Detection of DNA Methylation

In the burgeoning field of epigenetics, many methods are available to determine the methylation status of DNA samples. However, a method considered the "gold standard" to assess the differential methylation of specific regions of interest is bisulfite sequencing. Bisulfite treatment mediates deamination of cytosine residues into uracil, except for 5-methylcytosine (5mC) residues resistant to this conversion [224,225]. Comparing treated DNA samples with untreated samples via next-generation sequencing enables the detection of methylated cytosines across the entire genome. Although bulk methods have been used for locus-specific DNA methylation analysis, whole-genome examinations require novel approaches, such as single-cell reduced representation bisulfite sequencing (scRRBS) and single-cell bisulfite sequencing (scBS-seq). Additionally, scBS-seq was modified from the well-established bisulfite sequencing methodology to identify methylated cytosines in genomic DNA from single cells. Similar to the development of RRBS to reduce the cost of sequencing [226], single-cell RRBS uses restriction enzymes and bisulfite sequencing to enrich regions of the genome with abundant CpG content, reducing the number of nucleotides required for sequencing [227]. However, limitations such as partial enzyme digestion and bias during library preparation remain unresolved.

Multiple aspects of single-cell DNA methylation profiling in plants have been improved, including mapping rate, genomic coverage, and throughput. Changes in library preparation have increased the mapping rate, and advances in genomic coverage continue with methods such as bisulfite-converted randomly integrated fragment sequencing, which uses multiple displacement amplification instead of PCR to amplify converted DNA [228]. Additionally, technological refinement in throughput has also been achieved by single-cell combinatorial indexing for methylation analysis (sci-MET), which enables multiplex sequencing of single cells, decreasing the cost of sequencing, and further improving the mapping rate [229].

Bisulfite conversion severely degrades DNA, producing biases and unbalanced representation in sequencing libraries and limiting the thorough evaluation of methylomes. Enzymatic methyl-seq, a recent method tested in plants, uses the enzymatic conversion of unmethylated cytosines to uracils and does not harm the integrity of the DNA, achieving a higher yield than bisulfite sequencing [218]. This may provide advantages over bisulfite sequencing for single-cell applications by overcoming existing limitations in mapping efficiency and enabling the detection of methylated cytosine with few sequence reads. Another drawback of bisulfite sequencing is the generation of short reads that are insufficient for detecting DNA methylation at highly repetitive regions and long-range structures of the genome [230]. Epigenetic modifications can now be detected by electrical current signals in DNA using long-read nanopore sequencers [231]. This technology has been recently applied to plants, overcoming DNA degradation and amplification biases and allowing methylation profiling in more cytosines than bisulfite sequencing [232].

Development of Chromatin Accessibility Profiling Methods

Mapping highly accessible chromatin regions allow the localization of regulatory elements, improving our understanding of transcriptional regulation. An important technique, the Assay for Transposase-Accessible Chromatin with high-throughput sequencing (ATAC-seq), employs a hyperactive Tn5 transposase to induce DNA cleavage and insert sequencing adapters into open chromatin regions [233]. Here, sequencing reads allow the recognition of regions with increased accessibility, nucleosome positions, and transcription factor binding sites. Adaptation of ATAC-seq to plant systems has been challenging because of organellar DNA contamination during nuclei isolation. However, recent approaches for purifying plant cells have been developed to adequately isolate pure total nuclei; these include the Isolation of Nuclei TAgged in specific Cell Types (INTACT) method [234] and the more recent full-length RNA single-nucleus RNA-seq (FlsnRNA-seq)-based methods [235].

Although analysis of the bulk of chromatin accessibility has been performed in different plant species [236–238], examination at the single-cell level has only begun to emerge. Single-cell nucleosome-occupancy profiling methods being tested in plants include single-cell DNase sequencing (scDNase-seq), which utilizes DNase I in hypersensitive regions to capture accessible genomic sites [239]. Notably, the single-cell version of ATAC-seq (scATAC-seq) [240,241] has achieved high throughput in a split-pool method for combinatorial indexing (sci-ATAC-seq) [242].

Many improvements for mapping accessible chromatin regions in animal models have been developed and are being tested in plants. These include methods that capture more reads with a lower sequence bias with single-cell transposome hypersensitive site sequencing (scTHS-seq) [243] and alternative approaches using micrococcal nuclease that specifically cut nucleosome linker sequences in scMNase-seq [244]. Continued application of single-cell technologies to dissect the distinct modes of DNA accessibility in plant chromatin is likely to transform our view of molecular relationships and the evolution of cell type-specific gene regulatory networks [245].

ChIP-seq combines chromatin immunoprecipitation with massively parallel DNA sequencing and is a popular method for detecting genomic locations bound by proteins. However, to overcome the practical and economic limitations of ChIP-Seq, cleavage under targets and release using nuclease (CUT&RUN) was developed [246]. This method combines antibody-targeted controlled cleavage by micrococcal nuclease with DNA sequencing, allowing significant improvement in chromatin mapping and in situ identification of DNA-associated protein binding sites. ChIP-seq generally requires a large number of input cells, and several low-input approaches have been developed to solve this issue [247–250]. Notably, the development of ultra-low input (uli)-CUT&RUN, also based on the micrococcal nuclease-recruited antibody, uses as few as 100 cells for input [251]. This technique allows protein localization to chromatin with low input and high precision, permitting genome-wide profiling of chromatin-associated proteins in samples from single cells [252]. In the future, this technology could provide unprecedented insights into the epigenomic landscapes of important DNA-binding proteins, such as MADS-box TFs, from small plant samples at a resolution that has never been achieved before.

Mapping the Genome Structure of Single Cells

With the emergence of modern microscopic imaging techniques and chromosome conformation capture methods, our understanding of the 3D chromatin structure in plants has improved drastically [76]. Consecutive improvement in resolution and throughput, from chromosome conformation capture 3C, capture on-chip (4C), and capture carbon copy (5C) to Hi-C methods, have enabled unprecedented genome-wide resolution of the spatial organization and functional implications of chromatin conformation [253,254].

Applications of the Hi-C protocol to develop single-cell Hi-C (scHi-C) demonstrated the feasibility of retrieving chromatin structures within a single cell [255], facilitating the detection of chromosome conformational heterogeneity between animal cells [256,257] and more recently in plant cells [258–260]. As scHi-C has a low throughput, new strategies that increase throughput have been developed. Two substantial improvements are the single-cell combinational index Hi-C (sciHi-C), significantly expanding the sample size by employing a split-pool barcoding method [261], and the Dip-C, which can detect a greater number of contacts in single cells, allowing the 3D genome structure reconstruction of a single diploid cell [262].

The emergence of spatially resolved single-cell genomics and transcriptomics by imaging highlights how technological advances have expanded to provide biologists with extraordinary views of single-cell biology while preserving spatial context information. Remarkably, using microscopy through in situ sequencing or fluorescence in situ hybridization, it has been shown that genomic and transcriptomic sequences can be captured in a way that maintains spatial information while performing ex situ sequencing [263,264]. This opens a new path to tackle previously intractable problems in many biology fields. Therefore, spatially resolving functional profiles would address a wide range of issues, including identifying differences in expression patterns to address the spatial and functional organization of cells in plant tissues.

Advances in Epigenome Editing

The clustered, regularly interspaced short palindromic repeats and the catalytically inactive "dead" Cas9 system (CRISPR/dCas9) provide a robust manipulation tool to achieve targeted genome modifications [265,266]. Shortly after the emergence of this technology, researchers began incorporating regulatory proteins to edit the epigenome and modulate chromatin topology. The establishment of dCas9-based DNA-/chromatin-modifying enzymes has enabled this precise epigenome editing to control genome-wide gene regulation and chromatin status [267–270]. This includes the targeted positioning of histone modifiers, DNA methyltransferases, methylcytosine dioxygenases, and ubiquitin ligases [271,272]. New regulatory modules from synthetic and natural sources are currently being developed to enable the engineering of signaling, regulatory, and metabolic processes and achieve greater control of plant epigenomes.

Implementation of Transcriptional Regulators

The single-stranded guide RNA (sgRNA) is an essential component of the CRISPR/dCas9 system. Synthetic guide RNAs can be easily modified to obtain different targeting specificities, offering unparalleled simplicity and multiplexability. sgRNAs can simultaneously bind to different target loci and even increase the overall activity of transcriptional regulators when recruiting multiple sgRNAs to a sense strand [273,274]. sgRNA-led transcriptional effectors can stimulate gene expression by recruiting polymerases or transcription factors to gene sequences. Greater control over transcription can be achieved by employing multiple effectors simultaneously. This can be facilitated by an array of target sites for the RNA binding coat protein of phage Ms2 (MCP) that, when fused to sgRNAs, scaffolds are created for binding of transcriptional effectors [275,276]. Effector domains used in plants include the transcriptional activator VP64, acetyltransferase p300, methyltransferase KRYPTONITE, and strong repressors such as Kruppel-associated box (KRAB) [277,278]. Additionally, plant-specific short motifs, such as EDLL and SRDX, from the ethylene response factor, can also be utilized for transcriptional modulation [279,280].

The application of optogenetics, a method that uses light to control gene expression, faces experimental constraints in plants because of the fundamental role of light in shaping their development [281,282]. Nevertheless, optogenetics has been applied successfully in plants, including *Nicotiana benthamiana* and *A. thaliana* protoplasts [279]. Significant additions to the plant optogenetic toolbox include coupling PHYB, a red/far-red photoreceptor, to the VP16 transactivation domain to activate gene expression via red-light induction. A more recent system known as plant usable light-switch elements combines a blue-light-regulated repressor with a red-light-inducible switch to activate gene expression only under red light; expression remains inactive under white light or in the dark [283,284]. The creation and implementation of these optogenetic tools over the past few years have changed how we study the regulation of gene expression in plants. Consequently, the development of optogenetics for the spatiotemporal control of epigenetic modulators promises to provide unique insights into the epigenetic phenomena, chromatin dynamics, and gene expression mechanisms of plants.

Editing the Chromatin Topology of Plants

Multiple developmental and environmental factors, including light intensity, temperature, microbial infection, and signaling molecules, trigger global chromatin rearrangements [285]. Although this spatial and temporal organization is crucial for regulating cell functions, it remains largely uncharacterized in plants. The application of dCas9 to modify chromatin structure and DNA loop formation is emerging as an important tool for remodeling the genome and interrogating the functional roles of chromatin structure. Additionally, dCas9-based platforms have been used to target and manipulate chromatin structure and DNA loop formation. For example, the chromatin loop reorganization using CRISPR-dCas9 (CLOuD9) system was used to form artificial chromatin loops between distal genomic regions by tethering protein dimerization systems to dCas9 orthologs, resulting in the control of chromatin loop formation and selective modulation of gene expression [286]. In plants, monitoring the behavior of chromatin loci in the nucleus has been achieved by live imaging of visually trackable T-DNA insertions [287]. First developed in mammalian cells, advances in CRISPR/Cas-based chromatin-imaging methods that implement multicolor labeling for simultaneous tracking of different loci have now been applied to plant cells [288,289]. Further improvement of CRISPR imaging in plants has the potential to enable visualization of single genomic loci and significantly advance our understanding of chromatin dynamics in live cells.

FUTURE PERSPECTIVES

The rapid development of new sequencing technologies in other model systems and their applications to plant epigenome analyses would allow researchers to dissect epigenome dynamics at finer time scales under various conditions. In addition, higher-order chromatin analysis at single-cell resolutions, together with transcriptome information, could generate an integrated epigenome atlas of multiple tissues and specific cell lineages during plant development. At the cellular level, further studies of biochemical properties, organization, and functions of non-membrane nuclear structures formed by liquid-liquid phase separation would be essential for our comprehensive understanding of chromatin regulation in plants. In addition, the increase in high-quality genome data of non-model plant species will accelerate studies of epigenome dynamics, especially under natural environments, allowing us to further understand the epigenetic regulation of plant adaptation and stress responses under fluctuating natural environments. Finally, epigenome editing technology would be a promising application for breeding and improving traits in main crops; however, further technical breakthroughs would be required to precisely regulate targeted genes for practical application in agriculture.

References

[1] Deniz O, Frost JM, Branco MR. Regulation of transposable elements by DNA modifications. Nat Rev Genet 2019;20:417—31.

[2] Slotkin RK, Martienssen R. Transposable elements and the epigenetic regulation of the genome. Nat Rev Genet 2007; 8:272—85.

[3] Lisch D. How important are transposons for plant evolution? Nat Rev Genet 2013;14:49—61.

[4] Takuno S, Ran JH, Gaut BS. Evolutionary patterns of genic DNA methylation vary across land plants. Nat Plants 2016; 2:15222.

[5] Takuno S, Gaut BS. Gene body methylation is conserved between plant orthologs and is of evolutionary consequence. Proc Natl Acad Sci USA 2013;110:1797—802.

[6] Wang X, Hu L, Wang X, Li N, Xu C, Gong L, et al. DNA Methylation affects gene alternative splicing in plants: an example from rice. Mol Plant 2016;9:305—7.

[7] Horvath R, Laenen B, Takuno S, Slotte T. Single-cell expression noise and gene-body methylation in *Arabidopsis thaliana*. Heredity (Edinb) 2019;123:81—91.

[8] Choi J, Lyons DB, Kim MY, Moore JD, Zilberman D. DNA methylation and histone H1 jointly repress transposable elements and aberrant intragenic transcripts. Mol Cell 2020;77: 310—23 e317.

[9] Liang Z, Shen L, Cui X, Bao S, Geng Y, Yu G, et al. DNA N(6)-Adenine methylation in *Arabidopsis thaliana*. Dev Cell 2018; 45:406—16 e403.

[10] Zhou C, Wang C, Liu H, Zhou Q, Liu Q, Guo Y, et al. Identification and analysis of adenine N(6)-methylation sites in the rice genome. Nat Plants 2018;4:554—63.

[11] Law JA, Jacobsen SE. Establishing, maintaining and modifying DNA methylation patterns in plants and animals. Nat Rev Genet 2010;11:204—20.

[12] Matzke MA, Mosher RA. RNA-directed DNA methylation: an epigenetic pathway of increasing complexity. Nat Rev Genet 2014;15:394—408.

[13] Matzke MA, Kanno T, Matzke AJ. RNA-directed DNA methylation: the evolution of a complex epigenetic pathway in flowering plants. Annu Rev Plant Biol 2015;66:243—67.

[14] Hung YH, Slotkin RK. The initiation of RNA interference (RNAi) in plants. Curr Opin Plant Biol 2021;61:102014.

[15] Martienssen R, Moazed D. RNAi and heterochromatin assembly. Cold Spring Harb Perspect Biol 2015;7:a019323.

[16] Yaari R, Katz A, Domb K, Harris KD, Zemach A, Ohad N. RdDM-independent de novo and heterochromatin DNA methylation by plant CMT and DNMT3 orthologs. Nat Commun 2019;10:1613.

[17] Zhang H, Lang Z, Zhu JK. Dynamics and function of DNA methylation in plants. Nat Rev Mol Cell Biol 2018;19:489—506.

[18] Bouyer D, Kramdi A, Kassam M, Heese M, Schnittger A, Roudier F, et al. DNA methylation dynamics during early plant life. Genome Biol 2017;18:179.

[19] Kawakatsu T, Nery JR, Castanon R, Ecker JR. Dynamic DNA methylation reconfiguration during seed development and germination. Genome Biol 2017;18:171.

[20] Narsai R, Gouil Q, Secco D, Srivastava A, Karpievitch YV, Liew LC, et al. Extensive transcriptomic and epigenomic remodelling occurs during *Arabidopsis thaliana* germination. Genome Biol 2017;18:172.

[21] Lin JY, Le BH, Chen M, Henry KF, Hur J, Hsieh TF, et al. Similarity between soybean and Arabidopsis seed methylomes and loss of non-CG methylation does not affect seed development. Proc Natl Acad Sci USA 2017;114:E9730—9.

[22] Papareddy RK, Paldi K, Paulraj S, Kao P, Lutzmayer S, Nodine MD. Chromatin regulates expression of small RNAs to help maintain transposon methylome homeostasis in Arabidopsis. Genome Biol 2020;21:251.

[23] Baubec T, Finke A, Mittelsten Scheid O, Pecinka A. Meristem-specific expression of epigenetic regulators safeguards transposon silencing in Arabidopsis. EMBO Rep 2014;15:446—52.

[24] Ning YQ, Liu N, Lan KK, Su YN, Li L, Chen S, et al. DREAM complex suppresses DNA methylation maintenance genes and precludes DNA hypermethylation. Nat Plants 2020;6:942—56.

[25] Gutzat R, Rembart K, Nussbaumer T, Hofmann F, Pisupati R, Bradamante G, et al. Arabidopsis shoot stem cells display dynamic transcription and DNA methylation patterns. EMBO J 2020;39:e103667.

[26] Kawakatsu T, Stuart T, Valdes M, Breakfield N, Schmitz RJ, Nery JR, et al. Unique cell-type-specific patterns of DNA methylation in the root meristem. Nat Plants 2016;2:16058.

[27] Higo A, Saihara N, Miura F, Higashi Y, Yamada M, Tamaki S, et al. DNA methylation is reconfigured at the onset of reproduction in rice shoot apical meristem. Nat Commun 2020;11: 4079.

[28] Walker J, Gao H, Zhang J, Aldridge B, Vickers M, Higgins JD, et al. Sexual-lineage-specific DNA methylation regulates meiosis in Arabidopsis. Nat Genet 2018;50:130—7.

[29] Mosher RA, Melnyk CW. siRNAs and DNA methylation: seedy epigenetics. Trends Plant Sci 2010;15:204—10.

[30] Ono A, Kinoshita T. Epigenetics and plant reproduction: multiple steps for responsibly handling succession. Curr Opin Plant Biol 2021;61:102032.

[31] Ibarra CA, Feng X, Schoft VK, Hsieh TF, Uzawa R, Rodrigues JA, et al. Active DNA demethylation in plant companion cells reinforces transposon methylation in gametes. Science 2012;337:1360—4.

[32] Calarco JP, Borges F, Donoghue MT, Van F, Jullien Ex, PE, Lopes T, et al. Reprogramming of DNA methylation in pollen guides epigenetic inheritance via small RNA. Cell 2012;151: 194—205.

[33] Martinez G, Panda K, Kohler C, Slotkin RK. Silencing in sperm cells is directed by RNA movement from the surrounding nurse cell. Nat Plants 2016;2:16030.

[34] Park K, Kim MY, Vickers M, Park JS, Hyun Y, Okamoto T, et al. DNA demethylation is initiated in the central cells of Arabidopsis and rice. Proc Natl Acad Sci USA 2016;113:15138−43.

[35] Johnson L, Mollah S, Garcia BA, Muratore TL, Shabanowitz J, Hunt DF, et al. Mass spectrometry analysis of Arabidopsis histone H3 reveals distinct combinations of post-translational modifications. Nucleic Acids Res 2004;32:6511−18.

[36] Zhang K, Sridhar VV, Zhu J, Kapoor A, Zhu JK. Distinctive core histone post-translational modification patterns in *Arabidopsis thaliana*. PLoS One 2007;2:e1210.

[37] He K, Cao X, Deng X. Histone methylation in epigenetic regulation and temperature responses. Curr Opin Plant Biol 2021;61:102001.

[38] Jacob Y, Feng S, LeBlanc CA, Bernatavichute YV, Stroud H, Cokus S, et al. ATXR5 and ATXR6 are H3K27 monomethyltransferases required for chromatin structure and gene silencing. Nat Struct Mol Biol 2009;16:763−8.

[39] Pandey R, Muller A, Napoli CA, Selinger DA, Pikaard CS, Richards EJ, et al. Analysis of histone acetyltransferase and histone deacetylase families of *Arabidopsis thaliana* suggests functional diversification of chromatin modification among multicellular eukaryotes. Nucleic Acids Res 2002;30:5036−55.

[40] Hollender C, Liu Z. Histone deacetylase genes in Arabidopsis development. J Integr Plant Biol 2008;50:875−85.

[41] Lee K, Mas P, Seo PJ. The EC-HDA9 complex rhythmically regulates histone acetylation at the TOC1 promoter in Arabidopsis. Commun Biol 2019;2:143.

[42] Hung FY, Chen FF, Li C, Chen C, Chen JH, Cui Y, et al. The LDL1/2-HDA6 histone modification complex interacts with TOC1 and regulates the core circadian clock components in Arabidopsis. Front Plant Sci 2019;10:233.

[43] Ning YQ, Chen Q, Lin RN, Li YQ, Li L, Chen S, et al. The HDA19 histone deacetylase complex is involved in the regulation of flowering time in a photoperiod-dependent manner. Plant J 2019;98:448−64.

[44] Baumbusch LO, Thorstensen T, Krauss V, Fischer A, Naumann K, Assalkhou R, et al. The Arabidopsis thaliana genome contains at least 29 active genes encoding SET domain proteins that can be assigned to four evolutionarily conserved classes. Nucleic Acids Res 2001;29:4319−33.

[45] Jacob Y, Stroud H, Leblanc C, Feng S, Zhuo L, Caro E, et al. Regulation of heterochromatic DNA replication by histone H3 lysine 27 methyltransferases. Nature 2010;466:987−91.

[46] Pfluger J, Wagner D. Histone modifications and dynamic regulation of genome accessibility in plants. Curr Opin Plant Biol 2007;10:645−52.

[47] Sun Q, Zhou DX. Rice jmjC domain-containing gene JMJ706 encodes H3K9 demethylase required for floral organ development. Proc Natl Acad Sci USA 2008;105:13679−84.

[48] Lu F, Li G, Cui X, Liu C, Wang XJ, Cao X. Comparative analysis of JmjC domain-containing proteins reveals the potential histone demethylases in Arabidopsis and rice. J Integr Plant Biol 2008;50:886−96.

[49] Jiang D, Yang W, He Y, Amasino RM. Arabidopsis relatives of the human lysine-specific Demethylase1 repress the expression of FWA and flowering locus C and thus promote the floral transition. Plant Cell 2007;19:2975−87.

[50] Borg M, Jacob Y, Susaki D, LeBlanc C, Buendía D, Axelsson E, et al. Targeted reprogramming of H3K27me3 resets epigenetic memory in plant paternal chromatin. Nat Cell Biol 2020;22:621−9.

[51] Zheng S, Hu H, Ren H, Yang Z, Qiu Q, Qi W, et al. The Arabidopsis H3K27me3 demethylase JUMONJI 13 is a temperature and photoperiod dependent flowering repressor. Nat Commun 2019;10:1303.

[52] Crevillen P, Yang H, Cui X, Greeff C, Trick M, Qiu Q, et al. Epigenetic reprogramming that prevents transgenerational inheritance of the vernalized state. Nature 2014;515:587−90.

[53] Zhang X, Germann S, Blus BJ, Khorasanizadeh S, Gaudin V, Jacobsen SE. The Arabidopsis LHP1 protein colocalizes with histone H3 Lys27 trimethylation. Nat Struct Mol Biol 2007;14:869−71.

[54] Zhao S, Cheng L, Gao Y, Zhang B, Zheng X, Wang L, et al. Plant HP1 protein ADCP1 links multivalent H3K9 methylation readout to heterochromatin formation. Cell Res 2019;29:54−66.

[55] Zhang C, Du X, Tang K, Yang Z, Pan L, Zhu P, et al. Arabidopsis AGDP1 links H3K9me2 to DNA methylation in heterochromatin. Nat Commun 2018;9:4547.

[56] Du J, Zhong X, Bernatavichute YV, Stroud H, Feng S, Caro E, et al. Dual binding of chromomethylase domains to H3K9me2-containing nucleosomes directs DNA methylation in plants. Cell 2012;151:167−80.

[57] Li Z, Fu X, Wang Y, Liu R, He Y. Polycomb-mediated gene silencing by the BAH-EMF1 complex in plants. Nat Genet 2018;50:1254−61.

[58] Yang Z, Qian S, Scheid RN, Lu L, Chen X, Liu R, et al. EBS is a bivalent histone reader that regulates floral phase transition in Arabidopsis. Nat Genet 2018;50:1247−53.

[59] Qian S, Lv X, Scheid RN, Lu L, Yang Z, Chen W, et al. Dual recognition of H3K4me3 and H3K27me3 by a plant histone reader SHL. Nat Commun 2018;9:2425.

[60] Borg M, Jiang D, Berger F. Histone variants take center stage in shaping the epigenome. Curr Opin Plant Biol 2021;61:101991.

[61] Kumar SV. H2A.Z at the core of transcriptional regulation in plants. Mol Plant 2018;11:1112−14.

[62] Gomez-Zambrano A, Merini W, Calonje M. The repressive role of Arabidopsis H2A.Z in transcriptional regulation depends on AtBMI1 activity. Nat Commun 2019;10:2828.

[63] Kawashima T, Lorkovic ZJ, Nishihama R, Ishizaki K, Axelsson E, Yelagandula R, et al. Diversification of histone H2A variants during plant evolution. Trends Plant Sci 2015;20:419−25.

[64] Yelagandula R, Stroud H, Holec S, Zhou K, Feng S, Zhong X, et al. The histone variant H2A.W defines heterochromatin and promotes chromatin condensation in Arabidopsis. Cell 2014;158:98−109.

[65] Jacob Y, Bergamin E, Donoghue MT, Mongeon V, LeBlanc C, Voigt P, et al. Selective methylation of histone H3 variant H3.1 regulates heterochromatin replication. Science 2014;343:1249−53.

[66] Yan A, Borg M, Berger F, Chen Z. The atypical histone variant H3.15 promotes callus formation in *Arabidopsis thaliana*. Development 2020;147.

[67] Rutowicz K, Puzio M, Halibart-Puzio J, Lirski M, Kotlinski M, Kroten MA, et al. A specialized histone H1 variant is required for adaptive responses to complex abiotic stress and related DNA methylation in Arabidopsis. Plant Physiol 2015;169:2080−101.

[68] Zemach A, Kim MY, Hsieh PH, Coleman-Derr D, Eshed-Williams L, Thao K, et al. The Arabidopsis nucleosome remodeler DDM1 allows DNA methyltransferases to access H1-containing heterochromatin. Cell 2013;153:193−205.

[69] Bourguet P, Picard CL, Yelagandula R, Pelissier T, Lorkovic ZJ, Feng S, et al. The histone variant H2A.W and linker histone H1 co-regulate heterochromatin accessibility and DNA methylation. Nat Commun 2021;12:2683.

[70] Grob S, Schmid MW, Grossniklaus U. Hi-C analysis in Arabidopsis identifies the KNOT, a structure with similarities to the flamenco locus of Drosophila. Mol Cell 2014;55:678−93.

[71] Feng S, Cokus SJ, Schubert V, Zhai J, Pellegrini M, Jacobsen SE. Genome-wide Hi-C analyses in wild-type and mutants reveal

high-resolution chromatin interactions in Arabidopsis. Mol Cell 2014;55:694–707.
[72] Grob S, Grossniklaus U. Invasive DNA elements modify the nuclear architecture of their insertion site by KNOT-linked silencing in *Arabidopsis thaliana*. Genome Biol 2019;20:120.
[73] Sun L, Jing Y, Liu X, Li Q, Xue Z, Cheng Z, et al. Heat stress-induced transposon activation correlates with 3D chromatin organization rearrangement in Arabidopsis. Nat Commun 2020;11:1886.
[74] Liu C, Cheng YJ, Wang JW, Weigel D. Prominent topologically associated domains differentiate global chromatin packing in rice from Arabidopsis. Nat Plants 2017;3:742–8.
[75] Dong P, Tu X, Chu PY, Lu P, Zhu N, Grierson D, et al. 3D Chromatin architecture of large plant genomes determined by Local A/B compartments. Mol Plant 2017;10:1497–509.
[76] Ouyang W, Xiong D, Li G, Li X. Unraveling the 3D genome architecture in plants: present and future. Mol Plant 2020; 13:1676–93.
[77] Lu Z, Marand AP, Ricci WA, Ethridge CL, Zhang X, Schmitz RJ. The prevalence, evolution and chromatin signatures of plant regulatory elements. Nat Plants 2019;5:1250–9.
[78] Ricci WA, Lu Z, Ji L, Marand AP, Ethridge CL, Murphy NG, et al. Widespread long-range cis-regulatory elements in the maize genome. Nat Plants 2019;5:1237–49.
[79] Whittaker C, Dean C. The FLC locus: a platform for discoveries in epigenetics and adaptation. Annu Rev Cell Dev Biol 2017;33:555–75.
[80] Luo X, He Y. Experiencing winter for spring flowering: a molecular epigenetic perspective on vernalization. J Integr Plant Biol 2020;62:104–17.
[81] Sharma N, Geuten K, Giri BS, Varma A. The molecular mechanism of vernalization in Arabidopsis and cereals: role of Flowering Locus C and its homologs. Physiol Plant 2020; 170:373–83.
[82] Menon G, Schulten A, Dean C, Howard M. Digital paradigm for Polycomb epigenetic switching and memory. Curr Opin Plant Biol 2021;61:102012.
[83] Chandler VL, Stam M. Chromatin conversations: mechanisms and implications of paramutation. Nat Rev Genet 2004;5: 532–44.
[84] Chandler VL. Paramutation: from maize to mice. Cell 2007; 128:641–5.
[85] Hollick JB. Paramutation and related phenomena in diverse species. Nat Rev Genet 2017;18:5–23.
[86] Batista RA, Kohler C. Genomic imprinting in plants-revisiting existing models. Genes Dev 2020;34:24–36.
[87] Waters AJ, Bilinski P, Eichten SR, Vaughn MW, Ross-Ibarra J, Gehring M, et al. Comprehensive analysis of imprinted genes in maize reveals allelic variation for imprinting and limited conservation with other species. Proc Natl Acad Sci USA 2013;110:19639–44.
[88] Tamiru M, Hardcastle TJ, Lewsey MG. Regulation of genome-wide DNA methylation by mobile small RNAs. New Phytol 2018;217:540–6.
[89] Molnar A, Melnyk CW, Bassett A, Hardcastle TJ, Dunn R, Baulcombe DC. Small silencing RNAs in plants are mobile and direct epigenetic modification in recipient cells. Science 2010;328:872–5.
[90] Lewsey MG, Hardcastle TJ, Melnyk CW, Molnar A, Valli A, Urich MA, et al. Mobile small RNAs regulate genome-wide DNA methylation. Proc Natl Acad Sci USA 2016;113:E801–10.
[91] Kundariya H, Yang X, Morton K, Sanchez R, Axtell MJ, Hutton SF, et al. MSH1-induced heritable enhanced growth vigor through grafting is associated with the RdDM pathway in plants. Nat Commun 2020;11:5343.
[92] Kindgren P, Ard R, Ivanov M, Marquardt S. Transcriptional read-through of the long non-coding RNA SVALKA governs plant cold acclimation. Nat Commun 2018;9:4561.
[93] Kindgren P, Ivanov M, Marquardt S. Native elongation transcript sequencing reveals temperature dependent dynamics of nascent RNAPII transcription in Arabidopsis. Nucleic Acids Res 2020;48:2332–47.
[94] Deforges J, Reis RS, Jacquet P, Sheppard S, Gadekar VP, Hart-Smith G, et al. Control of cognate sense mRNA translation by cis-natural antisense RNAs. Plant Physiol 2019;180:305–22.
[95] Reis RS, Poirier Y. Making sense of the natural antisense transcript puzzle. Trends Plant Sci 2021;26:1104–15.
[96] Chen MX, Zhu FY, Gao B, Ma KL, Zhang Y, Fernie AR, et al. Full-length transcript-based proteogenomics of rice improves its genome and proteome annotation. Plant Physiol 2020; 182:1510–26.
[97] Ariel F, Jegu T, Latrasse D, Romero-Barrios N, Christ A, Benhamed M, et al. Noncoding transcription by alternative RNA polymerases dynamically regulates an auxin-driven chromatin loop. Mol Cell 2014;55:383–96.
[98] Ariel F, Lucero L, Christ A, Mammarella MF, Jegu T, Veluchamy A, et al. R-loop mediated transaction of the APOLO long noncoding RNA. Mol Cell 2020;77:1055–65 e1054.
[99] I.W.G.S. Consortium. A chromosome-based draft sequence of the hexaploid bread wheat (Triticum aestivum) genome. Science 2014;345.
[100] R.G.S.P. International. The map-based sequence of the rice genome. Nature 2005;436:793.
[101] Buisine N, Quesneville H, Colot V. Improved detection and annotation of transposable elements in sequenced genomes using multiple reference sequence sets. Genomics 2008;91: 467–75.
[102] Gill RA, Scossa F, King GJ, Golicz A, Tong C, Snowdon RJ, et al. On the role of transposable elements in the regulation of gene expression and subgenomic interactions in crop genomes. CRC Crit Rev Plant Sci 2021;40:157–89.
[103] Quadrana L, Silveira AB, Mayhew GF, LeBlanc C, Martienssen RA, Jeddeloh JA, et al. The Arabidopsis thaliana mobilome and its impact at the species level. elife 2016;5:e15716.
[104] Quesneville H. Twenty years of transposable element analysis in the *Arabidopsis thaliana* genome. Mobile DNA 2020;11:1–13.
[105] Zhu J, Kapoor A, Sridhar VV, Agius F, Zhu J-K. The DNA glycosylase/lyase ROS1 functions in pruning DNA methylation patterns in Arabidopsis. Curr Biol 2007;17:54–9.
[106] Yamamuro C, Miki D, Zheng Z, Ma J, Wang J, Yang Z, et al. Overproduction of stomatal lineage cells in Arabidopsis mutants defective in active DNA demethylation. Nat Commun 2014;5:4062.
[107] Halter T, Wang J, Amesefe D, Lastrucci E, Charvin M, Singla Rastogi M, et al. The Arabidopsis active demethylase ROS1 cis-regulates defence genes by erasing DNA methylation at promoter-regulatory regions. Elife 2021;10.
[108] Saze H, Shiraishi A, Miura A, Kakutani T. Control of genic DNA methylation by a jmjC domain-containing protein in *Arabidopsis thaliana*. Science 2008;319:462–5.
[109] Greenberg MV, Deleris A, Hale CJ, Liu A, Feng S, Jacobsen SE. Interplay between active chromatin marks and RNA-directed DNA methylation in *Arabidopsis thaliana*. PLoS Genet 2013;9:e1003946.
[110] Sigman MJ, Slotkin RK. The first rule of plant transposable element silencing: location, location, location. Plant Cell 2016; 28:304–13.
[111] Baduel P, Colot V. The epiallelic potential of transposable elements and its evolutionary significance in plants. Philos Trans R Soc B 2021;376:20200123.

[112] Banks JA, Masson P, Fedoroff N. Molecular mechanisms in the developmental regulation of the maize suppressor-mutator transposable element. Genes Dev 1988;2:1364—80.

[113] Iida S, Morita Y, Choi J-D, Park K-I, Hoshino A. Genetics and epigenetics in flower pigmentation associated with transposable elements in morning glories. Adv Biophys 2004;38:141—59.

[114] Wang Q, Wang Y, Sun H, Sun L, Zhang L. Transposon-induced methylation of the RsMYB1 promoter disturbs anthocyanin accumulation in red-fleshed radish. J Exp Bot 2020;71:2537—50.

[115] Quadrana L, Almeida J, Asís R, Duffy T, Dominguez PG, Bermúdez L, et al. Natural occurring epialleles determine vitamin E accumulation in tomato fruits. Nat Commun 2014;5:4027.

[116] Zhang X, Sun J, Cao X, Song X. Epigenetic mutation of RAV6 affects leaf angle and seed size in rice. Plant Physiol 2015;169:2118—28.

[117] He L, Wu W, Zinta G, Yang L, Wang D, Liu R, et al. A naturally occurring epiallele associates with leaf senescence and local climate adaptation in Arabidopsis accessions. Nat Commun 2018;9:1—11.

[118] Martin A, Troadec C, Boualem A, Rajab M, Fernandez R, Morin H, et al. A transposon-induced epigenetic change leads to sex determination in melon. Nature 2009;461:1135—8.

[119] Kinoshita T, Miura A, Choi Y, Kinoshita Y, Cao X, Jacobsen SE, et al. One-way control of FWA imprinting in Arabidopsis endosperm by DNA methylation. Science 2004;303:521—3.

[120] Lockton S, Gaut BS. The contribution of transposable elements to expressed coding sequence in *Arabidopsis thaliana*. J Mol Evol 2009;68:80—9.

[121] Le TN, Miyazaki Y, Takuno S, Saze H. Epigenetic regulation of intragenic transposable elements impacts gene transcription in *Arabidopsis thaliana*. Nucleic Acids Res 2015;43:3911—21.

[122] Li X, Guo K, Zhu X, Chen P, Li Y, Xie G, et al. Domestication of rice has reduced the occurrence of transposable elements within gene coding regions. BMC Genomics 2017;18:1—12.

[123] West PT, Li Q, Ji L, Eichten SR, Song J, Vaughn MW, et al. Genomic distribution of H3K9me2 and DNA methylation in a maize genome. PLoS One 2014;9:e105267.

[124] Espinas NA, Furci L, Shimajiri Y, Harukawa Y, Miura S, Takuno S, et al. Transcriptional regulation of genes bearing intronic heterochromatin in the rice genome. PLoS Genet 2020;16:e1008637.

[125] Do Kim K, El Baidouri M, Abernathy B, Iwata-Otsubo A, Chavarro C, Gonzales M, et al. A comparative epigenomic analysis of polyploidy-derived genes in soybean and common bean. Plant Physiol 2015;168:1433—47.

[126] Eulgem T, Tsuchiya T, Wang XJ, Beasley B, Cuzick A, Tör M, et al. EDM2 is required for RPP7-dependent disease resistance in Arabidopsis and affects RPP7 transcript levels. Plant J 2007;49:829—39.

[127] Saze H, Kitayama J, Takashima K, Miura S, Harukawa Y, Ito T, et al. Mechanism for full-length RNA processing of Arabidopsis genes containing intragenic heterochromatin. Nat Commun 2013;4:1—9.

[128] Wang X, Duan C-G, Tang K, Wang B, Zhang H, Lei M, et al. RNA-binding protein regulates plant DNA methylation by controlling mRNA processing at the intronic heterochromatin-containing gene IBM1. Proc Natl Acad Sci 2013;110:15467—72.

[129] Duan C-G, Wang X, Zhang L, Xiong X, Zhang Z, Tang K, et al. A protein complex regulates RNA processing of intronic heterochromatin-containing genes in Arabidopsis. Proc Natl Acad Sci 2017;114:E7377—84.

[130] Saze H. Epigenetic regulation of intragenic transposable elements: a two-edged sword. J Biochem 2018;164:323—8.

[131] Quadrana L. The contribution of transposable elements to transcriptional novelty in plants: the FLC affair. Transcription 2020;11:192—8.

[132] Fultz D, Choudury SG, Slotkin RK. Silencing of active transposable elements in plants. Curr Opin Plant Biol 2015;27:67—76.

[133] Liu J, He Y, Amasino R, Chen X. siRNAs targeting an intronic transposon in the regulation of natural flowering behavior in Arabidopsis. Genes Dev 2004;18:2873—8.

[134] Zhou J, Liu L, Li Q, Xu W, Li K, Wang ZW, et al. Intronic heterochromatin prevents cryptic transcription initiation in Arabidopsis. Plant J 2020;101:1185—97.

[135] Ito T, Nishio H, Tarutani Y, Emura N, Honjo MN, Toyoda A, et al. Seasonal stability and dynamics of DNA methylation in plants in a natural environment. Genes 2019;10:544.

[136] Niu X-M, Xu Y-C, Li Z-W, Bian Y-T, Hou X-H, Chen J-F, et al. Transposable elements drive rapid phenotypic variation in *Capsella rubella*. Proc Natl Acad Sci 2019;116:6908—13.

[137] Khan AR, Enjalbert J, Marsollier A-C, Rousselet A, Goldringer I, Vitte C. Vernalization treatment induces site-specific DNA hypermethylation at the VERNALIZATION-A1 (VRN-A1) locus in hexaploid winter wheat. BMC Plant Biol 2013;13:1—16.

[138] Saze H, Kakutani T. Heritable epigenetic mutation of a transposon-flanked Arabidopsis gene due to lack of the chromatin-remodeling factor DDM1. EMBO J 2007;26:3641—52.

[139] Tsuchiya T, Eulgem T. An alternative polyadenylation mechanism coopted to the Arabidopsis RPP7 gene through intronic retrotransposon domestication. Proc Natl Acad Sci 2013;110:E3535—43.

[140] Deremetz A, Le Roux C, Idir Y, Brousse C, Agorio A, Gy I, et al. Antagonistic actions of FPA and IBM2 regulate transcript processing from genes containing heterochromatin. Plant Physiol 2019;180:392—403.

[141] Parker MT, Knop K, Zacharaki V, Sherwood AV, Tome D, Yu X, et al. Widespread premature transcription termination of *Arabidopsis thaliana* NLR genes by the spen protein FPA. Elife 2021;10:e65537.

[142] Ong-Abdullah M, Ordway JM, Jiang N, Ooi S-E, Kok S-Y, Sarpan N, et al. Loss of Karma transposon methylation underlies the mantled somaclonal variant of oil palm. Nature 2015;525:533—7.

[143] Friedrich T, Faivre L, Bäurle I, Schubert D. Chromatin-based mechanisms of temperature memory in plants. Plant Cell Environ 2019;42:762—70.

[144] Verhoeven KJ, Jansen JJ, van Dijk PJ, Biere A. Stress-induced DNA methylation changes and their heritability in asexual dandelions. New Phytol 2010;185:1108—18.

[145] Song Y, Ci D, Tian M, Zhang D. Stable methylation of a non-coding RNA gene regulates gene expression in response to abiotic stress in Populus simonii. J Exp Bot 2016;67:1477—92.

[146] Liu J, Feng L, Li J, He Z. Genetic and epigenetic control of plant heat responses. Front Plant Sci 2015;6:267.

[147] Shen X, De Jonge J, Forsberg SK, Pettersson ME, Sheng Z, Hennig L, et al. Natural CMT2 variation is associated with genome-wide methylation changes and temperature seasonality. PLoS Genet 2014;10:e1004842.

[148] Lämke J, Brzezinka K, Altmann S, Bäurle I. A hit-and-run heat shock factor governs sustained histone methylation and transcriptional stress memory. EMBO J 2016;35:162—75.

[149] Yang C, Shen W, Chen H, Chu L, Xu Y, Zhou X, et al. Characterization and subcellular localization of histone deacetylases and their roles in response to abiotic stresses in soybean. BMC Plant Biol 2018;18:226.

[150] Sanchez DH, Paszkowski J. Heat-induced release of epigenetic silencing reveals the concealed role of an imprinted plant gene. PLoS Genet 2014;10:e1004806.

[151] Chang Y-N, Zhu C, Jiang J, Zhang H, Zhu J-K, Duan C-G. Epigenetic regulation in plant abiotic stress responses. J Integr Plant Biol 2020;62:563–80.

[152] Naydenov M, Baev V, Apostolova E, Gospodinova N, Sablok G, Gozmanova M, et al. High-temperature effect on genes engaged in DNA methylation and affected by DNA methylation in Arabidopsis. Plant Physiol Biochem 2015;87:102–8.

[153] Folsom JJ, Begcy K, Hao X, Wang D, Walia H. Rice Fertilization-Independent Endosperm1 regulates seed size under heat stress by controlling early endosperm development. Plant Physiol 2014;165:238.

[154] Kim JM, Sasaki T, Ueda M, Sako K, Seki M. Chromatin changes in response to drought, salinity, heat, and cold stresses in plants. Front Plant Sci 2015;6:114.

[155] Ashapkin VV, Kutueva LI, Aleksandrushkina NI, Vanyushin BF. Epigenetic mechanisms of plant adaptation to biotic and abiotic stresses. Int J Mol Sci 2020;21.

[156] Zhu J-K. Abiotic stress signaling and responses in plants. Cell 2016;167:313–24.

[157] Yang H, Berry S, Olsson TSG, Hartley M, Howard M, Dean C. Distinct phases of Polycomb silencing to hold epigenetic memory of cold in *Arabidopsis*. Science 2017;357:1142.

[158] Yang R, Hong Y, Ren Z, Tang K, Zhang H, Zhu J-K, et al. A Role for PICKLE in the regulation of cold and salt stress tolerance in Arabidopsis. Front Plant Sci 2019;10:900.

[159] Park J, Lim CJ, Shen M, Park HJ, Cha JY, Iniesto E, et al. Epigenetic switch from repressive to permissive chromatin in response to cold stress. Proc Natl Acad Sci USA 2018;115: e5400–9.

[160] Hu Y, Zhang L, Zhao L, Li J, He S, Zhou K, et al. Trichostatin A selectively suppresses the cold-induced transcription of the ZmDREB1 gene in maize. PLoS One 2011;6:e22132.

[161] Saraswat S, Yadav AK, Sirohi P, Singh NK. Role of epigenetics in crop improvement: water and heat stress. J Plant Biol 2017;60:231–40.

[162] Pavangadkar K, Thomashow MF, Triezenberg SJ. Histone dynamics and roles of histone acetyltransferases during cold-induced gene regulation in Arabidopsis. Plant Mol Biol 2010;74:183–200.

[163] Ferreira LJ, Azevedo V, Maroco J, Oliveira MM, Santos AP. Salt tolerant and sensitive rice varieties display differential methylome flexibility under salt stress. PLoS One 2015;10: e0124060.

[164] Kumar S, Beena AS, Awana M, Singh A. Salt-induced tissue-specific cytosine methylation downregulates expression of HKT genes in contrasting wheat (*Triticum aestivum* L.) genotypes. DNA Cell Biol 2017;36:283–94.

[165] Rus A, Yokoi S, Sharkhuu A, Reddy M, Lee BH, Matsumoto TK, et al. AtHKT1 is a salt tolerance determinant that controls Na(+) entry into plant roots. Proc Natl Acad Sci USA 2001;98:14150–5.

[166] Baek D, Jiang J, Chung JS, Wang B, Chen J, Xin Z, et al. Regulated AtHKT1 gene expression by a distal enhancer element and DNA methylation in the promoter plays an important role in salt tolerance. Plant Cell Physiol 2011;52:149–61.

[167] Sani E, Herzyk P, Perrella G, Colot V, Amtmann A. Hyperosmotic priming of Arabidopsis seedlings establishes a long-term somatic memory accompanied by specific changes of the epigenome. Genome Biol 2013;14:R59.

[168] Kumar S, Beena AS, Awana M, Singh A. Physiological, biochemical, epigenetic and molecular analyses of wheat (*Triticum aestivum*) genotypes with contrasting salt tolerance. Front Plant Sci 2017;8:1151.

[169] Xu R, Wang Y, Zheng H, Lu W, Wu C, Huang J, et al. Salt-induced transcription factor MYB74 is regulated by the RNA-directed DNA methylation pathway in Arabidopsis. J Exp Bot 2015;66:5997–6008.

[170] Yolcu S, Ozdemir F, Güler A, Bor M. Histone acetylation influences the transcriptional activation of POX in *Beta vulgaris* L. and *Beta maritima* L. under salt stress. Plant Physiol Biochem 2016;100:37–46.

[171] Kaldis A, Tsementzi D, Tanriverdi O, Vlachonasios KE. *Arabidopsis thaliana* transcriptional co-activators ADA2b and SGF29a are implicated in salt stress responses. Planta 2011;233: 749–62.

[172] Sokol A, Kwiatkowska A, Jerzmanowski A, Prymakowska-Bosak M. Up-regulation of stress-inducible genes in tobacco and Arabidopsis cells in response to abiotic stresses and ABA treatment correlates with dynamic changes in histone H3 and H4 modifications. Planta 2007;227:245–54.

[173] Chen LT, Luo M, Wang YY, Wu K. Involvement of Arabidopsis histone deacetylase HDA6 in ABA and salt stress response. J Exp Bot 2010;61:3345–53.

[174] Pandey G, Sharma N, Sahu PP, Prasad M. Chromatin-based epigenetic regulation of plant abiotic stress response. Curr Genomics 2016;17:490–8.

[175] Shinozaki K, Yamaguchi-Shinozaki K. Gene networks involved in drought stress response and tolerance. J Exp Bot 2007; 58:221–7.

[176] Zhu H, Xie W, Xu D, Miki D, Tang K, Huang C-F, et al. DNA demethylase ROS1 negatively regulates the imprinting of DOGL4 and seed dormancy in *Arabidopsis thaliana*. Proc Natl Acad Sci 2018;115:E9962.

[177] Mehdi S, Derkacheva M, Ramström M, Kralemann L, Bergquist J, Hennig L. The WD40 domain protein MSI1 functions in a histone deacetylase complex to fine-tune abscisic acid signaling. Plant Cell 2016;28:42.

[178] Ding Y, Avramova Z, Fromm M. The Arabidopsis trithorax-like factor ATX1 functions in dehydration stress responses via ABA-dependent and ABA-independent pathways. Plant J 2011;66:735–44.

[179] Kim JM, To TK, Ishida J, Morosawa T, Kawashima M, Matsui A, et al. Alterations of lysine modifications on the histone H3 N-tail under drought stress conditions in *Arabidopsis thaliana*. Plant Cell Physiol 2008;49:1580–8.

[180] Kim JM, To TK, Ishida J, Matsui A, Kimura H, Seki M. Transition of chromatin status during the process of recovery from drought stress in *Arabidopsis thaliana*. Plant Cell Physiol 2012;53:847–56.

[181] Ramirez-Prado JS, Latrasse D, Rodriguez-Granados NY, Huang Y, Manza-Mianza D, Brik-Chaouche R, et al. The Polycomb protein LHP1 regulates *Arabidopsis thaliana* stress responses through the repression of the MYC2-dependent branch of immunity. Plant J 2019;100:1118–31.

[182] Liang D, Zhang Z, Wu H, Huang C, Shuai P, Ye CY, et al. Single-base-resolution methylomes of Populus trichocarpa reveal the association between DNA methylation and drought stress. BMC Genet 2014;15(Suppl 1):S9.

[183] Wang W-S, Pan Y-J, Zhao X-Q, Dwivedi D, Zhu L-H, Ali J, et al. Drought-induced site-specific DNA methylation and its association with drought tolerance in rice (*Oryza sativa* L.). J Exp Bot 2011;62:1951–60.

[184] Gayacharan, Joel AJ. Epigenetic responses to drought stress in rice (*Oryza sativa* L.). Physiol Mol Biol Plants 2013;19: 379–87.

[185] Chen X, Xu X, Shen X, Li H, Zhu C, Chen R, et al. Genome-wide investigation of DNA methylation dynamics reveals a critical role of DNA demethylation during the early somatic embryogenesis of Dimocarpus longan Lour. Tree Physiol 2020; 40:1807–26.

[186] Ranf S. Pattern recognition receptors—versatile genetic tools for engineering broad-spectrum disease resistance in crops. Agronomy 2018;8:134.

[187] Lu Y, Tsuda K. Intimate association of PRR- and NLR-mediated signaling in plant immunity. Mol Plant Microbe Interact 2021;34:3–14.

[188] Berens ML, Berry HM, Mine A, Argueso CT, Tsuda K. Evolution of hormone signaling networks in plant defense. Annu Rev Phytopathol 2017;55:401–25.

[189] Zhang W, Zhao F, Jiang L, Chen C, Wu L, Liu Z. Different pathogen defense strategies in arabidopsis: more than pathogen recognition. Cells 2018;7:252.

[190] Snelders NC, Kettles GJ, Rudd JJ, Thomma BPHJ. Plant pathogen effector proteins as manipulators of host microbiomes? Mol Plant Pathol 2018;19:257–9.

[191] Nguyen Q-M, Iswanto ABB, Son GH, Kim SH. Recent advances in effector-triggered immunity in plants: new pieces in the puzzle create a different paradigm. Int J Mol Sci 2021;22:4709.

[192] Wilkinson SW, Magerøy MH, López Sánchez A, Smith LM, Furci L, Cotton TEA, et al. Surviving in a hostile world: plant strategies to resist pests and diseases. Annu Rev Phytopathol 2019;57:505–29.

[193] Hilker M, Schmülling T. Stress priming, memory, and signalling in plants. Plant Cell Environ 2019;42:753–61.

[194] Klessig DF, Choi HW, Dempsey DMA. Systemic acquired resistance and salicylic acid: past, present, and future. Mol Plant Microbe Interact 2018;31:871–88.

[195] Martinez-Medina A, Flors V, Heil M, Mauch-Mani B, Pieterse CMJ, Pozo MJ, et al. Recognizing plant defense priming. Trends Plant Sci 2016;21:818–22.

[196] Luna E, López A, Kooiman J, Ton J. Role of NPR1 and KYP in long-lasting induced resistance by β-aminobutyric acid. Front Plant Sci 2014;5:184.

[197] Rasmann S, De Vos M, Casteel CL, Tian D, Halitschke R, Sun JY, et al. Herbivory in the previous generation primes plants for enhanced insect resistance. Plant Physiol 2011;158:854–63.

[198] Luna E, Bruce TJ, Roberts MR, Flors V, Ton J. Next-generation systemic acquired resistance. Plant Physiol 2012;158:844–53.

[199] Quadrana L, Colot V. Plant transgenerational epigenetics. Annu Rev Genet 2016;50:467–91.

[200] Jaskiewicz M, Conrath U, Peterhänsel C. Chromatin modification acts as a memory for systemic acquired resistance in the plant stress response. EMBO Rep 2011;12:50–5.

[201] López A, Ramírez V, García-Andrade J, Flors V, Vera P. The RNA silencing enzyme RNA polymerase V is required for plant immunity. PLoS Genet 2011;7:e1002434.

[202] Schillheim B, Jansen I, Baum S, Beesley A, Bolm C, Conrath U. Sulforaphane modifies histone H3, unpacks chromatin, and primes defense. Plant Physiol 2018;176:2395–405.

[203] Baum S, Reimer-Michalski E-M, Bolger A, Mantai AJ, Benes V, Usadel B, et al. Isolation of open chromatin identifies regulators of systemic acquired resistance1. Plant Physiol 2019;181:817–33.

[204] Core L, Adelman K. Promoter-proximal pausing of RNA polymerase II: a nexus of gene regulation. Genes Dev 2019;33:960–82.

[205] Iwasaki M. Chromatin resetting mechanisms preventing transgenerational inheritance of epigenetic states. Front Plant Sci 2015;6.

[206] Chen W, Kong J, Lai T, Manning K, Wu C, Wang Y, et al. Tuning LeSPL-CNR expression by SlymiR157 affects tomato fruit ripening. Sci Rep 2015;5:7852.

[207] Miura K, Agetsuma M, Kitano H, Yoshimura A, Matsuoka M, Jacobsen SE, et al. A metastable DWARF1 epigenetic mutant affecting plant stature in rice. Proc Natl Acad Sci USA 2009;106:11218–23.

[208] Durand S, Bouché N, Perez Strand E, Loudet O, Camilleri C. Rapid establishment of genetic incompatibility through natural epigenetic variation. Curr Biol 2012;22:326–31.

[209] López Sánchez A, Stassen JHM, Furci L, Smith LM, Ton J. The role of DNA (de)methylation in immune responsiveness of Arabidopsis. Plant J 2016;88:361–74.

[210] Stassen JHM, López A, Jain R, Pascual-Pardo D, Luna E, Smith LM, et al. The relationship between transgenerational acquired resistance and global DNA methylation in Arabidopsis. Sci Rep 2018;8:14761.

[211] López Sánchez A, Pascual-Pardo D, Furci L, Roberts MR, Ton J. Costs and benefits of transgenerational induced resistance in Arabidopsis. Front Plant Sci 2021;12:644999.

[212] Reinders J, Wulff BBH, Mirouze M, Marí-Ordóñez A, Dapp M, Rozhon W, et al. Compromised stability of DNA methylation and transposon immobilization in mosaic Arabidopsis epigenomes. Genes Dev 2009;23:939–50.

[213] Johannes F, Porcher E, Teixeira FK, Saliba-Colombani V, Simon M, Agier N, et al. Assessing the impact of transgenerational epigenetic variation on complex traits. PLoS Genet 2009;5:e1000530.

[214] Latzel V, Allan E, Bortolini Silveira A, Colot V, Fischer M, Bossdorf O. Epigenetic diversity increases the productivity and stability of plant populations. Nat Commun 2013;4:2875.

[215] Furci L, Jain R, Stassen J, Berkowitz O, Whelan J, Roquis D, et al. Identification and characterisation of hypomethylated DNA loci controlling quantitative resistance in Arabidopsis. eLife 2019;8:e40655.

[216] Liégard B, Baillet V, Etcheverry M, Joseph E, Lariagon C, Lemoine J, et al. Quantitative resistance to clubroot infection mediated by transgenerational epigenetic variation in Arabidopsis. New Phytologist 2019;222:468–79.

[217] Krasileva KV. The role of transposable elements and DNA damage repair mechanisms in gene duplications and gene fusions in plant genomes. Curr Opin Plant Biol 2019;48:18–25.

[218] Feng S, Zhong Z, Wang M, Jacobsen SE. Efficient and accurate determination of genome-wide DNA methylation patterns in Arabidopsis thaliana with enzymatic methyl sequencing. Epigenetics Chromatin 2020;13:42.

[219] Cambiagno DA, Torres JR, Alvarez ME. Convergent epigenetic mechanisms avoid constitutive expression of immune receptor gene subsets. Front Plant Sci 2021;12:703667.

[220] Stuart T, Satija R. Integrative single-cell analysis. Nat Rev Genet 2019;20:257–72.

[221] Satterlee JW, Strable J, Scanlon MJ. Plant stem-cell organization and differentiation at single-cell resolution. Proc Natl Acad Sci USA 2020;117:33689–99.

[222] Shaw R, Tian X, Xu J. Single-cell transcriptome analysis in plants: advances and challenges. Mol Plant 2021;14:115–26.

[223] Zhang TQ, Chen Y, Wang JW. A single-cell analysis of the Arabidopsis vegetative shoot apex. Dev Cell 2021;56:1056–74 e1058.

[224] Cokus SJ, Feng S, Zhang X, Chen Z, Merriman B, Haudenschild CD, et al. Shotgun bisulphite sequencing of the Arabidopsis genome reveals DNA methylation patterning. Nature 2008;452:215–19.

[225] Lister R, O'Malley RC, Tonti-Filippini J, Gregory BD, Berry CC, Millar AH, et al. Highly integrated single-base resolution maps of the epigenome in Arabidopsis. Cell 2008;133:523–36.

[226] Meissner A, Gnirke A, Bell GW, Ramsahoye B, Lander ES, Jaenisch R. Reduced representation bisulfite sequencing for comparative high-resolution DNA methylation analysis. Nucleic Acids Res 2005;33:5868–77.

[227] Guo H, Zhu P, Wu X, Li X, Wen L, Tang F. Single-cell methylome landscapes of mouse embryonic stem cells and early

embryos analyzed using reduced representation bisulfite sequencing. Genome Res 2013;23:2126–35.

[228] Li X, Chen L, Zhang Q, Sun Y, Li Q, Yan J. BRIF-Seq: bisulfite-converted randomly integrated fragments sequencing at the single-cell level. Mol Plant 2019;12:438–46.

[229] Mulqueen RM, Pokholok D, Norberg SJ, Torkenczy KA, Fields AJ, Sun D, et al. Highly scalable generation of DNA methylation profiles in single cells. Nat Biotechnol 2018;36:428–31.

[230] Treangen TJ, Salzberg SL. Repetitive DNA and next-generation sequencing: computational challenges and solutions. Nat Rev Genet 2011;13:36–46.

[231] Laszlo AH, Derrington IM, Brinkerhoff H, Langford KW, Nova IC, Samson JM, et al. Detection and mapping of 5-methylcytosine and 5-hydroxymethylcytosine with nanopore MspA. Proc Natl Acad Sci USA 2013;110:18904–9.

[232] Ni HN, Nie P, Zhang F, Zhang J, Wu Z, Bai B, et al. Detection of cytosine methylations in plant from nanopore sequencing data using deep learning. bioRxiv 2021; 2021.02.07.430077.

[233] Buenrostro JD, Giresi PG, Zaba LC, Chang HY, Greenleaf WJ. Transposition of native chromatin for fast and sensitive epigenomic profiling of open chromatin, DNA-binding proteins and nucleosome position. Nat Methods 2013;10:1213–18.

[234] Deal RB, Henikoff S. The INTACT method for cell type-specific gene expression and chromatin profiling in *Arabidopsis thaliana*. Nat Protoc 2011;6:56–68.

[235] Long Y, Liu Z, Jia J, Mo W, Fang L, Lu D, et al. FlsnRNA-seq: protoplasting-free full-length single-nucleus RNA profiling in plants. Genome Biol 2021;22:66.

[236] Lu Z, Hofmeister BT, Vollmers C, DuBois RM, Schmitz RJ. Combining ATAC-seq with nuclei sorting for discovery of cis-regulatory regions in plant genomes. Nucleic Acids Res 2017;45:e41.

[237] Zhang W, Wu Y, Schnable JC, Zeng Z, Freeling M, Crawford GE, et al. High-resolution mapping of open chromatin in the rice genome. Genome Res 2012;22:151–62.

[238] Rodgers-Melnick E, Vera DL, Bass HW, Buckler ES. Open chromatin reveals the functional maize genome. Proc Natl Acad Sci USA 2016;113:E3177–84.

[239] Jin W, Tang Q, Wan M, Cui K, Zhang Y, Ren G, et al. Genome-wide detection of DNase I hypersensitive sites in single cells and FFPE tissue samples. Nature 2015;528:142–6.

[240] Buenrostro JD, Wu B, Litzenburger UM, Ruff D, Gonzales ML, Snyder MP, et al. Single-cell chromatin accessibility reveals principles of regulatory variation. Nature 2015;523:486–90.

[241] Chen X, Miragaia RJ, Natarajan KN, Teichmann SA. A rapid and robust method for single cell chromatin accessibility profiling. Nat Commun 2018;9:5345.

[242] Cusanovich DA, Daza R, Adey A, Pliner HA, Christiansen L, Gunderson KL, et al. Multiplex single cell profiling of chromatin accessibility by combinatorial cellular indexing. Science 2015;348:910–14.

[243] Lake BB, Chen S, Sos BC, Fan J, Kaeser GE, Yung YC, et al. Integrative single-cell analysis of transcriptional and epigenetic states in the human adult brain. Nat Biotechnol 2018;36:70–80.

[244] Lai B, Gao W, Cui K, Xie W, Tang Q, Jin W, et al. Principles of nucleosome organization revealed by single-cell micrococcal nuclease sequencing. Nature 2018;562:281–5.

[245] Farmer A, Thibivilliers S, Ryu KH, Schiefelbein J, Libault M. Single-nucleus RNA and ATAC sequencing reveals the impact of chromatin accessibility on gene expression in Arabidopsis roots at the single-cell level. Mol Plant 2021;14:372–83.

[246] Skene PJ, Henikoff S. An efficient targeted nuclease strategy for high-resolution mapping of DNA binding sites. Elife 2017;6.

[247] Dahl JA, Jung I, Aanes H, Greggains GD, Manaf A, Lerdrup M, et al. Broad histone H3K4me3 domains in mouse oocytes modulate maternal-to-zygotic transition. Nature 2016;537:548–52.

[248] Zhang B, Zheng H, Huang B, Li W, Xiang Y, Peng X, et al. Allelic reprogramming of the histone modification H3K4me3 in early mammalian development. Nature 2016;537:553–7.

[249] Liu X, Wang C, Liu W, Li J, Li C, Kou X, et al. Distinct features of H3K4me3 and H3K27me3 chromatin domains in pre-implantation embryos. Nature 2016;537:558–62.

[250] Ai S, Xiong H, Li CC, Luo Y, Shi Q, Liu Y, et al. Profiling chromatin states using single-cell itChIP-seq. Nat Cell Biol 2019;21:1164–72.

[251] Hainer SJ, Boskovic A, McCannell KN, Rando OJ, Fazzio TG. Profiling of pluripotency factors in single cells and early embryos. Cell 2019;177:1319–29 e1311.

[252] Patty BJ, Hainer SJ. Transcription factor chromatin profiling genome-wide using uliCUT&RUN in single cells and individual blastocysts. Nat Protoc 2021;16:2633–66.

[253] Padmarasu S, Himmelbach A, Mascher M, Stein N. In Situ Hi-C for plants: an improved method to detect long-range chromatin interactions. Methods Mol Biol 2019;1933:441–72.

[254] Pei L, Li G, Lindsey K, Zhang X, Wang M. Plant 3D genomics: the exploration and application of chromatin organization. New Phytol 2021;230:1772–86.

[255] Nagano T, Lubling Y, Stevens TJ, Schoenfelder S, Yaffe E, Dean W, et al. Single-cell Hi-C reveals cell-to-cell variability in chromosome structure. Nature 2013;502:59–64.

[256] Stevens TJ, Lando D, Basu S, Atkinson LP, Cao Y, Lee SF, et al. 3D structures of individual mammalian genomes studied by single-cell Hi-C. Nature 2017;544:59–64.

[257] Flyamer IM, Gassler J, Imakaev M, Brandao HB, Ulianov SV, Abdennur N, et al. Single-nucleus Hi-C reveals unique chromatin reorganization at oocyte-to-zygote transition. Nature 2017;544:110–14.

[258] Zhou S, Jiang W, Zhao Y, Zhou DX. Single-cell three-dimensional genome structures of rice gametes and unicellular zygotes. Nat Plants 2019;5:795–800.

[259] Liang Z, Zhang Q, Ji C, Hu G, Zhang P, Wang Y, et al. Reorganization of the 3D chromatin architecture of rice genomes during heat stress. BMC Biol 2021;19:53.

[260] Di Stefano M, Nutzmann HW, Marti-Renom MA, Jost D. Polymer modelling unveils the roles of heterochromatin and nucleolar organizing regions in shaping 3D genome organization in *Arabidopsis thaliana*. Nucleic Acids Res 2021;49:1840–58.

[261] Ramani V, Deng X, Qiu R, Gunderson KL, Steemers FJ, Disteche CM, et al. Massively multiplex single-cell Hi-C. Nat Methods 2017;14:263–6.

[262] Tan L, Xing D, Chang CH, Li H, Xie XS. Three-dimensional genome structures of single diploid human cells. Science 2018;361:924–8.

[263] Payne AC, Chiang ZD, Reginato PL, Mangiameli SM, Murray EM, Yao CC, et al. In situ genome sequencing resolves DNA sequence and structure in intact biological samples. Science 2021;371.

[264] Larsson L, Frisen J, Lundeberg J. Spatially resolved transcriptomics adds a new dimension to genomics. Nat Methods 2021;18:15–18.

[265] Qi LS, Larson MH, Gilbert LA, Doudna JA, Weissman JS, Arkin AP, et al. Repurposing CRISPR as an RNA-guided platform for sequence-specific control of gene expression. Cell 2013;152:1173–83.

[266] Zhu H, Li C, Gao C. Applications of CRISPR-Cas in agriculture and plant biotechnology. Nat Rev Mol Cell Biol 2020;21:661–77.

[267] Hilton IB, D'Ippolito AM, Vockley CM, Thakore PI, Crawford GE, Reddy TE, et al. Epigenome editing by a CRISPR-Cas9-based acetyltransferase activates genes from promoters and enhancers. Nat Biotechnol 2015;33:510–17.

[268] Zhang Y, Ma X, Xie X, Liu YG. CRISPR/Cas9-based genome editing in plants. Prog Mol Biol Transl Sci 2017;149:133–50.

[269] Liu C, Moschou PN. Phenotypic novelty by CRISPR in plants. Dev Biol 2018;435:170–5.

[270] Wada N, Ueta R, Osakabe Y, Osakabe K. Precision genome editing in plants: state-of-the-art in CRISPR/Cas9-based genome engineering. BMC Plant Biol 2020;20:234.

[271] Pulecio J, Verma N, Mejia-Ramirez E, Huangfu D, Raya A. CRISPR/Cas9-based engineering of the epigenome. Cell Stem Cell 2017;21:431–47.

[272] Moradpour M, Abdulah SNA. CRISPR/dCas9 platforms in plants: strategies and applications beyond genome editing. Plant Biotechnol J 2020;18:32–44.

[273] Didovyk A, Borek B, Tsimring L, Hasty J. Transcriptional regulation with CRISPR-Cas9: principles, advances, and applications. Curr Opin Biotechnol 2016;40:177–84.

[274] Li Z, Zhang D, Xiong X, Yan B, Xie W, Sheen J, et al. A potent Cas9-derived gene activator for plant and mammalian cells. Nat Plants 2017;3:930–6.

[275] Spille JH, Hecht M, Grube V, Cho WK, Lee C, Cisse II. A CRISPR/Cas9 platform for MS2-labelling of single mRNA in live stem cells. Methods 2019;153:35–45.

[276] Li C, Zong Y, Jin S, Zhu H, Lin D, Li S, et al. SWISS: multiplexed orthogonal genome editing in plants with a Cas9 nickase and engineered CRISPR RNA scaffolds. Genome Biol 2020;21:141.

[277] Lee JE, Neumann M, Duro DI, Schmid M. CRISPR-based tools for targeted transcriptional and epigenetic regulation in plants. PLoS One 2019;14:e0222778.

[278] Adli M. The CRISPR tool kit for genome editing and beyond. Nat Commun 2018;9:1911.

[279] Piatek A, Ali Z, Baazim H, Li L, Abulfaraj A, Al-Shareef S, et al. RNA-guided transcriptional regulation in planta via synthetic dCas9-based transcription factors. Plant Biotechnol J 2015;13:578–89.

[280] Lowder LG, Zhou J, Zhang Y, Malzahn A, Zhong Z, Hsieh TF, et al. Robust transcriptional activation in plants using multiplexed CRISPR-Act2.0 and mTALE-act systems. Mol Plant 2018;11:245–56.

[281] Christie JM, Zurbriggen MD. Optogenetics in plants. New Phytol 2021;229:3108–15.

[282] de Mena L, Rizk P, Rincon-Limas DE. Bringing light to transcription: the optogenetics repertoire. Front Genet 2018;9:518.

[283] Muller K, Siegel D, Rodriguez Jahnke F, Gerrer K, Wend S, Decker EL, et al. A red light-controlled synthetic gene expression switch for plant systems. Mol Biosyst 2014;10:1679–88.

[284] Ochoa-Fernandez R, Abel NB, Wieland FG, Schlegel J, Koch LA, Miller JB, et al. Optogenetic control of gene expression in plants in the presence of ambient white light. Nat Methods 2020;17:717–25.

[285] Liu C, Weigel D. Chromatin in 3D: progress and prospects for plants. Genome Biol 2015;16:170.

[286] Morgan SL, Mariano NC, Bermudez A, Arruda NL, Wu F, Luo Y, et al. Manipulation of nuclear architecture through CRISPR-mediated chromosomal looping. Nat Commun 2017;8:15993.

[287] Rosin FM, Watanabe N, Cacas JL, Kato N, Arroyo JM, Fang Y, et al. Genome-wide transposon tagging reveals location-dependent effects on transcription and chromatin organization in Arabidopsis. Plant J 2008;55:514–25.

[288] Ma H, Naseri A, Reyes-Gutierrez P, Wolfe SA, Zhang S, Pederson T. Multicolor CRISPR labeling of chromosomal loci in human cells. Proc Natl Acad Sci USA 2015;112:3002–7.

[289] Dreissig S, Schiml S, Schindele P, Weiss O, Rutten T, Schubert V, et al. Live-cell CRISPR imaging in plants reveals dynamic telomere movements. Plant J 2017;91:565–73.

SECTION V

Factors Influencing Epigenetic Changes

16. Dynamic changes in epigenetic modifications during mammalian early embryo development
17. Epigenetic biomarkers
18. Transposable elements shaping the epigenome
19. Dietary and metabolic compounds affecting covalent histone modifications
20. Epigenetics, stem cells, cellular differentiation, and associated neurological disorders and brain cancer
21. Epigenetic regulation of skeletal muscle regeneration
22. Epigenetics of X-chromosome inactivation
23. Epigenetics of memory processes
24. Transgenerational epigenetics
25. DNA methylation clocks in age-related disease

CHAPTER 16

Dynamic Changes in Epigenetic Modifications During Mammalian Early Embryo Development

Jie Yang and Wei Jiang

Frontier Science Center for Immunology and Metabolism, Medical Research Institute, Wuhan University, Wuhan, P.R. China

OUTLINE

Introduction	289
DNA methylation during mammalian early embryo development	291
Dynamic changes of DNA methylation in early embryo development	291
The function of DNA methylation during mammalian embryonic development	291
Dynamic changes and function of histone modifications in early embryo development	293
H3K4 methylation dynamic changes in early embryo development	293
H3K4 methylation function in early embryo development	294
H3K9 methylation dynamic changes in early embryo development	294
H3K9 methylation function in early embryo development	295
H3K27 methylation dynamic changes in early embryo development	295
H3K27me3 regulates XCI and imprinting	296
H3K27me3 is necessary for mammalian cell lineage commitment	296
Histone acetylation dynamic changes in early embryo development	296
Histone acetylation and XCI, imprinting	297
Histone acetylation and embryo development	297
Histone phosphorylation dynamic changes in early embryo development	298
Histone phosphorylation function in early embryo development	298
Summary	298
References	299

INTRODUCTION

Mammalian development originates from the zygote formed through paternal sperm entering the maternal oocyte for fertilization. The one-cell zygote cleavage generates two-cell, four-cell, and eight-cell embryos sequentially. After a series of cleavage, the multicellular embryo is compacted into a morula, and the morula continues to progress and undergoes further compaction and cavitation, which promotes the development of blastocyst. The early blastocyst includes inner cell mass (ICM), a fluid-filled blastocyst cavity, and trophectoderm (Fig. 16.1). Subsequently, the ICM segregates into epiblast and primitive endoderm (PE): epiblast differentiates into three germ layers (ectoderm, mesoderm, and endoderm), and PE forms the yolk sac. The trophectoderm finally contributes to the fetal placenta. After the blastocyst implants into the uterine wall, the embryo undergoes further development and generates an individual

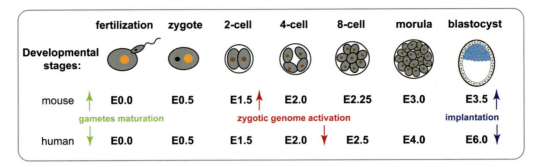

FIGURE 16.1 Schematic for early mammalian embryo development stages, mainly including the sperm and oocyte fusion, zygote, two-cell, four-cell, eight-cell embryo and blastocyst. The time of embryo development primarily refers to the literature [3].

organism [1,2]. During the early mammalian embryonic transition, there are a series of critical events to regulate embryo development properly, such as gametes maturation, fertilization, zygotic genome activation (ZGA), X chromosome inactivation (XCI), and genomic imprinting. Thus, it is vital to understand how these key events are regulated temporally and spatially during mammalian early development.

Epigenetic modifications are crucial in mammalian early embryonic development, from which DNA methylation and histone modifications are widely reported to participate in regulating developmental gene expression and the activation of the zygotic genome. DNA methylation takes place on the fifth position carbon of cytosine (5mC) in mammals and predominantly in CpG dinucleotides. DNA methylation is a heritable marker through semiconservation DNA replication [4,5]. Two different DNA methyltransferases, DNMT3 and DNMT1, are in charge of *de novo* DNA methylation and methylation maintenance during DNA replication, respectively. DNMT1 is the first reported mammalian DNA methyltransferase that maintains DNA methylation during cell division [6]. DNMT1 also possesses *de novo* DNA methylation ability, although this catalytic activity is repressed in the oocyte genome [7]. *De novo* DNA methylation is catalyzed mainly by DNMT3A and DNMT3B. DNMT3A and DNMT3B are functionally redundant and DNMT3B is mainly responsible for CpG islands methylation [8]. DNMT3L possesses a cysteine-rich region containing a C2C2 and an imperfect PHD zinc finger domain, showing high protein sequences similarity to DNMT3A and DNMT3B [9]. However, DNMT3L lacks the vital DNA cytosine catalytic motifs and is more likely to be a regulator of methylation and imprint establishment than to be a DNA methyltransferase [10]. Further study shows DNMT3L acts as a general stimulatory factor for *de novo* methylation by DNMT3A [11].

Moreover, multiple DNA methyltransferase isoforms are reported in different contexts. For instance, DNMT1s exists in somatic cells, DNMT1o and DNMT1p are specifically expressed in oocyte and pachytene spermatocytes separately [12]. DNA methylation is a dynamic transition utilizing methylation and demethylation. The ten-eleven translocation (TET) enzymes, TET1, TET2, and TET3, are widely expressed in vertebrates, which catalyzed 5mC oxidation to 5-hydroxymethyl cytosine (5hmc), and the 5hmC is further catalyzed and oxidized to 5-formylcytosin (5fC) and finally becomes 5-carboxylcytosin (5caC). Except for active demethylation by TET catalysis, passive demethylation is ongoing during DNA replication in cell division.

Histone modifications are also essential for regulating normal early embryo development and always in conjunction with DNA modification. Histone modifications occur at the histone tails of the nucleosome histone octamer and contain methylation, acetylation, phosphorylation, and other modifications. Histone methylation usually occurs at lysine (K) residues of H3 and H4 by histone methyltransferase and is erased by demethylase. There are three methylated states: mono-, di-, and tri-methylation, with active or repressive function in gene expression. H3K4me3, H3K36me3, and H3K79me3 are considered as active marks, while H3K27me3, H3K9me3, and H4K20me3 are repressive marks. Histone acetylation is added and erased by histone acetyltransferase (HAT) and histone deacetylase (HDAC) respectively. Well-studied HATs include at least five different subfamilies: KAT1, Gcn/PCAF, MYST, p300/CBP, and Rtt109 [13]. Histone acetylation creates open chromatin environment by carrying the negative charge to prevent interaction between histone tails and DNA [14,15]. Phosphorylation of histone occurs at serine (S) and threonine (T) residues of the histone tails. Histone phosphorylation regulates various biological processes, including cell cycle regulation, apoptosis, DNA repair, chromosome condensation, and gene expression. Phosphorylation of the Ser-10 of histone

H3 (H3S10ph) has primarily been reported in early mammalian development, which participates in mitotic chromosome condensation and decondensation. Aurora-B is a member of highly conserved serine/threonine protein kinase family responsible for H3S10ph during mitotic division [16]. Besides these histone modifications described above, histone ubiquitylation, sumoylation, and ADP-ribosylation are also functional for early embryo development and cellular processes regulation.

DNA METHYLATION DURING MAMMALIAN EARLY EMBRYO DEVELOPMENT

Dynamic changes of DNA methylation in early embryo development

DNA methylome with single-base resolution in mouse germ cells shows that sperm and oocyte are widely different in the extent and distribution of DNA methylation. Sperm genomes have almost complete coverage of DNA methylation (83.0%–89.4%) except CpG-rich regions, but the average methylation level of oocytes is less than half of sperm. The main difference between sperm and oocyte is largely contributed by specific retroviral element families [17,18]. However, both sperm and oocyte genomes are highly methylated compared with somatic cells.

The 5mC level of paternal and maternal chromosomes is almost equal at 3–6 hours of postfertilization. After 6–8 hours of fertilization, immunofluorescence results show that the mouse paternal genome undergoes significant and active demethylation, displaying a shallow methylation signal. The methylation of the maternal genome is maintained until the two-cell stage [19,20]. Single-cell epigenome sequencing results show that the median level of sperm methylation undergoes a significant decrease from 80.3% to 38.3% shortly after fertilization. Meanwhile, the maternal genome also shows mild demethylation from 32.4% to 27.8% [21]. After the four-cell stage, maternal methylation is gradually decreased by replication-dependent passive demethylation [19]. Loss of methylation during cleavage is mainly through passive demethylation in both parents. However, at least a significant proportion of maternal methylome also undergoes active demethylation during early embryonic development [22]. Further study indicates that Tet3 is enriched predominantly in the paternal pronucleus but also detectable in the maternal pronucleus, and thus Tet3-mediated DNA demethylation could partially contribute to the demethylation of the maternal genome [23].

During preimplantation development, genome-wide demethylation occurs, and the level of DNA methylation reaches the valley at the morula and blastocyst stage [12,24,25]. However, some specific loci are protected from demethylation, such as imprinted differentially methylated regions (DMRs) and some transposable elements [26–28]. During the blastocyst stage, remethylation takes place by *de novo* methylation, especially limited in ICM rather than trophectoderm [18,24]. Methylation after implantation restores as comparable to somatic cells. The global DNA methylation dynamics of the human embryo are very similar to mice [27–29]. The genome-wide demethylation occurs at the beginning of fertilization and completes at the two-cell stage in human early embryo development. The maternal genome is demethylated to a much lesser extent in human blastocysts than in mouse blastocysts because of increased imprinted DMRs in the human genome [26,28].

In brief, sperm genomes are almost completely methylated, and oocyte genomes are much less. After fertilization, paternal pronucleus undergoes significant active demethylation while the maternal pronucleus is mildly demethylated, and replication-dependent passive dilution also partially participates in the demethylation of both pronuclei during the zygotic stage. During embryo cleavage, the level of DNA methylation gradually decreases by replication-dependent passive demethylation and reaches the lowest level at the morula and blastocyst stage, but *de novo* methylation takes place at the blastocyst stage, especially in ICM. The degree of DNA methylation is comparable to somatic cells after implantation (Fig. 16.2A).

The function of DNA methylation during mammalian embryonic development

The dynamic of DNA methylation pattern during early embryonic development is related to gene expression, X chromosome activity and genomic imprinting. Abnormal development and embryonic lethality are found when introducing mutated methyltransferase genes into the germline of mice [8,39]: *Dnmt3a* homozygous mutant mice are runted and die around four weeks of age, and *Dnmt3b* homozygous mutant mice die before birth and the embryos show severe developmental defects [8].

X chromosome inactivation is a way to regulate developmental processes by dosage compensation of X-linked genes relative to XY males. The intact *Xist* gene is the master regulatory switch locus that initiates X chromosome inactivation [40]. Furthermore, alleles on the inactive X chromosome are tightly associated with methylation in promoter regions. *Xist* gene is highly expressed from the inactive X chromosome, and the 5′ end of *Xist* is completely demethylated on the

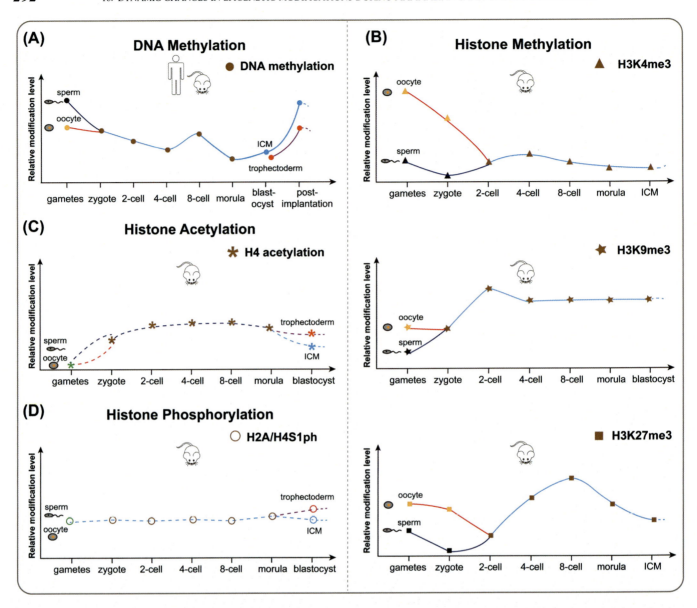

FIGURE 16.2 The dynamic changes of DNA methylation and histone modifications during mammalian early preimplantation development. The data is mainly from the DNA methylome sequencing results of Refs. [18,28,29] for DNA methylation, the ChIP-seq results of Refs. [30–32] for H3K4me3, the ChIP-seq results of Ref. [33] and the immunofluorescence results of Ref. [34] for H3K9me3, the ChIP-seq results of Ref. [35] and the immunofluorescence results of reference [34] for H3K27me3, the immunofluorescence results of Refs. [36,37] for H4 acetylation, and the immunofluorescence results of Ref. [38] for H2A/H4S1ph.

inactive X chromosome. DNA methyltransferase mutant male mouse embryo shows demethylation of *Xist* locus and increased *Xist* expression, which suggests DNA hypomethylation induces *Xist* expression [41,42]. In addition, *Xist* promotor is differentially methylated in oocytes but not sperm and maintained to the blastocyst stage, which is consistent with the paternal X chromosome inactivation that occurs at the blastocyst stage [43]. *Dnmt1*-deficient mouse embryo shows no effect on the imprinted X chromosome inactivation but an aberrant expression of other imprinted genes in the extraembryonic lineage. However, the random X chromosome inactivation is unstable in the *Dnmt1*-mutant embryonic lineage due to hypomethylation [44]. In addition, *Xist* RNA, histone deacetylation, and DNA methylation synergistically maintain the X chromosome inactive state [45].

Genomic imprinting is an epigenetic mark that causes a subset of mammalian genes to be expressed in a parental-origin-specific manner rather than from both chromosome homologs [46–48]. DNA methylation is involved in regulation of the expression of imprinting genes. In mouse embryos with mutated DNA methyltransferase, the paternal allele of imprinting gene *H19* is abnormally activated; whereas, typically expressed paternal allele of the *Igf-2* and maternal allele of the *Igf-2r* are repressed [49]. *Dnmt3l*, expressed during gametogenesis, is related to genomic imprinting. Bisulfite genomic sequencing of DNA from oocytes and embryos shows that *Dnmt3l* depletion prevents methylation of normally maternally methylated sequences. *Dnmt3l* homozygous mutant males are azoospermia, and homozygous females die before midgestation [10]. Furthermore, Dnmt3a but not Dnmt3b is involved in the initiation of genomic imprinting. *Dnmt3a* conditional mutant females are unable to establish the oocyte-specific imprintings at the maternally imprinted loci, and *Dnmt3a* conditional mutant males lack methylation at specific imprinting genes and show impaired spermatogenesis [50]. PGC7/Stella is a maternal factor required for protecting the maternal genome from DNA demethylation to maintain DNA methylation of imprinted loci. *PGC7/Stella*-deficient mice exhibit abnormal cleavage at two-four cell stages, and only 3% of embryos reach the blastocyst stage [51].

DNA methylation at the promoter region usually relates to repress gene expression, which is particularly important for imprinting genes expression, ZGA and retrotransposon silence. A study in mouse *Stella*-deficient oocyte shows that Stella controls the subcellular localization of UHRF1 (ubiquitin-like with PHD and RING finger domains 1), a critical protein in maintaining CG methylation by recruiting DNMT1 to hemi-methylated CG sites [52]. Stella safeguards the unique oocyte epigenome by regulating UHRF1 nuclear export to prevent interaction between UHRF1 and DNMT1. Loss of Stella in oocytes causes excessive DNA methylation at the whole genome. Consequently, a fraction of aberrant promoter hypermethylation is inherited by the two-cell-embryo stage and impairs ZGA [7]. Consistently, oocyte-specific *Uhrf1*-knockout mice show a significantly reduced level of CG methylation and developmental lethality at the blastocyst stage [52].

Tet3-deficient zygotes show blocked active demethylation in the paternal genome, hindering paternal *Oct4* and *Nanog* demethylation and delaying paternal *Oct4* expression in early embryos. *Tet3*-depleted germline in female mouse exhibits severely reduced reproductive capacity and increased abnormal development. However, maternal depletion of *Tet3* does not influence ZGA, and the embryos normally develop to the blastocyst stage [23,53]. The study of DNA methylome and transcriptomes in mouse germ cells indicates that sperm methylation on promoter negatively correlates with gene expression. Dnmt3l-dependent methylation on transcribed regions is required for maternal methylation imprinting and normal embryogenesis, and Dnmt3l-independent methylation on retroviral elements represses retrotransposons and is necessary for normal oogenesis [17]. By integrating reduced representation bisulfite sequencing (RRBS), whole-genome bisulfite sequencing (WGBS), and single-cell transcriptome data, further study shows DNA methylation level at promoter regions negatively correlated with gene expression, especially after ZGA at the eight-cell stage. Moreover, higher residual methylation retains at the evolutionarily younger and more active transposable elements [28].

In summary, DNA methylation is a common repressive modification to repress undesired gene expression, ensuring normal developmental process. DNA methylation in early mammalian development is involved in embryo development, X chromosome inactivation, imprinting gene expression, ZGA, and silencing of retrotransposons. The loss of methyltransferase in mouse germline causes abnormal development and embryonic lethality. Disruption of DNA methylation dynamic in zygote blocks paternal *Oct4* and *Nanog* expression and impairs ZGA. Hypermethylation in the germline of the female mouse causes severely reduced fertility and highly developmental failure probability.

DYNAMIC CHANGES AND FUNCTION OF HISTONE MODIFICATIONS IN EARLY EMBRYO DEVELOPMENT

H3K4 methylation dynamic changes in early embryo development

H3K4 mono- or tri-methylation (H3K4me1/H3K4me3) is ubiquitous in the maternal genome at all zygotic stages in mice, paternal pronucleus also stain H3K4me1, and both parental pronuclei show equal staining intensity at later pronucleus stage. H3K4me3 becomes detectable at 8–10 hours of postfertilization on paternal pronucleus, and shows no difference between paternal and maternal pronuclei at 12–16 hours of postfertilization [54].

With the optimization of chromatin immunoprecipitation followed by sequencing (ChIP-seq), H3K4me3 is further depicted genome-wide during early embryo development. Paternal H3K4me3 is widely reprogrammed upon fertilization, depleted in zygotes but reappears at the late two-cell stage, and H3K4me3 peaks on the promoter and global H3K4me3 of the paternal allele in zygote are different from sperm. Growing oocyte largely remains a canonical H3K4me3 pattern but full-grown and metaphase II (MII) oocytes show a non-canonical H3K4me3 pattern. Non-canonical H3K4me3 (ncH3K4me3) or broad H3K4me3 domain is first defined in metaphase II oocytes, which spans more than 10 kb DNA region and exists as broad peaks at promoters and a large number of distal loci. However, canonical H3K4me3 peaks are typically sharp and restricted to CpG-rich regions of the promoter. Non-canonical H3K4me3 accounts for ~22% of oocytes genome and passes on the maternal allele of zygote and two-cell embryo, and then is rapidly reduced in two-cell and almost erased by the four-cell stage owing to the active demethylation mediated by demethylases KDM5A and KDM5B. However, some of the H3K4me3 signals at transcriptional-start-site-containing domains are specifically maintained in the two-cell and eight-cell embryo. Meanwhile, canonical promoter H3K4me3 increases dramatically from the late two-cell stage and is maintained in the following stages until the blastocyst stage. The paternal H3K4me3 is weaker than the maternal in ICM but becomes comparable between parents in postimplantation embryos [30–32,35].

H3K4 methylation function in early embryo development

Kdm5A/B catalyzed active removal of broad H3K4me3 (non-canonical H3K4me3) in the two-cell stage is required for normal ZGA and also essential for early embryo development [31,32]. Stable H3K4me3, a subset of H3K4me3 that existed during all eight stages of embryo development (mouse MII oocyte, sperm, zygote, early two-cell, two-cell, four-cell, eight-cell embryo, and ICM), is enriched in higher chromatin architecture and related to higher gene expression level and transcriptional initiation during embryonic development [30]. Reduction of H3K4 methylation by *MLL2* knockdown in mouse oocytes results in anovulation and oocyte death, abnormal pro-apoptotic factors expression, and compromise of ZGA [55]. Loss of H3K4 methylation established by *MLL3/4* decreases the level of minor ZGA predominantly in the paternal pronucleus. Moreover, depletion of H3K4 methylation by overexpressing the H3K4M mutant (single K-M mutant) before fertilization significantly decreases the embryo survival rate [56]. Recent reports show that broad domains of H3K4me3 can be spread by aberrant H3K9me3 in *Kdm4A*-deficient oocytes, causing insufficient gene transcription activation, including endogenous retroviral elements and chimeric transcripts initiated from long terminal repeats during ZGA [57]. Remodeling of H3K4 methylation is also necessary to form paternal lamina-associated domains (LADs), which are thought to help organize chromosomes inside the nucleus and are associated with gene repression [58,59].

Taken together, H3K4me1 and H3K4me3 are ubiquitous in the maternal genome at all zygotic stages. H3K4me3 becomes detectable later than H3K4me1 at paternal pronucleus. However, the dynamic changes of H3K4me1 and H3K4me2 before preimplantation are still unclear and less reported. Paternal H3K4me3 is depleted in the zygote but reappears at the late two-cell stage. Maternal H3K4me3 is rapidly reduced in two-cell and eight-cell stage embryos. Non-canonical H3K4me3 reduces while canonical H3K4me3 takes place in the late two-cell embryos. Maternal H3K4me3 is stronger than paternal in ICM and becomes equal in both genomes in postimplantation embryos (Fig. 16.2B). H3K4me3 is a classical active marker, and deficiency of H3K4me3 causes insufficient gene transcription activation, abnormal pro-apoptotic factors expression, and compromise of ZGA in early mammalian development.

H3K9 methylation dynamic changes in early embryo development

Antibody assays of H3K9 di-methylation (H3K9me2) show that the oocyte is at a high level before fertilization and maintains such methylated state after fertilization, while the level of H3K9me2 on maternal pronucleus decreases passively during DNA replication at the first cleavage. The paternal pronucleus shows no staining of H3K9me2 and H3K9me3 but is positive for H3K9me1 immediately after fertilization. H3K9me2 is firstly detected in the paternal genome following full maturation of the pronuclei [60]. The asymmetric H3K9me2 in both parents persists until the late two-cell stage. At the four-cell stage, the asymmetric H3K9me2 is lost, and uniform methylation fluorescence is observed. The level of H3K9me2 further increases from the four-cell stage. At the blastocyst stage, the level of H3K9me2 is comparable between the ICM and the trophectoderm [36,61].

H3K9me3 staining shows that the H3K9me3 signal in mature sperm is detectable, and the metaphase II oocyte also stains strong H3K9me3. After fertilization, the paternal genome shows no immunocytologically detectable H3K9me3, while the polar body and maternal genome is H3K9me3-enriched. The level of H3K9me3 becomes indistinguishable on the parental chromosomes after the two-cell stage [62,63]. Genome-wide H3K9me3 profiling in mice shows that both parental genomes undergo large-scale reprogramming of H3K9me3 after fertilization. Only 14.4% of sperm H3K9me3 domains and 42.1% oocyte H3K9me3 domains are inherited during fertilization. The maternal and paternal H3K9me3 domains mostly are *de novo* established in the zygote, and there are more maternal-specific H3K9me3 regions than paternal from the zygote stage to the ICM stage. Global H3K9me3 dramatically increases at the two-cell stage, but a slight decline is observed at the four-cell stage. The level of H3K9me3 is stable from the four-cell stage to the morula stage. At the blastocyst, H3K9me3 in the trophectoderm is slightly higher than ICM. In addition, H3K9me3 is significantly enriched at long terminal repeats, increasing gradually after fertilization and persisting at a high level in preimplantation embryos. Of note, the H3K9me3 domains in the promoters disappear after fertilization and restore after implantation [33].

Recently, the dynamic change of H3K9me2 has been described in bovine early embryos. H3K9me2 is firstly detectable on female chromatin from 2 hours of postfertilization and keeps a similar intensity until 18 hours of postfertilization. The male H3K9me2 appears from 8 hours of postfertilization and becomes comparable to the female genome from 12 hours of postfertilization. The average level of H3K9me2 gradually increases from the two-cell stage to blastocysts. Both trophectoderm and ICM possess the equal intensity of H3K9me2 [64].

H3K9 methylation function in early embryo development

H3K9 methylation is a repressive marker and specifically marks facultative heterochromatin, which participates in gene silencing and assembly of heterochromatin. H3K9 methylation is also associated with X chromosome inactivation. H3K9 methylation occurs on the inactive X chromosome immediately after *Xist* RNA coating [65–67]. Except for the histone methyltransferase Suv39h family, G9a is identified as a candidate for H3K9 methylation in non-heterochromatic loci. *G9a*-deficient mice show strong embryonic lethality and no homozygous mutant embryo survived. A similar phenotype occurs in *GLP*-deficient mice, a *G9a*-related methyltransferase (GLP/Eu-HMTase1). *GLP*-deficiency in mice causes a severe reduction of H3K9me1 and H3K9me2 on euchromatin and shows embryonic lethality phenotype. These reports suggest euchromatin H3K9 methylation is essential for early embryogenesis and regulating transcriptional repression of developmental genes [68,69].

H3K9 methylation also participates in irreversible epigenetic inactivation of *Oct3/4* during early embryogenesis. Depleting *G9a* in oocytes reduces H3K9me2 enrichment and results in abnormal chromosome segregation, developmental delay, and frequent arrest of cleavage stage in embryos [70]. Following implantation, genes expressed during gametogenesis suffer a multi-step inactivation. Remarkably, the *Oct3/4* promoters undergo a pronounced increase in both H3K9me2 and H3K9me3 catalyzed by G9a, and H3K9 methylation can then bind HP1, and further recruit DNA methyltransferase with other effectors to *de novo* methylate *Oct3/4* promoters. Such epigenetic changes eventually inhibit *Oct3/4* re-expression and prevent reprogramming during early embryogenesis [71]. In addition, asymmetric H3K9 methylation may function as a precursor for asymmetric DNA methylation in early embryo development. Further mechanistic study shows that Stella can bind to H3K9me2-modified maternal chromatin and protect from 5mC to 5hmC catalyzed by Tet3. H3K9me2-modified imprinting loci are protected by Stella binding in mature sperm and early embryogenesis [61,72].

Collectively, the oocyte is highly enriched with H3K9me2 and H3K9me3 modification. Maternal H3K9me2 decreases during the first cleavage, and paternal H3K9me2 is undetectable until the late two-cell stage. H3K9me2 in both parents shows a consistent level from the four-cell stage and increases onward. However, H3K9me3 in both parental genomes suffers large-scale reconfiguration after fertilization, and the imbalance of H3K9me3 signal between maternal and paternal persists till blastocyst (Fig. 16.2B). Functionally, H3K9 methylation participates in XCI maintenance, early embryo development, irreversible epigenetic inactivation, chromosome segregation, and DNA methylation protection.

H3K27 methylation dynamic changes in early embryo development

During early mouse embryo development, mono-, di-, and tri-methylation on H3K27 (H3K27me1/2/3) are strongly associated with the maternal pronucleus. Immediately on fertilization, H3K27me1 is observed in the metaphase chromosomes of the oocyte and the decondensed sperm nucleus. H3K27me2 prominently stains in maternal pronucleus post gamete fusion, but paternal pronucleus is only weakly stained at the late zygote stage. The level of H3K27me2 is statistically increased from the two-cell to eight-cell stage and reaches the highest level in the early blastocysts [34].

H3K27me3 occurs in the paternal pronucleus at the late zygote stage, and maternal chromatin is detectable for H3K27me3 earlier than paternal chromatin [60]. H3K27me3 is detectable both in parental pronuclei, albeit weak in the two-cell mouse embryo, but increases in the four-cell stage, and then decreases from the eight-cell stage and becomes barely detectable in early blastocysts [34]. However, H3K27me3 is high at the oocytes and decreases gradually to reach the lowest level at the eight-cell stage but significantly increases at the blastocyst stage during bovine preimplantation [73]. H3K27me3 ChIP-seq data show that most H3K27me3 peaks in sperm are absent in zygotes, and the percentage of paternal H3K27me3 signal increases from zygotes to the two-cell stage. Broad H3K27me3 domains are observed on the paternal genome in zygotes and stages onward. H3K27me3 on the maternal genome is specifically erased from promoters but not distal regions after fertilization. In addition, H3K27me3 shows a massive loss in the two-cell stage embryos and a greater degree of dynamics in morula to ICM/trophectoderm transition. The absence of H3K27me3 in promoters of developmental genes persists to ICM, where weak enrichment emerges [35,74].

H3K27me3 regulates XCI and imprinting

The Polycomb group protein Eed-Ezh2 complex represses gene expression by marking H3K27me3. Eed-Ezh2 complex and H3K27me3 are recruited to the inactive X chromosome during the initiation of XCI in both extraembryonic and embryonic cells [75]. *Xist* locus shows a broad H3K27me3 domain which maintains from oocyte growth to preimplantation stage in the mouse. Reducing maternal H3K27me3 at the *Xist* locus by injecting *Kdm6b* mRNA induces maternal *Xist* depression and XCI [76]. Besides, maternal H3K27me3 also controls DNA methylation-independent imprinting. Ectopic removal of H3K27me3 specifically affects non-canonical imprinted genes [77]. Depleting maternal H3K27me3 in mouse oocyte by Eed knockout causes biallelic expression of all H3K27me3-mediated imprinted genes including *Xist*, and male-biased lethality [78]. During early mouse development, depletion of maternal *Ezh2* disrupts histone H3 methylation, at least for H3K27 and H3K9 methylation in oocyte and zygote. *Ezh2*-null mice are early embryonic lethal, but the loss of *Ezh2* in oocytes results in a long-term effect causing severe growth retardation of neonates, probably because the wild-type paternal allele partly rescues early lethal phenotype in maternal depletion of *Ezh2* [79,80].

H3K27me3 is necessary for mammalian cell lineage commitment

Moreover, H3K27me3 is essential for the first mammalian cell lineage commitment. The relative expression levels of *EED* and *KDM6B* mediate alternative recruitment of Polycomb repressor complex 2 (PRC2) to regulate H3K27me3 deposition at the key trophectoderm regulators *Cdx2* and *Gata3*, which are essential for suppression of ICM and preimplantation development and implantation of mouse embryos [81]. Recently, using cleavage under targets and releasing using nuclease (CUT&RUN) analysis and ultra-sensitive ChIP-Seq method, it is revealed that mono-ubiquitinated histone H2A at lysine 119 (H2AK119ub1) and H3K27me3 are highly co-localized in gametes [82,83]. H2AK119ub1 guides maternal inheritance and zygotic deposition of H3K27me3 in mouse embryos. In addition, H2AK119ub1 participates in repressing developmental genes without H3K27me3 but does not contribute to the maintenance of H3K27me3-dependent non-canonical imprinting in early development. Reduction of H2AK119ub1 by depletion of Polycomb group ring finger 1 (PCGF1) and PCGF6 leads to H3K27me3 loss at a subset of genes in oocytes, which is irreversibly inherited by embryos, causing loss of maternal H3K27me3-dependent imprinting, embryonic sublethal, and placental enlargement at term [82,83].

In short, the maternal pronucleus exhibits a strong H3K27me1/2/3 staining signal. From the two-cell to eight-cell stage embryos, the level of H3K27me2 is statistically increased and reaches its highest level in the early blastocysts. H3K27me3 is detectable in both pronuclei, although weak in the two-cell stage, but increases at the four-cell stage, and then decreases from the eight-cell stage and becomes barely detectable in the early blastocysts (Fig. 16.2B). Loss of H3K27me3 causes maternal *Xist* depression and XCI. Additionally, disruption of H3K27 methylation affects imprinted gene expression and the first mammalian cell lineage commitment.

Histone acetylation dynamic changes in early embryo development

Mature sperm and oocyte show an undetectable level of hyperacetylated H4. H4 acetylation immediately appears on the paternal but is only weakly associated with the metaphase plate of maternal chromatin after fertilization. The paternal pronucleus exhibit higher levels of H4 hyperacetylation than the maternal pronucleus throughout the G1 phase. During the S/G2 phase of the zygote, both pronuclei show equivalent levels of H4 hyperacetylation. In the two-cell and four-cell stage embryos, H4 acetylation is detectable throughout the nucleus with a varying degree of increased staining at the nuclear periphery, but this increasing tendency at the nuclear periphery reduces in eight-cell embryo and blastocyst. At the blastocyst, trophectoderm exhibits more H4 acetylation intensity than ICM. However, almost no staining of hyperacetylated H4 is observed in metaphase stage blastomeres, but a strong signal is detectable in the interphase stage at the four-cell and blastocyst stage [36–38,84,85].

Bovine oocyte shows negative staining for H4K5, K12, and K16 acetylation but weak H4K8 acetylation, while sperm lacks H4K5/12/16 acetylation. The paternal pronucleus acquires H4K5/12/16 acetylation faster than maternal but are both transiently hyperacetylated after fertilization, and both pronuclei are equally acetylated during S to G2-phase transition. H4K5 and H4K12 acetylation show similar patterns, decreasing from the zygote to the two-cell stage but gradually increasing and reaching the peak at the eight-cell stage. Reduced H4K5/12 acetylation is observed from the 16-cell to the morula stages, and trophectoderm is hyperacetylated rather than ICM at the blastocyst, while H4K8 and H4K16 acetylation are subtler during most stages. Interestingly, the peak of H4 acetylation at the eight-cell stage is coincident with ZGA, and H4 acetylation shows a significantly negative correlation with DNA methylation at different vital embryo stages [86].

In addition, the dynamic changes of histone H3 and H4 acetylation, including H3K9, H3K18, H4K5, and H4K8 acetylation have been carefully described in bovine oocytes, zygote, and preimplantation stage. H3 and H4 acetylation are quickly erased during bovine oocytes meiosis. During fertilization, H4K5 and H4K8 acetylation are detectable on paternal and maternal chromatin at postfertilization 8 and 12 hours, respectively, and increase after the four-cell stage. Nevertheless, H3K9 and H3K18 acetylation are resumed later than H4 acetylation. Then, H3 acetylation significantly decreases at the eight-cell stage but increases after the eight-cell stage, showing a similar dynamic pattern to H3K4me3 [64]. However, very few systematical studies of H3 acetylation dynamics in mouse or human early embryo development are reported.

Histone acetylation and XCI, imprinting

The inactive X chromosome lacks H4 acetylation in female mammals. However, H4 acetylation is involved in the process of X chromosome inactivation. A region of H4 hyperacetylation, located at 120 kb upstream from the *Xist* somatic promoter and containing the promoter P_0 (a promoter initiates the unstable *Xist* transcript), is gradually lost during differentiation of female mouse embryonic stem cells. Disruption of H4 deacetylation of hyperacetylated region prevents a normally inactivated X chromosome in differentiated cells [87,88]. In addition, histone hypoacetylation plays a crucial role in controlling the inactive X chromosome replication dynamics [89]. Histone acetylation is also necessary for genomic imprinting and genome integrity. Inhibition of meiotic histone deacetylation induces aneuploidy in fertilized mouse oocytes and further leads to embryonic death *in utero* at an early stage of development. Co-depletion of *Hdac1* and *Hdac2* in mouse oocytes results in arrest of oocyte development at the secondary follicle stage, accompanying increased histone acetylation, a global reduction of transcription and a pronounced decrease in H3K4 methylation [90]. Except for decreased H3K4 methylation, *Hdac1* and *Hdac2* double mutant displays global DNA hypomethylation and fails to establish imprinting markers for *Igf2r*, *Peg3*, and *Snrpn*. Further investigation shows that HDAC1/2 interacts with DNMT3A2 (an isoform of DNMT3A) through a SIN3A complex to establish maternal genomic imprints and maintain genomic integrity [91]. In addition, the older female embryos show a high incidence of aneuploidy, possibly due to inadequate meiotic histone deacetylation [92].

Histone acetylation and embryo development

Mammalian ortholog of *Drosophila M* of (males absent on the first) specifically acetylates H4K16. Ablation of the mouse *Mof* gene results in loss of H4K16 acetylation and eventually embryonic lethality and cell death. *Mof*-deficient mouse embryos are able of implantation but die before the onset of gastrulation [93]. Depleting H4K16 acetylation, by knocking down another H4K16 specifically and maternally derived acetyltransferase KAT8, significantly reduces blastocyst formation rate and arrests porcine embryos at the morula stage due to severe defects in lineage proliferation and genomic integrity [94]. During mouse preimplantation, co-knockdown *Hdac1* and *Hdac2* causes morula to blastocyst transition failure and disables to generate trophectoderm and pluripotent cells. Interestingly, co-knockdown of *Hdac1* and *Hdac2* increases the expression of DNA methyltransferases and thus promotes global DNA methylation [95]. These results also indicate that histone acetylation is essential for establishing the DNA methylation state during early development. In oocytes, the H4K12 acetylation levels statistically increase during mouse aging. Artificially increasing the level of H4K12 acetylation in oocytes of young mice by trichostatin A (TSA, an inhibitor targeting HDACs) treatment causes 38.9% of oocytes failing to form normal pronuclei after fertilization. These results show that acetylation patterns in mouse oocytes play a pivotal role in oocyte developmental potential [96]. In mouse zygotes, total protein acetylation increases, but histone H3 and H4 acetylation decrease in the two-cell stage embryos under 2% dimethyl sulfoxide treatment, in which epigenetic changes further perturb maternal-to-embryonic transition and arrest the two-cell or the four-cell stage development [97].

In summary, mature sperm and oocyte lack acetylation. After fertilization, the paternal pronucleus shows higher levels of H4 hyperacetylation than the maternal pronucleus at the G1 phase but is comparable during the S/G2 phase of the zygote. Along with development, the pattern of the H4 acetylation signal is detectable during interphase and mitosis. At the blastocyst, both mouse and bovine show higher H4 acetylation in trophectoderm than ICM (Fig. 16.2C). H3K9/18 acetylation is detectable after fertilization and significantly decreases at the eight-cell stage but increases after the eight-cell stage in bovine. Histone acetylation is essential for XCI, imprinting gene establishment, maternal-to-embryonic transition, and mammalian embryo development. The deficiency of histone acetylation causes embryonic lethality, developmental failure, and genomic integrity alteration.

Histone phosphorylation dynamic changes in early embryo development

Histone H4 and H2A N-terminal residues contain SGRGK sequences that can be phosphorylated. The dynamic change of histone H2A/H4 phosphoserine 1 (H2A/H4S1ph) in early mouse embryo development has been described. In the immature oocyte, H2A/H4S1ph surrounds the nucleolus and co-localizes with DNA. The sperm nuclei also stain positively for H2A/H4S1ph. In the zygote, both parental pronuclei are stained with equal intensity and maintained at subsequent stages. At the blastocyst stage, H2A/H4S1ph shows more intense in the trophectodermal blastomeres than ICM [38]. After fertilization, H3S10 is phosphorylated at mitosis from the zygote to the blastocyst stage. The signal of H3S10ph is less intense in the interphase than that at mitosis from the zygote to the morula stage but undetectable at blastocyst stage in the interphase [36]. In the one-cell and the two-cell mouse embryos, H3S10ph persists phosphorylated at anaphase and telophase of first and second embryonic mitoses but is dephosphorylated at late G1, and *de novo* phosphorylated at late S phase of the first and second meiotic division [98].

Histone phosphorylation function in early embryo development

H3S10ph plays a crucial regulatory role in mitotic chromosome condensation and decondensation. Interestingly, histone H3 phosphorylation shows a tight correlation with condensation of chromatin, but contradictory results are reported as well. Whether Aurora-B activity or histone H3 phosphorylation is essential for chromosome condensation in maturing porcine oocytes is still unclear [99,100]. In porcine embryos, H3S10ph is indicated to regulate the chromosome condensation in the first mitotic division based on the subcellular localization [101]. Aurora kinases catalyze histone H3 phosphorylation in mouse oocytes, which may be involved in the correct chromosome alignment and segregation during mouse oocyte maturation. Inhibition of Aurora kinases by chemical ZM447439 causes abnormalities in the first and second mitoses of the one- and the two-cell mouse embryos [98]. Another Aurora-B inhibitor treatment blocks mitotic division at the prophase stage and results in first mitotic division failure in porcine embryos [101]. In addition, the status of H3S10ph and H3T11ph also influences the degree and distribution of H3K9me2 and DNA modifications, which suggests crosstalk between H3S10/T11ph and H3K9me2 maintenance, and this crosstalk further affects the oxidative control of DNA methylation in the mouse zygote [102].

Collectively, H2A/H4S10ph are positive during mouse oocyte maturation, and paternal and maternal pronuclei are stained with equal intensity in the zygote and maintain at all subsequent development stages, while H2A/H4S1ph signals are more robust in trophectoderm than ICM at the blastocyst stage. H3S10ph signal exists at mitosis, but it is less intense in the interphase from the zygote to the morula stage and undetectable at blastocyst stage in the interphase (Fig. 16.2D). Histone phosphorylation shows a significant correlation with cell cycle and chromosomes condensation of mitotic division. Disruption of histone phosphorylation causes abnormal chromosome condensation, chromosome alignment, and segregation. However, there are still various unclear roles and dynamic changes about histone phosphorylation in early mammalian embryo development, such as if H3S10ph is necessary for chromosome condensation in maturing porcine oocytes, and the dynamic changes of histone phosphorylation in cleavage stages of mammalian embryo is not clarified yet.

SUMMARY

In summary, DNA methylation and histone modifications are dynamically changed during mammalian early development (Fig. 16.2). Such a dynamic pattern is correlated with pivotal embryo development events, including gametes maturation, zygotic genome activation, X chromosome inactivation, genomic imprinting, and normal embryo development. In this context, we summarize the dynamic changes and functions of classical DNA methylation and histone modifications during the early mammalian embryonic development. Extensive crosstalk among different modifications has been frequently proposed. For instance, repressive markers, including DNA methylation, H3K9me3, and H3K27me3, synergistically repress gene expression or are antagonistic to active markers in the same development event. The crosstalk among different modifications maintains and fine tunes the precise gene expression level and is thus necessary for normal embryo development. Unfortunately, although many efforts have been put into investigating epigenetic modification changes during development, the dynamics and functions of epigenetic modifications in early embryo development are still elusive and unclear. For example, except for classic epigenetic modifications mentioned above, other epigenetic modifications are less systematically reported during early mammalian embryo development. In addition, a few epigenetic modifications, such as H3K27me3 and H3K9 acetylation, are not functionally conserved in different species, which makes the study more complicated.

Moreover, contradictory results are frequently reported by different groups or using different detection methods. Along with developing many state-of-the-art technologies, such as ChIP-seq, WGBS, RRBS, and proteomics, we should integrate multiple layers of the dataset to systematically and comprehensively investigate the dynamics and functions of epigenetic modification in early embryo development in the near future. Furthermore, the unprecedented opportunity, that CRISPR/Cas9 technology offers for investigating and manipulating the epigenetic modifications and genome, facilitates further understanding of early mammalian embryo development and stem cell biology. Of note, CRISPR-off and CRISPR-on, two technologies for programmable writing and erasing epigenetic memories by fusing a single dead Cas9 protein to control DNA methylation and histone modifications [103], provide new ideas to investigate and manipulate epigenetic modifications during early mammalian embryo development.

References

[1] Niakan KK, et al. Human pre-implantation embryo development. Development 2012;139(5):829–41.

[2] Wamaitha SE, Niakan KK. Human pre-gastrulation development. Curr Top Dev Biol 2018;128:295–338.

[3] Cockburn K, Rossant J. Making the blastocyst: lessons from the mouse. J Clin Invest 2010;120(4):995–1003.

[4] Holliday R, Pugh JE. DNA modification mechanisms and gene activity during development. Science 1975;187(4173):226–32.

[5] Riggs AD. X inactivation, differentiation, and DNA methylation. Cytogenet Cell Genet 1975;14(1):9–25.

[6] Bestor T, et al. Cloning and sequencing of a cDNA encoding DNA methyltransferase of mouse cells. The carboxyl-terminal domain of the mammalian enzymes is related to bacterial restriction methyltransferases. J Mol Biol 1988;203(4):971–83.

[7] Li Y, et al. Stella safeguards the oocyte methylome by preventing de novo methylation mediated by DNMT1. Nature 2018;564(7734):136–40.

[8] Okano M, et al. DNA methyltransferases Dnmt3a and Dnmt3b are essential for de novo methylation and mammalian development. Cell 1999;99(3):247–57.

[9] Aapola U, et al. Isolation and initial characterization of a novel zinc finger gene, DNMT3L, on 21q22.3, related to the cytosine-5-methyltransferase 3 gene family. Genomics 2000;65(3):293–8.

[10] Bourc'his D, et al. Dnmt3L and the establishment of maternal genomic imprints. Science 2001;294(5551):2536–9.

[11] Chedin F, Lieber MR, Hsieh CL. The DNA methyltransferase-like protein DNMT3L stimulates de novo methylation by Dnmt3a. Proc Natl Acad Sci U S A 2002;99(26):16916–21.

[12] Zeng Y, Chen T. DNA methylation reprogramming during mammalian development. Genes (Basel) 2019;10(4).

[13] Marmorstein R, Zhou MM. Writers and readers of histone acetylation: structure, mechanism, and inhibition. Cold Spring Harb Perspect Biol 2014;6(7):a018762.

[14] Zhang YJ, et al. Overview of histone modification. Histone Mutations and Cancer 2021;1283:1–16.

[15] Hyun K, et al. Writing, erasing and reading histone lysine methylations. Exp Mol Med 2017;49(4):e324.

[16] Banerjee T, Chakravarti D. A peek into the complex realm of histone phosphorylation. Mol Cell Biol 2011;31(24):4858–73.

[17] Kobayashi H, et al. Contribution of intragenic DNA methylation in mouse gametic DNA methylomes to establish oocyte-specific heritable marks. PLoS Genet 2012;8(1):e1002440.

[18] Smith ZD, et al. A unique regulatory phase of DNA methylation in the early mammalian embryo. Nature 2012;484(7394):339–44.

[19] Mayer W, et al. Demethylation of the zygotic paternal genome. Nature 2000;403(6769):501–2.

[20] Oswald J, et al. Active demethylation of the paternal genome in the mouse zygote. Curr Biol 2000;10(8):475–8.

[21] Guo F, et al. Single-cell multi-omics sequencing of mouse early embryos and embryonic stem cells. Cell Res 2017;27(8):967–88.

[22] Wang L, et al. Programming and inheritance of parental DNA methylomes in mammals. Cell 2014;157(4):979–91.

[23] Shen L, et al. Tet3 and DNA replication mediate demethylation of both the maternal and paternal genomes in mouse zygotes. Cell Stem Cell 2014;15(4):459–71.

[24] Santos F, et al. Dynamic reprogramming of DNA methylation in the early mouse embryo. Dev Biol 2002;241(1):172–82.

[25] Li E, Zhang Y. DNA methylation in mammals. Cold Spring Harb Perspect Biol 2014;6(5):a019133.

[26] Okae H, et al. Genome-wide analysis of DNA methylation dynamics during early human development. PLoS Genet 2014;10(12):e1004868.

[27] Smith ZD, et al. DNA methylation dynamics of the human preimplantation embryo. Nature 2014;511(7511):611–15.

[28] Guo H, et al. The DNA methylation landscape of human early embryos. Nature 2014;511(7511):606–10.

[29] Zhu P, et al. Single-cell DNA methylome sequencing of human preimplantation embryos. Nat Genet 2018;50(1):12–19.

[30] Huang X, et al. Stable H3K4me3 is associated with transcription initiation during early embryo development. Bioinformatics 2019;35(20):3931–6.

[31] Zhang B, et al. Allelic reprogramming of the histone modification H3K4me3 in early mammalian development. Nature 2016;537(7621):553–7.

[32] Dahl JA, et al. Broad histone H3K4me3 domains in mouse oocytes modulate maternal-to-zygotic transition. Nature 2016;537(7621):548–52.

[33] Wang C, et al. Reprogramming of H3K9me3-dependent heterochromatin during mammalian embryo development. Nat Cell Biol 2018;20(5):620–31.

[34] Deng MT, et al. Exchanges of histone methylation and variants during mouse zygotic genome activation. Zygote 2020;28(1):51–8.

[35] Liu X, et al. Distinct features of H3K4me3 and H3K27me3 chromatin domains in pre-implantation embryos. Nature 2016;537(7621):558–62.

[36] Huang JC, et al. Comparison of histone modifications in in vivo and in vitro fertilization mouse embryos. Biochem Biophys Res Commun 2007;354(1):77–83.

[37] Thompson EM, et al. Progressive maturation of chromatin structure regulates HSP70.1 gene expression in the preimplantation mouse embryo. Development 1995;121(10):3425–37.

[38] Sarmento OF, et al. Dynamic alterations of specific histone modifications during early murine development. J Cell Sci 2004;117(Pt 19):4449−59.
[39] Li E, Bestor TH, Jaenisch R. Targeted mutation of the DNA methyltransferase gene results in embryonic lethality. Cell 1992;69(6):915−26.
[40] Penny GD, et al. Requirement for Xist in X chromosome inactivation. Nature 1996;379(6561):131−7.
[41] Panning B, Jaenisch R. DNA hypomethylation can activate Xist expression and silence X-linked genes. Genes Dev 1996;10(16): 1991−2002.
[42] Beard C, Li E, Jaenisch R. Loss of methylation activates Xist in somatic but not in embryonic cells. Genes Dev 1995;9(19): 2325−34.
[43] Zuccotti M, Monk M. Methylation of the mouse Xist gene in sperm and eggs correlates with imprinted Xist expression and paternal X-inactivation. Nat Genet 1995;9(3):316−20.
[44] Sado T, et al. X inactivation in the mouse embryo deficient for Dnmt1: distinct effect of hypomethylation on imprinted and random X inactivation. Dev Biol 2000;225(2):294−303.
[45] Csankovszki G, Nagy A, Jaenisch R. Synergism of Xist RNA, DNA methylation, and histone hypoacetylation in maintaining X chromosome inactivation. J Cell Biol 2001;153(4):773−84.
[46] da Rocha ST, Ferguson-Smith AC. Genomic imprinting. Curr Biol 2004;14(16):R646−9.
[47] Bartolomei MS. Genomic imprinting: employing and avoiding epigenetic processes. Genes Dev 2009;23(18):2124−33.
[48] Ferguson-Smith AC. Genomic imprinting: the emergence of an epigenetic paradigm. Nat Rev Genet 2011;12(8):565−75.
[49] Li E, Beard C, Jaenisch R. Role for DNA methylation in genomic imprinting. Nature 1993;366(6453):362−5.
[50] Kaneda M, et al. Role of de novo DNA methyltransferases in initiation of genomic imprinting and X-chromosome inactivation. Cold Spring Harb Symp Quant Biol 2004;69:125−9.
[51] Nakamura T, et al. PGC7/Stella protects against DNA demethylation in early embryogenesis. Nat Cell Biol 2007;9(1):64−71.
[52] Maenohara S, et al. Role of UHRF1 in de novo DNA methylation in oocytes and maintenance methylation in preimplantation embryos. PLoS Genet 2017;13(10):e1007042.
[53] Gu TP, et al. The role of Tet3 DNA dioxygenase in epigenetic reprogramming by oocytes. Nature 2011;477(7366):606−10.
[54] Lepikhov K, Walter J. Differential dynamics of histone H3 methylation at positions K4 and K9 in the mouse zygote. BMC Dev Biol 2004;4:12.
[55] Andreu-Vieyra CV, et al. MLL2 Is required in oocytes for bulk histone 3 lysine 4 trimethylation and transcriptional silencing. PLoS Biol 2010;8(8):e1000453.
[56] Aoshima K, et al. Paternal H3K4 methylation is required for minor zygotic gene activation and early mouse embryonic development. EMBO Rep 2015;16(7):803−12.
[57] Sankar A, et al. KDM4A regulates the maternal-to-zygotic transition by protecting broad H3K4me3 domains from H3K9me3 invasion in oocytes. Nat Cell Biol 2020;22(4):380−8.
[58] van Steensel B, Belmont AS. Lamina-associated domains: links with chromosome architecture, heterochromatin, and gene repression. Cell 2017;169(5):780−91.
[59] Borsos M, et al. Genome-lamina interactions are established de novo in the early mouse embryo. Nature 2019;569(7758): 729−33.
[60] Santos F, et al. Dynamic chromatin modifications characterise the first cell cycle in mouse embryos. Dev Biol 2005;280(1):225−36.
[61] Liu H, Kim JM, Aoki F. Regulation of histone H3 lysine 9 methylation in oocytes and early pre-implantation embryos. Development 2004;131(10):2269−80.
[62] Arney KL, et al. Histone methylation defines epigenetic asymmetry in the mouse zygote. Int J Dev Biol 2002;46(3):317−20.
[63] Cowell IG, et al. Heterochromatin, HP1 and methylation at lysine 9 of histone H3 in animals. Chromosoma 2002;111(1):22−36.
[64] Wu X, et al. Dynamic changes of histone acetylation and methylation in bovine oocytes, zygotes, and preimplantation embryos. J Exp Zool B Mol Dev Evol 2020;334(4):245−56.
[65] Boggs BA, et al. Differentially methylated forms of histone H3 show unique association patterns with inactive human X chromosomes. Nat Genet 2002;30(1):73−6.
[66] Peters AHFM, et al. Histone H3 lysine 9 methylation is an epigenetic imprint of facultative heterochromatin. Nat Genet 2002;30(1):77−80.
[67] Heard E, et al. Methylation of histone H3 at Lys-9 is an early mark on the X chromosome during X inactivation. Cell 2001;107 (6):727−38.
[68] Tachibana M, et al. G9a histone methyltransferase plays a dominant role in euchromatic histone H3 lysine 9 methylation and is essential for early embryogenesis. Genes Dev 2002;16(14): 1779−91.
[69] Tachibana M, et al. Histone methyltransferases G9a and GLP form heteromeric complexes and are both crucial for methylation of euchromatin at H3-K9. Genes Dev 2005;19(7):815−26.
[70] Yeung WKA, et al. Histone H3K9 methyltransferase G9a in oocytes is essential for preimplantation development but dispensable for CG methylation protection. Cell Rep 2019;27(1):282.
[71] Feldman N, et al. G9a-mediated irreversible epigenetic inactivation of Oct-3/4 during early embryogenesis. Nat Cell Biol 2006;8(2):188−94.
[72] Nakamura T, et al. PGC7 binds histone H3K9me2 to protect against conversion of 5mC to 5hmC in early embryos. Nature 2012;486(7403):415−19.
[73] Ross PJ, et al. Polycomb gene expression and histone H3 lysine 27 trimethylation changes during bovine preimplantation development. Reproduction 2008;136(6):777−85.
[74] Zheng H, et al. Resetting epigenetic memory by reprogramming of histone modifications in mammals. Mol Cell 2016;63(6):1066−79.
[75] Plath K, et al. Role of histone H3 lysine 27 methylation in X inactivation. Science 2003;300(5616):131−5.
[76] Inoue A, et al. Genomic imprinting of Xist by maternal H3K27me3. Genes Dev 2017;31(19):1927−32.
[77] Inoue A, et al. Maternal H3K27me3 controls DNA methylation-independent imprinting. Nature 2017;547(7664):419−24.
[78] Inoue A, et al. Maternal Eed knockout causes loss of H3K27me3 imprinting and random X inactivation in the extraembryonic cells. Genes Dev 2018;32(23−24):1525−36.
[79] Erhardt S, et al. Consequences of the depletion of zygotic and embryonic enhancer of zeste 2 during preimplantation mouse development. Development 2003;130(18):4235−48.
[80] O'Carroll D, et al. The Polycomb-group gene Ezh2 is required for early mouse development. Mol Cell Biol 2001;21(13):4330−6.
[81] Saha B, et al. EED and KDM6B coordinate the first mammalian cell lineage commitment to ensure embryo implantation. Mol Cell Biol 2013;33(14):2691−705.
[82] Chen Z, Djekidel MN, Zhang Y. Distinct dynamics and functions of H2AK119ub1 and H3K27me3 in mouse preimplantation embryos. Nat Genet 2021;53(4):551−63.
[83] Mei H, et al. H2AK119ub1 guides maternal inheritance and zygotic deposition of H3K27me3 in mouse embryos. Nat Genet 2021;53(4):539−50.
[84] Adenot PG, et al. Differential H4 acetylation of paternal and maternal chromatin precedes DNA replication and differential transcriptional activity in pronuclei of 1-cell mouse embryos. Development 1997;124(22):4615−25.

[85] Worrad DM, Turner BM, Schultz RM. Temporally restricted spatial localization of acetylated isoforms of histone H4 and RNA polymerase II in the 2-cell mouse embryo. Development 1995;121(9):2949−59.

[86] Maalouf WE, Alberio R, Campbell KH. Differential acetylation of histone H4 lysine during development of in vitro fertilized, cloned and parthenogenetically activated bovine embryos. Epigenetics 2008;3(4):199−209.

[87] Jeppesen P, Turner BM. The inactive X chromosome in female mammals is distinguished by a lack of histone H4 acetylation, a cytogenetic marker for gene expression. Cell 1993;74(2):281−9.

[88] O'Neill LP, et al. A developmental switch in H4 acetylation upstream of Xist plays a role in X chromosome inactivation. EMBO J 1999;18(10):2897−907.

[89] Casas-Delucchi CS, et al. Histone acetylation controls the inactive X chromosome replication dynamics. Nat Commun 2011;2:222.

[90] Ma P, et al. Compensatory functions of histone deacetylase 1 (HDAC1) and HDAC2 regulate transcription and apoptosis during mouse oocyte development. Proc Natl Acad Sci U S A 2012;109(8):E481−9.

[91] Ma P, et al. A DNMT3A2-HDAC2 complex is essential for genomic imprinting and genome integrity in mouse oocytes. Cell Rep 2015;13(8):1552−60.

[92] Akiyama T, Nagata M, Aoki F. Inadequate histone deacetylation during oocyte meiosis causes aneuploidy and embryo death in mice. Proc Natl Acad Sci U S A 2006;103(19):7339−44.

[93] Gupta A, et al. The mammalian ortholog of Drosophila MOF that acetylates histone H4 lysine 16 is essential for embryogenesis and oncogenesis. Mol Cell Biol 2008;28(1):397−409.

[94] Cao Z, et al. Maternal histone acetyltransferase KAT8 is required for porcine preimplantation embryo development. Oncotarget 2017;8(52):90250−61.

[95] Zhao P, et al. Essential roles of HDAC1 and 2 in lineage development and genome-wide DNA methylation during mouse preimplantation development. Epigenetics 2020;15(4):369−85.

[96] Suo L, et al. Changes in acetylation on lysine 12 of histone H4 (acH4K12) of murine oocytes during maternal aging may affect fertilization and subsequent embryo development. Fertil Steril 2010;93(3):945−51.

[97] Kang MH, et al. DMSO impairs the transcriptional program for maternal-to-embryonic transition by altering histone acetylation. Biomaterials 2020;230:119604.

[98] Teperek-Tkacz M, et al. Phosphorylation of histone H3 serine 10 in early mouse embryos: active phosphorylation at late S phase and differential effects of ZM447439 on first two embryonic mitoses. Cell Cycle 2010;9(23):4674−87.

[99] Jelinkova L, Kubelka M. Neither Aurora B activity nor histone H3 phosphorylation is essential for chromosome condensation during meiotic maturation of porcine oocytes. Biol Reprod 2006;74(5):905−12.

[100] Bui HT, Yamaoka E, Miyano T. Involvement of histone H3 (Ser10) phosphorylation in chromosome condensation without Cdc2 kinase and mitogen-activated protein kinase activation in pig oocytes. Biol Reprod 2004;70(6):1843−51.

[101] Chen C, et al. Phosphorylation of histone H3 on Ser-10 by Aurora B is essential for chromosome condensation in porcine embryos during the first mitotic division. Histochem Cell Biol 2017;148(1):73−83.

[102] Lan J, et al. Histone and DNA methylation control by H3 serine 10/threonine 11 phosphorylation in the mouse zygote. Epigenetics Chromatin 2017;10:5.

[103] Nunez JK, et al. Genome-wide programmable transcriptional memory by CRISPR-based epigenome editing. Cell 2021;184(9):2503−19.

CHAPTER

17

Epigenetic Biomarkers

Xiaotong Hu

Sir Run Run Shaw Hospital, Zhejiang University, Hangzhou, Zhejiang, P.R. China

OUTLINE

Introduction	303
Epigenetic biomarkers offer distinct advantages over genetic biomarkers	304
Minimally invasive tissues are suitable for detecting epigenetic biomarkers	305
Field cancerization and epigenetic biomarkers	306
Potential DNA methylation biomarkers in cancer and other diseases	307
Potential m6A methylation biomarkers in cancer and other diseases	308
Potential histone modification biomarkers in cancer and other diseases	309
Potential non-coding RNA biomarkers in diseases	310
Potential miRNA biomarkers in cancer and other diseases	310
Potential lncRNA biomarkers in cancer and other diseases	311
Epigenetic biomarker detection methods in the clinic	313
Challenges and future perspectives	316
Abbreviations	317
References	317

INTRODUCTION

An increasing amount of evidence has supported the key role epigenetic processes play in cell biology, physiology, and disease development, both at the individual-gene level and at the genome-wide level. Epigenetic mechanisms determine the phenotype without changes in the genotype. DNA methylation is the best studied epigenetic mechanism and acts together with other epigenetic entities: histone modification and non-coding RNAs (ncRNAs) to shape the chromatin structure according to its functional state. The epigenetic interface sits between the genetic blueprints stored in DNA sequences and phenotypes dictated by the pattern of gene expression. Epigenetic profiles, which are more susceptible to the changing environment, determine disease development and response to treatment. For example, recent studies in gastric cancer (GC) have strongly suggested the association of risk factors such as *Helicobacter pylori* infection and a unique aberrant DNA methylation pattern [1]. In addition, Mastroeni et al. determined the status of DNA methylation in monozygotic twins, where genetic similarities exist but discordant for Alzheimer disease (AD), proving that epigenetic mechanisms may mediate the effects of life events on AD risk [2].

Epigenetic biomarkers refer to the measurement of epigenetic alterations in tissues or peripheral fluids such as urine, blood, plasma, serum, and stool samples as markers of disease detection, progression, and therapy response. Although most research on epigenetic biomarkers has focused on cancer, epigenetic biomarkers have also been used as markers of other diseases such as neurological, metabolic and cardiovascular disorders, autoimmune diseases, and genomic imprinting disorders. Many diseases may remain asymptomatic and are always diagnosed at later stages and thus interventions are more likely ineffective. Epigenetic abnormalities can occur at all stages of disease progression such as dysplasia, local benign and malignant tumors, and finally, metastatic lesions. Moreover, epigenetic alterations are environmentally regulated and provide a plausible link between environmental epigenetics and disease susceptibility. In addition, epigenetic changes are reversible, and this makes them attractive therapeutic targets. Epigenetic biomarkers are valuable in disease screening, early detection, diagnosis, staging, risk stratification, and treatment monitoring. Information on epigenetic alterations in disease has rapidly increased due to the development of genome-wide techniques for their identification. Epigenetic biomarkers offer distinct advantages over genetic biomarkers and can be detected and quantified by numerous technologies, including genome-wide screening methods, as well as locus- or gene-specific high-resolution analysis in different tissue samples and body fluids obtained through minimally invasive procedures, making them highly suitable biomarkers. However, many studies have been conducted using small sample sizes and only a few results have been appropriately validated in independent cohorts or investigated in large prospective clinical trials. A number of limitations need to be resolved before these potential biomarkers may be used in the clinic. This chapter provides systemic information on recent research progress on epigenetic biomarkers and summarizes the advantages of epigenetic biomarkers over other biomarkers such as genetic biomarkers, the performance in liquid biopsy, and potential clinical application and challenges in cancer and other diseases.

EPIGENETIC BIOMARKERS OFFER DISTINCT ADVANTAGES OVER GENETIC BIOMARKERS

First, although both genetic and epigenetic biomarkers could be masked by normal cells, genetic biomarkers (such as loss of heterozygosity and mutation) are more likely to be masked by normal cells, while epigenetic biomarkers are not [3]. Non-malignant mutations harbored by hematopoietic cells (clonal hematopoiesis) could result in false-positive plasma genotyping results [4]. The readout of epigenetic biomarkers does not suffer from sample contamination and illegitimate expression.

Second, the genetic defects within a gene may involve several regions, not to mention its non-coding region, which have not or rarely been surveyed. For example, over 23,544 somatic mutations and 376 germline mutations have been reported for the p53 coding region in cancer. Currently, it is not possible to cover all the known mutations in genes analyzed even with the most powerful platform technology. On the contrary, epigenetic defects such as hypermethylation are always within a defined promoter region CpG island and are easier to be detected.

In addition, epigenetic defects are specific for diseases and occur at higher frequencies than genetic variations. For example, aberrant methylation occurs during the early stages of carcinogenesis, and distinct types of cancer exhibit specific patterns of methylation changes [5]. There are unique DNA methylation profiles that are stably maintained in each type and stage of tumor tissue and each normal tissue. Thus, methylation profile differences allow the identification of the origin tissue from which ctDNA is derived [6,7]. Moreover, cancer-deregulated miRNAs are significantly correlated to clinically relevant information such as disease aggressiveness and metastatic potential, patient response to therapy, time-to-relapse, and overall survival [8]. Diseases other than cancer reveal profound alterations in epigenetic defects, whereas genetic alterations such as mutations are rarely detected, which makes epigenetic analysis applicable to a broader range of diseases [9].

Furthermore, compared to DNA-, RNA-, and protein-based biomarkers, RNA- and protein-based biomarkers are biologically and biochemically unstable. This feature prevents their use in clinical practice, although they have been most intensively evaluated for more than a decade. DNA-based biomarkers are stable, and so the utility of DNA methylation biomarkers have been seriously considered [10].

Genome-wide analysis has revealed that about 70% of the genome is actively transcribed into ncRNAs, while <2% of the sequences encode proteins [11]. NcRNAs comprise a complex and diverse class that has been arbitrarily subdivided into two major groups: small ncRNAs, including microRNAs (miRNAs), and long non-coding RNAs (lncRNAs). One of the most remarkable aspects of miRNAs is their resistance to RNase digestion and other harsh conditions such as extreme pH, boiling, extended storage, and multiple freeze-thaw cycles outside the intracellular environment, and this aspect makes them potential biomarkers for clinical use. Their stability is due to several factors that prevent their degradation from both cellular and environmental degradation. First, miRNAs can be packaged in microvesicles such as exosomes and apoptotic bodies. Second, most miRNAs are bound to RNA-binding proteins such as AGO2 or lipoprotein complexes such as high density lipoproteins (HDLs) [12]. However, lncRNA expression is more specific to organs, tissues, cell types, developmental stages, and human diseases than protein-coding RNAs [13], making them promising candidates as diagnostic and prognostic biomarkers.

MINIMALLY INVASIVE TISSUES ARE SUITABLE FOR DETECTING EPIGENETIC BIOMARKERS

Many tissue types are suitable for the discovery of epigenetic biomarkers, including cell-based samples, such as tumor tissue biopsies, and cell-free samples, such as plasma. Biomarkers detected in tissue biopsies or postoperative tissues help identify high-risk patients who need advanced treatment or intensive follow-up. For example, stage I non-small cell lung cancer (NSCLC) patients with recurrence after surgery showed a higher frequency of DNA methylation in CDKN2A and cadherin 13 (CDH13) than patients without recurrences [14]. However, one of the standards of good disease biomarkers is that collection of the detected tissue should be simple and minimally invasive. For early diagnosis or curative effect monitoring of disease, a less invasive liquid biopsy is considered more effective. These minimally invasive tissues mainly include blood, urine, stools, ductal lavages, bronchial aspirate, sputum, and pleural effusions. The presence of cell-free DNA or circulating-free DNA (cfDNA) in human plasma has been known for some time [15]. cfDNA in blood is released from normal and cancer cells undergoing apoptosis or necrosis (Fig. 17.1).

cfDNA possesses tumor-related genetic alterations such as mutations and loss of heterozygosity. Similarly, epigenetic alterations, including DNA methylation are also detected in cfDNA of patients, and thus detecting tumor-specific epigenetic biomarkers in liquid biopsies is a potential tool for cancer diagnosis [16]. Analysis of DNA methylation in tumor-derived DNA from stools is an effective, non-invasive approach to the diagnosis of colorectal cancer (CRC). Analysis of DNA methylation patterns in pleural effusions is useful for the differential diagnosis of lung cancer and other thoracic malignancies.

Extensive studies have been conducted on the changes in histone modifications relative to disease. Additionally, histone modification alterations can be detected on nucleosomes released by apoptotic and necrotic cells in the blood of cancer patients. As nucleosomes are stable structures in the circulation [17], histone alterations such as methylation and acetylation on circulating nucleosomes may be potentially utilized as biomarkers.

miRNAs are released by apoptotic and necrotic cells, and can also be actively secreted into the blood in circulation by exosomes. Exosomes are endosome-derived membranous, nano-sized microvesicles that are actively secreted by a wide range of normal cells, as well as malignant cells and are also found in most biological fluids, including blood, urine, ascites, and cerebrospinal fluid. Exosomes contain cargoes such as miRNAs, proteins, and mRNA. Exosomal cargoes including miRNAs are thought to play an important role in cell-to-cell communication and have potential roles in disease development. There has been recent interest in identifying exosomal biomarkers from blood samples for both diagnosis and stratification of tumor patients. Pioneer studies have been conducted on exosomal miR-21 as a biomarker in hepatocellular carcinoma (HCC) [18] and exosomal miR-373 as a biomarker in breast cancer [19]. Furthermore,

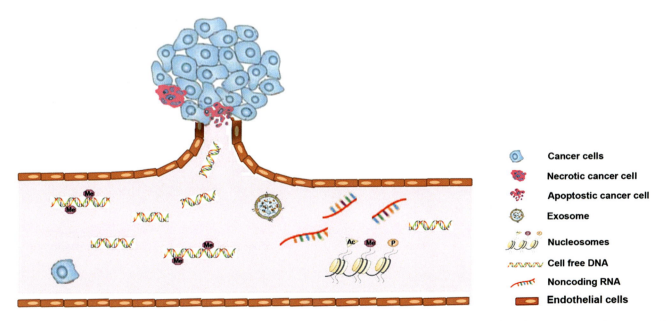

FIGURE 17.1 Tumor cells release cell-free DNA into the blood stream from normal and cancer cells undergoing apoptosis or necrosis. Circulating tumor cells, non-coding RNAs, exosomes, and nucleosomes can also be released. Epigenetic alterations can be detected using these non-invasive liquid biopsies. *Ac*, acetylation; *Me*, methylation; *P*, phosphorylase.

exosomes are robust and can be stored for extended periods without significantly affecting the integrity of encapsulated miRNAs. This feature of exosomes increases their potential applicability in the clinic. The detection of exosomal miRNAs in body fluids is a booming research field and could be a potential gold mine to advance diagnostics and monitor therapeutic efficiency with a higher sensitivity and specificity.

Similarly, an advantage of lncRNAs is that they may be used in the clinic as minimally invasive biomarkers due to their presence in body fluids such as serum/plasma and urine in a stable form protected from endogenous RNase, even though the precise mechanism of lncRNA release into the extracellular environment is not completely understood. Studies have shown that exosomes also contain lncRNAs. For example, Li et al. [20] analyzed the expression level of LINC00152 in plasma and plasma-derived exosomes from GC patients and found that there is no statistically expression difference between the two samples, suggesting that the majority of plasma LINC00152 is derived from exosomes. These disease-related epigenetic alterations could be detected in minimally invasive tissues, emphasizing its significance in clinical diagnosis, prognosis, and prediction of response to therapies.

FIELD CANCERIZATION AND EPIGENETIC BIOMARKERS

Epigenetic alterations are believed to be the early events in disease, particularly cancer development. It is generally accepted that the pathologically defined neighboring non-cancerous tissues represent the cells residing at the early stages of carcinogenesis. Previous studies have shown that normal adjacent cells to tumors frequently harbor genetic alterations such as loss of heterozygosity, microsatellite and chromosome instability, and gene mutations [21]. Epigenetic alterations have also been detected in premalignant tissues [22]. This phenomenon was first described in 1953 by Slaughter's group using the term "field cancerization" (also known as field effect or field defect), which refers to the presence of cancer-causing molecular abnormalities in histologically normal-appearing tissues surrounding a neoplasm, some of which are the same as those in the tumors [23,24]. Field cancerization reflects the susceptibility of normal tissues to undergo early cancer development due to exposure to carcinogens. Finally, development of second primary tumors and recurrences can be explained by the presence of residual abnormal tissue following surgery, which is pathologically normal (Fig. 17.2).

Epigenetic markers are also useful as a quantitative measure for field cancerization. Epigenetic alterations in field cancerization have recently been shown in various types of human cancers such as gastric, colorectal, esophageal, liver, and renal cancer [25,26]. For example, Issa et al. reported that four specific genes (*ESR*, *MYOD1*, *CDKN2A*, and *VCAN*) were highly methylated in the normal-appearing epithelia from ulcerative colitis patients with high-grade dysplasia or cancer than in non-ulcerative colitis individuals, indicating that methylation precedes dysplasia development [27]. Hypomethylation is also a marker of epigenetic field defects. Milicic et al. observed field cancerization in CRC, indicating that demethylation of P-cadherin might be an early event in the colorectal mucosa, prior to any morphological changes [28]. In addition, Sato et al. conducted miRNA expression profiling of paired tumor and non-tumor liver tissue samples from 73 HCC patients and found that the expression patterns of tumor-derived miRNAs tended to better predict early recurrence than late recurrence, whereas those of non-tumor-derived miRNAs tend to predict late recurrence after hepatic resection for HCC. These studies suggested that miRNA expression profiling of

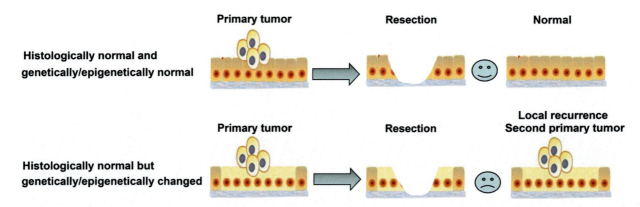

FIGURE 17.2 Field cancerization leads to disease recurrence after surgical resection and development of local recurrence or independent secondary tumors.

non-tumor liver tissues reflect the accumulation of genome abnormalities (field cancerization) in non-cancerous liver cells [29]. Because epigenetic changes always occur before histopathological changes, a better understanding of epigenetic field cancerization may improve cancer risk assessment.

POTENTIAL DNA METHYLATION BIOMARKERS IN CANCER AND OTHER DISEASES

Cancer is a leading cause of death worldwide, and largely due to the dearth of early detection methods, many cancers are diagnosed at advanced stages with poor prognosis. Current cancer biomarkers are based on proteins such as CEA, PSA, TPA, CA125, and CA19−9, and have low specificity. In addition, existing screening tests such as fecal occult-blood testing, colonoscopy, and stool DNA testing have low sensitivity and reproducibility. Therefore, the discovery of new biomarkers in cancer is particularly important to improve patient survival.

Cancer is a heterogeneous disease, and it is widely accepted that tumorigenesis is triggered by the accumulation of both genetic and epigenetic alterations. Epigenetics is an integral part of cancer initiation, development, and recurrence. The tumor microenvironment is mainly affected by all of the epigenomic components of cells. At the global level, DNA is often hypomethylated in cancer. Hypomethylation can cause activation of normally silent regions of the genome, particularly those related to development, including protooncogenes. Previous studies strongly support the hypothesis of early global hypomethylation in carcinogenesis. Hypomethylation and activation of long interspersed nucleotide element-1 (LINE-1), which comprises approximately 18% of the human genome, lead to chromosomal instability and is a biomarker of unfavorable prognosis in many cancers such as CRC [30] or lung neoplasia [31]. However, in clinical applications, the utilization of DNA hypomethylation as a biomarker is limited because a reduction in DNA methylation is technically more complicated to detect than a gain of a signal.

More important methylation changes in cancer are characterized by local hypermethylation of individual genes, which has been associated with tumor suppressor gene (TSG) downregulation [32]. In the past decade, many hypermethylated genes have been identified, and a number of putative DNA methylation markers have been evaluated in a variety of clinical specimens for cancer diagnosis, prognosis, and predictive information. Currently, DNA methylation is considered as one of the most promising epigenetic biomarkers [33,34]. Several commercial DNA methylation tests are now available for clinical application (Table 17.1).

DNA methylation of septin 9 (SEPT9) is a valuable marker for the diagnosis of CRC and is now used as a test in the clinic (Epi proColon) [35]. In a prospective study of SEPT9, the sensitivity and specificity of this marker to diagnose CRC were 48.2% and 91.5%, respectively [36]. The sensitivity of the SEPT9 test (73.3%) was not inferior to fecal immunochemical testing (FIT, 68.0%) for the detection of colorectal cancer; however, the specificity of SEPT9 was lower than FIT (81.5% vs 97.4%) [37].

Other commercially available DNA methylation tests for diagnosing CRC patients include DNA methylation of VIM and multitarget DNA testing using stool samples. The sensitivity and specificity of VIM methylation detection were respectively 46%−81% and 82%−90% [38], while those of the minimally invasive multitarget stool DNA test, which includes KRAS mutation, aberrant DNA methylation of NDRG4, BMP3, and β-actin (as reference for DNA quantification) with a hemoglobin immunoassay, reached 92.3% and 86.6%, respectively [39].

DNA methylation of short stature homeobox 2 (SHOX2) is a commercially available biomarker for the diagnosis of lung cancer. Using bronchial aspirates, SHOX2 methylation was found in 62% of lung cancers, which were negative by cytological analysis [40]. SHOX2 methylation was also detected in the plasma of lung cancer patients, with a sensitivity of 60% and specificity of 90% [41].

MGMT methylation is one of the first DNA methylation biomarkers identified. O^6-methylguanine DNA methyltransferase is a DNA repair protein that is encoded by the MGMT gene and is capable of

TABLE 17.1 Commercially Available Tests for DNA Methylation Detection in Clinic

DNA methylation biomarkers	Cancer type	Application	Source
SEPT9	Colorectal	Detection	Blood
NDRG4, BMP3, (KRAS mutation)	Colorectal	Detection	Stool
VIM	Colorectal	Detection	Stool
SHOX2	Lung	Detection	Bronchial fluid
MGMT	Brain	Prediction	Tumor

removing alkyl residues directly from the O6-position of guanines. Assessing the DNA methylation status of the MGMT gene is a tailored therapy for therapeutic choice in glioma, and MGMT methylation predicts survival benefit in glioma patients treated with alkylating agents [42].

A recent genome-wide assay called "plasma DNA tissue mapping" [43] is capable of tracing the tissue origin of cfDNA using methylation features in different tissues. In addition, a new approach [44] can identify the tissue origin of plasma cfDNA based on tissue-specific methylation biomarkers, which are focused on detecting a few markers for one tissue at a time rather than using a genome-wide approach. Another study [45] developed tissue-specific DNA methylation biomarkers for measuring the plasma concentrations of liver- and colon-derived DNA using a digital PCR-based method, and explored their clinical applications to metastatic CRC. The results confirmed the accuracy of the liver-specific marker in reflecting the concentration of liver-derived DNA in plasma (specificity: 93%; sensitivity: 60% in detecting HCC). The absolute concentration of liver-derived DNA better differentiates between CRC patients with and without liver metastasis compared with fractional concentration (AUCs, 0.85 vs 0.75; $P = .01$). These results demonstrate that ctDNA methylation is useful for differentiating patients with metastases to other organs and those without. More recently, Sadeh et al. [46] utilized chromatin immunoprecipitation of cell-free nucleosomes carrying active chromatin modifications followed by sequencing (cfChIP-seq) in assessing 268 human samples. During cell death, the genome is fragmented, and chromatin, mostly in the form of nucleosomes, is released into the circulation as cell-free nucleosomes (cf-nucleosomes), which retain some histone modifications. To detect the composition of cells/tissues that contribute to the cfDNA pool, they used published ChIP-seq data to define cell-type/tissue-specific signatures as promoters that have high signals only in one cell-type. Pathology-related changes in tumor transcriptional programs were then clearly identified. cfChIP-seq using low sequencing depth provides systemic and genome-wide information facilitates diagnosis and interrogation of physiological and pathological processes using blood samples.

The use of DNA methylation biomarkers in cancer thus looks very promising. Further validation studies of the potential cancer methylation biomarkers in prospective cohorts will find the most robust methylation biomarkers for clinical application.

Methylation biomarkers have been extensively characterized in cancer and may also be utilized in other diseases such as neurological, metabolic and cardiovascular disorders, autoimmune diseases, and genomic imprinting disorders. For example, assays using H19/IGF2 aberrant DNA methylation as a risk factor for colon cancer are commercially available for the neurodevelopmental disorders of genomic imprinting Prader–Willi syndrome (PWS) and Angelman syndrome (AS) [47]. AS is the most common age-dependent neurodegenerative disorder. Recent studies have shown that epigenetic modification is one of the pathogenic mechanisms of AD. Understanding of the epigenomes and transcriptomes of AD may warrant future identification of novel biological markers. The methylation frequency of HTERT, a mRNA component of telomerase, is higher in AD patients compared to elderly controls [48]. The association of HTERT methylation status with AD indicates that this gene may be involved in higher telomerase activity and immune dysfunctions in AD pathogenesis. In addition, other inflammatory genes such as iNOS, IL-1, and TNF-α are hypomethylated in the AD cortex [49]. This study suggests that many inflammatory genes are activated by epigenetic alterations and are involved in the pathogenesis of AD.

Genomic imprinting disorders include PWS, AS, Beckwith-Wiedemann syndrome (BWS), and Silver-Russell syndrome (SRS). To diagnose PWS or AS, the methylation status of the small nuclear ribonucleoprotein polypeptide N (SNRPN) gene locus or the methylation status and gene copy number changes at several sites across the region is commonly assessed [50]. BWS and SRS are disorders associated with imprinted genes on two differentially methylated regions (DMRs) of chromosome region 11p15.5. DRM1 consists of the imprinted insulin-like growth factor gene 2 (IGF2) and non-coding RNA H19. DRM2 contains several imprinted genes, including cyclin-dependent kinase inhibitor 1 C (CDKN1C), potassium voltage-gated channel, KQT-like subfamily member 1 (KCNQ1), and KCNQ1 opposite strand/antisense transcript 1 (KCNQ1OT1). Loss of methylation at DMR2 (KCNQ1OT1 hypomethylation) is the most frequent alteration in around 50% of BWS patients [51], whereas loss of methylation at DMR1 (H19 hypomethylation) is found in around 40% of SRS patients [52,53].

POTENTIAL M6A METHYLATION BIOMARKERS IN CANCER AND OTHER DISEASES

RNA epigenetics has received a great deal of attention in recent years and is an important mechanism for the regulation of RNA expression. The reversibility of epigenetics provides a scientific basis for early intervention and treatment of diseases. Five types of RNA methylation have been identified, and these involve N^1-methyladenosine (m1A), N^6-methyladenosine (m6A), 5-methylcytosine (m5C), N^7-methylguanine (m7G), and 2-O-methylation (2'OME) [54]. Among these, m6A is the most pervasive and highly conserved RNA modification in eukaryotic cells and affects almost all aspects of mRNA metabolism, including stability, alternative splicing, nuclear export, and translational efficiency [55,56]. Sequencing analysis showed that m6A modification is mainly concentrated in the common motif RRACH (R = G/A, H = A/C/U). In the RRACH motif, m6A modification is preferentially concentrated near the 3′-UTR, followed by the CDs and 5′-UTR regions [57,58]. The m6A modification is a highly dynamic and reversible process that involves enzymes responsible for installing modifications called "writers," removing methylation called "erasers," and recognizing modifications called "readers" (Fig. 17.3). Members of every class of regulators work together in a concerted manner to maintain a steady-state balance of m6A level inside the cell.

mRNA m6A modification has been demonstrated to play important roles in different physiological activities and human diseases. Mounting evidence has shown the importance of m6A methylation in embryonic development and stem cell regulation, including processes such as maintaining pluripotency and promoting differentiation [56,59,60]. Recent studies have demonstrated that m6A plays a critical role in controlling cancer cell proliferation, metastasis, immune evasion, and drug resistance [61–63]. It has a bifaceted role, i.e., acting as either an oncogene or a tumor suppressor in the pathogenesis of solid tumors and leukemia. m6A is also highly enriched in neural tissue and plays an important role in neurogenesis and neuronal disorders [64–66]. Other functional processes that involve m6A modification, include adipogenesis, development of obesity, and pathogenesis of type 2 diabetes [67,68]. Recent studies also demonstrate a critical role for m6A in the regulation of immune disorders and infections [69,70]. Huang et al. have demonstrated a significant increase in m6A RNA methylation in circulating tumor cells (CTCs) compared with whole blood cells, fueling subsequent investigations on RNA methylation in CTCs from lung cancer patients. This research facilitates early detection and characterization of upregulated m6A expression in CTCs, which may potentially be utilized in monitoring and preventing the development of overt metastatic diseases [71]. Pei et al. [72] also detected elevated m6A levels in peripheral blood leukocytes of non-small cell lung cancer patients by flow cytometry, indicating the potential use of m6A as a non-invasive biomarker. Furthermore, m6A regulators, including METTL3, WTAP, FTO, IGF2BPs, and YTHDFs, have been utilized in the prognosis of different types of cancers. m6A and its regulators may serve as biomarkers; however, its clinical application is hindered by the heterogeneity of m6A in patients and lack of assays to detect site-specific m6A from low-input clinical samples. Future single-cell sequencing techniques might be helpful in resolving such issues.

POTENTIAL HISTONE MODIFICATION BIOMARKERS IN CANCER AND OTHER DISEASES

Histone modifications regulate the structure of chromatin and play fundamental roles in gene regulation and expression. Thus, it is not surprising that aberrant patterns of histone marks are found in cancer and other diseases. For example, in CRC, two different studies have reported that global levels of acetylation of H4K12 (H4K12ac) and H3K18 (H3K18ac) increased in adenocarcinomas relative to normal tissues or adenoma [73,74]. Lower levels of dimethylation of H3K4

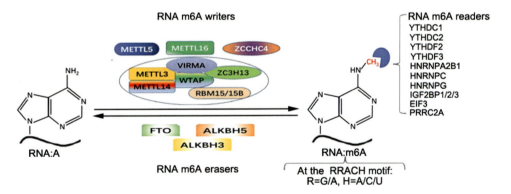

FIGURE 17.3 Dynamic reversible process and molecular composition of m6A. m6A modifications are catalyzed by m6A writers, erasers, and readers. The sixth N position of adenylate (A) can be methylated by m6A writers to form N6-methyladenosine (m6A). Writers include METTL3, METTL14, WTAP, RBM15/15B, VIRMA, and ZC3H13. METTL5, METTL16, and ZCCHC4 are independent m6A methyltransferases. Erasers are proteins with demethylase activity, and include FTO, ALKBH5, and ALKBH3. Readers are proteins that recognize m6A modifications and include the YTH family, HNRNP family, EIF3, IGF2BPs, and PRRC2A. m6A modifications are mainly concentrated in the common motif RRACH (R = G/A, H = A/C/U).

(H3K4me2) are correlated to poor survival rates. Multivariate survival analysis showed that H3K4me2 status is an independent prognostic factor in patients with CRC [75].

For clinical application, the detection of histone methylation in blood circulation was conducted [76]. Two histone methylation marks, H3K9me3 and H4K20me3, were investigated in blood circulating nucleosomes. H3K9me3 and H4K20me3 were found to be lower in patients with CRC in comparison with healthy controls [77]. Similar results were obtained in a subsequent study, including patients with CRC, breast, lung, and benign gastrointestinal disease [78]. These findings suggest that H3K9me3 and H4K20me3-related nucleosomes may be potentially utilized in CRC.

Changes in histone modifications have also been detected in other diseases such as neurodegenerative, autoimmune, and cardiovascular disorders. Many studies have shown that the aberrant expression of enzymes that induce covalent histone modifications, including histone deacetylases (HDACs), histone methylases, and demethylases, is often checked in these diseases [79].

However, the biomarker potential of histone marks faces more technical challenges compared with the diagnostic potential of DNA methylation, as histone modifications are less-stable modifications and have been shown to be more dynamic [80,81]. The assessment of histone marks is more difficult to employ and standardize because it relies on using antibodies, which vary in performance.

POTENTIAL NON-CODING RNA BIOMARKERS IN DISEASES

NcRNAs comprise a complex and diverse class that has been arbitrarily subdivided into two major groups, small ncRNAs and lncRNAs, according to their size (<200 nucleotides in length). Small ncRNAs include well-defined miRNAs, siRNAs, promoter- or antisense-associated short RNAs, and housekeeping RNAs. LncRNAs are classified into five general categories according to their location on the genome, namely, sense or antisense, intronic, bidirectional, and intergenic. Although the functions of most ncRNAs remain largely uncharacterized, numerous studies have provided evidence that ncRNAs are involved in many diseases and have illustrated their potential as novel biomarkers for various diseases [82,83]. However, only a few ncRNAs have been appropriately validated in independent cohorts or investigated for their function in disease development.

Potential miRNA biomarkers in cancer and other diseases

miRNAs are short (~22 nucleotides), endogenous non-protein-coding single-stranded RNA molecules that play role in posttranscriptional gene regulation through translational repression or inducing degradation of target messenger RNAs [84,85]. miRNAs contribute to various biological processes such as development, differentiation, and carcinogenesis [86]. Bioinformatics analyses indicate that each miRNA can control hundreds of gene targets, underscoring the potential influence of miRNAs on almost every genetic pathway. Extensive studies have shown that circulating miRNAs in a wide range of body fluids are useful biomarkers for the diagnosis and prognosis of disease [82–87].

In 2008, Chim et al. first identified the expression of miRNAs in circulation by qRT-PCR [88]. In the same year, upregulated miR-21 expression in serum was found to be linked to the relapse-free individuals by comparing B cell lymphoma patients with healthy controls [87]. MiR-21 is one of the first miRNAs that has been associated with a variety of cancers such as prostate cancer, pancreas cancer, liver cancer, GC, CRC, and NSCLC [89]. These studies have opened the door to the possibility of circulating miRNAs as potent, minimally invasive analytical, or diagnostics markers for different diseases, particularly cancer. These potential biomarker miRNAs have also been correlated to disease status, stage, aggressiveness, and response to therapy.

Subsequently, serum miR-141 has been reported to be capable of distinguishing prostate cancer patients from healthy controls, thereby providing evidence that tumor-restricted miRNAs may serve as blood-based biomarkers [90]. Recently, Chen et al. performed miRNA profiling of plasma from patients with prostate cancer or benign prostatic hyperplasia (BPH) and identified that a five-miRNA model (let-7c, let-7e, miR-30c, miR-622, and miR-1285) could differentiate prostate cancer from BPH, with an AUC of 0.924, and prostate cancer from healthy individuals, with AUC of 0.860. These miRNAs have been shown to improve the diagnostic performance of PSA assays [91].

In pancreatic ductal adenocarcinoma (PDAC), miRNAs were first reported as useful circulating biomarkers in 2009. Wang et al. reported a panel of four overexpressed plasma miRNAs (miR-21, -210, -155, and -196a) that could clearly distinguish patients from healthy controls, with a sensitivity of 64% and a specificity of 89% [92]. Recently, Kojima et al. detected 2555 human miRNAs using 100 serum samples from PDAC patients and 150 from healthy controls and found that a combination of eight miRNAs (miR-6075, -4294, -6880-5p, -6799-5p, -125a-3p, -4530, -6836-3p, and

-4476) had very high sensitivity (80.3%) and specificity (97.6%) [93].

Furthermore, Ng et al. [94] reported the combination of plasma miR-145 and -451 levels provided the best markers for breast cancer prediction, with a sensitivity of 90% and specificity of 92%, in discriminating breast cancer from control subjects, including all other types of cancers.

miRNAs in plasma could also serve as an early detection marker for GC and CRC. miR-17−5p, -21, -106a, and -106b were upregulated, while let-7a was downregulated in plasma of GC patients compared with healthy controls. The ratio of miR-106b/let-7a yielded the highest sensitivity and specificity, with an AUC value of 0.879 [95]. More interestingly, the diagnostic potential of another panel of five circulating miRNAs (miR-1, -20a, -27a, -34, and -423−5p) for GC was greater than standard testing with CEA and CA19−9 [96]. For CRC, elevated levels of plasma miR-29a and -92a were deemed potential early diagnostic markers for advanced adenoma patients [97].

Although most miRNAs studies have focused on cancer, current studies have revealed that a number of miRNAs may act as potential biomarkers for other diseases such as metabolic disorders, autoimmune, and inflammatory diseases. For instance, miR-19b, -106a, and -629 or -191 could distinguish Crohn's disease from ulcerative colitis [98,99]. Moreover, miRNA let-7d and let-7e were identified as possible therapeutic biomarkers in patients with Crohn's disease [100]. Circulating miRNAs may also represent good markers for heart failure (HF) and response to therapeutic treatment. In humans, urinary and plasma miR-21 have been associated with severe acute kidney injury and poor outcomes after adult cardiac surgery [101]. Circulating levels of miR-423−5p and miR-18b* were able to distinguish HF patients from controls, while miR-423−5p could differentiate HF patients from non-HF patients with other causes of dyspnea [102]. In addition, miR-1 is not only a potential indicator of acute myocardial infarction (AMI) but also a determinant of AMI size and recovery [103−105]. Other potential miRNA biomarkers are listed in Table 17.2.

Potential lncRNA biomarkers in cancer and other diseases

LncRNAs play important regulatory roles in transcription, translation, chromatin modification, and cellular organization. Misregulation of lncRNAs has been associated with various human diseases. Although lncRNAs have only recently been discovered, at least 321 experimentally verified lncRNAs have been associated with 221 various types of diseases, with most related to cancer [106]. lncRNAs represent another group of potential biomarkers for cancer diagnosis and prognosis.

Prostate cancer antigen 3 (PCA3 lncRNA or DD3) is a well-studied lncRNA and the most specific to prostate cancer (PCa) as it is not expressed in other normal human tissues. PROGENSA PCA3 test is the first urine-based molecular diagnostic test approved by the US Food and Drug Administration [107]. The sensitivity and specificity of urine PCA3 expression for PCa diagnosis reach 62% and 75%, respectively, supporting PCA3 as a reasonable marker for prostate cancer diagnosis [108].

Urothelial carcinoma-associated 1 (UCA1) is an lncRNA dysregulated in a broad range of human cancers [109]. UCA1 has been shown to be a promising biomarker for bladder cancer diagnosis and therapeutic monitoring. UCA1 levels in blood or urine of bladder cancer patients could be used to distinguish bladder cancer from other urinary tract diseases such as neurogenic bladder, renal cell carcinoma, and upper urinary tract restriction or reflux, with an overall AUC of 0.882 and an overall specificity of 91.8% [110,111]. UCA1 has also been shown to be overexpressed in the serum of HCC patients than in patients with chronic hepatitis C virus infection or healthy volunteers [112] and in gastric juice from GC patients than in normal subjects [113]. Recently, meta-analysis was conducted to assess the association between UCA1 and cancer prognosis. UCA1 levels were significantly and negatively associated with overall survival times in CRC, NSCLC, ovarian cancer, and GC [114].

HOX transcript antisense RNA (HOTAIR) is another well-studied lncRNA. In cervical cancers, high serum levels of HOTAIR were significantly correlated with tumor recurrence and shorter overall survival [115].

Studies have shown that the combination of several lncRNAs could considerably increase the biomarker efficiency. For example, Tang et al. showed that plasma RP11−160H22.5, XLOC_014172, and LOC149086 transcripts were also overexpressed in HCC. The sensitivity and specificity for HCC diagnosis reached 82% and 73%, with a merged AUC of 0.896. More interestingly, lncRNAs XLOC_014172 and LOC149086 also have prognostic value for metastasis prediction, with sensitivity and specificity of 91% and 90% [116].

Dong et al. found that the combination of CUDR, LSINCT-5, and PTENP1 provided the best diagnostic value in GC, with an AUC of 0.92, a sensitivity of 74.1%, and a specificity of 100%. They were also sufficiently sensitive and specific for early GC detection and distinguishing benign peptic ulcers from GC [117].

In addition, Shi et al. found a panel of three lncRNAs (XLOC_006844, LOC152578, and XLOC_000303) that were significantly upregulated in CRC plasma samples compared to those from healthy controls [118]. They

TABLE 17.2 Representative Potential MiRNA Biomarkers in Cancer and Other Diseases

MiRNAs	Related diseases	Application
miR-21	Non-small cell lung cancer, colorectal cancer, laryngeal squamous cell carcinoma, hepatocellular cancer, esophageal cancer, squamous cell carcinoma, gastric cancer, ovarian cancer, breast cancer, pancreatic denocarcinoma, cervical cancer, and glioblastoma, B cell lymphoma	Diagnostic and prognostic biomarker
miR-9	Osteosarcoma, nasopharyngeal carcinoma, non-small cell lung cancer, biliary tract cancer	Prognostic biomarker
miR-210	Colorectal cancer, cell renal cell carcinoma	Diagnostic and prognostic biomarker
miR-200c	Gastric cancer, colorectal cancer	Diagnostic and prognostic biomarker
miR-205	Bladder cancer, prostate cancer	Diagnostic and prognostic biomarker
miR-150	Colorectal cancer	Diagnostic and prognostic biomarker
miR-1825/ miR-484	prostate cancer	Diagnostic biomarker
miR-103	Breast cancer	Potential diagnostic biomarker
miR-16	Melanoma	Diagnostic and prognostic biomarker
miR-10a	Cervical cancer, breast cancer, bladder cancer, gastric cancer	Potential biomarkers of aggressive progression and poor prognosis
miR-18a	Esophageal cancer, pancreatic cancer, hepatocellular cancer, colorectal cancer	Potential diagnostic biomarker
miR-155	B-cell malignancies, chronic lymphocytic leukemia, breast cancer, esophageal cancer	Diagnostic and prognostic biomarker
miR-221	Larynx cancer, breast cancer	Diagnostic and prognostic biomarker
miR-20a	Cervical cancer, multiple myeloma, rectal cancer, prostate cancer, gastric cancer,	Potential biomarkers of aggressive progression and poor prognosis
miR-152	Bladder cancer, non-small-cell lung cancer	Diagnostic biomarker
let-7c	Non-small-cell lung cancer	Diagnostic biomarker
miR-125b	Ewing sarcoma, non-small-cell lung cancer, rectal adenocarcinoma	Diagnostic and prognostic biomarker
miR-126	Colorectal, esophageal cancer, hepatocellular carcinoma, acute myeloid leukemia, gastric cancer, thyroid cancer, breast cancer, oral cancer, and prostate cancer	Diagnostic and prognostic biomarker
miR-1260b	Colorectal cancer	Potential prognostic biomarker
miR-22	Osteosarcoma	Diagnostic and prognostic biomarker
miR-744	Pancreatic cancer	Diagnostic and prognostic biomarker
miR-25	Esophageal squamous cell carcinoma	Diagnostic and prognostic biomarker
miR-196a	Breast cancer	Prognostic biomarker
miR-193b	Ovarian cancer	Diagnostic and prognostic biomarker
miR-34a	Prostate cancer	Diagnostic and prognostic biomarker
miR-221	Larynx cancer, breast cancer	Diagnostic and prognostic biomarker
miR-19b/miR-106a/miR-629	Crohn disease	Diagnostic biomarker
miR-191	Ulcerative colitis, Crohn disease	Diagnostic biomarker
let-7c	Cervical intraepithelial lesions	Diagnostic biomarker
let-7d	Crohn disease	Prognostic biomarker
let-7e	Crohn disease	Prognostic biomarker

(Continued)

TABLE 17.2 (Continued)

MiRNAs	Related diseases	Application
miR-10a	Coronary artery disease	Potential diagnostic biomarker
miR-210	Preeclampsia	Diagnostic and prognostic biomarker
miR-155	Ankylosing spondylitis, male fertility	Diagnostic and prognostic biomarker
miR-150	Inflammatory active disease	Early diagnostic biomarker
miR-150–5p	Myasthenia gravis	Diagnostic biomarker
miR-23a	Pre-diabetes and type 2 diabetes	diagnostic biomarker
miR-125b	Alzheimer disease	Diagnostic biomarker

have a strong diagnostic value, with an AUC of 0.919 in the training set and an AUC of 0.975 in the validation set.

Dysregulation of lncRNAs is involved not only in cancer processes but also in a range of other disease types, in which the two top categories are cardiovascular diseases (including HF, atherosclerosis, and coronary artery disease) and neurodegenerative disorders (including schizophrenia, AD, and Huntington disease).

Serum lncRNA LIPCAR is considered a novel biomarker of cardiac remodeling and is predictive of mortality in HF patients [119]. In patients with ischemic HF, three lncRNAs (CDKN2B-AS1/ANRIL, HOTAIR, and LOC285194/TUSC7) showed similar modulation in peripheral blood mononuclear cells and heart tissues, suggesting a potential role as disease biomarkers [120]. In addition, circulating UCA1 could be potentially used as a novel biomarker for the diagnosis and/or prognosis of AMI [121].

In neurodegenerative diseases, lncRNAs FMR1-AS1 and FMR6 are expressed in peripheral blood leukocytes and may represent early diagnostic biomarkers for fragile X syndrome (FXS) and fragile X tremor/ataxia syndrome (FXTAS). Kraus et al. reported that five lncRNAs are significantly differentially expressed in Parkinson disease (PD), of which H19 upstream conserved 1 and 2 are significantly downregulated, while lincRNA-p21, MALAT1, SNHG1, and TncRNA are significantly upregulated. These may serve as potential new biomarkers even in early PD [122].

As for other diseases, the expression profile of lncRNAs in labial salivary glands of primary Sjögren syndrome (pSS) patients revealed that many novel lncRNA transcripts that contribute to the pathogenesis of pSS are dysregulated in pSS and aid in the development of new diagnostic biomarkers [123]. Xu et al. [124] found that the expression of two lncRNAs (Vax2os1 and Vax2os2) were significantly upregulated in the aqueous humor of choroidal neovascularization patients, making them predictive biomarkers for the diagnosis of ocular neovascular diseases [125]. In addition, MALAT1 is upregulated in peripheral blood samples from proliferative vitreoretinopathy (PVR) patients, implying that it may represent an easily detectable biomarker for minimally invasive diagnosis to identify high-risk PVR patients [126]. Moreover, recently a genome-wide long non-coding RNA analysis identified that a panel of five circulating lncRNAs (NR_038395, NR_038452, ENST00000482343, ENST00000544649, and ENST00000393610) was a potential minimally invasive biomarker for endometriosis [127]. Additional potential lncRNA biomarkers are listed in Table 17.3.

The future development of lncRNA applications relies on technologies to identify and validate their functions, structures, and mechanisms. Comprehensive understanding of genome-wide interaction networks of lncRNAs with proteins, chromatins, and other RNAs in regulating cellular processes will allow personalized medicine to use lncRNAs as highly specific biomarkers in diagnosis, prognosis, and therapeutic targets.

EPIGENETIC BIOMARKER DETECTION METHODS IN THE CLINIC

Methods in epigenetics have been discussed in the section "Methods in Epigenetics." In this chapter, we focus on epigenetic biomarker detection methods and challenges in their clinical application.

In general, these techniques can be divided into various types of pretreatment (enzyme digestion, affinity enrichment, and sodium bisulfite) followed by different analytical steps (locus-specific, gel-based, array-based, and next-generation sequencing-based analysis) [128].

Bisulfite treatment of DNA is widely used for the determination of DNA methylation status. During bisulfite treatment, unmethylated cytosines are converted into uracils,

TABLE 17.3 Representative Potential LncRNA Biomarkers in Cancer and Other Diseases

LncRNAs	Related diseases	Application
HOTAIR	Lung cancer, breast cancer, colorectal cancer, esophageal cancer, laryngeal cancer, nasopharyngeal cancer, hepatocellular cancer, gastric cancer, pancreatic carcinoma, non-small cell lung cancer, mesenchymal glioma	Prognostic biomarker of metastasis
SRA	Breast cancer, uterus cancer, ovary carcinomas	Co-activator of estrogen receptor alpha
PANDAR	Bladder cancer, gastric cancer, hepatocellular carcinoma, non-small cell lung cancer	Diagnostic and prognostic biomarker
MALAT1	Lung cancer, breast cancer, colorectal cancer, bladder cancer, kidney cancer, endometrium cancer, pancreatic carcinoma	Predict poor overall survival
LincRNA-p21	Lung cancer, colorectal cancer, leukemia, head and neck cancer	Predict the treatment response
H19	Prostate cancer, glioma cancer, gastric cancer, breast cancer, bladder cancer, kidney cancer	Putative susceptibility and diagnostic biomarker
GAS5	Prostate cancer, breast cancer, kidney cancer	Predict metastasis and poor prognosis
ANRIL	Breast cancer, melanoma, leukemia	Prognostic biomarker
MEG3	Breast cancer, lung cancer, bladder cancer, kidney cancer, leukemia, osteosarcoma	Predictor biomarker in progression and poor prognosis
XIST	Prostate cancer, breast cancer	Diagnostic and prognostic biomarker
GCAT1	Gastric cancer	Potential diagnostic and prognostic biomarker
SUMO1P3	Gastric cancer, bladder cancer	Potential diagnostic and prognostic biomarker
CCAT1	Gastric cancer, colorectal cancer, breast cancer, Hepatocellular carcinoma, gallbladder cancer	Potential diagnostic and prognostic biomarker
ENST00000435885.1/ XLOC 013014/ ENST00000547963.1	Esophagus cancer	Prognostic biomarker
uc.73	Colorectal cancer	Prognostic biomarker
HULC	Liver cancer, gastric cancer, Osteosarcoma	Diagnostic and prognostic biomarker
MALAT-1	Liver cancer, lung adenocarcinomas, bladder cancer	Predict poor prognosis
HOTTIP	Liver cancer, gastric cancer, prostate cancer, colorectal cancer, osteosarcoma, hepatocellular carcinoma	Predicts poor prognosis
HOXA13	Liver cancer, non-small cell lung cancer, gastric cancer, glioma	Diagnostic biomarker and promotes invasion
TUG1	Non-small cell lung cancer, colorectal cancer, gastric cancer	Poor prognosis biomarker
BANCR	Non-small cell lung cancer, melanoma	Associated with clinical progression and poor prognosis
LincRNA-RoR	Breast cancer, colorectal cancer, glioma, pancreatic cancer	Promotes invasion, metastasis and tumor growth
UCA1	Osteosarcoma, bladder cancer, renal cell carcinoma, epithelial ovarian cancer, gastric cancer	Poor prognostic biomarker

(Continued)

TABLE 17.3 (Continued)

LncRNAs	Related diseases	Application
linc-UBC1	Bladder cancer, gastric cancer	Negative prognostic biomarker
PCA3/ TMPRSS2-ERG	Prostate cancer	Diagnostic biomarkers
HOTAIRM1	Leukemia, colorectal cancer	Potential diagnostic biomarker
UCA1	Acute myocardial infarction	Diagnostic biomarker
LIPCAR	Heart failure patients	Diagnostic and prognostic biomarker
ANRIL/HOTAIR /TUSC7	Ischemic heart failure	Potential role as disease biomarkers
H19 upstream conserved 1 and 2/ lincRNA-p21/ Malat1/ SNHG1/TncRNA	Parkinson disease	Potential diagnostic biomarkers
MALAT1	Proliferative vitreoretinopathy	Diagnostic biomarker
Vax2os1/Vax2os2	Ocular neovascular diseases	Diagnostic biomarker
NR_038395/NR_038452/ ENST00000482343/ENST00000544649/ ENST00000393610	Endometriosis	Diagnostic biomarker

whereas methylated cytosines remain intact. However, after bisulfite conversion, 5-hydroxymethylcytosines also remain intact, indicating overestimation of the DNA methylation level in some cases. In addition, bisulfite treatment always leads to significant DNA degradation. Thus, methylation-sensitive restrictive enzyme—mediated methods are considered to be more sensitive than bisulfite treatment methods for detecting low-abundance methylated TSGs in liquid biopsy cfDNA.

Conventional methylation-specific PCR (MSP) is frequently used for the detection of DNA methylation. However, it involves qualitative analysis and is thus not suitable in clinical settings. In contrast, quantitative MSP analysis such as MethyLight, pyrosequencing, or SMART-MSP might be more suitable for clinical applications. However, the limit of this detection is as low as 5% for the each CpG; it might not have enough power to assess the low number of methylated DNA fragments in cfDNA. New quantitative DNA methylation assays have recently emerged such as Methyl-BEAMing based on emulsion PCR assay [129] and MethyLight coupled with digital PCR [130]. These have been more sensitive digital approaches, have less bias during sample preparation, and generate more accurate data, making them a revolutionary approach to DNA methylation analysis. There is another important issue that deserves attention when using PCR amplification analysis. Multiple splice variants of detected genes exist and specific primers should be designed accordingly. In addition, heterogeneity at the subject and tissue levels can also affect reproducibility in target gene-focused investigations.

Because research on DNA methylation has evolved from a locus-specific approach to genome-wide determination of the methylome at base pair resolution, array-based technology with high reproducibility and accuracy is widely used in the analysis of clinical samples [131]. Whole-genome shotgun bisulfite sequencing (WGBS-seq) is a more comprehensive approach to DNA methylation analysis. However, it is time-consuming and too expensive for analyzing multiple clinical samples [132]. To reduce the amount of sequencing, targeted bisulfite sequencing methods have been developed using array capture or affinity enrichment of methylated DNA using antibody for methylated cytosine or methyl-binding proteins before sequencing [129]. Reduced representation bisulfite sequencing (RRBS) is a semi-targeted bisulfite sequencing method, which analyze DNA fragments after methylation-insensitive restriction enzyme (MspI) digestion [133]. Today, next-generation sequencing (NGS) has greatly enhanced precision and resolution in preclinical and clinical epigenetic studies. It is combined with other techniques to identify epigenetic marks at the genomic scale. Targeted DNA methylation sequencing by NGS is reported as a novel DNA methylation analysis for cfDNA [134]. However, further validation studies may be still required when applying this method for clinical application.

In terms of histone modification biomarkers, antibodies to acetylated histone H3 and H4 have been used for ChIP to determine histone acetylation in specific regions of gene promoter and other regulatory regions. Genome-wide analysis was initially conducted using a technique known as ChIP-chip. ChIP-chip is now being superseded by ChIP-seq, which combines ChIP with NGS technologies involving high-throughput sequencing of all enriched DNA sequences. It provides the opportunity for increased sensitivity, higher resolution, and more comprehensive screening of genomic profiles while requiring less input material than ChIP-chip. However, because NGS generally produces a notoriously large quantity of data, more powerful bioinformatics support is essential for data processing and analysis [135].

Methylated RNA immunoprecipitation sequencing (MeRIP-Seq or m6A seq) is the mainstay for transcriptome-wide m6A profiling. This technique, analogous to ChIP-Seq in the mapping of histone modifications, relies on a specific anti-m6A antibody to pull down m6A-containing RNA fragments, which can then be mapped by NGS. However, this technique detects m6A-containing RNA fragments rather than specific m6A-modified sites. The resolution of this method is therefore limited by the size of the RNA fragment pulled down, typically 100−200 nt. The resolution of m6A profiling can be improved by combining antibody-based immunoprecipitation with photo-, as is seen with PA-m6A-Seq (photo-crosslinking-assisted m6A sequencing) and miCLIP (m6A individual nucleoside resolution and crosslinking immunoprecipitation). miCLIP is currently the most widely used technique for transcriptome-wide m6A mapping [136].

The recent development of third-generation single-molecule sequencing technology empowers direct detection of nucleotide sequence and modifications in RNA, which is emerging as an ideal platform for transcriptome-wide m6A profiling. Most recently, Lorenz et al. demonstrated the ability of nanopore-based sequencing to detect m6A modifications in endogenous mRNA transcripts [137]. This quickly evolving m6A detection method is expected to markedly accelerate the discovery and validation of m6A modification sites in the human transcriptome. This information will generate a more comprehensive m6A landscape for human cancers and eventually facilitate the development of new biomarkers for cancer diagnosis and molecular classifications.

miRNA or lncRNA biomarkers can be easily measured by qRT-PCR using the TaqMan PCR method with stem-looped RT primers or oligo-dT primers. Several novel detection methods have been developed to allow for rapid quantification of RNA and improve sensitivity and specificity, including the use of simple molecular beacons, enzymatic luminescence, nanoparticle-based probes, and different forms of electrophoresis [138]. High-throughput miRNA and lncRNA quantification technologies, including comprehensive RNA sequence analysis by NGS, are convenient clinical measuring tools. However, in terms of exosomal miRNA detection in the clinical setting, there remains a need for a rapid and inexpensive standard procedure that allows effective and pure isolation of exosomes and accurate genome-wide quantification [139].

Technical analyses of epigenetic alterations are currently under development, and researchers face a number of challenges when studying epigenetic changes. Technique selection depends on the quality and quantity of input DNA needed, purity and type of tissue or fluid DNA that is being extracted, extent of genome coverage, and overall assay reproducibility, sensitivity, specificity, accuracy, and quantification. Improvements in extraction and amplification procedures and growing availability of NGS technologies may allow the discovery of more specific and sensitive RNA markers in the near future.

CHALLENGES AND FUTURE PERSPECTIVES

Studies have revealed a number of epigenetic changes that could be used as disease biomarkers, thereby providing clinical value in diagnosis and prognosis of disease. However, only a few of these epigenetic marks have been appropriately validated in independent cohorts or investigated in large, prospective clinical trials. Many results are only from primary tissue biopsy and may totally differ from the liquid biopsy of the same patient. Moreover, for those minimally invasive epigenetic biomarker detection methods, insufficient material and technical shortcomings remain as the main challenges faced by researchers. For example, most current studies still use bisulfite-treated DNA methylation detection. A considerable loss of initial input DNA can occur, especially in those samples where only a limited amount of genomic DNA is available such as those from very small biopsies. In addition, tissue-specific methylation traits have not all been identified, and there are overlaps between the methylation signature among tissue types as well as cancers of different origin. Therefore, although methylation might serve as a lead to finding the origin of a tumor, it is currently difficult to be fully certain of the origin of a cfDNA sequence. Moreover, an agreed methodology as the gold standard for epigenetic alterations has yet to be established. A number of limitations need to be resolved before these biomarkers may be utilized in the clinic.

Advances in study and clinical trial designs will certainly help speed up validation and clinical implementation of potential epigenetic biomarkers. Standardization of specimen type and of sample preparation is also necessary when results from different groups or samples are to be meaningfully obtained and compared. Ideally, epigenetic biomarker detection methods for the clinical setting should be inexpensive, easy to use, automatable, and capable of processing many samples in parallel to minimize costs. There is an urgent need to develop cost-effective genome-wide technologies with high sensitivity and specificity that use minimally invasive samples. In addition, a panel of multiple disease-specific biomarkers will diagnose and predict disease prognosis more accurately than a single marker. Moreover, genetics and epigenetics act together and play important roles in disease development, and the combination of different epigenetic, genetic, or protein biomarkers will significantly increase sensitivity and specificity. In addition, powerful bioinformatics platforms will play a crucial role in this type of research as these have been used in all aspects of miRNA research.

It is fully anticipated that there will be a rapid improvement in technological capability and in gaining new insights into the role and mechanisms of epigenetic phenomena in development and disease. Additional epigenetic biomarkers will be successfully integrated into clinical practice in the near future.

Abbreviations

2'OME	2-O-methylation
Ac	Acetylation
AD	Alzheimer disease
AMI	Acute myocardial infarction
AS	Angelman syndrome
BPH	Benign prostatic hyperplasia
BWS	Beckwith-Wiedemann syndrome
CDKN1C	Cyclin-dependent kinase inhibitor 1C
cfDNA	Cell-free DNA or circulating free DNA
CRC	Colorectal cancer
DMRs	Differentially methylated regions
FXS	Fragile X syndrome
FXTAS	Fragile X tremor/ataxia syndrome
HCC	Hepatocellular carcinoma
HDACs	Histone deacetylases
HF	Heart failure
IGF2	Insulin-like growth factor gene 2
KCNQ1	KQT-like subfamily, member 1
KCNQ1OT1	KCNQ1 opposite strand/antisense transcript 1
lncRNAs	Long non-coding RNAs
m1A	N1-methyla-denosine
m5C	5-Methylcytosine
m6A	N6-methyladenosine
m7G	N7-methylguanine
Me	Methylation
MeRIP-Seq	Methylated RNA immunoprecipitation sequencing
MGMT	O6-methylguanine DNA methyltransferase
miCLIP	m6A individual nucleoside resolution and crosslinking immunoprecipitation
miRNAs	MicroRNAs
MSP	Methylation-specific PCR
NGS	Next-generation sequencing
NSCLC	Non-small cell lung cancer
P	Phosphorylase
PA-m6A-Seq	Photo-crosslinking-assisted m6A sequencing
PCa	Prostate cancer
PCA3	Prostate cancer antigen 3
PD	Parkinson disease
PDAC	Pancreatic ductal adenocarcinoma
pSS	Sjögren syndrome
PVR	Proliferative vitreoretinopathy
PWS	Prader–Willi syndrome
RRBS	Reduced representation bisulfite sequencing
SHOX2	Short stature homeobox 2
SNRPN	Small nuclear ribonucleoprotein polypeptide N
SRS	Silver-Russell syndrome
UCA1	Urothelial carcinoma-associated 1
WGBS-seq	Whole-genome shotgun bisulfite sequence

References

[1] Ushijima T, Nakajima T, Maekita T. DNA methylation as a marker for the past and future. J Gastroenterol 2006;41(5):401–7.

[2] Mastroeni D, McKee A, Rogers J, Coleman PD. Epigenetic differences in cortical neurons from a pair of monozygotic twins discordant for Alzheimer's disease. PLoS One 2009;4(8):e6617.

[3] Kolble K, Ullrich OM, Pidde H, Barthel B, Diermann J, Rudolph B, et al. Microsatellite alterations in serum DNA of patients with colorectal cancer. Lab Invest 1999;79(9):1145–50.

[4] Hu Y, Ulrich BC, Supplee J, Kuang Y, Lizotte PH, Feeney NB, et al. False-positive plasma genotyping due to clonal hematopoiesis. Clin Cancer Res 2018;24(18):4437–43.

[5] Paluszczak J, Baer-Dubowska W. Epigenetic diagnostics of cancer—the application of DNA methylation markers. J Appl Genet 2006;47(4):365–75.

[6] Dor Y, Cedar H. Principles of DNA methylation and their implications for biology and medicine. Lancet 2018;392(10149):777–86.

[7] Locke WJ, Guanzon D, Ma C, Liew YJ, Duesing KR, Fung KYC, et al. DNA methylation cancer biomarkers: translation to the clinic. Front Genet 2019;10:1150.

[8] Ferracin M, Negrini M. Micromarker 2.0: an update on the role of microRNAs in cancer diagnosis and prognosis. Expert Rev Mol Diagn 2015;15(10):1369–81.

[9] Volkmar M, Dedeurwaerder S, Cunha DA, Ndlovu MN, Defrance M, Deplus R, et al. DNA methylation profiling identifies epigenetic dysregulation in pancreatic islets from type 2 diabetic patients. EMBO J 2012;31(6):1405–26.

[10] Jubb AM, Quirke P, Oates AJ. DNA methylation, a biomarker for colorectal cancer: implications for screening and pathological utility. Ann NY Acad Sci 2003;983(1):251–67.

[11] Sanchez Y, Huarte M. Long non-coding RNAs: challenges for diagnosis and therapies. Nucleic Acid Ther 2013;23(1):15–20.

[12] Khoury S, Tran N. Circulating microRNAs: potential biomarkers for common malignancies. Biomark Med 2015;9(2):131–51.

[13] Djebali S, Davis CA, Merkel A, Dobin A, Lassmann T, Mortazavi A, et al. Landscape of transcription in human cells. Nature 2012;489(7414):101–8.

[14] Brock MV, Hooker CM, Ota-Machida E, Han Y, Guo M, Ames S, et al. DNA methylation markers and early recurrence in stage I lung cancer. N Engl J Med 2008;358(11):1118–28.

[15] Schwarzenbach H, Hoon DS, Pantel K. Cell-free nucleic acids as biomarkers in cancer patients. Nat Rev Cancer 2011;11(6):426–37.

[16] Laird PW. The power and the promise of DNA methylation markers. Nat Rev Cancer 2003;3(4):253–66.

[17] Holdenrieder S, Von Pawel J, Nagel D, Stieber P. Long-term stability of circulating nucleosomes in serum. Anticancer Res 2010;30(5):1613–15.

[18] Wang H, Hou L, Li A, Duan Y, Gao H, Song X. Expression of serum exosomal microRNA-21 in human hepatocellular carcinoma. BioMed Res Int 2014;2014:864894.

[19] Eichelser C, Stückrath I, Müller V, Milde-Langosch K, Wikman H, Pantel K, et al. Increased serum levels of circulating exosomal microRNA-373 in receptor-negative breast cancer patients. Oncotarget 2014;5(20):9650–63.

[20] Li Q, Shao Y, Zhang X, Zheng T, Miao M, Qin L, et al. Plasma long noncoding RNA protected by exosomes as a potential stable biomarker for gastric cancer. Tumor Biol 2015;36(3):2007–12.

[21] Braakhuis BJ, Tabor MP, Kummer JA, Leemans CR, Brakenhoff RH. A genetic explanation of Slaughter's concept of field cancerization: evidence and clinical implications. Cancer Res 2003;63(8):1727–30.

[22] Ushijima T. Epigenetic field for cancerization. J Biochem Mol Biol 2007;40(2):142–50.

[23] Dakubo GD, Jakupciak JP, Birch-Machin MA, Parr RL. Clinical implications and utility of field cancerization. Cancer Cell Int 2007;7:2.

[24] Slaughter DP, Southwick HW, Smejkal W. Field cancerization in oral stratified squamous epithelium; clinical implications of multicentric origin. Cancer 1953;6(5):963–8.

[25] Ramachandran K, Singal R. DNA methylation and field cancerization. Epigenomics 2012;4(3):243–5.

[26] Ushijima T. Epigenetic field for cancerization: its cause and clinical implications. BMC Proc 2013;7(Suppl. 2):K22.

[27] Issa JPJ, Ahuja N, Toyota M, Bronner MP, Brentnall TA. Accelerated age-related CpG island methylation in ulcerative colitis. Cancer Res 2001;61(9):3573–7.

[28] Milicic A, Harrison LA, Goodlad RA, Hardy RG, Nicholson AM, Presz M, et al. Ectopic expression of P-cadherin correlates with promoter hypomethylation early in colorectal carcinogenesis and enhanced intestinal crypt fission in vivo. Cancer Res 2008;68(19):7760–8.

[29] Sato F, Hatano E, Kitamura K, Myomoto A, Fujiwara T, Takizawa S, et al. MicroRNA profile predicts recurrence after resection in patients with hepatocellular carcinoma within the Milan Criteria. PLoS One 2011;6(1):e16435.

[30] Hur K, Cejas P, Feliu J, Moreno-Rubio J, Burgos E, Boland CR, et al. Hypomethylation of long interspersed nuclear element-1 (LINE-1) leads to activation of proto-oncogenes in human colorectal cancer metastasis. Gut 2014;63(4):635–46.

[31] Daskalos A, Nikolaidis G, Xinarianos G, Savvari P, Cassidy A, Zakopoulou R, et al. Hypomethylation of retrotransposable elements correlates with genomic instability in non-small cell lung cancer. Int J Cancer 2009;124(1):81–7.

[32] Park SY, Kim BH, Kim JH, Cho NY, Choi M, Yu EJ, et al. Methylation profiles of CpG islands loci in major types of human cancer. J Korean Med Sci 2007;22(2):311–17.

[33] Heyn H, Esteller M. DNA methylation profiling in the clinic: applications and challenges. Nat Rev Genet 2012;13(10):679–92.

[34] Paska AV, Hudler P. Aberrant methylation patterns in cancer: a clinical view. Biochem Med (Zagreb) 2015;25(2):161–76.

[35] Tetzner R, Model F, Weiss G, Schuster M, Distler J, Steiger KV, et al. Circulating methylated SEPT9 DNA in plasma is a biomarker for colorectal cancer. Clin Chem 2009;55(7):1337–46.

[36] Church TR, Wandell M, Lofton-Day C, Mongin SJ, Burger M, Payne SR, et al. Prospective evaluation of methylated SEPT9 in plasma for detection of asymptomatic colorectal cancer. Gut 2014;63(2):317–25.

[37] Johnson DA, Barclay RL, Mergener K, Weiss G, König T, Beck J, et al. Plasma Septin9 vs fecal immunochemical testing for colorectal cancer screening: a prospective multicenter study. PLoS One 2014;9(6):e98238.

[38] Ned RM, Melillo S, Marrone M. Fecal DNA testing for colorectal cancer screening: the ColoSureTM test. PLOS Curr 2011;3 RRN1220.

[39] Imperiale TF, Ransohoff DF, Itzkowitz SH, Levin TR, Lavin P, Lidgard GP, et al. Multitarget stool DNA testing for colorectal-cancer screening. N Engl J Med 2014;2014(370):1287–97.

[40] Schmidt B, Liebenberg V, Dietrich D, Schlegel T, Kneip C, Seegebarth A, et al. SHOX2 DNA methylation is a biomarker for the diagnosis of lung cancer based on bronchial aspirates. BMC Cancer 2010;10:600.

[41] Kneip C, Schmidt B, Seegebarth A, Weickmann S, Fleischhacker M, Liebenberg V, et al. SHOX2 DNA methylation is a biomarker for the diagnosis of lung cancer in plasma. J Thorac Oncol 2011;6(10):1632–8.

[42] Hegi ME, Diserens AC, Gorlia T, Hamou MF, de Tribolet N, Weller M, et al. MGMT gene silencing and benefit from temozolomide in glioblastoma. N Engl J Med 2005;352(10):997–1003.

[43] Kun S, Jiang P, Allen Chan KC, Wong J, Cheng YKY, Liang RHS, et al. Plasma DNA tissue mapping by genome wide methylation sequencing for noninvasive prenatal, cancer, and transplantation assessments. Proc Natl Acad Sci USA 2015;112:E5503–12.

[44] Lehmann-Werman R, Neiman D, Zemmour H, Moss J, Magenheim J, Vaknin-Dembinsky A, et al. Identification of tissuespecific cell death using methylation patterns of circulating DNA. Proc Natl Acad Sci USA 2016;113:E1826–34.

[45] GAi W, Ji L, Lam WKJ, Sun K, Jiang P, Chan AWH, et al. Liver- and colon-specific DNA methylation markers in plasma for investigation of colorectal cancers with or without liver metastases. Clin Chem 2018;64(8):1239–49.

[46] Wojdacz TK, Dobrovic A, Algar EM. Rapid detection of methylation change at H19 in human imprinting disorders using methylation-sensitive high-resolution melting. Hum Mutat 2008;29(10):1255–60.

[47] Silva PNO, Gigek CO, Leal MF, Bertolucci PHF, de Labio RW, Payao SLM, et al. Promoter methylation analysis of SIRT3, SMARCA5, HTERT and CDH1 genes in aging and Alzheimer's disease. J Alzheimer's Dis 2008;13(2):173–6.

[48] Akiyama H, Barger S, Barnum S, Bradt B, Bauer J, Cole GM, et al. Inflammation and Alzheimer's disease. Neurobiol Aging 2000;21(3):383–421.

[49] Ramsden SC, Clayton-Smith J, Birch R, Buiting K. Practice guidelines for the molecular analysis of Prader-Willi and Angelman syndromes. BMC Med Genet 2010;11:70.

[50] Smilinich NJ, Day CD, Fitzpatrick GV, Caldwell GM, Lossie AC, Cooper PR, et al. A maternally methylated CpG island in KvLQT1 is associated with an antisense paternal transcript and loss of imprinting in Beckwith-Wiedemann syndrome. Proc Natl Acad Sci USA 1999;96(14):8064–9.

[51] Eggermann T, Begemann M, Binder G, Spengler S. Silver-Russell syndrome: Genetic basis and molecular genetic testing. Orphanet J Rare Dis 2010;5:19.

[52] Horike S, Ferreira JC, Meguro-Horike M, Choufani S, Smith AC, Shuman C, et al. Screening of DNA methylation at the H19 promoter or the distal region of its ICR1 ensures efficient detection of chromosome 11p15 epimutations in Russell-Silver syndrome. Am J Med Genet A 2009;149A(11):2415–23.

[53] Roundtree IA, Evans ME, Pan T, He C. Dynamic RNA modifications in gene expression regulation. Cell 2017;169:1187–200.

[54] Wang X, Lu Z, Gomez A, Hon GC, Yue Y, Han D, et al. N6-methyladenosine-dependent regulation of messenger RNA stability. Nature 2014;505:117–20.

[55] Mauer J, Luo X, Blanjoie A, Jiao X, Grozhik AV, Patil DP, et al. Reversible methylation of m6Am in the 5′ cap controls mRNA stability. Nature 2017;541:371–5.

REFERENCES

[56] Bartosovic M, Molares HC, Gregorova P, Hrossova D, Kudla G, Vanacova S. N6-methyladenosine demethylase FTO targets pre-mRNAs and regulates alternative splicing and 3′-end processing. Nucleic Acids Res 2017;45(19):11356−70.

[57] Huang J, Yin P. Structural Insights into N 6-methyladenosine (m6A) Modification in the Transcriptome. Genom Proteom Bioinform 2018;16(2):85−98.

[58] Tang C, Klukovich R, Peng H, Wang Z, Yu T, Zhang Y, et al. ALKBH5-dependent m6A demethylation controls splicing and stability of long 3′-UTR mRNAs in male germ cells. Proc Natl Acad Sci USA 2018;115(2):E325−33.

[59] Motorin Y, Lyko F, Helm M. 5-methylcytosine in RNA: detection, enzymatic formation and biological functions. Nucleic Acids Res 2009;38(5):1415−30.

[60] Barbieri I, Kouzarides T. Role of RNA modifications in cancer. Nat Rev Cancer 2020;20(6):303−22.

[61] Haruehanroengra P, Zheng YY, Zhou Y, Huang Y, Sheng J. RNA modifications and cancer. RNA Biol 2020;17(11):1560−75.

[62] Wood S, Willbanks A, Cheng JX. The Role of RNA modifications and RNA-modifying proteins in cancer therapy and drug resistance. Curr Cancer Drug Targets 2021;21(4):326−52.

[63] Livneh I, Moshitch-Moshkovitz S, Amariglio N, Rechavi G, Dominissini D. The m6A epitranscriptome: transcriptome plasticity in brain development and function. Nat Rev Neurosci 2020;21(1):36−51.

[64] Widagdo J, Anggono V. The m6A-epitranscriptomic signature in neurobiology: from neurodevelopment to brain plasticity. J Neurochem 2018;147(2):137−52.

[65] Park CW, Lee SM, Yoon KJ. Epitranscriptomic regulation of transcriptome plasticity in development and diseases of the brain. BMB Rep 2020;53(11):551−64.

[66] Vissers C, Sinha A, Ming GL, Song H. The epitranscriptome in stem cell biology and neural development. Neurobiol Dis 2020;146:105139.

[67] Shi H, Wang X, Lu Z, Zhao BS, Ma H, Hsu PJ, et al. YTHDF3 facilitates translation and decay of N6-methyladenosine-modified RNA. Cell Res 2017;27(3):315−28.

[68] Trixl L, Lusser A. The dynamic RNA modification 5-methylcytosine and its emerging role as an epitranscriptomic mark. Wiley Interdiscip Rev RNA 2018;10(1):e1510.

[69] Shulman Z, Stern-Ginossar N. The RNA modification N6-methyladenosine as a novel regulator of the immune system. Nat Immunol 2020;21(5):501−12.

[70] Williams GD, Gokhale NS, Horner SM. Regulation of viral infection by the RNA modification N6-methyladenosine. Annu Rev Virol 2019;6(1):235−53.

[71] Huang W, Qi CB, Lv SW, Xie M, Feng YQ, Huang WH, et al. Determination of DNA and RNA methylation in circulating tumor cells by mass spectrometry. Anal Chem 2016;88(2):1378−84.

[72] Pei Y, Lou X, Li K, Xu X, Guo Y, Xu D, et al. Peripheral blood leukocyte N6-methyladenosine is a noninvasive biomarker for non-small-cell lung carcinoma. Onco Targets Ther 2020;13:11913−21.

[73] Ashktorab H, Belgrave K, Hosseinkhah F, Brim H, Nouraie M, Takkikto M, et al. Global histone H4 acetylation and HDAC2 expression in colon adenoma and carcinoma. Dig Dis Sci 2009;54(10):2109−17.

[74] Nakazawa T, Kondo T, Ma D, Niu D, Mochizuki K, Kawasaki T, et al. Global histone modification of histone H3 in colorectal cancer and its precursor lesion. Hum Pathol 2012;43(6):834−42.

[75] Tamagawa H, Oshima T, Shiozawa M, Morinaga S, Nakamura Y, Yoshihara M, et al. The global histone modification pattern correlates with overall survival in metachronous liver metastasis of colorectal cancer. Oncol Rep 2012;27(3):637−42.

[76] Deligezer U, Akisik EE, Erten N, Dalay N. Sequence specific histone methylation is detectable on circulating nucleosomes in plasma. Clin Chem 2008;54(7):1125−31.

[77] Deligezer U, Akisik EZ, Akisik EE, Kovancilar M, Bugra D, Erten N, et al. H3K9me3/H4K20me3 ratio in circulating nucleosomes as potential biomarker for colorectal cancer. In: Gahan PB, editor. Circulating nucleic acids in plasma and serum. Dordrecht: Springer; 2011. p. 97−103.

[78] Leszinski G, Gezer U, Siegele B, Stoetzer OJ, Holdenrieder S. Histone modifications H3K9me3 and H4K20me3 on circulating nucleosomes in cancer disease. Anticancer Res 2012;32(5):2199−206.

[79] Haluskova J. Epigenetic studies in human diseases. Folia Biol (Praha) 2010;56(3):83−96.

[80] Feinberg AP. Phenotypic plasticity and the epigenetics of human disease. Nature 2007;447(7143):433−40.

[81] Berdasco M, Esteller M. Aberrant epigenetic landscape in cancer: how cellular identity goes awry. Dev Cell 2010;19(5):698−711.

[82] Chen X, Ba Y, Ma L, Cai X, Yin Y, Wang K, et al. Characterization of microRNAs in serum: a novel class of biomarkers for diagnosis of cancer and other diseases. Cell Res 2008;18(10):997−1006.

[83] Gilad S, Meiri E, Yogev Y, Benjamin S, Lebanony D, Yerushalmi N, et al. Serum microRNAs are promising novel biomarkers. PLoS One 2008;3(9):e3148.

[84] Esquela-Kerscher A, Slack FJ. Oncomirs-microRNAs with a role in cancer. Nat Rev Cancer 2006;6(4):259−69.

[85] Chen K, Rajewsky N. The evolution of gene regulation by transcription factors and microRNAs. Nat Rev Genet 2007;8(2):93−103.

[86] Yan W, Qian L, Chen J, Chen W, Shen B. Comparison of prognostic microRNA biomarkers in blood and tissues for gastric cancer. J Cancer 2016;7(1):95−106.

[87] Lawrie CH, Gal S, Dunlop HM, Pushkaran B, Liggins AP, Pulford K, et al. Detection of elevated levels of tumour-associated microRNAs in serum of patients with diffuse large B-cell lymphoma. Br J Haematol 2008;141(5):672−5.

[88] Chim SS, Shing TK, Hung EC, Leung TY, Lau TK, Chiu RW, et al. Detection and characterization of placental microRNAs in maternal plasma. Clin Chem 2008;54(3):482−90.

[89] Wang W, Li J, Zhu W, Gao C, Jiang R, Li W, et al. MicroRNA-21 and the clinical outcomes of various carcinomas: a systematic review and *meta*-analysis. BMC Cancer 2014;14:819.

[90] Mitchell PS, Parkin RK, Kroh EM, Fritz BR, Wyman SK, Pogosova-Agadjanyan EL, et al. Circulating microRNAs as stable blood-based markers for cancer detection. Proc Natl Acad Sci USA 2008;105(30):10513−18.

[91] Chen ZH, Zhang GL, Li HR, Luo JD, Li ZX, Chen GM, et al. A panel of five circulating microRNAs as potential biomarkers for prostate cancer. Prostate 2012;72(13):1443−52.

[92] Wang J, Chen J, Chang P, LeBlanc A, Li D, Abbruzzesse JL, et al. MicroRNAs in plasma of pancreatic ductal adenocarcinoma patients as novel blood-based biomarkers of disease. Cancer Prev Res 2009;2(9):807−13.

[93] Kojima M, Sudo H, Kawauchi J, Takizawa S, Kondou S, Nobumasa H, et al. MicroRNA markers for the diagnosis of pancreatic and biliary-tract cancers. PLoS One 2015;10(2):e0118220.

[94] Ng EK, Li R, Shin VY, Jin HC, Leung CP, Ma ES, et al. Circulating microRNAs as specific biomarkers for breast cancer detection. PLoS One 2013;8(1):e53141.

[95] Tsujiura M, Ichikawa D, Komatsu S, Shiozaki A, Takeshita H, Kosuga T, et al. Circulating microRNAs in plasma of patients with gastric cancers. Br J Cancer 2010;102(7):1174−9.

[96] Liu R, Zhang C, Hu Z, Li G, Wang C, Yang C, et al. A five-microRNA signature identified from genomewide serum

microRNA expression profiling serves as a fingerprint for gastric cancer diagnosis. Eur J Cancer 2011;47(5):784–91.
[97] Huang Z, Huang D, Ni S, Peng Z, Sheng W, Du X. Plasma microRNAs are promising novel biomarkers for early detection of colorectal cancer. Int J Cancer 2010;127(1):118–26.
[98] Chen WX, Ren LH, Shi RH. Implication of miRNAs for inflammatory bowel disease treatment: systematic review. World J Gastrointest Pathophysiol 2014;5(2):63–70.
[99] Glinsky GV. Disease phenocode analysis identifies SNP-guided microRNA maps (MirMaps) associated with human "master" disease genes. Cell Cycle 2008;7(23):3680–94.
[100] Fujioka S, Nakamichi I, Esaki M, Asano K, Matsumoto T, Kitazono T. Serum microRNA levels in patients with Crohn's disease during induction therapy by infliximab. J Gastroenterol Hepatol 2014;29(6):1207–14.
[101] Du J, Cao X, Zou L, Chen Y, Guo J, Chen Z, et al. MicroRNA-21 and risk of severe acute kidney injury and poor outcomes after adult cardiac surgery. PLoS One 2013;8(5):e63390.
[102] Tijsen AJ, Creemers EE, Moerland PD, de Windt LJ, van der Wal AC, Kok WE, et al. MiR423-5p as a circulating biomarker for heart failure. Circ Res 2010;106(6):1035–9.
[103] Wang GK, Zhu JQ, Zhang JT, Li Q, Li Y, He J, et al. Circulating microRNA: a novel potential biomarker for early diagnosis of acute myocardial infarction in humans. Eur Heart J 2010;31(6):659–66.
[104] Cheng Y, Tan N, Yang J, Liu X, Cao X, He P, et al. A translational study of circulating cell-free microRNA-1 in acute myocardial infarction. Clin Sci (Lond) 2010;119(2):87–95.
[105] Ai J, Zhang R, Li Y, Pu J, Lu Y, Jiao J, et al. Circulating microRNA-1 as a potential novel biomarker for acute myocardial infarction. Bioch Biophys Res Commun 2010;391(1):73–7.
[106] Chen G, Wang Z, Wang D, Qiu C, Liu M, Chen X, et al. LncRNADisease: a database for long-non-coding RNA-associated diseases. Nucleic Acids Res 2013;41(Database issue):D983–6.
[107] Lee GL, Dobi A, Srivastava S. Prostate cancer: diagnostic performance of the PCA3 urine test. Nat Rev Urol 2011;8(3):123–4.
[108] Xue WJ, Ying XL, Jiang JH, Xu YH. Prostate cancer antigen 3 as a biomarker in the urine for prostate cancer diagnosis: a meta-analysis. J Cancer Res Ther 2014;10(7):218–21.
[109] Xue P, Chen W, Li X. Urothelial cancer associated 1: a long noncoding RNA with a crucial role in cancer. J Cancer Res Clin Oncol 2016;142(7):1407–19.
[110] Wang XS, Zhang Z, Wang HC, Cai JL, Xu QW, Li MQ, et al. Rapid identification of UCA1 as a very sensitive and specific unique marker for human bladder carcinoma. Clin Cancer Res 2006;12(16):4851–8.
[111] Zhang Q, Su M, Lu G, Wang J. The complexity of bladder cancer: long noncoding RNAs are on the stage. Mol Cancer 2013;12(1):101.
[112] Kamel MM, Matboli M, Sallam M, Montasser IF, Saad AS, El-Tawdi AH. Investigation of long noncoding RNAs expression profile as potential serum biomarkers in patients with hepatocellular carcinoma. Transl Res 2016;168:134–45.
[113] Zheng Q, Wu F, Dai WY, Zheng DC, Zheng C, Ye H, et al. Aberrant expression of UCA1 in gastric cancer and its clinical significance. Clin Transl Oncol 2015;17(8):640–6.
[114] Hong HH, Hou LK, Pan X, Wu CY, Huang H, Li B, et al. Long non-coding RNA UCA1 is a predictive biomarker of cancer. Oncotarget 2016;7(28):44442–7.
[115] Li J, Wang Y, Yu J, Dong R, Qiu H. A high level of circulating HOTAIR is associated with progression and poor prognosis of cervical cancer. Tumor Biol 2015;36(3):1661–5.
[116] Tang J, Jiang R, Deng L, Zhang X, Wang K, Sun B. Circulation long non-coding RNAs act as biomarkers for predicting tumorigenesis and metastasis in hepatocellular carcinoma. Oncotarget 2015;6(6):4505–15.
[117] Dong L, Qi P, Xu MD, Ni SJ, Huang D, Xu QH, et al. Circulating CUDR, LSINCT-5 and PTENP1 long noncoding RNAs in sera distinguish patients with gastric cancer from healthy controls. Int J Cancer 2015;137(5):1128–35.
[118] Shi J, Li X, Zhang F, Zhang C, Guan Q, Cao X, et al. Circulating lncRNAs associated with occurrence of colorectal cancer progression. Am J Cancer Res 2015;5(7):2258–65.
[119] Kumarswamy R, Bauters C, Volkmann I, Maury F, Fetisch J, Holzmann A, et al. Circulating long noncoding RNA, LIPCAR, predicts survival in patients with heart failure. Circ Res 2014;114(10):1569–75.
[120] Greco S, Zaccagnini G, Perfetti A, Fuschi P, Valaperta R, Voellenkle C, et al. Long noncoding RNA dysregulation in ischemic heart failure. J Transl Med 2016;14(1):183.
[121] Yan Y, Zhang B, Liu N, Qi C, Xiao Y, Tian X, et al. Circulating long noncoding RNA UCA1 as a novel biomarker of acute myocardial infarction. Biomed Res Int 2016;2016:8079372.
[122] Kraus TF, Haider M, Spanner J, Steinmaurer M, Dietinger V, Kretzschmar HA. Altered long noncoding RNA expression precedes the course of Parkinson's disease-a preliminary report. Mol Neurobiol 2016;2016:1–9.
[123] Shi H, Cao N, Pu Y, Xie L, Zheng L, Yu C. Long non-coding RNA expression profile in minor salivary gland of primary Sjogren's syndrome. Arthritis Res Ther 2016;18(1):109.
[124] Xu XD, Li KR, Li XM, Yao J, Qin J, Yan B. Long non-coding RNAs: new players in ocular neovascularization. Mol Biol Rep 2014;41(7):4493–505.
[125] Meola N, Pizzo M, Alfano G, Surace EM, Banfi S. The long noncoding RNA Vax2os1 controls the cell cycle progression of photoreceptor progenitors in the mouse retina. RNA 2012;18(1):111–23.
[126] Zhou RM, Wang XQ, Yao J, Shen Y, Chen SN, Yang H, et al. Identification and characterization of proliferative retinopathy-related long noncoding RNAs. Biochem Biophys Res Commun 2015;465(3):324–30.
[127] Wang WT, Sun YM, Huang W, He B, Zhao YN, Chen YQ. Genome-wide long non-coding RNA analysis identified circulating LncRNAs as novel non-invasive diagnostic biomarkers for gynecological disease. Sci Rep 2016;6:23343.
[128] Laird PW. Principles and challenges of genomewide DNA methylation analysis. Nat Rev Genet 2010;11(3):191–203.
[129] Li M, Chen WD, Papadopoulos N, Goodman SN, Bjerregaard NC, Laurberg S, et al. Sensitive digital quantification of DNA methylation in clinical samples. Nat Biotechnol 2009;27(9):858–63.
[130] Campan M, Moffitt M, Houshdaran S, Shen H, Widschwendter M, Daxenbichler G, et al. Genome-scale screen for DNA methylation-based detection markers for ovarian cancer. PLoS One 2011;6(12):e28141.
[131] Dedeurwaerder S, Defrance M, Calonne E, Denis H, Sotiriou C, Fuks F. Evaluation of the infinium methylation 450K technology. Epigenomics 2011;3(6):771–84.
[132] Lee EJ, Luo J, Wilson JM, Shi H. Analyzing the cancer methylome through targeted bisulfite sequencing. Cancer Lett 2013;340(2):171–8.
[133] Meissner A, Mikkelsen TS, Gu H, Wernig M, Hanna J, Sivachenko A, et al. Genomescale DNA methylation maps of pluripotent and differentiated cells. Nature 2008;454(7205):766–70.
[134] Vaca-Paniagua F, Oliver J, Nogueira da Costa A, Merle P, McKay J, Herceg Z, et al. Targeted deep DNA methylation analysis of

circulating cell-free DNA in plasma using massively parallel semiconductor sequencing. Epigenomics 2015;7(3):353–62.

[135] Jang H, Shin H. Current trends in the development and application of molecular technologies for cancer epigenetics. World J Gastroenterol 2013;19(7):1030–9.

[136] Linder B, Grozhik AV, Olarerin-George AO, Meydan C, Mason CE, Jaffrey SR. Single-nucleotide-resolution mapping of m6A and m6Am throughout the transcriptome. Nat Methods 2015;12(8):767–72.

[137] Garalde DR, Snell EA, Jachimowicz D, Sipos B, Lloyd JH, Bruce M, et al. Highly parallel direct RNA sequencing on an array of nanopores. Nat Methods 2018;15(3):201–6.

[138] Castoldi M, Schmidt S, Benes V, Hentze MW, Muckenthaler MU. miChip: an array-based method for microRNA expression profiling using locked nucleic acid capture probes. Nat Protoc 2008;3(2):321–9.

[139] Thind A, Wilson C. Exosomal miRNAs as cancer biomarkers and therapeutic targets. J Extracell Vesicles 2016;5:31292.

CHAPTER 18

Transposable Elements Shaping the Epigenome

Karen Giménez-Orenga[1] and Elisa Oltra[1,2]

[1]Centro de Investigación Traslacional San Alberto Magno, Universidad Católica de Valencia San Vicente Mártir, Valencia, Spain [2]Department of Pathology, School of Medicine and Health Sciences, Universidad Católica de Valencia San Vicente Mártir, Valencia, Spain

OUTLINE

Introduction	323	Nutriepigenetics	335
Classification and structure of transposable elements	324	Influence of transposable elements on host epigenetics	335
Class I transposable elements	324	Transposable elements influence on chromatin structure	335
Class II Transposable Elements	328	Transposable elements impact on genetics and epigenetics of gene regulatory regions	337
Transposable elements genomic annotation	328	Heterochromatin spreading	340
Epigenetic control of transposable elements	329	X-chromosome inactivation	341
DNA Methylation	329	Genomic Imprinting and Epialleles	342
Histone Modifications	329	Regulation of gene expression by TE Regulatory RNA Networks	342
Regulatory Mechanisms in trans	330	Concluding remarks	343
Posttranscriptional Modifications	331	References	343
Factors influencing transposable elements epigenetics	334		
Environmental stressors	334		
Viral infections	334		

INTRODUCTION

Following Barbara McClintock's pioneering experiments in the 1940s, revealing the existence of transposable elements (TEs) in maize [1,2], Britten and Davidson postulated a role for these repetitive sequences in controlling gene expression by providing regulatory binding sites for messenger RNA (mRNA) production [3,4]. However, several decades and the advance of technology were needed to abandon the concept that TEs are "molecular parasites" exclusively contributing to genetic instability. Attributes such as "dark matter" or "junk DNA" to describe TE's function as mere spacers of protein coding genes are commonly found in the old literature. Nowadays, the prescient early predictions of the above cited authors count with ample experimental support to recognize TEs as major contributors of eukaryotic genome plasticity and a main force for species adaptation and evolution [5–7]. Exaptation, a process by which TE sequences were co-opted to perform a non-TE function in the recipient host, allowed for the acquisition of new functions (e.g., V(D)J recombination, a process by which immunoglobulin genes are arranged with multiple variable (V), diversity (D) and joining (J) regions to create specific DNA arrangements, generating diverse antibodies and B and T cells in jawed vertebrates) [8]. Also, exaptation of a protein encoded by a TE, syncytin-1, with cell-cell fusion properties, mediates trophoblast fusion during mammalian placentation [9,10]. TEs are also involved in the fine tuning of preexisting host functions by adaptation [11,12]. However, when uncontrolled, TEs may lead to disease or even determine clinical treatment outcomes [13–15] which explains the increased interest for their study.

This chapter covers TE structure and classification, in light of their different transposition mechanisms; it describes epigenetic and posttranscriptional mechanisms that control TE expression under the influence of environmental factors; to end with an in depth description of the main mechanisms by which TEs impact host epigenetics, including their capacity to: (1) affect chromatin structure, (2) work as mediators for the spreading of repressive marks or (3) shape gene regulatory networks.

CLASSIFICATION AND STRUCTURE OF TRANSPOSABLE ELEMENTS

The unprecedented advance of biotechnology in recent decades allowing the complete sequencing of many living organism genomes [16–18] has uncovered that TEs are present in roughly all eukaryotes, representing a significant portion of their genomes, with values ranging from 12% in the model organism *Caenorhabditis elegans* to over 90% in some plants [19]. Moreover, according to the updated version of the human genome, around 50% of the human sequences are annotated as TEs [20].

The initial efforts for the classification of mobile DNA elements dates to 1989, when Finnegan proposed a division of TEs according to the requirement of an RNA intermediate for their mobility within the genome (class I, RNA transposons or retrotransposons, or "copy and paste" elements), or to the absence of such requirement (class II, DNA transposons, or "cut and paste" elements) [21]. Despite the perpetuation of this binary categorization through the years, still being recognized as an initial guideline for TE classification, species-specific TEs not fitting with either category were soon discovered. For example the non-autonomous DNA miniature inverted repeat transposable elements (MITEs). Non-autonomous elements rely on the transposition machinery of other autonomous elements that encode the necessary proteins for their mobilization, including a reverse transcriptase (RT) and integrase for class I TEs or a transposase in class II TEs [22].

In this context, Wicker proposed in 2007 a unified classification system adding detail to Finnegan's [23], which has been refined over time through a deeper understanding of TE's sequence and structure [22,24]. Accordingly, TEs are classified in hierarchical order, fitting with phylogenetics by class, subclass, order, superfamily, family, and subfamily. This classification system defines two classes, as previously mentioned, depending on the presence (class I) or absence (class II) of an RNA transposition intermediate. "Subclass," in the case of class II TEs and "order," in the case of class I TEs, distinguish TEs according to their insertion mechanism. Every subclass/order is composed of different superfamilies which share replication strategy but present distinct features, such as the structure of protein coding or non-coding domains or the presence and size of target site duplication (TSD). TSDs are 1–20 bp direct repeats flanking TEs after transposition derived from the target sequence [25]. Finally, superfamilies encompass several families which are defined by DNA sequence conservation and in turn comprise different subfamilies defined on the basis of phylogenetic data [23].

The relative abundance of different TE types varies largely across species. Fig. 18.1A shows the distribution of TEs in human, mice, *Drosophila*, and *Arabidopsis thaliana*, according to annotation from RepeatMasker. As it can be observed on Fig. 18.1A, the most abundant TEs in human are long interspersed nuclear elements (LINEs), followed by short interspersed nuclear elements (SINEs) and long terminal repeats (LTR) retrotransposons, with an overall of 94%–97%; being DNA transposons the least prevalent (only 3%–6%), while other clades present with different class I/class II relative abundance ratios [26–28].

Class I Transposable Elements

Class I TEs retrotranscribe *via* an RNA intermediate, i.e., they are transcribed from a genomic TE copy to generate an RNA intermediate, which then is reverse-transcribed into DNA by a TE-encoded reverse transcriptase (RT) [22,23]. Class I includes five orders: LTR retrotransposons, LINEs, SINEs, *Dictyostelium* intermediate repeat sequence (DIRS) or tyrosine recombinase-encoding retrotransposons [29] and Penelope like elements (PLE) [30] (Fig. 18.1B). These orders differ in their replication and insertion mechanisms (Fig. 18.2). In view of the content covered in this chapter, we will center our focus on the first three orders: LTR retrotransposons, LINEs and SINEs.

LTR Retrotransposons

LTR retrotransposons range in size from a few hundred pairs to 10 kb. Their name is given by the presence of LTRs at both ends of the element, ranging from 300 pb to 1 kb, that contain transcription initiation (e.g., RNA pol II promoter) and termination

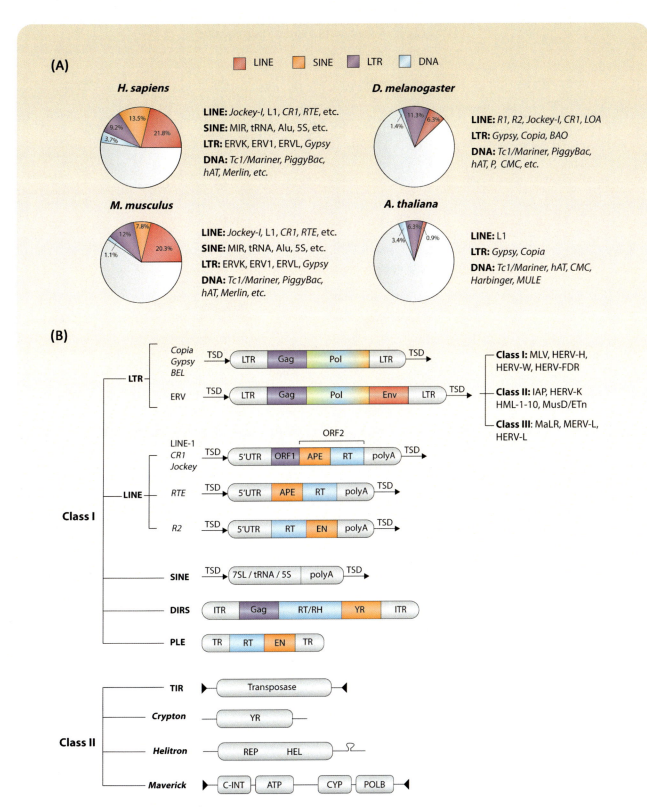

FIGURE 18.1 **Relative abundance and classification of eukaryotic TEs.** (A) Graphical representation of the genomic presence and composition of TEs in *Homo sapiens* (H. sapiens), *Mus musculus* (M. musculus), *Drosophila melanogaster* (D. melanogaster) and *Arabidopsis thaliana* (A. thaliana) retrieved from RepeatMasker. Superfamilies in each species are indicated next to the pie-diagrams. (B) Classification of eukaryotic TEs following the classification system proposed by Wicker et al. [23] by class (class I, class II), order (LTR retrotransposons (LTR), long interspersed nuclear element (LINE), short interspersed nuclear element (SINE), *Dictyostelium* intermediate repeat sequence (DIRS), Penelope like element (PLE), terminal inverted repeat element (TIR), *Crypton, Helitron, Maverick*) and superfamily (*Copia, Gypsy, BEL*, endogenous retrovirus (ERV), LINE-1, *CR1, Jockey*, RTE, R2), including the structure and coding regions of each element. Apurinic endonuclease (APE), packaging ATPase (ATP), c-integrase (C-INT), cysteine protease (CYP), endonuclease (EN), envelope (ENV), helicase (HEL), inverted terminal repeat (ITR), open reading frame (ORF), DNA polymerase B (POLB), replication initiator protein (REP), RNase H (RH), reverse transcriptase (RT), target site duplication (TSD), terminal repeat (TR), tyrosine recombinase (YR).

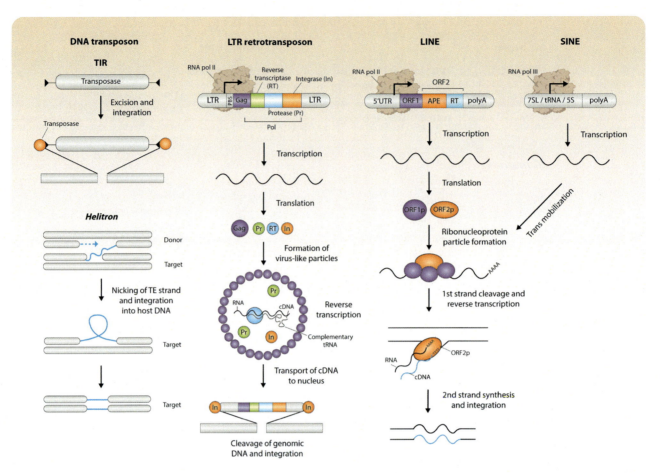

FIGURE 18.2 **Transposition mechanisms of main eukaryotic TE types.** DNA transposons mobilize as a DNA intermediate. Terminal inverted repeat elements (TIRs) encode a transposase that recognizes the terminal repeats that flank the element and excises the double stranded TE from the donor site to integrate it into another genomic site. Nonetheless, *Helitrons* form a rolling-circle *via* nicking and peeling one of its own DNA strands and integrating it at another site in the genome followed by DNA repair. Retrotransposons, however, mobilize as an RNA intermediate. LTR retrotransposons transpose through reverse transcription of their own mRNA, a process primed by a transfer RNA (tRNA) complementary to the primer binding site (PBS) located within its mRNA. This process occurs within a cytoplasmic viral-like particle encoded by the *gag* gene. Besides, LINE retrotransposons mobilize by a mechanism termed target-site primed reverse transcription (TPRT) in which its mRNA is exported to the cytoplasm, translated into ORF1p and ORF2p proteins and translocated to the nucleus forming a ribonucleoprotein with ORF1p and ORF2p that nicks one strand of host DNA to prime reverse transcription of LINE mRNA. Regarding SINEs, they are non-autonomous elements that do not encode the machinery necessary for self-mobilization. Instead, a LINE-like region located at their 3' parasitizes LINE's enzymatic machinery for their mobilization.

signals. Autonomous LTR retrotransposons encode at least *gag* and *pol* genes, flanked by LTR sequences. A protease, RT, ribonuclease H (RNase H) and integrase are coded by the *pol* gene while structural proteins are coded by *gag* [22,24]. LTR retrotransposons mobilize through reverse transcription of their own mRNA into double-stranded DNA (dsDNA), catalyzed by RT, which is then inserted into the host genome by an integrase [31]. This process occurs *via* a cytoplasmic viral-like particle encoded by the *gag* gene [32] and is primed by a transfer RNA (tRNA) complementary to the primer binding site (PBS), a region present in the viral RNA [33]. During their reverse transcription, the LTRs are entirely regenerated at both ends of the element (Fig. 18.2).

According to comparative analysis of the conserved domain of reverse transcriptases, LTR retrotransposons are subdivided into four superfamilies: *Copia*, *Gypsy*, *BEL* and endogenous retroviruses (ERVs) [22,24,34] (Fig. 18.1B). ERVs are a particular branch belonging to LTR retrotransposons that originated by retroviral infection of the germline millions of years ago. They encode an envelope gene, *env*, in addition to *gag* and *pol*, which generally lacks coding potential except for few cases [9,10]. In addition, ERVs contain signals for RNA polyadenylation and splicing, as well as the PBS and polypurine tract that define the boundaries of the LTRs [24]. However, over time, ERVs and TEs in general, have progressively diverged, accumulating mutations and other type of alterations sometimes becoming unique [35,36].

Nowadays, these unique TEs harbor internal SNPs (single nucleotide polymorphisms) with potential diagnostic value that are useful to classify TE families, or even lineages within a family [37,38]. Rearrangements of TEs, including recombination events between similar 5′ and 3′ LTRs in ERVs, have led to loss of some or even all coding regions, constituting in the latter case solitary LTRs (Solo-LTR) [35,36].

Despite resembling retroviruses in sequence and structure, ERVs lack an extracellular phase in their life cycle and depend on vertical transmission [24]. However, their homology to retroviruses allowed for their classification into three classes, based on phylogenetic relationships of the conserved regions of the *pol* gene. Class I LTR retrotransposons are most closely related to the exogenous gamma- and epsilon retroviruses, class II to the alpha- and beta-retroviruses and class III to the spuma-viruses (Fig. 18.1B) [24].

However, additional classification systems exist for them. For example, ERVs can also be classified according to the length of TSDs. ERVs with 4-bp TSDs are classified as ERV1 and correspond to class I, ERVs with 6-bp TSDs as ERV2 and correspond to class II while ERVs with 5-bp TSDs are classified as ERV3 and do not have a corresponding infectious retrovirus group [22].

Moreover, the amino acid one-letter code of the tRNA used to prime retrotranscription by binding the PBS defines different families [39]. For example, human ERV (HERV)-W family members bind tRNA$_{Trp}$. This system identifies the most probable tRNA priming reverse transcription according to PBS sequences, which are also subject of mutations and deletions, leading to inconsistencies, e.g., the HERV-K family is not primed by a tRNA$_{Lys}$ but most likely by a tRNA$_{Met}$ [35]. Thus, some HERVs have been named according to the symbol of a nearby gene (e.g., HERV-ADP) or to the presence of amino acid motifs (e.g., HERV-FRD) [35]. Other classification criteria include indication of the species in which the elements were initially identified. This is done by adding one or two letters before the ERV name. For example, letter "H" in HERV or "Mu" in MuERV, indicate that those elements were initially identified in human and mouse, respectively [35].

The existence of different classification and nomenclature systems for ERVs can lead to confusion. This is the case for some databases e.g., Repbase, which designates groups of ERVs and solo-LTRs with numerical names. For example, LTRs related to the primate HERV9 group are designated as LTR12. To facilitate the understanding of the varied classifications, some authors have published helpful conversion tables for human ERVs and associated LTRs [33,35].

LINEs

LINEs are 4–7 kb long elements characterized by the presence of a 5′ untranslated region (5′UTR) with an internal pol II active promoter and a repetitive tail, usually poly-A, at their 3′ [40]. They lack LTRs and encode between one to three proteins necessary for their mobilization, usually containing a RT and an endonuclease, which can be restriction-like endonuclease (RLE) or apurinic endonuclease (APE) [23,32]. LINEs span five major superfamilies: LINE-1 (L1), *CR1, Jockey, RTE*, and *R2* [23]. Whereas, *R2* elements have a unique ORF encoding an RT and an *RLE*; L1, *CR1*, and *Jockey* elements include two ORFs, the first, ORF1, encoding a 40 kDa protein with RNA binding and nucleic acid chaperone activities and the second, ORF2, encoding a RT and an APE (Fig. 18.1B) [32,41]. Interestingly, the Dualen element is extraordinary in that it encodes two endonucleases, a RLE and an APE [41].

L1 retrotransposition occurs by a mechanism termed target-site primed reverse transcription (TPRT) [22] (Fig. 18.2). L1 RNA is transcribed from an internal pol II promoter within its 5′UTR, located upstream the TSS, which ensures the preservation of the integrity of its promoter sequence [33]. L1 mRNA is exported to the cytoplasm where it is translated into ORF1p and ORF2p proteins, which associate with L1 to form ribonucleoprotein (RNP) particles. Then, L1 RNPs translocate to the nucleus and L1 endonuclease nicks one strand of host DNA at a loosely defined consensus site (5′-TTTT/A-3′), exposing a free 3′-OH group that the L1 RT can use to prime first strand L1 cDNA synthesis [22,33,36]. During this process, the 5′UTR is usually truncated and the L1 sequence inverted, leading to inactive L1 copies [40]. In addition, inserted L1s are flanked in most cases by variable length TSDs [42].

SINEs

SINEs range in size from 150 to 500 bp and have a 5′UTR with an internal pol III promoter, a repetitive tail, usually poly-A, at their 3′, and lack LTRs. They are derived from actively transcribed small RNA genes such as 5S ribosomal RNA (rRNA), 7SL RNA or tRNA [22,43] (Fig. 18.1B). However, SINEs do not encode the machinery necessary for self-mobilization, meaning that they are non-autonomous elements. Instead, SINEs possess a LINE-like region at their 3′ end which allows LINE's enzymatic machinery parasitization for their mobilization [32], which also generates 5–15 bp TSDs [42] (Fig. 18.2). Thus, SINEs and LINEs are often found in pairs [23] but unlike LINEs, de novo acquisition of SINEs is relatively common in mammals [32]. Novel SINEs have originated at several periods during the evolution of vertebrates, with unusual elements such as the primate-specific SVA

SINEs [44,45]. Other examples of recent SINEs are human *Alu* and mouse B1 and B2 [46].

Class II Transposable Elements

Class II TEs, also known as DNA transposons, mobilize as a DNA intermediate. They are classified into three major subclasses depending on the transposition mechanism: "cut and paste" or terminal inverted repeat (TIR) transposons (*hATs, PiggyBACs, and Mariners*), rolling-circle transposons (*Helitrons*) and self-synthesizing transposons (*Mavericks*) [47] (Figs. 18.1B and 18.2).

The most diverse and widespread group of DNA transposons are TIRs, which have a single ORF that encodes a DDE-motif transposase that recognizes the TIRs flanking the TE and excises the double-stranded TE from the donor site to integrate it into another genomic site [6,48]. Cut-and-paste transposons are characterized by their TIRs ranging in size from 10 bp to 1 kb [6]. The superfamily Crypton also uses a "cut and paste" transposition mechanism, but in this case, it is mediated by a tyrosine recombinase (YR) [6,22,49]. This transposition mechanism has the potential to increase DNA transposon copy numbers when transposition occurs during replication from a region which has already been replicated to another that has not yet done it [23]. It can also happen if the excised site is repaired by homologous recombination with another chromosome containing the element [6]. Besides, Helitron superfamily encodes a single protein with helicase and endonuclease, or rolling-circle replication initiator (Rep), enzymatic domains and its transposition is achieved *via* nicking and peeling one of its own DNA strands and integrating it at another site in the genome, followed by DNA repair (Fig. 18.2) [50]. Finally, Mavericks are very large elements that encode DNA polymerase B (Pol B), retroviral-like integrase, adenoviral-like protease (Pro), and putative ATPase (ATP), allowing their mobilization by a self-synthesizing mechanism involving direct synthesis of a DNA copy by a protein-primed polymerase B [22].

Although logical and comprehensive this classification system has not been widely implemented. Despite the urgent needs for some standards to address comparative genomic studies, TE classification continues to be a subject of debate with some classification systems completely forgoing original Finnegan's RNA-DNA transposon dichotomic system [25,51,52].

In addition to these classifications, TEs can be classified in accordance with their activity and evolutionary age. TE families currently or recently active are referred as "young" whereas "old" are those highly divergent, less active, or inactive families [53]. The number of progenitor or "master" copies and those remaining active for a given species is highly variable [53] (Fig. 18.1A). For instance, in human only the youngest L1, *Alu*, and SVA elements can still be actively mobilized [45,54]. However, among the hundreds of thousands of L1 copies only 80–100 are retrotransposition competent [55], and only about 20 of them are responsible for most ongoing L1 activity [55–57].

TRANSPOSABLE ELEMENTS GENOMIC ANNOTATION

Annotation of TEs has been challenged by their unique features:

- TEs are **repetitive and interspersed**. For example, newly acquired TEs are identical to their ancestors. Although ancient TEs have accumulated mutations, diverging and becoming unique, the recently incorporated TEs are almost identical to each other, complicating their identification and genome mapping.
- TEs are **polymorphic**. Their activity may render the presence or absence of an element at a given locus, as well as different copy numbers or be subject of mutations (SNPs), within a species or even for an individual. For example, 20% of inherited structural variants in human derive from new TE insertions [58], and insertional differences at 285 sites are found on average for L1 elements [59]. TE insertional polymorphisms are even more prevalent in mice or plants, the latter showing thousands of strongly active TEs [60,61].
- TE transcripts are **diverse**. Transcription mainly occurs from internal RNA pol II or pol III promoters leading to extended RNA species by a significant fraction of 3′ readthrough due to weak polyadenylation signals [62–64]. Thus, TE transcription would not be limited to its own mRNA but would extend to 3′ neighbor regions because of the lack of a strong polyadenylation signal. Also, some copies integrated into ribosomal DNA (rRNA) are co-transcribed with rRNA and sometimes self-cleaved by a ribozyme present on their 5′ end [65]. In addition to full-length, shorter TE transcript isoforms derived from alternative splicing or premature polyadenylation [64,66,67] are generated. Pervasive transcription, given TE abundance in eukaryotic genomes, and chimeric RNA species further enhances TE transcript repertoire. In fact, over one third of human protein coding transcripts and three quarters of long noncoding RNAs (lncRNAs) harbor an exon of TE origin [68–70].

All these features have led to numerous complications that still hinder their analysis qualitatively and quantitatively [71]. Reduced massive sequencing costs are rendering enlarged public collections of data, which together with long-read and single cell sequencing technologies, and the development of adequate computational tools, should lead to the refinement of current TE annotations and classifications, as well as to an improved understanding of TE roles in the different organisms.

Information on TEs is currently cataloged into three types of repositories: those attending to consensus sequences associating to each TE family (TE-centric), such as Repbase [72] or Dfam [73]; those cataloging individual TE instances within the corresponding reference genome (genome-centric), such as RepeatMasker [74]; and those registering individual insertions diverging from the reference genome (polymorphism-centric), such as dbVar [75]. For a list of additional repositories and computational tools, readers are referred to the recent comprehensive reviews by Goerner-Potvin and Bourque [76], Arkhipova [77], and Lanciano and Cristofari [71].

EPIGENETIC CONTROL OF TRANSPOSABLE ELEMENTS

Due to their ability to transpose, and to the presence of gene regulatory elements on their structure, most TEs could lead to mutations, genomic instability, or aberrant gene expression. Thus, the "genome defense" model proposes that silencing of TEs must be ensured in order to prevent disease, or even cell death [78,79]. Repression of TEs occurs at different levels, mainly relying on transcriptional silencing through epigenetic modifications that increase chromatin packaging, preventing the access of transcription activating factors and the binding of RNA polymerase [80,81]. The main epigenetic mechanisms involved include DNA methylation, histone tail posttranslational modifications and small-RNA based silencing, which work closely to repress TEs (Fig. 18.3).

DNA Methylation

DNA methylation consists in the transfer of a methyl group from S-adenyl methionine (SAM) onto carbon 5 (C5) of cytosines to form 5-methylcytosine (5mC) and it is mediated by different DNA methyltransferases (Dnmts) [81]. In mammals, 5mC patterns are established *de novo* during development by the Dnmt3 family consisting of Dnmt3a and Dnmt3b enzymes and one cofactor, Dnmt3l, which lacks catalytic activity. A *de novo* Dnmt gene (*Dnmt3c*) that evolved *via* duplication of Dnmt3b and is responsible for methylating the promoters of evolutionarily young retrotransposons in the rodent's male germline has also been identified [82]. Established 5mC patterns are maintained by Dnmt1, which copy the methylation pattern of the parental strand onto the newly synthesized during DNA replication [81] (Fig. 18.3).

In mammals, 5mC mainly locates on cytosines preceding a guanine nucleotide, called CpG sites, however, there is also evidence of DNA methylation on non-CpG sites [83–85]. Those CpG sites are generally methylated irrespective of their genomic context, counting for around 80% of the total in the mouse genome [86]. Yet higher CpG density genomic regions, known as CpG islands, are often associated with promoters and usually appear unmethylated. CpG density determines potential transcriptional effects of DNA methylation at promoter regions. Low CpG-density promoters are usually methylated but methylation does not correlate with transcriptional silencing; meanwhile, intermediate CpG-density promoters that can be either methylated or unmethylated and high-CpG density promoters , which are usually hypomethylated except for imprinted genes and the inactive X-chromosome, are more strongly affected by methylation with clear effects on gene silencing [87]. In contrast to mammals, plants cytosine methylation occurs at both symmetric CG and asymmetric CNN sites, where N = A, T or C [88]. The homolog of mammalian Dnmt1, Met1, and the plant specific Cmt3 drive maintenance of CG and CNG methylation, respectively, while Dnmt3 homologs Drm1 and Drm2 *de novo* methylate cytosines in all sequence contexts.

Like other epigenetic modifications, DNA methylation is dynamic and can be reversed *via* passive or active demethylation. The first occurs in the absence of Dnmt activity through the dilution of 5mC with cell divisions producing global genomic demethylation, while the second is specific to mammals and other metazoans and is mediated by the action of Ten-eleven translocation (TET) enzymes, targeting demethylation at individual CpG sites [89] (Fig. 18.3). TETs catalyze the iterative oxidation of 5mC to 5-hydroxymethylcytosine (5hmC), 5-formylcytosine (5fC) and 5-carboxylcytosine (5caC), which are poorly recognized by the maintenance enzyme Dnmt1 allowing their dilution through cell divisions. 5mC or its derivatives can also be replaced with unmodified cytosine by base excision repair mechanisms [90]. In contrast, TETs may also drive chromatin changes in a catalytically independent way by recruiting chromatin regulators with repressor activity such as OGT (O-linked β-d-N-acetylglucosamine transferase) [91], the SIN3 corepressor complex [92] and the Polycomb repressive complex 2 (PRC2) [93].

Histone Modifications

Regarding histone tail posttranslational modifications (PTM), other major players in TE silencing, the

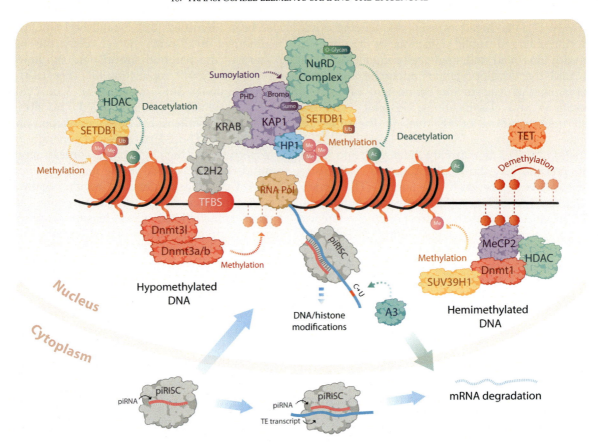

FIGURE 18.3 **Epigenetic and posttranscriptional mechanisms controlling class I TE expression in mammals.** The main epigenetic mechanisms involved include DNA methylation and demethylation by DNA methyltransferases (Dnmts, red) and Ten-eleven translocation (TET, orange) enzymes, respectively, histone tail posttranslational modifications such as methylation by histone methyltransferases SETBB1 and SUV39H1 (yellow) or deacetylation by histone deacetylases (HDACs, green) and small-RNA based silencing, which work closely to repress TEs. Epigenetic marks are specifically directed to TEs by KRAB-containing zinc finger proteins (KRAB-ZFPs) and Piwi-interacting RNAs-induced silencing complex (piRISC), the last also mediating TE transcript degradation. In addition, APOBEC3 (A3) proteins posttranscriptionally edit TE mRNAs introducing several cytosine to uracil (C-to-U) mutations during the minus first-strand DNA synthesis causing its degradation.

particular trimethylation of histone 3 at lysine 9 (H3K9me3) must be highlighted as the main histone mark found in TEs. This PTM is mediated by histone methyltransferases like SETDB1 (also known as ESET) or SUV39H1 in mammals [88–90], Su(var)3–9 in *Drosophila melanogaster* (*D. melanogaster*) [94], SUVH4, SUVH5 and SUVH6 in plants [95] or Clr4 in yeast [96]. Furthermore, biotinylation of lysine 12 in histone 4 (H4K12bio) and lysine 9 in histone H2A (H2AK9bio) mediated by holocarboxylase synthetase (HCS) also associates with TE repression, particularly of LTR transcripts, as reported in mouse and human cell cultures [97].

DNA methylation and histone modifications interplay to regulate TE expression, especially in mammals. For example, Dnmt3l binds to histone 3 tails in hypomethylated DNA and recruits Dnmt3a and Dnmt3b to establish *de novo* methylation [81,98]. Also, methyl-CpG-binding proteins (MBP), like MeCP2, bind hypomethylated DNA regions which are unprotected by transcription factors and not only recruits Dnmt1 to maintain DNA methylation [81,99] but also binds methyltransferases and histone deacetylases (HDAC) to establish repressive marks and to remove activating's [100] (Fig. 18.3).

Regulatory Mechanisms in trans

To achieve DNA methylation and/or histone PTMs, cells possess two main specific mediators: TE-derived small RNAs and transcriptional repressor KRAB-containing zinc finger proteins (KRAB-ZFPs), the last being specific to mammals.

Epigenetic repression of TEs by TE-derived small RNAs relies on Piwi-interacting RNAs (piRNAs) in animals, and small interfering RNAs (siRNAs) in plants and yeast. Regarding piRNAs, they are one of the components of the PIWI (P-element-Induced Wimpy testis)-piRNA pathway [101]. The PIWI-piRNA pathway consists of two components assembled as ribonucleoprotein complexes called piRNA-induced silencing complex (piRISC): a subfamily of Argonaute proteins with RNase-H activity, named PIWI, and 26–31 nucleotides long RNAs, known as piRNAs, that specifically bind TEs. piRNAs are generated from the expression of active TEs but also from piRNA clusters, which are loci that accumulate TEs sequences from previous invasions [102]. Those clusters are mainly located in pericentromeric and subtelomeric heterochromatin regions and their origin is believed to lie in eventual transpositions from previously active TE infections, working in a similar way to an "immune memory" system [103].

In contrast to piRNAs derived from clusters, which are antisense RNAs complementary to TEs, piRNAs derived from active TE transcripts must be processed from long single stranded transcripts in order to be produced [103]. Long TE transcripts are cleaved and their 5′ ends loaded into PIWI proteins, which then trim their 3′ ends to finally modify them by adding 2′ methyl groups. Interestingly, instead of being degraded, targeted transcripts are also processed to generate more piRNAs which are then loaded into PIWI proteins, generating the phenomenon called "ping pong" amplification loop [104]. This feed-back amplifying mechanism allows PIWI-mediated silencing to be more specific and effective towards targeted active elements. In this way, the PIWI pathway can mediate TE regulation at both, transcriptional and posttranscriptional levels. On one hand, by directly influencing TE epigenetic control in the nucleus, establishing heterochromatin within their loci and, on another hand, by targeting and destroying in the cytoplasm those TE transcripts that escaped epigenetic silencing [102–104] (Fig. 18.3).

Unlike piRNAs, siRNAs are generated from TE dsRNA precursors which are processed into 21–24 nucleotide siRNA complexes by a Dicer protein. siRNAs are then incorporated into an Argonaute (Ago) protein that degrades one of the strands. The ribonucleoprotein complex formed by Ago and siRNA, called siRNA-induced silencing complex (siRISC), directs the cleavage of TE transcripts which potentially function as templates for RNA-dependent RNA polymerase-mediated reverse strand synthesis and for the processing of additional siRNAs by Dicer. Similarly to piRNA-PIWI complexes in animals, siRNA-Ago complexes can target DNA and histone modifying complexes to TEs [105].

An alternative mechanism to impose heterochromatin packaging at TEs in mammals is through the KRAB-containing family of zinc finger proteins (KRAB-ZFPs). These proteins are transcription factors comprised by an array of C2H2 zinc fingers, which confers them with the capacity to bind specific DNA sequences lying within TEs, and a Krüppel-associated box (KRAB) domain that helps recruiting the corepressor KRAB-associated protein 1 (KAP1), also known as TRIM28 [106,107]. Three molecules of KAP1 bind to the KRAB domain of the KRAB-ZFPs through their N-terminal RING-B box-coiled coil (RBCC) region, while the C-terminal PHD (plant homeodomain) domain of KAP1 functions as an intramolecular E3 ligase that sumoylates several lysines in its bromodomain, enhancing its corepressor activity [108,109]. Sumoylation is a PTM that consists on the covalent attachment of a small ubiquitin-like modifier (SUMO) to a lysine residue. KAP1 mediates the assembly of heterochromatin epigenetic modifiers like the histone methyl transferase SETDB1 [110], the heterochromatin protein 1 (HP1) [111] or DNMTs, though the latter seems to be restricted to the embryo [106,112]. KAP1 interaction with SETDB1 establishes H3K9me through its C-terminal ubiquitinated SET domain [110]. However, SETDB1 can directly bind, in a KAP1-independent way, to H3K9me1/K14ac or H3K9me2/K14ac histone tails through its N-terminal triple Tudor domain (TTD) to deacetylate K14 via SETDB1-associated HDACs and further methylate K9 [113]. The introduction of H3K9me3 by SETDB1 allows the binding of HP1, which also interacts with KAP1 preventing transcription factor binding as well as maintaining the heterochromatic structure [114]. Since the axis KRAB/KAP1/SETDB1 localizes to restricted regions within TEs, and H3K9me3 marks must spread out all over to ensure their silencing, KAP1 may form a complex with HP1 and histone methyltransferases all over the TE region to mediate the spreading of heterochromatin [115,116]. In addition to SETDB1 and HP1, KAP1 also recruits the NuRD HDAC complex, which removes transcription-promoting histone acetylation modifications to repress TEs (Fig. 18.3) [117]. The inhibitory activity of KAP1-associated repressive chromatin factors, such as the HDAC1 of the NuRD complex, is enhanced by methylation-dependent O-linked glycosylation, a PTM introduced by the O-linked β-N-acetylglucosamine transferase (OGT) [118].

Posttranscriptional Modifications

Although epigenetic mechanisms are the main barrier controlling TE expression, posttranscriptional mechanisms repress those TEs that have escaped epigenetic control. For example, ERVs can be regulated by RNA-binding proteins, like the RNA-binding motif protein 4 (RBM4) [119] or by m6A RNA methylation of TE transcripts, protecting cell integrity by reducing the half-life of ERV mRNAs [120].

Nonetheless, the most important posttranscriptional mechanism limiting class I TE expression is that mediated by the APOBEC (apolipoprotein B mRNA editing enzyme, catalytic polypeptide-like) family of proteins.

The APOBECs are vertebrate-specific enzymes with the capacity to deaminate unmethylated cytosine residues into uracil (C-to-U) on single stranded DNA or RNA. The family includes APOBEC1, activation-induced cytidine deaminase (AID), APOBEC2, APOBEC3 and APOBEC4 in humans. Though all APOBEC family members participate in host protection against endogenous and exogenous pathogens, restriction of TE mobilization mainly relies on APOBEC3 (A3) proteins [121]. In this regard, the expansion of the *A3* gene cluster from one to three genes in pre-primate mammals to seven in primates coincided with the singular low levels of retrotransposon activity found in primates [121], suggesting that *A3* gene cluster evolved to protect hosts from the potential genomic instability caused by TE activation. Thus, the A3 family in primate lineage consists of seven members (A3A-A3H). A3 enzymes can restrict TE mobilization either through DNA editing-dependent or independent mechanisms. An example of DNA editing independence is the inhibition of the mouse LTR retrotransposon intracisternal A-particle (IAP) retrotransposition by human A3A [122]. For mechanisms dependent on the editing of DNA, A3s can introduce several cytosine to uracil (C-to-U) mutations during the minus first-strand DNA synthesis causing its degradation. In some cases, retrotransposition proceeds and a hypermutated DNA transposon bearing a series of guanosine to adenosine (G-to-A) mutations in the plus strand is integrated into the genome [123,124]. This may lead to preferential retention of mutated forms by positive selection and exaptation of that TE in benefit of the host [123]. Thus, A3 enzymes exert a dual inhibitory effect, decreasing the number of transposed cDNA copies and extensively editing them [125,126].

Class II TEs are posttranscriptionally controlled as well involving self-regulatory strategies [127,128]. One such mechanism was described as overproduction inhibition (OPI) for the class II DNA superfamily *Tc1/Mariner* by Lohe and Hartl [127]. The authors observed that the overall level of transposase activity was reduced by overproduction of the enzyme in *D. melanogaster*. Moreover, hypomorphic missense mutations in the transposase exhibited dominant negative complementation, antagonizing wildtype enzyme activity, indicating the assembly of mixed-subunit oligomeric forms with reduced activity as a way to control the mobilization of these TEs. The authors suggest that these regulatory mechanisms may broadly apply to *Mariner* TEs allowing to explain the inactivation of naturally acquired *Mariner* elements and the observed difference in copy numbers across species [127]. Although results about OPI in eukaryotes appear contradictory in the literature, it is reported that certain members of the *Mariner* and *PiggyBac* superfamilies use OPI to self-regulate their transposition in most eukaryotic cells. Furthermore, studies with Green fluorescent protein (GFP) fusions show that transposases are sequestered in cytoplasmic or nucleoplasmic foci; or within the nucleoli, depending on the type and expression levels which is interpreted as a way to protect the genome against the potentially genotoxic effects of these enzymes [129]. The mechanisms of DNA transposition and the autoregulation of eukaryotic DNA transposons have been recently comprehensively reviewed [130,131].

Regulatory Mechanisms Interplay

The extent to which these mechanisms regulate TE expression depends on the element to be regulated, the cell stage and the organism. Mammals and plants make use of both DNA methylation and histone modifications, whereas TE expression in *D. melanogaster* and yeast is controlled by DNA methylation [105]. While newly emerged elements in mammals are controlled by histone/DNA methylation-inducing small RNA-based mechanisms, older elements are preferentially silenced by KAP1-recruiting protein repressors [132,133]. This may be explained by the adaptation rate of the mechanism.

Since their appearance, KRAB-ZFPs have been engaged in a constant evolutionary arms race with TEs. As a consequence, KRAB-ZFPs have expanded and diversified in mammals correlating with the invasion of new TEs to maintain them silenced and even domesticating them [134]. Although highly specific, a long adaptation period required by KRAB-ZFPs leaves a temporary window for TEs to cause genomic instability. By contrast, small RNA-based mechanisms such as PIWI-piRNA pathway allow for a faster adaptation to the presence of new TEs albeit not as finely tuned if compared to KRAB-ZFPs based mechanisms.

In addition, the preponderance of a mechanism over the other depends on the cell's stage. While somatic cells bear a specific epigenetic landscape, it is erased in primordial germ cells (PGCs) and embryonic stem cells (ESC), in the course of embryogenesis, by a process called epigenetic reprogramming. During this process, two waves of DNA reprogramming take place, removing the epigenetic marks that control gene expression patterns to provide cells with the potency to differentiate into various cell types [135]. The first wave takes place right after the fusion of the gametes to form the zygote and the second throughout the formation of the PGCs. During this process most TEs become derepressed but although some level of expression is needed for embryogenesis to take place [136,137], their expression must be tightly controlled to prevent insertional mutagenesis by retrotransposition. In this context, the PIWI-piRNA pathway becomes fundamental repressing TEs in PGCs during reprogramming to prevent new TE insertions that would otherwise be inherited by future generations [138,139]. The control of TEs by the PIWIs seems also relevant in somatic cells [103,140].

DNA methylation and histone modifications target different TE families, such as ERVs, LINEs, SINEs, and SVAs, in a specific way [106,141,142] and even some TEs require the participation of both mechanisms in order to become silenced, such as IAPs [133,143].

In human and mouse ESCs, the formation of heterochromatin differs depending on evolutionary age of L1 elements, being KRAB-ZFP and KAP1 responsible for the formation of heterochromatin at older L1s [132,133] while small RNA-based repression controls younger L1s [144]. The PIWI-piRNA pathway contributes to establish genomic methylation patterns on L1s during re-methylation stage [145]. Moreover, it has been shown that evolutionary young full-length L1s located within introns of transcriptionally active genes are preferentially targeted by the HUSH (HUman Silencing Hub) complex, which cooperates with KAP1 for H3K9me3 deposition to ensure their silencing. This seems particularly important as young intact L1 elements pose the highest threat to genome integrity [146,147]. Young L1s can also be bound by TETs leading to DNA demethylation, though selective pressure to ensure genomic stability may have led to the evolution of TET-dependent host-silencing strategies to control L1 expression, especially during early embryonic development. Both TET1 and TET2 bind to genomic 5′ UTR of KAP1-unbound young L1s leading to low 5mC levels [148]. However, TET-dependent repressive activities, consisting in the recruitment of the SIN3A transcriptional repressor, counteract their oxidative DNA demethylation. Thus, TET enzymes may have a dual role in the regulation of young L1s that have escaped the KRAB-ZFP/KAP1 silencing machinery, on one side demethylate KRAB-ZFP/KAP1 L1s and on another recruit transcriptional repressors such as SIN3A.

In this sense, TETs also bind LTR elements such as IAPs [148] and loss of SETDB1 enables TET1 and TET2 binding and the accumulation of 5hmC [149–151]. As for L1s, a role for H3K9me3 in protecting ERVs from TET mediated DNA demethylation has also been proposed in ESCs [149].

Interestingly, TET2 does not contain a DNA-binding domain unlike the other members of TET family, and it is recruited to chromatin by the RNA-binding protein Paraspeckle component 1 (PSPC1) [152]. TETs bind LTR-derived elements co-opted by mammalian genomes as stage- and tissue-specific transcriptional regulatory modules to dynamically control neighbor gene expression such as pluripotency factors in ESCs [152]. Moreover, TETs also act on ERVs via non-catalytic pathways as observed in L1s, for example, by driving loss of the repressive H4R3me2 marks at IAPs [93,151]. Specifically, TET2 and PSPC1 bind LTR retrotransposons such as IAP and MusD and regulate neighbor gene expression by both transcriptional repression via HDACs and posttranscriptional destabilization of RNAs through 5hmC modification [148,152].

Additionally, DNA methylation enables stable silencing of some ERVs like IAPs [153,154]. In fact, the origin of DNA methylation has been repeatedly linked to TE silencing [155]. Importantly, CpG rich TEs such as IAPs [156,157] have the potential to become hypomethylated and recruit proteins that promote a permissive chromatin environment such as TETs, while CpG poor TEs remain methylated [158]. Thus, sequence-specific mechanisms drive DNA methylation at those TEs to ensure their methylation [159]. IAPs are silenced in a sequence-specific manner by KRAB-KZFPs [160] though PIWI-piRNA pathway contributes to establish genomic methylation patterns on IAP elements during re-methylation stage [145]. DNA methylation plays an essential role in ERV repression in post-implantation embryos and male germ cells [153,154] whilst during epigenetic reprogramming, and in ESCs, ERV silencing is mostly dependent on PTM of histones, particularly on SETDB1-mediated H3K9me3 [132,133,143]. Further, SETDB1′s role in ERV silencing is more pronounced in naïve ESCs than primed ESCs [151]. H3K9me3 is mainly targeted at young LTR and non-LTR transposons such as LINEs by SUV39H [141]; whereas, SETDB1-mediated H3K9me3 primarily suppresses ERVK transposons family, to which IAP elements belong [132,133,143]. Interestingly, IAP elements are quite resistant to DNA methylation during epigenetic reprogramming [153,160] as a result of the action of G9a/GLP complex, which recruits DNMTs de novo to maintain methylation at IAPs and imprinting control regions [161–163]. Furthermore, H3K9me2 enrichment impairs demethylation during epigenetic reprogramming by recruiting the DNMT1 chaperone NP95/UHRF1 [163–165].

Lastly, several A3 isoforms are capable of inactivating mouse IAP and MusD elements (hA3G) [122,125]. Both hA3G and hA3B induce C to T hypermutation of the IAP genome whilst hA3A seem to block IAP retrotransposition by deaminase-independent mechanisms [122,166,167]. Although no new HERV insertions have been identified in the human genome, there are polymorphic HERV-K (HML2) indicating potential activity [168,169], and a reconstructed pseudo-ancestral HERV-K sequence was shown to be susceptible of a particular A3 isoform (hA3F) [170], suggesting the participation of this family of enzymes during the evolutive control of ERVs in hominid.

But what about the non-LTR TEs that are still mobile in the human genome? A3 proteins effectively inhibit L1 retrotransposition and to a lesser extent L1-aid *Alu* elements mobility [122,171,172]. This indicates that APOBEC deaminases must enter the nucleus where reverse transcription takes place. The possibility that these deaminases interact with L1 ORF2p, additionally controlling transposition, has also been proposed [122,172,173]. Notably, an A3 isoform lacking the L1 ORF2p interaction domain (hA3G) cannot inhibit L1 retrotransposition but markedly decreases *Alu* retrotransposition [174]. Catalytically inactive hA3G mutants effectively inhibit *Alu* transposition. hA3G, in fact, sequesters *Alu* RNA in Staufen-containing high mobility complexes. The mechanism does not involve *Alu* RNA editing either, indicating that it differs from the hA3A and hA3B mediated inhibition of *Alu*s requiring interference of L1 machinery in the nucleus [174–176]. The lack of activity of the hA3G on L1 may reflect the blockade of its catalytic activity by L1 RNA, a phenomenon termed *cis*-preference. The relative high levels of APOBEC3 in ovary, testis and embryonic cells where most of the L1 and *Alu* transpositions take place [122,177–179], points at a physiologically relevant role for these enzymes to protect the human genome from potentially deleterious effects of retrotransposition.

In summary, the interplay of different epigenetic mechanisms and their varied influence on different types of retroelements enables the control of potentially harmful activities for the host and provides the host with new regulatory functions through cooption of retroviral elements.

FACTORS INFLUENCING TRANSPOSABLE ELEMENTS EPIGENETICS

Environmental Stressors

Environmental stressors are known to alter epigenetic modifications, including DNA methylation, histone PTMs, and the expression patterns of non-coding RNAs [180]. Activation of TEs by stress may provide the cell with the possibility to generate new stress-inducible genes or regulatory networks to protect itself from genome instability while adapting to survive under the newly encountered harmful conditions. This agrees with Barbara McClintock's hypothesis, which postulates that mobile elements allow the cell to respond to environmental "shocks" more flexibly through the mobilization of transposons from heterochromatic domains [181]. Thus, TEs may help the process of adaptation by increasing mutation rates and, therefore, genetic variability towards natural selection and organism positive fitness [182,183].

Exposure to natural or anthropogenic environmental physiological stressors such as endocrine disruptors, air pollution, ionizing or UV-radiation, extreme temperatures, heavy metals, or psychological abuse has been shown to lead to hypomethylation of L1s and ERVs, while showing hypermethylation of *Alu* sequences [184–187]. For example, prenatal exposure of mice to bisphenol A (BPA), considered as endocrine disruptor, led to DNA hypomethylation at nine CpG sites directly upstream of a cryptic promoter of an IAP retroelement [188]. However, chronic exposure to persistent organic pollutants showed hypomethylation of *Alu* retroelements [189]. Furthermore, stress-mediated changes in N(6)-methyladenine (6 mA) dynamics at the DNA level drive LINE expression with loss of 6 mA modifications [190]. Stress also increases H3K9me3 at TE loci, coinciding with the increased levels of the Suv39h2 methyltransferase in rat hippocampus [191].

Overall, it seems that although specificity between the stress type and the TEs affected exist [192,193] several open questions remain to be answered; for example, why some TE families are more prone to respond to stress, and what are the reasons for some TEs being upregulated by stress while others become inhibited [194].

Viral Infections

Another factor influencing TE's epigenetics are viral infections. Many viruses have been reported to influence TE expression in several species. For example, Sindbis virus (SINV) was found to decrease TE transcripts in somatic tissues of *Drosophila* by modulating the piRNA and siRNA repertoires, increasing TE-derived sRNA production that regulate TE expression [195]. In that sense, it has been reported that TE-derived RT makes DNA copies of RNA viruses upon viral infection to boost the production of antiviral siRNA, allowing a stronger immune response [196,197]. Also, upregulation of TEs near antiviral response genes has been observed in human and mouse, regardless of virus or host cell types [198].

Epigenetic dysregulation by viral infections has also been observed in several diseases such as multiple sclerosis (MS), amyotrophic lateral sclerosis (ALS) or schizophrenia (SZ), in which activation of some HERV families such as HERV-W and HERV-K has been linked to pathogenesis [14]. *Herpesviridae* infections are associated with a higher risk for developing MS [199,200] while influenza seems to contribute to SZ, especially when infections occur during the prenatal stages [201,202]. On another hand, herpes viruses like herpes simplex virus type 1 (HSV-1), human herpes virus type 6 (HHV6) and Epstein-Barr virus (EBV) are efficient at activating HERV-W elements, even on methylated promoters [203,204]. In fact, *in vitro* infection by HSV-1 was shown to transactivate HERV-W by stimulating LTR-directed transcription [205], inducing its expression in neuronal and brain endothelial cells [206]. Another HERV-W activator is HHV6 [207]. Also, the binding of the Epstein–Barr virus major envelope protein gp350 activates the transcription of HERV-W/ MSRV in peripheral blood mononuclear cells (PBMCs) and astrocytes of MS patients without requiring EBV entry [199]. Indeed, *in vivo* infection triggered higher MSRV activation in PBMCs when patients had or had passed symptomatic infectious mononucleosis, the clinical manifestation of EBV [208]. On the other hand, influenza virus has been shown to transactivate HERV-W elements by impairing the KRAB-ZFP/ KAP1/SETDB1 axis, which plays an important role in controlling HERV expression, as earlier described. *In vitro*, influenza virus triggers loss of sumoylation on KAP1 and lowers the levels of H3K9 trimethylation and SETDB1 without affecting promoter DNA methylation status [209,210]. Furthermore, it has been recently demonstrated that Sars-CoV-2 virus also has the ability to activate HERV-W expression on T cells of COVID patients, correlating with a worse disease prognosis [211,212].

Nutriepigenetics

After the pioneer findings by the groups of Wolf, Morgan, and Waterland [213–215], environmental epigenomics and epinutrigenomics (i.e., the study of the epigenetic changes produced by diet), have become prolific areas of research partially due to their potential for controlling or preventing detrimental retrotransposon activation. For example, as previously described, prenatal exposure of mice to bisphenol A (BPA) hypomethylated a cryptic IAP-derived promoter, leading to obesity, diabetes, and tumorigenesis. Interestingly, supplementation with genistein, a naturally occurring isoflavone, or other methyl donor supplements counteracted BPA exposure effects in the offspring [188].

As earlier mentioned, additional mechanisms towards retrotransposon repression in mouse and human LTRs include H4K12bio, a rare PTM in human [216] introduced by the enzyme HCS in presence of the vitamin biotin [188]. Another nutrient showing epigenetic control of TEs are the stilbenoids derived from grape's resveratrol which inhibit the MOS-1 eukaryote transposase, the enzyme mediating the mobility of the *Mariner* class II transposons, capable to inhibit HIV integrase as well [217]. Two additional examples worth mentioning are a study demonstrating associations between poor diets, hypomethylation of L1 retroelements and cancer, as well as association between methylation of the same element in participants adhering to healthy diets [218]; and another study showing repression of TEs, miRNAs, and lncRNAs in mice fed with lifespan extending diets. The authors importantly identified miRNA-34a, miRNA-107, and miRNA-212–3p targeting the chromatin remodeler gene Chd1 and suggested a crosstalk between these epigenetic regulators and TEs [219].

Activation of TEs by dysregulation of epigenetic marks may contribute to a variety of neurologic diseases and cancer [14]. For example, failure for SIRT6 (longevity regulating protein Sirtuin 6) to ribosylate KAP1, a silencer modification of L1 promoters, due to genotoxic stress or aging, leads to the activation of this abundant TE contributing to age-related diseases [220].

Although many questions remain to be addressed, the study of the lifestyle effects on the epigenetic landscape of TEs will certainly translate in future improved healthcare.

INFLUENCE OF TRANSPOSABLE ELEMENTS ON HOST EPIGENETICS

Since most TEs harbor gene expression regulatory sequences (transcription initiation and termination signals, RNA polyadenylation, splicing signals, etc.) and they are heavily controlled by several epigenetic mechanisms, it can be expected that TE-associated epigenetic machinery influences the epigenetics of genomic nearby regions. From the chromatin state of large chromosome regions to punctual chromatin changes, TEs have the ability to shape the cell's gene expression patterns as described next (Fig. 18.4).

Transposable Elements Influence on Chromatin Structure

In eukaryotes, nucleosomes are formed by the wrapping of short DNA stretches of about 147 bp around a core of eight histone subunits [221]. For this to occur, the double-stranded DNA has to bend at particular intervals, an event facilitated by the presence of certain dinucleotide motifs, meaning that genetic sequence may underlie nucleosome formation and thus histone epigenetic modification preferences [222,223]. As an overall, three-dimensional (3D) chromatin architecture not only reflects the packaging of DNA patterns in the nucleus of the cell, but also determines gene regulatory neighborhoods [224].

Topologically Associating Domains

Chromatin conformation capture techniques allowed the identification of topologically associating domains (TADs) which are self-associating domains or genomic

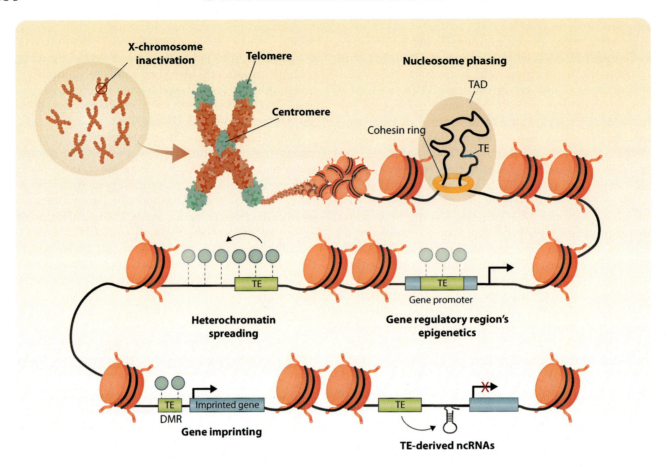

FIGURE 18.4 Epigenetic influence of TEs on eukaryotic cells. TEs influence host gene expression in many ways, from large chromatin changes to punctual epigenetic modifications. TEs mediate inactivation of X chromosome (XCI) in eutherian mammals. The presence and transcription of LINE elements facilitate the silencing of chromosomal regions prone to escape XCI. Further, due to their capacity to recruit the epigenetic silencing machinery, their enrichment in pericentromeric and subtelomeric chromosomal regions sustains cell's constitutive heterochromatin in those regions. Also, TEs structure containing regulatory sequences such as transcription factor binding sites or insulators has the capacity to determine nucleosome positioning and define topologically associating domains (TADs), greatly influencing host regulatory network. Going deeper, their insertion into host regulatory regions such as promoters or enhancers influences the epigenetic state of those elements, therefore affecting the expression of the genes under their control. Also, TEs are known to be donors of cryptic regulatory elements whose epigenetic state also influences gene expression. In this sense, when inserted into differentially methylated regions (DMR) their targeting by the epigenetic machinery may influence gene imprinting. At the least, TEs are a great source of non-coding RNAs regulating gene expression networks.

regions in which DNA sequences interact at a higher frequency with other DNA sequences within the TAD than with those outside, allowing coordinated regulation of gene expression and avoiding interference with neighbor regulatory elements [225]. Though 5 CCCTC 3′-binding factor (CTCF) sites are distributed throughout the genome, they appear enriched on TAD boundaries with a preference for a specific orientation to define the TAD. CTCF is a transcriptional regulator which can function as a transcriptional activator by binding histone acetyltransferase (HAT)-containing complexes, as a transcriptional repressor by binding HDAC-containing complexes, or as a transcriptional insulator by acting as buffers between euchromatin and heterochromatin, blocking the spread of active or repressive sites and communicating between enhancers and closely located promoters to establish genomic regulatory domains [226]. According to the "loop extrusion" model for TAD formation, TADs would arise by cohesin-mediated chromatin folding and CTCF-mediated insulation counteracting cohesin action to define TAD boundaries [225]. CTCFs located at TAD boundaries are oriented convergently, that is in opposite directions, to define a TAD. Otherwise, cohesin-mediated looping persists until encountering an inverted CTCF that dissociates cohesin from DNA, leading to long-range changes in gene expression [225].

TEs have been proposed to shape eukaryotic genome's architecture, connecting repetitive DNA elements, local chromatin environment, and epigenetic regulation of gene expression [227–229]. *De novo* heterochromatin formation has been observed upon insertions of TEs into euchromatin [227] and some TEs such as *Alu*/B1 and B2 elements in mice and *Alu* elements in humans are enriched at boundary regions of domains, indicating a role for TEs in the spatial organization of the genome [228]. Based on these observations and on the high relative abundance of TEs, it is tempting to speculate that perhaps repetitive elements may influence genome packaging and gene expression by their specific binding properties, as proposed by Huda and Jordan [229]. These same authors also propose that TE sequences could determine, for example by providing TFBSs (transcription factor binding sites) or TSSs, precise nucleosome positioning, a disposition receiving the name of nucleosome phasing, with downstream regulatory consequences (Fig. 18.4). In support of TE-biased nucleosome phasing, an enrichment of the insulator CTCF binding elements has been described [228,230]. SINEs are enriched in CTCF-sites seemingly promoting genome's organization and controlling transcription by forming cohesin-dependent loop structures at TAD boundaries [228,230]. In the case of human pluripotent stem cells (hPSCs), TAD boundaries were found to be devoid of CTCF clusters, and instead be dependent on the transcription of HERV-H, as silencing of this retroelement abolished its insulation effect, impacting on the gene regulatory landscape. The authors also observed that human ESC TADs weakened during cell differentiation towards cardiomyocytes, further supporting HERV-H's role in pluripotency [231]. Furthermore, bio-informatic screening and functional genomic data from CD4+ T cells identified a set of over 1000 predicted mammalian retrotransposon MIR insulators genome-wide, which were experimentally validated using *in vitro* and *in vivo* enhancer-blocking assays, as reported by Wang et al. [232]. Furthermore, in *Drosophila*, the *Gypsy* retrotransposon carries an insulator that is able to block the promoter's interactions with distal enhancers [233].

Centromere and Telomere Structures

TEs influence extends to centromere and telomere chromosomal regions (Fig. 18.4). Centromere and telomere are heterochromatin rich structures essential for chromosome functions and genome integrity. Centromere regulates chromosome segregation during cell division while telomere prevents chromosome shortening after replication [234].

In eukaryotes, centromere consists of long tandem arrays of satellite repeats free of TEs surrounded by a peripheral region rich in TEs. Interestingly, pericentromeric regions and subtelomeric regions are enriched in TEs, particularly in LINEs and SINEs, in humans [234]. However, in *Arabidopsis thaliana* satellite repeats are enriched in retrotransposons while pericentromeric region is enriched in DNA transposons [235]. Due to the abundance of TEs in those structures, it is believed that the epigenetic machinery repressing TEs might have a role in sustaining cell's constitutive heterochromatin [236]. Indeed, some satellite repeats in centromere regions have homology to known TEs [237]. Also, it is known that centromere protein B (CENPB), which helps to regulate centromeric heterochromatin, resembles the transposase of *Tc1/Mariner* transposon [237].

Regarding telomeres, they are maintained in eukaryotes by telomerase enzyme with the exception of *Drosophila*, which has no telomerase homolog, and non-LTR retrotransposons HeT-A, TAHRE, and TART have assumed the responsibility by transporting their Gag proteins into the nucleus where they recruit more copies to the chromosome end [236]. These particular elements share epigenetic control mechanisms with other TEs [237] and, interestingly, are found only in telomeres [238]. In humans, telomerase functions as a reverse transcriptase which uses telomere RNA template to add repeats of DNA for telomere elongation upon cell division [239]. In addition, the protein complex shelterin binds to telomeres to protect them from being "misinterpreted" as DNA double strand breaks, and therefore preventing the activation of DNA repair pathways. In this sense, inhibition of L1s has been associated with telomere shortening and L1 knock-down reduced mRNA telomerase levels in humans [240] as well as mRNA and protein levels of *klf-4* and *c-myc* [241], both known as important transcription factors for the expression of telomerase. Moreover, L1 knock-down reduced mRNA and protein shelterin levels, which cannot be achieved by knocking down KLF-4 or c-myc. Reduced shelterin levels correlated with an induction of double strand breaks, named telomere induced foci, specifically at telomeres and not any other genomic region, possibly as a result of reducing telomere protection by shelterin. This suggests that L1s influences telomere maintenance is mediated by those proteins.

Finally, it should be mentioned that L1s are able to directly influence global epigenetic mechanisms and lead to the loss of heterochromatin. On one hand, the formation of RNA:DNA hybrids by its encoded reverse transcriptase, complexes later processed by the Dicer endonuclease, prevents small regulatory RNA production leading to global altered transcription. On the other hand, L1s are able to modulate Dnmt1 or Dnmt3b enzymes involved in genome-wide epigenetic regulation [242].

Transposable Elements Impact on Genetics and Epigenetics of Gene Regulatory Regions

Transcription of eukaryotic genes is a highly complicated process requiring the ultra-precise cooperation of functionally diversified proteins and DNA sequences [243]. Transcription is regulated by *cis*-regulatory regions, which are DNA sequences that provide binding sites for trans-acting factors, including promoters, enhancers, silencers and insulators. Promoters are orientation dependent transcription factor docking sites nearby TSSs, while enhancers and silencers consist of clusters of protein binding sites cooperatively upregulating or inhibiting gene expression regardless of their location, distance, or orientation relative to gene promoters [244]. Insulators, nonetheless, are regions protecting genes from the regulatory influence of surrounding sequences.

Regulation of Gene Expression by Transposable Element-derived cis-Elements

A considerable contribution of cis-regulatory regions to the human genome is attributed to TEs, which is not surprising, as TEs contain *cis*-regulatory elements. For example, those found at the 5′ UTR of LINEs and the LTRs of ERVs [27,28]. Experimental evidence supports that TEs are a source of alternative promoters, enhancers, or sequences with insulator functions, also capable of tuning the activity of pre-existent promoters [192,245,246]. In fact, almost half of the open chromatin regions contain TE sequences and whole-genome analyses have shown that almost 25% of genes harbor TE sequences in their promoters or regulating regions [247], which suggests that they have a function regulating host gene expression [248].

LTR retroelements and LINES contain internal RNA pol II promoters, while SINEs contain RNA pol II or pol III promoters, and chimeric SVAs carry a core enhancer within the SINE-R moiety [53,249–251], allowing them to act as ectopic promoters [252]. Genome-wide analysis of the contribution of TEs to pol II promoters performed by combining RepeatMasker and ENCODE annotations for TEs and TFBS, respectively, found that 75% of the greater-than 35 K promoters evaluated (1500 bp TSS upstream regions) contained TE-derived sequences [253]. Also, since ERVs and L1s contain promoters to drive antisense transcripts, by extending transcription to nearby genes they may interfere with normal gene expression [254].

Moreover, long-range intra-chromosomal interaction studies has allowed the identification of TE-regulated genes by TE-derived enhancers. For example, it was found that the retroelement MER41B upregulated the interferon-inducible myeloid differentiation transcriptional activator IFI16 gene expression located about 20 kb away from the gene. Similarly, the technique identified the interaction between interferon induced transmembrane protein 1 and 3 genes (IFITM1 and IFITM3, respectively) and the distant retroelement MER41A, and the regulated expression of the striated-muscle gene MYPN [255]. Another example of LTR retrotransposon providing alternative enhancer elements is observed in acute myeloid leukemia (AML), where epigenetic silencing of an ERV-derived enhancer alters the expression of adjacent genes [256].

In addition to complete promoters and enhancers, TEs also regulate host gene expression by providing TFBS to which some host transcription factors (TFs) may bind, as it is the case for ESR1, TP53, SOX2, or CTCF [257]. In general, SINE elements largely contribute to regulatory regions, particularly *Alu* elements, being more frequently in promoter regions [253,258,259]. By contrast, LINE elements usually do not locate at regulatory regions, and this reflects in the amount of SINE-derived TFBSs in relation to LINE-derived TFBSs [260]. This is probably because of LINEs large size potentially disrupting genic regions [259]. Furthermore, LTRs are important regulators of gene expression by retaining their TFBS-enriched regulatory elements after being integrated [248,261].

Host TFs not only bind to TFBSs harbored by TEs but also new TFBS can be generated from postinsertional mutations. For example, a binding site for c-Myc could derive from the punctual cytosine to thymine mutation in *Alu* sequences [261]. Another example is provided by the creation of p53 binding sites by mutation of LTRs [261]. Contrary as it may seem, this process is not arbitrary, and TEs can provide clusters of TFBSs for TFs that work together to cooperatively regulate gene expression [262]. Systematic genome-wide expression analysis across 18 different tissues attributed tissue-specific gene expression to 62 different LTRs [263]. Moreover, evaluated data from the Genotype Tissue Expression (GTEx) Project and the Roadmap Epigenomics Project (REP) studies found that house-keeping genes rarely contain TE insertions, while genes with tissue-specific expression profiles contained TEs, particularly LTRs, within active chromatin regions, further supporting previous findings [263,264]. This suggests that TEs' role in gene regulation may be much more important than expected.

Another way in which TEs can influence host genes is by providing alternative splice sites and polyadenylation signals, hence increasing transcript diversity. Around 40%–60% of human genes produce different types of transcripts due to alternative splicing [17,265], giving a wide range of proteins. In fact, it explains the phenotypic difference between humans and mice having each around 30,000 protein-coding genes. Interestingly, some retroelement sequences inserted into a gene can be recruited as exons, by a process named exonization [266]. This further widens the variety of transcripts by employing the donor and acceptor splice sites harbored by those retroelements. In humans, the most exonizated retroelements are

Alus, constituting around a 5% of the alternative spliced exons [266]. Interestingly, *Alu* elements do not harbor splice sites but they arise from point mutations in their sequences. Most exonized *Alus* are alternatively spliced, i.e., they are only included in a fraction of synthetized transcripts, in contrast to constitutive exons which included at all times [265]. This is important since constitutively expressed *Alu* exons are associated with genetic disorders [265]. However, by being alternatively spliced, the wild-type version of the protein and its function are maintained, allowing the slow cooption i.e., slow adaptation with time through mutations of the exonized *Alu* until deriving a protein with novel functions that benefits the host [267]. In contrast to *Alu* elements, L1 harbor functional splice sites in their sequences prior to integration [268], so that they also influence host gene expression by alternative splicing.

A clear example of exonized L1 element is the human *ATRN* gene, encoding Attractin. It is a soluble protein with dipeptidyl peptidase IV activity found in human plasma. It is expressed and released by activated T lymphocytes, playing a fundamental role in the immune response. Attractin assists the formation of immune cell clusters by mediating the interaction between monocytes and T cells or by regulating the levels of chemokines that influence T-cell movement [269]. The soluble functional form of Attractin arose from the exonization of an L1 element, which became exon 25. Nonetheless, the complete gene contains 30 exons, the last five of which encode transmembrane and cytoplasmic domains. Alternative splicing of the gene regulates the availability of the soluble form with respect to the transmembrane's. The exonized L1 element contains a stop codon and a polyadenylation signal so that, when included in the transcript it forms a shorter version of the protein-lacking transmembrane and cytoplasmic domains [270]. This is an example of how TEs can lead to the generation of new gene products, regulating in this case the inflammatory response.

Regulation of Gene Expression by TE-associated Epigenetic Marks

In line with this, the capacity of TEs to recruit silencing factors mediating their own suppression makes them crucial players in controlling patterns of gene expression. As detailed earlier, TEs expression is tightly regulated by DNA methylation and repressive histone modifications, being, in most cases, strongly buried in heterochromatin. Thus, insertion of a TE introduces repressive epigenetic marks to the site where it is located, increasing DNA methylation levels or modifying histone tails of nearby regions [271]. Even when they do not directly affect protein coding regions, their epigenetic state can directly influence the expression of the genes where they were inserted, particularly when present into regulatory regions.

On one hand, since TE insertions into non-coding regions can act by themselves as alternative promoters or enhancers for genes located in their close vicinity, or further away, the epigenetic state of the cryptic TE-derived *cis* element directly regulates expression of genes under its control [252,272]. For example, in Hodgkin lymphoma, a disease presenting with aberrant overexpression of colony stimulating factor 1 receptor (CSF1R) correlating with worse prognosis for patient survival [273], the increase in CSF1R levels generally owes to hypomethylation of an LTR retrotransposon of the mammalian apparent LTR retrotransposons (MaLR) THE1B family located 6.2 kb upstream of the gene. This MaLR LTR element contains a promoter sequence with two CpG sites that are hypermethylated in non-Hodgkin cell lines. By contrast, those CpGs appear hypomethylated in a patient's Hodgkin lymphoma cells, enabling MaLR LTR sequences to act as alternative promoters of the *CSF1R* gene, leading to the production of additional transcript copies which characterizes the illness [274]. However, transcripts originated from alternative promoters usually express at lower levels to avoid deleterious effects. Nonetheless, there are few exceptions in which TE-derived promoters have been co-opted as constitutive, i.e., have become part of existing or new promoters for gene regulation, and most or all transcripts transcribe from them, such as the case of mouse neuronal apoptosis inhibitory protein (Naip), which is exclusively transcribed from an LTR-derived promoter [275]. Another example of promoter cooption is the ectopic regulation of the human *INSL4* gene. Rather than inducing a physiological disorder, the insertion of an HERV-derived LTR regulatory sequence into *INSL4* promoter allows the specific up-regulation of *INSL4* during human cytotrophoblast differentiation into syncytiatrophoblast [276]. In this case, the insertion of a TE sequence regulates the expression of the target gene in a tissue-specific manner contributing to speciation. In this line, functionally relevant TE-derived TFBSs are more conserved than adjacent ones [277] and their epigenetic state changes depending on cellular physiology. For example, cancerous cells harbor derepressed TE-derived TFBS for oncogenic TFs [278]. Indeed, TE-derived TFBSs could act as competitors of other TFBSs in binding to TFs, thus affecting overall gene expression.

On another hand, when TEs are inserted into promoters or other regulatory regions, their epigenetic state directly affects the native regulatory region of the gene (Fig. 18.4). A good example is the regulation of migration and invasion enhancer 1 (*MIEN1*), a gene involved in several types of cancer enhancing cell migration and invasion [279]. *MIEN1* promoter contains a SINE *Alu* region that directly regulates *MIEN1* expression. Under physiological conditions i.e., usual conditions of that organism and cell type, *MIEN1* promoter is hypermethylated by the DNA methylases Dnmt1, Dnmt3a, and Dnmt3b to prevent its expression. By contrast, hypomethylation of its promoter, specifically at the SINE *Alu* region, is observed in cancer cells [280]. This observation is in good agreement with others suggesting that SINE *Alu* regions are generally hypomethylated in cancer [281,282]. Mechanistically, loss of repressive DNA methylation at SINE *Alu* regions unmasks *MIEN1* promoter and allows the binding of transcriptional activators such as the upstream stimulatory factor (USF), facilitating cancer progression [280]. Furthermore, the influence of TEs on gene regulation should be even more noticeable when its insertion is located into the regulatory regions of genes containing epigenetic remodeling functions, such as those coding for zinc finger proteins [283].

Thus, insertion of TEs into coding or non-coding regions have had a huge impact on gene expression either by providing additional regulatory regions or by altering the epigenetic state of the site where they are located.

Heterochromatin Spreading

The assembly of the heterochromatic machinery locates at specific locus within a TE region and then spreads out all over in order to silence the entire TE. Heterochromatin spreading usually extends until encountering a chromatin barrier insulator that protects transcriptionally active regions, but sometimes it can extend further [284]. The effects of heterochromatin spreading on gene expression were first described in *D. melanogaster* under the name of "position effect variegation" (PEV). It describes the phenomenon by which genes located near to heterochromatic regions may give a variegated phenotype because of heterochromatin spreading from the assembly site [285] (Fig. 18.4). Yet, in addition to *Drosophila* [286], it has later been reported in a wide variety of species, including *A. thaliana* [287], rice [288], maize [289], and mice, [290] supporting the idea of heterochromatin spreading not being an isolated event. The strength of silencing inversely correlates with distance to TEs [291,292] and varies across species, being on average of about a hundred base pairs in *A. thaliana* [287] about 1 kb in maize [289], and up to 4–5 kb in mice and *D. melanogaster* [290,293,294]. In specific cases, heterochromatin can even spread beyond 20 kb from the TE insertion site [294]. Furthermore, the robustness of spreading mediated silencing and the type of epigenetic marks that are spread also vary between TE families.

Both SINE and LINE elements are highly enriched in DNA methylation and have been related to PEV events regarding DNA methylation spreading. In particular, *Alus* are enriched with CpG dinucleotides that mainly locate at gene-rich sub-chromosomal regions, where they can nucleate and spread DNA methylation [295,296]. When inserted in proximal or even in coding regions, *Alu* elements can regulate gene expression by methylating regulatory or coding sequences [295–298], which, in most cases, have a deleterious impact on host's physiology. An example of methylation spreading from an *Alu* element has been described for the Proopiomelanocortin (*POMC*) human gene [299]. *POMC* is constituted by three exons flanked by two introns, with two CpG sites located at the promoter region and at the intron 2-exon 3 boundary, respectively. *POMC* codes for a pre-proprotein that can generate different active peptides according to the sites posttranslationally cleaved. An epigenetic variant of *POMC* derived from an hypermethylated exon 3 CpG island has been associated with severe obesity in children [299]. This variant impairs the binding of the transcription enhancer P300 to its binding site in exon 3, reducing the levels of functional *POMC* long transcripts that contain the coding regions for melanocortin peptides which regulate body weight [300]. The trigger of this hypermethylation event resides in an *Alu* insertion at intron 2. Under normal conditions, *Alu*-induced heterochromatin spreading is restricted to intron 2; however, if spread beyond, it can hypermethylate exon 3 CpG's island giving rise to the described *POMC* epigenetic variant that increases the risk for obesity. In addition, *Alu*-induced methylation spreading has been evolutionarily co-opted to reshape methylation patterns around regulatory regions of key pathway-genes, conferring intra-species differences in gene expression. In particular, the epigenetic remodeling of key brain pathways by this mechanism contributes to the phenotypic differences between humans and primates [301].

With respect to LINE elements, they are mainly located within gene-poor sub-chromosomal regions [296], however, some copies are found near coding genes such as the β-globin gene (*HBB*). Severely reduced levels of β globin chain in hemoglobin, the protein encoded by *HBB*, gives rise to β + —thalassemia which can produce anemia [302]. It is known that a full-length L1 element is inserted in antisense orientation into intron 2 of the *HBB* and decreased rates of transcription of the *HBB* gene are caused by an altered chromatin state due to methylation spreading from the L1 insertion site into promoter and enhancer regions of the *HBB* gene [303]. Furthermore, since L1 elements contain potent CpG methylation sites [304] and are enriched in X chromosome [305], they have been shown to facilitate spreading of CpG methylation in the process of X chromosome inactivation [306].

LTR retrotransposons, however, are distributed homogeneously throughout the genome but, as LINEs, they are rarely located close to TSS such as *Alus* [307]. LTRs can also regulate flanking gene expression by providing cryptic sites for heterochromatin formation and spreading; *A. thaliana* and *D. melanogaster* LTR families show the most dominant spreading effects [287,290,294]. In mouse and humans, both DNA methylation and repressive histone PTMs are responsible for LTR retrotransposon silencing, though the second has a major role in this process; *i.e.*, LTR retrotransposons are more likely to spread repressive histone marks such as H3K9me3 than DNA methylation. Different LTRs present different potency for H3K9me3 spreading, being the mouse IAP elements among the LTRs with higher degree of H3K9me3 spreading, up to 5 kb [290]. Although LTRs are rarely located close to genes, an experimental approach on ESC lines showed that IAP copies are capable of disrupting beta 1,3-galactosyltransferase-like (*B3galtl*) gene expression by H3K9me3 spreading. A solitary IAP LTR located 369 bp upstream of the TSS of *B3galtl* in J1 ESCs robustly promotes spreading of H3K9me3 mark in the first 1 kb, affecting the CpG island promoter of that gene in *cis* and reducing its expression [290]. Since there are numerous situations where LTRs have been co-opted as constitutive, tissue-specific or developmental-specific promoters or enhancers for genes [286,308], LTRs may have an indirect impact on host adaptive potential by the spreading of epigenetic marks.

Additionally, in *D. melanogaster* and *A. thaliana* another spreading mechanism implying spreading of RNA targeting to neighboring sequences has been observed [309,310]. In *D. melanogaster* TEs inserted into the 3′ UTR region of actively transcribed genes in ovaries induce piRNA production towards the 3′ end of gene transcripts. For example, *pogoN1* insertion close to the *Madm* gene induces piRNA production towards the 3′ end of *Madm* transcripts independently of its orientation [309]. Also, intergenic and intronic TE insertion induces the generation of antisense small RNAs relative to neighboring exons. For example, insertion of the retrotransposon *Doc* in the intron of *CG42315* induced spreading of sRNA production into the mirtron that encodes the somatic microRNA *dme-miRNA-1011* (miRBase), which resulted in ectopic miRNA overexpression in ovaries where this miRNA is, normally, not expressed [309]. In plants, TE targeted by small RNAs can fuel the production of additional small RNAs from the neighboring sequences, a phenomenon known as "transitivity" [310,311]. The produced small RNAs not only regulate expression at the location where they are produced but can also drive silencing in neighboring cells by spreading from cell-to-cell or *via* the organism's vasculature [311,312].

X-chromosome Inactivation

Eutherian mammals are all mammals having a placenta, with the exception of monotremes and marsupials. Female eutherian mammal genomes include two X chromosomes, one of which must be inactivated to balance gene dosage between males and females by a process termed X chromosome inactivation (XCI) [313]. XCI is mediated by Xist lncRNA which is transcribed from the X chromosome that is going to be silenced. Xist contains six tandem repeats with sequence similarity to TEs, which has led to propose that it originated from TE insertions in eutherians [307]. Within these repeats, the A-repeat seems fundamental for the proper silencing and re-organization of the X chromosome [314]. To mediate XCI, Xist A-repeat is bound by Spen, an RNA binding corepressor protein that interacts with several heterochromatin complexes such as the NuRD complex and HDACs to mediate silencing in *cis* [315–317]. However, Spen also binds ERV sequences to mediate their heterochromatic silencing. In fact, Spen loss leads to ERV reactivation in mouse ESC [318]. Due to Xist RNA sequence similarity to TEs, it is thought that XCI derived from cooption of TE's silencing machinery to restrain X chromosome expression (Fig. 18.4). In this sense, Xist RNA would coat one of the two X chromosomes in females intending to spread the "viral infection" all over which would make Spen to bury the X chromosome in heterochromatin. This proposed mechanism is supported by the fact that partial X silencing is achieved when an A-repeat knockout is replaced by an ERV sequence, though the full XCI process does not take place [318].

The XCI process seems to start by localizing Xist RNA at different distal regions across the X chromosome, thereby creating a silent nuclear compartment depleted of RNA pol II and transcription factors. This compartment is rich in LINE and SINE sequences that are rapidly silenced within Xist RNA coated regions at the beginning of the XCI process, [319,320]. The compartment's function is to nucleate heterochromatin along the X chromosome and recruit genes located outside these regions to silence them [319]. LINE elements not only participate in the nucleation process but also their density determines silencing efficiency. In LINE-rich regions, genes are more effectively brought into Xist compartments to be silenced, whilst LINE-poor regions are found far from Xist-induced compartments resisting silencing by remaining outside. This is supported by the fact that genes escaping inactivation in human X chromosome are L1 poor [306]. Interestingly enough, despite a generalized depletion of LINEs within Xist compartments, some young full-length LINEs in the X chromosome are expressed in an XCI-induced manner, facilitating the silencing of chromosomal regions prone to escape XCI. Activated LINEs RNA first locate adjacent to Xist domains but then transcribe from within the Xist RNA compartment. Importantly, LINE expression is specific to the XCI process and cannot be achieved only by chromatin changes [321]. Furthermore, the underlying mechanisms for their activation remain unknown. However, it is interesting that Xist-mediated XCI is present in eutherian mammals but absent in marsupials whose X-chromosome is poor in LINEs [322].

Genomic Imprinting and Epialleles

Genomic imprinting is the epigenetic process whereby one of the two parentally inherited alleles of a gene is preferentially expressed over the other. In this process, whether an allele is expressed or silenced depends on the sex of the parent from which it has been inherited. In this sense, they can be differentiated into paternally expressed genes (PEGs) if the paternal allele is expressed and maternally expressed genes (MEGs) if it is the maternal allele the one being expressed [323]. Most imprinted genes are MEGs and locate at promoters, while few are PEGs and locate at intergenic regions [324]. Furthermore, most imprinted genes are placental, though post-natal imprinting has also been described in the brain [325]. Genomic imprinting can be established either through activation or suppression of expression of one of the parental alleles. It consists on methylation, either of the maternal or the paternal allele, of the CpG islands within a genomic element named imprinting control region (ICR), or differentially methylated region (DMR), that regulates the expression of multiple genes. Since this methylation is acquired in the germline during oogenesis or spermatogenesis, it is often referred to as germline DMRs (gDMRs) [324].

Gender differences in TEs epigenetics have been reported (e.g., hypomethylated IAPs and L1s in oocytes and hypomethylated SINEs in sperm) [326,327] and the epigenetic mechanisms underlying genomic imprinting are similar to those regulating TE expression e.g., imprinting control regions are maintained methylated by G9a/GLP complex as IAPs during epigenetic reprogramming [161–163], and SETDB1 maintains allele-specific DNA methylation at ICRs through ZFP57-KAP1 [149,328,329]. For this reason, it has been suggested that genomic imprinting arose in therian mammals as a host defense mechanism to the accumulation of TEs in their genomes [236,324,330,331] (Fig. 18.4). In fact, some gDMRs are TE-derived, such as the imprinted retrotransposon-like 1 (RLT1) element and the paternally expressed 10 (PEG10) genes [332–334]. Also, the paternally imprinted Rasgrf1 locus in mice is achieved through targeting TE sequences within a DMR adjacent to this locus by the PIWI-piRNA pathway, thus directing its methylation and silencing [335]. Moreover, it has been proposed that silencing of DNA after a retrotransposon insertion drives evolution of genomic imprinting in mammals, as insertion of PEG10 is shown to be sufficient to establish imprinting in the inserted regions [336].

Additionally, ERVs can establish metastable epialleles which are differentially expressed alleles driven by epigenetic modifications established during early development [337]. The best-known example of this phenomenon was described for the Agouti gene. A gene epigenetically controlled by a cryptic IAP LTR-retrotransposon promoter located upstream of its TSS. Thus, Agouti gene expression depends on the epigenetic state of the IAP promoter, activating its expression when it is hypomethylated. Furthermore, since some IAPs are not epigenetically reprogrammed in the early embryo, their epigenetic state can be inherited from ancestors [213,338]. The variable methylation of IAP loci is established as a result of their partial KZFP-mediated silencing which allows the binding of proteins, such as TETs, protecting those loci from methylation [339].

Regulation of gene expression by TE Regulatory RNA Networks

Well-known epiregulators of gene expression are non-coding RNAs (ncRNAs), which include regulatory RNAs such as small non-coding RNAs (sncRNAs) and lncRNAs. Their origin and functionality have been discovered to lie within TE sequence insertions, as it is the case of miRNAs, a kind of sncRNAs mediating gene silencing by cleaving and degrading target transcripts. To become functional, transcribed miRNA precursors' stem-loop structure must be processed (cleaved) to form mature miRNAs [340,341]. Interestingly, similar stem-loop structures are present on TE sequences flanking genomic ncRNA loci [342], thus being hypothesized that TEs are involved in the origin of miRNAs. In fact, not only has been reported a high sequence identity between MITE and the hsa-miR-548 family, but also MITE contain a stem-loop structure which can be processed to form a mature miRNA [343]. Indeed, the group of Yuan et al. [344] have shown that an entire miRNA family, the miR-1302, has arisen from TE sequences. TE-derived miRNAs seem to account for around 20% of human miRNAs, of which about 11% derived from inverted TE sequences, 52% partly overlap with non-inverted TE sequences and 37% completely arose from TE sequences [345]. Furthermore, with respect to lncRNAs, 65%–75% of them also harbor TE sequences, constituting about 27% of their transcript length [70,346]. They can be transcribed from different genomic regions such as intergenic (lincRNAs), intronic, antisense coding regions or even form exon shuffling circular lncRNAs [347]. In particular, lincRNAs constitute most of the lncRNAs [348] of which a huge percentage associate with TEs that contribute mainly to introns, followed by exons and promoters [349].

Regarding functionality, it has been found that TE-derived miRNAs target genes which harbor the same type of TE element from which they derive in their 3′ UTRs. For example, miR-28−5 and miR-15a derive from a LINE-2 (L2) element and target genes that harbor L2 sequences in their 3′ UTR such as the E2F transcription factor 6 (E2F6) [350]. The same has been described for lncRNAs but with *Alu* sequences. For example, a lncRNA harboring *Alu* sequences binds to the 3′ UTR of tissue plasminogen activator inhibitor type 1 (SERPINE1) transcript [351]. In both cases TE sequences influence the base pairing of the regulatory RNA required to form a dsRNA structure that is then degraded. Furthermore, *Alu* base pairing would be the source for lncRNAs circularization [352]. In addition, TEs provide regulatory RNAs with lineage and tissue-specificity. For example, the previously cited miR-1302 family are placental specific. This seems interesting since the TE from which it derived is eutherian-specific and placenta is characteristic of eutherian pregnancy [344]. One more example are the HERV-H sequences inserted in the TSS near lncRNA genes which can induce their expression in pluripotent cells [353]. In fact, Kang and coworkers showed that among those human lncRNAs with tissue-specific expression, around 70% derive from TEs [346].

Hence, TEs have greatly influenced RNA-mediated gene expression regulation by contributing to the origin, evolution and functionality of regulatory RNAs (Fig. 18.4). However, their impact is variable among TE families. Both SINE and LINE are the major contributors to miRNAs and lncRNAs; while DNA transposons have a considerable contribution to miRNAs, it is the lowest to lncRNAs [345,346]. Further, LTRs are over-represented in exonic and promoter regions of lncRNA genes in contrast to protein coding genes [70]. Intronic regions, however, show similar levels to coding-regions consistent with their emergence before the rodent and primate lineages [349].

CONCLUDING REMARKS

Since their discovery around the middle of the past century, our understanding of the role of TEs in the eukaryotic cell has drastically changed, from being considered as "spacer" functionless sequences to reveal their role as main drivers of evolution and key regulators of gene expression.

The progress of technology during the twentieth and current centuries has permitted the advance of genetics and molecular biology fields to unprecedent levels. The -omics, single cell analysis methods, and the bioinformatics have permitted to discern an ample variety of existing TEs and, thus, the requirement for more complex classification systems still pending of standarization.

Documented by a vast array of experimental data, it is now known that insertion of TEs into coding or non-coding regions has had a tremendous impact on the regulation of gene expression in eukaryotes. By providing additional *cis*-regulatory regions or by altering the epigenetic state of the site where they landed, TEs have been major drivers in the evolution of species and also in the emergence of diseases. The understanding of the genetic and epigenetic mechanisms in which TEs participate, which are starting to emerge, should allow for the growth of new scientific knowledge and for the development of improved future healthcare programs.

References

[1] McClintock B. Mutable loci in maize. Annual Report of the Director of the Department of Genetics; Carnegie Institution of Washington Yearbook. New York, NY: Cold Spring Harbor; 1947−8. p. 155−69.

[2] McCLINTOCK B. The origin and behavior of mutable loci in maize. Proc Natl Acad Sci U S A 1950;36(6):344−55. Available from: https://doi.org/10.1073/pnas.36.6.344 PMID: 15430309; PMCID: PMC1063197.

[3] Britten RJ, Davidson EH. Gene regulation for higher cells: a theory. Science 1969;165(3891):349−57. Available from: https://doi.org/10.1126/science.165.3891.349 PMID: 5789433.

[4] Davidson EH, Britten RJ. Regulation of gene expression: possible role of repetitive sequences. Science 1979;204(4397):1052−9. Available from: https://doi.org/10.1126/science.451548 PMID: 451548.

[5] Cordaux R, Batzer MA. The impact of retrotransposons on human genome evolution. Nat Rev Genet 2009;10(10):691−703. Available from: https://doi.org/10.1038/nrg2640 PMID: 19763152; PMCID: PMC2884099.

[6] Feschotte C, Pritham EJ. DNA transposons and the evolution of eukaryotic genomes. Annu Rev Genet 2007;41:331−68. Available from: https://doi.org/10.1146/annurev.genet.40.110405.090448 PMID: 18076328; PMCID: PMC2167627.

[7] Fambrini M, Usai G, Vangelisti A, Mascagni F, Pugliesi C. The plastic genome: the impact of transposable elements on gene functionality and genomic structural variations. Genesis 2020;58 (12):e23399. Available from: https://doi.org/10.1002/dvg.23399 Epub 2020 Nov 24; PMID: 33230956.

[8] Zhou L, Mitra R, Atkinson PW, Hickman AB, Dyda F, Craig NL. Transposition of hAT elements links transposable elements and V(D)J recombination. Nature 2004;432(7020):995−1001. Available from: https://doi.org/10.1038/nature03157 PMID: 15616554.

[9] Lavialle C, Cornelis G, Dupressoir A, Esnault C, Heidmann O, Vernochet C, et al. Paleovirology of 'syncytins', retroviral env genes exapted for a role in placentation. Philos Trans R Soc Lond B Biol Sci 2013;368(1626):20120507. Available from: https://doi.org/10.1098/rstb.2012.0507 PMID: 23938756; PMCID: PMC3758191.

[10] Cornelis G, Vernochet C, Carradec Q, Souquere S, Mulot B, Catzeflis F, et al. Retroviral envelope gene captures and syncytin exaptation for placentation in marsupials. Proc Natl Acad Sci U S A 2015;112(5):E487−96. Available from: https://doi.org/10.1073/pnas.1417000112 Epub 2015 Jan 20; PMID: 25605903.

[11] Song M, Pebworth MP, Yang X, Abnousi A, Fan C, Wen J, et al. Cell-type-specific 3D epigenomes in the developing human cortex. Nature 2020;587(7835):644–9. Available from: https://doi.org/10.1038/s41586-020-2825-4 Epub 2020 Oct 14; PMID: 33057195.

[12] Villanueva-Cañas JL, Horvath V, Aguilera L, González J. Diverse families of transposable elements affect the transcriptional regulation of stress-response genes in Drosophila melanogaster. Nucleic Acids Res 2019;47(13):6842–57. Available from: https://doi.org/10.1093/nar/gkz490 PMID: 31175824; PMCID: PMC6649756.

[13] Grundy EE, Diab N, Chiappinelli KB. Transposable element regulation and expression in cancer. FEBS J 2022;289:1160–1179;. Available from: https://doi.org/10.1111/febs.15722 Epub ahead of print; PMID: 33471418.

[14] Giménez-Orenga K, Oltra E. Human endogenous retrovirus as therapeutic targets in neurologic disease. Pharmaceuticals (Basel) 2021;14(6):495. Available from: https://doi.org/10.3390/ph14060495 PMID: 34073730; PMCID: PMC8225122.

[15] Petrizzo A, Ragone C, Cavalluzzo B, Mauriello A, Manolio C, Tagliamonte M, et al. Human endogenous retrovirus reactivation: implications for cancer immunotherapy. Cancers (Basel) 2021;13(9):1999. Available from: https://doi.org/10.3390/cancers13091999 PMID: 33919186; PMCID: PMC8122352.

[16] Goffeau A, Barrell BG, Bussey H, Davis RW, Dujon B, Feldmann H, et al. Life with 6000 genes. Science 1996;274 (5287):546. Available from: https://doi.org/10.1126/science.274.5287.546 563–7; PMID: 8849441.

[17] Lander ES, Linton LM, Birren B, Nusbaum C, Zody MC, Baldwin J, et al. International human genome sequencing consortium. Initial sequencing and analysis of the human genome. Nature 2001;409(6822):860–921. Available from: https://doi.org/10.1038/35057062 Erratum in: Nature 2001 Aug 2;412 (6846):565. Erratum in: Nature 2001 Jun 7;411(6838):720. Szustakowki, J [corrected to Szustakowski, J]; PMID: 11237011.

[18] Mouse Genome Sequencing ConsortiumWaterston RH, Lindblad-Toh K, Birney E, Rogers J, Abril JF, Agarwal P, et al. Initial sequencing and comparative analysis of the mouse genome. Nature 2002;420(6915):520–62. Available from: https://doi.org/10.1038/nature01262 PMID: 12466850.

[19] Chénais B, Caruso A, Hiard S, Casse N. The impact of transposable elements on eukaryotic genomes: from genome size increase to genetic adaptation to stressful environments. Gene 2012;509(1):7–15. Available from: https://doi.org/10.1016/j.gene.2012.07.042 Epub 2012 Aug 16; PMID: 22921893.

[20] Tang W, Mun S, Joshi A, Han K, Liang P. Mobile elements contribute to the uniqueness of human genome with 15,000 human-specific insertions and 14 Mbp sequence increase. DNA Res 2018;25(5):521–33. Available from: https://doi.org/10.1093/dnares/dsy022 PMID: 30052927; PMCID: PMC6191304.

[21] Finnegan DJ. Eukaryotic transposable elements and genome evolution. Trends Genet 1989;5(4):103–7. Available from: https://doi.org/10.1016/0168-9525(89)90039-5 PMID: 2543105.

[22] Kojima KK. Structural and sequence diversity of eukaryotic transposable elements. Genes Genet Syst 2020;94(6):233–52. Available from: https://doi.org/10.1266/ggs.18-00024 Epub 2018 Nov 9; PMID: 30416149.

[23] Wicker T, Sabot F, Hua-Van A, Bennetzen JL, Capy P, Chalhoub B, et al. A unified classification system for eukaryotic transposable elements. Nat Rev Genet 2007;8(12):973–82. Available from: https://doi.org/10.1038/nrg2165 PMID: 17984973.

[24] Mager DL, Stoye JP. Mammalian endogenous retroviruses. Microbiol Spectr 2015;3(1). Available from: https://doi.org/10.1128/microbiolspec.MDNA3-0009-2014 MDNA3–0009-2014; PMID: 26104559.

[25] Kapitonov VV, Jurka J. A universal classification of eukaryotic transposable elements implemented in Repbase. Nat Rev Genet 2008;9(5):411–12. Available from: https://doi.org/10.1038/nrg2165-c1 author reply 414; PMID: 18421312.

[26] Deniz Ö, Frost JM, Branco MR. Regulation of transposable elements by DNA modifications. Nat Rev Genet 2019;20(7):417–31. Available from: https://doi.org/10.1038/s41576-019-0106-6 Erratum in: Nat Rev Genet 2019 Mar 20; PMID: 30867571.

[27] Pappalardo AM, Ferrito V, Biscotti MA, Canapa A, Capriglione T. Transposable elements and stress in vertebrates: an overview. Int J Mol Sci 2021;22(4):1970. Available from: https://doi.org/10.3390/ijms22041970 PMID: 33671215; PMCID: PMC7922186.

[28] Pace 2nd JK, Feschotte C. Thpappe evolutionary history of human DNA transposons: evidence for intense activity in the primate lineage. Genome Res 2007;17(4):422–32. Available from: https://doi.org/10.1101/gr.5826307 Epub 2007 Mar 5; PMID: 17339369; PMCID: PMC1832089.

[29] Glöckner G, Szafranski K, Winckler T, Dingermann T, Quail MA, Cox E, et al. The complex repeats of Dictyostelium discoideum. Genome Res 2001;11(4):585–94. Available from: https://doi.org/10.1101/gr.162201 PMID: 11282973; PMCID: PMC311061.

[30] Arkhipova IR, Pyatkov KI, Meselson M, Evgen'ev MB. Retroelements containing introns in diverse invertebrate taxa. Nat Genet 2003;33(2):123–4. Available from: https://doi.org/10.1038/ng1074 Epub 2003 Jan 13; PMID: 12524543.

[31] Levin HL, Moran JV. Dynamic interactions between transposable elements and their hosts. Nat Rev Genet 2011;12 (9):615–27. Available from: https://doi.org/10.1038/nrg3030 PMID: 21850042; PMCID: PMC3192332.

[32] Platt 2nd RN, Vandewege MW, Ray DA. Mammalian transposable elements and their impacts on genome evolution. Chromosome Res 2018;26(1–2):25–43. Available from: https://doi.org/10.1007/s10577-017-9570-z Epub 2018 Feb 1; PMID: 29392473; PMCID: PMC5857283.

[33] de Parseval N, Heidmann T. Human endogenous retroviruses: from infectious elements to human genes. Cytogenet Genome Res 2005;110(1–4):318–32. Available from: https://doi.org/10.1159/000084964 PMID: 16093684.

[34] Nefedova L, Kim A. Mechanisms of LTR-retroelement transposition: lessons from Drosophila melanogaster. Viruses 2017;9(4):81. Available from: https://doi.org/10.3390/v9040081 PMID: 28420154; PMCID: PMC5408687.

[35] Bannert N, Kurth R. The evolutionary dynamics of human endogenous retroviral families. Annu Rev Genomics Hum Genet 2006;7:149–73. Available from: https://doi.org/10.1146/annurev.genom.7.080505.115700 PMID: 16722807.

[36] Deininger PL, Batzer MA. Mammalian retroelements. Genome Res 2002;12(10):1455–65. Available from: https://doi.org/10.1101/gr.282402 PMID: 12368238.

[37] Deininger PL, Batzer MA, Hutchison 3rd CA, Edgell MH. Master genes in mammalian repetitive DNA amplification. Trends Genet 1992;8(9):307–11. Available from: https://doi.org/10.1016/0168-9525(92)90262-3 PMID: 1365396.

[38] Bourgeois Y, Boissinot S. On the population dynamics of junk: a review on the population genomics of transposable elements. Genes (Basel) 2019;10(6):419. Available from: https://doi.org/10.3390/genes10060419 PMID: 31151307; PMCID: PMC6627506.

[39] Griffiths DJ. Endogenous retroviruses in the human genome sequence. Genome Biol 2001;2(6). Available from: https://doi.org/10.1186/gb-2001-2-6-reviews1017 REVIEWS1017; PMID: 11423012.

[40] Bodak M, Yu J, Ciaudo C. Regulation of LINE-1 in mammals. Biomol Concepts 2014;5(5):409–28. Available from: https://doi.org/10.1515/bmc-2014-0018 PMID: 25367621.

REFERENCES

[41] Kojima KK, Fujiwara H. An extraordinary retrotransposon family encoding dual endonucleases. Genome Res 2005;15(8):1106—17. Available from: https://doi.org/10.1101/gr.3271405 PMID: 16077010; PMCID: PMC1182223.

[42] Richardson SR, Doucet AJ, Kopera HC, Moldovan JB, Garcia-Perez JL, Moran JV. The influence of LINE-1 and SINE retrotransposons on mammalian genomes. Microbiol Spectr 2015;3(2). Available from: https://doi.org/10.1128/microbiolspec.MDNA3-0061-2014 MDNA3—0061-2014; PMID: 26104698.

[43] Ohshima K, Okada N. SINEs and LINEs: symbionts of eukaryotic genomes with a common tail. Cytogenet Genome Res 2005;110(1—4):475—90. Available from: https://doi.org/10.1159/000084981 PMID: 16093701.

[44] Wang H, Xing J, Grover D, Hedges DJ, Han K, Walker JA, et al. SVA elements: a hominid-specific retroposon family. J Mol Biol 2005;354(4):994—1007. Available from: https://doi.org/10.1016/j.jmb.2005.09.085 Epub 2005 Oct 19; PMID: 16288912.

[45] Mills RE, Bennett EA, Iskow RC, Devine SE. Which transposable elements are active in the human genome? Trends Genet 2007;23(4):183—91. Available from: https://doi.org/10.1016/j.tig.2007.02.006 Epub 2007 Feb 27; PMID: 17331616.

[46] Ichiyanagi K. Epigenetic regulation of transcription and possible functions of mammalian short interspersed elements, SINEs. Genes Genet Syst 2013;88(1):19—29. Available from: https://doi.org/10.1266/ggs.88.19 PMID: 23676707.

[47] Sotero-Caio CG, Platt 2nd RN, Suh A, Ray DA. Evolution and diversity of transposable elements in vertebrate genomes. Genome Biol Evol 2017;9(1):161—77. Available from: https://doi.org/10.1093/gbe/evw264 PMID: 28158585; PMCID: PMC5381603.

[48] Doak TG, Doerder FP, Jahn CL, Herrick G. A proposed superfamily of transposase genes: transposon-like elements in ciliated protozoa and a common "D35E" motif. Proc Natl Acad Sci U S A 1994;91(3):942—6. Available from: https://doi.org/10.1073/pnas.91.3.942 PMID: 8302872; PMCID: PMC521429.

[49] Kojima KK, Jurka J. Crypton transposons: identification of new diverse families and ancient domestication events. Mob DNA 2011;2(1):12. Available from: https://doi.org/10.1186/1759-8753-2-12 PMID: 22011512; PMCID: PMC3212892.

[50] Kapitonov VV, Jurka J. Helitrons on a roll: eukaryotic rolling-circle transposons. Trends Genet 2007;23(10):521—9. Available from: https://doi.org/10.1016/j.tig.2007.08.004 Epub 2007 Sep 11; PMID: 17850916.

[51] Seberg O, Petersen G. A unified classification system for eukaryotic transposable elements should reflect their phylogeny. Nat Rev Genet 2009;10(4):276. Available from: https://doi.org/10.1038/nrg2165-c3 PMID: 19238178.

[52] Piégu B, Bire S, Arensburger P, Bigot Y. A survey of transposable element classification systems—a *call* for a fundamental update to meet the challenge of their diversity and complexity. Mol Phylogenet Evol 2015;86:90—109. Available from: https://doi.org/10.1016/j.ympev.2015.03.009 Epub 2015 Mar 20; PMID: 25797922.

[53] Hancks DC, Kazazian Jr. HH. SVA retrotransposons: evolution and genetic instability. Semin Cancer Biol 2010;20(4):234—45. Available from: https://doi.org/10.1016/j.semcancer.2010.04.001 Epub 2010 Apr 21; PMID: 20416380.

[54] Beck CR, Collier P, Macfarlane C, Malig M, Kidd JM, Eichler EE, et al. LINE-1 retrotransposition activity in human genomes. Cell 2010;141(7):1159—70. Available from: https://doi.org/10.1016/j.cell.2010.05.021 PMID: 20602998; PMCID: PMC3013285.

[55] Gardner EJ, Lam VK, Harris DN, Chuang NT, Scott EC, Pittard WS, et al. 1000 genomes project consortium, devine SE. The mobile element locator tool (MELT): population-scale mobile element discovery and biology. Genome Res 2017;27(11):1916—29. Available from: https://doi.org/10.1101/gr.218032.116 Epub 2017 Aug 30; PMID: 28855259.

[56] Brouha B, Schustak J, Badge RM, Lutz-Prigge S, Farley AH, Moran JV, et al. Hot L1s account for the bulk of retrotransposition in the human population. Proc Natl Acad Sci U S A 2003;100(9):5280—5. Available from: https://doi.org/10.1073/pnas.0831042100 Epub 2003 Apr 7; PMID: 12682288.

[57] Huang CR, Burns KH, Boeke JD. Active transposition in genomes. Annu Rev Genet 2012;46:651—75. Available from: https://doi.org/10.1146/annurev-genet-110711-155616 PMID: 23145912; PMCID: PMC3612533.

[58] Rodriguez-Martin B, Alvarez EG, Baez-Ortega A, Zamora J, Supek F, Demeulemeester J, , et al.PCAWG Structural Variation Working GroupCampbell PJ, Tubio JMC PCAWG consortium. Pan-cancer analysis of whole genomes identifies driver rearrangements promoted by LINE-1 retrotransposition. Nat Genet 2020;52(3):306—19. Available from: https://doi.org/10.1038/s41588-019-0562-0 Epub 2020 Feb 5; PMID: 32024998.

[59] Ewing AD, Kazazian Jr. HH. High-throughput sequencing reveals extensive variation in human-specific L1 content in individual human genomes. Genome Res 2010;20(9):1262—70. Available from: https://doi.org/10.1101/gr.106419.110 Epub 2010 May 20; PMID: 20488934.

[60] Zhang Y, Maksakova IA, Gagnier L, van de Lagemaat LN, Mager DL. Genome-wide assessments reveal extremely high levels of polymorphism of two active families of mouse endogenous retroviral elements. PLoS Genet 2008;4(2):e1000007. Available from: https://doi.org/10.1371/journal.pgen.1000007 PMID: 18454193; PMCID: PMC2265474.

[61] Carpentier MC, Manfroi E, Wei FJ, Wu HP, Lasserre E, Llauro C, et al. Retrotranspositional landscape of Asian rice revealed by 3000 genomes. Nat Commun 2019;10(1):24. Available from: https://doi.org/10.1038/s41467-018-07974-5 PMID: 30604755; PMCID: PMC6318337.

[62] Thompson PJ, Macfarlan TS, Lorincz MC. Long terminal repeats: from parasitic elements to building blocks of the transcriptional regulatory repertoire. Mol Cell 2016;62(5):766—76. Available from: https://doi.org/10.1016/j.molcel.2016.03.029 PMID: 27259207; PMCID: PMC4910160.

[63] Conti A, Carnevali D, Bollati V, Fustinoni S, Pellegrini M, Dieci G. Identification of RNA polymerase III-transcribed Alu loci by computational screening of RNA-Seq data. Nucleic Acids Res 2015;43(2):817—35. Available from: https://doi.org/10.1093/nar/gku1361 Epub 2014 Dec 29; PMID: 25550429.

[64] Hermant C, Torres-Padilla ME. TFs for TEs: the transcription factor repertoire of mammalian transposable elements. Genes Dev 2021;35(1—2):22—39. Available from: https://doi.org/10.1101/gad.344473.120 PMID: 33397727; PMCID: PMC7778262.

[65] Eickbush DG, Eickbush TH. R2 retrotransposons encode a self-cleaving ribozyme for processing from an rRNA cotranscript. Mol Cell Biol 2010;30(13):3142—50. Available from: https://doi.org/10.1128/MCB.00300-10 Epub 2010 Apr 26; PMID: 20421411.

[66] Perepelitsa-Belancio V, Deininger P. RNA truncation by premature polyadenylation attenuates human mobile element activity. Nat Genet 2003;35(4):363—6. Available from: https://doi.org/10.1038/ng1269 Epub 2003 Nov 16; PMID: 14625551.

[67] Teixeira FK, Okuniewska M, Malone CD, Coux RX, Rio DC, Lehmann R. piRNA-mediated regulation of transposon alternative splicing in the soma and germ line. Nature 2017;552 (7684):268—72. Available from: https://doi.org/10.1038/nature25018 Epub 2017 Dec 6; PMID: 29211718.

[68] Jensen TH, Jacquier A, Libri D. Dealing with pervasive transcription. Mol Cell 2013;52(4):473—84. Available from: https://doi.org/10.1016/j.molcel.2013.10.032 PMID: 24267449.

[69] Rebollo R, Farivar S, Mager DL. C-GATE — catalogue of genes affected by transposable elements. Mob DNA 2012;3(1):9. Available from: https://doi.org/10.1186/1759-8753-3-9 PMID: 22621612; PMCID: PMC3472293.

[70] Kapusta A, Kronenberg Z, Lynch VJ, Zhuo X, Ramsay L, Bourque G, et al. Transposable elements are major contributors to the origin, diversification, and regulation of vertebrate long noncoding RNAs. PLoS Genet 2013;9(4):e1003470. Available from: https://doi.org/10.1371/journal.pgen.1003470 Epub 2013 Apr 25; PMID: 23637635.

[71] Lanciano S, Cristofari G. Measuring and interpreting transposable element expression. Nat Rev Genet 2020;21(12):721—36. Available from: https://doi.org/10.1038/s41576-020-0251-y Epub 2020 Jun 23; PMID: 32576954.

[72] Bao W, Kojima KK, Kohany O. Repbase update, a database of repetitive elements in eukaryotic genomes. Mob DNA 2015;6:11. Available from: https://doi.org/10.1186/s13100-015-0041-9 PMID: 26045719; PMCID: PMC4455052.

[73] Hubley R, Finn RD, Clements J, Eddy SR, Jones TA, Bao W, et al. The Dfam database of repetitive DNA families. Nucleic Acids Res 2016;44(D1):D81—9. Available from: https://doi.org/10.1093/nar/gkv1272 Epub 2015 Nov 26; PMID: 26612867.

[74] Tarailo-Graovac M, Chen N. Using RepeatMasker to identify repetitive elements in genomic sequences. Curr Protoc Bioinform 2009. Available from: https://doi.org/10.1002/0471250953.bi0410s25 Chapter 4:Unit 4.10; PMID: 19274634.

[75] Lappalainen I, Lopez J, Skipper L, Hefferon T, Spalding JD, Garner J, et al. DbVar and DGVa: public archives for genomic structural variation. Nucleic Acids Res 2013;41(Database issue):D936—41. Available from: https://doi.org/10.1093/nar/gks1213 Epub 2012 Nov 27; PMID: 23193291.

[76] Goerner-Potvin P, Bourque G. Computational tools to unmask transposable elements. Nat Rev Genet 2018;19(11):688—704. Available from: https://doi.org/10.1038/s41576-018-0050-x PMID: 30232369.

[77] Arkhipova IR. Using bioinformatic and phylogenetic approaches to classify transposable elements and understand their complex evolutionary histories. Mob DNA 2017;8:19. Available from: https://doi.org/10.1186/s13100-017-0103-2 PMID: 29225705; PMCID: PMC5718144.

[78] Ayarpadikannan S, Kim H-S. The impact of transposable elements in genome evolution and genetic instability and their implications in various diseases. Genomics Inform 2014;12(3):98—104.

[79] Klein SJ, O'Neill RJ. Transposable elements: genome innovation, chromosome diversity, and centromere conflict. Chromosome Res 2018;26(1—2):5—23. Available from: https://doi.org/10.1007/s10577-017-9569-5 Epub 2018 Jan 13; PMID: 29332159.

[80] He JP, Chen JK. Epigenetic control of transposable elements and cell fate decision. Yi Chuan 2021;43(9):822—34. Available from: https://doi.org/10.16288/j.yczz.21-113 PMID: 34702696; Groh S, Schotta G. Silencing of endogenous retroviruses by heterochromatin. Cell Mol Life Sci 2017;74(11):2055—65.

[81] Moore LD, Le T, Fan G. DNA methylation and its basic function. Neuropsychopharmacology 2013;38(1):23—38.

[82] Barau J, Teissandier A, Zamudio N, Roy S, Nalesso V, Hérault Y, et al. The DNA methyltransferase DNMT3C protects male germ cells from transposon activity. Science 2016;354(6314):909—12. Available from: https://doi.org/10.1126/science.aah5143 PMID: 27856912.

[83] Xie W, Barr CL, Kim A, Yue F, Lee AY, Eubanks J, et al. Base-resolution analyses of sequence and parent-of-origin dependent DNA methylation in the mouse genome. Cell 2012;148(4):816—31. Available from: https://doi.org/10.1016/j.cell.2011.12.035 PMID: 22341451; PMCID: PMC3343639.

[84] Ramsahoye BH, Biniszkiewicz D, Lyko F, Clark V, Bird AP, Jaenisch R. Non-CpG methylation is prevalent in embryonic stem cells and may be mediated by DNA methyltransferase 3a. Proc Natl Acad Sci U S A 2000;97(10):5237—42. Available from: https://doi.org/10.1073/pnas.97.10.5237 PMID: 10805783; PMCID: PMC25812.

[85] Ziller MJ, Müller F, Liao J, Zhang Y, Gu H, Bock C, et al. Genomic distribution and inter-sample variation of non-CpG methylation across human cell types. PLoS Genet 2011;7(12):e1002389. Available from: https://doi.org/10.1371/journal.pgen.1002389 Epub 2011 Dec 8; PMID: 22174693; PMCID: PMC3234221.

[86] Gruenbaum Y, Stein R, Cedar H, Razin A. Methylation of CpG sequences in eukaryotic DNA. FEBS Lett 1981;124(1):67—71. Available from: https://doi.org/10.1016/0014-5793(81)80055-5 PMID: 7215556.

[87] Hackett JA, Surani MA. DNA methylation dynamics during the mammalian life cycle. Philos Trans R Soc Lond B Biol Sci 2013;368(1609):20110328. Available from: https://doi.org/10.1098/rstb.2011.0328 PMID: 23166392; PMCID: PMC3539357.

[88] Henderson IR, Jacobsen SE. Epigenetic inheritance in plants. Nature 2007;447(7143):418—24. Available from: https://doi.org/10.1038/nature05917 PMID: 17522675.

[89] Caldwell BA. The role of active DNA demethylation in mammalian epigenetic reprogramming. ProQuest Dissertations Publishing; 2021. p. 28549127.

[90] Ito S, Kuraoka I. Epigenetic modifications in DNA could mimic oxidative DNA damage: a double-edged sword. DNA Repair (Amst) 2015;32:52—7. Available from: https://doi.org/10.1016/j.dnarep.2015.04.013 Epub 2015 May 1; PMID: 25956859.

[91] Vella P, Scelfo A, Jammula S, Chiacchiera F, Williams K, Cuomo A, et al. Tet proteins connect the O-linked N-acetylglucosamine transferase Ogt to chromatin in embryonic stem cells. Mol Cell 2013;49(4):645—56. Available from: https://doi.org/10.1016/j.molcel.2012.12.019 Epub 2013 Jan 24; PMID: 23352454.

[92] Williams K, Christensen J, Pedersen MT, Johansen JV, Cloos PA, Rappsilber J, et al. TET1 and hydroxymethylcytosine in transcription and DNA methylation fidelity. Nature 2011;473(7347):343—8. Available from: https://doi.org/10.1038/nature10066 Epub 2011 Apr 13; PMID: 21490601; PMCID: PMC3408592.

[93] Neri F, Incarnato D, Krepelova A, Rapelli S, Pagnani A, Zecchina R, et al. Genome-wide analysis identifies a functional association of Tet1 and Polycomb repressive complex 2 in mouse embryonic stem cells. Genome Biol 2013;14(8):R91. Available from: https://doi.org/10.1186/gb-2013-14-8-r91 PMID: 23987249; PMCID: PMC4053938.

[94] Ebert A, Lein S, Schotta G, Reuter G. Histone modification and the control of heterochromatic gene silencing in Drosophila. Chromosome Res 2006;14(4):377—92. Available from: https://doi.org/10.1007/s10577-006-1066-1 PMID: 16821134.

[95] Ebbs ML, Bender J. Locus-specific control of DNA methylation by the Arabidopsis SUVH5 histone methyltransferase. Plant Cell 2006;18(5):1166—76. Available from: https://doi.org/10.1105/tpc.106.041400 Epub 2006 Mar 31; PMID: 16582009.

[96] Verdel A, Vavasseur A, Le Gorrec M, Touat-Todeschini L. Common themes in siRNA-mediated epigenetic silencing pathways. Int J Dev Biol 2009;53(2—3):245—57. Available from: https://doi.org/10.1387/ijdb.082691av PMID: 19412884.

[97] Chew YC, West JT, Kratzer SJ, Ilvarsonn AM, Eissenberg JC, Dave BJ, et al. Biotinylation of histones represses transposable elements in human and mouse cells and cell lines and in *Drosophila melanogaster*. J Nutr 2008;138(12):2316—22.

[98] Ooi SKT, Qiu C, Bernstein E, Li K, Jia D, Yang Z, et al. DNMT3L connects unmethylated lysine 4 of histone H3 to de novo methylation of DNA. Nature 2007;448(7154):714—17.

[99] Kimura H, Shiota K. Methyl-CpG-binding protein, MeCP2, is a target molecule for maintenance DNA methyltransferase, Dnmt1. J Biol Chem 2003;278(7):4806–12.

[100] Fuks F, Hurd PJ, Wolf D, Nan X, Bird AP, Kouzarides T. The methyl-CpG-binding protein MeCP2 links DNA methylation to histone methylation. J Biol Chem 2003;278(6):4035–40.

[101] Ernst C, Odom DT, Kutter C. The emergence of piRNAs against transposon invasion to preserve mammalian genome integrity. Nat Commun 2017;8(1):1411.

[102] Vourekas A, Zheng Q, Alexiou P, Maragkakis M, Kirino Y, Gregory BD, et al. Mili and Miwi target RNA repertoire reveals piRNA biogenesis and function of Miwi in spermiogenesis. Nat Struct Mol Biol 2012;19(8):773–81.

[103] Tóth KF, Pezic D, Stuwe E, Webster A. The piRNA pathway guards the germline genome against transposable elements. Adv Exp Med Biol 2016;886:51–77.

[104] Russell SJ, Stalker L. LaMarre J. PIWIs, piRNAs and retrotransposons: complex battles during reprogramming in gametes and early embryos. Reprod Domest Anim 2017;52(Suppl 4):28–38.

[105] Girard A, Hannon GJ. Conserved themes in small-RNA-mediated transposon control. Trends Cell Biol 2008;18(3):136–48. Available from: https://doi.org/10.1016/j.tcb.2008.01.004 Epub 2008 Feb 20; PMID: 18282709.

[106] Turelli P, Castro-Diaz N, Marzetta F, Kapopoulou A, Raclot C, Duc J, et al. Interplay of TRIM28 and DNA methylation in controlling human endogenous retroelements. Genome Res 2014;24(8):1260–70.

[107] Ecco G, Imbeault M, Trono D. KRAB zinc finger proteins. Development 2017;144(15):2719–29.

[108] Stoll GA, Oda S-I, Chong Z-S, Yu M, McLaughlin SH, Modis Y. Structure of KAP1 tripartite motif identifies molecular interfaces required for retroelement silencing. Proc Natl Acad Sci U S A 2019;116(30):15042–51.

[109] Ivanov AV, Peng H, Yurchenko V, Yap KL, Negorev DG, Schultz DC, et al. PHD domain-mediated E3 ligase activity directs intramolecular sumoylation of an adjacent bromodomain required for gene silencing. Mol Cell 2007;28(5):823–37.

[110] Schultz DC, Ayyanathan K, Negorev D, Maul GG, Rauscher 3rd FJ. SETDB1: a novel KAP-1-associated histone H3, lysine 9-specific methyltransferase that contributes to HP1-mediated silencing of euchromatic genes by KRAB zinc-finger proteins. Genes Dev 2002;16(8):919–32.

[111] Ryan RF, Schultz DC, Ayyanathan K, Singh PB, Friedman JR, Fredericks WJ, et al. KAP-1 corepressor protein interacts and colocalizes with heterochromatic and euchromatic HP1 proteins: a potential role for Krüppel-associated box–zinc finger proteins in heterochromatin-mediated gene silencing. Mol Cell Biol 1999;19(6):4366–78.

[112] Quenneville S, Turelli P, Bojkowska K, Raclot C, Offner S, Kapopoulou A, et al. The KRAB-ZFP/KAP1 system contributes to the early embryonic establishment of site-specific DNA methylation patterns maintained during development. Cell Rep 2012;2(4):766–73.

[113] Jurkowska RZ, Qin S, Kungulovski G, Tempel W, Liu Y, Bashtrykov P, et al. H3K14ac is linked to methylation of H3K9 by the triple Tudor domain of SETDB1. Nat Commun 2017;8(1):2057.

[114] Machida S, Takizawa Y, Ishimaru M, Sugita Y, Sekine S, Nakayama J-I, et al. Structural basis of heterochromatin formation by human HP1. Mol Cell 2018;69(3):385–97 e8.

[115] Stewart MD, Li J, Wong J. Relationship between histone H3 lysine 9 methylation, transcription repression, and heterochromatin protein 1 recruitment. Mol Cell Biol 2005;25(7):2525–38.

[116] Groner AC, Meylan S, Ciuffi A, Zangger N, Ambrosini G, Dénervaud N, et al. KRAB-zinc finger proteins and KAP1 can mediate long-range transcriptional repression through heterochromatin spreading. PLoS Genet 2010;6(3):e1000869.

[117] Schultz DC, Friedman JR, Rauscher 3rd FJ. Targeting histone deacetylase complexes via KRAB-zinc finger proteins: the PHD and bromodomains of KAP-1 form a cooperative unit that recruits a novel isoform of the Mi-2alpha subunit of NuRD. Genes Dev 2001;15(4):428–43.

[118] Boulard M, Rucli S, Edwards JR, Bestor TH. Methylation-directed glycosylation of chromatin factors represses retrotransposon promoters. Proc Natl Acad Sci U S A 2020;117(25):14292–8.

[119] Foroushani AK, Chim B, Wong M, Rastegar A, Smith PT, Wang S, et al. Posttranscriptional regulation of human endogenous retroviruses by RNA-binding motif protein 4, RBM4. Proc Natl Acad Sci U S A 2020;117(42):26520–30.

[120] Chelmicki T, Roger E, Teissandier A, Dura M, Bonneville L, Rucli S, et al. m6A RNA methylation regulates the fate of endogenous retroviruses. Nature 2021;591(7849):312–16.

[121] Koito A, Ikeda T. Intrinsic immunity against retrotransposons by APOBEC cytidine deaminases. Front Microbiol 2013;4:28. Available from: https://doi.org/10.3389/fmicb.2013.00028 PMID: 23431045; PMCID: PMC3576619.

[122] Bogerd HP, Wiegand HL, Doehle BP, Lueders KK, Cullen BR. APOBEC3A and APOBEC3B are potent inhibitors of LTR-retrotransposon function in human cells. Nucleic Acids Res 2006;34(1):89–95. Available from: https://doi.org/10.1093/nar/gkj416 PMID: 16407327; PMCID: PMC1326241.

[123] Chiu YL, Greene WC. The APOBEC3 cytidine deaminases: an innate defensive network opposing exogenous retroviruses and endogenous retroelements. Annu Rev Immunol 2008;26:317–53. Available from: https://doi.org/10.1146/annurev.immunol.26.021607.090350 PMID: 18304004.

[124] Schumann GG, Gogvadze EV, Osanai-Futahashi M, Kuroki A, Münk C, Fujiwara H, et al. Unique functions of repetitive transcriptomes. Int Rev Cell Mol Biol 2010;285:115–88. Available from: https://doi.org/10.1016/B978-0-12-381047-2.00003-7 PMID: 21035099.

[125] Esnault C, Heidmann O, Delebecque F, Dewannieux M, Ribet D, Hance AJ, et al. APOBEC3G cytidine deaminase inhibits retrotransposition of endogenous retroviruses. Nature 2005;433(7024):430–3. Available from: https://doi.org/10.1038/nature03238 PMID: 15674295.

[126] Esnault C, Millet J, Schwartz O, Heidmann T. Dual inhibitory effects of APOBEC family proteins on retrotransposition of mammalian endogenous retroviruses. Nucleic Acids Res 2006;34(5):1522–31. Available from: https://doi.org/10.1093/nar/gkl054 PMID: 16537839; PMCID: PMC1401513.

[127] Lohe AR, Hartl DL. Autoregulation of mariner transposase activity by overproduction and dominant-negative complementation. Mol Biol Evol 1996;13(4):549–55. Available from: https://doi.org/10.1093/oxfordjournals.molbev.a025615 PMID: 8882498.

[128] Saha A, Mitchell JA, Nishida Y, Hildreth JE, Ariberre JA, Gilbert WV, et al. A trans-dominant form of Gag restricts Ty1 retrotransposition and mediates copy number control. J Virol 2015;89(7):3922–38. Available from: https://doi.org/10.1128/JVI.03060-14 Epub 2015 Jan 21. Erratum in: J Virol 2016 Apr 29;90(10):5210; PMID: 25609815.

[129] Bire S, Casteret S, Arnaoty A, Piégu B, Lecomte T, Bigot Y. Transposase concentration controls transposition activity: myth or reality? Gene 2013;530(2):165–71. Available from: https://doi.org/10.1016/j.gene.2013.08.039 Epub 2013 Aug 28; PMID: 23994686.

[130] Claeys Bouuaert C, Lipkow K, Andrews SS, Liu D, Chalmers R. The autoregulation of a eukaryotic DNA transposon. Elife 2013;2:e00668. Available from: https://doi.org/10.7554/eLife.00668 PMID: 23795293; PMCID: PMC3687335.

[131] Hickman AB, Dyda F. Mechanisms of DNA transposition. Microbiol Spectr 2015;3(2). Available from: https://doi.org/10.1128/microbiolspec.MDNA3-0034-2014 MDNA3-0034-2014; PMID: 26104718.

[132] Matsui T, Leung D, Miyashita H, Maksakova IA, Miyachi H, Kimura H, et al. Proviral silencing in embryonic stem cells requires the histone methyltransferase ESET. Nature 2010;464 (7290):927–31. Available from: https://doi.org/10.1038/nature08858 Epub 2010 Feb 17. Erratum in: Nature 2014 Sep 4;513(7516):128; PMID: 20164836.

[133] Rowe HM, Jakobsson J, Mesnard D, Rougemont J, Reynard S, Aktas T, et al. KAP1 controls endogenous retroviruses in embryonic stem cells. Nature 2010;463(7278):237–40. Available from: https://doi.org/10.1038/nature08674 PMID: 20075919.

[134] Bruno M, Mahgoub M, Macfarlan TS. The arms race between KRAB-Zinc finger proteins and endogenous retroelements and its impact on mammals. Annu Rev Genet 2019;53:393–416. Available from: https://doi.org/10.1146/annurev-genet-112618-043717 Epub 2019 Sep 13; PMID: 31518518.

[135] Saitou M, Kagiwada S, Kurimoto K. Epigenetic reprogramming in mouse pre-implantation development and primordial germ cells. Development 2012;139(1):15–31.

[136] Luo YB, Zhang L, Lin ZL, Ma JY, Jia J, Namgoong S, et al. Distinct subcellular localization and potential role of LINE1-ORF1P in meiotic oocytes. Histochem Cell Biol 2016;145 (1):93–104. Available from: https://doi.org/10.1007/s00418-015-1369-4 Epub 2015 Oct 14; PMID: 26464247.

[137] Grow EJ, Flynn RA, Chavez SL, Bayless NL, Wossidlo M, Wesche DJ, et al. Intrinsic retroviral reactivation in human pre-implantation embryos and pluripotent cells. Nature 2015;522 (7555):221–5. Available from: https://doi.org/10.1038/nature14308 Epub 2015 Apr 20; PMID: 25896322.

[138] Richardson SR, Gerdes P, Gerhardt DJ, Sanchez-Luque FJ, Bodea GO, Muñoz-Lopez M, et al. Heritable L1 retrotransposition in the mouse primordial germline and early embryo. Genome Res 2017;27(8):1395–405. Available from: https://doi.org/10.1101/gr.219022.116 Epub 2017 May 8; PMID: 28483779.

[139] Feusier J, Watkins WS, Thomas J, Farrell A, Witherspoon DJ, Baird L, et al. Pedigree-based estimation of human mobile element retrotransposition rates. Genome Res 2019;29 (10):1567–77. Available from: https://doi.org/10.1101/gr.247965.118 PMID: 31575651; PMCID: PMC6771411.

[140] Hyun S. Small RNA pathways that protect the somatic genome. Int J Mol Sci [Internet] 2017;18(5). Available from: https://doi.org/10.3390/ijms18050912. Available from.

[141] Bulut-Karslioglu A, De La Rosa-Velázquez IA, Ramirez F, Barenboim M, Onishi-Seebacher M, Arand J, et al. Suv39h-dependent H3K9me3 marks intact retrotransposons and silences LINE elements in mouse embryonic stem cells. Mol Cell 2014;55(2):277–90.

[142] Kondo Y, Issa J-PJ. Enrichment for histone H3 lysine 9 methylation at Alu repeats in human cells. J Biol Chem 2003;278 (30):27658–62.

[143] Karimi MM, Goyal P, Maksakova IA, Bilenky M, Leung D, Tang JX, et al. DNA methylation and SETDB1/H3K9me3 regulate predominantly distinct sets of genes, retroelements, and chimeric transcripts in mESCs. Cell Stem Cell 2011;8 (6):676–87. Available from: https://doi.org/10.1016/j.stem.2011.04.004 PMID: 21624812; PMCID: PMC3857791.

[144] Ciaudo C, Jay F, Okamoto I, Chen CJ, Sarazin A, Servant N, et al. RNAi-dependent and independent control of LINE1 accumulation and mobility in mouse embryonic stem cells. PLoS Genet 2013;9(11):e1003791. Available from: https://doi.org/10.1371/journal.pgen.1003791 Epub 2013 Nov 7. Retraction in: PLoS Genet 2015 Sep;11(9):e1005519. Erratum in: PLoS Genet 2015 May;11(5):e1005247; PMID: 24244175.

[145] Aravin AA, Sachidanandam R, Bourc'his D, Schaefer C, Pezic D, Toth KF, et al. A piRNA pathway primed by individual transposons is linked to de novo DNA methylation in mice. Mol Cell 2008;31(6):785–99. Available from: https://doi.org/10.1016/j.molcel.2008.09.003 PMID: 18922463.

[146] Liu N, Lee CH, Swigut T, Grow E, Gu B, Bassik MC, et al. Selective silencing of euchromatic L1s revealed by genome-wide screens for L1 regulators. Nature 2018;553(7687):228–32. Available from: https://doi.org/10.1038/nature25179 Epub 2017 Dec 6; PMID: 29211708.

[147] Robbez-Masson L, Tie CHC, Conde L, Tunbak H, Husovsky C, Tchasovnikarova IA, et al. The HUSH complex cooperates with TRIM28 to repress young retrotransposons and new genes. Genome Res 2018;28(6):836–45. Available from: https://doi.org/10.1101/gr.228171.117 Epub 2018 May 4; PMID: 29728366.

[148] de la Rica L, Deniz Ö, Cheng KC, Todd CD, Cruz C, Houseley J, et al. TET-dependent regulation of retrotransposable elements in mouse embryonic stem cells. Genome Biol 2016;17 (1):234. Available from: https://doi.org/10.1186/s13059-016-1096-8 PMID: 27863519.

[149] Leung D, Du T, Wagner U, Xie W, Lee AY, Goyal P, et al. Regulation of DNA methylation turnover at LTR retrotransposons and imprinted loci by the histone methyltransferase Setdb1. Proc Natl Acad Sci U S A 2014;111(18):6690–5. Available from: https://doi.org/10.1073/pnas.1322273111 Epub 2014 Apr 22; PMID: 24757056.

[150] Abdalla H, Yoshizawa Y, Hochi S. Active demethylation of paternal genome in mammalian zygotes. J Reprod Dev 2009;55 (4):356–60. Available from: https://doi.org/10.1262/jrd.20234 PMID: 19721335.

[151] Deniz Ö, de la Rica L, Cheng KCL, Spensberger D, Branco MR. SETDB1 prevents TET2-dependent activation of IAP retroelements in naïve embryonic stem cells. Genome Biol 2018;19(1):6. Available from: https://doi.org/10.1186/s13059-017-1376-y PMID: 29351814; PMCID: PMC5775534.

[152] Guallar D, Bi X, Pardavila JA, Huang X, Saenz C, Shi X, et al. RNA-dependent chromatin targeting of TET2 for endogenous retrovirus control in pluripotent stem cells. Nat Genet 2018;50 (3):443–51. Available from: https://doi.org/10.1038/s41588-018-0060-9 Epub 2018 Feb 26; PMID: 29483655.

[153] Walsh CP, Chaillet JR, Bestor TH. Transcription of IAP endogenous retroviruses is constrained by cytosine methylation. Nat Genet 1998;20(2):116–17. Available from: https://doi.org/10.1038/2413 PMID: 9771701.

[154] Bourc'his D, Bestor TH. Meiotic catastrophe and retrotransposon reactivation in male germ cells lacking Dnmt3L. Nature 2004;431(7004):96–9. Available from: https://doi.org/10.1038/nature02886 Epub 2004 Aug 18; PMID: 15318244.

[155] Yoder JA, Walsh CP, Bestor TH. Cytosine methylation and the ecology of intragenomic parasites. Trends Genet 1997;13 (8):335–40. Available from: https://doi.org/10.1016/s0168-9525(97)01181-5 PMID: 9260521.

[156] Elmer JL, Hay AD, Kessler NJ, Bertozzi TM, Ainscough EAC, Ferguson-Smith AC. Genomic properties of variably methylated retrotransposons in mouse. Mob DNA 2021;12(1):6. Available from: https://doi.org/10.1186/s13100-021-00235-1 PMID: 33612119; PMCID: PMC7898769.

[157] Kazachenka A, Bertozzi TM, Sjoberg-Herrera MK, Walker N, Gardner J, Gunning R, et al. Identification, characterization, and heritability of murine metastable epialleles: implications

[158] Long HK, King HW, Patient RK, Odom DT, Klose RJ. Protection of CpG islands from DNA methylation is DNA-encoded and evolutionarily conserved. Nucleic Acids Res 2016;44(14):6693–706. Available from: https://doi.org/10.1093/nar/gkw258 Epub 2016 Apr 15; PMID: 27084945.

[159] Coluccio A, Ecco G, Duc J, Offner S, Turelli P, Trono D. Individual retrotransposon integrants are differentially controlled by KZFP/KAP1-dependent histone methylation, DNA methylation and TET-mediated hydroxymethylation in naïve embryonic stem cells. Epigenetics Chromatin 2018;11(1):7. Available from: https://doi.org/10.1186/s13072-018-0177-1 PMID: 29482634; PMCID: PMC6389204.

[160] Liu S, Brind'Amour J, Karimi MM, Shirane K, Bogutz A, Lefebvre L, et al. Setdb1 is required for germline development and silencing of H3K9me3-marked endogenous retroviruses in primordial germ cells. Genes Dev 2014;28(18):2041–55. Available from: https://doi.org/10.1101/gad.244848.114 Erratum in: Genes Dev 2015 Jan 1;29(1):108; PMID: 25228647.

[161] Dong KB, Maksakova IA, Mohn F, Leung D, Appanah R, Lee S, et al. DNA methylation in ES cells requires the lysine methyltransferase G9a but not its catalytic activity. EMBO J 2008;27(20):2691–701. Available from: https://doi.org/10.1038/emboj.2008.193 Epub 2008 Sep 25; PMID: 18818693.

[162] Tachibana M, Matsumura Y, Fukuda M, Kimura H, Shinkai Y. G9a/GLP complexes independently mediate H3K9 and DNA methylation to silence transcription. EMBO J 2008;27(20):2681–90. Available from: https://doi.org/10.1038/emboj.2008.192 Epub 2008 Sep 25; PMID: 18818694.

[163] Zhang T, Termanis A, Özkan B, Bao XX, Culley J, de Lima Alves F, et al. G9a/GLP complex maintains imprinted DNA methylation in embryonic stem cells. Cell Rep 2016;15(1):77–85. Available from: https://doi.org/10.1016/j.celrep.2016.03.007 Epub 2016 Mar 24; PMID: 27052169.

[164] Rothbart SB, Krajewski K, Nady N, Tempel W, Xue S, Badeaux AI, et al. Association of UHRF1 with methylated H3K9 directs the maintenance of DNA methylation. Nat Struct Mol Biol 2012;19(11):1155–60. Available from: https://doi.org/10.1038/nsmb.2391 Epub 2012 Sep 30; PMID: 23022729.

[165] Liu X, Gao Q, Li P, Zhao Q, Zhang J, Li J, et al. UHRF1 targets DNMT1 for DNA methylation through cooperative binding of hemi-methylated DNA and methylated H3K9. Nat Commun 2013;4:1563. Available from: https://doi.org/10.1038/ncomms2562 PMID: 23463006.

[166] Dutko JA, Schäfer A, Kenny AE, Cullen BR, Curcio MJ. Inhibition of a yeast LTR retrotransposon by human APOBEC3 cytidine deaminases. Curr Biol 2005;15(7):661–6. Available from: https://doi.org/10.1016/j.cub.2005.02.051 PMID: 15823539.

[167] Schumacher AJ, Nissley DV, Harris RS. APOBEC3G hypermutates genomic DNA and inhibits Ty1 retrotransposition in yeast. Proc Natl Acad Sci U S A 2005;102(28):9854–9. Available from: https://doi.org/10.1073/pnas.0501694102 Epub 2005 Jul 6; PMID: 16000409.

[168] Belshaw R, Dawson AL, Woolven-Allen J, Redding J, Burt A, Tristem M. Genomewide screening reveals high levels of insertional polymorphism in the human endogenous retrovirus family HERV-K(HML2): implications for present-day activity. J Virol 2005;79(19):12507–14. Available from: https://doi.org/10.1128/JVI.79.19.12507-12514.2005 PMID: 16160178; PMCID: PMC1211540.

[169] Turner G, Barbulescu M, Su M, Jensen-Seaman MI, Kidd KK, Lenz J. Insertional polymorphisms of full-length endogenous retroviruses in humans. Curr Biol 2001;11(19):1531–5. Available from: https://doi.org/10.1016/s0960-9822(01)00455-9 PMID: 11591322.

[170] Lee YN, Bieniasz PD. Reconstitution of an infectious human endogenous retrovirus. PLoS Pathog 2007;3(1):e10. Available from: https://doi.org/10.1371/journal.ppat.0030010 PMID: 17257061; PMCID: PMC1781480.

[171] Muckenfuss H, Hamdorf M, Held U, Perković M, Löwer J, Cichutek K, et al. APOBEC3 proteins inhibit human LINE-1 retrotransposition. J Biol Chem 2006;281(31):22161–72. Available from: https://doi.org/10.1074/jbc.M601716200 Epub 2006 May 30; PMID: 16735504.

[172] Stenglein MD, Harris RS. APOBEC3B and APOBEC3F inhibit L1 retrotransposition by a DNA deamination-independent mechanism. J Biol Chem 2006;281(25):16837–41. Available from: https://doi.org/10.1074/jbc.M602367200 Epub 2006 Apr 28; PMID: 16648136.

[173] Chen H, Lilley CE, Yu Q, Lee DV, Chou J, Narvaiza I, et al. APOBEC3A is a potent inhibitor of adeno-associated virus and retrotransposons. Curr Biol 2006;16(5):480–5. Available from: https://doi.org/10.1016/j.cub.2006.01.031 PMID: 16527742.

[174] Chiu YL, Witkowska HE, Hall SC, Santiago M, Soros VB, Esnault C, et al. High-molecular-mass APOBEC3G complexes restrict Alu retrotransposition. Proc Natl Acad Sci U S A 2006;103(42):15588–93. Available from: https://doi.org/10.1073/pnas.0604524103 Epub 2006 Oct 9; PMID: 17030807.

[175] Hulme AE, Bogerd HP, Cullen BR, Moran JV. Selective inhibition of Alu retrotransposition by APOBEC3G. Gene 2007;390(1-2):199–205. Available from: https://doi.org/10.1016/j.gene.2006.08.032 Epub 2006 Sep 27; PMID: 17079095.

[176] Kazazian Jr. HH. Mobile elements: drivers of genome evolution. Science 2004;303(5664):1626–32. Available from: https://doi.org/10.1126/science.1089670 PMID: 15016989.

[177] Jarmuz A, Chester A, Bayliss J, Gisbourne J, Dunham I, Scott J, et al. An anthropoid-specific locus of orphan C to U RNA-editing enzymes on chromosome 22. Genomics 2002;79(3):285–96. Available from: https://doi.org/10.1006/geno.2002.6718 PMID: 11863358.

[178] Maksakova IA, Romanish MT, Gagnier L, Dunn CA, van de Lagemaat LN, Mager DL. Retroviral elements and their hosts: insertional mutagenesis in the mouse germ line. PLoS Genet 2006;2(1):e2. Available from: https://doi.org/10.1371/journal.pgen.0020002 PMID: 16440055; PMCID: PMC1331978.

[179] OhAinle M, Kerns JA, Malik HS, Emerman M. Adaptive evolution and antiviral activity of the conserved mammalian cytidine deaminase APOBEC3H. J Virol 2006;80(8):3853–62. Available from: https://doi.org/10.1128/JVI.80.8.3853-3862.2006 PMID: 16571802.

[180] Goyal D, Limesand SW, Goyal R. Epigenetic responses and the developmental origins of health and disease. J Endocrinol 2019;242(1):T105–19. Available from: https://doi.org/10.1530/JOE-19-0009 PMID: 31091503.

[181] McClintock B. The significance of responses of the genome to challenge. Science 1984;226(4676):792–801. Available from: https://doi.org/10.1126/science.15739260 PMID: 15739260.

[182] Zeh DW, Zeh JA, Ishida Y. Transposable elements and an epigenetic basis for punctuated equilibria. Bioessays 2009;31(7):715–26. Available from: https://doi.org/10.1002/bies.200900026 PMID: 19472370.

[183] Casacuberta E, González J. The impact of transposable elements in environmental adaptation. Mol Ecol 2013;22(6):1503–17. Available from: https://doi.org/10.1111/mec.12170 Epub 2013 Jan 7; PMID: 23293987.

[184] Miousse IR, Chalbot MC, Lumen A, Ferguson A, Kavouras IG, Koturbash I. Response of transposable elements to environmental stressors. Mutat Res Rev Mutat Res 2015;765:19–39.

Available from: https://doi.org/10.1016/j.mrrev.2015.05.003 Epub 2015 May 30; PMID: 26281766.

[185] Wright RO, Schwartz J, Wright RJ, Bollati V, Tarantini L, Park SK, et al. Biomarkers of lead exposure and DNA methylation within retrotransposons. Environ Health Perspect 2010;118 (6):790–5. Available from: https://doi.org/10.1289/ehp.0901429 Epub 2010 Jan 11; PMID: 20064768.

[186] Karimi A, Madjd Z, Habibi L, Akrami SM. Exposure of hepatocellular carcinoma cells to low-level As_2O_3 causes an extra toxicity pathway via L1 retrotransposition induction. Toxicol Lett 2014;229(1):111–17. Available from: https://doi.org/10.1016/j.toxlet.2014.05.025 Epub 2014 Jun 21; PMID: 24960058.

[187] Habibi L, Shokrgozar MA, Tabrizi M, Modarressi MH, Akrami SM. Mercury specifically induces LINE-1 activity in a human neuroblastoma cell line. Mutat Res Genet Toxicol Environ Mutagen 2014;759:9–20. Available from: https://doi.org/10.1016/j.mrgentox.2013.07.015 Epub 2013 Nov 13; PMID: 24240092.

[188] Dolinoy DC, Huang D, Jirtle RL. Maternal nutrient supplementation counteracts bisphenol A-induced DNA hypomethylation in early development. Proc Natl Acad Sci U S A 2007;104(32):13056–61. Available from: https://doi.org/10.1073/pnas.0703739104 Epub 2007 Aug 1; PMID: 17670942.

[189] Kim KY, Kim DS, Lee SK, Lee IK, Kang JH, Chang YS, et al. Association of low-dose exposure to persistent organic pollutants with global DNA hypomethylation in healthy Koreans. Environ Health Perspect 2010;118(3):370–4. Available from: https://doi.org/10.1289/ehp.0901131 Epub 2009 Nov 6; PMID: 20064773.

[190] Yao B, Cheng Y, Wang Z, Li Y, Chen L, Huang L, et al. DNA N6-methyladenine is dynamically regulated in the mouse brain following environmental stress. Nat Commun 2017;8 (1):1122. Available from: https://doi.org/10.1038/s41467-017-01195-y PMID: 29066820; PMCID: PMC5654764.

[191] Hunter RG, Murakami G, Dewell S, Seligsohn M, Baker ME, Datson NA, et al. Acute stress and hippocampal histone H3 lysine 9 trimethylation, a retrotransposon silencing response. Proc Natl Acad Sci U S A 2012;109(43):17657–62. Available from: https://doi.org/10.1073/pnas.1215810109 Epub 2012 Oct 4; PMID: 23043114.

[192] Franchini LF, López-Leal R, Nasif S, Beati P, Gelman DM, Low MJ, et al. Convergent evolution of two mammalian neuronal enhancers by sequential exaptation of unrelated retroposons. Proc Natl Acad Sci U S A 2011;108(37):15270–5. Available from: https://doi.org/10.1073/pnas.1104997108 Epub 2011 Aug 29. Erratum in: Proc Natl Acad Sci U S A 2011 Dec 20;108 (51):20850; PMID: 21876128.

[193] Simonti CN, Pavlicev M, Capra JA. Transposable element exaptation into regulatory regions is rare, influenced by evolutionary age, and subject to pleiotropic constraints. Mol Biol Evol 2017;34(11):2856–69. Available from: https://doi.org/10.1093/molbev/msx219 PMID: 28961735; PMCID: PMC5850124.

[194] Horváth V, Merenciano M, González J. Revisiting the relationship between transposable elements and the eukaryotic stress response. Trends Genet 2017;33(11):832–41. Available from: https://doi.org/10.1016/j.tig.2017.08.007 Epub 2017 Sep 22; PMID: 28947157.

[195] Roy M, Viginier B, Saint-Michel É, Arnaud F, Ratinier M, Fablet M. Viral infection impacts transposable element transcript amounts in Drosophila. Proc Natl Acad Sci U S A 2020;117(22):12249–57. Available from: https://doi.org/10.1073/pnas.2006106117 Epub 2020 May 20; PMID: 32434916.

[196] Goic B, Vodovar N, Mondotte JA, Monot C, Frangeul L, Blanc H, et al. RNA-mediated interference and reverse transcription control the persistence of RNA viruses in the insect model Drosophila. Nat Immunol 2013;14(4):396–403. Available from: https://doi.org/10.1038/ni.2542 Epub 2013 Feb 24; PMID: 23435119.

[197] Tassetto M, Kunitomi M, Andino R. Circulating immune cells mediate a systemic RNAi-based adaptive antiviral response in Drosophila. Cell 2017;169(2):314–25. Available from: https://doi.org/10.1016/j.cell.2017.03.033 e13; PMID: 28388413.

[198] Macchietto MG, Langlois RA, Shen SS. Virus-induced transposable element expression up-regulation in human and mouse host cells. Life Sci Alliance 2020;3(2):e201900536. Available from: https://doi.org/10.26508/lsa.201900536 PMID: 31964680; PMCID: PMC6977392.

[199] Mameli G, Poddighe L, Mei A, Uleri E, Sotgiu S, Serra C, et al. Expression and activation by Epstein Barr virus of human endogenous retroviruses-W in blood cells and astrocytes: inference for multiple sclerosis. PLoS One 2012;7(9):e44991. Available from: https://doi.org/10.1371/journal.pone.0044991 Epub 2012 Sep 27; PMID: 23028727.

[200] Perron H, Lang A. The human endogenous retrovirus link between genes and environment in multiple sclerosis and in multifactorial diseases associating neuroinflammation. Clin Rev Allergy Immunol 2010;39(1):51–61. Available from: https://doi.org/10.1007/s12016-009-8170-x PMID: 19697163.

[201] Limosin F, Rouillon F, Payan C, Cohen JM, Strub N. Prenatal exposure to influenza as a risk factor for adult schizophrenia. Acta Psychiatr Scand 2003;107(5):331–5. Available from: https://doi.org/10.1034/j.1600-0447.2003.00052.x PMID: 12752028.

[202] Ebert T, Kotler M. Prenatal exposure to influenza and the risk of subsequent development of schizophrenia. Isr Med Assoc J 2005;7(1):35–8 PMID: 15658144.

[203] Bhende PM, Seaman WT, Delecluse HJ, Kenney SC. The EBV lytic switch protein, Z, preferentially binds to and activates the methylated viral genome. Nat Genet 2004;36(10):1099–104. Available from: https://doi.org/10.1038/ng1424 Epub 2004 Sep 12; PMID: 15361873.

[204] Charvet B, Reynaud JM, Gourru-Lesimple G, Perron H, Marche PN, Horvat B. Induction of proinflammatory multiple sclerosis-associated retrovirus envelope protein by human herpesvirus-6A and CD46 receptor engagement. Front Immunol 2018;9:2803. Available from: https://doi.org/10.3389/fimmu.2018.02803 PMID: 30574140; PMCID: PMC6291489.

[205] Lee WJ, Kwun HJ, Kim HS, Jang KL. Activation of the human endogenous retrovirus W long terminal repeat by herpes simplex virus type 1 immediate early protein 1. Mol Cells 2003;15 (1):75–80 PMID: 12661764.

[206] Ruprecht K, Obojes K, Wengel V, Gronen F, Kim KS, Perron H, et al. Regulation of human endogenous retrovirus W protein expression by herpes simplex virus type 1: implications for multiple sclerosis. J Neurovirol 2006;12(1):65–71. Available from: https://doi.org/10.1080/13550280600614973 PMID: 16595376.

[207] Mameli G, Astone V, Arru G, Marconi S, Lovato L, Serra C, et al. Brains and peripheral blood mononuclear cells of multiple sclerosis (MS) patients hyperexpress MS-associated retrovirus/HERV-W endogenous retrovirus, but not Human herpesvirus 6. J Gen Virol 2007;88(Pt 1):264–74. Available from: https://doi.org/10.1099/vir.0.81890-0 PMID: 17170460.

[208] Mameli G, Madeddu G, Mei A, Uleri E, Poddighe L, Delogu LG, et al. Activation of MSRV-type endogenous retroviruses during infectious mononucleosis and Epstein-Barr virus latency: the missing link with multiple sclerosis? PLoS One 2013;8(11):e78474. Available from: https://doi.org/10.1371/journal.pone.0078474 PMID: 24236019; PMCID: PMC3827255.

[209] Nellåker C, Yao Y, Jones-Brando L, Mallet F, Yolken RH, Karlsson H. Transactivation of elements in the human endogenous retrovirus W family by viral infection. Retrovirology

2006;3:44. Available from: https://doi.org/10.1186/1742-4690-3-44 PMID: 16822326; PMCID: PMC1539011.

[210] Li F, Nellåker C, Sabunciyan S, Yolken RH, Jones-Brando L, Johansson AS, et al. Transcriptional derepression of the ERVWE1 locus following influenza A virus infection. J Virol 2014;88(8):4328−37. Available from: https://doi.org/10.1128/JVI.03628-13 Epub 2014 Jan 29; PMID: 24478419.

[211] Garcia-Montojo M, Nath A. HERV-W envelope expression in blood leukocytes as a marker of disease severity of COVID-19. EBioMedicine 2021;67:103363. Available from: https://doi.org/10.1016/j.ebiom.2021.103363 Epub 2021 May 13; PMID: 33993053.

[212] Balestrieri E, Minutolo A, Petrone V, Fanelli M, Iannetta M, Malagnino V, et al. Evidence of the pathogenic HERV-W envelope expression in T lymphocytes in association with the respiratory outcome of COVID-19 patients. EBioMedicine 2021;66:103341. Available from: https://doi.org/10.1016/j.ebiom.2021.103341 Epub 2021 Apr 15; PMID: 33867312.

[213] Morgan HD, Sutherland HG, Martin DI, Whitelaw E. Epigenetic inheritance at the agouti locus in the mouse. Nat Genet 1999;23(3):314−18.

[214] Wolff GL, Kodell RL, Moore SR, Cooney CA. Maternal epigenetics and methyl supplements affect agouti gene expression in Avy/a mice. FASEB J 1998;12(11):949−57 PMID: 9707167.

[215] Waterland RA, Jirtle RL. Transposable elements: targets for early nutritional effects on epigenetic gene regulation. Mol Cell Biol 2003;23(15):5293−300. Available from: https://doi.org/10.1128/MCB.23.15.5293-5300.2003 PMID: 12861015; PMCID: PMC165709.

[216] Kuroishi T, Rios-Avila L, Pestinger V, Wijeratne SS, Zempleni J. Biotinylation is a natural, albeit rare, modification of human histones. Mol Genet Metab 2011;104(4):537−45. Available from: https://doi.org/10.1016/j.ymgme.2011.08.030 Epub 2011 Sep 3; PMID: 21930408.

[217] Pflieger A, Waffo Teguo P, Papastamoulis Y, Chaignepain S, Subra F, Munir S, et al. Natural stilbenoids isolated from grapevine exhibiting inhibitory effects against HIV-1 integrase and eukaryote MOS1 transposase in vitro activities. PLoS One 2013;8(11):e81184. Available from: https://doi.org/10.1371/journal.pone.0081184 PMID: 24312275; PMCID: PMC3842960.

[218] Agodi A, Barchitta M, Quattrocchi A, Maugeri A, Canto C, Marchese AE, et al. Low fruit consumption and folate deficiency are associated with LINE-1 hypomethylation in women of a cancer-free population. Genes Nutr 2015;10(5):480. Available from: https://doi.org/10.1007/s12263-015-0480-4 Epub 2015 Jul 17; PMID: 26183162.

[219] Green CD, Huang Y, Dou X, Yang L, Liu Y, Han JJ. Impact of dietary interventions on noncoding RNA networks and mRNAs encoding chromatin-related factors. Cell Rep 2017;18 (12):2957−68. Available from: https://doi.org/10.1016/j.celrep.2017.03.001 PMID: 28329687.

[220] Van Meter M, Kashyap M, Rezazadeh S, Geneva AJ, Morello TD, Seluanov A, et al. SIRT6 represses LINE1 retrotransposons by ribosylating KAP1 but this repression fails with stress and age. Nat Commun 2014;5:5011. Available from: https://doi.org/10.1038/ncomms6011 PMID: 25247314; PMCID: PMC4185372.

[221] Kornberg RD, Lorch Y. Twenty-five years of the nucleosome, fundamental particle of the eukaryote chromosome. Cell 1999;98(3):285−94. Available from: https://doi.org/10.1016/s0092-8674(00)81958-3 PMID: 10458604.

[222] Segal E, Fondufe-Mittendorf Y, Chen L, Thåström A, Field Y, Moore IK, et al. A genomic code for nucleosome positioning. Nature 2006;442(7104):772−8. Available from: https://doi.org/10.1038/nature04979 Epub 2006 Jul 19; PMID: 16862119.

[223] Kaplan N, Moore I, Fondufe-Mittendorf Y, Gossett AJ, Tillo D, Field Y, et al. Nucleosome sequence preferences influence in vivo nucleosome organization. Nat Struct Mol Biol 2010;17 (8):918−20. Available from: https://doi.org/10.1038/nsmb0810-918 PMID: 20683473; PMCID: PMC2969171.

[224] Schmitt AD, Hu M, Ren B. Genome-wide mapping and analysis of chromosome architecture. Nat Rev Mol Cell Biol 2016;17 (12):743−55. Available from: https://doi.org/10.1038/nrm.2016.104 Epub 2016 Sep 1; PMID: 27580841.

[225] Tena JJ, Santos-Pereira JM. Topologically associating domains and regulatory landscapes in development, evolution and disease. Front Cell Dev Biol 2021;9:702787. Available from: https://doi.org/10.3389/fcell.2021.702787 PMID: 34295901; PMCID: PMC8290416.

[226] Klemm SL, Shipony Z, Greenleaf WJ. Chromatin accessibility and the regulatory epigenome. Nat Rev Genet 2019;20 (4):207−20. Available from: https://doi.org/10.1038/s41576-018-0089-8 PMID: 30675018.

[227] Lippman Z, Gendrel AV, Black M, Vaughn MW, Dedhia N, McCombie WR, et al. Role of transposable elements in heterochromatin and epigenetic control. Nature 2004;430(6998):471−6. Available from: https://doi.org/10.1038/nature02651 PMID: 15269773.

[228] Dixon JR, Selvaraj S, Yue F, Kim A, Li Y, Shen Y, et al. Topological domains in mammalian genomes identified by analysis of chromatin interactions. Nature 2012;485(7398):376−80. Available from: https://doi.org/10.1038/nature11082 PMID: 22495300; PMCID: PMC3356448.

[229] Huda A, Jordan IK. Epigenetic regulation of Mammalian genomes by transposable elements. Ann N Y Acad Sci 2009;1178:276−84. Available from: https://doi.org/10.1111/j.1749-6632.2009.05007.x PMID: 19845643.

[230] Kentepozidou E, Aitken SJ, Feig C, Stefflova K, Ibarra-Soria X, Odom DT, et al. Clustered CTCF binding is an evolutionary mechanism to maintain topologically associating domains. Genome Biol 2020;21(1):5. Available from: https://doi.org/10.1186/s13059-019-1894-x PMID: 31910870; PMCID: PMC6945661.

[231] Zhang Y, Li T, Preissl S, Amaral ML, Grinstein JD, Farah EN, et al. Transcriptionally active HERV-H retrotransposons demarcate topologically associating domains in human pluripotent stem cells. Nat Genet 2019;51(9):1380−8. Available from: https://doi.org/10.1038/s41588-019-0479-7 Epub 2019 Aug 19; PMID: 31427791.

[232] Wang J, Vicente-García C, Seruggia D, Moltó E, Fernandez-Miñán A, Neto A, et al. MIR retrotransposon sequences provide insulators to the human genome. Proc Natl Acad Sci U S A 2015;112(32):E4428−37. Available from: https://doi.org/10.1073/pnas.1507253112 Epub 2015 Jul 27; PMID: 26216945.

[233] Lei EP, Corces VG. RNA interference machinery influences the nuclear organization of a chromatin insulator. Nat Genet 2006;38(8):936−41. Available from: https://doi.org/10.1038/ng1850 Epub 2006 Jul 23; PMID: 16862159.

[234] Achrem M, Szućko I, Kalinka A. The epigenetic regulation of centromeres and telomeres in plants and animals. Comp Cytogen 2020;14(2):265−311. Available from: https://doi.org/10.3897/CompCytogen.v14i2.51895 PMID: 32733650; PMCID: PMC7360632.

[235] Arabidopsis Genome Initiative. Analysis of the genome sequence of the flowering plant Arabidopsis thaliana. Nature 2000;408(6814):796−815. Available from: https://doi.org/10.1038/35048692 PMID: 11130711.

[236] Slotkin RK, Martienssen R. Transposable elements and the epigenetic regulation of the genome. Nat Rev Genet 2007;8 (4):272−85.

[237] Wong LH, Choo KH. Evolutionary dynamics of transposable elements at the centromere. Trends Genet 2004;20(12):611−16. Available from: https://doi.org/10.1016/j.tig.2004.09.011 PMID: 15522456.

[238] George JA, DeBaryshe PG, Traverse KL, Celniker SE, Pardue ML. Genomic organization of the Drosophila telomere retrotransposable elements. Genome Res 2006;16(10):1231—40. Available from: https://doi.org/10.1101/gr.5348806 Epub 2006 Sep 8; PMID: 16963706.

[239] Liu K, Schoonmaker MM, Levine BL, June CH, Hodes RJ, Weng NP. Constitutive and regulated expression of telomerase reverse transcriptase (hTERT) in human lymphocytes. Proc Natl Acad Sci U S A 1999;96(9):5147—52. Available from: https://doi.org/10.1073/pnas.96.9.5147 PMID: 10220433; PMCID: PMC21831.

[240] Aschacher T, Wolf B, Enzmann F, Kienzl P, Messner B, Sampl S, et al. LINE-1 induces hTERT and ensures telomere maintenance in tumour cell lines. Oncogene 2016;35(1):94—104. Available from: https://doi.org/10.1038/onc.2015.65 Epub 2015 Mar 23; PMID: 25798839.

[241] Oricchio E, Sciamanna I, Beraldi R, Tolstonog GV, Schumann GG, Spadafora C. Distinct roles for LINE-1 and HERV-K retroelements in cell proliferation, differentiation and tumor progression. Oncogene 2007;26(29):4226—33. Available from: https://doi.org/10.1038/sj.onc.1210214 Epub 2007 Jan 22; PMID: 17237820.

[242] Mueller C, Aschacher T, Wolf B, Bergmann M. A role of LINE-1 in telomere regulation. Front Biosci (Landmark Ed) 2018;23:1310—19. Available from: https://doi.org/10.2741/4645 PMID: 28930601.

[243] Maston GA, Evans SK, Green MR. Transcriptional regulatory elements in the human genome. Annu Rev Genomics Hum Genet 2006;7:29—59. Available from: https://doi.org/10.1146/annurev.genom.7.080505.115623 PMID: 16719718.

[244] Ong CT, Corces VG. Enhancer function: new insights into the regulation of tissue-specific gene expression. Nat Rev Genet 2011;12(4):283—93. Available from: https://doi.org/10.1038/nrg2957 Epub 2011 Mar 1; PMID: 21358745.

[245] Conley AB, Piriyapongsa J, Jordan IK. Retroviral promoters in the human genome. Bioinformatics 2008;24(14):1563—7. Available from: https://doi.org/10.1093/bioinformatics/btn243 Epub 2008 Jun 5; PMID: 18535086.

[246] Hambor JE, Mennone J, Coon ME, Hanke JH, Kavathas P. Identification and characterization of an Alu-containing, T-cell-specific enhancer located in the last intron of the human CD8 alpha gene. Mol Cell Biol 1993;13(11):7056—70. Available from: https://doi.org/10.1128/mcb.13.11.7056-7070.1993 PMID: 8413295.

[247] Jordan IK, Rogozin IB, Glazko GV, Koonin EV. Origin of a substantial fraction of human regulatory sequences from transposable elements. Trends Genet 2003;19(2):68—72.

[248] Jacques PÉ, Jeyakani J, Bourque G. The majority of primate-specific regulatory sequences are derived from transposable elements. PLoS Genet 2013;9(5):e1003504. Available from: https://doi.org/10.1371/journal.pgen.1003504 Epub 2013 May 9; PMID: 23675311.

[249] Swergold GD. Identification, characterization, and cell specificity of a human LINE-1 promoter. Mol Cell Biol 1990;10(12):6718—29. Available from: https://doi.org/10.1128/mcb.10.12.6718-6729.1990 PMID: 1701022; PMCID: PMC362950.

[250] Roy AM, West NC, Rao A, Adhikari P, Alemán C, Barnes AP, et al. Upstream flanking sequences and transcription of SINEs. J Mol Biol 2000;302(1):17—25. Available from: https://doi.org/10.1006/jmbi.2000.4027 PMID: 10964558.

[251] Ono M, Kawakami M, Takezawa T. A novel human nonviral retroposon derived from an endogenous retrovirus. Nucleic Acids Res 1987;15(21):8725—37. Available from: https://doi.org/10.1093/nar/15.21.8725 PMID: 2825118; PMCID: PMC306401.

[252] Kines KJ. Expressing genes do not forget their LINEs: transposable elements and gene expression. Front Biosci 2012;17(1):1329.

[253] Kellner M, Makałowski W. Transposable elements significantly contributed to the core promoters in the human genome. Sci China Life Sci 2019;62(4):489—97. Available from: https://doi.org/10.1007/s11427-018-9449-0 Epub 2019 Mar 19; PMID: 30915629.

[254] Speek M. Antisense promoter of human L1 retrotransposon drives transcription of adjacent cellular genes. Mol Cell Biol 2001;21(6):1973—85. Available from: https://doi.org/10.1128/MCB.21.6.1973-1985.2001 PMID: 11238933; PMCID: PMC86790.

[255] Raviram R, Rocha PP, Luo VM, Swanzey E, Miraldi ER, Chuong EB, et al. Analysis of 3D genomic interactions identifies candidate host genes that transposable elements potentially regulate. Genome Biol 2018;19(1):216. Available from: https://doi.org/10.1186/s13059-018-1598-7 PMID: 30541598; PMCID: PMC6292174.

[256] Deniz Ö, Ahmed M, Todd CD, Rio-Machin A, Dawson MA, Branco MR. Endogenous retroviruses are a source of enhancers with oncogenic potential in acute myeloid leukaemia. Nat Commun 2020;11(1):3506.

[257] Bourque G, Leong B, Vega VB, Chen X, Lee YL, Srinivasan KG, et al. Evolution of the mammalian transcription factor binding repertoire via transposable elements. Genome Res 2008;18(11):1752—62.

[258] Ito J, Sugimoto R, Nakaoka H, Yamada S, Kimura T, Hayano T, et al. Systematic identification and characterization of regulatory elements derived from human endogenous retroviruses. PLoS Genet 2017;13(7):e1006883.

[259] Zeng L, Pederson SM, Cao D, Qu Z, Hu Z, Adelson DL, et al. Genome-wide analysis of the association of transposable elements with gene regulation suggests that Alu elements have the largest overall regulatory impact. J Comput Biol 2018;25(6):551—62.

[260] Nikitin D, Penzar D, Garazha A, Sorokin M, Tkachev V, Borisov N, et al. Profiling of human molecular pathways affected by retrotransposons at the level of regulation by transcription factor proteins. Front Immunol 2018;9:30.

[261] Zemojtel T, Kielbasa SM, Arndt PF, Behrens S, Bourque G, Vingron M. CpG deamination creates transcription factor-binding sites with high efficiency. Genome Biol Evol 2011;3:1304—11.

[262] Testori A, Caizzi L, Cutrupi S, Friard O, De Bortoli M, Cora' D, et al. The role of Transposable elements in shaping the combinatorial interaction of transcription factors. BMC Genomics 2012;13:400.

[263] Pavlicev M, Hiratsuka K, Swaggart KA, Dunn C, Muglia L. Detecting endogenous retrovirus-driven tissue-specific gene transcription. Genome Biol Evol 2015;7(4):1082—97. Available from: https://doi.org/10.1093/gbe/evv049 PMID: 25767249; PMCID: PMC4419796.

[264] Trizzino M, Kapusta A, Brown CD. Transposable elements generate regulatory novelty in a tissue-specific fashion. BMC Genomics 2018;19(1):468. Available from: https://doi.org/10.1186/s12864-018-4850-3 PMID: 29914366; PMCID: PMC6006921.

[265] Kreahling J, Graveley BR. The origins and implications of Aluternative splicing. Trends Genet 2004;20(1):1—4.

[266] Sorek R, Ast G, Graur D. Alu-containing exons are alternatively spliced. Genome Res 2002;12(7):1060—7.

[267] Makalowski W. Genomics. Not junk after all. Science 2003;300(5623):1246—7.

[268] Belancio VP, Hedges DJ, Deininger P. LINE-1 RNA splicing and influences on mammalian gene expression. Nucleic Acids Res 2006;34(5):1512—21.

[269] Duke-Cohan JS, Tang W, Schlossman SF. Attractin: a cub-family protease involved in T cell-monocyte/macrophage interactions. Adv Exp Med Biol 2000;477:173—85.

[270] Tang W, Gunn TM, McLaughlin DF, Barsh GS, Schlossman SF, Duke-Cohan JS. Secreted and membrane attractin result from alternative splicing of the human ATRN gene. Proc Natl Acad Sci U S A 2000;97(11):6025−30.

[271] Watkins WS, Feusier JE, Thomas J, Goubert C, Mallick S, Jorde LB. The simons genome diversity project: a global analysis of mobile element diversity. Genome Biol Evol 2020;12(6):779−94. Available from: https://doi.org/10.1093/gbe/evaa086 PMID: 32359137; PMCID: PMC7290288.

[272] Jönsson ME, Garza R, Sharma Y, Petri R, Södersten E, Johansson JG, et al. Activation of endogenous retroviruses during brain development causes an inflammatory response. EMBO J 2021;40(9):e106423.

[273] Koh YW, Park C, Yoon DH, Suh C, Huh J. CSF-1R expression in tumor-associated macrophages is associated with worse prognosis in classical Hodgkin lymphoma. Am J Clin Pathol 2014;141(4):573−83.

[274] Stacey KJ, Sagulenko V. A clear link between endogenous retroviral LTR activity and Hodgkin's lymphoma. Cell Res 2010;20(8):869−71.

[275] Romanish MT, Lock WM, van de Lagemaat LN, Dunn CA, Mager DL. Repeated recruitment of LTR retrotransposons as promoters by the anti-apoptotic locus NAIP during mammalian evolution. PLoS Genet 2007;3(1):e10.

[276] Bièche I, Laurent A, Laurendeau I, Duret L, Giovangrandi Y, Frendo J-L, et al. Placenta-specific INSL4 expression is mediated by a human endogenous retrovirus element. Biol Reprod 2003;68(4):1422−9.

[277] Polavarapu N, Mariño-Ramírez L, Landsman D, McDonald JF, Jordan IK. Evolutionary rates and patterns for human transcription factor binding sites derived from repetitive DNA. BMC Genomics 2008;9:226.

[278] Jiang J-C, Upton KR. Human transposons are an abundant supply of transcription factor binding sites and promoter activities in breast cancer cell lines. Mob DNA 2019;10(1):16.

[279] Kpetemey M, Dasgupta S, Rajendiran S, Das S, Gibbs LD, Shetty P, et al. MIEN1, a novel interactor of Annexin A2, promotes tumor cell migration by enhancing AnxA2 cell surface expression. Mol Cancer 2015;14(1):156.

[280] Rajendiran S, Gibbs LD, Van Treuren T, Klinkebiel DL, Vishwanatha JK. MIEN1 is tightly regulated by SINE Alu methylation in its promoter. Oncotarget 2016;7(40):65307−19.

[281] Gama-Sosa MA, Slagel VA, Trewyn RW, Oxenhandler R, Kuo KC, Gehrke CW, et al. The 5-methylcytosine content of DNA from human tumors. Nucleic Acids Res 1983;11(19):6883−94.

[282] Yegnasubramanian S, Haffner MC, Zhang Y, Gurel B, Cornish TC, Wu Z, et al. DNA hypomethylation arises later in prostate cancer progression than CpG island hypermethylation and contributes to metastatic tumor heterogeneity. Cancer Res 2008;68(21):8954−67.

[283] Landry JR, Medstrand P, Mager DL. Repetitive elements in the 5′ untranslated region of a human zinc-finger gene modulate transcription and translation efficiency. Genomics 2001;76(1−3):110−16.

[284] Allshire RC, Madhani HD. Ten principles of heterochromatin formation and function. Nat Rev Mol Cell Biol 2018;19(4):229−44.

[285] Girton JR, Johansen KM. Chromatin structure and the regulation of gene expression: the lessons of PEV in Drosophila. Adv Genet 2008;61:1−43.

[286] Cohen CJ, Lock WM, Mager DL. Endogenous retroviral LTRs as promoters for human genes: a critical assessment. Gene 2009;448(2):105−14. Available from: https://doi.org/10.1016/j.gene.2009.06.020 Epub 2009 Jul 3; PMID: 19577618.

[287] Quadrana L, Bortolini Silveira A, Mayhew GF, LeBlanc C, Martienssen RA, Jeddeloh JA, et al. The *Arabidopsis thaliana* mobilome and its impact at the species level. Elife 2016;5:e15716. Available from: https://doi.org/10.7554/eLife.15716 PMID: 27258693; PMCID: PMC4917339.

[288] Choi JY, Purugganan MD. Evolutionary epigenomics of retrotransposon-mediated methylation spreading in rice. Mol Biol Evol 2018;35(2):365−82. Available from: https://doi.org/10.1093/molbev/msx284 PMID: 29126199; PMCID: PMC5850837.

[289] Eichten SR, Ellis NA, Makarevitch I, Yeh CT, Gent JI, Guo L, et al. Spreading of heterochromatin is limited to specific families of maize retrotransposons. PLoS Genet 2012;8(12):e1003127. Available from: https://doi.org/10.1371/journal.pgen.1003127 Epub 2012 Dec 13; PMID: 23271981.

[290] Rebollo R, Karimi MM, Bilenky M, Gagnier L, Miceli-Royer K, Zhang Y, et al. Retrotransposon-induced heterochromatin spreading in the mouse revealed by insertional polymorphisms. PLoS Genet 2011;7(9):e1002301. Available from: https://doi.org/10.1371/journal.pgen.1002301 Epub 2011 Sep 29; PMID: 21980304.

[291] Sun FL, Haynes K, Simpson CL, Lee SD, Collins L, Wuller J, et al. cis-Acting determinants of heterochromatin formation on *Drosophila melanogaster* chromosome four. Mol Cell Biol 2004;24(18):8210−20. Available from: https://doi.org/10.1128/MCB.24.18.8210-8220.2004 PMID: 15340080; PMCID: PMC515050.

[292] Locke J, Kotarski MA, Tartof KD. Dosage-dependent modifiers of position effect variegation in Drosophila and a mass action model that explains their effect. Genetics 1988;120(1):181−98. Available from: https://doi.org/10.1093/genetics/120.1.181 PMID: 3146523; PMCID: PMC1203489.

[293] Lee YC. The role of piRNA-mediated epigenetic silencing in the population dynamics of transposable elements in *Drosophila melanogaster*. PLoS Genet 2015;11(6):e1005269. Available from: https://doi.org/10.1371/journal.pgen.1005269 PMID: 26042931; PMCID: PMC4456100.

[294] Lee YCG, Karpen GH. Pervasive epigenetic effects of Drosophila euchromatic transposable elements impact their evolution. Elife 2017;6:e25762. Available from: https://doi.org/10.7554/eLife.25762 PMID: 28695823; PMCID: PMC5505702.

[295] Chen L-L, Yang L. ALUternative regulation for gene expression. Trends Cell Biol 2017;27(7):480−90.

[296] Kang M-I, Rhyu M-G, Kim Y-H, Jung Y-C, Hong S-J, Cho C-S, et al. The length of CpG islands is associated with the distribution of Alu and L1 retroelements. Genomics 2006;87(5):580−90.

[297] Yates PA, Burman RW, Mummaneni P, Krussel S, Turker MS. Tandem B1 elements located in a mouse methylation center provide a target for de novo DNA methylation. J Biol Chem 1999;274(51):36357−61.

[298] Turker MS. Gene silencing in mammalian cells and the spread of DNA methylation. Oncogene 2002;21(35):5388−93.

[299] Kuehnen P, Mischke M, Wiegand S, Sers C, Horsthemke B, Lau S, et al. An Alu element-associated hypermethylation variant of the POMC gene is associated with childhood obesity. PLoS Genet 2012;8(3):e1002543.

[300] MacNeil DJ, Howard AD, Guan X, Fong TM, Nargund RP, Bednarek MA, et al. The role of melanocortins in body weight regulation: opportunities for the treatment of obesity. Eur J Pharmacol 2002;450(1):93−109.

[301] Prendergast JGD, Chambers EV, Semple CAM. Sequence-level mechanisms of human epigenome evolution. Genome Biol Evol 2014;6(7):1758−71.

[302] Needs T, Gonzalez-Mosquera LF, Lynch DT. Beta thalassemia. StatPearls. Treasure Island (FL): StatPearls Publishing; 2021.

[303] Lanikova L, Kucerova J, Indrak K, Divoka M, Issa J-P, Papayannopoulou T, et al. β-Thalassemia due to intronic LINE-1 insertion in the β-globin gene (HBB): molecular

mechanisms underlying reduced transcript levels of the β-globin(L1) allele. Hum Mutat 2013;34(10):1361—5.
[304] Hata K, Sakaki Y. Identification of critical CpG sites for repression of L1 transcription by DNA methylation. Gene 1997;189 (2):227—34. Available from: https://doi.org/10.1016/s0378-1119(96)00856-6 PMID: 9168132.
[305] Lyon MF. X-chromosome inactivation: a repeat hypothesis. Cytogenet Cell Genet 1998;80(1-4):133—7. Available from: https://doi.org/10.1159/000014969 PMID: 9678347.
[306] Bailey JA, Carrel L, Chakravarti A, Eichler EE. Molecular evidence for a relationship between LINE-1 elements and X chromosome inactivation: the Lyon repeat hypothesis. Proc Natl Acad Sci U S A 2000;97(12):6634—9. Available from: https://doi.org/10.1073/pnas.97.12.6634 PMID: 10841562; PMCID: PMC18684.
[307] Elisaphenko EA, Kolesnikov NN, Shevchenko AI, Rogozin IB, Nesterova TB, Brockdorff N, et al. A dual origin of the Xist gene from a protein-coding gene and a set of transposable elements. PLoS One 2008;3(6):e2521.
[308] Kunarso G, Chia NY, Jeyakani J, Hwang C, Lu X, Chan YS, et al. Transposable elements have rewired the core regulatory network of human embryonic stem cells. Nat Genet 2010;42 (7):631—4. Available from: https://doi.org/10.1038/ng.600 Epub 2010 Jun 6; PMID: 20526341.
[309] Shpiz S, Ryazansky S, Olovnikov I, Abramov Y, Kalmykova A. Euchromatic transposon insertions trigger production of novel Pi- and endo-siRNAs at the target sites in the drosophila germline. PLoS Genet 2014;10(2):e1004138. Available from: https://doi.org/10.1371/journal.pgen.1004138 PMID: 24516406; PMCID: PMC3916259.
[310] Van Houdt H, Bleys A, Depicker A. RNA target sequences promote spreading of RNA silencing. Plant Physiol 2003;131 (1):245—53. Available from: https://doi.org/10.1104/pp.009407 PMID: 12529532; PMCID: PMC166804.
[311] Vaistij FE, Jones L, Baulcombe DC. Spreading of RNA targeting and DNA methylation in RNA silencing requires transcription of the target gene and a putative RNA-dependent RNA polymerase. Plant Cell 2002;14(4):857—67. Available from: https://doi.org/10.1105/tpc.010480 PMID: 11971140; PMCID: PMC150687. Brosnan CA, Voinnet O. Cell-to-cell and long-distance siRNA movement in plants: mechanisms and biological implications. Curr Opin Plant Biol 2011 Oct;14(5):580-7. doi: 10.1016/j.pbi.2011.07.011. Epub 2011 Aug 19. PMID: 21862389.
[312] Himber C, Dunoyer P, Moissiard G, Ritzenthaler C, Voinnet O. Transitivity-dependent and -independent cell-to-cell movement of RNA silencing. EMBO J 2003;22(17):4523—33. Available from: https://doi.org/10.1093/emboj/cdg431 PMID: 12941703; PMCID: PMC202373.
[313] Chow J, Heard E. X inactivation and the complexities of silencing a sex chromosome. Curr Opin Cell Biol 2009;21(3):359—66.
[314] Wutz A, Rasmussen TP, Jaenisch R. Chromosomal silencing and localization are mediated by different domains of Xist RNA. Nat Genet 2002;30(2):167—74.
[315] Monfort A, Di Minin G, Postlmayr A, Freimann R, Arieti F, Thore S, et al. Identification of Spen as a crucial factor for Xist function through forward genetic screening in haploid embryonic stem cells. Cell Rep 2015;12(4):554—61.
[316] Chu C, Zhang QC, da Rocha ST, Flynn RA, Bharadwaj M, Calabrese JM, et al. Systematic discovery of Xist RNA binding proteins. Cell 2015;161(2):404—16.
[317] McHugh CA, Chen C-K, Chow A, Surka CF, Tran C, McDonel P, et al. The Xist lncRNA interacts directly with SHARP to silence transcription through HDAC3. Nature 2015;521(7551):232—6.
[318] Carter AC, Xu J, Nakamoto MY, Wei Y, Zarnegar BJ, Shi Q, et al. Spen links RNA-mediated endogenous retrovirus silencing and X chromosome inactivation. Elife [Internet] 2020;9. Available from: https://doi.org/10.7554/eLife.54508. Available from.
[319] Chaumeil J, Le Baccon P, Wutz A, Heard E. A novel role for Xist RNA in the formation of a repressive nuclear compartment into which genes are recruited when silenced. Genes Dev 2006;20(16):2223—37.
[320] Clemson CM, Hall LL, Byron M, McNeil J, Lawrence JB. The X chromosome is organized into a gene-rich outer rim and an internal core containing silenced nongenic sequences. Proc Natl Acad Sci U S A 2006;103(20):7688—93.
[321] Chow JC, Ciaudo C, Fazzari MJ, Mise N, Servant N, Glass JL, et al. LINE-1 activity in facultative heterochromatin formation during X chromosome inactivation. Cell 2016;166(3):782.
[322] Mikkelsen TS, Broad Institute Genome Sequencing PlatformWakefield MJ, Aken B, Amemiya CT, Chang JL, et al. Genome of the marsupial Monodelphis domestica reveals innovation in non-coding sequences. Nature 2007;447(7141):167—77.
[323] Kelsey G, Feil R. New insights into establishment and maintenance of DNA methylation imprints in mammals. Philos Trans R Soc Lond B Biol Sci 2013;368(1609):20110336.
[324] Renfree MB, Suzuki S, Kaneko-Ishino T. The origin and evolution of genomic imprinting and viviparity in mammals. Philos Trans R Soc Lond B Biol Sci 2013;368(1609):20120151.
[325] Kaneko-Ishino T, Kohda T, Ishino F. The regulation and biological significance of genomic imprinting in mammals. J Biochem 2003;133(6):699—711.
[326] Peaston AE, Evsikov AV, Graber JH, de Vries WN, Holbrook AE, Solter D, et al. Retrotransposons regulate host genes in mouse oocytes and preimplantation embryos. Dev Cell 2004;7 (4):597—606.
[327] Rubin CM, VandeVoort CA, Teplitz RL, Schmid CW. Alu repeated DNAs are differentially methylated in primate germ cells. Nucleic Acids Res 1994;22(23):5121—7.
[328] Li X, Ito M, Zhou F, Youngson N, Zuo X, Leder P, et al. A maternal-zygotic effect gene, Zfp57, maintains both maternal and paternal imprints. Dev Cell 2008;15(4):547—57. Available from: https://doi.org/10.1016/j.devcel.2008.08.014 PMID: 18854139; PMCID: PMC2593089.
[329] Quenneville S, Verde G, Corsinotti A, Kapopoulou A, Jakobsson J, Offner S, et al. In embryonic stem cells, ZFP57/KAP1 recognize a methylated hexanucleotide to affect chromatin and DNA methylation of imprinting control regions. Mol Cell 2011;44(3):361—72. Available from: https://doi.org/10.1016/j.molcel.2011.08.032 PMID: 22055183; PMCID: PMC3210328.
[330] McDonald JF, Matzke MA, Matzke AJ. Host defenses to transposable elements and the evolution of genomic imprinting. Cytogenet Genome Res 2005;110(1—4):242—9.
[331] Ondičová M, Oakey RJ, Walsh CP. Is imprinting the result of "friendly fire" by the host defense system? PLoS Genet 2020;16(4):e1008599. Available from: https://doi.org/10.1371/journal.pgen.1008599 PMID: 32271759; PMCID: PMC7144951.
[332] Ono R, Nakamura K, Inoue K, Naruse M, Usami T, Wakisaka-Saito N, et al. Deletion of Peg10, an imprinted gene acquired from a retrotransposon, causes early embryonic lethality. Nat Genet 2006;38(1):101—6.
[333] Sekita Y, Wagatsuma H, Nakamura K, Ono R, Kagami M, Wakisaka N, et al. Role of retrotransposon-derived imprinted gene, Rtl1, in the feto-maternal interface of mouse placenta. Nat Genet 2008;40(2):243—8.
[334] Youngson NA, Kocialkowski S, Peel N, Ferguson-Smith AC. A small family of sushi-class retrotransposon-derived genes in mammals and their relation to genomic imprinting. J Mol Evol 2005;61(4):481—90.
[335] Watanabe T, Tomizawa S, Mitsuya K, Totoki Y, Yamamoto Y, Kuramochi-Miyagawa S, et al. Role for piRNAs and

noncoding RNA in de novo DNA methylation of the imprinted mouse Rasgrf1 locus. Science 2011;332 (6031):848–52. Available from: https://doi.org/10.1126/science.1203919 PMID: 21566194; PMCID: PMC3368507.

[336] Suzuki S, Ono R, Narita T, Pask AJ, Shaw G, Wang C, et al. Retrotransposon silencing by DNA methylation can drive mammalian genomic imprinting. PLoS Genet 2007;3(4):e55.

[337] Bertozzi TM, Ferguson-Smith AC. Metastable epialleles and their contribution to epigenetic inheritance in mammals. Semin Cell Dev Biol 2020;97:93–105.

[338] Seisenberger S, Andrews S, Krueger F, Arand J, Walter J, Santos F, et al. The dynamics of genome-wide DNA methylation reprogramming in mouse primordial germ cells. Mol Cell 2012;48(6):849–62.

[339] Costello KR, Leung A, Trac C, Lee M, Basam M, Pospisilik JA, et al. Sequence features of retrotransposons allow for epigenetic variability. Elife 2021;10:e71104. Available from: https://doi.org/10.7554/eLife.71104 PMID: 34668484; PMCID: PMC8555987.

[340] Azlan A, Dzaki N, Azzam G. Argonaute: the executor of small RNA function. J Genet Genomics 2016;43(8):481–94.

[341] Peters L, Meister G. Argonaute proteins: mediators of RNA silencing. Mol Cell 2007;26(5):611–23.

[342] Hadjiargyrou M, Delihas N. The intertwining of transposable elements and non-coding RNAs. Int J Mol Sci 2013;14(7):13307–28.

[343] Piriyapongsa J, Jordan IK. A family of human microRNA genes from miniature inverted-repeat transposable elements. PLoS One 2007;2(2):e203. Available from: https://doi.org/10.1371/journal.pone.0000203 PMID: 17301878; PMCID: PMC1784062.

[344] Yuan Z, Sun X, Jiang D, Ding Y, Lu Z, Gong L, et al. Origin and evolution of a placental-specific microRNA family in the human genome. BMC Evol Biol 2010;10:346.

[345] Qin S, Jin P, Zhou X, Chen L, Ma F. The role of transposable elements in the origin and evolution of MicroRNAs in human. PLoS One 2015;10(6):e0131365.

[346] Kang D, Kim Y-J, Hong K, Han KTE. composition of human long noncoding RNAs and their expression patterns in human tissues. Genes Genomics 2015;37(1):87–95.

[347] Jalali S, Gandhi S, Scaria V. Navigating the dynamic landscape of long noncoding RNA and protein-coding gene annotations in GENCODE. Hum Genomics 2016;10(1):35.

[348] Cabili MN, Trapnell C, Goff L, Koziol M, Tazon-Vega B, Regev A, et al. Integrative annotation of human large intergenic noncoding RNAs reveals global properties and specific subclasses. Genes Dev 2011;25(18):1915–27.

[349] Kannan S, Chernikova D, Rogozin IB, Poliakov E, Managadze D, Koonin EV, et al. Transposable element insertions in long intergenic non-coding RNA genes. Front Bioeng Biotechnol 2015;3:71.

[350] Spengler RM, Oakley CK, Davidson BL. Functional microRNAs and target sites are created by lineage-specific transposition. Hum Mol Genet 2014;23(7):1783–93.

[351] Gong C, Maquat LE. "Alu"strious long ncRNAs and their role in shortening mRNA half-lives. Cell Cycle 2011;10(12):1882–3.

[352] Jeck WR, Sorrentino JA, Wang K, Slevin MK, Burd CE, Liu J, et al. Circular RNAs are abundant, conserved, and associated with ALU repeats. RNA 2013;19(2):141–57.

[353] Kelley D, Rinn J. Transposable elements reveal a stem cell-specific class of long noncoding RNAs. Genome Biol 2012;13(11):R107.

CHAPTER 19

Dietary and Metabolic Compounds Affecting Covalent Histone Modifications

Gareth W. Davison

Faculty of Life and Health Sciences, Ulster University, Northern Ireland, United Kingdom

OUTLINE

Introduction	357	Histone Deacetylation	371
Metabolic and Dietary Control of Histone and Transcriptional Dynamics	359	Other Histone Modifications	373
		Concluding Perspectives	374
Metabolism	359	Acknowledgments	375
Diet	360		
Regulation of Chromatin Dynamics	360	Abbreviations	375
One-Carbon Metabolism and Histone Methylation	360	References	375
Histone Demethylation	362	Further Reading	380
Histone Acetylation	367		
Histone Acetyltransferase Inhibitors and Activators	370		

INTRODUCTION

In eukaryotic nuclei, DNA and histones are efficiently packed together to contain heritable genetic information in the form of chromatin [1]. The basic repeating unit of chromatin, the nucleosome, consists of super-helical DNA (147 bp) intertwined around an octamer of core histone proteins. The octamer comprises two copies each of four core histones: an H2A-H2B tetramer, and two H3-H4 dimers positioned on both sides of the tetramer. An additional 60 bp histone, H1, sits on the outer surface of the nucleosome and binds to linker DNA joining nucleosomes together [2]. The core histones are structurally similar with a basic N-terminal domain, a globular domain organized by the histone fold, and a C-terminal unstructured tail [3], and while the terminal tails generally protrude from the nucleosome core particles, they are susceptible to a host of posttranslational modifications (PTMs). These positively charged amino acids at the 5′ tail, protruding from the core structure, leads to numerous PTMs including, but not limited to, methylation (1–3 methyl groups), acetylation, phosphorylation, ubiquitination, SUMOylation, lactylation, biotinylation, GlcNAcylation, citrullination, ADP-ribosylation, serotylation, hydroxylation, and oxidation. The majority of modifications are localized at the amino- and carboxyl-terminal histone tails, particularly in H3 and H4 histones. Only a select minority of modifications are localized to the histone globular domains.

Histone modifications and histone-modifying complexes are now commonly designated as the "histone code": a genomic network effectively designed to ensure an efficient silent state of chromatin and, from time-to-time and in a gene specific manner, to allow genes to be transcriptionally expressed [1]. The former, classically referred to as heterochromatin, operates through a state of chromatin compaction leading to transcriptional silencing, while at the other end of the continuum, euchromatin is a state of open chromatin promoting gene transcription (Fig. 19.1). Modifications of distinct amino acid residues in histones have unique functions. For example, trimethylation of lysine 4 in histone H3 (H3K4me3) leads to transcriptional activation of proximal DNA, whereas di-methylation of K9 (H3K9me) is allied to transcriptional silencing [2]. Other well characterized examples include H3K4, H3K36, and H3K79 methylation as gene activating marks, while H3K9, H3K27, and H4K20 methylation are silencing marks [4]. In general, histone acetylation and phosphorylation act as transcriptional activators, while histone deacetylation, biotinylation, and SUMOylation inhibit gene activation. Methylation and ubiquitination can either repress or activate gene expression, depending solely on the specific histone residue targeted. A degree of extra complexity partially originates from which methylation on lysine or arginine the marks sit and may be one of three different isoforms: mono-, di-, or tri-methyl on lysine, and mono- or di-methyl (asymmetric or symmetric) on arginine [5]. Collectively, these modifications not only control the transcriptional state of chromatin, but also the regulation of numerous downstream processes such as DNA replication, repair, recombination, and apoptosis (Fig. 19.1).

Histone modifications are generally written, read, and erased by numerous regulatory enzymes specific to the modification and target amino acid residue. To illustrate, human species have in the region of 50 different histone methyltransferases (HMTs: methylation writers), 28 different histone demethylases (HDMs: methylation erasers), and 147 different histone methylation readers [4]. Due to these unique characteristics, histone modifications are reversible, thus controlling

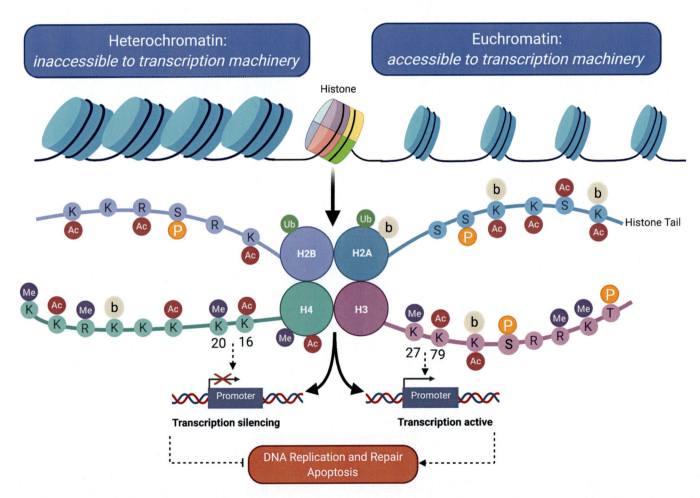

FIGURE 19.1 Chromatin and regulation of gene transcription. Gene transcription is regulated by chromatin accessibility. The nucleosome comprises 147 base pairs of tightly coiled DNA around each (two copies) of the four core histone proteins H2A, H2B, H3, and H4 [6]. *R*, arginine; *K*, lysine; *S*, serine; *Me*, methylation; *Ac*, acetylation; *Ub*, ubiquitination; *b*, biotinylation; *P*, phosphorylation. H3K27me and H3K79ac characterize an open chromatin state leading to active gene transcription (of DNA replication, DNA repair and apoptosis as an example), while H4K16ac and H4K20me represent a closed chromatin state leading to transcriptional repression.

and modulating chromatin structure through gene activation or repression. Each modification is derived from intermediates of cell metabolism, where intermediates such as S-adenosylmethionine (SAM) or acetyl-coenzyme A (acetyl-CoA) are used as substrates and cofactors for a variety of histone-modifying enzymes, allowing metabolism to directly communicate environmental changes to the chromatin state [7]. More specifically, histone methylation is directly linked to central metabolism through critical redox intermediates such as nicotinamide adenine dinucleotide (NAD^+) and SAM, and any fluctuation in these metabolic enzymes can directly affect intracellular signaling, and subsequently be determined on the histones via the appearance of marks and modifications, leading to a reprogramming of gene transcription [8].

Intracellular metabolic compounds are generally affected by a plethora of scenarios ranging from hypoxia, circadian rhythm, oxidative stress, redox balance, inflammation, disease, and hyperglycemia. The aberrant regulation of said molecular networks, perhaps due to gene mutations or environmental perturbations such as stress, smoking, or excess alcohol consumption, is also linked to changes across the life cycle, to embryonic development, cell identity, tumorigenesis, tumor progression, and microbiome-host commensalism. While these networks can be pharmacologically manipulated, diet and nutrition may also influence the metabolic cell state. Dietary compounds control the ability of nutrients to act as sensing molecules either to generate cell energy or regulate histones. Likewise, a deregulated nutrient sensing system can adversely affect the histone code by increasing the appearance of PTMs leading to a state of either eu- or hetero-chromatin, altering the phenotype and disease [7]. For example, early work ascertained the link between epigenetics and maternal and intrauterine malnutrition and growth retardation to the onset and progression of type 2 diabetes mellitus. The "thrifty phenotype hypothesis" proposed by Hales and Barker [9] proposed that undernutrition during fetal development, coupled with a low birthweight, can lead to adverse and permanent changes in glucose homeostasis in the offspring [10]. It is now known that environmentally linked epigenetic mechanisms play a principal role in the regulation of transcriptional activation and are involved in a plethora of relevant metabolic pathways (pancreatic and β-cell function, peripheral glucose uptake, insulin resistance) aligned to the susceptibility of the T2DM phenotype [11–13]. More subtle effects can be exerted by naturally occurring dietary patterns such as those involving an increased dose of substrate. In the offspring of pregnant mice fed a high-fat diet during pregnancy, higher histone H3 methylation at lysine 9 occurs in the promoter region of adiponectin in adipose tissue, suggesting that exposure to a high-fat diet in utero increases susceptibility to a metabolic syndrome-like phenomenon through histone modifications.

Nutrition, diet, and the metabolic state of the cell exchange many interacting factors with covalent histone modifications (Fig. 19.2). While such an interplay between chromatin and small molecule metabolites may be salient for the genome during organismal survival, growth, and development, elucidating various strategies by which chromatin can be manipulated or modified by metabolism and/or diet, may prove important for the complex understanding of the cellular mechanisms aligned to healthy and disease states. This chapter will examine the multifactorial interrelationship between the plethora of dietary and metabolic compounds and the regulation of posttranslational histone modifications. To that end, a selective focus will be afforded to the methylation and acetylation of histones H3 and H4, as these modifications correlate tightly with transcriptional activation and repression.

METABOLIC AND DIETARY CONTROL OF HISTONE AND TRANSCRIPTIONAL DYNAMICS

Metabolism

Endogenous metabolism occurs via a set of life-sustaining chemical modifications within cells, in which the modifications are catalyzed by specific enzymatic reactions in order to maintain cell homeostasis in response to environmental conditions [14]. Several aspects of epigenetic modifications require intermediates of cell metabolism for customary enzymatic function; conversely, and in a reciprocal biochemical arrangement, any diversity in cell metabolism can alter specific acetyltransferases and methyltransferases conferring widespread fluctuations in epigenetic modification patterns [6,15]. Specifically, while histone PTMs can transcriptionally regulate metabolism, biochemical reactions involving the disposition and removal of histone PTMs can also directly influence metabolism [8]. In addition to intracellular metabolic fluctuations, the epigenome may also be regulated by cell metabolite localization, for example, mitochondria (specifically the tricarboxylic acid [TCA] cycle) produce several regulatory molecules which are expended as co-substrates for numerous transcriptional and epigenetic processes [15,16]. To the contrary, TCA enzymes may block enzymatic action. This interplay between metabolism and epigenetics, termed *metaboloepigenetics* [17], and the prospect of metabolic intermediates controlling gene expression is no doubt exciting; however, mechanistic connections are not fully understood and limited to only a few disease states such as cancer [18] and type 2 diabetes mellitus [6].

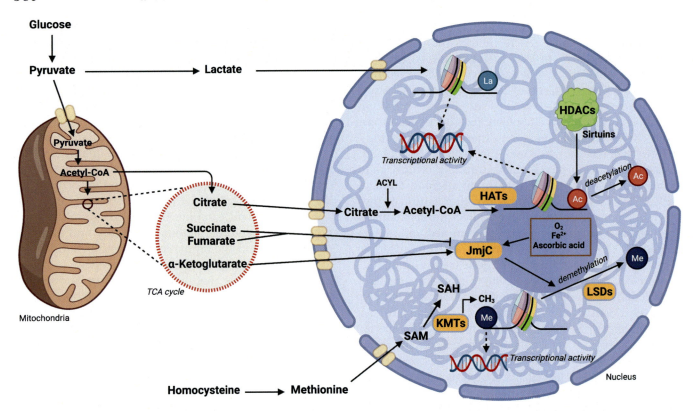

FIGURE 19.2 **The diet-metabolic interrelationship in histone modification and transcriptional regulation.** Dietary intake and metabolism, which can be influenced by environmental behaviors, regulate histone modifications which control an open or closed histone state, leading to active or repressed gene transcription and protein activity. *TF*, transcription factor.

Diet

An understanding of the complex nature of nutrient-gene interaction is important to not only ascertain how diet modulates the risk of disease development but also how diet may be used to manipulate early life nutrition and improve healthy aging. Cells generally respond to a diversity of nutrients through various signaling pathways, eventually converging onto chromatin where the expression of genes is either tightly fixed or modified, leading to physiological or pathological processes including embryonic development, aging, and carcinogenesis [19,20]. In the broadest sense, nutrients such as glucose, fatty acids, amino acids, and vitamins are utilized by specific metabolic pathways to generate metabolites that can be consumed as substrates or activity modulators of chromatin-modifying enzymes. Molecules such as SAM and acetyl-CoA (derived from glycolytic and oxidative phosphorylation pathways) regulate mainstream *canonical* modifications (inclusive of histone methylation and acetylation), while homocysteine, NAD^+, and lactate, control several *emerging* modifications such as homocysteinylation, ADP-ribosylation, and lactylation. Adjacent metabolic pathways and networks relating to one-carbon and methionine metabolism, ketogenesis, caloric overload and restriction, and redox metabolism/oxidative stress, also regulate the global epigenomic landscape in concert with chromatin modifiers, remodelers, and transcription factors [7]. Moreover, natural compounds found regularly in accessible food products (e.g., sulforaphane in broccoli and cabbage) can regulate the epigenome. While most bioactive food components work through a histone deacetylase (HDAC) inhibitory network, others can activate enzymes directly to cause histone PTMs and a reprogramming of gene transcription [19].

REGULATION OF CHROMATIN DYNAMICS

One-carbon Metabolism and Histone Methylation

One-carbon metabolism is a biochemical pathway that provides methyl groups for methylation reactions and nutrients for multiple biological functions [21,22]. The enzymes aligned to one-carbon metabolism are dependent on several vitamins or nutrients to serve as cofactors or methyl acceptors or donors, and encompass folate, cobalamin (vitamin B_{12}), pyridoxal-5′-phosphate (vitamin B_6), betaine, choline, and methionine [22]. Two major components of one-carbon metabolism comprise the folate and methionine cycles, that essentially transfer single carbon units to acceptor molecules. SAM is the major methyl (CH_3) donor in the cell, providing the activate CH_3 group as the main substrate for methyltransferase (e.g., histone methyltransferase) reactions on arginine and lysine residues of histones, leading to PTMs [6,15,21]. SAM has an estimated intracellular concentration of ~10 μM, but amounts vary from 10 to 100-fold under certain physiological conditions. The Michaelis-Menten constant (k_m) values for SAM for individual HMTs are in the low micromolar range, postulating that fluctuations of SAM can influence the rate of histone methylation [8]. Tetrahydrofolate (THF) when methylated on its N-5 atom (N^5-MTHF, methyltetrahydrofolate) also acts as a methyl donor. However, unlike SAM, the transfer potential of the methyl donor group of N^5-MTHF is not sufficiently high enough for the majority of methylation reactions [23]. SAM is produced by methionine adenosyltransferase (MAT) using methionine and adenosine triphosphate (ATP) as substrates. The methyl group of SAM is chemically activated by the positive charge on the adjacent sulfur atom, allowing the SAM methyl group to be more reactive compared to the methyl group on N^5-MTHF [6,23]. When SAM sheds a methyl group, it is converted to S-adenosylhomocysteine (SAH). Homocysteine can also be methylated to resynthesize methionine via the betaine-homocysteine methyltransferase reaction which absorbs choline and betaine in the process [22]. One caveat worth highlighting is the ability of SAH to act as a potent inhibitor of the HMTs. As such, the SAM/SAH ratio can greatly influence methyltransferase activity in vivo [24]. That said, SAH is physiologically controlled at low concentrations via hydrolysis to homocysteine (via S-adenosylhomocysteine hydrolase [SAHH]), which is recycled to methionine via the remethylation pathway by CH_3 transfer from N^5-MTHF, or alternatively catabolized to amino acids or glutathione. As such, the SAM/SAH ratio can serve as a salient biosensor of cell metabolic state, and any fluctuations in the SAM/SAH ratio by a change in nutrient availability, can thus modify the activity of the HMTs affecting histone methylation [6,14,25].

Methylation of histone H3 and H4 is aided by the recruitment of HMT enzymes to covalently add CH_3 from SAM onto the side-chain nitrogen atoms of mainly lysine and arginine residues [24,26]. Several HMTs exist belonging to three families of enzymes: SET-domain-containing protein family, DOT1-like protein family methylating lysine, and the protein arginine N-methyltransferase (PRMT) family which methylates arginine [27]. Moreover, the lysine methyltransferases contain a conserved domain Su(var)3–9, enhancer of zeste, and trithorax (SET) responsible for the disposition of CH_3 on several histone lysine residues [26]. The fundamental consequence of histone methylation is determined by the specific histone residue modified, the number of methyl groups added (mono-, di-, or tri-methylation), and the location within the N-terminal regions of either H3 or H4 [14,23]. The upshot of these methylation marks is a state of euchromatin (lightly packed DNA promoting transcription) or heterochromatin (condensed DNA suppressing transcription; Fig. 19.1) [6]. For instance, methylation of lysine is a relatively stable mark localized near active or poised transcriptional units and is represented by H3K4, H3K48, H3K36, and H3K79, leading to transcriptional activation, while methylation of H3K9, H3K27, and H4K20me1, cause compaction of chromatin and transcriptional repression. Histone arginine methylation is understood to be a more dynamic mark that correlates positively with the euchromatin state; whereas, a methyl loss from arginine on H3 and H4 is closely aligned to a state of heterochromatin. For example, H3R17me2 represents activation, while H3R2me2 is associated with transcriptional repression [5,14].

Nitric oxide (NO) is a free radical and commonly referred to as a reactive nitrogen species (RNS). NO is purported to be important for a plethora of endogenous cell signaling mechanisms, including epigenetic regulation of gene expression and cell phenotype. Active histone methylation is paralleled to a rise in NO activity. Cells cultured in methionine-free media show substantial global histone methylation following NO treatment, suggesting that active methylation in histones may also be controlled by a pathway other than active SAM [28,29]. H3K9 methyltransferase expression activity was ascertained by an NO dose response, and while SETDB2 and SUV39H2 seem to be transcriptionally regulated, protein methyltransferase G9a activity is adversely regulated following NO exposure [29]. This work suggests that although activation of methyltransferase activity by NO stimulation is unlikely, adjacent enzymatic down regulation could interfere with histone methylation patterns [28]. Further substantial investigative work is required to fully determine the role of NO in histone methylation.

Hyperglycemia is purported to regulate the chromatin state. Under conditions of high glucose, Sun et al. [30] examined the role of H3Kme in *TGF-β1*-induced gene expression of collagen-α1[I] and plasminogen activator inhibitor-1, showing an increase in chromatin marks (H3K4me1, H3K4me2, and H3K4me3), and a decrease in repressive marks (H3K9me2 and H3K9me3) at gene promoters. *TGF-β1* also enhanced the expression of the H3K4 methyltransferase SET7/9 and recruitment to these promoters. In contrast, glucose deprivation is known to induce an NAD^+/NAD ratio related in part to a redox-induced increase in the histone methyltransferase Suv39H1 and in H3K9me [31].

As demonstrated by Okabe et al. [32], vascular endothelial cells under transient stress by hyperglycemia accumulate SET7 in the nucleus, up-regulating pro-inflammatory gene expression. Aligned to diabetic pathology, H3K4me1 is enhanced at the promoter region of *Keap1* which is accompanied by an increase in SET7/9. This is notable, as in the hyperglycemic state, an increase in H3K4me1 and binding of Sp1 at the *Keap1* promoter in the retina, can block movement of the transcription factor *Nrf2* to the nucleus, thus impairing the transcriptional activity of a salient cell regulator [33,34]. Likewise, a glucose stimulated increase in H4K20me3 at the promoter/enhancer region of the antioxidant enzyme *SOD2*, can downregulate the expression of this mitochondrial protective protein leading to a compromised retina [6,35]. The role of protein arginine methyltransferase 5 (*Prmt5*) in pancreatic β-cells was recently investigated by Ma et al. [36], highlighting that islet-specific knockout of *Prmt5* decreases insulin gene expression, hypothesizing a contributory role in impaired glucose tolerance. Other data concerning insulin per se, has shown that insulin through a reactive oxygen species (ROS) mechanism enhances H3K4me1 and H3K9me1 [37], while insulin reduces methylation at H3R17 aligned to a down regulation in *PEPCK* and *G6Pase* [38].

Inflammation is an evolutionarily conserved process characterized by a diverse milieu of specialized immune and non-immune cells designed to protect from bacteria, viruses, and infections, by eliminating pathogens and promoting tissue repair and recovery [39]. Metabolic reprogramming is purported to control the proliferation and differentiation of immune cell populations, coupled with concurrent changes in chromatin state [7]. For example, antigen receptor engagement in T cells increase metabolic flux through the methionine cycle, supplying methyl donors for the purposes of histone methylation [40]. Methionine turnover also maintains SAM synthesis and H3K4me3 in $CD4^+$ T helper cells in culture [41]. Lipopolysaccharides (LPS) increase the binding of H3K4me3 and histone methyltransferases SET1 and myeloid/lymphoid leukemia to iNOS and TNFα gene promoters; this was unaffected by the addition of SAM to the murine macrophages [42]. TNFα was also shown to induce the recruitment of SET7/9 (H3K4 methyltransferase) at the MCP-1 and TNFα promoters, along with an increase in H3K4me3 [43]. Epigenetic and transcriptional mechanisms are also connected to dysregulated inflammatory pathways in aging, where an upregulation of IL-23p19 expression in dendritic cells is associated with differential transcription factor binding and selective changes in H3K4 methylation [44].

HISTONE DEMETHYLATION

Active histone demethylation (removal of methyl group) is tightly associated with products of cell metabolism and is carried out via two pathways: lysine-specific demethylases (LSD) and through the JumonjiC (JmjC) domain proteins [45]. In 2004, Shi and colleagues [46], first identified a nuclear lysine-specific histone demethylase 1 (LSD1 also known as KDMIA or AOF2), while demonstrating its ability to remove methyl groups from monomethylated or demethylated H3K4 and H3K9 [27]. LSD1 predominately operates via the co-factor flavin adenine dinucleotide (FAD^+), and through an oxidation reaction is reduced to $FADH_2$ in cell mitochondria [18] following demethylation of H3K4me1, H3K4me2, H3K9me1, and H3K9me2 [14]. This oxidant reaction catalyzed by LSD1 needs a protonated lysine epsilon amino group, thus restricting its activity to mono-methylated and di-methylated H3K4 [23]. The paralog LSD2 (also known as KDM1B or AOF1) also utilizes FAD^+ to demethylate H3K4me1 and H3K4me2 [47]. The LSD1 and 2 demethylases cleave the α-carbon bond of the methylated lysine to form an imine intermediate, which is hydrolyzed to formaldehyde, subsequently releasing a well-known ROS, hydrogen peroxide (H_2O_2) alongside the demethylated lysine [27]. It is conceivable that H_2O_2 produced through the LSD1 demethylation reaction may act as an intermediatory molecule to damage DNA in close proximity affecting gene transcription [23]. Moreover, although FAD is produced in mitochondria, there is clear crosstalk with nuclear LSD1 rendering the latter particularly sensitive to oscillations in FAD (FAD to $FADH_2$). These changes in FAD occur from other flavin-linked dehydrogenases and oxidases derived from fatty acid β-oxidation and TCA cycle activity. In this sense, LSD1 can regulate mitochondrial activity through binding to certain genes, thereby repressing transcription associated with the loss of H3K4 methylation [23].

Relative to the LSD family of demethylases, the majority of histone demethylation occurs through the JmjC-domain containing histone demethylases. The ~30 JmjC-domain contains histone demethylases (e.g., JHDM1, JHDM3A) operating via iron (Fe^{2+}), oxygen and α-ketoglutarate (α-KG) cofactors to remove methyl groups from arginine and mono-, di-, and tri-methylated lysine residues (H3K4, H3K9, H3K27, H3K36, and H4K20) in an oxidative reaction producing hydroxymethyl lysine, CO_2, and the TCA enzyme succinate. To the contrary, the TCA intermediate fumarate and the oncometabolite 2-hydroxyglutarate (2-HG) inhibits JmjC activity [14,48]. On the latter, and following 2-HG inhibition per se, tumors harboring gain of function mutations in isocitrate dehydrogenase (IDH) 1 and 2 can hypermethylate histones leading to transcriptional modifications and cancer proneness [49]. Vitamin C (L-ascorbic acid) is an additional co-factor of the JmjC domain-contain histone demethylases, where vitamin C is required for optimal JHDM1 activity [50], while demethylation mediated by JHDM3A is inhibited in the absence of vitamin C [51]. Vitamin C is also known to control epigenetic reprogramming by facilitating the histone demethylases to switch the pre-iPSC fate on or off by regulating the H3K9 methylation status of the core pluripotency loci. Moreover, cell vitamin C levels are normally under the control of redox biology, where the status of redox couples, including the $FAD/FADH_2$ ratio, can dictate the availability of vitamin C, and thus the ability of histones to demethylate [27]. In this sense, vitamin C can act as an important "redox switch" for enzymatic activity of JmjC histone lysine demethylases [52].

In keeping with the redox balance theme, a transient ROS (e.g., H_2O_2) generation during early development in Caenorhabditis elegans modulates gene transcription required for a prolonged lifespan by inhibiting the H3K4 methyltransferases MLL1–4, leading to a reduction in global H3K4me3. Thus, this work identifies the histone methyltransferases as redox sensitive units of the H3K4-trimethylating complex of proteins [53]. As outlined by Dai and colleagues [7], the notion that ROS production from metabolic reprogramming involving the oxidation and depletion of SAM requires further exploration, and another potential mechanism could involve the redox-responsiveness of vitamin B_{12}-utilizing enzyme methionine synthase. While hypoxia is also known to generate ROS in cells, a low oxygen tension can also inhibit numerous JHDMs including KDM5A, KDM6A, and KDM6B, to activate a state of histone hypermethylation [54,55]. Specifically, the oxygen sensing protein hypoxia-inducible factor (HIF) may target these histone demethylases. HIF is regulated by a hydroxylation reaction that possesses a similar Fe^{2}-dependent mechanism with JmjC-dependent demethylation, which also requires oxygen and α-KG. Additionally, HIF is known to directly up-regulate expression of KDM3A, KDM4B, KDM4C, and KDM6B, leading to demethylation of H3K9me2 and H3K27me3 [56]. The mechanism by which a low oxygen availability drives histone PTMs may include the effects of hypoxia on aspects of oxidative metabolism including the cell concentration of $FAD/FADH_2$, succinate, and α-KG [57].

As indicated, ROS can generate in cells leading to a state of oxidative stress including damage to DNA. In an interesting study aligned to certain TCA cycle intermediates, Sulkowski et al. [58] identified a connection between 2-HG, succinate, fumarate, and DNA repair. The TCA cycle metabolites were found to inhibit the lysine demethylase KDM4B, resulting in aberrant hypermethylation of H3K9 at loci in close proximity to DNA double-strand breaks. This interaction conceals a local H3K9 trimethylation signal that is crucial for the functioning of the DNA homology-dependent repair (HDR) pathway. Consequently, the recruitment of salient HDR transcriptional regulators, Tip60 and ATM, is impaired at DNA break sites leading to reduced end resection. It is thus plausible that a disruption in chromatin signaling and DNA repair in this way may further enhance susceptibility to disease. While facets of the aforementioned may be perceived as hypothetical, its contents should be subject to scientific inquiry through rigorous experimental testing in a disease orientated setting such as type 2 diabetes mellitus where enhanced DNA damage exists, and basic metabolism is compromised [6]. Moreover, and in keeping with the diabetic model, preserving H3K27 and H3K4 methylation using the lysine demethylase inhibitor GSK-J4 improves β-cell function, and by knockout of the H3K4 demethylase KDM5B, insulin sensitivity and glucose homeostasis may also be improved, highlighting the potential role of KDM5B in diabetes related metabolism [59,60].

Metabolic regulation of JHDM's can concomitantly regulate various aspects of immune cell biology [7]. αKG production via glutaminolysis is salient for alternative (M2) activation of macrophages and for clearance of endotoxin, inclusive of engagement of JHDM3-dependent histone reprogramming of M2 genes. This M2-promoting mechanism seems to be further modulated by a high αKG/succinate ratio [61]. In a similar line of investigation, Leone and colleagues [62], utilized a glutamine antagonist (JHU083) to promote T-cell activation by reducing αKG concentration and histone hypermethylation. Moreover, iron deficiency seems to affect the humoral immune response (part of the immune system responsible for controlling the antibody response to a pathogen through the activation and differentiation of B cells in plasma) via diminished activity aligned to the iron-dependent JHDMs and a H3K9 hypermethylation state [7,63].

Diet-derived macro- and micro-nutrients are essential for proper functioning and efficient cell networks, and many of these salient nutrients serve as metabolic intermediates involved in the initiation and/or maintenance of PTMs. Dietary impact on the architecture of histone molecules can be transient, reversible, and may persist in a positive or negative manner through the lifespan [64]. On a basic mechanistic level and as stated, histone methylation requires several cofactors from the diet, such as folate, methionine, choline, and betaine. Indeed, the first report in 1984 that diet could modify methylation per se, showed that rats fed a low methyl donor diet reduced cytosine methylation in hepatic nuclear DNA [65]. To the present day, a diet depleted of the primary methyl donor methionine, also leads to depletion of hepatic H3K27me3 and H3K4 di- and mono-methylation [25]. In keeping with the hypomethylated state, a low choline and betaine diet during pregnancy results in hypomethylation of H3K9 up-stream of the RE1 binding site in the calbindin 1 promoter of mice brain tissue [66]. However, in contrast, few studies have demonstrated that nutrient supplementation in isolation can cause hypermethylation; for example, choline alone supplementation enriches H3K9me2 and H3K27me3 alongside the expression of mRNA and protein SUV39H1 methyltransferases in a period of gestation in liver and brain rat tissue [67]. Choline supplementation (930 mg/day) also leads to H3K9me2 enrichment in the human placenta [68]. While folic acid supplementation is synonymous with DNA methylation, a growing body of work postulates that folate deficiency and/or supplementation affects global histone methylation. In a recent experiment, [69,70] shows that folic acid supplementation promotes H3K9me2 in the LCN2 promoter, thus providing a potential chemoprotective mechanism of folic acid supplementation on tumorigenesis.

There is a plethora of data showing that a lipid diet and fatty acid molecules can affect histone methylation (Tables 19.1 and 19.2), and perhaps this is unsurprising

TABLE 19.1 Typical Modifications in Histone Methylation (Lysine, K; Arginine, R) by Metabolic Model

Effect	Residue/methylation	Metabolic model	Reference
Increase	H3K4me1	Insulin, hyperglycemia, choline deficiency	[37,67]
	H3K4me2	Calorie restriction, methyl deficiency, hypoxia, hexavalent chromium, aging, glucose	[44,80–84]
	H3K4me3	Hexavalent chromium, arsenite, nickel, LPS, aging, hypoxia, glucose, ischemia, TNFα	[42,44,82–88]
	H3K9me1	Protein restriction, nickel, insulin via ROS with hyperglycemia, glucose deprivation, gestational choline supply, hypoxia, chromium, high glucose	[31,37,67,83,84,89–91]
	H3K9me2	Calorie restriction	[80]
	H3K9me3	Calorie restriction, methyl deficient model, hexavalent chromium, maternal high-fat diet	[92–96]
	H3K27me2	High-fat diet	[97]
	H3K27me3	Food restriction, hypoxia, gestational choline supply	[67,82,89]
	H3K36me3	Hypoxia	[83]
	H3K79me2	Hypoxia	[82]
	H4R3me2	Hypoxia	[82]
	H3R26me1	High-fat diet	[98]
	H3R49me1	High-fat diet	[98]
	H4R35me1	High-fat diet	[98]
Decrease	H3K4me1	Insulin via ROS with hyperglycemia	[37]
	H3K4me2	Calorie restriction, hyperglycemia, glucose induced cAMP, high-fat diet during gestation	[76,86,91,99,100]
	H3K4me3	Calorie restriction, chromium, SAM, LPS, high-fat diet, low-protein diet during gestation and lactation	[42,83,94,101,102]
	H3K9me1	Insulin via ROS	[37]
	H3K9me2	Protein restriction, choline/methionine restriction, high glucose	[81,89,91]
	H3K9me3	db/db mice, high glucose, low-protein diet during gestation	[103,104]
	H3K18me2	High-fat diet	[98]
	H3K27me3	Hypoxia, hexavalent chromium, methyl deficient diet	[82,84,92]
	H3R17	Insulin	[38]
	H3R26me1	High-fat diet	[98]
	H3R49me1	High-fat diet	[98]
	H4K20me3	Calorie restriction, methyl deficient diet, glucose-induced cAMP	[86,92–94]

Updated from Milagro, FI. Martinez, JA (2017) Dietary and metabolic compounds affecting covalent histone modifications. In: Tollefsbol, T.O. Handbook of Epigenetics. 2nd ed. Academic Press. U.S.A.

TABLE 19.2 Typical Aspects of Metabolism, Diet, and Phytogenic Compounds on Histone Methylation and Acetylation

	Effector	Effect
Metabolism	Nitric oxide	Increase in global methylation and H3K9 methyltransferase expression
	Lipopolysaccharides	Increase in binding of H3K4me3 and histone methyltransferases SET1
	TCA cycle intermediates	Fumarate and 2-hydroxyglutarate inhibit JmjC activity, inhibits KDM4B, interferes with DNA repair and associated H3K4me3
	Insulin	Increase in H3K4me1 and H3K9me1, decrease in H3R17
	Hyperglycemia and high glucose	Inhibits sirtuins and induces acetylation (e.g., at H3K9, H3K27, H4K5, H4K8, H4K12), increase in H3K4me1, H3K4me2, and H3K4me3, decrease in H3K9me2 and H3K9me3
	Glucose deprivation	Increase in Suv39H1 and H3K9me2, increase in demethylase activities
	Diabetes	Decrease in H3K9me3 and H3K4me2, increase in H3K9me2
	Hypoxia	Activates histone acetylation (i.e., H3K9), activates JmjC-containing proteins, increase in H3K9me2 and G9a, inhibits KDM5A, KDM6A and KDM6B, affects succinate and α-KG concentration
	Oxygen	Activates KDMs
	Exercise	Increase in H3K36, H3K9, and H3K14 acetylation
	Aging	Acetylates H4K16, increase in H3K4me2 and H3K4me3
	Inflammation	Activates histone acetylation, increase in H3K4me3 and H3K4, recruits 65, p300, and SET7/9, increase in H2A.Z and SET1 MT
	Oxidative stress (inclusive of ROS production)	Inhibits HDAC activity, inhibits H3K4 methyltransferases MLL1–4, decrease in global H3K4me3, increase in HAT activity, decrease in H3K9ac, H4K8ac and H4K16ac
	Estrogens	Decrease in H3 and H4 acetylation
	Cold exposure	Acetylates K56, K67, K73, K151, and H3K9 and H3K14 Decrease in HDAC activity
	Alcohol	Histone hyperacetylation
	Surrogate alcohols	Acetylates H3K9
Diet	Methyl deficient diet	Downregulates methyltransferases Loss of H4K20me3 and H3K9me3 Increase in macroH2A and H3K9me3 Reduces H3K27me3 and H4K20me3 Increase in histone H1 proportion
	Low dietary choline and betaine	H3K9 hypomethylation
	Folic acid supplementation	Promotes H3K9me2
	Choline deficiency	Increases H3K4Me2
	Choline supplementation	Up-regulates lysine methyltransferase activity Enriches H3K9Me2 and H3K27Me3
	Calorie restriction	Upregulates SIRT1, increase in H4 acetylation
	Low-protein diet	Increase in H3K27me3, H3K9me1/me3, decrease in H3K4me3, increase in H3K9 and H3 acetylation
	High-fat diet	Increase in hepatic H3K14ac and H3K9me3 in liver, increase in H3K9me in adipose tissue, increase in H3K20me in leptin promoter, increase in H3K27me3 and H3K27ac
	High carbohydrate/fat ratio	Acetylates H3K9
	Biotin deficiency	Activates histone acetylation
	High-salt diet	Inhibits class IIa HDACs

(Continued)

TABLE 19.2 (Continued)

	Effector	Effect
	Zinc deficiency	Inhibits class I and II HDACs
	Vitamin C	Regulates JmjC-domain containing histone demethylases
	Glutamine deficiency	Decrease in H3K4me3 with a reduced expression of SETD1 and ASH2L
	Glutamine supplementation	Global H3K4me3 hypomethylation
Phytogenic compounds	Genistein	Activates histone methylation, inhibits HDAC6
	Diallyl disulfide	Increase in H3 and H4 acetylation, inhibits HDAC activity
	Sulforaphane	Inhibits HDAC activity
	Gallic acid	Inhibits HDAC activity
	Apigenin	Inhibits HDAC activity
	Luteolin	Inhibits HDAC activity
	Butyrate	Inhibits HDAC activity
	Short-chain fatty acid composition	Inhibits HDAC activity
	Theophylline	Increases HDAC activity
	Quercetin	Activates histone acetylation
	Resveratrol	Activates SIRT1
	Luteolin	Activates SIRT1
	Butein and fisetin	Activates SIRT1
	Caffeic acid	Activates SIRT1
	Quercetin	Activates SIRT1
	Isoliquiritigenin	Activates SIRT1
	Piceatannol	Activates SIRT1
	Anacardic acid	Inhibits HAT activity
	Garcinol	Inhibits HAT activity
	Curcumin	Inhibits HAT activity
	Epigallocatechin-3-gallate	Increase in H3K9, H3K14 and H3 acetylation, inhibits HAT activity
	Delphinidin	Inhibits HAT activity
	Plumbagin	Inhibits lysine acetyltransferase activity
	Glycosaminoglycans	Inhibits HAT activity
	Retinoic acid	Acetylates H3K9

Updated from Milagro, FI. Martinez, JA (2017) Dietary and metabolic compounds affecting covalent histone modifications. In: Tollefsbol, T.O. Handbook of Epigenetics. 2nd ed. Academic Press. U.S.A.

given the inextricable link between the TCA cycle (β-oxidation) and histone modifications. A high-fat diet causes persistent changes in liver H3K4me3, aligned to enhanced chromatin accessibility and altered gene expression [71]. Other work shows enrichment of H3K9me at adiponectin and increased H3K20 methylation in the leptin promoter following a high-fat diet in mice adipose tissue [72], while Zheng and colleagues [73] fed mice a 58% kcal fat-rich diet and observed an increase in H3K27me3 and a decrease in H3K9me3 at *Pepck*. In contrast to this work, Inoue et al. [74] observed an induction of mono-, di-, and tri-methylation of H3K4 at promoter and transcribed regions of the *Si* (sucrase-isomaltase) and *Sglt1* (sodium-dependent glucose cotransporter) genes in rat jejunum in response to a high-starch/low fat-diet (71% energy as cornstarch/7%

energy as corn oil). Supplementation with the short chain fatty acids acetate, butyrate, and propionate, regulates global histone methylation and transcriptional responses in liver and colon of germ-free mice [75]. A diet low in protein also affects nutrient sensing pathways which can modify histones. Gestational protein restriction attenuates H3K4me3 and H3K20me3 and enhances H3K9me3 and H3K27me3 levels on *IGF2* in fetal liver tissue [76]. H3K9 trimethylation at the *Cyp7a1* gene promoter was also increased following a low-protein diet during gestation; this led to a transcriptional repression of *Cyp7a1* in the offspring of rats [77]. Glutamine is an essential and abundant amino acid in cells, and its deficiency is linked to a reduction in H3K4me3 in promoters of several cancer-related genes. This decrease in histone H3K4me3 correlates with a reduced expression of *SETD1* and *ASH2L*, genes encoding the histone H3K4 methyltransferase complex [78]. In a recent supplementation and metabolomic based study, dietary uptake of glutamine was shown to inhibit melanoma tumor growth (in a patient-derived tumor model) by increasing intra-tumoral αKG concentration, leading to a global state of H3K4me3 hypomethylation. These data provide evidence that glutamine can act as a potential dietary intervention to block tumor growth, thus highlighting a previously unrealized therapeutic avenue in cancer biology [79].

HISTONE ACETYLATION

Histone acetylation progresses the act of transcription by efficiently weakening electrostatic interactions between DNA and histones and between nucleosomes. Metabolic activity is tightly aligned to the remodeling of chromatin through histone acetylation [105], where the TCA cycle is a prominent controlling domain within cell mitochondria. Mitochondrial oxidative metabolism begins with the reaction that combines the two-carbon acetyl-CoA (generated from substrate oxidation) with a four-carbon oxaloacetate to generate the six-carbon citrate [106]. Non-metabolized acetyl-CoA is fluxed out of mitochondria into the cytosol after being converted to citrate, where in the cytosol the citrate is cleaved back into oxaloacetate and acetyl-CoA by ATP citrate lyase (ACLY). The most salient signaling function of acetyl-CoA per se, is its ability to transfer an acetyl group to lysine amino acids on the N-terminal tails of histone to yield acetylation (Fig. 19.3) [6,26,106]. Acetylation (half-life = 3 min) can occur in any of the histones controlled by a histone acetyltransferase (HAT) enzyme, and the activity of HAT is sensitive to fluxes in intracellular acetyl-CoA concentration ($\sim$10-fold under normal physiological conditions and within the k_m range of HATs) [23,24]. Acetyl-CoA can also freely diffuse from mitochondria into the cytosolic space and nucleus further affecting histone acetylation levels [105]. The HATs and corresponding acetylation sites may not be affected by acetyl-CoA fluctuations to the same extent, and it is unknown whether acetylation status of various lysine residues differs in responsiveness to acetyl-CoA fluctuations [5]. HATs belong to three main families: GNAT (Gcn5-related N-acetyltransferase), MYST (MOZ, Ybf2/Sas3, Sas2, Tip60), and p300/CBP. These enzymes typically lead to the abstraction of a proton from the ε-amino group of lysine, which is then primed for nucleophilic attack on the keto-carbon of acetyl-CoA; this intermediate structure decomposes, releasing free coenzyme A and acetyl-lysine. HAT enzymes are not specific for lysine residues on histone, but salient for a plethora of acetylation reactions [6,107].

As stated, histone acetylation controls the regulation of gene transcription. For example, acetylation at specific gene loci (i.e., H3K9ac, H3K27ac) allows the histones to be more accessible for transcriptional activation. Although the precise mechanism is debatable, it may feature a combination of (1) histone acetylation neutralizing their positive charge, thus weakening the interaction of the nucleosome with the DNA, leading to an opening of chromatin and active transcription and (2), histone acetylation acting as a docking station for the recruitment of transcription regulators [6,26,48]. Numerous stimuli can lead to histone acetylation inclusive of bacteria and viruses, cigarette smoke, hypoxia, biotin deficiency, quercetin, and caffeine [5]. From a metabolic perspective, glucose availability, inflammation and oxidative stress are important in acetylation. On glucose availability, there is a sense of inevitability as glucose is the primary cell substrate (alongside fatty acid biosynthesis) regulating the generation of acetyl-CoA within the glycolytic and mitochondrial pathways, from which acetylation stems. To this end, Gaikwad et al. [108] ascertained the effect of hyperglycemia on histone H3 modification in diabetic rats, and showed for the first time, an increase in histone H3 acetylation and gene expression of the *Fbn1* gene following hyperglycemia; suggesting that the change in expression of the *Fbn1* gene is epigenetically and metabolically regulated. By contrast, in conditions of glucose depletion such as that experienced during starvation, cells continue to demand mitochondrial acetyl-CoA causing fatty acid conversion to acyl-CoAs in the cytosol, the acyl-CoAs are then subsequently transported into mitochondria through carnitine and oxidized to replenish acetyl-CoA to further acetylate histone residues [8]. Inflammatory gene expression is regulated by histone acetylation. The induction of glycolysis (and lactate dehydrogenase) generates acetyl-CoA and histone acetylation (H3K9ac) in effector T cells, subsequently up-regulating the production of the pro-inflammatory

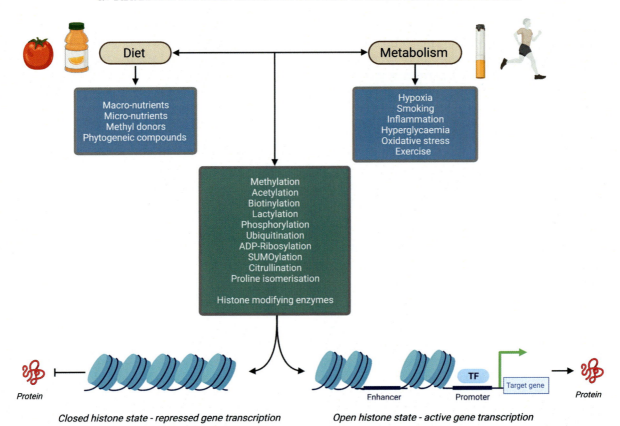

FIGURE 19.3 **Metabolic control of histone lactylation, methylation, and acetylation.** The mitochondria can flux metabolites into the nucleus to provide substrate for histone modifying enzymes, leading to histone modifications (histone methylation, acetylation, and lactylation shown here) and gene transcription. TCA cycle metabolites also block enzymatic activity. One-carbon metabolism provides substrate for histone methylation. *La*, lactate; *Ac*, acetylation; *Me*, methylation; *Acyl*, ATP citrate lyase; *SAM*, S-adenosylmethionine; *SAH*, S-adenosylhomocysteine; CH_3, methyl donor; *HDACs*, histone deacetylases; *HAT*, histone acetyltransferase; *LSDs*, lysine specific demethylases; *KMT*, lysine methyltransferase; *JmjC*, JumonjiC; O_2, oxygen; Fe^{2+}, iron.

cytokine interferon (IFN)-γ [109]. With regards to oxidative stress, there is an unambiguous link to increased HAT activity of p300/CBP and NFkB DNA binding in the promotion of pro-inflammatory gene expression [110,111]. In a diabetes setting, insulin leads to the production of ROS which is accompanied by histone H3 acetylation; interestingly, this effect was inhibited by the antioxidant enzyme catalase [112]. In line with another enzymatic antioxidant, *SOD1* deficiency was associated with a corresponding increase in the free radical species superoxide, alongside a decrease in H3 acetylation at the *peroxiredoxin1* gene promoter [113]. These contrasting experiments highlight that oxidative stress may potentially regulate the balance between an increase vs a decrease in histone acetylation [114]. To further compound this notion, alveolar epithelial cells treated with H_2O_2 increased H4 acetylation [115], while an increase in H_2O_2 levels in HL-60 leukemia cells decreased histone acetylation, probably by diminished activities of both *SOD* and catalase enzymes [116]. While the metabolic source of the ROS is open to debate, it is possible that mitochondrial leakage and activated leukocytes of primary free radical species are involved in the intracellular production of H_2O_2. As outlined by Garcia-Gimenez et al. [117], recent evidence highlights an intricate relationship between the effects of oxidative stress and histone acetylation. While Niu et al. [118] observed a decrease in H4K8ac and H3K9ac with H_2O_2, no paralleling changes were found in other acetylation sites in H4, such as K5, K9 and K12. Interesting other data presented by Niu and colleagues show that acetylation in histones H3 and H4 in bronchial epithelial cells are dependent on H_2O_2 dose and exposure time. For example, a 250 μM dose of H_2O_2 across a 3-hour time window correlates tightly with a decrease in H3K9ac and H4K8ac, while a low dose exposure of 25 μM H_2O_2 extended over a 3-week period leads to no change in the same histone residues. Cellular adaption to ROS in this instance may explain data variance [117], while further investigative work is required to interrogate the complex interrelationship between oxidative stress, antioxidant balance and histone acetylation.

Exercise can amplify TCA cycle turnover where acetyl-CoA production regulates histone acetylation. Indeed, it appears that a variety of exercise-induced substrates (i.e., glucose, fatty acids) provide the acetyl-CoA required for acetylation reactions across cell types aligned to different energetic contexts [119]. In a noteworthy study, acute (60 minutes) exercise performed at a moderate intensity, increases H3K36 acetylation in human skeletal muscle tissue [120]. Using an animal model, Smith et al. [121] demonstrates that rodent swimming exercise hyperacetylates H3K9 and H3K14 at *Glut4* promoter in triceps muscle. The role of exercise in epigenetic regulation is a burgeoning area in biology, and recently McGee and Hargreaves [119] outline several pertinent and unanswered questions. For example, it is currently unknown (1) how high-intensity exercise enhances histone acetylation at exercise response gene regions, (2) if exercise-induced histone acetylation is dependent on acetyl-CoA production or is it mainly dependent on acetate redistribution within distinct chromatin regions, and (3) what specific intracellular substrates supply acetyl-CoA? Further intensive investigative work is required to address these relevant questions.

Regarding dietary effects on histone acetylation per se, a moderate body of work demonstrates that caloric restriction and manipulation of macronutrients (fat, glucose, protein) can hyperacetylate histones (Fig. 19.4 and Table 19.2). Restricting overall calorie consumption is important for healthy aging, and the practice can stimulate several epigenetic effects in mammals. For example, Wheatley et al. [122] observed an increase in histone H4 acetylation on *Glut4* gene promoters in adipose tissue of obese mice; this change was accompanied by an increased expression of *Glut4* mRNA. Reducing overall

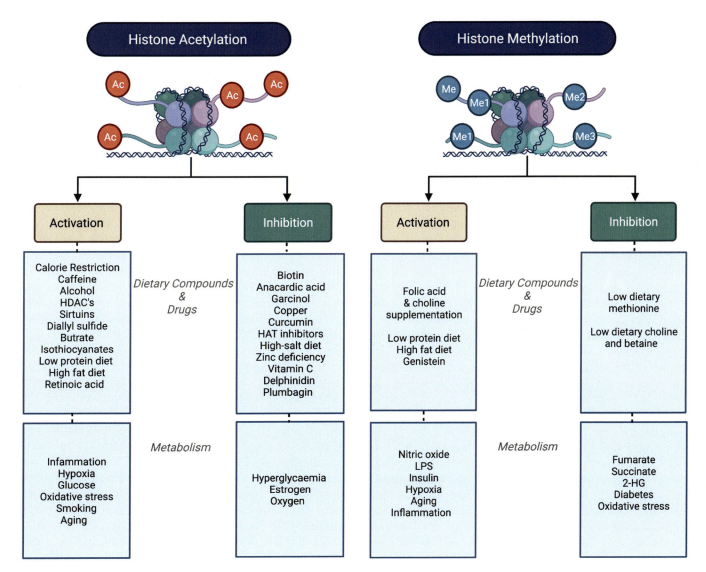

FIGURE 19.4 **Dietary and metabolic control of histone acetylation and methylation.** Dietary compounds, drugs, and aspects of metabolism (not designed to be an exhaustive list) controlling the activation and inhibition of histone acetylation and methylation. Ac on histone tail, histone acetylation; Me, Me1, Me2, Me3 on histone tails, histone methylation, mono-methylation, di-methylation, and tri-methylation, respectively.

calorie intake from glucose leads to an increase in H3 and H4 acetylation and *Tert* expression in fetal lung fibroblasts [99]. A low-protein diet can alter histones in the offspring. Pregnant pigs consuming a low-protein diet increased histone H3 acetylation on *G6PC* gene promoters in offspring male compared to offspring from control diet fed mothers [104]. In a later experiment by the same group, H3K9ac and muscle *MSTN* transcription was induced in male offspring following a low-protein diet in pregnant and lactating female pigs [102]. At the other end of the caloric intake continuum, a diet high in fat modifies acetylation across a range of gene promoters. Zhang and colleagues [97] observed hyperacetylation on histone H3 and H4 on *p21(Cip1)* and *p16(Ink4a)* respectively following a high-fat diet in an obese rat model. An increase in H3K27 acetylation at enhancers affects expression of multiple hepatic genes induced by a high-fat diet (40% calories from fat sources) in mice [123]. A high-fat diet during pregnancy leads to enhanced antioxidant genes (*Pon1* and *Pon3*) in male offspring, with the promoter of *Pon1* in liver tissue enriched with acetylation marks on histone H4 [100]. Supplementing with the fatty acid butyrate can hyperacetylate histones in pig alveolar macrophage [124] and breast cancer [125] cell lines. Moreover, phenethyl isothiocyanate found naturally in cruciferous vegetables [126], green tea polyphenols [127] and curcumin [128] are all known to hyperacetylate human histones through decreased HDAC activity (see following sections).

There is now emerging evidence demonstrating that maternal diet during pregnancy and in early life nutrition can affect adaptive immunity, ascertained in part through the study of respiratory allergies and metabolic diseases such as type 2 diabetes and obesity [129]. For example, Harb et al. [130] shows that an increase in protein kinase C (PKC) ζ (PKCζ) in blood cord T cells, following prenatal supplementation with fish oil, correlates with H3 acetylation at the PKCζ gene (*PRKCZ*) promoter, implying that PKCζ is amenable to regulation by prenatal nutritional exposures. In a follow-on experiment using 173 term placentas collected from a prospective birth cohort, Acevedo et al. [131] found that regular olive oil usage during pregnancy increased H3 acetylation on several immune-related gene promoters including *IL10RA* and *IL7R*. Moreover, maternal fish consumption was linked to an increase in H4 acetylation at the *CD14* gene in placentas of the female offspring. This data advocates that mothers who regularly ingest fish or olive oil as part of a diet can influence immune priming in newborns. Histone acetylation may also provide a nexus between vitamin consumption and allergy, although data is currently scant. In 1,25-dihydroxyvitamin D pre-treated monocytes from asthma patients, H4 acetylation is enhanced at the glucocorticoid response element of the MKP-1 promoter region [132], suggesting that vitamin D has corticosteroid-enhancing effects in patients with asthma.

In work demonstrating that raw cow's milk can reduce allergic symptoms, Abbring et al. [133] showed that acetylation of Th1, Th2, and regulatory T-cell-related genes of spenocyte-derived CD4[+] in C3H/HeOuJ mice were elevated after raw cow's milk compared to processed milk exposure. Following allergic induction (ovalbumin), the general immune stimulation was resolved and Th2 gene acetylation was lower compared to processed milk. Although epigenetics per se was not quantified, exposure to farm raw milk in the first year of life is aligned to modifications in gene expression of the innate immune receptors [134]. These data advocate for good early life nutrition (e.g., raw cow's milk) aligned to early life environmental exposures (e.g., farming environment) to reduce susceptibility of allergic diseases and sustain effects on health outcomes as the life journey progresses [129].

Histone Acetyltransferase Inhibitors and Activators

There is an emerging body of HAT inhibitors purporting to play an integral part in the treatment and management of a range of diseases such as cancer, diabetes, and inflammatory and neurological based diseases [111]. Related to this, Balasubramanyam and colleagues [135] show that anacardic acid (a 6-pentadecylsalicylic acid phenolic lipid) derived from the shell of cashew nuts inhibits p300 and p300/CBP-associated factor HAT activity. Anacardic acid can also attenuate MYST isoforms such as Tip60 [136]. Another polyphenol found in green tea, Epigallocatechin-3-gallate, abrogates p300-induced p65 acetylation and reduces p300 binding to the promoter region of *IL-6* [137]. In keeping with p300 targeting, the polyphenol curcumin disrupts the p300/GATA4 complex, inhibiting HAT activity [138], while delphinidin, an anthocyanidin extracted from pomegranate, inhibits p300/CBP in MH7A cells [139]. Other non-phenolic compound-based products can interfere with HAT activity. Plumbagin, a hydroxynaphthoquinone extracted from the root of *Plumbago rosea*, specifically inhibits lysine acetyltransferase KAT3B/p300 activity [140]. Several plant-based compounds such as alkaloids also possess inhibitory activity against p300 [141]. Finally, glycosaminoglycans are a sub-class of carbohydrates known to potently inhibit p300 and pCAF HAT activity in vitro [142]. Although the aforementioned outlines promising interactions as HAT inhibitors, several caveats regarding the underlying investigative approach remains. For instance, HAT inhibitors have undesired properties such as antioxidant activity (i.e., phenolic compounds), reactivity, instability, low in vivo potency, or lack selectively between HAT subtypes and other enzymes. Moreover, HATs are bi-substrate enzymes that catalyze reactions between two substrates: acetyl-CoA and a lysine-containing substrate, and this molecular characteristic may dictate the inhibitory potency of small molecule HAT inhibitors and may prove challenging for the reproducibility of enzyme inhibition experiments. A comprehensive characterization of the detailed molecular aspects of HATs and their inhibitors is recommended prior to any study inception; this will help generate a framework for further development towards therapeutic agents [111].

Environmentally driven HAT activators lead to changes in chromatin acetylation; best evidence currently exists for carbohydrate/glucose/lipids, ethanol, hypoxia and cold exposure. A high (70% energy as cornstarch) versus low (5%) carbohydrate diet alters H3K9 acetylation and expression of the SGLT1 gene in the jejunum of mice [143]. A glycolytic burst of glucose induces global (H3K9, H3K27, H4K5, H4K8, H4K12) acetylation in yeast cells via Gcn5 and Esa1 stimulation, suggesting that metabolic regulation of HAT's can modify cell acetylation in response to a physiological/dietary cue [144]. Concerning other macro substrates, Chung et al. [145] has recently shown that oleic-palmitic acid induces H3K9, H3K8, and H3K16 acetylation accompanied by an increase in lipogenic gene expression. Metabolism of alcohol contributes to neuron histone acetylation through direct deposition of acetyl-CoA (derived from alcohol) onto histones in an ACSS2-dependent manner, proposing a direct link between alcohol metabolism and gene regulation in the brain [146]. On other alcohol data, ethanol exposure leads to lysine hyperacetylation in several hepatic proteins (e.g., histone H3, p53, PGC-1α, SREPB-1c) spanning both cytosolic and mitochondrial cell regions, and possibly due to increased expression of p300/CBP, MYST, and PCAF [147]. Surrogate alcohols with an increased carbon chain length such as 1-propanol, 1-butanol, and isopentanol also modulate H3K9ac through a HAT dependent mechanism [148].

Atmospheric hypoxia and cold exposure regulate histone acetylation (Fig. 19.4). Regarding the former, hypoxic stress through an extremely low partial pressure of oxygen (1%) elevates H3 and H4 acetylation in neuroblastoma cells, with a corresponding increase in N-myc transcription [149]. In an earlier investigation, Watson et al. [150] used an identical hypoxic exposure (1% O_2) on prostate cells and observed enhanced H3K9 acetylation in the absence of any significant change in HAT and HIF-1α protein expression. The lack of inactivity in HIF-1α in this regard is important as CBP and p300 are linked to the HIF-1α complex, and further work is required to determine how H3K9ac occurs in the absence of induced HAT activity. Considering temperature, it has been shown that cold-induced thermogenesis increases the acetylation profile (K56, K67, K73, K151) of specific proteins (e.g., UCP-1) associated with brown adipose tissue; the mechanism is related to an alteration in mitochondrial acetyl-CoA flux [151]. Cold stress exacerbates asthma by H3K9 and H3K14 hyperacetylation in IL-4 gene promoter in CD_4^+ T cells via an up- and down-regulation in p300 and HDAC activity, respectively [152].

Finally, aside from HATs and HDACs regulating cell acetylation, other posttranslational modifications such as phosphorylation can also elevate HTA function and repress HDACs [5].

Histone Deacetylation

Histone deacetylation involves the removal of an acetyl group from a histone residue, and it usually precedes a closed chromatin configuration, leading to transcriptional repression. Histone demethylases (HDACs) are a highly conserved enzymatic family of proteins with responsibility for removing the acetyl group from the DNA-binding histone complex and may be subdivided into various classifications: class I (HDAC 1–3, 8); class IIa (HDAC 4, 5, 7, 9); class IIb (HDAC 6, 10); class III (sirtuins SIRT 1–7); class IV (HDAC 11). While class I, IIa/b, and IV are zinc-dependent, class III requires NAD^+ for catalysis [153]. On class III sirtuin proteins, there are seven isoforms distributed within mammalian cells, with SIRT1, 2, 6, and 7 residing in the nucleus [18,154]. Acetylation of H3K9 and H3K14 is generally targeted by SIRT1, H4K16 by SIRT2, H3K9, H3K18, and H3K56 by SIRT6, and H3K18 within various gene loci by SIRT7 [14]. The glycolytic substrate NAD^+ is an important redox cofactor in sirtuin activity, yielding deacetylated lysine and O-acetyl-ADP-ribose as by-products of the reaction [155]. Class III deacetylation reactions are metabolic responsive [48]. NAD^+ is known to fluctuate depending on specific metabolic cues, while NAD is subject to oxidation during normal cell metabolism, yielding NAD^+ and NADH, and any adjustment in the NAD^+/NADH ratio can subsequently affect sirtuin activity. For example, when cell energy is positively charged (e.g., increased glucose flux), there is a decrease in the NAD^+/NADH ratio, and this metabolic sensor attenuates sirtuin activity [6,24]. In contrast, when cell NAD^+ concentration is enhanced (increased NAD^+/NADH ratio) through adenosine monophosphate-activated protein kinase (AMPK) in a scenario such as a calorie restrictive state, sirtuin activation follows; when cells are deprived of ATP and other metabolic substrates, SIRT1 and six genes are up-regulated, leading to histone (H3K9ac and H3K14ac) deacetylation [6,14,156]. A slight caveat in data interpretation remains as most studies demonstrate that NAD^+ does not generally increase following starvation, and as such, the notion that cells produce higher levels of NAD^+ from a calorie restrictive state is perhaps counterintuitive, particularly when the need for an electron acceptor for hydrocarbon oxidation is generally limited [23]. SIRT1 upregulation and activity can also improve metabolic homeostasis through PGC-1α deacetylation, leading to gluconeogenesis and enhanced liver hepatic glucose output. Other work involving SIRT6, H3K9, and H3K56 deacetylation further links metabolism to chromatin remodeling (reviewed in [14]).

While NAD^+-dependent HDAC activity is controlled primarily by general metabolism, other HDAC (class I) enzymes may also be inhibited by certain metabolites such as butyrate and β-hydroxybutyrate arising from fatty acid oxidation or ketogenesis during prolonged exercise or starvation [23,157]. The work of Shimazu and colleagues [157] have shown that H3K9 and H3K14 acetylation occurs (alongside a corresponding upregulation in FOXO3A-regulated genes) following β-hydroxybutyrate administration in mice. Moreover, NAD^+ may be linked to circadian rhythm, and Nakahata et al. [158] demonstrates that HDAC activity of the NAD^+-dependent SIRT1 enzyme correlates with H3 Lys9/Lys14 acetylation at promoters of active circadian genes. Genetic ablation of SIRT1 also led to disturbances in the circadian cycle corresponding to a state of histone deacetylation. SIRT1 in this instance seems to function as an enzymatic rheostat of the circadian clock [158].

Oxidative stress inhibits HDAC activity, particularly in class III. As NAD^+ is prone to redox regulation, oxidative stress can upset the balance, by reducing cell NAD^+ causing a downregulation in SIRT1 activity [159]. HDAC and sirtuin inhibitors can induce apoptosis (via apoptosis-inducing genes) related to oxidative stress and DNA damage mechanisms [160]. A global loss of heterochromatin may be induced by an oxidative stress state; H_2O_2 treatment leads to a decrease in H3K9ac, H4K8ac, and H4K16ac in a HDAC-dependent manner [118]. Length of exposure time to oxidative stress may also be important in the regulation of chromatin. To this end, chronic oxidative stress (via H_2O_2 administration) modifies histones ascertained by a decrease in H3K9ac and H3K18ac in MCF7 cells [161]; this may be due in part to an over expression of HDAC1 as observed in immortalized human renal tubular epithelial cells [162]. Coupled with the notion that low dose oxidative stress can drive epigenetic reprogramming by ROS acting as intracellular signaling molecules, there seems to be an unambiguous link between oxidative stress/ROS production and histone (de)acetylation, thus paving the way for further investigative inquiry.

Inhibition of HDAC activity was first reported in 1978, when Candido and colleagues showed that the bacterially derived fatty acid butyrate, attenuates histone deacetylase activity in cultured cells [163]. Since then, numerous, recent investigations have ascertained that several bioactive dietary components inhibit HDAC activity. Most evidence exist for type I, II, and III HDAC enzymes [164], while cancer disease has been garnished with the most attention. Butyrate, sulforaphane and diallyl disulfide are popular agents in blocking type I and II HDAC activity, while resveratrol mainly interferes with type III deacetylases (Davies and Ross, 2007). On the former, short chain butyrate, found in fiber rich foods, is the most effective fatty acid in stimulating and repressing the expression of genes related to the cell proliferation complex. To this end, Davie [125] has shown that butyrate induces gene expression to halt cell proliferation by inhibiting HDAC activity recruited to the $p21^{Waf1/Cip1}$ promoter, leading to histone hyperacetylation and transcriptional activity of the $p21^{Waf1/Cip1}$ gene. Butyrate is also known to improve cardiac function through HDAC inhibition in diabetic mice through an increase of Glut1 acetylation and p38 phosphorylation [165]. Zou et al. [166] has recently shown that short chain fatty acids (including butyrate) inhibit HDAC activity and activates p38 MAPK to promote B10 cell generation and function; this work is important for a range of inflammatory and autoimmune diseases. In a related experiment, dietary manipulation of the gut microbiome using dietary fiber, which promotes the expansion of short chain fatty acid-producing bacteria Bifidobacterium and Prevotella, leads to HDAC inhibition and protects against acute kidney injury through GPR41 up-regulation [167]. Sulforaphane is an active isothiocyanate molecule found in cruciferous vegetables such as broccoli and inhibits HDAC activity by direct interaction with the HDAC active sites [168,169]. Sulforaphane-related HDAC inhibition was determined in a human cell line model showing an attenuation of tumorigenesis in Apc-minus mice via histone H3 acetylation in bax and P21 gene promoter regions [168,169]. Isothiocyanates with a comparable spacer length to that of sulforaphane, such as sulforaphene, erucin, and phenylbutyl isothiocyanate (plant compounds in the Brassicaceae family), all exhibit similiar HDAC inhibitory activities, and nutrient compounds with a longer or shorter spacer length, such as 6-erucin and phenethyl isothiocyanate, also display a similar inhibitory effect [5,168,169]. In an interesting study pitching sulforaphane against other phytochemicals (gallic acid and piceatannol) and the class II inhibitors MC1568 and TMP269, Choi and colleagues [170] showed that sulforaphane possesses only a moderate inhibitory effect on HDAC2 and HDAC class IIa, but stronger than piceatannol. Gallic acid proved a powerful inhibitor of HDAC8 and class IIa and b, but it only mildly inhibits activity in HDAC1 and 6; MC1568 does not have any inhibitory potential in HDAC class IIa.

Geinstein, found in soy, is traditionally shown to inhibit DNA methyltransferase activity, and when combined with sulforaphane, downregulates HDAC2 and 3 activity, alongside other molecules (*KLF4* and *hTERT*). These compounds were also shown to inhibit cell cycle progression to G2 phase in MDA-MB-231 and G1 phase in MCF-7 breast cancer cell lines. This salient work demonstrates that this combinational approach is more effective in preventing or treating mammary cancer than either of these dietary components administered alone, thus providing a new dietary and exploratory avenue for a potential preventive and therapeutic effect against breast cancer [171]. In keeping with this line of cancer related research, several other natural and dietary HDAC inhibitors (such as butein found in *Toxicodendron vernicifluum*, *Dahlia*, *Butea monosperma*, and *Coreopsis*; kaempferol found in grapes, green tea, tomatoes, potatoes, and onions; resveratrol found in grapes, red wine, blueberries and peanuts; sinapinic acid in wine and vinegar; zerumbone in ginger; and diallyl disulfide found in garlic) can exhibit an antitumor effect by activating cell cycle arrest, inducing apoptosis and autophagy, inhibiting angiogenesis, increasing ROS generation causing oxidative stress, and activating mitotic cell death in cancer cells [172]. On the latter dietary inhibitor, diallyl disulfide is an organosulfur compound from garlic that induces apoptosis in the MCF-7 breast-cancer cell line through HDAC inhibition and interfering with cell-cycle growth phases. HDAC inhibition is understood to lower the removal of the acetyl group from the acetylated substrate in MCF-7 cells, thus inducing histone H4 acetylation [173]. Flavones are a group of phytochemicals synthesized from flavanones via the enzyme flavone synthases, and the compounds apigenin (found in citrus fruits, onions, celery) and luteolin (in parsley broccoli, carrots apples) both inhibit HDAC activity associated with classes I, IIa, and IIb [174]. The polyphenol epigallocatechin gallate found mainly in green tea and chocolate increases H3K9, H3K14, and H3 acetylation via HDAC inhibition in human endothelial cells [175].

Resveratrol is from the stilbene phytochemical class and found mainly in red wine, berries, grapes, and other plant sources. It is active through the modulation of numerous signal transduction pathways aligned to cell division and growth, inflammation, angiogenesis, and metastasis [176]. The type III HDAC (in)hibitory effect of this polyphenolic compound is via SIRT1, however as resveratrol can bind to several proteins (more than 20) it can be difficult to separate epigenetic effects from effects on other cell structures [5]. SIRT1 activation enhances HDAC inhibition-mediated upregulation of GADD45G by repressing the binding of NF-κB/STAT3 complex to its promoter in malignant lymphoid cells [177], while Bagul et al. [178] has shown that resveratrol activates SIRT1, which subsequently deacetylates NFkB-p65 at H3K9. Other SIRT1-activating compounds include caffeic acid found in coffee, potatoes and sunflower seeds, the flavonol quercetin, found in peppers, onions, apples, and berries [174], the plant polyphenols butein and fisetin (a flavonoid abundant in strawberries, apples, persimmons, onions, and cucumbers) [179], the resveratrol metabolite piceatannol (found in passion fruit, grapes, and Japanese knotweed), the liquorice flavonoid isoliquiritigenin, and the flavonoid luteolin, found in several plants including celery, broccoli, carrots, green peppers, oranges, perilla, and olive oil [5,180]. Lastly, calorie restriction enhances SIRT1 in human peripheral blood mononuclear cells [181], fat, muscle, and liver [182], and as such may contribution to longevity.

OTHER HISTONE MODIFICATIONS

Other salient, but perhaps less-well known PTMs may play a role in the regulation of eu- and/or heterochromatin. To date, more than 200 PTMs in proteins have been reported [183], and several are enzymatic based which can be manipulated by metabolism per se or through dietary intervention. In a recent metabolic focused experiment, Zhang et al. [184] identified 28 lactylation sites on core histone lysine residues that can directly stimulate gene transcription. Lactate is an abundant metabolite emerging from glycolysis, where glucose is converted to two pyruvate molecules which are then fluxed into lactate [6,185]. As lysine lactylation is purported to depend solely on glucose flow through glycolysis, this emerging domain is ripe for further investigation involving dietary compounds that regulate intracellular glucose flow (i.e., carbohydrates) and lactate production (i.e., sodium bicarbonate). Sodium benzoate is a widely used chemical food preservative (an antimicrobial and flavoring agent to prevent canned fruit and fruit beverages from molding) and was recently ascertained to stimulate 22 lysine benzoylation (K_{bz}) marks on histones (H1, H2B, H2A, H3, H4) in HepG2 and RAW cells. These histone K_{bz} marks are enriched on promoters of active genes, particularly those related to glycerophospholipid metabolism, phospholipase D signaling, and insulin secretion. SIRT2 can also remove these K_{bz} marks [186]. That stated, the question whether dietary sodium benzoate can lead to tangible epigenetic changes remain opens [187]. Biotinylation (on K9, K13, K125, K127, K129 in histone H2A; K4, K9, K18 in histone H3; K8, K12 in histone H4) is a conserved and reversible modification and depends on exogenous biotin supply [2]. Ubiquitination operates through a ubiquitin ligase on a lysine residue. Histone H2B monoubiquitination at lysine 120 is necessary for trimethylation of H3K4 and H3K79 [188], while dietary restriction decreases the age-related increase in ubiquitination [43]. Carbonylation is also reduced by overall calorie restriction and aging [189]. As indicated by Molina-Serrano et al. [19], most histone modifications aligned to caloric restriction mainly affect stress-response genes; by inference implicating oxidative stress and redox regulation (particularly in carbonylation) in the aging process [190]. Moreover, oxidative stress can act as inductors of other histone modifications inclusive of phosphorylation, ADP-ribosylation, ubiquitination and biotinylation [5]. Protein phosphorylation is perhaps one of the most common PTMs. It's a reversible reaction which works through several kinases and phosphatases to add or remove a phosphate group (PO_4) to the polar group of R of various amino acids (e.g., H3S28; [191]). Phosphorylation can control gene transcription, and is regulated by several dietary compounds, such as genistein [192], sulforaphane [193], cryptotanshinone [194], curcumin [195], and capsaicin—the principal ingredient in chili plants [196]. SUMOylation is activated via several sumo-protein ligases (e.g., SUMO E1) and can be inhibited by ginkgolic acid and anacardin acid [197]. Finally, histone malonylation, specifically a decrease in H3K23mal, is induced by a high-fat diet in mice [98], while β-N-glycosylation (O-GlcNAc) is a highly dynamic modification, which responds to hormones (i.e., human growth hormone), nutrients (i.e., glucose), and stress [198]. Other PTMs exist, such as proline isomerization [199], homocysteinylation [200], serotonylation [201], ADP-ribosylation [202], citrullination [203], and lysine propionylation and butyrylation [204]; data is scare from the perspective of dietary manipulation and extensive work is required on basic biochemistry to ascertain their therapeutic potential.

CONCLUDING PERSPECTIVES

There is an inextricable link between diet and the metabolic state of the cell, and both as individual entities can influence the transcription of genes through histone modifications such as acetylation, methylation, biotinylation and lactylation. Nutrient quality can influence how metabolism functions, while metabolism drives cell nutrient need; this balance can subsequently dictate the level and type of histone modifications arising in cells. A comprehensive insight to the crosstalk between nutrition and metabolism and its effect on modifying histones is salient for the complex understanding of disease driven by environmental paradigms (e.g., high-fat diet, smoking, lack of exercise etc.). However, the broad domain of metabolic and nutrient sensing and its control and/or regulation of the chromatin state still requires a primary focus on basic biochemistry to ascertain the indubitable potential in this exciting area of epigenetics. For example, it is not overly clear as to the *extent* to which metabolites can categorically interfere with the chromatin state, as such, more evidence is needed for ROS, NAD^+, fumarate, succinate, and αKG. In particular, the link between ROS production, metabolism, and histone modification deserves further inquiry. ROS have the potential to play a significant role in shaping the epigenetic landscape as modulators of epigenetic mechanisms through redox signaling, with essential implications for localized gene transcription. To this author's knowledge, no data exists to determine if select ROS can oxidize the plethora of histone modifying enzymes; if key regulating enzymes and transcription factors are oxidized, then the effect on the eu- or hetero-chromatin state might be significant [6]. With the advent of targeted gene editing using CRISPR-Cas9, there is an opportunity to ascertain the effect of metabolism on histone modifications across a range of animal and human cell lines through novel in vitro experimentation. Whole gene knockout, or DNA prime editing directed at metabolic TCA cycle intermediates and epigenetic related enzymes is now possible. It is also now possible to determine chromatin accessibility and corresponding transcriptomics at a single-cell resolution using sc-assay for transposase-accessible chromatin (ATAC)-seq and scRNA-seq respectively [205]. On the latter, spatial transcriptomics can be examined to identify cell subpopulations and transcriptional variations related to human tissues in disease [6].

More work is required to examine how the many phytogenic compounds, the reduction of calories, lipids, and proteins, and the manipulation of methyl donors in the diet can modify posttranslational histones and subsequent gene transcription: arising from the perspective of decreasing the susceptible and management of disease such as cancer, and autoimmune and metabolic diseases. There is limited data interrogating the SAM pathway, and its control of methyl donor biochemistry, inclusive of methyltransferase enzyme activity. In addition to methionine, folic acid, choline, and betaine, consideration should also be afforded to vitamin B_2, B_6, and B_{12} as controllers of histone methylation. Additionally, supplementing with vitamin C to ascertain the effect on JmjC domain-contain histone demethylases should also be considered. On the dietary and nutrient supplementation aids known to influence histone modifications, there is a requirement to determine the dose, timing (acute vs chronic, gestation vs perinatal age vs across the lifespan) and the absorption kinetics to support understanding of pharmacodynamics and a healthy longevity. HDAC inhibitors hold promise as potential therapeutic drugs in several diseases (e.g., cancer and neurodegenerative disease), however, the potential safety limitations of these HDAC inhibitors cannot be ignored. For example, few studies have shown HDAC inhibitors to be neurotoxic rather than neuroprotective, and off-target actions of select inhibitors remains a problem [206].

Collectively, and with a greater emphasis on experimental work aligned to cell metabolism and nutrient interaction arising from a varied dietary intake, coupled with many new bioinformatic and in silico developments, the future direction of our understanding of the regulators and controllers of the many histone modifications that now exist is guaranteed. To this end, this narrative should be viewed as a continuation into the complex multidirectional connection between diet and metabolism affecting covalent histone modifications.

Acknowledgments

All Figures were produced using Biorender.com.

Abbreviations

Acetyl-CoA	Acetyl-coenzyme A
ACYL	ATP citrate lyase
AMPK	Adenosine monophosphate-activated protein kinase
ATP	Adenosine triphosphate
ADP	Adenosine ditriphosphate
CH_3	Methyl donor
FAD	Flavin adenine dinucleotide
Fe^{2+}	Iron
GLUT1/4	Glucose transporter 1 and 4
HAT	Histone acetyltransferase
HDAC(s)	Histone deacetylase(s)
HDR	DNA homology-dependent repair
HIF	Hypoxia-inducible factor
HMT	Histone methyltransferase
H_2O_2	Hydrogen peroxide
2-HG	2-hydroxyglutarate
IDH	Isocitrate dehydrogenase
IL	Interleukin
iPSC	Induced pluripotent stem cells
JMJD6	Bifunctional arginine demethylase and lysyl-hydroxylase
JHDM	JmjC domain-contain histone demethylases
JmjC	JumonjiC
KDM	Histone lysine demethylase
KDM/LSD	Lysine-specific histone demethylase or lysine (K)-specific demethylase
KMT	Histone lysine methyltransferase
α-KG	α-Ketoglutarate
K_{bz}	Lysine benzoylation
k_m	Michaelis-Menten constant
MAT	Methionine adenosyltransferase
MTHF	Methyltetrahydrofolate
NAD	Nicotinamide adenine dinucleotide
NFKB	Nuclear factor kappa B
NO	Nitric oxide
LPS	Lipopolysaccharide
PO_4	Phosphate group
PCAF	P300/CBP-associated factor
PRMT	Protein arginine N-methyltransferase
PTMs	Posttranslational modifications
RNS	reactive nitrogen species
ROS	Reactive oxygen species
SAH	S-adenosylhomocysteine
SAHH	S-adenosylhomocysteine hydrolase
SAM	S-adenosyl methionine
SGLT1	Sodium/glucose cotransporter 1
SIRT	Sirtuin
SOD	Superoxide dismutase
TCA	tricarboxylic acid cycle
TNFα	Tumor necrosis factor alpha

References

[1] Delage B, Dashwood RH. Dietary manipulation of histone structure and function. Ann Rev Nutr 2008;28:347−66.
[2] Hassan YI, Zempleni J. Epigenetic regulation of chromatin structure and gene function by biotin. J Nutr 2006;136:1763−5.
[3] Davie JR, Chadee DN. Regulation and regulatory parameters of histone modifications. J Cell Biochem 1998;72:203−13.
[4] Ideraabdullah FY, Zeisel SH. Dietary modulation of the epigenome. Physiol Rev 2018;98:667−95.
[5] Milagro FI, Martinez JA. Dietary and metabolic compounds affecting covalent histone modifications. In: Tollefsbol TO, editor. Handbook of Epigenetics. 2nd ed. USA: Academic Press; 2017.
[6] Davison GW, Irwin RE, Walsh CP. The metabolic-epigenetic nexus in type 2 diabetes mellitus. Free Radic Biol Med 2021; 170:194−206.
[7] Dai Z, Ramesh V, Locasale JW. The evolving metabolic landscape of chromatin biology and epigenetics. Nat Rev Genet 2020;21:737−53.
[8] Ye C, Tu BP. Sink into the epigenome: histones as repositories that influence cellular metabolism. Trends in Endocrinology and Metabolism 2018;29:626−37.
[9] Hales CN, Barker DJP. Type 2 (non-insulin-dependent) diabetes mellitus: the thrifty phenotype hypothesis. Diabetologia 1992;35: 595−601.
[10] Ling C, Rönn T. Epigenetics in human obesity and type 2 diabetes. Cell Metab 2019;7:1028−44.
[11] Bansal A, Simmons RA. Epigenetics and developmental origins of diabetes: correlation or causation? Am J Physiol Endocrinol Metab 2018;315:E15−28.
[12] Zhao J, Goldberg J, Bremner JD, Vaccarino V. Global DNA methylation is associated with insulin resistance: a monozygotic twin study. Diabetes 2012;61:542−6.
[13] Dhawan S, Natarajan R. Epigenetics and Type 2 Diabetes Risk. Curr Diab Rep 2019;19. Available from: https://doi.org/10.1007/s11892-019-1168-8.
[14] Etchegaray P, Mostoslavsky R. Interplay between metabolism and epigenetics: a nuclear adaptation to environmental changes. Mol Cell 2016;62:695−711.
[15] Keating ST, El-Osta A. Epigenetics and metabolism. Circ Res 2015;116:715−36.
[16] Shaughnessy DT, McAllister K, Worth L, et al. Mitochondria, energetics, epigenetics, and cellular responses to stress. Environ Health Perspect 2014;122:1271−8.
[17] Donohoe DR, Bultman SJ. Metaboloepigenetics: interrelationships between energy metabolism and epigenetic control of gene expression. J Cell Physiol 2012;227:3169−77.
[18] Miranda-Gonçalves V, Lameirinhas A, Henrique R, Jerónimo C. Metabolism and epigenetic interplay in cancer: regulation and putative therapeutic targets. Front Genet 2018;9:427.
[19] Molina-Serrano DI, Kyriakou DI, Kirmizis A. Histone modifications as an intersection between diet and longevity. Front Genet 2019;10:192.
[20] Choi SW, Friso S. Epigenetics: a new bridge between nutrition and health. Advances in Nutrition 2010;1:8−16.
[21] Mentch SJ, Locasale JW. One-carbon metabolism and epigenetics: understanding the specificity. Ann NY Acad Sci 2016;1363:91−8.
[22] Friso S, Udali S, De Santis D, Choi SW. One-carbon metabolism and epigenetics. Mol Aspects Med 2017;54:28−36.

[23] Kaelin WG, McKnight SL. Influence of metabolism on epigenetics and disease. Cell 2013;153:56—69.

[24] Wong CC, Qian Y, Yu J. Interplay between epigenetics and metabolism in oncogenesis: mechanisms and therapeutic approaches. Oncogene 2017;36:3359—74.

[25] Mentch SJ, Mehrmohamdi M, Huang L, et al. Histone methylation dynamics and gene regulation occur through the sensing of one-carbon metabolism. Cell Metab 2015;22:861—73.

[26] Vanzan L, Sklias A, Herceg Z, Murr R. Mechanisms of histone modifications. In: Tollefsbol TO, editor. Handbook of Epigenetics. 2nd ed. USA: Academic Press; 2017.

[27] Young JI, Züchner S, Wang G. Regulation of the epigenome by vitamin C. Annu Rev Nutr 2015;25:545—64.

[28] Socco S, Bovee RC, Palczewski MB, Hickok JR, Thomas DD. Epigenetics: the third pillar of nitric oxide signaling. Pharmacological Research 2017;121:52—8.

[29] Hickok JR, Vasudevan D, Antholine WE. Nitric oxide modifies global histonemethylation by inhibiting Jumonji C domain-containing demethylases. J Biol Chem 2013;288:16004—15.

[30] Sun G, Reddy MA, Yuan H, Lanting L, Kato M, Natarajan R. Epigenetic Histone Methylation Modulates Fibrotic Gene Expression. J Am Soc Nephrol 2010;21:2069—80.

[31] Murayama A, Ohmori K, Fujimura A, Minami H, Yasuzawa-Tanaka K, Kuroda T, et al. Epigenetic control of rDNA loci in response to intracellular energy status. Cell 2008;133:627—39.

[32] Okabe J, Orlowski C, Balcerczyk A, et al. Distinguishing hyperglycemic changes by Set7 in vascular endothelial cells. Circ Res 2012;110:1067—76.

[33] Mishra M, Zhong Q, Kowluru RA. Epigenetic modifications of Keap1 regulate its interaction the protective factor Nrf2 in the development of diabetic retinopathy. Invest Ophthalmol Vis Sci 2014;55:7256—65.

[34] Kowluru RA, Mishra M. Epigenetic regulation of redox signaling in diabetic retinopathy: role of Nrf2. Free Radic Biol Med 2017;103:155—64.

[35] Zhong Q, Kowluru RA. Epigenetic changes in mitochondrial superoxide dismutase in the retina and the development of diabetic retinopathy. Diabetes 2011;60:1304—13.

[36] Ma J, He X, Cao Y, et al. Islet-specific Prmt5 excision leads to reduced insulin expression and glucose intolerance in mice. J Endocrinol 2020;244:41—52.

[37] Kabra DG, Gupta J, Tikoo K. Insulin induced alteration in post-translational modifications of histone H3 under a hyperglycemic condition in L6 skeletal muscle myoblasts. Biochim Biophys Acta 2009;1792:574—83.

[38] Sun H, Zhou X, Chen H, Li Q, Costa M. Modulation of histone methylation and MLH1 gene silencing by hexavalent chromium. Toxicol Appl Pharmacol 2009;237:258—66.

[39] Furman D, Campisi J, Verdin E, Carrera-Bastos P, Targ S, Franceschi C, et al. Chronic inflammation in the etiology of disease across the life span. Nat Med 2019;25:1822—32.

[40] Sinclair LV, et al. Antigen receptor control of methionine metabolism in T cells. eLife 2019;8:e44210.

[41] Roy DG, et al. Methionine metabolism shapes T helper cell responses through regulation of epigenetic reprogramming. Cell Metab 2020;31:250—66 e259.

[42] Ara AI, Xia M, Ramani K, Mato JM, Lu SC. S-adenosylmethionine inhibits lipopolysaccharide-induced gene expression via modulation of histone methylation. Hepatology 2008;47:1655—66.

[43] Li F, Zhang L, Craddock J, Bruce-Keller A, Dasuri K, Nguyen AT, et al. Aging and dietary restriction effects on ubiquitination, sumoylation, and the proteasome in the heart. Mech Ageing Dev 2008;129:515—21.

[44] El Mezayen R, El Gazzar M, Myer R, High KP. Aging-dependent upregulation of IL-23p19 gene expression in dendritic cells is associated with differential transcription factor binding and histone modifications. Aging Cell 2009;8:553—65.

[45] Lu C, Thompson CB. Metabolic regulation of epigenetics. Cell Metab 2012;16:9—17.

[46] Shi Y, Lan F, Matson C, Mulligan P, Whetstine JR, Cole PA, et al. Histone demethylation mediated by the nuclear amine oxidase homolog LSD1. Cell 2004;119:941—53.

[47] Karytinos A, Forneris F, Profumo A, et al. A novel mammalian Flavin-dependent histone demethylase. J Biol Chem 2009;284: 17775—82.

[48] Wang J, Hodes GE, Zhang H, Zhang S, Zhao W, Golden SA, et al. Epigenetic modulation of inflammation and synaptic plasticity promotes resilience against stress in mice. Nat. Commun 2018;9:477.

[49] Schvartzman JM, Thompson CB, Finley LWS. Metabolic regulation of chromatin modifications and gene expression. J Cell Biol 2018;217:2247—59.

[50] Tsukada Y, Fang J, Erdjument-Bromage H, Warren ME, Borchers CH, Tempst P, et al. Histone demethylation by a family of JmjC domain-containing proteins. Nature 2006;439:811—16.

[51] Klose RJ, Yamane K, Bae Y, Zhang D, Erdjument-Bromage H, Tempst P, et al. The transcriptional repressor JHDM3A demethylates trimethyl histone H3 lysine 9 and lysine 36. Nature 2006;442:312—16.

[52] Mikhed Y, Görlach A, Knaus UG, Daiber A. Redox regulation of genome stability by effects on gene expression, epigenetic pathways and DNA damage/repair. Redox Biol 2015;5:275—89.

[53] Bazopoulou D, et al. Developmental ROS individualizes organismal stress resistance andlifespan. Nature 2019;576:301—5.

[54] Batie M, et al. Hypoxia induces rapid changes to histone methylation and reprograms chromatin. Science 2019;363:1222—6.

[55] Chakraborty AA, et al. Histone demethylase KDM6A directly senses oxygen to control chromatin and cell fate. Science 2019; 363:1217—22.

[56] Salminen A, Kaarniranta K, Kauppinen A. Hypoxia-inducible histone lysine demethylases: impact on the aging process and age related diseases. Aging Dis 2016;7:180—200.

[57] Fan J, Krautkramer KA, Feldman JL, Denu JM. Metabolic regulation of histone post-translational modifications. ACS Chem Biol 2015;10:95—101.

[58] Sulkowski PL, Oeck S, Dow J, et al. Oncometabolites suppress DNA repair by disrupting local chromatin signalling. Nature 2020;582:586—91.

[59] Backe MB, Andersson K, Bacos K, et al. Lysine demethylase inhibition protects pancreatic β cells from apoptosis and improves β-cell function. Mol Cell Endocrinol 2018;460:47—56.

[60] Backe MB, Jin C, Andreone L, Sankar A, Agger K, Helin K, et al. The lysine demethylase KDM5B regulates islet function and glucose homeostasis. J Diabetes Res 2019;2019:5451038 ID 5451038.

[61] Liu P-S, et al. α-Ketoglutarate orchestrates macrophage activation through metabolic and epigenetic reprogramming. Nat Immunol 2017;18:985—94.

[62] Leone RD, et al. Glutamine blockade induces divergent metabolic programs to overcome tumor immune evasion. Science 2019;366:1013.

[63] Jiang Y, et al. Iron-dependent histone 3 lysine 9 demethylation controls B cell proliferation and humoral immune responses. Nat Commun 2019;10:2935.

[64] Stover PJ, James WPT, Krook A, Garza C. Emerging concepts on the role of epigenetics in the relationships between nutrition and health. J Intern Med 2018;284:37—49.

[65] Wilson MJ, Shivapurkar N, Poirier LA. Hypomethylation of hepatic nuclear DNA in rats fed with a carcinogenic methyl-deficient diet. Biochem J 1984;218:987—90.

[66] Mehedint MG, Niculescu MD, Craciunescu CN, Zeisel SH. Choline deficiency alters global histone methylation and

epigenetic marking at the Re1 site of the calbindin 1 gene. FASEB J 2010;24:184—95.

[67] Davison JM, Mellott TJ, Kovacheva VP, Blusztajn JK. Gestational choline supply regulates methylation of histone H3, expression of histone methyltransferases G9a (Kmt1c) and Suv39h1 (Kmt1a), and DNA methylation of their genes in rat fetal liver and brain. J Biol Chem 2009;284:1982—9.

[68] Jiang X, Yan J, West AA, Perry CA, Malysheva OV, Devapatla S, et al. Maternal choline intake alters the epigenetic state of fetal cortisol-regulating genes in humans. FASEB J 2012;26:3563—74.

[69] Zhang Y, Xue G, Miao H, Zhou C, Sun S, Zhang Y. Folic acid supplementation acts as a chemopreventive factor in tumorigenesis of hepatocellular carcinoma by inducing H3K9Me2-dependent transcriptional repression of LCN2. Oncotarget 2021;12:366—78.

[70] Zhang Y, Gilmour A, Ahn YH, de la Vega L, Dinkova-Kostova A. The isothiocyanate sulforaphane inhibits mTOR in an NRF2-independent manner. Phytomedicine 2021;86:153062.

[71] Leung A, Trac C, Du J, Natarajan R, Schones DE. Persistent chromatin modifications induced by high-fat diet. J Biol Chem 2016;291:10446—55.

[72] Masuyama H, Hiramatsu Y. Effects of a high-fat diet exposure in utero on the metabolic syndrome-like phenomenon in mouse offspring through epigenetic changes in adipocytokine gene expression. Endocrinology 2012;153:2823—30.

[73] Zheng J, Zhang Q, Mul JD, Yu M, Xu J, Qi C, et al. Maternal high-calorie diet is associated with altered hepatic microRNA expression and impaired metabolic health in offspring at weaning age. Endocrine 2016;54:70—80.

[74] Inoue S, Honma K, Mochizuki K, Goda T. Induction of histone H3K4 methylation at the promoter, enhancer, and transcribed regions of the Si and Sglt1 genes in rat jejunum in response to a high-starch/low-fat diet. Nutrition 2015;31:366—72.

[75] Krautkramer KA, Kreznar JH, Romano KA, Vivas EI, Barrett-Wilt G, Rabaglia ME, et al. Diet-microbiota interactions mediate global epigenetic programming in multiple host tissues. Mol Cell 2016;64:982—92.

[76] Strakovsky RS, Zhang X, Zhou D, Pan Y. Gestational high-fat diet programs hepatic phosphoenolpyruvate carboxykinase gene expression and histone modification in neonatal offspring rats: programming gluconeogenesis by gestational high-fat diet. J Physiol 2011;589:2707—17.

[77] Sohi G, Marchand K, Revesz A, Arany E, Hardy DB. Maternal protein restriction elevates cholesterol in adult rat offspring due to repressive changes in histone modifications at the cholesterol 7alpha-hydroxylase promoter. Mol Endocrinol 2011;25:785—98.

[78] Simpson NE, Tryndyak VP, Pogribna M, Beland FA, Pogribny IP. Modifying metabolically sensitive histone marks by inhibiting glutamine metabolism affects gene expression and alters cancer cell phenotype. Epigenetics 2012;7:1413—20.

[79] Gabra MBI, Yang Y, Li H, et al. Dietary glutamine supplementation suppresses epigenetically-activated oncogenic pathways to inhibit melanoma tumour growth. Nat Commun 2020;11:3326.

[80] Jiménez-Chillarón JC, Díaz R, Martínez D, Pentinat T, Ramón-Krauel M, Ribó S, et al. The role of nutrition on epigenetic modifications and their implications on health. Biochimie 2012;94:2242—63.

[81] Dobosy JR, Fu VX, Desotelle JA, Srinivasan R, Kenowski ML, Almassi N, et al. A methyl-deficient diet modifies histone methylation and alters Igf2 and H19 repression in the prostate. Prostate 2008;68:1187—95.

[82] Johnson AB, Denko N, Barton MC. Hypoxia induces a novel signature of chromatin modifications and global repression of transcription. Mutat Res 2008;640:174—9.

[83] Xia X, Lemieux ME, Li W, Carroll JS, Brown M, Liu XS, et al. Integrative analysis of HIF binding and transactivation reveals its role in maintaining histone methylation homeostasis. Proc Natl Acad Sci USA 2009;106:4260—5.

[84] Brasacchio D, Okabe J, Tikellis C, Balcerczyk A, George P, Baker EK, et al. Hyperglycemia induces a dynamic cooperativity of histone methylase and demethylase enzymes associated with gene-activating epigenetic marks that coexist on the lysine tail. Diabetes 2009;58:1229—36.

[85] Villeneuve LM, Reddy MA, Lanting LL, Wang M, Meng L, Natarajan R. Epigenetic histone H3 lysine 9 methylation in metabolic memory and inflammatory phenotype of vascular smooth muscle cells in diabetes. Proc Natl Acad Sci 2008;105:9047—52.

[86] Zhou X, Li Q, Arita A, Sun H, Costa M. Effects of nickel, chromate, and arsenite on histone 3 lysine methylation. Toxicol Appl Pharmacol 2009;236:78—84.

[87] Li Y, Reddy MA, Miao F, Shanmugam N, Yee JK, Hawkins D, et al. Role of the histone H3 lysine 4 methyltransferase, SET7/9, in the regulation of NF-kappaB-dependent inflammatory genes. Relevance to diabetes and inflammation. J Biol Chem 2008;283:26771—81.

[88] Zager RA, Johnson AC. Renal ischemia-reperfusion injury upregulates histone-modifying enzyme systems and alters histone expression at proinflammatory/profibrotic genes. Am J Physiol Renal Physiol 2009;296:F1032—41.

[89] Lillycrop KA, Slater-Jefferies JL, Hanson MA, Godfrey KM, Jackson AA, Burdge GC. Induction of altered epigenetic regulation of the hepatic glucocorticoid receptor in the offspring of rats fed a protein-restricted diet during pregnancy suggests that reduced DNA methyltransferase-1 expression is involved in impaired DNA methylation and changes in histone modifications. Brit J Nutr 2007;97:1064—73.

[90] Miao F, Wu X, Zhang L, Yuan YC, Riggs AD, Natarajan R. Genome-wide analysis of histone lysine methylation variations caused by diabetic conditions in human monocytes. J Biol Chem 2007;282:13854—63.

[91] Hall RK, Wang XL, George L, Koch SR, Granner DK. Insulin represses phosphoenolpyruvate carboxykinase gene transcription by causing the rapid disruption of an active transcription complex: a potential epigenetic effect. Mol Endocrinol 2007;21:550—63.

[92] Pogribny IP, Tryndyak VP, Bagnyukova TV, Melnyk S, Montgomery B, Ross SA, et al. Hepatic epigenetic phenotype predetermines individual susceptibility to hepatic steatosis in mice fed a lipogenic methyl-deficient diet. J Hepatol 2009;51:176—86.

[93] Pogribny IP, Tryndyak VP, Muskhelishvili L, Rusyn I, Ross SA. Methyl deficiency, alterations in global histone modifications, and carcinogenesis. J Nutr 2007;137:216S—222SS.

[94] Sharif J, Nakamura M, Ito T, Kimura Y, Nagamune T, Mitsuya K, et al. Food restriction in pregnant mice can induce changes in histone modifications and suppress gene expression in fetus. Nucl Acids Symp Series 2007;51:125—6.

[95] Suter MA, Ma J, Vuguin PM, Hartil K, Fiallo A, Harris RA, et al. In utero exposure to a maternal high-fat diet alters the epigenetic histone code in a murine model. Am J Obstet Gynecol 2014;210:463 e11.

[96] Hancock RL, Dunne K, Walport LJ, Flashman E, Kawamura A. Epigenetic regulation by histone demethylases in hypoxia. Epigenomics 2015;7:791—811.

[97] Zhang X, Zhou D, Strakovsky R, Zhang Y, Pan X. Hepatic cellular senescence pathway genes are induced through histone modifications in a diet-induced obese rat model. Am J Physiol Gastrointest Liver Physiol 2012;302:G558—64.

[98] Nie L, Shuai L, Zhu M, Liu P, Xie ZF, Jiang S, et al. The landscape of histone modifications in a high-fat diet-induced obese (DIO) mouse model. Mol Cell Proteom 2017;16:1324—34.

[99] Daniel M, Tollefsbol TO. Epigenetic linkage of aging, cancer and nutrition. J Exp Biol 2015;218:59—70.

[100] Strakovsky RS, Zhang X, Zhou D, Pan Y-X. The regulation of hepatic Pon1 by a maternal high-fat diet is gender specific and may occur through promoter histone modifications in neonatal rats. J Nutr Biochem 2014;25:170–6.

[101] Suter MA, Takahashi D, Grove KL, Aagaard KM. Postweaning exposure to a high-fat diet is associated with alterations to the hepatic histone code in Japanese macaques. Ped Res 2013;74:252–8.

[102] Jia Y, Gao G, Song H, Cai D, Yang X, Zhao R. Low-protein diet fed to crossbred sows during pregnancy and lactation enhances myostatin gene expression through epigenetic regulation in skeletal muscle of weaning piglets. Eur J Nutr 2016;55:1307–14.

[103] Kikuno N, Shiina H, Urakami S, Kawamoto K, Hirata H, Tanaka Y, et al. Genistein mediated histone acetylation and demethylation activates tumor suppressor genes in prostate cancer cells. Int J Cancer 2008;123:552–60.

[104] Jia Y, Cong R, Li R, Yang X, Sun Q, Parvizi N, et al. Maternal low-protein diet induces gender-dependent changes in epigenetic regulation of the glucose-6-phosphatase gene in newborn piglet liver. J Nutr 2012;142:1659–65.

[105] Peleg S, Feller C, Ladurner AG, Imhof A. The metabolic impact on histone acetylation and transcription in ageing. Trends Biochem Sci 2016;41:700–11.

[106] Martinez-Reynes I, Chandel NS. Mitochondrial TCA cycle metabolites control physiology and disease. Nat Comm 2020;11:102.

[107] Cry AR, Domann FE. The redox basis of epigenetic modifications: from mechanisms to functional consequences. Antioxid Redox Signal 2011;15:551–89.

[108] Gaikwad A, Gupta J, Tikoo K. Epigenetic changes and alteration of Fbn1 and Col3A1 gene expression under hyperglycaemia and hyperinsulinaemic conditions. Biochem J 2010;432:333–41.

[109] Peng M, Yin N, Chhangawala S, Xu K, Leslie CS, Li MO. Aerobic glycolysis promotes T helper 1 cell differentiation through an epigenetic mechanism. Science 2016;354:481–4.

[110] Sundar IK, Yao H, Rahman I. Oxidative stress and chromatin remodeling in chronic obstructive pulmonary disease and smoking related diseases. Antioxid Redox Signal 2013;18:1956–71.

[111] Wapenaar H, Dekker FJ. Histone acetyltransferases: challenges in targeting bi-substrate enzymes. Clin Epigenetics 2016;8:59.

[112] Gupta J, Tikoo K. Involvement of insulin-induced reversible chromatin remodeling in altering the expression of oxidative stress-responsive genes under hyperglycemia in 3T3-L1 preadipocytes. Gene 2012;504:181–91.

[113] Wang X, Vatamaniuk MZ, Roneker CA, Pepper MP, Hu LG, Simmons RA, et al. Knockouts of SOD1 and GPX1 exert different impacts on murine islet function and pancreatic integrity. Antioxid Redox Signal 2011;14:391–401.

[114] Hervouet E, Zhang J, Boyer-Guittaut M. ROS and epigenetics. In: Ahmad SI, editor. Reactive oxygen species in biology and human health. CRC Press; 2016. p. 519–33.

[115] Tomita K, Barnes PJ, Adcock IM. The effect of oxidative stress on histone acetylation and IL-8 release. Biochem Biophys Res Commun 2003;301:572–7.

[116] Lin CJ, Kang J, Zheng R. Oxidative stress is involved in inhibition of copper on histone acetylation in cells. Chem Biol Interact 2005;151:167–76.

[117] García-Giménez JL, Garcés C, Romá-Mateo C, Pallardó FV. Oxidative stress-mediated alterations in histone post-translational modifications. Free Radic Biol Med 2021;170:6–18.

[118] Niu Y, DesMarais TL, Tong Z, Yao Y, Costa M. Oxidative stress alters global histone modification and DNA methylation. Free Radic Biol Med 2015;82:22–8.

[119] McGee SL, Hargreaves M. Epigenetics and exercise. Trends Endocrinol Metab 2019;30:636–45.

[120] McGee SL, Fairlie E, Garnham AP, Hargreaves M. Exercise-induced histone modifications in human skeletal muscle. J Physiol 2009;587:5951–8.

[121] Smith JA, et al. CaMK activation during exercise is required for histone hyperacetylation and MEF2A binding at the MEF2 site on the Glut4 gene. Am J Physiol Endocrinol Metab 2008;295:E698–704.

[122] Wheatley KE, Nogueira LM, Perkins SN, Hursting SD. Differential effects of calorie restriction and exercise on the adipose transcriptome in diet-induced obese mice. J Obes 2011;2011:1–13.

[123] Siersbæk M, Varticovski L, Yang S, Baek S, Nielsen R, Mandrup S, et al. High-fat diet-induced changes of mouse hepatic transcription and enhancer activity can be reversed by subsequent weight loss. Sci Rep 2017;7:40220.

[124] Xiong H, Guo B, Gan Z, Song D, Lu Z, Yi H, et al. Butyrate upregulates endogenous host defense peptides to enhance disease resistance in piglets via histone deacetylase inhibition. Sci Rep 2016;6:27070.

[125] Davie JR. Inhibition of histone deacetylase activity by butyrate. J Nutr 2003;133:2485S–93S.

[126] Wang LG, Beklemisheva A, Liu XM, Ferrari AC, Feng J, Chiao JW. Dual action on promoter demethylation and chromatin by an isothiocyanate restored GSTP1 silenced in prostate cancer. Mol Carcinog 2007;46:24–31.

[127] Pandey M, Shukla S, Gupta S. Promoter demethylation and chromatin remodeling by green tea polyphenols leads to re-expression of GSTP1 in human prostate cancer cells. Int J Cancer 2010;126:2520–33.

[128] Liu HL, Chen Y, Cui GH, Zhou JF. Curcumin, a potent antitumor reagent, is a novel histone deacetylase inhibitor regulating B-NHL cell line Raji proliferation. Acta Pharmacol Sin 2005;26:603–9.

[129] Esch BCAMv, Porbahaie M, Abbring S, Garssen J, Potaczek DP, Savelkoul HFJ, et al. The impact of milk and its components on epigenetic programming of immune function in early life and beyond: implications for allergy and asthma. Front Immunol 2020;11:2141.

[130] Harb H, Irvine J, Amarasekera M, Hii CS, Kesper DA, Ma Y, et al. The role of PKCζ in cord blood t-cell maturation towards th1 cytokine profile and its epigenetic regulation by fish oil. Biosci Rep 2017;37(2):BSR20160485.

[131] Acevedo N, Frumento P, Harb H, Alhamwe BA, Johansson C, Eick L, et al. Histone acetylation of immune regulatory genes in human placenta in association with maternal intake of olive oil and fish consumption. Int J Mol Sci 2019;20:1060.

[132] Zhang Y, Leung DYM, Goleva E. Anti-inflammatory and corticosteroid-enhancing actions of vitamin D in monocytes of patients with steroid-resistant and those with steroid-sensitive asthma. J Allergy Clin Immunol 2014;133:1744–52.

[133] Abbring S, Wolf J, Ayechu-Muruzabal V, Diks MAP, Alhamwe BA, Alhamdan F, et al. Raw cow's milk reduces allergic symptoms in a murine model for food allergy – a potential role for epigenetic modifications. Nutrients 2019;11:1721.

[134] Loss G, Bitter S, Wohlgensinger J, Frei R, Roduit C, Genuneit J, et al. Prenatal and early-life exposures alter expression of innate immunity genes: the pasture cohort study. J Allergy Clin Immunol 2012;130:523–30.

[135] Balasubramanyam K, Swaminathan V, Ranganathan A, Kundu TK. Small molecule modulators of histone acetyltransferase p300. J Biol Chem 2003;278:19134–40.

[136] Wu J, Xie N, Wu Z, Zhang Y, Zheng YG. Bisubstrate inhibitors of the MYST HATs Esa1 and Tip60. Bioorg Med Chem 2009;17:1381–6.

[137] Choi KC, Myung GJ, Lee YH, Joo CY, Seung HK, Kang HB, et al. Epigallocatechin-3-gallate, a histone acetyltransferase inhibitor, inhibits EBV-induced B lymphocyte transformation via suppression of RelA acetylation. Cancer Res 2009;69:583−92.

[138] Morimoto T, Sunagawa Y, Kawamura T, Takaya T, Wada H, Nagasawa A, et al. The dietary compound curcumin inhibits p300 histone acetyltransferase activity and prevents heart failure in rats. J Clin Investig 2008;118:868−78.

[139] Seong AR, Yoo JY, Choi K, Lee MH, Lee YH, Lee J, et al. Delphinidin, a specific inhibitor of histone acetyltransferase, suppresses inflammatory signaling via prevention of NF-kappaB acetylation in fibroblast-like synoviocyte MH7A cells. Biochem Biophys Res Commun 2011;410:581−6.

[140] Vasudevarao MD, Mizar P, Kumari S, Mandal S, Siddhanta S, Swamy MM, et al. Naphthoquinone-mediated inhibition of lysine acetyltransferase KAT3B/p300, basis for non-toxic inhibitor synthesis. J Biol Chem 2014;289:7702−17.

[141] Guo-Bo L, Lu-Yi H, Hui L, Sen J, Lin-Li L, Sheng-Yong Y. Identification of new p300 histone acetyltransferase inhibitors from natural products by a customized virtual screening method. RSC Adv 2016;6:61137−40.

[142] Buczek-Thomas J, Hsia E, Rich CB, Foster JA, Nugent MA. Inhibition of histone acetyltransferase by glycosaminoglycans. J Cell Biochem 2008;105:108−20.

[143] Honma K, Mochizuki K, Goda T. Inductions of histone H3 acetylation at lysine 9 on SGLT1 gene and its expression by feeding mice a high-carbohydrate/fat ratio diet. Nutrition 2009;25:40−4.

[144] Friis RM, Wu BP, Reinke SN, Hockman DJ, Sykes BD, Schultz MC. A glycolytic burst drives glucose induction of global histone acetylation by picNuA4 and SAGA. Nucleic Acids Res 2009;37:3969−80.

[145] Chung S, Hwang JT, Park JH, Choi HK. Free fatty acid-induced histone acetyltransferase activity accelerates lipid accumulation in HepG2 cells. Nutr Res Pract 2019;13:196−204.

[146] Mews P, Egervari G, Nativio R, Sidoli S, Donahue G, Lombroso SI, et al. Alcohol metabolism contributes to brain histone acetylation. Nature 2019;574:717−21.

[147] Shepard BD, Tuma PL. Alcohol-induced protein hyperacetylation: mechanisms and consequences. World J Gastroenterol 2009;15:1219−30.

[148] Choudhury M, Shukla SD. Surrogate alcohols and their metabolites modify histone H3 acetylation: involvement of histone acetyl transferase and histone deacetylase. Alcohol Clin Exp Res 2008;32:829−39.

[149] Poljaková J, Groh T, Gudino ŽO, Hraběta J, Bořek-Dohalská L, Kizek R, et al. Hypoxia-mediated histone acetylation and expression of N-myc transcription factor dictate aggressiveness of neuroblastoma cells. Oncol Rep 2014;31:1928−34.

[150] Watson JA, Watson CJ, McCrohan AM, Woodfine K, Tosetto M, McDaid J, et al. Generation of an epigenetic signature by chronic hypoxia in prostate cells. Hum Mol Genet 2009;18:3594−604.

[151] Entwisle SW, Sanchez-Gurmaches J, Lawrence RT, Pedersen DJ, Jung SM, Martin-Perez M, et al. Cold-Induced thermogenesis increases acetylation on the brown fat proteome and metabolome. BioRxiv 2018;. Available from: https://doi.org/10.1101/445718 preprint.

[152] Zhou J, Geng F, Xu J, Peng L, Ye X, Yang D, et al. PM2.5 exposure and cold stress exacerbates asthma in mice by increasing histone acetylation in IL-4 gene promoter in CD4 + T cells. Toxicol Lett 2019;316:147−53.

[153] Jänsch N, Meyners C, Muth M, Kopranovic A, Witt O, Oehme I, et al. The enzyme activity of histone deacetylase 8 is modulated by a redox-switch. Redox Biol 2019;20:60−7.

[154] Imai S, Armstrong CM, Kaeberlein M, et al. Transcriptional silencing and longevity Sir2 is an NAD- dependent histone deacetylase. Nature 2000;403:795−800.

[155] Rodriguez H, Rafehi H, Bhave M, El-Osta A. Metabolism and chromatin dynamics in health and disease. Mol Aspects Med 2017;54:1−15.

[156] Canto C, Gerhart-Hines Z, Feige JN, et al. AMPK regulates energy expenditure by modulating NAD^+ metabolism and SIRT1 activity. Nature 2009;458:1056−60.

[157] Shimazu T, Hirschey MD, Newman J, He W, Shirakawa K, Le Moan N, et al. Suppression of oxidative stress by β-hydroxybutyrate, an endogenous histone deacetylase inhibitor. Science 2013;339:211−14.

[158] Nakahata Y, Kaluzova M, Grimaldi B, Sahar S, Hirayama J, Chen D, et al. The NAD + -dependent deacetylase SIRT1 modulates CLOCK-mediated chromatin remodeling and circadian control. Cell 2008;134:329−40.

[159] Braidy N, Guillemin GJ, Mansour H, Chan-Ling T, Poljak A, Grant R. Age related changes in NAD + metabolism oxidative stress and Sirt1 activity in wistar rats. PLoS One 2011;6:e19194.

[160] Robert C, Rassool FV. HDAC inhibitors: roles of DNA damage and repair. Adv Cancer Res 2012;116:87−129.

[161] Mahalingaiah PK, Ponnusamy L, Singh KP. Chronic oxidative stress causes estrogen-independent aggressive phenotype, and epigenetic inactivation of estrogen receptor alpha in MCF-7 breast cancer cells. Breast Cancer Res Treat 2015;153:41−56.

[162] Mahalingaiah PK, Ponnusamy L, Singh KP. Oxidative stress-induced epigenetic changes associated with malignant transformation of human kidney epithelial cells. Oncotarget 2017;14:11127−43.

[163] Candido EPM, Reeves R, Davie JR. Sodium butyrate inhibits histone deacetylation in cultured cells. Cell 1978;14:105−13.

[164] Sedley L. Advances in nutritional epigenetics − A fresh perspective for an old idea. lessons learned, limitations, and future directions. Epigenetics Insights 2020;13:1−20.

[165] Chen Y, Du J, Zhao YT, Zhang L, Lv G, Zhuang S, et al. Histone deacetylase (HDAC) inhibition improves myocardial function and prevents cardiac remodeling in diabetic mice. Cardiovasc Diabetol 2015;14:99.

[166] Zou F, Qiu Y, Huang Y, Zou H, Cheng X, Niu Q, et al. Effects of short-chain fatty acids in inhibiting HDAC and activating p38 MAPK are critical for promoting B10 cell generation and function. Cell Death Dis 2021;12:582.

[167] Liu Y, Li YJ, Loh YW, Singer J, Zhu W, Macia L, et al. Fiber derived microbial metabolites prevent acute kidney injury through G-protein coupled receptors and HDAC inhibition. Front Cell Dev Biol 2021;9:648639.

[168] Myzak MC, Dashwood WM, Orner GA, Ho E, Dashwood RH. Sulforaphane inhibits histone deacetylase in vivo and suppresses tumorigenesis in Apc-minus mice. FASEB J 2006;20:506−8.

[169] Myzak MC, Ho E, Dashwood RH. Dietary agents as histone deacetylase inhibitors. Mol Carcinogen 2006;45:443−6.

[170] Choi SY, Kee HJ, Jin L, Ryu Y, Sun S, Kim GR, et al. Inhibition of class IIa histone deacetylase activity by gallic acid, sulforaphane, TMP269, and panobinostat. Biomed Pharmacother 2018;101:145−54.

[171] Paul B, Li Y, Tollefsbol TO. The effects of combinatorial genistein and sulforaphane in breast tumor inhibition: role in epigenetic regulation. Int J Mol Sci 2018;19:1754.

[172] Singh AK, Bishayee A, Pandey AK. Targeting histone deacetylases with natural and synthetic agents: an emerging anticancer strategy. Nutrients. 2018;10:731.

[173] Altonsy MO, Habib TN, Andrews SC. Diallyl disulfide-induced apoptosis in a breast-cancer cell line (MCF-7) may be caused by inhibition of histone deacetylation. Nutr Cancer 2012;64:1251–60.

[174] Evans LW, Ferguson BS. Food bioactive HDAC inhibitors in the epigenetic regulation of heart failure. Nutrients 2018;10:1120.

[175] Ciesielski O, Biesiekierska M, Balcerczyk A. Epigallocatechin-3-gallate (EGCG) alters histone acetylation and methylation and impacts chromatin architecture profile in human endothelial cells. Molecules 2020;25:2326.

[176] Meeran SM, Ahmed A, Tollefsbol TO. Epigenetic targets of bioactive dietary components for cancer prevention and therapy. Clin Epigenetics 2010;1:101–16.

[177] Scuto A, Kirschbaum M, Buettner R, Kujawski M, Cermak JM, Atadja P, et al. SIRT1 activation enhances HDAC inhibition-mediated upregulation of GADD45G by repressing the binding of NF-κB/STAT3 complex to its promoter in malignant lymphoid cells. Cell Death Dis 2013;4:e635.

[178] Bagul PK, Deepthi N, Sultana R, Banerjee SK. Resveratrol ameliorates cardiac oxidative stress in diabetes through deacetylation of NFkB-p65 and histone 3. J Nutr Biochem 2015;26:1298–307.

[179] Yang J, Kong X, Martins-Santos ME, Aleman G, Chaco E, Liu GE, et al. Activation of SIRT1 by resveratrol represses transcription of the gene for the cytosolic form of phosphoenolpyruvate carboxykinase (GTP) by deacetylating hepatic nuclear factor 4alpha. J Biol Chem 2009;284:27042–53.

[180] Adams J, Klaidman L. Sirtuins, nicotinamide and aging: a critical review. Lett Drug Design Discov 2007;4:44–8.

[181] Crujeiras AB, Parra D, Goyenechea E, Martínez JA. Sirtuin gene expression in human mononuclear cells is modulated by caloric restriction. Eur J Clin Invest 2008;38:672–8.

[182] Chen D, Bruno J, Easlon E, Lin SJ, Cheng HL, Alt FW, et al. Tissue-specific regulation of SIRT1 by calorie restriction. Genes Dev 2008;22:1753–7.

[183] Qausain S, Srinivasan H, Jamal S, Nasiruddin M, Khan KA. Phosphorylation and acetylation of proteins as posttranslational modification: implications in human health and associated diseases. In: Dar TA, Singh LR, editors. Protein modificomics. Academic Press; 2019. p. 69–86.

[184] Zhang Di, Tang Z, Huang H, et al. Metabolic regulation of gene expression by histone lactylation. Nature 2019;574:575–83.

[185] Izzo LT, Wellen KE. Histone lactylation links metabolism and gene regulation. Nature 2019;574:492–3.

[186] Huang H, Zhang D, Wang Y, Perez-Neut M, Han Z, Zheng YG, et al. Lysine benzoylation is a histone mark regulated by SIRT2. Nat Commun 2018;9:3374.

[187] Zhang Y, Kutateladze TG. Diet and the epigenome. Nat Commun 2018;9:3375.

[188] Nakanishi S, Lee JS, Gardner KE, Gardner JM, Takahashi Y, Chandrasekharan MB, et al. Histone H2BK123 monoubiquitination is the critical determinant for H3K4 and H3K79 trimethylation by COMPASS and Dot1. J Cell Biol 2009;186:371–7.

[189] Sharma R, Nakamura A, Takahashi R, Nakamoto H, Goto S. Carbonyl modification in rat liver histones: decrease with age and increase by dietary restriction. Free Rad Biol Med 2006;40:1179–84.

[190] Hauck AK, Zhou T, Upadhyay A, Sun Y, O'Connor MB, Chen Y, et al. Histone carbonylation is a redox-regulated epigenomic mark that accumulates with obesity and aging. Antioxidants 2020;9:1–13.

[191] Ardito F, Giuliani M, Perrone D, Troiano G, Muzio LL. The crucial role of protein phosphorylation in cell signaling and its use as targeted therapy. Int J Mol Med 2017;40:271–80.

[192] Yuseran H, Hartoyo E, Nurseta T, Kalim H. Genistein inhibits the proliferation of human choriocarcinoma cells via the downregulation of estrogen receptor-α phosphorylation at serine 118. Clin Nutr Open Sci 2021;35:77–85.

[193] Su X, Jiang X, Meng L, Dong X, Shen Y, Xin Y. Anticancer activity of sulforaphane: the epigenetic mechanisms and the Nrf2 signaling pathway. Oxid Med Cell Longev 2018;2018 ID 5438179.

[194] Shin DS, Kim HN, Shin KD, Yoon YJ, Kim SJ, Han DC, et al. Cryptotanshinone inhibits constitutive signal transducer and activator of transcription 3 function through blocking the dimerization in DU145 prostate cancer cells. Cancer Res 2009;69:193–202.

[195] Pany S, You Y, Das J. Curcumin inhibits protein kinase Cα activity by binding to its C1 domain. Biochemistry 2016;55:6327–36.

[196] Bhutani M, Pathak AK, Nair AS, Kunnumakkara AB, Guha S, Sethi G, et al. Capsaicin is a novel blocker of constitutive and interleukin-6 - Inducible STAT3 activation. Clin Cancer Res 2007;13:3024–32.

[197] Fukuda I, Ito A, Hirai G, Nishimura S, Kawasaki H, Saitoh H, et al. Ginkgolic acid inhibits protein SUMOylation by blocking formation of the E1-SUMO intermediate. Chem Biol 2009;16:133–40.

[198] Sadakierska-Chudy A, Filip M. A comprehensive view of the epigenetic landscape. Part II: histone post-translational modification, nucleosome level, and chromatin regulation by ncRNAs. Neurtox Res 2015;27:172–97.

[199] Nelson CJ, Santos-Rosa H, Kouzarides T. Proline isomerization of histone H3 regulates lysine methylation and gene expression. Cell 2006;126:905–16.

[200] Zhang Q, Bai B, Mei X, Wan C, Cao H, Li D, et al. Elevated H3K79 homocysteinylation causes abnormal gene expression during neural development and subsequent neural tube defects. Nat Commun 2018;9:3436.

[201] Farrelly LA, Thompson RE, Zhao S, Lepack AE, Lyu Y, Bhanu NV, et al. Histone serotonylation is a permissive modification that enhances TFIID binding to H3K4me3. Nature 2019;567:535–9.

[202] Chambon P, Weill JD, Mandel P. Nicotinamide mononucleotide activation of new DNA-dependent polyadenylic acid synthesizing nuclear enzyme. Biochem Biophys Res Commun 1963;11:39–43.

[203] Van Venrooij WJ, Pruijn GJ. How citrullination invaded rheumatoid arthritis research. Arthritis Res Ther 2014;16:103.

[204] Chen Y, Sprung R, Tang Y, Ball H, Sangras B, Kim SC, et al. Lysine propionylation and butyrylation are novel post-translational modifications in histones. Mol Cell Proteom 2007;6:812–19.

[205] Wang C, Sun D, Huang X, et al. Integrative analyses of single-cell transcriptome and regulome using MAES TRO. Genome Biol 2020;21:198.

[206] Shukla S, Tekwani BL. Histone deacetylases inhibitors in neurodegenerative diseases, neuroprotection and neuronal differentiation. Front Pharmacol 2020;11:537.

Further reading

Davis CD, Ross SA. Brief critical review dietary components impact histone modifications and cancer risk. Nutr Rev 2007;65:88–94.

CHAPTER

20

Epigenetics, Stem Cells, Cellular Differentiation, and Associated Neurological Disorders and Brain Cancer

Bhairavi Srinageshwar[1], Gary L. Dunbar[2] and Julien Rossignol[1]

[1]Central Michigan University, Mt. Pleasant, MI, United States
[2]Field Neurosciences Institute, Saginaw, MI, United States

OUTLINE

Introduction to Epigenetics	382
DNA Methylation	382
Histone Methylation	382
Histone Acetylation	382
Histone Phosphorylation	382
Histone Ubiquitination and SUMOylation	382
Epigenetics and the Human Brain	383
Stem Cells	383
Eukaryotic Chromosomal Organization	384
Histones and Their Structure	384
Epigenetics and Neurological Disorders	384
The Huntingtin Gene	385
Stem Cell Therapy for HD	386
Conclusions for Epigenetics in HD	387
Impairment of One-Carbon Metabolism in PD	387
The SNCA Gene	388
The PARK Gene	388
Stem Cell Therapy for PD	388
Conclusions for Epigenetics in PD	389
MeCP2 Gene and Its Function	389
iPSC Models of RTT	390
Conclusions for Epigenetics in RTT	390
DNA Methylation in MS	392
Micro-RNA	393
Histone Acetylation	393
Stem Cell Therapy for MS	393
Conclusions for Epigenetics in MS	394
Epigenetics and Glioblastoma	394
Glioblastoma Multiforme and Stem Cell Therapy	394
Epigenetic Changes in Glioma Stem Cells	394
Methylation	394
Acetylation	395
Genes Involved in Glioblastoma	395
Hypermethylation Status of IDH 1 and 2	395
Hypermethylated Status of MGMT Gene	396
Stem Cell Transplantations in Glioblastoma	396
MSC Transplantations for GB	396
TRAIL Producing MSCs	396
Neural Stem Cell Transplantation for GB	397
Conclusions for Epigenetics in GB	397
Conclusions	397
Abbreviations	398
References	398

INTRODUCTION TO EPIGENETICS

Epigenetics is defined as structural and functional changes occurring in histones and DNA, in the absence of alterations of the DNA sequence, which, in turn, has a significant impact on how gene expression is altered in a cell [1]. The term "epigenetics" was coined by the famous developmental biologist, Cornard Hal Waddington, as "the branch of biology that studies the causal interactions between genes and their products, which bring the phenotype into being"[2]. Epigenetics bridge the gap between the environment and gene expression, which was once believed to function independently [3]. Epigenetic changes can lead to increase or decrease in gene expression, thereby activating and/or deactivating genes, depending on the nature of the epigenetic control. Some of the most important histone modifications include: (1) methylation; (2) acetylation; (3) phosphorylation; (4) ubiquitination; and (5) SUMOylation and DNA modification, including DNA methylation. These changes are discussed in detail elsewhere [3,4]–"sumoylation on page 178–179, methylation on page 174–175", but are briefly described later as an overview for this chapter.

DNA Methylation

DNA methylation and some of the histone modifications are interdependent and play an important role in gene activation and repression during development [5]. DNA methylation reactions are catalyzed by a family of enzymes called DNA methyl transferases (DNMTs), which add methyl groups to a cytosine base of the DNA at the 5′-end, giving rise to the 5′-methyl cytosine. This reaction can either activate or repress gene expression, depending on the site of methylation and it can also determine how well the enzymes for gene transcription can access the DNA that is wrapped around the histone [6].

Another type of DNA methylation that was recently found in the human genome is the N6-methyladenine (6 mA) modification. Though this modification was extensively found in the prokaryotes and in some eukaryotes, the recent discovery in human genome showed that about 0.05% of adenine residue is methylation in the human genome. Alkylated DNA repair protein AlkB homolog (ALKBH1) and N-6 adenine-specific DNA methyltransferase 1 (N6AMT1) are mammalian DNA methylase that mediates the methylation of adenine in the human genome. This new DNA modification has shown to play a role in cancer and promote tumorigenesis, the details of which is described later in this chapter [7–9].

Histone Methylation

Trimethylation of lysine at position 4 on histone 3 (H3K4me3) promotes gene transcription (i.e., gene activation), whereas trimethylation of lysine at position 27 on histone 3 (H3K27me3) inhibits gene transcription (i.e., gene silencing). Alternate gene activation and repression promote a balanced dose of gene expression, which is required for those genes involved in overall development and maturation of organisms. This ensures that only appropriate genes are turned "ON" and "OFF" at any given point of time [10]. Huntington disease (HD), Parkinson disease (PD), and multiple sclerosis (MS) are some of the diseases that are caused by abnormal DNA methylation pattern. These are discussed in more detail in later sections of this chapter.

Histone Acetylation

Histone acetylation involves the acetylation of lysine residue, which is catalyzed by the enzymes, histone acetyltransferases (HATs), and histone deacetylases (HDACs), which have opposing effects on each other. HATs transfer acetyl groups to lysine, whereas HDACs remove acetyl groups from lysine. However, presence or absence of an acetyl group (CH_3CO) on a lysine residue alters the charge on the amino acid and can decrease the interaction of the N-terminal region of histones with the negatively charged phosphate groups of DNA. These events are involved in transformation of the condensed chromatin into a more relaxed structure, which can induce gene expression [11].

Histone Phosphorylation

Phosphorylation involves the addition of phosphate group to threonine, serine, and tyrosine residues. Phosphorylation of serine at position 139 on histone 2 (H2) occurs as a response to DNA damage during cell cycling. This relaxes the chromatin; thereby, the proteins or factors responsible for repairing the damaged regions of the DNA have greater access to the DNA, which aids in the recovery of the DNA damage. Moreover, phosphorylation of threonine and serine residues on histone 3 (H3) facilitates regulation of gene expression [12]. Histidine phosphorylation is one of the epigenetic modifications occurring in the prokaryotic cells and in lower eukaryotes that plays a major role in cell signaling. The histidine is phosphorylated at the imidazole ring, but occurs only on those nitrogen atoms that are unprotonated [13].

Histone Ubiquitination and SUMOylation

Ubiquitination and SUMOylation are associated with posttranslational modifications that regulate transcription of gene and protein translation activities. It is well established that the addition of ubiquitin molecules to proteins facilitate targeted protein degradation. In addition to ubiquitin, various small ubiquitin-like molecules (SUMO) are observed in cells, known as small ubiquitin-related modifier. These molecules have activities, which are similar to ubiquitin and can attach covalently to those proteins, which are involved in changing chromatin structure and gene expression [3].

EPIGENETICS AND THE HUMAN BRAIN

Generation of neurons and glial cells from progenitor cells involve epigenetic mechanisms that take place throughout the developmental stages of the brain. The outer layer of the embryo, called ectoderm, forms the central nervous system, and during the process of development, DNA methylation controls the epigenetic mechanisms of the embryonic cells. For example, to prevent the differentiation of nonneuronal cells into mature neurons, the proneural genes, as well as gene which are associated with proteins involved with neurogenesis, such as the brain-derived neurotrophic factor (*BDNF*) gene, are silenced by DNA methylation at their promoter region. However, DNA remethylation can take place on a neuronal gene, such as Sox2 [14], which allows for the selective initiation of neuronal development. Similarly, during initial development of the cortex, the genes involved in the formation of glial cells are suppressed by DNA methylation, promoting the formation of more neurons during the early stages of neuronal development. Eventually, during the later stages of cortical development, the DNA methylation is reversed, leading to the generation of glial cells [15,16]. Some of the genes involved in postnatal neurogenesis are *Sox2*, *Dlx2*, *Sp8*, and *Neurog2* and those involved in the formation of glial cells are *Sparcl1* and *Nkx2−2*. It has been shown that during the differentiation of neural stem cells (NSCs) or progenitor cells, DNA methylation is facilitated by DNMT3a, which silences the genes *Sparcl1* and *Nkx2−2*, leading to the inhibition of glial cell formation and promotion of mature neuron development from NSCs [17]. It is also believed that the normal aging process in humans is associated with modification of the epigenome in the brain, affecting certain genes related to neurogenesis, especially within the cortex [18]. This process leads to the disruption of synapses, abnormal neurotransmission, and is associated to age-related disorders [19]. However, a comprehensive description of the epigenetic mechanisms related to aging is beyond the scope of this chapter.

Stem Cells

Dysregulation of epigenetic mechanisms has a direct impact on gene expression patterns that lead to abnormal gene functions, which form the basis of most of the genetic disorders (monogenic or polygenic) in humans. Environmental stress can also alter epigenetic mechanisms, which could become a cause for the predisposition of certain diseases, such as autism, schizophrenia, and congenital heart disease [20]. This chapter focuses on the role of epigenetics and epigenetic changes that take place during stem cell differentiation, which can be used as a potential therapy for neurological diseases. Stem cells have a unique property of proliferation, differentiation, and self-renewal. Stem cell plasticity is an important characteristic and is based on the degree of pluripotency, which is the ability of a cell to differentiate into another cell lineage. Stem cells can be classified as: (1) totipotent, such as embryonic stem cells (ESCs) of the morula, which can be differentiated into any cell type, including placental cells; (2) pluripotent, such as induced pluripotent stem cells (iPSCs), which can be differentiated into any cell type, except for placental cells; and (3) multipotent, such as mesenchymal stem cells (MSCs) and neural stem cells (NSCs), which can be differentiated into many, but not all, cell types.

ESCs are highly unspecialized and can form any type of specialized cells under appropriate conditions and environments. ESCs divide and renew themselves, which make them appropriate candidates for regenerative medicine or cell replacement therapy [21]. Previously, we have reviewed the role of epigenetics in MSCs, NSCs, and iPSCs and their association with neurodegenerative diseases [22]. In addition, the basics of stem cells, and the various epigenetic mechanisms associated with them, are explained comprehensively in the previous edition of this book. Therefore, the aim of the present chapter is to discuss the role that epigenetics of stem cells might play in a subset of neurological diseases.

Some of the genetic disorders and glioblastoma that occur as a consequence of abnormal epigenetic mechanisms include: (1) HD; (2) PD; (3) Rett syndrome (RTT); (4) spinocerebellar ataxia (SCA); and (5) MS. Currently, stem cell-based therapies are being tested as a potential treatment for these diseases. A sampling of these treatment strategies are outlined in this chapter and include examples of: (1) transplants of bone marrow-derived mesenchymal stem cells (BM-MSCs) for treating HD; (2) transplants of neural stem cells (NSCs) for treating PD; (3) transplants of induced pluripotent stem cells (iPSCs) for treating RTT; (4) transplants of umbilical cord-derived mesenchymal stem cells (UC-MSCs) for treating SCA; (5) transplants of hematopoietic stem cells (HSCs) for treating MS; and (6) transplants of BM-MSCs, adipose-tissues-derived MSC (AT-MSCs) and induced neural stem cells (iNSCs) for treating glioblastoma. These treatment

TABLE 20.1 An Overview of Various Epigenetic Mechanisms Associated with Neurodegenerative Diseases and Glioblastoma

Neurodegenerative disease	Stem cell-based therapy	Epigenetic mechanism involved with disease
HD	BM-MSCs	Histone 3 (H3) methylation leading to reduced trophic factors [23]
PD	NSCs	DNA methylation leading to metabolic defects [24]
RTT	iPSCs	Point mutations in MeCP2 gene leading to defective epigenetic regulatory molecules [25]
SCA	UC-MSCs	Methylation and acetylation of histone leading to reduced RNA expression [26,27]
MS	HSCs	DNA methylation, histone acetylation and posttranscriptional modification by miRNA leading to compromised immune response [28–30]
GB	BM-MSCs, AT-MSCs iNSCs	Known epigenetic roles in glioma stem cells [31,32]

HD, Huntington disease; MS, multiple sclerosis; PD, Parkinson disease; RTT, Rett syndrome; SCA, spinocerebellar ataxias, GB, Glioblastoma; HSC, hematopoietic stem cell; NSC, Neural stem cell.

strategies and the epigenetic mechanisms involved with these disorders are summarized in Table 20.1.

To better understand the role of epigenetics in neurological disorders, a working knowledge about histones, chromosomes, and the general hierarchy in the chromosomal organization, is needed, which is briefly reviewed in the next section.

Eukaryotic Chromosomal Organization

The strands of DNA are about 2 nm wide and are packed around the core of four pairs of histone proteins (H2A, H2B, H3, and H4), forming the nucleosome, which is the first level of chromosomal organization (Fig. 20.1). These nucleosomes are the building blocks of the chromatin structure, which is about 30 nm in diameter. The formation of chromatin structure involves the fifth histone, H1, which is near the adjacent nucleosome, thereby compacting the nucleosome or chromatin to form chromatin coils, which are about 300 nm in diameter. These fibers are further condensed to make loops of 700 nm in diameter, which, in turn, form the intact metaphase chromosomes, which are about 1400 nm wide [33].

Histones and Their Structure

The genome consists of two molecules of each histone protein (H2A, H2B, H3, and H4) giving rise to an octamer, which is made of about 130 amino acids. The nucleosome, as discussed before, consists of DNA, that is, 146–147 nucleotides long, making about 1.65 turns around the octamer. The core histones are highly conserved in eukaryotes, having a "tail" at their N-terminal end where the epigenetic modifications, such as methylation, acetylation, and/or phosphorylation, take place. These regulate the chromatin structure, which has an impact on recruiting various proteins involving activation and repression of gene expression [33,34]. Defects in chromatin organization and deficiency of enzymes lead to various forms of human diseases, such as RTT, Rubinstein–Taybi syndrome, and Coffin–Lowry syndrome [35].

Epigenetics and Neurological Disorders

Huntington Disease and Mesenchymal Stem Cells

Huntington disease (OMIM #143100) is a devastating, fatal, autosomal dominant neurodegenerative disorder, which most profoundly affects the striatal region of the brain. The disease is characterized by cognitive, motor, and psychiatric disturbances [36]. There is atrophy and loss of the medium spiny GABAergic neurons in the caudate and putamen regions of the striatum, which leads to motor and cognitive impairment [37]. The disease is due to the expansion of CAG repeats in the Huntingtin gene (HTT), which produces mutant huntingtin protein (mHTT) that is highly toxic, leading to the signs and symptoms of the disease [38].

Currently our laboratory is investigating the role of transplanting BM-MSCs and UC-MSCs as a potential treatment for HD. We have shown that BM-MSCs create an optimal microenvironment in the striatum that slows the progression of neuronal loss and dysfunction by restoring the various neurotrophic factors, including BDNF, which is down regulated in HD. In our previous studies, we transplanted BM-MSCs, which were genetically altered to overexpress BDNF, into the striata of YAC128 mice (a slowly progressing, transgenic HD mouse model, which carries the entire human mHTT gene) and R6/2 mice, (fast-progressing, transgenic HD mouse model, carrying exon 1 of human mHTT gene), and observed profound neuroprotective effects, including a significant reduction in the motor symptoms of the disease. These observations suggest

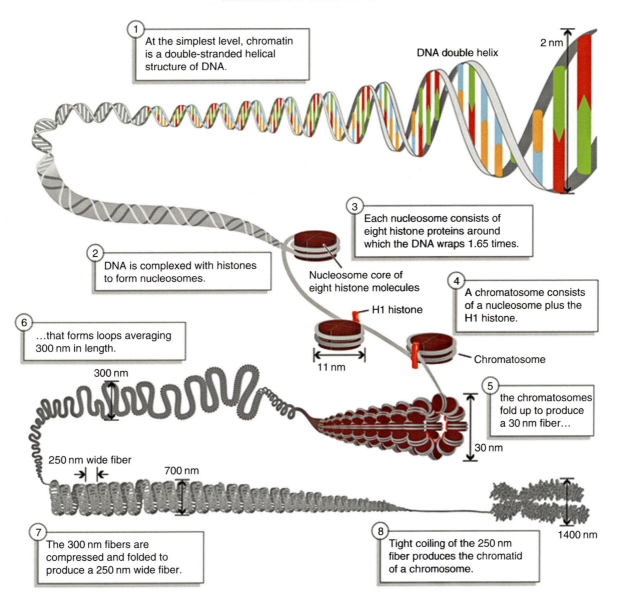

FIGURE 20.1 The eukaryotic chromosomal organization. The steps showing the chromosomal organization involved in condensation of a 2 nm wide DNA wrapping around the histone molecules to form the nucleosome, which in turn twists to form coils (30 nm) and loops (300 nm) that further condenses into a 1400 nm wide chromosome [33].

that manipulation of BM-MSCs could be a potential therapeutic for alleviating HD pathology [39–41].

The *Huntingtin* Gene

The normal HTT protein is localized in the cytoplasm, whereas mHTT is found in the nucleus, as well as in the cytoplasm. Though both wild type HTT and the mHTT can interact and inhibit acetyltransferase function, the actual presence of the mHTT in the nucleus was shown to be specifically responsible for interfering with the acetylation of histones [42]. The cognitive symptoms observed in HD are linked to the hypermethylated status of certain genes, such as *Sox2* and *Pax6*. Because these genes play vital roles in the proliferation and maintenance of neural stem cells, which eventually become neurons, alterations of their structure may contribute to subsequent dysfunctions, such as an impairment of hippocampal neurogenesis [43]. Similarly, the hypoacetylation of certain genes, such as *BDNF* gene, has been found in HD patients and animal models of HD. The CREB binding protein (also known as CREBBP) is a histone acetyltransferase and a transcriptional cofactor that regulates histone acetylation and gene activation. When the mHTT interacts with the CREB binding protein, it loses its transcriptional activator and HAT functions, which lead to hypoacetylation of histones. This in turn, leads to transcriptional dysregulation of certain genes in the neurons in HD brain [44]. Later, it was found that defects, other than DNA methylation and acetylation, are involved in HD. Lee and co-workers [45] investigated the role of microRNA molecules (miRNA) and found lower levels of 9 types of miRNA in 12-month-old YAC128 and 10-week-old R6/2 mice. Understanding the defective epigenetic mechanisms in HD has led to the use of HDAC inhibitors and miRNA as potential treatments for this disease.

HD and BDNF. BDNF is an important neurotropic factor, which is expressed abundantly in different brain regions. H3K4me3 is a histone involved in transcription of the *BDNF* gene, which is significantly reduced in the cortex of the HD brain. BDNF is expressed in the cortical neurons that project into the striatum and has been shown to be essential for the survival of striatal neurons. As mentioned earlier, histone methylation leads to gene silencing and histone acetylation leads to gene activation. However, H3K4 is an exception because either methylation or acetylation of this histone leads to gene activation. Modification of H3K4me3 is widely studied and is of interest to researchers because H3 is associated with the promoters of the genes that are actively transcribed (e.g., *BDNF*) [46]. Research using chromatin immunoprecipitation (ChIP) analysis has revealed a correlation between *BDNF* expression and H3K4me3 levels in HD [23]. In the brain, repressor element-1 silencing transcription factor/neuron-restrictive silencer factor (REST/NRSF) is the main factor that recruits other cofactors to regulate neuronal gene expression. REST, a transcriptional repressor, plays a major role in the regulation of the *BDNF* gene. Under normal circumstances, REST binds with the HTT in the cytoplasm, but in the case of mHTT, there is no interaction between the REST and HTT (Fig. 20.2). This leads to nuclear translocation of the REST, which then inhibits *BDNF* expression, resulting in neurodegeneration.

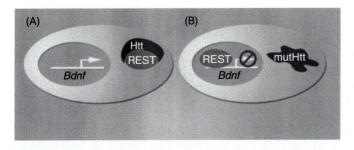

FIGURE 20.2 Role of REST in down-regulation of *BDNF* in HD. (A) The wild-type HTT protein binds to the REST protein in the cytoplasm, thereby preventing the REST molecule to bind to the *BDNF* promoter. (B) The mHTT fails to bind to the REST which causes REST to bind to the *BDNF* promoter and inhibits trophic factor transcription leading to reduced BDNF expression as seen in the HD brain [47].

REST not only targets the *BDNF* gene, but also influences other neuronal genes that are down regulated in HD [46].

These findings show that rescue of REST-regulated genes may prove to have a promising therapeutic effect on HD. H3K4me3 levels are significantly lower at the REST region of the *BDNF* gene in the cortex when comparisons are made between 8- and 12-week-old R6/2 mice. H3K4me3 not only regulates *BDNF* gene expression, but also has an impact on other genes, a postulate that was confirmed by transcriptome analysis and genome-wide analysis, which revealed that there are about 98 major genes showing differential expressions in the cortex and the striatum in 12-week-old R6/2 mice. In the cortex, the differentially expressed genes are associated with neurotransmission (e.g., *Grla3*, *Grm4*, and *Bagra5*), G-protein signaling (e.g., *Rgs9* and *Arpp21*), synaptic transmission (e.g., *Snap25* and *Rph3a*), inflammation (e.g., *C4a* and *Dusp6*), and calcium signaling (e.g., *Scn4b*, *Hpca*, and *Itpr1*). In the striatum, the differentially expressed genes are associated with neurotransmission (e.g., *Drd2*, *Grm3*, and *Gabrd*), G-protein signaling (e.g., *Rgs9* and *Arpp21*), synaptic transmission (e.g., *Snapr* and *Dlg4*), and calcium signaling (e.g., *Scn4b*, *Hpca*, and *Itpr1*) [23].

Stem Cell Therapy for HD

MSCs are adult stem cells that are abundantly present in bone-marrow (BM). MSCs can also be derived from umbilical cord (UC) and adipose tissue (AT). The MSCs, derived from BM and AT, have a greater survival rate, when compared to the MSCs derived from other sources [48]. BM-MSCs can differentiate into osteogenic, adipogenic, and chondrogenic lineages. However, by triggering an epigenetic mechanism, these MSCs can differentiate into a neuronal-like lineage. This can be achieved by passaging the MSCs and every time the MSCs get passaged, a series of methylation and acetylation reactions take place [49].

Recently, we have found that MSCs at higher passages (passaged about 40–50 times prior to transplantation) have more therapeutic efficacy for alleviating motor symptoms in the R6/2 mice, compared to lower-passaged MSCs (passaged about 3–8 times prior to transplantation). This study showed that passaging the cells up to 40–50 times produces a subpopulation of MSCs that have the potential to create an optimal environment within the transplantation site in the striatum, as well as generating BDNF, which is usually deficient in HD brains [41]. Genes that are associated with steering the MSCs toward osteogenic, adipogenic, and chondrogenic lineages, such as *osteopontin (OPN)*, *peroxisome proliferator-activated receptors gamma 2 (PPAR-γ2)*, and *fatty acid binding protein 4 (FABP4)*, undergo methylation and histone acetylation, as the cells are passaged, thereby reducing differentiation into osteogenic, adipogenic, or chondrogenic lineages [49,50]. The acetylation status of H3K9 at the promoter regions of these genes undergoes changes, thereby leading to less activation [51]. Therefore, when higher passaged MSCs are transplanted into animal models, the environment at the transplantation site is more favorable for increasing BDNF, which is required for neuronal survival and alleviating symptoms of HD [41].

Similarly, in another study, we genetically modified the BM-MSCs to overexpress BDNF and transplanted these cells into the striatum of the YAC128 mice. The mice, which received BM-MSCs that overexpressed BDNF showed improvement in motor coordination, compared to YAC128 mice which did not receive BM-MSCs that overexpressed BDNF [39]. Interestingly, one of our previous studies involved transplantation of UC-MSCs into the striata of the R6/2 mice and though these mice showed reduction in both spatial memory and motor deficits, the extent of behavioral sparing was slightly higher when BM-MSCs were used, suggesting that the source of the MSCs may affect their efficacy when transplanted [40].

Conclusions for Epigenetics in HD

The epigenetic alterations of genes, such as *OPN*, *PPAR-γ2*, and *FABP4*, steer MSCs away from osteogenic, adipogenic, and chondrogenic lineages as a function of cell passaging and may play an important role in driving these stem cells into a neuronal-like lineage. The aforementioned studies show that MSCs have a therapeutic effect on HD and by utilizing higher-passaged MSCs, the transplants appear to be more efficacious than using lower-passaged MSCs. However, the higher-passaged MSCs may have less clinical utility, because they are more susceptible to other epigenetic effects and tumor formation following transplantation, as they have been shown to carry chromosomal abnormalities that may adversely affect survivability and successful engraftment at the transplantation site [52]. Therefore, even though higher-passaged MSCs may have a stronger therapeutic effect for HD, it is important to study the epigenetic mechanisms of MSCs at different passages to select a subpopulation of cells that will show their maximum therapeutic effects, without adverse effects.

Epigenetic mechanisms play an important role in the dysregulation of genes that are associated with the secretion of trophic factors, such as BDNF, as described previously. Therefore, targeting epigenetic markers and improving the expression of *BDNF* may prove to be beneficial in alleviating the signs and symptoms of HD [23].

Although various studies have described histone modifications and histone variants that are associated with HD, including phosphorylation of histone 2 variants observed in HD cell line and in R6/2 mouse model [53], a detailed description of this and many other histone variations is beyond the scope of this chapter.

Parkinson Disease and Neural Stem Cells

Parkinson disease (OMIM #168600) is a late-onset neurodegenerative disease, mainly affecting individuals of about 65–85 years of age. The disease is characterized by impairment of both motor and non-motor symptoms, including rigidity, bradykinesia, tremor, postural instability, depression, abnormal sleep patterns, cognitive dysfunction, and autonomic insufficiency [24,54]. PD is due to the degeneration of dopaminergic neurons in substantia nigra, pars compacta (SNpc). The genetic factors, such as mutations in *PARK* genes and environmental factors, such as aging and exposure to neurotoxins, contribute to the disease [55,56]. As such, PD can be caused either by sporadic mutations or can be inherited. Major gene candidates that are associated with PD, include *PARK* genes, leucine-rich repeat kinase 2 gene (*LRRK2*), and the α-synuclein gene (*SNCA*). The mutations in *PARK* gene family (*PARK 1–15*) and *SNCA* show Mendelian inheritance patterns, suggesting familial PD. Sporadic PD is caused by variants found in *SNCA* and *LRRK 2* genes. PD also shows polygenic and complex inheritance patterns, combined with environmental factors [24].

Impairment of One-Carbon Metabolism in PD

The group of metabolic reactions consisting of various enzymes and coenzymes that are involved in various biological functions that involve lipid metabolism, redox reaction, and methylation reaction is known as one-carbon metabolism. These metabolic activities take place by utilizing glucose, amino acids, such as serine and glycine, and vitamins, such as B6 and B12 [57]. Therefore, impairment in DNA methylation is a part of one-carbon metabolism that is found in PD.

As mentioned earlier, DNA methylation is one of the major epigenetic modifications that ensures condensation or relaxation of chromatin structure, depending on the methylation status of the DNA that is wrapped around the histones. The chemical reaction is catalyzed by DNMT1, where cytosine gets methylated to form 5′methyl-cytosine, a reaction that involves two major molecules, S-adenosylmethionine (SAM), and S-adenosylhomocysteine (SAH). SAM is a universal methylating agent produced from folate and homocysteine (HCY), which methylates histones and DNA [24]. SAM, SAH, and HCY are some of the biomarkers of the metabolic pathway associated with the one-carbon metabolism [58]. Hence, the DNA methylation potential depends on the levels of SAM and the potential of DNMT1 to catalyze the reaction and transfer the methyl group to cytosine. This reaction was found to be impaired in many neurodegenerative diseases, especially in genes associated with Alzheimer disease (AD) and PD. Due to defects in one-carbon metabolism, the rate of methylation decreases, leading to increased gene expression. An example of such defects in metabolism was found during the analysis of methylation status of the *SNCA* gene in PD patients [24].

The SNCA Gene

The α-synuclein protein is important for dopaminergic neurogenesis during early embryonic development. The Lewy bodies found in PD consist of SNCA and protein inclusion bodies, which lead to disease pathogenesis. Lower levels of SNCA are associated with loss of dopaminergic neurons during embryonic stage, whereas increased expression of SNCA may be a risk factor or a threat during later stages [59]. The genetic abnormalities associated with *SNCA* that lead to PD are the point mutations and copy number variants. Previous studies have reported that the mutation of the *SNCA* genes found in dopaminergic neuron results in decreased methylation, leading to mono-allelic expression of the gene. However, mRNA levels from this single allele exceed that of the control subjects having normal biallelic expression. Methylation status of the intronic regions (human intron 1 having 66 CpG sites) of the *SNCA* gene was found to be similar to those reported in previous studies, especially in reference to the cortex and substantia nigra regions of the brain [60,61]. The miRNA also plays a role in regulating the gene expression of SNCA. Doxakis [62] has analyzed two of the major miRNA of the brain, mi-RNA7 and mi-RNA153, and found that overexpression of these miRNAs in neuronal culture lowered the SNCA levels.

The PARK Gene

The majority of *PARK* gene mutations are also associated with juvenile form of PD. Cai and co-workers [63] investigated about 33 CpG regions on the *PARK* gene promoter in three groups of individuals, including PD patients carrying *PARK* gene mutations, PD patients without *PARK* gene mutations, and age-matched controls. DNA methylation analysis revealed that there was no significant difference in methylation status between the three groups, which suggests that *PARK* gene methylation does not contribute in the PD pathogenesis.

Stem Cell Therapy for PD

More than half of the dopaminergic neurons are lost in SNpc before the actual onset of PD [64]. There are various studies and literature reviews that have investigated the role of neural stem cells in PD [54,59,65] and the importance of neurotrophic factors, particularly glial cell line-derived neurotrophic factor (GDNF). The NSCs are multipotent stem cells that are specifically found in the subventricular zone (SVZ), subgranular zone (SGZ), and the dentate gyrus (DG) of the hippocampus. The environment at these regions are the most favorable for the differentiation of NSCs into neurons [66]. Sanberg [65] has discussed the role of transplantation of undifferentiated NSCs into the striatum of a primate model of PD and indicated that NSCs were able to survive and migrate to the site of neurodegeneration and replace the lost neurons. The animals recovered from their behavioral deficits. Redmond and co-workers [67] also transplanted undifferentiated NSCs into the primate model of PD and found that though, these undifferentiated NSCs had therapeutic effect, only a small population of these NSCs partially differentiated into dopaminergic neurons, due to a less-than optimal microenvironment at the site of transplantation. These studies indicate that the partially differentiated NSCs migrate to the substantia nigra, through the nigrostriatal pathway, following their unilateral transplantation into the striatum. However, these finding suggest that cell replacement therapy provides only minimal neuroprotective effects.

Other approaches, such as dopamine replacement therapy and deep brain stimulation (which decrease tremor and rigidity), also failed to show neuroprotective effects. GDNF is well known to increase survivability of dopaminergic neurons, and delivering GDNF into the brain of PD animal models, such as MPTP (1-methyl-4-phenyl-1,2,3,6-tetrahydropyridine)-treated rhesus monkey [68] and 6-OHDA (6-hydroxydopamine) injected rats [69,70], were shown to be neuroprotective. Overall, GDNF is critical for establishing the nigrostriatal dopamine system during development and plays an important role in protecting dopaminergic neurons from degeneration by maintaining their morphology and neurochemical and biochemical reactions that are taking place in them, as well as ensuring proper neuronal differentiation and long-term survivability of neurons [71].

Open-labeled clinical trials using GDNF showed tolerance and clinical benefits in patients within 3 months of treatment, but randomized clinical trials failed to reveal significant benefits [72]. Deng and co-workers [73] have shown that the co-transplantation of dopaminergic neurons and NSCs can reduce motor symptoms in a rodent model of PD. However, to increase the number of NSCs that differentiate into dopaminergic neurons, the overexpression of *nuclear receptor (Nurr1)* and factors, derived from local type 1 astrocytes, are necessary. Wagner and co-workers [74] showed that NSCs having these factors were able to differentiate into dopaminergic neurons, compared to NSCs that did not express these factors. Similarly, *Nurr1* and *Pitx3* are needed to produce dopaminergic neurons from embryonic stem cells. Nurr1 is a nuclear hormone receptor that is involved in the dopaminergic neurogenesis, while Pitx3 is a transcription factor that is important for the differentiation and maintenance of dopaminergic neurons in the mid-brain. Nurr 1, along with Pitx3, influences the expression of some of the genes involved in production of dopamine, such as *tyrosine hydroxylase (TH)* and *dopamine transporter (DAT)*, which are associated with dopamine signaling [75]. Nurr1 is usually present in a silenced state when not combined with Pitx3. This is due to the binding of silencing mediator of retinoic acid and thyroid hormone receptor (SMRT) that leads to HDAC-mediated silencing of Nurr1. However, in the presence of Pitx3, the binding of SMRT is reduced and Nurr1 is activated, thereby the Nurr1-Pitx3 complex can bind to the promoter of *TH* and *DAT* genes and activate them. Therefore, Nurr1, on its own, cannot activate the target genes associated with dopamine production [76].

Conclusions for Epigenetics in PD

The familial form of PD involves defects in epigenetic mechanisms, such as one-carbon metabolism reaction, which is associated with defects in rate of DNA methylation that leads to PD. The rate of transfer of methyl group to the cytosine by DNMT1 has been highly reduced, thereby interfering with the one-carbon metabolism, which was confirmed by measuring SAM and SAH, the biomarkers of the one-carbon metabolism associated with DNA methylation. Similarly, abnormal increase in expression of certain miRNAs leads to decreased *SNCA* gene expression, which is also associated with this disease. Cell replacement therapy to increase production of GDNF through transplantation of partially differentiated NSCs has proven, thus far, to have only limited efficacy. To achieve conversion of NSCs into dopaminergic neurons, the expression of Nurr1 and Pitx3 are required. The complex formed between Nurr1 and Pitx3 is associated with increased gene expression of *TH* and *DAT*, which, in turn, increases the production of dopamine. Therefore, complexing Nurr1 and Pitx3 is necessary, since Nurr1, *per se*, is considered to be in a silenced state, and binding with Pitx3 leads to the activation of the complex.

Rett Syndrome and Induced Pluripotent Stem Cells

Rett syndrome (OMIM #312750) is a X-linked autosomal dominant progressive neurodevelopmental disorder affecting, predominantly, the female population and classified as one of the autism-spectrum disorders (ASDs). The classical symptoms of this disease include speech disability, stereotypic hand use, autistic characteristics, and seizures that gradually develop after 18 months of age [25]. The cause of the disease is associated with point mutations in *methyl-CpG binding protein 2 gene* (MeCP2), although there are other sets of genes and environmental factors that contribute to the onset of this disease. Most of the mutations observed in *MeCP2* gene are point mutations (missense or nonsense). Some of the hotspots include: (1) p. R133C and p. T158M, found in the methyl binding domain (MBD); and (2) p. R306C, p. R168X, p. R294X, and p. R255X located in transcriptional repression domain (TRD) [77–80]. The *MeCP2* gene has a major role to play in the epigenetic regulation of various gene expressions related to ASD [77]. The domains of *MeCP2* gene, such as the MBD and TRD, are involved in chromatin remodeling and protein interactions, respectively [81].

MeCP2 Gene and Its Function

MeCP2 gene is involved in coding an epigenetic regulatory molecule, and mutations or large-scale deletions, duplications, and insertions of *MeCP2* cause RTT syndrome. RTT can be classified into two categories, either atypical RTT or classical RTT. More than 95% of the patients having mutation in *MeCP2* gene are considered having classical RTT. In general, *MeCP2* binds to the DNA via the MBD and silences the gene. Similarly, the MeCP2 protein helps with chromatin remodeling by binding to the DNA via its TRD. The methylation of histone 3 at lysine position 9 (H3K9me) is achieved by *MeCP2*, thereby silencing the gene to which it is bound [25]. However, Yasui and co-workers [82] have extensively studied the MeCP2 binding sites on the genes and found that only about 6% of the CpG islands are bound by MeCP2. Their study indicated that: (1) the main function of MeCP2 is not associated with silencing the methylated regions of the gene and (2) the genes having the maximum methylation status are not bound by MeCP2. Previous publications have shown that the human genome has methylated cytosine as 5-hydroxymethylcytosine (5hmC) and 5-methylcytosine (5mC). It has been shown that

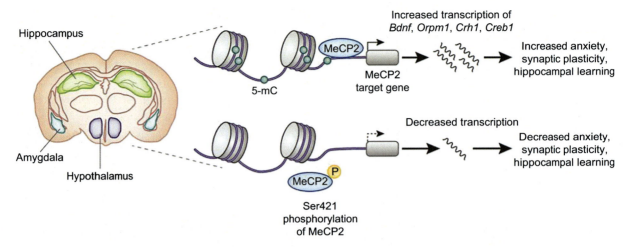

FIGURE 20.3 Function of MeCP2 gene. MeCP2 regulates the transcription of certain genes in the brain by binding to 5-methylcytosine (5mC). The main function of MeCP2 gene is not just silencing the gene to which it is bound, but also has the ability to increase the transcription of genes, such as BDNF, ORPM1, and CREB 1 [18].

5hmC is found abundantly in neuronal genes that are active and that MeCP2 has a very high affinity toward 5hmC compared to 5mC, which plays a major role in how the gene expressions are regulated in neurons (Fig. 20.3).

An interesting finding is that the MeCP2 competes with the histone, H1, to bind to the nucleosome, indicating that the levels of H1 and MeCP2 are not always corelated with each other, especially in neurons. The finding that *MeCP2* gene is associated with activating genes when bound to 5hmC, as well as with silencing the genes when linked with H3K9, underlies the dual nature of the protein. Therefore, there are some genes that are down regulated when MeCP2 is lost and upregulated with increased MeCP2 gene expression [83].

iPSC Models of RTT

Induced pluripotent stem cells have been used as a cell model for RTT [84]. Takahashi and Yamanaka [85] first reprogrammed the somatic cells into iPSCs by overexpressing four major genes, such as *Oct3/4*, *Sox2*, *Klf4*, and *c-Myc*, which are now collectively known as the Yamanaka factors. Reprogramming of iPSCs involves loss and gain of DNA methylation on H3 at lysine positions 27 (H3K27me3) and 4 (H3K4me3), as the cells get transformed from somatic stage, to pluripotent stage as has been discussed in detail in our previously work [22].

The iPSCs derived from the RTT patients have been reprogrammed to form neurons, which show significant pathological changes, such as reduced nuclear size, lower expression of neuronal markers, reduced dendritic spine density, loss of synapses, and lower levels of intracellular calcium. Electrophysiological analysis of neurons obtained from the RTT-derived iPSCs show decreased excitatory and inhibitory postsynaptic potential (EPSP and IPSP). Although cell replacement research for RTT using iPSC transplantation has not been translated to the clinic, the in vitro remodeling of fibroblasts derived from the RTT patients to form iPSCs have been successfully achieved by Marchetto and coworkers [86]. These findings are paving the way toward a better understanding of the pathophysiology of the disease and for identifying drugs or treatments that are patient- or mutation-specific [86,87].

Conclusions for Epigenetics in RTT

In general, it has been assumed that MeCP2 usually silences the gene to which it is associated, but subsequent studies have indicated that this might not always be the case. The *MeCP2* gene can increase, as well as decrease, the gene expression depending on the methylation status, thereby emphasizing the dual role of the gene. Although both RTT syndrome and iPSCs have well-known and strong epigenetic components, the use of iPSCs for treating RTT has not been investigated. The stem-cell-based model of RTT is very useful to study a specific mutation leading to a phenotype-genotype correlation and correcting the mutated RTT-iPSCs, in vitro, and then transplanting them may prove to be a future treatment for RTT. Given that highly specific and targeted cell replacement therapy can be achieved using the corrected iPSCs, utilizing epigenetically corrected iPSCs for treating RTT syndrome is worthy of further investigation as such an approach has significant promise.

The Spinocerebellar Ataxia and Mesenchymal Stem Cells

SCAs are a group of neurodegenerative disorders that are caused by trinucleotide repeat expansions (mainly CAG expansions that lead to elongated polyglutamine tracts). There are about 30 different genes responsible for the disease that is inherited in an autosomal dominant pattern. SCA patients have neuronal degeneration in cerebellum, brain stem, and spinal cord. The main characteristics of this disease are related to retinopathy, neuropathy, cognitive dysfunction, and dementia [88]. Unfortunately, there is no cure for SCAs. Approximately 28 different types of SCAs have been discovered so far, with the most common forms being SCA1, SCA2, SCA3, and SCA7. A detailed description of the different types of SCAs is discussed by Paulson [89]. Interestingly, SCA1, SCA2, SCA3, SCA6, SCA7, and SCA17 are due to the repeat expansion in the coding regions of the gene; whereas, SCA8, SCA10, SCA12, and SCA31 have repeat expansions in the non-coding regions of the gene [89].

SCA1 (OMIM #164400)

SCA1 mainly affects the brain stem and Purkinje cells of cerebellum and is characterized by ataxia of limbs and abnormal gait, leading to chorea. Assessments of expanded Ataxin1 gene having 82 CAG repeats in the cerebellar Purkinje cells in SCA1 animal models have shown reduced gene expression, which is involved in signal transduction and calcium homeostasis [90]. Ataxin 1 interacts with two major proteins, the retinoid acid receptor-related orphan receptor α (RORA, a type of HAT), and acetyltransferase tat-interactive protein 60 (Tip60, MW 60 kDa), which is a nuclear receptor coactivator that helps in the interaction of RORA (using ATXN1 with HMG-box protein 1 [AXH] domain of ATXN1). However, in the presence of the mutant Ataxin 1 gene or protein, the interactions with Tip60 and RORAα are disrupted, which lead to disease pathogenesis [91].

SCA7 (OMIM # 164500)

The symptoms associated with SCA7 include cerebellar neurodegeneration and retinal degeneration. Similar to SCA1, SCA7 also interacts with transcription factor cone-rod homeobox protein (CRX). CRX is a transcription activator of genes that are involved in formation of photoreceptors in the eyes. The interactions of mutant SCA7 with CRX cause abnormal formation of the photoreceptors, leading to retinal degeneration. Although the actual function of *Ataxin 7* gene is unknown, it has recently been shown that Ataxin 7 is a subunit of HATs [26]. Therefore, the mutant form of the protein is involved in the alteration of HATs, which eventually leads to the disruption of histone acetylation.

SCA8 (OMIM # 608768)

SCA8 is due to the combined repeat expansion of CTG and CAG in the *Ataxin 8* gene. Chen and co-workers [27] studied the SCA8 cell line, or transcript known as ATXN8OS. Epigenetic analysis of this transcript revealed increased levels of H3K9me2 and hypoacetylation of H3K14 that led to repression of ATXN8OS RNA in cell lines that have 157 repeats. Similarly, methylation of arginine residues and phosphorylation of serine or threonine were found in the cell lines with about 88 repeats, which eventually led to decreased RNA expression.

Stem cell therapy for SCA. Although preclinical and clinical trials have been conducted using drugs, antioxidants, and neurotrophic factors, none of these trials were successful in alleviating the symptoms in SCA patients [92]. However, UC-MSCs transplantations in a mouse model of SCA produced promising effects. The major advantages of using UC-MSCs are: (1) there are no ethical issues that arise from their use; (2) they are highly multipotent stem cells; and (3) they have immunosuppressive properties, resulting in reduced risk of tumor formation posttransplantation. Using SCA mice, Zhang and co-workers [93] showed that UC-MSCs provided a therapeutic effect on these animals, including the restoration of motor functions at 8 weeks posttransplantation. The results of this study also showed that the cerebellar atrophy and the number of cells undergoing apoptosis were reduced. There was also an increased production of growth factors, such as insulin-like growth factor-1 (IGF-1) and vascular endothelial growth factor (VEGF). Jin and co-workers [94] performed intravenous and intrathecal transplantation of UC-MSCs into patients affected with SCA and found that UC-MSCs are safe and have the capacity to alleviate the symptoms in SCA patients.

Another clinical study, conducted by Dongmei and co-workers [95] showed that transplantation of UC-MSCs could alleviate the symptoms of SCA, without any side effects, when evaluated using International Cooperative Ataxia Rating Scale (ICARS) and Activity of Daily Living Scale (ADL). Most importantly, these clinical findings provide converging evidence with the preclinical experimental results that UC-MSCs are safe and capable of alleviating the symptoms of SCA, indicating a potentially safe and efficacious therapy for SCA.

Conclusions for Epigenetics in SCA

There is a strong correlation between epigenetic defects in the genes *Ataxin 1, Ataxin 7, and Ataxin 8* and the neuropathological phenotypes associated with SCA in patients. Each ataxin type has its own associated epigenetic mechanism. For example, the Ataxin 1 is a complex that is associated with RORA and Tip60, which are acetyltransferases and the interaction with them is disrupted in the presence of mutant Ataxin 1. Similarly, Ataxin 7 is a part of HAT and presence of mutant Ataxin 7 leads to improper histone acetylation. Detailed analysis of SCA 8 cell line revealed hyper- and hypomethylation of H3K9 and H3K14, respectively [27].

To the best of our knowledge, there are no epigenetic mechanisms related to the UC-MSCs that would drive the cells to take on a specific neuronal phenotype that would confer significant therapeutic effects. However, previous studies have shown a favorable safety profile and beneficial therapeutic effects of UC-MSCs for SCA.

Multiple Sclerosis and Hematopoietic Stem Cells

Multiple sclerosis (OMIM #126200) is an autoimmune neurological disease characterized by the loss of myelin sheath, leading to demyelination and neurodegeneration of brain and spinal cord. The majority of the affected individuals are females between 20 and 40 years of age. In MS, the T-lymphocytes become stimulated by various factors, which, in turn, activate the inflammatory pathways, leading to the symptoms of the disease [96]. MS involves the genetic, epigenetic, and environmental factors (nutritional status). Epigenetic causes include: (1) DNA methylation; (2) posttranscriptional modification by miRNA; and (3) histone acetylation, such as what occurs with the HLA-DRB 1 (human leukocyte antigen having beta chain) gene on chromosome 6, which is responsible for the production of major histocompatibility complex class II (MHC class-II) antigen and which plays an important role in immune response mechanisms [28,97]. The environmental factors include vitamin D deficiency and frequent smoking which, in turn, leads to epigenetic changes observed in MS.

DNA Methylation in MS

Baranzini and co-workers [98] were the first to study the RNA transcriptome sequences and the epigenome sequences of CD4+ T-lymphocytes from three sets of MS-discordant, monozygotic twins, and found that there were no significant differences in the DNA methylation patterns. However, based on this study, alterations in DNA methylation patterns, as a cause of the disease cannot be ruled out, because the sample size was too small to make definitive conclusions. Other studies have shown that DNA methylation could be the cause of MS [99].

The methylation of CpG islands in some of the genes may be responsible for the disease, because the methylation pattern determines how the two different types of T-helper cells (Th1 and Th2) are formed, which, in turn, gives rise to cytokines, such as interferon-gamma (IFN-γ), interleukin-2 (IL-2), interleukin-4 (IL-4), and tumor necrosis factor-α (TNF-α). In MS, there is a dominance of IFN-γ expression associated with Th1, compared to IL-4 molecules that are associated with Th2. Therefore, abnormal DNA methylation pattern that is found in the promoter region of IFN-γ may explain why the immune response by Th1 is greater than Th2, leading to the pathophysiology of the disease (Fig. 20.4). Similarly, DNA methylation or histone deacetylation is associated with the IL-4 genes, leading to the gene silencing. HDAC inhibitors are given as a potential treatment for MS, because they lead to reduction in inflammation, demyelination, and neuronal degeneration [100].

Epigenetics not only control cytokine gene activation in MS, but also affects myelin structure, by regulating myelin basic protein (MBP). For example, hypomethylation of an enzyme, called peptidylarginine deiminase 2 (PAD 2), has been observed in MS [101]. This enzyme is responsible for conversion of arginine to citrulline and its level is increased in MS. Due to overactivation of PAD 2, the MBP is citrullinated and becomes vulnerable for degradation by the myelin-associated proteases, such as cathepsin D. This further leads to a reduced binding capacity of the MBP, causing lipid vesicle fragmentation, which, in turn, leads to the myelin breakdown observed in MS [101] (Fig. 20.5).

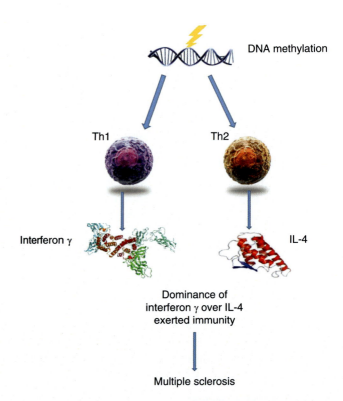

FIGURE 20.4 Impact of DNA methylation on T cells. Compromised immune response exerted by Th1 and Th2 cells (dominance of IFN-γ over IL-4) is due to DNA methylation that leads to MS.

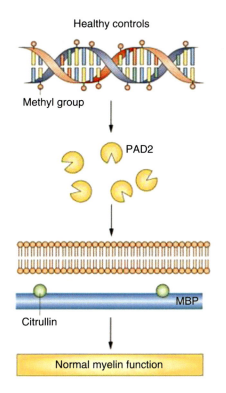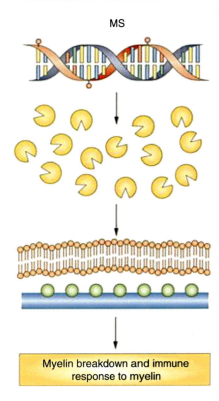

FIGURE 20.5 Comparison of citrullinated myelin in normal and MS patients. Hypomethylation causes increase in the production of enzyme peptidyl-arginine deiminase 2 (PAD 2) leading to increased citrullinated myelin basic protein, thus gives abnormal structure and thereby myelin breakdown in MS patients [97].

TABLE 20.2 Roles of Different Types of miRNA in MS

miRNA	Role in Multiple Sclerosis
mi-RNA155	Dysregulation of gene expression in CD4 + cells and peripheral blood mononuclear cells [30]
mi-RNA326	Dysregulation of gene expression in CD4 + [30]
mi-RNA18b, mi-RNA493, and mi-RNA599	Increased expression in relapse remitting MS [30]

Micro-RNA

In vitro analysis of MS induced lesions and pooled cells have shown that there were about ten miRNAs that were upregulated, especially mi-RNA155 and mi-RNA326. However, mi-RNA326 is more commonly associated with disease relapse, rather than being directly associated with the initiation of the disease. It was found that the patients have more severe form of disease; following relapse, and showed a high expression of mi-RNA326 (Table 20.2). This, in turn, leads to higher activation of T-cells and increased expression of inflammatory cytokines, which then results in abnormal immune response, leading to MS [29].

Histone Acetylation

In 2007, the International Multiple Sclerosis Genetics Consortium (IMSGC) conducted a genome-wide association study (GWAS) that included about 12,000 subjects. It was found that two genes, other than *HLA-DRB 1*, are associated with MS. Single nucleotide polymorphisms (SNPs) in *interleukin-2 receptor α gene (IL2RA)* and *interleukin-7 receptor α gene (IL7RA)* were found to be risk factors for MS. Again, in 2011, IMSGC conducted another GWAS study and found that abnormalities in some of the genes involved in cytokines and signal transduction pathways cause MS, reinforcing the fact that the most commonly affected gene by histone acetylation was the *HLA-DRB 1* gene [28].

Stem Cell Therapy for MS

Among different stem cell populations, HSCs have drawn a special attention for use as a potential therapy of MS, because of their multipotency. These stem cells differentiate into a very large population of cells, including all functional types of blood cells, B-cells, T-cells, and many other cell types, which make them a promising candidate for therapy of many diseases [102]. The first therapy trials using HSCs was started in 1995, followed by a second successful therapy trial on humans in 1998. These studies revealed that scores on the Expanded Disability Status Scale (EDSS) improved for patients who received the HSCs, confirming their use as a potential treatment for this disease [103,104]. However, subsequent clinical trials [105] revealed that patients who have a severe form of this disease (based on their EDSS score) and who received the transplants did not show significant benefits from the cell treatment. It is possible that these stem cells may exert their beneficial effects only during the early stages of disease and not after the disease has progressed to its severe stage [105]. Most of the transplants associated with the HSCs are autogenic in nature, because allogeneic or HLA-matched transplants cause an increase in mortality [96].

Atkins and Freedman [106] reviewed the pros and cons of using HSCs as a potential treatment for MS. One of the major advantages of using HSCs is that once they are extracted from bone-marrow and transplanted into MS patients, the B-cells, and, especially the T-cells, mature and become activated, which play a major role in boosting the immune levels. However, because it is important to ensure that other immune cells, such as macrophages, do not populate the regions near the graft, it is advisable to use immune ablative conditioning regimens to achieve the maximum benefit from the T-cells.

A phase II clinical trial conducted by Mancardi and co-workers [107] was reported recently in 2015, which showed that autologous HSC transplantation resulted in suppression of lesion-induced inflammation. These results are based on a 4-year follow up of patients having a progressive or a relapsing form of the disease.

Collectively, these studies show that HSCs, due to their multipotent and immunomodulatory properties, have a significant promise for producing an effective therapy for MS.

Conclusions for Epigenetics in MS

The compromised immune response seen in MS is due to the epigenetic mechanisms that affect the T-cells and the cytokines, which form the major molecules or proteins of the immune system. Hypomethylation of PAD 2 enzyme is observed in patients affected with MS, which leads to the breakdown of the MBP. Further analysis showed that miRNA, such as mi-RNA155, mi-RNA326, mi-RNA18b, mi-RNA493, and mi-RNA599 are associated with MS. The GWAS study and other studies have shown that *HLA-DRB 1* is the main candidate gene for MS. However, there are some SNPs associated other genes, such as *IL2RA* and *IL7RA*, that pose a risk factor for this disease. HSCs have been used as a stem-cell-based therapy for MS, whereby their activation of the T-cells improve immune responses, producing favorable outcomes.

EPIGENETICS AND GLIOBLASTOMA

Glioblastoma Multiforme and Stem Cell Therapy

Glioblastoma (GB; OMIM # 137800) is an aggressive grade IV astrocytoma that arises from glial cells in humans. There is no cure or effective treatment for this deadly brain cancer, with poor prognosis and resulting in a median survival of the diagnosed only 15–24 months for those with GB. Some of the treatment strategies include surgical resuscitation of the tumor, stereotactic radiation therapy, and chemotherapeutic drugs and/or anti-angiogenic agents, such as temozolomide (TMZ), lomustine and bevacizumab. GB develops twice as much in males than females, and the risk of developing GB increases with age [108,109]. Detailed description and the symptoms of GB are described by us and others elsewhere [110,111].

Even though some of the treatment strategies are effective in reducing the tumor volume and killing tumor cells, tumor recurrence is seen in almost all the patients. There is no standard care nor established protocols to combat the recurrent tumor [112]. One of the major candidates that contribute to tumor reoccurrence is the presence of glioma stem cells (GSCs). Like other stem cells, GSCs can self-renew and proliferate with specific genetic and epigenetic properties. These cancer stem cells are a sub-population of the tumor cells that are highly resistant to chemotherapeutic drugs and radiation therapy, they promote angiogenesis, tumor invasion and immunosuppression. These GSCs are widely used to explore potential therapeutic targets to eradicate glioblastoma tumors [113]. Interestingly, there are many research reports that support and validate the origin of GSCs to be the astrocytes that are derived from the neural stem cells (NSCs) having their niche at the subventricular zone (SVZ). However, there are many other studies which does not support this hypothesis. Some research indicates that mutations occurring in certain genes of NSCs during developmental stages can contribute to a late malignant transformation [114].

EPIGENETIC CHANGES IN GLIOMA STEM CELLS

Methylation

Disruption of DNA methylation (5mC) and DNA hypo-methylation (5hmC) are the major contributors in promoting glioma stem cells (GSCs) proliferation, invasion, and eventually, leading to the formation of tumors. Zhou and colleagues, in 2019, studied the genetic and epigenetic profiles of 22 patient-derived xenograft (PDX) and neural stem cells which were derived from fetal embryos. Methylation profiles showed that the GSCs of the PDX cell lines had lower levels of 5mC and 5hmC compared to the NSCs. Further analysis of glioma brain tissue (generated using T98G and A172 cell lines) showed similar patterns of lowered levels of 5mC and 5hmC compared to a healthy brain tissue. Similarly, specific PDX tissue also showed reduced staining intensity of tissue DNA methylation compared to a healthy tissue [31].

As discussed in the initial section of the chapter, a novel DNA methylation 6 mA was recently identified to play a major role in cancer, specifically in human glioblastoma. Xie and colleagues in 2019 were the first research group to study how this novel DNA methylation is associated with regulation of human glioblastoma. The GSCs were obtained from five males and four females and their analysis showed elevated levels of 6 mA in tumor cells compared to the control cells. There was about 100-fold change in 6 mA levels between the GSCs and astrocytes (controls). Similar results were obtained when analyzing the glioblastoma tissue samples and control samples [9]. However, conversely, another study showed elevated levels of 6 mA in healthy adult tissue as well as in cells [7].

Next, the gene ontology analysis showed that 6 mA was a repressive marker that is associated with H3K4me3, thereby it suppresses tumor suppressor genes. This in turn leads to the survival and proliferation of the GSCs resulting in tumor formation (Xie et al., 2018). It has to be noted here that though N6AMT1 is one of the DNA methyltransferase for 6 mA, this is not universally true, as at least in glioblastoma, another enzyme ALKBH1 was identified as the DNA methyltransferase [8].

There are some contracting results that were found among different studies about the levels of 6 mA in healthy and cancer samples and this needs further investigations. Also, there are some studies that are investigating the potential of targeting 6 mA and DNA methyltransferase to obtain a cure for glioblastoma.

Acetylation

In 2019, Was and colleagues analyzed the acetylation status of the brain tissue of about 500 patients with grade IV glioblastoma and about 600 patients with low grade glioma. Comparing the expression levels of *HDAC 1* through 11, it was found that expression of *HDAC 4, 5, 6, 8,* and *11* was lowered in glioma brain tissue, when compared to the healthy controls. In contrast, the expression levels of *HDAC 1, 2, 3,* and *7* was increased, specifically in the grade IV glioma brain. *HDAC 9* and *10* did not show any difference in their expression levels between the tumor tissue and healthy control samples. Current research is aiming to target and lower HDACs as a potential therapy to lower the proliferation and invasiveness of GB [32].

GENES INVOLVED IN GLIOBLASTOMA

Hypermethylation Status of *IDH 1* and *2*

The enzyme, isocitrate dehydrogenase (IDH), is an integral part of the Kerb's cycle/tricarboxylic acid (TCA) that plays a major role in cellular metabolism. There are five human genes that code for IDH1 and two that are very closely associated with cancer, especially glioblastoma. They serve to protect the cells against oxidative damage, and are involved in lipogenesis, glutamine metabolism, and regulation of glycolysis. IDH1 is found in the cytoplasm and in peroxisomes, whereas IDH2 is found in the mitochondria. The IDH1 and IDH2 (NADP dependent) catalyzes a reversible reaction of converting alpha-ketoglutarate to D-2-hydroxyglutarate, thereby reducing the NADP+ to NADPH in the TCA cycle. In a healthy individual with a wild-type IDH1/2, D-2-hydroxyglutarate dehydrogenase catalyzes the conversion of D-2-hydroxyglutarate back to alpha-ketoglutarate, thereby reducing the accumulation of D-2-hydroxyglutarate. However, a mutant form of *IDH1/2* leads to the accumulation of D-2-hydroxyglutarate in the blood serum, which is often detected and used as a biomarker in glioma patients [115]. Further, this leads to upregulation of hypoxia inducible factor-1alpha (HIF-1a) leading to angiogenesis and tumor vascularization, promoting tumor microenvironment and glioma invasion [116] (Fig. 20.6).

The common *IDH* mutations in glioma are p.R132H, found in IDH 1, and p.R172 as well as p.R140 found in IDH2. Glioma with these mutations show global CpG island hypermethylation, leading to arrest in cell differentiation. Bledea and colleagues, in 2019, studied the hypermethylation status of the chromosomes induced by IDH mutations in different cohorts. The study found that about 1.5% of the probes were hypermethylated in IDH mutant induced astrocytoma, and though this percentage is very small, the IDH mutations, per se, is highly associated with glioma, which determines the diagnosis and prognosis of the tumor [117]. The prognosis of patients with IDH mutant tumor is better than a IDH wild-type tumor, in terms of survival following diagnosis. Glioma patients with IDH mutant can survive for almost twice as longer as patients with IDH wildtype gene [118]. Though many studies have shown that IDH mutant patients have

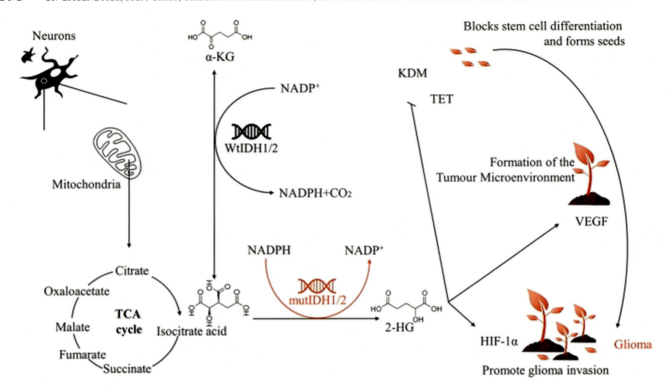

FIGURE 20.6 IDH mutations in Glioblastoma. The IDH is closely associated in the Kreb's cycle and the presence of IDH1/2 mutations in glioblastoma leads to better prognosis and better survival of GB patients [116].

better therapeutic outcomes with TMZ treatment, there are a few other studies that contradict this result.

Hypermethylated Status of MGMT Gene

The O6-methylguanine-DNA methyltransferase (MGMT) enzyme is involved in DNA repair mechanisms and causes DNA cross-links at the alkylated regions, thus halting DNA replication. TMZ, an FDA-approved drug for glioma, alkylates the DNA, thus enabling DNA cross-linking, leading to reduced cell proliferation. TMZ is highly effective for treating glioma patients with hypermethylated *MGMT* promotor, compared to those with unmethylated *MGMT* promotor.

From the studies above, it is clear that the prognosis of glioma is best when the patients are (1) *IDH1/2* mutant as well as (2) have hypermethylated *MGMT* promoters, which improve the overall survival of the patients.

STEM CELL TRANSPLANTATIONS IN GLIOBLASTOMA

MSC Transplantations for GB

Mesenchymal stem cells (MSCs) have been used as a potential stem-cell based therapy for GB, as MSCs are known to have anti-inflammatory and anti-tumoral properties that home in on at the tumor site. Mohme and colleagues in 2020, auto transplanted genetically modified mouse bone-marrow derived mesenchymal stem cells (BM-MSCs) that over-express pro-inflammatory cytokines (interleukins 7 and 12; IL7 and IL12) to achieve T cells and natural-killer-cell responses to combat tumors in Gl261 and CT-2A mouse models of GB. The study outcome showed that the MSCs that overexpressed IL7 and IL12 improved the survivability of the both tumor models in mice compared to the control mice injected with saline or native MSCs. Further, it was also shown that the treatment improved the T-cell infiltration, into the tumor thereby achieving immunomodulatory effects of MSCs that were over-expressing IL7 and IL12. Other outcomes include changes in the levels of CD11, CD86, CD49b natural killer cells, as well as CD4 and CD8, tumor infiltrating T-lymphocytes in the blood, in addition to altering the tumor environment [119].

TRAIL Producing MSCs

One of the most popular and widely used gene therapies that can be delivered by MSCs is the tumor-necrosis-factor-α-related apoptosis-inducing ligand (TRAIL). Choi and colleagues, in 2011, transplanted adipose-tissues-derived MSCs (AT-MSCs), intracranially, producing TRAIL into the F98 rat-tumor model of GB and found that the TRAIL reduced the tumor size and increased the number of apoptotic cells, compared to vehicle-treated animals with the tumor. Moreover, the rats that received the AT-MSC-TRAIL lived about 55 days longer than the rats treated with saline or only AT-MSCs. Tissue analysis revealed that the rats that received the AT-MSC-TRAIL treatment expressed more neuronal and glial lineage markers compared to the vehicle-treated rats, indicating regeneration of brain tissue following tumor treatment [120].

NEURAL STEM CELL TRANSPLANTATION FOR GB

In 2016, Bago and colleagues studied the therapeutic effects of transplanting induced neural stem cells (iNSCs) into a mouse tumor model. The iNSCs were transplanted into the brain tumor at 3 days following implantation of patient-derived GB cells (GBM8 cell line) in mice. Groups of mice were euthanized at three different time points and the brain tissue was analyzed for the migration and homing of the iNSCs to the tumor site. The study results indicated that the iNSCs showed tumor tropism and migrated towards the tumor site, and this was comparable to the tropism showed by the native NSCs.

Furthermore, the authors also over-expressed TRAIL in the iNSCs and transplanted them into the tumors of GB mice. Tissue analysis showed that the mice that received the iNSC-TRAIL cells had decreased tumor volume and survived almost twice as long as the vehicle-treated mice [121].

Sheets and colleagues, in 2020, used scaffolds and matrices to improve the NSC transplants in GB mice. They found that the scaffolded NSC transplants survived for 8 days in GB brain, compared to the NSC transplants without scaffold, which survived only for 3 days. However, the NSCs delivered using the matrix survived about eight times longer than those given scaffolds. The matrix NSCs were genetically modified to express thymidine kinase (TK), and were delivered to the resuscitated PDX tumor brain (postoperatively) in mice and the pro-drug, ganciclovir (GCV), was injected into the animals, intraperitoneally, to activate the gene. The study outcomes showed an overall reduction in tumor signal, as evidenced by in vivo imaging system, as well as an increase in median survival of the mice, compared to the vehicle-treated control animals [122].

Gutova and colleagues, in 2019, studied the effects of modified NSCs, which express and secrete human carboxylesterase (CE-NSCs) into a tumor mouse model created by transplanting cells from either U251 human GB cell line, PBT017 PDX cell line, or the Gl261 mouse GB cell line. The CE-NSCs were intraventricularly transplanted into the U251, PDX, and Gl261 GB mice at 10 days, 14 days or 7 days following tumor formation, respectively. Like any other previous studies mentioned, the modified NSCs migrated and homed in on the tumor site. However, in the case of Gl261 transplanted brain, there were some cells present in ipsilateral as well as in the contralateral hemispheres of the brain [123].

Conclusions for Epigenetics in GB

All the above-mentioned studies show that genetically modified stem cells can be potentially therapeutic in combating GB, compared to naïve stem cells. These modified stem cells, as well as epigenetic factors associated with GB, should be taken into consideration when planning treatment strategies. Moreover, the epigenetic status of glioma stem cells are also critical in determining the therapeutic outcomes in patients.

CONCLUSIONS

Epigenetics not only play a role in determining the stem cell fate, but also form the underlying bases of neurodegenerative diseases, such as HD, PD, RTT, SCA, and MS, as well as brain tumor/GB discussed in this chapter. Although the genetic bases of these diseases and cancer can vary, the epigenetic causes are similar. For example, alterations in the levels of DNA methylation and histone acetylation gives rise to SCA and MS. Understanding the epigenetic mechanisms of stem cells is an important aspect that needs to be carefully considered when designing strategies to achieve optimal efficacy and efficiency when cell replacement therapy for neurodegenerative diseases is being considered.

In order to translate and improve the outcomes of clinical trials, further research on how epigenetics drive stem cell fate is necessary. For example, because the use of MSCs for treating HD has been predominantly used in preclinical trials, researchers should be cognizant that the subpopulation of cells being used is the direct function of the number of passages these cells have undergone. Though higher passaged MSCs have proven to be highly beneficial in restoring the motor symptoms associated with HD, there is also a risk of chromosomal abnormalities that would lead to tumor formation or other adverse effects. Therefore, studying the epigenetic mechanism of MSCs to provide an effective cell replacement therapy is necessary. Some diseases, such as RTT have an epigenetic basis, which makes them excellent candidates for the yet-to-be-developed epigenetic-driven stem-cell replacement therapies. As such, a mutation corrected RTT-iPSC cell line for the transplantation may prove to be a promising therapeutic approach. Similarly, identifying the dysregulated epigenetic markers in each disease (such as H3K4me3 in HD and H3K9 and H3K14 in SCA8) may lead to designing a more targeted therapy for such neurological disorders. Similarly, characterization of the different types of methylation status and other epigenetic status of the glioma stem cells as well as the specific gene mutations associated with brain tumor can lead to better prognosis and diagnosis leading to improved therapeutic outcomes. Addressing these types of issues and furthering our knowledge about the epigenetic mechanisms of stem cells and the diseases associated with epigenetic alterations could pave way toward developing more effective and long-lasting therapeutic approaches.

Abbreviations

6 mA	N6-methyladenine
ADL	Activity of Daily Living Scale
ALKBH1	Alkylated DNA repair protein AlkB homolog
ASDs	Autism-spectrum disorders
AT	Adipose tissue
AT-MSCs	adipose-tissues-derived MSCs
BDNF	Brain derived neurotrophic factor
BM	Bone marrow
BM-MSCs	Bone marrow-derived MSCs
cAMP	Cyclic adenosine monophosphate
CBP	CREB-binding protein
CE	carboxylesterase
ChIP	Chromatin immunoprecipitation
CRE	cAMP response element
CREB	CRE-binding protein
CRX	Cone-rod homeobox protein
DAT	Dopamine transporter
DNMTs	DNA methyl transferases
EDSS	Expanded Disability Status Scale
EPSP	Excitatory postsynaptic potential
ESCS	Embryonic stem cells
GB	Glioblastoma
GCV	ganciclovir
GDNF	Glial derived neurotrophic factor
GSCs	Glioma stem cells
GWAS	Genome-wide association study
HATs	Histone acetyltransferases
HD	Huntington disease
HDACs	Histone deacetylases
HIF-1a	hypoxia inducible factor-1alpha
HLA-DRB	Human leukocyte antigen having beta chain
HSCs	Hematopoietic stem cells HSCs
6-OHDA	6-Hydroxydopamine
5hmC	5-Hydroxymethylcytosine
ICARS	International Cooperative Ataxia Rating Scale
IDH	isocitrate dehydrogenase
IFN α	Interferon α
IGF-1	Insulin-like growth factor-1
IL-2	Interleukin-2
IL-4	Interleukine-4
IL-7	Interleukine-7
IL-12	Interleukine-12
iNSCs	induced neural stem cells
iPSCs	Induced pluripotent stem cells
LRRK2	Leucine-rich repeat kinase 2 gene
MBD	Methyl binding domain
MBS	Myelin basic protein
MeCP2	Methyl-CpG binding protein 2 gene
MHC class-II	Histocompatibility complex class II
5mC	5-Methylcytosine
MGMT	O6-methylguanine-DNA methyltransferase
mHtt	Mutant huntingtin protein
miRNA	MicroRNA
MPTP	1-Methyl-4-phenyl-1,2,3,6-tetrahydropyridine
MS	Multiple sclerosis
MSC	Mesenchymal stem cells
N6AMT1	N-6 adenine-specific DNA methyltransferase 1
NRSF	Neuron-restrictive silencer factor
NSCs	Neural stem cells
PAD 2	Peptidyl-arginine deiminase 2
PD	Parkinson's disease
PDX	Patient-derived xenograft
PSP	Inhibitory post-synaptic potential
REST	Repressor element-1 silencing transcription factor
RTT	Rett syndrome
SAH	S-adenosylhomocysteine
SAM	S-adenosylmethionine
SCA	Spinocerebellar ataxia
SGZ	Subgranular zone
SMRT	Silencing mediator of retinoic acid and thyroid hormone receptor
SNCA	α-Synuclein
SNpc	Substantia nigra, pars compacta
SUMO	Small ubiquitin-like molecules
SVZ	Subventricular zone
TCA	tricarboxylic acid
Th	T-helper-cells
TH	Tyrosine hydroxylase
TK	thymidine kinase
TMZ	Temozolamide
TNF-α	Tumor necrosis factor-α
TRAIL	tumor-necrosis-factor-α-related apoptosis-inducing ligand
TRD	Transcriptional repression domain
UC	Umbilical cord
UC-MSCs	Umbilical cord derived MSCs
VEGF	Vascular endothelial growth factor

References

[1] Landgrave-Gómez J, Mercado-Gómez O, Guevara-Guzmán R. Epigenetic mechanisms in neurological and neurodegenerative diseases. Front Cell Neurosci 2015;9:58. Available from: http://www.ncbi.nlm.nih.gov/pmc/articles/PMC4343006/.

[2] Goldberg AD, Allis CD, Bernstein E. Epigenetics: a landscape takes shape. Cell 2007;128(4):635–8.

[3] Gill G. SUMO and ubiquitin in the nucleus: different functions, similar mechanisms? Genes Dev 2004;18(17):2046–59.

[4] Sadakierska-Chudy A, Filip M. A comprehensive view of the epigenetic landscape. Part II: histone post-translational modification, nucleosome level, and chromatin regulation by ncRNAs. Neurotox Res 2015;27:172–97.

[5] Cedar H, Bergman Y. Linking DNA methylation and histone modification: patterns and paradigms. Nat Rev Genet 2009;10(5):295–304.

[6] Jin B, Li Y, Robertson KD. DNA methylation. Genes Cancer 2011;2(6):607–17.

[7] Xiao CL, Zhu S, He M, et al. N^6-Methyladenine DNA modification in the human genome. Mol Cell 2018;71(2):306–18. Available from: https://doi.org/10.1016/j.molcel.2018.06.015 e7.

[8] Zhang M, Yang S, Nelakanti R, et al. Mammalian ALKBH1 serves as an N^6-mA demethylase of unpairing DNA. Cell Res 2020;30(3):197–210. Available from: https://doi.org/10.1038/s41422-019-0237-5.

[9] Xie Q, Wu TP, Gimple RC, et al. N^6-methyladenine DNA modification in glioblastoma. Cell 2018;175(5):1228–43. Available from: https://doi.org/10.1016/j.cell.2018.10.006 e20.

[10] Voigt P, Tee W-W, Reinberg D. A double take on bivalent promoters. Genes Dev 2013;27(12):1318–38.

[11] Bannister AJ, Kouzarides T. Regulation of chromatin by histone modifications. Cell Res 2011;21(3):381–95.

[12] Rossetto D, Avvakumov N, Côté J. Histone phosphorylation: a chromatin modification involved in diverse nuclear events. Epigenetics 2012;7(10):1098–108.

[13] Puttick J, Baker EN, Delbaere LTJ. Histidine phosphorylation in biological systems. Biochim Biophys Acta 2008;1784(1):100–5.

[14] Ellis P, Fagan BM, Magness ST, Hutton S, Taranova O, Hayashi S, et al. SOX2, a persistent marker for multipotential neural stem cells derived from embryonic stem cells, the embryo or the adult. Dev Neurosci 2004;26(2–4):148–65.

[15] Gapp K, Woldemichael BT, Bohacek J, Mansuy IM. Epigenetic regulation in neurodevelopment and neurodegenerative diseases. Neuroscience 2014;264:99–111.

[16] Hirabayashi Y, Gotoh Y. Epigenetic control of neural precursor cell fate during development. Nat Rev Neurosci 2010;11(6):377–88.

[17] Wu H, Coskun V, Tao J, Xie W, Ge W, Yoshikawa K, et al. Dnmt3a-dependent nonpromoter DNA methylation facilitates transcription of neurogenic genes. Science 2010;329(5990):444–8.

[18] Jakovcevski M, Akbarian S. Epigenetic mechanisms in neurological disease. Nat Med 2012;18(8):1194–204.

[19] Azpurua J, Eaton BA. Neuronal epigenetics and the aging synapse. Front Cell Neurosci 2015;9:208. Available from: http://www.ncbi.nlm.nih.gov/pmc/articles/PMC4444820/.

[20] Brookes E, Shi Y. Diverse epigenetic mechanisms of human disease. Annu Rev Genet 2014;48:237–68.

[21] Ul Hassan A, Hassan G, Rasool Z. Role of stem cells in treatment of neurological disorder. Int J Health Sci (Qassim) 2009;3(2):227–33.

[22] Srinageshwar B, Maiti P, Dunbar GL, Rossignol J. Role of epigenetics in stem cell proliferation and differentiation: implications for treating neurodegenerative diseases. Int J Mol Sci 2016;17(2):e199.

[23] Vashishtha M, Ng CW, Yildirim F, Gipson TA, Kratter IH, Bodai L, et al. Targeting H3K4 trimethylation in Huntington disease. Proc Natl Acad Sci USA 2013;110(32):E3027–36.

[24] Coppedè F. Genetics and epigenetics of Parkinson's disease. Sci World J 2012;2012:489830.

[25] Kubota T, Miyake K, Hirasawa T. Role of epigenetics in Rett syndrome. Epigenomics 2013;5(5):583–92.

[26] Helmlinger D, Hardy S, Sasorith S, Klein F, Robert F, Weber C, et al. Ataxin-7 is a subunit of GCN5 histone acetyltransferase-containing complexes. Hum Mol Genet 2004;13(12):1257–65.

[27] Chen I-C, Lin H-Y, Lee G-C, Kao S-H, Chen C-M, Wu Y-R, et al. Spinocerebellar ataxia type 8 larger triplet expansion alters histone modification and induces RNA foci. BMC Mol Biol 2009;10:9.

[28] Küçükali CI, Kürtüncü M, Çoban A, Çebi M, Tüzün E. Epigenetics of multiple sclerosis: an updated review. Neuromolecular Med 2015;17(2):83–96.

[29] Du C, Liu C, Kang J, Zhao G, Ye Z, Huang S, et al. MicroRNA miR-326 regulates TH-17 differentiation and is associated with the pathogenesis of multiple sclerosis. Nat Immunol 2009;10(12):1252–9.

[30] Huynh JL, Casaccia P. Epigenetic mechanisms in multiple sclerosis: implications for pathogenesis and treatment. Lancet Neurol 2013;12(2):195–206.

[31] Zhou D, Alver BM, Li S, et al. Distinctive epigenomes characterize glioma stem cells and their response to differentiation cues. Genome Biol 2018;19(1):43.

[32] Was H, Krol SK, Rotili D, et al. Histone deacetylase inhibitors exert anti-tumor effects on human adherent and stem-like glioma cells. Clin Epigenetics 2019;11(1):11.

[33] Alberts B, Johnson A, Lewis J, Raff M, Roberts K, Walter P. Chromosomal DNA and its packaging in the chromatin fiber. *Molecular biology of the cell* (4th edition) 2002. New York: Garland Science. Available from: http://www.ncbi.nlm.nih.gov/books/NBK26834/.

[34] Mariño-Ramírez L, Kann MG, Shoemaker BA, Landsman D. Histone structure and nucleosome stability. Expert Rev Proteom 2005;2(5):719–29.

[35] Hendrich B, Bickmore W. Human diseases with underlying defects in chromatin structure and modification. Hum Mol Genet 2001;10(20):2233–42.

[36] Roos RA. Huntington's disease: a clinical review. Orphanet J Rare Dis 2010;5:40.

[37] Walker FO. Huntington's disease. Lancet 2007;369(9557):218–28.

[38] The Huntington's disease collaborative research group. A novel gene containing a trinucleotide repeat that is expanded and unstable on Huntington's disease chromosomes. Cell 1993;72(6):971–83.

[39] Dey ND, Bombard MC, Roland BP, Davidson S, Lu M, Rossignol J, et al. Genetically engineered mesenchymal stem cells reduce behavioral deficits in the YAC 128 mouse model of Huntington's disease. Behav Brain Res 2010;214(2):193–200.

[40] Fink KD, Rossignol J, Crane AT, Davis KK, Bombard MC, Bavar AM, et al. Transplantation of umbilical cord-derived mesenchymal stem cells into the striata of R6/2 mice: behavioral and neuropathological analysis. Stem Cell Res Ther 2013;4(5):130.

[41] Rossignol J, Fink KD, Crane AT, Davis KK, Bombard MC, Clerc S, et al. Reductions in behavioral deficits and neuropathology in the R6/2 mouse model of Huntington's disease following transplantation of bone-marrow-derived mesenchymal stem cells is dependent on passage number. Stem Cell Res Ther 2015;6:9.

[42] Urdinguio RG, Sanchez-Mut JV, Esteller M. Epigenetic mechanisms in neurological diseases: genes, syndromes, and therapies. Lancet Neurol 2009;8(11):1056–72.

[43] Ng CW, Yildirim F, Yap YS, Dalin S, Matthews BJ, Velez PJ, et al. Extensive changes in DNA methylation are associated with expression of mutant huntingtin. Proc Natl Acad Sci USA 2013;110(6):2354–9.

[44] Lee J, Hwang YJ, Kim KY, Kowall NW, Ryu H. Epigenetic mechanisms of neurodegeneration in Huntington's disease. Neurotherapeutics 2013;10(4):664–76.

[45] Lee S-T, Chu K, Im W-S, Yoon H-J, Im J-Y, Park J-E, et al. Altered microRNA regulation in Huntington's disease models. Exp Neurol 2011;227(1):172–9.

[46] Moumné L, Betuing S, Caboche J. Multiple aspects of gene dysregulation in Huntington's disease. Front Neurol 2013;4:127.

[47] Bithell A, Johnson R, Buckley NJ. Transcriptional dysregulation of coding and non-coding genes in cellular models of Huntington's disease. Biochem Soc Trans 2009;37(Pt 6):1270–5.

[48] Kern S, Eichler H, Stoeve J, Klüter H, Bieback K. Comparative analysis of mesenchymal stem cells from bone marrow, umbilical cord blood, or adipose tissue. Stem Cells 2006;24(5):1294–301.

[49] Li Z, Liu C, Xie Z, Song P, Zhao RCH, Guo L, et al. Epigenetic dysregulation in mesenchymal stem cell aging and spontaneous differentiation. PLoS One 2011;6(6):e20526.

[50] Phinney DG. Functional heterogeneity of mesenchymal stem cells: implications for cell therapy. J Cell Biochem 2012;113(9):2806–12.

[51] Guo L, Zhou Y, Wang S, Wu Y. Epigenetic changes of mesenchymal stem cells in three-dimensional (3D) spheroids. J Cell Mol Med 2014;18(10):2009–19.

[52] Moll G, Rasmusson-Duprez I, von Bahr L, Connolly-Andersen A-M, Elgue G, Funke L, et al. Are therapeutic human mesenchymal stromal cells compatible with human blood? Stem Cells 2012;30(7):1565–74.

[53] Sadri-Vakili G, Cha J-HJ. Histone deacetylase inhibitors: a novel therapeutic approach to Huntington's disease (complex mechanism of neuronal death). Curr Alzheimer Res 2006;3(4):403–8.

[54] Han F, Baremberg D, Gao J, Duan J, Lu X, Zhang N, et al. Development of stem cell-based therapy for Parkinson's disease. Transl Neurodegener 2015;4:16.

[55] Klein C, Westenberger A. Genetics of Parkinson's disease. Cold Spring Harb Perspect Med 2012;2(1). Available from: http://www.ncbi.nlm.nih.gov/pmc/articles/PMC3253033/.

[56] Lesage S, Brice A. Parkinson's disease: from monogenic forms to genetic susceptibility factors. Hum Mol Genet 2009;18(R1):R48–59.

[57] Mentch SJ, Locasale JW. One-carbon metabolism and epigenetics: understanding the specificity. Ann NY Acad Sci 2016;1363:91–8.

[58] King WD, Ho V, Dodds L, Perkins SL, Casson RI, Massey TE. Relationships among biomarkers of one-carbon metabolism. Mol Biol Rep 2012;39(7):7805–12.

[59] Chou C-H, Fan H-C, Hueng D-Y, Chou C-H, Fan H-C, Hueng D-Y. Potential of neural stem cell-based therapy for Parkinson's disease. Parkinson's Dis 2015;2015:e571475 2015.

[60] Matsumoto L, Takuma H, Tamaoka A, Kurisaki H, Date H, Tsuji S, et al. CpG demethylation enhances alpha-synuclein expression and affects the pathogenesis of Parkinson's disease. PLoS One 2010;5(11):e15522.

[61] Jowaed A, Schmitt I, Kaut O, Wüllner U. Methylation regulates alpha-synuclein expression and is decreased in Parkinson's disease patients' brains. J Neurosci 2010;30(18):6355–9.

[62] Doxakis E. Post-transcriptional regulation of alpha-synuclein expression by mir-7 and mir-153. J Biol Chem 2010;285(17):12726–34.

[63] Cai M, Tian J, Zhao G, Luo W, Zhang B. Study of methylation levels of parkin gene promoter in Parkinson's disease patients. Int J Neurosci 2011;121(9):497–502.

[64] Fitzmaurice AG, Rhodes SL, Lulla A, Murphy NP, Lam HA, O'Donnell KC, et al. Aldehyde dehydrogenase inhibition as a pathogenic mechanism in Parkinson disease. Proc Natl Acad Sci USA 2013;110(2):636–41.

[65] Sanberg PR. Neural stem cells for Parkinson's disease: to protect and repair. Proc Natl Acad Sci USA 2007;104(29):11869–70.

[66] Ma DK, Marchetto MC, Guo JU, Ming G, Gage FH, Song H. Epigenetic choreographers of neurogenesis in the adult mammalian brain. Nat Neurosci 2010;13(11):1338–44.

[67] Redmond DE, Bjugstad KB, Teng YD, Ourednik V, Ourednik J, Wakeman DR, et al. Behavioral improvement in a primate Parkinson's model is associated with multiple homeostatic effects of human neural stem cells. Proc Natl Acad Sci USA 2007;104(29):12175–80.

[68] Gash DM, Zhang Z, Ovadia A, Cass WA, Yi A, Simmerman L, et al. Functional recovery in parkinsonian monkeys treated with GDNF. Nature 1996;380(6571):252–5.

[69] Sullivan AM, Opacka-Juffry J, Blunt SB. Long-term protection of the rat nigrostriatal dopaminergic system by glial cell line-derived neurotrophic factor against 6-hydroxydopamine in vivo. Eur J Neurosci 1998;10(1):57–63.

[70] Kirik D, Georgievska B, Björklund A. Localized striatal delivery of GDNF as a treatment for Parkinson disease. Nat Neurosci 2004;7(2):105–10.

[71] d'Anglemont de Tassigny X, Pascual A, López-Barneo J. GDNF-based therapies, GDNF-producing interneurons, and trophic support of the dopaminergic nigrostriatal pathway. Implications for Parkinson's disease. Front Neuroanat 2015;9:10.

[72] Patel NK, Gill SS. GDNF delivery for Parkinson's disease. Acta Neurochir Suppl 2007;97(Pt 2):135–54.

[73] Deng X, Liang Y, Lu H, Yang Z, Liu R'en, Wang J, et al. Cotransplantation of GDNF-overexpressing neural stem cells and fetal dopaminergic neurons mitigates motor symptoms in a rat model of Parkinson's disease. PLoS One 2013;8(12):e80880.

[74] Wagner J, Akerud P, Castro DS, Holm PC, Canals JM, Snyder EY, et al. Induction of a midbrain dopaminergic phenotype in Nurr1-overexpressing neural stem cells by type 1 astrocytes. Nat Biotechnol 1999;17(7):653–9.

[75] van Heesbeen HJ, Mesman S, Veenvliet JV, Smidt MP. Epigenetic mechanisms in the development and maintenance of dopaminergic neurons. Development 2013;140(6):1159–69.

[76] Jacobs FMJ, van Erp S, van der Linden AJA, von Oerthel L, Burbach JPH, Smidt MP. Pitx3 potentiates Nurr1 in dopamine neuron terminal differentiation through release of SMRT-mediated repression. Development 2009;136(4):531–40.

[77] Amir RE, Van den Veyver IB, Wan M, Tran CQ, Francke U, Zoghbi HY. Rett syndrome is caused by mutations in X-linked MECP2, encoding methyl-CpG-binding protein 2. Nat Genet 1999;23(2):185–8.

[78] Trappe R, Laccone F, Cobilanschi J, Meins M, Huppke P, Hanefeld F, et al. MECP2 mutations in sporadic cases of Rett syndrome are almost exclusively of paternal origin. Am J Hum Genet 2001;68(5):1093–101.

[79] Girard M, Couvert P, Carrié A, Tardieu M, Chelly J, Beldjord C, et al. Parental origin of de novo MECP2 mutations in Rett syndrome. Eur J Hum Genet 2001;9(3):231–6.

[80] Zhang X, Zhao Y, Bao X, Zhang J, Cao G, Wu X. Genetic features and mechanism of Rett syndrome in Chinese population. Zhonghua Yi Xue Yi Chuan Xue Za Zhi 2014;31(1):1–5.

[81] Bienvenu T, Chelly J. Molecular genetics of Rett syndrome: when DNA methylation goes unrecognized. Nat Rev Genet 2006;7(6):415–26.

[82] Yasui DH, Peddada S, Bieda MC, Vallero RO, Hogart A, Nagarajan RP, et al. Integrated epigenomic analyses of neuronal MeCP2 reveal a role for long-range interaction with active genes. Proc Natl Acad Sci USA 2007;104(49):19416–21.

[83] Mellen M, Ayata P, Dewell S, Kriaucionis S, Heintz N. MeCP2 binds to 5hmc enriched within active genes and accessible chromatin in the nervous system. Cell 2012;151(7):1417–30.

[84] Dajani R, Koo S-E, Sullivan GJ, Park I-H. Investigation of Rett syndrome using pluripotent stem cells. J Cell Biochem 2013;114(11):2446–53.

[85] Takahashi K, Yamanaka S. Induction of pluripotent stem cells from mouse embryonic and adult fibroblast cultures by defined factors. Cell 2006;126(4):663–76.

[86] Marchetto MCN, Carromeu C, Acab A, Yu D, Yeo GW, Mu Y, et al. A model for neural development and treatment of Rett syndrome using human induced pluripotent stem cells. Cell 2010;143(4):527–39.

[87] Kim K-Y, Hysolli E, Park I-H. Neuronal maturation defect in induced pluripotent stem cells from patients with Rett syndrome. Proc Natl Acad Sci USA 2011;108(34):14169–74.

[88] Teive HAG. Spinocerebellar ataxias. Arq Neuropsiquiatr 2009;67(4):1133–42.

[89] Paulson HL. The Spinocerebellar Ataxias. J Neuroophthalmol 2009;29(3):227–37.

[90] Lin X, Antalffy B, Kang D, Orr HT, Zoghbi HY. Polyglutamine expansion down-regulates specific neuronal genes before pathologic changes in SCA1. Nat Neurosci 2000;3(2):157–63.

[91] Gehrking KM, Andresen JM, Duvick L, Lough J, Zoghbi HY, Orr HT. Partial loss of Tip60 slows mid-stage neurodegeneration in a spinocerebellar ataxia type 1 (SCA1) mouse model. Hum Mol Genet 2011;20(11):2204–12.

[92] Ogawa M. Pharmacological treatments of cerebellar ataxia. Cerebellum 2004;3(2):107–11.

[93] Zhang M-J, Sun J-J, Qian L, Liu Z, Zhang Z, Cao W, et al. Human umbilical mesenchymal stem cells enhance the expression of neurotrophic factors and protect ataxic mice. Brain Res 2011;1402:122–31.

[94] Jin J-L, Liu Z, Lu Z-J, Guan D-N, Wang C, Chen Z-B, et al. Safety and efficacy of umbilical cord mesenchymal stem cell therapy in hereditary spinocerebellar ataxia. Curr Neurovasc Res 2013;10(1):11–20.

[95] Dongmei H, Jing L, Mei X, Ling Z, Hongmin Y, Zhidong W, et al. Clinical analysis of the treatment of spinocerebellar ataxia and multiple system atrophy-cerebellar type with umbilical cord mesenchymal stromal cells. Cytotherapy 2011;13(8):913–17.

[96] Bakhuraysah MM, Siatskas C, Petratos S. Hematopoietic stem cell transplantation for multiple sclerosis: is it a clinical reality. Stem Cell Res Ther 2016;7:12. Available from: http://www.ncbi.nlm.nih.gov/pmc/articles/PMC4715306/.

[97] Koch MW, Metz LM, Kovalchuk O. Epigenetic changes in patients with multiple sclerosis. Nat Rev Neurol 2013;9(1):35–43.

[98] Baranzini SE, Mudge J, van Velkinburgh JC, Khankhanian P, Khrebtukova I, Miller NA, et al. Genome, epigenome and RNA sequences of monozygotic twins discordant for multiple sclerosis. Nature 2010;464(7293):1351–6.

[99] Liggett T, Melnikov A, Tilwalli S, Yi Q, Chen H, Replogle C, et al. Methylation patterns of cell-free plasma DNA in relapsing-remitting multiple sclerosis. J Neurol Sci 2010;290(1–2):16–21.

[100] Kürtüncü M, Tüzün E. Multiple sclerosis: could it be an epigenetic disease? Med Hypotheses 2008;71(6):945–7.

[101] Mastronardi FG, Noor A, Wood DD, Paton T, Moscarello MA. Peptidyl argininedeiminase 2 CpG island in multiple sclerosis white matter is hypomethylated. J Neurosci Res 2007;85(9):2006–16.

[102] Seita J, Weissman IL. Hematopoietic stem cell: self-renewal vs differentiation. Wiley Interdiscip Rev Syst Biol Med 2010;2(6):640–53.

[103] Burt RK, Traynor AE, Cohen B, Karlin KH, Davis FA, Stefoski D, et al. T cell-depleted autologous hematopoietic stem cell transplantation for multiple sclerosis: report on the first three patients. Bone Marrow Transplant 1998;21(6):537–41.

[104] Burt RK, Burns W, Hess A. Bone marrow transplantation for multiple sclerosis. Bone Marrow Transplant 1995;16(1):1–6.

[105] Mandalfino P, Rice G, Smith A, Klein JL, Rystedt L, Ebers GC. Bone marrow transplantation in multiple sclerosis. J Neurol 2000;247(9):691–5.

[106] Atkins HL, Freedman MS. Hematopoietic stem cell therapy for multiple sclerosis: top 10 lessons learned. Neurotherapeutics 2013;10(1):68–76.

[107] Mancardi GL, Sormani MP, Gualandi F, Saiz A, Carreras E, Merelli E, et al. Autologous hematopoietic stem cell transplantation in multiple sclerosis: a phase II trial. Neurology 2015;84.

[108] Minniti G, Niyazi M, Alongi F, Navarria P, Belka C. Current status and recent advances in reirradiation of glioblastoma. Radiat Oncol 2021;16(1):36.

[109] Tan AC, Ashley DM, López GY, Malinzak M, Friedman HS, Khasraw M. Management of glioblastoma: state of the art and future directions. CA Cancer J Clin 2020;70(4):299–312.

[110] Rossignol J, Srinageshwar B, Dunbar GL. Current therapeutic strategies for glioblastoma. Brain Sci 2019;10(1):15 Published 2019 Dec 26.

[111] IJzerman-Korevaar M, Snijders TJ, de Graeff A. Teunissen SCCM, de Vos FYF. Prevalence of symptoms in glioma patients throughout the disease trajectory: a systematic review. J Neurooncol 2018;140(3):485–96.

[112] Djamel-Eddine YC, De Witte O, Mélot C, Lefranc F. Recurrent glioblastomas: should we operate a second and even a third time? Interdiscip Neurosurg 2019;18:100551.

[113] Valor LM, Hervás-Corpión I. The epigenetics of glioma stem cells: a brief overview. Front Oncol 2020;10:602378 Published 2020 Dec 2.

[114] Matarredona ER, Pastor AM. Neural stem cells of the subventricular zone as the origin of human glioblastoma stem cells. Therapeutic implications. Front Oncol 2019;9:779.

[115] Waitkus MS, Diplas BH, Yan H. Biological role and therapeutic potential of IDH mutations in cancer. Cancer Cell 2018;34(2):186–95.

[116] Huang J, Yu J, Tu L, Huang N, Li H, Luo Y. Isocitrate dehydrogenase mutations in glioma: from basic discovery to therapeutics development. Front Oncol 2019;9:506.

[117] Bledea R, Vasudevaraja V, Patel S, et al. Functional and topographic effects on DNA methylation in IDH1/2 mutant cancers. Sci Rep 2019;9(1):16830 Published 2019 Nov 14.

[118] Pandith AA, Qasim I, Baba SM, et al. Favorable role of *IDH*1/2 mutations aided with *MGMT* promoter gene methylation in the outcome of patients with malignant glioma. Future Sci OA 2020;7(3):FSO663.

[119] Mohme M, Maire CL, Geumann U, et al. Local intracerebral immunomodulation using interleukin-expressing mesenchymal stem cells in glioblastoma. Clin Cancer Res 2020;26(11):2626–39.

[120] Choi SA, Hwang SK, Wang KC, et al. Therapeutic efficacy and safety of TRAIL-producing human adipose tissue-derived mesenchymal stem cells against experimental brainstem glioma. Neuro Oncol 2011;13(1):61–9.

[121] Bagó JR, Alfonso-Pecchio A, Okolie O, et al. Therapeutically engineered induced neural stem cells are tumour-homing and inhibit progression of glioblastoma. Nat Commun 2016;7:10593.

[122] Sheets KT, Ewend MG, Mohiti-Asli M, et al. Developing implantable scaffolds to enhance neural stem cell therapy for post-operative glioblastoma. Mol Ther 2020;28(4):1056–67.

[123] Gutova M, Flores L, Adhikarla V, et al. Quantitative evaluation of intraventricular delivery of therapeutic neural stem cells to orthotopic glioma. Front Oncol 2019;9:68.

CHAPTER

21

Epigenetic Regulation of Skeletal Muscle Regeneration

Rodolfo Daniel Ávila-Avilés[1], Claudia Negrón-Lomas[1] and J. Manuel Hernández-Hernández[2]

[1]Laboratory of Epigenetics of Skeletal Muscle Regeneration, Department of Genetics and Molecular Biology, Centre for Research and Advanced Studies-IPN, Mexico City, Mexico [2]Instituto Politécnico Nacional 2508, San Pedro Zacatenco, Ciudad de México, México

OUTLINE

Introduction	403	Small Molecules as a Therapeutic Alternative in the Epigenetic Control of Regeneration	410
Epigenetic Control in the Maintenance of Quiescence	404	Conclusion	412
Epigenetic Control of the Activation and Proliferation of SCs	406	Acknowledgments	413
		References	413
Epigenetic Control of SCs Differentiation	408		

INTRODUCTION

Skeletal muscle (SkM) comprises nearly 40% of total body mass and plays a crucial role in moving, breathing, and regulating systemic energy metabolism [1]. This tissue is mainly composed of mature multinucleated, non-mitotic muscle cells called myofibers. Likewise, the SkM is composed of heterogeneous myofibers that can be of slow or rapid contraction, which differ in the composition of contractile proteins, oxidative capacity and substrate preference for the production of adenosine triphosphate (ATP).

Slow-twitch fibers show low fatigability, high oxidative capacity, and a preference for fatty acids as a substrate for ATP production. In contrast, fast-twitch fibers have a lower fatigue resistance, greater contraction strength, lower oxidative capacity, and a preference for glucose as a substrate for ATP production through anaerobic glycolysis [2]. Thus, fiber type composition of SkM profoundly impacts systemic energy consumption [3]. In order to upkeep with physical demands and tissue turnover, SkM is dynamically remodeled and repaired through life [4]. Satellite cells (SCs), a population of muscle resident stem cells, are responsible for the regeneration of injured muscles [5]. In the healthy resting muscle, SCs are in a quiescent state, under reversible mitotic arrest and slow metabolic activity. In response to muscle injury, SCs can be activated, resulting in proliferation, differentiation, and self-renewal through asymmetric divisions to produce quiescent satellite cells and proliferating myoblasts [6,7].

The process of myogenesis is controlled by several myogenic transcription factors that act as terminal effectors of signaling cascades and produce appropriate developmental stage-specific transcripts. Paired-box protein 7 (Pax7) maintains the population of quiescent satellite cells and, together with myogenic factor 5 (Myf5), plays a role in the expansion of activated myoblasts. Myoblast determination protein (MyoD) is believed to determine the differentiation potential of an activated myoblast, and acts together with myogenin and myocyte enhancer factor 2 (MEF2) to drive differentiation. Finally, muscle-specific regulatory factor 4 (MRF4) is required for hypertrophy, although it may have other roles as well. These transcription factors do not act alone, but exist as part of complex signaling cascades that control every stage of myogenesis (Fig. 21.1). Thus, a tightly regulated balance between self-renewal and differentiation is required to maintain the SC pool and generate appropriate numbers of transient amplifying progenitors. This regulation is necessary to support growth and regeneration of SkM [8].

However, in muscle-wasting diseases, such as Duchenne muscular dystrophy (DMD), reduced numbers of asymmetric divisions of muscle stem cells lead to a diminished population of myogenic progenitors and severe muscle degeneration, in addition to myofiber degeneration imposed by the absence of dystrophin and the consequent cell membrane fragility [9]. Understanding the mechanisms by which muscle is maintained, grows, or is lost is essential for combating muscle disease and may be useful for aiding muscle maintenance, allowing for better mobility and independence during later stages of life.

EPIGENETIC CONTROL IN THE MAINTENANCE OF QUIESCENCE

Quiescence is a state where cells enter a reversible cycle arrest in G_0 phase of the cell cycle [10]; it is a way to regulate stemness by preventing SCs exhaustion. Satellite cells (SCs) stemness is the ability to perform efficient cell fate decisions to self-renew and to generate myogenic progenitors for long-term regenerative potential. It has been demonstrated that loss of quiescence might cause an imbalance in the stem cell pool leading to stem cell depletion and impaired muscle regeneration [11]. Interdependent extrinsic and intrinsic signals coordinate the maintenance of SCs stemness and influence their ability to self-renew as well as their regenerative capacity [12].

Maintenance of the quiescent state requires repression of the genes encoding for both cell cycle proteins and permanent cell cycle exit. A comparative analysis of gene expression profiles revealed that around 500 genes are overexpressed in quiescent SCs, compared to activated SCs [13]. These include negative regulators

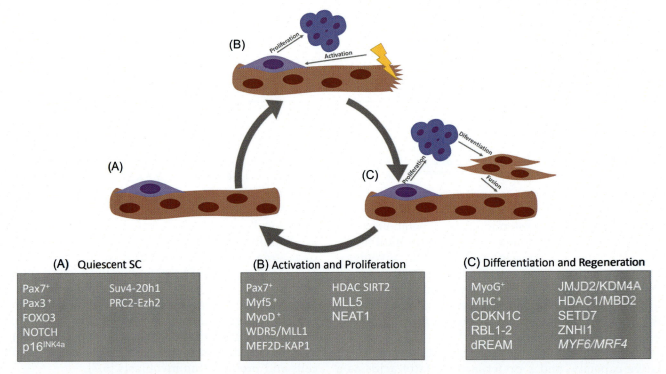

FIGURE 21.1 Transcription factors and chromatin modifiers regulating the different stages of myogenesis discussed throughout the text. A graphical representation of myogenesis is shown. (A-C) Upon activation induced by damage, quiescent satellite cells become activated to form proliferating myoblasts, which differentiate to form a multinucleated myotube. A fraction of satellite cells proliferates to renew the pool of stem cells required for future regeneration events. The lower portion of the figure shows the transcription factors required for this process and the stages for which they are required.

of cell cycle progression such as cyclin-dependent kinase inhibitor 1B (CDKN1B) and 1C (CDKN1C), tumor suppressor protein RB, and fibroblast growth factor inhibitor factor, sprouty 1 (SPRY1) [13,14]. For instance, the cellular tumor antigen p53 has been shown to maintain a reversible cell cycle arrest in quiescent SCs; whereas, the activation of tumor suppressor ARF (p16^{INK4a}) leads to a definitive cell cycle arrest and senescence [15–17]. Different pathways contribute to the quiescent SCs transcriptional network to maintain this balance. Specifically, the expression of forkhead box (FOXO) transcription factors, FOXO1, FOXO3A, and FOXO4, were reported recently by García-Prat et al. in 2020, to be required for the maintenance of the SCs quiescence [18]. Gopinath et al. in 2014, reported that FOXO3 maintains the expression of Notch pathway components [19]. It has also been reported that active Notch pathway leads to a decreased expression of E3 ubiquitin-protein ligase Mdm2 (MDM2) which allows accumulation of p53 to maintain cell cycle arrest and quiescence until injury [17–21]. On the other hand, p16^{INK4A} needs to be kept in a repressed state to prevent SCs from entering a definitive senescent cell cycle arrest [17].

As mentioned above, in the quiescent state, transcription levels are relatively low due to the condensed state of the chromatin [21]. For instance, many genes are expressed including the myogenic regulatory factors (MRFs) *MYOD, MYF5, MYOG and MRF4*, which show accumulation of mRNA but not of their encoded proteins. Boonsanay et al. suggested that heterochromatin-mediated MYOD repression may have an important role to maintain quiescence [21]. Also, de Morrée et al. demonstrated through high-resolution single-molecule and single-cell analyses that quiescent SCs express high levels of MYOD transcript in vivo, whereas MYOD protein is absent [22]. They also elucidated that MYOD mRNA interacts with Staufen1, a potent regulator of mRNA localization, translation, and stability, which prevents MYOD translation through its interaction with the MYOD 3′-UTR and that blocking MYOD translation maintains the quiescent phenotype of SCs. It has been proposed that SCs use additional mechanisms beyond transcriptional regulation to modulate their fate as it is the case for MYF5 and MYOD mRNA nuclear retention and that this must be controlled through epigenetic mechanisms. For example, expression of a H4K20me2 (dimethylation of lysine 20 of histone 4) methyltransferase, Suv4–20 h1, is required for condensation of chromatin and repression of MYOD expression in quiescent SCs [21,23]. On the other hand, there are studies showing that there is no difference in terms of the MYOD mRNA levels in the quiescent vs activated satellite cells [15].

PAX7 is a master regulator of satellite cell function as it controls the expression of genes promoting satellite cell survival and proliferation, while inhibiting their differentiation [24,25]. For example, *PAX7* methylation by CARM1 during asymmetric division has been shown to promote *PAX7* interaction with MLL1/2 to activate MYF5 transcription in the committed daughter cell [26–29]. Additionally, Addicks et al. found that *PAX7* recruits the H3K4 methyltransferases *MLL1/2* to epigenetically activate target genes. They also determined that *MLL1*, but not *MLL2*, is absolutely required for PAX7 expression and SC function in vivo. In fact, the deletion of *MLL1* in satellite cells reduced satellite cell proliferation and self-renewal, and significantly impaired skeletal muscle regeneration [30]. Recent findings from Sincennes et al. provided important insight into the mechanistic control of SC self-renewal through the regulation of posttranslational modifications on *PAX7*. They found that acetylation of PAX7 protein regulates its transcriptional activity and therefore, regulates muscle stem cell function and differentiation potential to facilitate metabolic adaptation of muscle tissue [29]. With these findings, the regulation of PAX7 as a master regulator of satellite cell identity is clear, however, its regulation during quiescence has yet to be fully explored.

To maintain quiescence, SCs prevent cell cycle entry through the expression of specific cell cycle inhibitors. As mentioned earlier, the choice of cell cycle inhibitors is indispensable, as the expression of p16^{INK4a} cell cycle inhibitor leads to senescence and permanent cell cycle exit. It has been reported that the repression of p16^{INK4a} in SCs is assured by the Polycomb PRC1 complex where the Ring1B E3 ubiquitin ligase mediates H2A monoubiquitination at lysine 119 (H2AUb) of the INK4a locus [17,31,32]. In addition, the Polycomb PRC2 complex containing the EZH2 subunit was shown to bind at the INK4a promoter to control its transcription in in vitro culture of mouse embryonic fibroblasts through the depositing of the repressive H3K27me3 mark [33]. It has also been suggested that the regulation of the INK4a locus by PRC2 is also likely to occur in SCs as a SCs-specific KO of EZH2 prevented the expansion of the stem cell population [34].

It is well known that MYOD has an important function in SCs commitment, and its expression is needed for their activation and proliferation. While the gene is expressed at low levels, a repressive chromatin environment is maintained at the *MYOD* locus to prevent high-level expression until activation. Boonsanay et al. elucidated that histone methyltransferase (HMT) Suv4–20 h1 adds H4K20me2 marks at *MYOD* promoter and at its distal regulatory region (DRR), 5 kb upstream of *MYOD* transcription starting site. Addition of H4K20me2 at these sites induces heterochromatin formation and decreased MYOD expression in early activated SCs [21]. The paper by Boonsanay et al. shows that Suv4—20h1 controls MYOD expression and satellite cell quiescence by promoting facultative heterochromatin formation and that loss of Suv4—20h1 leads to the activation of satellite cells. Since this mechanism is necessary to repress MYOD expression during quiescence, too [21,23], it could be inferred that these histone PTMs are already present at the quiescent state and maintained during the early activation.

Coupled with that mentioned above, the DRR and promoter of MYOD are also marked by the repressive H3K9me2 modification. While the enzyme responsible for H3K9me2 marking is not known, some mechanisms maintaining this mark have been discovered. For example, it has been elucidated that the E3 ubiquitin ligase Deltex2 is essential to maintain H3K9me2 mark through the inhibition of lysine demethylase Jumonji domain containing 1 C (JMJD1c) enzyme. Inhibition of JMJD1c functions prevents demethylation to allow maintenance of H3K9me2 marks at the MYOD promoter and DRR. It is noted that the removal of H3K9me2 at these regulatory regions is necessary for the increased expression of MYOD that drives SCs activation [35].

The expression of muscle-specific genes is also repressed both in quiescent and activated SCs. In this sense, Polycomb group genes code for chromatin multiprotein complexes that are responsible for maintaining gene silencing of transcriptional programs during differentiation and in adult tissues [36]. This complex mediates the addition of H3K27me3 marks to myosin heavy chain 2b (MYH4) and myogenin (MYOG) promoters and to muscle creatine kinase gene (MCK) enhancer, leading to their repression [37]. Stojic et al. showed later that wo distinct Polycomb repressive complex (PRC)2 complexes contribute to skeletal muscle cell differentiation: the PRC2-Ezh2 complex, which is bound to the myogenin (MYOG) promoter and muscle creatine kinase (mCK) enhancer in proliferating myoblasts, and the PRC2-Ezh1 complex, which replaces PRC2-Ezh2 on MyoG promoter in postmitotic myotubes. From this study, it could be inferred that the PRC2-Ezh1 complex is necessary for controlling the proper timing of MyoG transcriptional activation and thus, in contrast to PRC2-Ezh2 complex, is required for myogenic differentiation [36]. On the other hand, Luo et al. provided evidence for a role of focal adhesion kinase (FAK) in chromatin remodeling, myogenin induction, and muscle differentiation by interaction with MBD2. They found that FAK translocates into nuclei under specific circumstances (e.g., exposure to oxidative stress of muscle cells), in which it forms a complex with MBD2, attenuates MBD2 transcriptional repressor activity, activates gene (e.g., myogenin) expression, and promotes muscle-terminal differentiation [38].

Technical limitations have hindered the understanding of DNA methylation functions on the maintenance of SCs quiescent state. It could be inferred that the DNA is methylated at the MYOD locus to prevent transcription based on the original lineage conversion studies in fibroblast cell lines [39]. However, recent studies in primary fibroblasts have shown that the MYOD locus is not methylated in normal conditions and only becomes methylated as part of a genome-wide increase in CpG island methylation in response to stress [40]. Thus, the role for DNA methylation at MYOD and other genes in regulating the transition between satellite cell quiescence and activation is an area that still needs to be explored [12].

It is important to mention that most of the epigenetic information collected using isolated SCs does not represent quite the quiescent state but a transition between quiescence and activation known as early activation [15,16]. This is because most studies investigating the role of chromatin and epigenetic regulations in the maintenance of SCs quiescence have largely been performed on SCs isolated from uninjured muscle, with the assumption that cells retain the characteristics of a quiescent cell during the sorting procedure [10]. However, recent work has shown the importance of isolation protocols in the study of true quiescence SCs. Indeed, the development of in situ fixation techniques to lock cells in a quiescent state prior to isolation has exposed important differences in gene expression as well as histone post-transcriptional modifications.

EPIGENETIC CONTROL OF THE ACTIVATION AND PROLIFERATION OF SCS

Concomitant with SCs activation and cell cycle progression, RNA and DNA synthesis become more active [14]. Activated SCs present surface markers involved in activation several signal transduction pathways, including c-Met, FGFR1, FGFR4, syndecan-3, and syndecan-4 [41—44]. On the other hand, it has been described that the mechanical stress produced by exercise or myofibril rupture induces the release of TNF-α, nitric oxide (NO) and the hepatocyte growth factor (HGF) which act as activation signals. Specifically, the release of NO by nitric oxide synthase (NOS) triggers the release of HGF from the extracellular matrix, which associates with its c-Met receptor, improving the signaling of p38/MAPK to modulate cell proliferation and differentiation [5,45].

The expression of Myf5, a key transcription factor involved in regulating proliferation of SCs, is upregulated after activation [46]. This correlates with the enrichment of the H3K4me3 mark in the MYF5 promoter. The HMT complex (WDR5/ASH2L/MLL1) which is recruited by PAX7 through interaction with the CARM1 protein, is responsible for this modification [26,45]. Likewise, in activated SCs, the expression of Pax7 is regulated by the antagonism of the proteins of the PcG and TrxG to silence or activate their expression, respectively. KO mice for lysine methyltransferase MLL1 have been described to exhibit loss of PAX7 and MYF5 expression in activated and proliferating SCs but no effect in quiescent SCs [21]. Therefore, it is concluded that the expression of PAX7 must be maintained during the activation and proliferation of SCs.

Maintaining the open state of the PAX7 gene in activated cells is associated with the participation of the SWI/SNF remodeling complex. Specifically, the Brg1 subunit of the SWI/SNF complex is a phosphorylation target for casein kinase 2, which has been related to the activation of the complex and the proliferation of SCs [47]. This increase in the expression of PAX7 correlates with its positioning in chromatin regions with activation marks such as H3K4me1 and H3K27Ac [48]. MyoD regulatory elements are among regions enriched in PAX7.

It has been shown that in proliferating cells the central enhancer of MyoD (CER) undergoes a regulation by the transcription factor MSX1 (Msh homeobox 1) that binds to the histone isoform H1b, inducing compaction of chromatin in CER thus reducing MyoD's expression [49]. This repression of MyoD during the proliferation of SCs is controlled in response to signaling via the Notch1 pathway, by the transient expression of the inhibitor HES1 [50]. Also the repression of MyoD could be regulate by Pax7 [51]. Although the expression of MyoD is negatively regulated during proliferation, low levels of this protein allow the transcriptional repression of differentiation factors during proliferation. MyoD acts as a transcriptional repressor through interaction with MEF2D and KAP1. This interaction allows stabilizing the association with corepressors such as G9a and HDAC and coactivators such as p300 and LSD1. Therefore, these factors are recruited into the chromatin of genes associated with differentiation, which allows expression as a switch between proliferation and differentiation [52]. In addition to those already mentioned, the HP1α and β proteins interact with the MyoD-MEF2D-KAP1 complex, which establishes a heterochromatin state in their target genes, in cooperation with HDAC1 [46].

Furthermore, in C2C12 cells, Myog is repressed by the formation of a MyoD/P300/CBP(PCAF) complex, which in association with HDAC-SIRT2 promotes proliferation and downregulates differentiation [53]. Similarly, it has been described that the recruitment of Suv39h1 at the Myog locus allows its transcriptional repression by the histone mark H3K9me3; thus, inhibiting differentiation [54]. The recruitment of Suv39h1 is mediated by MyoD, which is phosphorylated at serines 199 and 200 via the p38/MAPK pathway [55]. Although, MyoD has been shown to play a role in the repression of the expression of genes associated with differentiation during the proliferation state, it also has a predominant role as a transcription factor for these genes by binding to the E boxes in the target genes. In this sense, the union of transcriptional repressors such as Snai1 together with HDAC1/2 in the E boxes has been described, which prevents the union of MyoD and transcriptionally represses these genes [56].

In addition to transcriptional control through repression of differentiation genes, activated cells must exert a positive control over proliferation. In this sense, the recruitment of histone acetyltransferases such as the p300/CBP and PCAF/GCN5 complexes recruited by the E2F family of transcription factors on the chromatin regions of cyclins has been reported [57,58]. The increase in the expression of E2F1 and its binding to PCAF allows the recruitment on the target genes of the transcription factor, which induces the passage from G1 to S [59]. Likewise, E2F recruits histone methylases of the KMT2 family, such as the MLL5 protein (which introduces the brand H3K4me3) on the locus of cylin A2, or SET2 (which introduces the brand H3K36me) that induces the increase of cylins D1 and E2, controlling the passage G1/S and G2/M [60].

In addition to this control, there are many lncRNAs associated with histone modifications by recruiting histone modifiers [61]. Nuclear paraspeckle assembly transcript 1 (NEAT1), is an lncRNA that accelerates myoblast proliferation and inhibits differentiation. The first effect by inducing the reduction of expression level of the cyclin dependent kinase inhibitor P21 gene; in addition, NEAT1 suppress the transcription of specific myogenic genes, such as Myog, Myh4, and Tnni2. Different studies show that NEAT1 interact with the enhancer region of Ezh2, recruiting Ezh2 to specific promoters [62,63], leading to repression of gene transcription [64,65].

Other lncRNAs can also regulate proliferation by recruiting chromosome modification complexes to target gene promoters. Overexpression of SYNPO2 intron sense-overlapping lncRNA (SYISL), induces myoblast proliferation and reduces differentiation. Whereas, its inhibition induces muscle differentiation and regeneration. Mechanistically, SYISL recruits PRC2 to promoter elements in the activation of several target genes such as p21, Myog, or Myh4 [66].

In addition to transcriptional control mediated by posttranslational modifications of histones, SC activation and proliferation is controlled by DNA methylation. This transcriptional control is mainly focused on the repression of the expression of inhibitors of the cell cycle, as has been shown for the inhibitor CDN1C (an inhibitor of cyclin C1) which is repressed by the action of the methyltransferase DNMT3A. For instance, knocking down of the DNA methyltransferase 3A (DNMT3A) in myoblasts results in a decrease in cell proliferation and an increase in CDN1C levels. Likewise, CDN1C knockdown rescues proliferation in myoblasts [67]. In an analogous manner, it has been shown that MyoD controls the expression of the ZFP238 protein, which recruits DNMT3A and HDAC1 to the promoter region of genes associated with differentiation, suggesting a CDN1C modulation via ZFP238 [68,69].

In conjunction with the repression of cell cycle inhibitors, DNA methylation has been shown to be important in promoters of differentiation genes such as Myog. It has been shown that the repression of Myog expression is achieved through the methylation of its promoter, as demonstrated by methylation-sensitive restriction endonuclease analysis and bisulfite sequencing revealing hypermethylation in proliferating C2C12 cells [70]. In addition to the methylation of the CpG islands in the Myog promoter, the binding of a methylation-dependent protein called ZBT38 is necessary, which functions as a transcriptional repressor, thus maintaining proliferation. Because of this, the ZBT38 knock out leads to differentiation in C2C12 cells [71].

EPIGENETIC CONTROL OF SCS DIFFERENTIATION

One of the first steps during the transition to differentiation is the stop of SCs proliferation by cell cycle arrest, which is mediated by the expression of the cell cycle inhibitors CDKN1C or CDKN1A (p21) [72]. Also, a progressive increase of MYOD expression is allowed by the reduction of Pax7 levels, leading to the expression of early differentiation markers such as MYOG. This also coincides with the abrogation of MYF5 [73]. Indeed, PRC2 interacts with Yin Yang 1 transcription factor, (YY1) via the phosphorylation of threonine 372 of EZH2, a subunit of PRC2 complex, by p38α. This interaction allows the replacement of H3K4me3 marks by H3K27me3 and leads to the formation of a repressive chromatin state at Pax7 promoter and to the repression of its expression [38,74,75].

Additional factors involved in cell differentiation include the E2F family of transcription factors. Among them, retinoblastoma-associated protein (RB), retinoblastoma-like protein 1 (RBL1) and 2 (RBL2) pocket proteins, are known to control gene expression of genes regulating cell cycle and promoting differentiation in various tissues [76,77] RB interacts with HDAC1 in order to maintain cell proliferation and can also regulate cell cycle exit through its interaction with E2F4. RB interacts with histone methyltransferases in order to increase the levels of H3K9me3 and H3K27me3 at promoters of genes involved in cell cycle. These events lead to cell cycle arrest and induction of cell differentiation. This repression is PRC1 and PRC2 dependent, where the addition of H2AK119Ub1 and H3K27me3 marks establish a silent state [78]. This inhibition is maintained by the recruitment of dimerization partner, RB-like, E2F, and MuvB (dREAM) chromatin compaction complex to target genes through interactions between its L3MBTL1 subunit, E2F4, and HP1γ heterochromatin protein [79,80].

MYOG expression is also essential for SCs to commit to differentiation and the ability of transcription factors such MYOD, MEF2D, SIX4, and FOXO3 to create a transcriptional competent state at the *MYOG* promoter depends upon the combined activity of multiple epigenetic enzymes. One of the initial events in activating the *MYOG* gene is the removal of the repressive H3K9me2 and H3K9me3 marks from the promoter by the action of the lysine demethylase JMJD2/KDM4A [81]. In addition, focal adhesion kinase (FAK) helps achieve the open chromatin state by facilitating the departure of HDAC enzymes from the promoter. In this case, FAK binds the methyl-CpG-binding protein MBD2, where it induced phosphorylation of HDAC1 that breaks up the HDAC1/MBD2 interaction and dissociates it from the promoter [82]. Once the repressive marks are cleared, the promoter can then be modified to accumulate transcriptionally permissive marks. One of the first marks to appear is the dimethylation at arginine 8 in histone 3, (H3R8me2) at the promoter. This is achieved through the PRMT5 protein, a type II arginine methyltransferase. Once the H3R8me2 mark is in place, the epigenetic mark acts to allow a stable association of the chromatin remodeling complex switch/sucrose non-fermentable (SWI/SNF) through recognition of modified histone tail by the Brg1 subunit of the complex. The association of SWI/SNF with the promoter then allows chromatin decompaction for RNA polymerase II to access the gene [83].

In addition, the histone methyltransferase SETD7 is targeted to the *MYOG* promoter by MyoD to introduce the H3K4me1 mark. SETD7 is required for differentiation as silencing of SETD7 leads to a reduced number of myotubes and loss of expression of MYOG [36,84,85]. The addition of H3K4me1 by SETD7 prevents the reintroduction of repressive H3K9me3 repressive marks by blocking Suv39h1 function. Without the presence of HDAC1 at *MYOG* promoter, the histone acetyltransferase P300/CAF leads to its enrichment of H3K9 and H3K14 acetylation and the expression of MYOG [86].

This association of the histone acetyltransferase P300 with the *MYOG* promoter is facilitated by chromatin-binding protein, NUPR1 (P8), which also recruits the RNA helicase DDX5 to the locus to promote high levels of gene expression [87]. The introduction of H3K36me3 marks is also essential to high level expression of MYOG as silencing of SETD2 blocked its expression and prevented myotube formation during differentiation [60]. Finally, histone exchanges within the nucleosome can alter the expression of the MYOG gene. During the differentiation process, the subunit ZNHI1 (p18Hamlet) of SNF2 complex is phosphorylated by P38α which allows its recruitment to MYOG promoter and allows the replacement of H2A histones by its less-stable variant H2A.Z [88].

For differentiation to proceed, cells must also begin to express functional genes that define the muscle lineage. Activation of these MYOD target genes requires the recruitment of SWI/SNF complex which is facilitated by the presence of histone 4 hyperacetylation that is recognized by the bromodomain of the transcription activator BRG1 (SMCA4), leading to chromatin decompaction and gene expression [89]. SWI/SNF recruitment is facilitated by the incorporation of the MYOD-associated SMRD3 (BAF60C) subunit into the chromatin remodeling complex [90]. In addition, MYOD recruits the histone acetyltransferase P300, the JDP2, the AP1 (JUN) and RUNX1 TFs, and SETD7 HMT, leading to active histone modification marks H3K27Ac, H3K18ac, and H3K4me1 that target RNA polymerase II to the promoter region of muscle differentiation genes [91]. Interestingly, SETD7 methylates non-histone proteins such as the TF SRF. SRF acetylation promotes its binding to its serum response element at the muscle-specific gene Acta1 to promote its expression. This regulation is necessary for differentiation and is revered by KDM2B [92]. These data suggest an indirect function of SETD7 in the regulation of differentiation. However, apart from Acta1, no other SRF targets have been identified in SCs so far.

On the other hand, while many repressive enzymes are removed from the MYOD target genes during differentiation, some complexes remain to permit repression of gene expression in response to a changing environment. An example of this is the MYH4, MYOG, and CKM genes which are marked by the PRC2 mediated H3K27me3 modifications in proliferating SCs to repress their expression and to prevent differentiation. As differentiation initiates, a switch occurs between subunits in the PRC2 complex where the EZH2 subunit in proliferation (PRC2-EZH2) is replaced by a functionally inactive EZH1-containing PRC2 complex that lacks the EED subunit (PRC2-EZH1) [34,93]. To allow gene expression during differentiation, removal of the H3K27me3 modifications is mediated by the H3K27 demethylase KDM6A (or UTX), a H3K27 demethylase that opens the chromatin to allow the expression of MYOG and the entrance in differentiation [94,95]. The stable association of PRC2-EZH1 complex that lacks methyltransferase activity helps to maintain the transcriptional permissive state. In response to cellular stresses such as muscle atrophy, the PRC2-EZH1 complex incorporates an EED subunit to form a functional HMT complex that can re-introduce H3K27me3 marks at these muscle genes to prevent their expression [34,93].

As MYOG begins to push the differentiation program forward, additional epigenetic events will lead to the expression of terminal differentiation-related genes and fusion of differentiated myocytes into myotubes. TrxG complexes, containing a H3K4me3 methyltransferase ASH2L, are recruited by MEF2D to muscle-specific genes such as a muscle cytoplasmic enzyme, the muscle creatine kinase, where they mediate the addition of H3K4me3 marks that promote gene expression [96]. It has been observed that the recruitment is modulated through phosphorylation of MEF2D at threonine 308 and 315 by P38α MAPK. Interestingly, the function of a histone arginine methyltransferase Prmt5 is critical for early differentiation but is dispensable for late differentiation while the type I arginine methyltransferase, CARM1/PRMT4 is necessary for late differentiation as it deposits dimethylation of arginine 17 at histone 3 to permit gene activation [83,97]. This reveals the refined roles of distinct epigenetic enzymes in ensuring the temporal expression of genes during muscle differentiation.

When fully differentiated, myocytes fuse thanks to Myomaker (a muscle-specific membrane protein required for fusion of embryonic myoblasts) and Myomerger (a muscle-specific membrane protein involved in formation of multinucleated muscle cells by mediating the transition from the early hemifusion stage to complete fusion to form multinucleated myofibers). While the epigenetic regulation of the Myomaker and Myomerger proteins has yet to be elucidated, the membrane protein CDON which positively regulates fusion was shown to be regulated by the Trithorax HMT ASH1L which deposits dimethylation of lysine 36 of histone 3 (H3K36me2) at the transcription start site to prevent Polycomb-mediated repression [98]. Interestingly, the absence of ASH1L provokes a diminution of fusion capacities of mouse and human myocytes in vitro without impairment of myosin heavy chain (MHC) protein expression. This suggests a direct control of fusion by a Trithorax complex and one can imagine that understanding of the regulation of fusion by epigenetic will emerge soon.

Several studies have revealed the important role of DNA methylation in regulating muscle differentiation. Indeed, during differentiation, the whole DNA methylation landscape was reported to decrease [99]. Several years ago, sodium arsenic treatment of C2C12 was reported to reduce differentiation capacities of the cells.

This was correlated with increased DNA methylation of the CpG site of *MYOG* promoter and a diminution of MYOG expression [100]. Hypermethylation of *MYOG* promoter in C2C12 cells decreases quickly after the induction of differentiation [70]. DNA methylation is known to repress TF binding. However, using a model of 293 T cells expressing luciferase reporter construct and different TFs, the binding of Sine oculis homeobox homolog 1 (SIX1) and MEF2A at *MYOG* promoter was confirmed. Silencing SIX1 in C2C12 leads to increased methylation of *MYOG* promoter, suggesting that SIX1 could have a role in the repression of the methylation [101,102].

Recently, ten-eleven translocation (TET) proteins have been shown to catalyze the conversion of DNA 5-methylcytosine into different oxidized forms, 5-hydroxymethylcytosine, 5-formylcytosine, and finally in 5-carboxylcytosine, demonstrating active demethylation capacities [100–104]. Supporting the idea of a decrease in DNA methylation during differentiation, TET1 and TET2 expression are significantly increased in myoblasts after induction of differentiation. Interestingly, inhibition of TET2 but not TET1 by siRNA in C2C12 results in increased DNA methylation of *MYOG*, *MYF6/MRF4*, and *MYMKy* (Myomaker) gene promoters. The increased DNA methylation at the promoter of these genes correlates with a decreased expression and an abrogation of C2C12 differentiation [105]. The presence of CpG methylation at Myog promoter is necessary for the binding of ZBT38 protein and the decreased expression of Myog. The diminution of methylation at *MYOG* promoter may lead to ZBT38 removal and abrogation of its repression [71]. In addition to *MYOG* promoter demethylation, a general diminution of DNA methylation at the CpG sites of *MYOD* promoter occurs after 3 days of differentiation in C2C12 cells [106].

DNA methylation changes during differentiation are not only attributed to demethylation of the promoter of genes required for myogenic differentiation. The addition of DNA methylation at CpG sites on the promoter of specific genes is also necessary to allow differentiation of myoblasts. In C2C12 cells, after 3 days of differentiation, an increase in DNA methylation of CpG sites of Pax7 and Myf5 promoters was shown. Interestingly, after 5 days of differentiation, the DNA methylation is still higher as compared with the proliferative state but slightly reduced as compared to day 3 of differentiation.

Genetic deletion of DNMT3A in mice resulted in fibrosis and reduction of cross-section area of the muscle after regeneration from an acute injury. The diminution of DNMT3A is correlated with a diminution of promoter DNA methylation and the expression of *GDF5*, an important muscle gene. Interestingly, GDF5 increased expression does not change proliferation or differentiation capacities. However, a diminution of myofiber size, length, and nuclei number and a decreased expression of differentiation-related genes was reported, suggesting that DNA methylation of *GDF5* promoter is necessary to avoid undesired muscle atrophy [107].

Modification of DNA methylation during differentiation is also correlated with histone modifications. In particular, heterochromatin protein HP1γ recognizes and binds H3K9me3 over the genome, interacts with DNMT1, and recruits HMTs [108,109]. During this phenomenon, the protein level of HP1γ does not change, but its spatial localization does and is correlated with the presence of a methyl-CpG-binding protein, MECP2. These functions allow the maintenance of specific gene silencing by the addition of DNA methylation during differentiation and suggest interactions between DNA methylation and changes in histone PTMs [110].

SMALL MOLECULES AS A THERAPEUTIC ALTERNATIVE IN THE EPIGENETIC CONTROL OF REGENERATION

Enzymatic activities mediating a plethora of post-translational modifications of histones and non-histone proteins is vital for many cellular and biological processes. Because these modifications mediated by epigenetic factors are reversible, modulating its activity is a promising strategy to control muscle differentiation and regeneration. A growing list of different compounds and small molecules with activity for the modulation of epigenetic modifiers have been identified and designed (Table 21.1).

As discussed before, DNA methylation has a relevant role in muscle differentiation. Inhibition of DNMT by 5-azacitidine in conjunction with treatment of VEGF induce differentiation of human primary granulosa cells (GC) to muscle cells [147]. The treatment with 5-azacitidine and VEGF allows a temporary upregulation of pluripotency markers, accompanied by the loss of specific ultrastructural characteristics of GC by autophagocytosis. This was associated with a temporary chromatin decondensation. Following exposure to VEGF, induction of muscle-specific genes with the appearance of multinucleated cells was observed, along with a considerable amount of non-spatially organized filaments [147].

In contrast, Jun Lee et al. in 2007 [148] showed that myoblast cells can be induced to transdifferentiate into smooth muscle cells by modulating their epigenetic program using Zebularin, a DMNT1 inhibitor that induces the morphological transformation of C2C12

TABLE 21.1 Summary of Compiled Modifiers [111].

Type	Name	Epigenetic Factor	Reference
DNMT inhibitors	5-Azacytidine	DNMT1	[112]
	Decitabine	DNMT1	[112]
	Zebularine	DNMT1	[113]
	3-DeazaneplanocinA	Global demethylation	[114]
	Epigallocatechin-3-gallate EGCG	DNMT1	[115]
	S110	DNMT1	[116]
	RG108	DNMT1	[117]
HMT inhibitors	GSK126	EZH2	[118]
	Tazemetostat	EZH2	[119]
	BIX-01294	G9a	[120]
	Pyrazole amide compounds	PRMT4	[121]
	Ellagic acid (TBBD)	PRMT4	[122]
	GSK3235025	PRMT5	[123]
	GSK3326595	PRMT5	[123]
	AMI-1	PRMTs	[124]
	RM65	PRMTs	[125]
HDAC inhibitors	Trichostatin A	HDAC1	[126,127]
	Valproic acid	HDAC1	[128]
	Phenyl butyrate	HDAC1	[129]
	Sodium butyrate	HDAC1	[130]
	Ms-275	HDAC1 and HDAC3	[131]
	CI-994	HDAC1 and HDAC3	[132]
	Romidepsin (FK228)	HDAC1	[133]
	Compound 60	HDAC1 and HDAC2	[134]
	RG2833	HDAC1 and HDAC3	[135]
KAT inhibitors	Lys CoA	p300	[135]
	H3CoA20	PCAF	[135]
	Anacardic acid	p300, PCAF	[136]
	Garcinol	p300, PCAF	[137]
	Curcumin	p300/CBP	[138]
	γ-Butyrolactones	CBP, Gcn5	[139]
	Isothiazolones	p300/PCAF	[140]
	LTK-14	p300	[115]
	Plumbagin	p300/CBP, PCAF, Tip60	[141]
	Iso-garcinol	p300/CBP, PCAF, Tip60, MOF, MOZ	[142]
	PTK1	P300/CBP	[143]
	Embelin	PCAF	[144]
	C646	p300/CBP	[145]
	A-485	p300/CBP	[146]

myoblasts into smooth muscle cells, accompanied by de novo synthesis of smooth muscle markers such as α-actin and smooth muscle transgelin. Furthermore, an increase in p21 and a decrease in cyclinD1 mRNA was observed after treatment, indicating an inhibition of cell cycle progression [148].

Also, it has been shown that (-)-epigallocatechin-3-gallate (EGCG), increased mRNA levels of genes associated with the oxidation of fatty acids in skeletal muscle, such as nuclear respiratory factor (nrf) 1, medium chain acyl coA decarboxylase (MCAD), uncoupling protein (ucp) 3 and peroxisome proliferator response element (ppar) α [149]. And it has been reported that EGCG can reduce signaling for apoptosis and promote skeletal muscle recovery in the fast plantaris muscle and the slow soleus muscle after hindlimb suspension (HLS) in senescent animals [141].

Other strategy is the inhibition of histone methyltransferase proteins. EZH2, a subunit of the PCR2 complex has been shown to be inhibited by small molecules such as GSK126 or Tazemetostat. Exposition of primary myoblast to GSK126 enhance proliferation, resulting in upregulation of genes like Ccnd1, Ccne1, Cdk2, Cdk4, Pcna and cyclin D1 [150]. On the other hand, an inhibition of the histone methyltransferase G9a by the molecule BIX-01294, led to a significant rescue of myogenic differentiation [52,151] as reported Mei Tze et al., in 2012.

Protein arginine methylation catalyzed by PRMT4 and PRMT5, regulates multiple biological processes such as chromatin remodeling and RNA splicing, and are implicated on muscle regeneration [111]. Inhibitors of PRMTs include, Ellagic acid (TBBD) and pyrazole-containing derivatives that specifically inhibits H3R17me2 (dimethylation on arginine17 of histone H3), and H3R26me2 mediated by PRMT4. Specifically, TBBD and its derivatives have been reported to function by binding to histone in the KAPRK motif altering Proline 16 for inhibition upon enzyme interaction [122]. GSK3235025, and its derivative GSK3326595 are other small molecules that are suggested to be specific inhibitor of PRMT5 that catalyze H3R8me2 and H4R3me2, marks for transcriptional repression [111].

It has previously been discussed how different HDACs control development, homeostasis and regenerative potential of specific tissues and organs. Thus, HDACs inhibition have been used to treat several aspects of degenerative diseases including neuromuscular, neurodegenerative, and cardiovascular conditions [152], and to correct phenotypes in experimental models. In particular, treatment with HDAC inhibitors (HDACi) as a new pharmacological intervention in the context of Duchenne muscular dystrophy (DMD) has promising potential leading to include different protocols in preclinical studies and early clinical studies [153–155].

The route of action of HDACi is mainly based on its effect on fibroadipogenic progenitors (FAP). FAPs support SC-mediated repair with acute injuries, but they become cellular effectors of fibrotic and adipogenic degeneration of muscles exposed to conditions of chronic damage, in DMD and other neuromuscular disorders. It has been shown in murine model for DMD, the mdx mice in early stages of the disease, that FAPs can promote SC-mediated compensatory regeneration and are susceptible to HDACi-mediated enhancement of their pro-regenerative activity and inhibition of their fibroadipogenic potential [156,157]. Additional observations have revealed that exposure to HDACi promotes the formation and release of pro-regenerative and antifibrotic extracellular vesicles (VE) from FAPs of DMD muscles, in the early stages of the disease [158].

Protein acetyltransferases (KAT) form a ternary complex with the lysine substrate and acetyl-coA within their active site during the acetylation process. Due to this, the design and generation of KAT inhibitors (KATi) was based on molecules that mimic the ternary complex. As an example, Philip Cole et al. in 2000, described Lys-CoA and H3-CoA-20, as inhibitors of the KAT activity of p300 and PCAF, respectively [135]. Along with the design of inhibitors, a screening of natural products with KAT modulating activity has been carried out; as an example, anacardic acid, which was isolated from cashew shells. Initially, it was found to be a non-specific inhibitor of p300/CBP and PCAF [136]. Another natural compound is curcumin, one of the main curcuminoids in turmeric (*Curcuma longa*). This compound is a p300 inhibitor without having an effect on PCAF activity [138]. A third compound that has KAT inhibitory activity is arcinol, this inhibits the KAT activity of p300 and PCAF KAT [137]. A fourth natural compound is Embelin isolated from *Embelia ribes*. This targets the blocking of PCAF KAT activity and has been shown to block differentiation into C2C12 cells through inhibition of PCAF-mediated MyoD acetylation [144].

CONCLUSION

As elucidation of the epigenetic mechanisms mediating skeletal muscle regeneration are being solved, advances in the design of molecules focused on increasing some of the multiple functional properties of muscle, in the context of either diseases or in health conditions, is of the utmost need. This chapter has highlighted the molecules and factors involved in the regulation of critical epigenetic processes taking place during skeletal muscle regeneration. Of particular interest are skeletal muscle satellite cells, since they could prove to be useful for restoring metabolic and wider homeostatic processes not only in the skeletal muscle, but in the whole organism. We look forward to future research establishing clinically and therapeutically relevant roles for satellite cells in combination with small molecules and new drugs, combating conditions to skeletal muscle.

Acknowledgments

Work in our research group is supported by Consejo Nacional de Ciencia y Tecnología (CONACYT), grants CF-2019/140637 and CF-2472263.

References

[1] Baskin KK, Winders BR, Olson EN. Muscle as a "mediator" of systemic metabolism. Cell Metab 2015;21(2).
[2] Bassel-Duby R, Olson EN. Signaling pathways in skeletal muscle remodeling. Annu Rev Biochem 2006;75(1).
[3] Schiaffino S, Reggiani C. Fiber types in mammalian skeletal muscles. Physiol Rev 2011;91(4).
[4] Wang YX, Rudnicki MA. Satellite cells, the engines of muscle repair. Nat Rev Mol Cell Biol 2012;13(2).
[5] Dumont NA, Wang YX, Rudnicki MA. Intrinsic and extrinsic mechanisms regulating satellite cell function. Development 2015;142(9).
[6] Bentzinger CF, Wang YX, Dumont NA, Rudnicki MA. Cellular dynamics in the muscle satellite cell niche. EMBO Rep 2013;14(12).
[7] Kuang S, Kuroda K, le Grand F, Rudnicki MA. Asymmetric self-renewal and commitment of satellite stem cells in muscle. Cell 2007;129(5).
[8] Seale P, Sabourin LA, Girgis-Gabardo A, Mansouri A, Gruss P, Rudnicki MA. Pax7 is required for the specification of myogenic satellite cells. Cell 2000;102(6).
[9] Dumont NA, Wang YX, von Maltzahn J, Pasut A, Bentzinger CF, Brun CE, et al. Dystrophin expression in muscle stem cells regulates their polarity and asymmetric division. Nat Med 2015;21(12).
[10] Machado L, Esteves de Lima J, Fabre O, Proux C, Legendre R, Szegedi A, et al. In situ fixation redefines quiescence and early activation of skeletal muscle stem cells. Cell Rep 2017;21(7).
[11] Orford KW, Scadden DT. Deconstructing stem cell self-renewal: genetic insights into cell-cycle regulation. Nat Rev Genet 2008;9(2).
[12] Ancel S, Stuelsatz P, Feige JN. Muscle stem cell quiescence: controlling stemness by staying asleep. Trends Cell Biol 2021;31(7).
[13] Wozniak AC, Pilipowicz O, Yablonka-Reuveni Z, Greenway S, Craven S, Scott E, et al. C-met expression and mechanical activation of satellite cells on cultured muscle fibers. J Histochem Cytochem 2003;51(11).
[14] Ávila-Avilés RD, López-Méndez LC, Hernández-Hernández JM. Células satélite y su contribución a la regeneración muscular en salud y enfermedad. Revista Mexicana de Industria y Salud 2019.
[15] Liu L, Cheung TH, Charville GW, Hurgo BMC, Leavitt T, Shih J, et al. Chromatin modifications as determinants of muscle stem cell quiescence and chronological aging. Cell Rep 2013;4(1).
[16] Liu L, Cheung TH, Charville GW, Rando TA. Isolation of skeletal muscle stem cells by fluorescence-activated cell sorting. Nat Protoc 2015;10(10).
[17] Sousa-Victor P, Gutarra S, García-Prat L, Rodriguez-Ubreva J, Ortet L, Ruiz-Bonilla V, et al. Geriatric muscle stem cells switch reversible quiescence into senescence. Nature 2014;506(7488).
[18] García-Prat L, Perdiguero E, Alonso-Martín S, Dell'Orso S, Ravichandran S, Brooks SR, et al. FoxO maintains a genuine muscle stem-cell quiescent state until geriatric age. Nat Cell Biol 2020;22(11).
[19] Gopinath SD, Webb AE, Brunet A, Rando TA. FOXO3 promotes quiescence in adult muscle stem cells during the process of self-renewal. Stem Cell Rep 2014;2(4).
[20] Bjornson CRR, Cheung TH, Liu L, Tripathi PV, Steeper KM, Rando TA. Notch signaling is necessary to maintain quiescence in adult muscle stem cells. Stem Cells 2012;30(2).
[21] Boonsanay V, Zhang T, Georgieva A, Kostin S, Qi H, Yuan X, et al. Regulation of skeletal muscle stem cell quiescence by Suv4-20h1-dependent facultative heterochromatin formation. Cell Stem Cell 2016;18(2).
[22] de Morrée A, van Velthoven CTJ, Gan Q, Salvi JS, Klein JDD, Akimenko I, et al. Staufen1 inhibits MyoD translation to actively maintain muscle stem cell quiescence. Proc Natl Acad Sci 2017;114(43).
[23] Li Y, Dilworth FJ. Compacting chromatin to ensure muscle satellite cell quiescence. Cell Stem Cell 2016;18(2).
[24] Soleimani VD, Punch VG, Kawabe Y, Jones AE, Palidwor GA, Porter CJ, et al. Transcriptional dominance of Pax7 in adult myogenesis is due to high-affinity recognition of homeodomain motifs. Dev Cell 2012;22(6).
[25] Sincennes M-C, Brun CE, Rudnicki MA. Concise review: epigenetic regulation of myogenesis in health and disease. Stem Cells Transl Med 2016;5(3).
[26] McKinnell IW, Ishibashi J, le Grand F, Punch VGJ, Addicks GC, Greenblatt JF, et al. Pax7 activates myogenic genes by recruitment of a histone methyltransferase complex. Nat Cell Biol 2008;10(1).
[27] Kawabe Y, Wang YX, McKinnell IW, Bedford MT, Rudnicki MA. Carm1 regulates Pax7 transcriptional activity through MLL1/2 recruitment during asymmetric satellite stem cell divisions. Cell Stem Cell 2012;11(3).
[28] Chang NC, Sincennes M-C, Chevalier FP, Brun CE, Lacaria M, Segalés J, et al. The dystrophin glycoprotein complex regulates the epigenetic activation of muscle stem cell commitment. Cell Stem Cell 2018;22(5).
[29] Sincennes M-C, Brun CE, Lin AYT, Rosembert T, Datzkiw D, Saber J, et al. Acetylation of PAX7 controls muscle stem cell self-renewal and differentiation potential in mice. Nat Commun 2021;12(1).
[30] Addicks GC, Brun CE, Sincennes M-C, Saber J, Porter CJ, Francis Stewart A, et al. MLL1 is required for PAX7 expression and satellite cell self-renewal in mice. Nat Commun 2019;10(1).
[31] Cao R, Tsukada Y, Zhang Y. Role of Bmi-1 and Ring1A in H2A ubiquitylation and hox gene silencing. Mol Cell 2005;20(6).
[32] Wang H, Wang L, Erdjument-Bromage H, Vidal M, Tempst P, Jones RS, et al. Role of histone H2A ubiquitination in Polycomb silencing. Nature 2004;431(7010).
[33] Agherbi H, Gaussmann-Wenger A, Verthuy C, Chasson L, Serrano M, Djabali M. Polycomb mediated epigenetic silencing and replication timing at the INK4a/ARF locus during senescence. PLoS One 2009;4(5).
[34] Juan AH, Derfoul A, Feng X, Ryall JG, Dell'Orso S, Pasut A, et al. Polycomb EZH2 controls self-renewal and safeguards the transcriptional identity of skeletal muscle stem cells. Genes Dev 2011;25(8).
[35] Luo D, de Morree A, Boutet S, Quach N, Natu V, Rustagi A, et al. Deltex2 represses MyoD expression and inhibits myogenic

[36] Tao Y, Neppl RL, Huang Z-P, Chen J, Tang R-H, Cao R, et al. The histone methyltransferase Set7/9 promotes myoblast differentiation and myofibril assembly. J Cell Biol 2011;194(4).

[35] differentiation by acting as a negative regulator of Jmjd1c. Proc Natl Acad Sci 2017;114(15).

[37] Caretti G. The Polycomb Ezh2 methyltransferase regulates muscle gene expression and skeletal muscle differentiation. Genes Dev 2004;18(21).

[38] Asp P, Blum R, Vethantham V, Parisi F, Micsinai M, Cheng J, et al. Genome-wide remodeling of the epigenetic landscape during myogenic differentiation. Proc Natl Acad Sci 2011;108(22).

[39] Jones PA, Wolkowicz MJ, Rideout WM, Gonzales FA, Marziasz CM, Coetzee GA, et al. De novo methylation of the MyoD1 CpG island during the establishment of immortal cell lines. Proc Natl Acad Sci 1990;87(16).

[40] Diede SJ, Yao Z, Keyes CC, Tyler AE, Dey J, Hackett CS, et al. Fundamental differences in promoter CpG island DNA hypermethylation between human cancer and genetically engineered mouse models of cancer. Epigenetics 2013;8(12).

[41] Templeton TJ, Hauschka SD. FGF-mediated aspects of skeletal muscle growth and differentiation are controlled by a high affinity receptor, FGFR1. Dev Biol 1992;154(1).

[42] Sheehan SM, Allen RE. Skeletal muscle satellite cell proliferation in response to members of the fibroblast growth factor family and hepatocyte growth factor. J Cell Physiol 1999;181(3).

[43] Cornelison DDW, Filla MS, Stanley HM, Rapraeger AC, Olwin BB. Syndecan-3 and Syndecan-4 specifically mark skeletal muscle satellite cells and are implicated in satellite cell maintenance and muscle regeneration. Dev Biol 2001;239(1).

[44] Ono Y, Urata Y, Goto S, Nakagawa S, Humbert PO, Li T-S, et al. Muscle stem cell fate is controlled by the cell-polarity protein scrib. Cell Rep 2015;10(7).

[45] Troy A, Cadwallader AB, Fedorov Y, Tyner K, Tanaka KK, Olwin BB. Coordination of satellite cell activation and self-renewal by par-complex-dependent asymmetric activation of p38α/β MAPK. Cell Stem Cell 2012;11(4).

[46] Massenet J, Gardner E, Chazaud B, Dilworth FJ. Epigenetic regulation of satellite cell fate during skeletal muscle regeneration. Skeletal Muscle 2021;11(1).

[47] Padilla-Benavides T, Nasipak BT, Imbalzano AN. Brg1 controls the expression of Pax7 to promote viability and proliferation of mouse primary myoblasts. J Cell Physiol 2015;230(12).

[48] Lilja KC, Zhang N, Magli A, Gunduz V, Bowman CJ, Arpke RW, et al. Pax7 remodels the chromatin landscape in skeletal muscle stem cells. PLoS One 2017;12(4).

[49] Lee H. Msx1 cooperates with histone H1b for inhibition of transcription and myogenesis. Science 2004;304(5677).

[50] Lahmann I, Bröhl D, Zyrianova T, Isomura A, Czajkowski MT, Kapoor V, et al. Oscillations of MyoD and Hes1 proteins regulate the maintenance of activated muscle stem cells. Genes Dev 2019;33(9—10).

[51] Olguin HC, Yang Z, Tapscott SJ, Olwin BB. Reciprocal inhibition between Pax7 and muscle regulatory factors modulates myogenic cell fate determination. J Cell Biol 2007;177(5).

[52] Ling BMT, Bharathy N, Chung T-K, Kok WK, Li S, Tan YH, et al. Lysine methyltransferase G9a methylates the transcription factor MyoD and regulates skeletal muscle differentiation. Proc Natl Acad Sci 2012;109(3).

[53] Fulco M, Schiltz RL, Iezzi S, King MT, Zhao P, Kashiwaya Y, et al. Sir2 regulates skeletal muscle differentiation as a potential sensor of the redox state. Mol Cell 2003;12(1).

[54] Mal AK. Histone methyltransferase Suv39h1 represses MyoD-stimulated myogenic differentiation. EMBO J 2006;25(14).

[55] Gillespie MA, le Grand F, Scimè A, Kuang S, von Maltzahn J, Seale V, et al. p38-γ—dependent gene silencing restricts entry into the myogenic differentiation program. J Cell Biol 2009;187(7).

[56] Soleimani VD, Yin H, Jahani-Asl A, Ming H, Kockx CEM, van Ijcken WFJ, et al. Snail regulates MyoD binding-site occupancy to direct enhancer switching and differentiation-specific transcription in myogenesis. Mol Cell 2012;47(3).

[57] Takahashi Y, Rayman JB, Dynlacht BD. Analysis of promoter binding by the E2F and pRB families in vivo: distinct E2F proteins mediate activation and repression. Genes Dev 2000;14(7).

[58] Taubert S, Gorrini C, Frank SR, Parisi T, Fuchs M, Chan H-M, et al. E2F-dependent histone acetylation and recruitment of the Tip60 acetyltransferase complex to chromatin in late G_1. Mol Cell Biol 2004;24(10).

[59] Rao VK, Ow JR, Shankar SR, Bharathy N, Manikandan J, Wang Y, et al. G9a promotes proliferation and inhibits cell cycle exit during myogenic differentiation. Nucleic Acids Res 2016;44(17).

[60] Yi X, Tao Y, Lin X, Dai Y, Yang T, Yue X, et al. Histone methyltransferase Setd2 is critical for the proliferation and differentiation of myoblasts. Biochim Biophys Acta Mol Cell Res 2017;1864(4).

[61] Arredondo-Robles Av, Rodríguez-López KP, Ávila-Avilés RD. Long non-coding RNAs in cervical cancer. J Appl Genet 2020;405 Springer.

[62] Wang S, Zuo H, Jin J, Lv W, Xu Z, Fan Y, et al. Long noncoding RNA Neat1 modulates myogenesis by recruiting Ezh2. Cell Death Dis 2019;10(7).

[63] Alpizar-Reyes B, Mr B-T, Contreras-Gómora I, Ma M-A, Rd Á-A. Relationship of lncRNA to Breast Cancer. J Cancer Immunol Rev 2020.

[64] Viré E, Brenner C, Deplus R, Blanchon L, Fraga M, Didelot C, et al. The Polycomb group protein EZH2 directly controls DNA methylation. Nature 2006;439(7078).

[65] Shen X, Liu Y, Hsu Y-J, Fujiwara Y, Kim J, Mao X, et al. EZH1 mediates methylation on Histone H3 Lysine 27 and Complements EZH2 in maintaining stem cell identity and executing pluripotency. Mol Cell 2008;32(4).

[66] Jin JJ, Lv W, Xia P, Xu ZY, Zheng AD, Wang XJ, et al. Long noncoding RNA SYISL regulates myogenesis by interacting with polycomb repressive complex 2. Proc Natl Acad Sci 2018;115(42).

[67] Naito M, Mori M, Inagawa M, Miyata K, Hashimoto N, Tanaka S, et al. Dnmt3a regulates proliferation of muscle satellite cells via p57Kip2. PLoS Genet 2016;12(7).

[68] Yokoyama S, Ito Y, Ueno-Kudoh H, Shimizu H, Uchibe K, Albini S, et al. A systems approach reveals that the myogenesis genome network is regulated by the transcriptional repressor RP58. Dev Cell 2009;17(6).

[69] Fuks F. Dnmt3a binds deacetylases and is recruited by a sequence-specific repressor to silence transcription. EMBO J 2001;20(10).

[70] Fuso A, Ferraguti G, Grandoni F, Ruggeri R, Scarpa S, Strom R, et al. Early demethylation of non-CpG, CpC-rich elements in the myogenin 5′-flanking region. Cell Cycle 2010;9(19).

[71] Oikawa Y, Omori R, Nishii T, Ishida Y, Kawaichi M, Matsuda E. The methyl-CpG-binding protein CIBZ suppresses myogenic differentiation by directly inhibiting myogenin expression. Cell Res 2011;21(11).

[72] Zhang P, Wong C, Liu D, Finegold M, Harper JW, Elledge SJ. p21CIP1 and p57KIP2 control muscle differentiation at the myogenin step. Genes Dev 1999;13(2).

[73] Berkes CA, Tapscott SJ. MyoD and the transcriptional control of myogenesis. Semin Cell Dev Biol 2005;16(4—5).

[74] Bhanu Nv, Sidoli S, Yuan Z-F, Molden RC, Garcia BA. Regulation of proline-directed kinases and the trans-histone code H3K9me3/H4K20me3 during human myogenesis. J Biol Chem 2019;294(20).

[75] Palacios D, Mozzetta C, Consalvi S, Caretti G, Saccone V, Proserpio V, et al. TNF/p38α/Polycomb signaling to Pax7 locus in satellite cells links inflammation to the epigenetic control of muscle regeneration. Cell Stem Cell 2010;7(4).

[76] Balciunaite E, Spektor A, Lents NH, Cam H, te Riele H, Scime A, et al. Pocket protein complexes are recruited to distinct targets in quiescent and proliferating cells. Mol Cell Biol 2005;25(18).

[77] Dimova DK. Cell cycle-dependent and cell cycle-independent control of transcription by the Drosophila E2F/RB pathway. Genes Dev 2003;17(18).

[78] Schwartz YB, Pirrotta V. Polycomb silencing mechanisms and the management of genomic programmes. Nat Rev Genet 2007;8(1).

[79] Trojer P, Li G, Sims RJ, Vaquero A, Kalakonda N, Boccuni P, et al. L3MBTL1, a Histone-methylation-dependent chromatin lock. Cell 2007;129(5).

[80] Peng XL, So KK, He L, Zhao Y, Zhou J, Li Y, et al. MyoD- and FoxO3-mediated hotspot interaction orchestrates super-enhancer activity during myogenic differentiation. Nucleic Acids Res 2017;45(15).

[81] Verrier L, Escaffit F, Chailleux C, Trouche D, Vandromme M. A new isoform of the histone demethylase JMJD2A/KDM4A is required for skeletal muscle differentiation. PLoS Genet 2011;7(6).

[82] Luo S-W, Zhang C, Zhang B, Kim C-H, Qiu Y-Z, Du Q-S, et al. Regulation of heterochromatin remodelling and myogenin expression during muscle differentiation by FAK interaction with MBD2. EMBO J 2009;28(17).

[83] Dacwag CS, Ohkawa Y, Pal S, Sif S, Imbalzano AN. The protein arginine methyltransferase Prmt5 Is required for myogenesis because it facilitates ATP-dependent chromatin remodeling. Mol Cell Biol 2007;27(1).

[84] Nishioka K. Set9, a novel histone H3 methyltransferase that facilitates transcription by precluding histone tail modifications required for heterochromatin formation. Genes Dev 2002;16(4).

[85] Wang H, Cao R, Xia L, Erdjument-Bromage H, Borchers C, Tempst P, et al. Purification and functional characterization of a Histone H3-Lysine 4-specific methyltransferase. Mol Cell 2001;8(6).

[86] Mal A, Harter ML. MyoD is functionally linked to the silencing of a muscle-specific regulatory gene prior to skeletal myogenesis. Proc Natl Acad Sci 2003;100(4).

[87] Sambasivan R, Cheedipudi S, Pasupuleti N, Saleh A, Pavlath GK, Dhawan J. The small chromatin-binding protein p8 coordinates the association of anti-proliferative and pro-myogenic proteins at the myogenin promoter. J Cell Sci 2009;122(19).

[88] Cuadrado A, Corrado N, Perdiguero E, Lafarga V, Muñoz-Canoves P, Nebreda AR. Essential role of p18Hamlet/SRCAP-mediated histone H2A.Z chromatin incorporation in muscle differentiation. EMBO J 2010;29(12).

[89] de la Serna IL, Ohkawa Y, Berkes CA, Bergstrom DA, Dacwag CS, Tapscott SJ, et al. MyoD targets chromatin remodeling complexes to the myogenin locus prior to forming a stable DNA-bound complex. Mol Cell Biol 2005;25(10).

[90] Forcales Sv, Albini S, Giordani L, Malecova B, Cignolo L, Chernov A, et al. Signal-dependent incorporation of MyoD-BAF60c into Brg1-based SWI/SNF chromatin-remodelling complex. EMBO J 2012;31(2).

[91] Blum R, Dynlacht BD. The role of MyoD1 and histone modifications in the activation of muscle enhancers. Epigenetics 2013;8(8).

[92] Kwon D-H, Kang J-Y, Joung H, Kim J-Y, Jeong A, Min H-K, et al. SRF is a nonhistone methylation target of KDM2B and SET7 in the regulation of skeletal muscle differentiation. Exp Mol Med 2021;53(2).

[93] Stojic L, Jasencakova Z, Prezioso C, Stützer A, Bodega B, Pasini D, et al. Chromatin regulated interchange between polycomb repressive complex 2 (PRC2)-Ezh2 and PRC2-Ezh1 complexes controls myogenin activation in skeletal muscle cells. Epigenetics Chromatin 2011;4(1).

[94] Faralli H, Wang C, Nakka K, Benyoucef A, Sebastian S, Zhuang L, et al. UTX demethylase activity is required for satellite cell–mediated muscle regeneration. J Clin Investig 2016;126(4).

[95] Seenundun S, Rampalli S, Liu Q-C, Aziz A, Palii C, Hong S, et al. UTX mediates demethylation of H3K27me3 at muscle-specific genes during myogenesis. EMBO J 2010;29(8).

[96] Rampalli S, Li L, Mak E, Ge K, Brand M, Tapscott SJ, et al. p38 MAPK signaling regulates recruitment of Ash2L-containing methyltransferase complexes to specific genes during differentiation. Nat Struct Mol Biol 2007;14(12).

[97] Dacwag CS, Bedford MT, Sif S, Imbalzano AN. Distinct protein arginine methyltransferases promote ATP-dependent chromatin remodeling function at different stages of skeletal muscle differentiation. Mol Cell Biol 2009;29(7).

[98] Castiglioni I, Caccia R, Garcia-Manteiga JM, Ferri G, Caretti G, Molineris I, et al. The Trithorax protein Ash1L promotes myoblast fusion by activating Cdon expression. Nat Commun 2018;9(1).

[99] Tsumagari K, Baribault C, Terragni J, Varley KE, Gertz J, Pradhan S, et al. Early de novo DNA methylation and prolonged demethylation in the muscle lineage. Epigenetics 2013;8(3).

[100] Steffens AA, Hong G-M, Bain LJ. Sodium arsenite delays the differentiation of C2C12 mouse myoblast cells and alters methylation patterns on the transcription factor myogenin. Toxicol Appl Pharmacol 2011;250(2).

[101] Palacios D, Summerbell D, Rigby PWJ, Boyes J. Interplay between DNA methylation and transcription factor availability: implications for developmental activation of the mouse *Myogenin* gene. Mol Cell Biol 2010;30(15).

[102] Liu Y, Chu A, Chakroun I, Islam U, Blais A. Cooperation between myogenic regulatory factors and SIX family transcription factors is important for myoblast differentiation. Nucleic Acids Res 2010;38(20).

[103] Pfaffeneder T, Hackner B, Truß M, Münzel M, Müller M, Deiml CA, et al. The discovery of 5-formylcytosine in embryonic stem cell DNA. Angew Chem Int Ed 2011;50(31).

[104] Tahiliani M, Koh KP, Shen Y, Pastor WA, Bandukwala H, Brudno Y, et al. Conversion of 5-methylcytosine to 5-hydroxymethylcytosine in mammalian DNA by MLL partner TET1. Science 2009;324(5929).

[105] Zhong X, Wang Q-Q, Li J-W, Zhang Y-M, An X-R, Hou J. Ten-eleven translocation-2 (Tet2) Is involved in myogenic differentiation of skeletal myoblast cells in vitro. Sci Rep 2017;7(1).

[106] Chao Z, Zheng X-L, Sun R-P, Liu H-L, Huang L-L, Cao Z-X, et al. Characterization of the methylation status of Pax7 and myogenic regulator factors in cell myogenic differentiation. Asian-Australas J of Anim Sci 2015;29(7).

[107] Hatazawa Y, Ono Y, Hirose Y, Kanai S, Fujii NL, Machida S, et al. Reduced Dnmt3a increases Gdf5 expression with suppressed satellite cell differentiation and impaired skeletal muscle regeneration. FASEB J 2018;32(3).

[108] Ait-Si-Ali S, Guasconi V, Fritsch L, Yahi H, Sekhri R, Naguibneva I, et al. A Suv39h-dependent mechanism for silencing S-phase genes in differentiating but not in cycling cells. EMBO J 2004;23(3).

[109] Smallwood A, Esteve P-O, Pradhan S, Carey M. Functional cooperation between HP1 and DNMT1 mediates gene silencing. Genes Dev 2007;21(10).

[110] Agarwal N, Hardt T, Brero A, Nowak D, Rothbauer U, Becker A, et al. MeCP2 interacts with HP1 and modulates its heterochromatin association during myogenic differentiation. Nucleic Acids Res 2007;35(16).

[111] Singh AK, Halder-Sinha S, Clement JP, Kundu TK. Epigenetic modulation by small molecule compounds for neurodegenerative disorders. Pharmacol Res 2018;132.

[112] Christman JK. 5-Azacytidine and 5-aza-2′-deoxycytidine as inhibitors of DNA methylation: mechanistic studies and their implications for cancer therapy. Oncogene 2002;21(35).

[113] Zhou L, Cheng X, Connolly BA, Dickman MJ, Hurd PJ, Hornby DP, et al. Methylation inhibitor that forms a covalent complex with DNA methyltransferases. J Mol Biol 2002;321(4).

[114] Tan J, Yang X, Zhuang L, Jiang X, Chen W, Lee PL, et al. Pharmacologic disruption of Polycomb-repressive complex 2-mediated gene repression selectively induces apoptosis in cancer cells. Genes Dev 2007;21(9).

[115] Lee W, Lee TH, Park B-J, Chang J-W, Yu J-R, Koo H-S, et al. Caenorhabditis elegans calnexin is N-glycosylated and required for stress response. Biochem Biophys Res Commun 2005;338(2).

[116] Castillo-Aguilera O, Depreux P, Halby L, Arimondo P, Goossens L. DNA methylation targeting: the DNMT/HMT crosstalk challenge. Biomolecules 2017;7(4).

[117] Brueckner B, Garcia Boy R, Siedlecki P, Musch T, Kliem HC, Zielenkiewicz P, et al. Epigenetic reactivation of tumor suppressor genes by a novel small-molecule inhibitor of human DNA methyltransferases. Cancer Res 2005;65(14).

[118] McCabe MT, Ott HM, Ganji G, Korenchuk S, Thompson C, van Aller GS, et al. EZH2 inhibition as a therapeutic strategy for lymphoma with EZH2-activating mutations. Nature 2012;492(7427).

[119] Knutson SK, Kawano S, Minoshima Y, Warholic NM, Huang K-C, Xiao Y, et al. Selective inhibition of EZH2 by EPZ-6438 leads to potent antitumor activity in *EZH2* -mutant non-hodgkin lymphoma. Mol Cancer Ther 2014;13(4).

[120] Kubicek S, O'Sullivan RJ, August EM, Hickey ER, Zhang Q, Teodoro ML, et al. Reversal of H3K9mc2 by a small-molecule inhibitor for the G9a Histone methyltransferase. Mol Cell 2007;25(3).

[121] Purandare Av, Chen Z, Huynh T, Pang S, Geng J, Vaccaro W, et al. Pyrazole inhibitors of coactivator associated arginine methyltransferase 1 (CARM1). Bioorg Med Chem Lett 2008;18(15).

[122] Selvi BR, Batta K, Kishore AH, Mantelingu K, Varier RA, Balasubramanyam K, et al. Identification of a novel inhibitor of coactivator-associated arginine methyltransferase 1 (CARM1)-mediated methylation of Histone H3 Arg-17. J Biol Chem 2010;285(10).

[123] Chan-Penebre E, Kuplast KG, Majer CR, Boriack-Sjodin PA, Wigle TJ, Johnston LD, et al. A selective inhibitor of PRMT5 with in vivo and in vitro potency in MCL models. Nat Chem Biol 2015;11(6).

[124] Cheng D, Yadav N, King RW, Swanson MS, Weinstein EJ, Bedford MT. Small molecule regulators of protein arginine methyltransferases. J Biol Chem 2004;279(23).

[125] Spannhoff A, Machmur R, Heinke R, Trojer P, Bauer I, Brosch G, et al. A novel arginine methyltransferase inhibitor with cellular activity. Bioorg Med Chem Lett 2007;17(15).

[126] Vigushin DM, Ali S, Pace PE, Mirsaidi N, Ito K, Adcock I, et al. Trichostatin A is a histone deacetylase inhibitor with potent antitumor activity against breast cancer in vivo. Clin Cancer Res 2001;7(4).

[127] Yoshida M, Kijima M, Akita M, Beppu T. Potent and specific inhibition of mammalian histone deacetylase both in vivo and in vitro by trichostatin A. J Biol Chem 1990;265(28).

[128] Gottlicher M. Valproic acid defines a novel class of HDAC inhibitors inducing differentiation of transformed cells. EMBO J 2001;20(24).

[129] Daosukho C, Chen Y, Noel T, Sompol P, Nithipongvanitch R, Velez JM, et al. Phenylbutyrate, a histone deacetylase inhibitor, protects against Adriamycin-induced cardiac injury. Free Radi Biol Med 2007;42(12).

[130] Candido E. Sodium butyrate inhibits histone deacetylation in cultured cells. Cell 1978;14(i).

[131] Saito A, Yamashita T, Mariko Y, Nosaka Y, Tsuchiya K, Ando T, et al. A synthetic inhibitor of histone deacetylase, MS-27−275, with marked in vivo antitumor activity against human tumors. Proc Natl Acad Sci 1999;96(8).

[132] Loprevite M, Tiseo M, Grossi F, Scolaro T, Semino C, Pandolfi A, et al. In vitro study of CI-994, a histone deacetylase inhibitor, in non-small cell lung cancer cell lines. Oncol Res 2005;15(1).

[133] Nakajima H, Kim YB, Terano H, Yoshida M, Horinouchi S. FR901228, a potent antitumor antibiotic, is a novel histone deacetylase inhibitor. Exp Cell Res 1998;241(1).

[134] Schroeder FA, Lewis MC, Fass DM, Wagner FF, Zhang Y-L, Hennig KM, et al. A selective HDAC 1/2 inhibitor modulates chromatin and gene expression in brain and alters mouse behavior in two mood-related tests. PLoS One 2013;8(8).

[135] Lau OD, Kundu TK, Soccio RE, Ait-Si-Ali S, Khalil EM, Vassilev A, et al. HATs off. Mol Cell 2000;5(3).

[136] Balasubramanyam K, Swaminathan V, Ranganathan A, Kundu TK. Small molecule modulators of histone acetyltransferase p300. J Biol Chem 2003;278(21).

[137] Balasubramanyam K, Altaf M, Varier RA, Swaminathan V, Ravindran A, Sadhale PP, et al. Polyisoprenylated benzophenone, garcinol, a natural histone acetyltransferase inhibitor, represses chromatin transcription and alters global gene expression. J Biol Chem 2004;279(32).

[138] Balasubramanyam K, Varier RA, Altaf M, Swaminathan V, Siddappa NB, Ranga U, et al. Curcumin, a novel p300/CREB-binding protein-specific inhibitor of acetyltransferase, represses the acetylation of histone/nonhistone proteins and histone acetyltransferase-dependent chromatin transcription. J Biol Chem 2004;279(49).

[139] Biel M, Kretsovali A, Karatzali E, Papamatheakis J, Giannis A. Design, synthesis, and biological evaluation of a small-molecule inhibitor of the histone acetyltransferase Gcn5. Angew Chem Int Ed 2004;43(30).

[140] Stimson L, Rowlands MG, Newbatt YM, Smith NF, Raynaud FI, Rogers P, et al. Isothiazolones as inhibitors of PCAF and p300 histone acetyltransferase activity. Mol Cancer Ther 2005;4(10).

[141] Ravindra KC, Selvi BR, Arif M, Reddy BAA, Thanuja GR, Agrawal S, et al. Inhibition of Lysine Acetyltransferase KAT3B/p300 Activity by a Naturally Occurring Hydroxynaphthoquinone, Plumbagin. J Biol Chem 2009;284(36).

[142] Mantelingu K, Reddy BAA, Swaminathan V, Kishore AH, Siddappa NB, Kumar GVP, et al. Specific inhibition of p300-HAT alters global gene expression and represses HIV replication. Chem Biol 2007;14(6).

[143] Vasudevarao MD, Mizar P, Kumari S, Mandal S, Siddhanta S, Swamy MM, et al. Naphthoquinone-mediated inhibition of lysine acetyltransferase KAT3B/p300, basis for non-toxic inhibitor synthesis. J Biol Chem 2014;289(11).

[144] Modak R, Basha J, Bharathy N, Maity K, Mizar P, Bhat AV, et al. Probing p300/CBP associated factor (PCAF)-dependent pathways with a small molecule inhibitor. ACS Chem Biol 2013;8(6).

[145] Bowers EM, Yan G, Mukherjee C, Orry A, Wang L, Holbert MA, et al. Virtual ligand screening of the p300/CBP histone acetyltransferase: identification of a selective small molecule inhibitor. Chem Biol 2010;17(5).

[146] Lasko LM, Jakob CG, Edalji RP, Qiu W, Montgomery D, Digiammarino EL, et al. Discovery of a selective catalytic p300/CBP inhibitor that targets lineage-specific tumours. Nature 2017;550(7674).

[147] Brevini TAL, Pennarossa G, Rahman MM, Paffoni A, Antonini S, Ragni G, et al. Morphological and molecular changes of human granulosa cells exposed to 5-azacytidine and addressed toward muscular differentiation. Stem Cell Rev Rep 2014;10(5).

[148] Lee WJ, Kim HJ. Inhibition of DNA methylation is involved in transdifferentiation of myoblasts into smooth muscle cells. Mol Cells 2007;24(3).

[149] Sae-tan S, Grove KA, Kennett MJ, Lambert JD. (−)-Epigallocatechin-3-gallate increases the expression of genes related to fat oxidation in the skeletal muscle of high fat-fed mice. Food Funct 2011;2(2).

[150] Adhikari A, Davie JK. The PRC2 complex directly regulates the cell cycle and controls proliferation in skeletal muscle. Cell Cycle 2020;19(18).

[151] Ling BMT, Gopinadhan S, Kok WK, Shankar SR, Gopal P, Bharathy N, et al. G9a mediates Sharp-1–dependent inhibition of skeletal muscle differentiation. Mol Biol Cell 2012;23(24).

[152] Puri PL, Sartorelli V. HDACs and sirtuins: targets for new pharmacological interventions in human diseases. Pharmacol Res 2010;62(1).

[153] Minetti GC, Colussi C, Adami R, Serra C, Mozzetta C, Parente V, et al. Functional and morphological recovery of dystrophic muscles in mice treated with deacetylase inhibitors. Nat Med 2006;12(10).

[154] Consalvi S, Mozzetta C, Bettica P, Germani M, Fiorentini F, del Bene F, et al. Preclinical studies in the mdx mouse model of duchenne muscular dystrophy with the histone deacetylase inhibitor givinostat. Mol Med 2013;19(1).

[155] Bettica P, Petrini S, D'Oria V, D'Amico A, Catteruccia M, Pane M, et al. Histological effects of givinostat in boys with Duchenne muscular dystrophy. Neuromuscul Disord 2016;26(10).

[156] Mozzetta C, Consalvi S, Saccone V, Tierney M, Diamantini A, Mitchell KJ, et al. Fibroadipogenic progenitors mediate the ability of HDAC inhibitors to promote regeneration in dystrophic muscles of young, but not old Mdx mice. EMBO. Mol Med 2013;5(4).

[157] Saccone V, Consalvi S, Giordani L, Mozzetta C, Barozzi I, Sandona M, et al. HDAC-regulated myomiRs control BAF60 variant exchange and direct the functional phenotype of fibroadipogenic progenitors in dystrophic muscles. Genes Dev 2014;28(8).

[158] Sandonà M, Consalvi S, Tucciarone L, de Bardi M, Scimeca M, Angelini DF, et al. HDAC inhibitors tune miRNAs in extracellular vesicles of dystrophic muscle-resident mesenchymal cells. EMBO Rep 2020;21(9).

CHAPTER 22

Epigenetics of X-chromosome Inactivation

Cíntia Barros Santos-Rebouças

Department of Genetics, Institute of Biology Roberto Alcantara Gomes, State University of Rio de Janeiro, Rio de Janeiro, Rio de Janeiro, Brazil

OUTLINE

Introduction	419	Physiological and Pathogenic XCI Skewing	429
Brief Historical Perspective of XCI	420	X-chromosome, Sex Bias and Diseases	432
X-chromosome Evolution and the Incomplete Nature of XCI	420	XCI Plasticity: Opportunities for Epigenetic Therapeutics	434
Imprinted and Random XCI	422	Concluding Remarks	434
XCI Regulation and Main Epigenetic Steps	423	Acknowledgments	434
XCI Differences Between Mice and Humans	427	Glossary	435
X-autosome Dosage Compensation	427	Abbreviations	436
Methods for Exploring XCI Status	429	References	436

INTRODUCTION

In eutherian mammals, such as mice and humans, the male is usually the one with two different sex chromosomes (XY, heterogametic sex), while the female has two of the same sex chromosome type (XX, homogametic sex). Even though the current X and Y chromosomes of mammals have arisen from the same ancestral homologous pair of autosomes, their cumulative evolutionary changes made them readily distinguishable concerning size, gene content, structure and function. Therefore, X-chromosome inactivation (XCI) emerged as a multi-layered and synergistic epigenetic phenomenon to ensure the dosage balance of X-linked gene expression between the hetero- and homogametic sexes at early embryogenesis, transcriptionally silencing all but one X-chromosome per diploid set. Thereby, XCI serves to equalize the expression output of X-linked genes in females and males, ensuring that only a single X-chromosome functions in both sexes. As a consequence, failure to persuade XCI in XX embryos leads to early lethality during development [1,2].

Epigenetic players involved in XCI dynamics include long non-coding RNAs (lncRNAs), DNA methylation, posttranslational histone modifications, histone variants, chromatin remodelers, chromosome ultrastructure, and nuclear localization [3]. The complexity of XCI epigenetic regulation is still enhanced by a considerable number of genes on the inactive X-chromosome (Xi) that are able to fully or partially escape from XCI, *locus* by *locus* fine-tuning regulation, and the remarkable reversibility reprogramming of XCI. Additionally, a second epigenetic dosage compensation is responsible for balancing the expression between the single expressed dose of most X-linked genes and the double expressed dose of autosomal genes that occurs through the upregulation of genes on the active X-chromosome (Xa) in both sexes [4,5].

Although much progress about XCI has been achieved over the past 60 years, there are still many questions to be solved in this puzzle landscape. With this regard, ongoing researches have focused on unraveling the epigenetic mechanisms behind XCI process as a fascinating model not only for understanding gene expression regulation in general, but also to decipher its repercussions on physiological/pathological differences between sexes concerning X-linked genes. This chapter provides an overview of the main epigenetic regulatory events surrounding XCI and their intricacies in health and disease.

BRIEF HISTORICAL PERSPECTIVE OF XCI

Sex dosage compensation is directly related to sex determination at early development and the evolutionary history of an organism. Thus, to compensate for differences in sexual chromosomes dosage, species evolved to distinctly adjust the regulatory mechanisms inherent to sex-linked genes transcription. Whereas, female worms decrease the transcription levels of both X-chromosomes by half to equalize with males, in flies this dosage balance is accomplished by the twofold overexpression of the single X-chromosome in males, since both female X-chromosomes are transcribed at normal levels. Moreover, monotremes acquired another mechanism for sex dosage compensation. They have multiple X and Y chromosomes that seem to have arisen independently of eutherian and marsupial sex chromosomes, in a manner that dosage compensation of X-linked genes in such organisms occurs on a gene-by-gene basis, rather than through chromosome-wide silencing. Contrariwise, placental mammals and pouched marsupials acquired a unique dosage compensation mechanism, in which all but one X-chromosome are silenced in female somatic cells [6–9].

The first evidence of a functional dissimilarity between the two X-chromosomes on mammalian female cells was found by Barr and Bertram [10], who reported the existence of a deeply stained body in the nucleus of feline nerve cells, only observed in females. This "Barr body," that was postulated to derive from the heterochromatin of the sex chromosomes, was subsequently found in different female mammals' cells [11], with evidences that only one of the two X-chromosomes exhibited this singular heteropyknotic feature, possibly reflecting a more compact conformation than its homolog [12]. Aware of these findings, Mary Lyon [13] noted a variegated pattern in the coat of female mice heterozygous for a mutation in an X-linked gene that controls coat color and interpreted it as a consequence of a random clonal growth of cells with different X-chromosome being expressed (normal or mutant). Lyon hypothesized that (1) the peculiar heteropyknotic structure of the "Barr body" represented its inactive function, with its X-chromosome homologous being transcriptionally active; (2) the inactivation occurs at early embryonic development and (3) it could be either of maternal or paternal origin in different cells of the same organism [14]. A noticeable example of this parental mosaic-like pattern in females can be seen in calico cats. These female cats have an alternating coat in white, orange, and black spots, which correspond to two functionally different parental cell populations, one that has X-chromosomes, in which the allele responsible for the orange color is active and another, in which the allele responsible for the black color is active. Males, on the other hand, do not exhibit an alternating coat color, because they have only one X-chromosome. Lyon also suggested that X-chromosome genes with Y homologs would escape from XCI [14] and that there might be some center(s) from which XCI spreads [15]. Besides, long interspersed nuclear elements (LINEs), for which X-chromosome is particularly rich, should act as booster elements to promote the spread of silencing [16,17]. Lyon still proposed that DNA methylation would be part of the mechanism for stabilizing XCI, after spreading has occurred [18]. All the initial foresights of Lyon, that are deemed as the Lyon Law, were further experimentally confirmed and magnified by her and others, serving as the fundamental basis for the current knowledge about XCI.

X-CHROMOSOME EVOLUTION AND THE INCOMPLETE NATURE OF XCI

The mammalian X and Y chromosomes originated from a pair of ancestral autosomes some 300 million years ago. Throughout the evolutionary process, a partial barrier to recombination has been established between the two chromosomes due to subsequent inversions and, although many original functional elements have been conserved on the X-chromosome, the Y-chromosome has degenerated, losing much of the traces of the ancestral autosome. Besides great discrepancies in size and in gene co854ntent, the pair of sex chromosomes in humans, unlike the other chromosomes, now have their own genetic/epigenetic mechanisms of regulation [19]. So, two different sexes were created, the homogametic sex, in which the two sex chromosomes are equivalent (46,XX) and can fully recombine like any pair of autosomes during gametogenesis and the heterogametic sex (46,XY), in which pairing and meiotic recombination will involve only the ends of the sexual chromosomes that maintained a high degree of similarity. These major homologous regions, called pseudoautosomal regions (PARs), are divided into PAR1 and PAR2. The genes in the PARs encode proteins necessary for both sexes, participating in signal transduction, bone growth, energy metabolism, and synthesis of hormones and receptors [20]. The molecular evolution of the X-chromosome was accompanied by a disproportional enrichment of genes involved with brain development and function, either by recruitment of new genes to the X-chromosome or by acquisition of a new function by ancient X-borne genes. Indeed, X-linked genes are highly expressed in brain in comparison to autosomal genes, and from all the protein-coding genes located at X-chromosome, roughly 40% are known to be expressed in the brain [21]. Furthermore, X-chromosome has a higher proportion of essential brain-expressed microRNAs than would be expected [22]. The higher accumulation of cognitive genes on the X-chromosome rather than on autosomes might be a result of a rapid fixation of recessive alleles that conferred advantage to hemizygous males, which could explain the huge enlargement of the hominid brain over the past million years [23]. Besides cognitive genes, the X-chromosome is also enriched for crucial regulatory imune-related genes [24].

The first sequence of the human X-chromosome was disclosed in 2005 [25,26] and after almost two decades of methodological improvements, the X-chromosome recently became the first human chromosome to be sequenced in a gapless, telomere-to-telomere assembly from high-coverage ultra-long nanopore reads [27]. Our X-chromosome is 156 Mb long and carries some 1.547 genes (858 protein-coding and 689 non-coding), which represents about 3.4% of the genes encoded by the whole genome (GRCh38.p13; Ensembl, [28]). However, the X-chromosome is overrepresented by genes related to Mendelian diseases, being responsible for 6% of all entries with phenotype description and molecular basis known described in the Online Mendelian Inheritance in Man database [29]. The largest PAR (PAR1) has 2.6 Mb and is located in the terminal region of the short arms of X and Y, while PAR2 has only 320 Kb and is at the opposite end of the long arms. The approximately 24 genes contained in the PAR1 have a high rate of recombination and are not subject to XCI in females. However, intriguingly, there is some dosage imbalance even in the PAR1, with a greater expression of some genes in males, suggesting that the combined Xa and Xi expression in females fails to reach the expression arising from X and Y chromosomes in males [30]. On the other hand, the five genes present in PAR2 originated from duplication events from the X- to the Y-chromosome throughout evolution. Part of these genes may undergo XCI, probably as a reflection of the status of the X-chromosome, before the duplication events. Curiously, some PAR2 genes are inactivated on both Xi and in the Y chromosome, achieving dosage compensation in a unique manner [31].

Apart from the PARs, cumulative evidence confirmed the initial assumptions of Lyon that human XCI is typically incomplete, since 15%–23% of the whole X-chromosome genes in female cells are bi-allelic expressed from both the Xa and Xi, known as XCI escape genes [30,32,33]. So, this biallelic expression across the Xi results in a higher expression of these genes in females compared with males, which causes a sex-bias in gene expression. Even that in aneuploidies only one X-chromosome per diploid cell is active, expression of genes escaping from XCI is directly dependent on the total number of X-chromosomes in a cell. Physiologically, the degree of escape from XCI can vary at some extent among genes, individuals and tissues, some of them with tissue-specific dosage effects of the escape [30,34,35]. When the XCI status of a gene differs among individuals or among cells or tissues within an individual, the gene is referred as a variable escape gene. Conventionally, genes are considered as escapees if they are expressed at a level that is at least 10% of the Xa allele expression [32].

The clustered distribution of escape and variable escape genes along the X-chromosome is a consequence of sex chromosome evolution and stretches insight into their epigenetic regulation. The number of XCI escape genes is higher on the more evolutionarily recent strata of the X-chromosome, located at the distal short arm [25,36]. Besides, not all genes that escape from XCI need to be dosage-sensitive. Some of these bi-allelic expressed genes have a paralogue on the Y chromosome with a similar function, and highly conserved dosage-sensitive X/Y paralogs that escape from XCI in females are candidates for being responsible for embryo survival [37]. Other escape genes, however, have lost their Y paralogues, or their Y paralogues have evolved a distinct, often testis-specific, role [35]. Escape genes that reside outside the PARs and have not retained a Y-linked paralog are putative candidates for sex-specific phenotypes and can also modulate sex-related disease susceptibility. Furthermore, the misexpression of escape genes could explain why individuals with sex aneuploidies, like Turner (45,X) or Klinefelter (47,XXY) syndromes, may have some clinical manifestations, even in the presence of only one Xa [33]. In mice, fewer genes escape from XCI in comparison to humans [38] and 7 of the 17 constitutive escape genes also escape in humans, suggesting that conserved

V. FACTORS INFLUENCING EPIGENETIC CHANGES

cis-acting sequences regulates Xi expression [36]. Recently, an evaluation of XCI status across 12 species showed that XCI status is largely conserved across species, highlighting the existence of domains of escape from silencing [38].

As new high throuput methodologies have been expanded with more individuals and cell types being studied, XCI landscape has emerged, becoming evident that the number of genes that constitutive escape or variable escape from XCI is higher than initially thought [39]. Thereby, escape from XCI might be understood as an additional epigenetic layer for regulating gene expression and maintain tissues/phenotypic differences between sexes. The observation that escape genes are evolutionarily conserved and are a stronger purifying selection than X-inactivated genes highlights a noteworthy role of these genes and suggests a potential importance of them to human diseases [40], having an central role in sexually dimorphic traits. Besides, escaping from XCI represents an exciting opportunity to artificially reactivate alleles for treating females with X-linked diseases caused by heterozygous pathogenic variants.

IMPRINTED AND RANDOM XCI

Much of our understanding about XCI dynamics comes from mouse studies, in which embryonic development is easily accessible *in vivo*. Two types of XCI that are vital for embryo viability can be recognized in mice: imprinted and random. At fertilization, the female zygote has both X-chromosomes (maternal and paternal) active. The first inactivation takes place still during pre-implantation development (4- to 8-cell stage female embryos) and is imprinted de novo, resulting in the inactivation exclusive of the paternal X-chromosome (Xi^P) [39]. Maternal trimethylation of H3 on lysine 27 (H3K27me3) appears to have an important role in the paternal imprinted XCI through repression of maternal *Xist* [41]. In the trophectoderm and the primitive endoderm that originate the placenta and yolk sac, respectively, the Xi^P silencing is stably maintained throughout development. Contrariwise, in the epiblast precursor cells within the inner cell mass of the blastocyst, the transcriptional gene silencing is erased in Xi^P (by 60–64-cell stage), a process known as X-chromosome reactivation (XCR) [42,43]. So for a brief period, these cells carry two Xa. The extent and nature of this reactivation reprogramming remain enigmatic, but it was shown to be linked to pluripotency factors, such as Nanog and Prdm14 and is correlated with the epiblast differentiation [44,45]. However, different genes are reactivated by distinct mechanisms at different times, with some of them having a minimal epigenetic memory in the inner cell mass, whereas others may require active erasure of chromatin marks [46].

After paternal imprinted XCI has been erased by X-chromosome reactivation in the epiblast, a second wave of inactivation, at this moment random, takes place in each cell of the embryo-proper. At this stage, both maternal and paternal X-chromosomes have the same probability for being silenced, but once established, XCI is steadily transmitted to cell progenies through subsequent mitosis. This randomness and stable inheritance give rise to an overall 1:1 ratio of cells expressing either the maternal or the paternal X-chromosome, resulting in a mosaicism pattern in female somatic cells and tissues with different functional X-chromosomes (maternal or paternal) being expressed independently in each set of cells. Another stage during mouse development at which XCR takes place is in the germline preceding meiosis to erase random XCI [47]. Indeed, the reversal of XCI patterns during female gametogenesis is obligatory to the beginning of a new XCI cycle at the following generation. In the same way, XCI marks also need to be completely erased in reprogrammed cells to accomplish pluripotency (Fig. 22.1).

In humans, the existence of imprinted XCI remains controversial. In term placentas, an analysis of allele-specific expression of 22 X-linked genes, using single nucleotide polymorphisms (SNPs) in transcribed regions, showed that the parental expression of X-linked genes was heterogeneous, indicating that XCI in human placenta is random [48]. Subsequent allele-specific methylome and transcriptome analysis from human placentas, however, demonstrated maternal imprinting, although the authors suggest that this could be a consequence of an incomplete erasure of epigenetic marks, needing further confirmation [49]. In the same way, no one X-linked imprinted gene has been identified in the human genome until now, although some evidence suggests that they could exist as in Turner syndrome the degree of the social cognition deficit is dependent on the parent of origin of the missing X-chromosome [50].

Both imprinted and random XCI relies on the X-inactivation center (Xic), a *locus* on the X-chromosome containing a number of mostly non-coding RNA genes and on the upregulation of an X-linked lncRNA within the Xic (lncRNA, 15,000–17,000 nt), called X inactive-specific transcript (Xist for mice; XIST for humans), mapped to Xq13.2 [51]. During XCI, Xist/XIST RNA is monoallelic expressed from only one X-chromosome (the future Xi) and capable of coating this X-chromosome in *cis*, triggering a cascade of *trans*-acting factors and subjacent events that lead to transcriptional silencing. Xic, which is defined as the minimal critical region necessary

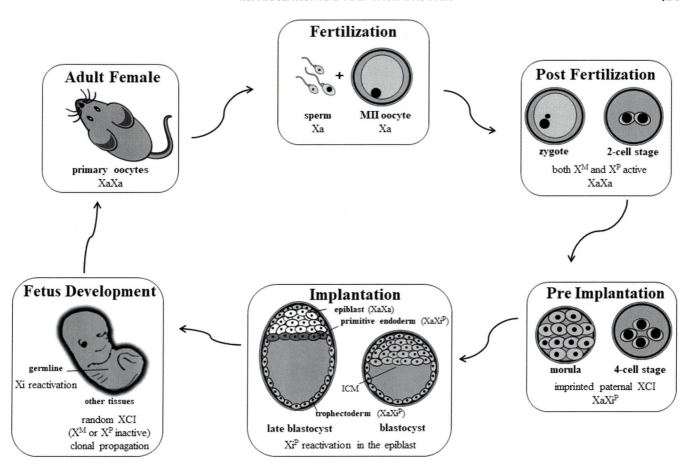

FIGURE 22.1 Summarized kinetics of inactivation and reactivation of the X-chromosome during the mice embryonic development and adult life. At fertilization, both X-chromosomes on sperm and meiosis II (MII) oocyte are active. During female mouse development, embryonic cells at 4-cell stage inactivate the paternally inherited X-chromosome (imprinted XCI; Xi^P). Once stablished, the Xi^P remains inactive in extra-embryonic tissues (trophectoderm and placenta). In contrast, the Xi^P is reactivated in the inner cell mass (ICM) in epiblast cells that gives rise to the embryo. Around implantation, a random XCI occurs at epiblast cells and either the maternal or the paternal X-chromosome have the same probability to become inactive in each cell. The inactive state is maintained by clonal propagation through cell divisions in a parental proportion of 1:1. In the fetus primordial germ cells, however, the Xi is reactivated so that the newly formed maternal X-chromosomes are active at the time of oocyte fertilization.

and sufficient to trigger XCI, is relatively poor in protein-coding genes and enriched for non-coding genes and repetitive sequences [25,52]. Besides, Xic is portioned into megabase-sized local domains of increased chromatin self-interactions, termed topologically associating domains (TADs) that rely on the architectural binding protein CTCF [53]. TADs segregate *Xist cis*-regulation into two domains: the positive one contains *Xist* promoter and its known activators, whereas the adjacent domain includes its negative regulators for the timely and efficient XCI during female development [54].

XCI REGULATION AND MAIN EPIGENETIC STEPS

The three major steps in XCI include: XCI initiation from Xic with a variety of *cis*- and *trans*-acting players, XCI spreading mediated by the protagonist Xist/XIST RNA, and stable maintenance of the inactive state during the subsequent cell divisions. Previous to XCI initiation, the X-chromosomes need to be counted and, thereby, only one X-chromosome is kept active by a diploid cell, while all other X-chromosomes are silenced (n-1 rule). Thus, X-chromosome counting in humans enables that X-chromosome aneuploidies lead to syndromes with a relatively moderate clinical impact (e.g., Turner and Klinefelter syndromes), conversely to autosomal monosomies and most of the autosomal trisomies that are incompatible with life. The knowledge about the counting mechanism is still evolving, but the two X-chromosomes get close nuclear proximity before and at the onset of the differentiation process and this X-X pairing is thought to mediate the *trans*-sensing of the two X-chromosomes, participating in the counting process [55]. However, more recent functional evidence suggested that positioning at the lamina and homologous pairing between the Xic *loci* are not critical determinants during initiation of random XCI, although a role of homologous pairing at later XCI stages cannot be ruled out [56]. Also, X-linked and autosomal regulatory elements are believed to interact and determine the X:autosomes ratio. An X-linked U3 ubiquitin ligase gene present in mice and conserved in humans, upstream of *Xist/XIST*, called *Rlim/RLIM* (or *Rnf12/RNF12*), was identified as at least one of the responsible for this function. Moreover, the lncRNA Jpx, transcribed from the mouse Xic, has also been implicated in the counting process, since a deletion of a single *Jpx* precludes the upregulation of *Xist* [57,58].

But which element protects the Xa from being silenced by its own *Xist locus*? Thus, to have an Xa in each cell, it is necessary to choose the X-chromosome that will be maintained active and repress the *Xist locus* in this X-chromosome. In mice, during imprinted XCI, the *Xist* allele from the maternal X-chromosome is prevented from being expressed in earliest differentiating tissues by the antisense noncoding transcript Tsix, transcribed from the same X-chromosome. Besides, the regions corresponding to the Xist and Tsix TADs exhibit synteny across mammals [52,59], suggesting evolutionary constraints to maintain these *loci* close. Mutations of either *Xist* or *Tsix* cause complete XCI skewing [60], but even that a chromosome with a *Xist* mutation cannot be silenced, an X-chromosome with a *Tsix* mutation is generally the one that will be preferentially inactivated [61]. Furthermore, homozygosity for *Tsix* deletion in mice revealed extremely low fertility, and in the few surviving offspring, the XCI curiously returns to a random pattern [62]. Indeed, Gayen and colleagues demonstrated that Tsix is not compulsory at the onset of random XCI, but it protects the Xa from ectopic silencing as soon as X-inactivation has begun [63]. Although *Tsix* is functional in rodents, it is truncated in humans [61]. Exclusive to humans XCI is the expression of a non-coding RNA from the Xa, known as XACT, which has been proposed to antagonistically compete with XIST RNA [64].

After choosing the Xa, the XCI initiation takes place by silencing all the rest of the X-chromosomes in the cell, which is marked by *Xist* upregulation. In mice, the non-coding genes *Jpx* and *Ftx*, located upstream and in the same TAD of Xist, act as Xist activators, contributing to its accumulation by different mechanisms. While Jpx RNA activates *Xist* expression through its *trans*-acting ability to evict CTCF from *Xist* promoter, Ftx promotes *Xist* expression in a transcription-dependent manner in *cis* (reviewed in [65]). The *XIST* upregulation in humans is less understood. *Jpx* and its human homolog *JPX* have deep divergence in their nucleotide sequences and RNA secondary structures. However, upregulation of the human *JPX* expression also shows a positive correlation with *XIST* upregulation in female pre-implantation embryos and tissues. JPX RNA also interacts with CTCF protein [66,67].

The human *XIST* is expressed as early as in mice. However, instead of being restricted to the paternal X-chromosome, the human *XIST* is expressed in both females and males, but at that time, as a little amount of transcripts is made, it is not enough to silence the X-chromosomes expressing *XIST* [68,69]. Afterwards, *XIST* is upregulated and becomes limited to the future Xi(s) on females. Upregulation of *XIST* was speculated to be preceded by some dosage compensation independent of XIST RNA. Thus, single-cell RNA sequencing (scRNA-Seq) data analysis indicated that human females' embryos achieve preimplantation dosage compensation in some genes by downregulation of both X-chromosomes, which is known as dampening [69]. Nonetheless, a more stringent evaluation of the same data showed decrease of biallelic and increase of monoallelic X-linked genes expression from 4-cell to the blastocyst state, suggesting that XCI takes place during human preimplantation development, questioning the existence of dampening [4]. However, the idea of X-dampening and the role of *XIST* in this process needs further considerations [70]. Moreover, as the XCI onset walk together with the tissue differentiation, the *Xist* upregulation was previous linked to the downregulation of critical pluripotency factors (Nanog, Oct4/Pou5F1, Sox2) [71]. However, in mouse embryonic stem cells the increased resolution of single cell RNA-seq has further showed that XCI progression is not closely synchronized with loss of pluripotency and increase of differentiation at single-cell level, although these processes are globally correlated [72]. It is worth noting that while human X-chromosome aneuploidies (47,XXX, 48,XXXX, …) cells have only one Xa, in triploid cells (69,XXX and 69,XXY) both females and males most often have two Xa, even that the number of X-chromosomes is the same in the two situations (e.g., 47,XXX vs 69,XXX). Besides, in tetraploids, there are two Xi and two Xa. This suggests that the number of Xa is dependent on chromosome ploidy. Moreover, as the number of autosomal sets differs between X trisomy and triploid cells, it could point to a possible role of dosage-sensitive *XIST* repressor (s), encoded by autosomal gene(s) with likely candidates on chromosome 19 [61].

Upon *Xist/XIST* overexpression, it coats the future Xi in *cis* and the propagation of the XCI to the rest of the chromosome, starting the XCI spreading stage that occurs by rapid depletion of RNA polymerase II and induction of chromatin modifications. At this stage, genes along the Xi start to be progressively silenced with different kinetics (early, middle or late inactivation) [73], but genes in the vicinity of the Xic are often firstly inactivated during differentiation [74]. Thenceforth, different epigenetic modifications take place on the future Xi. The major epigenetic modifications include depletion of active chromatin marks (e.g., histone H3 lysine 4 trimethylation H3K4me3, H3 and H4 acetylation), and recruitment of epigenetic modifiers, such as Polycomb repressive complexes (PRC1 and PRC2), resulting in H2A ubiquitination (H2AK119ub1) and histone H3 lysine 27 di-/tri-methylation (H3K27me3), respectively. Moreover, the Xi is also enriched for histone H4 lysine K20 mono-methylation, histone H3 lysine K9 di-methylation and the macroH2A histone variant [46,75]. The exact hierarchy of XCI spreading events triggered by Xist/XIST RNA and the causal relationships of these processes are not fully deciphered. However, histone deacetylation and H2AK119 ubiquitination appear to be the earliest chromatin modifications. The histone deacetylation enzyme HDAC3 is pre-bound on the X-chromosome and upon *Xist* coating, it likely mediates histone deacetylation and facilitates transcriptional silencing [76]. It should be noted, however, that although HDAC3 comes up as a major mediator of gene silencing, it is only one of many pathways involved in this process. Afterwards, PRC1-associated H2AK119Ub and then PRC2-associated H3K27me3 accumulate at large intergenic domains that can then spread into genes [76].

Among placental mammals, *Xist* is poorly conserved, except for a series of unique repeat regions, known as the A-to-F repeats, whose size and number vary in different species. From these repeats, the most conserved is the A-repeat located at the 5' region of *Xist*, that was shown to be crucial for XCI [77], affecting the capacity of Xist RNA to coat the future Xi or to recruit chromatin-modifying players, as PRC2 [78]. In human cells, XIST is highly methylated with at least 78 N^6-methyladenosine (m^6A) residues than other RNAs. m^6A formation in XIST is mediated by the RNA-binding motif protein 15 (RBM15) and its paralogue RBM15B, which binds the m^6A-methylation complex and recruits it to specific sites in RNA. Besides, knockdown of *RBM15* and *RBM15B*, or knockdown of the methyltransferase like 3 (*METTL3*) gene, an m^6A methyltransferase, impairs XIST-mediated gene silencing, revealing that RNA methylation is also required for XIST-mediated transcriptional silencing [79]. Moreover, Wilms tumor 1 associated protein (WTAP), also involved in N^6-methyladenosine modification on RNA, interacts with Xist and was postulated to have a role in its stability or acting as a guide for its target sites [80]. On the other hand, the Rbm15/m6A-methyltransferase complex was recently shown to have only minor contributions to induce gene silencing [81].

Xist interactome has revealed that Xist acts as a scaffold to recruit crucial Xist-interacting factors (RNA-binding proteins, RBPs) to the Xi. A substantial number of RBPs have intrinsically disordered regions and are predicted to be prone to phase separation, a mechanism that has been hypothesized to be important for Xist RNA function, possibly impacting on Xist-mediated silencing or Xist RNA localization [82]. From the recognized Xist-interacting RBPs, a common identified factor is known as SPEN (Split ends homolog or SMART/HDAC1-associated repressor protein, SHARP). SPEN is thought to be recruited by the A-repeat region of Xist, promoting silencing by interaction with transcriptional corepressors as NCOR1 and SMRT/NCOR2 and recruitment of histone deacetylases, and leading to the loss of active chromatin marks such as H3K27ac [83–86]. Moreover, some studies have shown that interactions between the SHARP/SMART/HDAC3 complex and Xist is required to expel RNA polymerase II from the Xi [83,85,87,88]. In pre-implantation mouse embryos and in embryonic stem cells, the role of SPEN in gene silencing on the Xi was confirmed [89]. In parallel, components of PRC1, but not PRC2, were also recognized as direct Xist partners [84,85]. Subsequent studies showed, however, that Polycomb recruitment is mediated by the PCGF3/5-PRC1 complex, which catalyzes chromosome-wide H2A lysine 119 ubiquitylation, signaling for the recruitment of other PRC1 complexes and PRC2 [90,91]. Xist RNA-binding protein hnRNPK is the principal Xist RNA Polycomb interaction domain binding factor required to recruit PCGF3/5-PRC1, in a manner that hnRNPK recognizes the B/C-repeat region on Xist and directs recruit PRC1 and PRC2 to Xi [90,91]. Indeed, B-repeat represents the stronger binding region for hnRNPK within Xist, but this protein also interacts with the C-repeat [81,92]. Currently, both A-repeat/SPEN and B/C-repeat/hnRPNK/PRC1 pathways are recognized as central for establishment of XCI, gene silencing, and Polycomb recruitment, with A-repeat initiating silencing and low levels of Polycomb recruitment in early differentiation stages of mouse embryonic stem cells and B/C-repeat subsequently functioning to enhance and stabilize high levels of Polycomb occupancy. Once established, XCI no longer requires A-repeat [93,94]. Recently [95] argue that Polycomb recruitment in the absence of B-repeat should be attributed strictly to C-repeat, opening contributions also for C-repeat. Additionally, the nuclear matrix protein Cip1-interacting zinc finger protein 1 (CIZ1), another Xist RNA interactor, binds to the E-repeat of Xist and is enriched on the Xi of mouse and human female cells, with a crucial role in anchoring Xist to the nuclear matrix in specific somatic lineages [96].

3D nuclear organization and X-chromosome conformation also change during XCI spreading. It is historically known that Xi is located at the periphery of the nucleus or at the nucleolus [10,97]. In this context, lamin B receptor, a component of nuclear lamina, was identified by proteomic screens as a direct Xist partner [83,84]. Once inactivated, the Xi is tethered to the nuclear periphery via Xist interaction with lamin B receptor [83] and this interaction changes the 3D nuclear structure and facilitates XCI by enabling Xist and its silencing proteins to spread to actively transcribed genes across the X-chromosome [72]. However, the role of a lamin B receptor in XCI was less evident in a subsequent systematic analysis of Xist-mediated allelic silencing in mouse embryonic stem cell-based models [81]. TADs in the Xi are also lost (in exception of escape genes of the Xi) [80], accompanied by decrease of binding CTCF and cohesion proteins on the Xi [84,98]. However, it is still unclear if the TADs lost are necessary for silencing genes or if it is a consequence of it. Indeed, a more likely possibility is that loss of TADs and gene silencing are independent events, since even in the absence of TADs (CTCF or cohesion degron), genes are mostly normally expressed [99–101].

Xist RNA accumulation leads to a quick X-chromosome conformation change with the establishment of a repressive compartment in the Xi, devoid of RNA polymerase II, in which genes become relocated as they undergo silencing. On the other hand, escape genes keep on the periphery of this Xist RNA silent compartment, where they are accessible to the transcription machinery. Detailed structure analysis of the Xi in mice and humans reveals two megadomains encompassing superloops separated by a border containing the macrosatellite Dxz4/DXZ4 in both mice and humans, respectively [98]. DXZ4 binds CTCF exclusively in the Xi [102]. Besides, the *Firre locus* is vital for the chromatin superloop formation [103] and the lncRNA Firre anchors the Xi to the nucleolus by binding CTCF and maintains H3K27me3 methylation [104]. The silent nuclear compartment consists mostly by repeat-rich regions. A specific class of these repeats, LINEs, was proposed to work as "way stations" for efficient XCI spreading and silencing of the Xi, thereby boosting the inactivation signal through the X-chromosome [16,17]. Indeed, LINEs are retrotransposable elements enriched on both human and mouse X-chromosomes relative to autosomes and there is a correlation between gene silencing and the presence of LINEs [105]. LINE-enriched regions seem to be mutually exclusive with Xist RNA coated regions, suggesting that LINE elements might have a role in propagating silence into regions less associated with Xist [54]. Another difference between the two X-chromosomes, as a reflection of their different chromatin states, is the time when the replication initiates prior to cell division, with the Xi being the last chromosome to initiate replication [106].

As we saw, the Xi has its own epigenetic code, different not only from its homolog, but also from all the other chromosomes with respect to transcriptional regulation, chromatin composition, 3D nuclear organization, structure conformation, and DNA replication timing at interphase. However, how some genes fully or partially evade from Xist/XIST-mediated silencing remains to be solved. It appears that genes could employ different mechanisms to escape from XCI. While some genes resist from the XCI initiation (constitutive escapees, such as *Kdm5c* or *Utx*), others are initially silenced and subsequently reactivated in a tissue-specific manner (facultative escapees) [54]. It is known that escape genes are usually depleted in repressive marks related to XCI and enriched for epigenetic marks of active transcription. Furthermore, they are commonly found in the periphery of the silent domain of the Xi, where they interact with each other [5,98]. Besides, although TADs are lost in the Xi, escape genes retain some TAD-like structure and also show enrichment for CTCF and YY1, factors that mediate long-range DNA interactions, suggesting that interactions with regulatory elements could maintain their transcriptional activity [54]. Also, LINEs are generally depleted around escape genes, in comparison to the rest of the genome [107] and mice that have fewer escape genes also have a greater density of LINEs [108]. So, the peculiar 3D organization of escape genes with their TADs maintenance and LINEs distribution could have a role in their ability of escape.

Once XCI spreading has finished, the last step in XCI consists in the acquisition of further epigenetic modifications to ensure the transcriptional silencing of the Xi during subsequent cell progenies. Maintenance of XCI depends on the synergy among Xist RNA, CpG island DNA methylation and histone modifications. CpG island DNA methylation is enriched in the Xi and has long been described to play a role in XCI maintenance [109]. DNA methylation is deposited on the Xi in a Xist-dependent manner by the de novo DNA methyltransferase DNMT3B [110] and is stably associated with it across cell divisions [111]. Although DNA demethylation in mouse somatic cells lead to Xi reactivation in specific genes [112], treatment of human somatic cells with demethylation drugs (e.g., azacytidine) does not reverse XCI in most Xi genes [113]. Besides, individuals with immunodeficiency centromeric instability-facial anomalies syndrome (ICF syndrome), characterized by pathogenic variants in the *DNMT3B* DNA methyltransferase gene maintain XCI, even with global hypomethylation across the Xi [114]. These findings point out that although DNA methylation is indeed a crucial step in maintaining transcriptional silencing in the Xi, other players are also needed for this role.

Therefore, heterochromatin marks, such as the histone H2A variant (macroH2A), and SMCHD1 protein are enriched on Xi and are thought to be recruited later on XCI to maintain the stability of XCI silencing [115]. Moreover, both PRC1 and PRC2 seem to be related to the spreading and maintenance of silencing, but dispensable for XCI initiation [54,116,117]. Indeed, the localization of the architectural protein SmcHD1 to Xi occurs several days after the XCI initiation and depends on the PRC1 activity [118]. On the other hand, while Xist is essential for the initiation of XCI, once the inactive state is maintained by epigenetic mechanisms, Xist RNA has long be thought to become dispensable [112]. However, high-level Xist expression keep on throughout female life and some recent studies with more sensitive assays are against this observation, since partial Xi reactivation has been demonstrated when Xist is deleted in post-XCI mice and humans' cells, with an apparent shift in X-to-autosome dosage balance [119]. As many Xist-interacting RBPs (e.g., SHARP, SAF-A, Ptbp1) contain intrinsically disordered regions and undergo RNA-mediated phase-separation *in vitro*, it was suggested that protein compartmentalization and phase separation could be also involved in the XCI maintenance. Deletion of *Ptbp1* does not disrupt XCI initiation, but have an impact on the compartment formation and maintenance of silencing [120].

Although the predominant model is that, once established XCI is maintained across somatic cells, recent studies have found XCI diversity in somatic cells and some of them have unique and dynamic XCI maintenance mechanisms, such as those of the immune system. As an example, resting mature T and B cells from mice and humans lack Xist/XIST RNA signals and enrichment for the heterochromatin modifications H3K27me3, H2A-ubiquitin, H420me1, and macroH2A on the Xi [121,122]. Furthermore, even that dosage compensation through XCI is a global process affecting transcriptional levels in almost the whole X-chromosome, fine-tuning adjustments can be made *locus* by *locus*, regulating one gene at a time. Fig. 22.2 summarizes the main epigenetic steps involved in XCI.

XCI DIFFERENCES BETWEEN MICE AND HUMANS

XCI in humans has been challenging to study, since it initiates early in the embryo development, besides ethical matters. Thereby, as previously mentioned, much of our understanding about XCI epigenetics came from mice. However, cumulative evidence has demonstrated that XCI is remarkably distinct in rodents, compared to human and other mammals, mainly regarding the onset of the silent event, parental imprinting of the Xi, the nature of the lncRNAs involved, the epigenetic stability of the inactivation once it occurs, and evolutionary differences in the genes escaping from XCI [123,124] (Table 22.1). Indeed, species-specific staging of embryogenesis and distinct evolutionary courses in genes within the Xic can justify such differences [61]. Yet considering that pluripotent cells of the blastocyst have two Xa, reprogramming of somatic cells to an embryonic state took off as a valuable system to elucidate XCI in human and mainly the reactivation of genes along Xi. So, besides human embryonic stem cells (hESCs) and human induced pluripotent stem cells (iPSCs) are contouring the ethical issues associated with embryo cells, these approaches have opened up new opportunities to explore XCI, XCR and their epigenetic modulation during embryogenesis, at least *in vitro*.

X-AUTOSOME DOSAGE COMPENSATION

In addition to the dosage compensation in females achieved by the XCI process, Susumu Ohno proposed that the single X-chromosome gene content should be expressed twice in male mammals to compensate for gene dosage between sex chromosomes and autosomes [125]. However, succeeding studies have found contradictory findings. While early microarray studies supported the Ohno's hypothesis [126,127], more recent RNA-seq data have been inconsistent, with at most weak or infrequent X-chromosome upregulation [128–130]. Nevertheless, an X:autosomes expression ratio close to 1 was reported in both male and female embryonic stem cells, consistent with a transcriptional upregulation dosage compensation between the single Xa and the autosomes in humans [4], as initially proposed by Ohno [125]. Much of these discrepancies among studies might be influenced by the filtering of the genes investigated (reviewed in Posynick and Brown, [131]). Moreover, the lack of dosage compensation at the transcriptome level does not necessarily refute the Ohno's hypothesis, as X-chromosome upregulation could happen at the translational and/or posttranslational levels. In this sense, Chen and colleagues [132], analyzed a human proteomic database from 22 tissues and found that X-chromosome upregulation is also absent at protein level. As new genomic, transcriptomic, proteomic, and posttranslational methodologies become available, these contradictory findings need better exploration.

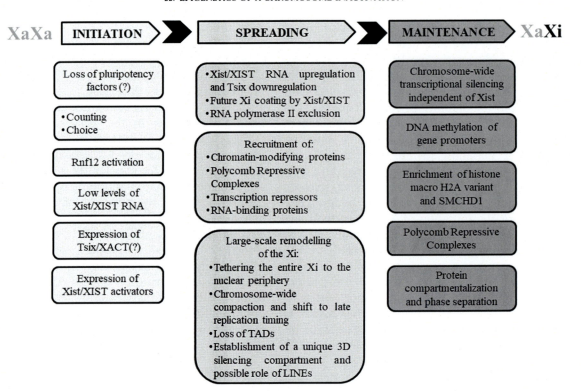

FIGURE 22.2 Global XCI regulation steps, divided in three main sequential phases: initiation, spreading, and maintenance. At fertilization, both X-chromosomes in females are active and characterized by low levels of *Xist/XIST* expression and higher levels of their negative regulators (Tsix in mice and possibly XACT in humans). At the XCI onset, counting and choosing which X-chromosome will keep active occurs, with loss of pluripotency factors (recently questioned) and upregulation of *Rnf12/RNF12* factor. Further, upregulation of Xist/XIST and Xist/XIST RNA coating of the future Xi happens with RNA polymerase II exclusion and recruitment of chromatin modifying proteins, DNA methylation enzymes, transcription repressors and Xist/XIST-interacting RNA-binding proteins. Concomitantly, large-scale remodeling of the Xi takes place to ensure XCI spreading, including tethering the Xi to the nuclear periphery, Xi compaction and shift to late replication timing, loss of TADs, and establishment of a silencing compartment in the Xi with relocation of silenced genes into this Xist repressive nuclear compartment. Later on, in the maintenance step, for propagating XCI across mitotic divisions, the promoter gene regions on Xi become methylated and Xi gets enriched for macro H2A histone variant and SMCHD1, with a potential role of the accumulation of PRC1/PRC2 (initiated at the XCI spreading) and protein compartmentalization/phase separation in this step. Note that these general steps can vary between mice and humans (see the text for details).

TABLE 22.1 Main XCI Differences Between Mice and Humans

Features	Mice	Humans
Xist/XIST expression at XCI initiation	Only from the future Xi in females (exclusively the paternal X-chromosome)	From all X-chromosomes in both females and males
Xist/XIST upregulation on Xi	Selective, only the future Xi is upregulated; at 4 to 8 cells	All X-chromosomes but one is upregulated; not before day 7
Imprinted XCI	In trophectoderm	Contradictory existence
Random XCI	In embryo proper	In placenta and embryo proper
Repression of *Xist* on the future Xa	*Tsix* early. Later, yet unknown	Some other dosage-sensitive *XIST* repressors (e.g., XACT)
Functional Tsix	In imprinted XCI	Not functional
XACT	Absent	Might inhibit *XIST* upregulation early, but not later
Genes escaping from XCI	Fewer escape genes compared to humans	15%–23% of all X-linked genes; 7 of the 17 constitutive escape genes in mice also escape in humans

METHODS FOR EXPLORING XCI STATUS

Different large-scale approaches have been employed to explore the XCI status of the X-chromosome genes, including the use of somatic cell hybrids to quantify Xi gene expression [32]; measurement of allele-specific expression in heterozygous transcribed SNPs, from which genes subject to XCI are monoallelically expressed, while escape genes have bi-allelic expression [30,32,133]; comparison of male and female expression [30]; quantification of DNA methylation on CpG islands [134] and the evaluation of X-linked expression in females with complete XCI skewing [30]. Nonetheless, the random nature of XCI, resulting in parental functional mosaicism within tissues, usually offers an additional challenge for delineating XCI status and from the 1.547 coding and non-coding genes located at the X-chromosome [28], only about 639 genes (41%) have already a XCI status call [30,33], sometimes without a consensus status among the different studies. Also, in cases of XCI investigation, one of the most critical issues to be addressed in the analysis of XCI status is the selection of the tissue or cell type, which should be as closely related as possible to the pathological or physiological condition evaluated.

The advent of allele-specific analysis of single-cell RNA-seq has enabling to assess the Xi expression directly, overpassing some of these previous limitations [3,39]. Moreover, recently, high-coverage ultra-long nanopore sequencing, used to sequence the whole X-chromosome from telomere-to-telomere, appeared to be also sensitive to methylated bases, in a manner that precisely anchored ultra-long reads provided also methylation patterns over repetitive regions across the X-chromosome that were difficult to detect with short-read sequencing [27]. Although the X-chromosome methylation profile found in this study needs further considerations, this new methodology reinforced a decreased methylation across the majority of the PAR1 and PAR2 pseudoautosomal regions [27] and could be an alternative tool to explore global methylation patterns and XCI status in regions yet unexplored across the X-chromosome.

Among the methods developed to quantitatively define the XCI status for individual genes, the most cost-effective and historically used is the DNA-methylation-based approach known as the human androgen receptor (HUMARA) assay. This methyl-CpG-sensitive restriction-endonuclease-based PCR approach targets a highly polymorphic short tandem repeat in the androgen receptor gene (*AR*), located at Xq12, and the methylation status of its alleles on the Xi usually correlates with the whole XCI [135] and is vastly used for exploring XCI in females with X-linked conditions. The combination of the HUMARA assay with other polymorphic regions on the X-chromosome has been used to enhance the informativeness mainly in females homozygous for the *AR* alleles [136]. These targeted DNA-methylation-based approaches are very efficient to detect non-random XCI, making possible genotype correlations with clinical phenotypes.

PHYSIOLOGICAL AND PATHOGENIC XCI SKEWING

Non-random XCI or XCI skewing is defined as a preferential inactivation of the same X-chromosome in over than 80% of cells, leading to a disproportional functional ratio between the Xi and Xa. XCI skewing can arise by chance in a small percentage of females, or due to primary non-random choice or to secondary stochastic or genetic events. In primary skewing, variants in genes acting in the XCI process (ex. *XIST*) prevent the cell from silencing the X-chromosome carrying a variant and even variants in autosomal genes, such those encoding factors such as CTCF or YY1, which interact to some of the XCI key regulators might be able to modify the overall 1:1 ratio of cells expressing either the maternal or the paternal X-chromosome [55]. On the other hand, secondary XCI skewing happens as a strong postinactivation cell selection, working for or against cells carrying one particular X-chromosome (Fig. 22.3).

A compelling illustration of postinactivation cell selection is the Lesch-Nyhan syndrome, in which blood cells expressing the mutated allele show a clear growth disadvantage and gradually disappear from the blood [137]. In contrast, in adrenal leukodystrophy, some pathogenic variants confer a proliferative advantage, possible explaining the increase of the disease severity with ageing in women [138]. Non-random XCI can also be responsible for phenotypic discordance between monozygotic female twins who are carriers of an X-linked disease. As an example, XCI analysis in peripheral blood cells from female monozygotic twins harboring a 26-bp deletion who were clinically discordant for the Rett syndrome identified a favorable skewing in the asymptomatic twin (ratio 99:1), and random XCI in the affected twin [139]. Similar XCI profiles were additionally demonstrated for other conditions with phenotype discordance between female twins, including Fragile X syndrome, Duchenne muscular dystrophy, Hunter syndrome, Fabry disease, color blindness, hemophilia B, and Aicardi syndrome [55,140]. Besides X-linked conditions, high frequency of XCI skewing was also reported in breast and ovarian cancers [141,142]. Table 22.2 illustrates some of the X-linked conditions exhibiting XCI skewing in females.

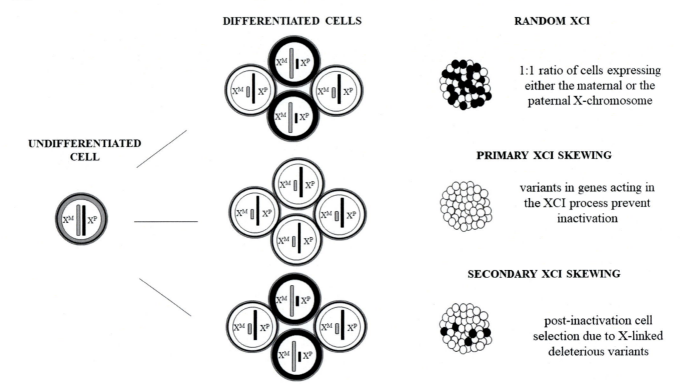

FIGURE 22.3 Random and skewed XCI. In female undifferentiated cells, both the maternal (X^M; gray) and the paternal (X^P; black) X-chromosomes are initially active. The expected random XCI leads to a mosaic population of cells after clonal propagation, with a 1:1 ratio of cells expressing either the maternal or the paternal X-chromosome. Primary skewed XCI can arise by the presence of variants in genes acting in the XCI process, such as variants in *XIST*, leading in general to a complete skewed pattern. Secondary skewed XCI results from cell selection after a random XCI initiation step, usually caused by a deleterious X-linked variant in one of the X-chromosomes.

The degree of XCI skewing and if it is favorable or not to the expression of a given pathogenic variant can directly influence the penetrance and the clinical variability of an X-linked phenotype in females, acting as a modulator of phenotype outcomes [159]. Heterozygous females for X-linked pathogenic variants usually show XCI skewing against the X-chromosome that harbors the damaging allele, keeping the normal X-chromosome and autosomal dosage. Nonetheless, XCI skewing against the abnormal X- chromosome does not always take place. In these later cases, the cause for the XCI skewing could be the existence of another damaging allele on the second X-chromosome, thereby obliging one of the detrimental alleles to be active. Contrariwise, in balanced X/autosome rearrangements, the normal X-chromosome is usually preferentially silenced, in order to ensure functional euploidy, while in unbalanced X/autosome translocation, the consequences are generally mitigated by the preferential inactivation of the abnormal X-chromosome [146,159]. Indeed, XCI skewing can also modify the inheritance mode of X-driven phenotypes. Whereas, heterozygous females for an X-linked dominant condition can be phenotypically healthy or have a milder phenotype, XCI skewing can functionally transform an X-linked recessive condition into a dominant trait, if the normal allele has been turned off in a greater proportion of cells [164]. So, in heterozygous females with XCI skewing, the allelic dosages of genes subject to XCI are unbalanced and can be functionally homozygous, if the XCI skewed is complete with 100% of the cells expressing the same parental X-chromosome.

Although XCI skewing toward one parental X-chromosome has been observed in many human traits, the extent to which genetics and also environment influence XCI skewing is still unknown. XCI patterns analysis in epithelial and hematopoietic cells of over 500 healthy female mother-neonate pairs indicated no correlation between the XCI patterns of mothers and their respective neonates [165]. However, twin studies reinforce that genetic factors can contribute to XCI skewing [166–168]. Moreover, it was demonstrated that XCI skewing correlates positively with age in blood tissues, with an increase of XCI skewing rates after 55–60 years of age [169,170]. An XCI study based on multi-tissue transcriptomic data from a large cohort of female twins showed that XCI patterns are tissue-specific and that blood-derived tissues exhibited the highest prevalence of XCI skewing, with is proportional to age. The consequences of age-related XCI skewing in blood remain unknow, but may impact the

TABLE 22.2 Examples of Human X-linked Diseases that Exhibit Skewed XCI in Carrier Females as a Basis for Modulating Phenotype Variability

X-linked Condition	OMIM Number	Inheritance	Gene	Location	XCI Status	References
Alpha-Thalassemia/Mental Retardation syndrome	301040	XLD	ATRX	Xp21.1	inactive	Badens et al. [143]
Borjeson-Forssman-Lehmann syndrome	301900	XLR	PHF6	Xq26.2	inactive	Carter et al. [144]
Coffin-Lowry syndrome	303600	XLD	RPS6KA3	Xp22.12	inactive	Wang et al. [145]
Cornelia de Lange syndrome 2	300590	XLD	SMC1A	Xp11.22	escape	Fieremans et al. [146]
Cornelia de Lange syndrome 5	300882	XLD	HDAC8	Xq13.1	inactive	Fieremans et al. [146]
Duchenne Muscular dystrophy	310200	XLR	DMD	Xp21.2-p21.1	inactive	[147]
Fanconi anemia, complementation group B	300514	XLR	FANCB	Xp22.2	inactive	Holden et al. [148]
Focal dermal hypoplasia	305600	XLD	PORCN	Xp11.23	inactive	Brady et al. [149]
Hemolytic anemia, G6PD deficient	300908	XLD	G6PD	Xq28	inactive	Wang et al. [150]
Hemophilia A	306700	XLR	F8	Xq28	inactive	Radic et al. [151]
Incontinentia pigmenti	308300	XLD	IKBKG	Xq28	escape	Fusco et al. [152]
Intellectual developmental disorder, X-linked, syndrome, Snijders Blok type	300958	XLD, XLR	DDX3X	Xp11.4	escape	Snijders Blok et al. [153]
Keipert syndrome	301026	XLR	GPC4	Xq26.2	inactive	Amor et al. [154]
Lesch-Nyhan syndrome	300322	XLR	HPRT1	Xq26.2-q26.3	inactive	Torres and Puig [155]
Lymphoproliferative syndrome, X-linked, 2	300635	XLR	XIAP	Xq25	inactive	Woon et al. [156]
Menkes syndrome	309350	XLD	FLNA	Xq28	inactive	Desai et al. [157]
Mental retardation, X-linked 12/35	300395	XLR	CUL4B	Xq25	inactive	Zou et al. [158]
Mental retardation, X-linked 99, syndromic, female-restricted	300968	XLD	USP9X	Xp11.4	escape	Vianna et al. [159]
Mental retardation, X-linked, syndromic 33	300966	XLR	TAF1	Xq13.1	inactive	Vianna et al. [159]
Oculofaciocardiodental syndrome	300166	XLD	BCOR	Xp11.4	inactive	Hedera and Gorski [160]
Orofaciodigital syndrome I	311200	XLD	OFD1	Xp22.2	escape	Morleo and Franco [161]
Rett syndrome	312750	XLD	MECP2	Xq28	inactive	Knudsen et al. [162]
Wiskott-Aldrich syndrome	301000	XLR	WAS	Xp11.23	variable escape	Lacout et al. [163]

XLD, X-linked dominant condition; XLR, X-linked recessive condition; OMIM number, Online Mendelian Inheritance in Man phenotype number.
Adapted from XCI according to the combined status described in T. Tukiainen, A.C. Villani, A. Yen, M.A. Rivas, J.L. Marshall, R. Satija, et al. Landscape of X chromosome inactivation across human tissues. Nature 2017, 550(7675), 244–248.

immune system of healthy ageing women, influencing the epigenetic predisposition to age-related traits. Increased XCI skewing with age was also associated with smoking, with is consistent with the increase of clonal hematopoiesis observed in smokers [167].

Extreme XCI skewing (>90%) can be also used as a valuable, cost-effective starting tool to predict X-linked sequence or structural pathogenic variants in females, mainly those involved in brain development and function disturbances, such as intellectual disability, due to the enrichment of cognitive genes on the X-chromosome [146,159]. Furthermore, another application of XCI skewing screening in females concerns the identification of healthy heterozygous carriers of X-linked recessive pathogenic variants, which is of extreme importance for genetic counseling, family reproductive planning and potential early treatment (when available, e.g., lysosomal storage disorders like Fabry disease and Hunter syndrome) of potentially affected offspring. XCI skewing does not always affect all tissues in the same way, in a manner that it is crucial to ascertain different cell types, particularly from accessible tissues related to the given condition. Indeed, as XCI profiles are clonal, XCI status screening can be used in tumor biopsies to access cancer propagation and metastases origin. In tumors affecting multiple organs, a similar XCI profile could inform that the tumors likely originated from the same initial transformation, whereas different XCI profiles can be suggestive of independent mutational events, influencing clinical treatments [55].

Besides XCI skewing, other X-linked compensation mechanisms may protect females for expressing an X-linked pathogenic variant. Recently, we reported an asymptomatic female, who harbors a large Xq25-q28 deletion (>32 Mb) on the X-chromosome. Regardless of XCI extreme skewing, the deletion forced the structural hemizygosis of a number of critical XCI escape and variable escape genes, but only three blood-expressed genes spanning the deletion (*GPR112*, *SLC6A8*, *FUNDC2*) showed abnormal gene expression on transcriptome-wide analysis. These findings provide evidence that an additional protective gene-by-gene mechanism occurs at the transcriptional level in the Xa to reduce the deleterious dosage imbalance of escape and variable escape genes and preclude major clinical symptoms in large Xq deletions [171].

X-CHROMOSOME, SEX BIAS AND DISEASES

Sexual dimorphism concerning human diseases is usually attributed to sex-hormonal specificities induced by gonadal differentiation, sex-specific life experiences or distinct health care between men and women that could differentially modulate susceptible genes across the genome. Nonetheless, X-linked genes and their XCI statuses can have a crucial but still underappreciated modulation role in dissimilar phenotypes/presentations between sexes involving both rare, as well as common disorders [172].

Pathogenic X-linked variants in genes subject to XCI have different phenotype consequences in males and females. As males are hemizygous for genes on the X-chromosome, X-linked recessive conditions have often a significant male bias. For genes subject to XCI, heterozygous females will have two populations of cells, one with the pathogenic variant on the Xa and one with the silenced pathogenic variant on Xi. So, as previously mentioned, phenotype on females will be related to the proportion of cells carrying the variant on the Xa and XCI skewing could favor or not the expression of this pathogenic allele [115]. However, some X-linked pathogenic variants are lethal for males still in utero and the survivors are exclusively females or mosaic males, as seen in *incontinentia pigmenti* and orofacial digital syndrome type 1 [173]. Moreover, a pathogenic variant in an escape gene can cause a clinical phenotype irrespective of being located on the Xa or Xi [146,159]. As an example, truncated variants in the escape gene *USP9X* are lethal in male fetus, and the phenotype consequences are confined to females; whereas, non-truncated *USP9X* variants cause hippocampal-related intellectual disability, hypotonia and aggressivity in males, with no abnormalities in female carriers [159,174]. Besides *USP9X*, current literature has also reinforced that females with variants in other intellectual disability-related escape genes, such as *DDX3X* and *SMC1A* undergo extreme XCI skewing [146,153,159,174] (Table 22.2), although the mechanisms underlying XCI skewing in females harboring variants in escape genes remain to be elucidated.

In humans, males and females also show historical distinctions concerning innate and adaptive immune responses [175]. Females have more robust immune responses and a better protection against infections (bacterial, viral, and parasitic pathogens). In immune responses to viruses, sex differences in viral cell entry, recognition of viral motifs and immune cell activation could result in different clinical outcomes between sexes, with males presenting more severe phenotypes. The COVID-19 global pandemic caused by the coronavirus SARS-CoV-2 is a clearly example of a viral infection, in which males are more susceptible to infection, severe disease, and mortality. Previous outbreaks in SARS (SARS-CoV virus) and MERS (MERS-CoV virus) exhibited a similar male bias for severity [176,177]. Curiously, cell tropism in both SARS-CoV and SARS-CoV-2 relies on the angiotensin converting enzyme 2 (ACE-2) and on the serine protease TMPRSS2, that facilitate the integration of the coronavirus to the human host. Both genes encoding these proteins have sex-related expression differences, with hormone response elements in their promoters [178,179]. Additionally, *ACE-2*, located at Xp22.2, is supposed to escape from XCI [33] and is more expressed in females than males in a variety of tissues [180]. So it should be expected that increased *ACE2* levels in females would enable more viral entry and have a subsequent increased risk for COVID-19, which was not clinically evidenced. One of the possible explanations for this unexpected observation is that elevated *ACE2* levels could have a protective effect, decreasing the susceptibility for severe lung injury, as well as other vascular-related injuries in consequence of SARS-CoV-2 infection [176,181].

Unlike what happens with infections, females are three to seven times more susceptible to develop autoimmune diseases compared to males, including systemic lupus erythematosus, rheumatoid arthritis, multiple sclerosis, Sjogren syndrome, and scleroderma [182]. Autoimmune conditions are also increased in patients with Klinefelter syndrome (47,XXY), while females with Turner syndrome (45,X), even those taking estrogens, have the same risk as 46,XY males. So, it is evident that the number of X-chromosomes can influence the risk for developing autoimmune diseases. In this context, although the excessive prevalence of autoimmune diseases in females are thought to be due to multiple issues, the expression of Toll-like receptors, an integral part of the innate immunity, from the Xi is partially responsible for the sex differences in some autoimmune conditions, such as Sjogren disease and lupus [183]. Particularly in lupus, 90% of the patients are female and several X-linked genes are overexpressed and/or hypomethylated in them, including escape or variable escape genes, such as *CD40LG* [3,184]. Other X-linked genes escaping from XCI that could influence autoimmunity include *CXCR3*, *KDM6A*, and *CXorf21*, which were shown to be higher expressed in females and in Klinefelter individuals compared to 46,XY males [172]. Some cumulative findings also suggest that abnormal epigenetic modifications in the Xi and consequent perturbation of XCI may be responsible for abnormal X-linked genes expression in lupus disease. In this context, circulating T and B cells display non-canonical features of XCI, with Xi lacking Xist RNA localization and enrichment for heterochromatin modifications [176]. Understanding the X-chromosome mechanisms that influence the sex-related differential immune responses has direct implications in the treatments of these autoimmune disorders, but also can help to identify sex-biased protections or harmful elements and improve infection vaccine strategies and efficacy.

Most cancers have predominance in males than females and male cancer mortality rates are higher than those in females. Pathogenic variants in X-linked tumor suppressor genes that escape from XCI could underlie a significant fraction of this male excess in different cancer types, as males would need only a single deleterious variant to manifest the phenotype, while females would require two. Thus, the "Escape from X-Inactivation Tumor-Suppressor" (EXITS) hypothesis argues that biallelic expression of X-linked tumor suppressor genes gives females an enhanced cancer protection and contributes to the higher prevalence of some tumors in males. Dunford et al. [185] investigated putative EXITS genes from >4100 cancers across 21 tumor types for sex bias, and have found that many genes with an increased incidence of loss-of-function variants in males across cancer types are known to escape from XCI. Thus, EXITS genes challenges the vastly known "two-hits" theory for suppressor genes and presents the new concept that a single hit can also result in the loss of tumor suppressor function [186]. Examples of potential EXITS genes that inhibit tumorigenesis include *KDM5C* (lysine demethylase 5C), *KDM6A* (lysine demethylase 6A), *MAGEC3* (MAGE family member C3), *DDX3X* (DEAD-box helicase 3, X-linked), *CNKSR2* (connector enhancer of kinase suppressor of Ras 2), and *ATRX* (ATRX chromatin remodeler) [185,187]. It is noteworthy, however, that EXITS *loci* may interact with hormonal and/or environmental factors to modulate cancer risk and sex bias. Sex immunological differences have been also postulated to contribute to the male bias in different cancer types [188]. Distinction between hormonal influences and genes escaping from XCI consequences to sexual dimorphism is challenging, but the study of X-chromosome aneuploidies could be beneficial.

XCI/XCR can be also widely used as epigenetic biomarkers for diseases. Breast tumors and cell lines regularly show epigenetic instability of the Xi, with highly abnormal 3D nuclear organization and global heterochromatin perturbations, including gain of euchromatic marks and aberrant distributions of repressive marks, such as H3K27me3 and promoter DNA methylation [189,190]. Moreover, given that the X-chromosome harbors vital cell differentiation and proliferation genes, and that Xist silences hundreds of genes (including oncogenes), a direct causal relationship between the absence of Xist RNA and tumor development has been demonstrated [191]. *Xist* downregulation and XCI loss are also noted in basal-like cancer, breast cancer, susceptibility gene 1-null triple-negative breast cancer, and ovarian cancer cells [186]. Multiple disturbances in the epigenetics of the X-chromosome, including aberrations in *XIST* expression, were also postulated to promote Alzheimer disease pathogenesis [192,193].

XCR occurs physiologically in the inner cells mass of mice and in primordial cells of mice and humans, as well as in different reprogramming settings. Methodological advances accomplished by single cell RNA-seq, allele-resolution transcriptomics, and epigenomics, as well as cellular reprogramming has revealed that XCR is a gradual and a multiple pathways process involving opposite epigenetic marks from XCI, such as *Xist* repression, H3K27me3 loss, DNA hypomethylation of X-linked promoters, and other likely active marks, such as acquisition of TADs, loss of the megadomains, and activation of Polycomb targets (reviewed in Talon et al. [194]). Nonetheless, reactivation of some Xi genes can also occur in somatic cells as a consequence of ageing or due to some congenital or acquired diseases, such as autoimmune disease and cancer [124]. As an example, epigenetic instability of the Xi and loss of XCI are observed in some cancers in females, resulting in the disappearance of the Barr body and reactivation of X-linked genes [190,195]. Besides, breast tumors and cell lines exhibit abnormal 3D nuclear organization and global heterochromatin perturbations, with the gain of euchromatic marks and abnormal distribution of repressive marks, such as H3K27me3 and promoter DNA methylation [189,190].

XCI PLASTICITY: OPPORTUNITIES FOR EPIGENETIC THERAPEUTICS

The facultative nature of the Xi heterochromatin together with the XCI reversibility in differentiated cells, and the ability of some genes escaping from XCI, open up an exciting opportunity to treat X-linked diseases caused by heterozygous pathogenic variants in females. The best exemplified successful case of Xi targeted reactivation is the Rett syndrome mouse model, in which the inducible expression of the wild type Mecp2 allele was sufficient to reverse the advanced neurological symptoms, even in adult mice [196,197]. The same strategy can be extended to targeted reactivation of silenced tumor supressor genes in cancer. Contrariwise, a stimulating demonstration of the power of XIST in chromosome therapy by inducing gene silencing in *cis* is the correction of trisomy 21 by insertion of a highly expressed *XIST* transgene into the *DYRK1A locus* on chromosome 21, in Down syndrome pluripotent stem cells. The XIST RNA-coated chromosome 21 triggered stable heterochromatin modifications, chromosome-wide transcriptional silencing and DNA methylation, correcting chromosome 21 expression to near normal disomic levels and restoring cellular phenotypes [198]. Fascinatingly, immature hematopoietic precursor cells conserve their aptitude to initiate ectopic XCI, suggesting that their X-chromosomes structure is still permissive to changes in XCI profile [199]. So collecting cells from the patient, forcing the inactivation of the pathogenic allele, or recreation of a balanced mosaicism among hematopoietic stem cells before reimplantation in the patient could represent an important treatment tool for autoimmune conditions [55]. Furthermore, given that *XIST* expression loss might drive tumorigenesis, *XIST* reactivation might also be thought as a possible epigenetic therapy for different types of cancer. Indeed, XIST RNA levels are strongly associated with disease-free periods in ovarian cancer patients, suggesting that *XIST* expression may be a potential marker for chemotherapeutic responses in such types of cancer and maybe others [200]. Nonetheless, the outcomes from studies into the reversibility of the XCI or silencing genes need to be interpreted carefully, due to the huge number of XCI-interacting players and possible genome-wide consequences [119]. In this sense, due to the multiplicity of epigenetic silencing mechanisms concerning XCI, the major limitations/pitfalls of the Xi reactivation strategies include partial reactivation of Xi with extremely low expression levels of target gene restoration, which has been ameliorated in past years through the use of combined effective drugs [201] and reactivation of other *loci* across the X-chromosome with untold physiological side effects [202]. Besides, X-chromosome dosage is known to be essential throughout development, a field that still needs more investigation, mainly concerning targeted drug delivery to the central nervous system.

CONCLUDING REMARKS

The X-chromosome is truly unique in its fascinating biological portrait and evolutionary history, which made it emerge as a mechanistic paradigm for studying epigenetics. Thanks to the methodological advances, the continuous discovery of key players in XCI dynamics has bringing new insights in this fascinating field. In this sense, single-cell multiomic methodologies focusing on XCI, although yet challenging, promises a lot and should allow a better delineation of the epigenetic marks surrounding XCI and the order of their establishment. Besides, allele-specific analysis of single-cell RNA-seq has arisen as a potent methodology to assess XCI, passing over the complications with mosaicism. Deciphering the fine epigenetic modulation subjacent to XCI, the peculiarities of escape genes, the role of XCI/XCR skewing in the pathogenicity of different conditions and understanding how to reactivate genes along the Xi will certainly have essential implications towards the development of efficient genetic/epigenetic therapies. Nonetheless, there are many open issues that need to be addressed.

Acknowledgments

C.B.S.R is supported by Conselho Nacional de Desenvolvimento Científico e Tecnológico (CNPq), Fundação de Amparo à Pesquisa do Estado do Rio de Janeiro (FAPERJ), Coordenação de Aperfeiçoamento de Pessoal de Nível Superior (CAPES) - Finance Code 001 and Centro de Produção da Universidade do Estado do Rio de Janeiro (CEPUERJ) - Brazil.

Glossary

Barr body — Dense, dark-staining structure observed at the periphery of the nucleus of interphase cells with more than one X-chromosome, representing the inactive X-chromosome. Also called sex chromatin.

Blastocyst — Early embryogenesis stage, in which the embryo is capable of giving rise to all tissues. The internal cells of the blastocyst form the inner cell mass that will generate the embryo proper and the outer layer consists of cells collectively called the trophectoderm, which originate the placenta.

Carrier — the term usually used for heterozygous females, harboring two distinct alleles for a referred X-linked locus, one of which has a mutation.

Chromatin — The complex of DNA and proteins (mainly histones) that forms the chromosome scaffold, protecting our genes from damage and playing an essential role in expression regulation, through the modulation of the DNA accessibility to transcriptional machinery. Regions containing actively transcribed genes are known as euchromatin, whereas silenced regions are called heterochromatin.

Dampening — In the dampening model, dosage compensation in preimplantation embryos is achieved by downregulating the expression from the two X-chromosomes in females. Following the biallelic *Xist* expression at the 8-cell stage, the expression of X-linked genes is reduced, preserving their biallelic expression until the late blastocyst stage.

Dosage compensation — The process by which organisms equalize the expression of genes between distinct biological sexes across species. Another kind of dosage compensation involves the equalization of genes expression between X-chromosome and autosomes.

Embryonic stem cells — Pluripotent cells lines derived from the inner cell mass of the early blastocyst stage during embryo development that could differentiate into the three germ layers.

Escape gene — X-chromosome genes in female cells that are bi-allelic expressed from both the active and inactive X-chromosomes. These genes can escape X-chromosome constitutively or have variable expression profiles among different cells, tissues or individuals.

Hemizygote — A diploid individual with a single allele copy at a given gene instead of two. The term is usually used for males that have only one copy of X-linked genes.

Heterogametic sex — The sex of a species, in which the sex chromosomes are different (ex., human males—46,XY).

Heteropyknotic feature — the state of chromosomes or of parts of a chromosome in a nucleus of retaining more or less stain or have a more or less coiled, in comparison to other chromosomes or regions.

Homogametic sex — The sex of a species, in which the sex chromosomes are identical (ex., human females—46,XX).

Induced pluripotent stem cells — A type of stem cell from differentiated somatic tissues that can be reprogrammed *in vitro* to induce cellular pluripotency.

Inner cell mass — The mass of cells inside the blastocyst that originates the embryo proper.

Mosaicism — The condition of possessing more than one cell line with distinct genetic constitutions originated from the same zygote in a tissue or individual.

Pseudoautosomal regions — Homologous nucleotide sequences on the tips of both X and Y chromosomes, that exchange during meiosis and are inherited as autosomal genes. Genes on these regions are expressed from X and Y chromosomes in males and from both X-chromosomes in females.

Random X Inactivation — Usual inactivation process, in which the maternal and the paternal X-chromosomes have the same probability to became inactive in a mammalian embryo.

Sexual dimorphism — When different sexes of the same species are distinctly affected by a condition, concerning prevalence and/or severity.

Skewed X Inactivation — Non-random X-chromosome Inactivation, in which a preferential inactivation of the same parental X-chromosome occurs, leading to a disproportional functional ratio between the inactive and active X-chromosomes.

X-inactivation center (Xic)	*locus* on the X- chromosome essential for the X-chromosome inactivation in *cis*.
X-inactive specific transcript (XIST/Xist)	Long non-coding RNA gene within the X-inactivation center that stars the X-chromosome inactivation, coating the future inactive X-chromosome in *cis*.

Abbreviations

hESCs	human embryonic stem cells
iPSCs	induced pluripotent stem cells
LINEs	long interspersed nuclear elements
OMIM	Online mendelian inheritance in man
PAR1/PAR2	Pseudoautosomal regions 1 and 2
PRC1/PRC2	Polycomb repressive complexes 1 and 2
TAD	Topologically associating domains
TSIX	XIST antisense RNA
Xa	active X-chromosome
XACT	X active specific transcript
XCI	X-chromosome inactivation
Xi	inactive X-chromosome
Xic	X-inactivation center
Xi^P	inactivation of the paternal X-chromosome
XIST	human X inactive-specific transcript (RNA or protein)
XIST	human X inactive-specific transcript gene
Xist	mouse X inactive-specific transcript (RNA or protein)
Xist	mouse X inactive-specific transcript gene

References

[1] Borensztein M, Syx L, Ancelin K, Diabangouaya P, Picard C, Liu T, et al. Xist-dependent imprinted X inactivation and the early developmental consequences of its failure. Nat Struct Mol Biol 2017;24(3):226–33.

[2] Wang F, Shin J, Shea JM, Yu J, Bošković A, Byron M, et al. Regulation of X-linked gene expression during early mouse development by Rlim. Elife 2016;5:e19127.

[3] Carrel L, Brown CJ. When the Lyon (ized chromosome) roars: ongoing expression from an inactive X chromosome. Philos Trans R Soc B: Biol Sci 2017;372(1733):20160355.

[4] Moreira de Mello JC, Fernandes GR, Vibranovski MD, Pereira LV. Early X chromosome inactivation during human preimplantation development revealed by single-cell RNA-sequencing. Sci Rep 2017;7(1):10794.

[5] Disteche CM, Berletch JB. X-chromosome inactivation and escape. J Genet 2015;94(4):591–9.

[6] Meyer BJ. Sex in the worm counting and compensating X-chromosome dose. Trends Genet 2000;16(6):247–53.

[7] Park Y, Kuroda MI. Epigenetic aspects of X-chromosome dosage compensation. Science 2001;293(5532):1083–5.

[8] Whitworth DJ, Pask AJ. The X factor: X chromosome dosage compensation in the evolutionarily divergent monotremes and marsupials. Semin Cell Dev Biol 2016;56:117–21.

[9] Deng X, Meller VH. Non-coding RNA in fly dosage compensation. Trends Biochem Sci 2006;31(9):526–32.

[10] Barr ML, Bertram EG. A morphological distinction between neurones of the male and female, and the behaviour of the nucleolar satellite during accelerated nucleoprotein synthesis. Nature 1949;163(4148):676.

[11] Ohno S, Kaplan WD, Kinosita R. Somatic association of the positively heteropycnotic x-chromosomes in female mice (Mus musculus). Exp Cell Res 1958;15(3):616–18.

[12] Ohno S, Hauschka TS. Allocycly of the X-chromosome in tumors and normal tissues. Cancer Res 1960;20:541–5.

[13] Lyon MF. Gene action in the X-chromosome of the mouse (Mus musculus L.). Nature 1961;190:372–3.

[14] Lyon MF. Sex chromatin and gene action in the mammalian X-chromosome. Am J Hum Genet 1962;14(2):135–48.

[15] Lyon MF, Searle AG, Ford CE, Ohno S. A mouse translocation suppression sex-linked variegation. Cytogenetics 1964;3:306–23.

[16] Lyon MF. Do LINEs have a role in X-chromosome inactivation? J Biomed Biotechnol 2006;2006(1):59746.

[17] Lyon MF. X-chromosome inactivation: a repeat hypothesis. Cytogenet Cell Genet 1998;80(1–4):133–7.

[18] Lyon MF. Some milestones in the history of X-chromosome inactivation. Annu Rev Genet 1992;26:16–28.

[19] Carrel L, Cottle AA, Goglin KC, Willard HF. A first-generation X-inactivation profile of the human X chromosome. Proc Natl Acad Sci USA 1999;96(25):14440–4.

[20] Helena Mangs A, Morris BJ. The human pseudoautosomal region (PAR): origin, function and future. Curr Genomics 2007;8(2):129–36.

[21] Ropers HH. X-linked mental retardation: many genes for a complex disorder. Curr Opin Genet Dev 2006;16(3):260–9.

[22] Gonçalves TF, Piergiorge RM, Dos Santos JM, Gusmão J, Pimentel MMG, Santos-Rebouças CB. Network profiling of brain-expressed X-chromosomal microRNA genes implicates shared key microRNAs in intellectual disability. J Mol Neurosci 2019;67(2):295–304.

[23] Santos-Rebouças CB, Pimentel MMG. New insights into X-linked mental retardation: from clinical to molecular approaches. New York: Nova Science Publishers; 2007.

[24] Bianchi I, Lleo A, Gershwin ME, Invernizzi P. The X chromosome and immune associated genes. J Autoimmun 2012;38(2–3):J187–92.

[25] Ross MT, Grafham DV, Coffey AJ, Scherer S, McLay K, Muzny D, et al. The DNA sequence of the human X chromosome. Nature 2005;434(7031):325–37.

[26] Harsha HC, Suresh S, Amanchy R, Deshpande N, Shanker K, Yatish AJ, et al. A manually curated functional annotation of the human X chromosome. Nat Genet 2005;37(4):331–2.

[27] Miga KH, Koren S, Rhie A, Vollger MR, Gershman A, Bzikadze A, et al. Telomere-to-telomere assembly of a complete human X chromosome. Nature 2020;585(7823):79–84.

[28] Ensembl, Available from: https://www.ensembl.org/Homo_sapiens/Location/Chromosome?r = X:126321265-156040895 [accessed 30.05.22].

[29] OMIM, Online Mendelian Inheritance in Man, Available from: https://www.omim.org/statistics/entry [Updated 27.05.22].

[30] Tukiainen T, Villani AC, Yen A, Rivas MA, Marshall JL, Satija R, et al. Landscape of X chromosome inactivation across human tissues. Nature 2017;550(7675):244–8.

[31] Ciccodicola A, D'Esposito M, Esposito T, Gianfrancesco F, Migliaccio C, Miano MG, et al. Differentially regulated and evolved genes in the fully sequenced Xq/Yq pseudoautosomal region. Hum Mol Genet 2000;9(3):395–401.

[32] Carrel L, Willard HF. X-inactivation profile reveals extensive variability in X-linked gene expression in females. Nature 2005;434(7031):400–4.

REFERENCES

[33] Balaton BP, Cotton AM, Brown CJ. Derivation of consensus inactivation status for X-linked genes from genome-wide studies. Biol Sex Differ 2015;6:35.

[34] Berletch JB, Ma W, Yang F, Shendure J, Noble WS, Disteche CM, et al. Escape from X inactivation varies in mouse tissues. PLoS Genet 2015;11(3):e1005079.

[35] Deng X, Berletch JB, Nguyen DK, Disteche CM. X chromosome regulation: diverse patterns in development, tissues and disease. Nat Rev Genet 2014;15(6):367−78.

[36] Balaton BP, Brown CJ. Escape Artists of the X Chromosome. Trends Genet 2016;32(6):348−59.

[37] Bellott DW, Hughes JF, Skaletsky H, Brown LG, Pyntikova T, Cho TJ, et al. Mammalian Y chromosomes retain widely expressed dosage-sensitive regulators. Nature 2014;508(7497):494−9.

[38] Balaton BP, Fornes O, Wasserman WW, Brown CJ. Cross-species examination of X-chromosome inactivation highlights domains of escape from silencing. Epigenetics Chromatin 2021;14(1):12.

[39] Keniry A, Blewitt ME. Studying X chromosome inactivation in the single-cell genomic era. Biochem Soc Trans 2018;46(3):577−86.

[40] Slavney A, Arbiza L, Clark AG, Keinan A. Strong constraint on human genes escaping X-inactivation is modulated by their expression level and breadth in both sexes. Mol Biol Evol 2016;33(2):384−93.

[41] Chen Z, Zhang Y. Maternal H3K27me3-dependent autosomal and X chromosome imprinting. Nat Rev Genet 2020;21(9):555−71.

[42] Ohhata T, Wutz A. Reactivation of the inactive X chromosome in development and reprogramming. Cell Mol Life Sci 2013;70(14):2443−61.

[43] Pasque V, Plath K. X chromosome reactivation in reprogramming and in development. Curr Opin Cell Biol 2015;37:75−83.

[44] Payer B, Rosenberg M, Yamaji M, Yabuta Y, Koyanagi-Aoi M, Hayashi K, et al. Tsix RNA and the germline factor, PRDM14, link X reactivation and stem cell reprogramming. Mol Cell 2013;52(6):805−18.

[45] Silva J, Nichols J, Theunissen TW, Guo G, van Oosten AL, Barrandon O, et al. Nanog is the gateway to the pluripotent ground state. Cell 2009;138(4):722−37.

[46] Borensztein M, Okamoto I, Syx L, Guilbaud G, Picard C, Ancelin K, et al. Contribution of epigenetic landscapes and transcription factors to X-chromosome reactivation in the inner cell mass. Nat Commun 2017;8(1):1297.

[47] Payer B. Developmental regulation of X-chromosome inactivation. Semin Cell Dev Biol 2016;56:88−99.

[48] Moreira de Mello JC, de Araújo ES, Stabellini R, Fraga AM, de Souza JE, Sumita DR, et al. Random X inactivation and extensive mosaicism in human placenta revealed by analysis of allele-specific gene expression along the X chromosome. PLoS One 2010;5(6):e10947.

[49] Hamada H, Okae H, Toh H, Chiba H, Hiura H, Shirane K, et al. Allele-specific methylome and transcriptome analysis reveals widespread imprinting in the human placenta. Am J Hum Genet 2016;99(5):1045−58.

[50] Knickmeyer RC. Turner syndrome: advances in understanding altered cognition, brain structure and function. Curr Opin Neurol 2012;25(2):144−9.

[51] Marahrens Y, Panning B, Dausman J, Strauss W, Jaenischm R. Xist-deficient mice are defective in dosage compensation but not spermatogenesis. Genes Dev 1997;11(2):156−66.

[52] Chureau C, Prissette M, Bourdet A, Barbe V, Cattolico L, Jones L, et al. Comparative sequence analysis of the X-inactivation center region in mouse, human, and bovine. Genome Res 2002;12(6):894−908.

[53] Dixon JR, Selvaraj S, Yue F, Kim A, Li Y, Shen Y, et al. Topological domains in mammalian genomes identified by analysis of chromatin interactions. Nature 2012;485(7398):376−80.

[54] Galupa R, Heard E. X-chromosome inactivation: a crossroads between chromosome architecture and gene regulation. Annu Rev Genet 2018;52:535−66.

[55] Morey C, Avner P. Genetics and epigenetics of the X chromosome. Ann N Y Acad Sci 2010;1214:E18−33.

[56] Pollex T, Heard E. Nuclear positioning and pairing of X-chromosome inactivation centers are not primary determinants during initiation of random X-inactivation. Nat Genet 2019;51(2):285−95.

[57] Sun S, Del Rosario BC;, Szanto A, Ogawa Y, Jeon Y, Lee JT. Jpx RNA activates Xist by evicting CTCF. Cell 2013;153(7):1537−51.

[58] Tian D, Sun S, Lee JT. The long noncoding RNA, Jpx, is a molecular switch for X chromosome inactivation. Cell 2010;143(3):390−403.

[59] Nesterova TB, Slobodyanyuk SY, Elisaphenko EA, Shevchenko AI, Johnston C, Pavlova ME, et al. Characterization of the genomic Xist locus in rodents reveals conservation of overall gene structure and tandem repeats but rapid evolution of unique sequence. Genome Res 2001;11(5):833−49.

[60] Jégu T, Aeby E, Lee JT. The X chromosome in space. Nat Rev Genet 2017;18(6):377−89.

[61] Migeon BR. Choosing the active X: the human version of X inactivation. Trends Genet 2017;33(12):899−909.

[62] Lee JT. Homozygous Tsix mutant mice reveal a sex-ratio distortion and revert to random X-inactivation. Nat Genet 2002;32(1):195−200.

[63] Gayen S, Maclary E, Buttigieg E, Hinten M, Kalantry S. A primary role for the Tsix lncRNA in maintaining random X-chromosome inactivation. Cell Rep 2015;11(8):1251−65.

[64] Vallot C, Patrat C, Collier AJ, Huret C, Casanova M, Liyakat Ali TM, et al. XACT noncoding RNA competes with XIST in the control of X chromosome activity during human early development. Cell Stem Cell 2017;20(1):102−11.

[65] Patrat C, Ouimette JF, Rougeulle C. X chromosome inactivation in human development. Development 2020;147(1):dev183095.

[66] Oh HJ, Lee JT. Long noncoding RNA functionality beyond sequence: the Jpx model: commentary on "Functional conservation of lncRNA JPX despite sequence and structural divergence" by Karner et al. (2019). J Mol Biol 2020;432(2):301−4.

[67] Karner H, Webb CH, Carmona S, Liu Y, Lin B, Erhard M, et al. Functional conservation of LncRNA JPX despite sequence and structural divergence. J Mol Biol 2020;432(2):283−300.

[68] Okamoto I, Patrat C, Thépot D, Peynot N, Fauque P, Daniel N, et al. Eutherian mammals use diverse strategies to initiate X-chromosome inactivation during development. Nature 2011;472(7343):370−4.

[69] Petropoulos S, Edsgärd D, Reinius B, Deng Q, Panula SP, Codeluppi S, et al. Single-cell RNA-Seq reveals lineage and X chromosome dynamics in human preimplantation embryos. Cell 2016;167(1):285.

[70] Saiba R, Arava M, Gayen S. Dosage compensation in human pre-implantation embryos: X-chromosome inactivation or dampening? EMBO Rep 2018;19(8):e46294.

[71] Deuve JL, Avner P. The coupling of X-chromosome inactivation to pluripotency. Annu Rev Cell Dev Biol 2011;27:611−29.

[72] Chen G, Schell JP, Benitez JA, Petropoulos S, Yilmaz M, Reinius B, et al. Single-cell analyses of X Chromosome inactivation dynamics and pluripotency during differentiation. Genome Res 2016;26(10):1342−54.

[73] Patrat C, Okamoto I, Diabangouaya P, Vialon V, Le Baccon P, Chow J, et al. Dynamic changes in paternal X-chromosome activity during imprinted X-chromosome inactivation in mice. Proc Natl Acad Sci USA 2009;106(13):5198−203.

[74] Lin H, Gupta V, Vermilyea MD, Falciani F, Lee JT, O'Neill LP, et al. Dosage compensation in the mouse balances up-regulation and silencing of X-linked genes. PLoS Biol 2007;5(12):e326.

[75] da Rocha ST, Heard E. Novel players in X inactivation: insights into Xist-mediated gene silencing and chromosome conformation. Nat Struct Mol Biol 2017;24(3):197−204.

[76] Żylicz JJ, Bousard A, Žumer K, Dossin F, Mohammad E, da Rocha ST, et al. The implication of early chromatin changes in X chromosome inactivation. Cell 2019;176(1−2):182−97 e23.

[77] Wutz A, Rasmussen TP, Jaenisch R. Chromosomal silencing and localization are mediated by different domains of Xist RNA. Nat Genet 2002;30(2):167−74.

[78] Zhao J, Sun BK, Erwin JA, Song JJ, Lee JT. Polycomb proteins targeted by a short repeat RNA to the mouse X chromosome. Science 2008;322(5902):750−6.

[79] Patil DP, Chen CK, Pickering BF, Chow A, Jackson C, Guttman M, et al. m(6)A RNA methylation promotes XIST-mediated transcriptional repression. Nature 2016;537(7620):369−73.

[80] Pinheiro I, Heard E. X chromosome inactivation: new players in the initiation of gene silencing. F1000Res 2017;6 F1000 Faculty Rev-344.

[81] Nesterova TB, Wei G, Coker H, Pintacuda G, Bowness JS, Zhang T, et al. Systematic allelic analysis defines the interplay of key pathways in X chromosome inactivation. Nat Commun 2019;10(1):3129.

[82] Cerase A, Armaos A, Neumayer C, Avner P, Guttman M, Tartaglia GG. Phase separation drives X-chromosome inactivation: a hypothesis. Nat Struct Mol Biol 2019;26(5):331−4.

[83] McHugh CA, Chen CK, Chow A, Surka CF, Tran C, McDonel P, et al. The Xist lncRNA interacts directly with SHARP to silence transcription through HDAC3. Nature 2015;521(7551):232−6.

[84] Minajigi A, Froberg J, Wei C, Sunwoo H, Kesner B, Colognori D, et al. Chromosomes. A comprehensive Xist interactome reveals cohesin repulsion and an RNA-directed chromosome conformation. Science 2015;349(6245). Available from: https://doi.org/10.1126/science.aab2276.

[85] Chu C, Zhang QC, da Rocha ST, Flynn RA, Bharadwaj M, Calabrese JM, et al. Systematic discovery, of Xist RNA binding proteins. Cell 2015;161(2):404−16.

[86] Giaimo BD, Robert-Finestra T, Oswald F, Gribnau J, Borggrefe T. Chromatin regulator SPEN/SHARP in X inactivation and disease. Cancers (Basel) 2021;13(7):1665.

[87] Moindrot B, Cerase A, Coker H, Masui O, Grijzenhout A, Pintacuda G, et al. A pooled shRNA screen identifies Rbm15, Spen, and Wtap as factors required for Xist RNA-mediated silencing. Cell Rep 2015;12(4):562−72.

[88] Monfort A, Di Minin G, Postlmayr A, Freimann R, Arieti F, Thore S, et al. Identification of Spen as a crucial factor for Xist function through forward genetic screening in haploid embryonic stem cells. Cell Rep 2015;12(4):554−61.

[89] Dossin F, Pinheiro I, Żylicz JJ, Roensch J, Collombet S, Le Saux A, et al. SPEN integrates transcriptional and epigenetic control of X-inactivation. Nature 2020;578(7795):455−60.

[90] Pintacuda G, Young AN, Cerase A. Function by structure: spotlights on Xist long non-coding RNA. Front Mol Biosci 2017;4:90.

[91] Almeida M, Pintacuda G, Masui O, Koseki Y, Gdula M, Cerase A, et al. PCGF3/5-PRC1 initiates Polycomb recruitment in X chromosome inactivation. Science 2017;356(6342):1081−4.

[92] Bousard A, Raposo AC, Żylicz JJ, Picard C, Pires VB, Qi Y, et al. The role of Xist-mediated Polycomb recruitment in the initiation of X-chromosome inactivation. EMBO Rep 2019;20(10):e48019.

[93] Colognori D, Sunwoo H, Wang D, Wang CY, Lee JT. Xist repeats A and B account for two distinct phases of X inactivation establishment. Dev Cell 2020;54(1):21−32 e5.

[94] Colognori D, Sunwoo H, Wang D, Wang CY, Lee JT. Xist repeat A contributes to early recruitment of Polycomb complexes during X-chromosome inactivation. Dev Cell 2021;56(9):1236−7.

[95] Wei G, Almeida M, Bowness JS, Nesterova TB, Brockdorff N. Xist repeats B and C, but not repeat A, mediate de novo recruitment of the Polycomb system in X chromosome inactivation. Dev Cell 2021;56(9):1234−5.

[96] Ridings-Figueroa R, Stewart ER, Nesterova TB, Coker H, Pintacuda G, Godwin J, et al. The nuclear matrix protein CIZ1 facilitates localization of Xist RNA to the inactive X-chromosome territory. Genes Dev 2017;31(9):876−88.

[97] Rego A, Sinclair PB, Tao W, Kireev I, Belmont AS. The facultative heterochromatin of the inactive X chromosome has a distinctive condensed ultrastructure. J Cell Sci 2008;121(Pt 7):1119−27.

[98] Deng X, Ma W, Ramani V, Hill A, Yang F, Ay F, et al. Bipartite structure of the inactive mouse X chromosome. Genome Biol 2015;16(1):152.

[99] Nora EP, Goloborodko A, Valton AL, Gibcus JH, Uebersohn A, Abdennur N, et al. Targeted degradation of CTCF decouples local insulation of chromosome domains from genomic compartmentalization. Cell 2017;169(5):930−44 e22.

[100] Rao SSP, Huang SC, Glenn, St Hilaire B;, Engreitz JM;, Perez EM, et al. Cohesin loss eliminates all loop domains. Cell 2017;171(2):305−20 e24.

[101] Kriz AJ, Colognori D, Sunwoo H, Nabet B, Lee JT. Balancing cohesin eviction and retention prevents aberrant chromosomal interactions, Polycomb-mediated repression, and X-inactivation. Mol Cell 2021;81(9):1970−87 e9.

[102] Horakova AH, Calabrese JM, McLaughlin CR, Tremblay DC, Magnuson T, Chadwick BP. The mouse DXZ4 homolog retains Ctcf binding and proximity to Pls3 despite substantial organizational differences compared to the primate macrosatellite. Genome Biol 2012;13(8):R70.

[103] Barutcu AR, Maass PG, Lewandowski JP, Weiner CL, Rinn JL. A TAD boundary is preserved upon deletion of the CTCF-rich Firre locus. Nat Commun 2018;9(1):1444.

[104] Yang F, Deng X, Ma W, Berletch JB, Rabaia N, Wei G, et al. The lncRNA Firre anchors the inactive X chromosome to the nucleolus by binding CTCF and maintains H3K27me3 methylation. Genome Biol 2015;16(1):52.

[105] Loda A, Brandsma JH, Vassilev I, Servant N, Loos F, Amirnasr A, et al. Genetic and epigenetic features direct differential efficiency of Xist-mediated silencing at X-chromosomal and autosomal locations. Nat Commun 2017;8(1):690.

[106] Chadwick BP, Willard HF. Barring gene expression after XIST: maintaining facultative heterochromatin on the inactive X. Semin Cell Dev Biol 2003;14(6):359−67.

[107] Wang Z, Willard HF, Mukherjee S, Furey TS. Evidence of influence of genomic DNA sequence on human X chromosome inactivation. PLoS Comput Biol 2006;2(9):e113.

[108] Ngamphiw C, Tongsima S, Mutirangura A. Roles of intragenic and intergenic L1s in mouse and human. PLoS One 2014;9(11):e113434.

[109] Norris DP, Brockdorff N, Rastan S. Methylation status of CpG-rich islands on active and inactive mouse X chromosomes. Mammalian Genome 1991;1(2):78−83.

[110] Gendrel AV, Apedaile A, Coker H, Termanis A, Zvetkova I, Godwin J, et al. Smchd1-dependent and -independent pathways determine developmental dynamics of CpG island methylation on the inactive X chromosome. Dev Cell 2012;23(2):265−79.

[111] Hermann A, Gowher H, Jeltsch A. Biochemistry and biology of mammalian DNA methyltransferases. Cell Mol Life Sci 2004;61(19−20):2571−87.

[112] Csankovszki G, Nagy A, Jaenisch R. Synergism of Xist RNA, DNA methylation, and histone hypoacetylation in maintaining X chromosome inactivation. J Cell Biol 2001;153(4):773–84.

[113] Wolf SF, Migeon BR. Studies of X chromosome DNA methylation in normal human cells. Nature 1982;295(5851):667–71.

[114] Hansen RS, Stöger R, Wijmenga C, Stanek AM, Canfield TK, Luo P, et al. Escape from gene silencing in ICF syndrome: evidence for advanced replication time as a major determinant. Hum Mol Genet 2000;9(18):2575–87.

[115] Balaton BP, Dixon-McDougall T, Peeters SB, Brown CJ. The eXceptional nature of the X chromosome. Hum Mol Genet 2018;27(R2):R242–9.

[116] Kalantry S, Magnuson T. The Polycomb group protein EED is dispensable for the initiation of random X-chromosome inactivation. PLoS Genet 2006;2(5):e66.

[117] Leeb M, Wutz A. Ring1B is crucial for the regulation of developmental control genes and PRC1 proteins but not X inactivation in embryonic cells. J Cell Biol 2007;178(2):219–29.

[118] Jansz N, Nesterova T, Keniry A, Iminitoff M, Hickey PF, Pintacuda G, et al. Smchd1 targeting to the inactive X Is dependent on the Xist-HnrnpK-PRC1 pathway. Cell Rep 2018;25(7):1912–23 e9.

[119] Yang L, Yildirim E, Kirby JE, Press W, Lee JT. Widespread organ tolerance to Xist loss and X reactivation except under chronic stress in the gut. Proc Natl Acad Sci USA 2020;117(8):4262–72.

[120] Strehle M, Guttman M. Xist drives spatial compartmentalization of DNA and protein to orchestrate initiation and maintenance of X inactivation. Curr Opin Cell Biol 2020;64:139–47.

[121] Sierra I, Anguera MC. Enjoy the silence: X-chromosome inactivation diversity in somatic cells. Curr Opin Genet Dev 2019;55:26–31.

[122] Wang J, Syrett CM, Kramer MC, Basu A, Atchison ML, Anguera MC. Unusual maintenance of X chromosome inactivation predisposes female lymphocytes for increased expression from the inactive X. Proc Natl Acad Sci USA 2016;113(14):E2029–38.

[123] Migeon BR. An overview of X inactivation based on species differences. Semin Cell Dev Biol 2016;56:111–16.

[124] Cantone I, Fisher AG. Human X chromosome inactivation and reactivation: implications for cell reprogramming and disease. Philos Trans R Soc B: Biol Sci 2017;372(1733):20160358.

[125] Ohno S. Sex chromosomes and sex-linked genes. Berlin: Springer; 1967.

[126] Nguyen DK, Disteche CM. Dosage compensation of the active X chromosome in mammals. Nat Genet 2006;38(1):47–53.

[127] Gupta V, Parisi M, Sturgill D, Nuttall R, Doctolero M, Dudko OK, et al. Global analysis of X-chromosome dosage compensation. J Biol 2006;5(1):3.

[128] Pessia E, Engelstädter J, Marais GA. The evolution of X chromosome inactivation in mammals: the demise of Ohno's hypothesis? Cell Mol Life Sci 2014;71(8):1383–94.

[129] Xiong Y, Chen X, Chen Z, Wang X, Shi S, Wang X, et al. RNA sequencing shows no dosage compensation of the active X-chromosome. Nat Genet 2010;42(12):1043–7.

[130] Disteche CM. Dosage compensation of the sex chromosomes and autosomes. Semin Cell Dev Biol 2016;56:9–18.

[131] Posynick BJ, Brown CJ. Escape From X-Chromosome inactivation: an evolutionary perspective. Front Cell Dev Biol 2019;7:241.

[132] Chen X, Zhang J. No X-chromosome dosage compensation in human proteomes. Mol Biol Evol 2015;32(6):1456–60.

[133] Cotton AM, Ge B, Light N, Adoue V, Pastinen T, Brown CJ. Analysis of expressed SNPs identifies variable extents of expression from the human inactive X chromosome. Genome Biol 2013;14(11):R122.

[134] Cotton AM, Price EM, Jones MJ, Balaton BP, Kobor MS, Brown CJ. Landscape of DNA methylation on the X chromosome reflects CpG density, functional chromatin state and X-chromosome inactivation. Hum Mol Genet 2015;24(6):1528–39.

[135] Allen RC, Zoghbi HY, Moseley AB, Rosenblatt HM, Belmont JW. Methylation of HpaII and HhaI sites near the polymorphic CAG repeat in the human androgen-receptor gene correlates with X chromosome inactivation. Am J Hum Genet 1992;51(6):1229–39.

[136] Machado FB, Machado FB, Faria MA, Lovatel VL, Alves da Silva AF;, Radic CP, et al. 5meCpG epigenetic marks neighboring a primate-conserved core promoter short tandem repeat indicate X-chromosome inactivation. PLoS One 2014;9(7):e103714.

[137] Nyhan WL, Bakay B, Connor JD, Marks JF, Keele DK. Hemizygous expression of glucose-6-phosphate dehydrogenase in erythrocytes of heterozygotes for the Lesch-Nyhan syndrome. Proc Natl Acad Sci USA 1970;65(1):214–18.

[138] Migeon BR, Moser HW, Moser AB, Axelman J, Sillence D, Norum RA. Adrenoleukodystrophy: evidence for X linkage, inactivation, and selection favoring the mutant allele in heterozygous cells. Proc Natl Acad Sci USA 1981;78(8):5066–70.

[139] Hoffbuhr K, Devaney JM, LaFleur B, Sirianni N, Scacheri C, Giron J, et al. MeCP2 mutations in children with and without the phenotype of Rett syndrome. Neurology 2001;56(11):1486–95.

[140] Gringras P, Chen W. Mechanisms for differences in monozygous twins. Early Hum Dev 2001;64(2):105–17.

[141] Buller RE, Sood AK, Lallas T, Buekers T, Skilling JS. Association between nonrandom X-chromosome inactivation and BRCA1 mutation in germline DNA of patients with ovarian cancer. J Natl Cancer Inst 1999;91(4):339–46.

[142] Kristiansen M, Langerød A, Knudsen GP, Weber BL, Børresen-Dale AL, Orstavik KH. High frequency of skewed X inactivation in young breast cancer patients. J Med Genet 2002;39(1):30–3.

[143] Badens C, Martini N, Courrier S, DesPortes V, Touraine R, Levy N, et al. ATRX syndrome in a girl with a heterozygous mutation in the ATRX Zn finger domain and a totally skewed X-inactivation pattern. Am J Med Genet A 2006;140(20):2212–15.

[144] Carter MT, Picketts DJ, Hunter AG, Graham GE. Further clinical delineation of the Börjeson-Forssman-Lehmann syndrome in patients with PHF6 mutations. Am J Med Genet A 2009;149A(2):246–50.

[145] Wang Y, Martinez JE, Wilson GL, He XY, Tuck-Muller CM, Maertens P, et al. A novel RSK2 (RPS6KA3) gene mutation associated with abnormal brain MRI findings in a family with Coffin-Lowry syndrome. Am J Med Genet A 2006;140(12):1274–9.

[146] Fieremans N, Van Esch H, Holvoet M, Van Goethem G, Devriendt K, Rosello M, et al. Identification of intellectual disability genes in female patients with a Skewed X-inactivation pattern. Hum Mutat 2016;37(8):804–11.

[147] Viggiano E, Picillo E, Ergoli M, Cirillo A, Del Gaudio S, Politano L. Skewed X-chromosome inactivation plays a crucial role in the onset of symptoms in carriers of Becker muscular dystrophy. J Gene Med 2017;19(4):e2952.

[148] Holden ST, Cox JJ, Kesterton I, Thomas NS, Carr C, Woods CG. Fanconi anaemia complementation group B presenting as X linked VACTERL with hydrocephalus syndrome. J Med Genet 2006;43(9):750–4.

[149] Brady PD, Van Esch H, Fieremans N, Froyen G, Slavotinek A, Deprest J, et al. Expanding the phenotypic spectrum of PORCN variants in two males with syndromic microphthalmia. Eur J Hum Genet 2015;23(4):551–4.

[150] Wang J, Xiao QZ, Chen YM, Yi S, Liu D, Liu YH, et al. DNA hypermethylation and X chromosome inactivation are major determinants of phenotypic variation in women heterozygous for G6PD mutations. Blood Cells Mol Dis 2014;53(4):241−5.

[151] Radic CP, Rossetti LC, Abelleyro MM, Tetzlaff T, Candela M, Neme D, et al. Phenotype-genotype correlations in hemophilia A carriers are consistent with the binary role of the phase between F8 and X-chromosome inactivation. J Thromb Haemost 2015;13(4):530−9.

[152] Fusco F, Bardaro T, Fimiani G, Mercadante V, Miano MG, Falco G, et al. Molecular analysis of the genetic defect in a large cohort of IP patients and identification of novel NEMO mutations interfering with NF-kappaB activation. Hum Mol Genet 2004;13(16):1763−73.

[153] Snijders Blok L, Madsen E, Juusola J, Gilissen C, Baralle D, Reijnders MR, et al. Mutations in DDX3X are a common cause of unexplained intellectual disability with gender-specific effects on Wnt signaling. Am J Hum Genet 2015;97(2):343−52.

[154] Amor DJ, Stephenson SEM, Mustapha M, Mensah MA, Ockeloen CW, Lee WS, et al. Pathogenic variants in GPC4 cause keipert syndrome. Am J Hum Genet 2019;104(5):914−24.

[155] Torres RJ, Puig JG. Skewed X inactivation in Lesch-Nyhan disease carrier females. J Hum Genet 2017;62(12):1079−83.

[156] Woon ST, Ameratunga R, Croxson M, Taylor G, Neas K, Edkins E, et al. Follicular lymphoma in a X-linked lymphoproliferative syndrome carrier female. Scand J Immunol 2008;68(2):153−8.

[157] Desai V, Donsante A, Swoboda KJ, Martensen M, Thompson J, Kaler SG. Favorably skewed X-inactivation accounts for neurological sparing in female carriers of Menkes disease. Clin Genet 2011;79(2):176−82.

[158] Zou Y, Liu Q, Chen B, Zhang X, Guo C, Zhou H, et al. Mutation in CUL4B, which encodes a member of cullin-RING ubiquitin ligase complex, causes X-linked mental retardation. Am J Hum Genet 2007;80(3):561−6.

[159] Vianna EQ, Piergiorge RM, Gonçalves AP, Dos Santos JM, Calassara V, Rosenberg C, et al. Understanding the landscape of X-linked variants causing intellectual disability in females through extreme X chromosome inactivation skewing. Mol Neurobiol 2020;57(9):3671−84.

[160] Hedera P, Gorski JL. Oculo-facio-cardio-dental syndrome: skewed X chromosome inactivation in mother and daughter suggest X-linked dominant Inheritance. Am J Med Genet A 2003;123A(3):261−6.

[161] Morleo M, Franco B. Dosage compensation of the mammalian X chromosome influences the phenotypic variability of X-linked dominant male-lethal disorders. J Med Genet 2008;45 (7):401−8.

[162] Knudsen GP, Neilson TC, Pedersen J, Kerr A, Schwartz M, Hulten M, et al. Increased skewing of X chromosome inactivation in Rett syndrome patients and their mothers. Eur J Hum Genet 2006;14(11):1189−94.

[163] Lacout C, Haddad E, Sabri S, Svinarchouk F, Garçon L, Capron C, et al. A defect in hematopoietic stem cell migration explains the nonrandom X-chromosome inactivation in carriers of Wiskott-Aldrich syndrome. Blood 2003;102(4):1282−9.

[164] Vacca M, Della Ragione F, Scalabrì F, D'Esposito M. X inactivation and reactivation in X-linked. diseases. Semin Cell Dev Biol 2016;56:78−87.

[165] Bolduc V, Chagnon P, Provost S, Dubé MP, Belisle C, Gingras M, et al. No evidence that skewing of X chromosome inactivation patterns is transmitted to offspring in humans. J Clin Invest 2008;118(1):333−41.

[166] Vickers MA, McLeod E, Spector TD, Wilson IJ. Assessment of mechanism of acquired skewed X inactivation by analysis of twins. Blood 2001;97(5):1274−81.

[167] Zito A, Davies MN, Tsai PC, Roberts S, Andres-Ejarque R, Nardone S, et al. Heritability of skewed X-inactivation in female twins is tissue-specific and associated with age. Nat Commun 2019;10(1):5339.

[168] Kristiansen M, Knudsen GP, Bathum L, Naumova AK, Sørensen TI, Brix TH, et al. Twin study of genetic and aging effects on X chromosome inactivation. Eur J Hum Genet 2005;13(5):599−606.

[169] Busque L, Mio R, Mattioli J, Brais E, Blais N, Lalonde Y, et al. Nonrandom X-inactivation patterns in normal females: lyonization ratios vary with age. Blood 1996;88(1):59−65.

[170] Knudsen GP, Pedersen J, Klingenberg O, Lygren I, Ørstavik KH. Increased skewing of X chromosome inactivation with age in both blood and buccal cells. Cytogenet Genome Res 2007;116(1−2):24−8.

[171] Santos-Rebouças CB, Boy R, Vianna EQ, Gonçalves AP, Piergiorge RM, Abdala BB, et al. Skewed X-chromosome inactivation and compensatory upregulation of escape genes precludes major clinical symptoms in a female with a large Xq deletion. Front Genet 2020;11:101.

[172] Migeon BR. X-linked diseases: susceptible females. Genet Med 2020;22(7):1156−74.

[173] Franco B, Ballabio A. X-inactivation and human disease: X-linked dominant male-lethal disorders. Curr Opin Genet Dev 2006;16(3):254−9.

[174] Reijnders MR, Zachariadis V, Latour B, Jolly L, Mancini GM, Pfundt R, et al. De Novo loss-of-function mutations in USP9X cause a female-specific recognizable syndrome with developmental delay and congenital malformations. Am J Hum Genet 2016;98(2):373−81.

[175] Klein SL, Flanagan KL. Sex differences in immune responses. Nat Rev Immunol 2016;16(10):626−38.

[176] Forsyth KS, Anguera MC. Time to get ill: the intersection of viral infections, sex, and the X chromosome. Curr Opin Physiol 2021;19:62−72.

[177] Jin JM, Bai P, He W, Wu F, Liu XF, Han DM, et al. Gender differences in patients with COVID-19: focus on severity and mortality. Front in Public Health 2020;8:152.

[178] Lucas JM, Heinlein C, Kim T, Hernandez SA, Malik MS, True LD, et al. The androgen-regulated protease TMPRSS2 activates a proteolytic cascade involving components of the tumor microenvironment and promotes prostate cancer metastasis. Cancer Discov 2014;4(11):1310−25.

[179] Stelzig KE, Canepa-Escaro F, Schiliro M, Berdnikovs S, Prakash YS, Chiarella SE. Estrogen regulates the expression of SARS-CoV-2 receptor ACE2 in differentiated airway epithelial cells. Am J Physiol-Lung Cell Mol Physiol 2020;318(6):L1280−1.

[180] Chen J, Jiang Q, Xia X, Liu K, Yu Z, Tao W, et al. Individual variation of the SARS-CoV-2 receptor ACE2 gene expression and regulation. Aging Cell 2020;19(7):e13168.

[181] Imai Y, Kuba K, Rao S, Huan Y, Guo F, Guan B, et al. Angiotensin-converting enzyme 2 protects from severe acute lung failure. Nature 2005;436(7047):112−16.

[182] Billi AC, Kahlenberg JM, Gudjonsson JE. Sex bias in autoimmunity. Curr Opin Rheumatol 2019;31(1):53−61.

[183] Syrett CM, Anguera MC. When the balance is broken: X-linked gene dosage from two X chromosomes and female-biased autoimmunity. J Leukoc Biol 2019;106(4):919−32.

[184] Hewagama A, Gorelik G, Patel D, Liyanarachchi P, McCune WJ, Somers E, et al. Overexpression of X-linked genes in T cells from women with lupus. J Autoimmun 2013;41:60−71.

[185] Dunford A, Weinstock DM, Savova V, Schumacher SE, Cleary JP, Yoda A, et al. Tumor-suppressor genes that escape from X-inactivation contribute to cancer sex bias. Nat Genet 2017;49 (1):10−16.

[186] Wang D, Tang L, Wu Y, Fan C, Zhang S, Xiang B, et al. Abnormal X chromosome inactivation and tumor development. Cell Mol Life Sci 2020;77(15):2949–58.

[187] Clocchiatti A, Cora E, Zhang Y, Dotto GP. Sexual dimorphism in cancer. Nat Rev Cancer 2016;16(5):330–9.

[188] Natri H, Garcia AR, Buetow KH, Trumble BC;, Wilson MA. The pregnancy pickle: evolved immune compensation due to pregnancy underlies sex differences in human diseases. Trends Genet 2019;35(7):478–88.

[189] Sun Z, Prodduturi N, Sun SY, Thompson EA, Kocher JP. Chromosome X genomic and epigenomic aberrations and clinical implications in breast cancer by base resolution profiling. Epigenomics 2015;7(7):1099–110.

[190] Chaligné R, Popova T, Mendoza-Parra MA, Saleem MA, Gentien D, Ban K, et al. The inactive X chromosome is epigenetically unstable and transcriptionally labile in breast cancer. Genome Res 2015;25(4):488–503.

[191] Yildirim E, Kirby JE, Brown DE, Mercier FE, Sadreyev RI, Scadden DT, et al. Xist RNA is a potent suppressor of hematologic cancer in mice. Cell 2013;152(4):727–42.

[192] Chanda K, Mukhopadhyay D. LncRNA Xist, X-chromosome Instability and Alzheimer's Disease. Curr Alzheimer Res 2020;17(6):499–507.

[193] Bajic VP, Essack M, Zivkovic L, Stewart A, Zafirovic S, Bajic VB, et al. The X files: "The mystery of X chromosome instability in Alzheimer's disease". Front Genet 2020;10:1368.

[194] Talon I, Janiszewski A, Chappell J, Vanheer L, Pasque V. Recent advances in understanding the reversal of gene silencing during X chromosome reactivation. Front Cell Dev Biol 2019;7:169.

[195] Moore KL, Barr ML. The sex chromatin in human malignant tissues. Br J Cancer 1957;11(3):384–90.

[196] Robinson L, Guy J, McKay L, Brockett E, Spike RC, Selfridge J, et al. Morphological and functional reversal of phenotypes in a mouse model of Rett syndrome. Brain 2012;135(Pt 9):2699–710.

[197] Guy J, Gan J, Selfridge J, Cobb S, Bird A. Reversal of neurological defects in a mouse model of Rett syndrome. Science 2007;315(5815):1143–7.

[198] Jiang J, Jing Y, Cost GJ, Chiang JC, Kolpa HJ, Cotton AM, et al. Translating dosage compensation to trisomy 21. Nature 2013;500(7462):296–300.

[199] Savarese F, Flahndorfer K, Jaenisch R, Busslinger M, Wutz A. Hematopoietic precursor cells transiently reestablish permissiveness for X inactivation. Mol Cell Biol 2006;26(19):7167–77.

[200] Huang KC, Rao PH, Lau CC, Heard E, Ng SK, Brown C, et al. Relationship of XIST expression and responses of ovarian cancer to chemotherapy. Mol Cancer Ther 2002;1(10):769–76.

[201] Carrette LLG, Wang CY, Wei C, Press W, Ma W, Kelleher 3rd RJ, et al. A mixed modality approach towards Xi reactivation for Rett syndrome and other X-linked disorders. Proc Natl Acad Sci USA 2018;115(4):E668–75.

[202] Sripathy S, Leko V, Adrianse RL, Loe T, Foss EJ, Dalrymple E, et al. Screen for reactivation of MeCP2 on the inactive X chromosome identifies the BMP/TGF-β superfamily as a regulator of XIST expression. Proc Natl Acad Sci USA 2017;114(7):1619–24.

CHAPTER 23

Epigenetics of Memory Processes

Sravani Pulya and Balaram Ghosh

Epigenetic Research Laboratory, Department of Pharmacy, Birla Institute of Technology and Science-Pilani Hyderabad Campus, Shamirpet, Hyderabad, Telangana, India

OUTLINE

Introduction	443	DNA Methylation and Long-term Memory	452
Histone Posttranslational Modifications: Long-term Memory Regulation	444	Histone Variant Exchange	456
		Epitranscriptomics: RNA Epigenetics	457
Histone Acetylation and HATs	445	Summary	458
Histone Deacetylation and Histone Deacetylases	447	Acknowledgments	458
Histone Methylation	448	Abbreviations	458
Histone Modifications: Gene Expression	449	References	458
Manipulating Histone Modifications	450		

INTRODUCTION

Memory formation is a complex process involving both short-term and long-term memories depending on both synapse-synapse and synapse-nucleus signaling in the CNS. Short-term memory relies on the dynamic regulation of underlying gene activity; whereas, long-term memory is characterized by the long-lasting alterations in gene activity states followed by transcriptional regulation in the nucleus. Memory consolidation is a time-dependent process of transforming recently learned memories to long-term memories, lasting for a longer time period, that require changes in the protein synthesis and gene expression. Neuronal stimulation leads to increased levels of protein synthesis and gene expression following the activation of neuronal dendrites, synapses and proteins required for the transcription and translational regulation for synaptic functions and neuroplasticity. The gene expression modifications are regulated by epigenetic mechanisms involving chromatin modifications. Epigenetic mechanisms involve the dynamic regulation of histone protein modifications and covalent modifications of DNA, such as DNA methylation, in response to the environmental stimuli governing the regulation of learning and memory as represented in the Fig. 23.1. Recently, miRNAs and glial cells have received attention for their role in regulating memory processes. Over the past decade, lot of research has been carried out that established epigenetic mechanisms and their role in CNS. Epigenetics in neurons differ from that of traditional epigenetics in that they are non-heritable as adult neurons are non-dividing in nature. This led to a new sub-field of epigenetics, known as neuroepigenetics, that studies the different roles of epigenetic mechanisms in regard to synaptic plasticity, learning and memory regulation, neurotoxicity, and CNS disorders [1].

This chapter reviews in detail the epigenetic mechanisms governing the memory processes including posttranslational modifications of histones, covalent modifications

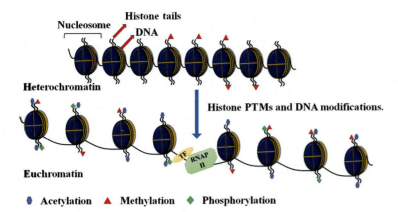

FIGURE 23.1 General representation of epigenetic modifications and chromatin remodeling. DNA wrapped around histone octamer constituting the nucleosome, the fundamental unit of chromatin. Two different transcriptional states of chromatin; heterochromatin, closed and transcriptionally repressed state, and euchromatin, open and transcriptionally active state. Chromatin remodeling is regulated by the modifications on the histone tail residues and covalent DNA modifications catalyzed by their respective enzymes. *TF*, Transcription factor; *RNAP II*, RNA polymerase II.

on DNA, and the emerging role of histone variants and epitranscriptomics. Further, the manipulations of these modifications are also discussed in detail, including the recent literature regarding their critical role in learning and memory processes.

HISTONE POSTTRANSLATIONAL MODIFICATIONS: LONG-TERM MEMORY REGULATION

Eukaryotic nuclei consist of approximately 1.7 turns of DNA or about ~147 bp of DNA wrapped around the core of histone proteins containing 2 dimers of H2A and H2B and a tetramer of H3-H4 together forming a complex called nucleosome, which is the fundamental unit of chromatin. Each nucleosome is joined to another by small segments of DNA together with linker histone H1. Positively charged histones are tightly bound to the negatively charged DNA with high affinity, resulting in a compact chromatin structure leading to a transcriptionally unfavorable state. Hence, structural modifications of the chromatin regulate gene expression. Chromatin exists in different forms such as heterochromatin, which is the transcriptionally inactive and highly compact structure of nucleosome, and another being euchromatin which forms the open nucleosome associated with transcriptional activation. These altered structures of chromatin are regulated by several factors that involve the presence of activators, repressors, and chromatin remodeling complexes, along with and also due to the covalent modifications of DNA as well as PTMs (posttranslational modifications) of the histone proteins [2]. The accessibility of DNA towards regulatory proteins, and thus gene expression, depends on the PTMs of the N-terminal of histone tails consisting of lysine residues. Several PTMs of histone proteins are known that include acetylation, phosphorylation, methylation, SUMOylation, and ubiquitination. Recently, serotonylation, dopaminylation, deamination, propionylation,

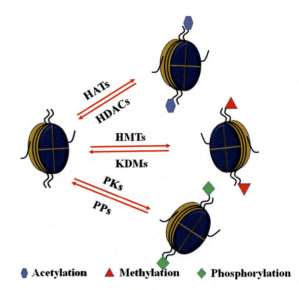

FIGURE 23.2 Posttranslational modifications of the histones. Histone acetylation: catalyzed by histone acetyltransferases (HATs) and histone deacetylases (HDACs). Histone methylation: catalyzed by histone methyltransferases (HMTs) and lysine specific demethylases (KDMs). Histone phosphorylation: catalyzed by protein kinases (PKs) and protein phosphatases (PPs).

butyrylation, and histone globular domain modifications were also reported [3–5]. These PTMs affect the interactions between nucleosomes, thus affecting the overall chromatin structure and remodeling.

Among the PTMs, acetylation, methylation, and phosphorylation of histone proteins have been studied in relation to the epigenetic regulation of memory and learning (Fig. 23.2) [6]. Histone acetylation is primarily associated with gene activation; whereas, histone methylation can be either an activating/repressing gene expression, depending on the modified histone amino acid residue. Histone acetylation or deacetylation catalyzes the addition or removal of an acetyl group on ϵ-amino lysine residues and is catalyzed by two enzymes, namely histone acetyl transferases (HATs) and histone deacetylases (HDACs) respectively [7]. Histone methyltransferases (HMT) and demethylases are the enzymes

TABLE 23.1 Brief Summary of the Most Extensively Studied Posttranslational Modifications of Histones, Their Target Residues Modified by the Enzymes Catalyzing Those Modifications

PTMs	Target Residues	Target Histones Affected	Function	Enzymes Involved
Acetylation	Lysine	H3K9/27 H4K5/K8/K12/K16	DNA repair, replication and transcriptional activation	HATs/HDACs
Methylation	Lysine Arginine	H3K4/K9/K27/K36/K79/R2/R17 H4K12/K20/R3	Transcriptional activation/repression	HMTs/KDMs
Phosphorylation	Serine Threonine	H3S10/S28, H4S1	Transcriptional activation/DNA repair	PKs/PPs

Functional roles of the modified histone residues in transcriptional regulation and gene expression are given.

that catalyze multiple levels (e.g., mono-, di-, or tri-) of methylation/demethylation on the specific sites of either lysine and arginine residues on H3 and H4 histones respectively, thus regulating the efficiency of gene expression differently. On the other hand, histone phosphorylation is majorly associated with transcriptional activation of gene expression. Histone phosphorylation occurs on threonine 3, serine 10, and serine 28 of histone H3 and serine 1 of histone H4 residues catalyzed by nuclear kinases and phosphatases [8]. All these modifications of histones in correlation with other covalent modifications on DNA together govern the regulation of gene expression and activity in CNS (Table 23.1).

HISTONE ACETYLATION AND HATS

One of the most widely and extensively studied PTM of histones is histone acetylation which is widely implicated in memory and learning [7,9–16]. Histone acetyl transferases (HATs) are the enzymes that catalyze the reaction by the transfer of an acetyl group from acetyl coenzyme A onto the N-terminal amino group on the histone tail. Upon the addition of an acetyl group onto the positively charged histones, the affinity between histone and DNA decreases, leading to the permissive state of chromatin towards transcriptional activation. Some evidence also suggests that the acetylated lysine residues function as markers that are recognized by various other transcription regulators and chromatin remodeling proteins, RNA polymerase II, thus facilitating transcriptional activation [7]. Mammalian HATs consist of five different families namely, (1) the GNAT (GCN5-related acetyltransferases; GCN5, PCAF, ELP3) family; (2) the MYST (MOZ, YBF2/SAS3, SAS2 and TIP60 protein) family; (3) the transcription factor related HATs; [TFIIIC (Transcription Factor IIIC), TAF1]; (4) the nuclear receptor coactivator family namely, SRC (steroid receptor coactivator), ACTR/NCOA3 (nuclear receptor coactivator 3); and (5) the cAMP response element-binding protein (CREB)-binding protein (CBP) and p300 family [17]. Another type B HAT, known as HAT1, located in the cytoplasm, functions in acetylating the newly synthesized histones before chromatin assembly [17]. Interestingly, HATs are usually found in high levels in hippocampal and prefrontal cortex areas of the brain that are involved in learning and memory.

Hippocampal histone acetylation was found to positively correlate with the contextual fear conditioning (CFC), novel object recognition (NOR) and spatial memory formation studies. Hence, disorders with reduced cognitive abilities such as Alzheimer disease were associated with hypoacetylation as reported in several studies including pre-clinical models and human postmortem brain samples [18,19]. In this context, one of the earliest and most extensively studied HAT enzyme is CBP. Early studies suggested CBP-deficient mice have normal short-term memory (STM) but deficient long-term memory (LTM). The early embryonic lethality of CBP knockout mice due to the failure of neural tube closure has demonstrated its essential role during embryogenesis [20]. Bousiges et al., for the first time reported that the CBP is induced upon spatial training and 1 h post-Morris water maze spatial training; bulk levels of H2B/H4 acetylation were increased at the promoter regions of memory-related genes *Zif268* and *Bdnf* in rat dorsal hippocampus [21]. On the other hand, studies in CBP knockout mice in mature neurons in the postnatal brain have not reported any neurological defects or neurodegeneration but exhibited deficits in the formation of spatial, associative, and object-recognition memory along with impaired LTM [22,23]. For example, studies in the mouse models for RSTS (Rubinstein-Taybi syndrome) suggested a direct role for CBP and histone acetylation in cognition and memory. The specific deletion of CBP gene in postmitotic principal neurons of the forebrain displayed no abnormalities and neurodegeneration. This study identified CBP acetyltransferase domain prefers the acetylation of H2A and H2B dimers, and the loss of CBP does not affect neuronal viability. The study further confirmed that CBP deficiency caused object recognition memory (ORM) in the postnatal brain of adult neurons [24]. Also, studies in aplysia demonstrated that the switch from short to long term memory facilitation is induced by CREB-like proteins [25] and several studies in Drosophila also reaffirm the involvement of CREB in LTM formation [26]. A recent study in CBP knockdown mice indicated that the acetylation of H3K23 catalyzed by CBP contributed to learning, likely by controlling specific learning-related genes, suggesting a novel epigenetic regulatory mechanism [27]. Mice with a truncated form of CBP when studied through passive avoidance tests and CFC, were deficient in LTM with normal STM, thus suggesting the crucial role of CBP in mammalian LTM as in the case of Rubinstein-Taybi syndrome (RTS) [28,29]. Similar results were also found in CBP deficient mice with significant impairments in NOR, spatial and fear memory, including deficits in hippocampal long-term potentiation (LTP), demonstrating that CBP acetyltransferase activity is critical for gene activation regulating memory consolidation [30]. Another study demonstrated by object recognition tasks indicated that the acute hypo functioning of the CBP in the medial prefrontal cortex (mPFC) resulted in normal information acquisition and normal STM; whereas, LTM for object location was disrupted [31]. Literature reports established that CBP is required for the correct differentiation and integration of different neurons into hippocampus, cortical, cerebellar neuronal and spinal cord circuits. Recently, it was found that even postdifferentiation, CBP and its interaction with SRF (serum response factor) is required to activate the downstream genes and govern the developmental processes of dendritic growth, spine maturation, and activity-dependent synaptic changes in newborn neurons [32,33]. Structurally, CBP and p300 contain a lysine acetyltransferase (KAT) domain and multiple independent protein-binding domains that interact with many other proteins, promoting the acetylation of the target lysine residues, thus playing critical roles during the development of the nervous system and also regulating the stimuli-driven transcription and plasticity in postmitotic neurons [32]. Though CBP/P300 belong to the same family and are often demonstrated interchangeably for their intervention in intellectual development and cognitive function, few selective inhibition studies suggest otherwise. Both are found to acetylate H3K14, H3K18, H3K23, H3K27, H4K5, H4K8, H4K12, and H4K16 with varied degrees of specificity. CBP overexpression resulted in increased AcH4K12 but not AcH4K8; whereas, p300 functions otherwise. Studies reported with single cell RNA-seq data show that CBP is more expressed than p300 in neurons of both developing and adult brain functions [32,34,35]. Also, the deregulation of acetylated H4K12 levels in aged mice is associated with the impairment of their cognitive abilities which after restoration reinstated the hippocampal gene expression associated with memory consolidation [36]. Treatment with SAHA, a pan-HDAC inhibitor when delivered orally has shown to be promising probable therapy in age-related cognitive disorders [37]. PCAF, a transcriptional coactivator with intrinsic histone deacetylase activity, is found to be long-lasting in the chromatin remodeling necessary for memory formation and stress responses as studied in the case of PCAF knockout mice. Studies established that p300/CBP acetylation activity in the lateral amygdala is critical for fear conditioning and fear memory consolidation [32]. Compound C646, a selective p300/CBP inhibitor (also called as HAT inhibitor) when locally infused into the lateral nucleus of amygdala significantly impaired consolidation and reconsolidation of pavlovian fear memories [38]. A study was conducted involving the use of selective HAT inhibitors and short-interference RNAs (siRNAs) when injected into hippocampus or perirhinal cortex of the rats, it was found that in the hippocampus, CBP/p300 inhibition impaired LTM but not STM of the object, whereas p300 specific inhibition in the perirhinal cortex had no effect. On the other hand, PCAF specific inhibition impaired both LTM and STM [39]. In another recent study, the function of CBP KIX domain was studied using KIX domain mutant mice, preventing phospho-CREB binding, suggesting its potential role in understanding the effects of circadian and memory impairments due to aging in hippocampal memory and gene expression [40].

Apart from CBP, the most widely studied HAT, several other HATs were also found to play critical roles in learning and memory. A MYST family HAT, Tip60 is found to play a critical role in nervous system development, function, aging, and neurodegenerative diseases. Tip60 is known to acetylate the target substrates H2A, H3, and H4 histone proteins. Studies have reported the loss of Tip60 in Alzheimer disease Drosophila model and also in human postmortem Alzheimer disease brain, and the possibility of Tip60 hypoacetylation to be a common event that contributes to the epigenetic mediated gene mis-regulation, and cognition deficits in other neurodegenerative diseases as well [41,42]. H2A.Z modification by Tip60 HAT has contributed to the fear memory consolidation in mice [43]. Mouse deficient in Tip60, indicated the hypoacetylation of H4K12 leading to the downregulation of genes in CA1 demonstrating neuronal loss and mild memory deficit [44]. Tip60 variants were found to cause histone hypoacetylation and gene expression dysregulation leading to neurodevelopmental syndrome, sleep disorders and a disruption in circadian clock [45].

Several studies have also reported the use of HAT inhibitors implicating memory and learning processes.

In an interesting study, two HAT inhibitors, garcinol (p300/CBP and PCAF family inhibitor) and C646 (specific p300/CBP family inhibitor), were injected in the honeybee and found that both inhibitors impaired acetylation-mediated memory processes with either transcription-dependent/independent mechanisms, respectively [46]. Garcinol, when systemically injected was found to impair the reconsolidation of cocaine-cue memories in a temporally constrained, cue-specific, and persistent manner [47]. Studies indicated the key role of garcinol in the reconsolidation of auditory fear memories through modulating epigenetic processes in the lateral amygdala [46,48]. Recently, the treatment of C646 decreased the acetylation of H3K9 in the adult optic tectum of zebrafish and the EP300 inhibition has downregulated the notch target genes her4 and her6 thus promoting neuroepithelial-like stem cell proliferation and also adult neurogenesis [49]. In a recent report, Nickel exposed to mouse hippocampal and cultured neurons, led to the hypoacetylation at H3K9 thus repressing the expression of H3K9 modulated neural genes leading to impairment of neural function without causing neuron death. Further, treatment with sodium butyrate, pan-HDACis attenuated the affect caused by Nickel, suggesting that the chronic exposure to Nickel caused impaired memory through reduced acetylation [50]. One of the first evidence using TTK21, a small molecule CBP/p300 activator, in conjugation to glucose-based carbon nanosphere, acetylated histones in the hippocampus and frontal cortex in-vivo leading to the maturation and differentiation of adult neuronal progenitors contributing to LTM [51]. Another study, for the first time identified CSP-TTK21 treatment has restored plasticity, spatial memory and behavioral deficits in the diseased brain of mice with Alzheimer's disease like Tau pathology [52].

HISTONE DEACETYLATION AND HISTONE DEACETYLASES

Histone deacetylases (HDACs) possess transcriptional repressive activity by catalyzing the deacetylation on the N-terminal end of acetylated lysine residues thus restoring their positive charge and condensation of DNA onto the histone core. Four classes of HDAC isoforms are known, classified according to their mechanism of action, including class I, IIA, IIB, and IV are the classic zinc dependent HDACs, the class III being sirtuins that are NAD+ dependent [53]. Several HDAC inhibitors (HDACis) have been identified involving both pan-HDACis and isoform specific HDACis which are widely used to study the implications involving histone PTMs in adult memory formation and learning [19,54]. Studies using isoform selective and pan-HDACis have reported their significant therapeutic intervention in cognitive and memory enhancement [55–60].

HDACs, particularly class I HDACs are found to be potent negative regulators of LTM formation [9,61,62]. Several pan-HDACis enhance synaptic plasticity as well as long-term memory by acetylation mechanisms [63]. Literature findings suggest that under normal conditions, the activity of transcriptional repressors dominate transcriptional activators i.e., the activity of HDACs dominate that of HATs due to which long-term synaptic enhancement and memory formation are impaired. One of the earliest studies reported that the use of HDACis TSA or NaB enhanced the induction of LTP at Schaffer-collateral synapses in CA1 area of hippocampus in-vitro prior to contextual fear conditioning, also enhancing the LTM formation in-vivo [9]. Studies using HDACis have demonstrated CBP and CREB interactions mediating the hippocampus dependent memory and synaptic plasticity [64]. HDAC inhibition in CBP mutant mice led to persistent forms of LTM for object recognition and explored the role of HDACs as potential memory suppressor genes. In a study, the hippocampal delivery of a GABA agonist, muscimol caused the inactivation of the hippocampus, and blocked the LTM for object location with no effect on object recognition, thus indicating the role of hippocampus in object location memory (OLM) [55]. Further, using location-dependent object recognition tasks and engaging the hippocampus along with HDAC inhibition, it is demonstrated that CBP is essential for the long term memory modulation in the hippocampus and also established that differential roles of HDAC isoforms in learning and memory [55]. Among the HDAC isoforms, class I HDACs were widely studied in relation to memory, learning, and cognitive functions [56,60]. HDAC1 is found to be associated with the promotion of fear and extinction learning; whereas, HDAC2 is associated with the suppression of fear and extinction learning. A study suggested that hippocampal HDAC1 in adult mice promotes H3K9 deacetylation followed by trimethylation of target genes thus affecting the extinction of contextual fear memories [65]. HDAC1 is also found to be upregulated in the prefrontal cortex in the case of patients with schizophrenia particularly those suffering from early life stress [66]. Recently, a study reported the region-specific expression of HDAC1 and HDAC2, in which the acetylation of H3/H4 and the neuronal activation were found to be associated differentially in fear memory extinction and its consolidation [67]. In centromedial amygdala (CeM) of the central amygdala, HDAC1 was downregulated following fear conditioning; whereas, it is upregulated following extinction. In the contrary, HDAC2 was downregulated in CeM following conditioning and not extinction. Upregulation of HDAC1 and downregulation of HDAC2 were found in centrolateral amygdala

(CeL) following both fear conditioning and extinction [67]. Exifone, a selective HDAC1 activator was associated with neuroprotective activity in human neuronal models of neurodegeneration derived from a tauopathy patient [68]. One of the first studies revealed that HDAC2 overexpression led to reduced AcH4K12 thus impaired hippocampus-dependent associative and special learning in pavlovian fear conditioning paradigm and Morris water maze test [60]. Moreover, HDAC2 knockout mice have exhibited an increase in spine density and presynaptic terminals, further suggesting HDAC2 as a negative regulator of memory [62]. A study reported by McQuown and group in genetically modified mice, with HDAC3 focal deletions in the CA1 area of dorsal hippocampus together with inhibiting HDAC3 using RGFP136, significantly enhanced LTM in a persistent manner through increased *Nr4a2* and *c-fos* gene expression in the hippocampus, thus demonstrating the critical role of HDAC3 in the epigenetic mechanisms underlying LTM formation [61]. Further, HDAC3 was found to negatively regulate cocaine induced CPP acquisition in mice, which is correlated with increased gene expression of *c-fos* and *Nr4a2* with increased AcH4K8 levels during the consolidation phase of acquisition [69]. Thus, HDAC3 was found to be potentially involved in LTM formation, but not STM which is different from HDAC2 which is involved in the enhancement of short-time memory. Thus, the beneficial effect of various HDAC inhibitors and genetic deletion of HDACs in improving learning and memory, synaptic plasticity in various non-diseased rodent models together with other animal models of neurodegeneration and brain injury implies the significance of epigenetic manipulations to alter memory capacity [56,57,70].

Though several reports discussed the role of class IIa HDACs mainly HDAC4 and HDAC5 in synaptic plasticity and memory, little is established unlike class I HDACs. However, loss of HDAC4 in brain resulted in hippocampal memory and learning impairments, suggesting HDAC4 to be a crucial positive regulator of memory, in contrast to class I HDACs [71]. On the contrary, dysregulation of HDAC4 was found to modulate LTM functions in *Drosophila* when overexpression of HDAC4 catalytic mutant abolished LTM; this was further supported by a recent study indicating that HDAC4 abundance in the nucleus causes impairment in the neuronal development and LTM [72,73]. Class IIb HDAC, HDAC6, was found to be translocating both in the cytoplasm and nucleus, thus impacting synaptic function both positively and negatively. The nuclear localization of HDAC6 leads to decreased expression of *Bdnf*, thus negatively regulating synaptic functions followed by memory impairment. HDAC6 inhibition or knockdown caused a decrease in the glutamatergic signaling, thus impacting synaptic functions during stress conditions [74]. Further, several studies established the role of HDAC6 in neurological disorders due to the impairment of synaptic function and plasticity [53,75,76].

HISTONE METHYLATION

Another important histone PTM i.e., histone methylation in particular methylated H3K4, is also found to be widely regulated in the hippocampus and striatum-dependent memory formation in mice, thus leading to transcriptional activation. Since its first description in 1964, H3K9 methylation was found to be a highly conserved chromatin modification. Like acetylation, histone methylation is also found to be regulated by six of each H3K4 methyltransferases and demethylases [77]. Since then, several studies reported the essential role of histone methylation in synaptic plasticity and memory. In the one hour postcontextual fear conditioning in mice, bulk levels of trimethylated H3K4 (H3K4Me3) at the promoter regions of memory-related genes *Zif268* and *Bdnf*, transcriptional activation marker and H3K9me2 (a marker for transcriptional silencing) were found to be increased. Mice deficient in Mll, a H3K4-specific methyltransferase, exhibited deficits in contextual fear memory compared to wild-type mice, thus affirming the complicated mechanisms regulated by gene expression as well as gene silencing that occurs during memory formation in the hippocampus [78]. HDAC inhibition with NaB increased H3K4me3 and decreased H3K9me2 in the hippocampus following CFC [78]. Studies have reported the importance and role of neuronal KMT2a (Mll1) and KMT2b (Mll2) in the memory formation that was demonstrated in KMT2a/KMT2b knockdown mice, which lead to differential gene expression and the decreased H3K4 methylation in the hippocampal neurons of mice [79]. On the other hand, histone demethylation is known to be regulated by a family of lysine specific demethylases LSD1/KDM1 (KDM1A and KDM1B), and KDM5 (KDM5A, KDM5B, KDM5C, and KDM5D) characterized by different demethylating functions such as KDM1 family demethylates mono- and di-methyl H3K4; whereas, the KDM5 family removes methyl groups from mono-, di-, tri-methylated H3K4 [77]. In a recent study, the analysis of Chip-seq datasets of H3K4me3 revealed the role of broad domains in CA1 neurons of hippocampus to be associated with the transcriptional activation during memory and learning, which are regulated by KMT2a and KMT2b genes; they also identified that the widening of the domains occur during contextual fear conditioning in mice [80]. Contextual fear conditioning has increased the levels of H3K9me2 in the area of CA1 hippocampus and also in the entorhinal cortex (EC) 1 h after training in a water maze competition task. Whereas, the inhibition of G9a/GLP, a H3K9 dimethyl transferase immediately after learning, increased H3K9 acetylation levels while H3K9me2 levels were unchanged in dorsal hippocampus or gradually decreased in the dorsal striatum suggesting region specific outcomes of the same modification on memory posttraining [81]. Rats trained for intracranial self-stimulation (ICSS) induced reward seeking behavior in the lateral hypothalamus-medial forebrain bundle (LH-MFB) lead to increased H3K4me2 levels at *Bdnf* IV and IX promoter regions; whereas, H3K9me2 was decreased due to increased LSD 1 (lysine specific demethylase1) expression suggesting the critical role of LSD 1 driven *Bdnf* expression in LH [82].

HISTONE MODIFICATIONS: GENE EXPRESSION

Several studies established the role of specific histone modifications regulating gene expression, leading to LTM formations. Initial studies investigated the role of histone modifications in CA1 of the hippocampus following contextual fear conditioning (CFC) in mice. It was found that CFC in mice increased H3 acetylation and phosphorylation but not H4, indicating the LTM formation was regulated by the extracellular signal-regulated kinase/mitogen activated protein kinase (ERK/MAPK) pathway and mitogen- and stress-activated protein kinase 1 (MSK1) [83]. During the initial stages of LTM consolidation in a contextual fear conditioning paradigm, H3 acetylation in CA1 was found to be regulated by the activation of NMDA receptors and ERK signaling activated by PKC/PKA pathways. Further, the use of TSA or NaB enhanced the induction of LTP, thus also contributing to LTM formation and consolidation in vivo [9]. Interestingly, latent inhibition was found to be associated with H4 hyperacetylation, indicating that different histone codes and modifications may underlie different kinds of memories [84]. MSK1 and MSK2 are known to phosphorylate CREB at S133, H3, and also play a critical role in the regulation of immediate early genes, also regulating *Bdnf* signaling and basal synaptic transmission in CA1 area of the hippocampus [85,86]. More recently, the contrasting roles of MSK1 and MSK2 in cognitive processes have been reported. Loss of MSK1 in knockout mice was found to affect neuronal activity and hippocampal-dependent spatial recognition memory consolidation; whereas, in contrast MSK2 deletion led to increased CREB and EGR1 basal levels in dentate gyrus (DG) [87]. Studies investigated the role of PKCα-mediated phosphorylation of CBP at S436, regulating age dependent hippocampal neurogenesis and memory with no alterations in CREB binding and contextual fear memory [88]. Also, the phosphorylation of LSD1, by PKCα was found to be involved in regulating presynaptic gene expression, a thus regulating the hippocampus-dependent memory in phosphorylation-dependent manner when studied through LSD 1 knock-in mice with a mutated PKCα phosphorylation site [89]. Studies demonstrated the Ser/Thr protein phosphatase 1 (PP1) to be a potent negative regulator of hippocampus-dependent LTM and synaptic plasticity. Due to its abundance in brain structures like LA (amygdala), it is also involved in the regulation of fear memory, related to posttraumatic stress disorder (PTSD) in humans [90].

Several studies have implicated the involvement of various immediate early genes (IEG) in LTM formation. *Nr4a1* and *Nr4a2* are two CREB-dependent IEGs whose expression was found to be involved in regulating learning and memory. The increased expression of these genes were found in the hippocampus and cortex regions of brain following learning [91]. Histone acetylation was found to be necessary for the upregulation of *Nr4a1* and *Nr4a2* when studied with the HDAC inhibitor TSA [64]. HDAC3 deletion in the dorsal hippocampus region found to be modulating LTM formation in an *Nr4a2* dependent manner also contributing to age-related cognitive decline in vivo [61,92]. Blocking the expression of *Nr4a1* or *Nr4a2* using siRNA, showed that *Nr4a2* is necessary for both LTM for object location as well as recognition; whereas, *Nr4a1* is required for only object location in correlation with their expression in specific regions of brain [93]. A very recent study reported the longest known transcriptional changes detected in the hippocampus after CFC, where *Fosl2* and *Nfil3* genes were upregulated for almost a week after learning, thus contributing to LTM storage [94]. *Bdnf* is one of the most widely studied genes, as it has been shown to possess a critical role in long-term potentiation (LTP) underlying LTM formation and neurogenesis [95]. It is found abundantly and is widely expressed together with its major receptor TrkB in all the key areas of brain including the hippocampus, neocortex, cerebellum, striatum, and amygdala, making it primarily responsible for cognitive functions. Many studies in in vivo animal models has established the increased expression of *Bdnf* mRNA in the hippocampus following training in the Morris water maze, radial arm maze and CFC, thus indicating the regulation of *Bdnf* activity in hippocampal as well as parietal cortex in learning and memory [96–98]. Differential regulation of exon-specific *Bdnf* mRNAs (I, IV, VI, IX) was found to be associated with LTM consolidation following contextual fear learning.

Upregulation of both *Bdnf* exons I and IV was found in amygdala and only *Bdnf* IV exon was found in the hippocampus, suggesting region specific regulation of *Bdnf* exons during fear memory consolidation. This is also found to be associated with the histone modifications particularly the increase in the H3 acetylation levels and also the binding of active phosphorylated CREB at promoters IV and I of *Bdnf* exons [99]. A recent study using the Morris water maze test, the neonatal propofol treatment of rats demonstrated hippocampal NeuN (neuronal nuclei antigen) positive cell loss with long-lasting neuronal and synaptic abnormalities and also suggesting a possible role of PKA, CREB, and *Bdnf* expression in adult hippocampus, leading to propofol induced neuroapoptosis, thus impairing LTM and learning in adulthood [100]. Several studies have suggested the direct correlation of *Bdnf* mRNA expression levels with the NMDA receptor activity in the hippocampal cultured neurons [101,102]. All these evidences strongly suggest that the NMDA receptor and CREB- dependent mechanisms are involved in regulating the *Bdnf* specific variants required for LTM formation. The recently proposed theory of cognitive epigenetic priming (CEP) describes the enhanced memory formation when HDAC inhibition is paired with an external stimulus such as learning (Fig. 23.3). According to this theory, HDAC inhibition leads to a primed state, thus enhancing histone acetylation globally including brain areas, where the genes in the brain areas are then specifically targeted by any neuronal activity, thus further inducing histone acetylation and leading to gene transcriptional activation and enhanced cognition [19]. Another recent investigation using an activity-dependent tagging system in mice for the determination of epigenetic state, 3D genome architecture and transcriptional landscape of engram cells over the life-span of memory formation and recall, revealed that the process of memory encoding leads to a priming event that increases the accessibility of enhancers with no marked transcriptional changes in the genome. This further resulted in spatial reorganization of a larger chromatin segments with promoter-enhancer interactions. When engram neurons get reactivated, the promoter-enhancer interactions upregulate the genes involved in the local protein translation events. This study as a whole details the transcriptional and epigenomic landscape across the lifespan of memory formation and recall in the hippocampal engram ensemble [15].

MANIPULATING HISTONE MODIFICATIONS

The understanding of the mechanisms underlying long term memory formation through posttranslational modifications of histones have largely been implicated by demonstrating the pharmacological manipulation using HDAC inhibitors towards various modifications (Fig. 23.4) [30,55,56,103]. Most of the animal models have utilized HDAC inhibitors to demonstrate the increased acetylation following HDAC inhibition affecting the cognitive abilities. Most widely used are *pan*-HDACis such as TSA, NaB, valproic acid, and SAHA differentially inhibiting various HDAC isoforms, thus resulting in histone hyperacetylation [9,63,64,70]. This ability of HDAC*i*s modulating histone modifications via histone hyperacetylation made them potential therapeutic targets towards the treatment of various cognitive disorders, fear-related disorders such as posttraumatic stress disorder (PTSD), and also several pathologies involving memory and learning deficits such as Alzheimer disease and neural atrophy. HDAC inhibition using *pan*-HDACis, NaB and TSA, has led to increased H3 acetylation levels followed by increased long term memory, LTP, and also facilitating

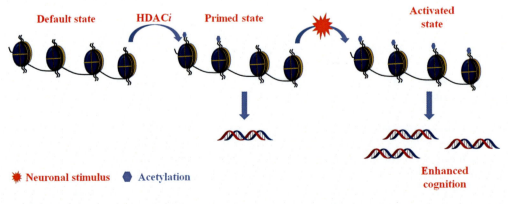

FIGURE 23.3 **Cognitive epigenetic priming.** Treatment with HDAC*i* enhances histone acetylation inducing a primed state, followed by neuronal stimulation that further enhances histone acetylation levels thus inducing gene transcription leading to an activated state causing enhanced cognition.

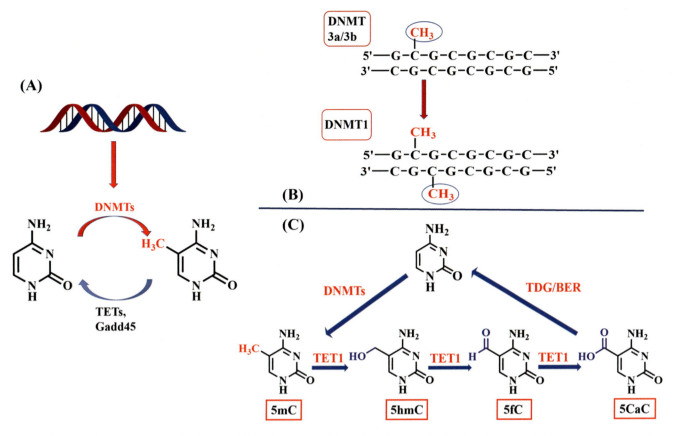

FIGURE 23.4 **(A) Covalent modifications on DNA involves DNA methylation and DNA demethylation.** (B) DNA methylation, a transcriptionally repressive state, is catalyzed by enzyme DNA methyltransferases (DNMTs), involving the addition of -CH3 (methyl) group at 5'-position of cytosine ring. DNMT3a/3b, catalyzes the de novo methylation at unmodified cytosine, and DNMT1 catalyzes the maintenance methylation on previously methylated cytosine. (C) Active DNA demethylation is carried out by the enzymatic mechanism catalyzed by TET1 and non-enzymatic mechanism carried out by Gadd45 followed by TDG/BER pathway. The conversion of 5mC to 5hmC to 5fC and finally to 5CaC was catalyzed by TET1 followed by base excision repair-thymine DNA glycosylase pathway to obtain demethylated cytosine. The detailed description of the mechanisms and their role in memory and processes is given in the text description. 5mC, 5-methylcytosine; 5hmC, 5-hydroxy methylcytosine; 5fC, 5-formylcytosine; 5CaC, 5-carboxycytosine; DNMT, DNA methyltransferases; TET1, Ten-eleven translocation methyl cytosine dioxygenases; *Gadd45*, growth arrest and DNA damage 45; TDG, thymine DNA glycosylase; BER, base-excision repair.

fear extinction in the CA1 of hippocampus in vivo [9,104,105]. Studies demonstrated the enhanced LTM in both acquisition and extinction of cued-fear memory and also in contextual fear conditioning following the intra-hippocampal administration of VPA and NaB [63,105,106]. A very recent report studied the selective enhancement of weak context fear memory, but not strong context fear memory in rats, by the use of NaB that was impaired by cycloheximide [70]. Early induction of LTP led to increased HDAC activity, that was found with the HDAC inhibition using TSA in vivo, suggesting that HDAC activity may negatively regulate plasticity-related gene expression involved in the initial or intermediate stabilization of LTP but not its long-term maintenance [107]. Given the wide application of *pan-HDACis* in the regulation of LTM, and reports suggesting the role of different HDAC isoforms, studies later have been focused on the roles of specific HDAC isoforms using isoform selective HDAC inhibitors.

Class I HDAC inhibition with Ms-275, deacetylates H3K9, thus impairing fear extinction memory [65]. Recent studies have also identified the contrast role of Ms-275 in enhancing the consolidation of recent fear memory extinction by increased acetylation on H4 in the cortex and amygdala [108]. Further, class I HDAC inhibition using Ms-275 was found to enhance object recognition memory consolidation in the insular cortex followed by the activation of *Bdnf/TrkB* pathway at 4 h after acquisition, suggesting the temporal expression of *Bdnf* in cortical circuits in ORM [56]. Given the abundant expression of HDAC3 in different regions of brain, and its role in the negative regulation of LTM, several studies have reported the use of HDAC3 specific inhibitors. HDAC3 inhibition was found to increase H4K8 acetylation further, enhancing long-term object recognition memory in mice [61,109]. Class I HDAC inhibition using RGFP963 was found to enhance the consolidation of cued fear extinction, whereas RGFP966, a HDAC3 specific inhibitor, does not significantly enhance the fear memory consolidation, suggesting the varied roles of class I HDACs in regulating fear memory extinction [58,110]. HDAC3 inhibition with RGFP136 regulated the CBP-dependent transcription mechanism in hippocampus dependent LTM function [61]. Combining HDAC inhibition with learning paradigms known as cognitive epigenetic priming (CEP), has found to increase histone acetylation levels in various conditions thus ameliorating memory formation [10,19,111]. Crebinostat, a class I HDAC inhibitor reported by Fass et al., was identified as a probe from a library of small molecules that was found to activate CREB-mediated transcription [112]. It was found to enhance long-term memory, in a hippocampus dependent, CFC paradigm, when crebinostat was administered systematically in vivo. Also, crebinostat treated cultures mouse primary neurons, revealed the upregulation of CREB-target genes *Bdnf*, *Grn*, and downregulation of *Mapt* (tau) gene thus, implicating in age-related cognitive decline and also in cognitive disorders. CI-994, a class I HDAC inhibitor has shown increase in the H3 acetylation levels in hippocampus only when combined with extinction training [111]. Treatments with CI994 and RGFP136 has demonstrated the persistence of long term memory beyond the point of normal memory failure [104,109]. Tianeptaline, a class I HDAC*i* was shown to enhance hippocampal dependent CREB-mediated transcriptional regulation and the expression of Arc (activity-regulated cytoskeleton-associated protein) leading to enhanced CFC, thus enhancing memory and modulating neuroplasticity in vivo [113]. Recently, our group also identified an HDAC3 specific inhibitor, PT3, and demonstrated the upregulation of long-term memory-associated genes in the hippocampus tissue as well as in mouse *neuro-2a* cells and also demonstrated its ability of long-term memory enhancement in the novel object recognition mouse model [57]. Several mouse models of neurodegenerative diseases, in particular AD, demonstrated the protective effects when treated with HDAC*is* and improved learning and memory formation, and also increased dendritic spine density [114]. In a recent study, in the case of AD patients, it was found that the F2 area was more affected than the hippocampus. The decrease in the levels of CBP, PCAF, HDAC1, and HDAC2 with a substantial increase in the total H3 and H2B levels were found in the frontal cortex whereas only HDAC1 and CBP were found to decrease in the hippocampal region of the brain [115]. In various mouse models of AD, NaB treatment increased the acetylation levels of H3 and H4, thus improving learning abilities and restoring access to previously deleted long-term memories in both early and advanced stage disease progression in mice [116,117]. BG45 administration in exogenous amyloid β-treated cells and mice was found to upregulate the levels of three synapse-related genes, *grik2*, *scn3b*, and *synpr*, thus improving the expression of synaptic and cytoskeletal proteins following the modulation of class I HDACs in vitro and in vivo [118].

In brief, various experiments and data till now indicated the role of specific histone modifications in the CNS with regard to the formation and consolidation of memories in adult brain. Several memory modulators, including HAT activators and HDAC inhibitors, were found to function as cognitive enhancers thus, restoring long-term memories when tested in various animal models. Histone PTMs have been extensively studied and gathered most attention for their role in learning and memory. In particular, histone acetylation, methylation, and phosphorylation are reported so far in majority of literature and studies; whereas, the role of ubiquitination and sumoylation are still unexplored. Also, the currently available HDAC inhibitors are not fully understood in capacity to their inhibiting various other non-specific sites leading to unwanted effects other than those reported so far. Hence, recent research has largely been focused on HDAC isoform specific inhibitors and their implications in various mechanisms. All these observations suggest other roles of non-histone sites working in correlation with the histone modifications for their activity in memory consolidation and synaptic plasticity.

DNA METHYLATION AND LONG-TERM MEMORY

Covalent modification of DNA, DNA methylation, catalyzed by DNA methyltransferases, in addition to histone modifications, has been found to play an active role in the mechanisms involved in learning and memory processes. Brief representation of the covalent modifications of DNA is represented in the Fig. 23.5A [119]. DNA methylation catalyzed by DNMTs, is the addition of a -CH3 group by a methyl donor, S-adenosyl methionine through covalent bonding at 5′ position of cytosine next to guanine (CpG), which occurs in high frequencies in and around gene regulatory regions known as CpG islands. DNA methylation is basically transcriptional repressive activity that inhibits the binding of transcriptional complexes to the binding sites and also recruits HDACs and methyl CpG-binding protein 2 (MeCP2) thus leading to condensed chromatin states. Two categories of DNA methyltransferases are known. The de novo DNMTs—DNMT3a and DNMT3b—characterized by methylating unmethylated DNA at the embryonic stage. Another category is maintenance DNMT; DNMT1 that methylates on the newly synthesized complementary DNA strand following DNA replication and damage (Fig. 23.5B). All three DNMT genes display different

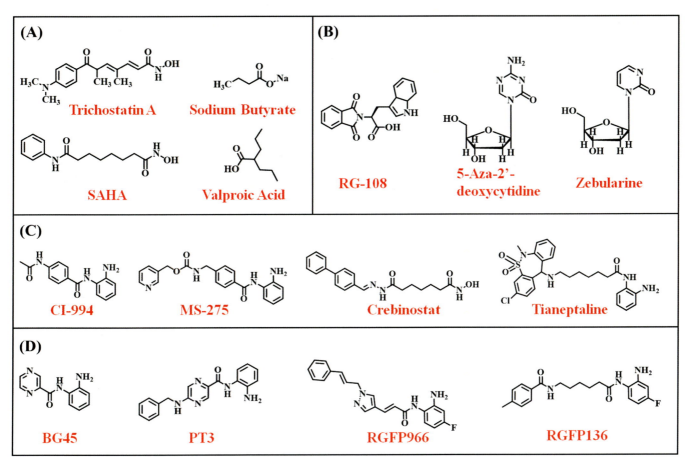

FIGURE 23.5 **Structures of reported memory modulators.** (A) Pan-HDAC inhibitors; (B) DNMT inhibitors; (C) Class I HDAC selective inhibitors; (D) HDAC3 selective inhibitors.

expression profiles in the CNS. DNMT1 is found to be expressed in neurons right from embryogenesis through the adult neurons. DNMT3a is expressed from late embryogenesis to adulthood and is also found in the postmitotic neurons of the adult brain; whereas, DNMT3b is only found in the neural progenitor tissue during early embryogenesis. It is also observed that DNMT1 methylates the unmodified DNA possessing methyl groups thus acting together with DNMT3a in the de novo methylation of DNA. One of the very first studies reported the increased expression of DNMT 3a and 3b in the CA1 area of the hippocampus following contextual fear conditioning. The studies also found that pharmacologically inhibiting DNMT using 5-AZA, the zebularine blocks normal memory consolidation leading to the demethylation of the positive regulators of memory, reelin, Bdnf, which are further acetylated by HATs, leading to transcriptional activation and other memory enhancing genes [120]. Whereas, DNMTs also act on negative regulators of memory such as PP1. Increased methylation of PP1 recruits methyl-CpG binding domain proteins (MBD), thus leading to transcriptional silencing of PP1 and other memory-suppressing genes [121]. Several animal models have established the role of methylated DNA-binding transcription regulator MeCP2 in regulating memory and synaptic plasticity and its mechanism by either activating or repressing gene transcription [122]. In a very recent study, the locus specific methylation at the TSS of MeCP2 in *Neuro-2a* cells and in mice led to decreased MeCP2 expression; it induced alterations in behavioral changes in mice such as reduced social interaction, enhanced anxiety/depression, along with poor performance in the memory tasks, thus suggesting the role of MeCP2 in ASD (autism specific disorder) [123]. Given the critical role of *Bdnf* in neurogenesis, neuronal maturation, synaptic plasticity, memory and learning, and also several neurological disorders, it is found to be associated with the DNA methylation and demethylation mechanism in its gene locus regulated by MeCP2. It was found that neural depolarization of MeCP2 led to its binding to the *Bdnf* promoter III. This caused transcriptional repression of *Bdnf* leading to decreased *Bdnf* protein expression and mRNA in the brains of mutant mice when compared to wild type mice, suggesting its role in the pathology of RTT [124]. It was also found that the NMDA receptor blockade is involved in DNA demethylation at the *Bdnf* gene loci correlating with the altered *Bdnf* gene expression during memory consolidation [99]. Methylation changes at memory-regulated genes such as *Bdnf*, *arc*, and *calcineurin*, are also reported for their role in memory formation and maintenance [99,125]. Hippocampal DNA methylation modifications tend to be stable and disappear within 24 h of training; whereas, changes in the cortex persist in association with remote memory following CFC. The study conducted by Miller et al. [125] reported the rapid methylation of *reelin* 1 h after CFC and its hypermethylation persisted for 7 days. On the other hand, no methylation of *calcineurin* was found until 1 day and this persisted for 30 days. Further, pharmacological manipulation of DNA methylation in dorsomedial prefrontal cortex (dmPFC) after 1 month disrupted remote memory and also increased *calcineurin* expression. This suggested the involvement of DNA methylation in both the hippocampus and cortex with regard to long-lasting memory persistence plays a critical role in gene expression regulation in the adult brain in both hippocampus and cortex in order to preserve long-lasting memories [125]. Also, studies involving DNMT inhibition with RG-108 in the mouse hippocampus and perirhinal cortex, and another study in mutant mice with double knockout DNMT1, DNMT3a in forebrain excitatory neurons showed abnormal long-term plasticity along with learning and memory deficits in object, spatial, and fear memory tasks [126]. The same group recently reported that the selective knockdown of DNMT1 and DNMT3a in primary cultured hippocampal neurons derived from Dnmt1,3a$^{2flox/2flox}$ mice led to both pre- and postsynaptic dysfunction caused by the presence of longer dendrites with less branches and low density of excitatory synapses together with reduced frequency and amplitude of miniature excitatory postsynaptic currents (mEPSCs) [127]. Also, the suppression of intracellular [Ca^{2+}] increase triggered by glutamate was observed due to DNMT1 and DNMT3a deletion. In another recent study, selective DNMT1 knockdown in induced pluripotent stem cell (iPSC)-derived excitatory cortical neurons lowered the number of actively firing neurons, thus decreasing synaptic activity of human glutamatergic neurons further pointing out the correlation between DNMT1 and synaptic transmission [128]. Rats pre-treated with 5-AZA and exposed to sevoflurane were found to be associated with increased expression of DNMT1, DNMT3a, and 5-methylcytosine levels along with decreases in TET1 and 5-hydroxymethylcytosine levels. It was also found that sevoflurane treatment induced hypermethylation of *Shank2*, *Psd95*, *Syn1*, and *Syp* genes, and downregulated expression of synaptic proteins, thus establishing evidence that repeated neonatal sevoflurane exposure impaired cognitive and spatial memory through DNA methylation of synaptic genes [129]. Halder et al. [130] reported a genome-wide study of changes in DNA methylation and the role of chromatin modifications in learning and memory processed in two different brain regions i.e., hippocampus and cortex, in two different cell populations and at three different time points. Fear conditioning-induced changes in the DNA methylation in the hippocampus showed both an increase and decrease in the methylation levels at 1 h, not 4 weeks post-training; whereas, the contrast was observed in anterior cingulate cortex (ACC) [130]. It was found that extensively over 6250 differentially methylated genes were found in the cortex when exposed to both context and shock when compared to 1206 differentially methylated genes in CA1 of hippocampus. Whereas, hippocampal changes were more specific than cortical changes at 1 h to the context-shock association with over 3000 differentially methylated regions in area CA1 in mice exposed to context and shock together over context only exposure [130]. Subsequently, remote memory was found to be associated with only 1223 differentially methylated genes in the cortex of mice, which is lower compared to 1 h posttraining. These results suggest that the number of genes involved in the maintenance of memory trace in the ACC were much smaller than those involved in the initial processing of the stimulus [130]. Recent study reported the impact of DNA methylation on memory traces or engrams that are specifically activated during learning processes. The upregulation of DNMT3a2 was found to stabilize engrams, thus leading to successful memory recall [131]. Further, comparing the mice that were exposed to context only against context-shock, only 154 genes were found to be differentially methylated. All these data indicate that only a certain set of genes are altered from the initial training and exhibit persistent changes in the methylation thus correlating with the memory maintenance [130]. Altogether, these findings suggest that the cellular mechanism involving covalent modifications of DNA might be actively involved in the LTM consolidation in adult CNS. It is understood that memory formation involves hypermethylation at memory suppressor genes and hypomethylation at memory enhancing genes caused by DNMT activity together in correlation with histone modifications in adult rat hippocampus and cortex [99,121,122,125,132].

Although, a direct correlation between DNA methylation and demethylation in memory formation has been established, little is known about DNA demethylation mechanisms so far. DNA demethylation can occur either through active or passive mechanisms. Historically, DNA demethylation is regarded as a passive process in differentiated cells, that refers to the loss of methyl group from 5mC either by DNMT1 inhibition or by multiple rounds of cell division. Whereas, active DNA demethylation has been studied as a controversial process that involves the enzymatic removal of methyl group from 5mC moiety by breaking the C-C bond [133,134]. However, several reports mention the critical role of active methylation and demethylation in mature cells in CNS. Early reports suggested the indirect evidence of active DNA demethylation in the adult mouse CNS after contextual fear conditioning using nonquantitative methods [99,119,132,133]. In the past decade, several mechanisms of demethylating the 5mC have been reported through modifying the methylated cytosine either by hydroxylation, deamination, oxidation, or together catalyzed by three distinct families of enzymes: the TET (Ten-eleven translocation [TET] methyl cytosine dioxygenases) family that modifies the 5mC first by hydroxylation followed by oxidation; the APOBEC/AID family that deaminates the 5mC or 5hmC; and finally, by the glycosylases (BER/TDG) family that mediate DNA repair [135,136]. Another mechanism involving the repair based DNA demethylation carried out by growth arrest and DNA damage 45 (Gadd45) that function by recruiting the enzymatic machinery responsible for TDG/BER pathway [137].

Gadd family of proteins are stress-inducible and are widely implicated in cell cycle regulation, apoptosis, DNA damage and repair. It is known that Gadd45 excise and repair methylated cytosines during DNA strand break repair. Mice deficit in Gadd45b exhibited impaired neural activity resulting in impaired DNA demethylation, thus failing to activate critical memory-related genes such as *Bdnf* and (fibroblast growth factor) *Fgf1*, and modulating adult neurogenesis in response to external stimuli by active DNA demethylation [138]. Gadd45 family isoforms such as Gadd45b and Gadd45g were found to be upregulated in the hippocampus following CFC in mice. Gadd45b knockout mice exhibited a significant deficit specifically in long-term contextual fear conditioning, with no differences observed during cued fear conditioning, suggesting the role of Gadd45b in the long-term hippocampus dependent fear and spatial memory storage [139]. Another study, reported the role of NF-kB activity in the regulation of Gadd45b expression in a fear conditioning paradigm. This is demonstrated by increase Gadd45b gene expression and *Bdnf* demethylation, which is prevented by pharmacologically inhibiting NF-kB [140]. Another study reported the downstream signaling of *Bdnf* in the nucleus accumbens (NAc) to be mediated by Gadd45b, thus contributing to susceptibility to social stress in mice [141]. Another Gadd45 isoform, Gadd45γ, is identified to be involved actively in DNA repair and DNA demethylation, influencing the learning-induced IEG expression in pre-limbic prefrontal cortex (PLPFC) [142]. Moreover, recently it was found that Gadd45γ expression regulates age-related cognitive abilities and synaptic plasticity in the hippocampus in adult mice and humans; it is also required for CREB activation in response to neuronal activity and memory-related gene expression [143].

Accumulating evidence now implicate the demethylation pathway involving TET enzymes (Fig. 23.5C) [144,145]. Three isoforms of TET enzymes were identified and all three were found to possess oxidase activity. One of the earliest evidence suggests the accumulation of 5-hydorxy methyl cytosine (5hmC) abundantly in brain tissues. Further, human TET1, initially identified as a fusion partner of the *Mll* gene in acute myeloid leukemia (AML) and later characterized as an ortholog of JBP1 and JBP2 (J-binding protein 1 and 2, thymidine hydroxylases class of enzymes that catalyze the oxidation of thymine to 5hmU) is involved in the conversion of 5mC to 5hmC, thus suggesting its role in the DNA demethylation [146]. TET1 is expressed in the dentate gyrus of adult mouse brain and is involved in the neuronal-activity induced, region-specific active DNA demethylation. TET1 knockout mice were found to downregulate neuronal activity-regulated genes *Npas4*, *c-Fos*, and *Arc*, and also exhibited hippocampal long-term depression, impaired memory extinction, and synaptic plasticity [147]. TET1 was found to be downregulated followed by neuronal activity in the hippocampus. Overexpression of TET1 or TET1m (catalytically inactive mutant of TET1) in CA1 was found to regulate the levels of 5mC and 5hmC, thus increasing the expression of immediate early genes involved in memory formation and impaired contextual fear memory. In contrast to these studies, TET1 knockout mice showed normal fear memory after 24 h and 30 days after CFC [148]. These discrepancies in the results indicate the different functions of TET1 in different types of memory. Further, studies have also reported the role of TET in the conversion of 5mC to generate 5hmC to 5fC (5 formyl cytosine) to 5caC (5 carboxyl cytosine) in an enzymatic dependent manner. And finally, the base—excision repair (BER) initiated by thymine DNA glycosylase (TDG), thus regulating the complete pathway of active DNA demethylation [145]. Recent discovery suggests the differential expression of TET1 gene isoforms, *Tet1s* and *Tet1FL*, in adult mouse brain from early development to adulthood. Individual repression of each gene encodes their differential expression and contrast changes in the basal synaptic transmission. *Tet1s*, being a more expressed isoform in neurons and its repression was found to enhance hippocampal-dependent memory whereas, the *Tet1FL* expressed at lower levels impairs the same [144]. Recent studies also suggest novel mechanisms involved in the role of a TET-TDG pathway in regulating the active DNA demethylation in both CpG and non-CpG sites [149].

HISTONE VARIANT EXCHANGE

Histone variant exchange is one such branch that was suggested to be a key regulator of neural plasticity and cognitive function. Apart from histone PTMs and DNA methylation, histone variants were also recently studied for their role in memory formation and regulation (Table 23.2). Histone variants are the non-allelic counterparts of histones that get replaced with each other in the chromatin [150]. All the histones except H4 possess their counterpart histone variants with the H2A family, which is the most diverse with H2A.Z, H2A.X, H2A.B variants and histone H3 and only one histone variant H3.3 [151]. Histone variants are highly diverse in function by generating diversified transcriptional outcomes thus introducing regulatory diversity in neurons. These histone variants are replication independent and form the major source of histones in postmitotic neurons [151].

In one of the very first studies, Maze and the group reported the dynamic exchange of H3.3 into the chromatin core, thus identifying histone turnover as a novel regulator or neural plasticity and memory in both embryonic and adult neurons. They reported the H3.3 accumulates across the life-span of both rodents and humans throughout the neuronal genome from embryonic development to the adolescence stage along with retaining the highly dynamic gene expression patterns through life [150]. They showed that H3.3, through the protein knockdown approach, reduced both excitatory and inhibitory neuronal synapses. Further, H3.3 blockade turnover caused impairment of the memory formation in the hippocampus, suggesting the critical role of H3.3 in memory consolidation [150]. Another study revealed the importance of H3.3 in contextual fear learning and memory by incorporating hemagglutinin (HA)-tagged H3.3 in the excitatory neurons of the forebrain in a transgenic mouse line that preferentially accumulated in the promoter regions of actively transcribed neuronal genes [152]. Another recent interesting study found a non-canonical H3.3 pattern in mature oocytes and early embryos, which gradually switches to its canonical form with duly incorporation of H3.1/H3.2 mediated by Caf-1 at the two-cell stage, in a transcription independent and replication dependent manner, thus suggesting the transition of non-canonical to canonical form of H3.3 in the developmental program of early embryos [153].

A histone variant H2A.B, from the H2A family, found to be the most abundant one in the hippocampus, was inhibited followed by the contextual fear conditioning of mice, suggesting its role in memory formation [151]. From the H2A family, H2A.Z variant was extensively studied till date and has emerged as a negative regulator of the recent and remote memory consolidation, following fear conditioning in the hippocampus and cortex, thus possessing a suppressive effect on memory [154]. Fear conditioning resulted in the removal of H2A.Z from the transcription start site (TSS), 30 minutes after training, subsequently increasing the expression of several memory-related genes. This process was transient, with H2A.Z binding returning to baseline levels 2 h posttraining in the hippocampus. Whereas, H2A.Z eviction in the prefrontal cortex (mPFC) 2 h posttraining was transient and the H2A.Z binding on plasticity related genes retuned to baseline 30 days after CFC, thus improving only remote memory. These data suggest the role of H2A.Z in remote memory maintenance but not initial memory consolidation [154]. A recent report studied the specific function of H2A.Z temporally by targeting its regulator Tip60 (part of H2A.Z deposition complex) [43]. Nu9056, a Tip60 inhibitor was found to reduce H2A.Z binding. On the other hand, use of TSA increased H2A.Z acetylation with no influence on the total H2A.Z in area CA1. Tip60 inhibition 23 h after training was shown to improve remote memory with no influence on recent memory. In contrast, Tip 60 inhibition with the same drug 30 days after learning was found to impair the recall of remote memory after 1 h but improved memory when the mice were tested 1 day later. These data suggest the role of H2A.Z role in recalling remote memory but not recent memory [43]. The neurons of H2A.Z knockout mice exhibited

TABLE 23.2 Summary of the Known Histone Variants Corresponding to the Canonical Histones and Their Functions in the Active Transcriptional Regulation and Chromatin Remodeling

Canonical Histone Proteins	Histone Variants	Function
H2A	H2A.X H2A.Z H2A.B	DNA damage repair, sex chromosome remodeling, transcription
H2B	Numerous variants known	—
H3	H3.3	Transcriptional activation in response to stress stimuli and gene expression
H4	—	—

abnormal dendrites that resulted in the enhanced neural progenitors proliferation with reduced differentiation. This lead to decreased exploratory activity and impaired learning and memory, resulting in cortical neurogenesis defects and neurodevelopmental disorder [155]. In another study using conditional-inducible H2A.Z knockout mice, regulation of androgen receptor-mediated fear memory was indicated, suggesting the role of the androgen receptor as H2A.Z regulator in neurons, thus implicating sex differences in fear-related disorders [156]. H2A.Z manipulation effects are also reported to be sex-specific and task-specific, wherein H2A.Z deletion in mice was found to improve fear memory only in males, with improving memory on a non-aversive object in place task in both genders suggesting stress induced H2A.Z effects [157].

EPITRANSCRIPTOMICS: RNA EPIGENETICS

Most of the studies have focused on the epigenetic regulation of gene expression through covalent modifications on DNA and histone PTMs till date. Gene translation was also found to play a vital role in learning and memory. Towards this, mRNA modifications (epitranscriptomics) have recently been studied more for the implication in learning and memory [158,159]. The most prominent and widely studied modification is N6 methyl adenine (m6A) mRNA i.e., methylation of adenine at 6th position of RNA and its role in fear memory [160]. The m6A mRNA modification is highly dynamic and reversibly regulated by methyltransferases and demethylases [161]. METTL3 and METTL14 are the known transferases that catalyze methylation leading to m6A and two demethylases are reported, which includes fat mass and obesity associated protein (FTO) and α-ketoglutarate dependent dioxygenase alkB homolog 5 (ALKBH5). Due to its high abundance in brain, m6A was found to be well regulated in the memory formation. One of the earliest demonstrations reported the upregulation of m6A in mouse medial prefrontal cortex (mPFC) following training. Simultaneously, knockdown of FTO in mPFC led to the increase of m6A levels, thus resulting in the enhanced consolidation of fear memory [162]. Later, another study reported the decreased FTO levels at the synapses shortly after fear conditioning in the dorsal CA1, further suggesting its critical role in LTM formation [163]. A recent study, characterized by depleting METTL3 or YTHDF1 (binding protein of m6A) in mouse hippocampus, was found to reduce memory consolidation followed by impaired learning and long-term potentiation [164,165]. Another study involving proteomic and functional analyses indicated the role of the neural

m6A/YTHDF1 pathway in translational activation, suggesting its role in learning and memory in *Drosophila* [166]. One study demonstrating cocaine-induced conditioned place preference (CPP) in mice, significantly downregulated the levels of FTO, further leading to increased levels of m6A, thus uncovering the role of m6A RNAs modification in cocaine reward [167]. Several recent studies have also established the roles of miRNAs in long-term synaptic plasticity and memory in various animal models of learning and memory [168–170]. A few recent studies suggested an interesting crosstalk between different epigenetic mechanisms, in particular, involvement of micro RNAs. One such interesting study reported the perinatal exposure of bisphenol-A (BPA), an estrogenic endocrine disruptor that impaired spatial memory and induced upregulation of synaptic proteins neurexin1 and neuroligin3 in male mice. This study evidenced that perinatal BPA exposure reduced 5mC DNA and increased acetylated H3 levels in the cerebral cortex and hippocampus in the postnatal stage of the male mice. It was also found that BPA exposure caused changes in the mRNA levels of DNMT1 and DNMT3a in both cerebral cortex and hippocampus at different ages of male mice. The results altogether demonstrated the role of BPA exposure in brain function and behavior that might due to the underlying epigenetic changes [171]. In another study, the increase in the levels of Tet3 upon CFC were found to be regulated by *SAM-68* and *miR-29b*, thus suggesting an interesting crosstalk between TETs and micro RNAs in adult brain for memory formation [172]. An interesting study reported that *Lym*-miR-137 was upregulated after single-trail conditioning followed by the downregulation of *Lym*-CREB2 mRNA, thus establishing its role in LTM [173]. Another study involving the genome wide screening followed by gene expression profiling, identified miRNA-138 as a potential regulator of memory function in humans [174]. Very recently, the role of glial cells such as astrocytes and microglia have been reported and studied in detail in regulating the memory processes [175,176]. Astrocytic ApoE was found to be associated with a variety of miRNAs, which further silence the genes involved in cholesterol biosynthesis in neurons. This promotes the acetyl-CoA substrate that acetylates histones and leads to transcriptional activation. A study demonstrated that enhanced levels of acH3K27 in the promoters of multiple IEGs increases memory consolidation in mice, thus identifying the novel epigenetic mechanism mediated by ApoE modulating brain function and memory processes [175]. Given the regulation of FTO preferably at synapses suggests that epitranscriptomics together with DNA modifications might regulate memory formation through long-term potentiation at the synaptic terminals.

SUMMARY

The findings discussed in this chapter further evidenced the role of epigenetic mechanisms governing neural plasticity, thus regulating learning and memory. A great deal of research has been carried out in a very short period of time in the field of neuroepigenetics, trying to understand the underlying mechanisms influencing gene expression in region specific areas of the brain at different time points, regulating different memory processes. It is now well established that histone posttranslational modifications, along with DNA methylation, forms the primary mechanism for chromatin remodeling, thus regulating transcriptional outcomes. Further, novel mechanisms involving histone variant exchange and mRNA epigenetics are also shown to be working in correlation with the chromatin modifications, thus bridging the gap in understanding the epigenetic events in the brain affecting memory. Pharmacological inhibition and manipulations of the epigenetic modifications using selective inhibitors or activators were found to restore memory and learning processes in several animal models of neurological disorders, further indicating their therapeutic potential in treating age-related cognitive disorders, neuropsychiatric disorders, posttraumatic stress disorders, and neurotoxicity. Thus, continued extensive research in this field could help us in further elucidating the complex mechanisms involved in the memory formation and maintenance and exploiting these mechanisms in the treatment of neurological disorders through epigenetic based therapy in the future.

Acknowledgments

The research in the epigenetic research laboratory of Dr. Balaram Ghosh has been supported by Council of Scientific and Industrial Research (CSIR-37(1722)/19/EMR-II). Sravani Pulya acknowledges CSIR for providing Senior research fellowship (File no: 09/1026 (0024)/2018-EMR-I). Authors sincerely acknowledge the Department of Pharmacy, Birla Institute of Technology and Science-Pilani, Hyderabad campus, India.

Abbreviations

5-Aza	5-Aza-2′-deoxycytidine
5hmC	5-Hydroxylmethyl cytosine
5mC	5-Methyl cytosine
AD	Alzheimer's disease
BDNF	Brain-derived neurotrophic factor
BER	Base excision repair
CAF-1	Chromatin assembly factor-1
CBP	CREB-binding protein
CFC	Contextual fear conditioning
CNS	Central nervous system
CpG	Cytosine–guanine dinucleotide
CPP	Conditional place preference
CREB	cAMP response element–binding protein
DNMT	DNA methyltransferase
ERK/MAPK	extracellular signal–related kinase/mitogen-activated protein kinase
ERK	Extracellular regulated kinase
FTO	Fat mass and obesity associated protein
Gadd45	Growth arrest and DNA damage inducible
Grn	Granulin
H	Histone
HAT	Histone acetyltransferase
HDAC	Histone deacetylases
HMT	Histone methyltransferases
IEG	Immediate early genes
LH-MFB	Lateral hypothalamus-medial forebrain bundle
LSD1	Lysine specific demethylase
LTM	Long term memory
LTP	long-term potentiation
Mapt	Microtubule associated protein Tau.
MeCP2	Methyl Cpg–binding protein 2
MSK1	Mitogen- and stress-activated kinase 1
METTL3	Methyltransferase like 3
METTL14	Methyltransferase like 14
NaB	Sodium butyrate
NMDA	N-methyl-D-aspartate receptor
NOR	Novel object recognition
OLM	object location memory
ORM	object recognition memory
PCAF	p300/CBP–associated factor
PKC	Protein kinase C
PP1	Protein phosphatase1
PTM	Posttranslational modifications.
PTSD	Posttraumatic stress disorder
SRF	Serum response factor
STM	Short term memory
TDG	Thymine DNA glycosylase
TET	Ten-eleven translocation methylcytosine dioxygenase
TrkB	Tropomyosin receptor kinase B
TSA	Trichostatin A
TSS	Transcription start site
YTHDF1	YTH N6-methyladenosine RNA binding protein 1

References

[1] Bird A. Perceptions of epigenetics. Nature 2007;447:396–8. Available from: https://doi.org/10.1038/nature05913.

[2] Quina AS, Buschbeck M, Di Croce L. Chromatin structure and epigenetics. Biochem Pharmacol 2006;72:1563–9. Available from: https://doi.org/10.1016/j.bcp.2006.06.016.

[3] Lawrence M, Daujat S, Schneider R. Lateral thinking: how histone modifications regulate gene expression. Trends Genet 2016;32:42–56. Available from: https://doi.org/10.1016/j.tig.2015.10.007.

[4] Farrelly LA, Thompson RE, Zhao S, Lepack AE, Lyu Y, Bhanu NV, et al. Histone serotonylation is a permissive modification that enhances TFIID binding to H3K4me3. Nature 2019;567:535–9. Available from: https://doi.org/10.1038/s41586-019-1024-7.

[5] Lepack AE, Werner CT, Stewart AF, Fulton SL, Zhong P, Farrelly LA, et al. Dopaminylation of histone H3 in ventral tegmental area regulates cocaine seeking. Science 2020;368:197–201. Available from: https://doi.org/10.1126/science.aaw8806.

[6] Sen N. Epigenetic regulation of memory by acetylation and methylation of chromatin: implications in neurological disorders, aging, and addiction. Neuromol Med 2015;17:97–110. Available from: https://doi.org/10.1007/s12017-014-8306-x.

REFERENCES

[7] Turner BM. Histone acetylation and an epigenetic code. Bioessays 2000;22:836–45. Available from: https://doi.org/10.1002/1521-1878(200009)22:9<836::AID-BIES9>3.0.CO;2-X.

[8] Grinkevich LN. Epigenetics and the formation of long-term memory. Neurosci Behav Physiol 2014;44:200–13. Available from: https://doi.org/10.1007/s11055-014-9897-2.

[9] Levenson JM, O'Riordan KJ, Brown KD, Trinh MA, Molfese DL, Sweatt JD. Regulation of histone acetylation during memory formation in the hippocampus. J Biol Chem 2004;279:40545–59. Available from: https://doi.org/10.1074/jbc.M402229200.

[10] Gräff J, Tsai L-H. Histone acetylation: molecular mnemonics on the chromatin. Nat Rev Neurosci 2013;14:97–111. Available from: https://doi.org/10.1038/nrn3427.

[11] Federman N, Fustiñana MS, Romano A. Histone acetylation is recruited in consolidation as a molecular feature of stronger memories. Learn Mem 2009;16:600–6. Available from: https://doi.org/10.1101/lm.1537009.

[12] Lopez-Atalaya JP, Barco A. Can changes in histone acetylation contribute to memory formation? Trends Genet 2014;30:529–39. Available from: https://doi.org/10.1016/j.tig.2014.09.003.

[13] Bisagno V, Cadet JL. Histone deacetylases and immediate early genes: key players in psychostimulant-induced neuronal plasticity. Neurotox Res 2021;39:2134–40. Available from: https://doi.org/10.1007/s12640-021-00420-3.

[14] Geng H, Chen H, Wang H, Wang L. The histone modifications of neuronal plasticity. Neural Plast 2021;2021. Available from: https://doi.org/10.1155/2021/6690523 6690523.

[15] Marco A, Meharena HS, Dileep V, Raju RM, Davila-Velderrain J, Zhang AL, et al. Mapping the epigenomic and transcriptomic interplay during memory formation and recall in the hippocampal engram ensemble. Nat Neurosci 2020;23:1606–17. Available from: https://doi.org/10.1038/s41593-020-00717-0.

[16] Marco A. Activity-dependent remodeling of genome architecture in engram cells facilitates memory formation and recall. Neural Regen Res 2022;17(5):991–3. Available from: https://www.nrronline.org/article.asp?issn=1673-5374;year=2022;volume=17;issue=5;spage=991;epage=993;aulast=Marco accessed 24.10.21.

[17] Selvi BR, Cassel J-C, Kundu TK, Boutillier A-L. Tuning acetylation levels with HAT activators: therapeutic strategy in neurodegenerative diseases. Biochim Biophys Acta 2010;1799:840–53. Available from: https://doi.org/10.1016/j.bbagrm.2010.08.012.

[18] Stilling RM, Fischer A. The role of histone acetylation in age-associated memory impairment and Alzheimer's disease. Neurobiol Learn Mem 2011;96:19–26. Available from: https://doi.org/10.1016/j.nlm.2011.04.002.

[19] Burns AM, Gräff J. Cognitive epigenetic priming: leveraging histone acetylation for memory amelioration. Curr Opin Neurobiol 2021;67:75–84. Available from: https://doi.org/10.1016/j.conb.2020.08.011.

[20] Yao T-P, Oh SP, Fuchs M, Zhou N-D, Ch'ng L-E, Newsome D, et al. Gene dosage–dependent embryonic development and proliferation defects in mice lacking the transcriptional integrator p300. Cell 1998;93:361–72. Available from: https://doi.org/10.1016/S0092-8674(00)81165-4.

[21] Bousiges O, Vasconcelos AP, de, Neidl R, Cosquer B, Herbeaux K, Panteleeva I, et al. Spatial memory consolidation is associated with induction of several lysine-acetyltransferase (Histone Acetyltransferase) expression levels and H2B/H4 acetylation-dependent transcriptional events in the rat hippocampus. Neuropsychopharmacol 2010;35:2521–37. Available from: https://doi.org/10.1038/npp.2010.117.

[22] Chen G, Zou X, Watanabe H, van Deursen JM, Shen J. CREB binding protein is required for both short-term and long-term memory formation. J Neurosci 2010;30:13066–77. Available from: https://doi.org/10.1523/JNEUROSCI.2378-10.2010.

[23] Barrett RM, Malvaez M, Kramar E, Matheos DP, Arrizon A, Cabrera SM, et al. Hippocampal focal knockout of CBP affects specific histone modifications, long-term potentiation, and long-term memory. Neuropsychopharmacology 2011;36:1545–56. Available from: https://doi.org/10.1038/npp.2011.61.

[24] Valor LM, Pulopulos MM, Jimenez-Minchan M, Olivares R, Lutz B, Barco A. Ablation of CBP in forebrain principal neurons causes modest memory and transcriptional defects and a dramatic reduction of histone acetylation but does not affect cell viability. J Neurosci 2011;31:1652–63. Available from: https://doi.org/10.1523/JNEUROSCI.4737-10.2011.

[25] Bartsch D, Ghirardi M, Skehel PA, Karl KA, Herder SP, Chen M, et al. Aplysia CREB2 represses long-term facilitation: relief of repression converts transient facilitation into long-term functional and structural change. Cell 1995;83:979–92. Available from: https://doi.org/10.1016/0092-8674(95)90213-9.

[26] Tubon TC, Zhang J, Friedman EL, Jin H, Gonzales ED, Zhou H, et al. dCREB2-mediated enhancement of memory formation. J Neurosci 2013;33:7475–87. Available from: https://doi.org/10.1523/JNEUROSCI.4387-12.2013.

[27] Li K-L, Zhang L, Yang X-M, Fang Q, Yin X-F, Wei H-M, et al. Histone acetyltransferase CBP-related H3K23 acetylation contributes to courtship learning in Drosophila. BMC Dev Biol 2018;18:20. Available from: https://doi.org/10.1186/s12861-018-0179-z.

[28] Oike Y, Hata A, Mamiya T, Kaname T, Noda Y, Suzuki M, et al. Truncated CBP protein leads to classical Rubinstein—Taybi syndrome phenotypes in mice: implications for a dominant-negative mechanism. Hum Mol Genet 1999;8:387–96. Available from: https://doi.org/10.1093/hmg/8.3.387.

[29] Petrif F, Giles RH, Dauwerse HG, Saris JJ, Hennekam RCM, Masuno M, et al. Rubinstein-Taybi syndrome caused by mutations in the transcriptional co-activator CBP. Nature 1995;376:348–51. Available from: https://doi.org/10.1038/376348a0.

[30] Korzus E, Rosenfeld MG, Mayford M. CBP histone acetyltransferase activity is a critical component of memory consolidation. Neuron 2004;42:961–72. Available from: https://doi.org/10.1016/j.neuron.2004.06.002.

[31] Vieira PA, Korzus E. CBP-dependent memory consolidation in the prefrontal cortex supports object-location learning. Hippocampus 2015;25:1532–40. Available from: https://doi.org/10.1002/hipo.22473.

[32] Lipinski M, del Blanco B, Barco A. CBP/p300 in brain development and plasticity: disentangling the KAT's cradle. Curr Opin Neurobiol 2019;59:1–8. Available from: https://doi.org/10.1016/j.conb.2019.01.023.

[33] Blanco B, del, Guiretti D, Tomasoni R, Lopez-Cascales MT, Muñoz-Viana R, Lipinski M, et al. CBP and SRF co-regulate dendritic growth and synaptic maturation. Cell Death Differ 2019;26:2208–22. Available from: https://doi.org/10.1038/s41418-019-0285-x.

[34] Telley L, Govindan S, Prados J, Stevant I, Nef S, Dermitzakis E, et al. Sequential transcriptional waves direct the differentiation of newborn neurons in the mouse neocortex. Science 2016;351:1443–6. Available from: https://doi.org/10.1126/science.aad8361.

[35] Tie F, Banerjee R, Stratton CA, Prasad-Sinha J, Stepanik V, Zlobin A, et al. CBP-mediated acetylation of histone H3 lysine 27 antagonizes Drosophila Polycomb silencing. Development 2009;136:3131–41. Available from: https://doi.org/10.1242/dev.037127.

[36] Peleg S, Sananbenesi F, Zovoilis A, Burkhardt S, Bahari-Javan S, Agis-Balboa RC, et al. Altered histone acetylation is associated with age-dependent memory impairment in mice. Science 2010;328:753–6. Available from: https://doi.org/10.1126/science.1186088.

[37] Benito E, Urbanke H, Ramachandran B, Barth J, Halder R, Awasthi A, et al. HDAC inhibitor–dependent transcriptome and memory reinstatement in cognitive decline models. J Clin Invest 2015;125:3572–84. Available from: https://doi.org/10.1172/JCI79942.

[38] Maddox SA, Watts CS, Schafe GE. p300/CBP histone acetyltransferase activity is required for newly acquired and reactivated fear memories in the lateral amygdala. Learn Mem 2013;20:109–19. Available from: https://doi.org/10.1101/lm.029157.112.

[39] Mitchnick KA, Creighton SD, Cloke JM, Wolter M, Zaika O, Christen B, et al. Dissociable roles for histone acetyltransferases p300 and PCAF in hippocampus and perirhinal cortex-mediated object memory. Genes Brain Behav 2016;15:542–57. Available from: https://doi.org/10.1111/gbb.12303.

[40] Chatterjee S, Angelakos CC, Bahl E, Hawk JD, Gaine ME, Poplawski SG, et al. The CBP KIX domain regulates long-term memory and circadian activity. BMC Biol 2020;18:155. Available from: https://doi.org/10.1186/s12915-020-00886-1.

[41] Beaver M, Bhatnagar A, Panikker P, Zhang H, Snook R, Parmar V, et al. Disruption of Tip60 HAT mediated neural histone acetylation homeostasis is an early common event in neurodegenerative diseases. Sci Rep 2020;10:18265. Available from: https://doi.org/10.1038/s41598-020-75035-3.

[42] Xu S, Wilf R, Menon T, Panikker P, Sarthi J, Elefant F. Epigenetic control of learning and memory in Drosophila by Tip60 HAT action. Genetics 2014;198:1571–86. Available from: https://doi.org/10.1534/genetics.114.171660.

[43] Narkaj K, Stefanelli G, Wahdan M, Azam AB, Ramzan F, Steininger CFD, et al. Blocking H2A.Z incorporation via Tip60 inhibition promotes systems consolidation of fear memory in mice. ENeuro 2018;5. Available from: https://doi.org/10.1523/ENEURO.0378-18.2018.

[44] Urban I, Kerimoglu C, Sakib MS, Wang H, Benito E, Thaller C, et al. TIP60/KAT5 is required for neuronal viability in hippocampal CA1. Sci Rep 2019;9:16173. Available from: https://doi.org/10.1038/s41598-019-50927-1.

[45] Humbert J, Salian S, Makrythanasis P, Lemire G, Rousseau J, Ehresmann S, et al. De Novo KAT5 variants cause a syndrome with recognizable facial dysmorphisms, cerebellar atrophy, sleep disturbance, and epilepsy. Am J Hum Genet 2020;107:564–74. Available from: https://doi.org/10.1016/j.ajhg.2020.08.002.

[46] Merschbaecher K, Hatko L, Folz J, Mueller U. Inhibition of different histone acetyltransferases (HATs) uncovers transcription-dependent and -independent acetylation-mediated mechanisms in memory formation. Learn Mem 2016;23:83–9. Available from: https://doi.org/10.1101/lm.039438.115.

[47] Monsey MS, Sanchez H, Taylor JR. The naturally occurring compound garcinia indica selectively impairs the reconsolidation of a cocaine-associated memory. Neuropsychopharmacology 2017;42:587–97. Available from: https://doi.org/10.1038/npp.2016.117.

[48] Monsey MS, Ruiz SG, Taylor JR. Regulation of garcinol on histone acetylation in the amygdala and on the reconsolidation of a cocaine-associated memory. Front Behav Neurosci 2020;13. Available from: https://doi.org/10.3389/fnbeh.2019.00281.

[49] Shimizu Y, Kawasaki T. Histone acetyltransferase EP300 regulates the proliferation and differentiation of neural stem cells during adult neurogenesis and regenerative neurogenesis in the zebrafish optic tectum. Neurosci Lett 2021;756:135978. Available from: https://doi.org/10.1016/j.neulet.2021.135978.

[50] Zhou C, Liu M, Mei X, Li Q, Zhang W, Deng P, et al. Histone hypoacetylation contributes to neurotoxicity induced by chronic nickel exposure in vivo and in vitro. Sci Total Environ 2021;783:147014. Available from: https://doi.org/10.1016/j.scitotenv.2021.147014.

[51] Chatterjee S, Mizar P, Cassel R, Neidl R, Selvi BR, Mohankrishna DV, et al. A Novel activator of CBP/p300 acetyltransferases promotes neurogenesis and extends memory duration in adult mice. J Neurosci 2013;33:10698–712. Available from: https://doi.org/10.1523/JNEUROSCI.5772-12.2013.

[52] Chatterjee S, Cassel R, Schneider-Anthony A, Merienne K, Cosquer B, Tzeplaeff L, et al. Reinstating plasticity and memory in a tauopathy mouse model with an acetyltransferase activator. EMBO Mol Med 2018;10:e8587. Available from: https://doi.org/10.15252/emmm.201708587.

[53] Haggarty SJ, Tsai L-H. Probing the role of HDACs and mechanisms of chromatin-mediated neuroplasticity. Neurobiol Learn Mem 2011;96:41–52. Available from: https://doi.org/10.1016/j.nlm.2011.04.009.

[54] Villain H, Florian C, Roullet P. HDAC inhibition promotes both initial consolidation and reconsolidation of spatial memory in mice. Sci Rep 2016;6:27015. Available from: https://doi.org/10.1038/srep27015.

[55] Haettig J, Stefanko DP, Multani ML, Figueroa DX, McQuown SC, Wood MA. HDAC inhibition modulates hippocampus-dependent long-term memory for object location in a CBP-dependent manner. Learn Mem 2011;18:71–9. Available from: https://doi.org/10.1101/lm.1986911.

[56] Ramirez-Mejia G, Gil-Lievana E, Urrego-Morales O, Soto-Reyes E, Bermúdez-Rattoni F. Class I HDAC inhibition improves object recognition memory consolidation through BDNF/TrkB pathway in a time-dependent manner. Neuropharmacology 2021;187:108493. Available from: https://doi.org/10.1016/j.neuropharm.2021.108493.

[57] Pulya S, Mahale A, Bobde Y, Routholla G, Patel T, Swati, et al. PT3: a Novel Benzamide class histone deacetylase 3 inhibitor improves learning and memory in novel object recognition mouse model. ACS Chem Neurosci 2021;12(5):883–92. Available from: https://doi.org/10.1021/acschemneuro.0c00721.

[58] Rumbaugh G, Sillivan SE, Ozkan ED, Rojas CS, Hubbs CR, Aceti M, et al. Pharmacological selectivity within class I histone deacetylases predicts effects on synaptic function and memory rescue. Neuropsychopharmacology 2015;40:2307–16. Available from: https://doi.org/10.1038/npp.2015.93.

[59] Sharma S, Sarathlal KC, Taliyan R. Epigenetics in neurodegenerative diseases: the role of histone deacetylases. CNS Neurol Disord Drug Targets 2019;18:11–18. Available from: https://doi.org/10.2174/1871527317666181004155136.

[60] Schmauss C. The roles of class I histone deacetylases (HDACs) in memory, learning, and executive cognitive functions: a review. Neurosci Biobehav Rev 2017;83:63–71. Available from: https://doi.org/10.1016/j.neubiorev.2017.10.004.

[61] McQuown SC, Barrett RM, Matheos DP, Post RJ, Rogge GA, Alenghat T, et al. HDAC3 is a critical negative regulator of long-term memory formation. J Neurosci 2011;31:764–74. Available from: https://doi.org/10.1523/JNEUROSCI.5052-10.2011.

[62] Guan J-S, Haggarty SJ, Giacometti E, Dannenberg J-H, Joseph N, Gao J, et al. HDAC2 negatively regulates memory formation and synaptic plasticity. Nature 2009;459:55–60. Available from: https://doi.org/10.1038/nature07925.

[63] Bredy TW, Barad M. The histone deacetylase inhibitor valproic acid enhances acquisition, extinction, and reconsolidation of conditioned fear. Learn Mem 2008;15:39–45. Available from: https://doi.org/10.1101/lm.801108.

[64] Vecsey CG, Hawk JD, Lattal KM, Stein JM, Fabian SA, Attner MA, et al. Histone deacetylase inhibitors enhance memory and synaptic plasticity via CREB: CBP-dependent transcriptional activation. J Neurosci 2007;27:6128. Available from: https://doi.org/10.1523/JNEUROSCI.0296-07.2007.

REFERENCES

[65] Bahari-Javan S, Maddalena A, Kerimoglu C, Wittnam J, Held T, Bähr M, et al. HDAC1 regulates fear extinction in mice. J Neurosci 2012;32:5062–73. Available from: https://doi.org/10.1523/JNEUROSCI.0079-12.2012.

[66] Bahari-Javan S, Varbanov H, Halder R, Benito E, Kaurani L, Burkhardt S, et al. HDAC1 links early life stress to schizophrenia-like phenotypes. PNAS 2017;114:E4686–94. Available from: https://doi.org/10.1073/pnas.1613842114.

[67] Siddiqui SA, Singh S, Ugale R, Ranjan V, Kanojia R, Saha S, et al. Regulation of HDAC1 and HDAC2 during consolidation and extinction of fear memory. Brain Res Bull 2019;150:86–101. Available from: https://doi.org/10.1016/j.brainresbull.2019.05.011.

[68] Patnaik D, Pao P-C, Zhao W-N, Silva MC, Hylton NK, Chindavong PS, et al. Exifone is a potent HDAC1 activator with neuroprotective activity in human neuronal models of neurodegeneration. ACS Chem Neurosci 2021;12:271–84. Available from: https://doi.org/10.1021/acschemneuro.0c00308.

[69] Rogge GA, Singh H, Dang R, Wood MA. HDAC3 is a negative regulator of cocaine-context-associated memory formation. J Neurosci 2013;33:6623–32. Available from: https://doi.org/10.1523/JNEUROSCI.4472-12.2013.

[70] Vinarskaya AKh, Balaban PM, Roshchin MV, Zuzina AB. Sodium butyrate as a selective cognitive enhancer for weak or impaired memory. Neurobiol Learn Mem 2021;180:107414. Available from: https://doi.org/10.1016/j.nlm.2021.107414.

[71] Kim M-S, Akhtar MW, Adachi M, Mahgoub M, Bassel-Duby R, Kavalali ET, et al. An essential role for histone deacetylase 4 in synaptic plasticity and memory formation. J Neurosci 2012;32:10879–86. Available from: https://doi.org/10.1523/JNEUROSCI.2089-12.2012.

[72] Main P, Tan WJ, Wheeler D, Fitzsimons HL. Increased abundance of nuclear HDAC4 impairs neuronal development and long-term memory. Front Mol Neurosci 2021;14. Available from: https://doi.org/10.3389/fnmol.2021.616642.

[73] Fitzsimons HL, Schwartz S, Given FM, Scott MJ. The histone deacetylase HDAC4 regulates long-term memory in Drosophila. PLoS One 2013;8:e83903. Available from: https://doi.org/10.1371/journal.pone.0083903.

[74] Lee JB, Wei J, Liu W, Cheng J, Feng J, Yan Z. Histone deacetylase 6 gates the synaptic action of acute stress in prefrontal cortex. J Physiol 2012;590:1535–46. Available from: https://doi.org/10.1113/jphysiol.2011.224907.

[75] Pulya S, Amin SKA, Adhikari N, Biswas S, Jha T, Ghosh B. HDAC6 as privileged target in drug discovery: a perspective. Pharmacol Res 2020;105274. Available from: https://doi.org/10.1016/j.phrs.2020.105274.

[76] Francelle L, Outeiro TF, Rappold GA. Inhibition of HDAC6 activity protects dopaminergic neurons from alpha-synuclein toxicity. Sci Rep 2020;10:1–14. Available from: https://doi.org/10.1038/s41598-020-62678-5.

[77] Hyun K, Jeon J, Park K, Kim J. Writing, erasing and reading histone lysine methylations. Exp Mol Med 2017;49:e324. Available from: https://doi.org/10.1038/emm.2017.11 –e324.

[78] Gupta S, Kim SY, Artis S, Molfese DL, Schumacher A, Sweatt JD, et al. Histone methylation regulates memory formation. J Neurosci 2010;30:3589–99. Available from: https://doi.org/10.1523/JNEUROSCI.3732-09.2010.

[79] Kerimoglu C, Sakib MS, Jain G, Benito E, Burkhardt S, Capece V, et al. KMT2A and KMT2B mediate memory function by affecting distinct genomic regions. Cell Rep 2017;20:538–48. Available from: https://doi.org/10.1016/j.celrep.2017.06.072.

[80] Collins BE, Sweatt JD, Greer CB. Broad domains of histone 3 lysine 4 trimethylation are associated with transcriptional activation in CA1 neurons of the hippocampus during memory formation. Neurobiol Learn Mem 2019;161:149–57. Available from: https://doi.org/10.1016/j.nlm.2019.04.009.

[81] Nicolay-Kritter K, Lassalle J, Guillou J-L, Mons N. The histone H3 lysine 9 methyltransferase G9a/GLP complex activity is required for long-term consolidation of spatial memory in mice. Neurobiol Learn Mem 2021;179:107406. Available from: https://doi.org/10.1016/j.nlm.2021.107406.

[82] Sagarkar S, Choudhary AG, Balasubramanian N, Awathale SN, Somalwar AR, Pawar N, et al. LSD1-BDNF activity in lateral hypothalamus-medial forebrain bundle area is essential for reward seeking behavior. Prog Neurobiol 2021;202:102048. Available from: https://doi.org/10.1016/j.pneurobio.2021.102048.

[83] Chwang WB, Arthur JS, Schumacher A, Sweatt JD. The nuclear kinase mitogen- and stress-activated protein kinase 1 regulates hippocampal chromatin remodeling in memory formation. J Neurosci 2007;27:12732–42. Available from: https://doi.org/10.1523/JNEUROSCI.2522-07.2007.

[84] Levenson JM, Sweatt JD. Epigenetic mechanisms in memory formation. Nat Rev Neurosci 2005;6:108–18. Available from: https://doi.org/10.1038/nrn1604.

[85] Reyskens KMSE, Arthur JSC. Emerging roles of the mitogen and stress activated kinases MSK1 and MSK2. Front Cell Dev Biol 2016;4. Available from: https://doi.org/10.3389/fcell.2016.00056.

[86] Daumas S, Hunter CJ, Mistry RB, Morè L, Privitera L, Cooper DD, et al. The kinase function of MSK1 regulates BDNF signaling to CREB and basal synaptic transmission, but is not required for hippocampal long-term potentiation or spatial memory. ENeuro 2017;4. Available from: https://doi.org/10.1523/ENEURO.0212-16.2017.

[87] Morice E, Enderlin V, Gautron S, Laroche S. Contrasting functions of mitogen- and stress-activated protein kinases 1 and 2 in recognition memory and in vivo hippocampal synaptic transmission. Neuroscience 2021;463:70–85. Available from: https://doi.org/10.1016/j.neuroscience.2021.03.004.

[88] Syal C, Seegobin M, Sarma SN, Gouveia A, Hsu K, Niibori Y, et al. Ectopic expression of aPKC-mediated phosphorylation in p300 modulates hippocampal neurogenesis, CREB binding and fear memory differently with age. Sci Rep 2018;8:13489. Available from: https://doi.org/10.1038/s41598-018-31657-2.

[89] Lim C-S, Nam HJ, Lee J, Kim D, Choi JE, Kang SJ, et al. PKCα-mediated phosphorylation of LSD1 is required for presynaptic plasticity and hippocampal learning and memory. Sci Rep 2017;7:1–15. Available from: https://doi.org/10.1038/s41598-017-05239-7.

[90] Koshibu K, Gräff J, Mansuy IM. Nuclear protein phosphatase-1: an epigenetic regulator of fear memory and amygdala long-term potentiation. Neuroscience 2011;173:30–6. Available from: https://doi.org/10.1016/j.neuroscience.2010.11.023.

[91] Bridi MS, Hawk JD, Chatterjee S, Safe S, Abel T. Pharmacological activators of the NR4A nuclear receptors enhance LTP in a CREB/CBP-dependent manner. Neuropsychopharmacol 2017;42:1243–53. Available from: https://doi.org/10.1038/npp.2016.253.

[92] Kwapis JL, Alaghband Y, López AJ, Long JM, Li X, Shu G, et al. HDAC3-mediated repression of the Nr4a family contributes to age-related impairments in long-term memory. J Neurosci 2019;39:4999–5009. Available from: https://doi.org/10.1523/JNEUROSCI.2799-18.2019.

[93] McNulty SE, Barrett RM, Vogel-Ciernia A, Malvaez M, Hernandez N, Davatolhagh MF, et al. Differential roles for Nr4a1 and Nr4a2 in object location vs. object recognition long-term memory. Learn Mem 2012;19:588–92. Available from: https://doi.org/10.1101/lm.026385.112.

[94] Mizuno K, Jeffries AR, Abel T, Giese KP. Long-lasting transcription in hippocampal area CA1 after contextual fear conditioning. Neurobiol Learn Mem 2020;172:107250. Available from: https://doi.org/10.1016/j.nlm.2020.107250.

[95] Liu PZ, Nusslock R. Exercise-mediated neurogenesis in the hippocampus via BDNF. Front Neurosci 2018;12: 52. Available from: https://doi.org/10.3389/fnins.2018.00052.

[96] Alonso M, Bekinschtein P, Cammarota M, Vianna MRM, Izquierdo I, Medina JH. Endogenous BDNF is required for long-term memory formation in the rat parietal cortex. Learn Mem 2005;12:504–10. Available from: https://doi.org/10.1101/lm.27305.

[97] Hall J, Thomas KL, Everitt BJ. Rapid and selective induction of BDNF expression in the hippocampus during contextual learning. Nat Neurosci 2000;3:533–5. Available from: https://doi.org/10.1038/75698.

[98] Ma YL, Wang HL, Wu HC, Wei CL, Lee EHY. Brain-derived neurotrophic factor antisense oligonucleotide impairs memory retention and inhibits long-term potentiation in rats. Neuroscience 1997;82:957–67. Available from: https://doi.org/10.1016/S0306-4522(97)00325-4.

[99] Lubin FD, Roth TL, Sweatt JD. Epigenetic regulation of bdnf gene transcription in the consolidation of fear memory. J Neurosci 2008;28:10576–86. Available from: https://doi.org/10.1523/JNEUROSCI.1786-08.2008.

[100] Zhong Y, Chen J, Li L, Qin Y, Wei Y, Pan S, et al. PKA-CREB-BDNF signaling pathway mediates propofol-induced long-term learning and memory impairment in hippocampus of rats. Brain Res 2018;1691:64–74. Available from: https://doi.org/10.1016/j.brainres.2018.04.022.

[101] Crozier RA, Black IB, Plummer MR. Blockade of NR2B-Containing NMDA receptors prevents BDNF enhancement of glutamatergic transmission in hippocampal neurons. Learn Mem 1999;6:257–66. Available from: https://doi.org/10.1101/lm.6.3.257.

[102] Crozier RA, Bi C, Han YR, Plummer MR. BDNF modulation of NMDA receptors is activity dependent. J Neurophysiol 2008;100:3264–74. Available from: https://doi.org/10.1152/jn.90418.2008.

[103] Amin SKA, Adhikari N, Kotagiri S, Jha T, Ghosh B. Histone deacetylase 3 inhibitors in learning and memory processes with special emphasis on benzamides. Eur J Med Chem 2019;166:369–80. Available from: https://doi.org/10.1016/j.ejmech.2019.01.077.

[104] Stefanko DP, Barrett RM, Ly AR, Reolon GK, Wood MA. Modulation of long-term memory for object recognition via HDAC inhibition. Proc Natl Acad Sci U S A 2009;106:9447–52. Available from: https://doi.org/10.1073/pnas.0903964106.

[105] Lattal KM, Barrett R, Wood MA. Systemic or intrahippocampal delivery of histone deacetylase inhibitors facilitates fear extinction. Behav Neurosci 2007;121:1125–31. Available from: https://doi.org/10.1037/0735-7044.121.5.1125.

[106] Wilson CB, McLaughlin LD, Ebenezer PJ, Nair AR, Francis J. Valproic acid effects in the hippocampus and prefrontal cortex in an animal model of post-traumatic stress disorder. Behav Brain Res 2014;268:72–80. Available from: https://doi.org/10.1016/j.bbr.2014.03.029.

[107] Kyrke-Smith M, Logan B, Abraham WC, Williams JM. Bilateral histone deacetylase 1 and 2 activity and enrichment at unique genes following induction of long-term potentiation in vivo. Hippocampus 2021;31:389–407. Available from: https://doi.org/10.1002/hipo.23297.

[108] Whittle N, Maurer V, Murphy C, Rainer J, Bindreither D, Hauschild M, et al. Enhancing dopaminergic signaling and histone acetylation promotes long-term rescue of deficient fear extinction. Transl Psychiatry 2016;6:e974. Available from: https://doi.org/10.1038/tp.2016.231 –e974.

[109] Malvaez M, McQuown SC, Rogge GA, Astarabadi M, Jacques V, Carreiro S, et al. HDAC3-selective inhibitor enhances extinction of cocaine-seeking behavior in a persistent manner. Proc Natl Acad Sci USA 2013;110:2647–52. Available from: https://doi.org/10.1073/pnas.1213364110.

[110] Bowers ME, Xia B, Carreiro S, Ressler KJ. The Class I HDAC inhibitor RGFP963 enhances consolidation of cued fear extinction. Learn Mem 2015;22:225–31. Available from: https://doi.org/10.1101/lm.036699.114.

[111] Gräff J, Joseph NF, Horn ME, Samiei A, Meng J, Seo J, et al. Epigenetic priming of memory updating during reconsolidation to attenuate remote fear memories. Cell 2014;156:261–76. Available from: https://doi.org/10.1016/j.cell.2013.12.020.

[112] Fass DM, Reis SA, Ghosh B, Hennig KM, Joseph NF, Zhao W-N, et al. Crebinostat: a novel cognitive enhancer that inhibits histone deacetylase activity and modulates chromatin-mediated neuroplasticity. Neuropharmacology 2013;64:81–96. Available from: https://doi.org/10.1016/j.neuropharm.2012.06.043.

[113] Zhao W-N, Ghosh B, Tyler M, Lalonde J, Joseph NF, Kosaric N, et al. Class I histone deacetylase inhibition by tianeptinaline modulates neuroplasticity and enhances memory. ACS Chem Neurosci 2018;9:2262–73. Available from: https://doi.org/10.1021/acschemneuro.8b00116.

[114] De Simone A, Milelli A. Histone deacetylase inhibitors as multitarget ligands: new players in Alzheimer's disease drug discovery? ChemMedChem 2019;14:1067–73. Available from: https://doi.org/10.1002/cmdc.201900174.

[115] Schueller E, Paiva I, Blanc F, Wang X-L, Cassel J-C, Boutillier A-L, et al. Dysregulation of histone acetylation pathways in hippocampus and frontal cortex of Alzheimer's disease patients. Eur Neuropsychopharmacol 2020;33:101–16. Available from: https://doi.org/10.1016/j.euroneuro.2020.01.015.

[116] Fernando WMADB, Martins IJ, Morici M, Bharadwaj P, Rainey-Smith SR, Lim WLF, et al. Sodium butyrate reduces brain amyloid-β levels and improves cognitive memory performance in an Alzheimer's disease transgenic mouse model at an early disease stage. J Alzheimer's Dis 2020;74:91–9. Available from: https://doi.org/10.3233/JAD-190120.

[117] Govindarajan N, Agis-Balboa RC, Walter J, Sananbenesi F, Fischer A. Sodium butyrate improves memory function in an Alzheimer's disease mouse model when administered at an advanced stage of disease progression. J Alzheimer's Dis 2011;26:187–97. Available from: https://doi.org/10.3233/JAD-2011-110080.

[118] Han Y, Chen L, Guo Y, Wang C, Zhang C, Kong L, et al. Class I HDAC inhibitor improves synaptic proteins and repairs cytoskeleton through regulating synapse-related genes in vitro and in vivo. Front Aging Neurosci 2021;12. Available from: https://doi.org/10.3389/fnagi.2020.619866.

[119] Levenson JM, Roth TL, Lubin FD, Miller CA, Huang I-C, Desai P, et al. Evidence that DNA (cytosine-5) methyltransferase regulates synaptic plasticity in the hippocampus. J Biol Chem 2006;281:15763–73. Available from: https://doi.org/10.1074/jbc.M511767200.

[120] Sui L, Wang Y, Ju L-H, Chen M. Epigenetic regulation of reelin and brain-derived neurotrophic factor genes in long-term potentiation in rat medial prefrontal cortex. Neurobiol Learn Mem 2012;97:425–40. Available from: https://doi.org/10.1016/j.nlm.2012.03.007.

[121] Miller CA, Campbell SL, Sweatt JD. DNA methylation and histone acetylation work in concert to regulate memory formation and synaptic plasticity. Neurobiol Learn Mem 2008;89:599–603. Available from: https://doi.org/10.1016/j.nlm.2007.07.016.

[122] Robinson HA, Pozzo-Miller L. The role of MeCP2 in learning and memory. Learn Mem 2019;26:343–50. Available from: https://doi.org/10.1101/lm.048876.118.

[123] Lu Z, Liu Z, Mao W, Wang X, Zheng X, Chen S, et al. Locus-specific DNA methylation of Mecp2 promoter leads to autism-

like phenotypes in mice. Cell Death Dis 2020;11:1–11. Available from: https://doi.org/10.1038/s41419-020-2290-x.

[124] Chang Q, Khare G, Dani V, Nelson S, Jaenisch R. The disease progression of Mecp2 mutant mice is affected by the level of BDNF expression. Neuron 2006;49:341–8. Available from: https://doi.org/10.1016/j.neuron.2005.12.027.

[125] Miller CA, Gavin CF, White JA, Parrish RR, Honasoge A, Yancey CR, et al. Cortical DNA methylation maintains remote memory. Nat Neurosci 2010;13:664–6. Available from: https://doi.org/10.1038/nn.2560.

[126] Mitchnick KA, Creighton S, O'Hara M, Kalisch BE, Winters BD. Differential contributions of de novo and maintenance DNA methyltransferases to object memory processing in the rat hippocampus and perirhinal cortex – a double dissociation. Eur J Neurosci 2015;41:773–86. Available from: https://doi.org/10.1111/ejn.12819.

[127] Sun W, Kong Q, Zhang M, Mi X, Sun X, Yu M, et al. Virus-mediated Dnmt1 and Dnmt3a deletion disrupts excitatory synaptogenesis and synaptic function in primary cultured hippocampal neurons. Biochem Biophys Res Commun 2020;526:361–7. Available from: https://doi.org/10.1016/j.bbrc.2020.03.094.

[128] Bachmann S, Linde J, Bell M, Spehr M, Zempel H, Zimmer-Bensch G. DNA methyltransferase 1 (DNMT1) shapes neuronal activity of human iPSC-derived glutamatergic cortical neurons. Int J Mole Sci 2021;22:2034. Available from: https://doi.org/10.3390/ijms22042034.

[129] Fan X-Y, Shi G, Zhao P. Neonatal sevoflurane exposure impairs learning and memory by the hypermethylation of hippocampal synaptic genes. Mol Neurobiol 2021;58:895–904. Available from: https://doi.org/10.1007/s12035-020-02161-4.

[130] Halder R, Hennion M, Vidal RO, Shomroni O, Rahman R-U, Rajput A, et al. DNA methylation changes in plasticity genes accompany the formation and maintenance of memory. Nat Neurosci 2016;19:102–10. Available from: https://doi.org/10.1038/nn.4194.

[131] Gulmez Karaca K, Kupke J, Brito DVC, Zeuch B, Thome C, Weichenhan D, et al. Neuronal ensemble-specific DNA methylation strengthens engram stability. Nat Commun 2020;11:639. Available from: https://doi.org/10.1038/s41467-020-14498-4.

[132] Webb WM, Sanchez RG, Perez G, Butler AA, Hauser RM, Rich MC, et al. Dynamic association of epigenetic H3K4me3 and DNA 5hmC marks in the dorsal hippocampus and anterior cingulate cortex following reactivation of a fear memory. Neurobiol Learn Mem 2017;142:66–78. Available from: https://doi.org/10.1016/j.nlm.2017.02.010.

[133] Miller CA, Sweatt JD. Covalent modification of DNA regulates memory formation. Neuron 2007;53:857–69. Available from: https://doi.org/10.1016/j.neuron.2007.02.022.

[134] Wu SC, Zhang Y. Active DNA demethylation: many roads lead to Rome. Nat Rev Mol Cell Biol 2010;11:607–20. Available from: https://doi.org/10.1038/nrm2950.

[135] Heyward FD, Sweatt JD. DNA methylation in memory formation: emerging insights. Neuroscientist 2015;21:475–89. Available from: https://doi.org/10.1177/1073858415579635.

[136] Kohli RM, Zhang Y. TET enzymes, TDG and the dynamics of DNA demethylation. Nature 2013;502:472–9. Available from: https://doi.org/10.1038/nature12750.

[137] Niehrs C, Schäfer A. Active DNA demethylation by Gadd45 and DNA repair. Trends Cell Biol 2012;22:220–7. Available from: https://doi.org/10.1016/j.tcb.2012.01.002.

[138] Ma DK, Jang M-H, Guo JU, Kitabatake Y, Chang M, Powanpongkul N, et al. Neuronal activity–induced Gadd45b promotes epigenetic DNA demethylation and adult neurogenesis. Science 2009;323:1074–7. Available from: https://doi.org/10.1126/science.1166859.

[139] Leach PT, Poplawski SG, Kenney JW, Hoffman B, Liebermann DA, Abel T, et al. Gadd45b knockout mice exhibit selective deficits in hippocampus-dependent long-term memory. Learn Mem 2012;19:319–24. Available from: https://doi.org/10.1101/lm.024984.111.

[140] Jarome TJ, Butler AA, Nichols JN, Pacheco NL, Lubin FD. NF-κB mediates Gadd45β expression and DNA demethylation in the hippocampus during fear memory formation. Front Mol Neurosci 2015;8. Available from: https://doi.org/10.3389/fnmol.2015.00054.

[141] Labonté B, Jeong YH, Parise E, Issler O, Fatma M, Engmann O, et al. Gadd45b mediates depressive-like role through DNA demethylation. Sci Rep 2019;9:4615. Available from: https://doi.org/10.1038/s41598-019-40844-8.

[142] Li X, Marshall PR, Leighton LJ, Zajaczkowski EL, Wang Z, Madugalle SU, et al. The DNA repair-associated protein Gadd45γ regulates the temporal coding of immediate early gene expression within the prelimbic prefrontal cortex and is required for the consolidation of associative fear memory. J Neurosci 2019;39:970–83. Available from: https://doi.org/10.1523/JNEUROSCI.2024-18.2018.

[143] Brito DVC, Gulmez Karaca K, Kupke J, Mudlaff F, Zeuch B, Gomes R, et al. Modeling human age-associated increase in Gadd45γ expression leads to spatial recognition memory impairments in young adult mice. Neurobiol Aging 2020;94:281–6. Available from: https://doi.org/10.1016/j.neurobiolaging.2020.06.021.

[144] Greer CB, Wright J, Weiss JD, Lazarenko RM, Moran SP, Zhu J, et al. Tet1 isoforms differentially regulate gene expression, synaptic transmission, and memory in the mammalian brain. J Neurosci 2021;41:578–93. Available from: https://doi.org/10.1523/JNEUROSCI.1821-20.2020.

[145] Wu X, Zhang Y. TET-mediated active DNA demethylation: mechanism, function and beyond. Nat Rev Genet 2017;18:517–34. Available from: https://doi.org/10.1038/nrg.2017.33.

[146] Kriaucionis S, Heintz N. The nuclear DNA Base 5-hydroxymethylcytosine is present in Purkinje neurons and the brain. Science 2009;324:929–30. Available from: https://doi.org/10.1126/science.1169786.

[147] Rudenko A, Dawlaty MM, Seo J, Cheng AW, Meng J, Le T, et al. Tet1 is critical for neuronal activity-regulated gene expression and memory extinction. Neuron 2013;79:1109–22. Available from: https://doi.org/10.1016/j.neuron.2013.08.003.

[148] Kumar D, Aggarwal M, Kaas GA, Lewis J, Wang J, Ross DL, et al. Tet1 oxidase regulates neuronal gene transcription, active DNA hydroxymethylation, object location memory, and threat recognition memory. Neuroepigenetics 2015;4:12–27. Available from: https://doi.org/10.1016/j.nepig.2015.10.002.

[149] DeNizio JE, Dow BJ, Serrano JC, Ghanty U, Drohat AC, Kohli RM. TET-TDG active DNA demethylation at CpG and Non-CpG sites. J Mol Biol 2021;433:166877. Available from: https://doi.org/10.1016/j.jmb.2021.166877.

[150] Maze I, Wenderski W, Noh K-M, Bagot RC, Tzavaras N, Purushothaman I, et al. Critical role of histone turnover in neuronal transcription and plasticity. Neuron 2015;87:77–94. Available from: https://doi.org/10.1016/j.neuron.2015.06.014.

[151] Zovkic IB. Epigenetics and memory: an expanded role for chromatin dynamics. Curr Opin Neurobiol 2021;67:58–65. Available from: https://doi.org/10.1016/j.conb.2020.08.007.

[152] McNally AG, Poplawski SG, Mayweather BA, White KM, Abel T. Characterization of a novel chromatin sorting tool reveals importance of histone variant H3.3 in contextual fear memory and motor learning. Front Mol Neurosci 2016;9. Available from: https://doi.org/10.3389/fnmol.2016.00011.

[153] Ishiuchi T, Abe S, Inoue K, Yeung WKA, Miki Y, Ogura A, et al. Reprogramming of the histone H3.3 landscape in the

[154] Zovkic IB, Paulukaitis BS, Day JJ, Etikala DM, Sweatt JD. Histone H2A.Z subunit exchange controls consolidation of recent and remote memory. Nature 2014;515:582–6. Available from: https://doi.org/10.1038/nature13707.

[155] Shen T, Ji F, Wang Y, Lei X, Zhang D, Jiao J. Brain-specific deletion of histone variant H2A.z results in cortical neurogenesis defects and neurodevelopmental disorder. Nucleic Acids Res 2018;46:2290–307. Available from: https://doi.org/10.1093/nar/gkx1295.

[156] Ramzan F, Baumbach J, Monks AD, Zovkic IB. Histone H2A.Z is required for androgen receptor-mediated effects on fear memory. Neurobiol Learn Mem 2020;175:107311. Available from: https://doi.org/10.1016/j.nlm.2020.107311.

[157] Ramzan F, Creighton SD, Hall M, Baumbach J, Wahdan M, Poulson SJ, et al. Sex-specific effects of the histone variant H2A.Z on fear memory, stress-enhanced fear learning and hypersensitivity to pain. Sci Rep 2020;10:14331. Available from: https://doi.org/10.1038/s41598-020-71229-x.

[158] Li J, Yang X, Qi Z, Sang Y, Liu Y, Xu B, et al. The role of mRNA m6A methylation in the nervous system. Cell Biosci 2019;9:66. Available from: https://doi.org/10.1186/s13578-019-0330-y.

[159] Dermentzaki G, Lotti F. New insights on the role of N6-Methyladenosine RNA methylation in the physiology and pathology of the nervous system. Front Mol Biosci 2020;7. Available from: https://doi.org/10.3389/fmolb.2020.555372.

[160] Lee M, Kim B, Kim VN. Emerging roles of RNA modification: m6A and U-Tail. Cell 2014;158:980–7. Available from: https://doi.org/10.1016/j.cell.2014.08.005.

[161] Jia G, Fu Y, He C. Reversible RNA adenosine methylation in biological regulation. Trends Genet 2013;29:108–15. Available from: https://doi.org/10.1016/j.tig.2012.11.003.

[162] Widagdo J, Zhao Q-Y, Kempen M-J, Tan MC, Ratnu VS, Wei W, et al. Experience-dependent accumulation of N6-methyladenosine in the prefrontal cortex is associated with memory processes in mice. J Neurosci 2016;36:6771–7.

[163] Walters BJ, Mercaldo V, Gillon CJ, Yip M, Neve RL, Boyce FM, et al. The role of the RNA demethylase FTO (Fat Mass and Obesity-Associated) and mRNA methylation in hippocampal memory formation. Neuropsychopharmacol 2017;42:1502–10. Available from: https://doi.org/10.1038/npp.2017.31.

[164] Zhang Z, Wang M, Xie D, Huang Z, Zhang L, Yang Y, et al. METTL3-mediated N 6 -methyladenosine mRNA modification enhances long-term memory consolidation. Cell Res 2018;28:1050–61. Available from: https://doi.org/10.1038/s41422-018-0092-9.

[165] Shi H, Zhang X, Weng Y-L, Lu Z, Liu Y, Lu Z, et al. m 6 A facilitates hippocampus-dependent learning and memory through YTHDF1. Nature 2018;563:249–53. Available from: https://doi.org/10.1038/s41586-018-0666-1.

[166] Kan L, Ott S, Joseph B, Park ES, Dai W, Kleiner RE, et al. A neural m 6 A/Ythdf pathway is required for learning and memory in Drosophila. Nat Commun 2021;12:1458. Available from: https://doi.org/10.1038/s41467-021-21537-1.

[167] Xue A, Huang Y, Li M, Wei Q, Bu Q. Comprehensive analysis of differential m6A RNA methylomes in the hippocampus of cocaine-conditioned mice. Mol Neurobiol 2021;. Available from: https://doi.org/10.1007/s12035-021-02363-4.

[168] Sillivan SE, Jamieson S, de Nijs L, Jones M, Snijders C, Klengel T, et al. MicroRNA regulation of persistent stress-enhanced memory. Mol Psychiatry 2020;25:965–76. Available from: https://doi.org/10.1038/s41380-019-0432-2.

[169] Konopka W, Schütz G, Kaczmarek L. The microRNA contribution to learning and memory. Neuroscientist 2011;17:468–74. Available from: https://doi.org/10.1177/1073858411411721.

[170] Wang W, Kwon EJ, Tsai L-H. MicroRNAs in learning, memory, and neurological diseases. Learn Mem 2012;19:359–68. Available from: https://doi.org/10.1101/lm.026492.112.

[171] Kumar D, Thakur MK. Effect of perinatal exposure to Bisphenol-A on DNA methylation and histone acetylation in cerebral cortex and hippocampus of postnatal male mice. J Toxicol Sci 2017;42:281–9. Available from: https://doi.org/10.2131/jts.42.281.

[172] Kremer EA, Gaur N, Lee MA, Engmann O, Bohacek J, Mansuy IM. Interplay between TETs and microRNAs in the adult brain for memory formation. Sci Rep 2018;8:1678. Available from: https://doi.org/10.1038/s41598-018-19806-z.

[173] Korneev SA, Vavoulis DV, Naskar S, Dyakonova VE, Kemenes I, Kemenes G. A CREB2-targeting microRNA is required for long-term memory after single-trial learning. Sci Rep 2018;8:3950. Available from: https://doi.org/10.1038/s41598-018-22278-w.

[174] Schröder J, Ansaloni S, Schilling M, Liu T, Radke J, Jaedicke M, et al. MicroRNA-138 is a potential regulator of memory performance in humans. Front Hum Neurosci 2014;8:501. Available from: https://doi.org/10.3389/fnhum.2014.00501.

[175] Li X, Zhang J, Li D, He C, He K, Xue T, et al. Astrocytic ApoE reprograms neuronal cholesterol metabolism and histone-acetylation-mediated memory. Neuron 2021;109:957–70. Available from: https://doi.org/10.1016/j.neuron.2021.01.005 e8.

[176] Ayata P, Badimon A, Strasburger HJ, Duff MK, Montgomery SE, Loh Y-HE, et al. Epigenetic regulation of brain region-specific microglia clearance activity. Nat Neurosci 2018;21:1049–60. Available from: https://doi.org/10.1038/s41593-018-0192-3.

CHAPTER 24

Transgenerational Epigenetics

James P. Curley[1], Rahia Mashoodh[2] and Frances A. Champagne[1]

[1]Department of Psychology, University of Texas, Austin, TX, United States [2]Department of Zoology, University of Cambridge, Cambridge, United Kingdom

OUTLINE

Introduction	465	Experience-dependent Epigenetic Inheritance	471
Epigenetic Consequences of Prenatal Maternal Exposures	466	Epigenetics, Plasticity, and Evolving Concepts of Inheritance	472
Postnatal Maternal Regulation of the Epigenome	468	Summary	473
Paternal Influence on Offspring Development	469	Acknowledgments	473
Transgenerational Effects of Parental Influence	469	References	473
Germline-mediated Transgenerational Inheritance	470		

INTRODUCTION

The regulation of gene expression through epigenetic modifications provides a dynamic route through which environmental experiences can lead to persistent changes in cellular phenotype. This plasticity plays an important role in mediating cellular differentiation and the potential stability of these modifications can lead to stable and heritable variations in gene expression [1]. Environmentally induced changes in DNA methylation, posttranslational histone modifications, and expression of non-coding RNAs have been observed following a broad range of environmental exposures. The process of DNA methylation whereby cytosine is converted to 5-methylcytosine is mediated by methyltransferases (i.e., DNA methyltransferase, DNMT1, DNMT3) which promote maintenance and de novo DNA methylation [2]. The process of DNA methylation is dependent on the presence of methyl donors (provided by nutrients such as folic acid, methionine, and choline [3,4]) and the transcriptional repression associated with DNA methylation is sustained through methyl-binding proteins such as methyl CpG binding protein 2 (MeCP2, [5]). Oxidation of methylated cytosines through the activity of Ten-eleven translocation (Tet) proteins can lead to DNA hydroxymethylation which may be a critical step in the process of active demethylation of DNA [6]. Histone proteins, which form the core of the nucleosome, also significantly alter gene expression through interactions with DNA [7]. Histones can undergo multiple post-translational modifications, including methylation (di- and tri-), acetylation, and ubiquitination, which can alter the accessibility of DNA and chromatin density [7]. The prediction of transcriptional activation vs suppression in response to histone modifications is dependent on the type and location of modification [8]. For example, tri-methylation (me3) of histone 3 (H3) at the lysine 4 (K4) position within the histone tail is associated with transcriptional activation; whereas, H3K27me3 is associated with both increased and decreased transcriptional activity [9,10]. Non-coding RNAs (RNA molecules that do not encode for a protein, for example, microRNAs, piRNAs [piwi-interacting RNAs], lncRNAs [long non-coding RNAs]), play a critical role in gene regulation through inhibition of translation and interplay with DNA methylation and chromatin [11,12]. Importantly, there is crosstalk between these epigenetic mechanisms that contributes to dynamic yet potentially stable levels of transcriptional activity [13,14].

Experiences across the lifespan can induce modifications to the epigenome [15]. Moreover, these epigenetic effects can have implications for neurobiology, physiology, and behavior of an organism leading to divergent developmental outcomes. Thus, the molecular mechanisms that regulate gene expression can contribute to the "epigenesis" of phenotype as described by Waddington in the 1940s, in which the term "epigenetics" has its roots [16]. Within mammals, the experience of offspring during the earliest periods of development are largely shaped by interactions with parents, with the broader characteristics of the social and physical environment influencing developing organisms via parent-offspring interactions [17]. Maternal effects on offspring development occurring prenatally or postnatally are well established [18,19]. In addition to paternal effects occurring among biparental species, there is evidence for paternal pre-conception influences among species where there is no postnatal contact between fathers and offspring [20]. These parental effects are associated with epigenetic variation and in some cases can be observed to influence descendants across multiple generations [20]. In this review, we will discuss evidence of maternal and paternal epigenetic influence on offspring development (intergenerational effects), with particular focus on studies indicating an association between parental experiences/environmental exposures and epigenetic alterations in offspring (Fig. 24.1). An emerging theme within these studies is the multigenerational and transgenerational implications of these environmentally induced effects. Though both multigenerational and transgenerational effects typically involve altered phenotypes that can be observed across several generations (i.e., grand-offspring generations or later), these effects can be distinguished based on the persistence and transmission of the phenotype in the absence of direct exposure (a phenomenon considered transgenerational). Here we will explore the pathways through which parental influences may persist across multiple generations leading to the stable inheritance of an epigenetically-mediated phenotype. These epigenetic effects may be a mechanism of adaptive plasticity that confers the ability of an organism to be prepared for the challenges of current and future environments [21].

EPIGENETIC CONSEQUENCES OF PRENATAL MATERNAL EXPOSURES

A defining feature of mammalian development is the in utero gestation of offspring. During this gestational period, the growth and development of offspring is dependent on maternal physiology and sensitive to a wide range of maternal environmental exposures. In this section, we will highlight studies that focus on maternal nutrition, toxicological exposures, and stress during pregnancy and the epigenetic and phenotypic effects in offspring.

The quality of the maternal nutritional environment during pregnancy can have a significant impact on the

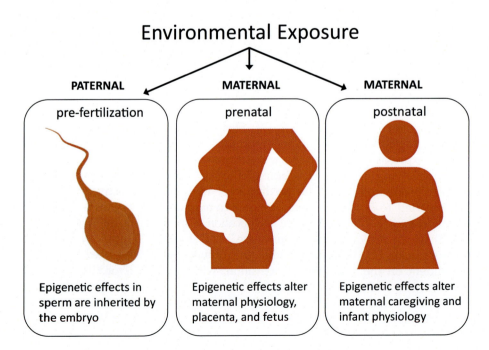

FIGURE 24.1 A summary figure of the parental targets of environmental exposure at different stages in reproduction/development and the nature of the epigenetic effects in each stage that shape offspring outcomes.

growth and development of the fetus, with long-term consequences for brain development and metabolism. Epidemiological studies of cohorts exposed prenatally to conditions of famine, suggest a heightened risk of metabolic disease, schizophrenia, and other neurodevelopmental abnormalities with the specific consequences dependent on the timing of exposure to maternal undernutrition [22,23]. Analysis of blood samples from siblings gestated during periods with or without maternal famine indicates an impact of early gestational exposure on genome-wide DNA methylation patterns, particularly within gene regulatory regions [24]. Though multiple biological pathways exhibit an epigenetic impact of prenatal famine, the decreased DNA methylation of the insulin-like growth factor II (*IGF2*) gene as a consequence of maternal periconceptual exposure to famine may be implicated in growth, metabolic, and neurodevelopmental outcomes [25]. Genome-wide analyses of DNA methylation in adults who were exposed gestationally to famine indicate that variation in DNA methylation within several genomic regions is a mediator between famine exposure and specific metabolic outcomes, including increased body mass index (BMI), lipid metabolism and triglyceride levels [26].

Laboratory studies in rodents have identified specific nutritional deficits, such as prenatal protein restriction or folic acid/choline deficiency, as having similar epigenetic consequences. Offspring of female rats placed on a protein deficient diet throughout gestation were found to have altered hepatic gene expression and DNA methylation profiles, including elevated glucocorticoid receptor (encoded by the *NR3C1* gene) and peroxisomal proliferator-activated receptor alpha (*PPARα*) gene expression associated with decreased DNA methylation of these genes [27–29]. Moreover, these transcriptional and epigenetic effects could be largely reversed when gestational protein restriction is accompanied by folic acid supplementation [27,28]. Dietary effects on levels of DNMT1 may account for these observed modifications in global and gene-specific methylation, as *DNMT1* expression is altered in hepatic [27,30] and brain tissue [31] as a function of protein/choline restriction. The impact of dietary supplementation with methyl-donors during fetal development is also demonstrated by the consequences for phenotype among mice with the A^{vy} allele of the Agouti gene. The expression of the A^{vy} allele is epigenetically regulated through levels of DNA methylation, with decreased methylation associated with yellow coat color and obesity among A^{vy} mice [32]. Though the maternal Agouti phenotype is typically inherited by offspring, gestational exposure to methyl donors through dietary supplementation of the mother can effectively silence the expression of the A^{vy} allele, inducing a pseudo wild-type phenotype [33]. Thus, the maternal nutritional environment can have a sustained impact on development through alterations in gene expression that are maintained through DNA methylation. These effects may manifest in response to both nutrient restriction [27,30] or supplementation (i.e., high fat diet [34]) directly within developing fetal tissues or within the placenta [35,36] with consequences for the fetal environment. Beyond DNA methylation, the epigenetic effects of prenatal dietary manipulation are increasingly evident on measures of histone acetylation and tri-methylation [34] and expression of microRNAs [37] in offspring tissues.

The rapid period of cellular proliferation and differentiation that occurs during fetal development provides a critical temporal window during which maternal gestational exposure to toxins may lead to long-term disruptions in offspring and there is increasing evidence for the epigenetic basis of these effects. In utero methyl mercury exposure in mice leads to DNA hypermethylation, increased histone trimethylation and decreased histone acetylation within the IV promoter of the brain derived neurotrophic factor (*BDNF*) gene in the hippocampus of offspring [38]. Exposure of pregnant mice to inhaled diesel exhaust particles results in decreased DNA methylation within *PPARγ* in offspring adipose tissue [39] and increased DNA methylation within *BDNF IV* in offspring hippocampus [40]. In humans, comparable levels of exposure to inhaled pollutants is associated with reduced global DNA methylation in white blood cells derived from cord blood samples [41]. In rats, prenatal exposure to the antiandrogenic fungicide vinclozolin results in increased rates of prostate disease, kidney disease, testis abnormalities, and tumor development [42]. This prenatal exposure to an endocrine disrupting chemical is associated with altered DNA methylation patterns in sperm, altered expression of microRNAs in primordial germ cells and impairments in reproduction in male offspring [43,44]. In utero exposure to the endocrine disruptor bisphenol-A (BPA) has been demonstrated to induce genome-wide changes in promoter DNA methylation in the fetal mouse brain [45], increased hippocampal DNA methylation within *BDNF IV* [46] and the promoter region of the gene encoding estrogen receptor alpha (*ESR1*) within the cortex of juvenile offspring [47] and altered hepatic levels of histone acetylation, di-methylation and trimethylation of the carnitine palmitoyltransferase I gene (*CPT1*) in male offspring [48]. Obesity-related phenotypes are also observed in association with prenatal BPA exposure and there is evidence that BPA-induced alterations in DNA methylation within growth-promoting genes, such as insulin-like growth factor II receptor (*IGF2R*) and mesoderm-specific transcript (*MEST*) are a mediator of these effects [49,50]. Though prevention or reduction of exposure to toxins during prenatal development may be an optimal strategy for shifting developmental outcomes, there is evidence that exposure-associated phenotypic and epigenetic effects can be altered by dietary interventions during pregnancy [51]. For example, BPA-induced hypomethylation of the A^{vy} allele in mice leads to metabolic abnormality and obesity in adulthood and similar to the case of prenatal protein restriction, BPA-induced effects can be reversed through folate supplementation in the mother's diet [52]. In humans, the effects of prenatal smoking on DNA methylation in newborns is reduced in association with high folate intake [53]. Other prenatal dietary supplements, such as vitamin C and B may also be able to counteract the epigenetic effects of toxin exposure and highlight the complexity of the prenatal "exposome" [51].

Evidence for the epigenetic influence of prenatal maternal mood and psychosocial stress has emerged from human cohort studies and animal models—providing further support for the role of epigenetic mechanisms in mediating developmental outcomes. These maternal experiences may also exacerbate the effects of prenatal toxins on developmental outcomes [54]. Among infants born to mothers with elevated ratings of depression during pregnancy, there is significant differential DNA methylation within the genome [55], elevated NR3C1 promoter DNA methylation levels [56], and decreased DNA methylation within the oxytocin receptor gene (OXTR) [57]. Though most epigenetic studies in human subjects have been dependent on the use of blood or buccal samples, epigenetic variation in human postmortem hippocampal tissue have also been observed in relationship to exposure to maternal depression, with some overlap with epigenetic markers detected in blood [55]. While ameliorating maternal depression during pregnancy using pharmacological approaches may be necessary to reduce developmental risk in offspring, concerns regarding the developmental consequences of prenatal exposure to antidepressants have emerged [58] and there is evidence for epigenetic effects in offspring associated with use of antidepressants during pregnancy [59].

In rodents, chronic gestational stress is associated with decreased hypothalamic DNA methylation of the corticotrophin-releasing-factor (CRF) gene promoter [60], increased hypothalamic DNA methylation of the NR3C1 promoter region [60], increased DNMT1 expression in the cortex and hippocampus [61], decreased hippocampal histone acetylation [62], and modified expression of microRNAs [63]. Prenatal stress can exacerbate the epigenetic effects of other prenatal exposures, such as exposure to lead [64], and it is evident that the effects of stress and other exposures occurring during fetal development are sex-specific, with males and females exhibiting differential epigenetic responses to prenatal adversity [36,47,60,64]. These sex-specific effects extend to the placenta, where epigenetic variation has been found to be induced by stress in studies of humans [65,66] and rodents [60,67,68] and may serve as a critical mediator of prenatal effects on offspring development.

POSTNATAL MATERNAL REGULATION OF THE EPIGENOME

Though dynamic epigenetic modifications were once thought to be limited to the very early stages of development, evidence for continued epigenetic influence of parents beyond the prenatal period has challenged this view. Studies of the effects of natural variations in postnatal care in rodents have established the mediating role of epigenetic factors in shaping individual differences in brain and behavior [69]. Reduced levels of postnatal maternal licking/grooming (LG) behavior in rats, in particular, has been found to alter hippocampal gene expression, histone acetylation and DNA methylation globally, with specific increases in DNA methylation and decreases in H3K9 acetylation within the NR3C1 [70], glutamate decarboxylase 1 (GAD1) [71], and GRM1 (encoding metabotropic glutamate receptor 1) [72] gene promoters in adult male offspring. In female rat offspring, the experience of low levels of LG during postnatal development leads to increased DNA methylation and decreased H3K4me3 within the hypothalamus at the ESR1 promoter region [73,74]. Cross-fostering manipulations have been used to illustrate the link between the postnatal experience of high vs. low levels of LG and these epigenetic outcomes [70]. In addition, pharmacological manipulations of histone acetylation and DNA methylation can be used to reverse the epigenetic and neurobehavioral consequences of variation in postnatal maternal LG [70,73,75]. Epigenetic effects within the brain have also been observed in rodents as a consequence of postnatal exposure to maternal separation [76], abusive caregiving [77], and communal rearing [78]. In humans, analyses of postmortem hippocampal tissue reveal similar epigenetic signatures in response to childhood maltreatment that have been observed in rodent studies of variation in parental care, with a significant parallel in the finding of increased hippocampal NR3C1 DNA methylation in response to a history of abusive caregiving [79,80]. Exposure to high versus low levels of postnatal maternal touch is associated with differential DNA methylation across the genome [81] and there is also an association between breastfeeding and offspring DNA methylation [82]. Though the pathways through which these postnatal effects are achieved have yet to be elucidated, it is likely that activation of transcription factors in response to variation in the quality of postnatal mother–infant interactions leads to a cascade of cellular/molecular changes with consequences for epigenetic profiles [83].

PATERNAL INFLUENCE ON OFFSPRING DEVELOPMENT

Mammalian development is characterized by intense prenatal and postnatal mother–infant interactions and thus studies of parental influence have primarily focused on maternal rather than paternal effects. However, even among species in which biparental care is not typical, significant paternal modulation of offspring development has been observed [20]. In rodents, premating exposure of males to alcohol is associated with reduced offspring birth weight, increased mortality, and numerous cognitive and behavioral abnormalities [84–86]. Likewise, offspring of cocaine-exposed males perform poorly on tests of spatial attention/working memory and have a reduced cerebral volume [87,88]. In rats, premating paternal exposure to delta-9-tetrohydrocannabinol (THC) is associated with impaired attention in offspring [89]. Variation in the dietary environment of fathers appears to be transmissible to offspring. For instance, reduced serum glucose and altered levels of corticosterone and IGF1 are found among offspring of male mice that undergo a 24-h complete fast two weeks before mating [90] and premating chronic paternal caloric restriction results in reduced serum leptin and altered behavior in offspring [91]. Studies across a diverse range of species [92–96], including humans [97], indicate that paternal exposure to stress at a broad range of developmental time points can induce sex-specific alterations in the behavior and neurobiology of offspring. Finally, epidemiological studies in humans have demonstrated increased risk of autism and schizophrenia that emerge as a function of increased paternal age [98]. Laboratory studies of paternal age effects in isogenic rodents also indicate that offspring of "old" fathers have reduced longevity, increased social deficits, and perform more poorly on learning and memory tasks [99–101]. The transmission of these paternal effects to offspring in the absence of any postnatal contact with fathers suggests that these exposures may lead to alterations in the male germ cells with consequences for postfertilization embryonic development [102].

Investigation of the role of epigenetic mechanisms in mediating paternal effects suggests that environmentally induced epigenetic changes within sperm cells (including DNA methylation, histone modifications and expression of non-coding RNAs) may propagate the effects of paternal experiences on development [102–104]. In males, chronic exposure to alcohol or cocaine can induce chromatin remodeling and changes in DNA methylation within numerous genes in both the brain and periphery [105,106]. In particular, alcohol exposure has been shown to decrease *DNMT* mRNA levels in the sperm cells of adult male rats [107], chronic cocaine exposure in adult male mice has been shown to decrease *DNMT1* while increasing *DNMT3* mRNA expression in the germ cell-rich cells of the seminiferous tubules of the testes [88], and cocaine self-administration by males is associated with increased H3 acetylation within the *BDNF* promoter [108]. These epigenetic alterations may impact genomic imprinting—the parent-of-origin epigenetic effects on gene expression—as analysis of sperm DNA methylation levels in human males that are heavy drinkers indicates reduced DNA methylation in the normally hypermethylated *H19* and *IG* regulatory regions [106]. In the case of paternal age, overall reductions in DNA methylation are observed in the sperm of "old" (12–14 months) compared to "young" (3 months) male mice [109] and hypermethylation of ribosomal DNA has been found in the sperm and liver cells of "old" (21–28 months) compared to "young/adult" (6 months) male rats [110]. Though there are many genetic and morphological abnormalities in sperm associated with aging, these epigenetic modifications may contribute to the aberrant developmental outcomes associated with increasing paternal age.

Male stress exposure occurring during postnatal development and in adulthood can also impact epigenetic outcomes in sperm. Postnatal maternal separation in male mice is associated with increased expression of several microRNAs, downregulation of piRNAs and increased DNA methylation within the regulatory region of *MECP2* [111,112]. Chronic stress exposure in adulthood is also associated with increased expression of several microRNAs in sperm [96] and fear conditioning to a specific odorant molecule (acetophenone) results in reduced DNA methylation in sperm within the gene encoding the odorant receptor responsive to that molecule [113]. Thus, environmental exposures may lead to altered epigenetic marks in male gametes with both broad and gene-specific consequences [103]. It has also been determined that direct manipulation of these environmentally-sensitive epigenetic marks in sperm can recapitulate the predicted phenotypic consequences in offspring, providing strong support for the role of paternal epigenetic transmission [111,114].

TRANSGENERATIONAL EFFECTS OF PARENTAL INFLUENCE

The stability of epigenetic modifications within an individual's own lifespan and evidence suggestive of a transmission of parental epigenetic changes to offspring provide a new perspective on the stable inheritance of traits. Moreover, there is increasing evidence that this non-genomic inheritance can be maintained over multiple generations, such that in addition to the developmental effects of parental experiences on F1 generation offspring (intergenerational effects), there may be observed influences of parental (F0) experiences on grand-offspring (F2) and possibly great-grand-offspring (F3). In general, there may be two distinct routes through which these types of epigenetic inheritance patterns can occur: germline-mediated vs experience-dependent/non-germline-mediated (Fig. 24.2). Within germline-mediated transgenerational effects, environmental exposures are thought to induce epigenetic alterations within the developing gametes that persist in the germline of descendants in the absence of continued exposure with consequences for F1, F2, and F3 generations. In contrast, experience-dependent/non-germline mediated epigenetic transmission requires that an experience or environmental exposure be present in each generation to re-establish the epigenetic modifications which permit the trait to persist in subsequent generations. This process results in a multigenerational continuity of a phenotype. The distinction between these two routes can be difficult to establish experimentally, particularly in the case of prenatal exposures in which F1 offspring and the F1 offspring's germline, which will give rise to the F2 generation, are exposed to the inducing environmental factor. Though both of these processes can lead to the stable inheritance of phenotype, there is certainly divergence in the routes through which this is achieved [115,116].

GERMLINE-MEDIATED TRANSGENERATIONAL INHERITANCE

There is growing evidence for the transgenerational impact of nutrition, toxins, and stress exposure that provides support for an inheritance pattern that is likely germline-mediated. Analysis of archival records from Sweden in which crop success (used as a proxy for food intake) and longevity can be determined in multiple generations, suggests that in humans, a high level of nutrition during the slow growth period that precedes puberty is associated with diabetes and cardiovascular disease mortality of grand-offspring [117]. Interestingly, these effects are sex-specific, with paternal grandfather nutrition predicting grandson mortality and paternal grandmother nutrition predicting granddaughter longevity. Laboratory studies in rodents have confirmed the transgenerational impact of nutrition and indicate that prenatal protein restriction can exert effects on growth and metabolism of offspring and grand-offspring through changes in DNA methylation status of *NR3C1* [118]. When F0 female mice are exposed to caloric restriction during late gestation, F2 grand-offspring are found to have impaired glucose tolerance and this effect is maintained even when the F1 generation is provided with ad libitum food throughout their lifetime. Among descendants of female mice (F0) placed on a high-fat diet throughout pregnancy, offspring (F1) and grand-offspring (F2) exhibit increased body length [119]. This diet-induced effect can also be observed in the F3 generation, but only when that generation is derived from the F1 exposed patriline [120]. In human cohort studies, paternal pre-conception consumption of betel nuts

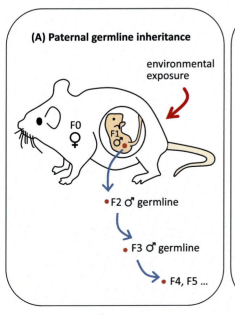 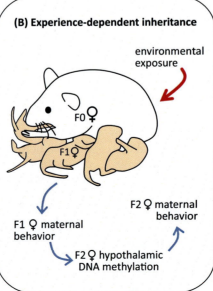

FIGURE 24.2 Illustration of the distinction between a paternal germline epigenetic inheritance (A) and an experience-dependent inheritance of an epigenetic effect (B). In an example of a paternal germline inheritance, an environmental exposure occurring during prenatal development results in an epigenetic alteration within the male F1 germline that is transmitted to F2 and F3 generation male offspring. In contrast, experience-dependent inheritance, such as the transmission of maternal behavior across generations, requires that each generation is exposed to different maternal care in infancy.

(which contain nitrosamines) leads to dose-dependent increases in offspring risk of metabolic syndrome [121] and in transgenerational studies of mice, 2–6 days of pre-conception betel nut consumption by F0 generation males is associated with increased glucose intolerance amongst F1, F2, and F3 generation offspring [122]. Similar metabolic effects are observed when males are exposed in utero to dexamethasone, with increased glucose intolerance observed among the offspring of these males when mated with non-exposed females [123]. However, in the case of prenatal dexamethasone exposure, these metabolic phenotypes do not persist beyond the F2 generation indicating that there is either compensation for the germline effects or that the effect is mediated by experience-dependent transmission.

The consequences of in utero exposure to endocrine-disrupting compounds has also been explored within a transgenerational model and provides evidence for the pervasive effects on epigenetic profiles of these early life exposures. In utero exposure to vinclozolin in rats has been demonstrated to disrupt DNA methylation in sperm and increase rates of infertility and risk of prostrate and kidney disease in F1, F2, and F3 offspring with the transmission though the patriline [43]. Prenatal vinclozolin-induced alterations in gene expression within the hippocampus and amygdala have also been observed for up to three generations postexposure with sex-specific effects on anxiety-like behavior [124]. Interestingly, mate-choice studies suggest that females presented with F3 vinclozolin-exposed or non-exposed males show a significant partner preference for non-exposed males, indicating an additional measure of decreased reproductive success as a consequence of treatment with endocrine disruptors [125]. The persistence of these disruptions beyond the F2 generation (with the F3 generation being the first "non-exposed" generation) suggests that the effects of these exposures have become incorporated into the germline and there is incomplete erasure of the associated epigenetic marks during the process of gametogenesis, fertilization, and embryogenesis [115]. Similar effects have been observed following F0 gestational exposure to a mixture of BPA and other endocrine disrupting compounds, with reproductive and sperm epigenome consequences observed in the F3 generation [126]. Understanding the mechanisms through which these toxin exposures lead to alterations in the epigenome will have significant implications for our understanding of environmental health issues.

The profound effects of stress exposure on development can be transmitted to offspring and grand-offspring and there is increasing evidence for the role of stress-induced epigenetic variation in the paternal germline in mediating the inheritance of these effects. Increased anxiety-like and depressive-like behavioral indices are observed in the female offspring (F2) of maternally separated males (F1) and though these behavioral effects are not observed in F2 male offspring, the grand-offspring (F3) generated from F2 exposed males exhibit these behavioral traits [112]. Moreover, the increased DNA methylation observed within the *MECP2* gene in the sperm of exposed males are recapitulated in the cortex of F2 male offspring and microRNAs differentially expressed in the sperm of exposed F1 males are similarly observed in the F2 hippocampus [111,112]. Chronic social stress in juvenile male mice results in behavioral deficits (social deficits and anxiety) that persist to the F3 generation—though only when this generation is derived from the exposed patriline—further supporting a paternal germline inheritance [95]. Finally, the impact of fear conditioning of F0 generation males to acetophenone results in olfactory sensitivity to this specific odorant in subsequent F1 and F2 generations [113]. Altered DNA methylation of the olfactory receptor responsive to acetophenone is observed in the sperm of both F0 and F1 males suggesting a germline basis to the transmission of memory of fear-related stimuli.

Though the stability and inheritance of epigenetic characteristics of the genome during the process of mitosis is well-established and supported by specific enzymes, such as DNMT1, the mechanisms accounting for the persistence of epigenetic effects during gametogenesis, fertilization, and postfertilization genomic reorganization are yet to be fully understood. The phenomenon of genomic imprinting establishes a pathway for the erasure and re-establishment of epigenetic marks within the genome and, in particular, within the germline [127]. However, it is unclear whether a similar process can be utilized following de novo epigenetic changes. Environmental exposures can impact the retention of histones, expression of non-coding RNAs and DNA methylation in sperm and these epigenetic changes can be observed in subsequent generations [103]. A critical question to address is the degree to which the induced epigenetic modification is directly inherited or influences the general molecular processes of epigenomic organization that occur during the pre- and postfertilization period resulting in a recapitulation of environmentally induced effects in non-exposed generations.

EXPERIENCE-DEPENDENT EPIGENETIC INHERITANCE

Across species, there is evidence for the transmission of individual differences in maternal behavior from mother to offspring and grand-offspring. In humans, mother–infant attachment classifications (secure, anxious/resistant, avoidant, disorganized) are similar across generations of female offspring [128] as are levels of parental bonding [129]. In rhesus and pigtail macaques, rates of postnatal maternal rejection and infant abuse are transmitted across matrilines and cross-fostering studies conducted between abusive and non-abusive macaques females indicates that the transmission of abusive behavior from mother to daughter is dependent on the experience of abuse during the postnatal period [130,131]. This matrilineal transmission is also evident in laboratory rodents. Natural variations in postnatal maternal LG observed in the F0 generation are associated with similar levels of LG in F1 and F2 generation females [132]. As such, under stable environmental conditions, offspring (F1) and grand-offspring (F2) of low-LG females display low levels of LG; whereas, offspring (F1) and grand-offspring (F2) of high-LG females display high levels of LG [132]. Similar to the multigenerational effects of abuse in macaques, cross-fostering studies have demonstrated that the transmission of maternal LG from mother to female offspring is dependent on the level of maternal LG received in infancy [133]. Communal postnatal rearing in mice results in increased postpartum maternal behavior in F0 females, in F1 females, and in F2 females that have not been communally reared but are the offspring of communally reared females [134]. A similar behavioral transmission can occur when the F0 maternal behavior is altered through a genetic mutation that is not inherited by offspring [135]. The experience-dependent nature of the transmission of maternal behavior is further highlighted in studies where environmental conditions of mothers are altered through chronic exposure to stress [136] or manipulation of the juvenile environment [132]. These environmental exposures impact maternal behavior (particularly LG) leading to a disruption of the inheritance of the predicted maternal phenotype. Since postnatal maternal LG can impact a broad range of behavioral and neurobiological outcomes, the transmission of LG maternal phenotypes can act as a vector for the transmission of F0 environmental exposures on the development of subsequent generations [137].

Epigenetic mechanisms may be critical in mediating the transmission of maternal behavior across generations and for the recapitulation of behavioral and neurobiological phenotypes that emerge as a consequence of this multigenerational transmission. Female offspring of low-LG mothers exhibit a reduced sensitivity to estrogen and have reduced levels of hypothalamic *ESR1* expression within the medial preoptic area (MPOA) of the hypothalamus, which likely accounts for the reduced level of postpartum maternal behavior observed in these females [138]. Female offspring reared by a low-LG dam have increased DNA methylation, decreased H3K4me3, and increased H3K9me3 at the *ESR1* gene promoter [73,74]. This epigenetic variation results in reduced binding of signal transducer and activator of transcription (Stat)5 to the *ESR1* promoter with consequences for the transcriptional activity of this gene [73]. Thus, epigenetic modifications to a gene that regulates several aspects of reproduction, including postpartum maternal behavior, results in differential levels of expression of *ESR1* in offspring in adulthood, which alters estrogen sensitivity and consequently leads to variations in the level of maternal care that these females (F1) provide to their own offspring (F2). The transmission from mother to daughter of variations in maternal LG within this multigenerational framework is mediated by the stability of brain region-specific epigenetic modifications that occur in infancy and influence behavior in adulthood. Similar experience-dependent effects of the postnatal environment in rats have been induced through exposure to abuse. Increase in DNA methylation in the *BDNF* IV promoter and consequent decrease in *BDNF* mRNA in the prefrontal cortex has been found in association with exposure to periods of postnatal abusive maternal care (dragging, burying, etc.) [77]. Moreover, these effects on *BDNF* IV promoter DNA methylation are perpetuated to the F1 offspring of abused females suggesting a role for epigenetic mechanisms in this transmission. Overall, these studies highlight the stable inheritance of traits that can be achieved through a behavioral transmission of epigenetic modifications. Moreover, the consequences of this transmission extend to all systems that are impacted by variation in maternal behavior, including social behavior, stress responsivity, and cognition, and involve epigenetic modification of genes within the brain regions that regulate these phenotypic outcomes.

EPIGENETICS, PLASTICITY, AND EVOLVING CONCEPTS OF INHERITANCE

Though the study of mechanisms of inheritance and the origins of divergent developmental trajectories has traditionally been the domain of the field of genetics, there is increasing evidence for the role of epigenetic modifications in maintaining environmentally induced variations in phenotype both within and across generations. The dynamic nature of these epigenetic effects provides a mechanism through which a single genotype can give rise to multiple phenotypic outcomes, conferring a heightened level of developmental plasticity to an organism. In contrast to environmentally induced genetic alterations/mutations, which are thought to be nondirected, there may be adaptive consequences associated with experience-dependent epigenetic modifications. For example, nutritional "programming" of fetal metabolism has been explored as an adaptive consequence of early life experience [139], and there is clearly a role for epigenetic mechanisms in mediating the effects of variations in prenatal food intake. When the prenatal period is characterized by undernutrition, a "thrifty phenotype" may result, which allows an individual to be conservative with regard to energy use and which promotes storage of glucose—with adverse health consequences associated with a mismatch between the quality of the prenatal and postnatal nutritional environment [140]. Similar adaptive consequences may be relevant to the development of heightened stress reactivity. Though elevated stress responses are typically considered to be a negative outcome and associated with increased susceptibility to physical and psychiatric disease, within an evolutionary perspective, the ability to respond rapidly to threat would be particularly advantageous under conditions of high predation/low resource availability [141]. Laboratory studies of maternal care in rodents suggest that chronic stress and social impoverishment can lead to reduced LG with consequences for the increased stress response of offspring via differential hippocampal *NR3C1* DNA methylation [70,132,136]. Though this environmentally induced phenotype is associated with impaired cognitive performance under standard testing conditions [142], synaptic plasticity is enhanced in offspring of low-LG mothers when corticosterone levels are elevated [143]. Thus, the consequences of early life experience can be considered as adaptive or maladaptive dependent on the consistency or "match" between early and later environmental conditions, and epigenetic mechanisms may play a critical role in shaping these phenotypic adaptations. The match between parental and offspring environmental conditions may also be an important predictor of whether transgenerational effects occur [21], further highlighting the adaptive role played by within and across generation plasticity that has been described

The concept that experience-induced characteristics can be transmitted across generations is reminiscent of Lamarckian theories of use/disuse and the inheritance of acquired characteristics [144]. Though the role of heritable epigenetic modifications in evolutionary processes certainly remains a topic of debate, both germline and experience-dependent epigenetic transmission may be important processes to be considered within an extended evolutionary synthesis and a more dynamic and interactive view of development [145]. Importantly, though there is growing support for multigenerational and transgenerational epigenetic consequences of environmental exposures, our understanding of the molecular, cellular, and behavioral pathways through which these outcomes are achieved is still evolving. These processes occur within the context of genetic variation and other inheritance systems and there are likely interactions between inheritance systems that make the prediction of individual-level outcomes challenging. Moreover, these epigenetic effects have often been explored from the perspective of pathology, yet the potential for multigenerational improvements in health and behavioral outcomes or resistance to subsequent environmental exposures are being increasingly observed [108,146,147]. Thus, broadening our concept of inheritance to include both genetic and epigenetic mechanisms and the interplay between these pathways [148] may provide insights into effective therapeutic approaches and lead to a greater appreciation of the benefits that can be achieved through intervention in parental and grandparental generations.

SUMMARY

In this chapter, we explore the epigenetic and developmental effects of maternal and paternal environmental exposures in offspring and emerging evidence for the role of epigenetics in transmitting environmental effects across generations. Evidence for the multigenerational and transgenerational impact of the environment is described and the chapter explores the unique routes through which mothers and fathers mediate this generational transmission. We differentiate between germline and experience-dependent routes of epigenetic inheritance and discuss the role of these mechanisms of inheritance in adaptation and plasticity.

Acknowledgments

The authors wish to acknowledge funding received from Grant Number DP2OD001674 from the Office of the Director, National Institutes of Health.

References

[1] Moore LD, Le T, Fan G. DNA methylation and its basic function. Neuropsychopharmacology 2013;38:23—38. Available from: https://doi.org/10.1038/npp.2012.112.

[2] Feng J, Fan G. The role of DNA methylation in the central nervous system and neuropsychiatric disorders. Int Rev Neurobiol 2009;89:67—84. Available from: https://doi.org/10.1016/S0074-7742(09)89004-1.

[3] Anderson OS, Sant KE, Dolinoy DC. Nutrition and epigenetics: an interplay of dietary methyl donors, one-carbon metabolism and DNA methylation. J Nutr Biochem 2012;23:853—9. Available from: https://doi.org/10.1016/j.jnutbio.2012.03.003.

[4] Pauwels S, Ghosh M, Duca RC, Bekaert B, Freson K, Huybrechts I, et al. Maternal intake of methyl-group donors affects DNA methylation of metabolic genes in infants. Clin Epigenetics 2017;9:16. Available from: https://doi.org/10.1186/s13148-017-0321-y.

[5] Fan G, Hutnick L. Methyl-CpG binding proteins in the nervous system. Cell Res 2005;15:255—61. Available from: https://doi.org/10.1038/sj.cr.7290294.

[6] Richa R, Sinha RP. Hydroxymethylation of DNA: an epigenetic marker. EXCLI J 2014;13:592−610.

[7] Cheung P, Allis CD, Sassone-Corsi P. Signaling to chromatin through histone modifications. Cell 2000;103:263−71.

[8] Jenuwein T, Allis CD. Translating the histone code. Science 2001;293:1074−80. Available from: https://doi.org/10.1126/science.1063127.

[9] Vakoc CR, Sachdeva MM, Wang H, Blobel GA. Profile of histone lysine methylation across transcribed mammalian chromatin. Mol Cell Biol 2006;26:9185−95. Available from: https://doi.org/10.1128/MCB.01529-06.

[10] Young MD, Willson TA, Wakefield MJ, Trounson E, Hilton DJ, Blewitt ME, et al. ChIP-seq analysis reveals distinct H3K27me3 profiles that correlate with transcriptional activity. Nucleic Acids Res 2011;39:7415−27. Available from: https://doi.org/10.1093/nar/gkr416.

[11] Sato F, Tsuchiya S, Meltzer SJ, Shimizu K. MicroRNAs and epigenetics. FEBS J 2011;278:1598−609. Available from: https://doi.org/10.1111/j.1742-4658.2011.08089.x.

[12] Statello L, Guo C-J, Chen L-L, Huarte M. Gene regulation by long non-coding RNAs and its biological functions. Nat Rev Mol Cell Biol 2021;22:96−118. Available from: https://doi.org/10.1038/s41580-020-00315-9.

[13] Molina-Serrano D, Schiza V, Kirmizis A. Cross-talk among epigenetic modifications: lessons from histone arginine methylation. Biochem Soc Trans 2013;41:751−9. Available from: https://doi.org/10.1042/BST20130003.

[14] Denis H, Ndlovu MN, Fuks F. Regulation of mammalian DNA methyltransferases: a route to new mechanisms. EMBO Rep 2011;12:647−56. Available from: https://doi.org/10.1038/embor.2011.110.

[15] Champagne FA. Epigenetic influence of social experiences across the lifespan. Dev Psychobiol 2010;52:299−311. Available from: https://doi.org/10.1002/dev.20436.

[16] Jablonka E, Lamb MJ. The changing concept of epigenetics. Ann N Y Acad Sci 2002;981:82−96.

[17] Champagne FA, Francis DD, Mar A, Meaney MJ. Variations in maternal care in the rat as a mediating influence for the effects of environment on development. Physiol Behav 2003;79:359−71.

[18] Monk C, Spicer J, Champagne FA. Linking prenatal maternal adversity to developmental outcomes in infants: the role of epigenetic pathways. Dev Psychopathol 2012;24:1361−76. Available from: https://doi.org/10.1017/S0954579412000764.

[19] Hane AA, Henderson HA, Reeb-Sutherland BC, Fox NA. Ordinary variations in human maternal caregiving in infancy and biobehavioral development in early childhood: a follow-up study. Dev Psychobiol 2010;52:558−67. Available from: https://doi.org/10.1002/dev.20461.

[20] Curley JP, Mashoodh R, Champagne FA. Epigenetics and the origins of paternal effects. Horm Behav 2011;59:306−14. Available from: https://doi.org/10.1016/j.yhbeh.2010.06.018.

[21] Colicchio JM, Herman J. Empirical patterns of environmental variation favor adaptive transgenerational plasticity. Ecol Evol 2020;10:1648−65. Available from: https://doi.org/10.1002/ece3.6022.

[22] Susser ES, Lin SP. Schizophrenia after prenatal exposure to the Dutch Hunger Winter of 1944-1945. Arch Gen Psychiatry 1992;49:983−8.

[23] Hulshoff Pol HE, Hoek HW, Susser E, Brown AS, Dingemans A, Schnack HG, et al. Prenatal exposure to famine and brain morphology in schizophrenia. Am J Psychiatry 2000;157:1170−2. Available from: https://doi.org/10.1176/appi.ajp.157.7.1170.

[24] Tobi EW, Goeman JJ, Monajemi R, Gu H, Putter H, Zhang Y, et al. DNA methylation signatures link prenatal famine exposure to growth and metabolism. Nat Commun 2014;5:5592. Available from: https://doi.org/10.1038/ncomms6592.

[25] Heijmans BT, Tobi EW, Stein AD, Putter H, Blauw GJ, Susser ES, et al. Persistent epigenetic differences associated with prenatal exposure to famine in humans. Proc Natl Acad Sci USA 2008;105:17046−9. Available from: https://doi.org/10.1073/pnas.0806560105.

[26] Tobi EW, Slieker RC, Luijk R, Dekkers KF, Stein AD, Xu KM, et al. DNA methylation as a mediator of the association between prenatal adversity and risk factors for metabolic disease in adulthood. Sci Adv 2018;4:eaao4364. Available from: https://doi.org/10.1126/sciadv.aao4364.

[27] Altobelli G, Bogdarina IG, Stupka E, Clark AJL, Langley-Evans S. Genome-wide methylation and gene expression changes in newborn rats following maternal protein restriction and reversal by folic acid. PLoS One 2013;8:e82989. Available from: https://doi.org/10.1371/journal.pone.0082989.

[28] Lillycrop KA, Phillips ES, Jackson AA, Hanson MA, Burdge GC. Dietary protein restriction of pregnant rats induces and folic acid supplementation prevents epigenetic modification of hepatic gene expression in the offspring. J Nutr 2005;135:1382−6.

[29] Lillycrop KA, Phillips ES, Torrens C, Hanson MA, Jackson AA, Burdge GC. Feeding pregnant rats a protein-restricted diet persistently alters the methylation of specific cytosines in the hepatic PPAR alpha promoter of the offspring. Br J Nutr 2008;100:278−82. Available from: https://doi.org/10.1017/S0007114507894438.

[30] Lillycrop KA, Slater-Jefferies JL, Hanson MA, Godfrey KM, Jackson AA, Burdge GC. Induction of altered epigenetic regulation of the hepatic glucocorticoid receptor in the offspring of rats fed a protein-restricted diet during pregnancy suggests that reduced DNA methyltransferase-1 expression is involved in impaired DNA methylation and changes in histone modifications. Br J Nutr 2007;97:1064−73. Available from: https://doi.org/10.1017/S000711450769196X.

[31] Kovacheva VP, Mellott TJ, Davison JM, Wagner N, Lopez-Coviella I, Schnitzler AC, et al. Gestational choline deficiency causes global and Igf2 gene DNA hypermethylation by upregulation of Dnmt1 expression. J Biol Chem 2007;282:31777−88. Available from: https://doi.org/10.1074/jbc.M705539200.

[32] Morgan HD, Sutherland HG, Martin DI, Whitelaw E. Epigenetic inheritance at the agouti locus in the mouse. Nat Genet 1999;23:314−18. Available from: https://doi.org/10.1038/15490.

[33] Waterland RA, Jirtle RL. Transposable elements: targets for early nutritional effects on epigenetic gene regulation. Mol Cell Biol 2003;23:5293−300.

[34] Suter MA, Ma J, Vuguin PM, Hartil K, Fiallo A, Harris RA, et al. In utero exposure to a maternal high-fat diet alters the epigenetic histone code in a murine model. Am J Obstet Gynecol 2014;210. Available from: https://doi.org/10.1016/j.ajog.2014.01.045 463.e1-463.e11.

[35] Reamon-Buettner SM, Buschmann J, Lewin G. Identifying placental epigenetic alterations in an intrauterine growth restriction (IUGR) rat model induced by gestational protein deficiency. Reprod Toxicol 2014;45:117−24. Available from: https://doi.org/10.1016/j.reprotox.2014.02.009.

[36] Gabory A, Ferry L, Fajardy I, Jouneau L, Gothié J-D, Vigé A, et al. Maternal diets trigger sex-specific divergent trajectories of gene expression and epigenetic systems in mouse placenta. PLoS One 2012;7:e47986. Available from: https://doi.org/10.1371/journal.pone.0047986.

[37] Casas-Agustench P, Fernandes FS, Tavares do Carmo MG, Visioli F, Herrera E, Dávalos A. Consumption of distinct dietary lipids during early pregnancy differentially modulates the expression of microRNAs in mothers and offspring. PLoS One 2015;10:e0117858. Available from: https://doi.org/10.1371/journal.pone.0117858.

[38] Onishchenko N, Karpova N, Sabri F, Castrén E, Ceccatelli S. Long-lasting depression-like behavior and epigenetic changes of BDNF gene expression induced by perinatal exposure to methylmercury. J Neurochem 2008;106:1378–87. Available from: https://doi.org/10.1111/j.1471-4159.2008.05484.x.

[39] Yan Z, Zhang H, Maher C, Arteaga-Solis E, Champagne FA, Wu L, et al. Prenatal polycyclic aromatic hydrocarbon, adiposity, peroxisome proliferator-activated receptor (PPAR) γ methylation in offspring, grand-offspring mice. PLoS One 2014;9:e110706. Available from: https://doi.org/10.1371/journal.pone.0110706.

[40] Miller RL, Yan Z, Maher C, Zhang H, Gudsnuk K, McDonald J, et al. Impact of prenatal polycyclic aromatic hydrocarbon exposure on behavior, cortical gene expression and DNA methylation of the Bdnf gene. Neuroepigenetics 2016;5:11–18. Available from: https://doi.org/10.1016/j.nepig.2016.02.001.

[41] Herbstman JB, Tang D, Zhu D, Qu L, Sjödin A, Li Z, et al. Prenatal exposure to polycyclic aromatic hydrocarbons, benzo[a]pyrene-DNA adducts, and genomic DNA methylation in cord blood. Environ Health Perspect 2012;120:733–8. Available from: https://doi.org/10.1289/ehp.1104056.

[42] Anway MD, Leathers C, Skinner MK. Endocrine disruptor vinclozolin induced epigenetic transgenerational adult-onset disease. Endocrinology 2006;147:5515–23. Available from: https://doi.org/10.1210/en.2006-0640.

[43] Anway MD. Epigenetic transgenerational actions of endocrine disruptors and male fertility. Science 2005;308:1466–9. Available from: https://doi.org/10.1126/science.1108190.

[44] Brieño-Enríquez MA, García-López J, Cárdenas DB, Guibert S, Cleroux E, Děd L, et al. Exposure to endocrine disruptor induces transgenerational epigenetic deregulation of microRNAs in primordial germ cells. PLoS One 2015;10:e0124296. Available from: https://doi.org/10.1371/journal.pone.0124296.

[45] Yaoi T, Itoh K, Nakamura K, Ogi H, Fujiwara Y, Fushiki S. Genome-wide analysis of epigenomic alterations in fetal mouse forebrain after exposure to low doses of bisphenol A. Biochem Biophys Res Commun 2008;376:563–7. Available from: https://doi.org/10.1016/j.bbrc.2008.09.028.

[46] Kundakovic M, Gudsnuk K, Herbstman JB, Tang D, Perera FP, Champagne FA. DNA methylation of BDNF as a biomarker of early-life adversity. Proc Natl Acad Sci USA 2015;112:6807–13. Available from: https://doi.org/10.1073/pnas.1408355111.

[47] Kundakovic M, Gudsnuk K, Franks B, Madrid J, Miller RL, Perera FP, et al. Sex-specific epigenetic disruption and behavioral changes following low-dose in utero bisphenol A exposure. Proc Natl Acad Sci USA 2013;110:9956–61. Available from: https://doi.org/10.1073/pnas.1214056110.

[48] Strakovsky RS, Wang H, Engeseth NJ, Flaws JA, Helferich WG, Pan Y-X, et al. Developmental bisphenol A (BPA) exposure leads to sex-specific modification of hepatic gene expression and epigenome at birth that may exacerbate high-fat diet-induced hepatic steatosis. Toxicol Appl Pharmacol 2015;284:101–12. Available from: https://doi.org/10.1016/j.taap.2015.02.021.

[49] Junge KM, Leppert B, Jahreis S, Wissenbach DK, Feltens R, Grützmann K, et al. MEST mediates the impact of prenatal bisphenol A exposure on long-term body weight development. Clin Epigenet 2018;10:58. Available from: https://doi.org/10.1186/s13148-018-0478-z.

[50] Choi Y-J, Lee YA, Hong Y-C, Cho J, Lee K-S, Shin CH, et al. Effect of prenatal bisphenol A exposure on early childhood body mass index through epigenetic influence on the insulin-like growth factor 2 receptor (IGF2R) gene. Environ Int 2020;143:105929. Available from: https://doi.org/10.1016/j.envint.2020.105929.

[51] Li S, Chen M, Li Y, Tollefsbol TO. Prenatal epigenetics diets play protective roles against environmental pollution. Clinical Epigenetics 2019;11:82. Available from: https://doi.org/10.1186/s13148-019-0659-4.

[52] Dolinoy DC, Huang D, Jirtle RL. Maternal nutrient supplementation counteracts bisphenol A-induced DNA hypomethylation in early development. Proc Natl Acad Sci USA 2007;104:13056–61. Available from: https://doi.org/10.1073/pnas.0703739104.

[53] Zhang B, Hong X, Ji H, Tang W-Y, Kimmel M, Ji Y, et al. Maternal smoking during pregnancy and cord blood DNA methylation: new insight on sex differences and effect modification by maternal folate levels. Epigenetics 2018;13:505–18. Available from: https://doi.org/10.1080/15592294.2018.1475978.

[54] Perera FP, Wheelock K, Wang Y, Tang D, Margolis AE, Badia G, et al. Combined effects of prenatal exposure to polycyclic aromatic hydrocarbons and material hardship on child ADHD behavior problems. Environ Res 2018;160:506–13. Available from: https://doi.org/10.1016/j.envres.2017.09.002.

[55] Nemoda Z, Massart R, Suderman M, Hallett M, Li T, Coote M, et al. Maternal depression is associated with DNA methylation changes in cord blood T lymphocytes and adult hippocampi. Transl Psychiatry 2015;5:e545. Available from: https://doi.org/10.1038/tp.2015.32.

[56] Oberlander TF, Weinberg J, Papsdorf M, Grunau R, Misri S, Devlin AM. Prenatal exposure to maternal depression, neonatal methylation of human glucocorticoid receptor gene (NR3C1) and infant cortisol stress responses. Epigenetics 2008;3:97–106.

[57] Unternaehrer E, Bolten M, Nast I, Staehli S, Meyer AH, Dempster E, et al. Maternal adversities during pregnancy and cord blood oxytocin receptor (OXTR) DNA methylation. Soc Cogn Affect Neurosci 2016;16:1460–1470. Available from: https://doi.org/10.1093/scan/nsw051

[58] Brown AS, Gyllenberg D, Malm H, McKeague IW, Hinkka-Yli-Salomäki S, Artama M, et al. Association of selective serotonin reuptake inhibitor exposure during pregnancy with speech, scholastic, and motor disorders in offspring. JAMA Psychiatry 2016;73:1163–70. Available from: https://doi.org/10.1001/jamapsychiatry.2016.2594.

[59] Cardenas A, Faleschini S, Cortes Hidalgo A, Rifas-Shiman SL, Baccarelli AA, DeMeo DL, et al. Prenatal maternal antidepressants, anxiety, and depression and offspring DNA methylation: epigenome-wide associations at birth and persistence into early childhood. Clinical Epigenetics 2019;11:56. Available from: https://doi.org/10.1186/s13148-019-0653-x.

[60] Mueller BR, Bale TL. Sex-specific programming of offspring emotionality after stress early in pregnancy. J Neurosci 2008;28:9055–65. Available from: https://doi.org/10.1523/JNEUROSCI.1424-08.2008.

[61] Dong E, Dzitoyeva SG, Matrisciano F, Tueting P, Grayson DR, Guidotti A. Brain-derived neurotrophic factor epigenetic modifications associated with schizophrenia-like phenotype induced by prenatal stress in mice. Biol Psychiatry 2015;77:589–96. Available from: https://doi.org/10.1016/j.biopsych.2014.08.012.

[62] Benoit JD, Rakic P, Frick KM. Prenatal stress induces spatial memory deficits and epigenetic changes in the hippocampus indicative of heterochromatin formation and reduced gene expression. Behav Brain Res 2015;281:1–8. Available from: https://doi.org/10.1016/j.bbr.2014.12.001.

[63] Monteleone MC, Adrover E, Pallarés ME, Antonelli MC, Frasch AC, Brocco MA. Prenatal stress changes the glycoprotein GPM6A gene expression and induces epigenetic changes in rat offspring brain. Epigenetics 2014;9:152–60. Available from: https://doi.org/10.4161/epi.25925.

[64] Schneider JS, Anderson DW, Kidd SK, Sobolewski M, Cory-Slechta DA. Sex-dependent effects of lead and prenatal stress on post-translational histone modifications in frontal cortex and hippocampus in the early postnatal brain. Neurotoxicology

[65] Appleton AA, Armstrong DA, Lesseur C, Lee J, Padbury JF, Lester BM, et al. Patterning in placental 11-B hydroxysteroid dehydrogenase methylation according to prenatal socioeconomic adversity. PLoS One 2013;8:e74691. Available from: https://doi.org/10.1371/journal.pone.0074691.

[66] Monk C, Feng T, Lee S, Krupska I, Champagne FA, Tycko B. Distress during pregnancy: epigenetic regulation of placenta glucocorticoid-related genes and fetal neurobehavior. Am J Psychiatry 2016;173:705−13. Available from: https://doi.org/10.1176/appi.ajp.2015.15091171.

[67] Jensen Peña C, Monk C, Champagne FA. Epigenetic effects of prenatal stress on 11β-hydroxysteroid dehydrogenase-2 in the placenta and fetal brain. PLoS One 2012;7:e39791. Available from: https://doi.org/10.1371/journal.pone.0039791.

[68] Howerton CL, Morgan CP, Fischer DB, Bale TL. O-GlcNAc transferase (OGT) as a placental biomarker of maternal stress and reprogramming of CNS gene transcription in development. Proc Natl Acad Sci USA 2013;110:5169−74. Available from: https://doi.org/10.1073/pnas.1300065110.

[69] Curley JP, Champagne FA. Influence of maternal care on the developing brain: mechanisms, temporal dynamics and sensitive periods. Front Neuroendocrinol 2016;40:52−66. Available from: https://doi.org/10.1016/j.yfrne.2015.11.001.

[70] Weaver ICG, Cervoni N, Champagne FA, D'Alessio AC, Sharma S, Seckl JR, et al. Epigenetic programming by maternal behavior. Nat Neurosci 2004;7:847−54. Available from: https://doi.org/10.1038/nn1276.

[71] Zhang T-Y, Hellstrom IC, Bagot RC, Wen X, Diorio J, Meaney MJ. Maternal care and DNA methylation of a glutamic acid decarboxylase 1 promoter in rat hippocampus. J Neurosci 2010;30:13130−7. Available from: https://doi.org/10.1523/JNEUROSCI.1039-10.2010.

[72] Bagot RC, Zhang T-Y, Wen X, Nguyen TTT, Nguyen H-B, Diorio J, et al. Variations in postnatal maternal care and the epigenetic regulation of metabotropic glutamate receptor 1 expression and hippocampal function in the rat. Proc Natl Acad Sci USA 2012;109(Suppl 2):17200−7. Available from: https://doi.org/10.1073/pnas.1204599109.

[73] Champagne FA, Weaver ICG, Diorio J, Dymov S, Szyf M, Meaney MJ. Maternal care associated with methylation of the estrogen receptor-alpha1b promoter and estrogen receptor-alpha expression in the medial preoptic area of female offspring. Endocrinology 2006;147:2909−15. Available from: https://doi.org/10.1210/en.2005-1119.

[74] Peña CJ, Neugut YD, Champagne FA. Developmental timing of the effects of maternal care on gene expression and epigenetic regulation of hormone receptor levels in female rats. Endocrinology 2013;154:4340−51. Available from: https://doi.org/10.1210/en.2013-1595.

[75] Weaver ICG, Champagne FA, Brown SE, Dymov S, Sharma S, Meaney MJ, et al. Reversal of maternal programming of stress responses in adult offspring through methyl supplementation: altering epigenetic marking later in life. J Neurosci 2005;25:11045−54. Available from: https://doi.org/10.1523/JNEUROSCI.3652-05.2005.

[76] Suri D, Veenit V, Sarkar A, Thiagarajan D, Kumar A, Nestler EJ, et al. Early stress evokes age-dependent biphasic changes in hippocampal neurogenesis, BDNF expression, and cognition. Biol Psychiatry 2013;73:658−66. Available from: https://doi.org/10.1016/j.biopsych.2012.10.023.

[77] Roth TL, Lubin FD, Funk AJ, Sweatt JD. Lasting epigenetic influence of early-life adversity on the BDNF gene. Biol Psychiatry 2009;65:760−9. Available from: https://doi.org/10.1016/j.biopsych.2008.11.028.

[78] Branchi I, Karpova NN, D'Andrea I, Castrén E, Alleva E. Epigenetic modifications induced by early enrichment are associated with changes in timing of induction of BDNF expression. Neurosci Lett 2011;495:168−72. Available from: https://doi.org/10.1016/j.neulet.2011.03.038.

[79] Suderman M, McGowan PO, Sasaki A, Huang TCT, Hallett MT, Meaney MJ, et al. Conserved epigenetic sensitivity to early life experience in the rat and human hippocampus. Proc Natl Acad Sci USA 2012;109(Suppl 2):17266−72. Available from: https://doi.org/10.1073/pnas.1121260109.

[80] McGowan PO, Sasaki A, D'Alessio AC, Dymov S, Labonté B, Szyf M, et al. Epigenetic regulation of the glucocorticoid receptor in human brain associates with childhood abuse. Nat Neurosci 2009;12:342−8. Available from: https://doi.org/10.1038/nn.2270.

[81] Moore SR, McEwen LM, Quirt J, Morin A, Mah SM, Barr RG, et al. Epigenetic correlates of neonatal contact in humans. Dev Psychopathol 2017;29:1517−38. Available from: https://doi.org/10.1017/S0954579417001213.

[82] Hartwig FP, Davey Smith G, Simpkin AJ, Victora CG, Relton CL, Caramaschi D. Association between breastfeeding and DNA methylation over the life course: findings from the avon longitudinal study of parents and children (ALSPAC). Nutrients 2020;12:3309. Available from: https://doi.org/10.3390/nu12113309.

[83] Hellstrom IC, Dhir SK, Diorio JC, Meaney MJ. Maternal licking regulates hippocampal glucocorticoid receptor transcription through a thyroid hormone-serotonin-NGFI-A signalling cascade. Philos Trans R Soc Lond, B, Biol Sci 2012;367:2495−510. Available from: https://doi.org/10.1098/rstb.2012.0223.

[84] Abel E. Paternal contribution to fetal alcohol syndrome. Addict Biol 2004;9:127−33. Available from: https://doi.org/10.1080/13556210410001716980 discussion 135-136.

[85] Meek LR, Myren K, Sturm J, Burau D. Acute paternal alcohol use affects offspring development and adult behavior. Physiol Behav 2007;91:154−60. Available from: https://doi.org/10.1016/j.physbeh.2007.02.004.

[86] Ledig M, Misslin R, Vogel E, Holownia A, Copin JC, Tholey G. Paternal alcohol exposure: developmental and behavioral effects on the offspring of rats. Neuropharmacology 1998;37:57−66.

[87] Abel EL, Moore C, Waselewsky D, Zajac C, Russell LD. Effects of cocaine hydrochloride on reproductive function and sexual behavior of male rats and on the behavior of their offspring. J Androl 1989;10:17−27.

[88] He F, Lidow IA, Lidow MS. Consequences of paternal cocaine exposure in mice. Neurotoxicol Teratol 2006;28:198−209. Available from: https://doi.org/10.1016/j.ntt.2005.12.003.

[89] Levin ED, Hawkey AB, Hall BJ, Cauley M, Slade S, Yazdani E, et al. Paternal THC exposure in rats causes long-lasting neurobehavioral effects in the offspring. Neurotoxicol Teratol 2019;74:106806. Available from: https://doi.org/10.1016/j.ntt.2019.04.003.

[90] Anderson LM, Riffle L, Wilson R, Travlos GS, Lubomirski MS, Alvord WG. Preconceptional fasting of fathers alters serum glucose in offspring of mice. Nutrition 2006;22:327−31. Available from: https://doi.org/10.1016/j.nut.2005.09.006.

[91] Govic A, Penman J, Tammer AH, Paolini AG. Paternal calorie restriction prior to conception alters anxiety-like behavior of the adult rat progeny. Psychoneuroendocrinology 2016;64:1−11. Available from: https://doi.org/10.1016/j.psyneuen.2015.10.020.

[92] Dietz DM, Laplant Q, Watts EL, Hodes GE, Russo SJ, Feng J, et al. Paternal transmission of stress-induced pathologies. Biol Psychiatry 2011;70:408−14. Available from: https://doi.org/10.1016/j.biopsych.2011.05.005.

REFERENCES

[93] Hoyer C, Richter H, Brandwein C, Riva MA, Gass P. Preconceptional paternal exposure to a single traumatic event affects postnatal growth of female but not male offspring. Neuroreport 2013;24:856–60. Available from: https://doi.org/10.1097/WNR.0000000000000007.

[94] Kinnally EL, Capitanio JP. Paternal early experiences influence infant development through non-social mechanisms in Rhesus Macaques. Front Zool 2015;12:S14. Available from: https://doi.org/10.1186/1742-9994-12-S1-S14.

[95] Saavedra-Rodríguez L, Feig LA. Chronic social instability induces anxiety and defective social interactions across generations. Biol Psychiatry 2013;73:44–53. Available from: https://doi.org/10.1016/j.biopsych.2012.06.035.

[96] Rodgers AB, Morgan CP, Bronson SL, Revello S, Bale TL. Paternal stress exposure alters sperm microRNA content and reprograms offspring HPA stress axis regulation. J Neurosci 2013;33:9003–12. Available from: https://doi.org/10.1523/JNEUROSCI.0914-13.2013.

[97] Yehuda R, Daskalakis NP, Lehrner A, Desarnaud F, Bader HN, Makotkine I, et al. Influences of maternal and paternal PTSD on epigenetic regulation of the glucocorticoid receptor gene in Holocaust survivor offspring. Am J Psychiatry 2014;171:872–80. Available from: https://doi.org/10.1176/appi.ajp.2014.13121571.

[98] McGrath JJ, Petersen L, Agerbo E, Mors O, Mortensen PB, Pedersen CB. A comprehensive assessment of parental age and psychiatric disorders. JAMA Psychiatry 2014;71:301–9. Available from: https://doi.org/10.1001/jamapsychiatry.2013.4081.

[99] Janecka M, Manduca A, Servadio M, Trezza V, Smith R, Mill J, et al. Effects of advanced paternal age on trajectories of social behavior in offspring. Genes Brain Behav 2015;14:443–53. Available from: https://doi.org/10.1111/gbb.12227.

[100] García-Palomares S, Pertusa JF, Miñarro J, García-Pérez MA, Hermenegildo C, Rausell F, et al. Long-term effects of delayed fatherhood in mice on postnatal development and behavioral traits of offspring. Biol Reprod 2009;80:337–42. Available from: https://doi.org/10.1095/biolreprod.108.072066.

[101] García-Palomares S, Navarro S, Pertusa JF, Hermenegildo C, García-Pérez MA, Rausell F, et al. Delayed fatherhood in mice decreases reproductive fitness and longevity of offspring. Biol Reprod 2009;80:343–9. Available from: https://doi.org/10.1095/biolreprod.108.073395.

[102] Jenkins TG, Carrell DT. The sperm epigenome and potential implications for the developing embryo. Reproduction 2012;143:727–34. Available from: https://doi.org/10.1530/REP-11-0450.

[103] Ben Maamar M, Nilsson EE, Skinner MK. Epigenetic transgenerational inheritance, gametogenesis and germline development. Biol Reprod 2021;105(3):570–92. Available from: https://doi.org/10.1093/biolre/ioab085.

[104] Carrell DT, Hammoud SS. The human sperm epigenome and its potential role in embryonic development. Mol Hum Reprod 2010;16:37–47. Available from: https://doi.org/10.1093/molehr/gap090.

[105] Novikova SI, He F, Bai J, Cutrufello NJ, Lidow MS, Undieh AS. Maternal cocaine administration in mice alters DNA methylation and gene expression in hippocampal neurons of neonatal and prepubertal offspring. PLoS One 2008;3:e1919. Available from: https://doi.org/10.1371/journal.pone.0001919.

[106] Ouko LA, Shantikumar K, Knezovich J, Haycock P, Schnugh DJ, Ramsay M. Effect of alcohol consumption on CpG methylation in the differentially methylated regions of H19 and IG-DMR in male gametes: implications for fetal alcohol spectrum disorders. Alcohol Clin Exp Res 2006;33:1615–27. Available from: https://doi.org/10.1111/j.1530-0277.2009.00993.x.

[107] Bielawski DM, Zaher FM, Svinarich DM, Abel EL. Paternal alcohol exposure affects sperm cytosine methyltransferase messenger RNA levels. Alcohol Clin Exp Res 2002;26:347–51.

[108] Vassoler FM, White SL, Schmidt HD, Sadri-Vakili G, Pierce RC. Epigenetic inheritance of a cocaine-resistance phenotype. Nat Neurosci 2013;16:42–7. Available from: https://doi.org/10.1038/nn.3280.

[109] Milekic MH, Xin Y, O'Donnell A, Kumar KK, Bradley-Moore M, Malaspina D, et al. Age-related sperm DNA methylation changes are transmitted to offspring and associated with abnormal behavior and dysregulated gene expression. Mol Psychiatry 2015;20:995–1001. Available from: https://doi.org/10.1038/mp.2014.84.

[110] Oakes CC, Smiraglia DJ, Plass C, Trasler JM, Robaire B. Aging results in hypermethylation of ribosomal DNA in sperm and liver of male rats. Proc Natl Acad Sci USA 2003;100:1775–80. Available from: https://doi.org/10.1073/pnas.0437971100.

[111] Gapp K, Jawaid A, Sarkies P, Bohacek J, Pelczar P, Prados J, et al. Implication of sperm RNAs in transgenerational inheritance of the effects of early trauma in mice. Nat Neurosci 2014;17:667–9. Available from: https://doi.org/10.1038/nn.3695.

[112] Franklin TB, Russig H, Weiss IC, Gräff J, Linder N, Michalon A, et al. Epigenetic transmission of the impact of early stress across generations. Biol Psychiatry 2010;68:408–15. Available from: https://doi.org/10.1016/j.biopsych.2010.05.036.

[113] Dias BG, Ressler KJ. Parental olfactory experience influences behavior and neural structure in subsequent generations. Nat Neurosci 2014;17:89–96. Available from: https://doi.org/10.1038/nn.3594.

[114] Rodgers AB, Morgan CP, Leu NA, Bale TL. Transgenerational epigenetic programming via sperm microRNA recapitulates effects of paternal stress. Proc Natl Acad Sci USA 2015;112:13699–704. Available from: https://doi.org/10.1073/pnas.1508347112.

[115] Skinner MK. What is an epigenetic transgenerational phenotype? F3 or F2. Reprod Toxicol 2008;25(2–6). Available from: https://doi.org/10.1016/j.reprotox.2007.09.001.

[116] Champagne FA. Epigenetic legacy of parental experiences: dynamic and interactive pathways to inheritance. Dev Psychopathol 2016;28:1219–28. Available from: https://doi.org/10.1017/S0954579416000808.

[117] Pembrey ME, Bygren LO, Kaati G, Edvinsson S, Northstone K, Sjöström M, et al. Sex-specific, male-line transgenerational responses in humans. Eur J Hum Genet 2006;14:159–66. Available from: https://doi.org/10.1038/sj.ejhg.5201538.

[118] Zambrano E, Martínez-Samayoa PM, Bautista CJ, Deás M, Guillén L, Rodríguez-González GL, et al. Sex differences in transgenerational alterations of growth and metabolism in progeny (F2) of female offspring (F1) of rats fed a low protein diet during pregnancy and lactation. J Physiol (Lond) 2005;566:225–36. Available from: https://doi.org/10.1113/jphysiol.2005.086462.

[119] Dunn GA, Bale TL. Maternal high-fat diet promotes body length increases and insulin insensitivity in second-generation mice. Endocrinology 2009;150:4999–5009. Available from: https://doi.org/10.1210/en.2009-0500.

[120] Dunn GA, Bale TL. Maternal high-fat diet effects on third-generation female body size via the paternal lineage. Endocrinology 2011;152:2228–36. Available from: https://doi.org/10.1210/en.2010-1461.

[121] Chen TH-H, Chiu Y-H, Boucher BJ. Transgenerational effects of betel-quid chewing on the development of the metabolic syndrome in the Keelung Community-based Integrated Screening Program. Am J Clin Nutr 2006;83:688–92.

[122] Boucher BJ, Ewen SW, Stowers JM. Betel nut (Areca catechu) consumption and the induction of glucose intolerance in adult CD1 mice and in their F1 and F2 offspring. Diabetologia 1994;37:49–55.

[123] Drake AJ, Walker BR, Seckl JR. Intergenerational consequences of fetal programming by in utero exposure to glucocorticoids in rats. Am J Physiol Regul Integr Comp Physiol 2005;288:R34–8. Available from: https://doi.org/10.1152/ajpregu.00106.2004.

[124] Skinner MK, Anway MD, Savenkova MI, Gore AC, Crews D. Transgenerational epigenetic programming of the brain transcriptome and anxiety behavior. PLoS One 2008;3:e3745. Available from: https://doi.org/10.1371/journal.pone.0003745.

[125] Crews D, Gore AC, Hsu TS, Dangleben NL, Spinetta M, Schallert T, et al. Transgenerational epigenetic imprints on mate preference. Proc Natl Acad Sci USA 2007;104:5942–6. Available from: https://doi.org/10.1073/pnas.0610410104.

[126] Manikkam M, Tracey R, Guerrero-Bosagna C, Skinner MK. Plastics derived endocrine disruptors (BPA, DEHP and DBP) induce epigenetic transgenerational inheritance of obesity, reproductive disease and sperm epimutations. PLoS One 2013;8:e55387. Available from: https://doi.org/10.1371/journal.pone.0055387.

[127] Ferguson-Smith AC. Genomic imprinting: the emergence of an epigenetic paradigm. Nat Rev Genet 2011;12:565–75. Available from: https://doi.org/10.1038/nrg3032.

[128] Benoit D, Parker KC. Stability and transmission of attachment across three generations. Child Dev 1994;65:1444–56.

[129] Miller L, Kramer R, Warner V, Wickramaratne P, Weissman M. Intergenerational transmission of parental bonding among women. J Am Acad Child Adolesc Psychiatry 1997;36:1134–9. Available from: https://doi.org/10.1097/00004583-199708000-00022.

[130] Maestripieri D. Early experience affects the intergenerational transmission of infant abuse in rhesus monkeys. Proc Natl Acad Sci USA 2005;102:9726–9. Available from: https://doi.org/10.1073/pnas.0504122102.

[131] Maestripieri D, Wallen K, Carroll KA. Infant abuse runs in families of group-living pigtail macaques. Child Abuse Negl 1997;21:465–71.

[132] Champagne FA, Meaney MJ. Transgenerational effects of social environment on variations in maternal care and behavioral response to novelty. Behav Neurosci 2007;121:1353–63. Available from: https://doi.org/10.1037/0735-7044.121.6.1353.

[133] Francis D, Diorio J, Liu D, Meaney MJ. Nongenomic transmission across generations of maternal behavior and stress responses in the rat. Science 1999;286:1155–8.

[134] Curley JP, Davidson S, Bateson P, Champagne FA. Social enrichment during postnatal development induces transgenerational effects on emotional and reproductive behavior in mice. Front Behav Neurosci 2009;3:25. Available from: https://doi.org/10.3389/neuro.08.025.2009.

[135] Curley JP, Champagne FA, Bateson P, Keverne EB. Transgenerational effects of impaired maternal care on behaviour of offspring and grandoffspring. Anim Behav 2008;75:1551–61. Available from: https://doi.org/10.1016/j.anbehav.2007.10.008.

[136] Champagne FA, Meaney MJ. Stress during gestation alters postpartum maternal care and the development of the offspring in a rodent model. Biol Psychiatry 2006;59:1227–35. Available from: https://doi.org/10.1016/j.biopsych.2005.10.016.

[137] Meaney. Maternal care, gene expression, and the transmission of individual differences in stress reactivity across generations. Annu Rev Neurosci 2001;24:1161–92. Available from: https://doi.org/10.1146/annurev.neuro.24.1.1161.

[138] Champagne FA, Weaver ICG, Diorio J, Sharma S, Meaney MJ. Natural variations in maternal care are associated with estrogen receptor alpha expression and estrogen sensitivity in the medial preoptic area. Endocrinology 2003;144:4720–4. Available from: https://doi.org/10.1210/en.2003-0564.

[139] McMillen IC, Robinson JS. Developmental origins of the metabolic syndrome: prediction, plasticity, and programming. Physiol Rev 2005;85:571–633. Available from: https://doi.org/10.1152/physrev.00053.2003.

[140] Wells JCK. The thrifty phenotype: an adaptation in growth or metabolism? Am J Hum Biol 2011;23:65–75. Available from: https://doi.org/10.1002/ajhb.21100.

[141] Cameron NM, Champagne FA, Parent C, Fish EW, Ozaki-Kuroda K, Meaney MJ. The programming of individual differences in defensive responses and reproductive strategies in the rat through variations in maternal care. Neurosci Biobehav Rev 2005;29:843–65. Available from: https://doi.org/10.1016/j.neubiorev.2005.03.022.

[142] Liu D, Diorio J, Day JC, Francis DD, Meaney MJ. Maternal care, hippocampal synaptogenesis and cognitive development in rats. Nat Neurosci 2000;3:799–806. Available from: https://doi.org/10.1038/77702.

[143] Champagne DL, Bagot RC, van Hasselt F, Ramakers G, Meaney MJ, de Kloet ER, et al. Maternal care and hippocampal plasticity: evidence for experience-dependent structural plasticity, altered synaptic functioning, and differential responsiveness to glucocorticoids and stress. J Neurosci 2008;28:6037–45. Available from: https://doi.org/10.1523/JNEUROSCI.0526-08.2008.

[144] Lamarck J-B. *Philosophie zoologique*. Musée d'Histoire Naturelle; 1809.

[145] Laland KN, Uller T, Feldman MW, Sterelny K, Müller GB, Moczek A, et al. The extended evolutionary synthesis: its structure, assumptions and predictions. Proc Biol Sci 2015;282:20151019. Available from: https://doi.org/10.1098/rspb.2015.1019.

[146] Zeybel M, Hardy T, Wong YK, Mathers JC, Fox CR, Gackowska A, et al. Multigenerational epigenetic adaptation of the hepatic wound-healing response. Nat Med 2012;18:1369–77. Available from: https://doi.org/10.1038/nm.2893.

[147] Arai JA, Li S, Hartley DM, Feig LA. Transgenerational rescue of a genetic defect in long-term potentiation and memory formation by juvenile enrichment. J Neurosci 2009;29:1496–502. Available from: https://doi.org/10.1523/JNEUROSCI.5057-08.2009.

[148] Danchin É, Charmantier A, Champagne FA, Mesoudi A, Pujol B, Blanchet S. Beyond DNA: integrating inclusive inheritance into an extended theory of evolution. Nat Rev Genet 2011;12:475–86. Available from: https://doi.org/10.1038/nrg3028.

CHAPTER 25

DNA Methylation Clocks in Age-related Disease

Peter D. Fransquet, Jo Wrigglesworth and Joanne Ryan
School of Public Health and Preventive Medicine, Monash University, Melbourne, VIC, Australia

OUTLINE

Aging	479	Cancer	485
Biomarkers of Aging	480	Dementia	487
DNA Methylation Changes with Age	480	Cardiovascular Disease	487
Epigenetic Clocks	480	Type 2 Diabetes	488
		Frailty and Physical Disability	488
First Generation Epigenetic Clocks	480	Epigenetic Clocks Without an Age-related Disease Focus	488
Initial Studies	480		
Horvath's Epigenetic Clock	483	Can Epigenetic Age be Modified?	489
Hannum's Epigenetic Clock	483	Biology of Epigenetic Age	489
Performance Comparisons	484	Genetics of Epigenetic Clocks	490
Second Generation Epigenetic Clocks	484	Links with Lifestyle and Health Factors	490
PhenoAge	484	Summary	491
GrimAge	485	References	491
Epigenetic Clocks and Age-related Disease	485		

AGING

The population is aging globally. Advances in medicine and healthcare have led to declining mortality rates earlier in life, and a major consequence of this is the rapidly aging population. Over nearly two centuries, life expectancy has been steadily increasing around the world [1], and predictions indicate this will continue to rise [2]. Aging is characterized by a progressive loss of physiological integrity and subsequent functional impairment, and eventually death [3]. There is a raft of molecular changes that occur within the cells and tissues, which includes cellular decay, damage to organ structures, shortening and alternations of telomeres, epigenetic drift and changes in gene expression. Aging represents the accumulation of biological changes that depends on an individual's genetic background and the influence of different exposures throughout their lifetime. Some individuals appear to evade major physiological deterioration until very late in life, for others, physiological decline manifests in late middle age. Chronological age (time since birth) is therefore an imprecise marker of these biological processes, and thus not a perfect measure of an individual's functional capacity. The considerable inter-individual variation in the rate of aging has spurned interest in identifying exposures that regulate this process; determine what drives longevity, and how this could predict the length of an individual's lifespan. Individual differences in aging are also thought to influence an individual's risk of many diseases [4]. There are numerous chronic conditions which are strongly linked to the aging process, becoming more prevalent with advanced age. This includes Alzheimer disease (AD) and related dementias [5], frailty [6] and physical disability [7], cancer [8,9], type 2 diabetes [10,11], and cardiovascular disease (CVD) [12]. Given their substantial societal and economic costs, the shift in an aging population is thus creating a global challenge [13]. This has driven interest to better our understanding of aging, and to identify robust markers of its biological process. New insights into the underlying mechanisms of aging would further increase our understanding of why some individuals can remain physically and cognitively robust as they age, and in developing new options for targeted interventions to slow aging and reduce the risk of age-related disease.

Biomarkers of Aging

An ideal biomarker of aging is a biological parameter that reflects an underlying aging process, and can predict later functional capability and lifespan better than chronological age [14]. Various biomarkers have been proposed, which capture different biological aspects of aging, and enable an accurate estimate of "age" across a range of tissues and at different stages of the life. These have also permitted the calculation of so-called age acceleration, which is the difference between an individual's estimated biological age and their (actual) chronological age. Individuals whose biological age is substantially older than their chronical age are said to have accelerated aging trajectory (greater physiological deterioration than expected for their age), while those with younger biological age compared to their chronological age, have negative age acceleration (relative preservation of physiological function). Well accepted biomarkers of aging include inflammation, immunosenescence (cellular senescence), genomic instability, somatic mutations, mitochondrial dysfunction, and telomere length/attrition. These all appear to be pivotal aspects of aging and are associated with a number of age-related diseases, conditions, and mortality [3]. Indeed, a recent development includes "iAge," a marker of aging developed from a collection of inflammatory measures, which are associated with age-related outcomes such as cardiovascular aging, frailty, immunosenescence, and multimorbidity [15]. Thus far however, one of the most promising and extensively investigated biomarkers of aging are based on DNA methylation.

DNA Methylation Changes with Age

Epigenetic modifications regulate normal cellular development and differentiation and are necessary for the long-term regulation of gene function. They may also be a fundamental mechanism that drives human aging [16]. Epigenetic patterns change with age; with a systematic increase in DNA methylation over the first year of life [17], which then remain relatively stable before a global decline in DNA methylation after reaching adulthood [18]. In contrast to this global decline in methylation, there is also an increase in intra- and interindividual variability in DNA methylation with age, known as epigenetic drift [19], and hypermethylation in specific gene regions associated with aging [20]. Genes involved include tumor-suppressors and other important regulators of growth, differentiation, angiogenesis, and DNA-repair genes. These age-related methylation changes could promote the increase in age-related disease. Epigenetic changes thus appear to be markers of the overall health and lifespan of the cell [21] and are considered one of the hallmarks of aging [3].

EPIGENETIC CLOCKS

A number of studies have characterized epigenetic changes that result from aging, and demonstrated that methylation patterns at CpGs in various tissues are predictors of chronological age. This has led to the development of DNA methylation algorithms of age which are considered among the most promising biomarkers of aging [4]. To be consistent with the most common terminology, we will refer to these as "epigenetic clocks," with their measure being referred to as "epigenetic age." The epigenetic clocks are not only highly correlated with chronological age, but they have permitted the identification of individuals whose epigenetic age and chronological age is discordant. Individuals with an older epigenetic age than expected compared to the population with the same chronological age are said to exhibit accelerated epigenetic aging, commonly termed "age acceleration." This measure is considered more informative than absolute epigenetic age, and is superior in its association with mortality and longevity [22,23]. Age acceleration has also been associated with a number of age-related diseases. It has been observed however, that age acceleration slows down or may even reverse in very old individuals, such that epigenetic age estimates may show a poorer correlation with chronological age in these individuals [24,25].

A vast and increasing number of epigenetic clocks have been developed in the last eight years which have spurned great interest in the field. An overview of just some of these epigenetic clocks is presented in Table 25.1.

FIRST GENERATION EPIGENETIC CLOCKS

Initial Studies

The first epigenome-wide study for correlations between DNA methylation at particular CpGs and

TABLE 25.1 Some of the Most Commonly Use DNA Methylation Clocks to Predict Aging and Risk of Age-Related Disease

Ref	Name	Year	Platform	Tissue	Estimated Against	Method	Characteristics of Training Sample N. Age range/mean ± SD, Ethnicity	CpGs	Correlation (R), MAD/MAE/MeAD (if stated)
[26]	Bocklandt	2011	Illumina 27 K	Saliva	Chronological age	Penalized regression	N = 66 male twins, unrelated males and females, 18–70 years, ethnicity NS	3	R = 0.83 MeAD = 5.2 years
[27]	Koch-Wagner	2011	Illumina 27 K	Multiple	Chronological age	Linear regression	N = 130 (3 datasets), 16–72 years, ethnicity NS	5	R^2 = 0.68 MeAD = 11 years
[28]	Garagnani	2012	Illumina 450 K	Blood	Chronological age	Spearman correlation	N = 64 (32 mother [42–83 years] and offspring [9–52 years]), ethnicity NS	3	R = 0.92, 0.80, 0.63
[29]	Hannum	2013	Illumina 450 K	Blood	Chronological age	Penalized regression	N = 482, 19–101 years, European Caucasian, Hispanic	71	R = 0.96
[30]	Horvath	2013	Illumina 27 K/450 K	Multiple (51 tissues)	Chronological age	Penalized regression	N = 3931 (39 cohorts), 43 ± 25 years Various ethnicities	353	R = 0.97 MAD = 3.6 years
[31]	Florath	2014	Illumina 450 K	Blood	Chronological age	Linear regression	N = 249, 50–75 years. Predominantly German nationals	17	R = 0.84, MAD = 2.6 years
[32]	Weidner	2014	Bisulfite pyrosequencing	Blood	Chronological age	Linear regression	N = 82 (two cohorts), 0–78 years ethnicity NS	3	MAD = 5.4 years
[33]	Zbiec-Piekarska	2015	Bisulfite pyrosequencing	Blood	Chronological age	Linear regression	N = 300, 2–75 years, ethnicity NS	5	R^2 = 0.94, MAD = 4.5 years
[34]	Bohlin	2016	Illumina 450 K	Cord blood	Gestational age	Penalized regression	N = 1068 newborns, 279.6 and 282.3 days, Norwegian nationals	96, 58	R^2 = 0.66, 0.55 respectively
[35]	EpiTOC	2016	Illumina 450 K	Blood	Cancer risk (total no. cell divisions)	Penalized Regression	N = 656, 19–101, European Caucasian, Hispanic	385	Not stated
[36]	Knight	2016	Illumina 27 K/450 K	Cord blood/blood spots	Gestational Age	Penalized Regression	N = 207, 24–42 weeks, Caucasian, African American	148	R = 0.99 MAD = 1.24 weeks
[37]	Vidal-Bralo	2016	Illumina 27 K (training)	Blood	Chronological age	Multiple regression	N = 390, 20–78 years, Caucasian	8	R^2 = 0.60 MAD = 7.3 years
[38]	Hong	2017	Illumina 450 K	Saliva	Chronological age	Linear regression	N = 113, 19–64 years. From South Korea but ethnicity NS	7	R = 0.945, MAD = 3.13 years
[39]	Zhang (2017)	2017	Illumina 450 K	Blood	All-Cause mortality	Weighted Cox regression	N = 954, 50–75 years. Participants from Germany, but ethnicity NS	58	Not stated
[40]	MiAge	2018	Illumina 450 K	8 cancer types	Mitotic age (total number of cell divisions)	Linear regression	N = 4020 tumor and normal tissue, age, and ethnicity NS	286	Not stated
[41]	PhenoAge	2018	Illumina 450 K	Blood	Lifespan/health span	Penalized regression	N = 456, 21–91 years, ethnicity NS	513	Various

(Continued)

TABLE 25.1 (Continued)

Ref	Name	Year Platform	Tissue	Estimated Against	Method	Characteristics of Training Sample N. Age range/mean ± SD, Ethnicity	CpGs	Correlation (R), MAD/MAE/MeAD (if stated)
[42]	Skin & Blood	2018 Illumina 450 K/EPIC	Multiple	Chronological age	Penalized regression	N = 896 (10 datasets), 0–94 years, ethnicity NS	391	Various
[43]	DNAmTL	2019 Illumina 450 K/EPIC	Blood	Telomere length	Penalized regression	N = 2256 (2 cohorts), 22–93 years, European, African ancestry	140	Various
[44]	CancerClock	2019 Illumina 450 K	Cancer tissues	Chronological Age	Penalized Regression	N = 704, 14–89 years, ethnicity NS	282	R = 0.94
[45]	GrimAge	2019 Illumina 450 K/EPIC	Blood	Time to death	Penalized Cox regression	N = 1731, 66 ± 8.6 years, various ethnicities from multiple cohorts	1030	R = 0.82
[46]	Falick Michaeli	2019 RRBS	Cord blood and placenta	Gestational age	Linear regression	N = 10 newborns, 29–40 weeks, Israelian nationals	743 and 275	R = 0.77, -0.55 respectively
[47]	Jung	2019 Multiplex Methylation SNaPshot	Blood, saliva, buccal	Chronological age	Linear regression	N = 300, 18–74 years Korean	5	R = 0.94 MAD = 3.8 years
[48]	Lee (2019)	2019 Illumina 450 K/EPIC	Placental tissues	Gestational age	Penalized regression model	N = 1102 and N = 963 from 14 datasets, 5–42 weeks; various ethnicities from multiple cohorts	546–558	R = 0.99 MAE = 0.96
[49]	Zhang (2019)	2019 Illumina 450 K/EPIC	Blood and saliva	Chronological age	Penalized regression	N = 13,566 (13 cohorts), 2–104 years, various ethnicities	514	R = 0.99
[50]	DunedinPoAm	2020 Illumina 450 K	Blood	Pace of aging (biological systems)	Penalized regression	N = 810, 38 years, predominantly Caucasian	46	R = 0.56
[51]	Lee (2020)	2020 Illumina 450 K/EPIC	Blood	Chronological age, three different cohort clocks	Penalized regression	N = 1592, 19–49 years and N = 2227, 18–88 years. From Norway and publicly available data. Ethnicity NS	1695, 1791, 1892	Training R = 0.999 MAD = 0.14 years
[52]	Meat	2020 Illumina 27 K/450 K/EPIC	Skeletal muscle	Chronological age	Penalized regression	N = 682, 18–89 years, Caucasian (99%)	200	Population R = 0.62 MAD = 7 years

MAD, mean absolute difference between predicted and observed age; *MAE*, mean absolute error; *MeAD*, mean absolute difference; *NS*, not stated; *RRBS*, reduced representation bisulfite sequencing; *SD*, standard deviation.

chronological age was carried out by Rakyan et al. in 2010 [53]. Whole blood from twin pair and singleton women (n = 93), aged 49–75 years, was used to derive a 360 CpG signature of aging, with 213 CpGs positively correlated and 147 negatively correlated with age. A majority (but not all) of these correlations were replicated in a separate sample aged between 20–69 (n = 25, now including five men), in specific blood cells (CD14+ monocytes and CD4+ T-cells) rather than whole blood. This finding provided evidence for both cell intrinsic and extrinsic methylation patterns of aging across both sexes. The correlation between aging and DNA methylation at specific sites across the epigenome was explored again in a subsequent study by Bocklandt et al. in 2011 [26]. In this study, CpG sites were analyzed from the saliva (primarily white blood and epithelial cells) of 34 monozygotic twin pairs, aged between 21 and 55. After adjustments for multiple comparisons, they found 88 CpGs which were correlated with age (69 positive, 19 negative), 10 of which were also found in Rakyan et al.'s previous study. Bocklandt et al. were able to further replicate their finding in saliva using three CpGs from 22 twins and 50 unrelated male and females aged 18–70 years. The average difference between estimated and actual age was 5.3 years (3.5 years in the twins).

In 2013, publications which greatly advanced the method in deriving accurate age estimation from DNA methylation at specific CpG sites came from two independent groups, detailing their development of the "first-generation" epigenetic clocks [29,30]. These epigenetic clocks, developed from varying tissue types, are both highly correlated with chronological age. Until recently, these were the most widely validated and recognized markers of DNA methylation aging, and the majority of research in this field has applied one or both of these calculators.

Horvath's Epigenetic Clock

Horvath's epigenetic clock is a multi-tissue age predictor developed on DNA methylation levels at 353 CpG sites, with 193 being positively correlated with age [30]. It was designed using data from 7844 noncancer samples of all ages across 51 different tissue and cell types, from several different cohorts. When all tissue types were considered, there was a very high correlation between Horvath's epigenetic clock and chronological age (r = 0.96), and the median difference between the predicted (DNA methylation) and actual (chronological) age was only 3.6 years. The rate of change of epigenetic age (ticking rate of epigenetic clock) is faster during periods of growth and development, in particular the first year of life, but with a slow constant rate after the age of 20 years until death. This pan tissue clock has two unique properties: (1) it appears to apply to all nucleated cells in the body and (2) it applies to the entire age spectrum: from children to centenarians, Horvath's pan tissue clock is relatively uncorrelated with blood cell composition [11], and is thought to reflect the innate aging process, capturing cell-intrinsic changes. While predicting age from molecular mechanisms is of great interest, determining the utility of epigenetic age acceleration as a predictor of age-related disease is of greater importance. A derivative of this Horvath's age acceleration is the "intrinsic epigenetic age acceleration," which adjusts for blood cell counts and is regressed on chronological age. It thus provides a measure that is independent of changes in cell composition [54], which is a variable component in blood samples, and relates to properties of hematopoietic stem cells.

Hannum's Epigenetic Clock

In comparison to Horvath's multi-tissue predictor, Hannum's epigenetic clock is based on a linear combination of blood specific DNA methylation at 71 CpG sites [29]. These sites are largely distinct from the 353 CpGs identified by Horvath, with only six overlapping between the two methods. Correlations were derived from DNA methylation in whole blood samples, collected from two cohorts of individuals of Caucasian or Hispanic ethnicity (n = 482 and 172) aged 19–101 years. Like the Horvath clock, there was a very high correlation between epigenetic age and chronological age in their sample (r = 0.91), with an error similar to that of Horvath's clock of 3.9 years. However, unlike Horvath, this biomarker was tailored for blood tissue and in adults; thus, may not be as applicable to other tissues and age groups. Hannum's clock may also be confounded by age-related changes in blood composition, but has the added benefit of providing an accurate prediction of life expectancy [55,56]. An additional measure was developed from Hannum's original clock, which considers age-related changes in cell composition an important consideration as changes in blood cell composition occur normally in aging [55]. The measure incorporates a weighted average of age-associated cell counts and is called the extrinsic epigenetic age acceleration. It provides an advantage over Horvath's clock in its ability to estimate a measure of age acceleration extrinsic of blood cell type composition. It has been shown to predict risk of mortality beyond chronological age and independent of known risk factors, including education level, smoking, physical activity, and body mass index [55].

Another clock to be considered first generation is the Weidner clock developed in 2014 [32]. The Weidner clock showed that using three CpGs, one of which (cg25809905) is also found in the Horvath clock, could estimate age with the mean absolute deviation from chronological age of 4.5 years (REF). In this study, methylation was initially analyzed across blood and cord blood cells (leukocytes, CD14+ monocytes, CD4+ T cells, lymphocytes and mononuclear cells) from four different studies totaling 575 participants aged 0–78 years. From this, they found 102 CpGs that correlated with age, and had a mean absolute deviation from chronological age of 3.34 years. For three of the CpGs that were highly correlated with age in a separate set of 82 blood samples, they validated the findings using a different technique to measure DNA methylation (pyrosequencing), resulting a mean absolute deviation from chronological age of 5.4 years. They finally validated in yet another separate sample of 69 participants, reaching a mean absolute deviation from chronological age of 4.5 years.

Performance Comparisons

A number of early studies measured both the Horvath and Hannum clocks, and compared their performance as biomarkers of aging. The correlation between these clocks has been variable, ranging from moderate ($r = 0.37$) [57], to strong ($r = 0.76$) [55]. This is likely largely due to Horvath's clock being built as a multi-tissue biomarker, and Hannum's clock being built only on blood. However, the fact that these clocks were developed separately, and have only six CpGs in common, suggests they are measuring slightly different components of aging, which may be tissue specific.

By its development, the intrinsic epigenetic age acceleration measure of Horvath's clock exhibits greater consistency across tissues, suggesting it is under stronger genetic control, and measures cell-intrinsic aging. On the other hand, Hannum's extrinsic epigenetic age acceleration measures immune system aging and incorporates a cell count weighted average. As there are changes in the cell composition of blood with age, extrinsic epigenetic age acceleration may be a better predictor of decline with age [58]. Although intrinsic measures seem to exhibit greater consistency across cell types and organs, extrinsic measures seem to be better suited for assessing age-related decline of tissue performance as they exhibit stronger predictive associations with time to death than intrinsic measures of age acceleration (based on multi-tissue epigenetic age) [55]. As such, it has been suggested that Hannum's clock may be a better predictor of mortality risk [55] and has been shown to be superior in estimating chronological age in long lived individuals [59]. However, this has not been found consistently across all studies [60].

Some studies have used a consensus measure which amalgamates the estimation methods of Horvath and Hannum clocks into one, and indicate that this measure might have superior estimation capacity [61,62]. Although to date, such a measure has only been used in a few specific contexts.

Both the Horvath and Hannum clocks were originally generated using data from Illumina methylation arrays, with Horvath using probes for CpGs available on both the Illumina 27 and 450 K arrays, and Hannum's being based on 450 K array probes. They can be applied to newer data obtained with the Illumina EPIC array (850 K), however some of the CpG sites are missing; 19 of 353 for Horvath's clock and 6 of 71 for Hannum's. A study investigating the effect of this, found that missing this small number of CpGs did not have a substantially impact effect on the estimates [63].

SECOND GENERATION EPIGENETIC CLOCKS

Although the Hannum and Horvath clocks are markers of chronological age, built based on CpGs that together are highly correlated with chronological age, newer clocks have been specifically designed to predict age-related health/disease phenotypes, incorporating age and DNA methylation estimates of specific proteomic measures.

PhenoAge

One of the most popular of these second generation clocks, "DNAm PhenoAge" (or simply PhenoAge), is founded on physiological markers of age and health-related dysregulation, representing a "phenotypic" as opposed to chronological age [41].

Levine et al. first identified a composite measure from levels of nine multi-system clinical chemistry biomarkers [41]. Together, these reflected a biomarker of phenotypic aging, which was predictive of age-associated outcomes such as specific and all-cause mortality, cognitive performance, and physical functioning. The nine biomarkers included albumin, creatinine, serum glucose, C-reactive protein, mean cell volume, red cell distribution width, alkaline phosphatase, and white blood cell percentage. Levine et al. then tested these biomarkers, together with chronological age, for their association with blood-based DNA methylation from 456 adults aged 21–100. They discovered methylation at 513 CpGs was associated with the Phenotypic Age biomarker, the composite of which gave rise to PhenoAge, a DNA methylation-based marker of phenotypic aging. Interestingly, this clock has five CpGs in common with both the Horvath and Hannum clocks, another one just with Hannum's and 36 others in common with Horvath's clock. This reflects the PhenoAge clock's representation of multiple aspects of aging, being derived from both chronological age-related CpGs (in line with the development of the Horvath and Hannum clocks), as well as physiological biomarkers of age. On a molecular level, PhenoAge was thus shown to capture age-related decline of biological systems, reflecting an increased activation of pro-inflammatory signaling, and a decrease in repair pathways and damage recognition.

PhenoAge is associated with life expectancy, physical and cognitive function, and even facial aging. It was found to be strongly correlated with various measures of multimorbidity and is considered a stronger predictor of cardiovascular disease and mortality than the first generation clocks. PhenoAge was also positively associated with lung cancer incidence in both smokers and non-smokers [41]. Evidence of PhenoAge's strong prediction of mortality was also shown in a follow-up study by Liu et al. [64]. After adjusting for chronological age, they demonstrated in a wide range of populations, varying in age, ethnicity, socioeconomic backgrounds, and different causes of death, that PhenoAge was still highly predictive of mortality. In fact, PhenoAge was shown to be associated with both all-cause mortality, and disease specific mortality, such as heart disease, cancer, chronic lower respiratory disease, diabetes, influenza or pneumonia, and nephritis/nephrosis. Interestingly though it was not associated with the risk of cerebrovascular disease [64].

GrimAge

Another second generation clock of biological age by Lu et al., was developed by considering DNA methylation markers of plasma proteins that increase with age, and a methylation mark of smoking pack-years, sex, and chronological age [45]. It was specifically developed to measure lifespan by regressing time-to-death onto these biomarkers to produce a single composite measure, GrimAge. GrimAge was created from measurements of blood plasma and DNA methylation levels of 2356 individuals, resulting in a 1030 CpG signature. However these CpGs are not publicly available so they cannot be compared for potential overlap with the other epigenetic clocks. The GrimAge measure incorporates DNA methylation-based estimated smoking pack years, as well as DNA methylation estimation of specific proteomic measures of seven plasma-proteins: leptin, Cystatin C, adrenomedullin, beta-2 microglobulin, plasminogen activation inhibitor 1, tissue inhibitor metalloproteinase, and growth differentiation factor-15 (GDF-15). GDF-15 alone has also been strongly associated with aging (r = 0.82) [65,66].

Currently, GrimAge, is the strongest predictor of overall mortality [67]. Compared to other clocks, including PhenoAge and Horvath, it is also a stronger predictor of time to death, cancer, and heart disease, and has been associated with lifespan and other comorbid health outcomes [45].

EPIGENETIC CLOCKS AND AGE-RELATED DISEASE

Age is the primary risk factor for a number of diseases, and epigenetic clocks have thus been the focus of many recent age-related disease studies. A predictor of chronological age which incorporates biological processes related to aging, may improve understanding of age-related physiological and pathological changes. While some studies have investigated these clock measures in target tissue related to the disease (e.g., brain, heart, tumor tissue for dementia, cardiovascular disease and cancer respectively), many studies have focused on epigenetic clocks measured in blood. The accessibility of blood makes it ideal for early screening and thus a blood predictive marker would have specific utility in this context. Further, commonly used epigenetic clocks (Horvath, Hannum, PhenoAge, and GrimAge) as well as measures of age acceleration can be calculated using blood methylation data though a publicly available website (https://dnamage.genetics.ucla.edu/home), making it relatively easy to include in studies. Below we synthesize evidence for the association between epigenetic clocks and some common age-related diseases. Fig. 25.1 shows these common diseases and lists the epigenetic clocks currently found to be most predictive of these outcomes.

Cancer

Cancer-associated DNA methylation changes are common and well documented, being involved in the initiation and development of cancers, [68] and thus have been a target for new therapeutic strategies [69].

Some studies have measured the epigenetic clock in normal and cancer tissue, and compared these. For example, aberrant epigenetic aging, as measured by Horvath's clock, has been found in brain cancer (glioma) tissue, compared to the high correlation between epigenetic age and chronological age in normal brain tissue [70]. Further, Horvath's age acceleration has

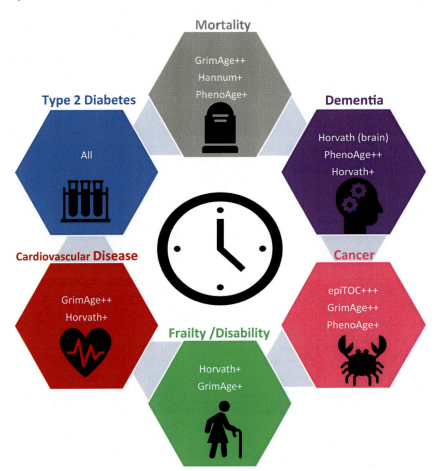

FIGURE 25.1 Common age-associated diseases and the most predictive DNA methylation clocks. DNA methylation measured in blood, unless indicated otherwise. ++ is highly predictive and + is more predictive than other clocks.

been associated with glioma molecular subtypes, and interestingly, increased age acceleration is associated with patient survival [70] and better gastric cancer prognosis [71]. In a similar manner, as measured in normal colon mucosa, PhenoAge age deceleration, rather than acceleration, was observed in a high-risk colon cancer group compared to the low-risk, possibly reflecting dysfunctional epigenetic aging [72].

In blood, early studies demonstrated that these epigenetic clocks were associated with cancer incidence and mortality, although were not always consistent. For example, a study of cause-specific mortality found that an increase in Horvath's age acceleration was a good predictor of cancer mortality, but they found no significant association with Hannum's clock [60]. However, another study which focused specifically on the Hannum clock found that accelerated aging was associated with higher risk of cancer incidence and mortality [73].

Focusing on specific cancers, intrinsic epigenetic age acceleration has been shown to predict incident lung cancer in women [74] and breast cancer risk [75]. One study which compared the performance of three of the clocks (Horvath, Hannum, and PhenoAge) in terms of 5-year breast cancer risk, found that PhenoAge acceleration had the strongest association (hazard ratio 1.15), compared to the others (1.08 and 1.10 respectively) [76]. PhenoAge, as well as GrimAge have also been associated with several other cancers. One study of several different cancer types found an association between these biological age measures and risk of colorectal, kidney, lung, mature B-cell, and urothelial cancers [77]. The greatest risk ratio from this study was GrimAge predicting lung cancer risk.

Aside from these aging clocks, there have also been specific epigenetic clocks developed for cancer risk prediction. One example is the new mitotic clocks, which are built on the premise that enhanced mitosis in a stem cell is a factor in cancer risk [35]. These clocks thus measure the acceleration in the stem cell mitotic rate of "ticking." The "epiTOC" was developed using peripheral blood and integrating mathematical modeling with a known model to estimate tissue-specific stem cell division rates [78]. It has been shown to be accelerated in cancer, in carcinogen-exposed normal epithelial cells, and in pre-cancerous lesions [35]. More recently, the epiTOC2 has been developed which can directly estimate, within a given tissue, the cumulative number of stem cell divisions using DNA methylation at 163 CpGs. This clock has been shown to be a strong predictor of cancer risk in pre-neoplastic lesions [79]. Another recent clock is a breast tissue specific aging clock, based on the methylation of 286 CpGs. Age acceleration of this clock is associated with breast tumor tissue compared to aging in normal breast tissue [80].

Dementia

Dementia is a clinical syndrome characterized by substantial decline in cognitive functioning, resulting in significant impairments in social or occupational functioning [81]. Alzheimer disease is the most common neuropathology resulting in dementia, affecting nearly one third of people over age 85 [82], although in older adults, mixed dementia is common [83].

To date there have been no epigenetic clocks specifically developed to predict risk of dementia or Alzheimer disease. Levine et al. examined the association between Alzheimer disease pathology and PhenoAge in brain tissue (prefrontal cortex) and found that it was one year older for individuals with pathology, namely neurofibrillary tangles, amyloid load, and neuritic plaques [41]. Higher brain PhenoAge was also associated with a decline in global cognition, as well as working and episodic memory in Alzheimer disease patients, but did not predict cognitive function in individuals without dementia.

Studies which have measured epigenetic clocks in blood tissue and examined associations with dementia, have been mixed. For example, two studies of community-based individuals from the Lothian Birth cohort 1921 and Aspirin in Reducing Events in the Elderly (ASPREE) studies, respectively, found no association between blood epigenetic accelerated aging (Horvath, Hannum, PhenoAge, and GrimAge) and incident all-cause dementia over a median 9 and 5 years, respectively [84,85]. However, another study found that higher GrimAge was associated with brain vascular lesions and lower cognitive ability at age 73 years, and with a trend for an association with cognitive decline from 70 to 79 years [86]. A study of 335 patients with Parkinson disease and 257 controls reported that Horvath age acceleration and extrinsic epigenetic age acceleration, but not intrinsic epigenetic age acceleration, were associated with disease risk; the second generation epigenetic clock measures were not examined [87].

A recently developed, tissue specific "Cortical clock," built on methylation from nearly 1400 human cortex samples (aged 1–108 years), has been shown to more accurately predict chronological age compared to other clocks, including Horvath's and PhenoAge [88]. Clocks such as these may be of use for brain specific age-related disease such as dementia. Evidence of this is seen in the 2021 study by Grodstein et al. [89]. Here they observed that clocks that had been trained using blood methylation were generally not predictive of neurologic phenotypes. In contrast, the cortical clock could predict the likelihood of neurodegenerative phenotypes, including Alzheimer disease diagnosis, as well as Alzheimer disease pathology. However, to truly be of use as an aging biomarker for age-related disease, epigenetic brain age clocks also need to be measured in easy to access tissue such as blood, given that brain biopsy is highly invasive and associated with several risks to the individual's health [90].

Cardiovascular Disease

Cardiovascular disease (CVD) is the overarching term for several pathologies, including but not limited to peripheral arterial disease, coronary heart disease, and cerebrovascular disease and stroke [91]. Age is the primary risk factor for CVD and associated diseases, which is one of the leading causes of mortality worldwide [92].

There is evidence that DNA methylation measured in blood can predict risk of heart disease and heart attack [93], however there are fewer studies which have assessed the association between epigenetic aging and incident cardiovascular disease outcomes. A study of 1863 participants aged 50–75 found that both the Horvath and Hannum clocks were associated with the risk of cardiovascular disease mortality, however they were weaker with the Hannum clock and became nonsignificant after adjustment [60]. Likewise, a 10-year longitudinal study of 832 participants aged 70 years old, who were free from CVD at inclusion, Horvath age acceleration, but not Hannum's clock, was associated with incident CVD even after adjusting for CVD risk factors [94]. In contrast, another study reported both clocks were associated with increased risk of fatal coronary heart disease [95]. A study (n = 827, mean age: 69.8) which used a separate 7 CpG clock [96] also found increased age acceleration in those with diagnosed coronary artery disease compared to controls [97]. More recently, a study of around 2500 people aged over 60 years old found GrimAge acceleration to be associated with increased heart attack risk, as well as stroke [98].

Similar to cancer studies, most epigenetic aging studies have focused on heart disease risk rather than heart disease outcomes themselves. One of the first examples of this was shown in a study, also by Horvath et al., which showed a weak association with extrinsic epigenetic age acceleration and cardiometabolic disease risk factors such as triglyceride levels,

C-reactive protein, and creatinine [54]. However, this was only in one of seven populations examined, and no associations were found with incident coronary heart disease. Another study of adolescences/young adults (n = 995, mean age of 17.3), found that a 5-year increase in extrinsic epigenetic age acceleration was associated with a 4% increased risk of cardiovascular disease endpoints (coronary death, heart attack, and stroke) by age 47 [99]. Other studies which have investigated risk factors for cardiovascular disease, but also other age-related diseases, are discussed in the "Links with lifestyle and health factors" section below.

Type 2 Diabetes

Type 2 diabetes (T2D) is a disease primarily defined as a resistance to peripheral insulin, as well as beta cell function and glucose metabolism impairment [11]. Prevalence of T2D greatly increases with age [100].

To date, relatively few studies have looked at the association between T2D and epigenetic age. One of the first of these studies in 2818 participants (mean age 59 years), observed that Horvath age acceleration was associated with a history of diabetes [101]. Another study of cardiovascular disease in African Americans (n = 2543, mean age 57), which also looked at T2D as a risk factor, found that both increased Horvath and Hannum age acceleration were associated with T2D [95]. GrimAge and PhenoAge have also been shown to be strong predictors of T2D risk and incidence. [45]. A separate study, including a discovery (n = 4450, mean age 51.4) and replication (n = 2578, mean age 50) cohort, found GrimAge (HR:1.52), and PhenoAge (HR:1.54) to be predictive of incidence of T2D [102]. GrimAge was also associated with incidence risk of T2D in a separate study of over 1000 individuals aged around 40, particularly in those that were obese [103].

There have been several studies related to diabetes risk factors also, in particular fasting glucose levels. The first example of which is a small longitudinal study of 43 women, which found an association between Horvath age acceleration and fasting glucose [104]. A more recent study including many risk factors for cardiometabolic disease, specifically in an African American population (n = 1100, mean age 57.1 years), found extrinsic epigenetic age acceleration, PhenoAge acceleration ($\beta = 4 \times 10^{-3}$, $P = .004$) and GrimAge acceleration ($\beta = 8 \times 10^{-3}$, $P = 4 \times 10^{-4}$) to be associated with fasting glucose levels, both observations being adjusted for age, sex, familial relatedness and multiple testing [105]. This trend was also observed in a separate population of 736 Ghanaians aged around 51 years, which found a positive association between intrinsic epigenetic age acceleration, extrinsic epigenetic age acceleration, PhenoAge, and GrimAge, with fasting glucose levels [106].

Frailty and Physical Disability

Frailty is a geriatric syndrome of progressive physical decline that is characterized as a heightened state of vulnerability to physiological stressors [107]. Physical disability is the inability to perform tasks of daily living which are necessary to maintain independent functioning [7,108]. Frailty is a risk factor for physical disability, and both are closely tied to the aging process [109].

A recent *meta*-analysis including 3092 participants from three studies found that epigenetic age was positively associated with frailty status [110]. Two of the included studies used the Horvath clock [111,112] and the third used extrinsic epigenetic age acceleration [113]. A more recent study compared performance of the four common epigenetic clocks (Horvath, Hannum, PhenoAge and GrimAge) as a marker of frailty in 490 participants from the Irish Longitudinal Study on Ageing. Here they found that only GrimAge was predictive of walking speed and frailty after adjusting for age, sex, and white blood cell counts [114].

Specific objective measures of age-related physical disabilities, which are also criteria in frailty diagnosis, have also been associated with epigenetic aging. For example, weaker grip strength [115,116] and slower walking speed [115] have been associated with Horvath age acceleration. Inversely, it has been reported that a decrease in epigenetic age measures are associated with improvements in functional age measures. One small (n = 16) study of older adults around 70 years of age, found that a decrease in Hannum age acceleration, and in particular PhenoAge, was associated with increased gait speed; and a decrease in Hannum age acceleration was also associated with improved grip strength [117].

In contrast however, more recent studies have found that there was no association between epigenetic age and frailty [118]. One example comes from a German cohort of 1649 participants aged 22–84 that found no association with a different measure of epigenetic age and frailty [96]. Another newer study of 31 Italian semi-supercentenarians also found there was no association between any of the four primary clocks and frailty (or interestingly, with chronological age) [119].

EPIGENETIC CLOCKS WITHOUT AN AGE-RELATED DISEASE FOCUS

Although epigenetic age is of great interest and importance for studies of age-related disease, there is also utility of epigenetic clocks in other fields. One field of which is forensic science [33,120]. Epigenetic clocks have been used to help narrow down the list of suspects by determining the approximate age of the culprit who left DNA at the scene of a crime [121]. Given the cost of assaying multiple CpGs, and the need for sufficient quantities of DNA, age estimates based on fewer CpGs have been the preferred clocks. In particular, clocks of greater interest can be applied across multiple relevant tissue types and body fluids in the forensic setting [122], as well as saliva, which is the most frequently encountered body fluid at crime scenes [38]. A number of specific clocks are being developed for this purpose, the earliest of which were proposed in 2015, with clocks still being developed today. More recent developments have focused on highly sensitive methods, given the small samples generally available in the forensic setting [123,124].

Another series of clocks have been developed for placenta, cord-blood, and blood spot samples to estimate gestational age [34,36,46,125]. While initially appearing less relevant for the study of age-related disease, according to the developmental origins of health and disease paradigm, early-life plays a crucial role in the risk of diseases in later-life [126]. Indeed, gestational age has been associated with high blood pressure [127] and cardiovascular disease [128], age-related cognitive dysfunction [129], and mortality risk [130]. Thus, DNA methylation-based estimates of gestational age could play an important role in predicting later risk of age-related disease. Other recent developments highlight the dynamic nature of epigenetics during embryogenesis, with the multi-tissue clock demonstrating that epigenetic age is decreased during early embryogenesis, but increases again at later stages of prenatal development [131].

CAN EPIGENETIC AGE BE MODIFIED?

Beyond the ability to estimate future risk of health and disease is the long-standing question of how people age and whether aging can be slowed down, reversed, or even prevented.

Biology of Epigenetic Age

Despite aging clocks being very good predictors of an individual's aging trajectory and their risk of age-related disease, the biological interpretation remains less clear. Understanding the underlying biological mechanisms could help explain different trajectories of age-related decline. For the first generation clocks, it is not clear what physiological or cellular process they were reflecting. Nor is it clear what effect DNA methylation at these specific CpGs has on gene expression; thus, the biological explanation for their association with aging trajectories. Are changes in methylation driving aging, or merely the consequence of the aging process?

Horvath's clock has characteristics that resemble a mitotic clock given that it is highly correlated with chronological age and cell passage number, it is close to zero in embryonic stem cells, and has the highest ticking rate during periods of growth. However, it also tracks with chronological age in brain tissue, which is non-proliferative. Thus, it has been speculated that it represents the epigenetic maintenance system, which mitigates epigenomic instability that increases with age [30]. This is supported by a later study showing it is an intrinsic feature of the cells distinct from cell senescence [132]. Other mechanisms include circadian rhythmicity, with overlapping observed between cytosines exhibiting age-related changes and those with circadian epigenetic oscillations [133]. A 2019 study by Søraas et al., which followed allogeneic hematopoietic stem cell transplantation recipients over 17 years, interestingly found that the Horvath age of reconstituted blood was not influenced by the age of the recipient [134]. This suggests that human hematopoietic stem cells have an intrinsic DNAm age which is unaffected by the environment.

The second generation clocks with a heavier focus on proteomic measures are targeting processes where the biological mechanisms are clearer. PhenoAge for example exhibits strong correlations with gene expression levels. However, it remains that epigenetic age in blood (rather than just the target/disease tissue) can predict disease outcomes. In terms of age-related diseases, it may be attributed to the role of inflammatory mechanisms and thus the immune system in many of these.

Another consideration in the field is that epigenetic aging in different tissues, although sometimes overlapping, can be quite different, likely due to each tissue's underlying biology. Clocks designed using specific tissues have been shown to be generally more accurate when measuring epigenetic age in said tissue, as is seen in the cortical clock mentioned above, as well as the "MEAT" skeletal muscle clock [52]. To date however, clocks are generally developed using blood, brain, or multiple tissues, and there yet to be specific aging clocks based on specific tissues, which may be a primary area of age-related disease, including heart, lung, kidney, and liver. More work is needed to understand the mechanisms underpinning the links between epigenetic clocks, aging and age-related diseases, and importantly, to determine whether the rate of aging, as measured by these epigenetic clocks, can be modified.

Given that epigenetic changes are reversible, this opens up the potential to modify aging trajectories and interventions to promote longevity and years lived free of disease and disability.

Genetics of Epigenetic Clocks

It remains unclear the extent to which the way we age is heritable, and what component is due to our lifetime experiences and exposures. It has been shown that the offspring of semi-supercentenarians are epigenetically younger than age-matched controls [23]. This highlights the likely involvement of common genetic factors underlying these epigenetic clocks, although common shared environment could also be partly involved. Other studies have demonstrated race/ethnic differences in epigenetic aging rates, but these may not be consistent across all measures. For example, compared to Caucasians, African Americans had lower but Hispanics and Tsimane Amerindians had higher extrinsic aging [54]. In terms of intrinsic measures, no difference was observed between Caucasians and African Americans, but Hispanics and Tsimane Amerindians had lower epigenetic aging.

A newer genome-wide association study (GWAS) of the Horvath and Hannum age acceleration estimates in blood, using a sample of 13,493 individuals of European ancestry, revealed novel single nucleotide variants associated with these aging biomarkers [135]. In particular, metabolic and immune pathways were found to share genetic underpinnings with these epigenetic clocks. However, the identified genes were largely distinct from those reported in a previous meta-analysis involving 9907 individuals [136]. This indicates that there is a need for further studies on the genetics of these, as well as the newer epigenetic clocks.

Links with Lifestyle and Health Factors

In addition to the capacity of these epigenetic clocks to predict aging and the risk of age-related disease, there is increasing interest in interventions which could help halt or reverse the age acceleration, and thus potentially lower disease risk or prevent disease. It is well established that DNA methylation patterns are influenced by external stimuli [137] and that epigenetic changes are potentially reversible. Determining the extent to which these different factors are associated with epigenetic age, could help identify modifiable factors which could delay aging or reduce the risk of age-related disease. Thus, a key focus to date has been investigating the extent to which environmental, lifestyle, and health factors influence epigenetic age.

One of the first studies to assess this was carried out in 2017 by Quach et al. [138]. Primarily a study of over 4000 postmenopausal women, they found that factors such as diet (fish, and a blood carotenoid level proxy for vegetable and fruit consumption), education, moderate alcohol consumption, and BMI were associated with lower extrinsic epigenetic age acceleration. Lifestyle and health factors were also the focus of a systematic review published in 2020, but the findings indicate currently insufficient evidence for a strong association between any particular factor and epigenetic age [110]. However, they did find in their meta-analysis that body mass index (BMI) was significantly positively associated with both intrinsic epigenetic age acceleration and extrinsic epigenetic age acceleration measures. This finding was also observed in a more recent systematic review of biological, social, and environmental factors associated with age acceleration, which also found associations between BMI and increased age acceleration of Horvath, PhenoAge and GrimAge clocks [139]. Also in line with modifiable age related disease risk factors, a study of metabolic and inflammatory biomarkers in 830 participants with a mean age of approximately 48.7 years, observed an association between Horvath's age acceleration and increased postprandial high-density lipoproteins (HDL), and decreased soluble interleukin 2 receptor subunit alpha (both markers of heart disease risk) [140]. They also found extrinsic epigenetic age acceleration to be associated with other risk factors such as postprandial triglycerides, interleukin-6, C-reactive protein, and tumor necrosis factor (TNF) alpha, as well as being inversely associated with fasting HDL. It is well known that higher adiposity is a risk factor for age related diseases such as heart disease [141]. An aforementioned study which found an association between epigenetic aging and fasting glucose, also found associations with other poorer health risk factors, for example BMI and waist circumference [104]. However, after adjustment for multiple testing only the association with BMI remained, and findings could not be replicated in a separate study of 157 participants. A newer study observed that, using blood samples of 2758 women (mean age 57 years), age acceleration measured by all four of the primarily used clocks (Horvath, Hannum, PhenoAge, GrimAge) was associated with measures of adiposity (body mass index, waist-to-hip ratio, and waist circumference) [142]. Other modifiable lifestyle factors, such as alcohol and drug use, have also been observed to associate with increased blood PhenoAge [143] and GrimAge acceleration [144].

There is also body of evidence that shows socioeconomic and sociodemographic factors are associated with epigenetic aging. One study of 1834 women found that PhenoAge was higher in non-Hispanic black women who had lower education levels [145]. Another study found that adults without a qualification had increased GrimAge acceleration compared to people with a higher education [146], and adults who had fathers in an unskilled occupational social class had a higher PhenoAge acceleration compared to adults with fathers in professional social class [146]. Likewise, a less advantaged childhood and having parents in semiskilled/unskilled occupations has been associated with higher Hannum age [147]. Neighborhood deprivation has also been associated with Hannum, PhenoAge, and GrimAge clocks [148], but not Horvath.

Of note, mental health factors such as stress, posttraumatic stress disorder, and other psychiatric symptoms have been associated with differences in epigenetic age, however thus far findings have been less consistent [110]. The contribution that these environmental, lifestyle, and health factors make to chronic disease is substantial, and because of their ability to change through modifying behaviors, they are primary points for prevention of chronic disease. Because of the close relationship between risk and epigenetic aging, some studies are using the epigenetic clock to measure the mitigation of modifiable risk—the theory being that the reversal of epigenetic aging may result in cellular rejuvenation [149], and thus could reverse the risk of age-related disease. The first of these studies found a decrease in GrimAge compared to chronological age after undergoing treatment for thymus regeneration, with the aim to prevent or reverse signs of immunosenescence [150]. A recent animal study found that expression of certain genes in the retinal ganglion cells of mice, not only induce youthful epigenetic reprogramming as measured by a separate biological aging clock [151], but also reverses glaucoma vision loss and promotes axon regeneration after injury [152]. The most current research shows that changes in diet and lifestyle factors (such as exercise, deep breathing, and optimized sleep) significantly decreases Horvath epigenetic aging both compared to control group as well as within individuals over an 8-week trial period [153].

However, for many of these factors it remains unclear whether they are the cause, or consequence of the rate if the epigenetic clock. A lack of longitudinal studies which can help establish temporality, combined with the challenges of testing causation in human studies, means that there is still much to be learnt.

SUMMARY

In this chapter we have described the development of DNA methylation biomarkers of biological age, namely epigenetic clocks. We detailed the characteristics of the first generation of epigenetic clocks and provided a comparison between commonly used measures. We then summarized the newer epigenetic clocks which have been developed, and which incorporate proteomic measures, as well as more specific clocks for individual diseases. This leads to the section of age-related disease where we describe the predictive capacity of the clocks to estimate the risk of cancer, dementia, cardiovascular disease, diabetes, and frailty. We then discuss the possibility of modifying these epigenetic clocks as a potential to reverse biological age, describing studies that have investigated the association with environmental exposures and lifestyle factors, and developments in the area of underlying mechanisms. Our aim was to provide the reader with a thorough summary of the area of DNA methylation measures of aging, commonly referred to as "epigenetic clocks," their association with age-related disease, and future avenues of research in this field.

References

[1] Oeppen J, Vaupel JW. Demography. Broken limits life expectancy. Science 2002;296(5570):1029–31.

[2] Kontis V, Bennett JE, Mathers CD, Li G, Foreman K, Ezzati M. Future life expectancy in 35 industrialised countries: projections with a Bayesian model ensemble. Lancet 2017;389(10076):1323–35.

[3] López-Otín C, Blasco MA, Partridge L, Serrano M, Kroemer G. The hallmarks of aging. Cell 2013;153(6):1194–217.

[4] Jylhävä J, Pedersen NL, Hägg S. Biological age predictors. EBioMedicine 2017;21:29–36.

[5] Fratiglioni L, Ahlbom A, Viitanen M, Winblad B. Risk factors for late-onset Alzheimer's disease: a population-based, case-control study. Ann Neurol 1993;33(3):258–66.

[6] Fried LP, Tangen CM, Walston J, Newman AB, Hirsch C, Gottdiener J, et al. Frailty in older adults: evidence for a phenotype. J Gerontol A Biol Sci Med Sci 2001;56(3):M146–56.

[7] Guralnik JM, LaCroix AZ, Branch LG, Kasl SV, Wallace RB. Morbidity and disability in older persons in the years prior to death. Am J Public Health 1991;81(4):443–7.

[8] Aunan JR, Cho WC, Søreide K. The biology of aging and cancer: a brief overview of shared and divergent molecular hallmarks. Aging Dis 2017;8(5):628–42.

[9] White MC, Holman DM, Boehm JE, Peipins LA, Grossman M, Henley SJ. Age and cancer risk: a potentially modifiable relationship. Am J Prev Med 2014;46(3 Suppl 1):S7–15.

[10] Selvin E, Parrinello CM. Age-related differences in glycaemic control in diabetes. Diabetologia 2013;56(12):2549–51.

[11] Vaiserman A, Lushchak O. Developmental origins of type 2 diabetes: focus on epigenetics. Ageing Res Rev 2019;55:100957.

[12] Niccoli T, Partridge L. Ageing as a risk factor for disease. Curr Biol 2012;22(17):R741–52.

[13] Partridge L, Deelen J, Slagboom PE. Facing up to the global challenges of ageing. Nature 2018;561(7721):45–56.

[14] Johnson TE. Recent results: biomarkers of aging. Exp Gerontol 2006;41(12):1243–6.

[15] Sayed N, Huang Y, Nguyen K, Krejciova-Rajaniemi Z, Grawe AP, Gao T, et al. An inflammatory aging clock (iAge) based on deep learning tracks multimorbidity, immunosenescence, frailty and cardiovascular aging. Nat Aging 2021;1(7):598–615.

[16] Slieker RC, van Iterson M, Luijk R, Beekman M, Zhernakova DV, Moed MH, et al. Age-related accrual of methylomic variability is linked to fundamental ageing mechanisms. Genome Biol 2016;17(1):191.

[17] Martino DJ, Tulic MK, Gordon L, Hodder M, Richman TR, Metcalfe J, et al. Evidence for age-related and individual-specific changes in DNA methylation profile of mononuclear cells during early immune development in humans. Epigenetics. 2011;6(9):1085—94.

[18] Jones MJ, Goodman SJ, Kobor MS. DNA methylation and healthy human aging. Aging Cell 2015;14(6):924—32.

[19] Teschendorff AE, West J, Beck S. Age-associated epigenetic drift: implications, and a case of epigenetic thrift? Hum Mol Genet 2013;22(R1):R7—r15.

[20] Johansson A, Enroth S, Gyllensten U. Continuous aging of the human DNA methylome throughout the human lifespan. PLoS One 2013;8(6):e67378.

[21] Benayoun BA, Pollina EA, Brunet A. Epigenetic regulation of ageing: linking environmental inputs to genomic stability. Nat Rev Mol Cell Biol 2015;16(10):593—610.

[22] Fransquet PD, Wrigglesworth J, Woods RL, Ernst ME, Ryan J. The epigenetic clock as a predictor of disease and mortality risk: a systematic review and *meta*-analysis. Clin Epigenetics 2019;11(1):62.

[23] Horvath S, Pirazzini C, Bacalini MG, Gentilini D, Di Blasio AM, Delledonne M, et al. Decreased epigenetic age of PBMCs from Italian semi-supercentenarians and their offspring. Aging (Albany NY) 2015;7(12):1159—70.

[24] Bell CG, Lowe R, Adams PD, Baccarelli AA, Beck S, Bell JT, et al. DNA methylation aging clocks: challenges and recommendations. Genome Biol 2019;20(1):249.

[25] El Khoury LY, Gorrie-Stone T, Smart M, Hughes A, Bao Y, Andrayas A, et al. Systematic underestimation of the epigenetic clock and age acceleration in older subjects. Genome Biol 2019;20(1):283.

[26] Bocklandt S, Lin W, Sehl ME, Sánchez FJ, Sinsheimer JS, Horvath S, et al. Epigenetic predictor of age. PLoS One 2011;6(6):e14821.

[27] Koch CM, Wagner W. Epigenetic-aging-signature to determine age in different tissues. Aging (Albany N Y) 2011;3(10):1018—27.

[28] Garagnani P, Bacalini MG, Pirazzini C, Gori D, Giuliani C, Mari D, et al. Methylation of ELOVL2 gene as a new epigenetic marker of age. Aging Cell 2012;11(6):1132—4.

[29] Hannum G, Guinney J, Zhao L, Zhang L, Hughes G, Sadda S, et al. Genome-wide methylation profiles reveal quantitative views of human aging rates. Mol Cell 2013;49(2):359—67.

[30] Horvath S. DNA methylation age of human tissues and cell types. Genome Biol 2013;14(10):R115.

[31] Florath I, Butterbach K, Müller H, Bewerunge-Hudler M, Brenner H. Cross-sectional and longitudinal changes in DNA methylation with age: an epigenome-wide analysis revealing over 60 novel age-associated CpG sites. Hum Mol Genet 2014;23(5):1186—201.

[32] Weidner CI, Lin Q, Koch CM, Eisele L, Beier F, Ziegler P, et al. Aging of blood can be tracked by DNA methylation changes at just three CpG sites. Genome Biol 2014;15(2):R24.

[33] Zbieć-Piekarska R, Spólnicka M, Kupiec T, Parys-Proszek A, Makowska Ż, Pałeczka A, et al. Development of a forensically useful age prediction method based on DNA methylation analysis. Forensic Sci Int Genet 2015;17:173—9.

[34] Bohlin J, Håberg SE, Magnus P, Reese SE, Gjessing HK, Magnus MC, et al. Prediction of gestational age based on genome-wide differentially methylated regions. Genome Biol 2016;17(1):207.

[35] Yang Z, Wong A, Kuh D, Paul DS, Rakyan VK, Leslie RD, et al. Correlation of an epigenetic mitotic clock with cancer risk. Genome Biol 2016;17(1):205.

[36] Knight AK, Craig JM, Theda C, Bækvad-Hansen M, Bybjerg-Grauholm J, Hansen CS, et al. An epigenetic clock for gestational age at birth based on blood methylation data. Genome Biol 2016;17(1):206.

[37] Vidal-Bralo L, Lopez-Golan Y, Gonzalez A. Simplified assay for epigenetic age estimation in whole blood of adults. Front Genet 2016;7:126.

[38] Hong SR, Jung SE, Lee EH, Shin KJ, Yang WI, Lee HY. DNA methylation-based age prediction from saliva: high age predictability by combination of 7 CpG markers. Forensic Sci Int Genet 2017;29:118—25.

[39] Zhang Y, Wilson R, Heiss J, Breitling LP, Saum K-U, Schöttker B, et al. DNA methylation signatures in peripheral blood strongly predict all-cause mortality. Nat Commun 2017;8(1):14617.

[40] Youn A, Wang S. The MiAge calculator: a DNA methylation-based mitotic age calculator of human tissue types. Epigenetics. 2018;13(2):192—206.

[41] Levine ME, Lu AT, Quach A, Chen BH, Assimes TL, Bandinelli S, et al. An epigenetic biomarker of aging for lifespan and healthspan. Aging (Albany NY) 2018;10(4):573—91.

[42] Horvath S, Oshima J, Martin GM, Lu AT, Quach A, Cohen H, et al. Epigenetic clock for skin and blood cells applied to Hutchinson Gilford Progeria Syndrome and ex vivo studies. Aging (Albany N Y) 2018;10(7):1758—75.

[43] Lu AT, Seeboth A, Tsai P-C, Sun D, Quach A, Reiner AP, et al. DNA methylation-based estimator of telomere length. Aging. 2019;11(16):5895—923.

[44] Zhu T, Gao Y, Wang J, Li X, Shang S, Wang Y, et al. CancerClock: a DNA methylation age predictor to identify and characterize aging clock in pan-cancer. Front Bioeng Biotechnol 2019;7:388.

[45] Lu AT, Quach A, Wilson JG, Reiner AP, Aviv A, Raj K, et al. DNA methylation GrimAge strongly predicts lifespan and healthspan. Aging (Albany NY) 2019;11(2):303—27.

[46] Falick Michaeli T, Spiro A, Sabag O, Karavani G, Yagel S, Eventov-Friedman S, et al. Determining gestational age using genome methylation profile: a novel approach for fetal medicine. Prenat Diagn 2019;39(11):1005—10.

[47] Jung SE, Lim SM, Hong SR, Lee EH, Shin KJ, Lee HY. DNA methylation of the ELOVL2, FHL2, KLF14, C1orf132/MIR29B2C, and TRIM59 genes for age prediction from blood, saliva, and buccal swab samples. Forensic Sci Int Genet 2019;38:1—8.

[48] Lee Y, Choufani S, Weksberg R, Wilson SL, Yuan V, Burt A, et al. Placental epigenetic clocks: estimating gestational age using placental DNA methylation levels. Aging. 2019;11(12):4238—53.

[49] Zhang Q, Vallerga CL, Walker RM, Lin T, Henders AK, Montgomery GW, et al. Improved precision of epigenetic clock estimates across tissues and its implication for biological ageing. Genome Med 2019;11(1):54.

[50] Belsky DW, Caspi A, Arseneault L, Baccarelli A, Corcoran DL, Gao X, et al. Quantification of the pace of biological aging in humans through a blood test, the DunedinPoAm DNA methylation algorithm. Elife. 2020;9:e54870.

[51] Lee Y, Haftorn KL, Denault WRP, Nustad HE, Page CM, Lyle R, et al. Blood-based epigenetic estimators of chronological age in human adults using DNA methylation data from the Illumina MethylationEPIC array. BMC Genomics 2020;21(1):747.

[52] Voisin S, Harvey NR, Haupt LM, Griffiths LR, Ashton KJ, Coffey VG, et al. An epigenetic clock for human skeletal muscle. J Cachexia Sarcopenia Muscle 2020;11(4):887—98.

[53] Rakyan VK, Down TA, Maslau S, Andrew T, Yang TP, Beyan H, et al. Human aging-associated DNA hypermethylation

occurs preferentially at bivalent chromatin domains. Genome Res 2010;20(4):434—9.

[54] Horvath S, Gurven M, Levine ME, Trumble BC, Kaplan H, Allayee H, et al. An epigenetic clock analysis of race/ethnicity, sex, and coronary heart disease. Genome Biol 2016;17(1):171.

[55] Chen BH, Marioni RE, Colicino E, Peters MJ, Ward-Caviness CK, Tsai PC, et al. DNA methylation-based measures of biological age: meta-analysis predicting time to death. Aging (Albany NY) 2016;8(9):1844—65.

[56] Horvath S, Raj K. DNA methylation-based biomarkers and the epigenetic clock theory of ageing. Nat Rev Genet 2018;19 (6):371—84.

[57] Belsky DW, Moffitt TE, Cohen AA, Corcoran DL, Levine ME, Prinz JA, et al. Eleven telomere, epigenetic clock, and biomarker-composite quantifications of biological aging: do they measure the same thing? Am J Epidemiol 2018;187 (6):1220—30.

[58] Marioni RE, Shah S, McRae AF, Chen BH, Colicino E, Harris SE, et al. DNA methylation age of blood predicts all-cause mortality in later life. Genome Biol 2015;16(1):25.

[59] Armstrong NJ, Mather KA, Thalamuthu A, Wright MJ, Trollor JN, Ames D, et al. Aging, exceptional longevity and comparisons of the Hannum and Horvath epigenetic clocks. Epigenomics. 2017;9(5):689—700.

[60] Perna L, Zhang Y, Mons U, Holleczek B, Saum KU, Brenner H. Epigenetic age acceleration predicts cancer, cardiovascular, and all-cause mortality in a German case cohort. Clin Epigenetics 2016;8:64.

[61] Gross AM, Jaeger PA, Kreisberg JF, Licon K, Jepsen KL, Khosroheidari M, et al. Methylome-wide analysis of chronic HIV infection reveals five-year increase in biological age and epigenetic targeting of HLA. Mol Cell 2016;62(2):157—68.

[62] Proskovec AL, Rezich MT, O'Neill J, Morsey B, Wang T, Ideker T, et al. Association of epigenetic metrics of biological age with cortical thickness. JAMA Netw Open 2020;3(9):e2015428.

[63] McEwen LM, Jones MJ, Lin DTS, Edgar RD, Husquin LT, MacIsaac JL, et al. Systematic evaluation of DNA methylation age estimation with common preprocessing methods and the Infinium MethylationEPIC BeadChip array. Clin Epigenetics 2018;10(1):123.

[64] Liu Z, Kuo PL, Horvath S, Crimmins E, Ferrucci L, Levine M. A new aging measure captures morbidity and mortality risk across diverse subpopulations from NHANES IV: a cohort study. PLoS Med 2018;15(12):e1002718.

[65] Lehallier B, Gate D, Schaum N, Nanasi T, Lee SE, Yousef H, et al. Undulating changes in human plasma proteome profiles across the lifespan. Nat Med 2019;25(12):1843—50.

[66] Xie S, Lu L, Liu L. Growth differentiation factor-15 and the risk of cardiovascular diseases and all-cause mortality: a meta-analysis of prospective studies. Clin Cardiol 2019;42(5):513—23.

[67] Föhr T, Waller K, Viljanen A, Sanchez R, Ollikainen M, Rantanen T, et al. Does the epigenetic clock GrimAge predict mortality independent of genetic influences: an 18 year follow-up study in older female twin pairs. Clin Epigenetics 2021;13(1):128.

[68] Michalak EM, Burr ML, Bannister AJ, Dawson MA. The roles of DNA, RNA and histone methylation in ageing and cancer. Nat Rev Mol Cell Biol 2019;20(10):573—89.

[69] Locke WJ, Guanzon D, Ma C, Liew YJ, Duesing KR, Fung KYC, et al. DNA methylation cancer biomarkers: translation to the clinic. Front Genet 2019;10:1150.

[70] Zheng C, Berger NA, Li L, Xu R. Epigenetic age acceleration and clinical outcomes in gliomas. PLoS One 2020;15(7): e0236045.

[71] Zhou YJ, Lu XF, Meng JL, Wang QW, Chen JN, Zhang QW, et al. Specific epigenetic age acceleration patterns among four molecular subtypes of gastric cancer and their prognostic value. Epigenomics. 2021;13(10):767—78.

[72] Wang T, Maden SK, Luebeck GE, Li CI, Newcomb PA, Ulrich CM, et al. Dysfunctional epigenetic aging of the normal colon and colorectal cancer risk. Clin Epigenetics 2020;12(1):5.

[73] Zheng Y, Joyce BT, Colicino E, Liu L, Zhang W, Dai Q, et al. Blood epigenetic age may predict cancer incidence and mortality. EBioMedicine. 2016;5:68—73.

[74] Levine ME, Hosgood HD, Chen B, Absher D, Assimes T, Horvath S. DNA methylation age of blood predicts future onset of lung cancer in the women's health initiative. Aging (Albany NY) 2015;7(9):690—700.

[75] Ambatipudi S, Horvath S, Perrier F, Cuenin C, Hernandez-Vargas H, Le Calvez-Kelm F, et al. DNA methylome analysis identifies accelerated epigenetic ageing associated with postmenopausal breast cancer susceptibility. Eur J Cancer 2017;75: 299—307.

[76] Kresovich JK, Xu Z, O'Brien KM, Weinberg CR, Sandler DP, Taylor JA. Methylation-based biological age and breast cancer risk. J Natl Cancer Inst 2019;111(10):1051—8.

[77] Dugué PA, Bassett JK, Wong EM, Joo JE, Li S, Yu C, et al. Biological aging measures based on blood DNA methylation and risk of cancer: a prospective study. JNCI Cancer Spectr 2021;5(1):pkaa109.

[78] Christensen BC, Kelsey KT. A new timepiece: an epigenetic mitotic clock. Genome Biol 2016;17(1):216.

[79] Teschendorff AE. A comparison of epigenetic mitotic-like clocks for cancer risk prediction. Genome Med 2020;12(1):56.

[80] Castle JR, Lin N, Liu J, Storniolo AMV, Shendre A, Hou L, et al. Estimating breast tissue-specific DNA methylation age using next-generation sequencing data. Clin Epigenetics 2020;12(1):45.

[81] Livingston G, Huntley J, Sommerlad A, Ames D, Ballard C, Banerjee S, et al. Dementia prevention, intervention, and care: 2020 report of the Lancet Commission. Lancet. 2020;396(10248):413—46.

[82] Jack Jr. CR, Bennett DA, Blennow K, Carrillo MC, Dunn B, Haeberlein SB, et al. NIA-AA research framework: toward a biological definition of Alzheimer's disease. Alzheimers Dement 2018;14(4):535—62.

[83] Neuropathology Group of the Medical Research Council Cognitive Function and Ageing Study (MRC CFAS). Pathological correlates of late-onset dementia in a multicentre, community-based population in England and Wales. Neuropathology Group of the Medical Research Council Cognitive Function and Ageing Study (MRC CFAS). Lancet 2001;357(9251):169—75.

[84] Fransquet PD, Lacaze P, Saffery R, Shah RC, Vryer R, Murray A, et al. Accelerated epigenetic aging in peripheral blood does not predict dementia risk. Curr Alzheimer Res 2021;18(5):443—51.

[85] Sibbett RA, Altschul DM, Marioni RE, Deary IJ, Starr JM, Russ TC. DNA methylation-based measures of accelerated biological ageing and the risk of dementia in the oldest-old: a study of the Lothian Birth Cohort 1921. BMC Psychiatry 2020;20(1):91.

[86] Hillary RF, Stevenson AJ, Cox SR, McCartney DL, Harris SE, Seeboth A, et al. An epigenetic predictor of death captures multi-modal measures of brain health. Mol Psychiatry 2019;26 (8):3806—16.

[87] Horvath S, Ritz BR. Increased epigenetic age and granulocyte counts in the blood of Parkinson's disease patients. Aging (Albany NY) 2015;7(12):1130—42.

[88] Shireby GL, Davies JP, Francis PT, Burrage J, Walker EM, Neilson GWA, et al. Recalibrating the epigenetic clock: implications for assessing biological age in the human cortex. Brain. 2020;143(12):3763—75.

[89] Grodstein F, Lemos B, Yu L, Klein HU, Iatrou A, Buchman AS, et al. The association of epigenetic clocks in brain tissue with

brain pathologies and common aging phenotypes. Neurobiol Dis 2021;157:105428.

[90] Serrano GE, Intorcia A, Carew J, Chiarolanza G, Hidalgo JA, Sue LI, et al. Feasibility study: comparison of frontal cortex needle core vs open biopsy for detection of characteristic proteinopathies of neurodegenerative diseases. J Neuropathol Exp Neurol 2015;74(9):934—42.

[91] Stewart J, Manmathan G, Wilkinson P. Primary prevention of cardiovascular disease: a review of contemporary guidance and literature. JRSM Cardiovasc Dis 2017;6 2048004016687211.

[92] Dhingra R, Vasan RS. Age as a risk factor. Med Clin North Am 2012;96(1):87—91.

[93] Agha G, Mendelson MM, Ward-Caviness CK, Joehanes R, Huan T, Gondalia R, et al. Blood leukocyte DNA methylation predicts risk of future myocardial infarction and coronary heart disease. Circulation. 2019;140(8):645—57.

[94] Lind L, Ingelsson E, Sundström J, Siegbahn A, Lampa E. Methylation-based estimated biological age and cardiovascular disease. Eur J Clin Invest 2018;48(2).

[95] Roetker NS, Pankow JS, Bressler J, Morrison AC, Boerwinkle E. Prospective study of epigenetic age acceleration and incidence of cardiovascular disease outcomes in the ARIC Study (Atherosclerosis Risk in Communities). Circ Genom Precis Med 2018;11(3):e001937.

[96] Vetter VM, Meyer A, Karbasiyan M, Steinhagen-Thiessen E, Hopfenmüller W, Demuth I. Epigenetic clock and relative telomere length represent largely different aspects of aging in the Berlin Aging Study II (BASE-II). J Gerontol A Biol Sci Med Sci 2019;74(1):27—32.

[97] Banszerus VL, Vetter VM, König M, Landmesser U, Demuth I. Evidence of accelerated epigenetic aging in patients diagnosed with coronary artery disease: results of the LipidCardio study. medRxiv 2020; 2020.02.23.20026906.

[98] Wang C, Ni W, Yao Y, Just A, Heiss J, Wei Y, et al. DNA methylation-based biomarkers of age acceleration and all-cause death, myocardial infarction, stroke, and cancer in two cohorts: the NAS, and KORA F4. EBioMedicine. 2021;63:103151.

[99] Huang RC, Lillycrop KA, Beilin LJ, Godfrey KM, Anderson D, Mori TA, et al. Epigenetic age acceleration in adolescence associates With BMI, inflammation, and risk score for middle age cardiovascular disease. J Clin Endocrinol Metab 2019;104(7):3012—24.

[100] Ogrin R, Neoh S, Aylen T, Audehm R, Churilov L, Thurgood L, et al. Older people with type 2 diabetes-individualizing management with a specialized (OPTIMISE) community team: protocol for a safety and feasibility mixed methods study. JMIR Res Protoc 2019;8(6):e13986.

[101] Dugué PA, Bassett JK, Joo JE, Baglietto L, Jung CH, Wong EM, et al. Association of DNA methylation-based biological age with health risk factors and overall and cause-specific mortality. Am J Epidemiol 2018;187(3):529—38.

[102] Hillary RF, Stevenson AJ, McCartney DL, Campbell A, Walker RM, Howard DM, et al. Epigenetic measures of ageing predict the prevalence and incidence of leading causes of death and disease burden. Clin Epigenetics 2020;12(1):115.

[103] Kim K, Joyce BT, Zheng Y, Schreiner PJ, Jacobs Jr. DR, Catov JM, et al. DNA methylation GrimAge and incident diabetes: The Coronary Artery Risk Development in Young Adults (CARDIA) Study. Diabetes. 2021;70(6):1404—13.

[104] Grant CD, Jafari N, Hou L, Li Y, Stewart JD, Zhang G, et al. A longitudinal study of DNA methylation as a potential mediator of age-related diabetes risk. Geroscience. 2017;39(5—6):475—89.

[105] Ammous F, Zhao W, Ratliff SM, Mosley TH, Bielak LF, Zhou X, et al. Epigenetic age acceleration is associated with cardiometabolic risk factors and clinical cardiovascular disease risk scores in African Americans. Clin Epigenetics 2021;13(1):55.

[106] Chilunga FP, Henneman P, Elliott HR, Cronje HT, Walia GK, Meeks KAC, et al. Epigenetic-age acceleration in the emerging burden of cardiometabolic diseases among migrant and non-migrant African populations: a population-based cross-sectional RODAM substudy. Lancet Healthy Longev 2021;2(6):E327—39.

[107] Mitnitski AB, Mogilner AJ, Rockwood K. Accumulation of deficits as a proxy measure of aging. ScientificWorldJournal. 2001;1:323—36.

[108] Katz S, Ford AB, Moskowitz RW, Jackson BA, Jaffe MW. Studies of illness in the aged. the index of adl: a standardized measure of biological and psychosocial function. JAMA. 1963;185:914—19.

[109] Hoogendijk EO, Afilalo J, Ensrud KE, Kowal P, Onder G, Fried LP. Frailty: implications for clinical practice and public health. Lancet. 2019;394(10206):1365—75.

[110] Ryan J, Wrigglesworth J, Loong J, Fransquet PD, Woods RL. A systematic review and meta-analysis of environmental, lifestyle, and health factors associated with DNA methylation age. J Gerontol A Biol Sci Med Sci 2020;75(3):481—94.

[111] Breitling LP, Saum KU, Perna L, Schöttker B, Holleczek B, Brenner H. Frailty is associated with the epigenetic clock but not with telomere length in a German cohort. Clin Epigenetics 2016;8:21.

[112] Kim S, Myers L, Wyckoff J, Cherry KE, Jazwinski SM. The frailty index outperforms DNA methylation age and its derivatives as an indicator of biological age. Geroscience. 2017;39(1):83—92.

[113] Gale CR, Marioni RE, Harris SE, Starr JM, Deary IJ. DNA methylation and the epigenetic clock in relation to physical frailty in older people: the Lothian Birth Cohort 1936. Clin Epigenetics 2018;10(1):101.

[114] McCrory C, Fiorito G, Hernandez B, Polidoro S, O'Halloran AM, Hever A, et al. GrimAge outperforms other epigenetic clocks in the prediction of age-related clinical phenotypes and all-cause mortality. J Gerontol A Biol Sci Med Sci 2021;76(5):741—9.

[115] Marioni RE, Shah S, McRae AF, Ritchie SJ, Muniz-Terrera G, Harris SE, et al. The epigenetic clock is correlated with physical and cognitive fitness in the Lothian Birth Cohort 1936. Int J Epidemiol 2015;44(4):1388—96.

[116] Simpkin AJ, Cooper R, Howe LD, Relton CL, Davey Smith G, Teschendorff A, et al. Are objective measures of physical capability related to accelerated epigenetic age? Findings from a British birth cohort. BMJ Open 2017;7(10):e016708.

[117] Petersen CL, Christensen BC, Batsis JA. Weight management intervention identifies association of decreased DNA methylation age with improved functional age measures in older adults with obesity. Clin. Epigenetics. 2021;13(1):46.

[118] Vetter VM, Spira D, Banszerus VL, Demuth I. Epigenetic clock and leukocyte telomere length are associated with vitamin d status but not with functional assessments and frailty in the Berlin Aging Study II. J Gerontol A Biol Sci Med Sci 2020;75(11):2056—63.

[119] Bacalini MG, Gentilini D, Monti D, Garagnani P, Mari D, Cesari M, et al. No association between frailty index and epigenetic clocks in Italian semi-supercentenarians. Mech Ageing Dev 2021;197:111514.

[120] Zbieć-Piekarska R, Spólnicka M, Kupiec T, Makowska Ż, Spas A, Parys-Proszek A, et al. Examination of DNA methylation status of the ELOVL2 marker may be useful for human age prediction in forensic science. Forensic Sci Int Genet 2015;14:161—7.

[121] Lee HY, Lee SD, Shin KJ. Forensic DNA methylation profiling from evidence material for investigative leads. BMB Rep 2016;49(7):359–69.

[122] Jung SE, Shin KJ, Lee HY. DNA methylation-based age prediction from various tissues and body fluids. BMB Rep 2017;50(11):546–53.

[123] Fleckhaus J, Schneider PM. Novel multiplex strategy for DNA methylation-based age prediction from small amounts of DNA via Pyrosequencing. Forensic Sci Int Genet 2020;44:102189.

[124] Thong Z, Tan JYY, Loo ES, Phua YW, Chan XLS, Syn CK-C. Artificial neural network, predictor variables and sensitivity threshold for DNA methylation-based age prediction using blood samples. Sci Rep 2021;11(1):1744.

[125] Lee Y, Choufani S, Weksberg R, Wilson SL, Yuan V, Burt A, et al. Placental epigenetic clocks: estimating gestational age using placental DNA methylation levels. Aging (Albany NY) 2019;11(12):4238–53.

[126] Heindel JJ, Vandenberg LN. Developmental origins of health and disease: a paradigm for understanding disease cause and prevention. Curr Opin Pediatr 2015;27(2):248–53.

[127] Cooper R, Atherton K, Power C. Gestational age and risk factors for cardiovascular disease: evidence from the 1958 British birth cohort followed to mid-life. Int J Epidemiol 2009;38(1):235–44.

[128] Barker DJ. The fetal and infant origins of adult disease. Br Med J 1990;301(6761):1111.

[129] Mosing MA, Lundholm C, Cnattingius S, Gatz M, Pedersen NL. Associations between birth characteristics and age-related cognitive impairment and dementia: a registry-based cohort study. PLoS Med 2018;15(7):e1002609.

[130] Risnes K, Bilsteen JF, Brown P, Pulakka A, Andersen AN, Opdahl S, et al. Mortality among young adults born preterm and early term in 4 nordic nations. JAMA Netw Open 2021;4(1):e2032779.

[131] Kerepesi C, Zhang B, Lee SG, Trapp A, Gladyshev VN. Epigenetic clocks reveal a rejuvenation event during embryogenesis followed by aging. Sci Adv 2021;7(26).

[132] Lowe D, Horvath S, Raj K. Epigenetic clock analyses of cellular senescence and ageing. Oncotarget. 2016;7(8):8524–31.

[133] Oh G, Ebrahimi S, Carlucci M, Zhang A, Nair A, Groot DE, et al. Cytosine modifications exhibit circadian oscillations that are involved in epigenetic diversity and aging. Nat Commun 2018;9(1):644.

[134] Søraas A, Matsuyama M, de Lima M, Wald D, Buechner J, Gedde-Dahl T, et al. Epigenetic age is a cell-intrinsic property in transplanted human hematopoietic cells. Aging Cell 2019;18(2):e12897.

[135] Gibson J, Russ TC, Clarke TK, Howard DM, Hillary RF, Evans KL, et al. A meta-analysis of genome-wide association studies of epigenetic age acceleration. PLoS Genet 2019;15(11):e1008104.

[136] Lu AT, Xue L, Salfati EL, Chen BH, Ferrucci L, Levy D, et al. GWAS of epigenetic aging rates in blood reveals a critical role for TERT. Nat Commun 2018;9(1):387.

[137] Martin EM, Fry RC. Environmental Influences on the epigenome: exposure-associated DNA methylation in human populations. Annu Rev Public Health 2018;39:309–33.

[138] Quach A, Levine ME, Tanaka T, Lu AT, Chen BH, Ferrucci L, et al. Epigenetic clock analysis of diet, exercise, education, and lifestyle factors. Aging (Albany NY) 2017;9(2):419–46.

[139] Oblak L, van der Zaag J, Higgins-Chen AT, Levine ME, Boks MP. A systematic review of biological, social and environmental factors associated with epigenetic clock acceleration. Ageing Res Rev 2021;69:101348.

[140] Irvin MR, Aslibekyan S, Do A, Zhi D, Hidalgo B, Claas SA, et al. Metabolic and inflammatory biomarkers are associated with epigenetic aging acceleration estimates in the GOLDN study. Clin Epigenetics 2018;10:56.

[141] Goossens GH. The metabolic phenotype in obesity: fat mass, body fat distribution, and adipose tissue function. Obes Facts 2017;10(3):207–15.

[142] Kresovich JK, Garval EL, Martinez Lopez AM, Xu Z, Niehoff NM, White AJ, et al. Associations of body composition and physical activity level with multiple measures of epigenetic age acceleration. Am J Epidemiol 2021;190(6):984–93.

[143] Luo A, Jung J, Longley M, Rosoff DB, Charlet K, Muench C, et al. Epigenetic aging is accelerated in alcohol use disorder and regulated by genetic variation in APOL2. Neuropsychopharmacology. 2020;45(2):327–36.

[144] Shu C, Sosnowski DW, Tao R, Deep-Soboslay A, Kleinman JE, Hyde TM, et al. Epigenome-wide study of brain DNA methylation following acute opioid intoxication. Drug Alcohol Depend 2021;221:108658.

[145] Liu Z, Chen BH, Assimes TL, Ferrucci L, Horvath S, Levine ME. The role of epigenetic aging in education and racial/ethnic mortality disparities among older U.S. Women. Psychoneuroendocrinology. 2019;104:18–24.

[146] George A, Hardy R, Castillo Fernandez J, Kelly Y, Maddock J. Life course socioeconomic position and DNA methylation age acceleration in mid-life. J Epidemiol Community Health 2021;75(11):1084–90.

[147] Hughes A, Smart M, Gorrie-Stone T, Hannon E, Mill J, Bao Y, et al. Socioeconomic position and DNA methylation age acceleration across the life course. Am J Epidemiol 2018;187(11):2346–54.

[148] Lawrence KG, Kresovich JK, O'Brien KM, Hoang TT, Xu Z, Taylor JA, et al. Association of neighborhood deprivation with epigenetic aging using 4 clock metrics. JAMA Netw Open 2020;3(11):e2024329.

[149] Mendelsohn AR, Larrick JW. Epigenetic age reversal by cell-extrinsic and cell-intrinsic means. Rejuvenation Res 2019;22(5):439–46.

[150] Fahy GM, Brooke RT, Watson JP, Good Z, Vasanawala SS, Maecker H, et al. Reversal of epigenetic aging and immunosenescent trends in humans. Aging Cell 2019;18(6):e13028.

[151] Wang M, Lemos B. Ribosomal DNA harbors an evolutionarily conserved clock of biological aging. Genome Res 2019;29(3):325–33.

[152] Lu Y, Brommer B, Tian X, Krishnan A, Meer M, Wang C, et al. Reprogramming to recover youthful epigenetic information and restore vision. Nature. 2020;588(7836):124–9.

[153] Fitzgerald KN, Hodges R, Hanes D, Stack E, Cheishvili D, Szyf M, et al. Potential reversal of epigenetic age using a diet and lifestyle intervention: a pilot randomized clinical trial. Aging (Albany NY) 2021;13(7):9419–32.

SECTION VI

Evolutionary Epigenetics

26. Evolution of epigenetic mechanisms in plants: insights from H3K4 and H3K27 methyltransferases
27. Evolution, dynamics and functions of epigenetic mechanisms in animals
28. Adaptive evolution and epigenetics

CHAPTER 26

Evolution of Epigenetic Mechanisms in Plants: Insights from H3K4 and H3K27 Methyltransferases

J. Armando Casas-Mollano[1], Ericka Zacarias[1] and Juliana Almeida[2]

[1]BioTechnology Institute, University of Minnesota, Twin-Cities, Saint Paul, MN, United States [2]Centre of Natural Sciences and Humanities, Federal University of ABC, Sao Bernardo do Campo, Brazil

OUTLINE

Introduction	499
Histone Lysine Methylation	500
Histone Lysine Methyltransferases in Plants	500
Evolution of Plant Histone Lysine Methyltransferases Methylating H3K4	502
Phylogenetic Relationships Between Histone Lysine Methyltransferases Methylating H3K4	502
ATX1-like Histone Lysine Methyltransferases in Plants	502
ATX5-like Histone Lysine Methyltransferases in Plants	505
SET1-like Histone Lysine Methyltransferases in Plants	506
ATXR3-like Histone Lysine Methyltransferases in Plants	507
Interplay Between Plant Histone Lysine Methyltransferases in Stablishing H3K4 Methylation	508
Evolution of Plant Histone Lysine Methyltransferases Methylating H3K27	508
Phylogenetic Relationship Between Histone Lysine Methyltransferases Methylating H3K27	508
Histone Lysine Methyltransferases Methylating H3K27 in Early-diverging Plant Lineages	509
Histone Lysine Methyltransferases Methylating H3K27 in Eudicotyledonous Plants	511
Histone Lysine Methyltransferases Methylating H3K27 in Monocotyledonous Plants	513
Perspectives	514
Acknowledgments	515
References	515

INTRODUCTION

Histones are subject to a multitude of posttranslational modifications (PTMs) that occur preferentially at their N-terminal tails and include methylation, acetylation, phosphorylation, ubiquitination, among others [1–3]. These PTMs can be added or removed at specific amino acids by histone modifying enzymes [1,2].

Modification of histones with different functional groups may affect chromatin in two ways; whereas, some histone PTMs can alter chromatin accessibility by changing histone-DNA and histone-histone interactions, others may result in the recruitment and/or exclusion of effector proteins that alter the chromatin structure [1,4,5]. Together, histone PTMs and their modifying enzymes influence the organization and function of chromatin, and play a crucial role in all DNA-templated processes such as transcription, replication, and DNA-repair [1,4].

A hypothesis put forward by Strahl and Allis postulates that histone modifications index the chromatin so that, alone or in combination, they may form a "histone code" possibly recognized by effector proteins, which, in turn, direct a specific biological readout [6]. However, the meaning of histone PTMs does not appear to strictly adhere to a code as individual chromatin marks do not always predict a single biological outcome [1,2,5,7]. Moreover, while many histone PTMs appear to be conserved, the functions of some marks may differ in some organisms, suggesting that the histone code may not be universal [8–10]. Elucidating these organismal "dialects" of the "language" of histone modifications will require the characterization of histone marks and their modifying enzymes in different eukaryotic lineages, including plants [8].

Plants and animals not only share many conserved histone PTMs but the histone-modifying systems responsible for their implementation are also alike [9,10]. However, scrutiny of the ever-increasing data has revealed differences in biological functions for some PTMs. Novel functions and even modifying activities that have arisen since these two eukaryotic lineages diverged have also been integrated into the histone code. Thus, to illustrate how plant histone-modifying systems have been shaped during evolution and how they impact the histone code and the regulation of several processes in plants, here we follow the evolutionary path and functions acquired along the way by the two most studied PTMs in plants, histone H3 lysine 4 and lysine 27 methylation, and the enzymes responsible for these modifications.

HISTONE LYSINE METHYLATION

Histone lysine methylation is a modification whose particular outcome, either transcriptional repression, or activation, depends on the residue being modified [1,11]. Lysine can accept more than one methyl group producing monomethylated, dimethylated, or trimethylated residues [12]. Methylation at specific residues and at different levels creates two layers of complexity whose combination results in a large coding potential for lysine methylation. For instance, trimethylation of histone H3 lysine 4 (H3K4me3) is associated with transcriptional activation and is usually placed at the 5′ end of active genes, whereas H3K27me3 correlates with gene repression most often at euchromatic, and also heterochromatic, regions [10,11,13–15]. Yet other histone methylation marks, H3K9me2 or H3K9me3, can be associated with the highly condensed chromatin that constitutes heterochromatic regions, gene silencing and suppression of transposable elements. In plants, H3K9me2 is particularly linked to a self-reinforcing loop for the maintenance of non-CG DNA methylation [10,11,16].

Within the flowering plants, analysis of histone PTMs carried out in grass-like monocot (*e.g.*, sugarcane) and eudicot plants (*e.g.*, *Arabidopsis* and soybean) has shown conservation in lysine methylation of several residues, including H3K4, K9, K18, K23, K27, and K36, and histone H4 lysine 20 (H4K20) [17–20]. Additional analysis using modification-specific antibodies revealed conservation of methylation at H3K4, K9, K27, and K36 in other plant species, although their functions may not always be conserved [9]. In the unicellular alga *Chlamydomonas reinhardtii*, H3K9me2 is not present and then H3K9me1 is the mark found associated with silent heterochromatin; in higher plants, such as *Arabidopsis thaliana*, H3K9me2 is the predominant heterochromatic mark though [9,10,21]. Moreover, species-specific changes in the association of histone PTMs to chromatin domains imply that their function may have also diverged. For example, H3K27me2 has been found exclusively associated with euchromatin in barley (*Hordeum vulgare*), while in *Arabidopsis* it is co-localized with highly condensed heterochromatin [9]. Furthermore, H3K27me3, normally a euchromatin-related PTM, has been found to have a distribution typical for heterochromatic marks in the fava bean *Vicia faba* [9].

HISTONE LYSINE METHYLTRANSFERASES IN PLANTS

Histone lysine methyltransferases (HKMTs) can actively add methyl groups to lysine residues based on the catalytic activity of the SET domain. Common to all HKMTs, the SET domain is a conserved sequence of ~130 to 140 amino acids that was originally recognized as a homologous region encoded by three fruit fly (*Drosophila melanogaster*) genes: the strongest *Suppressor of variegation* [Su(var)3–9], the Polycomb group (PcG) gene *enhancer of zeste* [E(Z)], and the Trithorax group (TrxG) gene *Trithorax (Trx)* [12]. SET domain-containing proteins are found not only in all domains of life, including eukaryotes and prokaryotes, but also in viruses, and may have originated from an ancient bacterial protein [22]. Furthermore, the widespread distribution of SET-domain proteins in eukaryotes suggests that chromatin regulation by histone lysine methylation evolved in the last common ancestor of all eukaryotes.

In plants, a large number of genes encoding SET-domain group (SDGs) proteins have been identified in the available genomes of several species, including bryophytes represented by the common liverwort

(*Marchantia polymorpha*, 32 SDGs), eudicot plants such as *A. thaliana* (47 SDGs), *Brassica rapa* (49 SDGs), grapevine (*Vitis vinifera*, 33 SDGs), tomato (*Solanum lycopersicum*, 44 SDGs), wild cotton (*Gossypium raimondii*, 52 SDGs), poplar (*Populus trichocarpa*, 59 SDGs), woodland strawberry (*Fragaria vesca*, 45 SDGs), apple (*Malus × domestica*, 67 SDGs), litchi (*Litchi chinensis*, 37 SDGs), as well as monocots, such as maize (*Zea mays*, 43 SDGs), rice (*Oryza sativa*, 43 SDGs), *Brachypodium distachyon* (41 SDGs), foxtail millet (*Setaria italica*, 53 SDGs), bread wheat (*Triticum aestivum*, 59 SDGs) and two orchids (*Dendrobium catenatum*, 44 SDGs and *Phalaenopsis equestris*, 42 SDGs) (Table 26.1) [23–38,40]. Importantly, the large number of SDGs, 32–67 SDGs, observed in plants, in stark contrast to the homologs found in yeast (4 SDGs) and metazoans—mouse (17 SDGs), *Drosophila* (14 SDGs)—is believed to be the result of gene duplication events adopting new functions during plant evolution [35].

Gene duplication is the primary source of evolutionary innovation including the origin of novel functions and the generation of new expression patterns [41]. A gene can be duplicated by tandem duplication, segmental duplication of chromosomes, and whole genome duplications (WGD) [42]. Tandem duplications tend to generate copies in genomic clusters; whereas, segmental duplications and WGD result in dispersion of gene copies [42]. Transposable elements and retrotransposition-like events may also mediate gene duplication [24,42]. The immediate result of gene duplication is functional redundancy between the new pair. Thus, it is possible for one copy to retain the original function while the second is prone to accumulate mutations. If the mutations lead to a new function then the new copy may be maintained by natural selection (neofunctionalization), otherwise it may be lost by deleterious mutations (non-functionalization) [41]. Neofunctionalization can be achieved by acquiring novel biochemical functions or by adopting differential expression patterns. In an alternative scenario, both copies can accumulate mutations that affect their function in such a way that their added capacity is

TABLE 26.1 Number of SET-domain Genes in Several Plants

Groups/species	Total SDGs	Class I	Class II	Class III	Class IV	Class V	Class VI	Class VII	Genome Size (Mbp)	References
Bryophytes										
Marchantia polymorpha	32	3	5	4	1	5	14	n.d.	~280	[23]
Eudicots										
Arabidopsis thaliana	47	3	5	7	2	15	5	10	135	[24–26]
Brassica rapa	49	4	7	9	4	20	5	n.d.	485	[27]
Vitis vinifera	33	2	4	6	2	12	5	2	487	[28]
Solanum lycopersicum	44	3	10	6	2	14	7	2	~900	[29]
Gossypium raimondii	52	3	5	7	2	18	5	12	880	[30]
Populus trichocarpa	59	4	6	11	4	18	5	11	~500	[31]
Fragaria vesca	45	2	5	7	2	15	7	7	240	[32]
Litchi chinensis	37	3	5	5	2	13	5	4	~550	[33]
Malus × domestica	67	5	5	6	3	25	12	11	742	[34]
Monocots										
Zea mays	43	3	6	6	2	18	5	3	2300	[25,26,35]
Oryza sativa	43	2	6	6	2	16	5	6	466	[36]
Brachypodium distachyon	41	2	6	7	2	20	4	n.d.	271	[27]
Setaria italica	53	1	6	6	2	21	6	11	485	[37]
Dendrobium catenatum	44	2	5	7	3	14	6	7	1110	[38]
Phalaenopsis equestris	42	2	4	7	2	13	6	8	1160	[38]
Triticum aestivum[a]	59 (160)	3 (9)	6 (16)	7 (19)	3 (7)	26 (73)	4 (12)	10 (24)	~17000	[40]

[a]The numbers refers to the genes in a subgenome while the numbers in parenthesis indicate sum of homeologs found in each category.
n.d., Not determined in the study.

equivalent to the single-copy ancestral gene (subfunctionalization) [41].

Analysis of the physical localization of SDG proteins in monocot and eudicot plants (*Arabidopsis*, poplar, wild cotton, grapevine, woodland strawberry, apple, rice, maize, foxtail millet, and bread wheat) reveals a dispersed chromosomal distribution with many SDGs found in large duplicated regions throughout their respective genomes [24,26,28,30–32,34–37,40]. Thus, segmental duplication and WGD (polyploidization) appear to be the main mechanisms responsible for the expansion of the SET-domain gene family in plants. Furthermore, phylogenetic analysis of SDGs between the model flowering plant *Arabidopsis* (eudicot) and maize (monocot) identified at least 19 orthology groups with members of these two species, leading to the conclusion that most of the duplications and subsequent functional diversification of the SET domain family predate monocot-eudicot divergence. However, WGD events do not always lead to large increases in the number of SET-domain genes. For instance, a recent polyploidization event in *B. rapa* compared to *Arabidopsis* added only few more SDGs to the family (Table 26.1) suggesting that most of the SET-domain genes were lost after duplication [27]. Thus, a pattern of ancient and lineage-specific duplications, and also losses of duplicated genes during diploidization, have likely been the main determinants of the SDG numbers in different plant species.

Plant SET-domain proteins have been divided into seven classes; five (I-V) primarily assigned according to their domain architecture and differences in enzymatic specificity [24,35], while the other two classes (VI and VII) were added later to accommodate proteins with a more divergent SET-domain [25]. Class I includes E(Z) homologs that catalyze H3K27me3; class II, homologs of the *Drosophila* protein Absent, Small, or Homeotic discs 1 (ASH1) with specificity for H3K36; class III, Trx-like proteins that methylate H3K4; class IV, homologs of the plant-specific proteins ARABIDOPSIS TRITHORAX RELATED5 and 6, (ATXR5 and ATXR6) implicated in H3K27me1; class V, homologs of the *Drosophila* Su(var) 3–9 that methylate H3K9. Class VI includes proteins with a split SET domain and, finally, class VII contains proteins that methylate residues on non-histone targets. Although several proteins from each class have been studied in some detail, SDGs from class I and III remain the best studied due to their function in multiple plant developmental pathways.

EVOLUTION OF PLANT HISTONE LYSINE METHYLTRANSFERASES METHYLATING H3K4

Methylation of H3K4 has been observed in its three levels, mono-, di- and trimethylation, in most eukaryotes investigated so far. Although H3K4me1 has been associated with different chromatin configurations, H3K4me2 and H3K4me3 have been almost invariably correlated with transcriptional activation. In the yeast *Saccharomyces cerevisiae*, a single HKMT (Set1) is responsible for all H3K4 methylation [43–46]. By contrast, multiple methyltransferases have evolved to methylate H3K4 in metazoans. For instance, in *Drosophila*, three orthologs of the yeast Set1 have been identified; dSet1, Trx, and Trithorax-related (Trr) [47–50]. Mammals also possess several H3K4 methyltransferases namely SET1A, SET1B, mixed-lineage leukemia 1 (MLL1), MLL2, MLL3, and MLL4 [51–56].

Phylogenetic Relationships Between Histone Lysine Methyltransferases Methylating H3K4

While the yeast and metazoan SET1 proteins are similar at the SET domain and overall protein structure, the Trx/MLL1 related-genes encode proteins with more complex domain architecture and are more divergent than their SET1 paralogs (Fig. 26.1). Indeed, phylogenetic analysis including H3K4-specific HKMTs from yeast, *Drosophila* and humans revealed three orthology groups. The first one contains SET1 from yeast, human, and flies, while the second group clusters together *Drosophila* Trx and human MLL1 and MLL4. The third clade includes *Drosophila* Trr and human MLL2 and MLL3 [50].

In the plant kingdom, phylogenetic analysis allowed the identification of four subgroups of HKMTs, ATX1-like, ATX5-like, SET1-like and ATXR3-like, that methylate H3K4 (Fig. 26.2). The subgroups are all monophyletic and with representatives from chlorophyte unicellular algae, mosses, eudicot and monocot plants, indicating that their origins may predate the emergence of the green lineage. In addition, three of these subgroups (ATX1-like, ATX5-like, and SET1-like) share homologs with animal proteins, but the fourth subgroup, ATXR3-like, appears to have evolved specifically in the plant lineage, suggesting additional evolutionary novelty in the regulation of H3K4 methylation in plants. Thus, the presence of four subgroups of HKMTs targeting H3K4 points to a rather complex regulation of this chromatin modification already present in the common ancestor of all plants.

ATX1-like Histone Lysine Methyltransferases in Plants

ATX1-like, together with the ATX5-like subgroup, correspond to homologs of the *Drosophila* Trx protein. The *A. thaliana* genome encodes five Trx homologs properly named ARABIDOPSIS TRITHORAX-LIKE1–5. However, phylogenetic analysis showed that while ATX1-like subgroup harbors ATX1 and ATX2, ATX5-like subgroup is

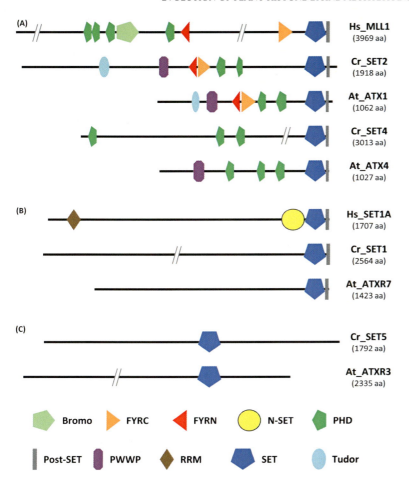

FIGURE 26.1 Composition and domain architecture of histone lysine methyltransferases (HKMTs) involved in H3K4 methylation from representative animals and plants. (A) Schematic representation of Trithorax (Trx)-like proteins from animals and plants. (B) Diagram of SET1–like proteins. (C) Illustration of proteins from the plant-specific *Arabidopsis* TRITHORAX RELATED3 (ATXR3)-like subgroup. Domains present in each protein were identified using the SMART database. Proteins are drawn to scale. The name of each protein is preceded by a two-letter abbreviation of the species as follows: *At*, *Arabidopsis thaliana*; *Cr*, *Chlamydomonas reinhardtii*; *Hs*, *Homo sapiens*. aa, Amino acid; Bromo, Bromodomain; FYRC, F/Y-rich C-terminus; FYRN, F/Y-rich N-terminus; PHD, plant homeodomain finger; PWWP, Pro-Trp-Trp-Pro; RRM, RNA recognition motif; SET, Su(var)3–9, enhancer of zeste, Trithorax domain.

comprised of ATX3–5 homologs (Fig. 26.2) [13,24,57]. The ATX1-like/ATX5-like divergence may have occurred before green lineage evolution, as proteins belonging to both subgroups were already present as early as in the common ancestor between green unicellular algae and land plants (Fig. 26.2).

Proteins belonging to the ATX1-like subgroup consist of (1) an N-terminal Tudor domain; (2) a Pro-Trp-Trp-Pro (PWWP; named after its central motif) domain; (3) F/Y-rich N-terminus (FYRN) and F/Y-rich C-terminus (FYRC) domains, collectively called domain associated with SET in Trx (DAST); (4) an extended plant homeodomain finger (ePHD); and (5) a C-terminal SET domain with its Post-SET companion domain (Fig. 26.1A) [13,57,58]. Tudor is a binding domain specific for different histones methylation marks [59]. PWWP is involved in chromatin-targeting to lysine modified nucleosomes [60]. DAST mediates protein-protein interactions, especially heterodimer formation [61]. The ePHD regulates ATX1 sub-cellular localization by binding to the lipid messenger molecule phosphatidylinositol 5-phosphate [62].

The functions of *Arabidopsis* ATX1 and ATX2 have been characterized in some detail (Table 26.2). A knockout mutant in *ATX1* exhibited pleiotropic phenotypes, affecting the activation of genes involved in secondary cell wall formation in the inflorescence stem, the response to environmental changes such as drought stresses, and developmental programs, including flowering time and leaf, flower and root development [63–67]. During flower development, ATX1 regulates floral organ identity by maintaining normal expression levels of the floral homeotic genes *APETALA1* (*AP1*), *AP2*, *PISTILLATA* (*PI*), and *AGAMOUS* (*AG*) [63]. ATX1 also participates in the regulation of flowering time by activating the expression of the floral repressor *FLOWERING LOCUS C* (*FLC*) and two other *FLC* paralogs, *MADS AFFECTING FLOWERING4* (*MAF4*) and *MAF5* [64,74]. In the *atx1* mutant, downregulation of *FLC* correlates with the loss of H3K4me3 in the chromatin of this gene [64]. Similar observations were made for other two ATX1 targets, the *LIPID TRANSFER PROTEIN* (*LTP*) and the *nine-cis-epoxycarotenoid dioxygenase* (*NCED*), related to abscisic acid synthesis and drought response [66,87]. Interestingly, H3K4me2 levels in the chromatin of *FLC* and *LTP* were unaffected in the *atx1* mutant, suggesting that ATX1 is required for H3K4me3 but not H3K4me2.

ATX1 and *ATX2* originated from a segmental chromosomal duplication [24]. The restricted expression

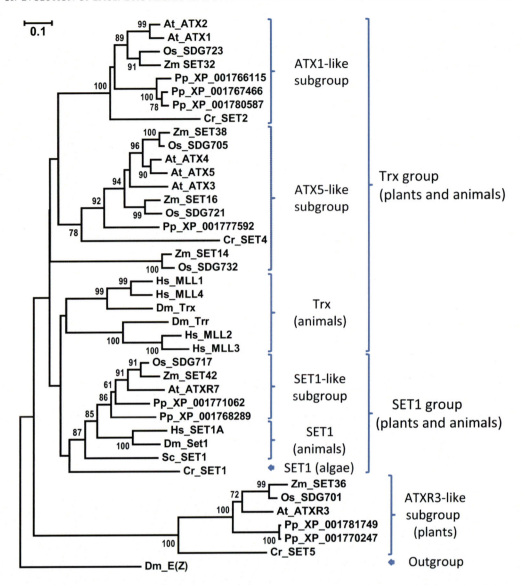

FIGURE 26.2 Phylogenetic relationships among HKMTs methylating H3K4 from plants and animals. Clustering of plant and/or animal proteins, forming major subgroups, is indicated to the right of the tree. The phylogeny was reconstructed using amino acid sequences corresponding to the SET domain obtained with the SMART database. Sequences were then aligned using the ClustalX program and MEGA version 6.0 was used to construct a Neighbor-joining tree. Reliability of the inferred tree was tested by bootstrap analysis based on 1000 pseudoreplicates. Only bootstrap values bigger than 60% are shown. The SET domain of the *Drosophila* enhancer of zeste [Dm_E(Z)] protein was used as an outgroup to place the root of the tree. The name of each protein is preceded by a two-letter abbreviation of the species as follows: At, *Arabidopsis thaliana*; Cr, *Chlamydomonas reinhardtii*; Dm, *Drosophila melanogaster*; Hs, *Homo sapiens*; Os, *Oryza sativa*; Pp, *Physcomitrella patens*; Sc, *Saccharomyces cerevisiae*; Zm, *Zea mays*.

pattern of *ATX2* relative to *ATX1* together with the lack of a visible *atx2*-related phenotype suggests that, after duplication, *ATX1* likely retained the ancestral function, while *ATX2* might have acquired novel, although limited, functions [68]. In fact, ATX2 is necessary for H3K4me2, but not H3K4me3, at its target loci [68]. Despite these differences, *atx2-1* null mutation enhances the early-flowering phenotype of *atx1-2*, suggesting that ATX1 and ATX2 likely play a redundant role in *FLC* transcriptional activation [64]. Thus, these observations indicate that ATX1 and ATX2 may share some common targets, perhaps through the same transcription factor set for target binding, yet the underlying mechanisms of activation may differ, as other HKMTs, such as ATXR7 and ATXR3, are also involved in *FLC* activation (see further).

A rice homolog of the ATX1/ATX2 pair, Os_SDG723, has been identified as a promoter of flowering, under long-day and short-day conditions, suggesting an evolutionarily conserved role of the ATX1-like subgroup in flowering time regulation [69]. Mutants in *Os_SDG723*, *ostrx1-2* and *ostrx1-3*, displayed reduced expression of

TABLE 26.2 Functions of Plant H3K4 HKMTs

Subgroup	Gene	Species	H3K4 Methylation[a]	Function	References
ATX1-like	ATX1	Arabidopsis	H3K4me3	Activation of genes involved in secondary cell wall formation, drought stress responses, developmental programs, including flowering time and leaf, flower and root development	[63–67]
	ATX2	Arabidopsis	H3K4me2	Regulation of flowering time	[64,68]
	Os_SDG723	rice	H3K4me3	Flowering time promotion	[69,70]
ATX5-like	ATX5	Arabidopsis	H3K4me2, H3K4me3	Sugar response, abscisic acid (ABA) signaling, stress response, plant growth and reproductive development	[39,71,72]
	ATX4	Arabidopsis	H3K4me2, H3K4me3	ABA signaling, stress response, plant growth and reproductive development	[39,72]
	ATX3	Arabidopsis	H3K4me2, H3K4me3	Plant growth and reproductive development	[72]
	Os_SDG705, Os_SDG721	rice	H3K4me2, H3K4me3	Gibberellic acid synthesis and signaling	[73]
SET1-like	ATXR7	Arabidopsis	H3K4me1, H3K4me2, H3K4me3	Regulation of flowering time, seed dormancy, lipid accumulation, cuticle integrity, plant immunity, and response to environmental stress	[74–79]
	Cr_SET1	Chlamydomonas	H3K4me1	Repression of single-copy transgenes and transposable elements	[80]
ATXR3-like	ATXR3	Arabidopsis	H3K4me3	Flowering time regulation, plant growth and sporophytic and gametophytic development	[81–85]
	Os_SDG701	rice	H3K4me2, H3K4me3	Flowering time, gametophyte development, and seed production	[86]

[a]Refers to in vivo H3K4 methylation mediated by each enzyme at individual loci or at genome-wide levels.

the flowering promoting genes RICE FLOWERING TIME LOCUS T1 (RFT1), Early heading date 1 (Ehd1) and Heading date 3a (Hd3a) [69,70]. Os_SDG723 can also bind the promoter of Ehd1 through its interaction with the transcription factor SIP (SDG723/OsTrx1 Interaction Protein 1). ostrx1-2 and ostrx1-3 also displayed reduced levels of H3K4me3 at RFT1, Ehd1 and Hd3a suggesting that Os_SDG723 is necessary for the deposition of H3K4me3, required for proper gene expression of its target genes [70].

Notably, the presence of a single rice and maize homolog of ATX1, Os_SDG723 and Zm_SET32, respectively, suggests the duplication event that gave origin to the pair ATX1/ATX2 may have occurred after monocot-eudicot divergence (Fig. 26.2) [57]. Thus, neofunctionalization of the dimethyl H3K4 activity of ATX2 has probably evolved as a lineage-specific feature not shared by the other members of the ATX1-like subgroup in which di- and trimethylation of H3K4 may be carried out by a single enzyme or by an alternative HKMT.

Although the function of ATX1 is remarkably similar to that of its animal homolog Trx, whose activity is also required to maintain the homeotic gene expression in Drosophila, the origin of these two regulatory activities is a case of convergent evolution [63]. Multicellularity originated independently in animals and plants, after which homeotic genes evolved in parallel in these two groups, likely as a selective pressure for functions conferring segment identity in multicellular organisms undergoing development [88]. Thus, a similar requirement for mechanisms that maintain gene expression patterns during development may have led to the convergent use of Trx-like proteins for the regulation of homeotic genes in animals and plants.

ATX5-like Histone Lysine Methyltransferases in Plants

Proteins belonging to ATX5-like subgroup are readily distinguished from ATX1-like by the lack of the Tudor and DAST domains, and the presence of an extra PHD finger (Fig. 26.1A) [57]. In spite of the single ATX5-like representative found in the green alga Chlamydomonas and the moss Physcomitrella patens, monocots, and eudicots possess two ATX5-like paralogs, which may have been originated likely before the monocot-eudicot split and diverged over time experiencing additional lineage-specific duplications (Fig. 26.2). Indeed, the pair ATX4/ATX5 appears to have been derived from a lineage-specific duplication, whereas the monocot proteins, Zm_SET14 and Os_SDG732, have diverged from the subgroup. Whether Zm_SET14 and Os_SDG732 have acquired novel functions or have been rapidly evolving remains to be explored.

Genes belonging to the ATX5-like subgroup have been characterized in *Arabidopsis* and rice where they have been shown to be involved in the response to environmental cues (glucose response, dehydration stress), phytohormone signaling (abscisic acid, gibberellic acid), and plant growth and development (Table 26.2) [39,71–73,89]. In *Arabidopsis*, ATX5 modulates glucose response by targeting *HEME OXYGENASE 1* (*HY1*), a repressor of the transcriptional regulator of sugar-responsive genes, *ABSCISIC ACID-INSENSITIVE4* (*ABI4*). ATX5 maintains the expression of *HY1* by catalyzing H3K4me3 deposition at the chromatin of this gene. However, increased concentrations of glucose impair ATX5 activity, resulting in *HY1* downregulation and the subsequent activation of the *ABI4*-mediated glucose response [71]. ATX5, together with ATX4, also play a role in abscisic acid (ABA)-signaling and drought stress responses [39]. Indeed, null mutants for *ATX4* and *ATX5* are both, hypersensitive to ABA treatment and more stress tolerant than the wildtype. ATX4 and ATX5 regulate these processes, at least partly, by directly binding to the negative regulator of the ABA-response, *ABA-HYPERSENSITIVE GERMINATION 3* (*AHG3*), in an ABA-dependent manner. Upon binding, the activity of both proteins increases H3K4me2 and H3K4me3 levels at the *AHG3* gene and induces its expression [39]. Interestingly, ATX4 and ATX5 appears to have partly redundant roles in the regulation ABA signaling since the double mutant for both genes displays exacerbated phenotypes [39]. Indeed, the target genes regulated by both proteins largely overlap, though a small subset of genes is uniquely modulated by each protein [39]. Thus, it appears that after duplication ATX4 and ATX5 remained extensively redundant for some functions but have been individually coopted for some others.

Genetic analysis of null mutants has shown that ATX4 and ATX5 also have redundant functions with ATX3. The lack of obvious phenotypes observed in single and double mutant combinations for *ATX3*, *ATX4* or *ATX5* contrast with the severe phenotype of the triple mutant, *atx3/atx4/atx5*, which display retarded growth and reduced seed set [72]. More recently, it has been shown that the *atx4* and *atx5* single mutants present slight growing and developmental phenotypes, which are exacerbated in the double mutant *atx4/atx5* [39]. However, despite the inconsistencies observed in the phenotypes of single and double mutants, likely due to the differences in growing conditions, the functional redundancy observed between ATX3, ATX4 and ATX5 remains true.

Similarly to *Arabidopsis*, ATX5-like genes in rice, *Os_SDG705* and *Os_SDG721*, were demonstrated to be partly redundant. Null mutants for *Os_SDG705* or *Os_SDG721* are semi-dwarf and display reduced panicle branching whereas double mutant plants grow even shorter, and with severely impaired panicle branching, than any of the single mutants [73]. The dwarf phenotype in the *sdg705* and *sdg721* mutants is explained by the reduced expression of genes involved in gibberellic acid (GA) synthesis and signaling, triggered by the loss of H3K4me3 [73]. Thus, even though some biological processes are regulated by individual ATX5-like genes, the majority of the pathways are partly modulated by more than one member. This suggests that subfunctionalization, by large, appears to be the main result of the duplication in the ATX5-like subgroup.

The proteins encoded by ATX5-like genes are HKMTs that catalyze di- and trimethylation of H3K4. In vitro experiments showing the HKMT activity of any of the ATX5-like proteins have proven unsuccessful; however, mutant analysis and transgene overexpression have demonstrated that ATX3/4/5 and Os_SDG705/721 are responsible for H3K4me2 and H3K4me3 at a substantial number of loci in the genome [39,72,73]. For instance, *Arabidopsis* ATX3/4/5 are responsible for 12% and 16% of total H3K4me2 and H3K4me3 methylation sites in the genome, respectively [72]. The contribution of individual ATX5-like genes to the global levels of H3K4me2 and H3K4me3 remains unclear. In *Arabidopsis*, Li and colleagues shown that *atx4* and *atx5* null-mutants exhibit reduced levels of H3K4me2 and H3K4me3, whereas another group found reduced levels of the same histone modifications at some loci only in *atx5* mutants [39,72]. In contrast, in rice reduced levels of H3K4me2 and H3K4me3 were only observed in the double mutant *sdg705/sdg721* [73]. Thus, subfunctionalization of ATX5-like genes appears to have occurred at the histone methylation level in which different members contribute to stablish proper levels of H3K4me2 and H3K4me3 at a given loci.

SET1-like Histone Lysine Methyltransferases in Plants

The SET1-like subgroup of plant HKMTs corresponds to proteins most related to the yeast and metazoan SET1. In this subgroup, the occurrence of a single homolog predominates across most plant species studied, apart from two paralogs found in the moss *P. patens* (Fig. 26.2). Plant SET1-like proteins contain both SET and Post-SET domains and lack the N-terminal RNA recognition motif (RRM) domain characteristic of their animal orthologs (Fig. 26.1B). Two proteins of this subgroup have been characterized so far. The first one, ARABIDOPSIS TRITHORAX-RELATED7 (ATXR7) participates in the regulation of flowering time, seed dormancy, lipid accumulation, cuticle integrity, plant immunity and response to environmental stress (Table 26.2) [74–79]. A loss-of-function mutation in ATXR7 resulted in lower expression of FLC, and its paralogs MAF4 and MAF5, leading to early flowering [74]. The atxr7 mutant also displayed reduced accumulation of lipids and defects in cuticle permeability, which were associated with enhanced susceptibility to pathogens [75]. The observed phenotypes were explained, at least in part, by reduced expression of CAROTENOID ISOMERASE2 (CCR2) and ECERIFERUM3 (CER3), encoding enzymes in the biosynthesis of carotenoids and cuticular wax, respectively [75]. ATXR7 also interacts with the plant-specific nuclear protein MODIFIER OF SNC1, 9 (MOS9) and both are required for the transcriptional activity of the NLR-encoding genes, SUPPRESSOR OF NPR1-1, CONSTITUTIVE1 (SNC1), and RECOGNITION OF PERONOSPORA PARASITICA4 (RPP4) [77]. In the atxr7 mutant, loss of transcriptional activation correlated with changes in H3K4 methylation. In the FLC gene, depletion of methyl H3K4 was observed at all levels, but was more dramatic in H3K4me1 [74]. CCR2 and CER3 also showed significantly reduced levels of H3K4me1, H3K4me2, and H3K4me3 in the atrx7 mutant [75]. Thus, in contrast to ATX1, ATXR7 is an HKMT that regulates multiple levels of H3K4 methylation at its target genes. Yet both ATX1 and ATXR7 can control the levels of H3K4me3 at a subset of genes related to the thermotolerance to high temperatures and in the transcriptional memory of heat stress in Arabidopsis. Together with ATX1, ATXR7 regulate the H3K4me3 levels not only required for stress-responsive genes during heat stress but also for preventing DNA methylation at loci associated with heat stress transcriptional memory [79].

Contrary to the activating function of ATXR7 and the other Trx-like HKMTs, the SET1-like protein from Chlamydomonas, Cr_SET1, is necessary to repress the expression of single-copy aadA transgenes (conferring resistance to spectinomycin) and the euchromatic TOC1 retrotransposon-like element [80]. In Chlamydomonas, single-copy transgenes and transposons integrated into euchromatic regions are associated with high levels of H3K4me1. Interestingly, downregulation of Cr_SET1 expression promoted the reactivation of these transcriptional units, which is correlated with decreased levels of H3K4me1, but not H3K4me2 [80]. Indeed, Cr_SET1 is not only required for H3K4me1 at these loci, but also for global H3K4me1 levels in Chlamydomonas. Neofunctionalization into a repressor protein may help explain the divergence of Cr_SET1 with respect to the other proteins of this subgroup. Nonetheless, ATXR7 also affects global levels of H3K4me1 [75], suggesting a conserved function for SET1-like subgroup of plant HKMTs in the implementation of H3K4me1, even though their biological outcomes are contrasting.

ATXR3-like Histone Lysine Methyltransferases in Plants

The ATXR3-like subgroup of H3K4 HKMTs cluster together ARABIDOPSIS TRITHORAX-RELATED3 (ATXR3) and its homologs (Fig. 26.2). This subgroup appears to correspond to novel SET-domain proteins that may have evolved exclusively in the plant lineage, as no homologs could be found in any other eukaryotic groups analyzed. The SET domain of ATXR3-like proteins is more divergent being not strictly located to the C-terminal end as in SET1-like or Trx-like proteins (Fig. 26.1C). Orthologs of ATXR3-like proteins could be already identified in unicellular algae indicating an early origin during plant evolution (Fig. 26.2) [84]. In addition, the ATXR3-like subgroup is represented by a single gene in Chlamydomonas, Arabidopsis, rice and maize, meaning that, with the exception of mosses that have two paralogs, this subgroup has not undergone expansion during plant evolution (Fig. 26.2).

ATXR3 and its rice homolog, Os_SDG701, participate in transcriptional activation in plants. A knockout mutant of Arabidopsis ATXR3 displayed pleiotropic developmental phenotypes, including smaller and curled leaves, short roots, early flowering, and defects in pollen and embryo sac development (Table 26.2) [81–85]. Consistent with these phenotypes, a large number of genes were differentially regulated in the atxr3 mutant, with some of the downregulated genes thought to be responsible for the phenotypes observed, such as the floral regulator FLC, genes essential for gametophyte development (SPOROCYTELESS/NOZZLE, SPL/NZZ and BTB AND TAZ DOMAIN PROTEIN3, BT3) and genes preferentially expressed in egg cells (EGG CELL1, EC1) and the endosperm (MEA) [81,82]. Moreover, the atxr3 mutant showed depleted global levels of H3K4me3 and, most important, at several of its targets, indicating the involvement of ATXR3 in the deposition of H3K4me3 on these genes [78,82]. ATXR3 is also the main HKMT catalyzing H3K4me3 during epigenetic reprogramming of the male gametophyte. H3K4me3 deposited by ATXR3 is necessary for proper chromatin decondensation in the microspore and the vegetative cell nucleus of the pollen. ATXR3 activity is also required for the transcriptional reactivation of transposable elements (TEs) in the vegetative cell of the pollen, thereby promoting the biosynthesis of mobile siRNAs from TEs that direct gene silencing in the sperm cell nucleus [85].

Os_SDG701, the rice homolog of ATXR3, is an essential gene also responsible for maintaining gene expression at a multitude of loci associated to several biological functions such as flowering time, gametophyte development, and seed production [86]. Os_SDG701 is a positive regulator of flowering since its overexpression and downregulation results in early and late flowering phenotypes, respectively. Os_SDG701 functions by depositing H3K4me3 in the chromatin of the flowering promoter genes *Heading date 3a* (*Hd3a*) and *RICE FLOWERING LOCUS T 1*, which results in their increased expression [86]. In addition, Os_SDG701 is required for H3K4me3 at a multitude of other target genes [86,90]. However, in contrast to ATXR3, Os_SDG701 is also necessary for H3K4me2 in vivo [90].

Interplay Between Plant Histone Lysine Methyltransferases in Stablishing H3K4 Methylation

As mentioned before, in yeast, Set1 is the only methyltransferase responsible for all levels of H3K4 methylation [44–46]. In metazoans, the increased number of HKMTs has led to a specialization of the methylation level specified by each enzyme. In *Drosophila* and mammalian cells, the bulk of H3K4me1, especially at enhancers, is carried out by the HKMTs Trr/MLL3/MLL4; whereas, dSET1 and SET1A/B are responsible for the majority of H3K4me2/3, leaving Trx/MLL1/MLL2 to mediate H3K4me3 at more specific targets and pathways [50,91–93]. In plants, there is also an apparent specialization of methyltransferases for different levels of H3K4 methylation, although it seems to have evolved slightly different than in animals. The SET1-like subgroup of HKMTs, represented by ATXR7 and Cr_SET1, is mainly responsible for H3K4me1 levels while global levels of H3K4me3 are frequently established by the plant specific ATXR3 protein and to a lesser extent by ATX5-like proteins [72,75,80–82]. The functions of plant ATX1-like HKMTs, similar to their animals counterparts, appears to be limited to specific target loci because its disruption caused only a slight decrease in the global levels of H3K4me3 [87].

At individual loci, the contributions of all HKMTs responsible for the establishment of proper levels of H3K4 methylation may vary. The lack of overlap between phenotypes and transcriptional profiles in the mutants for ATXR3 and ATX1, and ATX1 and ATX2, indicates these HKMTs may play distinct roles in gene regulation [68,82]. However, these three methyltransferases as well as ATXR7, have been shown to play a role in the proper activation of *FLC* in the *FRIGIDA*-Col-0 background [64,74,84]. Regarding flowering phenotype, *atxr3* is epistatic to *atxr7* while the phenotype of the double mutants, *atx1*/*atxr7* and *atx1*/*atx2*, are additive relative to that of their single mutants [64,74,84]. These observations suggest that ATXR3 and ATXR7 act together in *FLC* activation, but the activities of the other HKMTs are redundant. A model that can account for these observations is one in which ATXR7 is required for H3K4me1, and perhaps H3K4me2, that in turn serves as a substrate for ATX2 that can convert it to H3K4me2. Then, the combined activity of both ATX1 and ATXR3 might be required to produce the levels of H3K4me3 necessary to reach full transcriptional activation of *FLC*. Whether this model is correct or not will require further genetic analysis of the mutants, but it reflects the complex interactions occurring between HKMTs necessary to establish H3K4me3.

EVOLUTION OF PLANT HISTONE LYSINE METHYLTRANSFERASES METHYLATING H3K27

HKMTs involved in H3K27 methylation include homologs of the *Drosophila* E(Z) and the human EZH1 and EZH2. These enzymes are the catalytic component of a conserved multisubunit complex, Polycomb repressive complex 2 (PRC2), that catalyzes H3K27 trimethylation, a PTM associated with transcriptional gene silencing [94–98]. E(Z) and homologs of the PRC2 core subunits have been found across species from the major eukaryotic lineages. This supports the hypothesis that these proteins existed in the common unicellular ancestor of all eukaryotes likely functioning as a silencing mechanism to combat transposons and other genomic parasites [14].

Phylogenetic Relationship Between Histone Lysine Methyltransferases Methylating H3K27

Drosophila E(Z) and human EZH1 and EZH2 have a conserved domain structure that consists of two SANT domains, an EZD2 domain, a CXC cysteine rich region, and a C-terminal SET domain (Fig. 26.3A). The SANT domain [switching-defective protein 3 (Swi3), Adaptor 2 (Ada2), nuclear receptor corepressor (N-CoR), and transcription factor (TF)IIIB] is a histone-binding module that couples histone tail binding to catalytic activity [99]. EZD2 is a small region of ~50 amino acids containing five conserved cysteines present in plant and animal E(Z) proteins [100]. The CXC domain is a ~65-residue cysteine-rich region preceding the catalytic SET domain that may be necessary for enhanced HKMT activity of E(Z) [101]. Genes coding E(Z) proteins are represented by single members in *Drosophila* [E(Z)] and nematodes (maternal-effect sterile 2 or

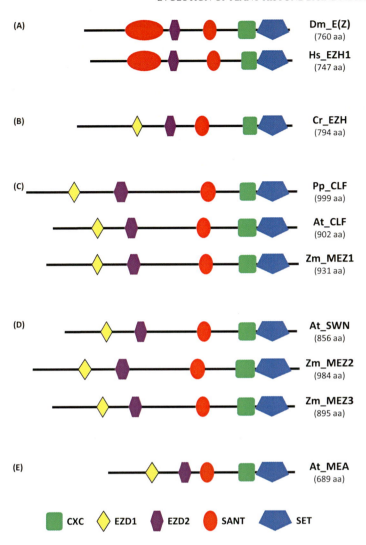

FIGURE 26.3 Domain architecture of E(Z) homologs from several species of plants and animals. Representative E(Z) homologs from (A) animals, (B) algae, and plants from the (C) CURLY LEAF (CLF)-like, (D) SWINGER (SWN)-like, and (E) MEA-like subgroups are depicted. Switching-defective protein 3 (Swi3), Adaptor 2 (Ada2), nuclear receptor corepressor (N-CoR), and transcription factor (TF)IIIB (SANT), CXC, and SET domains were identified using the SMART database, whereas the EZD1 and EZD2 domains were identified by aligning conserved plant and animal E(Z) homologs. Proteins are drawn to scale. The name of each protein is preceded by a two-letter abbreviation of the species as follows: At, *Arabidopsis thaliana*; Cr, *Chlamydomonas reinhardtii*; Dm, *Drosophila melanogaster*; Hs, *Homo sapiens*; Pp, *Physcomitrella patens*; Zm, *Zea mays*.

MES-2) and have likely undergone a single duplication event in mammals giving rise to EZH1 and EZH2 (Fig. 26.4) [98].

Plant E(Z) proteins have a domain structure fairly conserved with that of their animal orthologs (Fig. 26.3). They consist of (1) an N-terminal EZD1 domain, a conserved acidic region of ~70 amino acids present only in plant E(Z) proteins [102], (2) an EZD2 motif, (3) a single SANT domain, (4) a CXC domain and (5) a C-terminal SET domain (Fig. 26.3). This domain structure is conserved in most plants, although some proteins from the chlorophyte algae, such as *Ostreococcus tauri*, may lack the SANT domain [14]. Phylogenetic analysis reveals a monophyletic origin for all plant E(Z) homologs, an observation supported by the presence of single E(Z) proteins in the genomes of the early-divergent chlorophyte green algae (Fig. 26.4). In land plants, similarly to mammals, E(Z) proteins have experienced limited expansion. A single duplication event, predating the split between monocots and eudicots, gave origin to two subgroups, CLF-like and SWN-like, represented by the *Arabidopsis* CURLY LEAF (CLF) and SWINGER (SWN) proteins (Fig. 26.4). Additional lineage-specific duplications may have created novel E(Z) copies in either subgroup. For example, a single duplication in the CLF-like subgroup, early during Solanaceae evolution, gave origin to the tomato Sl_EZ2 and Sl_EZ3, and, in maize, MEZ2 and MEZ3 originated from a duplication in the SWN-like subgroup (Fig. 26.4). A relatively recent duplication, 35–85 million years ago, gave origin to a divergent subgroup, MEA-like, whose archetype, *Arabidopsis* MEDEA (MEA) originated from SWN during a WGD event within the Brassicaceae lineage [103]. The rapid evolution of MEA driven by positive selection has led to a rapid divergence from SWN and other E(Z) proteins [103,104].

Histone Lysine Methyltransferases Methylating H3K27 in Early-diverging Plant Lineages

E(Z) homologs have been characterized to different levels in few species that diverged earlier during plant

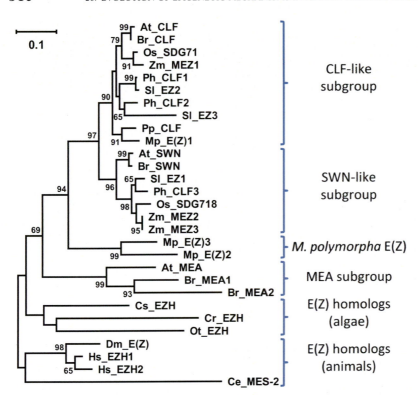

FIGURE 26.4 Unrooted neighbor-joining tree showing the relationship among E(Z) homologs from plants and animals. Clustering of E(Z) homologs from animals, algae, and land plants is indicated to the right of the tree. The tree was constructed using amino acid sequences corresponding to the SET domain, aligned with ClustalX, and MEGA version X was used to draw the tree. Bootstrap analysis based on 1000 pseudoreplicates was used to determine the robustness of the tree. Only bootstrap values bigger than 60% are shown. The name of each protein is preceded by a two-letter abbreviation of the species as follows: At, *Arabidopsis thaliana*; Br, *Brassica rapa*; Ce, *Caenorhabditis elegans*; Cr, *Chlamydomonas reinhardtii*; Cs, *Coccomyxa subellipsoidea* C-169; Dm, *Drosophila melanogaster*; Hs, *Homo sapiens*; Mp, *Marchantia polymorpha*; Os, *Oryza sativa*; Ot, *Ostreococcus tauri*; Ph, *Petunia hybrida*; Pp, *Physcomitrella patens*; Sl, *Solanum lycopersicum*; Zm, *Zea mays*.

evolution, including the chlorophyte green alga *Chlamydomonas* and the bryophytes *Physcomitrella patens* and *Marchantia polymorpha* (Table 26.3). In *Chlamydomonas*, a single *Enhancer of zeste* homolog (*Cr_EZH*) was identified [14]. Downregulation of *Cr_EZH* caused the release of silencing of the repetitive *aadA* transgenes, found in a heterochromatic silenced state, and of the *TOC1* retrotransposon present in dispersed copies throughout the *Chlamydomonas* genome, pointing a role for Cr_EZH in the modulation of repressive chromatin states. Even though H3K27me3 could not be assayed, silencing of *Cr_EZH* resulted in increased global levels of H3K4me3 and H4 acetylation, modifications generally associated with transcriptional activation, further implicating Cr_EZH in gene silencing [14]. In agreement with these observations, analysis of the genomic distribution of H3K27me3 in the red algae *Cyanidioschizon merolae* and the marine diatom *Phaeodactylum tricornutum* has revealed that H3K27me3 is enriched in silent genes, transposons, and other repetitive elements [125,126]. Thus, an early function for plant E(Z) homologs was possibly the generation of repressive chromatin as part of a defense mechanism against genomic parasites such as transposable elements [14,127]. Later in plant evolution, with the advent of multicellularity and the increased in complexity of gene regulation, E(Z) appears to have acquired specialized functions, such as regulation of genes involved in developmental programs and cell identity [14]. Nevertheless, association between H3K27me3 and transposable elements still occurs in *M. polymorpha*, even though to a less extent than in diatoms and red algae, suggesting that E(Z) homologs still played a role, yet limited, in the silencing of repeats elements in the common accessor of bryophytes and vascular plants [127].

E(Z) homologs have been characterized from the early-diverging bryophytes *P. patens* and *M. polymorpha*. The genome of the moss *P. patens* encodes a single copy of E(Z), Pp_CLF, and the liverwort *M. polymorpha* encodes three homologs, Mp_E(Z)1, Mp_E(Z)2, and Mp_E(Z)3 [23,105]. Interestingly, Pp_CLF and Mp_E(Z)1 cluster together with members of the CLF-like subgroup whereas MpE(Z)2, and MpE(Z)3 do not belong to either CLF-like or SWN-like subgroups (Fig. 26.4). Despite the limited information available from the two bryophyte representatives, the clustering pattern of E(Z) homologs indicates that Pp_CLF and Mp_E(Z)1 are canonical E(Z) proteins. MpE(Z)2, and MpE(Z)3 have diverged from CLF-like and SWN-like proteins instead, likely the result of rapid evolution after a lineage-specific duplication in *M. polymorpha*.

In land plants, the life cycle is characterized by an "alternation of generations" between the diploid sporophyte and the haploid gametophyte [128]. In mosses like *P. patens*, the gametophyte is the dominant form in the life cycle, whereas the sporophyte is reduced and parasitizes on the gametophyte [129]. For the two generations to strictly alternate, sporophyte development must be repressed in the gametophyte until

TABLE 26.3 Plant H3K27 HKMTs and Their Functions

Gene	species	Plant Subgroup	function	references
Cr_EZH	Chlamydomonas	none	Silencing of multi-copy transgenes and retrotransposons	[14]
Pp_CLF	Physcomitrellapatens	CLF-like	Repression of sporophyte development in the gametophyte	[105,106]
Mp_E(Z)1,	Marchantia polymorpha	CLF-like	repression of the sporophyte-specific genes in the gametophyte	[23,107]
Mp_E(Z)2, Mp_E(Z)3	Marchantia polymorpha	none	Not determined	[23]
CLF	Arabidopsis	CLF-like	Regulation of developmental programs such as those of leaves and flowers, flowering time regulation, stem cell maintenance, root meristem activity, and organ size control and lipid biosynthesis during postfertilization development	[100,108–113]
SWN	Arabidopsis	SWN-like	Repression of embryo development program in vegetative tissues	[114–116]
MEA	Arabidopsis	MEA-like	Repression of central cell proliferation and endosperm development in the female gametophyte before fertilization, and regulation of cell proliferation in the endosperm and the embryo after fertilization	[117–119]
Sl_EZ1	tomato	SWN-like	Flower development and carpel number determination	[120]
Sl_EZ2	tomato	CLF-like	Plant growth, development of leaves, flowers and fruits, and fruit ripening and cuticle formation	[121]
Sl_EZ3	tomato	CLF-like	Not determined	[120,121]
Os_SDG711	rice	CLF-like	Repression of flowering under long day conditions, flower and panicle development, gene repression during shoot meristem to leaf transition, gene silencing and DNA methylation maintenance.	[122–124]
Os_SDG718	rice	SWN-like	Flowering promotion under short day conditions	[122]

fertilization. Interestingly, in Pp_CLF deletion mutants, and mutants that are catalytically inactive, the gametophytic vegetative cells produced sporophyte-like bodies, expressing sporophyte-specific markers, in the absence of fertilization [105,106]. Furthermore, ectopic expression of Pp_CLF ceased the proliferation of the sporophyte-like apical cell and led to the formation of a sporangium-like (a structure in which spores are formed) organ [105]. Intriguingly, recent findings show that in the liverwort M. polymorpha, Mp_E(Z)1, a Pp_CLF ortholog, is involved in the repression of the sporophyte-specific genes, MpKNOX2 and MpBELL1, in the gametophyte [107]. Together these observations suggest that E(Z) homologs in bryophytes function in the repression of the sporophyte developmental program and cell proliferation in gametophytic vegetative cells [105,107]. Thus, even early during land plant evolution, E(Z) proteins had already been recruited for the control of developmental programs and the regulation of cell identity.

Deletion of Pp_CLF and mutations in conserved sites that render the SET domain catalytically inactive resulted in the complete depletion of H3K27me3 in P. patens histones [106]. In addition, genes associated with H3K27me3 in the wildtype, BELL1, BELL2, JAGGED1, JAGGED2 and AIB, showed increased expression and loss of the H3K27me3 mark in Pp_CLF mutants [106]. Similarly, in M. polymorpha downregulation of Mp_E(Z)1 resulted in global reduction of H3K27me3 levels [107]. Furthermore, MpKNOX2 and MpBELL1, two genes normally repressed in gametophyte tissues, become reactivated with the concomitant loss of H3K27me3 upon downregulation of Mp_E(Z)1 [107]. Thus, these results demonstrate that Pp_CLF and Mp_E(Z)1 are the major H3K27me3 HKMTs in bryophytes and that this modification regulates silencing of orthologs of transcription factors known to regulate a multitude of developmental processes in higher plants [106,107].

Histone Lysine Methyltransferases Methylating H3K27 in Eudicotyledonous Plants

In eudicot plants, E(Z) proteins have been characterized at some level in the model plant Arabidopsis, tomato, and petunia. Arabidopsis E(Z) homologs are encoded by three genes CLF, SWN, and MEA. CLF and SWN functions are partially redundant during sporophyte development whereas MEA act specifically during gametophyte and seed development [100,114,117]. Interestingly, CLF, MEA, and SWN are undetectable during sperm development, a crucial feature in the resetting of the epigenetic memory in plant paternal chromatin, which in plants involves the loss of global, and specifically H3K27me3, marks in the sperm, unlike in animals for which germline reprogramming involves mainly DNA methylation [130].

CLF participates in the regulation of several developmental programs including leaf and flower morphology, flowering time, stem cell maintenance, root meristem activity, and organ size control and lipid biosynthesis during postfertilization development (Table 26.3) [100,108–113]. Abnormal leaf and flower phenotypes in the *clf* mutant are caused by the ectopic expression of floral homeotic genes *AGAMOUS* (*AG*), *APETALA3* (*AP3*), and *AGAMOUS LIKE17* (*AGL17*); and the KNOTTED-like homeobox genes *KNOTTED1-LIKE2* (*KNAT2*) and *SHOOT MERISTEMLESS* (*STM*) [100,110]. CLF is also necessary for the repression of several flowering time genes. *clf* mutants exhibited higher expression of the floral repressor gene *FLC* and its paralogs *MAF4* and *MAF5*, as well as the floral promoter genes *AGL19* and *FLOWERING LOCUS T* (*FT*) [108,111]. CLF also acts during flower development to repress the expression of the stem cell regulator *WUSCHEL* (*WUS*) resulting in floral meristem termination after production of a defined number of floral organs [113]. In the root, CLF negatively regulates meristematic activity by repressing root stem cell and meristem genes, such as *AGL42*, *AGL21*, *WUSCHEL RELATED HOMEOBOX5* (*WOX5*), and *SHORT-ROOT* (*SHR*) [109].

In contrast to CLF, loss-of-function *swn* mutations produced no phenotype [114]. However, the *swnxclf* double mutant showed a much more severe phenotype than either single mutant, pinpointing the redundant functions between SWN and CLF. *swnxclf* double mutant spontaneously forms callus-like structures that produce somatic embryos, likely as a result of ectopic expression of embryonic regulators such as *LEAFY COTYLEDON1* (*LEC1*), *LEC2* and *FUSCA3* (*FUS3*) [114,115]. Furthermore, analysis of H3K27me3 profiles in seedlings from different mutants showed that CLF is the sole HKMT responsible for silencing and H3K27me3 at flower developmental genes, whereas it has a redundant function with SWN in the repression of embryonic traits [116]. Thus, SWN and CLF act redundantly in the repression of the embryonic program in vegetative tissues. The imprinted *MEA* gene is required for repression of central cell proliferation and endosperm development in the female gametophyte before fertilization, and also for proper proliferation of cells in the embryo and in the endosperm after fertilization [117–119]. Interestingly, the *swn* mutant enhances the autonomous seed formation phenotype of the *mea* mutant, indicating that SWN has partially redundant roles with MEA in repressing endosperm proliferation in the female gametophyte [131]. Not only different *Arabidopsis* E(Z) homologs have redundant functions, but they also seem to cooperate in the repression of the same targets, even though they act at different stages of plant development [115]. Thus, complex interactions between the paralogs CLF, SWN and MEA occur at different genomic regions and at different times during development with each enzyme contributing differently to the repression of individual loci.

Like other E(Z) homologs, CLF, SWN, and MEA are the catalytic engine responsible for the deposition of the H3K27me3 mark that mediates gene repression. Loss of CLF function led to reduced global levels of H3K27me3 [111]. Furthermore, transcriptional repression failure in several direct CLF targets (i.e., *AG*, *STM*, *FLC*, *AGL19*, and *FT*) was associated with the loss of the H3K27me3 mark from the chromatin of these genes in *clf* mutants [108,111,132]. Similarly, redundant regulation of *STM* by SWN and CLF requires either enzyme to mediate H3K27me3 and gene silencing [132]. In flower buds, MEA activity is responsible for mediating repression and H3K27me3 at its direct target *PHERES1* (*PHE1*) [115,133]. Thus, CLF, SWN, and MEA are HKMTs that mediate H3K27me3 and transcriptional repression of developmental programs ensuring appropriate gene expression in *Arabidopsis*.

Three tomato (*S. lycopersicum*) homologs of E(Z) have been identified, Sl_EZ1, Sl_EZ2, and Sl_EZ3 (Table 26.3) [120,121]. Sl_EZ2 and Sl_EZ3 cluster together in the CLF-like subgroup, while Sl_EZ1 belongs to the SWN-like subgroup (Fig. 26.4) [120]. Downregulation of *Sl_EZ2* produced pleiotropic phenotypes including reduced plant size, shorter internodes, altered leaf and flower morphology, and defects in fruit development, ripening and cuticle formation [121]. The multiple roles of Sl_EZ2 during plant development, as well as its phylogenetic relationship with *Arabidopsis* CLF, indicate that it may be its functional ortholog. Similar to *CLF*, *Sl_EZ2* knockdown decreased global levels of H3K27me3, suggesting a conserved role of Sl_EZ2 in the implementation of this modification. Yet expression analysis of some orthologs of CLF targets in the Sl_EZ2-deficient tomato plants indicated that these genes may not be regulated by Sl_EZ2 [121]. Indeed, some mutant phenotypes suggest that Sl_EZ2 may have diversified their functions and/or acquired new ones, especially during fleshy fruit development.

In contrast to *Sl_EZ2*, downregulation of *Sl_EZ1* resulted in defects only in flower development and fruits with increased carpel numbers. Phenotypes arising from the individual downregulation of *Sl_EZ1* and *Sl_EZ2* suggest that both proteins may have distinct, at least partly, nonoverlapping functions during tomato development [120,121]. Consistently, specific expression patterns were observed for the three tomato E(Z) homologs [121]. The functional separation of Sl_EZ1 and Sl_EZ2, in contrast to the redundancy of CLF-SWN function, indicates that the roles played by different E(Z) paralogs in *Arabidopsis* and tomato may have changed during evolution [129].

Three E(Z) homologs were also found in another member of the Solanaceae family, petunia (*Petunia hybrida*), namely Ph_CLF1, Ph_CLF2 and Ph_CLF3. Phylogenetic analysis showed that the duplication in the CLF-like orthologs of tomato and petunia likely predates the split between these closely-related species suggesting that it may have occurred early during speciation within Solanaceae lineage (Fig. 26.4) [120,134]. Although the function of petunia genes has not been characterized yet, the divergence in expression patterns and differences in regulation of splicing between *Ph_CLF1* and *Ph_CLF2* indicates likely neo- or subfunctionalization. Further analysis may yield additional clues on the functional diversification and emergence of new roles for the E(Z) genes in the Solanaceae family.

Histone Lysine Methyltransferases Methylating H3K27 in Monocotyledonous Plants

E(Z) homologs have been characterized in two monocot species, maize and rice. Three maize enhancer of zeste-like genes, *MEZ1*, *MEZ2*, and *MEZ3*, were the first to be identified [102]. MEZ1 belongs to the CLF-like subgroup, whereas MEZ2 and MEZ3 are more closely related and cluster with the SWN-like subgroup (Fig. 26.4) [102]. MEZ2 and MEZ3 may have been originated from a duplication later in monocot evolution, as the genome of rice and barley encode a single SWN-like ortholog (Fig. 26.4). To our knowledge, functional characterization of the maize E(Z) genes has not been carried out to date. However, the three genes were shown to be expressed throughout plant development [102].

Similar to *Arabidopsis MEA*, *MEZ1* is imprinted in the endosperm, although both alleles are expressed in other tissues [135]. Furthermore, activity of the maternally inherited allele of *MEA* and *MEZ1* is necessary for silencing of the paternal allele [136,137]. However, MEZ1 and *Arabidopsis* MEA appear to have evolved independently. *MEA* was probably originated from a duplication from *SWN* in the cruciferous plant lineage, whilst *MEZ1* belongs to the CLF-like subgroup that diverged from the SWN-like subgroup early during the evolution of land plants [135]. Independent origins of MEA and MEZ1 not only suggest that convergent evolution of imprinting may have occurred in the E(Z) family, but also indicate that there may be a requirement for maternal-specific expression in endosperm tissue in plants [135].

In contrast to maize, rice has two E(Z) paralogs, Os_SDG711 and Os_SDG718 belonging to the CLF-like and SWN-like subgroups, respectively (Fig. 26.4) (Table 26.3) [138]. Gene expression analysis showed that *Os_SDG711* was widely expressed, while the expression of *Os_SDG718* was more predominant in leaves. Furthermore, expression of *Os_SDG711* was higher in long days (14 h light/10 h dark) than in short days (10 h light/14 h dark), whereas *Os_SDG718* exhibited the opposite pattern [122]. Os_SDG711 is a repressor of flowering under long days by targeting *early heading date 1* (*Ehd1*), which, in turn, controls the expression of *Hd3a*, a floral promoter, ortholog of *Arabidopsis FT* [122]. Though indirectly, Os_SDG711 also controls the expression of *heading date 1* (*Hd1*), a repressor of flowering in long days [122]. Thus, repression of flowering under long day conditions requires the coordinated regulation of two different pathways by Os_SDG711. Similarly, Os_SDG718 is a promoter of flowering, but under short day conditions. Os_SDG718 represses the expression of *Os_LF* (a repressor of *Hd1*) which in turn leads to activation of *Hd1* that behaves as a flowering promoter under short day conditions [122]. Thus, Os_SDG711 and Os_SDG718 help to achieve a precise photoperiod control of flowering time by regulating key flowering genes in rice.

In contrast to *Arabidopsis*, downregulation of neither *Os_SDG711* nor *Os_SDG718* led to developmental and/or morphological phenotypes in vegetative or reproductive tissues, suggesting that these proteins may act redundantly or not at all in the regulation of developmental programs. The latter is likely not to be true, as several lines of evidence point to Os_SDG711 and Os_SDG718 having functions in rice development. Os_SDG718 directly interacts with DEFORMED FLORAL ORGAN1 (DFO1) an ortholog of the *Arabidopsis* repressor protein EMBRYONIC FLOWER1 (EMF1) [139]. In the *dfo1* mutant, loss of H3K27me3-mediated repression of the floral identity gene *Os_MADS58*, the ortholog of *Arabidopsis AG*, leads to abnormal floral morphology [139]. Thus, DFO and likely Os-SDG718 are regulators of the floral identity in rice. Overexpression of *Os_SDG711* resulted in higher number of stamens and reduced pollen viability, further implicating rice E(Z) homologs in the regulation of flower development [122]. Os_SDG711 has also been shown to be involved in the regulation of inflorescence meristem activity. Downregulation of Os_SDG711 resulted in reduced panicle (the rice branched inflorescences) size with fewer primary branches whereas its overexpression led to the opposite phenotype [140].

Furthermore, silencing of Os_SDG711 led to increased expression of many hormone and panicle developmental genes [140]. Similarly, Os_SDG711 was shown to be important for regulating gene expression during the transition from shoot apical meristem to leaf [123].

Similar to other E(Z) homologs, Os_SDG711 and Os_SDG718 encode HKMTs responsible for implementing H3K27me3 and for silencing their target genes. Os_SDG711, belonging to the CLF-like subgroup, has, so far, the largest effect in H3K27me3 levels in rice. Downregulation and overexpression of Os_SDG711 was shown to result in decreased and increased global levels of H3K27me3, respectively [122]. Furthermore, many of the genes upregulated in the inflorescence meristem (genes associated with panicle development, auxin and cytokinin synthesis or signaling, metabolism) of Os_SDG711-knockdown plants were found to lose H3K27me3 [140]. Similarly, downregulation of Os_SDG711 led to reduced levels of H3K27me3 and activation of starch synthesis and amylase genes in developing seeds [141]. Also, Os_SDG711 overexpression led to an increase in genes, gaining H3K27me3 during the transition from shoot apical meristem to leaf [123]. Os_SDG718, on the other hand, has been shown to be responsible for H3K27me3 levels solely at its target gene Os_LF [122]. These observations suggest that despite the fact that both Os_SDG711 and Os_SDG718 are HKMTs responsible for H3K27me3, Os_SDG711 is likely the predominant methyltransferase in rice.

Os_SDG711 has been shown to interact with Os_DRM2, a DNA methyltransferase responsible for genome wide CHH (where H equals A, T, or C) and CHG methylation, suggesting a functional interaction between H3K27me3 and DNA methylation in rice [123,142]. Indeed, a mutation in Os_DRM2 reduced H3K27me3 levels of a subset of genes, particularly developmental genes, marked with CHG and CHH methylation, while Os_SDG711 was shown to target many genes, enriched in CHG and CHH methylation, for H3K27 methylation [123]. Thus, Os_DRM2 not only mediates CHG and CHH methylation but also recruits Os_SDG711 to mediate H3K27me3 at a subset of genes in rice. In addition, the lower expression of genes marked by these two modifications, compared with those only associated to H3K27me3, indicates that they cooperate to repress developmental genes [123].

Os_SDG711 was also implicated in the maintenance of stable repression of the 03 g locus (Os03g02470). In wildtype plants, 03 g is a transcriptionally inactive gene associated with high levels of DNA methylation and H3K27me3 marks, and low levels of H3K4me3. Upon callus culture, the 03 g gene becomes activated (gaining H3K4me3 and losing H3K27me3 and DNA methylation) and remains as such in regenerated plants and for at least two generations [124]. However, in the gain-of-function mutant sdg711 and plants overexpressing Os_SDG711, stable activation of 03 g was impaired while histone and DNA methylation marks were brought back to their wildtype levels in the regenerated plants. Furthermore, stably expressed 03 g was resilenced by crossing with alleles overexpressing Os_SDG711 [124]. Taken together, these observations suggest that H3K27me3 mediated by Os_SDG711 is sufficient to maintain and/or rescue DNA methylation and to mediate stable repression of 03g, and potentially implicates this protein as a novel player in the generation and/or resetting of other epialleles in rice [124].

The characterization of E(Z) homologs reveals a conserved function in coordinating gene repression during plant development and regulating flowering time in rice, and perhaps in all monocot plants. Furthermore, additional functions for these proteins in transgenerational gene repression, the generation of epigenetic variation, and the coordinated regulation of developmental genes with DNA methylation has been discovered [123,124]. Interestingly, these two new features involve the interaction of DNA methylation with H3K27me3 in ensuring the stable repression of common targets.

PERSPECTIVES

Among the known chromatin-regulating modules, the SET domain is an ancient motif that may have been recruited for histone modification, and thereby chromatin regulation, in the dawn of eukaryotic evolution. Indeed, given their conservation throughout many eukaryotic supergroups, E(Z) homologs and some proteins of the PRC2 complex may have been already present in the last common ancestor of all eukaryotes [14]. In addition, it is interesting to observe remarkable similarities in the structure and function between other SET-domain proteins from distant eukaryotic lineages, such as plants and animals.

Early eukaryotes may have already had a complex set of HKMTs as supported by the presence of animal homologs in four out of five plant SDGs classes with HKMT activity [24,25,35]. In plants, the number of SET-domain containing proteins has been further increased during evolution from usually one representative in eukaryotic unicellular green algae to several paralogs in higher plants in each SDG class. Many of these paralogs may have their origins early in plant evolution, even before the divergence between monocots and eudicots [35]. Because Trx-like, E(Z) homologs and other plant SDGs have been recruited for the regulation of developmental pathways, as in the vertebrate lineage, we can speculate that increased complexity of gene regulation patterns associated with the adoption of multicellularity may have driven the expansion of SET-domain genes early in plant history. Other landmarks in plant evolution, such as the embrace of a sessile lifestyle and an open growth system (the continuous formation of organs throughout the life cycle) may have taken advantage of additional SDGs to tightly regulate growth patterns in response to environmental cues and the maintenance of stem cells for the constant generation of new organs. These same factors might have contributed to the expansion of other families of chromatin regulators, such as histone demethylases, histone acetyltransferases (HATs), and deacetylases (HDACs) [143,144]. Indeed, the evolutionary history of these chromatin regulators resembles the evolution of the SDG family itself. For instance, ancient WGDs during early angiosperm evolution may have contributed to the expansion of some members of the histone demethylase family [144]. Similarly, the plant-specific type-2 HDAC family (HD2), which likely originated as a single-copy gene in the Streptophyta green algae, appears to have increased its copy number in monocot and eudicot species [145]. Duplicated HD2 genes may have been the product of an ancient polyploidization event, predating the origin of all seed plants, and also from successive duplication events in different groups, particularly in eudicot species [145,146]. Thus, functional diversification of SDGs and other chromatin regulators during early plant evolution indicates that modulation of gene expression by histone modification may have been an important factor in the evolutionary success of plants [144].

The majority of evolutionary innovation during evolution of H3K27me3/H3K4me-based mechanisms is associated with the recruitment of the HMKTs to regulate novel processes. For instance, Sl_EZ2 has been coopted for transcriptional control underlying fleshy fruit development in tomato; whereas, neither CLF, SWN, nor MEA play a role during silique development in *Arabidopsis* [121]. Hence, while the mechanism of chromatin regulation remains the same, that is H3K4me3 mediating gene activation and H3K27me3 mediating gene repression, neofunctionalization of duplicated genes appears to be attained by adopting the regulation of new genomic targets or regulatory networks. Besides that, neofunctionalization may also come from new interactions with other histone modifying systems that may help to reinforce specific chromatin states or lead to the regulation of new targets. As an example, a novel interaction found in rice between Os_SDG711 and Os_DRM2 led to H3K27me3 and DNA methylation to cooperate in the silencing a subset of their target genes [123]. However, in a few cases, new enzymatic activities and even different functional meaning of the modification in question may arise. In *Arabidopsis*, ATX2 has diverged from ATX1 into an HKMT that specifically converts H3K4me1 to H3K4me2, whereas ATX1 is able to catalyze H3K4me3 [68]. Transcription from an internal promoter produced a solo-SET ATX1-derived protein that methylated nonhistone targets in the cytoplasm [147]. Although *Chlamydomonas* SET1 and *Arabidopsis* ATXR7 are both proteins mediating H3K4me1, only SET1-mediated H3K4me1 functions as a silencing mark in euchromatic transposons and transgenes in *Chlamydomonas*, whereas ATXR7 is involved in gene activation in *Arabidopsis* [74,80].

Novel functions resulting in changes of catalytic specificity may add novel layers of complexity to the regulation of a single modification, could also lead to the modification of novel histone residues, thereby increasing the complexity of the histone code. In addition, the emergence of novel functions for conserved histone modifications in some lineages, as in SET1-mediated H3K4me1, adds to mounting evidence against a universal histone code. Thus, to understand the language of covalent modifications in plants and all its "dialects" the function of HKMTs and other modifying enzymes should be investigated in as many lineages of the plant kingdom as possible. Comparative analysis between lineages will also help us to understand how the histone code and all its dialects evolved.

Acknowledgments

JA acknowledges the support of the São Paulo Research Foundation (FAPESP), grant 2020/06560−1.

References

[1] Kouzarides T. Chromatin modifications and their function. Cell 2007;128(4):693−705.
[2] Berger SL. The complex language of chromatin regulation during transcription. Nature 2007;447(7143):407−12.
[3] Huang H, Sabari BR, Garcia BA, Allis CD, Zhao Y. SnapShot: histone modifications. Cell 2014;159(2):458 e1.
[4] Tessarz P, Kouzarides T. Histone core modifications regulating nucleosome structure and dynamics. Nat Rev Mol Cell Biol 2014;15(11):703−8.
[5] Taverna SD, Li H, Ruthenburg AJ, Allis CD, Patel DJ. How chromatin-binding modules interpret histone modifications: lessons from professional pocket pickers. Nat Struct Mol Biol 2007;14(11):1025−40.
[6] Strahl BD, Allis CD. The language of covalent histone modifications. Nature 2000;403(6765):41−5.
[7] Sims 3rd RJ, Reinberg D. Is there a code embedded in proteins that is based on post-translational modifications? Nat Rev Mol Cell Biol 2008;9(10):815−20.
[8] Cerutti H, Casas-Mollano JA. Histone H3 phosphorylation: universal code or lineage specific dialects? Epigenetics 2009;4(2):71−5.

[9] Fuchs J, Demidov D, Houben A, Schubert I. Chromosomal histone modification patterns—from conservation to diversity. Trends Plant Sci 2006;11(4):199—208.

[10] Feng S, Jacobsen SE. Epigenetic modifications in plants: an evolutionary perspective. Curr Opin Plant Biol 2011;14(2):179—86.

[11] Lachner M, Jenuwein T. The many faces of histone lysine methylation. Curr Opin Cell Biol 2002;14(3):286—98.

[12] Zhang Y, Reinberg D. Transcription regulation by histone methylation: interplay between different covalent modifications of the core histone tails. Genes Dev 2001;15(18):2343—60.

[13] Alvarez-Venegas R. Regulation by polycomb and trithorax group proteins in Arabidopsis. Arabidopsis Book 2010;8:e0128.

[14] Shaver S, Casas-Mollano JA, Cerny RL, Cerutti H. Origin of the polycomb repressive complex 2 and gene silencing by an E(z) homolog in the unicellular alga Chlamydomonas. Epigenetics 2010;5(4):301—12.

[15] Weinhofer I, Hehenberger E, Roszak P, Hennig L, Kohler C. H3K27me3 profiling of the endosperm implies exclusion of polycomb group protein targeting by DNA methylation. PLoS Genet 2010;6(10):e1001152.

[16] Du J, Johnson LM, Jacobsen SE, Patel DJ. DNA methylation pathways and their crosstalk with histone methylation. Nat Rev Mol Cell Biol 2015;16(9):519—32.

[17] Johnson L, Mollah S, Garcia BA, Muratore TL, Shabanowitz J, Hunt DF, et al. Mass spectrometry analysis of Arabidopsis histone H3 reveals distinct combinations of post-translational modifications. Nucleic Acids Res 2004;32(22):6511—18.

[18] Moraes I, Yuan ZF, Liu S, Souza GM, Garcia BA, Casas-Mollano JA. Analysis of Histones H3 and H4 Reveals Novel and Conserved Post-Translational Modifications in Sugarcane. PLoS One 2015;10(7):e0134586.

[19] Zhang K, Sridhar VV, Zhu J, Kapoor A, Zhu JK. Distinctive core histone post-translational modification patterns in Arabidopsis thaliana. PLoS One 2007;2(11):e1210.

[20] Wu T, Yuan T, Tsai SN, Wang C, Sun SM, Lam HM, et al. Mass spectrometry analysis of the variants of histone H3 and H4 of soybean and their post-translational modifications. BMC Plant Biol 2009;9:98.

[21] Casas-Mollano JA, van Dijk K, Eisenhart J, Cerutti H. SET3p monomethylates histone H3 on lysine 9 and is required for the silencing of tandemly repeated transgenes in Chlamydomonas. Nucleic Acids Res 2007;35(3):939—50.

[22] Alvarez-Venegas R, Sadder M, Tikhonov A, Avramova Z. Origin of the bacterial SET domain genes: vertical or horizontal? Mol Biol Evol 2007;24(2):482—97.

[23] Bowman JL, Kohchi T, Yamato KT, Jenkins J, Shu S, Ishizaki K, et al. Insights into Land Plant Evolution Garnered from the Marchantia polymorpha Genome. Cell 2017;171(2):287—304 e15.

[24] Baumbusch LO, Thorstensen T, Krauss V, Fischer A, Naumann K, Assalkhou R, et al. The *Arabidopsis thaliana* genome contains at least 29 active genes encoding SET domain proteins that can be assigned to four evolutionarily conserved classes. Nucleic Acids Res 2001;29(21):4319—33.

[25] Ng DW, Wang T, Chandrasekharan MB, Aramayo R, Kertbundit S, Hall TC, et al. domain-containing proteins: structure, function and regulation. Biochim Biophys Acta 2007;1769(5—6):316—29.

[26] Qian Y, Xi Y, Cheng B, Zhu S, Kan X. Identification and characterization of the SET domain gene family in maize. Mol Biol Rep 2014;41(3):1341—54.

[27] Huang Y, Liu C, Shen WH, Ruan Y. Phylogenetic analysis and classification of the Brassica rapa SET-domain protein family. BMC Plant Biol 2011;11:175.

[28] Aquea F, Vega A, Timmermann T, Poupin MJ, Arce-Johnson P. Genome-wide analysis of the SET DOMAIN GROUP family in grapevine. Plant Cell Rep 2011;30(6):1087—97.

[29] Aiese Cigliano R, Sanseverino W, Cremona G, Ercolano MR, Conicella C, Consiglio FM. Genome-wide analysis of histone modifiers in tomato: gaining an insight into their developmental roles. BMC Genomics 2013;14:57.

[30] Huang Y, Mo Y, Chen P, Yuan X, Meng F, Zhu S, et al. Identification of SET domain-containing proteins in gossypium raimondii and their response to high temperature stress. Sci Rep 2016;6:32729.

[31] Lei L, Zhou SL, Ma H, Zhang LS. Expansion and diversification of the SET domain gene family following whole-genome duplications in *Populus trichocarpa*. BMC Evol Biol 2012;12:51.

[32] Gu T, Han Y, Huang R, McAvoy RJ, Li Y. Identification and characterization of histone lysine methylation modifiers in Fragaria vesca. Sci Rep 2016;6:23581.

[33] Peng M, Ying P, Liu X, Li C, Xia R, Li J, et al. Genome-wide identification of histone modifiers and their expression patterns during fruit abscission in litchi. Front Plant Sci 2017;8:639.

[34] Li W, Yan J, Wang S, Wang Q, Wang C, Li Z, et al. Genome-wide analysis of SET-domain group histone methyltransferases in apple reveals their role in development and stress responses. BMC Genomics 2021;22(1):283.

[35] Springer NM, Napoli CA, Selinger DA, Pandey R, Cone KC, Chandler VL, et al. Comparative analysis of SET domain proteins in maize and Arabidopsis reveals multiple duplications preceding the divergence of monocots and dicots. Plant Physiol 2003;132(2):907—25.

[36] Lu Z, Huang X, Ouyang Y, Yao J. Genome-wide identification, phylogenetic and co-expression analysis of OsSET gene family in rice. PLoS One 2013;8(6):e65426.

[37] Yadav CB, Muthamilarasan M, Dangi A, Shweta S, Prasad M. Comprehensive analysis of SET domain gene family in foxtail millet identifies the putative role of SiSET14 in abiotic stress tolerance. Sci Rep 2016;6:32621.

[38] Chen DH, Qiu HL, Huang Y, Zhang L, Si JP. Genome-wide identification and expression profiling of SET Domain Group family in *Dendrobium catenatum*. BMC Plant Biol 2020;20(1):40.

[39] Liu Y, Zhang A, Yin H, Meng Q, Yu X, Huang S, et al. Trithorax-group proteins ARABIDOPSIS TRITHORAX4 (ATX4) and ATX5 function in abscisic acid and dehydration stress responses. N Phytol 2018;217(4):1582—97.

[40] Batra R, Gautam T, Pal S, Chaturvedi D, Rakhi, Jan I, et al. Identification and characterization of SET domain family genes in bread wheat (*Triticum aestivum* L.). Sci Rep 2020;10(1):14624.

[41] Lynch M, Conery JS. The evolutionary fate and consequences of duplicate genes. Science 2000;290(5494):1151—5.

[42] Cannon SB, Mitra A, Baumgarten A, Young ND, May G. The roles of segmental and tandem gene duplication in the evolution of large gene families in *Arabidopsis thaliana*. BMC Plant Biol 2004;4:10.

[43] Briggs SD, Bryk M, Strahl BD, Cheung WL, Davie JK, Dent SY, et al. Histone H3 lysine 4 methylation is mediated by Set1 and required for cell growth and rDNA silencing in *Saccharomyces cerevisiae*. Genes Dev 2001;15(24):3286—95.

[44] Krogan NJ, Dover J, Khorrami S, Greenblatt JF, Schneider J, Johnston M, et al. COMPASS, a histone H3 (Lysine 4) methyltransferase required for telomeric silencing of gene expression. J Biol Chem 2002;277(13):10753—5.

[45] Nagy PL, Griesenbeck J, Kornberg RD, Cleary ML. A trithorax-group complex purified from Saccharomyces cerevisiae is required for methylation of histone H3. Proc Natl Acad Sci U S A 2002;99(1):90—4.

[46] Roguev A, Schaft D, Shevchenko A, Pijnappel WW, Wilm M, Aasland R, et al. The *Saccharomyces cerevisiae* Set1 complex includes an Ash2 homologue and methylates histone 3 lysine 4. EMBO J 2001;20(24):7137—48.

[47] Mohan M, Herz HM, Smith ER, Zhang Y, Jackson J, Washburn MP, et al. The COMPASS family of H3K4 methylases in Drosophila. Mol Cell Biol 2011;31(21):4310–18.

[48] Mazo AM, Huang DH, Mozer BA, Dawid IB. The trithorax gene, a trans-acting regulator of the bithorax complex in Drosophila, encodes a protein with zinc-binding domains. Proc Natl Acad Sci U S A 1990;87(6):2112–16.

[49] Sedkov Y, Cho E, Petruk S, Cherbas L, Smith ST, Jones RS, et al. Methylation at lysine 4 of histone H3 in ecdysone-dependent development of Drosophila. Nature 2003;426(6962):78–83.

[50] Ardehali MB, Mei A, Zobeck KL, Caron M, Lis JT, Kusch T. Drosophila Set1 is the major histone H3 lysine 4 trimethyltransferase with role in transcription. EMBO J 2011;30(14):2817–28.

[51] Goo YH, Sohn YC, Kim DH, Kim SW, Kang MJ, Jung DJ, et al. Activating signal cointegrator 2 belongs to a novel steady-state complex that contains a subset of trithorax group proteins. Mol Cell Biol 2003;23(1):140–9.

[52] Hughes CM, Rozenblatt-Rosen O, Milne TA, Copeland TD, Levine SS, Lee JC, et al. Menin associates with a trithorax family histone methyltransferase complex and with the hoxc8 locus. Mol Cell 2004;13(4):587–97.

[53] Lee JH, Skalnik DG. CpG-binding protein (CXXC finger protein 1) is a component of the mammalian Set1 histone H3-Lys4 methyltransferase complex, the analogue of the yeast Set1/COMPASS complex. J Biol Chem 2005;280(50):41725–31.

[54] Lee JH, Tate CM, You JS, Skalnik DG. Identification and characterization of the human Set1B histone H3-Lys4 methyltransferase complex. J Biol Chem 2007;282(18):13419–28.

[55] Gu Y, Nakamura T, Alder H, Prasad R, Canaani O, Cimino G, et al. The t(4;11) chromosome translocation of human acute leukemias fuses the ALL-1 gene, related to *Drosophila trithorax*, to the AF-4 gene. Cell 1992;71(4):701–8.

[56] Tkachuk DC, Kohler S, Cleary ML. Involvement of a homolog of *Drosophila trithorax* by 11q23 chromosomal translocations in acute leukemias. Cell 1992;71(4):691–700.

[57] Avramova Z. Evolution and pleiotropy of TRITHORAX function in Arabidopsis. Int J Dev Biol 2009;53(2–3):371–81.

[58] Fromm M, Avramova Z. ATX1/AtCOMPASS and the H3K4me3 marks: how do they activate Arabidopsis genes? Curr Opin Plant Biol 2014;21:75–82.

[59] Lu R, Wang GG. Tudor: a versatile family of histone methylation 'readers'. Trends Biochem Sci 2013;38(11):546–55.

[60] Qin S, Min J. Structure and function of the nucleosome-binding PWWP domain. Trends Biochem Sci 2014;39(11):536–47.

[61] Pless B, Oehm C, Knauer S, Stauber RH, Dingermann T, Marschalek R. The heterodimerization domains of MLL-FYRN and FYRC—are potential target structures in t(4;11) leukemia. Leukemia 2011;25(4):663–70.

[62] Alvarez-Venegas R, Sadder M, Hlavacka A, Baluska F, Xia Y, Lu G, et al. The Arabidopsis homolog of trithorax, ATX1, binds phosphatidylinositol 5-phosphate, and the two regulate a common set of target genes. Proc Natl Acad Sci U S A 2006;103(15):6049–54.

[63] Alvarez-Venegas R, Pien S, Sadder M, Witmer X, Grossniklaus U, Avramova Z. ATX-1 an Arabidopsis homolog of trithorax, activates flower homeotic genes. Curr Biol 2003;13(8):627–37.

[64] Pien S, Fleury D, Mylne JS, Crevillen P, Inze D, Avramova Z, et al. ARABIDOPSIS TRITHORAX1 dynamically regulates FLOWERING LOCUS C activation via histone 3 lysine 4 trimethylation. Plant Cell 2008;20(3):580–8.

[65] Napsucialy-Mendivil S, Alvarez-Venegas R, Shishkova S, Dubrovsky JG. Arabidopsis homolog of trithorax1 (ATX1) is required for cell production, patterning, and morphogenesis in root development. J Exp Bot 2014;65(22):6373–84.

[66] Ding Y, Avramova Z, Fromm M. The Arabidopsis trithorax-like factor ATX1 functions in dehydration stress responses via ABA-dependent and ABA-independent pathways. Plant J 2011;66(5):735–44.

[67] Wang X, Wang D, Xu W, Kong L, Ye X, Zhuang Q, et al. Histone methyltransferase ATX1 dynamically regulates fiber secondary cell wall biosynthesis in Arabidopsis inflorescence stem. Nucleic Acids Res 2021;49(1):190–205.

[68] Saleh A, Alvarez-Venegas R, Yilmaz M, Le O, Hou G, Sadder M, et al. The highly similar Arabidopsis homologs of trithorax ATX1 and ATX2 encode proteins with divergent biochemical functions. Plant Cell 2008;20(3):568–79.

[69] Choi SC, Lee S, Kim SR, Lee YS, Liu C, Cao X, et al. Trithorax group protein Oryza sativa Trithorax1 controls flowering time in rice via interaction with early heading date3. Plant Physiol 2014;164(3):1326–37.

[70] Jiang P, Wang S, Zheng H, Li H, Zhang F, Su Y, et al. SIP1 participates in regulation of flowering time in rice by recruiting OsTrx1 to Ehd1. N Phytol 2018;219(1):422–35.

[71] Liu Y, Wang J, Yin H, Zhang A, Huang S, Wang TJ, et al. Trithorax-group protein ATX5 mediates the glucose response via impacting the HY1-ABI4 signaling module. Plant Mol Biol 2018;98(6):495–506.

[72] Chen LQ, Luo JH, Cui ZH, Xue M, Wang L, Zhang XY, et al. ATX3, ATX4, and ATX5 encode putative H3K4 methyltransferases and are critical for plant development. Plant Physiol 2017;174(3):1795–806.

[73] Jiang P, Wang S, Ikram AU, Xu Z, Jiang H, Cheng B, et al. SDG721 and SDG705 are required for rice growth. J Integr Plant Biol 2018;60(7):530–5.

[74] Tamada Y, Yun JY, Woo SC, Amasino RM. ARABIDOPSIS TRITHORAX-RELATED7 is required for methylation of lysine 4 of histone H3 and for transcriptional activation of FLOWERING LOCUS C. Plant Cell 2009;21(10):3257–69.

[75] Lee S, Fu F, Xu S, Lee SY, Yun DJ, Mengiste T. Global regulation of plant immunity by histone lysine methyl transferases. Plant Cell 2016;28(7):1640–61.

[76] Liu Y, Geyer R, van Zanten M, Carles A, Li Y, Horold A, et al. Identification of the Arabidopsis REDUCED DORMANCY 2 gene uncovers a role for the polymerase associated factor 1 complex in seed dormancy. PLoS One 2011;6(7):e22241.

[77] Xia S, Cheng YT, Huang S, Win J, Soards A, Jinn TL, et al. Regulation of transcription of nucleotide-binding leucine-rich repeat-encoding genes SNC1 and RPP4 via H3K4 trimethylation. Plant Physiol 2013;162(3):1694–705.

[78] Berr A, Xu L, Gao J, Cognat V, Steinmetz A, Dong A, et al. SET DOMAIN GROUP25 encodes a histone methyltransferase and is involved in FLOWERING LOCUS C activation and repression of flowering. Plant Physiol 2009;151(3):1476–85.

[79] Song ZT, Zhang LL, Han JJ, Zhou M, Liu JX. Histone H3K4 methyltransferases SDG25 and ATX1 maintain heat-stress gene expression during recovery in Arabidopsis. Plant J 2021;105(5):1326–38.

[80] van Dijk K, Marley KE, Jeong BR, Xu J, Hesson J, Cerny RL, et al. Monomethyl histone H3 lysine 4 as an epigenetic mark for silenced euchromatin in Chlamydomonas. Plant Cell 2005;17(9):2439–53.

[81] Berr A, McCallum EJ, Menard R, Meyer D, Fuchs J, Dong A, et al. Arabidopsis SET DOMAIN GROUP2 is required for H3K4 trimethylation and is crucial for both sporophyte and gametophyte development. Plant Cell 2010;22(10):3232–48.

[82] Guo L, Yu Y, Law JA, Zhang X. SET DOMAIN GROUP2 is the major histone H3 lysine 4 trimethyltransferase in Arabidopsis. Proc Natl Acad Sci U S A 2010;107(43):18557–62.

[83] Yao X, Feng H, Yu Y, Dong A, Shen WH. SDG2-mediated H3K4 methylation is required for proper Arabidopsis root growth and development. PLoS One 2013;8(2):e56537.

[84] Yun JY, Tamada Y, Kang YE, Amasino RM. Arabidopsis trithorax-related3/SET domain GROUP2 is required for the winter-annual habit of Arabidopsis thaliana. Plant Cell Physiol 2012;53(5):834–46.

[85] Pinon V, Yao X, Dong A, Shen WH. SDG2-Mediated H3K4me3 is crucial for chromatin condensation and mitotic division during male gametogenesis in Arabidopsis. Plant Physiol 2017;174(2):1205–15.

[86] Liu K, Yu Y, Dong A, Shen WH. SET DOMAIN GROUP701 encodes a H3K4-methytransferase and regulates multiple key processes of rice plant development. N Phytol 2017;215(2):609–23.

[87] Alvarez-Venegas R, Avramova Z. Methylation patterns of histone H3 Lys 4, Lys 9 and Lys 27 in transcriptionally active and inactive Arabidopsis genes and in atx1 mutants. Nucleic Acids Res 2005;33(16):5199–207.

[88] Meyerowitz EM. Plants and the logic of development. Genetics 1997;145(1):5–9.

[89] Lee K, Park OS, Choi CY, Seo PJ. ARABIDOPSIS TRITHORAX 4 facilitates shoot identity establishment during the plant regeneration process. Plant Cell Physiol 2019;60(4):826–34.

[90] Liu Y, Liu K, Yin L, Yu Y, Qi J, Shen WH, et al. H3K4me2 functions as a repressive epigenetic mark in plants. Epigenetics Chromatin 2019;12(1):40.

[91] Wu M, Wang PF, Lee JS, Martin-Brown S, Florens L, Washburn M, et al. Molecular regulation of H3K4 trimethylation by Wdr82, a component of human Set1/COMPASS. Mol Cell Biol 2008;28(24):7337–44.

[92] Hu D, Gao X, Morgan MA, Herz HM, Smith ER, Shilatifard A. The MLL3/MLL4 branches of the COMPASS family function as major histone H3K4 monomethylases at enhancers. Mol Cell Biol 2013;33(23):4745–54.

[93] Herz HM, Mohan M, Garruss AS, Liang K, Takahashi YH, Mickey K, et al. Enhancer-associated H3K4 monomethylation by Trithorax-related, the Drosophila homolog of mammalian Mll3/Mll4. Genes Dev 2012;26(23):2604–20.

[94] Cao R, Wang L, Wang H, Xia L, Erdjument-Bromage H, Tempst P, et al. Role of histone H3 lysine 27 methylation in Polycomb-group silencing. Science 2002;298(5595):1039–43.

[95] Kuzmichev A, Nishioka K, Erdjument-Bromage H, Tempst P, Reinberg D. Histone methyltransferase activity associated with a human multiprotein complex containing the Enhancer of Zeste protein. Genes Dev 2002;16(22):2893–905.

[96] Czermin B, Melfi R, McCabe D, Seitz V, Imhof A, Pirrotta V. Drosophila enhancer of Zeste/ESC complexes have a histone H3 methyltransferase activity that marks chromosomal Polycomb sites. Cell 2002;111(2):185–96.

[97] Muller J, Hart CM, Francis NJ, Vargas ML, Sengupta A, Wild B, et al. Histone methyltransferase activity of a Drosophila Polycomb group repressor complex. Cell 2002;111(2):197–208.

[98] Margueron R, Reinberg D. The Polycomb complex PRC2 and its mark in life. Nature 2011;469(7330):343–9.

[99] Boyer LA, Latek RR, Peterson CL. The SANT domain: a unique histone-tail-binding module? Nat Rev Mol Cell Biol 2004;5(2):158–63.

[100] Goodrich J, Puangsomlee P, Martin M, Long D, Meyerowitz EM, Coupland G. A Polycomb-group gene regulates homeotic gene expression in Arabidopsis. Nature 1997;386(6620):44–51.

[101] Ketel CS, Andersen EF, Vargas ML, Suh J, Strome S, Simon JA. Subunit contributions to histone methyltransferase activities of fly and worm polycomb group complexes. Mol Cell Biol 2005;25(16):6857–68.

[102] Springer NM, Danilevskaya ON, Hermon P, Helentjaris TG, Phillips RL, Kaeppler HF, et al. Sequence relationships, conserved domains, and expression patterns for maize homologs of the polycomb group genes E(z), esc, and E(Pc). Plant Physiol 2002;128(4):1332–45.

[103] Spillane C, Schmid KJ, Laoueille-Duprat S, Pien S, Escobar-Restrepo JM, Baroux C, et al. Positive darwinian selection at the imprinted MEDEA locus in plants. Nature 2007;448(7151):349–52.

[104] Miyake T, Takebayashi N, Wolf DE. Possible diversifying selection in the imprinted gene, MEDEA, in Arabidopsis. Mol Biol Evol 2009;26(4):843–57.

[105] Okano Y, Aono N, Hiwatashi Y, Murata T, Nishiyama T, Ishikawa T, et al. A polycomb repressive complex 2 gene regulates apogamy and gives evolutionary insights into early land plant evolution. Proc Natl Acad Sci U S A 2009;106(38):16321–6.

[106] Pereman I, Mosquna A, Katz A, Wiedemann G, Lang D, Decker EL, et al. The Polycomb group protein CLF emerges as a specific tri-methylase of H3K27 regulating gene expression and development in Physcomitrella patens. Biochim Biophys Acta 2016;1859(7):860–70.

[107] Dierschke T, Flores-Sandoval E, Rast-Somssich MI, Althoff F, Zachgo S, Bowman JL. Gamete-specific expression of TALE class HD genes activates the diploid sporophyte program in Marchantia polymorpha. bioRxiv. 2020.

[108] Schonrock N, Bouveret R, Leroy O, Borghi L, Kohler C, Gruissem W, et al. Polycomb-group proteins repress the floral activator AGL19 in the FLC-independent vernalization pathway. Genes Dev 2006;20(12):1667–78.

[109] Aichinger E, Villar CB, Di Mambro R, Sabatini S, Kohler C. The CHD3 chromatin remodeler PICKLE and polycomb group proteins antagonistically regulate meristem activity in the Arabidopsis root. Plant Cell 23(3):1047–1060.

[110] Katz A, Oliva M, Mosquna A, Hakim O, Ohad N. FIE and CURLY LEAF polycomb proteins interact in the regulation of homeobox gene expression during sporophyte development. Plant J 2004;37(5):707–19.

[111] Jiang D, Wang Y, Wang Y, He Y. Repression of FLOWERING LOCUS C and FLOWERING LOCUS T by the Arabidopsis Polycomb repressive complex 2 components. PLoS One 2008;3(10):e3404.

[112] Liu J, Deng S, Wang H, Ye J, Wu HW, Sun HX, et al. CURLY LEAF regulates gene sets coordinating seed size and lipid biosynthesis. Plant Physiol 2016;171(1):424–36.

[113] Liu X, Kim YJ, Muller R, Yumul RE, Liu C, Pan Y, et al. AGAMOUS terminates floral stem cell maintenance in Arabidopsis by directly repressing WUSCHEL through recruitment of Polycomb Group proteins. Plant Cell 2011;23(10):3654–70.

[114] Chanvivattana Y, Bishopp A, Schubert D, Stock C, Moon YH, Sung ZR, et al. Interaction of Polycomb-group proteins controlling flowering in Arabidopsis. Development 2004;131(21):5263–76.

[115] Makarevich G, Leroy O, Akinci U, Schubert D, Clarenz O, Goodrich J, et al. Different Polycomb group complexes regulate common target genes in Arabidopsis. EMBO Rep 2006;7(9):947–52.

[116] Wang H, Liu C, Cheng J, Liu J, Zhang L, He C, et al. Arabidopsis flower and embryo developmental genes are repressed in seedlings by different combinations of polycomb group proteins in association with distinct sets of Cis-regulatory elements. PLoS Genet 2016;12(1):e1005771.

[117] Grossniklaus U, Vielle-Calzada JP, Hoeppner MA, Gagliano WB. Maternal control of embryogenesis by MEDEA, a polycomb group gene in Arabidopsis. Science 1998;280(5362):446–50.

[118] Kiyosue T, Ohad N, Yadegari R, Hannon M, Dinneny J, Wells D, et al. Control of fertilization-independent endosperm development by the MEDEA polycomb gene in Arabidopsis. Proc Natl Acad Sci U S A 1999;96(7):4186–91.

[119] Kinoshita T, Yadegari R, Harada JJ, Goldberg RB, Fischer RL. Imprinting of the MEDEA polycomb gene in the Arabidopsis endosperm. Plant Cell 1999;11(10):1945—52.

[120] How Kit A, Boureau L, Stammitti-Bert L, Rolin D, Teyssier E, Gallusci P. Functional analysis of SlEZ1 a tomato enhancer of zeste (E(z)) gene demonstrates a role in flower development. Plant Mol Biol 2010;74(3):201—13.

[121] Boureau L, How-Kit A, Teyssier E, Drevensek S, Rainieri M, Joubes J, et al. A CURLY LEAF homologue controls both vegetative and reproductive development of tomato plants. Plant Mol Biol 2016;90(4—5):485—501.

[122] Liu X, Zhou C, Zhao Y, Zhou S, Wang W, Zhou DX. The rice enhancer of zeste [E(z)] genes SDG711 and SDG718 are respectively involved in long day and short day signaling to mediate the accurate photoperiod control of flowering time. Front Plant Sci 2014;5:591.

[123] Zhou S, Liu X, Zhou C, Zhou Q, Zhao Y, Li G, et al. Cooperation between the H3K27me3 chromatin marker and non-CG methylation in epigenetic regulation. Plant Physiol 2016;172(2):1131—41.

[124] Chen X, Liu X, Zhao Y, Zhou DX. Histone H3K4me3 and H3K27me3 regulatory genes control stable transmission of an epimutation in rice. Sci Rep 2015;5:13251.

[125] Mikulski P, Komarynets O, Fachinelli F, Weber APM, Schubert D. Characterization of the Polycomb-Group Mark H3K27me3 in unicellular algae. Front Plant Sci 2017;8:607.

[126] Veluchamy A, Rastogi A, Lin X, Lombard B, Murik O, Thomas Y, et al. An integrative analysis of post-translational histone modifications in the marine diatom Phaeodactylum tricornutum. Genome Biol 2015;16:102.

[127] Montgomery SA, Tanizawa Y, Galik B, Wang N, Ito T, Mochizuki T, et al. Chromatin organization in early land plants reveals an ancestral association between H3K27me3, transposons, and constitutive heterochromatin. Curr Biol 2020;30(4):573—88 e7.

[128] Bennici A. Origin and early evolution of land plants: Problems and considerations. Commun Integr Biol 2008;1(2):212—18.

[129] Butenko Y, Ohad N. Polycomb-group mediated epigenetic mechanisms through plant evolution. Biochim Biophys Acta 2011;1809(8):395—406.

[130] Borg M, Jacob Y, Susaki D, LeBlanc C, Buendia D, Axelsson E, et al. Targeted reprogramming of H3K27me3 resets epigenetic memory in plant paternal chromatin. Nat Cell Biol 2020;22(6):621—9.

[131] Wang D, Tyson MD, Jackson SS, Yadegari R. Partially redundant functions of two SET-domain polycomb-group proteins in controlling initiation of seed development in Arabidopsis. Proc Natl Acad Sci U S A 2006;103(35):13244—9.

[132] Schubert D, Primavesi L, Bishopp A, Roberts G, Doonan J, Jenuwein T, et al. Silencing by plant Polycomb-group genes requires dispersed trimethylation of histone H3 at lysine 27. EMBO J 2006;25(19):4638—49.

[133] Kohler C, Hennig L, Spillane C, Pien S, Gruissem W, Grossniklaus U. The Polycomb-group protein MEDEA regulates seed development by controlling expression of the MADS-box gene PHERES1. Genes Dev 2003;17(12):1540—53.

[134] Mayama T, Ohtsubo E, Tsuchimoto S. Isolation and expression analysis of petunia CURLY LEAF-like genes. Plant Cell Physiol 2003;44(8):811—19.

[135] Haun WJ, Laoueille-Duprat S, O'Connell MJ, Spillane C, Grossniklaus U, Phillips AR, et al. Genomic imprinting, methylation and molecular evolution of maize Enhancer of zeste (Mez) homologs. Plant J 2007;49(2):325—37.

[136] Gehring M, Huh JH, Hsieh TF, Penterman J, Choi Y, Harada JJ, et al. DEMETER DNA glycosylase establishes MEDEA polycomb gene self-imprinting by allele-specific demethylation. Cell 2006;124(3):495—506.

[137] Haun WJ, Danilevskaya ON, Meeley RB, Springer NM. Disruption of imprinting by mutator transposon insertions in the 5′ proximal regions of the Zea mays Mez1 locus. Genetics 2009;181(4):1229—37.

[138] Luo M, Platten D, Chaudhury A, Peacock WJ, Dennis ES. Expression, imprinting, and evolution of rice homologs of the polycomb group genes. Mol Plant 2009;2(4):711—23.

[139] Zheng M, Wang Y, Wang Y, Wang C, Ren Y, Lv J, et al. DEFORMED FLORAL ORGAN1 (DFO1) regulates floral organ identity by epigenetically repressing the expression of OsMADS58 in rice (Oryza sativa). N Phytol 2015;206(4):1476—90.

[140] Liu X, Zhou S, Wang W, Ye Y, Zhao Y, Xu Q, et al. Regulation of histone methylation and reprogramming of gene expression in the rice inflorescence meristem. Plant Cell 2015;27(5):1428—44.

[141] Liu X, Luo J, Li T, Yang H, Wang P, Su L, et al. SDG711 is involved in rice seed development through regulation of starch metabolism gene expression in coordination with other histone modifications. Rice (N Y) 2021;14(1):25.

[142] Tan F, Zhou C, Zhou Q, Zhou S, Yang W, Zhao Y, et al. Analysis of chromatin regulators reveals specific features of rice DNA methylation pathways. Plant Physiol 2016;171(3):2041—54.

[143] Chua YL, Gray JC. Histone modifications and transcription in plants. Annual Plant Reviews 29: Regulation of Transcription in Plants. Blackwell Publishing Ltd; 2007. p. 79—111.

[144] Qian S, Wang Y, Ma H, Zhang L. Expansion and functional divergence of Jumonji c-containing histone demethylases: significance of duplications in ancestral angiosperms and vertebrates. Plant Physiol 2015;168(4):1321—37.

[145] Bourque S, Jeandroz S, Grandperret V, Lehotai N, Aime S, Soltis DE, et al. The evolution of HD2 proteins in green plants. Trends Plant Sci 2016;21(12):1008—16.

[146] Pandey R, Muller A, Napoli CA, Selinger DA, Pikaard CS, Richards EJ, et al. Analysis of histone acetyltransferase and histone deacetylase families of *Arabidopsis thaliana* suggests functional diversification of chromatin modification among multicellular eukaryotes. Nucleic Acids Res 2002;30(23):5036—55.

[147] Ndamukong I, Lapko H, Cerny RL, Avramova Z. A cytoplasm-specific activity encoded by the Trithorax-like ATX1 gene. Nucleic Acids Res 2011;39(11):4709—18.

CHAPTER 27

Evolution, Functions and Dynamics of Epigenetic Mechanisms in Animals

Günter Vogt
Faculty of Biosciences, University of Heidelberg, Heidelberg, Germany

OUTLINE

Introduction	521	Phenotypic Plasticity	531
Evolution of Epigenetic Mechanisms in The Animal Kingdom	521	Relevance of Epigenetics for Animal Ecology	534
		Asexually Reproducing Populations	534
DNA Methylation	522	Sessile Taxa	536
Histone Modifications	523	Invasions and Adaptive Radiations	536
Non-coding RNAs	523	Parasites	538
Features of Epigenetic Mechanisms and Role in Gene Regulation	524	Relevance of Epigenetics for Animal Evolution	538
		Domestication	538
Dynamics of Epigenetic Marks During Development	526	Polyploid Speciation	540
		Transgenerational Inheritance of Epigenetically Determined Phenotypes	542
Embryonic Development	526		
Cell and Tissue Differentiation	528	Conclusions and Perspectives	543
Aging	528	Summary	545
Generation of Phenotypic Variation in Populations by Epigenetic Mechanisms	530	References	545
Polyphenism in Insects	530		

INTRODUCTION

This chapter deals with the evolution of epigenetic mechanisms in the animal kingdom and their main functions in animal biology, including regulation of gene expression, development, generation of phenotypic variation from the same genome, and contribution to environmental adaptation and evolution. The main focus is on DNA methylation, the best investigated epigenetic mechanism in animals, and to a lower extent on histone modifications and non-coding RNAs. Animals in the modern sense are a branch of the Opisthokonta that includes the multicellular animals (Metazoa) and the unicellular Choanozoa, Filasterea, and Mesomycetozoa [1]. The closest relatives of the Animalia (Holozoa) are the Fungi (Holomycota).

EVOLUTION OF EPIGENETIC MECHANISMS IN THE ANIMAL KINGDOM

In this section, I will describe the evolutionary ancestry of DNA methylation, histone modifications, and non-coding RNAs and their diversification in the animal kingdom.

DNA Methylation

Methylation of the DNA is a widespread epigenetic mark in prokaryotes and eukaryotes and therefore predates the first animals [2]. DNA methylation is rather well investigated in animals and allows comparison across phyla and examination of correlations with phylogenetic position (evolutionary niveau) and genome size. DNA methylation data have been collected since the 1970s with different methods such as paper chromatography, capillary electrophoresis, ELISA-based colorimetric assays, immunodetection, MSAP (methylation sensitive amplification polymorphism), HPLC, mass spectrometry, and whole-genome bisulfite sequencing. Table 27.1 shows global DNA methylation (5-methylcytosine per total cytosine) values for 255 animal species from 21 major groups, varying from 0% to 12.6%. Although obtained with different methods and not of the same quality, they allow a rough comparison. As an example, in the marbled crayfish, *Procambarus virginalis*, capillary electrophoresis revealed a mean 5mC/total C level of 1.9% [3], whereas mass spectrometry revealed a moderately higher level of 2.4% [4].

The various clades of the multicellular animals have quite different patterns of DNA methylation (Table 27.1). The basal Porifera (sponges) and Cnidaria (corals and jellyfish) have well methylated genomes. Unmethylated and lowly methylated genomes are typical of Platyhelminthes (flatworms), Nematoda (roundworms) and Rotatoria (rotifers) but were also found in highly derived arthropod groups such as Acari (mites), Coleoptera (beetles), Diptera (flies) and Hymenoptera (bees, ants, and wasps). The genomes of the Annelida (ringed worms) and the species-rich Mollusca (snails, bivalves, and cephalopods) are well methylated. The Arthropoda (chelicerates, crustaceans, and insects), the largest animal group on earth, have variable

TABLE 27.1 Relationship of Genome Size and DNA Methylation in Animals

Group	DNA (1C) Mean (pg)	DNA (1C) Range (pg)	5mC/C Mean (%)	5mC/C Range (%)	Dnmt1 (copies)	Dnmt3 (copies)
Choanozoa	0.05 (2)	0.04–0.06	?	?	0	0
Porifera	0.24 (67)	0.04–1.8	3.78 (15)	3.0–9.4	1	1
Placozoa	0.05 (1)	0.05	?	?	0	0
Cnidaria	1.01 (10)	0.23–1.85	2.13 (5)	0–4.6	0–1	0–1
Ctenophora	1.74 (2)	0.31–3.16	0.48 (3)	0.1–1.2	1	0
Platyhelminthes	2.10 (61)	0.06–20.5	+ (7)	0– +	0–?	0–?
Nematoda	0.20 (41)	0.03–2.5	0.19 (8)	0–1.5	0–1	0–1
Rotatoria	0.60 (7)	0.25–1.22	0 (1)	0–0	0	0
Mollusca	1.80 (183)	0.4–5.9	2.77 (5)	1.95–4.9	1	0–1
Annelida	1.50 (126)	0.06–7.6	8.40 (1)	8.4–8.4	1–2	1–2
Chelicerata	2.40 (117)	0.08–7.5	+ (23)	0– +	0–2	0–1
Myriapoda	0.74 (15)	0.30–2.14	+ (1)	+	4	1
Crustacea	3.10 (227)	0.16–38.0	0.87 (3)	0–2.4	1	1
Insecta	1.60 (433)	0.1–16.9	0.43 (20)	0–9.3	0–3	0–1
Echinodermata	1.30 (46)	0.5–4.4	5.19 (3)	3.29–6.6	1	1
Urochordata	0.12 (4)	0.06–0.20	4.50 (2)	4.07–4.93	0–1	0–2
Pisces	1.86 (1527)	0.4–133	7.97 (65)	4.95–12.6	1	2–8
Amphibia	16.7 (463)	0.95–120.1	8.98 (4)	6.34–11.82	1	1
Reptilia	2.30 (309)	1.1–5.4	4.82 (43)	2.89–6.53	1	2
Aves	1.50 (205)	1.0–2.2	4.59 (7)	2.87–5.97	1	2
Mammalia	3.50 (432)	1.7–8.4	4.18 (39)	1.78–6.65	1	3

Genome size is given for haploid genomes (1C-values). Figures in brackets indicate species numbers; +, methylation present but not quantified; ?, no data available. Based on Ref. [5]. Updated with Ref. [6–9].

methylation levels ranging from 0% to 9.3%. The mean of insects is only 0.43% but this value is strongly biased by the preferred investigation of Hymenoptera, Coleoptera, and Diptera, which have low or zero methylation [10,11]. The Echinodermata (starfish and sea urchins) and Chordata (tunicates and vertebrates) have mean methylation levels of 5.19% and 4.75%, respectively. The range of vertebrates, the most derived group of the Chordata, spans from 1.78% to 12.6%. Unfortunately, little is known about DNA methylation in the three unicellular Holozoa branches, but genome-wide profiling in the choanozoan *Salpingoeca rosetta* revealed the absence of genes coding for DNA methylation enzymes, suggesting lack of DNA methylation [12].

Comparison among higher taxa demonstrates that there is no positive correlation between the degree of DNA methylation and evolutionary niveau or complexity of body plan. The basal Porifera that lack complex tissues and organs have mean methylation levels comparable to those of the highly complex mammals (3.78% vs 4.18%). High methylation levels close to 10% have been found in representatives of both basal and derived phyla. Examples are the sponge *Mycale massa* (9.4%) [13], the lepidopteran insect *Mamestra brassicae* (9.3%) [14], the fish *Liparis tunicatus* (12.6%) [15], and the frog *Rana temporaria* (11.8%) [16].

Literature often implied that small animal genomes are lowly methylated and large genomes are highly methylated [17]. However, comparison of the mean DNA methylation levels of 255 species from 21 higher taxa with the mean haploid genome sizes of 4278 species from the same taxa shows that there is no such a relationship (Table 27.1). For example, mean global DNA methylation levels of the taxonomically very distant sponges and mammals are similar (3.78% vs 4.18%) despite enormous differences in mean genome sizes (0.24 pg vs 3.5 pg). Respective examples are also found between different species of the same group. For instance, chum salmon *Oncorhynchus keta* has a genome size of 2.49 pg and a methylation level of 6.5% but kelp snailfish *Liparis tunicatus* has a genome size of only 0.88 pg and a methylation level of 12.1% [15].

The DNA methylation marks are established during embryogenesis and modified throughout life by a DNA methylation machinery consisting of DNA (cytosine-5) methyltransferases (Dnmts), Ten-eleven-translocation enzymes (Tets) and other proteins. DNA methyltransferases are evolutionarily conserved enzymes that have been classified into three groups, Dnmt1, Dnmt2, and Dnmt3 [18]. Dnmt1 serves for maintaining methylation patterns after DNA replication and Dnmt3 is responsible for de novo methylation during development. Dnmt2 is structurally similar to Dnmt1 and Dnmt3 but acts primarily as a tRNA methyltransferase. Tet catalyzes demethylation of 5-methylcytosine [19].

The ground pattern of the DNA methylation machinery in animals is probably single copies of Dnmt1, Dnmt3, and Tet as shown in Fig. 27.1A for the basal Chordata. Dnmt1 and Dnmt3 are shared by most animals but were lost in some lineages like platyhelminths and dipteran insects and expanded in others (Table 27.1). The number of Dnmt1 was increased in some species of the poriferans, annelids, chelicerates, crustaceans, myriapods, and insects, reaching a maximum of five copies in water flea *Daphnia pulex* [20,21]. Dnmt3 was expanded in some poriferans, annelids, chelicerates, insects, and chordates, reaching a maximum of eight copies in rainbow trout *Oncorhynchus mykiss* [21,22]. Tet genes occur in single copies in most invertebrates but in two copies in some poriferan, chelicerate, crustacean and insect species and in three copies with different functions in most vertebrates [21]. Tet is absent in the nematode *Caenorhabditis elegans* and some further nematodes [21,23].

Histone Modifications

Posttranslational modifications at the N-terminal tails of the histones such as acetylation and methylation of lysines and arginines exist in all eukaryotic crown phyla and in protists with chromatin structures and must therefore be considered evolutionary very old [26]. In the animal kingdom, conservation of methylation and acetylation of lysines and arginines is virtually absolute, but some variability may exist in phosphorylation of serine, threonine, and tyrosine [26]. Gaiti and colleagues principally confirmed the old ancestry of histone modifications in animals by investigating the sponge *Amphimedon queenslandica*, a morphologically simple, early branching animal [27]. The authors revealed that the regulatory landscape for gene expression used by the complex bilaterians was already in place at the beginning of animal multicellularity, including distal enhancers and marking of active promoters by H3K4me3. Presently, there is no evidence for the diversification of the ancestral set of histone modifications in the animal kingdom, making this epigenetic mechanism the most conserved one in animals.

Non-coding RNAs

In past years, tens of thousands of ncRNA genes, including short, long and circular forms, have been discovered in fully sequenced animal genomes. Comparison of the genomes of non-bilaterian metazoans, unicellular holozoans, and fungi revealed that the evolution of Piwi-interacting RNAs correlates with the emergence of metazoan multicellularity, while microRNAs, circular RNAs, and long ncRNAs

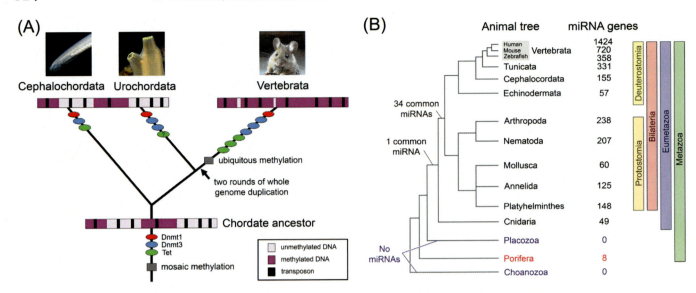

FIGURE 27.1 **Evolution of epigenetic mechanisms in animals.** (A) Evolution of DNA methylation patterns and the DNA methylation machinery in Chordata. Cephalochordates and urochordates show mosaic methylation, single copies of Dnmt1, single or duplicate copies of Dnmt3, and single copies of Tet. In mammals (representatives of vertebrates), DNA methylation is ubiquitous and Dnmt3 and Tet are expanded to three copies. These features probably go back to two rounds of whole genome duplication in the stem vertebrates. Transposons are heavily methylated in vertebrates but only sparsely methylated in cephalochordates and urochordates. (B) Evolution of miRNA genes in the animal kingdom (groups in tree except of vertebrates are represented by single species). The Porifera and Eumetazoa have independently evolved specific miRNAs. The Eumetazoa share one common miRNA gene, namely *mir-100*, and the Bilateria share 34 miRNA genes. A particularly high degree of miRNA diversification occurred in the vertebrates. The Choanozoa and Placozoa have no miRNA genes. *A based on R. Albalat, J. Martí-Solans, C. Cañestro. DNA methylation in amphioxus: from ancestral functions to new roles in vertebrates. Brief Funct Genom 2012;11 (2):142–155. B based on E. Berezikov. Evolution of microRNA diversity and regulation in animals. Nat Rev Genet 2011;12(12):846–860.*

predate the metazoans [28]. The ncRNAs diversified enormously during animal evolution. Differential expansion of the ncRNAs and the emergence of novel cis-regulatory DNA elements in bilaterians apparently accounts for their increased regulatory and morphological complexity relative to non-bilaterians.

miRNAs are rather well investigated for several animal phyla and have already been subjected to evolutionary analysis (Fig. 27.1B). Wheeler and colleagues studied the gains and losses of miRNA families in the animal kingdom and found continuous acquisition during the 650 million year history of the Metazoa with minimal secondary loss [29]. The Eumetazoa share only a single miRNA, but there was a considerable expansion to 34 miRNAs in the ancestor of the Bilateria [25]. The various bilaterian clades have evolved quite different numbers of specific miRNAs, and a particularly intense expansion has occurred in vertebrates (Fig. 27.1B). The Porifera, the most basal clade of the Metazoa, share no miRNAs with the Eumetazoa but have independently evolved some specific miRNAs. The multicellular Placozoa and Ctenophora and the unicellular Choanozoa have no miRNAs. The ancient origin of miRNAs, their dramatic expansion in bilaterian animals, and their function in providing robustness to transcriptional programs suggest that they have been crucial for the evolution of animal complexity [25]. However, the ctenophores can produce different cell types including neurons and muscles in the absence of miRNAs [30].

Mohammed and colleagues made a detailed analysis of miRNAs in the dipteran genus *Drosophila* and identified 649 novel, evolutionarily restricted variants in the 12 investigated species [31]. Penso-Dolphin and colleagues studied the role of miRNAs in the evolution of gene regulation in four tissues of five mammalian species [32]. They identified 413 novel miRNAs and highlighted the important contributions of intronic sequences, duplication events and repetitive elements in the emergence of the new miRNA loci. The novel miRNAs tended to be more tissue specific than the conserved ones. They seemed to also play an important role in the evolution of brain plasticity.

FEATURES OF EPIGENETIC MECHANISMS AND ROLE IN GENE REGULATION

Epigenetic mechanisms exert the biological functions discussed below by modulating gene expression. DNA methylation, histone modifications, ncRNAs, and other components like proteins of the methyl-CpG-binding domain (MBD) family usually interact in a complex manner to regulate the genome [33–36]. The interaction of these processes are difficult to assess and understudied in animals, but there are numerous papers on the effects of single epigenetic mechanisms on gene expression.

The involvement of DNA methylation in gene expression can best be investigated by analyzing genome-wide methylomes, which are meanwhile available for several dozen animal species, mainly insects and vertebrates, but there are also examples from poriferans, cnidarians, ctenophores, nematodes, bivalves, spiders, crustaceans, and lower chordates. Comparison of these methylomes revealed shared features like confinement of methylation to cytosines, methylation of gene bodies, and hypomethylation of promoters in active genes, but also some differences regarding transposon methylation.

Methylation in animals is mainly found in the context of CpG dinucleotides but can also occur in CHG and CHH, where H equals A, T, or C. In sea anemone *Nematostella vectensis*, bivalve *Crassostrea gigas*, ant *Camponotus floridanus*, and sea squirt *Ciona intestinalis*, more than 99% methylation modification was found in CpG contexts and the rest in CHG and CHH [37]. In humans, the corresponding value is 92.4% [37], but in the basal ctenophores it is less than 50% [38]. The degree to which CpGs are methylated varies considerably between species and higher taxa. In mouse and zebrafish it is 74% and 80%, respectively, but in honey bee it is only 1% [39].

Broadly speaking, four types of DNA methylation patterns can be distinguished in animal genomes: unmethylated, sparsely methylated, mosaically methylated, and ubiquitously methylated. Invertebrate species often show sparse methylation and mosaic methylation, comprising domains of methylated DNA interspersed with domains of unmethylated DNA. In contrast, vertebrates generally have ubiquitous DNA methylation, which is the genome-wide methylation of CpG sites (Fig. 27.1A). The CpG islands of promoter regions remain largely unmethylated [40]. Transition from mosaic to ubiquitous methylation in chordates has probably occurred some 465 million years ago in the early evolution of fishes as inferred from the presence of mosaic methylation in Cephalochordata and Urochordata (Fig. 27.1A) and ubiquitous methylation in cartilaginous fish [24,41]. Interestingly, the various patterns of DNA methylation can also occur in the genome of the same species. For example, in the marbled crayfish *Procambarus virginalis*, 41% of genes were heavily methylated, 33% were moderately methylated, and 26% were unmethylated. About 25% of the 20 longest scaffolds were ubiquitously methylated, 70% were mosaically methylated, and 5% were sporadically methylated [42].

Gene bodies are methylated in all animals with methylated genomes from Porifera to Mammalia, suggesting that it is a phylogenetically old and conserved feature [37,43,44]. Even in insects with low methylation levels like the honey bee, most methylation marks are enriched in gene bodies [43,45]. In the moderately methylated sea squirt *Ciona intestinalis*, the methylation level is almost two times higher in genes than in intergenic regions, but in the highly methylated genomes of zebrafish and mouse, there is only slightly higher CpG methylation in the gene bodies than in intergenic regions [39,43]. In the oyster *Crassostrea gigas*, methylated CpGs were particularly frequent in moderately expressed genes [37]. In the spider *Stegodyphus dumicola*, methylated genes were more stably expressed than unmethylated genes [9]. In crayfish *Procambarus virginalis*, higher methylated genes like housekeeping genes were commonly found in chromatin with limited accessibility and were stably expressed at various levels, whereas low-methylated genes often resided in chromatin with higher accessibility and showed higher variation in expression [46]. Gene body methylation also seems to reduce transcriptional noise and help in splicing [40,47,48].

Expression and repression of gene activity via promoter methylation is apparently widespread in animals. Most invertebrates and vertebrates have lower levels of CpG methylation in promoter regions when compared with gene bodies [40,43,44,49]. In the highly methylated fish and mouse, the genes exhibit a prominent dip in methylation just upstream of the transcription start site [43]. The respective sequence is called CpG island and is particularly rich in unmethylated CpGs [39]. CpG islands destabilize nucleosomes and attract proteins that create a transcriptionally permissive chromatin state [49].

Hypermethylation of CpGs in the promoter region is usually associated with silencing of the corresponding gene as shown for the distantly related protostomian mud crab *Scylla paramamosain* and the deuterostomian chicken *Gallus gallus* [50,51].

Transposons are heavily methylated in vertebrates but largely unmethylated in some insects [43,45,52]. These differences gave rise to the hypothesis that in invertebrates DNA methylation is not used as a defense mechanism against intragenomic parasites and that transposon defense may have evolved in vertebrates [43]. However, in the desert locust *Schistocerca gregaria*, the stick insect *Medauroidea extradentata*, the bivalve *Crassostrea gigas*, and the chordate *Ciona intestinalis* a significant fraction of transposons and repetitive elements is methylated [37,43,53,54], suggesting that transposon inactivation by methylation is much more widespread in animals than previously assumed.

Like DNA methylation, the histone modifications can be dynamically altered during development and in response to environmental signals as will be demonstrated below. They play a crucial role in regulating the chromatin state (open or closed), thus influencing all chromatin dependent processes including gene expression. Histone acetylation and H3K4 methylation often stimulates gene expression, whereas H3K27 methylation often represses gene expression [55,56]. The changes of histone modifications are catalyzed by numerous enzymes [57].

Methylation is currently one of the best characterized histone modifications and occurs on arginine and lysine residues. Histone methylation can regulate other modifications like acetylation, phosphorylation and ubiquitination in order to define a precise functional chromatin environment [58]. Jambhekar and colleagues reviewed the role of histone modifications in animal development and emphasized that the proper regulation of histone methylation is essential for ensuring the coordinated expression of gene networks that govern pluripotency, body patterning, tissue differentiation and organogenesis [59].

Small to long ncRNAs are further important epigenetic regulators of gene expression. Similar to protein-coding genes, their expression is controlled by genetic and epigenetic mechanisms. They regulate gene expression either at the transcriptional or posttranscriptional level. For example, they can repress key enzymes that drive epigenetic remodeling or bind to complementary sequences in promoters, recruiting specific protein complexes that modulate chromatin structure and gene expression [60,61]. Examples for posttranscriptional regulators are miRNAs that inhibit translation or cause mRNA degradation [62]. Small interfering RNAs (siRNAs) can regulate gene transcription through transposable element silencing and the interaction with DNA methylation marks and histone modifications [62,63]. Long ncRNAs (lncRNAs) are involved in transcriptional regulation and classical epigenetic processes such as X-chromosome inactivation and genomic imprinting [61,64]. Depending on their localization and interactions with DNA, RNA and proteins, lncRNAs can modulate chromatin function, regulate the assembly and function of nuclear bodies, alter the stability and translation of mRNAs, and interfere with signaling pathways [61,64]. The lncRNA Xist helps to inactivate X-chromosomes in mammals by interacting with methylated CpG islands and hypoacetylated histone H4 [65].

Proteins of the methyl-CpG-binding domain (MBD) family are important readers of the epigenome [36] as they recruit chromatin remodelers, histone deacetylases, and methylases to methylated DNA associated with gene repression. MBD protein binding usually requires both functional MBD domains and methyl-CpGs, but some MBD proteins also bind unmethylated DNA and active regulatory regions. In humans, mutations in MBD domains prevent specific binding to methylated sites resulting in gene deregulation and the onset of diseases including cancer.

DYNAMICS OF EPIGENETIC MARKS DURING DEVELOPMENT

Epigenetic signatures can considerably vary during an animal's lifetime. This dynamics is particularly obvious in embryonic development, tissue differentiation, and aging.

Embryonic Development

Development in animals is either direct or indirect. Direct developers show rather continuous phenotypic alterations from the zygote to the adult, whereas indirect developers have larval stages and sometimes quiescent pupal stages with different morphologies, behaviors, and ecologies interspersed between the embryonic and adult stages. Vertebrate and invertebrate examples of direct developers are mammals and freshwater crayfish, respectively, and examples of indirect developers are fishes and holometabolous insects.

In mouse, development from the zygote to birth lasts 18–21 days. 5mC/C in the 2.6 Gb genome of adult mice is ~5% and varies with sex, tissue, and condition [66]. The DNA methylation marks are globally erased and re-established a first time in the zygote and the following pre-implantation stages (Fig. 27.2A) and a second time in the primordial germ cells (PGCs) [67,68]. Expression of the methylation and demethylation enzymes varies accordingly in this period of time (Fig. 27.2A). After fertilization, DNA methylation of the sperm pronucleus in the zygote is actively reduced by Tet3. The DNA methylation marks in the maternal genome are passively lost over subsequent cell divisions because the oocyte derived Dnmt1o is largely excluded from the nucleus and, therefore, maintenance methylation is inefficient. The global DNA methylation level reaches a minimum around the blastocyst stage (32–140 cells) at day 4 of development. After implantation, the DNA marks are re-established by Dnmt3A and Dnmt3B, particularly in the inner cell mass that develops into the mouse. Removal of the DNA methylation marks in the PGCs occurs at days 12–14 of embryonic development. They are re-established during further development of the PGCs to either sperm or oocytes, restricting developmental potency.

In the parthenogenetic marbled crayfish, *Procambarus virginalis*, that develops from unfertilized eggs embryonic development lasts 17–26 days [71]. Based on morphological criteria, embryonic development was subdivided into

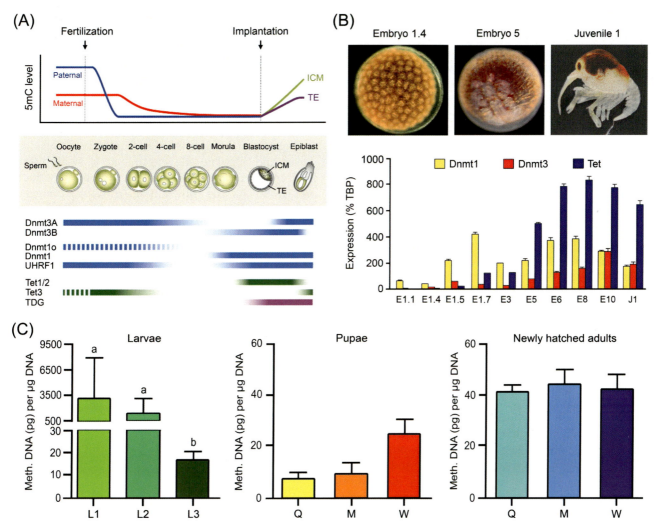

FIGURE 27.2 **Dynamics of DNA methylation during embryonic and larval development.** (A) Changes in cytosine methylation and expression of the DNA methylation and demethylation enzymes in early embryonic development of mouse, *Mus musculus*. The paternal 5mC is more rapidly reduced after fertilization than the maternal 5mC. In the blastocyst stage, DNA methylation reaches a minimum. After implantation, the DNA methylation pattern is re-established, particularly in the inner cell mass (ICM) that develops into the mouse. The trophoectoderm (TE) becomes part of the placenta. The methylation enzymes Dnmt1o (oocyte derived variant of Dnmt1), Dnmt1, UHRF1, Dnmt3A and Dnmt3B and the demethylation enzymes Tet 1/2, Tet3 and TDG are correspondingly downregulated and upregulated. (B) Dynamics of the DNA methylation and demethylation enzymes during embryonic development in the parthenogenetic marbled crayfish, *Procambarus virginalis*. mRNA expression levels are given relative to TBP (Tata box binding protein) expression. Dnmt1, Dnmt3, and Tet show very low expression until the 128 nucleus stage (embryo 1.4) and quite different expression dynamics thereafter. Embryo 5 corresponds to about 50% embryonic development, in which tissues are not yet discernible, and juvenile 1 is the hatching stage. Bars indicate standard deviation from three measurements. (C) Quantity of methylated cytosines in larvae, pupae and adults in stingless bee *Melipona scutellaris*. There is a dramatic reduction of DNA methylation during early larval development and a moderate increase between pupa and hatching adults. 1–3, subsequent larval stages; M, male; Q, queen; W, worker. Show are means ± SEM (n = 3). *A based on H. Wu, Y. Zhang. Reversing DNA methylation: mechanisms, genomics, and biological functions. Cell. 2014;156(1-2):45–68. B based on F. Gatzmann, C. Falckenhayn, J. Gutekunst, K. Hanna, G. Raddatz, V. Coutinho Carneiro, et al. The methylome of the marbled crayfish links gene body methylation to stable expression of poorly accessible genes. Epigenetics Chromatin 2018;11(1):57. C based on C.A.M. Cardoso-Júnior, P.T. Fujimura, C.D. Santos-Júnior, N. Araújo Borges, C. Ueira-Vieira, K. Hartfelder, et al. Epigenetic modifications and their relation to caste and sex determination and adult division of labor in the stingless bee Melipona scutellaris. Genet Mol Biol 2017;40(1):61–68.*

10 stages. Stage 1 that spans from the spawned egg to the beginning of gastrulation was further divided into 8 substages (1.1 to 1.8). Global DNA methylation of the 3.7 Gb genome is about 2.5% in adults [4], corresponding to about 50% of the mouse value. The DNA was already well methylated from embryonic stage 5 (Fig. 28.3D), the earliest stage where we could reliably determine global DNA methylation by mass spectrometry. In this stage, the tissues are beginning to be formed. Marbled crayfish have single copies of Dnmt1, Dnmt3, and Tet. Expression of

these enzymes was generally low until the 124 nucleus stage (Fig. 27.2B, embryo 1.4) [46], in which cell membranes between the nuclei are still lacking. These are only formed in the 256 cell stage. Dnmt1 was strongly upregulated in embryonic stage 1.5 (256–512 cell stage), while Dnmt3 expression increased continuously from this stage until stage 10. Tet mRNA levels increased strongly during mid embryogenesis and remained high until hatching (Fig. 27.2B) [46]. The expression levels and dynamics of the methylation and demethylation enzymes suggest that the DNA methylation pattern is intensely remodeled in the second half of embryonic development, in which the tissues are formed.

Holometabolous insects produce morphologically, functionally, and behaviorally very distinct life stages from the same genome, namely the larva, pupa and adult or imago. These diverse life stages allow insects to partition life history to feeding and growth (larva), quiescence and metamorphosis (pupa), and reproduction and dispersal (adult). Their expression is controlled by a hormone-mediated developmental program [72]. The involvement of epigenetic mechanisms in this indirect mode of development is only sparsely investigated. Using an ELISA-based methodology in the stingless bee Melipona scutellaris, Cardoso-Júnior and colleagues measured a dramatic reduction of DNA methylation during early larval development followed by relatively low values in the pupa and a moderate increase between pupa and adults (Fig. 27.2C) [70]. Using western blot assays, the authors also found significant differences in histone methylation and phosphorylation between the queens and workers.

Histone modifications are also involved in the regulation of embryonic development in mammals. Using immunofluorescence and confocal microscopy, Sarmento and colleagues measured changes in global levels of histone modifications during oocyte maturation and preimplantation of mouse [73]. They distinguished two strikingly distinct categories of histone modifications. The first category contained stable modifications including H3 lysine 9 methylation, H3 lysine 4 methylation and H4/H2A serine 1 phosphorylation. The second group contained dynamic and reversible marks including H4 acetylation, H3 arginine 17 methylation and H4 arginine 3 methylation. The removal of these marks in eggs and early embryos occurred during metaphase, suggesting that the responsible enzymes are cytoplasmic in nature.

ncRNAs are further important regulators of animal embryogenesis as shown by research with knock-out mice [74]. For example, mice lacking miRNAs showed a depletion of multipotent stem cells and died at the eighth day of embryonic development. According to Pauli and colleagues, ncRNAs are involved in maintenance of pluripotency, patterning of body axes, specification and differentiation of cell types, and organogenesis [75]. They control embryonic gene expression by several means, ranging from miRNA-induced degradation of mRNAs to lncRNA-mediated modification of chromatin. In fruit fly Drosophila melanogaster, lncRNAs were significantly upregulated in late embryonic and larval stages [76]. Li and colleagues have reviewed the regulatory functions of ncRNAs in insect development [64]. miRNAs can silence coding genes through translation repression and mRNA decay and regulate the expression level of numerous lncRNAs through the lncRNA decay pathway. lncRNAs and circRNAs can function as miRNA sponges, thus derepressing target gene expression. piRNAs mainly silence transposable elements in the germline at the transcriptional and post-transcriptional levels.

Cell and Tissue Differentiation

The various cell types in the body of eumetazoans originate from a single pluripotent cell, the zygote, and contain the same DNA sequence with the exception of a few random mutations. In humans, there are about 400 different cell types [77]. Despite genetic uniformity, they are morphologically and functionally highly diverse. Blake and colleagues performed a comparative study of gene expression and DNA methylation in livers, kidneys, hearts, and lungs of primates, humans, chimpanzees, and rhesus macaques [69]. They found a high degree of conservation in gene expression levels when considering the same tissue across species. They also measured significant differences in DNA methylation between tissues (Fig. 27.3A) and identified tissue-specific differentially methylated regions (DMRs).

In marbled crayfish, the expression of the methylation and demethylation enzymes varied considerably between tissues (Fig. 27.3B). Dnmt3, which is responsible for the establishment of the DNA methylation patterns, seems to represent the most tissue specific enzyme of the methylation machinery [46].

Histones and ncRNAs are also involved in tissue differentiation. Zhang and Zhang reported that histone modification profiles vary between human tissues and cells and considered them causative for cell-type specific expression of both protein-coding genes and miRNA genes [79]. Isakova and colleagues investigated the role of ncRNAs in the development of 11 tissues in mouse and revealed that ~30% of the total ncRNA transcriptome is tissue-specific [80].

Aging

Growth in animals is either determinate or indeterminate. Determinate growers stop growth in a certain period of life, e.g., humans at about 20 years, but

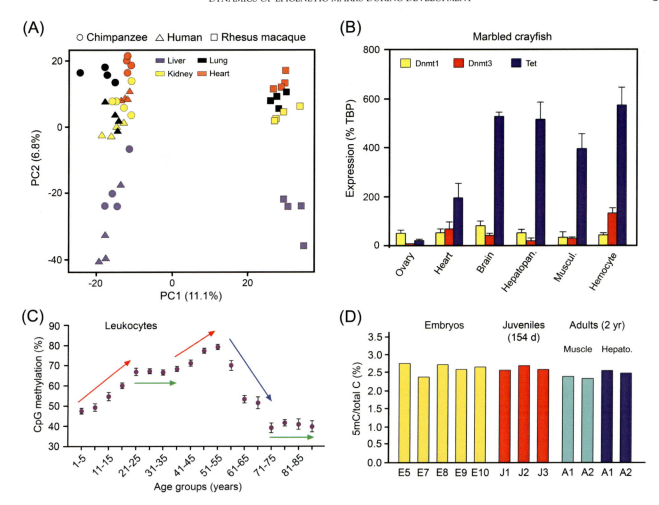

FIGURE 27.3 Differences of DNA methylation between tissues and dynamic alterations during aging. (A) Principal component analysis (PCA) of average methylation levels in 47 tissue samples from four humans (*Homo sapiens*), 4 chimpanzees (*Pan troglodytes*) and 4 rhesus macaques (*Macaca mulatta*), showing conservation of tissue specific DNA methylation in the primate species. (B) Expression of Dnmt1, Dnmt3, and Tet in various tissues and organs of the marbled crayfish *Procambarus virginalis* relative to TBP expression. The graph shows marked differences between tissues for all enzymes. Dnmt3, which is responsible for the establishment of the DNA methylation patterns, seems to represent the most tissue specific enzyme of the methylation machinery. Bars indicate standard deviation of three measurements. (C) Variation of CpG methylation in healthy humans between 2 and 97 years of age (n = 2116). Shown are the mean CpG methylation levels of 5-year age groups. Bars indicate standard error of the group means. The methylation level increases until 55 years and decreases thereafter, but there are also periods of stasis. (D) Variation of 5mC/total C levels measured by mass spectrometry in marbled crayfish from mid-embryonic development to adults of rather old age. Data for embryonic stages were obtained from pooled samples of many embryos, data for the juvenile stage from three whole clutch-mates (J1-J3) and data for the adult stage from the hepatopancreas and abdominal musculature of two adults (A1 and A2). The methylation level seems to decrease slightly with age. *A based on L.E. Blake, J. Roux, J. Hernando-Herraez, N.E. Banovich, R. Garcia Perez, C.J. Hsiao, et al. A comparison of gene expression and DNA methylation patterns across tissues and species. Genome Research. 2020;30(2):1–13, Graph in B based on F. Gatzmann, C. Falckenhayn, J. Gutekunst, K. Hanna, G. Raddatz, V. Coutinho Carneiro, et al. The methylome of the marbled crayfish links gene body methylation to stable expression of poorly accessible genes. Epigenetics Chromatin 2018;11(1):57. C based on S.-Y. Tsang, T. Ahmad, F.W.K. Mat, C. Zhao, S. Xiao, K. Xia, et al. Variation of global DNA methylation levels with age and in autistic children. Hum Genom 2016;10(1):31.*

continue to live on for longer periods of time. Indeterminate growers have no fixed size limit and grow until old age. Examples of determinate growers are mammals and insects and examples of indeterminate growers are many fishes and crustaceans. Determinate growth is generally associated with gradual senescence resulting in age-related deteriorations and diseases, whereas indeterminate growth is associated with negligible senescence. The epigenetic signatures may develop differently during aging in these growth formats.

Fraga and colleagues investigated the development of DNA methylation and histone modification patterns in monozygotic human twin pairs of different age [81]. They found that twins are epigenetically indistinguishable during the early years of life, but develop remarkable differences with increasing age in their overall content and genomic distribution of 5-methylcytosine and histone acetylation, affecting their gene-expression profile. For example, global DNA methylation increased from ~2.5% in 3-year-old children to ~4.5% in 50-year-old adults.

Tsang and colleagues investigated genome-wide CpG methylation in leukocytes from the peripheral blood of 2116 healthy Chinese ranging from 2 to 97 years [78]. They found a more complex pattern of methylation variation with age than earlier authors, including a steady increase of the methylation level from 2 to 25 years, a static phase between 26 and 40 years, another rise from 41 to 55 years reaching a peak at ~80%, followed by a sharp decrease to ~40% between 56 and 75 years and a static phase thereafter (Fig. 27.3C). A gender effect was observed only for the 41–55 age group in which methylation was significantly higher in females than in males.

Stubbs and colleagues identified 329 unique, tissue independent CpG sites in mouse, in which the methylation level increased continuously across 40 weeks [82]. They suggested using this epigenetic feature as a clock for aging research. Experimental modulation of its ticking rate and resetting the clock may help to better understand the biology of ageing.

In the indeterminately growing marbled crayfish, DNA methylation was on average 2.62% in the second half of embryonic development, 2.58% in 154 days old juveniles and 2.41% in adults of about 2 years (Fig. 27.3D), corresponding to a reduction of ~8% during lifetime. Most marbled crayfish that reach adulthood die at an age of 2–3 years [71]. Unlike in determinately growing humans and mice, there was no rise of the methylation level in marbled crayfish. This may be a consequence of indeterminate growth in which stem cells remain active, regenerating tissues and appendages until old age.

GENERATION OF PHENOTYPIC VARIATION IN POPULATIONS BY EPIGENETIC MECHANISMS

The examples of tissue differentiation and holometabolous development have shown that animals can generate different phenotypes from the same DNA sequence by modulating gene expression with the help of epigenetic mechanisms. This also holds for the production of polyphenism and phenotypic plasticity in populations.

Polyphenism in Insects

Polyphenism describes the ability to produce two or more discrete phenotypes from the same DNA sequence. It is widespread in insects and best exemplified by the different castes in social species. The different phenotypes can better exploit resources, cope with temporally heterogeneous environments and partition labor in the colony [83]. Polyphenism is triggered by environmental cues that modulate hormonal pathways via epigenetic marks [83–85].

The honey bee *Apis mellifera* is here presented as an example of the association of epigenetic modifications and polyphenism. Honey bees produce morphologically, behaviorally and reproductively different queens and workers from the same genome by differential feeding of the larvae. Presumptive queens are fed with royal jelly and presumptive workers with pollen. Both morphs are diploid but the workers are considerably smaller (Fig. 27.4A) and sterile. Queens produce the entire offspring and regulate life in the hive by pheromones, whereas workers act as foragers or nurses. Longevity is about 2 years in queens but only 3–6 weeks in workers.

Several papers demonstrated possible roles of DNA methylation and histone modifications in producing the queen and worker phenotypes in honey bee. Lyko and colleagues reported that the DNA of the brains of queens and workers differ in methylation of more than 550 genes, including genes involved in metabolism, brain development and neural functions [45]. An example is shown in Fig. 27.4B. Herb and colleagues found substantial differences in DNA methylation between nurse and forager subcastes of workers [87]. Reverting foragers back to nurses re-established methylation signatures for a majority of genes. Foret and colleagues sequenced the larval and adult methylomes in both queens and workers [88]. They measured a significantly higher number of differentially methylated genes (DMGs) in the larval head relative to the adult brain (2399 vs 560) with 82% of DMGs hypermethylated in worker larvae and 18% hypermethylated in queen larvae. Several highly conserved metabolic and signaling pathways were enriched in methylated genes, including genes coding for juvenile hormone and the insulin signaling pathway that regulate caste determination.

Wojciechowski and colleagues produced genome-wide maps of the chromatin structure in honey bee at a key larval stage in which development into queen or worker was virtually irreversible [86]. Using ChIP-seq, which combines chromatin immunoprecipitation with DNA sequencing to identify the binding sites of DNA-associated proteins, they found extensive genome-wide differences in histone modifications (H3K4me3, H3K27ac and H3K36me3), many of which correlated with caste-specific transcription. The authors identified H3K27ac as a key chromatin modification with a pronounced caste-specific distribution in exons, introns and intergenic regions (Fig. 27.4C). In 96 h queens, an increase in enrichment of H3K27ac was almost exclusively located within 0–1 kb

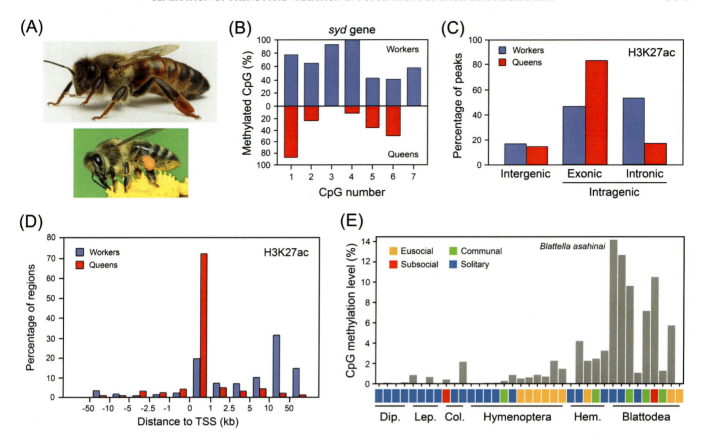

FIGURE 27.4 **Involvement of epigenetic mechanisms in polyphenism and sociality of insects.** (A) Dimorphism of reproducing queen (upper panel) and foraging worker (lower panel) in honey bee, *Apis mellifera*. (B) Different methylation of CpGs of *syd* gene (encodes the catalytic component of the chromatin structure-remodeling complex) in workers and queens. (C) Different enrichment of unique H3K27ac ChIP-seq regions in intergenic and intragenic locations of 96 h workers and 96 h queens. (D) Different location of unique intronic H3K27ac ChIP-seq regions relative to the nearest transcription start site (TSS) in 96 h queens and 96 h workers. In queens, an enrichment of H3K27ac is almost exclusively observed close to the TSS, but in workers it is located more downstream. (E) Extensive variation of DNA methylation throughout insects revealed by WGBS. The genomic level of DNA methylation ranges from 0% in Diptera (Dip.) to 14% in the cockroach *Blattella asahinai*. Overall, levels are highest in the phylogenetically basal Blattodea. There is no obvious correlation between DNA methylation level and social behavior. Col., Coleoptera; Hem., Hemiptera; Lep., Lepidoptera. *Photos in A: queen by Alex Wild, worker by Andreas Trepte; B based on F. Lyko, S. Foret, R. Kucharski, S Wolf, C. Falckenhayn, R. Maleszka. The honey bee epigenomes: differential methylation of brain DNA in queens and workers. PLoS Biol 2010;8(11):e1000506. C and D based on M. Wojciechowski, R. Lowe, J. Maleszka, D. Conn, M. Maleszka, P.J. Hurd. Phenotypically distinct female castes in honey bees are defined by alternative chromatin states during larval development. Genome Res 2018;28(10):1532–1542. E based on A.J. Bewick, K.J. Vogel, A.J. Moore, R.J. Schmitz. Evolution of DNA methylation across insects. Mol Biol Evol 2017;34(3):654–665.*

downstream of the transcription start sites, whereas in 96 h workers enrichment was mostly located outside this region (Fig. 27.4D).

Apparently, there is no simple cause-effect relationship between epigenetic mechanisms and social behavior in the species-rich and highly diverse insects. In the primitively eusocial wasp *Polistes dominula*, only seven genes are methylated and Dnmt3 is absent [89], calling into question the general importance of DNA methylation in social behavior. Bewick and colleagues came to the same conclusion by whole genome bisulfite sequencing (WGBS) of 41 species from several insect orders and investigation of the bimodality of CpG$_{o/e}$ values in 123 social and asocial insect species (Fig. 27.4E) [10].

Phenotypic Plasticity

All animals are probably able to produce a continuous spectrum of phenotypes from the same genome, which is called phenotypic plasticity. This phenotypic diversity can be achieved either by stochastic developmental or environmentally induced changes of the epigenetic marks on the DNA and chromatin [3,90–93]. Both mechanisms differ in function but usually act together. Environmentally induced phenotypic variation (EIPV) is predominant in predictable environments and strengthens adaptation to the prevailing conditions and stochastic developmental phenotypic variation (SDPV) is predominant in unpredictable environments and contributes to evolutionary bet-hedging.

SDPV is well documented for bacteria, protists, fungi, plants, and animals and is therefore considered a general biological principle [90]. It is mainly caused by stochastic alterations of epigenetic marks on the DNA and chromatin during development [3,94]. Such epimutations are reversible and do not change the DNA sequence. In the model plant *Arabidopsis thaliana* epimutations are about five orders of magnitude more frequent than genetic mutations (10^{-4} vs 10^{-9} per base pair and generation) [95]. Therefore, epimutations have the potential to generate phenotypic variation much more rapidly than genetic mutations. Since SDPV produces a range of epigenotypes and related phenotypes from the same genotype a priori without knowing the future conditions, it is a risk-spreading strategy that enhances the chance of survival when the environmental conditions change.

EIPV is also ubiquitous in all living organisms as documented by the extensive literature on phenotypic plasticity [96,97]. It is induced by environmental signals that likely cause modifications of the epigenetic marks on genes and chromatin via sense organs and neurohormones, leading to directional changes of gene expression. Probably, only relatively strong and longer lasting environmental cues like starvation, predator pressure, toxicants, high and low temperature, salinity, and oxygen are capable of producing EIPV in animals [98].

SDPV and EIPV can best be distinguished and quantified in the laboratory by raising genetically identical clutch-mates in the same or in different environments. A good experimental animal for this purpose is the obligatory parthenogenetic marbled crayfish, *Procambarus virginalis* (Fig. 27.5). This species was detected in 1995 in the German aquarium trade; it was then spread throughout the world and has meanwhile invaded diverse habitats on three continents [71]. The marbled crayfish is a monoclonal all-female species that produces offspring genetically identical to the mother and among each other. Genetic identity was repeatedly demonstrated, first by the microsatellite technique and later by comparison of whole genome sequences [4,99].

Marbled crayfish have a maximum body length without chelae of ca. 12 cm. They are directly developing and mostly reproduce twice a year. Clutch size is ~50 to 730 offspring, depending on female size, providing an extraordinarily large source of genetically identical clutch-mates for experimentation. The maximum age is 4.5 years. Marbled crayfish can be raised in very simple laboratory settings and all life stages can be fed with the same pellet food. The haploid genome size is 3.7 Gb comprising 21,772 predicted genes [99]. Global DNA methylation is about 2.5% as shown by mass spectrometry [4]. A reference methylome established by WGBS revealed that 74% of genes are methylated and 26% are unmethylated [42,46]. Releases of marbled crayfish from aquarium stocks in the last 20 years have led to the establishment of numerous wild populations in Europe (17 countries), Africa (Madagascar), and Asia (Israel, Japan, China, and Taiwan). It now lives in tropical to cold-temperate biomes in a broad range of habitats, including rivers, lakes, ponds, rice fields, and thermal and polluted waters [71,103].

The degree of SDPV in laboratory raised *P. virginalis* was lowest for morphological characters, higher for biochemical and life history traits such as growth, reproduction, and longevity, and highest for behavioral traits and the coloration pattern [3]. For example, life span among members of my laboratory colony that had reached adulthood varied between 312 and 1610 days, despite genetic identity and identical rearing and feeding. A behavioral example is shown in Fig. 27.5B. When five size-matched stage-6 juveniles with neutral agonistic behavior were placed together in a culture vessel with only a nylon net as shelter, a social hierarchy was gradually established in the following 34 days. At the end of the experiment, the group consisted of 1 dominant, 1 subdominant, and 3 subordinates. Interestingly, growth of the dominant speeded up compared to the subdominant and subordinates, although all specimens had unlimited access to the food and fed regularly. SDPV of coloration was extremely high so that each marbled crayfish was individually recognized. The color pattern differed markedly between clutch-mates and between mother and offspring (Fig. 27.5A), indicating that it is not inherited.

Measurement of global DNA methylation with capillary electrophoresis in communally raised clutch-mates revealed differences between tissues, specimens (Fig. 27.5C) and age groups. Adults of 626 days had on average ~10% lower 5mC/C levels when compared to 188 days old juveniles [3]. The methylation differences between individuals could not yet be mechanistically linked to variations of morphological, behavioral and life history traits.

Research with other clonal animals like polyembryonic armadillos, highly inbred rats, and cloned pigs and goats confirmed the tendency observed in marbled crayfish that SDPV is lowest for morphological characters, higher for biochemical and life history traits, and particularly high for behavioral traits and coloration. Spotted colors in cloned cats and cattle and inbred guinea pigs were extremely variable, identifying each specimen individually as in the marbled crayfish [90].

EIPV in marbled crayfish was studied by comparison of laboratory raised and wild specimens and comparison of differently adapted wild populations. Comparing 20% of whole genome sequences of specimens from different laboratory lineages and wild populations from Germany and Madagascar that are separated from each other since about 20–40 generations revealed only very small differences of 129–219 single nucleotide variants (SNVs) between samples [99]. The vast majority of these SNVs were silent mutations. The maximum number of

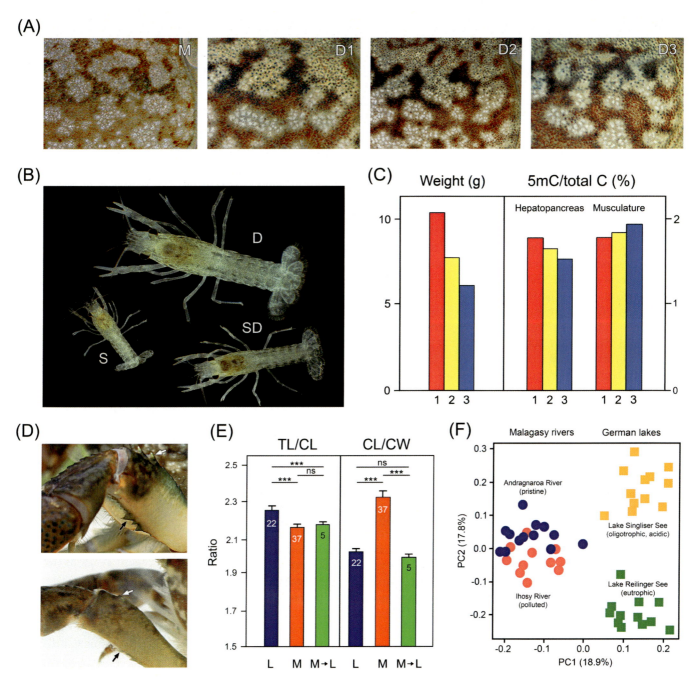

FIGURE 27.5 Generation of epigenetically mediated phenotypic variation in genetically identical marbled crayfish, *Procambarus virginalis*. A-C: Stochastic developmental phenotypic variation (SDPV); D-F: Environmentally induced phenotypic variation (EIPV). (A) Comparison of color patterns of posterio-lateral cephalothorax area of a mother (M) and three adult daughters (D1-D3) from the same clutch and raised together in the same narrowly controlled laboratory environment, showing striking differences in marmoration between all individuals. (B) Establishment of social hierarchy and growth differences in clutch-mates kept for 34 days under social stress conditions. The experiment was started with five size-matched juvenile clutch-mates of indifferent agonistic behavior and ended with one dominant (D), one subdominant (SD) and three subordinates (S) of remarkably different sizes. (C) Variation of body weight and global DNA methylation (measured by capillary electrophoresis) in the hepatopancreas and abdominal musculature of three 626 days old, communally reared clutch-mates (1–3). The graph shows methylation differences between individuals and tissues. (D) Chelipeds of laboratory-raised (upper panel) and wild specimen (lower panel) from Lake Moosweiher, Germany, showing much bigger and sharper spines (arrows) in the wild specimen. (E) Body proportions of marbled crayfish from my laboratory colony (L) and Lake Moosweiher (M), showing significant differences in total length (TL)/carapace length (CL) ratio and CL/carapace width (CW) ratio. The adult offspring of a specimen transferred from Lake Moosweiher to the laboratory (M→L) had a TL/CL ratio similar to the wild population but a CL/CW ratio similar to the laboratory population, indicating partial adaptation to the new environment. Figures in columns give numbers of specimens investigated; ***, significantly different ($P < 0.001$); ns, not significantly different. (F) Principal component analysis of DNA methylation of 122 genes in the hepatopancreas of specimens from Andragnaroa River and Ihosy River (Madagascar) and Lake Singliser See and Lake Reilinger See (Germany), showing clear separation of the populations. $P < 0.05$. *A-C from G. Vogt, M. Huber, M. Thiemann, G. van den Boogaart, O.J. Schmitz, C.D. Schubart. Production of different phenotypes from the same genotype in the same environment by developmental variation. J Exp Biol 2008;211(4):510–523. D from G. Vogt, C. Lukhaup, M. Pfeiffer, N.J. Dorn, B. W. Williams, R. Schulz, et al. Morphological and genetic characterization of the marbled crayfish, including a determination key. Zootaxa 2018;4524(3):329–350. E based on G. Vogt. Epigenetic variation in animal populations: sources, extent, phenotypic implications and ecological and evolutionary relevance. J Biosci 2021;46:24. F based on S. Tönges, G. Venkatesh, R. Andriantsoa, K. Hanna, F. Gatzmann, G. Raddatz, et al. Location-dependent DNA methylation signatures in a clonal invasive crayfish. Front Cell Dev Biol 2021;9:794506.*

non-synonymous SNVs that change the amino acid sequence of proteins was only 4 compared to the reference genome. These data suggest that any phenotypic differences between laboratory-raised and wild populations must have been caused by epigenetic mechanisms rather than DNA polymorphism.

Comparison of isogenic marbled crayfish populations grown under strikingly different conditions revealed considerable phenotypic differences. For example, the wild specimens from German lakes all had prominent sharp spines on their carapaces and chelipeds [100], but laboratory raised specimens of the same size lacked these spines and had only small blunt knobs instead (Fig. 27.5D). Specimens transferred from Lake Moosweiher to the laboratory maintained their spines through the following molts until the end of life, but in the F1 generation they were considerably reduced resembling members of the laboratory colony.

The laboratory specimens also had significantly longer pleons and broader carapaces when compared to equal-sized specimens from Lake Moosweiher (Fig. 27.5E) [101]. Interestingly, the adult offspring of a specimen that was transferred from Lake Moosweiher to the laboratory and reproduced there one year later had a total length/carapace length ratio similar to their mother and the wild population, but a carapace length/carapace width ratio different from their mother and more similar to the laboratory population (Fig. 27.5E), indicating partial adaptation to the new environment. There were also marked differences in population structure between wild populations from different bioclimatic regions. Andriantsoa and colleagues analyzed populations from a pond, a lake, two rivers, and a rice field in humid, subarid, and dry regions of Madagascar and revealed significant differences in population density and size-frequency distribution among populations despite genetic identity [103].

To investigate the association between environmental adaptation and epigenetic signatures in more detail, Tönges and colleagues [102] have recently compared two Malagasy populations of marbled crayfish from pristine Andragnaroa River (clear mountain river with relatively low pH and temperature) and polluted Ihosy river (higher pH and temperature, very high contents of aluminium due to nearby mining activities) with two German lake populations living in acidic, oligotrophic Lake Singliser See (former lignite mining site, no fishes present) and slightly basic, eutrophic Lake Reilinger See (predatory fishes present) (Fig. 27.5F). They identified specific and highly localized DNA methylation signatures for each of these populations that remained stable over consecutive years, providing evidence for the efficacy of EIPV and the existence of epigenetic ecotypes.

RELEVANCE OF EPIGENETICS FOR ANIMAL ECOLOGY

Epigenetic mechanisms are probably effective in all animal populations, but their role is particularly evident in asexual lineages, invasive populations, and sessile taxa. Asexually reproducing populations have no or little genetic variation, which in sexual populations obscures the contribution of epigenetic variation to phenotypic variation. Invasive groups are often small and genetically impoverished but manage to adapt to previously inexperienced conditions despite this handicap. Sessile taxa cannot evade unfavorable environmental conditions by migration and must, therefore, possess a rapidly responding compensatory system aside of long-term genetic adaptation.

Asexually Reproducing Populations

Several asexually reproducing invertebrates and vertebrates have been used in the past years to examine the role of epigenetics in environmental adaptation. One of them is the marbled crayfish discussed above. Another interesting example is the New Zealand mud snail, *Potamopyrgus antipodarum*, one of only few mollusks capable of asexual reproduction. Genetic and karyotypic data suggest that in their native range clonal lineages emerged repeatedly from diploid sexual females by spontaneous transition from diploidy to triploidy and parallel change from gonochorism to parthenogenesis [104].

Potamopyrgus antipodarum has a shell size of 5–12 mm and reaches sexual maturity after 3–6 months. There are 1–6 generations per year and longevity is 18 months. The mud snail is ovoviviparous and each female produces between 20 and 120 juveniles per clutch. They live in streams, lakes, and reservoirs in fresh and brackish water, feed on periphyton, macrophytes, and detritus, and survive dry and cold periods buried in the mud. Population density can be extremely high, amounting to many thousands of individuals per m^2. Mud snail populations, even clonal ones, show great differences for size and fecundity that are linked to environmental variables such as water temperature, salinity, and current [105,106].

In order to determine the epigenetic proportion of phenotypic variation, Thorson and colleagues compared morphological traits and DNA methylation in *P. antipodarum* from different sites in Oregon and Washington, US [105]. These populations originated from a single clone that was introduced in the western United States about 30 years ago. Dybdahl and Drown analyzed genetic markers in these populations and demonstrated the near absence of genotypic variation within and among populations [107]. Thorson and colleagues revealed habitat specific differences in shell

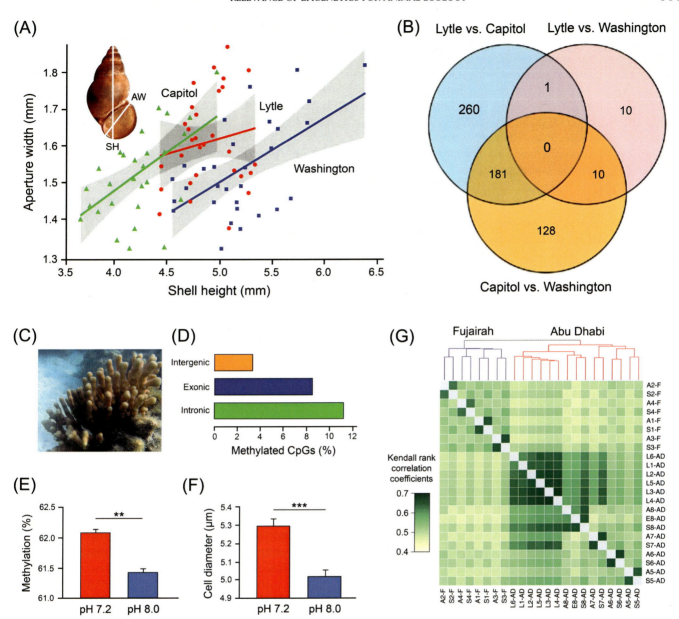

FIGURE 27.6 **Relationship between environmental adaptation and epigenetic variation.** A, B: asexual populations; C-G: sessile animals. (A) Differences in allometry of shell height (SH) and aperture width (AW) between populations of monoclonal mud snail *Potamopyrgus antipodarum* from pristine Lake Lytle, urban Capitol Lake and polluted Lake Washington. Shaded areas indicate 95% confidence intervals. (B) Venn diagram showing differently methylated regions (DMR) and their overlap between lake comparisons. (C) Coral *Stylophora pistillata* in natural environment. (D) Distribution of DNA methylation in the genome of *S. pistillata* showing that introns are more intensely methylated than exons and intergenic regions. (E) Comparison of the effect of pH on DNA methylation in laboratory raised, genetically identical *S. pistillata* showing significantly higher mean methylation levels at stressful pH 7.2 ($P < 0.01$). (F) Comparison of the effect of pH on cell size in the same genets demonstrating significantly larger cells at pH 7.2 ($P < 0.001$). (G) Clustering performed on pair wise correlation of methylation data from *Platygyra daedalea* sampled from unstressed population at Fujairah, Gulf of Oman (F, specimens 1–4) and temperature-stressed population at Abu Dhabi, Arabian-Persian Gulf (AD, specimens 5–8). Within each population, samples were collected from adults (A), their spawned sperm (S) and eggs (E), and larval offspring from reciprocal crosses between E7 and S8 (L1–L3) and S7 and E8 (L4–L6). The graph demonstrates grouping of samples by environmental origin and a strong effect of transgenerational inheritance on methylation patterns: gametes cluster best with respective adults and larvae with their parents. *A and B based on J.L.M. Thorson, M. Smithson, I. Sadler-Riggleman, D. Beck, M. Dybdahl, M.K. Skinner. Regional epigenetic variation in asexual snail populations among urban and rural lakes. Environ Epigenet 2019;5(4):dvz020. C from https://en.wikipedia.org/wiki/Stylophora_pistillata; D-F based on Y.J. Liew, D. Zoccola, Y. Li, E. Tambutte, A.A. Venn, C.T. Michell, et al. Epigenome-associated phenotypic acclimatization to ocean acidification in a reef-building coral. Sci Adv 2018;4(6):eaar8028. G based on Y.J. Liew, E.J. Howells, X. Wang, C.T. Michell, J.A. Burt, Y. Idaghdour, et al. Intergenerational epigenetic inheritance in reef-building corals. Nat Clim Change 2020;10:254–259.*

shape that were correlated with water current speed [105]. Using methylated DNA immunoprecipitation (MeDIP) and Illumina sequencing, the authors also measured significant genome wide DNA methylation differences in the foot pad tissue between lake and river populations, suggesting that environmentally induced epigenetic changes may have caused adaptive phenotypic changes in less than 100 generations.

Thorson and colleagues also compared genetically identical *P. antipodarum* populations between a rural lake (Lake Lytle, OR) and two polluted urban lakes (Capitol Lake, WA, and Lake Washington, WA) [108]. They measured differences in shell shape and allometric growth (Fig. 27.6A) between lakes and identified numerous DMRs (Fig. 27.6B). A higher number of DMRs was shared between rural Lake Lytle and Capitol Lake characterized by high water temperature and high levels of phosphorous and fecal bacteria. A similar relationship was also observed between the two urban lakes, but not between the rural lake and Lake Washington heavily polluted by heavy metals and organic xenobiotics. The presence of site-specific differences in DNA methylation between genetically identical lake populations suggests an epigenetic response to the specific environmental factors.

Sessile Taxa

Sessile animals are the Porifera, Cirripedia, Bryozoa, and Pelmatozoa, and most Cnidaria, Bivalvia and Tunicata. Some of them, particularly corals, can form large clonal assemblages by asexual reproduction. In contrast to the vagile animals, they cannot evade unfavorable environmental conditions. Therefore, epigenetic variation could play a similarly big role in environmental adaptation of these animals as for the immobile plants.

Liew and colleagues investigated DNA methylation in the scleractinian coral *Stylophora pistillata* (Fig. 27.5C) and showed that introns have proportionally more methylated positions (11.3%) than exons (8.6%) or intergenic regions (3.3%) (Fig. 27.6D) [109]. Gene body methylation significantly reduced transcriptional noise by fine-tuning the expression of highly expressed genes. Exposure of isogenic coral fragments from the same genet to long-term pH-stress (pH 7.2) significantly increased mean methylation levels (Fig. 27.6E) when compared to the control (pH 8.0). Widespread methylation changes were observed in genes regulating cell cycle and body size. Enhanced DNA methylation at lower pH was accompanied by an increase in cell size (Fig. 27.6F) and polyp size resulting in more porous skeletons. The authors concluded that DNA methylation seems to help corals buffering the impacts of environmental changes by fine-tuning gene expression.

Liew and colleagues also analyzed DNA methylation patterns and their environmental implications in the reef-building brain coral *Platygyra daedalea* using WGBS [110]. They identified 1.42 million CpG positions (3.2% of all CpGs) that were consistently methylated in the ~800 Mb genome. Adults and their gametes and larval offspring were sampled from two distant populations in the Arabian Peninsula. The Abu Dhabi population lives inside the Arabian-Persian Gulf under extreme temperatures (winter <19°C and summer >35°C) and salinities (40–46 psu) and has persisted through several major thermal stress events (coral bleaching) during the past two decades. The Fujairah population lives south of the strait of Hormus under comparatively milder conditions (22°C–33°C, 36–39 psu) and has not experienced coral bleaching in recent years.

Liew and colleagues showed that the DNA methylation patterns in *P. daedalea* are determined by genotype, developmental stage, and the parental environment (Fig. 27.6G) [110]. Interestingly, genome-wide CpG methylation was inherited from adults to their sperm and larvae (Fig. 27.6G). Variation in hypermethylation of genes in adults and their sperm from distinct environments suggests intergenerational epigenetic acclimatization to local temperature and salinity. Furthermore, genotype-independent adjustments of methylation levels in stress-related genes were strongly correlated with offspring survival rates under heat stress, suggesting that epigenetic mechanisms could help corals to adapt to climate change.

Invasions and Adaptive Radiations

Invaders can be ecologically and evolutionarily very successful even when the founding population in the new environment is small and genetically rather uniform. This phenomenon is well known as "invasion paradox", but the underlying mechanisms are poorly studied [111]. Particularly the first steps of invasion, the survival of the invading specimens and the establishment of a founder population, may depend much on the generation of epigenetic variation [91,112,113]. Good models for studying this topic in animals are the invasive *Procambarus virginalis* and *Potamopyrgus antipodarum* discussed above and some relatively recent and ongoing invasions of sexually reproducing species.

An illustrative example of the latter is the house sparrow, *Passer domesticus* that was introduced to Mombasa (Kenya) in the 1950s and since has evolved significant phenotypic variation [114]. Analysis of MSAP and microsatellite loci revealed that the Kenyan house sparrow populations have a low genetic diversity but a relatively high epigenetic diversity. There was a significant negative correlation between genetic and epigenetic diversity (Fig. 27.7A) and a positive correlation between epigenetic diversity and the inbreeding coefficient,

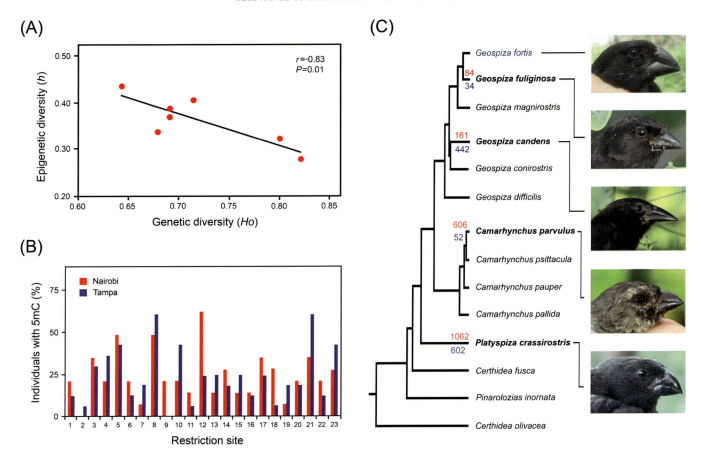

FIGURE 27.7 **Genetic and epigenetic variation in invasions and adaptive radiations of birds.** (A) Negative correlation of genetic and epigenetic diversity in invasive populations of house sparrow, *Passer domesticus*, from seven Kenyan cities. Genetic diversity was determined by microsatellite analysis and epigenetic diversity by MSAP. *h*, haplotype diversity; *Ho*, heterozygosity; *P*, probability value; *r*, Pearson correlation coefficient. (B) Comparison of percentage of house sparrows with methylated CpGs at 23 restriction sites between the invasive populations of Tampa, Florida (n = 16) and Nairobi, Kenya (n = 14). There are marked differences between the ~150 year old Floridan invasion and the ~65 year old Kenyan invasion. (C) Correlation of genetic and epigenetic changes with phylogenetic distance in five species of Darwin's finches. Blue numbers on branches indicate genetic mutations (CNVs) and red numbers indicate epimutations (DMRs) in four species compared to a reference species (*Geospiza fortis*). The graph shows that the number of DMRs increases consistently with phylogenetic distance, whereas the number of CNVs does not. A based on A.L. Liebl, A.W. Schrey, C.L. Richards, L.B. Martin. Patterns of DNA methylation throughout a range expansion of an introduced songbird. Integr Comp Biol 2013;53(2):351–358. B based on A.W. Schrey, C.A.C. Coon, M.T. Grispo, M. Awad, T. Imboma, E.D. McCoy, et al. Epigenetic variation may compensate for decreased genetic variation with introductions: a case study using house sparrows (*Passer domesticus*) on two continents. Genet Res Int 2012;2012:979751. C based on M.K. Skinner, C. Gurerrero-Bosagna, M. Muksitul Haque, E.E. Nilsson, J.A.H. Koop, S.A. Knutie, et al. Epigenetics and the evolution of Darwin's finches. Genome Biol Evol 2014;6(8):1972–1989.

suggesting that DNA methylation helps to overcome genetic barriers typically associated with invasions such as bottlenecks and inbreeding.

Schrey and colleagues compared the house sparrow invasion in Kenya (~65 years ago) with an older invasion in Florida (~150 years ago) and found that samples from Kenya had less genetic diversity at multiple microsatellite loci than samples from Florida [115]. Overall, DNA methylation was higher in the Kenyan population, but some restriction fragments were more methylated in the Floridian population (Fig. 27.7B). The authors concluded that epigenetic variation may have compensated for relatively low genetic variation in both invasions of house sparrow.

Founder populations often show rapid phenotypic divergence from source populations after colonizing new environments. To investigate the involvement of epigenetic mechanisms in this early step of invasion, Hu and colleagues experimentally colonized eight small Caribbean islands with brown anole lizard, *Anolis sagrei*, deriving from a common source population [117]. Four days after colonization they measured genome-wide DNA methylation in recaptured specimens from islands with high versus low habitat quality. Habitat quality explained a significant proportion of the recorded epigenetic variation. Differentially methylated cytosines mapped to genes involved in signal transduction, immune response and circadian rhythm likely to be relevant to habitat change.

A classic example of invasion and subsequent adaptive radiation is the Darwin's finches that evolved over a period of 2–3 million years from a single invader and radiated into 14 species occupying distinct ecological niches on Galapagos. In five species investigated, epimutations (DMRs) were more common than genetic copy number variants (CNVs) (Fig. 27.7C). Moreover, the number of DMRs increased with phylogenetic distance, whereas the number of CNVs did not [116]. The number, chromosomal location, regional clustering, and lack of overlap of epimutations and CNVs suggest that epigenetic changes are distinct and probably have contributed to evolutionary change independently from genetic change.

Parasites

There is some evidence that DNA methylation is also used by parasites to adapt to their unique life style and by the hosts to respond to the parasitic infections. Gao and colleagues demonstrated the presence of DNA methylation in the human nematode parasite *Trichinella spiralis* [7], but free-living nematodes like *Caenorhabditis elegans* have unmethylated genomes. The authors observed a drastic increase in DNA methylation of parasitism-related genes during the development from newborn larvae to encapsulated larvae to infectious adults. Liu and colleagues compared the capsule-forming *Trichinella spiralis* with the non-encapsulated *Trichinella pseudospiralis* and recorded considerable differential expansion and methylation of parasitism-related multicopy gene families, especially for the DNase II members and glutathione S-transferases [118].

Geyer and colleagues found evidence for cytosine methylation in the genome of *Schistosoma mansoni*, the causative agent of schistosomiasis, and demonstrated its importance for egg production and egg development [119]. Geyer and colleagues also characterized the DNA methylation machinery of the intermediate host of schistosomiasis, the gastropod *Biomphalaria glabrata* [6], and found that it is responsive to soluble products of the parasite.

RELEVANCE OF EPIGENETICS FOR ANIMAL EVOLUTION

The significance of epigenetic mechanisms for animal evolution can be studied on the example of adaptive radiations as shown for Darwin's finches but also on the examples of domestication and polyploid speciation. Domestication is evolution in time laps and polyploid speciation is characterized by particularly intense rearrangements of the chromatin and genome with the help of epigenetic mechanisms.

Domestication

In domesticated animals, the breeding history is very short in terms of evolutionary times and rather well documented. Moreover, since selection was done by humans the selective forces and their direction are known. Domestication of most animals started a few thousand years ago with a maximum of c. 30,000 years in dogs [120]. In several fish species, domestication has just begun. Domesticated species, particularly dogs and fowl, have the additional advantage of exceptionally broad variation of phenotypic traits like body size, body shape, and coloration, and their genetics is rather well investigated [121].

In dog, which descended from the gray wolf, *Canis lupus*, some of the phenotypic changes related to domestication have already been linked to specific genes. For example, different alleles involved in the fight-or-flight response have been subject to strong selection and resulted in behavioral differences between dogs and wolves [122]. Janowitz Koch and colleagues established that domestication of dogs has also been associated with epigenetic alterations [123]. They analyzed methylation differences across the genomes and revealed species-specific patterns of DMRs at 68 sites. The authors concluded that selection may have not only acted on genes but also on methylation patterns.

Bélteky and colleagues investigated differences in hypothalamic DNA methylation between two selected lines of red jungle fowl, *Gallus gallus* (ancestor of chicken) that were bred for either high or low fear of humans over five generations [124]. The authors found 22 DMRs between the two lineages in genes involved in cellular metabolism and neural signaling. They concluded that selection for tameness can cause divergent epigenetic patterns within only five generations and that these changes may have had an important role in chicken domestication.

Anastasiadi and Piferrer investigated domestication in European sea bass, *Dicentrarchus labrax*, one of the main farmed fish species in the Mediterranean [125]. A good quality reference genome is available and selective breeding programs are now applied to this species, which is at the beginning of domestication. Anastasiadi and Piferrer detected DNA methylation differences between early domesticates and their wild counterparts despite the absence of genetic differences. About one-fifth of the methylation differences that persisted into adulthood were established by the time of gastrulation and affected genes involved in developmental processes. For example, the *adamts9* gene that encodes an

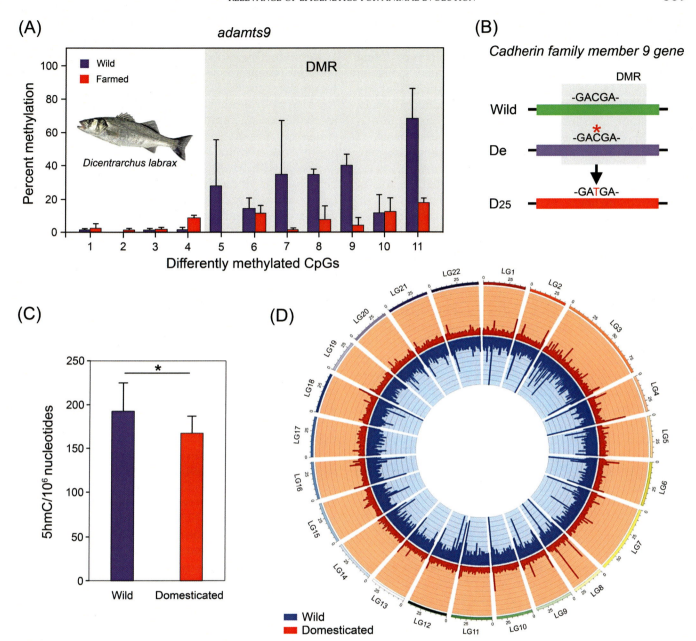

FIGURE 27.8 Changes of DNA methylation and hydroxymethylation patterns during early domestication of fish. (A) Mean methylation of CpGs in DMR of *adamts9* gene in the muscle of sea bass, *Dicentrarchus labrax*, showing significantly lower methylation in farmed than wild specimens, which is associated with higher expression levels. This matrix metalloproteinase remodels the extracellular matrix during development. (B) CpG in DMR of *cadherin family member 9* gene of sea bass that got methylated (asterisk) in early domesticates (De) and was converted into a TpG after 25 years of selective breeding (D_{25}). (C) Difference of hydroxymethylated cytosines (5hmC) in fast muscles between wild tilapia, *Oreochromis niloticus*, and their offspring reared in captivity. Domesticates have a significantly lower 5mhC level. (D) Circular representation of the tilapia nuclear genome established from fast muscles showing sites with substantial levels of 5hmC in wild specimens and their domesticated offspring and differences between the two. *A and B based on D. Anastasiadi and F. Piferrer. Epimutations in developmental genes underlie the onset of domestication in farmed European Sea Bass. Mol Biol Evol 2019;36(10):2252–2264. C and D based on J. Konstantinidis, P. Sætrom, R. Mjelle, A.V. Nedoluzhko, D. Robledo, J.M.O. Fernandes. Major gene expression changes and epigenetic remodelling in Nile tilapia muscle after just one generation of domestication. Epigenetics 2020;15(10):1052–1067.*

extracellular matrix metalloproteinase was hypomethylated in embryos of early domesticates and this pattern was maintained in the adult muscle (Fig. 27.8A), resulting in higher *adamts9* expression levels.

Some early domesticated sea bass showed lower jaw malformation, a key feature of the domestication syndrome. Interestingly, associated DNA methylation patterns significantly overlapped with 5-methylcytosine-to-thymine mutations after 25 years of selective breeding (Fig. 27.8B), suggesting that epimutations with phenotypic effect have been converted into phenotype fixing genetic mutations. Furthermore, epimutated genes in sea bass coincided with corresponding genes that are under positive selection in other domesticated animals. The authors concluded from their work that epimutations in developmental genes underlie the onset of domestication and cause Darwin's domestication syndrome, which might be genetically fixed later on.

Konstantinidis and colleagues investigated the role of hydroxymethylation in domestication of Nile tilapia, *Oreochromis niloticus*, at a genome-wide level and single nucleotide resolution and found that the muscle hydroxymethylome changed already after a single generation of domestication [126]. Hydroxymethylcytosine is a metabolite of methylcytosine but is meanwhile regarded as a stable epigenetic mark involved in development, aging and diseases [127]. The overall decrease in hydroxymethylcytosine levels in domesticated tilapia (Fig. 27.8C,D) was accompanied by the downregulation of 2015 genes, mainly immune genes, whereas several myogenic and metabolic genes that affect growth were upregulated when compared to the wild specimens. There were 126 differentially hydroxymethylated cytosines (DHMCs) between groups, which were not due to genetic variation. They were associated with genes involved in immune defense, growth and neuronal pathways. The DHMCs were mostly located within gene bodies suggesting a functional role in gene expression.

Polyploid Speciation

New animal species mostly arise by the divergence of allele frequencies between or within populations and the establishment of reproductive barriers. Less frequently, species originate by the duplication of entire genomes (autopolyploidy) or interspecific hybridization associated with genome duplication (allopolyploidy) [128]. Both types of speciation are accompanied or followed by reproductive isolation, chromatin remodeling, alteration of gene expression, and changes of life history features, but these effects are usually much more pronounced in polyploid speciation. Apparently, epigenetic mechanisms contribute to all of these processes.

The generation of new species and new higher taxa by polyploidy has played a considerable role in animal evolution [129,130]. A well-established higher taxon example is the vertebrates, which have experienced two rounds of polyploidy in their stem line and a further round in the stem line of the teleost fishes. These events led to the expansion of the DNA methylation machinery (Fig. 27.1A) [24]. Extant polyploid species are relatively frequent in water fleas, insects, fishes and amphibians, but extremely rare in mammals.

Reproductive isolation, the absence of gene flow between parent species and neospecies, is an important requirement for the separate evolution of a new species. An animal example for the involvement of epigenetic mechanisms in reproductive isolation is the North American tessellated darter fish, *Etheostoma olmste*. Smith and colleagues investigated DNA methylation variation with MSAP and found that the strength of behavioral reproductive isolation was associated with epigenetic divergence [131]. Laporte and colleagues compared differently methylated transposable elements (TE) between the "dwarf" and "normal" whitefish, two nascent species of the *Coregonus clupeaformis* species complex, that diverged some 15,000 generations ago [132]. Their data support a role of DNA methylation reprogramming and associated derepression of TEs in postzygotic isolation.

New polyploid genomes are usually unstable and require chromatin rearrangement. One common feature is the reduction of the amount of DNA. DNA loss mostly concerns redundant genes, and in extreme cases the polyploid genome can be downsized to the diploid state. For example, after doubling of the genome in the stem line of fishes 70%—80% of duplicated genes have been lost [133].

The DNA methylation level of polyploids can either be higher or lower when compared to the parent species. For example, hybrids of kangaroos *Macropus eugenii* x *Wallabia bicolor* were characterized by a genome-wide hypomethylation [134]. Removal of DNA methylation from retrotransposons in the hybrids facilitated their amplification and caused gross changes in genome structure. An increase of DNA methylation compared to the parent species was observed in hybrids of red crucian carp *Carassius gibelio auratus* x common carp *Cyprinus carpio* [135].

Polyploids often have life history traits that are different from those of the parent species. Growth, number of offspring, and other quantitative traits can either decrease or increase when compared to the diploid ancestors. In allopolyploids, the increase of life history traits is usually explained as the result of heterozygosity (hybrid vigor). In autopolyploids, trait alteration is caused by changes of gene dosage, rearrangement of gene-networks and modulation of gene expression. All of these alterations obviously require the contribution of epigenetic mechanisms.

In the triploid, obligatory parthenogenetic marbled crayfish, *Procambarus virginalis*, which originated from the diploid, sexually reproducing slough crayfish, *Procambarus fallax*, some 30 years ago, speciation was accompanied by reproductive isolation and marked alterations in DNA content, global DNA methylation,

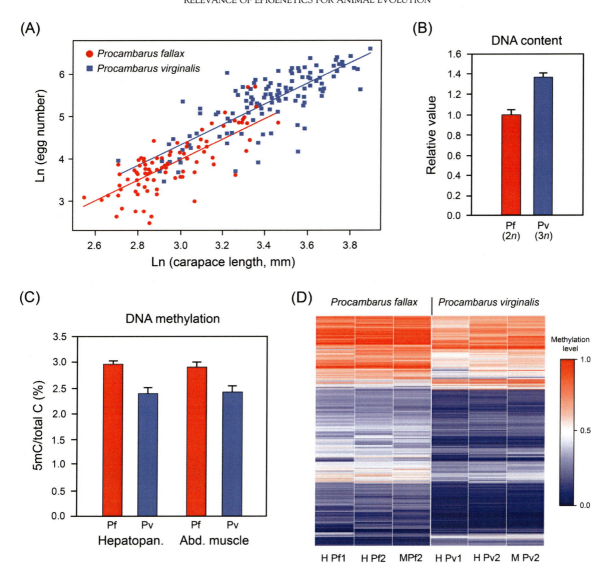

FIGURE 27.9 Changes of fecundity, DNA content, and DNA methylation during autopolyploid speciation of *Procambarus virginalis* from *Procambarus fallax*. (A) Comparison of pleopodal egg numbers per female and clutch, demonstrating that fecundity is much higher in the triploid neospecies when compared to its diploid parent species. Lines are linear model prediction lines. (B) Comparison of DNA content in hemocytes. Flow cytometry of two biological and three technical replicates demonstrates a significant, ~1.4-fold higher DNA content in *P. virginalis* (Pv) than in *P. fallax* (Pf) ($P = 1.33 \times 10^{-7}$), suggesting that some DNA has been lost after triploidization. (C) Comparison of global DNA methylation in the hepatopancreas and abdominal musculature of both species. Mass spectrometry of the organs from three laboratory raised females per species revealed a significant, ~20% lower DNA methylation level in *P. virginalis* ($P = 1.48 \times 10^{-7}$). (D) Comparison of gene body methylation in 2357 genes between two *P. fallax* (Pf1 and Pf2) and two *P. virginalis* (Pv1 and Pv2) specimens. The heatmap of the hepatopancreases (H) and abdominal muscles (M) shows numerous differentially methylated genes between both species, most of them being hypomethylated in *P. virginalis*. A based on G. Vogt, N.J. Dorn, M. Pfeiffer, C. Lukhaup, B.W. Williams, R. Schulz, et al. *The dimension of biological change caused by autotriploidy: a meta-analysis with the triploid* Procambarus virginalis *and its diploid parent* Procambarus fallax. Zool Anz 2019;281:53–67. B and C based on G. Vogt, C. Falckenhayn, A. Schrimpf, K. Schmid, K. Hanna, J. Panteleit, et al. *The marbled crayfish as a paradigm for saltational speciation by autopolyploidy and parthenogenesis in animals.* Biology Open 2015;4(11):1583–1594. D based on F. Gatzmann, C. Falckenhayn, J. Gutekunst, K. Hanna, G. Raddatz, V. Coutinho Carneiro, et al. *The methylome of the marbled crayfish links gene body methylation to stable expression of poorly accessible genes.* Epigenetics Chromatin 2018;11(1):57.

genome-wide DNA methylation patterns, and life history traits [4,46,136]. In laboratory experiments, marbled crayfish females readily copulated with males of *P. fallax*, but the offspring was always pure marbled crayfish as demonstrated by microsatellite analysis [4], indicating that reproductive isolation occurs on the cytological or genetic rather than the behavioral level.

Body size and fecundity are significantly enhanced in marbled crayfish [136] indicating superior fitness. Marbled crayfish grow to a maximum body length of

~12 cm corresponding to a body weight of 40 g, whilst *P. fallax* grow to a maximum of ~9 cm and 18 g. Fecundity is on average much higher in *P. virginalis* due to bigger body size, but it is also ca. 40% higher in specimens of equal sizes (Fig. 27.9A).

The triploid marbled crayfish has a 1.4-fold instead of an expected 1.5-fold increased DNA content when compared to its diploid parent species (Fig. 27.9B), suggesting loss of some DNA after triploidization [4]. Global DNA methylation is about 20% lower in marbled crayfish (Fig. 27.9C), arguing for a considerable rearrangement of the DNA methylation pattern and gene expression during speciation [4]. Differences in gene body methylation in hundreds of genes between both species support this view (Fig. 27.9D) [46]. Most of these genes were hypomethylated in the marbled crayfish, which seems to be associated with higher phenotypic plasticity and higher invasiveness when compared to its parent species.

Transgenerational Inheritance of Epigenetically Determined Phenotypes

The polyploid speciation examples suggest that epigenetic changes follow genetic changes helping to reorganize the new genome. On the other hand, the invasion and domestication examples suggest that epigenetic mechanisms can also trigger evolutionary change. This possibility would require selection, transgenerational inheritance and final genetic integration of epigenetic modifications and associated phenotypic traits.

The idea of transgenerational epigenetic inheritance (TEI) is still controversial, although there are convincing examples published for bacteria, plants, and animals, suggesting that it is a universal phenomenon of life [137–139]. Good examples are published for animals diverse as mouse, brine shrimps, and corals [110,140,141]. For example, in parthenogenetic brine shrimp *Artemia* raised from a single female, the transgenerational inheritance of acquired resistance to heat shock and bacterial infection was accompanied by TEI of certain patterns of DNA methylation and histone acetylation [141]. In animals, the transfer of histone modifications and ncRNAs via sperm seems to be particularly important for the inheritance of epigenetic marks and related phenotypic traits across generations [110,142].

The main argument against transgenerational epigenetic inheritance in animals has come from the erasure of epigenetic signatures in the germline and zygote of mammals, which was generalized for all animals. However, in zebrafish *Danio rerio*, the paternal methylome is largely maintained throughout early embryogenesis, whereas the maternal methylome is maintained until the 16-cell stage and then progressively reprogrammed by parallel losses and gains of methylation marks [143]. In the honey bee, there is apparently no significant DNA methylation erasure in the gametes and zygote, and DNA methylation marks are stably transferred from fathers to daughters [144]. In the phylogenetically basal corals, many CpG methylation signatures are transmitted from parents to offspring via the sperm, too [110]. And even the mammalian genome can bypass epigenetic reprogramming during development and transmit information from parents to offspring via DNA methylation marks, histone modification marks, and miRNAs [145].

Experimental evidence suggests that only some epigenetically mediated phenotypes are transgenerationally inherited. For example, neither the status of being a honey bee queen nor the distinct marmoration pattern of a marbled crayfish is inherited. In these cases, it is the ability to generate such different phenotypes from the same genome that is inherited, and this ability enables subsequent generations to produce similar patterns of epigenetic and phenotypic variation de novo. Whether an epigenetic pattern is inherited or not seems to depend on trait and conditions. In a well-adapted population living in a stable environment, it makes little sense to inherit epigenetic variants over many generations because it would incur costs but provide no advantage. However, in a changing environment new and better suited epigenetic variants may be selected and transgenerationally inherited to better cope with the new conditions.

Deterministic selection models showed that newly arising epimutations are principally stable enough to respond effectively to long-term selection yielding epimutation-selection equilibria that are close to those expected for DNA sequence mutation rates [95]. Kronholm and Collins revealed that the long-term effects of epimutations depend crucially on their stability and fitness effects relative to genetic mutations [146].

In sexually reproducing species, beneficial epigenetically determined phenotypes are thought to be fixed on the long range by genetic assimilation, a process by which a phenotype originally produced in response to environmental signals later becomes genetically encoded via random mutation and selection [147,148]. An alternative, more directional mechanism, which would also be applicable to asexually reproducing species, is the conversion of epimutations with phenotypic effects to corresponding genetic mutations, e.g., by deamination of methylated cytosines to thymines as indicated in Fig. 27.8B. The C-to-T transition is by far the most common single nucleotide mutation in living organisms and it is promoted by methylation of the cytosine [149]. Methylated CpGs have a mutation rate that is at least 10–50 times greater than that of unmethylated cytosines [150]. Methylated CpGs can even have mutational effects on the surrounding DNA [149].

Hydrolytic deamination of 5mC is a major source of human mutations, and in bacteria, methylated cytosines were identified as hot spots for cytosine-to-thymine mutations that modulate antibiotics susceptibility [151,152]. Most animals display lower observed than expected densities of CpG dinucleotides ($C_{o/e}$ values) in their genomes [153], suggesting that they have repeatedly experienced transitions from mCpG to TpG. This feature may reflect multiple genetic integrations of advantageous epimutations throughout evolution.

Yi and Goodisman have recently reviewed the strong and pervasive effects of epigenetic mechanisms on DNA sequence evolution in eukaryotic organisms [154]. They discussed how the physical interactions between histones, their modifications and the DNA affect DNA mutation and repair, how DNA methylation suppresses the movement of transposable elements, and how genome expansion through gene duplication is controlled by epigenetic information. The authors concluded that epigenetic mechanisms have widespread indirect effects on DNA sequences and represent a potent cause and constraint of genome evolution.

CONCLUSIONS AND PERSPECTIVES

Surveys across the animal kingdom revealed that DNA methylation, histone modifications, and ncRNA pathways are ubiquitous in animals, with a few exceptions. Histone modifications are apparently similar in all groups, but DNA methylation and ncRNAs have undergone considerable diversification during animal evolution. DNA methylation, the best investigated epigenetic mechanism in animals, is quite variable between and within phyla with respect to both 5mC/C levels and genomic methylation patterns. In some species and clades, DNA methylation was secondarily lost. There is obviously no consistent correlation between methylation level and evolutionary niveau or genome size as previously assumed. The ancestral set of single copies of DNA methylation and demethylation enzymes was expanded in some taxa but lost in others.

There is considerable evidence that DNA methylation, histone modifications and ncRNAs are all involved in regulation of gene expression, production of phenotypic variation, and transgenerational inheritance of acquired characters. Regulation of gene expression by methylation of promoters and gene bodies is apparently conserved in animals with methylated genomes but silencing of repetitive sequences by methylation is variable. Hypomethylation and hypermethylation of promoters is usually associated with gene expression and silencing, respectively. Hypermethylation of gene bodies promotes stable expression of genes, particularly of housekeeping genes, and hypomethylation of gene bodies leads to more variable expression. Interestingly, hypomethylation of the entire genome in a given species seems to increase phenotypic plasticity and invasiveness.

It is a long-running discussion whether the epigenetic variation observed in animal populations is just a consequence of genetic variation (DNA polymorphism) or whether it can arise independently. Experiments with clonal animals revealed that epigenetic variation can be surprisingly broad despite the absence of genetic variation, and therefore, it must be considered as a source of phenotypic variation in its own right. In sexually reproducing populations, where the amount and location of modifiable CpGs usually varies among population members, variation of DNA methylation is to some degree dependent on genetic variation. Of course, independence from genetic variation does not mean independence from the genome because epigenetic mechanisms require genes encoding Dnmts and Tets, histone modifying enzymes, other chromatin related proteins, and ncRNAs and need to interact with genes to modify their expression.

Obviously, all animals are capable of producing phenotypic variation from a single DNA sequence by epigenetic mechanisms, resulting in different cell types, polyphenism and phenotypic plasticity. The generation of different cells types and tissues from the single DNA sequence of the zygote is a key feature of all Eumetazoa. Polyphenism is typical of holometabolous insects with morphologically and functionally distinct life stages and castes. Phenotypic plasticity is probably effective in all animal populations, but it is of particular relevance in populations with no or little genetic variation, small invasive groups and sessile taxa. It helps to survive changing, adverse and previously inexperienced environmental conditions.

The production of different cell types and morphologically distinct life stages with the help of epigenetic mechanisms is triggered by developmental programs, but different morphs and castes are often induced by environmental factors. Epigenetically based phenotypic variation in populations can be produced either by stochastic developmental or environmentally induced changes of the epigenetic marks on the DNA and chromatin. Usually, both occur together but with different weighing: environmentally induced phenotypic variation is directional and predominant in predictable environments, strengthening adaptation to the prevailing conditions, and stochastic developmental phenotypic variation is predominant in unpredictable environments and contributes to evolutionary bet-hedging.

Epigenetic variation has apparently a different function in animal populations when compared to genetic variation, because it is responsive to environmental cues, can be rapidly established and is principally

> **BOX 27.1**
>
> **Overview of epigenetics in animals**
>
> **Animal groups used for epigenetics research:** Most frequently: Insecta and Vertebrata; less frequently: Porifera, Ctenophora, Cnidaria, Nematoda, Crustacea, Chelicerata, Mollusca, and Tunicata.
>
> **Epigenetic mechanisms investigated:** DNA methylation, histone modifications, and ncRNAs.
>
> **Biological role of epigenetics in animals:** Gene regulation, transposon silencing, imprinting, X-chromosome inactivation. Involved in development, cell and tissue differentiation, aging, growth, sex determination, reproduction, response to adverse environmental conditions and toxicants, immune defense, diseases, polyphenism, phenotypic plasticity, domestication, speciation, and transgenerational inheritance of acquired characters.
>
> **Generation of phenotypic variation from the same DNA sequence:** By developmental programs (deterministic), developmental stochasticity (probabilistic), and environmental induction (directional). All variants are mediated by modulation of gene expression via epigenetic mechanisms.
>
> **Dependence of epigenetic variation on genetic variation:** Semi-independent in sexually reproducing, genetically polymorphic populations and independent in asexually reproducing clonal populations.
>
> **Relevance of epigenetic variation for animal ecology:** Probably effective in all animal populations underpinning phenotypic plasticity. Particularly important for asexually reproducing and genetically impoverished populations, invasive groups, sessile taxa, and long-lived species, helping them to stay in the game of life when the conditions change. Enables rapid and simultaneous adaptation of many population members to environmental stressors and can be reversed when the stressor disappears.
>
> **Relevance of epigenetics for animal evolution:** Epigenetic changes can follow and consolidate genetic changes or elicit evolutionary change. Epigenetic marks and associated phenotypes can be inherited across generations. Depending on condition, they may either be reversed to the native state or selected and genetically integrated on the long term, resulting in new evolutionary trajectories.

reversible. Combined with transgenerational epigenetic inheritance, it is a perfect means to cope with transient environmental stressors and environmental changes. If the adverse conditions should disappear in the lifetime of the exposed generation or the subsequent generation, the epigenetic marks and related phenotypes could easily be reverted to the old state, but when the adverse conditions should become permanent the epigenetic variants could persist and be selected and genetically integrated in the long term.

Polyploid speciation demonstrates that epigenetic mechanisms can help consolidating evolutionary processes triggered by genetic changes. Nascent species arisen by a macro-mutation can be stabilized by epigenetic solidification of reproductive isolation, chromatin rearrangement and alteration of gene expression. However, since epigenetically triggered phenotypes are principally selectable, epigenetic variation could also be the starting point of evolutionary change. This possibility is supported by animal domestication, in which the early availability of epigenetic variants enables selection of desired traits long before sufficient genetic variation is available for selection. Therefore, epigenetic changes can be both, followers and leaders in animal evolution, depending on context.

The latter possibility may help to explain why animal evolution is often much faster than expected from the mere combination of random genetic mutation and natural selection. Epigenetically triggered evolutionary changes seem to be particularly effective when the environmental conditions change. Therefore, one might speculate that they considerably accelerate phenotypic diversification and speciation in transitional phases between periods of geological stasis, contributing to the understanding of the punctuated equilibrium theory of evolution proposed by Eldredge and Gould [155] (Box 27.1).

Transgenerational epigenetic inheritance is certainly the most controversial topic of evolutionary epigenetics in animals, but evidence for its prevalence is steadily increasing. Obviously, only a minor fraction of epigenetic variants are passed on to the next generation, depending on many factors like trait, environmental condition, selective advantage, etc. Preliminary experimental data give an idea on how principally reversible epimutations with phenotypic effects could be converted into phenotype fixing genetic mutations, e.g., by deamination of mCpG into TpG.

The dynamics of epigenetic signatures is apparently very variable depending on location in the genome and functional context, but this issue is greatly understudied in animals and other organisms. Epigenetic modifications that are involved in on-off-switching of inducible genes are assumed to have short lifetimes (hours to days), whilst epigenetic marks involved in tissue differentiation, environmental adaptation and evolutionary change must have long lifetimes. Future experiments that compare epigenetic patterns between inducible and constitutive genes and between genes specifically expressed in different life periods (e.g., embryogenesis and reproductive period) could shed more light on this topic. Long-lived, metastable epigenetic signatures can be identified by comparing different tissues and populations adapted to different environments. Understanding the relationships between stable epigenetic fingerprints and their eliciting cues (food, environmental factors, toxicants, pathogens, etc) would pave the way to the development of epigenetic biomarkers in husbandry, veterinary medicine and ecology.

Epigenetic signatures in animals depend on many factors including genome, tissue, developmental stage, life history, life style, environmental conditions, and health status. Future studies should consider this aspect more thoroughly and standardize as much influencing factors as possible to obtain meaningful results. For example, pooling of different tissues and individuals must be avoided. This is now possible by using recently developed omics techniques that enable relatively fast analysis of larger numbers of small-sized tissue samples at reasonable costs. Future work should also be extended to understudied taxonomic units and unconventional life styles and life histories. Of particular interest are members of the same clade that differ significantly in genome size, body size, life span, and growth format.

Most papers on epigenetic variation in animals demonstrated associations between epigenetic marks and phenotypes, while cause-effect relationships between the two remained largely elusive. Therefore, more laboratory experiments should be directed towards identifying the mechanistic links between epigenome and phenotype. Cause-effect relationships can be investigated by using pharmacological blockers of DNA methylation and histone modifications, RNAi, and engineering of genes and epigenetic marks of interest by CRISPR-Cas [156]. Alternatively, the epigenotype-to-phenotype relationships could be investigated by association studies with genetically identical clutch-mates including genome-wide epigenetic modifications, transcriptomes, proteomes and phenomes.

SUMMARY

This chapter discusses the evolution of epigenetic mechanisms and their main functions in animals. Histone modifications are apparently conserved throughout the animal kingdom, but DNA methylation and ncRNA pathways have undergone considerably diversification during animal evolution. There is no consistent correlation between DNA methylation level and evolutionary niveau or genome size as previously assumed. Research with asexually reproducing populations and genetically impoverished invasive groups revealed that the production of epigenetic variation is independent or semi-independent from genetic variation, depending on the mode of reproduction. Epigenetic mechanisms develop their effects through the activation and repression of genes and modulation of gene expression. This way, they can produce phenotypic variation from the same DNA sequence, resulting in different cell types, polyphenism and phenotypic plasticity in populations. The generation of different cells and tissues from the DNA sequence of the zygote with the help of epigenetic mechanisms is a key feature of all Eumetazoa. Polyphenism is typical of many insects and optimizes exploitation of resources and labor division in social colonies. Phenotypic plasticity is a feature of all animals and helps populations, particularly asexual, sessile and invasive ones, to survive fluctuating, adverse and previously inexperienced environmental conditions. The provision of epigenetic variation in populations is a means for short-term and medium-term adaptation to environmental changes. It can be rapidly established in numerous population members in response to environmental cues and reversed, when the triggering cues disappear, which is a qualitative difference to adaptation by selection on genetic variation. Epigenetic mechanisms also contribute significantly to speciation and domestication, two prominent aspects of evolution. These examples demonstrate that epigenetic mechanisms can do both, consolidate genetically caused change and trigger evolutionary change, or, with other words, be followers and leaders in evolution. There are convincing examples of transgenerational inheritance of acquired phenotypic traits and associated epigenetic signatures in animals, and there are new hypotheses on how epigenetically triggered phenotypes could be genetically integrated in the long term.

References

[1] Torruella G, Derelle R, Paps J, Lang BF, Roger AJ, Shalchian-Tabrizi K, et al. Phylogenetic relationships within the Opisthokonta based on phylogenomic analyses of conserved single-copy protein domains. Mol Biol Evol 2012;29(2):531–44.

[2] Iyer LM, Abhiman S, Aravind L. Natural history of eukaryotic DNA methylation systems. Prog Mol Biol Transl Sci 2011;101:25–104.

[3] Vogt G, Huber M, Thiemann M, van den Boogaart G, Schmitz OJ, Schubart CD. Production of different phenotypes from the same genotype in the same environment by developmental variation. J Exp Biol 2008;211(4):510–23.

[4] Vogt G, Falckenhayn C, Schrimpf A, Schmid K, Hanna K, Panteleit J, et al. The marbled crayfish as a paradigm for

saltational speciation by autopolyploidy and parthenogenesis in animals. Biol Open 2015;4(11):1583—94.

[5] Vogt G. Evolution of epigenetic mechanisms in animals and their role in speciation. In: Tollefsbol T, editor. Handbook of epigenetics. The new molecular and medical genetics. 2nd edition. Cambridge, MA: Academic Press; 2017. p. 409—26.

[6] Geyer KK, Niazi UH, Duval D, Cosseau C, Tomlinson C, Chalmers IW, et al. The *Biomphalaria glabrata* DNA methylation machinery displays spatial tissue expression, is differentially active in distinct snail populations and is modulated by interactions with *Schistosoma mansoni*. PLoS Negl Trop Dis 2017;11(5):e0005246.

[7] Gao F, Liu X, Wu X-P, Wang X-L, Gong D, Lu H, et al. Differential DNA methylation in discrete developmental stages of the parasitic nematode *Trichinella spiralis*. Genome Biol 2012;13:R100.

[8] Kyger R, Luzuriaga-Neira A, Layman T, Sandberg TOM, Singh D, Huchon D, et al. Myxosporea (Myxozoa, Cnidaria) lack DNA cytosine methylation. Mol Biol Evol 2021;38(2):393—404.

[9] Liu S, Aageaard A, Bechsgaard J, Bilde T. DNA methylation patterns in the social spider, *Stegodyphus dumicola*. Genes 2019;10(2):137.

[10] Bewick AJ, Vogel KJ, Moore AJ, Schmitz RJ. Evolution of DNA methylation across insects. Mol Biol Evol 2017;34(3):654—65.

[11] Provataris P, Meusemann K, Niehuis O, Grath S, Misof B. Signatures of DNA methylation across insects suggest reduced DNA methylation levels in Holometabola. Genome Biol Evol 2018;10(4):1185—97.

[12] Fairclough SR, Chen Z, Kramer E, Zeng Q, Young S, Robertson HM, et al. Premetazoan genome evolution and the regulation of cell differentiation in the choanoflagellate *Salpingoeca rosetta*. Genome Biol 2013;14(2):R15.

[13] Regev A, Lamb MJ, Jablonka E. The role of DNA methylation in invertebrates: developmental regulation or genome defense? Mol Biol Evol 1998;15(7):880—91.

[14] Mandroli M, Volpi N. The genome of the lepidopteran *Mamestra brassicae* has a vertebrate-like content of methylcytosine. Genetica 2003;119(2):187—91.

[15] Varriale A, Bernardi G. DNA methylation and body temperature in fishes. Gene 2006;385:111—21.

[16] Jabbari K, Cacciò S, Païs de Barros JP, Desgrès J, Bernardi G. Evolutionary changes in CpG and methylation levels in the genome of vertebrates. Gene 1997;205(1—2):109—18.

[17] Lechner M, Marz M, Ihling C, Sinz A, Stadler PF, Krauss V. The correlation of genome size and DNA methylation rate in metazoans. Theory Biosci 2013;132(1):47—60.

[18] Lyko F. The DNA methyltransferase family: a versatile toolkit for epigenetic regulation. Nat Rev Genet 2018;19(2):81—92.

[19] Ross SE, Bogdanovic O. TET enzymes, DNA demethylation and pluripotency. Biochem Soc Trans 2019;47(3):875—85.

[20] Glastad KM, Hunt BG, Yi SV, Goodisman MAD. DNA methylation in insects: on the brink of the epigenomic era. Insect Mol Biol 2011;20(5):553—65.

[21] Planques A, Kerner P, Ferry L, Grunau C, Gazave E, Vervoort M. DNA methylation atlas and machinery in the developing and regenerating annelid *Platynereis dumerilii*. BMC Biol 2021;19:148.

[22] Liu J, Hu H, Panserat S, Marandel L. Evolutionary history of DNA methylation related genes in chordates: new insights from multiple whole genome duplications. Sci Rep 2020;10(1):970.

[23] Raddatz G, Guzzardo PM, Olova N, Fantappié MR, Rampp M, Schaefer M, et al. Dnmt2-dependent methylomes lack defined DNA methylation patterns. Proc Natl Acad Sci USA 2013;110(21):8627—31.

[24] Albalat R, Martí-Solans J, Cañestro C. DNA methylation in amphioxus: from ancestral functions to new roles in vertebrates. Brief Funct Genom 2012;11(2):142—55.

[25] Berezikov E. Evolution of microRNA diversity and regulation in animals. Nat Rev Genet 2011;12(12):846—60.

[26] Waterborg JH. Evolution of histone H3: emergence of variants and conservation of posttranslational modification sites. Biochem Cell Biol 2012;90(1):79—95.

[27] Gaiti F, Jindrich K, Fernandez-Valverde SL, Roper KE, Degnan BM, Tanurdžić M. Landscape of histone modifications in a sponge reveals the origin of animal cis-regulatory complexity. eLife 2017;6:e22194.

[28] Gaiti F, Calcino AD, Tanurdžića M, Degnana BM. Origin and evolution of the metazoan non-coding regulatory genome. Dev Biol 2017;427(2):193—202.

[29] Wheeler BM, Heimberg AM, Moy VN, Sperling EA, Holstein TW, Heber S, et al. The deep evolution of metazoan microRNAs. Evol Dev 2009;11(1):50—68.

[30] Moran Y, Agron M, Praher D, Technau U. The evolutionary origin of plant and animal microRNAs. Nat Ecol Evol 2017;1(3):27.

[31] Mohammed J, Flynt AS, Panzarino AM, Mondal MH, DeCruz M, Siepel A, et al. Deep experimental profiling of microRNA diversity, deployment, and evolution across the *Drosophila* genus. Genome Res 2018;28(1):52—65.

[32] Penso-Dolfin L, Moxon S, Haerty W, Di Palma F. The evolutionary dynamics of microRNAs in domestic mammals. Sci Rep 2018;8:17050.

[33] Cedar H, Bergman Y. Linking DNA methylation and histone modification: patterns and paradigms. Nat Rev Genet 2009;10(5):295—304.

[34] O'Leary VB, Hain S, Maugg D, Smida J, Azimzadeh O, Tapio S, et al. Long non-coding RNA *PARTICLE* bridges histone and DNA methylation. Sci Rep 2017;7:1790.

[35] Rose NR, Klose RJ. Understanding the relationship between DNA methylation and histone lysine methylation. Biochim Biophys Acta 2014;1839(12):1362—72.

[36] Du Q, Luu P-L, Stirzaker C, Clark SJ. Methyl-CpG-binding domain proteins: readers of the epigenome. Epigenomics 2015;7(6):1051—73.

[37] Wang X, Li Q, Lian J, Li L, Jin L, Cai H, et al. Genome-wide and single-base resolution DNA methylomes of the Pacific oyster *Crassostrea gigas* provide insight into the evolution of invertebrate CpG methylation. BMC Genom 2014;15:1119.

[38] Dabe EC, Sanford RS, Kohn AB, Bobkova Y, Moroz LL. DNA methylation in basal metazoans: insights from ctenophores. Integr Comp Biol 2015;55(6):1096—110.

[39] Feng S, Cokus SJ, Zhang X, Chen PY, Bostick M, Goll MG, et al. Conservation and divergence of methylation patterning in plants and animals. Proc Natl Acad Sci USA 2010;107(19):8689—94.

[40] Jiang N, Wang L, Chen J, Wang L, Leach L, Luo Z. Conserved and divergent patterns of DNA methylation in higher vertebrates. Genome Biol Evol 2014;6(11):2998—3014.

[41] Peat JR, Ortega-Recalde O, Kardailsky O, Hore TA. The elephant shark methylome reveals conservation of epigenetic regulation across jawed vertebrates. F1000Research 2017;6:526.

[42] Falckenhayn C. *The methylome of the marbled crayfish Procambarus virginalis*. Doctoral thesis. Heidelberg, Germany: University of Heidelberg. 2016.

[43] Zemach A, McDaniel IE, Silva P, Zilberman D. Genome-wide evolutionary analysis of eukaryotic DNA methylation. Science 2010;238(5980):916—19.

[44] Kenny NJ, Francis WR, Rivera-Vicéns RF, Juravel X, de Mendoza A, Díez-Vives C, et al. Tracing animal genomic evolution with the chromosomal-level assembly of the freshwater sponge *Ephydatia muelleri*. Nat Commun 2020;11:3676.

[45] Lyko F, Foret S, Kucharski R, Wolf S, Falckenhayn C, Maleszka R. The honey bee epigenomes: differential methylation of brain DNA in queens and workers. PLoS Biol 2010;8(11):e1000506.

References

[46] Gatzmann F, Falckenhayn C, Gutekunst J, Hanna K, Raddatz G, Coutinho Carneiro V, et al. The methylome of the marbled crayfish links gene body methylation to stable expression of poorly accessible genes. Epigenetics Chromatin 2018;11(1):57.

[47] Neri F, Rapelli S, Krepelova A, Incarnato D, Parlato C, Basile G, et al. Intragenic DNA methylation prevents spurious transcription initiation. Nature 2017;543(7643):72–7.

[48] Maor GL, Yearim A, Ast G. The alternative role of DNA methylation in splicing regulation. Trends Genet 2015;31(5):274–80.

[49] Deaton AM, Bird A. CpG islands and the regulation of transcription. Genes Dev 2011;25(10):1010–22.

[50] Lim W-J, Kim KH, Kim J-Y, Jeong S, Kim N. Identification of DNA-methylated CpG islands associated with gene silencing in the adult body tissues of the Ogye chicken using RNA-seq and reduced representation bisulfite sequencing. Front Genet 2019;10:346.

[51] Jiang Q, Lin D, Huang H, Wang G, Ye H. DNA methylation inhibits the expression of CFSH in mud crab. Front Endocrinol 2020;11:163.

[52] Wang X, Wheeler D, Avery A, Rago A, Choi J-H, Colbourne JK, et al. Function and evolution of DNA methylation in *Nasonia vitripennis*. PLoS Genet 2013;9(10):e1003872.

[53] Krauss V, Eisenhardt C, Unger T. The genome of the stick insect *Medauroidea extradentata* is strongly methylated within genes and repetitive DNA. PLoS One 2009;4(9):e7223.

[54] Falckenhayn C, Boerjan B, Raddatz G, Frohme M, Schoofs L, Lyko F. Characterization of genome methylation patterns in the desert locust *Schistocerca gregaria*. J Exp Biol 2013;216(8):1423–9.

[55] Lennartsson A, Ekwall K. Histone modification patterns and epigenetic codes. Biochim Biophys Acta 2009;1790(9):863–8.

[56] Bannister AJ, Kouzarides T. Regulation of chromatin by histone modifications. Cell Res 2011;21(3):381–95.

[57] Marmorstein R, Trievel RC. Histone modifying enzymes: structures, mechanisms, and specificities. Biochim Biophys Acta 2008;1789(1):58–68.

[58] Izzo A, Schneider R. Chatting histone modifications in mammals. Brief Funct Genom 2011;9(5–6):429–43.

[59] Jambhekar A, Dhall A, Shi Y. Roles and regulation of histone methylation in animal development. Nat Rev Mol Cell Biol 2019;20(10):625–41.

[60] Frias-Laserre D, Villagra CA. The importance of ncRNAs as epigenetic mechanisms in phenotypic variation and organic evolution. Front Microbiol 2017;8:2483.

[61] Statello L, Guo C-J, Chen L-L, Huarte M. Gene regulation by long non-coding RNAs and its biological functions. Nat Rev Mol Cell Biol 2021;22(2):96–118.

[62] Yao Q, Chen Y, Zhou X. The roles of microRNAs in epigenetic regulation. Curr Opin Chem Biol 2019;51:11–17.

[63] Holoch D, Moazed D. RNA-mediated epigenetic regulation of gene expression. Nat Rev Genet 2015;16(2):71–84.

[64] Li M-Z, Xiao H-M, He K, Li F. Progress and prospects of non-coding RNAs in insects. J Integr Agric 2019;18(4):729–47.

[65] Csankovszki G, Nagy A, Jaenisch R. Synergism of Xist RNA, DNA methylation, and histone hypoacetylation in maintaining X chromosome inactivation. J Cell Biol 2001;153(4):773–83.

[66] Nohara K, Baba T, Murai H, Kobayashi Y, Suzuki T, Tateishi Y, et al. Global DNA methylation in the mouse liver is affected by methyl deficiency and arsenic in a sex-dependent manner. Arch Toxicol 2011;85(6):653–61.

[67] Seisenberger S, Peat JR, Hore TA, Santos F, Dean W, Reik W. Reprogramming DNA methylation in the mammalian life cycle: building and breaking epigenetic barriers. Philos Trans R Soc Lond B 2012;68(1609):20110330.

[68] Wu H, Zhang Y. Reversing DNA methylation: mechanisms, genomics, and biological functions. Cell 2014;156(1–2):45–68.

[69] Blake LE, Roux J, Hernando-Herraez I, Banovich NE, Garcia Perez R, Hsiao CJ, et al. Comparison of gene expression and DNA methylation patterns across tissues and species. Genome Res 2020;30(2):1–13.

[70] Cardoso-Júnior CAM, Fujimura PT, Santos-Júnior CD, Araújo Borges N, Ueira-Vieira C, Hartfelder K, et al. Epigenetic modifications and their relation to caste and sex determination and adult division of labor in the stingless bee *Melipona scutellaris*. Genet Mol Biol 2017;40(1):61–8.

[71] Vogt G. Biology, ecology, evolution, systematics and utilization of the parthenogenetic marbled crayfish, *Procambarus virginalis*. In: Ribeiro FB, editor. Crayfish: evolution, habitat and conservation strategies. Hauppauge, NY: Nova Science Publishers; 2020. p. 137–227.

[72] Rolff J, Johnston PR, Reynolds S. Complete metamorphosis of insects. Philos Trans R Soc Lond B 2019;374(1783):20190063.

[73] Sarmento OF, Digilio LC, Wang Y, Perlin J, Herr JC, Allis CD. Dynamic alterations of specific histone modifications during early murine development. J Cell Sci 2004;117(19):4449–59.

[74] Beermann J, Piccoli M-T, Viereck J, Thum T. Non-coding RNAs in development and disease: background, mechanisms, and therapeutic approaches. Physiol Rev 2016;96(4):1297–325.

[75] Pauli A, Rinn JL, Schier AF. Non-coding RNAs as regulators of embryogenesis. Nat Rev Genet 2011;12(2):136–49.

[76] Chen MJ, Chen LK, Lai YS, Lin Y-Y, Wu D-C, Tung Y-A. Integrating RNA-seq and ChIP-seq data to characterize long non-coding RNAs in *Drosophila melanogaster*. BMC Genom 2016;17:220.

[77] Vickaryous MK, Hall BK. Human cell type diversity, evolution, development, and classification with special reference to cells derived from the neural crest. Biol Rev 2006;81(3):425–55.

[78] Tsang S-Y, Ahmad T, Mat FWK, Zhao C, Xiao S, Xia K, et al. Variation of global DNA methylation levels with age and in autistic children. Hum Genom 2016;10(1):31.

[79] Zhang Z, Zhang MQ. Histone modification profiles are predictive for tissue/cell-type specific expression of both protein-coding and microRNA genes. BMC Bioinform 2011;12:155.

[80] Isakova A, Fehlmann T, Keller A, Quake SR. A mouse tissue atlas of small noncoding RNA. Proc Natl Acad Sci USA 2020;117(41):25634–45.

[81] Fraga MF, Ballestar E, Paz MF, Ropero S, Setien F, Ballestar ML, et al. Epigenetic differences arise during the lifetime of monozygotic twins. Proc Natl Acad Sci USA 2005;102(30):10604–9.

[82] Stubbs TM, Bonder MJ, Stark A-K, Krueger F, BI Ageing Clock Teamvon Meyenn F, et al. Multi-tissue DNA methylation age predictor in mouse. Genome Biol 2017;18(1):68.

[83] Simpson SJ, Sword GA, Lo N. Polyphenism in insects. Curr Biol 2011;21(18):R738–49.

[84] Glastad KM, Hunt BG, Goodisman MAD. Epigenetics in insects: genome regulation and the generation of phenotypic diversity. Annu Rev Entomol 2018;64:185–203.

[85] Yang C-H, Pospisilik JA. Polyphenism – a window into gene-environment interactions and phenotypic plasticity. Front Genet 2019;10:132.

[86] Wojciechowski M, Lowe R, Maleszka J, Conn D, Maleszka M, Hurd PJ. Phenotypically distinct female castes in honey bees are defined by alternative chromatin states during larval development. Genome Res 2018;28(10):1532–42.

[87] Herb BR, Wolschin F, Hansen KD, Aryee MJ, Langmead B, Irizarry R, et al. Reversible switching between epigenetic states in honeybee behavioral subcastes. Nat Neurosci 2012;15(10):1371–3.

[88] Foret S, Kucharski R, Pellegrini M, Feng S, Jacobsen SE, Robinson GE, et al. DNA methylation dynamics, metabolic fluxes, gene splicing, and alternative phenotypes in honey bees. Proc Natl Acad Sci USA 2012;109(13):4968–73.

[89] Standage DS, Berens AJ, Glastad KM, Severin AJ, Brendel VP, Toth AL. Genome, transcriptome and methylome sequencing of a primitively eusocial wasp reveal a greatly reduced DNA methylation system in a social insect. Mol Ecol 2016; 25(8):1769–84.

[90] Vogt G. Stochastic developmental variation, an epigenetic source of phenotypic diversity with far-reaching biological consequences. J Biosci 2015;40(1):159–204.

[91] Vogt G. Facilitation of environmental adaptation and evolution by epigenetic phenotype variation: insights from clonal, invasive, polyploid, and domesticated animals. Environ Epigenet 2017;3(1):dvx002.

[92] Vogt G. Disentangling the environmentally induced and stochastic developmental components of phenotypic variation. In: Levine H, Jolly MK, Kulkarni P, Nanjundiah V, editors. Phenotypic switching: implications in biology and medicine. San Diego, CA: Academic Press; 2020. p. 207–51.

[93] Leung C, Breton S, Angers B. Facing environmental predictability with different sources of epigenetic variation. Ecol Evol 2016;6(15):5234–45.

[94] Feinberg AP, Irizarry RA. Stochastic epigenetic variation as a driving force of development, evolutionary adaptation, and disease. Proc Natl Acad Sci USA 2010;107(Suppl. 1):1757–64.

[95] Van der Graaf A, Wardenaar R, Neumann DA, Taudt A, Shaw RG, Jansen RC, et al. Rate, spectrum, and evolutionary dynamics of spontaneous epimutations. Proc Natl Acad Sci USA 2015;112(21):6676–81.

[96] DeWitt T, Scheiner SM. editors Phenotypic plasticity: functional and conceptual approaches. New York, NY: Oxford University Press; 2004.

[97] Fox RJ, Donelson JM, Schunter C, Ravasi T, Gaitán-Espitia JD. Beyond buying time: the role of plasticity in phenotypic adaptation to rapid environmental change. Philos Trans R Soc Lond B 2019;374(1768):20180174.

[98] Strader ME, Kozal LC, Leach TS, Wong JM, Chamorro JD, Housh MJ, et al. Examining the role of DNA methylation in transcriptomic plasticity of early stage sea urchins: developmental and maternal effects in a kelp forest herbivore. Front Mar Sci 2020;7:205.

[99] Gutekunst J, Andriantsoa R, Falckenhayn C, Hanna K, Stein W, Rasamy JR, et al. Clonal genome evolution and rapid invasive spread of the marbled crayfish. Nat Ecol Evol 2018;2(3):567–73.

[100] Vogt G, Lukhaup C, Pfeiffer M, Dorn NJ, Williams BW, Schulz R, et al. Morphological and genetic characterization of the marbled crayfish, including a determination key. Zootaxa 2018;4524(3):329–50.

[101] Vogt G. Epigenetic variation in animal populations: sources, extent, phenotypic implications and ecological and evolutionary relevance. J Biosci 2021;46:24.

[102] Tönges S, Venkatesh G, Andriantsoa R, Hanna K, Gatzmann F, Raddatz G, et al. Location-dependent DNA methylation signatures in a clonal invasive crayfish. Front Cell Dev Biol 2021;9:794506.

[103] Andriantsoa R, Tönges S, Panteleit J, Theissinger K, Coutinho Carneiro V, Rasamy J, et al. Ecological plasticity and commercial impact of invasive marbled crayfish populations in Madagascar. BMC Ecol 2019;19:8.

[104] Neiman M, Jokela J, Lively CM. Variation in asexual lineage age in *Potamopyrgus antipodarum*, a New Zealand snail. Evolution 2005;59(9):1945–52.

[105] Thorson JLM, Smithson M, Beck D, Sadler-Riggleman I, Nilsson E, Dybdahl M, et al. Epigenetics and adaptive phenotypic variation between habitats in an asexual snail. Sci Rep 2017;7:14139.

[106] Verhaegen G, Neiman M, Haase M. Ecomorphology of a generalist freshwater gastropod: complex relations of shell morphology, habitat, and fecundity. Org Divers Evol 2018;18:425–41.

[107] Dybdahl MF, Drown DM. The absence of genotypic diversity in a successful parthenogenetic invader. Biol Invasions 2011;13:1663–72.

[108] Thorson JLM, Smithson M, Sadler-Riggleman I, Beck D, Dybdahl M, Skinner MK. Regional epigenetic variation in asexual snail populations among urban and rural lakes. Environ Epigenet 2019;5(4):dvz020.

[109] Liew YJ, Zoccola D, Li Y, Tambutte E, Venn AA, Michell CT, et al. Epigenome-associated phenotypic acclimatization to ocean acidification in a reef-building coral. Sci Adv 2018;4(6): eaar8028.

[110] Liew YJ, Howells EJ, Wang X, Michell CT, Burt JA, Idaghdour Y, et al. Intergenerational epigenetic inheritance in reef-building corals. Nat Clim Change 2020;10:254–9.

[111] Fridley JD, Stachowicz JJ, Naeem S, Sax DF, Seabloom EW, Smith MD, et al. The invasion paradox: reconciling pattern and process in species invasions. Ecology 2007;88(1):3–17.

[112] Hawes NA, Fidler AE, Tremblay LA, Pochon X, Dunphy BJ, Smith KF. Understanding the role of DNA methylation in successful biological invasions: a review. Biol Invasions 2018;20:2285–300.

[113] Coutinho Carneiro V, Lyko F. Rapid epigenetic adaptation in animals and its role in invasiveness. Integr Comp Biol 2020;60(2):267–74.

[114] Liebl AL, Schrey AW, Richards CL, Martin LB. Patterns of DNA methylation throughout a range expansion of an introduced songbird. Integr Comp Biol 2013;53(2):351–8.

[115] Schrey AW, Coon CAC, Grispo MT, Awad M, Imboma T, McCoy ED, et al. Epigenetic variation may compensate for decreased genetic variation with introductions: a case study using house sparrows (*Passer domesticus*) on two continents. Genet Res Int 2012;2012:979751.

[116] Skinner MK, Gurerrero-Bosagna C, Muksitul Haque M, Nilsson EE, Koop JAH, Knutie SA, et al. Epigenetics and the evolution of Darwin's finches. Genome Biol Evol 2014;6(8):1972–89.

[117] Hu J, Askary AM, Thurman TJ, Spiller DA, Palmer TM, Pringle RM, et al. The epigenetic signature of colonizing new environments in anolis lizards. Mol Biol Evol 2019;36(10):2165–70.

[118] Liu X, Feng Y, Bai X, Wang X, Qin R, Tang B, et al. Comparative multi-omics analyses reveal differential expression of key genes relevant for parasitism between non-encapsulated and encapsulated *Trichinella*. Commun Biol 2021;4:134.

[119] Geyer KK, Rodríguez López CM, Chalmers IW, Munshi SE, Truscott M, Heald J, et al. Cytosine methylation regulates oviposition in the pathogenic blood fluke *Schistosoma mansoni*. Nat Commun 2011;2:424.

[120] Larson G, Fuller DQ. The evolution of animal domestication. Annu Rev Ecol Evol Syst 2014;45:115–36.

[121] Wright D. The genetic architecture of domestication in animals. Bioinform Biol Insights 2015;9(Suppl.4):11–20.

[122] Cagan A, Blass T. Identification of genomic variants putatively targeted by selection during dog domestication. BMC Evol Biol 2016;16:10.

[123] Janowitz Koch I, Clark MM, Thompson MJ, Deere-Machemer KA, Wang J, Duarte L, et al. The concerted impact of domestication and transposon insertions on methylation patterns between dogs and grey wolves. Mol Ecol 2016;25(8):1838–55.

[124] Bélteky J, Agnvall B, Bektic L, Höglund A, Jensen P, Guerrero-Bosagna C. Epigenetics and early domestication: differences in hypothalamic DNA methylation between red junglefowl divergently selected for high or low fear of humans. Genet Sel Evol 2018;50(1):13.

[125] Anastasiadi D, Piferrer F. Epimutations in developmental genes underlie the onset of domestication in farmed European Sea Bass. Mol Biol Evol 2019;36(10):2252−64.

[126] Konstantinidis J, Sætrom P, Mjelle R, Nedoluzhko AV, Robledo D, Fernandes JMO. Major gene expression changes and epigenetic remodelling in Nile tilapia muscle after just one generation of domestication. Epigenetics 2020;15(10):1052−67.

[127] Richa R, Rajeshwar PS. Hydroxymethylation of DNA: an epigenetic marker. EXCLI J 2014;13:592−610.

[128] Faria R, Navarro A. Chromosomal speciation revisited: rearranging theory with pieces of evidence. Trends Ecol Evol 2010;25(11):660−9.

[129] Gregory TR, Mable BK. Polyploidy in animals. In: Gregory TR, editor. The evolution of the genome. Burlington, MA: Academic Press; 2005. p. 427−517.

[130] Abbott R, Albach D, Ansell S, Arntzen JW, Baird SJE, Bierne N, et al. Hybridization and speciation. J Evol Biol 2013;26(2):229−46.

[131] Smith TA, Martin M, Nguyen M, Mendelson TC. Epigenetic divergence as a potential first step in darter speciation. Mol Ecol 2016;25(8):1883−94.

[132] Laporte M, Le Luyer J, Rougeux C, Dion-Côté A-M, Krick M, Bernatchez L. DNA methylation reprogramming, TE derepression, and postzygotic isolation of nascent animal species. Sci Adv 2019;5(10):eaaw1644.

[133] Inoue J, Sato Y, Sinclair R, Tsukamoto K, Nishida M. Rapid genome reshaping by multiple-gene loss after whole-genome duplication in teleost fish suggested by mathematical modeling. Proc Natl Acad Sci USA 2015;112(48):14918−23.

[134] O'Neill RJW, O'Neill MJ, Graves JAM. Undermethylation associated with retroelement activation and chromosome remodelling in an interspecific mammalian hybrid. Nature 1998;393(6680):68−72.

[135] Xiao J, Song C, Liu S, Tao M, Hu J, Wang J, et al. DNA methylation analysis of allotetraploid hybrids of red crucian carp (*Carassius auratus* red var.) and common carp (*Cyprinus carpio* L.). PLoS One 2013;8(2):e56409.

[136] Vogt G, Dorn NJ, Pfeiffer M, Lukhaup C, Williams BW, Schulz R, et al. The dimension of biological change caused by autotriploidy: a meta-analysis with the triploid *Procambarus virginalis* and its diploid parent *Procambarus fallax*. Zool Anz 2019;281:53−67.

[137] Jablonka E. The evolutionary implications of epigenetic inheritance. Interface Focus 2017;7(5):20160135.

[138] Skvortsova K, Iovino N, Bogdanović O. Functions and mechanisms of epigenetic inheritance in animals. Nat Rev Mol Cell Biol 2018;19(12):775−90.

[139] Anastasiadi D, Venney CJ, Bernatchez L, Wellenreuther M. Epigenetic inheritance and reproductive mode in plants and animals. Trends Ecol Evol 2021;36(12):1124−40.

[140] Daxinger L, Whitelaw E. Understanding transgenerational epigenetic inheritance via the gametes in mammals. Nat Rev Genet 2012;13(3):153−62.

[141] Norouzitallab P, Baruah K, Vandegehuchte M, Van Stappen G, Catania F, Vanden Bussche J, et al. Environmental heat stress induces epigenetic transgenerational inheritance of robustness in parthenogenetic *Artemia* model. FASEB J 2014;28(8):3552−63.

[142] Lismer A, Siklenka K, Lafleur C, Dumeaux V, Kimmins S. Sperm histone H3 lysine 4 trimethylation is altered in a genetic mouse model of transgenerational epigenetic inheritance. Nucleic Acids Res 2020;48(20):11380−93.

[143] Jiang L, Zhang J, Wang JJ, Wang L, Zhang L, Li G, et al. Sperm, but not oocyte, DNA methylome is inherited by zebrafish early embryos. Cell 2013;153(4):773−84.

[144] Yagound B, Remnant EJ, Buchmann G, Oldroyd BP. Intergenerational transfer of DNA methylation marks in the honey bee. Proc Natl Acad Sci USA 2020;117(51):32519−27.

[145] Baxter FA, Drake AJ. Non-genetic inheritance via the male germline in mammals. Philos Trans R Soc Lond B 2018;374(1770):20180118.

[146] Kronholm I, Collins S. Epigenetic mutations can both help and hinder adaptive evolution. Mol Ecol 2016;25(8):1856−68.

[147] Waddington CH. Genetic assimilation of an acquired character. Evolution 1953;7(2):118−26.

[148] Pigliucci M, Murren CJ, Schlichting CD. Phenotypic plasticity and evolution by genetic assimilation. J Exp Biol 2006;209(12):2362−7.

[149] Walser JC, Furano AV. The mutational spectrum of non-CpG DNA varies with CpG content. Genome Res 2010;20(7):875−82.

[150] Lutsenko E, Bhagwat AS. Principal causes of hot spots for cytosine to thymine mutations at sites of cytosine methylation in growing cells. A model, its experimental support and implications. Mutat Res 1999;437(1):11−20.

[151] Shen JC, Rideout III WM, Jones PA. The rate of hydrolytic deamination of 5-methylcytosine in double-stranded DNA. Nucleic Acids Res 1994;22(6):972−6.

[152] Ghosh D, Veeraraghavan B, Elangovan R, Vivekanandan P. Antibiotic resistance and epigenetics: more to it than meets the eye. Antimicrob Agents Chemother 2020;64(2):e02225−19.

[153] Yi SV, Goodisman MA. Computational approaches for understanding the evolution of DNA methylation in animals. Epigenetics 2009;4(8):551−6.

[154] Yi SV, Goodisman MAD. The impact of epigenetic information on genome evolution. Philos Trans R Soc Lond B 2021;376(1826):20200114.

[155] Eldredge N, Gould SJ. Punctuated equilibria: an alternative to phyletic gradualism. In: Schopf TJM, editor. Models in paleobiology. San Francisco, CA: Freeman Cooper; 1972. p. 82−115.

[156] Liu XS, Wu H, Ji X, Stelzer Y, Wu X, Czauderna S, et al. Editing DNA methylation in the mammalian genome. Cell 2016;167(1):233−47.

CHAPTER 28

Adaptive evolution and epigenetics

Ilkka Kronholm

Department of Biological and Environmental Science, University of Jyväskylä, Jyväskylä, Finland

OUTLINE

Introduction	551	Modeling Spontaneous Epigenetic Variation	555
Epigenetic Variation	551	Limits of Epigenetic Contributions to Adaptation	559
Modeling Evolution	552	Conclusions and Future Directions	560
Induced Epigenetic Changes	552	Acknowledgments	561
Spontaneous Epigenetic Variation	554	References	561

INTRODUCTION

Adaptive evolution by natural selection requires that there are differences in reproductive success among individuals and that these differences are heritable. Individuals that are better matched to their environment will leave more offspring and, thus, the properties that these individuals have will become more common in the population as a result of natural selection. The mathematical foundations of modern evolutionary theory were laid by R. A. Fisher in 1930 with the publication of *The Genetical Theory of Natural Selection* [1], long before it was shown that in the 1940s and 50s that DNA was the genetic material. Indeed, the chemical basis of heredity does not matter in evolutionary theory. If epigenetic variation is heritable there is no *a priori* reason to doubt it could be under selection in the same manner as genetic variation. The question then becomes, does epigenetic variation have the necessary properties (e.g., sufficient stability) to contribute to adaptive evolution? Here, I will examine the properties of epigenetic variation and review what is known about adaptation as it relates to epigenetics. First, I will discuss different types of epigenetic variation, then by focusing on spontaneous changes I will examine the parameters that determine how it can drive an evolutionary response. Second, I will consider models of adaptation and explore the limits of epigenetic variation with respect to possible effects on evolutionary change. Given that the importance of epigenetic variation in evolution is far from clear and many questions remain, suggestions for future research are provided.

EPIGENETIC VARIATION

Currently there is no consensus on how the term *epigenetics* should be defined and, historically, the term has had different meanings [2]. Today, the term epigenetics is typically applied to inherited changes in transcription that are caused by chromatin modifications rather than changes in DNA sequence [3]. From an evolutionary perspective, this definition is unsatisfactory because changes to chromatin that are inherited but do not affect transcription, those that are functionally neutral, would not fall under this definition. Furthermore, while some heritable RNA molecules are often considered epigenetic [4] it is unclear if they affect chromatin in cases where these molecules do not act in the RNA-directed DNA methylation pathway of plants [5]. For the purposes of this chapter, *epigenetics* will refer to chromatin states that are inherited and *epigenetic change* to a change in such states, regardless of the molecular mechanism responsible. There are several different types of non-genetic inheritance [6], such as cultural inheritance [7], that could potentially affect evolution. However, this chapter will focus on changes in chromatin modifications, such as DNA methylation, histone modifications, or inherited RNA molecules. While this distinction between epigenetics and other forms of non-genetic inheritance is somewhat arbitrary, it seems reasonable to deal with different mechanisms one at a time.

In order for any type of change to contribute to evolution it must be inherited, but perfect inheritance by all offspring is not required. There are important taxonomic differences in epigenetic inheritance: the mechanisms of epigenetic inheritance are most clear for plants, nematodes, and fungi. For mammals, outside of imprinted loci, there is some evidence of non-genetic inheritance at the phenotypic level, but mechanisms are not well established. Mammalian germ cells undergo extensive epigenetic reprogramming during development, and while some loci can escape this reprogramming, the potential for epigenetic inheritance seems smaller in mammals [8]. For different types of epigenetic changes, the inheritance and mechanism of DNA methylation across cell divisions is well established (see chapter 2 about DNA methylation). For histone modifications, there is some evidence for inheritance of histone methylation at certain residues, particularly methylation of histone 3 lysine 9 (H3K9me), and histone 3 lysine 27 [9–13], and perhaps also histone 3 lysine 4, as disruption of this mark causes transgenerational effects [14]. Recent studies have highlighted the importance of small RNAs in the inheritance of histone modifications [13,15,16]. In fission yeast, H3K9me and small RNAs form a positive feedback loop that maintains the silenced epigenetic state and allows epigenetic inheritance [13]. It should be noted that while there are a large number of possible histone modifications [17], there is evidence of inheritance for only a few. It has also been established that in some cases, epigenetic inheritance is mediated by small RNA molecules [18–22].

For epigenetic changes to play a role in adaptive evolution, they must affect the phenotype and be at least partially independent of the genome. Epigenetic changes exert their effect on phenotype via gene expression changes. While transcription factors are ultimately in control of transcription [23], certain chromatin modifications can prevent transcription-factor binding and thus transcription itself [17]. Furthermore, a large proportion of chromatin modifications are influenced by genomic context [24] and the function of many others are not yet known. From an evolutionary perspective, the most interesting epigenetic marks and genomic locations are those where variation in the epigenetic state can exist, even if the DNA sequence remains identical. Regardless of the molecular mechanism, two different types of epigenetic variation can be distinguished [3] with different effects on adaptive evolution: (1) inherited epigenetic changes that are induced by the environment, and (2) spontaneous epigenetic changes.

MODELING EVOLUTION

In the following sections, I'll describe how evolutionary models have been applied to both induced and spontaneous epigenetic variation and what can we learn from these models. However, first I'll address the question, why use mathematical models? Sometimes modeling approaches are criticized for not being realistic. This is by design. Models are not supposed to mimic biological reality to every detail. Indeed, that would be counterproductive to what the models attempt to do. Think of an analogy to a map, maps do not represent the real world in a 1:1 scale. A map is an abstraction of the real world that shows some general features of the landscape that will helps us get from A to B. By analogy, the same thing applies to mathematical models, they are an abstraction that will help us learn how the complicated biological system functions. The reason we should frame our model in terms of mathematics rather than verbal arguments, is that this will force us to be precise and make our assumptions about the system transparent.

There are different ways to model evolutionary change, and population or quantitative genetic models are often used [25]. As mentioned earlier, these models do not require or make any assumptions about the molecular basis of heredity since they operate at the level of phenotypic variation [26]. Thus, epigenetic variation can be incorporated into different types of evolutionary models [27–33]. Modeling studies have revealed that certain parameters are important in determining how epigenetic variation influences adaptive evolution. Next, I will discuss these parameters, what their realistic empirical values could be, and then illustrate the influence of epigenetic variation on adaptation for induced epigenetic changes and spontaneous epigenetic changes.

INDUCED EPIGENETIC CHANGES

In the ecological and evolutionary literature, it has been established that the environment experienced by the parents can influence the phenotype of their offspring. These have been referred to as maternal, paternal, intergenerational, and transgenerational effects and have been observed in animals [34–39], plants [34,40–43], and fungi [44]. Parental effects can be mediated by a variety of mechanisms, such as maternal investment in eggs. However, in certain cases it has been shown that these effects are mediated by epigenetic mechanisms. When organisms perceive or experience certain stimuli from the environment, such as herbivores or pathogens [19,43], epigenetic changes, that can be inherited by subsequent generations, are induced [45,46]. In plants, induction of defense mechanisms in the offspring have been linked to epigenetic mechanisms; responses to herbivory and pathogens require small RNAs [47] and a specific RNA-directed methylation pathway [48,49], respectively. Wibowo et al. [50] identified an intergenerational response to salt stress in *Arabidopsis*, the survival of offspring was enhanced under salt stress if the maternal plants had experienced salt stress. This response was mediated by DNA methylation changes transmitted via seeds but not pollen—where methylation marks were reset due to high DNA-glycosylase activity. In animals, the best characterized cases can be found in nematodes [19,51–54]. Transmitted responses have also been investigated in fruit flies, where the paternal diet influenced the triglyceride content of their offspring and the response was dependent on histone 3 lysine 9 and lysine 27 methylation [55].

It should be stressed that it is the genetic program itself that ultimately creates the ability to be phenotypically plastic and epigenetic changes are a mechanism by which the environmental signal is transmitted to the offspring. Regulation of transmitted responses are under genetic control [56], and the resetting of epigenetic markers is an active process [57,58]. While in many cases it is not known if or how the transgenerational responses are adaptive, it is unlikely that such mechanisms would have evolved by neutral processes alone.

Phenotypic plasticity, where environment influences the development of an organism that only happens within generation is common. Organisms must respond to their environment all the time to maintain a favorable internal state. Then what kind of environments favor the evolution of induced epigenetic changes? I'll approach this question using the modeling framework of Botero et al. [59]. They modeled the evolution of phenotypic plasticity using an individual based simulation model, where individuals could evolve reversible or developmental plasticity, a bet-hedging strategy, where individuals increased the variability of their offspring, or population could track the environment by genetic adaptation. Which strategy was favored by evolution depended on how often environmental fluctuations occurred relative to generation time, and how useful environmental cues were to organisms for predicting the environment. I've extended the model of Botero et al. to include the possibility of evolution between generation plasticity that is transmitted via induced epigenetic changes [60]. To summarize the model results: transmission of plastic effects by epigenetic changes evolved when the environment changed every 3–32 generations, when environmental cues predicted environmental changes with a probability of 0.6 or greater (Fig. 28.1), and when costs of plasticity were small or moderate. So environmental changes need to occur in the short term, and there has to be environmental cues available that predict the offspring environment with moderate probability. See [60] for details of the model.

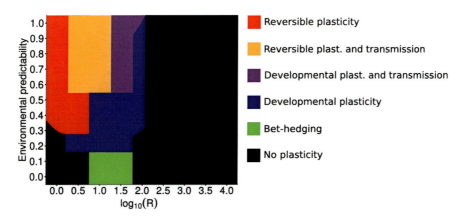

FIGURE 28.1 Evolution favors different strategies to cope with fluctuating environments, depending on rate of environmental fluctuation, R, and environmental predictability. R is expressed in $\log_{10}$ units of generations, so $R = 1.5$ corresponds to environmental fluctuations that happen approximately every 32 generations. Reversible plasticity means that indiduals can adjust their phenotype based on an environmental cue multiple times during their lifetime, developmental plasticity allows only one adjustment, bet-hedging means that individuals increase varibility in their offspring, and transmission means that there is an epigenetic change in adults, based on the current environmental cue, and juveniles will adjust their phenotype based on this epigenetic state. Plasticity is costly in these simulations. *Simplified and redrawn from Fig. 5 in I. Kronholm, Evolution of anticipatory effects mediated by epigenetic changes, Environmental Epigenetics, 8: 1-12 (2022).*

Similar predictions are also made by other models [61–65]. There is also some experimental evidence that transmitted plasticity evolves this way, as experiments with nematodes have shown that transmitted effects do evolve in predictably fluctuating environments [66,67].

Another question concerns how epigenetic regulation of phenotypic plasticity might affect subsequent adaptive evolution. Suppose a population is presented with a large and permanent environmental change that requires an evolutionary response. We know that phenotypic plasticity can speed up adaptation by keeping population size large [68] and, thus, keeping mutational supply and the amount of phenotypic variation high that selection can act upon. Phenotypic plasticity can also speed up the process of adaptation by moving the population closer to the new optimum, so that the population can start adapting closer to new optimum. This phenomenon is called the Baldwin effect [69]. Transmitted plastic effects can function in a similar manner if there is a positive covariance between parental and offspring phenotypes and the phenotype is adaptive [70]. In the model described above, when environment changes slowly, hundreds of generations between environmental fluctuations, the population tracks the optimum mainly by genetic adaptation. If costs of plasticity are minimal, populations can experience transient periods of plasticity and genetic assimilation, as they evolve plasticity during periods of environmental change, which is again lost when the environment is not changing. Overall, the theory that plasticity promotes adaptation has some support from experiments, as plastic genotypes tend to adapt faster [71,72].

While induced epigenetic changes are likely an adaptation to environmental fluctuations and may influence adaptive evolution indirectly via phenotypic plasticy, they are ultimately controlled by a genetic program and thus depend on the current capabilities of the organism. Epigenetic changes themselves are not responsible for adaptation but can be the mechanism that mediates it. Next, I will discuss epigenetic changes that are random with respect to fitness and which can be directly adaptive.

SPONTANEOUS EPIGENETIC VARIATION

All genetic variation is generated by spontaneous mutations. While mutations are random, it is not the case that all mutations are equally likely to happen or that all phenotypic effects are equally likely. However, mutations are random with respect to fitness and decades of research have supported the view that they occur randomly [73]. Genetic mutations happen because of errors in DNA replication, as no biochemical system is error free. In an analogous manner, spontaneous epigenetic changes or epimutations can happen due to errors in their copying, deletion or propagation within the genome. For example, methylation or demethylation of a gene promoter can happen as a result of a random change in the propagation of an epigenetic mark. The best evidence that epigenetic marks experience random changes similar to DNA has been obtained from mutation accumulation experiments. These are experiments where initially clonal lines are propagated for many generations in such way that efficiency of natural selection is greatly reduced allowing the accumulation of even slightly deleterious mutations [74]. Such experiments have been used to study genetic mutation rates, but they can be used to examine epigenetic changes as well. This has been done in the plant *Arabidopsis thaliana* where changes in DNA methylation have been examined in mutation accumulation lines [75–77]. These studies have revealed that spontaneous DNA methylation gains and losses happen throughout the genome an order of magnitude faster than the genetic mutation rate, especially with respect to losses or reversions. Moreover, spontaneous changes in DNA methylation over longer regions, called differentially methylated regions (DMRs), which are more likely to be biologically functional than single cytosine changes, happen at a rate comparable to single cytosine changes [78]. A study of somatic rates of DNA methylation change in poplar trees found similar results [79]. In animals, epigenetic mutations have been studied in nematodes, and small RNA dependent epimutations happen at a rate that is much faster than genetic mutations, but most of these epimutations are quite unstable [80]. However, they may still have the potential to affect phenotype, as one study found appreciable amounts of phenotypic variation were likely generated by these epigenetic changes [81]. Rates of DNA methylation are yet to be studied for animals with similar experiments, but somatic DNA methylation changes indicate a similar pattern [82]. Similarly, measurements in human twins have shown that DNA methylation patterns diverge over time [83,84]. Reversion rates for genetic mutations are often so low that they can effectively be ignored in evolutionary models, this is not the case for epigenetic changes.

It is possible to generate *Arabidopsis* lines that are genetically nearly identical, but where variation in DNA methylation has been artificially induced [85]. Some of these methylation changes segregate in a stable manner, and research has shown that epigenetic variation can contribute extensively to phenotypic variation [85–88] and that epiQTLs are stable enough to be selected upon [87].

Evidence of spontaneous epigenetic variation also comes from natural epigenetic variants or epialleles.

These are alleles that have the same DNA sequence but one has a different epigenetic state. Most examples of naturally occuring epialleles can be found in plants. Prominent examples include an epiallele in toadflax *cycloidea* gene that affects flower morphology [89], where the methylated allele of the gene is silenced. Another example is the *Cnr* locus affecting fruit ripening in tomato, where the promoter of one of the *Cnr* alleles is methylated [90]. In rice, an epiallele of the *DWARF1* gene causes short plant stature, the epiallele is silenced by DNA methylation and repressive chromatin marks [91]. The *DWARF1* epiallele is unstable with a forward mutation rate of 0.04 and reverse mutation rate of 0.008 [91], these rates are even higher if chimeric individuals are taken into account. Both the toadflax *cycloidea* and the tomato *Cnr* epialleles also exhibit some somatic instability [89,90]. In barley, an epiallele of the *cly1* gene is responsible for floret closing [92]. Epialleles have been observed in other organisms than plants. For example, in the fungus *Mucor circinelloides* RNAi-dependent epimutations can confer resistance to a fungicide [22] and in the oomycete *Phytophora sojae*, a plant pathogen, gene silencing mediated by small RNAs is transmitted between generations in a non-Mendelian fashion [21].

The previous examples concern alleles that were initially identified phenotypically. Many more examples have been revealed when variation in the epigenetic marks themselves, such as DNA methylation, has been examined. Lauria et al. [93] investigated variation in DNA methylation in several inbred lines of maize and showed that 12% of polymorphisms were meiotically inherited and that new polymorphisms occurred at appreciable rates. Meiotically stable epialleles in maize have also been observed by Regulski et al. [94], who also observed paramutation like behavior for some epialleles. In addition, DMRs have been shown to be transmitted in outcrossing events in *Arabidopsis* [95]. In a soybean mapping study, many but not all methylated regions were associated with genetic variants [96]. In outbreeding species it is difficult to decouple epigenetic and genetic variation, but in inbreeding species or species with very low genetic variation, epigenetic variation is common. In *Arabidopsis*, extensive variation in methylation patterns has been observed [97,98], and in apomictic dandelions variation in DNA methylation is common and while most of it is correlated with genetic variation, some of it seems independent [99]. In a clonally reproducing fish, variation in DNA methylation appears to be the only source of variation [100]. Furthermore, in plants from salt marsh environments, epigenetic variation seems to be correlated with the environment but not with genetic variation [101].

While there is no doubt that spontaneous epigenetic variation in DNA methylation exists, an important question is, how much of the observed variation in DNA methylation is under genetic control and how much is spontaneous epigenetic variation? By genetic control I mean that genetic variants in either *cis* or *trans* determine the methylation status of the locus in question. Recent high resolution studies of *Arabidopsis* suggest that much of the natural variation in DNA methylation can be explained by genetic variants [102,103] and environmental influences were small. Gene expression studies in human cell lines have suggested that most allele-specific expression patterns are associated with genetic variants [104], and twin studies show that DNA methylation differences are mainly under genetic control [84,105]. This suggests that functional epigenetic variation is rare in humans. However, in the case of a genetic variant controlling methylation in *cis*, it is difficult to determine if the linked genetic variants are causal because a methylation change can be in strong linkage disequilibrium with a nearby genetic variant.

Research into the natural variation of chromatin marks other than DNA methylation is in the early exploratory stages. Pioneering studies with human cell lines have revealed that variation among individuals exists [106], and that variation in chromatin marks is due to genetic polymorphism [106–108]. When these kinds of studies are performed in other species it will be revealed if genetic control is the norm. In yeast, preliminary studies of histone modifications have shown that many of the chromatin marks co-vary to some extent, but this co-variation is not complete as they also vary independently [109]. The genetic control of these epigenetic changes in yeast remains an open question. An investigation in epigenetic changes in the parasitic helminth *Schistosoma mansoni* found both environmentally induced and spontaneous changes in histone modifications. In roughly half of the cases, spontaneous histone modifications were mitotically inherited [110]. While much more is known about the natural variation of DNA methylation, it will be interesting to see if spontaneous variation in other chromatin marks is more common in non-human organisms.

Despite the many unresolved issues surrounding spontaneous epigenetic changes, it is clear that epigenetic variation, and DNA methylation in particular, fulfill the criteria that are required for it to be under selection and contribute to evolution. It seems that epigenetic variation has different properties than genetic variation and next I will examine how these differences influence adaptation.

MODELING SPONTANEOUS EPIGENETIC VARIATION

Here, I'll examine how to model spontaneous epigenetic variation and its effect on adaptation. From standard evolutionary theory, previous models, and from the molecular biology of epigenetic changes, we know that the important parameters for adaptive evolution are:

- Supply of epigenetic and genetic mutations (forward mutation rate).
- Stability of epigenetic mutations (backmutation rate).
- Distribution of mutational effects.

Mutational supply, $N\mu$, is a compound parameter that includes population size, N, multiplied by the mutation rate, μ. If mutations are rare, the rate of adaptive evolution is limited by their supply, as a population must wait for beneficial mutations to occur. It is well known that adaptive evolution is faster when $N\mu$ increases [111,112], but eventually saturates when $N\mu$ is high enough and the rate of adaptation becomes determined by the rate at which beneficial mutations increase in frequency. A common assumption is that epigenetic changes, such as DNA methylation changes, have a much higher mutation rate than genetic mutations and this seems to be the case. A study of spontaneous methylation changes at individual CpG sites in mutation accumulation lines of *Arabidopsis* revealed that the forward mutation rate (i.e., unmethylated to methylated cytosine) was 2.56×10^{-4} per site per generation [77]. For DMR changes, the forward mutation rate was 1.2×10^{-4} [78]. Compared to a genetic mutation rate of 7×10^{-9} per site per generation [113], these epimutation rates are several orders of magnitude higher and can differ according to genomic context [77,78]. Furthermore, individual alleles can have even higher mutation rates of up to 0.04 in the case of the *DWARF1* epiallele in rice [91]. Very little is known about the rates of spontaneous changes to histone modification, but rates are known to vary from at least 10^{-3}–10^{-4} [110].

The high supply of epigenetic mutations means that epigenetic variation will always exceed genetic variation and that for any evolutionary problem that can be solved by epigenetic mutations, selection is likely, at least at first, to favor a beneficial epigenetic allele to increase in frequency. This should be the case especially in small populations, where the supply of genetic mutations is low.

Another crucial aspect is the stability of epigenetic mutations, or the backmutation rate. In *Arabidopsis* CpG sites, an analysis of mutation-accumulation data has shown that methylated cytosines demethylate (i.e., backmutate) at a rate of 6.30×10^{-4} per site per generation [77] for single sites and at a rate of 4.6×10^{-4} for DMRs [78]. Higher rates have been reported for epialleles affecting the phenotype, with *DWARF1* values ranging from 0.008 to 0.27 [91]. Typically, backmutation rates of genetic loci are so low that they can safely be ignored in evolutionary models. However, this is not the case for epigenetic mutations. Backmutation decelerates the rate of adaptation relative to a stable allele that is increasing in frequency [33,114]. Instability does not prevent adaptation per se, as the selection coefficient of the allele greatly influences its rate of change. Thus, strong selection is required for unstable epigenetic alleles to play a role in adaptation [31,33]. Observed backmutation rates for epigenetic alleles are within the range that can support adaptation with an epigenetic basis [33]. Besides a lower rate of adaptation due to backmutation, the final fitness of the population will be lower because a beneficial epiallele with a high backmutation rate will never be fully fixed but rather maintained in a mutation-selection balance.

Adaptation from new mutations can be modeled by assuming that there is a statistical distribution of mutational effects, such as one where most mutations are deleterious and a small portion are beneficial. When a new mutation occurs, its phenotypic effect is drawn from this distribution. We have some evidence concerning the actual shape of this distribution for genetic mutations [115], but no empirical data are available for epigenetic mutations. However, there are reasons to suspect these distributions are different. From a molecular perspective, epigenetic changes can activate, modify or silence the expression of a given gene. It should be kept in mind that, all of these effects can be achieved with genetic mutations, but genetic mutations can also change the amino acid sequence and thus protein structure, genes can be duplicated or deleted, and large genomic rearrangements can occur. It would then stand to reason that, genetic mutations have a wider variety of possible phenotypic effects than epigenetic mutations, and this is true at least in simulated regulatory networks [33]. However, whether the phenotypic effects that epigenetic mutations have are smaller or greater than genetic mutations on average is unknown. If epigenetic mutations have, on average, smaller phenotypic effects than genetic mutations, this could alleviate the negative effects of their instability [33].

I will now illustrate adaptation with epigenetic variation using a simple population genetic model. The models are parameterized mainly based on *Arabidopsis* DNA methylation data. Other species, such as fruit flies or baker's yeast, can have very low levels of DNA methylation or lack it entirely. However, the idea is to illustrate general principles, and currently the best data available are from *Arabidopsis* DNA methylation. For simplicity, I assume a sexually reproducing diploid species with a very large population size, non-overlapping generations, and random mating. A biological example of an organism with these assumptions can be found among annual plants. I also assume that the same phenotype can be achieved with both epigenetic and genetic changes and that epialleles are inherited in a Mendelian fashion. Consider one locus with two alleles, one of which is in a different epigenetic state, for example a methylated and a nonmethylated allele. The standard equation for allele frequency change is:

$$\Delta q = pqs - \mu_1 q + \mu_2 p \quad (28.1)$$

where q is the frequency of the epigenetic allele, p is the frequency of the genetic allele, s is the selection coefficient, μ_1 is the backmutation rate of the epigenetic allele, and μ_2 is the forward mutation rate of the epigenetic allele. It is assumed here that other mutations are sufficiently rare to be safely ignored.

From Eq. (28.1) it can be seen that the epiallele increases in frequency when $\Delta q > 0$, which happens when the combined effect of natural selection and input of new mutations $(1-q)qs + \mu_2(1-q)$ is greater than the effects of backmutation $\mu_1 q$. Starting from a low frequency of allele q and iterating the difference equation $q_{t+1} = q_t + \Delta q$, where q_t is the frequency of allele q in generation t, it can be seen that the epiallele initially increases in frequency but has a lower equilibrium frequency when in mutation-selection balance than a normal genetic mutation (Fig. 28.2).

In any realistic scenario we must take into account that epigenetic mutations never occur in a vacuum, and that genetic mutations are always present, albeit at a lower frequency. This can result in a phenomenon where an adaptation initially produced by an abundant epiallele is eventually replaced by a genetic mutation as a consequence of epigenetic instability. This has been observed in different models [31,33] that make different assumptions regarding the fitness landscape and the mapping of genotype to phenotype. A simple version of this can be illustrated by considering one locus with multiple alleles. I have adapted this example from a similar one given in the supplementary material of Kronholm and Collins [33]. Let us assume an environmental change after which the expression of a constitutively expressed gene is now slightly deleterious and a silenced allele is now beneficial. The alleles in this model are: the ancestral allele that is expressed and now poorly adapted to the new environment; an epigenetically silenced allele that is methylated; and an allele that contains a genetic mutation that silences the gene. The terms in the model are given in Table 28.1.

Following Rice [25], the equations for selection on multiple alleles can be written in matrix form. Changes in allele frequencies are due to selection and mutation:

$$\Delta \vec{p} = \Delta \vec{p}_{sel} + \Delta \vec{p}_{mut} \quad (28.2)$$

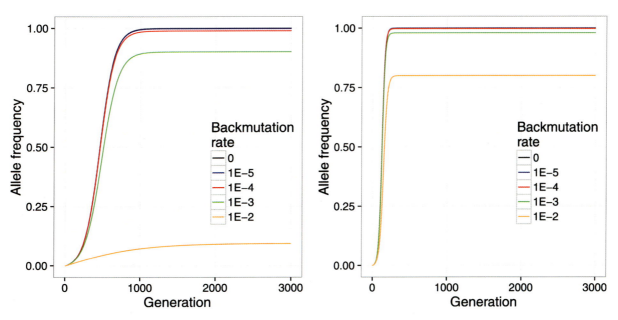

FIGURE 28.2 Allele frequency dynamics of mutation-selection balance for an epiallele. The forward epigenetic mutation rate, μ_2 is 1×10^{-4}, lines of different colors correspond to different backmutation rates from 0 to 1×10^{-2}. In the left panel, selection coefficient, $s = 0.01$, and in the right panel $s = 0.05$. Black and blue lines completely overlap in the figure.

TABLE 28.1 Parameters of the Model, s Is the Selection Coefficient Againts the Ancestral Allele

Allele	A_1	A_2	A_3			
Frequency	p_1	p_2	p_3			
Type	Ancestral	Epigenetic	Genetic			
Genotype	A_1A_1	A_1A_2	A_1A_3	A_2A_2	A_2A_3	A_3A_3
Fitness	$1-s$	$1-1/2s$	$1-1/2s$	1	1	1
Mutation	$A_1 \to A_2$	$A_1 \to A_3$	$A_2 \to A_3$	$A_2 \to A_1$		
Mutation rate	μ_1	μ_2	μ_3	μ_4		

where $\vec{p}$ is a vector of allele frequencies for the three alleles. Change due to selection is:

$$\Delta \vec{p}_{sel} = \Delta \begin{pmatrix} p_1 \\ p_2 \\ p_3 \end{pmatrix}_{sel}$$

$$= \frac{1}{2\bar{w}} \begin{pmatrix} p_1(1-p_1) & -p_1p_2 & -p_1p_3 \\ -p_2p_1 & p_2(1-p_2) & -p_2p_3 \\ -p_3p_1 & -p_3p_2 & p_3(1-p_3) \end{pmatrix} \begin{pmatrix} 2w_1^* \\ 2w_2^* \\ 2w_3^* \end{pmatrix} \quad (28.3)$$

where w_i^* is the marginal fitness of allele A_i. Marginal fitness is calculated from:

$$w_i^* = \sum_{j=1}^{j} p_j w_{ij} \quad (28.4)$$

where w_{ij} is the fitness for genotype A_iA_j. The population mean fitness is:

$$\bar{w} = \sum_{i=1}^{i} p_i w_i^*. \quad (28.5)$$

Using the mutational pathways and their rates from Table 28.1, change in allele frequencies due to mutation is

$$\Delta \vec{p}_{mut} = \Delta \begin{pmatrix} p_1 \\ p_2 \\ p_3 \end{pmatrix}_{mut} = \begin{pmatrix} -p_1\mu_1 - p_1\mu_2 + p_2\mu_4 \\ p_1\mu_1 - p_2\mu_3 - p_2\mu_4 \\ p_2\mu_3 + p_1\mu_3 \end{pmatrix}. \quad (28.6)$$

Substituting Eqs. (28.3 and 28.6) for Eq. (28.2) provides the change in allele frequencies for one generation. Evolutionary dynamics of the different alleles can be obtained by iterating the difference equation:

$$\vec{p}_{t+1} = \vec{p}_t + \Delta \vec{p}. \quad (28.7)$$

For this example, I parameterize the model based on empirical data from *Arabidopsis* mutation accumulation experiment, that investigated CpG methylation [77].

In the model, the forward mutation rate to a methylated state for a CpG site was $\mu_1 = 2.56 \times 10^{-4}$, the backmutation rate was $\mu_4 = 6.30 \times 10^{-4}$, and the genetic mutation rate was $\mu_2 = \mu_3 = 7 \times 10^{-9}$ [113]. Results from running the model with these values are shown in Fig. 28.3. It can be seen that the epiallele initially increases to a high frequency because the forward mutation rate is much faster than for the genetic allele and initially only epialleles are exposed to selection. Since the epiallele is unstable, it is slowly replaced by the genetic allele but this takes a considerable amount of time (Fig. 28.3). Replacement of the epiallele is driven by its backmutation. Note that the mean fitness of the population does not change considerably during this process. Compared to the situation where only the genetic allele is available, the initial stage of adaptation is faster when the epiallele is available (Fig. 28.3).

This model implies that if there exists such two mutation systems then the epigenetic system is used at the initial stage of adaptation, and afterwards the same phenotype is fixed using genetic mutations. Similar results have been obtained with more complex models that make different assumptions on how the (epi)genotype maps to the phenotype [31,33]. While obviously the exact pattern of adaptation depends on the parameter values, values derived from empirical data make this pattern at least plausible.

There are now a few studies that have examined the role of epigenetic variation in adaptation directly, using experimental evolution. A study by Wang et al. [116] used an artificial system in fission yeast with mutants that had uncontrolled heterochromatin spreading. In populations of these mutants, an adaptive epigenetic change that restored original fitness was rapidly selected, showing that benificial epigenetic changes can arise and spread, just like genetic mutations. Another study used the single celled algae *Chlamydomonas reinhardtii* that had multiple different evolutionary treatments, including manipulations of the epigenetic system either using chemicals or mutants [117]. In general, adaptation was slower in treatments where the epigenetic system was manipulated, and in non-

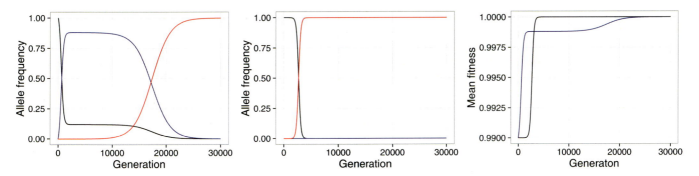

FIGURE 28.3 Left panel: evolutionary dynamics with epigenetic mutation. Black line tracks the frequency of the ancestral allele, blue is the epiallele, and red is the genetic allele. Middle panel: evolutionary dynamics with only genetic mutation. Black line is the ancestral allele and red line is the genetic allele. Right panel: trajectories of population mean fitness. Black line is fitness of a population adapting with only genetic mutations from the middle panel and blue line is the fitness of a population adapting with epigenetic and genetic mutations from the left panel.

manipulated treatments DNA-methylation changes occurred that were at least consistent with adaptive changes. Overall, the results of this study support the idea that epigenetic changes can contribute to adaptation. Furthermore, an excellent study was performed by Stajic et al. [118] with budding yeast that used an artificial system and some clever tricks of yeast genetics. They inserted a transgene into different regions of heterochromatin that exhibited different amounts of spontaneous silencing. By evolving the populations in an environment where silencing of the transgene was favored, they observed that the evolutionary dynamics were qualitatively similar as the model described above. In particular, the populations initially adapted using epigenetic changes and this accelerated adaptation and they fixed the advantageous phenotype later using genetic mutations that abolished the function of this transgene. The time scale of this process was much faster than in the model, possibly indicating some additional benefit of silencing the gene genetically. Finally, another excellent study in fission yeast showed that caffeine resistance was conferred by unstable epigenetic changes and the authors were able to show that it was H3K9 methylation at particular genes that made the cells resistant to caffeine [119], proving that epigenetic changes were contributing to the first step of adaptation.

The model also makes some indirect predictions: if adaptation from using epigenetic changes happens first, then epigenetically encoded traits should be more common in new populations that have recently colonized new areas and that are in the process of adaptation or have recently adapted to a new environment. In empirical studies it has been observed that epigenetic variation is more common in recent orchid hybrids [120], similarly artificial hybrids of *Arabidopsis* are more susceptible to DNA demethylating agents than their natural counterparts [121]. Alternatively, such patterns could also be considered at the genomic level, where the prediction is that newly duplicated genes should have more epigenetic variation than older duplicates. Indeed, this is what has been observed when Keller and Yi looked at methylation data from duplicated genes in humans. They observed that duplicate gene pairs were initially methylated but gradually lost methylation as they aged and diverged in their methylation patterns [122]. Similarly, a new gene in *Arabidopsis* exhibited extensive variation in DNA methylation that was correlated with gene expression [123]. These observations are at least consistent with the prediction that epigenetic variation is important in the initial stages of adaptation and which is then gradually replaced by genetic changes that fix the phenotype permanently. While these observations do not prove that the model is correct, they do suggest that it will be interesting to look in different systems whether this pattern holds in general.

LIMITS OF EPIGENETIC CONTRIBUTIONS TO ADAPTATION

While epigenetic variation seems to be common, genetic changes are undoubtedly the main driver of evolutionary change. In artificial selection experiments, selected phenotypes seem rather stable indicating that a large part of the observed phenotypic variation has a genetic basis or that at least a majority of the epigenetic variation contributing to the phenotype is stable. While many epialleles exist, one could still ask the question: why have decades of research on laboratory mutants detected so few epialleles? Confirmation bias may play a role, where researchers have discarded mutants that were not stable or have not behaved in strict Mendelian fashion, since no clear mechanism to explain such behavior has been apparent until now. Alternatively, it may be simply that epialleles affecting the phenotype are rare. Data pertaining to the phenotypic effects of epigenetic variation might provide more insight into this question.

Another issue regarding the phenotypic effects of epimutation is that a large part of the research in this field in the past has focused on methylation of single CpG sites and discovering differentially methylated positions. It is unclear what are the phenotypic effects of methylation of single cytosines. In many of the characterized epialleles, larger regions such as promoters have been methylated at several positions [89–91]. This is also observed in lines where epigenetic variation has been artificially induced, methylation changes responsible are longer contiguous stretches of methylation [87]. Differentially methylated regions may be more often responsible for phenotypic effects than single site changes. Furthermore, environmentally induced epigenetic changes that have phenotypic effects appear to be mainly DMRs [50]. In *Arabidopsis*, DMRs appear to have mutation rates that are in the same order of magnitude as for single sites [78], while surveys of natural populations have indicated that DMRs tend to be stable [103]. The reasons for this contradiction are yet to be resolved.

The high mutation rate of epigenetic variation raises a further issue. Similar to genetic mutations, there is little doubt that the majority of spontaneous epigenetic mutations that effect phenotype are deleterious in an organism that is adapted reasonably well to its environment [115]. If all epigenetic changes that happen at high rates would contribute to phenotype, such a high epimutation rate would cause a large load of deleterious epimutations. Furthermore, in some cases it has been reported that epigenetic variation has properties resembling paramutation [94]. Paramutation is a phenomenon where the paramutagenic allele converts the other to the same state when they both are present in a heterozygous genotype [124]. The paramutagenic allele always spreads in a population via the paramutagenic process, and mathematically this is equivalent to models of meiotic drive [125]. Beneficial paramutagenic alleles spread rapidly as selection and paramutation work in synergy to increase allele frequency, but deleterious paramutagenic alleles will also spread. As such, it is unlikely that paramutagenic alleles are common, or if they are, they cannot affect the phenotype very often. Otherwise common paramutagenic alleles would result in a large mutational load that would likely lead to a severe decrease in fitness and possibly extinction.

Our knowledge of the taxonomic spread of different epigenetic mechanisms is also poor. Understandably, our mechanistic understanding of different epigenetic mechanisms is based on a few model organisms, but how common all of the different mechanisms are can have a large effect on how important epigenetic variation turns out to be in general. We know that that in some taxa there can be quite a bit of variation whether DNA methylation is present and how much DNA is methylated; for example, there is lot of variation in the amount of DNA methylated in flowering plants [126], fungi [127], and insects [128]. Many yeasts lack DNA methylation all together [129], and DNA methylation seems to be absent in flies, including the fruit fly *Drosophila melanogaster*. However, extremely low levels of DNA methylation in *Drosophila* has been reported [129]. While it may be that different species may use different pathways to silence genes epigenetically, variation in epigenetic mechanisms may also allow some species to have more epigenetic variation than others.

CONCLUSIONS AND FUTURE DIRECTIONS

For spontaneous epigenetic variation it is now clear that it can in principle contribute to adaptation. The question remains, how important is epigenetic variation? The contribution of spontaneous epigenetic variation depends on three parameters mentioned earlier: mutational supply of epigenetic mutations, their stability, and the distribution of their phenotypic effects. Based on data from mutation accumulation experiments, we can now say something about supply and stability [75–78]. However, the data is limited to cytosine methylation in *Arabidopsis*; to know if rates of spontaneous DNA methylation changes reflect general trends, we need more data for different organisms. The amount of DNA methylation in different species is highly variable [130], even among vascular plants [126], and *Arabidopsis* does not have a typical plant genome. The data suggest that levels of DNA methylation in *Arabidopsis* may be at or close to an evolutionary equilibrium for the observed forward and backmutation rates [77], and if this is the case for other species they may have very different epimutation rates. However, estimates of somatic epimutation rates in poplar suggest that the per generation epimutation rate is similar to that of *Arabidopsis* [79]. Data from more species is needed to resolve this issue.

The stability of epialleles is the second important parameter to understand their role in evolution. Data indicate that in *Arabidopsis*, single CpG sites and DMRs have rates of changes that are similar orders of magnitude and are stable enough to contribute to adaptation [77,78]. However, it seems that stability of DMRs varies depending on their genomic context and other properties, and it is important to characterize what makes some epialles stable and some unstable. Furthermore, instability of an epiallele should not be taken to mean that it is irrelevant to adaptive evolution. By using evolutionary models, one can check what is the magnitude of the selection coefficient that would be required for those alleles to increase in frequency. Again, we need more data from different species and epialleles to establish what is typical for epialleles.

The third important parameter is the distribution of phenotypic effects for epigenetic mutations. Estimating this distribution is a hard empirical problem. However, it is also perhaps one of the most important in terms of evolution. While some epialleles do affect the phenotype, many of them seem neutral. When they affect the phenotype, epigenetic mutations are likely to change regulation of gene expression, so they could have similar effects as *cis*-regulatory mutations [131]. Progress may be made by making use of the artificially induced epigenetic variation present in the *Arabidopsis* epiRIL population [85]. Given that some DMRs that affect the phenotype have already been characterized in this population [87], the construction of lines harboring only single DMR change in an otherwise uniform genetic background could provide insights into their distribution of effects. Another possibility is to make use of the lines characterized in mutation accumulation studies, for instance if the DMR change that happened can be crossed into a uniform background.

Given that its mechanism is well understood, the majority of studies on epigenetic marks in an evolutionary context have focused on cytosine methylation. Consequently, mutation accumulation studies of epigenetic marks other than cytosine methylation are needed. Certain sequencing technologies now allow looking at other modifications than cytosine methylation, and it has become clear that many eukaryotes also undergo adenosine methylation [132–134], but very little is known about its role, inheritance, and ability to spontaneously change. Some studies in bacteria suggest that there is variation in adenosine methylation and it can be involved in adaptation [135]. Histone modifications, such as histone H3 lysine 9 methylation and H3 lysine 27 methylation, that are showing some evidence of transmission to the next generation, are also interesting modifications to study.

For spontaneous epigenetic changes, we now have evidence that epigenetic changes can contribute to adaptation in principle [116,117]. Evolutionary models predict that epigenetic changes should happen first, and they are later replaced by genetic changes to fix the new phenotype [31,33]. Moreover, we now have experimental support for this prediction [118]. The next steps should involve better characterizing epigenetic changes that happen during adaptation naturally, whether in lab populations or in nature. Further tests of the model predictions could also be done in natural populations, such as comparing younger and older populations, on whether the former have more epigenetic variation than the latter. Similar studies for duplicated genes could also be helpful. Such studies would also benefit from population genetic theory developed for markers with high mutation rates [136] and test for finding the patterns left by natural selection on epigenetic variation [137].

While we have a good understanding of which environmental conditions favor the evolution of transmitted plastic responses, there are some gaps in our understanding regarding the evolution of transmitted induced responses. One important parameter that has a large influence on whether induced responses evolve is whether organisms pay any cost of plasticity. Costs of plasticity are incorporated in many theoretical models, but our empirical understanding of costs is very limited. While we certainly have good examples of cases where plastic responses are transmitted via induced epigenetic changes, it seems that, in general, transmitted plastic effects are quite weak [138]. This could be because usually people have not looked for them, or that conditions that favor their evolution are rare in nature. Especially for effects that last for multiple generations. As more data from different taxa are accumulated, we will see how common transmitted plasticity ultimately is.

Since the publication of the previous edition of this book, there have been a number of new discoveries, most notably we now have direct empirical evidence, albeit in laboratory conditions, that epigenetic variation can contribute to adaptation. While there are many details to be worked out, the big question remaining is, how important epigenetic variation is for adapation? No single study can answer this question, and the answer will only come as data accumulates over a long period, and of course we cannot know the outcome in advance. Nevertheless, the field of evolutionary epigenetics is now more exiting than ever.

Acknowledgments

This research is supported by the Academy of Finland grant no. 321584. I would like to thank Tarmo Ketola and Joannes van Cann for comments on the previous edition of this chapter.

References

[1] Fisher RA. The Genetical Theory of Natural Selection. Oxford: Clarendon Press; 1930.
[2] Haig D. Weismann rules! OK? Epigenetics and the Lamarckian temptation. Biol Philos 2007;22:415–28.
[3] Richards EJ. Inherited epigenetic variation — revisiting soft inheritance. Nat Rev Genet 2006;7:395–401.

[4] Holoch D, Moazed D. RNA-mediated epigenetic regulation of gene expression. Nat Rev Genet 2015;16:71–84.

[5] Matzke MA, Mosher RA. RNA-directed DNA methylation: an epigenetic pathway of increasing complexity. Nat Rev Genet 2014;15:394–408.

[6] Jablonka E, Raz G. Transgenerational epigenetic inheritance: prevalence, mechanisms, and implications for the study of heredity and evolution. Q Rev Biol 2009;84:131–76.

[7] Feldman MW, Cavalli-Sforza LL. Models for cultural inheritance: a general linear model. Ann Hum Biol 1975;2:215–26.

[8] Baxter FA, Drake AJ. Non-genetic inheritance via the male germline in mammals. Phil Trans R Soc B 2019;374:20180118.

[9] Gaydos LJ, Wang W, Strome S. H3K27me and PRC2 transmit a memory of repression across generations and during development. Science 2014;345:1515–18.

[10] Ragunathan K, Jih G, Moazed D. Epigenetic inheritance uncoupled from sequence-specific recruitment. Science 2014;348:1256899.

[11] Audergon PNCB, Catania S, Kagansky A, Tong P, Shukla M, Pidoux AL, et al. Epigenetics. Restricted epigenetic inheritance of H3K9 methylation. Science 2015;348:132–5.

[12] van de Werken C, van der Heijden GW, Eleveld C, Teeuwssen M, Albert M, Baarends WM, et al. Paternal heterochromatin formation in human embryos is H3K9/HP1 directed and primed by sperm-derived histone modifications. Nat Commun 2014;5:5868.

[13] Yu R, Wang X, Moazed D. Epigenetic inheritance mediated by coupling of RNAi and histone H3K9 methylation. Nature 2018;558:615–19.

[14] Siklenka K, Erkek S, Godmann M, Lambrot R, McGraw S, Lafleur C, et al. Disruption of histone methylation in developing sperm impairs offspring health transgenerationally. Science 2015;350:aab2006.

[15] Jih G, Iglesias N, Currie MA, Bhanu NV, Paulo JA, Gygi SP, et al. Unique roles for histone H3K9me states in RNAi and heritable silencing of transcription. Nature 2017;547:463–7.

[16] Shipkovenska G, Durango A, Kalocsay M, Gygi SP, Moazed D. A conserved rna degradation complex required for spreading and epigenetic inheritance of heterochromatin. eLife 2020;9:e54341.

[17] Bannister AJ, Kouzarides T. Regulation of chromatin by histone modifications. Cell Res 2011;21:381–95.

[18] Rassoulzadegan M, Grandjean V, Gounon P, Vincent S, Gillot I, Cuzin F. RNA-mediated non-mendelian inheritance of an epigenetic change in the mouse. Nature 2006;441:469–74.

[19] Rechavi O, Minevich G, Hobert O. Transgenerational inheritance of an acquired small RNA-based antiviral response in C. elegans. Cell 2011;147:1248–56.

[20] Ashe A, Sapetschnig A, Weick E-M, Mitchell J, Bagijn M, Cording A, et al. piRNAs can trigger a multigenerational epigenetic memory in the germline of C. elegans. Cell 2012;150:88–99.

[21] Qutob D, Patrick Chapman B, Gijzen M. Transgenerational gene silencing causes gain of virulence in a plant pathogen. Nat Commun 2013;4:1349.

[22] Calo S, Shertz-Wall C, Lee SC, Bastidas RJ, Nicolas FE, Granek JA, et al. Antifungal drug resistance evoked via RNAi-dependent epimutations. Nature 2014;513:555–8.

[23] Davidson EH. The regulatory genome: gene regulatory networks in development and evolution. Academic Press; 2006.

[24] Teixeira FK, Heredia F, Sarazin A, Roudier F, Boccara M, Ciaudo C, et al. A role for RNAi in the selective correction of DNA methylation defects. Science 2009;323:1600–4.

[25] Rice SH. Evolutionary theory: mathematical and conceptual foundations. Sinauer Associates, Inc; 2004.

[26] Lynch M, Walsh B. Genetics and analysis of quantitative traits. Sunderland: Sinauer Associates, Inc; 1998.

[27] Bonduriansky R, Day T. Nongenetic inheritance and its evolutionary implications. Ann Rev Ecol Evol Syst 2009;40:103–25.

[28] Day T, Bonduriansky R. A unified approach to evolutionary consequences of genetic and nongenetic inheritance. Am Nat 2011;178:E18–36.

[29] Johannes F, Colomé-Tatché M. Quantitative epigenetics through epigenomic perturbation of isogenic lines. Genetics 2011;188:215–27.

[30] Geoghegan JL, Spencer HG. Population-epigenetic models of selection. Theor Popul Biol 2012;81:232–42.

[31] Klironomos F, Berg J, Collins S. How epigenetic mutations can affect genetic evolution: Model and mechanism. Bioessays 2013;35:571–8.

[32] Furrow RE. Epigenetic inheritance, epimutation, and the response to selection. PLoS One 2014;9:e101559.

[33] Kronholm I, Collins S. Epigenetic mutations can both help and hinder adaptive evolution. Mol Ecol 2016;25:1856–68.

[34] Agrawal AA, Laforsch C, Tollrian R. Transgenerational induction of defences in animals and plants. Nature 1999;401:60–3.

[35] Salinas S, Munch SB. Thermal legacies: transgenerational effects of temperature on growth in a vertebrate. Ecol Lett 2012;15:159–63.

[36] Eggert H, Kurtz J, Diddens-de Buhr MF. Different effects of paternal trans-generational immune priming on survival and immunity in step and genetic offspring. Proc R Soc B Biol Sci 2014;281:20142089.

[37] Shama LNS, Wegner KM. Grandparental effects in marine sticklebacks: transgenerational plasticity across multiple generations. J Evol Biol 2014;27:2297–307.

[38] Zajitschek S, Hotzy C, Zajitschek F, Immler S. Short-term variation in sperm competition causes sperm-mediated epigenetic effects on early offspring performance in the zebrafish. Proc R Soc B Biol Sci 2014;281 20140422.

[39] Chen B, Li S, Ren Q, Tong X, Zhang X, Kang L. Paternal epigenetic effects of population density on locust phase-related characteristics associated with heat-shock protein expression. Mol Ecol 2015;24:851–62.

[40] Galloway LF, Etterson JR. Transgenerational plasticity is adaptive in the wild. Science 2007;318:1134–6.

[41] Holeski LM. Within and between generation phenotypic plasticity in trichome density of Mimulus guttatus. J Evol Biol 2007;20:2092–100.

[42] Slaughter A, Daniel X, Flors V, Luna E, Hohn B, Mauch-Kani B. Descendants of primed Arabidopsis plants exhibit resistance to biotic stress. Plant Physiol 2012;158:835–43.

[43] Verhoeven KJ, van Gurp TP. Transgenerational effects of stress exposure on offspring phenotypes in apomictic dandelion. PLoS One 2012;7:e38605.

[44] Zimmerman KCK, Levitis DA, Pringle A. Beyond animals and plants: dynamic maternal effects in the fungus Neurospora crassa. J Evol Biol 2016;1379–93.

[45] Kou H, Li Y, Song X, Ou X, Xing S, Ma J, et al. Heritable alteration in DNA methylation induced by nitrogen-deficiency stress accompanies enhanced tolerance by progenies to the stress in rice (Oryza sativa L.). J Plant Physiol 2011;168:1685–93.

[46] Ou X, Zhang Y, Xu C, Lin X, Zang Q, Zhuang T, et al. Transgenerational inheritance of modified DNA methylation patterns and enhanced tolerance induced by heavy metal stress in rice (Oryza sativa L.). PLoS One 2012;7:e41143.

[47] Rasmann S, De Vos M, Casteel CL, Tian D, Halitschke R, Sun JY, et al. Herbivory in the previous generation primes plants for enhanced insect resistance. Plant Physiol 2012;158:854–63.

[48] Luna E, Bruce TJ, Roberts MR, Flors V, Ton J. Next-generation systemic acquired resistance. Plant Physiol 2012;158:844–53.

[49] Luna E, Ton J. The epigenetic machinery controlling transgenerational systemic acquired resistance. Plant Signal Behav 2012;7:615–18.

REFERENCES

[50] Wibowo A, Becker C, Marconi G, Durr J, Price J, Hagmann J, et al. Hyperosmotic stress memory in *Arabidopsis* is mediated by distinct epigenetically labile sites in the genome and is restricted in the male germline by DNA glycosylase activity. eLife 2016;5:e13546.

[51] Rechavi O, Houri-Ze'evi L, Anava S, Goh W, Kerk S, Hannon G, et al. Starvation-induced transgenerational inheritance of small rnas in *C. elegans*. Cell 2014;158:277−87.

[52] Jobson MA, Jordan JM, Sandrof MA, Hibshman JD, Lennox AL, Baugh LR. Transgenerational effects of early life starvation on growth, reproduction and stress resistance in *Caenorhabditis elegans*. Genetics 2015;201:201−12.

[53] Kishimoto S, Uno M, Okabe E, Nono M, Nishida E. Environmental stresses induce transgenerationally inheritable survival advantages via germline-to-soma communication in *Caenorhabditis elegans*. Nat Commun 2017;8:14031.

[54] Ivimey-Cook ER, Sales K, Carlsson H, Immler S, Chapman T, Maklakov AA. Transgenerational fitness effects of lifespan extension by dietary restriction in *Caenorhabditis elegans*. bioRxiv 2020.

[55] Öst A, Lempradl A, Casas E, Weigert M, Tiko T, Deniz M, et al. Paternal diet defines offspring chromatin state and intergenerational obesity. Cell 2014;159:1352−64.

[56] Houri-Zeèvi L, Korem Y, Sheftel H, Faigenbloom L, Toker I, Dagan Y, et al. A tunable mechanism determines the duration of the transgenerational small RNA inheritance in *C. elegans*. Cell 2016;165:88−99.

[57] Yamaguchi S, Shen L, Liu Y, Sendler D, Zhang Y. Role of Tet1 in erasure of genomic imprinting. Nature 2013;504:460−4.

[58] Crevillen P, Yang H, Cui X, Greeff C, Trick M, Qiu Q, et al. Epigenetic reprogramming that prevents transgenerational inheritance of the vernalized state. Nature 2014;515:587−90.

[59] Botero CA, Weissing FJ, Wright J, Rubenstein DR. Evolutionary tipping points in the capacity to adapt to environmental change. Proc Natl Acad Sci 2015;112:184−9.

[60] Kronholm I. Evolution of anticipatory effects mediated by epigenetic changes. Environ Epigenet 2022;8:1−12. Available from: https://doi.org/10.1093/eep/dvac007.

[61] Jablonka E, Oborny B, Molnár I, Kisdi E, Hofbauer J, Czárán T. The adaptive advantage of phenotypic memory in changing environments. Philos Trans R Soc Lond B Biol Sci 1995; 350:133−41.

[62] Lachmann M, Jablonka E. The inheritance of phenotypes: an adaptation to fluctuating environments. J Theor Biol 1996;181:1−9.

[63] Kuijper B, Hoyle RB. When to rely on maternal effects and when on phenotypic plasticity? Evolution 2015;69:950−68.

[64] Leimar O, McNamara JM. The evolution of transgenerational integration of information in heterogeneous environments. Am Nat 2015;185:E55−69 PMID: 25674697.

[65] Greenspoon PB, Spencer HG. The evolution of epigenetically mediated adaptive transgenerational plasticity in a subdivided population. Evolution 2018;72:2773−80.

[66] Dey S, Proulx SR, Teotónio H. Adaptation to temporally fluctuating environments by the evolution of maternal effects. PLoS Biol 2016;14:1−29.

[67] Lind MI, Zwoinska MK, Andersson J, Carlsson H, Krieg T, Larva T, et al. Environmental variation mediates the evolution of anticipatory parental effects. Evol Lett 2020;4:371−81.

[68] Chevin L-M, Lande R, Mace GM. Adaptation, plasticity, and extinction in a changing environment: towards a predictive theory. PLoS Biol 2010;8:e1000357.

[69] Lande R. Adaptation to an extraordinary environment by evolution of phenotypic plasticity and genetic assimilation. J Evol Biol 2009;22:1435−46.

[70] Hoyle RB, Ezard THG. The benefits of maternal effects in novel and in stable environments. J R Soc Interface 2012;9:2403−13.

[71] Schaum CE, Collins S. Plasticity predicts evolution in a marine alga. Proc R Soc B Biol Sci 2014;281 20141486.

[72] Lind MI, Yarlett K, Reger J, Carter MJ, Beckerman AP. The alignment between phenotypic plasticity, the major axis of genetic variation and the response to selection. Proc R Soc Lond B: Biol Sci 2015;282:20151651.

[73] Lenski RE, Mittler JE. The directed mutation controversy and Neo-Darwinism. Science 1993;259:188−94.

[74] Halligan DL, Keightley PD. Spontaneous mutation accumulation studies in evolutionary genetics. Ann Rev Ecol Evol Syst 2009;40:151−72.

[75] Becker C, Hagmann J, Müller J, Koenig D, Stegle O, Borgwardt K, et al. Spontaneous epigenetic variation in the *Arabidopsis thaliana* methylome. Nature 2011;480:245−9.

[76] Schmitz RJ, Schultz MD, Lewsey MG, O'Malley RC, Urich MA, Libiger O, et al. Transgenerational epigenetic instability is a source of novel methylation variants. Science 2011;334: 369−73.

[77] van der Graaf A, Wardenaar R, Neumann DA, Taudt A, Shaw RG, Jansen RC, et al. Rate, spectrum, and evolutionary dynamics of spontaneous epimutations. Proc Natl Acad Sci 2015;112:6676−81.

[78] Denkena J, Johannes F, Colomé-Tatché M. Region-level epimutation rates in *Arabidopsis thaliana*. Heredity 2021;127: 190−202.

[79] Hofmeister BT, Denkena J, Colomé-Tatché M, Shahryary Y, Hazarika R, Grimwood J, et al. The somatic genetic and epigenetic mutation rate in a wild long-lived perennial *Populus trichocarpa*. Genome Biol 2020;21:259.

[80] Beltran T, Shahrezaei V, Katju V, Sarkies P. Epimutations driven by small RNAs arise frequently but most have limited duration in *Caenorhabditis elegans*. Nat Ecol Evol 2020;4:1539−48.

[81] Johnson LM, Smith OJ, Hahn DA, Baer CF. Short-term heritable variation overwhelms 200 generations of mutational variance for metabolic traits in *Caenorhabditis elegans*. Evolution 2020;74:2451−64.

[82] Bennett-Baker PE, Wilkowski J, Burke DT. Age-associated activation of epigenetically repressed genes in the mouse. Genetics 2003;165:2055−62.

[83] Fraga MF, Ballestar E, Paz MF, Ropero S, Setien F, Ballestar ML, et al. Epigenetic differences arise during the lifetime of monozygotic twins. Proc Natl Acad Sci U S A 2005;102:10604−9.

[84] van Dongen J, Nivard MG, Willemsen G, Hottenga J-J, Helmer Q, Dolan CV, et al. Genetic and environmental influences interact with age and sex in shaping the human methylome. Nat Commun 2016;7:11115.

[85] Johannes F, Porcher E, Teixeira FK, Saliba-Colombani V, Simon M, Agier N, et al. Assessing the impact of transgenerational epigenetic variation on complex traits. PLoS Genet 2009;5:e1000530.

[86] Latzel V, Allan E, Bortolini Silveira A, Colot V, Fischer M, Bossdorf O. Epigenetic diversity increases the productivity and stability of plant populations. Nat Commun 2013;4:2875.

[87] Cortijo S, Wardenaar R, Colomé-Tatché M, Gilly A, Etcheverry M, Labadie K, et al. Mapping the epigenetic basis of complex traits. Science 2014;343:1145−8.

[88] Kooke R, Johannes F, Wardenaar R, Becker F, Etcheverry M, Colot V, et al. Epigenetic basis of morphological variation and phenotypic plasticity in *Arabidopsis thaliana*. Plant Cell 2015; 27:337−48.

[89] Cubas P, Vincent C, Coen E. An epigenetic mutation responsible for natural variation in floral symmetry. Nature 1999;401:157−61.

[90] Manning K, Mahmut T, Poole M, Hong Y, Thompson AJ, King GJ, et al. A naturally occurring epigenetic mutation in a gene encoding an SBP-box transcription factor inhibits tomato fruit ripening. Nat Genet 2006;38:948−52.

[91] Miura K, Agetsuma M, Kitano H, Yoshimura A, Matsuoka M, Jacobsen SE, et al. A metastable *DWARF1* epigenetic mutant affecting plant stature in rice. Proc Natl Acad Sci 2009;106: 11218–23.

[92] Wang N, Ning S, Wu J, Tagiri A, Komatsuda T. An epiallele at *cly1* affects the expression of floret closing (cleistogamy) in barley. Genetics 2015;199:95–104.

[93] Lauria M, Piccinini S, Pirona R, Lund G, Viotti A, Motto M. Epigenetic variation, inheritance, and parent-of-origin effects of cytosine methylation in maize (*Zea mays*). Genetics 2014;196:653–66.

[94] Regulski M, Lu Z, Kendall J, Donoghue MT, Reinders J, Llaca V, et al. The maize methylome influences mRNA splice sites and reveals widespread paramutation-like switches guided by small RNA. Genome Res 2013;23:1651–62.

[95] Hofmeister BT, Lee K, Rohr NA, Hall DW, Schmitz RJ. Stable inheritance of DNA methylation allows creation of epigenotype maps and the study of epiallele inheritance patterns in the absence of genetic variation. Genome Biol 2017;18:155.

[96] Schmitz RJ, He Y, Valdés-López O, Khan SM, Joshi T, Urich MA, et al. Epigenome-wide inheritance of cytosine methylation variants in a recombinant inbred population. Genome Res 2013;23:1663–74.

[97] Vaughn MW, Tanurd M, Lippman Z, Jiang H, Carrasquillo R, Rabinowicz PD, et al. Epigenetic natural variation in *Arabidopsis thaliana*. PLoS Biol 2007;5:e174.

[98] Schmitz RJ, Schultz MD, Urich MA, Nery JR, Pelizzola M, Libiger O, et al. Patterns of population epigenomic diversity. Nature 2013;495:193–8.

[99] Preite V, Snoek LB, Oplaat C, Biere A, van der Putten WH, Verhoeven KJF. The epigenetic footprint of poleward range-expanding plants in apomictic dandelions. Mol Ecol 2015;24: 4406–18.

[100] Massicotte R, Whitelaw E, Angers B. DNA methylation: a source of random variation in populations. Epigenetics 2011;6: 421–7.

[101] Foust CM, Preite V, Schrey AW, Alvarez M, Robertson MH, Verhoeven KJF, et al. Genetic and epigenetic differences associated with environmental gradients in replicate populations of two salt marsh perennials. Mol Ecol 2016;25:1639–52.

[102] Dubin MJ, Zhang P, Meng D, Remigereau M-S, Osborne EJ, Paolo Casale F, et al. DNA methylation in *Arabidopsis* has a genetic basis and shows evidence of local adaptation. eLife 2015;4:e05255.

[103] Hagmann J, Becker C, Müller J, Stegle O, Meyer RC, Wang G, et al. Century-scale methylome stability in a recently diverged *Arabidopsis thaliana* lineage. PLoS Genet 2015;11:e1004920.

[104] Lappalainen T, Sammeth M, Friedlander MR, 't Hoen PAC, Monlong J, Rivas MA, et al. Transcriptome and genome sequencing uncovers functional variation in humans. Nature 2013;501:506–11.

[105] McRae AF, Powell JE, Henders AK, Bowdler L, Hemani G, Shah S, et al. Contribution of genetic variation to transgenerational inheritance of DNA methylation. Genome Biol 2014;15:R73.

[106] Kasowski M, Kyriazopoulou-Panagiotopoulou S, Grubert F, Zaugg JB, Kundaje A, Liu Y, et al. Extensive variation in chromatin states across humans. Science 2013;342:750–2.

[107] Waszak SM, Delaneau O, Gschwind AR, Kilpinen H, Raghav SK, Witwicki RM, et al. Population variation and genetic control of modular chromatin architecture in humans. Cell 2015;162: 1039–50.

[108] Grubert F, Zaugg JB, Kasowski M, Ursu O, Spacek DV, Martin AR, et al. Genetic control of chromatin states in humans involves local and distal chromosomal interactions. Cell 2015; 162:1051–65.

[109] Filleton F, Chuffart F, Nagarajan M, Bottin-Duplus H, Yvert G. The complex pattern of epigenomic variation between natural yeast strains at single-nucleosome resolution. Epigenetics Chromatin 2015;8:26.

[110] Roquis D, Rognon A, Chapparo C, Boissier J, Arancibia N, Cosseau C, et al. Frequency and mitotic heritability of epimutations in *Schistosoma mansoni*. Mol Ecol 2016;25:1741–58.

[111] Orr HA. The rate of adaptation in asexuals. Genetics 2000;155: 961–8.

[112] de Visser JAGM, Zeyl CW, Gerrish PJ, Blanchard JL, Lenski RE. Diminishing returns from mutation supply rate in asexual populations. Science 1999;283:404–6.

[113] Ossowski S, Schneeberger K, Lucas-Lledó JI, Warthmann N, Clark RM, Shaw RG, et al. The rate and molecular spectrum of spontaneous mutations in *Arabidopsis thaliana*. Science 2010;327:92–4.

[114] Geoghegan JL, Spencer HG. Exploring epiallele stability in a population-epigenetic model. Theor Popul Biol 2013;83: 136–44.

[115] Eyre-Walker A, Keightley PD. The distribution of fitness effects of new mutations. Nat Rev Genet 2007;8:610–18.

[116] Wang J, Reddy BD, Jia S. Rapid epigenetic adaptation to uncontrolled heterochromatin spreading. eLife 2015;4: e06179.

[117] Kronholm I, Bassett A, Baulcombe D, Collins S. Epigenetic and genetic contributions to adaptation in *Chlamydomonas*. Mol Biol Evol 2017;34:2285–306.

[118] Stajic D, Perfeito L, Jansen LET. Epigenetic gene silencing alters the mechanisms and rate of evolutionary adaptation. Nat Ecol Evol 2019;3:491–8.

[119] Torres-Garcia S, Yaseen I, Shukla M, Audergon PNCB, White SA, Pidoux AL, et al. Epigenetic gene silencing by heterochromatin primes fungal resistance. Nature 2020;585:453–8.

[120] Paun O, Bateman RM, Fay MF, Hedren M, Civeyrel L, Chase MW. Stable epigenetic effects impact adaptation in allopolyploid orchids (Dactylorhiza: Orchidae). Mol Biol Evol 2010;27:2465–73.

[121] Madlung A, Masuelli RW, Watson B, Reynolds SH, Davison J, Comai L. Remodeling of DNA methylation and phenotypic and transcriptional changes in synthetic *Arabidopsis* allotetraploids. Plant Physiol 2002;129:733–46.

[122] Keller TE, Yi SV. DNA methylation and evolution of duplicate genes. Proc Natl Acad Sci 2014;111:5926–31.

[123] Silveira AB, Trontin C, Cortijo S, Barau J, Del Bem LEV, Loudet O, et al. Extensive natural epigenetic variation at a *de novo* originated gene. PLoS Genet 2013;9:e1003437.

[124] Chandler VL. Paramutation: from maize to mice. Cell 2007; 128:641–5.

[125] Unckless RL, Messer PW, Connallon T, Clark AG. Modeling the manipulation of natural populations by the mutagenic chain reaction. Genetics 2015;201:425–31.

[126] Alonso C, Pérez R, Bazaga P, Herrera CM. Global DNA cytosine methylation as an evolving trait: phylogenetic signal and correlated evolution with genome size in angiosperms. Front Genet 2015;6.

[127] Nai Y-S, Huang Y-C, Yen M-R, Chen P-Y. Diversity of fungal DNA methyltransferases and their association with DNA methylation patterns. Front Microbiol 2021;11:3614.

[128] Bewick AJ, Vogel KJ, Moore AJ, Schmitz RJ. Evolution of DNA methylation across insects. Mol Biol Evol 2016;34: 654–65.

[129] Capuano F, Mülleder M, Kok R, Blom HJ, Ralser M. Cytosine dna methylation is found in *Drosophila melanogaster* but absent in *Saccharomyces cerevisiae, Schizosaccharomyces pombe,* and other yeast species. Anal Chem 2014;86:3697–702.

[130] Feng S, Cokus SJ, Zhang X, Chen P-Y, Bostick M, Goll MG, et al. Conservation and divergence of methylation patterning in plants and animals. Proc Natl Acad Sci 2010;107:8689—94.

[131] Wittkopp PJ, Kalay G. *Cis*-regulatory elements: molecular mechanisms and evolutionary processes underlying divergence. Nat Rev Genet 2012;13:59—69.

[132] Fu Y, Luo G-Z, Chen K, Deng X, Yu M, Han D, et al. N^6-methyldeoxyadenosine marks active transcription start sites in *Chlamydomonas*. Cell 2015;161:879—92.

[133] Greer E, Blanco M, Gu L, Sendinc E, Liu J, Aristizábal-Corrales D, et al. DNA methylation on N^6-adenine in *C. elegans*. Cell 2015;161:868—78.

[134] Koziol MJ, Bradshaw CR, Allen GE, Costa ASH, Frezza C, Gurdon JB. Identification of methylated deoxyadenosines in vertebrates reveals diversity in DNA modifications. Nat Struct Mol Biol 2016;23:24—30.

[135] Bruneaux M, Kronholm I, Ashrafi R, Ketola T. Roles of adenine methylation and genetic mutations in adaptation to different temperatures in Serratia marcescens. Epigenetics 2021. Available from: https://doi.org/10.1080/15592294.2021.1966215 In press.

[136] Charlesworth B, Jain K. Purifying selection, drift, and reversible mutation with arbitrarily high mutation rates. Genetics 2014;198:1587—602.

[137] Wang J, Fan C. A neutrality test for detecting selection on DNA methylation using single methylation polymorphism frequency spectrum. Genome Biol Evol 2015;7:154—71.

[138] Uller T, Nakagawa S, English S. Weak evidence for anticipatory parental effects in plants and animals. J Evol Biol 2013;26:2161—70.

SECTION VII

Epigenetic Epidemiology

29. Epigenetics of livestock health, production, and breeding
30. Nutritional epigenome and fetal metabolic programming
31. Epigenetics of drug addiction
32. Environmental influence of epigenetics
33. The gut microbiome influence on human epigenetics, health, and disease
34. Population pharmacoepigenomics

CHAPTER 29

Epigenetics of Livestock Health, Production, and Breeding

Eveline M. Ibeagha-Awemu[1] and Hasan Khatib[2]

[1]Sherbrooke Research and Development Centre, Agriculture and Agri-Food Canada, Sherbrooke, QC, Canada
[2]Department of Animal and Dairy Sciences, University of Wisconsin, Madison, WI, United States

OUTLINE

Introduction	569	Application of Genome and Epigenome Editing Technologies in Livestock Breeding	595
Development of Animal Breeding	570	Conclusion	595
Epigenetic Source of Phenotypic Variation in Livestock Health and Production	571	Summary	595
Epigenetic Mechanisms Regulate Livestock Reproduction and Growth	573	Livestock Epigenetics Case Study	595
Parental Nutritional Impacts on Livestock Epigenome and Production	575	Objective	596
Epigenetic Mechanisms Regulate Livestock Immune Response and Health Outcomes	576	Scope	596
		Audience	596
Epigenetic Mechanisms Impact Lipid Metabolism and Livestock Products	583	Rationale	596
		Expected Results and Deliverables	597
Revisiting Animal Breeding Planning and Management: The Role of Epigenetic Mechanisms	587	Safety Considerations	597
		Results	598
Transgenerational Epigenetic Inheritance and Genomic Imprinting	589	Challenges and Solutions	598
		Learning and Knowledge Outcomes	598
Biomarker and Therapeutic Potentials of Epigenetic Marks in Livestock Management	590	Acknowledgments	598
The Role of Epigenetic Mechanisms in Livestock Breeding	591	Glossary	598
		References	599

INTRODUCTION

A substantial challenge currently facing the food animal industry is developing strategies to secure sustainable global food supply and safety, and to meet the demands of an ever-increasing population. To overcome this challenge, solutions must be devised to improve the production of food, including animal products globally. Thus, new knowledge on factors impacting food production and quality, such as deposition of adipose tissue in beef cattle, milk quality in dairy cattle, egg quality in chickens, and so on, would be valuable to animal industries to improve production efficiency and meet consumer demands. Current breeding strategies account for only a portion of the phenotypic variance in a trait, while the elusive portion could be due to other factors, including epigenetics. Adequate ascertainment of this elusive portion of variation will ensure continuous gains in livestock traits of economic importance. Changes in epigenetic mechanisms, including DNA methylation, RNA methylation, chromatin remodeling, histone modification, and noncoding RNA regulation impact human health, livestock well-being, reproduction, health and productivity which is an indication that these factors contribute to the phenotypic outcome. Therefore, continued improvement of genetic gain of economically important traits in livestock will require detailed characterization of functional genomic and epigenomic variations, which will ultimately link all manner of DNA sequence variations, epigenetic modifications/variations, and their interaction with the genome and environment, and application in livestock improvement programs.

DEVELOPMENT OF ANIMAL BREEDING

Present-day advanced animal breeding and animal care, and exploitation for human purpose started as a simple act with animal domestication thousands of years ago. Before the birth of the subject area "genetics" and the development of advanced breeding tools, humans used intuitive selection criteria of visually favoring animals showing best performance in desired traits largely without knowledge of the underlying mechanisms. Consequently, different breeds of farm animals with distinct traits were developed in diverse environments. Notably, Robert Bakewell (1725–1795), a British farmer, was the first to practice modern animal breeding through careful record keeping and selection of superior animals as parents of the next generation [1]. The science behind animal breeding started to take shape with the rediscovery of the work of the Austrian Monk, Gregor Mendel (1822–1884), in 1900 and a broader understanding of the inheritance of distinctly visible traits [2]. At the beginning of the 20th century, conflicting ideas regarding simple Mendelian inheritance and inheritance of traits showing continuous variation (e.g., body weight, milk production), as well as the theory of inheritance of acquired phenotypes through nongenetic inheritance proposed by Jean-Baptiste Lamarck (1777–1829) raged on for several decades [3]. Modern quantitative genetics started with the works of Ronald A. Fisher (1890–1962), Sewall Wright (1889–1988), and Jay L. Lush (1896–1982) as exemplified when Fisher statistically demonstrated a correlation between Mendel's principles of inheritance and accumulative factors on a continuous scale between relatives [4,5]. This led to the development and application of animal breeding methods in the 1930s and 1940s [6,7]. In 1942, Waddington introduced the term "epigenetics" [8] to signify the whole complex of developmental processes that lie between the genotype and the expression of the phenotype, which denotes the nongenetic component of variation but was completely ignored by animal breeders.

The desire to increase productivity by rapid dispersal of valuable genes from top genetic merit males led to the development of artificial insemination in the 1950s, which has improved the genetic quality of livestock breeds worldwide [9]. To interpret and apply data that was being generated, statistical methods that accounted for environmental and genetic factors were developed, including BLUP (best linear unbiased prediction) and REML (residual or restricted maximum likelihood), in the 1960s–1980s (reviewed in [5]). Using these approaches and their developments, estimated breeding values (EBVs) of sires were predicted based on measurements in own performance, pedigree, and sibling and application in breeding programs.

The development of molecular genetic techniques from the 1970s enabled the application of genetic markers in genetic improvement and relied on the ability to genotype individuals for specific or limited loci; starting the notion of marker assisted selection (MAS) [5,10]. MAS was limited to markers with major effects which prompted linkage mapping studies following the development of microsatellite markers and detection of quantitative trait loci (QTL) from the 1980s onwards. Today, with the application of single nucleotide polymorphisms (SNPs) and other genomic variations in genome wide association studies (GWAS), thousands of QTLs have been detected for many livestock traits (https://www.animalgenome.org/).

Advances in next generation sequencing and genotyping techniques, development of bioinformatics tools and the publication of the genome sequences of chicken [11], cow [12], horse [13], pig [14], goat [15], sheep [16], salmon [17], catfish [18] and buffalo [19,20], etc., started a new era in livestock genetics and breeding. The generation of thousands of genetic markers mainly in the form of SNPs provided the possibility to predict breeding values from dense markers, unlike previous traditional data used in evaluations. Thus, genomically best linear unbiased prediction (GBLUP) was developed under the principle that quantitative traits are controlled by a large number of loci that contribute equally to the trait implying homogeneous variance among SNPs [21]. Genomic estimated breeding value (GEBV), which is the sum of the effects of dense genetic markers or their haplotypes across the entire genome, were predicted and applied in selection programs [22]. Thus, genomic selection, which refers to selection decisions based on GEBV, has revolutionized livestock breeding; with wide application in dairy cattle [23,24]. Gains resulting from genomic selection include reductions in generation interval and cost of progeny testing as GEBVs are available early in life [25–27] (Box 29.1). The theory of nongenetic and more specifically epigenetic inheritance in all these developments was however completely ignored, but recognition of the contribution of epigenetics to phenotypic variation is increasing in recent years [28–36]. The historical development of the animal breeding act has been discussed extensively by several authors [2,3,5,9,10,23,26,27] and summarized, as well as the evolution of the science of genetics and epigenetics in Fig. 29.1. It is expected that, further management and technological advances including developments in gene editing approaches will continue to support increased gains in animal breeding programs.

Today, data for genomic selection is drawn from associations between dense makers and complex traits (milk production, health, reproduction and growth traits, etc.) mostly generated through large genome-wide association studies (GWAS). Mammalian genomes are complex and phenotypic expression of complex traits is under the control of many genes contributing small effects and environmental factors. Most GWAS data (considers mainly SNP effects) are able to detect only a part of the phenotypic variance or heritability in traits. To ensure continuous gains in livestock traits of economic importance, the elusive portion of variation must be identified and adequately accounted for. The elusive portion of variation is probably due to a high number of variants of smaller effects yet to be identified: rare or low frequency variants with possibly large effects which are poorly represented in available genotyping arrays (genotyping arrays focus on common variants with a population frequency of $\geq 5\%$); copy number variants; gene $\times$ gene interactions; inadequate documentation of shared environment among relatives; epigenetic variation/effects and noncoding RNA regulation of phenotypic expression [41–43]. Furthermore, it has been opined that epigenetic mechanisms form part of the uncovered players in the expression of complex animal production traits and disease etiology, and are likely responsible for a portion of the missing variation in production traits and also for a portion of the heritability that is not accounted for in existing genetic assessment schemes [30]. The detailed characterization of livestock functional genomic and epigenomic variation will ultimately link all manner of DNA sequence variations, epigenetic modifications/variations, and their interaction with the environment for a holistic view of information for improved decisions and application.

EPIGENETIC SOURCE OF PHENOTYPIC VARIATION IN LIVESTOCK HEALTH AND PRODUCTION

The epigenetic landscape of the genome is influenced by an interaction of genetic and environmental factors resulting in varying phenotypic outcomes. The heritable basis underlying epigenomic variation in complex traits will pave the way toward thoughtful exploitation of the epigenetic source of phenotypic variation in livestock traits. The epigenetic sources of variation are due to epigenetic mechanisms (DNA methylation or hydroxymethylation, RNA methylation, histone modifications, chromatin remodeling, and noncoding RNA) [44–49] (Box 29.2), which respond to different types of internal and external environmental cues as directed by the underlying genetic composition (Fig. 29.2). It is well accepted that epigenetic mechanisms are major players in developmental processes, such as gametogenesis, embryogenesis, cellular differentiation, X-chromosome inactivation, telomere length regulation, genomic imprinting, and health outcomes, among others.

The epigenome of the cell, unlike the genome, is fairly unstable, as well as highly dynamic throughout life and is governed by a multifaceted interaction between genetic and environmental factors [30,34,44]. Furthermore, the epigenome is cell- and tissue-specific and is subject to change with age and other factors, including acquired factors. Moreover, the epigenome captures specific environmental circumstances and growth stages to direct phenotypic outcomes at specific moments in time [50]. The preservation of genomic and epigenomic homeostasis and a dynamic

---

BOX 29.1

Gains of genomic selection [25,27].

- Higher rate of genetic gain.
- Increased reliability of predicting breeding values.
- Higher intensity of selection.
- Shortened generation interval.
- Selection of animals possible at early age.
- Rapid genetic improvement in lowly heritable traits (e.g., fertility, lifespan, health, etc.).

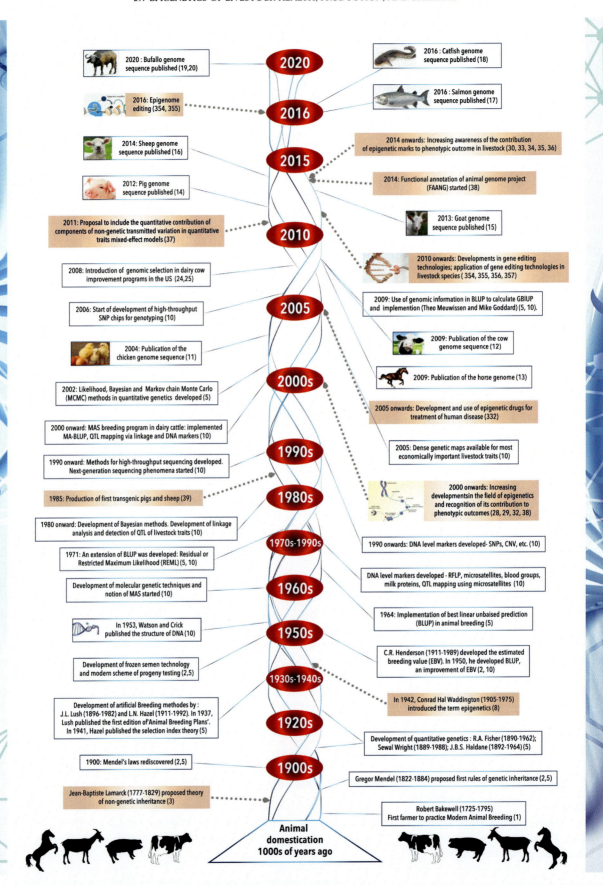

FIGURE 29.1 Historical development of the animal breeding act and millstones in genome sequencing and epigenetics research. *Further details have been presented or reviewed in References* [1–3,5,8,10–20,22,24–26,28–30,32–40].

> **BOX 29.2**
>
> **Epigenetic mechanisms [44–49].**
>
> **DNA methylation:** Involves covalent addition of a methyl or hydroxymethyl group to the 5′ position of cytosine base in the DNA sequence in a reaction catalyzed by a class of enzymes known as DNA methyltransferases. DNA methylation occurs mostly at cytosine-phosphate-guanosine (CpG) dinucleotides and to a lesser extent at CpA, CpT, or CpC dinucleotides.
>
> **RNA methylation:** This is a posttranscriptional modification involving transfer of a methyl group from an active methyl compound to another compound. More than 160 modifications have been found in mRNA and noncoding RNA species. The most common RNA methylation modifications include m^6A (N6-methyladenosine), m^5C (5-methylcytidine), m^7G (N7-methylguanosine) and 2-O-methylation. m^6A is the most abundant RNA modification in eukaryotes. Reported roles of RNA methylation include RNA splicing, stability, nuclear export, translation, initiation of miRNA biogenesiss, immunogenicity, DNA damage repair, perturbs cellular differentiation, sex determination, spermatogenesis, embryonic development, learning, memory, immune response, and several human and animal diseases.
>
> **Chromatin remodeling:** The structure of chromatin is under strict regulation by several mechanisms that encompass chromatin remodeling, histone modification, histone variant incorporation, and histone eviction, which imposes major impediments on aspects of transcription. Chromatin may exist in a relaxed state (euchromatin) which is associated with active gene transcription and expression or in a highly compacted (heterochromatin) and silenced state which is associated with repressed gene expression through hindered access of transcription factors to genes.
>
> **Histone modification:** The N-terminal tails of histones are subject to posttranslational modifications and today, over 100 distinct histone modifications including lysine methylation, lysine acetylation, serine/threonine phosphorylation, ubiquitination, sumoylation, and crotonylation have been described.
>
> **Noncoding RNA (ncRNA):** Are a class of RNA species that are not translated into proteins and mediate their functions as RNA. ncRNA generally regulate gene expression at the transcriptional and posttranscriptional levels and are involved in all biological processes investigated till date. Epigenetic related ncRNA interfere with transcription, mRNA stability, or translation and include long noncoding RNA (lncRNA), small interfering RNA (siRNAs), piwi-interacting RNA (piRNA), small nucleolar RNA (snoRNA) and microRNA (miRNA), etc.
>
> **MicroRNA (miRNA):** are a class of small noncoding RNA (18–24 nucleotides in length) which act to regulate gene expression through an RNA interference mechanism.

balance of stability and reversibility in gene expression patterns is required to ensure cell identity, maintain growth, development and health, and enable cellular response to stimuli. Deviance from this balance is highlighted by numerous associations between aberrant epigenomic perturbations and human/animal diseases (e.g., cancer, metabolic, and mental disorders) and developmental disorders [34,51]. Increasingly, it is evident that epigenetic mechanisms profoundly influence livestock health, growth and development, productivity, and phenotypic outcomes.

Epigenetic Mechanisms Regulate Livestock Reproduction and Growth

Genetic diversity and dynamic epigenetic modifications are essential elements for normal cellular growth and development, through response to environmental stimulus and involvement in numerous biological processes, organ development, and consequently supporting adaptation to various conditions and productivity. Recent advances have supported profiling of the epigenetic alterations or involvement in livestock reproduction and growth processes in response to environmental factors, paternal environment, underlying genetic composition, and reproduction technologies. The important impacts of epigenetic regulatory mechanisms on placental and embryo development of various livestock species have been discussed in detail in many reviews [52–55].

A compendium of research results on DNA methylome pattern screening identified that; (1) compared with somatic tissues, sperm DNA methylome was characterized by low methylation levels [56,57]; (2) dysregulated sperm DNA methylation and histone modifications impact male fertility and related traits [56,58–60]; (3) methylome variations between high and low motile bull sperms are enriched for genes involved in chromatin remodeling and repetitive element activities in pericentric regions [61]; (4) methylation differences is a potential regulatory mechanism underlying sperm fertility differences due to age and

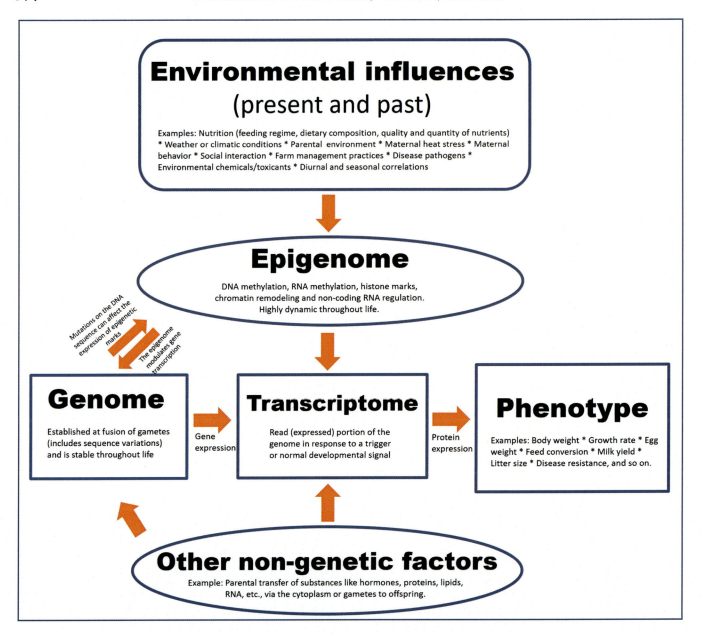

FIGURE 29.2 Phenotypic outcome is a multilevel interaction between the genome, epigenome, environmental factors, as well as other nongenetic factors. The phenotype is not only a function of an individual's DNA sequence (genome) but also a consequence of how the genome is read or regulated by the epigenotype taking into consideration both past and present environmental information.

other factors including heat and oxidative stress [62–65]; and (5) DNA methylation is involved in the regulation of repressed development of the reproduction system of inbred chickens [66]. The sperm DNA methylome patterns of 28 bulls enabled classification into conversed, variable, and highly variable methylated regions and association of highly variable methylated regions (enriched for genes with important functions in spermatogenesis and fertilization) with reproduction traits, and the variable methylated regions (enriched with genes with roles in sperm motility) with functions in sperm motility [67].

Moreover, embryo DNA methylation screening at different stages of development associated DNA methylation alterations with important regulatory roles in embryo viability and fetus development [68–71]. Maternal stressors such as metabolic disorder, heat stress and negative energy balance cause DNA methylation alterations which are associated with poor offspring performance [72,73].

In general, the use of assisted reproduction technologies to solve declining fertility problems in livestock species [74] has been linked to a portion of the epigenetic disturbance during development [31,75]. For example, the majority of embryos derived through nuclear transfer die during postimplantation development [76]. Investigations into this condition established that demethylation and de novo methylation processes are not properly accomplished in cloned embryos shortly after fertilization, as well as passive demethylation of the maternal genome during normal embryonic development [77,78]. Differential histone 4 (H4) acetylation in the blastomeres and failure of histone modification reprogramming in cloned bovine embryos has been demonstrated [79,80]. Furthermore, subtle DNA methylation abnormalities in cloned fetuses in midgestation have been observed [81].

Epigenetic impacts are not only limited to the gametes and reproductive organs but also extend to the somatic cells at all growth stages. Profiling of epigenetic processes in different porcine, bovine and chicken somatic tissues such as liver, brain, mammary gland tissues, tooth, small intestine, longissimus dorsi muscle, retina, cornea, liver, and muscle revealed that epigenetic alterations impact the growth, development, health, and productivity of these species [82–86]. Moreover, comparison of the histone H1 subtypes of five avian species (chicken, quail, gray partridge, pheasant, and duck) uncovered roles for histone modification in the development and evolution of poultry [87]. The identification of epigenomic alterations in different tissues and involvement in various processes will further understanding of the involvement of epigenetic processes in livestock reproduction, growth and development.

Parental Nutritional Impacts on Livestock Epigenome and Production

A compendium of recent data indicates that xenobiotics and dietary exposures, and deficiencies at critical periods of development modify the epigenome resulting in the incidence and severity of many human diseases including obesity, heart disease, hypertension, diabetes, and mental illness [88]. Therefore, developmental programming is the concept that parental stimulus or insult at critical periods in fetal or early postnatal development could have long-term impacts on the offspring. This perception is based on human epidemiological evidence implicating intrauterine growth restriction as an important risk factor for adult onset of diseases including hypertension and type II diabetes [89]. In food animal species, a suboptimal condition in utero during gestation is also well documented to have postnatal impacts of compromised health, slower growth rate, increased fat deposition, reduced muscle mass, and reduced meat quality in progeny [90]. Indeed, early studies in the 1960s have shown that severe undernutrition or heavy feeding in early pregnancy can lead to a decline in embryo survival [91].

Recent studies have begun to build a body of evidence of an association between maternal nutrient intake and progeny's postnatal body composition and glucose metabolism, as well as that timing and duration of nutrient modification during gestation differentially impacts outcomes. For example, it has been reported that late gestation maternal nutrition can impact postnatal body composition, insulin sensitivity, and growth rate in ruminants [92–94]. In addition, considerable evidence indicates that prepartum plane of nutrition, parental age and metabolism at time of conception impacts calf birthweight [95,96], milk production [97], calf immunoglobulin production [98], decrease protein synthesis [99], and alters offspring epigenetic marks and gene expression [100]. Although these factors may influence the survival, health, and growth rate of offspring, few studies have evaluated the long-term postnatal consequences related to livestock production traits of economic importance.

Studies in sheep and cattle have shown that maternal nutrition during different stages of pregnancy can impact embryo development, the epigenome of fetuses, postnatal progeny body composition, insulin sensitivity, and growth rate. Our data show that maternal nutrition of corn or hay in sheep had substantial effects on gene expression and DNA methylation in the fetus [101,102]. In these studies, mature pregnant multiparous ewes were naturally bred to a single sire, and from day 67 of gestation until necropsy (day 130), they were individually fed different primary energy sources. Interestingly, a maternal starch-based diet mainly altered fetal muscle development while a maternal diet with high fiber, protein, and fat concentrations mainly impacted fetal subcutaneous and perirenal adipose tissues. Pathway analysis revealed that several gene ontology (GO) terms and Kyoto Encyclopedia of Genes and Genomes (KEGG) pathways were enriched with differentially expressed genes associated with different metabolic processes and tissue and organ development. In a different study in dairy cattle, we assessed the effects of differential supplementation of one single amino acid on cows' diet before pregnancy on embryo quality and embryo gene expression profiles [103]. Two diet treatments differing slightly in methionine level were supplemented to two groups of Holstein cows from calving to embryo flushing at day 6 after synchronized ovulations. Although embryos in both treatments showed similar development and morphological appearance, their transcriptomic profiles were significantly different. Out of 10,662 genes analyzed for differential expression, 276 genes showed at least a twofold expression difference between diet treatments. In another study, poor maternal nutrition through diet restriction (60% National Research Council [NRC] recommendation) or overnutrition (140% NRC) from day 30 of gestation until day 45 or day 90 or day 135 of gestation or until parturition resulted in altered fetuses/offspring longissimus dorsi mRNA expression compared to controls [104]. Moreover, 2205 genes were found to be differentially expressed over time between days 90 and 135, and birth, corroborating observation of changes in muscle function, increased protein accretion, increased metabolic activity during myogenesis, as well as changes in the expression of genes involved in metabolic processes, protein synthesis, and cell cycle during fetal myogenesis [104]. Data generated on a Copenhagen sheep model demonstrates negative implications of fetal overnutrition and particularly late gestation undernutrition which impaired a wide range of endocrine functions regulating growth and consequently also reproductive traits in offspring [105]. Overall, these findings provide evidence that maternal diet supplementation before or during pregnancy can modulate gene expression and epigenetic changes in the offspring and that these changes can even have transgenerational effects.

Recently, increasing evidence supports that parental diets affect offspring phenotypes via epigenetic modifications such as DNA methylation. Although many epigenetic modifications that exist in sperm (e.g., chromatin structure, DNA methylation) are erased after fertilization, some modifications persist through embryonic development, supporting the hypothesis that paternal epigenetic marks can be heritable and transmitted to the next generations [106]. Interestingly, Ng and colleagues reported that high-fat diet consumption in rat sires induced increased body weight and body fat content and caused impaired glucose tolerance and insulin resistance in F1 female offspring [107]. The paternal high-fat diet also altered the expression of 642 genes in pancreatic islet cells in adult female offspring.

In a recent study, Ryan and colleagues reported the effects of increased paternal dietary intake of methyl donors, supplemented for 6 weeks before mating, on cognitive and neural functions of the offspring [108]. They found that mice born to fathers fed methyl donor supplements expressed hippocampus-dependent spatial learning deficits, decreased hippocampal synaptic plasticity, and impaired contextual fear conditioning. Interestingly, mice born to fathers fed the methyl-donor supplements also expressed increased DNA methylation patterns and decreased expression of the *KCNMB2* gene involved in neural learning and memory regulation [108]. Furthermore, this study revealed that the susceptibility to diet-mediated epigenetic modifications increased during specific critical windows of development, such as the period before mating. Morgan et al. [109] reported that a low protein diet fed to male mice 7 weeks before insemination impaired F1 neonatal and adult offspring renin-angiotensin system activity. The effects of the paternal diet were also observed in the growth of the F2 generation. Interestingly, the authors found that the expression profiles of several epigenetic regulators of RNA methylation, DNA methylation, and histone modifications were altered in the testes of adult F1 male offspring [109]. These studies have established that paternal diet influences epigenetic modification in the sperm, which alters DNA methylation and gene expression and leads to altered phenotypes in the offspring. However, most of these studies have been conducted in mice and rats. There are a few reports in sheep and cattle on the impact of paternal nutrition on sperm traits, and those are limited to one generation [110,111]. In a recent study, we evaluated the impact of pre-pubertal diet in Polypay rams (a composite sheep breed constituted in the United States in the late 1960s) on production and reproduction traits, DNA methylation, and transmission to offspring [112]. The paternal genome's susceptibility to epigenetic modifications increased during specific critical windows of development, such as puberty. Therefore, from weaning until puberty, rams were separated once daily for diet administration. Half of the rams (12 twin rams) were fed the control diet (basal concentrate diet), and the other half (12 twin rams) were fed a treatment diet (basal concentrate diet plus 0.22% added rumen-protected methionine). Methionine-treated rams reached puberty at 22 weeks of age, significantly different from control rams that reached puberty at 24.8 weeks ($P < .0001$). A total of 10 rams were chosen for breeding to produce the F1 generation. For these rams, a total of 824 differentially methylated cytosines (DMCs) and 216 differentially methylated regions (DMRs) were found in the sperm DNA. The diet of F0 rams had a significant association with scrotal circumference and weight at puberty of F1 offspring. Overall, we demonstrated through this study that paternal methionine supplementation altered DNA methylation at different genomic regions in gametes and affected the offspring's phenotypes [112].

Epigenetic Mechanisms Regulate Livestock Immune Response and Health Outcomes

The findings of numerous investigations have demonstrated a role for epigenetic modifications in the immune response to disease pathogens and other stressors, as well as involvement of the epigenetic state of cells and organs in the development of livestock diseases [34,113]. Experimental challenge of the bovine mammary gland with pathogenic *Escherichia coli* (*E. coli*) bacteria resulted in the remethylation of a hypomethylated region of the upper promoter of alpha S1 casein gene and consequently shutdown of alpha S1 casein synthesis [114]. In the peripheral blood cells of clinical mastitic Chinese Holstein dairy cows, hypermethylation of the CD4 gene promoter, which also repressed its expression, was demonstrated in cows with clinical mastitis [115,116]. During an infection of Mycobacterium bovis, DNA methylation was reported to alter gene expression in CD4 + T cells [117]. Lipopolysaccharide (LPS) stimulation of peripheral blood mononuclear cells from healthy calves resulted in differential expression of HDAC6, HDAC7, and DNMT3A genes while treatment with trichostatin A, a histone deacetylase inhibitor, significantly inhibited the expression of three proinflammatory cytokines (TNF, IL2, and IFN); thus, suggesting an important role for the measured epigenetic enzymes in the regulation of bovine innate immune gene expression [118]. Stimulation of endometrial or mammary epithelial cells with bacterial LPS resulted in altered expression of genes related to the immune response and milk production (IL-6, IL-8, ACACA, ACSS2, and S6K1) [119,120]. Moreover, synergistic effects of co-stimulation of mammary epithelial cells with LPS, peptidoglycan and lipoteichoic acid resulted in decreased methylation levels (DNA hypomethylation) and consequently increased gene expression (transcriptome changes) and inflammatory responses, and decrease in lactation performance compared to stimulation with LPS only [121].

Several reports have identified a plethora of DMRs in cow mammary gland tissues, peripheral blood, and porcine mammary epithelial cells in response to mastitis caused by Staphylococcus aureus (S. aureus) and or E. coli [84,122–124], indicative of important regulatory functions of DNA methylation in mammary gland immunity. In goats, promoter hypomethylation was found to regulate the expression of genes with roles in inflammation and apoptosis in the liver during subacute ruminal acidosis precipitated by high-concentrate diets [125]. Meanwhile, altered DNA methylation state of genes with roles in transportation and signaling and scrapie pathogenesis were reported in sheep naturally infected with scrapie [126]. DNA methylation has been suggested as a possible mechanism regulating immune response differences between two chicken lines differing in genetic resistance to multiple pathogens, following enrichment results of DMR genes identified when the two lines were compared [127]. Examining DNA methylation patterns in chicken Marek disease (MD)-resistant line 6$_3$ and MD-susceptible line 7$_2$ at 21 days after Marek disease virus type 1 infection, Tian et al. [128] identified abundant infection-induced DMRs between the two groups, with more DMRs in line 7$_2$ than in line 6$_3$. Similarly, a high number of DMRs were detected in chickens immunized with infectious laryngotracheitis vaccine compared with unvaccinated birds [128]. Avian influenza virus infection was found to increase the global DNA methylation levels of immune organs (e.g., thymus and bursa) of chickens with avian influenza virus infection [129]. Meanwhile, differentially methylated peaks in blood methylome were found in the promoter regions of immune related genes in Salmonella enterica-infected chickens [130]. The level of promoter methylation of the BPI gene in three breeds of pigs (Yorkshire, Sutai, and Meishan) associated negatively with its gene expression, contributing to intestinal immunity and disease susceptibility [131]. Moreover, promoter methylation of PACSIN1 down-regulated its expression, which indirectly promoted the production of IL6, IL8, and TNFα, demonstrating potential roles in mediating porcine responses to disease pathogens [132].

RNA methylation, especially m^{6}A modification known to play roles in various immune and inflammatory responses to disease pathogens has been extensively studied in humans, while limited reports are available in livestock. Studying m^{6}A RNA methylation modification in porcine intestinal epithelial (IPEC-J2) cells exposed to Clostridium perfringens beta2 (CPB2) toxin, Zhang et al. [133] identified 1448 differential m^{6}A RNA modifications (437 upregulated and 1011 down-regulated) between CPB2 toxin-exposed and nonexposed cells. Combining m^{6}A sequencing and RNA sequencing results, the authors [133] found that genes with up-regulated m^{6}A peaks but with down-regulated expression were mainly enriched in Wnt signaling pathways suggesting that this pathway may be impacted by m^{6}A methylation in CPB2-induced IPEC-J2 cells. In another study, 16,691 m^{6}A peaks within 7066 mRNA transcripts and 10,029 peaks within 4891 mRNA transcripts were found in nontreated bovine mammary epithelial cells (BMEC) and in BMEC treated with inactivated E. coli for 24 hours, respectively [134]. Comparing the two groups, the authors [134] observed that 474 hypermethylated mRNAs and 2101 hypomethylated mRNAs in the E. coli group were mainly enriched in NF-κB, MAPK, and TGF-β signaling pathways, indicative of a role for m^{6}A RNA modification in bovine mastitis.

In addition to DNA and RNA methylation, histone modifications and miRNAs also regulate the host response to disease pathogens. Using a genome-wide approach to determine histone H3K27me3 modifications on blood lymphocytes in lactating Holsteins, He et al. [135] reported a blueprint of bovine H3K27me3 marks that mediate gene silencing, as well as indications that H3K27me3 plays its repressed role mainly in the regulatory regions of genes in bovine lymphocyte. Assessing livers from cows with experimentally induced E. coli mastitis, it was established that the chromatin at the promoters of genes encoding TLR2, TLR4, LBP, and HP were opened up 24 h after infection, which led to their enhanced mRNA expression [136]. Integration of RNA sequencing (mRNA-Seq and miRNA-Seq) and chromatin immunoprecipitation sequencing (ChIP-seq) data from *Mycobacterium bovis*-infected macrophage indicated that discrepant distribution of H3K4me3 and RNA polymerase II at key immune genes was responsible for alveolar macrophage transcriptional reprogramming [137]. Altered gene expression in porcine alveolar macrophages following stimulation with poly I:C or LPS was found to associate significantly with H3K27ac alteration at active regulatory regions enriched for transcription factor binding motifs of transcription factor genes with functions in the inflammation response [138]. In chickens, Marek disease virus was responsible for temporal chromatin signatures in the bursa of Fabricius chickens at different stages of Marek disease development, including enrichment of altered H3K27me3 signatures in pathways related to the immune response [139,140]. These data demonstrate that the presence of pathogens or its ligands changed the DNA methylation status or chromatin modification of cells or organs with consequences on the immune response and productivity. Specific examples of altered epigenetic modifications in response to disease pathogens and effects on the immune response and health were summarized recently [34] (Table 29.1).

There is growing evidence that epigenetic mechanisms can also change genome activity under some environmental conditions like heat stress, transportation,

TABLE 29.1 Epigenetic Changes due to Disease Pathogens and Other Stress Factors Impacting Livestock Health

Stimulus or Disease	Breed	Changed Epigenetic Mark and Site	Reference
Disease Pathogens			
NDV infection under heat stress condition	Fayoumi and Leghorn chickens	Greater differences in histone modification (H3K27ac and H3K4me1) levels in bursa of Leghorns than Fayoumis, associated genes enriched in biological processes gene ontology terms related to cell cycle and receptor signaling of lymphocytes	[141]
E. coil and *S. aureus* induced mastitis	Holstein cows	Genome wide DNA methylation profiles of mammary gland tissues and peripheral blood revealed a plethora of DMRs	[84,122,124,142]
Bacteria colonization in intestine	Danish Landrace × Large White × Duroc pigs	Changed genome wide DNA methylation in immune genes at distal small intestine	[143,144]
Mycobacterium avium subsp *paratuberculosis* (Johne's disease)	Holstein cows	A total of 11,263 DMCs and 1259 DMRs in ileum tissue, 62,459 DMCs and 8086 DMRs in ileum lymph node tissues reported in response to *Mycobacterium avium* subsp *paratuberculosis*	[145]
Mycobacterium bovis infection	Alveolar macrophages (in vitro)	H3K4me3 is more prevalent in chromatin, at a genome-wide level in alveolar macrophages of infected group	[137]
E. coli challenge	Sows mammary epithelial cells (in vitro)	561 and 898 DMCs identified at 3 h and 24 h after *E. coli* challenge of mammary epithelial cells	[123]
LPS challenge	Mammary epithelial cell (in vitro)	High LPS dose induced hypomethylation of immune-related genes, while low LPS dose induced hypermethylation of lactation-related genes	[119]
Porcine epidemic diarrhea virus	Large white piglets	Higher H3K4me3 enrichment and expression levels of some antiviral genes in jejunum of infected group	[146]
Clinical mastitis	Holstein cows	Changed DNA methylation regulated *IL6R* transcription in mammary gland tissues	[147]
LPS challenge	Cows	Decreased methylation of specific CpG sites in IL-6; increased expression level of IL-6, IL-8 and DNMTs in endometrial cells	[120]
Subacute ruminal acidosis	Dairy goat	Reduced promoter methylation level of LOC101896713 and *CASP8* in liver	[148]
Bovine viral diarrhea virus	Madin-darby bovine kidney cell (in vitro)	*DNMT1* silencing induced decreased methylation level of miR-29b and repressed bovine viral diarrhea virus replication in Madin-darby bovine kidney cell	[149]
Poly I:C stimulation	Dapulian and Landrace pigs	5827 DMRs, and 70 DM and DE genes identified in peripheral blood mononuclear cell	[150]

(Continued)

TABLE 29.1 (Continued)

Stimulus or Disease	Breed	Changed Epigenetic Mark and Site	Reference
Porcine reproductive and respiratory syndrome virus	Transgenic pigs	*HDAC6* overexpression enhances resistance to PRRSV infection in serum (in vitro) and lung tissues (in vivo)	[151]
Salmonella enterica infection	Domestic chickens	879 DMRs identified in blood, including 135 DMRs in the promoter regions	[130]
Mycobacterium bovis infection	Holstein-Friesian cattle	Genome wide DNA methylation profile of CD4 + T cells revealed 760 DMRs	[117]
Clinical mastitis	Holstein cattle	Altered methylation of *CD4* in blood	[115,116]
E. coli induced mastitis	Holstein cows	H3K27me3 levels in silent genes were higher in lymphocytes from subclinical *S. aureus* mastitis cattle than in healthy cows	[152]
H5N1 influenza virus	BWEL-SPF chicken	Significantly upregulated total DNA methylation levels in the thymus and bursa of Fibricius of infected group compared with noninfected group	[129]
Marek disease	Two specific-pathogen-free inbred lines of White Leghorn chicken	Different H3K27me3 marks associated with immune-related genes identified in bursa of Fibricius	[139,140]
Clostridium perfringens beta2 (CPB2) toxin	Porcine intestinal epithelial cells (IPEC-J2)	A total of 1448 m^6A RNA modification sites (437 upregulated and 1011 down-regulated) were significantly different between CPB2 toxin exposed cells and nonexposed cells	[133]
E. coli challenge	Bovine mammary epithelial cells	A total of 2508 genes with m^6A RNA modifications were differentially expressed between bovine mammary epithelial cells challenged with *E. coli* compared with nonchallenged cells.	[134]

Other Stress Factors

Weaning stress	Piglet	2674 differentially methylated regions identified in peripheral blood mononuclear cells between two groups of weaned piglets exhibiting high and low stress responses (measured by cortisol concentration and lesion scores)	[153]
Heat stress through embryonic thermal manipulation	Broiler chickens	785 H3K4me3 and 148 H3K27me3 differential peaks identified in hypothalamus and muscle tissues	[154]
Heat stress	Hu sheep	Increased m6A methylation of RNA of *HSP* in liver	[155]
Hypoxic stress (different breeds and high [H]/low [L] altitude)	Tibetan (T) and Yorkshire (Y) pigs	6829 (TH vs YH), 11997 (TL vs YL), 2828 (TH vs TL), and 1286 (YH vs YL) DMRs identified in heart tissue	[156]
Hypoxic stress (different species and highest plateau)	Tibetan goat, Tibetan sheep, Chuanzhong goat, and Small-tailed Han sheep	Higher methylation in *HIF-1α*, *HIF-3α*, and *EPO*, and lower methylation rate in *FIH-1* observed in various tissues (heart, liver, lungs, kidney, muscle, and brain)	[157]
Prenatal transportation stress	Brahman cows and their offspring	Detected were 7407 hyper- and 8721 hypo- methylated CpG sites, including 1205 DMCs located within promoter regions in blood of calves from prenatal stressed cows	[158]
Heat stress through embryonic thermal manipulation	Chicken; PB-2 and NN lines	Altered methylation at *HSPs* in brain tissues (hypothalamus)	[159,160]
Heat stress	Male DLY pigs	57147 DMRs corresponding to 1422 DMGs identified in *longissimus dorsi* muscles	[161]

H3K4me3, tri-methylation at the 4th lysine residue of histone H3 protein; *H3K27me3*, tri-methylation at the 27th lysine residue of histone H3 protein; *DMR*, differential methylated region; *DMG*, differential methylated gene; *HSP*, heat-shock protein; *DMC*, differential methylated CpG; *TH*, Tibetan pigs grown in the highland; *TL*, Tibetan pigs grown in the lowland; *YH*, Yorkshire pigs grown in the highland; *YL*, Yorkshire pigs grown in the lowland; *E. coli*, Escherichia coli; *S. aureus*, Staphylococcus aureus; *LPS*, lipopolysaccharide; *PRRSV*, Porcine reproductive and respiratory syndrome virus; *DM*, differentially methylated; *DE*, differentially expressed; *NDV*, Newcastle disease virus; *H3K27ac*, acetylation at the 27th lysine residue of histone H3 protein; *H3K4me1*, the mono-methylation at the 4th lysine residue of the histone H3 protein.

Source: Modified from Wang M, Ibeagha-Awemu EM. Impacts of epigenetic processes on the health and productivity of livestock. Front Genet. 2021;11:1812.

and toxicants. With changing global temperatures, heat stress is amongst factors that negatively impact animal production and health, and important roles of epigenetic factors in the regulation of host response to heat stress have emerged. DMRs in genes with functions in cellular defense, energy and lipid metabolism, and stress responses were reported in longissimus dorsi muscles of heat-stressed pigs [161]. DNA methylation and histone H3K27me3 and H3K4me3 alterations were identified as regulatory factors of chickens' adaptation to embryonic thermal manipulation, which could improve thermal adaptability to heat stress in later or postnatal life [154,159,160]. An important environmental stressor, such as hypoxic stress, affects porcine growth, especially in high-altitude regions. An epigenetic connection to hypoxic stress response was found when some genes containing DMRs in heart tissues of Tibetan pigs from high and low altitude regions were significantly enriched in hypoxia-inducible factor (*HIF*) 1 signaling pathway indicative of impact on hypoxia-related processes [156]. Moreover, higher methylation levels in the hypoxia genes, *HIF-1a*, *HIF-3a*, and *EPO* and lower levels in *HIF-1*, were reported in the liver, kidney, heart, lungs, brain, and muscle tissues of plateau goat and sheep, indicative of a modulatory role of epigenetic processes in hypoxia resistance of plateau animals [157]. In cattle, transportation was shown to induce maternal stress, which in turn induced methylome changes in calves [158]. The methylome changes were elevated DNA methylation levels at the promoter regions of genes enriched in pathways related to stress response, behavior, immune response, and metabolism in blood from prenatally stressed bull calves compared to non-prenatally stressed calves [158].

Recent investigations have highlighted the important roles of noncoding RNA regulation in farm animal diseases, as well as their biomarker and therapeutic potentials, which have been summarized in several reviews [162–165]. Recently, Do et al. [162] cataloged several miRNAs of potential importance in bovine, swine, chicken and small ruminant diseases (Table 29.2). Specifically, the Table 29.2 summarized the association of specific

TABLE 29.2 Potential miRNA Biomarkers for Various Livestock Diseases[a]

Disease	Pathogen	Phenotype or Tissue or Cell Studied	Changed or Potential miRNA Biomarkers	References
Cattle Diseases				
Mastitis	*Streptococcus uberis*	BMEC[b]	miR-200c, miR-210, miR-193a, miR-29b-2, miR-130a, miR-98, let-7b, miR-24-2, miR-128-2, let-7d, miR-128-1, let-7e, miR-185, miR-652, miR-494, miR-2342, miR-29c, miR-29e, miR-29b-2, miR-100, and miR-130	[166]
		BMEC	miR-181a, miR-16, and miR-31	[167]
		Blood	miR-25, miR30e-5p, miR-342, miR-191, miR-399b, miR-451, and miR-486	[168]
	Staphylococcus aureus and *Escherichia coli*	BMEC	miR-2339, miR-21–3p, miR-423–5p, miR-499, miR-92a, miR-193a-3p, miR-23a, miR-99b, miR-21–3p, miR-193a-3p, miR-365–3p, miR-30c, and miR-30b-5p	[169]
	Streptococcus agalactiae	Mammary gland tissue	miR-223, miR-16, miR-136, miR-136, miR-3660, miR-335, and miR-378	[170]
	E. coli and *S. aureus*	Mammary gland tissue	miR-144, miR-451, and miR-7863	[171]
	Streptococcus agalactiae	Milk	miR-21, miR-146a, miR-155, miR-222, and miR-383	[172]
	CMT[c]	Milk	let-7i, miR-21, miR-27, miR-99b, miR-146, miR-147, miR-155, and miR-223	[172]
Bovine tuberculosis	*Mycobacterium bovis*	Lung	miR-142–5p, miR-146a, and miR-423–3p	[173]
Johne disease	*M. avium* subsp. *paratuberculosis*	Blood	miR-19b, miR-19b-2, miR-1271, miR-100, miR-301a, miR-32, miR-6517, and miR-7857	[174]
		Ileum	miR-146b, miR-196b, miR-2483–5p, miR-133b, miR-1247–5p, miR-184, miR-202, miR-105a, novel-53, miR-433, miR-2400, miR-137, miR-424–3p, and miR-138	[175]
		Serum	miR-1976, miR-873–3p, miR-520f-3p, and miR-126–3p	[176]
		Feces	miR-223, miR-19b, miR-27b, miR-30d, miR-24, and miR-16	[177]

(Continued)

TABLE 29.2 (Continued)

Disease	Pathogen	Phenotype or Tissue or Cell Studied	Changed or Potential miRNA Biomarkers	References
Diarrhea	Bovine viral diarrhea virus	Serum	miR-423–5p, and miR-151–3p	[178]
Foot and mouth disease	Foot and mouth disease virus	Serum	miR-17–5p, miR-31, and miR-1281	[179]
Swine diseases				
Porcine reproductive and respiratory syndrome	Porcine reproductive and respiratory syndrome virus	Porcine alveolar macrophages	miR-30a-3p, miR-132, miR-27b, miR-29b, miR-146a, and miR-9–2	[180]
		Blood monocytes and porcine alveolar macrophages	miR-181	[181]
			miR-125b	[182]
			miR-23, miR-378, and miR-505	[183]
		MARC-145 cell	miR-145, and miR-127	[184]
		Lung	miR-183, miR-219, miR-28–3p, and miR-143–3p	[185]
		Lung	miR-26	[186]
		Lung	miRNA-30c	[187]
		Lung	miR-22	[188]
		Lung	miR-373	[189]
		Alveolar macrophages	miR-140, miR-92b, miR-545, miR-1306, miR-374b, and miR-199b	[190]
		Alveolar macrophages	miR-10a-5p	[191]
		Blood	miR-125b and miR-145–5p	[192]
Swine influenza infection	Influenza A virus	*In silico*	miR-124a and miR-145	[193]
		Influenza A virus subtype H1N2	miRNAs miR-15a, miR-21, miR-146, miR-206, miR-223, and miR-451	[194]
Multiple diseases	*Salmonella* species	Whole blood	miR-155	[195]
		Intestines	miR-29a	[196]
	Lawsonia intracellularis	Intestines	miR-486, miR-500, miR-127, miR-215, miR-194b-5p, and miR-122	[197]
	E. coli F18	Intestines	miR-143, let-7f, miR-30e, miR-148a, miR-148b, miR-181a, miR-192, miR-27b, miR-15b, miR-21, miR-215, and miR-152	[198]
		Duodenum	miR-196b, miR-499–5p, and miR-218–3p	[199]
	Trichuris suis	Serum	let-7d-3p	[200]
	Actinobacillus pleuropneumoniae	Lung	miR-664–5p, miR-451, and miR-15a	[201]
	Porcine cytomegalovirus	Macrophages	miR101, miR-7, miR-128, miR155–5p, miR-196–5p, miR-18a, miR-19b, and miR-24–3p	[202]
	African swine fever virus	Spleen and submandibular lymph node	miR-126–5p, miR-92c, miR-92a, miR-30e-5p miR-500a-5p, miR-125b, miR-451, and miR-125a	[203]
	Influenza A virus	Lung	miR-15a, miR-18a, miR-21, miR-29b, and miR-590–3p	[204]

(Continued)

TABLE 29.2 (Continued)

Disease	Pathogen	Phenotype or Tissue or Cell Studied	Changed or Potential miRNA Biomarkers	References
Chicken Diseases				
Marek disease	Gallid herpesvirus 2	Spleen and liver	miR-221, miR−140, miR−199, miR-181a, miR−146b, miR−146c, miR−26a, and miR-21	[205]
		Spleen	miR-15, miR-456, and let-7i	[206]
		Spleen and liver	miR-103 and miR-219b	[207,208]
	Marek disease virus	Bursa samples	miR-30a, miR-1662, miR-9−1, miR-9−2, miR-499, miR-193b, and miR-1684a	[209]
Avian leukosis	Avian leukosis virus	Liver	miR-221, miR-222, miR-1456, miR-1704, miR-1777, miR-1790, miR-2127, let-7b, let-7i, miR-125b, miR-375, and miR-458	[210]
		Liver	miR-221, miR-193a, miR-193b and miR-125b, miR-221, and miR-222	[211,212]
		Liver	mir-34b-5p, let-7b and let-7i	[213,214]
		Chicken embryo fibroblasts	miR-184−3p, miR-146a-3p, miR-146a-5p, miR-3538 and miR-155	[215]
Bursal disease	Bursal disease virus	DF-1 cells	miR-9, miR-2127, and miR-130b	[216−218]
Avian influenza	Avian influenza viruses	Lung and trachea	miR-146, miR-15, and miR-21	[219]
		Lung	miR-34a, miR-122−1, miR-122−2, miR-146a, miR-155, miR-206, miR-1719, miR-1594, miR-1599, and miR-451	[220]
		Embryo fibroblasts	miR-146c, miR-181a, miR-181b, miR-30b, miR-30c, miR-30e, miR-455, miR-1599, and miR-1416	[221]
Chronic respiratory diseases	*Mycoplasma gallisepticum*	Lung	miR-8 family, miR-499 family, and miR-17 family	[222]
		Cell (DF-1)	miR-99a	[223]
		Chicken embryonic lungs and DF-1 cells,	miR-19a	[224]
Small Ruminants Diseases				
Cystic Echinococcosis	*Echinococcus granulosus*	Intestine	miR-21−3p, miR-542−5p, miR-671, miR-134−5p, miR-26b, and miR-27a	[225]
Enzootic nasal adenocarcinoma	Enzootic nasal tumor virus	Tumor and para-carcinoma nasal	miR-449b-3p, miR-449a-3p, miR-133a-3p, miR-449c, miR-133b, miR-9−5p, mi miR-148a-3p, miR-296−3p, miR-873−3p, and miR-331−3p	[226]
Bluetongue virus infection	Bluetongue virus	Testis	let-7d, let-7f, miR-106b, miR-10a, miR-10b, miR-136, miR-148a, miR-17−5p, miR-191, miR-194, miR-29a, miR-29b, miR-30a-3p, miR-30b, miR-362, miR-369−3p, miR-369−5p, miR-379−5p, miR-3958−3p, miR-409−3p, miR-412−3p, miR-432, miR-493−5p, miR-541−5p, and miR-758−3p	[227]
Peste des petits ruminants	Peste des petits ruminants virus	Spleen and lung	miR-21−3p, miR-1246, miR-27a-5p, miR-760−3p, miR-320a, and miR-363	[228]
Prion diseases	Prion virus	Plasma	miR-342−3p, let-7b, and miR-21−5p	[229]
Peste des petits ruminant disease	Peste des petits ruminants virus	Peripheral blood lymphocyte	miR-150, miR-370−3p, and miR-411b-3p	[230]

(Continued)

TABLE 29.2 (Continued)

Disease	Pathogen	Phenotype or Tissue or Cell Studied	Changed or Potential miRNA Biomarkers	References
Lung infection	Small ruminant lentiviruses	Lung	miR-21, miR-148a, let-7f, let-7b, miR-99a, and miR-125b	[231]
Peste des petits ruminants virus infection	Peste des petits ruminants virus	Peripheral blood mononuclear cells	miR-204−3p, miR-338−3p, miR-30b-3p, miR-199a-5p, miR-199a-3p, and miR-1	[232]
Peste des petits ruminants virus infection	Peste des petits ruminants virus	Peripheral blood mononuclear cells	miR-218 and miR-1	[233,234]

[a]This table is modified from Do et al. [162].
[b]BMEC, bovine mammary epithelial cells.
[c]CMT, California mastitis test.

miRNAs and their potential to serve as biomarkers for specific livestock pathogens and/or diseases. These varied sources of data demonstrate the regulatory roles of miRNA in the etiology of livestock diseases unequivocally. Thus, a deeper understanding of how different disease pathogens of relevance in livestock production direct epigenetic modifications and phenotypic outcomes may guide well-versed efforts toward control, complete eradication, or treatment of livestock diseases.

Epigenetic Mechanisms Impact Lipid Metabolism and Livestock Products

Lipid metabolism and livestock products of economic importance, including milk, milk fat, milk proteins, beef, wool, eggs, etc., are under the regulation of many factors (genetics, nutrition, health, age, parity, breed, environmental conditions, management practices, etc.) including epigenetics, which can be exploited for improved productivity and or quality.

Lipid Metabolism, Milk and Its Constituents

The mammary glands of ruminants produce milk and fat which contribute enormously to human nutrition. Producing milk of desired nutritional composition (e.g., increased protein and unsaturated fatty acid contents) has generated research and commercial interests. Strategies employed to improve milk production and quality include nutrition and genetics. In addition to identified genetic markers for these traits, regulatory roles of epigenetic mechanisms on lipid metabolism and milk production have been demonstrated [30,34,235−237].

Following the initial report that factors like mastitis and milking times altered DNA methylation level around the STAT5-binding enhancer region in the CSN1S1 promoter followed by negative regulation of αS1-casein synthesis in milk during lactation [114,238], many reports of altered epigenetic marks in relation to lipid metabolism and milk production traits have emerged. For example, altered DNA methylation levels in blood or mammary gland tissues have been reported between lactating dairy cows with high and low milk yield or high and low milk protein yields [239−241]. In dairy goats, altered DNA methylation levels of genes involved in milk production, including RXRa and NPY, have been reported [242]. Involvement of m^{6}A RNA methylation in the regulation of lipid metabolism by curcumin following LPS-induced hepatic lipid metabolism disorder was reported in piglets [243]. In another study, it was found that adipocyte and preadipocyte contents of m^{6}A and m^{6}A associated enzymes were different during yak adipocyte differentiation [244]. Moreover, enrichment analysis of genes with significantly different m^{6}A peaks showed participation of some of the genes (e.g., KLF9, FOXO1, ZNF395 and UHRF1) in pathways associated with adipogenic metabolism [244].

Epigenetic factors are also known to interact in their regulatory roles. Impaired fatty acid synthesis in goat milk, due to inhibition of miR-145 expression, increased methylation levels of some lipid-related genes (FASN, SCD1, PPARG, and SREBF1) [245]. Moreover, increased promoter DNA methylation of ACACA and SCD genes decreased their expression resulting in reduced milk fat concentrations in goats in response to a high grain diet [246]. The mRNA expression levels of some DNA methyltransferase enzymes like DNMT1, DNMT3A, and DNMT3B were downregulated by increased expression of miR-148a, miR-152, and miR-29s, respectively, which also impacted bovine mammary gland epithelial cell activities and milk synthesis [247−249]. Inhibition of miR-29s expression in cow mammary gland epithelial cells triggered increased global DNA hypermethylation, and upregulated the methylation levels of the promoters of important lactation-related genes, including CSN1S1, EIF5, SREBP1, PPARG, and GLUT1, and decreased secretion of triglycerides, lactose, and lactoprotein [250]. Moreover, decreased methylation levels of CpG islands in the promoter regions of genes (e.g., *FTO*, *INS*, *IGF1*, *CAV1*, etc.), which function in the regulation of a plethora of genes with roles in metabolism and milk synthesis was due to miRNAs targeting *DNMTs* [248]. Additional examples of altered DNA methylation marks or histone modifications involved in the regulation of milk processes were summarized recently (Table 29.3) [34].

MiRNA expression in the mammary gland of farm animals also have roles in the development and maintenance of mammary lipid synthesis and lipid metabolism [280–284]. A number of highly expressed miRNAs (e.g., miR-148a, miR-26a, miR-30a-p, Let-7a-5p, miR-99a-5p, miR-21–5p, Let-7b, miR-30d, miR-200c, miR-200a, miR-191, miR-186, miR-92a, miR-24–3p, let-7f, let-7g, miR-27b, and miR-27b) reported in bovine milk, mammary gland tissues, and at

TABLE 29.3 Epigenetic alterations underlying different livestock product phenotypes[a]

Context	Organ Studied	Epigenetic Alteration/Modification	Production Trait	Reference
Milk and Component Traits				
High and low milk yielding cows	Blood	A total of 10,877 and 6617 differentially methylated regions identified between high and low milk yielding cows at dry and lactating periods, respectively. Potential candidate genes for milk traits identified were *DOCK1*, *PTK2* and *PIK3R1*.	Milk yield	[251]
High and low milk yielding Holstein cows	Jugular venous blood	DNA methylation rates in the lower-yielding cows were significantly higher than those in the higher-yielding animals	Milk and protein yield	[240]
High and low milk yielding Holstein cows	Peripheral blood mononuclear cells	72 DMRs reported between high and low milk yielding cows 252 DMRs across herd environments	Milk yield	[239]
Holstein cows at different lactation stages	Blood	DNA methylation of *EEF1D* was lower in the dry period than the early stage of lactation	Milk production	[252]
High and low milk protein producing Holstein cows	Mammary gland tissue	A total of 2420 DMCs including 708 DMCS co-located with QTLs for milk traits reported between cows producing milk with very high protein content compared with cows producing milk with very low protein content	Milk protein content	[241]
High and low milk fat producing Holstein cows	Mammary gland tissue	A total of 706 DMCs including 83 DMCS co-located with QTLs for milk traits reported between cows producing milk with very high fat content compared with cows producing milk with very low fat content	Milk fat content	[241]
Xinong Saanen goats at different lactation stages	Mammary gland tissue	Methylation levels of 95 and 54 genes in the lactation period were up- or downregulated, respectively, relative to the dry period	Milk production	[242]
Xinong Saanen goats at different lactation stages	Mammary gland tissue	Inhibition of miR-145 increased methylation levels of *FASN*, *SCD1*, *PPARG* and *SREBF1*	Milk fat synthesis	[245]
Guanzhong goat fed high or low concentrate diets	Mammary gland tissue	Increased DNA methylation level in the promoter regions of the *ACACA* and *SCD* genes	Milk fat production and composition	[246]
Holstein cows milked once or twice daily	Liver and mammary gland tissue	Altered methylation of four CpG sites within the distal upstream regulatory region of the *CSN1S1* gene	Milk production (yield and milking frequency)	[238]
Escherichia coli and *Staphylococcus aureus* induced mastitis in Holstein cows	Liver and mammary gland tissues	Remethylation of upstream promoter of *CSN1S1* gene in response to *E. coli* infection	αS1-casein synthesis	[114]

(Continued)

TABLE 29.3 (Continued)

Context	Organ Studied	Epigenetic Alteration/Modification	Production Trait	Reference
Meat Production and Quality Traits				
Chicken	Intramuscular preadipocytes and matured preadipocytes	Several differentially methylated regions (7580) and genes identified between the two groups in vitro	Intramuscular fat deposition	[253]
Angus cattle with divergent beef tenderness	Longissimus dorsi muscles	DNA methylation profiles related to beef tenderness, and 7215 DMRs between tender and tough beef	Beef tenderness	[254]
Simmental, Yunling, and Wenshan cattle breeds differing in meat-production abilities	*Longissimus dorsi* muscles	18 DM and DE genes between Simmental and Wenshan cattle, 14 DM and DE genes between Simmental and Yunling cattle, 28 DM genes between Wenshan and Yunling cattle	Meat quality	[255]
Polled Yak at three growth stages	Longissimus dorsi muscles	1344, 822, and 420 genes with DM CCGG sites, 2282, 3056, and 537 genes with DM CCWGG sites between 6 months old vs 90-day-old, 6 months old vs 3 years old and 3 years old vs 90-day-old fetuses, respectively	Muscle development and growth	[256]
Highest and lowest pH littermates of Duroc	Longissimus dorsi muscles	3468 DMRs, including 44 and 21 protein-coding genes with hyper- and hypo methylated regions in their gene bodies	Postmortem energy metabolism and pH	[257]
Pigs with high or low boar taint	Testis	32 DE genes with DMCs	Boar taint	[258]
Duroc and Pietrain (differs in metabolic characters)	Longissimus dorsi muscles	More than 2000 DMCs identified	Muscle metabolism	[259]
Qinchuan cattle (fetal and adult)	Longissimus dorsi muscles	3 DMCs in the core promoter region of *SIX1*; Histone H4 and E2F2 bind to *SLX1*	Muscle development	[260]
Small Tailed Han and Dorper × Small Tailed Han crossbred sheep differing in meat production ability	Longissimus dorsi muscles	808 DMRs and global loss of DNA methylation in the DMRs in the crossbred sheep, 12 potential DMGs	Meat production	[261]
Japanese black and Chinese Red Steppes cattle exhibiting different meat-production ability	Longissimus dorsi muscles	23,150 DMRs identified, 331 DMRs correlated negatively with expression of DE genes, 21 DMRs located in promoter regions	Muscle development and related meat quality traits	[262]
Maternal nutrient restriction in mid to late gestation of Angus x Simmental crossbred cows	Longissimus dorsi and semitendinosus muscle	Altered methylation status of a DMR in *IGF2* and differential expression of miR-1 and miR-133a	Muscle function	[263]
Castrated and noncastrated pigs Huainan pigs	Liver and adipose tissues	*GHR* methylation rate in the liver of castrated and noncastrated pigs were 93.33% and 0%, respectively	Castration-induced fat deposition	[264]
Polish Large White, Duroc and Pietrain (differs in metabolic traits)	Subcutaneous fat, visceral fat, and Longissimus dorsi muscle	H3K9ac and H3K4me3 correlated to expression level of selected genes	Adipose tissue accumulation	[265]
Gushi hens at different growth stages	Breast muscle	2714 DMRs and 378 DMGs identified	Intramuscular fat deposition and water holding capacity	[266]
Obese (Tongcheng), lean (Landrace), and miniature (Wuzhishan) pig breeds	Blood leukocytes	2807 (Tongcheng vs Landrace), 2969 (Tongcheng vs Wuzhishan), and 5547 (Landrace vs Wuzhishan) DMGs identified	Fat-related phenotype variance	[267]

(Continued)

TABLE 29.3 (Continued)

Context	Organ Studied	Epigenetic Alteration/Modification	Production Trait	Reference
Lean (Landrace) and obese (Rongchang) pig breeds	Backfat	483 DMRs identified in the promoter regions of genes	Fat deposition and fatty acid composition	[268]
Daninghe and Qingjiaoma chickens under different feeding conditions	Breast muscle	46 CpG sites and 3 CpG islands in *UCP3*, different methylation level of *UCP* and *FATP1* between groups	Breast muscle (intramuscular fat content)	[269,270]
Egg Production				
Hy-Line Brown commercial female chickens before and after reproductive maturation	Ovaries	Increased methylation of two CpG sites in ERα; Increased H3K27ac and decreased H3K36me3 related to increased ERα mRNA transcript	Reproductively mature, egg production	[271]
Betaine supplementation of laying hens	Liver	Hypomethylation of promoter in *GR*	Egg production	[272,273]
Wool Yield and Quality				
Coarse type and fine type cashmere goats	Skin	9085 DM N6-methyladenosine sites,	Cashmere fiber growth	[274]
Cashmere goats at anagen and telogen stages	Skin	Altered promoter methylation of *HOTAIR* gene	Cashmere fiber growth	[275]
Different generations of cashmere goats	Skin	336 hyper- and 753 hypo- methylated 5mC, corresponding to 214 hyper- and 560 hypo-methylated genes	Cashmere traits	[276]
Cashmere goats at anagen and telogen stages	Skin	1311 DMRs corresponding to 493 DMGs (269 hyper- DMGs and 224 hypo-methylated DMGs)	Hair cycling and cashmere growth	[277]
Cashmere goats and other goat species	Skin	Altered methylation degree of *HOXC8* exon 1	Cashmere fiber growth	[278]
Fine-type Liaoning cashmere goat and coarse-type Liaoning Cashmere goat	Skin	A total of 9085 differential RNA m^{6}A sites found between fine-type Liaoning cashmere goat and coarse-type Liaoning Cashmere goat skin samples	Cashmere fiber growth	[279]

[a]This table was modified from Wang and Ibesagha-Awemu [34]. DM: differentially methylated, DE: differentially expressed; DMR: differential methylated region; DMG: differential methylated gene; DMC: differential methylated CpG; H3K9ac: histone H3 lysine 9 acetylation; H3K4me3: tri-methylation at the 4th lysine residue of histone H3 protein; H3K27ac: acetylation at the 27th lysine residue of histone H3 protein; H3K36me3: tri-methylation at the 36th lysine residue of histone H3 protein.

different stages of the entire bovine lactation curve, and results of gene ontology and pathways analyses suggest housekeeping regulatory roles in cellular growth and development, lipid/protein metabolism, milk synthesis (e.g., fatty acid metabolism, protein synthesis, and amino acid metabolism), tissue morphology and organ development [281,285,286]. Furthermore, functional analyses have directly linked these miRNAs with regulatory roles in milk synthesis, fatty acid metabolism, and adipocyte differentiation. Down-regulation of miR-26a/b and their host genes (*CTDSPL, CTDSP2,* and *CTDSP1*) in goat mammary epithelial cells decreased the expression of genes related to fatty acid synthesis (*PPARG, LXRA, SREBF1, FASN, ACACA, GPAM, LPIN1, DGAT1,* and *SCD1*), triacylglycerol accumulation and unsaturated fatty acid synthesis, as well as a functional association between miR-26a/b, their host genes, and *INSIG1* [287]. MiRNA-24 was shown to control triacylglycerol synthesis in goat mammary epithelial cells by targeting the fatty acid synthase gene [288]. MiR-26a and miR-26b are upregulated in early adipogenesis and their inhibition prevented lipid accumulation while their overexpression accelerated it thus establishing them as key regulators of human white and brite adipocyte differentiation [289]. MiR-103 was shown to control milk fat accumulation in goat mammary gland during lactation [290], and a role for miRNA-143 in the differentiation of bovine intramuscular fat has been demonstrated [291]. These lines of evidence demonstrate unequivocally the regulatory roles of miRNA in lipid metabolism and milk traits in livestock species.

Adipogenesis and Meat

Meat quality is a trait of high interest in the beef, swine, poultry, and small ruminant industries. DNA methylation and noncoding RNA have been identified as important regulatory mechanisms modulating important meat traits such as intramuscular fat deposition and skeletal muscle development in chickens, cattle, goats, sheep, and pigs [253,255,266,292].

The whole genome-wide DNA methylation patterns of longissimus, latissimus dorsi muscles and adipose tissues from different breeds of cattle, sheep, and pigs provided insight on the epigenetic regulatory mechanisms modulating the expression of genes involved in the regulation of muscle development, obesity development, lipid translocation, lipid transport, meat tenderness and fatty acid metabolism [254,261,262,293–295]. For instance, DMRs identified in Angus cattle concerning beef tenderness were significantly enriched in ATP binding cassette subfamily and myosin-related genes (e.g., *ABCA1*, *ABCA7* and *ABCG1*) with roles in beef tenderness and fatty acid metabolism [254]. In addition, numerous DMRs and DMCs were significantly associated with differences in muscle development and meat quality in several cattle (beef) breeds [255,256] and chickens [266] or differentiated between obese and lean pigs, revealing vital roles of DNA methylation in lipogenesis [267,268] as well as roles in intramuscular fat deposition in chickens [253,266,269]. DNA methylation alterations are also reported to potentially regulate other meat quality traits like meat color, pH, and carcass traits in cattle [257,296], and boar taint in pigs [258]. Specific examples of altered epigenetic marks in relation to these traits were reported recently [34] (Table 29.3). Moreover, miRNA profiling in skeletal muscles has associated miRNAs with diverse carcass and muscle quality and development traits in various livestock species including sheep, cattle and pig [297–299].

Wool and Egg Production

The profitability of the egg industry relies on the reproductive maturation of the ovaries. A plethora of investigations have reported significant involvement of epigenetic regulatory processes in the reproductive maturation of the ovaries and egg laying performance, amongst others. Recently, epigenetic modifications identified during ovarian development and maturation in estrogen receptor alpha (*ERα*) gene with important roles in egg laying, associated altered CpG sites, higher histone H3K27ac, and lower H3K36me3 with *ERα* expression [271]. Meanwhile, altered DNA methylation patterns in response to betaine supplementation, and association with enhanced egg laying performance in hens have been reported [300]. In another study, promoter region methylation in response to betaine supplementation possibly regulated altered expression of liver transport-related and lipid synthesis genes, which consequently supported the synthesis and release of yolk precursor substances in the liver and egg laying performance [273].

The economic importance of wool is of high value in the goat industries and efforts to improve wool production includes understanding the molecular mechanisms and specifically the involvement of epigenetic factors. Several reports have implicated altered DNA methylation patterns and histone acetylation in the regulation of goat fetal fibroblast cells, which are critical for cashmere wool production and hair follicle morphogenesis as well as in wool fiber development and transformation of furs with special characters [301–304]. Further examples of changed epigenetic marks in relation to various egg and wool phenotypes are listed in Table 29.3. In addition to DNA methylation and histone modifications, RNA m^6A methylation, miRNAs and circRNAs have been implicated in the growth and development of cashmere fibers in cashmere goats [305–307]. For example, studying m^6A methylation in relation to fiber growth, Wang et al. [279] identified 9085 differential RNA m^6A sites between fine-type Liaoning cashmere goat and coarse-type Liaoning Cashmere goat skin samples, and involvement of differently methylated genes in intermediate filament and keratin filament; thus implicating m^6A modification during growth of cashmere fiber.

REVISITING ANIMAL BREEDING PLANNING AND MANAGEMENT: THE ROLE OF EPIGENETIC MECHANISMS

As discussed in the section "Development of Animal Breeding" aforementioned, developments in animal breeding have relied on systematic selection principles that considered the underlying genotype, as well as information on relatives. Advanced breeding models were developed based on these principles which have led to tremendous improvement in livestock traits. In particular, genomic breeding has accelerated dairy cow productivity, and the goal of animal breeders is to continue to improve gains. With increasing knowledge as outlined in section "Epigenetic Source of Phenotypic Variation in Livestock Health and Production" aforementioned, it is no longer debatable that sources of variation embedded in the epigenome are an untapped resource in animal breeding and management. The limitations of the inclusion of epigenomics information in livestock breeding plans as well as the research gaps were recently discussed (Box 29.3) [30,34]. To allow for applicability of epigenetics, epigenetic marks and sources of epigenetic variation must be accurately determined and analyzed. As with DNA variations, epigenetic marks may have enhancing, deleterious or neutral attributes and the potential to acclimatize and respond to environmental cues with profound effects on heredity and breeding [30].

> **BOX 29.3**
>
> **Limitations of inclusion of epigenetic information in livestock breeding plans [30].**
>
> - Limited epigenetic research activities on farm animal species as compared to extensive work on humans and model organisms.
> - Insufficient recognition of the importance of epigenome contributions to the emergence of livestock phenotypes of economic importance and disease traits.
> - Limited tools specific for livestock species. In humans, availability of epigenome maps (DNA methylation, chromatin modification state, and chromatin structures) has enabled the development of assays that support both small, large scale and genome wide epigenetic studies.
> - Limited funding for livestock epigenetic research.
> - Involvement of a limited number of researchers on a global scale.
>
> Research gaps and future perspectives of livestock epigenetics research: [34]
>
> - Develop tools for livestock epigenetics research.
> - Expand epigenetics exploration in livestock organs and tissues under varying conditions.
> - Recognize epigenetic contribution to livestock phenotype diversity.
> - Examine, document and exploit epigenetic inheritance in livestock.
> - Explore potential application of epigenetic biomarkers in livestock production and health.

Informed application of epigenetic data for breeding purposes will require a clear distinction between the nongenetic and genetic sources of variation. Largely, the principles of modern animal breeding are based on transmissible characters coded in the DNA sequence and interaction with the environment from generation to generation otherwise known as genetic inheritance. Nowadays, various lines of evidence suggest that Lamarck's theory of the transmission of nongenetic factors (i.e., not determined by the DNA sequence) from one generation to another, otherwise known as nongenetic inheritance is also a contributing factor [37,308]. Nongenetic inheritance has been defined as any effect on offspring phenotype brought about by the transmission of factors other than DNA sequences from parents or more remote ancestors [308]. Other connotations of nongenetic inheritance include soft inheritance, Lamarckian inheritance, transgenerational epigenetic effects, and non-Mendelian inheritance. Genetic inheritance encompasses mechanisms whereby the environment (internal and external) within which an ancestor's genes reside influences development in descendants. Examples include transmission of epigenetic state not limited by parent of origin, transmission of epigenetic state limited by parent of origin (genomic imprinting), transmission of cytoplasmic/somatic factors, transgenerational epigenetic inheritance, ecological inheritance, and cultural inheritance. Thus, nongenetic inheritance ensues from various interacting mechanisms including epigenetics, parental effects, ecological, and cultural inheritance (reviewed by Ref. [37]). This implies that all forms of genetic and nongenetic inheritance contribute to phenotypic outcomes.

Epigenetic inheritance has been referred to as the manner in which the environment affects the expression of the genome of an individual during development and the development of its descendants [29]. Epigenetic effects could be intragenerational (represents development and phenotypic modifications of an organism within a single generation) or intergenerational (refers to heritable states of gene expression from one generation to the next generation without alterations in the DNA sequence) or transgenerational (refers to heritable states of gene expression from one generation to the next and subsequent generations without alterations in the DNA sequence).

Parental effects have been defined as the effects that parents have on the phenotype of their offspring that are unrelated to the offspring's own genotype [309]. Thus, the parental environment can affect the phenotype of offspring through the transmission of various substances, such as proteins, lipids, hormones, RNA, and so on, which could be via gametic substances or the cytoplasm. In this case, the phenotype of the offspring does not only depend on its own genotype and environmental exposures but also its parental environment and genotype. A typical example is in birds whereby both cellular (yolk cell) and extracellular (albumin) signals loaded during egg formation directly guide gene expression of the developing embryo and are shown to modify both morphological and physiological phenotypes [310,311]. Transmission of RNA through eggs/sperms has been shown to affect the phenotype of offspring in many ways, as well as a role in transgenerational epigenetic inheritance [312,313].

In livestock, prenatal maternal conditions such as dam's body condition score, metabolic status and health, and bull's environment or condition and metabolic status have been reported to influence offspring fertility and milk production (reviewed in [99]). For example, Swartz et al. [314] used somatic cell score (SCS) as a proxy for mammary gland health to study the lactation records of 15,992 (4366 herds, first lactation) or 3570 (1554 herds, second lactation) daughters and reported a decline in fat yield of daughters born to first lactation cows with elevated SCS. Amongst the suggested underlying mechanisms, changes in the epigenome due to the presence of mastitis pathogens were suggested to play a role in controlling milk fat synthesis in the daughters [314], and therefore portray an intergenerational effect on fat yield. In livestock breeding therefore, parental nutritional imbalances and exposure of breeding animals to disease pathogens and environmental toxicants orchestrate epigenetic changes that consequently affect phenotypic outcomes in the offspring.

Transgenerational Epigenetic Inheritance and Genomic Imprinting

Transgenerational epigenetic inheritance has been discussed in detail in Chapter 25 of this volume, and the underlying mechanisms have been elaborated upon [315]. Transgenerational epigenetic inheritance occurs when germline reprogramming is perturbed enabling the stable transmission of epigenetic marks (e.g., DNA methylation, histone modification, etc.) acquired during development or imposed by the environment from one generation to the next. In livestock breeding, it is important to identify sources of variation due to transgenerational epigenetics inheritance. While several examples of transgenerational epigenetics phenomena have been reported in model organisms and humans, only a few reports abound in livestock species. An investigation of the nutritional effects of methylating micronutrients down a male large white pig line (F0, F1, and F2 generations) indicated significant differences in gene expression, carcass traits, as well as in DNA methylation at the promoter of the *IYD* gene between F0 and F2 generations, suggesting transgenerational epigenetic inheritance [316]. In dairy cows, females born to cows that were lactating while pregnant were shown to produce less milk (52 kg), have a shorter lifespan (16 days less) and were metabolically less efficient than females whose fetal life was established in the absence of maternal lactation suggesting a probable intergenerational effect on the offspring that resulted from an interaction between pregnancy and lactation [317]. It is not clear if these effects persisted beyond the first generation. It is thought that competition for nutrients for milk production and fetal needs during the peak of lactation, a period characterized by negative energy balance, could

have affected nutrient availability during a critical window of fetal germline reprogramming and consequently, the effects in later life. However, it has been observed that true transgenerational epigenetic inheritance can only be detected in F3 generation of sporadic or environmentally (toxins, nutrition, and stress) induced epigenetic changes in females or F2 of males [315].

As outlined in section "Epigenetic Source of Phenotypic Variation in Livestock Health and Production," environmental influences or the exposome, such as exposure to chemicals, nutrition, maternal behavior, pathogens, environment, management practices, climatic conditions, and so on, cause modifications in gene expression and phenotypic outcomes that persist throughout life and may be transmitted to future generations. However, the transgenerational nature of these effects have not been systematically investigated in livestock species. The occurrence of transgenerational epigenetic inheritance is important for animal breeding purposes [318], and it is expedient that these effects be documented for application in improvement breeding.

Genomic imprinting (discussed in detail in Chapter 40) is the preferential expression of one of two gene alleles in the offspring based on parental origin and is strictly regulated by epigenetic processes, notably DNA methylation, histone modifications, chromatin remodeling, noncoding RNA, and protein insulators. Genomic imprinting plays physiological roles in metabolism and body composition throughout life and as such contributes to the typical disparity and architecture of complex traits. The effects of epigenetic mechanisms on imprinted gene control of livestock growth and development, and productivity and the epigenetic consequences of artificial reproductive technologies on bovine imprinted genes have been discussed in details by several authors [319–323]. Moreover, genomic imprinting has been established in cattle, sheep, goat, pig, and hinny (horse × donkey hybrid) [319–322,324].

Imprinted genes are regulated by epigenetic modifications, such as DNA methylation, histone modification, and chromatin remodeling and the process of reprogramming the differentially methylated regions thought to regulate these genes occurs at gametogenesis and at fertilization. Thus, by reprogramming the genome, the imprints of the parental chromosomes are erased, and in each generation, new imprints are made within the offspring's genome. Naturally occurring genetic variation at imprinted gene loci and epigenetic dysregulation associated with reproductive technologies have been shown to influence economically important traits in livestock species [320,321]. Some notable effects of imprinted gene loci on livestock traits include the callipyge locus (DLK1–DIO3 imprinted domain) in sheep which is due to a single nucleotide polymorphism in the *DLK1-DIO3* imprinted gene cluster, the *IGF2* locus due to a mutation in a conserved CpG island within intron 3 of the IGF2 gene in the pig. These imprinted loci have been associated with growth, carcass, and reproduction traits in the pig and sheep (reviewed in [320,321]). SNPs in the *IGF2* gene have been associated with milk yield, milk protein and fat yield, milk fat, and protein percentage, progeny carcass conformation, and progeny carcass weight in Irish Holstein Friesian cows [325]. It has also been observed that genomic imprinting contributes meaningfully to the genetic variance of many complex traits as evidenced by estimated proportions between 8% and 25% of the total additive genetic variance [326,327]. Dysregulated expression, epigenetic mechanisms and genetic mutations at imprinted gene clusters are known to lead to disease conditions in humans such as Beckwith-Wiedemann syndrome, Prader-Willi syndrome, Angelman syndrome, transient neonatal diabetes mellitus, Cushing syndrome, McCune-Albright syndrome, progressive osseous heteroplasia, ovarian cancer, etc. [328]. Disorders are not limited to humans as *IGF2R*, *IGF2*, and *H19* are all imprinted genes that have shown irregular expression in animals suffering from Large Offspring syndrome which causes fetal deformation or death in livestock [329]. Large Offspring syndrome has been observed in animals produced by assisted reproduction technology which includes in vitro fertilization or intracytoplasmic sperm injection [329]. Therefore, due to the roles of imprinted genes in embryonic growth and development, they serve as suitable candidates for studying the effects of assisted reproduction technology on embryonic development and fetal programming. Current breeding practices do not account for the imprinting status of genes. Meanwhile, overwhelming evidence demonstrates epigenetic influences on gene imprinting and resultant phenotypic effects.

Biomarker and Therapeutic Potentials of Epigenetic Marks in Livestock Management

Integrating new technologies to enhance disease resistance and reduce disease transmission is desirable given that conventional management methods fall short of adequately tackling these challenges. Upon infection, some livestock diseases (e.g., mastitis, JD, etc.) can develop into a clinical or subclinical type of infection, depending on the causal pathogen. For most subclinical or chronic conditions, currently available diagnostic or therapeutic tools cannot adequately address the situation, resulting in economic losses due to disease persistence on animal farms. For instance, JD, a chronic incurable disease of ruminants, progresses slowly upon infection by *M. avium* spp. *paratuberculosis* and can be divided into three stages including silent infection, subclinical, and clinical/advanced disease and last throughout the life of the animal. The subclinical stage of the infection is characterized by a long incubation period lasting 4 to 6 years, depending on the host immunological status. JD diagnosis, whether based on the presence of bacterial or host antigens at the subclinical stage, is particularly challenging as currently available tests show low sensitivity and specificity [330,331]. Therefore, specific biomolecule patterns such as gene expression/variation, proteins, metabolites, and epigenetic marks (miRNA, DNA methylation, etc.) can be developed as effective biomarkers to improve disease diagnosis and therapy. The biomarker potential of epigenetic markers in livestock diseases was recently summarized [34,162].

An epigenetic biomarker, which represents a measurable epigenetic mark in body fluids or different tissues, can be used for disease diagnosis (delineate a disease condition), prognosis (predict disease outcome), prediction (response to treatment or therapy), therapy monitoring (monitor the response to treatment) and as a disease risk biomarker (forecast tool for future disease development). As shown in sections above, epigenetic mechanisms respond to various types of internal (e.g., maternal environment, etc.) and external environmental factors (e.g., disease pathogens, nutrition, management practices, etc.) and therefore represent the evolution of individual phenotype variation under specific conditions and consequently have the potential to contribute to enhanced livestock disease management. Examples of changed epigenetic marks listed in Table 29.2 can be developed and used in livestock disease diagnosis and therapy. Although there is no currently developed epigenetic test (diagnosis) or drug (therapy) for livestock disease management, several examples abound for human health management. Several first, second and third generation epidrugs (epigenetic drugs) containing DNMT inhibitors (e.g., azacytidine, zebularine and guadecitabine), histone deacetylase inhibitors (e.g., hydroxamic acid, vorinostat, belinostat, panobinostat, tucidinostat, and valproic acid), histone methyltransferase inhibitors (e.g., EPZ004777, EPZ-5676, DZNep, pinometostat, and tazemetostat), histone demethylase inhibitors (e.g., pargyline, phenelzine, and tranylcypromine), and bromodomain and extra-terminal domain inhibitors (e.g., thienotriazolodiazepines and benzodiazepines) have been used to manage several human conditions including several types of cancers and myelodysplastic syndrome [332,333]. Meanwhile, several miRNA targets are presently undergoing clinical trials for oncology therapy and therapy for other human disease conditions such as osteoporosis, liver fibrosis, Alzheimer disease and Parkinson disease, etc. [334]. An overview of the steps for miRNA biomarker development for livestock was recently proposed [162].

Recent reports show that miRNAs could be promising noninvasive biomarkers in the diagnosis of cancer and fertility. MiRNAs are secreted into the extracellular environment and are present in serum, urine, and saliva with relatively high stability in serum as they withstand freezing, thawing and pH changes [335,336]. MiRNAs are present within exosomes in the extracellular environment with the potential to act as signaling molecules [337]. As these miRNAs are dynamically expressed throughout mammalian embryonic development and have numerous mRNA targets for one miRNA, it is of great importance to better understand their roles.

Given that many other cell types have been found to secrete miRNAs, we investigated and demonstrated that several miRNAs were present in culture media of human and bovine embryos and their expression levels were correlated with embryo quality or developmental competence [338]. Assessing the milieu of miRNA in in vitro culture media from embryos of differing quality (conditioned media was collected from both blastocyst and degenerate embryo which failed to develop from the morula to blastocyst stage) by miRNA-sequencing revealed 11 miRNAs that were differentially expressed between degenerate and blastocyst embryos [339]. Several of the miRNAs identified as differentially expressed regulate genes associated with cellular proliferation. For example, overexpression of miR-146a is associated with an increase in cell proliferation in a p-ERK-dependent manner [340], and both miR-24 and miR-148a overexpression act to repress p27 thereby increasing cell proliferation [341,342]. MiRNAs present in the media may reflect developmental abnormalities due to altered embryo genetics. For example, miR-146a which was more highly expressed in the degenerate media in the study of Kropp and Khatib [339]; the CC genotype of this miRNA was reported to be more frequent in chromosomally normal spontaneously aborted human fetuses [343]. Differential secretion into culture media of miR-191 by aneuploid embryos compared to normal euploid embryos has been reported [344]. Moreover, there is ample evidence of the roles of miRNAs in cell–cell signaling [337,345]. Sohel et al. [345] demonstrated uptake of exosome-coupled miRNA from bovine follicular fluid into granulosa cells with a corresponding increase in endogenous miRNA in transfected cells and down-regulation of target genes of the exosome-coupled miRNA. Also, mimic supplementation experiments suggest that miRNAs are taken up by the embryo and that extracellular miRNAs affect embryonic development [339]. Thus, the presence of miRNAs in the culture media of in vitro fertilized embryos and the roles they play in early embryonic development and cell communication make them favorable candidates as promising noninvasive biomarkers of embryo quality and pregnancy.

Therefore, careful exploitation of epigenetic events could be useful for the prevention and treatment of livestock diseases and for enhancing livestock reproduction, growth and development.

The Role of Epigenetic Mechanisms in Livestock Breeding

The contribution of epigenetic mechanisms to phenotypic variation in livestock is gaining prominence, and it is being advocated for consideration in breeding programs [30,31,33–36]. Aided by developments in sequencing technologies and bioinformatics tools, initial examination of epigenetic variation at the individual gene level, cell/tissue type has progressed to whole epigenome and population epigenome levels. In fact, many recent studies on an array of species, from yeast to humans and livestock, have uncovered the regulatory mechanisms (*cis* and *trans*) that direct patterns of individual and population epigenome variation at the level of single epigenetic marks and epigenome maps of different cell types as well as in response to environmental factors [32,34,346,347]. An integrative analysis of 111 reference human epigenomes (histone modification patterns, chromatin accessibility, DNA methylation and RNA expression) established global maps of regulatory elements, define regulatory modules of coordinated activity (activators and repressors) which demonstrated the central role of epigenomics information in understanding gene regulation, cellular differentiation and human disease [346]. In livestock, efforts by the Functional Annotation of Animal Genome Consortium (FAANG project, http://www.faang.org) and many other research groups have reported the epigenome maps of different cell types and tissues under various environmental conditions [34,38]. The maps of functional elements will fill the genotype-to-phenotype gap that is necessary to fully understand how the genome is read and translated into complex phenotypes, and will provide epigenomics data for use in the improved management of livestock species.

Epigenome-wide association studies (EWAS), the equivalent of GWAS would be necessary to quantify the source of variation attributable to epigenetic mechanisms. EWAS has provided new insights into the pathogenesis of common diseases and complex disorders in humans [348] and its application in livestock breeding is expected to achieve comparable results. Following the initial use of the term EWAS [348], about

2698 EWAS based studies have been conducted in humans (PubMedCentral search with key words "epigenome wide association study," June 6, 2022) and associations of DNA methylation sites and histone modifications with various human disease conditions (e.g., cancer, diabetes, asthma, obesity, Alzheimer, schizophrenia, asthma, Parkinson, etc.), smoking, glucose metabolism, growth disparity between identical twins, aging, nutrition, maternal care and heavy metal exposure, etc. As of today, no EWAS has been conducted in livestock due to the nonavailability of livestock specific DNA methylation genotyping assays.

For the detailed characterization of livestock genetic and epigenetic functional variation, a multidimensional approach that links DNA sequence variations, epigenetic modifications/marks, other nongenetic sources of variation and environmental interaction, and effects on phenotypic outcome is necessary. It is known that the phenotypic expression of a trait could be directly/indirectly influenced by DNA sequence variation and regulated by epigenetic changes triggered by environmental influences. Thus, the epigenome modulates the transcriptome as detected by the genome (including residing DNA variations), resulting in varying protein expression levels and phenotypic expression (Fig. 29.2). The implication is that phenotypic variance of a trait is a contribution of both the genetic, epigenetic, and other nongenetic variances modulated by environmental influences. For example, it has been demonstrated that DNA-methylation is a contributing factor to the nongenetic component of the etiology of type 1 diabetes (T1D) in monozygotic twins in an EWAS which showed that T1D-associated methylation variable positions arise very early in the etiological process in T1D discordant monozygotic twins as compared to controls [349].

Farm management, feeding practices, and disease pathogens are factors that have greatly enhanced or limited livestock productivity. The quality of feed (nutrient composition, pelleted or mashed, grain size, etc.) directly affects animal productivity. In dairy cows, maternal environment, mammary gland development, milking frequency, length of the daylight period, and disease pathogens directly influence productivity. It was revealed that the prenatal environment of dam and grand-dam influenced milk production in subsequent daughters by contributing minor effects, which were believed to be mediated by epigenetic mechanisms [314,350]. These small positive effects were passed onto daughters, an indication of a pathogen-induced or management-induced effect that was independent of the underlying genome which could be exploited for enhanced productivity. Nutritionally induced compensatory growth at different stages of development before first parturition positively influences mammary development and lactation performance throughout life and a probable link between this phenomenon and epigenetic control of metabolic imprinting and mammary gene expression has been proposed [351]. These effects should be known and exploited at any given production circle.

Epigenomics data or the nongenetic component of variation can be exploited for livestock improvement management and breeding. To do so, the following should be considered: (1) transient (intragenerational) or heritable (transgenerational) nature of the epigenetic variation; (2) whether the epigenetic effect is independent of DNA variant effect; and (3) whether the epigenetic effect is causal or consequential. For animal breeding purposes, sources of epigenetic variation should be clearly defined and used appropriately (Fig. 29.3). It was stated recently that epigenetic data on how epigenetic mechanisms influence livestock traits would complement genomic information and provide a better understanding of the factors that shape livestock phenotypes and directional application in breed improvement and management practices [30]. Therefore, calculations for GEBVs, which considers the sum of the effects of dense genetic markers or their haplotypes across the entire genome, should be expanded to include epigenomics and other nongenetic sources of variation. It has been proposed that the effects of genetic and nongenetic inheritance should be dissected and included in the estimation of trait heritability [37]. Thus, the quantitative contribution of the components of inclusive heritability (the percentage of variation in a trait that is transmitted between generations, whatever the mechanism of transmission and should be greater than or equal to heritability) [37] are summarized in Fig. 29.4.

Mixed-effect models or animal model approaches which are based on pedigrees can allow the estimation of several variance components simultaneously [37]. Using animal models to estimate genetic parameters in the analysis of quantitative traits requires a decomposition of the phenotypic variance—covariance matrix (P matrix) into a matrix of additive genetic variance—covariance (G matrix) and a residual defined by $G = A \times V_A$, where V_A is the additive genetic variance and A is the additive relationship matrix with individual elements $A_{ij} = 2\Theta_{ij}$ (where Θ_{ij} is the coefficient of coancestry between individuals i and j) (reviewed in [37]). Therefore, the construction of an additive genetic relationship matrix, A, based on pedigree information is required to solve an animal model equation. Similarly, nongenetic heritability can be partitioned by building corresponding matrices and inclusion in the animal model equation. For animal breeding purposes, the nongenetic component of variation due to epigenetic inheritance and paternal effects could be built into the animal model equations, as there is yet no documented

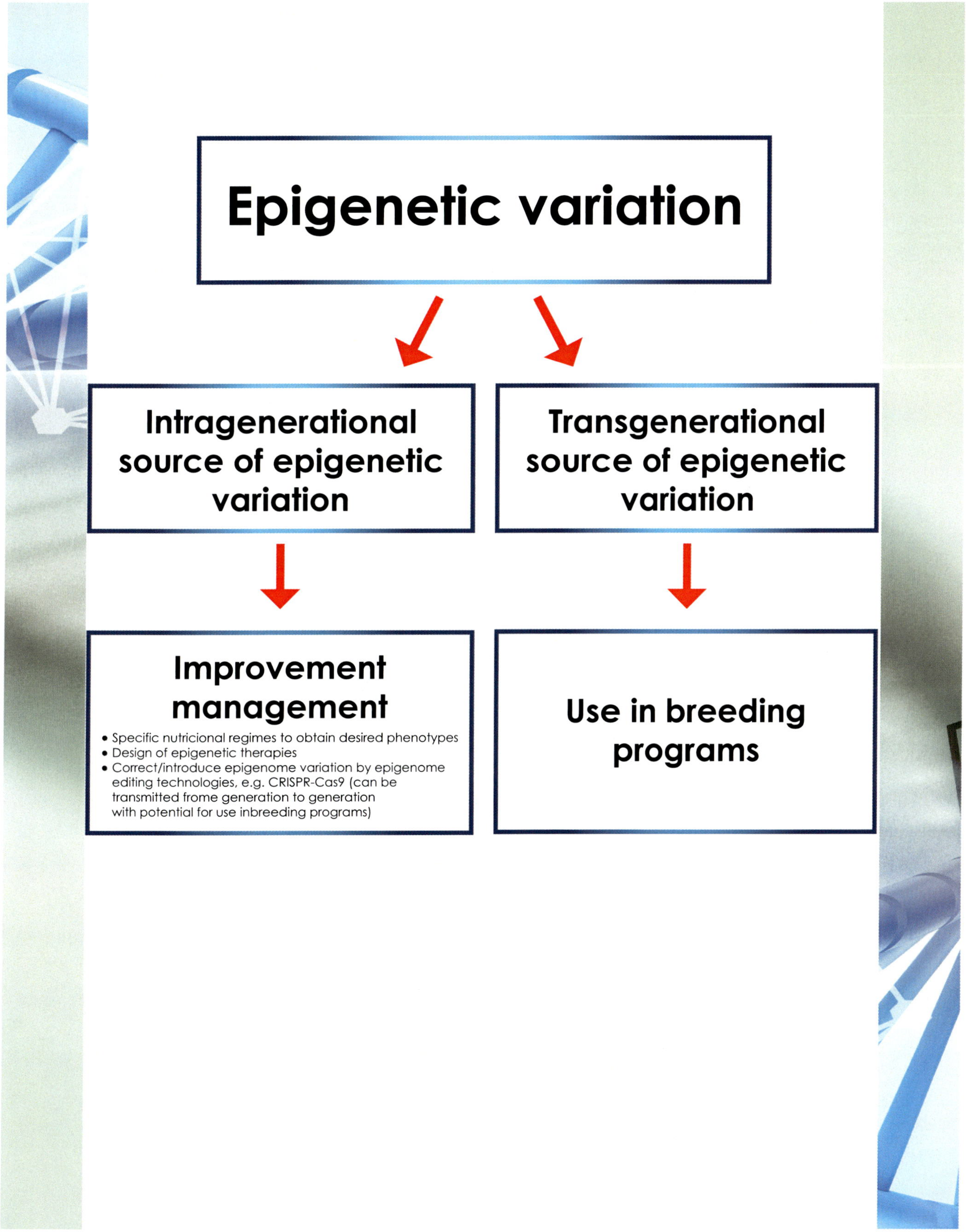

FIGURE 29.3 Sources of epigenetic variation for appropriate use in livestock improvement.

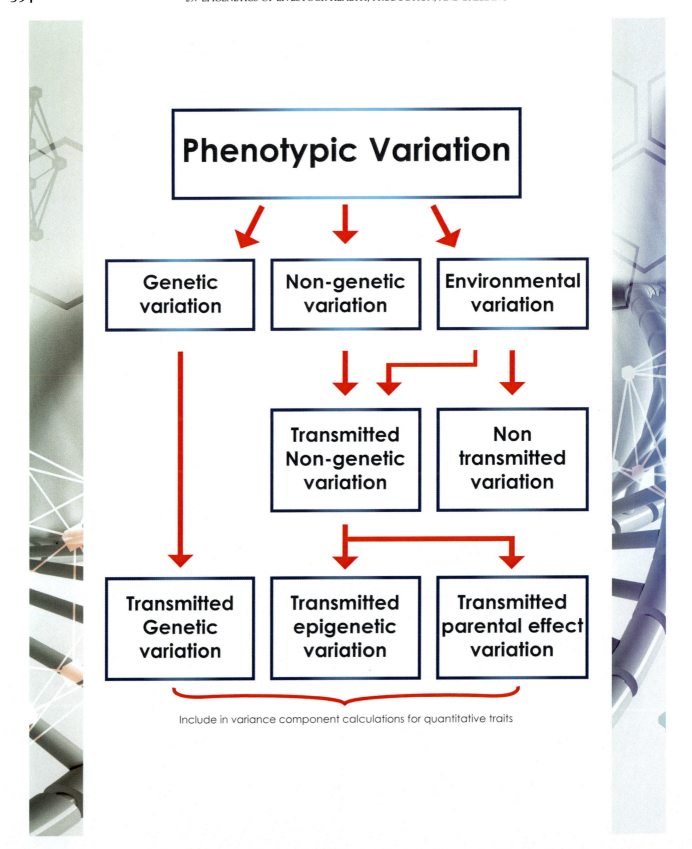

FIGURE 29.4 Contribution of the components of inclusive heritability in variance component calculations for livestock quantitative traits.

evidence of the transmission of ecological and social variation in livestock species (Fig. 29.4). Furthermore, novel expression for covariance between relatives was introduced through a model that takes epigenetic transmissibility or the probability of transmission of ancestral phenotypes and environmental induction into account [352].

Application of Genome and Epigenome Editing Technologies in Livestock Breeding

The feasibility of adding or removing epigenetic marks by different compounds has made it appealing to introduce a wide range of epigenetic drugs for cancer treatment, among other applications [332]. For example, the FDA-approved DNMTi 5-aza-2'-deoxycytidine, a DNA methylation inhibitor, has been known as a cancer drug since 1968 because of its cytotoxic effects on cancer cells [333]. However, this compound has a wide range of side effects such as anemia, neutropenia, thrombocytopenia, and kidney toxicity [353]. Importantly, epigenetic drugs, such as 5-aza-2'-deoxycytidine, are global epigenetic agents that alter both target and nontarget genes [353]. With the discovery of the clustered regularly interspaced short palindromic repeats (CRISPR) system, epigenome editing aimed at adding or removing epigenetic modifications to specific genes is now feasible [354]. Fusion proteins composed of an effector domain (e.g., DNA methyltransferase or histone acetyltransferase) and a deactivated Cas9 (dCas9), along with guide RNAs (gRNA), can activate or silence specific target genes. The dCas9 does not cut the DNA; its role is to direct the effector domain to the target gene through gRNA binding.

Generally, technologies like zinc finger nucleases (ZFN), transcription activator-like effector nucleases (TALEN), and CRISPR/Cas9 have been used to achieve efficient genome editing by introducing genes/alleles of interest into a recipient genome, by editing a section of the genome or by switching genes on or off [355,356]. Genome editing technologies have been used to target certain genes in livestock species like the *MSTN* or double muscling gene in cattle, sheep and pigs; the polled allele in cattle; the *RELA* gene that confers resistance to African Swine fever in pigs; the *LDLR* gene in a pig model of atherosclerosis; lysostaphin (*LSS*) gene for mastitis resistance in cattle; *SKI* gene for muscle development in chickens; and the *CD163* gene that is thought to confer resistance to porcine reproductive and respiratory syndrome virus in pig, etc. to the production of transgenic animals (reviewed in [39]). Further examples of gene editing to address environmental adaptation, disease, and product quality in livestock and plants of agricultural importance were reviewed recently [40].

Epigenetic editing at loci of interest represents an innovative method that might selectively and heritably alter gene expression. Thus, in contrast to genome editing, epigenome editing technology changes the chromatin state without changing the DNA sequence [357]. A proof-of-principle of this technology has been demonstrated in human cell lines, where targeted acetylation of both promoters and enhancers resulted in successful transcriptional activation of several genes [358]. For animal breeding purposes, these technologies can be used to introduce/fix favorable alleles and epigenetic marks for increased productivity. Although epigenome editing has not been performed in livestock, a recent simulation study [359] showed that promotion of alleles for polygenic traits by genome editing could double genetic gains when compared to conventional genomic selection. Therefore, the application of epigenome editing could significantly impact animal production, reproduction, and health using targeted manipulation of candidate genes affecting traits of economic importance.

CONCLUSION

Animal breeding art has evolved to a science-based activity and exploitation of all sources of trait variation will ensure continued improvement in livestock traits of economic importance. This will, however, depend on the adequate characterization of epigenomics sources of variation of livestock traits. Tools to facilitate livestock epigenetics research and increased research funding are factors that will ensure optimal exploitation of epigenomics data in livestock management and improvement.

SUMMARY

In this chapter we have provided a summary of the development of the animal breeding art and how epigenetics could play a role in animal management and improvement breeding. We have presented an overview of the current state of livestock epigenetic research and summarized data on the associations of epigenetic marks and livestock health, reproduction, growth and development traits, as well as livestock products. We have also outlined the potential application of epigenetic information in livestock health management and breeding. Our general aim was to present an overview of livestock epigenetic research and potential application in livestock managemnet and breeding.

LIVESTOCK EPIGENETICS CASE STUDY

Investigating the role of DNA methylation in subclinical Johne disease (JD) in cattle [145].

Objective

The objective of the study was to investigate the role of DNA methylation in cow's (ileal [IL] and ileal lymph node [ILLN] tissues) response to persistent subclinical *Mycobacterium avium* subsp *paratuberculosis* (MAP) infection. Cows studied included those diagnosed with persistent MAP infection over a period of 1–4 years and healthy control (MAP-ve) cows from the same farms.

Scope

1. **Research workflow (Fig. 29.5):** (1) Selection of participating commercial farms; (2) selection of experimental animals (animals diagnosed for the presence or absence of MAP using two tests); (3) selected MAP positive (MAP + ve) and MAP negative (MAP-ve) cows purchased and transported to research center; (4) animals euthanized and intestinal tissues collected; (5) intestinal tissues tested for the presence or absence of MAP bacteria; (6) final groups of cows selected, based on the results of the three tests, for subsequent analyses; (7) DNA isolation and library preparation for whole genome bisulfite sequencing; (8) next generating sequencing of prepared libraries; (9) bioinformatics processing of generated reads; (10) identification of DNA methylated sites and genes associated with JD; and (11) functional analysis to predict the functions of genes.
2. **Cost:** The entire project cost from cow selection until publication of one manuscript and one PhD student stipend for 2 years was $90,000 CAN.
3. **Project limitation:** (1) Diagnostic tests to identify MAP + ve cows have their limitations requiring the use of more than one test to identify cows. This is laborious and expensive. (2) Small sample size that was due to the difficulty of identifying naturally infected cows with certainty. Animals were at different physiological stages and age, which could compound the results generated.

Audience

The research team included two government research scientists/university professors with expertise in genomics/epigenomics and immunology; two university professors with expertise in genomics/epigenomics and lactation biology, a research technician, and one PhD student. This research is of importance to dairy farmers, veterinarians, researchers, farmers, students, policy makers, and drug companies.

Rationale

Mycobacterium avium subsp. *paratuberculosis* (MAP) causes Johne disease (JD) in bovine and other ruminants. JD is a major threat to cattle health and the economic profitability of the dairy industry. The economic impact of JD is estimated at ~$90 M annually in

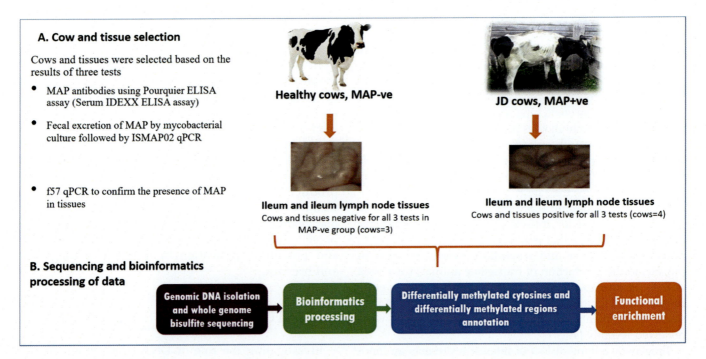

FIGURE 29.5 The experimental workflow and actions taken.

Canada and at $200–250 M annually in the United States. JD is a chronic enteric disease characterized by a long incubation period, imbalance in gut immunity, and ability of the pathogen to subvert host cell immune responses to ensure its intracellular survival and replication. A long subclinical period of 2–5 years (phase 2) between the infection (phase I) and the clinical (phase III) periods of the disease provides the pathogen (MAP) an opportunity to adapt to its environment for survival. Numerous investigations have characterized the host response to MAP infection, but the mechanisms used by MAP to subvert host immune responses are still unclear. Also, potential association between JD and human disease like Crohn disease and arthritis rheumatoid [360–362] underscores the importance of controlling JD.

In our recent studies, we observed that MAP impacted the expression of epigenetics genes, lncRNA pattern of macrophage (the first host immune cell defense encountered by MAP while invading cows) and the miRNA expression patterns of small intestinal tissues from MAP infected cows [363–367]. Furthermore, comparing the transcriptome landscape of small intestinal tissues and draining lymph nodes from MAP infected (MAP + ve) cows with MAP negative (MAP − ve) cows, a heighten immune response by the IL and a depressed immune response by the ILLN was observed, suggesting involvement of epigenetic factors [145]. This study, therefore, aimed to investigate the involvement of DNA methylation in cow's response to subclinical MAP infection, using the whole genome bisulfite sequencing technique.

Hypothesis of the Research: The DNA methylation patterns of the ileum and ileum lymph node tissues of cows are altered during subclinical MAP infection.

Expected Results and Deliverables

- Characterize the DNA methylation patterns of ileum and ileum lymph node tissues from cows with subclinical MAP infection and corresponding tissues from healthy (MAP-ve) cows.
- Identify altered CpG sites and differentially methylated regions associated with JD.
- Identify differentially methylated CpG sites between MAP + ve and MAP-ve IL and ILLN tissues.
- Identify biological processes and pathways impacted by DNA methylation during subclinical MAP infection.
- Identify immune processes and genes with altered DNA methylation states during subclinical MAP infection.
- Identify potential epigenetic biomarkers of JD.

Safety Considerations

MAP is a potential zoonotic disease so appropriate safety gears were used during animal sampling. Collected samples were processed in a biosafety level 2 laboratory.

The process, workflow and actions taken Tools and materials used

Process	Tools and Equipment
Animal selection	16 commercial dairy farms with a history of JD and about 1500 cows sampled
Blood	Pourquier serum ELISA assay (IDEXX Laboratories, Markham, Ontario, Canada) to detect MAP antigens
Feces	MAP fecal PCR to diagnose for the presence of MAP
Feces	Mycobacterial fecal culture (Laboratoire d'épidémiosurveillance animale du Québec, Saint-Hyacinthe, Québec, Canada) to confirm the excretion of live MAP bacteria
Ileum and ileum lymph node tissues	f57 real time qPCR to target the MAP-specific, single-copy DNA element F57
DNA isolation	DNeasy Blood and Tissue Kit (Qiagen)
DNA quantification	Quant-IT PicoGreen dsDNA Assay Kit (Life Technologies)
Bisulfite conversion of DNA	EZ DNA Methylation-Lightning Kit (Zymo)
Library preparation	NEBNext Ultra II DNA Library Prep Kit for Illumina (New England BioLabs)
Library size selection	SPRIselect beads (Beckman Coulter)
Library quantification	Kapa Illumina GA with Revised Primers-SYBR Fast Universal kit (Kapa Biosystems)
Fragment size determination	LabChip GX (PerkinElmer)
Next generation sequencing	Illumina NovaSeq S4 system (Illumina)
Read quality check and adaptor trimming	TrimGalore v0.4.4 and and CutAdapt v1.13
Read quality matrix visualization	FastQC v0.11.3
Index reference genome (ARS-UCD1.2)	BSseeker2 v2.1.7

(Continued)

(Continued)

Process	Tools and Equipment
Alignment of reads to reference genome	Bowtie2 aligner within BSseeker2
Reads fixmated, sorted and duplicates removed	SAMtools v1.7
Methylation level determination	BSseeker2 v2.1.7
Methylation site coverage and distribution, and correlation analysis	MethylKit v1.12.0
Methylation density plots	Circlize, an R package
DNA methylation dynamic plots	Beanplot program in R package
Differential methylation analysis	Radmeth function of Methpipe v3.4.3
Gene ontology and pathways enrichment	ClueGO, a Cytoscape plug-in

Results

- The DNA methylation patterns of the IL and ILLN tissues of cows at the subclinical stage of JD and healthy cows were described.
- Higher numbers of differentially methylated cytosines (DMCs) and regions (DMRs) (FDR < 0.1) were found in ILLN tissues than in IL tissues.
- A total of 11,263 DMCs and 1259 DMRs were found between MAP + ve and MAP − ve IL tissues.
- A total of 62,459 DMCs and 8086 DMRs were found between MAP + ve and MAP − ve ILLN tissues.
- DMCs were annotated to 503 genes in the IL and to 1717 genes in the ILLN.
- DMRs were annotated to 394 genes in the IL and to 1305 genes in the ILLN.
- More gene ontology terms and pathways were found in ILLN than in IL tissues.
- Many genes with hypo/hypermethylated DMRs in their promoters or 5′UTRs and having roles in multiple processes such as cellular, metabolic processes, and immune regulation (e.g., *CCL21, PLAGL1, BARHL2, CHRNG, BTC, GLA, BLK, SLC18A1, TEX101, PPA1, SLC18A1, BLK, TDGF1, EIF4E, CCDC56*, etc.) were found.
- More immune pathways were enriched for genes in the ILLN that in the IL.
- Important immune pathways found in the ILLN included Cell adhesion molecules (CAMs), Bacterial invasion of epithelial cells, Platelet activation and Inflammatory mediator regulation of TRP channels.
- The ILLN was more impacted by DNA methylation than the IL.

Challenges and Solutions

The process of identifying cows infected with MAP bacteria over a long period of time and cows completely free of the pathogen was very challenging, expensive, and also time consuming. To overcome the challenge of wrongly classifying cows, three diagnostic tests were used.

Learning and Knowledge Outcomes

This is the first study to describe the role of DNA methylation in JD. The study confirmed initial suggestions of the involvement of epigenetic processes in the host response to MAP infection. Several identified CpG sites and genes associated to MAP infection are potential biomarkers of MAP infection, which can support development of improved diagnostic and therapeutic solutions for JD management. For subsequent experiments, a higher number of infected and noninfected cows should be used. Sequencing of samples to identify methylated sites should be performed at higher depth (e.g., 60x and above).

Acknowledgments

We acknowledge financial support by Agriculture and Agri-Food Canada (grant no. J-002223) to Eveline M. Ibeagha-Awemu and the United States of America Department of Agriculture—Agriculture and Food Research Initiative Competitive Grant no. 2020−67015−30828 to Hasan Khatib.

Glossary

ACACA	Acetyl-CoA carboxylase alpha
ABCA1	ATP binding cassette subfamily A member 1
ABCA7	ATP binding cassette subfamily A member 7
ABCG1	ATP binding cassette subfamily G member 1
BARHL2	BarH like homeobox 2
BLK	BLK proto-oncogene, Src family tyrosine kinase
BLUP	Best linear unbiased prediction
BPI	Bctericidal permeability increasing protein
BTC	Betacellulin
CAND1	Cullin associated and neddylation dissociated 1
Cas9	Type II CRISPR RNA-guided endonuclease Cas9
CAV1	Caveolin 1
CCDC56	Coiled-coil domain containing 56
CCL21	C-C motif chemokine ligand 21
CD4	Cluster of differentiation 4
CD163	CD163 molecule
CDKN1C	Cyclin dependent kinase inhibitor 1C
CHRNG	Cholinergic receptor nicotinic gamma subunit,
CRISPR	Clustered regularly interspaced short palindromic repeats
CSN1S1	Casein alpha s1
CTDSP1	CTD (carboxy-terminal domain) RNA polymerase II polypeptide A small phosphatase 1
CTDSP2	CTD RNA polymerase II polypeptide A small phosphatase 2

CTDSPL	CTD small phosphatase like
CYP4F29P	Cytochrome P450 family 4 subfamily F member 29, pseudogene
DGAT1	Diacylglycerol O-acyltransferase 1
DLK1	Delta like noncanonical Notch ligand 1
DIO3	Deiodinase, iodothyronine, type III
DMC	Differentially methylated cytosine
DMR	Differentially methylated region
DNA	Deoxyribonucleic acid
DNAH12	Dynein axonemal heavy chain 12
DNMT1	DNA methyltransferase 1
DNMT3A	DNA methyltransferase 3 alpha
DNMT3B	DNA methyltransferase 3 beta
EBV	Estimated breeding values
EIF5	E74 like ETS transcription factor 5
EIF4E	Eukaryotic translation initiation factor 4E
ELOVL6	Elongation of very long chain fatty acids protein 6
EPO	Erythropoietin
ERα	Estrogen receptor alpha
EWAS	Epigenome wide association study
FASN	Fatty acid synthase
FTO	FTO alpha-ketoglutarate dependent dioxygenase
GBLUP	Genomically best linear unbiased prediction
GEBV	Genomic estimated breeding values
GHR	Growth hormone receptor
GLA	Galactosidase alpha
GLUT1	Glucose transporter 1
GO	Gene ontology
GPAM	Glycerol-3-phosphate acyltransferase, mitochondrial
GWAS	Genome wide association study
H3K4me3	Trimethylation at the fourth lysine residue of the histone H3 protein
H3K27ac	Acetylation at the 27th lysine residue of the histone H3 protein
H3K27me3	Trimethylation at the 27th lysine residue of the histone H3 protein
H3K36me3	Trimethylation at the 36th lysine residue of the histone H3 protein
H19	H19, imprinted maternally expressed transcript
HB	Haptoglobin
HDAC6	Histone deacetylase 6
HDAC7	Histone deacetylase 7
HIF1	Hypoxia-inducible factor 1
HIF-1a	Hypoxia-inducible factor 1 alpha
HIF-3a	Hypoxia-inducible factor 3 alpha
iDMRs	infection-induced differential methylation regions
IFN	Interferon gamma
IGF1	Insulin-like growth factor 1
IGF2	Insulin-like growth factor 2
IGF2R	Insulin-like growth factor 2 receptor
IL	Ileum
ILLN	Ileum lymph node
INS	Insulin
INSIG1	Insulin induced gene 1
IL2, IL6, IL8	interleukin-2, -6, -8
IYD	Iodotyrosine deiodinase
JD	Johne disease
KCNMB2	Potassium calcium-activated channel subfamily M regulatory beta subunit 2
KEGG	Kyoto Encyclopedia of Genes and Genomes
LDLR	Low density lipoprotein receptor
LPIN1	Lipin 1
LPS	Lipopolysaccharide
LXRA (NR1H3)	Nuclear receptor subfamily 1 group H member 3
m^6A	N6-methyladenosine
MAGEL2	MAGE family member L2
MAPK	Mitogen-activated protein kinase
MAS	Marker assisted selection
MD	Marek disease
MSTN	Myostatin
NF-κB	Nuclear Factor Kappa B
NPY	Neuropeptide Y
PACSNI1	Protein kinase C and casein kinase substrate in neurons 1
PLAGL1	PLAG1 like zinc finger 1
PPA1	Inorganic pyrophosphatase 1
PPARA	Peroxisome proliferator activated receptor alpha
PPARG	Peroxisome proliferator activated receptor gamma
QTL	Quantitative trait loci
RELA	V-rel reticuloendotheliosis viral oncogene homolog
REML	Residual or restricted maximum likelihood
RNA	Ribonucleic acid
RXRA	Retinoid X receptor alpha
SAA3	Serum amyloid A3
SCD	Stearoyl-coenzyme A desaturase
SCD1	Stearoyl-coenzyme A desaturase 1
SCS	Somatic cell score
SLC18A1	Solute carrier family 18 member A1
SNP	Single nucleotide polymorphism
SP1	Sp1 transcription factor
SREBF1	Sterol regulatory element binding transcription factor 1
STAT5	Sgnal transducer and activator of transcription 5
TALEN	Transcription activator-like effector nucleases
TDGF1	Teratocarcinoma-derived growth factor 1
TEX101	Testis expressed gene 101
TGF-β	Transforming growth factor-beta,
TLR2, TRL4	Toll-like receptor 2, toll-like receptor 4
TNFα	Tumor necrosis factor alpha
T1D	Type 1 diabetes
UBE3A	Ubiquitin protein ligase E3A
ZFN	Zinc finger nucleases
ZFP64	Zinc finger protein 64

References

[1] Wood RJ, Orel V. Genetic prehistory in selective breeding: a prelude to mendel. Oxford: Oxford University Press; 2001.

[2] Blasco A. Animal breeding methods and sustainability animal breeding sustainability. In: Christou P, Savin R, Costa-Pierce BA, Misztal I, Whitelaw CBA, editors. Sustainable food production. New York: Springer; 2013. p. 41−57.

[3] Price WB. The origins of theoretical population genetics. University of Chicago Press; 2001. p. 211.

[4] Fisher RA. The correlation between relatives on the supposition of mendelian inheritance. Trans R Soc Edinb 1918;52:399−433.

[5] Hill WG. Applications of population genetics to animal breeding, from wright, fisher and lush to genomic prediction. Genetics. 2014;196(1):1−16.

[6] Lush JL. Animal breeding plans. Lowa State Coll Press; 1943. p. 457.

[7] Hazel LN. The genetic basis for constructing selection indexes. Genetics. 1943;28(6):476−90.

[8] Waddington CH. The epigenotype. Int J Epidemiol 2012;41(1):10−13.

[9] Brassley P. Cutting across nature? The history of artificial insemination in pigs in the United Kingdom. Stud Hist Philos Biol Biomed Sci 2007;38(2):442–61.

[10] Weller JI. Genomic selection in animals. John Wiley & Sons, Inc; 2016. p. 171.

[11] ICGSC. Sequence and comparative analysis of the chicken genome provide unique perspectives on vertebrate evolution. Nature. 2004;432(7018):695–716.

[12] Elsik CG, Tellam RL, Worley KC, Gibbs RA, Muzny DM, Weinstock GM, et al. The genome sequence of taurine cattle: a window to ruminant biology and evolution. Science. 2009;324 (5926):522–8.

[13] Wade CM, Giulotto E, Sigurdsson S, Zoli M, Gnerre S, Imsland F, et al. Genome sequence, comparative analysis, and population genetics of the domestic horse. Science. 2009;326(5954):865–7.

[14] Groenen MAM, Archibald AL, Uenishi H, Tuggle CK, Takeuchi Y, Rothschild MF, et al. Analyses of pig genomes provide insight into porcine demography and evolution. Nature. 2012;491(7424):393–8.

[15] Dong Y, Xie M, Jiang Y, Xiao N, Du X, Zhang W, et al. Sequencing and automated whole-genome optical mapping of the genome of a domestic goat (*Capra hircus*). Nat Biotech 2013;31(2):135–41.

[16] Jiang Y, Xie M, Chen W, Talbot R, Maddox JF, Faraut T, et al. The sheep genome illuminates biology of the rumen and lipid metabolism. Science. 2014;344(6188):1168–73.

[17] Lien S, Koop BF, Sandve SR, Miller JR, Kent MP, Nome T, et al. The Atlantic salmon genome provides insights into rediploidization. Nature. 2016;533(7602):200–5.

[18] Liu Z, Liu S, Yao J, Bao L, Zhang J, Li Y, et al. The channel catfish genome sequence provides insights into the evolution of scale formation in teleosts. Nat Commun 2016;7(1):11757.

[19] Dutta P, Talenti A, Young R, Jayaraman S, Callaby R, Jadhav SK, et al. Whole genome analysis of water buffalo and global cattle breeds highlights convergent signatures of domestication. Nat Commun 2020;11(1):4739.

[20] El-Khishin DA, Ageez A, Saad ME, Ibrahim A, Shokrof M, Hassan LR, et al. Sequencing and assembly of the Egyptian buffalo genome. PLoS One 2020;15(8):e0237087.

[21] Goddard M. Genomic selection: prediction of accuracy and maximisation of long term response. Genetica. 2009;136 (2):245–57.

[22] Meuwissen T, Hayes B, Goddard M. Accelerating improvement of livestock with genomic selection. Annu Rev Anim Biosci 2013;1:221–37.

[23] Miglior F, Fleming A, Malchiodi F, Brito LF, Martin P, Baes CF. A 100-year review: identification and genetic selection of economically important traits in dairy cattle. J Dairy Sci 2017;100 (12):10251–71.

[24] Wiggans GR, VanRaden PM, Cooper TA. The genomic evaluation system in the United States: past, present, future. J Dairy Sci 2011;94(6):3202–11.

[25] García-Ruiz A, Cole JB, VanRaden PM, Wiggans GR, Ruiz-López FJ, Van Tassell CP. Changes in genetic selection differentials and generation intervals in US Holstein dairy cattle as a result of genomic selection. Proc Natl Acad Sci 2016;113(28):E3995–4004.

[26] Pryce JE, Daetwyler HD. Designing dairy cattle breeding schemes under genomic selection: a review of international research. Ani Prod Sci 2012;52(3):107–14.

[27] Wiggans GR, Cole JB, Hubbard SM, Sonstegard TS. Genomic selection in dairy cattle: the USDA experience. Annu Rev Anim Biosci 2017;5(1):309–27.

[28] Burggren WW. Epigenetics as a source of variation in comparative animal physiology - or - Lamarck is lookin' pretty good these days. J Exp Biol 2014;217(Pt 5):682–9.

[29] Burggren WW, Crews D. Epigenetics in comparative biology: why we should pay attention. Integr Comp Biol 2014;54(1):7–20.

[30] Ibeagha-Awemu EM, Zhao X. Epigenetic marks: regulators of livestock phenotypes and conceivable sources of missing variation in livestock improvement programs. Front Genet 2015;6:302.

[31] Meirelles FV, Bressan FF, Smith LC, Perecin F, Chiaratti MR, Ferraz JBS. Cytoplasmatic inheritance, epigenetics and reprogramming DNA as tools in animal breeding. Livest Sci 2014;166 (1):199–205.

[32] Taudt A, Colome-Tatche M, Johannes F. Genetic sources of population epigenomic variation. Nat Rev Genet 2016;17 (6):319–32.

[33] Triantaphyllopoulos KA, Ikonomopoulos I, Bannister AJ. Epigenetics and inheritance of phenotype variation in livestock. Epigenetics Chromatin 2016;9:31.

[34] Wang M, Ibeagha-Awemu EM. Impacts of epigenetic processes on the health and productivity of livestock. Front Genet 2021;11:1812.

[35] Doherty R, Farrelly CO, Meade KG. Comparative epigenetics: relevance to the regulation of production and health traits in cattle. Anim Genet 2014;45(Suppl.1):3–14.

[36] Goddard ME, Whitelaw E. The use of epigenetic phenomena for the improvement of sheep and cattle. Front Genet 2014;5(JUL).

[37] Danchin E, Charmantier A, Champagne FA, Mesoudi A, Pujol B, Blanchet S. Beyond DNA: integrating inclusive inheritance into an extended theory of evolution. Nat Rev Genet 2011;12 (7):475–86.

[38] Andersson L, Archibald AL, Bottema CD, Brauning R, Burgess SC, Burt DW, et al. Coordinated international action to accelerate genome-to-phenome with FAANG, the functional annotation of animal genomes project. Genome Biol 2015;16(1):1–6.

[39] Van Eenennaam AL, De Figueiredo Silva F, Trott JF, Zilberman D. Genetic engineering of livestock: the opportunity cost of regulatory delay. Annu Rev Anim Biosci 2021;9:453–78.

[40] Karavolias NG, Horner W, Abugu MN, Evanega SN. Application of gene editing for climate change in agriculture. Front SustaFood Syst 2021;5:296.

[41] Eichler EE, Flint J, Gibson G, Kong A, Leal SM, Moore JH, et al. Missing heritability and strategies for finding the underlying causes of complex disease. Nat Rev Genet 2010;11(6):446–50.

[42] Kilpinen H, Dermitzakis ET. Genetic and epigenetic contribution to complex traits. Hum Mol Genet 2012;21(R1):R24–8.

[43] Manolio TA, Collins FS, Cox NJ, Goldstein DB, Hindorff LA, Hunter DJ, et al. Finding the missing heritability of complex diseases. Nature. 2009;461(7265):747–53.

[44] Bernstein BE, Meissner A, Lander ES. The mammalian epigenome. Cell. 2007;128(4):669–81.

[45] Bird A. DNA methylation patterns and epigenetic memory. Genes Dev 2002;16(1):6–21.

[46] Kouzarides T. Chromatin modifications and their function. Cell. 2007;128(4):693–705.

[47] Morris KV, Mattick JS. The rise of regulatory RNA. Nat Rev Genet 2014;15(6):423–37.

[48] Zhang M, Song J, Yuan W, Zhang W, Sun Z. Roles of RNA methylation on tumor immunity and clinical implications. Front Immunol 2021;12(600):600.

[49] Zhou Y, Kong Y, Fan W, Tao T, Xiao Q, Li N, et al. Principles of RNA methylation and their implications for biology and medicine. Biomed Pharmacother 2020;131:110731.

[50] Ibeagha-Awemu EM, Yu Y. Consequence of epigenetic processes on animal health and productivity: is additional level of regulation of relevance? Anim Front 2021;11(6):7–18.

[51] Zhang L, Lu Q, Chang C. Epigenetics in health and disease. Adv Exp Med Biol 2020;1253:3–55.

REFERENCES

[52] Franco MM. Epigenetics in genetic improvement and animal reproduction. Archivos Latinoamericanos de Producción Anim 2017;25:1–2.

[53] Wang X, Qu J, Li J, He H, Liu Z, Huan Y. Epigenetic reprogramming during somatic cell nuclear transfer: recent progress and future directions. Front. Genet. 2020;11:205.

[54] Zhu L, Marjani SL, Jiang Z. The epigenetics of gametes and early embryos and potential long-range consequences in livestock species—Filling in the picture with epigenomic analyses. Front Genet 2021;12(223):223.

[55] Das L, Parbin S, Pradhan N, Kausar C, Patra SK. Epigenetics of reproductive infertility. Front Biosci (Schol Ed). 2017;9(4):509–35.

[56] Perrier JP, Sellem E, Prézelin A, Gasselin M, Jouneau L, Piumi F, et al. A multi-scale analysis of bull sperm methylome revealed both species peculiarities and conserved tissue-specific features. BMC Genomics 2018;19(1):404.

[57] Zhou Y, Connor EE, Bickhart DM, Li C, Baldwin RL, Schroeder SG, et al. Comparative whole genome DNA methylation profiling of cattle sperm and somatic tissues reveals striking hypomethylated patterns in sperm. GigaScience. 2018;7(5):giy039.

[58] Gross N, Peñagaricano F, Khatib H. Integration of whole-genome DNA methylation data with RNA sequencing data to identify markers for bull fertility. Anim Genet 2020;51(4):502–10.

[59] Kropp J, Carrillo JA, Namous H, Daniels A, Salih SM, Song J, et al. Male fertility status is associated with DNA methylation signatures in sperm and transcriptomic profiles of bovine preimplantation embryos. BMC Genomics 2017;18(1):280.

[60] Kutchy NA, Menezes ESB, Chiappetta A, Tan W, Wills RW, Kaya A, et al. Acetylation and methylation of sperm histone 3 lysine 27 (H3K27ac and H3K27me3) are associated with bull fertility. Andrologia 2018;50(3).

[61] Capra E, Lazzari B, Turri F, Cremonesi P, Portela AMR, Ajmone-Marsan P, et al. Epigenetic analysis of high and low motile sperm populations reveals methylation variation in satellite regions within the pericentromeric position and in genes functionally related to sperm DNA organization and maintenance in Bos taurus. BMC Genomics 2019;20(1):940.

[62] Lambert S, Blondin P, Vigneault C, Labrecque R, Dufort I, Sirard MA. Spermatozoa DNA methylation patterns differ due to peripubertal age in bulls. Theriogenology. 2018;106:21–9.

[63] Rahman MB, Schellander K, Luceño NL, Van, Soom A. Heat stress responses in spermatozoa: mechanisms and consequences for cattle fertility. Theriogenology. 2018;113:102–12.

[64] Takeda K, Kobayashi E, Nishino K, Imai A, Adachi H, Hoshino Y, et al. Age-related changes in DNA methylation levels at CpG sites in bull spermatozoa and in vitro fertilization-derived blastocyst-stage embryos revealed by combined bisulfite restriction analysis. J Reprod Dev 2019;65(4):305–12.

[65] Wyck S, Herrera C, Requena CE, Bittner L, Hajkova P, Bollwein H, et al. Oxidative stress in sperm affects the epigenetic reprogramming in early embryonic development. Epigenetics Chromatin 2018;11(1):60.

[66] Han W, Xue Q, Li G, Yin J, Zhang H, Zhu Y, et al. Genome-wide analysis of the role of DNA methylation in inbreeding depression of reproduction in Langshan chicken. Genomics. 2020;112(4):2677–87.

[67] Liu S, Fang L, Zhou Y, Santos DJA, Xiang R, Daetwyler HD, et al. Analyses of inter-individual variations of sperm DNA methylation and their potential implications in cattle. BMC Genomics 2019;20(1):888.

[68] Ispada J, de Lima CB, Sirard MA, Fontes PK, Nogueira MFG, Annes K, et al. Genome-wide screening of DNA methylation in bovine blastocysts with different kinetics of development. Epigenetics Chromatin 2018;11(1):1.

[69] Salilew-Wondim D, Saeed-Zidane M, Hoelker M, Gebremedhn S, Poirier M, Pandey HO, et al. Genome-wide DNA methylation patterns of bovine blastocysts derived from in vivo embryos subjected to in vitro culture before, during or after embryonic genome activation. BMC Genomics 2018;19(1):424.

[70] Duan JE, Jiang ZC, Alqahtani F, Mandoiu I, Dong H, Zheng X, et al. Methylome dynamics of bovine gametes and in vivo early embryos. Front Genet 2019;10:512.

[71] Cao P, Li H, Zuo Y, Nashun B. Characterization of DNA methylation patterns and mining of epigenetic markers during genomic reprogramming in SCNT embryos. Front Cell Dev Biol 2020;8:877.

[72] Akbarinejad V, Gharagozlou F, Vojgani M. Temporal effect of maternal heat stress during gestation on the fertility and anti-Müllerian hormone concentration of offspring in bovine. Theriogenology. 2017;99:69–78.

[73] Desmet KL, Van Hoeck V, Gagné D, Fournier E, Thakur A, O'Doherty AM, et al. Exposure of bovine oocytes and embryos to elevated non-esterified fatty acid concentrations: integration of epigenetic and transcriptomic signatures in resultant blastocysts. BMC Genomics 2016;17(1):1004.

[74] Garnsworthy PC, Sinclair KD, Webb R. Integration of physiological mechanisms that influence fertility in dairy cows. Animal. 2008;2(08):1144–52.

[75] Urrego R, Rodriguez-Osorio N, Niemann H. Epigenetic disorders and altered gene expression after use of assisted reproductive technologies in domestic cattle. Epigenetics. 2014;9(6):803–15.

[76] Cibelli JB, Campbell KH, Seidel GE, West MD, Lanza RP. The health profile of cloned animals. Nat Biotech 2002;20(1):13–14.

[77] Dean W, Santos F, Stojkovic M, Zakhartchenko V, Walter J, Wolf E, et al. Conservation of methylation reprogramming in mammalian development: aberrant reprogramming in cloned embryos. Proc Natl Acad Sci U S A 2001;98(24):13734–8.

[78] Yang X, Smith SL, Tian XC, Lewin HA, Renard J-P, Wakayama T. Nuclear reprogramming of cloned embryos and its implications for therapeutic cloning. Nat Genet 2007;39(3):295–302.

[79] Maalouf WE, Alberio R, Campbell KH. Differential acetylation of histone H4 lysine during development of in vitro fertilized, cloned and parthenogenetically activated bovine embryos. Epigenetics. 2008;3(4):199–209.

[80] Santos F, Zakhartchenko V, Stojkovic M, Peters A, Jenuwein T, Wolf E, et al. Epigenetic marking correlates with developmental potential in cloned bovine preimplantation embryos. Curr Biol 2003;13(13):1116–21.

[81] Couldrey C, Lee RS. DNA methylation patterns in tissues from mid-gestation bovine foetuses produced by somatic cell nuclear transfer show subtle abnormalities in nuclear reprogramming. BMC Dev Biol 2010;10.

[82] Lee I, Rasoul BA, Holub AS, Lejeune A, Enke RA, Timp W. Whole genome DNA methylation sequencing of the chicken retina, cornea and brain. Sci Data 2017;4:170148.

[83] Liu Z, Han S, Shen X, Wang Y, Cui C, He H, et al. The landscape of DNA methylation associated with the transcriptomic network in layers and broilers generates insight into embryonic muscle development in chicken. Int J Biol Sci 2019;15(7):1404–18.

[84] Wang M, Liang Y, Ibeagha-Awemu EM, Li M, Zhang H, Chen Z, et al. Genome-wide DNA methylation analysis of mammary gland tissues from chinese holstein cows with staphylococcus aureus induced mastitis. Front Genet 2020;11:1295.

[85] Zhou Y, Xu L, Bickhart DM, Abdel Hay EH, Schroeder SG, Connor EE, et al. Reduced representation bisulphite sequencing of ten bovine somatic tissues reveals DNA methylation patterns and their impacts on gene expression. BMC Genomics 2016;17(1):779.

[86] Su Y, Fan Z, Wu X, Li Y, Wang F, Zhang C, et al. Genome-wide DNA methylation profile of developing deciduous tooth germ in miniature pigs. BMC Genomics 2016;17(1):134.

[87] Kowalski A, Pałyga J. Distribution of non-allelic histone H1 subtypes in five avian species. Ann Anim Sci 2017;17(2):385−98.

[88] Skinner M, Lumey LH, Fleming TP, Sapienza C, Hoyo C, Aronica L, et al. RW-2018-research workshop: the effect of nutrition on epigenetic status, growth, and health. JPEN J Parenter Enter Nutr 2019;43(5):627−37.

[89] Barker DJ. In utero programming of chronic disease. Clin Sci (London, Engl: 1979) 1998;95(2):115−28.

[90] Wu G, Bazer FW, Wallace JM, Spencer TE. Board-invited review: intrauterine growth retardation: implications for the animal sciences. J Anim Sci 2006;84(9):2316−37.

[91] Bellows RA, Pope AL, Chapman AB, Casida LE. Effect of level and sequence of feeding and breed on ovulation rate, embryo survival and fetal growth in the mature Ewe. J Anim Sci 1963;22(1):101−8.

[92] Radunz AE, Fluharty FL, Relling AE, Felix TL, Shoup LM, Zerby HN, et al. Prepartum dietary energy source fed to beef cows: II. Effects on progeny postnatal growth, glucose tolerance, and carcass composition. J Anim Sci 2012;90(13):4962−74.

[93] Radunz AE, Fluharty FL, Susin I, Felix TL, Zerby HN, Loerch SC. Winter-feeding systems for gestating sheep II. Effects on feedlot performance, glucose tolerance, and carcass composition of lamb progeny. J Anim Sci 2011;89(2):478−88.

[94] Radunz AE, Fluharty FL, Zerby HN, Loerch SC. Winter-feeding systems for gestating sheep I. Effects on pre- and post-partum ewe performance and lamb progeny preweaning performance. J Anim Sci 2011;89(2):467−77.

[95] Boyd GW, Kiser TE, Lowrey RS. Effects of prepartum energy intake on steroids during late gestation and on cow and calf performance. J Anim Sci 1987;64(6):1703−9.

[96] Collier RJ, Doelger SG, Head HH, Thatcher WW, Wilcox CJ. Effects of heat stress during pregnancy on maternal hormone concentrations, calf birth weight and postpartum milk yield of Holstein cows. J Anim Sci 1982;54(2):309−19.

[97] Holst P, Killeen I, Cullis B. Nutrition of the pregnant ewe and its effect on gestation length, lamb birth weight and lamb survival. Aust J Agric Res 1986;37(6):647−55.

[98] Blecha F, Bull RC, Olson DP, Ross RH, Curtis S. Effects of prepartum protein restriction in the beef cow on immunoglobin content in blood and colostral whey and subsequent immunoglobin absorption by the neonatal calf. J Anim Sci 1981;53(5):1174−80.

[99] Wu C, Sirard M-A. Parental effects on epigenetic programming in gametes and embryos of dairy cows. Front Genet 2020;11:1234.

[100] Altmann S, Murani E, Schwerin M, Metges CC, Wimmers K, Ponsuksili S. Maternal dietary protein restriction and excess affects offspring gene expression and methylation of non-SMC subunits of condensin I in liver and skeletal muscle. Epigenetics. 2012;7(3):239−52.

[101] Lan X, Cretney EC, Kropp J, Khateeb K, Berg MA, Penagaricano F, et al. Maternal diet during pregnancy induces gene expression and DNA methylation changes in fetal tissues in sheep. Front Genet 2013;4:49.

[102] Penagaricano F, Wang X, Rosa GJ, Radunz AE, Khatib H. Maternal nutrition induces gene expression changes in fetal muscle and adipose tissues in sheep. BMC Genomics 2014;15:1034.

[103] Penagaricano F, Souza AH, Carvalho PD, Driver AM, Gambra R, Kropp J, et al. Effect of maternal methionine supplementation on the transcriptome of bovine preimplantation embryos. PLoS One 2013;8(8):e72302.

[104] Gauvin MC, Pillai SM, Reed SA, Stevens JR, Hoffman ML, Jones AK, et al. Poor maternal nutrition during gestation in sheep alters prenatal muscle growth and development in offspring. J Anim Sci 2020;98(1):skz388.

[105] Khanal P, Nielsen MO. Impacts of prenatal nutrition on animal production and performance: a focus on growth and metabolic and endocrine function in sheep. J Anim Sci Biotechnol 2017;8(1):75.

[106] Slyvka Y, Zhang Y, Nowak FV. Epigenetic effects of paternal diet on offspring: emphasis on obesity. Endocrine. 2015;48(1):36−46.

[107] Ng SF, Lin RC, Laybutt DR, Barres R, Owens JA, Morris MJ. Chronic high-fat diet in fathers programs β-cell dysfunction in female rat offspring. Nature. 2010;467(7318):963−6.

[108] Ryan DP, Henzel KS, Pearson BL, Siwek ME, Papazoglou A, Guo L, et al. A paternal methyl donor-rich diet altered cognitive and neural functions in offspring mice. Mol Psychiatry 2018;23(5):1345−55.

[109] Morgan HL, Paganopoulou P, Akhtar S, Urquhart N, Philomin R, Dickinson Y, et al. Paternal diet impairs F1 and F2 offspring vascular function through sperm and seminal plasma specific mechanisms in mice. J Physiol 2020;598(4):699−715.

[110] Gholami H, Chamani M, Towhidi A, Fazeli MH. Effect of feeding a docosahexaenoic acid-enriched nutriceutical on the quality of fresh and frozen-thawed semen in Holstein bulls. Theriogenology. 2010;74(9):1548−58.

[111] Samadian F, Towhidi A, Rezayazdi K, Bahreini M. Effects of dietary n-3 fatty acids on characteristics and lipid composition of ovine sperm. Animal. 2010;4(12):2017−22.

[112] Gross N, Taylor T, Crenshaw T, Khatib H. The intergenerational impacts of paternal diet on DNA methylation and offspring phenotypes in sheep. Front Genet 2020;11:597943.

[113] Emam M, Livernois A, Paibomesai M, Atalla H, Mallard B. Genetic and epigenetic regulation of immune response and resistance to infectious diseases in domestic ruminants. Vet Clin North Am Food Anim Pract 2019;35(3):405−29.

[114] Vanselow J, Yang W, Herrmann J, Zerbe H, Schuberth HJ, Petzl W, et al. DNA-remethylation around a STAT5-binding enhancer in the αS1-casein promoter is associated with abrupt shutdown of a αS1-casein synthesis during acute mastitis. J Mol Endocrinol 2006;37(3):463−77.

[115] Wang XS, Zhang Y, He YH, Ma PP, Fan LJ, Wang YC, et al. Aberrant promoter methylation of the CD4 gene in peripheral blood cells of mastitic dairy cows. Genet Mol Res 2013;12(4):6228−39.

[116] Usman T, Yu Y, Wang Y. P2001 CD4 promoter hyper methylation is associated with lower gene expression in clinical mastitis cows and vice versa in the healthy controls. J Anim Sci 2016;94(suppl_4):38.

[117] Doherty R, Whiston R, Cormican P, Finlay EK, Couldrey C, Brady C, et al. The CD4(+) T cell methylome contributes to a distinct CD4(+) T cell transcriptional signature in Mycobacterium bovis-infected cattle. Sci Rep 2016;6:31014.

[118] Doherty R, O'Farrelly C, Meade KG. Epigenetic regulation of the innate immune response to LPS in bovine peripheral blood mononuclear cells (PBMC). Vet Immunol Immunopathol 2013;154(3−4):102−10.

[119] Chen J, Wu Y, Sun Y, Dong X, Wang Z, Zhang Z, et al. Bacterial lipopolysaccharide induced alterations of genome-wide DNA methylation and promoter methylation of lactation-related genes in bovine mammary epithelial cells. Toxins. 2019;11(5):298.

[120] Wang J, Yan X, Nesengani LT, Ding H, Yang L, Lu W. LPS-induces IL-6 and IL-8 gene expression in bovine endometrial cells "through DNA methylation. Gene 2018;677:266−72.

[121] Wu Y, Sun Y, Dong X, Chen J, Wang Z, Chen J, et al. The synergism of PGN, LTA and LPS in inducing transcriptome changes, inflammatory responses and a decrease in lactation as well as the associated epigenetic mechanisms in bovine mammary epithelial cells. Toxins. 2020;12(6):387.

[122] Ju Z, Jiang Q, Wang J, Wang X, Yang C, Sun Y, et al. Genome-wide methylation and transcriptome of blood neutrophils reveal the roles of DNA methylation in affecting transcription of protein-coding genes and miRNAs in E. coli-infected mastitis cows. BMC Genomics 2020;21(1):102.

[123] Sajjanar B, Trakooljul N, Wimmers K, Ponsuksili S. DNA methylation analysis of porcine mammary epithelial cells reveals differentially methylated loci associated with immune response against Escherichia coli challenge. BMC Genomics 2019;20(1):623.

[124] Song M, He Y, Zhou H, Zhang Y, Li X, Yu Y. Combined analysis of DNA methylome and transcriptome reveal novel candidate genes with susceptibility to bovine Staphylococcus aureus subclinical mastitis. Sci Rep 2016;6:29390.

[125] Chang G, Liu X, Ma N, Yan J, Dai H, Roy AC, et al. Dietary addition of sodium butyrate contributes to attenuated feeding-induced hepatocyte apoptosis in dairy goats. J Agric Food Chem 2018;66(38):9995—10002.

[126] Hernaiz A, Sentre S, Bolea R, López-Pérez O, Sanz A, Zaragoza P, et al., editors. Epigenetic changes in the central nervous system of sheep naturally infected with scrapie. Proceedings of the 18th Jornadas Sobre Producción Animal; 2019; Zaragoza: Asociación Interprofesional para el Desarrollo Agrario.

[127] Li J, Li R, Wang Y, Hu X, Zhao Y, Li L, et al. Genome-wide DNA methylome variation in two genetically distinct chicken lines using MethylC-seq. BMC Genomics 2015;16:851.

[128] Tian F, Zhan F, VanderKraats ND, Hiken JF, Edwards JR, Zhang H, et al. DNMT gene expression and methylome in Marek's disease resistant and susceptible chickens prior to and following infection by MDV. Epigenetics. 2013;8(4):431—44.

[129] Zhang YH, Meng JL, Gao Y, Zhang JY, Niu SL, Yu XZ, et al. Changes in methylation of genomic DNA from chicken immune organs in response to H5N1 influenza virus infection. Genet Mol Res 2016;15(3).

[130] Wang F, Li J, Li Q, Liu R, Zheng M, Wang Q, et al. Changes of host DNA methylation in domestic chickens infected with Salmonella enterica. J Genet 2017;96(4):545—50.

[131] Wang H, Wu J, Wu S, Wu S, Bao W. DNA methylation differences of the BPI promoter among pig breeds and the regulation of gene expression. RSC Adv 2017;7(76):48025—30.

[132] Feng W, Zhou L, Wang H, Hu Z, Wang X, Fu J, et al. Functional analysis of DNA methylation of the PACSIN1 promoter in pig peripheral blood mononuclear cells. J Cell Biochem 2019;120(6):10118—27.

[133] Zhang J, Yang Q, Yang J, Gao X, Luo R, Huang X, et al. Comprehensive analysis of transcriptome-wide m(6)A methylome upon clostridium perfringens beta2 toxin exposure in porcine intestinal epithelial cells by m(6)A sequencing. Front Genet 2021;12:689748.

[134] Li T, Lin C, Zhu Y, Xu H, Yin Y, Wang C, et al. Transcriptome profiling of m6A mRNA modification in bovine mammary epithelial cells treated with Escherichia coli. Int J Mol Sci 2021;22(12):6254.

[135] He Y, Yu Y, Zhang Y, Song J, Mitra A, Zhang Y, et al. Genome-wide bovine H3K27me3 modifications and the regulatory effects on genes expressions in peripheral blood lymphocytes. PLoS One 2012;7(6):e39094.

[136] Chang G, Petzl W, Vanselow J, Gunther J, Shen X, Seyfert HM. Epigenetic mechanisms contribute to enhanced expression of immune response genes in the liver of cows after experimentally induced Escherichia coli mastitis. Vet J (London, Engl: 1997) 2015;203(3):339—41.

[137] Hall TJ, Vernimmen D, Browne JA, Mullen MP, Gordon SV, MacHugh DE, et al. Alveolar macrophage chromatin is modified to orchestrate host response to mycobacterium bovis infection. Front Genet 2019;10:1386.

[138] Herrera-Uribe J, Liu H, Byrne KA, Bond ZF, Loving CL, Tuggle CK. Changes in H3K27ac at gene regulatory regions in porcine alveolar macrophages following LPS or PolyIC exposure. Front Genet 2020;11(817):817.

[139] Mitra A, Luo J, He Y, Gu Y, Zhang H, Zhao K, et al. Histone modifications induced by MDV infection at early cytolytic and latency phases. BMC Genomics 2015;16(1):311.

[140] Song J. P2020 Genome-wide assessment of inbred chicken lines indicates epigenetics signatures of resistance to Marek's disease. J Anim Sci 2016;94(suppl_4):47.

[141] Chanthavixay G, Kern C, Wang Y, Saelao P, Lamont SJ, Gallardo RA, et al. Integrated transcriptome and histone modification analysis reveals NDV infection under heat stress affects bursa development and proliferation in susceptible chicken line. Front Genet 2020;11:567812.

[142] Wu Y, Chen J, Sun Y, Dong X, Wang Z, Chen J, et al. PGN and LTA from staphylococcus aureus induced inflammation and decreased lactation through regulating DNA methylation and histone H3 acetylation in bovine mammary epithelial cells. Toxins. 2020;12(4):238.

[143] Pan X, Gong D, Nguyen DN, Zhang X, Hu Q, Lu H, et al. Early microbial colonization affects DNA methylation of genes related to intestinal immunity and metabolism in preterm pigs. DNA Res 2018;25(3):287—96.

[144] Pan X, Thymann T, Gao F, Sangild PT. Rapid gut adaptation to preterm birth involves feeding-related DNA methylation reprogramming of intestinal genes in pigs. Front Immunol 2020;11:565.

[145] Ibeagha-Awemu EM, Bissonnette N, Bhattarai S, Wang M, Dudemaine P-L, McKay SD, et al. Whole genome methylation analysis reveals role of DNA methylation in cow's ileal and ileal lymph node responses to Mycobacterium avium subsp. paratuberculosis infection. Front Genet 2021;12:797490.

[146] Wang H, Yang L, Qu H, Feng H, Wu S, Bao W. Global mapping of H3K4 trimethylation (H3K4me3) and transcriptome analysis reveal genes involved in the response to epidemic diarrhea virus infections in pigs. Animals (Basel) 2019;9(8):523.

[147] Zhang Y, Wang X, Jiang Q, Hao H, Ju Z, Yang C, et al. DNA methylation rather than single nucleotide polymorphisms regulates the production of an aberrant splice variant of IL6R in mastitic cows. Cell Stress Chaperones 2018;23(4):617—28.

[148] Chang G, Zhang K, Xu T, Jin D, Guo J, Zhuang S, et al. Epigenetic mechanisms contribute to the expression of immune related genes in the livers of dairy cows fed a high concentrate diet. PLoS One 2015;10(4):e0123942.

[149] Fu Q, Shi H, Chen C. Roles of bta-miR-29b promoter regions DNA methylation in regulating miR-29b expression and bovine viral diarrhea virus NADL replication in MDBK cells. Arch Virol 2017;162(2):401—8.

[150] Wang H, Wang J, Ning C, Zheng X, Fu J, Wang A, et al. Genome-wide DNA methylation and transcriptome analyses reveal genes involved in immune responses of pig peripheral blood mononuclear cells to poly I:C. Sci Rep 2017;7(1):9709.

[151] Lu T, Song Z, Li Q, Li Z, Wang M, Liu L, et al. Overexpression of histone deacetylase 6 enhances resistance to porcine reproductive and respiratory syndrome virus in pigs. PLoS One 2017;12(1):e0169317.

[152] He Y, Song M, Zhang Y, Li X, Song J, Zhang Y, et al. Whole-genome regulation analysis of histone H3 lysin 27 trimethylation in subclinical mastitis cows infected by *Staphylococcus aureus*. BMC Genomics 2016;17:565.

[153] Corbett RJ, Luttman AM, Wurtz KE, Siegford JM, Raney NE, Ford LM, et al. Weaning induces stress-dependent DNA methylation and transcriptional changes in piglet PBMCs. Front Genet 2021;12, 633564.

[154] David S-A, Vitorino Carvalho A, Gimonnet C, Brionne A, Hennequet-Antier C, Piégu B, et al. Thermal manipulation during embryogenesis impacts H3K4me3 and H3K27me3 histone marks in chicken hypothalamus. Front Genet 2019;10:1207.

[155] Lu Z, Ma Y, Li Q, Liu E, Jin M, Zhang L, et al. The role of N (6)-methyladenosine RNA methylation in the heat stress response of sheep (Ovis aries). Cell Stress Chaperones 2019;24 (2):333–42.

[156] Zhang B, Ban D, Gou X, Zhang Y, Yang L, Chamba Y, et al. Genome-wide DNA methylation profiles in Tibetan and Yorkshire pigs under high-altitude hypoxia. J Anim Sci Biotechnol 2019;10:25.

[157] Wang Y, Luo M, Jiang M, Lin Y, Zhu J. Comparative analysis of tissue expression and methylation reveals the crucial hypoxia genes in hypoxia-resistant animals. Can J Anim Sci 2017;98(1):204–12.

[158] Littlejohn BP, Price DM, Neuendorff DA, Carroll JA, Vann RC, Riggs PK, et al. Prenatal transportation stress alters genome-wide DNA methylation in suckling Brahman bull calves. J Anim Sci 2018;96(12):5075–99.

[159] Kisliouk T, Cramer T, Meiri N. Methyl CpG level at distal part of heat-shock protein promoter HSP70 exhibits epigenetic memory for heat stress by modulating recruitment of POU2F1-associated nucleosome-remodeling deacetylase (NuRD) complex. J Neurochem 2017;141(3):358–72.

[160] Vinoth A, Thirunalasundari T, Shanmugam M, Uthrakumar A, Suji S, Rajkumar U. Evaluation of DNA methylation and mRNA expression of heat shock proteins in thermal manipulated chicken. Cell Stress Chaperones 2018;23(2):235–52.

[161] Hao Y, Cui Y, Gu X. Genome-wide DNA methylation profiles changes associated with constant heat stress in pigs as measured by bisulfite sequencing. Sci Rep 2016;6:27507.

[162] Do DN, Dudemaine P-L, Mathur M, Suravajhala P, Zhao X, Ibeagha-Awemu EM. MiRNA regulatory functions in farm animal diseases, and biomarker potentials for effective therapies. Int J Mol Sci 2021;22(6):3080.

[163] Dong H, Gao Q, Peng X, Sun Y, Han T, Zhao B, et al. Circulating microRNAs As potential biomarkers for veterinary infectious diseases. Front Vet Sci 2017;4:186.

[164] Szczepanek J, Pareek C, Tretyn A, editors. The role of microRNAs in animal physiology and pathology. Transl Anim Sci 2018;.

[165] Taxis TM, Casas E. MicroRNA expression and implications for infectious diseases in livestock. CAB Rev 2017;12:26.

[166] Lawless N, Foroushani AB, McCabe MS, O'Farrelly C, Lynn DJ. Next generation sequencing reveals the expression of a unique miRNA profile in response to a gram-positive bacterial infection. PLoS One 2013;8(3):e57543.

[167] Naeem A, Zhong K, Moisá SJ, Drackley JK, Moyes KM, Loor JJ. Bioinformatics analysis of microRNA and putative target genes in bovine mammary tissue infected with Streptococcus uberis. J Dairy Sci 2012;95(11):6397–408.

[168] Li Z, Wang H, Chen L, Wang L, Liu X, Ru C, et al. Identification and characterization of novel and differentially expressed microRNAs in peripheral blood from healthy and mastitis Holstein cattle by deep sequencing. Anim Genet 2014;45(1):20–7.

[169] Jin W, Ibeagha-Awemu EM, Liang G, Beaudoin F, Zhao X, Guan le L. Transcriptome microRNA profiling of bovine mammary epithelial cells challenged with Escherichia coli or Staphylococcus aureus bacteria reveals pathogen directed microRNA expression profiles. BMC Genomics 2014;15:181.

[170] Pu J, Li R, Zhang C, Chen D, Liao X, Zhu Y, et al. Expression profiles of miRNAs from bovine mammary glands in response to *Streptococcus agalactiae*-induced mastitis. The. J Dairy Res 2017;84(3):300–8.

[171] Luoreng Z-M, Wang X-P, Mei C-G, Zan L-S. Comparison of microRNA profiles between bovine mammary glands infected with *Staphylococcus aureus* and *Escherichia coli*. Int J Biol Sci 2018;14(1):87–99.

[172] Lai YC, Fujikawa T, Maemura T, Ando T, Kitahara G, Endo Y, et al. Inflammation-related microRNA expression level in the bovine milk is affected by mastitis. PLoS One 2017;12(5): e0177182.

[173] Vegh P, Magee DA, Nalpas NC, Bryan K, McCabe MS, Browne JA, et al. MicroRNA profiling of the bovine alveolar macrophage response to Mycobacterium bovis infection suggests pathogen survival is enhanced by microRNA regulation of endocytosis and lysosome trafficking. Tuberculosis (Edinburgh, Scotl) 2015;95(1):60–7.

[174] Malvisi M, Palazzo F, Morandi N, Lazzari B, Williams JL, Pagnacco G, et al. Responses of bovine innate immunity to Mycobacterium avium subsp. paratuberculosis infection revealed by changes in gene expression and levels of microRNA. PLoS One 2016;11(10):e0164461.

[175] Liang G, Malmuthuge N, Guan Y, Ren Y, Griebel PJ, Guan le L. Altered microRNA expression and pre-mRNA splicing events reveal new mechanisms associated with early stage Mycobacterium avium subspecies paratuberculosis infection. Sci Rep 2016;6:24964.

[176] Gupta SK, Maclean PH, Ganesh S, Shu D, Buddle BM, Wedlock DN, et al. Detection of microRNA in cattle serum and their potential use to diagnose severity of Johne's disease. J Dairy Sci 2018;101(11):10259–70.

[177] Shaughnessy RG, Farrell D, Stojkovic B, Browne JA, Kenny K, Gordon SV. Identification of microRNAs in bovine faeces and their potential as biomarkers of Johne's Disease. Sci Rep 2020;10(1):5908.

[178] Taxis TM, Bauermann FV, Ridpath JF, Casas E. Circulating microRNAs in serum from cattle challenged with bovine viral diarrhea virus. Front Genet 2017;8:91.

[179] Stenfeldt C, Arzt J, Smoliga G, LaRocco M, Gutkoska J, Lawrence P. Proof-of-concept study: profile of circulating microRNAs in Bovine serum harvested during acute and persistent FMDV infection. Virol J 2017;14(1):71.

[180] Yoo JAH, Liu D. HC. Characterization of the microRNAome in porcine reproductive and respiratory syndrome virus infected macrophages. PLoS One 2013;8(12):e82054.

[181] Guo XK, Zhang Q, Gao L, Li N, Chen XX, Feng WH. Increasing expression of microRNA 181 inhibits porcine reproductive and respiratory syndrome virus replication and has implications for controlling virus infection. J Virol 2013;87 (2):1159–71.

[182] Wang D, Cao L, Xu Z, Fang L, Zhong Y, Chen Q, et al. MiR-125b reduces porcine reproductive and respiratory syndrome virus replication by negatively regulating the NF-κB pathway. PLoS One 2013;8(2):e55838.

[183] Zhang Q, Guo XK, Gao L, Huang C, Li N, Jia X, et al. MicroRNA-23 inhibits PRRSV replication by directly targeting PRRSV RNA and possibly by upregulating type I interferons. Virology. 2014;450–451:182–95.

[184] Zhou A, Li S, Zhang S. miRNAs and genes expression in MARC-145 cell in response to PRRSV infection. Infect Genet Evol 2014;27:173—80.

[185] Li J, Chen Z, Zhao J, Fang L, Fang R, Xiao J, et al. Difference in microRNA expression and editing profile of lung tissues from different pig breeds related to immune responses to HP-PRRSV. Sci Rep 2015;5:9549.

[186] Li L, Wei Z, Zhou Y, Gao F, Jiang Y, Yu L, et al. Host miR-26a suppresses replication of porcine reproductive and respiratory syndrome virus by upregulating type I interferons. Virus Res 2015;195:86—94.

[187] Zhang Q, Huang C, Yang Q, Gao L, Liu HC, Tang J, et al. MicroRNA-30c modulates type I IFN responses to facilitate porcine reproductive and respiratory syndrome virus infection by targeting JAK1. J Immunol 2016;196(5):2272—82.

[188] Xiao S, Du T, Wang X, Ni H, Yan Y, Li N, et al. MiR-22 promotes porcine reproductive and respiratory syndrome virus replication by targeting the host factor HO-1. Vet Microbiol 2016;192:226—30.

[189] Chen J, Shi X, Zhang X, Wang A, Wang L, Yang Y, et al. MicroRNA 373 facilitates the replication of porcine reproductive and respiratory syndrome virus by its negative regulation of Type I interferon induction. J Virol 2017;91(3) e01311-16.

[190] Zhou X, Michal JJ, Jiang Z, Liu B. MicroRNA expression profiling in alveolar macrophages of indigenous Chinese Tongcheng pigs infected with PRRSV in vivo. J Appl Genet 2017;58(4):539—44.

[191] Zhao G, Hou J, Xu G, Xiang A, Kang Y, Yan Y, et al. Cellular microRNA miR-10a-5p inhibits replication of porcine reproductive and respiratory syndrome virus by targeting the host factor signal recognition particle 14. The. J Gen Virol 2017;98(4):624—32.

[192] Fleming DS, Miller LC. Identification of small non-coding RNA classes expressed in swine whole blood during HP-PRRSV infection. Virology. 2018;517:56—61.

[193] He T, Feng G, Chen H, Wang L, Wang Y. Identification of host encoded microRNAs interacting with novel swine-origin influenza A (H1N1) virus and swine influenza virus. Bioinformation. 2009;4(3):112—18.

[194] Skovgaard K, Cirera S, Vasby D, Podolska A, Breum S, Dürrwald R, et al. Expression of innate immune genes, proteins and microRNAs in lung tissue of pigs infected experimentally with influenza virus (H1N2). Innate Immun 2013;19(5):531—44.

[195] Huang TH, Uthe JJ, Bearson SM, Demirkale CY, Nettleton D, Knetter S, et al. Distinct peripheral blood RNA responses to Salmonella in pigs differing in Salmonella shedding levels: intersection of IFNG, TLR and miRNA pathways. PLoS One 2011;6(12):e28768.

[196] Hoeke L, Sharbati J, Pawar K, Keller A, Einspanier R, Sharbati S. Intestinal *Salmonella typhimurium* infection leads to miR-29a induced caveolin 2 regulation. PLoS One 2013;8(6):e67300.

[197] Li H, Zhang M, Zheng E. Comprehensive miRNA expression profiles in the ilea of Lawsonia intracellularis-infected pigs. J Vet Med Sci 2017;79(2):282—9.

[198] Ye L, Su X, Wu Z, Zheng X, Wang J, Zi C, et al. Analysis of differential miRNA expression in the duodenum of Escherichia coli F18-sensitive and -resistant weaned piglets. PLoS One 2012;7(8):e43741.

[199] Wu Z, Qin W, Wu S, Zhu G, Bao W, Wu S. Identification of microRNAs regulating Escherichia coli F18 infection in Meishan weaned piglets. Biol Direct 2016;11(1):59.

[200] Hansen EP, Kringel H, Thamsborg SM, Jex A, Nejsum P. Profiling circulating miRNAs in serum from pigs infected with the porcine whipworm, Trichuris suis. Vet Parasitol 2016;223:30—3.

[201] Podolska A, Anthon C, Bak M, Tommerup N, Skovgaard K, Heegaard PM, et al. Profiling microRNAs in lung tissue from pigs infected with *Actinobacillus pleuropneumoniae*. BMC Genomics 2012;13:459.

[202] Wang J, Xie H, Ling Q, Lu D, Lv Z, Zhuang R, et al. Coding-noncoding gene expression in intrahepatic cholangiocarcinoma. Transl Res 2016;168:107—21.

[203] Núñez-Hernández F, Pérez LJ, Muñoz M, Vera G, Accensi F, Sánchez A, et al. Differential expression of porcine microRNAs in African swine fever virus infected pigs: a proof-of-concept study. Virol J 2017;14(1):198.

[204] Brogaard L, Larsen LE, Heegaard PMH, Anthon C, Gorodkin J, Dürrwald R, et al. IFN-λ and microRNAs are important modulators of the pulmonary innate immune response against influenza A (H1N2) infection in pigs. PLoS One 2018;13(4):e0194765.

[205] Lian L, Qu L, Chen Y, Lamont SJ, Yang N. A systematic analysis of miRNA transcriptome in Marek's disease virus-induced lymphoma reveals novel and differentially expressed miRNAs. PLoS One 2012;7(11):e51003.

[206] Tian F, Luo J, Zhang H, Chang S, Song J. MiRNA expression signatures induced by Marek's disease virus infection in chickens. Genomics. 2012;99(3):152—9.

[207] Han B, Lian L, Li X, Zhao C, Qu L, Liu C, et al. Chicken gga-miR-103-3p targets CCNE1 and TFDP2 and inhibits MDCC-MSB1 cell migration. G3 (Bethesda, Md) 2016;6(5):1277—85.

[208] Zhao C, Li X, Han B, You Z, Qu L, Liu C, et al. Gga-miR-219b targeting BCL11B suppresses proliferation, migration and invasion of Marek's disease tumor cell MSB1. Sci Rep 2017;7(1):4247.

[209] Heidari M, Zhang L, Zhang H. MicroRNA profiling in the bursae of Marek's disease virus-infected resistant and susceptible chicken lines. Genomics. 2020;112(3):2564—71.

[210] Li H, Ji J, Xie Q, Shang H, Zhang H, Xin X, et al. Aberrant expression of liver microRNA in chickens infected with subgroup J avian leukosis virus. Virus Res 2012;169(1):268—71.

[211] Dai Z, Ji J, Yan Y, Lin W, Li H, Chen F, et al. Role of gga-miR-221 and gga-miR-222 during tumour formation in chickens infected by subgroup J avian leukosis virus. Viruses. 2015;7(12):6538—51.

[212] Wang Q, Gao Y, Ji X, Qi X, Qin L, Gao H, et al. Differential expression of microRNAs in avian leukosis virus subgroup J-induced tumors. Vet Microbiol 2013;162(1):232—8.

[213] Ji J, Shang H, Zhang H, Li H, Ma J, Bi Y, et al. Temporal changes of microRNA gga-let-7b and gga-let-7i expression in chickens challenged with subgroup J avian leukosis virus. Vet Res Commun 2017;41(3):219—26.

[214] Li Z, Luo Q, Xu H, Zheng M, Abdalla BA, Feng M, et al. MiR-34b-5p suppresses melanoma differentiation-associated Gene 5 (MDA5) signaling pathway to promote avian leukosis virus subgroup J (ALV-J)-infected cells proliferaction and ALV-J replication. Front Cell Infect Microbiol 2017;7:17.

[215] Zhou D, Xue J, He S, Du X, Zhou J, Li C, et al. Reticuloendotheliosis virus and avian leukosis virus subgroup J synergistically increase the accumulation of exosomal miRNAs. Retrovirology. 2018;15(1):45.

[216] Fu M, Wang B, Chen X, He Z, Wang Y, Li X, et al. MicroRNA gga-miR-130b suppresses infectious bursal disease virus replication via targeting of the viral genome and cellular suppressors of cytokine signaling 5. J Virol 2018;92(1) e01646-17.

[217] Ouyang W, Wang YS, Du XN, Liu HJ, Zhang HB. gga-miR-9* inhibits IFN production in antiviral innate immunity by targeting interferon regulatory factor 2 to promote IBDV replication. Vet Microbiol 2015;178(1—2):41—9.

[218] Ouyang W, Wang YS, Meng K, Pan QX, Wang XL, Xia XX, et al. gga-miR-2127 downregulates the translation of chicken

p53 and attenuates chp53-mediated innate immune response against IBDV infection. Vet Microbiol 2017;198:34–42.

[219] Wang Y, Brahmakshatriya V, Zhu H, Lupiani B, Reddy SM, Yoon BJ, et al. Identification of differentially expressed miRNAs in chicken lung and trachea with avian influenza virus infection by a deep sequencing approach. BMC Genomics 2009;10:512.

[220] Wang Y, Brahmakshatriya V, Lupiani B, Reddy SM, Soibam B, Benham AL, et al. Integrated analysis of microRNA expression and mRNA transcriptome in lungs of avian influenza virus infected broilers. BMC Genomics 2012;13:278.

[221] Peng X, Gao QS, Zhou L, Chen ZH, Lu S, Huang HJ, et al. MicroRNAs in avian influenza virus H9N2-infected and non-infected chicken embryo fibroblasts. Genet Mol Res 2015;14(3):9081–91.

[222] Zhao Y, Hou Y, Zhang K, Yuan B, Peng X. Identification of differentially expressed miRNAs through high-throughput sequencing in the chicken lung in response to Mycoplasma gallisepticum HS. Comparative biochemistry and physiology Part D. Genomics Proteom 2017;22:146–56.

[223] Zhao Y, Wang Z, Hou Y, Zhang K, Peng X. gga-miR-99a targets SMARCA5 to regulate Mycoplasma gallisepticum (HS strain) infection by depressing cell proliferation in chicken. Gene. 2017;627:239–47.

[224] Hu Q, Zhao Y, Wang Z, Hou Y, Bi D, Sun J, et al. Chicken gga-miR-19a targets ZMYND11 and plays an important role in host defense against Mycoplasma gallisepticum (HS Strain) infection. Front Cell Infect Microbiol 2016;6:102.

[225] Jiang S, Li X, Wang X, Ban Q, Hui W, Jia B. MicroRNA profiling of the intestinal tissue of Kazakh sheep after experimental Echinococcus granulosus infection, using a high-throughput approach. Parasite (Paris, Fr) 2016;23:23.

[226] Wang B, Ye N, Cao SJ, Wen XT, Huang Y, Yan QG. Identification of novel and differentially expressed microRNAs in goat enzootic nasal adenocarcinoma. BMC Genomics 2016;17(1):896.

[227] Du J, Gao S, Tian Z, Xing S, Huang D, Zhang G, et al. MicroRNA expression profiling of primary sheep testicular cells in response to bluetongue virus infection. Infect Genet Evol 2017;49:256–67.

[228] Pandey A, Sahu AR, Wani SA, Saxena S, Kanchan S, Sah V, et al. Modulation of host miRNAs transcriptome in lung and spleen of peste des petits ruminants virus infected sheep and goats. Front Microbiol 2017;8:1146.

[229] Sanz Rubio D, López-Pérez Ó, de Andrés Pablo Á, Bolea R, Osta R, Badiola JJ, et al. Increased circulating microRNAs miR-342-3p and miR-21-5p in natural sheep prion disease. The. J Gen virology 2017;98(2):305–10.

[230] Yang Y, Qin X, Meng X, Zhu X, Zhang X, Li Y, et al. MicroRNA expression profile in peripheral blood lymphocytes of sheep vaccinated with Nigeria 75/1 peste des petits ruminants virus. Viruses. 2019;11:11.

[231] Bilbao-Arribas M, Abendaño N, Varela-Martínez E, Reina R, de Andrés D, Jugo BM. Expression analysis of lung miRNAs responding to ovine VM virus infection by RNA-seq. BMC Genomics 2019;20(1):62.

[232] Qi X, Wang T, Xue Q, Li Z, Yang B, Wang J. MicroRNA expression profiling of goat peripheral blood mononuclear cells in response to peste des petits ruminants virus infection. Vet Res 2018;49(1):62.

[233] Qi X, Li Z, Li H, Wang T, Zhang Y, Wang J. MicroRNA-1 negatively regulates peripheral NK cell function via tumor necrosis factor-like weak inducer of apoptosis (TWEAK) Signaling pathways during PPRV infection. Front Immunol 2019;10:3066.

[234] Qi X, Wang T, Li Z, Wan Y, Yang B, Zeng W, et al. MicroRNA-218 regulates signaling lymphocyte activation molecular (SLAM) mediated peste des petits ruminants virus infectivity in goat peripheral blood mononuclear cells. Front Immunol 2019;10:2201.

[235] Devinoy E, Rijnkels M. Epigenetics in mammary gland biology and cancer. J Mammary Gland Biol Neoplasia 2010;15(1):1–4.

[236] Singh K, Erdman RA, Swanson KM, Molenaar AJ, Maqbool NJ, Wheeler TT, et al. Epigenetic regulation of milk production in dairy cows. J Mammary Gland Biol Neoplasia 2010;15(1):101–12.

[237] Wu J, Frazier K, Zhang J, Gan Z, Wang T, Zhong X. Emerging role of m6A RNA methylation in nutritional physiology and metabolism. Obes Rev 2020;21(1):e12942.

[238] Nguyen M, Boutinaud M, Pétridou B, Gabory A, Pannetier M, Chat S, et al. DNA methylation and transcription in a distal region upstream from the bovine AlphaS1 casein gene after once or twice daily milking. PLoS One 2014;9(11):e111556.

[239] Dechow CD, Liu WS. DNA methylation patterns in peripheral blood mononuclear cells from Holstein cattle with variable milk yield. BMC Genomics 2018;19(1):744.

[240] Wang L, Sun HZ, Guan LL, Liu JX. Short communication: relationship of blood DNA methylation rate and milk performance in dairy cows. J Dairy Sci 2019;102(6):5208–11.

[241] Wang M, Bissonnette N, Dudemaine P-L, Zhao X, Ibeagha-Awemu EM. Whole genome DNA methylation variations in mammary gland tissues from holstein cattle producing milk with various fat and protein contents. Genes. 2021;12(11):11.

[242] Zhang X, Zhang S, Ma L, Jiang E, Xu H, Chen R, et al. Reduced representation bisulfite sequencing (RRBS) of dairy goat mammary glands reveals DNA methylation profiles of integrated genome-wide and critical milk-related genes. Oncotarget 2017;8:70.

[243] Lu N, Li X, Yu J, Li Y, Wang C, Zhang L, et al. Curcumin attenuates lipopolysaccharide-induced hepatic lipid metabolism disorder by modification of m6A RNA methylation in piglets. Lipids. 2018;53(1):53–63.

[244] Zhang Y, Liang C, Wu X, Pei J, Guo X, Chu M, et al. Integrated study of transcriptome-wide m 6 A methylome reveals novel insights into the character and function of m 6 A methylation during yak adipocytes differentiation. Front Cell Dev Biol 2021;9:689067.

[245] Wang H, Shi H, Luo J, Yi Y, Yao D, Zhang X, et al. MiR-145 regulates lipogenesis in goat mammary cells via targeting INSIG1 and epigenetic regulation of lipid-related genes. J Cell Physiol 2017;232(5):1030–40.

[246] Tian P, Luo Y, Li X, Tian J, Tao S, Hua C, et al. Negative effects of long-term feeding of high-grain diets to lactating goats on milk fat production and composition by regulating gene expression and DNA methylation in the mammary gland. J Anim Sci Biotechnol 2017;8:74.

[247] Liang S, Nie ZW, Guo J, Niu YJ, Shin KT, Ock SA, et al. Overexpression of microRNA-29b decreases expression of DNA methyltransferases and improves quality of the blastocysts derived from somatic cell nuclear transfer in cattle. Microsc Microanal 2018;24(1):29–37.

[248] Melnik BC, Schmitz G. Milk's role as an epigenetic regulator in health and disease. Diseases. 2017;5(1):12.

[249] Wang J, Bian Y, Wang Z, Li D, Wang C, Li Q, et al. MicroRNA-152 regulates DNA methyltransferase 1 and is involved in the development and lactation of mammary glands in dairy cows. PLoS One 2014;9(7):e101358.

[250] Bian Y, Lei Y, Wang C, Wang J, Wang L, Liu L, et al. Epigenetic regulation of miR-29s affects the lactation activity of dairy cow mammary epithelial cells. J Cell Physiol 2015;230(9):2152–63.

[251] Dong W, Yang J, Zhang Y, Liu S, Ning C, Ding X, et al. Integrative analysis of genome-wide DNA methylation and gene expression profiles reveals important epigenetic genes related to milk production traits in dairy cattle. J Anim Breed Genet 2021;138(5):562–73.

[252] Liu X, Yang J, Zhang Q, Jiang L. Regulation of DNA methylation on EEF1D and RPL8 expression in cattle. Genetica. 2017;145(4–5):387–95.

[253] Zhang M, Li D, Zhai Y, Wang Z, Ma X, Zhang D, et al. The landscape of DNA methylation associated with the transcriptomic network of intramuscular adipocytes generates insight into intramuscular fat deposition in chicken. Front Cell Dev Biol 2020;8:206.

[254] Zhao C, Ji G, Carrillo JA, Li Y, Tian F, Baldwin RL, et al. The profiling of DNA methylation and its regulation on divergent tenderness in angus beef cattle. Front Genet 2020;11:939.

[255] Chen Z, Chu S, Xu X, Jiang J, Wang W, Shen H, et al. Analysis of longissimus muscle quality characteristics and associations with DNA methylation status in cattle. Genes. Genomics. 2019;41(10):1147–63.

[256] Ma X, Jia C, Chu M, Fu D, Lei Q, Ding X, et al. Transcriptome and DNA methylation analyses of the molecular mechanisms underlying with longissimus dorsi muscles at different stages of development in the polled yak. Genes. 2019;10(12):970.

[257] Park H, Seo K-S, Lee M, Seo S. Identification of meat quality-related differentially methylated regions in the DNA of the longissimus dorsi muscle in pig. Anim Biotechnol 2020;31(3):189–94.

[258] Wang X, Kadarmideen HN. Genome-wide DNA methylation analysis using next-generation sequencing to reveal candidate genes responsible for boar taint in pigs. Anim Genet 2019;50(6):644–59.

[259] Ponsuksili S, Trakooljul N, Basavaraj S, Hadlich F, Murani E, Wimmers K. Epigenome-wide skeletal muscle DNA methylation profiles at the background of distinct metabolic types and ryanodine receptor variation in pigs. BMC Genomics 2019;20(1):492.

[260] Wei D, Li A, Zhao C, Wang H, Mei C, Khan R, et al. Transcriptional regulation by CpG sites methylation in the core promoter region of the bovine SIX1 gene: roles of histone H4 and E2F2. Int J Mol Sci 2018;19(1).

[261] Cao Y, Jin HG, Ma HH, Zhao ZH. Comparative analysis on genome-wide DNA methylation in longissimus dorsi muscle between Small Tailed Han and Dorper × Small Tailed Han crossbred sheep. Asian-Australas J Anim Sci 2017;30(11):1529–39.

[262] Fang X, Zhao Z, Yu H, Li G, Jiang P, Yang Y, et al. Comparative genome-wide methylation analysis of longissimus dorsi muscles between Japanese black (Wagyu) and Chinese Red Steppes cattle. PLoS One 2017;12(8):e0182492.

[263] Paradis F, Wood KM, Swanson KC, Miller SP, McBride BW, Fitzsimmons C. Maternal nutrient restriction in mid-to-late gestation influences fetal mRNA expression in muscle tissues in beef cattle. BMC Genomics 2017;18(1):632.

[264] Wang J, Chen J, Zhang J, Gao B, Bai X, Lan Y, et al. Castration-induced changes in the expression profiles and promoter methylation of the GHR gene in Huainan male pigs. Anim Sci J 2017;88(8):1113–19.

[265] Kociucka B, Stachecka J, Szydlowski M, Szczerbal I. Rapid communication: the correlation between histone modifications and expression of key genes involved in accumulation of adipose tissue in the pig. J Anim Sci 2017;95(10):4514–19.

[266] Zhang M, Yan FB, Li F, Jiang KR, Li DH, Han RL, et al. Genome-wide DNA methylation profiles reveal novel candidate genes associated with meat quality at different age stages in hens. Sci Rep 2017;7:45564.

[267] Yang Y, Zhou R, Mu Y, Hou X, Tang Z, Li K. Genome-wide analysis of DNA methylation in obese, lean, and miniature pig breeds. Sci Rep 2016;6:30160.

[268] Zhang S, Shen L, Xia Y, Yang Q, Li X, Tang G, et al. DNA methylation landscape of fat deposits and fatty acid composition in obese and lean pigs. Sci Rep 2016;6:35063.

[269] Gao G, Wang H, Zhao X, Li Q, Wang C, Li J, et al. Effect of feeding conditions on the methylation status of Fatp1 gene in chicken breast muscle. Braz J Poult Sci 2017;19:55–8.

[270] Gao GL, Wang HW, Zhao XZ, Li Q, Li J, Li QR, et al. Feeding conditions and breed affect the level of DNA methylation of the mitochondrial uncoupling protein 3 gene in chicken breast muscle1. J Anim Sci 2015;93(4):1522–34.

[271] Guo M, Chen Y, Chen Q, Guo X, Yuan Z, Kang L, et al. Epigenetic changes associated with increased estrogen receptor alpha mRNA transcript abundance during reproductive maturation in chicken ovaries. Anim Reprod Sci 2020;214:106287.

[272] Omer NA, Hu Y, Hu Y, Idriss AA, Abobaker H, Hou Z, et al. Dietary betaine activates hepatic VTGII expression in laying hens associated with hypomethylation of GR gene promoter and enhanced GR expression. J Anim Sci Biotechnol 2018;9:2.

[273] Omer NA, Hu Y, Idriss AA, Abobaker H, Hou Z, Yang S, et al. Dietary betaine improves egg-laying rate in hens through hypomethylation and glucocorticoid receptor-mediated activation of hepatic lipogenesis-related genes. Poult Sci 2020;99(6):3121–32.

[274] Wang Y, Zheng Y, Guo D, Zhang X, Guo S, Hui T, et al. m6A methylation analysis of differentially expressed genes in skin tissues of coarse and fine type liaoning cashmere goats. Front Genet 2020;10:1318.

[275] Jiao Q, Yin RH, Zhao SJ, Wang ZY, Zhu YB, Wang W, et al. Identification and molecular analysis of a lncRNA-HOTAIR transcript from secondary hair follicle of cashmere goat reveal integrated regulatory network with the expression regulated potentially by its promoter methylation. Gene. 2019;688:182–92.

[276] Dai B, Zhang M, Yuan J-L, Ren L-Q, Han X-Y, Liu D-J. Integrative analysis of methylation and transcriptional profiles to reveal the genetic stability of cashmere traits in the Tβ4 overexpression of cashmere goats. Animals (Basel) 2019;9(12):1002.

[277] Li C, Li Y, Zhou G, Gao Y, Ma S, Chen Y, et al. Whole-genome bisulfite sequencing of goat skins identifies signatures associated with hair cycling. BMC Genomics 2018;19(1):638.

[278] Bai WL, Wang JJ, Yin RH, Dang YL, Wang ZY, Zhu YB, et al. Molecular characterization of HOXC8 gene and methylation status analysis of its exon 1 associated with the length of cashmere fiber in Liaoning cashmere goat. Genetica. 2017;145(1):115–26.

[279] Wang Y, Zheng Y, Guo D, Zhang X, Guo S, Hui T, et al. m6A methylation analysis of differentially expressed genes in skin tissues of coarse and fine type liaoning cashmere goats. Front Genet 2020;10(1318):1318.

[280] Do DN, Ibeagha-Awemu EM. Non-coding RNA roles in ruminant mammary gland development and lactation. In: Gigli I, editor. Current topics in lactation. Rijeka: InTech; 2017. Ch. 05.

[281] Do DN, Li R, Dudemaine P-L, Ibeagha-Awemu EM. MicroRNA roles in signalling during lactation: an insight from differential expression, time course and pathway analyses of deep sequence data. Sci Rep 2017;7:44605.

[282] Li HY, Xi QY, Xiong YY, Liu XL, Cheng X, Shu G, et al. Identification and comparison of microRNAs from skeletal

muscle and adipose tissues from two porcine breeds. Anim Genet 2012;43(6):704—13.
[283] Cui X, Zhang S, Zhang Q, Guo X, Wu C, Yao M, et al. Comprehensive microRNA expression profile of the mammary gland in lactating dairy cows with extremely different milk protein and fat percentages. Front Genet 2020;11(1504):1504.
[284] Wang J, Hao Z, Hu J, Liu X, Li S, Wang J, et al. Small RNA deep sequencing reveals the expressions of microRNAs in ovine mammary gland development at peak-lactation and during the non-lactating period. Genomics. 2021;113(1, Part 2):637—46.
[285] Li R, Beaudoin F, Ammah A, Bissonnette N, Benchaar C, Zhao X, et al. Deep sequencing shows microRNA involvement in bovine mammary gland adaptation to diets supplemented with linseed oil or safflower oil. BMC Genomics 2015;16(1):884.
[286] Li R, Dudemaine P-L, Zhao X, Lei C, Ibeagha-Awemu EM. Comparative analysis of the miRNome of bovine milk fat, whey and cells. PLoS One 2016;11(4):e0154129.
[287] Wang H, Luo J, Zhang T, Tian H, Ma Y, Xu H, et al. MicroRNA-26a/b and their host genes synergistically regulate triacylglycerol synthesis by targeting the INSIG1 gene. RNA Biol 2016;1—11.
[288] Wang H, Luo J, Chen Z, Cao WT, Xu HF, Gou DM, et al. MicroRNA-24 can control triacylglycerol synthesis in goat mammary epithelial cells by targeting the fatty acid synthase gene. J Dairy Sci 2015;98(12):9001—14.
[289] Karbiener M, Pisani DF, Frontini A, Oberreiter LM, Lang E, Vegiopoulos A, et al. MicroRNA-26 family is required for human adipogenesis and drives characteristics of brown adipocytes. Stem Cell 2014;32(6):1578—90.
[290] Lin X, Luo J, Zhang L, Wang W, Gou D. MiR-103 controls milk fat accumulation in goat (*Capra hircus*) mammary gland during lactation. PLoS One 2013;8(11):e79258.
[291] Li H, Zhang Z, Zhou X, Wang Z, Wang G, Han Z. Effects of microRNA-143 in the differentiation and proliferation of bovine intramuscular preadipocytes. Mol Biol Rep 2011;38(7):4273—80.
[292] Li M, Wang T, Wu H, Zhang J, Zhou C, Jiang A, et al. Genome-wide DNA methylation changes between the superficial and deep backfat tissues of the pig. Int J Mol Sci 2012;13(6):7098—108.
[293] Couldrey C, Brauning R, Bracegirdle J, Maclean P, Henderson HV, McEwan JC. Genome-wide DNA methylation patterns and transcription analysis in sheep muscle. PLoS One 2014;9:7.
[294] Fan Y, Liang Y, Deng K, Zhang Z, Zhang G, Zhang Y, et al. Analysis of DNA methylation profiles during sheep skeletal muscle development using whole-genome bisulfite sequencing. BMC Genomics 2020;21(1):327.
[295] Li M, Wu H, Luo Z, Xia Y, Guan J, Wang T, et al. An atlas of DNA methylomes in porcine adipose and muscle tissues. Nature. Communications. 2012;3:850.
[296] Te Pas MF, Lebret B. Oksbjerg NJAAB. Invited review: measurable biomarkers linked to meat quality from different pig production systems. Arch Tierz 2017;60:271—83.
[297] Kaur M, Kumar A, Siddaraju NK, Fairoze MN, Chhabra P, Ahlawat S, et al. Differential expression of miRNAs in skeletal muscles of Indian sheep with diverse carcass and muscle traits. Sci Rep 2020;10(1):16332.
[298] Kappeler BIG, Regitano LCA, Poleti MD, Cesar ASM, Moreira GCM, Gasparin G, et al. MiRNAs differentially expressed in skeletal muscle of animals with divergent estimated breeding values for beef tenderness. BMC Mol Biol 2019;20(1):1.
[299] Iqbal A, Ping J, Ali S, Zhen G, Juan L, Kang JZ, et al. Role of microRNAs in myogenesis and their effects on meat quality in pig — A review. Asian-Australas J Anim Sci 2020;33(12):1873—84.

[300] Xing J, Jiang Y. Effect of dietary betaine supplementation on mRNA level of lipogenesis genes and on promoter CpG methylation of fatty acid synthase (FAS) gene in laying hens. Afr J Biotechnol 2012;11(24):6633—40.
[301] Palazzese L, Czernik M, Iuso D, Toschi P, Loi P. Nuclear quiescence and histone hyper-acetylation jointly improve protamine-mediated nuclear remodeling in sheep fibroblasts. PLoS One 2018;13(3):e0193954.
[302] Wang S, Li F, Liu J, Zhang Y, Zheng Y, Ge W, et al. Integrative analysis of methylome and transcriptome reveals the regulatory mechanisms of hair follicle morphogenesis in cashmere goat. Cells. 2020;9(4):969.
[303] Wang X, Wang Z, Wang Q, Wang H, Liang H, Liu D. Epigenetic modification differences between fetal fibroblast cells and mesenchymal stem cells of the Arbas Cashmere goat. Res Vet Sci 2017;114:363—9.
[304] Xiao P, Zhong T, Liu Z, Ding Y, Guan W, He X, et al. Integrated analysis of methylome and transcriptome changes reveals the underlying regulatory signatures driving curly wool transformation in Chinese Zhongwei Goats. Front Genet 2019;10:1263.
[305] Zheng Y, Hui T, Yue C, Sun J, Guo D, Guo S, et al. Comprehensive analysis of circRNAs from cashmere goat skin by next generation RNA sequencing (RNA-seq). Sci Rep 2020;10(1):516.
[306] Han W, Yang F, Wu Z, Guo F, Zhang J, Hai E, et al. Inner mongolian cashmere goat secondary follicle development regulation research based on mRNA-miRNA co-analysis. Sci Rep 2020;10(1):4519.
[307] Wang Y, Zheng Y, Guo D, Zhang X, Guo S, Hui T, et al. m6A methylation analysis of differentially expressed genes in skin tissues of coarse and fine type liaoning cashmere goats. Front Genet 2019;10:1318.
[308] Bondurianksy R, Day T. Nongenetic inheritance and its evolutionary implications. Annu Rev 2009;40(1):103—25.
[309] Mousseau TA, Fox CW. The adaptive significance of maternal effects. Trends Ecol Evol 1998;13(10):403—7.
[310] Ho DH. Transgenerational epigenetics: the role of maternal effects in cardiovascular development. Integr Comp Biol 2014;54(1):43—51.
[311] Ho DH, Reed WL, Burggren WW. Egg yolk environment differentially influences physiological and morphological development of broiler and layer chicken embryos. J Exp Biol 2011;214(Pt 4):619—28.
[312] Rassoulzadegan M, Grandjean V, Gounon P, Vincent S, Gillot I, Cuzin F. RNA-mediated non-mendelian inheritance of an epigenetic change in the mouse. Nature. 2006;441(7092):469—74.
[313] Youngson NA, Whitelaw E. Transgenerational epigenetic effects. Annu Rev 2008;9:233—57.
[314] Swartz TH, Bradford BJ, Clay JS. Intergenerational cycle of disease: maternal mastitis is associated with poorer daughter performance in dairy cattle. J Dairy Sci 2021;104(4):4537—48.
[315] Heard E, Martienssen Robert A. Transgenerational epigenetic inheritance: myths and mechanisms. Cell. 2014;157(1):95—109.
[316] Braunschweig M, Jagannathan V, Gutzwiller A, Bee G. Investigations on transgenerational epigenetic response down the male line in F2 pigs. PLoS One 2012;7(2):e30583.
[317] González-Recio O, Ugarte E, Bach A. Trans-generational effect of maternal lactation during pregnancy: a holstein cow model. PLoS One 2012;7(12):e51816.
[318] Feeney A, Nilsson E, Skinner M. Epigenetics and transgenerational inheritance in domesticated farm animals. J Anim Sci Biotechnol 2014;5(1):48.
[319] Imumorin IG, Peters SO, De Donato M. Genomic imprinting and imprinted gene clusters in the bovine genome. Livestock epigenetics. Wiley-Blackwell; 2012. p. 89—111.

[320] Magee DA, Spillane C, Berkowicz EW, Sikora KM, MacHugh DE. Imprinted loci in domestic livestock species as epigenomic targets for artificial selection of complex traits. Anim Genet 2014;45(Suppl 1):25–39.

[321] O'Doherty A, MacHugh D, Spillane C, Magee D. Genomic imprinting effects on complex traits in domesticated animal species. Front Genet 2015;6:156.

[322] Smith LC, Therrien J, Filion F, Bressan F, Meirelles FV. Epigenetic consequences of artificial reproductive technologies to the bovine imprinted genes SNRPN, H19/IGF2, and IGF2R. Front Genet 2015;6:58.

[323] Tian XC. Genomic imprinting in farm animals. Annu Rev Anim Biosci 2014;2:23–40.

[324] Wang X, Miller DC, Harman R, Antczak DF, Clark AG. Paternally expressed genes predominate in the placenta. Proc Natl Acad Sci U S A 2013;110(26):10705–10.

[325] Berkowicz EW, Magee DA, Sikora KM, Berry DP, Howard DJ, Mullen MP, et al. Single nucleotide polymorphisms at the imprinted bovine insulin-like growth factor 2 (IGF2) locus are associated with dairy performance in Irish Holstein-Friesian cattle. J Dairy Res 2011;78(1):1–8.

[326] Magee DA, Berry DP, Berkowicz EW, Sikora KM, Howard DJ, Mullen MP, et al. Single nucleotide polymorphisms within the bovine DLK1-DIO3 imprinted domain are associated with economically important production traits in cattle. J Heredity 2011;102(1):94–101.

[327] Neugebauer N, Rader I, Schild HJ, Zimmer D, Reinsch N. Evidence for parent-of-origin effects on genetic variability of beef traits. J Anim Sci 2010;88(2):523–32.

[328] Kalish JM, Jiang C, Bartolomei MS. Epigenetics and imprinting in human disease. Int J Dev Biol 2014;58(2–4):291–8.

[329] Young LE, Fairburn HR. Improving the safety of embryo technologies: possible role of genomic imprinting. Theriogenology. 2000;53(2):627–48.

[330] Park H-E, Yoo HS. Biomarkers as diagnostic tools for mycobacterial infections in cattle. Anim Health Res Rev 2021;22:72–84.

[331] Karuppusamy S, Kirby GM, Mutharia L, Tripathi BN. An update on mycobacterium avium subspecies paratuberculosis antigens and their role in the diagnosis of Johne's disease. World J Microbiol Biotechnol 2019;35(8):120.

[332] Montalvo-Casimiro M, González-Barrios R, Meraz-Rodriguez MA, Juárez-González VT, Arriaga-Canon C, Herrera LA. Epidrug repurposing: discovering new faces of old acquaintances in cancer therapy. Front Oncol 2020;10(2461):2461.

[333] Heerboth S, Lapinska K, Snyder N, Leary M, Rollinson S, Sarkar S. Use of epigenetic drugs in disease: an overview. Genet Epigenetics 2014;6:9–19.

[334] Chakraborty C, Sharma AR, Sharma G, Lee S-S. Therapeutic advances of miRNAs: a preclinical and clinical update. J Adv Res 2021;28:127–38.

[335] Chen X, Ba Y, Ma L, Cai X, Yin Y, Wang K, et al. Characterization of microRNAs in serum: a novel class of biomarkers for diagnosis of cancer and other diseases. Cell Res 2008;18(10):997–1006.

[336] Mitchell PS, Parkin RK, Kroh EM, Fritz BR, Wyman SK, Pogosova-Agadjanyan EL, et al. Circulating microRNAs as stable blood-based markers for cancer detection. Proc Natl Acad Sci U S A 2008;105(30):10513–18.

[337] Valadi H, Ekstrom K, Bossios A, Sjostrand M, Lee JJ, Lotvall JO. Exosome-mediated transfer of mRNAs and microRNAs is a novel mechanism of genetic exchange between cells. Nat Cell Biol 2007;9(6):654–9.

[338] Kropp J, Salih SM, Khatib H. Expression of microRNAs in bovine and human pre-implantation embryo culture media. Front Genet 2014;5:91.

[339] Kropp J, Khatib H. Characterization of microRNA in bovine in vitro culture media associated with embryo quality and development. J Dairy Sci 2015;98(9):6552–63.

[340] Wu S, Li RW, Li W, Li CJ. Transcriptome characterization by RNA-seq unravels the mechanisms of butyrate-induced epigenomic regulation in bovine cells. PLoS One 2012;7(5):e36940.

[341] Giglio S, Cirombella R, Amodeo R, Portaro L, Lavra L, Vecchione A. MicroRNA miR-24 promotes cell proliferation by targeting the CDKs inhibitors p27Kip1 and p16INK4a. J Cell Physiol 2013;228(10):2015–23.

[342] Guo SL, Peng Z, Yang X, Fan KJ, Ye H, Li ZH, et al. miR-148a promoted cell proliferation by targeting p27 in gastric cancer cells. Int J Biol Sci 2011;7(5):567–74.

[343] Jeon YJ, Kim SY, Rah H, Choi DH, Cha SH, Yoon TK, et al. Association of the miR-146aC>G, miR-149T>C, miR-196a2T>C, and miR-499A>G polymorphisms with risk of spontaneously aborted fetuses. Am J Reprod Immunol 2012;68(5):408–17.

[344] Rosenbluth EM, Shelton DN, Wells LM, Sparks AE, Van Voorhis BJ. Human embryos secrete microRNAs into culture media—a potential biomarker for implantation. Fertil Steril 2014;101(5):1493–500.

[345] Sohel MM, Hoelker M, Noferesti SS, Salilew-Wondim D, Tholen E, Looft C, et al. Exosomal and non-exosomal transport of extra-cellular microRNAs in follicular fluid: implications for bovine oocyte developmental competence. PLoS One 2013;8(11):e78505.

[346] Roadmap Epigenomics C, Kundaje A, Meuleman W, Ernst J, Bilenky M, Yen A, et al. Integrative analysis of 111 reference human epigenomes. Nature. 2015;518(7539):317–30.

[347] Rivera CM, Ren B. Mapping human epigenomes. Cell. 2013;155(1). Available from: https://doi.org/10.1016/j.cell.2013.09.011.

[348] Rakyan VK, Down TA, Balding DJ, Beck S. Epigenome-wide association studies for common human diseases. Nat Rev Genet 2011;12(8):529–41.

[349] Rakyan VK, Beyan H, Down TA, Hawa MI, Maslau S, Aden D, et al. Identification of Type 1 diabetes–associated dna methylation variable positions that precede disease diagnosis. PLoS Genet 2011;7(9):e1002300.

[350] Gudex B, Johnson D, Singh K. Prenatal maternal and possible transgenerational epigenetic effects on milk production. PLoS One 2014;9(6):e98928.

[351] Park CS. Role of compensatory mammary growth in epigenetic control of gene expression. FASEB J 2005;19(12):1586–91.

[352] Tal O, Kisdi E, Jablonka E. Epigenetic contribution to covariance between relatives. Genetics. 2010;184(4):1037–50.

[353] Jones PA, Ohtani H, Chakravarthy A, De, Carvalho DD. Epigenetic therapy in immune-oncology. Nat Rev Cancer 2019;19(3):151–61.

[354] Kungulovski G, Jeltsch A. Epigenome editing: state of the art, concepts, and perspectives. Trends Genet 2016;32(2):101–13.

[355] Randhawa S, Sengar S. Chapter One - The evolution and history of gene editing technologies. In: Ghosh D, editor. Progress in molecular biology and translational science, 178. Academic Press; 2021. p. 1–62.

[356] Nakamura M, Gao Y, Dominguez AA, Qi LS. CRISPR technologies for precise epigenome editing. Nat Cell Biol 2021;23(1):11–22.

[357] Lau CH, Suh Y. In vivo epigenome editing and transcriptional modulation using CRISPR technology. Transgenic Res 2018;27(6):489–509.

[358] Hilton IB, D'Ippolito AM, Vockley CM, Thakore PI, Crawford GE, Reddy TE, et al. Epigenome editing by a CRISPR-Cas9-based acetyltransferase activates genes from promoters and enhancers. Nat Biotech 2015;33(5):510–17.

[359] Jenko J, Gorjanc G, Cleveland MA, Varshney RK, Whitelaw CBA, Woolliams JA, et al. Potential of promotion of alleles by genome editing to improve quantitative traits in livestock breeding programs. Genet Sel Evol 2015;47(1):55.

[360] Bo M, Jasemi S, Uras G, Erre GL, Passiu G, Sechi LA. Role of infections in the pathogenesis of rheumatoid arthritis: focus on mycobacteria. Microorganisms. 2020;8(10):1459.

[361] McNees AL, Markesich D, Zayyani NR, Graham DY. Mycobacterium paratuberculosis as a cause of Crohn's disease. Expert Rev Gastroenterol Hepatol 2015;9(12):1523—34.

[362] Over K, Crandall PG, O'Bryan CA, Ricke SC. Current perspectives on Mycobacterium avium subsp. paratuberculosis, Johne's disease, and Crohn's disease: a review. Crit Rev Microbiol 2011;37(2):141—56.

[363] Ariel O, Gendron D, Dudemaine PL, Gevry N, Ibeagha-Awemu EM, Bissonnette N. Transcriptome profiling of bovine macrophages infected by Mycobacterium avium spp. paratuberculosis depicts foam cell and innate immune tolerance phenotypes. Front Immunol 2019;10:2874.

[364] Gupta P, Peter S, Jung M, Lewin A, Hemmrich-Stanisak G, Franke A, et al. Analysis of long non-coding RNA and mRNA expression in bovine macrophages brings up novel aspects of Mycobacterium avium subspecies paratuberculosis infections. Sci Rep 2019;9(1):1571.

[365] Marete A, Ariel O, Ibeagha-Awemu E, Bissonnette N. Identification of long non-coding RNA isolated from naturally infected macrophages and associated with bovine Johne's disease in Canadian holstein using a combination of neural networks and logistic regression. Front Vet Sci 2021;8:639053.

[366] Wang M, Bissonnette N, Griebel P, Dudemaine P-L, Do DN, Ibeagha-Awemu EM. PSVI-15 Transcriptome analysis of ileal lymph nodes identifies key microRNAs affecting disease progression in Holstein cows with subclinical Johne's disease. J Anim Sci 2019;97(Suppl_3):207—8.

[367] Wang M, Bissonnette N, Griebel P, Dudemaine P-L, Do DN, Mao Y, et al. PSVI-14 Differentially expressed microRNAs with potential regulatory roles in ileum of Holstein cows with subclinical Johne's disease. J Anim Sci 2019;97(Suppl_3):206—7.

CHAPTER

30

Nutritional Epigenetics and Fetal Metabolic Programming

Ho-Sun Lee[1,2]

[1]Interdisciplinary Program in Bioinformatics and Department of Statistics, Seoul National University, Seoul, Republic of Korea [2]Toxicology Division, National Forensic Service Daegu Institute, Republic of Korea

OUTLINE

Introduction	611	Vitamin D	617
Metabolic Sensing by Epigenetic Mechanisms	612	High Fat/high Sugar Diet	617
DNA Methylation	612	Ketogenic Diet	618
Histone Modification	614	Dietary Fiber and SCFA	618
Non-coding RNA Modification	615	Summary and Perspectives	619
Impact of Prenatal Nutrition on Fetal Reprogramming	615	Glossary	620
Maternal Diet and Metabolic Epigenome	616	Acknowledgments	620
B Vitamins	616	References	620
Choline and Betaine	616		

INTRODUCTION

Metabolic syndrome (MetS) is defined by a cluster of interconnected components characterized by abdominal obesity, hypertension, dyslipidemia, insulin resistance, and glucose intolerance that directly increase the risk of coronary heart disease, other forms of cardiovascular atherosclerotic disease, and diabetes mellitus type 2 (T2D). The most widely accepted definitions for MetS have been produced by the World Health Organization (WHO), the European Group for the Study of Insulin Resistance (EGIR) and the National Cholesterol Education Program—Third Adult Treatment Panel (NCEP ATP III) [1–3]. While all groups agree on the core five components of the MetS, as well as its definition as the presence of at least three of the factors, they provide different clinical criteria to identify a cluster and the diagnosis of MetS [2]. In general, the International Diabetes Federation estimates that around a quarter of the world's adult population has MetS [4]. Therefore, the prevalence of main outcomes of MetS, such as T2D and cardiovascular disease, is a worldwide epidemic and an acute societal problem.

Because recent evidence suggests a link between metabolic disorders and the epigenome regulation, the current global epidemic of obesity and metabolic disorders underlines the need for a better understanding of epigenome deregulation by changing diet, lifestyle and environmental cues. Epigenetics refers to changes in gene function that cannot be explained by changes in the underlying DNA sequence. Epigenetic modifications such as DNA methylation patterns and histone modifications make important contributions to epigenetic memory (Table 30.1). It is increasingly accepted that nutrients and their metabolites influence long-term epigenetic

TABLE 30.1 Epigenetic Modifications and Their Function

Epigenetic Mechanism	Residue	Modification	Major Function	Relationship to Metabolism
DNA	CpG	Methylation	Transcriptional repression	Folate intake influences global DNA methylation Production of inhibitor metabolites such as R-2-hydroxyglutarate and S-2HG may increase DNA methylation by competitively inhibiting Tet demethylase *IDH1* or *IDH2* mutations influence DNA methylation and gene expression profiles
Histone H3 Histone H4	H3K9 H3K4 H3K27 H3K36 H3R17 H3R26 H4K20 H4KR3	me1 me2 me3	Transcriptional activation transcriptional repression poised to transcriptional activation	Production of R-2-hydroxyglutarate and S-2HG by mutant isocitrate dehydrogenase and lactate dehydrogenase may increase histone methylation by competitively inhibiting the histone demethylases Methylation of histones is related to the availability of dietary methionine and intracellular production of SAM, linking metabolism to epigenetic regulation
Histone H3 Histone H4	H3K9 H3K4 H3K14 H3K36 H4K4 H4K8 H4K12	ac	transcriptional activation	Acetylation is promoted by the acetyl group donor acetyl-CoA and inhibited by CoA

ac, acetylation; *me*, methylation (1: mono; 2: bi; 3:tri; a: asymmetric; s: symmetric).

alterations, and may be involved in the development of metabolic disorders especially through alteration of MetS components. Altered epigenetic signatures may even potentially result in transgenerational alterations in the phenotype [5]. In animal models and epidemiological evidence, dietary imbalance, such as a high fat or low protein diet, has been shown to have impacts on the epigenome of metabolic organs or susceptibility to metabolic disorders [6]. Therefore, the epigenome can be considered an interface between the metabolism and the environment that is central to the generation of metabolic phenotypes and their stability throughout life [5].

In this chapter, we will focus on understanding how maternal dietary factors influence the epigenome during pregnancy, and how epigenetic factors may contribute to the regulation of metabolic disorders in later life. We will also discuss how these advances could provide the bases for future preventive strategies for metabolic disorders via the nutrition-based epigenetic modification.

METABOLIC SENSING BY EPIGENETIC MECHANISMS

DNA Methylation

Epigenetic markers are enzyme-mediated chemical modifications of DNA and of their associated chromatin proteins. The underlying mechanisms include chromatin remodeling, DNA methylation (occurring at the fifth position of cytosine, which is largely confined to CpG dinucleotides), histone modifications (acetylation, methylation, phosphorylation, ubiquitination, sumoylation), chromatin remodeling, and non-coding regulatory RNA. A growing body of evidence supports the hypothesis that histone modifications and DNA methylation are linked to the regulation of the principal adipogenic transcription factor and its target genes, resulting in obesity and hyperlipidemia [7,8].

DNA methylation is the most extensively studied epigenetic mechanism, and occurs within the one-carbon metabolism pathway, which is a complex set of biochemical reactions, involved in the transfer of single-carbon moieties (Fig. 30.1). Sites of DNA methylation are occupied by various proteins, including methyl-CpG binding domain (MBD) proteins, which recruit the enzymatic machinery to establish silent chromatin. All cytosines [CpGs and CpHs (H = A, T, C)] in DNA can potentially become methylated. Most of the attention on DNA methylation has focused on CpG sites. However, there are more CpH sites in the genome than CpG sites, and non-CpG methylations, which are highly cell-type specific and associated with mammalian embryonic stem cells and neurons, have emerged as important epigenetic markers [9,10]. Dynamic erasure and re-establishment of DNA methylation occur in different stages of development. The mammalian genome undergoes two extensive waves of reprogramming of CpG methylation patterns during early embryogenesis and after germline development. Maintenance of DNA methylation is mediated by DNA methyltransferases

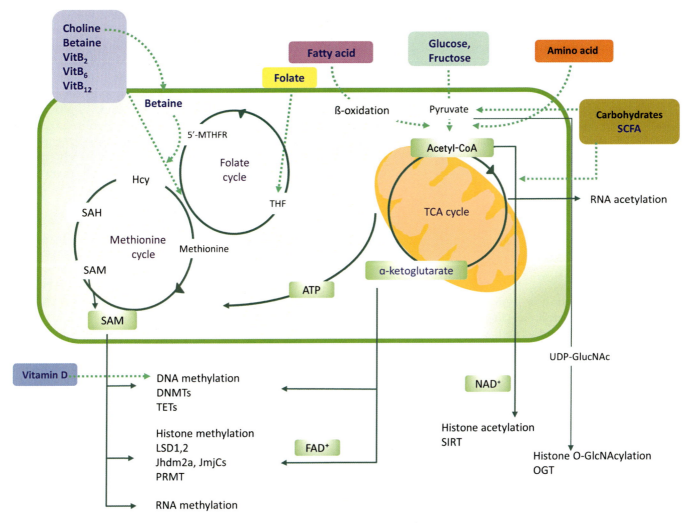

FIGURE 30.1 The interaction between one-carbon metabolism and nutrients on epigenetic modification. One-carbon metabolism is driven by the folate and methionine cycles responds to changes in the TCA cycle. Nutrients enter the metabolic pathways, which participate in epigenetic modification. *Hcy*, homocysteine; *JHDM2A*, JmjC-domain-containing histone dimethyl; *LSD*, lysine demethylase; *MTHFR*, methylenetetrahydrofolate reductase; *SAH*, S-adenosyl-l-homocysteine; *SAM*, S-adenosyl-methionine; *TCA cycle*, tricarboxylic acid cycle; *THF*, tetrahydrofolate.

(DNMTs), and this process is relatively well understood at a biochemical level. In particular, DNA or histone methylation depends on S-adenosyl-methionine (SAM), whose synthesis is dependent on diet, and on different enzymes whose activities, in turn, rely on the one-carbon cycle. Therefore, the importance of dietary methyl donors, including folate, choline, and vitamin B12, in epigenetic regulation has been examined in rodent models and in humans [5]. Based on the role of methyl donors in the development of diseases, it has been suggested that disruption of the one-carbon cycle may have an important role in the development of MetS. In addition, dietary availability of methyl donors has an impact on reprogramming developmental plasticity [5].

DNA demethylation is the process of removal of a methyl group from the nucleotide in a DNA residue that occurs either via oxidation or deamination. Although locus-specific DNA methylation pattern in somatic cells is stably maintained, genome-wide loss of 5mC, or DNA demethylation at specific sites, has been observed in specific developmental stages such as preimplantation embryos and developing primordial germ cells [11]. Active DNA demethylation is carried out by Ten-eleven translocations (Tets) family of methylcytosine dioxygenases that can convert 5mC into 5-hydroxymethylcytosine, 5-formylcytosine, and 5-carboxylcytosine [11]. Tet1-dependent and Tet2-dependent demethylation, which mostly affects imprinted loci and germline-specific genes, and Tet3 is active in the fertilized zygotes, leading to hydroxymethylation and active DNA demethylation [12]. Tet enzymes in collaboration with regulatory elements, including the AICDA deaminase, the MBD4 glycosylase, and the GADD45-alpha BER, lead to DNA demethylation and enhancer activation [13]. Recent study indicates the role of non-CpG methylation for neurological development that methylated CAC sequences (CpA methylation) in gene bodies after birth are recognized by methyl-CpG-binding protein 2, the binding of which establishes a lifelong epigenetic memory of gene silencing in neuron [14]. Tet-mediated oxidation of 5mC requires the metabolite α-ketoglutarate from the tricarboxylic acid (TCA) cycle; therefore, Tets have the potential to regulate DNA methylation patterns and thereby diverse physiological functions, including metabolic signaling [15].

Histone Modification

Histone methylation or demethylation has also been associated with metabolic disorders (Fig. 30.1). For example, protein arginine methyltransferases (PRMTs) catalyze the transfer of methyl groups from SAM to the guanidine nitrogen of arginine residues. PRMT1 catalyzes the addition of two methyl groups to the same residues, resulting in methylation of peroxisome proliferator-activated receptor gamma coactivator-1a (PGC-1α), which is involved in energy metabolism, including mitochondrial fatty acid oxidation. The histone H3K9 demethylase, jhdm2a, was identified as a crucial regulator of metabolic imbalance and hyperlipidemia [16]. Lysine-specific demethylases (LSD) 1 and 2 comprise the amine oxidase family, which require flavin adenosine dinucleotide (FAD) as a coenzyme. It was reported that FAD-dependent LSD1 epigenetically represses energy expenditure, which facilitates energy storage in white adipocytes [6]. Sirtuin (SIRT)1, a member of the histone deacetylase (HDAC) family, is a nuclear NAD^+-dependent protein deacetylase that has emerged as a key metabolic sensor, and is considered a key regulator of glucose homeostasis, insulin sensitivity, oxidative stress, and anti-inflammatory activity [17]. Liu et al., reported that high-fat diet induced metabolic damage via SREBP-1c expression and improvement of anti-inflammation and anti-oxidation are mediated through SIRT1 mechanism [18]. On the other hand, SIRT1, activated by the $NAD^+/NADH$ ratio increases under low-nutrient conditions, promotes fat metabolism in adipocytes by repressing peroxisome proliferator-activated receptor (PPAR)γ, which means that upregulating SIRT1 increases lipolysis, thereby inducing fat loss [19].

A number of studies provided the evidence for an unsuspected role for chromatin as a metabolic sensor [20,21]. Histone acetylation is the key mechanism of nutrient-sensitive epigenetic modification [22]. It involves the transfer of an acetyl-CoA derivative acetyl group to the ε-amino group of N-terminal lysine by histone acetyltransferases (HAT). There are several nuclear metabolic enzymes that supply a local source of acetyl-CoA for histone acetylation, including ATP-citrate lyase (ACLY) and acetyl-CoA synthetase 2 (ACSS2). Acetyl-CoA is closely linked to histone acetylation in response to glucose and growth factor stimulation and for controlling of a gene expression program implicated in lipid metabolism [23]. Lysine acetylation of proteins not only triggers gene transcription but is also a critical posttranslational modification that regulates the activity of core metabolic enzymes [24].

Histone acylation is also dynamically regulated upon various metabolic perturbations [25]. In addition to acetylation, histone lysine residues can be posttranslationally modified by other acyl groups, resulting in propionylation, butyrylation, 2-hydroxybutyrylation, ß-hydroxyisobutyrylation, crotonylation, malonylation, succinylation, and glutarylation [26]. O-GlcNAcylation is a posttranslational modification consisting of the addition of O-linked N-acetylglucosamine residue (GlcNAc) to specific serine/threonine residues of proteins [27]. For example, histone O-GlcNAcylation has been shown to be an epigenetic mark, controlling gene expression during embryonic development, and cell differentiation, in the etiology of epigenetic-related diseases. Chromatin remodeling is susceptible to a number of post-translational modifications by histone proteins, which impact their interactions with DNA [28]. GlcNAcylation involves the attachment of O-linked N-acetylglucosamine moieties to Ser and the residues of cytoplasmic, nuclear, and mitochondrial proteins [29]. Intracellular levels of the UDP-GlcNAc precursor molecule and the degree of global chromatin GlcNAcylation depend on the energetic state of the cell, making O-GlcNAcylation for a functional link between nutrition and regulation of gene expression [30]. Histones become reversibly mono- or poly- ADP-ribosylated resulting from transfer of an ADP-ribose moiety from NAD^+ to specific residues such as lysine and arginine [30]. ADP-ribosylation is a reaction catalyzed by most of members of the poly ADP-ribose polymerase (PARP) family. PARP enzymes (especially PARP1 and PARP2) can be activated in response to various metabolic regulations, including obesity, hyperlipidemia, T2D, and nonalcoholic fatty liver disease (NAFLD) [31].

Histone homocysteinylation is an emerging modification whereby homocysteine can be activated by methionyl-tRNA synthetase to produce Hcy—thiolactone, which may response with the ε-amino group of a protein lysine residue. Elevated H3K79 homocysteinylation has been observed from human fetal brains during pregnancy [32]. Cultured neuronal cells and developing mouse brain indicate that H3K4me3Q5ser exerts a permissive transcriptional activity [33]. Lactylation of lysines is strictly associated with the glycolysis pathway [34]. Lactate, an end product of glycolysis, generates lactyl-CoA, which contributes a lactyl group to lysine residues of histone proteins through histone acetyltransferases p300, generating a novel modification called lysine lactylation [35]. Histone lactylation is also emerging epigenetic hallmark with high glycolytic activity, which drives metabolic reprogramming and works as a regulator of immune reactions and intercellular communication [36].

Non-coding RNA Modification

Many recent studies have also implicated non-coding RNAs, a distinct epigenetic mechanism, in the regulation of metabolic function. Non-coding RNAs can be divided by their size into small non-coding RNAs (<200 nt length, microRNAs [miRNAs]), and small nucleolar RNAs, and long non-coding RNAs (>200 nt length). MicroRNAs play significant roles in the posttranscriptional regulation of gene expression in many biological processes including regulation of metabolic function. Mir-143 [37], let-7 [38], mir-200a [39] induced by a high fat diet were shown to be involved in metabolic disorders such as insulin resistance. In mice fed a high-fat diet, the effect of hepatic miR-27 levels was shown to regulate the expression of numerous miR-27 target genes associated with lipid metabolism, including angiopoietin-like 3 (Angptl3) and glycerol-phosphate acyltransferase 1 (Gpam) [40]. Long non-coding RNA (lncRNA) is a newly emerging regulatory mechanism in metabolic control [41]. Insulin-like growth factor 1 signaling was shown to be triggered by lncRNA (i.e., CRNDE) expression; hence, lncRNAs may also be implicated in the metabolic effects of insulin and the development of insulin resistance [42]. A recent study reported that lncRNAs regulate cholesterol metabolism; lnc-HC (lncRNA derived from hepatocytes) binds hnRNPA2B1 to regulate expression of Cyp7a1 and Abca1 in hepatocytes [43], and increases in nuclear SREBP-1c protein are caused by lnc-MALT1 [44]. Pioglitazone, which is a drug with hypoglycemic action used to treat T2D, was shown to regulate HDAC7 expression in MetS by up-regulating lnc-MEG3 [45]. More recently, sperm-derived transfer RNAs (tRNAs) resulted from paternal diet have been implicated as a regulator of metabolic phenotype in the F1 generation [46,47]. However, the role of lnc RNA in the development of metabolic disorders through diet remains largely unknown.

Emerging interests in the modification of mRNA have shown that the N6 methyladenosine (m^6A) modification is involved in lipid metabolism and the progression of metabolic disorders [48,49]. The m^6A modification is a widespread and reversible process, catalyzed by methyltransferase-like (METTL)3, METTL14, and WTAP that act as methylases to transfer the methyl-group onto the sixth carbon atom of substrate mRNA from SAM. Meanwhile, FTO and ALKBH5 are demethylases that eliminate the m^6A methylation from mRNA strand. A high-fat diet was significantly associated with increased m^6A levels combined with downregulated FTO and overexpression of METTL3, which in turn results in metabolic disorders [50]. It has been demonstrated that acetylation of cytidine in mRNAs promotes translation efficiency. It is catalyzed by N-acetyltransferase 10 (NAT10), whose activity is dependent upon the abundance of acetyl-CoA and ATP. Acetate from carbohydrate may affect the acetylation of mRNA and its further translation [51].

IMPACT OF PRENATAL NUTRITION ON FETAL REPROGRAMMING

In 1986, David Barker proposed the Developmental Origins of Health and Disease (DOHaD) hypothesis, denoted fetal re-programming, that the environmental stimuli such as nutrients during vulnerable development stages may permanently program change of organ structure and function of the offspring toward the development of non-communicable diseases [52]. He reported that infants with the lowest weight at birth had the highest death rates from coronary heart disease. Based on epidemiological studies, researchers hypothesized that fetal and perinatal events, including maternal undernutrition, were central to determine one's risk of developing chronic metabolic diseases [53,54]. Later, increasing evidence from epidemiological and animal studies suggested the importance of the intrauterine and early postnatal environment in programming of the fetus to adverse metabolic outcomes, and supported the notion of the thrifty phenotype hypothesis and DOHaD paradigm [55,56]. Emerging evidence further highlights the role of epigenetic mechanisms in mediating the effects of environmental exposure, which in certain instances may emphasize the mechanism underlying intergenerational transmission. Maternal inductions of intergenerational responses have been reported, and a variety of macronutrient and timing interventions have been used, including fasting, calorie restriction, dietary protein, fat, and methyl donor content, amongst others [57]. For instance, maternal undernutrition or a high-fat diet predisposes the first generation (F1) to metabolic disorders via epigenetic regulation. Interestingly, the F1 can transmit similar phenotypes across generations, despite normal nutrition in the F1 generation [58,59].

MATERNAL DIET AND METABOLIC EPIGENOME

An adequate diet of the fetus requires a balanced supply of nutrients in the maternal diet, and altered intakes of various nutrients have been involved in fetal development problems. There is mounting evidence supporting that maternal nutrients may modify the pattern of DNA methylation and histone modification. Changing activity of the enzymes regulating folate, methionine and TCA cycles is one way that nutrients influence the epigenome.

B Vitamins

In particular, undernutrition and overnutrition of folate and B vitamins lead to similar long-term phenotypic results in the offspring. High levels of folate, vitamin B12 and choline during pregnancy caused epigenetic changes and resulted in cardiovascular disease in offspring. Higher maternal folate during pregnancy predicted higher adiposity and insulin resistance in children at 6 years of age [60]. Folate depletion in human tissues led to not only in global hypomethylation but also in targeted hypermethylation of specific loci such as TFAP2A, STX11, and CYS1 [61].

One-carbon nutrients, especially methyl donors, and its metabolites are thought to be of particular importance in the periconceptional period and during embryonic development [62]. Folate is a methyl-donor, formerly known as vitamin B9, that has been extensively investigated for maternal nutrient and DNA methylation. Periconceptual supplementation with folate is known to significantly reduce the risk of pregnancies complicated by neural tube abnormalities [63]. Folate exposure during the periconceptional period was associated with global DNA methylation with DNMT activity in brain tissue of offspring [64]. In addition, periconceptional folate supplementation has also been associated with decreased DNA methylation at H19/IGF2 with birthweight [65,66], and hypomethylation at LEP and DNMT1 [67], and hypermethylation at RXRA in buccal cells [67]. Intervention studies also suggest folic acid supplementation beyond the first trimester of pregnancy or early life leads to significantly lower overall DNA methylation levels at LINE1, IGF2, and BDNF [68]. With multiple nutrients, folate supplementation was also involved in DNA hypomethylation in GNASAS, MEG3, IGF2R and MEST [69].

In particular, undernutrition and overnutrition of folate and B vitamins lead to similar long-term phenotypic results in the offspring. High levels of folate, vitamin B12 and choline during pregnancy caused epigenetic changes and resulted in cardiovascular disease in offspring. Higher maternal folate during pregnancy predicted higher adiposity and insulin resistance in children at 6 years of age [60]. Folate depletion in human tissues led to not only in global hypomethylation but also in targeted hypermethylation of specific loci such as TFAP2A, STX11, and CYS1 [61].

Vitamin B2, known as riboflavin, is an essential component of two major coenzymes, FMN (Flavin mononucleotide) and FAD (Flavin adenosine dinucleotide), and is found in some foods including egg, meats, and green vegetables. Dietary factors like vitamins B2, B6, and B12 are important cofactors of folate and methionine cycles, and involved in availability of methyl donor SAM for HMTs and DNMTs. Vitamin B2 intake from the periconceptional period to the third trimester of pregnancy increases ZAX1 DNA methylation, which controls fatal growth [70]. Maternal vitamin B6 (pyridoxal phosphate) concentration was associated with higher weight gain in offspring with methylation at the MEG3 DMR, which is also linked to chronic diseases [71].

Vitamin B12 is naturally found in foods such as fish, meat, poultry, eggs, and dairy products. Vitamin B12 (cobalamin) deficiency limits the enzymatic conversion of homocysteine to methionine and increases homocysteine levels, where hyperhomocysteine may appear in vitamin B12 deficiency. Several studies have found an association between maternal vitamin B12 and lipid metabolism. For example, women of child-bearing age with vitamin B12 deficiency exhibited hypomethylation in the promoter regions of the SREBF1 and LDLR genes which are essential to the regulation of cholesterol biosynthesis [72]. Besides, there are evidences to support the causative association between low maternal vitamin B12 status and high risk of neural tube defects, and lower birthweight and other adverse health outcome in the offspring [73]. From the observations of animal and human studies, the association of maternal B vitamins on one-carbon metabolism and negative long-term outcomes of offspring remain a field of clinical and public health significance.

Choline and Betaine

Choline and betaine are important methyl-donors associated with one carbon metabolism and epigenetic mechanisms that modify regulation of gene expression. The oxidation of choline serves as a methyl donor for homocysteine-methionine conversion via betaine homocysteine methyltransferase (BHMT) in the liver and kidney. Choline is an abundant essential nutrient in the diet, particularly in foods containing high-protein such as red meat, large eggs, soybeans, and wheat germ [74]. Choline has been demonstrated in studies where choline deficiency leads to abnormalities in lipid metabolism [75,76]. Choline is a precursor to phosphatidylcholine, which is abundant phospholipid in cell membranes [75]. Choline deficient diet resulted in abnormality of fatty acid oxidation and decreased ATP production in rat [77,78]. It has been shown that maternal choline supplementation during pregnancy modifies fetal histone and DNA methylation, suggesting that coordinated epigenetic mechanisms contribute to the long-term developmental effects of varied choline intake *in utero* [79]. Betaine is a methyl derivative of glycine and is found in foods such as shrimp, beetroots, and whole grains. Maternal betaine induced hypermethylation at *DNMT1*, *RXRA*, and *POMC* genes that are associated with body weight regulation, and hypomethylation at *LEP*, which is involved in the development of obesity and insulin resistance [80]. Higher levels of maternal betaine were found to be associated with lower birthweight, decreased subcutaneous adipose tissue, and small for gestational age [81]. On the other hands, betaine supplementation during pregnancy normalized fetal growth in diabetic pregnancy [82]. Maternal dietary supplementation can be a nutritional strategy to improve lifelong health of children. Future research is required to further elucidate the impact of nutrients on the fetal epigenome and to determine the level of maternal nutrients necessary for the optimal physiological function of offspring.

Vitamin D

Vitamin D is an essential nutrient that is involved in the formation of bone and regulation of several functions throughout the body including immune system [83]. This vitamin from the skin and diet is metabolized in the liver to its major circulation form, 25-hydroxyvitamin D (25(OH)D), and subsequently converted in the kidneys to its active form, 1,25(OH)$_2$D. During pregnancy, the fetus is entirely dependent on the maternal supply of 25(OH)D. Low maternal vitamin D levels during pregnancy have been associated with preterm birth and detrimental effects on infant bone development and infectious diseases [84].

Maternal vitamin D deficiency is linked to adverse health outcomes of offspring via epigenetically regulated fetal organs and tissue. The reduced availability of maternal 25(OH)D leads to decrease in the 25(OH)D entering the placenta and thus a decrease in the availability of the 1,25(OH)$_2$D required for both genomic and non-genomic signaling responses in the fetus. Maternal vitamin D supplementation (3800 IU/day vs 400 IU/day) during pre- and postnatal fetal development resulted in numerous DNA methylation changes of genes for collagen metabolic process and apoptosis in mother and infant leukocytes using 450 k DNA methylation microarray [85]. However, there was no association between imprinted genes, including IGF2/H19 and vitamin D deficiency, even though low vitamin D status led to low birthweight, and high BMI at 1 and 3 years after birth [86]. In an animal study, Sprague–Dawley rats showed that offspring of vitamin D deficient mothers exhibited differential methylation in adipose tissue with expression changes of lipid metabolism genes for *Vldlr* and *Hif1α*: *Vldlr* was hypermethylated and low expressed, and *Hif1α* was demethylated and high expressed [87], and imprinted gene for *H19/Igf2ICR* and *Snrpn*, *Dlk1/Gtl2*, and *Grb10* in the liver [88]. Both human and animal studies support the role of vitamin D as a modulator of DNA methylation not only on fetal development but also on metabolic process.

High Fat/High Sugar Diet

The overall increase in fat-rich diet or fructose (high sugar) intake is linked to the incidence of metabolic disorders, such as insulin resistance, T2D, hypertension, and obesity. Maternal fructose may induce epigenetic alterations in stem cells, leading to adverse effects during fetal development and children's health later in life [89].

Classically, the high-fat diets used in animal models contain a high fraction of saturated fatty acid, with a low fraction of monounsaturated and polyunsaturated fatty acid [90] or high dietary fat (a combination of different types for fatty acids) intake as a percentage of total dietary energy [91]. Maternal high-fat diet results in significantly increased lipid transfer across the placenta, which leads to organ specific effects in the liver, muscle, and pancreas, and caused changes in hypothalamic neuronal function and immune response [5]. Maternal high-fat diet induces circadian dysregulation to impair the CLOCK:BMAL1 chromatin recruitment [92] and pronounced transcriptional regulator PPARγ [93,94]. High fat or western diet during pregnancy induces circulating immune cell activation through Toll-like receptor 4 (TLR4)- c-Jun N-terminal kinase (JNK)/ nuclear factor-kappa B (NF-κB) signaling, resulting in TNF-α, IL-1β, IL-6, and other pro-inflammatory cytokine secretion. It contributes to insulin resistance [95] and the development of neurodegenerative diseases. From 10% to 60% fructose consumption during gestation and lactation increased hypertension and obesity in animal studies [96]. In previous study, maternal high fructose intake (20% fructose water) during gestation and lactation decreased serum high-density lipoprotein cholesterol (HDL-C) level in offspring by modifying miR-206 expression level in the liver that suppresses the expression of *Lxr*a [97]. On the other hands, male offspring from fructose-fed mothers (10% fructose water) had elevated plasma HDL-C levels and induced LXRα gene promoter methylation [98]. Maternal fructose intake (20%) affects hippocampal function in offspring through decreased brain-derived neurotrophic factor (*BDNF*) gene expression and increased DNA methylation at the *BDNF* promoter [99]. In fructose (10%) fed groups during pregnancy, the maternal microbiome modulation was observed with significantly reduced *Lactobacillus* and *Bacteroides* [100]. Paternal diet also has been affecting the metabolic status of offspring over multiple generations. In the Dutch famine study, the offspring of prenatally undernourished fathers, but not mothers, were heavier and more obese than those of fathers and mothers who had not been prenatally [101]. Paternal high-fat diet exposure significantly decreased the acetyl H3K9 level and increased the dimethyl H3K9 level in the adiponectin promoter region, and the monomethyl H4K20 level in the leptin promoter region was increased in adipose tissue of the offspring [102]. In addition, in a fly model of paternal intergenerational metabolic reprogramming, a paternal high sugar diet was shown to alter chromatin de-repression in the F1 generation [103]. Specifically, susceptibility to obesity was a consequence of reduced stage-specific epigenetic regulation of histone H3K27me3- and H3K9me3-defined domains both in mature sperm and in offspring embryos [103].

Ketogenic Diet

The ketogenic diet is a low-carbohydrate, fat-rich diet that has been used to treat epilepsy for a century [28,104,105]. Despite the anti-seizure efficacy of the ketogenic diet, the underlying mechanism is not completely understood. Ketogenesis takes place in the mitochondrial matrix of hepatocytes, where free fatty acids (FFAs) are released from adipose tissue during lipolysis. FFAs are broken down via β-oxidation to acetyl-CoA, which is used as a precursor for the production of β-hydroxybutyrate and acetoacetate and acetone (the three major ketone bodies). Ketone bodies coordinate cellular functions via β-hydroxybutyrylation that integrate the classic DNA methylation and histone modifications, including histone lysine acetylation, methylation, and histone phosphorylation and ubiquitination. Ketogenic diet has been shown to promote the formation of β-hydroxybutyrate and lead to decreased H3ac levels and increased H3K4me3 levels for lysine-specific demethylase 2D in the hippocampus and rescue of the neurogenesis in a mouse model [106]. Ketogenesis is intensified under conditions characterized by insufficient or inaccessible availability of glucose [107]. Ketogenesis is the pathway that is active under conditions caused by prolonged exercise, starvation, carbohydrate restriction, ketogenic diet or insulin deficiency. In rats, β-hydroxybutyrate adequately replaces the glucose deficit in the placenta, fetal brain, and liver during fasting hypoglycemia [108].

The maternal ketone body can cross the placenta and be used as fuels and lipogenic substrates by the fetus. Animal experiments and clinical studies have shown that exposure to high-ketone environments during pregnancy was closely related to adverse outcome in the offspring. The maternal ketogenic diet results in alterations in fetal growth, as embryo volume and organ volume percentage [109], and associated with an risk of congenital malformations in the offspring [110]. Therefore, changes in maternal ketone body during pregnancy may control the availability of lipid metabolites to the fetus through epigenetic mechanism, which could effectively counteract the early-life metabolic programming of adult risk.

Dietary Fiber and SCFA

Pregnant mothers have microbiomes in their gastrointestinal tract, vagina, skin, oral cavity, and placenta. While the pathway from the microbiome to the fetus is not fully understood, it may involve translocation of the microbiome through the membrane of the vaginal route, or of the mother's intestinal tract and may exist within the feto-placental unit.

Early development of the infant microbiome is related to the delivery mode, gestational age, and maternal probiotic or prebiotic uses [111]. Maternal diets have an impact on the early establishment of the fetal and neonatal microbiome and can lead to specific signatures that predispose to the development of obesity later in life. Maternal high-fat diets have been shown to affect the gut microbiome profiles of the offspring, including *Coprococcus*, *Coriobacteriace*, *Helicobacterioceae* and *Allobaculum* in mice [112] and *Bacteroides* in humans [113].

Short-chain fatty acids (SCFAs) are the main metabolites produced by fermentation of dietary fiber and resistant starch through gut microbiota. The maternal gut microbiome also produces B vitamins, vitamin K, and polyunsaturated fatty acids [114], which can subsequently be passed to the developing fetus by placenta or while breastfeeding. As mentioned above, B vitamins (riboflavin, pyridoxine, folate, and cyanocobalamin) are important cofactors for the enzymes in the folate cycle where the conversion from homocysteine to methionine, which is required for the availability of SAM. The other metabolites such as niacin, pantothenic acid, and SCFAs are sources of acetyl-CoA or NAD+, which affect histone acetylation via SIRT inhibition or HAT activation [115]. Histone acylation modification is also derived from SCFA by acyl-CoA syntheses (Fig. 30.2) [116].

Recent papers illustrated the critical role of SCFA in the fetal development and metabolic programming

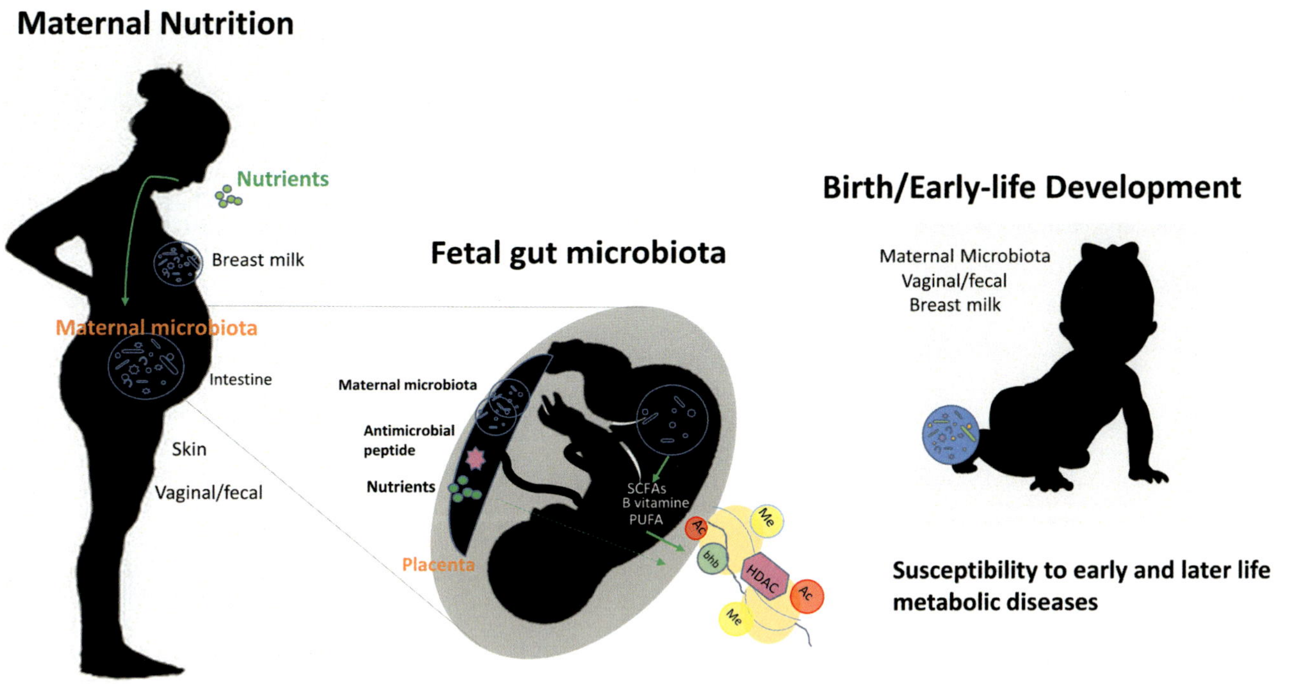

FIGURE 30.2 Fetal metabolic programming by maternal gut microbiota and nutrition. There is maternal microbiota from various maternal sites, including the intestine, vagina, and breast milk. Molecular signals originating from the maternal microbiota and its products such as antimicrobial peptide and during pregnancy can be transferred to the fetus during in utero development via the placenta and through the mode of delivery, and through breast milk during the postnatal period. The molecular mechanism of epigenetics may involve development of metabolic disorders and obesity later in life. *Ac*, Acetylation; *HDAC*, histone deacetylase; *Me*, methylation; *PUFA*, polyunsaturated fatty acid; *SCFA*, short-chain fatty acid.

during pregnancy [117]. Maternal gut microbial metabolites can change the host cellular levels of important epigenetic modifiers like DNMTs and DNA demethylases [115]. Maternal SCFA acetate, which suppresses asthma by promoting acetylation at the Foxp3 promoter and inducing Treg cells, through HDAC9 inhibition in offspring [118]. Besides, the SCFA butyrate induces histone acetylation in promoters of oxidative stress resistance factors *FOXO3A* and *MT2* by inhibiting HDAC [119]. Butyrate has been shown to induce M2 macrophage with the activation of the H2K9/STAT6 signaling pathway, therefore improving insulin sensitivity [120,121]. While growing evidence supports beneficial effects of maternal butyrate, including its role in insulin resistance [122], its effect remains controversial with opposite results [123]. Maternal microbiome-derived propionate mediates both insulin level and sympathetic nervous system by GPR43 and GPR41 signals, respectively [124]. Besides, microbiome metabolite trimethylamine N-oxide (TMAO) by dietary choline induces aortic stiffness and increases systolic arterial pressure with aging [125]. The SCFA pentanoate has an immune modulating effect that inhibits HDAC1 and HDAC8 in $CD4^+$ T cells, thereby reducing interleukin (IL)-17A production and enhancing IL-10 production toward elevated glucose oxidation [115,126]. Therefore, maternal fiber induced SCFA can be a preventive strategy for metabolic disorders and obesity in the offspring.

SUMMARY AND PERSPECTIVES

Epigenetic process is now regarded as an important mechanism that is involved in "fetal development reprogramming" for both early and later life outcomes. It is now increasingly accepted that nutrients, and their metabolites, influence long-term epigenetic alterations, and may be involved in the development of metabolic disorders. While it has been widely recognized that suboptimal maternal nutrition during pregnancy may have a marked effect on the maternal-fetal epigenome which constitutes a risk for metabolic disorders, intervention strategies aimed at modulating epigenetic mechanisms are challenging. This review explored the recent findings on epigenetic mechanisms for metabolic disorders with maternal nutrients. Several nutrients can work together to impact the fetal epigenome. Further efforts are expected to evaluate the optimal diet depending on maternal condition and critical time window during the prenatal or perinatal period to elucidate a better understanding of the role of multiple nutrition, and prevention of metabolic and other disorders. In addition, future studies are expected to provide new advances that should constitute a basis for developing a novel preventive strategy for metabolic reprogramming.

Acknowledgments

This work was supported by the National Research Foundation of Korea (NRF) grant funded by the Korea government (MSIT) (No.2022R1F1A1074097).

Glossary

MetS	metabolic syndrome
T2D	diabetes mellitus type 2
WHO	World Health Organization
EGIR	European Group for the Study of Insulin Resistance
NCEP ATP III	National Cholesterol Education Program—Third Adult Treatment Panel
DNMT	DNA methyltransferase
SAM	S-adenosyl-methionine
Tets	Ten-eleven translocations
TCA	tricarboxylic acid
PRMTs	protein arginine methyltransferases
PGC-1α	proliferator-activated receptor gamma coactivator-1a
LSD	lysine-specific demethylases
FAD	flavin adenosine dinucleotide
SIRT	sirtuin
HDAC	histone deacethylase
PPAR	peroxisome proliferator-activated receptor
ACLY	ATP-citrate lyase
ACSS2	acetyl-CoA synthetase 2
GlcNAc	O-linked N-acetylglucosamine residue
PARP	poly ADP-ribose polymerase
NAFLD	non-alcoholic fatty liver disease
lncRNA	long non-coding RNA
Angptl3	angiopoietin-like 3
Gpam	glycerol-phosphate acyltransferase 1
tRNAs	transfer RNAs
METTL	methyltransferase-like
m6A	N6 methyladenosine
NAT10	N-acetyltransferase 10
DOHaD	Developmental Origins of Health and Disease
HDL-C	high-density lipoprotein cholesterol
BDNF	brain-derived neurotrophic factor
SCFAs	short-chain fatty acids
TMAO	trimethylamine N-oxide
HPA	hypothalamic-pituitary-adrenal
GCs	glucocorticoids
Pomc	pro-opiomelanocortin
ENS	enteric nervous system
5-HT	5-hydroxytryptophan
TNF	tumor necrosis factor
GLP-1	glucagon-like peptide-1
TLR4	toll-like receptor 4
JNK	c-Jun N-terminal kinase
NF-κB	nuclear factor-kappa B

References

[1] Alberti KG, Zimmet PZ. Definition, diagnosis and classification of diabetes mellitus and its complications. Part 1: diagnosis and classification of diabetes mellitus provisional report of a WHO consultation. Diabet Med 1998;15(7):539–53.

[2] Grundy SM, Brewer Jr. HB, Cleeman JI, Smith Jr. SC, Lenfant C, American Heart A, et al. Definition of metabolic syndrome: report of the National Heart, Lung, and Blood Institute/American Heart Association conference on scientific issues related to definition. Circulation. 2004;109(3):433–8.

[3] Zimmet P, Magliano D, Matsuzawa Y, Alberti G, Shaw J. The metabolic syndrome: a global public health problem and a new definition. J Atheroscler Thromb 2005;12(6):295–300.

[4] Liu S, Song J, Peng J, Tang Z, Zhang J, Zhang L. Association of serum uric acid/creatinine ratio and metabolic syndrome in euthyroid population. Wei Sheng Yan Jiu 2020;49(3):374–80.

[5] Lee HS. Impact of maternal diet on the epigenome during in utero life and the developmental programming of diseases in childhood and adulthood. Nutrients. 2015;7(11):9492–507.

[6] Hino S, Nagaoka K, Nakao M. Metabolism-epigenome crosstalk in physiology and diseases. J Hum Genet 2013;58(7):410–15.

[7] Carson C, Lawson HA. Epigenetics of metabolic syndrome. Physiol Genomics 2018;50(11):947–55.

[8] Pant R, Firmal P, Shah VK, Alam A, Chattopadhyay S. Epigenetic regulation of adipogenesis in development of metabolic syndrome. Front Cell Dev Biol 2020;8:619888.

[9] Unnikrishnan A, Freeman WM, Jackson J, Wren JD, Porter H, Richardson A. The role of DNA methylation in epigenetics of aging. Pharmacol Ther 2019;195:172–85.

[10] Price AJ, Collado-Torres L, Ivanov NA, Xia W, Burke EE, Shin JH, et al. Divergent neuronal DNA methylation patterns across human cortical development reveal critical periods and a unique role of CpH methylation. Genome Biol 2019;20(1):196.

[11] Zeng Y, Chen T. DNA methylation reprogramming during mammalian development. Genes (Basel) 2019;10(4):257.

[12] Greenberg MVC, Bourc'his D. The diverse roles of DNA methylation in mammalian development and disease. Nat Rev Mol Cell Biol 2019;20(10):590–607.

[13] Pastor WA, Aravind L, Rao A. TETonic shift: biological roles of TET proteins in DNA demethylation and transcription. Nat Rev Mol Cell Biol 2013;14(6):341–56.

[14] de Mendoza A, Poppe D, Buckberry S, Pflueger J, Albertin CB, Daish T, et al. The emergence of the brain non-CpG methylation system in vertebrates. Nat Ecol Evol 2021;5(3):369–78.

[15] Cardaci S, Ciriolo MR. TCA cycle defects and cancer: when metabolism tunes redox state. Int J Cell Biol 2012;2012:161837.

[16] Inagaki T, Tachibana M, Magoori K, Kudo H, Tanaka T, Okamura M, et al. Obesity and metabolic syndrome in histone demethylase JHDM2a-deficient mice. Genes Cell 2009;14(8):991–1001.

[17] Liang F, Kume S, Koya D. SIRT1 and insulin resistance. Nat Rev Endocrinol 2009;5(7):367–73.

[18] Liu X, Gao Y, Li M, Geng C, Xu H, Yang Y, et al. Sirt1 mediates the effect of the heme oxygenase inducer, cobalt protoporphyrin, on ameliorating liver metabolic damage caused by a high-fat diet. J Hepatol 2015;63(3):713–21.

[19] Nogueiras R, Habegger KM, Chaudhary N, Finan B, Banks AS, Dietrich MO, et al. Sirtuin 1 and Sirtuin 3: physiological modulators of metabolism. Physiol Rev 2012;92(3):1479–514.

[20] Navarro E, Funtikova AN, Fito M, Schroder H. Prenatal nutrition and the risk of adult obesity: long-term effects of nutrition on epigenetic mechanisms regulating Gene expression. J Nutr Biochem 2016;39:1–14.

[21] Li X. SIRT1 and energy metabolism. Acta Biochim Biophys Sin (Shanghai) 2013;45(1):51–60.

[22] Zhang JJ, Fan TT, Mao YZ, Hou JL, Wang M, Zhang M, et al. Nuclear dihydroxyacetone phosphate signals nutrient sufficiency and cell cycle phase to global histone acetylation. Nat Metab 2021;3(6):859–75.

[23] McDonnell E, Crown SB, Fox DB, Kitir B, Ilkayeva OR, Olsen CA, et al. Lipids reprogram metabolism to become a major carbon source for histone acetylation. Cell Rep 2016;17(6): 1463–72.

[24] Xu W, Li Y, Liu C, Zhao S. Protein lysine acetylation guards metabolic homeostasis to fight against cancer. Oncogene. 2014;33(18):2279–85.

[25] Sabari BR, Zhang D, Allis CD, Zhao Y. Metabolic regulation of gene expression through histone acylations. Nat Rev Mol Cell Biol 2017;18(2):90–101.

[26] Jo C, Park S, Oh S, Choi J, Kim EK, Youn HD, et al. Histone acylation marks respond to metabolic perturbations and enable cellular adaptation. Exp Mol Med 2020;52(12):2005–19.

[27] Yang X, Qian K. Protein O-GlcNAcylation: emerging mechanisms and functions. Nat Rev Mol Cell Biol 2017;18(7):452–65.

[28] Zhang S, Roche K, Nasheuer HP, Lowndes NF. Modification of histones by sugar beta-N-acetylglucosamine (GlcNAc) occurs on multiple residues, including histone H3 serine 10, and is cell cycle-regulated. J Biol Chem 2011;286(43):37483–95.

[29] Hardiville S, Hart GW. Nutrient regulation of gene expression by O-GlcNAcylation of chromatin. Curr Opin Chem Biol 2016;33:88–94.

[30] O'Sullivan J, Tedim Ferreira M, Gagne JP, Sharma AK, Hendzel MJ, Masson JY, et al. Emerging roles of eraser enzymes in the dynamic control of protein ADP-ribosylation. Nat Commun 2019;10(1):1182.

[31] Szanto M, Bai P. The role of ADP-ribose metabolism in metabolic regulation, adipose tissue differentiation, and metabolism. Genes Dev 2020;34(5–6):321–40.

[32] Zhang Q, Bai B, Mei X, Wan C, Cao H, Dan L, et al. Elevated H3K79 homocysteinylation causes abnormal gene expression during neural development and subsequent neural tube defects. Nat Commun 2018;9(1):3436.

[33] Farrelly LA, Thompson RE, Zhao S, Lepack AE, Lyu Y, Bhanu NV, et al. Histone serotonylation is a permissive modification that enhances TFIID binding to H3K4me3. Nature. 2019;567 (7749):535–9.

[34] Cavalieri V. The expanding constellation of histone post-translational modifications in the epigenetic landscape. Genes (Basel) 2021;12(10).

[35] Zhang D, Tang Z, Huang H, Zhou G, Cui C, Weng Y, et al. Metabolic regulation of gene expression by histone lactylation. Nature. 2019;574(7779):575–80.

[36] Chen AN, Luo Y, Yang YH, Fu JT, Geng XM, Shi JP, et al. Lactylation, a novel metabolic reprogramming code: current status and prospects. Front Immunol 2021;12:688910.

[37] Takanabe R, Ono K, Abe Y, Takaya T, Horie T, Wada H, et al. Up-regulated expression of microRNA-143 in association with obesity in adipose tissue of mice fed high-fat diet. Biochem Biophys Res Commun 2008;376(4):728–32.

[38] Frost RJ, Olson EN. Control of glucose homeostasis and insulin sensitivity by the Let-7 family of microRNAs. Proc Natl Acad Sci U S A 2011;108(52):21075–80.

[39] Chartoumpekis DV, Zaravinos A, Ziros PG, Iskrenova RP, Psyrogiannis AI, Kyriazopoulou VE, et al. Differential expression of microRNAs in adipose tissue after long-term high-fat diet-induced obesity in mice. PLoS One 2012;7(4):e34872.

[40] Vickers KC, Shoucri BM, Levin MG, Wu H, Pearson DS, Osei-Hwedieh D, et al. MicroRNA-27b is a regulatory hub in lipid metabolism and is altered in dyslipidemia. Hepatology. 2013;57(2):533–42.

[41] Zhao XY, Lin JD. Long noncoding RNAs: a new regulatory code in metabolic control. Trends Biochem Sci 2015;40(10): 586–96.

[42] Ellis BC, Graham LD, Molloy PL. CRNDE, a long non-coding RNA responsive to insulin/IGF signaling, regulates genes involved in central metabolism. Biochim Biophys Acta 2014;1843(2):372–86.

[43] Lan X, Yan J, Ren J, Zhong B, Li J, Li Y, et al. A novel long non-coding RNA Lnc-HC binds hnRNPA2B1 to regulate expressions of Cyp7a1 and Abca1 in hepatocytic cholesterol metabolism. Hepatology. 2016;64(1):58–72.

[44] Yan C, Chen J, Chen N. Long noncoding RNA MALAT1 promotes hepatic steatosis and insulin resistance by increasing nuclear SREBP-1c protein stability. Sci Rep 2016;6:22640.

[45] Liu HZ, Wang QY, Zhang Y, Qi DT, Li MW, Guo WQ, et al. Pioglitazone up-regulates long non-coding RNA MEG3 to protect endothelial progenitor cells via increasing HDAC7 expression in metabolic syndrome. Biomed Pharmacother 2016;78:101–9.

[46] Sharma U, Conine CC, Shea JM, Boskovic A, Derr AG, Bing XY, et al. Biogenesis and function of tRNA fragments during sperm maturation and fertilization in mammals. Science. 2016;351 (6271):391–6.

[47] Chen Q, Yan M, Cao Z, Li X, Zhang Y, Shi J, et al. Sperm tsRNAs contribute to intergenerational inheritance of an acquired metabolic disorder. Science. 2016;351(6271):397–400.

[48] Li Y, Wang J, Huang C, Shen M, Zhan H, Xu K. RNA N6-methyladenosine: a promising molecular target in metabolic diseases. Cell Biosci 2020;10:19.

[49] Yang C, Hu Y, Zhou B, Bao Y, Li Z, Gong C, et al. The role of m(6)A modification in physiology and disease. Cell Death Dis 2020;11(11):960.

[50] Li Y, Zhang Q, Cui G, Zhao F, Tian X, Sun BF, et al. m(6)A regulates liver metabolic disorders and hepatogenous diabetes. Genom Proteom Bioinform 2020;18(4):371–83.

[51] Shu XE, Swanda RV, Qian SB. Nutrient Control of mRNA Translation. Annu Rev Nutr 2020;40:51–75.

[52] Barker DJ. The intrauterine origins of cardiovascular disease. Acta Paediatr Suppl 1993;82(Suppl 391):93–9.

[53] Ravelli AC, van der Meulen JH, Michels RP, Osmond C, Barker DJ, Hales CN, et al. Glucose tolerance in adults after prenatal exposure to famine. Lancet. 1998;351(9097):173–7.

[54] He L, Zeng H, Li F, Feng J, Liu S, Liu J, et al. Homocysteine impairs coronary artery endothelial function by inhibiting tetrahydrobiopterin in patients with hyperhomocysteinemia. Am J Physiol Endocrinol Metab 2010;299(6):E1061–5.

[55] Ross SA, Milner JA. Epigenetic modulation and cancer: effect of metabolic syndrome? The. Am J Clin Nutr 2007;86(3):s872–7.

[56] Hales CN, Barker DJ. The thrifty phenotype hypothesis. Br Med Bull 2001;60:5–20.

[57] Barres R, Zierath JR. The role of diet and exercise in the transgenerational epigenetic landscape of T2DM. Nat Rev Endocrinol 2016;12(8):441–51.

[58] Burdge GC, Slater-Jefferies J, Torrens C, Phillips ES, Hanson MA, Lillycrop KA. Dietary protein restriction of pregnant rats in the F0 generation induces altered methylation of hepatic gene promoters in the adult male offspring in the F1 and F2 generations. Br J Nutr 2007;97(3):435–9.

[59] Lane N, Dean W, Erhardt S, Hajkova P, Surani A, Walter J, et al. Resistance of IAPs to methylation reprogramming may provide a mechanism for epigenetic inheritance in the mouse. Genesis. 2003;35(2):88–93.

[60] Deshmukh U, Katre P, Yajnik CS. Influence of maternal vitamin B12 and folate on growth and insulin resistance in the offspring. Nestle Nutr Inst Workshop Ser 2013;74:145–54 discussion 54-6.

[61] Gonseth S, Roy R, Houseman EA, de Smith AJ, Zhou M, Lee ST, et al. Periconceptional folate consumption is associated with neonatal DNA methylation modifications in neural crest regulatory and cancer development genes. Epigenetics. 2015;10 (12):1166–76.

[62] James P, Sajjadi S, Tomar AS, Saffari A, Fall CHD, Prentice AM, et al. Candidate genes linking maternal nutrient exposure to offspring health via DNA methylation: a review of existing evidence in humans with specific focus on one-carbon metabolism. Int J Epidemiol 2018;47(6):1910–37.

[63] Liu J, Li Z, Ye R, Liu J, Ren A. Periconceptional folic acid supplementation and sex difference in prevention of neural tube defects and their subtypes in China: results from a large prospective cohort study. Nutr J 2018;17(1):115.

[64] Li W, Li Z, Li S, Wang X, Wilson JX, Huang G. Periconceptional folic acid supplementation benefit to development of early sensory-motor function through increase DNA methylation in rat offspring. Nutrients. 2018;10(3):292.

[65] Hoyo C, Murtha AP, Schildkraut JM, Jirtle RL, Demark-Wahnefried W, Forman MR, et al. Methylation variation at IGF2 differentially methylated regions and maternal folic acid use before and during pregnancy. Epigenetics. 2011;6(7):928–36.

[66] Steegers-Theunissen RP, Obermann-Borst SA, Kremer D, Lindemans J, Siebel C, Steegers EA, et al. Periconceptional maternal folic acid use of 400 microg per day is related to increased methylation of the IGF2 gene in the very young child. PLoS One 2009;4(11):e7845.

[67] Pauwels S, Ghosh M, Duca RC, Bekaert B, Freson K, Huybrechts I, et al. Maternal intake of methyl-group donors affects DNA methylation of metabolic genes in infants. Clin Epigenetics 2017;9:16.

[68] Caffrey A, Irwin RE, McNulty H, Strain JJ, Lees-Murdock DJ, McNulty BA, et al. Gene-specific DNA methylation in newborns in response to folic acid supplementation during the second and third trimesters of pregnancy: epigenetic analysis from a randomized controlled trial. Am J Clin Nutr 2018;107(4):566–75.

[69] Cooper WN, Khulan B, Owens S, Elks CE, Seidel V, Prentice AM, et al. DNA methylation profiling at imprinted loci after periconceptional micronutrient supplementation in humans: results of a pilot randomized controlled trial. FASEB J 2012;26(5):1782–90.

[70] Azzi S, Sas TC, Koudou Y, Le Bouc Y, Souberbielle JC, Dargent-Molina P, et al. Degree of methylation of ZAC1 (PLAGL1) is associated with prenatal and post-natal growth in healthy infants of the EDEN mother child cohort. Epigenetics. 2014;9(3):338–45.

[71] McCullough LE, Miller EE, Mendez MA, Murtha AP, Murphy SK, Hoyo C. Maternal B vitamins: effects on offspring weight and DNA methylation at genomically imprinted domains. Clin Epigenetics 2016;8:8.

[72] Adaikalakoteswari A, Finer S, Voyias PD, McCarthy CM, Vatish M, Moore J, et al. Vitamin B12 insufficiency induces cholesterol biosynthesis by limiting s-adenosylmethionine and modulating the methylation of SREBF1 and LDLR genes. Clin Epigenetics 2015;7:14.

[73] Behere RV, Deshmukh AS, Otiv S, Gupte MD, Yajnik CS. Maternal vitamin B12 status during pregnancy and its association with outcomes of pregnancy and health of the offspring: a systematic review and implications for policy in India. Front Endocrinol (Lausanne) 2021;12 619176.

[74] Zhu J, Wu Y, Tang Q, Leng Y, Cai W. The effects of choline on hepatic lipid metabolism, mitochondrial function and antioxidative status in human hepatic C3A cells exposed to excessive energy substrates. Nutrients. 2014;6(7):2552–71.

[75] da Silva RP, Kelly KB, Lewis ED, Leonard KA, Goruk S, Curtis JM, et al. Choline deficiency impairs intestinal lipid metabolism in the lactating rat. J Nutr Biochem 2015;26(10):1077–83.

[76] Jack-Roberts C, Joselit Y, Nanobashvili K, Bretter R, Malysheva OV, Caudill MA, et al. Choline supplementation normalizes fetal adiposity and reduces lipogenic gene expression in a mouse model of maternal obesity. Nutrients. 2017;9(8):899.

[77] Serviddio G, Giudetti AM, Bellanti F, Priore P, Rollo T, Tamborra R, et al. Oxidation of hepatic carnitine palmitoyl transferase-I (CPT-I) impairs fatty acid beta-oxidation in rats fed a methionine-choline deficient diet. PLoS One 2011;6(9):e24084.

[78] James SJ, Cross DR, Miller BJ. Alterations in nucleotide pools in rats fed diets deficient in choline, methionine and/or folic acid. Carcinogenesis. 1992;13(12):2471–4.

[79] Romano KA, Martinez-Del Campo A, Kasahara K, Chittim CL, Vivas EI, Amador-Noguez D, et al. Metabolic, epigenetic, and transgenerational effects of gut bacterial choline consumption. Cell Host Microbe 2017;22(3):279–90 e7.

[80] Pauwels S, Ghosh M, Duca RC, Bekaert B, Freson K, Huybrechts I, et al. Dietary and supplemental maternal methyl-group donor intake and cord blood DNA methylation. Epigenetics. 2017;12(1):1–10.

[81] Molto-Puigmarti C, Obeid R, Mommers M, Eussen SJ, Thijs C. Maternal plasma choline and betaine in late pregnancy and child growth up to age 8 years in the KOALA Birth Cohort Study. Am J Clin Nutr 2021;114(4):1438–46.

[82] Salahi P, Gharabaghi M, Rocky A, Alirezaei M. In vivo: maternal betaine supplementation normalized fetal growth in diabetic pregnancy. Arch Gynecol Obstet 2020;302(4):837–44.

[83] Fetahu IS, Hobaus J, Kallay E. Vitamin D and the epigenome. Front Physiol 2014;5:164.

[84] Karras SN, Fakhoury H, Muscogiuri G, Grant WB, van den Ouweland JM, Colao AM, et al. Maternal vitamin D levels during pregnancy and neonatal health: evidence to date and clinical implications. Ther Adv Musculoskelet Dis 2016;8(4):124–35.

[85] Anderson CM, Gillespie SL, Thiele DK, Ralph JL, Ohm JE. Effects of maternal vitamin D supplementation on the maternal and infant epigenome. Breastfeed Med 2018;13(5):371–80.

[86] Benjamin Neelon SE, White AJ, Vidal AC, Schildkraut JM, Murtha AP, Murphy SK, et al. Maternal vitamin D, DNA methylation at imprint regulatory regions and offspring weight at birth, 1 year and 3 years. Int J Obes (Lond) 2018;42(4):587–93.

[87] Ideraabdullah FY, Belenchia AM, Rosenfeld CS, Kullman SW, Knuth M, Mahapatra D, et al. Maternal vitamin D deficiency and developmental origins of health and disease (DOHaD). J Endocrinol 2019; JOE-18-0541.R2.

[88] Xue J, Schoenrock SA, Valdar W, Tarantino LM, Ideraabdullah FY. Maternal vitamin D depletion alters DNA methylation at imprinted loci in multiple generations. Clin Epigenetics 2016;8:107.

[89] Harvey A, Caretti G, Moresi V, Renzini A, Adamo S. Interplay between metabolites and the epigenome in regulating embryonic and adult stem cell potency and maintenance. Stem Cell Rep 2019;13(4):573–89.

[90] Sears B, Perry M. The role of fatty acids in insulin resistance. Lipids Health Dis 2015;14:121.

[91] Winzell MS, Ahren B. The high-fat diet-fed mouse: a model for studying mechanisms and treatment of impaired glucose tolerance and type 2 diabetes. Diabetes. 2004;53(Suppl 3):S215–19.

[92] Cleal JK, Bruce KD, Shearer JL, Thomas H, Plume J, Gregory L, et al. Maternal obesity during pregnancy alters daily activity and feeding cycles, and hypothalamic clock gene expression in adult male mouse offspring. Int J Mol Sci 2019;20(21):5408.

[93] Eckel-Mahan KL, Patel VR, de Mateo S, Orozco-Solis R, Ceglia NJ, Sahar S, et al. Reprogramming of the circadian clock by nutritional challenge. Cell. 2013;155(7):1464–78.

[94] Murakami M, Tognini P, Liu Y, Eckel-Mahan KL, Baldi P, Sassone-Corsi P. Gut microbiota directs PPARgamma-driven reprogramming of the liver circadian clock by nutritional challenge. EMBO Rep 2016;17(9):1292—303.

[95] Shi J, Fan J, Su Q, Yang Z. Cytokines and abnormal glucose and lipid metabolism. Front Endocrinol (Lausanne) 2019;10:703.

[96] Tain YL, Chan JY, Hsu CN. Maternal fructose intake affects transcriptome changes and programmed hypertension in offspring in later life. Nutrients. 2016;8(12):757.

[97] Yamazaki M, Munetsuna E, Yamada H, Ando Y, Mizuno G, Fujii R, et al. Maternal fructose consumption down-regulates Lxra expression via miR-206-mediated regulation. J Nutr Biochem 2020;82:108386.

[98] Rodrigo S, Fauste E, de la Cuesta M, Rodriguez L, Alvarez-Millan JJ, Panadero MI, et al. Maternal fructose induces gender-dependent changes in both LXRalpha promoter methylation and cholesterol metabolism in progeny. J Nutr Biochem 2018;61:163—72.

[99] Yamazaki M, Yamada H, Munetsuna E, Ishikawa H, Mizuno G, Mukuda T, et al. Excess maternal fructose consumption impairs hippocampal function in offspring via epigenetic modification of BDNF promoter. FASEB J 2018;32(5):2549—62.

[100] Astbury S, Song A, Zhou M, Nielsen B, Hoedl A, Willing BP, et al. High fructose intake during pregnancy in rats influences the maternal microbiome and gut development in the offspring. Front Genet 2018;9:203.

[101] Veenendaal MV, Painter RC, de Rooij SR, Bossuyt PM, van der Post JA, Gluckman PD, et al. Transgenerational effects of prenatal exposure to the 1944-45 Dutch famine. BJOG 2013;120(5):548—53.

[102] Masuyama H, Mitsui T, Eguchi T, Tamada S, Hiramatsu Y. The effects of paternal high-fat diet exposure on offspring metabolism with epigenetic changes in the mouse Adiponectin and Leptin gene promoter. Am J Physiol Endocrinol Metab 2016;311(1):E236—45 ajpendo 00095 2016.

[103] Ost A, Lempradl A, Casas E, Weigert M, Tiko T, Deniz M, et al. Paternal diet defines offspring chromatin state and intergenerational obesity. Cell. 2014;159(6):1352—64.

[104] Kossoff EH. More fat and fewer seizures: dietary therapies for epilepsy. Lancet Neurol 2004;3(7):415—20.

[105] McDonald TJW, Cervenka MC. The expanding role of ketogenic diets in adult neurological disorders. Brain Sci 2018;8(8):148.

[106] Benjamin JS, Pilarowski GO, Carosso GA, Zhang L, Huso DL, Goff LA, et al. A ketogenic diet rescues hippocampal memory defects in a mouse model of Kabuki syndrome. Proc Natl Acad Sci U S A 2017;114(1):125—30.

[107] Dabek A, Wojtala M, Pirola L, Balcerczyk A. Modulation of cellular biochemistry, epigenetics and metabolomics by ketone bodies. implications of the ketogenic diet in the physiology of the organism and pathological states. Nutrients. 2020;12(3):788.

[108] Emilio, H., Miguel, A.L. Maternal-fetal transfer of lipid metabolites. 2018.

[109] Sussman D, van Eede M, Wong MD, Adamson SL, Henkelman M. Effects of a ketogenic diet during pregnancy on embryonic growth in the mouse. BMC Pregnancy Childbirth 2013;13:109.

[110] Horton Jr. WE, Sadler TW. Effects of maternal diabetes on early embryogenesis. Alterations in morphogenesis produced by the ketone body, B-hydroxybutyrate. Diabetes 1983;32(7):610—16.

[111] Mesa MD, Loureiro B, Iglesia I, Fernandez Gonzalez S, Llurba Olive E, Garcia Algar O, et al. The evolving microbiome from pregnancy to early infancy: a comprehensive review. Nutrients. 2020;12(1):133.

[112] Wankhade UD, Zhong Y, Kang P, Alfaro M, Chintapalli SV, Thakali KM, et al. Enhanced offspring predisposition to steatohepatitis with maternal high-fat diet is associated with epigenetic and microbiome alterations. PLoS One 2017;12(4):e0175675.

[113] Chu DM, Antony KM, Ma J, Prince AL, Showalter L, Moller M, et al. The early infant gut microbiome varies in association with a maternal high-fat diet. Genome Med 2016;8(1):77.

[114] Yoshii K, Hosomi K, Sawane K, Kunisawa J. Metabolism of dietary and microbial vitamin B family in the regulation of host immunity. Front Nutr 2019;6:48.

[115] Lee HS. The interaction between gut microbiome and nutrients on development of human disease through epigenetic mechanisms. Genomics Inf 2019;17(3):e24.

[116] Chen XF, Chen X, Tang X. Short-chain fatty acid, acylation and cardiovascular diseases. Clin Sci (Lond) 2020;134(6):657—76.

[117] Zietek M, Celewicz Z, Szczuko M. Short-chain fatty acids, maternal microbiota and metabolism in pregnancy. Nutrients. 2021;13(4):1244.

[118] Thorburn AN, McKenzie CI, Shen S, Stanley D, Macia L, Mason LJ, et al. Evidence that asthma is a developmental origin disease influenced by maternal diet and bacterial metabolites. Nat Commun 2015;6:7320.

[119] Shimazu T, Hirschey MD, Newman J, He W, Shirakawa K, Le Moan N, et al. Suppression of oxidative stress by beta-hydroxybutyrate, an endogenous histone deacetylase inhibitor. Science. 2013;339(6116):211—14.

[120] Nakajima A, Nakatani A, Hasegawa S, Irie J, Ozawa K, Tsujimoto G, et al. The short chain fatty acid receptor GPR43 regulates inflammatory signals in adipose tissue M2-type macrophages. PLoS One 2017;12(7):e0179696.

[121] Canfora EE, Jocken JW, Blaak EE. Short-chain fatty acids in control of body weight and insulin sensitivity. Nat Rev Endocrinol 2015;11(10):577—91.

[122] Liu H, Wang J, He T, Becker S, Zhang G, Li D, et al. Butyrate: a double-edged sword for health? Adv Nutr 2018;9(1):21—9.

[123] Huang Y, Gao S, Chen J, Albrecht E, Zhao R, Yang X. Maternal butyrate supplementation induces insulin resistance associated with enhanced intramuscular fat deposition in the offspring. Oncotarget. 2017;8(8):13073—84.

[124] Kimura I, Miyamoto J, Ohue-Kitano R, Watanabe K, Yamada T, Onuki M, et al. Maternal gut microbiota in pregnancy influences offspring metabolic phenotype in mice. Science. 2020;367(6481):eaaw8429.

[125] Brunt VE, LaRocca TJ, Bazzoni AE, Sapinsley ZJ, Miyamoto-Ditmon J, Gioscia-Ryan RA, et al. The gut microbiome-derived metabolite trimethylamine N-oxide modulates neuroinflammation and cognitive function with aging. Geroscience. 2021;43(1):377—94.

[126] Luu M, Pautz S, Kohl V, Singh R, Romero R, Lucas S, et al. The short-chain fatty acid pentanoate suppresses autoimmunity by modulating the metabolic-epigenetic crosstalk in lymphocytes. Nat Commun 2019;10(1):760.

CHAPTER 31

Epigenetics of Drug Addiction

Ryan D. Shepard[1,2] and Fereshteh S. Nugent[1]

[1]Department of Pharmacology & Molecular Therapeutics, F. Edward Hebert School of Medicine, Uniformed Services University of the Health Sciences, Bethesda, MD, United States [2]Synapse and Neural Circuit Research Section, National Institute of Neurological Disorders and Stroke, National Institutes of Health, Bethesda, MD, United States

OUTLINE

Introduction	625
What Is Addiction?	626
Dopamine and Reward Circuits	626
Synaptic Plasticity, Learning and Memory and Addiction	628
Epigenetic Processes in Drug Addiction	630
Histone Modifications	630
DNA Methylation	631
Micro-RNAs	631
Synergistic and Opposing Effects of Intersectional Signaling in Epigenetic Regulation	632
Utilizing Epigenetic Targets for Diagnosis and Treatments to Combat Addiction and SUDs	632
Conclusion	634
Summary	634
Acknowledgments	634
References	634
Further Reading	637

INTRODUCTION

Drug addiction and substance use disorders (SUDs) are health concerns that detrimentally affect societies globally. Not only does addiction come at a large personal cost to both families and afflicted individuals, it can also strain economic and medical resources within communities. Although psychological and pharmacological treatment strategies do provide some therapeutic utility for individuals with SUD, there are varying degrees of efficacy which stems from an incomplete understanding of processes underlying the induction and maintenance of addictive behaviors.

As of 2021, it has been estimated that 275 million people engage in drug use worldwide and approximately 36 million people suffer from SUDs [1]. The use of illicit substances can have many negative health consequences that are both directly and indirectly due to the drug of choice. Unfortunately, overdose can be a common cause of death which is sometimes due to prolonged use [2,3]. Additionally, long-term usage of an illicit substance can impair both mental and physical health [4,5]. Moreover, in the case where needles are used for drug delivery, there is additional concern for contracting infectious diseases such as hepatitis B and human immunodeficiency virus (HIV) [6,7]. It is also important to note that patients diagnosed with SUD could use a variety of illicit substances and are often not confined to one drug of choice [8]. Because there is a high degree of risk with chronic drug usage following the first exposure, it is important to acknowledge and investigate mechanisms of drug addiction that reinforce the rewarding aspects of the drug and contribute to compulsive drug-seeking behaviors.

WHAT IS ADDICTION?

In general, addiction is viewed as a chronic relapsing brain disease in which susceptible individuals develop compulsive drug seeking and taking behaviors in spite of serious negative consequences. According to the Diagnostic and Statistical Manual of Mental Disorders (DSM-5) [9], SUD diagnosis requires that a person meet at least two of the following criteria within a 12-month period:

- Hazardous use.
- Social/Interpersonal problems related to use.
- Neglect of major roles to use.
- Withdrawal.
- Tolerance.
- Use of larger amounts or for longer periods.
- Repeated attempts to quit or control usage.
- Large amounts of time devoted towards usage.
- Physical/mental problems associated with usage.
- Activities given up due to usage.
- Craving for substance.

Moreover, SUDs can be further assessed by the type of substance such as: alcohol, stimulant use disorder (including amphetamines, cocaine, and other stimulants), caffeine, cannabis, phencyclidine (and other hallucinogens), inhalants, tobacco, opioids, sedatives, hypnotics, and anxiolytics. In addition to these diagnostic criteria, the addiction process can be conceptualized as having three major components: (1) binge/intoxication, (2) withdrawal/negative affect, and (3) preoccupation/anticipation [10,11]. Over time, reinforcing patterns in these behaviors make the drug association stronger and more difficult to disrupt (Fig. 31.1).

In order to better understand the addiction process, different theories and working frameworks on addiction have been developed [12], which center around positive reinforcement where the excessively rewarding effects of drugs and drug cues reshape behaviors to promote habitual or compulsive drug-seeking. Conversely, negative reinforcement also impacts addictive behaviors in which negative consequences of drug usage, such as dysphoria and negative effects following withdrawal, can also promote drug seeking behavior. Both positive and negative reinforcement are major drivers of persistent drug seeking and taking behavior. On a biological level, one of the primary ways in which addiction can occur is due to the buildup of tolerance—a state in which a person needs to administer increasing quantities of a drug to achieve the initial euphoric effect. This can also lead to drug dependence. An important note is that the development of tolerance and/or dependence does not equate to addiction, as medically directed usage of a drug overtime can also lead to tolerance. Although drugs of abuse utilize a wide variety of pharmacological mechanisms, the canonical neurotransmitter associated with reward and motivation is dopamine (DA).

DOPAMINE AND REWARD CIRCUITS

Dopamine (DA) is a monoamine utilized as a neurotransmitter by the central nervous system (CNS) and is critical for processes associated with reward, motivation, and movement [13]. Although DA neurons can be found throughout the CNS, discrete brain structures release DA as a potent

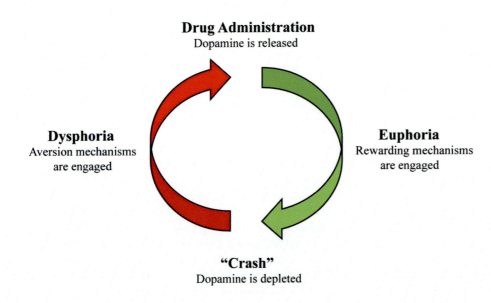

FIGURE 31.1 Drugs of abuse "hijack" reward circuitry which result in enhanced dopamine (DA) release and leads to a euphoric state. Following administration, plasticity processes that "re-wire" the reward circuitry are engaged which can then lead to long-term adaptations. As DA is depleted, this promotes a state of dysphoria and results in neural processes regulating aversion to become activated. In aversive states, individuals can be motivated to utilize drugs to bring themselves back to a state of euphoria.

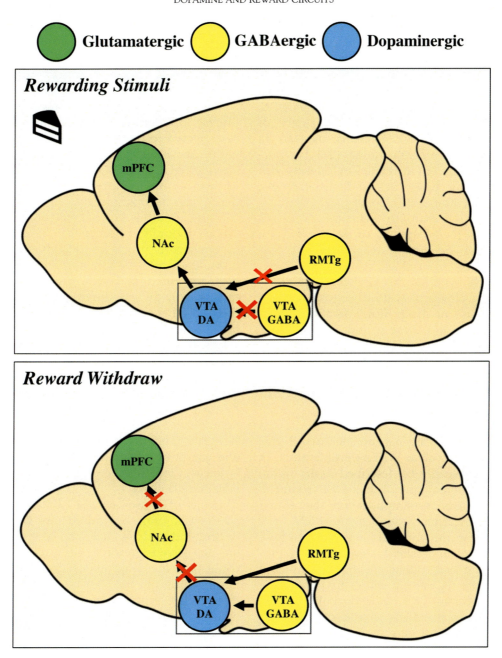

FIGURE 31.2 General Schematic of the Brain Reward Pathway. In response to reward (top panel), DA is released from the VTA to activate the reward pathway and inhibition from GABA neurons is restricted. In the absence of reward (bottom panel), GABA neurons dampen activation of the reward pathway by restricting DA release from the VTA.

regulator of circuit function. For example, the mesocorticolimbic dopaminergic pathway (also referred to as the brain reward pathway) consists of a set of interconnected brain structures which originates within the ventral tegmental area (VTA), the source of DA neurons for this pathway [14,15]. In response to natural rewarding stimuli, VTA DA neurons are stimulated and release DA into the nucleus accumbens (NAc) which further modulate distal brain regions such as the medial prefrontal cortex (mPFC, although direct VTA DA projections to mPFC can also be engaged) (Fig. 31.2). There are multiple circuit mechanisms and upstream brain structures which regulate DA release from VTA DA neurons. Here, we take a reductionist approach to explain how VTA DA neurons are regulated by providing a few well-documented examples of critical brain regions in the reward pathway that regulate DA release. For example, glutamatergic (excitatory) inputs from mPFC can stimulate VTA DA neurons

[16,17] whereas GABAergic (inhibitory) neurons from within the VTA [18], the rostromedial tegmental nucleus (RMTg; also known as "tail" of the VTA) [19,20] and NAc [21] can inhibit VTA DA neurons.

It is important to note that although DA neurons are the primary neuronal cell type within the VTA, recent studies have demonstrated the existence of glutamatergic and GABAergic neurons, as well as those that co-express neurotransmitters [22]. For purposes of this chapter, we will relate DA release with respect to rewarding properties and mechanisms, but a caveat is that literature also suggests that DA release can also contribute to aversion processes [23,24]. A common theme among all drugs of abuse is that they utilize the brain reward pathway and promote DA release from the VTA into the NAc, therefore usurping synaptic mechanisms that mediate the reinforcing effects of naturally rewarding stimuli. Long-term use of drugs fundamentally alters this reward pathway and these maladaptations contribute to the addiction cycle.

SYNAPTIC PLASTICITY, LEARNING AND MEMORY AND ADDICTION

Processes underlying learning and memory (L&M) are critical for the survival, as well as proliferation, of all species and can broadly be described as an organism's response to external stimuli in the environment. Although this is a gross oversimplification, the cellular and physiological mechanisms underlying L&M have been studied for centuries and still have not been fully elucidated. To better understand these complex processes, many researchers utilize the hippocampus, a critical structure that regulates a wide variety of memory encoding processes [25]. Indeed, both clinical observations and animal models have widely demonstrated that damage and/or perturbation to hippocampal function ultimately leads to memory deficits [26].

Synaptic plasticity is considered to be the cellular basis of L&M derived from the Hebbian learning rule "neurons that wire together, fire together," meaning the magnitude of the association between synapses (specialized junctions between neurons) dictates how closely neurons will communicate with one another [27]. For example, in Hebbian plasticity, long-term potentiation (LTP) can be defined as a strengthening of synaptic communication whereas long-term depression (LTD) is a weakening of communication between two neurons. Traditionally when describing LTP and LTD phenomenon, these processes are considered to be input-specific and the "rules" underlying these mechanisms may differ at different synapses and/or cell types. In contrast, homeostatic plasticity is less about input specificity and more so about how a neuron can "tune" its activity relative to the network of neurons it communicates with [28]. Although important to L&M, many processes underlying synaptic plasticity have yet to be fully understood.

With respect to plasticity and L&M, a multitude of diverse cellular and transcriptional processes have been shown to be involved. More specifically, an area within the hippocampus known as the dentate gyrus has garnered interest due to long-term neurogenesis (growth of new neurons) in the subventricular zone that occurs throughout life, although recent evidence has suggested this process may not be as robust as previously thought in humans [29,30]. There are a variety of epigenetic processes including histone modifications, DNA methylation, and microRNAs that have all been identified as critical regulators that control the rate of growth and differentiation of neurons within this and other structures [31,32]. As a result, the formation of new memories requires these epigenetic processes. For example, LTP in the hippocampus has been found to rely on DNA methyl-transferases (DNMTs) [33,34]. Additionally, histone posttranslational modifications (PTMs) also impact L&M in the hippocampus [35–37]. As a result, it is not surprising that deficits in neurogenesis and/or perturbation of L&M processes have been shown to contribute to memory impairment and cognitive decline [26]. Thus, epigenetic regulation is keenly tied to processes involved in L&M such as synaptic plasticity.

Drugs of abuse can also induce and alter synaptic plasticity in brain reward circuits. Of note, some forms of drug-induced plasticity can be transient while others are long-lived, forming drug-related memories that could promote drug taking, drug seeking and relapse. Earlier studies in the VTA were the first to demonstrate that a single exposure to any addictive drugs including opioids, alcohol and psychostimulants can induce glutamatergic (excitatory) LTP in VTA DA neurons regardless of their initial mechanism of action [38,39]. Later on, it was shown that LTP at inhibitory GABAergic synapses onto VTA DA neurons can also be inhibited by a single exposure to morphine [40]. These seminal studies have formed the foundation for the current perspective of synaptic basis for drug addiction introducing a pathological learning model of addiction where addictive drugs usurp synaptic mechanisms that underlie reward-based learning and motivated behaviors. For example, drug-induced glutamatergic and GABAergic plasticity in VTA DA neurons, following prolonged exposures to drugs of abuse, could result in persistent changes in VTA DA signaling that trigger synaptic plasticity (as well as epigenetic modifications) in VTA projections areas such as the NAc, which can then promote reinforcing, motivational, and aversive properties of addictive drugs in the three stages of addiction cycle [10,41]. Given the considerable amount of heterogeneity among neurons that have the same neurochemical identity (e.g., VTA DA neurons) based on their afferent and efferent connections and how they respond to rewarding and aversive stimuli, future studies are necessary to verify whether distinct inputs onto different subpopulation of neurons in reward circuits are able to exhibit different forms of activity-dependent plasticity and how drugs of abuse would modulate the plasticity in a cell- and pathway-specific manner.

Fortunately, circuit-based studies of different forms of drug-evoked synaptic plasticity in critical addiction-related brain circuits, specifically the mesocorticolimbic DA circuits, have become possible through the use of advanced transgenic technology and optogenetics where the precise contribution of distinct synaptic inputs in drug-evoked plasticity and behavior can be interrogated [42]. However, it is also important to note that drug-induced plasticity seems to be transient when the route of administration is passive even in response to repeated drug exposures [43]. Different animal models and modes of studying the rewarding aspect/abuse potential associated with drugs of abuse are often more complex and can create a mosaic of effects when considering the broad range of addiction literature [21,22]. For example, some experimental designs incorporate acute administration which involves a single drug exposure by the experimenter.

Additionally, there are multiple ways in which drug can be administered in animals by either volitional (self-administration) or passive means (orally, subcutaneously, intraperitoneal, intravenously, etc.) which can influence the onset of the drug effect, as well as activation/inhibition of downstream processes. Moreover, the duration of drug exposure can also impact neuroadaptations and changes at the cellular level. Lastly, more complex behavioral research paradigms can be used to assess how learning, motivation, and cue-induced reinforcement using environmental stimuli affect drug use and relapse. Therefore, synaptic studies of addiction need to incorporate robust and translationally relevant animal models of addiction such as operant models of drug self-administration where addicted animals transition from volitional to compulsive drug taking and seeking behaviors (Table 31.1). This is also a valid point in epigenetic studies of addiction in which drug-evoked epigenetic modifications of plasticity could differ or sometimes be opposite in experimenter-administered studies vs those that use operant models of addiction [44]. For more comprehensive discussions of drug-induced plasticity, see these review articles [45–48]. Given the role of epigenetics in L&M, it is not surprising that epigenetic processes are engaged by drugs of abuse to induce persistent drug-induced plasticity. We will now discuss some of the examples of epigenetic dysregulation by addictive drugs.

TABLE 31.1 Listed Are Basic Behavioral Paradigms That Are Utilized in Addiction Research, as well as Pros and Cons of Utilizing Each Research Model

Method of Drug Administration	Description	Strengths	Limitations
Acute intraperitoneal (i.p.) injection	Single time systemic administration of a drug by the experimenter	Useful for identifying acute molecular, cellular, and behavioral effects Relatively easy to implement	Does not address long-term adaptations from drug exposure Time points after the drug administration can create different effects Provides no context in terms of motivation and drug learning
Repeated i.p. injection	Multiple systemic administrations of a drug done by the experimenter	Good at looking at potentially long-term mechanisms associated with drug use	Different lengths of administration can result in different effects Provides no context in motivation and drug learning
Conditioned place preference (CPP)	Drug exposure is combined with an environmental context	Can examine effects of a drug post-hoc Relatively easy way to assess drug reward learning	Does not provide information on active drug seeking behavior
2 bottle selection/ preference	Rodents are offered a two-bottle choice and have *ad libitum* access for a defined period	Can examine effects of a drug post-hoc Can minimally assess drug seeking behavior and reward learning	Dependent on a drug being soluble Relies on animal passively identifying and not having to "work" for the drug
Self-Administration	Rodents operantly learn to receive drug by pairing a task	Allows for assessment of complex drug-seeking behaviors and motivation	Complex technique which requires additional equipment and staff skilled at animal surgeries

EPIGENETIC PROCESSES IN DRUG ADDICTION

In addition to neurophysiological processes underlying drug-induced synaptic plasticity, recent data from both clinical observations and preclinical animal models of addiction have demonstrated that drugs of abuse can alter the epigenome which subsequently affects synaptic mechanisms underlying drug reward, drug-taking and drug-seeking behaviors. Of note, these epigenetic processes have been shown to regulate key genes associated with plasticity. Given that epigenetic modifications can influence the transcriptional rate of genes, research on epigenetic processes involved in the addiction cycle has provided mechanistic insight into how drugs of abuse induce persistent alterations in the reward circuitry independent of their canonical pharmacological actions.

Histone Modifications

Histone modifiers regulate the chromatin structure dynamically by regulating the access to genes by altering how "tightly" DNA can associate with histones. Alterations to histones occur on the "tails" of the histone and so far a multitude of sites have been identified that are amenable to enzymatic alterations. Although there are various mechanistic processes that regulate rates of gene transcription by modifying histones (acetylation, methylation, phosphorylation, palmitoylation, sumoylation, etc.) [49], only certain processes have been observed within the context of addiction and SUDs.

One of the most studied epigenetic mechanisms in the field of addiction is the role of histone modifications—specifically, consequences associated with alterations to histone acetylation and methylation. With respect to histone acetylation, preclinical models involving drug exposure to rodents have pinpointed that acetylation at specific histone residues are critical for gene regulation in L&M that are subsequently modified following drug exposure [50,51]. In fact, a large body of literature points to how histone deacetylases (HDACs) become aberrant in drug addiction that results in drug-induced alterations to the acetylation dynamics of histones. Indeed, alterations of epigenetic marks through HDACs within brain reward structures such as the VTA and NAc have been identified. In the context of cocaine, many studies have found an overall upregulation of histone acetylation marks [52–55] where one of the potential HDACs that could be involved is HDAC5 [56]. Indeed, this has also been the case for opioid use [57,58]. In the case of HDAC5 and cocaine, an earlier report found that chronic cocaine injections increase the export of HDAC5 from the nucleus, resulting in an overall decrease in HDAC5-mediated deacetylation. Consistently, viral-mediated overexpression of HDAC5 in mouse NAc diminished cocaine preference in CPP while knockdown of HDAC5 did the opposite [56]. However, a recent cocaine self-administration study reports that maintaining HDAC5 nuclear localization and its deacetylase activity in the NAc during cocaine exposures only reduces cue-induced reinstatement while other cocaine-related behaviors remain unchanged [59]. In another report, chronic cocaine administration is found to increase the expression of some of downstream HDAC3-target genes important in cocaine-induced plasticity in the NAc by elevating the acetylation of H4K8 in promoter regions. More importantly, disruption of HDAC3 activity only in NAc D1R-MSNs, but not D2R-MSNs, enhanced cocaine CPP while reducing cocaine-seeking following a 30-d withdrawal in cocaine self-administration paradigm in mice [60].

In spite of these studies that are reporting examples of drug-induced decreases in HDAC activity and histone hyperacetylation in drug-induced plasticity and behavior, some studies have reported the opposite. For example, hypoacetylation of H3K9 and upregulation of HDAC2 in the VTA following acute morphine administration as well as a severe early life stress (as a risk factor for addiction) has been observed [50,61]. Blocking HDACs and deacetylation in the VTA have been shown to affect alcohol consumption as well [62,63]. Importantly, changes in acetylation have been shown to alter the expression of critical genes such as FosB, CREB, BDNF, and MEF2 [44,64]. Differences observed in studies of drug-induced changes in histone acetylation could be due to a variety of factors such as dosage, route of administration, acute vs chronic drug exposures and addiction paradigms, as well as possible differences in HDAC-mediated histone acetylation at specific genes in a single cell type or brain region. Furthermore, this highlights the importance and necessity of the use of highly relevant animal models of addiction when manipulating epigenetic marks in drug-induced plasticity and behavior in cell-type- but also gene locus-specific manner.

The use of histone deacetylase (HDAC) inhibitors have been shown to ameliorate these drug-induced acetylation changes indicating a potentially critical role for how drugs mobilize HDACs specifically to alter the epigenome which strengthen neurophysiological processes and plasticity that reinforce the rewarding properties of drugs [50,65]. Importantly, recent studies looking at patients with substance abuse history have identified similar alterations in histone acetylation, which appears to be dose-dependent in that the magnitude of the alteration is correlated with the length of usage. Specifically, the level of H3, as well as H3K27 acetylation specifically, within the dorsal striatum positively correlated to the length of use and resulted in transcriptional alteration of genes which control glutamatergic transmission [66].

In contrast to histone acetylation, which involves only one PTM to residues which bi-directionally regulate transcription (i.e., acetylation increases transcription, deacetylation decreases transcription), histone methylation can involve multiple methyl groups, and depending on the residue that is being modified, the impact on transcription can be different. Broadly, actions of histone methylation can either be silencing or activating. For example, the activity of methyltransferase G9 in the NAc has been studied in the context of cocaine [67,68] and opioids [69] where important transcription factors, such as FosB expression, are impacted and differential effects at both the acute and chronic usage level contribute to drug-induced plasticity.

More recently, intriguing findings have demonstrated that DA can epigenetically regulate histones themselves though dopaminylation [70]; this process is important for cocaine-mediated alterations to transcription. Initially, the motivation behind the investigation to identify dopaminylation as an epigenetic process was based on a recent finding that serotonin, another important monoamine neurotransmitter that regulates mood and is also involved in addiction, could covalently bind to H3 glutamine 5 (H3Q5) through a process called serotonylation [71,72]. Using both postmortem samples from heroin addicts and a rodent model of cocaine self-administration, it was identified that H3Q5 [H3 glutamine 5 dopaminylation (H3Q5dop)] served as a target for transglutaminase 2 (TGM2) to covalently attach dopamine [70]. Importantly, dopaminylation of H3Q5 was altered by cocaine use and subsequent withdrawal time that resulted in dysregulated VTA physiology and transcription of genes associated with addiction, as well as synaptic plasticity. This adds an interesting layer to how transcription of critical genes can be regulated by neurotransmitters, and potentially points to how activity-dependent release of neurotransmitters can epigenetically impact genes outside of the canonical histone modifiers. In the future, it will be interesting to examine whether dopaminylation is a process utilized by other drugs of abuse to regulate transcription, as well as whether other environmental influences such as stress also affect this process.

DNA Methylation

Unlike histone PTMs which alter the association of chromatin (i.e., how closely associated DNA is to the histone), methylation of DNA occurs as a direct modification to the DNA base, specifically cytosine, which affects the direct transcription of downstream genes. The main role of DNA methylation is to silence genes which occur via two main processes: either blocking transcription by making the promoter inaccessible to the corresponding transcription factors or by recruiting proteins which can bind to the methylated residue making chromatin less accessible for transcription. In general, DNA methylation is considered to be more stably expressed than other epigenetic modifiers. Within the context of epigenetic modifications that are engaged during SUDs, DNMT3 has received the most attention because it is the only DNA methyltransferase expressed in adulthood that can produce de novo DNA methylation [73] for which there are many isoforms [74]. Indeed, DNMT3a has been shown to increase in the NAc during chronic injections of cocaine, but later to substantially increase in the absence of cocaine [75]. This role of DNMT3a in mediating the reinforcing effects of cocaine on NAc circuitry has also been demonstrated during cocaine self-administration that potentially deals with how intermediate early genes (Arc, FosB, Egr2) are regulated [76]. Although many addiction studies have focused on repression of transcription by DNA methylation, loss of cystine-methylation has also been shown to be involved in addiction. Using a chronic alcohol administration model, the NMDAR receptor subunit NR2B expression becomes enhanced in cortical neurons which is due to demethylation of CpG islands in the NR2B gene [77]. Clinically, evidence has pointed to hypermethylation of the OPRM1 (encodes for the μ opioid receptor [MOR]) promoter and has been associated with decreased MOR expression [78–80]. Additionally, increased methylation has also been observed in patients with alcoholism [81], although DNMT3b is decreased.

Micro-RNAs

MicroRNAs (miRNA) are a class of non-coding RNAs that are the most understood of RNAs that can epigenetically regulate transcription. Mechanistically, mi-RNA-mediated suppression of gene expression occurs through mi-RNA association with the 3' UTR region on mRNA that results in the formation of the RNA-induced silencing (RISC) complex. In general, the formation of the RISC complex results in mRNA degradation or blocked transcription which decreases the abundance of targeted mRNA transcripts which subsequently decreases protein expression of the selected gene. Interestingly, solely affecting formation of the RISC complex alone can impact addictive phenotypes in rodents. Indeed, knockout of Argonaute 2 (Ago2) is sufficient to reduce the rewarding effects of cocaine and thus decrease cocaine self-administration [82]. Although less understood than histone modifications and DNA methylation, growing literature has revealed that miRNA and other ncRNAs epigenetically regulate transcription as a posttranslational event which are recruited in the addiction process [83]. So far, miR-181, miR-212, miR-124, miR-9, and Let-7 have been identified as critical regulators involved in addiction [84,85].

First, cocaine-induced effects on miRNA as an epigenetic regulator have been the most extensively studied and have well-documented impacts on striatal circuitry. Extended access to cocaine in a self-administration paradigm has been found to involve mirR-212 which either through overexpression or inhibition affects cocaine-seeking behavior by altering CREB activation which is involved in cocaine reinforcement [86]. This phenomenon involving miR-212 and cocaine seeking has been observed in other studies [87,88] and pinpoints that although changes to miR-212 expression might potentially be universal, site-specific expression should be considered. Second, alcohol use disorder (AUD) has also been shown to utilize miRNAs to reinforce alcohol consumption, namely through miR-9. Interestingly, upregulation of miR-9 post-transcriptionally regulates ion channel function which increases neuronal firing, resulting in changes to alcohol tolerance [89]. Additionally, increased expression of miR-214 and miR-206 have also been implicated in alcohol seeking [90,91]. Last, given that opioids have been shown to induce a variety of epigenetic changes, it is not surprising that part of the rewarding properties of opioids also rely on miRNA-dependent processes. For example, agonism-dependent stimulation of MORs affect miR-190 expression [92] which has implications for long-term MOR-dependent stimulation on miR-190 activity. Additionally, MOR expression itself has also been shown to be regulated by Let-7 which serves as a negative regulator of MORs and potentially contributes to the development of tolerance [93].

Given that the discovery of unique miRNAs are increasing with time, it is likely that the list of miRNAs associated with addiction and SUDs will continue to grow. Additionally, discrimination between specific miRNAs within the context of specific drug usage will provide valuable insight into how drugs of abuse utilize discrete mechanisms to reinforce drug-induced reward adaptations.

Synergistic and Opposing Effects of Intersectional Signaling in Epigenetic Regulation

Although in the above sections we describe epigenetic alterations within the context of single studies and by drug classes, crosstalk signaling between epigenetic regulators have been documented and add another level of complexity to regulation of critical transcription factors. For example, methyl CpG binding protein 2 (MeCP2) has been shown to be affected by cocaine-dependent [94,95] and heroin-dependent regulation of miR-212 [96], respectively. This demonstrates an important concept that although research might focus on identifying specific epigenetic alterations in response to drugs of abuse, interaction between epigenetic processes occurs and this is an aspect of studying the role of epigenetic regulators in the future. Additionally, newer discoveries of serotonylation and dopaminylation have also pinpointed that some epigenetic processes are synergistic in nature. For example, serotonylation of H3Q5 (H3Q5ser) can coexist with trimethylation of H3K4 (H3K4me3) as a dual modification (H3K4me3Q5ser). Interestingly, serotonylation can stabilize the histone trimethylation and seems to be specific to serotonergic neurons [72]. Moreover, this motif of dual modification has also been observed with dopaminylation (H3K4me3Q5dop) [70]. However, it is important to note that the identification of these dual histone modifications has, so far, only been observed in vitro. Therefore, it remains unknown whether this is a bona fide in vivo epigenetic alteration. Lastly, another important consideration is that although a majority of the studies described here are with respect to a single drug, polydrug use is observed in SUDs [8], and could lead to many drug-dependent epigenetic processes being engaged that require investigators to re-examine how addiction is modeled experimentally in animals.

UTILIZING EPIGENETIC TARGETS FOR DIAGNOSIS AND TREATMENTS TO COMBAT ADDICTION AND SUDS

In order to create better interventional treatments for SUDs, research has been devoted to understanding how we can more efficiently utilize biomarkers for diagnosis to complement psychological assessment [97]. Currently, one major limitation of examining epigenetic biomarkers with respect to brain region-specific effects in addition, is that a majority of the clinical observations are done postmortem. Even though postmortem studies can yield insight by examining epigenetic markers by brain region, there are issues that still need to be considered such as individual variability, duration of drug usage, sample collection, and other confounding variables.

Another limitation for examining epigenetic marks are due to both the temporal and spatial expression throughout not just the CNS, but the entire body as well. Nevertheless, using peripheral sample collection, such as from the blood, has still yielded insight. For example, examining miRNAs in the periphery is potentially a possibility and evidence has suggested that this might be a feasible avenue. Indeed, serum miRNA composition in patients diagnosed with SUD (heroin and methamphetamine) are different from patients without SUD diagnosis [98]. Moreover, alterations in miRNA composition have also been seen following acute alcohol consumption [99] and in methamphetamine users [100]. Another example of how peripheral blood samples have been used was by examining levels of the MOR in lymphocytes from Caucasian subjects with histories of heroin abuse and addiction that were currently on methadone treatment. Interestingly, decreased MOR expression was observed in the lymphocytes which was associated with increased DNA methylation [79]. Indeed, lymphocytes have also been assayed to detect changes in the DNA methylation which has been shown to potentially implicate a variety of genes in alcohol users [81,101–104]. These observations have been made possible by advancements in genetic sequencing (DNA methylation microarrays and bisulfite sequencing) that allow DNA methylation and other marks to be evaluated as potential diagnostic criteria [103,104], as well as allowing scientists to conduct epigenetic-wide association studies (EWAS) [105,106]. In addition, as genome-wide association studies (GWAS) have become critical for understanding how certain genes are related to diseases, EWAS studies take the same approach by attempting to examine potentially common epigenetic marks across populations. With respect to epigenetic markers within a population, studies have yielded insight into how common epigenetic-mediated mechanisms might contribute to addiction-related behaviors within a population [107–109]. Although human studies on epigenetic changes due to drug usage are critical for understanding the complete psychopathology surrounding addiction, studies have many caveats which can hinder the interpretation and overall applicability of observations. For example, one major consideration is the inherent difference of demographics that are used in studies and must be taken into consideration. Additionally, although samples can be taken from the periphery, it is unknown whether alterations in epigenetic marks from these samples are reminiscent of changes occurring within the brain and therefore must be interpreted with caution. Indeed, there are also potential issues of cellular mosaicism whereby epigenetic changes are not expressed in all cells whether this is due to change or is a change in response to the environment. Lastly, it is important to remember that EWAS studies are undertaken by selectively studying disease phenotypes, such as addiction, and therefore observed epigenetic marks that are concomitantly observed may not be related to the disease phenotype. Likewise, other epigenetic modifications could be involved which may not necessarily be as commonly observed in these studies. Given the numerous potential confounds that could happen in EWAS studies, caution should be taken both when designing and interpreting EWAS studies [110]. Nevertheless, taken together, there is promise for the development of more advanced diagnostics that can potentially be used to help identify at-risk people and allow for earlier intervention to combat SUDs.

A majority of epigenetic studies attempt to demonstrate causality by showing epigenetic mechanisms are directly tied to physiology and behavior. Pre-clinical evidence has suggested there could be opportunities to target epigenetic modifiers pharmacologically. Although there are no currently approved therapies for treating addiction or any psychiatric illness, HDAC inhibitors are approved for use in cancer treatments which suggests they are well-tolerated [111]. This has prompted speculation that HDAC inhibitors might be useful as antidepressants and for use in other mental health disorders due to our understanding of how histone acetylation is involved in L&M [112–114]. However, one limitation in many studies has been demonstrating direct manipulation of enzymes which mediate epigenetic processes and this slows down the development of more selective drugs that can target epigenetic modifiers. Now, the advent of more advanced genetic editing using CRISPR (clustered regularly interspaced short palindromic repeats) has allowed for more extensive research into the mechanistic roles of genetic regulation. Indeed, the use of CRISPR as well as viral-mediated delivery of engineered zinc finger proteins (ZFP) for targeting of histone marks has now been demonstrated as useful tools to probe and manipulate regulation of histone acetylation and methylation [115–118], as well as DNA methylation [119,120]. For example, bidirectional regulation of cyclin-dependent kinase 5 (Cdk5) gene expression in mouse NAc has been achieved through locus-specific histone manipulations using a viral-mediated delivery of ZFP, which was sufficient to induce opposing cocaine-related behaviors [118].

The greatest challenge remains in how to selectively target epigenetic modifications in a site- and cell-specific manner. For example, systemic administration of compounds that target epigenetic modifiers could have detrimental consequences by affecting epigenetic processes in unwanted tissues and cells. One potential way to mitigate concerns with pharmacotherapy is to design drugs that have enhanced selectivity for discrete enzymes that carry out epigenetic processes. Additionally, it is possible that advancements in neurosurgery will allow for delivery of agents to specific sites within the brain. This has most recently been accomplished with the use of optogenetics to combat blindness [121] and approval of Luxturna by the Food and Drug Administration which involves viral delivery of RPE65 to combat hereditary blindness [122]. It is possible in the future that gene therapies could be used to target maladaptive epigenetic changes due to addiction or other environmental influences.

CONCLUSION

Technological advancements in genetic sequencing and screening are now giving us greater insight into how environmental factors affect gene expression which is critical for elucidating the pathophysiology underlying mental health disorders that potentially have multiple contributing factors. In addition to canonical mechanisms that reinforce the rewarding properties of drugs, there is now greater appreciation for how drug-induced epigenetic alterations play a key role in addiction. Although a multitude of epigenetic mechanisms have been identified, more research is required both to understand lesser understood epigenetic modifications and to gain greater mechanistic insight into established epigenetic processes. Indeed, this also requires ascertaining the role of critical epigenetic regulators and discriminating selectively between different enzymes in order to create more selective therapeutic treatments. It still remains of high importance to understand alterations to the epigenome by drugs in specific brain regions, as well as whether these changes synergize, based on the connectivity of related structures such as the brain reward pathway. Lastly, new cutting edge research on the inheritance of epigenetic markers will help identify whether certain populations are more vulnerable based on epigenetic inheritance due to epigenetic modifications accrued by the parents [80,123,124]. Based on preclinical research attempting to ameliorate changes in the epigenome due to drug exposure through both pharmacological and gene editing methods, it is possible that future treatment strategies for SUDs will involve interventions which target epigenetic modifications.

SUMMARY

In this chapter, we have provided a general introduction into the addiction field by addressing the addiction process, brain reward circuitry, and how drugs of abuse utilize DA for drug-induced reward and motivation. Additionally, we briefly describe epigenetic processes involved in basic processes of L&M to bridge how understanding of these processes informs us about potential epigenetic roles in addiction. We then provide extensive examples as to the epigenetic modifications (histone modifications, DNA methylation, and miRNA) that have been observed both in laboratory and clinical studies. Lastly, we synthesize how, by understanding drug-induced epigenetic modifications, it is possible in the future that diagnostic methods and/or treatments will be used to target these maladaptive alterations in the epigenome.

Acknowledgments

The opinions and assertions contained herein are the private opinions of the authors and are not to be construed as official or reflecting the views of the Uniformed Services University of the Health Sciences or the Department of Defense or the National Institute of Neurological Disorders and Stroke (NINDS) or the Government of the United States. This work was supported by the National Institute of Drugs of Abuse (NIH/NIDA) Grant#R01 DA039533 to FN. The funding agency did not contribute to writing this chapter or deciding to submit it.

References

[1] Crime, U.N.O.O.D.&. and United Nations Office On Drugs & Crime World Drug Report 2021. CrimRxiv, 2021.
[2] Wilson N, et al. Drug and opioid-involved overdose deaths — United States, 2017—2018. Morb Mortal Wkly Rep 2020;69:290—7.
[3] Hedegaard H, et al. *Drug overdose deaths in the United States, 1999—2019*, U.S. Department of Health and Human Services, Centers for Disease Control and Prevention. Natl Cent Health Stat 2020.
[4] Farrell M. Physical complications of drug abuse. Routledge Library Editions: Addictions. Routledge; 2021.
[5] Newcomb MD, Locke T. Health, social, and psychological consequences of drug use and abuse. In: Sloboda Z, editor. Epidemiology of drug abuse. US: Springer; 2005. p. 45—59.
[6] Degenhardt L, et al. Global prevalence of injecting drug use and sociodemographic characteristics and prevalence of HIV, HBV, and HCV in people who inject drugs: a multistage systematic review. Lancet Glob Health 2017;5:e1192—207.
[7] Degenhardt L, et al. Estimating the burden of disease attributable to injecting drug use as a risk factor for HIV, hepatitis C, and hepatitis B: findings from the Global Burden of Disease Study 2013. Lancet Infect Dis 2016;16:1385—98.
[8] Crummy EA, et al. One is not enough: understanding and modeling polysubstance use. Front Neurosci 2020;14:569.
[9] American Psychiatric Association. Diagnostic and statistical manual of mental disorders (DSM-5®). American Psychiatric Pub; 2013.
[10] Koob GF, Volkow ND. Neurobiology of addiction: a neurocircuitry analysis. Lancet Psychiatry 2016;3:760—73.
[11] Kwako LE, Koob GF. Neuroclinical framework for the role of stress in addiction. *Chronic.* Stress (Thousand Oaks) 2017;1 2470547017698140.
[12] Hogarth L. Addiction is driven by excessive goal-directed drug choice under negative affect: translational critique of habit and compulsion theory. Neuropsychopharmacology 2020;45:720—35.
[13] Wise RA. Dopamine, learning and motivation. Nat Rev Neurosci 2004;5:483—94.
[14] Baik J-H. Stress and the dopaminergic reward system. Exp Mol Med 2020;52:1879—90.
[15] Haber SN, Knutson B. The reward circuit: linking primate anatomy and human imaging. Neuropsychopharmacology 2010;35:4—26.
[16] Au-Young SM, et al. Medial prefrontal cortical output neurons to the ventral tegmental area (VTA) and their responses to burst-patterned stimulation of the VTA: neuroanatomical and in vivo electrophysiological analyses. Synapse 1999;34:245—55.

[17] Gao M, et al. Functional coupling between the prefrontal cortex and dopamine neurons in the ventral tegmental area. J Neurosci 2007;27:5414−21.

[18] Bouarab C, et al. VTA GABA Neurons at the Interface of Stress and Reward. Front Neural Circuits 2019;13:78.

[19] Jhou TC, et al. The rostromedial tegmental nucleus (RMTg), a GABAergic afferent to midbrain dopamine neurons, encodes aversive stimuli and inhibits motor responses. Neuron 2009;61:786−800.

[20] Bourdy R, et al. Control of the nigrostriatal dopamine neuron activity and motor function by the tail of the ventral tegmental area. Neuropsychopharmacology 2014;39:2788−98.

[21] Edwards NJ, et al. Circuit specificity in the inhibitory architecture of the VTA regulates cocaine-induced behavior. Nat Neurosci 2017;20:438−48.

[22] Morales M, Margolis EB. Ventral tegmental area: cellular heterogeneity, connectivity and behaviour. Nat Rev Neurosci 2017;18:73−85.

[23] de Jong JW, et al. A neural circuit mechanism for encoding aversive stimuli in the mesolimbic dopamine system. Neuron 2019;101:133−51 e7.

[24] Yuan L, et al. Topography of reward and aversion encoding in the mesolimbic dopaminergic system. J Neurosci 2019;39:6472−81.

[25] Bird CM, Burgess N. The hippocampus and memory: insights from spatial processing. Nat Rev Neurosci 2008;9:182−94.

[26] Abrous DN, Wojtowicz JM. Interaction between neurogenesis and hippocampal memory system: new vistas. Cold Spring Harb Perspect Biol 2015;7:a018952.

[27] Stanton PK. LTD, LTP, and the sliding threshold for long-term synaptic plasticity. Hippocampus 1996;6:35−42.

[28] Turrigiano GG, Nelson SB. Homeostatic plasticity in the developing nervous system. Nat Rev Neurosci 2004;5:97−107.

[29] Moreno-Jiménez EP, et al. Evidences for adult hippocampal neurogenesis in humans. J Neurosci 2021;41:2541−53.

[30] Franjic D, et al. Transcriptomic taxonomy and neurogenic trajectories of adult human, macaque, and pig hippocampal and entorhinal cells. Neuron 2021;110(3):452−469.e14. Available from: https://doi.org/10.1016/j.neuron.2021.10.036.

[31] Covic M, et al. Epigenetic regulation of neurogenesis in the adult hippocampus. Heredity 2010;105:122−34.

[32] Yao B, et al. Epigenetic mechanisms in neurogenesis. Nat Rev Neurosci 2016;17:537−49.

[33] Levenson JM, et al. Evidence that DNA (cytosine-5) methyltransferase regulates synaptic plasticity in the hippocampus. J Biol Chem 2006;281:15763−73.

[34] Feng J, et al. Dnmt1 and Dnmt3a maintain DNA methylation and regulate synaptic function in adult forebrain neurons. Nat Neurosci 2010;13:423−30.

[35] Vecsey CG, et al. Histone deacetylase inhibitors enhance memory and synaptic plasticity via CREB:CBP-dependent transcriptional activation. J Neurosci 2007;27:6128−40.

[36] Levenson JM, et al. Regulation of histone acetylation during memory formation in the hippocampus. J Biol Chem 2004;279:40545−59.

[37] Guan Z, et al. Integration of long-term-memory-related synaptic plasticity involves bidirectional regulation of gene expression and chromatin structure. Cell 2002;111:483−93.

[38] Saal D, et al. Drugs of abuse and stress trigger a common synaptic adaptation in dopamine. neurons, Neuron 2003;37:577−82.

[39] Ungless MA, et al. Single cocaine exposure in vivo induces long-term potentiation in dopamine neurons. Nature 2001;411:583−7.

[40] Nugent FS, et al. Opioids block long-term potentiation of inhibitory synapses. Nature 2007;446:1086−90.

[41] Volkow ND, et al. The neuroscience of drug reward and addiction. Physiol Rev 2019;99:2115−40.

[42] Bocklisch C, et al. Cocaine disinhibits dopamine neurons by potentiation of GABA transmission in the ventral tegmental area. Science 2013;341:1521−5.

[43] Chen BT, et al. Cocaine but not natural reward self-administration nor passive cocaine infusion produces persistent LTP in the VTA. Neuron 2008;59:288−97.

[44] Stewart AF, et al. Epigenetics of drug addiction. Cold Spring Harb Perspect Med 2021;11:a040253.

[45] Langlois LD, Nugent FS. Opiates and plasticity in the ventral tegmental area. ACS Chem Neurosci 2017;8:1830−8.

[46] Nestler EJ, Lüscher C. The molecular basis of drug addiction: linking epigenetic to synaptic and circuit mechanisms. Neuron 2019;102:48−59.

[47] Lüscher C, Malenka RC. Drug-evoked synaptic plasticity in addiction: from molecular changes to circuit remodeling. Neuron 2011;69:650−63.

[48] Koob GF, Le Moal M. Plasticity of reward neurocircuitry and the "dark side" of drug addiction. Nat Neurosci 2005;8:1442−4.

[49] Jenuwein T, Allis CD. Translating the histone code. Science 2001;293:1074−80.

[50] Shepard RD, Nugent FS. Early life stress- and drug-induced histone modifications within the ventral tegmental area. Front Cell Dev Biol 2020;8:588476.

[51] Renthal W, Nestler EJ. Histone acetylation in drug addiction. SemCell Dev Biol 2009;20:387−94.

[52] Kumar A, et al. Chromatin remodeling is a key mechanism underlying cocaine-induced plasticity in striatum. Neuron 2005;48:303−14.

[53] Wang L, et al. Chronic cocaine-induced H3 acetylation and transcriptional activation of CaMKIIα in the nucleus accumbens is critical for motivation for drug reinforcement. Neuropsychopharmacology 2010;35:913−28.

[54] Malvaez M, et al. CBP in the nucleus accumbens regulates cocaine-induced histone acetylation and is critical for cocaine-associated behaviors. J Neurosci 2011;31:16941−8.

[55] Schmidt HD, et al. Increased brain-derived neurotrophic factor (BDNF) expression in the ventral tegmental area during cocaine abstinence is associated with increased histone acetylation at BDNF exon I-containing promoters. J Neurochem 2012;120:202−9.

[56] Renthal W, et al. Histone deacetylase 5 epigenetically controls behavioral adaptations to chronic emotional stimuli. Neuron 2007;56:517−29.

[57] Sheng J, et al. Histone H3 phosphoacetylation is critical for heroin-induced place preference. Neuroreport 2011;22:575−80.

[58] Wang Z, et al. Epigenetic upregulation of PSD-95 contributes to the rewarding behavior by morphine conditioning. Eur J Pharmacol 2014;732:123−9.

[59] Taniguchi M, et al. HDAC5 and Its target gene, Npas4, function in the nucleus accumbens to regulate cocaine-conditioned behaviors. Neuron 2017;96:130−44 e6.

[60] Campbell RR, et al. HDAC3 activity within the nucleus accumbens regulates cocaine-induced plasticity and behavior in a cell-type-specific manner. J Neurosci 2021;41:2814−27.

[61] Authement ME, et al. Morphine-induced synaptic plasticity in the VTA is reversed by HDAC inhibition. J Neurophysiol 2016;116:1093−103.

[62] You C, et al. Histone deacetylase inhibitor suberanilohydroxamic acid treatment reverses hyposensitivity to γ-aminobutyric acid in the ventral tegmental area during ethanol withdrawal. Alcohol Clin Exp Res 2018;42:2160−71.

[63] Arora DS, et al. Hyposensitivity to gamma-aminobutyric acid in the ventral tegmental area during alcohol withdrawal: reversal by histone deacetylase inhibitors. Neuropsychopharmacology 2013;38:1674–84.

[64] Robison AJ, Nestler EJ. Transcriptional and epigenetic mechanisms of addiction. Nat Rev Neurosci 2011;12:623–37.

[65] Browne CJ, et al. Epigenetic mechanisms of opioid addiction. Biol Psychiatry 2020;87:22–33.

[66] Egervari G, et al. Striatal H3K27 acetylation linked to glutamatergic gene dysregulation in human heroin abusers holds promise as therapeutic target. Biol Psychiatry 2017;81:585–94.

[67] Maze I, et al. Essential role of the histone methyltransferase G9a in cocaine-induced plasticity. Science 2010;327:213–16.

[68] Renthal W, et al. Delta FosB mediates epigenetic desensitization of the c-fos gene after chronic amphetamine exposure. J Neurosci 2008;28:7344–9.

[69] Sun H, et al. Morphine epigenomically regulates behavior through alterations in histone H3 lysine 9 dimethylation in the nucleus accumbens. J Neurosci 2012;32:17454–64.

[70] Lepack AE, et al. Dopaminylation of histone H3 in ventral tegmental area regulates cocaine seeking. Science 2020;368:197–201.

[71] Farrelly LA, et al. Histone serotonylation is a permissive modification that enhances TFIID binding to H3K4me3. Nature 2019;567:535–9.

[72] Zhao S, et al. Histone H3Q5 serotonylation stabilizes H3K4 methylation and potentiates its readout. Proc Natl Acad Sci U S A 2021;118.

[73] Feng J, et al. Dynamic expression of de novo DNA methyltransferases Dnmt3a and Dnmt3b in the central nervous system. J Neurosci Res 2005;79:734–46.

[74] Gujar H, et al. The roles of human DNA methyltransferases and their isoforms in shaping the epigenome. Genes 2019;10, 172.

[75] LaPlant Q, et al. Dnmt3a regulates emotional behavior and spine plasticity in the nucleus accumbens. Nat Neurosci 2010;13:1137–43.

[76] Cannella N, et al. Dnmt3a2 in the nucleus accumbens shell is required for reinstatement of cocaine seeking. J Neurosci 2018;38:7516–28.

[77] Marutha Ravindran CR, Ticku MK. Role of CpG islands in the up-regulation of NMDA receptor NR2B gene expression following chronic ethanol treatment of cultured cortical neurons of mice. Neurochem Int 2005;46:313–27.

[78] Ebrahimi G, et al. Elevated levels of DNA methylation at the OPRM1 promoter region in men with opioid use disorder. Am J Drug Alcohol Abuse 2018;44:193–9.

[79] Nielsen DA, et al. Increased OPRM1 DNA methylation in lymphocytes of methadone-maintained former heroin addicts. Neuropsychopharmacology 2009;34:867–73.

[80] PhD VMC, et al. Elevated levels of DNA methylation at the OPRM1 promoter in blood and sperm from male opioid addicts. J Opioid Manag 2011;7:258–64.

[81] Bönsch D, et al. Lowered DNA methyltransferase (DNMT-3b) mRNA expression is associated with genomic DNA hypermethylation in patients with chronic alcoholism. J Neural Transm 2006;113:1299–304.

[82] Schaefer A, et al. Argonaute 2 in dopamine 2 receptor-expressing neurons regulates cocaine addiction. J Exp Med 2010;207:1843–51.

[83] Kenny PJ. Epigenetics, microRNA, and addiction. Dialogues Clin Neurosci 2014;16:335–44.

[84] Bali P, Kenny PJ. Transcriptional mechanisms of drug addiction. Dialogues Clin Neurosci 2019;21:379–87.

[85] Smith ACW, Kenny PJ. MicroRNAs regulate synaptic plasticity underlying drug addiction. Genes Brain Behav 2018;17:e12424.

[86] Hollander JA, et al. Striatal microRNA controls cocaine intake through CREB signalling. Nature 2010;466:197–202.

[87] Quinn RK, et al. Distinct miRNA expression in dorsal striatal subregions is associated with risk for addiction in rats. Transl Psychiatry 2015;5:e503.

[88] Quinn RK, et al. Temporally specific miRNA expression patterns in the dorsal and ventral striatum of addiction-prone rats. Addict Biol 2018;23:631–42.

[89] Pietrzykowski AZ, et al. Posttranscriptional regulation of BK channel splice variant stability by miR-9 underlies neuroadaptation to alcohol. Neuron 2008;59:274–87.

[90] Bahi A, Dreyer J-L. Striatal modulation of BDNF expression using microRNA124a-expressing lentiviral vectors impairs ethanol-induced conditioned-place preference and voluntary alcohol consumption. Eur J Neurosci 2013;38:2328–37.

[91] Tapocik JD, et al. microRNA-206 in rat medial prefrontal cortex regulates BDNF expression and alcohol drinking. J Neurosci 2014;34:4581–8.

[92] Zheng H, et al. Yin Yang 1 phosphorylation contributes to the differential effects of mu-opioid receptor agonists on microRNA-190 expression. J Biol Chem 2010;285:21994–2002.

[93] He Y, et al. Regulation of opioid tolerance by let-7 family microRNA targeting the mu opioid receptor. J Neurosci 2010;30:10251–8.

[94] Im H-I, et al. MeCP2 controls BDNF expression and cocaine intake through homeostatic interactions with microRNA-212. Nat Neurosci 2010;13:1120–7.

[95] Viola TW, et al. Increased cocaine-induced conditioned place preference during periadolescence in maternally separated male BALB/c mice: the role of cortical BDNF, microRNA-212, and MeCP2. Psychopharmacology 2016;233:3279–88.

[96] Yan B, et al. MiR-218 targets MeCP2 and inhibits heroin seeking behavior. Sci Rep 2017;7:40413.

[97] Kwako LE, et al. Addiction biomarkers: dimensional approaches to understanding addiction. Trends Mol Med 2018;24:121–8.

[98] Gu W-J, et al. Altered serum microRNA expression profile in subjects with heroin and methamphetamine use disorder. Biomed Pharmacother 2020;125:109918.

[99] McCrae JC, et al. Ethanol consumption produces a small increase in circulating miR-122 in healthy individuals. Clin Toxicol 2016;54:53–5.

[100] Zhao Y, et al. Decreased expression of plasma microrna in patients with Methamphetamine (MA) use disorder. J Neuroimmune Pharmacol 2016;11:542–8.

[101] Bönsch D, et al. DNA hypermethylation of the alpha synuclein promoter in patients with alcoholism. Neuroreport 2005;16:167–70.

[102] Heberlein A, et al. Epigenetic down regulation of nerve growth factor during alcohol withdrawal. Addict Biol 2013;18:508–10.

[103] Philibert RA, et al. A pilot examination of the genome-wide DNA methylation signatures of subjects entering and exiting short-term alcohol dependence treatment programs. Epigenetics 2014;9:1212–19.

[104] Brückmann C, et al. Validation of differential GDAP1 DNA methylation in alcohol dependence and its potential function as a biomarker for disease severity and therapy outcome. Epigenetics 2016;11:456–63.

[105] Heyn H, Esteller M. DNA methylation profiling in the clinic: applications and challenges. Nat Rev Genet 2012;13:679–92.

[106] Lax E, Szyf M. The role of dna methylation in drug addiction: implications for diagnostic and therapeutics. Prog Mol Biol Transl Sci 2018;157:93–104.

[107] Montalvo-Ortiz J, et al. Genomewide study of epigenetic biomarkers of opioid dependence in European-American women.

[108] Kozlenkov A, et al. DNA methylation profiling of human prefrontal cortex neurons in heroin users shows significant difference between genomic contexts of hyper- and hypomethylation and a younger epigenetic age. Genes 2017;8152. Available from: https://doi.org/10.3390/genes8060152.

[109] Lohoff F, et al. Epigenome-wide association study and multitissue replication of individuals with alcohol use disorder: evidence for abnormal glucocorticoid signaling pathway gene regulation. Mol Psychiatry 2021;26:2224–37. Available from: https://doi.org/10.1038/s41380-020-0734-4.

[110] Birney E, et al. Epigenome-wide association studies and the interpretation of disease -omics. PLoS Genet 2016;126. Available from: https://doi.org/10.1371/journal.pgen.1006105.

[111] West AC, Johnstone RW. New and emerging HDAC inhibitors for cancer treatment. J Clin Investigation 2014;124:30–9.

[112] Machado-Vieira R, et al. Histone deacetylases and mood disorders: epigenetic programming in gene-environment interactions. CNS Neurosci Ther 2011;17:699–704.

[113] Park H-S, et al. Epigenetic targeting of histone deacetylases in diagnostics and treatment of depression. Int J Mol Sci 2021;22:5398.

[114] Covington 3rd HE, et al. Antidepressant actions of histone deacetylase inhibitors. J Neurosci 2009;29:11451–60.

[115] Hamilton PJ, et al. Neuroepigenetic editing. Methods Mol Biol 2018;1767:113–36.

[116] Xu S-J, Heller EA. Recent advances in neuroepigenetic editing. Curr Opin Neurobiol 2019;59:26–33.

[117] Yim YY, et al. In vivo locus-specific editing of the neuroepigenome. Nat Rev Neurosci 2020;21:471–84.

[118] Heller EA, et al. Targeted epigenetic remodeling of the Cdk5 gene in nucleus accumbens regulates cocaine- and stress-evoked behavior. J Neurosci 2016;36:4690–7.

[119] Kang JG, et al. Regulation of gene expression by altered promoter methylation using a CRISPR/Cas9-mediated epigenetic editing system. Sci Rep 2019;9:11960.

[120] Xu X, et al. A CRISPR-based approach for targeted DNA demethylation. Cell Discov 2016;2:16009.

[121] Sahel J-A, et al. Partial recovery of visual function in a blind patient after optogenetic therapy. Nat Med 2021;27:1223–9.

[122] Prado DA, et al. Gene therapy beyond luxturna: a new horizon of the treatment for inherited retinal disease. Curr Opin Ophthalmol 2020;31:147–54.

[123] Novikova SI, et al. Maternal cocaine administration in mice alters DNA methylation and gene expression in hippocampal neurons of neonatal and prepubertal offspring. PLoS One 2008;3:e1919.

[124] He F, et al. Consequences of paternal cocaine exposure in mice. Neurotoxicol Teratol 2006;28:198–209.

Further Reading

Christopher, E.J. et al., *Animal Models of Addiction*. Oxford Medicine Online, 2017.

Kuhn BN, et al. Understanding addiction using animal models. Front Behav Neurosci 2019;13, 262.

Pascoli V, et al. Stochastic synaptic plasticity underlying compulsion in a model of addiction. Nature 2018;564:366–71.

CHAPTER

32

Environmental Influence on Epigenetics

Marisol Resendiz[1],*, Darryl S. Watkins[1],*, Nail Can Öztürk[2] and Feng C. Zhou[1,3]

[1]Stark Neuroscience Research Institute Indiana University School of Medicine, Indianapolis, IN, United States
[2]Department of Anatomy, Mersin University, Mersin, Turkey [3]Department of Anatomy, Cell Biology and Physiology, IUSM, Indianapolis, IN, Uinted States

OUTLINE

Introduction	639	Conclusion and Future Direction	658
The Extent (timeline) of Environmental Influence	641	Research case study [18]	660
Parental and Ancestral Predisposition	641	Objective	660
Prenatal Exposure	641	Scope	660
Postnatal and Lifetime Influence	642	Audience	660
Exerting Environment	642	Rationale	660
Mental or Physiological Environment	643	Expected Results and Deliverables	661
Enriched Environment	643	Safety Considerations	661
Activity and Experience-dependent Changes	645	The Process, Workflow and Actions Taken	661
Stress	646	Tools and Materials Used	661
Exercise	649	Results	661
Nutrition	650	Challenges and Solutions	661
Hazardous Environmental Pollutants and Chemicals	650	Learning and Knowledge Outcomes	662
Heavy Metals and Pollutants	651	Acknowledgments	662
Prescription Medicine	654	Glossary	662
Addictive Substances	654	References	663

INTRODUCTION

Epigenetics, contrary to earlier understanding, is dynamic throughout life and generations (as described in this chapter and Chapter 25). The acquisition of epigenetic modifications after the initial mass erasure and re-establishment at the early implantation stage [1–5] alludes to a fascinatingly important phenomenon, covered extensively herein. In addition, acquired heritable epigenetics, such as with imprinting genes (see Chapter 25), are also essential in shaping organismal life. The overall epigenetic landscape, collectively termed the *epigenome* (e.g., the sum of the entire DNA methylation profile, histone code, and microRNA landscape), is primarily formed in each species and in each individual as

*Co-first authors.

well as in each tissue as a combination of genetic factors, epigenetic memory, and acquired epigenetic inputs over the individual's lifetime (and even prior to birth). Although the formation of the histone code is still largely unclear, much has been uncovered regarding the dynamism of DNA methylation. Following the initial global DNA demethylation of blastocysts and germ lines, most subsequent DNA methylation is differentially programmed over the developmental stages according to cell, tissue, and organ type [1]. After reaching maturity, a methylome and chromatin histone code are presumably established in each tissue type of the individual. Although epigenetic features were thought to be relatively stable thereafter, many new findings are challenging the established dogma. Today, the methylome, histone code, and other epigenetic marks are known to be affected by a myriad of environmental influences and modifications beyond the inheritable and programmed epigenome occurring during development. This chapter will lay out how several environmental factors can influence the epigenome (though the breadth of epigenetic modifications is vast, this chapter will limit the scope to more commonly reported modifications such as DNA methylation, histone modifications, miRNA profiles and epigenetic enzyme alterations), including transient and fixed or "permanent" modifications. Some epigenetic codes change in level and strategic position to alter transcription and affect genetic fates, functions, and cellular behavior, while other epigenetic changes occurring under a threshold or not at a strategic position, make no immediate transcriptional impact.

Environmental effects on the epigenome result not only from life experience but also the life experiences of parents who contribute their sperm and eggs. Some environmental influences may be traced generations back, referred to as "multigenerational" ("intergenerational" is used interchangeably) or "transgenerational" if achieved by exposure three generations prior (or older). While extensive and complementary studies of intergenerational epigenetic inheritance have been published in both animal and human studies, the majority of transgenerational studies performed in animals await validation and those suspected in humans (particularly in families with rare disease inheritance) have been redirected toward underlying genetic variation rather than persistent, heritable epigenetic modifications [6]. Additionally, the influence of aging and oxidative stress can largely impact heritable transmission of epigenetic modifications carried through the germline, complicating the singularity of the environmental exposure studied [7].

All told, the *in utero* development of the embryo can be profoundly affected by the prenatal environment. The remaining environmental influence on the epigenome occurs during the lifetime of the individual and accumulates such that it may alter gene transcription, protein production, physiological function, and behaviors of the individual. The temporal environmental inputs will be discussed first in the section, The Extent of Environmental Influence.

The Environmental Factors are discussed in Section 32.3. In addition to physical agents, this chapter includes the category—mental or physiological environmental factors. These refer to intangible environmental factors e.g., quality of living, maternal care, learning and experience, stress, aging, etc. which influence epigenetics and potentially impact cellular, tissue, organ function, or behaviors. In the second category—hazardous environmental pollutants and chemicals—we include hazardous materials such as heavy metals, pollutants, prescriptive medicine, and addictive substances, which have been increasingly found to alter various epigenetic elements. These are considered more immediate and tractable environmental factors exerting inputs into the epigenome.

Lastly, certain environmental factors can be recorded as epigenetic memory in susceptible regions of the DNA, such as on differentially (DNA) methylated regions (DMR) or regions which may serve as a marker or a coded trace of the environmental past [8,9]. If the DMR leads to a consequential change in transcription and function, we refer to this as a "Functional DMR." The presence of a functional DMR may allow for early diagnosis of diseases or disorders that otherwise appear unsuspicious during early stages. This potentiality will be discussed in the section, Conclusion and Future Direction. Understanding functional DMRs can better elucidate the mechanisms contributing to the potential treatment of diseases or disorders. Furthermore, reversing environmental impacts through the normalization of hazardous exposures and provision of "anti-agents" such as folic acid, choline, or histone deacetylase inhibitors (HDACi) is an emerging horizon in the field [10,11].

An important note in the discussion of environmental factors is that many studies begin with animal observation, some are reinforced by different epigenetic marks, e.g., DNA methylation, histone modification, or microRNA, and some are further observed in human data. A couple of caveats should be noted such as the dose, length, and type of environmental exposures which can further the extent of the epigenetic changes and may not always be comparable between human and animal models; the resistance between species to various environmental factors may not be identical. In general, animal models are designed in consideration of human parameters e.g., body weight, time course, lifespan, etc. However, translational or reverse translational challenes persist and are represented throughout the discussion in this chapter. Overall, rich observations have been documented in animal models, particularly with regard to DNA methylation. While clinical cases lag, they are increasingly confirmatory of preclinical reports. Finally, epigenetic marks beyond DNA methylation while less documented, are increasingly appearing as explorations deepen.

THE EXTENT (TIMELINE) OF ENVIRONMENTAL INFLUENCE

It is important to be aware that environmental exposures which impose epigenetic changes may not be contemporary, but also extend to the period of time prior to birth or even generations prior. An individual's epigenetic make-up can be influenced by an environment traced back through three continuous but chronologically distinct stages of an individual: ancestral (including parents, grandparents, and beyond), prenatal, and postnatal stages, each profoundly different. Thus, the transcription of the genome may ultimately be a function of *environment x epigenetics x gene interactions* spanning several stages. This is significant from a clinical point of view due to the potential timeline of disease etiology expanding from environmental factors during the months or years concurrent to the illness to molecular factors. These factors can be traced as far back as perinatal life during developmental formation or to the individuals' parents and ancestors (imprinting and transgenerational transmission) (e.g., some cases of autism [12] in experimental animal model and clinic correlation [13]), which began laying the epigenetic profile prior to conception through the germline. The three stages are further outlined below.

Parental and Ancestral Predisposition

The definition of parental predisposition includes not only maternal but also paternal epigenetic alterations occurring prior to pregnancy or conception of an individual. As such, some affected individuals are not exposed directly but rather inherit the epigenetic characteristics or marks from their parents' epigenomes which escaped germ line mass erasure [3,5]. Studies of human populations and animal models suggest that a father's experiences or lifestyle factors such as diet or environmental stress can influence development and the health of his descendants. For example, the epigenetic mark designated as H3K27me3 modifies histone methylation and DNA accessibility in *Caenorhabditis elegans*. This mark is demonstrably transmitted from sperm to embryos [14]. Similarly, H3K27me3 in maternal fruit-fly eggs is maintained in the embryo after fertilization, corroborating maternal transmission of epigenetic marks to the offspring [15].

Ancestral epigenetic alterations may be precursors to the environmental influence placed upon the paternal and maternal epigenome, the majority of which would mostly be eliminated during the two stages of germ line maturation—soon after fertilization and during the formation of sperm and eggs [1,2]. Nevertheless, increasing reports show that some of the ancestral epigenetic code can escape global erasure events and survive into the epigenome of the offspring. These transgenerational epigenetic changes are rare, but when substantiated, have been known to be carried over 3–4 generations. Some instances of multi- or transgenerational influences on the epigenome include climate, radiation levels, oxygen levels, and paternal smoking and drinking [16–18]. These long range ancestral environmental influences may not be random, rather, they may entail an epigenetic memory that contributes to the characteristic epigenome among a species or population heterogeneity within a species across time and geographic distributions. The parental or grandparental environmental inputs conceivably not only provide inheritable characteristics but can also affect the offspring in a way that makes the individual disease prone or resistant to future environmental catastrophe.

A second source of ancestral epigenetic influence is derived from the inherited genomic structure. The impact of the genome on the epigenome is manifested through DNA methylation patterns of imprinted regions [19]. Concordant gains or loss of methylation within individual twin pairs are more common in monozygotic than dizygotic twin pairs, indicating that *de novo* and/or maintenance methylation is influenced by the underlying DNA sequence [19]. Many polygenic differences within populations occurring at C and T nucleotides or even at adjacent G to C can also conceivably influence the landscape of DNA methylation, implying that DNA methylation states are tightly maintained according to genetic sequences.

Prenatal Exposure

Prenatal environmental influence occurs during pregnancy from fertilization to birth, at which time maternal exposure to environmental factors which alter the epigenetic make-up of the fetal epigenome are at the forefront. The epigenetic modifications occurring in the prenatal stage include dynamic phases of demethylation and remethylation [1–5] around the blastocyst stage, followed by cumulative remethylation through the remainder of gestation. It is well-known that the uterine environment can contribute to the diversity of monozygotic twins [20]. Many prenatal environmental influences critical to development not only determine health and normality but may also plant seeds of developmental disorders and syndromes in later life [21]. How the adverse environmental exposure of the embryo in the womb might increase vulnerability to developmental deficiency through alteration of the epigenome remains a compelling question and warrants further study. Many of these prenatal epigenetic-altering environmental factors are discussed in the forthcoming sections.

Postnatal and Lifetime Influence

Postnatal to Adulthood

The influence of the environment on the individual epigenome continues after birth and persists throughout the lifespan, as evidenced by genome-wide DNA methylation variabilities found in adolescent monozygotic twins followed since birth [20]. While some modifications may be a transient response to the plasticity of the organism such as in the case of neuronal synapse formation in the brain, other changes may be lasting and are perhaps carried throughout the lifespan via somatic cells. It remains an important endeavor to establish lasting epigenetic modifications which alter functionality. More recent findings demonstrate that experience-dependent epigenetic alterations lead to behavioral changes (See Section 32.4).

Ageing

In a simple definition, chronological age is the amount of time that has passed from birth to the given date for a person. However, chronological age or the passing of time may not always align with an individual's biological age. In 1988, Baker and Sprott proposed the identification of biomarkers to predict the functional capability of an individual or organ in relation to chronological time [22]. Advances in laboratory techniques and biostatistics during and after the Human Genome Project have further progressed the efforts of developing biological markers of ageing [23,24]. Although the general theme of biological ageing and its molecular markers is beyond the scope of the current review, in recent years, the notion of an "epigenetic clock" has gained traction as a promising molecular indicator of biological age [25].

DNA methylation dynamics have become recognizable as one of the hallmarks of biological ageing [26]. Some of the earliest findings exemplify traces of significant methylation changes in human saliva which occurred with aging [27]. Later, pronounced efforts employing machine learning tools helped to accumulate sound evidence that patterns of certain CpG sequences are significantly correlated with accelerated aging. This along with other evidence substantiates the concept of the epigenetic clock in general [28–30]. A recent systematic review highlighted 156 published works that reported on factors relevant to accelerated epigenetic aging [31]. While numerous versions of an "epigenetic clock" exist currently (GrimAge, Hannum, Horvath and Levine clocks), there tends to be agreement with regard to environmental effects. Further, a meta-analysis of influential factors (36 factors out of 57 evaluated), including alcohol use, lung function, cancer, smoking, and physical activity in association with the aforementioned epigenetic clocks demonstrated significant relevance with epigenetic age acceleration [31].

Many of these findings will be discussed below as we introduce specific environmental factors in Section Exerting Environment. The extent (timeline) of the three stages of environmental influence is illustrated in Fig. 32.1. A point worthy of mentioning in this Section is that the deposit and potential functions of epigenetic marks throughout the three chronological stages are cumulative and dynamic. Their density, landscape (e.g., relative to the regulatory regions or gene bodies), and pattern (e.g., CpG island and C vs C adjacent to G) are all potential contributors to the regulatory functions of the epigenetic machinery. Many inherited epigenetic modifications occur at sub-threshold (non-functional) levels but may upon further environmental impacts, across generations or across lifetime stages, accumulate into meaningful (functional) levels capable of triggering transcriptional change and leading to diseases or disorders.

EXERTING ENVIRONMENT

Living organisms and their individual cells are surrounded by environmental factors which can influence their functions as well as the formation and maintenance of their structures. The mechanisms by which the environment influences the individual's cells, tissues, and organs, and their functions in health and disease are major endeavors in biomedical research. A compelling question remains how the environment penetrates to reach the chromatin inside the nucleus to alter transcription.

Environmental factors may directly change the physiology and subsequently structure, cell signaling, cell cycling, leading to different biological states, growth dynamics or pathogenesis (Fig. 32.2). To date, the environmental factors described in this chapter have been demonstrated to affect epigenetics through alteration of the methyl or ethyl donors which fuel the epigenetic changes, altering or

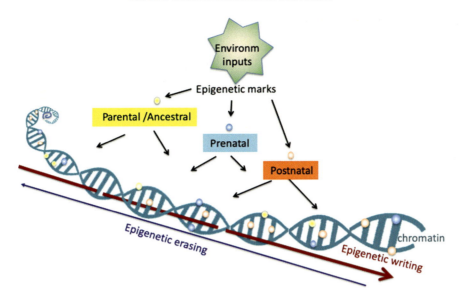

FIGURE 32.1 The extent of the environment. The environmental exposures may lead to epigenetic alterations on chromatin (DNA and histone including methylation, acetylation, ubiquitination, etc.). Three major stages by which the environment can change an individual's epigenome. Exposures at the ancestral and parental stage may result in transgenerational epigenetic change including imprinting, which passes through germ lines. Exposures at the prenatal stage are often carried into life and have profound effects on epigenetic make-up contributing to development and postnatal life. Recent evidence demonstrated that many environmental exposures continue to impact postnatal and adult epigenetics which add to the prenatal epigenetic portfolio. These epigenetic buildups are also constitutionally erased around the fertilization and blastocyst stages followed by remethylation, and some are erased in all above mentioned stages in different scales under different mechanisms. Environm: environment.

translocating (from cytoplasm to nucleus) the epigenetic enzymes such as DNA methyltransferases (DNMTs), DNA demethylation enzymes Ten-eleven translocases (Tets), histone methyltransferases (HMT), histone acetylases (HA), and histone deacetylases (HDAC), or through modifications of the non-coding RNA to influence gene transcription (Fig. 32.3).

In this chapter, a broader sense of the environment includes mental or physiological events, as well as hazardous chemical exposures, including air pollutants. The exerting environment is most likely a comparative term, as the reactivity is relative and incremental. Many environmental elements have influence on the living organism at various levels. The review below is not exclusive but covers recent major findings where environmental influences, acting via epigenetic mechanisms, have been the focus of research. We also try to frame the literature in the context of animal versus human research, gender-based effects, and correlative versus causal links.

MENTAL OR PHYSIOLOGICAL ENVIRONMENT

Mental (Sections 32.4.1–32.4.3) and physiological (Sections 32.4.4 and 32.4.5) environments include nonphysical or intangible experiences which have been found to provide epigenetic inputs in the organism, many leading to altered gene transcription and correlated with lasting behavioral changes. This category includes the following items.

Enriched Environment

Besides the genetic blueprint, three types of epigenetics also participate in the design of early life, the inherited (as a template) epigenetics, the intrinsic epigenetic program (for recent review see [33]), and the environmental related epigenetics [21] which together affect development and behavior [34]. To the latter, during early postnatal life, the environmental setting and early social experiences such as mother-offspring bonding and peer interaction can affect epigenetics. First, environmental enrichment (EE) in animal studies, defined by rich and colorful living spaces with play-toys and litter-mate interaction, has been known to improve cognitive function in young as well as aged mice [35]. Enriched odor exposure increases the number of newborn neurons in the adult olfactory bulb as detected by the proliferation marker bromodeoxyuridine (BrdU), improving odor memory [36]. Recent environmental enrichment studies highlight the role of histone 3 acetylation (an activating mark) in conferring improvements in spatial working memory [37] through promoter binding of choline acetyltransferase (ChAT), plasticity markers like nerve growth factor (NGF) and brain-derived neurotrophic factor (BDNF), both of which are known to enhance nerve growth, neuronal connections and synaptic formation

The epigenetic modification of the *BDNF* gene in particular has been demonstrably affected in the hippocampus by an enriched environment in mice [41]. A significant increase in histone H3 lysine 4 trimethylation (H3K4me3) at the *BDNF* promoter regions 3 and 6, a decrease in histone H3 lysine 9 (H3K9me3, favors transcription) at the *BDNF* promoter region 4 and H3K27me3 (favors inhibition of transcription) at *BDNF* promoter regions 3 and 4 in the hippocampus, likely facilitate *BDNF* transcription.

Environmental enrichment also modulates hydroxymethylation (5hmC) of genes known to be involved in axon guidance in the hippocampus. Interestingly, the alteration of 5hmC mostly occurs at the gene body [35], likely corresponding to the conversion and translocation of 5mC to 5hmC in the promoter as a prerequisite of gene activation (for review see [42]). Similar studies have focused on the ability of environmental enrichment to shape neuroplasticity, such as enhanced hippocampal long-term potentiation (LTP) mediated at least partially by miR-132 suppression of HDAC3 [43]. This epigenetic path has been implicated in intergenerational effects of environmental enrichment on neuroplasticity as well [44]. Environmental enrichment-dependent downregulation of HDAC2 and HDAC5 have also demonstrated a role in neuroplasticity. Specifically, environmental enrichment-related corticospinal tract rewiring was observed in a mouse model of stroke, mediated by Hdac2 downregulation and subsequent induction of the pro-plasticity proteins NGF, BDNF, PSD95, synapsin, and GAP43 [45]. In a model of environmentally enriched postnatal rearing, enhanced neuroplasticity in the visual cortex was attributed to decreased HDAC5 on the *Mef2c*, *Arc*, and *Egr1* promoters and simultaneous enrichment of H3 acetylation at the *Mef2c* promoter [46].

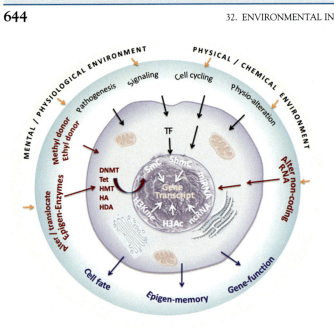

FIGURE 32.2 Environmental factors including mental/physiological events and physical/chemical substrates influence life down to the cellular levels. They may directly change the physiology and subsequently its structure, cell signaling, and cell cycling, leading to different biological states, growth dynamics, or pathogenesis. If these cellular changes are functions of gene alteration, the bridge between the environmental factors and gene transcription is not readily apparent. Besides some direct influence on functional proteins which allow access to chromatin in the nucleus, the environment also instigates epigenetic events which may also reach the chromatin and can regulate gene transcription. To date, environmental factors which are described in the chapter are known to affect the methyl or ethyl donors which fuel epigenetic changes, alter or translocate the epigenetic enzymes (such as DNMT, Tet, HMT, HA, and HAD) from cytoplasm to nucleus, or alter the non-coding RNA to influence gene transcription and mRNA. The outcome of the environmental impacts through cellular and epigenetic mechanisms change gene function, cell fate, or deposit an epigenetic memory awaiting future reinforcement of environmental inputs. *DNMTs*, DNA methyltransferases; *Tets*, DNA demethylation enzymes Ten-eleven translocation; *HMT*, histone methyltransferase; *HA*, histone acetylase; *HDAC*, histone deacetylase.

and in food preference memory via orbitofrontal upregulation [38]

Enriched environments with group caging and toy play also increased neural stem/progenitor cell proliferation and neurogenesis in the subventricular zone of stroke-lesioned adult rats [39]. Also in rodents, a communal nest where two or more mothers keep their pups together and share care-giving duties was used as strategy for the study of the long-term effects of mother-offspring bonding and peer interaction (for review see [40]). Mice reared in a communal nest display relevant changes in brain function and behavior, including enhanced generation of new brain cells and elaborate adult social competencies. This development is closely associated with high levels of neural

Finally, the enriched environment has demonstrated disease-modifying potential. For example, the maternal enriched environment can alter maternal body weight as well as metabolic gene expression in mouse offspring through modulation of epigenetic marks of metabolic genes in both oocytes and offspring [47]. Similarly, in a model of fetal alcohol spectrum disorders (FASD), a disease characterized by life-long CNS dysfunction, learning disability, and low IQ, an enriched social environment has been reported to alleviate behavioral abnormality [48]. Some of the induced changes in gene expression were mediated by epigenetic processes including DNA methylation and miRNAs in an animal model [49],

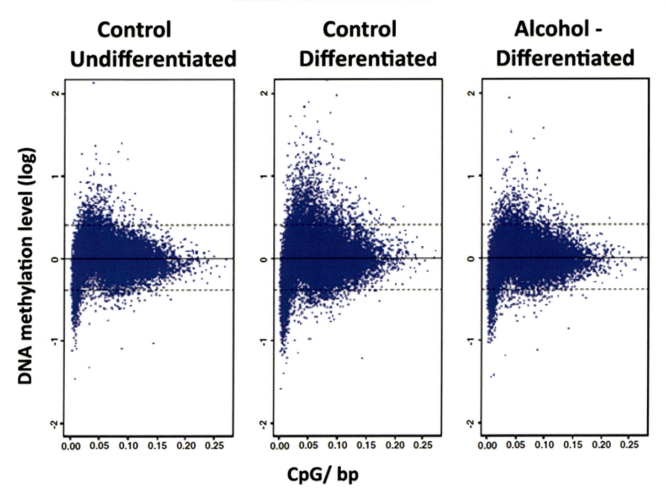

FIGURE 32.3 Alcohol induced epigenomic change on DNA methylation which influences neural stem cell differentiation. Genome-wide distribution of neural stem cell DNA methylation during differentiation and upon alcohol treatment. Scatter-plotting of promoter methylation levels against CpG content in neural stem cells of three treatment groups demonstrated the pattern of DNA methylation. Each dot represents the methylation level of a gene at the promoter. During differentiation, a characteristic dynamic global DNA methylation pattern showed an increased number of genes with hypermethylation as well as an increased number of genes with hypomethylation (middle panel as compared with left panel), while alcohol prevented that dynamic change and prevented many neural stem cells from differentiation. Graft note: X-axis denotes the observed-to-expected CpG ratio, and Y-axis represents the average log 2 transformation of the methylation signal level through all the probes in −1300 to +500-bp from the transcription start site. The dashed lines at y = 0.4 and −0.4 are defined as cutoffs for hypermethylation (above upper line) and hypomethylation (below lower line), which represents DNA-methylation levels >1.3-folds of the genome-wide median methylation levels. The genes with DNA methylation levels between the two lines are defined as moderately methylated [32]. *Adopted from Fig. 32.3 of F.C. Zhou et al. Alcohol Alters DNA Methylation Patterns and Inhibits Neural Stem Cell Differentiation, Alcohol Clin Exp Res, 2011. w35(4): p. 735–746.*

ultimately ameliorating impaired cognitive ability [50]. These findings highlight how early environmental influences determine developmental trajectories through epigenetic mediation and how a positive environmental factors can normalize or even reverse the disease course.

Activity and Experience-dependent Changes

Activity Related Neuronal Plasticity

Mounting evidence demonstrates that a significant scale of external stimuli such as sensory inputs, active cortical motor activities, or cognitive comprehension (here defined as "activity") can reshape neuron plasticity in various degrees such as dendritic and synaptic remodeling and receptor reorganization to sustain adaptive function. This remodeling is likely a result of revised gene expression. Increasing evidence has pointed out that activity-dependent plasticity is mediated by gene expression modifications in the adult CNS. The activity-dependent neuronal plasticity encompassing profound changes in brain wiring and information processing is supported by gene-directed alteration of synapse numbers or synaptic strength (for review see [51]). Activity or experience-dependent neuronal activity has been found to modify the DNA methylation landscape in the adult hippocampus [52]. This activity-modified CpG methylation shows a broad genomic distribution with significant enrichment in brain-specific genes related to neuronal plasticity, synaptic function, protein phosphorylation, neuronal differentiation, calcium signaling, and genes encoding Notch signaling. Methylation changes require active DNMT or demethylation enzymes such as Growth Arrest and DNA Damage gene (*Gadd45*), thymine DNA glycosylase, and Ten-eleven translocation (*Tet*) oxygenases in the adult brain. GADD45a, which erases DNA methylation and is marked by DNA repair-mediated DNA demethylation [44], was deemed essential for activity-dependent induction of DNA demethylation [53]. Recently, GADD45b was also found to be important for the activity-induced demethylation and transcription of genes, such as *Bdnf* and *Fibrous growth factor 1* (*Fgf1*), in addition to its critical role in adult neurogenesis [54].

Understanding of how epigenetics mediates transcriptional change of genes regulating neuronal plasticity has progressed. First, it was found that activity-dependent glutamate channel N-methyl-D-aspartate (NMDA) receptor activation (a part of canonical pathway for neuronal plasticity) modified the epigenetics of the *Bdnf* gene promoters in hippocampal neurons [55] (a key brain region for functional plasticity). Second, DNA methylation-CpG binding-protein 2 (MeCP2) was found to bind selectively to *Bdnf* gene promoter 3 and function to repress expression of the *Bdnf* gene. In contrast, re-activation of *Bdnf* transcription requires calcium-dependent phosphorylation of MeCP2 [56]. Activity-dependent membrane depolarization triggers the calcium-dependent phosphorylation and release of MeCP2 from the *Bdnf* promoter 3 and decreases CpG methylation within the regulatory region of the *Bdnf* gene, through which transcription is facilitated [56]. The involvement of these *Bndf* epigenetic dynamics showcased an epigenetic mediation of activity-dependent neuronal plasticity [57].

To date, environmental enrichment studies have been limited to animal (rodent) studies. While it is advantageous from an epigenetic perspective that environmental enrichment (particularly social interaction) be precisely controlled and quantified, human studies may be much more difficult to define. For one, the complexity of early human factors and social interactions is greater and therefore harder to standardize. Additionally, the modifications and gene/functional reach of environmental enrichment (or lack thereof) have yet to be substantiated in humans. Finally, the causal links in animal studies are sparse and require extensive validation. What is certain is that the epigenetic mechanisms which internalize environmental inputs in animals are present in humans, particularly for predictive adaptive responses in the presence of early life stressors (mainly documented in cortisol related genes/pathways) such as maternal separation [58].

Experience and Memory Formation

Experience-induced changes in DNA methylation are also associated with the formation and maintenance of memory (Review see [59]). DNA methylation actively mediates memory formation, through regulation of both memory-promoting and memory-suppressing genes [60]. Regulation of histone acetylation also occurs during memory formation in the hippocampus [61]. Further, DNA methylation and histone acetylation have been shown to work in concert to regulate memory formation and synaptic plasticity [62]. BDNF also regulates plasticity-related processes underlying memory and learning. BDNF may be so essential for working memory that a single base mutation resulting in the amino acid substitution Val66Met results in impaired dendritic mRNA trafficking and associated memory and cognitive deficits in humans [63]. Epigenetic alterations of the *Bdnf* gene have also been linked with memory, stress, and neuropsychiatric disorders. In the hippocampus, *Bdnf* methylation is a molecular mechanism underlying the persistent cognitive deficits featured in posttraumatic stress disorder (PTSD) [64]. Further, it has been reported that memory deficiency induced by neonatal anesthesia in rats can be improved by increasing histone acetylation of the *Bdnf* gene through environmental enrichment [65]. As with environmental enrichment, the limitations of many studies include the limited generalizability of animal-generated outcomes due to the controlled and strictly defined "experience" compared to humans.

Stress

Stress or stressors are described here as negative mental or environmental factors which have been shown to lead to physiological disturbance, developmental abnormality, among many other disorders. Stress affects epigenetics across a wide range of the lifespan from prenatal to postnatal periods and into adult life, as discussed below.

Prenatal Stress

Stress during gestation can exert powerful impacts on maternal mental health and on fetal brain development in animals and humans. Maternal stress can induce epigenetic signatures of psychiatric and neurological diseases in the offspring. A series of examples of epigenetic mediation were identified through microRNAs (miR), which affect a wide range of genes and their functions. Prenatal stress altered a large spectrum of miRs, e.g., elevation of miR-103 down-regulated one of its gene target *Ptplb* which is involved in brain pathologies [66]. Elevation of miR-323 and miR-98 may alter inflammatory responses and down-regulation of miR-145, a marker of multiple sclerosis in humans, may mediate neurological disease.

A small but meaningful study of human sibling pairs discordant for intrauterine hyperglycemia due to maternal gestational diabetes mellitus (GDM) demonstrated global hypermethylation in GDM offspring according to peripheral blood DNA examination. Specifically, 12 differentially methylated CpG sites were identified on diabetes and obesity-related genes. mRNA expression of the hepatocyte nuclear factor-4 alpha (*HNF-4α*) gene in particular was correlated with CpG methylation status [67]. Genetic variants of *HNF-4α* have been shown to contribute to the risk and pathophysiology of diabetes mellitus and related risk factors. Maternal obesity has similarly been shown to reprogram the placental epigenome in a study of 42 "obese" versus "lean" normative pregnancies. Placental villi DNA of offspring from obese mothers exhibited a general shift toward increased 5mC and decreased 5hmC methylation. This was concurrent with decreases in alpha ketoglutarate (α-KG), a cofactor of the TET enzymes which facilitate conversion of 5mC to 5hmC. Increased 5mC was specifically profiled <100 bp near the transcription start site (TSS) of the pregnancy-specific glycoprotein (PSG) cluster on chromosome 19q13 and the *growth hormone-chorionic somatomammotropin hormone* (*GH-CSH*) gene cluster on chromosome 17q24. Indeed, hypermethylation of these regions was correlated with decreased mRNA expression of *Gh2, Cshl1, Psg2*, and the *Mia* gene [68].

Maternal exposure to stresses (e.g., chronic restraint in animal models) during pregnancy presents elevated maternal glucocorticoid levels, which is implicated in the etiology of developmental deficits in offspring e.g., hypertension, hyperglycemia, autism [69], and anxiety behaviors [70]. Placental 11b-hydroxysteroid dehydrogenase type 2 (HSD11B2) functions to block the impact of maternal glucocorticoid effect on the fetus, however, epigenetic mechanisms have been identified by which prenatal stress alters *Hsd11b2* gene expression to compromise its buffering effect [71]. In humans, maternal stressors, particularly, late-pregnancy depressive symptoms, have been associated with the methylation signature of the glucocorticoid receptor gene *Nr3c1* in newborns and infant cord blood [72]. A landmark study of objective maternal stress was carried out from women pregnant during a 1998 Quebec ice storm [73]. The offspring of these mothers were profiled 13 years later and the epigenetic landscape of the children's T cells revealed many immune system gene pathways were affected by prenatal maternal stress-specific (PNMS) changes in DNA methylation. Among these, the *lymphotoxin alpha* (*Lta*) gene, involved in the regulation of the innate and adaptive immune system, exhibited 18 differentiated methylated CGs that correlated with objective PNMS. Additionally, the secretory granule neuroendocrine protein 1 (*Sgne1*) gene was functionally dependent on CpG methylation downstream of the TSS, the area overlapping differentially methylated CpGs correlated with objective PNMS. Recently, in a cohort of 117 pregnant females, late term subjective maternal stress (including anxiety, depressive-like behaviors, and stressful life events) predicted peripheral mononuclear cell methylation in 3-month-old offspring and depressive symptoms predicted the DNA transcription epigenetic component [74], reiterating the influence of prenatal maternal stress on the immune system epigenetics of offspring.

Prenatal stress is not limited to maternal influence. Paternal stress may also affect the epigenetic profile of the offspring. For example, paternal stress experience across the lifespan was found to induce epigenetic changes in germ cells (specifically via sperm miR alterations) and impact offspring hypothalamic-pituitary-adrenal (HPA) stress axis regulation [75]. Another animal study showcased the ability of miRs to confer paternal stress intergenerationally. This corticosterone (CORT)-supplementation study reported a myriad of sperm-altered miRNA coinciding with transient increases in early-life fear conditioning, reduced hippocampal *Igf2* mRNA and elevated *Bdnf* exon IV transcript levels in female offspring. In contrast, F1 males demonstrated enhanced anxiety and depression-like behaviors in later life which coincided with increased hippocampal *Igf2* mRNA. *Igf2* and *Bdnf* contain predicted binding sites for miR-98 and miR-190-b, respectively, two validated CORT-sperm altered miRs [76]. These miRs and other epigenetic changes may contribute to the epigenome of the fertilized egg and subsequent offspring if they escape the global erasure at conception. Though the existing body of literature fairly represents human and animal prenatal stress, it is important to note that findings are not interchangeable at this stage. This is particularly due to the robust, complex, and sometimes intangible nature of maternal stress (e.g., depression, life-threatening events) compared to the laboratory-generated models of prenatal stress (e.g., chronic restraint) which are relatively straightforward.

Childhood Adversity or Early Life Stress

Stress continues to affect epigenetics at all stages of postnatal life, which can be cumulative. A wide range of maternal stresses affecting growth of children have been found to be linked to epigenetic causes. Maternal depression, quality of maternal care, and maternal separation have all been found to affect development of individuals through the influence of epigenetics [77]. Mother—infant interactions may lead to epigenetic modifications at the promoter regions of steroid receptor genes, as was demonstrated in a mouse model of maternal care. For example, levels of offspring hippocampal glucocorticoid receptors (GR) regulating HPA function were found to respond to stress upon the level of maternal care. Epigenetic response to such stress, e.g., DNA methylation of $GR1_7$, was found decreased, corresponding to elevated levels of GR expression in the offspring [78]. Further, the binding site for the transcription factor *NGFI* at the *GR* promoter, was found to be differentially methylated. The differential methylation of the *GR* promoter during the postpartum period was sustained into adulthood. Indeed, this observation is extended as humans with a history of childhood abuse similarly experience-altered HPA stress responses and an increased risk of suicide. Epigenetic differences with increased DNA methylation were found in a neuron-specific GR nuclear receptor subfamily 3 group C member 1 (*NR3C1*), at the promoter in postmortem hippocampus obtained from suicide victims with a history of childhood abuse as compared with suicide victims with no childhood abuse [79]. Increased DNA methylation at the *Nr3c1* promoter corresponded to decreased levels of the GR mRNA. Another human study of childhood emotional abuse severity further pinpoints methylation of the *Nr3c1* exon 1 F as a predictor of cortisol levels in a depressed cohort [80].

Other indicators of early life stress include hypermethylation of the *Bdnf* exon IV in the prefrontal cortex and whole blood tissue in a rodent model of infant maltreatment [81]. On the other hand, hypomethylation of the GR sensitivity modulator *Fkbp5* at two CpGs at intron 7 has been reported in preschool-aged children with documented history of mild-severe maltreatment [82] and in a population of children in foster care (age 12), where time spent institutionalized was negatively correlated with CpG1 methylation [83]. Interestingly, in a study of patients with psychotic disorders who experienced various degrees of early life stress, *Fkbp5* hypomethylation at CpG2 and CpG4 (intron 7) was associated with history of adverse events and improved cognitive ability in adulthood [84], suggesting a protective role against elevated cortisol in early psychosis. Interestingly, early life exposure to intimate partner violence (prenatal and postnatal) appears to mitigate offspring anxiety and depressive-like behaviors in contrast to offspring exposed to prenatal stress only. In this cohort, *Fkbp5* was hypermethylated at a plausible binding site for the AP-1 complex, an activator of gene transcription, suggesting a mechanism for mitigating the hormonal stress response [85].

A major category of early life stress in the literature is maternal separation and subsequent effects on offspring. In a mouse model of maternal deprivation and social isolation, female offspring (but not males) demonstrated reduced reward-seeking behavior and decreased expression of the *dopamine D1 receptor* gene *Drd1a* in the VTA in adulthood [86]. Decreased *Drd1a* expression was correlated with hypermethylation at 29/31 CpG sites. In contrast, a juvenile rat model of maternal separation observed dopaminergic hyperexcitability in the VTA, likely due to GABAergic LTD mediated by histone deacetylation [77]. In a two-hit rat model of early life stress (maternal separation) and adult stress (contextual fear conditioning, CRC), maternal separation enhanced contextual fear conditioning compared to maternally reared offspring. Molecular investigation showed that *Reln* methylation was decreased in the maternally deprived weanling hippocampus in addition to Reelin protein, the latter of which was detectable into adulthood. Further, through pharmacological treatment, the contribution of *Reln* methylation was evident in CRC [87,88]. In another two-hit model of maternal separation and subsequent single prolonged stress, only offspring who experienced early life stress demonstrated increased anxiety, depression, and contextual fear memory. Adulthood stress-induced reductions in *Bdnf* were dependent on the severity of early life stress and correlated with increased HDAC2 and decreased H3K9 acetylation. Interestingly male offspring demonstrated a degree of second stress resilience [89]. The concept of sex-dependent effects of early life stress is re-iterated by a recent direct and cross-generational assessment of maternally deprived rats, wherein *Bdnf* hypermethylation and decreased mRNA after maternal separation were observed in females up to the F2 generation, despite behavioral indications of a blunted stress response [90].

Finally, chronic social stress in adult animals has been shown to induce long-term hypomethylation in the promoter region of the gene encoding the *corticotropin-releasing factor* (*Crf*), accompanied by an increase in mRNA expression in animals that display social avoidance [91]. In a human study of 7th grade children, hypermethylation of the *Nr3c1* exon 1 F transcript, particularly at CpG2, were associated with social stressors, particularly being bullied or lacking friends [92]. Similarly, in a study of 4- to 5-month old term infants exposed to infantile social stress via the face-to-face still-face paradigm, buccal cell DNA analysis revealed a hypermethylated *Nr3c1* promoter which was correlated with heart rate variability (a measure of stress responsivity) [93]. A major limitation of animal studies of early life stress is that the paradigms for "chronic" stress may not accurately reflect the human experience. For example, most rodent models of maternal separation are limited to isolation of pups for a few hours daily from birth to young adulthood for the purposes of pup survival. Human maternal separation, on the other hand, occurs predominantly on a greater scale, suggesting incongruence of epigenetic, transcriptional, and behavioral outcomes.

Exercise

Exercise is commonly known to affect muscle and adipose tissue, and in recent years has been further recognized to impact adult brain neural stem cells [94]. Exercise is a good model to showcase the change of gene transcription of an individual during lifestyle changes. The question remains how exercise as an environmental factor influences gene transcription to sustain a phenotype, and whether epigenetics plays a role in this transition. Acute exercise has been shown to remodel promoter methylation in human skeletal muscle [95]. On the other hand, a 6-month exercise routine influences genome-wide DNA methylation patterns in human adipose tissue [96]. In the brain, physical exercise improves cognitive responses to psychological stress through enhancement of epigenetic mechanisms and gene expression in the dentate gyrus [97]. For example, exercised rats showed a significantly higher number of dentate granule neurons expressing the histone 3 acetylation (H3Ac; favors activation of transcription) and induction of the immediate-early gene *c-Fos*. Similarly, four exercise modalities improved survival and aversive memory in aged rats. This occurred in conjunction with exercise-induced mitigation of *c-Fos* promoter reductions in H3K9 acetylation and H3K4me3 in the hippocampus [98].

Both acute and chronic exercise significantly impact DNA methylation, in a highly tissue- and gene-specific manner (for review see [99]). As mentioned earlier, BDNF affects synaptic structure and function. Exercise has also been found to promote the expression of *Bdnf* by increasing beta-hydroxybutyrate which activates the *Bdnf* promoter 1 and transmitter release in the hippocampus [100]. Exercise also induces changes in class I histone deacetylase (HDAC) expression and its binding to the hippocampal *Bdnf* promoter to affect *Bdnf* transcription [100]. In senescence-accelerated mice, a 6-month exercise regimen increased *Bdnf* expression and increased the hippocampal HAT to HDAC activity ratio, a crucial component of histone acetylation [101]. In a pilot study of institutionalized older adults, a multimodal 8-week exercise protocol further demonstrated that H3 hyperacetylation was correlated with improved cognition and quality of life, though *Bdnf* mRNA expression was not impacted in that study [102].

Parental exercise has been shown to impact not just the parent but the offspring as well. For example in female rats, strength training during pregnancy revealed that male offspring observed DNA hypomethylation in the hippocampus with parallel increases in the number of cells in the hilus of the dentate gyrus [103]. A similar rodent study additionally recorded increases in the spatial cognitive ability of male offspring along with improvements in neuroplasticity indicators in the dentate gyrus [104]. Spatial learning and hippocampal hypomethylation of offspring were similarly observed in a preconception exercise program of paternal rats. However, no changes in markers of neuroplasticity were observed with paternal exercise [105]. Parental exercise has additionally shown ameliorating potential for various offspring insults, including some of alcohol's neurobehavioral deficiencies in an animal model of fetal alcohol spectrum disorders [48], maternal signatures of gestational diabetes and GDM-associated signatures in the cord blood of human infants [106], and cognitive deficits and BDNF inhibition in a rodent model of *in utero* sevoflurane exposure (the latter of which was p300 histone acetyltransferase dependent) [107]. For a systematic review of intergenerational exercise-derived epigenetic modifications see [108].

Finally, exercise displays disease-modifying potential utilizing various epigenetic mechanisms. Notably, differentially methylated regions profiled in humans observing various degrees of alcohol misuse and exercise showed 32 overlapping regions, 15 of which were enriched in genes related to synaptic plasticity [109]. These opposite epigenetic associations suggest exercise as an epigenetically driven intervention for alcohol use disorders. Corroborating evidence can be taken from rodent studies of chronic alcohol dosing and adolescent intermittent alcohol. A 12-week treadmill protocol in chronically dosed mice ameliorated markers of oxidative and endoplasmic reticulum stress, *Dnmt* expression, hyperhomocysteinemia, and blood brain barrier dysfunction [110]. In adolescent rats intermittently dosed with ethanol, voluntary wheel running from P56–95 restored cholinergic neuron and neuroimmune markers, as well as learning deficits. Promoter hypermethylation (both DNA and H3K9me2) of *ChAT* and increased H3K9me2 of the *Trka* promoter in response to intermittent alcohol were reversed by voluntary exercise [111]. Finally, a pilot randomized controlled trial of postmenopausal women found that after a 6-month intervention to reduce sedentary behavior, DNA methylation at 12 loci (including the weight associated genes *Nampt* and *Runx3*) were correlated with changes in percent body weight, suggesting a link between exercise-associate weight loss and epigenetic regulation [112].

Nutrition

Since epigenetic donors such as methyl and acetyl donors are derived from diet, nutritional imbalance can affect the supply chain. Dietary and nutrition disparity during early life has long been considered as a critical environmental factor, which may exert long lasting effects on development led by altered gene transcription. Increasing evidence shows alteration of DNA methylation globally and in specific genes. A major entry point to epigenetics is the involvement of these dietary factors in pathways that interact with methyl donors [113]. These factors include folic acid, methionine [114], vitamin B complex, polyphenols, flavonoids, and phytoestrogen genistein. Apparently, many natural products including medicinal herbs may have impacts on epigenetics, though studies have yet to be done to fill these knowledge gaps. This discussion is mainly covered in Chapter 31. Briefly, animal studies of methyl-donor supplementation diets (i.e., folic acid, betaine, choline) variably demonstrate effects on global DNA methylation. For example, animals studies of the intergenerational effects of methyl-donor supplementation/deficiency have reported methylation-associated upregulation of important developmental and imprinting genes in maternal models [115,116], and to a lesser extent in paternal supplementation [117]. On the other hand, there is limited evidence from supplementation studies at this time that methyl donors other than folic acid can alter DNA methylation patterns in humans [118]. This may be due to lack of homogeneity in study protocols and lack of convergence of genomic loci.

Mounting evidence corroborates the influence of maternal diets on offspring, particularly with regard to high fat diets. Animal studies have implicated the role of hypermethylation at the promoter, enhancer, and transcription factor binding sites of lipid metabolism genes like pro-opiomelanocortin (*Pomc*) in the arcuate nucleus and hepatic *Lipin1* [119,120]. Additional evidence demonstrates the role of the maternal high fat diet in dysregulation of leptin by decreased binding of the repressive histone mark H3K9me3 at the *Lepr* promoter in the hippocampus of female offspring only [121], echoing the sex-dependent pattern of maternal inheritance seen across other intergenerational studies.

The importance of the gut microbiome in maintaining intestinal homeostasis has been increasingly acknowledged. Evidence from mice studies employing genetic manipulation have demonstrated the importance of microbiota-induced epigenetic programming for maintenance of gut health *in vivo* [122]. Additionally, diet-related changes in gut microbiota were found to reprogram histone acetylation at enhancers and binding sites of colorectal cancer genes and lipid metabolism-associated transcription factors [123]. While limited, evidence from human studies implicates differential DNA methylation as a correlate of microbiota diversity [124,125]. For a more thorough review of diet-related microbiome epigenomics see [126]. A caveat of exercise and nutrition epigenetic studies is the myriad of potentially confounding lifestyle factors in human studies which are more difficult to assess and control compared to animals. On the other hand, basal metabolic differences limit the generalizability of animal studies.

The mental or physiological environmental factors mentioned above are certainly a small fraction of a yet to be identified larger picture due to the relative age of the field and the still-developing methodology. Current studies have yet to deeply investigate some of the more emerging epigenetic elements, including 5-hydroxymethylation (5hmC, another form of DNA methylation with an opposing role on transcription), emerging histone codes, Polycomb and trithorax complexes [127]. Current research in these aspects is in the midst of moving from cellular epigenetics (e.g., average changes of 5mC or H3K9me3 in a tissue) towards epigenomics and single gene epigenetics. In light of analysis tools such as high throughput methylation and chromatin chip analyses, more exciting findings are expected on the horizon.

What we have learned so far is that intangible factors such as profound experiences (mental) and physical exercise are capable of provoking neuroendo-inflammatory states, such that life experiences find their way into the organismal chromatin via modification of epigenetic codes, ultimately influencing gene transcriptions. Of the above described categories, other potential factors would be compelling for future investigation, including psychiatric conditions, e.g., anxiety, and depressive mood, intensive skill training, or disease-prone diets, e.g. high sugar and high fat diets. What is important of these recent findings is that intangible environmental factors have certainly demonstrated the capacity for altering epigenetic coding to regulate transcription, such as in the learning activity ocurring through the epigenetic regulation of *Bdnf* and *Nmda* transcription which underlie neuronal plasticity and brain function. On the other hand, the the same mechanisms lend themselves to facilitate the transference of detrimental environmental factors, such as in how epigenetics may alter the HPA axis to induce stress/inflammatory factors affecting long-term brain and endocrine functions.

HAZARDOUS ENVIRONMENTAL POLLUTANTS AND CHEMICALS

The following section will examine environmental exposures such as hazardous substances or chemicals which pose health risks, facilitate myriad disease states, and have also been shown to alter epigenetic profiles. This includes but is not limited to heavy metal toxicants and pollutants such as: arsenic, cadmium, lead, mercury, endocrine disrupting chemicals, pesticides, prescription medicines, air pollution, and addictive substances. Literature on the epigenetic effects exerted by many of the chemicals listed above is constantly being updated. The following section will give an overview of some of the seminal and more recent studies that corroborate how hazardous environmental pollutants and chemicals impact the epigenome, leaving distinct altered epigenetic signatures.

Heavy Metals and Pollutants

The epigenome is malleable and susceptible to environmental assault. Heavy metal toxicants such as lead, arsenic, cadmium, mercury, chromium, nickel, etc., have all been reported to impact the epigenome in different capacities. In recent years, the list has broadened in scope to include non-metal pollutants such as bisphenol A, phthalates, air pollution, pesticides, and prescription medications like valproic acid and g-hydroxybutyrate.

Lead

Lead (Pb) toxicity represents a significant health threat worldwide. Exposure to Pb is not uncommon, as Pb is found in many household products, public water supply via Pb-containing water pipes, and industrial waste (for recent review see [128]). The deleterious effects of Pb can be dose dependent; however, it is important to note that Pb concentrations as low as (BPb <10 µg/dL) detected in the human blood can interfere with appropriate brain development and neuroplasticity, cause damage to vital organs such as the kidneys, and drive perturbations in the reproductive and cardiovascular systems [129]. There is epidemiologic evidence that chronic Pb exposure to humans result in higher instances of aging-associated diseases such as Alzheimer and Parkinson disease, and although specific mechanisms underlying the causative nature remain unclear, an emerging role for epigenetics is beginning to be uncovered [130,131] (for recent review see [132]). In rodent models, perinatal and/or early life Pb exposure has been shown to increase DNMT3 and H3K27me3 protein expression, while decreasing DNMT1, MECP2, H3K9Ac and H3K4me2 protein expression within the hippocampus [133,134]. An earlier review on Pb exposure and essential epigenetic biomarkers for appropriate fetal brain development points to multiple modifications within the epigenome, including differential DNA methylation changes, histone code and non-coding RNA modifications [135]. A recent human study using umbilical cord blood of Pb exposed children indicated changes in genomic 5hmC and 5mC expression, common epigenetic biomarkers [136]. Pb exposure in a dose-dependent manner is additionally associated with hyperactivity, a prominent component of attention-deficit/hyperactivity disorder (ADHD) [137,138]. Interestingly, Pb exposure leads to increases in histone acetylation by elevating p300 (a histone acetyltransferase) enzymatic activity within the hippocampus, suggesting a possible epigenetic role in ADHD pathology [139].

Arsenic and Cadmium

Arsenic (As) is one of the most concerning pollutants found in contaminated drinking water globally and is considered a human carcinogen [140]. As-induced cancer outcomes are well documented and new studies continue to produce novel insight and therapeutic strategies to combat As-mediated cancer risks. Non-cancer related outcomes such as altered immune responses, neurodevelopmental perturbations, and vascular disease (VD) are also associated with As exposure [141]. Mechanisms by which As exposure dysregulates specific epigenetic signatures within the epigenome to give rise to non-cancer related health outcomes are yet to be fully understood; however, new and exciting research investigating this perplexing problem are starting to develop. Methylation sequencing analysis has shown that prenatal As exposure induced CpG 5mC alterations in newborn cord blood and is associated with changes in gene expression and subsequent birth outcomes [142]. In addition, another study using cord blood samples from As-exposed subjects *in utero* with low-level concentrations between 0.10 to 0.15 µg/L, revealed aberrant DNA methylation changes in a gene specific manner [143]. Furthermore, increased expression of miRNAs associated with innate and adaptive immune responses were detected in individuals exposed to higher levels (up to 236.0 µg/L) of As in drinking water [144]. The neurotoxicity of As has been shown to be transgenerational, in that ancestral exposure to As produced deficits in motor and cognitive function in subsequent generations of zebrafish via decreases in *Bdnf* mRNA expression and increases in H3K4me3 methylation [145]. As exposure is also associated with VD. A recent study using newborn cord blood identified five CpGs associated with low-density lipoprotein (LDL), a biomarker for cardiovascular disease (CVD), to have differential methylation alterations when exposed to As *in utero* [146]. These data suggest a causal relationship between As exposure, lipid metabolism and CVD. Taken together, As exposure gives rise to long-lasting epigenetic reprogramming.

Cadmium (Cd) exposure occurs mostly through consumption of vegetables, seafood and tobacco smoking [147]. Cd is an established human lung carcinogen and is known to mediate aspects of CVD, but the underlying mechanisms remain unclear. Recent studies have demonstrated Cd-induced epigenetic changes, which may contribute to Cd-mediated toxicity and its pathogenic risks. Cd treated human bronchial epithelial cells (BEAS-2B) demonstrated increased global H3K4me3 and H3K9me2 by inhibiting the activities of H3K4 and H3K9 demethylases, which persisted after Cd removal [148]. In a human study, a population exposed to Cd contamination in Thailand indicated hypomethylation of epigenetic genes O-6-methylguanine DNA methyltransferase (*Mgmt*), Metallothionein 2 A (*Mt2a*) and *Dnmt3b* in blood samples of women, and the hypermethylation of *Line-1* in men were associated with increasing levels of Cd in their urine [149]. In Cd-linked preeclampsia pregnancy disorder, characterized by high blood pressure and proteinuria, miRNAs that target the transforming growth factor (TGF)-beta pathway were significantly altered within the preeclamptic placenta and Cd-treated trophoblasts [150].

Furthermore, an epigenomic DNA methyl-sequencing study in mother–baby pairs showed that maternal exposure to Cd with a blood concentration of (0.2 ug/L) differentially altered DNA methylation of specific transcription factor binding sites within the CpG islands of the gene sets between the mother and the infant [151]. In addition, research suggests specific imprinting control regions (ICRs) may be more vulnerable to Cd-mediated epigenetic alterations when probing DNA methylation changes using mother–baby cord blood samples and whole genome bisulfite sequencing assays [152]. Interestingly, chronic Cd exposure has been demonstrated to impact over 1788 methylation sites in male rodent spermatozoa, impacting transcription initiation and splicing, and suggesting a role for Cd-mediated aberrant methylome changes that impact fertility and future offspring in mammalians [153].

Methylmercury

The largest source of exposure to methylmercury (MeHg) is through the consumption of ocean fish. MeHg poison is most harmful with high exposure during early life, which leads to neurological disorders including mental retardation, cerebral palsy, hyperkinesia, and dysarthria [154]. It is known that MeHg can be transferred from the mother to the fetus via neutral amino acids across the placenta [155]. Neonates and infants can be continuously exposed to MeHg through maternal breast milk. The epigenetic mechanisms regulating MeHg toxicity that lead to neurodevelopmental perturbations have just begun to be explored. Prenatal exposure to MeHg has been associated with lower cognitive performance and lower DNA methylation levels on paraoxonase 1 gene (*Pon1*) in early childhood males [156]. An animal study in which young adult mice previously exposed to MeHg at the dose of 0.5 mg/kg/day from gestation day 7 (E7) to postnatal day 7 (P7) exhibited depression-like behavior, increased repressive chromatin marks such as DNA hypermethylation, reduced H3 acetylation, and increased H3K27me3 at the *Bdnf* promoter region [157], and was also associated with decreased BDNF protein expression. Interestingly, treatment with the antidepressant fluoxetine increased H3 acetylation, restored BDNF protein expression, and ameliorated depressive behavior [157].

In human studies, cervical swabs from pregnant women with toenail mercury concentrations (0.17 ± 0.09 μg/g) demonstrated decreased expression of microRNA miR-575 and miR-4286 [158]. A methylation sequencing study showed that pregnancy exposure led to hypermethylation in a panel of CpG islands in newborn cord blood cells [159]. Another analysis of cord blood from a cohort of newborns from maternal mercury indicated that mothers had DMRs associated with genes *Angpt2*, *Prpf18*, *Foxd2*, and *Tceanc2* [160], which are potentially associated with neurological development. Although the animal studies indicated above and the recent clinical findings have progressed the study of MeHg's effects on the epigenome, compelling translational studies are needed to fill the gaps and give rise to new therapeutic strategies for individuals affected by MeHg toxicity.

Endocrine Disrupting Chemicals

Endocrine disrupting chemicals (EDCs) like bisphenol A (BPA) and phthalates are chemicals capable of interfering with appropriate hormonal-related processes, and has become an increasingly disconcerting environmental hazard due to the wide use of plastic bottles, protective linings of beverages, food containers, cosmetics, and basic personal care commodities (for recent review see BPA [161] and phthalates [162]). Studies show that exposure to EDCs such as BPA can affect neurodevelopment, sexually dimorphic behaviors, and mediate hyperactivity disorder [163]. Moreover, exposure to phthalates have also been correlated with hyperactivity, sex-specific perturbations, allergies, and asthma [164,165]. Prenatal/early-life exposure to EDCs alter appropriate methylation programming in both rodents and humans [166–168]. Prenatal exposure to low doses of BPA leads to the acceleration of neurogenesis and aberrant neuronal migration [169]. In fact, BPA was found to induce MECP2 binding and reduced H3K9 acetylation at the promoter of potassium chloride cotransporter 2 (*Kcc2*), along with decreased *Kcc2* expression, which led to a delayed perinatal chloride shift in cortical neurons [170]. This could be a potential epigenetic mechanism of BPA-induced neurotoxicity. Early-life exposure to BPA significantly decreased methylation levels at repetitive elements *LINE-1* and *IAP*, and increased methylation at imprinted genes *Esr1* and *Igf2* in mouse tail tissue [171]. Interestingly, maternal dietary exposure to different mixtures of phthalates from gestation through lactation periods show alterations in tail *IAP* DNA methylation in a sex-specific manner in the weanlings of the exposed dam [172]. In addition, human studies have shown that maternal exposure to phthalates decreases *LINE-1* methylation in infant cord blood [173,174]. Newly emerging evidence suggests that mixtures of EDCs such as BPA and phthalates can modify epigenetic programming within the hypothalamus of female rodents altering maternal behaviors across multiple generations, implicating long-lasting deleterious effects of EDCs [175]. These data offer new insight into how EDCs may alter epigenetic profiles that regulate their negative health impact.

Pesticides

Pesticides including insecticides and fungicides provide a very interesting paradigm for epigenetics research. One aspect is how they may be toxic and/or carcinogenic hazards to humans; another is how some species through epigenetic mechanisms rather than genetic mutations develop epigenetic disruption resistance. In addition, researchers are starting to investigate the transgenerational effects of pesticides to the epigenome.

As an environmental hazard, a well-documented case reported that Vinclozolin (trade name Ronilan, Curalan, Vorlan, Touche), a fungicide commonly used to control blights, rots, and molds in fruits and vegetables, has long lasting effects on epigenetic codes. Vinclozolin is an antiandrogen which can cause sex organ malformations, transgenerational endocrine-related perturbations, and is a possible human carcinogen. Embryonic exposure can influence sexual differentiation, gonadal formation, and reproductive functions. In a rat model, Vinclozolin-mediated epigenetic modifications were transmitted across 3–4 generations, demonstrating altered DNA methylation carried through sperm [176]. Vinclozolin exposure as low as (3 mg/L) in the ancestral generation led to an alteration of global DNA methylation levels in exposed individuals and the two subsequent generations (F1 and F2) not exposed. In addition, a significant decrease in cross-sensitivity to the insecticide, Imidacloprid, in F1 and F2 was also identified [177]. Transgenerational epigenetic inheritance was first confirmed in the studies of pesticides on rodents, which led to the studies of transgenerational effects on prenatal malnutrition in a Dutch population exposed to famine during World War II [16]. The studies also identified an epigenetic link across generations [178]. Further study of this area is warranted to better understand the etiology of pesticide-linked toxicity which might have been seeded generations prior. In more recent studies, researchers have started to sample blood from individuals who have varying occupational exposure to a myriad of pesticides. Data suggest genome-wide DNA methylation changes using Illumina 450 K arrays. Furthermore, there 31 CpGs annotated to 29 genes (several of which were previously associated with pesticide exposure), when compared to controls were differentially methylated [179].

Air Pollution

Air pollution is undoubtedly one of the most alarming forms of pollutant exposures as the World Health Organization estimates approximately seven million deaths per year globally are related specifically to air pollution [180]. Air pollution can be defined as any particulates that modify the natural atmospheric characteristics that are hazardous to living organisms. Exposure to air pollution gives rise to respiratory disorders such as asthma, causes cardiovascular disease and has been correlated with neurodegenerative diseases such as Alzheimer disease (AD) (see recent review [181]). Many of the toxicants and pollutants discussed previously have been found in atmospheric air; thus, it is not surprising that compelling evidence postulates air pollution as a hazardous exposure capable of dysregulation of normative epigenetic programming. Recently, a study demonstrated that 4-hour and 7-day exposure to traffic-related air pollution (TRAP) causes hypomethylation in *LINE-1* and inducible nitrous oxide (*iNOS*) promoter along with hypermethylation in adenomatous polyposis coli (*APC*) and $p16^{CDKN2A}$ promoter regions [182]. Coincidentally, one of the initial reasons for the study mentioned above was that *iNOS*, *APC* and $p16^{CDKN2A}$ were previously found to experience changes in methylation in blood samples collected from lung cancer patients [183–185]. A human methylome-wide association study (MWAS) using peripheral blood leukocyte DNA found that air pollution was correlated with DNA methylation changes in specific genomic regions that were associated with myriad disease states, including neurological, pulmonary and cardiovascular [186]. Furthermore, researchers have also demonstrated that both humans and rodents exposed to air pollution display a decrease in H3K9me2 and H3K9me3 expression within the frontal cortex [187]. The overview on air pollution in this section is not exhaustive; however, the data suggests that air pollution may mediate precise epigenetic modifications that have deleterious health outcomes.

Heavy metal toxicants and pollutants facilitate health hazards impacting various biological systems. However, the long-lasting effects from fetal to adult reprogramming of functional gene cohorts can only be demystified after a better mechanistic understanding of their causal relationship to the epigenetic events described above. The field also warns as emerging evidence continues to suggest the potential carry-over or transfer of specific epigenetic signatures in the germ line that may contribute to distinct transgenerational and multigenerational epigenome changes. Although there are rigorous human studies that correlate these deleterious agents to changes in epigenetic machinery and programming, highly controlled causative studies are lacking. Currently, animal research allows for more invasive and controlled paradigms, giving rise to basic mechanistic understanding of environmental epigenomic impacts. As research continues, higher-order mammalian models such as primate and human studies along with stringent translational methodologies will be needed to fully elucidate how heavy metal toxicants and various pollutants mediate epigenomic changes that propagate myriad disease states.

Prescription Medicine

Epigenetic mechanisms for disease modifying therapeutics as well as negative side effects of pharmaceutical agents are major new fields of study. Pharmaceutical agents with epigenetic mechanisms are often a double-edged sword. A list of drugs with side effects postulated to occur through epigenetic alterations is further discussed elsewhere (for review, see [188]). A few examples (e.g., valproic acid, and γ-hydroxybutyrate) are showcased below.

Valproic Acid

Valproic acid (VPA) (commonly known as valproate, convulex, divalproex and depakote) is used as an antiepileptic drug for the treatment of seizures and bipolar disorder [189]. Currently, VPA is active in over 100 clinical trials globally [190]. In spite of the therapeutic uses, several studies have demonstrated that VPA caused higher risks of infant growth defects during early pregnancy [191]. Developing research suggests VPA may disrupt specific epigenetic machinery that facilitates such negative outcomes. VPA acts as an inhibitor of histone deacetylase (HDACi) [192]. VPA inhibits HDAC to suppress *Cyp19* gene expression and inhibits the growth and survival of endometrial stromal cells [193]. As such, VPA treatment in human embryonic cells delayed cell differentiation into neuronal precursor cells and was linked to a genomic increase of H3 acetylation [194].

In contrast, there are also positive VPA-mediated epigenetic correlates. VPA was used to reduce sensorimotor gating deficits and HDAC2 overexpression in an animal model of schizophrenia [195]. Treating human fibroblast cells with relevant therapeutic doses of VPA illustrated marked changes in histone modifications and impacted the deacetylation of NFkB-associated genes [190]. Interestingly, the inhibition of HDAC suppresses NFkB-associated genes and signaling, which have been linked with comorbidities of epilepsy and mild depression, suggesting early treatment of a HDACi like VPA may be advantageous [196]. Delineating the deleterious versus the efficacious VPA-mediated acetylation/deacetylation balance will provide a clearer path for creating disease-modifying therapeutics.

γ-hydroxybutyrate

γ-hydroxybutyrate GHB is an endogenous metabolite and precursor to the major human brain inhibitory neurotransmitter γ-aminobutyric acid (GABA). GHB is used in anesthesia and as a treatment for insomnia and clinical depression [197]. Several studies have revealed its function in regulating epigenetic modifications [198]. GHB has been shown to inhibit HDAC, leading to accumulation of acetylated histones in the adult rat brain [198]. In addition, GHB inhibits TET activity, driving decreases in 5hmC levels in adult glioblastomas and pediatric pontine gliomas cells, suggesting a role in mediating cell-proliferative states via epigenetic mechanisms [199]. Despite its medical applications, GHB has been abused for more than 20 years and is commonly known as a "date-rape drug" or a "party drug" [200]. To date, teratogenic effects have not been reported for GHB. A deeper mechanistic understanding of the therapeutic benefits of GHB vs the molecular underlying risks that drive abuse potential will need to be delineated.

Addictive Substances

Many mind-altering substances, in addition to changing the sensitivity and function of transmission in the brain, alter epigenetics to reinforce and precipitate their addictive behaviors. Meanwhile, drugs of abuse also damage tissues and organs, particularly the brain through disruption of the normal epigenetic program. In the developing stages, organs such as the heart and brain are also sensitive and vulnerable to many substances of abuse, including alcohol. The alcohol induced epigenetic alterations are illustrated and show-cased as just one example. This can be extended in a general sense to other susbtances of abuse.

Alcohol

According to the National Institute on Alcohol Abuse and Alcoholism (NIAAA), in 2019, 25.8% of people aged 18 and older (29.7% of men and 22.2% of women both in this age group) reportedly engaged in binge drinking in the month prior to survey, increasing the risk of developing fetal alcohol spectrum disorders (FASD). Since alcohol drinking problems may start from early adolescence and persist into late adulthood, alcohol associated diseases (e.g., addiction) may preserve the continuous state of alcohol consumption which may subsequently result in a broad-spectrum health and socioeconomic concerns varying from liver disease to cancer, mental dysfunctions to FASD, and even death. In this sense, the magnitude of alcohol use disorder-related research is enormous. Besides genetic factors, epigenetics has been found to mediate the brain adaptation underlying alcohol addiction and withdrawal, and potential organ damage. In this sense, the field of epigenetics and intersection of genetics and the environment, occupies a marked portion under alcohol use disorder related research. When consumed, alcohol and its metabolites modulate the epigenome not only by serving as a methyl and acetyl donor but by directly interacting with enzymes involved in methylation and acetylation (for reviews see [201,202]).

One of the earliest efforts in epigenetic related alcohol research demonstrated that alcohol alters neural stem cell differentiation and their genomic DNA methylation patterns [32]. Gradually, it became more apparent that alcohol dysregulates epigenetic dynamics including DNA methylation, histone modifications and microRNA, affecting transcription of many genes in key neuroanatomical structures and contributing to the neurobiological architecture of alcohol use related pathology [203]. More recently in 2018, alcohol exposure has been shown to augment the levels of *Dnmt3a*, *Dnmt3b*, and *Dnmt3l* transcripts as well as DNMT3a protein levels, experimented on a neural precursor cell line and primary mouse embryonic fibroblasts [204]. A prominent rodent study in 2019 showed that intermittent alcohol exposure to adolescent rats increased DNMT activity and DNA methylation at neuropeptide-Y (*Npy*) and exon IV of *Bdnf* genes in the adult amygdala, accompanying increased anxiety and alcohol consumption, while 5-azacytidine treatment attenuated the increased anxiety and alcohol consumption, and restored hypermethylation observed in *Npy* and *Bdnf* genes [205]. Another study performed in adult rats provided strong evidence of DNA methylation dependent behavior of mRNA expression of *Glial cell line-derived neurotrophic factor* in the nucleus accumbens and ventral tegmental areas in response to alcohol exposure and withdrawal [206]. Sound evidence of alcohol dependent epigenetic perturbation was shown in an animal model mimicking intermittent ethanol exposure during adolescence in rats. This investigation showed that subjecting rats to an intermittent alcohol exposure paradigm accumulated long-lasting deficits in CREB signaling in the amygdala via histone acetylation mechanisms [207]. Another investigation found a strong interplay between the histone methylation of promoter regions of both the *ChAT* and *Trka* genes and cholinergic neuron loss in the basal forebrain of rats induced by intermittent alcohol exposure during adolescence [111]. A recent investigation providing an elaborate set of analyses of *in vitro* and *in vivo* alcohol exposure, as well as in postmortem brain samples from human alcoholics showed an important link between histone demethylase KDM6B and alcohol dependence. Apparently, KDM6B protein was upregulated both in the brain tissue of human alcoholics as well as alcohol dependent rats, the latter accompanied by a decrease of trimethylation levels at H3K27me3 [208].

Based on a 2018 study supported by NIAAA and covering more than 6000 first graders across four US communities, it was estimated that as many as 1%−5% of first-grade children possess some form of fetal alcohol spectrum disorder [209]. For a comprehensive review the recent book chapter on Fetal Alcohol Spectrum Disorders in Cassidy and Allanson's *Management of Genetic Syndromes* can be viewed [210]. Excessive alcohol exposure induces bi-directional changes of 5mC DNA methylation (hyper and hypomethylation) [32,211], particularly during embryonic and brain development [212]. Prenatal alcohol has also been shown to interfere with the intrinsic DNA methylation program [33]. Alcohol can also increase histone 3 acetylation globally [213] and alter miRNA expression in neural stem cells in a cell-type and stage specific manner [49]. Transgenerational epigenetic changes leading to hypomethylation were also found in rats with parental alcohol exposure. For instance the DNA methylation of the *Pomc* gene was altered over F2 and F3 generations, specifically in the imprinting gene *H19* in F1 sperm and transgenerationally in the offspring brain [214]. Such findings indicate that the effects of alcohol exposure may be carried beyond expectation in subsequent generations. Alcohol induced FASD and accompanied mental and cognitive deficits are also likely to impose their action through epigenetics not only at the prenatal level, but also continuously in postnatal life (Fig. 32.4).

Cocaine

According to the results of 2018 National Survey on Drug Use and Health (NSDUH), 874,000 new drug users (aged 12 or older) were recent initiates of cocaine. As a psychostimulant, cocaine has a strong impact on the mesolimbic reward pathway and when repeatedly consumed, casual drug use status can easily turn into a drug craving and compulsive drug taking state, which is linked to molecular and cellular adaptations across multiple brain structures (for review see [215]). The neuronal adaptations mediating mind alteration and the evolution of the addictive behavior have been recently accumulating evidence of epigenetic mediation. Cocaine self-administration persistently

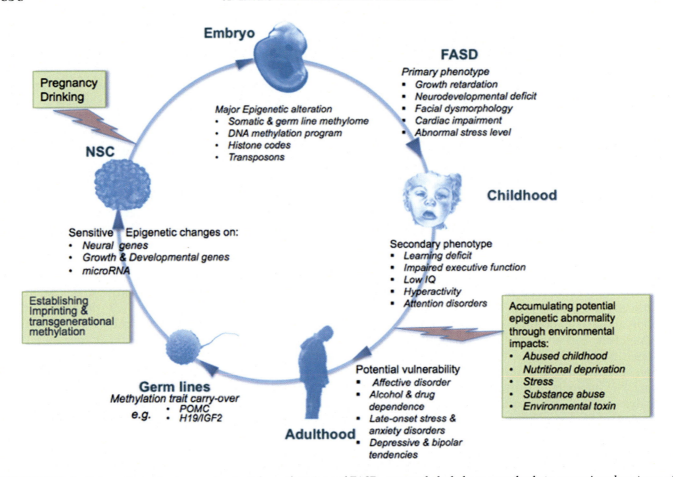

FIGURE 32.4 Epigenetic mechanisms and potential manifestations of FASD—maternal alcohol exposure leads to expansive phenotypes of FASD, but the mechanisms remain elusive. Besides immediate cellular effects, it is now understood that alcohol extensively alters epigenetics during fetal and neural stem cell (NSC) development through genomic DNA methylation, cellular DNA methylation programming (see text), histone modification, transposons, and microRNA. These expansive epigenetic changes are likely a major upstream disrupter of gene transcription leading to primary phenotypes of FASD (e.g., growth retardation, neurodevelopmental deficit) and collectively compromising brain function and mental faculty as secondary phenotypes in early life. It is not expected that all epigenetic changes lead to transcriptional and phenotypic changes, but increasing evidence suggests that continuous environmental insults may lead to increased epigenetic abnormality. The primary seeding of epigenetic errors and secondary, cumulative epigenetic abnormality via abrasive environments (e.g., childhood abuse, stress) may result in a potential manifestation of FASD beyond the classical diagnosis in adulthood. Moreover, epigenetic errors (e.g., POMC and H19 intergenic regulator) carried in the germ lines that pass the checkpoints of global demethylation, may influence the next generations. Epigenetic medicine challenges clinicians to take a deeper as well as a broader view on FASD—using both the magnifying and telescopic lens of epigenetics. *Adopted from Fig. 32.4 of M. Resendiz et al., Epigenetic medicine and fetal alcohol spectrum disorders. Epigenomics, 2013. 5(1): p. 73–86.*

increases DMRs in the genome, with additional DMRs being found during abstinence in the neurons of the mouse frontal cortex. Interestingly, altered DNA methylation associated with isoform-specific regions e.g., *Cdh13, Cpeb4* and *Mctp1* were also associated with changes in the expression of specific isoforms [216]. Another study performed methylome analysis in the prefrontal cortex of rats following self-cocaine administration, observing increased DNA methylation patterns in gene promoters, but not in gene bodies, that are inversely correlated with transcriptional activation [217]. Tedious work by Engmann and colleagues revealed important cocaine induced alterations of 3D architecture of chromatin at genes (*Auts2* and *Caln1*) showing epigenetic regulation through histone and DNA methylation [218]. Cocaine-induced transcriptional changes through histone modifications are also under investigation, for example expression changes in *FosB*, an important early immediate gene, was associated with incremental enrichment of H3K4Me3 [219]. While limited to a handful of animal studies such as those listed above, there is a strong association between the cocaine-induced deviation of neuronal transcription and epigenetic modifications. Although the epigenetic perspective in cocaine-related research is still in its infancy in general, such relation in the human brain is particularly understudied. Of interest, a recent study performed on postmortem brain (caudate nucleus) of patients who had long-term histories of cocaine dependence found cocaine-related hypomethylation of the *Irx2* gene upon a detailed methylome wide profiling and reproduced similar results in the dorsal striatum from cocaine self-administration in mice [220]. Cocaine also triggers epigenetic alterations in other reward related circuits including the corticostriatal [17] and nucleus accumbens pathways [221]. Moreover, cocaine addiction affects MeCP2 homeostasis and its expression [222]. A causal effect has been recently supported by a study demonstrating that methyl supplementation attenuates cocaine-seeking behaviors and cocaine-induced *c-Fos* activation in a DNA methylation-dependent manner [11].

Cocaine use during pregnancy was found to impair cognitive function and social behavior of the offspring. The elevated DMR2 region of *Igf2* in the hippocampus decreased *Igf2* mRNA and protein expressions in a mouse model [223]. The *Igf2* gene has a critical role for embryonic brain development and memory consolidation and enhancement. Intra-hippocampal injection of recombinant *Igf2* reactivated the repressed *calcium calmodulin kinase II* alpha and reversed cognitive deficits in prenatal cocaine treated offspring [223]. *Bdnf* is also a target of cocaine induced epigenetic modifiers. A self-administration cocaine rat model displayed increased H3 acetylation and decreased MeCP2 association with *Bdnf* promoter IV in the medial prefrontal cortex (mPFC) of cocaine treated rats, a phenomenon which was associated with increased *Bdnf* expression [224].

Cannabinoids

Cannabis, the major ingredient in marijuana, is the most used illicit substance in the United States and worldwide. Based on the public data of NSDUH report in 2019, among the adolescents aged from 12 to 17, young adults aged 18 to 25, and adults aged 26 or older, the percentage reporting marijuana use disorder in the previous year were estimated as 2.8% (or 699,000 people), 5.8% (or 2.0 million people) and 1.0% (or 2.2 million people), respectively [225]. The major psychoactive intoxicant within cannabis, Delta-9-tetrahydrocannabinol (THC), targets the endocannabinoid system. Despite the impact of cannabis exposure on the neurobiology of the adult brain being well defined, the knowledge regarding epigenetic consequences of cannabis exposure to adolescent and adult neuronal processes is still limited. Studies in the last decade, examining different adult brain regions of rodents that were exposed to THC or cannabinoid agonists during adolescence or adulthood have demonstrated various epigenetic alterations including DNA methylation, histone modifications, and microRNA (for review see [226]).

Maternal cannabis use is also a serious concern. A recent study examining the cannabis use during pregnancy in the United States (via analyzing the public-use data files covering 2005–2018 from the NSDUH) highlighted that over one in ten pregnant women with depression reported past-month cannabis use [227]. Cannabis intoxication during pregnancy has been linked with adverse outcomes that include fetal growth restrictions, lower birthweight, and long-term impact on brain development [228,229]. THC in maternal circulation can readily cross the placental membrane. THC binds to CB receptors of the fetal cannabinoid signaling system, altering neurodevelopment and possibly rewiring cannabinoid signaling circuitry [230]. In humans, it was shown that maternal cannabis exposure reduced dopamine receptor D2 (*Drd2*) expression in the ventral striatum of fetuses. A rat model study similarly showed that decreased *Drd2* is associated with reduced [transcription activating] H3K4 and increased [transcription repressive] H3K9, which together decreased Polymerase II binding. Since *Drd2* is involved in the regulation of addiction, this study also provides a link between prenatal drug exposure and later life addiction [231]. A recent study mimicking cannabidiol exposure from 2 weeks to mating and through gestation and lactation found thousands of DMRs in the cerebral cortex and hippocampus of F1 offspring and sex-specific behavioral outcomes [232].

Opioids

Opium from the opium poppy has been one of the most addictive substances since ancient times and to this day. A wide range of derivative drugs e.g., hydrocodone, oxycodone, morphine, tramadol with similar chemical natures are included under the category of opioids and are often prescribed for pain control. In contrast to prescriptive use, chronic opioid use is associated with increased DNA methylation which in turn is correlated with increased clinical pain [233]. Increased DNA methylation was found in opioid-treated patients compared to non-opioid-treated patients for pain control [233]. Opium use has been further associated with an increased risk of cancers of the larynx, lung, and stomach [234]. It has also been shown that prenatal opioid exposure causes more severe phenotypes than cocaine exposure on the offspring nervous system [235]. Although the epigenetic effect on prenatal opioid exposure is not clear, it has been reported that sperm DNA from opioid addicts has increased methylation at the promoter region of *Opioid mu1 receptor* [236] an important regulator of drug dependence.

Nicotine

Consumption of nicotine through any form of tobacco products can induce nicotine addiction. Although its harmful effects have been largely publicly recognized, the prevalence of maternal smoking during pregnancy is still estimated as 8% in the Western world [237]. While *in utero* exposure to maternal tobacco smoking is known to disrupt fetal development and cause adverse birth outcomes, its long-term consequences on offspring have also become evident [237]. A considerable number of studies designed to analyze various human tissues associated with maternal tobacco support the idea that alterations in epigenetic mechanisms, including DNA methylation, histone modifications, and miRNA expression are some of the possible links between maternal smoking during pregnancy and related clinical outcomes (see review [238]). A considerable portion of the studies investigating the link between DNA methylation and maternal smoking during pregnancy in humans have utilized cord blood DNA of offspring. Several of these studies report increased expression and decreased methylation of the *Ahrr* gene, a gene associated with smoking and lung cancer [239–241].

CONCLUSION AND FUTURE DIRECTION

Although insights on the environmental factors capable of epigenetic modification have been rapidly amassing in recent years, to date, the understanding of the environmental impacts that lead to transcriptional change via epigenetic mechanisms are still rudimentary. In this chapter we have identified that mental, physiological, chemical, and hazardous environments can cause changes to DNA methylation (e.g., 5mC and 5hmC), histone codes (e.g., H3K9me3, H3K27me3, H3Ac) and miRs through alteration of methyl or acetyl donors or their transferring enzymes. Although parallel transcriptional changes are often cited along with the epigenetic changes, the causal relationships are far from conclusive. Major gaps in the field include that many studies are examined at cellular levels, where epigenetic changes including their transferring enzymes are reported as the average changes in the cells, restricting interpretation of any singular gene regulation. A number of studies have thus moved towards gene-based analysis using high-throughput sequencing or pyrosequencing analyses for DNA methylation and histone identification, by which the epigenetics of a single gene/loci are being identified to offer a more precise interpretation of gene function. Nevertheless, many of these improved approaches are continually conducted in tissues with mixed cell types. This is a problem since epigenetics are of a cell-specific repertoire. As such, epigenetic data acquired from tissues of mixed vascular cells or glial cells and neurons will produce an average of the heterogeneity of cells. Consequently, the use of cell sorting is on the rise, and bioinformatic algorithms are increasingly being used to estimate cell type composition. Furthermore, studies directly comparing epigenetic profiles of various tissue types (i.e., blood, liver, etc.) to infer gene-specific epigenetic regulation patterns across tissues are being explored. (TarGET consortium, https://pubmed.ncbi.nlm.nih.gov/29509741/).

The current state of DNA methylation study is also often incomprehensive. This is mainly due to the fact that separation of the two major forms of DNA methylation, 5mC and 5hmC, from small amounts of sample are technically demanding. The 5hmC mark has recently been identified to play a bivalent role in favoring transcription to oppose the repressive role of 5mC (for review see [42,242]). As such, there remains a need for distinctive DNA methylation profiling to fully characterize the role of the most widely studied epigenetic modification. Additionally, while many epigenetic studies zero in on DNA methylation due to the abundance of comparative data in the literature, emphasis must be given to the other modification classes in order to more robustly build the epigenetic profile of a phenomenon (e.g., environmental input, disease). Finally, the majority of environmental epigenetic studies to date have singularly focused on one or more epigenetic modifications occurring in parallel. This is fundamentally flawed as we know that epigenetic modifications "talk" with one another (e.g., Suv39h1 methylates H3K9 and mediates localization of Dnmt3b to pericentromeric repeats, Dnmt1/3b mutations show H3K9 methylation changes) [243]. Thus, increasing need has arisen for the concurrent study of multiple epigenetic modifications that are sparse and largely correlative at the moment. Only then can we work toward a more complete understanding of the epigenetic language which guides transcriptional and functional outcomes to shape the state of the cell, organ, and individual across many contexts.

Another direction, mentioned earlier, is the reversal of harmful environmental exertions. For instance, while environmental factors can confer epigenetic modifications to facilitate detrimental outcomes, new evidence demonstrates the ameliorating potential of environmental and lifestyle influences like nutrition, exercise, and an enriched environment. Through extensive investigation, utilizing pharmacological, transgenic, and whole genome/epigenome sequencing, human and animal models of disease including stroke, Alzheimer disease, alcoholism, spinal cord injury, and obesity/gestational diabetes have demonstrated sensitivity to disease modifying environmental interventions, mediated at least in part by critical epigenetic modifications (Table 32.1). In environmental enrichment animal studies, the ameliorating potential of EE was demonstrated in a model of stroke, SCI, and Aβ-oligomer-related decreases in hippocampal LTP. In these studies, parallel use of pharmacological inhibitors and KO mice outlined the critical nature of HDAC2 repression, H4K8ac upregulation, and miR-132 targeting of HDAC3

TABLE 32.1 Disease-related Reprogrammable Epigenetic Modifications

	Phenotype	Gene Expression	Epigenetics	LOF	GOF	References
Environmental Enrichment	Improved stroke-induced motor function and neuroplasticity deficits	normalized expression of NGF/BDNF, PSD95, synapsin, and GAP43	normalized increases in HDAC2 protein and activity	HDAC2 inhibition mimicked functional stroke recovery and normalization of synaptic plasticity markers	HDAC2 overexpression inhibited functional recovery after stroke	[45]
	Enhanced neurite outgrowth, sciatic nerve regeneration, and increased muscle reinnervation after injury	upregulated genes involved in neuronal activity, calcium signaling and cytoskeletal rearrangements	upregulation of H3K27ac and H4K8ac by calcium-dependent Cpb-activation	Egr3 KO abolished EE-dependent DRG outgrowth	CSP-TTK21 promoted sensory axon regeneration, synapse formation, sensorimotor function, and upregulation of H4K8ac	[244]
	persistent hippocampal LTP	normalized upregulation of HDAC3	normalized Aβ-related decreases in miRNA-132–3p	HDAC inhibitors induced hippocampal LTP, prevented Aβ-impaired LTP; HDAC3 inhibition mimics EE neuroprotection	overexpression of miR-132 overcomes Aβ-related impairment of hippocampal LTP	[43]
Nutrition	Increased glucose tolerance of HFD mice of PPAR alpha-supplemented dams	HFD elevated Fgf21 mRNA expression in PPARα-supplemented dams	Enhanced Fgf21 demethylation of offspring	Fgf21 methylation is PPAR alpha-dependent according to PPAR alpha KO mice		[245]
	*2 year weight change/fat loss in overweight humans	reduced NFATC2IP	cis-meQTL rs11150675 (determines DNAm of NFATC2IP) positively correlated with fat loss			[246]
Exercise	*Upregulation of genes involved in glycogen storage and metabolism, glycolysis, TCA cycle and oxidative stress	PDHA1 and citrate synthase, ANT1/2; SOD2 and CAT were increased by regular exercise group	Hypomethylation of 714 promoters-including genes related to the phenotype			[247]
	*Increased step count, reduced weight and improved diastolic blood pressure		NAMPT intron hypomethylation and RUNX3 hypermethylation inversely correlated with weight loss			[112]

LOF, Loss of function; *GOF*, Gain of function; *KO*, Knockout; *HFD*, High-fat diet; *EE*, Environmental enrichment; *LTP*, Long-term potentiation. * indicates a human study.

[45,89,244] in reversing disease-generated deficits in neuroplasticity. Lifestyle factors like diet and exercise possess similar potential, even prenatally. For example, the cis-meQTL rs11150675 genotype which determines the DNA methylation of *Nfat2cip* was positively correlated with 2-year fat loss in a trial of overweight or obese humans [246]. Similarly, *Fgf21* methylation was responsible for conferring the increased glucose tolerance of mouse offspring in response to a high-fat diet [245]. Regarding exercise interventions in response to aging and obesity, epigenome-wide studies in humans have implicated DNA methylation of a variety of sites integral to metabolism, glycolysis, and weight-associated genes [112,247]. As more consequential epigenetic marks are identified in human disease, environmental factor detection along with targeted pharmacological and epigenetic editing tools hold promise for the mitigation of disease risk, course, and outcomes set in motion or exacerbated by environment-associated epigenomics.

Aside from regulating gene transcription, the exerting environment may often impact chromatin through epigenetic memory. The newly created, unique epigenetic marks or landscapes associated with specific genes or intergenic regions at specific loci may serve as a biomarker reflecting unique environmental impacts. The differential DMRs with predictive capacity may serve as a diagnostic tool for hazardous environmental exposures or toxicity in the future. A number of examples have revealed that potential. Lead-associated changes with distinctive 5hmC and 5mC profiles in genomic loci, if specific to the lead-exposure, might have some potential as early biomarkers of prenatal lead exposure [136]. Prenatal arsenic exposure may be identified in newborn cord blood through DNA methylation levels of a panel of genes, possibly allowing the prediction of birth outcomes and head circumference [142]. A variety of epigenetic modifications, including DNA methylation, histone modification, and non-coding RNA expression, may be used to profile risk assessment of toxic metals, e.g., arsenic, lead, and mercury in the future [248]. More DMRs are likely to be identified for diagnostic purposes in the future, like forensic detection of an environmental insult, or the history of alcohol exposure during pregnancy. The challenges, as indicated above, are the current characterization of epigenetic features in cell and tissue specific manners. While human sampling is often limited to blood cells, buccal smears, saliva, and urine due to ethical reasons, the research findings may not reflect the environmental impact on the appropriate target cells used for diagnostic purposes. In short, the surrogate analyses will have to be cross-checked with postmortem tissue analysis and supported by animal studies. Undoubtedly, epigenetic diagnosis will require further polishing. As the analytical tools and understanding become more mature, epigenetics may return favorable outcomes for prediction efforts of a variety of environmental impacts.

RESEARCH CASE STUDY [18]

Transgenerational inheritance of the adverse effects of fetal alcohol exposure on immune gene interferon-Υ.

Objective

The goal of this study was to investigate whether fetal alcohol exposure has long-term (transgenerational) epigenetic effects on the production of the immune gene coding interferon-Υ (IFN-Υ), a cytokine known to regulate both innate and adaptive immunity.

Scope

Although very challenging in scope, this study used well-controlled conditions in rodents to test whether drinking during pregnancy affects the immune gene and associated epigenetic markers in the offspring, over extended generations. Humans with similar alcohol use disorders are not as easily controlled. The rodent model facilitated maternal drinking in a predetermined amount of alcohol over a well-known gestation stage and studied changes in the immune gene and protein altered via changes in DNA methylation. The study evaluated the F1–F3 generations, legitimizing the observation of transgenerational epigenetic changes. Currently, this design cannot be studied in humans.

Humans have longer lifespans which require different genetic and cellular mechanisms for innate and adaptive immunity maintenance. Thus it is unclear if drinking during pregnancy affects immune function via epigenetics in rodent models to the same extent as it would in humans. Drugs of abuse like alcohol and a myriad of environmental factors are capable of altering DNA methylation in the current generation, while the mechanisms by which they exert those changes and whether those changes are inheritable across multiple generations are mostly unknown. The current study, however, presents a compelling case for future investigation.

Audience

This study should generally interest basic scientists and human researchers whose focus is alcohol use disorders. This study may also appeal to individuals who study epigenetics, immunology, and endocrinology. Moreover, the primary audience will be those who are interested in how maternal alcohol consumption during pregnancy drives aberrant epigenetic changes within their progeny. The study may also appeal to substance abuse and sexual dimorphism researchers who are interested in trans-generational epigenetic changes.

Rationale

Fetal alcohol exposure produces a myriad of deleterious effects in infants such as higher instances of pneumonia. Alcohol-induced epigenetic effects can be ubiquitous across a wide range of genes often erased during early implantation; the current study focuses on a single gene evaluated across multiple generations (F1–F3) which were not directly exposed to alcohol. This design allows for clear observation of how maternal alcohol ingestion during pregnancy alters the epigenetic programming, driving alteration of a critical immune-related protein (sex-specifically) across generations.

Expected Results and Deliverables

The key finding of this study demonstrated that maternal drinking altered not only the Ifn-ϒ mRNA and IFN-ϒ protein but also the methylation level of individual CpGs within the promoter region of *Ifn-ϒ* gene (in the spleen tissue) of the offspring. This study showed these changes persist sex-specifically to the F3 generation. The F3 finding is very important since maternal pregnancy exposure may also expose the fetal tissue (F1) and its germ line (F2). On the other hand, the F3 generation serves as a clean observation free of direct exposure of the environmental factor (alcohol). This is a star example that some ancestral environmental factors may have lasting effects on future generations.

The second important result from this study was the lasting nature of the alterations into the adult stage of the offspring. Furthermore, in this study sex differences were considered, concluding that only the male germ line's progeny transmitted the epigenetic changes across all generations observed.

Safety Considerations

This was an animal study which followed the guidelines of animal use protocols approved by the Association for the Assessment and Accreditation of Laboratory Animal Care at Rutgers University.

The Process, Workflow and Actions Taken

Acknowledging the findings and scientific novelty of this research first depends on understanding the rigorous and tightly controlled experimental design. A schematic of the detailed breeding paradigm can be reviewed in Fig. 32.2 of the article. Briefly, offspring of the alcohol fed (AF), pair fed (PF) (isocaloric liquid control diet) and chow ad libitum control (AD) groups of pregnant rats primed the first generation (F1) of the study. To avoid possible poor nursing, litters of the AF and PF groups were cross fostered. Male and female germ lines of second and third generations were obtained through a breeding paradigm in which AF, PF, or AD male or female rats were mated with control animals of the opposite gender. Gene and protein expression analysis and CpG methylation assessments were utilized on the spleen tissue of 60–90 days old F1–3 rats. To generate the gene expression level data set, spleen tissues of eight animals per group were used for real-time PCR analysis. *Ifn-ϒ* mRNA expression was normalized with the mean of three reference gene expression levels, which strengthened the relative expression. Ifn-ϒ protein expression was observed via Western-blots in which the relative abundance of Ifn-ϒ was normalized to actin protein expression within each sample. Eight animals per group were included in the methylation analysis of four individual CpG dinucleotides in the proximal region of the Ifn-ϒ gene using pyrosequencing along with appropriate commercial assays. Only one pup from each litter was used in an experimental measure. The quantity of specimens used in each molecular analysis were optimal.

Tools and Materials Used

Scientifically standard molecular protocols were applied by using high quality commercial tools and materials which yield well controlled and reliable data.

Results

Maternal alcohol consumption during the second and third trimesters by Isogenic Fisher 344 rats resulted in significant reduction in both *Ifn-ϒ* mRNA and IFN-ϒ protein expression levels along with elevated promoter methylation of the corresponding gene in the spleen tissue of the male and female adult offspring. This inversely correlation persisted into the F2 and F3 generations within the male germline. Rats exposed to an isocaloric liquid control diet did not generate significant alternations compared to the control group. Since the cross generational effects of fetal alcohol exposure on the expression (gene/protein) and DNA methylation dynamics of an immune gene were largely unexplored previously, there was no clear expectation. It is interesting that distortion of fetal development by prenatal alcohol exposure exerts its effect on a canonical immune element transgenerationally in a sexually dymorphic fashion.

Challenges and Solutions

This study provides clear and interesting evidence of sex-specific, cross generational transmission of an epigenetic influence on an important immune gene in the spleen tissue. Regardless of the specific immune-related gene focus, this tediously performed animal experimentation lacks immune-specific functional measurements within the groups. For example, such work could be reinforced by measuring pro-inflammatory and inflammatory cytokines in the blood, showing systemic changes or using other organ tissues in concert to corroborate those exclusive epigenetic findings.

Learning and Knowledge Outcomes

This study highlights that the fetal alcohol exposure transmits its epigenetic influence on the expression of *Ifn-Υ* gene with no sex discrimination in the first generation, but persists through the F2-F3 generations via the male germ line exclusively. Labor intensive and ethically concerned, this animal model is a well-designed work and further studies could adapt the same model to produce convenient comparisons. Explicit immune-system related molecular findings in the spleen could be better extrapolated within the context of fetal alcohol syndrome (FAS) when supported by various other immune-specific assays.

Acknowledgments

While writing this commentary, FCZ is supported by National Institute of Health AA016698 and P50AA07611, AA024216, and by M. W. Keck Foundation.

Glossary

5hmC	5-hydroxymethylcytosine
5mC	5-methylcytosine
Adcyl	adenylate cyclase 1 gene
ADHD	attention-deficit/hyperactivity disorder
Angpt2	angiopoietin 2 gene
As	arsenic
Avp	arginine vasopressin
BDNF	brain derived neurotrophic factor
BEAS-2B	human bronchial epithelial cells
BrdU	bromodeoxyuridine
BPA	bisphenol A
Cbep4	cytoplasmic polyadenylation element binding protein 4
Cd	cadmium
Cdh13	cadherin 13
ChAT	choline acetyltransferase
Chst14	carbohydrate sulfotransferase 14 gene
CNS	central nervous system
CREB	cAMP response element-binding protein
Creb3l4	cAMP response element-binding protein 3-like 4
CVD	cardiovascular disease
Cyp19	cytochrome p450 family 19 gene
DEHP	diethylhexyl phthalate
DMR	differentially methylated region
DNA	genetic code / Deoxyribonucleic acid
DNMT1	DNA methyltransferase 1
DNMT3	DNA methyltransferase 3
DNMT3B	DNA methyltransferase 3B
DRD2	dopamine receptor D2
EDC	endocrine disrupting chemical
EE	environmental enrichment
Esr1	estrogen receptor 1 gene
F1	the generation resulting immediately from a cross of the first set of parents
FASD	fetal alcohol spectrum disorders
Fgf	fibroblast growth factor gene
Foxd2	forkhead box protein D2 gene
Gadd45	growth arrest and DNA damage 45 gene
GDM	gestational diabetes mellitus
GHB	γ-hydroxybutyrate
GR	flucocorticoid receptor
HA	histone acetylase
HDAC	histone deacetylase
HDACi	histone deacetylase inhibitor
H3K9Ac	histone H3 acetylation
H3K4me3	trimethylated histone H3 on lysine 4
H3K9me2	dimethylated histone H3 on lysine 9
H3K27me3	trimethylated histone H3 on lysine 27
HMT	histone methyltransferase
HPA	hypothalamic-pituitary-adrenal (axis)
HSD11B2	11b-hydroxysteroid dehydrogenase type 2
IAP	inhibitor of apoptosis protein
ICR	imprinting control region
Igf2	insulin growth factor 2
Kcc2	potassium chloride cotransporter 2
KDM6B	Lysine demethylase 6B
LDL	lol-density lipoprotein
LTP	long term potentiation
Line-1	long interspersed nuclear element 1 gene
Mctp1	Multiple C2 And transmembrane domain containing 1
MeCP2	methylation CpG binding protein 2
MeHg	methylmercury
Mgmt	O-6-methylguanine DNA methyltransferase
miRNA (miR)	microRNA
mPFC	medial prefrontal cortex
Mt2a	metallothionein 2 A
NIAAA	National Institute on Alcohol Abuse and Alcoholism
NGF	nerve growth factor
Ngf	nerve growth factor gene
NMDA	N-methyl-D-aspartate glutamate receptor
NPY	neuropeptide-Y
NR3C1	glucocorticoid receptor nuclear receptor subfamily 3 group C member 1
NSDUH	National Survey on Drug Use and Health
Paqr4	progestin and adipoQ receptor family member 4 gene
Pb	lead
Pitx3	pituitary homeobox 3 gene
PNMS	prenatal maternal stress
Pon1	paraoxonase 1 gene
Prpf18	pre-mRNA processing factor 18 gene
Ptplb	protein tyrosine phosphatase-like member b gene
Sox2	SRY-box transcription factor 2 gene
Tceanc2	transcription elongation factor a N-terminal and central domain containing 2 gene
Tet	Ten-eleven translocation
THC	Delta-9-tetrahydrocannabinol
TGF-β	transforming growth factor beta
Tpd52	tumor protein D52 gene
TrkA	tropomyosin receptor kinase A
TSS	transcriptional start site
VD	vascular disease
VPA	valproic acid
Wnt10b	wnt family member 10B gene

References

[1] Mayer W, et al. Demethylation of the zygotic paternal genome. Nature 2000;403(6769):501–2.

[2] Popp C, et al. Genome-wide erasure of DNA methylation in mouse primordial germ cells is affected by AID deficiency. Nature 2010;463(7284):1101–5.

[3] Reik W, Dean W, Walter J. Epigenetic reprogramming in mammalian development. Science 2001;293(5532):1089–93.

[4] Rideout 3rd WM, Eggan K, Jaenisch R. Nuclear cloning and epigenetic reprogramming of the genome. Science 2001;293(5532):1093–8.

[5] Surani MA. Reprogramming of genome function through epigenetic inheritance. Nature 2001;414(6859):122–8.

[6] Horsthemke B. A critical view on transgenerational epigenetic inheritance in humans. Nat Commun 2018;9(1) 2973-2973.

[7] Xavier MJ, et al. Transgenerational inheritance: how impacts to the epigenetic and genetic information of parents affect offspring health. Hum Reprod Update 2019;25(5):519–41.

[8] Zhang Y, Zhang J. Identification of functionally methylated regions based on discriminant analysis through integrating methylation and gene expression data. Mol Biosyst 2015;11(7):1786–93.

[9] Fan H, et al. Systematically prioritizing functional differentially methylated regions (fDMRs) by integrating multi-omics data in colorectal cancer. Sci Rep 2015;5(1):12789.

[10] Rucker FG, et al. Molecular dissection of valproic acid effects in acute myeloid leukemia identifies predictive networks. Epigenetics 2016;11(7):517–25.

[11] Wright KN, et al. Methyl supplementation attenuates cocaine-seeking behaviors and cocaine-induced c-Fos activation in a DNA methylation-dependent manner. J Neurosci 2015;35(23):8948–58.

[12] Choi CS, et al. The transgenerational inheritance of autism-like phenotypes in mice exposed to valproic acid during pregnancy. Sci Rep 2016;6:36250.

[13] Garrido N, et al. Sperm DNA methylation epimutation biomarker for paternal offspring autism susceptibility. Clinical. Epigenetics 2021;13(1):6.

[14] Tabuchi TM, et al. Caenorhabditis elegans sperm carry a histone-based epigenetic memory of both spermatogenesis and oogenesis. Nat Commun 2018;9(1):4310.

[15] Zenk F, et al. Germ line-inherited H3K27me3 restricts enhancer function during maternal-to-zygotic transition. Science 2017;357(6347):212–16.

[16] Painter RC, et al. Transgenerational effects of prenatal exposure to the Dutch famine on neonatal adiposity and health in later life. BJOG 2008;115(10):1243–9.

[17] Sadri-Vakili G. Cocaine triggers epigenetic alterations in the corticostriatal circuit. Brain Res 2015;1628(Pt A):50–9.

[18] Gangisetty O, Palagani A, Sarkar DK. Transgenerational inheritance of fetal alcohol exposure adverse effects on immune gene interferon-ϒ. Clin Epigenetics 2020;12(1):70.

[19] Coolen MW, et al. Impact of the genome on the epigenome is manifested in DNA methylation patterns of imprinted regions in monozygotic and dizygotic twins. PLoS One 2011;6(10):e25590.

[20] Levesque ML, et al. Genome-wide DNA methylation variability in adolescent monozygotic twins followed since birth. Epigenetics 2014;9(10):1410–21.

[21] Lo CL, Zhou FC. Environmental alterations of epigenetics prior to the birth. Int Rev Neurobiol 2014;115:1–49.

[22] Baker GT, Sprott RL. Biomarkers of aging. Exp Gerontol 1988;23(4):223–39.

[23] Tibshirani R. Regression shrinkage and selection via the lasso. J R Stat Soc B Stat Methodol 1996;58(1):267–88.

[24] Zou H, Hastie T. Regularization and variable selection via the elastic net. J R Stat Soc B Stat Methodol 2005;67(2):301–20.

[25] Jylhävä J, Pedersen NL, Hägg S. Biological age predictors. EBioMedicine 2017;21:29–36.

[26] López-Otín C, et al. The hallmarks of aging. Cell 2013;153(6):1194–217.

[27] Bocklandt S, et al. Epigenetic predictor of age. PLoS One 2011;6(6):e14821.

[28] Garagnani P, et al. Methylation of ELOVL2 gene as a new epigenetic marker of age. Aging Cell 2012;11(6):1132–4.

[29] Horvath S. DNA methylation age of human tissues and cell types. Genome Biol 2013;14(10):3156.

[30] Hannum G, et al. Genome-wide methylation profiles reveal quantitative views of human aging rates. Mol Cell 2013;49(2):359–67.

[31] Oblak L, et al. A systematic review of biological, social and environmental factors associated with epigenetic clock acceleration. Ageing Res Rev 2021;69:101348.

[32] Zhou FC, et al. Alcohol alters DNA methylation patterns and inhibits neural stem cell differentiation. Alcohol Clin Exp Res 2011;w35(4):735–46.

[33] Zhou FC. DNA methylation program during development. Front Biol 2012;6:485–94.

[34] Fagiolini M, Jensen CL, Champagne FA. Epigenetic influences on brain development and plasticity. Curr Opin Neurobiol 2009;19(2):207–12.

[35] Irier H, et al. Environmental enrichment modulates 5-hydroxymethylcytosine dynamics in hippocampus. Genomics 2014;104(5):376–82.

[36] Rochefort C, et al. Enriched odor exposure increases the number of newborn neurons in the adult olfactory bulb and improves odor memory. J Neurosci 2002;22(7):2679–89.

[37] Wang X, et al. Enriched environment improves working memory impairment of mice with traumatic brain injury by enhancing histone acetylation in the prefrontal cortex. Peer J 2018;6:e6113.

[38] Cintoli S, et al. Environmental enrichment induces changes in long-term memory for social transmission of food preference in aged mice through a mechanism associated with epigenetic processes. Neural Plasticity 2018;2018:3725087.

[39] Komitova M, et al. Enriched environment increases neural stem/progenitor cell proliferation and neurogenesis in the subventricular zone of stroke-lesioned adult rats. Stroke 2005;36(6):1278–82.

[40] Branchi I. The mouse communal nest: investigating the epigenetic influences of the early social environment on brain and behavior development. Neurosci Biobehav Rev 2009;33(4):551–9.

[41] Kuzumaki N, et al. Hippocampal epigenetic modification at the brain-derived neurotrophic factor gene induced by an enriched environment. Hippocampus 2011;21(2):127–32.

[42] Resendiz M, et al. Epigenetic medicine and fetal alcohol spectrum disorders. Epigenomics 2013;5(1):73–86.

[43] Wei Z, et al. Environmental enrichment prevents Aβ oligomer-induced synaptic dysfunction through mirna-132 and hdac3 signaling pathways. Neurobiol Dis 2020;134:104617.

[44] Benito E, et al. RNA-dependent intergenerational inheritance of enhanced synaptic plasticity after environmental enrichment. Cell Rep 2018;23(2):546–54.

[45] Lin Y-H, et al. HDAC2 (Histone deacetylase 2): a critical factor in environmental enrichment-mediated stroke recovery. J Neurochem 2020;155(6):679–96.

[46] Puang Shu J, et al. MEF2C and HDAC5 regulate Egr1 and Arc genes to increase dendritic spine density and complexity in early enriched environment. Neuronal Signal 2020;4(3).

[47] Wei Y, et al. Enriched environment-induced maternal weight loss reprograms metabolic gene expression in mouse offspring. J Biol Chem 2015;290(8):4604–19.

[48] Kelly SJ, Goodlett CR, Hannigan JH. Animal models of fetal alcohol spectrum disorders: impact of the social environment. Dev Disabil Res Rev 2009;15(3):200–8.

[49] Miranda RC. MicroRNAs and fetal brain development: implications for ethanol teratology during the second trimester period of neurogenesis. Front Genet 2012;3:77.

[50] Chokroborty-Hoque A, Alberry B, Singh SM. Exploring the complexity of intellectual disability in fetal alcohol spectrum disorders. Front Pediatr 2014;2:90.

[51] Carulli D, Foscarin S, Rossi F. Activity-dependent plasticity and gene expression modifications in the adult CNS. Front Mol Neurosci 2011;4:50.

[52] Guo JU, et al. Neuronal activity modifies the DNA methylation landscape in the adult brain. Nat Neurosci 2011;14(10):1345–51.

[53] Barreto G, et al. Gadd45a promotes epigenetic gene activation by repair-mediated DNA demethylation. Nature 2007;445(7128):671–5.

[54] Ma DK, et al. Neuronal activity-induced Gadd45b promotes epigenetic DNA demethylation and adult neurogenesis. Science 2009;323(5917):1074–7.

[55] Tian F, Marini AM, Lipsky RH. NMDA receptor activation induces differential epigenetic modification of Bdnf promoters in hippocampal neurons. Amino Acids 2010;38(4):1067–74.

[56] Chen WG, et al. Derepression of BDNF transcription involves calcium-dependent phosphorylation of MeCP2. Science 2003;302(5646):885–9.

[57] Karpova NN. Role of BDNF epigenetics in activity-dependent neuronal plasticity. Neuropharmacology 2014;76(Pt C):709–18.

[58] Bergman NJ. Birth practices: maternal-neonate separation as a source of toxic stress. Birth Defects Res 2019;111(15):1087–109.

[59] Sweatt JD. Experience-dependent epigenetic modifications in the central nervous system. Biol Psychiatry 2009;65(3):191–7.

[60] Heyward FD, Sweatt JD. DNA methylation in memory formation: emerging insights. Neuroscientist 2015;21(5):475–89.

[61] Levenson JM, et al. Regulation of histone acetylation during memory formation in the hippocampus. J Biol Chem 2004;279(39):40545–59.

[62] Miller CA, Campbell SL, Sweatt JD. DNA methylation and histone acetylation work in concert to regulate memory formation and synaptic plasticity. Neurobiol Learn Mem 2008;89(4):599–603.

[63] Baj G, et al. Toward a unified biological hypothesis for the BDNF Val66Met-associated memory deficits in humans: a model of impaired dendritic mRNA trafficking. Front Neurosci 2013;7:188.

[64] Roth TL, et al. Epigenetic modification of hippocampal Bdnf DNA in adult rats in an animal model of post-traumatic stress disorder. J Psychiatr Res 2011;45(7):919–26.

[65] Wu J, Bie B, Naguib M. Epigenetic manipulation of brain-derived neurotrophic factor improves memory deficiency induced by neonatal anesthesia in rats. Anesthesiology 2016;124(3):624–40.

[66] Zucchi FC, et al. Maternal stress induces epigenetic signatures of psychiatric and neurological diseases in the offspring. PLoS One 2013;8(2):e56967.

[67] Kim E, et al. DNA methylation profiles in sibling pairs discordant for intrauterine exposure to maternal gestational diabetes. Epigenetics 2017;12(10):825–32.

[68] Mitsuya K, et al. Alterations in the placental methylome with maternal obesity and evidence for metabolic regulation. PLoS One 2017;12(10):e0186115.

[69] Mehler MF, Purpura DP. Autism, fever, epigenetics and the locus coeruleus. Brain Res Rev 2009;59(2):388–92.

[70] Holmes MC, et al. The mother or the fetus? 11beta-hydroxysteroid dehydrogenase type 2 null mice provide evidence for direct fetal programming of behavior by endogenous glucocorticoids. J Neurosci 2006;26(14):3840–4.

[71] Jensen Pena C, Monk C, Champagne FA. Epigenetic effects of prenatal stress on 11beta-hydroxysteroid dehydrogenase-2 in the placenta and fetal brain. PLoS One 2012;7(6):e39791.

[72] Cao-Lei L, Laplante DP, King S. Prenatal maternal stress and epigenetics: review of the human research. Curr Mol Biol Rep 2016;2(1):16–25.

[73] Cao-Lei L, et al. DNA methylation signatures triggered by prenatal maternal stress exposure to a natural disaster: project ice storm. PLoS One 2014;9(9):e107653.

[74] Letourneau N, et al. Prenatal maternal distress and immune cell epigenetic profiles at 3-months of age. Dev Psychobiol 2021;63(5):973–84.

[75] Rodgers AB, et al. Paternal stress exposure alters sperm microRNA content and reprograms offspring HPA stress axis regulation. J Neurosci 2013;33(21):9003–12.

[76] Short AK, et al. Elevated paternal glucocorticoid exposure alters the small noncoding RNA profile in sperm and modifies anxiety and depressive phenotypes in the offspring. Translational. Psychiatry 2016;6(6) e837-e837.

[77] Champagne FA, Curley JP. Epigenetic mechanisms mediating the long-term effects of maternal care on development. Neurosci Biobehav Rev 2009;33(4):593–600.

[78] Weaver IC, et al. Early environmental regulation of hippocampal glucocorticoid receptor gene expression: characterization of intracellular mediators and potential genomic target sites. Ann N Y Acad Sci 2004;1024:182–212.

[79] McGowan PO, et al. Epigenetic regulation of the glucocorticoid receptor in human brain associates with childhood abuse. Nat Neurosci 2009;12(3):342–8.

[80] Farrell C, et al. DNA methylation differences at the glucocorticoid receptor gene in depression are related to functional alterations in hypothalamic–pituitary–adrenal axis activity and to early life emotional abuse. Psychiatry Res 2018;265:341–8.

[81] Duffy HBD, Roth TL. Increases in Bdnf DNA methylation in the prefrontal cortex following aversive caregiving are reflected in blood tissue. Front Hum. Neurosci 2020;14(501):594244.

[82] Tyrka AR, et al. Childhood maltreatment and methylation of FK506 binding protein 5 gene (FKBP5). Dev Psychopathol 2015;27(4pt2):1637–45.

[83] Non AL, et al. DNA methylation at stress-related genes is associated with exposure to early life institutionalization. Am J Phys Anthropol 2016;161(1):84–93.

[84] Misiak B, et al. Adverse childhood experiences and methylation of the FKBP5 gene in patients with psychotic disorders. J Clin Med 2020;9(12):3792.

[85] Serpeloni F, et al. Does prenatal stress shape postnatal resilience? – An epigenome-wide study on violence and mental health in humans. Front Genet 2019;10:269.

[86] Sasagawa T, et al. Long-term effects of maternal separation coupled with social isolation on reward seeking and changes in dopamine D1 receptor expression in the nucleus accumbens via DNA methylation in mice. Neurosci Lett 2017;641:33–9.

[87] Shepard RD, et al. Histone deacetylase inhibition reduces ventral tegmental area dopamine neuronal hyperexcitability involving AKAP150 signaling following maternal deprivation in juvenile male rats. J Neurosci Res 2020;98(7):1457–67.

[88] Wang R-H, et al. Maternal deprivation enhances contextual fear memory via epigenetically programming second-hit stress-induced reelin expression in adult rats. Int J Neuropsychopharmacol 2018;21(11):1037–48.

[89] Sun H, et al. Maternal separation-induced histone acetylation correlates with BDNF-programmed synaptic changes in an

[89 cont.] animal model of PTSD with sex differences. Mol Neurobiol 2021;58(4):1738–54.

[90] Coley EJL, et al. Cross-generational transmission of early life stress effects on HPA regulators and bdnf are mediated by sex, lineage, and upbringing. Front Behav Neurosci 2019;13:101.

[91] Elliott E, et al. Resilience to social stress coincides with functional DNA methylation of the Crf gene in adult mice. Nat Neurosci 2010;13(11):1351–3.

[92] Efstathopoulos P, et al. NR3C1 hypermethylation in depressed and bullied adolescents. Transl Psychiatry 2018;8(1):121.

[93] Aghagoli G, et al. Social stress-related epigenetic changes associated with increased heart rate variability in infants. Front. Behav Neurosci 2020;13:294.

[94] Blackmore DG, et al. Exercise increases neural stem cell number in a growth hormone-dependent manner, augmenting the regenerative response in aged mice. Stem Cell 2009;27(8): 2044–52.

[95] Barres R, et al. Acute exercise remodels promoter methylation in human skeletal muscle. Cell Metab 2012;15(3):405–11.

[96] Ronn T, et al. A six months exercise intervention influences the genome-wide DNA methylation pattern in human adipose tissue. PLoS Genet 2013;9(6):e1003572.

[97] Collins A, et al. Exercise improves cognitive responses to psychological stress through enhancement of epigenetic mechanisms and gene expression in the dentate gyrus. PLoS One 2009;4(1):e4330.

[98] de Meireles LCF, et al. Exercise modalities improve aversive memory and survival rate in aged rats: role of hippocampal epigenetic modifications. Mol Neurobiol 2019;56(12):8408–19.

[99] Voisin S, et al. Exercise training and DNA methylation in humans. Acta Physiol (Oxf) 2015;213(1):39–59.

[100] Sleiman SF, et al. Exercise promotes the expression of brain derived neurotrophic factor (BDNF) through the action of the ketone body beta-hydroxybutyrate. Elife 2016;5:e15092.

[101] Li X, et al. Exercise enhances the expression of brain-derived neurotrophic factor in the hippocampus accompanied by epigenetic alterations in senescence-accelerated mice prone 8. Neurosci Lett 2019;706:176–81.

[102] Fraga I, et al. Effects of a multimodal exercise protocol on functional outcomes, epigenetic modulation and brain-derived neurotrophic factor levels in institutionalized older adults: a quasi-experimental pilot study. Neural Regeneration Res 2021;16(12):2479–85.

[103] Meireles ALF, et al. Strength training during pregnancy influences hippocampal plasticity but not body development in neonatal rats. J Musculoskelet Neuronal Interact 2021;21(2): 279–86.

[104] Segabinazi E, et al. Effects of maternal physical exercise on global DNA methylation and hippocampal plasticity of rat male offspring. Neuroscience 2019;418:218–30.

[105] Spindler C, et al. Paternal physical exercise modulates global DNA methylation status in the hippocampus of male rat offspring. Neural Regen Res 2019;14(3):491–500.

[106] Antoun E, et al. Maternal dysglycaemia, changes in the infant's epigenome modified with a diet and physical activity intervention in pregnancy: secondary analysis of a randomised control trial. PLOS Med 2020;17(11):e1003229.

[107] Wu Z, et al. Maternal treadmill exercise reduces the neurotoxicity of prenatal sevoflurane exposure in rats via activation of p300 histone acetyltransferase. Neurochem Res 2020;45 (7):1626–35.

[108] Axsom JE, Libonati JR. Impact of parental exercise on epigenetic modifications inherited by offspring: a systematic review. Physiological Rep 2019;7(22):e14287.

[109] Chen J, et al. Opposite epigenetic associations with alcohol use and exercise intervention. Front Psychiatry 2018;9:594.

[110] George AK, et al. Exercise mitigates alcohol induced endoplasmic reticulum stress mediated cognitive impairment through ATF6-Herp signaling. Sci Rep 2018;8(1):5158.

[111] Vetreno RP, et al. Neuroimmune and epigenetic involvement in adolescent binge ethanol-induced loss of basal forebrain cholinergic neurons: restoration with voluntary exercise. Addiction Biol 2020;25(2):e12731.

[112] McEwen LM, et al. DNA methylation signatures in peripheral blood mononuclear cells from a lifestyle intervention for women at midlife: a pilot randomized controlled trial. Appl Physiol Nutr Metab 2018;43(3):233–9.

[113] Choi SW, Friso S. Epigenetics: a new bridge between nutrition and health. Adv Nutr 2010;1(1):8–16.

[114] Zeisel SH. Epigenetic mechanisms for nutrition determinants of later health outcomes. Am J Clin Nutr 2009;89(5):1488S–93S.

[115] Yan Z, et al. Maternal chronic folate supplementation ameliorates behavior disorders induced by prenatal high-fat diet through methylation alteration of BDNF and Grin2b in offspring hippocampus. Mol Nutr Food Res 2017;61(12):1700461.

[116] Kwan ST, et al. Maternal choline supplementation during normal murine pregnancy alters the placental epigenome: results of an exploratory study. Nutrients 2018;10(4):417.

[117] Ly L, et al. Intergenerational impact of paternal lifetime exposures to both folic acid deficiency and supplementation on reproductive outcomes and imprinted gene methylation. Mol Hum Reprod 2017;23(7):461–77.

[118] ElGendy K, et al. Effects of dietary interventions on DNA methylation in adult humans: systematic review and meta-analysis. Br J Nutr 2018;120(9):961–76.

[119] Gali Ramamoorthy T, et al. Maternal overnutrition programs epigenetic changes in the regulatory regions of hypothalamic Pomc in the offspring of rats. Int J Obes 2018;42(8):1431–44.

[120] Rouschop SH, et al. Gene expression and DNA methylation as mechanisms of disturbed metabolism in offspring after exposure to a prenatal HF diet [S]. J Lipid Res 2019;60(7):1250–9.

[121] Glendining KA, et al. Maternal obesity modulates sexually dimorphic epigenetic regulation and expression of leptin receptor in offspring hippocampus. Brain Behav Immun 2020;88:151–60.

[122] Ansari I, et al. The microbiota programs DNA methylation to control intestinal homeostasis and inflammation. Nat Microbiol 2020;5(4):610–19.

[123] Qin Y, et al. An obesity-associated gut microbiome reprograms the intestinal epigenome and leads to altered colonic gene expression. Genome Biol 2018;19(1):7.

[124] Remely M, et al. Therapeutic perspectives of epigenetically active nutrients. Br J Pharmacol 2015;172(11):2756–68.

[125] Barcik W, et al. The role of lung and gut microbiota in the pathology of asthma. Immunity 2020;52(2):241–55.

[126] Cuevas-Sierra A, et al. Diet, gut microbiota, and obesity: links with host genetics and epigenetics and potential applications. Adv Nutr 2019;10(suppl_1):S17–30.

[127] Schuettengruber B, et al. Genome regulation by polycomb and trithorax proteins. Cell 2007;128(4):735–45.

[128] Wani AL, Ara A, Usmani JA. Lead toxicity: a review. Interdiscip Toxicol 2015;8(2):55–64.

[129] Sanders T, et al. Neurotoxic effects and biomarkers of lead exposure: a review. Rev Env Health 2009;24(1):15–45.

[130] Bakulski KM, et al. Alzheimer's disease and environmental exposure to lead: the epidemiologic evidence and potential role of epigenetics. Curr Alzheimer Res 2012;9(5):563–73.

[131] Wu S, et al. Environmental lead exposure aggravates the progression of Alzheimer's disease in mice by targeting on blood brain barrier. Toxicol Lett 2020;319:138–47.

[132] Bihaqi SW. Early life exposure to lead (Pb) and changes in DNA methylation: relevance to Alzheimer's disease. Rev Env Health 2019;34(2):187—95.

[133] Schneider JS, Kidd SK, Anderson DW. Influence of developmental lead exposure on expression of DNA methyltransferases and methyl cytosine-binding proteins in hippocampus. Toxicol Lett 2013;217(1):75—81.

[134] Eid A, et al. Developmental lead exposure and lifespan alterations in epigenetic regulators and their correspondence to biomarkers of Alzheimer's disease. Alzheimers Dement (Amst) 2016;2:123—31.

[135] Senut MC, et al. Epigenetics of early-life lead exposure and effects on brain development. Epigenomics 2012;4(6):665—74.

[136] Sen A, et al. Lead exposure induces changes in 5-hydroxymethylcytosine clusters in CpG islands in human embryonic stem cells and umbilical cord blood. Epigenetics 2015;10(7):607—21.

[137] Donzelli G, et al. Particulate matter exposure and attention-deficit/hyperactivity disorder in children: a systematic review of epidemiological studies. Int J Environ Res Public Health 2020;17(1).

[138] Ji Y, et al. A prospective birth cohort study on early childhood lead levels and attention deficit hyperactivity disorder: new insight on sex differences. J Pediatrics 2018;199:124—31 e8.

[139] Luo M, et al. Epigenetic histone modification regulates developmental lead exposure induced hyperactivity in rats. Toxicol Lett 2014;225(1):78—85.

[140] Kuivenhoven M, Mason K. StatPearls, Editor Arsenic toxicity. Treasure Island, FL: StatPearls Publishing; 2021.

[141] Howe CG, Gamble MV. Influence of arsenic on global levels of histone posttranslational modifications: a review of the literature and challenges in the field. Curr Env Health Rep 2016;3(3):225—37.

[142] Rojas D, et al. Prenatal arsenic exposure and the epigenome: identifying sites of 5-methylcytosine alterations that predict functional changes in gene expression in newborn cord blood and subsequent birth outcomes. Toxicol Sci 2015;143(1):97—106.

[143] Koestler DC, et al. Differential DNA methylation in umbilical cord blood of infants exposed to low levels of arsenic in utero. Env Health Perspect 2013;121(8):971—7.

[144] Rager JE, et al. Prenatal arsenic exposure and the epigenome: altered microRNAs associated with innate and adaptive immune signaling in newborn cord blood. Env Mol Mutagen 2013;55(3):196—208.

[145] Valles S, et al. Exposure to low doses of inorganic arsenic induces transgenerational changes on behavioral and epigenetic markers in zebrafish (Danio rerio). Toxicol Appl Pharmacol 2020;396:115002.

[146] Kaushal A, et al. Genome-wide DNA methylation at birth in relation to in utero arsenic exposure and the associated health in later life. Environ Health 2017;16(1):50.

[147] Olsson IM, et al. Cadmium in blood and urine—impact of sex, age, dietary intake, iron status, and former smoking—association of renal effects. Env Health Perspect 2002;110(12):1185—90.

[148] Xiao C, et al. Cadmium induces histone H3 lysine methylation by inhibiting histone demethylase activity. Toxicol Sci 2015;145(1):80—9.

[149] Virani S, et al. DNA methylation is differentially associated with environmental cadmium exposure based on sex and smoking status. Chemosphere 2016;145:284—90.

[150] Brooks SA, et al. miRNAs as common regulators of the transforming growth factor (TGF)-beta pathway in the preeclamptic placenta and cadmium-treated trophoblasts: links between the environment, the epigenome and preeclampsia. Food Chem Toxicol 2016;98(Pt A):50—7.

[151] Sanders AP, et al. Cadmium exposure and the epigenome: exposure-associated patterns of DNA methylation in leukocytes from mother-baby pairs. Epigenetics 2014;9(2):212—21.

[152] Cowley M, et al. Effects of cadmium exposure on DNA methylation at imprinting control regions and genome-wide in mothers and newborn children. Env Health Perspect 2018;126(3):037003.

[153] Saintilnord WN, et al. Chronic exposure to cadmium induces differential methylation in mice spermatozoa. Toxicological Sci 2021;180(2):262—76.

[154] Myers GJ, Davidson PW. Prenatal methylmercury exposure and children: neurologic, developmental, and behavioral research. Env Health Perspect 1998;106(Suppl 3):841—7.

[155] Kajiwara Y, et al. Methylmercury transport across the placenta via neutral amino acid carrier. Arch Toxicol 1996;70(5):310—14.

[156] Cardenas A, et al. Persistent DNA methylation changes associated with prenatal mercury exposure and cognitive performance during childhood. Sci Rep 2017;7(1):288.

[157] Onishchenko N, et al. Long-lasting depression-like behavior and epigenetic changes of BDNF gene expression induced by perinatal exposure to methylmercury. J Neurochem 2008;106(3):1378—87.

[158] Sanders AP, et al. Altered miRNA expression in the cervix during pregnancy associated with lead and mercury exposure. Epigenomics 2015;7(6):885—96.

[159] Cardenas A, et al. Differential DNA methylation in umbilical cord blood of infants exposed to mercury and arsenic in utero. Epigenetics 2015;10(6):508—15.

[160] Bakulski KM, et al. Prenatal mercury concentration is associated with changes in DNA methylation at TCEANC2 in newborns. Int J Epidemiol 2015;44(4):1249—62.

[161] Ma Y, et al. The adverse health effects of bisphenol A and related toxicity mechanisms. Env Res 2019;176:108575.

[162] Giuliani A, et al. Critical review on the presence of phthalates in food and evidence of their biological impact. Int J Env Res Public Health 2020;17(16):5655.

[163] Palanza P, et al. Effects of developmental exposure to bisphenol A on brain and behavior in mice. Env Res 2008;108(2):150—7.

[164] Braun JM, Sathyanarayana S, Hauser R. Phthalate exposure and children's health. Curr Oppediatrics 2013;25(2):247—54.

[165] Praveena SM, et al. Phthalates exposure and attention-deficit/hyperactivity disorder in children: a systematic review of epidemiological literature. Env Sci Pollut Res Int 2020;27(36):44757—70.

[166] Chen CH, et al. Association between fetal exposure to phthalate endocrine disruptor and genome-wide DNA methylation at birth. Env Res 2018;162:261—70.

[167] Taylor JA, et al. Prenatal exposure to bisphenol A disrupts naturally occurring bimodal DNA methylation at proximal promoter of fggy, an obesity-relevant gene encoding a carbohydrate kinase, in gonadal white adipose tissues of CD-1 mice. Endocrinology 2018;159(2):779—94.

[168] Moody L, et al. Tissue-specific changes in Srebf1 and Srebf2 expression and DNA methylation with perinatal phthalate exposure. Environ epigenetics 2019;5(2). p. dvz009-dvz009.

[169] Nakamura K, et al. Murine neocortical histogenesis is perturbed by prenatal exposure to low doses of bisphenol A. J Neurosci Res 2006;84(6):1197—205.

[170] Yeo M, et al. Bisphenol A delays the perinatal chloride shift in cortical neurons by epigenetic effects on the Kcc2 promoter. Proc Natl Acad Sci U S A 2013;110(11):4315—20.

[171] Kochmanski J, et al. Longitudinal effects of developmental bisphenol A and variable diet exposures on epigenetic drift in mice. Reprod Toxicol 2016;68:154—63.

[172] Neier K, et al. Perinatal exposures to phthalates and phthalate mixtures result in sex-specific effects on body weight,

organ weights and intracisternal A-particle (IAP) DNA methylation in weanling mice. J Dev Orig Health Dis 2019;10(2):176−87.

[173] Montrose L, et al. Maternal levels of endocrine disrupting chemicals in the first trimester of pregnancy are associated with infant cord blood DNA methylation. Epigenetics 2018;13(3):301−9.

[174] Huen K, et al. Maternal phthalate exposure during pregnancy is associated with DNA methylation of LINE-1 and Alu repetitive elements in Mexican-American children. Environ Res 2016;148:55−62.

[175] López-Rodríguez D, et al. Multi- and transgenerational outcomes of an exposure to a mixture of endocrine-disrupting chemicals (EDCs) on puberty and maternal behavior in the female rat. Environ Health Perspect 2021;129(8) 87003-87003.

[176] Skinner MK, et al. Transgenerational epigenetic programming of the brain transcriptome and anxiety behavior. PLoS One 2008;3(11):e3745.

[177] Oppold A, et al. Epigenetic alterations and decreasing insecticide sensitivity of the Asian tiger mosquito Aedes albopictus. Ecotoxicol Env Saf 2015;122:45−53.

[178] Heijmans BT, et al. Persistent epigenetic differences associated with prenatal exposure to famine in humans. Proc Natl Acad Sci U S A 2008;105(44):17046−9.

[179] van der Plaat DA, et al. Occupational exposure to pesticides is associated with differential DNA methylation. Occup Env Med 2018;75(6):427−35.

[180] Organization, W.H. *Air Pollution*. 2021; Available from: https://www.who.int/health-topics/air-pollution#tab = tab_1.

[181] Manisalidis I, et al. Environmental and health impacts of air pollution: a review. Front. Public Health 2020;8(14).

[182] Ding R, et al. Characteristics of DNA methylation changes induced by traffic-related air pollution. Mutat Res Genet Toxicol Env Mutagen 2016;796:46−53.

[183] Suga Y, et al. Quantitative p16 and ESR1 methylation in the peripheral blood of patients with non-small cell lung cancer. Oncol Rep 2008;20(5):1137−42.

[184] Fujiwara K, et al. Identification of epigenetic aberrant promoter methylation in serum DNA is useful for early detection of lung cancer. Clin Cancer Res 2005;11(3):1219−25.

[185] Hsu HS, et al. Characterization of a multiple epigenetic marker panel for lung cancer detection and risk assessment in plasma. Cancer 2007;110(9):2019−26.

[186] Gondalia R, et al. Methylome-wide association study provides evidence of particulate matter air pollution-associated DNA methylation. Environ Int 2019;132 104723-104723.

[187] Calderón-Garcidueñas L, et al. Reduced repressive epigenetic marks, increased DNA damage and Alzheimer's disease hallmarks in the brain of humans and mice exposed to particulate urban air pollution. Env Res 2020;183:109226.

[188] Csoka AB, Szyf M. Epigenetic side-effects of common pharmaceuticals: a potential new field in medicine and pharmacology. Med Hypotheses 2009;73(5):770−80.

[189] Loscher W. Basic pharmacology of valproate: a review after 35 years of clinical use for the treatment of epilepsy. CNS Drugs 2002;16(10):669−94.

[190] Lunke S, et al. Epigenetic evidence of an Ac/Dc axis by VPA and SAHA. Clin. Epigenet 2021;13(1):58.

[191] Wyszynski DF, et al. Increased rate of major malformations in offspring exposed to valproate during pregnancy. Neurology 2005;64(6):961−5.

[192] Gottlicher M, et al. Valproic acid defines a novel class of HDAC inhibitors inducing differentiation of transformed cells. EMBO J 2001;20(24):6969−78.

[193] Chen Y, et al. Valproic acid-induced histone acetylation suppresses CYP19 gene expression and inhibits the growth and survival of endometrial stromal cells. Int J Mol Med 2015;36(3):725−32.

[194] Balmer NV, et al. Epigenetic changes and disturbed neural development in a human embryonic stem cell-based model relating to the fetal valproate syndrome. Hum Mol Genet 2012;21(18):4104−14.

[195] Bator E, et al. Valproic acid (VPA) reduces sensorimotor gating deficits and HDAC2 overexpression in the MAM animal model of schizophrenia. Pharmacol Rep 2015;67(6):1124−9.

[196] Citraro R, et al. Effects of histone deacetylase inhibitors on the development of epilepsy and psychiatric comorbidity in WAG/Rij rats. Mol Neurobiol 2020;57(1):408−21.

[197] Mamelak M, Scharf MB, Woods M. Treatment of narcolepsy with gamma-hydroxybutyrate. A review of clinical and sleep laboratory findings. Sleep 1986;9(1 Pt 2):285−9.

[198] Klein C, et al. Pharmacological doses of gamma-hydroxybutyrate (GHB) potentiate histone acetylation in the rat brain by histone deacetylase inhibition. Neuropharmacology 2009;57(2):137−47.

[199] El-Habr EA, et al. A driver role for GABA metabolism in controlling stem and proliferative cell state through GHB production in glioma. Acta Neuropathologica 2017;133(4):645−60.

[200] O'Connell T, Kaye L, Plosay 3rd JJ. Gamma-hydroxybutyrate (GHB): a newer drug of abuse. Am Fam Physician 2000;62(11):2478−83.

[201] Resendiz M, Mason S, Lo C, Badin J, Chiu YJ, Zhou FC. Alcohol metabolism and epigenetic methylation and acetylation, in molecular aspects of alcohol and nutritionIn: Patel VB, editor. Maryland Hdights, MO, USA: Academic Press; 2016. p. 287−304.

[202] Pandey SC, Bohnsack JP. Alcohol makes its epigenetic marks. Cell Metab 2020;31(2):213−14.

[203] Ponomarev I. Epigenetic control of gene expression in the alcoholic brain. Alcohol Res 2013;35(1):69−76.

[204] Miozzo F, et al. Alcohol exposure promotes DNA methyltransferase DNMT3A upregulation through reactive oxygen species-dependent mechanisms. Cell Stress Chaperones 2018;23(1):115−26.

[205] Sakharkar AJ, et al. Altered amygdala DNA methylation mechanisms after adolescent alcohol exposure contribute to adult anxiety and alcohol drinking. Neuropharmacology 2019;157:107679.

[206] Maier HB, et al. Alcohol consumption alters Gdnf promoter methylation and expression in rats. J Psychiatr Res 2020;121:1−9.

[207] Zhang H, et al. Adolescent alcohol exposure epigenetically regulates CREB signaling in the adult amygdala. Sci Rep 2018;8(1):10376.

[208] Johnstone AL, et al. Dysregulation of the histone demethylase KDM6B in alcohol dependence is associated with epigenetic regulation of inflammatory signaling pathways. Addiction Biol 2021;26(1):e12816.

[209] May PA, et al. Prevalence of fetal alcohol spectrum disorders in 4 US communities. JAMA 2018;319(5):474−82.

[210] Hoyme HE, Shah PE. Fetal alcohol spectrum disorders. Cassidy Allanson's Manag Genet Syndromes 2021;405−24.

[211] Liu Y, et al. Alcohol exposure alters DNA methylation profiles in mouse embryos at early neurulation. Epigenetics 2009;4(7).

[212] Chen Y, Ö NC, Zhou FC. DNA methylation program in developing hippocampus and its alteration by alcohol. PLoS One 2013;8(3):e60503.

[213] Shukla SD, et al. Acetaldehyde alters MAP kinase signalling and epigenetic histone modifications in hepatocytes. Novartis Found Symp 2007;285:217−24 discussion 224-8.

[214] Knezovich JG, Ramsay M. The effect of preconception paternal alcohol exposure on epigenetic remodeling of the h19 and rasgrf1 imprinting control regions in mouse offspring. Front Genet 2012;3:10.

[215] Pierce RC, et al. Environmental, genetic and epigenetic contributions to cocaine addiction. Neuropsychopharmacology 2018;43(7):1471–80.

[216] Baker-Andresen D, et al. Persistent variations in neuronal DNA methylation following cocaine self-administration and protracted abstinence in mice. Neuroepigenetics 2015;4:1–11.

[217] Fonteneau M, et al. Inhibition of DNA methyltransferases regulates cocaine self-administration by rats: a genome-wide DNA methylation study. Genes, Brain Behav 2017;16(3):313–27.

[218] Engmann O, et al. Cocaine-induced chromatin modifications associate with increased expression and three-dimensional looping of *Auts2*. Biol Psychiatry 2017;82(11):794–805.

[219] Cates HM, et al. Fosb induction in nucleus accumbens by cocaine is regulated by E2F3a. eNeuro 2019;6(2).

[220] Vaillancourt K, et al. Cocaine-related DNA methylation in caudate neurons alters 3D chromatin structure of the IRXA gene cluster. Mol Psychiatry 2020;26(7):3134–51.

[221] Massart R, et al. Role of DNA methylation in the nucleus accumbens in incubation of cocaine craving. J Neurosci 2015;35(21):8042–58.

[222] Ausio J. MeCP2 and the enigmatic organization of brain chromatin. Implic Depression Cocaine Addiction Clin Epigenetics 2016;8:58.

[223] Zhao Q, et al. Prenatal cocaine exposure impairs cognitive function of progeny via insulin growth factor II epigenetic regulation. Neurobiol Dis 2015;82:54–65.

[224] Sadri-Vakili G, et al. Cocaine-induced chromatin remodeling increases brain-derived neurotrophic factor transcription in the rat medial prefrontal cortex, which alters the reinforcing efficacy of cocaine. J Neurosci 2010;30(35):11735–44.

[225] SAMHSA. *Key substance use and mental health indicators in the United States: Results from the 2019 National Survey on Drug Use and Health NSDUH Series H-55 2020*, Available from: https://www.samhsa.gov/data/.

[226] Bara A, et al. Cannabis and synaptic reprogramming of the developing brain. Nat Rev Neurosci 2021;22(7):423–38.

[227] Goodwin RD, et al. Cannabis use during pregnancy in the United States: the role of depression. Drug Alcohol Dependence 2020;210:107881.

[228] Metz TD, Borgelt LM. Marijuana use in pregnancy and while breastfeeding. Obstet Gynecol 2018;132(5):1198–210.

[229] Bolhuis K, et al. Maternal and paternal cannabis use during pregnancy and the risk of psychotic-like experiences in the offspring. Schizophr Res 2018;202:322–7.

[230] Richardson KA, Hester AK, McLemore GL. Prenatal cannabis exposure – The "first hit" to the endocannabinoid system. Neurotoxicol Teratol 2016.

[231] DiNieri JA, et al. Maternal cannabis use alters ventral striatal dopamine D2 gene regulation in the offspring. Biol Psychiatry 2011;70(8):763–9.

[232] Wanner NM, et al. Developmental cannabidiol exposure increases anxiety and modifies genome-wide brain DNA methylation in adult female mice. Clin Epigenetics 2021;13(1):4.

[233] Doehring A, et al. Chronic opioid use is associated with increased DNA methylation correlating with increased clinical pain. Pain 2013;154(1):15–23.

[234] Kamangar F, et al. Opium use: an emerging risk factor for cancer? Lancet Oncol 2014;15(2):e69–77.

[235] Das A, Poole WK, Bada HS. A repeated measures approach for simultaneous modeling of multiple neurobehavioral outcomes in newborns exposed to cocaine in utero. Am J Epidemiol 2004;159(9):891–9.

[236] Chorbov VM, et al. Elevated levels of DNA methylation at the OPRM1 promoter in blood and sperm from male opioid addicts. J Opioid Manag 2011;7(4):258–64.

[237] Lange S, et al. National, regional, and global prevalence of smoking during pregnancy in the general population: a systematic review and meta-analysis. Lancet Glob Health 2018;6(7):e769–76.

[238] Nakamura A, François O, Lepeule J. Epigenetic alterations of maternal tobacco smoking during pregnancy: a narrative review. Int J Environ Res Public Health 2021;18(10):5083.

[239] Joubert BR, et al. 450K epigenome-wide scan identifies differential DNA methylation in newborns related to maternal smoking during pregnancy. Env Health Perspect 2012;120(10):1425–31.

[240] Novakovic B, et al. Postnatal stability, tissue, and time specific effects of AHRR methylation change in response to maternal smoking in pregnancy. Epigenetics 2014;9(3):377–86.

[241] Knopik VS, et al. Prenatal substance exposure and offspring development: does DNA methylation play a role? Neurotoxicol Teratol 2019;71:50–63.

[242] Chen Y, et al. Diversity of two forms of DNA methylation in the brain. Front Genet 2014;5:46.

[243] Winter S, Fischle W. Epigenetic markers and their cross-talk. Essays Biochem 2010;48:45–61.

[244] Hutson TH, et al. Cbp-dependent histone acetylation mediates axon regeneration induced by environmental enrichment in rodent spinal cord injury models. Sci Transl Med 2019;11(487):eaaw2064.

[245] Yuan X, et al. Epigenetic modulation of Fgf21 in the perinatal mouse liver ameliorates diet-induced obesity in adulthood. Nat Commun 2018;9(1):636.

[246] Sun D, et al. Genetic, epigenetic and transcriptional variations at NFATC2IP locus with weight loss in response to diet interventions: the POUNDS lost trial. Diabetes Obes Metab 2018;20(9):2298–303.

[247] Sailani MR, et al. Lifelong physical activity is associated with promoter hypomethylation of genes involved in metabolism, myogenesis, contractile properties and oxidative stress resistance in aged human skeletal muscle. Sci Rep 2019;9(1):3272.

[248] Ray PD, Yosim A, Fry RC. Incorporating epigenetic data into the risk assessment process for the toxic metals arsenic, cadmium, chromium, lead, and mercury: strategies and challenges. Front Genet 2014;5:201.

CHAPTER

33

Gut Microbiome Influence on Human Epigenetics, Health, and Disease

Martin M. Watson[1], Mark van der Giezen[1,2] and Kjetil Søreide[3,4]

[1]Department of Chemistry, Bioscience and Environmental Engineering, University of Stavanger, Stavanger, Norway
[2]Biosciences, University of Exeter, Exeter, United Kingdom [3]Department of Gastrointestinal Surgery, Stavanger University Hospital, Stavanger, Norway [4]Department of Clinical Medicine, University of Bergen, Bergen, Norway

OUTLINE

Introduction	669	Pathogenic Infections and Epigenetic Modifications	680
Early Microbiome Exposure and Epigenetic Influence	673	Bacterial Infections	680
		Viral Infections	681
Human Gut Microflora	675	Parasitic Infections	682
Microbiome, Epigenetics, and Effect on Metabolism	675	Summary	683
		Abbreviations	683
Gut microbiota, Inflammation, and Colorectal Carcinogenesis	677	References	683
Pathogenic Infections and Cancer	677		

INTRODUCTION

As human beings, our physiology is influenced by the surrounding environment in every way. How we live and commute, what we eat and drink influence us more than we realized and understood. Importantly, we live together with trillions of microorganisms—bacteria, viruses, archaea, and eukaryotes-collectively known as microbiota [1]. With the broader term microbiome, both microbiota and its "theater of activity," the collection of interacting factors such as DNA/RNA, proteins, metabolites, etc., are intended. Microbiota have been shown to interact with one another and also with the host immune and organ systems in ways that influence the development of disease or maintain health [2–4]. The human host has coevolved with these microorganisms over hundreds of thousands of years and developed complex mechanisms that monitor and control this ecosystem, with a profound impact on host physiology (Fig. 33.1). More importantly, because of their symbiotic relationship in terms of fitness, biology, and evolution, the human body and the microorganisms inhabiting it can be viewed as a singular entity, often termed "the holobiont" [5,6].

Many sites on our body display a distinctive set of microbiota, but in likely no other place is the interaction between human cells and microorganisms as important as in the gastrointestinal tract (Fig. 33.2). Here microbes may influence both physiological and pathogenetic processes through regulation of various molecular mechanisms [2,7]. Such cellular mechanisms have homeostatic roles beyond the traditional concept

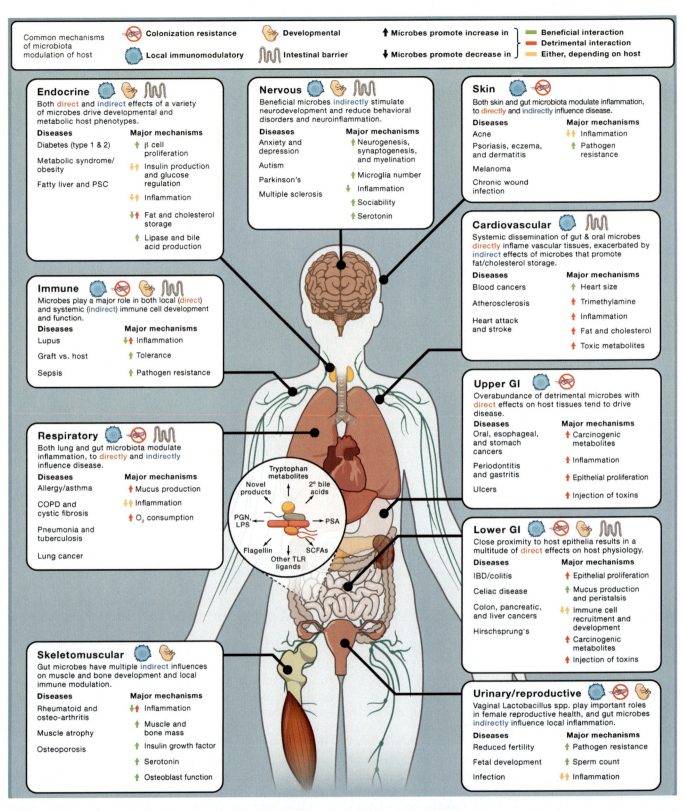

FIGURE 33.1 The body's major organ systems and the homeostatic defects that lead to disease. Different microbiota patterns reside on all environmentally exposed surfaces of the body (such as on the skin, mouth, lung, and reproductive and urinary tracts); however, the greatest density and diversity of microbes is found within the gastrointestinal tract. Through both local and systemic mechanisms, the microbiota influences development and function of mammalian's physiology. For example, our commensal microbes can regulate the development of tissue-specific cells such as the microglia in the brain, intestinal epithelial cells, or β cells in the pancreas. Disruptions to the microbiota, or dysbiosis, perturb these homeostatic processes and can lead to organ specific diseases. In recent years, specific group of microbes and pleiotropic molecules associated with them have emerged as important in their mammalian hosts homeostatic processes, in multiple organ systems. In many cases, specific members of the microbiota can enhance a process while others inhibit it, and the resulting outcome of these interactions on the host are often context dependent. Many diseases stem from either a loss or an imbalance of healthy host-microbe interactions, thus highlighting the importance of these pathways to overall health. *Adapted from J.H. Hill and J.L. Round, SnapShot: microbiota effects on host physiology. Cell, 2021; 184(10): 2796-2796.e1, with permission. Copyright 2021 by Elsevier.*

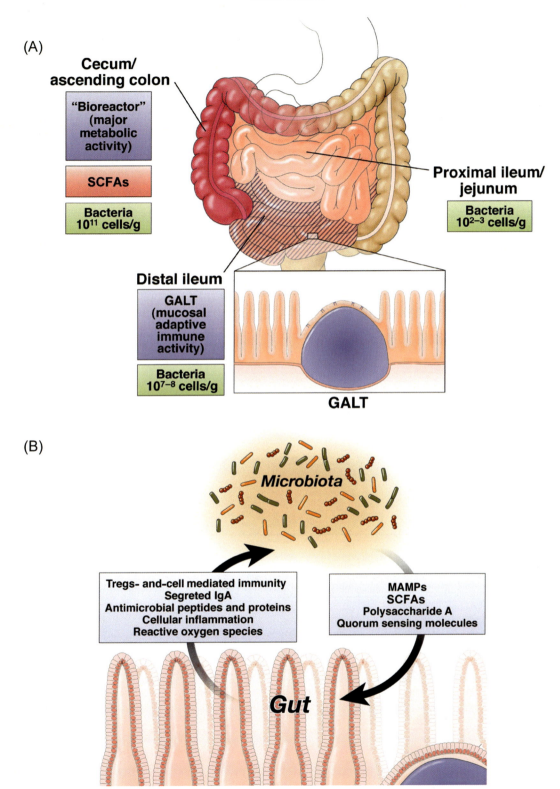

FIGURE 33.2 Microbiota as a "bioreactor" and gut-microbiota "crosstalk." (A) Preferred sites of commensal/probiotic interaction with the gut. Cecum/ascending colon is a "bioreactor" with the greatest amounts of bacteria, metabolic activity, and short-chain fatty acid (SCFA) fermentation. Concentration of SCFA diminishes along the colon. The distal ileum is enriched in GALT (Peyer's patches) and is the dominant site of luminal sampling and mucosal adaptive immune activity. (B) Mechanisms of microbiota and gut crosstalk. Both parties in the symbiotic dyad possess means to alter and shape each other, resulting in a "negotiated settlement" at equilibrium. A breakdown on this crosstalk may result in a "dysbiotic" microbiota and clinical consequences. *Adapted from A.S. Neish, Microbes in gastrointestinal health and disease. Gastroenterology, 2009; 136(1): 65–80, with permission. Copyright 2009 by Elsevier.*

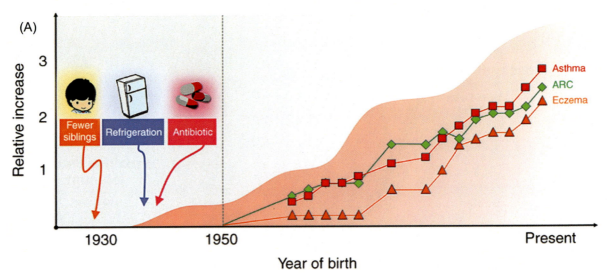
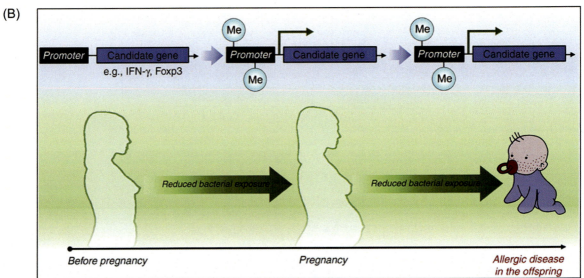
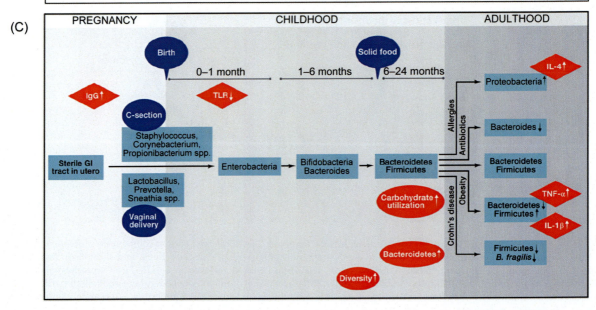

of defense against potential pathogens, suggesting that these pathways contribute directly to the well-being of the gut [7]. The microbes of the large bowel, for example, provide us with genetic and metabolic attributes we have not been required to evolve on our own (Fig. 33.2), including the ability to harvest otherwise inaccessible nutrients [8].

Microbiota may however influence the host's epigenetic makeup at an earlier stage, even before birth. Indeed, the microbiota population changes during the course of a lifetime, affects several aspects of health and disease, and may even influence the process of aging.

This chapter will review some associations between both commensal and pathogenic infective agents and their (known or potential) influence on human health through direct or epigenetic mechanisms. Inclusion of all aspects, associations and mechanisms of this emerging field is beyond the scope of this chapter, and recent and updated reviews on the related topics are therefore referred to, wherever applicable. The microbiome and its (epi)genetics is today explored through a vast field of anatomic locations (from immune system to parenchymal organs) and types of diseases (from autoimmune disease to neoplasia). The field is ever-changing, and has a geographical bias that recently became apparent, with the discovery of thousands of microbial species yet to be named, often enriched in non-westernized regions and populations [9,10]. Thus, this chapter may provide some examples and educational points to link microbiota, epigenetics and human health that may already be outdated at the time of publication.

EARLY MICROBIOME EXPOSURE AND EPIGENETIC INFLUENCE

It is now generally understood that exposure to microbes can influence development and epigenetic make-up of different tissues, and that the quality of early microbiota establishment may in part explain later disease patterns and disease development [11]. Mode of delivery, breastfeeding, geographical location, and household exposures such as pet animals and siblings are all covariates that significantly correlate with the early establishment of intestinal microbial diversity [12—17]. Microbiota exposure as early as during fetal life may have important roles for development, and some suggest that even maternal microbiota composition may be influential on later disease risk long before gestation [18—20]. Some examples are autoimmune diseases (including allergies and asthma), obesity, and cancer. Improvements in socioeconomic standards, advances in technology (e.g., refrigeration, water filtration, antibiotic treatments), and the increased rate of cesarean births have possibly reduced overall exposure to microorganisms (Fig. 33.3A) and are often argued as causal to the subsequently recorded increase in autoimmune diseases in society [18,21]. Likewise, changes in diet and caloric intake may influence the gut microbiota and its genetic make-up (the so-called microbiome), which again affects the landscape of epigenetic modifications in the host, and the risk of various diseases, including cancer [17,22].

Previously, it was believed that the fetus resides within a sterile environment and that the newborn attains its intestinal microbiota only during and after birth. Numerous studies in humans as well as other species appear to contradict this dogma [19,20,23—26]; however, a recent study showed no detectable microbiota in fetal meconium, pointing to birth as the event initiating colonization of the gut [27]. While the child is exposed to skin, oral, vaginal, and gut microflora during and after birth, there appear to be important and substantial links between mother, the microbiome composition, and the fetus [18,19]. Indeed, exposure to microorganisms during vaginal delivery establishes the type of gut intestinal flora

FIGURE 33.3 Lifetime microbial exposure and risk/benefit in health and disease. (A) There is a temporal dissociation between the major changes in microbial exposure to children in westernized countries, which occurred during the first half of the century, and the onset of the accelerating increase in allergic diseases that happened during the second half of the century. (B) Epigenetic changes of the promoter region of immunoregulatory genes important for allergy development could explain how the effect of microbial exposure to the mother during or even before pregnancy is transferred to the next generation. (C). Development of the microbiota. The gastrointestinal tract of the fetus is sterile until birth, after which the newborn is initially colonized. Depending on delivery mode, the initial communities tend toward a skin-like (cesarean section) or a vaginal-like (vaginal delivery) configuration. During the first weeks of life, there is a reduced activity of Toll-like receptor (TLRs), potentially allowing the necessary formation of a stable bacterial community in the gut. As the infant grows, and with the introduction of solid foods, the microbiota diversity increases, and the community converges toward an adult-like state. At the same time, the immune system "learns" to differentiate between commensal and pathogenic bacteria. By adulthood, a relatively stable community composition (but varying between different individuals) is achieved, dominated mostly by Bacteroidetes and Firmicutes. Different diseases are characterized by significant changes in the microbiota and associated changes in the production of cytokines. *Adapted from Parts A and B, T.R. Abrahamsson, R.Y. Wu, and M.C. Jenmalm, Gut microbiota and allergy: the importance of the pregnancy period. Pediatr Res, 2015; 77(1—2): 214—219, with permission from Macmillan Publishers Ltd., Copyright 2015 by Elsevier. Part C, J.C. Clemente, et al., The impact of the gut microbiota on human health: an integrative view. Cell, 2012; 148(6): 1258—1270, with permission. Copyright 2012 by Elsevier.*

and often leads to a higher abundance of Bacteroidetes vs Firmicutes bacterial phyla, which in turn associates with a faster development of the child and a higher taxonomic diversity [15]. This knowledge has led to concerns about the rates and consequences of cesarean deliveries (which prevents exposure of the newborn to the vaginal canal and are performed in a sterile surgical environment). Further, this data has raised the question of whether a vaginal swab to introduce maternal vaginal microbes into cesarean-sectioned infants' gastrointestinal tracts may be beneficial. On the other side of the age spectrum is the knowledge that the gut microbiota may change with age (Fig. 33.3) and even speed up some of the processes known to be associated with biological aging, such as cognitive function [28–30].

Advances in understanding and in analytical techniques of microbial studies have enabled progress in characterizing the taxonomic composition, metabolic capacity, and immunomodulatory activity of the human microbiota, further confirming its role in human health [31,32]. The microbiota has established multiple mechanisms to influence the eukaryotic host, generally in a beneficial fashion, and maintain their stable niche [33,34]. As their microbial genomes allow for a number of metabolic processes beyond the capabilities of the host genome, a diverse and abundant microbiome is seen as an essential part of normal physiology in humans (Fig. 33.4).

Gaining a fuller understanding of both partners in the normal gut–microbiota interaction may thus shed light on how the relationship can go awry and contribute to disease. Further, increased understanding may reveal mechanisms through which these relationships could be manipulated toward therapeutic ends [7,8,35].

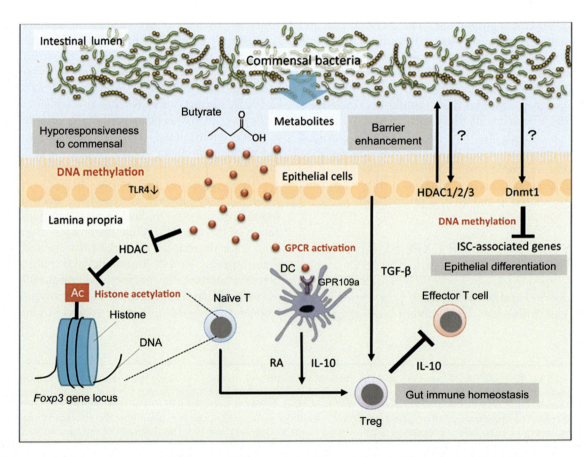

FIGURE 33.4 Commensal microbiota-dependent regulation of intestinal immune system. Commensal bacteria actively consume indigestible materials and produce a diversity of metabolites, including short-chain fatty acids (SCFAs). Among them, butyrate upregulates histone H3 acetylation at the Foxp3 gene locus. Butyrate also binds to GPR109a on dendritic cells to induce production of retinoic acids and IL-10. Butyrate together with acetate and propionate induce transforming growth factor-β (TGF-β) secretion by intestinal epithelial cells (IECs). Collectively, these biological effects of SCFAs facilitate differentiation of naïve CD4+ T cells into Treg cells. Dnmt1-dependent DNA methylation contributes to repression of ISC-associated genes in IECs, critical for differentiation and proliferation of IECs. In addition, IEC-intrinsic histone deacetylases (HDACs) regulate epithelial composition and barrier functions. Commensal bacteria epigenetically downregulate the expression of epithelial TLR4 to maintain hyporesponsiveness to commensal bacteria. *Adapted from Y. Obata, Y. Furusawa, and K. Hase, Epigenetic modifications of the immune system in health and disease. Immunol Cell Biol, 2015; 93(3): 226–232, with permission from Macmillan Publishers Ltd. Copyright 2015 by Elsevier.*

HUMAN GUT MICROFLORA

Humans, as any other animal, live in constant direct interaction with an intricate microbial environment. The total number of bacteria, viruses, protozoa, and archaea normally inhabiting our mucosal surfaces exceeds the quantity of cells ($\sim 10^{13}$) in our bodies. Bacteria, which alone constitute more than 99% in volume and genome of the microbiota inhabiting the human host, were until very recently thought to outnumber our cells in a 10:1 ratio. This ratio is nowadays estimated to be more likely 1.3:1 [1]. Perhaps the greatest habitat for microbes is the gastrointestinal (GI) tract, with increasing recorded concentrations of bacteria from the stomach ($\sim 10^1 - 10^4$ cells/g or mL) to the small ($\sim 10^4 - 10^8$ cells/g or mL) and large bowel ($10^{10} - 10^{12}$ cells/g or mL) [36,37]. Interestingly, it was suggested that a single defecation event could momentarily reverse the order of bacterial-to-human cells ratio [1].

Gut microbiota concentration and composition is largely influenced by diet [38], type of birth (as mentioned earlier), age [35], and geographical location [39,40]. Molecular analysis has demonstrated different communities of bacteria from individual to individual [41–43]. In a study on 124 European individuals, between 1000 and 1150 different species of bacteria were discovered inhabiting the gastrointestinal tract, with each individual harboring at least 160 species, with a great overlap [44]. While the upper GI tract (stomach and small intestines) have a relatively simple microbiota composition and density, the large intestine is home to a dynamic microbial ecosystem, densely populated by living microorganisms. Among them are bacteria, which achieve estimated concentrations $>10^{11}$ cells/g of wet content in the colon [36,37], or about 60% of fecal mass.

Conversely, very little is known about the distribution of microbial eukaryotes, both unicellular (protists) and multicellular like helminths, in the human gut [45,46].

The great bioreactor that the gut microflora represents in the human body produces a plethora of low-molecular weight, biologically active molecules as a result of both endogenous protein expression and food digestion. These include, but are not limited to, neurotransmitters (serotonin, dopamine, etc.), short-chain fatty acids (SCFAs) such as butyrate, propionate, or acetate, chemical moieties (acetyl-, methyl-, etc.), and simple gases (hydrogen sulfide, H_2S; nitrous oxide, NO; carbon mono- and dioxide, CO/CO_2; ammonia, NH_3, methane, CH_4; etc.). Such molecules are able to signal in auto-, para-, and endocrine fashion, exerting well-established functions as for example digestion and oxygen regulation inside the gut [47], inflammation [48,49], neuroendocrine signaling [50,51], and epigenetic mediation [52,53]. Bacteria are able to influence the host epigenome either directly by producing modulating metabolites, or indirectly by activating signaling cascades that lead to epigenetic regulation of genes. Similarly to viruses, bacteria provoke histone modifications and chromatin remodeling in infected cells, thereby altering the host's transcriptional program and, in most cases, dampening the host innate immune response [6]. Still, our current knowledge is likely to represent just the "tip of the iceberg" of the full spectrum of microbiota influence on the human body.

Furthermore, microorganisms inhabiting us do not only yield molecules as a by-product of the digestion of food we ingest; they also produce a great deal of their own. The microbiome involves the collection of over 3 million genes [44] expressed by the totality of microbes inhabiting the human body. Estimates in number of bacterial genes alone reach 2 million genes, 100 times the human estimates of 20,000 [54].

Scattered amidst the commensal microflora are potential pathogens intrinsically capable of both producing and preventing symptomatic infectious diseases [7,8,55–57]. The long coexistence of microbial pathogens with their eukaryotic hosts, and their coevolution, have provided pathogens with an amazing capacity to exploit host cell functions for survival, replication inside or outside cells, and escape from early innate immune responses [58]. The fact that microbes are so well adapted to their host has been of great benefit for cell biologists who are increasingly using them to study fundamental cell processes.

Living within the colonic lumen are thus hundreds of grams of bacteria, which have a great influence on the host's homoeostatic processes, and important health benefits. When the fine-tuned symbiosis that we and our "forgotten organ" evolved into is thrown out of balance, though, the dysbiotic endogenous microflora can negatively affect our health.

MICROBIOME, EPIGENETICS, AND EFFECT ON METABOLISM

The influence of the gut microbiota state in human health and disease has been increasingly recognized and investigated in recent years. There is a clear epigenetic link, strengthened by recent evidence in animal studies, between the gut microbiome and metabolic diseases such as obesity and diabetes (reviewed in Ref. [59]). Differences in microbiota composition, their relative gene expression, and metabolic programming have been described extensively between obese and lean individuals, pointing to a contribution of the gut microbiota to these metabolic phenotypes [60–63] (Fig. 33.5). As such, the microbiota is considered a potential tool in therapy and diagnostics.

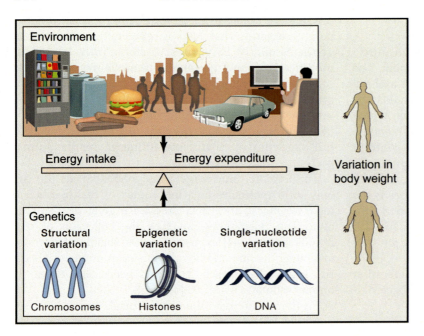

FIGURE 33.5 Contribution of genes and environmental factors to weight gain. Human adiposity is influenced by complex interactions between genetic and environmental influences. The current environment potently facilitates the development of obesity. Abundance of highly processed food has a major impact on energy intake; whereas, numerous other environmental factors, such as television watching, leisure activities, and transport, negatively affect energy expenditure. In any environment, there is a variation in body fat and BMI in large part influenced by genetic variation disrupting energy homeostasis by either decreasing energy expenditure or increasing energy intake [64]. *Adapted from A.A. van der Klaauw, and I.S. Farooqi. The hunger genes: pathways to obesity. Cell 2015;161(1):119–32, with permission. Copyright 2015 by Elsevier.*

In the human gut, the dominant bacterial phylae are Firmicutes, Bacteroidetes, Proteobacteria, and Actinobacteria. Their relative proportion and overall host health are believed to be mutually influenced according to diet, metabolic state, and weight, particularly obesity [38,59,61,65–67]. Obese individuals often exhibit increased proportions of Firmicutes and decreased proportions of Bacteroidetes in the gut [68], and this imbalance is also often associated with a western diet. Whole-genome methylation analysis significantly correlated high levels of the Firmicutes phylum with aberrant methylation pattern in genes associated with metabolism, obesity, and obesity-related diseases, in a pilot study [69]. There is, however, still debate over the link between a particular phylum and energy absorption from the diet.

SCFAs, such as butyrate, acetate, and propionate are end by-products of the fermentation of intestinal bacteria in the presence of complex carbohydrates or other indigestible foods [63]. These fatty acids can diffuse through the epithelial cells and into the intestinal lamina propria and are known to interact with the host cells they come across both directly (e.g., by direct ligand-receptor signaling) and indirectly (by interfering with chromatin structure) [70]. SCFAs are found at varying concentrations in obese and lean individuals, and are known to interact with metabolic processes such as lipogenesis, glucose homeostasis, and insulin sensitivity [47,70].

Different microbiota composition was shown to affect epigenetic regulation of free fatty-acid receptors (FFARs) by decreased promoter methylation in obese and diabetes type 2 patients [71]. Promoter methylation increased in obese patients following nutritional counseling and decrease in body mass index. The molecular mechanisms of the microbiota–host gene methylation observation remain elusive. It is however known that bacteria are important contributors of methyl group donor molecules, such as folate and cofactors needed in DNA methylation processes, and imbalances in the gut microbiota composition can lead to hypo- or hypermethylation in the host [22,52].

Low levels of diffused inflammation in the gut are also associated with both obesity and type 2 diabetes. Butyrate, of which Firmicutes are the main contributors in the human gut, is thought to have a protective effect towards low-grade inflammation and therefore obesity. This is believed to happen through inhibition of histone deacetylases (HDACs) and consequent upregulation of key genes that prevent immune cells infiltration [3,53,72]. Further, a recent genome-wide analysis of DNA methylation in obese patients showed differential methylation patterns in patients with a low versus high Bacteroidetes-to-Firmicute ratio [73]. This was particularly true in adipose tissue, for genes implicated in glucose and energy metabolism, such as *HDAC7* and *IGF2BP2*.

The epigenome is influenced by environmental factors throughout life [22]. For example, nutritional factors can have profound effects on the expression of specific genes via epigenetic modification, and these may be passed on to subsequent generations with potentially detrimental effects.

Many cancers are also associated with altered epigenetic profiles, leading to perturbations in expression of the genes involved in cell growth, differentiation, or DNA damage repair [74]. Autoimmune and neoplastic diseases increase in frequency with increasing age alongside with epigenetic dysregulation [28], as also seen in studies in monozygotic twins. Differences in methylation status of CpG sites, monoallelic silencing, and other epigenetic regulatory mechanisms have been observed in key inflammatory response genes.

GUT MICROBIOTA, INFLAMMATION, AND COLORECTAL CARCINOGENESIS

The thought that dietary habits contribute to colorectal cancer (CRC) development through modulation of the gut microbiota is not new and CRC development has been related to infections of viruses, bacteria, and parasites [75]. Nevertheless, accumulating evidence points to a dual role of the gut microbiota, both promoting and protective, in the context of colorectal carcinogenesis.

Food fermentation is one of the principal modes of active metabolites production by gut-residing microbes. Hydrogen sulfide, for example, a secondary metabolite produced by digestion of red meat by sulfate-reducing bacteria, is a known oncometabolite [68,76].

On the other side, bacterial fermentation is known to produce a multifaceted tumor-suppressive effect via production of metabolites, such as equol, urolithins, and more importantly SCFAs, from digestion of dietary fibers and plants.

The ever-mentioned butyrate (possibly the most studied among bacterial active metabolites) was shown to drive differentiation of monocytes into macrophages through inhibition of histone deacetylase 3 (HDAC3) [77]. The study highlighted the importance of microbiota-derived butyrate in regulating metabolic and transcriptional program of macrophages, enhancing protective antimicrobial activity in the gut. A thorough review of the protective and detrimental metabolites of bacterial fermentation is given by Hullar et al. [78], and Bhat and Kapila reviewed their role as epigenetic modulators in depth [70].

Due to the Warburg effect [79] endured by tumor cells to survive in anaerobic conditions, butyrate is no longer efficiently metabolized in mitochondria. Consequently, it is free to accumulate in the nucleus, where it actively inhibits HDACs, effectively regulating gene expression involved in pathways such as proliferation, and inducing apoptosis of cancer cells in an epigenetic fashion [80]. The protective effect of dietary fiber and butyrate on CRC was recently confirmed in a gnotobiotic mouse model [81]. Butyrate was also shown to regulate protein expression indirectly by modulating posttranscriptional regulation, specifically through up- and downregulation of certain miRNAs (reviewed in [82]), however the exact mechanisms of miRNA interference is currently unknown.

Bacterial effector proteins have also shown to epigenetically interfere with cellular pathways in the context of cancers [83]. An example is Salmonella's acetyltransferase AvrA, which is believed to promote proliferation of epithelial cells in the colon by stabilization of β-catenin, a key component in the Wnt signaling pathway [84].

One of the most-studied contexts in which the mediator role of microbiota is recognized is the inflammatory process. Several mechanisms of tumor promotion have been suggested, including the ability of certain components of the microbial flora of the colon to produce and maintain a state of chronic low-grade inflammation, as reviewed by Irrazábal et al. [68]. The genotoxic stress that results may contribute to colon carcinogenesis and influence genetic stability [68,85] (Figs. 33.6 and 33.7). In a cellular model, it was demonstrated that Enterococcus faecalis promotes chromosomal instability in mammalian cells by producing a superoxide able to oxidatively damage DNA [86].

Conversely, the SCFAs butyrate and propionate produced by microbial fermentation were repeatedly shown in mice to activate FoxP3 expression in naïve CD4 + T-lymphocytes, which results in expansion of the Th17 Tregs cells, thus playing an immunosuppressive role [76,86–88].

Historically, patients with inflammatory bowel disease (IBD) have a higher risk to develop CRC due to the chronic inflammation of their bowel, resulting from (and fueled by) a perturbed immunologic response to the residing microbial community [89,90]. Whether the alterations in intestinal flora are the cause or the consequence of the associated chronic inflammation is still unclear; however, it holds potential as a non-invasive monitoring biomarker of Crohn disease [91]. Although this response is thought to be due at least in part to a genetic predisposition, an abnormal intestinal microflora has been reported numerous times in IBD patients [91–93]. What appears important is the role of proinflammatory cytokines like tumor necrosis factor (TNF-α)—for which nowadays therapeutic inhibitor drugs are available [94,95]—in IBD development, as they may alter the microbial composition, restructure the local immune cell populations, enhance virulence, and increase invasion [93].

While dysbiosis of gut microbiota may cause instability at the chromosomal level in colonic epithelium directly, via inflammation, oxidative DNA damage and production of oncometabolites, symbiosis, and certain taxonomic profiles produce a protective effect by interfering with proliferation of tumor cell populations.

PATHOGENIC INFECTIONS AND CANCER

Of notice, several other cancers have been linked to viral, bacterial, and parasitic infections [75,83]. Infective agents are thought to be responsible for almost

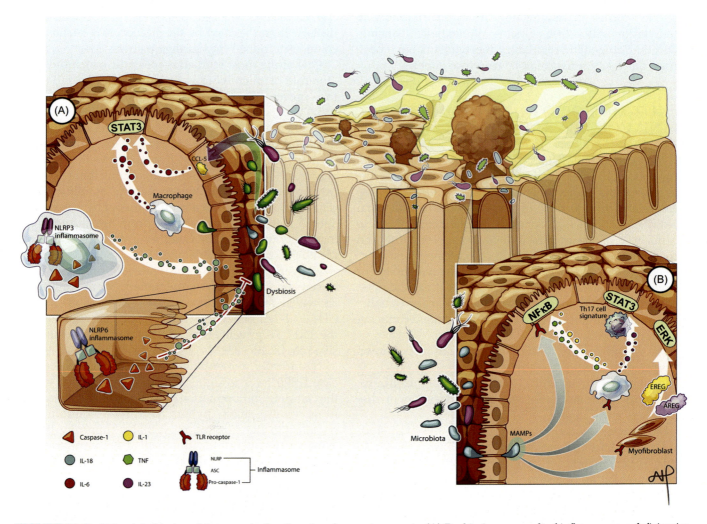

FIGURE 33.6 Role of dysbiosis and immune dysfunctions in colon carcinogenesis. (A) Dysbiosis, as a result of inflammasome deficiencies, could promote tumorigenesis. Inflammasome-derived interleukin (IL)-18 is necessary for tissue repair, protection against tumors, and the maintenance of the microbial ecology equilibrium. In turn, this phenotype associated with the lack of IL-18 could be exacerbated by a dysbiotic microbiota that could lead to chronic inflammation, increased IL-6 signaling, and tumorigenesis. In intestinal epithelial cells, IL-6 activates STAT3 signaling, protecting normal and premalignant cells from apoptosis. (B) Dysbiosis and immune dysfunctions may allow increased bacterial translocation due to altered barrier function. Microorganism-associated molecular patterns (MAMPs) are detected by Toll-like receptors (TLRs) present in epithelial cells, macrophages, and myofibroblasts, leading to the activation of different pathways that promote cancer development. Epiregulin (EREG) and amphiregulin (AREG) are epidermal growth factor receptor (EGFR) ligands and therefore induce proliferation through MAPK/ERK pathway activation. Th17 cytokines mark the early stages of colorectal cancer (CRC) by induction of STAT3. *Adapted from T. Irrazabal, et al., The multifaceted role of the intestinal microbiota in colon cancer. Mol Cell, 2014; 54(2): 309–320 with permission. Copyright 2014 by Elsevier.*

one-fifth of all cancers, with an estimated total of infection-attributable cancer at 2.2 million cases of the global cancer burden in 2018 [96]. The principal agents involved have been estimated to be (in decreasing order): the bacterium *Helicobacter pylori* (810,000 cases, mainly non-cardia gastric cancers), the human papilloma viruses (HPV; 690,000 cases, mostly cervical cancers), the hepatitis B and C viruses (520,000 cases combined, mainly hepatocellular carcinoma). The remaining 210,000 new infection-associated cancers were estimated to be related to Epstein–Barr virus, human T-cell lymphotropic virus type 1 (HTLV-1), human herpesvirus type 8 (HHV8, or Kaposi sarcoma herpesvirus), and parasitic infections such as schistosomes and liver flukes.

Notably, alteration of gut microbiota homeostasis is linked to extra-colonic diseases as well, such as in breast and liver [97–99]. *H. pylori* is a gram-negative bacterium that colonizes the stomach persistently in over half of the world's population. *H. pylori* is linked to various diseases of the stomach, such as peptic ulcer disease, gastric lymphoma, and gastric cancer [100]. In addition, *H. pylori* may also play a role in IBD, irritable bowel disease, and potentially also in CRC.

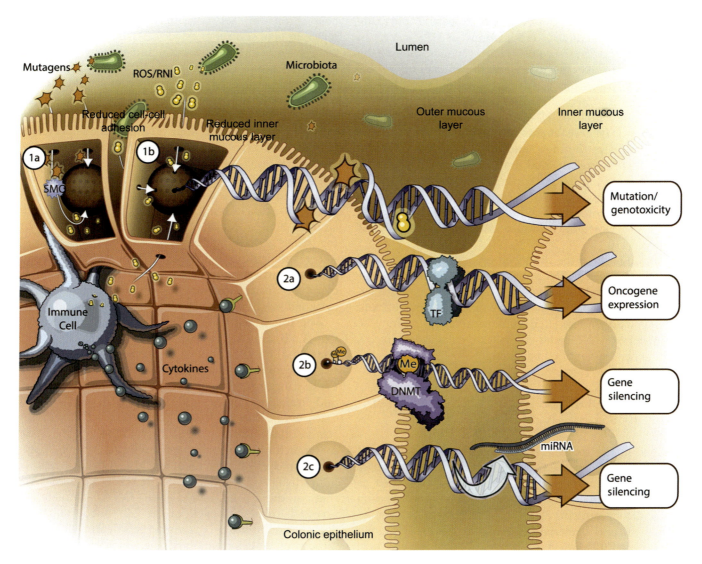

FIGURE 33.7 Inflammation-induced mechanisms of cancer development in the colon. During dysbiosis, specific mutagens could reach colon epithelial cells. (1a) Mutagens from food or released by luminal microbiota enter into the cell, where they can directly cause DNA damage or induce the expression of enzymes like spermine oxidase (SMO) that lead to increased reactive oxygen species (ROS). (1b) ROS and reactive nitrogen intermediates species released by either microbes in the lumen or immune cells within the lamina propria can also lead to genotoxicity. (2) Inflammation leads to the activation of different pathways that could lead to gene dysregulation and tumorigenesis. (2a) Certain cytokines activate transcription factors that drive the expression of genes that control survival, proliferation, and angiogenesis. (2b) Cytokines, including IL-6 and IL-1β, have been associated with the induction of DNA methyltransferases (DNMT), which silence tumor suppressor genes. (2c) Other cytokine inflammatory pathways induce miRNA expression, which downregulates the expression of tumor suppressor genes. *Adapted from T. Irrazabal, et al., The multifaceted role of the intestinal microbiota in colon cancer. Mol Cell, 2014; 54(2): 309–320, with permission. Copyright 2014 by Elsevier.*

Aberrant expression of cell cycle control proteins has been demonstrated in *H. pylori* infected gastric epithelial cells, suggesting that perturbation of the cell cycle plays a role in the pathogenesis of various *H. pylori* associated diseases. Downregulation of E-cadherin (an adhesion molecule involved in tumor invasion and metastasis) in gastric cancers associated to *H. pylori* infection has been observed for more than a decade [101,102], and is caused by silencing of E-cadherin by promoter CpG methylation [102]. Further, in a study of the modulation of the cell cycle control protein $p21^{WAF1}$ by *H. pylori* in a gastric carcinoma cell line and in primary gastric cells derived from healthy tissue [103], the investigators observed an upregulation of $p21^{WAF1}$ in both gastric cancer cells and primary cells. Analysis revealed that the increased expression of $p21^{WAF1}$ induced by *H. pylori* is associated with the release of HDAC-1 from the $p21^{WAF1}$ promoter and hyperacetylation of histone H4 [103].

Decline in gastric cancer risk is at least in part due to antibiotic-mediated eradication of *H. pylori* in asymptomatic individuals [104], which has been shown to reverse the methylation patterns of important tumor promoters in the gastric mucosa [105].

On the contrary, the incidence of CRC is still the third-highest globally [106]. Although it is not directly related to infectious agents, it is tempting to speculate that changes in the diets, and also in the way we prepare and process our food, introduced novel strains of bacteria allowing them unprecedented contact with the gut epithelium [107,108], where they might enter in competition with the residing commensal population [109], and potentially induce carcinogenic molecular changes [110].

On the basis of the described epigenetic changes that may be induced by *H. pylori* infection, such as hypermethylation and silencing of *hMLH1* (a major mechanism of carcinogenesis in mismatch repair-deficient CRC carcinogenesis) [105], similar mechanisms may be inferred in the colorectal epithelium, either by *H. pylori*, or by other unidentified bacterial species.

PATHOGENIC INFECTIONS AND EPIGENETIC MODIFICATIONS

Once inside the infected cells, both the genomes of certain viruses and the proviral genomes of retroviruses are regularly subject to epigenetic regulation such as DNA methylation, histone modifications, and binding of regulatory proteins [111]. In turn, proteins translated from viral and bacterial genomes are routinely shown to epigenetically affect protein expression in the host [112]. These microbial-host interactions often involve promoter silencing of key genes by hypermethylation at CpG islands, as in the case of *Campylobacter rectus* in the placenta or *H. pylori* in gastric mucosa. The resulting perturbation of epigenetic balances may contribute to cellular dysfunctions and, eventually, to the development of pathological conditions.

Bacterial Infections

Bacterial pathogens have evolved various strategies to avoid innate immune surveillance and establish or maintain their environments within the human body [113,114]. The identification of few bacterial effectors capable of entering the nucleus and modifying chromatin structure in the host raises the questions of how pathogens modulate chromatin structure and why. Chromatin is a dynamic structure that maintains the stability and accessibility (or lack thereof) of the host DNA genome to the transcription machinery. Arbibe [113] reviewed the various strategies used by pathogens to interface with host chromatin. Microorganisms might use chromatin injury to take control of major cellular functions, such as the cell cycle of infected cells, and thereby secure their own existence or evade immune surveillance. Further, by modulating epigenetic modifications of chromatin structure at specific genomic locations, the pathogen becomes able to exploit transcriptional activity belonging to host cells for its own genetic code. The mammalian cell cycle, with its network of signaling proteins is involved in processes that reach beyond the simple life and death of a cell, and a feature that many bacterial pathogens have evolved to interfere with. Bacterial toxins like cyclomodulins and nucleomodulins are examples of bacterial-secreted compounds that directly impact the pathways regulating host cell cycle.

Cyclomodulins include cytolethal distending toxins (CDTs), vacuolating cytotoxin, cycle-inhibiting factor, cytotoxic necrotizing factors, dermonecrotic toxin, and cytotoxin-associated antigen A.

The functional CDT is composed of three proteins; Cdt-A, -B, and -C, are secreted by a number of pathogens and may initiate a eukaryotic cell cycle block prior to mitosis, at the G2 stage [115,116]. CdtA and CdtC function as dimeric subunits able to bind CdtB and transport intracellularly. Once inside the cell, CdtB enters the nucleus, and provokes DNA double-strand breaks in a similar fashion to the DNAse I enzymatic activity, inducing phosphorylation of histone H2AX and ultimately resulting in cell cycle arrest, cellular distension, and cell death [117]. Epithelial cells that undergo CDT-mediated DNA damage are arrested in the cell cycle at the G2/M boundary, and eventually initiate programmed cell death.

Several issues remain to be elucidated regarding CDT biology, including the molecular mechanisms of CDT interaction with DNA [118].

The recently discovered bacterial nucleomodulins are toxins capable of epigenetically modify the hosts' chromatin structure by modifying DNA and histones, as well as disturbing protein ubiquitination and localization in the nucleus [119,120]. A well-studied example is the eukaryote-like SET domain protein group, histone-methylating proteins encoded by bacteria, such as *Legionella* and *Chlamydia* that are able to modify chromatin structure in the host.

Knowledge on how bacteria may influence the host environment through genetic and epigenetic manipulation is increasing [42,100,121,122]. Clearly, upon infection, pathogens may reprogram host gene expression. In eukaryotic cells, genetic reprogramming is induced by the concerted activation/repression of transcription factors and various histone modifications that control DNA accessibility in chromatin [123]. One study exploring the microbe-host interaction during infection reported that *Listeria monocytogenes* produces listeriolysin O, a toxin able to induce dephosphorylation of histone H3 and deacetylation of histone H4 during early phases of infection. Perfringolysin and pneumolysin, two other molecules belonging to the same family of bacterial toxins and produced by Clostridium perfringens and Streptococcus pneumoniae, respectively, produced similar epigenetic changes [58,123]. The loss of histone phosphorylation and acetylation correlates with downregulation of cytokine CXCL2 and phosphatase DUSP4, which are two proteins heavily involved in host immune response to pathogens. These findings highlighting epigenetic regulation as a strategy common to intra- and extracellular pathogens to modulate the host response during infection [123].

Innate immunity is the first line of defense against a bacterial infection, and most organisms are able to mount an efficient early, nonspecific response leading to the recruitment of cellular effectors and inflammation. Microbial-associated molecular patterns (MAMPs) are bacterial components able to induce an immune response, and include LPS, bacterial flagellin, lipoteichoic acid, peptidoglycan, and nucleic acids [7,58]. Host cells recognize MAMPs through pattern recognition receptors (PRRs) such as Toll-like receptors (TLRs) present either on their cell surface or on endosomes, or nucleotide-binding oligomerization domain proteins (NODs) and NOD-like receptors (NLRs), in the cytoplasm. These receptors activate signaling cascades leading to transcriptional activation of immunity genes, such as cytokine genes.

Viral Infections

Viruses are essentially protein-coated genetic material, lacking any organelle or molecular structure enabling a self-sufficient life cycle. As such, they rely on infection of a host cell and the exploitation of its components to replicate and infect other cells and hosts. Human-infecting viruses are therefore able to influence and interact with the human genome, and multiple mechanisms have been described that regulate cycling between latent and lytic phases, as well as disease development.

Epstein-Barr virus (EBV) is a human herpesvirus infecting memory B cells, present in a latent form in most of the world population. In immunocompetent individuals, primary EBV infection is usually asymptomatic or causing infectious mononucleosis, a self-limiting disease. In immunocompromised patients, EBV is associated with a wide variety of neoplasms, sometimes developing also in patients with an apparently intact immune system [124–126]. DNA methylation and histone modifications strictly regulate expression of latent viral oncogenes and RNAs [127], resulting in a silencing of the EBV genome in memory B cells, or in a cell-type dependent reactivation of latent promoters as, for example, in tumor cells. In immortalized oral keratinocytes, EBV infection produced genome-wide gene methylation and recapitulated the methylation patterns seen in EBV-associated neoplasia [128]. The viral oncoprotein latent membrane protein 1 (LMP1) was shown to upregulate DNA methyltransferases (in particular DNMT3b) via NF-κB signaling [129]. This in turn produced an increase in methylation at CpG islands in the promoter region of tumor suppressor gene PTEN, effectively silencing its expression [129]. In nasopharyngeal carcinoma cell lines, the hypermethylation of the e-cadherin promoter was attributed to LMP1 expression via JNK/AP1-signaling [130].

A further example of epigenetic regulation of viral latency/lytic phase and host epigenome modifications is human cytomegalovirus (HCMV) infection. HCMV is a very common (estimated at 83% of the general population) herpes virus, albeit generally asymptomatic in immunocompetent subjects. In immunocompromised individuals such as congenitally infected infants, HIV and organ transplant patients, however, HCMV infection can result in serious and sometimes lethal complications [131]. Specifically in the gastrointestinal tract, HCMV can cause complications like gastroenteritis, ulcers, hepatitis, and graft rejection. After infection, recruitment of histone proteins to the virus' major immediate early promoter (MIEP, which contains binding sites for multiple transcription factors) causes abundant promoter methylation, and therefore transcriptional inactivation and latency [132]. Successive histone acetylation reactivates immediate early gene expression pathways, and the virus re-enters the lytic phase. All phases of HCMV infection are thus tightly regulated by epigenetic modifications of chromatin structure.

It is interesting to note how HCMV seems to become re-activated in 10%−43% (depending on diagnostic method) of IBD patients, with a peak at 20%−67% in steroid-refractory patients [133].

It was indeed shown how glucocorticoids such as dexamethasone were able to induce reactivation of transcriptional activity of the MIEP in both pregnant women and transplant recipients [134,135].

A combination of long-term immunomodulatory therapy and chronic inflammation may therefore be the cause for HCMV's exit from the latent phase in IBD patients, which is increasingly reported although the issue is still a matter of debate.

Cell division and proliferation is part of homeostatic management of tissue regeneration. Each cell division is however associated with a progressive shortening of telomeres, eventually reaching a threshold that triggers mechanisms of cellular senescence and irreversible cell cycle arrest. Therefore, the ability of virus-infected cells to evade such senescence program is important to guarantee proliferation of infected cells and thus viral long-term survival. EBV, Kaposi sarcoma-associated herpesvirus, HPV, hepatitis B virus (HBV), hepatitis C virus, and human T-cell leukemia virus-1 each appear to mediate activation of telomerase reverse transcriptase hTERT by cis-activation, posttranscriptional or epigenetic [136,137]. It is likely that, as future studies further expose the strategies used by viruses to deregulate telomerase activity and control of telomere length, novel mechanisms will emerge and underscore the importance of increased telomerase activity in sustaining virus-infected cells, and its potential in therapeutic targeting.

Epigenetic transformations during virus-host infection do not exclusively affect the host. Fernandez et al. [111] investigated the whole DNA methylomes of the papilloma (16 and 18) and hepatitis B viruses, and all the transcription start sites of EBV by bisulfite sequencing of multiple time points during the progression of diseases associated with them. The methylation status of the viral DNA was shown to evolve from an unmethylated to a highly methylated one alongside disease progression, from asymptomatic carriers, through premalignant lesions, to invasive cancers [111]. One interpretation of this finding given by the authors is the possibility that DNA methylation might serve to downregulate expression of viral antigen, effectively "hiding" the virus from the human immune system. Further, the investigators suggest that the DNA methylomes found in the study could be used for further research to understand how the viral proteins are able to use the human methyltransferases to achieve immune evasion [111]. Also, these findings have potential clinical applications as detection of methylated viral genomes in biological fluids such as serum and blood is relatively non-invasive [111].

As viral infections are often systemic and hardly localized specifically in the gut, it is interesting to review how other components of the gut microbiome (and their dysbiosis) affect viral immunity in the host. The microbiota of specific, virus-targeted mucosa such as that of the GI tract (but also of the lungs or the vagina), plays important roles in both physical and molecular defenses. The layer of epithelium, as well as the mucus coating it, provide a physical barrier preventing viral attachment and infection, as demonstrated in animal models [138]. In vivo studies comparing germ-free with conventionally raised animals showed that the gut microbiota contributes to the maintenance of the mucus layer [139–141]. It is unclear whether this process is mediated by epigenetic mechanisms; however, it was shown how specific bacteria groups induce expression of distinct mucin genes [139,142], and regulate mucin release from goblet cells into the lumen by triggering a process relying on the enzyme meprin β [143]. In the specific case of Rotavirus infection, one of the most common in the intestines, certain species of bacteria such as Lactobacillus and Bifidobacterium were shown to be able to reduce expression of NSP4, a toxin rotavirus uses to disrupt and penetrate epithelial tight junctions during infection [144,145].

Parasitic Infections

Despite their huge disease burden, mainly in developing nations [146], parasites are relatively little studied when it comes to the gut microbiome. Intestinal parasites such as *Giardia intestinalis*, *Entamoeba histolytica*, and *Cryptosporidium* species do receive attention; however, their role in the gut microbiome and effect on human host beyond disease are somewhat understudied. In addition, the realization that (microbial) eukaryotes found in the human intestines are not necessarily parasitic or pathogenic is relatively new [46,147]. Large-scale studies are now reporting the presence of microorganisms, previously considered parasites (*Blastocystis* and *Dientamoeba fragilis*), in healthy subjects [148,149]. Whether these "parasites" are actual commensals with possible beneficial effect on the human host, perhaps modulated via the immune system, is currently being investigated [46]. The unusual anaerobic biochemistry of these microbial intestinal eukaryotes [150] clearly links to potential positive effects due to the production of SCFAs such as acetate and propionate. *Giardia*, *Entamoeba*, and *Blastocystis* are known acetate producers while the latter, and nematodes such as *Ascaris lumbricoides* and the trematode *Fasciola hepatica* produce propionate [150].

As mentioned above, Tregs play an immunosuppressive role and are important to distinguish self from nonself and prevent autoimmune diseases [76,86,87]. SCFAs, as produced by the gut microbiome, can induce differentiation naïve CD4 + T cells into Treg lineage, via activation of FoxP3 expression. However, several intestinal parasites stimulate Treg activity to evade the immune system [151]. The murine intestinal roundworm *Heligmosomoides polygyrus* excretes a transforming growth factor-β (TGF-β) mimic to induce FoxP3 expression [152]. The *H. polygyrus* TGF-β mimic (Hp-TGM) can induce Tregs differentiation from human naïve CD4 + T cells in a similar fashion as TGF-β. Hp-TGM is actually capable of epigenetic changes in the Treg-specific demethylated region (TSDR) by reducing methylation in the FoxP3 locus [153]. Levels of H3K27 acetylation was similar between TGF-β and Hp-TGM [153], but it is clear that *H. polygyrus* epigenetically regulates its host's immune system to aid its survival in the gut.

The intestinal parasite *Cryptosporidium parvum*, of the large Apicomplexa phylum, causes diarrhea in humans and livestock worldwide and is a major cause of parasite death [154]. There are no drugs for vulnerable patients and infected patients (and livestock) shed large numbers of infectious parasites when infected. This often leads to contaminated drinking water sources and spread of disease [155]. In immune competent hosts, the innate immune system controls parasite growth and initiates the adaptive immune response [154,155]. However, *C. parvum* can also excrete non-coding RNAs (ncRNAs) that modulate host gene expression [156]. For example, *C. parvum* ncRNAs have been shown to lead to suppression of Interleukin-33 (IL33) [157], an IL-1 cytokine family member that is important in innate immunity via activation of multiple essential components of allergic inflammation [158]. Suppression of IL33 and other genes is linked to H3K9 histone methylation, a clear indicator of gene silencing [159]. In the case of *C. parvum*, the ncRNAs are recruited to the promoter regions with the histone methyltransferase G9a [157] leading to their epigenetic effect on host gene expression.

SUMMARY

Modulation of host transcription by microbial pathogens is now a well-accepted concept. Microorganisms induce epigenetic regulation of gene expression in their hosts, by means of DNA methylation, histone modifications and post-translational interference, to sustain their own life cycle and existence. Epigenetic mechanisms are an integral part of how microorganisms in our gut multiply, compete with other strains, evade the immune system and ultimately contribute to the health or disease state of the human host.

There is, however, a need to further determine the molecular mechanisms involved in epigenetic modifications induced by bacteria, viruses, and eukaryotic parasites. In recent times, exciting new knowledge and gut microbiome manipulation began to yield promising results in clinical settings.

Epigenetic changes in the context of symbiosis and dysbiosis will have to be further investigated for specific diseases and conditions to determine how these phenomena develop and contribute to health and disease processes, and to capitalize on their emerging potential new therapeutic implications evolving from this research field.

Abbreviations

BCAFF	B-cell activating factor
CDTs	Cytolethal distending toxins
Cif	Cycle inhibiting factor
CIN	Chromosmal instability
CRC	Colorectal cancer
EBV	Epstein–Barr virus
FFARs	Free fatty acid receptors
FMT	Fecal microbiota transplantation
HDACs	Histone deacetylases
HPV	Human papilloma viruses
HIV	Human immunodeficiency virus
IBD	Inflammatory bowel disease
IL	Interleukine
LMP1	Latent membrane protein 1
MAMP	Microbial-associated molecular pattern
NLR	Nod-like receptor
NODs	Nucleotide-binding oligomerization domain proteins
PRR	Pattern recognition receptor
ROS	Reactive oxygen species
SCFA	Short-chain fatty acid
TLR	Toll-like receptor
TGF-β	Transforming growth factor-β

References

[1] Sender R, Fuchs S, Milo R. Revised estimates for the number of human and bacteria cells in the body. PLoS Biol 2016;14(8):e1002533.

[2] Clemente JC, et al. The impact of the gut microbiota on human health: an integrative view. Cell 2012;148(6):1258–70.

[3] Zheng D, Liwinski T, Elinav E. Interaction between microbiota and immunity in health and disease. Cell Res 2020;30(6):492–506.

[4] Hill JH, Round JL. SnapShot: microbiota effects on host physiology. Cell 2021;184(10) 2796-2796.e1.

[5] van de Guchte M, Blottière HM, Doré J. Humans as holobionts: implications for prevention and therapy. Microbiome 2018;6(1):81.

[6] Miro-Blanch J, Yanes O. Epigenetic regulation at the interplay between gut microbiota and host metabolism. Front Genet 2019;10:638.

[7] Neish AS. Microbes in gastrointestinal health and disease. Gastroenterology 2009;136(1):65–80.

[8] Backhed F, et al. Host-bacterial mutualism in the human intestine. Science 2005;307(5717):1915–20.

[9] Pasolli E, et al. Extensive unexplored human microbiome diversity revealed by over 150,000 genomes from metagenomes spanning age, geography, and lifestyle. Cell 2019;176(3):649–62 e20.

[10] Nayfach S, et al. A genomic catalog of Earth's microbiomes. Nat Biotechnol 2021;39(4):499–509.

[11] Holmes E, et al. Gut microbiota composition and activity in relation to host metabolic phenotype and disease risk. Cell Metab 2012;16(5):559–64.

[12] Yuan C, et al. Association between cesarean birth and risk of obesity in offspring in childhood, adolescence, and early adulthood. JAMA Pediatrics 2016;170(11). p. e162385-e162385.

[13] Koplin JJ, et al. Environmental and demographic risk factors for egg allergy in a population-based study of infants. Allergy 2012;67(11):1415–22.

[14] Klement E, et al. Breastfeeding and risk of inflammatory bowel disease: a systematic review with meta-analysis. Am J Clin Nutr 2004;80(5):1342–52.

[15] Stewart CJ, et al. Temporal development of the gut microbiome in early childhood from the TEDDY study. Nature 2018;562(7728):583–8.

[16] Virtanen SM, et al. Microbial exposure in infancy and subsequent appearance of type 1 diabetes mellitus—associated autoantibodies: a cohort study. JAMA Pediatrics 2014;168(8):755–63.

[17] Aagaard K, Stewart CJ, Chu D. Una destinatio, viae diversae. EMBO Rep 2016;17(12):1679–84.

[18] Abrahamsson TR, Wu RY, Jenmalm MC. Gut microbiota and allergy: the importance of the pregnancy period. Pediatr Res 2015;77(1–2):214–19.

[19] Neu J. The microbiome during pregnancy and early postnatal life. Semin Fetal Neonatal Med 2016;21(6):373–9.

[20] Milani C, et al. The first microbial colonizers of the human gut: composition, activities, and health implications of the infant gut microbiota. Microbiology Mol Biol Rev 2017;81(4). p. e00036-17.

[21] Neu J, Rushing J. Cesarean vs vaginal delivery: long term infant outcomes and the hygiene hypothesis. Clperinatology 2011;38(2):321–31.

[22] Paul B, et al. Influences of diet and the gut microbiome on epigenetic modulation in cancer and other diseases. Clin Epigenetics 2015;7:112.

[23] Nagpal R, et al. Sensitive quantitative analysis of the meconium bacterial microbiota in healthy term infants born vaginally or by cesarean section. Front Microbiol 2016;7:1997.

[24] Hansen R, et al. First-pass meconium samples from healthy term vaginally-delivered neonates: an analysis of the microbiota. PLoS One 2015;10(7):e0133320.

[25] Del Chierico F, et al. Phylogenetic and metabolic tracking of gut microbiota during perinatal development. PLoS One 2015;10(9):e0137347.

[26] Collado MC, et al. Human gut colonisation may be initiated in utero by distinct microbial communities in the placenta and amniotic fluid. Sci Rep 2016;6(1):23129.

[27] Kennedy KM, et al. Fetal meconium does not have a detectable microbiota before birth. Nat Microbiology 2021;6(7):865–73.

[28] Aunan JR, et al. Molecular and biological hallmarks of ageing. Br J Surg 2016;103(2):e29–46.

[29] O'Toole PW, Jeffery IB. Gut microbiota and aging. Science 2015;350(6265):1214–15.

[30] Amamoto R, et al. Yearly changes in the composition of gut microbiota in the elderly, and the effect of lactobacilli intake on these changes. Sci Rep 2021;11(1):12765.

[31] Zhang X, et al. Advancing functional and translational microbiome research using meta-omics approaches. Microbiome 2019;7(1):154.

[32] Fricker AM, Podlesny D, Fricke WF. What is new and relevant for sequencing-based microbiome research? A mini-review. J Adv Res 2019;19:105–12.

[33] Markandey M, et al. Gut microbiota: sculptors of the intestinal stem cell niche in health and inflammatory bowel disease. Gut Microbes 2021;13(1):1990827.

[34] Zhou Y, Hu G, Wang MC. Host and microbiota metabolic signals in aging and longevity. Nat Chem Biol 2021;17(10):1027–36.

[35] Tamburini S, et al. The microbiome in early life: implications for health outcomes. Nat Med 2016;22(7):713–22.

[36] Derrien M, van Hylckama JE. Vlieg, fate, activity, and impact of ingested bacteria within the human gut microbiota. Trends Microbiol 2015;23(6):354–66.

[37] O'Hara AM, Shanahan F. The gut flora as a forgotten organ. EMBO Rep 2006;7(7):688–93.

[38] Wu GD, et al. Linking long-term dietary patterns with gut microbial enterotypes. Science 2011;334(6052):105–8.

[39] Yatsunenko T, et al. Human gut microbiome viewed across age and geography. Nature 2012;486(7402):222–7.

[40] Dong TS, Gupta A. Influence of early life, diet, and the environment on the microbiome. Clin Gastroenterol Hepatol 2019;17(2):231–42.

[41] Green GL, et al. Molecular characterization of the bacteria adherent to human colorectal mucosa. J Appl Microbiol 2006;100(3):460–9.

[42] Lozupone CA, et al. Diversity, stability and resilience of the human gut microbiota. Nature 2012;489(7415):220–30.

[43] Kuczynski J, et al. Direct sequencing of the human microbiome readily reveals community differences. Genome Biol 2010;11(5):210.

[44] Qin J, et al. A human gut microbial gene catalogue established by metagenomic sequencing. Nature 2010;464(7285):59–65.

[45] Andersen LO, Vedel Nielsen H, Stensvold CR, et al. Waiting for the human intestinal Eukaryotome. ISME J 2013;7(7):1253–5.

[46] Stensvold CR, van der Giezen M. Associations between gut microbiota and common luminal intestinal parasites. Trends Parasitol 2018;34(5):369–77.

[47] Byndloss MX, et al. Microbiota-activated PPAR-γ signaling inhibits dysbiotic Enterobacteriaceae expansion. Science 2017;357(6351):570–5.

[48] Obata Y, Furusawa Y, Hase K. Epigenetic modifications of the immune system in health and disease. Immunol Cell Biol 2015;93(3):226–32.

[49] Meijer K, de Vos P, Priebe MG. Butyrate and other short-chain fatty acids as modulators of immunity: what relevance for health? Curr Opin Clin Nutr Metab Care 2010;13(6):715–21.

[50] Clarke G, et al. Minireview: gut microbiota: the neglected endocrine organ. Mol Endocrinol 2014;28(8):1221–38.

[51] Galland L. The gut microbiome and the brain. J Med Food 2014;17(12):1261–72.

[52] Shenderov BA. Gut indigenous microbiota and epigenetics. Microb Ecol Health Dis 2012;23. Available from: https://doi.org/10.3402/mehd.v23i0.17195.

[53] Berni Canani R, Di Costanzo M, Leone L. The epigenetic effects of butyrate: potential therapeutic implications for clinical practice. Clin Epigenetics 2012;4(1). p. 4-4.

[54] Gilbert JA, et al. Current understanding of the human microbiome. Nat Med 2018;24(4):392–400.

[55] Guarner F, Malagelada JR. Gut flora in health and disease. Lancet 2003;361(9356):512–19.

[56] Hecht G. In the beginning was Helicobacter pylori: roles for microbes in other intestinal disorders. Gastroenterology 2007;132(2):481–3.

[57] Tan CY, Ramirez ZE, Surana NK. A modern-world view of host–microbiota–pathogen interactions. J Immunol 2021;207(7):1710.

[58] Hamon MA, Cossart P. Histone modifications and chromatin remodeling during bacterial infections. Cell Host Microbe 2008;4(2):100–9.

[59] Sharma M, et al. The epigenetic connection between the gut microbiome in obesity and diabetes. Front Genet 2019;10:1329.

[60] Gerard P. Gut microbiota and obesity. Cell Mol Life Sci 2016;73(1):147–62.

[61] Pigeyre M, et al. Recent progress in genetics, epigenetics and metagenomics unveils the pathophysiology of human obesity. Clin Sci (Lond) 2016;130(12):943–86.

[62] Nieuwdorp M, et al. Role of the microbiome in energy regulation and metabolism. Gastroenterology 2014;146(6):1525–33.

[63] Canfora EE, Jocken JW, Blaak EE. Short-chain fatty acids in control of body weight and insulin sensitivity. Nat Rev Endocrinol 2015;11(10):577–91.

[64] van der Klaauw AA, Farooqi IS. The hunger genes: pathways to obesity. Cell 2015;161(1):119–32.

[65] Turnbaugh PJ, et al. An obesity-associated gut microbiome with increased capacity for energy harvest. Nature 2006;444(7122):1027–31.

[66] Ley RE, et al. Microbial ecology: human gut microbes associated with obesity. Nature 2006;444(7122):1022–3.

[67] Faith JJ, et al. The long-term stability of the human gut microbiota. Science 2013;341(6141):1237439.

[68] Irrazabal T, et al. The multifaceted role of the intestinal microbiota in colon cancer. Mol Cell 2014;54(2):309–20.

[69] Kumar H, et al. Gut microbiota as an epigenetic regulator: pilot study based on whole-genome methylation analysis. mBio 2014;5(6):e02113–14.

[70] Bhat MI, Kapila R. Dietary metabolites derived from gut microbiota: critical modulators of epigenetic changes in mammals. Nutr Rev 2017;75(5):374–89.

[71] Remely M, et al. Effects of short chain fatty acid producing bacteria on epigenetic regulation of FFAR3 in type 2 diabetes and obesity. Gene 2014;537(1):85–92.

[72] Chang PV, et al. The microbial metabolite butyrate regulates intestinal macrophage function via histone deacetylase inhibition. Proc Natl Acad Sci 2014;111(6):2247.

[73] Ramos-Molina B, et al. Gut microbiota composition is associated with the global DNA methylation pattern in obesity. Front Genet 2019;10:613.

[74] Gao D, Herman JG, Guo M. The clinical value of aberrant epigenetic changes of DNA damage repair genes in human cancer. Oncotarget 2016;.

[75] Collins D, Hogan AM, Winter DC. Microbial and viral pathogens in colorectal cancer. Lancet Oncol 2011;12(5):504–12.

REFERENCES

[76] Bultman SJ. Interplay between diet, gut microbiota, epigenetic events, and colorectal cancer. Mol Nutr Food Res 2017;61(1).

[77] Schulthess J, et al. The short chain fatty acid butyrate imprints an antimicrobial program in macrophages. Immunity 2019;50 (2):432−45 e7.

[78] Hullar MA, Burnett-Hartman AN, Lampe JW. Gut microbes, diet, and cancer. Cancer Treat Res 2014;159:377−99.

[79] Vaupel P, Multhoff G. Revisiting the Warburg effect: historical dogma vs current understanding. J Physiol 2021;599(6):1745−57.

[80] Sebastián C, Mostoslavsky R. Untangling the fiber yarn: butyrate feeds Warburg to suppress colorectal cancer. Cancer Discov 2014;4(12):1368−70.

[81] Donohoe DR, et al. A gnotobiotic mouse model demonstrates that dietary fiber protects against colorectal tumorigenesis in a microbiota- and butyrate-dependent manner. Cancer Discov 2014;4(12):1387−97.

[82] Bishop KS, Xu H, Marlow G. Epigenetic regulation of gene expression induced by butyrate in colorectal cancer: involvement of microRNA. Genet & epigenetics 2017;9 1179237X17729900-1179237X17729900.

[83] van Elsland D, Neefjes J. Bacterial infections and cancer. EMBO Rep 2018;19(11):e46632.

[84] Lu R, et al. Enteric bacterial protein AvrA promotes colonic tumorigenesis and activates colonic beta-catenin signaling pathway. Oncogenesis 2014;3(6):e105.

[85] Abreu MT, Peek Jr. RM. Gastrointestinal malignancy and the microbiome. Gastroenterology 2014;146(6):1534−46 e3.

[86] Wang X, Huycke MM. Extracellular superoxide production by Enterococcus faecalis promotes chromosomal instability in mammalian cells. Gastroenterology 2007;132(2):551−61.

[87] Smith PM, et al. The microbial metabolites, short-chain fatty acids, regulate colonic Treg cell homeostasis. Science 2013;341 (6145):569−73.

[88] Arpaia N, et al. Metabolites produced by commensal bacteria promote peripheral regulatory T-cell generation. Nature 2013;504(7480):451−5.

[89] Dubinsky M, Braun J. Diagnostic and prognostic microbial biomarkers in inflammatory bowel diseases. Gastroenterology 2015;149(5):1265−74 e3.

[90] Tilg H, Moschen AR. Food, immunity, and the microbiome. Gastroenterology 2015;148(6):1107−19.

[91] Marlicz W, et al. Emerging concepts in non-invasive monitoring of Crohn's disease. Ther Adv Gastroenterol 2018;11. p. 1756284818769076.

[92] Pascal V, et al. A microbial signature for Crohn's disease. Gut 2017;66(5):813−22.

[93] Hecht GA. Inflammatory bowel disease−live transmission. N Engl J Med 2008;358(5):528−30.

[94] Ben Ghezala I, et al. Small molecule drugs in inflammatory bowel diseases. Pharmaceuticals (Basel) 2021;14(7).

[95] Sanchis-Artero L, et al. Evaluation of changes in intestinal microbiota in Crohn's disease patients after anti-TNF alpha treatment. Sci Rep 2021;11(1):10016.

[96] de Martel C, et al. Global burden of cancer attributable to infections in 2018: a worldwide incidence analysis. Lancet Glob Health 2020;8(2):e180−90.

[97] Erdman SE, Poutahidis T. Gut bacteria and cancer. Biochim Biophys Acta 2015;1856(1):86−90.

[98] Kwa M, et al. The intestinal microbiome and estrogen receptor-positive female breast cancer. J Natl Cancer Inst 2016;108(8):djw029.

[99] Xie G, et al. Distinctly altered gut microbiota in the progression of liver disease. Oncotarget 2016;7(15):19355−66.

[100] Amieva M, Peek Jr. RM. Pathobiology of helicobacter pylori-induced gastric cancer. Gastroenterology 2016;150(1):64−78.

[101] Terres AM, et al. H pylori infection is associated with downregulation of E-cadherin, a molecule involved in epithelial cell adhesion and proliferation control. J Clin Pathol 1998;51(5):410−12.

[102] Chan AO, et al. Promoter methylation of E-cadherin gene in gastric mucosa associated with Helicobacter pylori infection and in gastric cancer. Gut 2003;52(4):502−6.

[103] Xia G, et al. Helicobacter pylori regulates p21(WAF1) by histone H4 acetylation. Biochem Biophys Res Commun 2008;369 (2):526−31.

[104] Lee Y-C, et al. Association between helicobacter pylori eradication and gastric cancer incidence: a systematic review and meta-analysis. Gastroenterology 2016;150(5):1113−24 e5.

[105] Perri F, et al. Aberrant DNA methylation in non-neoplastic gastric mucosa of H. pylori infected patients and effect of eradication. Am J Gastroenterol 2007;102(7):1361−71.

[106] Sung H, et al. Global cancer statistics 2020: GLOBOCAN estimates of incidence and mortality worldwide for 36 cancers in 185 countries. Cancer J Clin 2021;71(3):209−49.

[107] Kadhum HJ, et al. Characteristics of cytotoxic necrotizing factor and cytolethal distending toxin producing Escherichia coli strains isolated from meat samples in Northern Ireland. Food Microbiol 2006;23(5):491−7.

[108] Zheng J, et al. Adherence to and invasion of human intestinal epithelial cells by Campylobacter jejuni and Campylobacter coli isolates from retail meat products. J Food Prot 2006;69 (4):768−74.

[109] Kau AL, et al. Human nutrition, the gut microbiome and the immune system. Nature 2011;474(7351):327−36.

[110] Dolara P, et al. Red wine polyphenols influence carcinogenesis, intestinal microflora, oxidative damage and gene expression profiles of colonic mucosa in F344 rats. Mutat Res 2005;591(1−2):237−46.

[111] Fernandez AF, et al. The dynamic DNA methylomes of double-stranded DNA viruses associated with human cancer. Genome Res 2009;19(3):438−51.

[112] Minarovits J. Microbe-induced epigenetic alterations in host cells: the coming era of patho-epigenetics of microbial infections. A review. Acta Microbiol Immunol Hung 2009;56(1):1−19.

[113] Arbibe L. Immune subversion by chromatin manipulation: a 'new face' of host-bacterial pathogen interaction. Cell Microbiol 2008;10(8):1582−90.

[114] Brodsky IE, Medzhitov R. Targeting of immune signalling networks by bacterial pathogens. Nat Cell Biol 2009;11(5):521−6.

[115] Smith JL, Bayles DO. The contribution of cytolethal distending toxin to bacterial pathogenesis. Crit Rev Microbiol 2006;32(4):227−48.

[116] Nougayrede JP, et al. Escherichia coli induces DNA double-strand breaks in eukaryotic cells. Science 2006;313(5788):848−51.

[117] Pons BJ, et al. Cell transfection of purified cytolethal distending toxin B subunits allows comparing their nuclease activity while plasmid degradation assay does not. PLoS One 2019;14(3): e0214313.

[118] Nesic D, Hsu Y, Stebbins CE. Assembly and function of a bacterial genotoxin. Nature 2004;429(6990):429−33.

[119] Hanford HE, Von Dwingelo J, Abu Kwaik Y. Bacterial nucleomodulins: a coevolutionary adaptation to the eukaryotic command center. PLOS Pathog 2021;17(1):e1009184.

[120] Khan AA, Khan Z. Bacterial nucleomodulins and cancer: an unresolved enigma. Transl Oncol 2021;14(1):100922.

[121] Rolando M, Gomez-Valero L, Buchrieser C. Bacterial remodelling of the host epigenome: functional role and evolution of effectors methylating host histones. Cell Microbiol 2015;17(8):1098−107.

[122] Kamada N, Nunez G. Regulation of the immune system by the resident intestinal bacteria. Gastroenterology 2014;146(6):1477−88.

[123] Hamon MA, et al. Histone modifications induced by a family of bacterial toxins. Proc Natl Acad Sci U S A 2007;104 (33):13467−72.

[124] Niller HH, Wolf H, Minarovits J. Epigenetic dysregulation of the host cell genome in Epstein-Barr virus-associated neoplasia. Semin Cancer Biol 2009;19(3):158–64.
[125] Niller HH, et al. EBV-associated neoplasms: alternative pathogenetic pathways. Med Hypotheses 2004;62(3):387–91.
[126] Niller HH, Wolf H, Minarovits J. Regulation and dysregulation of Epstein-Barr virus latency: implications for the development of autoimmune diseases. Autoimmunity 2008;41(4):298–328.
[127] Tempera I, Lieberman PM. Epigenetic regulation of EBV persistence and oncogenesis. SemCancer Biol 2014;26:22–9.
[128] Birdwell CE, et al. Genome-wide DNA methylation as an epigenetic consequence of Epstein-Barr Virus infection of immortalized keratinocytes. J Virol 2014;88(19):11442–58.
[129] Peng H, et al. Higher methylation intensity induced by EBV LMP1 via NF-κB/DNMT3b signaling contributes to silencing of PTEN gene. Oncotarget 2016;7(26):40025–37.
[130] Tsai CL, et al. Activation of DNA methyltransferase 1 by EBV LMP1 Involves c-Jun NH(2)-terminal kinase signaling. Cancer Res 2006;66(24):11668–76.
[131] Wang Y-Q, Zhao X-Y. Human cytomegalovirus primary infection and reactivation: insights from virion-carried molecules. Front Microbiol 2020;11:1511.
[132] Reeves MB. Chromatin-mediated regulation of cytomegalovirus gene expression. Virus Res 2011;157(2):134–43.
[133] Rahier JF, et al. Second European evidence-based consensus on the prevention, diagnosis and management of opportunistic infections in inflammatory bowel disease. J Crohn's Colitis 2014;8(6):443–68.
[134] Inoue-Toyoda M, et al. Glucocorticoids facilitate the transcription from the human cytomegalovirus major immediate early promoter in glucocorticoid receptor- and nuclear factor-I-like protein-dependent manner. Biochem Biophys Res Commun 2015;458(1):180–5.
[135] Van Damme E, et al. Glucocorticosteroids trigger reactivation of human cytomegalovirus from latently infected myeloid cells and increase the risk for HCMV infection in D+ R+ liver transplant patients. J Gen Virol 2015;96(Pt 1):131.
[136] Bellon M, Nicot C. Regulation of telomerase and telomeres: human tumor viruses take control. J Natl Cancer Inst 2008;100(2):98–108.
[137] Bellon M, Nicot C. Telomerase: a crucial player in HTLV-I-induced human T-cell leukemia. Cancer Genomics Proteom 2007;4(1):21–5.
[138] Lieleg O, et al. Mucin biopolymers as broad-spectrum antiviral agents. Biomacromolecules 2012;13(6):1724–32.
[139] Schroeder BO. Fight them or feed them: how the intestinal mucus layer manages the gut microbiota. Gastroenterol Rep 2019;7(1):3–12.
[140] Earle KA, et al. Quantitative imaging of gut microbiota spatial organization. Cell Host Microbe 2015;18(4):478–88.
[141] Desai MS, et al. A dietary fiber-deprived gut microbiota degrades the colonic mucus barrier and enhances pathogen susceptibility. Cell 2016;167(5):1339–53 e21.
[142] Bergström A, et al. Nature of bacterial colonization influences transcription of mucin genes in mice during the first week of life. BMC Res Notes 2012;5:402.
[143] Schütte A, et al. Microbial-induced meprin β cleavage in MUC2 mucin and a functional CFTR channel are required to release anchored small intestinal mucus. Proc Natl Acad Sci 2014;111(34):12396–401.
[144] Gonzalez-Ochoa G, et al. Modulation of rotavirus severe gastroenteritis by the combination of probiotics and prebiotics. Arch Microbiol 2017;199(7):953–61.
[145] Olaya Galán NN, et al. In vitro antiviral activity of Lactobacillus casei and Bifidobacterium adolescentis against rotavirus infection monitored by NSP4 protein production. J Appl Microbiol 2016;120(4):1041–51.
[146] Speich B, et al. Effect of sanitation and water treatment on intestinal protozoa infection: a systematic review and meta-analysis. Lancet Infect Dis 2016;16(1):87–99.
[147] Nash AK, et al. The gut mycobiome of the human microbiome project healthy cohort. Microbiome 2017;5(1):153.
[148] Lhotska Z, et al. A study on the prevalence and subtype diversity of the intestinal protist Blastocystis sp. in a gut-healthy human population in the Czech Republic. Front Cell Infect Microbiol 2020;10:544335.
[149] Stensvold CR. Pinning down the role of common luminal intestinal parasitic protists in human health and disease – status and challenges. Parasitology 2019;146(6):695–701.
[150] Müller M, et al. Biochemistry and evolution of anaerobic energy metabolism in eukaryotes. Microbiol Mol Biol Rev 2012;76:444–95.
[151] Maizels RM, Smits HH, McSorley HJ. Modulation of host immunity by helminths: the expanding repertoire of parasite effector molecules. Immunity 2018;49(5):801–18.
[152] Grainger JR, et al. Helminth secretions induce de novo T cell Foxp3 expression and regulatory function through the TGF-beta pathway. J Exp Med 2010;207(11):2331–41.
[153] Cook L, et al. Induction of stable human FOXP3(+) Tregs by a parasite-derived TGF-beta mimic. Immunol Cell Biol 2021;99 (8):833–47.
[154] Borad A, Ward H. Human immune responses in cryptosporidiosis. Future Microbiol 2010;5(3):507–19.
[155] Crawford CK, Kol A. The mucosal innate immune response to cryptosporidium parvum, a global one health issue. Front Cell Infect Microbiol 2021;11:689401.
[156] Villares M, Berthelet J, Weitzman JB. The clever strategies used by intracellular parasites to hijack host gene expression. Semin Immunopathol 2020;42:215–26.
[157] Wang Y, et al. Delivery of parasite Cdg7_Flc_0990 RNA transcript into intestinal epithelial cells during Cryptosporidium parvum infection suppresses host cell gene transcription through epigenetic mechanisms. Cell Microbiol 2017;19(11). Available from: https://doi.org/10.1111/cmi.12760.
[158] Chan BCL, et al. IL33: roles in allergic inflammation and therapeutic perspectives. Front Immunol 2019;10:364.
[159] Barski A, et al. High-resolution profiling of histone methylations in the human genome. Cell 2007;129(4):823–37.

CHAPTER 34

Population Pharmacoepigenomics

Jacob Peedicayil

Department of Pharmacology & Clinical Pharmacology, Christian Medical College, Vellore, Tamil Nadu, India

OUTLINE

Introduction	687
General Aspects of Population Pharmacoepigenomics	688
Population Variations of Epigenetic Patterns and Population Pharmacoepigenomics	688
Human Epigenome Projects and Population Pharmacoepigenomics	688
Population Pharmacoepigenomics in Relation to Pharmacokinetics	689
Population Pharmacoepigenomics in Relation to Pharmacodynamics	689
3D and 4D Chromatin Structures in the Nuclei of Cells in the Liver and Other Body Organs and Population Pharmacoepigenomics	690
Population Pharmacoepigenomics in Relation to Adverse Drug Reactions and Drug Interactions	691
Conclusions	691
Population Pharmacoepigenomics Case Study	691
Objective	691
Scope	692
Audience	692
Rationale	692
Expected Results and Deliverables	692
Safety Considerations	692
Process, Workflow and Actions Taken	692
Tools and Materials Used	692
Results	692
Learning and Knowledge Outcomes	692
Abbreviations	693
References	693

INTRODUCTION

Pharmacoepigenetics refers to the study of the epigenetic basis of variation in response to drugs [1,2]. Pharmacoepigenomics refers to the application of pharmacoepigenetics on a genome-wide scale [1,3]. Together, these sub-disciplines of pharmacology involve studying the role of epigenetics in the variation in response to drugs within and between individuals, in the effects of drugs on gene expression profiles, in the mechanisms of action of drugs and adverse drug reactions (ADRs), and in the discovery of new drug targets [1,3–5]. Population epigenetics and epigenomics, which address issues concerning the prevalence and importance of epigenetic variation within and between different populations, is an active area of research [6–10]. Population pharmacoepigenomics, the study of how the epigenetic basis of variation in response to drugs varies between individuals and within and between populations, was discussed in the second edition of this book [11]. Since then, considerable developments have taken place in this field. The current chapter discusses the topic of population pharmacoepigenomics from the latest perspectives.

GENERAL ASPECTS OF POPULATION PHARMACOEPIGENOMICS

Population epigenetics and epigenomics are active areas of current research at the interface of molecular genetics, genomics, and population biology, and address questions concerning the prevalence and importance of epigenetic variation within and across populations [6]. Genetic variations in populations are known to influence epigenetic variations [12]. For example, DNA variants can influence epigenetic mechanisms through the sequence-specific activity of transcription factors [12]. Recent studies in many epigenomes from yeast to humans have begun to dissect the cis- and trans-regulatory genetic mechanisms that shape patterns of population epigenomic variation at the level of integrated chromatin state maps. In this context, factors beyond cis- and trans-acting genetic control are also important. Various aspects of the environment, including exposure to mutagens and other compounds; diet; pathogens; physical conditions, such as abnormal temperatures; and psychosocial factors such as maternal nurturing can also induce epigenetic variation [6]. Epigenetic patterns also vary in the population according to sex and age of individuals [13]. Variations in epigenetic patterns across populations can have biological significance. In mammals, the vast majority of epigenetic information appears to be reset during gametogenesis or during early development.

POPULATION VARIATIONS OF EPIGENETIC PATTERNS AND POPULATION PHARMACOEPIGENOMICS

The field of epigenomics has rapidly progressed from the study of individual reference epigenomes to surveying epigenomic variations in populations. Studies of epigenomic variation within populations suggest high levels of phenotypically relevant variation, with the patterns of epigenetic regulation varying between individuals, genome regions, and with relation to the environment. Of the three major mechanisms of epigenetic regulation, DNA methylation, histone modifications, and non-coding RNA—mediated gene regulation, the best whole-genome information presently available pertains to DNA methylation patterns. A study by Heyn et al. [14] comparing genome-wide DNA methylation in three different ethnic populations (Caucasian American, African American, and Han Chinese American), generated 493 differentially methylated CpG regions (DMRs). The authors found that these DMRs influence phenotypic differences between the three ethnic groups, such as appearance, susceptibility to certain diseases and pathogens, drug metabolism, sensory perception, and susceptibility to disease. Many of these DMRs were associated with underlying genetic changes, suggesting a direct relationship between the genetic and epigenetic codes. However, one-third of the DMRs were not associated with any genetic variation, suggesting that the variation might be independent of the DNA sequence. In another study, Kasowski et al. [15] studied five histone modifications in 19 people of diverse ethnic origins (5 European, 7 Yoruban, 2 Asian, 1 non-European Caucasian, and 4 San,). The Yoruban people are a native West African ethnic group, and the San people are speakers of three distinct languages in Southern Africa. The authors found extensive variation in these modifications in regulatory regions of the genome. This variation correlated with genetic variation and population divergence, and was associated with disruption of binding of transcription factor—binding motifs. A lot of population variation has also been found to occur in non-coding RNAs, such as microRNAs (miRNAs). In this context, Carbonell et al. [16] analyzed the data on 14 populations from the 1000 Genomes Project along with data obtained from 60 exomes of healthy individuals from a population from Southern Spain, sequenced in the context of the Medical Genome Project.

Analysis of the miRNA sequences of the 1152 individuals found a total of 527 variants in miRNAs, affecting not only preprocessed miRNAs, but also the mature miRNAs. Of these 527 variants, 45 variants affected the recognition region of the corresponding miRNAs and were found in 43 different miRNAs, 26 of which are known to contribute to the pathogenesis of 57 diseases. Hence, this study showed that 44% of human miRNAs contain variants. The average number of variants per individual in miRNAs was 28. Thus, this study found a level of variability in miRNA genes comparable to that observed in coding regions of the genome.

HUMAN EPIGENOME PROJECTS AND POPULATION PHARMACOEPIGENOMICS

Soon after the completion of the Human Genome Project, it was realized that an epigenome, a map of the genome-wide changes to the DNA, and its protein scaffold would be required in addition to the DNA sequence for a proper understanding of human biology and disease. With recent technological advancements, such as advances in DNA sequencing technology and high-resolution measurements of various epigenetic marks at a genome-wide scale, the field of epigenomics has developed rapidly [9]. Large-scale projects have been initiated to collect reference epigenomes for different tissues, developmental stages, disease states, and environmental influences.

Such projects include the Encyclopedia of DNA Elements (ENCODE) Project, the NIH Roadmap Epigenome Project, the International Human Epigenome Project, and the Genotype Tissue Expression (GTEx) Project [17–23]. The data from these projects include DNA methylation, histone modifications, chromatin accessibility, and RNA transcripts in stem cells and primary ex vivo tissues selected to represent the normal counterparts of tissues and organ systems frequently involved in human disease [18]. The data from these projects is relevant to pharmacology, as it could help in the development of new drugs, such as anticancer drugs [24]. These data are being made freely available online. Online programs, portals, tools, and data servers, such as the Encode Portal [22], EpiMINE [25], ePIANNO [26], DeepBlue [27], and the University of California Santa Cruz (UCSC) Genome Browser Database [28] are available to help researchers access and utilize the available data. The data from these projects are proving to be very valuable in annotation of the transcription units, cis-regulatory sequences, and other genomic features in the human genome. Integrative analysis of epigenomic maps is facilitating the study of the gene regulatory programs involved in the pathogenesis of disease [20]. For example, the vast majority of trait- and disease-associated single-nucleotide polymorphisms (SNPs) identified in genome-wide association studies (GWAS) lie outside of any protein-coding sequence. By correlating these SNPs with accessible chromatin regions identified in epigenome projects, it is possible to correlate the SNPs with regulatory regions of the genome [29].

POPULATION PHARMACOEPIGENOMICS IN RELATION TO PHARMACOKINETICS

Pharmacokinetics is the branch of pharmacology dealing with the absorption, distribution, metabolism, and excretion (or elimination) (ADME) of drugs. ADME of drugs is known to vary from individual to individual and within and between populations. Population pharmacokinetics refers to the study of the variability in plasma drug concentrations between individuals when standard dosage regimens of drugs are administered [30–32]. The variability involved in population pharmacokinetics is accounted for in terms of patient variables, such as age, sex, body weight, and disease status [30].

Many proteins, such as hepatic enzymes as well as membrane transporters are known to be involved in the ADME of drugs [33]. About 300 genes are known to encode proteins involved in the ADME of drugs [34]. Research on pharmacoepigenetics is demonstrating that epigenetic mechanisms regulate such genes [34–37]. Many studies have shown that aberrant DNA methylation or histone acetylation in the promoter regions of these genes alters their expression [35]. Moreover, there is growing evidence that miRNAs influence cellular ADME processes via post-translational silencing of ADME genes [35]. Several variables in normal individuals, as well as patients, can influence pharmacokinetics. These include age [38], sex [39], body weight [40,41], racial and ethnic background [42,43], diet and nutrition [44], and disease status [33]. It is likely that epigenetics plays a major role in the ways by which all these variables modify the ADME of drugs in the body [45–53]. However, at present, little is known about how the epigenetic regulation of genes underlying ADME varies between individuals and within and between populations, and how interindividual differences in the epigenetic regulation of ADME genes contribute to interindividual differences in the pharmacokinetics of drugs [36,54]. Some examples of an epigenetic basis for patient variables affecting pharmacokinetics are age [37], gender [38], body weight [40,41], race and ethnicity [42,43], diet and nutrition [44], and presence of disease [32] (Table 34.1).

TABLE 34.1 Evidence for an Epigenetic Basis for Patient Variables Affecting Pharmacokinetics

Age	Age-related changes in DNA methylation and histone acetylation may influence rates of hepatic drug metabolism	[55]
Sex	Sex-specific tissue expression of genes can occur due to changes in transcriptional control due to changes in chromatin accessibility (as in conversion of euchromatin to heterochromatin)	[56]
Race and ethnicity	Ethnicity-associated CpG methylation of genes like those underlying ADME may have caused differences in ADME	[57]
Disease status	hsa-miR-148a-3p-mediated CYP2B6 down-regulation in alcoholic hepatitis	[58]

ADME, absorption, distribution, metabolism, excretion.

POPULATION PHARMACOEPIGENOMICS IN RELATION TO PHARMACODYNAMICS

Pharmacodynamics is the branch of pharmacology dealing with the mechanisms of action of drugs. Pharmacodynamics involves the study of the biochemical and physiological changes produced by drugs in the body during the prevention and treatment of disease. It is well known that the major way by which drugs act is via drug receptors [59]. Like pharmacokinetics, pharmacodynamics is known to vary between individuals and populations. Population pharmacodynamics refers to the study of the

TABLE 34.2 Evidence for an Epigenetic Basis for Patient Variables Affecting Pharmacodynamics

Variables	Finding	References
Age	Epigenetic drugs like those increasing DNA methylation, HDAC inhibitors, and drugs targeting miRNAs may help in the treatment of Alzheimer disease	[66]
Sex	Epigenetic mechanisms like DNA methylation and histone acetylation could play a role in the sexually dimorphic pharmacodynamic effects of drugs	[67]
Race and ethnicity	DNA methylation signatures associated with interferon-related genes can be influenced by ethnicity	[68]
Diet and nutrition	Diet and nutrition impact mood disorders by affecting DNA methylation	[69]
Disease status	DNA methylation pattern of the T-cell differentiation factor forkhead box P1 (FOXP1) gene predicts response to drug treatment in NSCLC	[70]

NSCLC, nonsmall cell lung cancer; SLE, systemic lupus erythematosus.

variability in pharmacological effects of drugs between individuals when standard dosage regimens of drugs are administered [60]. Like in the case of pharmacokinetics, many variables in patients can influence pharmacodynamics. These include age [37,61], sex [62], racial and ethnic factors [59,60], and disease status [58]. It is likely that epigenetics plays a major role in the ways by which all these patient variables modify the mechanisms of action of drugs in the body [2,47,63–65]. However, at present, like in the case of pharmacokinetics, very little is known about how the epigenetic regulation of genes underlying drug receptors varies between individuals and within and between populations, and how interindividual differences in the epigenetic regulation of genes encoding drug receptors contribute to interindividual differences in pharmacodynamics [2,49]. Some examples of an epigenetic basis for patient variables affecting pharmacodynamics are given in Table 34.2.

3D AND 4D CHROMATIN STRUCTURES IN THE NUCLEI OF CELLS IN THE LIVER AND OTHER BODY ORGANS AND POPULATION PHARMACOEPIGENOMICS

In eukaryotes, the genome adopts in the nucleus a 3D configuration of chromatin. This varies between cell types, developmental stages, environmental conditions, and disease states [71]. The 3D configuration of chromatin influences gene expression and cell functions [72,73].

The 3D chromatin structure also varies with time (the fourth dimension) resulting in the 4D structure of chromatin [73]. Chromosome conformation capture (3 C), first at the level of a single locus or a set of loci, and then across the genome, has made it possible to link chromatin structure with gene regulation, DNA replication timing, and somatic copy number alterations [68]. The 3D and 4D chromatin structures are influenced by a number of factors, including DNA sequence, epigenetic mechanisms such as DNA methylation, histone modifications, non-coding RNA-mediated regulation of gene expression, and nuclear localization [71]. The resulting topology impacts many cellular functions, such as gene transcription, RNA processing, and DNA repair and replication [71]. Recently, it was proposed that there should be an international consortium to coordinate the elucidation of the 3D genome that varies with cell types, developmental stages, environmental conditions, and disease [71].

In recent years, there have been new developments regarding the use of 3 C. For example, Hi-C is an extension of 3 C that is capable of identifying long range chromosomal interactions in an unbiased genome-wide fashion [72]. In this regard, Xu et al. [74] used Hi-C to study the liver of normal and nonalcoholic fatty liver disease (NAFLD) mice. The authors identified variations in thousands of genome-wide regions with respect to 3D chromatin organization and genomic rearrangements between normal and NAFLD mice. The authors revealed gene dysregulation frequently accompanied by these variations. Candidate target genes were identified in NAFLD affected by genetic rearrangements and spatial organization disruption. The authors inferred that their data offer insights into novel mechanisms of NAFLD pathogenesis and can provide a new conceptual framework for the treatment of NAFLD.

With regard to population pharmacoepigenomics, the 3D and 4D chromatin structures are of relevance because they vary in the population [75,76]. 3D and 4D chromatin structures are also relevant to pharmacology since they can also influence the actions of drugs [77]. In this context, they can influence pharmacodynamics, due to effects in different body tissues, depending on the main site of action of the concerned drug. For example, drugs can be developed to intercalate with DNA, alkylate or bind to the DNA groove, and either block transcription factor binding or change the 3D conformation of the consensus DNA strand [77]. The 3D genome of cancer cells is also a target for anticancer drugs [78]. "Mining" the topography and dynamics of the 4D genome could also help identify novel central nervous system (CNS) drug pathways [79]. Perturbations in 3D genome organization can also promote acquired drug resistance in cancer cells [80]. The 3D and 4D chromatin structures are also known to influence ADR and toxic effects of drugs. For example,

TABLE 34.3 Evidence for the Importance of 3D and 4D Chromatin Structures in Population Pharmacoepigenomics

Chromatin Structure	Importance	References
3D	DNA-intercalating, alkylating, or DNA groove-binding small molecules can target transcription factors by changing the 3D conformation of the consensus DNA strand	[75]
3D	3D genome of cancer cells can be targeted by anticancer drugs	[78]
4D	Mining the topography and dynamics of the 4D genome could help identify novel CNS drugs	[79]
3D	Changes in the 3D genome organization can promote cancer evolution and drug resistance in cancer cells	[80]
3D and 4D	DNA methylation topology is a potential indicator of causal impacts on chromatin distribution with a possible application in the study of toxic effects of drugs	[81]

ADR, adverse drug reactions; *CNS*, central nervous system.

it has been found that the 3D DNA methylation phenotypes of the genome correlate with cytotoxicity levels of the DNA hypomethylating drugs, zebularine and 5-azacytidine, in prostate and liver cancer cell models [81]. More details of the role of 3D and 4D chromatin on population pharmacoepigenomics are given in Table 34.3.

POPULATION PHARMACOEPIGENOMICS IN RELATION TO ADVERSE DRUG REACTIONS AND DRUG INTERACTIONS

Adverse drug reactions (ADRs), also referred to as adverse drug effects (ADEs), are common and important problems associated with the use of drugs in patients [82,83]. ADRs are thought to contribute to 100,000 deaths annually in the United States and to millions of episodes of discomfort or injury worldwide [84]. ADRs are usually due to the use of excessive doses of drugs or allergic responses to drugs. ADRs can also be the result of drug interactions that are thought to contribute to at least 4% of preventable ADRs [84]. Some ADRs due to drug interactions are due to pharmacodynamic reasons, such as in the case of two drugs administered together that have similar effects on a drug receptor, resulting in a synergistic effect. More often, drug interaction–induced ADRs are due to pharmacokinetic reasons, such as when one drug alters the metabolism of another drug, leading to toxic blood levels of the second drug [85]. In this context, drug interactions are of two major types: pharmacodynamic interactions where the interaction takes place at the drug receptor; and pharmacokinetic interactions where the drug interactions involve drug ADME [86]. Epigenetic mechanisms of gene expression are known to underlie ADRs [34,64,87]. As discussed earlier, as epigenetic mechanisms of gene expression can influence the pharmacodynamics and pharmacokinetics of drugs, epigenetic mechanisms of gene expression are likely to underlie drug interactions. However, at present we are in the early stages of understanding the roles of epigenetics in ADRs and drug interactions, and how these roles vary between individuals and within and between populations.

CONCLUSIONS

Epigenetics plays a major role in pharmacology, affecting pharmacokinetics (ADME), pharmacodynamics, ADRs, and drug interactions. Epigenetic mechanisms regulating ADME, pharmacodynamics, ADRs, and drug interactions vary between individuals and within and between populations. At present we are in the early stages of studying these epigenetic variations.

Much more research needs to be performed on how epigenetics influences ADME of drugs, the pharmacodynamic aspects of drugs, ADRs, and interactions of drugs, and how such influences vary between individuals and within and between populations. More research is also required on the relevance of population variation of epigenetic patterns, the human epigenome projects, and the 3D and 4D chromatin structures in the nuclei of cells in the liver and other body organs to population pharmacoepigenomics. The resulting knowledge could greatly improve our understanding of the mechanisms of action of drugs, ADME of drugs, ADRs, and drug interactions.

POPULATION PHARMACOEPIGENOMICS CASE STUDY

Given below is a case study that demonstrates an epigenetic basis for how the epigenetic profile in a population (in this instance, patients with advanced nonsmall-cell lung cancer), can be associated with response to treatment with drugs.

Objective

Antiprogrammed death-1 (PD-1) treatment for advanced nonsmall-cell lung cancer (NSCLC) has improved the survival of patients with this disorder. However, a high percentage of patients with advanced NSCLC do not respond to this treatment. This study investigated the use of DNA methylation profiles to determine the efficacy of anti-PD-1 treatment in patients with advanced NSCLC.

Scope

In a multicenter study the authors recruited adult patients from 15 hospitals in France, Spain, and Italy, who had histologically proven advanced NSCLC and had been subjected to PD-1 blockade during the course of the disease. The study population comprised a discovery cohort to determine the correlation between epigenetic features and clinical benefit from PD-1 blockade and two validation cohorts to assess the validity of the authors' assumptions.

Audience

This case study will be useful for medical oncologists, respiratory physicians and surgeons, clinical pharmacologists and epigeneticists.

Rationale

The rationale for the study was to use DNA methylation profiles to determine the efficacy of two drugs that block PD-1, an immune checkpoint receptor expressed in activated B and T cells.

Expected Results and Deliverables

From this study it can be learnt that DNA methylation patterns vary between individuals and between populations and that these variations can be associated with pharmacological properties of drugs.

Safety Considerations

Possible adverse drug reactions due to the drugs used for anti-PD-1 treatment should be looked out for.

Process, Workflow and Actions Taken

The authors first established an epigenomic profile on a microarray DNA methylation signature of 301 differentially methylated CpG sites (EPIMMUNE) in a discovery set of tumor samples from patients treated with nivolumab or pembrolizumab. The EPIMMUNE signature was validated in an independent set of patients. A derived DNA methylation marker was validated by a single-methylation assay in a validation cohort of patients. The main study outcomes were progression-free survival and overall survival.

Tools and Materials Used

The authors used the Kaplan-Meier method to estimate progression-free and overall survival, and estimated the differences between the groups using the log-rank test. The authors constructed a multivariate Cox model to identify the variables independently associated with progression-free and overall survival.

Results

The authors obtained samples from 142 patients between 23 June 2014 and 18 May 2017. There were 34 in the discovery cohort, 47 in the EPIMMUNE validation cohort, and 61 in the derived methylation marker cohort (the T-cell differentiation factor forkhead box P1 [FOXP1]). The FOXP1 was the best single predictive DNA methylation marker from the EPIMMUNE signature. The EPIMMUNE signature in patients with advanced NSCLC treated with anti-PD-1 agents was associated with improved progression-free survival and overall survival. The EPIMMUNE-positive signature was not associated with PD-L1 expression, the presence of CD8+ cells, or mutational load. EPIMMUNE-negative tumors were enriched in tumor-associated macrophages and neutrophils, cancer-associated fibroblasts, and senescent endothelial cells. The EPIMMUNE-positive signature was associated with improved progression-free survival in the EPIMMUNE validation cohort. The unmethylated status of FOXP1 was associated with improved progression-free survival and overall survival in the FOXP1 validation cohort. The EPIMMUNE signature and unmethylated FOXP1 were not associated with clinical benefit in patients who did not receive immunotherapy.

Learning and Knowledge Outcomes

This study demonstrates that the epigenetic status of NSCLC tumors indicates which patients are most likely to benefit from nivolumab or pembrolizumab treatments. The methylation status of FOXP1 can be correlated with validated predictive biomarkers like PD-L1 staining and mutational load to better choose patients who will demonstrate clinical benefit from PD-1 blockade. The predictive power of this should be investigated in prospective clinical trials.

Reference

The above case study was taken from Duruisseaux M, et al. (2018). Epigenetic prediction of response to anti-PD-1 treatment in nonsmall-cell lung cancer: A multicenter, retrospective analysis. *Lancet Respir Med* 2018; 6: 771–81.

Abbreviations

ADEs adverse drug effect
ADME absorption, distribution, metabolism, and excretion (or elimination)
ADRs adverse drug reactions
CNS central nervous system
DMRs differentially methylated regions
FOXP1 Forkhead box P1
miRNAs MicroRNAs
NAFLD nonalcoholic fatty liver disease
PD-1 antiprogrammed- death-1
SNPs single-nucleotide polymorphisms

References

[1] Peedicayil J. Pharmacoepigenetics and pharmacoepigenomics: an overview. Curr Drug Discov Technol 2019;16:392–9.
[2] Gomez A, Ingelman-Sundberg M. Pharmacoepigenetics: its role in interindividual differences in drug response. Clin Pharmacol Ther 2009;85:426–30.
[3] Szyf M. Toward a discipline of pharmacoepigenomics. Curr Pharmacogenomics 2004;2:357–77.
[4] Cressman AM, Piquette-Miller M. Epigenetics: a new link toward understanding human disease and drug response. Clin Pharmacol Ther 2012;92:669–73.
[5] Cascorbi I. Overlapping effects of genetic variation and epigenetics on drug response: challenges of pharmacoepigenomics. Pharmacogenomics 2013;14:1807–9.
[6] Richards EJ. Population epigenetics. Curr Opin Genet Dev 2008;18:221–6.
[7] Johnson LJ, Tricker PJ. Epigenomic plasticity within populations: its evolutionary significance and potential. Heredity 2010;105:113–21.
[8] Schmitz RJ, Schultz MD, Urich MA, Nery JR, Pelizzola M, Libiger O, et al. Patterns of population epigenomic diversity. Nature 2013;495:193–8.
[9] Taudt A, Colomé-Tatché M, Johannes F. Genetic sources of population epigenomic variation. Nat Rev Genet 2016;17:319–32.
[10] Greally JM. Population epigenetics. Curr Opin Syst Biol 2017;1:84–9.
[11] Peedicayil J. Population pharmacoepigenomics. In: Tollefsbol T, editor. Handbook of epigenetics: the new molecular and medical genetics. Waltham, MA: Elsevier; 2017. p. 511–6.
[12] Furey TS, Sethupathy P. Genetics driving epigenetics. Science 2013;342:705–6.
[13] van Dongen J, Nivard MG, Willemsen G, Hottenga J-J, Helmer Q, Dolan CV. Genetic and environmental influences interact with age and sex in shaping the human methylome. Nat Commun 2016;7:11115.
[14] Heyn H, Moran S, Hernando-Herraez I, Sayols S, Gomez A, Sandoval J, et al. DNA methylation contributes to natural human variation. Genome Res 2013;23:1363–72.
[15] Kasowski M, Kyriazopoulou-Panagiotopoulou S, Grubert F, Zaugg JB, Kundaje A, Liu Y, et al. Extensive variation in chromatin states across humans. Science 2013;342:750–2.
[16] Carbonell J, Alloza E, Arce P, Borrego S, Santoyo J, Ruiz-Ferrer M, et al. A map of human microRNA variation uncovers unexpectedly high levels of variability. Genome Med 2012;4:62.
[17] Satterlee JS, Schübeler D, Ng H-H. Tackling the epigenome: challenges and opportunities for collaboration. Nat Biotechnol 2010;10:1039–44.
[18] Bernstein BE, Stamatoyannopoulos JA, Costello JF, Ren B, Milosavljevic A, Meissner A, et al. The NIH roadmap epigenomics mapping consortium. Nat Biotechnol 2010;28:1045–8.
[19] Wylie C. NIH Epigenomics program. Creating a wealth of reference data. Epigenetics 2010;5:670–84.
[20] Rivera CM, Ren B. Mapping human epigenomes. Cell 2013;155:39–55.
[21] Roadmap Epigenomics Consortium. Integrative analysis of 111 reference human epigenomes. Nature 2015;518:317–30.
[22] Sloan CA, Chan ET, Davidson JM, Malladi VS, Strattan JS, Hitz BC, et al. ENCODE data at the ENCODE portal. Nucleic Acids Res 2016;44:D726–32.
[23] GTEx Consortium. The genotype-tissue expression (GTEx) project. Nat Genet 2013;45:580–5.
[24] Kulis M, Esteller M. DNA methylation and cancer. Adv Genet 2010;70:27–56.
[25] Jammula S, Pasini D. EpiMINE, a computational program for mining epigenomic data. Epigenetics Chromatin 2016;9:42.
[26] Liu C-H, Ho B-C, Chen C-L, Chang Y-H, Hsu Y-C, Li Y-C, et al. ePIANNO: ePIgenomics ANNOtation tool. PLoS One 2016;11:e0148321.
[27] Albrecht F, List M, Bock C, Lengauer T. DeepBlue epigenomic data server: programmatic data retrieval and analysis of epigenome region sets. Nucleic Acids Res 2016;44:W581–6.
[28] Lee BT, Barber GP, Benet-Pagès A, Casper J, Clawson H, Diekhans M, et al. The UCSC genome browser database: 2022 Update. Nucleic Acids Res 2021.
[29] Perkel JM. A guide to the epigenome. Scientist 2012;26:60–2.
[30] Aarons L. Population pharmacokinetics: theory and practice. Br J Clin Pharmacol 1991;32:669–70.
[31] Sheiner LB, Ludden TM. Population pharmacokinetics/dynamics. Annu Rev Pharmacol Toxicol 1992;32:185–209.
[32] Duffull SB, Wright DF, Winter HR. Interpreting population pharmacokinetic-pharmacodynamic analyses—a clinical viewpoint. Br J Clin Pharmacol 2011;71:807–14.
[33] Buxton ILO, Benet LZ. Pharmacokinetics: the dynamics of drug absorption, distribution, metabolism, and elimination. In: Brunton LL, Chabner B, Knollman B, editors. The pharmacological basis of therapeutics. New York, NY: McGraw-Hill; 2011. p. 17–39.
[34] Kacevska M, Ivanov M, Ingelman-Sundberg M. Perspectives on epigenetics and its relevance to adverse drug reactions. Clin Pharmacol Ther 2011;89:902–7.
[35] Nuzziello N, Liguori M. Pharmacoepigenomics in neurodegenerative diseases. In: Tollefsbol TO, editor. Medical epigenetics. Cambridge, MA: Elsevier; 2021. p. 559–81.
[36] Ingelman-Sundberg M, Cascorbi I. Pharmacogenomic or epigenomic biomarkers in drug treatment: two sides of the same medal? Clin Pharmacol Ther 2016;99:478–80.
[37] Fisel P, Schaeffeler E, Schwab M. DNA methylation of ADME genes. Clin Pharmacol Ther 2016;99:512–27.
[38] Mangoni AA, Jackson SHD. Age-related changes in pharmacokinetics and pharmadynamics: basic principles and practical applications. Br J Clin Pharmacol 2004;57:6–14.
[39] Gandhi M, Aweeka F, Greenblatt RM, Blaschke TF. Sex differences in pharmacokinetics and pharmacodynamics. Annu Rev Pharmacol Toxicol 2004;44:499–523.
[40] Cheymol G. Effects of obesity on pharmacokinetics: implications for drug therapy. Clin Pharmacokinet 2000;39:215–31.
[41] Hanley MJ, Abernethy DR, Greenblatt DJ. Effect of obesity on the pharmacokinetics of drugs in humans. Clin Pharmacokinet 2010;49:71–87.
[42] Kalow W. Interethnic variation of drug metabolism. Trends Pharmacol Sci 1991;12:102–7.
[43] Johnson JA. Predictability of the effects of race or ethnicity on pharmacokinetics of drugs. Int J Clin Pharmacol Ther 2000;38:53–60.
[44] Anderson KE. Influences of diet and nutrition on clinical pharmacokinetics. Clin Pharmacokinetics 1988;14:325–46.

[45] Fraga MF. Esteller. Epigenetics and aging: the targets and the marks. Trends Genet 2007;23:413—18.
[46] Szyf M. Implications of a life-long dynamic epigenome. Epigenomics 2009;1:9—12.
[47] Ingelman-Sundberg M, Gomez A. The past, present and future of pharmacoepigenomics. Pharmacogenomics 2010;11:625—7.
[48] Adkins RM, Krushkal J, Tylavsky FA, Thomas F. Racial differences in gene-specific DNA methylation levels are present at birth. Birth Defects Res A 2011;91:728—36.
[49] Zhong X-B, Leeder JS. Epigenetic regulation of ADME-related genes: focus on drug metabolism and transport. Drug Metab Dispos 2013;41:1721—4.
[50] Piferrer F. Epigenetics of sex determination and gonadogenesis. Dev Dyn 2013;242:360—70.
[51] Cordero P, Li J, Oben JA. Epigenetics of obesity: beyond the genome sequence. Curr Opin Clin Nutr Metab Care 2015;18:361—6.
[52] Remely M, Stefanska B, Lovrecic L, Magnet U, Haslberger AG. Nutriepigenomics: the role of nutrition in epigenetic control of human diseases. Curr Opin Clin Nutr Metab Care 2015;18:328—33.
[53] Egger G, Liang G, Aparicio A, Jones PA. Epigenetics in human disease and prospects for epigenetic therapy. Nature 2004;429:457—63.
[54] Ivanov M, Barragan I, Ingelman-Sundberg M. Epigenetic mechanisms of importance for drug treatment. Trends Pharmacol Sci 2014;35:384—96.
[55] Kronfol MM, Jahr FM, Dozmorov MG, Phansalker PS, Xie LY, Aberg KA, et al. DNA methylation and histone acetylation changes to cytochrome P450 2E1 regulation in normal aging and impact on rates of drug metabolism in the liver. Geroscience 2020;42:819—32.
[56] Waxman DJ, Holloway MG. Sex differences in the expression of hepatic drug metabolizing enzymes. Mol Pharmacol 2009;76:215—28.
[57] Chu SK, Yang H-C. Interethnic DNA methylation difference and its implications in pharmacoepigenetics. Epigenomics 2017;9:1437—54.
[58] Luo J, Xie M, Hou Y, Ma W, Jin Y, Chen J, et al. A novel epigenetic mechanism unravels hsa-miR-148a-3p-mediated CYP2B6 downregulation in alcoholic hepatitis disease. Biochem Pharmacol 2021;188:114582.
[59] Blumenthal DK, Garrison JC. Pharmacodynamics: molecular mechanisms of drug action. In: Brunton LL, Chabner B, Knollman B, editors. The pharmacological basis of therapeutics. New York, NY: McGraw-Hill; 2011. p. 41—72.
[60] Minto C, Schnider T. Expanding clinical applications of population pharmacodynamic modelling. Br J Clin Pharmacol 1998;46:321—33.
[61] Klotz U. Effect of age on pharmacokinetics and pharmacodynamics in man. Int J Clin Pharmacol Ther 1998;36:581—5.
[62] Rang HP, Ritter JM, Flower RJ, Henderson G. Rang & Dale's pharmacology. 8th ed. London: Churchill Livingstone; 2016.
[63] Ivanov M, Kacevska M, Ingelman-Sundberg M. Epigenomics and interindividual differences in drug response. Clin Pharmacol Ther 2012;92:727—36.
[64] Priestley CC, Anderton M, Doherty AT, Duffy P, Mellor HR, Powell H, et al. Epigenetics-relevance to drug safety science. Toxicol Res 2012;1:23—31.
[65] Cascorbi I, Schwab M. Epigenetics in drug response. Clin Pharmacol Ther 2016;99:468—70.
[66] Adwan L, Zawia NH. Epigenetics: a novel therapeutic approach for the treatment of Alzheimer's disease. Pharmacol Ther 2013;139:41—50.
[67] Mauvais- Jarvis F, Berthold HK, Campesi I, Carrero J-J, Dakal S, Franconi F, et al. Sex-and gender-based pharmacological response to drugs. Pharmacol Rev 2021;73:730—62.
[68] Joseph S, George NI, Green-Knox B, Treadwell EL, Word B, Yim S, et al. Epigenome-wide association study of peripheral blood mononuclear cells in systemic lupus erythematousus: identifying DNA methylation signatures associated with interferon-related genes based on ethnicity and SLEDAI. J Autoimmunity 2019;96:147—57.
[69] Hoepner CT, McIntyre RS, Papakostas GI. Impact of supplementation and nutritional interventions on pathogenic processes of mood disorders: a review of the evidence. Nutrients 2021;13:767.
[70] Duruisseaux M, Martinez-Cardus A, Calleja-Cervantes ME, Moran S, Castro de Moura M, Davalos V, et al. Epigenetic prediction of response to anti-PD-1 treatment in non-small-cell lung cancer: a multicentre, retrospective analysis. Lancet Respir Dis 2018;6:771—81.
[71] Tashiro S, Lanctôt C. The international nucleosome consortium. Nucleus 2015;6:89—92.
[72] van Berkum NL, Lieberman-Aiden E, Williams L, Imakaev M, Gnirke A, Mirny LA, et al. Hi-C: a method to study the three dimensional architecture of genomes. J Vis Exp 2010;39:e1869.
[73] Belton J-M, McCord RP, Gibcus JH, Naumova N, Zhan Y, Dekker J. Hi-C: a comprehensive technique to capture the conformation of genomes. Methods 2012;58:268—76.
[74] Xu L, Yin L, Qi Y, Tan X, Gao M, Peng J. 3D organization and rearrangement of genome provide insights into pathogenesis of NAFLD by integrated Hi-C, Nanopore, and RNA sequencing. Acta Pharm Sin B 2021;11:3150—64.
[75] Gorkin DU, Qiu Y, Hu M, Fietez-Brant K, Liu T, Schmitt AD, et al. Common DNA sequence variation influences 3-dimensional conformation of the human genome. Genome Biol 2019;20:255.
[76] Sadowski M, Kraft A, Szalaj P, Wlasnowuski M, Tang Z, Ruan Y, et al. Spatial chromatin architecture alteration by structural variations in human genomes at the population scale. Genome Biol 2019;20:148.
[77] Hagenbuchner J, Ausserlechner MJ. Targeting transcription factors by small compounds—current strategies and future implications. Biochem Pharmacol 2016;107:1—13.
[78] Kantidze OL, Gurova KV, Studitsky VM, Razin SV. The 3D genome as a target for anticancer therapy. Trends Mol Med 2020;26:141—9.
[79] Higgins GA, Allyn-Feuer A, Georgoff P, Nikolian V, Allam HB, Athey BD, et al. Mining the topography and dynamics of the 4D nucleosome to identify novel CNS drug pathways. Methods 2017;123:102—18.
[80] Manjon A.G., Peric-hupkes D., Liu N.Q., Friskes A., Jousten S., Teunissen H., et al. Perturbations in 3D genome organization can promote acquired drug resistance. In Press. https://doi.org/10.1101/2021.02.02.429315.
[81] Gertych A, Oh JH, Wawrowsky KA, Weisenberger DJ, Tajbakhsh J. 3-D DNA methylation phenotypes correlate with cytotoxicity levels in prostate and liver cancer cell models. BMC Pharmacol Toxicol 2013;14:11.
[82] Pirmohamed M, Breckenridge AM, Kitteringham NR, Park BK. Adverse drug reactions. BMJ 1998;316:1295—8.
[83] Aronson JK, Ferner RE. Joining the DoTS: new approach to classifying adverse drug reactions. BMJ 2003;327:1222—5.
[84] Stricker BH, Psaty BM. Detection, verification, and quantification of adverse drug reactions. BMJ 2004;329:44—7.
[85] Huang S-M, Lesko LJ, Temple R. Adverse drug reactions and interactions. In: Waldman SA, Terzic A, editors. Pharmacology and therapeutics: principles to practice. Philadelphia, PA: Saunders Elsevier; 2009. p. 265—73.
[86] Osterhoudt KC, Penning TM. Drug toxicity and poisoning. In: Brunton LL, Chabner B, Knollman B, editors. The pharmacological basis of therapeutics. New York, NY: McGraw-Hill; 2011. p. 73—87.
[87] Cacabelos R, Cacabelos N, Carril JC. The role of pharmacogenomics in adverse drug reactions. Expert Rev Clin Pharmacol 2019;12:407—42.

SECTION VIII

Epigenetics and Human Disease

36. Cancer epigenetics
37. The role of epigenetics in autoimmune disorders
38. Epigenetics of brain disorders
39. Epigenetics of metabolic diseases
40. Imprinting disorders in humans

CHAPTER 35

Cancer Epigenetics

Marina Alexeeva[1,2], Marcus Roalsø[1,2,3] and Kjetil Søreide[1,2,4]

[1]Department of Gastrointestinal Surgery, Stavanger University Hospital, Stavanger, Norway [2]Gastrointestinal Translational Research Unit, Laboratory for Molecular Biology, Stavanger University Hospital, Stavanger, Norway [3]Department of Quality and Health Technology, University of Stavanger, Stavanger Norway [4]Department of Clinical Medicine, University of Bergen, Bergen, Norway

OUTLINE

Introduction	697
Epigenetic Influences Over a Lifetime and Cancer Risk	698
Epigenetics in Cancer: Remodelers, Writers, Readers, and Erasers	699
DNA Methylation	700
Histone Modification	702
Noncoding RNA	703
Importance of Epigenetic Changes Together with Genetic Alterations	703
Epigenetics in Intratumoral Heterogeneity and Therapeutic Resistance	704
Genetic and Epigenetic Classification of Cancer	705
Limitations to Epigenetic Classification and Prognostication	708
Epigenetic Biomarkers in Cancer	708
Cancer Epigenome in Epidemiology	708
Epigenetics as Cancer Therapeutic Targets	709
Future Perspectives	710
Abbreviations	710
References	711

INTRODUCTION

Cancer develops through the accumulated changes at the chromosomal level or genetic base pair-level, potentially affecting oncogenes ("drivers") or tumor suppressor genes ("brakes") that eventually lead to uncontrolled cell division, invasion, and metastasis in various solid organ cancers [1,2]. The mechanisms involved in the hallmarks to cancer development are well described and continues to be explored [3]. Fundamentally, cancer is a genomic disease that alters cellular information flow to modify cellular homeostasis and promote growth, described through mechanisms outlined as the hallmarks of cancer [1,2]. The discovery of a universal genetic code for protein-coding genes produced countless breakthroughs in understanding how such mutations drive cancer, establishing the scientific principles on which the development of targeted therapies for malignancies are based. Hence, the discovery that genes were altered in and through noncoding regions as well as by epigenetic mechanisms came as a game changing discovery. Epigenetics refers to—heritable or developed—traits that are not attributable to changes in the DNA sequence itself, but rather in the mechanisms around the DNA control as such. In more specific terms, it can be used to describe how chromatin-associated proteins and reversible chemical modifications of DNA and histone proteins maintain transcriptional programs by regulating chromatin structure.

Epigenetic changes are influenced by the exogenous factors throughout a lifetime, are present in all human cancers, and are known to cooperate with genetic alterations to drive the cancer phenotype through the specific hallmarks of cancer (Fig. 35.1) [3,4].

These epigenetic changes involve mechanisms such as DNA methylation, histone modifiers and readers, chromatin remodelers, noncoding RNAs, and other components of chromatin [5]. Cancer genetics and epigenetics are inextricably linked in generating the malignant phenotype; epigenetic changes can cause mutations in genes, and, conversely, mutations are frequently observed in genes that modify the epigenome [4].

Notably, epigenetics is not a new field in basic science; it has been explored since the 1940s [6,7]. However, only more recently has the potential for use as biomarkers and therapeutic options made cancer epigenetics a closer ally for mainstream clinical cancer care [8,9]. Epigenetic therapies, in which the goal is to reverse epigenetic changes, are now entering as novel modes of medical care for some types of malignant disease, including solid tumors [10–13].

In this chapter we will explore some areas where cancer epigenetics is increasingly playing a role in the understanding of cancer as a disease. Further, as an example and a well-described public health challenge, colorectal cancer will be specifically mentioned as a cancer disease model when suitable, as coverage of all cancer types will be beyond the scope of the chapter. Due to the vast and developing field of research going beyond the scope of this introductory chapter, the references are chosen mainly for their updated, comprehensive expert review insight. Thus, for the interested reader who wishes to pursue further in-depth knowledge to certain themes, the references provide a direction to sources of detailed knowledge. While specific mechanisms of epigenetics are explored in previous chapters in this book, this chapter overview will largely focus on the areas of immediate or future importance for understanding epigenetics and the role in cancer.

EPIGENETIC INFLUENCES OVER A LIFETIME AND CANCER RISK

Epigenetic studies include the investigation of DNA methylation, histone modifications, chromatin remodeling and gene regulation by noncoding RNAs

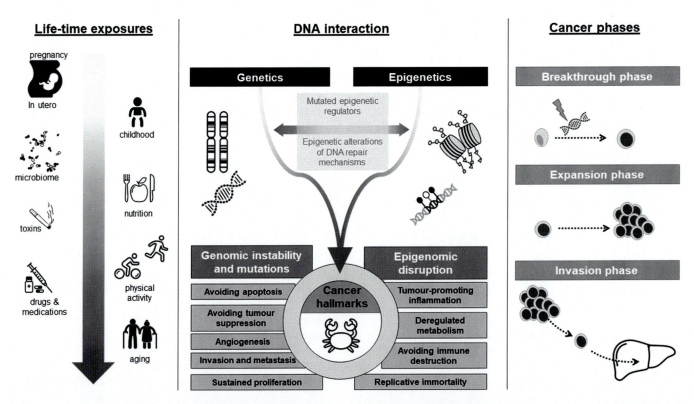

FIGURE 35.1 Interplay between genomic and epigenetic changes acquired during lifetime influence the hallmarks of cancer and facilitate the progress of cancer development. Epigenetic could be changed by exogenous and endogenous factors through a lifetime of exposure (left panel). Acquired epigenetic changes interplay with mutations in DNA and influence the cancer hallmarks for cancer risk and progression (middle panel). Accumulation of epigenetic disruptions and genomic instabilities contribute to cancer development occurring in three phases, breakthrough phase, expansion, and invasion (left panel) [2]. *Reproduced with permission from Elsevier; Drake TM, Soreide K. Cancer epigenetics in solid organ tumors: A primer for surgical oncologists. Eur J Surg Oncol. 2019;45(5):736-46.*

(ncRNAs). Epigenetic alterations are critical for early developmental processes, the silencing of the inactive X-chromosome and tissue-specific gene regulation [14]. Epigenetic marks changes during fetal development, through adult life, and with aging [15]. Age-associated changes in epigenome are well documented and thought to promote diseases of aging [16,17]. Epigenetic changes directly contributing to aging-related diseases such as cancer include the DNA hypo- or hypermethylation, accumulation of histone variants, changes in chromatin accessibility, loss of histones and heterochromatin, aberrant histone modifications, and deregulated expression/activity of miRNAs [18]. One of the known biomarkers for aging is an epigenetic (methylation) age, or Horvath's clock, a set of CpG methylation representing a mathematical model, allows to measure chronological age [19,20]. Epigenetic age acceleration, i.e., the difference between DNA methylation age and chronological age, has been associated with increased risk for colorectal cancer, and may be used as a biomarker for cancer risk [21,22].

Chemical and physical environmental stressors, diet, life habits, and pharmacological treatments can affect the epigenome through different mechanisms [23,24]. An important future challenge in the field of epigenetics will be to describe how the environment affects epigenetic change and to learn if interaction between various epigenetic influences can determine healthy from those attributed to risk of disease phenotypes during a lifetime. This may give the opportunity for potential risk-reducing interventions, lifestyle changes or even epigenetic preventive therapy [25].

The way in which energy is used in cells is determined under the influence of environmental factors such as nutritional availability [26]. Metabolic adaptation is mainly achieved through the modulation of metabolic gene expression [27,28] and may also involve epigenetic mechanisms that enable long-term regulation. Recent studies have identified that nutrients and their metabolites exert an important influence on the epigenome, as they serve as substrates and/or coenzymes for epigenetic-modifying enzymes. Some epigenetic factors have been shown to regulate metabolic genes leading to a shift in energy flow. These findings suggest the concept of metabolism-epigenome crosstalk that may contribute to the formation of a long-term metabolic phenotype [26]. This is particularly relevant to the pathogenesis of obesity and associated metabolic disorders, in which pre- and post-natal nutritional conditions affect disease risks in adulthood.

Investigating tissue specific alterations, the aging human colorectal mucosa develops aberrant patterns of DNA methylation that may contribute to its increasing vulnerability to cancer [29]. Evidence suggests that age-dependent loss of global methylation, together with hypermethylation of CpG islands associated with cancer-related genes, may be influenced by nutritional and metabolic factors [30,31]. Several compounds of nutrition, such as folates and vitamins, are essential for the maintenance of normal DNA methylation. Folate metabolism is known to modify epigenetic mechanisms under experimental conditions, and more recent findings has explored the important roles of vitamin C and D in maintenance of the epigenome [32,33]. Human intervention trials and cross-sectional studies suggest a role for folates and other nutritional and metabolic factors as determinants of colorectal mucosal DNA methylation [29]. Moreover, most cancer cells exploit metabolic pathways for their hyperproliferative activity [34], while metabolic dysregulation leads to aberrant epigenetic regulation in some cancers [35].

EPIGENETICS IN CANCER: REMODELERS, WRITERS, READERS, AND ERASERS

Genetic mechanisms of mutation, copy number alteration, insertions, deletions, and recombination are particularly well suited as vehicles of persistent phenotypic changes in cancer. Several past cancer studies investigated this and developed sequential progression models, such as that presented for colorectal cancer [36]. For this reason, cancer has long been viewed as a genetic disease, although different methylation profiles where noted to occur in different genes in the same tumor model [37].

Knowingly, genetic events occur at low frequency and are thus not a particularly efficient means for malignant transformation [38]. However, some cancer cells overcome this bottleneck by acquiring DNA repair defects, thus boosting the mutation rate, such as seen in mismatch repair deficient tumors that lead to high occurrence of microsatellite instability (MSI) throughout the genome [39,40]. Mechanisms of epigenetic control offer an alternative path to acquiring stable oncogenic traits [4,41]. Epigenetic states are flexible yet persist through multiple cell divisions and exert clear-cut effects on the cellular phenotype. Although cancer cells have long been known to undergo epigenetic changes, genome-scale genomic and epigenomic analyses have only recently revealed the widespread occurrence of mutations in (epi)genetic regulators and the breadth of alterations to the epigenome in cancer cells. There are several classes of epigenetic regulators (Fig. 35.2), those that write the marks (dubbed *writers*), DNA methyltransferases, histone methyltransferases, and histone acetyltransferases; those that read the marks (dubbed *readers*), the methyl binding domain proteins (MBDs); those that erase the

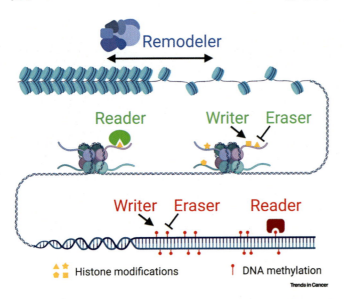

FIGURE 35.2 Epigenetic remodelers, writers, erasers, and readers. The basic functional unit of chromatin is the nucleosome, which is composed of DNA wrapped around histones. Core histone tails are projected from nucleosomes and are subject to post-translational modifications (yellow icons). These include methylation, acetylation, phosphorylation and ubiquitination. The main epigenetic regulators can be categorized as remodelers, writers, erasers and readers of epigenetic marks. The packaging state of chromatic is dynamic and regulated by chromatic-remodeling complexes (*remodelers*). Epigenetic *writers* are responsible for the addition of chemical modifications. Epigenetic *erasers* catalyze the removal of the covalent modifications. Epigenetic *readers* are proteins with specific domains that recognize and bind to particular modifications [42]. *Reproduced with permission from Cao J, Jan Q. Cancer Epigenetics, Tumor Immunity, and Immunotherapy. Trends Cancer, 2020. 6(7): p. 580–592, with permission from* Trends of Cancer.

marks (dubbed *erasers*), histone deacetylases (HDAC) and histone demethylases (HDM), active DNA demethylation (TET, AID); and those that remodel (dubbed *remodelers*) the chromatin, such as components of the SWI/SNF complex [12,43]. The frequent, recurrent mutation of specific epigenetic modifiers in a variety of cancers demonstrates that altered epigenetic regulation plays an important role in driving tumorigenesis. While epigenetics represents the wide range of changes that regulate gene expression, yet does not cause changes in the primary base pair sequence of the DNA in itself, these can be broadly defined as DNA methylation, histone modifications, genome imprinting, and noncoding RNA (ncRNA) [44]. Previous chapters have detailed mechanistic insights to each of these mechanisms, so only a brief presentation related to cancerogenesis will be given here.

DNA Methylation

The best-known and most explored epigenetic marker/mechanism is DNA methylation [6]. DNA methylation occurs in the context of chemical modifications by the addition of a methyl group to DNA at the 5-carbon of the cytosine pyrimidine ring to form a 5-methylcytosine (5mC). DNA methylation has critical roles in the control of gene activity and the architecture of the chromatin of the cell. In humans, DNA methylation occurs in cytosines (C) that precede guanines (G); these are called dinucleotide "CpGs." Regions in the DNA that contain many adjacent cytosine and guanine nucleotides are called "CpG islands." CpG islands are not randomly distributed in the genome; instead, there are CpG-rich regions, which span the 5' end of the regulatory region of many genes [7,45]. These islands occur in approximately 40% of the promoters of human genes [46]. DNA methylation of cytosines in the CpG-islands is associated with transcriptional silencing because 5mC constitutes a steric hindrance interfering with binding of some transcriptional factors, as well as recruitment of proteins that selectively bind to methylated DNA, the MBD family of proteins. MBDs indirectly represses transcription via interaction with co-repressors that modify chromatin. In addition to CpG sites, hypermethylation occurs at the repetitive genomic sequences, like pericentromeric, centromeric and telomeric areas, and contributes to maintenance of chromosomic stability [47]. The levels of DNA methylation are regulated by writes and erasers. The DNA methyltransferases (writers) modify more than 28 million CpG bases across the human genome, while the TET enzymes (erasers) are involved in 5mC removal by active deamination. Hypo- and hypermethylation will consequently alter the access to and expression of certain genes or DNA parts [48], as depicted in Fig. 35.3A and B.

For many years epigenetic dysregulation has been demonstrated in a cancer. Regulators of DNA methylation are often disturbed in cancer and that leads to an aberrant DNA methylation pattern [50,51]. Generally, typical features of cancer are a global loss of DNA methylation with focal hypermethylation as observed for tumor-suppressor gene promoters (Fig. 35.3A and B). Global loss of DNA methylation gives a rise to deletions, translocation, and chromosomal rearrangements in tumors. The CpG islands are usually unmethylated in normal cells, and the genes downstream of these unmethylated promoters are transcribed in the presence of transcriptional activators. Genomic platforms have confirmed that almost 10% of normally unmethylated CpG islands in the promoters, many of them belonging to tumor suppressor genes, become abnormally methylated and thus silenced in cancer (Fig. 35.3) [7,45]. Moreover, cancer associated gene silencing due to the hypermethylated CpG islands at promoter regions affect the genes involved in all malignant capability acquisitions [52].

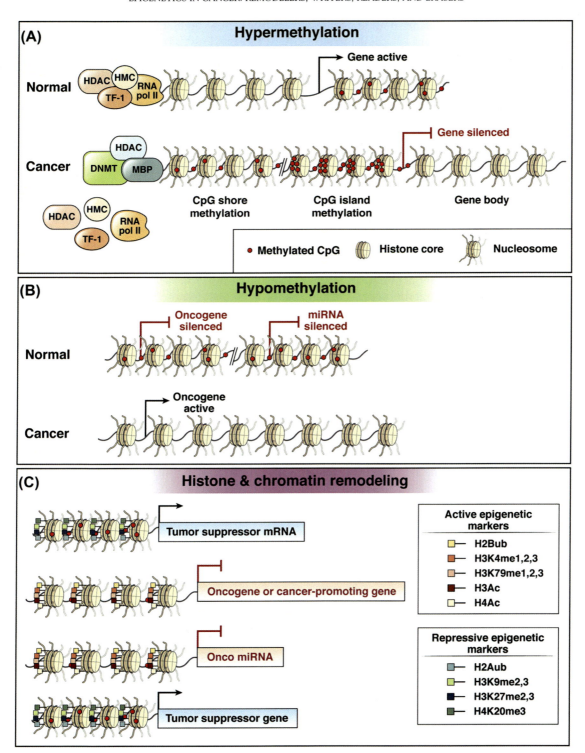

FIGURE 35.3 An illustration of various epigenetic alterations in cancer. The concept of aberrant DNA hypermethylation (A) and DNA hypomethylation (B) in the context of a "cancer cell" vs "normal cell" are shown. TF-1, transcription factor 1; RNA pol II, RNA polymerase II; HDAC, histone deacetylase complex; DNMT, DNA methyltransferase; MBP, methyl binding protein. CpG islands associated with promoters are generally unmethylated in normal cells that allow the transcription factors to bind and activate the expression of genes. However, in the cancer cell (A) the CpG sites (red circles) of tumor suppressor genes are usually hypermethylated leading to gene silencing and closed chromatin hypomethylation of CpG sites (B) can lead to the activation of oncogenes contributing to cancer formation. C. Structure of DNA chromatin and histone modification states that affect conformation of chromatin in cancer. *Ac*, acetylated; *H2A*, histone 2A; *H2B*, histone 2B; *H3*, histone 3; *H4*, histone 4; *K*, lysine; *me*, methylated; *Onco miRNA*, oncogenic microRNA; *ub*, ubiquitylated [49]. *Reproduced from Grady W, Yu M, Markowitz SD. Epigenetic Alterations in the Gastrointestinal Tract: Current and Emerging Use for Biomarkers of Cancer.* Gastroenterology, 2021. 160 (3): p. 690–709 with permission from Gastroenterology.

In colorectal cancer, a CpG-island methylator phenotype (CIMP) has been recognized as a separate molecular group (besides microsatellite instable and chromosomal instable tumors) that has distinct clinical and pathological characteristics [40,53]. For one, prognosis is demonstrated to be consistently poorer with the CIMP group [54]. It was also shown that CIMP-positive status is more representative in multiple colorectal cancers [55].

Histone Modification

In addition to DNA methylation, N-terminal tails of core histone (H2A, H2B, H3 and H4) undergo posttranscriptional modification including acetylation, methylation, phosphorylation, and ubiquitination (Fig. 35.3C). The enzymes responsible for such modifications throughout histone tails include histone acetyltransferases (HATs), histone deacetylases (HDACs) and sirtuins, histone methyltransferases (HMTs), histone demethylases (HDMs), histone kinases and phosphatases, histone ubiquitin ligases, and deubiquitinases (as reviewed in detail in [7]). Modifications of histone N-terminal tails that usually occur at lysine, arginine, and create so-called "histone code" which dictates the dynamics of chromatin. Heterochromatin is a closed chromatin conformation that is often associated with DNA methylation and inactive gene transcription. In contrast, the euchromatin state is in an open conformation and associates with active gene transcription, presumably secondary to increased transcription factor binding. DNA methyltransferases and MBDs work with histone-modifying enzymes for the purpose of regulating all DNA-templated processes including transcription, repair, replication, and recombination [6,7,56]. Methylation of histone at lysine is catalyzed by different enzymes, and one of the profound one is enhancer of zeste homolog 2 (EZH2), a catalytic subunit of Polycomb repressive complex 2 (PRC2). Depending on the position of lysine in the histone tail and a level of its methylation (mono-, di-, and trimethylation) the methylated lysine can serve either as activation mark or repressive mark for transcription. PRC2 cooperation on the trimethylation of lysine 27 (K27) at the H3 is emerged as one of the most influential epigenetic controls in colorectal cancer [46]. Mutation in H3K27 to methionine or isoleucine leads to that the K27 cannot be longer methylated casing the aberrant methylation pattern. The K27M/I histones are also called "oncohistone" and found in glioblastoma together with others mutated histones [57,58].

Importantly, this histone code is not static but rather is changing in a context-dependent manner to facilitate or repress, for instance, gene transcription. New histone modifications are still being discovered and are found in new combinations. The influence of these marks on the deposition, interpretation and erasure of other histone modifications is known as histone crosstalk (Fig. 35.4) and is of great importance for the transcriptional readout of a gene [7,43]. Consistently, with the fact that disruption of normal patterns of covalent histone modifications is a defined hallmark of human cancer many such enzymes that add or remove these chemical groups, and also those that recognize them (the readers), are mutated or misregulated in cancer.

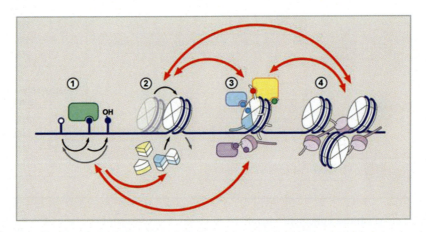

FIGURE 35.4 Hierarchical organization of epigenetic regulation. Dynamic and reversible epigenetic processes generate diverse regulatory environment. DNA methylation (1) (mC, methylcytosine, represented by closed circle) may result in eviction of DNA-binding proteins, or recruitment of methyl-binding factors. mC oxidation (shown as hydroxymethylcytosine, hmC) generates additional diversity. Core and linker histone exchange, including variant histone incorporation (2), regulates local DNA accessibility and, together with histone modifications (3), introduces local variations to chromatin structure. These are interpreted by the "reader" machinery (3) and drive higher-order chromatin organization and nuclear topology (4). Red arrows above and below indicate crosstalk between regulatory layers. *Reproduced with permission from Elsevier in Soshnev AA, Josefowicz SZ, Allis CD. Greater Than the Sum of Parts: Complexity of the Dynamic Epigenome. Mol Cell. 2016 Jun 2;62 (5):681-94.*

Noncoding RNA

Although the most thoroughly studied sequences in the human genome are those of protein-coding genes, these translated regions account for only 2% of the entire DNA. The remaining transcribed but nonprotein-coding part of the genome is made up of noncoding RNAs (ncRNAs) [44,59,60], which are increasingly recognized as being fundamental to embryogenesis and development and are also disrupted in diseases such as cancer [2,7,44,61]. The ncRNAs include microRNAs (miRNAs), long noncoding RNAs (lncRNAs) and circular RNA (circRNA). The involvement of long noncoding RNAs in progression of colorectal cancer is given in Fig. 35.5. Extensive reviews are presented elsewhere [44,63–65].

Importance of Epigenetic Changes Together with Genetic Alterations

Notably, tumors evolve in three broad phases, as reviewed by Vogelstein and Kinzler [66]. In the *breakthrough phase*, a cell acquires a driver-gene mutation and begins to proliferate abnormally [66]. It takes many cell divisions, and many years, for the cells

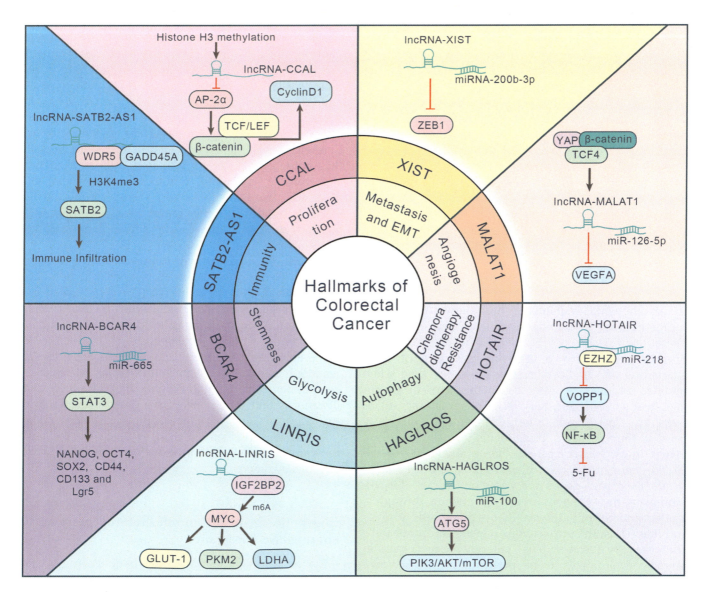

FIGURE 35.5 **The role of lncRNAs in hallmarks of colorectal cancer.** Representative LncRNAs and related oncogenic mechanisms in the proliferation, metastasis, angiogenesis, chemoradiotherapy, autophagy, glycolysis, stemness and immunity of colorectal cancer [62]. *Reproduced from Yang Y, Yan X, Li X, Ma Y, Goel A. Long non-coding RNAs in colorectal cancer: Novel oncogenic mechanisms and promising clinical applications. Cancer Letters 504 (2021) 67–80 with permission from Cancer Letters.*

resulting from this proliferation to be observable clinically, if they ever are. For example, nevi on the skin and small adenomas in the intestine are detected only because they can be easily observed on visual inspection or colonoscopy. Similarly sized lesions in internal organs (e.g., kidneys or the pancreas) would generally be undetectable clinically. The *expansion phase* is driven by a second driver-gene mutation enabling the cell to thrive in its local environment despite low concentrations of growth factors, nutrients, oxygen, and appropriate cell-to-cell contacts [66]. Known cancer mutation rates suggest that this second mutation is unlikely to occur absent a large increase in cell number during the breakthrough phase [38]. The likelihood of progression thus depends on both the mutation rate and the number of cells at risk for acquiring another mutation. The first two mutations lead to the abnormal proliferation and disordered cellular architecture defining benign tumors. Subsequent mutations enable cells to invade normal tissues and grow in otherwise hostile environments. Such cells are, by definition, malignant.

The mutation initiating the breakthrough phase is often very specific, a limited number of growth-regulating pathways seem able to initiate neoplasia in a given cell type. As tumors progress, this specificity seems to be progressively lost, so a greater number of driver genes can transform a cell from the expansion phase to the *invasive phase* (see Ref. [66] for full overview). The fact that so few genetic mutations are required in neoplastic transformation leaves several unanswered questions. This "dark matter", as alluded to [38], is most likely controlled and influenced by epigenetic mechanisms.

Genome-wide sequencing has shown that every tumor harbors thousands of genetic (and epigenetic) alterations that are not present in the patient's germline. Of notice, only a very small fraction of these alterations are in "driver genes"—genes that when mutated endow the tumor cell with a growth advantage over surrounding cells [38]. The remaining alterations are "passengers," found in tumor cells only because they occurred coincidentally during the long march toward tumorigenesis. Only about 200 of the 20,000 genes in the human genome have been shown to act as driver genes for common cancers [38,66]. Moreover, these genes appear to function through a limited number of pathways that regulate cells' growth and fate. Thus, the implications of genes in tumor initiation may be reduced to a few general principles in cancer development from preneoplasia to invasive cancer.

How epigenetics may involve genetics in carcinogenesis (and vice versa) is not fully understood, other than the two being mutually included [4,67]. Notably, techniques in genetics research have expanded tremendously over the past decades, enabling a more rapid and widespread investigation into genetic disruptions. At the same time, epigenetics research has largely focused on DNA methylation, as this feature is the most stable marker that survives various sample processing, including DNA extraction and even formalin fixation and paraffin embedding [68]. However, mechanisms and relations related to epigenetic reprogramming in cancer cells, roles in cell plasticity, and roles for intratumor heterogeneity are being explored, of which some relations are briefly mentioned here.

Epigenetic reprogramming of neoplastic cells have been proposed [69], in that by activating specific transcription factor drivers and modulating collaborating chromatin regulators, cancer cells may dynamically regulate their epigenetic circuits to rewire differentiated cancer cells into stem-like cells, thus refueling cancer growth [69]. Epigenetic reprogramming of neoplastic cells have been proposed [69], in that by activating specific transcription factor drivers and modulating collaborating chromatin regulators, cancer cells may dynamically regulate their epigenetic circuits to rewire differentiated cancer cells into stem-like cells, thus refueling cancer growth [69]. While the idea of stem-cells or "stemness" remains a controversial topic in and of itself, the idea and theory of this cellular plasticity model is highly entertained as a mechanism for both understanding cancer development and as a mode for potential intervention [63,70–73].

An epigenetic plasticity of tumor cells have been proposed to be fundamental for tumor cells to be able to metastasize [73]. Cancer stem cells (CSCs) have been identified in various tumors and are defined by their potential to initiate tumors upon transplantation, self-renewal, and reconstitution of tumor heterogeneity [74]. Modifications of the epigenome can favor tumor initiation by affecting genome integrity, DNA repair, and tumor cell plasticity. Importantly, an in-depth understanding of the epigenomic alterations underlying neoplastic transformation may open new avenues for chromatin-targeted cancer treatment, as these epigenetic changes could be inherently more amenable to inhibition and reversal than hard-wired genomic alterations.

Epigenetics in Intratumoral Heterogeneity and Therapeutic Resistance

By the time of diagnosis, any patient's tumor may consist of tens of thousands or even millions of cells that have differentiated and created several subclones within the tumor. This heterogeneous mixture of functionally distinct cancer cells can be caused by varying levels of receptor activity, differentiation, and distinct metabolic and epigenetic states. Several classes of epigenetic regulators (Fig. 35.2) of writers, readers, and erasers have been implicated in mechanisms leading to intratumoral heterogeneity and chemotherapy resistance [72,75,76]. The clinical implications are starting to unfold with current epigenetic mechanisms being explored for its particular role in chemotherapy resistance [77] and for epigenetic regulation of immune response in tumors [42,78]. Knowledge and the ability to biopsy tumor changes in subclones at time of diagnosis, and as well during follow-up, may help tailor the appropriate therapy to the epigenetic and genetic makeup of the tumor. Further, epigenetic marks may be used as biomarker of treatment success; either for measure of response or as indicator of recurrence. Lastly, specific epigenetic drugs may be used together with conventional drugs in order to achieve a more tailored and specific response to the tumor at hand. Several principles are reviewed in detail elsewhere [42,44,78–80].

GENETIC AND EPIGENETIC CLASSIFICATION OF CANCER

Tumors have historically been classified based on their (likely) tissue of origin and differentiation and grade. This still represents the mainstay for diagnosis and prognosis, but increasingly the molecular features help classify tumors into distinct therapeutic and prognostic classes. In solid organ cancers, the tumor-node-metastasis (TNM) system is still used for prognosis and therapeutic decisions, but it is recognized to have several limitations with risk of over- and undertreatment [81,82]. Also, the TNM system does not capture the specific underlying genetic pathways involved, although these are to some degree incorporated into routine testing for a few select mutations (e.g., RAS (rat sarcoma virus) mutation status for EGFR (epidermal growth factor receptor) treatment) [83].

The CRC could be classified based on the consensus molecular subtypes (CMS) that reflects the difference in tumor biology and captures the intrinsic heterogeneity of CRC. CMS classification of CRC possesses a strong potential for prediction of response to therapies and prognosis [84].

Commonly observed alterations across sporadic CRCs have now allowed classification of CRC into at least three possible distinct groups [40], and emerging markers of good, bad, and ugly disease behavior are proposed for advanced disease (e.g., liver metastases) [85]. For the assessment of primary colorectal cancers (Fig. 35.6) there are at least three suggested classifications that come into consideration.

One is a hypermutated group that includes defective DNA mismatch repair with microsatellite instability and POLE mutations (about ~15% of CRC patients) [86], containing multiple frameshifted genes and $BRAF^{V600E}$. The second is a nonhypermutated group with multiple somatic copy number alterations and aneuploidy, previously known as chromosomal instability (CIN) type of tumors (in ~85%), containing oncogenic activation of KRAS and PIK3CA and mutation and loss of heterozygosity of tumor suppressor genes, such as APC and TP53. A third group is named CpG island methylator phenotype (CIMP) type CRCs (in ~20%) that overlap greatly with microsatellite instability CRCs and some nonhypermutated CRCs. CIMP tumors have methylated CpG islands and epigenetic alterations are essential in these cancers. Lastly a fourth group (or, maybe more appropriate, a modifier group or trait occurring across all other groups) is named after elevated microsatellite alterations at selected tetranucleotide (EMAST) repeats (found in up to ~60% of CRC, but also in a range of other cancer types [39]) that associates with metastatic behavior in both hypermutated and nonhypermutated groups.

Components from these classifications are now used as diagnostic, prognostic, and treatment biomarkers [40,84,87], yet universal agreement on such a new classification has not been reached with alternative proposals published [88,89]. It should also be noted that evidence from the EMAST classification is not yet clear [39,90], although studies are emerging [91–96]. Additional common biomarkers may come from genome-wide association studies and microRNAs, among other sources, as well as from the unique alteration profile of an individual CRC to apply a precision medicine approach to modern cancer care. The delineation of molecular markers in CRC metastasis (Fig. 35.7) is also becoming of age and may play an even greater role with advancement in epigenetic therapies in the future [97]. As an example, clinical behavior of colorectal cancer is determined by several factors, including demographic data (age, gender, race) and tumor presentation (location, stage) and timing of presentation of metastasis (synchronous or metachronous), as depicted in Fig. 35.7A. Embedded in the cancer cells are the molecular pathways, which follows distinct forms of genomic instability yet with partly overlapping areas. Hypermutated cancers belong to the microsatellite instable (MSI) cancers and in part the CpG-island methylator phenotype (CIMP) cancers. Nonhypermutated cancers follow in large parts the chromosomal instability (CIN)–driven pathways, often involving KRAS mutations from an early stage. The propensity to develop metastasis may possibly be modified through the elevated microsatellite alterations at selected tetranucleotide repeat (EMAST) and associated mechanisms, such as regulation of

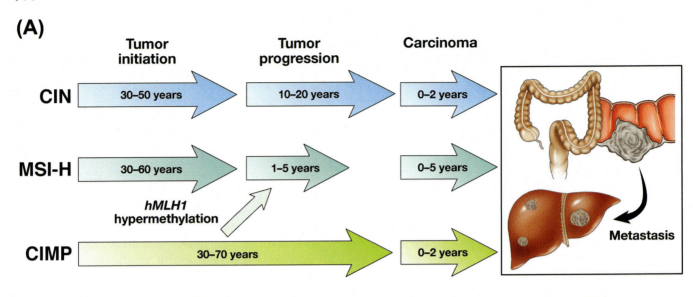
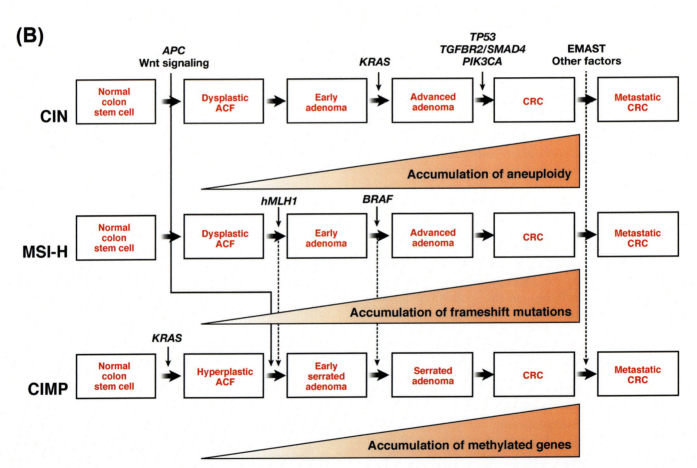

FIGURE 35.6 **Timeline for sporadic CRC pathogenesis and its characterized molecular pathways.** (A) The timelines are based on the mean age of CRC for each type. Tumorigenesis can be broken into tumor initiation (development of an adenoma), tumor progression that culminates in a malignancy (carcinoma) that can spread as metastasis. MSI-H tumors are known to have a shortened progression stage. (B) Each pathway has its feature of moving from normal to cancer and potentially metastasis, with varying histology. Wnt signaling is the gatekeeper for all three pathways. The CIMP pathway contributes to both the MSI-H (through hypermethylation of *hMLH1*) and CIN pathways, and specifically characterizes a serrated pathway. EMAST can modulate all three pathways. *Reproduced from Carethers JM, Jung BH. Genetics and Genetic Biomarkers in Sporadic Colorectal Cancer. Gastroenterology. 2015 Oct;149(5):1177–1190.e3. with permission from Gastroenterology.*

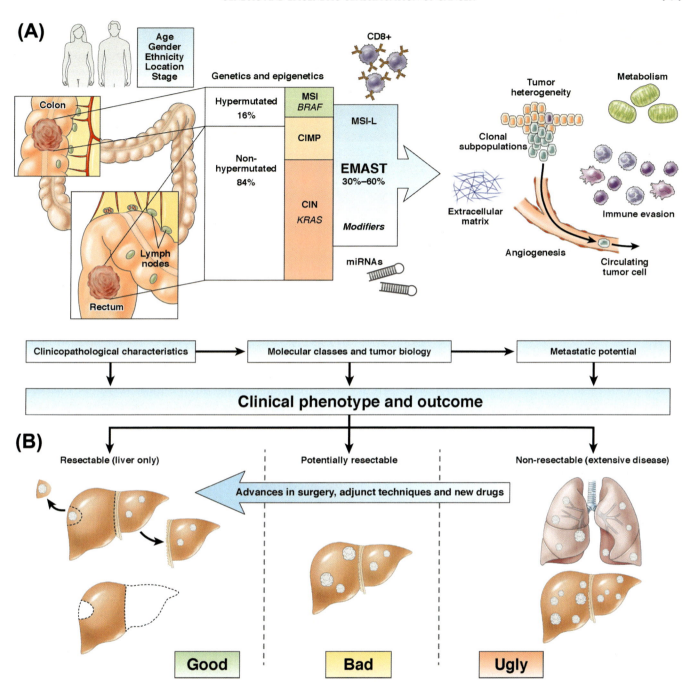

FIGURE 35.7 **Clinical and molecular influence on aggressiveness of colorectal liver metastasis.** (A) Clinical behavior of colorectal cancer is determined by several factors, including demographic data (age, gender, race) and tumor presentation (location, stage) and timing of presentation of metastasis (synchronous or metachronous). Embedded in the cancer cells are the molecular pathways, which follows distinct forms of genomic instability yet with partly overlapping areas. Hypermutated cancers belong to the microsatellite instable (MSI) cancers and in part the CpG-island methylator phenotype (CIMP) cancers. Non-hypermutated follow in large parts the chromosomal instability (CIN)–driven pathways, often involving KRAS mutations from an early stage. The propensity to develop metastasis may possibly be modified through the elevated microsatellite alterations at selected tetranucleotide repeat (EMAST) and associated epigenetic mechanisms, such as regulation of microRNAs or activity and numbers and activity of $CD8^+$ immune cells. Finally, the microenvironment contains numerous factors that may facilitate or propagate metastasis to invade, spread and settle in a new organ sites, particularly the liver and the lungs. (B) Determined by the clinical presentation, the genetic traits and molecular mechanisms, the prognosis in colorectal liver metastasis is related to resectabilty for long-term survival. "Good" cases amenable for surgery have fewer bad genetic traits. Patients with concomitant liver and lung metastases have an "ugly" tumor biology and are more likely to have higher frequencies of both KRAS and BRAF mutations and respond poorly to any line of treatment. The "bad" cases are in between, and the shift from "nonresectable" to "resectable" experiences a positive drift with time and where changing practice in surgical strategy, novel techniques and use of conversion chemotherapy regimens improve outcomes. *Reproduced with permission from Søreide K, Watson MM, Hagland HR. Deciphering the Molecular Code to Colorectal Liver Metastasis Biology Through Microsatellite Alterations and Allelic Loss: The Good, the Bad, and the Ugly.* Gastroenterology. *2016 Apr;150(4):811-4.*

microRNAs or activity and numbers of CD8$^+$ immune cells. Finally, the microenvironment contains numerous factors that may facilitate or propagate metastasis to invade, spread and settle in a new organ sites, particularly the liver and the lungs. Determined by the clinical presentation (Fig. 35.7B), the genetic traits and molecular mechanisms, the prognosis in colorectal liver metastasis is related to resectabilty for long-term survival. Novel biomarkers may aid in understanding aggressiveness of liver metastasis, assist in clinical decision making and help to find new and more efficient therapies.

Limitations to Epigenetic Classification and Prognostication

Generalizability and reproducibility in using epigenetic markers for classification and diagnosis has been hampered by the lack of standardized and unified protocols and analytical designs. For example, we found in a study that the call of CIMP classification would deviate substantially between cases depending on what definitions, genes and panels where used for defining CIMP status [98]. A systematic review examined all published studies on CRC prognosis according to the different definitions of CIMP and identified 36 studies [99]. Among these, 30 (83%) studies reported the association of CIMP and CRC prognosis and 11 (31%) studies reported the association of CIMP with survival after chemotherapy. Overall, 16 different definitions of CIMP were identified [99]. The majority of studies reported a poorer prognosis for patients with CIMP-positive CRC than with CIMP-negative CRC. Inconsistent results or varying effect strengths could not be explained by different CIMP definitions used. No consistent variation in response to specific therapies according to CIMP status was found. As the authors conclude, comparative analyses of different CIMP panels in the same large study populations are needed to further clarify the role of CIMP definitions and to find out how methylation information can best be used to predict CRC prognosis and response to specific CRC therapies. From both studies [98,99] goes the notion that better standardization and agreement between studies are needed.

EPIGENETIC BIOMARKERS IN CANCER

The ability to identify high- and low-risk patients in cancer beyond the regularly used Tumor-Node-Metastasis (TNM) system, is still somewhat poor [81]. Molecular biology has, over the years, given insight into basic principles of cancer initiation and development; and increasingly these are being exploited as biomarkers, and several epigenetic markers or tools have been proposed in various tumors. This includes aberrations increasing risk of tumor development, (epi-)genetic changes associated with the stepwise progression of the disease, and errors predicting response to a specific treatment. Because several epigenetic changes occur before histopathological changes are present, they can serve as biomarkers for cancer diagnosis and risk assessment [100]. Many cancers may remain asymptomatic until relatively late stages; in managing the disease, efforts should be focused on early detection, accurate prediction of disease progression, and frequent monitoring [48]. Based on epigenomic information, several biomarkers have been identified that may serve as diagnostic tools; some such biomarkers also may be useful in identifying individuals who will respond to therapy and, potentially, live longer [40,45,48,85,100–102].

As techniques develop and investigations expand, the putative list of potential epigenetic biomarkers grow longer. However, methylation status is a (if not the most) frequently investigated marker [103], as indicated by the CpG-island methylation status explored for CRC. Further, noncoding RNAs may serve as biomarkers, exemplified as microRNA profiles. However, as the average CRC methylome has hundreds to thousands of abnormally methylated genes and dozens of altered miRNAs [48,104], it is hard to define any given marker as valid. Further, biomarker validity and robustness depends on stage of disease (e.g., preneoplasia, early stage or, metastatic situation), will likely be dependent on a set of markers (e.g., related to other genetic mutations or, for example, as set of miRNA alterations) and, is dependent on the outcome as endpoint (i.e., whether it is suited for diagnosis, prognosis, or therapy prediction) [105]. Thus, much is still to be learned for the best use and exact role of epigenetic markers in cancer disease outcome prediction. Much like the Genome-wide association studies (GWAS) studies, epigenome-wide association studies (EWAS) are likely to require large international consortium-based approaches to reach the numbers of subjects, and statistical and scientific rigor, required for robust findings.

Cancer Epigenome in Epidemiology

Epigenetic epidemiology includes the study of variation in epigenetic traits and the risk of disease in populations. Its application to the field of cancer has provided insight into how lifestyle and environmental factors influence the epigenome and how epigenetic events may be involved in carcinogenesis [106]. Furthermore, it has the potential to bring benefit to patients through the identification of diagnostic markers that enable the early detection of disease and prognostic markers that can inform upon appropriate treatment strategies. However, there are a number of challenges associated with the conduct of such studies and with the identification of biomarkers that can be applied to the clinical setting [106].

Completion of the human genome a decade ago laid the foundation for using genetic information in assessing risk to identify individuals and populations that are likely to develop cancer, and designing treatments tailored to a person's genetic profile (coining the term "precision medicine"). GWAS, completed during the past several years, have identified risk-associated single nucleotide polymorphisms (SNPs) that can be used as screening tools in epidemiologic studies of a variety of tumor types. Just as GWAS grew from the field of genetic epidemiology, so too do EWAS derive from the burgeoning field of epigenetic epidemiology, with both aiming to understand the molecular basis for disease risk [107]. EWAS to identify regulatory mechanisms and to predict their causal link to common diseases currently make use of a crude proxy by examining percentage methylation differences at ~2% of CpG sites in the genome, sites selected in part for their accessibility using available detection methods. Now, novel pipelines in which candidate functional SNPs are first evaluated by fine mapping, epigenomic profiling, and epigenome editing, and then interrogated for causal function by using genome editing to create isogenic cell lines followed by phenotypic characterization are reported [108], which may generate new insight into causality and relations between genetics and epigenetics and relation to disease risk.

While genetic risk of disease is currently unmodifiable, there is, at least in theory, a potential for that epigenetic risk may be reversible and or modifiable through modifiers. There have been numerous diseases, exposures and lifestyle factors investigated with EWAS, with several significant associations now identified (reviewed in [107,109]). Currently, it has been difficult for researchers to understand disease risk from GWAS results [110], as most GWAS-identified SNPs are located in noncoding regions of the genome. Thus, the GWAS field has been left with the conundrum as to how a single-nucleotide change in a noncoding region could confer increased risk for a specific disease. One possible answer to the causality puzzle is that the variant SNPs cause changes in gene expression levels rather than causing changes in protein function [110]. The current hypothesis is that one or more of these risk-associated noncoding SNPs cause changes in gene expression of a critical gene [110]. However, functional follow-up experiments are both expensive and time-consuming and one cannot test each possible candidate SNP for causality. Many of the SNPs found in noncoding regions are in relation to promoters and enhancers. While investigators can bioinformatically identify promoter SNPs, it is more difficult to identify SNPs within enhancers. Unlike promoters, the enhancers do not occur at a defined distance from a transcription start site (TSS). However, they can be identified by specific epigenomic profiles. Several methodological and biological implications are drawn from past findings with the need for improved technology, analyses, and interpretations (as discussed in detail in [110]). However, it is hoped that information gained from GWAS and EWAS will have potential for applications in cancer control and treatment in the future [107,109]. However, much like the GWAS studies, EWAS are likely to require large international consortium-based approaches to reach the numbers of subjects, and statistical and scientific rigor, required for robust findings.

EPIGENETICS AS CANCER THERAPEUTIC TARGETS

Genetic mutations and gross structural defects in the DNA sequence permanently alter genetic loci in ways that significantly disrupt gene function. In contrast, genes modified by aberrant epigenetic modifications remain structurally intact and are thus subject to partial or complete reversal of modifications that restore the original (i.e., nondiseased) state. Such reversibility makes epigenetic modifications ideal targets for therapeutic intervention [12]. For a comprehensive presentation of targets of epigenetic writers, readers, and erasers, we refer to comprehensive overviews on this topic [11,12].

Epigenetic therapy is currently entertained in comparatively few malignant diseases, although the field is expanding. At present there are nine drugs approved for clinical use by the US Food and Drug Administration (FDA) [111,112]. The two cytidine analogs azacitidine and decitabine were the first to be approved for use in a preleukemic disorder called myelodysplastic syndrome (MDS) and in acute myelogenic leukemia (AML) [113,114]. The two drugs lead to reversal of DNA hypermethylation by inactivation of DNA methyltransferase (DNMT). Response rates vary and most responders relapse within two years [115].

For those with further refractory disease, targeted therapies are now available against mutations in isocitrate dehydrogenase 1 and 2 (IDH1 and IDH2) [116]. While not epigenetic modulators, both are critical enzymes in the citric acid cycle, and are examples of indirect inhibition of epigenetic dysregulation. Mutated variants increase levels of the metabolite 2-hydroxyglutarate, promoting tumor growth, by inhibition of αKG dioxygenases, including histone demethylases and DNA hydroxylase, which in turn interfere with the cells ability to remove methyl groups from DNA. This impedes transcription, preventing cellular differentiation, which is a defining feature of leukemia. Notably, the two drugs enasidenib and ivosidenib, targeting IDH2 and IDH1 respectively, were the result of biomarker driven research [117,118], and the first approval of drugs targeting oncometabolites.

Four histone deacetylase (HDAC) inhibitors are currently in clinical use, namely vorinostat and romidepsin (cutaneous T-cell lymphoma), and belinostat, and panobinostat (peripheral T-cell lymphoma) [119]. Their mechanism of action is not fully elucidated and appears to vary depending on tumor origin and biology, drug type, and dosing. Chromatin relaxation ultimately leads to transcription of genes that promote cellular differentiation or inhibit cell growth, impeding tumor progression.

The last group of drugs currently FDA approved target histone lysine methyltransferases (KMTs), in the end leading to transcriptional repression [120]. The enhancer of zeste homolog 2 (EZH2) is targeted by tazemetostat, providing favorable responses in patients with follicular lymphomas, with both wildtype and mutated *EZH2* [121,122].

While the role of epigenetics on solid organ cancers is still in its early days and awaits further trials [97,123,124], EZH2 inhibition by tazemetostat has proven efficacious in epithelioid sarcoma in patients not eligible for complete resection [125,126]. Why solid tumors appear to be less sensitive to epigenetic agents, compared to hematological malignancies, is a conundrum as the solid tumors harbor significant epigenetic aberrations [127]. Awaiting further progress in drug therapeutics, epigenetic biomarkers may provide important benefits in early diagnosis and prediction of drug responses in solid tumors [128].

While first generation epigenetic drugs were unspecific, broadly targeting, with challenging safety profiles influencing efficacy, several other avenues are being explored, such as epigenetic sensitization to radiotherapy or cytotoxic chemotherapy, and epigenetic immune modulation and priming for immune therapy [97,129–132]. The latter have moved into clinical trials, where epigenetic drugs combined with immune checkpoint therapy (ICT) have reported promising results in both liquid and solid tumors [133–137]. In addition, epigenetics is promising to improve chimeric antigen receptor (CAR) T-cell therapy [138–140] Further, several new mechanisms and possible early phase trials are discussed in more detail elsewhere [12,97,123,124,141–145].

One of the intricacies of cancer treatments involve unintended epigenetic modifications that may exacerbate tumor progression—a side effect that clearly would contradict the use of epigenetic therapy for curative purposes. The specificity restrictions (i.e., tumor specific effect not involving the normal epigenome) posed by epigenetic therapies and ways to address such limitations is presented in detail elsewhere [146]. As stated, with the next generation of targets and drugs, there is hope that novel epigenetic therapies may improve drug targeting and drug delivery, optimize dosing schedules, and improve the efficacy of preexisting treatment modalities [143].

FUTURE PERSPECTIVES

For cancer as a genetic disease, it is becoming clear that a wide range of epigenetic pathways, affecting nucleic acids and histone proteins, are dynamic and reversible, offering considerable promise for a better understanding of cancer processes and having the potential for treatment [2,79,114,147–149]. These insights have had a paradigm-shift impact on our understanding of normal and perturbed development. Indeed, the dynamic epigenome adds an additional layer of complexity to the function of our genome, leading to "a sum greater than its parts" [43]. As such, advances in genome editing, visualization technology, and genome-wide analyses have revealed unprecedented complexity of chromatin pathways, offering explanations to long-standing questions and presenting new challenges [43]. The combined investigation of GWAS and EWAS studies may yield better determination of risk factors for improved cancer control [109]. Novel understanding of epigenetic therapies may render new therapy options understanding and overcoming chemotherapy refractory neoplasia [150]. Epigenetics in cancer may yield a wider range of options tailored towards personalized cancer therapy [150–152], in particular, the rapidly emerging field of immune-oncology looks exciting [42].

Abbreviations

CIMP	CpG island methylator phenotype
CIN	chromosomal instability
ncRNA	noncoding RNA
CR	cromatin regulator
TF	transcription factor
CSC	cancer stem cells
GWAS	genome-wide association studies
EWAS	epigenome-wide association studies
SNP	single nucleotide polymorphisms
miRNA	microRNAs
piRNa	PIWI-interacting RNA
snoRNA	small nucleolar RNA
siRNA	small interfering RNA
lncRNA	long noncoding RNA
HAT	histone acetyltransferases

HDAC	histone deacetylases
HMT	histone methyltransferases
HDM	histone demethylases
MSI	microsatellite instability
CRC	colorectal cancer
TNM	tumor-node-metastasis system
TET	Ten-eleven translocation
PTM	post-translational modification

References

[1] Makohon-Moore A, Iacobuzio-Donahue CA. Pancreatic cancer biology and genetics from an evolutionary perspective. Nat Rev Cancer 2016.

[2] Drake TM, Søreide K. Cancer epigenetics in solid organ tumours: a primer for surgical oncologists. Eur J Surg Oncol 2019;45(5):736−46.

[3] Hanahan D, Weinberg RA. Hallmarks of cancer: the next generation. Cell 2011;144(5):646−74.

[4] Shen H, Laird PW. Interplay between the cancer genome and epigenome. Cell 2013;153(1):38−55.

[5] Baylin SB, Jones PA. Epigenetic determinants of cancer. Cold Spring Harb Perspect Biol 2016;8(9):a019505.

[6] Esteller M. Epigenetics in cancer. N Engl J Med 2008;358(11): 1148−59.

[7] Simo-Riudalbas L, Esteller M. Cancer genomics identifies disrupted epigenetic genes. Hum Genet 2014;133(6):713−25.

[8] Rodriguez-Paredes M, Esteller M. Cancer epigenetics reaches mainstream oncology. Nat Med 2011;17(3):330−9.

[9] Dawson MA, Kouzarides T. Cancer epigenetics: from mechanism to therapy. Cell 2012;150(1):12−27.

[10] Song Y, Wu F, Wu J. Targeting histone methylation for cancer therapy: enzymes, inhibitors, biological activity and perspectives. J Hematol Oncol 2016;9(1):49.

[11] Cheng Y, et al. Targeting epigenetic regulators for cancer therapy: mechanisms and advances in clinical trials. Signal Transduct Target Ther 2019;4:62.

[12] Simo-Riudalbas L, Esteller M. Targeting the histone orthography of cancer: drugs for writers, erasers and readers. Br J Pharmacol 2015;172(11):2716−32.

[13] Brien GL, Valerio DG, Armstrong SA. Exploiting the epigenome to control cancer-promoting gene-expression programs. Cancer Cell 2016;29(4):464−76.

[14] Weichenhan D, Plass C. The evolving epigenome. Hum Mol Genet 2013;22(R1) R1−6.

[15] Torano EG, et al. The impact of external factors on the epigenome: in utero and over lifetime. Biomed Res Int 2016;2016:2568635.

[16] Field AE, et al. DNA methylation clocks in aging: categories, causes, and consequences. Mol Cell 2018;71(6):882−95.

[17] Chiavellini P, et al. Aging and rejuvenation − a modular epigenome model. Aging (Albany NY) 2021;13(4):4734−46.

[18] Saul D, Kosinsky RL. Epigenetics of aging and aging-associated diseases. Int J Mol Sci 2021;22(1):401.

[19] Horvath S, Raj K. DNA methylation-based biomarkers and the epigenetic clock theory of ageing. Nat Rev Genet 2018;19(6):371−84.

[20] Aunan JR, et al. Molecular and biological hallmarks of ageing. Br J Surg 2016;103(2):e29−46.

[21] Zheng C, Li L, Xu R. Association of epigenetic clock with consensus molecular subtypes and overall survival of colorectal cancer. Cancer Epidemiol Biomarkers Prev 2019;28(10):1720−4.

[22] Dugue PA, et al. DNA methylation-based biological aging and cancer risk and survival: pooled analysis of seven prospective studies. Int J Cancer 2018;142(8):1611−19.

[23] Ideraabdullah FY, Zeisel SH. Dietary modulation of the epigenome. Physiol Rev 2018;98(2):667−95.

[24] Romani M, Pistillo MP, Banelli B. Environmental epigenetics: crossroad between public health, lifestyle, and cancer prevention. Biomed Res Int 2015;2015:587983.

[25] Arora I, Sharma M, Tollefsbol TO. Combinatorial epigenetics impact of polyphenols and phytochemicals in cancer prevention and therapy. Int J Mol Sci 2019;20(18):4567.

[26] Hino S, Nagaoka K, Nakao M. Metabolism-epigenome crosstalk in physiology and diseases. J Hum Genet 2013;58(7):410−15.

[27] Hagland HR, Soreide K. Cellular metabolism in colorectal carcinogenesis: influence of lifestyle, gut microbiome and metabolic pathways. Cancer Lett 2015;356(2 Pt A):273−80.

[28] Soreide K, Sund M. Epidemiological-molecular evidence of metabolic reprogramming on proliferation, autophagy and cell signaling in pancreas cancer. Cancer Lett 2015;356(2 Pt A):281−8.

[29] Johnson IT, Belshaw NJ. The effect of diet on the intestinal epigenome. Epigenomics 2014;6(2):239−51.

[30] Boukouris AE, Zervopoulos SD, Michelakis ED. Metabolic enzymes moonlighting in the nucleus: metabolic regulation of gene transcription. Trends Biochem Sci 2016;41(8):712−30.

[31] Aunan JR, Cho WC, Søreide K. The biology of aging and cancer: a brief overview of shared and divergent molecular hallmarks. Aging Dis 2017;8(5):628−42.

[32] Camarena V, Wang G. The epigenetic role of vitamin C in health and disease. Cell Mol Life Sci 2016;73(8):1645−58.

[33] Fetahu IS, Hobaus J, Kallay E. Vitamin D and the epigenome. Front Physiol 2014;5:164.

[34] Pavlova NN, Thompson CB. The emerging hallmarks of cancer metabolism. Cell Metab 2016;23(1):27−47.

[35] Hagland HR, et al. Molecular pathways and cellular metabolism in colorectal cancer. Dig Surg 2013;30(1):12−25.

[36] Vogelstein B, et al. Genetic alterations during colorectal-tumor development. N Engl J Med 1988;319(9):525−32.

[37] Feinberg AP, Vogelstein B. Hypomethylation distinguishes genes of some human cancers from their normal counterparts. Nature 1983;301(5895):89−92.

[38] Vogelstein B, et al. Cancer genome landscapes. Science 2013;339 (6127):1546−58.

[39] Watson MM, Berg M, Soreide K. Prevalence and implications of elevated microsatellite alterations at selected tetranucleotides in cancer. Br J Cancer 2014;111(5):823−7.

[40] Carethers JM, Jung BH. Genetics and genetic biomarkers in sporadic colorectal cancer. Gastroenterology 2015;149(5): 1177−90 e3.

[41] Kumar R, et al. Epigenomic regulation of oncogenesis by chromatin remodeling. Oncogene 2016;35(34):4423−36.

[42] Cao J, Yan Q. Cancer epigenetics, tumor immunity, and immunotherapy. Trends. Cancer 2020;6(7):580−92.

[43] Soshnev AA, Josefowicz SZ, Allis CD. Greater than the sum of parts: complexity of the dynamic epigenome. Mol Cell 2016;62 (5):681−94.

[44] Dragomir MP, et al. Non-coding RNAs in GI cancers: from cancer hallmarks to clinical utility. Gut 2020;69(4):748−63.

[45] Berg M, Soreide K. Genetic and epigenetic traits as biomarkers in colorectal cancer. Int J Mol Sci 2011;12(12):9426−39.

[46] Lorzadeh A, et al. Epigenetic regulation of intestinal stem cells and disease: a balancing act of DNA and histone methylation. Gastroenterology 2021;160(7):2267−82.

[47] Vukic M, Daxinger L. DNA methylation in disease: immunodeficiency, centromeric instability, facial anomalies syndrome. Essays Biochem 2019;63(6):773−83.

[48] Okugawa Y, Grady WM, Goel A. Epigenetic alterations in colorectal cancer: emerging biomarkers. Gastroenterology 2015;149 (5):1204−25 e12.

[49] Grady WM, Yu M, Markowitz SD. Epigenetic alterations in the gastrointestinal tract: current and emerging use for biomarkers of cancer. Gastroenterology 2021;160(3):690−709.

[50] Bray JK, et al. Roles and regulations of TET enzymes in solid tumors. Trends. Cancer 2021;7(7):635−46.

[51] Nishiyama A, Nakanishi M. Navigating the DNA methylation landscape of cancer. Trends Genet 2021;37(11):1012−27.

[52] Ortiz-Barahona V, Joshi RS, Esteller M. Use of DNA methylation profiling in translational oncology. Semin Cancer Biol 2020;83:523−35.

[53] Nazemalhosseini Mojarad E, et al. The CpG island methylator phenotype (CIMP) in colorectal cancer. Gastroenterol Hepatol Bed Bench 2013;6(3):120−8.

[54] Juo YY, et al. Prognostic value of CpG island methylator phenotype among colorectal cancer patients: a systematic review and meta-analysis. Ann Oncol 2014;25(12):2314−27.

[55] Tapial S, et al. Cimp-positive status is more representative in multiple colorectal cancers than in unique primary colorectal cancers. Sci Rep 2019;9(1):10516.

[56] Sandoval J, Esteller M. Cancer epigenomics: beyond genomics. Curr Opin Genet Dev 2012;22(1):50−5.

[57] Deshmukh S, et al. Oncohistones: a roadmap to stalled development. FEBS J 2022;289(5):1315−28.

[58] Nacev BA, et al. The expanding landscape of 'oncohistone' mutations in human cancers. Nature 2019;567(7749):473−8.

[59] Gutschner T, Diederichs S. The hallmarks of cancer: a long non-coding RNA point of view. RNA Biol 2012;9(6):703−19.

[60] Xue B, He L. An expanding universe of the non-coding genome in cancer biology. Carcinogenesis 2014;35(6):1209−16.

[61] Liz J, Esteller M. lncRNAs and microRNAs with a role in cancer development. Biochim Biophys Acta 2016;1859(1):169−76.

[62] Yang Y, et al. Long non-coding RNAs in colorectal cancer: novel oncogenic mechanisms and promising clinical applications. Cancer Lett 2021;504:67−80.

[63] Dehghani H. Regulation of chromatin organization in cell stemness: the emerging role of long non-coding RNAs. Stem Cell Rev Rep 2021;17(6):2042−53.

[64] Garg M, Sethi G. Emerging role of long non-coding RNA (lncRNA) in human malignancies: a unique opportunity for precision medicine. Cancer Lett 2021;519:1.

[65] Slack FJ, Chinnaiyan AM. The role of non-coding RNAs in oncology. Cell 2019;179(5):1033−55.

[66] Vogelstein B, Kinzler KW. The path to cancer − Three strikes and you're out. N Engl J Med 2015;373(20):1895−8.

[67] Huidobro C, Fernandez AF, Fraga MF. The role of genetics in the establishment and maintenance of the epigenome. Cell Mol Life Sci 2013;70(9):1543−73.

[68] Laird PW. Principles and challenges of genomewide DNA methylation analysis. Nat Rev Genet 2010;11(3):191−203.

[69] Suva ML, Riggi N, Bernstein BE. Epigenetic reprogramming in cancer. Science 2013;339(6127):1567−70.

[70] Hammoud SS, Cairns BR, Jones DA. Epigenetic regulation of colon cancer and intestinal stem cells. Curr Opin Cell Biol 2013;25(2):177−83.

[71] Ravasio R, Ceccacci E, Minucci S. Self-renewal of tumor cells: epigenetic determinants of the cancer stem cell phenotype. Curr Opin Genet Dev 2016;36:92−9.

[72] Easwaran H, Tsai HC, Baylin SB. Cancer epigenetics: tumor heterogeneity, plasticity of stem-like states, and drug resistance. Mol Cell 2014;54(5):716−27.

[73] Bedi U, et al. Epigenetic plasticity: a central regulator of epithelial-to-mesenchymal transition in cancer. Oncotarget 2014;5(8):2016−29.

[74] Avgustinova A, Benitah SA. The epigenetics of tumour initiation: cancer stem cells and their chromatin. Curr Opin Genet Dev 2016;36:8−15.

[75] Mazor T, et al. Intratumoral heterogeneity of the epigenome. Cancer Cell 2016;29(4):440−51.

[76] Pribluda A, de la Cruz CC, Jackson EL. Intratumoral heterogeneity: from diversity comes resistance. Clin Cancer Res 2015;21(13):2916−23.

[77] Lv JF, et al. Epigenetic alternations and cancer chemotherapy response. Cancer Chemother Pharmacol 2016;77(4):673−84.

[78] Topper MJ, et al. The emerging role of epigenetic therapeutics in immuno-oncology. Nat Rev Clin Oncol 2020;17(2):75−90.

[79] Duan YC, et al. Research progress of dual inhibitors targeting crosstalk between histone epigenetic modulators for cancer therapy. Eur J Med Chem 2021;222:113588.

[80] Liu M, et al. Understanding the epigenetic regulation of tumours and their microenvironments: opportunities and problems for epigenetic therapy. J Pathol 2017;241(1):10−24.

[81] Lea D, et al. Accuracy of TNM staging in colorectal cancer: a review of current culprits, the modern role of morphology and stepping-stones for improvements in the molecular era. Scand J Gastroenterol 2014;49(10):1153−63.

[82] Roalsø M, Aunan JR, Søreide K. Refined TNM-staging for pancreatic adenocarcinoma − Real progress or much ado about nothing? Eur J Surg Oncol 2020;46(8):1554−7.

[83] Berg M, Soreide K. EGFR and downstream genetic alterations in KRAS/BRAF and PI3K/AKT pathways in colorectal cancer: implications for targeted therapy. Discov Med 2012;14(76):207−14.

[84] Ten Hoorn S, et al. Clinical value of consensus molecular subtypes in colorectal cancer: a systematic review and meta-analysis. J Natl Cancer Inst 2022;114(4):503−16.

[85] Soreide K, Watson MM, Hagland HR. Deciphering the molecular code to colorectal liver metastasis biology through microsatellite alterations and allelic loss: the good, the bad, and the ugly. Gastroenterology 2016;150(4):811−14.

[86] Kanani A, Veen T, Søreide K. Neoadjuvant immunotherapy in primary and metastatic colorectal cancer. Br J Surg 2021;108(12):1417−25.

[87] Martini G, et al. Molecular subtypes and the evolution of treatment management in metastatic colorectal cancer. Ther Adv Med Oncol 2020;12:1758835920936089.

[88] Sadanandam A, et al. A colorectal cancer classification system that associates cellular phenotype and responses to therapy. Nat Med 2013;19(5):619−25.

[89] Guinney J, et al. The consensus molecular subtypes of colorectal cancer. Nat Med 2015;21(11):1350−6.

[90] Ranjbar R, et al. EMAST frequency in colorectal cancer: a meta-analysis and literature review. Biomark Med 2020;14(11):1021−30.

[91] Watson MM, et al. Elevated microsatellite alterations at selected tetranucleotides in early-stage colorectal cancers with and without high-frequency microsatellite instability: same, same but different? Cancer Med 2016;5(7):1580−7.

[92] Carethers JM, Koi M, Tseng-Rogenski SS. EMAST is a form of microsatellite instability that is initiated by inflammation and modulates colorectal cancer progression. Genes (Basel) 2015;6(2):185−205.

[93] Venderbosch S, et al. EMAST is associated with a poor prognosis in microsatellite instable metastatic colorectal cancer. PLoS One 2015;10(4):e0124538.

[94] Watson MM, et al. Elevated microsatellite alterations at selected tetranucleotides (EMAST) in colorectal cancer is associated with an elderly, frail phenotype and improved recurrence-free survival. Ann Surg Oncol 2020;27(4):1058−67.

[95] Watson MM, et al. Prevalence of PD-L1 expression is associated with EMAST, density of peritumoral T-cells and recurrence-free survival in operable non-metastatic colorectal cancer. Cancer Immunol Immunother 2020;69(8):1627−37.

[96] Watson MM, et al. Elevated microsatellite alterations at selected tetranucleotides (EMAST) is not attributed to MSH3 loss in stage I-III colon cancer: an automated, digitalized assessment by immunohistochemistry of whole slides and hot spots. Transl Oncol 2019;12(12):1583–8.

[97] Abdelfatah E, et al. Epigenetic therapy in gastrointestinal cancer: the right combination. Ther Adv Gastroenterol 2016;9(4):560–79.

[98] Berg M, Hagland HR, Soreide K. Comparison of CpG island methylator phenotype (CIMP) frequency in colon cancer using different probe- and gene-specific scoring alternatives on recommended multi-gene panels. PLoS One 2014;9(1):e86657.

[99] Jia M, et al. Different definitions of CpG island methylator phenotype and outcomes of colorectal cancer: a systematic review. Clin Epigenetics 2016;8:25.

[100] Verma M. The role of epigenomics in the study of cancer biomarkers and in the development of diagnostic tools. Adv Exp Med Biol 2015;867:59–80.

[101] Meseure D, et al. Long noncoding RNAs as new architects in cancer epigenetics, prognostic biomarkers, and potential therapeutic targets. Biomed Res Int 2015;2015:320214.

[102] Dai X, et al. Cancer hallmarks, biomarkers and breast cancer molecular subtypes. J Cancer 2016;7(10):1281–94.

[103] Lin X, Barton S, Holbrook JD. How to make DNA methylome wide association studies more powerful. Epigenomics 2016;8(8):1117–29.

[104] Reis AH, Vargas FR, Lemos B. Biomarkers of genome instability and cancer epigenetics. Tumour Biol 2016;37(10):13029–38.

[105] Søreide K. Receiver-operating characteristic curve analysis in diagnostic, prognostic and predictive biomarker research. J Clin Pathol 2009;62(1):1–5.

[106] Barrow TM, Michels KB. Epigenetic epidemiology of cancer. Biochem Biophys Res Commun 2014;455(1–2):70–83.

[107] Flanagan JM. Epigenome-wide association studies (EWAS): past, present, and future. Methods Mol Biol 2015;1238:51–63.

[108] Spisak S, et al. CAUSEL: an epigenome- and genome-editing pipeline for establishing function of noncoding GWAS variants. Nat Med 2015;21(11):1357–63.

[109] Verma M. Genome-wide association studies and epigenome-wide association studies go together in cancer control. Future Oncol 2016;12(13):1645–64.

[110] Tak YG, Farnham PJ. Making sense of GWAS: using epigenomics and genome engineering to understand the functional relevance of SNPs in non-coding regions of the human genome. Epigenetics Chromatin 2015;8:57.

[111] Bates SE. Epigenetic therapies for cancer. N Engl J Med 2020;383(7):650–63.

[112] Mohammad HP, Barbash O, Creasy CL. Targeting epigenetic modifications in cancer therapy: erasing the roadmap to cancer. Nat Med 2019;25(3):403–18.

[113] Yun S, et al. Targeting epigenetic pathways in acute myeloid leukemia and myelodysplastic syndrome: a systematic review of hypomethylating agents trials. Clin Epigenetics 2016;8:68.

[114] Paluch BE, et al. Epigenetics: a primer for clinicians. Blood Rev 2016;30(4):285–95.

[115] Wei AH, et al. Oral azacitidine maintenance therapy for acute myeloid leukemia in first remission. N Engl J Med 2020;383(26):2526–37.

[116] Stein EM, et al. Enasidenib in mutant IDH2 relapsed or refractory acute myeloid leukemia. Blood 2017;130(6):722–31.

[117] Welch JS, et al. The origin and evolution of mutations in acute myeloid leukemia. Cell 2012;150(2):264–78.

[118] Lu C, et al. Induction of sarcomas by mutant IDH2. Genes Dev 2013;27(18):1986–98.

[119] Eckschlager T, et al. Histone deacetylase inhibitors as anticancer drugs. Int J Mol Sci 2017;18(7):1414.

[120] Husmann D, Gozani O. Histone lysine methyltransferases in biology and disease. Nat Struct Mol Biol 2019;26(10):880–9.

[121] Knutson SK, et al. Selective inhibition of EZH2 by EPZ-6438 leads to potent antitumor activity in EZH2-mutant non-Hodgkin lymphoma. Mol Cancer Ther 2014;13(4):842–54.

[122] Italiano A, et al. Tazemetostat, an EZH2 inhibitor, in relapsed or refractory B-cell non-Hodgkin lymphoma and advanced solid tumours: a first-in-human, open-label, phase 1 study. Lancet Oncol 2018;19(5):649–59.

[123] Valdespino V, Valdespino PM. Potential of epigenetic therapies in the management of solid tumors. Cancer Manag Res 2015;7:241–51.

[124] Ramachandran S, et al. Epigenetic therapy for solid tumors: highlighting the impact of tumor hypoxia. Genes (Basel) 2015;6(4):935–56.

[125] Stacchiotti S, et al. Safety and efficacy of tazemetostat, a first-in-class EZH2 inhibitor, in patients (pts) with epithelioid sarcoma (ES)(NCT02601950). 2019, American Society of Clinical Oncology.

[126] Hoy SM. Tazemetostat: first approval. Drugs 2020;80(5):513–21.

[127] Jin N, et al. Advances in epigenetic therapeutics with focus on solid tumors. Clin Epigenetics 2021;13(1):83.

[128] Berdasco M, Esteller M. Clinical epigenetics: seizing opportunities for translation. Nat Rev Genet 2019;20(2):109–27.

[129] Jonsson M, et al. Hypoxia-independent gene expression signature associated with radiosensitisation of prostate cancer cell lines by histone deacetylase inhibition. Br J Cancer 2016;115(8):929–39.

[130] Min HY, et al. Essential role of DNA methyltransferase 1-mediated transcription of insulin-like growth factor 2 in resistance to histone deacetylase inhibitors. Clin Cancer Res 2016;23(5):1299–311.

[131] Beg AA, Gray JE. HDAC inhibitors with PD-1 blockade: a promising strategy for treatment of multiple cancer types? Epigenomics 2016;8(8):1015–17.

[132] Griffin GK, et al. Epigenetic silencing by SETDB1 suppresses tumour intrinsic immunogenicity. Nature 2021;595(7866):309–14.

[133] Levy BP, et al. Randomised phase 2 study of pembrolizumab plus CC-486 versus pembrolizumab plus placebo in patients with previously treated advanced non-small cell lung cancer. Eur J Cancer 2019;108:120–8.

[134] Hellmann MD, et al. Entinostat plus pembrolizumab in patients with metastatic NSCLC previously treated with anti-PD-(L)1 therapy. Clin Cancer Res 2021;27(4):1019–28.

[135] Gray JE, et al. Phase I/Ib study of pembrolizumab plus vorinostat in advanced/metastatic non-small cell lung cancer. Clin Cancer Res 2019;25(22):6623–32.

[136] Daver N, et al. Efficacy, safety, and biomarkers of response to azacitidine and nivolumab in relapsed/refractory acute myeloid leukemia: a nonrandomized, open-label, Phase II study. Cancer Discov 2019;9(3):370–83.

[137] Ny L, et al. The PEMDAC phase 2 study of pembrolizumab and entinostat in patients with metastatic uveal melanoma. Nat Commun 2021;12(1):5155.

[138] Akbari B, et al. Epigenetic strategies to boost CAR T cell therapy. Mol Ther 2021;29(9):2640–59.

[139] Wang Y, et al. Low-dose decitabine priming endows CAR T cells with enhanced and persistent antitumour potential via epigenetic reprogramming. Nat Commun 2021;12(1):409.

[140] You L, et al. Decitabine-mediated epigenetic reprograming enhances anti-leukemia efficacy of CD123-targeted chimeric antigen receptor T-cells. Front Immunol 2020;11:1787.

[141] Cramer SA, Adjei IM, Labhasetwar V. Advancements in the delivery of epigenetic drugs. Expert Opin Drug Deliv 2015;12(9):1501–12.

[142] Ganesan A. Multitarget drugs: an epigenetic epiphany. Chem Med Chem 2016;11(12):1227–41.

[143] Hamm CA, Costa FF. Epigenomes as therapeutic targets. Pharmacol Ther 2015;151:72–86.

[144] Morera L, Lubbert M, Jung M. Targeting histone methyltransferases and demethylases in clinical trials for cancer therapy. Clin Epigenetics 2016;8:57.

[145] Isaacs JS. Hsp90 as a "Chaperone" of the epigenome: insights and opportunities for cancer therapy. Adv Cancer Res 2016;129:107–40.

[146] Bojang Jr. P, Ramos KS. The promise and failures of epigenetic therapies for cancer treatment. Cancer Treat Rev 2014;40(1):153–69.

[147] Soto J, et al. The impact of next-generation sequencing on the DNA methylation-based translational cancer research. Transl Res 2016;169:1–18 e1.

[148] Ning B, et al. Targeting epigenetic regulations in cancer. Acta Biochim Biophys Sin (Shanghai) 2016;48(1):97–109.

[149] Zhou WM, et al. Methylation landscape: targeting writer or eraser to discover anti-cancer drug. Front Pharmacol 2021;12:690057.

[150] Morales La Madrid A, Hashizume R, Kieran MW. Future clinical trials in DIPG: bringing epigenetics to the clinic. Front Oncol 2015;5:148.

[151] Falahi F, Sgro A, Blancafort P. Epigenome engineering in cancer: fairytale or a realistic path to the clinic? Front Oncol 2015;5:22.

[152] Miozzo M, Vaira V, Sirchia SM. Epigenetic alterations in cancer and personalized cancer treatment. Future Oncol 2015;11(2):333–48.

CHAPTER

36

The Role of Epigenetics in Autoimmune Disorders

Kerstin Klein[1,2]

[1]Department of BioMedical Research, University of Bern, Bern, Switzerland [2]Department of Rheumatology and Immunology, Bern University Hospital, Bern, Switzerland

OUTLINE

Epigenetic Mechanisms Influence Autoimmune Processes	715
Mechanisms of Autoimmunity	716
Unraveling the Cell-type Specific Effects of Genetic Risk Factors in Autoimmune Diseases	717
Epigenetics of Immune Cells	719
T Cells	719
B Cells	719
Macrophages	720
Dendritic Cells	720
Epigenetics of Systemic Lupus Erythematosus	720
Epigenetics of Rheumatoid Arthritis	722
Multiple Sclerosis	723
Type 1 Diabetes Mellitus	724
Systemic Sclerosis	725
Use of Epigenetic Modifications as Potential Biomarkers	726
Use of Epigenetic Modifiers for Potential Diagnosis and Therapy in Autoimmune Diseases	727
Summary	728
Multiomics and machine learning accurately predict clinical response to adalimumab and etanercept therapy in patients with rheumatoid arthritis [203] [ex: Autoimmunity Epigenetics Case Study]	728
Objective	728
Scope	728
Audience	728
Rationale	728
Expected Results and Deliverables	728
Safety Considerations	729
The Process, Workflow and Actions Taken	729
Tools and Materials Used	729
Results	729
Challenges	729
Learning and Knowledge Outcomes	729
Abbreviations	730
References	730

EPIGENETIC MECHANISMS INFLUENCE AUTOIMMUNE PROCESSES

Epigenetic mechanisms influence autoimmune responses at different levels. On the one hand, the differentiation of cells of the innate and the adaptive immune system is controlled by epigenetic mechanisms. On the other hand, many epigenetic modifications were shown to be altered in specific cell types in disease. However, it has not yet been fully understood whether these epigenetic changes are inflammation-induced or predispose individuals to autoimmune diseases.

Posttranslational histone modifications or methyl groups are added to histone side chains or the DNA, respectively by "writer" proteins. These epigenetic marks are recognized by "reader" proteins that serve as platforms and docking sites for effector proteins mediating fundamental processes, such as transcription, DNA replication and recombination, DNA damage response, and chromatin remodeling. Epigenetic "eraser" proteins are capable of removing epigenetic marks [1]. Expression levels of enzymes corresponding to all three classes of these epigenetic modifiers were described to be altered in autoimmune diseases and their inhibitors are future promising therapies for patients [2–4]. In addition to DNA methylation and posttranslational histone modifications, short non-coding miRNAs and long non-coding RNAs contribute to epigenetic factors controlling gene expression.

In this chapter, I focus on epigenetic modification on selected immune disorders. I restricted the description of epigenetic mechanisms in autoimmunity to DNA methylation, and posttranslational histone modifications. Furthermore, I discuss the potential of targeting epigenetic enzymes in autoimmunity.

MECHANISMS OF AUTOIMMUNITY

Self-tolerance is essential for the immune system to function normally. Aberrations in any part of the innate or the adaptive immune system can lead to misdirected reactions resulting in autoimmunity where pathological immune responses are directed against host antigens (see Fig. 36.1).

In autoimmune disorders, antigen-presenting cells, belonging to the adaptive immunity, recognize self-antigens and propagate a self-directed autoimmune reaction. Consequently, activated B-cells

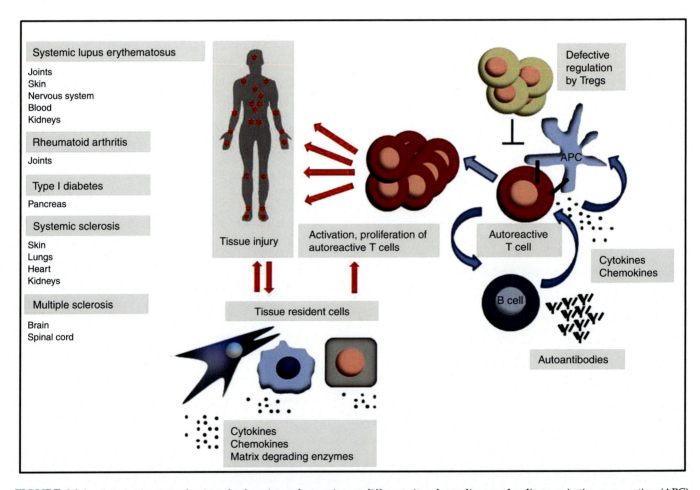

FIGURE 36.1 Autoimmune mechanisms lead to tissue destruction at different sites depending on the disease. Antigen presenting (APC) recognizes self-antigens and propagates a self-directed immune reaction. Autoantibodies, directed against the organisms own structures are produced by activated B cells. A defective regulation by regulatory T cells (Treg), as well as the lack of thymic deletion enables the proliferation and further activation of autoreactive T cells, tissue damage, and the occurrence of new antigenic epitopes that activate more lymphcytes. Tissue resident cells, such as fibroblasts, macrophages, and others contribute to the ongoing inflammation and tissue destruction by the production of inflammatory mediators and matrix degrading enzymes. The ongoing autoimmune reaction creates an inflammatory loop, where multiple cells of the immune system together with tissue resident cells further amplify the tissue destruction.

produce auto-antibodies, which are directed against the organisms own tissues, and depending on the localization of the immune reaction, result in diverse complications. In addition, a failure of control mechanisms occurs. The lack of the thymic deletion of autoreactive lymphocytes or the disturbed control by regulatory T cells (Treg) leads to the activation of autoreactive T cells and tissue damage. This leads to the occurrence of new antigenic epitopes, which activate more lymphocytes [5]. Furthermore, tissue resident cells, such as fibroblasts, macrophages, and others contribute to the inflammatory milieu and to the tissue destruction by secretion of cytokines, chemokines, and matrix degrading enzymes [6]. The ongoing autoimmune reaction creates an inflammatory environment in which multiple immune and tissue resident cells interact, creating an inflammatory loop due to the massive production of cytokines and other mediators [5].

Due to the chronic nature and their prevalence in young populations, autoimmune disorders are not only a significant clinical problem but also a socioeconomic burden due to associated healthcare costs. The majority of patients with autoimmune diseases develop symptoms well after the abnormal immune reactions begin, making it difficult to pinpoint the factors responsible for the initiation of the disease [5]. Most of the autoimmune diseases have an unclear pathogenesis, however, a combination of genetic, environmental, and endogenous factors are known to play a role (see Fig. 36.2).

A generally higher disease concordance rate in monozygotic (MZ) twins relative to dizygotic twins or other family members indicates a genetic contribution to autoimmune diseases. However, disease concordance in MZ twins is incomplete, suggesting additional factors playing a role [7]. Whereas, human leukocyte antigen (HLA) variants are well known genetic risk factors for autoimmune diseases, genome-wide association studies (GWAS) have identified in the past years more than hundred susceptibility loci, so called single nucleotide polymorphisms (SNP), contributing to disease risk. Most of these risk loci are in non-coding regions of the DNA, making it difficult to connect them to disease mechanisms [8].

Many environmental factors, including infections, ultraviolet (UV) light, smoking, dietary factors, drug intake or silica dust, and mineral oil exposure were suggested to contribute to the etiology of autoimmune disorders. Additionally, a role of the microbiome as influencing factor of local and systemic immune response was discovered in recent years. Differences in the microbiome alter the composition of bacterial metabolites in the gut. Decreased levels of butyrate, a bacterial fermentation product, were associated with a disturbed integrity of the gut epithelial layer, enabling the bacterial antigen

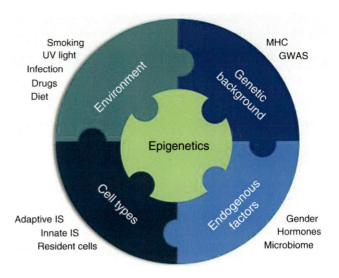

FIGURE 36.2 The puzzle of autoimmunity. The etiology of autoimmune diseases is multifactorial, with genetic, environmental, endogenous, and epigenetic factors playing a role. Many cell types, including cells of the adaptive and innate immune system, as well as resident cells in disease affected target tissues, contribute to the pathogenic mechanisms in autoimmune diseases. Risk factors and epigenetic mechanisms can have cell-type specific effects. As risk and genetic factors have the potential to modify the epigenetic landscape, epigenetic mechanisms provide the link between environmental, endogenous, and genetic risk factors and are central to the understanding of autoimmune processes. *GWAS*, genome-wide association studies; *IS*, immune system; *MHC*, major histocompatibility complex.

exposure to the immune system [9]. Most autoimmune diseases have a higher prevalence in women compared to men, such as systemic lupus erythematosus (SLE) [10], rheumatoid arthritis (RA) [11,12], multiple sclerosis (MS) [13], and systemic sclerosis (SSc) [14], pointing to gender and hormones as important endogenous factors in autoimmunity. On the other hand, a higher prevalence of type 1 diabetes mellitus (T1D) was described in males than females [15] (Table 36.1).

The understanding of the interplay between immunity, life style, as well as genetic, endogenous, and environmental factors is poor but key to the understanding and resolving of the autoimmune etiology.

Unraveling the Cell-type Specific Effects of Genetic Risk Factors in Autoimmune Diseases

The integration of cell type-specific epigenetic changes into genetic susceptibility loci has revolutionized the understanding about genetic factors as causes of disease [36,37,46–52]. Genetic fine mapping, the assessment of cell-type specific chromatin landscapes using assay for transposase-accessible chromatin using sequencing (ATACseq), chromatin immunoprecipitation sequencing (ChIPseq), and DNA methylation arrays, the analysis of 3D chromatin interactions,

TABLE 36.1 Overview on Prevalence, XCI, and Epigenetic Factors Among Autoimmune Diseases and Different Cell Types

	SLE	RA	MS	T1D	SSc
Prevalence (female:male)	9:1 [16]	3:1 [17]	3.5:1 [13]	1:1.5 [15]	3–8:1 [14]
Skewed XCI	Present [10]	Present [11]	Not present [18]	—	Present [14]
CD40L (females, T cells)	Demethylated → increased expression levels [10]	Demethylated → increased expression levels [12]	—	—	Demethylated → increased expression levels [19]
DNA methylation					
CD4+ T cells	Hypomethylation [20,21] (Lei, 2009 #1393) Differential methylation [22]	Hypomethylation [23]	Differential methylation [24] [25]	Differential methylation [26]	Hypomethylation [27–29]
CD8+ T cells	Hypomethylation [30,31]	Hypomethylation [23]	Differential methylation [25,32]	—	Hypomethylation [28]
B cells	Differential methylation [22]	Differential methylation [33,34]	Differential methylation [25,35]	Differential methylation [26,36]	—
CD14+ monocytes	Differential methylation [22]	Differential methylation [23,37]	Differential methylation [25]	Differential methylation [26,38]	—
Oligodendrocytes	—	—	Differential methylation [38]	—	—
Neurons	—	—	Hypomethylation [39]	—	—
Synovial fibroblasts	—	Differential methylation [40,41]	—	—	—
Skin fibroblasts	—	—	—	—	Hypomethylation [42,43]
LINE-1					
CD4+ T cells	Hypomethylation [31]	—	—	—	—
CD8+ T cells	Hypomethylation [31]	—	—	—	—
B cells	Hypomethylation [31]	—	—	—	—
Neutrophils	Hypomethylation [31]	—	—	—	—
Synovial fibroblasts	—	Hypomethylation [44]	—	—	—
5-Hydroxymethylation					
CD4+ T cells	Increased [45]	—	—	—	—
Neurons	—	—	Hyperhydroxymethylation [39]	—	—
Skin fibroblasts	—	—	—	—	Increased [42]

XCI, X Chromosome inactivation.

together with a subsequent correlation with gene expression profiles enables to link genetic variants to disease-relevant genes and their enrichment in relevant pathways [8,46,50–53]. This attempt leads eventually to new drug development or to the identification of disease specific biomarkers. Several of the biological candidate genes identified in this manner are known drug targets [48] and therefore, serve as a proof of principle that underscore the potential of integrative omics analysis. In addition, these studies have identified new potential drug targets [48,49]. The first landmark study by Farh et al. was based on the analysis of

21 autoimmune disease GWAS, transcriptomic data sets, cis-regulatory element annotation and a series of epigenetic analysis in resting and stimulated immune cell subsets. The authors showed that ~60% of likely causal genetic variants are located in stimulus and cell-type specific enhancers, enriched for both histone modifications, and the transcription of non-coding RNA (16). More recent studies that included cell types that were initially omitted, have highlighted the role of tissue-resident cells in carrying a substantial genetic risk for specific diseases, such as skin-resident dendritic cells (DC) in SSc [50], joint-resident synovial fibroblasts (SF) in RA [52,54], oligodendrocytes in MS [55], and acinar and ductal cells of the exocrine pancreas in T1D [51]. Studying the functional consequences of affected candidate genes in disease-relevant cell types is only at the beginning and the future challenge for a better understanding of the etiology of autoimmune diseases.

EPIGENETICS OF IMMUNE CELLS

T Cells

$CD4^+$ and $CD8^+$ T cells are generated in the thymus, where a set of repressive histone modifications and DNA methylation control and manifest lineage decision [56]. During the differentiation of naïve $CD8^+$ T cells into memory and effector T cells activation-associated histone modifications are lost, and repressive histone marks and DNA methylation marks are gained in gene loci whose expression is reduced in effector cells. In contrast, decreased repressive and an increased activating epigenetic modifications were associated with effector cell-associated transcription factors and functional effector genes. Naïve, memory, and effector cells are characterized by distinct DNA methylation and enhancer profiles that underlie cell type-specific responses [57].

$CD4^+$ T cells have been implicated as potent inducers of autoimmune diseases. T helper (Th) cells can be divided into several subsets, such as Th1, Th2, Th17, Th9, Th22, follicular T (Tfh), and Treg cells, based on the expression and function of unique transcription factors and their cytokine profiles [58]. T-box transcription factor 21 (Tbet), GATA binding protein 3 (Gata3), and RAR-related orphan receptor C (Rorγt) are master transcriptional regulators of Th1, Th2, and Th17 lineages, respectively [59]. In naïve $CD4^+$ T cells, the chromatin at the genetic loci of key transcription factors and cytokines is in an inactive and poised state, characterized by high levels of DNA methylation and repressive histone modifications. The differentiation into T-cell subsets is triggered by T-cell receptor signaling and the appropriate cytokines, leading to the activation of master transcription factors that recruit chromatin remodeling enzymes [4]. A partially primed, pre-existing chromatin landscape has been described that is then utilized by signal-dependent master transcription factors to direct lineage polarization [60,61]. Some differentiated Th1- and Th2-specific loci (*IFNG* and *TBX21* for Th1 cells, *IL4*, *IL13*, and *Gata3* for Th2 cells) are associated with the gain of active histone marks and the loss of repressive marks [4]. The deficiency of enhancer of zeste homolog 2 (EZH2) in T cells resulted in an expansion of Th1, Th2, and Th17 cells, but a reduced differentiation of Treg [62]. In Th17 cells, only few epigenetic changes depend on the master transcription factor Rorγt [59], and the hyperacetylation of *IL17a* and *IL17f* promoters are dependent on the transcription factor signal transducer and activator of transcription (STAT) 3 [63,64]. Also, Forkhead Box P3 (Foxp3), the master transcription factor of immunosuppressive Treg, is under tight epigenetic control. DNA hypomethylation combined with the formation of characteristic histone modifications establish an open chromatin structure, thereby imprinting Foxp3 in Treg [65]. Subsequently, Foxp3 itself interacts with the histone acetyltransferase Tat-interactive protein 60 (TIP60) and the histone deacetylases (HDAC) 7 and 9 to at specific gene loci to actively repress their transcription in Treg [66]. The thymus-induced epigenetic signature remains stable once Tregs leave the circulation and immigrate into peripheral tissues [67].

B Cells

B cell lineage programming and differentiation are orchestrated by a set of lineage- defining transcription factors and signal-dependent transcription factors. These factors are controlled by different epigenetic mechanisms [68]. Mechanistic studies in conditional knockout mice have uncovered the role of several epigenetic enzymes, including DNA and histone modifiers, in B-cell development [69]. Recent single cell technologies, in particular single cell ATACseq, have enabled the studying on open chromatin states associated with developmental and differentiation states even in rare cell populations [70]. B-cell maturation and differentiation were associated with a massive re-configuration of the DNA methylome, with changes in approximately 30% of all measured CpG sites. Early stages of B-cell differentiation were characterized by enhancer demethylation which were enriched for binding sites of key B-cell transcription factors, such as paired box 5 (PAX5), EBF transcription factor 1 (EBF1), interferon regulator factor (IRF) 4, and others. The expression of these transcription factors inversely correlated with the methylation levels of their binding sites. At late differentiation stages, demethylation of heterochromatin and methylation gain at polycomb-repressed regions were common [71].

EZH2 is the catalytic subunit of the polycomb repressive complex 2, the writer complex of repressive histone 3 lysine 27 trimethylation (H3K27me3) [72]. EZH2 has been shown to be involved in the epigenetic regulation of B-cell differentiation at multiple levels [68], from the differentiation of pro- to pre-B cells to the formation of germinal center B cells [73–76].

Histone acetylation plays a major role in the generation of B-cell receptors [77] and in the regulation of B lymphocyte-induced maturation protein 1 (Blimp-1), the master regulator of plasma cell differentiation [78,79]. The lysine-specific histone demethylase 1 A (LSD1) was shown to interact with Blimp-1 to remove histone 3 lysine 4 monomethylation (H3K4me1) marks and to maintain the repression of B-cell lineage genes. Furthermore, LSD1 was recruited and interacted with several transcription factors to fine tune multiple cell fate stages during B-cell differentiation [68].

Macrophages

Macrophages are myeloid cells of the immune system that are specialized in phagocytosis, a cellular activity critical for tissue remodeling and debris clearing, as well as the innate immunological defense against pathogens. Most tissue macrophages are established prenatally and persist through adulthood without substantial input from adult hematopoiesis [80]. The lineage determination of macrophages is regulated by the expression of the transcription factor Pu.1 that facilitates a stable opening of the chromatin and the recruitment of additional transcription factors. Regulation of gene expression in macrophages occurs mainly on the levels of enhancers [81,82]. Enhancers shared by many tissue macrophages are active and contain H3K4me1 marks and the activating histone mark histone 3 lysine 27 acetylation (H3K27ac) that prepare the cell to respond to a future stimulus. In contrast, enhancers that are unique for a specific set of tissue macrophages are poised, which might indicate the potential of the respective population to respond to local challenges. Signals from the local environment have a dominant role in shaping tissue-specific regulatory regions and gene activity independent of development [80], leading to the loss of suppressive H3K27me3 marks and the gain of H3K27ac in enhancers and promoters [81].

Epigenetic mechanisms are the basis for an immune memory that enable macrophages to adjust their response upon a repeated trigger with inflammatory stimuli. This can lead to either tolerance, with a reduced response to prevent inflammation-induced tissue damage, or training, resulting in a stronger immune response [83]. A tolerant state in monocyte-derived macrophages has been characterized by the loss of histone acetylation and of histone 3 lysine 4 trimethylation (H3K4me3) marks in the promoters of pro-inflammatory genes after repeated stimulation with Toll-like receptor (TLR) ligands such as lipopolysaccharide (LPS) [84]. In contrast, training with β-glucans, cell wall components of *Candida albicans*, induced a genome-wide gain of H3K27ac in promoters and enhancers, and an increase of H3K4me3 at promotors of inflammatory cytokines [82,85].

Dendritic Cells

DC are professional antigen-presenting cells that reside in nearly every tissue and are crucial for primary T cell responses. DC sense a vast number of antigens through a broad spectrum of pattern recognition receptors, such as TLR, and migrate to draining lymph nodes where they initiate the systemic immune response [86]. DC develop from progenitor cells into common DC progenitors and further into either conventional DC (cDC) or plasmacytoid DC (pDC). The majority of studies on human DC are derived from cultured monocytes (moDC) that were induced with GM-CSF and IL-4 [86,87]. This differentiation process was shown to be regulated by DNA methylation, DNA hydroxymethylation and histone modifications [87–90].

Pu.1 and C/EBP have been identified as the pioneer transcription factors that regulate DC lineage commitment by orchestrating enhancer regions [91,92]. H3K4me1- and H3K27ac-positive enhancers determine the DC subset specification and are likely established early in DC differentiation [87]. Afterwards, the global chromatin landscape across different DC subsets was shown to be remarkably stable after stimulation [92–94]. However, chromatin remodeling has been described in selected gene promoters in LPS-stimulated DC [86]. Immune tolerance and training have, similar to macrophages, been described for DC, where similar epigenetic mechanisms underlie these effects [86].

EPIGENETICS OF SYSTEMIC LUPUS ERYTHEMATOSUS

SLE is a chronic, systemic autoimmune disease with diverse clinical manifestation, and affecting multiple organs, including the joints, skin, nervous system, blood, and kidneys. Depending on the type of lupus, the symptoms can include rashes, hair loss, aching and swelling of joints, fever, anemia, and abnormal blood clotting. The disease can take a mild course when it affects only a few organs; nevertheless severe forms can be life-threatening [95].

SLE is characterized by autoreactive B and T lymphocytes together with the presence of diverse sets of autoantibodies in the circulation of affected patients, such as anti-nuclear antibodies (ANAs), anti-double-stranded DNA (dsDNA) antibodies, and anti-Smith (anti-Sm) antibodies, which have been used as classical serological markers in SLE. The incidence of SLE varies among geographic regions and ethnicities [96]. SLE is much more prevalent in women compared to men, with a ratio of 9:1 [16].

Beside a genetic factor driving SLE [97], resulting in a concordance rate of ~20% in heterozygotic (HZ) twins, environmental factors, such as smoking, UV light exposure, viruses, and hormones have been suggested to contribute to the pathogenesis [95].

First evidence for the involvement of epigenetic alterations in the pathogenesis of autoimmune diseases arose in 1986 when Richardson et al. demonstrated that $CD4^+$ T cells became autoreactive upon treatment with the demethylating agent 5-azacytidin (5-aza) [98]. Meanwhile, numerous studies have shown that $CD4^+$ T cells in SLE are globally hypomethylated [20,21,27]. Changes in DNA methylation patterns in $CD4^+$ T cells correlated with diverse clinical manifestations [99–101], and occurred early in lupus flares [102]. Hypomethylation in $CD4^+CD28^+KIR^+CD11a^{hi}$ T cells correlated with disease activity and was associated with a pro-inflammatory gene expression profile [103].

Inhibitors of DNA methylation, such as procainamide and hydralazine induce lupus-like disease in mice [104] and approximately 12% of patients under high dose hydralazine treatment over long periods were shown to develop lupus-like symptoms, which usually disappeared upon discontinuing the medication [105]. Experimental evidence suggested a link of DNA hypomethylation and the extracellular signal-regulated kinase (ERK) pathway. Hydralazine decreased levels of DNA methylation via inhibition of the ERK pathway [106], while a decreased signaling via the ERK pathway contributed to DNA hypomethylation in T cells of SLE patients [107]. In a mouse model of SLE, decreased ERK signaling in T cells resulted in the down regulation of DNA methyltransferase 1 (DNMT1) and overexpression of the methylation-sensitive genes CD11a and CD70, characteristic for T cells in human lupus. Moreover, these animals also developed anti-dsDNA antibodies [108].

Genome-wide DNA methylation profiling revealed a hypomethylation in $CD8^+$ T cells derived from patients with SLE compared to healthy controls. Hypomethylation particularly affected the *HLA-DRB1* locus and genes associated with a type-I interferon response, including STAT1 [30].

Beside reports of altered DNA methylation levels in SLE, recently, $CD4^+$ T cells were shown to have globally elevated levels of 5-hydroxymethylcytosine (5-hmC), an epigenetic modification which associated with active transcription and the over expression of selected immune-related genes, such as SOCS1, NR2F6, and IL15RA. In addition, expression levels of Ten–eleven translocation (TET) enzymes TET2 and TET3, enzymes responsible for converting 5-methylcytosine (5-mC) to 5-hmC were increased compared to healthy controls [45].

Aberrant DNA methylation profiles and an SLE-specific landscape of chromatin accessibility in naïve B cells provided evidence for epigenetic mechanisms underlying B-cell abnormalities in SLE [22,109]. Scharer et al. have unraveled the molecular basis of the B-cell dysfunction in SLE by integrating the DNA methylomes, chromatin accessibility profiles and transcriptomes of five human B-cell subsets derived from patients with SLE and healthy controls. Molecular signatures associated with SLE were already present in resting naïve cells and epigenome alterations were then transferred to effector B cells [110]. Recently, an ethnicity-dependency of DNA methylation patterns has been described in B cell lineages from African Americans and European Americans. An enrichment of CpG hypomethylation of interferon-regulated genes was specifically associated with African American female SLE patients [111].

Also histone modifications were studied as contributors to the pathogenesis of SLE. Hypoacetylation of the histones 3 (H3) and H4 was first described in the autoimmune field in splenocytes isolated from lupus prone mice [112]. Inactivation of the histone acetyl transferase p300 leads to the development of severe lupus-like disease in mice, the formation of anti-dsDNA antibodies, as well as glomerulonephritis and premature death [113]. The data from animal models are reflected by human data; in $CD4^+$ T cells, a global histone H3 and H4 hypoacetylation, as well as H3K9 hypomethylation was observed. The degree of H3 hypoacetylation correlated negatively with an increased disease activity [114]. Increased levels of EZH2 and H3K27me3 were reported in $CD4^+$ T cells and B cells of SLE patients [115,116]. In $CD4^+$ T cells, EZH2 activity contributed to the hypomethylation of regulatory regions of genes involved in leukocyte adhesion and migration. This mechanism led to an increased adherence of T cells to endothelial cells [115]. In B cells, EZH2 promoted the differentiation towards plasmablasts. EZH2 over expression in B cells of SLE patients correlated with diseases activity and production of autoantibodies [116].

The transcription factor regulatory factor X1 (RFX1) was shown to be down regulated in $CD4^+$ T cells of patients with SLE by a STAT3-dependent process. RFX1 recruited a repressor complex consisting of DNMT1, HDAC1, and SUV39H1 to the IL-17A promoter. Hence, RFX1 deficiency promoted the differentiation of naïve $CD4^+$ cells towards Th17 cells, by an increase of H3ac and H3K9me3, and a decreased DNA methylation at the IL-17A promoter, followed by the over production of IL-17A [117].

EPIGENETICS OF RHEUMATOID ARTHRITIS

RA is a chronic, systemic autoimmune disease, characterized by chronic inflammation of the joints with severe pain and swelling, joint damage, and disability, which leads ultimately to joint destruction and loss of function [118]. The destructive arthritis primarily and typically affects the small joints of the hands and feet, but also larger joints and extra-articular structures get involved during the course of the disease. Patients with RA are more prone to atherosclerosis and have generally an increased risk for cardiovascular events, such as heart attacks and strokes [119].

RA affects about 0.5%—1% of the population worldwide, being the most common inflammatory joint disease. The disease onset is most commonly observed in individuals aged between 30 and 60. Women have a three times increased risk of RA compared to men [17] and there is evidence that an altered DNA methylation contributes to this gender bias. Female RA patients exhibited an increased skewed X-chromosome inactivation (XCI) pattern compared to controls [11]. The promoter of the X-chromosome encoded gene *CD40L* was demethylated in RA patients and overexpressed in CD4$^+$ T cells in female but not male RA patients [12], similarly to what is seen in SLE patients [10]. Even though a variety of treatment options exist, RA therapy is not curative and not always induces remission. Even more, about one third of the patients do not respond satisfactorily to therapy and primary or secondary resistance to disease-modifying anti-rheumatic drugs is one of the major unsolved issues in RA therapy [120].

The pathogenic phenomena in RA occur mainly in the synovial tissues. The main function of the synovial tissue in a healthy joint is the supply with synovial fluid, which lubricates the cartilage and minimizes the friction of the cartilage covering the bones. In RA, the synovial tissue overgrows as a result of the increased survival rate of the synovial resident cells, the increased proliferation of sublining SF and the infiltration by immune cells [121,122]. Activation of the infiltrating immune cells leads to a self-directed attack localized to joints and is followed by joint inflammation (arthritis). Activation of resident synovial cells parallels the inflammation and results in an aggressive invasion of SF and macrophages into cartilage and bone, leading to their damage [121].

Studies on epigenetics have contributed to a better understanding of the pathogenesis of RA. RA SF are intrinsically activated and resemble an activated phenotype [123]. RA SF produce large amounts of cytokines and chemokines and therefore promote the ongoing inflammation. On the other hand, RA SF produce matrix degrading metalloproteinases, enabling them to actively invade and destroy cartilage [6]. Furthermore, the joint-specific transcriptional diversity of SF, that translates into joint-specific functions, might predispose certain joints for specific forms of arthritis [124]. DNA methylation and histone modifications maintain the joint-specific patterns in SF [124,125]. This joint-specific diversity of SF might be also relevant in patients' response to therapy, underlined by a study providing evidence for a joint-specific epigenetic regulation of the Janus kinase (JAK)-STAT signaling in SF that led to differences of knee and hip SF in their sensitivity to the JAK inhibitor tofacitinib [126].

Investigations to which extent epigenetic changes are responsible for the activated phenotype of RA SF are ongoing; however, there is already substantial evidence that several diverse epigenetic factors control the inflammatory behavior, the invasiveness, and matrix-degrading properties, as well as the apoptosis resistance of RA SF. A comprehensive epigenomic analysis, integrating activating and repressive histone modifications, open chromatin states, whole-genome DNA methylation and RNA expression in RA compared to osteoarthritis (OA) SF has identified active RA-associated regulatory regions in enhancers and promoters [127]. A single or repeated exposure of SF with inflammatory stimuli, such as tumor necrosis factor (TNF) or LPS, was shown to induce sustained open chromatin states characterized by activating histone marks in promotors of inflammatory genes, leading to their continuous expression [128,129]. This is in stark contrast to monocytes and macrophages that induced tolerance under the same conditions.

Regarding the role of histone modifications in RA, the majority of data is still based on the use of inhibitors against HDAC (HDACi) [130] or bromodomain and extra-terminal domain (BET) proteins [129,131,132] and less data are available on the individual role of specific enzymes or histone modifications. Different members of HDAC were shown to have both pro- and anti-inflammatory functions in RA, underscoring the importance of performing functional studies of individual enzymes [130]. Increased levels of the histone deacetylases sirtuin 1 (SIRT1) expression in synovial tissues and synovial cells of RA patients were shown to have pro-inflammatory properties [133]. Furthermore, inflammatory cytokines were shown to suppress HDAC5 in RA SF, leading to an increased nuclear accumulation of the transcription factor IRF1 and the induction of type I interferon response genes [134]. On the other hand, HDAC3 silencing in RA SF resembled effects of pan-HDACi, leading to the suppression of inflammatory genes and the production of type I IFN [135].

Smoking, the main risk factor for RA [136], was not only shown to induce changes in lung tissues of patients [137] but also affects the epigenetic landscape directly in synovial tissues and RA SF by modifying the expression levels of SIRT1 [138] and SIRT6 [139]. Increased levels of SIRT6 in collagen-induced arthritis (CIA) mice and synovial tissues of RA smokers, and RA SF after exposure with cigarette smoke extract were described to have anti-destructive and anti-inflammatory roles [139,140]. Furthermore, nicotine was shown to modify the microRNA profile of $CD8^+$ T cells, affecting the FOXO-signaling pathway and leading to a high prevalence of $CD8^+$ T cells with a naïve-memory phenotype [141].

RA SF exhibit profound changes in their DNA methylation patterns compared to OA SF. CpG islands in the promoter of the retrotransposable element long interspersed nuclear element-1 (LINE-1) are hypomethylated in RA SF, leading to the reactivation of LINE-1 transcription [44]. Nakano et al. identified hypo- and hypermethylated genes in RA SF that cluster to pathways relevant to cell movement, adhesion, and trafficking [40]. In addition, RA SF have decreased levels of a global DNA methylation [41] probably due to an increased recycling of polyamines in RA SF, leading to reduced amounts of the methyldonor S-adenosyl methionine (SAM) [142]. Based on these findings, a new therapeutic concept was suggested, aiming to reverse the global hypomethylation by inhibiting the polyamine recycling pathway and/or a supplementation of SAM. This treatment concept was sufficient to reduce the invasiveness of RA SF engrafted in the severe immune deficient (SCID) mouse model by 70% [143]. Changes in DNA methylation signatures have already been detected in SF from early and very early RA patients [144,145]. A global hypomethylation was also found in peripheral blood $CD4^+$ and $CD8^+$ T cells [23]. A further sub analysis revealed that differentially methylated regions in $CD4^+$ memory T cells exceeded those of $CD4^+$ naïve T cells compared to health controls and were affected by disease activity, diseases duration and treatment with methotrexate [146].

MULTIPLE SCLEROSIS

MS is a chronic neurodegenerative autoimmune disease occurring in 0.1% of the population and being characterized by the autoimmune destruction of myelin and neurons and the formation of focal demyelinated plaques in the white matter of the central nervous system [147,148]. Although T cells are considered as the key mediators of MS nearly every cell type of the innate and the adaptive immune system has been implicated in the immunopathology [149,150].

There are different types and stages of MS, ranging from slowly progressing to rare malignant forms leading to progressive disability. The majority of disability in MS relates to spinal cord dysfunction [148]. The genetic heritability of MS was estimated to be up to 54%, in which HLA variants contribute to a threefold increased risk to develop MS. MS is more often transmitted to the next generation by mothers than fathers, suggesting epigenetic influences. Among the best-established environmental factors for MS are infections with Epstein-Barr virus, low vitamin D and sun exposure, high BMI, and smoking [150].

Global transcriptome and DNA methylome patterns in bronchoalveolar lavage cells were shown to be profoundly changed in MS patients that smoked compared to non-smokers [151]. Genome-wide DNA methylation analyses revealed a hypomethylation of the HLA-DRB1 risk locus in $CD4^+$ T cells in MS patients compared to healthy controls [24,152,153]. This might have a great significance in the MS pathogenesis since class II HLA molecules are crucially involved in presentation of myelin epitopes to T cells and subsequent formation of anti-myelin immunity [154]. Interestingly, this epigenetic change was not found in $CD8^+$ T cells in the same set of patients, highlighting the importance of studying distinctive cell subtypes when investigating epigenetic changes [32].

There is growing evidence that epigenetic changes are capable of causing myelin damage. Huynh et al. studied genome-wide DNA methylation changes in combination with RNA sequencing (RNAseq) in normal appearing white matter (NAWM) of MS patients, which is characterized by the lack of inflammatory infiltrates and the absence of demyelination. This group found that genes regulating oligodendrocyte survival, a cell type capable of at least partially remyelinating lesion sites, were hypermethylated and expressed at lower levels in MS-affected brains compared to controls, whereas genes related to proteolytic processing were hypomethylated and expressed at higher levels. This indicates that epigenetic changes in genes affecting oligodendrocyte susceptibility to damage can already be detected in pathology-free areas of MS-affected brains [155]. Furthermore, hypomethylation of the peptidyl argininedeiminase 2 (PAD2) promoter leads to the overexpression of PAD2 in peripheral blood mononuclear cells (PBMC) and the white matter of MS patients. PAD2 catalyzes the citrullination of myelin basic protein (MBP) in the white matter of brains of patients with MS, leading to the loss of myelin stability in the MS brain [156]. Furthermore, an increased citrullination of histones by PAD4 in the white matter of MS patients and animal models of demyelination was shown. The authors concluded that the TNF-induced PAD4 nuclear translocation, followed by histone citrullination might be part of the process leading to the irreversible changes and apoptosis in oligodendrocytes [157]. In rodents, histone deacetylation favors oligodendrocyte differentiation, while acetylation is associated with increased levels of transcriptional inhibitors of oligodendrocyte differentiation. Interestingly, there is also a shift from histone deacetylation to histone acetylation in the white matter of MS patients with disease duration [158].

Betaine depletion is a hallmark of MS brains, leading to decreased levels of H3K4me3 and mitochondrial impairment in neurons. Betaine supplementation in a mouse model of MS, increased the ratio of SAM/SAH and restored levels of H3K4me3, and leading to a subsequent protection from axonal damage [159]. Betaine administration to oligodendrocytes modulated their maturation by increasing their histone- and DNA methyl-transferase activities [160]. In contrast, the restriction of methionine followed by a subsequent reduction of intracellular SAM and decreased levels of H3K4me3, limited the expansion of Th17 cells. These mechanisms delayed the onset and severity of experimental allergic encephalomyelitis (EAE), a model of brain inflammation and MS [161].

TYPE 1 DIABETES MELLITUS

T1D is an autoimmune disorder hallmarked by the organ-specific immune destruction of the insulin-producing β-cells in the islet of the Langerhans within in the pancreas. Once these cells are destroyed, patients with T1D loose blood glucose control, which can result in acute conditions, such as ketoacidosis and severe hyperglycemia and secondary complications including heart disease, blindness, and kidney failure [162]. The onset of T1D usually occurs during childhood or adolescence. More than 80% of β-cells has already been destroyed by the time the first clinical symptoms become apparent, since the autoimmune processes remain subclinical for many years in most patients. Once diagnosed, patients require lifelong insulin treatment [163]. Studies in non-obese diabetic (NOD) mice, the best characterized animal model for autoimmune diabetes [164], have demonstrated that the disease occurs as a consequence of a breakdown in immune regulation, resulting in the expansion of autoreactive $CD4^+$ and $CD8^+$ T cells, autoantibody producing B lymphocytes, and activation of the innate immune system that collaborate to destroy the insulin-producing β-cells. Autoreactive T cells specific for β-cell proteins including insulin itself, as well as autoantibodies against β-cell proteins can be detected in very young NOD mice, as well as in peripheral blood of susceptible individuals before disease onset [162]. A recent study showed that beta cell-specific $CD8^+$ T cells from patients with T1D retained a stem cell-like epigenetic state that resemble a hybrid of naïve and effector associated epigenetic programs and led to sustained autoreactivity [165]. Rui et al. have shown that the inflammatory milieu in NOD mice or the in vitro stimulation of human β-cells with inflammatory cytokines induce the expression of methyltransferases in pancreatic islet or β-cells, respectively. This leads to a subsequent methylation of the insulin gene and a decreased expression of insulin over time [166]. Interferon-α-induced changes in the DNA methylation in β-cells were suggested as a potential mechanism underlying the initial β-cell loss in T1D [167].

Studies in MZ twin pairs identified a discordance rate of ~50% for childhood onset of T1D, implicating roles for genetic and non-genetic factors in the etiology of the disease [38]. Environmental factors, such as population hygiene, sun exposure, viral infections, dietary factors, and the gut microbiome were postulated to play a role in T1D [9,162].

Several studies compared global DNA methylation patterns in MZ twins discordant for T1D, a powerful method to analyze the contribution of epigenetic mechanisms to disease etiology. A study in a small group of discordant MZ twins identified a global hypomethylation of CpG sites within promoter regions in peripheral blood of MZ twins with T1D compared to twins without T1D [168]. Genome-wide DNA methylation profiles of purified $CD14^+$ monocytes in MZ twins discordant for T1D revealed several T1D-associated methylation variable positions (MVPs) that significantly correlated with the diabetic state. Interestingly, these T1D-MVPs raise early in the etiological disease process. [38]. Another study has compared the genome-wide methylation profiles in three immune effector cell types, B cells, $CD4^+$ T cells and monocytes, in MZ twins discordant for T1D, revealing changes in immune cell metabolism and cell cycle pathways. The analysis of DNA methylation profiles in cord blood of newborns that later progressed to overt T1D suggested that T1D-associated patterns were likely acquired after birth. [26]. Another study pointed to very early changes of DNA methylation at specific loci in peripheral blood of pre-diseased patients, with some of them already present at birth [169]. Global methylation analysis of B cell lines from MZ twins discordant for T1D identified significant methylation changes in genes functionally annotated to immune response and defense response pathways between affected and unaffected twins. The integration of differentially methylated CpG sites of affected and unaffected twins with GWAS data mapped several known T1D associated genes, such as *HLA*, *INS*,

IL-2RB, and *CD226* [36]. Differential methylated CpG sites were also identified in DNA from peripheral blood of T1D patients at high risk for diabetic nephropathy compared to patients with no signs of renal disease [170], and differential methylated loci were suggested as potential biomarkers for end-stage kidney disease [171]. A genome-wide analysis of DNA methylation in subjects with T1D identified differentially methylated CpG sites that could be used a prospective biomarker for the development of proliferative diabetic retinopathy, the main cause of vision loss in subjects with diabetes [172].

Epigenetic mechanisms are also involved in a phenomenon known as metabolic memory in which an early exposure to hyperglycemia predisposed T1D patients to the development of long-term diabetic complications. Levels of H3K9ac were shown to be enriched in human monocytes of T1D patients with microvascular complications and associated with a history of near-normal levels of glycemia (HbA_{1c} level). Hyperacetylated promoters were found in genes encoding signaling molecules of the nuclear factor-κB pathway and were enriched in known genes related to diabetes complications further emphasizing the role of long-term epigenetic changes in microvascular complications in T1D [173]. In addition to long lasting hyperglycemia-induced epigenetic changes in T1D patients, a genome-wide DNA methylation study also identified DNA methylation patterns in leukocytes that were associated with kidney function of nondiabetic adult offspring exposed in utero to T1D [174]. The Diabetes Control and Complications Trial, in which patients with and without complications were followed over many years, persistent DNA methylation patterns in whole blood and in monocytes were identified that underlined metabolic memory [175,176].

SYSTEMIC SCLEROSIS

SSc is a complex multisystem autoimmune disease with a high morbidity and mortality. Pathologic features include the dysregulation of the immune system, endothelial dysfunction, and activation of fibroblasts, resulting in a progressive tissue fibrosis [177]. Deposition of collagen occurs in the skin but can also occur in internal organs, including the lungs, heart, gastrointestinal tract, and kidneys. Two subtypes of SSc, limited cutaneous SSc (lSSc), in which skin fibrosis affects only the distal areas of hands and fingers, and diffuse cutaneous SSc (dSSc), in which skin fibrosis affects the entire body, can be distinguished [178]. By synthesizing excessive amounts of collagen and other components of the extracellular matrix (ECM), activated myofibroblasts mediate the tissue fibrosis in SSc. Together with a reduced expression of matrix metalloproteinases this leads to an excessive matrix accumulation [179]. Furthermore, SSc fibroblasts are characterized by a hyperresponsiveness to cytokines and chemokines that further increases their production of ECM. Activated antigen-specific T cells are the major cells infiltrating skin and lungs [180], however, many different cell types, including tissue-resident cells and innate and adaptive immune cells, contribute to the pathophysiology in SSc [177].

The exact etiology of SSc is undetermined. A concordance rate of only 4.2% in MZ twins suggests, that the genetic predisposition plays only a moderate role and contributes to a much lower rate to the pathogenesis compared to other autoimmune diseases [181]. Environmental factors, such as silica exposure, bacterial and virus infections, drugs, and others might influence genetically susceptible individuals through epigenetic mechanisms and result in autoimmune disorders [182].

Global changes in DNA methylation have been described in $CD4^+$ and $CD8^+$ T cells and fibroblasts of patients with SSc compared to controls [27–29,42,43]. The most common genes affected by DNA hypomethylation in T-cell subsets were type I interferon related genes [28], a signature that is shared across different autoimmune diseases [183]. Conflicting results have been reported on the differential expression of DNMT in SSc patients. One study reported that the mRNA expression of DNMT1 was reduced and associated with a global DNA hypomethylation in T cells and fibroblasts of SSc patients [42]. Another study provided evidence for an increased protein expression of DNMT1 and DNMT3a in skin tissues and dermal fibroblasts of SSc patients compared to controls in in larger sample sizes [184]. In addition, the expression of TET1, the enzyme that further converts 5-mC into 5-hmC, was increased in skin of patients with SSc. In fibroblasts, TET1 was induced by hypoxia via HIF-1α-independent pathways [42]. Furthermore, methyl-CpG-binding protein 2 (MECP2) was increased in dermal fibroblasts from SSc patients [185,186].

Activated Wnt signaling with decreased expression of endogenous inhibitors has been described to play a key role in the pathomechanism in SSc. Dees et al. showed that promoter hypermethylation down regulated the Wnt antagonists Dickkopf-related protein 1 (DKK1) and secreted frizzled-related protein 1 (SFRP1) in SSc fibroblasts. In addition, inhibition of DNMT effectively reduced Wnt signaling. These data suggest a novel link between epigenetic alterations and increased Wnt signaling in SSc. This mechanism suggests the use of DNMT inhibitors for the treatment of SSc, which are already approved for a clinical use in cancer [187].

The methylation status of extracellular matrix proteins such as collagen regulates fibrotic processes in

SSc. A genome-wide DNA methylation study identified common and subset-specific differentially methylated regions in skin fibroblasts from patients with dSSc and lSSc. Commonly affected pathways were associated with ECM-receptor interaction and focal adhesion [43]. Furthermore, a hypermethylation of the regulatory region of the transcription factor Fli-1 proto-oncogene (FLI1) resulted in decreased levels of FLI1 in SSc fibroblasts compared to healthy controls. This was associated with an increased collagen expression. The addition of the demethylating agent 5-aza resulted in increased FLI1 expression and reduced expression of collagen in SSc fibroblasts [188]. The expression of DNMT1 and DNMT3 was regulated by TGFβ, leading to a subsequent hypermethylation of the SOCS3 promoter, the down regulation of the SOCS3 expression, and STAT3 activation. This mechanism promoted the transition of fibroblasts towards myofibroblasts. In line with these findings, administration of 5-aza reversed the activation of dermal SSc fibroblasts and ameliorated bleomycin-induced fibrosis in mice [184].

In line with the findings from global DNA methylation studies, an integrative approach using RNAseq and ATACseq of fibroblasts derived from lesional skin of patients with dSSc and healthy controls, has identified an increased signature of pro-fibrotic genes in lesional skin fibroblasts. A distal enhancer of TGFβ2 exhibited a higher chromatin accessibility of chromatin, leading to its continuous over expression in SSc skin fibroblasts, providing an autocrine signal for the ongoing pro-fibrotic gene expression [189].

In addition to global changes in DNA methylation patterns, changes in histone modifications have been assessed. Aberrant patterns of H3K4me3 and H3K27ac in monocytes of SSc patients were associated with genes corresponding to immune, interferon and antiviral pathways and were enriched for binding sites of IRF and STAT transcription factors [190]. H3K27me3 and the methyltransferase EZH2 have been shown to be over expressed in skin fibroblasts and endothelial cells of SSc patients compared to healthy controls [115,191]. However, there are conflicting results regarding the benefits of targeting enzymes associated with the increased levels of H3K27me3. Tsou et al. have identified EZH2 of being pro-fibrotic and anti-angiogenic and provided evidence of beneficial effects of targeting EZH2 in vitro and in vivo by using the inhibitor 3-deazaneplanocin A hydrochloride (DZNep). On the other hand, Krämer et al. showed an enhanced fibrosis using DZNep. The same group showed that also the histone demethylase Jumonji domain-containing protein 3 (JMJD3) was over expressed in SSc skin fibroblast of SSc patients compared to controls. Knockdown of JMJD3 or pharmacological inhibition using GSKJ4 reversed fibroblast activation in a Fos-related antigen 2 (FRA2)-dependent manner [192].

Endothelial dysfunction is an early hallmark of SSc [177]. HDAC5 was shown to be over expressed in endothelial cells of patients with dSSc compared to controls, where it seemed to have anti-angiogenic function. Knockdown of HDAC5 and subsequent ATACseq has identified several HDAC5 target genes associated with angiogenesis and fibrosis, providing a new link between epigenetic regulation and pathomechanisms in SSc [193].

USE OF EPIGENETIC MODIFICATIONS AS POTENTIAL BIOMARKERS

The knowledge on epigenetic aberrations in autoimmune diseases is growing fast. Research in this field is important since the current diagnosis for most of these disorders is far from satisfactory and often delayed. Current diagnosis is often possible only after the appearance of symptoms, missing the chance for an early intervention before disease onset or substantial tissue damage. Moreover, the prediction of disease progress is difficult. Even though there has been substantial progress in the treatment of some of these diseases, there is no efficient therapy available for most autoimmune disorders and none of them described in this chapter can be cured.

Due to the high and easy accessibility of peripheral blood and other body fluids, researchers hoped to identify epigenetic changes that can be used as biomarkers for disease risk, prediction of disease outcome, and associated complications, as well as for prediction of therapy response. DNA methylation is the best characterized epigenetic mark and many studies have examined DNA methylation profiles in PBMC. However, any disease-related signatures in PBMC identified might just display differences in the relative abundance of individual cell types as each cell subset generates a unique methylation profile. Therefore, it is important to identify cell- or tissue-specific changes in DNA-methylation [194]. However, there are also some exceptions to this rule. In SLE, a robust hypomethylation of interferon-regulated genes was described to be independent of the cell type analyzed. Based on this, Zhao et al. identified two hypomethylated CpG sites in the *IFI44L* promoter in PBMC that can distinguish patients with SLE from healthy and patients with RA or the primary Sjögren syndrome with high sensitivity and specificity. Furthermore, *IFI44L* promoter methylation further decreased in patients with renal damage, and increased in SLE patients in remission [195]. There is a need for similar studies in other autoimmune diseases, where changes in methylation patterns are not only compared to healthy individuals but also to patients suffering from distinct diseases to increase the diagnostic value [196].

In recent years, many studies focused on addressing changes in DNA methylation in specific cell subsets. Monitoring epigenetic changes in DC of patients with SLE were suggested for assessing kidney involvement. Whereas, a hypermethylation of DC was associated with severe lupus nephritis, decreased levels of H3K4me3 and H3K27me3 correlated with early stages of lupus nephritis [197].

Interferon β treatment of patients with MS changed specifically the global DNA methylation pattern of monocytes [198]. Whether this finding is useful for the development of a biomarker for treatment response needs further evaluation.

Methotrexate treatment of RA patients was shown to alter DNA methylation levels in blood samples [23,199]. However, initial results reporting an association of DNA methylation patterns in leukocytes with the response of RA patients treated with methotrexate [200,201] were not confirmed in another study [202]. Differential DNA methylation was also used to predict treatment response of RA patients treated with adalimumab or etanercept [203,204]. These data entered a machine learning model, with an overall predictive accuracy of 84.7% and 88% for adalimumab and etanercept, respectively [204]. Machine learning based on differentially methylated regions in PBMC was also used to predict the differentiation of undifferentiated arthritis towards RA [205]. Data from such approaches need further validation, but together they highlight the potential of an additive value of including DNA methylation profiles into machine learning models to predict treatment response or disease outcome.

In T1D, differentially methylation loci were suggested as biomarkers for renal disease and diabetic retinopathy [170–172]. β-cell derived circulating unmethylated insulin DNA was suggested as an early and non-invasive biomarker for the detection of T1D [206].

Since DNA hypomethylation was observed in different cell types and several autoimmune diseases (Table 36.1), drugs increasing the activity of DNMT could be of therapeutic value. Unfortunately, the detailed analysis of drugs regulating DNA methylation has revealed them to have numerous "off target" effects and some of those do not affect methylation at all [207].

USE OF EPIGENETIC MODIFIERS FOR POTENTIAL DIAGNOSIS AND THERAPY IN AUTOIMMUNE DISEASES

Given the numerous epigenetic aberrations in autoimmune and chronic inflammatory conditions identified, targeting enzymes involved in writing, reading, and erasing these modifications is a therapeutic opportunity.

The use of HDACi as therapeutic drugs for different autoimmune diseases has been intensively studied over more than 20 years and HDACi therapy has shown promising results in pre-clinical models of many autoimmune and inflammatory diseases, including SLE, RA, MS, SSc, and T1D. The use of pan-HDACi was associated with many deleterious side effects during clinical trials for cancer treatment, including fatigue, nausea, thrombocytopenia, and electrocardiograph abnormalities [2]. In recent years, several class- and isoform-specific HDACi have been developed that are now under investigation [208]. Since they target, in contrast to the first generation of pan-HDACi, only selected members of the HDAC family, an improved safety profile is expected. Since most of the beneficial effects of HDACi were shown to be chromatin-independent and non-epigenetic [209], we do not discuss them here in detail.

The inhibition of epigenetic reader proteins, such as the BET proteins, is another therapeutic concept for autoimmune diseases. In 2013, Mele and coworkers showed that therapeutic dosing of the pan-BET inhibitor JQ-1 was efficacious in two mouse models of autoimmunity, the models for CIA and EAE. In these models BET inhibition suppressed the differentiation and activation of Th17 cells [3]. Numerous studies in mouse models for arthritis and fibrosis followed, providing evidence for the beneficial use of BET inhibitors in these conditions [132]. BET inhibition was also reported to have beneficial effects in a model of murine lupus [210]. Anti-inflammatory properties of BET inhibitors were initially described in murine LPS-stimulated bone marrow-derived macrophages [211], which were later confirmed in human monocytes [210,212]. In line with these effects, BET inhibitors restored the aberrant expression of interferon pathway genes found in monocytes of SSc patients [190]. In RA SF, the pan-inhibitor I-BET suppressed the expression of a broad range of inflammatory cytokines, chemokines, and matrix metalloproteinases in presence of inflammatory stimuli and reduced proliferation rates and chemotactic properties towards PBMC [129,131]. I-BET also exhibited beneficial effects on the modulation of T1D in NOD mice and human islet cells by dampening islet inflammation, enhancing β-cell regeneration, and attenuating the senescence-associated secretory phenotype of islet cells [213,214]. Whereas, therapeutic effects in CIA and EAE seemed to be dependent on T cells [3], the effects in NOD mice were independent of T-cell function [214]. So far, no specific inhibitors of BET proteins are available and data on the individual function of BET proteins autoimmunity are limited. However, first clinical trials for BET inhibitors in different types of cancer, the treatment of type 2 diabetes mellitus, cardiovascular and coronary artery disease are ongoing, underscoring the therapeutic potential of these drugs [132]. As potential side effects, reversible hematological toxicities, including thrombocytopenia, anemia, and neutropenia, and a few non-hematological events such as gastrointestinal events and fatigue were reported [215,216].

Another therapeutic concept that has been investigated in several autoimmune conditions is the pharmacological inhibition of the histone-lysine N-methyltransferase EZH2. Whereas conflicting results have been reported regarding the therapeutic use of EZH2 inhibitors in mouse models of SSc [115,191], beneficial effects of EZH2 inhibitors have been reported in lupus-like disease in mice [217,218]. DZNep improved the survival of mice, reduced the renal involvement, splenomegaly, and lymphadenopathy, and reduced the production of anti-dsDNA auto-antibodies [217]. Furthermore, EZH2 deficiency restricted disease progression in an animal model of MS by impairing the trans-endothelial migration of neutrophils and DC, by regulating their integrin signaling and adhesion [219]. Since EZH2 deficiency was reported to suppress the differentiation of Treg [220], the use of EZH2 inhibitors has to be considered with care and would need a more in depth investigation in autoimmune conditions.

SUMMARY

New high throughput methods and large-scale approaches have significantly improved the knowledge of altered epigenetic patterns in autoimmune conditions. The integration of genetic and epigenetic data sets with 3D chromatin interactions and gene expression profiles in different cell types have provided new insights into disease pathogenesis and provide the basis for new therapies. Several epigenetic modifications have been detected, many of them already at early stages of autoimmune diseases, suggesting that targeting of associated enzymes would eventually enable to interfere with aberrant cell activation in autoimmune diseases. To achieve this, a better knowledge on the contribution of individual enzymes in different cell types is still necessary. Furthermore, more isoform-specific inhibitors are needed. The integration of epigenetic alterations, in particular altered DNA methylation, into machine learning approaches is one possibility to predict treatment response or disease outcome in the future. However, more confirmatory studies are needed to validate such biomarker studies.

MULTIOMICS AND MACHINE LEARNING ACCURATELY PREDICT CLINICAL RESPONSE TO ADALIMUMAB AND ETANERCEPT THERAPY IN PATIENTS WITH RHEUMATOID ARTHRITIS [203] [EX: AUTOIMMUNITY EPIGENETICS CASE STUDY]

Objective

To predict the response to treatment with anti-tumor necrosis factor (anti-TNF), namely adalimumab or etanercept, in patients with rheumatoid arthritis (RA) prior to treatment initiation.

Scope

Clinical data and peripheral blood of patients included in the BioCURA cohort were collected over 5 years (2009–2014). RNA sequencing was performed on peripheral blood mononuclear cells (PBMC) as well as on sorted cell populations (CD4$^+$ T cells, CD14$^+$ monocytes). DNA methylation profiles were only assessed in PBMC. Sorted cell population for DNA methylation data would have provided an additional value for the study, since epigenetic modifications are cell type-specific. Molecular profiles assessed in the study entered a machine learning approach, which needs a comprehensive bioinformatics expert knowledge. Molecular profiles were assessed on baseline samples. It would be interesting to see how the markers identified change over treatment and whether some of them can be used to monitor treatment response.

Audience

This case study, published by Tao et al. in *Arthritis & Rheumatology*, 2021 [203], is of interest for physicians with an interest in predictive biomarkers for treatment response, as well as basic translational researchers involved in performing biomarker studies including epigenetic data sets.

Rationale

Approximately 30% of patients with RA do not respond sufficiently to treatment with their initial anti-TNF therapy. Factors that would allow a prediction of therapy response are highly needed in clinical practice.

Expected Results and Deliverables

The study provides evidence for the benefits of using molecular signatures obtained by RNA sequencing and DNA methylation profiling into machine learning models for personalized medicine.

Safety Considerations

Good bio banking, including sample and clinical data collection over several years and following patients over several visits, is a pre-requisite for performing such a study. RNA sequencing and DNA methylation profiling needs to be performed in a large set of patient samples that provides sufficient statistical power in the analysis. For bioinformatics analysis, expert knowledge is needed.

The Process, Workflow and Actions Taken

- Patients included were selected from the BioCURA conhort, and followed over 12 months after initiation of therapy with adalimumab or etanercept. Treatment response was evaluated after 6 months or earlier in non-responders, based on changes in the disease activity scores.
- Before initiation of anti-TNF therapy, PBMC were isolated from peripheral blood samples. In a substudy, $CD4^+$ T cells and $CD14^+$ monocytes were isolated from PBMC by positive and negative selection to assess if different cell populations underlined the molecular patterns identified in adalimumab and etanercept responses.
- Total RNA was isolated and sent for library preparation and RNA sequencing (HiSeq 4000).
- DNA was isolated from frozen PBMC, bisulfite treated, amplified, and hybridized to the Infinium MethylationEPIC BeadChip array (Illumina).
- Gene expression profiles and DNA methylation signatures were analyzed using standard bioinformatics procedures for these types of data.
- Each of the data sets was then used separately for a machine learning approach based on random forest algorithms.

Tools and Materials Used

For the assessment of DNA methylation patterns, the Infinium MethylationEPIC BeadChip array (Illumina) was used that allows for simultaneous interrogation of more than 850,000 CpG sites, providing a high genome-wide coverage of RefSeq genes. This array contains over 90% of the probes of HumanMethylation450 BeadChip that is widely used in other studies.

Results

Distinct sets of differentially expressed genes and hypermethylation patterns in PBMC underlined the response to etanercept and adalimumab. To test whether different cell types underlined the distinct signatures in etanercept and adalimumab responders, RNA sequencing was additionally performed in sorted $CD4^+$ T cells and $CD14^+$ monocytes. There was little overlap in differentially expressed genes in $CD4^+$ T cells and $CD14^+$ monocytes in responders to both treatments, supporting the need for performing such studies in specific cell types. Expression profiles and DNA methylation signatures entered machine learning models based on random forest algorithms. The best model for predicting the response to adalimumab was based on the gene expression in PBMC, with an overall accuracy of 85.9%. The prediction model based on DNA methylation patterns in PBMC reached an accuracy of 84.7% and was superior over the prediction model using differential gene expression in specific cell types.

The best model for predicting the response to etanercept was based on DNA methylation signatures with an overall accuracy of 88%, which was far superior to the best model based on gene expression in $CD4^+$ T cells, with an overall accuracy of 79.0%.

All models used predicted that approximately 30% of patients will not respond to their initial anti-TNF therapy what is in line with current clinical observations.

A small number of patients that entered validation of the prediction models did not respond to their first anti-TNF therapy and were then switched to the other one. In this set-up, accurate prediction was achieved in 88.9% and 77.8%, using gene expression models and DNA methylation models, respectively.

Challenges

The biggest challenge is the transfer of such studies into clinical practice.

Learning and Knowledge Outcomes

The study shows that different prediction models, including different data sets derived from diverse cell types have to be tested for different drugs in order to evaluate the best model for each drug.

Data obtained by RNA sequencing and DNA methylation profiling should be integrated into machine learning models instead of handling the sets independently, to once test whether the prediction models could be further improved. The assessment of DNA methylation should also be performed in sorted cell types. It would be interesting to see, whether the set of differentially expressed genes and differentially methylated regions in responders and non-responders could be further scaled down. The integration of clinical data and molecular profiles into machine learning models might help to scale down the molecular parameters needed in the prediction model that enable a more ease transfer into clinical practice. The predictive models need further evaluation in more cohorts.

Abbreviations

ATACseq	Accessible chromatin using sequencing
5-aza	5-azacytidin
BET	Bromodomain and extra terminal domain
Blimp-1	B lymphocyte-induced maturation protein 1
ChIPseq	Chromatin Immunoprecipitation sequencing
CIA	Collagen induced arthritis
DC	Dendritic cell
DNMT	DNA methyltransferase
dsDNA	double-stranded DNA
dSSc	diffuse cutaneous SSc
DZNep	3-deazaneplanocin A hydrochloride
EAE	Experimental allergic encephalomyelitis
ECM	Extracellular matrix
ERK	Extracellular signal-regulated kinase
EZH2	Enhancer of zeste homolog 2
FLI1	Fli-1 proto-oncogene
FOXP3	Forkhead box P3
Gata3	GATA binding protein 3
GWAS	Genome-wide association study
H3, H4	Histone 3, histone 4
H3ac	Histone 3 acetylation
H3K27ac	Histone 3 lysine 27 acetylation
H3K27me3	Histone 3 lysine 27 trimethylation
H3K4me1	Histone 3 lysine 4 monomethylation
H3K4me3	Histone 3 lysine 4 trimethylation
H3K9ac	Histone 3 lysine 9 acetylation
HAT	Histone acetyltransferase
HDAC	Histone deacetylase
HDACi	Histone deacetylase inhibitor
HLA	Human leukocyte antigen
5-hmC	5-Hydroxymethylcytosine
IL	Interleukin
IRF	interferon regulator factor
JAK	Janus kinase
JMJD3	Jumonji domain-containing protein 3
LINE-1	long interspersed nuclear element-1
LPS	lipopolysaccharide
LSD1	histone demethylase 1 A
lSSc	limited cutaneous SSc
Ms	Multiple sclerosis
5-mC	5-Methylcytosine
MVP	Methylation variable positions
MZ	Monozygotic
NOD	Non-obese diabetic
PBMC	Peripheral blood derived mononuclear cells
RA	Rheumatoid arthritis
RFX1	Regulatory factor X1
RNAseq	RNA sequencing
Rorγt	RAR-related orphan receptor C
SAM	S-adenosyl methionine
SF	Synovial fibroblast
SIRT	Sirtuin
SLE	Systemic lupus erythematosus
SSc	Systemic sclerosis
STAT	Signal transducer and activator of transcription
T1D	Type 1 diabetes mellitus
TET	Ten–eleven translocation
Th	T helper
Tbet	T-box transcription factor 21
TLR	Toll-like receptor
TNF	tumor necrosis factor
Treg	Regulatory T cells
XCI	X-chromosome inactivation

References

[1] Musselman CA, Lalonde ME, Cote J, Kutateladze TG. Perceiving the epigenetic landscape through histone readers. Nat Struct Mol Biol 2012;19(12):1218–27.

[2] Regna NLAR, Christopher M. Isoform-selective HDAC inhibition in autoimmune disease. J Clin Cell Immunol 2014;5(2):206–13.

[3] Mele DA, Salmeron A, Ghosh S, Huang HR, Bryant BM, Lora JM. BET bromodomain inhibition suppresses TH17-mediated pathology. J Exp Med 2013;210(11):2181–90.

[4] Rodriguez RM, Lopez-Larrea C, Suarez-Alvarez B. Epigenetic dynamics during CD4(+) T cells lineage commitment. Int J Biochem Cell Biol 2015;67:75–85.

[5] Rosenblum MD, Remedios KA, Abbas AK. Mechanisms of human autoimmunity. J Clin Invest 2015;125(6):2228–33.

[6] Neumann E, Lefevre S, Zimmermann B, Gay S, Muller-Ladner U. Rheumatoid arthritis progression mediated by activated synovial fibroblasts. Trends Mol Med 2010;16(10):458–68.

[7] Hewagama A, Richardson B. The genetics and epigenetics of autoimmune diseases. J Autoimmun 2009;33(1):3–11.

[8] Eyre S, Worthington J. Take your PICS: moving from GWAS to immune function. Immunity 2014;41(6):883–5.

[9] Davis-Richardson AG, Triplett EW. A model for the role of gut bacteria in the development of autoimmunity for type 1 diabetes. Diabetologia 2015;58(7):1386–93.

[10] Lu Q, Wu A, Tesmer L, Ray D, Yousif N, Richardson B. Demethylation of CD40LG on the inactive X in T cells from women with lupus. J Immunol 2007;179(9):6352–8.

[11] Chabchoub G, Uz E, Maalej A, Mustafa CA, Rebai A, Mnif M, et al. Analysis of skewed X-chromosome inactivation in females with rheumatoid arthritis and autoimmune thyroid diseases. Arthritis Res Ther 2009;11(4):R106.

[12] Liao J, Liang G, Xie S, Zhao H, Zuo X, Li F, et al. CD40L demethylation in CD4(+) T cells from women with rheumatoid arthritis. Clin Immunol 2012;145(1):13–18.

[13] Harbo HF, Gold R, Tintore M. Sex and gender issues in multiple sclerosis. Ther Adv Neurol Disor 2013;6(4):237–48.

[14] Ozbalkan Z, Bagislar S, Kiraz S, Akyerli CB, Ozer HT, Yavuz S, et al. Skewed X chromosome inactivation in blood cells of women with scleroderma. Arthritis Rheum 2005;52(5):1564–70.

[15] Diaz-Valencia PA, Bougneres P, Valleron AJ. Global epidemiology of type 1 diabetes in young adults and adults: a systematic review. BMC Public Health 2015;15:255.

[16] Soto ME, Vallejo M, Guillen F, Simon JA, Arena E, Reyes PA. Gender impact in systemic lupus erythematosus. Clin Exp Rheumatol 2004;22(6):713–21.

[17] Symmons DP. Epidemiology of rheumatoid arthritis: determinants of onset, persistence and outcome. Best Pract Res Clin Rheumatol 2002;16(5):707–22.

[18] Knudsen GP. Gender bias in autoimmune diseases: X chromosome inactivation in women with multiple sclerosis. J Neurol Sci 2009;286(1–2):43–6.

[19] Lian X, Xiao R, Hu X, Kanekura T, Jiang H, Li Y, et al. DNA demethylation of CD40l in CD4 + T cells from women with systemic sclerosis: a possible explanation for female susceptibility. Arthritis Rheum 2012;64(7):2338–45.

References

[20] Richardson B, Scheinbart L, Strahler J, Gross L, Hanash S, Johnson M. Evidence for impaired T cell DNA methylation in systemic lupus erythematosus and rheumatoid arthritis. Arthritis Rheum 1990;33(11):1665−73.

[21] Coit P, Jeffries M, Altorok N, Dozmorov MG, Koelsch KA, Wren JD, et al. Genome-wide DNA methylation study suggests epigenetic accessibility and transcriptional poising of interferon-regulated genes in naive CD4 + T cells from lupus patients. J Autoimmun 2013;43:78−84.

[22] Ulff-Moller CJ, Asmar F, Liu Y, Svendsen AJ, Busato F, Gronbaek K, et al. Twin DNA methylation profiling reveals flare-dependent interferon signature and B cell promoter hypermethylation in systemic lupus erythematosus. Arthritis Rheumatol 2018;70(6):878−90.

[23] de Andres MC, Perez-Pampin E, Calaza M, Santaclara FJ, Ortea I, Gomez-Reino JJ, et al. Assessment of global DNA methylation in peripheral blood cell subpopulations of early rheumatoid arthritis before and after methotrexate. Arthritis Res Ther 2015;17:233.

[24] Graves MC, Benton M, Lea RA, Boyle M, Tajouri L, Macartney-Coxson D, et al. Methylation differences at the HLA-DRB1 locus in CD4 + T-Cells are associated with multiple sclerosis. Multiple Sclerosis 2014;20(8):1033−41.

[25] Ewing E, Kular L, Fernandes SJ, Karathanasis N, Lagani V, Ruhrmann S, et al. Combining evidence from four immune cell types identifies DNA methylation patterns that implicate functionally distinct pathways during Multiple Sclerosis progression. EBioMedicine 2019;43:411−23.

[26] Paul DS, Teschendorff AE, Dang MA, Lowe R, Hawa MI, Ecker S, et al. Increased DNA methylation variability in type 1 diabetes across three immune effector cell types. Nat Commun 2016;7:13555.

[27] Lei W, Luo Y, Lei W, Luo Y, Yan K, Zhao S, et al. Abnormal DNA methylation in CD4 + T cells from patients with systemic lupus erythematosus, systemic sclerosis, and dermatomyositis. Scand J Rheumatol 2009;38(5):369−74.

[28] Ding W, Pu W, Wang L, Jiang S, Zhou X, Tu W, et al. Genome-wide DNA methylation analysis in systemic sclerosis reveals hypomethylation of IFN-associated genes in CD4(+) and CD8(+) T cells. J Invest Dermatol 2018;138(5):1069−77.

[29] Li T, Ortiz-Fernandez L, Andres-Leon E, Ciudad L, Javierre BM, Lopez-Isac E, et al. Epigenomics and transcriptomics of systemic sclerosis CD4 + T cells reveal long-range dysregulation of key inflammatory pathways mediated by disease-associated susceptibility loci. Genome Med 2020;12(1):81.

[30] Miller S, Tsou PS, Coit P, Gensterblum-Miller E, Renauer P, Rohraff DM, et al. Hypomethylation of STAT1 and HLA-DRB1 is associated with type-I interferon-dependent HLA-DRB1 expression in lupus CD8 + T cells. Ann Rheum Dis 2019;78(4):519−28.

[31] Nakkuntod J, Avihingsanon Y, Mutirangura A, Hirankarn N. Hypomethylation of LINE-1 but not Alu in lymphocyte subsets of systemic lupus erythematosus patients. Clin Chim Acta 2011;412(15−16):1457−61.

[32] Maltby VE, Graves MC, Lea RA, Benton MC, Sanders KA, Tajouri L, et al. Genome-wide DNA methylation profiling of CD8 + T cells shows a distinct epigenetic signature to CD4 + T cells in multiple sclerosis patients. Clin Epigenetics 2015;7:118.

[33] Glossop JR, Emes RD, Nixon NB, Packham JC, Fryer AA, Mattey DL, et al. Genome-wide profiling in treatment-naive early rheumatoid arthritis reveals DNA methylome changes in T and B lymphocytes. Epigenomics 2016;8(2):209−24.

[34] Julia A, Absher D, Lopez-Lasanta M, Palau N, Pluma A, Waite Jones L, et al. Epigenome-wide association study of rheumatoid arthritis identifies differentially methylated loci in B cells. Hum Mol Genet 2017;26(14):2803−11.

[35] Maltby VE, Lea RA, Graves MC, Sanders KA, Benton MC, Tajouri L, et al. Genome-wide DNA methylation changes in CD19(+) B cells from relapsing-remitting multiple sclerosis patients. Sci Rep 2018;8(1):17418.

[36] Stefan M, Zhang W, Concepcion E, Yi Z, Tomer Y. DNA methylation profiles in type 1 diabetes twins point to strong epigenetic effects on etiology. J Autoimmun 2014;50:33−7.

[37] Liu Y, Aryee MJ, Padyukov L, Fallin MD, Hesselberg E, Runarsson A, et al. Epigenome-wide association data implicate DNA methylation as an intermediary of genetic risk in rheumatoid arthritis. Nat Biotechnol 2013;31(2):142−7.

[38] Rakyan VK, Beyan H, Down TA, Hawa MI, Maslau S, Aden D, et al. Identification of type 1 diabetes-associated DNA methylation variable positions that precede disease diagnosis. PLoS Genet 2011;7(9):e1002300.

[39] Kular L, Needhamsen M, Adzemovic MZ, Kramarova T, Gomez-Cabrero D, Ewing E, et al. Neuronal methylome reveals CREB-associated neuro-axonal impairment in multiple sclerosis. Clin Epigenetics 2019;11(1):86.

[40] Nakano K, Whitaker JW, Boyle DL, Wang W, Firestein GS. DNA methylome signature in rheumatoid arthritis. Ann Rheum Dis 2013;72(1):110−17.

[41] Karouzakis E, Gay RE, Michel BA, Gay S, Neidhart M. DNA hypomethylation in rheumatoid arthritis synovial fibroblasts. Arthritis Rheum 2009;60(12):3613−22.

[42] Hattori M, Yokoyama Y, Hattori T, Motegi S, Amano H, Hatada I, et al. Global DNA hypomethylation and hypoxia-induced expression of the ten eleven translocation (TET) family, TET1, in scleroderma fibroblasts. Exp Dermatol 2015;24(11):841−6.

[43] Altorok N, Tsou PS, Coit P, Khanna D, Sawalha AH. Genome-wide DNA methylation analysis in dermal fibroblasts from patients with diffuse and limited systemic sclerosis reveals common and subset-specific DNA methylation aberrancies. Ann Rheum Dis 2015;74(8):1612−20.

[44] Neidhart M, Rethage J, Kuchen S, Kunzler P, Crowl RM, Billingham ME, et al. Retrotransposable L1 elements expressed in rheumatoid arthritis synovial tissue: association with genomic DNA hypomethylation and influence on gene expression. Arthritis Rheum 2000;43(12):2634−47.

[45] Zhao M, Wang J, Liao W, Li D, Li M, Wu H, et al. Increased 5-hydroxymethylcytosine in CD4(+) T cells in systemic lupus erythematosus. J Autoimmun 2016;69:64−73.

[46] Farh KK, Marson A, Zhu J, Kleinewietfeld M, Housley WJ, Beik S, et al. Genetic and epigenetic fine mapping of causal autoimmune disease variants. Nature 2015;518(7539):337−43.

[47] Okada Y, Wu D, Trynka G, Raj T, Terao C, Ikari K, et al. Genetics of rheumatoid arthritis contributes to biology and drug discovery. Nature 2014;506(7488):376−81.

[48] Whitaker JW, Boyle DL, Bartok B, Ball ST, Gay S, Wang W, et al. Integrative omics analysis of rheumatoid arthritis identifies non-obvious therapeutic targets. PloS One 2015;10(4):e0124254.

[49] Okada Y, Muramatsu T, Suita N, Kanai M, Kawakami E, Iotchkova V, et al. Significant impact of miRNA-target gene networks on genetics of human complex traits. Sci Rep 2016;6:22223.

[50] Liu Q, Zaba LC, Satpathy AT, Longmire M, Zhang W, Li K, et al. Chromatin accessibility landscapes of skin cells in systemic sclerosis nominate dendritic cells in disease pathogenesis. Nat Commun 2020;11(1):5843.

[51] Chiou J, Geusz RJ, Okino ML, Han JY, Miller M, Melton R, et al. Interpreting type 1 diabetes risk with genetics and single-cell epigenomics. Nature 2021;594(7863):398−402.

[52] Ge X, Frank-Bertoncelj M, Klein K, Mcgovern A, Kuret T, Houtman M, et al. Functional genomics atlas of synovial fibroblasts defining rheumatoid arthritis heritability. medRxiv 2020;2020.12.16.20248230.

[53] Martin P, McGovern A, Orozco G, Duffus K, Yarwood A, Schoenfelder S, et al. Capture Hi-C reveals novel candidate genes and complex long-range interactions with related autoimmune risk loci. Nat Commun 2015;6:10069.

[54] Tsuchiya H, Ota M, Fujio K. Multiomics landscape of synovial fibroblasts in rheumatoid arthritis. Inflammation and regeneration 2021;41(1):7.

[55] Factor DC, Barbeau AM, Allan KC, Hu LR, Madhavan M, Hoang AT, et al. Cell type-specific intralocus interactions reveal oligodendrocyte mechanisms in MS. Cell 2020;181 (2):382−95 e21.

[56] Gialitakis M, Sellars M, Littman DR. The epigenetic landscape of lineage choice: lessons from the heritability of CD4 and CD8 expression. Curr Top Microbiol Immunol 2012;356:165−88.

[57] Henning AN, Roychoudhuri R, Restifo NP. Epigenetic control of CD8(+) T cell differentiation. Nat Rev Immunol 2018;18 (5):340−56.

[58] Murphy KM, Stockinger B. Effector T cell plasticity: flexibility in the face of changing circumstances. Nat Immunol 2010;11 (8):674−80.

[59] Wang C, Collins M, Kuchroo VK. Effector T cell differentiation: are master regulators of effector T cells still the masters? Curr Opin Immunol 2015;37:6−10.

[60] Hashimoto S, Ogoshi K, Sasaki A, Abe J, Qu W, Nakatani Y, et al. Coordinated changes in DNA methylation in antigen-specific memory CD4 T cells. J Immunol 2013;190(8):4076−91.

[61] Vahedi G, Takahashi H, Nakayamada S, Sun HW, Sartorelli V, Kanno Y, et al. STATs shape the active enhancer landscape of T cell populations. Cell 2012;151(5):981−93.

[62] Yang XP, Jiang K, Hirahara K, Vahedi G, Afzali B, Sciume G, et al. EZH2 is crucial for both differentiation of regulatory T cells and T effector cell expansion. Sci Rep 2015;5:10643.

[63] Akimzhanov AM, Yang XO, Dong C. Chromatin remodeling of interleukin-17 (IL-17)-IL-17F cytokine gene locus during inflammatory helper T cell differentiation. J Biol Chem 2007;282(9):5969−72.

[64] Wei L, Laurence A, Elias KM, O'Shea JJ. IL-21 is produced by Th17 cells and drives IL-17 production in a STAT3-dependent manner. J Biol Chem 2007;282(48):34605−10.

[65] Huehn J, Beyer M. Epigenetic and transcriptional control of Foxp3 + regulatory T cells. Semin Immunol 2015;27(1):10−18.

[66] Li B, Samanta A, Song X, Iacono KT, Bembas K, Tao R, et al. FOXP3 interactions with histone acetyltransferase and class II histone deacetylases are required for repression. Proc Natl Acad Sci U S A 2007;104(11):4571−6.

[67] Delacher M, Imbusch CD, Weichenhan D, Breiling A, Hotz-Wagenblatt A, Trager U, et al. Genome-wide DNA-methylation landscape defines specialization of regulatory T cells in tissues. Nat Immunol 2017;18(10):1160−72.

[68] Wiggins KJ, Scharer CD. Roadmap to a plasma cell: epigenetic and transcriptional cues that guide B cell differentiation. Immunol Rev 2021;300(1):54−64.

[69] Zhang Y, Good-Jacobson KL. Epigenetic regulation of B cell fate and function during an immune response. Immunol Rev 2019;288(1):75−84.

[70] Martin-Subero JI, Oakes CC. Charting the dynamic epigenome during B-cell development. Semin Cancer Biol 2018;51:139−48.

[71] Kulis M, Merkel A, Heath S, Queiros AC, Schuyler RP, Castellano G, et al. Whole-genome fingerprint of the DNA methylome during human B cell differentiation. Nat Genet 2015;47(7):746−56.

[72] Cao R, Wang L, Wang H, Xia L, Erdjument-Bromage H, Tempst P, et al. Role of histone H3 lysine 27 methylation in Polycomb-group silencing. Science 2002;298(5595):1039−43.

[73] Su IH, Basavaraj A, Krutchinsky AN, Hobert O, Ullrich A, Chait BT, et al. Ezh2 controls B cell development through histone H3 methylation and Igh rearrangement. Nat Immunol 2003;4(2):124−31.

[74] Mandal M, Powers SE, Maienschein-Cline M, Bartom ET, Hamel KM, Kee BL, et al. Epigenetic repression of the Igk locus by STAT5-mediated recruitment of the histone methyltransferase Ezh2. Nat Immunol 2011;12(12):1212−20.

[75] Jacobsen JA, Woodard J, Mandal M, Clark MR, Bartom ET, Sigvardsson M, et al. EZH2 regulates the developmental timing of effectors of the pre-antigen receptor checkpoints. J Immunol 2017;198(12):4682−91.

[76] Beguelin W, Popovic R, Teater M, Jiang Y, Bunting KL, Rosen M, et al. EZH2 is required for germinal center formation and somatic EZH2 mutations promote lymphoid transformation. Cancer Cell 2013;23(5):677−92.

[77] Xu CR, Feeney AJ. The epigenetic profile of Ig genes is dynamically regulated during B cell differentiation and is modulated by pre-B cell receptor signaling. J Immunol 2009;182(3):1362−9.

[78] Lee SC, Bottaro A, Insel RA. Activation of terminal B cell differentiation by inhibition of histone deacetylation. Mol Immunol 2003;39(15):923−32.

[79] Tanaka H, Muto A, Shima H, Katoh Y, Sax N, Tajima S, et al. Epigenetic regulation of the Blimp-1 Gene (Prdm1) in B cells involves Bach2 and histone deacetylase 3. J Biol Chem 2016;291 (12):6316−30.

[80] Amit I, Winter DR, Jung S. The role of the local environment and epigenetics in shaping macrophage identity and their effect on tissue homeostasis. Nat Immunol 2016;17 (1):18−25.

[81] Kapellos TS, Iqbal AJ. Epigenetic control of macrophage polarisation and soluble mediator gene expression during inflammation. Mediators Inflamm 2016;2016:6591703.

[82] Saeed S, Quintin J, Kerstens HH, Rao NA, Aghajanirefah A, Matarese F, et al. Epigenetic programming of monocyte-to-macrophage differentiation and trained innate immunity. Science 2014;345(6204):1251086.

[83] Hoeksema MA, de Winther MP. Epigenetic regulation of monocyte and macrophage function. Antioxid Redox Signal 2016;25 (14):758−74.

[84] Foster SL, Hargreaves DC, Medzhitov R. Gene-specific control of inflammation by TLR-induced chromatin modifications. Nature 2007;447(7147):972−8.

[85] Quintin J, Saeed S, Martens JHA, Giamarellos-Bourboulis EJ, Ifrim DC, Logie C, et al. Candida albicans infection affords protection against reinfection via functional reprogramming of monocytes. Cell Host Microbe 2012;12(2):223−32.

[86] Boukhaled GM, Corrado M, Guak H, Krawczyk CM. Chromatin architecture as an essential determinant of dendritic cell function. Front Immunol 2019;10:1119.

[87] Tian Y, Meng L, Zhang Y. Epigenetic regulation of dendritic cell development and function. Cancer J 2017;23(5):302−7.

[88] Zhang X, Ulm A, Somineni HK, Oh S, Weirauch MT, Zhang HX, et al. DNA methylation dynamics during ex vivo differentiation and maturation of human dendritic cells. Epigenetics Chromatin 2014;7:21.

[89] Klug M, Schmidhofer S, Gebhard C, Andreesen R, Rehli M. 5-Hydroxymethylcytosine is an essential intermediate of active DNA demethylation processes in primary human monocytes. Genome Biol 2013;14(5):R46.

[90] Pacis A, Tailleux L, Morin AM, Lambourne J, MacIsaac JL, Yotova V, et al. Bacterial infection remodels the DNA methylation landscape of human dendritic cells. Genome Res 2015;25 (12):1801−11.

[91] Paul F, Arkin Y, Giladi A, Jaitin DA, Kenigsberg E, Keren-Shaul H, et al. Transcriptional heterogeneity and lineage commitment in myeloid progenitors. Cell 2015;163(7):1663−77.

REFERENCES

[92] Garber M, Yosef N, Goren A, Raychowdhury R, Thielke A, Guttman M, et al. A high-throughput chromatin immunoprecipitation approach reveals principles of dynamic gene regulation in mammals. Mol Cell 2012;47(5):810–22.

[93] Lin Q, Chauvistre H, Costa IG, Gusmao EG, Mitzka S, Hanzelmann S, et al. Epigenetic program and transcription factor circuitry of dendritic cell development. Nucleic Acids Res 2015;43(20):9680–93.

[94] Bornstein C, Winter D, Barnett-Itzhaki Z, David E, Kadri S, Garber M, et al. A negative feedback loop of transcription factors specifies alternative dendritic cell chromatin States. Mol Cell 2014;56(6):749–62.

[95] Tsokos GC. Systemic lupus erythematosus. N Engl J Med 2011;365(22):2110–21.

[96] Wu H, Zhao M, Tan L, Lu Q. The key culprit in the pathogenesis of systemic lupus erythematosus: aberrant DNA methylation. Autoimmun Rev 2016;15(7):684–9.

[97] Crispin JC, Hedrich CM, Tsokos GC. Gene-function studies in systemic lupus erythematosus. Nat Rev Rheumatol 2013;9(8):476–84.

[98] Richardson B. Effect of an inhibitor of DNA methylation on T cells. II. 5-Azacytidine induces self-reactivity in antigen-specific T4+ cells. Hum Immunol 1986;17(4):456–70.

[99] Zhao M, Liu S, Luo S, Wu H, Tang M, Cheng W, et al. DNA methylation and mRNA and microRNA expression of SLE CD4+ T cells correlate with disease phenotype. J Autoimmun 2014;54:127–36.

[100] Renauer P, Coit P, Jeffries MA, Merrill JT, McCune WJ, Maksimowicz-McKinnon K, et al. DNA methylation patterns in naive CD4+ T cells identify epigenetic susceptibility loci for malar rash and discoid rash in systemic lupus erythematosus. Lupus Sci Med 2015;2(1):e000101.

[101] Coit P, Renauer P, Jeffries MA, Merrill JT, McCune WJ, Maksimowicz-McKinnon K, et al. Renal involvement in lupus is characterized by unique DNA methylation changes in naive CD4+ T cells. J Autoimmun 2015;61:29–35.

[102] Coit P, Dozmorov MG, Merrill JT, McCune WJ, Maksimowicz-McKinnon K, Wren JD, et al. Epigenetic reprogramming in naive CD4+ T cells favoring T cell activation and non-Th1 effector T cell immune response as an early event in lupus flares. Arthritis Rheumatol 2016;68(9):2200–9.

[103] Gensterblum E, Renauer P, Coit P, Strickland FM, Kilian NC, Miller S, et al. CD4+CD28+KIR+CD11a(hi) T cells correlate with disease activity and are characterized by a pro-inflammatory epigenetic and transcriptional profile in lupus patients. J Autoimmun 2018;86:19–28.

[104] Scheinbart LS, Johnson MA, Gross LA, Edelstein SR, Richardson BC. Procainamide inhibits DNA methyltransferase in a human T cell line. J Rheumatol 1991;18(4):530–4.

[105] Vedove CD, Del Giglio M, Schena D, Girolomoni G. Drug-induced lupus erythematosus. Arch Dermatol Res 2009;301(1):99–105.

[106] Deng C, Lu Q, Zhang Z, Rao T, Attwood J, Yung R, et al. Hydralazine may induce autoimmunity by inhibiting extracellular signal-regulated kinase pathway signaling. Arthritis Rheum 2003;48(3):746–56.

[107] Oelke K, Richardson B. Decreased T cell ERK pathway signaling may contribute to the development of lupus through effects on DNA methylation and gene expression. Int Rev Immunol 2004;23(3–4):315–31.

[108] Sawalha AH, Jeffries M, Webb R, Lu Q, Gorelik G, Ray D, et al. Defective T-cell ERK signaling induces interferon-regulated gene expression and overexpression of methylation-sensitive genes similar to lupus patients. Genes Immun 2008;9(4):368–78.

[109] Scharer CD, Blalock EL, Barwick BG, Haines RR, Wei C, Sanz I, et al. ATAC-seq on biobanked specimens defines a unique chromatin accessibility structure in naive SLE B cells. Sci Rep 2016;6:27030.

[110] Scharer CD, Blalock EL, Mi T, Barwick BG, Jenks SA, Deguchi T, et al. Epigenetic programming underpins B cell dysfunction in human SLE. Nat Immunol 2019;20(8):1071–82.

[111] Breitbach ME, Ramaker RC, Roberts K, Kimberly RP, Absher D. Population-specific patterns of epigenetic defects in the B cell lineage in patients with systemic lupus erythematosus. Arthritis Rheumatol 2020;72(2):282–91.

[112] Garcia BA, Busby SA, Shabanowitz J, Hunt DF, Mishra N. Resetting the epigenetic histone code in the MRL-lpr/lpr mouse model of lupus by histone deacetylase inhibition. J Proteome Res 2005;4(6):2032–42.

[113] Forster N, Gallinat S, Jablonska J, Weiss S, Elsasser HP, Lutz W. p300 protein acetyltransferase activity suppresses systemic lupus erythematosus-like autoimmune disease in mice. J Immunol 2007;178(11):6941–8.

[114] Hu N, Qiu X, Luo Y, Yuan J, Li Y, Lei W, et al. Abnormal histone modification patterns in lupus CD4+ T cells. J Rheumatol 2008;35(5):804–10.

[115] Tsou PS, Coit P, Kilian NC, Sawalha AH. EZH2 modulates the DNA methylome and controls T Cell adhesion through junctional adhesion molecule a in lupus patients. Arthritis Rheumatol 2018;70(1):98–108.

[116] Zhang M, Iwata S, Hajime M, Ohkubo N, Todoroki Y, Miyata H, et al. Methionine commits cells to differentiate into plasmablasts through epigenetic regulation of BTB and CNC Homolog 2 by the Methyltransferase EZH2. Arthritis Rheumatol 2020;72(7):1143–53.

[117] Zhao M, Tan Y, Peng Q, Huang C, Guo Y, Liang G, et al. IL-6/STAT3 pathway induced deficiency of RFX1 contributes to Th17-dependent autoimmune diseases via epigenetic regulation. Nat Commun 2018;9(1):583.

[118] Ferrari M, Onuoha SC, Pitzalis C. Trojan horses and guided missiles: targeted therapies in the war on arthritis. Nat Rev Rheumatol 2015;11(6):328–37.

[119] Van Doornum S, McColl G, Wicks IP. Accelerated atherosclerosis: an extraarticular feature of rheumatoid arthritis? Arthritis Rheum 2002;46(4):862–73.

[120] Klarenbeek NB, Allaart CF, Kerstens PJ, Huizinga TW, Dijkmans BA. The BeSt story: on strategy trials in rheumatoid arthritis. Curr Opin Rheumatol 2009;21(3):291–8.

[121] Karouzakis E, Neidhart M, Gay RE, Gay S. Molecular and cellular basis of rheumatoid joint destruction. Immunol Lett 2006;106(1):8–13.

[122] Mizoguchi F, Slowikowski K, Wei K, Marshall JL, Rao DA, Chang SK, et al. Functionally distinct disease-associated fibroblast subsets in rheumatoid arthritis. Nat Commun 2018;9(1):789.

[123] Muller-Ladner U, Kriegsmann J, Franklin BN, Matsumoto S, Geiler T, Gay RE, et al. Synovial fibroblasts of patients with rheumatoid arthritis attach to and invade normal human cartilage when engrafted into SCID mice. Am J Pathol 1996;149(5):1607–15.

[124] Frank-Bertoncelj M, Trenkmann M, Klein K, Karouzakis E, Rehrauer H, Bratus A, et al. Epigenetically-driven anatomical diversity of synovial fibroblasts guides joint-specific fibroblast functions. Nat Commun 2017;8:14852.

[125] Ai R, Hammaker D, Boyle DL, Morgan R, Walsh AM, Fan S, et al. Joint-specific DNA methylation and transcriptome signatures in rheumatoid arthritis identify distinct pathogenic processes. Nat Commun 2016;7:11849.

[126] Hammaker D, Nygaard G, Kuhs A, Ai R, Boyle DL, Wang W, et al. Joint location-specific JAK-STAT signaling in rheumatoid arthritis fibroblast-like synoviocytes. ACR Open Rheumatol 2019;1(10):640–8.

[127] Ai R, Laragione T, Hammaker D, Boyle DL, Wildberg A, Maeshima K, et al. Comprehensive epigenetic landscape of

[128] Klein K, Frank-Bertoncelj M, Karouzakis E, Gay RE, Kolling C, Ciurea A, et al. The epigenetic architecture at gene promoters determines cell type-specific LPS tolerance. J Autoimmun 2017;83:122–33.

[129] Loh C, Park SH, Lee A, Yuan R, Ivashkiv LB, Kalliolias GD. TNF-induced inflammatory genes escape repression in fibroblast-like synoviocytes: transcriptomic and epigenomic analysis. Ann Rheum Dis 2019;78(9):1205–14.

[130] Klein K, Gay S. Epigenetics in rheumatoid arthritis. Curr Opin Rheumatol 2015;27(1):76–82.

[131] Klein K, Kabala PA, Grabiec AM, Gay RE, Kolling C, Lin LL, et al. The bromodomain protein inhibitor I-BET151 suppresses expression of inflammatory genes and matrix degrading enzymes in rheumatoid arthritis synovial fibroblasts. Ann Rheum Dis 2016;75(2):422–9.

[132] Klein K. Bromodomain protein inhibition: a novel therapeutic strategy in rheumatic diseases. RMD Open 2018;4(2):e000744.

[133] Niederer F, Ospelt C, Brentano F, Hottiger MO, Gay RE, Gay S, et al. SIRT1 overexpression in the rheumatoid arthritis synovium contributes to proinflammatory cytokine production and apoptosis resistance. Ann Rheum Dis 2011;70(10):1866–73.

[134] Angiolilli C, Grabiec AM, Ferguson BS, Ospelt C, Malvar Fernandez B, van Es IE, et al. Inflammatory cytokines epigenetically regulate rheumatoid arthritis fibroblast-like synoviocyte activation by suppressing HDAC5 expression. Ann Rheum Dis 2016;75(2):430–8.

[135] Angiolilli C, Kabala PA, Grabiec AM, Van Baarsen IM, Ferguson BS, Garcia S, et al. Histone deacetylase 3 regulates the inflammatory gene expression programme of rheumatoid arthritis fibroblast-like synoviocytes. Ann Rheum Dis 2016;76(1):277–85.

[136] Klareskog L, Gregersen PK, Huizinga TW. Prevention of autoimmune rheumatic disease: state of the art and future perspectives. Ann Rheum Dis 2010;69(12):2062–6.

[137] Reynisdottir G, Karimi R, Joshua V, Olsen H, Hensvold AH, Harju A, et al. Structural changes and antibody enrichment in the lungs are early features of anti-citrullinated protein antibody-positive rheumatoid arthritis. Arthritis Rheumatol 2014;66(1):31–9.

[138] Engler A, Tange C, Frank-Bertoncelj M, Gay RE, Gay S, Ospelt C. Regulation and function of SIRT1 in rheumatoid arthritis synovial fibroblasts. J Mol Med 2016;94(2):173–82.

[139] Engler A, Niederer F, Klein K, Gay RE, Kyburz D, Camici GG, et al. SIRT6 regulates the cigarette smoke-induced signalling in rheumatoid arthritis synovial fibroblasts. J Mol Med 2014;92(7):757–67.

[140] Lee HS, Ka SO, Lee SM, Lee SI, Park JW, Park BH. Overexpression of sirtuin 6 suppresses inflammatory responses and bone destruction in mice with collagen-induced arthritis. Arthritis Rheum 2013;65(7):1776–85.

[141] Wasen C, Ospelt C, Camponeschi A, Erlandsson MC, Andersson KME, Silfversward ST, et al. Nicotine changes the microRNA profile to regulate the FOXO memory program of CD8(+) T cells in rheumatoid arthritis. Front Immunol 2020;11:1474.

[142] Karouzakis E, Gay RE, Gay S, Neidhart M. Increased recycling of polyamines is associated with global DNA hypomethylation in rheumatoid arthritis synovial fibroblasts. Arthritis Rheum 2012;64(6):1809–17.

[143] Neidhart M, Karouzakis E, Jungel A, Gay RE, Gay S. Inhibition of spermidine/spermine n1-acetyltransferase activity: a new therapeutic concept in rheumatoid arthritis. Arthritis Rheumatol 2014;66(7):1723–33.

[144] Ai R, Whitaker JW, Boyle DL, Tak PP, Gerlag DM, Wang W, et al. DNA methylome signature in synoviocytes from patients with early rheumatoid arthritis compared to synoviocytes from patients with longstanding rheumatoid arthritis. Arthritis Rheumatol 2015;67(7):1978–80.

[145] Karouzakis E, Raza K, Kolling C, Buckley CD, Gay S, Filer A, et al. Analysis of early changes in DNA methylation in synovial fibroblasts of RA patients before diagnosis. Sci Rep 2018;8(1):7370.

[146] Guderud K, Sunde LH, Flam ST, Maehlen MT, Mjaavatten MD, Lillegraven S, et al. Rheumatoid arthritis patients, both newly diagnosed and methotrexate treated, show more DNA methylation differences in CD4(+) memory than in CD4(+) naive T cells. Front Immunol. 2020;11:194.

[147] Weiner HL. The challenge of multiple sclerosis: how do we cure a chronic heterogeneous disease? Ann Neurol. 2009;65(3):239–48.

[148] Lassmann H, Bruck W, Lucchinetti CF. The immunopathology of multiple sclerosis: an overview. Brain Pathol. 2007;17(2):210–18.

[149] Carrithers MD, Visintin I, Kang SJ, Janeway Jr. CA. Differential adhesion molecule requirements for immune surveillance and inflammatory recruitment. Brain 2000;123(Pt 6):1092–101.

[150] Ruhrmann S, Stridh P, Kular L, Jagodic M. Genomic imprinting: a missing piece of the multiple sclerosis puzzle? Int J Biochem Cell Biol 2015;67:49–57.

[151] Ringh MV, Hagemann-Jensen M, Needhamsen M, Kullberg S, Wahlstrom J, Grunewald J, et al. Methylome and transcriptome signature of bronchoalveolar cells from multiple sclerosis patients in relation to smoking. Multiple Sclerosis 2021;27(7):1014–26.

[152] Maltby VE, Lea RA, Sanders KA, White N, Benton MC, Scott RJ, et al. Differential methylation at MHC in CD4(+) T cells is associated with multiple sclerosis independently of HLA-DRB1. Clin Epigenetics 2017;9:71.

[153] Kular L, Liu Y, Ruhrmann S, Zheleznyakova G, Marabita F, Gomez-Cabrero D, et al. DNA methylation as a mediator of HLA-DRB1*15:01 and a protective variant in multiple sclerosis. Nat Commun 2018;9(1):2397.

[154] Kucukali CI, Kurtuncu M, Coban A, Cebi M, Tuzun E. Epigenetics of multiple sclerosis: an updated review. Neuromolecular Med 2015;17(2):83–96.

[155] Huynh JL, Garg P, Thin TH, Yoo S, Dutta R, Trapp BD, et al. Epigenome-wide differences in pathology-free regions of multiple sclerosis-affected brains. Nat Neurosci 2014;17(1):121–30.

[156] Mastronardi FG, Noor A, Wood DD, Paton T, Moscarello MA. Peptidyl argininedeiminase 2 CpG island in multiple sclerosis white matter is hypomethylated. J Neurosci Res 2007;85(9):2006–16.

[157] Mastronardi FG, Wood DD, Mei J, Raijmakers R, Tseveleki V, Dosch HM, et al. Increased citrullination of histone H3 in multiple sclerosis brain and animal models of demyelination: a role for tumor necrosis factor-induced peptidylarginine deiminase 4 translocation. J Neurosci 2006;26(44):11387–96.

[158] Pedre X, Mastronardi F, Bruck W, Lopez-Rodas G, Kuhlmann T, Casaccia P. Changed histone acetylation patterns in normal-appearing white matter and early multiple sclerosis lesions. J Neurosci 2011;31(9):3435–45.

[159] Singhal NK, Sternbach S, Fleming S, Alkhayer K, Shelestak J, Popescu D, et al. Betaine restores epigenetic control and supports neuronal mitochondria in the cuprizone mouse model of multiple sclerosis. Epigenetics 2020;15(8):871–86.

[160] Sternbach S, West N, Singhal NK, Clements R, Basu S, Tripathi A, et al. The BHMT-betaine methylation pathway

epigenetically modulates oligodendrocyte maturation. PloS One 2021;16(5):e0250486.

[161] Roy DG, Chen J, Mamane V, Ma EH, Muhire BM, Sheldon RD, et al. Methionine metabolism shapes T helper cell responses through regulation of epigenetic reprogramming. Cell Metabolism 2020;31(2):250−66 e9.

[162] Bluestone JA, Herold K, Eisenbarth G. Genetics, pathogenesis and clinical interventions in type 1 diabetes. Nature 2010;464 (7293):1293−300.

[163] Reimann M, Bonifacio E, Solimena M, Schwarz PE, Ludwig B, Hanefeld M, et al. An update on preventive and regenerative therapies in diabetes mellitus. Pharmacol Ther 2009;121(3):317−31.

[164] Anderson MS, Bluestone JA. The NOD mouse: a model of immune dysregulation. Ann Rev Immunol 2005;23:447−85.

[165] Abdelsamed HA, Zebley CC, Nguyen H, Rutishauser RL, Fan Y, Ghoneim HE, et al. Beta cell-specific CD8(+) T cells maintain stem cell memory-associated epigenetic programs during type 1 diabetes. Nat Immunol 2020;21(5):578−87.

[166] Rui J, Deng S, Lebastchi J, Clark PL, Usmani-Brown S, Herold KC. Methylation of insulin DNA in response to proinflammatory cytokines during the progression of autoimmune diabetes in NOD mice. Diabetologia 2016;59(5):1021−9.

[167] Stefan-Lifshitz M, Karakose E, Cui L, Ettela A, Yi Z, Zhang W, et al. Epigenetic modulation of beta cells by interferon-alpha via PNPT1/mir-26a/TET2 triggers autoimmune diabetes. JCI Insight 2019;4(5):e126663.

[168] Elboudwarej E, Cole M, Briggs FB, Fouts A, Fain PR, Quach H, et al. Hypomethylation within gene promoter regions and type 1 diabetes in discordant monozygotic twins. J Autoimmun 2016;68:23−9.

[169] Johnson RK, Vanderlinden LA, Dong F, Carry PM, Seifert J, Waugh K, et al. Longitudinal DNA methylation differences precede type 1 diabetes. Sci Rep 2020;10(1):3721.

[170] Bell CG, Teschendorff AE, Rakyan VK, Maxwell AP, Beck S, Savage DA, et al. Methylation analysis for diabetic nephropathy in type 1 diabetes mellitus. BMC Med Genom 2010;3:33.

[171] Smyth LJ, Kilner J, Nair V, Liu H, Brennan E, Kerr K, et al. Assessment of differentially methylated loci in individuals with end-stage kidney disease attributed to diabetic kidney disease: an exploratory study. Clin Epigenetics 2021;13(1):99.

[172] Agardh E, Lundstig A, Perfilyev A, Volkov P, Freiburghaus T, Lindholm E, et al. Genome-wide analysis of DNA methylation in subjects with type 1 diabetes identifies epigenetic modifications associated with proliferative diabetic retinopathy. BMC Med 2015;13:182.

[173] Miao F, Chen Z, Genuth S, Paterson A, Zhang L, Wu X, et al. Evaluating the role of epigenetic histone modifications in the metabolic memory of type 1 diabetes. Diabetes 2014;63 (5):1748−62.

[174] Gautier JF, Porcher R, Abi Khalil C, Bellili-Munoz N, Fetita LS, Travert F, et al. Kidney dysfunction in adult offspring exposed in utero to type 1 diabetes is associated with alterations in genome-wide DNA methylation. PloS One 2015;10(8): e0134654.

[175] Chen Z, Miao F, Paterson AD, Lachin JM, Zhang L, Schones DE, et al. Epigenomic profiling reveals an association between persistence of DNA methylation and metabolic memory in the DCCT/EDIC type 1 diabetes cohort. Proc Natl Acad Sci U S A 2016;113(21):E3002−11.

[176] Chen Z, Miao F, Braffett BH, Lachin JM, Zhang L, Wu X, et al. DNA methylation mediates development of HbA1c-associated complications in type 1 diabetes. Nat Metab 2020;2(8):744−62.

[177] Cutolo M, Soldano S, Smith V. Pathophysiology of systemic sclerosis: current understanding and new insights. Expert Rev Clin Immunol 2019;15(7):753−64.

[178] Gabrielli A, Avvedimento EV, Krieg T. Scleroderma. N Engl J Med 2009;360(19):1989−2003.

[179] Korman B. Evolving insights into the cellular and molecular pathogenesis of fibrosis in systemic sclerosis. Transl Res 2019;209:77−89.

[180] Gu YS, Kong J, Cheema GS, Keen CL, Wick G, Gershwin ME. The immunobiology of systemic sclerosis. Semin Arthritis Rheum 2008;38(2):132−60.

[181] Feghali-Bostwick C, Medsger Jr. TA, Wright TM. Analysis of systemic sclerosis in twins reveals low concordance for disease and high concordance for the presence of antinuclear antibodies. Arthritis Rheum 2003;48(7):1956−63.

[182] Altorok N, Kahaleh B. Epigenetics and systemic sclerosis. Semin Immunopathol 2015;37(5):453−62.

[183] Chen S, Pu W, Guo S, Jin L, He D, Wang J, et al. Methylation profiles reveal common epigenetic patterns of interferon-related genes in multiple autoimmune diseases. Front Genet 2019;10:223.

[184] Dees C, Potter S, Zhang Y, Bergmann C, Zhou X, Luber M, et al. TGF-beta-induced epigenetic deregulation of SOCS3 facilitates STAT3 signaling to promote fibrosis. J Clin Invest 2020;130(5):2347−63.

[185] O'Reilly S, Ciechomska M, Fullard N, Przyborski S, van Laar JM. IL-13 mediates collagen deposition via STAT6 and microRNA-135b: a role for epigenetics. Sci Rep 2016;6:25066.

[186] He Y, Tsou PS, Khanna D, Sawalha AH. Methyl-CpG-binding protein 2 mediates antifibrotic effects in scleroderma fibroblasts. Ann Rheum Dis 2018;77(8):1208−18.

[187] Dees C, Schlottmann I, Funke R, Distler A, Palumbo-Zerr K, Zerr P, et al. The Wnt antagonists DKK1 and SFRP1 are downregulated by promoter hypermethylation in systemic sclerosis. Ann Rheum Dis 2014;73(6):1232−9.

[188] Wang Y, Fan PS, Kahaleh B. Association between enhanced type I collagen expression and epigenetic repression of the FLI1 gene in scleroderma fibroblasts. Arthritis Rheum 2006;54 (7):2271−9.

[189] Shin JY, Beckett JD, Bagirzadeh R, Creamer TJ, Shah AA, McMahan Z, et al. Epigenetic activation and memory at a TGFB2 enhancer in systemic sclerosis. Sci Transl Med 2019;11 (497).

[190] van der Kroef M, Castellucci M, Mokry M, Cossu M, Garonzi M, Bossini-Castillo LM, et al. Histone modifications underlie monocyte dysregulation in patients with systemic sclerosis, underlining the treatment potential of epigenetic targeting. Ann Rheum Dis 2019;78(4):529−38.

[191] Kramer M, Dees C, Huang J, Schlottmann I, Palumbo-Zerr K, Zerr P, et al. Inhibition of H3K27 histone trimethylation activates fibroblasts and induces fibrosis. Ann Rheum Dis 2013;72 (4):614−20.

[192] Bergmann C, Brandt A, Merlevede B, Hallenberger L, Dees C, Wohlfahrt T, et al. The histone demethylase Jumonji domain-containing protein 3 (JMJD3) regulates fibroblast activation in systemic sclerosis. Ann Rheum Dis 2018;77(1):150−8.

[193] Tsou PS, Wren JD, Amin MA, Schiopu E, Fox DA, Khanna D, et al. Histone deacetylase 5 is overexpressed in scleroderma endothelial cells and impairs angiogenesis via repression of proangiogenic factors. Arthritis Rheumatol 2016;68(12):2975−85.

[194] Dang MN, Bradford CM, Pozzilli P, Leslie RD. Methylation analysis in distinct immune cell subsets in type 1 diabetes. Methods Mol Biol 2016;1433:143−51.

[195] Zhao M, Zhou Y, Zhu B, Wan M, Jiang T, Tan Q, et al. IFI44L promoter methylation as a blood biomarker for systemic lupus erythematosus. Ann Rheum Dis 2016;75(11):1998−2006.

[196] Ospelt C. Epigenetic biomarkers in rheumatology - the future? Swiss Med Wkly 2016;146:w14312.

[197] Wardowska A, Komorniczak M, Bullo-Piontecka B, Debska-Slizien MA, Pikula M. Transcriptomic and epigenetic alterations in dendritic cells correspond with chronic kidney disease in lupus nephritis. Front Immunol 2019;10:2026.

[198] Diniz SN, da Silva CF, de Almeida IT, da Silva Costa FE, de Oliveira EML. INFbeta treatment affects global DNA methylation in monocytes of patients with multiple sclerosis. J Neuroimmunol 2021;355:577563.

[199] Cribbs AP, Kennedy A, Penn H, Amjadi P, Green P, Read JE, et al. Methotrexate restores regulatory T cell function through demethylation of the FoxP3 upstream enhancer in patients with rheumatoid arthritis. Arthritis Rheumatol 2015;67(5):1182—92.

[200] Gosselt HR, van Zelst BD, de Rotte M, Hazes JMW, de Jonge R, Heil SG. Higher baseline global leukocyte DNA methylation is associated with MTX non-response in early RA patients. Arthritis Res Ther 2019;21(1):157.

[201] Nair N, Plant D, Verstappen SM, Isaacs JD, Morgan AW, Hyrich KL, et al. Differential DNA methylation correlates with response to methotrexate in rheumatoid arthritis. Rheumatology (Oxford) 2020;59(6):1364—71.

[202] Gosselt HR, Vallerga CL, Mandaviya PR, Lubberts E, Hazes JMW, de Jonge R, et al. Epigenome wide association study of response to methotrexate in early rheumatoid arthritis patients. PloS One 2021;16(3):e0247709.

[203] Plant D, Webster A, Nair N, Oliver J, Smith SL, Eyre S, et al. Differential methylation as a biomarker of response to etanercept in patients with rheumatoid arthritis. Arthritis Rheumatol 2016;68(6):1353—60.

[204] Tao W, Concepcion AN, Vianen M, Marijnissen ACA, Lafeber F, Radstake T, et al. Multiomics and machine learning accurately predict clinical response to adalimumab and etanercept therapy in patients with rheumatoid arthritis. Arthritis Rheumatol 2021;73(2):212—22.

[205] de la Calle-Fabregat C, Niemantsverdriet E, Canete JD, Li T, van der Helm-van Mil AHM, Rodriguez-Ubreva J, et al. The DNA methylation profile of undifferentiated arthritis patients anticipates their subsequent differentiation to rheumatoid arthritis. Arthritis Rheumatol 2021;.

[206] Zhang K, Lin G, Han Y, Xie J, Li J. Circulating unmethylated insulin DNA as a potential non-invasive biomarker of beta cell death in type 1 diabetes: a review and future prospect. Clin Epigenetics 2017;9:44.

[207] Flotho C, Claus R, Batz C, Schneider M, Sandrock I, Ihde S, et al. The DNA methyltransferase inhibitors azacitidine, decitabine and zebularine exert differential effects on cancer gene expression in acute myeloid leukemia cells. Leukemia 2009;23(6):1019—28.

[208] Ghiboub M, Elfiky AMI, de Winther MPJ, Harker NR, Tough DF, de Jonge WJ. Selective targeting of epigenetic readers and histone deacetylases in autoimmune and inflammatory diseases: recent advances and future perspectives. J Pers Med 2021;11(5):336.

[209] Klein K, Gay S. Epigenetic modifications in rheumatoid arthritis, a review. Curr Opin Pharmacol 2013;13(3):420—5.

[210] Wei S, Sun Y, Sha H. Therapeutic targeting of BET protein BRD4 delays murine lupus. Int Immunopharmacol 2015;29(2):314—19.

[211] Nicodeme E, Jeffrey KL, Schaefer U, Beinke S, Dewell S, Chung CW, et al. Suppression of inflammation by a synthetic histone mimic. Nature 2010;468(7327):1119—23.

[212] Chan CH, Fang C, Yarilina A, Prinjha RK, Qiao Y, Ivashkiv LB. BET bromodomain inhibition suppresses transcriptional responses to cytokine-Jak-STAT signaling in a gene-specific manner in human monocytes. Eur J Immunol 2015;45(1):287—97.

[213] Thompson PJ, Shah A, Apostolopoulou H, Bhushan A. BET proteins are required for transcriptional activation of the senescent islet cell secretome in type 1 diabetes. Int J Mol Sci 2019;20(19).

[214] Fu W, Farache J, Clardy SM, Hattori K, Mander P, Lee K, et al. Epigenetic modulation of type-1 diabetes via a dual effect on pancreatic macrophages and beta cells. Elife 2014;3:e04631.

[215] Amorim S, Stathis A, Gleeson M, Iyengar S, Magarotto V, Leleu X, et al. Bromodomain inhibitor OTX015 in patients with lymphoma or multiple myeloma: a dose-escalation, open-label, pharmacokinetic, phase 1 study. Lancet Haematol 2016;3(4):e196—204.

[216] Berthon C, Raffoux E, Thomas X, Vey N, Gomez-Roca C, Yee K, et al. Bromodomain inhibitor OTX015 in patients with acute leukaemia: a dose-escalation, phase 1 study. Lancet Haematol 2016;3(4):e186—95.

[217] Rohraff DM, He Y, Farkash EA, Schonfeld M, Tsou PS, Sawalha AH. Inhibition of EZH2 ameliorates lupus-like disease in MRL/lpr Mice. Arthritis Rheumatol 2019;71(10):1681—90.

[218] Zhen Y, Smith RD, Finkelman FD, Shao WH. Ezh2-mediated epigenetic modification is required for allogeneic T cell-induced lupus disease. Arthritis Res Ther 2020;22(1):133.

[219] Gunawan M, Venkatesan N, Loh JT, Wong JF, Berger H, Neo WH, et al. The methyltransferase Ezh2 controls cell adhesion and migration through direct methylation of the extranuclear regulatory protein talin. Nat Immunol 2015;16(5):505—16.

[220] Xiao XY, Li YT, Jiang X, Ji X, Lu X, Yang B, et al. EZH2 deficiency attenuates Treg differentiation in rheumatoid arthritis. J Autoimmun 2020;108:102404.

CHAPTER 37

Epigenetics of Brain Disorders

Ali Jawaid[1,2,3], Eloïse A. Kremer[1,3], Nancy V.N. Carullo[1,3] and Isabelle M. Mansuy[1,3]

[1]Laboratory of Neuroepigenetics, Brain Research Institute, Medical Faculty of the University of Zurich, Zurich, Switzerland [2]Department of Health Sciences and Technology, Institute for Neuroscience, ETH Zurich, Zurich, Switzerland [3]Zurich Neuroscience Center, ETH and University of Zurich, Zurich, Switzerland

OUTLINE

Introduction	738
Important Epigenetic Mechanisms for the Brain	738
DNA Methylation	738
Histone Modifications	738
Epigenetic Dysregulation in Neurodevelopmental Disorders: The Example of Rett Syndrome	739
DNA Methylation	739
MeCP2 Mutations	740
MeCP2 Interactions with Chromatin Modifiers	740
Non-coding RNA in RS	740
Epigenetic Dysregulation in Neurodegenerative Disorders: The Example of Alzheimer's Disease	742
Histone Modifications and Histone Modifier in AD	742
DNA Methylation in AD	744
Non-coding RNA in AD	745
Epigenetic Dysregulation in Psychiatric Disorders: The Example of Depression	745
Histone Modifications and Histone Modifier in Depression	746
DNA Methylation in Depression	747
Non-coding RNA in Depression	747
Epigenetic Dysregulation by Environmental Stress: The Example of Early-life Stress	748
DNA Methylation in Early-life Stress	748
Non-coding RNA in Early-life Stress	749
Intergenerational and Transgenerational Epigenetic Effects of Early-life Stress	749
Conclusions and Outlook	749
Summary	750
Research case study	751
Objective	751
Scope	751
Audience	751
Rationale	751
Expected Results and Deliverables	751
Safety Considerations	751
The Process, Workflow and Actions Taken	751
Tools and Materials Used	751
Results	751
Challenges and Solutions	752
Learning and Knowledge Outcomes	752
Acknowledgments	752
References	752

INTRODUCTION

The brain is a complex organ, which is highly plastic during development and keeps some plasticity and responsiveness throughout life. In mammals, this plasticity is particularly high during prenatal, postnatal, and pubertal periods. It is characterized by the formation and maintenance of synapses, the formation of new neurons, their integration into existing neuronal networks, and the reorganization of neuronal networks. This plasticity is essential for brain processes, such as learning and memory formation, and requires active changes in gene expression. Several brain disorders result from dysfunctional or impaired regulation of the molecular processes governing brain plasticity, underscoring their potential role in the pathogenesis of brain disorders.

Gene expression in the brain can be modulated in a cell, region, and context-specific manner by epigenetic mechanisms. These mechanisms operate at the transcriptional and posttranscriptional level in response to activity. They can implicate several processes involving DNA methylation (reviewed in Ref. [1]) histone modifications (reviewed in Ref. [2]), and non-coding RNAs (ncRNAs), such as long non-coding RNAs (lncRNAs) or microRNAs (miRNAs) (reviewed in Refs. [3,4]). Additional processes like nucleosome positioning [5], histone turnover [6], competitive endogenous RNAs (reviewed in Refs. [7,8]), prion-like seeding and phase separation (reviewed in Refs. [9,10]) also constitute active nongenetic regulation in the brain. This chapter describes the contribution of DNA methylation, histone modifications, and miRNAs to brain disorders using a neurodevelopmental, neurodegenerative, and psychiatric disorder as examples. It also indicates how epigenetic processes contribute to the effects of early postnatal life adversity on brain functions and their transmission to subsequent generations.

IMPORTANT EPIGENETIC MECHANISMS FOR THE BRAIN

DNA Methylation

DNA methylation most commonly occurs at cytosine-guanine dinucleotides (CpG), although CpH methylation (H is A, C, or T) is also present in the human and mouse brain postnatally [11–13]. CpG methylation has traditionally been viewed as a fairly stable epigenetic mark mostly responsible for gene silencing. Silencing takes place by direct inhibition of the binding of transcription factors, or by recruitment of methyl CpG—binding domain (MBD) proteins and associated repressive chromatin-remodeling components (reviewed in [14–16]). However, since CpG methylation is also present at the promoter and coding region of actively transcribed genes, it can also be associated with transcriptional activity [17,18]. Recent evidence has shown that in many cells including postmitotic cells like adult neurons, DNA methylation is dynamically regulated and CpGs can be actively methylated and demethylated. While DNA methylation is mediated by DNA methyltransferases (DNMTs), DNA demethylation involves a succession of biochemical steps. It can occur in a passive DNA replication-dependent process or through active demethylation implicating several enzymes. During this process, 5-methylcytosine (5-mC) is successively oxidized into 5-hydroxymethylcytosine (5-hmC), 5-formylcytosine (5-fC), and 5-carboxylcytosine (5-caC) by Ten—eleven translocation (TET) proteins. 5-fC and 5-cAC can ultimately be removed and repaired by thymine DNA glycosylase (TDG) generating an unmodified cytosine [19]. In contrast to 5-fC and 5-caC, 5-hmC is enriched in the adult brain, especially in the hippocampus, cortex, and cerebellum [20–22]. The high abundance and stability of 5-hmC in the brain [23,24], suggest that this cytosine modification may not only be a transient by-product of 5-mC metabolism, but may also serve as an independent epigenetic mark thought to play an important role in brain processes. As both DNA methylation and hydroxymethylation patterns are modulated during development and upon neuronal activity and memory processes in different brain regions [25–27], alterations in their profile may be relevant for the development of brain disorders. Importantly, some proteins like the methyl-DNA-binding protein MeCP2 and the protein phosphatase PP1 can regulate multiple epigenetic pathways in a tissue- and context-specific manner [28–30]. Perturbation of such epigenetic regulators, experimentally or in disease states, affects many core brain processes, suggesting their implication in brain disorders [28–32].

Histone Modifications

Covalent posttranslational modifications (PTMs) of histone proteins are important epigenetic marks that primarily involve acetylation, methylation, phosphorylation, ubiquitination, SUMOylation, and monoaminylation (reviewed in Ref. [33,34]). These modifications control the accessibility of the chromatin to the transcriptional machinery, in a sequence and activity-dependent manner [33]. Generally, acetylation and phosphorylation, occurring respectively on lysine (K) and serine (S), threonine (T) or tyrosine (Y) residues, are associated with transcriptional activation [35]. Histone methylation on K is associated with both actively transcribed and silenced genes [36,37].

Similarly, histone ubiquitination is associated with both transcriptional silencing and activation depending on whether it occurs on H2A or H2B, respectively (reviewed in Ref. [38]). Finally, histone SUMOylation, which requires addition of small-ubiquitin like modifier (SUMO) to histones, negatively regulates gene transcription in the brain, often in association with other epigenetic regulators [39,40]. An extensive review of histone modifications discovered in the past decade, including monoaminylation, is provided in [34]. Importantly, histone PTMs are integral to the regulation of important physiological brain processes. For instance, synaptic mechanisms underlying hippocampal memory formation require dynamic changes in histone acetylation and methylation [41–44]. Similarly, histone acetylation is also essential for neuronal resistance against ischemic and oxidative insult. In these contexts, H3 and H4 acetylation, and HAT binding level, specifically CREB binding protein (CBP) directly contribute to neuronal sensitivity [45,46].

Enzymes important for regulation of major histone PTMs, such as histone acetyl transferases (HATs), histone deacetylases (HDACs), and histone methyltransferases (HMTs) are abundant in the brain. Owing to this and the important roles for histone PTMs in brain processes, their critical involvement in brain disorders has become an important topic of current investigations [31,47]. In this chapter, we will discuss the implications of specific histone modifiers that are important for brain functions and disease including HATs like CBP, and HDACs like HDAC1/2 and SIRT1 (silent mating type information regulation 2 homolog 1).

Non-coding RNA

Epigenetic regulation of gene expression can also be achieved by ncRNA such as lncRNA and miRNAs. The family of lncRNA (non-coding RNAs longer than 200 bases) can have a wide range of functions including their actions as scaffold, decoys, or guides for transcription machinery and epigenetic modifiers. Furthermore, lncRNA can aide spatiotemporal expression specificity [48]. They are key regulators of gene expression and brain function in healthy and diseased states (reviewed in [49]) and, therefore, play an important role in an animal's behavior and memory formation [31,50]. In contrast, miRNAs are 20–22 nucleotide long RNAs that act by translational repression or degradation of mRNA targets (reviewed in Ref. [51]). MiRNA biogenesis involves a succession of steps in the cell regulated by the RNAses, Drosha and its partner DeGeorge syndrome critical region 8 (DGCR8), in the nucleus, followed by further processing by another RNAse, Dicer and its partner TAR RNA-binding protein (TRBP), in the cytoplasm. Once formed, miRNAs associate with Argonaute proteins in the cytoplasm to form an RNA-induced silencing complex (RISC). A number of RNA-binding proteins can additionally modulate the efficiency of the microprocessor, Dicer, or RISC (reviewed in Ref. [51]). Many miRNAs and components of miRNA biogenesis machinery are abundantly expressed in the brain, mostly in a region-specific manner. They have been linked to important brain processes like neurogenesis (reviewed in Ref. [52]), neuronal activity (reviewed in Ref. [53,54]), and memory formation (reviewed in Ref. [55,56]). Animal models with experimental manipulation of global miRNA biogenesis or functional inhibition/overexpression of specific miRNAs can lead to pathological phenotypes, highlighting their important contribution to brain disorders (reviewed in Ref. [4,57]).

EPIGENETIC DYSREGULATION IN NEURODEVELOPMENTAL DISORDERS: THE EXAMPLE OF RETT SYNDROME

Neurodevelopmental disorders are characterized by impaired functions of the central nervous system that appear early in development and often persist into adulthood. The impairments may manifest early in life as in the case of fetal alcohol syndrome, or later. Several genetically determined neurodevelopmental disorders have been documented to also involve epigenetic dysregulation that alter physiological functions (reviewed in Ref. [58,59]). This chapter focuses on Rett syndrome (RS), a disorder associated with aberrant DNA methylation, histone PTMs, and miRNAs (reviewed in Refs. [60–62]).

RS is a relatively common (worldwide prevalence of 1:10,000) and progressive neurological disorder characterized by an arrest of central nervous system (CNS) development and intellectual disability. It is generally caused by loss-of-function mutations in the X-linked *methyl-CpG-binding protein 2* (*MeCP2*) gene [63], which is lethal when hemizygous in males, and therefore affects exclusively females. MeCP2 is normally abundant in the brain, in particular in neurons where its level is comparable to that of histone octamers [64]. It is a member of the MBD protein family that binds methylated CpGs with high affinity to regulate gene transcription in a bidirectional fashion. MeCP2 has three domains, a CpG-binding domain (CBD), a transcriptional repressor domain (TRD), and a WW domain [65].

DNA Methylation

MeCP2 not only binds to methylated CpG, but also recognizes methylated CpH (H = A, C, or T) [12,66], and hydroxymethylated CpGs [67], suggesting that it occupies a substantial part of DNA. MeCP2 deficiencies e.g., in induced *MeCP2* mutant mice, therefore, strongly perturb regions of 5-mC and 5-hmC [68]. MeCP2 binding contributes to transcriptional regulation through the recruitment of HDACs and other transcriptional co-repressor complexes, such as Sin3a/HDACI and ski/NcoR/HDACII. This results in chromatin compaction and gene silencing [65,69]. In addition to acting as a transcriptional repressor, MeCP2 can also function as transcriptional activator. This occurs by interacting with transcription factors like CREB1 at regulatory regions of target genes [70], and/or through binding to 5-hmC which is associated with active genes. Specifically, MeCP2 recognizes and binds to CpA repeats with high affinity and demonstrates a selectivity for hmCA sites [71]. Binding to hmCA sites excludes nucleosomes and, therefore, protects bound regions from nucleosome invasion ultimately regulating gene transcription. Interestingly, a Rett syndrome-associated MeCP2 mutation inhibits MeCP2 binding to hmC sites, and loss of MeCP2 causes increased nucleosome density and affects chromatin accessibility [71,72]. Hence, MeCP2 appears to be an adapter molecule, acting as DNA-binding protein that helps recruiting transcriptional regulators.

MeCP2 Mutations

In humans, RS can be caused by loss-of-function mutations in either *MeCP2* CBD or TRD [63]. Missense mutations are generally milder while nonsense mutations cause more severe symptoms. Similarly, CBD mutations generally lead to a more severe phenotype [65]. In mice, neuron-specific *MeCP2* deletion has been shown to recapitulate RS symptoms. MeCP2-deficient mice have reduced brain weight and abnormal brain circuitry [73], an overall decrease in exploratory activity [73,74], cognitive deficits, and impaired synaptic plasticity [75]. MeCP2 ablation in forebrain GABAergic neurons also impairs sensory information processing [76], a symptom present in RS patients. Some of the cognitive deficits induced by MeCP2 deficiency could be reversed by overexpression of wild-type human MeCP2 in young mutant mice [77], confirming that they resulted from a lack of MeCP2. However, at the same time, MeCP2 overexpression induces seizures, suggesting that an excess in MeCP2 is also deleterious for brain functions, and may lead to neurological functions related to RS.

Since MeCP2 functions as a transcriptional regulator, dysregulation of its target genes likely contributes to RS symptoms. Global transcriptome analyses in human RS brain and MeCP2-deficient mice have shown dysregulation of a multitude of genes. Genes with long sequences, in particular, appear to be prime targets of MeCP2-mediated transcriptional repression in the mouse brain [78,79]. Many of these long genes encode proteins implicated in neuronal physiology, axon guidance, and synapse formation, such as the calcium/calmodulin-dependent kinase *Camk2*, and the voltage-gated potassium channel *Kcnh7*, and may explain neuronal dysfunctions in RS [79]. Importantly, these genes contain a high density of mCA required for length-dependent gene repression by MeCP2.

An important target is the brain-derived neurotrophic factor (*BDNF*) gene, where expression is reduced in *MeCP2-null* mice. Overexpression of BDNF in *MeCP2-null* mice ameliorates the phenotype; it reduces neuronal atrophy and improves survival [80]. In addition, delivery of exogenous BDNF restores synaptic dysfunctions in *MeCP2-null* mice [81], suggesting that changes in gene expression in MeCP2 mutants have an impact on RS phenotype. Further to *BDNF*, many other genes including myelin-associated proteins and dopamine decarboxylase, have been identified as direct-binding targets of MeCP2 in the mouse brain [79,82].

MeCP2 Interactions with Chromatin Modifiers

MeCP2 also interacts with other proteins, including chromatin-modifying factors [17,83,84], epigenetic regulators, such as DNMT1 and TET1 [85], and transcriptional modulators, such as coREST, suv39H1, cSK1, and so on [61]. It can therefore influence the epigenetic profile (DNA methylation) of target genes. It can also influence histone PTMs, in particular histone acetylation and methylation (see Fig. 37.1). For *BDNF*, these changes are mediated by the formation of a complex between MeCP2 and HDACI, reducing H3 and H4 acetylation. This is paralleled by increased H3K9 dimethylation, which inhibits gene transcription, but decreased H3K4 dimethylation, which promotes gene transcription [86]. This indicates complex and possibly self-compensatory mechanisms between transcriptional programs regulated by MeCP2 as a result of its dual role in DNA methylation and histone modifications [60,62,78,87]. Notably, the activity and expression of MeCP2 itself is under the control of PTMs. Activity-dependent phosphorylation of MeCP2 at T 308 regulates its transcriptional repressor activity by modulating its interaction with the nuclear receptor co-repressor complex (NCoR) [88]. Similarly, MeCP2 SUMOylation is an important regulator of transcriptional repressor activity, which can rescue the behavioral deficits in mutant-MeCP2 model of RS [89].

Non-coding RNA in RS

Further to DNA methylation and histone modifications, miRNAs are also involved in RS symptoms

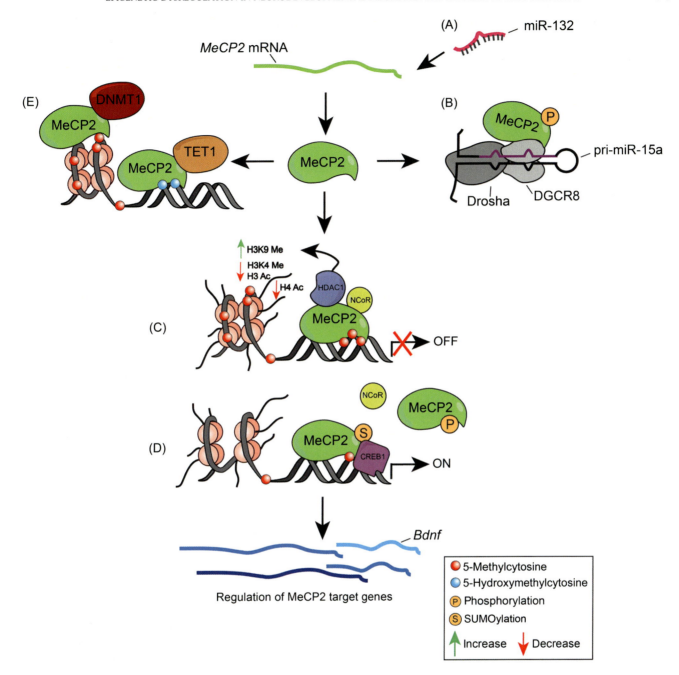

FIGURE 37.1 Roles of MeCP2 as regulator of the epigenetic code. (A) Posttranscriptional regulation of MeCP2 by miRNAs. MeCP2 translation is regulated by miRNAs. For example, miR-132 targets MeCP2 mRNA in neurons, which ultimately leads to increased *Bdnf* transcript levels [89]. (B) MeCP2 modulates the microprocessor complex. MeCP2 regulates gene expression posttranscriptionally by modulating miRNA processing via interfering with the assembly of Drosha and DGCR8 complex [90]. MeCP2 can selectively modulate the biogenesis of certain miRNAs, in particular miR-15a. This miRNA is implicated in neuronal maturation by controlling *BDNF* expression levels [91]. (C) MeCP2 functions as a transcriptional repressor. MeCP2 functions as a transcriptional repressor by recruiting transcriptional co-repressor complexes (e.g., NCoR) and/or HDACs (e.g., HDAC1). Recruitment of HDAC1 results in reduced acetylation of H3 and H4 at *Bdnf* promoter, as well as increased H3K9 dimethylation and decreased H3K4 dimethylation. This leads to chromatin condensation, thus limited access to the transcriptional machinery and mediates decreased *Bdnf* transcription. (D) MeCP2 functions as a transcriptional activator. MeCP2 also functions as a transcriptional activator by interacting with transcription factors, such as CREB1 at the promoter of target genes [67]. Recruitment of CREB1 results in chromatin remodeling permissible for transcription. In addition, phosphorylation of MeCP2 can result in the inability of MeCP2 to bind to its methylated-binding sites and interferes with the recruitment of NCoR [87]. Further, MeCP2 SUMOylation increases CREB DNA binding, thus enhancing *Bdnf* mRNA expression [88]. (E) MeCP2 influences the DNA methylation profile of target genes. MeCP2 binds to both methylated and hydroxymethylated cytosines [64], and interacts with epigenetic regulators, such as DNMT1 and TET1 [83]. MeCP2 could therefore influence the epigenetic profile (DNA methylation) of target genes, such as *Bdnf*. *Ac*, Acetyl; *BDNF*, brain-derived neurotrophic factor; *CREB1*, cAMP responsive element–binding protein 1; *DGCR8*, DiGeorge syndrome chromosomal region 8; *HDAC*, histone deacetylase; *Me*, methyl; *MeCP2*, methyl CpG–binding protein gene; *miR*, miRNA; *NCoR*, nuclear receptor co-repressor 1; *pri-miR*, primary miRNA.

[61,62,92]. Some miRNAs mediate their effect by regulating MeCP2 expression, for instance, MeCP2 is a putative target of miR-132, a CREB-dependent miRNA [93]. Similarly, miR-22 promotes the differentiation of smooth muscle cells from stem cells by controlling posttranscriptional expression of MeCP2 [90]. MiRNAs may also be downstream effectors of MeCP2 dysfunction in RS, since MeCP2 was recently shown to regulate the microprocessor-mediated biogenesis of miRNAs by interfering with Drosha-DGCR8 binding [94]. Through this interaction and possibly others, MeCP2 can selectively modulate the microprocessor-mediated biogenesis of miR-137 [95], miR-199a [91], and miR-15a [96]. These miRNAs regulate important molecular pathways, such as neuronal maturation [84,95], mammalian targets of rapamycin (mTOR) signaling [91], and BDNF expression [96] respectively, and hence likely contribute to RS etiology and/or phenotype.

In conclusion, a large body of evidence supports the involvement of epigenetic mechanisms in RS, primarily related to a multidimensional and intricate role of MeCP2 as regulator of the epigenetic code. Future studies should examine the potential crosstalk of different epigenetic mechanisms regulated by MeCP2 to identify safe and potent therapeutic targets for RS.

EPIGENETIC DYSREGULATION IN NEURODEGENERATIVE DISORDERS: THE EXAMPLE OF ALZHEIMER'S DISEASE

Neurodegenerative diseases are pathological conditions characterized by a gradual loss of cells in the nervous system. In CNS, such loss usually has devastating consequences on cognition and locomotion. A hallmark feature of most neurodegenerative conditions is intracellular or extracellular deposition of pathologically aggregated proteins. These pathological deposits include beta-amyloid (β-amyloid) in the case of Alzheimer's disease (AD), Tau in the case of AD and frontotemporal lobar degeneration (FTLD), TAR DNA-binding protein of 43 kDa (TDP-43) in the case of amyotrophic lateral sclerosis (ALS), FTLD, fused in sarcoma (FUS) in the case of FTLD, and Huntingtin (HTT) in the case of Huntington disease. Emerging evidence suggests that epigenetic mechanisms contribute to the deposition of these pathological aggregates, and to pathways leading to neuronal death and/or cognitive and motor dysfunction downstream of the pathological deposits (reviewed in Refs. [97,98]). Here, AD is used as a prototype neurodegenerative condition to discuss the contribution of different epigenetic mechanisms in brain degeneration (see Fig. 37.2).

AD is one of the most common neurodegenerative diseases worldwide with an estimated prevalence of 5%–7% above the age of 60 years in most world regions [117]. AD pathology is characterized by two major hallmarks in the brain: extracellular amyloid plaques and intracellular neurofibrillary tangles (NFTs) [111]. Amyloid plaques are deposits of the amyloid β (Aβ) peptide, produced through enzymatic cleavage of the amyloid precursor protein (APP) by β- and γ-secretases. NFTs are intraneuronal aggregates of hyperphosphorylated Tau, a microtubule-binding protein (reviewed in [118]). E4 polymorphism of apolipoprotein E, a cholesterol carrier protein with additional roles in Aβ metabolism and transport across the blood brain barrier, is the largest known genetic risk factor for late-onset AD [111]. The second leading genetic risk locus is the *BIN1* gene which has been proposed to be involved in TAU pathology [119]. Interestingly, many of the AD associated SNVs at the *BIN1* locus fall outside of the gene body and concentrate in an upstream regulatory element instead. These variants have been proposed to alter enhancer activity which results in altered *BIN1* transcription levels [120]. Similarly, mutations in presenilin-1 (PS1) and presenilin-2 (PS2), two genes that encode the γ-secretase complex that cleaves APP to Aβ, can lead to familial forms of AD and are discussed in more detail below.

Epigenetic mechanisms have been extensively studied in the CdK-p25 mouse model of AD. Cyclin dependent kinase 5 (Cdk5) and its regulatory subunit p35 are important for CNS development. p25, a truncated form of p35, increases Cdk5 activity, leading to Tau hyperphosphorylation, formation of NFTs, astrogliosis, and neurodegeneration [112]. Studies on CdK-p25 AD model implicated altered histone PTMs in AD [113,121–123].

Histone Modifications and Histone Modifier in AD

Histone PTMs, especially aberrant histone acetylation may contribute to AD through an interaction of the components of APP-Aβ pathway and histone acetylation machinery in neurons. Both hyper- and hypoacetylation have been reported in AD. Hyperacetylation is in part related to APP-Aβ dependent regulation of HATs. APP intracellular domain (AICD), which is produced from APP by γ secretase, forms a complex with the nuclear adaptor protein Fe65 and the HAT TIP60 to activate transcription of genes [113]. Further, *presenilin 1 (PS1)*, a gene coding for the γ secretase complex, may itself contribute to histone hyperacetylation in AD pathology. Loss-of-function mutations in *PS1* or

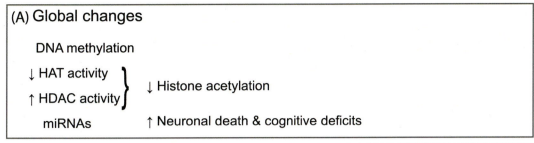
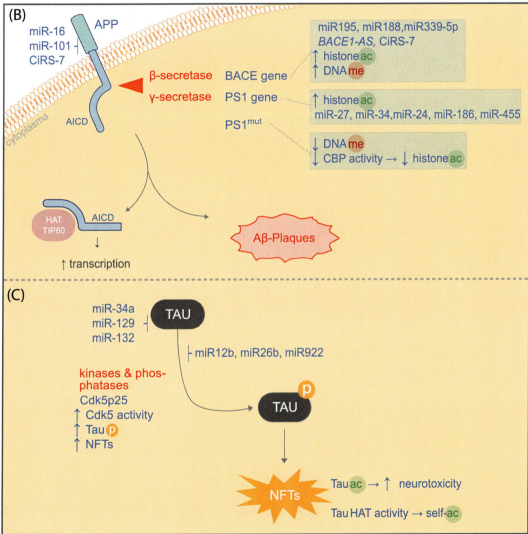

FIGURE 37.2 Epigenetic regulation and dysregulation in AD. The disease has been associated with broad alterations of epigenetic gene regulation, as well as specific changes contributing to the formation of two major hallmarks of AD: Aβ-plaques and neurofibrillary tangles (NFTs) [97,98]. (A) Global changes in DNA methylation, posttranscriptional histone modifications, and miRNA mediated transcriptional regulation have been associated AD [99,100]. These changes include altered DNA methylation patterns and decreased histone acetylation resulting from decreased "writer" HAT activity and increased "eraser" HDAC activity. Additionally, a number of miRNAs that are differentially expressed in AD patients and AD model organisms can be linked to increased neuronal death and cognitive deficits. (B) Epigenetic mechanisms regulate APP processing and can contribute to Aβ-plaque formation. miRNAs and circular RNA CiRS-7 regulate APP expression [101]. The transmembrane protein APP is cleaved by secretases, releasing the intracellular domain AICID. AICID interacts with HATs and other proteins to induce transcription programs. CiRS-7 has also been associated with β-secretase gene BACE1 expression. BACE1 is further regulated by miRNAs and lncRNA BACE1-AS [56,96,102–105]. In AD, the promoter locus can be hypermethylated and show increased histone acetylation. The PS1 gene locus codes for the γ-secretase protein. Its expression is regulated by several miRNAs and, in the context of AD, shows increased occupancy of acetylated histones [101,106–110]. AD associated mutations in PS1 further show decreases in DNA methylation and decreased histone acetylation. (C) TAU expression and phosphorylation are regulated by many miRNAs. Additionally, kinase Cdk5p25 demonstrated increased Cdk5 activity leading to increased phosphorylated TAU and increased NFT formation in AD [111–114]. TAU itself possesses HAT activity and the capacity to self-acetylate. Importantly, acetylation of TAU has been shown to increase the toxicity of NFTs [115,116].

mutations associated with familial AD inhibit the proteasomal degradation of the HAT CREB-binding protein (CBP), and result in increased CREB-mediated gene expression in cultured neurons [121]. Consistent with the involvement of hyperacetylation, overexpression of the HDAC SIRT1 in CdK-p25 mice confers substantial protection against AD-related pathologies and memory loss [122].

In contrast, other lines of evidence suggest that AD is associated with histone hypoacetylation related to a decreased activity of HATs or an increase in HDACs. In cultured cortical neurons, the overexpression of APP decreases H3 and H4 acetylation by reducing the level of CBP [123]. Similarly, loss-of-function mutations in *PS1* and *PS2* genes in mice reduce the expression of CBP, as well as CBP/CREB target genes *c-fos* and *BDNF* [114]. Further, HDAC2 is significantly increased in hippocampal CA1 neurons in an AD mouse model, which correlated with decreased promoter acetylation and expression of many plasticity related genes. Virus-mediated knock-down of HDAC2 increases the expression of these genes, and improves cognitive functions in these mice, indicating that HDAC2 induces an epigenetic blockade on cognitive functions in AD. This blockade can be induced not only by Aβ oligomers but also by other neurotoxic stimuli, such as hydrogen peroxide, and depends on the binding of stress elements to HDAC2 promoter [124]. Moreover, in CdK-p25 AD mice, the intracerebroventricular injection of sodium butyrate, a potent class I/II HDAC inhibitor, rescues memory loss, and reinstates synaptic connectivity [125]. Further, components of APP-Aβ pathway can themselves be regulated by histone PTMs. For example, cellular models of AD show increased histone acetylation of PS1 and BACE1, and increased expression of HAT p300 [126]. Similarly, apolipoprotein E4, which increases the risk of AD, favors the nuclear translocation of histones in human neurons, and reduces BDNF expression through low-density lipoprotein receptor related protein (LRP-1) [127].

Changes in histone in AD are not limited to acetylation but also include altered methylation and ubiquitination, as observed in the frontal cortex of AD patients [128]. These changes may contribute to AD by regulating Tau deposition, or by mediating neurotoxicity downstream to pathological Tau aggregation. Interestingly, targeting of the ubiquitin-proteosome system by inhibition of de-ubiquitylation can alleviate some neurogenerative pathologies and decrease α-synuclein levels [129].

Phosphorylation and acetylation of Tau inhibit its physiological binding and stabilization of neuronal microtubules, and instead promote Tau insolubility and aggregation [115,130]. Further, mimic of Tau Lys 280 acetylation exacerbates human Tau-mediated neurotoxicity in Drosophila [116]. In contrast, histone K methylation occurs endogenously in normal human brains, and resists Tau aggregation in vitro [131]. Interestingly, a recent study showed that Tau protein has intrinsic HAT activity, and can self-acetylate [132]. This observation may explain why overexpressing Tau leads to Tau pathological aggregation, and questions the utility of HDAC inhibitors in Tau models of AD and other Tauopathies.

DNA Methylation in AD

Further to histone PTMs, DNA methylation and hydroxymethylation might also be involved in the pathology of AD. In general, global levels of DNMT and TET enzymes in AD brain regions, and the presence of their reaction products, 5-mC and 5-hmC, have been investigated in several studies. AD patients have decreased levels of 5-mC and DNMT1 in neurons of the entorhinal cortex [133]. Similarly, a global reduction of 5-mC and 5-hmC is observed in the hippocampus of AD patients [134]. Another study however, reported opposite results with increased level of 5-mC, 5-hmC, and TET1 in hippocampal tissue [99]. In the frontal and temporal cortex of AD patients, both 5-mC and 5-hmC are elevated [135]. Further analyses are therefore required to understand these discrepancies. Gene-specific alterations in DNA methylation have also been identified in AD patients. A postmortem study in humans reported hypomethylation of the *PS1* promoter region in late-onset AD patients when compared to age-matched healthy subjects [136]. Furthermore, in vitro hypomethylation of the promoter region of *PS1* increases presenilin expression, which enhances β-amyloid formation [137]. This effect can be reversed by application of the methyl donor S-adenosylmethionine (SAM) that rescues methylation, decreases presenilin expression, and reduces β-amyloid formation. These observations suggest that methyl donors or drugs targeting the methyl metabolism may be potential therapeutic agents to treat AD [138].

In contrast, other AD-related susceptibility genes, such as *BACE1*, which codes for β-secretase, or the gene coding for apolipoprotein E are *hyper*methylated in late-onset AD [136]. This suggests that DNA methylation is presumably altered bidirectionally and in a gene-specific manner in AD, similar to histone acetylation. Recently, new evidence emerged, including two independent epigenome-wide association studies, for altered genome-wide DNA methylation profiles in AD [100,139,140]. These studies uncovered site-specific methylation alterations and expression changes in genes and enhancers not previously associated with AD. In particular, four novel loci were independently identified, underpinning their likely association with AD risk. Furthermore, AD-associated genes and genes important for neuronal functions are differentially hydroxymethylated in the hippocampus of a mouse model of AD as demonstrated via 5-hmC genome-wide profiling [141]. This study further showed that treatment of culture hippocampal neurons with toxic Aβ (1−42) peptide decreases global 5-hmC expression. Further studies are, however, needed to identify the alterations in DNA methylation and hydroxymethylation at specific genes and their possible effect on gene expression to better characterize the mechanisms underlying AD pathogenesis.

Non-coding RNA in AD

ncRNAs may contribute to AD pathology by post-transcriptionally regulating the expression of proteins involved in the production or clearance of amyloid-β. For example, β secretase gene *BACE1* expression can be regulated by miR-107 [102], miR-195 [103], miR-188 [104], miR-339−5p [105], the miR-29 cluster (reviewed in [101]), as well as lncRNA *BACE1-AS* (reviewed in [101]) and CiRS-7 [106]. Several miRNAs have been proposed to target and regulate the different components of γ-secretase, including miR-27a, miR-24, miR-34a, miR-186, miR-455 (reviewed in [101]). Similarly, miR-153, miR-101, miR-16 and CiRS-7 regulate the expression of APP [106−110] while APP trafficking has been shown to be regulated by several lncRNA (reviewed in [101]). In addition to amyloid production, proteins important for amyloid clearance across the blood-brain barrier, such as receptor for advanced glycosylation end products (RAGE) can also be regulated by miRNAs [142]. Several miRNAs also regulate PS1 expression in AD [101,106−110,143]. Further, miRNAs are also involved in the production of Tau, the major component of NFTs. Tau expression is under the control of miR-219 [144]. Similarly, miR-125b [145], miR-26b [146], and miR-922 [90] induce Tau phosphorylation by regulating Tau-related kinases and phosphatases, which correlates with cognitive deficits in AD models. Finally, miRNAs may link amyloid/Tau deposition to pathways responsible for neuronal death and/or cognitive deficits in AD. Cognitive deficits in an AD transgenic model were related to miR-206-dependent dysregulation of BDNF [147]. Similarly, another study showed that oxidative damage produced by soluble Aβ peptide is likely mediated by altered expression of miR-145 and miR-210 [148]. Finally, miR-146a arbitrates the inflammatory circuits in AD models and stressed human brain cells through nuclear factor kappa-light-chain-enhancer of activated B cells (NFκB) [149].

In conclusion, it is becoming increasingly clear that epigenetic processes are implicated in AD at multiple levels. They mediate the effects of AD risk factors, regulate the expression of many proteins involved in amyloid production and/or clearance, block cognitive functioning in AD, and are responsible for neuronal death downstream from pathological Aβ/Tau aggregation. Some epigenetic marks like histone acetylation are currently being investigated as biomarkers in AD [150] and HDAC inhibitors for potential therapeutic approaches. However, the widely used non-specific HDAC inhibitor valproic acid could not reverse cognitive decline or neuropsychiatric symptoms associated with AD in a double-blind place-controlled trial [151]. More selective inhibitors may have better outcomes [152]. The involvement of epigenetic mechanisms in mediating the risk of AD also highlights their implication in AD prevention, particularly in selected groups. Of particular note here are dietary factors, environment enrichment, and exercise, as they all have potential to change gene expression epigenetically, and have been shown to delay/ameliorate the symptoms of AD. Emerging evidence suggests that certain diets and exercise can have protective effects in AD through possible epigenetic regulation of BDNF. Similarly, polyphenols in diet have the potential to counteract miRNAs alterations induced by ApoE knock-out, indicating their potential for delaying AD in individuals with ApoE4 polymorphism [153−156].

EPIGENETIC DYSREGULATION IN PSYCHIATRIC DISORDERS: THE EXAMPLE OF DEPRESSION

Psychiatric diseases, such as depression and personality, anxiety, or psychotic disorders are multifaceted conditions with complex etiology, and are often difficult to treat. A substantial body of evidence suggests that depressive phenotypes are partly caused by epigenetic mechanisms (see Fig. 37.3). According to DSM-V, major depressive disorder (MDD) is characterized by low mood, markedly diminished interest or pleasure in activities, fatigue or lethargy, feelings of worthlessness or excessive guilt, psychomotor agitation or retardation, and abnormal weight or sleep changes (American Psychiatric Association). This chronic illness affects a sizable population with about 1.5%−19% lifetime prevalence worldwide [157,158]. Depression is usually difficult to treat, and only half of depressed patients show complete remission [159−161]. A major issue with most treatments is that the symptoms are usually ameliorated only after a few weeks of therapy. The reasons for such delay are not entirely known but could reflect the contribution of epigenetic

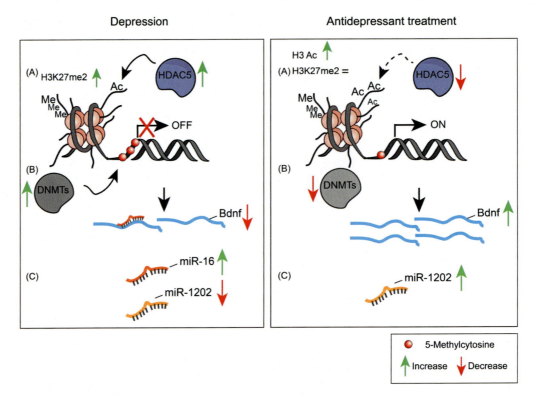

FIGURE 37.3 The importance of PTMs, DNA methylation, and miRNAs in depression. (A) Alterations in histone PTMs. Upon chronic social defeat stress, a rodent model of depression, H3K27 dimethylation is increased at *Bdnf* promoter. This leads to chromatin condensation, which shuts down *Bdnf* gene expression. In parallel, global H3 acetylation levels are decreased, accompanied by increased HDAC5 expression. Chronic antidepressant (imipramine) treatment reduces HDAC5 levels, leading to increased H3 acetylation, whereas H3K27 dimethylation remains unaffected. Nonetheless, the increase in H3 acetylation is sufficient to reinstate *Bdnf* gene expression [157]. (B) Alterations in DNA methylation. Depression in humans is associated with increased DNMT expression in the brain and differential methylation levels at several candidate gene promoters, such as *Bdnf*. The increase in 5-mC levels at the *Bdnf* promoter correlates with low *Bdnf* expression in patients suffering from depression [163]. Treatment of blood cells from depressed individuals with an antidepressant (paroxetine) results in reduced DNMT1 activity, decreased *Bdnf* methylation, which leads to increased *Bdnf* expression [164]. (C) Alterations in miRNA profiles. Depression-like behaviors in rats induced by maternal deprivation are associated with increased miR-16 expression, a miRNA that targets *Bdnf*, in the hippocampus, as well as, reduced *Bdnf* transcripts [165]. In the blood of individuals with depression, the level of miR-1202 is reduced, while antidepressant treatment reverses this effect [166]. *Bdnf*, Brain-derived neurotrophic factor; *DNMT*, DNA methyltransferase; *HDAC*, histone deacetylase; *5-mC*, 5-methylcytosine; *miR*, miRNA; *PTMs*, posttranslational histone modifications.

mechanisms to the etiology of depression. Epigenetic mechanisms are also implicated in the appearance of depression after stress exposure, and in the effect of antidepressants [162].

Histone Modifications and Histone Modifier in Depression

The importance of histone PTMs and BDNF in depression can be appreciated from a rodent model of social defeat. Brain-derived neurotrophic factor (BDNF) is a brain neurotrophin with critical functions in neuronal development and function, that has been linked to depression-associated phenotypes. Social defeat stress induces symptoms of depression that can be reversed with long-term antidepressant treatment, mimicking the observation in humans [160]. In mice, chronic social defeat decreases the expression of two splice variants of BDNF in the hippocampus and induces H3K27 dimethylation, a mark for transcriptional repression [86], in their respective promoter regions [167]. In contrast, hippocampal H3K27 trimethylation, a mark of transcriptional repression, decreases after chronic stress [168]. Similarly, While behavioral anomalies induced by social defeat are reversed by chronic antidepressant treatment, the increase in H3K27 dimethylation is not, suggesting that this histone hypermethylation acts as lasting stress-induced mark [167]. Instead, H3 acetylation and H3K4 methylation, marks of transcriptional activation [86], are increased at the same promoters leading to a reversal of BDNF downregulation [167,169]. Long-term antidepressant treatment also down regulates the expression of HDAC5 in animals exposed to chronic stress [170]. Thus, changes in histone acetylation and methylation by chronic stress at the *BDNF* gene in the hippocampus are likely important for the development of depressive behaviors, and histone acetylation is a target of antidepressant treatments. Further, chronic unpredictable mild stress (CUMS) also decreases H3K9 methylation at the promoter of *corticotrophin-releasing hormone receptor 1* (*CHRR1*) in the rat hippocampus, and is associated with altered hypothalamic-pituitary-adrenal (HPA) axis, a characteristic of depression [171]. Further to the hippocampus, other neuroanatomical regions have changes in histone PTMs at genes involved in behavioral response to chronic stress. In ventral striatum, histone PTMs at glial-derived neurotrophic factor (GDNF) gene have been associated with the susceptibility and adaptation to chronic stress in mice [163].

DNA Methylation in Depression

Alterations in DNA methylation have been reported upon antidepressant administration in preclinical and clinical studies. In a rat model of depression, three-week treatment with the selective serotonin reuptake inhibitor (SSRI) escitalopram reduced mRNA levels of *DNMT1* and *DNMT3a* in forebrain neurons associated with decreased global DNA methylation and partial reversal of pro-depressant behaviors [164]. Consistently, hypomethylation of the serotonin transporter gene, a main target of SSRIs, affects SSRI treatment response in humans [172]. Intraperitoneal administration of the SSRI fluoxetine leads to increased MeCP2 and methyl CpG-binding domain protein 1 (MBD1) in the adult rat brain [173].

Interestingly, disruption of *BDNF* IV activity or decreased *BDNF* IV promoter methylation can also lead to aberrant neuronal activity and depression-like behaviors [174–176]. In line with the observed epigenetic changes in response to SSRI treatment, cognitive behavior therapy (CBT) can also lead to hypomethylation. Following psychotherapy, patients with borderline personality disorder showed decreased CpG methylation at two exons of *BDNF*, *BDNF CpG exon I* and *IV*, in peripheral blood leukocytes; whereas, methylation was increased in non-responders [177].

Typically, both methyl-binding proteins are enriched in neurons, as compared to glial cells [178,179]. Upon fluoxetine administration, both proteins were found to accumulate in neurons, including GABA-ergic interneurons. This is of particular interest because abnormal GABAergic transmission and anomalies in GABA-related gene methylation are linked to major depression and suicide. Depressed patients who committed suicide have higher level of methylation in the GABA-A α1 receptor subunit promoter, and increased DNMT3b mRNA and protein in the prefrontal cortex when compared to control individuals who died of other causes [180]. This suggests the interesting possibility that antidepressant treatments can target the epigenetic machinery in cell types affected by depression. Strikingly, TET1, an enzyme important for DNA demethylation, is protective against depression, further highlighting the importance of precise DNA methylation dynamics in depression [181].

Several studies further reported differential DNA methylation at candidate genes, such as *BDNF* in different tissues of depressed individuals. Increased 5-mC level at *BDNF* promoter IV is associated with decreased *BDNF* transcript levels in Wernicke's area of suicide victims, some of whom suffering from MDD [175]. This suggests that changes in DNA methylation levels could be responsible for *BDNF* downregulation in depression. *BDNF* promoter hypermethylation is observed in both buccal tissue and blood of depressed patients, and is correlated with suicidal behavior. The correlation between BDNF level and its promoter methylation in blood cells and brain suggests that blood BDNF is an easily accessible biomarker of depressive disorders. Treatment of peripheral blood cells isolated from depressed patients with the antidepressant paroxetine reduces DNMT1 activity and *BDNF* methylation, and correlates with increased *BDNF* expression [182]. Acquiring blood samples to assess *BDNF* methylation is less invasive and may serve as diagnostic tool for MDD [176,183,184] and potentially even be used for treatment monitoring. Genome-wide profiling of DNA methylation may also be useful since profiling in blood of patients suffering from depression is distinct from non-depressed individuals [185]. But DNA methylation profiling can be variable as revealed by analyses in monozygotic twins from depression-discordant pairs [166,186–188], suggesting the possibility that the methylome of patients with MDD may be more sensitive to environmental influence.

Non-coding RNA in Depression

Further, miRNAs have also been associated with the long-term effects of chronic stress, an important inducer of depression. Chronic mild stress alters the expression of miR-186 and miR-709 in the hippocampus and prefrontal cortex in mice [189]. miR-19 levels have been associated with pro-depressive effects of chronic stress by decreasing the expression of a number of target genes [190] including adrenergic receptor b1 (*adrb1*) in BLA [191] and *FKBP5* in the amygdala [192]. A possible mode of action of miRNAs is to regulate the molecular pathways underlying stress resilience. MiR-135 is considered an endogenous "antidepressant," which has been shown to prevent the development of pro-depressive phenotype in mice by modifying serotonergic activity [193]. Further, miRNAs can contribute to the regulation of neurogenesis (miR-124 through targeting Sox9; miR-128 through targeting LRRC4C), neurotrophic signaling (miR-26 a and b through targeting BDNF) and inflammation (miR-155 and miR-181), which have all been implicated in depression [194–196]. Not surprisingly, clinically efficient antidepressants lead to notable changes in brain miRNA profiles. For example, fluoxetine promotes the biogenesis of miR-16 in raphe nuclei [197], and imipramine and citalopram regulate miRNA miR-1202 [198].

In conclusion, the contribution of epigenetic mechanisms to molecular pathways affected by stress, their role in determining the susceptibility to depression, and their alteration by antidepressants, highlight their potential role in depressive prophylaxis and therapeutics. It is also possible that non-pharmacological strategies like CBT and psychotherapy used to treat depression also implicate epigenetic changes. This supports the idea that epigenetic marks are amenable to non-pharmacological treatment approaches in psychiatry, and could serve as therapeutic targets and prognostic markers in the future. Off note again here is the potential of dietary factors and exercise in prevention and/or treatment of psychiatric disorders. Changes in serotonergic signaling by tryptophan-rich diet, exercise-induced changes in BDNF, and induction of neurogenesis are a few ways through which diet and exercise could alter the development or clinical course of depression [153,154].

EPIGENETIC DYSREGULATION BY ENVIRONMENTAL STRESS: THE EXAMPLE OF EARLY-LIFE STRESS

It has long been recognized that detrimental experiences in early postnatal life have long-lasting consequences for brain functions, especially in the affective domains. The quality of the social and parental environment early, postnatal in particular, is a critical determinant of an organism's neuropsychological development. In humans, childhood trauma or prolonged separation from the mother results in deviant behaviors, such as drug abuse, in adult life [165]. In mice, unpredictable maternal separation combined with unpredictable maternal stress was also shown to lead to severe and persistent emotional and cognitive dysfunctions [199]. Such adverse experiences in early postnatal life are known to induce multiple epigenetic modifications in the brain, resulting in altered gene expression (reviewed in Ref. [200]). Different molecular factors including glucocorticoids and glucocorticoid receptor (GR), mineralocorticoids, and mineralocorticoid receptor (MR), arginine vasopressin (avp) and oxytocin, are noteworthy as they are known to be altered by early traumatic stress.

DNA Methylation in Early-life Stress

Maternal care is indeed an important environmental factor for behavioral responses later in life. The offspring of high-nurturing female rats have decreased DNA methylation and increased H3K9 acetylation at *NR3C1*, a gene coding for GR, in the brain [201]. In contrast, the offspring of low nurturing females have decreased brain *NR3C1* methylation, suggesting that differential epigenetic marking underlies changes in GR expression by maternal care. In humans, childhood maltreatment and adversity also lead to alterations of DNA methylation at *NR3C1* as observed in the hippocampal autopsy specimens of suicide victims with a history of childhood abuse [202]. Importantly, methylation profile at *NR3C1* promoter in response to early-life trauma is comparable in rats and humans [203]. It is not surprising that epigenetic regulation of GR is under stringent regulation considering its critical role in mediating the effect of stress on the brain, and serving as an important determinant of neuro-endocrine integration [194]. Likewise, disrupted maternal care in early childhood decreases hippocampal Mr expression in adults in association with changes in histone PTMs [204]. Parental contact in early postnatal life also determines DNA methylation of the oxytocin gene in the brain and the periphery of the offspring in rodents [205]. Finally, periodic mother-infant separation in mice leads to DNA hypomethylation in the promoter region of *avp*, which correlates with an increased and persistent expression of avp resulting in increased vulnerability to stress in adulthood [205,206]. Such persistent increase in avp could have pleiotropic effects on adult brain functions related to its role in regulation of social behaviors, pair bonding, sexual preferences, and its noted dysregulation in anxiety and depression [207].

Genome-wide CpG promoter methylation screen revealed differentially methylated regions in postmortem hippocampal tissue of suicide victims with a history of childhood abuse. These changes are associated with the dysregulation of a network of genes involved in neuronal plasticity [208]. In adult monkey, early-life maternal deprivation is associated with changes in DNA hydroxymethylation at the promoter of genes related to psychiatric disorders and/or perinatal adversity in cortex [209]. In maternally deprived animals, hydroxymethylation was decreased at the promoter of the dopaminergic receptor 3, the adrenoreceptor alpha 1, and the serotonergic transporter. In contrast, hydroxymethylation was increased at the promoter of the GABAA receptor alpha 2, the transcriptional repressor REST and the MAP kinase 1. Hydroxymethylation within regulatory regions has been hypothesized to constitute an epigenetic mark presumably associated with transcriptional modulators [24,67,85]. It should be noted however, that further studies are required to fully understand its functional implications. Altogether, these findings suggest that early life adversity triggers variations in both DNA methylation and hydroxymethylation that persist into adulthood.

Non-coding RNA in Early-life Stress

Besides DNA methylation, miRNAs may also be important mediators of the effects of early postnatal environmental influence, since several of them are affected by early adversity. miR-16 expression is increased in the rat hippocampus after early-life trauma, leading to decreased hippocampal BDNF [210]. Maternal deprivation in early childhood also alters miR-504 expression in nucleus accumbens in rats [211].

Intergenerational and Transgenerational Epigenetic Effects of Early-life Stress

Epigenetic modifications can carry the effects of early-life trauma not only through life but also across generations. Exposure to early-life trauma induced by unpredictable maternal separation and unpredictable maternal stress has been shown to alter behavior and metabolism across generations, even in the absence of any traumatic exposure in the offspring. Transmission of these effects implicates sperm RNAs, since sperm RNAs from traumatized males injected into naïve control fertilized oocytes recapitulate the effects of trauma in the resulting animals, as well as in their progeny. These RNAs include miRNAs, which expression is altered by trauma not only in sperm, but also in the serum and brain of exposed mice, and in the brain of the offspring [204]. Apart from miRNAs, early-life trauma also alters DNA methylation in sperm and brain across generations. It alters DNA methylation in several genes, such as corticotropin-releasing factor receptor 2 (CRFR2), Mecp2 and cannabinoid receptor 1 (CB1), some associated with depressive-like behaviors, in the brain and the sperm of exposed males and their offspring [199], suggesting another potential means of transgenerational epigenetic inheritance. Such epigenetic alterations were recently shown to be reversed by environmental or pharmacological manipulations.

Crossfostering of pups, or treatment with the HDAC inhibitor trichostatin A leads to NR3C1 DNA hypomethylation and histone hyperacetylation in the offspring of low nurturing mothers [201]. Likewise, methyl supplementation via administration of L-methionine, a SAM precursor, can reverse maternal programming of stress responses via GR [212]. Importantly, suboptimal maternal care also decreases histone acetylation of a large number of genes in the mouse hippocampus, including ATRX and Reelin, and these changes are reversed by treatment with trichostatin A. [213]. Finally, paternal environmental enrichment also reverses the transmission of behavioral traits, and corrects aberrant DNA methylation of NR3C1 in a mouse model of early-life trauma [214].

Future research on this immensely intriguing topic should focus on the effectors of these epigenetic changes, and on identifying the mechanisms underlying the effects of early-life trauma on the brain and the germline. Circulating hormones and cytokines could be relevant as they can readily access the brain and germ cells. Indeed, maternal immune activation mediates the transgenerational effects of prenatal stress [215]. Further, the observation that environmental enrichment can reverse some of the long-term adverse consequences of early postnatal trauma through epigenetic modulation also raises the question if dietary factors could have similar preventive or therapeutic benefits. Influence of folate-containing diet to modulate DNA methylation, tryptophan-rich diet to modify serotonergic signaling in the brain, and antioxidant and anti-inflammatory properties of certain foods are important to consider [153–156]. Although the field is extremely dynamic and the subject of considerable research, more basic, as well as preclinical, and clinical research will be needed for applying these findings to the clinic.

CONCLUSIONS AND OUTLOOK

It is becoming increasingly clear that epigenetic mechanisms play a pivotal role in higher-order brain functions, in both physiological and pathological conditions. Epigenetic mechanisms contribute to the etiopathology of brain disorders, determine an individual's disease susceptibility, and underlie the therapeutic efficacy of treatments. In some cases, epigenetic mechanisms serve as vectors carrying the effects of environmental stressors to subsequent generations. A precise delineation of these processes is necessary for a better understanding of the mechanisms underlying brain diseases, and for the development of potential treatments and preventive strategies. Future research should therefore explore how epigenetic mechanisms contribute to the risk to or protection against brain disorders. Possible crosstalk between epigenetic mechanisms, such as regulation of HDACs or DNMTs binding by ncRNAs, or potential master epigenetic regulators and functional "epigenetic codes" should be identified. Further to DNA methylation, histone PTMs, and miRNAs, other forms of epigenetic regulation, such as alterations in regulatory elements, lncRNAs, piwi RNAs (piRNAs), RNA methylation, or prion-like mechanisms may also be involved. LncRNAs were found to be altered in mouse models of AD [216] and depression [217]. Similarly, piRNAs have emerged as critical vectors of intergenerational inheritance in *Caenorhabditis elegans* and *Drosophila*, and as regulators of long-term memory in mice [218,219]. Similarly, mutations and variants in regulatory elements such as enhancers and their binding partners have been linked to a variety of neurodevelopmental and neuropsychiatric diseases (reviewed in [32]). These processes merit attention in the future.

It is also important to consider the possible solution for the issue of the use of epigenetic treatments for brain disorders. The complexity of the epigenetic landscape in the brain, and of its regulation in different brain cell types during development and adulthood, with constant interaction with the environment needs to be carefully studied [220]. Additional challenges include the possibility that manipulating an epigenetic effector could have contrasting effects in different brain regions and/or functions, or in the periphery [220]. A relevant example is that of the miRNA cluster miR183−96−182, which promotes memory formation when overexpressed in the mouse hippocampus [221], but has also been linked with depressive-like behavior in rat [222]. An additional challenge is the low permeability of the blood-brain barrier. Some of the challenges of using epigenetic drugs to treat brain diseases need to be addressed in the future, and consider the mode of regulation of epigenetic enzymes and their targets in different brain cells, neuroanatomical regions, and functional contexts. It would also be important to determine if epigenetic therapies are more likely to benefit selected groups based on genetic features like the presence or absence of metabolizing enzymes, in which case those groups would need to be carefully isolated through genetic screening and/or clinical histories.

SUMMARY

Normal brain functions build on a unique challenge to constantly adapt to new stimuli while maintaining information about previous experiences for long periods of time. At the core of this dynamic yet stable encoding are multiple gene expression networks that can be regulated on various levels. Non-coding RNAs and epigenetic modifications to DNA and histones provide exquisite control over transcriptional programs. Dysregulation of these epigenetic control mechanisms can lead to aberrant gene expression and contribute to disease states. The neurodevelopmental disorder Rett syndrome for example, is characterized by dysfunction of transcriptional regulator and methyl-binding protein MeCP2. MeCP2 itself interacts with a number of transcription factors and epigenetic modifiers including DNMTs, TETs, and HDACs (see Fig. 37.1). In Rett syndrome, a multitude of altered PTMs on histones or MeCP2, as well as altered miRNA levels have been identified and are likely to contribute to the disease phenotype.

Pathological aggregation of proteins is a hallmark of many neurodegenerative diseases including Alzheimer disease. In the case of AD, two of the main hallmarks of the disease are amyloid plaques and neurofibrillary tangles. While global changes in miRNA levels, DNA methylation, and histone acetylation have been associated with AD, there are several specific changes that contribute to the aggregation of Aβ-plaques and hyperphosphorylated Tau into NFTs (see Fig. 37.2). APP and Tau processing is directly affected by the levels of miRNAs, lncRNAs, and histone acetylation and DNA methylation at genes coding for key processing enzymes. An imbalance in these regulatory mechanisms can contribute to protein aggregation and disease. Further to neurological conditions, psychiatric diseases such as depression also involve epigenetic mechanisms and factors for their etiology and expression. Importantly, antidepressant treatment has been shown to reverse depression-associated epigenetic changes such as increased H3K23me2, DNA methylation, DNMT and HDAC activity, as well as correct the levels of a subset of miRNAs (see Fig. 37.3).

Pathologies involving adverse experiences such as early-life stress are also characterized by lasting epigenetic changes that affect gene expression and behavior in adults and in some case, also their offspring. Changes in DNA methylation, hydroxymethylation, and miRNAs have been associated with early-life stress and can persist into adulthood and be transmitted across generations. Interestingly, in rodent models, environmental enrichment can alleviate some of the early-life stress associated epigenetic alterations and behavioral phenotypes.

To summarize, the importance of epigenetic dysfunction in brain diseases is now fully appreciated. With a better understanding of their modes of regulation and the identification of their specific downstream targets, and their region- and cell-type specific effects, safe and efficient epigenetic treatments can be envisaged for treatment of brain disorders in future.

RESEARCH CASE STUDY

A case report: Transcriptional and epigenetic processes in peripheral blood cells of monozygotic twins discordant for Alzheimer disease.

Objective

The goal of the study is to evaluate the respective influence of genes vs nongenetic factors in AD development by exploiting the comparable genetic makeup of monozygotic twins. It also aims at identifying potential new biomarkers.

Scope

The study is based on one pair of twins that was followed across time for several years. During the study, biological fluids including cerebral spinal fluid (CSF) and peripheral blood were repeatedly collected for analysis, and both twins also underwent extensive behavioral, neurological, and neuropsychological tests. Although the study encompasses only two subjects, the advantage of focusing on such a low subject number is, that it allows to better control confounding factors like genetic background, age, exposure during early development. Each parameter could be examined extremely carefully for each twin, and the analyses allow to dissociate every difference between the twins in relation to key target genes.

Audience

This study suggests that epigenetic factors play a role in the molecular underpinning of AD, and provides potential biomarkers for the disease with possibly new leads for therapeutic consideration. It is a preliminary study that may help set the basis for future in depth studies of epigenetic factors in AD.

Rationale

There is a large variety of potential AD-associated target genes and pathways, making the study of AD complex. To circumvent this complexity, this study focuses on specific gene targets and examines differences in their expression level and epigenetic state, to identify individual signatures associated with the disease. It compares genetic and epigenetic parameters in homozygotic twins who differ in their manifestation of the disease but have comparable genetic make-up and similar environments.

Expected Results and Deliverables

Detect differences in target genes expression and/or epigenetic state in peripheral blood samples, and identify potential new biomarkers for AD that may also represent new leads to therapeutic development.

Safety Considerations

Both participants and care givers gave their informed consent to participate to the study.

The Process, Workflow and Actions Taken

Two monozygotic twins were evaluated over the course of 3–4 years starting at age 70 and 71 respectively with one initial assessment and two follow-ups after 2 and 4 years for the affected twin and after 2 and 3 years for the unaffected twin (three assessments per subject in total). One of the twins was first diagnosed with mild cognitive impairment at the age of 70 with progression to dementia at a follow-up evaluation 2 years later. Cognitive functioning, affective, and functional states were assessed in both twins, as well as genetic testing for Alzheimer's associated risk gene apolipoprotein E (APOE) variants, complete blood tests (for *Treponema pallidum*, TSH and thyroid hormones, vitamin B12, and folate), and CT-scans and/or MRI. Lumbar punctures were also conducted to screen for Amyloid-β (Aβ), Tau, and pTau in CSF. The affected twin was also subjected to FDG-PET.

Tools and Materials Used

Neuropsychological assessment included a test battery to assess different aspects of cognitive functioning. Neuropsychiatric Inventory (NPI), means of the Activities of Daily Living (ADL) and Instrumental ADL (IADL) scales were used to assess affective and functional states. CSF from lumbar punctures and peripheral blood were tested for a list of AD-associated biomarkers.

Results

The twin affected with MCI progressed to dementia and AD diagnosis within 2 years after the first assessment. At the initial evaluation, the patient presented with cerebral atrophy and moderate hypometabolism, but normal blood test. At the time of the first follow-up and AD diagnosis, the affected patient had Aβ, Tau, and pTau in CSF, as well as global cognitive deficits. Compared to the unaffected twin, elevated mRNA levels of sirtuin 1 (SIRT1), amyloid precursor protein (APP), and peptidyl prolyl isomerase 1 (PIN1) were observed, while the level of APOE and presenilin 1

(PSEN1) were not different. The affected twin further showed DNA hypermethylation of the APOE promoter, and DNA hypomethylation at promoter regions of APP, PIN1, SIRT1, and PSEN1. There was however no difference in DNMT mRNA expression levels. There was also an overall decrease in total histone level and histone methylation in the affected twin. Likewise, the level of histone acetylation was reduced but HDAC2 and HDAC9 was increased in the affected twin.

Challenges and Solutions

One of the assets of the study is that there was no significant difference in the environment of the twins, suggesting that it is unlikely that environmental factors contribute to disease development. Instead, the similarity of the environment allowed focus on a few key genes and blood samples without too many confounding factors.

Learning and Knowledge Outcomes

The study found differences in the epigenetic regulation of several AD-associated genes. These findings highlight the relevance and importance for future studies with larger cohorts to investigate epigenetic mechanisms in AD in depth. The study further suggests HDACs as possible target for therapeutic and diagnostic in AD. These findings support the relevance of recent and ongoing clinical trials that focus on HDAC inhibitors to treat neurological disorders and AD (reviewed in [223]).

Acknowledgments

We thank the anonymous reviewers for their useful and constructive comments. The lab of IM Mansuy is funded by the University of Zürich, the Swiss Federal Institute of Technology Zürich, the Swiss National Science Foundation, the European Commission H2020.

References

[1] Tognini P, Napoli D, Pizzorusso T. Dynamic DNA methylation in the brain: a new epigenetic mark for experience-dependent plasticity. Front Cell Neurosci 2015;9:331. Available from: https://doi.org/10.3389/fncel.2015.00331.

[2] Rudenko A, Tsai LH. Epigenetic modifications in the nervous system and their impact upon cognitive impairments. Neuropharmacology 2014;80:70−82. Available from: https://doi.org/10.1016/j.neuropharm.2014.01.043.

[3] Follert P, Cremer H, Béclin C. MicroRNAs in brain development and function: a matter of flexibility and stability. Front Mol Neurosci 2014;7:5. Available from: https://doi.org/10.3389/fnmol.2014.00005.

[4] Wang W, Kwon EJ, Tsai LH. MicroRNAs in learning, memory, and neurological diseases. Learn Mem 2012;19:359−68. Available from: https://doi.org/10.1101/lm.026492.112.

[5] Brown AN, Vied C, Dennis JH, Bhide PG. Nucleosome repositioning: a novel mechanism for nicotine- and cocaine-induced epigenetic changes. PLoS One 2015;10:e0139103. Available from: https://doi.org/10.1371/journal.pone.0139103.

[6] Maze I, Wenderski W, Noh KM, Bagot RC, Tzavaras N, Purushothaman I, et al. Critical role of histone turnover in neuronal transcription and plasticity. Neuron 2015;87:77−94. Available from: https://doi.org/10.1016/j.neuron.2015.06.014.

[7] Thomson DW, Dinger ME. Endogenous microRNA sponges: evidence and controversy. Nat Rev Genet 2016;17:272−83. Available from: https://doi.org/10.1038/nrg.2016.20.

[8] Lin W, Liu H, Tang Y, Wei Y, Wei W, Zhang L, et al. The development and controversy of competitive endogenous RNA hypothesis in non-coding genes. Mol Cell Biochem 2021;476:109−23. Available from: https://doi.org/10.1007/s11010-020-03889-2.

[9] Arzate-Mejía RG, Lottenbach Z, Schindler V, Jawaid A, Mansuy IM. Long-term impact of social isolation and molecular underpinnings. Front Genet 2020;11:621. Available from: https://doi.org/10.3389/fgene.2020.589621.

[10] March ZM, King OD, Shorter J. Prion-like domains as epigenetic regulators, scaffolds for subcellular organization, and drivers of neurodegenerative disease. Brain Res 2016;1647:9−18. Available from: https://doi.org/10.1016/j.brainres.2016.02.037.

[11] Lister R, Mukamel EA, Nery JR, Urich M, Puddifoot CA, Johnson ND, et al. Global epigenomic reconfiguration during mammalian brain development. Science (80-) 2013;341:1237905. Available from: https://doi.org/10.1126/science.1237905.

[12] Guo JU, Su Y, Shin JH, Shin J, Li H, Xie B, et al. Distribution, recognition and regulation of non-CpG methylation in the adult mammalian brain. Nat Neurosci 2014;17:215−22. Available from: https://doi.org/10.1038/nn.3607.

[13] Jeong H, Mendizabal I, Berto S, Chatterjee P, Layman T, Usui N, et al. Evolution of DNA methylation in the human brain. Nat Commun 2021;12:1−12. Available from: https://doi.org/10.1038/s41467-021-21917-7.

[14] Bird A. DNA methylation patterns and epigenetic memory. Genes Dev 2002;16:6−21. Available from: https://doi.org/10.1101/gad.947102.

[15] Klose RJ, Bird AP. Genomic DNA methylation: the mark and its mediators. Trends Biochem Sci 2006;31:89−97. Available from: https://doi.org/10.1016/j.tibs.2005.12.008.

[16] Rausch C, Hastert FD, Cardoso MC. DNA modification readers and writers and their interplay. J Mol Biol 2020;432:1731−46. Available from: https://doi.org/10.1016/j.jmb.2019.12.018.

[17] Jones PL, Veenstra GJC, Wade PA, Vermaak D, Kass SU, Landsberger N, et al. Methylated DNA and MeCP2 recruit histone deacetylase to repress transcription. Nat Genet 1998;19:187−91. Available from: https://doi.org/10.1038/561.

[18] Bahar Halpern K, Vana T, Walker MD. Paradoxical role of DNA methylation in activation of FoxA2 gene expression during endoderm development. J Biol Chem 2014;289:23882−92. Available from: https://doi.org/10.1074/jbc.M114.573469.

[19] Kohli RM, Zhang Y. TET enzymes, TDG and the dynamics of DNA demethylation. Nature 2013;502:472−9. Available from: https://doi.org/10.1038/nature12750.

[20] Globisch D, Münzel M, Müller M, Michalakis S, Wagner M, Koch S, et al. Tissue distribution of 5-hydroxymethylcytosine and search for active demethylation intermediates. PLoS One 2010;5:e15367. Available from: https://doi.org/10.1371/journal.pone.0015367.

[21] Münzel M, Globisch D, Brückl T, Wagner M, Welzmiller V, Michalakis S, et al. Quantification of the sixth DNA base hydroxymethylcytosine in the brain. Angew Chemie Int Ed 2010;49:5375−7. Available from: https://doi.org/10.1002/anie.201002033.

[22] Song CX, Szulwach KE, Fu Y, Dai Q, Yi C, Li X, et al. Selective chemical labeling reveals the genome-wide distribution of 5-hydroxymethylcytosine. Nat Biotechnol 2011;29:68−75. Available from: https://doi.org/10.1038/nbt.1732.

[23] Bachman M, Uribe-Lewis S, Yang X, Williams M, Murrell A, Balasubramanian S. 5-Hydroxymethylcytosine is a predominantly stable DNA modification. Nat Chem 2014;6:1049–55. Available from: https://doi.org/10.1038/nchem.2064.

[24] Kozlenkov A, Li J, Apontes P, Hurd YL, Byne WM, Koonin EV, et al. A unique role for DNA (hydroxy)methylation in epigenetic regulation of human inhibitory neurons. Sci Adv 2018;4:eaau6190. Available from: https://doi.org/10.1126/sciadv.aau6190.

[25] Lister R, Mukamel EA. Turning over DNA methylation in the mind. Front Neurosci 2015;9:252. Available from: https://doi.org/10.3389/fnins.2015.00252.

[26] Stroud H, Su SC, Hrvatin S, Greben AW, Renthal W, Boxer LD, et al. Early-life gene expression in neurons modulates lasting epigenetic states. Cell 2017;171:1151–64. Available from: https://doi.org/10.1016/j.cell.2017.09.047 e16.

[27] Day JJ, Sweatt JD. DNA methylation and memory formation. Nat Neurosci 2010;13:1319–23. Available from: https://doi.org/10.1038/nn.2666.

[28] Jenuwein T, Allis CD. Translating the histone code. Science (80-) 2001;293:1074–80. Available from: https://doi.org/10.1126/science.1063127.

[29] Turner BM. Cellular memory and the histone code. Cell 2002;111:285–91. Available from: https://doi.org/10.1016/S0092-8674(02)01080-2.

[30] Gräff J, Mansuy IM. Epigenetic codes in cognition and behaviour. Behav Brain Res 2008;192:70–87. Available from: https://doi.org/10.1016/j.bbr.2008.01.021.

[31] Mews P, Calipari ES, Day J, Lobo MK, Bredy T, Abel T. From circuits to chromatin: the emerging role of epigenetics in mental health. J Neurosci 2021;41:873–82. Available from: https://doi.org/10.1523/JNEUROSCI.1649-20.2020.

[32] Carullo NVN, Day JJ. Genomic enhancers in brain health and disease. Genes (Basel) 2019;10:43. Available from: https://doi.org/10.3390/genes10010043.

[33] Bannister AJ, Kouzarides T. Regulation of chromatin by histone modifications. Cell Res 2011;21:381–95. Available from: https://doi.org/10.1038/cr.2011.22.

[34] Chan JC, Maze I. Nothing is yet set in (Hi)stone: novel post-translational modifications regulating chromatin function. Trends Biochem Sci 2020;45:829–44. Available from: https://doi.org/10.1016/j.tibs.2020.05.009.

[35] Li B, Carey M, Workman JL. The role of chromatin during transcription. Cell 2007;128:707–19. Available from: https://doi.org/10.1016/j.cell.2007.01.015.

[36] Klose RJ, Zhang Y. Regulation of histone methylation by demethylimination and demethylation. Nat Rev Mol Cell Biol 2007;8:307–18. Available from: https://doi.org/10.1038/nrm2143.

[37] Peters AHFM, Schübeler D. Methylation of histones: playing memory with DNA. Curr Opin Cell Biol 2005;17:230–8. Available from: https://doi.org/10.1016/j.ceb.2005.02.006.

[38] Cao J, Yan Q. Histone ubiquitination and deubiquitination in transcription, DNA damage response, and cancer. Front Oncol 2012;2:26. Available from: https://doi.org/10.3389/fonc.2012.00026.

[39] Cheng J, Huang M, Zhu Y, Xin Y-J, Zhao Y-K, Huang J, et al. SUMOylation of MeCP2 is essential for transcriptional repression and hippocampal synapse development. J Neurochem 2014;128:798–806. Available from: https://doi.org/10.1111/jnc.12523.

[40] Stielow C, Stielow B, Finkernagel F, Scharfe M, Jarek M, Suske G. SUMOylation of the polycomb group protein L3MBTL2 facilitates repression of its target genes. Nucleic Acids Res 2014;42:3044–58. Available from: https://doi.org/10.1093/nar/gkt1317.

[41] Kerimoglu C, Agis-Balboa RC, Kranz A, Stilling R, Bahari-Javan S, Benito-Garagorri E, et al. Histone-methyltransferase mll2 (kmt2B) is required for memory formation in mice. J Neurosci 2013;33:3452–64. Available from: https://doi.org/10.1523/JNEUROSCI.3356-12.2013.

[42] Gräff J, Woldemichael BT, Berchtold D, Dewarrat G, Mansuy IM. Dynamic histone marks in the hippocampus and cortex facilitate memory consolidation. Nat Commun 2012;3:1997. Available from: https://doi.org/10.1038/ncomms1997.

[43] Mahgoub M, Monteggia LM. A role for histone deacetylases in the cellular and behavioral mechanisms underlying learning and memory. Learn Mem 2014;21:564–8. Available from: https://doi.org/10.1101/lm.036012.114.

[44] Marco A, Meharena HS, Dileep V, Raju RM, Davila-Velderrain J, Zhang AL, et al. Mapping the epigenomic and transcriptomic interplay during memory formation and recall in the hippocampal engram ensemble. Nat Neurosci 2020;23:1606–17. Available from: https://doi.org/10.1038/s41593-020-00717-0.

[45] Yildirim F, Ji S, Kronenberg G, Barco A, Olivares R, Benito E, et al. Histone acetylation and CREB binding protein are required for neuronal resistance against ischemic injury. PLoS One 2014;9:e95465. Available from: https://doi.org/10.1371/journal.pone.0095465.

[46] Ryu H, Lee J, Olofsson BA, Mwidau A, Deodoglu A, Escudero M, et al. Histone deacetylase inhibitors prevent oxidative neuronal death independent of expanded polyglutamine repeats via an Sp1-dependent pathway. Proc Natl Acad Sci U S A 2003;100:4281–6. Available from: https://doi.org/10.1073/pnas.0737363100.

[47] Volmar CH, Wahlestedt C. Histone deacetylases (HDACs) and brain function. Neuroepigenetics 2015;1:20–7. Available from: https://doi.org/10.1016/j.nepig.2014.10.002.

[48] Mercer TR, Mattick JS. Structure and function of long noncoding RNAs in epigenetic regulation. Nat Struct Mol Biol 2013;20:300–7. Available from: https://doi.org/10.1038/nsmb.2480.

[49] Bhatti GK, Khullar N, Sidhu IS, Navik US, Reddy AP, Reddy PH, et al. Emerging role of non-coding RNA in health and disease. Metab Brain Dis 2021;36:1119–34. Available from: https://doi.org/10.1007/s11011-021-00739-y.

[50] Kyzar EJ, Zhang H, Pandey SC. Adolescent alcohol exposure epigenetically suppresses amygdala arc enhancer RNA expression to confer adult anxiety susceptibility. Biol Psychiatry 2019;85(11):904–14. Available from: https://doi.org/10.1016/j.biopsych.2018.12.021.

[51] Ha M, Kim VN. Regulation of microRNA biogenesis. Nat Rev Mol Cell Biol 2014;15:509–24. Available from: https://doi.org/10.1038/nrm3838.

[52] Wakabayashi T, Hidaka R, Fujimaki S, Asashima M, Kuwabara T. MicroRNAs and epigenetics in adult neurogenesis. Adv Genet 2014;86:27–44. Available from: https://doi.org/10.1016/B978-0-12-800222-3.00002-4.

[53] Elramah S, Landry M, Favereaux A. MicroRNAs regulate neuronal plasticity and are involved in pain mechanisms. Front Cell Neurosci 2014;8, 31. Available from: https://doi.org/10.3389/fncel.2014.00031.

[54] Kiltschewskij D, Cairns MJ. Temporospatial guidance of activity-dependent gene expression by microRNA: mechanisms and functional implications for neural plasticity. Nucleic Acids Res 2019;47:533–45. Available from: https://doi.org/10.1093/nar/gky1235.

[55] Saab BJ, Mansuy IM. Neuroepigenetics of memory formation and impairment: the role of microRNAs. Neuropharmacology 2014;80:61–9. Available from: https://doi.org/10.1016/j.neuropharm.2014.01.026.

[56] Kremer EA, Gaur N, Lee MA, Engmann O, Bohacek J, Mansuy IM. Interplay between TETs and microRNAs in the adult brain for memory formation. Sci Rep 2018;8:1678. Available from: https://doi.org/10.1038/s41598-018-19806-z.

[57] Kanach C, Blusztajn JK, Fischer A, Delalle I. MicroRNAs as candidate biomarkers for alzheimer's disease. Non-Coding RNA 2021;7:1—9. Available from: https://doi.org/10.3390/ncrna7010008.

[58] Millan MJ. An epigenetic framework for neurodevelopmental disorders: from pathogenesis to potential therapy. Neuropharmacology 2013;68:2—82. Available from: https://doi.org/10.1016/j.neuropharm.2012.11.015.

[59] Coda DM, Gräff J. Neurogenetic and neuroepigenetic mechanisms in cognitive health and disease. Front Mol Neurosci 2020;13:205. Available from: https://doi.org/10.3389/fnmol.2020.589109.

[60] Kubota T, Miyake K, Hirasawa T. Role of epigenetics in Rett syndrome. Epigenomics 2013;5:583—92. Available from: https://doi.org/10.2217/epi.13.54.

[61] Lyst MJ, Bird A. Rett syndrome: a complex disorder with simple roots. Nat Rev Genet 2015;16:261—74. Available from: https://doi.org/10.1038/nrg3897.

[62] Lavery LA, Zoghbi HY. The distinct methylation landscape of maturing neurons and its role in Rett syndrome pathogenesis. Curr Opin Neurobiol 2019;59:180—8. Available from: https://doi.org/10.1016/j.conb.2019.08.001.

[63] Amir RE, Van Den Veyver IB, Wan M, Tran CQ, Francke U, Zoghbi HY. Rett syndrome is caused by mutations in X-linked MECP2, encoding methyl- CpG-binding protein 2. Nat Genet 1999;23:185—8. Available from: https://doi.org/10.1038/13810.

[64] Skene PJ, Illingworth RS, Webb S, Kerr ARW, James KD, Turner DJ, et al. Neuronal MeCP2 Is expressed at near histone-octamer levels and globally alters the chromatin state. Mol Cell 2010;37:457—68. Available from: https://doi.org/10.1016/j.molcel.2010.01.030.

[65] Weaving LS, Ellaway CJ, Gécz J, Christodoulou J. Rett syndrome: clinical review and genetic update. J Med Genet 2005;42:1—7. Available from: https://doi.org/10.1136/jmg.2004.027730.

[66] Tillotson R, Cholewa-Waclaw J, Chhatbar K, Connelly JC, Kirschner SA, Webb S, et al. Neuronal non-CG methylation is an essential target for MeCP2 function. Mol Cell 2021;81:1260—75. Available from: https://doi.org/10.1016/j.molcel.2021.01.011 e12.

[67] Mellén M, Ayata P, Dewell S, Kriaucionis S, Heintz N. MeCP2 binds to 5hmC enriched within active genes and accessible chromatin in the nervous system. Cell 2012;151:1417—30. Available from: https://doi.org/10.1016/j.cell.2012.11.022.

[68] Szulwach KE, Li X, Li Y, Song CX, Wu H, Dai Q, et al. 5-hmC-mediated epigenetic dynamics during postnatal neurodevelopment and aging. Nat Neurosci 2011;14:1607—16. Available from: https://doi.org/10.1038/nn.2959.

[69] Ito-Ishida A, Baker SA, Sillitoe RV, Sun Y, Zhou J, Ono Y, et al. MeCP2 levels regulate the 3d structure of heterochromatic foci in mouse neurons. J Neurosci 2020;40:8746—66. Available from: https://doi.org/10.1523/JNEUROSCI.1281-19.2020.

[70] Chahrour M, Sung YJ, Shaw C, Zhou X, Wong STC, Qin J, et al. MeCP2, a key contributor to neurological disease, activates and represses transcription. Science (80-) 2008;320:1224—9. Available from: https://doi.org/10.1126/science.1153252.

[71] Ibrahim Abdulkhaleg, Papin Christophe, Mohideen-Abdul Kareem, Le Gras Stéphanie, Stoll Isabelle, Bronner Christian, et al. MeCP2 is a microsatellite binding protein that protects CA repeats from nucleosome invasion. Science 2021;372:eabd5581. Available from: https://doi.org/10.1126/SCIENCE.ABD5581.

[72] Mellén M, Ayata P, Heintz N. 5-hydroxymethylcytosine accumulation in postmitotic neurons results in functional demethylation of expressed genes. Proc Natl Acad Sci 2017;114:E7812—21. Available from: https://doi.org/10.1073/PNAS.1708044114.

[73] Chen RZ, Akbarian S, Tudor M, Jaenisch R. Deficiency of methyl-CpG binding protein-2 in CNS neurons results in a Rett-like phenotype in mice. Nat Genet 2001;27:327—31. Available from: https://doi.org/10.1038/85906.

[74] Guy J, Hendrich B, Holmes M, Martin JE, Bird A. A mouse Mecp2-null mutation causes neurological symptoms that mimic rett syndrome. Nat Genet 2001;27:322—6. Available from: https://doi.org/10.1038/85899.

[75] Gemelli T, Berton O, Nelson ED, Perrotti LI, Jaenisch R, Monteggia LM. Postnatal loss of methyl-CpG binding protein 2 in the forebrain is sufficient to mediate behavioral aspects of Rett syndrome in mice. Biol Psychiatry 2006;59:468—76. Available from: https://doi.org/10.1016/j.biopsych.2005.07.025.

[76] Goffin D, Brodkin ES, Blendy JA, Siegel SJ, Zhou Z. Cellular origins of auditory event-related potential deficits in Rett syndrome. Nat Neurosci 2014;17:804—6. Available from: https://doi.org/10.1038/nn.3710.

[77] Collins AL, Levenson JM, Vilaythong AP, Richman R, Armstrong DL, Noebels JL, et al. Mild overexpression of MeCP2 causes a progressive neurological disorder in mice. Hum Mol Genet 2004;13:2679—89. Available from: https://doi.org/10.1093/hmg/ddh282.

[78] Sugino K, Hempel CM, Okaty BW, Arnson HA, Kato S, Dani VS, et al. Cell-type-specific repression by methyl-CpG-binding protein 2 is biased toward long genes. J Neurosci 2014;34:12877—83. Available from: https://doi.org/10.1523/JNEUROSCI.2674-14.2014.

[79] Gabel HW, Kinde B, Stroud H, Gilbert CS, Harmin DA, Kastan NR, et al. Disruption of DNA-methylation-dependent long gene repression in Rett syndrome. Nature 2015;522:89—93. Available from: https://doi.org/10.1038/nature14319.

[80] Chang Q, Khare G, Dani V, Nelson S, Jaenisch R. The disease progression of Mecp2 mutant mice is affected by the level of BDNF expression. Neuron 2006;49:341—8. Available from: https://doi.org/10.1016/j.neuron.2005.12.027.

[81] Kline DD, Ogier M, Kunze DL, Katz DM. Exogenous brain-derived neurotrophic factor rescues synaptic dysfunction in Mecp2-null mice. J Neurosci 2010;30:5303—10. Available from: https://doi.org/10.1523/JNEUROSCI.5503-09.2010.

[82] Urdinguio RG, Lopez-Serra L, Lopez-Nieva P, Alaminos M, Diaz-Uriarte R, Fernandez AF, et al. Mecp2-null mice provide new neuronal targets for rett syndrome. PLoS One 2008;3:e3669. Available from: https://doi.org/10.1371/journal.pone.0003669.

[83] Nan X, Ng HH, Johnson CA, Laherty CD, Turner BM, Eisenman RN, et al. Transcriptional repression by the methyl-CpG-binding protein MeCP2 involves a histone deacetylase complex. Nature 1998;393:386—9. Available from: https://doi.org/10.1038/30764.

[84] Rastegar M, Kaufmann WE, Berube N, Ausió J, Good KV, Vincent JB. MeCP2: the genetic driver of rett syndrome epigenetics. Front Genet 2021;12:620859. Available from: https://doi.org/10.3389/fgene.2021.620859.

[85] Cartron P-F, Nadaradjane A, Lepape F, Lalier L, Gardie B, Vallette FM. Identification of TET1 partners that control its DNA-demethylating function. Genes Cancer 2013;4:235—41. Available from: https://doi.org/10.1177/1947601913489020.

[86] Kouzarides T. Chromatin modifications and their function. Cell 2007;128:693—705. Available from: https://doi.org/10.1016/j.cell.2007.02.005.

[87] Shahbazian MD, Young JI, Yuva-Paylor LA, Spencer CM, Antalffy BA, Noebels JL, et al. Mice with truncated MeCP2 recapitulate many Rett syndrome features and display hyperacetylation of histone H3. Neuron 2002;35:243—54. Available from: https://doi.org/10.1016/S0896-6273(02)00768-7.

[88] Ebert DH, Gabel HW, Robinson ND, Kastan NR, Hu LS, Cohen S, et al. Activity-dependent phosphorylation of MeCP2 threonine 308 regulates interaction with NCoR. Nature 2013;499:341—5. Available from: https://doi.org/10.1038/nature12348.

[89] Tai DJC, Liu YC, Hsu WL, Ma YL, Cheng SJ, Liu SY, et al. MeCP2 SUMOylation rescues Mecp2-mutant-induced behavioural deficits

in a mouse model of Rett syndrome. Nat Commun 2016;7:10552. Available from: https://doi.org/10.1038/ncomms10552.

[90] Zhao H, Wen G, Huang Y, Yu X, Chen Q, Afzal TA, et al. MicroRNA-22 regulates smooth muscle cell differentiation from stem cells by targeting methyl CpG−binding protein 2. Arterioscler Thromb Vasc Biol 2015;35:918−29. Available from: https://doi.org/10.1161/ATVBAHA.114.305212.

[91] Tsujimura K, Irie K, Nakashima H, Egashira Y, Fukao Y, Fujiwara M, et al. MiR-199a links MeCP2 with mTOR signaling and its dysregulation leads to Rett syndrome phenotypes. Cell Rep 2015;12:1887−901. Available from: https://doi.org/10.1016/j.celrep.2015.08.028.

[92] Good KV, Vincent JB, Ausió J. MeCP2: the genetic driver of Rett syndrome epigenetics. Front Genet 2021;12:620859. Available from: https://doi.org/10.3389/fgene.2021.620859.

[93] Klein ME, Lioy DT, Ma L, Impey S, Mandel G, Goodman RH. Homeostatic regulation of MeCP2 expression by a CREB-induced microRNA. Nat Neurosci 2007;10:1513−14. Available from: https://doi.org/10.1038/nn2010.

[94] Cheng TL, Wang Z, Liao Q, Zhu Y, Zhou WH, Xu W, et al. MeCP2 suppresses nuclear microRNA processing and dendritic growth by regulating the DGCR8/Drosha complex. Dev Cell 2014;28:547−60. Available from: https://doi.org/10.1016/j.devcel.2014.01.032.

[95] Smrt RD, Szulwach KE, Pfeiffer RL, Li X, Guo W, Pathania M, et al. MicroRNA miR-137 regulates neuronal maturation by targeting ubiquitin ligase mind bomb-1. Stem Cells 2010;28:1060−70. Available from: https://doi.org/10.1002/stem.431.

[96] Gao Y, Su J, Guo W, Polich ED, Magyar DP, Xing Y, et al. Inhibition of miR-15a promotes BDNF expression and rescues dendritic maturation deficits in MeCP2-deficient neurons. Stem Cells 2015;33:1618−29. Available from: https://doi.org/10.1002/stem.1950.

[97] Landgrave-Gómez J, Mercado-Gómez O, Guevara-Guzmán R. Epigenetic mechanisms in neurological and neurodegenerative diseases. Front Cell Neurosci 2015;9:58. Available from: https://doi.org/10.3389/fncel.2015.00058.

[98] Jakovcevski M, Akbarian S. Epigenetic mechanisms in neurological disease. Nat Med 2012;18:1194−204. Available from: https://doi.org/10.1038/nm.2828.

[99] Bradley-Whitman MA, Lovell MA. Epigenetic changes in the progression of Alzheimer's disease. Mech Ageing Dev 2013;134:486−95. Available from: https://doi.org/10.1016/j.mad.2013.08.005.

[100] Li P, Marshall L, Oh G, Jakubowski JL, Groot D, He Y, et al. Epigenetic dysregulation of enhancers in neurons is associated with Alzheimer's disease pathology and cognitive symptoms. Nat Commun 2019;10:1−14. Available from: https://doi.org/10.1038/s41467-019-10101-7.

[101] Laura Idda M, Munk R, Abdelmohsen K, Gorospe M. Noncoding RNAs in Alzheimer's disease. Wiley Interdiscip Rev RNA 2018;9: e1463. Available from: https://doi.org/10.1002/wrna.1463.

[102] Wang WX, Rajeev BW, Stromberg AJ, Ren N, Tang G, Huang Q, et al. The expression of microRNA miR-107 decreases early in Alzheimer's disease and may accelerate disease progression through regulation of β-site amyloid precursor protein-cleaving enzyme 1. J Neurosci 2008;28:1213−23. Available from: https://doi.org/10.1523/JNEUROSCI.5065-07.2008.

[103] Zhu HC, Wang LM, Wang M, Song B, Tan S, Teng JF, et al. MicroRNA-195 downregulates Alzheimer's disease amyloid-β production by targeting BACE1. Brain Res Bull 2012;88:596−601. Available from: https://doi.org/10.1016/j.brainresbull.2012.05.018.

[104] Zhang J, Hu M, Teng Z, Tang YP, Chen C. Synaptic and cognitive improvements by inhibition of 2-AG metabolism are through upregulation of microRNA-188-3p in a mouse model of Alzheimer's disease. J Neurosci 2014;34:14919−33. Available from: https://doi.org/10.1523/JNEUROSCI.1165-14.2014.

[105] Long JM, Ray B, Lahiri DK. MicroRNA-339-5p down-regulates protein expression of β-site amyloid precursor protein-cleaving enzyme 1 (BACE1) in human primary brain cultures and is reduced in brain tissue specimens of alzheimer disease subjects. J Biol Chem 2014;289:5184−98. Available from: https://doi.org/10.1074/jbc.M113.518241.

[106] Lukiw WJ. Circular RNA (circRNA) in Alzheimer's disease (AD). Front Genet 2013;4:307. Available from: https://doi.org/10.3389/fgene.2013.00307.

[107] Long JM, Ray B, Lahiri DK. MicroRNA-153 physiologically inhibits expression of amyloid-β precursor protein in cultured human fetal brain cells and is dysregulated in a subset of Alzheimer disease patients. J Biol Chem 2012;287:31298−310. Available from: https://doi.org/10.1074/jbc.M112.366336.

[108] Zhang B, Chen CF, Wang AH, Lin QF. MiR-16 regulates cell death in Alzheimer's disease by targeting amyloid precursor protein. Eur Rev Med Pharmacol Sci 2015;19:4020−7.

[109] Zhong Z, Yuan K, Tong X, Hu J, Song Z, Zhang G, et al. MiR-16 attenuates β-amyloid-induced neurotoxicity through targeting β-site amyloid precursor protein-cleaving enzyme 1 in an Alzheimer's disease cell model. Neuroreport 2018;29:1365−72. Available from: https://doi.org/10.1097/WNR.0000000000001118.

[110] Barbato C, Pezzola S, Caggiano C, Antonelli M, Frisone P, Ciotti MT, et al. A lentiviral sponge for miR-101 regulates RanBP9 expression and amyloid precursor protein metabolism in hippocampal neurons. Front Cell Neurosci 2014;8:37. Available from: https://doi.org/10.3389/fncel.2014.00037.

[111] LaFerla FM, Kitazawa M. Antipodal effects of p25 on synaptic plasticity, learning, and memory − Too much of a good thing is bad. Neuron 2005;48:711−12. Available from: https://doi.org/10.1016/j.neuron.2005.11.022.

[112] Patrick GN, Zukerberg L, Nikolic M, De La Monte S, Dikkes P, Tsai LH. Conversion of p35 to p25 deregulates Cdk5 activity and promotes neurodegeneration. Nature 1999;402:615−22. Available from: https://doi.org/10.1038/45159.

[113] Cao X, Südhof TC. A transcriptivety active complex of APP with Fe65 and histone acetyltransferase Tip60. Science (80-) 2001;293:115−20. Available from: https://doi.org/10.1126/science.1058783.

[114] Saura CA, Choi SY, Beglopoulos V, Malkani S, Zhang D, Rao BSS, et al. Loss of presenilin function causes impairments of memory and synaptic plasticity followed by age-dependent neurodegeneration. Neuron 2004;42:23−36. Available from: https://doi.org/10.1016/S0896-6273(04)00182-5.

[115] Cohen TJ, Guo JL, Hurtado DE, Kwong LK, Mills IP, Trojanowski JQ, et al. The acetylation of tau inhibits its function and promotes pathological tau aggregation. Nat Commun 2011;2:1255. Available from: https://doi.org/10.1038/ncomms1255.

[116] Gorsky MK, Burnouf S, Dols J, Mandelkow E, Partridge L. Acetylation mimic of lysine 280 exacerbates human Tau neurotoxicity in vivo. Sci Rep 2016;6:22685. Available from: https://doi.org/10.1038/srep22685.

[117] Prince M, Bryce R, Albanese E, Wimo A, Ribeiro W, Ferri CP. The global prevalence of dementia: a systematic review and metaanalysis. Alzheimer's Dement 2013;9:63−75. Available from: https://doi.org/10.1016/j.jalz.2012.11.007 e2.

[118] Serrano-Pozo A, Frosch MP, Masliah E, Hyman BT. Neuropathological alterations in Alzheimer disease. Cold Spring Harb Perspect Med 2011;1:a006189. Available from: https://doi.org/10.1101/cshperspect.a006189.

[119] Chapuis J, Hansmannel F, Gistelinck M, Mounier A, Van Cauwenberghe C, Kolen KV, et al. Increased expression of BIN1 mediates Alzheimer genetic risk by modulating tau

[120] Hnisz D, Abraham BJ, Lee TI, Lau A, Saint-André V, Sigova AA, et al. XSuper-enhancers in the control of cell identity and disease. Cell 2013;155:934. Available from: https://doi.org/10.1016/j.cell.2013.09.053.

[121] Marambaud P, Wen PH, Dutt A, Shioi J, Takashima A, Siman R, et al. A CBP binding transcriptional repressor produced by the PS1/ε-cleavage of N-Cadherin is inhibited by PS1 FAD mutations. Cell 2003;114:635−45. Available from: https://doi.org/10.1016/j.cell.2003.08.008.

[122] Kim D, Nguyen MD, Dobbin MM, Fischer A, Sananbenesi F, Rodgers JT, et al. SIRT1 deacetylase protects against neurodegeneration in models for Alzheimer's disease and amyotrophic lateral sclerosis. EMBO J 2007;26:3169−79. Available from: https://doi.org/10.1038/sj.emboj.7601758.

[123] Rouaux C, Jokic N, Mbebi C, Boutillier S, Loeffler JP, Boutillier AL. Critical loss of CBP/p300 histone acetylase activity by caspase-6 during neurodegeneration. EMBO J 2003;22:6537−49. Available from: https://doi.org/10.1093/emboj/cdg615.

[124] Gräff J, Rei D, Guan JS, Wang WY, Seo J, Hennig KM, et al. An epigenetic blockade of cognitive functions in the neurodegenerating brain. Nature 2012;483:222−6. Available from: https://doi.org/10.1038/nature10849.

[125] Fischer A, Sananbenesi F, Wang X, Dobbin M, Tsai LH. Recovery of learning and memory is associated with chromatin remodelling. Nature 2007;447:178−82. Available from: https://doi.org/10.1038/nature05772.

[126] Lu X, Deng Y, Yu D, Cao H, Wang L, Liu L, et al. Histone acetyltransferase p300 mediates histone acetylation of PS1 and BACE1 in a cellular model of Alzheimer's disease. PLoS One 2014;9:e103067. Available from: https://doi.org/10.1371/journal.pone.0103067.

[127] Sen A, Nelson TJ, Alkon DL. ApoE4 and Aβ oligomers reduce BDNF expression via HDAC nuclear translocation. J Neurosci 2015;35:7538−51. Available from: https://doi.org/10.1523/JNEUROSCI.0260-15.2015.

[128] Anderson KW, Turko IV. Histone post-translational modifications in frontal cortex from human donors with Alzheimer's disease. Clin Proteomics 2015;12:26. Available from: https://doi.org/10.1186/s12014-015-9098-1.

[129] Liu X, Balaraman K, Lynch CC, Hebron M, Wolf C, Moussa C. Novel ubiquitin specific protease-13 inhibitors alleviate neurodegenerative pathology. Metabolites 2021;11:622. Available from: https://doi.org/10.3390/METABO11090622.

[130] Gong CX, Lidsky T, Wegiel J, Zuck L, Grundke-Iqbal I, Iqbal K. Phosphorylation of microtubule-associated protein tau is regulated by protein phosphatase 2A in mammalian brain. Implications for neurofibrillary degeneration in Alzheimer's disease. J Biol Chem 2000;275:5535−44. Available from: https://doi.org/10.1074/jbc.275.8.5535.

[131] Funk KE, Thomas SN, Schafer KN, Cooper GL, Liao Z, Clark DJ, et al. Lysine methylation is an endogenous post-translational modification of tau protein in human brain and a modulator of aggregation propensity. Biochem J 2014;462:77−88. Available from: https://doi.org/10.1042/BJ20140372.

[132] Cohen TJ, Friedmann D, Hwang AW, Marmorstein R, Lee VMY. The microtubule-associated tau protein has intrinsic acetyltransferase activity. Nat Struct Mol Biol 2013;20:756−62. Available from: https://doi.org/10.1038/nsmb.2555.

[133] Mastroeni D, Grover A, Delvaux E, Whiteside C, Coleman PD, Rogers J. Epigenetic changes in Alzheimer's disease: decrements in DNA methylation. Neurobiol Aging 2010;31:2025−37. Available from: https://doi.org/10.1016/j.neurobiolaging.2008.12.005.

[134] Chouliaras L, Mastroeni D, Delvaux E, Grover A, Kenis G, Hof PR, et al. Consistent decrease in global DNA methylation and hydroxymethylation in the hippocampus of Alzheimer's disease patients. Neurobiol Aging 2013;34:2091−9. Available from: https://doi.org/10.1016/j.neurobiolaging.2013.02.021.

[135] Coppieters N, Dieriks BV, Lill C, Faull RLM, Curtis MA, Dragunow M. Global changes in DNA methylation and hydroxymethylation in Alzheimer's disease human brain. Neurobiol Aging 2014;35:1334−44. Available from: https://doi.org/10.1016/j.neurobiolaging.2013.11.031.

[136] Wang SC, Oeize B, Schumacher A. Age-specific epigenetic drift in late-onset Alzheimer's disease. PLoS One 2008;3:2698. Available from: https://doi.org/10.1371/journal.pone.0002698.

[137] Scarpa S, Fuso A, D'Anselmi F, Cavallaro RA. Presenilin 1 gene silencing by S-adenosylmethionine: a treatment for Alzheimer disease? FEBS Lett 2003;541:145−8. Available from: https://doi.org/10.1016/S0014-5793(03)00277-1.

[138] Scarpa S, Cavallaro RA, D'Anselmi F, Fuso A. Gene silencing through methylation: an epigenetic intervention on Alzheimer disease. J Alzheimer's Dis 2006;9:407−14. Available from: https://doi.org/10.3233/JAD-2006-9406.

[139] Lunnon K, Smith R, Hannon E, De Jager PL, Srivastava G, Volta M, et al. Methylomic profiling implicates cortical deregulation of ANK1 in Alzheimer's disease. Nat Neurosci 2014;17:1164−70. Available from: https://doi.org/10.1038/nn.3782.

[140] De Jager PL, Srivastava G, Lunnon K, Burgess J, Schalkwyk LC, Yu L, et al. Alzheimer's disease: early alterations in brain DNA methylation at ANK1, BIN1, RHBDF2 and other loci. Nat Neurosci 2014;17:1156−63. Available from: https://doi.org/10.1038/nn.3786.

[141] Shu L, Sun W, Li L, Xu Z, Lin L, Xie P, et al. Genome-wide alteration of 5-hydroxymenthylcytosine in a mouse model of Alzheimer's disease. BMC Genomics 2016;17. Available from: https://doi.org/10.1186/s12864-016-2731-1.

[142] Mercado-Pimentel ME, Onyeagucha BC, Li Q, Pimentel AC, Jandova J, Nelson MA. The S100P/RAGE signaling pathway regulates expression of microRNA-21 in colon cancer cells. FEBS Lett 2015;589:2388−93. Available from: https://doi.org/10.1016/j.febslet.2015.07.010.

[143] Lukiw WJ. microRNA-146a signaling in Alzheimer's Disease (AD) and Prion Disease (PrD). Front Neurol 2020;11:462. Available from: https://doi.org/10.3389/fneur.2020.00462.

[144] Santa-Maria I, Alaniz ME, Renwick N, Cela C, Fulga TA, Van Vactor D, et al. Dysregulation of microRNA-219 promotes neurodegeneration through post-transcriptional regulation of tau. J Clin Invest 2015;125:681−6. Available from: https://doi.org/10.1172/JCI78421.

[145] Banzhaf-Strathmann J, Benito E, May S, Arzberger T, Tahirovic S, Kretzschmar H, et al. Micro RNA -125b induces tau hyperphosphorylation and cognitive deficits in Alzheimer's disease. EMBO J 2014;33:1667−80. Available from: https://doi.org/10.15252/embj.201387576.

[146] Absalon S, Kochanek DM, Raghavan V, Krichevsky AM. MiR-26b, upregulated in Alzheimer's disease, activates cell cycle entry, Tau-phosphorylation, and apoptosis in postmitotic neurons. J Neurosci 2013;33:14645−59. Available from: https://doi.org/10.1523/JNEUROSCI.1327-13.2013.

[147] Lee ST, Chu K, Jung KH, Kim JH, Huh JY, Yoon H, et al. MiR-206 regulates brain-derived neurotrophic factor in Alzheimer disease model. Ann Neurol 2012;72:269−77. Available from: https://doi.org/10.1002/ana.23588.

[148] Li JJ, Dolios G, Wang R, Liao FF. Soluble beta-amyloid peptides, but not insoluble fibrils, have specific effect on neuronal MicroRNA expression. PLoS One 2014;9:e90770. Available from: https://doi.org/10.1371/journal.pone.0090770.

[149] Lukiw WJ, Zhao Y, Jian GC. An NF-κB-sensitive micro RNA-146a-mediated inflammatory circuit in alzheimer disease and in stressed human brain cells. J Biol Chem 2008;283:31315—22. Available from: https://doi.org/10.1074/jbc.M805371200.

[150] Wey HY, Gilbert TM, Zürcher NR, She A, Bhanot A, Taillon BD, et al. Insights into neuroepigenetics through human histone deacetylase PET imaging. Sci Transl Med 2016;8. Available from: https://doi.org/10.1126/scitranslmed.aaf7551 351ra106-351ra106.

[151] Tariot PN, Schneider LS, Cummings J, Thomas RG, Raman R, Jakimovich LJ, et al. Chronic divalproex sodium to attenuate agitation and clinical progression of Alzheimer disease. Arch Gen Psychiatry 2011;68:853—61. Available from: https://doi.org/10.1001/archgenpsychiatry.2011.72.

[152] Falkenberg KJ, Johnstone RW. Histone deacetylases and their inhibitors in cancer, neurological diseases and immune disorders. Nat Rev Drug Discov 2014;13:673—91. Available from: https://doi.org/10.1038/nrd4360.

[153] Joy Dauncey M. Genomic and epigenomic insights into nutrition and brain disorders. Nutrients 2013;5:887—914. Available from: https://doi.org/10.3390/nu5030887.

[154] Dauncey MJ. Nutrition, the brain and cognitive decline: insights from epigenetics. Eur J Clin Nutr 2014;68:1179—85. Available from: https://doi.org/10.1038/ejcn.2014.173.

[155] Dauncey MJ. Nutrition, genes, and neuroscience: implications for development, health, and disease. Diet and exercise in cognitive function and neurological diseases. Wiley Blackwell; 2015. p. 1—13. Available from: https://doi.org/10.1002/9781118840634.ch1.

[156] Pizzorusso T, Tognini P. Interplay between metabolism, nutrition and epigenetics in shaping brain dna methylation, neural function and behavior. Genes (Basel) 2020;11:1—18. Available from: https://doi.org/10.3390/genes11070742.

[157] Weissman MM. Cross-national epidemiology of major depression and bipolar disorder. JAMA J Am Med Assoc 1996;276:293. Available from: https://doi.org/10.1001/jama.1996.03540040037030.

[158] Kessler RC, Bromet EJ. The epidemiology of depression across cultures. Annu Rev Public Health 2013;34:119—38. Available from: https://doi.org/10.1146/annurev-publhealth-031912-114409.

[159] Berton O, Nestler EJ. New approaches to antidepressant drug discovery: beyond monoamines. Nat Rev Neurosci 2006;7:137—51. Available from: https://doi.org/10.1038/nrn1846.

[160] Tsankova N, Renthal W, Kumar A, Nestler EJ. Epigenetic regulation in psychiatric disorders. Nat Rev Neurosci 2007;8:355—67. Available from: https://doi.org/10.1038/nrn2132.

[161] Kronman H, Torres-Berrío A, Sidoli S, Issler O, Godino A, Ramakrishnan A, et al. Long-term behavioral and cell-type-specific molecular effects of early life stress are mediated by H3K79me2 dynamics in medium spiny neurons. Nat Neurosci 2021;24:667—76. Available from: https://doi.org/10.1038/s41593-021-00814-8.

[162] Menke A, Binder EB. Epigenetic alterations in depression and antidepressant treatment. Dialogues Clin Neurosci 2014;16:395—404. Available from: https://doi.org/10.31887/dcns.2014.16.3/amenke.

[163] Uchida S, Hara K, Kobayashi A, Otsuki K, Yamagata H, Hobara T, et al. Epigenetic status of Gdnf in the ventral striatum determines susceptibility and adaptation to daily stressful events. Neuron 2011;69:359—72. Available from: https://doi.org/10.1016/j.neuron.2010.12.023.

[164] Melas PA, Rogdaki M, Lennartsson A, Björk K, Qi H, Witasp A, et al. Antidepressant treatment is associated with epigenetic alterations in the promoter of P11 in a genetic model of depression. Int J Neuropsychopharmacol 2012;15:669—79. Available from: https://doi.org/10.1017/S1461145711000940.

[165] Khoury L, Tang YL, Bradley B, Cubells JF, Ressler KJ. Substance use, childhood traumatic experience, and posttraumatic stress disorder in an urban civilian population. Depress Anxiety 2010;27:1077—86. Available from: https://doi.org/10.1002/da.20751.

[166] Dempster EL, Wong CCY, Lester KJ, Burrage J, Gregory AM, Mill J, et al. Genome-wide methylomic analysis of monozygotic twins discordant for adolescent depression. Biol Psychiatry 2014;76:977—83. Available from: https://doi.org/10.1016/j.biopsych.2014.04.013.

[167] Tsankova NM, Berton O, Renthal W, Kumar A, Neve RL, Nestler EJ. Sustained hippocampal chromatin regulation in a mouse model of depression and antidepressant action. Nat Neurosci 2006;9:519—25. Available from: https://doi.org/10.1038/nn1659.

[168] Hunter RG, McCarthy KJ, Milne TA, Pfaff DW, McEwen BS. Regulation of hippocampal H3 histone methylation by acute and chronic stress. Proc Natl Acad Sci U S A 2009;106:20912. Available from: https://doi.org/10.1073/PNAS.0911143106.

[169] Wilkinson MB, Xiao G, Kumar A, LaPlant Q, Renthal W, Sikder D, et al. Imipramine treatment and resiliency exhibit similar chromatin regulation in the mouse nucleus accumbens in depression models. J Neurosci 2009;29:7820—32. Available from: https://doi.org/10.1523/JNEUROSCI.0932-09.2009.

[170] Renthal W, Maze I, Krishnan V, Covington HE, Xiao G, Kumar A, et al. Histone deacetylase 5 epigenetically controls behavioral adaptations to chronic emotional stimuli. Neuron 2007;56:517—29. Available from: https://doi.org/10.1016/j.neuron.2007.09.032.

[171] Wan Q, Gao K, Rong H, Wu M, Wang H, Wang X, et al. Histone modifications of the Crhr1 gene in a rat model of depression following chronic stress. Behav Brain Res 2014;271:1—6. Available from: https://doi.org/10.1016/j.bbr.2014.05.031.

[172] Domschke K, Tidow N, Schwarte K, Deckert J, Lesch KP, Arolt V, et al. Serotonin transporter gene hypomethylation predicts impaired antidepressant treatment response. Int J Neuropsychopharmacol 2014;17:1167—76. Available from: https://doi.org/10.1017/S146114571400039X.

[173] Cassel S, Carouge D, Gensburger C, Anglard P, Burgun C, Dietrich JB, et al. Fluoxetine and cocaine induce the epigenetic factors MeCP2 and MBD1 in adult rat brain. Mol Pharmacol 2006;70:487—92. Available from: https://doi.org/10.1124/mol.106.022301.

[174] Colucci-D'Amato L, Speranza L, Volpicelli F. Neurotrophic factor BDNF, physiological functions and therapeutic potential in depression, neurodegeneration and brain cancer. Int J Mol Sci 2020;21:1—29. Available from: https://doi.org/10.3390/IJMS21207777.

[175] Keller S, Sarchiapone M, Zarrilli F, Videtič A, Ferraro A, Carli V, et al. Increased BDNF promoter methylation in the Wernicke area of suicide subjects. Arch Gen Psychiatry 2010;67:258—67. Available from: https://doi.org/10.1001/archgenpsychiatry.2010.9.

[176] Kang HJ, Kim JM, Lee JY, Kim SY, Bae KY, Kim SW, et al. BDNF promoter methylation and suicidal behavior in depressive patients. J Affect Disord 2013;151:679—85. Available from: https://doi.org/10.1016/j.jad.2013.08.001.

[177] Perroud N, Salzmann A, Prada P, Nicastro R, Hoeppli ME, Furrer S, et al. Response to psychotherapy in borderline personality disorder and methylation status of the BDNF gene. Transl Psychiatry 2013;3:207. Available from: https://doi.org/10.1038/tp.2012.140.

[178] Zachariah RM, Olson CO, Ezeonwuka C, Rastegar M. Novel MeCP2 isoform-specific antibody reveals the endogenous MeCP2E1 expression in murine brain, primary neurons and astrocytes. PLoS One 2012;7:49763. Available from: https://doi.org/10.1371/journal.pone.0049763.

[179] Zhao X, Ueba T, Christie BR, Barkho B, McConnell MJ, Nakashima K, et al. Mice lacking methyl-CpG binding protein

1 have deficits in adult neurogenesis and hippocampal function. Proc Natl Acad Sci U S A 2003;100:6777–82. Available from: https://doi.org/10.1073/pnas.1131928100.

[180] Poulter MO, Du L, Weaver ICG, Palkovits M, Faludi G, Merali Z, et al. GABAA receptor promoter hypermethylation in suicide brain: implications for the involvement of epigenetic processes. Biol Psychiatry 2008;64:645–52. Available from: https://doi.org/10.1016/j.biopsych.2008.05.028.

[181] Feng J, Pena CJ, Purushothaman I, Engmann O, Walker D, Brown AN, et al. Tet1 in nucleus accumbens opposes depression- and anxiety-like behaviors. Neuropsychopharmacology 2017;42:1657–69. Available from: https://doi.org/10.1038/npp.2017.6.

[182] Gassen NC, Fries GR, Zannas AS, Hartmann J, Zschocke J, Hafner K, et al. Chaperoning epigenetics: FKBP51 decreases the activity of DNMT1 and mediates epigenetic effects of the antidepressant paroxetine. Sci Signal 2015;8:ra119. Available from: https://doi.org/10.1126/scisignal.aac7695 –ra119.

[183] Fuchikami M, Morinobu S, Segawa M, Okamoto Y, Yamawaki S, Ozaki N, et al. DNA methylation profiles of the brain-derived neurotrophic factor (BDNF) gene as a potent diagnostic biomarker in major depression. PLoS One 2011;6:23881. Available from: https://doi.org/10.1371/journal.pone.0023881.

[184] Januar V, Ancelin ML, Ritchie K, Saffery R, Ryan J. BDNF promoter methylation and genetic variation in late-life depression. Transl Psychiatry 2015;5:619. Available from: https://doi.org/10.1038/tp.2015.114.

[185] Uddin M, Koenen KC, Aiello AE, Wildman DE, De Los Santos R, Galea S. Epigenetic and inflammatory marker profiles associated with depression in a community-based epidemiologic sample. Psychol Med 2011;41:997–1007. Available from: https://doi.org/10.1017/S0033291710001674.

[186] Byrne EM, Carrillo-Roa T, Henders AK, Bowdler L, McRae AF, Heath AC, et al. Monozygotic twins affected with major depressive disorder have greater variance in methylation than their unaffected co-twin. Transl Psychiatry 2013;3:269. Available from: https://doi.org/10.1038/tp.2013.45.

[187] Davies MN, Krause L, Bell JT, Gao F, Ward KJ, Wu H, et al. Hypermethylation in the ZBTB20 gene is associated with major depressive disorder. Genome Biol 2014;15:1–12. Available from: https://doi.org/10.1186/gb-2014-15-4-r56.

[188] Córdova-Palomera A, Fatjó-Vilas M, Gastó C, Navarro V, Krebs MO, Fañanás L. Genome-wide methylation study on depression: differential methylation and variable methylation in monozygotic twins. Transl Psychiatry 2015;5:e557. Available from: https://doi.org/10.1038/tp.2015.49 –e557.

[189] Babenko O, Golubov A, Ilnytskyy Y, Kovalchuk I, Metz GA. Genomic and epigenomic responses to chronic stress involve miRNA-mediated programming. PLoS One 2012;7:e29441. Available from: https://doi.org/10.1371/journal.pone.0029441.

[190] Mazzelli M, Maj C, Mariani N, Mora C, Begni V, Pariante CM, et al. The long-term effects of early life stress on the modulation of miR-19 levels. Front Psychiatry 2020;11. Available from: https://doi.org/10.3389/fpsyt.2020.00389.

[191] Volk N, Paul ED, Haramati S, Eitan C, Fields BKK, Zwang R, et al. MicroRNA-19b associates with Ago2 in the amygdala following chronic stress and regulates the adrenergic receptor beta 1. J Neurosci 2014;34:15070–82. Available from: https://doi.org/10.1523/JNEUROSCI.0855-14.2014.

[192] Volk N, Pape JC, Engel M, Zannas AS, Cattane N, Cattaneo A, et al. Amygdalar microRNA-15a is essential for coping with chronic stress. Cell Rep 2016;17:1882–91. Available from: https://doi.org/10.1016/j.celrep.2016.10.038.

[193] Issler O, Haramati S, Paul ED, Maeno H, Navon I, Zwang R, et al. MicroRNA 135 is essential for chronic stress resiliency, antidepressant efficacy, and intact serotonergic activity. Neuron 2014;83:344–60. Available from: https://doi.org/10.1016/j.neuron.2014.05.042.

[194] Woodbury ME, Freilich RW, Cheng CJ, Asai H, Ikezu S, Boucher JD, et al. miR-155 is essential for inflammation-induced hippocampal neurogenic dysfunction. J Neurosci 2015;35:9764–81. Available from: https://doi.org/10.1523/JNEUROSCI.4790-14.2015.

[195] Hutchison ER, Kawamoto EM, Taub DD, Lal A, Abdelmohsen K, Zhang Y, et al. Evidence for miR-181 involvement in neuroinflammatory responses of astrocytes. Glia 2013;61:1018–28. Available from: https://doi.org/10.1002/glia.22483.

[196] Dwivedi Y. Emerging role of microRNAs in major depressive disorder: diagnosis and therapeutic implications. Dialogues Clin Neurosci 2014;16:43–61. Available from: https://doi.org/10.31887/dcns.2014.16.1/ydwivedi.

[197] Baudry A, Mouillet-Richard S, Schneider B, Launay JM, Kellermann O. MiR-16 targets the serotonin transporter: a new facet for adaptive responses to antidepressants. Science (80-) 2010;329:1537–41. Available from: https://doi.org/10.1126/science.1193692.

[198] Lopez JP, Lim R, Cruceanu C, Crapper L, Fasano C, Labonte B, et al. MiR-1202 is a primate-specific and brain-enriched microRNA involved in major depression and antidepressant treatment. Nat Med 2014;20:764–8. Available from: https://doi.org/10.1038/nm.3582.

[199] Franklin TB, Russig H, Weiss IC, Grff J, Linder N, Michalon A, et al. Epigenetic transmission of the impact of early stress across generations. Biol Psychiatry 2010;68:408–15. Available from: https://doi.org/10.1016/j.biopsych.2010.05.036.

[200] Kundakovic M, Champagne FA. Early life experience, epigenetics, and the developing brain. Neuropsychopharmacology 2015;40:141–53. Available from: https://doi.org/10.1038/npp.2014.140.

[201] Weaver ICG, Cervoni N, Champagne FA, D'Alessio AC, Sharma S, Seckl JR, et al. Epigenetic programming by maternal behavior. Nat Neurosci 2004;7:847–54. Available from: https://doi.org/10.1038/nn1276.

[202] McGowan PO, Sasaki A, D'Alessio AC, Dymov S, Labonté B, Szyf M, et al. Epigenetic regulation of the glucocorticoid receptor in human brain associates with childhood abuse. Nat Neurosci 2009;12:342–8. Available from: https://doi.org/10.1038/nn.2270.

[203] Suderman M, McGowan PO, Sasaki A, Huang TCT, Hallett MT, Meaney MJ, et al. Conserved epigenetic sensitivity to early life experience in the rat and human hippocampus. Proc Natl Acad Sci U S A 2012;109:17266–72. Available from: https://doi.org/10.1073/pnas.1121260109.

[204] Gapp K, Jawaid A, Sarkies P, Bohacek J, Pelczar P, Prados J, et al. Implication of sperm RNAs in transgenerational inheritance of the effects of early trauma in mice. Nat Neurosci 2014;17:667–9. Available from: https://doi.org/10.1038/nn.3695.

[205] Hammock EAD. Developmental perspectives on oxytocin and vasopressin. Neuropsychopharmacology 2015;40:24–42. Available from: https://doi.org/10.1038/npp.2014.120.

[206] Kember RL, Dempster EL, Lee THA, Schalkwyk LC, Mill J, Fernandes C. Maternal separation is associated with strain-specific responses to stress and epigenetic alterations to Nr3c1, Avp, and Nr4a1 in mouse. Brain Behav 2012;2:455–67. Available from: https://doi.org/10.1002/brb3.69.

[207] Meyer-Lindenberg A, Domes G, Kirsch P, Heinrichs M. Oxytocin and vasopressin in the human brain: social neuropeptides for translational medicine. Nat Rev Neurosci 2011;12:524–38. Available from: https://doi.org/10.1038/nrn3044.

[208] Labonté B, Suderman M, Maussion G, Navaro L, Yerko V, Mahar I, et al. Genome-wide epigenetic regulation by early-

life trauma. Arch Gen Psychiatry 2012;69:722−31. Available from: https://doi.org/10.1001/archgenpsychiatry.2011.2287.

[209] Massart R, Suderman M, Provencal N, Yi C, Bennett AJ, Suomi S, et al. Hydroxymethylation and DNA methylation profiles in the prefrontal cortex of the non-human primate rhesus macaque and the impact of maternal deprivation on hydroxymethylation. Neuroscience 2014;268:139−48. Available from: https://doi.org/10.1016/j.neuroscience.2014.03.021.

[210] Bai M, Zhu X, Zhang Y, Zhang S, Zhang L, Xue L, et al. Abnormal hippocampal BDNF and miR-16 expression is associated with depression-like behaviors induced by stress during early life. PLoS One 2012;7:e46921. Available from: https://doi.org/10.1371/journal.pone.0046921.

[211] Zhang Y, Zhu X, Bai M, Zhang L, Xue L, Yi J. Maternal deprivation enhances behavioral vulnerability to stress associated with miR-504 expression in nucleus accumbens of rats. PLoS One 2013;8:69934. Available from: https://doi.org/10.1371/journal.pone.0069934.

[212] Weaver ICG, Champagne FA, Brown SE, Dymov S, Sharma S, Meaney MJ, et al. Reversal of maternal programming of stress responses in adult offspring through methyl supplementation: altering epigenetic marking later in life. J Neurosci 2005;25:11045−54. Available from: https://doi.org/10.1523/JNEUROSCI.3652-05.2005.

[213] Weaver ICG, Meaney MJ, Szyf M. Maternal care effects on the hippocampal transcriptome and anxiety-mediated behaviors in the offspring that are reversible in adulthood. Proc Natl Acad Sci U S A 2006;103:3480−5. Available from: https://doi.org/10.1073/pnas.0507526103.

[214] Gapp K, Bohacek J, Grossmann J, Brunner AM, Manuella F, Nanni P, et al. Potential of environmental enrichment to prevent transgenerational effects of paternal trauma. Neuropsychopharmacology 2016;41:2749−58. Available from: https://doi.org/10.1038/npp.2016.87.

[215] Weber-Stadlbauer U, Richetto J, Labouesse MA, Bohacek J, Mansuy IM, Meyer U. Transgenerational transmission and modification of pathological traits induced by prenatal immune activation. Mol Psychiatry 2017;22:102−12. Available from: https://doi.org/10.1038/mp.2016.41.

[216] Lee DY, Moon J, Lee ST, Jung KH, Park DK, Yoo JS, et al. Distinct expression of long non-coding RNAs in an Alzheimer's disease model. J Alzheimer's Dis 2015;45:837−49. Available from: https://doi.org/10.3233/JAD-142919.

[217] Huang X, Li LY, Shi MY, Lin JJ. The link between long non-coding RNAs and depression. Prog Neuropsychopharmacol Biol Psychiatry 2017;73:73−8. Available from: https://doi.org/10.1016/j.pnpbp.2016.06.004.

[218] Rajasethupathy P, Antonov I, Sheridan R, Frey S, Sander C, Tuschl T, et al. A role for neuronal piRNAs in the epigenetic control of memory-related synaptic plasticity. Cell 2012;149:693−707. Available from: https://doi.org/10.1016/j.cell.2012.02.057.

[219] Ashe A, Sapetschnig A, Weick EM, Mitchell J, Bagijn MP, Cording AC, et al. PiRNAs can trigger a multigenerational epigenetic memory in the germline of C. elegans. Cell 2012;150:88−99. Available from: https://doi.org/10.1016/j.cell.2012.06.018.

[220] Szyf M. Prospects for the development of epigenetic drugs for CNS conditions. Nat Rev Drug Discov 2015;14:461−74. Available from: https://doi.org/10.1038/nrd4580.

[221] Woldemichael BT, Jawaid A, Kremer EA, Gaur N, Krol J, Marchais A, et al. The microRNA cluster miR-183/96/182 contributes to long-term memory in a protein phosphatase 1-dependent manner. Nat Commun 2016;7, 12594. Available from: https://doi.org/10.1038/ncomms12594.

[222] Li Y, Li S, Yan J, Wang D, Yin R, Zhao L, et al. MiR-182 (microRNA-182) suppression in the hippocampus evokes antidepressant-like effects in rats. Prog Neuropsychopharmacol. Biol Psychiatry 2016;65:96−103. Available from: https://doi.org/10.1016/j.pnpbp.2015.09.004.

[223] Bondarev AD, Attwood MM, Jonsson J, Chubarev VN, Tarasov VV, Schiöth HB. Recent developments of HDAC inhibitors: emerging indications and novel molecules. Br J Clin Pharmacol 2021;. Available from: https://doi.org/10.1111/BCP.14889.

CHAPTER 38

Epigenetics of Metabolic Diseases

Linn Gillberg[1] and Line Hjort[2]

[1]Department of Biomedical Sciences, University of Copenhagen, Copenhagen, Denmark [2]Department of Endocrinology, Copenhagen University Hospital, Copenhagen, Denmark

OUTLINE

Introduction	761	Epigenetic Features of Metabolic Diseases	769
Impact of Age and Lifestyle Factors on the Epigenome	762	Metabolic Syndrome	769
		Obesity	769
Age and the Epigenome	762	Type 2 Diabetes	770
Diet, Metabolism, and the Epigenome	763	Conclusions	771
Physical Activity, Inactivity, and the Epigenome	764	Abbreviations	772
Epigenetic Memory, Prenatal Exposure, and Risk of Metabolic Disease	766	References	772
Thrifty Phenotype and Birthweight	766		
Diabetes in Pregnancy	767		

INTRODUCTION

A growing amount of literature suggests that epigenetic mechanisms, in a complex interplay between genetic and environmental factors, play a role in the development of metabolic diseases, such as obesity, metabolic syndrome, type 2 diabetes (T2D), and cardiovascular disease (CVD). Some of the first data to indicate that both environmental exposures and age could have an effect on the epigenetic profile of our genome was obtained from a study of monozygotic twins. This study showed that despite sharing identical genomes, the patterns of DNA methylation and histone acetylation in several tissues were more diverse in twin pairs who were older, had different lifestyles, and had spent less of their lives together [1]. Other early studies suggested that epigenetic mechanisms are involved in the process of aging by affecting genes important for mitochondrial oxidative phosphorylation, including *COX7A1* [2]. Thereafter, numerous studies show highly significant associations between epigenetic modifications and age or lifestyle factors. Some of the most convincing evidence of how epigenetic modifications provide a link between adverse environmental factors early in life, and increased risk of metabolic disease later in life, comes from studies of adult individuals who were born by mothers exposed to famine or gestational diabetes during pregnancy [3–5].

The prevalence of T2D and the metabolic syndrome has risen globally in parallel with the obesity epidemic over the past decades, and are major threats to global health [6–8]. The complexity of metabolic disorders, such as T2D is reflected by effects on various tissues and organs including the liver, intestines, the brain, skeletal muscle, adipose tissue, and endocrine organs, such as the pancreas. Moreover, metabolic diseases have heterogeneous etiologies involving interactions between a wide range of factors of genetic, developmental, and environmental origin [7,9]. However, much of the etiology and pathogenesis of metabolic diseases remain unknown. Consequently, further understanding of the complex interplay between tissue-specific epigenetic mechanisms, environmental factors and the genome may provide new clues as to why some individuals develop metabolic diseases whereas others do not.

In this chapter we aim to provide an overview of: [1] how our age and lifestyle influences metabolic traits and the epigenome, [2] how adverse fetal exposures can affect epigenetic modifications and thereby increase the risk of metabolic disease later in life, and, [3] how the epigenome is affected in some of the most common metabolic diseases including the metabolic syndrome, obesity, and T2D.

IMPACT OF AGE AND LIFESTYLE FACTORS ON THE EPIGENOME

Metabolic syndrome, obesity, T2D, and CVD are slowly progressive metabolic diseases that increase in prevalence with age, excess weight, and lifestyle related factors, such as diet, smoking, stress, toxins, and physical inactivity. During the last years, those risk factors for metabolic diseases have been associated with altered epigenomes in several human tissues (Fig. 38.1).

Age and the Epigenome

Aging is characterized by a gradual decline in cellular functions and associated with increased risk of metabolic disorders, neurodegenerative diseases, and several cancers. Epigenetic remodeling is closely related to aging, and it is believed that epigenetic information integrates environmental input with the genetic background in the process of aging [10].

The first study to demonstrate that age and environmental factors can have a lifelong impact on the phenotype by altering epigenetic patterns across multiple human tissues was conducted in 2005 by Fraga et al. [1]. Today, numerous studies demonstrate highly significant correlations between age and methylation at specific cytosines in the human genome of various tissues relevant to metabolic disease, suggesting that epigenetics are indeed involved in the age-related gradual decline of cellular functions [11–16]. One of the

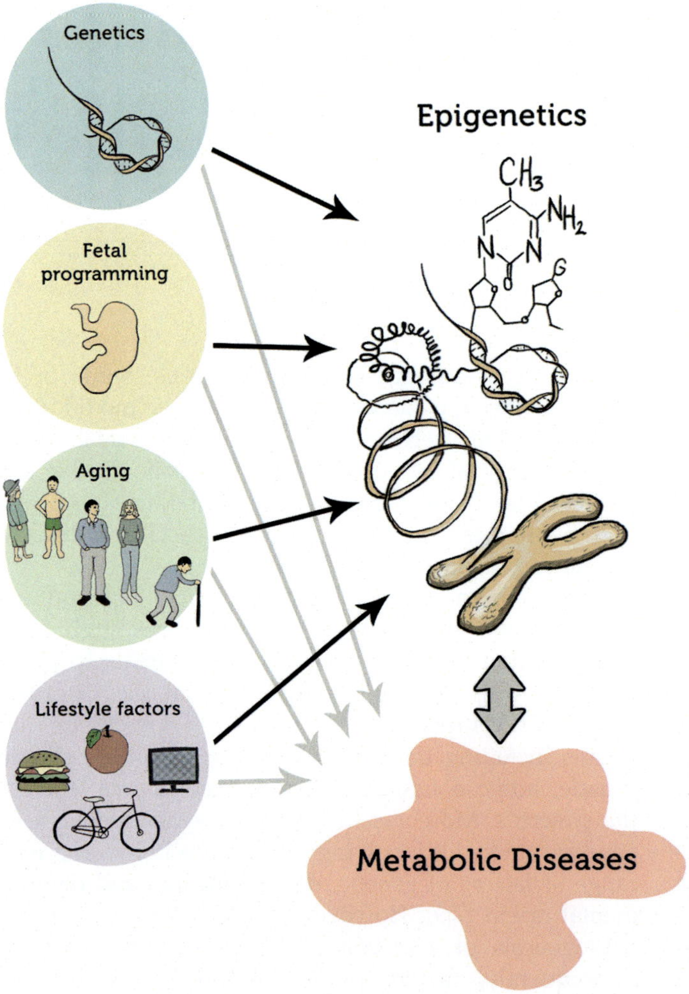

FIGURE 38.1 Epigenetic involvement in the development of metabolic diseases. Epigenetics as a central co-actor in the complex interplay between genetics, fetal programming, age, and lifestyle-related factors in the development of metabolic diseases, such as metabolic syndrome, obesity, and T2D.

studies supporting this was performed by Bigot et al. who showed that the increased methylation of the muscle stem-cell pool regulator SPRY1 with age might be a way for the body to decrease the regenerative response and thereby inhibit the replenishment of the muscle stem-cell pool with old age [15].

Several independent but highly overlapping predictive models of biological aging and prediction of lifespan have been developed from age-associated epigenomic data derived from several human cell types [11,14,17–22]. These "epigenetic age clocks" may be promising markers of individual (biological) aging and increase our understanding of why some individuals develop complex age-related diseases. Furthermore, many age-associated alterations in DNA methylations are positioned in CpG islands of genes important for human metabolism. One example is the Krüppel-like factor 14 gene (KLF14) in which highly significant associations between age and site-specific DNA methylation have been identified in human blood [11,14,23], adipose tissue [13], pancreatic islets [12], saliva [23], buccal swab [23,24], brain and muscle tissue [24,25]. In this case, age-associated DNA methylation of disease-relevant genes in blood, saliva, and buccal swab reflects the DNA methylation in other tissues and might, besides being potential biomarker candidates, be directly involved in age-related pathologies that increase the risk of metabolic diseases.

Age-related changes in cell and organ functions have also been linked to microRNAs (miRNAs) and histone modifications, especially the ones connected with the sirtuin (SIRT) family of proteins and histone methylation [26]. The multi-target deacetylases SIRT1 and SIRT6 are located in the nucleus where they modulate chromatin structure and the activity of regulatory proteins involved in aging [26]. SIRT6 functions as a histone H3K9ac deacetylase that modulates telomeric chromatin and thereby prevents the cells from telomere dysfunction and premature cellular senescence [27]. Patients with the accelerated aging disease Werner syndrome have abnormal telomere structures that resembles the ones of SIRT6 depleted cells [27]. Overexpression of Sirt6 in transgenic male, but not female, mice caused a significant increase in the average lifespan together with deceased serum levels of insulin-like growth factor 1 (IGF1) and altered phosphorylation levels of components in IGF1 signaling, which is a key pathway in the regulation of lifespan [28].

In a study where more than 800 miRNAs were profiled in human peripheral blood mononuclear cells, the expression of nine miRNAs were identified to be significantly lower in older individuals [29]. Among these were miR-221 whose potential target PI3-kinase, which is an integrator of signaling pathways promoting tumorigenesis, was upregulated with advancing age, supporting a possible role for miRNAs in human aging and age-related diseases.

Diet, Metabolism, and the Epigenome

Metabolic regulation ultimately functions at a molecular level. Chromatin, which is under genetic and epigenetic control, facilitates transcriptional regulation across cell types, and increasing evidence point to a regulatory interface between chromatin and energy homeostasis [30]. The influence of dietary factors on epigenetic regulation provides an explanation of how lifestyle interacts with the epigenome to affect human metabolism in health and disease.

Dietary nutrients provide substrates for epigenetic building blocks (i.e., methyl and acetyl groups) and serve as cofactors for epigenetic enzymes. Thus, our diet can influence epigenetic patterns both by changing the availability of epigenetic substrates and by altering the activity of the enzymes that are responsible for addition or removal of epigenetic marks. For example, S-adenosyl methionine (SAM) is the methyl donor substrate that provides methyl groups to both DNA and histone methyltransferases. Some of the most important methyl donors that provide methyl groups to SAM are the vitamins, folate and cobalamin (vitamin B12), the nutrient choline, and the amino acid methionine [31]. Interestingly, methionine restriction without calorie restriction has been associated with longevity in rodents. In a study of Wistar rats a 7-week isocaloric diet with 40% methionine restriction caused significantly decreased global methylation of genomic DNA from the rat heart as well as decreased mitochondrial ROS production and lowered oxidative damage to DNA and proteins of the heart [32]. The authors suggested that beneficial effects caused by the methionine dietary restriction on the heart might be involved in the longevity of rats. In addition to methyl donors bioactive food components, such as the antioxidant isoflavone genistein have the potential to alter DNA methylation, partially through a direct inactivation of DNA methyltransferases [33]. Similarly, the well-known antioxidant vitamin C, which has been found lower in plasma from individuals with diabetes [34], is a cofactor for the dioxygenase enzymes that catalyze active demethylation of both DNA (Ten—eleven translocation enzymes) and histones (Jumonji C domain-containing histone demethylases) [35–37].

Many of the aging-related functions altered by calorie restriction are regulated by the sirtuin SIRT1, which has been shown to silence transcription via deacetylation of histones (e.g., H4K16ac, H3K9ac, H1K26ac) [38]. It also causes decreased levels of the transcriptionally active chromatin mark H3K79me2 and increases levels of transcriptionally repressed chromatin marks H3K9me3 and H4K20me1 [39]. Interestingly, SIRT1 affects differentiation of adipocytes, myocytes and metabolism via modulation of PGC-1α and PPARγ activity, and its activity might thereby provide a link to calorie restricted diets and increased lifespan [26].

The impact of methyl donors, bioactive food components, and nutrients on epigenetic modifications has frequently been studied in a setting where the impact of maternal diet on DNA methylation in the offspring is investigated. For example, dietary supplementation with the omega-3 fatty acid docosahexaenoic acid during pregnancy caused increased DNA methylation levels of the imprinted gene insulin-like growth factor 2 (*IGF2*), which is important for growth and development, in cord blood from the infants [40]. The most visual example of how methyl donor availability during gestation is crucial for the establishment of DNA methylation patterns in the offspring comes from Agouti^{vy} mice. Offspring of mothers fed a diet rich in methyl donors received a wild-type coat color (brown), whereas offspring of mothers fed a standard diet developed a yellow coat color [41]. The brown coat color was due to DNA hypermethylation of a long terminal repeat controlling expression of the *agouti* gene that caused decreased gene expression of *agouti* and thereby altered coat pigmentation.

One of the hitherto most consistent example of epigenetic alterations in adulthood caused by environmental exposures during prenatal, childhood, or adult life, is DNA promoter methylation of the metabolic master regulator peroxisome proliferator-activated receptor gamma coactivator 1 alpha (gene, *PPARGC1A*; protein, PGC-1α). The broad role of PGC-1α comprises its regulation of genes encoding proteins required for energy metabolism, mitochondrial biogenesis, and adaptive thermogenesis [42]. In 2003, the *PPARGC1A* expression was found to be significantly downregulated in skeletal muscle from prediabetic and diabetic subjects [43]. Site-specific DNA promoter methylation of *PPARGC1A* has been studied in both adipose tissue and skeletal muscle from healthy young men in response to a 5-day high-fat overfeeding (HFO) diet [44,45]. Interestingly, in both tissues the promoter methylation of *PPARGC1A* increased significantly with the HFO on two of the four CpG sites studied, whereas the *PPARGC1A* mRNA expression did not change significantly (Table 38.1). From the same study cohort, the plasticity of DNA methylation in adipose tissue in response to 5-days HFO was investigated on a genome-wide scale using Illumina's 450 K arrays [59]. Here, the insulin responsive glucose transporter 4 (*GLUT4*) and the obesity-related gene *CDK5* were among the genes that changed both DNA methylation and its corresponding gene expression in response to overfeeding [59]. Although the genome-wide changes of adipose tissue DNA methylation were not particularly dynamic in response to the short-term HFO in this study, more extensive DNA methylation changes have been reported in skeletal muscle after short-term HFO [60]. This could indicate that DNA methylation in skeletal muscle is more plastic in response to dietary changes than DNA methylation in adipose tissue [59,60].

Beneficial effects of intermittent fasting and diet restriction on healthy aging and metabolic disease prevention might be related to epigenetic mechanisms [61]. In a study investigating the effect of 36 h fasting on DNA methylation in muscle and fat tissue in healthy young men, only nominal methylation differences were observed in skeletal muscle and adipose tissue after 36 h fasting compared to after a control diet, in genes with significant mRNA changes [62]. In the promoter region of the important adipokines leptin (*LEP*) and adiponectin (*ADIPOQ*), methylation levels were slightly but significantly increased after 36 h fasting compared to after a control diet in subcutaneous adipose tissue (SAT) from young men born with normal birthweights [63].

In a high-fat diet-induced obesity mouse model, altered histone methylation was found in paternal spermatozoa at genes important for embryogenesis, suggesting that dietary exposures can modulate histone modifications of genes involved in developmental processes [64]. Our diet may also have an influence on the miRNA expression. In the liver from high-fat diet-fed mice compared to chow fed mice, expression of miR-21 was decreased which, by targeting fatty acid–binding protein 7, was related to intracellular lipid accumulation in the liver [65].

In liver from rats exposed to 8 weeks control or high-fat sucrose diets with or without methyl donor supplementation, the global DNA methylation was surprisingly *decreased* in liver from the rats exposed to methyl donor supplementation [66]. This finding, together with the disassociation between diet-induced epigenetic plasticity in adipose tissue and skeletal muscle from the same subjects, indicate that focused studies (i.e., well-matched study groups, well-considered target tissues and standardized interventions) with clear hypotheses are needed when investigating the influence of diet on global, genome-wide, and site-specific epigenetic patterns in humans and animals.

The basic understanding of how susceptible the human epigenome is to dietary changes is largely unknown. Further studies aimed at elucidating the impact of diet on the epigenome may provide optimized dietary guidelines and thereby improve the prevention of metabolic disorders.

Physical Activity, Inactivity, and the Epigenome

It is well-known that regular exercise has widespread beneficial effects for general health by regulating metabolic processes including the oxidative capacity, the

TABLE 38.1 Lifestyle Factors and Metabolic Disease Affects Epigenetics as Exemplified by PPARGC1A Promoter Methylation

Exposure	Cohort	Tissue	Sample Size	Method	Results	References
High-fat diet	Healthy young men	Skeletal muscle and subcutaneous adipose tissue	$n = 26$	BSseq.	Increased promoter methylation after 5 days high-fat overfeeding	[44–46]
	Mice, maternal	Skeletal muscle	$n = 4–8$, in each group	Pyroseq.	Increased promoter methylation and decreased gene expression in offspring	[47]
Physical activity	Healthy, sedentary men	Skeletal muscle [48,49]	$n = 28$	MeDIP-chip	Decreased promoter methylation, increased gene expression	[48,49]
	Healthy, sedentary men and women	Skeletal muscle [48,49]	$n = 14$	MeDIP-chip, BSseq.	Decreased promoter methylation, increased gene expression	[48,49]
	Healthy, trained men	Leukocytes	$n = 8$	Pyroseq.	Decreased promoter methylation, increased gene expression	[50]
Physical inactivity	Healthy young men	Skeletal muscle	$n = 20$	BSseq.	Increased promoter methylation and decreased gene expression after 9 days bed rest	[51]
Low birth weight	Healthy young men born with LBW	Skeletal muscle and subcutaneous adipose tissue	$n = 17–19$ cases, 16–26 controls	BSseq. and Pyroseq.	Increased promoter methylation	[44–46]
Maternal BMI	Neonatal offspring of obese women	Umbilical cord blood	$n = 88$	Ms-PCR	Positive correlation between maternal BMI and promoter methylation in newborns	[52]
Adiposity	Preadolescent children	Peripheral blood	$n = 40$	Pyroseq.	Promoter methylation in children predicted adiposity 5–7 years later in life	[53]
GDM and T1D in pregnancy	Adult male and female offspring	Skeletal muscle	$n = 64$ cases, 34 controls	Pyroseq. and qPCR	Decreased gene expression in GDM offspring with no methylation changes	[54]
T2D	T2D, men and women	Pancreatic islets	$n = 12$ cases, 48 controls	Seq.	Increased promoter methylation, decreased gene expression	[55]
	T2D, men and women	Skeletal muscle	$n = 17$ NGT, 8 IGT, 17 T2D	MeDIP-chip, BSseq.	Increased promoter methylation	[56,57]
	Twins discordant for T2D, men and women	Skeletal muscle	$n = 11$ pairs	BSseq.	Increased promoter methylation	[58]

The table shows a selection of studies where DNA methylation and gene expression of the metabolic master regulator peroxisome proliferator-activated receptor gamma coactivator 1 alpha (PPARGC1A) is associated with metabolic diseases and its predisposing factors.
BMI, body mass index; GDM, gestational diabetes mellitus; LBW, low birth weight; T1D, type 1 diabetes; T2D, type 2 diabetes, Pyroseq, pyro-sequencing; BSseq, Bisulfite-sequencing; MeDIP-chip, Methylated DNA Immunoprecipitation Chip; Ms-PCR, Methylation-specific PCR.

cardiovascular function, and whole-body glucose and fatty acid homeostasis [67]. Besides restructuring the muscular and skeletal system to compensate for mechanical stress, physical exercise affects the expression of genes important for metabolism.

Genome-wide DNA methylation changes have been observed after regular exercise in several tissues spanning from peripheral blood cells [68] to tissues important for whole-body metabolism, including skeletal muscle and adipose tissue [48,49,69]. Even an acute exercise intervention has been shown to change DNA promoter methylation in skeletal muscle from healthy sedentary men and women [49]. Physical activity has also been shown to regulate histone deacetylases in human skeletal muscle that induced expression of genes involved in metabolic pathways through the transcription factor, myocyte enhancer factor 2 [70] and PPARGC1A [71]. In two independent studies, PPARGC1A promoter methylation decreased significantly with exercise in human skeletal muscle [48,49]. Similarly, PPARGC1A promoter methylation decreased and mRNA expression increased after aerobic exercise in leukocytes from healthy trained men [50]. In a study by Alibegovic et al., exposure to 9 days bed rest in healthy young men led to the opposite of what was found during exercise; PPARGC1A DNA methylation increased and gene expression decreased in skeletal muscle (Table 38.1) [51]. Overall, there is evidence to suggest that a sedentary lifestyle can lead to a changed epigenome, and conversely that physical exercise may induce mechanisms whereby a worsened epigenetic phenotype can be reversed.

The epigenetic effects of physical activity are genome-wide, and e.g., affect genes involved in muscle growth and metabolism [72]. In general, even a small amount of exercise seem to induce global hypomethylation in muscle cells [48,49]. If situated in regulatory regions, exercise-induced epigenetic changes may alter the expression of genes important in muscle repair, differentiation and growth pathways [72,73]. Furthermore, the intensity of the exercise has been shown to correlate directly with the level of promoter demethylation and gene expression with more strenuous exercise activating a higher number of genes [72].

The skeletal muscle mass as well as its adaptability to environmental changes is believed to be programmable, so that it is more capable to respond to a stimuli that has been encountered before [74]. In this regard, skeletal muscle may have an "epigenetic memory." This was supported by a study of mouse muscle cells where a TNF-α-induced increase in DNA methylation of the muscle differentiation regulator *Myod* was maintained throughout muscle proliferation and lifespan [75]. More recently, studies have shown that parental exercise is associated with improved metabolic health in the offspring via epigenetic signals. In mice, both exercise training of male mice before contraception [76] and exercise training of female mice during pregnancy [77] are associated with improvements of metabolic health and changed DNA methylation or miRNA levels in male and female offspring.

miRNAs are believed to be involved in the regulation of inflammation and angiogenesis. Indeed, aerobic exercise has been shown to reduce the overall number of miRNAs that produce negative inflammatory effects in skeletal muscle [78]. Stimuli that cause the body to enter an anabolic state, such as resistance training have also been shown to generate a reduction in inflammatory miRNAs [79]. This reduction may play a role in the growth of the muscle cell (hypertrophy) but needs to be further studied.

Physical activity has also been shown to have beneficial and protective effects in cancer patients [80]. Hypermethylation in the promoter regions of the tumor suppressing genes *APC*, *RASSF1A*, and *L3MBTL1* are common epigenetic markers for breast cancer and physical exercise has been shown to decrease the methylation at these sites which has been suggested to lower the risk of developing breast cancer [81,82].

EPIGENETIC MEMORY, PRENATAL EXPOSURE, AND RISK OF METABOLIC DISEASE

It is well established that suboptimal environmental exposures during fetal life may have a permanent effect on human tissue function and the risk of metabolic diseases. This is often referred to as developmental programming. Even though the precise mechanism of how events during prenatal life can be "remembered" and affect susceptibility to metabolic diseases in adult life remain unknown, the concept of an epigenetic "memory" that is established in utero is an attractive explanation [83] (Fig. 38.1).

Thrifty Phenotype and Birthweight

More than 25 years ago, Hales and Barker proposed "the thrifty phenotype hypothesis" with which they suggested that poor nutritional conditions that give rise to reduced fetal growth might be responsible for an adult phenotype that is beneficial for survival under poor nutritional conditions but detrimental when nutrition is abundant [84]. Since then, their observations that developmental programming plays an important role in the development of metabolic abnormalities, including T2D and CVD, have been confirmed and expanded by numerous studies [3,9,85]. Some of the most convincing evidence is represented by studies of adult individuals who were born by mothers exposed to famine during pregnancy. In these studies, individuals born at term with a low birthweight (LBW; usually <2500 g), most likely due to prenatal undernutrition, had impaired glucose tolerance and were at increased risk for diabetes later in life [3,4]. Importantly, Harder et al. examined the relationship between birthweight and risk of T2D in a large meta-analysis in 2007, and discovered an U-shaped relationship [86]. Thus, not only a low birthweight but also a high birthweight (usually >4000 g) is related to an increased risk of T2D. High birthweight is primarily associated with overnutritional and hyperglycemic pregnancies and will be covered in the subsequent section.

The causal relationship between environmental conditions in pregnancy and epigenome changes that are linked to offspring metabolic health has been shown in animal studies. In a study of C57BL/6 mice it was demonstrated that bisphenol A (BPA) exposure during pregnancy disrupts metabolic health across multiple generations. Furthermore, the metabolic changes (including increased body fat and disturbed glucose homeostasis) were directly linked to changes of *Igf2* mRNA expression and DNA methylation at a differentially methylated region in the *Igf2* promoter in embryonic tissues [87]. Additional studies have shown that developmental exposure of male mice to the endocrine disruptor 2,3,7,8-tetrachlorodibenzo-p-dioxin is associated with placental dysfunction and intrauterine growth restriction of the further generations, suggesting that epigenetic modifications within the sperm have occurred [88].

By identifying lower DNA methylation levels of the *IGF2* gene in adult individuals who were prenatally exposed to famine during the Dutch Hunger Winter of 1944–45, Heijmans et al. were the first to show that in uterine environmental conditions can cause epigenetic changes that persist throughout human life [3]. Since then, alterations of epigenetic patterns and plasticity in LBW compared to normal birthweight (NBW) individuals have been identified in multiple human tissues that are relevant to metabolic diseases [44,45,59,89,90]. Also in stem cells, individuals born with LBW seems to have altered epigenetic patterns. In cultured adipose tissue stem cells derived from LBW subject, DNA methylation of the gene encoding the adipose-derived hormone leptin (*LEP*) was increased; whereas, the mRNA expression and the secreted leptin protein was downregulated in LBW compared to NBW subjects [89]. Also, in muscle stem cells (satellite cells) it has been demonstrated that impaired muscle development predisposing LBW individuals to T2D is linked to and potentially caused by distinct DNA methylation and transcriptional changes including down regulation of *HDAC7* and *FYN* [91]. These studies indicate that individuals exposed to restricted growth intrauterine have dysfunctional stem cells, which may be epigenetically programmed.

The promoter methylation of *PPARGC1A* has been studied in skeletal muscle in two independent populations of young healthy LBW and NBW men, and in both populations the methylation levels were significantly increased in men born with low birthweights compared normal (Table 38.1) [44,46]. Similar to *LEP*, DNA methylation of the *PPARGC1A* promoter was significantly increased in adipose tissue from the LBW men [45]. In 2009, Gemma et al. found a positive correlation between maternal body mass index (BMI) and *PPARGC1A* promoter methylation in the umbilical cord [52]. More recently Clarke-Harris et al. showed that DNA methylation at four CpG sites in the *PPARGC1A* promoter in whole blood from 5- to 7-year-old children predicted adiposity in the same children aged 9–14 years [53]. These results suggest that *PPARGC1A* promoter methylation in whole blood and umbilical cord may serve as a biomarker for estimates of the intrauterine milieu, and could thereby predict the future risk of obesity and metabolic diseases. On the other hand, DNA methylation of *PPARGC1A* in skeletal muscle and adipose tissue may be involved in the progression of insulin resistance, and T2D.

Altered posttranscriptional regulation, such as mRNA degradation, mRNA translation, and expression of long non-coding RNAs and miRNAs has also been suggested to link fetal programming, and LBW with increased risk of T2D. Indeed, a miRNA profile related to an adverse adipocyte phenotype has been observed in adult LBW individuals and in prediabetic rats that had been exposed to suboptimal nutrition in early life [92]. Also in skeletal muscle, LBW in twins and undernutrition during pregnancy in rats has been associated with altered expression of specific miRNAs that potentially target insulin signaling [93].

Taken together, these studies suggest that an adverse intrauterine environment that causes restrictive fetal growth is memorized in adult life as alterations in the epigenetic patterns and plasticity, and thereby increase the risk of metabolic disorders later in life [3,9,44,59,85,89,90]. However, it is also of importance to note, that genetic components regulating birthweight do play a large role, and when possible, these should be accounted for in epigenetic studies related to birthweight. Indeed, it was recently shown using Mendelian randomization that both indirect maternal and direct fetal genetic effects drive the observational relationship between lower birthweight and higher blood pressure in the offspring later in life [94]. Furthermore, by using maternal birthweight-lowering genotypes as a proxy for an adverse intrauterine environment no evidence was found that an adverse intrauterine environment causally raises offspring blood pressure. This indicates that the inverse birthweight-blood pressure association is attributable to genetic effects and not to intrauterine programming.

Diabetes in Pregnancy

A hyperglycemic intrauterine environment caused by maternal type 1 diabetes (T1D), T2D, or gestational diabetes mellitus (GDM) can result in a high birthweight and an increased risk of metabolic diseases of the offspring later in life. In addition to studies of undernutrition and growth restriction during pregnancy, studies on overnutrition exemplified by GDM have supported epigenetics as a link between fetal programming and development of metabolic disease. Along with the epidemic proportions of people with T2D worldwide, the prevalence of diabetes in pregnancy is escalating as well mainly as a result of obesity, physical inactivity, increased aging, and urbanization [95,96].

GDM affects 1%–10% of all births worldwide, but is likely progressing more rapidly in developing countries as in India with 10%–17% of pregnancies affected [97,98]. Generally, offspring of mothers with GDM, T1D, or T2D are affected by the increased nutrition supply causing an accelerated growth of the fetus that cause a higher birthweight, which in undiagnosed cases can be life threatening for both mother and child. In addition to acute effects comprising complicated pregnancies and births, numerous observational human studies have shown that a hyperglycemic (high blood sugar) intrauterine environment caused by maternal diabetes has long-term effects on offspring health, including an increased risk of T2D and obesity, already at an early age [9,54,99,100]. However, the underlying molecular mechanisms are still unknown. Data strongly indicate that the intrauterine hyperglycemic and hypermetabolic environment in diabetic mothers place the offspring at an increased long-term risk for T2D at a degree exceeding what is known for LBW. In a Danish study, it was shown that GDM offspring exhibit up to an eightfold elevated risk of T2D, fourfold higher risk of metabolic syndrome, and a twofold higher risk of overweight [99,100]. A possible mechanism behind the increased disease risk in offspring of diabetic mothers could be tissue-wide epigenetic changes of genes important for metabolic and developmental pathways [101]. There is evidence to suggest that an "epigenetic memory" in skeletal muscle and fat stem cells are mediating these effects [89,102,103]. Also fetal exposition to T1D has been associated with adulthood pre-diabetes, including renal dysfunction that seems to be mediated by epigenetic mechanisms [104].

Another small-scale study reported that global methylation levels were reduced in the placenta from GDM women; whereas, no major differences in methylation levels were detected in cord blood sample [105]. Furthermore, the maternally imprinted *MEST* gene, the non-imprinted glucocorticoid receptor *NR3C1* gene, and interspersed ALU repeats showed significantly decreased methylation levels in both placenta tissue and cord blood from GDM mothers [106]. At the epigenome-wide level, a discovery study on a subpopulation (n = 188) from the Danish National Birth Cohort, recently identified 76 differentially methylated CpG sites in 9–16-year-old offspring of women with gestational diabetes, relative to controls, none of which were attenuated following adjustment for offspring BMI [107]. However, most of this epigenetic variation was due to confounding by maternal pre-pregnancy BMI. Nevertheless, 13 methylation variants were independently associated with maternal gestational diabetes. Also, elevated levels of miRNAs (miR-16–5p, -29a-3p, and -134–5p) were recently observed in women, who had normal glucose tolerance at baseline and later in pregnancy developed GDM, compared with controls who had normal glucose tolerance throughout their pregnancy [108]. Elevated levels of a combination of the three miRNAs could distinguish later GDM from pregnant women with normal glucose tolerance, and thereby propose a new panel of biomarkers that may be used for prediction of later GDM development. These studies together support the idea that intrauterine exposure to diabetes change the maternal and fetal epigenome and may potentially have long-lasting effects on the offspring epigenome later in life.

Diabetes in pregnancy is typically associated with increased inflammation, which also increases the risk of developing metabolic disease in the offspring. Changing the maternal lifestyle by, for example, physical activity might improve metabolic health in both the mother and the child. One study examined this in mice exposed to high-fat diet with or without voluntary exercise before and during pregnancy [47]. They found increased methylation of the *Ppargc1a* promoter with a high-fat diet and corresponding decreased gene expression in muscle from both neonatal and 12-month-old offspring (Table 38.1). When the mice exercised, the hypermethylation was prevented and an increase in *Ppargc1a* expression was observed, which seemed to have a protective effect of the otherwise worsened metabolic profile in the offspring [47]. A recent human study investigated whether a lifestyle intervention (physical activity with or without dietary advice) in pregnant women with obesity also were associated with epigenetic variation in cord blood and body composition in the offspring. In the cord blood, DNA methylation was altered at 379 CpG sites, annotated to 370 genes, following a lifestyle intervention vs control subjects [109]. These 370 genes were significantly overrepresented in 15 biological processes including response to fatty acids and adipose tissue development. The offspring of mothers included in a lifestyle intervention were born with more lean mass compared with control subjects. DNA methylation at 17 CpG sites (annotated to following genes amongst others: *DISC1*, *GBX2*, *HERC2*, and *HUWE1*), partially mediated the effect of the lifestyle intervention on lean mass in the offspring. Moreover, 22 CpG sites were associated with offspring BMI z scores (measures of relative weight adjusted for child age and sex) during the first 3 years of life [109]. Indeed, further knowledge of how the underlying molecular mechanisms by which exercise during pregnancy is affecting health and risk of metabolic disease in the offspring are needed to further improve the prevention strategies of metabolic diseases.

EPIGENETIC FEATURES OF METABOLIC DISEASES

Epigenetics has been established as one of the underlying causes of cancer development [110], but the link between epigenetics and metabolic diseases has only recently been recognized. Numerous studies have found altered epigenetic fingerprints in patients with metabolic diseases compared to healthy controls. Interestingly, some of the disease-related epigenetic modifications, including promoter methylation of *PPARGC1A*, have also been associated with risk factors for metabolic diseases, including aging, overweight and adverse fetal programming [12,13,44,55].

Metabolic Syndrome

Metabolic syndrome is a group of conditions, including central obesity, hypertension, and dyslipidemia that, when occurring together, culminate in an increased risk of T2D and CVD. The major risk factors for the development of metabolic syndrome are physical inactivity and a high-calorie diet, which contributes to central (abdominal, visceral) obesity, and insulin resistance. Other factors that increase the risk of metabolic syndrome are age, family history of T2D, or a past medical history of GDM or CVD [111]. In individuals with the metabolic syndrome or any of its components, it is possible to delay and even prevented the development of serious health problems with lifestyle changes (i.e., physical activity and weight loss).

A single precise definition of the metabolic syndrome is of much debate, but the defining components that cluster together are central obesity, hypertension, insulin resistance, and elevated fasting triglycerides level in the blood (dyslipidemia) [111]. Other components that are included in some of the diagnostic criteria are a decreased level of fasting high-density lipoprotein cholesterol and increased levels of fasting glucose or insulin in the blood. Due to the diverse defining conditions and the various tissues affected in the metabolic syndrome, it is not trivial to link epigenetic mechanisms to the pathogenesis of this complex metabolic disease. Studies of epigenetic modifications in relevant target tissues of well-characterized individuals with the metabolic syndrome vs healthy controls are therefore scarce.

DNA methylation of a CpG site in the body region of the suppressor of cytokine signaling 3 (*SOCS3*) gene, that is upregulated in obesity and has been associates with both insulin and leptin resistance [112], was analyzed in peripheral blood mononuclear cells from 839 individuals of whom one-third had the metabolic syndrome [113]. Interestingly, the methylation of *SOCS3* was inversely associated with the gene expression and with the presence of metabolic syndrome and obesity, as well as with several of the conditions related to metabolic syndrome including fasting levels of triglycerides, low-density lipoprotein cholesterol and glucose in the blood, waist to hip ratio, and insulin resistance. The results from this study suggest that lower methylation of *SOCS3* may lead to increased gene expression that suppresses cytokine signaling, thereby increasing the risk of metabolic syndrome [113].

In opposition to metabolic syndrome per se, numerous extensive epigenetic studies have been performed on some of its defining components. For example, dyslipidemia in patients with metabolic syndrome (typically high levels of triglycerides and low levels of high-density lipoprotein cholesterol in the blood) has been associated with altered regulation of epigenetic modifications. In vitro cell studies of human skeletal muscle, and pancreatic islets indicate that the fatty acids palmitate and oleate can affect DNA methylation of metabolically important genes [56,114,115]. Also, it is suggested that miRNAs, including miR-33, -34a, and -21, are critical regulators of lipid metabolism, and thus may be involved in the development of dyslipidemia and metabolic syndrome [116]. Hypertension is another important component of the metabolic syndrome that increases the risk of CVD and that has been associated with epigenetics mechanisms [117]. In a study of whole blood from 60 patients with essential hypertension, the average level of DNA methylation was significantly lower in the cases compared to the 30 control subjects with the patients affected by severe hypertension showing the lowest global methylation levels [118]. Thus, by being associated with different defining components of metabolic syndrome, the epigenome may be a central co-actor in the development of metabolic syndrome at many different levels. As discussed by Carson and Lawson [119], future integration of big epigenomic, genomic, and phenotypic data with a systems biology approach might move us closer to predicting an individual's risk of developing metabolic syndrome, to step further towards precision medicine, and to understanding how associations among metabolic traits evolved.

Obesity

Obesity affects more than 500 million people worldwide and up to one-third of the population in the western countries; it is associated with several comorbidities, including T2D and CVD [8]. Besides being a metabolic disorder, obesity is recognized as a risk factor for other metabolic diseases, primarily through the underlying insulin resistance that accompanies the accumulation of adipose tissue together with the lipid accumulation in the vascular system (atherosclerosis).

Impact of body fat percentage and BMI on DNA methylation has been reported in human tissues and blood. These studies suggest that obesity to some extent may be epigenetically regulated [13,120–124]. Global DNA methylation in human SAT was significantly associated with fat distribution (waist measurement and waist–hip ratio) in a study by Keller et al. [121]. Significant associations between BMI and DNA methylation at three CpG sites in intron 1 of the gene encoding hypoxia inducible factor 3 alpha (*HIF3A*), in both whole blood (from three cohorts) and SAT (from a fourth female cohort), were demonstrated by Dick et al. [120], and has since been replicated in several other studies. In one study, the association between BMI and *HIF3A* methylation was significant at all three CpG sites in SAT from females, but not males, indicating a gender-specific effect of BMI on *HIF3A* DNA methylation [13]. Also in skeletal muscle cells, lipid exposure has been shown to induce differential DNA methylation and gene expression responses in cells obtained from obese women compared to lean controls [115]. Interestingly, the weight-loss induced surgery Roux-en-Y gastric bypass (RYGB) has also been shown to cause a dynamic effect on the skeletal muscle epigenome in obese individuals [125].

The epigenetic regulation of the multi-target deacetylases *SIRT1* and *SIRT7* was investigated in a study where the gene expression was significantly different in adipose tissue from obese vs normal-weight subjects [126]. Here, DNA methylation was similar between the groups, but the expression of several miRNAs including miR-125a-5p and miR-193a-3p were significantly altered in obese individuals, and correlated negatively with body weight and BMI [126].

Long non-coding RNAs have also been found differentially expressed in adipose tissue from metabolically unhealthy (dysglycemic) vs metabolically healthy (normoglycemic) obese individuals [58,127]. Using a genome-wide approach, Shi et al. found *GAS5* to be downregulated in adipocytes from dysglycemic vs normoglycemic obese individuals, and demonstrated that *GAS5* positively regulates *INSR* gene expression in human adipocytes and preadipocytes in vitro [127].

In addition to long non-coding RNAs, differential levels of specific histone modifications have been observed both globally [128] and at promoters of genes important in adipogenesis, inflammation, and lipid metabolism [129] in adipose tissue from dysglycemic versus normoglycemic obese individuals. Castellano-Castillo et al. demonstrated associations between H3K4me3 levels at the promoters of several genes important for lipid metabolism (e.g., *LPL*, *SREBF2*, *PPARG* and *IL6*) and levels of glucose, insulin, and homeostatic index of insulin resistance [129]. This suggests that histone methylation may have an effect on various pathways associated with metabolic disease and thereby in the development of obesity and T2D [130].

Recent findings suggest that obesity may increase the risk of passing on metabolic dysfunction to the next generation through transgenerational epigenetic inheritance [131]. For example, Donkin et al. found unaltered spermatozoal histone positioning in obese compared to lean men, but markedly different DNA methylation patterns, suggesting paternal transmission of epigenetic modifications [132]. This study indicates that obesity may propagate metabolic dysfunction to the next generation through remodeling of the epigenome in gametes.

Type 2 Diabetes

T2D is a serious metabolic disease characterized by hyperglycemia due to multiple organ defects including insulin resistance and insufficient insulin secretion. The global prevalence of diabetes is estimated to 9% and the number of people with diabetes in the world exceeds 400 million [7]. Diabetes patients are at high risk of developing serious chronic complications and increased mortality from the long-term effects of hyperglycemia, including both microvascular damage (diabetic retinopathy, nephropathy, and neuropathy) and CVD (stroke, ischemic heart disease, and amputation) [133].

Because of the heterogeneous etiology and pathogenesis of T2D integrative approaches are needed to understand its multilayered complexity, and thereby optimize the prevention and treatment. Epigenetics has been implicated in the pathogenesis of T2D because of its ability to integrate genetic factors with environmental exposures. Studies have especially focused on tissues that are relevant to the pathogenic state of disease, i.e., in skeletal muscle, adipose tissue, and the liver, which are the main sites of insulin resistance in people with T2D or at risk of developing the disease, and in pancreatic islands where the beta-cells produce and secrete insufficient amounts of insulin in T2D individuals.

Epigenetic alterations of numerous genes have been identified in pancreatic islets from subjects with T2D versus controls [55,134–138]. In 2008 Ling et al. found that the *PPARGC1A* promoter was hypermethylated in pancreatic islets from T2D compared with nondiabetic subjects [55]. Moreover, they found a reduction of *PPARGC1A* mRNA levels in T2D subjects and a borderline significant inverse correlation between the promoter methylation and mRNA expression of *PPARGC1A*. Interestingly, the same sites in the *PPARGC1A* promoter were later demonstrated to be hypermethylated in skeletal muscle from prediabetic (insulin resistant) and T2D individuals (Table 38.1) [56,57]. Increased promoter methylation and decreased gene expression have also been identified in the promoter region of the gene that encodes insulin (INS) as well as in PDX1, a key transcription factor for beta-cell development and function, in pancreatic islands from T2D individuals [136,137]. More recent genome-wide DNA methylation studies in pancreatic islets from diabetic subjects and nondiabetic controls have identified multiple other genes with altered methylation, and expression in diabetic islets [134,135,139].

In human adipose tissue and liver biopsies, altered epigenetic, and transcriptional patterns of metabolically important genes have been identified in T2D subjects compared to healthy controls [140,141]. Nilsson et al. investigated genome-wide DNA methylation in SAT from 28 diabetic and 28 nondiabetic individuals identifying 15,627 CpG sites with differential DNA methylation in T2D individuals [141]. Another study investigated if DNA methylation levels at the 15,627 CpG sites that were differential in SAT from T2D patients, also were significantly associated with age, BMI, or HbA1c in SAT from another cohort of 96 healthy males [13,141]. Interestingly, DNA methylation at 8% of the T2D-associated CpG sites were significantly associated with the age in this study [13]. Most of these sites (90%) changed positively with age and were increased in T2D individuals. Moreover, SAT DNA methylation in 6% of the sites that were differentially methylated in SAT from T2D individuals were associated with BMI. Also, despite a narrow range in HbA1c levels, 30 of the sites that were differentially methylated in T2D individuals correlated with HbA1c levels; methylation of 26 of those sites showed a negatively correlation with HbA1c in the healthy men and were lower in T2D individuals compared with controls [13,141]. The consistency between site-specific DNA methylation differences in adipose tissue from T2D patients and significant correlations age, BMI, or HbA1c at the same sites in adipose tissue from healthy individuals suggests that epigenetic mechanisms interact with lifestyle factors in the development of T2D (Fig. 38.1).

Recently, VPS39 was identified as an important regulator of myoblast differentiation and muscle glucose uptake, and it was shown that VPS39 is downregulated in myoblasts and myotubes from individuals with T2D [142]. The downregulation of VPS39 seems to be driven by a pathway connecting VPS39-deficiency in human myoblasts to impaired autophagy, abnormal epigenetic reprogramming, dysregulation of myogenic regulators, and perturbed differentiation. VPS39 knockdown in human myoblasts has profound effects on autophagic flux, insulin signaling, epigenetic enzymes, DNA methylation and expression of myogenic regulators. Overall, VPS39-deficiency contributes to impaired muscle differentiation and reduced glucose uptake, and epigenetic regulation of VPS39 thereby offers a therapeutic target for T2D [142].

Besides DNA methylation, T2D have been associated with distinctive miRNA profiles in the blood and in T2D target tissues, such as skeletal muscle [93,143]. These miRNA profiles are sometimes detected several years before the manifestation of disease and may thereby be central players of pathological processes [93,143]. Moreover, distinct miRNAs appear to be predictive of long-term microvascular and macrovascular complications of diabetes [143].

The chronic hyperglycemia of diabetes patients causes damage of the vascular system throughout the body including the cells lining blood vessels (endothelial cells and vascular smooth muscle cells), the heart (cardiac cells), and the kidneys (renal cells) [133]. Cardiovascular complications of diabetes are, in addition to hyperglycemia, characterized by inflammation, oxidative stress, and endothelial dysfunction that per se have been shown to induce various epigenetic changes that can become irreversible (reviewed in Ref. [133]). In this regard, an "epigenetic memory" may be involved in the progression of cardiovascular complications from T2D. This was supported by Miao et al. who found enrichment in H3 lysine-9 acetylation of promoter regions in monocytes from diabetes patients with microangiopathy compared to diabetic patients without microvascular complications [144]. Notably, the epigenetic changes induced by hyperglycemia and inflammation in this study were detected in peripheral blood monocytes. Thus, there is a potential of using epigenetic markers of blood cells and circulating miRNAs in the blood to estimate health status, and thereby early detection of diabetes progression and its associated complications [133,143]. Additionally, epigenetic patterns may indicate optimal treatments for T2D patients. In blood cells from drug-naïve T2D patients, DNA methylation of specific CpG sites differed between glycemic responders and non-responders to metformin (11 sites), and between metformin-tolerant and -intolerant patients (four sites) [145]. Based on these data, methylation risk scores were calculated to be able to discriminate glycemic responders/non-responders and patients tolerant/intolerant to metformin in drug-naïve T2D patients. Thereby, blood-borne epigenetic markers may help patients with T2D receive optimal therapy.

CONCLUSIONS

We are just beginning to understand how, and to which extent, epigenetic mechanisms interact with the genome, transcriptional machinery, and lifestyle exposures to affect the risk of slowly progressive metabolic diseases later in life and in future generations (Fig. 38.1). Specific epigenetic alterations have been detected in blood samples as well as in stem cells and primary metabolic tissues from people with metabolic disease and in individuals at increased disease risk (i.e., due to being exposed to adverse intrauterine environments).

During the past years, genome-wide studies of human blood and tissues have identified increasing amount of epigenetic modifications that are influenced by lifestyle factors (i.e., physical activity, diet, and smoking) or are associated with age and clinical characteristics (i.e., BMI and insulin resistance). Further insight into these associations will ideally lead to improved predictive biomarkers of metabolic diseases and thereby increase the possibilities to postpone, or prevent, metabolic diseases, such as T2D [16]. However, although present at numerous genomic sites in relevant tissues, the magnitude of most epigenetic changes associated with metabolic disease are, unlike cancer, relatively small. In this respect, it might be advantageous to identify risk-related epigenetic marks in well-characterized human studies where levels of DNA methylation, histone modifications or miRNAs in disease-related tissues are correlated with quantitative metabolic traits, such as BMI, waist-to-hip ratio, blood pressure, HbA1c, insulin sensitivity or fasting blood levels of glucose, insulin, cholesterol or triglycerides in the blood. Also, further descriptive as well as mechanistic studies are required to reveal how epigenetic mechanisms are inherited and, in interaction with the genome, the prenatally established epigenetic memory and postnatal environmental exposures influence human health, and disease. In support of functional and disease-related interactions between genetic and epigenetic variations, the NIH Roadmap Epigenomics Mapping Consortium published some of the most comprehensive human epigenomic data in 2015, where they demonstrate epigenomic enrichments in more than 100 types of cells and tissues related to a trait which had previously been associated with a genetic variant [146].

Most of the knowledge we have concerning DNA methylation, histone modifications and miRNAs in the pathogenesis of metabolic diseases has emerged during the last ten years, and the research field holds infinite possibilities. Future challenges in the field of epigenetics of metabolic diseases include an improved understanding of how environmental factors and genetic variation influence epigenetic modifications, how epigenetic factors affect gene transcription and disease-related phenotypes, if epigenetic changes induced by today's lifestyle are reversible, and if they are inherited to future generations. Based on recent research and progress in development of epigenetic biomarkers, it is our belief that epigenetic modifications will serve as markers for the diagnosis, prognosis, and individualized treatment of metabolic diseases in the future.

Abbreviations

CVD	Cardiovascular disease
GDM	Gestational diabetes mellitus
HIF3A	Hypoxia inducible factor 3 alpha
IGF2	Insulin growth factor 2
KLF14	Krüppel-like factor 14
LBW	Low birthweight
miRNA	microRNA
NBW	Normal birthweight
NAFLD	Non-alcoholic fatty liver disease
PPARGC1A	Peroxisome proliferator-activated receptor gamma coactivator 1 alpha
SAM	S-Adenosyl methionine
SAT	Subcutaneous adipose tissue
T1D	Type 1 diabetes
T2D	Type 2 diabetes

References

[1] Fraga MF, Ballestar E, Paz MF, Ropero S, Setien F, Ballestar ML, et al. Epigenetic differences arise during the lifetime of monozygotic twins. Proc Natl Acad Sci U S A 2005;102(30):10604−9.

[2] Rönn T, Poulsen P, Hansson O, Holmkvist J, Almgren P, Nilsson P, et al. Age influences DNA methylation and gene expression of COX7A1 in human skeletal muscle. Diabetologia 2008;51(7):1159−68. Available from: http://link.springer.com/10.1007/s00125-008-1018-8.

[3] Heijmans BT, Tobi EW, Stein AD, Putter H, Blauw GJ, Susser ES, et al. Persistent epigenetic differences associated with prenatal exposure to famine in humans. Proc Natl Acad Sci U S A 2008;105(44):17046−9. Available from: http://www.ncbi.nlm.nih.gov/pubmed/18955703.

[4] Ravelli AC, van der Meulen JH, Michels RP, Osmond C, Barker DJ, Hales CN, et al. Glucose tolerance in adults after prenatal exposure to famine. Lancet (London, Engl) 1998;351(9097):173−7. Available from: http://www.ncbi.nlm.nih.gov/pubmed/9449872.

[5] Finer S, Mathews C, Lowe R, Smart M, Hillman S, Foo L, et al. Maternal gestational diabetes is associated with genome-wide DNA methylation variation in placenta and cord blood of exposed offspring. Hum Mol Genet 2015;24(11):3021−9. Available from: http://www.ncbi.nlm.nih.gov/pubmed/25634562.

[6] Ford ES, Giles WH, Mokdad AH. Increasing prevalence of the metabolic syndrome among u.s. Adults. Diabetes Care 2004;27(10):2444−9. Available from: http://www.ncbi.nlm.nih.gov/pubmed/15451914.

[7] International Diabetes Federation. *IDF Diabetes Atlas − 7th Edition.* 2015 [cited 2016 Nov 11]. Available from: http://www.diabetesatlas.org.

[8] World health organization. *Obesity and Overweight.* 2016 [cited 2016 Aug 30]. Available from: http://www.who.int/mediacentre/factsheets/fs311/en/.

[9] Vaag A, Brøns C, Gillberg L, Hansen NS, Hjort L, Arora GP, et al. Genetic, nongenetic and epigenetic risk determinants in developmental programming of type 2 diabetes. Acta Obstetricia et Gynecologica Scandinavica 2014;93(11):1099−108.

[10] Benayoun BA, Pollina EA, Brunet A. Epigenetic regulation of ageing: linking environmental inputs to genomic stability. Nat Rev Mol Cell Biol 2015;16(10):593−610. Available from: http://doi.org/10.1038/nrm4048%5Cn, http://www.ncbi.nlm.nih.gov/pubmed/26373265%5Cn, http://www.pubmedcentral.nih.gov/articlerender.fcgi?artid = PMC4736728.

REFERENCES

[11] Horvath S. DNA methylation age of human tissues and cell types. Genome Biol 2013;14(10):R115. Available from: http://www.pubmedcentral.nih.gov/articlerender.fcgi?artid=4015143&tool=pmcentrez&rendertype=abstract.

[12] Bacos K, Gillberg L, Volkov P, Olsson AH, Hansen T, Pedersen O, et al. Blood-based biomarkers of age-associated epigenetic changes in human islets associate with insulin secretion and diabetes. Nat Commun 2016;7:11089. Available from: http://www.ncbi.nlm.nih.gov/pubmed/27029739.

[13] Rönn T, Volkov P, Gillberg L, Kokosar M, Perfilyev A, Jacobsen AL, et al. Impact of age, BMI and HbA1c levels on the genome-wide DNA methylation and mRNA expression patterns in human adipose tissue and identification of epigenetic biomarkers in blood. Hum Mol Genet 2015;24(13):3792–813. Available from: http://www.ncbi.nlm.nih.gov/pubmed/25861810.

[14] Hannum G, Guinney J, Zhao L, Zhang L, Hughes G, Sadda S, et al. Genome-wide methylation profiles reveal quantitative views of human aging rates. Mol Cell 2013;49(2):359–67. Available from: http://doi.org/10.1016/j.molcel.2012.10.016.

[15] Bigot A, Duddy WJ, Zamalou G, Butler-browne G, Mouly V, Bigot A, et al. Age-associated methylation suppresses SPRY1, leading to a failure of re-quiescence and loss of the reserve stem cell pool in elderly muscle. Cell Rep 2015;13:1172–82. Available from: http://www.sciencedirect.com/science/article/pii/S2211124715011092.

[16] Gillberg L, Ling C. The potential use of DNA methylation biomarkers to identify risk and progression of type 2 diabetes. Front Endocrinol (Lausanne) 2015;6:43. Available from: http://www.ncbi.nlm.nih.gov/pubmed/25870586.

[17] Zhang Y, Wilson R, Heiss J, Breitling LP, Saum K-U, Schöttker B, et al. DNA methylation signatures in peripheral blood strongly predict all-cause mortality. Nat Commun 2017;8(1):14617. Available from: http://www.nature.com/articles/ncomms14617.

[18] Levine ME, Lu AT, Quach A, Chen BH, Assimes TL, Bandinelli S, et al. An epigenetic biomarker of aging for lifespan and healthspan. Aging (Albany NY) 2018;10(4):573–91. Available from: https://www.aging-us.com/lookup/doi/10.18632/aging.101414.

[19] Zhang Y, Saum K-U, Schöttker B, Holleczek B, Brenner H. Methylomic survival predictors, frailty, and mortality. Aging (Albany NY) 2018;10(3):339–57. Available from: https://www.aging-us.com/lookup/doi/10.18632/aging.101392.

[20] Lu AT, Quach A, Wilson JG, Reiner AP, Aviv A, Raj K, et al. DNA methylation GrimAge strongly predicts lifespan and healthspan. Aging (Albany NY) 2019;11(2):303–27. Available from: https://www.aging-us.com/lookup/doi/10.18632/aging.101684.

[21] Fransquet PD, Wrigglesworth J, Woods RL, Ernst ME, Ryan J. The epigenetic clock as a predictor of disease and mortality risk: a systematic review and meta-analysis. Clin Epigenetics 2019;11(1):62. Available from: https://clinicalepigeneticsjournal.biomedcentral.com/articles/10.1186/s13148-019-0656-7.

[22] Gao X, Colicino E, Shen J, Just AC, Nwanaji-Enwerem JC, Wang C, et al. Comparative validation of an epigenetic mortality risk score with three aging biomarkers for predicting mortality risks among older adult males. Int J Epidemiol 2019;48(6):1958–71. Available from: https://academic.oup.com/ije/article/48/6/1958/5481894.

[23] Jung S-E, Lim SM, Hong SR, Lee EH, Shin K-J, Lee HY. DNA methylation of the ELOVL2, FHL2, KLF14, C1orf132/MIR29B2C, and TRIM59 genes for age prediction from blood, saliva, and buccal swab samples. Forensic Sci Int Genet 2019;38:1–8. Available from: http://www.ncbi.nlm.nih.gov/pubmed/30300865.

[24] Naue J, Sänger T, Hoefsloot HCJ, Lutz-Bonengel S, Kloosterman AD, Verschure PJ. Proof of concept study of age-dependent DNA methylation markers across different tissues by massive parallel sequencing. Forensic Sci Int Genet 2018;36:152–9. Available from: http://www.ncbi.nlm.nih.gov/pubmed/30031222.

[25] Sillanpää E, Heikkinen A, Kankaanpää A, Paavilainen A, Kujala UM, Tammelin TH, et al. Blood and skeletal muscle ageing determined by epigenetic clocks and their associations with physical activity and functioning. Clin Epigenetics 2021;13(1):110. Available from: https://clinicalepigeneticsjournal.biomedcentral.com/articles/10.1186/s13148-021-01094-6.

[26] Ashapkin VV, Kutueva LI, Vanyushin BF. Aging as an epigenetic phenomenon. Curr Genom 2017;18(5). Available from: http://www.eurekaselect.com/151530/article.

[27] Michishita E, McCord RA, Berber E, Kioi M, Padilla-Nash H, Damian M, et al. SIRT6 is a histone H3 lysine 9 deacetylase that modulates telomeric chromatin. Nature 2008;452(7186):492–6. Available from: http://www.ncbi.nlm.nih.gov/pubmed/18337721.

[28] Kanfi Y, Naiman S, Amir G, Peshti V, Zinman G, Nahum L, et al. The sirtuin SIRT6 regulates lifespan in male mice. Nature 2012;483(7388):218–21. Available from: http://www.ncbi.nlm.nih.gov/pubmed/22367546.

[29] Hooten NN, Abdelmohsen K, Gorospe M, Ejiogu N, Zonderman AB, Evans MK. microRNA expression patterns reveal differential expression of target genes with age. PLoS One 2010;5(5):e10724.

[30] Lempradl A, Pospisilik JA, Penninger JM. Exploring the emerging complexity in transcriptional regulation of energy homeostasis. Nat Rev Genet 2015;16(11):665–81.

[31] Parle-McDermott A, Ozaki M. The impact of nutrition on differential methylated regions of the genome. Adv Nutr An Int Rev J 2011;2(6):463–71. Available from: http://www.pubmedcentral.nih.gov/articlerender.fcgi?artid=3226384&tool=pmcentrez&rendertype=abstract.

[32] Sanchez-roman I, Gomez A, Gomez J, Suarez H, Sanchez C, Pamplona R, et al. Forty percent methionine restriction lowers DNA methylation, complex I ROS generation, and oxidative damage to mtDNA and mitochondrial proteins in rat heart. J Bioenerg Biomembr 2011;1:699–708.

[33] Fang MZ, Chen D, Sun Y, Jin Z, Christman JK, Yang CS. Reversal of hypermethylation and reactivation of p16INK4a, RARβ, and MGMT genes by genistein and other isoflavones from soy. Clin Cancer Res 2005;11(19 I):7033–41.

[34] Sinclair AJ, Taylor PB, Lunec J, Girling AJ, Barnett AH. Low plasma ascorbate levels in patients with type 2 diabetes mellitus consuming adequate dietary vitamin C. Diabet Med 1994;11(9):893–8. Available from: http://www.ncbi.nlm.nih.gov/pubmed/7705029.

[35] Blaschke K, Ebata KT, Karimi MM, Zepeda-Martínez JA, Goyal P, Mahapatra S, et al. Vitamin C induces Tet-dependent DNA demethylation and a blastocyst-like state in ES cells. Nature. 2013;500(7461):222–6.

[36] Klose RJ, Yamane K, Bae Y, Zhang D, Erdjument-Bromage H, Tempst P, et al. The transcriptional repressor JHDM3A demethylates trimethyl histone H3 lysine 9 and lysine 36. Nature. 2006;442(7100):312–16.

[37] Gillberg L, Ørskov AD, Liu M, Harsløf LBS, Jones PA, Grønbæk K. Vitamin C – A new player in regulation of the cancer epigenome. Semin Cancer Biol 2018;51:59–67.

[38] Vaquero A, Reinberg D. Calorie restriction and the exercise of chromatin. Genes Dev 2009;23(16):1849–69. Available from: http://www.ncbi.nlm.nih.gov/pubmed/19608767.

[39] Vaquero A, Scher M, Lee D, Erdjument-Bromage H, Tempst P, Reinberg D. Human SirT1 interacts with histone H1 and promotes formation of facultative heterochromatin. Mol Cell

[40] Lee H-S, Barraza-Villarreal A, Biessy C, Duarte-Salles T, Sly PD, Ramakrishnan U, et al. Dietary supplementation with polyunsaturated fatty acid during pregnancy modulates DNA methylation at IGF2/H19 imprinted genes and growth of infants. Physiol Genomics 2014;46(23):851–7. Available from: http://www.ncbi.nlm.nih.gov/pubmed/25293351.

2004;16(1):93–105. Available from: http://www.ncbi.nlm.nih.gov/pubmed/15469825.

[41] Cooney CA, Dave AA, Wolff GL. Maternal methyl supplements in mice affect epigenetic variation and DNA methylation of offspring. J Nutr 2002;132(8 Suppl):2393S–400S. Available from: http://www.ncbi.nlm.nih.gov/pubmed/12163699.

[42] Puigserver P, Wu Z, Park CW, Graves R, Wright M, Spiegelman BM. A cold-inducible coactivator of nuclear receptors linked to adaptive thermogenesis. Cell 1998;92(6):829–39. Available from: http://www.ncbi.nlm.nih.gov/pubmed/9529258.

[43] Patti ME, Butte AJ, Crunkhorn S, Cusi K, Berria R, Kashyap S, et al. Coordinated reduction of genes of oxidative metabolism in humans with insulin resistance and diabetes: potential role of PGC1 and NRF1. Proc Natl Acad Sci U S A 2003;100(14):8466–71. Available from: http://www.ncbi.nlm.nih.gov/pubmed/12832613.

[44] Brøns C, Jacobsen S, Nilsson E, Rönn T, Jensen CB, Storgaard H, et al. Deoxyribonucleic acid methylation and gene expression of PPARGC1A in human muscle is influenced by high-fat overfeeding in a birth-weight-dependent manner. J Clin Endocrinol Metab 2010;95(6):3048–56. Available from: http://www.ncbi.nlm.nih.gov/pubmed/20410232.

[45] Gillberg L, Jacobsen SC, Rönn T, Brøns C, Vaag A. PPARGC1A DNA methylation in subcutaneous adipose tissue in low birth weight subjects—impact of 5 days of high-fat overfeeding. Metabolism 2014;63(2):263–71. Available from: http://www.ncbi.nlm.nih.gov/pubmed/24262291.

[46] Jørgensen SW, Brøns C, Bluck L, Hjort L, Færch K, Thankamony A, et al. Metabolic response to 36 hours of fasting in young men born small vs appropriate for gestational age. Diabetologia. 2014;58(1):178–87.

[47] Laker RC, Lillard TS, Okutsu M, Zhang M, Hoehn KL, Connelly JJ, et al. Exercise prevents maternal high-fat diet-induced hypermethylation of the Pgc-1α gene and age-dependent metabolic dysfunction in the offspring. Diabetes 2014;63(5):1605–11. Available from: http://www.ncbi.nlm.nih.gov/pubmed/24430439.

[48] Nitert MD, Dayeh T, Volkov P, Elgzyri T, Hall E, Nilsson E, et al. Impact of an exercise intervention on DNA methylation in skeletal muscle from first-degree relatives of patients with type 2 diabetes. Diabetes 2012;61(12):3322–32. Available from: http://diabetes.diabetesjournals.org/cgi/doi/10.2337/db11-1653.

[49] Barrès R, Yan J, Egan B, Treebak JT, Rasmussen M, Fritz T, et al. Acute exercise remodels promoter methylation in human skeletal muscle. Cell Metab 2012;15(3):405–11. Available from: http://linkinghub.elsevier.com/retrieve/pii/S15504131120-00058.

[50] Hunter DJ, James L, Hussey B, Wadley AJ, Lindley MR, Mastana SS. Impact of aerobic exercise and fatty acid supplementation on global and gene-specific DNA methylation. Epigenetics 2019;14(3):294–309. Available from: https://www.tandfonline.com/doi/full/10.1080/15592294.2019.1582276.

[51] Alibegovic AC, Sonne MP, Hojbjerre L, Bork-Jensen J, Jacobsen S, Nilsson E, et al. Insulin resistance induced by physical inactivity is associated with multiple transcriptional changes in skeletal muscle in young men. AJP Endocrinol Metab 2010;299(5):E752–63. Available from: http://ajpendo.physiology.org/cgi/doi/10.1152/ajpendo.00590.2009.

[52] Gemma C, Sookoian S, Alvariñas J, García SI, Quintana L, Kanevsky D, et al. Maternal pregestational BMI Is associated with methylation of the PPARGC1A promoter in newborns. Obesity 2009;17(5):1032–9. Available from: http://doi.wiley.com/10.1038/oby.2008.605.

[53] Clarke-Harris R, Wilkin TJ, Hosking J, Pinkney J, Jeffery AN, Metcalf BS, et al. PGC1 promoter methylation in blood at 5–7 years predicts adiposity from 9 to 14 years (EarlyBird 50). Diabetes 2014;63(7):2528–37. Available from: http://diabetes.diabetesjournals.org/cgi/doi/10.2337/db13-0671.

[54] Kelstrup L, Hjort L, Houshmand-Oeregaard A, Clausen TD, Hansen NS, Broholm C, et al. Gene expression and DNA methylation of PPARGC1A in muscle and adipose tissue from adult offspring of women with diabetes in pregnancy. Diabetes 2016;65(10):2900–10. Available from: http://www.ncbi.nlm.nih.gov/pubmed/27388218.

[55] Ling C, Del Guerra S, Lupi R, Rönn T, Granhall C, Luthman H, et al. Epigenetic regulation of PPARGC1A in human type 2 diabetic islets and effect on insulin secretion. Diabetologia 2008;51(4):615–22. Available from: http://link.springer.com/10.1007/s00125-007-0916-5.

[56] Barrès R, Osler ME, Yan J, Rune A, Fritz T, Caidahl K, et al. Non-CpG methylation of the PGC-1α promoter through DNMT3B controls mitochondrial density. Cell Metab 2009;10(3):189–98. Available from: http://linkinghub.elsevier.com/retrieve/pii/S1550413109002290.

[57] Ribel-Madsen R, Fraga MF, Jacobsen S, Bork-Jensen J, Lara E, Calvanese V, et al. Genome-wide analysis of DNA methylation differences in muscle and fat from monozygotic twins discordant for Type 2 diabetes. PLoS One 2012;7(12):e51302. Available from: http://dx.plos.org/10.1371/journal.pone.0051302.

[58] Gao H, Kerr A, Jiao H, Hon C-C, Rydén M, Dahlman I, et al. Long non-coding RNAs associated with metabolic traits in human white adipose tissue. EBioMedicine 2018;30:248–60. Available from: http://www.ncbi.nlm.nih.gov/pubmed/29580841.

[59] Gillberg L, Perfilyev A, Brøns C, Thomasen M, Grunnet LG, Volkov P, et al. Adipose tissue transcriptomics and epigenomics in low birthweight men and controls: role of high-fat overfeeding. Diabetologia 2016;59(4):799–812.

[60] Jacobsen SC, Brøns C, Bork-Jensen J, Ribel-Madsen R, Yang B, Lara E, et al. Effects of short-term high-fat overfeeding on genome-wide DNA methylation in the skeletal muscle of healthy young men. Diabetologia 2012;55(12):3341–9. Available from: http://www.ncbi.nlm.nih.gov/pubmed/22961225.

[61] Li Y, Daniel M, Tollefsbol TO. Epigenetic regulation of caloric restriction in aging. BMC Med 2011;9(1):98. Available from: http://bmcmedicine.biomedcentral.com/articles/10.1186/1741-7015-9-98.

[62] Gillberg L, Rönn T, Jørgensen SW, Perfilyev A, Hjort L, Nilsson E, et al. Fasting unmasks differential fat and muscle transcriptional regulation of metabolic gene sets in low vs normal birth weight men. EBioMedicine 2019;47:341–51. Available from: https://linkinghub.elsevier.com/retrieve/pii/S2352396-41930533X.

[63] Hjort L, Jørgensen SW, Gillberg L, Hall E, Brøns C, Frystyk J, et al. 36 h fasting of young men influences adipose tissue DNA methylation of LEP and ADIPOQ in a birth weight-dependent manner. Clin Epigenetics 2017;9:40. Available from: http://www.ncbi.nlm.nih.gov/pubmed/28439315.

[64] Terashima M, Barbour S, Ren J, Yu W, Han Y, Muegge K. Effect of high fat diet on paternal sperm histone distribution and male offspring liver gene expression. Epigenetics 2015;10(9):861–71. Available from: http://www.tandfonline.com/doi/full/10.1080/15592294.2015.1075691.

[65] Ahn J, Lee H, Jung CH, Ha T. Lycopene inhibits hepatic steatosis via microRNA-21-induced downregulation of fatty acid-binding protein 7 in mice fed a high-fat diet. Mol Nutr Food Res 2012;56(11):1665–74. Available from: http://doi.wiley.com/10.1002/mnfr.201200182.

[66] Cordero P, Campion J, Milagro FI, Martinez JA. Transcriptomic and epigenetic changes in early liver steatosis associated to obesity: effect of dietary methyl donor supplementation. Mol Genet Metab 2013;110(3):388–95. Available from: http://www.ncbi.nlm.nih.gov/pubmed/24084163.

[67] Hawley JA, Hargreaves M, Joyner MJ, Zierath JR. Integrative biology of exercise. Cell 2014;159(4):738–49. Available from: http://www.ncbi.nlm.nih.gov/pubmed/25417152.

[68] Denham J, O'Brien BJ, Marques FZ, Charchar FJ. Changes in the leukocyte methylome and its effect on cardiovascular-related genes after exercise. J Appl Physiol 2015;118(4):475–88. Available from: http://www.ncbi.nlm.nih.gov/pubmed/25539938.

[69] Rönn T, Volkov P, Davegårdh C, Dayeh T, Hall E, Olsson AH, et al. A six months exercise intervention influences the genome-wide DNA methylation pattern in human adipose tissue. PLoS Genet 2013;9(6):e1003572. Available from: http://www.ncbi.nlm.nih.gov/pubmed/23825961.

[70] McGee SL, Fairlie E, Garnham AP, Hargreaves M. Exercise-induced histone modifications in human skeletal muscle. J Physiol 2009;587(Pt 24):5951–8. Available from: http://www.ncbi.nlm.nih.gov/pubmed/19884317.

[71] Egan B, Carson BP, Garcia-Roves PM, Chibalin AV, Sarsfield FM, Barron N, et al. Exercise intensity-dependent regulation of peroxisome proliferator-activated receptor γ coactivator-1α mRNA abundance is associated with differential activation of upstream signalling kinases in human skeletal muscle. J Physiol 2010;588(10):1779–90. Available from: http://doi.wiley.com/10.1113/jphysiol.2010.188011.

[72] Ntanasis-Stathopoulos J, Tzanninis JG, Philippou A, Koutsilieris M. Epigenetic regulation on gene expression induced by physical exercise. J Musculoskelet Neuronal Interact 2013;13(2):133–46. Available from: http://www.ncbi.nlm.nih.gov/pubmed/23728100.

[73] Barrès R, Zierath JR. The role of diet and exercise in the transgenerational epigenetic landscape of T2DM. Nat Rev Endocrinol 2016;12(8):441–51. Available from: http://www.ncbi.nlm.nih.gov/pubmed/27312865.

[74] Sharples AP, Stewart CE, Seaborne RA. Does skeletal muscle have an 'epi'-memory? The role of epigenetics in nutritional programming, metabolic disease, aging and exercise. Aging Cell 2016;15(4):603–16.

[75] Sharples AP, Polydorou I, Hughes DC, Owens DJ, Hughes TM, Stewart CE. Skeletal muscle cells possess a 'memory' of acute early life TNF-α exposure: role of epigenetic adaptation. Biogerontology. 2016;17(3):603–17.

[76] Stanford KI, Rasmussen M, Baer LA, Lehnig AC, Rowland LA, White JD, et al. Paternal exercise improves glucose metabolism in adult offspring. Diabetes 2018;67(12):2530–40. Available from: http://diabetes.diabetesjournals.org/lookup/doi/10.2337/db18-0667.

[77] Kusuyama J, Alves-Wagner AB, Conlin RH, Makarewicz NS, Albertson BG, Prince NB, et al. Placental superoxide dismutase 3 mediates benefits of maternal exercise on offspring health. Cell Metab 2021;33(5):939–56. Available from: https://linkinghub.elsevier.com/retrieve/pii/S155041312100111X.

[78] Baggish AL, Hale A, Weiner RB, Lewis GD, Systrom D, Wang F, et al. Dynamic regulation of circulating microRNA during acute exhaustive exercise and sustained aerobic exercise training. J Physiol 2011;589(Pt 16):3983–94. Available from: http://www.ncbi.nlm.nih.gov/pubmed/21690193.

[79] Güller I, Russell AP. MicroRNAs in skeletal muscle: their role and regulation in development, disease and function. J Physiol 2010;588(Pt 21):4075–87. Available from: http://www.ncbi.nlm.nih.gov/pubmed/20724363.

[80] Pedersen L, Idorn M, Olofsson GH, Lauenborg B, Nookaew I, Hansen RH, et al. Voluntary running suppresses tumor growth through epinephrine- and IL-6-dependent NK cell mobilization and redistribution. Cell Metab 2016;23(3):554–62. Available from: http://www.ncbi.nlm.nih.gov/pubmed/26895752.

[81] Coyle YM, Xie X-J, Lewis CM, Bu D, Milchgrub S, Euhus DM. Role of physical activity in modulating breast cancer risk as defined by APC and RASSF1A promoter hypermethylation in nonmalignant breast tissue. Cancer Epidemiol Biomarkers Prev 2007;16(2):192–6. Available from: http://www.ncbi.nlm.nih.gov/pubmed/17301249.

[82] Zeng H, Irwin ML, Lu L, Risch H, Mayne S, Mu L, et al. Physical activity and breast cancer survival: an epigenetic link through reduced methylation of a tumor suppressor gene L3MBTL1. Breast Cancer Res Treat 2012;133(1):127–35. Available from: http://www.ncbi.nlm.nih.gov/pubmed/21837478.

[83] Fernandez-Twinn DS, Hjort L, Novakovic B, Ozanne SE, Saffery R. Intrauterine programming of obesity and type 2 diabetes. Diabetologia 2019;62(10):1789–801. Available from: http://link.springer.com/10.1007/s00125-019-4951-9.

[84] Hales CN, Barker DJ. Type 2 (non-insulin-dependent) diabetes mellitus: the thrifty phenotype hypothesis. Diabetologia 1992;35(7):595–601. Available from: http://www.ncbi.nlm.nih.gov/pubmed/1644236.

[85] Ozanne SE. Epigenetics and metabolism in 2014: metabolic programming—knowns, unknowns and possibilities. Nat Rev Endocrinol 2015;11(2):67–8. Available from: http://doi.org/10.1038/nrendo.2014.218%5Cnpapers3://publication/doi/10.1038/nrendo.2014.218.

[86] Harder T, Rodekamp E, Schellong K, Dudenhausen JW, Plagemann A. Birth weight and subsequent risk of type 2 diabetes: a meta-analysis. Am J Epidemiol 2007;165(8):849–57. Available from: http://www.ncbi.nlm.nih.gov/pubmed/17215379.

[87] Susiarjo M, Xin F, Bansal A, Stefaniak M, Li C, Simmons RA, et al. Bisphenol a exposure disrupts metabolic health across multiple generations in the mouse. Endocrinology 2015;156(6):2049–58. Available from: http://www.ncbi.nlm.nih.gov/pubmed/25807043.

[88] Ding T, Mokshagundam S, Rinaudo PF, Osteen KG, Bruner-Tran KL. Paternal developmental toxicant exposure is associated with epigenetic modulation of sperm and placental Pgr and Igf2 in a mouse model. Biol Reprod 2018;99(4):864–76. Available from: http://www.ncbi.nlm.nih.gov/pubmed/29741588.

[89] Schultz NS, Broholm C, Gillberg L, Mortensen B, Jørgensen SW, Schultz HS, et al. Impaired leptin gene expression and release in cultured preadipocytes isolated from individuals born with low birth weight. Diabetes 2014;63(1):111–21.

[90] Jacobsen SC, Gillberg L, Bork-Jensen J, Ribel-Madsen R, Lara E, Calvanese V, et al. Young men with low birthweight exhibit decreased plasticity of genome-wide muscle DNA methylation by high-fat overfeeding. Diabetologia 2014;57(6):1154–8. Available from: http://www.ncbi.nlm.nih.gov/pubmed/24570141.

[91] Broholm C, Ribel-Madsen R, Hjort L, Olsson AH, Ahlers JMD, Hansen NS, et al. Epigenome- and transcriptome-wide changes in muscle stem cells from low birth weight men. Endocr Res 2020;45(1):58–71. Available from: https://www.tandfonline.com/doi/full/10.1080/07435800.2019.1669160.

[92] Fernandez-twinn DS, Cannell IG, David H, Warner M, Vaag AA, Brøns C, et al. Programming of adipose tissue miR-483-3p and GDF-3 expression by maternal diet in type 2 diabetes. Diabetologia 2012;44:1003–12.

[93] Bork-jensen J, Scheele C, Christophersen DV, Nilsson E, Friedrichsen M, Fernandez-twinn DS, et al. Glucose tolerance is associated with differential expression of microRNAs in skeletal muscle: results from studies of twins with and without type 2 diabetes. Diabetologia 2015;363–73.

[94] Warrington NM, Beaumont RN, Horikoshi M, Day FR, Helgeland Ø, Laurin C, et al. Maternal and fetal genetic effects on birth weight and their relevance to cardio-metabolic risk factors. Nat Genet 2019;51(5):804–14. Available from: http://www.nature.com/articles/s41588-019-0403-1.

[95] Ben-Haroush A, Yogev Y, Hod M. Epidemiology of gestational diabetes mellitus and its association with Type 2 diabetes. Diabet Med 2004;21(2):103–13. Available from: http://www.ncbi.nlm.nih.gov/pubmed/14984444.

[96] Buchanan TA, Xiang AH. Gestational diabetes mellitus. J Clin Invest 2005;115(3):485–91. Available from: http://www.ncbi.nlm.nih.gov/pubmed/15765129.

[97] Seshiah V, Balaji V, Balaji MS, Paneerselvam A, Arthi T, Thamizharasi M, et al. Prevalence of gestational diabetes mellitus in South India (Tamil Nadu)—a community based study. J Assoc Physicians India 2008;56:329–33. Available from: http://www.ncbi.nlm.nih.gov/pubmed/18700640.

[98] Lawrence JM, Contreras R, Chen W, Sacks DA. Trends in the prevalence of preexisting diabetes and gestational diabetes mellitus among a racially/ethnically diverse population of pregnant women, 1999–2005. Diabetes Care 2008;31(5):899–904. Available from: http://www.ncbi.nlm.nih.gov/pubmed/18223030.

[99] Clausen TD, Mathiesen ER, Hansen T, Pedersen O, Jensen DM, Lauenborg J, et al. High prevalence of type 2 diabetes and pre-diabetes in adult offspring of women with gestational diabetes mellitus or type 1 diabetes: the role of intrauterine hyperglycemia. Diabetes Care 2008;31(2):340–6. Available from: http://www.ncbi.nlm.nih.gov/pubmed/18000174.

[100] Clausen TD, Mathiesen ER, Hansen T, Pedersen O, Jensen DM, Lauenborg J, et al. Overweight and the metabolic syndrome in adult offspring of women with diet-treated gestational diabetes mellitus or type 1 diabetes. J Clin Endocrinol Metab 2009;94(7):2464–70. Available from: http://www.ncbi.nlm.nih.gov/pubmed/19417040.

[101] Fernandez-Twinn DS, Ozanne SE. Early life nutrition and metabolic programming. Ann N Y Acad Sci 2010;1212:78–96. Available from: http://www.ncbi.nlm.nih.gov/pubmed/21070247.

[102] Hansen NS, Hjort L, Broholm C, Gillberg L, Schrölkamp M, Schultz HS, et al. Metabolic and transcriptional changes in cultured muscle stem cells from low birth weight subjects. J Clin Endocrinol Metab 2016;2015–4214. Available from: http://press.endocrine.org/doi/10.1210/jc.2015-4214.

[103] Vaag AA, Grunnet LG, Arora GP, Brøns C. The thrifty phenotype hypothesis revisited. Diabetologia 2012;55(8):2085–8. Available from: http://www.ncbi.nlm.nih.gov/pubmed/22643933.

[104] Gautier JF, Porcher R, Khalil CA, Bellili-Munoz N, Fetita LS, Travert F, et al. Kidney dysfunction in adult offspring exposed in utero to type 1 diabetes is associated with alterations in genome-wide DNA methylation. PLoS One 2015;10(8):1–14.

[105] Nomura Y, Lambertini L, Rialdi A, Lee M, Mystal EY, Grabie M, et al. Global methylation in the placenta and umbilical cord blood from pregnancies with maternal gestational diabetes, preeclampsia, and obesity. Reprod Sci 2014;21(1):131–7. Available from: http://www.ncbi.nlm.nih.gov/pubmed/23765376.

[106] El Hajj N, Pliushch G, Schneider E, Dittrich M, Müller T, Korenkov M, et al. Metabolic programming of MEST DNA methylation by intrauterine exposure to gestational diabetes mellitus. Diabetes 2013;62(4):1320–8. Available from: http://www.ncbi.nlm.nih.gov/pubmed/23209187.

[107] Hjort L, Novakovic B, Grunnet LG, Maple-Brown L, Damm P, Desoye G, et al. Diabetes in pregnancy and epigenetic mechanisms—how the first 9 months from conception might affect the child's epigenome and later risk of disease. Lancet Diabetes Endocrinol 2019;7(10):796–806. Available from: https://linkinghub.elsevier.com/retrieve/pii/S2213858719300786.

[108] Sørensen A, van Poppel M, Desoye G, Damm P, Simmons D, Jensen D, et al. The predictive value of miR-16, -29a and -134 for early identification of gestational diabetes: a nested analysis of the DALI cohort. Cell 2021;10(1):170. Available from: https://www.mdpi.com/2073-4409/10/1/170.

[109] Jönsson J, Renault KM, García-Calzón S, Perfilyev A, Estampador AC, Nørgaard K, et al. Lifestyle intervention in pregnant women with obesity impacts cord blood DNA methylation, which associates with body composition in the offspring. Diabetes 2021;70(4):854–66. Available from: http://diabetes.diabetesjournals.org/lookup/doi/10.2337/db20-0487.

[110] Baylin SB, Jones PA. A decade of exploring the cancer epigenome—biological and translational implications. Nat Rev Cancer 2011;11(10):726–34. Available from: http://www.nature.com/doifinder/10.1038/nrc3130.

[111] Neill SO, Driscoll LO. Metabolic syndrome: a closer look at the growing epidemic and its associated pathologies. Obes Rev 2015;16:1–12.

[112] Pedroso JAB, Ramos-Lobo AM, Donato J. SOCS3 as a future target to treat metabolic disorders. Hormones 2019;18(2):127–36. Available from: http://link.springer.com/10.1007/s42000-018-0078-5.

[113] Ali O, Cerjak D, Jr JWK, James R, Blangero J, Carless MA, et al. Methylation of SOCS3 is inversely associated with metabolic syndrome in an epigenome-wide association study of obesity. Epigenetics 2016;11:699–707.

[114] Hall E, Volkov P, Dayeh T, Bacos K, Ronn T, Nitert MD, et al. Effects of palmitate on genome-wide mRNA expression and DNA methylation patterns in human pancreatic islets. BMC Med 2014;12:103. Available from: http://www.ncbi.nlm.nih.gov/pubmed/24953961.

[115] Maples JM, Brault JJ, Shewchuk BM, Witczak CA, Zou K, Rowland N, et al. Lipid exposure elicits differential responses in gene expression and DNA methylation in primary human skeletal muscle cells from severely obese women. Physiol Genomics 2015;47(5):139–46.

[116] Yang Z, Cappello T, Wang L. Emerging role of microRNAs in lipid metabolism. Acta Pharm Sin B 2015;5(2):145–50. Available from: https://doi.org/10.1016/j.apsb.2015.01.002.

[117] Khalil CA. The emerging role of epigenetics in cardiovascular disease. Ther Adv Chronic Dis 2014;5:178–87.

[118] Smolarek I, Wyszko E, Barciszewska AM, Nowak S, Gawronska I, Jablecka A, et al. Global DNA methylation changes in blood of patients with essential hypertension. Med Sci Monit 2010;16(3) CR149-R155.

[119] Carson C, Lawson HA. Epigenetics of metabolic syndrome. Physiol Genomics 2018;50(11):947–55. Available from: https://www.physiology.org/doi/10.1152/physiolgenomics.00072.2018.

[120] Dick KJ, Nelson CP, Tsaprouni L, Sandling JK, Aïssi D, Wahl S, et al. DNA methylation and body-mass index: a genome-wide analysis. Lancet. 2014;383(9933):1990–8.

[121] Keller M, Kralisch S, Rohde K, Schleinitz D, Dietrich A, Schön MR, et al. Global DNA methylation levels in human adipose tissue are related to fat distribution and glucose homeostasis. Diabetologia. 2014;57(11):2374–83.

[122] Wang X, Pan Y, Zhu H, Hao G, Huang Y, Barnes V, et al. An epigenome-wide study of obesity in African American youth

[122] and young adults: novel findings, replication in neutrophils, and relationship with gene expression. Clin Epigenetics 2018;10(1):3. Available from: https://clinicalepigeneticsjournal.biomedcentral.com/articles/10.1186/s13148-017-0435-2.

[123] Wahl S, Drong A, Lehne B, Loh M, Scott WR, Kunze S, et al. Epigenome-wide association study of body mass index, and the adverse outcomes of adiposity. Nature 2017;541(7635):81−6. Available from: http://www.nature.com/articles/nature20784.

[124] Ling C, Rönn T. Epigenetics in human obesity and type 2 diabetes. Cell Metab 2019;29(5):1028−44. Available from: https://linkinghub.elsevier.com/retrieve/pii/S1550413119301378.

[125] Barres R, Kirchner H, Rasmussen M, Yan J, Kantor FR, Krook A, et al. Weight loss after gastric bypass surgery in human obesity remodels promoter methylation. Cell Rep 2013;3(4):1020−7. Available from: http://www.ncbi.nlm.nih.gov/pubmed/23583180.

[126] Kurylowicz A, Owczarz M, Polosak J, Jonas MI, Lisik W, Jonas M, et al. SIRT1 and SIRT7 expression in adipose tissues of obese and normal-weight individuals is regulated by microRNAs but not by methylation status. Int J Obes 2016;40(August):1−28. Available from: https://doi.org/10.1038/ijo.2016.131.

[127] Shi Y, Parag S, Patel R, Lui A, Murr M, Cai J, et al. Stabilization of lncRNA GAS5 by a small molecule and its implications in diabetic adipocytes. Cell. Chem Biol 2019;26(3):319−30. Available from: http://www.ncbi.nlm.nih.gov/pubmed/30661991.

[128] Jufvas A, Sjödin S, Lundqvist K, Amin R, Vener AV, Strålfors P. Global differences in specific histone H3 methylation are associated with overweight and type 2 diabetes. Clin Epigenetics 2013;5(1):15. Available from: http://www.ncbi.nlm.nih.gov/pubmed/24004477.

[129] Castellano-Castillo D, Denechaud P-D, Fajas L, Moreno-Indias I, Oliva-Olivera W, Tinahones F, et al. Human adipose tissue H3K4me3 histone mark in adipogenic, lipid metabolism and inflammatory genes is positively associated with BMI and HOMA-IR. PLoS One 2019;14(4):e0215083. Available from: http://www.ncbi.nlm.nih.gov/pubmed/30958852.

[130] Andrade S, Morais T, Sandovici I, Seabra AL, Constância M, Monteiro MP. Adipose tissue epigenetic profile in obesity-related dysglycemia − a systematic review. Front Endocrinol (Lausanne) 2021;12:681649. Available from: http://www.ncbi.nlm.nih.gov/pubmed/34290669.

[131] Donkin I, Barrès R. Sperm epigenetics and influence of environmental factors. Mol Metab 2018;14:1−11. Available from: https://linkinghub.elsevier.com/retrieve/pii/S2212877818301042.

[132] Donkin I, Versteyhe S, Ingerslev LR, Qian K, Mechta M, Nordkap L, et al. Obesity and bariatric surgery drive epigenetic variation of spermatozoa in humans. Cell Metab 2016;23(2):369−78. Available from: http://www.ncbi.nlm.nih.gov/pubmed/26669700.

[133] Pasquier J, Hoarau-véchot J, Fakhro K, Rafii A, Khalil CA. Epigenetics and cardiovascular disease in diabetes. Curr Diab Rep 2015;15:108.

[134] Dayeh T, Volkov P, Salö S, Hall E, Nilsson E, Olsson AH, et al. Genome-wide DNA methylation analysis of human pancreatic islets from type 2 diabetic and non-diabetic donors identifies candidate genes that influence insulin secretion. PLoS Genet 2014;10(3):e1004160. Available from: http://dx.plos.org/10.1371/journal.pgen.1004160.

[135] Volkmar M, Dedeurwaerder S, Cunha DA, Ndlovu MN, Defrance M, Deplus R, et al. DNA methylation profiling identifies epigenetic dysregulation in pancreatic islets from type 2 diabetic patients. EMBO J 2012;31(6):1405−26. Available from: http://emboj.embopress.org/cgi/doi/10.1038/emboj.2011.503.

[136] Yang BT, Dayeh TA, Volkov PA, Kirkpatrick CL, Malmgren S, Jing X, et al. Increased DNA methylation and decreased expression of PDX-1 in pancreatic islets from patients with type 2 diabetes. Mol Endocrinol 2012;26(7):1203−12. Available from: http://press.endocrine.org/doi/abs/10.1210/me.2012-1004.

[137] Yang BT, Dayeh TA, Kirkpatrick CL, Taneera J, Kumar R, Groop L, et al. Insulin promoter DNA methylation correlates negatively with insulin gene expression and positively with HbA(1c) levels in human pancreatic islets. Diabetologia 2011;54(2):360−7. Available from: http://link.springer.com/10.1007/s00125-010-1967-6.

[138] Hall E, Jönsson J, Ofori JK, Volkov P, Perfilyev A, Dekker Nitert M, et al. Glucolipotoxicity alters insulin secretion via epigenetic changes in human islets. Diabetes 2019;68(10):1965−74. Available from: http://diabetes.diabetesjournals.org/lookup/doi/10.2337/db18-0900.

[139] Ouni M, Saussenthaler S, Eichelmann F, Jähnert M, Stadion M, Wittenbecher C, et al. Epigenetic changes in islets of langerhans preceding the onset of diabetes. Diabetes 2020;69(11):2503−17. Available from: http://diabetes.diabetesjournals.org/lookup/doi/10.2337/db20-0204.

[140] Nilsson E, Matte A, Perfilyev A, de Mello VD, Käkelä P, Pihlajamäki J, et al. Epigenetic alterations in human liver from subjects with type 2 diabetes in parallel with reduced folate levels. J Clin Endocrinol Metab 2015;100(11):E1491−501. Available from: http://press.endocrine.org/doi/10.1210/jc.2015-3204.

[141] Nilsson E, Jansson PA, Perfilyev A, Volkov P, Pedersen M, Svensson MK, et al. Altered DNA methylation and differential expression of genes influencing metabolism and inflammation in adipose tissue from subjects with type 2 diabetes. Diabetes 2014;63(9):2962−76. Available from: http://diabetes.diabetesjournals.org/cgi/doi/10.2337/db13-1459.

[142] Davegårdh C, Säll J, Benrick A, Broholm C, Volkov P, Perfilyev A, et al. VPS39-deficiency observed in type 2 diabetes impairs muscle stem cell differentiation via altered autophagy and epigenetics. Nat Commun 2021;12(1):2431. Available from: http://www.nature.com/articles/s41467-021-22068-5.

[143] Guay C, Regazzi R. Circulating microRNAs as novel biomarkers for diabetes mellitus. Nat Rev Endocrinol 2013;9(9):513−21. Available from: https://doi.org/10.1038/nrendo.2013.86.

[144] Miao F, Chen Z, Genuth S, Paterson A, Zhang L, Wu X, et al. Evaluating the role of epigenetic histone modifications in the metabolic memory of type 1 diabetes. Diabetes 2014;63(5):1748−62.

[145] García-Calzón S, Perfilyev A, Martinell M, Ustinova M, Kalamajski S, Franks PW, et al. Epigenetic markers associated with metformin response and intolerance in drug-naïve patients with type 2 diabetes. Sci Transl Med 2020;12(561):eaaz1803. Available from: https://stm.sciencemag.org/lookup/doi/10.1126/scitranslmed.aaz1803.

[146] Roadmap Epigenomics Consortium Kundaje A, Meuleman W, Ernst J, Bilenky M, Yen A, et al. Integrative analysis of 111 reference human epigenomes. Nature 2015;518(7539):317−30. Available from: http://www.ncbi.nlm.nih.gov/pubmed/4530010.

CHAPTER 39

Imprinting Disorders in Humans

Thomas Eggermann

Institut für Humangenetik, RWTH Aachen, Aachen, Germany

OUTLINE

Introduction	779
Clinical Findings in Imprinting Disorders	779
Molecular Findings in Imprinting Disorders	781
Causes of Disturbed Imprinting	782
Cis-acting Factors	783
Trans-acting Factors	784
Maternal Effect Mutations and Multilocus Imprinting Disturbance	784
Translational Use of New Findings in ImpDis and New Methodologies	785
Molecular Diagnostics	*785*
Genetic and Reproductive Counseling	787
Clinical Management	787
Concluding Remarks	787
References	787

INTRODUCTION

Genomic imprinting describes the monoallelic and parent-of-origin dependent expression of more than 100 human genes. Molecularly, the phenomenon of genomic imprinting is based on epigenetic mechanisms and, therefore, comprises regulation by DNA modifications (e.g., 5-methylcytosine), non-coding RNAs (ncRNA) and chromatin structuring (e.g., by histone modifications). Imprinted genes tend to cluster, and each cluster is regulated by at least one differentially methylated region (DMR). Imprinted regions often harbor several DMRs, and these DMRs are then organized hierarchically with a germline DMR (gDMR) as the key regulator and subordinated somatic DMRs. An example is the imprinted region in 14q32 with the MEG3/DLK1:IG-DMR as the gDMR and the subordinated MEG3:TSS-DMR and MEG8:Int2-DMR (see [1]). As this hierarchical organization requires a balanced regulation, it is not surprising that the fine-tuned expression of imprinted genes is the prerequisite of proper development, and its disturbance has a severe impact on the phenotype.

CLINICAL FINDINGS IN IMPRINTING DISORDERS

The majority of imprinted genes play a role in growth and development, and accordingly molecular alterations within imprinting domains result in aberrant intrauterine and/or postnatal growth, metabolic disturbances, puberty, asymmetry, dysmorphisms and/or cognitive impairment. Additionally, some disorders are characterized by an increased risk to develop (embryonal) tumors.

Currently, 13 imprinting disorders (ImpDis) have been defined (Table 39.1). In fact, each ImpDis is characterized by specific features, and patients with the typical phenotypes often exhibit molecular changes at specific imprinted loci. However, many of these features are unspecific, and there is a broad clinical over-

TABLE 39.1 Overview on the Currently Known 13 ImpDis, the Associated Molecular Disturbances, and the Major Clinical Findings

Imprinting Disorder OMIM#	Prevalence	Chromosome	Molecular Defects (Frequency)	MLID[a]	Main Clinical Features
Transient neonatal diabetes mellitus (TNDM) 601410	1/300.000	6q24	upd(6)pat (41%)		IUGR, transient diabetes mellitus, hyperglycemia without ketoacidosis, macroglossia, abdominal wall defects
			dup(6q24)pat (33%)		
			PLAGL1:alt-TSS-DMR, LOM (26%)	50%	
Silver-Russell syndrome (SRS) 180860	1/75.000–1/100.000	7 11p15.15	upd(7)mat (7%–10%) upd(11p15)mat (1 case) dup(11p15)mat (<1%)	1 case	IUGR, PNGR, relative macrocephaly at birth, body asymmetry, prominent forehead, feeding difficulties
			H19/IGF2:IG:DMR, LOM (40%)	10%	
			CDKN1C, IGF2, HMGA2, PLAG1 point mutations		
Birk-Barel syndrome 612292	Unknown	8q24.3	KCNK9 point mutations		Intellectual disability, hypotonia, dysmorphism
Beckwith–Wiedemann syndrome (BWS) 30650	1/15.000	11p15.5	upd(11p15)pat (20%) dup(11p15)pat (2%–4%)		Macroglossia, exomphalos, lateralized overgrowth, Wilms tumor or nephroblastomatosis, hyperinsulinism, adrenal cortex cytomegaly, placental mesenchymal dysplasia
			H19/IGF2:IG-DMR, GOM (4%)		
			KCNQ1OT1:TSS-DMR, LOM (50%)	25%	
			CDKN1C point mutations (5% sporadic; familial 25%–50%)		
Temple syndrome (TS14) 616222	Unknown	14q32	upd(14)mat (70%)		IUGR, PNGR, neonatal hypotonia, feeding difficulties in infancy, truncal obesity, scoliosis, precocious puberty
			del(14q32)pat (10%)		
			MEG3/DLK1:IG-DMR, LOM (20%)	Single cases	
Kagami–Ogata syndrome (KOS14) 608149	Unknown	14q32	upd(14)pat (70%)		IUGR, polyhydramnion, abdominal wall defects, bell-shaped thorax, coat-hanger ribs
			del(14q32)mat (20%)		
			MEG3/DLK1:IG-DMR, GOM (10%)		
(familial) Central Precocious Puberty (CPPB)	Unknown	14q32	DLK1 point mutations		Central precocious puberty
Prader–Willi syndrome (PWS) 176270	1/25.000–1/10.000	15q11q13	del(15q11q13)pat (70%–75%) upd(15)mat (25%–30%)		PNGR, Intellectual disability, neonatal hypotonia, hypogenitalism, hypopigmentation, obesity, hyperphagia
			SNURF:TSS-DMR, GOM (1%)	1 case	

(Continued)

TABLE 39.1 (Continued)

Imprinting Disorder OMIM#	Prevalence	Chromosome	Molecular Defects (Frequency)	MLID[a]	Main Clinical Features
Angelman syndrome (AS) 105830	1/20.000–1/12.000	15q11q13	del(15q11q13)mat (75%)		Severe intellectual disability, microcephaly, no speech, unmotivated laughing, ataxia, seizures, scoliosis
			upd(15)pat (1%–2%)		
			SNURF:TSS-DMR, LOM (1%)		
			UBE3A point mutations (10%)		
Central precocious puberty 2 (CPPB2) 615356	Unknown	15q11.2	*MKRN3* point mutations		Early activation of the hypothalamic–pituitary–gonadal axis resulting in gonadotropin-dependent precocious puberty
Schaaf–Yang syndrome (SYS) 615547	Unknown	15q11.2	*MAGEL2* point mutations		Delayed psychomotor development, intellectual disability, hypotonia
upd(16)mat	Unknown	16p13.3	*ZNF597*:TSS-DMR, LOM	1 case	Similar to SRS
Pseudohypoparathyroidism 1B (PHP1B) 603233	Unknown	20q13	del(20q13)mat (single cases)		Resistance to PTH and other hormones, Albright hereditary osteodystrophy, subcutaneous ossifications, feeding behavior anomalies, abnormal growth patterns
			GNAS DMRs: LOM (>60%)	12.5%	
			upd(20)pat (2%–20%)		
			GNAS point mutations		
Mulchandani–Bhoj–Conlin syndrome (MBCS) 617352	Unknown	20	upd(20)mat		IUGR, PNGR, feeding difficulties

[a]The percentage of MLID prevalence is only estimated from different studies but is probably underestimated as multilocus testing is not standardized and not implemented in all laboratories.

IUGR, intrauterine growth retardation; *PNGR*, postnatal growth retardation; *PTH*, parathyroid hormone; *upd*, uniparental disomy; *pat*, paternal; *mat*, maternal; *del*, deletion; *dup*, duplication; *LOM*, loss of methylation; *GOM*, gain of methylation.

lap between several disorders (see [2]). This clinical overlap also refers to a broad spectrum of other monogenetically caused inborn disorders. As a result, the diagnostic workup of imprinting disorders is complex and requires a careful medical history. Thus, patients with a molecular ImpDis (Table 39.1) might remain undetected or misdiagnosed, with major consequences for their therapeutic management. Accordingly, the application of genetic tests targeting a broad spectrum of imprinted loci and molecular changes is currently in discussion.

MOLECULAR FINDINGS IN IMPRINTING DISORDERS

ImpDis are not a clinically defined group of entities, but they are characterized by the same patterns of molecular alterations affecting DMRs. Four different types have been defined (Fig. 39.1), and three of them represent changes of the DNA (copy number variations [CNVs], single nucleotide variants [SNVs], uniparental disomies [UPDs]). The fourth class comprises epimutations, and they consist of modified methylation marks without a change of the DMR's DNA sequence. For the majority of epimutations, the molecular cause is currently unknown (primary epimutations), but there is a growing number epimutations which are associated with genomic alterations (secondary epimutations, Fig. 39.2) [3]. Epimutations at specific imprinted loci are often associated with specific ImpDis, but the number of reports on patients with overlapping molecular and/or clinical alterations is increasing. An example is Silver-Russell syndrome (SRS) which is classically associated with 11p15.5 and chromosome 7 disturbances. However, due to the clinical overlap with the 14q32-associated Temple syndrome (TS14), patients with SRS features sometimes exhibit molecular changes in 14q32 [4]. Furthermore, in several epimutation carriers, more than one imprinted locus is affected in addition to the disease-specific DMR, a molecular condition which has been called Multilocus Imprinting Disturbance (MLID) [5].

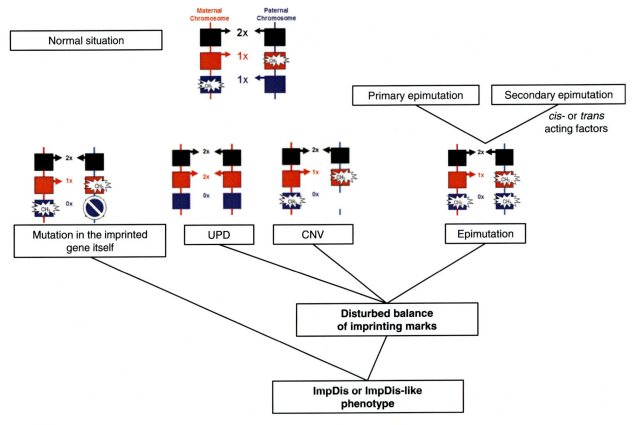

FIGURE 39.1 The four types of molecular alterations which have been identified to disturb the balanced expression of imprinted genes. *UPD*, uniparental disomy; *CNV*, copy number variant.

MLID particularly occurs in transient neonatal diabetes mellitus (TNDM), Beckwith–Wiedemann syndrome (BWS), SRS and Pseudohypoparathyroidism 1B (PHP1b), in these cohorts it accounts for 10%–50% of cases (Table 39.1).

The frequencies of the four molecular changes differ among the 13 ImpDis (Table 39.1). In Prader–Willi and Angelman syndromes (PWS, AS), deletions are the preponderant molecular subgroups; whereas, epimutations mainly occur in BWS and SRS. UPD appears to be frequent in TNDM and the chromosome 14 associated disorders, TS14 and Kagami–Ogata syndrome (KOS14). In contrast, with the exception of *UBE3A* and *CDKN1C* point mutations in AS and BWS, SNVs are rare or have never been observed in the well-established ImpDis. Nevertheless, there is a growing number of ImpDis with SNVs as the only molecular change (e.g., Schaaf–Yang syndrome).

Functionally, all four types of (epi)mutations lead to the imbalanced expression of the imprinted genes regulated by the affected DMR, and a reduced or increased dosage can be expected (Fig. 39.1). Depending on their physiological function, the respective gene products can either have an indirect physiological effect as they serve as regulators of the expression of protein-coding genes (e.g., *KCNQ1OT1* regulates the expression of *CDKN1C* in 11p15 [6]), or they are protein-coding with an immediate effect on the phenotype (e.g., *IGF2*). In fact, many imprinted genes are involved in fetal and neonate growth and metabolism, either by directly affecting the growth rate or indirectly by mediating the prenatal exchange of nutrients between mother and fetus or by modulating maternal care (see [7]). However, disturbances of genomic imprinting do not only affect early life but have life-long health consequences.

The etiology of UPDs, CNVs, and SNVs correspond to that of other chromosomal and monogenetic disorders, e.g., UPDs are the result of chromosomal nondisjunction. In contrast, the causes of epimutations are manifold, though they remain currently unclear in the majority of cases.

CAUSES OF DISTURBED IMPRINTING

Though the understanding of the etiology of epimutations is just at the beginning, there is a growing number of studies indicating that both environmental as

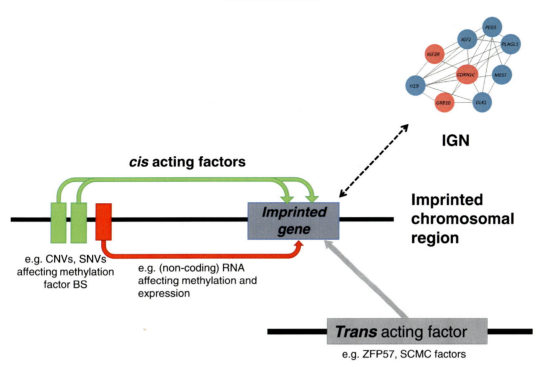

FIGURE 39.2 Functional interaction of *cis* and *trans* acting factors with imprinted genes. *BS*, binding sites; *SNV*, single nucleotide variant; *CNV*, copy number variant; *SCMC*, subcortical maternal complex.

well as genetic factors have an impact on genomic imprinting.

Currently, the majority of epimutations are classified as primary, probably due to the inadequacy of the currently applied diagnostic test approaches which do not address the genomic alterations behind them. Nevertheless, the sporadic occurrence of primary epimutations in the vast majority of cases indicate that non-genetic factors play a relevant role in the proper establishment and maintenance of imprinting marks. In fact, several environmental factors like the maternal nutritional situation, drugs and endocrine-disturbing substances have been identified to affect the methylation imprint in model systems [8–12]. The potential of exogenic factors to disturb the setting of imprinting marks suits to the vulnerability of this process in oocyte maturation and early embryogenesis as the most important ontogenetic window of resetting the genomic imprints. In humans, there is an increasing number of epidemiological studies indicating transgenerational effects of the exposure to environmental challenges on the methylation status of later generations. Assisted reproductive technologies (ART) have already been identified as another environmental risk factor for ImpDis. In a meta-analyses of more than 350 publications, Lazaraviciute and colleagues [13] could recently confirm an increased frequency of children with ImpDis born after ART. Nevertheless, the causal relationship between ART and altered imprinting could not be elucidated as the clinical indications for infertility treatment and the ART protocols differed markedly.

Whereas, the molecular links between environmental factors and aberrant imprinting has not yet been identified, there is an increasing number of reports on DNA sequence alterations with an impact on DMRs. These variants can be localized close to the DMR (*cis*-acting), or in another chromosomal region (*trans*-acting) (Fig. 39.2). Furthermore, in early development, *trans*-acting variants can even be localized in the maternal genome and affect DNA methylation of the later embryo, prior to of the embryonic genome activation (maternal effect mutations).

CIS-ACTING FACTORS

In the recent years, *cis*-acting factors have been identified for several clinically relevant germline DMRs. They consist of genomic DNA sequences close to the DMR which might either serve as binding sites for proteins required for the correct establishment and maintenance of imprinting marks, or which contribute to the parent-of-origin specific chromatin folding as the prerequisite for methylation and transcription factor binding. *Cis*-acting factors also include (non-coding) RNA transcripts encoded in imprinted regions which mediate the proper DMR methylation as well.

The functional relevance of methylation factor binding sites in imprinted regions has been illustrated by BWS and SRS patients carrying deletions within the imprinting center 1 (IC1, H19/IGF2:IG-DMR) in 11p15. At least two genes are under the control of the H19/IGF2:IG-DMR in this imprinting center, the paternally expressed growth factor IGF2 and the maternally expressed H19 gene. Disturbances of their expression by one of the aforementioned molecular alterations result in either overgrowth (BWS) or growth restriction (SRS). The IC1 harbors binding sites for methylation-specific DNA binding factors (see [14]). CCCTC-binding factor (CTCF) binds to the unmethylated maternal IC1 allele and serves as an insulator for enabling enhancers to interact with the appropriate gene (here: the maternal H19 copy); whereas, the maintenance of the paternal IC1 methylation is attributable to binding of the KRAB zinc finger protein ZFP57. This cluster of repetitive CTCF/ZFP57 binding sites (BS) as well as BS for further factors involved in the life cycle of imprinting (e.g., YY1, OCT4/SOX, ZBTB33 [15–17]) is therefore needed for the correct function of the IC1, and copy number variations of the IC1 disturb spatial organization and methylation of the H19/IGF2:IG-DMR.

For several imprinted regions, the *cis*-acting role of (non-coding) RNA transcripts in the regulation of methylation has been hypothesized: Beygo and coworkers [1] reported a KOS14 patient with a deletion in 14q32 between the germline MEG3/DLK1:IG-DMR and the MEG8:Int2-DMR organized in a hierarchical manner. Whereas, the MEG3/DLK1:IG-DMR showed normal methylation patterns, the MEG8:Int2-DMR was hypomethylated. The authors suggested that this loss of methylation might be caused by the loss of a transcript encoded by the region deleted on the maternal allele of the patient, which is required to maintain the methylation of the MEG8-DMR on the maternal chromosome. This assumption is based on similar observations from BWS patients carrying KCNQ1 mutations on the maternal allele and LOM of the imprinting center 2 in 11p15.5 on the same chromosome [18]. Here, the lack of an intact KCNQ1 transcript probably has an impact in *cis* on the hypermethylation status of the IC2.

TRANS-ACTING FACTORS

Due to its complexity, the imprinting cycle of life is prone to disturbances, and these can occur during all steps, from establishment to maintenance and erasure. In fact, *trans*-acting genomic mutations affecting the maintenance of methylation in the oocyte and early embryo have already been identified, but disturbances of further stages of development can be assumed (see [9]).

In the rare imprinting disorder transient neonatal diabetes mellitus (TNDM), mutations in the ZFP57 gene were the first *trans*-acting alterations identified [19]. Germline mutations in ZFP57 have been identified to cause LOM and are associated with transient neonatal diabetes mellitus [19], but so far have not been linked to SRS or BWS [20,21]. ZFP57 variants are mainly associated with LOM of the PLAGL1:alt-TSS-DMR, GRB10:alt-TSS-DMR and PEG3:TSS-DMR, by interaction with the KAP1 co-repressor complex it protects these loci from demethylation during early embryogenesis (see [22]). In fact, further factors are involved in the complex maintenance of imprinting marks, and their disturbances might contribute to aberrant imprinting as well [23,24].

The functional interaction between imprinted genes in the so-called "imprinted gene network" (IGN) has been suggested [25] more than 15 years ago, and could be confirmed by numerous studies (see [26]). Some examples are the results of Stelzer et al. [27] in induced pluripotent stem cells (iPSCs) from PWS patients and human parthenogenetic IPSCs, in which the upregulation of 14q32 encoded maternally expressed genes could be observed. They could identify a long ncRNA in the PWS region (IPW) in 15q11q13 as a regulator of imprinted genes in 14q32. Clinically, this observation fits with the overlap between PWS and TS14.

MATERNAL EFFECT MUTATIONS AND MULTILOCUS IMPRINTING DISTURBANCE

Maternal effect variants are another group of *trans*-acting factors, but in contrast to the aforementioned factors, trans means that variants in the maternal genome have an impact on the methylation status of the embryonic genome. These maternal effect mutations occur in genes encoding members of the subcortical maternal membrane (SCMC; see [28]) (Fig. 39.3, Table 39.2). As the SCMC in the (maturing) oocyte is required for the proper meiotic division in the oocyte and later in the first mitotic division prior to the zygote genome activation, maternal SCMC mutations can cause preimplantation embryonic arrest (for example: [29]) and infertility, as well as aneuploidy in the offspring [30]. Additionally, the proper rearrangement of cell compartments as well as establishment and maintenance of imprinting marks can be affected. Altogether, disturbances of this key structure in oocyte and early embryo maturation generally result in infertility, hydatidiform moles, miscarriages, pregnancy-associated disorders, aneuploidies, and disturbed

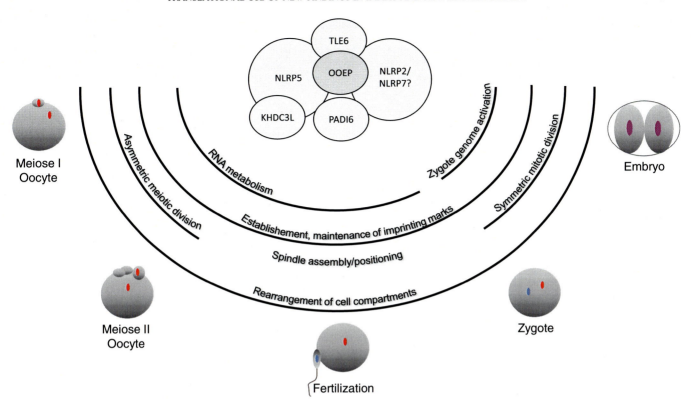

FIGURE 39.3 The SCMC comprises at least seven proteins with different roles in oocyte maturation and early embryonic development.

imprinting in the embryonic genome (see [31]). Though the members of the SCMC play different roles in the oocyte maturation process, a correlation of their disturbances with specific clinical outcomes is currently not obvious. In respect to epimutations in the offspring of maternal effect mutations carrying mothers, both specific epimutations as well as a heterogeneous MLID pattern have been identified [30,32], but a correlation between affected SCMC gene, epimutation pattern, or clinical impact is currently not obvious. Additional functional, molecular, and clinical studies are needed to further enlighten the relevance of maternal-effect variants, and the understanding of the regulation of genomic imprinting is the basis of a more personalized treatment and counseling.

TRANSLATIONAL USE OF NEW FINDINGS IN IMPDIS AND NEW METHODOLOGIES

Molecular Diagnostics

The last decade has seen a remarkable advance of knowledge on genomic imprinting disturbances, as well as an enormous progress in molecular diagnostic methods. Next generation sequencing-based analysis of the whole exome is meanwhile well implemented in daily human genetic routine, and it is only a question of time that datasets from further omic approaches (e.g., whole genome sequencing, epigenomics, transcriptomics) are used to comprehensively decipher the molecular cause of a congenital disorder.

The identification of new ImpDis, their clinical and molecular overlap as well as the agreement on consensus guidelines for diagnostics and treatment (e.g., [33–35]) has had a significant impact on clinical management, and thereby constantly contributes to an improvement of care of the patients and their families.

In the field of molecular diagnosis of ImpDis, several methods have been developed to address the different molecular subtypes. In particular, the implementation of methylation-specific multiplex ligation-dependent probe amplification (Ms-MLPA) [36] has given a remarkable impetus in genetic testing of ImpDis, as it enables the detection of the major three molecular classes of variants (UPD, CNVs, epimutations) in one run, is relatively easy to handle, and commercially available. Furthermore, the different assays Ms-MLPA assay target all currently known and clinically relevant DMRs. The standardised definition and nomenclature of these DMRs [37] has additionally facilitated and harmonized

TABLE 39.2 Factors of the SCMC and Clinical Outcome in Case of Mutations

Human Gene	Mouse Gene	Known Functions	Effect on Female Fertility	HYDM	Miscarriages	Aneuploidy	MLID in Offspring	Representative References
NLRP2 (NLR Family Pyrin domain-containing 2)	Nlrp2	– Proper methylation at imprinted loci – Spindle assembly			Yes	Yes	Yes	[30,41]
NLRP5 (NLR Family Pyrin domain-containing 5)	Mater	– Interacts with KHDC3L – Interacts with OOEP/TE6 – Spindle assembly/position	Yes	Yes	Yes		Yes	[42–44]
NLRP7 (NLR Family Pyrin domain-containing 7)	No orthologue	– Compensatory to NLRP2		Yes	Yes	Yes	Yes	[30,45]
OOEP (Oocyted permitting embryonic development)	Floped	– Interacts with NLRP5/TLE6 – Spindle assembly/position					Yes	[30]
TLE6 (Transducing-like enhancer of split 6)	Tle6	– Interacts with NLRP5/OOEP – Spindle assembly/position	Yes					[46]
PADI6 (Peptidyl arginine deiminase 6)	Padi6	– Spindle assembly/position – Rearrangement of organelles	Yes	Yes	Yes	Yes	Yes	[30]
KHDC3L (KH Domain containing 3-like ECTA1, C6ORF221)	Filia	– Interacts with NLRP5 – Spindle assembly		Yes	Yes		Yes	[47]

HYDM, hydatidiform mole; MLID, multilocus imprinting disturbance.

molecular testing of ImpDis. Quality assessment schemes (organized by European Molecular Quality Network—EMQN) for the most frequent ImpDis furthermore contribute to an increasing quality of diagnostic testing and reporting. Testing for ImpDis should be, as genetic test in general, targeted which means that only those chromosomal regions should be analyzed for which evidence for an association with specific phenotypes exist. This targeted testing avoids the detection of unintended and incidental findings or variants of uncertain significance (VUS).

However, the broad spectrum of molecular alterations detectable at least in some ImpDis (like SRS [38]) leads to the discussions whether a set of all clinically relevant DMRs should be analyzed in patients suspicious to suffer from an ImpDis. The application of such multilocus testing approaches should be considered carefully, and probably restricted to those ImpDis with a broad spectrum of molecular alterations; whereas, specific phenotypes and ImpDis like KOS14 probably do not require multilocus testing for diagnostic purposes.

Despite the challenges and limitations of multilocus testing in ImpDis, new (epi)genome-wide approaches like whole exome/whole genome sequencing and long-read sequencing generally provide the chance to identify new pathomechanism, and first studies have already proven their potential to significantly increase the diagnostic rate [39].

GENETIC AND REPRODUCTIVE COUNSELING

In any case, the benefit of testing for the patient and his family and its commensurability have to be considered, and unneeded uncertainty of the family by incidental findings or VUS has to be avoided. Nevertheless, the identification of risk factors like pathogenic SNVs or CNVs predisposing to an increased recurrence risk for ImpDis is the basis of a targeted genetic counseling. In specific situations and for research issues multilocus testing is a very valuable tool to identify MLID, as MLID contributes to an understanding of genomic imprinting and its disturbances. The clinical use of multilocus testing regarding a more personalized treatment is currently in discussions, but it has already been shown that the detection of MLID and underlying maternal effect variants can contribute to a more precise genetic and reproductive counseling. In fact, the proof of maternal effect variants explains recurrent miscarriages and ImpDis in the offspring, which might be circumvented by oocyte donation (see [31]).

CLINICAL MANAGEMENT

The determination of the precise molecular cause of an ImpDis is not only required to confirm the clinical diagnosis in this clinically heterogeneous group of congenital disorders, but in several entities it is needed for a more personalized clinical management. In BWS for example, the risk for tumor development and decision on monitoring is related to the molecular alterations, as these are associated with different risks to develop (embryonal) tumors (for review: [33]). In SRS the knowledge about the molecular cause helps to decide on growth hormone treatment, in the chromosome 7-associated SRS subgroup the management of cognitive impairment should be pursued [34]. On the other hand, the identification of the disease-causing mutation also avoids harms for the patients by medical malpractice.

The molecular mechanism causing ImpDis—loss of the only active gene copy or its double dosage—are promising targets for novel therapeutic approaches. Site specific methylation or demethylation should allow the activation of a silenced allele or to correct aberrant imprinting marks. In fact, such a targeted approach is hampered by several limitations: Imprinted genes are localized in chromosomal clusters; therefore, it is not always clear which gene in the affected gene cluster contributes to the disease. Furthermore, the targeted activation of specific DMRs requires specific methylation factors, which are often not specific for one DMR. However, due to the life-long handicap of some ImpDis, they are candidates for genetic therapy, and in mice with an AS phenotype first promising approaches have been reported [40]. However, genetic therapy will only be considered as an option in ImpDis with severe phenotypes, with the aim to significantly improve the clinical feature. Finally, the benefits of the therapy for the patient and his family must outweigh the side effects.

CONCLUDING REMARKS

ImpDis are a well-known group of congenital disorders since several years, but the recent years have shown a significant increase in knowledge on their molecular basis, spectrum of (epi)mutations and clinical spectrum. As a result, new ImpDis have been defined, and the molecular diagnostic strategy could be adapted to the broadening spectrum of molecular alterations. With the identification of cis- and trans-acting factors as well as maternal effect mutations which cause aberrant methylation at clinically relevant DMRs, new insights in the complex regulation of genomic imprinting and epigenetics in general have been obtained; therefore, ImpDis can be regarded as a model for epigenetic regulation and its disturbances in humans. Finally, the improved and more precise diagnostic strategies allow the a more personaliszd medicine, and even the step from curative to causal therapeutic strategies becomes imaginable.

References

[1] Beygo J, Kuchler A, Gillessen-Kaesbach G, Albrecht B, Eckle J, Eggermann T, et al. New insights into the imprinted MEG8-DMR in 14q32 and clinical and molecular description of novel patients with Temple syndrome. Eur J Hum Genet 2017;25(8):935−45.
[2] Eggermann T, Davies JH, Tauber M, van den Akker E, Hokken-Koelega A, Johansson G, et al. Growth restriction and genomic imprinting-overlapping phenotypes support the concept of an imprinting network. Genes (Basel) 2021;12(4).
[3] Horsthemke B. Mechanisms of imprint dysregulation. Am J Med Genet C Semin Med Genet 2010;154C(3):321−8.
[4] Geoffron S, Abi Habib W, Chantot-Bastaraud S, Dubern B, Steunou V, Azzi S, et al. Chromosome 14q32.2 imprinted region disruption as an alternative molecular diagnosis of Silver-Russell syndrome. J Clin Endocrinol Metab 2018;103(7):2436−46.
[5] Sanchez-Delgado M, Riccio A, Eggermann T, Maher ER, Lapunzina P, Mackay D, et al. Causes and consequences of multi-locus imprinting disturbances in humans. Trends Genet 2016;32(7):444−55.
[6] Chiesa N, De Crescenzo A, Mishra K, Perone L, Carella M, Palumbo O, et al. The KCNQ1OT1 imprinting control region and non-coding RNA: new properties derived from the study of Beckwith-Wiedemann syndrome and Silver-Russell syndrome cases. Hum Mol Genet 2012;21(1):10−25.
[7] Cassidy FC, Charalambous M. Genomic imprinting, growth and maternal-fetal interactions. J Exp Biol 2018;221(Pt Suppl 1).

[8] Kitsiou-Tzeli S, Tzetis M. Maternal epigenetics and fetal and neonatal growth. Curr Opin Endocrinol Diabetes Obes 2017;24(1):43–6.

[9] Monk D, Mackay DJG, Eggermann T, Maher ER, Riccio A. Genomic imprinting disorders: lessons on how genome, epigenome and environment interact. Nat Rev Genet 2019;20(4):235–48.

[10] Xavier MJ, Salas-Huetos A, Oud MS, Aston KI, Veltman JA. Disease gene discovery in male infertility: past, present and future. Hum Genet 2021;140(1):7–19.

[11] Murphy SK, Itchon-Ramos N, Visco Z, Huang Z, Grenier C, Schrott R, et al. Cannabinoid exposure and altered DNA methylation in rat and human sperm. Epigenetics. 2018;13(12):1208–21.

[12] Dunford AR, Sangster JM. Maternal and paternal periconceptional nutrition as an indicator of offspring metabolic syndrome risk in later life through epigenetic imprinting: a systematic review. Diabetes Metab Syndr 2017;11(Suppl 2):S655–62.

[13] Lazaraviciute G, Kauser M, Bhattacharya S, Haggarty P, Bhattacharya S. A systematic review and meta-analysis of DNA methylation levels and imprinting disorders in children conceived by IVF/ICSI compared with children conceived spontaneously. Hum Reprod Update 2015;21(4):555–7.

[14] Sparago A, Cerrato F, Riccio A. Is ZFP57 binding to H19/IGF2:IG-DMR affected in Silver-Russell syndrome? Clin Epigenetics 2018;10:23.

[15] Zimmerman DL, Boddy CS, Schoenherr CS. Oct4/Sox2 binding sites contribute to maintaining hypomethylation of the maternal igf2/h19 imprinting control region. PLoS One 2013;8(12):e81962.

[16] Bohne F, Langer D, Martine U, Eider CS, Cencic R, Begemann M, et al. Kaiso mediates human ICR1 methylation maintenance and H19 transcriptional fine regulation. Clin Epigenetics 2016;8:47.

[17] Kim JD, Hinz AK, Bergmann A, Huang JM, Ovcharenko I, Stubbs L, et al. Identification of clustered YY1 binding sites in imprinting control regions. Genome Res 2006;16(7):901–11.

[18] Valente FM, Sparago A, Freschi A, Hill-Harfe K, Maas SM, Frints SGM, et al. Transcription alterations of KCNQ1 associated with imprinted methylation defects in the Beckwith-Wiedemann locus. Genet Med 2019;21(8):1808–20.

[19] Mackay DJ, Callaway JL, Marks SM, White HE, Acerini CL, Boonen SE, et al. Hypomethylation of multiple imprinted loci in individuals with transient neonatal diabetes is associated with mutations in ZFP57. Nat Genet 2008;40(8):949–51.

[20] Boonen SE, Hahnemann JM, Mackay D, Tommerup N, Brondum-Nielsen K, Tumer Z, et al. No evidence for pathogenic variants or maternal effect of ZFP57 as the cause of Beckwith-Wiedemann syndrome. Eur J Hum Genet 2012;20(1):119–21.

[21] Spengler S, Gogiel M, Schonherr N, Binder G, Eggermann T. Screening for genomic variants in ZFP57 in Silver-Russell syndrome patients with 11p15 epimutations. Eur J Med Genet 2009;52(6):415–16.

[22] Monteagudo-Sanchez A, Hernandez Mora JR, Simon C, Burton A, Tenorio J, Lapunzina P, et al. The role of ZFP57 and additional KRAB-zinc finger proteins in the maintenance of human imprinted methylation and multi-locus imprinting disturbances. Nucleic Acids Res 2020;48(20):11394–407.

[23] Kagami M, Hara-Isono K, Matsubara K, Nakabayashi K, Narumi S, Fukami M, et al. ZNF445: a homozygous truncating variant in a patient with Temple syndrome and multilocus imprinting disturbance. Clin Epigenetics 2021;13(1):119.

[24] Takahashi N, Coluccio A, Thorball CW, Planet E, Shi H, Offner S, et al. ZNF445 is a primary regulator of genomic imprinting. Genes Dev 2019;33(1–2):49–54.

[25] Varrault A, Gueydan C, Delalbre A, Bellmann A, Houssami S, Aknin C, et al. Zac1 regulates an imprinted gene network critically involved in the control of embryonic growth. Dev Cell 2006;11(5):711–22.

[26] Patten MM, Cowley M, Oakey RJ, Feil R. Regulatory links between imprinted genes: evolutionary predictions and consequences. Proc Biol Sci 2016;283(1824).

[27] Stelzer Y, Sagi I, Yanuka O, Eiges R, Benvenisty N. The noncoding RNA IPW regulates the imprinted DLK1-DIO3 locus in an induced pluripotent stem cell model of Prader-Willi syndrome. Nat Genet 2014;46(6):551–7.

[28] Monk D, Sanchez-Delgado M, Fisher R. NLRPs, the subcortical maternal complex and genomic imprinting. Reproduction 2017;154(6):R161–70.

[29] Zheng W, Chen L, Dai J, Dai C, Guo J, Lu C, et al. New biallelic mutations in PADI6 cause recurrent preimplantation embryonic arrest characterized by direct cleavage. J Assist Reprod Genet 2020;37(1):205–12.

[30] Begemann M, Rezwan FI, Beygo J, Docherty LE, Kolarova J, Schroeder C, et al. Maternal variants in NLRP and other maternal effect proteins are associated with multilocus imprinting disturbance in offspring. J Med Genet 2018;55(7):497–504.

[31] Elbracht M, Mackay D, Begemann M, Kagan KO, Eggermann T. Disturbed genomic imprinting and its relevance for human reproduction: causes and clinical consequences. Hum Reprod Update 2020;26(2):197–213.

[32] Soellner L, Kraft F, Sauer S, Begemann M, Kurth I, Elbracht M, et al. Search for cis-acting factors and maternal effect variants in Silver-Russell patients with ICR1 hypomethylation and their mothers. Eur J Hum Genet 2019;27(1):42–8.

[33] Brioude F, Kalish JM, Mussa A, Foster AC, Bliek J, Ferrero GB, et al. Expert consensus document: clinical and molecular diagnosis, screening and management of Beckwith-Wiedemann syndrome: an international consensus statement. Nat Rev Endocrinol 2018;14(4):229–49.

[34] Wakeling EL, Brioude F, Lokulo-Sodipe O, O'Connell SM, Salem J, Bliek J, et al. Diagnosis and management of Silver-Russell syndrome: first international consensus statement. Nat Rev Endocrinol 2017;13(2):105–24.

[35] Mantovani G, Bastepe M, Monk D, de Sanctis L, Thiele S, Usardi A, et al. Diagnosis and management of pseudohypoparathyroidism and related disorders: first international consensus statement. Nat Rev Endocrinol 2018;14(8):476–500.

[36] Eijk-Van Os PG, Schouten JP. Multiplex Ligation-dependent Probe Amplification (MLPA(R)) for the detection of copy number variation in genomic sequences. Methods Mol Biol 2011;688:97–126.

[37] Monk D, Morales J, den Dunnen JT, Russo S, Court F, Prawitt D, et al. Recommendations for a nomenclature system for reporting methylation aberrations in imprinted domains. Epigenetics. 2018;13(2):117–21.

[38] Eggermann T, Bruck J, Knopp C, Fekete G, Kratz C, Tasic V, et al. Need for a precise molecular diagnosis in Beckwith-Wiedemann and Silver-Russell syndrome: what has to be considered and why it is important. J Mol Med (Berl) 2020;98(10):1447–55.

[39] Wright CF, FitzPatrick DR, Firth HV. Paediatric genomics: diagnosing rare disease in children. Nat Rev Genet 2018;19(5):325.

[40] Silva-Santos S, van Woerden GM, Bruinsma CF, Mientjes E, Jolfaei MA, Distel B, et al. Ube3a reinstatement identifies distinct developmental windows in a murine Angelman syndrome model. J Clin Invest 2015;125(5):2069–76.

[41] Meyer E, Lim D, Pasha S, Tee LJ, Rahman F, Yates JR, et al. Germline mutation in NLRP2 (NALP2) in a familial imprinting disorder (Beckwith-Wiedemann syndrome). PLoS Genet 2009;5(3):e1000423.

[42] Docherty LE, Rezwan FI, Poole RL, Turner CL, Kivuva E, Maher ER, et al. Mutations in NLRP5 are associated with reproductive wastage and multilocus imprinting disorders in humans. Nat Commun 2015;6:8086.

[43] Rezaei M, Suresh B, Bereke E, Hadipour Z, Aguinaga M, Qian J, et al. Novel pathogenic variants in NLRP7, NLRP5, and PADI6 in patients with recurrent hydatidiform moles and reproductive failure. Clin Genet 2021;99(6):823—8.

[44] Li M, Jia M, Zhao X, Shi R, Xue X. A new NLRP5 mutation causes female infertility and total fertilization failure. Gynecol Endocrinol 2021;37(3):283—4.

[45] Akoury E, Gupta N, Bagga R, Brown S, Dery C, Kabra M, et al. Live births in women with recurrent hydatidiform mole and two NLRP7 mutations. Reprod Biomed Online 2015;31(1):120—4.

[46] Alazami AM, Awad SM, Coskun S, Al-Hassan S, Hijazi H, Abdulwahab FM, et al. TLE6 mutation causes the earliest known human embryonic lethality. Genome Biol 2015;16:240.

[47] Demond H, Anvar Z, Jahromi BN, Sparago A, Verma A, Davari M, et al. A KHDC3L mutation resulting in recurrent hydatidiform mole causes genome-wide DNA methylation loss in oocytes and persistent imprinting defects post-fertilisation. Genome Med 2019;11(1):84.

SECTION IX

Epigenetic Therapy

40. Clinical applications of histone deacetylase inhibitors
41. Combination epigenetic therapy

CHAPTER

40

Clinical Applications of Histone Deacetylase Inhibitors

Romain Pacaud*, Jose Garcia, Scott Thomas* and Pamela N. Munster*

Division of Hematology and Oncology, University of California, San Francisco, CA, United States

OUTLINE

Introduction	793	Entinostat (Ms-275, SND-275)	805
HDACi for the Treatment of Hematological Malignancies	795	Valproic Acid	806
		Belinostat (Beleodaq, PXD101)	806
		Others HDACi	807
Vorinostat (Zolinza, Suberoylanilide Hydroxamic Acid)	796	Clinical Applications of HDACi for Noncancer Diseases	807
Panobinostat (Farydak, LBH589)	799	Current Therapeutic Applications	807
Romidepsin (Istodax, Depsipeptide, FK228)	799	Clinical Evaluation of Valproic Acid and Phenylbutyrate	808
Entinostat (Ms-275, SND-275)	800	Others HDACi	808
Valproic Acid	800	HDACi Dependent Reactivation of Latent HIV	809
Belinostat (Beleodaq, PXD101)	800		
Others HDACi	801	Conclusions and the Future Directions of the Clinical Applications of HDACi	809
HDACi in the Treatment of Solid Tumors	801		
Vorinostat (Zolinza, Suberoylanilide Hydroxamic Acid)	802	References	810
Panobinostat (Farydak, LBH589)	804		
Romidepsin (Istodax, Depsipeptide, FK228)	805		

INTRODUCTION

Histone acetyltransferases (HATs) and histone deacetylases (HDACs) play a crucial role in gene regulation. HATs catalyze the transfer of acetyl groups to lysine residues of histones, resulting in the relaxation of chromosomal DNA [1]. In general, this promotes transcription in the affected chromosomal regions. On the other hand, HDACs reverse histone acetylation, causing chromosomal DNA condensation and altered transcription. HDACs have also been shown to directly modify a variety of nonhistone substrates, such as p53, Ku70, pRB, and E2F-1 [2–4]. There are currently 18 known human HDACs that are separated into 4 classes (Table 40.1) [5]. Class I, II, and IV HDACs are zinc-dependent proteins, while class III HDACs require NAD^+ [6]. Most of the clinically relevant work to date involves class I, II, and IV HDACs, and thus they will be the focus of this chapter.

HDACi have shown early promise as a treatment option for a variety of cancer and noncancer diseases

*These authors contributed equally to the writing of this chapter.

TABLE 40.1 Histone Deacetylases (HDACs) and Their Inhibitors

	Class I	Class IIa	Class IIb	Class III	Class IV
Members	HDAC1 HDAC2 HDAC3 HDAC8	HDAC4 HDAC5 HDAC7 HDAC9	HDAC6 HDAC10	Sirt1–7	HDAC11
Location	Nuclear	Nuclear/cytoplasmic	Nuclear/cytoplasmic	Nuclear/cytoplasmic/mitochondrial	Nuclear
Substrates	Histones Stat3 Bcl-6 P53 E2F-1 LSD1 YY1 PCNA	Histones HP-1 GATA-1 Smad7 Hsc70 PGC1a MSH2	Histones α-Tubulin Hsp90 Smad7 MSH2	p53 NF-κB DNA pol B RNA pol I TAF Tubulin	Related to Rpd3 protein SHMT2
HDACi	Abexinostat Belinostat Entinostat Mocetinostat Panobinostat Phenylbutyrate Romidepsin Valproic acid Vorinostat K560[a] Chidamide Chromopeptide A[a] CHR-3996[a] Largazole[a] Givinostat Sodium butyrate[a] Resminostat TSA Quisinostat Pracinostat	Abexinostat Belinostat Panobinostat Phenylbutyrate Valproic Acid Vorinostat Givinostat TSA Phenylbutyrate Pracinostat	Abexinostat Belinostat Panobinostat Ricolinostat Vorinostat Quisinostat Givinostat TSA Sodium butyrate Resminostat	Nicotinamides	Abexinostat Belinostat Panobinostat Vorinostat Pracinostat

[a]Still experimental, not studied in cancer, or too few studies.
HDACs are divided into four main classes based on their similarity to yeast proteins [6]. HDACs have a variety of histone and non-histone substrates.

and several of them have been approved over the years, mainly in hematological malignancies. Despite the presumed global effect, pharmacological inhibition of HDACs alters the expression of ~0.5%–20% of genes [7]. Preclinical work has demonstrated that HDACi appear to have a differential effect on transformed cells compared to normal cells [8]. Depending on the class and subtype of a specific HDACi, its concentration, and the cell type being treated, it can broadly modulate a wide range of cellular functions including cell cycle arrest, or the up and downregulation of pro- and antiapoptotic genes. HDACi have been implicated in the regulation of angiogenesis, invasion, and metastases genes. An emerging area of interest is their effects on the modulation of DNA repair genes, as well as their interference in the immune response to cancer therapy [1,5,9–14].

Clinically relevant HDACi are subdivided based on their chemical structure and include hydroxamic acids, short-chain fatty acids, cyclic peptides, and benzamides. In vitro, each of them presents different activity toward HDAC inhibition, depending on their targets and range of activities (Table 40.2). The potency of hydroxamic acids lies within the micro- to nanomolar range in vitro [5,15]. These drugs include vorinostat [suberoylanilide hydroxamic acid (SAHA)], panobinostat (LBH589), belinostat (PXD101), trichostatin A (TSA), quisinostat (JNJ-26481585), givinostat (ITF2357), rocilinostat (ACY-1215), and abexinostat (PCI-24781). Short-chain fatty acids, which include butyrate and valproic acid (VPA), typically have weaker potencies in the millimolar range in vitro and shorter half-lives, however higher clinical concentrations can be achieved [15]. VPA causes the specific degradation of HDAC2 [16]. Cyclic peptides are a group of potent HDACi that function in the nanomolar range [15], with romidepsin (depsipeptide) as the most notable. Romidepsin is quite particular compared to the other HDACi. Indeed,

TABLE 40.2 Histone Deacetylases IC$_{50}$ (nM) per Histone Deacetylases Inhibitors

HDACi	HDAC1	HDAC2	HDAC3	HDAC8	HDAC4	HDAC5	HDAC7	HDAC9	HDAC6	HDAC10	HDAC11	Ref.
Abexinostat	21	63.4	148	370	60.4	48.2	350	168	11.5	52	13.6	[22]
Belinostat	41	125	30	216	115		67	128	82			[23]
Chidamide	95	160	67	733						78	432	[24]
Entinostat	181	1160	2310					505				[23]
Givinostat	28	56	21		52		163		27			[23]
Mocetinostat	9.26	35.6	192		810	899				5800	195	[22]
Panobinostat	0.31	1.51	1.69	38.9	0.42	0.84	38.6	1.65	1.98	0.61	0.32	[22]
Pracinostat	49	96	43	140	56	47	137	70	1000	40	93	[25]
Quisinostat	0.11	0.33	4.86	4.26	0.64	3.69	119	32.1	76.8	0.46	0.37	[22]
Resminostat	42.5		50.1	877					71.8			[26]
Ricolinostat	58	48	51	100	7000	5000	1400		4.7			[27]
Romidepsin	36	47			510				1400			[21]
Trichostatin A	2	3	4	456	6		5	6	3			[23]
Valproic acid	700000	800000	1000000		1500000	1000000	1300000					[28]
Vorinostat	61.8	241	257	212	154	101	2490	182	20	152	100	[22]

All the values were obtained through cell-free assays.

this molecule is a natural compound that was isolated from the bacteria *Chromobacterium violaceum*. First isolated as a potential antibiotic, it was more remarkable for its antitumor activity, even though the mechanism of action was not fully understood yet [17,18]. Benzamides are generally effective in the micromolar range in vitro and are represented by mocetinostat (MGCD0103) and entinostat (Ms-275) [5]. The mechanism underlying the function of individual HDACi remains largely unknown. Several cocrystal structures with HDAC8 suggest the hydroxamic acid moiety of TSA and vorinostat chelate with the zinc atom within the catalytic pocket of the protein [19,20]. The cyclic peptide romidepsin is thought to function similarly, perturbing the coordination of the zinc atom by HDAC8 and thus inhibiting its catalytic activity [21]. Depending on their structure, some HDACi (e.g., vorinostat and panobinostat) can indifferently inhibit all the HDAC isoforms and are thus called pan-inhibitors. Others can instead act specifically against some HDACs and are considered HDAC isoform-selective inhibitors (e.g., entinostat and rocilinostat) [13]. The clinical impact of select versus pan-HDACi has not clearly been established. Both have some advantages and disavantages. Pan-HDACi is usually considered the most potent inhibitors of HDAC in vitro (Table 40.2), but they can be way more toxic due to their broad range of action.

.The limited understanding of the mechanisms underpinning the HDACi mode of action renders the understanding of clinical on and off-target toxicities more challenging. Adjustment of dosing and schedules throughout multiple studies have allowed their clinical integration as monotherapy and combination regimens [29,30]. Data compiled from phase I and II clinical trials have demonstrated that HDACi are clinically efficacious for the treatment of various hematological malignancies. Furthermore, HDACi has been shown to potentiate the effects of existing treatments for solid and hematological malignancies and are currently under investigation in multiple combinatorial treatments. HDACi is also being evaluated for the treatment of numerous noncancer diseases, including neurological, viral, neuromuscular, pulmonary, and inflammatory diseases. In this chapter, the clinical use of HDACi will be discussed, with an emphasis on the treatment options for hematological and solid malignancies.

HDACI FOR THE TREATMENT OF HEMATOLOGICAL MALIGNANCIES

HDACi have demonstrated clinical efficacy in patients with hematological malignancies, including cutaneous T-cell lymphoma (CTCL), peripheral T-cell lymphoma (PTCL), B-cell lymphoma, acute myeloid leukemia (AML), myelodysplastic syndromes (MDS), and multiple myeloma (MM), at least in part due to their effects on common translocations (e.g., AML1/ETO or PLZF-RARα fusion proteins) [31,32]. Four HDACi are approved by the Food and Drug Administration (FDA) for the treatment of hematological malignancies. Vorinostat was approved for the treatment of CTCL in 2006, followed by romidepsin (CTCL, 2009; PTCL, 2011), belinostat (PTCL, 2014), and panobinostat (MM, 2015) [33]. Several other HDACi were granted the orphan drug status like mocetinostat for MDS and Diffuse Large B-cell Lymphoma (DLBCL) [34,35], as well as givinostat for polycythemia vera (PV) [36], VPA for DLBCL [37], and pracinostat for Acute Myeloid Leukemia (AML) [38].

Despite promising results from clinical trials of HDACi in select hematological malignancies, their efficacy as single agents has been modest. Numerous clinical trials are underway to evaluate their therapeutic potential when used in combination with other treatments.

Vorinostat (Zolinza, Suberoylanilide Hydroxamic Acid)

Vorinostat is a hydroxymate HDACi that targets class I, II, and IV HDACs. It is well tolerated at an oral dose of 400 mg/d and leads to a substantial increase in acetylated biomarkers in peripheral blood mononuclear cells (PBMCs) [13,39–41]. In 2006, vorinostat was approved as a therapeutic agent for patients who had undergone at least two systemic therapies for recurrent, progressive, or persistent CTCL [40]. Early clinical trials in patients with various solid and hematological malignancies demonstrated that vorinostat was generally well tolerated through either intravenous or oral dosing (Table 40.3) [42,43]. In contrast, vorinostat as a single agent is not effective in AML [44] and DLBCL [45]. Early studies showed vorinostat may enhance the antitumor activity of conventional and investigational drugs in hematological cancers. For instance, a phase I trial of vorinostat combined with a CHOP (cyclophosphamide, doxorubicin, vincristine, prednisone) regimen in patients with PTCL, showed a 2-year progression-free survival (PFS) rate of 79% [46]. The same regimen (plus rituximab) was tested in DLBCL and resulted in a 73% 2-years PFS, however excessive toxicity precluded further development [47]. In addition, a phase II trial of vorinostat, idarubicin, and cytarabine in patients with AML or MDS demonstrated an overall objective response of 85%, with 76% achieving complete response (CR) and 9% in complete remission with incomplete platelet recovery [48]. Unfortunately, a phase III trial where patients were randomized into two groups (both groups received idarubicin and cytarabine, the first one received also vorinostat and the second one did not) failed to confirm the promising results from the phase II trial, by showing no advantage to vorinostat adding (and more toxicity in the vorinostat group) (NCT01802333) [49]. Vorinostat was also explored in combination with the CDK9 inhibitor alvocidib in a phase I trial. Although 13 patients out of 26 total displayed stable disease progression, no objective response was achieved, and the combination was not further pursued [50].

Both histone deacetylation and DNA hypermethylation are crucial steps in the epigenetic regulation of cancer, thus combinations of HDACi and DNA methyltransferase (DNMT) inhibitors (e.g., decitabine, azacitidine) have been explored in MDS and leukemia where DNMT inhibitors (DNMTi) have been approved as single agents and other heme malignancies [91]. Notably, decitabine and vorinostat followed by chemotherapy showed an overall response rate (ORR) of 46.2% in heavily pretreated patients with relapsed/refractory acute lymphoblastic leukemia. Correlative studies of these trials suggested that decitabine triggered a global DNA hypomethylation which differed for responders versus nonresponders [51]. Despite encouraging efficacy results, these combinations were burdened with excess toxicity [92]. Sequential and concurrent administration of decitabine and vorinostat in patients with AML yielded an ORR of 23%, favoring the concurrent treatment [52]. This combination was further explored in a phase II trial with the addition of interleukin-2/activated natural-killer cell infusion in high-risk MDS patients. The regimen was found to be active (ORR: 5/9 patients), yet 77.8% of patients experienced grade 3 toxicities and the 1-year overall survival (OS) was only 33.3% [53]. Similarly, azacitidine, a second approved DNMTi underwent extensive testing with vorinostat in hematological malignancies. A phase II study focusing on patients with poor performance, organ dysfunction, or comorbidities carrying an AML or MDS suggested a benefit in adding vorinostat to azacitidine (ORR is 41% and CR of 23%) with an improvement in the 60 days survival (82% vs 67%) [54], yet could not be reproduced in another phase II trial showing no additional benefit of adding vorinostat to azacitidine in AML and MDS [55] upfront or upon progression [93]. Thus overall, it appears that the addition of vorinostat to DNMTi may not robustly add a therapeutic benefit but increases the risk of toxicity. Currently, vorinostat in combination with azacitidine is studied after allogeneic hematopoietic cell transplantation to prevent any relapse [94].

The ability of vorinostat to inhibit multiple pathways by degrading HIF-1α, inactivating Hsp90, and degrading Akt have made vorinostat an attractive agent to test in combination with bortezomib in patients with MM. In previously treated MM patients, 16 of 142 patients had a partial response (PR) or better response during a phase IIb study, yielding an ORR of 11% [56] and a 0.8-month increase in PFS compared to the treatment with placebo and bortezomib [95]. Newer strategies are now focused on combining checkpoint inhibitors and HDACi [96].

TABLE 40.3 Clinical Trials of HDACi for the Treatment of Hematological Malignancies

Hematological tumors	HDACi	Other drugs	Complete or partial response	Stable disease	References
AML	Vorinostat		1/37		Schaefer [44]
Diffuse large B-cell lymphoma	Vorinostat		1/18	1/18	Crump [45]
PTCL	Vorinostat	CHOP	12/14		Oki [46]
DLBCL	Vorinostat	Rituximab, CHOP	73% PFS (/72)		Persky [47]
AML/MDS	Vorinostat	Idarubicin, Cytarabine	64/75		Garcia-Manero [48]
AML/MDS	Vorinsotat	Idarubicin, Cytarabin	37% EFS (/216) Without vorinostat: 41% EFS (/261)		Clinicaltrials.gov [49]
Relapsed/ref acute leukemia	Vorinostat	Alvocidib		13/26	Holkova [50]
Relapsed/ref acute lymphoblastic leukemia	Vorinostat	Decitabine, vincristine, prednisone, doxorubicin, PEG-asparaginase	4/13 (CR) 2/13 (PR)	1/13	Burke [51]
AML	Vorinostat	Decitabine	8/36		How [52]
High-risk MDS	Vorinostat	Decitabine, Interleukin-2	5/9 (ORR)		Clinicaltrials.gov [53]
AML/MDS	Vorinostat	Azacitidine	20/81		Montalban-Bravo [54]
AML/MDS	Vorinostat	Azacitidine	42% ORR (vs. 41% Aza. Alone)		Craddock [55]
Relapsed/refMM	Vorinostat	Bortezomib	16/142		Siegel [56]
Hodgkin lymphoma/Non-Hodgkin Lymphoma	Vorinostat	Azacytidine Busulfan Gemcitabine Melphalan Rituximab	45/60 EFS		Clinicaltrials.gov [57]
CTCL	Panobinostat		2/10 (CR) 4/10 (PR)	1/10	Ellis [58]
Hodgkin's lymphoma RR to ASCT	Panobinostat		30/129 (CR) 5/129 (PR)		Younes [59]
Refractory CTCL	Panobinostat		24/139		Duvic [60]
Relapsed/ref MM	Panobinostat	Bortezomib, Dexamethasone	1/55 (CR) 18/55 (PR)	10/55	Richardson [61]
Relapsed/ref MM	Panobinostat	Bortezomib, Dexamethasone	235/387		San-Miguel [62]
Relapsed/ref MM	Panobinostat	Carfilzomib	63% ORR (/32)		Kaufman [63]
Relapsed MM	Panobinostat	Lenalidomide, Dexamethasone	2/27 CR 9/27 PR		Chari [64]
MM, Lymphomas	Panobinostat	Everolimus	20%–39% 0rr in lymphomas 0%–7% ORR in MM		Clinicaltrials.org [65]
AML/MDS	Panobinostat	Azacitidine	27.5% CR vs 14.3% CR with Aza alone		Garcia-Manero [66]
AML/MDS	Panobinostat	Decitabine	6/51 CR 11/51 PR	2/51	Uy [67]
CTCL	Romidepsin		4/71 (CR) 20/71 (PR)	26/71	Piekarz [68]

(Continued)

TABLE 40.3 (Continued)

Hematological tumors	HDACi	Other drugs	Complete or partial response	Stable disease	References
CTCL	Romidepsin		6/96 (CR) 27/96 (PR)	45/96	Whittaker [69]
PTCL	Romidepsin		8/45 (CR) 9/45 (PR)	5/45	Piekarz [70]
CTCL/PTCL	Romidepsin	Doxorubicin	4/21 CR 6/21 PR		Vu [71]
T-cell/B-cell lymphomas	Romidepsin	Pralatrexate	4/33 CR 7/33 PR	4/33	Amengual [72]
AML/MDS	Romidepsin		1/11 CR	6/11	Klimek [73]
Relapsed/ref acute leukemia	Entinostat		0/39		Gojo [74]
Relapsed/ref Hodgkin lymphoma	Entinostat		6/49	6/49 (≥6 months) 13/49 (<6 months)	Batlevi [75]
Relapsed/ref Hodgkin Lymphoma	Entinostat	Pembrolizumab	92% ORR (/13)		Sermer [76]
Myeloid neoplasm	Entinostat	Azacitidine	6 months OS vs, 13 months with Aza. alone		Prebet [77]
MDS/AML	Valproic acid	ATRA	8/18	4/18	Kuendgen [78]
MDS/AML	Valproic acid	ATRA	6/20		Pilatrino [79]
AML	Valproic acid	ATRA, Cytarabine	11/36		Fredly [80]
Advanced hematological neoplasia	Belinostat		0/16	5/16	Gimsing [81]
Relapsed/ref PTCL or CTCL	Belinostat		6/24 (PTCL) 4/24 (CTCL)	4/29 (PTCL) 10/29 (CTCL)	Foss [82]
Relapsed/ref PTCL	Belinostat		13/120 (CR) 18/120 (PR)	18/120	O'Connor [83]
PTCL	Belinostat	Cyclophosphamide, Doxorubicin Vincristine, Prednisone	86% ORR (/23)		Johnston [84]
Relapsed/ref follicular lymphoma	Abexinostat		56% ORR		Targeted Oncology [85]
PTCL	Chidamide	CHOP	26/43 CR 11/43 PR	1/43	Shi [86]
Relapsed/ref MM	Ricolinostat	Bortezomib, Dexamethasone	37% ORR		Vogl [87]
Myelofibrosis	Pracinostat	Ruxolitinib	80% ORR		Clinicaltrials.gov [88]
High-risk MDS	Pracinostat	Azacitidine	33% ORR		Atallah [89]
AML	Pracinostat	Azacitidine	42% CR		Garcia-Manero [90]

AML, acute myeloid leukemia; *ATRA*, all-trans retinoic acid; *CHOP*, cyclophosphamide, doxorubicin, vincristine, prednisone; *CTCL*, cutaneous T-cell lymphoma; *MDS*, myelodysplastic syndromes; *MM*, multiple myeloma; *PTCL*, peripheral T-cell lymphoma; *RR to ASCT*, relapsed or refractory to autologous stem-cell transplantation.

Panobinostat (Farydak, LBH589)

Panobinostat is a hydroxamic acid HDACi with high potency approved for the treatment of MM. It falls in the pan-HDACi category, by targeting class I, II, and IV HDACs (Tables 40.1 and 40.2). In an initial large phase I trial in various solid tumors and CTCL, panobinostat showed a response in 7 out of 10 CTCL patients but no objective responses in other malignancies [58,97].

The promising single-agent activity of panobinostat was noted in heavily pretreated patients with Hodgkin's lymphoma who relapsed or were refractory to autologous stem-cell transplantation [59]. Tumor reductions occurred in 96 of 129 patients, and 35 patients achieved an objective response. Panobinostat monotherapy also demonstrated antitumor activity in a phase II trial of patients with refractory CTCL regardless of prior bexarotene treatment [60].

The effects of panobinostat in modulating several cellular mechanisms involving cancer cell survival, angiogenesis, and immunity [98] prompted its exploration in combination with specific other agents to enhance antitumor activity. In a phase II trial of panobinostat in combination with bortezomib and dexamethasone in heavily pretreated bortezomib-refractory MM patients, the ORR was 34.5%, including one near-CR and 18 PR [61]. In a larger phase III trial including relapsed/refractory MM patients, the three-drugs combination led to a clinically relevant and significant increase in PFS compared with placebo, bortezomib, and dexamethasone [62]. While the proportion of patients achieving an OR was similar in both groups, a greater proportion of patients achieved a near-complete CR in the panobinostat group. These two studies confirmed the efficacy of panobinostat, bortezomib, and dexamethasone in patients with the bortezomib-refractory disease, further validating the benefit of HDAC inhibition in patients with MM. Thus, panobinostat in combination with bortezomib received approval for patients with MM who have received at least two prior standard therapies, including bortezomib and an immunomodulatory agent [99]. A new generation of proteasome inhibitor (PI), carfilzomib, displays similar results in combination with panobinostat, yet with fewer side effects [63]. Another approach aimed to reduce toxicity was to combine panobinostat and lenalidomide with dexamethasone in a phase II trial, leading to encouraging outcomes [64]. Further combinations with promising results include the addition of everolimus (mTOR inhibitor) to escalating doses of panobinostat in recurrent MM or lymphoma patients. At 20 mg of panobinostat (with a steady dose of everolimus), the ORR for lymphoma was 95% and 50% for MM. At a higher dose (30/40 mg), the ORR for lymphoma was 100%, but 0% for MM. However, the OS was significantly better at the higher dose (35.4 and 21.7 months for lymphoma and MM respectively) than at the lower dose (17.1 and 16.6 months respectively), but at the cost of more pronounced toxicity [65]. Similar to other HDACi, the addition of panobinostat to DNMTi did not demonstrate benefit to single-agent DNMTi therapy [66,67].

Romidepsin (Istodax, Depsipeptide, FK228)

Romidepsin, one of the first HDACi, had raised great interest due to its broad and potent in vitro and in vivo cytotoxic activity against human cancer cell lines, as well as human tumors in mice (xenografts), with minimal cytotoxicity in normal cells. On the opposite of a hydroxamic acid-like panobinostat, romidepsin is a selective HDACi, with a particular potency against HDAC1 and 2, followed by HDAC4, and HDAC6 to a lesser extent (Table 40.2) [100]. In a case report of a phase I trial, a CR was noted in one patient with PTCL, and PR was noted in three patients with CTCL after treatment with romidepsin [101], creating much hope for this disease compared to standard chemotherapy. This case report prompted a larger phase II study of romidepsin in patients with CTCL [68], confirming its activity by a response rate of 34%, including 4 CR, 20 PR, and 26 patients experiencing disease stabilizations [68]. A second phase II study of romidepsin monotherapy in patients with pretreated CTCL further solidified the data leading to the FDA approval of romidepsin in this disease context [69]. Similarly, a phase II trial of romidepsin in patients with PTCL who had received prior therapy demonstrated an ORR of 38% with a median response duration of 8.9 months [70]. As such, romidepsin has been FDA approved for patients with CTCL who received at least one prior systemic therapy, as well as patients with PTCL after one prior therapy. Due to its effect as a single agent, combination strategies have been tested to improve the therapeutic response. In relapsed or refractory T-cell lymphomas, romidepsin was combined with liposomal doxorubicin in a phase I dose-escalation study. The ORR for CTCL was 70% (and not as efficient in PTCL with 27%) [71]. The safety profile of this combination was acceptable and further studies, focused on CTCL, are being conducted. Romidepsin also displayed an interesting effect in another type of T-cell lymphoma, peripheral T-cell lymphoma (PTCL) in combination with pralatrexate. Indeed, the combination reached an ORR of 71% in T-cell lymphoma, compared to 33% in non-T-cell lymphoma. These data allowed us to proceed to a multicenter phase II study currently ongoing [72]. Romidepsin did not show any significant benefit as monotherapy with MDS or AML patients [73].

Other combinations are currently studied in different clinical trials, like azacitidine, dexamethasone, lenalidomide, or pembrolizumab [102,103].

Entinostat (Ms-275, SND-275)

Entinostat, a benzamide that specifically inhibits class I HDAC, has a relatively long half-life [104,105]. Trials have shown significant biological activity in patients with hematological malignancies receiving entinostat treatment [74,104]. In a phase II study of entinostat monotherapy in relapsed/refractory Hodgkin lymphoma, the ORR was modest (12%) [75]. Nonetheless, entinostat provided clinical benefits (CR + PR + stable disease) in 51% of this heavily pretreated population. While entinostat demonstrated a well-tolerated clinical profile suitable for combination, dose reductions were necessary to mitigate hematologic toxicities. In all, a well-tolerated clinical profile, mechanism of action, and the potential synergy with immunotherapy support further exploration of entinostat in combination with PD-1/PD-L1 inhibitors. A phase II study of entinostat and pembrolizumab in relapsed and refractory Hodgkin lymphoma suggested a promising ORR of 92%, including 3 patients (out of 13 patients) who progressed on prior anti-PD-1 therapy [76]. DNMTi were also tested in combination with entinostat. However, significant benefits were difficult to reach, due to a high toxicity profile [77].

Valproic Acid

VPA is an aliphatic short-branched chain fatty acid that was only identified later as an HDACi [106]. It selectively inhibits class I HDACs (and to a lesser extent class IIa HDACs) but only at high concentrations (IC_{50} in the millimolar range in vitro) (Table 40.2). Due to that, it is one of the least potent HDACi currently. Nevertheless, it has been widely studied because of its well-tolerated characteristic known through its use as an antiepileptic drug. VPA has been studied in patients with MDS and AML as a single therapy or in combination with other treatments. A phase II trial of VPA alone or in combination with all-*trans* retinoic acid (ATRA) in patients with MDS or AML secondary to MDS reported a response rate of nearly 44% in the VPA monotherapy cohort [78]. Interestingly, none of the patients who received VPA and ATRA concurrently from the beginning of the trial demonstrated a clinical response, suggesting that pretreatment with VPA may be necessary for a synergistic effect of both drugs [78]. A separate phase II study of sequential VPA and ATRA therapy in patients with AML or MDS, however, opposed this concept. In this trial, all the hematologic effects had to be attributed to VPA because they appeared before ATRA was added to the scheduled treatment and the withdrawal of ATRA did not change hematologic parameters [79].

A subsequent phase II study of VPA with intermittent ATRA and low-dose cytarabine in AML patients not eligible for conventional chemotherapy demonstrated that the combination treatment was well tolerated. Moreover, disease stabilization with the improvement of normal peripheral blood cell counts was observed for a subset of patients. VPA serum levels did not differentiate between responders and nonresponders indicating that it may not be a good biomarker. While remission is historically very uncommon in AML patients, this study had two patients with complete remission, suggesting that the combination with cytotoxic drugs may increase the antileukemic effects of VPA and ATRA [80].

Like with other HDACi, much emphasis has been placed on exploring the aberrant DNA methylation that is frequently seen in leukemia, with a combination of DNMTi and HDACi. As such, VPA was also studied in combination with azacitidine for the treatment of MDS and AML, often with the addition of ATRA [107]. In the absence of randomized controlled studies, the superiority of the combination over monotherapy remains unclear, and the combination is undergoing testing as maintenance treatment after allogeneic stem cell transplantation in patients with high-risk MDS/AML [108]. Similarly, a preclinical study assessing the effects of the combination of decitabine with VPA in leukemia cell lines demonstrated this combination to have synergistic activity [109], which did not translate into clinical efficacy in a small phase II study of decitabine with or without VPA in patients with MDS or AML. Adding VPA did not improve outcome [110], suggesting that HDACi may not improve DNMTi monotherapy in clinical settings.

Belinostat (Beleodaq, PXD101)

Numerous clinical trials examined the efficacy of belinostat, a hydroxamic acid, either alone or in conjunction with other modalities in the treatment of patients with hematological malignancies. In the first phase I study of belinostat in patients with advanced hematological neoplasia, no complete or partial responses were noted in the heavily pretreated patient population. Nonetheless, out of 16 patients, disease stabilization was achieved in five patients, including two patients with DLBCL, one patient with MM, and two patients with chronic myelogenous leukemia [81].

The focus of belinostat development was shifted early on towards establishing the efficacy and safety of belinostat monotherapy in patients with relapsed/refractory PTCL or CTCL [82]. The ORR was 25% (PTCL) and 13.8% (CTCL), with the treatment being well-tolerated in both cohorts. The favorable safety profile of belinostat prompted a larger phase II trial in patients with relapsed/refractory PTCL which showed an ORR of 25.8% and a median duration response of 13.6 months [83]. Its safety profile suggested that it could be combined with other regimens to further improve the outcome for patients with PTCL. Complete and durable responses with manageable toxicity have led to the FDA approval of belinostat for this indication in 2014 [82]. After this approval, belinostat was tested in multiple combinations, for example, concurrently given with a cyclophosphamide, doxorubicin hydrochloride (hydroxydaunorubicin), vincristine sulfate (Oncovin), and prednisone (CHOP) regimen. Some grade 3 Adverse Effects (AE) were noticed but nothing over the toxicity end point. On 33 patients total, the ORR was 86% with an impressive 71% CR [84]. Other studies explored the combination with azacitidine, bortezomib, or pevonedistat, but no results are yet available.

Others HDACi

Abexinostat, a hydroxamic acid, received a fast-track designation by the FDA in follicular lymphoma (with three lines of pretreatment), after demonstrating in a phase II trial an ORR of 56% [85]. In addition, a multicenter phase II study is evaluating the safety and efficacy of oral abexinostat in patients with relapsed or refractory follicular lymphoma, as well as another phase II trial in relapsed or refractory DLBCL [111,112]. In DLBCL, a combination with ibrutinib (a tyrosine kinase inhibitor specific to B cells) was also explored [113]. However, further, development was halted for strategic decisions. Mocetinostat, a benzamide HDACi, was studied in Hodgkin's lymphomas, demonstrating a tolerable treatment and promising effects as monotherapy in a phase II trial [114]. It is also studied in MDS and other blood cancers. Trichostatin A (TSA) has been widely studied preclinically but challenging pharmacokinetic (PK) profiles halted early development. Chidamide, another benzamide HDACi, much more focused on solid tumors, has been evaluated to a lesser degree in heme malignancies. One phase Ib/II trial got interesting results in a study of chidamide and CHOP regimen combination in untreated PTCL. The ORR reached in this study was 82%, with manageable AE [86]. Ricolinostat, an HDAC6 inhibitor, also shows a promising effect for relapsed or refractory MM in phase I/II trials in combination with bortezomib and dexamethasone [87]. Finally, pracinostat is also studied, particularly in MDS or other myeloid-related cancers. A phase II study combining pracinostat and ruxolitinib (a JAK inhibitor) shows an ORR of 80% with a safe toxicity profile [88]. It also has been combined at low doses with azacitidine in a phase II trial and demonstrated a tolerable dose for patients with an interesting 33% CR in MDS patients and also in another phase II study with older AML patients newly diagnosed with an ORR of 52% and more importantly a CR of 42% [89,90].

In summary, multiple HDACi have shown efficacy in heme malignancies, however, the very competitive landscape in lymphomas, considerable unaddressed cumulative toxicities and the high response rate required to show a benefit has relegated HDACi to the treatment of CTCL and PTCL. Panobinostat and others with activity and approval in myeloma are less frequently used due to adverse effects. The interest in pan-HDACi uses has been brought up multiple times, especially because of their toxicity (mostly hydroxamate-based HDACi). Indeed, HDACs are more or more prevalent depending on the tissue origin. For example, class II HDACs are highly represented in cardiac tissue, and their inhibition was correlated to severe cardio-toxicity [115]. They are also considered mutagenic agents in certain circumstances but are still widely tested due to their high HDAC inhibition potency [116]. The inability to clearly show superiority over DNMTi has precluded their use in MDS, and their place in leukemias remains to be established. Foremost, the reasons for hesitations in combinations are added toxicities, requiring dose adjustments, thus there are opportunities for the development of less toxic HDACi. In this optic, selective HDACi are currently in development, offering potentially a better-tuned treatment for patients, with less toxicity. However, the HDAC's mode of action redundancies may undercut the antitumor effect of these drugs [117]. Strategies that favor tailored combinations may be the key to successful HDACi action, but this requires tedious experimental work to find the right conditions [118].

HDACI IN THE TREATMENT OF SOLID TUMORS

The approval of HDACi in heme malignancies brought much interest to these agents for solid tumor malignancies. HDACi show cell cycle arrest, differentiation, and apoptosis at clinically feasible concentrations across many solid tumor cancer cell lines [15].

In mouse xenograft models, treatment with HDACi has been shown to reduce tumor size, and inhibit angiogenesis and metastasis [15]. Most recently, HDACi and epigenetic modifiers have gained interest for their effects on immunomodulatory cells and their potential use in combination with immunotherapy and checkpoint inhibitors [13].

Similarly, to their use in hematological malignancies, HDACi is reasonably well tolerated in patients with solid tumors. The most common toxicities include nausea, vomiting, anorexia, diarrhea, fatigue, neutropenia, and thrombocytopenia. Toxicities are more pronounced with cumulative drug exposure and often off target. Cardiac toxicities, particularly long QT intervals have been reported with romidepsin [119] and other HDACi early on; however, a study aimed to determine the risk of QT-prolongation with super-therapeutic doses of vorinostat (800 mg) did not demonstrate an increased risk of cardiac toxicity (such as QT-prolongations or arrhythmias) [120,121]. The occurrence of uncorrected electrolyte abnormalities with HDACi may have contributed to QT prolongation.

The efficacy of HDACi in solid tumors is more modest than in hematological malignancies. Data from phase 1 trials suggest that treatment-induced acetylation in PBMCs occurs at clinically feasible levels and acetylation in PBMCs correlates with acetylation in tumors [39]. However, acetylation may be a necessary pharmacological marker and is not a clear predictor of tumor response [23,122]. Further studies will be needed to determine whether acetylation is necessary and sufficient to induce target gene and protein modification required for tumor response and whether the presence of select HDAC enzymes further predicts response as already being reported in some articles [30,39,123−125].

The modest activity of HDACi in solid tumors may be further explained by cumulative toxicities often limiting extended exposure without interruptions and dose modifications. Additionally, most HDACi exhibit a short pharmacological half-life, further influencing the therapeutic window of currently available agents. Several preclinical in vitro and xenograft studies have suggested that the effects of HDACi on growth and differentiation were reversible upon drug withdrawal; hence, continuous drug exposure maybe in fact be necessary [126]. Some or all of these factors may contribute to the limited clinical efficacy of these drugs when given as single agents (Table 40.4).

Preclinical data have shown increased efficacy in solid tumors when HDACi are used in combination with other anticancer therapies. As such, several clinical trials are currently investigating the benefit of HDACi in combination with other therapeutics for the treatment of solid tumors.

Vorinostat (Zolinza, Suberoylanilide Hydroxamic Acid)

The efficacy of vorinostat in solid tumors is under investigation in different settings and tumor types. A phase II trial of vorinostat to treat recurrent glioblastoma multiforme reported moderate single-agent efficacy, with 9 out of 52 patients having stable disease (SD) status at 6 months [127]. However, treatment with vorinostat in a single agent setting did not achieve clinical response in the treatment of solid tumors in breast cancer [128], head and neck cancer [129], nonsmall cell lung cancer (NSCLC) [130], and prostate cancer [131] phase II trials.

Preclinical data have demonstrated that HDACi, specifical vorinostat, were shown to decondensed chromatin, allowing for increased access to DNA-damaging chemotherapies, such as topoisomerase inhibitors, nucleosides, cisplatin, and cisplatin-derivatives [148]. Results from clinical trials that use many of these combinations have been promising. A phase I trial combining vorinostat with 5-fluorouracil (5-FU), leucovorin, and oxaliplatin (FOLFOX) in refractory colon cancer patients established the maximum tolerated dose of vorinostat (300 mg oral, twice daily for 1 week every 2 weeks) in combination with FOLFOX [149]. A phase II trial combining vorinostat with the same drugs (except oxaliplatin) displayed interesting results with high ORR (100% in the highest dose group). However, the OS did not change between the low and high doses of vorinostat, and toxicity increased with serious AE when the dose is high [150]. A phase I trial combining vorinostat and capecitabine, a 5-FU prodrug, in patients with advanced solid tumors reported PR in 4 out of 28 patients (14%), 1 unconfirmed CR, and 18 patients (64%) with SD [151].

Preclinical data have also shown that the combination of HDACi with capecitabine results in synergistic antiproliferative and proapoptotic effects in colorectal and breast cancer [152,153]. A phase I trial of vorinostat in combination with capecitabine and cisplatin completed in patients with advanced gastric cancer showed a median PFS and OS of 7.1 and 18.0 months, respectively, with an ORR of 56%, suggesting that this combination is feasible and recommended for further development in advanced gastric cancer [154]. However, this combination did not show any significant beneficial effect for relapsed or metastatic head and neck cancer patients with only 2 patients out of 25 reaching the response criteria evaluation threshold [155].

TABLE 40.4 Clinical Trials of HDACi as Single Agents for the Treatment of Solid Tumors

Solid tumors	HDACi	Complete or partial response	Stable disease	References
GBM	Vorinostat	2/66	10/66	Galanis [127]
MBC	Vorinostat	0/14	2/14	Luu [128]
SCCHN	Vorinostat	1/13	3/13	Blumenschein [129]
NSCLC	Vorinostat	0/16	8/16	Traynor [130]
Prostate cancer	Vorinostat	0/27	2/27	Bradley [131]
Solid tumors	Panobinostat	0/14	6/14	Morita [132]
Refractory renal carcinoma	Panobinostat	0/20		Hainsworth [133]
Metastatic breast cancer		1/44 CR 1/44 PR	17/44	Clinicaltrials.gov [134]
Metastatic hormone ref. prostate cancer	Panobinostat	11.4% PFS (/35)		Clinicaltrials.gov [135]
Metastatic CRC	Romidepsin	0/25	4/25	Whitehead [136]
Lung cancer	Romidepsin	0/19	9/19	Schrump [120]
Metastatic RCC	Romidepsin	2/29		Stadler [137]
CRPC	Romidepsin	2/35	11/35	Molife [138]
Solid tumors	Romidepsin	0/28	9/28	Amiri-Kordestani [139]
Solid tumors	Entinostat	0/28	15/28	Ryan [104]
Solid tumors	Entinostat	0/22	1/22	Kummar [105]
Solid tumors	Entinostat	2/27	6/27	Gore [140]
Metastatic melanoma	Entinostat	0/28	7/28	Hauschild [141]
Solid tumors	Valproic acid	0/18	2/18	Atmaca [142]
Prostate cancer	Valproic acid	4/6 CR		Clinicaltrials.gov [143]
Solid tumors	Belinostat	0/48	18/48	Steele [144]
Thymic epithelial tumors	Belinostat	2/41	25/41	Giaccone [145]
Advanced solid tumors	Belinostat		13/47	Takebe [146]
Metastatic urothelial carcinoma	Mocetinostat	1/9 PR		Grivas [147]

CRPC, Castration-resistant prostate cancer; *GBM*, glioblastoma multiforme; *MBC*, metastatic breast cancer; *NSCLC*, non-small cell lung cancer; *RCC*, renal cell carcinoma; *SCCHN*, squamous cell carcinoma of the head and neck.

HDACi with efficacy against HDAC6 led to acetylation of tubulin and thus sensitize cells to mitotic inhibitors, such as paclitaxel [156]. Notably, a phase I study combining vorinostat with paclitaxel and carboplatin in patients with advanced solid tumors saw a PR in 11 out of 25 patients (44%), with 10 in NSCLC, and SD for 7 patients [157]. A subsequent randomized, double-blind phase II trial in NSCLC reported an increased response rate in patients receiving the HDACi combination over the chemotherapy alone [158]. The addition of vorinostat increased toxicity and the extended randomized phase III trial was halted, as the study did not meet its efficacy endpoint during a planned interim analysis (NCT00473889). Toxicities prompted the discontinuation of a phase II trial in ovarian cancer with vorinostat, carboplatin, and gemcitabine combination treatment [159].

HDACi have been shown to reverse resistance to hormonal therapy. Vorinostat showed encouraging activity in phase II clinical trial, reversing resistance to tamoxifen in breast cancer patients who progressed on aromatase inhibitors. The combination was well tolerated, with an objective response of 19% and stabilized disease for 9 out of 43 additional patients (21%) for greater than 6 months [124]. Notable findings from the trial were the strong correlation between HDAC2 expression with the response and the histone acetylation of PBMCs, not at the C_{max} but the trough level. A second phase II study also exploring hormone-resistance reversing in ER-positive breast cancer context with vorinostat, tamoxifen, and the anti-PD-1 pembrolizumab explored the efficacy of this combination. The limited ORR in 34 patients concluded the termination of the study. However, an analysis of a subgroup of 5 responders reveals an immunological signature of T-cell exhaustion ($CD8^{PD-1+/CTLA-4+}$) and treatment-induced depletion of regulatory T-cells ($CD4^{Foxp3+/CTLA-4+}$). These findings advocated for a preselection of the population by analysis of these biomarkers before treatment [160]. An ongoing phase II trial is focusing on stage IV breast cancer and will test the importance of vorinostat by adding or retracting the drug in the combination regimen [161]. Preclinical data have shown that vorinostat reverses resistance to the tyrosine kinase inhibitors (TKI) gefitinib and erlotinib by increasing reactive oxygen species in an in vitro model of NSCLC [162]. A phase I/II trial in NSCLC patients with epidermal growth factor receptor (EGFR) mutations following TKI progression showed that although the combination of vorinostat and erlotinib was well tolerated, the low percentage of response did not encourage further investigation of this combination [163,164].

A phase I/II trial demonstrated that the combination of vorinostat with gefitinib was well tolerated in previously treated NSCLC. While the combination therapy did not improve outcomes in an unselected patient population, biomarker analysis suggested that this combination may be more effective for EGFR-mutant NSCLC patients. Indeed, higher response rates and longer survival were observed in this group of patients and could be attributed to the addition of vorinostat [164]. Since further studies are required to confirm the combination benefit, a phase I trial (NCT02151721) focused on patients with EGFR-mutant NSCLC has been initiated, hoping to reverse the EGFR-TKI resistance with the addition of vorinostat. However, as of today, the trial status is unknown [165].

HDACi also targets HIF-1 transcription factors and synergizes with angiogenesis inhibitors, such as bevacizumab (Avastin) and sorafenib [166]. In a phase I trial, the combination of vorinostat and bevacizumab in patients with advanced clear cell renal cancer was associated with increased acetylated biomarkers in PBMCs. Three patients (12%) had SD [167]. A phase II trial combining vorinostat with paclitaxel and bevacizumab in patients with metastatic breast cancer (MBC) reported an objective response of 55% and SD for at least 24 weeks in 30% of the patients [168]. A phase I trial of vorinostat combined with bevacizumab in recurrent glioblastoma showed encouraging results with improved survival and PFS, with two patients remaining alive for nearly 2 years after enrollment. Although patient enrollment was modest, results suggested that this combination should have been further investigated to clarify its efficacy [169]. A phase II trial comparing bevacizumab alone and in combination with vorinostat did not show a benefit and ended the investigations [170]. Similarly, a phase II trial testing vorinostat in combination with bortezomib (a proteasome inhibitor) in patients with recurrent glioblastoma failed to improve PFS and eventually was shut down [171].

HDACi as epigenetic drugs has a wide range of effects, including alteration in tumor immunity, tumor-associated antigens, and regulation of MHC molecules [13]. Vorinostat can induce tumor cell apoptosis that stimulates the uptake of dead tumor cells by antigen-presenting cells (APCs), leading to immunogenic cell death and eradication of solid tumors, when combined with antibodies that augment host antitumor immune responses and promote APCs function. Moreover, HDACi has been shown to modulate the PD-1/PD-L1 pathway [13]. Based on this, several clinical trials are evaluating the combination of HDACi and immunotherapy with a specific focus on checkpoint inhibitors. Phase I/II trials are currently combining vorinostat and the anti-PD-1 pembrolizumab in patients with recurrent squamous cell head and neck cancer (HN), or salivary gland cancer (SGC) with proof of activity. Indeed, in HN, 32% showed PR and 20% SD, while in SGC, 16% showed PR and 56% SD. Toxicity was a problem and doses have to be adjusted [172]. Also tested in patients with advanced renal or urothelial cell carcinoma, results showed modest benefits but demonstrated a safe toxicity level and also highlighted a possible need to preselect patients depending on their checkpoint naive or resistant status [173]. This combination was also tested in other solid tumors, in the breast as previously reported, and also in glioblastoma [160,174].

Panobinostat (Farydak, LBH589)

As a single agent, panobinostat has shown moderate effects in solid tumors. In a phase I trial, intravenous panobinostat was evaluated in solid tumors. 6 out of 14 patients achieved SD for at least 4 months, but no complete or partial responses were observed. Thrombocytopenia was seen in all patients and was rapidly resolved, with severity depending on the dose of panobinostat and baseline platelet count [132]. Several phase II studies integrated panobinostat into the standard of care regimens in metastatic or locally unresectable clear cell renal carcinoma, metastatic breast cancer patients, metastatic hormone-refractory prostate cancer, or glioblastoma with very limited or no activity [133–135,175].

Promising results have been shown, even in a phase, I trial that combined panobinostat with fractionated stereotactic reirradiation therapy in the treatment of glioma [175].

In a phase I trial with an expansion cohort in sarcoma, panobinostat was combined with epirubicin in patients with solid tumors. The combined treatment was well tolerated, and 12 of 20 patients with sarcoma who progressed on prior anthracycline treatment benefited from the combination [176]. The immune-modulatory efficacy of panobinostat was tested in combination with the anti-CTLA-4 antibody ipilimumab, but no benefit was observed [177]. A phase I ongoing trial is testing the safety and efficacy of panobinostat and PDR001 (anti-PD-1) in various solid tumors [178], and results are awaited.

Romidepsin (Istodax, Depsipeptide, FK228)

Similar to other agents of this class, treatment with romidepsin has resulted in minimal clinical response when used as a single agent in phase II trials for metastatic colon cancer [136], lung cancer [120], renal cancer [137], and prostate cancer [138]. A romidepsin dose of 7 mg/m^2 was found tolerable and resulted in histone acetylation in PBMCs in a phase I trial involving solid tumors. Twenty-eight patients were enrolled, and although there were no objective responses with romidepsin alone, nine patients achieved SD. This suggests that this schedule may be useful for developing combination studies in solid tumors [139].

The combination of romidepsin with gemcitabine in solid tumors resulted in additive hematologic toxicities with 7% of the patients achieving PR, 52% of the patients had SD with encouraging results obtained in breast cancer patients [179].

A preclinical combination of DNMT and HDACi has shown to synergistically reexpress tumor suppressor genes and inhibit tumor growth in xenograft models. More recently, epigenetic modifiers have been shown to modulate genes involved in viral response [13]. On this basis, a phase I dose-escalation trial of oral 5-azacitidine and romidepsin was evaluating the combination efficacy in advanced solid tumors with an expansion cohort in virally mediated cancers and liposarcoma, but the combination was not active [180]. In addition, a phase I trial is currently investigating whether epigenetic modulators, such as romidepsin and oral 5-azacitidine, can enhance the response of pembrolizumab in microsatellite stable advanced colorectal cancer. Indeed, preclinical in vivo data have recently demonstrated romidepsin's ability to increase chemokines expression in tumors, T cells, and macrophages, increase T cell infiltration, and T-cell-dependent tumor regression. When combined with PD-1 blockade, the combination increased the activation of tumor-infiltrating T cells [181,182]. Another ongoing multicenter phase I/II trial is focused on the combination of romidepsin, azacitidine, paclitaxel, and gemcitabine, and was completed by a study of a posttreatment with durvalumab (anti-PD-L1) and lenalidomide in pancreatic cancer [183].

Entinostat (Ms-275, SND-275)

Entinostat, one of the first HDACi under development has shown early promising therapeutic potential in both solid and hematologic malignancies, associated with long-term tolerability (a patient with metastatic melanoma has been treated over 5 years with entinostat) [74,104,105,140]. However as seen in several studies, while SD was frequent, responses were rarer. In a phase II study evaluating entinostat in pretreated metastatic melanoma, SD was seen in 7 patients out of 28 [141]. Like with other HDACi, the strong experimental preclinical synergy of combining HDACi and DNMT inhibitors was reported prompting the clinical exploration of this combination. Early clinical data showed prolonged disease control in patients with extensively pretreated recurrent metastatic NSCLC. The dose of azacitidine administered was well below the maximal tolerated dose to allow epigenetic activity of the drug while avoiding toxicity from a long treatment [184]. Nonetheless, more extended Phase II data suggested that benefits were limited to a few patients and further studies were terminated. A multicenter phase II clinical trial evaluated this combination in women with advanced breast cancer but showed no clinical activity in advanced triple-negative breast cancer (TNBC) and limited benefits in estrogen receptor-positive (ER$^+$) breast cancer [185]. Likewise, the combination was inactive in metastatic colorectal cancer [186] and previously treated NSCLC, even in the presence of erlotinib [187].

In 2013, entinostat received a breakthrough designation from the FDA in combination with exemestane, for the treatment of hormonal therapy refractory breast cancer. Based on results from the phase II ENCORE 301 study, entinostat showed the potential to reverse resistance to hormonal therapies in patients, with advanced ER$^+$ breast cancer [123]. Yet like many other HDACi trials, this was not confirmed in the phase III trial E2112. The combination failed to achieve the primary endpoint by not showing a significant difference in OS compared to exemestane alone [188].

Similar to other HDACi, there has been an increased interest in entinostat immune-modulatory activity [13]. In the correlative analysis of the randomized phase II ENCORE 301 trial, entinostat was shown to induce myeloid-derived suppressor cells (MDSCs) reduction and increase HLA-Dr expression on monocytes, without altering the CD8/CD4 T cell ratio when combined with the aromatase inhibitor exemestane in hormone receptor-positive breast cancer [189]. In vitro experiments showed that low doses of entinostat transcriptionally reduced regulatory T-cell (Treg) Foxp3 expression [190]. Moreover, impressive tumor suppression in CT26 colorectal tumors and a 4T1 MBC mouse model was observed when entinostat was combined, with azacitidine and the anti-CTLA-4 and anti-PD-1 antibodies. Epigenetic modifiers decreased FoxP3$^+$ Treg, and entinostat in particular was able to reduce circulating granulocytic MDSCs that directly inhibit the function of CD8$^+$ T cells [191]. Currently, three clinical trials are testing entinostat in combination, with immunotherapy. A phase I trial (NCT02453620) in unresectable HER2-negative MBC combines entinostat, with nivolumab and ipilimumab; a phase II study (NCT01928576) is combining entinostat, with azacitidine and nivolumab in NSCLC; and a phase I/II trial (NCT02437136) is combining pembrolizumab and entinostat in NSCLC, melanoma, and microsatellite stable metastatic colorectal cancer. Results are awaited for the first trials. The third one, despite any strong benefits, shows interesting activity for the colorectal part of the study [192].

Valproic Acid

Many trials have evaluated valproic acid's efficacy in various solid tumor settings as a single agent, but there is little reason to think that VPA will have much efficacy as a single agent in solid tumors. Efficacy even in preclinical models requires high, not clinically achievable concentrations.

A phase I trial evaluated its safety in advanced cancer and showed that no objective responses were observed and stable disease status was obtained only in a low percentage of patients [142]. A phase II trial evaluating the biochemical nonmetastatic prostate cancer progression showed positive results with VPA treatment and hormonal therapy. Prostate-specific antigen (PSA) doubling time was longer on the VPA arm than with the standard of care. Also, 68% of the patients treated with VPA achieved CR, against 17% only in the standard of care arm, which did not appear to hold sufficient promise and the trial was terminated by the investigators [143].

Mainly evaluated in combination therapies, VPA has shown promising results in a different types of solid tumors. Sequence-specific combination of the HDACi VPA and the topoisomerase II inhibitor epirubicin in patients with advanced solid tumors resulted in a PR for 9 out of 41 patients (22%) and SD for 16 patients (39%) [193]. A dose-expansion study in breast cancer showed a 50% response rate [39]. Pulse dosing allowed the administration of high levels of VPA without cumulative toxicities. Responses were strongly correlated with the ability to acetylate histones and demonstrate HDAC2 expression. Impressive results were shown when different classes of epigenetic modifiers, such as HDACi and hypomethylating agents, were combined. In a small phase II study where VPA was administered, with hydralazine, doxorubicin, and cyclophosphamide in locally advanced breast cancer, of the 16 total patients enrolled, 31% achieved CR and 50% achieved PR [194]. Another small study evaluated the addition of VPA and hydralazine to chemotherapy to overcome chemo-resistance in refractory solid tumors. This study in predominantly ovarian cancer showed clinical benefit in 12 out of the 15 patients enrolled, with 4 PR and 8 SD [195]. It is important to mention that, in both studies, VPA was combined with hydralazine, a relatively weak indirect nonnucleoside DNMT inhibitor [194,195]. A phase III trial was performed but has not been reported (NCT00533299). In advanced platinum-resistant cancer, the combination of azacitidine, VPA, and carboplatin showed only SD in three patients with prostate cancer, cervical cancer, and colorectal cancer [196]. A subsequent phase I trial involving 66 patients with advanced cancer, combined VPA with 5-azacitidine and showed strong histone H3 acetylation and a global decrease in DNA methylation of PBMCs. In this trial, 14 patients (25%) achieved SD for 4–6 months [197]. Several phases I clinical trials have incorporated HDACi with decitabine. VPA and decitabine in NSCLC showed strong serum demethylation and reexpression of hypermethylated genes of interest. However, the toxicity, even at low doses, was unacceptable [198].

VPA also shows synergistic effects in combination with capecitabine, promoting upregulation of thymidine phosphorylase, the key enzyme responsible for capecitabine conversion in the active 5FU metabolite [153]. A phase I/II trial in combination with short-course radiotherapy and capecitabine in colorectal cancer demonstrated a safe dose of the combination between VPA and capecitabine, as well as encouraging responses in the all cohort of patients [199]. While VPA has shown to be an interesting and likely effective HDACi, the lack of sponsorship has prevented the conduct of larger trials.

Belinostat (Beleodaq, PXD101)

While mainly being investigated in hematological diseases, a phase I trial of belinostat in patients, with advanced refractory solid tumors resulted in SD for 18 out of 46 patients. The study included patients with colorectal, renal, melanoma, and various other solid tumors. Disease stabilization was seen in 50% of the patients at the highest tolerated dose [144]. The moderate activity of belinostat was found even in patients, with recurrent or refractory advanced thymic epithelial tumors [145]. In a group of advanced cancers with liver dysfunctions, a phase I trial determined that increased doses of belinostat could be found in bad liver function cancers, but no direct correlation between toxicity and belinostat could be made. They also reported that 28% of the patients were showing an SD status [146] When combined with carboplatin in resistant ovarian cancer, belinostat showed minimal activity in this population of patients [200]. However, when it was combined with doxorubicin in soft tissue sarcoma, the combined agents showed a response in 12 out of 20 patients, warranting further investigation [201]. Worth noting are two recent phase I trials, involving belinostat, one in combination with ribociclib (CDK inhibitor) in metastatic TNBC and recurrent ovarian cancer, and another one with talazoparib in metastatic breast, ovary, and prostate cancer (NCT04315233 and NCT04703920 respectively).

Others HDACi

Several other HDACi have been tested for solid tumor malignancies. The benzamide type HDACi, chidamide (formerly known as HB-8000 or CS055Chidamide) is approved by the Chinese FDA for PTCL. Chidamide was evaluated in a phase III trial in combination with exemestane compared to placebo and exemestane in ER-positive, HER2-negative breast cancers after progression on endocrine therapy, and showed a significant improvement in PFS (7.4 vs 3.8 months) [202]. Thirty-six studies involving chidamide in breast cancer, NSCL, and other diseases are currently recruiting. Results are eagerly awaited as this is the only HDACi so far with efficacy in Phase III trials in solid tumor malignancies.

Mocetinostat, also a benzamide type HDACi, has been studied in locally advanced/metastatic urothelial carcinoma as a single agent, however, unlike its efficacy in lymphomas, it did not show enough potential [147]. Mocetinostat is further studied as an immune modulator combination with immune checkpoint inhibitors.

A third HDACi of interest, the hydroxamic acid quisinostat, administered with paclitaxel and carboplatin in the treatment of ovarian cancer showed an ORR of 50% and a tolerable toxicity profile in the phase II study [203]. However, no phase III studies are being pursued to this day, (the company making the drug did not display any new progression about clinical assays).

CLINICAL APPLICATIONS OF HDACI FOR NONCANCER DISEASES

HDACs have elicited growing interest in the treatment of various nononcological diseases and a substantial body of preclinical work demonstrates the promise of HDACi for the treatment of noncancer disorders, including neuromuscular, cardiovascular, hematological, pulmonary, infectious, and inflammatory diseases [204–207]. Early on, HDACi has been investigated for their ability to regulate the human immunodeficiency virus (HIV) latency, directly and indirectly by inducing histone deacetylation at HIV integrated sites, and by causing nonhistone protein modification, such as NF-κB [208,209]. In this section, we will review results from clinical trials treating noncancer diseases, with HDACi, focusing mainly on VPA and phenylbutyrate, which have undergone extensive clinical testing.

Current Therapeutic Applications
Valproic Acid

For more than 30 years, VPA has been used to treat seizure disorders. In addition to its use as an anticonvulsant, VPA is commonly prescribed as a first-line treatment for the prevention and maintenance of mania in patients with bipolar disorder and migraine headaches [210–213]. A major factor complicating the use of VPA is its teratogenicity. Children of pregnant mothers treated with VPA are at greater risk for major congenital malformations, such as neural tube defects, midline abnormalities, and the development of spina bifida, as well as neurodevelopmental delay [214]. Despite its common use and broad applications, the cellular mechanisms underpinning the therapeutic benefits of VPA are not clear, nor is the extent to which its HDAC inhibition activity contributes to these benefits. The effectiveness of VPA for the treatment of these varied neurological disorders argues for the modulation of several target pathways. Furthermore, the response can vary significantly depending on the duration of treatment. For example, immediate short-term administration of VPA elicits rapid temporary control of seizure disorders, while several days of treatment are required to achieve full benefit for mood stabilization. Additionally, following the cessation of prolonged treatment, antiseizure, antimigraine, and antimanic benefits have been found to persist in some patients whereas no persistent effects on depression were found [215]. VPA may provoke autism during the early phases of development [216]. VPA has also been used off-label for schizophrenia [217]. Other reports suggest neuroprotective abilities, such as the prevention of cerebral ischemia and the promotion of hippocampal neurogenesis in rats [218]. While many of the findings are still without a clear mechanism, a compelling hypothesis asserts that the rapid therapeutic actions of VPA are the result of direct interaction with components of key biochemical pathways, while the long-term and persistent effects are the result of HDAC inhibition and consequent epigenetic modulation, and altered protein expression [215].

Phenylbutyrate

Phenylbutyrate (PB) is a fatty-acid HDACi that targets class I and IIa HDACs, which for more than a decade has been applied in the clinic for the treatment of urea cycle disorders [219]. However, for this treatment, phenylbutyrate's efficacy does not lie with its HDAC inhibition activity. Rather, for patients unable to metabolize glutamine to urea, phenylbutyrate provides an alternate pathway for nitrogen excretion. PB is metabolized to phenylacetyl-CoA and conjugated to glutamine in the liver, which can then be excreted in the urine [219]. The effects of histones acetylation (targeted or globally) via HDAC2 on SOD1 and other mRNA expression patterns have created interest to test such agents for amyotrophic lateral sclerosis which led to studies of PB in this disease [220]. Early clinical data suggest better symptom control and improvement of survival when initiated shortly after symptoms first manifest [221,222].

Clinical Evaluation of Valproic Acid and Phenylbutyrate

The established safety profile and clinical utility of VPA and PB render them ideal HDACi for proof of principle studies in other disorders arising from dysregulation of acetylation and genetic mutations, including cystic fibrosis, spinal muscular atrophy, hemoglobinopathies, neurodegenerative diseases, and HIV infection [223–225].

Cystic fibrosis is an autosomal recessive disorder with mutations in the chloride ion-channel protein, cystic fibrosis transmembrane conductance regulator (CFTR), which primarily manifests in mucus overproduction and impaired pulmonary function [226]. In two small randomized, double-blind, placebo-controlled studies, cystic fibrosis patients, with homozygous F508del-CFTR mutations were treated with PB. In both the studies, patients exhibited a modest but statistically significant improvement in chloride transport, indicating improved CFTR function [227,228]. However, while PB is still in phase I/II trial concerning cystic fibrosis treatment (NCT02323100), most of the patients are now treated with more mutation-specific drugs, like Kalydeco (ivacaftor) for G551D mutation and Orkambi (lumacaftor) for F508del mutation, that correct the structure and function of the misfolded proteins.

Spinal muscular atrophy (SMA) is a severe hereditary neuromuscular disease caused by motor neuron degeneration, resulting in weakness and muscle wasting. The most common form of SMA is due to mutations in the survival motor neuron 1 (SMN1) gene. A second gene, SMN2, is nearly identical to SMN1 [229]. It was hypothesized that increased expression of full-length SMN from the SMN2 gene could be induced with HDAC inhibition and thus compensate for mutant SMN1. In an initial pilot study, children with SMA treated with PB for 3 to 9 weeks demonstrated increased motor function [230]. Using leukocytes as a surrogate for motor neurons, PB was shown to elevate SMN mRNA levels [222]. However, in a randomized, double-blind, placebo-controlled trial enrolling 107 children with SMA, no significant improvement over placebo was observed with PB treatment [231]. With VPA treatment, increased levels of SMN mRNA and protein were observed in PBMCs in 7 out of 10 SMA mutant carriers, and increased levels of SMN2 mRNA in 7 out of 20 SMA patients [232]. In a small study regrouping adults, with milder SMA (type III/IV), increased muscle strength was observed, with 8 months of VPA treatment [233]. In a phase II trial, tolerability and efficacy of VPA treatment were examined in a heterogeneous group of subjects, with varying degrees of disease severity [234]. Of 42 patients, 27 showed an improvement in gross motor function. However, no significant increase in SMN mRNA was observed from whole blood purification. Benefits were primarily seen in nonambulatory patients, with type II SMA who were 5 years old or younger. A phase III trial is testing the combination of VPA and levocarnitine, a carnitine supplement that was started in 2013 but has not resulted (NCT01671384).

Due to preclinical work suggesting that VPA promotes neuronal protection [235], several trials have been initiated to evaluate the drug's use in the treatment of neurodegenerative diseases, such as Alzheimer's, Huntington's, and amyotrophic lateral sclerosis as previously described. The results are still awaited.

Others HDACi

Like VPA and PB, several other HDACi are studied in neuromuscular diseases. The hydroxamate HDACi givinostat has been explored in multiple trials on Duchenne muscular dystrophy (DMD), a genetic disorder (present on chromosome X) leading to muscle degeneration due to a mutation in the dystrophin gene. Dystrophin is essential for muscle cell maintenance, and its absence leads to cell damage and loss, with no curative options [236]. A phase I/II trial tested givinostat in 18 boys suffering from DMD. Biopsies before and 12 months after therapy showed a significant increase in the muscle tissue with a decrease in fibrotic tissue formation [237]. A follow-up on the phase III trial (NCT0285179) and a phase 2/3 trial (NCT03373968) are ongoing and one of the trials has reported a noticeable delay in symptoms development observed in patients treated with givinostat [238].

These findings raise hope to define new strategies for mostly untreatable and to date incurable diseases but are a long way from being clinically established options.

HDACi Dependent Reactivation of Latent HIV

Viral reactivation from latently infected cells has been one of the therapeutic strategies to eradicate HIV as its eradication is hampered by latency in infected $CD4^+$ T cells. Class, I HDAC activity has been linked to the repression of the HIV virus and depletion of the viral reservoir [239] through HDAC2 and HDAC3 [240]. In a pilot study, VPA was administered with highly active antiretroviral therapy in patients infected with HIV. Infected $CD4^+$ T cells were substantially reduced in three of four patients treated, with a mean decrease of 75%, and the treatment was well tolerated [241].

Various clinical trials have evaluated a potential role for HDACi in combination with other strategies to express latent virus as part of an HIV cure. In phase I/II clinical trial, 15 patients with HIV were treated with panobinostat for 8 weeks while maintaining combination antiretroviral therapy. Panobinostat effectively disrupted HIV latency in vivo by increasing cell-associated HIV RNA, but it did not reduce the number of latently infected cells [242]. Further studies explore the effects of panobinostat in combination with interferon-alpha2a on the virus residual reservoir composed of infected cells (NCT02471430). Similarly, a significant increase in cell-associated RNA was observed in 90% of patients receiving vorinostat and suppressive combination antiretroviral therapy [243]. Nevertheless, vorinostat like panobinostat did not control latently infected cells in a larger phase II trial [244]. Romidepsin was also explored as a means to increase HIV-1 transcription [245]. However, a similar study could not confirm any change in the viremia after treatment with romidepsin despite showing an effect on CD4 T-cells altogether suggesting that the HDACi effects may not be sufficiently translated into clinical benefits [246]. Further mechanistic studies and the development of more selective HDACi may be needed to define the role of HDACi in HIV and other viral diseases. Yet, in the setting of more specific anti-HIV therapy, epigenetic modifiers may be too nonspecific to have a major role in this disease.

CONCLUSIONS AND THE FUTURE DIRECTIONS OF THE CLINICAL APPLICATIONS OF HDACI

HDACi provides an exciting means to treat a variety of human diseases with acceptable side effects. The most advanced development of these agents is in the treatment of CTCL, where several HDACi have shown clinical efficacy in patients and other treatment options remain limited. HDACi have shown efficacy in other hematological malignancies. However, as single agents for the treatment of solid tumor malignancies, the efficacy of HDACi has been more modest.

Work to develop more specific HDACi with an improved therapeutic window and better target identification is underway to increase their efficacy. Furthermore, multiple preclinical and clinical studies suggest that HDACi may have an additive or synergistic role in the target modulation of many novel biological therapeutics and chemotherapy, in both hematological and solid tumor malignances. These strategies include the combined epigenetic modulation of histone and nonhistone targets by HDAC and DNMT inhibitors, VEGF inhibitors, and antiestrogens among many others. Promising clinical evidence using HDACi also comes from clinical trials targeting noncancer diseases, such as spinal muscle atrophy and HIV infection. The promising early efficacy of VPA, PB, givinostat, and others, clinically demonstrated their limited toxicity profile and thus they provide attractive options for evaluating the benefit of HDAC inhibition for the treatment of various diseases.

One of the limitations of current HDACi is their tissue and drug-specific action and relatively narrow therapeutic window. Many on-target and off-target effects are dependent on still poorly understood host factors. The role of individual HDACs as therapeutic targets as well as the potential for mutual redundancy of HDACs needs further exploration. Many HDACi effects on biological targets require sustained dosing, as effects are mostly reversible upon drug removal. Careful specific patient selection will also have a significant impact on future clinical trials of HDACi, especially as related to cancer treatment. In addition, the mechanisms of resistance to HDACi need to be further explored to maximize the clinical efficacy of these drugs [247] and much more research is needed on markers that could correlate with resistance to HDACi. Determining the specific type and grade of cancer that is most susceptible to HDACi treatment will also aid in selecting patients most likely to respond.

Further insight into the molecular mechanisms that govern the inhibition of HDACs could help to design more specific and potent drugs, and the creation of more selective HDACi could improve their effectiveness in the future. Recognizing the specific roles of select HDACs involved in exerting the desired effects of HDACi and their synergy with other therapeutics will allow the development of better and more potent drugs. A potential strategy to fine-tune HDACi uses is to develop drugs not targeting a specific HDAC, but the protein complex it is part of. Indeed, it is known that certain HDACs (mostly class I) do not act as a monomer (except HDAC8) and are part of corepressor complexes like Sin3, NuRD, or CoREST. Proteins present in those complexes are essential for HDAC activity. The absence of those proteins decrease drastically HDAC activity and it is also proven that they were activator of their deacetylation role [248,249]. Those complexes which contain HDACs may differ from one type of tissue to another. Targeting only one of the members of those complexes could offer a similar effect on HDAC inhibition, but with tissue specificity, which could help to manage unwanted toxicity. They are also, by themselves, known to be involved in some cancer development. For example, in Ewing sarcoma, the oncogenic fusion protein EWSR1-FLI1 can bind to the NuRD complex, increasing the recruitment of this complex onto gene promoter regions, leading to expression repression [250]. Understanding and developing therapies targeting those complexes are avenues that must be explored deeper.

Altogether, the approval of several HDACi has supported their use in the treatment of cancer, and their applicability as novel treatment strategies either alone or in combination with a large variety of other therapeutics will undoubtedly expand in the future.

References

[1] Mottet D, Castronovo V. Histone deacetylases: target enzymes for cancer therapy. Clin Exp Metastasis 2008;25:183–9. Available from: https://doi.org/10.1007/s10585-007-9131-5.

[2] Chan HM, Krstic-Demonacos M, Smith L, Demonacos C, La NB. Thangue, Acetylation control of the retinoblastoma tumour-suppressor protein. Nat Cell Biol 2001;3:667–74. Available from: https://doi.org/10.1038/35083062.

[3] Kawaguchi Y, Kovacs JJ, McLaurin A, Vance JM, Ito A, Yao T-P. The deacetylase HDAC6 regulates aggresome formation and cell viability in response to misfolded protein stress. Cell. 2003;115:727–38. Available from: https://doi.org/10.1016/S0092-8674(03)00939-5.

[4] Marzio G, Wagener C, Gutierrez MI, Cartwright P, Helin K, Giacca M. E2F family members are differentially regulated by reversible acetylation. J Biol Chem 2000;275:10887–92. Available from: https://doi.org/10.1074/jbc.275.15.10887.

[5] Xu WS, Parmigiani RB, Marks PA. Histone deacetylase inhibitors: molecular mechanisms of action. Oncogene. 2007;26:5541–52. Available from: https://doi.org/10.1038/sj.onc.1210620.

[6] Carew JS, Giles FJ, Nawrocki ST. Histone deacetylase inhibitors: mechanisms of cell death and promise in combination cancer therapy. Cancer Lett 2008;269:7–17. Available from: https://doi.org/10.1016/j.canlet.2008.03.037.

[7] Chueh AC, Tse JWT, Tögel L, Mariadason JM. Mechanisms of histone deacetylase inhibitor-regulated gene expression in cancer cells. Antioxid Redox Signal 2014;23:66–84. Available from: https://doi.org/10.1089/ars.2014.5863.

[8] Ungerstedt JS, Sowa Y, Xu W-S, Shao Y, Dokmanovic M, Perez G, et al. Role of thioredoxin in the response of normal and transformed cells to histone deacetylase inhibitors. Proc Natl Acad Sci U S A 2005;102:673–8. Available from: https://doi.org/10.1073/pnas.0408732102.

[9] Wallace DM, Cotter TG. Histone deacetylase activity in conjunction with E2F-1 and p53 regulates Apaf-1 expression in 661W cells and the retina. J Neurosci Res 2009;87:887–905. Available from: https://doi.org/10.1002/jnr.21910.

[10] Yang QC, Zeng BF, Shi ZM, Dong Y, Jiang ZM, Huang J, et al. Inhibition of hypoxia-induced angiogenesis by trichostatin A via suppression of HIF-1a activity in human osteosarcoma. J Exp Clin Cancer Res CR 2006;25:593–9.

[11] Marchion DC, Bicaku E, Daud AI, Richon V, Sullivan DM, Munster PN. Sequence-specific potentiation of topoisomerase II inhibitors by the histone deacetylase inhibitor suberoylanilide hydroxamic acid. J Cell Biochem 2004;92:223–37. Available from: https://doi.org/10.1002/jcb.20045.

[12] Zhang Y, Carr T, Dimtchev A, Zaer N, Dritschilo A, Jung M. Attenuated DNA damage repair by trichostatin A through BRCA1 suppression. Radiat Res 2007;168:115–24. Available from: https://doi.org/10.1667/RR0811.1.

[13] Terranova-Barberio M, Thomas S, Munster PN. Epigenetic modifiers in immunotherapy: a focus on checkpoint inhibitors. Immunotherapy. 2016;8:705–19. Available from: https://doi.org/10.2217/imt-2016-0014.

[14] Park J, Thomas S, Munster PN. Epigenetic modulation with histone deacetylase inhibitors in combination with immunotherapy. Epigenomics. 2015;7:641–52. Available from: https://doi.org/10.2217/epi.15.16.

[15] Dokmanovic M, Marks PA. Prospects: histone deacetylase inhibitors. J Cell Biochem 2005;96:293–304. Available from: https://doi.org/10.1002/jcb.20532.

[16] Krämer OH, Zhu P, Ostendorff HP, Golebiewski M, Tiefenbach J, Peters MA, et al. The histone deacetylase inhibitor valproic acid selectively induces proteasomal degradation of HDAC2. EMBO J 2003;22:3411–20. Available from: https://doi.org/10.1093/emboj/cdg315.

[17] Ueda H, Manda T, Matsumoto S, Mukumoto S, Nishigaki F, Kawamura I, et al. FR901228, a novel antitumor bicyclic depsipeptide produced by Chromobacterium violaceum No. 968 III. Antitumor activities on experimental tumors in mice. J Antibiot (Tokyo) 1994;47:315–23. Available from: https://doi.org/10.7164/antibiotics.47.315.

[18] Bertino EM, Otterson GA. Romidepsin: a novel histone deacetylase inhibitor for cancer. Expert Opin Investig Drugs 2011;20:1151–8. Available from: https://doi.org/10.1517/13543784.2011.594437.

[19] Vannini A, Volpari C, Filocamo G, Casavola EC, Brunetti M, Renzoni D, et al. Crystal structure of a eukaryotic zinc-dependent histone deacetylase, human HDAC8, complexed with a hydroxamic acid inhibitor. Proc Natl Acad Sci 2004;101:15064–9. Available from: https://doi.org/10.1073/pnas.0404603101.

REFERENCES

[20] Somoza JR, Skene RJ, Katz BA, Mol C, Ho JD, Jennings AJ, et al. Structural snapshots of human HDAC8 provide insights into the class I histone deacetylases. Struct Lond Engl 2004;1993(12):1325–34. Available from: https://doi.org/10.1016/j.str.2004.04.012.

[21] Furumai R, Matsuyama A, Kobashi N, Lee K-H, Nishiyama M, Nakajima H, et al. FK228 (depsipeptide) as a natural prodrug that inhibits class I histone deacetylases. Cancer Res 2002;62:4916–21.

[22] Arts J, King P, Mariën A, Floren W, Beliën A, Janssen L, et al. JNJ-26481585, a novel "second-generation" oral histone deacetylase inhibitor, shows broad-spectrum preclinical antitumoral activity. Clin Cancer Res 2009;15:6841–51. Available from: https://doi.org/10.1158/1078-0432.CCR-09-0547.

[23] Khan N, Jeffers M, Kumar S, Hackett C, Boldog F, Khramtsov N, et al. Determination of the class and isoform selectivity of small-molecule histone deacetylase inhibitors. Biochem J 2008;409:581–9. Available from: https://doi.org/10.1042/BJ20070779.

[24] Ning Z-Q, Li Z-B, Newman MJ, Shan S, Wang X-H, Pan D-S, et al. Chidamide (CS055/HBI-8000): a new histone deacetylase inhibitor of the benzamide class with antitumor activity and the ability to enhance immune cell-mediated tumor cell cytotoxicity. Cancer Chemother Pharmacol 2012;69:901–9. Available from: https://doi.org/10.1007/s00280-011-1766-x.

[25] Novotny-Diermayr V, Sangthongpitag K, Hu CY, Wu X, Sausgruber N, Yeo P, et al. SB939, a novel potent and orally active histone deacetylase inhibitor with high tumor exposure and efficacy in mouse models of colorectal cancer. Mol Cancer Ther 2010;9:642–52. Available from: https://doi.org/10.1158/1535-7163.MCT-09-0689.

[26] Mandl-Weber S, Meinel FG, Jankowsky R, Oduncu F, Schmidmaier R, Baumann P. The novel inhibitor of histone deacetylase resminostat (RAS2410) inhibits proliferation and induces apoptosis in multiple myeloma (MM) cells. Br J Haematol 2010;149:518–28. Available from: https://doi.org/10.1111/j.1365-2141.2010.08124.x.

[27] Santo L, Hideshima T, Kung AL, Tseng J-C, Tamang D, Yang M, et al. Preclinical activity, pharmacodynamic, and pharmacokinetic properties of a selective HDAC6 inhibitor, ACY-1215, in combination with bortezomib in multiple myeloma. Blood. 2012;119:2579–89. Available from: https://doi.org/10.1182/blood-2011-10-387365.

[28] Gurvich N, Tsygankova OM, Meinkoth JL, Klein PS. Histone deacetylase is a target of valproic acid-mediated cellular differentiation. Cancer Res 2004;64:1079–86. Available from: https://doi.org/10.1158/0008-5472.can-03-0799.

[29] Subramanian S, Bates SE, Wright JJ, Espinoza-Delgado I, Piekarz RL. Clinical toxicities of histone deacetylase inhibitors. Pharmaceuticals. 2010;3:2751–67. Available from: https://doi.org/10.3390/ph3092751.

[30] Aggarwal R, Thomas S, Pawlowska N, Bartelink I, Grabowsky J, Jahan T, et al. Inhibiting histone deacetylase as a means to reverse resistance to angiogenesis inhibitors: Phase I study of abexinostat plus pazopanib in advanced solid tumor malignancies. J Clin Oncol 2017;35:1231–9. Available from: https://doi.org/10.1200/JCO.2016.70.5350.

[31] Blum W, Marcucci G. Targeting epigenetic changes in acute myeloid leukemia. Clin Adv Hematol Oncol HO 2005;3:855–65.

[32] Hussain L, Maimaitiyiming Y, Islam K, Naranmandura H. Acute promyelocytic leukemia and variant fusion proteins: PLZF-RARα fusion protein at a glance. Semin Oncol 2019;46:133–44. Available from: https://doi.org/10.1053/j.seminoncol.2019.04.004.

[33] Eckschlager T, Plch J, Stiborova M, Hrabeta J. Histone deacetylase inhibitors as anticancer drugs. Int J Mol Sci 2017;18. Available from: https://doi.org/10.3390/ijms18071414.

[34] New Drugs/Drug News. Pharm Ther 2014;39:539–78.

[35] Nguyen T, Parker R, Hawkins E, Holkova B, Yazbeck V, Kolluri A, et al. Synergistic interactions between PLK1 and HDAC inhibitors in non-Hodgkin's lymphoma cells occur in vitro and in vivo and proceed through multiple mechanisms. Oncotarget. 2017;8:31478–93. Available from: https://doi.org/10.18632/oncotarget.15649.

[36] Chifotides HT, Bose P, Verstovsek S. Givinostat: an emerging treatment for polycythemia vera. Expert Opin Investig Drugs 2020;29:525–36. Available from: https://doi.org/10.1080/13543784.2020.1761323.

[37] Search Orphan Drug Designations and Approvals. (n.d.). https://www.accessdata.fda.gov/scripts/opdlisting/oopd/detailedIndex.cfm?cfgridkey = 563116 [accessed 06.05.21].

[38] Shafer D, Grant S. Update on rational targeted therapy in AML. Blood Rev 2016;30:275–83. Available from: https://doi.org/10.1016/j.blre.2016.02.001.

[39] Munster P, Marchion D, Bicaku E, Lacevic M, Kim J, Centeno B, et al. Clinical and biological effects of valproic acid as a histone deacetylase inhibitor on tumor and surrogate tissues: phase I/II trial of valproic acid and epirubicin/FEC. Clin Cancer Res J Am Assoc Cancer Res 2009;15:2488–96. Available from: https://doi.org/10.1158/1078-0432.CCR-08-1930.

[40] Mann BS, Johnson JR, Cohen MH, Justice R, Pazdur R. FDA approval summary: vorinostat for treatment of advanced primary cutaneous T-cell lymphoma. Oncologist 2007;12:1247–52. Available from: https://doi.org/10.1634/theoncologist.12-10-1247.

[41] Ree AH, Saelen MG, Kalanxhi E, Østensen IHG, Schee K, Røe K, et al. Biomarkers of histone deacetylase inhibitor activity in a phase 1 combined-modality study with radiotherapy. PLoS One 2014;9:e89750. Available from: https://doi.org/10.1371/journal.pone.0089750.

[42] Kelly WK, O'Connor OA, Krug LM, Chiao JH, Heaney M, Curley T, et al. Phase I study of an oral histone deacetylase inhibitor, suberoylanilide hydroxamic acid, in patients with advanced cancer. J Clin Oncol J Am Soc Clin Oncol 2005;23:3923–31. Available from: https://doi.org/10.1200/JCO.2005.14.167.

[43] Kelly WK, Richon VM, O'Connor O, Curley T, MacGregor-Curtelli B, Tong W, et al. Phase I clinical trial of histone deacetylase inhibitor: suberoylanilide hydroxamic acid administered intravenously. Clin Cancer Res J Am Assoc Cancer Res 2003;9:3578–88.

[44] Schaefer EW, Loaiza-Bonilla A, Juckett M, DiPersio JF, Roy V, Slack J, et al. Mayo P2C Phase II Consortium, A phase 2 study of vorinostat in acute myeloid leukemia. Haematologica. 2009;94:1375–82. Available from: https://doi.org/10.3324/haematol.2009.009217.

[45] Crump M, Coiffier B, Jacobsen ED, Sun L, Ricker JL, Xie H, et al. Phase II trial of oral vorinostat (suberoylanilide hydroxamic acid) in relapsed diffuse large-B-cell lymphoma. Ann Oncol J Eur Soc Med Oncol 2008;19:964–9. Available from: https://doi.org/10.1093/annonc/mdn031.

[46] Oki Y, Younes A, Copeland A, Hagemeister F, Fayad LE, McLaughlin P, et al. Phase I study of vorinostat in combination with standard CHOP in patients with newly diagnosed peripheral T-cell lymphoma. Br J Haematol 2013;162:138–41. Available from: https://doi.org/10.1111/bjh.12326.

[47] Persky DO, Li H, Rimsza LM, Barr PM, Popplewell LL, Bane CL, et al. I/II trial of Vorinostat (SAHA) in combination with Rituximab-CHOP in patients with newly diagnosed advanced stage diffuse large B-cell Lymphoma (DLBCL): SWOG S0806. Am J Hematol 2018;93:486–93. Available from: https://doi.org/10.1002/ajh.25010.

[48] Garcia-Manero G, Tambaro FP, Bekele NB, Yang H, Ravandi F, Jabbour E, et al. Phase II trial of Vorinostat with idarubicin and

cytarabine for patients with newly diagnosed acute myelogenous leukemia or myelodysplastic syndrome. J Clin Oncol 2012;30:2204–10. Available from: https://doi.org/10.1200/JCO.2011.38.3265.

[49] National Cancer Institute (NCI), A Randomized Phase III Study of Standard Cytarabine Plus Daunorubicin (7 + 3) Therapy or Idarubicin With High Dose Cytarabine (IA) vs IA With Vorinostat (NSC-701852) (IA + V) in Younger Patients With Previously Untreated Acute Myeloid Leukemia (AML), clinicaltrials.gov, 2018. https://clinicaltrials.gov/ct2/show/NCT01802333 [accessed 05.05.21].

[50] Holkova B, Supko JG, Ames MM, Reid JM, Shapiro GI, Tombes MB, et al. A phase I trial of vorinostat and alvocidib in patients with relapsed, refractory or poor prognosis acute leukemia, or refractory anemia with excess blasts-2. Clin Cancer Res J Am Assoc Cancer Res 2013;19:1873–83. Available from: https://doi.org/10.1158/1078-0432.CCR-12-2926.

[51] Burke MJ, Lamba JK, Pounds S, Cao X, Ghodke-Puranik Y, Lindgren BR, et al. A therapeutic trial of decitabine and vorinostat in combination with chemotherapy for relapsed/refractory acute lymphoblastic leukemia. Am J Hematol 2014;89:889–95. Available from: https://doi.org/10.1002/ajh.23778.

[52] How J, Minden MD, Brian L, Chen EX, Brandwein J, Schuh AC, et al. A phase I trial of two sequence-specific schedules of decitabine and vorinostat in patients with acute myeloid leukemia. Leuk Lymphoma 2015;56:2793–802. Available from: https://doi.org/10.3109/10428194.2015.1018248.

[53] Decitabine and Vorinostat Conditioning Followed by CD3-/CD19- NK Cells Infusion for High Risk Myelodysplastic Syndromes - Study Results - ClinicalTrials.gov (n.d.). https://clinicaltrials.gov/ct2/show/results/NCT01593670 [accessed 15.07.21].

[54] Montalban-Bravo G, Huang X, Jabbour E, Borthakur G, DiNardo CD, Pemmaraju N, et al. A clinical trial for patients with acute myeloid leukemia or myelodysplastic syndromes not eligible for standard clinical trials. Leukemia. 2017;31:318–24. Available from: https://doi.org/10.1038/leu.2016.303.

[55] Craddock CF, Houlton AE, Quek LS, Ferguson P, Gbandi E, Roberts C, et al. Outcome of azacitidine therapy in acute myeloid leukemia is not improved by concurrent vorinostat therapy but is predicted by a diagnostic molecular signature. Clin Cancer Res J Am Assoc Cancer Res 2017;23:6430–40. Available from: https://doi.org/10.1158/1078-0432.CCR-17-1423.

[56] Siegel DS, Dimopoulos M, Jagannath S, Goldschmidt H, Durrant S, Kaufman JL, et al. VANTAGE 095: an international, multicenter, open-label study of Vorinostat (MK-0683) in combination with bortezomib in patients with relapsed and refractory multiple myeloma. Clin Lymphoma Myeloma Leuk 2016;16:329–34. Available from: https://doi.org/10.1016/j.clml.2016.02.042 e1.

[57] Nieto Y, Valdez BC, Thall PF, Jones RB, Wei W, Myers A, et al. Double epigenetic modulation of high-dose chemotherapy with azacitidine and vorinostat for patients with refractory or poor-risk relapsed lymphoma. Cancer. 2016;122:2680–8. Available from: https://doi.org/10.1002/cncr.30100.

[58] Ellis L, Pan Y, Smyth GK, George DJ, McCormack C, Williams-Truax R, et al. Histone deacetylase inhibitor panobinostat induces clinical responses with associated alterations in gene expression profiles in cutaneous T-cell lymphoma. Clin Cancer Res Off J Am Assoc Cancer Res 2008;14:4500–10. Available from: https://doi.org/10.1158/1078-0432.CCR-07-4262.

[59] Younes A, Sureda A, Ben-Yehuda D, Zinzani PL, Ong T-C, Prince HM, et al. Panobinostat in patients with relapsed/refractory Hodgkin's lymphoma after autologous stem-cell transplantation: results of a phase II study. J Clin Oncol J Am Soc Clin Oncol 2012;30:2197–203. Available from: https://doi.org/10.1200/JCO.2011.38.1350.

[60] Duvic M, Dummer R, Becker JC, Poulalhon N, Ortiz Romero P, Grazia Bernengo M, et al. Panobinostat activity in both bexarotene-exposed and -naïve patients with refractory cutaneous T-cell lymphoma: results of a phase II trial. Eur J Cancer 2013;49:386–94. Available from: https://doi.org/10.1016/j.ejca.2012.08.017.

[61] Richardson PG, Schlossman RL, Alsina M, Weber DM, Coutre SE, Gasparetto C, et al. PANORAMA 2: panobinostat in combination with bortezomib and dexamethasone in patients with relapsed and bortezomib-refractory myeloma. Blood. 2013;122:2331–7. Available from: https://doi.org/10.1182/blood-2013-01-481325.

[62] San-Miguel JF, Hungria VTM, Yoon S-S, Beksac M, Dimopoulos MA, Elghandour A, et al. Panobinostat plus bortezomib and dexamethasone vs placebo plus bortezomib and dexamethasone in patients with relapsed or relapsed and refractory multiple myeloma: a multicentre, randomised, double-blind phase 3 trial. Lancet Oncol 2014;15:1195–206. Available from: https://doi.org/10.1016/S1470-2045(14)70440-1.

[63] Kaufman JL, Mina R, Jakubowiak AJ, Zimmerman TL, Wolf JJ, Lewis C, et al. Combining carfilzomib and panobinostat to treat relapsed/refractory multiple myeloma: results of a Multiple Myeloma Research Consortium Phase I Study. Blood Cancer J 2019;9:3. Available from: https://doi.org/10.1038/s41408-018-0154-8.

[64] Chari A, Cho HJ, Dhadwal A, Morgan G, La L, Zarychta K, et al. A phase 2 study of panobinostat with lenalidomide and weekly dexamethasone in myeloma. Blood Adv 2017;1:1575–83. Available from: https://doi.org/10.1182/bloodadvances.2017007427.

[65] Mayo Clinic, A Phase I/II Study of the Histone Deacetylase (HDAC) Inhibitor LBH589 (Panobinostat) in Combination With mTOR Inhibitor RAD001 (Everolimus) in Patients With Relapsed Multiple Myeloma or Lymphoma, clinicaltrials.gov, 2020. https://clinicaltrials.gov/ct2/show/results/NCT00918333 [accessed 14.07.21].

[66] Garcia-Manero G, Sekeres MA, Egyed M, Breccia M, Graux C, Cavenagh JD, et al. A phase 1b/2b multicenter study of oral panobinostat plus azacitidine in adults with MDS, CMML or AML with ≤30% blasts. Leukemia 2017;31:2799–806. Available from: https://doi.org/10.1038/leu.2017.159.

[67] Uy GL, Duncavage EJ, Chang GS, Jacoby MA, Miller CA, Shao J, et al. Dynamic changes in the clonal structure of MDS and AML in response to epigenetic therapy. Leukemia 2017;31:872–81. Available from: https://doi.org/10.1038/leu.2016.282.

[68] Piekarz RL, Frye R, Turner M, Wright JJ, Allen SL, Kirschbaum MH, et al. Phase II multi-institutional trial of the histone deacetylase inhibitor romidepsin as monotherapy for patients with cutaneous T-cell lymphoma. J Clin Oncol 2009;27:5410–17. Available from: https://doi.org/10.1200/JCO.2008.21.6150.

[69] Whittaker SJ, Demierre M-F, Kim EJ, Rook AH, Lerner A, Duvic M, et al. Final results from a multicenter, international, pivotal study of romidepsin in refractory cutaneous T-cell lymphoma. J Clin Oncol J Am Soc Clin Oncol 2010;28:4485–91. Available from: https://doi.org/10.1200/JCO.2010.28.9066.

[70] Piekarz RL, Frye R, Prince HM, Kirschbaum MH, Zain J, Allen SL, et al. Phase 2 trial of romidepsin in patients with peripheral T-cell lymphoma. Blood. 2011;117:5827–34. Available from: https://doi.org/10.1182/blood-2010-10-312603.

[71] Vu K, Wu C-H, Yang C-Y, Zhan A, Cavallone E, Berry W, et al. Romidepsin plus liposomal doxorubicin is safe and effective in patients with relapsed or refractory T-cell lymphoma: results of

[72] Amengual JE, Lichtenstein R, Lue J, Sawas A, Deng C, Lichtenstein E, et al. A phase 1 study of romidepsin and pralatrexate reveals marked activity in relapsed and refractory T-cell lymphoma. Blood. 2018;131:397−407. Available from: https://doi.org/10.1182/blood-2017-09-806737.

[73] Klimek VM, Fircanis S, Maslak P, Guernah I, Baum M, Wu N, et al. Tolerability, pharmacodynamics, and pharmacokinetics studies of depsipeptide (romidepsin) in patients with acute myelogenous leukemia or advanced myelodysplastic syndromes. Clin Cancer Res J Am Assoc Cancer Res 2008;14:826−32. Available from: https://doi.org/10.1158/1078-0432.CCR-07-0318.

[74] Gojo I, Jiemjit A, Trepel JB, Sparreboom A, Figg WD, Rollins S, et al. Phase 1 and pharmacologic study of MS-275, a histone deacetylase inhibitor, in adults with refractory and relapsed acute leukemias. Blood 2006;109:2781−90. Available from: https://doi.org/10.1182/blood-2006-05-021873.

[75] Batlevi CL, Kasamon Y, Bociek RG, Lee P, Gore L, Copeland A, et al. ENGAGE- 501: phase II study of entinostat (SNDX-275) in relapsed and refractory Hodgkin lymphoma. Haematologica. 2016;101:968−75. Available from: https://doi.org/10.3324/haematol.2016.142406.

[76] Sermer DJ, Vardhana SA, Ames A, Biggar E, Moskowitz AJ, Batlevi CL, et al. Early data from a phase II trial investigating the combination of pembrolizumab (PEM) and entinostat (ENT) in relapsed and refractory (R/R) Hodgkin lymphoma (HL). J Clin Oncol 2020;38. Available from: https://doi.org/10.1200/JCO.2020.38.15_suppl.e20018 e20018−e20018.

[77] Prebet T, Sun Z, Ketterling RP, Zeidan A, Greenberg P, Herman J, et al.Eastern Cooperative Oncology Group and North American Leukemia intergroup Azacitidine with or without Entinostat for the treatment of therapy-related myeloid neoplasm: further results of the E1905 North American Leukemia Intergroup study. Br J Haematol 2016;172:384−91. Available from: https://doi.org/10.1111/bjh.13832.

[78] Kuendgen A, Strupp C, Aivado M, Bernhardt A, Hildebrandt B, Haas R, et al. Treatment of myelodysplastic syndromes with valproic acid alone or in combination with all-trans retinoic acid. Blood. 2004;104:1266−9. Available from: https://doi.org/10.1182/blood-2003-12-4333.

[79] Pilatrino C, Cilloni D, Messa E, Morotti A, Giugliano E, Pautasso M, et al. Increase in platelet count in older, poor-risk patients with acute myeloid leukemia or myelodysplastic syndrome treated with valproic acid and all-trans retinoic acid. Cancer. 2005;104:101−9. Available from: https://doi.org/10.1002/cncr.21132.

[80] Fredly H, Ersvær E, Kittang AO, Tsykunova G, Gjertsen BT, Bruserud Ø. The combination of valproic acid, all-trans retinoic acid and low-dose cytarabine as disease-stabilizing treatment in acute myeloid leukemia. Clin Epigenetics 2013;5:13. Available from: https://doi.org/10.1186/1868-7083-5-13.

[81] Gimsing P, Hansen M, Knudsen LM, Knoblauch P, Christensen IJ, Ooi CE, et al. A phase I clinical trial of the histone deacetylase inhibitor belinostat in patients with advanced hematological neoplasia. Eur J Haematol 2008;81:170−6. Available from: https://doi.org/10.1111/j.1600-0609.2008.01102.x.

[82] Foss F, Advani R, Duvic M, Hymes KB, Intragumtornchai T, Lekhakula A, et al. A Phase II trial of Belinostat (PXD101) in patients with relapsed or refractory peripheral or cutaneous T-cell lymphoma. Br J Haematol 2015;168:811−19. Available from: https://doi.org/10.1111/bjh.13222.

[83] O'Connor OA, Horwitz S, Masszi T, Van Hoof A, Brown P, Doorduijn J, et al. Belinostat in patients with relapsed or refractory peripheral T-cell lymphoma: results of the pivotal phase II BELIEF (CLN-19) study. J Clin Oncol J Am Soc Clin Oncol 2015;33:2492−9. Available from: https://doi.org/10.1200/JCO.2014.59.2782.

[84] Johnston PB, Cashen AF, Nikolinakos PG, Beaven AW, Barta SK, Bhat G, et al. Belinostat in combination with standard cyclophosphamide, doxorubicin, vincristine and prednisone as first-line treatment for patients with newly diagnosed peripheral T-cell lymphoma. Exp Hematol Oncol 2021;10:15. Available from: https://doi.org/10.1186/s40164-021-00203-8.

[85] FDA Grants Fast Track Designation to Abexinostat for Fourth-Line Follicular Lymphoma, Target. Oncol. (n.d.). https://www.targetedonc.com/view/fda-grants-fast-track-designation-to-abexinostat-for-fourthline-follicular-lymphoma [accessed 16.07.21].

[86] Shi Y, Gui L, Cao J, Zhang H, Zhu J, Ji D, et al. A phase 1b/2 study of chidamide combined with CHOP in previously untreated patients with peripheral T-cell lymphoma (PTCL). Blood. 2019;134. Available from: https://doi.org/10.1182/blood-2019-128078 2840−2840.

[87] Vogl DT, Raje N, Jagannath S, Richardson P, Hari P, Orlowski R, et al. Ricolinostat, the first selective histone deacetylase 6 inhibitor, in combination with bortezomib and dexamethasone for relapsed or refractory multiple myeloma. Clin Cancer Res J Am Assoc Cancer Res 2017;23:3307−15. Available from: https://doi.org/10.1158/1078-0432.CCR-16-2526.

[88] Ruxolitinib and Pracinostat Combination Therapy for Patients With Myelofibrosis (MF) - Study Results - ClinicalTrials.gov, (n.d.). https://clinicaltrials.gov/ct2/show/results/NCT02267278 [accessed 16.07.21].

[89] Atallah EL, Khaled SK, Cooper BW, Warlick ED, Ramies DA, Li J, et al. Phase II study of lower-dose pracinostat plus azacitidine safety and efficacy in patients with high/very high-risk myelodysplastic syndromes. J Clin Oncol 2020;38. Available from: https://doi.org/10.1200/JCO.2020.38.15_suppl.7556 7556−7556.

[90] Garcia-Manero G, Abaza Y, Takahashi K, Medeiros BC, Arellano M, Khaled SK, et al. Pracinostat plus azacitidine in older patients with newly diagnosed acute myeloid leukemia: results of a phase 2 study. Blood Adv 2019;3:508−18. Available from: https://doi.org/10.1182/bloodadvances.2018027409.

[91] Sato T, Issa J-PJ, Kropf P. DNA hypomethylating drugs in cancer therapy. Cold Spring Harb Perspect Med 2017;7:a026948. Available from: https://doi.org/10.1101/cshperspect.a026948.

[92] Burke MJ, Kostadinov R, Sposto R, Gore L, Kelley SM, Rabik C, et al. Decitabine and vorinostat with chemotherapy in relapsed pediatric acute lymphoblastic leukemia: a TACL pilot study. Clin Cancer Res Off J Am Assoc Cancer Res 2020;26:2297−307. Available from: https://doi.org/10.1158/1078-0432.CCR-19-1251.

[93] Prebet T, Delaunay J, Wattel E, Braun T, Cony-Makhoul P, Dimicoli S, et al. A phase II add-on study of vorinostat (VOR) in higher risk myelodysplastic syndrome with failure of hypomethylating agents (HMA): the GFM Azavor Study. Blood. 2015;126:2900. Available from: https://doi.org/10.1182/blood.V126.23.2900.2900.

[94] B. Oshrine, Epigenetic Modification for Relapse Prevention: a Dose-finding Study of Vorinostat Used in Combination With Low-dose Azacitidine in Children Undergoing Allogeneic Hematopoietic Cell Transplantation for Myeloid Malignancies, clinicaltrials.gov, 2021. https://clinicaltrials.gov/ct2/show/NCT03843528 [accessed 14.07.21].

[95] Dimopoulos M, Siegel DS, Lonial S, Qi J, Hajek R, Facon T, et al. Vorinostat or placebo in combination with bortezomib in patients with multiple myeloma (VANTAGE 088): a multicentre, randomised, double-blind study. Lancet Oncol 2013;14:1129−40. Available from: https://doi.org/10.1016/S1470-2045(13)70398-X.

[96] City of Hope Medical Center, A Phase 1 Study of Pembrolizumab Plus Vorinostat for Relapsed or Refractory Diffuse Large B-Cell Lymphoma, Follicular Lymphoma, and Hodgkin Lymphoma, clinicaltrials.gov, 2021. https://clinicaltrials.gov/ct2/show/NCT03150329 [accessed 14.07.21].

[97] Prince HM, George DJ, Johnstone R, Williams-Truax R, Atadja P, Zhao C, et al. LBH589, a novel histone deacetylase inhibitor (HDACi), treatment of patients with cutaneous T-cell lymphoma (CTCL). Changes in skin gene expression profiles related to clinical response following therapy. J Clin Oncol 2006;24. Available from: https://doi.org/10.1200/jco.2006.24.18_suppl.7501 7501−7501.

[98] Moore DC, Arnall JR, Harvey RD. Incidence and management of adverse events associated with panobinostat in the treatment of relapsed/refractory multiple myeloma. J Oncol Pharm Pract 2019;25:613−22. Available from: https://doi.org/10.1177/1078155218788706.

[99] Raedler LA. Farydak (Panobinostat): first HDAC inhibitor approved for patients with relapsed multiple myeloma. Am Health Drug Benefits 2016;9:84−7.

[100] Bieliauskas AV, Pflum MKH. Isoform-selective histone deacetylase inhibitors. Chem Soc Rev 2008;37:1402−13. Available from: https://doi.org/10.1039/b703830p.

[101] Piekarz RL, Robey R, Sandor V, Bakke S, Wilson WH, Dahmoush L, et al. Inhibitor of histone deacetylation, depsipeptide (FR901228), in the treatment of peripheral and cutaneous T-cell lymphoma: a case report. Blood. 2001;98:2865−8. Available from: https://doi.org/10.1182/blood.v98.9.2865.

[102] National Cancer Institute (NCI), A Phase 1 Study of Romidepsin, CC-486 (5-azacitidine), Dexamethasone, and Lenalidomide (RAdR) for Relapsed/Refractory T-cell Malignancies, clinicaltrials.gov, 2021. https://clinicaltrials.gov/ct2/show/NCT04447027 [accessed 14.07.21].

[103] M.D. Anderson Cancer Center, A Phase I/II Study of Pembrolizumab (MK-3475) in Combination With Romidepsin in Patients With Relapsed or Refractory Peripheral T-Cell Lymphoma, clinicaltrials.gov, 2020. https://clinicaltrials.gov/ct2/show/NCT03278782 [accessed 14.07.21].

[104] Ryan QC, Headlee D, Acharya M, Sparreboom A, Trepel JB, Ye J, et al. Phase I and pharmacokinetic study of MS-275, a histone deacetylase inhibitor, in patients with advanced and refractory solid tumors or lymphoma. J Clin Oncol 2005;23:3912−22. Available from: https://doi.org/10.1200/JCO.2005.02.188.

[105] Kummar Shivaani, Gutierrez Martin, Gardner Erin R, Donovan Erin, Hwang Kyunghwa, Chung Eun Joo, et al. Phase I trial of MS-275, a histone deacetylase inhibitor, administered weekly in refractory solid tumors and lymphoid malignancies. Clin Cancer Res J Am Assoc Cancer Res 2007;13:5411−17. Available from: https://doi.org/10.1158/1078-0432.ccr-07-0791.

[106] Phiel CJ, Zhang F, Huang EY, Guenther MG, Lazar MA, Klein PS. Histone deacetylase is a direct target of valproic acid, a potent anticonvulsant, mood stabilizer, and teratogen. J Biol Chem 2001;276:36734−41. Available from: https://doi.org/10.1074/jbc.M101287200.

[107] Ornstein MC, Mukherjee S, Sekeres MA. More is better: combination therapies for myelodysplastic syndromes. Best Pract Res Clin Haematol 2015;28:22−31. Available from: https://doi.org/10.1016/j.beha.2014.11.002.

[108] P. Stiff, Maintenance Therapy With Azacitidine and Valproic Acid After Allogeneic Stem Cell Transplant in Patients With High-Risk Acute Myelogenous Leukemia (AML) or Myelodysplastic Syndrome (MDS)(Version 1_06 Jan 2012), clinicaltrials.gov, 2021. https://clinicaltrials.gov/ct2/show/NCT02124174 [accessed 14.07.21].

[109] Yang H, Hoshino K, Sanchez-Gonzalez B, Kantarjian H, Garcia-Manero G. Antileukemia activity of the combination of 5-aza-2′-deoxycytidine with valproic acid. Leuk Res 2005;29:739−48. Available from: https://doi.org/10.1016/j.leukres.2004.11.022.

[110] Issa J-P, Garcia-Manero G, Huang X, Cortes J, Ravandi F, Jabbour E, et al. Results of phase 2 randomized study of low-dose decitabine with or without valproic acid in patients with myelodysplastic syndrome and acute myelogenous leukemia. Cancer. 2015;121:556−61. Available from: https://doi.org/10.1002/cncr.29085.

[111] Xynomic Pharmaceuticals, Inc., An Open-label, Single-arm, Multi-center Phase 2 Study to Evaluate the Efficacy and Safety of Oral Histone Deacetylase (HDAC)-Inhibitor Abexinostat, as Monotherapy in Patients With Relapsed or Refractory Follicular Lymphoma (FL), clinicaltrials.gov, 2021. https://clinicaltrials.gov/ct2/show/NCT03934567 [accessed 14.07.21].

[112] Xynomic Pharmaceuticals, Inc., An Open-label, Single-arm, Multi-center Phase 2 Study to Evaluate the Efficacy and Safety of Abexinostat as Monotherapy in Patients With Relapsed or Refractory Diffuse Large B-cell Lymphoma (DLBCL), clinicaltrials.gov, 2021. https://clinicaltrials.gov/ct2/show/NCT03936153 [accessed 14.07.21].

[113] Memorial Sloan Kettering Cancer Center, Phase I Study of Abexinostat and Ibrutinib in Diffuse Large B-cell Lymphoma and Mantle Cell Lymphoma, clinicaltrials.gov, 2021. https://clinicaltrials.gov/ct2/show/NCT03939182 [accessed 14.07.21].

[114] Younes A, Oki Y, Bociek RG, Kuruvilla J, Fanale M, Neelapu S, et al. Mocetinostat for relapsed classical Hodgkin's lymphoma: an open-label, single-arm, phase 2 trial. Lancet Oncol 2011;12:1222−8. Available from: https://doi.org/10.1016/S1470-2045(11)70265-0.

[115] Benedetti R, Conte M, Altucci L. Targeting histone deacetylases in diseases: where are we? Antioxid Redox Signal 2015;23:99−126. Available from: https://doi.org/10.1089/ars.2013.5776.

[116] Shen S, Kozikowski AP. Why hydroxamates may not be the best histone deacetylase inhibitors—what some may have forgotten or would rather forget? ChemMedChem. 2016;11:15−21. Available from: https://doi.org/10.1002/cmdc.201500486.

[117] Milazzo G, Mercatelli D, Di Muzio G, Triboli L, De Rosa P, Perini G, et al. Histone deacetylases (HDACs): evolution, specificity, role in transcriptional complexes, and pharmacological actionability. Genes. 2020;11:556. Available from: https://doi.org/10.3390/genes11050556.

[118] Zhang L, Han Y, Jiang Q, Wang C, Chen X, Li X, et al. Trend of histone deacetylase inhibitors in cancer therapy: isoform selectivity or multitargeted strategy. Med Res Rev 2015;35:63−84. Available from: https://doi.org/10.1002/med.21320.

[119] Shah MH, Binkley P, Chan K, Xiao J, Arbogast D, Collamore M, et al. Cardiotoxicity of histone deacetylase inhibitor depsipeptide in patients with metastatic neuroendocrine tumors. Clin Cancer Res 2006;12:3997−4003. Available from: https://doi.org/10.1158/1078-0432.CCR-05-2689.

[120] Schrump DS, Fischette MR, Nguyen DM, Zhao M, Li X, Kunst TF, et al. Clinical and molecular responses in lung cancer patients receiving romidepsin. Clin Cancer Res 2008;14:188−98. Available from: https://doi.org/10.1158/1078-0432.CCR-07-0135.

[121] Munster PN, Rubin EH, Van Belle S, Friedman E, Patterson JK, Van Dyck K, et al. A single supratherapeutic dose of vorinostat does not prolong the QTc interval in patients with advanced cancer. Clin Cancer Res Off J Am Assoc Cancer Res 2009;15:7077−84. Available from: https://doi.org/10.1158/1078-0432.CCR-09-1214.

[122] Li S, Shi B, Liu X, An H-X. Acetylation and deacetylation of DNA repair proteins in cancers. Front Oncol 2020;0. Available from: https://doi.org/10.3389/fonc.2020.573502.

[123] Yardley DA, Ismail-Khan RR, Melichar B, Lichinitser M, Munster PN, Klein PM, et al. Randomized phase II, double-blind, placebo-controlled study of exemestane with or without entinostat in postmenopausal women with locally recurrent or metastatic estrogen receptor-positive breast cancer progressing on treatment with a nonsteroidal aromatase inhibitor. J Clin Oncol J Am Soc Clin Oncol 2013;31:2128—35. Available from: https://doi.org/10.1200/JCO.2012.43.7251.

[124] Munster PN, Thurn KT, Thomas S, Raha P, Lacevic M, Miller A, et al. A phase II study of the histone deacetylase inhibitor vorinostat combined with tamoxifen for the treatment of patients with hormone therapy-resistant breast cancer. Br J Cancer 2011;104:1828—35. Available from: https://doi.org/10.1038/bjc.2011.156.

[125] Munster PN, Marchion D, Thomas S, Egorin M, Minton S, Springett G, et al. Phase I trial of vorinostat and doxorubicin in solid tumours: histone deacetylase 2 expression as a predictive marker. Br J Cancer 2009;101:1044—50. Available from: https://doi.org/10.1038/sj.bjc.6605293.

[126] Munster PN, Troso-Sandoval T, Rosen N, Rifkind R, Marks PA, Richon VM. The histone deacetylase inhibitor suberoylanilide hydroxamic acid induces differentiation of human breast cancer cells. Cancer Res 2001;61:8492—7.

[127] Galanis E, Jaeckle KA, Maurer MJ, Reid JM, Ames MM, Hardwick JS, et al. Phase II trial of vorinostat in recurrent glioblastoma multiforme: a north central cancer treatment group study. J Clin Oncol J Am Soc Clin Oncol 2009;27:2052—8. Available from: https://doi.org/10.1200/JCO.2008.19.0694.

[128] Luu TH, Morgan RJ, Leong L, Lim D, McNamara M, Portnow J, et al. A phase II trial of vorinostat (suberoylanilide hydroxamic acid) in metastatic breast cancer: a california cancer consortium study. Clin Cancer Res 2008;14:7138—42. Available from: https://doi.org/10.1158/1078-0432.CCR-08-0122.

[129] Blumenschein GR, Kies MS, Papadimitrakopoulou VA, Lu C, Kumar AJ, Ricker JL, et al. Phase II trial of the histone deacetylase inhibitor vorinostat (Zolinza™, suberoylanilide hydroxamic acid, SAHA) in patients with recurrent and/or metastatic head and neck cancer. Invest N Drugs 2008;26:81—7. Available from: https://doi.org/10.1007/s10637-007-9075-2.

[130] Traynor AM, Dubey S, Eickhoff JC, Kolesar JM, Schell K, Huie MS, et al. Vorinostat (NSC# 701852) in patients with relapsed non-small cell lung cancer: a wisconsin oncology network phase II study. J Thorac Oncol 2009;4:522—6. Available from: https://doi.org/10.1097/JTO.0b013e3181952478.

[131] Bradley D, Rathkopf D, Dunn R, Stadler WM, Liu G, Smith DC, et al. Vorinostat in advanced prostate cancer patients progressing on prior chemotherapy (National Cancer Institute Trial 6862). Cancer. 2009;115:5541—9. Available from: https://doi.org/10.1002/cncr.24597.

[132] Morita S, Oizumi S, Minami H, Kitagawa K, Komatsu Y, Fujiwara Y, et al. Phase I dose-escalating study of panobinostat (LBH589) Administered intravenously to Japanese patients with advanced solid tumors. Invest N Drugs 2012;30:1950—7. Available from: https://doi.org/10.1007/s10637-011-9751-0.

[133] Hainsworth JD, Infante JR, Spigel DR, Arrowsmith ER, Boccia RV, Burris HA. A phase II trial of panobinostat, a histone deacetylase inhibitor, in the treatment of patients with refractory metastatic renal cell carcinoma. Cancer Invest 2011;29:451—5. Available from: https://doi.org/10.3109/07357907.2011.590568.

[134] Study of Panobinostat Monotherapy in Women With HER2-negative Locally Recurrent or Metastatic Breast Cancer - Study Results - ClinicalTrials.gov (n.d.). https://clinicaltrials.gov/ct2/show/results/NCT00777049 [accessed 16.07.21].

[135] Efficacy and Safety Study of Panobinostat in Participants With Metastatic Hormone Refractory Prostate Cancer - Study Results - ClinicalTrials.gov (n.d.). https://clinicaltrials.gov/ct2/show/results/NCT00667862 [accessed 16.07.21].

[136] Whitehead RP, Rankin C, Hoff PMG, Gold PJ, Billingsley KG, Chapman RA, et al. Phase II trial of romidepsin (NSC-630176) in previously treated colorectal cancer patients with advanced disease: a Southwest Oncology Group study (S0336). Invest N Drugs 2008;27:469. Available from: https://doi.org/10.1007/s10637-008-9190-8.

[137] Stadler WM, Margolin K, Ferber S, McCulloch W, Thompson JA. A phase II study of depsipeptide in refractory metastatic renal cell cancer. Clin Genitourin Cancer 2006;5:57—60. Available from: https://doi.org/10.3816/CGC.2006.n.018.

[138] Molife LR, Attard G, Fong PC, Karavasilis V, Reid AHM, Patterson S, et al. Phase II, two-stage, single-arm trial of the histone deacetylase inhibitor (HDACi) romidepsin in metastatic castration-resistant prostate cancer (CRPC). Ann Oncol J Eur Soc Med Oncol 2010;21:109—13. Available from: https://doi.org/10.1093/annonc/mdp270.

[139] Amiri-Kordestani L, Luchenko V, Peer CJ, Ghafourian K, Reynolds J, Draper D, et al. Phase I trial of a new schedule of romidepsin in patients with advanced cancers. Clin Cancer Res J Am Assoc Cancer Res 2013;19:4499—507. Available from: https://doi.org/10.1158/1078-0432.CCR-13-0095.

[140] Gore L, Rothenberg ML, O'Bryant CL, Schultz MK, Sandler AB, Coffin D, et al. A phase I and pharmacokinetic study of the oral histone deacetylase inhibitor, MS-275, in patients with refractory solid tumors and lymphomas. Clin Cancer Res J Am Assoc Cancer Res 2008;14:4517—25. Available from: https://doi.org/10.1158/1078-0432.CCR-07-1461.

[141] Hauschild A, Trefzer U, Garbe C, Kaehler KC, Ugurel S, Kiecker F, et al. Multicenter phase II trial of the histone deacetylase inhibitor pyridylmethyl-N-{4-[(2-aminophenyl)-carbamoyl]-benzyl}-carbamate in pretreated metastatic melanoma. Melanoma Res 2008;18:274—8. Available from: https://doi.org/10.1097/CMR.0b013e328307c248.

[142] Atmaca A, Al-Batran S-E, Maurer A, Neumann A, Heinzel T, Hentsch B, et al. Valproic acid (VPA) in patients with refractory advanced cancer: a dose escalating phase I clinical trial. Br J Cancer 2007;97:177—82. Available from: https://doi.org/10.1038/sj.bjc.6603851.

[143] Sidney Kimmel Comprehensive Cancer Center at Johns Hopkins, Randomized, Controlled Phase II Study of Valproic Acid in Patients With Non-metastatic Biochemical Progression of Prostate Cancer, clinicaltrials.gov, 2018. https://clinicaltrials.gov/ct2/show/results/NCT00670046 [accessed 15.07.21].

[144] Steele NL, Plumb JA, Vidal L, Tjørnelund J, Knoblauch P, Rasmussen A, et al. A phase 1 pharmacokinetic and pharmacodynamic study of the histone deacetylase inhibitor belinostat in patients with advanced solid tumors. Clin Cancer Res J Am Assoc Cancer Res 2008;14:804—10. Available from: https://doi.org/10.1158/1078-0432.CCR-07-1786.

[145] Giaccone G, Rajan A, Berman A, Kelly RJ, Szabo E, Lopez-Chavez A, et al. Phase II study of belinostat in patients with recurrent or refractory advanced thymic epithelial tumors. J Clin Oncol J Am Soc Clin Oncol 2011;29:2052—9. Available from: https://doi.org/10.1200/JCO.2010.32.4467.

[146] Takebe N, Beumer JH, Kummar S, Kiesel BF, Dowlati A, O'Sullivan Coyne G, et al. A phase I pharmacokinetic study of belinostat in patients with advanced cancers and varying degrees of liver dysfunction. Br J Clin Pharmacol 2019;85:2499—511. Available from: https://doi.org/10.1111/bcp.14054.

[147] Grivas P, Mortazavi A, Picus J, Hahn NM, Milowsky MI, Hart LL, et al. Mocetinostat for patients with previously treated, locally advanced/metastatic urothelial carcinoma and inactivating alterations of acetyltransferase genes. Cancer 2019;125:533–40. Available from: https://doi.org/10.1002/cncr.31817.

[148] Marchion DC, Bicaku E, Daud AI, Sullivan DM, Munster PN. Valproic acid alters chromatin structure by regulation of chromatin modulation proteins. Cancer Res 2005;65:3815–22. Available from: https://doi.org/10.1158/0008-5472.CAN-04-2478.

[149] Fakih MG, Pendyala L, Fetterly G, Toth K, Zwiebel JA, Espinoza-Delgado I, et al. A phase I, pharmacokinetic and pharmacodynamic study on vorinostat in combination with 5-fluorouracil, leucovorin, and oxaliplatin in patients with refractory colorectal cancer. Clin Cancer Res J Am Assoc Cancer Res 2009;15:3189–95. Available from: https://doi.org/10.1158/1078-0432.CCR-08-2999.

[150] Vorinostat, Fluorouracil, and Leucovorin Calcium in Treating Patients With Metastatic Colorectal Cancer That Has Not Responded to Previous Treatment - Study Results - ClinicalTrials.gov (n.d.). https://clinicaltrials.gov/ct2/show/results/NCT00942266 [accessed 16.07.21].

[151] Townsley C, Oza AM, Tang P, Siu LL, Pond GR, Sarveswaran P, et al. Expanded phase I study of vorinostat (VOR) in combination with capecitabine (CAP) in patients (pts) with advanced solid tumors. J Clin Oncol 2008;26. Available from: https://doi.org/10.1200/jco.2008.26.15_suppl.11096 11096–11096.

[152] Di Gennaro E, Piro G, Chianese MI, Franco R, Di Cintio A, Moccia T, et al. Vorinostat synergises with capecitabine through upregulation of thymidine phosphorylase. Br J Cancer 2010;103:1680–91. Available from: https://doi.org/10.1038/sj.bjc.6605969.

[153] Terranova-Barberio M, Roca MS, Zotti AI, Leone A, Bruzzese F, Vitagliano C, et al. Valproic acid potentiates the anticancer activity of capecitabine in vitro and in vivo in breast cancer models via induction of thymidine phosphorylase expression. Oncotarget 2016;7:7715–31. Available from: https://doi.org/10.18632/oncotarget.6802.

[154] Yoo C, Ryu M-H, Na Y-S, Ryoo B-Y, Lee C-W, Maeng J, et al. Phase I and pharmacodynamic study of vorinostat combined with capecitabine and cisplatin as first-line chemotherapy in advanced gastric cancer. Invest N Drugs 2014;32:271–8. Available from: https://doi.org/10.1007/s10637-013-9983-2.

[155] Capecitabine and Vorinostat in Treating Patients With Recurrent and/or Metastatic Head and Neck Cancer - Study Results - ClinicalTrials.gov (n.d.). https://clinicaltrials.gov/ct2/show/results/NCT01267240 [accessed 16.07.21].

[156] Dowdy SC, Jiang S, Zhou XC, Hou X, Jin F, Podratz KC, et al. Histone deacetylase inhibitors and paclitaxel cause synergistic effects on apoptosis and microtubule stabilization in papillary serous endometrial cancer cells. Mol Cancer Ther 2006;5:2767–76. Available from: https://doi.org/10.1158/1535-7163.MCT-06-0209.

[157] Ramalingam SS, Parise RA, Ramanathan RK, Ramananthan RK, Lagattuta TF, Musguire LA, et al. Phase I and pharmacokinetic study of vorinostat, a histone deacetylase inhibitor, in combination with carboplatin and paclitaxel for advanced solid malignancies. Clin Cancer Res J Am Assoc Cancer Res 2007;13:3605–10. Available from: https://doi.org/10.1158/1078-0432.CCR-07-0162.

[158] Ramalingam SS, Maitland M, Frankel P, Argiris AE, Koczywas M, Gitlitz B, et al. Randomized, double-blind, placebo-controlled phase II study of carboplatin and paclitaxel with or without vorinostat, a histone deacetylase inhibitor (HDAC), for first-line therapy of advanced non-small cell lung cancer (NCI 7863). J Clin Oncol 2009;27. Available from: https://doi.org/10.1200/jco.2009.27.15_suppl.8004 8004–8004.

[159] U.A.M. MD, Phase Ib/II Study of Combination of Vorinostat, Carboplatin and Gemcitabine + Vorinostat Maintenance in Women With Recurrent, Platinum-Sensitive Epithelial Ovarian, Fallopian Tube, or Peritoneal Cancer, clinicaltrials.gov, 2018. https://clinicaltrials.gov/ct2/show/results/NCT00910000 [accessed 15.07.21].

[160] Terranova-Barberio M, Pawlowska N, Dhawan M, Moasser M, Chien AJ, Melisko ME, et al. Exhausted T cell signature predicts immunotherapy response in ER-positive breast cancer. Nat Commun 2020;11:3584. Available from: https://doi.org/10.1038/s41467-020-17414-y.

[161] University of California, San Francisco, Epigenetic Priming for Immune Therapy in ER-Positive Breast Cancer in Biomarker Select Population, clinicaltrials.gov, 2021. https://clinicaltrials.gov/ct2/show/NCT04190056 [accessed 15.07.21].

[162] Leone A, Roca MS, Ciardiello C, Terranova-Barberio M, Vitagliano C, Ciliberto G, et al. Vorinostat synergizes with EGFR inhibitors in NSCLC cells by increasing ROS via upregulation of the major mitochondrial porin VDAC1 and modulation of the c-Myc-NRF2-KEAP1 pathway. Free Radic Biol Med 2015;89:287–99. Available from: https://doi.org/10.1016/j.freeradbiomed.2015.07.155.

[163] Reguart N, Rosell R, Cardenal F, Cardona AF, Isla D, Palmero R, et al. Phase I/II trial of vorinostat (SAHA) and erlotinib for non-small cell lung cancer (NSCLC) patients with epidermal growth factor receptor (EGFR) mutations after erlotinib progression. Lung Cancer Amst Neth 2014;84:161–7. Available from: https://doi.org/10.1016/j.lungcan.2014.02.011.

[164] Han J-Y, Lee SH, Lee GK, Yun T, Lee YJ, Hwang KH, et al. Phase I/II study of gefitinib (Iressa®) and vorinostat (IVORI) in previously treated patients with advanced non-small cell lung cancer. Cancer Chemother Pharmacol 2015;75:475–83. Available from: https://doi.org/10.1007/s00280-014-2664-9.

[165] S.Y. Ph.D M.D., Phase I of Vorinostat-Iressa Combined Therapy on Resistance by BIM Polymorphysim in EGFR Mutant Lung Cancer, clinicaltrials.gov, 2018. https://clinicaltrials.gov/ct2/show/NCT02151721 [accessed 15.07.21].

[166] Ellis L, Hammers H, Pili R. Targeting tumor angiogenesis with histone deacetylase inhibitors. Cancer Lett 2009;280:145–53. Available from: https://doi.org/10.1016/j.canlet.2008.11.012.

[167] Hammers HJ, Verheul H, Wilky B, Salumbides B, Holleran J, Egorin MJ, et al. Phase I safety and pharmacokinetic/pharmacodynamic results of the histone deacetylase inhibitor vorinostat in combination with bevacizumab in patients with kidney cancer. J Clin Oncol 2008;26. Available from: https://doi.org/10.1200/jco.2008.26.15_suppl.16094 16094–16094.

[168] Ramaswamy B, Fiskus W, Cohen B, Pellegrino C, Hershman DL, Chuang E, et al. Phase I-II study of vorinostat plus paclitaxel and bevacizumab in metastatic breast cancer: evidence for vorinostat-induced tubulin acetylation and Hsp90 inhibition in vivo. Breast Cancer Res Treat 2012;132:1063–72. Available from: https://doi.org/10.1007/s10549-011-1928-x.

[169] Chinnaiyan P, Chowdhary S, Potthast L, Prabhu A, Tsai Y-Y, Sarcar B, et al. Phase I trial of vorinostat combined with bevacizumab and CPT-11 in recurrent glioblastoma. Neuro-Oncol 2012;14:93–100. Available from: https://doi.org/10.1093/neuonc/nor187.

[170] Puduvalli VK, Wu J, Yuan Y, Armstrong TS, Groves MD, Raizer JJ, et al. Brain tumor trials collaborative bayesian adaptive randomized phase II trial of bevacizumab plus vorinostat vs bevacizumab alone in adults with recurrent glioblastoma (BTTC-1102). J Clin Oncol 2015;33. Available from: https://doi.org/10.1200/jco.2015.33.15_suppl.2012 2012–2012.

[171] Friday BB, Anderson SK, Buckner J, Yu C, Giannini C, Geoffroy F, et al. Phase II trial of vorinostat in combination with bortezomib in recurrent glioblastoma: a north central cancer treatment group study. Neuro-Oncol 2012;14:215–21. Available from: https://doi.org/10.1093/neuonc/nor198.

[172] Rodriguez CP, Wu QV, Voutsinas J, Fromm JR, Jiang X, Pillarisetty VG, et al. A phase II Trial of pembrolizumab and vorinostat in recurrent metastatic head and neck squamous cell carcinomas and salivary gland cancer. Clin Cancer Res J Am Assoc Cancer Res 2020;26:837–45. Available from: https://doi.org/10.1158/1078-0432.CCR-19-2214.

[173] Pili R, Quinn DI, Albany C, Adra N, Logan TF, Greenspan A, et al. Immunomodulation by HDAC inhibition: results from a phase Ib study with vorinostat and pembrolizumab in metastatic urothelial, renal, and prostate carcinoma patients. J Clin Oncol 2019;37. Available from: https://doi.org/10.1200/JCO.2019.37.15_suppl.2572 2572–2572.

[174] H. Lee Moffitt Cancer Center and Research Institute, A Phase I Trial of Pembrolizumab and Vorinostat Combined With Temozolomide and Radiation Therapy for Newly Diagnosed Glioblastoma, clinicaltrials.gov, 2021. https://clinicaltrials.gov/ct2/show/NCT03426891 [accessed 15.07.21].

[175] Shi W, Palmer JD, Werner-Wasik M, Andrews DW, Evans JJ, Glass J, et al. Phase I trial of panobinostat and fractionated stereotactic re-irradiation therapy for recurrent high grade gliomas. J Neurooncol 2016;127:535–9. Available from: https://doi.org/10.1007/s11060-016-2059-3.

[176] Thomas S, Aggarwal R, Jahan T, Ryan C, Troung T, Cripps AM, et al. A phase I trial of panobinostat and epirubicin in solid tumors with a dose expansion in patients with sarcoma. Ann Oncol 2016;27:947–52. Available from: https://doi.org/10.1093/annonc/mdw044.

[177] Khushalani NI, Markowitz J, Eroglu Z, Giuroiu I, Ladanova V, Reiersen P, et al. A phase I trial of panobinostat with ipilimumab in advanced melanoma. J Clin Oncol 2017;35. Available from: https://doi.org/10.1200/JCO.2017.35.15_suppl.9547 9547–9547.

[178] Novartis Pharmaceuticals, Phase Ib, Open-label, Multi-center Study to Characterize the Safety, Tolerability and Pharmacodynamics (PD) of PDR001 in Combination With LCL161, Everolimus (RAD001) or Panobinostat (LBH589), clinicaltrials.gov, 2021. https://clinicaltrials.gov/ct2/show/NCT02890069 [accessed 15.07.21].

[179] Jones SF, Infante JR, Spigel DR, Peacock NW, Thompson DS, Greco FA, et al. Phase 1 results from a study of romidepsin in combination with gemcitabine in patients with advanced solid tumors. Cancer Invest 2012;30:481–6. Available from: https://doi.org/10.3109/07357907.2012.675382.

[180] Gaillard SL, Zahurak M, Sharma A, Reiss KA, Sartorius-Mergenthaler S, Downs M, et al. A phase I trial of oral DNA methyltransferase inhibitor CC-486 and the histone deacetylase inhibitor romidepsin in advanced solid tumors. Cancer. 2019;125:2837–45. Available from: https://doi.org/10.1002/cncr.32138.

[181] Murphy AG, Walker R, Lutz ER, Parkinson R, Ahuja N, Zheng L, et al. Epigenetic priming prior to pembrolizumab in mismatch repair-proficient advanced colorectal cancer. J Clin Oncol 2019;37. Available from: https://doi.org/10.1200/JCO.2019.37.4_suppl.591 591–591.

[182] Zheng H, Zhao W, Yan C, Watson CC, Massengill M, Xie M, et al. HDAC inhibitors enhance T-cell chemokine expression and augment response to PD-1 immunotherapy in lung adenocarcinoma. Clin Cancer Res J Am Assoc Cancer Res 2016;22:4119–32. Available from: https://doi.org/10.1158/1078-0432.CCR-15-2584.

[183] GWT-TUD GmbH, A Multicenter, Phase I/II Study of Sequential Epigenetic and Immune Targeting in Combination With Nab-Paclitaxel/Gemcitabine in Patients With Advanced Pancreatic Ductal Adenocarcinoma., clinicaltrials.gov, 2021. https://clinicaltrials.gov/ct2/show/NCT04257448 [accessed 15.07.21].

[184] Juergens RA, Wrangle J, Vendetti FP, Murphy SC, Zhao M, Coleman B, et al. Combination epigenetic therapy has efficacy in patients with refractory advanced non-small cell lung cancer. Cancer Discov 2011;1:598–607. Available from: https://doi.org/10.1158/2159-8290.CD-11-0214.

[185] Connolly RM, Jankowitz RC, Zahnow CA, Zhang Z, Rudek MA, Jeter SC, et al. Abstract 4666: a phase 2 study investigating the safety, efficacy and surrogate biomarkers of response of 5-azacitidine (5-AZA) andentinostat (MS-275) in patients with triple-negative advanced breast cancer. Cancer Res 2013;73. Available from: https://doi.org/10.1158/1538-7445.AM2013-4666 4666–4666.

[186] Azad NS, El-Khoueiry AB, Mahoney MR, Adkins D, Flynn PJ, Bahary N, et al. A phase II study of combination epigenetic therapy in metastatic colorectal cancer (mCRC): a phase II consortium (P2C)/Stand Up 2 Cancer (SU2C) study. J Clin Oncol 2013;31. Available from: https://doi.org/10.1200/jco.2013.31.15_suppl.3539 3539–3539.

[187] Witta SE, Jotte RM, Konduri K, Neubauer MA, Spira AI, Ruxer RL, et al. Randomized phase II trial of erlotinib with and without entinostat in patients with advanced non-small-cell lung cancer who progressed on prior chemotherapy. J Clin Oncol J Am Soc Clin Oncol 2012;30:2248–55. Available from: https://doi.org/10.1200/JCO.2011.38.9411.

[188] ECOG-ACRIN Provides Syndax Pharmaceuticals With Results of Phase 3 E2112 Trial of Entinostat Plus Exemestane in Patients with HR+, HER2- Breast Cancer | Syndax Pharmaceuticals, Inc., (n.d.). https://ir.syndax.com/news-releases/news-release-details/ecog-acrin-provides-syndax-pharmaceuticals-results-phase-3-e2112 [accessed 16.07.21].

[189] Chumsri S, Lee M-J, Tomita Y, Lee S, Tomita S, Cruickshank S, et al. Abstract P2-11-10: epigenetic immune modulation by entinostat in breast cancer: correlative analysis of ENCORE 301 trial. Cancer Res 2016;76. Available from: https://doi.org/10.1158/1538-7445.SABCS15-P2-11-10 P2-P2-11–10.

[190] Shen L, Ciesielski M, Ramakrishnan S, Miles KM, Ellis L, Sotomayor P, et al. Class I histone deacetylase inhibitor entinostat suppresses regulatory T cells and enhances immunotherapies in renal and prostate cancer models. PLoS One 2012;7: e30815. Available from: https://doi.org/10.1371/journal.pone.0030815.

[191] Kim K, Skora AD, Li Z, Liu Q, Tam AJ, Blosser RL, et al. Eradication of metastatic mouse cancers resistant to immune checkpoint blockade by suppression of myeloid-derived cells. Proc Natl Acad Sci U S A 2014;111:11774–9. Available from: https://doi.org/10.1073/pnas.1410626111.

[192] Azad NS, Shirai K, McRee AJ, Opyrchal M, Johnson DB, Ordentlich P, et al. ENCORE 601: a phase 2 study of entinostat in combination with pembrolizumab in patients with microsatellite stable metastatic colorectal cancer. J Clin Oncol 2018;36. Available from: https://doi.org/10.1200/JCO.2018.36.15_suppl.3557 3557–3557.

[193] Münster P, Marchion D, Bicaku E, Schmitt M, Lee JH, DeConti R, et al. Phase I trial of histone deacetylase inhibition by valproic acid followed by the topoisomerase II inhibitor epirubicin in advanced solid tumors: a clinical and translational study. J Clin Oncol J Am Soc Clin Oncol 2007;25:1979–85. Available from: https://doi.org/10.1200/JCO.2006.08.6165.

[194] Arce C, Pérez-Plasencia C, González-Fierro A, de la Cruz-Hernández E, Revilla-Vázquez A, Chávez-Blanco A, et al. A proof-of-principle study of epigenetic therapy added to neoadjuvant doxorubicin cyclophosphamide for locally advanced breast cancer. PLoS One 2006;1:e98. Available from: https://doi.org/10.1371/journal.pone.0000098.

[195] Candelaria M, Gallardo-Rincón D, Arce C, Cetina L, Aguilar-Ponce JL, Arrieta O, et al. A phase II study of epigenetic therapy with hydralazine and magnesium valproate to overcome chemotherapy resistance in refractory solid tumors. Ann Oncol J Eur Soc Med Oncol 2007;18:1529–38. Available from: https://doi.org/10.1093/annonc/mdm204.

[196] Falchook GS, Fu S, Naing A, Hong DS, Hu W, Moulder S, et al. Methylation and histone deacetylase inhibition in combination with platinum treatment in patients with advanced malignancies. Invest N Drugs 2013;31. Available from: https://doi.org/10.1007/s10637-013-0003-3 10.1007/s10637-013-0003-3.

[197] Braiteh F, Soriano AO, Garcia-Manero G, Hong D, Johnson MM, De Padua Silva L, et al. Phase I study of epigenetic modulation with 5-azacytidine and valproic acid in patients with advanced cancers. Clin Cancer Res J Am Assoc Cancer Res 2008;14:6296–301. Available from: https://doi.org/10.1158/1078-0432.CCR-08-1247.

[198] Chu BF, Karpenko MJ, Liu Z, Aimiuwu J, Villalona-Calero MA, Chan KK, et al. Phase I study of 5-aza-2′-deoxycytidine in combination with valproic acid in non-small-cell lung cancer. Cancer Chemother Pharmacol 2013;71:115–21. Available from: https://doi.org/10.1007/s00280-012-1986-8.

[199] Budillon A, Delrio P, Pecori B, Tatangelo F, Gennaro ED, Romano C, et al. Phase I/II study of valproic acid (VPA) and short-course radiotherapy (SCRT) plus capecitabine (CAP) as preoperative treatment in low-moderate risk rectal cancer (V-shoRT-R3). Ann Oncol 2018;29:viii167–8. Available from: https://doi.org/10.1093/annonc/mdy281.045.

[200] Dizon DS, Blessing JA, Penson RT, Drake RD, Walker JL, Johnston CM, et al. A phase II evaluation of belinostat and carboplatin in the treatment of recurrent or persistent platinum-resistant ovarian, fallopian tube, or primary peritoneal carcinoma: a Gynecologic Oncology Group study. Gynecol Oncol 2012;125:367–71. Available from: https://doi.org/10.1016/j.ygyno.2012.02.019.

[201] Vitfell-Rasmussen J, Lind-Hansen M, Safwat A, Rossen P, Knoblauch P, Jones RL, et al. A phase I/II clinical trial of belinostat (PXD101) in combination with doxorubicin in patients with soft tissue sarcomas (STS). J Clin Oncol 2015;33. Available from: https://doi.org/10.1200/jco.2015.33.15_suppl.10516 10516–10516.

[202] Jiang Z, Li W, Hu X, Zhang Q, Sun T, Cui S, et al. Tucidinostat plus exemestane for postmenopausal patients with advanced, hormone receptor-positive breast cancer (ACE): a randomised, double-blind, placebo-controlled, phase 3 trial. Lancet Oncol 2019;20:806–15. Available from: https://doi.org/10.1016/S1470-2045(19)30164-0.

[203] Tjulandin S, Fedyanin M, Vladimirov VI, Kostorov V, Lisyanskaya AS, Krikunova L, et al. A multicenter phase II study of the efficacy and safety of quisinostat (an HDAC inhibitor) in combination with paclitaxel and carboplatin chemotherapy (CT) in patients (pts) with recurrent platinum resistant high grade serous epithelial ovarian, primarily peritoneal or fallopian tube carcinoma cancer (OC). J Clin Oncol 2017;35. Available from: https://doi.org/10.1200/JCO.2017.35.15_suppl.5541 5541–5541.

[204] Kazantsev AG, Thompson LM. Therapeutic application of histone deacetylase inhibitors for central nervous system disorders. Nat Rev Drug Discov 2008;7:854–68. Available from: https://doi.org/10.1038/nrd2681.

[205] Rotili D, Simonetti G, Savarino A, Palamara AT, Migliaccio AR, Mai A. Non-cancer uses of histone deacetylase inhibitors: effects on infectious diseases and beta-hemoglobinopathies. Curr Top Med Chem 2009;9:272–91. Available from: https://doi.org/10.2174/156802609788085296.

[206] Blanchard F, Chipoy C. Histone deacetylase inhibitors: new drugs for the treatment of inflammatory diseases? Drug Discov Today 2005;10:197–204. Available from: https://doi.org/10.1016/S1359-6446(04)03309-4.

[207] Haberland M, Montgomery RL, Olson EN. The many roles of histone deacetylases in development and physiology: implications for disease and therapy. Nat Rev Genet 2009;10:32–42. Available from: https://doi.org/10.1038/nrg2485.

[208] Shirakawa K, Chavez L, Hakre S, Calvanese V, Verdin E. Reactivation of latent HIV by histone deacetylase inhibitors. Trends Microbiol 2013;21:277–85. Available from: https://doi.org/10.1016/j.tim.2013.02.005.

[209] Martínez-Bonet M, Clemente MI, Serramía MJ, Muñoz E, Moreno S, Muñoz-Fernández MÁ. Synergistic activation of latent HIV-1 expression by novel histone deacetylase inhibitors and bryostatin-1. Sci Rep 2015;5:16445. Available from: https://doi.org/10.1038/srep16445.

[210] Löscher W. Basic pharmacology of valproate. CNS Drugs 2002;16:669–94. Available from: https://doi.org/10.2165/00023210-200216100-00003.

[211] Haddad PM, Das A, Ashfaq M, Wieck A. A review of valproate in psychiatric practice. Expert Opin Drug Metab Toxicol 2009;5:539–51. Available from: https://doi.org/10.1517/17425250902911455.

[212] Mathew NT. Antiepileptic drugs in migraine prevention. Headache. 2001;41(Suppl 1):S18–24. Available from: https://doi.org/10.1046/j.1526-4610.2001.01154-4.x.

[213] Lee S-Y, Wang T-Y, Chen S-L, Chang Y-H, Chen P-S, Huang S-Y, et al. Combination of dextromethorphan and memantine in treating bipolar spectrum disorder: a 12-week double-blind randomized clinical trial. Int J Bipolar Disord 2020;8:11. Available from: https://doi.org/10.1186/s40345-019-0174-8.

[214] Duncan S. Teratogenesis of sodium valproate. Curr Opin Neurol 2007;20:175–80. Available from: https://doi.org/10.1097/WCO.0b013e32805866fb.

[215] Rosenberg G. The mechanisms of action of valproate in neuropsychiatric disorders: can we see the forest for the trees? Cell Mol Life Sci CMLS 2007;64:2090–103. Available from: https://doi.org/10.1007/s00018-007-7079-x.

[216] Chaliha D, Albrecht M, Vaccarezza M, Takechi R, Lam V, Al-Salami H, et al. Review of the valproic-acid-induced rodent model of autism. Dev Neurosci 2020;42:12–48. Available from: https://doi.org/10.1159/000509109.

[217] Horowitz E, Bergman LC, Ashkenazy C, Moscona-Hurvitz I, Grinvald-Fogel H, Magnezi R. Off-label use of sodium valproate for schizophrenia. PLoS One 2014;9:e92573. Available from: https://doi.org/10.1371/journal.pone.0092573.

[218] Kim HJ, Leeds P, Chuang D-M. The HDAC inhibitor, sodium butyrate, stimulates neurogenesis in the ischemic brain. J Neurochem 2009;110:1226–40. Available from: https://doi.org/10.1111/j.1471-4159.2009.06212.x.

[219] Feillet F, Leonard JV. Alternative pathway therapy for urea cycle disorders. J Inherit Metab Dis 1998;21(Suppl 1):101–11. Available from: https://doi.org/10.1023/a:1005365825875.

[220] Bennett SA, Tanaz R, Cobos SN, Torrente MP. Epigenetics in amyotrophic lateral sclerosis: a role for histone post-translational modifications in neurodegenerative disease. Transl Res 2019;204:19–30. Available from: https://doi.org/10.1016/j.trsl.2018.10.002.

[221] Janssen C, Schmalbach S, Boeselt S, Sarlette A, Dengler R, Petri S. Differential histone deacetylase mRNA expression patterns in amyotrophic lateral sclerosis. J Neuropathol Exp Neurol 2010;69:573—81. Available from: https://doi.org/10.1097/NEN.0b013e3181ddd404.

[222] Brahe C, Vitali T, Tiziano FD, Angelozzi C, Pinto AM, Borgo F, et al. Phenylbutyrate increases SMN gene expression in spinal muscular atrophy patients. Eur J Hum Genet EJHG 2005;13:256—9. Available from: https://doi.org/10.1038/sj.ejhg.5201320.

[223] Anglès F, Hutt DM, Balch WE. HDAC inhibitors rescue multiple disease-causing CFTR variants. Hum Mol Genet 2019;28:1982—2000. Available from: https://doi.org/10.1093/hmg/ddz026.

[224] Shukla S, Tekwani BL. Histone deacetylases inhibitors in neurodegenerative diseases, neuroprotection and neuronal differentiation. Front Pharmacol 2020;0. Available from: https://doi.org/10.3389/fphar.2020.00537.

[225] Margolis DM. Histone deacetylase inhibitors and HIV latency. Curr Opin HIV AIDS 2011;6:25—9. Available from: https://doi.org/10.1097/COH.0b013e328341242d.

[226] Rowe SM, Clancy JP. Advances in cystic fibrosis therapies. Curr Opin Pediatr 2006;18:604—13. Available from: https://doi.org/10.1097/MOP.0b013e3280109b90.

[227] Rubenstein RC, Zeitlin PL. A pilot clinical trial of oral sodium 4-phenylbutyrate (Buphenyl) in deltaF508-homozygous cystic fibrosis patients: partial restoration of nasal epithelial CFTR function. Am J Respir Crit Care Med 1998;157:484—90. Available from: https://doi.org/10.1164/ajrccm.157.2.9706088.

[228] Zeitlin PL, Diener-West M, Rubenstein RC, Boyle MP, Lee CKK, Brass-Ernst L. Evidence of CFTR function in cystic fibrosis after systemic administration of 4-phenylbutyrate. Mol Ther J Am Soc Gene Ther 2002;6:119—26. Available from: https://doi.org/10.1006/mthe.2002.0639.

[229] Lunn MR, Wang CH. Spinal muscular atrophy. Lancet 2008;371:2120—33. Available from: https://doi.org/10.1016/S0140-6736(08)60921-6.

[230] Mercuri E, Bertini E, Messina S, Pelliccioni M, D'Amico A, Colitto F, et al. Pilot trial of phenylbutyrate in spinal muscular atrophy. Neuromuscul Disord 2004;14:130—5. Available from: https://doi.org/10.1016/j.nmd.2003.11.006.

[231] Mercuri E, Bertini E, Messina S, Solari A, D'Amico A, Angelozzi C, et al. Randomized, double-blind, placebo-controlled trial of phenylbutyrate in spinal muscular atrophy. Neurology. 2007;68:51—5. Available from: https://doi.org/10.1212/01.wnl.0000249142.82285.d6.

[232] Brichta L, Holker I, Haug K, Klockgether T, Wirth B. In vivo activation of SMN in spinal muscular atrophy carriers and patients treated with valproate. Ann Neurol 2006;59:970—5. Available from: https://doi.org/10.1002/ana.20836.

[233] Weihl CC, Connolly AM, Pestronk A. Valproate may improve strength and function in patients with type III/IV spinal muscle atrophy. Neurology. 2006;67:500—1. Available from: https://doi.org/10.1212/01.wnl.0000231139.26253.d0.

[234] Swoboda KJ, Scott CB, Reyna SP, Prior TW, LaSalle B, Sorenson SL, et al. Phase II open label study of valproic acid in spinal muscular atrophy. PLoS One 2009;4:e5268. Available from: https://doi.org/10.1371/journal.pone.0005268.

[235] Morrison BE, Majdzadeh N, D'Mello SR. Histone deacetylases: focus on the nervous system. Cell Mol Life Sci CMLS 2007;64:2258—69. Available from: https://doi.org/10.1007/s00018-007-7035-9.

[236] Duan D, Goemans N, Takeda S, Mercuri E, Aartsma-Rus A. Duchenne muscular dystrophy. Nat Rev Dis Prim 2021;7:1—19. Available from: https://doi.org/10.1038/s41572-021-00248-3.

[237] Bettica P, Petrini S, D'Oria V, D'Amico A, Catteruccia M, Pane M, et al. Histological effects of givinostat in boys with Duchenne muscular dystrophy. Neuromuscul Disord 2016;26:643—9. Available from: https://doi.org/10.1016/j.nmd.2016.07.002.

[238] F.R. PhD, Givinostat Continues to Delay DMD Progression in Long-term Study, (n.d.). https://musculardystrophynews.com/2021/02/23/givinostat-continues-to-delay-duchenne-muscular-dystrophy-dmd-progression-new-data-long-term-study/ [accessed 18.07.21].

[239] Demonté D, Quivy V, Colette Y, Van Lint C. Administration of HDAC inhibitors to reactivate HIV-1 expression in latent cellular reservoirs: implications for the development of therapeutic strategies. Biochem Pharmacol 2004;68:1231—8. Available from: https://doi.org/10.1016/j.bcp.2004.05.040.

[240] Keedy KS, Archin NM, Gates AT, Espeseth A, Hazuda DJ, Margolis DM. A limited group of class I histone deacetylases acts to repress human immunodeficiency virus type 1 expression. J Virol 2009;83:4749—56. Available from: https://doi.org/10.1128/JVI.02585-08.

[241] Lehrman G, Hogue IB, Palmer S, Jennings C, Spina CA, Wiegand A, et al. Depletion of latent HIV-1 infection in vivo: a proof-of-concept study. Lancet Lond Engl 2005;366:549—55. Available from: https://doi.org/10.1016/S0140-6736(05)67098-5.

[242] Rasmussen TA, Tolstrup M, Brinkmann CR, Olesen R, Erikstrup C, Solomon A, et al. Panobinostat, a histone deacetylase inhibitor, for latent-virus reactivation in HIV-infected patients on suppressive antiretroviral therapy: a phase 1/2, single group, clinical trial. Lancet HIV 2014;1:e13—21. Available from: https://doi.org/10.1016/S2352-3018(14)70014-1.

[243] Elliott JH, Wightman F, Solomon A, Ghneim K, Ahlers J, Cameron MJ, et al. Activation of HIV transcription with short-course vorinostat in HIV-infected patients on suppressive antiretroviral therapy. PLoS Pathog 2014;10:e1004473. Available from: https://doi.org/10.1371/journal.ppat.1004473.

[244] Fidler S, Stöhr W, Pace M, Dorrell L, Lever A, Pett S, et al. RIVER trial study group, Antiretroviral therapy alone vs antiretroviral therapy with a kick and kill approach, on measures of the HIV reservoir in participants with recent HIV infection (the RIVER trial): a phase 2, randomised trial. Lancet Lond Engl 2020;395:888—98. Available from: https://doi.org/10.1016/S0140-6736(19)32990-3.

[245] Søgaard OS, Graversen ME, Leth S, Olesen R, Brinkmann CR, Nissen SK, et al. The depsipeptide romidepsin reverses HIV-1 latency in vivo. PLoS Pathog 2015;11:e1005142. Available from: https://doi.org/10.1371/journal.ppat.1005142.

[246] McMahon DK, Zheng L, Cyktor JC, Aga E, Macatangay BJ, Godfrey C, et al. ACTG A5315 team, A phase 1/2 randomized, placebo-controlled trial of romidespin in persons with HIV-1 on suppressive antiretroviral therapy. J Infect Dis 2020;. Available from: https://doi.org/10.1093/infdis/jiaa777.

[247] Lee J-H, Choy ML, Marks PA. Mechanisms of resistance to histone deacetylase inhibitors. Adv Cancer Res 2012;116:39—86. Available from: https://doi.org/10.1016/B978-0-12-394387-3.00002-1.

[248] Ayer DE. Histone deacetylases: transcriptional repression with SINers and NuRDs. Trends Cell Biol 1999;9:193—8. Available from: https://doi.org/10.1016/S0962-8924(99)01536-6.

[249] Watson PJ, Fairall L, Santos GM, Schwabe JWR. Structure of HDAC3 bound to corepressor and inositol tetraphosphate. Nature 2012;481:335—40. Available from: https://doi.org/10.1038/nature10728.

[250] Sankar S, Bell R, Stephens B, Zhuo R, Sharma S, Bearss DJ, et al. Mechanism and relevance of EWS/FLI-mediated transcriptional repression in Ewing sarcoma. Oncogene 2016;35:6155—6. Available from: https://doi.org/10.1038/onc.2016.142.

CHAPTER

41

Combination Epigenetic Therapy

Rūta Navakauskienė
Life Sciences Center, Vilnius University, Vilnius, Lithuania

OUTLINE

Introduction	821	Histone Methylation/Demethylation and Epigenetic Modifiers	828
Chromatin-remodeling Agents, Combined Treatment, and Targeted Therapy	822	Histone Acetylation/Deacetylation and Epigenetic Modifiers	829
DNA Methylation/Demethylation and Combined Treatment	823	Conclusions and Future Perspectives	833
DNMT Inhibitors	826	References	837
Histone Modifications and Histone-modifying Enzymes for Epigenetic Treatment	828		

INTRODUCTION

The importance of chromatin structure in gene regulation and cellular functioning is stupendous. Changes in chromatin remodeling affect the transcriptional regulation more than changes in the activation of transcription factors and co-factors. Accordingly, histone acetylation may facilitate the binding of transcription factors and co-factors to the gene regulatory sequences. On the other hand, DNA and histone some specific methylation could inhibit the access of transcription factors and co-factors the regulatory DNA sequences. In eukaryotes, chromatin is organized from DNA and proteins (histone proteins and numerous non-histone proteins). This complex provides a three-dimensional genome organization. Chromatin is a dynamic, constantly volatile structure. Due to various modifications of DNA and histones, certain chromatin regions change from inactive to active and vice versa, accordingly such changes influence and regulate the expression of certain genes and determine important processes of cell development and functioning. Chromatin as a dynamic macromolecular structure can be reconfigured depending on the incoming physiological and environmental signals. Chromatin modifications affect the regulation of gene expression without altering the DNA sequence. Such modifications are attributed to chromatin protein posttranslational modifications (PTMs), nucleosome composition changes depending on DNA methylation, and antisense microRNAs operation. Identification of DNA and histone modifications has a major impact on gene expression regulatory processes. Most PTM histones are dynamic and each PTM needs specific enzymes for the chemical group connection and/or removal. Most of the known effectors are protein complexes containing several domains and recognizing different histone modifications which ensure the chromatin-remodeling and gene expression [1].

The bromodomain and extraterminal (BET) proteins recognize acetylated lysines in histones and in such way regulate gene expression. So, this epigenetic reader could be a target for chromatin remodeling and treatment of various diseases. Therefore, over the past few years, the development of BET inhibitors has begun and they were clinically tested [2–4].

Gene expression modulation dependent on chromatin modifications is known as an epigenetic regulation. For example, active chromatin regions may be determined by histone hyperacetylation. On the contrary, heterochromatin is determined as deacetylated chromatin with methylation of CpG islands of DNA [5]. In DNA modification the Ten-eleven translocation (TET) proteins play an important role as well and TET inhibitors could be chosen as a novel anticancer drug [6]. Inappropriate and unbalanced chromatin structure can lead to cell malignancy and various forms of cancer development. Chromatin modifications alter the noncovalent interaction between histones and DNA resulting in chromatin remodeling and changes in gene expression [1].

Nowadays, researchers are working profoundly in the field of chemotherapy and pay great attention to the search of natural and synthetic compounds to counteract the uncontrolled proliferation of cancer cells. Epigenetic therapy is a promising strategy in oncology based on return of the transformed cells to a normal way of differentiation after using a variety of chemical and biological agents. The aberrant chromosome remodeling causes the appearance of oncogenic chimeric protein complexes with repressive proteins, such as HDACs, HMTs, and DNMTs that are involved in gene activity regulation. Most chemotherapy is accompanied by negative side effects and resistance to the effects of drugs after relapse. New epigenetic modifiers HDAC, HMT, DNMT, BET, TET inhibitors (HDACi, HMTi, DNMTi, BETi, TETi) are constantly tested both in laboratories and in clinical trials. Some of them become a promising chemotherapeutic agent and are involved in cancer treatment that stop the growth of cancer cells and reinforce the efficiency of other anticancer agents without toxicity to normal cells. HDAC inhibition results in the acetylation of chromatin proteins, histones, and cause widespread changes in the expression of genes responsible for normal cell development. DNMTi—DNA methylation inhibitors activate silenced genes in cancer cells and work together with HMTi. Recently emerging evidences demonstrated that balanced metabolic processes along with epigenetic regulation are important for normal cellular functioning [7].

In some cancers, such as leukemia, differentiation therapy is applied, for example, HMT and / or HDAC inhibitors are used with all-*trans* retinoic acid (ATRA). Leukemia is a case of malignancies in which the development of normal cells is stopped at the intermediate differentiation stage. The aim of such therapies is to force cancer cells to differentiate and reenter the normal development way. Thus, in this chapter the combined epigenetic therapy for leukemia will be presented as well, analyzing how differentiation inducer all-*trans* retinoic acid together with epigenetic modifiers influence on the human acute myeloid/promyelocytic leukemia (AML/APL) cell differentiation process and effectively contribute to apoptosis promotion. Preclinical and clinical studies are underway, the search for new epigenetic inhibitors and their mechanistic studies are ongoing. The most important challenge in cancer treatment these days is overcoming resistance to therapy. For this purpose, various therapies are being tested, in particular, the simultaneous use of traditional and new epigenetic drugs and two epigenetic drugs with different effects (e.g., DNMTi and HDACi / HMTi, etc.). To make cancer cells more sensitive to chemotherapy, epigenetic drugs are given before chemotherapy, or both epigenetic drugs and chemotherapy are used together to achieve the maximum efficacy of synergistic treatment. Combinations of epigenetic modifiers with immunotherapy and metabolic drugs are also being investigated. All this would contribute to a more effective and specific epigenetic therapeutic treatment of different diseases. This is a quite a new area, but the initial data provide the hope that pharmacological chromatin-remodeling will develop into the new epigenetic strategy in cancer therapy.

CHROMATIN-REMODELING AGENTS, COMBINED TREATMENT, AND TARGETED THERAPY

Chromatin remodeling is a key regulator of gene expression. DNA structure dynamically changes during cell development and gene transcription. Proteins that regulate chromatin structure are involved in chromosome rearrangements, and in this respect, they are important for both chromatin structure surveys and chromatin remodeling. Changes of chromatin structure are often linked to the altered composition of chromatin-modifying enzymes and cause significant changes in cellular processes. Any abnormal alterations in epigenetic control mechanisms could contribute to various human diseases including cancer.

Tumor cells undergo aberrant epigenetic reorganization: DNA methylation (except at gene bodies), histone deacetylation, and histone methylation that can lead to both transcriptional activation and repression. In normal cells, the chromatin activation is related to active transcription markers—modified histones: H3K4me2, H3K4me3, H3K9me, H3K36me3, H3K79me2, H3S10P, H3Ac K9, -18, -56, H4Ac K5, -8, -12, -16, -20, H2BAc K6, -7, 16, 17, while repressed chromatin regions are enriched with repressive histone markers: H3K9me2, H3K27me3, H4K20me3. In tumor cells, the promoters of tumor suppressor genes lose nearly all acetylation and acquire repressive marks including DNA methylation, H3K27me3, H3K9me, and H4 deacetylation. Therefore, cancer therapies targeting epigenetic-modifying enzymes (e.g., DNMTs, HATs, HDACs, kinases, HMTs, and HDMs) are developing and bring a new era in anticancer therapeutics [8–10]. Also, the cytosine-based TET enzymes were presented as important epigenetic regulators in DNA modification. TET have various roles in gene expression by converting 5-methylcytosine to 5-hydroxymethylcytosine in DNA and in this way cause the opposite or alternative epigenetic effect-deciding biological processes. Recently efforts have been made to synthesize and analyze TET inhibitors to target DNA modification and chromatin remodeling [11]. DNA demethylation by TET is initiated through oxidation of 5mC, 5hmC, 5fc, 5caC. In some cases, especially in myeloid neoplasms, TET2 is frequently mutated that leads to defective myeloid differentiation and clonal expansion. It was demonstrated that treatment with TET inhibitor suppresses the clonal evolution of TET2-mutant cells. Therefore, TET inhibitors may constitute a new class of targeted agents in TET2-mutant cancer [12]. The other potential agents for chromatin remodeling are bromodomain and extraterminal (BET) proteins' inhibitors [13]. BET proteins via bromodomains (BRD) interact with acetylated lysines in histones and regulate the transcriptional program. Recent studies demonstrated that pharmacologic BET inhibition has clear preclinical antitumor activity in a variety of solid tumors and hematologic cancers [4,14]. BET inhibitors (BETi) have been also shown to be effective in the treatment of aggressive triple-negative breast cancer. However in such cases, when BETi is used at effective doses, the effects of BETi are highly toxic, therefore this limits their clinical implementation [15].

The schematic presentation of chromatin modifiers action is presented in Fig. 41.1. Most of these inhibitors reach preclinical studies; several of these compounds are in clinical trials (Table 41.1) with significant activity against a broad spectrum of both hematological and solid tumors. Therefore, a lot of efforts have been directed to develop epigenetic drugs for the treatment of a variety of malignancies. The detailed description of clinical trial, A Trial of Epigenetic Priming in Patients with Newly Diagnosed Acute Myeloid Leukemia, is presented at the end of the chapter.

Acute promyelocytic leukemia (APL) is a unique subtype of acute myeloid leukemia (AML) in which bone marrow cells that produce blood cells do not develop and function normally. APL is cytogenetically characterized by a balanced reciprocal translocation between chromosomes t(15;17)(q22;q12), t(11;17)(q23;q12), t(11;17)(q13;q12), t(5;17)(q23;p12) resulting in fusion of promyelocytic leukemia (PML), PLZF, NuMa, NPM1 and retinoic acid receptor alpha (RARα) genes. Therapeutic advances, including anthracyclines, cytarabine chemotherapy, and all-*trans* retinoic acid (ATRA) therapy, have significantly improved outcomes in APL patients. APL is currently the most curable form of adult AML [16]. The treatment of APL includes induction therapy, consolidation therapy, and maintenance therapy [17]. The major component of induction therapy is ATRA, which is commonly combined with other agents. For instance, the combination of arsenic trioxide with ATRA may be alternated with chemotherapy in newly diagnosed APL considered for patients who are not candidates for anthracycline-based therapy [18]. Though, epigenetic modulation of chromatin represents a promising new impact in drug discovery. The US Food and Drug Administration (FDA) has approved the use of an HDAC inhibitor, suberoylanilide hydroxamic acid (SAHA, marketed by Merck as Zolinza), for the treatment of cutaneous T-cell lymphoma (CTCL), belinostat, romidepsin, and chidamide against peripheral T-cell lymphoma (PTCL) and panobinostat, against multiple myeloma. Numerous HDAC inhibitors are under I–III clinical trials against more common solid tumors, including lung, breast, and prostate cancer [19]. Azacitidine and decitabine were approved for the treatment of high-risk myelodysplastic syndrome (MDS) [9].

The most promising use, especially for solid tumors, may be in combination of DNMTi and HDACi agents. Treatment of MDS and acute myeloid leukemia (AML) by HDAC inhibitors (valproic acid, sodium phenylbutyrate, entinostat) after low doses of 5-azacitidine or 5-aza-2′-deoxycytidine can enrich the effect of DNMT inhibitor on expression of genes with cancer-specific DNA hypermethylation [20–22]. Low combined doses of 5-azacitidine and the HDACi entinostat induced remarkable responses in non-small cell lung cancer (NSCLC). It is reported that the combination of epigenetic therapies with other cancer treatments (e.g., panobinostat with dexamethasone for relapsed multiple myeloma; belinostat in combination with carboplatin for NSCLC) is emerging strategy for cancer therapy as well [23].

DNA METHYLATION/DEMETHYLATION AND COMBINED TREATMENT

DNA methylation at the transcriptional start sites of genes is one of the most important epigenetic phenomena that control the gene expression and other biological processes [24]. Following methylation of DNA sequences, they attract various protein complexes (Polycomb repressive complexes [PRCs], HDAC, DNMT) that inhibit gene expression. It is known that methylation disorders are determined by many human diseases. The development of cancer is affected by DNA hypermethylation. In this case, the hypermethylation of promoter CpG islands decreases tumor

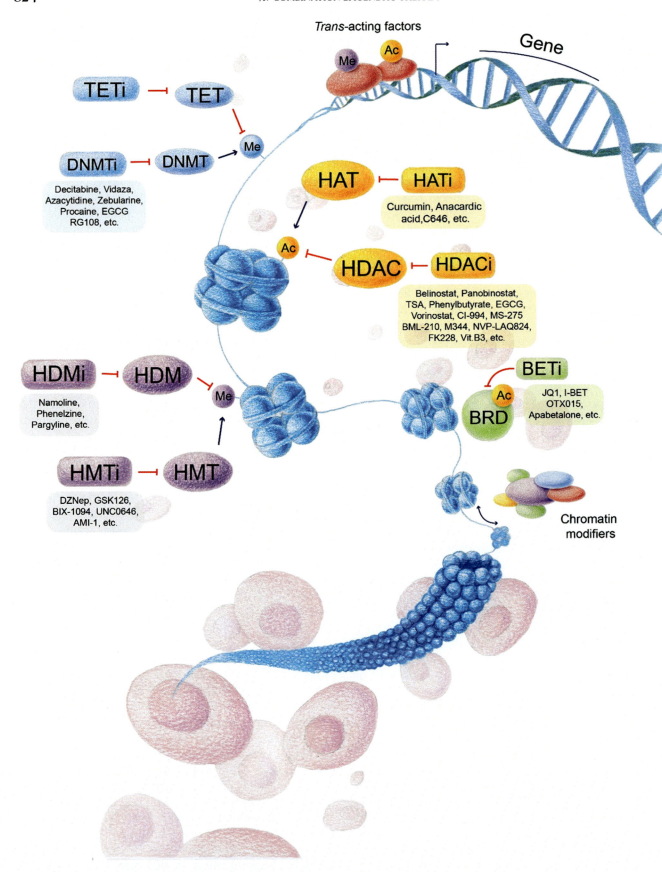

FIGURE 41.1 Chromatin structure and remodeling via combined epigenetic treatment.

TABLE 41.1 Clinical Trials in Epigenetic Combined Treatment

Drug	Combined Treatment Drug	Status (With Results)	Conditions
5-Azacytidine	Drug: Docetaxel		Non-small cell lung carcinoma
	Drug: Entinostat	Terminated	
	Drug: Pembrolizumab	Terminated	Solid tumors, advanced malignancies, metastatic cancer
	Drug: Epacadostat		
	Drug: CC-486	Terminated	Breast neoplasms
	Drug: Fulvestrant		
	• Drug: Nab-paclitaxel IV • Drug: CC-486 • Drug: Duravalumab	Active, not recruiting	• Carcinoma, non-small cell lung
	Drug: Entinostat	Terminated	• Stage IA non-small cell lung carcinoma • Stage IB non-small cell lung carcinoma
	• Drug: Lenalidomide	Terminated	• Follicular lymphoma, • Marginal zone lymphoma
	• Biological: Avelumab • Biological: Utomilumab • Biological: Rituximab	Terminated	• Diffuse large b-cell lymphoma
	• Biological: Avelumab • Biological: Utomilumab • Biological: Rituximab	Terminated	• Diffuse large B-cell lymphoma
	• Drug: Vorinostat	Completed	• Lymphoma
	• Drug: Clofarabine	Terminated	• Myelodysplastic syndromes
	• Drug: Decitabine • Drug: Idarubicin • Drug: Cytarabine	Terminated	Adult acute megakaryoblastic leukemia (M7), adult Acute monoblastic leukemia (M5a), adult acute monocytic leukemia (M5b), and others
	• Drug: Fludarabine • Drug: Cytarabine	Completed	Lymphoblastic leukemia, acute, childhood, myelogenous leukemia, acute, childhood
	• Drug: Cytarabine • Drug: Decitabine • Drug: Etoposide	Completed	Previously treated myelodysplastic syndrome, recurrent adult acute myeloid leukemia
Vorinostat (SAHA)	• Drug: Tamoxifen • Drug: Vorinostat • Drug: Pembrolizumab	Terminated	Breast neoplasms
	• Drug: Azacytidine	Completed	• Lymphoma
Decitabine	• Drug: Nivolumab • Drug: Tetrahydrouridine	Active, not recruiting	• Lung cancer • Non-small cell lung cancer
	• Drug: LBH589 Drug: Tamoxifen	Terminated	• Breast cancer • Breast tumors • Breast neoplasms
	• Drug: Cytarabine • Other: Supportive Care	Terminated	• Acute myeloid leukemia
	• Drug: Temozolomide, • Drug: Panobinostat	Terminated	• Metastatic melanoma
	• Drug: Azacitidine • Drug: Entinostat • Other: Laboratory biomarker analysis	Terminated	• Stage IA non-small cell lung carcinoma • Stage IB non-small cell lung carcinoma

(Continued)

TABLE 41.1 (Continued)

Drug	Combined Treatment Drug	Status (With Results)	Conditions
	• Drug: Idarubicin • Drug: Cytarabine	Terminated	• Adult acute megakaryoblastic leukemia (M7) • Adult acute monoblastic leukemia (M5a) • Adult acute monocytic leukemia (M5b), and others
	• Drug: Cytarabine • Drug: Etoposide	Completed	• Previously treated myelodysplastic syndrome • Recurrent adult acute myeloid leukemia
ATRA	• Drug: tranylcypromine • Drug: Cytarabine	Recruiting	• Acute myeloid leukemia • Myelodysplastic syndrome

Source: *According to clinicaltrials.gov 2021 June.*

suppressor genes expression and promotes the expansion of uncontrollably malignant cells [25].

Another methylation disorder, hypomethylation, is usually associated with prostate, liver, cervical tumors, and chronic lymphocytic leukemia development and progression. DNA hypomethylation biological significance for the development of cancer is less understood than hypermethylation. Researchers have found that a variety of repetitive sequences (e.g., LINE-1) replace the hypomethylation of the genome stability and cause cancer. It was also found that expression of hypomethylation activated oncogenes, such as H-RAS, Boris/CTCFL, FGFR1, and c-MYC leads to cancer progression [24].

DNMT1 is the most common methyltransferase in somatic cells. DNMT1 is responsible for the maintenance of methylation after DNA replication. This is carried out through two mechanisms: (1) directly interfacing with the replication fork, (2) interacting with UHRF1 protein [26]. DNMT3—methyltransferases (DNMT3A and DNMT3B) conduct de novo DNA methylation in embryogenesis and germ cells during development [27].

Mutations in the gene encoding DNMT3A were detected in patients with various hematological malignancies likewise in patients with AML and in acute monocytic leukemia [28–30]. DNMT3A mutations appear in AML patients from the disease initiation to relapse. This demonstrates the importance of DNMT3A as a tumor suppressor. However, the signaling pathways that are influenced by the DNMT3A mutations are still unclear. It was also reported that DNMT3A mutations do not influence the global methylation of DNA promoter regions [28,31]. Indeed, even if it is known that promoter methylation is associated to the gene silencing, the function of gene body methylation still has to be defined. It was shown [32] that DNMT inhibitor 5-aza-2′-deoxycytidine not only induces the gene expression, but also decreases the expression of oncogenes and metabolic genes commonly overexpressed in various cancers. Therefore, gene body methylation might be a therapeutic target for the normalization of gene overexpression induced during carcinogenesis [33].

DNMT Inhibitors

DNA methylation is a reversible process. Therefore, the use of special DNMT inhibitors can block the activity of DNMTs, activate silenced tumor suppressor genes, and by preventing the proliferation of cancer cells promote apoptosis. There are two families of DNMT inhibitors: nucleoside analog of cytidine and non-nucleoside DNMTs inhibitors [34]. 5-Azacytidine (Vidaza), decitabine (Dacogen or 5-aza-2′-deoxycytidine), and zebularine are assigned to the first family of DNMTi. The second family of DNMT inhibitors include hydralazine, procainamide (Pronestyl, Procan, Procanbid), RG108, epigallocatechin gallate (also known as epigallocatechin-3-gallate, EGCG; it could be assigned as HDACi as well).

5-Azacytidine and decitabine have been used clinically in the treatment of hematological malignancies, including the MDS and acute leukemia [35–38]. Azacytidine and decitabine are approved by FDA as suitable for MDS, AML, and chronic myelomonocytic leukemia (CMML) [39]. For elderly patients with newly diagnosed AML the combined epigenetic therapy is suggested, that is, decitabine, azacytidine with novel agents as capecitabine, volasertib, and low-dose cytarabine [40]. Combined treatments with azacytidine have been evaluated: azacytidine with lenalidomide, vorinostat, cytarabine, anti-CD33-gemtuzumab ozogamicin in AML and high-risk MDS [41–43]. It was reported that in order to achieve a second complete remission for relapsed patients with AML after the hematopoietic cell transplantation, a combination of decitabine and valproic acid [44] or decitabine alone were used prior to standard induction chemotherapy, which was given for patients with AML [45]. A combined treatment of MDA-MB-231 cells with decitabine and suramin decreases the invasive potential of these highly invasive breast cancer cells [46].

Zebularine, a novel nucleoside inhibitor with increased stability and low toxicity, may be used as a demethylating agent that preferentially targets tumor cells. Growth inhibitory and demethylation effects of zebularine in different bladder cancer, AML cell lines, and primary AML samples have been shown as well [47,48]. It was demonstrated that APL cell pre-treatment with zebularine before the induction of granulocytic differentiation with ATRA had the additional differentiation effects on leukemia cells and this was associated with early changes in modification status of histones (hyperAcH4 and H3K4me3) [49]. Zebularine cotreatment with ATRA at pharmacological dose (1 μM for NB4 cells) and higher (3 μM for KG1 cells) has increased the granulocytic differentiation in both cell lines. Pretreatment with zebularine before the treatment with different doses of ATRA alone or ATRA with HDAC inhibitors, such as phenylbutyrate (PB) and BML-210 [N-(2-aminophenyl)-N9 phenyloctanol diamine], resulted in significant acceleration and enhancement of differentiation and cell cycle arrest at G0/G1. The combined treatment of zebularine with ATRA decreased expression of DNMT1 in leukemia cells [50]. It was reported that zebularine may potentiate inhibitory effects of classical chemotherapeutic agents, such as oxaliplatin and 5-fluorouracil (5-FU), commonly used in colorectal cancer therapy [51].

RG108 is a novel small-molecule DNMT inhibitor that blocks the active site of human DNMTs. It is an attractive candidate as an epigenetic drug effectively blocking the DNMTs active center without causing any cytotoxic or genotoxic effects on human cells [52]. It was found that RG108, in combination with ATRA and HDAC inhibitors (PB and BML-210), reinforces promyelocytic leukemia cell differentiation into granulocytes with re-expression of methylation-silenced E-cadherin and changes in histone H4 and H3 modifications (H4 hyperacetylation, H3K4me3) [53]. RG108-pretreatment before the treatment by ATRA alone or ATRA together with PB has caused a remarkable effect on the enhancement or acceleration of histone modifications. A combined treatment with suramin and RG108 impairs the invasive potential of highly invasive breast cancer cell lines MDA-MB-231, BT-20, and HCC1954, whereas suramin or RG108 alone have no visible effect [46].

Green tea (Camellia sinensis) is high in EGCG, that is, catechin polyphenols, which is identified as a potent antioxidant and in vitro may have healing properties in treatment of many disorders, including cancer [54]. Various studies have shown that EGCG acts in various signal transduction pathways, blocks the carcinogenesis and inhibits tumor metastasis to the area of other organs [55,56]. The treatment of cervical cancer cells (HeLa) with EGCG inhibits DNMT activity and reduces DNMT3B mRNA expression [57].

EGCG forms hydrogen bonds with amino acids (Pro, Glu, Cys, Ser, and Arg) in catalytic DNMT pocket, directly inhibits DNMT as well as DNA methylation and activates the silenced gene transcription [57]. Recently it was found that EGCG can act not only as an inhibitor of DNMTs, but also as a HDACi. The green tea catechin EGCG inhibits RNA expression of HDAC1–HDAC3. Also, EGCG inhibits Polycomb-group proteins that are key epigenetic regulators [58]. EGCG is generally available in the form of tea, it is inexpensive and can be taken orally. Also, the compound has selectivity for cancer cells [59]. The combinations of EGCG and anticancer compounds, such as sulindac, celecoxib, erlotinib, luteolin, trichostatin A (TSA) have induced the similar synergistic anticancer effects in cell lines of human head, neck, lung, and breast cancer. The treatment with EGCG inhibited self-renewal of cancer stem cells [60]. The synergistic enhancements of apoptosis and GADD153 gene expression in human NSCLC cells by the combination of EGCG and celecoxib was reported [61]. It was observed that at low and physiologically achievable concentrations, combinatorial EGCG and sodium butyrate are effective in promoting apoptosis, inducing cell cycle arrest, and DNA-damage in colorectal cancer cells [62]. The effect of EGCG alone or in combination with TSA inhibit human malignant lymphoma CA46 cell proliferation through epigenetic modification of p16INK4a [63]. EGCG in combination with HDACi TSA or vorinostat demonstrate synergistic effects of reactivation of ERα expression in ERα-negative breast cancer cells or antimelanoma respectively [64,65]. Tyagi et al. demonstrated the therapeutic efficacy of a combination of non-toxic, low dose of 5-aza-2′-deoxycytidine with EGCG on growth inhibition of human breast cancer cells (MCF-7, MDA-MB 231). This combination has no significant toxicity to nontumorigenic MCF-10A cells [66].

Even hypomethylating agents such as azacytidine or decitabine represent one of the main therapeutic alternatives for AML patients, the additional effective therapeutic therapies are still being sought. Recently, the use of DNMTi with isocitrate dehydrogenase inhibitors (IDH) is in progress. IDH constitute another therapeutic class with DNA methylation effects in AML [67]. IDH inhibitors also showed encouraging results for relapsed/refractory AML (R/R AML) patients harboring an IDH mutation and are seen as promising chemotherapeutic drugs. The enasidenib and ivosidenib, the IDH2 and IDH1 inhibitors have been approved by FDA for the treatment of adult R/R AML with IDH2 and IDH1 mutations, respectively. It was demonstrated that IDH inhibitor monotherapy for R/R AML is effective and safe [68]. Recent studies have led to the emergence of new therapeutic options using DNMT and IDH inhibitors for specific groups of AML patients.

HISTONE MODIFICATIONS AND HISTONE-MODIFYING ENZYMES FOR EPIGENETIC TREATMENT

Histone Methylation/Demethylation and Epigenetic Modifiers

HMT

Histone methylation in the past few years has become a key component of chromatin states. So, histone methylation disorder is often associated with various human diseases, because many chromatin reorganization and regulatory mechanisms depend on the different histone methylation pattern. About 50% of the histone lysine methyltransferases alterations in their expression and/or activity are associated with certain disorders [69]. Methylation of some lysine and arginine residues of histones determines transcriptional activation. Histones H3 and H4 are methylated specifically and influence differently the chromatin status. Histone H3 can be methylated on positions R2, K4, K9, K14, R17, K27, K36, K79, and histone H4—on R3, K12, K20 and K59 positions. The most common methylation of histone H3 on K4, K36, and K79 positions is associated with chromatin activation and gene expression, and methylation of histone H3K9, K27, and K20 positions and histone H4 methylation—with heterochromatin and gene expression repression [70]. Thus, depending on the histone methylation site, this modification can activate or inactivate gene expression. Histone H3K4me3 is associated with activation of gene expression. H3K4 may be mono-, di- and trimethylated, each means the same transcriptional regulation. It was found that H3K4 dimethylation (H3K4me2) deals with the active and potentially active genes, and trimethylation of H3K4 (H3K4me3) is found in the transcription start regions, i.e., both modifications are associated with chromatin activation [71].

HMT Inhibitors

Although histone methylation of lysine residues was discovered 3 decades ago, the importance of this epigenetic modification is clarified only in recent years. Then HMTs inhibitors have been developed as potential cancer therapy agents [72]. HMT inhibitors are chemical compounds which have an ability to inhibit HMTs catalytic activity. H3K9me2 and H3K9me3 are classical heterochromatin tags with immense significance of gene expression inhibition. Chaetocin is the first identified HMT inhibitor that specifically targets the Suv39 histone H3K9 methyltransferases [73]. Combination of chaetocin with HDACi SAHA showed synergistic cytotoxicity on HL-60, U937, and KG-1a cells [74].

H3K9 mono- and dimethylation are catalyzed respectively by G9a and GLP lysine methyltransferases. Histone H3K9me1 and H3K9me2 are the only markers that are lost when the tumor suppressor genes are reactivated after treatment with demethylation agents. HMT G9a is specifically suppressed by inhibitor BIX-01294. It was shown that after exposure of various cancer cell lines with BIX-01294 the demethylation of H3K9me2 is observed. A small molecule compound BIX-01294 (a diazepine-quinazolin-amine derivative) specifically inhibits enzymatic activity of EHMT2, reduces H3K9me2 levels and significantly suppresses cancer cell growth [75–77]. BIX-01294 in combination with ATRA and HDACs (belinostat, BML-210) or DNMTs (zebularine, RG108) inhibitors increased leukemia cell differentiation into granulocytes, efficiently induced the expression of early myeloid differentiation marker CD11b [78]. BIX-01294-mediated depletion of HMT2 and the H3K9me2 marker inhibits cell-cycle progression and enhances ATRA-mediated granulocytic differentiation of promyelocytic leukemia HL-60 and NB4 cells. BIX-01294-induced a dose-dependent growth inhibition and non-apoptotic cell death might be due to cell inability to divide after arrest at the G1/S phase that may lead to chromosome instability and cellular senescence [76]. BIX-01294 as a single drug induced changes in the expression of two G1/S phase regulating proteins, cyclin E1 and p27, depletion of H3K9me2 paralleled with the reduction of the levels of chromatin modifiers, HDAC1 and DNMT1. The addition of ATRA and either HDAC inhibitors (belinostat, BML-210) or DNMT inhibitors (RG108, zebularine) resulted in changes of histone H4 and H3K9 modification status, leading to the reduction of HMT2 and DNMT1, components of the repressive complex, during DNA replication [77]. Those perturbations may trigger the local chromatin environment to activate the transcriptions of differentiation-related genes. Previous studies have also shown that treatment with BIX-01294 of several cancer cells [75] or combined treatment with HDAC inhibitors [79] caused transcriptional stimulation correlated with removal of H3K9me2 at the promoters of tumor suppressor or several HMT2 target genes with the marker restoration after removal of the EHMT2 inhibitor [80]. These findings indicate that BIX-01294 has anticancer effects and can be a useful target for the development of epigenetic therapies.

Another example could be a competitive inhibitor of DOT1L—adenosine derivative—EPZ004777. After leukemia cell treatment with the chemical compound the noticeable decrease in H3K79 mono- and dimethylation was observed. Pharmacological inhibition of H3K79 methylation causes the changes of gene expression, tumor cell differentiation and apoptosis of leukemia cells [69,81]. It was demonstrated that DOT1L inhibition was effective against primary patient DNMT3A-mutant AML samples, reducing colony forming capacity and inducing terminal differentiation in vitro. In vivo studies showed the efficacy of the DOT1L inhibitor EPZ5676 in a nude rat xenograft model of DNMT3A-mutant AML. These studies suggest that DOT1L could be a promising therapeutic target for the treatment of AML poor prognosis disease [82].

Most of HMT EZH2 inhibition studies were performed using carbocyclic adenosine analog 3-Deazaneplanocin A (DZNep). DZNep acts as both an S-adenosylhomocysteine synthesis inhibitor and a HMT EZH2 inhibitor [83]. DZNep effect on histone methylation is more global than specific. EZH2 inhibition by using DZNep or RNA interference leads to reduced breast cancer cell proliferation and decreased proliferation of prostate cancer cell lines [84]. The treatment of these cells with DZNep increases the expression of cell cycle regulator's p21, p27. This results in cell cycle arrest and apoptosis induction of cancer cells. This compound is capable of inducing apoptosis of breast cancer and colon cancer cells as well. DZNep affects many HMTs and thus can determine both repressing and activating effects on transcription [83]. It was also shown that DZNep may synergistically interact with HDAC and DNMTi and thereby activate silenced genes [85]. Treatment of DZNep, combined with the DNA hypomethylating agent 5-azacitidine, decreased levels of EZH2, suppressed methylation of H3K27me2 and -me3, and increased p16 expression associated with cell proliferation. DZNep might be able to enhance the therapeutic effects of DNMTi such as 5-azacitidine for patients with MDS [86]. The effect of DZNep and the combination of gemcitabine and DZNep in cholangiocarcinoma cells was determined [87]. Cotreatment of AML cells with DZNep and the pan-histone deacetylase inhibitor panobinostat caused depletion of EZH2, induced apoptosis of AML, and significantly improved survival of immunodeficiency mice with HL-60 leukemia [88]. The exact mechanism of DZNep action is not yet known, but this compound could be a promising drug for epigenetic treatment of cancer. To propose new possibilities for leukemia treatment, the potential of HDACi belinostat and HMTi DZNep were studied to enhance conventional therapy in vitro (immortalized cell lines) and ex vivo (APL patient bone marrow mononuclear cells). It was demonstrated that the combined treatment of HDACi, HMTi with conventional treatment (all-*trans* retinoic acid and idarubicin) used in the study had a stronger effect on cell proliferation inhibition, induction of apoptosis, and suppression of metabolism. The treatment accelerated granulocytic cell differentiation and caused chromatin remodeling (increased H3K14 and H4 acetylation levels). Therefore, 3-Deazaneplanocin A and belinostat enhanced conventional AML treatment could be considered for further investigations for clinical use [89].

Histone Acetylation/Deacetylation and Epigenetic Modifiers

Histone acetylation modeling is important for the general state of chromatin and, most importantly, for the regulation and phenotype of the entire cell. Histone acetylation can alter both the nucleosome position and gene activity. Some non-histone proteins including oncoproteins and tumor inhibitors (MYC, p53, and others) are acetylated as well. This shows that acetylation of histone and non-histone proteins is a very important process that affects both the normal cell regulation and the development of various diseases, especially cancer.

HAT and HDAC

Alterations in catalytic activity of histone acetyltransferase (HAT) and HDAC could affect the deviations from the normal cell development to pathological and cause cancer development. For example, during hematopoiesis, lineage specific transcription factors attract some HAT or HDAC and regulate targeted gene expression, and the alterations of this entire process could affect the pathogenesis. Some types of leukemia are characterized by chromosomal translocations and expression of fusion proteins (PML-RARα, PLZF-RARα, AML1-ETO, etc.) that are diagnostic for the disease. Determined fusion proteins have the ability to attract HDAC and/or DNMT and thereby contribute to gene silencing, which is directly related to the development of cancer. HDAC also interferes with oncoproteins, such as BCL6, whose activity is controlled by dynamic of acetylation/deacetylation. The influence of HAT on carcinogenesis appears to be dependent on the complex effect of many factors, as HAT can act both as oncogenes and as tumor inhibitors. Mutations in HAT have been identified in some tumors. These mutations may affect tumor progression. Therefore, HAT inhibitors are also being investigated as potential cancer therapy agents [90].

HDAC Inhibitors

HDACs are associated with the occurrence of cancer and have become targets of anticancer treatment. HDAC inhibitors have recently emerged as potential cancer therapeutic agents. They bind to the HDAC active center and cause an increase in histone acetylation. Thus, HDAC regulates tumor cell biological processes, such as cell cycle, differentiation and apoptosis [91]. It is observed that abnormal HDAC activity may be associated with cancer development. There are studies that suggest the changed HDAC expression of tumor samples. For example, in prostate cancer cells the expression of HDAC1 is significantly increased. On the contrary, the decreased HDAC1 expression levels are observed in gastric cancer cells. The researchers also found that proteins associated with oncogenesis process (p53, HSP90, E2F, PRB, and BCL6) may be HDAC substrates [92]. These findings suggest that HDAC plays an important role, for example, in tumor emergence and the development of functional HDAC activity associated not only with histones but also with non-histone proteins, such as transcription factors and other proteins that are involved in DNA replication, metabolism, apoptosis, protein folding, cellular signaling, and regulation [93].

HDAC inhibitors which are a new class of agents those are involved in chromatin structure remodeling and gene expression. Interest in HDAC inhibitors began almost 30 years ago and in recent years, an increasing number of various HDAC inhibitors have been defined to inhibit proliferation and induce differentiation and/or apoptosis of cancer cells. The first HDACi approved by FDA was vorinostat (Zolinza, Suberoylanilide hydroxamic acid (SAHA)) to treat the rare cancer cutaneous T-cell lymphoma (CTCL). Currently, about 12 different HDACis are in clinical trials as monotherapy or combination therapy in patients with hematologic and solid tumors. In the regulation of histone acetylation, HDACi have an antiproliferative, differentiating or proapoptotic properties. HDACs are the enzymes that catalyze the removal of acetyl groups in proteins from the lysine side groups, including those which form nucleosome histone chromatin protein—H2A, H2B, H3, and H4. Recently, HDAC inhibitors have been used as epigenetic drugs in the treatment of various hematological diseases.

HDAC inhibitors are very similar in structure, which is characterized by three main parts: zinc-binding domain, a hydrophobic group, and an intermediate enzyme in joining the group, which is usually an aromatic group [94]. HDACi synergistically act with other different anticancer agents, including radiation therapy, chemotherapy, differentiation agents, epigenetic therapy, and new targeted agents [95–98].

Currently, FDA have approved three major HDAC inhibitors used in cancer therapy: vorinostat, romidepsin (Istodax), and belinostat. Vorinostat and romidepsin are used for CTCL and belinostat for PTCL (fda.gov). It is important that the HDAC inhibitors can eliminate or at least decrease the abnormal gene expression inhibition, which is characteristic for the development of cancer. Numerous HDAC inhibitors [TSA, vorinostat (SAHA), belinostat, panabinostat, valproic acid, PB] were used for the treatment of cancer cells. HDAC inhibitors in combination with ATRA significantly improved granulocytic differentiation efficiency [48,50,53].

Belinostat (PXD 101) is a hydroxamate-type HDAC inhibitor with antitumor activity. It exhibits HDAC class I and II enzyme activity and thus increases histone acetylation [57]. Inhibition of HDAC occurs when the inhibitor reacts directly with the catalytic center of the enzyme [99]. This leads to cell cycle retardation, apoptosis and reduced cancer cell proliferation. Belinostat increases not only histone acetylation, but also non-histone protein acetylation levels by changing the cell microtubule dynamics and increasing cell death. Belinostat is well tolerated and the use of this compound in combination with other chemotherapeutic agents does not cause bone marrow intoxication. This compound is evaluated in phase I and II clinical trials in both solid and hematological tumors [99]. Currently, belinostat is used as drug in PTCL, carcinoma, and other hematological tissues or solid tumors [100]. There were studies of it as a single agent or in combination with chemotherapeutic agent carboplatin [90]. It was demonstrated [101] that belinostat-induced cancer cell death is accompanied by apoptosis: caspase 3 activation, nuclear protein PARP-1 degradation, changes in expression of cell cycle regulatory proteins' survivin, p27, cyclin E1 (G1/S phase), and cyclin A (G2/M phase). Belinostat affected PRC2 (Polycomb repressive complex 2) components—SUZ12 and the catalytic subunit of EZH2 (responsible for the transcriptional repression by methylation of histone H3K27me3). Belinostat, depending on the dose, induced histone H4 hyperacetylation, active chromatin specific histone H3K9 and H3K16 acetylation. Belinostat alone did not induce granulocytic differentiation of leukemia cells, but in combination with ATRA it increased differentiation of HL-60 and NB4 cells. HDAC inhibitor belinostat demonstrates anticancer properties: leukemia cell growth inhibition and induction of differentiation under the influence of epigenetic path through chromatin structure changes [102]. By using in vitro and in vivo models, the effective combined treatment of APL by choosing two epigenetic drugs, belinostat and DZNep, was reported. These observations indicate that APL xenograft mice somatic tissues are subjected to changes in specific histone modifications in response to epigenetic agents-derived treatment [103]. The combination of HDACI belinostat and the proteasome inhibitor bortezomib induced apoptosis and inhibited renal cancer growth synergistically [104]. Belinostat and panobinostat were active against the thyroid cancer cell lines, and belinostat was effective in preventing growth of human thyroid cancer xenografts in immunodeficient mice. Both

HDAC inhibitors induced apoptosis, elevated acetylation of H3, p21, and PARP [105]. The clinical activity of belinostat, carboplatin, and paclitaxel, a phase 1b/2 study was evaluated. It was demonstrated that combined treatment was well tolerated and clinical benefit in heavily pretreated patients with epithelial ovarian cancer was demonstrated [106]. The combination of belinostat and Bortezomib against pancreatic cancer and hepatocellular cancer cell lines has displayed significant synergistic antiproliferative and proapoptotic activity [107].

A new generation of HDAC inhibitors—depsipeptide FK228, sodium phenylbutyrate (PB), niacin (vitamin B3), belinostat, and BML-210 alone and in combination with ATRA show their impact on different leukemia cell lines proliferation, differentiation, apoptosis induced processes, and their molecular mechanisms [49,53,78,108–111]. It was shown that HDACi BML-210 inhibits various leukemia cell lines (NB4, HL-60, THP-1, K562) proliferation and can cause apoptosis depending on dose and time of exposure. For the induction of apoptosis of the apoptosis resistant cell line K562 demanded a longer treatment [112]. BML-210 in combination with ATRA increased antiproliferative and proapoptotic effects on cancer cells. It was found that BML-210 is partially (up to 30%) capable of inducing granulocytic differentiation. Leukemia cell combined treatment with BML-210 and ATRA has accelerated and increased cancer cells differentiation into granulocytes [113] and erythrocytes [110]. The aggressive T-cell lymphomas (TCL) are driven by epigenetic defects and combined epigenetic treatment using HDAC inhibitors (panobinostat, belinostat, romidepsin and vorinostat) and DNMT inhibitor (decitabine) in vitro and in vivo models of TCL demonstrated inhibition of tumor growth in the combination cohort compared to the single agent [114].

FK228 is a HDAC inhibitor with therapeutic effectiveness, which is already being used in clinical trials for the treatment of lymphoma patients. There were demonstrated the positive effects of FK228 on the differentiation induction of human promyelocytic leukemia cell lines NB4, HL-60, and KG-1 [109,115]. FK228 inhibits the growth of leukemia cells by the induction of apoptosis depending on dose of the compound. Differentiation resistant KG-1 cells FK228, together with an increased dose of ATRA did not prevent differentiation, but increased the apoptotic response beneficial effects of FK228 on human promyelocytic leukemia, which may be exerted through the induction of differentiation or apoptosis via histone modification and selective involvement of transcription factors, such as NF-κB and p53 [109].

The HDAC (HDAC1 and Sir2) inhibitors, sodium phenylbutyrate and vitamin B3, respectively were used alone and in combination with the differentiation agent, ATRA, in human promyelocytic leukemia cells. The most promising combination for differentiation therapy defined by 6-h pre-treatment with PB and vitamin B3 before exposition to ATRA alone or with vitamin B3 was demonstrated. Such treatment significantly accelerated and increased cancer cell differentiation (up to 95%) during the treatment [108,109]. Sodium phenylbutyrate and its combinations with ATRA and vitamin B3 causes histone H4 acetylation in the p21 promoter regions corresponding to p53 and/or Sp1 sites [109]. That indicates the possibility of using the combination of agents in ATRA-resistant AML treatment to produce both differentiation and apoptosis. The Table 41.2 summarizes the combined epigenetic treatments on AML cells giving positive effects on cancer cells proliferation inhibition, differentiation and/or apoptosis initiation caused certain gene and protein expression via epigenetic chromatin remodeling.

Other Combined Therapies and Possible Combination with Epigenetics

In many cases, when patients may be or become resistant to standard therapy during treatment and relapse after remission, it is necessary to develop new therapies. If relapse is observed after remission, it means that a certain population of cancer cells has been resistant to the treatment applied. It was shown that chemotherapy-resistant AML cells have been shown to have more intense oxidative phosphorylation. By inhibiting mitochondrial oxidative phosphorylation, cells have been shown to become more sensitive to chemotherapy in vitro and in vivo [122,123]. For decades, glycolysis has been considered the main source of energy for cancer cells, but recently, oxidative phosphorylation of mitochondria has become an important factor in the chemotherapeutic resistance of some types of cancer cells, and mitochondrial oxidative phosphorylation has attracted great interest as a potential and valuable clinical target in the treatment of cancer.

Because leukemic cells from different R/R AML patients have different metabolic activities, it is important to understand the effect of differences in cellular metabolism on treatment efficacy. Cancer cells with different energy status have recently been studied (higher oxidative phosphorylation/more intense glycolysis/relatively lower metabolic activity). It was demonstrated that metformin decreased R/R AML cell oxidative phosphorylation ex vivo, cotreatment with cytarabine and venetoclax slightly increased the effect. However, treatment with atovaquone did not have a marked effect on R/R AML, even though the other studies have shown that atovaquone significantly inhibits oxidative phosphorylation in various types of cancer cell lines [124]. Cell treatment had a slight effect on cell proliferation inhibition; combination of metformin, cytarabine,

TABLE 41.2 Combined Epigenetic Treatment of AML (Acute Myeloid Leukemia) Cells

Epigenetic Modifier	Combined Treatment	Reference
DNMT inhibitors	Zebularine Zebularine + ATRA	[49]
	Decitabine Decitabine + ATRA	[50,116]
	Procaine Procaine + ATRA	[116]
	RG108 RG108 + ATRA	[49]
HDAC inhibitors	FK228 FK228 + ATRA	[109,110,115]
	BML-210 BML-210 + ATRA	[110,113,115]
	TSA + ATRA	[115]
	EGCG EGCG + ATRA	[117,118]
	Phenylbutyrate + ATRA	[115]
	Belinostat Belinostat + ATRA	[101,102,119,120]
	VitamineB3 + ATRA	[108,115]
	Phenylbutyrate + vitamineB3 + ATRA	[108]
	FK228 + vitamineB3 + ATRA	[109,115,121]
HMT inhibitors	BIX-1094 BIX-1094 + ATRA	[118]
DNMTi + HDACi	Zebularine + Phenylbutyrate + ATRA	[49]
	Zebularine + BML-210 + ATRA	[49,50]
	RG108 + Phenylbutyrate RG108 + Phenylbutyrate + ATRA	[49,53]
	RG108 + BML-210 RG108 + BML-210 + ATRA	[49,53]
DNMTi + HMTi	Zebularine + BIX-1094 + ATRA	[78]
	RG108 + BIX-1094 + ATRA	[78]
HMTi + HDACi	BIX-1094 + EGCG	[118]
	BIX-1094 + Belinostat + ATRA BIX-1094 + BML-210 + ATRA	[78,101]
	DZNep + Belinostat + ATRA	[103,119,120]
	DZNep + Belinostat + IDA + ATRA	[89]

and the apoptosis-inducing agent venetoclax (BCL-2 inhibitor) had the strongest effect. Moreover, a slightly greater effect on cell proliferation and cell cycle regulation was demonstrated in cells with a higher initial rate of oxidative phosphorylation, as demonstrated by gene and protein expression analysis [123].

Thus, inhibition of oxidative phosphorylation induced molecular changes associated with inhibition of cancer processes, but these were not sufficient to treat R/R AML patients. The combined treatment of metformin that acts as an inhibitor of oxidative phosphorylation and MCL-1 (Bcl-2 family protein) inhibitor S63845 in AML cells was tested, and it was identified that combined treatment had stronger inhibitory effects on cellular metabolism and ATP levels consequently induced apoptosis in AML cells [125]. It was demonstrated that OXPHOS inhibitors

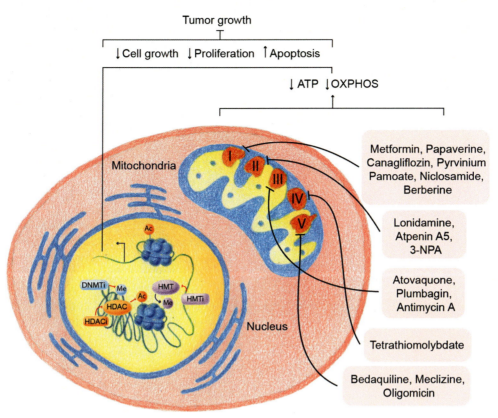

FIGURE 41.2 Combined OXPHOS and epigenetic treatment impact on tumor growth suppression.

improved anti-AML efficacy of IDH mutant inhibitors in vivo as IDH mutation displayed an enhanced mitochondrial oxidative metabolism. Therefore, the combinatory mitochondrial-targeted therapies can be applied together with IDHi to treat IDH mutant AML patients, especially those unresponsive to treatment or experiencing relapsing [126]. Based on the studies carried out, it is advisable to conduct further research in search of new effective combinations of therapeutic agents. Combined treatment using oxidative phosphorylation inhibitors with epigenetic modifiers can also have a potentially effective impact on therapy (Fig. 41.2).

CONCLUSIONS AND FUTURE PERSPECTIVES

Epigenetic modifications play a key role in the regulation of pathogenesis, age-related diseases, and especially carcinogenesis. Epigenetic therapy will be more effective if the molecular factors (epigenetic markers, transcription factors, etc.) involved in aging, cancer, and its progression are identified. Treatment of cancer depends on age of the patient, his/her specific cytogenetic disorders, and other changes. Therefore, cancer therapy should be personalized based on the patient cytogenetic and epigenetic changes, biological age, drug tolerance, disease stage, and so on. Thus, it is necessary to select a treatment strategy with maximum efficiency and minimal toxicity to the patient. The use of epigenetic factors in cancer therapy is vital, but not fully developed yet. Targeted epigenetic treatment is a new promising therapeutic approach in medical strategies. The synergistic effects of combined treatment of epigenetic modifiers and anticancer drugs are the most important for combined epigenetic treatment. The therapeutic potential of DNMT, HMT, and HDAC inhibitors alone and in the combination, which was assessed on cell lines and clinical trials plays a key role in the development of targeted epigenetic therapies.

In summary, chemical modulators, epigenetic modifiers, and their combinations are effective in inducing anticancer molecular changes in cancer cells. However, due to the different changes in the cancer cells, each patient needs an individually chosen therapy.

CLINICAL TRIAL

A Trial of Epigenetic Priming in Patients with Newly Diagnosed Acute Myeloid Leukemia

Study number/id number

NCT03164057

Synopsis

No Study Results Posted on ClinicalTrials.gov for this Study.

Condition

Acute Myeloid Leukemia
Myelodysplastic Syndromes

Study Phase

Phase 2
Study type: Interventional
Recruitment Status: Recruiting
Actual Study Start Date: June 15, 2017
Estimated Primary Completion Date: June 2025
Estimated Study Completion Date: June 2027
Last Update Posted: May 7, 2021

Study Drug

Drug: Azacitidine
Drug: Decitabine
Drug: Cytarabine
Drug: Daunorubicin
Drug: Etoposide
Combination Product: ITMHA
Drug: Idarubicin
Drug: Fludarabine
Drug: Mitoxantrone
Drug: Erwinia asparaginase
Drug: Sorafenib
Drug: G-CSF
Drug: Dexrazoxane
Biological: Stem Cell Transplant

Ethics

INCLUSION CRITERIA

- Diagnostic criteria: patients must have one of the following diagnoses:
  o Acute myeloid leukemia fulfilling the criteria of the WHO Classification (see Appendix I), or
  o >5% but <20% marrow myeloblasts and evidence of a clonal de novo AML genetic abnormality [e.g., t(8;21), inv(16), t(9;11)], or
  o Myeloid sarcoma (also referred to as extramedullary myeloid tumor, granulocytic sarcoma, or chloroma), with or without evidence of a leukemia process in the bone marrow or peripheral blood, with confirmation of myeloid differentiation, or
  o High grade myelodysplastic syndrome (MDS) with greater than 5% blasts, or
  o Patients with treatment related myeloid neoplasms including AML and MDS, provided their cumulative anthracycline dose has not exceeded 230 mg/m2 doxorubicin equivalents.
- Other criteria - Patients must meet all the following criteria:
  o Age >28 days and <22 years at time of study entry inclusive, and
  o No prior therapy for this malignancy except for one dose of intrathecal therapy and the use of hydroxyurea or low-dose cytarabine (100–200 mg/m2 per day for one week or less for hyperleukocytosis), and
  o Written informed consent according to institutional guidelines, and
  o Female patients of childbearing potential must have a negative pregnancy test within 2 weeks prior to enrollment, and
  o Male and female participants of reproductive potential must use an effective contraceptive method during the study and for a minimum of 6 months after study treatment.

EXCLUSION CRITERIA

- Down syndrome.
- Acute promyelocytic leukemia (APL).
- BCR-ABL1 chronic myeloid leukemia in blast crisis (CML-BC).
- Juvenile myelomonocytic leukemia (JMML).
- Fanconi anemia (FA).
- Kostmann syndrome.
- Shwachman syndrome.
- Other bone marrow failure syndromes or low grade (<5% bone marrow blasts) MDS.
- Use of concomitant chemotherapy, radiation therapy, or immunotherapy other than as specified in the protocol.
- Use of investigational agents within 30 days or any anticancer therapy for this malignancy within 2 weeks before study entry with the exception of IT therapy, hydroxyurea, or low-dose cytarabine as specified in the protocol document. The patient must have recovered from all acute toxicities from any previous therapy.
- Systemic fungal, bacterial, viral, or other infection not controlled (defined as exhibiting ongoing signs/symptoms related to the infection and without improvement, despite appropriate antibiotics or other treatment).

- Pregnant or lactating.
- Any significant concurrent disease, illness, or psychiatric disorder that would compromise patient safety or compliance, interfere with consent, study participation, follow up, or interpretation of study results.
- Prior chemotherapy, with the exception of hydroxyurea or low-dose cytarabine as specified in the protocol document. The patient must have recovered from all acute toxicities from any previous therapy.
- Patients with treatment related myeloid neoplasms with cumulative anthracyclines greater than 230 mg/m2 doxorubicin equivalents.
- Sexes eligible for study: All
- Ages: up to 21 years (child, adult)
- Accepts healthy volunteers: No

Study Administrative Structure

Principal Investigator and Contact: Jeffrey E. Rubnitz, MD, PhD 866−278−5833, jeffrey.rubnitz@stjude.org

Contact: Referral Office 866−278−5833 referralinfo@stjude.org

Listed Location Countries: United States

Responsible Party: St. Jude Children's Research Hospital

Study Sponsor: St. Jude Children's Research Hospital

Collaborators: Not provided

Introduction

The overall aim of this study is to determine if epigenetic priming with a DNA methyltransferase inhibitor (DMTi) prior to chemotherapy blocks is tolerable and carries evidence of a clinical efficacy signal as determined by minimal residual disease (MRD), event-free survival (EFS), and overall survival (OS). Tolerability for each of the agents, as well as total reduction in DNA methylation and outcome assessments will be done to simultaneously obtain preliminary biological and clinical data for each DMTi in parallel.

Study Objectives

PRIMARY OBJECTIVES

- Evaluate the tolerability of 5 days of epigenetic priming with azacitidine and decitabine as a single *agent DMTi* prior to standard AML chemotherapy blocks.
- Evaluate the change in genome-wide methylation burden induced by 5 days of epigenetic priming and the association of postpriming genome-wide methylation burden with event-free survival among pediatric AML patients.

SECONDARY OBJECTIVES

- Describe minimal residual disease levels following Induction I chemotherapy in patients that receive DMTi.
- Estimate the event-free survival and overall survival of patients receiving a DMTi prior to chemotherapy courses.

Detailed Description

To determine tolerability, priming with DMTi (azacitidine or decitabine) will be limited to Induction I and II during Part 1 of the study. If DMTi treatment is tolerated during Part 1, the investigators will go on to an Expansion Phase (Part 2) that includes DMTi priming prior to all chemotherapy blocks.

Treatment will consist of 5 blocks of conventional chemotherapy: Induction I, Induction II, Intensification I, Intensification II, and Intensification III over approximately 5 months.

RANDOMIZATION: Patients will be randomized to receive one of two DMTi (azacitidine or decitabine) for 5 days prior to Induction I. Intrathecal (ITHMA) treatments will be given right before treatment on this study or on day 1 of Induction I treatment. Leucovorin will be given 24−30 hours following ITHMA.

INDUCTION I CHEMOTHERAPY: Patients receive cytarabine, daunorubicin, and etoposide.

INDUCTION II CHEMOTHERAPY; Patients receive their assigned DMTi for 5 days followed by fludarabine, cytarabine, G-CSF, and idarubicin.

Patients are then evaluated and assigned to either the low-risk arm, intermediate-risk arm, or the high-risk arm for Intensification therapy.

Patients with ≥5% blasts following Induction II will be considered refractory and will go off therapy. The rare high risk patient with an MRD <0.1% following Induction I may proceed directly to stem cell transplant (SCT) after Induction II—if a suitable donor is available and the transplant can be performed without delay. MDS patients may proceed to SCT once they have achieved MRD <0.1% irrespective of the number of chemotherapy courses received.

INTENSIFICATION I CHEMOTHERAPY—LOW-RISK AML, INTERMEDIATE-RISK AML, and HIGH-RISK AML with no donor: Patients receive cytarabine and etoposide. After administration of 5 days of a DMTi prior to Inductions I and II satisfies a tolerability determination criterion, patients will also receive their randomly assigned DMTi for 5 days prior to cytarabine and etoposide.

INTENSIFICATION II CHEMOTHERAPY—LOW RISK AML, INTERMEDIATE-RISK AML, and HIGH-RISK AML with no donor: Patients receive mitoxantrone and cytarabine. After administration of 5 days of a DMTi prior to Inductions I and II satisfies a tolerability determination criterion, patients will also receive their randomly assigned DMTi for 5 days prior to mitoxantrone and cytarabine.

INTENSIFICATION I CHEMOTHERAPY—HIGH-RISK AML with a donor: Patients receive mitoxantrone and cytarabine followed by stem cell transplant (SCT). Treatment related AML patients and patients with treatment related MDS who have a donor but are not able to receive a SCT without delay will proceed to HR Intensification III and receive erwinia asparaginase and cytarabine. After administration of 5 days of a DMTi prior to earlier courses satisfies a tolerability criterion, patients will also receive their randomly assigned DMTi for 5 days prior to mitoxantrone and cytarabine or erwinia asparaginase and cytarabine.

Treatment related AML patients and treatment related MDS patients that are not able to receive a SCT should go off treatment following Intensification II.

INTENSIFICATION III CHEMOTHERAPY—INTERMEDIATE-RISK AML and HIGH-RISK AML with no donor: Patients receive erwinia asparaginase and cytarabine. After administration of 5 days of a DMTi prior to earlier courses satisfies a tolerability criterion, patients will also receive their randomly assigned DMTi for 5 days prior to erwinia asparaginase and cytarabine.

Investigational Plan

A Trial of Epigenetic Priming in Patients With Newly Diagnosed Acute Myeloid Leukemia—Full Text View—ClinicalTrials.gov

Experimental: AZA + ADE | AZA + FLAG + Ida | AE | MA

Part 1 Tolerability with AZA—Low Risk

Patients are randomized to receive 5 days of single agent azacitidine as part of Inductions I and II and then receive low-risk intensifications I and II without azacitidine.

Interventions: azacitidine, cytarabine, daunorubicin, etoposide, dexrazoxane, fludarabine, idarubicin, G-CSF, mitoxantrone, ITMHA.

Experimental: DAC + ADE | DAC + FLAG + Ida | AE | MA

Part 1 Tolerability with DAC—Low Risk

Patients are randomized to receive 5 days of single agent decitabine as part of Inductions I & II and then receive low-risk Intensifications I & II without decitabine.

Interventions: decitabine, cytarabine, daunorubicin, etoposide, dexrazoxane, fludarabine, idarubicin, G-CSF, mitoxantrone, ITMHA.

Experimental: AZA + ADE | AZA + FLAG + Ida + Sor | AZA + AE + Sor | AZA + MA + Sor

Part 2 Dose Expansion with AZA—Low Risk

Patients are randomized to receive 5 days of single agent azacitidine as part of Inductions I & II and low-risk Intensifications I & II. Sorafenib will be given to patients with FLT3-ITD. For these patients, AZA will be limited to the first two courses of Induction chemotherapy. They will not receive AZA with Intensification therapy.

Interventions: azacitidine, cytarabine, daunorubicin, etoposide, dexrazoxane, fludarabine, idarubicin, G-CSF, sorafenib, mitoxantrone, ITMHA.

Experimental: DAC + ADE | DAC + FLAG + Ida + Sor | DAC + AE + Sor | DAC + MA + Sor

Part 2 Dose Expansion with DAC—Low Risk

Patients are randomized to receive 5 days of single agent decitabine as part of Inductions I & II and low-risk Intensifications I & II. Sorafenib will be given to patients with FLT3-ITD. For these patients, DAC will be limited to the first two courses of Induction chemotherapy. They will not receive DAC with Intensification therapy.

Interventions: decitabine, cytarabine, daunorubicin, etoposide, dexrazoxane, fludarabine, idarubicin, G-CSF, sorafenib, mitoxantrone, ITMHA.

Experimental: AZA + ADE | AZA + FLAG + Ida | AE | MA | Asp + AraC

Part 1 Tolerability with AZA—Intermediate Risk

Patients are randomized to receive 5 days of single agent azacitidine as part of Inductions I & II and then receive intermediate risk Intensifications I, II, & III without azacitidine.

Interventions: azacitidine, cytarabine, daunorubicin, etoposide, dexrazoxane, fludarabine, idarubicin, G-CSF, mitoxantrone, erwinia asparaginase, ITMHA.

Experimental: DAC + ADE | DAC + FLAG + Ida | AE | MA | Asp + AraC

Part 1 Tolerability with DAC—Intermediate Risk

Patients are randomized to receive 5 days of single agent decitabine as part of Inductions I & II and then receive intermediate-risk Intensifications I, II & III without decitabine.

Interventions: decitabine, cytarabine, daunorubicin, etoposide, dexrazoxane, fludarabine, idarubicin, G-CSF, mitoxantrone, erwinia asparaginase, ITMHA.

Recruitment Information

Recruitment Status: Recruiting
Estimated Enrollment: (submitted: May 19, 2017) 200
Original Estimated Enrollment: Same as current
Estimated Study Completion Date: June 2027
Estimated Primary Completion Date: June 2025 (Final data collection date for primary outcome measure).

References

[1] Dawson MA, Kouzarides T. Cancer epigenetics: from mechanism to therapy. Cell 2012;150(1):12−27.

[2] Bowry A, Kelly RDW, Petermann E. Hypertranscription and replication stress in cancer. Trends Cancer 2021;7(9):863−77 S2405-8033(21)00098-4.

[3] Napoli C, Coscioni E, de Nigris F, Donatelli F. Emergent expansion of clinical epigenetics in patients with cardiovascular diseases. Curr Opin Cardiol 2021;36(3):295−300.

[4] Shorstova T, Foulkes WD, Witcher M. Achieving clinical success with BET inhibitors as anti-cancer agents. Br J Cancer 2021;124(9):1478−90.

[5] Rice JC, Briggs SD, Ueberheide B, Barber CM, Shabanowitz J, Hunt DF, et al. Histone methyltransferases direct different degrees of methylation to define distinct chromatin domains. Cell 2003;12(6):1591−8.

[6] Mikkelsen SU, Gillberg L, Lykkesfeldt J, Grønbæk K. The role of vitamin C in epigenetic cancer therapy. Free Radic Biol Med 2021;170:179−93 S0891-5849(21)00169-6.

[7] Morrison AJ. Cancer cell metabolism connects epigenetic modifications to transcriptional regulation. FEBS J 2022;;289(5):1302−14. Available from: https://doi.org/10.1111/febs.16032 Online ahead of print.

[8] Bojang Jr P, Ramos KS. The promise and failures of epigenetic therapies for cancer treatment. Cancer Treat Rev 2016;40(1):153−69.

[9] Pollock RM, Richon VM. Epigenetic approaches to cancer therapy. Drug Discov Today 2009;6(2):71−9.

[10] Hillyar C, Rallis KS, Varghese J. Advances in epigenetic cancer therapeutics. Cureus 2020;12(11):e11725.

[11] Chua GNL, Wassarman KL, Sun H, Alp JA, Jarczyk EI, Kuzio NJ, et al. Cytosine-based TET enzyme inhibitors. ACS Med Chem Lett 2019;10(2):180−5.

[12] Guan Y, Tiwari AD, Phillips JG, Hasipek M, Grabowski DR, Pagliuca S, et al. Therapeutic strategy for preferential targeting of TET2-mutant and TET dioxygenase−deficient cells in myeloid neoplasms. Blood Cancer Disco 2021;2(2):146−62. Available from: https://doi.org/10.1158/2643-3230.BCD-20-0173.

[13] Suarez-Alvarez B, Rodriguez RM, Ruiz-Ortega M, Lopez-Larrea C. BET proteins: an approach to future therapies in transplantation. Am J Transpl 2017;17(9):2254−62.

[14] Stathis A, Bertoni F. BET proteins as targets for anticancer treatment. Cancer Discov 2018;8(1):24−36.

[15] Mustafi S, Camarena V, Qureshi R, Yoon H, Volmar CH, Huff TC, et al. Vitamin C supplementation expands the therapeutic window of BETi for triple negative breast cancer. EBioMedicine 2019;43:201−10. Available from: https://doi.org/10.1016/j.ebiom.2019.04.006.

[16] Navakauskienė R, Navakauskas D, Borutinskaitė V, Matuzevičius D. Epigenetics and proteomics of leukemia − A synergy of experimental biology and computational informatics. Published by Springer; 2021p. 406. ISBN: 978-3-030-68707-6. Available from: https://doi.org/10.1007/978-3-030-68708-3.

[17] Kotiah SD. *Acute promyelocytic leukemia treatment protocols.* Available from: http://emedicine.medscape.com/article/2005126-overview.

[18] Liu CC, Liu CC, Wang H, Wang WD, Zhu MY, Geng QR, et al. Consolidation therapy of arsenic trioxide alternated with chemotherapy achieves remarkable efficacy in newly diagnosed acute promyelocytic leukemia. Onco Targets Ther 2015;8:3297−303.

[19] Halsall JA, Turner BM. Histone deacetylase inhibitors for cancer therapy: an evolutionarily ancient resistance response may explain their limited success. Bioessays 2016;38:1102−10.

[20] Maslak P, Chanel S, Camacho LH, et al. Pilot study of combination transcriptional modulation therapy with sodium phenylbutyrate and 5-azacytidine in patients with acute myeloid leukemia or myelodysplastic syndrome. Leukemia 2006;20:212−17.

[21] Voso MT, Santini V, Finelli C, et al. Valproic acid at therapeutic plasma levels may increase 5-azacytidine efficacy in higher risk myelodysplastic syndromes. Clin Cancer Res 2009;15:5002−7.

[22] Prebet T, Sun Z, Figueroa ME, et al. Prolonged administration of azacitidine with or without entinostat for myelodysplastic syndrome and acute myeloid leukemia with myelodysplasia-related changes: results of the US leukemia intergroup trial E1905. J Clin Oncol 2014;32:1242−8.

[23] Ahuja N, Sharma AR, Baylin SB. Epigenetic therapeutics: a new weapon in the war against cancer. Annu Rev Med 2016;67:73−89.

[24] Sadikovic B, Al-Romaih K, Squire J, Zielenska M. Cause and consequences of genetic and epigenetic alterations in human cancer. Curr Genom 2008;9(6):394−408.

[25] Jin B, Robertson KD. DNA methyltransferases, DNA damage repair, and cancer. Adv Exp Med Biol 2013;754:3−29.

[26] Sharif J, Koseki H. Recruitment of Dnmt1 roles of the SRA protein Np95 (Uhrf1) and other factors. Prog Mol BiolTransl Sci 2011;101:289−310.

[27] Kulis M, Esteller M. DNA methylation and cancer. Adv Genet 2010;70:27−56.

[28] Yang L, Rau R, Goodell MA. DNMT3A in haematological malignancies. Nat Rev Cancer 2015;15:152−65.

[29] Ley TJ, Ding L, Walter MJ, McLellan MD, Lamprecht T, Larson DE, et al. DNMT3A mutations in acute myeloid leukemia. N Engl J Med 2010;363(25):2424−33.

[30] Yan XJ, Xu J, Gu ZH, Pan CM, Lu G, Shen Y, et al. Exome sequencing identifies somatic mutations of DNA methyltransferase gene DNMT3A in acute monocytic leukemia. Nat Genet 2011;43(4):309−15.

[31] Challen GA, Sun D, Jeong M, Luo M, Jelinek J, Berg JS, et al. Dnmt3a is essential for hematopoietic stem cell differentiation. Nat Genet 2012;44(1):23−31.

[32] Yang X, Han H, De Carvalho DD, Lay FD, Jones PA, Liang G. Gene body methylation can alter gene expression and is a therapeutic target in cancer. Cancer Cell 2014;26(4):577−90.

[33] Yan W, Herman JG, Guo M. Epigenome-based personalized medicine in human cancer. Epigenomics 2016;8(1):119−33.

[34] Yang X, Lay F, Han H, Jones PA. Targeting DNA methylation for epigenetic therapy. Trends Pharmacol Sci 2010;31(11):536−46.

[35] Silverman LR, Holland JF, Weinberg RS, Alter BP, Davi RB, Ellison RR, et al. Effect of treatment with 5-aza cytidine on the in vivo and in vitro hematopoiesis in patients with myelodysplastic syndromes. Leukemia 1993;7:21−9.

[36] Issa JP, Garcia-Manero G, Giles FJ, Mannari R, Thomas D, Faderl S, et al. Phase I study of low-dose prolonged exposure schedules of the hypomethylating agent 5-aza-2'-deoxycytidine (decitabine) in hematopoietic malignancies. Blood 2004;103:1635−40.

[37] Bug G, Ottmann OG. The HDAC system and association with acute leukemias and myelodysplastic syndromes. Invest N Drugs 2010;28(Suppl. 1):S36−49.

[38] Müller S, Krämer OH. Inhibitors of HDACs-effective drugs against cancer? Curr Cancer Drug Targets 2010;10:210−28.

[39] Diesch J, Zwick A, Garz AK, Palau A, Buschbeck M, Götze KS. A clinical-molecular update on azanucleoside-based therapy for the treatment of hematologic cancers. Clin Epigenet 2016;8:71.

[40] Erba HP. Finding the optimal combination therapy for the treatment of newly diagnosed AML in older patients unfit for intensive therapy. Leuk Res 2015;39(2):183−91.

[41] Zeidan AM, Linhares Y, Gore SD. Current therapy of myelodysplastic syndromes. Blood Rev 2013;27:243–59.

[42] Bayraktar UD, Domingo GC, Schmit J, Pereira D. Azacitidine combined with gemtuzumab ozogamicin in patients with relapsed/refractory acute myeloid leukemia. Leuk Lymphoma 2011;52:913–15.

[43] Ishikawa T. Novel therapeutic strategies using hypomethylating agents in the treatment of myelodysplastic syndrome. Int J Clin Oncol 2014;19(1):10–15.

[44] Blum W, Klisovic RB, Hackanson B, Liu Z, Liu S, Devine H, et al. Phase I study of decitabine alone or in combination with valproic acid in acute myeloid leukemia. J Clin Oncol 2007;25 (25):3884–91.

[45] Scandura JM, Roboz GJ, Moh M, Morawa E, Brenet F, Bose JR, et al. Phase 1 study of epigenetic priming with decitabine prior to standard induction chemotherapy for patients with AML. Blood 2011;118(6):1472–80.

[46] Borges S, Döppler HR, Storz P. A combination treatment with DNA methyltransferase inhibitors and suramin decreases invasiveness of breast cancer cells. Breast Cancer Res Treat 2014;144 (1):79–91.

[47] Scott SA, Lakshimikuttysamma A, Sheridan DP, Sanche SE, Geyer CR, DeCoteau JF. Zebularine inhibits human myeloid leukemia cell growth in vitro in association with p15INK4B demethylation and reexpression. Exp Hematol 2007;5:263–73.

[48] Veerla S, Panagopoulos I, Jin Y, Lindgren D, Höglund M. Promoter analysis of epigenetically controlled genes in bladder cancer. Genes Chromosomes Cancer 2008;47:368–78.

[49] Savickiene J, Treigyte G, Borutinskaite VV, Navakauskiene R. Antileukemic activity of combined epigenetic agents, DNMT inhibitors zebularine and RG108 with HDAC inhibitors, against promyelocytic leukemia HL-60 cells. Cell Mol Biol Lett 2012;17 (4):501–25.

[50] Savickiene J, Treigyte G, Jonusiene V, Bruzaite R, Borutinskaite V, Navakauskiene R. Epigenetic changes by zebularine leading to enhanced differentiation of human promyelocytic leukemia NB4 and KG1 cells. Mol Cell Biochem 2012;359:245–61.

[51] Flis S, Gnyszka A, Flis K. DNA methyltransferase inhibitors improve the effect of chemotherapeutic agents in SW48 and HT-29 colorectal cancer cells. PLoS One 2014;9(3):e92305.

[52] Brueckner B, Garcia Boy R, Siedlecki P, Musch T, Kliem HC, Zielenkiewicz P, et al. Epigenetic reactivation of tumor suppressor genes by a novel small-molecule inhibitor of human DNA methyltransferases. Cancer Res 2005;65(14):6305–11.

[53] Savickiene J, Treigyte G, Jazdauskaite A, Borutinskaite VV, Navakauskiene R. DNA methyltransferase inhibitor RG108 and histone deacetylase inhibitors cooperate to enhance NB4 cell differentiation and E-cadherin re-expression by chromatin remodelling. Cell Biol Int 2012;36(11):1067–178.

[54] Katiyar S, Elmets CA, Katiyar SK. Green tea and skin cancer: photoimmunology, angiogenesis and DNA repair. J Nutr Biochem 2007;18:287–96.

[55] Chang CW, Hsieh YH, Yang WE, Yang SF, Chen Y, Hu DN. Epigallocatechin gallate inhibits migration of human uveal melanoma cells via downregulation of matrix metalloproteinase-2 activity and ERK1/2 pathway. Biomed Res Int 2014;2014:141582.

[56] AlQathama A, Prieto JM. Natural products with therapeutic potential in melanoma metastasis. Nat Prod Rep 2015;32 (8):1170–82.

[57] Khan MA, Hussain A, Sundaram MK, Alalami U, Gunasekera D, Ramesh L, et al. Epigallocatechin-3-gallate reverses the expression of various tumor-suppressor genes by inhibiting DNA methyltransferases and histone deacetylases in human cervical cancer cells. Oncol Rep 2015;33(4):1976–84.

[58] Rajendran P, Ho E, Williams DE, Dashwood RH. Dietary phytochemicals, HDAC inhibition, and DNA damage/repair defects in cancer cells. Clin Epigenet 2011;3(1):4.

[59] Sartippour MR, Shao ZM, Heber D, Beatty P, Zhang L, Liu C, et al. Green tea inhibits vascular endothelial growth factor (VEGF) induction in human breast cancer cells. J Nutr 2002;132(8):2307–11.

[60] Fujiki H, Sueoka E, Watanabe T, Suganuma M. Synergistic enhancement of anticancer effects on numerous human cancer cell lines treated with the combination of EGCG, other green tea catechins, and anticancer compounds. J Cancer Res Clin Oncol 2015;141(9):1511–22.

[61] Suganuma M, Saha A, Fujiki H. New cancer treatment strategy using combination of green tea catechins and anticancer drugs. Cancer Sci 2011;102(2):317–23.

[62] Saldanha SN, Kala R, Tollefsbol TO. Molecular mechanisms for inhibition of colon cancer cells by combined epigenetic-modulating epigallocatechin gallate and sodium butyrate. Exp Cell Res 2014;324(1):40–53.

[63] Wu DS, Shen JZ, Yu AF, Fu HY, Zhou HR, Shen SF. Epigallocatechin-3-gallate and trichostatin A synergistically inhibit human lymphoma cell proliferation through epigenetic modification of p16INK4a. Oncol Rep 2013;30(6):2969–75.

[64] Li Y, Yuan YY, Meeran SM, Tollefsbol TO. Synergistic epigenetic reactivation of estrogen receptor-α (ERα) by combined green tea polyphenol and histone deacetylase inhibitor in ERα-negative breast cancer cells. Mol Cancer 2010;9:274.

[65] Nihal M, Roelke CT, Wood GS. Anti-melanoma effects of vorinostat in combination with polyphenolic antioxidant (-)-epigallocatechin-3-gallate (EGCG). Pharm Res 2010;27(6):1103–14.

[66] Tyagi T, Treas JN, Mahalingaiah PK, Singh KP. Potentiation of growth inhibition and epigenetic modulation by combination of green tea polyphenol and 5-aza-2'-deoxycytidine in human breast cancer cells. Breast Cancer Res Treat 2015;149(3):655–68.

[67] Contieri B, Duarte BKL, Lazarini M. Updates on DNA methylation modifiers in acute myeloid leukemia. Ann Hematol 2020;99(4):693–701.

[68] Liu X, Gong Y. Isocitrate dehydrogenase inhibitors in acute myeloid leukemia. Biomark Res 2019;7:22.

[69] Fierz B, Muir TW. Chromatin as an expansive canvas for chemical biology. Nat Chem Biol 2012;8(5):417–27.

[70] Lee DY, Teyssier C, Strahl BD, Stallcup MR. Role of protein methylation in regulation of transcription. Endocr Rev 2005;2:147–70.

[71] Eissenberg JC, Shilatifard A. Histone H3 lysine 4 (H3K4) methylation in development and differentiation. Dev Biol 2010;339 (2):240–9.

[72] Zagni C, Chiacchio U, Rescifina A. Histone methyltransferase inhibitors: novel epigenetic agents for cancer treatment. Curr Med Chem 2013;20(2):167–85.

[73] Greiner D, Bonaldi T, Eskeland R, Roemer E, Imhof A. Identification of a specific inhibitor of the histone methyltransferase SU(VAR)3–9. Nat Chem Biol 2005;1(3):143–5.

[74] Lai YS, Chen JY, Tsai HJ, Chen TY, Hung WC. The SUV39H1 inhibitor chaetocin induces differentiation and shows synergistic cytotoxicity with other epigenetic drugs in acute myeloid leukemia cells. Blood. Cancer J 2015;5:e313.

[75] Cho HS, Kelly JD, Hayami S, Toyokawa G, Takawa M, Yoshimatsu M, et al. Enhanced expression of EHMT2 is involved in <!--> the proliferation of cancer cells through negative regulation of SIAH1. Neoplasia 2011;13:676–84.

[76] Kondo Y, Shen L, Ahmed S, Boumber Y, Sekido Y, Haddad BR, et al. Downregulation of histone H3 lysine 9 methyltransferase G9a induces centrosome disruption and chromosome instability in cancer cells. PLoS One 2008;3(4):e2037.

[77] Chang Y, Zhang X, Horton JR, Upadhyay AK, Spannhoff A, Liu J, et al. Structural basis for G9a-like protein lysine methyltransferase inhibition by BIX-01294. Nat Struct Mol Biol 2009;16:312–17.

[78] Savickiene J, Treigyte G, Stirblyte I, Valiuliene G, Navakauskiene R. Euchromatic histone methyltransferase 2 inhibitor, BIX-01294, sensitizes human promyelocytic leukemia HL-60 and NB4 cells to growth inhibition and differentiation. Leuk Res 2014;38(7):822–9.

[79] McGarvey KM, Fahrner JA, Greene E, Martens J, Jenuwein T, Baylin SB. Silenced tumor suppressor genes reactivated by DNA demethylation do not return to a fully euchromatic chromatin state. Cancer Res 2006;66:3541–9.

[80] Kubicek S, O'Sullivan RJ, August EM, Hickey ER, Zhang Q, Teodoro ML, et al. Reversal of H3K9me2 by a small-molecule inhibitor for the G9a histone methyl-transferase. Mol Cell 2007;25:473–81.

[81] Tian X, Zhang S, Liu HM, Zhang YB, Blair CA, Mercola D, et al. Histone lysine-specific methyltransferases and demethylases in carcinogenesis: new targets for cancer therapy and prevention. Curr Cancer Drug Targets 2013;13(5):558–79.

[82] Rau RE, Rodriguez B, Luo M, Jeong M, Rosen A, Rogers JH, et al. DOT1L as a therapeutic target for the treatment of DNMT3A-mutant acute myeloid leukemia. Blood 2016;128(7):971–81.

[83] Miranda TB, Cortez CC, Yoo CB, Liang G, Abe M, Kelly TK, et al. DZNep is a global histone methylation inhibitor that reactivates developmental genes not silenced by dna methylation. Mol Cancer Ther 2009;8(6):1579–88.

[84] Chase A, Cross NC. Aberrations of EZH2 in cancer. Clin Cancer Res 2011;17(9):2613–18.

[85] Tan JZ, Yan Y, Wang X, Jiang Y, Xu HE. EZH2: biology, disease, and structure-based drug discovery. Acta Pharmacol Sin 2014;35:161–74.

[86] Karoopongse E, Yeung C, Byon J, Ramakrishnan A, Holman ZJ, Jiang PY, et al. The KDM2B- let-7b -EZH2 axis in myelodysplastic syndromes as a target for combined epigenetic therapy. PLoS One 2014;9(9):e107817.

[87] Nakagawa S, Sakamoto Y, Okabe H, Hayashi H, Hashimoto D, Yokoyama N, et al. Epigenetic therapy with the histone methyltransferase EZH2 inhibitor 3-deazaneplanocin A inhibits the growth of cholangiocarcinoma cells. Oncol Rep 2014;31(2):983–8.

[88] Fiskus W, Wang Y, Sreekumar A, Buckley KM, Shi H, Jillella A, et al. Combined epigenetic therapy with the histone methyltransferase EZH2 inhibitor 3-deazaneplanocin A and the histone deacetylase inhibitor panobinostat against human AML cells. Blood 2009;114(13):2733–43.

[89] Vitkevičienė A, Skiauterytė G, Žučenka A, Stoškus M, Gineikienė E, Borutinskaitė V, et al. HDAC and HMT inhibitors in combination with conventional therapy: a novel treatment option for acute promyelocytic leukemia. J Oncol 2019;2019:6179573.

[90] Wu D, Qiu Y, Jiao Y, Qiu Y, Liu D. Small molecules targeting HATs, HDACs, and BRDs in cancer therapy. Front Oncol 2020;10:560487. Available from: https://doi.org/10.3389/fonc.2020.560487.

[91] Mottamal M, Zheng S, Huang TL, Wang G. Histone deacetylase inhibitors in clinical studies as templates for new anticancer agents. Molecules 2015;20(3):3898–941.

[92] Biancotto C, Frigè G, Minucci S. Histone modification therapy of cancer. Adv Genet 2010;70:341–86.

[93] Glozak MA, Sengupta N, Zhang X, Seto E. Acetylation and deacetylation of non-histone proteins. Gene 2005;363:15–23.

[94] Finnin MS, Donigian JR, Cohen A, Richon VM, Rifkind RA, Marks PA, et al. Structures of a histone deacetylase homologue bound to the TSA and SAHA inhibitors. Nature 1999;401 (6749):188–93.

[95] Dokmanovic M, Clarke C, Marks PA. Histone deacetylase inhibitors: overview and perspectives. Mol Cancer Res 2007;5 (10):981–9.

[96] Dokmanovic M, Marks PA. Prospects: histone deacetylase inhibitors. J Cell Biochem 2005;96(2):293–304.

[97] Bondarev AD, Attwood MM, Jonsson J, Chubarev VN, Tarasov VV, Schiöth HB. Recent developments of HDAC inhibitors: emerging indications and novel molecules. Br J Clin Pharmacol 2021. Available from: https://doi.org/10.1111/bcp.14889 Online ahead of print.

[98] Shetty MG, Pai P, Deaver RE, Satyamoorthy K, Babitha KS. Histone deacetylase 2 selective inhibitors: a versatile therapeutic strategy as next generation drug target in cancer therapy. Pharmacol Res 2021;170:105695.

[99] Grassadonia A, Cioffi P, Simiele F, Iezzi L, Zilli M, Natoli C. Role of hydroxamate-based histone deacetylase inhibitors (Hb-HDACIs) in the treatment of solid malignancies. Cancers 2013;5:919–42.

[100] Copeland A, Buglio D, Younes A. Histone deacetylase inhibitors in lymphoma. Curr Opin Oncol 2010;22:431–6.

[101] Savickiene J, Treigyte G, Valiuliene G, Stirblyte I, Navakauskiene R. Epigenetic and molecular mechanisms underlying the antileukemic activity of the histone deacetylase inhibitor belinostat in human acute promyelocytic leukemia cells. Anticancer Drugs 2014;25(8):938–49.

[102] Valiuliene G, Stirblyte I, Cicenaite D, Kaupinis A, Valius M, Navakauskiene R. Belinostat, a potent HDACi, exerts antileukaemic effect in human acute promyelocytic leukaemia cells via chromatin remodelling. J Cell Mol Med 2015;19(7):1742–55.

[103] Valiulienė G, Treigytė G, Savickienė J, Matuzevičius D, Alksnė M, Jarašienė-Burinskaja R, et al. Histone modifications patterns in tissues and tumours from acute promyelocytic leukemia xenograft model in response to combined epigenetic therapy. Biomed Pharmacother 2016;79:62–70.

[104] Asano T, Sato A, Isono M, Okubo K, Ito K, Asano T. Bortezomib and belinostat inhibit renal cancer growth synergistically by causing ubiquitinated protein accumulation and endoplasmic reticulum stress. Biomed Rep 2015;3 (6):797–801.

[105] Chan D, Zheng Y, Tyner JW, Chng WJ, Chien WW, Gery S, et al. Belinostat and panobinostat (HDACI): in vitro and in vivo studies in thyroid cancer. J Cancer Res Clin Oncol 2013;139(9):1507–14.

[106] Dizon DS, Damstrup L, Finkler NJ, Lassen U, Celano P, Glasspool R, et al. Phase II activity of belinostat (PXD-101), carboplatin, and paclitaxel in women with previously treated ovarian cancer. Int J Gynecol Cancer 2012;22(6):979–86.

[107] Spratlin JL, Pitts TM, Kulikowski GN, Morelli MP, Tentler JJ, Serkova NJ, et al. Synergistic activity of histone deacetylase and proteasome inhibition against pancreatic and hepatocellular cancer cell lines. Anticancer Res 2011;31(4):1093–103.

[108] Merzvinskyte R, Treigyte G, Savickiene J, Magnusson KE, Navakauskiene R. Effects of histone deacetylase inhibitors, sodium phenyl butyrate and vitamin B3, in combination with retinoic acid on granulocytic differentiation of human promyelocytic leukemia HL-60 cells. Ann N Y Acad Sci 2006;1091:356–67.

[109] Savickiene J, Treigyte G, Borutinskaite V, Navakauskiene R, Magnusson KE. The histone deacetylase inhibitor FK228 sensitizes distinctly the human leukemia cells to retinoic acid-induced differentiation. Ann N Y Acad Sci 2006;1091:368–84.

[110] Savickiene J, Borutinskaite VV, Treigyte G, Magnusson KE, Navakauskiene R. The novel histone deacetylase inhibitor BML-210 exerts growth inhibitory, proapoptotic and differentiation stimulating effects on the human leukemia cell lines. Eur J Pharmacol 2006;549(1–3):9–18.

[111] Borutinskaite VV, Magnusson KE, Navakauskiene R. Histone deacetylase inhibitor BML-210 induces growth inhibition and apoptosis and regulates HDAC and DAPC complex expression levels in cervical cancer cells. Mol Biol Rep 2012;39(12):10179—86.

[112] Savickiene J, Treigyte G, Magnusson KE, Navakauskiene R. p21 (Waf1/Cip1) and FasL gene activation via Sp1 and NFkappaB is required for leukemia cell survival but not for cell death induced by diverse stimuli. Int J Biochem Cell Biol 2005;37(4):784—96.

[113] Borutinskaitė V, Navakauskienė R. The histone deacetylase inhibitor BML-210 influences gene and protein expression in human promyelocytic leukemia NB4 cells via epigenetic reprogramming. Int J Mol Sci 2015;16(8):18252—69.

[114] Marchi E, Zullo KM, Amengual JE, Kalac M, Bongero D, McIntosh CM, et al. The combination of hypomethylating agents and histone deacetylase inhibitors produce marked synergy in preclinical models of T-cell lymphoma. Br J Haematol 2015;171:215—26.

[115] Savickiene J, Treigyte G, Magnusson KE, Navakauskiene R. Response of retinoic acid-resistant KG1 cells to combination of retinoic acid with diverse histone deacetylase inhibitors. Ann N Y Acad Sci 2009;1171:321—33.

[116] Borutinskaite V, Bauraite-Akatova J, Navakauskiene R. Anti-leukemic activity of DNA methyltransferase inhibitor procaine targeted on human leukaemia cells. Open Life Sci 2016;11:322—30.

[117] Borutinskaitė V, Virkšaitė A, Gudelytė G, Navakauskienė R. Green tea polyphenol EGCG causes anti-cancerous epigenetic modulations in acute promyelocytic leukemia cells. Leuk Lymphoma 2018;59(2):469—78. Available from: https://doi.org/10.1080/10428194.2017.1339881.

[118] Vitkeviciene A, Baksiene S, Borutinskaite V, Navakauskiene R. Epigallocatechin-3-gallate and BIX-01294 have different impact on epigenetics and senescence modulation in acute and chronic myeloid leukemia cells. Eur J Pharmacol 2018;838:32—40. Available from: https://doi.org/10.1016/j.ejphar.2018.09.005.

[119] Valiulienė G, Stirblytė I, Jasnauskaitė M, Borutinskaitė V, Navakauskienė R. Anti-leukemic effects of HDACi Belinostat and HMTi 3-Deazaneplanocin A on human acute promyelocytic leukemia cells. Eur J Pharmacol 2017;799:143—53. Available from: https://doi.org/10.1016/j.ejphar.2017.02.014.

[120] Valiulienė G, Vitkevičienė A, Navakauskienė R. The epigenetic treatment remodel genome-wide histone H4 hyper-acetylation patterns and affect signaling pathways in acute promyelocytic leukemia cells. Eur J Pharmacol 2020;889:173641. Available from: https://doi.org/10.1016/j.ejphar.2020.173641.

[121] Savickiene J, Treigyte G, Vistartaite G, Tunaitis V, Magnusson KE, Navakauskiene R. C/EBPα and PU.1 are involved in distinct differentiation responses of acute promyelocytic leukemia HL-60 and NB4 cells via chromatin remodeling. Differentiation 2011;81(1):57—67. Available from: https://doi.org/10.1016/j.diff.2010.08.003.

[122] Farge T, Saland E, de Toni F, Aroua N, Hosseini M, Perry R, et al. Chemotherapy-resistant human acute myeloid leukemia cells are not enriched for leukemic stem cells but require oxidative metabolism. Cancer Discov 2017;7(7):716—35.

[123] Vitkevičienė A, Janulis V, Žučenka A, Borutinskaitė V, Kaupinis A, Valius M, et al. Oxidative phosphorylation inhibition induces anticancerous changes in therapy-resistant-acute myeloid leukemia patient cells. Mol Carcinog 2019;58(11):2008—16.

[124] Ashton TM, McKenna WG, Kunz-Schughart LA, Higgins GS. Oxidative phosphorylation as an emerging target in cancer therapy. Clin Cancer Res 2018;24(11):2482—90.

[125] Valiulienė G, Vitkevičienė A, Skliutė G, Borutinskaitė V, Navakauskienė R. Pharmaceutical drug metformin and MCL1 inhibitor S63845 exhibit anticancer activity in myeloid leukemia cells via redox remodeling. Molecules 2021;26(8):2303.

[126] Stuani L, Sabatier M, Saland E, Cognet G, Poupin N, Bosc C, et al. Mitochondrial metabolism supports resistance to IDH mutant inhibitors in acute myeloid leukemia. J Exp Med 2021;218(5):e20200924.

SECTION X

The Future of Epigenetics

42. New directions for epigenetics: application of engineered DNA-binding molecules to locus-specific epigenetic research

CHAPTER 42

New Directions for Epigenetics: Application of Engineered DNA-binding Molecules to Locus-specific Epigenetic Research

Toshitsugu Fujita and Hodaka Fujii

Department of Biochemistry and Genome Biology, Hirosaki University Graduate School of Medicine, Aomori, Japan

OUTLINE

Introduction	843
General information on engineered DNA-binding molecules	844
ZF proteins	844
TAL proteins	844
CRISPR/Cas	844
Locus-specific epigenome editing	845
Proof-of-concept research on locus-specific epigenome editing	846
Locus-specific epigenome editing aimed at obtaining mechanistic insight into disorders and therapeutic applications	850
Potential problems with locus-specific epigenome editing	852
Advanced epigenome editing tools	854
Characteristics of engineered DNA-binding molecules in locus-specific epigenome editing	857
Locus-specific identification of epigenetic molecules that interact with target genomic regions	858
Development of locus-specific ChIP: iChIP and enChIP	858
Identification of epigenetic and related molecules in a locus-specific manner by enChIP	859
In vitro enChIP	860
Other methods	862
Conclusions	862
References	862

INTRODUCTION

Epigenetic mechanisms are regulated by DNA and histone modifications, and these modifications are finely tuned by diverse molecules in a locus-specific manner. Multiple biochemical technologies have been developed for the analysis of epigenetic modifications, including chromatin immunoprecipitation (ChIP), methylation-sensitive restriction enzyme digestion, methylation-specific affinity purification, and bisulfite treatment [1,2]. Applications of these technologies have revealed the important relationship between locus-specific epigenetic modifications and their biological output in the contexts of development and disease. To date, however, analyses of epigenetic regulation have mostly been descriptive endeavors, and the mechanistic details of locus-specific epigenetic regulation remain largely unclear. Therefore, novel approaches are required to obtain mechanistic insight into molecular mechanisms of epigenetic regulation, especially in a locus-specific manner.

Recently, engineered DNA-binding molecules such as zinc finger (ZF) proteins, transcription activator-like effector (TAL or TALE) proteins, and the clustered regularly interspaced short palindromic repeats (CRISPR) and

CRISPR-associated protein (Cas) (CRISPR/Cas) system have been widely used as genome editing tools [3–5]. Moreover, the applications of these engineered DNA-binding molecules are not restricted to genome editing. For example, taking advantage of these systems' abilities to recognize specified DNA sequences, one can target the desired molecules to loci of interest by fusing them with engineered DNA-binding molecules [6–11]. This localization strategy can be used to artificially control epigenetic modifications in a locus-specific manner, which is subsequently reflected in alteration of biological outputs such as transcriptional activity and cell fate. Therefore, it might be possible to adapt this strategy to elucidate the mechanistic details of locus-specific epigenetic regulation, as well as to develop novel epigenetic therapies through artificial alteration of epigenetic information in a locus-specific manner.

To obtain mechanistic insight into epigenetic regulation of genomic regions of interest, it is also essential to identify the epigenetic regulators and modifications associated with these regions. Methods for this purpose were limited until recently, when the advent of engineered DNA-binding molecules made it easier to target specific genomic loci while retaining their molecular interactions [12]. Combining such approaches with downstream analyses (e.g., mass spectrometry [MS], microarray, and next-generation sequencing [NGS]) makes it possible to identify locus-specific epigenetic regulators and modifications.

In the second edition of this book, to highlight new directions for epigenetic research, we described applications of engineered DNA-binding molecules to locus-specific artificial epigenome editing, as well as locus-specific identification of epigenetic regulators. In the third edition, we update the contents of the previous edition.

GENERAL INFORMATION ON ENGINEERED DNA-BINDING MOLECULES

Genome editing tools are required for knock-in or knock-out strategies used in developmental biology, molecular biology, and medical research. Until two decades ago, genome editing was dependent on the natural occurrence of homologous recombination, and was therefore restricted to certain cell types (e.g., embryonic stem [ES] cells). However, the advent of engineered DNA-binding molecules such as ZF proteins, TAL proteins, and CRISPR/Cas enabled us to perform genome editing more easily in cells of many more types. First, we will briefly describe the general properties of these molecules.

ZF Proteins

ZF proteins contain ZF motif(s), which consist of about 30 amino acid residues (a.a.). Each ZF motif binds to a three-base sequence of double-helical DNA (Fig. 42.1A) [13]. By sequentially linking ZF motifs, one can construct a ZF protein that recognizes a given DNA sequence. ZF proteins fused with functional domains of *Fok*I, an endonuclease, have been used as ZF-nucleases (ZFNs) to artificially introduce double-strand breaks in a sequence-specific manner [3,4]. Although ZF proteins represent the prototype of engineered DNA-binding molecules, technical expertise and experience are required to construct an effective protein, in particular for optimal design and assembly of DNA-binding motifs. For example, Carroll et al. reported a protocol for construction of an effective ZF protein by assembling ZF modules [14].

TAL Proteins

TAL proteins are transcription factors of plant pathogenic bacteria, such as *Xanthomonas* sp. A TAL protein generally recognizes about 20 bases of DNA in a sequence-specific manner via a DNA-recognition domain consisting of DNA-binding modules, each of which recognizes a single nucleotide (Fig. 42.1A) [3,6]. A DNA-binding module consists of a conserved repetitive region (33–35 a.a.); the 12th and 13th residues are decisive in recognition of the corresponding nucleotide. One can construct a TAL protein that recognizes a given DNA sequence by assembling the corresponding DNA-binding modules [15,16]. TAL proteins linked to *Fok*I have been used as locus-specific genome editing tools called TALENs [3,6]. The binding specificity of a TAL domain depends solely on the constituent TAL module, whereas that of a ZF protein is determined not only by each ZF motif but also by the combinations of the modules. Therefore, in contrast to the situation for ZF proteins, it is not necessary to consider combinations of TAL modules. Consequently, the design and combination of DNA-binding domains to construct a TAL protein are more straightforward.

CRISPR/Cas

The CRISPR/Cas9 system, which consists of the DNA-binding and cleavage protein Cas9 and a guide RNA (gRNA), was initially characterized as a bacterial adaptive immune system [8,17,18]. In this context, foreign DNA such as a plasmid or phage genome is first digested and integrated at the CRISPR array locus in the bacterial genome. Next, the integrated DNA is transcribed from the locus to yield a mature transcript called the CRISPR RNA (crRNA), which forms the CRISPR/Cas complex

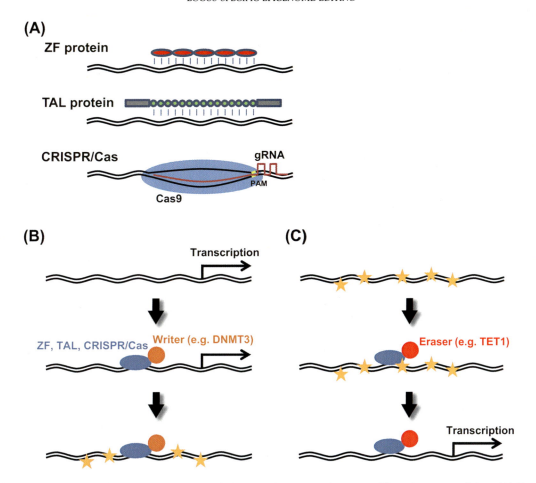

FIGURE 42.1 Engineered DNA-binding molecules and their application to locus-specific epigenome editing. (A) Engineered DNA-binding molecules. A zinc finger (ZF) protein consists of ZF motifs (red ellipses); one motif recognizes three bases. A transcription activator-like effector (TAL) protein possesses a DNA-binding domain consisting of an assembly of DNA-binding modules (green circles); each module recognizes one nucleotide. Flanking regions are shown in gray. CRISPR/Cas consists of Cas9 and gRNA; gRNA determines recognition of a 23 bp target including the PAM site (yellow). In this figure, sgRNA is shown as gRNA. For locus-tagging without DNA cleavage, a catalytically inactive form of Cas9 (dCas9) is used instead of wild-type Cas9. (B, C) Locus-specific epigenome editing. Engineered DNA-binding molecules (blue circle) fused with an epigenetic writer (orange circle, B) or eraser (red circle, C) can be used for locus-specific addition or removal of selected epigenetic marks (yellow stars).

along with trans-activating crRNA (tracrRNA) and the Cas9 protein. The complex recognizes invading foreign DNA complementary to the crRNA and cleaves it via the intrinsic nuclease activity of Cas9 (Fig. 42.1A). The RNA complex of crRNA and tracrRNA, called gRNA, is decisive for DNA targeting in this system. The CRISPR/Cas9 system derived from *Streptococcus pyogenes* (*S. pyogenes*) has been most widely applied to genome editing. The CRISPR/Cas9 complex recognizes a ~23 bp sequence containing the protospacer adjacent motif (PAM) sequence (5′-NGG-3′) at the 3′ position [8–10]. A single chimeric gRNA (sgRNA) mimicking the structure of the annealed crRNA/tracrRNA is generally used in place of the two separate RNAs (Fig. 42.1A), enabling use of a simplified two-component CRISPR/Cas9 system. Because gRNA (or sgRNA) expression vectors can be constructed time- and cost-effectively by standard genetic engineering, this system has been widely adopted for the purpose of genome editing. When it is necessary to tag a locus without introducing a DNA double-strand break, a nuclease-deficient mutant of Cas9 (dCas9) is used. The *S. pyogenes* dCas9 protein possesses two mutations in the RuvC1 and HNH nuclease domains (D10A and H841A) that disrupt nuclease activity but not the binding activity of Cas9 [19,20]. In addition to CRISPR/Cas9 from *S. pyogenes*, other orthologues have been used in genome editing and other applications [8–10].

LOCUS-SPECIFIC EPIGENOME EDITING

Artificial epigenetic modifications in loci of interest might introduce targeted changes in biological output, and are thus useful for investigating the biological significance of specific epigenetic marks. For example, intentional removal of CpG methylation is a promising approach for re-inducing transcription of suppressed target genes. To this end, epigenetic drugs, such as inhibitors of DNA methyltransferases (MTases) in replicative cells, are useful and effective. Alternatively, exogenous expression of epigenetic modifiers may introduce desired epigenetic modifications at target loci. However, because such approaches can affect epigenetic status on a genome-wide scale, technologies for locus-specific epigenome editing are required.

The DNA-recognition properties of engineered DNA-binding molecules have been utilized to locus-specifically tether various functional molecules such as the transcriptional regulators VP64 and KRAB, and fluorescent proteins, to achieve locus-specific transcriptional modification or visualization of target loci in living cells [11]. In addition, one can artificially edit epigenetic information in a locus-specific manner using fusions of engineered DNA-binding molecules and the enzymes directly responsible for epigenetic modifications. Alternatively, by analyzing the biological output of artificial epigenome editing, one can elucidate functions of the fused molecules, whose epigenetic roles remain unclear at this stage. In this section, we describe various applications of ZF proteins, TAL proteins, and CRISPR/Cas to artificial and locus-specific epigenome editing and focus on the use of epigenetic enzymes, but not transcriptional activators/repressors such as VP64 or KRAB (Fig. 42.1B and C and Table 42.1).

Proof-of-concept Research on Locus-specific Epigenome Editing

DNA Methylation

As the first proof-of-concept of locus-specific artificial CpG methylation, the ZF motifs of two ZF proteins, Zif268 (a mouse immediate early protein) and Zip53 (an engineered derivative of Zif268), were C-terminally fused to M.SssI, a CpG MTase from *Spiroplasma* [21]. The resultant fusion proteins could catalyze locus-specific CpG methylation of purified DNA in test tubes, demonstrating the feasibility of locus-specific CpG methylation and, in a broad sense, locus-specific epigenome editing. In addition, Zif268 and Zip53 fused with M.SssI or M.CviPI, a GpC MTase, were also used for targeted DNA methylation in yeast [22]. As expected, those fusion proteins increased CpG or GpC methylation around target sites ~20-fold relative to negative controls, suggesting that locus-specific DNA methylation can be introduced even in cells.

In mammals, CpG methylation is established de novo during early embryogenesis by DNMT3A and DNMT3B, and subsequently maintained by DNMT1. To achieve locus-specific DNA methylation, a ZF protein was fused with the catalytic domain of DNMT3A (ZF-DNMT3A) [27]. When the constructed protein targeted the *IE175k* promoter of Herpes Simplex Virus type 1 (HSV-1) in COS-7 monkey kidney cells infected with HSV-1, CpG sites around the target sequence were highly methylated, effectively suppressing transcription from the viral gene promoter. ZF-DNMT3A was also used for artificial DNA methylation in the promoter region of *VEGF-A* in SKOV3 human ovarian cells [28]. These results demonstrated that locus-specific DNA methylation around a target site is feasible, even in mammalian cells.

DNMT3L (DNMT3-Like), another member of the DNMT3 family, is homologous to DNMT3A and DNMT3B [88]. Although DNMT3L does not have CpG MTase activity, it can increase the activities of DNMT3A and DNMT3B via molecular interactions [89]. Taking advantage of this property, Siddique et al. generated a ZF protein linked sequentially to the catalytic domains of DNMT3A and DNMT3L (DNMT3A-3Lsc) [29]. This fusion protein induced CpG methylation and subsequent transcriptional suppression of *VEGF-A* more effectively than ZF-DNMT3A alone in SKOV3 cells. DNMT3A-3Lsc was also fused with a TAL protein to induce CpG methylation in the *CDKN2A* promoter region [30]. Although targeted CpG methylation was introduced by the fusion protein, it was not confined to the target *CDKN2A* gene, but extended toward surrounding genes (*p14ARF*, *CDKN2B*, and *MTAP*). In this regard, targeted epigenetic modifications can expand into regions surrounding a target locus, depending on chromosomal conformation (for more details of this "spreading" of epigenetic modifications, please see the first paragraph in the section, Potential Problems with Locus-specific Epigenome Editing). Sequential linking of two (or more) epigenetic modifiers to an engineered DNA-binding molecule is a promising strategy for promoting locus-specific epigenome editing.

CRISPR/Cas has also been used to achieve locus-specific DNA methylation. An advantage of CRISPR/Cas over the other two types of engineered DNA-binding proteins is that it can be used to target multiple sites using pooled sgRNAs. For locus-specific DNA methylation, dCas9 was fused with the catalytic domain of DNMT3A (dCas9-DNMT3A) [31]. Simultaneous targeting of multiple sites using pooled sgRNAs (four target sites in one locus) synergistically increased the levels of CpG methylation in the CpG island of the *IL6ST* or *BACH2*

TABLE 42.1 Molecules Fused to Engineered DNA/RNA-Binding Molecules for Locus-Specific Epigenome/Epitranscriptome Editing

Targeted Output	Fused Proteins	References[a]
DNA methylation (cytosine)	M.SssI	[21–25]
	M.CviPI	[22]
	M.HhaI	[26]
	DNMT3A	[27–52]
	DNMT3B	[48]
	DNMT3L	[29,30,33,42,46,47,51]
DNA methylation (adenine)	Dam	[53]
DNA demethylation	TDG	[54]
	TET1	[34,35,43,55–63]
	TET2	[55,64]
	TET3	[55,65]
	GADD45A	[63]
	NEIL2	[63]
H3K4 methylation	PRDM9	[66]
H3K4 demethylation	LSD1	[67–69]
H3K9 methylation	SUV39H1	[70,71]
	G9a	[70–73]
	GLP	[28]
	SUVDEL76	[73]
	HP1	[74]
	SS18	[74]
H3K27 methylation	EZH2	[33,75,76]
	FOG1	[75]
H3K27 acetylation	p300	[69,77–84]
	CBP	[85]
H3K27 deacetylation	HDAC3	[86]
H3K79 methylation	DOT1L	[66]
H2B monoubiquitination	UBE2A	[66]
RNA methylation (adenine)	METTL3	[87]
	METTL14	[87]
RNA demethylation (adenine)	ALKBH5	[87]
	FTO	[87]

[a]Major references are listed.

promoter in human HEK293 cells [31]. Synergistic effects were also observed when pooled sgRNAs targeting three to eight sites of the *CDKN2A* gene promoter were used in human 293 T cells, although the output (i.e., the reduction in transcription) was marginal [32]. By contrast, when *Cdkn1a* was simultaneously targeted with two sgRNAs in the mouse IL3-dependent myeloid progenitor cell line 32D, CpG methylation of the locus was more prominent, sufficient to induce a high degree of transcriptional suppression (~50%) [32]. DNMT3A and DNMT3L form a tetramer [90]. Hence, dCas9-DNMT3L recruits endogenous DNMT3A/DNMT3L to methylate CpG at the target gene *HER2* in HCT116 cells [33].

DNA Demethylation

Active CpG demethylation is initiated by oxidation of methylated cytosine (5-mC) by TET family proteins to yield 5-hydroxymethylcytosine (5-hmC). Subsequently, 5-hmC is converted in a stepwise manner to 5-formylcytosine (5-fC) and then 5-carboxylcytosine (5-caC); 5-fC and 5-caC can be subsequently removed by thymine-DNA-glycosylase (TDG) in base excision repair[91]. Targeted erasure of CpG methylation patterns around target gene promoters represents an attractive approach for re-expressing suppressed genes, as well as for studying the detailed molecular mechanisms that define how the loss of DNA methylation increases promoter activity (e.g., by promoting or reducing binding of particular transcription factors). Accordingly, the enzymes involved in active CpG demethylation have been employed for locus-specific CpG demethylation (Fig. 42.1C).

To evaluate the physiological significance of CpG methylation in mammals, TDG was linked to a ZF protein [54]. The resultant single ZF-TDG construct exerted modest effects on CpG demethylation of the target *Nos2* promoter, as well as on transcription of the gene; whereas, multiple constructs (specifically, four constructs targeting different sites in the promoter region) significantly increased CpG demethylation and restored induction of transcription upon stimulation with lipopolysaccharide or interferon γ (IFNγ) in NIH 3T3 cells. In another study, the catalytic domains of TET family proteins (TET1, 2, and 3) were fused with ZF proteins to remove CpG methylation in the promoter regions of epigenetically silenced *ICAM-1* and *EpCAM* [55]. Among the three TET family members, TET2 exhibited the strongest demethylation efficiency and the most robust re-induction of transcription in the fused form, whereas TET1 or TET3 exerted no or negligible effects.

The application of TAL proteins fused to full-length TET1 or its catalytic domain (TAL-TET1) was reported prior to the aforementioned study using ZF proteins [56]. TAL-TET1 targeting the *KLF4*, *RHOXF2*, or *HBB* gene significantly reduced CpG methylation on the target regions and restored robust expression of their gene products (~100-fold induction) in human K562, HeLa, and 293 T cells. ZF-TET1 targeting the *KLF4* or *HBB* gene induced CpG demethylation on the target regions with comparable efficiency [56].

CRISPR/Cas was also used to induce locus-specific CpG demethylation [57]. dCas9 was fused with the catalytic domain of TET1 (dCas9-TET1) to induce CRISPR/dCas9-mediated locus-specific CpG demethylation of the *BRCA1* promoter region, which is CpG-hypermethylated in ovarian and breast cancers [58]. This system induced modest but significant CpG demethylation around the target sites, as well as upregulation of *BRCA1* transcription, in HeLa cervical cancer and MCF7 breast cancer cells [57]. In addition, dCas9-TET1 demethylated the *BDNF* promoter region, resulting in upregulation of gene expression in cultured mouse cortical neurons and in the brain of a mouse model [34].

For CRISPR/dCas9-mediated locus-specific recruitment, effector molecules can be linked not only to dCas9, but also to the gRNA. In one study, RNA stem-loop structures that interact with the bacteriophage protein MS2 were fused with sgRNA to recruit MS2-linked transcriptional activators to target loci [92]. Several variations of this scaffold RNA system have been reported [93–95]. In their application of a scaffold RNA system to locus-specific epigenome editing, Xu et al. fused the catalytic domain of TET1 to MS2 (MS2-TET1) [59] (Fig. 42.2, left panels). Combinatorial use of MS2-TET1 and dCas9-TET1 yielded stronger CpG demethylation on target loci (*RANKL*, *MAGEB2*, and *MMP2*) than either fusion protein alone in HEK293FT, HeLa, or SH-SY5Y cells. In accordance with the reduction of CpG methylation levels, the transcription of target genes was upregulated.

To tether multiple copies of transcriptional regulators or fluorescent proteins to target loci, Tanenbaum et al. fused SunTag, an epitope tag consisting of multiple tandem repeats of the recognition sequence of the anti-GCN4 antibody, to dCas9 (dCas9-SunTag) [96], allowing proteins fused to the single-chain variable fragment of the antibody (scFV-GCN4) to be recruited to the loci where dCas9-SunTag interacts. Morita et al. applied dCas9-SunTag along with scFV-GCN4 fused to the catalytic domain of TET1 to locus-specific epigenome editing [60] (Fig. 42.2, right panels). In mouse ES cells, a modified version of the SunTag system exhibited stronger effects on targeted CpG demethylation (> 90%) of the *Gfap* locus than dCas9-TET1. Targeted CpG demethylation and the resultant transcriptional upregulation were also observed for other loci, such as *H19*. In addition, using in utero electroporation, Morita et al. successfully induced targeted CpG demethylation in mouse fetal brain [60]. dCas9-SunTag-mediated demethylation using TET1 was also successful in CHO cells [35]. In addition to TET1, the dCas9-SunTag system was also combined with DNMT3A to induce CpG methylation at various loci in human cell lines [36,37].

Histone Modification

Methylation of histone H3 lysine 9 (H3K9) is an epigenetic mark that correlates with gene silencing in a variety

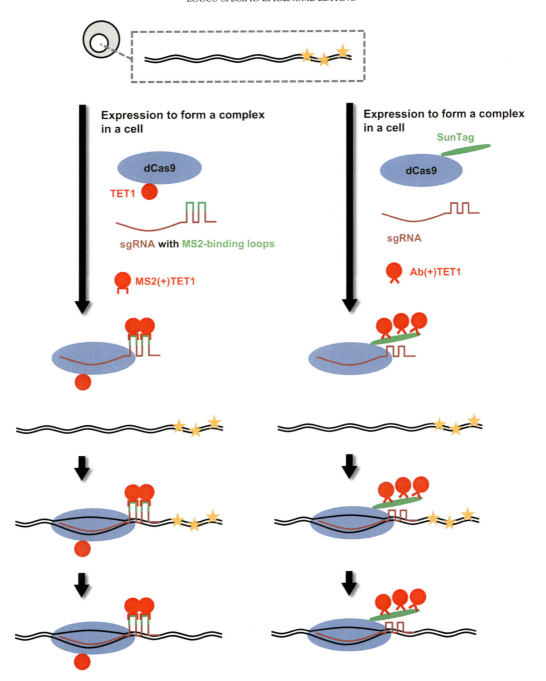

FIGURE 42.2 Different methods for locus-specific epigenome editing. (Left) Scaffold RNA-mediated epigenome editing system. The desired epigenetic regulators (e.g., TET1) fused with RNA-binding proteins (e.g., MS2) can be recruited to target loci via interaction with the modified sgRNA, which possesses RNA loop structures that interact with RNA-binding proteins. (Right) The SunTag-mediated epigenome editing system. Desired epigenetic regulators (e.g., TET1) fused to an antibody (Ab) against SunTag can be recruited to target loci via interaction with SunTag-fused dCas9.

of organisms [97]. Accordingly, the enzymes responsible for H3K9 methylation promote transcriptional suppression. Snowden et al. fused ZF proteins to the catalytic domains of the histone MTases SUV39H1 and G9a [70]. The resultant fusion proteins increased H3K9 methylation levels in the *VEGF-A* promoter and suppressed *VEGF-A* transcription in HEK293 cells. In contrast to the catalytic domain, full-length SUV39H1 did not effectively repress transcription in fusion form [70]. Simultaneous targeting by ZF proteins fused to G9a and v-ErbA, a viral protein that recruits the NCoR/SMRT corepressor complex, enforced transcriptional suppression of the *VEGF-A* gene [70]. In addition, the H3K9 MTase GLP was also fused to ZF proteins to induce H3K9 methylation in the promoter region of *VEGF-A* [28]. Interestingly, the induced epigenetic marks spread widely, at least ~5 kb upstream and ~15 kb downstream from the ZF-binding site.

LSD1, a histone demethylase that catalyzes removal of H3K4 methylation, an active histone mark [97], was used to induce locus-specific removal of histone methylation [67]. In K562 cells, LSD1 linked to TAL proteins targeting enhancer regions decreased the level of histone H3 lysine 4 dimethylation (H3K4me2) in a locus-specific manner. Consistent with the finding that LSD1 interacts with other chromatin-modifying enzymes, including a histone deacetylase [98], H3 lysine 27 acetylation (H3K27ac) levels were also reduced at these loci [67]. This enhancer-targeting strategy resulted in transcriptional suppression of proximal genes that were regulated by enhancers, suggesting that not only promoter regions but also distal enhancer regions can be considered as targets for artificial epigenome editing. This type of targeting strategy was also used to engineer locus-specific changes in CpG methylation [34,99]. A genome-wide study revealed that 26 of 40 TAL-LSD1 constructs substantially decreased levels of the epigenetic modifications of interest at their target loci [67].

To achieve CRISPR/dCas9-mediated removal of H3K4 methylation, LSD1 was fused to dCas9 of *Neisseria meningitidis* (dCas9-LSD1) [68]. dCas9-LSD1/sgRNA targeting the distal enhancer regions, but not the proximal promoter regions, of *Oct4* and *Tbox3* effectively suppressed transcription of these genes in mouse ES cells. Consistent with this, the levels of H3K4me2 and H3K27ac were reduced. In addition, these alterations resulted in genome-wide transcriptomic and morphological changes of the cells. Together, these results demonstrate that artificial locus-specific epigenome editing can lead to phenotypic alteration.

To develop another locus-specific histone modifier, dCas9 was fused to the catalytic core domain of histone acetyltransferase p300 to promote formation of H3K27ac marks in 293 T cells [77]. When used to target the promoter regions, as well as distal regulatory regions up to ~50 kbp away from the targets, the fusion protein activated transcription of the target genes (*IL1RN, MYOD, OCT4*, and *globins*). When distal regulatory regions were targeted, enrichment of H3K27ac was observed not only around the target regions but also in the regulated promoter regions [77]. Therefore, targeting of distal regulatory regions, in combination with genome-wide analysis of specific histone marks, may be used to identify the genes regulated by these regions. Locus-specific epigenome editing using p300 was also achieved using TAL and ZF proteins targeting the *IL1RN* and *ICAM1* genes, respectively, in 293 T cells [77]. O'Geen et al. showed that dCas9 fused to full-length H3K27 methyltransferase EZH2 (the catalytic subunit of the PRC2 complex) induces H3K27 trimethylation on the *HER2* gene in HCT116 cells [75]. They also demonstrated that dCas9 fused to the N-terminus of FOG1, which binds to the NuRD complex, can recruit the complex to a target *HER2* gene, which is followed by enhancement of H3K27 trimethylation by the PRC2 complex recruited via NuRD [75]. Wang et al. showed that dCas9 fused to the catalytic domain of EZH2 can induce the placement of histone marks on the *GRN* promoter in Hep3B cells [76].

Locus-specific Epigenome Editing Aimed at Obtaining Mechanistic Insight into Disorders and Therapeutic Applications

Cancer Biology

CpG methylation is a well-known epigenetic mark associated with silencing of tumor suppressor genes in cancer [100]. Accordingly, artificial and locus-specific removal of CpG methylation represents an attractive approach for development of cancer therapies and obtaining mechanistic insight into the roles of epigenetic regulation in carcinogenesis and cancer progression. For example, *Maspin*, a tumor suppressor gene associated with metastatic progression and poor prognosis in breast cancer [101], was targeted to evaluate the roles of the encoded protein in cancer cells [38]. In that study, ZF-DNMT3A targeting *Maspin* induced stable and heritable DNA methylation and suppressed transcription of the gene in human breast cancer cell lines. Moreover, this transcriptional suppression induced colony formation by breast cancer cells [38]. Interestingly, although the ZF-DNMT3A-mediated induction of DNA methylation was modest, transcriptional suppression was dramatic and sufficient to induce the expected biological outcome. Therefore, in some cases, modest levels of DNA methylation around a target site are sufficient to achieve epigenetic suppression of the target gene.

SOX2, an ES cell marker, is expressed in early-stage breast carcinoma [102]. Epigenetic suppression by ZF-DNMT3A targeting the *SOX2* promoter decreased *SOX2* expression and proliferation of breast cancer cells [38]. In another study focusing on *SOX2* [39], induced DNA methylation was maintained even after expression of ZF-DNMT3A ceased. Moreover, xenograft studies demonstrated stable *SOX2* repression and long-term (~100 days) inhibition of breast tumor growth [39], suggesting the potential of locus-specific epigenome editing as a novel epigenetic therapy.

EpCAM, a positive regulator of cellular proliferation [103], is highly expressed in ovarian cancer cells [104]. Stable expression of ZF-DNMT3A targeting the *EpCAM* promoter increased CpG methylation by ~50% around the target site [40]. The induction of CpG methylation resulted in suppression of *EpCAM* transcription as well as cellular proliferation.

CpG methylation at the *p16* tumor suppressor gene, especially in the CpG island in the promoter region, is correlated with reduced *p16* expression and metastasis of many cancers [105–111]. To examine the causal relationship between *p16* CpG methylation and cancer metastasis, ZF-DNMT3A was expressed in the human gastric cancer cell line BGC-823 and the gastric mucosa cell line GES-1, in which *p16* is still unmethylated and transcribed [41]. The resultant targeted CpG methylation suppressed *p16* expression and increased the cell motility of both cell lines in transwell migration assays, as well as pneumonic metastasis of BGC-823 in xenograft studies. These results confirmed that CpG methylation in the *p16* gene promoter plays significant roles in promoting cancer metastasis. CpG methylation of the *p16* gene by dCas9-DNMT3A-3L in primary human myoepithelial cells prevented senescence, increased cellular proliferation, and induced cancer-like gene expression [42]; these results also demonstrate the importance of *p16* as an anticancer gene.

ZF-TET2 was designed to manipulate the expression of several tumor suppressor genes (*C13ORF18*, *CCNA1*, *TFPI2*, and *Maspin*) [64]. ZF-TET2 decreased the level of CpG methylation, induced re-expression of a subset of its target genes in human cervical cancer cell lines, and decreased tumor cell growth.

TAL-TET1 and TAL fused to DNMT3A (TAL-DNMT3A) were used to manipulate expression of *CRMP4*, a suppressor of cancer metastasis [43]. In xenograft studies addressing cell migratory and tissue invasive activities, TAL-TET1 decreased the level of CpG methylation on the *CRMP4* promoter in metastatic PC3 cells, thereby inhibiting metastasis. Inversely, TAL-DNMT3A induced CpG methylation of the unmethylated *CRMP4* promoter in non-metastatic 22Rv1 cells, causing metastasis to occur [43].

HER2/neu/ERBB2 is overexpressed in several types of cancer [112–114]. To suppress HER2/neu expression, epigenome editing with ZF proteins fused with G9a or SUV39H1 was performed in breast cancer cell lines and an ovarian cancer cell line [71]. ZF-G9a targeting the *HER2/neu* promoter increased incorporation of H3K9me2 and decreased H3 acetylation in the target region. In accordance with the alterations in histone marks, *HER2/neu* expression was significantly suppressed, and cell proliferation and clonogenicity were inhibited. Moreover, the combination of ZF-G9a and lapatinib, a small-molecule inhibitor of HER2/neu, increased the inhibition of cell proliferation, suggesting that the combined use of chemical compounds and locus-specific epigenome editing represents an effective strategy for epigenetic therapy.

Neurological Diseases

Using ZF-G9a, Heller et al. investigated molecular mechanisms linking chromatin dynamics and neurobiological phenomena in a discrete brain region [72]. Using mice as a model animal, they locus-specifically modified epigenetic marks at *FosB*, which is involved in addiction to drugs (e.g., cocaine) and depression in both rodents and humans [115]. ZF-G9a enriched the H3K9me2 mark across the *FosB* promoter in the nucleus accumbens, a key brain reward region. The histone methylation was sufficient to control drug- and stress-evoked transcriptional and behavioral responses through coordination with the endogenous transcriptional machinery.

ZF-G9a and ZF fused with the histone methyltransferase SUVDEL76 (ZF-SUVDEL76) were also used to rewrite the H3K9me2 and H3K9me3 modifications on the *Del4/PSD95* gene in mouse hippocampus [73]. Epigenome editing by ZF-SUVDEL76 suppressed expression of the *Del4/PSD95* gene, which was followed by a decrease in the evoked excitatory postsynaptic currents mediated by NMDA, whose synaptic expression is regulated by PSD95. Interestingly, upregulation of the *Del4/PSD95* gene by ZF-VP64 rescued memory deficits in aged and Alzheimer disease mouse models [73].

The *SNCA* gene, which encodes α-synuclein, is a genetic risk factor for Parkinson disease (PD). PD patients have lower levels of CpG methylation at the *SNCA* intron 1 region, which is linked to upregulation of gene expression. Locus-specific targeting of dCas9-DNMT3A induced CpG methylation at the *SNCA* intron 1 region, and subsequent suppression of gene expression was observed in induced pluripotent stem cell (iPSC)-derived dopaminergic neurons from a PD patient [44]. Interestingly, PD phenotypes such as mitochondrial superoxide production or cell viability were rescued. Taken together, these results suggest that locus-specific epigenome editing is also useful for elucidating the molecular mechanisms underlying neuropsychiatric disease and is a potential tool that can be used to design novel therapies to treat intractable neuronal diseases.

Other Diseases

CpG methylation of the *RASAL1* and *KLOTHO* promoters correlates with progression of kidney fibrosis in patients and in fibrosis mouse models [116–118]. To ameliorate kidney fibrosis in these mice, dCas9 fused to the TET3 catalytic domain (dCas9-TET3) and sgRNA targeted to the *Rasal1* and *Klotho* promoters were expressed in the kidney for targeted CpG demethylation [65]. This system successfully initiated CpG demethylation at those promoters, induced protein expression, and lessened the severity of kidney fibrosis. These results provide the basis for a potential application for locus-specific epigenome editing to attenuate tissue fibrosis.

Fragile X syndrome (FXS) is the most common genetic form of intellectual disability in humans. A CGG trinucleotide repeat (>200 repeats) expansion mutation at the 5′ UTR of *FMR1*, which is accompanied by CpG hypermethylation, promotes CpG hypermethylation of the *FMR1* promoter and silencing of gene expression [119–121]. Using dCas9-TET1, Liu et al. succeeded in demethylating the repeats and the promoter region in patient-derived iPS cells [61]. Neuronal cells differentiated from the methylation-edited cells exhibited relatively normal neuronal phenotypes. Hence, targeted CpG demethylation of the CGG repeats is a potential therapeutic approach for the treatment of FXS.

Regulatory T cells (Tregs) play important roles in regulating immune responses and maintaining immunological tolerance [122]. Foxp3 is a master transcription factor that regulates Treg differentiation and function. Okada et al. showed that dCas9-p300 targeted to the *Foxp3* promoter induces and stabilizes expression of the gene in primary mouse T cells and reinforces in vitro suppression of Treg activity [78]. Therefore, locus-specific epigenome editing may be useful for regulation of immunological tolerance.

The dCas9-SunTag system, which recruits p300, was used to induce expression of the *Oct4* gene in mouse MEF cells, resulting in reprograming resembling that of iPSCs [79]. In addition, targeted CpG methylation by dCas9-DNMT3A enabled differentiation of ES cells to motor neurons [45]. The combination of dCas9-TET1 and dCas9-VP64, but not dCas9-VP64 alone, strongly increases the expression of master transcription factors for neural development, such as *Sox1*, in neuronal progenitor cells, which alters cell fate to favor neuronal cells [62]. Hence, locus-specific epigenome editing can stabilize or reprogram cellular lineage commitment.

Other Applications

H3K4me3 near TSSs is associated with active transcription. To examine the dynamics between histone crosstalk and the sustainability of gene re-expression, Cano-Rodriguez et al. used dCas9/ZF proteins fused to the histone writers PRDM9 (H3K4 methyltransferase), DOT1L (H3K79 methyltransferase), and UBE2A (related to H2B monoubiquitination) [66]. Using a combination of those fused proteins and 5-aza-2′-deoxycytidine (5-aza-dC), they found that H3K4 methyltransferases are sufficient to induce gene re-expression from promoters with no or low DNA methylation. In addition, they also showed that the combination of H3K4 and H3K79 methyltransferases in the presence of 5-aza-dC is required for stable gene re-expression. In this regard, locus-specific epigenome editing can synergically induce a desirable biological output when combined with epigenetic drugs such as 5-Aza-dC or trichostatin A [34,123].

Chromatin structure is finely organized and compartmentalized by topological associated domains (TADs), in which cell type-specific enhancers and promoters modulate lineage-specific gene expression [124,125]. Binding to unmethylated insulator motifs, the insulator protein CTCF plays an important role in modulating TAD boundaries [126,127]. To examine the importance of CTCF in TAD boundary formation, CTCF-binding motifs were artificially CpG-methylated using dCas9-DNMT3A [34]. CpG methylation inhibited CTCF binding, which resulted in alterations in locus topology and neighbor gene expression. Combinatorial use of dCas9-DNMT3A, dCas9-DNMT3L, and dCas9-KRAB targeting CTCF-binding motifs also produced similar results [46]. CpG methylation of CTCF-binding motifs was also successfully performed in mouse zygotes using dCas9-MQ1^{Q147L} (a mutant of M.SssI methyltransferase), and the resultant mouse pups retained the targeted CpG methylation [23].

Taken together, these findings demonstrate that epigenome editing tools can provide mechanistic insight into epigenetic regulation in various disorders, and could also be used to develop therapeutic approaches. In addition, they might be used to generate model animals that mimic pathological phenotypes, such as drug addiction and depression. For these strategies to be applied in the clinic, it will also be necessary to develop an appropriate means of delivering the relevant components to the target regions of the body. In this chapter, we focused on locus-specific epigenome editing in mammalian cells. However, locus-specific epigenome editing is widely applicable to various cells and organisms, including plants [128–131] and fungi [132].

Potential Problems with Locus-specific Epigenome Editing

Most of the aforementioned cases of epigenome editing involved introduction of specific epigenetic modifications around target sites, and these modifications propagated extensively (~ ± 1 kbp). In some cases, introduced epigenetic marks expanded more than 10 kb from target sites [28,30,36]. Such expansion can be attributed to the 3D structure of chromatin: physical chromosomal organization, including chromatin looping, may bring distant genomic regions into the vicinity of the target site, where they can interact with an engineered DNA-binding molecule fused to an epigenetic modifier. Alternatively, endogenous epigenetic modifiers may recognize specific epigenetic modifications introduced by artificial epigenome editing and subsequently drive expansion of the modifications around target regions.

On the other hand, detailed biochemical analyses demonstrated that, when dCas9-DNMT3A was employed, the highest CpG methylation activity was centered 27 bp downstream from the PAM sequence, where the C-terminally fused DNMT3A physically interacted [31]. However, CpG in the dCas9 binding site was not methylated, probably because the region was inaccessible to the fused DNMT3A [31]. Similar effects were observed when dCas9-DNMT3A-3L [47] or a ZF protein alone were used [55].

Off-target Effects

It is especially important to manage off-target effects caused by unintended association of engineered DNA-binding molecules with DNA sequences similar to the target site, or even dissimilar sequences [30,32,56]. A genome-wide analysis showed that locus-specific epigenome editing achieved relatively high specificity in terms of transcriptional output [77], whereas, in another study, modest CpG methylation was detectable at endogenous loci [32] even when irrelevant loci were targeted by the CRISPR/Cas-mediated CpG methylation system. Other genome-wide analyses revealed that global off-target CpG methylation occurs after expression of dCas9-DNMT3A or -DNMT3B in ES, 293T, and MCF7 cells [48,49]. In addition, although ZF proteins targeted to the *Dlg4* gene were specifically recruited to the target site [73], those targeted to the *SOX2* promoter bound to thousands of off-target sites [133]. The SunTag system, which uses DNMT3A, produced less off-target CpG methylation than dCas9-DNMT3 [36,37]. This difference may be caused by more intensive recruitment of DNMT3A to the target site compared with irrelevant genomic regions. The dCas9-SunTag-DNMT3A system methylated some CpHs (H = A, C, or T) in target regions [36], which might be considered undesirable off-target modification in some cases. Hence, the severity of off-target effects is dependent on the system used for locus-specific epigenome editing.

At this stage, such off-target effects might be minimized by controlling the expression levels of epigenome editing factors, although on-target effects also often decreased [48]. In this regard, the efficiency and specificity of genome editing using CRISPR/Cas can be optimized adjusting the dosage of CRISPR/Cas-expression plasmids [134–136]. Alternatively, because it is difficult to control gene expression from a plasmid that contains a strong promoter, the use of recombinant dCas9 proteins and gRNA ribonucleoproteins (RNPs) may make it easier to adjust the dosages for epigenome editing. In addition, because various high-fidelity variants of Cas9 have been recently developed [137], the use of those variants is a reasonable strategy to decrease off-target binding. In fact, a high-fidelity version of dCas9 (dHFCas9) fused to TET3 exhibited stronger on-target effects and weaker off-target effects [65]. Moreover, because the full-length proteins retain functional domains that interact with their endogenous binding partners, potentially resulting in mislocalization of the engineered fusion molecules, it would be preferable to utilize only the catalytic domains of epigenetic modifiers for epigenome editing. Hofacker et al. introduced mutations in the effector domains of fused DNMT3A, which reduces its DNA-binding ability [50]. Such mutants exhibited a drastic reduction in off-target methylation with only a slight loss in on-target modification. Other recent approaches (see the section, Advanced Epigenome Editing Tools) may also be used to manage off-target effects.

Long-term Effects of Locus-specific Epigenome Editing

A high level of CpG methylation generally occurs between 5 and 15 days after expression of dCas9-DNMT3A, although this modification is unstable and diminishes within 20–40 days due to loss of the expression construct from cells [31]. Therefore, positive feedback is required for inheritance of the induced epigenetic modifications [28]. In this regard, it is surprising that transient demethylation of CpG sites on a target promoter can induce stable gene expression for more than 80 days in CHO cells [35]. To promote long-term effects from locus-specific epigenome editing, the combinatorial use of DNMT3A, DNMT3L, and KRAB, which can recruit regulators of suppressive histone modifications, enables inheritable locus-specific epigenetic modification and transcriptional silencing of the *B2M* promoter (more than 60 days) [51]. Combinatorial use of DNMT3A, DNMT3L, and EZH2 produced similar results [33]. Therefore, combinatorial use of epigenetic modifiers for histone and DNA modification can be used to promote inheritable locus-specific epigenetic editing. When fused to engineered DNA-binding molecules, epigenetic modifiers had a stronger influence on transcriptional output than activators/repressors such as VP64 or KRAB [68,77], but they also exhibited a weaker or negligible influence in some cases [64,123]. This variability in influence may be dependent on the target positions or cell type. To induce dual effects, Li et al. used CRISPR/dCas9 systems to recruit p300 and VP64 (enhancer activation) or LSD1 and KRAB (enhancer repression) [69].

Other Factors Affecting Locus-specific Epigenome Editing

The efficiency of epigenome editing depends on various factors, including the target region, number of target sites per locus, identities of the epigenetic modifiers and engineered DNA-binding molecules, cell type, time course, and turnover of the expressed epigenome editing tools. For example, the desired epigenetic modifications cannot be obtained when the target sites are not appropriate for fused epigenetic modifiers; for example, locus-specific targeting of distal enhancers, but not proximal promoter regions, by dCas9-LSD1 induces desirable outputs [68], suggesting that the specific combination of fused epigenetic modifiers and their target regions is a critical issue. Targeting an edge, but not the body, of a H3K27ac peak, dCas9 fused to HDAC3 significantly reduced the H3K27ac levels [86]. Interestingly, the decrease in H3K27ac levels on the *Mecp2* gene promoter downregulated its transcription in N2a cells but upregulated its transcription in MC3T3-E1 cells [86], suggesting that epigenome editing can produce cell type-specific effects. Chromatin environments may be an important variable because nucleosomes can block recruitment of CRISPR/Cas9 complexes [138]. In addition, recruitment of CRISPR/Cas9 complexes can alter the native chromatin organization to favor an open chromatin state [139], which may stimulate binding of DNA-binding molecules around CRISPR/Cas9-binding sites.

In some cases, only a single sgRNA, as opposed to multiple sgRNAs, is sufficient to acquire desirable outputs of locus-specific epigenome editing. Whether single or multiple sgRNAs are required is most likely dependent on the chromatin environment and the target position of the gRNA(s) within the target locus, i.e., this requirement is highly context-dependent.

Advanced Epigenome Editing Tools

Site-specific CpG Methylation

As mentioned above, it is important to manage off-target effects caused by locus-specific epigenome editing. Because an epigenetic enzyme fused to an engineered DNA-binding molecule retains its activity even when not bound to a target locus, excessive expression may cause undesired effects. To manage such off-target effects, Chaikind et al. linked the N-terminal and C-terminal parts of M.HhaI, a CpG MTase, to two different ZF proteins [26]. In this system, M.HhaI exhibits CpG methylation activity only when its functional form is reconstituted on the juxtaposed target sites of the ZF proteins (Fig. 42.3A). In *Escherichia coli*, this system successfully induced methylation at the target region, with negligible methylation at non-target sites.

To dissect the site-specific effects of epigenome editing, tools with a narrower action range (e.g., capable of introduction of methylation solely at a single target CpG site) are required. Chaikind et al. also linked the N-terminal and C-terminal parts of M.SssI to two different ZF proteins [24]. As with M.HhaI [26], when the functional form of M.SssI was reconstituted, the enzyme could introduce CpG methylation at single CpG sites. By modifying the enzymatic activity, those authors succeeded in inducing methylation of a single CpG target site in *E. coli* [24] (Fig. 42.3A). Moreover, the same group fused dCas9 to the N-terminal half of M.SssI and co-expressed it with the C-terminal half and sgRNAs [25]. This split-expression system resulted in the ability to confine CpG methylation to regions adjacent to target PAM sites in *E. coli* and 293T cells, and drastically reduced off-target effects compared with systems that fused full-length M.SssI. dCas9-MQ1^{Q147L} also exhibited a relatively confined methylation pattern with a width of about 30–50 bp without considerable off-target CpG methylation [23]. Hence, M.SssI, which functions as a monomeric enzyme, may be more suitable for site-specific CpG methylation, enabling dissection of the effects of CpG methylation with single-base resolution. By contrast, DNMT3A fused to an engineered DNA-binding molecule recruits endogenous DNMT3A/DNMT3L to target sites to form an enzyme complex [90], which may expand CpG methylation to include regions other than target sites through sequential recruitment of DNMT3A/DNMT3L.

Tethering of Non-coding RNAs

Non-coding RNAs (ncRNAs) play various roles in fundamental processes in biology, including transcription, genome imprinting, chromatin remodeling, and epigenetics, in a locus-specific manner [140,141]. Therefore, artificial locus-specific tethering of ncRNAs would be useful in a wide range of biological research, such as induction of desirable biological outputs derived from ncRNA tethering or examination of the molecular mechanisms controlled by tethered ncRNAs. The CRISPR/dCas9-mediated scaffold RNA system is suitable for this purpose [92–95]. Moreover, Shechner et al. expanded the utility of this CRISPR/dCas9-mediated scaffold RNA system (named CRISPR-Display) and succeeded in tethering natural long ncRNAs (lncRNAs) up to 4.8 kb long [95] (Fig. 42.3B). Specifically, they linked various human natural ncRNA domains, such as the repressive NoRC-binding pRNA stem-loop, *Xist* A-repeat (RepA), enhancer-transcribed RNAs (eRNAs), and the 4799 nt lncRNA *HOTTIP*, to sgRNAs and evaluated their effects in reporter gene assays. Thus, CRISPR-Display is potentially useful for functional analysis of ncRNAs, including lncRNAs. In addition, it will be interesting to apply these systems to locus-specific epigenome editing using ncRNAs [140,141] (Fig. 42.3B).

Tethering Molecules on sgRNAs

The *Casilio* system [85], which is a tethering method similar to the MS2 system (Fig. 42.2, left panels), consists of dCas9, an sgRNA appended with one or more Pumilio/FBF (PUF)-binding site(s) (sgRNA-PBS),

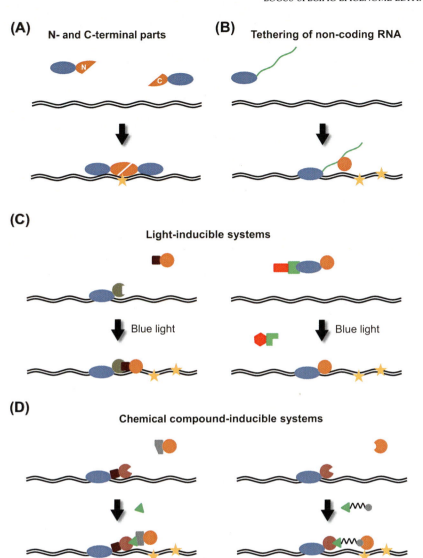

FIGURE 42.3 Advanced methods for locus-specific epigenome editing. (A) N-terminal and C-terminal parts of MTase (orange) are linked to two different ZF proteins (blue). Locus-specific epigenome editing (yellow star) occurs when the enzyme is reconstituted at the target locus. (B) CRISPR/dCas9 (blue)-mediated localization of specific RNAs, such as lncRNAs, at the target locus. The localized RNA (green), which is fused to a gRNA, may introduce a desirable output (e.g., epigenome editing, yellow star) through recruitment of epigenetic regulators (e.g., epigenetic writer, orange circle). (C) Light-inducible epigenome editing. (Left panel) A TAL protein (blue) fused with the light-sensitive CRY2 protein (dark yellow) can interact with the CIB1 protein (dark brown), which is fused to a histone modifier (orange), after exposure to blue light. (Right panel) Alternatively, the LOV2 photosensor domain (red), which is appropriately folded in the absence of light, is fused to a phage-derived anti-CRISPR protein (green). This fusion protein suppresses activity of the CRISPR/dCas9 system. After exposure to blue light, the LOV2 photosensor domain, together with ArcII4, becomes misfolded, leading to release of suppression and activation of the CRISPR/dCas9 system. (D) Chemical compound-inducible epigenome editing. (Left panel) The MS2 protein (dark brown) fused to FKBP (brown) was recruited to dCas9/gRNA_MS2_loop (blue). In the presence of rapamycin (green), the FRB protein (gray) fused to an epigenetic modifier (orange) binds to the CRISPR/dCas9 complex. (Right panel) Alternatively, FK506 (green) is linked to a specific chemical molecule (gray) that binds to an endogenous epigenetic reader (orange). In the presence of modified FK506, an endogenous epigenetic reader is recruited to a dCas9-FKBP target site (blue and brown).

which interact with a PUF RNA-binding domain [142] and an effector fused with the PUF domain (a PUF fusion molecule). This system successfully recruited multiple molecules of the HAT domain of CREB-binding protein (CBP) to the *OCT4* promoter/enhancer and activated gene expression. Recently, the *Casilio* system was combined with TET1 and proteins enhancing conversion of 5-mC to cytosine (GADD45A and NEIL2) to induce CpG demethylation of target genes (the *Casilio-ME* system) [63]. The *Casilio-ME* system can be customized into an inducible system, and results are superior to the TAL-TET1 system, dCas9/sgRNA_MS2-mediated TET1-tethering system, and the dCas9-SunTag-mediated TET1-tethering system.

Light- or Chemical Compound-mediated Inducible Systems

Light-inducible epigenome editing systems have also been developed. Transcriptional regulators or histone modifiers were fused to the cryptochrome-interacting basic-helix-loop-helix 1 (CIB1) protein while TAL proteins were linked with the light-sensitive cryptochrome 2 (CRY2) protein, an interacting partner of CIB1 [143]. Linking with CIB1, various histone modifiers showed light-inducible histone modifications on a TAL-targeted gene (Fig. 42.3C, left panel). Interestingly, applying AcrIIA4, a phage-derived anti-CRISPR protein inhibiting CRISPR/Cas9 activities [144], Bubeck et al. developed another optogenetic control of the CRISPR/Cas9 system [80]. In this system, ArcIIA4 was fused to the LOV2 photosensor domain, which is misfolded under light, so that activity of the CRISPR/Cas9 system can be optogenetically controlled (Fig. 42.3C, right panel). They demonstrated that this system, which uses dCas9-p300, upregulates transcription of the target gene *IL1RN* in 293 T cells in the absence of light [80]. Liu et al. also utilized ArcIIA4 to evaluate target epigenome editing with dCas9-TET1 [61].

Inducible locus-specific epigenome editing can also be achieved with chemical compounds. For example, the MS2 protein was fused with FK506-binding protein (FKBP), which dimerizes with FRB (FKBP-rapamycin-binding domain of mTOR) in the presence of rapamycin. FRB was fused to HP1 and SS18 to recruit the H3K9 methyltransferase SUV39H1 or the mSWI/SNF (BAF) chromatin regulatory complex (Fig. 42.3D, left panel). Co-expression of dCas9/sgRNA-MS2-loop, MS2-FKBP, and FRB-HP1 or FRB-SS18 induced desirable epigenetic outputs in the presence of rapamycin, which were followed by modification of target gene expression (Oct4 or Nkx2.9) in mouse ES cells [74]. Induced epigenome editing could be reversed within 24 hours by adding the FKBP-binding molecule FK506, a competitive inhibitor of rapamycin. Chiarella et al. used FK506 fused to specific chemical compounds that bind to endogenous epigenetic readers, such as bromodomain-containing proteins (Fig. 42.3D, right panel) [145]. An endogenous epigenetic reader can be recruited to a dCas9-FKBP/sgRNA target site in the presence of modified FK506. dCas9/gRNA-MS2-loop and MS2-FKBP can also be used instead of dCas9-FKBP/sgRNA. ABA-mediated linking between dCas9-PYL and ABI-p300 proteins was used to induce recruitment of p300 to the IL1RN promoter for enrichment of H3K27ac, which was followed by induction of gene expression [81]. The auxin-induced degradation system was utilized to control locus-specific epigenome editing [82]. Auxin-induced degradation of dCas9-p300 was observed in mammalian cells [82].

Locus-specific Chromatin Reorganization

Distal enhancers can interact with target gene promoters via chromatin looping to activate their transcription. In erythroid cells, the locus control region (LCR) controls transcription of β-globin genes in a developmental stage-specific manner. Chromatin looping between LCR and the target globin gene is mediated by Ldb1 proteins, which form homodimers. Deng et al. applied a ZF protein fused with Ldb1 to form chromatin loops between LCR and the targeted β-globin genes in mouse erythroid cells [146,147]. The forced formation of chromatin loops resulted in recruitment of RNA polymerase II and activation of gene transcription. Therefore, induction of chromatin loops by artificial DNA-binding molecules represents a novel approach to controlling gene transcription in a locus-specific manner.

Morgan et al. developed a method called chromatin loop reorganization using CRISPR-dCas9 (CLOuD9) to selectively and reversibly establish de novo chromatin loops [148]. The authors used abscisic acid (ABA)-mediated linkage between the pyrabactin resistance 1 (PYR1)/PYR1-like (PYL) and ABA insensitive (ABI) proteins [149]. They successfully dimerized Streptococcus pyogenes dCas9-PYL and Staphylococcus aureus dCas9-ABI in the presence of ABA. Therefore, each dCas9 fusion protein/sgRNA complex at target genomic regions can be brought into close enough proximity to make chromatin loops (Fig. 42.4A). CLOuD9 induced reversible looping between the β-globin promoter and a LCR, which activated gene expression through enrichment of H3K4me3. Wang et al. used dCas9-ABI and a nuclear membrane protein fused to PYL in the presence of ABA to artificially localize a CRISPR/dCas9-tagged locus to the nuclear membrane (Fig. 42.4B) [150]. Kim et al. took advantage of the heterodimerization ability of the N-terminal domain of CIB1 (CIBN) and CRY2 under blue light illumination to generate chromatin looping between CRISPR/dCas-CIBN-tagged loci in ES cells expressing CRY2 (Fig. 42.4C) [151].

Other Applications of Epigenome Editing

There are additional applications of artificial epigenome editing, other than the modification of epigenetic markers discussed above. Katayama et al. attempted to replace CpG-methylated promoter regions of OLIG and NANOG with unmethylated donor DNA by microhomology arm-mediated homologous recombination using CRISPR/Cas9 in 293T cells [152]. The desired effects of replacement (e.g., erasure of DNA methylation and transcriptional upregulation) were observed and maintained for ~21 days after induction of homologous recombination. Shrimp et al. utilized dCas9-p300-mediated transcriptional activation to screen for inhibitors of p300 [83]. Kuscu et al. induced H3K27ac on non-regulatory genomic regions using dCas9-p300, which reprogramed the chromatin state of these regions into enhancer-like elements [82]. Broche et al. used the promiscuous off-target binding activity of ZF-DNMT3A to investigate the global dynamics of randomly introduced DNA methylation and the subsequent changes to the epigenome network [52].

In most applications of artificial epigenome editing, engineered DNA-binding molecules such as CRISPR/dCas9 have been expressed from transgenes in target cells. Liszczak et al. used recombinant dCas9 fused with a pharmacologic agent to recruit epigenetic proteins to a target locus [153]. They utilized the split intein-mediated protein trans-splicing (PTS) system [154] to site-specifically link a recombinant dCas9 to epigenetic probes JQ1 (a small molecule) and UNC3866 (a modified peptide). Because JQ1 and UNC3866 bind to BET family proteins and the PRC1 complex, respectively, the authors succeeded in recruiting those proteins to the target sites of recombinant dCas9-JQ1/gRNA and dCas9-UNC3866/gRNA. Their results indicate that the combination of the PTS system with transduction of resultant

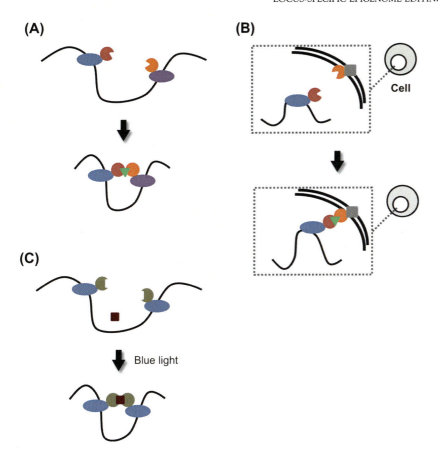

FIGURE 42.4 **Locus-specific chromatin reorganization.** (A) The CLOuD9 system. *Streptococcus pyogenes* dCas9-PYL (blue and brown) and *Staphylococcus aureus* dCas9-ABI (purple and orange) interact with each other in the presence of ABA (green) so that targeted loci come into proximity. (B) dCas9-ABI (blue and brown) binds to a PYL-fused protein (orange and gray) in nuclear compartments, such as a nuclear membrane, in the presence of ABA (green), which can artificially recruit a targeted locus to a specific nuclear compartment. (C) dCas9-CIBN (blue and dark yellow) forms heterodimers with the CRY2 protein (dark brown) after illumination with blue light, which can be used to induce chromosomal looping.

recombinant dCas9/gRNA molecules represents a useful approach for locus-specific epigenome editing.

Epigenetic networks are regulated by a circuit of epigenetic writers and readers. Park et al. developed an epigenetic regulatory system using engineered epigenetic writers and readers for N^6-methyladenine (m^6A), a DNA modification rarely found in metazoan genomes [53]. In this circuit system, the following three components were employed: (I) ZF or CRISPR/dCas9 fused to Dam (*E. coli* DNA adenine methyltransferase), which catalyzes methylation of adenines in GATC motifs; (II) Dam fused to an m^6A reader domain (RD, the binding domain of *Dpn* I); and (III) artificial DNA containing multiple copies of GATC. After induction of m^6A by ZF or CRISPR/dCas9 fused to Dam, m^6A was expanded on the artificial DNA by RD-Dam. Using this system, the authors examined how an epigenetic mark, such as m^6A, propagates and is maintained through multiple rounds of cell division in human cells. The authors suggest that this circuit system is a useful toolkit for investigating models of epigenetic regulation.

Epitranscriptome Editing

Recently, the CRISPR/Cas13a system was used to target and cleave RNA (RNA editing) [155,156].

In addition to RNA editing, the nuclease-dead form of Cas13a (dCas13a) can be used to tag and localize RNA in cells [157]. The CRISPR/Cas9 system can also recognize an RNA sequence in the presence of PAM-presenting oligonucleotides (PAMer), which are short single-stranded DNA oligos that target a PAM site [158,159]. RNA contains posttranscriptional chemical modifications, called RNA epigenetics, which are important for gene expression [160]. Liu et al. fused dCas9 to m^6A methyltransferases (METTL3 and METTL14) or demethylases (ALKBH5 or FTO) [87]. In the presence of PAMers, they succeeded in artificially modifying m^6A on target RNAs (*Hspa1a*, *Actb*, and *Malat1*) in mammalian cells. This type of RNA targeting system makes it possible to artificially edit and analyze the epitranscriptome, and to produce desirable biological outputs.

Characteristics of Engineered DNA-binding Molecules in Locus-specific Epigenome Editing

Among ZF proteins, TAL proteins, and CRISPR/Cas, the third class of molecules may be easiest to handle in the context of locus-specific epigenome editing. For example, to construct expression vectors for CRISPR/Cas systems, it is only necessary to clone synthesized DNA sequences corresponding to the target site into gRNA expression vectors. By contrast, although TAL or ZF protein expression vectors can be constructed by established systems [15,16,161–163], the process takes longer time and more effort. To target multiple loci, dCas9 fused with an epigenetic modifier can be used with multiple gRNAs in the CRISPR/dCas9 system. By contrast, for ZF or TAL proteins, an epigenetic modifier has to be fused to each DNA-binding protein. In addition, in the CRISPR/dCas9 system, epigenetic regulators can be linked not only to dCas9, but also to gRNA.

On the other hand, for successful locus targeting, CRISPR/Cas needs a PAM sequence adjacent to the target sequences. Although the specificity of PAM recognition by CRISPR/Cas can be altered intentionally [164], the requirement for a PAM sequence represents a limitation at this stage. Because ZF and TAL proteins do not have such strict sequence limitations, their target sites can be selected more flexibly.

Another potential problem is the size of the open reading frame of *S. pyogenes* Cas9, which might be too large to be efficiently packaged into routinely used delivery vehicles such as adeno-associated viruses (AAVs). To overcome this obstacle, several groups have investigated the use of smaller Cas9 orthologues (e.g., Cas9 from *S. aureus*) or the CRISPR/Cpf1 (recently renamed Cas12) system [165–168]. In this context, dCpf1-p300 was tested for its ability to promote locus-specific epigenome editing [84]. More recently, the CRISPR/CasΦ system was reported [169]. Because the molecular weight of the CasΦ protein is approximately half that of the Cas9 protein, the CRISPR/CasΦ system is attractive from the viewpoint of reduction in size.

TAL proteins are sensitive to CpG methylation in their binding sites [170,171]. Although some TAL DNA-binding modules can recognize 5-mC [170,171], this property might still prevent TAL proteins from efficiently targeting CpG-methylated loci for epigenome editing. By contrast, CRISPR/Cas and ZF proteins are unlikely to be affected by CpG methylation [134,172,173]. Therefore, it may be unnecessary to consider the direct effects of CpG methylation of target sites when CRISPR/Cas or ZF proteins are used for epigenome editing on CpG-methylated target loci, or on target loci where levels of CpG methylation remain unclear. However, in the context of allele-specific targeting of a locus in which one allele is allele-specifically CpG-methylated (e.g., genome imprinting), the sensitivity of TAL proteins to CpG methylation could be utilized for preferential targeting of the non-CpG-methylated allele. In addition, the sensitivity of TAL proteins to CpG methylation was used to develop an assay to detect CpG methylation in a test tube [174].

ZF modules can be artificially designed to distinguish CpG methylation or m^6A at target sites [175,176], which can also be useful for preferential targeting of modified alleles.

For in vivo delivery of these engineered DNA binding molecules, adeno-associated virus could be widely used, as mentioned above. In this regard, recent efforts led to generation of serotypes capable of crossing the blood-brain barrier, enabling delivery of AAVs to the central or peripheral nervous systems, or visceral organs, through intravenous administration [177]. Therefore, body-wide gene transfer could be achieved via intravenous or tissue injection of the AAV delivery system.

Herein, we have presented multiple examples of locus-specific epigenome editing. We anticipate that use of these tools will increase as new systems and their respective applications are developed.

LOCUS-SPECIFIC IDENTIFICATION OF EPIGENETIC MOLECULES THAT INTERACT WITH TARGET GENOMIC REGIONS

To elucidate the molecular mechanisms underlying epigenetic regulation, it is important to identify epigenetic writers/erasers as well as related regulators. The major enzymatic components involved in epigenetic modification (e.g., DNMTs, TETs, and HDACs) have been identified. However, because epigenetic mechanisms are regulated locus-specifically, it is also important to identify epigenetic molecules in a locus-specific manner. To this end, we developed a novel biochemical technology, engineered DNA-binding molecule-mediated ChIP (enChIP), which uses engineered DNA-binding proteins, such as TAL proteins and CRISPR/Cas, to target specific genomic loci. In combination with downstream biochemical analyses, this technology can be used for locus-specific identification of epigenetic regulators. In this section, we describe enChIP and its applications to the elucidation of locus-specific epigenetic mechanisms.

Development of Locus-specific ChIP: iChIP and enChIP

To identify molecules that interact with genomic regions of interest, our group has been developing locus-specific ChIP technologies, which are biochemical techniques for isolating target genomic regions in a locus-specific manner while retaining their molecular interactions [178]. In locus-specific ChIP, a target locus is tagged with DNA-binding molecules (and their recognition sequences) and isolated from cells by affinity purification. After isolation, the molecules interacting with the isolated genomic regions can be identified in a non-biased manner by downstream analyses such as MS, microarray, and NGS. Thus, locus-specific ChIP enables biochemical elucidation of the molecular mechanisms of genome functions, including epigenetic regulation.

The first form of locus-specific ChIP that we developed was insertional ChIP (iChIP) [178–180]. In iChIP, a locus is tagged with exogenous DNA-binding molecules, such as the bacterial DNA-binding protein LexA and its binding DNA elements, and the tagged locus is then affinity-purified. Using iChIP in combination with MS or RT-PCR, Our research group identified various proteins and RNA that interact with the insulator, *Pax5* promoter, and immunoglobulin (Ig) switch regions in vertebrate cells [181–184]. Bui et al. identified factors assembled on the silenced provirus, such as components of inhibitors of the histone acetyltransferase (INHAT), using iChIP [185]. In addition, McCullagh et al. analyzed chromosomal interactions in *Saccharomyces cerevisiae* using iChIP combined with microarrays [186]. In general, locus-tagging in iChIP requires insertion of exogenous DNA sequences around a target locus. Although the advent of genome editing tools made such gene targeting much easier, this step remains time-consuming. Therefore, to allow researchers to skip the insertion step, we subsequently developed another form of locus-specific ChIP, enChIP, in which engineered DNA-binding molecules such as TAL and CRISPR/Cas are employed for locus-tagging [12,187,188]. The general scheme of enChIP is as follows (details are shown in Fig. 42.5): (1) design of engineered DNA-binding molecules, (2) expression of the designed molecule for locus-tagging, (3) preparation of fragmented chromatin, and (4) affinity purification to isolate the tagged locus. Proteins, DNAs, or RNAs that interact with the isolated locus can be purified and identified in a non-biased manner by downstream analyses (e.g., MS, NGS, or microarray). In the next section, we describe the practical applications of enChIP.

Identification of Epigenetic and Related Molecules in a Locus-specific Manner by enChIP

We first used enChIP with CRISPR/Cas to identify the proteins that interact with the promoter region of *interferon regulatory factor 1* (*IRF-1*) in human cell lines [187,189]. Because transcription of *IRF-1* is upregulated by IFNγ stimulation, we identified the proteins whose interaction with the promoter region was increased in response to IFNγ stimulation using enChIP, followed by stable isotope labeling using amino acids in cell culture (SILAC) MS (enChIP-SILAC). The identified proteins included transcriptional regulators, proteins involved in histone deacetylation and/or corepressor functions, and histone variants [189]. Our research group also succeeded in identifying the *EPAS1* promoter-associated proteins, including histone variants, using enChIP-SILAC [190]. The identification of epigenetic regulators and specific histone variants was reasonable because physiological alteration of nucleosome occupancy or chromosome positioning is linked to changes in promoter activity. We believe that the enChIP using the CRISPR/Cas system is a powerful and useful method because other groups have used enChIP combined with MS (enChIP-Ms) or other modified systems to identify a variety of proteins related to transcription of the *globin* genes [191], the *DUX4* retrogene [192], the *WNT5A* gene [193], the *Sox9* gene [194], the *Ucp1* gene [195], the *MACC1* gene [196], and erythroid enhancers [191] in mammalian cells. In addition, enChIP-MS was used to identify proteins associated with the *var* gene family in *Plasmodium falciparum* [197]. González-Rico et al. identified a histone methyltransferase and chromatin assembly factors as important proteins for chromatin loop formation at the *Nanog* locus using enChIP-MS [198]. Morgan et al. identified the RNA helicases DDX5 and DDX17 as proteins important for de novo chromatin loop formation [148].

We also used enChIP with a TAL protein to identify proteins that interact with telomere regions in a mouse hematopoietic cell line [188]. In particular, in that study we used enChIP-MS to identify known and novel telomere-binding proteins. The identified proteins may be involved in maintenance of characteristic properties of chromatin structures in telomere regions [199]. After we published our first paper describing the enChIP system, another group also reported a locus-specific ChIP method that used a TAL protein to identify proteins and histone modifications on a galactose-inducible promoter in budding yeast [200]. enChIP using a TAL protein was also used to identify proteins associated with the *OCT4* enhancer in human ES cells [201]. Taken together, enChIP is useful for elucidating the molecular components involved in epigenetic regulation.

We also used enChIP in combination with RT-PCR (enChIP-RT-PCR) or RNA-sequencing (enChIP-RNA-Seq) to directly identify chromatin-associated RNAs [188,202], and succeeded in non-biased identification of ncRNAs that interact with telomere regions. Because ncRNAs are involved in maintenance of chromatin structures [141], the identified ncRNAs might function similarly in telomere regions. Other groups succeeded

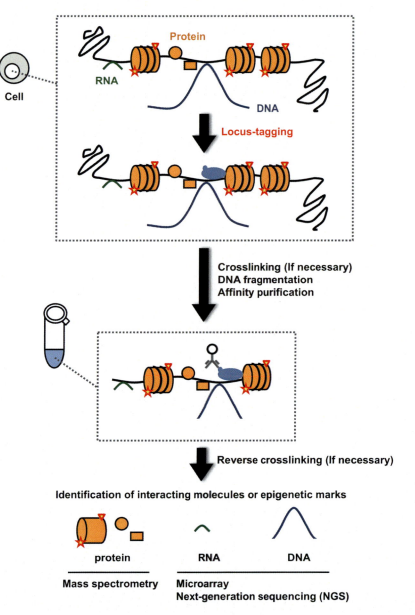

FIGURE 42.5 Scheme of enChIP. First, an engineered DNA-binding molecule (blue circle) such as a TAL protein or CRISPR/Cas, which consists of dCas9 plus gRNA (or sgRNA), is expressed in cells for the purpose of locus-tagging. Second, molecular interactions are fixed by crosslinking (if necessary), and chromatin DNA is fragmented by sonication or enzymatic digestion. Third, the tagged locus is affinity-purified. After reverse crosslinking (if necessary), molecules interacting with the locus can be identified by MS, microarray, NGS, or other methods. ZF proteins are also suitable for use in enChIP.

in identifying an oncogenic intronic miRNA and a circular RNA as transcriptional regulators that interact with the target promoters *IGF2* and *FLI1*, respectively, in cancer cells [203,204]. Thus, these results demonstrated the utility of enChIP as a non-biased research tool for identifying proteins as well as multiple types of RNAs bound to target loci in cells.

It would also be possible to use enChIP in combination with NGS (enChIP-Seq) to identify chromosomal interactions. Indeed, we used enChIP-Seq to identify genomic regions that interact with insulator regions in the human leukemia cell line K562 [205]. The same approach was successfully used to identify a functional distant enhancer of the *Sox9* and *Ccl5* genes in mouse cells [194,206]. In addition, enChIP-PCR can be used to confirm interactions between genomic regions [82].

In vitro enChIP

As discussed above, enChIP is a useful technique for epigenetic research. However, expression of locus-tagging molecules in cells may still be time-consuming in some cases (e.g., when using primary cells or tissues isolated from animals). Accordingly, we developed another enChIP method in which locus-tagging by engineered DNA-binding molecules in cells is dispensable [182,207].

In this method, in vitro enChIP, locus-tagging is performed after fragmentation of chromatin DNA, while the other steps are basically the same as in conventional enChIP (Fig. 42.6). First, an engineered DNA-binding molecule is prepared as a recombinant or synthetic molecule, bearing a tag or tags if necessary. Second, the cells to be analyzed are crosslinked with formaldehyde or other crosslinkers (if necessary) and lysed, and the chromatin DNA is fragmented by sonication or other methods. In a test tube, the prepared recombinant molecule is incubated with the fragmented chromatin DNA to tag the target locus. The tagged locus is then affinity-purified using an antibody against the tag or the recombinant molecule itself.

We first attempted in vitro enChIP using a recombinant TAL protein [182]. However, the yields of target loci were low, probably due to degradation of the recombinant protein. In subsequent efforts, we employed recombinant CRISPR/Cas ribonucleoproteins (RNPs) [207]. In that case, the recombinant 3xFLAG-dCas9 protein was not detectably degraded despite its large molecular weight (~160 kDa), and in vitro enChIP returned high yields of the target locus (~1% of input). Notably, the biotin–avidin purification system can be used for in vitro enChIP; specifically, the locus targeted by biotinylated gRNA plus recombinant 3xFLAG-dCas9 is affinity-purified by streptavidin-coated carriers [207]. In vitro enChIP should be applicable to identification of molecules that interact with genomic regions of interest, with the goal of elucidating the molecular mechanisms underlying genome functions. In fact, we succeeded in identifying novel chromosomal interactions that regulate *Pax5* expression in the chicken cell line DT40 [208]. In addition, Tsui et al. identified potential regulators of the histone gene cluster in *Drosophila* cells [209].

Taken together, these results indicate that in-cell enChIP and in vitro enChIP are useful tools for identifying genome-interacting molecules and epigenetic

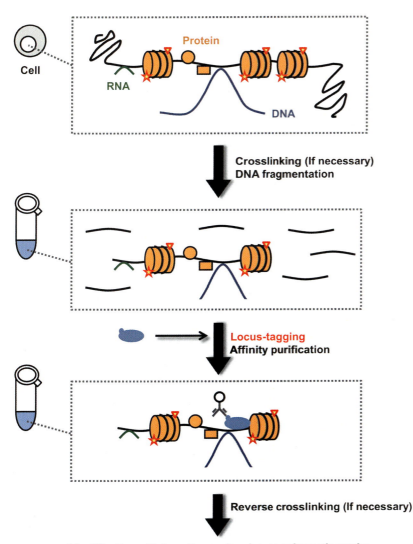

FIGURE 42.6 Scheme of in vitro enChIP. First, recombinant engineered DNA-binding molecules (blue circle) are prepared; for example, dCas9 fused with a tag(s) is purified as recombinant protein, and gRNA (or sgRNA) is chemically synthesized or prepared by in vitro transcription. Alternatively, a recombinant TAL protein is prepared. Second, fragmented chromatin DNA is prepared from intact cells; molecular interactions are fixed by crosslinking (if necessary), and chromatin DNA is fragmented by sonication or enzymatic digestion. Third, the recombinant engineered DNA-binding molecule is incubated with the fragmented chromatin DNA in a test tube, and the tagged locus is affinity-purified. ZF proteins are also suitable for use in in vitro enChIP.

marks in a locus-specific manner, as well as for dissecting the molecular mechanisms underlying genome functions, including epigenetic regulation. enChIP could also be used to identify epigenetic regulators as candidate molecules for development of clinical drugs. Although the 3xFLAG-tag has been widely used for enChIP, various additional epitope tags are compatible with CRISPR and TAL molecules [187,191,210,211]. AM-tag, a peptide tag from Active Motif, Inc., is compatible with enChIP (http://www.activemotif.jp/catalog/1172/enchip), and enChIP kits using this tag are available from the vendor. It is of note that we have a patent on enChIP [212], which covers any affinity purification schemes including engineered DNA-binding molecules, tags fused with engineered DNA-binding molecules such as dCas9 and gRNA, and IP with antibodies against the tags or engineered DNA-binding molecules such as dCas9 themselves, or other forms of affinity purification.

Other Methods

Recently, two groups independently developed a biochemical method using dCas9 fused to APEX2 (named C-BERST or GLoPro) to identify proteins interacting with a target genomic locus [213,214]. In the presence of hydrogen peroxide (H_2O_2), APEX2 oxidizes the phenol moiety of biotin–phenol (BP) compounds to phenoxyl radicals, which react with surface-exposed tyrosine residues in nearby proteins; this results in covalent biotinylation of those proteins. With this method, a locus is targeted by the CRISPR/dCas9–APEX2 complex, and proteins bound near the locus are biotinylated in the presence of BP and H_2O_2. Biotinylated proteins can be isolated using a biotin-streptavidin system and identified by MS analysis. These systems were used to identify proteins interacting with repetitive loci (telomeres and α-satellites) in U2OS cells and single-copy loci (*TERT* or *MYC*) in 293 T cells. In addition, Escobar et al. fused the *E. coli* biotin ligase (BirA) to dCas9 to biotinylate histones at a target locus [215]. They used this biotinylating system to track parental nucleosome segregation at target loci. Hence, the CRISPR/dCas9-based biotinylating systems are useful for identifying proteins and analyzing their epigenetic properties at target loci. The CRISPR/dCas13-based biotinylating system can also be used for identification and analysis of epigenetic proteins at target RNAs [216]. One potential drawback to these systems might be the presence of background biotinylation of fusion proteins not bound to target loci. Modifying this system to allow for selective induction of biotinylation only when these enzymes bind to their target loci might solve this problem.

CONCLUSIONS

In this chapter, we described applications of engineered DNA-binding molecules to epigenetic research. These molecules can be used for locus-specific epigenome editing and subsequent induction of desired biological output. Furthermore, locus-specific epigenome editing can be used in combination with multiple types of epigenetic modifiers. In this context, CRISPR/Cas is the optimal choice due to the convenience of its molecular design and its ability to perform simultaneous editing of multiple loci. In addition to the information provided in this chapter, interested readers may also benefit from recent reviews describing locus-specific epigenome editing [217–221]. Moreover, locus-specific ChIP is a powerful tool for isolating genomic regions of interest and biochemically identifying their interacting molecules, including epigenetic regulators. Our recent review articles are also beneficial for readers interested in locus-specific ChIP [222,223]. These research tools will facilitate elucidation of the molecular mechanisms underlying epigenetic regulation, and in the future may be applicable to clinical medicine.

References

[1] Collas P. The current state of chromatin immunoprecipitation. Mol Biotechnol 2010;45(1):87–100.
[2] Zilberman D, Henikoff S. Genome-wide analysis of DNA methylation patterns. Development 2007;134(22):3959–65.
[3] Gaj T, Gersbach CA, Barbas 3rd. CF. ZFN, TALEN, and CRISPR/Cas-based methods for genome engineering. Trends Biotechnol 2013;31(7):397–405.
[4] Wijshake T, Baker DJ, van de Sluis B. Endonucleases: new tools to edit the mouse genome. Biochim Biophys Acta 2014;1842(10):1942–50.
[5] Pauwels K, Podevin N, Breyer D, Carroll D, Herman P. Engineering nucleases for gene targeting: safety and regulatory considerations. N Biotechnol 2014;31(1):18–27.
[6] Sun N, Zhao H. Transcription activator-like effector nucleases (TALENs): a highly efficient and versatile tool for genome editing. Biotechnol Bioeng 2013;110(7):1811–21.
[7] Mali P, Esvelt KM, Church GM. Cas9 as a versatile tool for engineering biology. Nat Methods 2013;10(10):957–63.
[8] Doudna JA, Charpentier E. Genome editing. The new frontier of genome engineering with CRISPR-Cas9. Science 2014;346:1258096.
[9] Hsu PD, Lander ES, Zhang F. Development and applications of CRISPR-Cas9 for genome engineering. Cell 2014;157(6):1262–78.
[10] Harrison MM, Jenkins BV, O'Connor-Giles KM, Wildonger J. A CRISPR view of development. Genes Dev 2014;28(17):1859–72.
[11] Fujita T, Fujii H. Applications of engineered DNA-binding molecules such as TAL proteins and the CRISPR/Cas system in biology research. Int J Mol Sci 2015;16(10):23143–64.
[12] Fujii H, Fujita T. Isolation of specific genomic regions and identification of their associated molecules by engineered DNA-binding molecule-mediated chromatin immunoprecipitation (enChIP) using the CRISPR system and TAL proteins. Int J Mol Sci 2015;16(9):21802–12.

REFERENCES

[13] Durai S, Mani M, Kandavelou K, Wu J, Porteus MH, Chandrasegaran S. Zinc finger nucleases: custom-designed molecular scissors for genome engineering of plant and mammalian cells. Nucleic Acids Res 2005;33(18):5978–90.

[14] Carroll D, Morton JJ, Beumer KJ, Segal DJ. Design, construction and in vitro testing of zinc finger nucleases. Nat Protoc 2006;1(3):1329–41.

[15] Cermak T, Doyle EL, Christian M, Wang L, Zhang Y, Schmidt C, et al. Efficient design and assembly of custom TALEN and other TAL effector-based constructs for DNA targeting. Nucleic Acids Res 2011;39(12):e82.

[16] Reyon D, Tsai SQ, Khayter C, Foden JA, Sander JD, Joung JK. FLASH assembly of TALENs for high-throughput genome editing. Nat Biotechnol 2012;30(5):460–5.

[17] Ishino Y, Shinagawa H, Makino K, Amemura M, Nakata A. Nucleotide sequence of the iap gene, responsible for alkaline phosphatase isozyme conversion in *Escherichia coli*, and identification of the gene product. J Bacteriol 1987;169(12):5429–33.

[18] Barrangou R, Fremaux C, Deveau H, Richards M, Boyaval P, Moineau S, et al. CRISPR provides acquired resistance against viruses in prokaryotes. Science 2007;315(5819):1709–12.

[19] Jinek M, Chylinski K, Fonfara I, Hauer M, Doudna JA, Charpentier E. A programmable dual-RNA-guided DNA endonuclease in adaptive bacterial immunity. Science 2012;337(6096):816–21.

[20] Qi LS, Larson MH, Gilbert LA, Doudna JA, Weissman JS, Arkin AP, et al. Repurposing CRISPR as an RNA-guided platform for sequence-specific control of gene expression. Cell 2013;152(5):1173–83.

[21] Xu GL, Bestor TH. Cytosine methylation targetted to predetermined sequences. Nat Genet 1997;17(4):376–8.

[22] Carvin CD, Parr RD, Kladde MP. Site-selective in vivo targeting of cytosine-5 DNA methylation by zinc-finger proteins. Nucleic Acids Res 2003;31(22):6493–501.

[23] Lei Y, Zhang X, Su J, Jeong M, Gundry MC, Huang YH, et al. Targeted DNA methylation in vivo using an engineered dCas9-MQ1 fusion protein. Nat Commun 2017;8:16026.

[24] Chaikind B, Ostermeier M. Directed evolution of improved zinc finger methyltransferases. PLoS One 2014;9(5):e96931.

[25] Xiong T, Meister GE, Workman RE, Kato NC, Spellberg MJ, Turker F, et al. Targeted DNA methylation in human cells using engineered dCas9-methyltransferases. Sci Rep 2017;7(1):6732.

[26] Chaikind B, Kilambi KP, Gray JJ, Ostermeier M. Targeted DNA methylation using an artificially bisected M.HhaI fused to zinc fingers. PLoS One 2012;7(9):e44852.

[27] Li F, Papworth M, Minczuk M, Rohde C, Zhang Y, Ragozin S, et al. Chimeric DNA methyltransferases target DNA methylation to specific DNA sequences and repress expression of target genes. Nucleic Acids Res 2007;35(1):100–12.

[28] Kungulovski G, Nunna S, Thomas M, Zanger UM, Reinhardt R, Jeltsch A. Targeted epigenome editing of an endogenous locus with chromatin modifiers is not stably maintained. Epigenetics Chromatin 2015;8:12.

[29] Siddique AN, Nunna S, Rajavelu A, Zhang Y, Jurkowska RZ, Reinhardt R, et al. Targeted methylation and gene silencing of VEGF-A in human cells by using a designed Dnmt3a-Dnmt3L single-chain fusion protein with increased DNA methylation activity. J Mol Biol 2013;425(3):479–91.

[30] Bernstein DL, Le Lay JE, Ruano EG, Kaestner KH. TALE-mediated epigenetic suppression of CDKN2A increases replication in human fibroblasts. J Clin Invest 2015;125(5):1998–2006.

[31] Vojta A, Dobrinić P, Tadić V, Bočkor L, Korać P, Julg B, et al. Repurposing the CRISPR-Cas9 system for targeted DNA methylation. Nucleic Acids Res 2016;44(12):5615–28.

[32] McDonald JI, Celik H, Rois LE, Fishberger G, Fowler T, Rees R, et al. Reprogrammable CRISPR/Cas9-based system for inducing site-specific DNA methylation. Biol Open 2016;5(6):866–74.

[33] O'Geen H, Bates SL, Carter SS, Nisson KA, Halmai J, Fink KD, et al. Ezh2-dCas9 and KRAB-dCas9 enable engineering of epigenetic memory in a context-dependent manner. Epigenetics Chromatin 2019;12(1):26.

[34] Liu XS, Wu H, Ji X, Stelzer Y, Wu X, Czauderna S, et al. Editing DNA methylation in the mammalian genome. Cell 2016;167(1):233–47 e17.

[35] Marx N, Grünwald-Gruber C, Bydlinski N, Dhiman H, Ngoc Nguyen L, Klanert G, et al. CRISPR-based targeted epigenetic editing enables gene expression modulation of the silenced beta-galactoside alpha-2,6-sialyltransferase 1 in CHO cells. Biotechnol J 2018;13(10):e1700217.

[36] Huang YH, Su J, Lei Y, Brunetti L, Gundry MC, Zhang X, et al. DNA epigenome editing using CRISPR-Cas SunTag-directed DNMT3A. Genome Biol 2017;18(1):176.

[37] Pflueger C, Tan D, Swain T, Nguyen T, Pflueger J, Nefzger C, et al. A modular dCas9-SunTag DNMT3A epigenome editing system overcomes pervasive off-target activity of direct fusion dCas9-DNMT3A constructs. Genome Res 2018;28(8):1193–206.

[38] Rivenbark AG, Stolzenburg S, Beltran AS, Yuan X, Rots MG, Strahl BD, et al. Epigenetic reprogramming of cancer cells via targeted DNA methylation. Epigenetics 2012;7(4):350–60.

[39] Stolzenburg S, Beltran AS, Swift-Scanlan T, Rivenbark AG, Rashwan R, Blancafort P. Stable oncogenic silencing in vivo by programmable and targeted de novo DNA methylation in breast cancer. Oncogene 2015;34(43):5427–35.

[40] Nunna S, Reinhardt R, Ragozin S, Jeltsch A. Targeted methylation of the Epithelial Cell Adhesion Molecule (EpCAM) promoter to silence its expression in ovarian cancer cells. PLoS One 2014;9(1):e87703.

[41] Cui C, Gan Y, Gu L, Wilson J, Liu Z, Zhang B, et al. P16-specific DNA methylation by engineered zinc finger methyltransferase inactivates gene transcription and promotes cancer metastasis. Genome Biol 2015;16:252.

[42] Saunderson EA, Stepper P, Gomm JJ, Hoa L, Morgan A, Allen MD, et al. Hit-and-run epigenetic editing prevents senescence entry in primary breast cells from healthy donors. Nat Commun 2017;8(1):1450.

[43] Li K, Pang J, Cheng H, Liu WP, Di JM, Xiao HJ, et al. Manipulation of prostate cancer metastasis by locus-specific modification of the CRMP4 promoter region using chimeric TALE DNA methyltransferase and demethylase. Oncotarget 2015;6(12):10030–44.

[44] Kantor B, Tagliafierro L, Gu J, Zamora ME, Ilich E, Grenier C, et al. Downregulation of SNCA expression by targeted editing of DNA methylation: a potential strategy for precision therapy in PD. Mol Ther 2018;26(11):2638–49.

[45] Ziller MJ, Ortega JA, Quinlan KA, Santos DP, Gu H, Martin EJ, et al. Dissecting the functional consequences of de novo DNA methylation dynamics in human motor neuron differentiation and physiology. Cell Stem Cell 2018;22(4):559–74 e9.

[46] Tarjan DR, Flavahan WA, Bernstein BE. Epigenome editing strategies for the functional annotation of CTCF insulators. Nat Commun 2019;10(1):4258.

[47] Stepper P, Kungulovski G, Jurkowska RZ, Chandra T, Krueger F, Reinhardt R, et al. Efficient targeted DNA methylation with chimeric dCas9-Dnmt3a-Dnmt3L methyltransferase. Nucleic Acids Res 2017;45(4):1703–13.

[48] Lin L, Liu Y, Xu F, Huang J, Daugaard TF, Petersen TS, et al. Genome-wide determination of on-target and off-target characteristics for RNA-guided DNA methylation by dCas9 methyltransferases. Gigascience 2018;7(3):1–19.

[49] Galonska C, Charlton J, Mattei AL, Donaghey J, Clement K, Gu H, et al. Genome-wide tracking of dCas9-methyltransferase footprints. Nat Commun 2018;9(1):597.

[50] Hofacker D, Broche J, Laistner L, Adam S, Bashtrykov P, Jeltsch A. Engineering of effector domains for targeted DNA methylation with reduced off-target effects. Int J Mol Sci 2020;21(2):502.

[51] Amabile A, Migliara A, Capasso P, Biffi M, Cittaro D, Naldini L, et al. Inheritable silencing of endogenous genes by hit- and-run targeted epigenetic editing. Cell 2016;167(1):219–32 e14.

[52] Broche J, Kungulovski G, Bashtrykov P, Rathert P, Jeltsch A. Genome-wide investigation of the dynamic changes of epigenome modifications after global DNA methylation editing. Nucleic Acids Res 2021;49(1):158–76.

[53] Park M, Patel N, Keung AJ, Khalil AS. Engineering epigenetic regulation using synthetic read-write modules. Cell 2019;176(1–2):227–38 e20.

[54] Gregory DJ, Zhang Y, Kobzik L, Fedulov AV. Specific transcriptional enhancement of inducible nitric oxide synthase by targeted promoter demethylation. Epigenetics 2013;8(11):1205–12.

[55] Chen H, Kazemier HG, De Groote ML, Ruiters MHJ, Xu GL, Rots MG. Induced DNA demethylation by targeting ten-eleven translocation 2 to the human ICAM-1 promoter. Nucleic Acids Res 2014;42(3):1563–74.

[56] Maeder ML, Angstman JF, Richardson ME, Linder SJ, Cascio VM, Tsai SQ, et al. Targeted DNA demethylation and activation of endogenous genes using programmable TALE-TET1 fusion proteins. Nat Biotechnol 2013;31(12):1137–42.

[57] Choudhury SR, Cui Y, Lubecka K, Stefanska B, Irudayaraj J. CRISPR-dCas9 mediated TET1 targeting for selective DNA demethylation at BRCA1 promoter. Oncotarget 2016;7(29):46545–56.

[58] Esteller M, Silva JM, Dominguez G, Bonilla F, Matias-Guiu X, Lerma E, et al. Promoter hypermethylation and BRCA1 inactivation in sporadic breast and ovarian tumors. J Natl Cancer Inst 2000;92(7):564–9.

[59] Xu X, Tao Y, Gao X, Zhang L, Li X, Zou W, et al. A CRISPR-based approach for targeted DNA demethylation. Cell Discov 2016;2:16009.

[60] Morita S, Noguchi H, Horii T, Nakabayashi K, Kimura M, Okamura K, et al. Targeted DNA demethylation in vivo using dCas9-peptide repeat and scFv-TET1 catalytic domain fusions. Nat Biotechnol 2016;34(10):1060–5.

[61] Liu XS, Wu H, Krzisch M, Wu X, Graef J, Muffat J, et al. Rescue of fragile X syndrome neurons by DNA methylation editing of the FMR1 gene. Cell 2018;172(5):979–92 e6.

[62] Baumann V, Wiesbeck M, Breunig CT, Braun JM, Köferle A, Ninkovic J, et al. Targeted removal of epigenetic barriers during transcriptional reprogramming. Nat Commun 2019;10(1):2119.

[63] Taghbalout A, Du M, Jillette N, Rosikiewicz W, Rath A, Heinen CD, et al. Enhanced CRISPR-based DNA demethylation by Casilio-ME-mediated RNA-guided coupling of methylcytosine oxidation and DNA repair pathways. Nat Commun 2019;10(1):4296.

[64] Huisman C, Van Der Wijst MGP, Schokker M, Blancafort P, Terpstra MM, Kok K, et al. Re-expression of selected epigenetically silenced candidate tumor suppressor genes in cervical cancer by TET2-directed demethylation. Mol Ther 2016;24(3):536–47.

[65] Xu X, Tan X, Tampe B, Wilhelmi T, Hulshoff MS, Saito S, et al. High-fidelity CRISPR/Cas9- based gene-specific hydroxymethylation rescues gene expression and attenuates renal fibrosis. Nat Commun 2018;9(1):3509.

[66] Cano-Rodriguez D, Gjaltema RAF, Jilderda LJ, Jellema P, Dokter-Fokkens J, Ruiters MHJ, et al. Writing of H3K4Me3 overcomes epigenetic silencing in a sustained but context-dependent manner. Nat Commun 2016;7:12284.

[67] Mendenhall EM, Williamson KE, Reyon D, Zou JY, Ram O, Joung JK, et al. Locus-specific editing of histone modifications at endogenous enhancers. Nat Biotechnol 2013;31(12):1133–6.

[68] Kearns NA, Pham H, Tabak B, Genga RM, Silverstein NJ, Garber M, et al. Functional annotation of native enhancers with a Cas9-histone demethylase fusion. Nat Methods 2015;12(5):401–3.

[69] Li K, Liu Y, Cao H, Zhang Y, Gu Z, Liu X, et al. Interrogation of enhancer function by enhancer-targeting CRISPR epigenetic editing. Nat Commun 2020;11(1):485.

[70] Snowden AW, Gregory PD, Case CC, Pabo CO. Gene-specific targeting of H3K9 methylation is sufficient for initiating repression in vivo. Curr Biol 2002;12(24):2159–66.

[71] Falahi F, Huisman C, Kazemier HG, van der Vlies P, Kok K, Hospers GAP, et al. Towards sustained silencing of HER2/neu in cancer by epigenetic editing. Mol Cancer Res 2013;11(9):1029–39.

[72] Heller EA, Cates HM, Peña CJ, Sun H, Shao N, Feng J, et al. Locus-specific epigenetic remodeling controls addiction- and depression-related behaviors. Nat Neurosci 2014;17(12):1720–7.

[73] Bustos FJ, Ampuero E, Jury N, Aguilar R, Falahi F, Toledo J, et al. Epigenetic editing of the Dlg4/PSD95 gene improves cognition in aged and Alzheimer's disease mice. Brain 2017;140(12):3252–68.

[74] Braun SMG, Kirkland JG, Chory EJ, Husmann D, Calarco JP, Crabtree GR. Rapid and reversible epigenome editing by endogenous chromatin regulators. Nat Commun 2017;8(1):560.

[75] O'Geen H, Ren C, Nicolet CM, Perez AA, Halmai J, Le VM, et al. dCas9-based epigenome editing suggests acquisition of histone methylation is not sufficient for target gene repression. Nucleic Acids Res 2017;45(17):9901–16.

[76] Wang H, Guo R, Du Z, Bai L, Li L, Cui J, et al. Epigenetic targeting of granulin in hepatoma cells by synthetic CRISPR dCas9 epi-suppressors. Mol Ther Nucleic Acids 2018;11:23–33.

[77] Hilton IB, Ippolito AMD, Vockley CM, Thakore PI, Crawford GE, Reddy TE, et al. Epigenome editing by a CRISPR-Cas9-based acetyltransferase activates genes from promoters and enhancers. Nat Biotechnol 2015;33(5):510–17.

[78] Okada M, Kanamori M, Someya K, Nakatsukasa H, Yoshimura A. Stabilization of Foxp3 expression by CRISPR-dCas9-based epigenome editing in mouse primary T cells. Epigenetics Chromatin 2017;8(10):24.

[79] Liu P, Chen M, Liu Y, Qi LS, Ding S. CRISPR-based chromatin remodeling of the endogenous Oct4 or Sox2 locus enables reprogramming to pluripotency. Cell Stem Cell 2018;22(2):252–61 e4.

[80] Bubeck F, Hoffmann MD, Harteveld Z, Aschenbrenner S, Bietz A, Waldhauer MC, et al. Engineered anti-CRISPR proteins for optogenetic control of CRISPR–Cas9. Nat Methods 2018;15(11):924–7.

[81] Chen T, Gao D, Zhang R, Zeng G, Yan H, Lim E, et al. Chemically controlled epigenome editing through an inducible dCas9 system. J Am Chem Soc 2017;139(33):11337–40.

[82] Kuscu C, Mammeadov R, Czikora A, Unlu H, Tufan T, Fischer NL, et al. Temporal and spatial epigenome editing allows precise gene regulation in mammalian cells. J Mol Biol 2019;431(1):111–21.

[83] Shrimp JH, Grose C, Widmeyer SRT, Thorpe AL, Jadhav A, Meier JL. Chemical control of a CRISPR-Cas9 acetyltransferase. ACS Chem Biol 2018;13(2):455–60.

[84] Zhang X, Wang W, Shan L, Han L, Ma S, Zhang Y, et al. Gene activation in human cells using CRISPR/Cpf1-p300 and CRISPR/Cpf1-SunTag systems. Protein Cell 2018;9(4):380–3.

[85] Cheng AW, Jillette N, Lee P, Plaskon D, Fujiwara Y, Wang W, et al. Casilio: a versatile CRISPR-Cas9-Pumilio hybrid for gene regulation and genomic labeling. Cell Res 2016;26(2):254–7.

[86] Kwon DY, Zhao YT, Lamonica JM, Zhou Z. Locus-specific histone deacetylation using a synthetic CRISPR-Cas9-based HDAC. Nat Commun 2017;8:15315.

[87] Liu XM, Zhou J, Mao Y, Ji Q, Qian SB. Programmable RNA N 6-methyladenosine editing by CRISPR-Cas9 conjugates. Nat Chem Biol 2019;15(9):865−71.

[88] Subramaniam D, Thombre R, Dhar A, Anant S. DNA methyltransferases: a novel target for prevention and therapy. Front Oncol 2014;4:80.

[89] Kareta MS, Botello ZM, Ennis JJ, Chou C, Chédin F. Reconstitution and mechanism of the stimulation of de novo methylation by human DNMT3L. J Biol Chem 2006;281(36):25893−902.

[90] Jia D, Jurkowska RZ, Zhang X, Jeltsch A, Cheng X. Structure of Dnmt3a bound to Dnmt3L suggests a model for de novo DNA methylation. Nature 2007;449(7159):248−51.

[91] Meng H, Cao Y, Qin J, Song X, Zhang Q, Shi Y, et al. DNA methylation, its mediators and genome integrity. Int J Biol Sci 2015;11(5):604−17.

[92] Mali P, Aach J, Stranges PB, Esvelt KM, Moosburner M, Kosuri S, et al. CAS9 transcriptional activators for target specificity screening and paired nickases for cooperative genome engineering. Nat Biotechnol 2013;31(9):833−8.

[93] Zalatan JG, Lee ME, Almeida R, Gilbert LA, Whitehead EH, La Russa M, et al. Engineering complex synthetic transcriptional programs with CRISPR RNA scaffolds. Cell 2015;160(1−2):339−50.

[94] Konermann S, Brigham MD, Trevino AE, Joung J, Abudayyeh OO, Barcena C, et al. Genome-scale transcriptional activation by an engineered CRISPR-Cas9 complex. Nature 2015;517(7536):583−8.

[95] Shechner DM, Hacisuleyman E, Younger ST, Rinn JL. Multiplexable, locus-specific targeting of long RNAs with CRISPR-Display. Nat Methods 2015;12(7):664−70.

[96] Tanenbaum ME, Gilbert LA, Qi LS, Weissman JS, Vale RD. A protein-tagging system for signal amplification in gene expression and fluorescence imaging. Cell 2014;159(3):635−46.

[97] Zhang T, Cooper S, Brockdorff N. The interplay of histone modifications − writers that read. EMBO Rep 2015;16(11):1467−81.

[98] Lee MG, Wynder C, Bochar DA, Hakimi MA, Cooch N, Shiekhattar R. Functional interplay between histone demethylase and deacetylase enzymes. Mol Cell Biol 2006;26(17):6395−402.

[99] Mkannez G, Gagné-Ouellet V, Nsaibia MJ, Boulanger MC, Rosa M, Argaud D, et al. DNA methylation of a PLPP3 MIR transposon-based enhancer promotes an osteogenic programme in calcific aortic valve disease. Cardiovasc Res 2018;114(11):1525−35.

[100] Esteller M. CpG island hypermethylation and tumor suppressor genes: a booming present, a brighter future. Oncogene 2002;21(35):5427−40.

[101] Bodenstine TM, Seftor REB, Khalkhali-Ellis Z, Seftor EA, Pemberton PA, Hendrix MJC. Maspin: molecular mechanisms and therapeutic implications. Cancer Metastasis Rev 2012;31(3−4):529−51.

[102] Lengerke C, Fehm T, Kurth R, Neubauer H, Scheble V, Müller F, et al. Expression of the embryonic stem cell marker SOX2 in early-stage breast carcinoma. BMC Cancer 2011;11:42.

[103] Münz M, Kieu C, Mack B, Schmitt B, Zeidler R, Gires O. The carcinoma-associated antigen EpCAM upregulates c-myc and induces cell proliferation. Oncogene 2004;23(34):5748−58.

[104] Kim JH, Herlyn D, Wong KK, Park DC, Schorge JO, Lu KH, et al. Identification of epithelial cell adhesion molecule autoantibody in patients with ovarian cancer. Clin Cancer Res 2003;9(13):4782−91.

[105] Belinsky SA, Nikula KJ, Palmisano WA, Michels R, Saccomanno G, Gabrielson E, et al. Aberrant methylation of p16(INK4a) is an early event in lung cancer and a potential biomarker for early diagnosis. Proc Natl Acad Sci U S A 1998;95(20):11891−6.

[106] Matsuda Y, Ichida T, Matsuzawa J, Sugimura K, Asakura H. p16(INK4) is inactivated by extensive CpG methylation in human hepatocellular carcinoma. Gastroenterology 1999;116(2):394−400.

[107] Nuovo GJ, Plaia TW, Belinsky SA, Baylin SB, Herman JG. In situ detection of the hypermethylation-induced inactivation of the p16 gene as an early event in oncogenesis. Proc Natl Acad Sci U S A 1999;96(22):12754−9.

[108] Herman JG. p16(INK4): involvement early and often in gastrointestinal malignancies. Gastroenterology 1999;116(2):483−5.

[109] Belinsky SA, Liechty KC, Gentry FD, Wolf HJ, Rogers J, Vu K, et al. Promoter hypermethylation of multiple genes in sputum precedes lung cancer incidence in a high-risk cohort. Cancer Res 2006;66(6):3338−44.

[110] Luo D, Zhang B, Lv L, Xiang S, Liu Y, Ji J, et al. Methylation of CpG islands of p16 associated with progression of primary gastric carcinomas. Lab Invest 2006;86(6):591−8.

[111] Liu H, Liu XW, Dong G, Zhou J, Liu Y, Gao Y, et al. P16 methylation as an early predictor for cancer development from oral epithelial dysplasia: a double-blind multicentre prospective study. EBioMedicine 2015;2(5):432−7.

[112] Camilleri-Broët S, Hardy-Bessard AC, Le Tourneau A, Paraiso D, Levrel O, Leduc B, et al. HER-2 overexpression is an independent marker of poor prognosis of advanced primary ovarian carcinoma: a multicenter study of the GINECO group. Ann Oncol 2004;15(1):104−12.

[113] Tapia C, Savic S, Wagner U, Schönegg R, Novotny H, Grilli B, et al. HER2 gene status in primary breast cancers and matched distant metastases. Breast Cancer Res 2007;9(3):R31.

[114] Hechtman JF, Polydorides AD. HER2/neu gene amplification and protein overexpression in gastric and gastroesophageal junction adenocarcinoma: a review of histopathology, diagnostic testing, and clinical implications. Arch Pathol Lab Med 2012;136(6):691−7.

[115] Robison AJ, Nestler EJ. Transcriptional and epigenetic mechanisms of addiction. Nat Rev Neurosci 2011;12(11):623−37.

[116] Azuma M, Koyama D, Kikuchi I, Yoshizawa H, Thasinas D, Shiizaki K, et al. Promoter methylation confers kidney-specific expression of the Klotho gene. FASEB J 2012;26(10):4264−74.

[117] Sugiura H, Yoshida T, Shiohira S, Kohei J, Mitobe M, Kurosu H, et al. Reduced klotho expression level in kidney aggravates renal interstitial fibrosis. Am J Physiol Renal Physiol 2012;302(10):F1252−64.

[118] Satoh M, Nagasu H, Morita Y, Yamaguchi TP, Kanwar YS, Kashihara N. Klotho protects against mouse renal fibrosis by inhibiting Wnt signaling. Am J Physiol Renal Physiol 2012;303(12):F1641−51.

[119] Verkerk AJ, Pieretti M, Sutcliffe JS, Fu YH, Kuhl DP, Pizzuti A, et al. Identification of a gene (FMR-1) containing a CGG repeat coincident with a breakpoint cluster region exhibiting length variation in fragile X syndrome. Cell 1991;65(5):905−14.

[120] De Esch CEF, Ghazvini M, Loos F, Schelling-Kazaryan N, Widagdo W, Munshi ST, et al. Epigenetic characterization of the FMR1 promoter in induced pluripotent stem cells from human fibroblasts carrying an unmethylated full mutation. Stem Cell Rep 2014;3(4):548−55.

[121] Avitzour M, Mor-Shaked H, Yanovsky-Dagan S, Aharoni S, Altarescu G, Renbaum P, et al. FMR1 epigenetic silencing commonly occurs in undifferentiated fragile X-affected embryonic stem cells. Stem Cell Rep 2014;3(5):699−706.

[122] Sakaguchi S, Yamaguchi T, Nomura T, Ono M. Regulatory T cells and immune tolerance. Cell 2008;133(5):775−87.

[123] Huisman C, van der Wijst MGP, Falahi F, Overkamp J, Karsten G, Terpstra MM, et al. Prolonged re-expression of the hypermethylated gene EPB41L3 using artificial transcription factors and epigenetic drugs. Epigenetics 2015;10(5):384−96.

[124] Rao SSP, Huntley MH, Durand NC, Stamenova EK, Bochkov ID, Robinson JT, et al. A 3D map of the human genome at kilobase resolution reveals principles of chromatin looping. Cell 2014;159(7):1665−80.

[125] Dixon JR, Selvaraj S, Yue F, Kim A, Li Y, Shen Y, et al. Topological domains in mammalian genomes identified by analysis of chromatin interactions. Nature 2012;485(7398):376−80.

[126] Bell AC, Felsenfeld G. Methylation of a CTCF-dependent boundary controls imprinted expression of the Igf2 gene. Nature 2000;405(6785):482−5.

[127] Hark AT, Schoenherr CJ, Katz DJ, Ingram RS, Levorse JM, Tilghman SM. CTCF mediates methylation-sensitive enhancer-blocking activity at the H19/Igf2 locus. Nature 2000;405 (6785):486−9.

[128] Johnson LM, Du J, Hale CJ, Bischof S, Feng S, Chodavarapu RK, et al. SRA-and SET-domain-containing proteins link RNA polymerase V occupancy to DNA methylation. Nature 2014;507(7490):124−8.

[129] Gallego-Bartolomé J, Gardiner J, Liu W, Papikian A, Ghoshal B, Kuo HY, et al. Targeted DNA demethylation of the arabidopsis genome using the human TET1 catalytic domain. Proc Natl Acad Sci U S A 2018;115(9):E2125−34.

[130] Gallego-Bartolomé J, Liu W, Kuo PH, Feng S, Ghoshal B, Gardiner J, et al. Co-targeting RNA polymerases IV and V promotes efficient de novo DNA methylation in Arabidopsis. Cell 2019;176(5):1068−82 e19.

[131] Fal K, Tomkova D, Vachon G, Chabouté ME, Berr A, Carles CC. Chromatin manipulation and editing: challenges, new technologies and their use in plants. Int J Mol Sci 2021;22 (2):512.

[132] Li X, Huang L, Pan L, Wang B, Pan L. CRISPR/dCas9-mediated epigenetic modification reveals differential regulation of histone acetylation on Aspergillus niger secondary metabolite. Microbiol Res 2021;245:126694.

[133] Grimmer MR, Stolzenburg S, Ford E, Lister R, Blancafort P, Farnham PJ. Analysis of an artificial zinc finger epigenetic modulator: widespread binding but limited regulation. Nucleic Acids Res 2014;42(16):10856−68.

[134] Hsu PD, Scott DA, Weinstein JA, Ran FA, Konermann S, Agarwala V, et al. DNA targeting specificity of RNA-guided Cas9 nucleases. Nat Biotechnol 2013;31(9):827−32.

[135] Fu Y, Foden JA, Khayter C, Maeder ML, Reyon D, Joung JK, et al. High-frequency off-target mutagenesis induced by CRISPR-Cas nucleases in human cells. Nat Biotechnol 2013;31 (9):822−6.

[136] Pattanayak V, Lin S, Guilinger JP, Ma E, Doudna JA, Liu DR. High-throughput profiling of off-target DNA cleavage reveals RNA-programmed Cas9 nuclease specificity. Nat Biotechnol 2013;31(9):839−43.

[137] Kleinstiver BP, Pattanayak V, Prew MS, Tsai SQ, Nguyen NT, Zheng Z, et al. High-fidelity CRISPR-Cas9 nucleases with no detectable genome-wide off-target effects. Nature 2016;529 (7587):490−5.

[138] Horlbeck MA, Witkowsky LB, Guglielmi B, Replogle JM, Gilbert LA, Villalta JE, et al. Nucleosomes impede Cas9 access to DNA in vivo and in vitro. Elife 2016;5:e12677.

[139] Barkal AA, Srinivasan S, Hashimoto T, Gifford DK, Sherwood RI. Cas9 functionally opens chromatin. PLoS One 2016;11(3): e0152683.

[140] Lee JT. Epigenetic regulation by long noncoding RNAs. Science 2012;338(6113):1435−9.

[141] Cao J. The functional role of long non-coding RNAs and epigenetics. Biol Proced Online 2014;16:11.

[142] Chen Y, Varani G. Engineering RNA-binding proteins for biology. FEBS J 2013;280(16):3734−54.

[143] Konermann S, Brigham MD, Trevino A, Hsu PD, Heidenreich M, Cong L, et al. Optical control of mammalian endogenous transcription and epigenetic states. Nature 2013;500(7463): 472−6.

[144] Rauch BJ, Silvis MR, Hultquist JF, Waters CS, McGregor MJ, Krogan NJ, et al. Inhibition of CRISPR-Cas9 with bacteriophage proteins. Cell 2017;168(1−2):150−8 e10.

[145] Chiarella AM, Butler KV, Gryder BE, Lu D, Wang TA, Yu X, et al. Dose-dependent activation of gene expression is achieved using CRISPR and small molecules that recruit endogenous chromatin machinery. Nat Biotechnol 2020;38(1):50−5.

[146] Deng W, Lee J, Wang H, Miller J, Reik A, Gregory PD, et al. Controlling long-range genomic interactions at a native locus by targeted tethering of a looping factor. Cell 2012;149 (6):1233−44.

[147] Deng W, Rupon JW, Krivega I, Breda L, Motta I, Jahn KS, et al. Reactivation of developmentally silenced globin genes by forced chromatin looping. Cell 2014;158(4):849−60.

[148] Morgan SL, Mariano NC, Bermudez A, Arruda NL, Wu F, Luo Y, et al. Manipulation of nuclear architecture through CRISPR-mediated chromosomal looping. Nat Commun 2017;8:15993.

[149] Liang FS, Ho WQ, Crabtree GR. Engineering the ABA Plant stress pathway for regulation of induced proximity. Sci Signal 2011;4(164):rs2.

[150] Wang H, Xu X, Nguyen CM, Liu Y, Gao Y, Lin X, et al. CRISPR-mediated programmable 3D genome positioning and nuclear organization. Cell 2018;175(5):1405−17 e14.

[151] Kim JH, Rege M, Valeri J, Dunagin MC, Metzger A, Titus KR, et al. LADL: light-activated dynamic looping for endogenous gene expression control. Nat Methods 2019;16(7):633−9.

[152] Katayama S, Moriguchi T, Ohtsu N, Kondo T. A powerful CRISPR/Cas9-based method for targeted transcriptional activation. Angew Chem Int Ed Engl 2016;55(22):6452−6.

[153] Liszczak GP, Brown ZZ, Kim SH, Oslund RC, David Y, Muir TW. Genomic targeting of epigenetic probes using a chemically tailored Cas9 system. Proc Natl Acad Sci U S A 2017;114 (4):681−6.

[154] Shah NH, Dann GP, Vila-Perelló M, Liu Z, Muir TW. Ultrafast protein splicing is common among cyanobacterial split inteins: implications for protein engineering. J Am Chem Soc 2012;134 (28):11338−41.

[155] Abudayyeh OO, Gootenberg JS, Konermann S, Joung J, Slaymaker IM, Cox DBT, et al. C2c2 is a single-component programmable RNA-guided RNA-targeting CRISPR effector. Science 2016;353(6299):aaf5573.

[156] East-Seletsky A, O'Connell MR, Knight SC, Burstein D, Cate JHD, Tjian R, et al. Two distinct RNase activities of CRISPR-C2c2 enable guide-RNA processing and RNA detection. Nature 2016;538(7624):270−3.

[157] Abudayyeh OO, Gootenberg JS, Essletzbichler P, Han S, Joung J, Belanto JJ, et al. RNA targeting with CRISPR-Cas13. Nature 2017;550(7675):280−4.

[158] O'Connell MR, Oakes BL, Sternberg SH, East-Seletsky A, Kaplan M, Doudna JA. Programmable RNA recognition and cleavage by CRISPR/Cas9. Nature 2014;516(7530):263−6.

[159] Nelles DA, Fang MY, O'Connell MR, Xu JL, Markmiller SJ, Doudna JA, et al. Programmable RNA tracking in live cells with CRISPR/Cas9. Cell 2016;165(2):488−96.

[160] Liu N, Pan T. RNA epigenetics. Transl Res 2015;165(1):28–35.
[161] Wright DA, Thibodeau-Beganny S, Sander JD, Winfrey RJ, Hirsh AS, Eichtinger M, et al. Standardized reagents and protocols for engineering zinc finger nucleases by modular assembly. Nat Protoc 2006;1(3):1637–52.
[162] Maeder ML, Thibodeau-Beganny S, Sander JD, Voytas DF, Joung JK. Oligomerized pool engineering (OPEN): an "open-source" protocol for making customized zinc-finger arrays. Nat Protoc 2009;4(10):1471–501.
[163] Sander JD, Dahlborg EJ, Goodwin MJ, Cade L, Zhang F, Cifuentes D, et al. Selection-free zinc-finger-nuclease engineering by context-dependent assembly (CoDA). Nat Methods 2011;8(1):67–9.
[164] Kleinstiver BP, Prew MS, Tsai SQ, Topkar VV, Nguyen NT, Zheng Z, et al. Engineered CRISPR-Cas9 nucleases with altered PAM specificities. Nature 2015;523(7561):481–5.
[165] Shmakov S, Abudayyeh OO, Makarova KS, Wolf YI, Gootenberg JS, Semenova E, et al. Discovery and functional characterization of diverse class 2 CRISPR-Cas systems. Mol Cell 2015;60(3):385–97.
[166] Dong D, Ren K, Qiu X, Zheng J, Guo M, Guan X, et al. The crystal structure of Cpf1 in complex with CRISPR RNA. Nature 2016;532(7600):522–6.
[167] Yamano T, Nishimasu H, Zetsche B, Hirano H, Slaymaker IM, Li Y, et al. Crystal structure of Cpf1 in complex with guide RNA and target DNA. Cell 2016;165(4):949–62.
[168] Ran FA, Cong L, Yan WX, Scott DA, Gootenberg JS, Kriz AJ, et al. In vivo genome editing using *Staphylococcus aureus* Cas9. Nature 2015;520(7546):186–91.
[169] Pausch P, Al-Shayeb B, Bisom-Rapp E, Tsuchida CA, Li Z, Cress BF, et al. CRISPR-CasΦ from huge phages is a hypercompact genome editor. Science 2020;369(6501):333–7.
[170] Deng D, Yin P, Yan C, Pan X, Gong X, Qi S, et al. Recognition of methylated DNA by TAL effectors. Cell Res 2012;22(10):1502–4.
[171] Hu J, Lei Y, Wong WK, Liu S, Lee KC, He X, et al. Direct activation of human and mouse Oct4 genes using engineered TALE and Cas9 transcription factors. Nucleic Acids Res 2014;42(7):4375–90.
[172] Chen S, Oikonomou G, Chiu CN, Niles BJ, Liu J, Lee DA, et al. A large-scale in vivo analysis reveals that TALENs are significantly more mutagenic than ZFNs generated using context-dependent assembly. Nucleic Acids Res 2013;41(4):2769–78.
[173] Fujita T, Yuno M, Fujii H. Allele-specific locus binding and genome editing by CRISPR at the p16INK4a locus. Sci Rep 2016;6:30485.
[174] Kubik G, Schmidt MJ, Penner JE, Summerer D. Programmable and highly resolved in vitro detection of 5-methylcytosine by TALEs. Angew Chem Int Ed Engl 2014;53(23):6002–6.
[175] Isalan M, Choo Y. Engineered zinc finger proteins that respond to DNA modification by HaeIII and HhaI methyltransferase enzymes. J Mol Biol 2000;295(3):471–7.
[176] Maier JAH, Möhrle R, Jeltsch A. Design of synthetic epigenetic circuits featuring memory effects and reversible switching based on DNA methylation. Nat Commun 2017;8:15336.
[177] Chan KY, Jang MJ, Yoo BB, Greenbaum A, Ravi N, Wu WL, et al. Engineered AAVs for efficient noninvasive gene delivery to the central and peripheral nervous systems. Nat Neurosci 2017;20(8):1172–9.
[178] Fujita T, Fujii H. Biochemical analysis of genome functions using locus-specific chromatin immunoprecipitation technologies. Gene Regul Syst Bio 2016;10(Suppl 1):1–9.
[179] Hoshino A, Fujii H. Insertional chromatin immunoprecipitation: a method for isolating specific genomic regions. J Biosci Bioeng 2009;108(5):446–9.
[180] Fujita T, Fujii H. Efficient isolation of specific genomic regions by insertional chromatin immunoprecipitation (iChIP) with a second-generation tagged LexA DNA-binding domain. Adv Biosci Biotechnol 2012;3(5):626–9.
[181] Fujita T, Fujii H. Direct identification of insulator components by insertional chromatin immunoprecipitation. PLoS One 2011;6(10):e26109.
[182] Fujita T, Fujii H. Efficient isolation of specific genomic regions retaining molecular interactions by the iChIP system using recombinant exogenous DNA-binding proteins. BMC Mol Biol 2014;15:26.
[183] Fujita T, Kitaura F, Fujii H. A critical role of the Thy28-MYH9 axis in B cell-specific expression of the Pax5 gene in chicken B cells. PLoS One 2015;10(1):e0116579.
[184] Husain A, Xu J, Fujii H, Nakata M, Kobayashi M, Wang J, et al. SAMHD1-mediated dNTP degradation is required for efficient DNA repair during antibody class switch recombination. EMBO J 2020;39(15):e102931.
[185] Bui PL, Nishimura K, Seminario Mondejar G, Kumar A, Aizawa S, Murano K, et al. Template activating factor-I α regulates retroviral silencing during reprogramming. Cell Rep 2019;29(7):1909–22 e5.
[186] McCullagh E, Seshan A, El-Samad H, Madhani HD. Coordinate control of gene expression noise and interchromosomal interactions in a MAP kinase pathway. Nat Cell Biol 2010;12(10):954–62.
[187] Fujita T, Fujii H. Efficient isolation of specific genomic regions and identification of associated proteins by engineered DNA-binding molecule-mediated chromatin immunoprecipitation (enChIP) using CRISPR. Biochem Biophys Res Commun 2013;439(1):132–6.
[188] Fujita T, Asano Y, Ohtsuka J, Takada Y, Saito K, Ohki R, et al. Identification of telomere-associated molecules by engineered DNA-binding molecule-mediated chromatin immunoprecipitation (enChIP). Sci Rep 2013;3:3171.
[189] Fujita T, Fujii H. Identification of proteins associated with an IFNγ-responsive promoter by a retroviral expression system for enChIP using CRISPR. PLoS One 2014;9(7):e103084.
[190] Hamidian A, Vaapil M, von Stedingk K, Fujita T, Persson CU, Eriksson P, et al. Promoter-associated proteins of EPAS1 identified by enChIP-MS — A putative role of HDX as a negative regulator. Biochem Biophys Res Commun 2018;499(2):291–8.
[191] Liu X, Zhang Y, Chen Y, Li M, Zhou F, Li K, et al. In situ capture of chromatin interactions by biotinylated dCas9. Cell 2017;170(5):1028–43 e19.
[192] Campbell AE, Shadle SC, Jagannathan S, Lim JW, Resnick R, Tawil R, et al. NuRD and CAF-1-mediated silencing of the D4Z4 array is modulated by DUX4-induced MBD3L proteins. Elife 2018;7:e31023.
[193] Han B, Zhou B, Qu Y, Gao B, Xu Y, Chung S, et al. FOXC1-induced non-canonical WNT5A-MMP7 signaling regulates invasiveness in triple-negative breast cancer. Oncogene 2018;37(10):1399–408.
[194] Mochizuki Y, Chiba T, Kataoka K, Yamashita S, Sato T, Kato T, et al. Combinatorial CRISPR/Cas9 approach to elucidate a far-upstream enhancer complex for tissue-specific Sox9 expression. Dev Cell 2018;46(6):794–806 e6.
[195] Shamsi F, Xue R, Huang TL, Lundh M, Liu Y, Leiria LO, et al. FGF6 and FGF9 regulate UCP1 expression independent of brown adipogenesis. Nat Commun 2020;11(1):1421.
[196] Huang Y, Xiang Y, Xie Z, Cai Y, Yang Q, Huang H, et al. Mass spectrometry-based proteomic capture of proteins bound to the MACC1 promoter in colon cancer. Clin Exp Metastasis 2020;37(4):477–87.

[197] Bryant JM, Baumgarten S, Dingli F, Loew D, Sinha A, Claës A, et al. Exploring the virulence gene interactome with CRISPR/dCas9 in the human malaria parasite. Mol Syst Biol 2020;16(8):e9569.

[198] González-Rico FJ, Vicente-García C, Fernández A, Muñoz-Santos D, Montoliu L, Morales-Hernández A, et al. Alu retrotransposons modulate Nanog expression through dynamic changes in regional chromatin conformation via aryl hydrocarbon receptor. Epigenetics Chromatin 2020;13(1):15.

[199] Galati A, Micheli E, Cacchione S. Chromatin structure in telomere dynamics. Front Oncol 2013;3:46.

[200] Byrum SD, Taverna SD, Tackett AJ. Purification of a specific native genomic locus for proteomic analysis. Nucleic Acids Res 2013;41(20):e195.

[201] Fang F, Xia N, Angulo B, Carey J, Cady Z, Durruthy-Durruthy J, et al. A distinct isoform of ZNF207 controls self-renewal and pluripotency of human embryonic stem cells. Nat Commun 2018;9(1):4384.

[202] Fujita T, Yuno M, Okuzaki D, Ohki R, Fujii H. Identification of non-coding RNAs associated with telomeres using a combination of enChIP and RNA sequencing. PLoS One 2015;10(4):e0123387.

[203] Zhang Y, Hu JF, Wang H, Cui J, Gao S, Hoffman AR, et al. CRISPR Cas9-guided chromatin immunoprecipitation identifies miR483 as an epigenetic modulator of IGF2 imprinting in tumors. Oncotarget 2017;8(21):34177—90.

[204] Chen N, Zhao G, Yan X, Lv Z, Yin H, Zhang S, et al. A novel FLI1 exonic circular RNA promotes metastasis in breast cancer by coordinately regulating TET1 and DNMT1. Genome Biol 2018;19(1):218.

[205] Fujita T, Yuno M, Suzuki Y, Sugano S, Fujii H. Identification of physical interactions between genomic regions by enChIP-Seq. Genes Cells 2017;22(6):506—20.

[206] Seo W, Shimizu K, Kojo S, Okeke A, Kohwi-Shigematsu T, Fujii S, et al. Runx-mediated regulation of CCL5 via antagonizing two enhancers influences immune cell function and antitumor immunity. Nat Commun 2020;11(1):1562.

[207] Fujita T, Yuno M, Fujii H. Efficient sequence-specific isolation of DNA fragments and chromatin by in vitro enChIP technology using recombinant CRISPR ribonucleoproteins. Genes Cells 2016;21(4):370—7.

[208] Fujita T, Kitaura F, Yuno M, Suzuki Y, Sugano S, Fujii H. Locus-specific ChIP combined with NGS analysis reveals genomic regulatory regions that physically interact with the Pax5 promoter in a chicken B cell line. DNA Res 2017;24(5):537—48.

[209] Tsui C, Inouye C, Levy M, Lu A, Florens L, Washburn MP, et al. dCas9-targeted locus-specific protein isolation method identifies histone gene regulators. Proc Natl Acad Sci U S A 2018;115(12):E2734—41.

[210] Waldrip ZJ, Byrum SD, Storey AJ, Gao J, Byrd AK, Mackintosh SG, et al. A CRISPR-based approach for proteomic analysis of a single genomic locus. Epigenetics 2014;9(9):1207—11.

[211] Fujita T, Yuno M, Fujii H. enChIP systems using different CRISPR orthologues and epitope tags. BMC Res Notes 2018;11(1):154.

[212] Fujii H, Fujita T. *Method for isolating specific genomic region using molecule binding specifically to endogenous DNA sequence.* WO2014/125668; JP5954808.

[213] Gao XD, Tu LC, Mir A, Rodriguez T, Ding Y, Leszyk J, et al. C-BERST: defining subnuclear proteomic landscapes at genomic elements with dCas9-APEX2. Nat Methods 2018;15(6):433—6.

[214] Myers SA, Wright J, Peckner R, Kalish BT, Zhang F, Carr SA. Discovery of proteins associated with a predefined genomic locus via dCas9-APEX-mediated proximity labeling. Nat Methods 2018;15(6):437—9.

[215] Escobar TM, Oksuz O, Saldaña-Meyer R, Descostes N, Bonasio R, Reinberg D. Active and repressed chromatin domains exhibit distinct nucleosome segregation during DNA replication. Cell 2019;179(4):953—63 e11.

[216] Han S, Zhao BS, Myers SA, Carr SA, He C, Ting AY. RNA-protein interaction mapping via MS2- or Cas13-based APEX targeting. Proc Natl Acad Sci U S A 2020;117(36):22068—79.

[217] Thakore PI, Black JB, Hilton IB, Gersbach CA. Editing the epigenome: technologies for programmable transcription and epigenetic modulation. Nat Methods 2016;13(2):127—37.

[218] Kungulovski G, Jeltsch A. Epigenome editing: state of the art, concepts, and perspectives. Trends Genet 2016;32(2):101—13.

[219] Jurkowski TP, Ravichandran M, Stepper P. Synthetic epigenetics-towards intelligent control of epigenetic states and cell identity. Clin Epigenetics 2015;7(1):18.

[220] Gomez JA, Beitnere U, Segal DJ. Live-animal epigenome editing: convergence of novel techniques. Trends Genet 2019;35(7):527—41.

[221] Nakamura M, Gao Y, Dominguez AA, Qi LS. CRISPR technologies for precise epigenome editing. Nat Cell Biol 2021;23(1):11—22.

[222] Fujita T, Fujii H. Purification of specific DNA species using the CRISPR system. Biol Methods Protoc 2019;4(1):bpz008.

[223] Fujita H, Fujita T, Fujii H. Locus-specific genomic DNA purification using the CRISPR system: methods and applications. CRISPR J 2021;4(2):290—300.

Index

Note: Page numbers followed by "*f*" and "*t*" refer to figures and tables, respectively.

A

ABA-Hypersensitive Germination 3 (*AHG3*), 506
Abca1 gene, 615
Aberrant histone modifications role in disease, 45
Aberrant methylation, 304
Abexinostat, 794–795
Abiotic stress, 271–273
Abscisic acid (ABA), 273, 856
Abscisic Acid-Insensitive4 (*ABI4*), 506
Absorption, distribution, metabolism, and excretion (ADME), 689, 691
Acetate, 675–676
Acetyl coenzyme A (AcCoA), 203
Acetyl donors, 650
acetyl-CoA. *See* Acetyl-coenzyme A (acetyl-CoA)
Acetyl-CoA synthetase (Acs), 204
Acetyl-CoA synthetase 2 (ACSS2), 614
Acetyl-coenzyme A (acetyl-CoA), 28–30, 358–360, 367
Acetyl-phosphate (AcP), 203
Acetylation, 28–36, 112–114, 137, 305, 384, 395
ACLY. *See* ATP-citrate lyase (ACLY)
AcP. *See* Acetyl-phosphate (AcP)
Acquired heritable epigenetics, 639–640
Acs. *See* Acetyl-CoA synthetase (Acs)
ACSS2. *See* Acetyl-CoA synthetase 2 (ACSS2)
ACT-seq. *See* Antibody-guided chromatin tagmentation sequencing (ACT-seq)
ACT-seq method, 143
Activation-induced cytidine deaminase (AID), 331–332
Active DNA demethylation, 18–20, 613–614, 699–700
Active histone demethylation, 362
Active promoter regions, 138
Activity of Daily Living Scale (ADL), 391
Activity related neuronal plasticity, 645–646
Activity-dependent histone acetylation and cognitive function, 232–233
Activity-regulated cytoskeleton associated protein (Arc), 451–452
Acute lymphoblastic leukemia (ALL), 60–61
Acute myeloid leukemia (AML), 16, 61, 338, 455, 709, 795–796, 823
  combined epigenetic treatment of, 832*t*
  trial of epigenetic priming in patients with newly diagnosed, 833–836
Acute myocardial infarction (AMI), 311

Acute promyelocytic leukemia (APL), 823
  treatment of, 823
    consolidation therapy, 823
    induction therapy, 823
    maintenance therapy, 823
Acylation, 145–148
  2-hydroxyisobutyrylation and β-hydroxybutyrylation, 147–148
  benzoylation, 147
  butyrylation, crotonylation, and propionylation, 147
  glutarylation, 147
  lactylation, 147
AD. *See* Alzheimer's disease (AD)
ADA2b, 273
Adalimumab, 727
  multiomics and machine learning accurately predict clinical response to, 728–730
adamts9 gene, 538
Adaptive immune system, 716
ADCA-DN. *See* Autosomal dominant cerebellar ataxia, deafness, and narcolepsy (ADCA-DN)
Addiction, 628–629
Addictive substances, 640, 654–658
Adenocarcinomas, 309–310
Adenomatous polyposis coli (APC), 653
Adenosine monophosphate activated protein kinase (AMPK), 371–372
Adenosine triphosphate (ATP), 360–361
Adenosine-to-inosine (A-to-I), 62
S-adenosyl methionine (SAM), 3–4, 66, 329, 358–359, 387–388, 612–613, 723, 744, 763
ADEs. *See* Adverse drug effects (ADEs)
ADHD. *See* Attention-deficit/hyperactivity disorder (ADHD)
Adiponectin (*ADIPOQ*), 764
Adipose tissue (AT), 386
Adipose-tissues-derived MSC (AT-MSCs), 383–384, 396–397
ADL. Activity of Daily Living Scale (ADL)
ADLD. *See* Autosomal dominant adult-onset demyelinating leukodystrophy (ADLD)
ADME. *See* Absorption, distribution, metabolism, and excretion (ADME)
AdoMet, 12, 204
ADP-ribosylation, 44, 360
Adrenergic receptor b1 (*adrb1*), 747–748
ADRs. *See* Adverse drug reactions (ADRs)

Adulthood, postnatal to, 642
Advanced epigenome editing tools, 854–857
  epitranscriptome editing, 857
  light-or chemical compound-mediated inducible systems, 855–856
  locus-specific chromatin reorganization, 856, 857*f*
  other applications of epigenome editing, 856–857
  site-specific CpG methylation, 854
  tethering molecules on sgRNAs, 854–855
  tethering of non-coding RNAs, 854
Adverse drug effects (ADEs), 691
  population pharmacoepigenomics in relation to, 691
Adverse drug reactions (ADRs), 687, 691
  population pharmacoepigenomics in relation to, 691
AF. *See* Alcohol fed (AF)
AG. *See* AGAMOUS (AG)
AGAMOUS (*AG*), 503
Age, 762–763
  acceleration, 480
  age-related changes, 763
  factors on epigenome, 762–766
Age-related disease, 485–489
  epigenetic clocks without, 488–489
Aging, 479–480, 528–530, 698–699, 762
  aging-associated diseases, 651
  biomarkers of, 480
AGM. *See* Aorta-gonad-mesonephros (AGM)
Ago. *See* Argonaute (Ago)
AGO proteins. *See* ARGONAUTE proteins (AGO proteins)
Ago-loading, 174
AGO2, 304
Ago2. *See* Argonaute 2 (Ago2)
Agouti gene, 251–252, 342
Agouti viable yellow allele (A^{vy} allele), 251–252
AHG3. *See* *ABA-Hypersensitive Germination 3* (*AHG3*)
AICD. *See* APP intracellular domain (AICD)
AID. *See* Activation-induced cytidine deaminase (AID)
AIPP1. *See* ASI1-IMMUNOPRECIPITATED PROTEIN 1 (AIPP1)
Air pollution, 651, 653–654
Alcohol consumption, 490, 626, 654–655
Alcohol fed (AF), 661
Alcohol use disorder (AUD), 632
Alkaloids, 370–371

ALKBH1. *See* Alkylated DNA repair protein AlkB homolog (ALKBH1)
Alkylated DNA repair protein AlkB homolog (ALKBH1), 382
ALL. *See* Acute lymphoblastic leukemia (ALL)
All-trans retinoic acid (ATRA), 800, 822
Allelic gene expression, 257
Allobaculum, 618
α-ketoglutarate (α-KG), 362–363, 647
α-ketoglutarate dependent dioxygenase alkB homolog 5 (ALKBH5), 457
Alphaproteobacteria, 206
ALS. *See* Amyotrophic lateral sclerosis (ALS)
Alu/B1 elements, 336–337
Alu/B2 elements, 336–337
Alzheimer disease *Drosophila* model, 446
Alzheimer's disease (AD), 7, 234–235, 303, 387–388, 445–446, 479–480, 487, 651, 653, 742–745
  histone modifications and histone modifier in, 742–744
    DNA methylation in, 744–745
    epigenetic regulation and dysregulation in AD, 743f
    non-coding RNA in, 745
  pathogenesis, 235
AMI. *See* Acute myocardial infarction (AMI)
Amino acids, 45, 360
Amino-terminal (N-terminal), 28
γ-aminobutyric acid (GABA), 654
AML. *See* Acute myeloid leukemia (AML)
Amphimedon queenslandica, 523
AMPK. *See* Adenosine monophosphate activated protein kinase (AMPK)
Amyloid fibers, 77
Amyloid precursor protein (APP), 234–235, 742
Amyloid prions
  cis elements important for formation, 76
  genetic control of formation, 76–77
  of *Saccharomyces cerevisiae*, *Podospora anserina*, and organisms, 74–76
  variants, 77
Amyloid β (Aβ), 742
Amyotrophic lateral sclerosis (ALS), 234, 334–335, 742
Anacardic acid, 370–371, 412
ANAs. *See* Anti-nuclear antibodies (ANAs)
Ancestral allele, 557
Ancestral epigenetic alterations, 641
Ancestral predisposition, 641
Androgen receptor gene (AR gene), 429
Angelman syndrome (AS), 7, 308, 589–590, 782
Angiopoietin-like 3 (Angptl3), 615
Angiotensin converting enzyme 2 (ACE-2), 432
Angptl3. *See* Angiopoietin-like 3 (Angptl3)
Animal breeding
  development of, 570–571, 572f
  planning and management, 587–595
    biomarker and therapeutic potentials of epigenetic marks in livestock management, 590–591

epigenetic mechanisms role in livestock breeding, 591–595
genome and epigenome editing technologies in livestock breeding, 595
transgenerational epigenetic inheritance and genomic imprinting, 589–590
Animal care, 570
Animal models, 592–595, 611–612
  of addiction, 629
Animals, epigenetic mechanisms in
  dynamics of epigenetic marks during development, 526–530
  evolution in animal kingdom, 521–524
  features and role in gene regulation, 524–526
  generation of phenotypic variation in populations by epigenetic mechanisms, 530–534
  relevance of epigenetics
    for animal ecology, 534–538
    for animal evolution, 538–543
ANRIL, 166
Ant (*Camponotus floridanus*), 525
Antennapedia complexes, 223
Anterior cingulate cortex (ACC), 452–454
"Anti-agents", 640
Anti-double stranded DNA (dsDNA), 720–721
Anti-nuclear antibodies (ANAs), 720–721
anti-Sm antibodies. *See* Anti-Smith antibodies (anti-Sm antibodies)
Anti-Smith antibodies (anti-Sm antibodies), 720–721
anti-TNF. *See* Anti-tumor necrosis factor (anti-TNF)
Anti-tumor necrosis factor (anti-TNF), 728
Antibody-guided chromatin tagmentation sequencing (ACT-seq), 143
Anticancer compounds, 827
Antigen-presenting cells (APC), 716–717, 804
Antisense transcripts, 174
Aorta-gonad-mesonephros (AGM), 117
AP1 gene. *See* *APETALA1* gene (*AP1* gene)
APC. *See* Adenomatous polyposis coli (APC); Antigen-presenting cells (APC)
APC gene, 766
APE. *See* Apurinic endonuclease (APE)
APETALA1 gene (*AP1* gene), 503
Apis mellifera, 530
APL. *See* Acute promyelocytic leukemia (APL)
Aplysia protein, 74–76
APOBEC-coupled epigenetic sequencing (ACE-Seq), 131
APOBEC3A. *See* Apolipoprotein mRNA editing enzyme, catalytic polypeptide-like 3 A (APOBEC3A)
Apolipoprotein mRNA editing enzyme, catalytic polypeptide-like 3 A (APOBEC3A), 129
Apoptosis, 41
APP. *See* Amyloid precursor protein (APP)
APP intracellular domain (AICD), 742–744
Apurinic endonuclease (APE), 327

AR gene. *See* Androgen receptor gene (AR gene)
Arabidopsis, 502, 506–507, 515, 552–555
  CpG sites, 556
  DNA methylation, 556–557
  genome, 266–267
  plants, 274
Arabidopsis hallerii, 270–271
Arabidopsis thaliana, 127–128, 264–265, 272, 324, 337, 500, 532, 554
ARABIDOPSIS TRITHORAX-RELATED3 (*ATXR3*), 507
ARABIDOPSIS TRITHORAX-RELATED7 (*ATXR7*), 506–507
Arc. *See* Activity-regulated cytoskeleton associated protein (Arc)
Arc gene, 452–454
Archaea, 669–673
ARG kinases, 202
Arginine vasopressin (avp), 748
Arginines, 36
Argonaute (Ago), 331
Argonaute 2 (Ago2), 631–632
ARGONAUTE proteins (AGO proteins), 264
Argonaute proteins (Ago proteins), 174
Arsenic (As), 651–652
ART. *See* Assisted reproductive technologies (ART)
Artificial epigenetic modifications, 845–846
Artificial hypomethylated epialleles, 275
AS. *See* Angelman syndrome (AS)
Ascaris lumbricoides, 682
ASDs. *See* Autism-spectrum disorders (ASDs)
Asexually reproducing populations, 534–536
ASI1-IMMUNOPRECIPITATED PROTEIN 1 (AIPP1), 270
Asparagines (N), 76
Assay for transposase-accessible chromatin sequencing (ATAC-seq), 139
Assisted reproductive technologies (ART), 783
AT. *See* Adipose tissue (AT)
AT-MSCs. *See* Adipose-tissues-derived MSC (AT-MSCs)
ATAC-seq, 277
ATAC-seq. *See* Assay for transposase-accessible chromatin sequencing (ATAC-seq)
Ataxia-telangiectasia mutated (ATM), 35
Ataxin 1, 391–392
Ataxin 7, 391–392
ATM. *See* Ataxia-telangiectasia mutated (ATM)
Atmospheric hypoxia, 371
ATP. *See* Adenosine triphosphate (ATP)
ATP-citrate lyase (ACLY), 614
ATP-dependent chromatin remodeling, 272
ATRA. *See* All-trans retinoic acid (ATRA)
ATRX-DNMT3-DNMT3L (ADD), 15
Attention-deficit/hyperactivity disorder (ADHD), 651
ATX1 gene, 503–504
ATX1-like histone lysine methyltransferases in Plants, 502–505
  plant H3K4 HKMTs, 505t

ATX2 gene, 503–504
ATX5-like histone lysine methyltransferases in plants, 505–506
ATXR3, 507
  ATXR3-like histone lysine methyltransferases in Plants, 507–508
ATXR3. *See* ARABIDOPSIS TRITHORAX-RELATED3 (*ATXR3*)
ATXR7. *See* ARABIDOPSIS TRITHORAX-RELATED7 (*ATXR7*)
Atypical Ser kinases, 202
AUD. *See* Alcohol use disorder (AUD)
Aurora-B, 290–291
Autism, 383
Autism-spectrum disorders (ASDs), 389
Autoimmune disorder, 310–311, 677, 716–717, 724
  epigenetic mechanisms influence autoimmune processes, 715–716
  epigenetic modifications use as potential biomarkers, 726–727
  epigenetic modifiers use for potential diagnosis and therapy in, 727–728
  genetic risk factors in, 717–719
  immune cells, epigenetics of, 719–720
  mechanisms of autoimmunity, 716–719
  multiomics and machine learning, 728–730
  multiple sclerosis, 723–724
  rheumatoid arthritis, epigenetics of, 722–723
  systemic lupus erythematosus, epigenetics of, 720–722
  systemic sclerosis, 725–726
  type 1 diabetes mellitus, 724–725
Autoimmunity, mechanisms of, 716–719
  antigen presenting, 716f
  puzzle of autoimmunity, 717f
  unraveling cell-type specific effects of genetic risk factors in autoimmune diseases, 717–719
Autosomal dominant adult-onset demyelinating leukodystrophy (ADLD), 97–98
Autosomal dominant cerebellar ataxia, deafness, and narcolepsy (ADCA-DN), 13
avp. *See* Arginine vasopressin (avp)
Axin fused allele (*Axin^{Fu}* allele), 252
Axin^{Fu} allele. *See* *Axin fused* allele (*Axin^{Fu}* allele)
5-aza-2′-deoxycytidine, 595
Azacitidine, 823
5-azacytidin (5-aza), 721, 826–827
Azacytidine, 826–828
Azelaic acid, 274
Aβ. *See* Amyloid β (Aβ)

B

B cells, 719–720
B lymphocyte-induced maturation protein 1 (Blimp-1), 720
B Vitamins, 616
B-cell lymphoma, 795–796

B3galtl gene expression. *See* Beta 1,3-galactosyltransferase-like gene expression (B3galtl gene expression)
BACE1, 744–745
Bacillus anthracis, 204, 208
Bacillus cereus, 208–209
Back-splicing, 170
Bacteria, 669–673, 675, 680–682
  DNA methylation in, 204–207
  protein acetylation in, 203–204
Bacterial infections, 681–682
Bacterial pathogens, 680
Bacteroides, 617–618
BAH. *See* Bromo-Adjacent Homology (BAH)
Balanced cytogenetic abnormalities (BCA), 93
Baldwin effect, 554
Bar gene, 90
Barbara McClintock's hypothesis, 334
Barley (*Hordeum vulgare*), 500
Base excision repair (BER), 18–20, 44, 455
BCA. *See* Balanced cytogenetic abnormalities (BCA)
BDNF. *See* Brain-derived neurotrophic factor (BDNF)
Bdnf DNA methylation, 646
Bdnf gene, 387, 449–450, 452–454, 746–747
Beckwith–Wiedemann syndrome (BWS), 308, 589–590, 781–782
Beleodaq. *See* Belinostat
Belinostat, 794–795, 800–801, 806–807, 830–831
Benign prostatic hyperplasia (BPH), 310
Benzamides, 794–795
Benzothiadiazole-primed wild-type *Arabidopsis* plants (BTH-primed wild-type *Arabidopsis* plants), 274–275
Benzoylation, 145
BER. *See* Base excision repair (BER)
Best linear unbiased prediction (BLUP), 570
BET inhibitor (BETi), 822
BET proteins, 727–728
BET proteins. *See* Bromodomain and extraterminal domain proteins (BET proteins)
Beta 1,3-galactosyltransferase-like gene expression (B3galtl gene expression), 340–341
Betaine, 363–364, 616–617
  depletion, 724
Betaine homocysteine methyltransferase (BHMT), 616–617
BETi. *See* BET inhibitor (BETi)
Bevacizumab, 394
BHMT. *See* Betaine homocysteine methyltransferase (BHMT)
Bifidobacterium, 372–373, 682
BIN1 gene, 742
Bioinformatics analyses, 310
Biological age, 480
Biomarker(s), 484–485
  of aging, 480
  epigenetic modifications use as potential biomarkers, 726–727

and therapeutic potentials of epigenetic marks in livestock management, 590–591
Biotic stress, plant epigenetics and, 273–276
Biotic stressors, 275
Biotin-labeled NTPs (biotin-NTPs), 173
biotin-NTPs. *See* Biotin-labeled NTPs (biotin-NTPs)
Biotrophic pathogen *Pseudomonas syringae*, 275
Bisphenol A (BPA), 334–335, 457, 467–468, 651–653, 766–767
Bisulfite
  bisulfite-converted randomly integrated fragment sequencing method, 276
  bisulfite-treated RNAs, 167
  conversion, 276
  genomic sequencing of DNA, 292–293
  sequencing, 125–126
  treatment, 843
  of DNA, 313–315
Bithorax complexes, 223
Bivalent chromatin, 223
Bivalve (*Crassostrea gigas*), 525
Blastocyst, 289–290
Blimp-1. *See* B lymphocyte-induced maturation protein 1 (Blimp-1)
BLUP. *See* Best linear unbiased prediction (BLUP)
BM. *See* Bone-marrow (BM)
BM-MSCs. *See* Bone marrow-derived mesenchymal stem cells (BM-MSCs); Bone-marrow derived mesenchymal stem cells (BM-MSCs)
BMEC. *See* Bovine mammary epithelial cells (BMEC)
BMI. *See* Body mass index (BMI)
BMI1 gene, 117–118
BNS. *See* BONSAI (BNS)
Body fluids, 306
Body mass index (BMI), 466–467, 490, 767
Bone marrow-derived mesenchymal stem cells (BM-MSCs), 383–384
Bone-marrow (BM), 386
Bone-marrow derived mesenchymal stem cells (BM-MSCs), 396
BONSAI (BNS), 270–271
Boundary elements, 97
Bovine mammary epithelial cells (BMEC), 577
Bovine oocyte, 296–297
BPA. *See* Bisphenol A (BPA)
BPH. *See* Benign prostatic hyperplasia (BPH)
Brachypodium distachyon, 500–501
Brain cancer, 485–486
Brain disorders, epigenetics of
  environmental stress, epigenetic dysregulation by, 748–749
  important epigenetic mechanisms for brain, 738–739
  neurodegenerative disorders, epigenetic dysregulation in, 742–745
  neurodevelopmental disorders, epigenetic dysregulation in, 739–742
  psychiatric disorders, epigenetic dysregulation in, 745–748

Brain-derived neurotrophic factor (BDNF), 383, 467–468, 617–618, 644, 740, 746–747
Brassica rapa, 500–501
BrdU. See Bromodeoxyuridine (BrdU)
Breakthrough phase, genetic alterations, 703–704
BRIC-seq. See 5′-bromo-uridine (BrU) immunoprecipitation chase-deep sequencing analysis (BRIC-seq)
Bromo-Adjacent Homology (BAH), 12–13, 267
5′-bromo-uridine (BrU) immunoprecipitation chase-deep sequencing analysis (BRIC-seq), 172–173
5′-bromo-uridine (BrU), 172
Bromodeoxyuridine (BrdU), 643–644
Bromodomain and extraterminal domain proteins (BET proteins), 722, 821–823
Bromodomain containing protein (4BRD4), 34–35
BrU-labeled RNAs (BrURNA), 172
BrURNA. See BrU-labeled RNAs (BrURNA)
BTH-primed wild-type *Arabidopsis* plants. See Benzothiadiazole-primed wild-type *Arabidopsis* plants (BTH-primed wild-type *Arabidopsis* plants)
Burkitt lymphoma, 94
1-butanol, 371
Butea monosperma, 372–373
Butyrate, 372–373, 675–676
Butyrylation, 145, 147
BWS. See Beckwith–Wiedemann syndrome (BWS)

C

C Repeat Binding Factor (CBF)-Cold-Responsive (COR) pathway, 272
C-5-methylcytosine (5mC), 204
c-Myc gene, 390
C-reactive protein, 487–488
C-terminal. See Carboxy-termina (C-terminal)
C2C12 cells, 407
C^5 position (m5C), 64
Cadherin 13 (CDH13), 305
Cadmium (Cd), 651–652
Caenorhabditis elegans, 38, 169, 324, 363, 523, 538, 641, 749–750
CAGE. See Cap analysis gene expression (CAGE)
Calcineuri gene, 452–454
Calcium calmodulin kinase II alpha, 657
Camellia sinensis. See Green tea (*Camellia sinensis*)
Camk2 gene, 740
cAMP binding protein (CBP), 445
cAMP response element binding protein (CREB), 232–233, 445
Campomelic dysplasia, 97
Camponotus floridanus. See Ant (*Camponotus floridanus*)
Campylobacter rectus, 680
Cancer, 7, 307, 359, 433, 479–480, 485–487, 698–699, 703, 727–728
  biology, 850–851
  cancer-associated DNA methylation, 485
  drug, 595
  epigenetic biomarkers in, 708–709
  genetic and epigenetic classification of, 705–708
  genetics, 698
  pathogenic infections and, 680–683
  PcG-mediated control of gene expression in cancer cells, 117–118
  potential DNA methylation biomarkers in, 307–308
  potential histone modification biomarkers in, 309–310
  potential lncRNA biomarkers in, 311–313
  potential M6A methylation biomarkers in, 308–309
  potential miRNA biomarkers in, 310–311
Cancer cutaneous T-cell lymphoma (CTCL), 830
Cancer epigenetics, 698–705
  cancer epigenome in epidemiology, 708–709
  DNA methylation, 700–702
  epigenetic biomarkers in cancer, 708–709
  epigenetic influences over lifetime and cancer risk, 698–699
  epigenetics as cancer therapeutic targets, 709–710
  epigenetics in intratumoral heterogeneity and therapeutic resistance, 704–705
  genetic and epigenetic classification of cancer, 705–708
  histone modification, 702
  importance of epigenetic changes together with genetic alterations, 703–704
  limitations to epigenetic classification and prognostication, 708
  noncoding RNA, 703
Cancer epigenome in epidemiology, 708–709
Cancer stem cells (CSCs), 704
Cancer therapeutic targets, epigenetics as, 709–710
Candida albicans, white/opaque switch of, 81–82
Cannabinoid receptor 1 (CB1), 749
Cannabinoids, 657
Cannabis, 657
Cap analysis gene expression (CAGE), 173
Capecitabine, 826–827
CAPS. See Chemical-assisted pyridine borane sequencing (CAPS)
CapSTARR-seq, 139
CaptureSeq, 167
CAR T-cell therapy. See Chimeric antigen receptor T-cell therapy (CAR T-cell therapy)
Carbon 5 (C5), 329
Carbonylation, 373–374
Carboplatin, 830–831
Carboxy-termina (C-terminal), 28
5-carboxylcytosine (5caC), 3–4, 12, 130, 290, 329, 455, 613–614, 738
Carcinogenesis, 704, 708–709
Cardiovascular atherosclerotic disease, 611
Cardiovascular disease (CVD), 186, 310, 479–480, 487–488, 651, 761
Carnitine palmitoyltransferase I gene (CPT1), 467–468
Carotenoid Isomerase2 (*CCR2*), 506–507
Casein kinase 2, 407
Casein kinase II (CK2), 41
Casilio system, 854–855
Cathepsin D, 392
Caulobacter, 206
CB1. See Cannabinoid receptor 1 (CB1)
CBD. See CpG-binding domain (CBD)
CBP. See cAMP binding protein (CBP); CREB-binding protein (CBP)
CBT. See Cognitive behavior therapy (CBT)
CBX6 protein, 116–117
CBX7 protein, 116–117
5′ CCCTC 3′-binding factor (CTCF), 335–336, 784
CD226 gene, 724–725
cDC. See Either conventional DC (cDC)
Cdk5. See Cyclin dependent kinase 5 (Cdk5)
CDKN1B. See Cyclin-dependent kinase inhibitor 1B (CDKN1B)
CDKN1C. See Cyclin-dependent kinase inhibitor 1C (CDKN1C)
cDNA library construction, 164–169
  depletion of ribosomal RNA from cDNA libraries, 165–166
  detection of low-abundance RNA species, 167–169
  library preparation for strand-specific RNA-seq, 166–167
CDTs. See Cytolethal distending toxins (CDTs)
CE-NSCs. See Human carboxylesterase (CE-NSCs)
CeL. See Centrolateral amygdale (CeL)
Celecoxib, 827
Cell and tissue differentiation, 528
Cell-based samples, 305
Cell-free nucleosomes (cf-nucleosomes), 308
Cell-free samples, 305
Cell-specific imprinting, 256
Cell-to-cell movement of siRNAs, 268–269
Cells, 111
Cellular RNA, 174
CeM. See Centromedial amygdala (CeM)
CENPB. See Centromere protein B (CENPB)
Central enhancer of MyoD (CER), 407
Central nervous system (CNS), 98, 235, 626–628, 690–691, 739
Centrolateral amygdale (CeL), 447–448
Centromedial amygdala (CeM), 447–448
Centromere protein B (CENPB), 337
Centromere structures, 337
CEP. See Cognitive epigenetic priming (CEP)
CER. See Central enhancer of MyoD (CER)
ceRNA network, 186–187
ceRNAs network analysis. See Competitive endogenous RNAs network analysis (ceRNAs network analysis)
CES. See Chromatin entry sites (CES)
cf-nucleosomes. See Cell-free nucleosomes (cf-nucleosomes)

CFC. *See* Contextual fear conditioning (CFC)
cfDNA. *See* Circulating-free DNA (cfDNA)
CFTR. *See* Cystic fibrosis transmembrane conductance regulator (CFTR)
CG. *See* Crippled growth (CG)
ChAT. *See* Choline acetyltransferase (ChAT)
CheA. *See* Chemotaxis protein A (CheA)
Chemical compound-mediated inducible systems, 855–856
Chemical environmental stressors, 699
Chemical moieties, 675
Chemical-assisted pyridine borane sequencing (CAPS), 130
Chemotaxis, 204
Chemotaxis protein A (CheA), 202–203
Chemotaxis protein Y (CheY), 202–203
Chemotherapy, 822
cheRNAs. *See* Chromatin-enriched RNAs (cheRNAs)
CheY. *See* Chemotaxis protein Y (CheY)
ChIC. *See* Chromatin immunocleavage (ChIC)
Chidamide, 801, 805
ChIL. *See* Chromatin integration labeling (ChIL)
Chimeric antigen receptor T-cell therapy (CAR T-cell therapy), 710
ChIP. *See* Chromatin immunoprecipitation (ChIP)
ChIP-chip, 315–316
ChIPseq. *See* Chromatin immunoprecipitation sequencing (ChIPseq)
Chlamydia, 680–681
Chlamydomonas, 78, 505–507, 515
Chlamydomonas reinhardtii, 500, 558–559
Choline, 363–364, 616–617, 640
Choline acetyltransferase (ChAT), 643–644
ChOR-seq. *See* Chromatin occupancy after replication and sequencing (ChOR-seq)
Chrom-seq. *See* Chromatin-associated RNA-seq (Chrom-seq)
Chromatin, 89, 335, 357, 444, 499–500, 642, 650, 680, 763
  accessibility profiling methods, development of, 277
  boundary elements, 90
  conformation capture techniques, 335–336
  one-carbon metabolism and histone methylation, 360–362
  regulation of chromatin dynamics, 360–362
  remodelling, 614
Chromatin entry sites (CES), 230–231
Chromatin histone code, 639–640
Chromatin immunocleavage (ChIC), 140
Chromatin immunoprecipitation (ChIP), 386, 843
Chromatin immunoprecipitation sequencing (ChIPseq), 4–5, 140, 149, 152–153, 173, 277, 293–294, 530–531, 577–578, 717–719
Chromatin integration labeling (ChIL), 140
Chromatin loop reorganization using CRISPR-dCas9 (CLOuD9), 856

Chromatin modification, 274–275, 552, 821
Chromatin occupancy after replication and sequencing (ChOR-seq), 137–138
Chromatin organization, 27–28, 137–138
Chromatin regulator, 138, 153
Chromatin remodeling, 569–570, 573
Chromatin structure, 821, 852
Chromatin topology of plants, 278
Chromatin-associated RNA-seq (Chrom-seq), 167
Chromatin-binding protein (BRD4), 143
Chromatin-enriched RNAs (cheRNAs), 167
Chromatin-linked adaptor for MSL proteins (CLAMP), 230–231
Chromatin-remodeling agents, 822–823
Chromium, 651
Chromobox2 (CBX2), 226
CHROMOMETHYLASES (CMTs), 264
Chromosomal instability (CIN), 705
  CIN-driven pathways, 705–708
Chromosomal position effect (CPE), 90
  Bar gene, 90
  CPE-associated pathogenesis, 93–94
  in flies, 90
  genome topology and long-distance interactions, 92–93
  in human pathologies, 93–98
    aniridia, 95t
    autosomal recessive neurodegenerative, 96–97
    Burkitt lymphoma, 94
    campomelic dysplasia, 97
    chromosomal rearrangements, 93
    constitutional syndromes, 95t
    CPE-associated pathogenesis, 93–94
    developmental syndromes, 94
    DNA copy number variation, 94
    FMR1 gene, 96–97
    malignant hemopathies, 94
    mesomelic dysplasia, 97–98
    Pelizaeus–Merzbacher disease, 98
    repeat expansions disease, 96
    split-hand/foot malformation, 97
    structural variants, 93
    structural variations associated with disease phenotype, 93f
    translocations of genes, 94
  in model organisms, 90–92
    chromatin configuration modulates gene, 91f
  telomeric position effect, 98–101
  White gene, 90
Chromosome, 419, 501–502
Chromosome conformation capture (3C), 690
Chronic lymphocytic leukemia (CLL), 60
Chronic myelomonocytic leukemia (CMML), 826–827
Chronic social stress, 471
Chronic stress, 469
Chronic unpredictable mild stress (CUMS), 746–747
Chronological age, 642
CHRR1. *See* Corticotrophin-releasing hormone receptor 1 (CHRR1)
CIA. *See* Collagen-induced arthritis (CIA)

CIB1. *See* Cryptochrome-interacting basic-helix-loop-helix 1 (CIB1)
CIMP. *See* CpG island methylator phenotype (CIMP); CpG-island methylator phenotype (CIMP)
CIN. *See* Chromosomal instability (CIN)
Ciona intestinalis. *See* Sea squirt (*Ciona intestinalis*)
Cip1-interacting zinc finger protein 1 (CIZ1), 425
circRNA. *See* Circular RNA (circRNA)
Circular RNA (circRNA), 55, 163–164, 703
  algorithm for circular RNAs detection, 170–171, 171t
Circular RNAs, 58–59
Circulating tumor cells (CTCs), 309
Circulating-free DNA (cfDNA), 305
Cis elements important for amyloid prion formation, 76
Cis-acting factors, 783–784
Cis-regulators, 90
Cis-regulatory elements (CREs), 138, 268
Citrullination, 44–45
CIZ1. *See* Cip1-interacting zinc finger protein 1 (CIZ1)
CK2. *See* Casein kinase II (CK2)
CKM gene, 409
CLAMP. *See* Chromatin-linked adaptor for MSL proteins (CLAMP)
Class I transposable elements, 324–328
  LINEs, 327
  LTR retrotransposons, 324–327
  SINEs, 327–328
Class II transposable elements, 332
Class II transposable elements, 328
Classification system, 324
ClassIIa HDACs, 232–233
Cleavage under targeted accessible chromatin (CUTAC), 143
Cleavage under targets and release using nuclease (CUT&RUN), 140–143
Cleavage under targets and tagmentation (CUT&Tag), 140–143
Clinical trials of HDACi, 795–796
  for hematological malignancies treatment, 797t
  phase I trial, 805–806
  phase II study, 805–806
  as single agents for treatment of solid tumors, 803t
CLL. *See* Chronic lymphocytic leukemia (CLL)
Clostridium perfringens, 681
Clostridium perfringens beta2 (CPB2), 577
CLOuD9. *See* Chromatin loop reorganization using CRISPR-dCas9 (CLOuD9)
Clustered regularly interspaced short palindromic repeats (CRISPR), 7–8, 595, 633
Clustered regularly interspaced short palindromic repeats and CRISPR-associated protein (CRISPR/Cas), 843–845
CMML. *See* Chronic myelomonocytic leukemia (CMML)

CMS. *See* Consensus molecular subtypes (CMS)
CMTs. *See* CHROMOMETHYLASES (CMTs)
CNS. *See* Central nervous system (CNS)
CNV. *See* Copy number variation (CNV)
CoBATCH method. *See* Combinatorial barcoding and targeted chromatin release (CoBATCH method)
Cocaine, 655–657
Codon substitution frequency (CSF), 169
Coefficient of variation, 189–190
Coffin–Lowry syndrome, 384
Cognitive behavior therapy (CBT), 747
Cognitive epigenetic priming (CEP), 449–452
Cognitive function, activity-dependent histone acetylation and, 232–233
Cognitive function, epigenetic language in postmitotic neurons underlying, 232–238
Cognitive function, histone deacetylase interplay in, 233–234
"Cold acclimation", 272
Cold stress, 272
Collagen-induced arthritis (CIA), 723
Colonoscopy, 307
Colony stimulating factor 1 receptor (CSF1R), 339
Colorectal cancer (CRC), 305–307, 677
Colorectal carcinogenesis, 677–680
Comamonas testosteroni, 209
Combination epigenetic therapy
  chromatin structure and remodeling via combined epigenetic treatment, 824f
  chromatin-remodeling agents, combined treatment, and targeted therapy, 822–823
  clinical trial, 833–836
  clinical trials in epigenetic combined treatment, 825t
  DNA methylation/demethylation and combined treatment, 823–828
  histone modifications and histone-modifying enzymes for epigenetic treatment, 828–833
Combinatorial barcoding and targeted chromatin release (CoBATCH method), 143
Combined assay of transcriptome and enriched chromatin binding (CoTECH), 144–145
Combined treatment, 822–823
Competitive endogenous RNAs network analysis (ceRNAs network analysis), 185–186
  basic principles of, 185–186
  comparison of mathematical models for, 190
  computational epigenetic approaches for, 190–195
  mathematical models for, 187–190
  network, 186–187
Computational epigenetics
  computational epigenetic approaches for competitive endogenous RNAs network analysis, 190–195

databases and Tools for ceRNA network analysis, 191t
and diseases, 190
Conditional-based mutual information approach, 195
Conditioned place preference (CPP), 457
Cone-rod homeobox protein (CRX), 391
Congenital heart disease, 383
Consensus molecular subtypes (CMS), 705
Context-based ceRNA prediction approach, 190, 194
Context-based hypergeometric test-coexpression approach, 194
Contextual fear conditioning (CFC), 445–446, 449, 648
Conventional methylation-specific PCR (MSP), 315
Cooption of retroviral elements, 334
Coprococcus, 618
Copy number variation (CNV), 94, 781–782
Coregonus clupeaformis, 540
Coreopsis, 372–373
Coriobacteriace, 618
CORT-supplementation study. *See* Corticosterone-supplementation study (CORT-supplementation study)
"Cortical clock", 487
Corticosterone-supplementation study (CORT-supplementation study), 647
Corticotrophin-releasing hormone receptor 1 (CHRR1), 746–747
Corticotrophin-releasing-factor (CRF), 468, 648
Corticotropin-releasing factor receptor 2 (CRFR2), 749
CoTECH. *See* Combined assay of transcriptome and enriched chromatin binding (CoTECH)
Cotton (*Gossypium raimondii*), 500–501
Covalent histone modification, 373–374
  histone acetylation, 367–373
    histone acetyltransferase inhibitors and activators, 370–371
    histone deacetylation, 371–373
  histone demethylation, 362–367
  metabolism, diet, and phytogenic compounds on histone methylation and acetylation, 365t
  typical modifications in histone methylation, 364t
  metabolic and dietary control of histone and transcriptional dynamics, 359–360
  diet, 360
  metabolism, 359
  regulation of chromatin dynamics, 360–362
Covalent histone modifications, 220
CPB2. *See Clostridium perfringens* beta2 (CPB2)
CPE. *See* Chromosomal position effect (CPE)
CpG. *See* Cytosine-phosphate-guanosine (CpG)
CpG island methylator phenotype (CIMP), 705

CpG-binding domain (CBD), 739
CpG-density promoters (IPC), 329
CpG-island methylator phenotype (CIMP), 702, 705–708
CPP. *See* Conditioned place preference (CPP)
CPT1. *See* Carnitine palmitoyltransferase I gene (CPT1)
Crassostrea gigas. *See* Bivalve (*Crassostrea gigas*)
CRC. *See* Colorectal cancer (CRC)
CREB. *See* cAMP response element binding protein (CREB)
CREB-binding protein (CBP), 232–233, 385–386, 738–739, 742–744, 854–855
CREBBP. *See* CREB-binding protein (CBP)
CREs. *See* Cis-regulatory elements (CREs)
CRF. *See* Corticotrophin-releasing-factor (CRF)
CRFR2. *See* Corticotropin-releasing factor receptor 2 (CRFR2)
Cri-Du-Chat Syndrome, 100
Crippled growth (CG), 80–81
CRISPR. *See* Clustered regularly interspaced short palindromic repeats (CRISPR)
CRISPR RNA (crRNA), 844–845
CRISPR-dCas9 system (CLOuD9 system), 278
CRISPR/Cas. *See* Clustered regularly interspaced short palindromic repeats and CRISPR-associated protein (CRISPR/Cas)
Crotonylation, 145, 147
crRNA. *See* CRISPR RNA (crRNA)
CRX. *See* Cone-rod homeobox protein (CRX)
CRY2. *See* Cryptochrome 2 (CRY2)
Cryptochrome 2 (CRY2), 855
Cryptochrome-interacting basic-helix-loop-helix 1 (CIB1), 855
Cryptosporidium, 682
Cryptosporidium parvum, 683
Crystallography, 16
CSCs. *See* Cancer stem cells (CSCs)
CSF. *See* Codon substitution frequency (CSF)
CSF1R. *See* Colony stimulating factor 1 receptor (CSF1R)
CTCF. *See* CCCTC-binding factor (CTCF)
CTCs. *See* Circulating tumor cells (CTCs)
CUMS. *See* Chronic unpredictable mild stress (CUMS)
Curcuma longa. *See* Turmeric (*Curcuma longa*)
Cushing syndrome, 589–590
CUT&RUN. *See* Cleavage under targets and release using nuclease (CUT&RUN)
CUT&Tag. *See* Cleavage under targets and tagmentation (CUT&Tag)
CUTAC. *See* Cleavage under targeted accessible chromatin (CUTAC)
Cutaneous T-cell lymphoma (CTCL), 795–796, 823
CVD. *See* Cardiovascular disease (CVD)
CXCR3 gene, 433
CXorf21 gene, 433
Cyclic peptides, 794–795
Cyclin dependent kinase 5 (Cdk5), 633, 742
Cyclin-dependent kinase inhibitor 1B (CDKN1B), 404–405

Cyclin-dependent kinase inhibitor 1C (CDKN1C), 308, 404–405
Cycloidea gene, 554–555
Cyclomodulins, 680
Cyp7a1 gene, 615
Cystic fibrosis, 796
Cystic fibrosis transmembrane conductance regulator (CFTR), 808
Cytolethal distending toxins (CDTs), 680
Cytoplasmic symbionts, 73–74
Cytosine to uracil (C-to-U), 331–332
Cytosine-5-methylenesulfonate, 130
Cytosine-guanine dinucleotides, 738
Cytosine-phosphate-guanosine (CpG), 452–454, 573, 738
  island, 452–454, 525, 700
    arrays, 123
  methylation, 738, 850
    site-specific CpG methylation, 854
  rich TEs, 333
  sites, 329
Cytosines (C), 125–126, 465–466, 700
Cytotaxis, 78

D

D4Z4 gene, 101
DA. *See* Dopamine (DA)
Dahlia gene, 372–373
DAM. *See* DNA adenine methyltransferase (DAM)
DamID, 92
Danio rerio. *See* Zebrafish (*Danio rerio*)
Daphnia pulex, 523
Darwin's domestication syndrome, 539–540
DAT. *See* Dopamine transporter (DAT)
"Date-rape drug", 654
DC. *See* Dendritic cells (DC)
dCas9-SunTag system, 852
dCas9. *See* DeactivatedM Cas9 (dCas9)
DCC. *See* Dosage compensation complex (DCC)
ddm1-based epiRILs, 275
DDR. *See* DNA damage response (DDR)
DDX3X gene, 433
De novo methyltransferases, 62
DeactivatedM Cas9 (dCas9), 595
Deazaneplanocin A (DZNep), 829
3-deazaneplanocin A hydrochloride (DZNep), 726
Decitabine, 823, 826–828
Deep brain stimulation, 388
DeepBlue, 689
DEFICIENS gene, 270–271
DeGeorge syndrome critical region 8 (DGCR8), 739
Deinococcus radiodurans, 205–206
Delta-9-tetrahydrocannabinol (THC), 469, 657
Dementia, 487
Demethylases, 444–445
Demethylation, 614
Dendritic cells (DC), 717–720
Dendrobium catenatum, 500–501
Dentate gyrus (DG), 388, 449
Deoxy-UTP (dUTP), 167

Depression, 186
  DNA Methylation in, 747
  histone modifications and histone modifier in, 746–747
  non-coding RNA in, 747–748
Depsipeptide. *See* Romidepsin
Deterministic selection models, 542
Deubiquitinating enzymes (DUBs), 41–42
Developmental Origins of Health and Disease (DOHaD), 615
Developmental programming, 766
Developmental syndromes, 94
Dexamethasone, 681
DG. *See* Dentate gyrus (DG)
DGCR8. *See* DeGeorge syndrome critical region 8 (DGCR8)
DHMCs. *See* Differentially hydroxymethylated cytosines (DHMCs)
DHS. *See* DNase I hypersensitive sites (DHS)
DHU. *See* Dihydrouracil (DHU)
Diabetes, 676
  in pregnancy, 767–768
Diabetes mellitus type 2 (T2D), 611
Diagnosis and therapy in autoimmune diseases, 727–728
Diagnostic and Statistical Manual of Mental Disorders (DSM-5), 626
Diagnostics, molecular, 785–786
Dicentrarchus labrax, 538
DICER-LIKE proteins, 264
Dickkopf-related protein 1 (DKK1), 725
Dictyostelium intermediate repeat sequence (DIRS), 324
Diet, 360, 763–764
Diet-derived macro-nutrients, 363
Diet-derived micro-nutrients, 363
Dietary factors, 6–7
Dietary fiber, 618–619
Dietary HDACi, 372–373
Dietary nutrients, 763
Dietary patterns, 359
Differentially hydroxymethylated cytosines (DHMCs), 540
Differentially methylated cytosines (DMCs), 576
Differentially methylated genes (DMGs), 530
Differentially methylated region (DMR), 291, 308, 342, 528, 554, 576, 640, 688, 779
Diffuse cutaneous SSc (dSSc), 725
Diffuse Large B-cell Lymphoma (DLBCL), 795–796
Dihydrouracil (DHU), 130
Dimerization partner RB-like, E2F, and MuvB (dREAM), 408
Dinucleotide "CpGs", 700
Diploid embryos, 253
DIRS. *See* Dictyostelium intermediate repeat sequence (DIRS)
Disease, 186, 310
  aberrant histone modifications role in, 45
  alteration of histone modifications in, 153
  pathogens, 592
  potential DNA methylation biomarkers in cancer and, 307–308

  potential histone modification biomarkers in cancer and, 309–310
  potential lncRNA biomarkers in, 311–313
  potential M6A methylation biomarkers in cancer and, 308–309
  potential miRNA biomarkers in, 310–311
  potential non-coding RNA biomarkers in, 310–313
Disruptor of telomeric silencing-1 (Dot1), 36, 42–43
Distal regulatory region (DRR), 405–406
Distal SAR signaling, 274
DKK1. *See* Dickkopf-related protein 1 (DKK1)
DLBCL. *See* Diffuse Large B-cell Lymphoma (DLBCL)
DM. *See* Steinert myotonic dystrophy (DM)
DMAP1. *See* DNA methyltransferase associated protein 1 (DMAP1)
DMCs. *See* Differentially methylated cytosines (DMCs)
DMD. *See* Duchenne muscular dystrophy (DMD)
DME. *See* DNA demethylases (DME)
DMGs. *See* Differentially methylated genes (DMGs)
DMR. *See* Differentially methylated region (DMR)
DMRs. *See* Differentially methylated CpG regions (DMRs)
DMTi. *See* DNA methyltransferase inhibitor (DMTi)
DNA, 276, 843
DNA adenine methyltransferase (DAM), 204
DNA cytosine methylation, 264
DNA cytosine modifications
  5-hydroxymethylcytosine mapping methodologies, 130–131
  ACE-Seq, 131
  DNA methylation and hydroxymethylation profiles, 124f
  enzymatic methyl-sequencing and pyrimidine borane-based methods, 129–130
  infinium and epic methylation bead chips, 127
  infinium EPIC arrays, 127f
  RRBS, 126f
  MBD protein-based affinity pulldown, 124–125
  methylated DNA immunoprecipitation, 124
  methylated-CPG island recovery assay, 125
  OxBS-Seq, 131
  single cell whole genome methylation analysis, 132–133
  SMRT-seq and nanopore sequencing, 131–132
  sodium bisulfite-based approaches, 128–129
  TAB-seq, 131
  targeted bisulfite sequencing, 125–127
  WGBS, 127–128
DNA Damage gene, 645–646

DNA damage response (DDR), 35
  histone phosphorylation role in, 40–41
  histone ubiquitination role in, 43
DNA demethylases (DME), 265
  activities, 264
  enzymes ten-eleven translocation (Tets), 642–643
DNA demethylation, 17–20, 19f, 613–614, 848
  active, 18–20
    TET proteins, 18
    TET proteins in development, 20
    TET-dependent demethylation pathways, 18–20
  passive, 17–18
DNA fluorescent in-situ hybridization experiments (DNA-FISH experiments), 232–233
DNA homology-dependent repair pathway (HDR pathway), 363
DNA hydroxymethylation, 748–749
DNA hypermethylation, 652
DNA methylation, 3–4, 11–16, 38–39, 59–60, 123–124, 206, 219–220, 264–266, 271–272, 275, 290, 293, 303, 330, 333, 382, 409–410, 426–427, 465–466, 469, 480, 490, 521–523, 522t, 525–526, 538, 551–552, 554–555, 569–570, 573, 611–614, 631, 639–640, 680–681, 698, 700–702, 721, 726–727, 738, 749, 761, 764, 769, 846–848
  in AD, 744–745
  advances in detection of, 276
  arrays, 717–719
  in bacteria, 204–207
  brain, 738
  changes with age, 480
  clocks in age-related disease
    aging, 479–480
    biomarkers of aging, 480
    DNA methylation changes with age, 480
    epigenetic clocks, 480
    first generation epigenetic clocks, 480–484
    second generation epigenetic clocks, 484–485
  de novo, 15–16
    DNMT3A and DNMT3B, 15–16
    DNMT3L, 16
  in depression, 747
  DNA methylation/demethylation and combined treatment, 823–828
    DNMT inhibitors, 826–828
  dynamic changes in early embryo development, 291
  dynamics during plant life cycle, 264–266
  in early-life stress, 748–749
  epigenetic age modified, 489–491
  epigenetic clocks and age-related disease, 485–488
    cancer, 485–487
    CVD, 487–488
    dementia, 487
    frailty and physical disability, 488
    T2D, 488
  epigenetic clocks without age-related disease focus, 488–489
  in eukaryotes and prokaryotes, 205t
  and long-term memory, 452–455, 453f
  maintenance DNA methylation, 12–15
    DNMT1, 12–13
    UHRF1, 13–15
  major proteins involved in, 14f
  during mammalian embryonic development, function of, 291
  mechanisms of, 13f
  in MS, 392
  neurodevelopmental disorders, 739–740
  and regulation, 264
DNA methylome, 125, 291, 573–574, 723
DNA methyltransferase (DNMT), 382, 709
DNA methyltransferase 1 (DNMT1), 12–13, 250, 290, 721, 725–726
DNA methyltransferase 2 (Dnmt2), 219–220
DNA methyltransferase 3A (DNMT3A), 15–16, 255, 407–408, 452–454
DNA methyltransferase associated protein 1 (DMAP1), 12–13
DNA methyltransferase inhibitor (DMTi), 835
DNA methyltransferases (DNMTs), 12, 204, 272, 612–613, 628, 642–643, 738
DNA modifications, 217–218
DNA repair
  histone acetylation role in, 35–36
  histone methylation role in, 39
DNA transposons, 220–221
DNA-binding molecules, 844–845
  CRISPR/Cas9 system, 844–845
  locus-specific epigenome editing, 850–852
  TAL proteins, 844
  ZF proteins, 844
DNA-binding proteins, 277
DNA-binding response regulator (DNA-binding RR), 202
DNA-binding RR. See DNA-binding response regulator (DNA-binding RR)
DNA-FISH experiments. See DNA fluorescent in-situ hybridization experiments (DNA-FISH experiments)
DNA-methylation, 592
DNA-RNA hybrid, 269
"DNAm PhenoAge", 484
DNase hypersensitivity assay, 257
DNase I hypersensitive sites (DHS), 139
DNMT. See DNA methyltransferase (DNMT)
Dnmt gene (Dnmt3c), 329
DNMT inhibitor (DNMTi), 822, 826–828
DNMT1 gene, 12–13
Dnmt1 knockout (KO), 13
DNMT1. See DNA methyltransferase 1 (DNMT1)
Dnmt1/3a/3b triple KO (TKO), 13
Dnmt2. See DNA methyltransferase 2 (Dnmt2)
DNMT3, 725–726
DNMT3 mRNA, 469
DNMT3-like (DNMT3L), 15–16, 258
  Dnmt3l-dependent methylation, 293
  Dnmt3l-independent methylation, 293

DNMT3A. See DNA methyltransferase 3A (DNMT3A)
DNMT3A gene, 15–16, 292–293
Dnmt3a/3b double KO (DKO), 13
DNMT3A2, 15
DNMT3B gene, 12, 15–16, 292–293, 652
Dnmt3c. See Dnmt gene (Dnmt3c)
DNMT3L. See DNMT3-like (DNMT3L)
Dnmt3L KO, 16
DNMTi. See DNMT inhibitor (DNMTi)
DNMTs. See DNA methyltransferases (DNMTs)
DOHaD. See Developmental Origins of Health and Disease (DOHaD)
Domain associated with SET in Trx (DAST), 503
Domain Rearranged Methyltransferases, 264
Domesticated animals, 538
Domestication of animal, 538–540
Dopamine (DA), 626–628
Dopamine D1 receptor gene Drd1a, 648
Dopamine receptor D2 (Drd2), 657
Dopamine replacement therapy, 388
Dopamine to H3 at glutamine 5 to form H3Q5dop (H3Q5dop), 148
Dopamine transporter (DAT), 388–389
Dosage compensation, 229–232
Dosage compensation complex (DCC), 229, 231
  male-specific lethal complex/dosage compensation complex, 229–230
  regulation of dosage compensation complex targeting to X Chromosome, 230–231
  spreading and target recognition, 231
  transcriptional activation of active x-linked genes by, 231–232
Dot1. See Disruptor of telomeric silencing-1 (Dot1)
Double fertilization, 266
Double strand break (DSB), 35
Double-stranded DNA (dsDNA), 324–326
Double-stranded β-helix (DSBH), 18
Drd2. See Dopamine receptor D2 (Drd2)
dREAM. See Dimerization partner RB-like, E2F, and MuvB (dREAM)
Drosophila, 3, 5, 37, 90, 92, 111–112, 216, 222–223, 226–227, 229, 233, 324, 340, 445–446, 502, 749–750
  Aβ, 235
  connectome, 216
  disease models to study dysregulation of Hat activity, 234–235
  genetics, 238
  genome, 216, 221–222
    heterochromatin within, 220–221
  MOF, 297
  S2 cells, 174
  Tip60 mediated restoration of histone acetylation homeostasis, 236f
Drosophila MB. See Drosophila mushroom body (Drosophila MB)
Drosophila melanogaster, 99, 216, 329–330, 500, 528, 560
  Antennapedia complexes, 223

Bithorax complexes, 223
chromatin packaging and gene control regulation, 217–220
  DNA methylation, 219–220
  histone acetylation, 218–219
  histone methylation, 219
  noncoding RNAs, 219
development
  Hox gene expression, epigenetic modifications maintain patterns of, 223
  during M-Phase, 228–229
  PcG proteins and gene repression, 223–226
  PREs and PcG-TrxG-mediated gene expression, 226–227
  role of epigenetics, 223–229
  during S-Phase, 227–228, 228f
  silencing and activation, mechanisms of, 223
  transgenerational epigenetic inheritance, 227
  TrxG proteins and gene activation, 226
dosage compensation, 229–232
  dosage compensation complex spreading and target recognition, 231
  male-specific lethal complex/dosage compensation complex, 229–230
  regulation of dosage compensation complex targeting to X Chromosome, 230–231
  transcriptional activation of active x-linked genes by dosage compensation complex, 231–232
epigenetic language in postmitotic neurons underlying cognitive function, 232–238
  activity-dependent histone acetylation and cognitive function, 232–233
  Drosophila disease models to study dysregulation of HAT activity, 234–235
  histone acetylation targeted therapeutics for neurodegenerative disorders, 235–237
  histone acetyltransferase, 233–234
  learning and memory, environmental enrichment improves, 237–238
Eu-Het transition zone, 220
homeotic (Hox genes), 216
Hox clusters organization, 224f
kenyon cells, 217
methyl/phos switch, 228–229
model organism in epigenetic research, 216–217
mushroom body (MB), 237–238
position-effect variegation, 220–223
  epigenetic histone modifications regulate position-effect variegation in, 221–223
  heterochromatin Within *Drosophila* Genome, 220–221
  protein regulators of position-effect variegation, 221
Drosophila mushroom body (*Drosophila* MB), 233

Drosophila neurodegenerative models, 235
Drosophila neuromuscular junction, 216–217
Drosophila visual system, 216–217
Drought stress, 273
DRR. *See* Distal regulatory region (DRR)
Drug, 490
Drug addiction (DA), 625
  epigenetics of
    epigenetic processes in, 630–632
    epigenetic targets for diagnosis and treatments to combat addiction and SUDs, 632–634
    and reward circuits, 626–628
    synaptic plasticity, learning and memory and addiction, 628–629
    synergistic and opposing effects of intersectional signaling, 632
Drugs of abuse, 6–7
DSB. *See* Double strand break (DSB)
DSBH. *See* Double-stranded β-helix (DSBH)
dsDNA. *See* Anti-double stranded DNA (dsDNA); Double-stranded DNA (dsDNA)
DSM-5. *See* Diagnostic and Statistical Manual of Mental Disorders (DSM-5)
DSN treatment. *See* Duplex-specific nuclease treatment (DSN treatment)
dSSc. *See* Diffuse cutaneous SSc (dSSc)
DUBs. *See* Deubiquitinating enzymes (DUBs)
Duchenne muscular dystrophy (DMD), 404, 412, 808–809
Duplex-specific nuclease treatment (DSN treatment), 165
dUTP. *See* Deoxy-UTP (dUTP)
DUX4 gene, 101
DWARF1 gene, 554–556
Dyrec-seq, 174
Dysregulation of lncRNAs, 313
Dystrophin, 808–809
DZNep. *See* Deazaneplanocin A (DZNep)

E
E(var) genes, 221
E(var)3-93E genes, 221
E2F transcription factor 6 (E2F6), 343
E2F6. *See* E2F transcription factor 6 (E2F6)
E3 ubiquitin-protein ligase Mdm2 (MDM2), 404–405
EAE. *See* Experimental allergic encephalomyelitis (EAE)
EARLY FLOWERING 6 (ELF6), 267
Early heading date 1 (Ehd1), 504–505
Early microbiome exposure and epigenetic influence, 675
Early-life stress, 748–749
  DNA methylation in, 748–749
  intergenerational and transgenerational epigenetic effects of, 749
  non-coding RNA in, 749
Early-life trauma, 749
EBF transcription factor 1 (EBF1), 719
EBF1. *See* EBF transcription factor 1 (EBF1)
EBV. *See* Epstein-Barr virus (EBV)
EBVs. *See* Estimated breeding values (EBVs)
EC. *See* Entorhinal cortex (EC)

ECM. *See* Extracellular matrix (ECM)
Ecotypes, 270–271
Ectoderm, 383
EDCs. *See* Endocrine disrupting chemicals (EDCs)
EDM2. *See* Enhanced Downy Mildew 2 (EDM2)
EDSS. *See* Expanded Disability Status Scale (EDSS)
EE. *See* Environmental enrichment (EE)
Effector-triggered immunity (ETI), 274
EFS. *See* Event-free survival (EFS)
EGIR. *See* European Group for the Study of Insulin Resistance (EGIR)
Ehd1. *See* *Early heading date 1 (Ehd1)*
EIPV. *See* Environmentally induced phenotypic variation (EIPV)
Either conventional DC (cDC), 720
Elaeis guineensis, 270–271
Electron microscopy (EM), 226
Elevated microsatellite alterations at selected tetranucleotide (EMAST), 705
ELF6. *See* EARLY FLOWERING 6 (ELF6)
EM. *See* Electron microscopy (EM)
EM-seq method. *See* Enzymatic methyl-sequencing method (EM-seq method)
EMAST. *See* Elevated microsatellite alterations at selected tetranucleotide (EMAST)
Embryo development, Histone acetylation and, 297–298
Embryo transfer experiments, 257
Embryogenesis, 251
  epigenetic re-programming during, 251
Embryonic development, 526–528
  DNA methylation during embryonic and larval development, 527f
Embryonic exposure, 653
Embryonic stem cells (ESCs), 11, 38, 62–63, 114, 332–333, 383, 844
  PcG-mediated control of gene expression in, 116–117
Emergent properties, 79–80
EMQN. *See* European Molecular Quality Network (EMQN)
EMT. *See* Epithelial to mesenchymal transition (EMT)
enChIP. *See* Engineered DNA-binding molecule-mediated ChIP (enChIP)
enChIP in combination with RNA-sequencing (enChIP-RNASeq), 859–860
enChIP in combination with RT-PCR (enChIP-RT-PCR), 859–860
Encode Portal, 689
ENCODE Project. *See* Encyclopedia of DNA Elements Project (ENCODE Project)
Encoding estrogen receptor alpha gene (ESR1 gene), 467–468
Encyclopedia of DNA Elements Project (ENCODE Project), 689
Endocrine disrupting chemicals (EDCs), 652–653
Endocrine disruptors, 334
Endogenous metabolism, 359

Endogenous retroviruses (ERVs), 326–327, 333
Endothelial dysfunction, 726
Engineered DNA-binding molecule-mediated ChIP (enChIP), 858–859
　identification of epigenetic and related molecules in locus-specific manner by, 859–860
　other methods, 862
　scheme of, 860f
　in vitro enChIP, 860–862
Engineered DNA-binding molecules, 843–845
　and application to locus-specific epigenome editing, 845f
　characteristics in locus-specific epigenome editing, 857–858
　CRISPR/Cas9 system, 844–845
　TAL proteins, 844
　ZF proteins, 844
Enhanced Downy Mildew 2 (EDM2), 270
Enhancer, 138–139
　adoption, 97–98
　dual activity of, 139
Enhancer of variegation (E(var)), 90
Enhancer of zeste homolog 2 (EZH2), 702, 710, 719
Enhancer of zestes, 221
Enhancer RNAs (eRNAs), 139, 154–155, 163–164, 854
Entamoeba histolytica, 682
Enterococcus faecalis, 677
Entinostat, 794–795, 800, 805–806
Entorhinal cortex (EC), 448–449
Environment x epigenetics x gene interactions, 641
Environmental enrichment (EE), 237–238, 643–644
Environmental exposures, 650–651
Environmental hazard, 653
Environmental influence, extent of, 641–642
　parental and ancestral predisposition, 641
　postnatal and lifetime influence, 642
　　ageing, 642
　　postnatal to adulthood, 642
　　prenatal exposure, 641–642
　research case study, 660–662
　　audience, 660
　　challenges and solutions, 661–662
　　expected results and deliverables, 661
　　learning and knowledge outcomes, 662
　　objective, 660
　　process, workflow and actions taken, 661
　　rationale, 660–661
　　results, 661
　　safety considerations, 661
　　scope, 660
　　tools and materials used, 661
Environmental stress, 383
　epigenetic dysregulation by, 748–749
Environmental stressors, 334
Environmentally induced phenotypic variation (EIPV), 531–532
Enzymatic methyl-sequencing method (EM-seq method), 129–130, 276

EpCAM gene, 850
ePHD. See Extended plant homeodomain finger (ePHD)
(epi) genome-wide approaches, 786
Epialleles, 342
ePIANNO tool, 689
EPIC methylation bead chips, 127
Epidemiology, cancer epigenome in, 708–709
Epigallocatechin gallate, 412, 826
Epigallocatechin-3-gallate (EGCG). See Epigallocatechin gallate
Epigenetic abnormalities, 303
Epigenetic age, 480, 489–491
　biology of, 489–490
　genetics of epigenetic clocks, 490
　links with lifestyle and health factors, 490–491
Epigenetic aging, 480, 485–486
Epigenetic biomarker, 590–591
　in cancer, 708–709
　detection methods in clinic, 313–316
　distinct advantages over genetic biomarkers, 304
　field cancerization and, 306–307
　minimally invasive tissues are suitable for detecting epigenetic biomarkers, 305–306
　　available tests for DNA methylation detection in clinic, 307t
　　potential DNA methylation biomarkers in cancer and diseases, 307–308
　　potential histone modification biomarkers in cancer and other diseases, 309–310
　　potential M6A methylation biomarkers in cancer and diseases, 308–309
　　potential non-coding RNA biomarkers in diseases, 310–313
　　　potential lncRNA biomarkers in cancer and diseases, 311–313
　　　potential miRNA biomarkers in cancer and diseases, 310–311
Epigenetic changes, 552–554
　cope with fluctuating environments, 553f
　in glioma stem cells, 394–395
　　acetylation, 395
　　methylation, 394–395
Epigenetic classification and prognostication, limitations to, 708
Epigenetic clock, 6, 480, 485–488, 642
　without age-related disease focus, 488–489
　genetics of, 490
Epigenetic contributions to adaptation, limits of, 559–560
Epigenetic control mechanisms, 217
Epigenetic diagnosis, 658
Epigenetic donors, 650
Epigenetic drift, 480
Epigenetic dynamic, 655
Epigenetic dysregulation
　in neurodegenerative disorders, 742–745
　in neurodevelopmental disorders, 739–742
　in psychiatric disorders, 745–748
Epigenetic enzymes, 639–640, 642–643
Epigenetic epidemiology, 708–709

Epigenetic histone modifications regulate position-effect variegation in Drosophila, 221–223
Epigenetic influences over lifetime and cancer risk, 698–699
Epigenetic inheritance, 73–74, 552, 588
　amyloid prions
　　of Saccharomyces cerevisiae, Podospora anserina, and organisms, 74–76
　　variants, 77
　of Avy metastable epiallele, 253f
　cis elements important for amyloid prion formation, 76
　crippled growth, 80–81
　cytotaxis of cilia and complex structures, 78
　environmental modulation of, 252–253
　genetic control of amyloid prion formation and propagation, 76–77
　lactose operon and positive feedback loop, 79–80
　mixed heredity, 78–79
　regulatory inheritance, 79
　self-driven assembly of hsp60 mitochondrial chaperonin, 77–78
　structural heredity, 74
　white/opaque switch of Candida albicans, 81–82
Epigenetic language in postmitotic neurons underlying cognitive function, 232–238
Epigenetic markers, 250–251, 306–307, 553, 612
　dynamics during development, 526–530
　　aging, 528–530
　　cell and tissue differentiation, 528
　　embryonic development, 526–528
　in livestock management, 590–591
Epigenetic mechanisms, 271–272, 274, 359, 362, 387, 569–570, 690, 720, 725
　for brain, 738–739
　　DNA Methylation, 738
　　histone modifications, 738–739
　crosstalks between epigenetic modifications, 209
　for disease, 654
　DNA methylation in bacteria, 204–207
　evolution in animal kingdom, 521–524, 524f
　　DNA methylation, 522–523
　　histone modifications, 523
　　non-coding RNAs, 523–524
　　and role in gene regulation, features of, 524–526
　generation of phenotypic variation in populations by, 530–534, 531f
　　phenotypic plasticity, 531–534
　　polyphenism in insects, 530–531
　impact lipid metabolism and livestock products, 583–587
　influence autoimmune processes, 715–716
　metabolic sensing by, 612–615
　　DNA methylation, 612–614
　　histone modification, 614–615
　　non-coding RNA modification, 615

in plants
  evolution of plant histone lysine methyltransferases methylating H3K4, 502–508
  histone lysine methylation, 500
  histone lysine methyltransferases in plants, 500–502
protein methylation, 207–209
protein phosphorylation in prokaryote, 201–204
protocols for molecular modeling, 210–211
regulate livestock immune response and health outcomes, 576–583
  disease pathogens and stress factors impacting livestock health, 578t
  potential miRNA biomarkers, 580t
regulate livestock reproduction and growth, 573–575
role in animal breeding planning and management, 587–595
role in livestock breeding, 591–595
Epigenetic medicine, 655–657, 656f
Epigenetic memory, 223, 227, 268, 274, 766, 771
of metabolic diseases, 766–768
Epigenetic modifications, 272, 384
of BDNF gene, 644
crosstalks between, 209
DNA methylation during mammalian early embryo development, 291–293
  dynamic changes of DNA methylation in early embryo development, 291
  function of DNA methylation during mammalian embryonic development, 291
dynamic changes and function of histone modifications in early embryo development, 293–298
of histone proteins regulate chromatin packaging and gene control in *Drosophila*, 217–220
pathogenic infections and, 680–681
use as potential biomarkers, 726–727
Epigenetic modifiers
histone acetylation/deacetylation and, 829–833
histone methylation/demethylation and, 828–829
use for potential diagnosis and therapy in autoimmune diseases, 727–728
Epigenetic phenomena in plants, 268–269
genome imprinting, 268
mobile sRNAs, 268–269
noncoding RNAs, 269
paramutation, 268
vernalization, 268
Epigenetic priming trial in patients with newly diagnosed AML, 833–836
Epigenetic processes, 446–447
Epigenetic profiles, 303
Epigenetic programming, 251
in gametogenesis, 250–251
  epigenetic re-programming in mouse development, 250f

Epigenetic re-programming, 251
during embryogenesis, 251
of neoplastic cells, 704
Epigenetic reader proteins, 727–728
Epigenetic regulation
of lncRNAs, 64–65
of miRNA expression, 60–62
Epigenetic regulation, 822
Epigenetic remodelling, 762
Epigenetic therapeutics, opportunities for, 434
Epigenetic therapy, 698, 709, 822
Epigenetic transformations, 682
Epigenetic variation, 534, 551–552
modeling spontaneous epigenetic variation, 555–559
  left panel, 559f
  mutation-selection balance, 557f
spontaneous epigenetic variation, 554–555
Epigenetic-wide association studies (EWAS), 632–633
Epigenetically determined phenotypes, transgenerational inheritance of, 542–543
Epigenetics, 3, 59–60, 216, 307, 382–383, 443, 551–552, 570, 611–612, 639–640, 769, 843
for animal ecology, 534–538
  asexually reproducing populations, 534–536
  invasions and adaptive radiations, 536–538
  parasites, 538
  sessile taxa, 536
for animal evolution, 538–543
  domestication, 538–540
  polyploid speciation, 540–542
  transgenerational inheritance of epigenetically determined phenotypes, 542–543
in cancer, 699–705
as cancer therapeutic targets, 709–710
CircRNAs and, 67
combined therapies and possible combination with, 831–833
DA and reward circuits, 626–628
development of animal breeding, 570–571
DNA methylation, 382
in MS, 392
in drug addiction, 630–632
  DNA methylation, 631
  histone modifications, 630–631
  miRNA, 631–632
and effect on metabolism, 677
engineered DNA-binding molecules, 844–845
environmental influence on
  exerting environment, 642–643
  extent of, 641–642
  mental or physiological environment, 643–650
epigenetic changes in glioma stem cells, 394–395
epigenetic epidemiology, 6–7
epigenetic therapy, 7

eukaryotic chromosomal organization, 384
evolutionary epigenetics, 6
factors influencing epigenetic changes, 5–6
future of, 7–8
and glioblastoma, 394
in HD, 387
histone acetylation, 382, 393
histone methylation, 382
and histone modifications, 28
histone phosphorylation, 382
histone ubiquitination and SUMOylation, 382–383
histones and structure, 384
and human brain, 383–394
and human disease, 7
Huntingtin Gene, 385–386
identification and related molecules in locus-specific manner by enChIP, 859–860
of immune cells, 719–720
impairment of one-carbon metabolism in PD, 387–388
in intratumoral heterogeneity and therapeutic resistance, 704–705
iPSC Models of RTT, 390
livestock epigenetics case study, 595–598
locus-specific epigenome editing, 845–858
locus-specific identification of epigenetic molecules interact with target genomic regions, 858–862
MeCP2 Gene and Function, 389–390
of memory processes
  DNA methylation and long-term memory, 452–455
  epitranscriptomics, 457
  histone acetylation and hats, 445–447
  histone deacetylation and histone deacetylases, 447–448
  histone methylation, 448–449
  histone modifications, 449–450
  histone posttranslational modifications, 444–445
  histone variant exchange, 456–457
  manipulating histone modifications, 450–452
methods in, 4–5
micro-RNA, 393
by miRNAs, 62–64
model organisms of, 5
molecular mechanisms of, 3–4
in MS, 394
and neurological disorders, 384–385
in PD, 389
of phenotypic variation in livestock health and production, 571–587
piRNAs and, 66
plasticity, and evolving of inheritance, 472–473
regulation by lncRNAs, 65–66
regulation of transposable elements in genic regions, 269–270
revisiting animal breeding planning and management, 587–595
of rheumatoid arthritis, 722–723

Epigenetics (Continued)
  in RTT, 390–392
  in SCA, 391–392
  stem cell therapy
    for HD, 386–387
    for MS, 393–394
  stem cells, 383–384
  synaptic plasticity, learning and memory and addiction, 628–629, 629t
  synergistic and opposing effects of intersectional signaling in epigenetic regulation, 632
  of systemic lupus erythematosus, 720–722
  transposable elements impact on epigenetics of gene regulatory regions, 337–340
  utilizing epigenetic targets for diagnosis and treatments to combat addiction and SUDs, 632–634
Epigenetics inheritance, 589
Epigenome, 639–640, 762–763
  from Bulk to Single-Cell, 276–278
    advances in detection of DNA methylation, 276
    advances in epigenome editing, 278
    development of chromatin accessibility profiling methods, 277
    editing chromatin topology of plants, 278
    implementation of transcriptional regulators, 278
    mapping genome structure of single cells, 277–278
  classification and structure of TE, 324–328
  emerging technologies for epigenome studies in plants, 276–278
  epigenetic control of transposable elements, 329–334
  factors influencing transposable elements epigenetics, 334–335
  influence of transposable elements on host epigenetics, 335–343
  postnatal maternal regulation of, 468–469
  transposable elements genomic annotation, 328–329
Epigenome editing
  advances in, 278
  technologies application in livestock breeding, 595
Epigenome-wide association studies (EWAS), 591–592, 708
EpiMINE tool, 689
Epimutation, 781–782
Epithelial to mesenchymal transition (EMT), 117
Epitranscriptome editing, 857
Epitranscriptomics, 457
EPSP. See Excitatory postsynaptic potential (EPSP)
Epstein-Barr virus (EBV), 334–335, 681
Equol, 677
EricScript tool, 171–172
ERK pathway. See Extracellular signal-regulated kinase pathway (ERK pathway)

ERK/MAPK. See Extracellular signal-regulated kinase/mitogen activated protein kinase (ERK/MAPK)
Erlotinib, 827
eRNAs. See Enhancer RNAs (eRNAs)
6-erucin, 372–373
ERVs. See Endogenous retroviruses (ERVs)
Escape from X-Inactivation Tumor-Suppressor (EXITS), 433
Escape genes, 421–422
Escherichia coli, 73–74, 140, 202–204, 576–577
ESCs. See Embryonic stem cells (ESCs)
Esophageal cancer, 306–307
ESR1 gene. See Encoding estrogen receptor alpha gene (ESR1 gene)
EST. See Expressed sequence tag (EST)
Estimated breeding values (EBVs), 570
Estrogen receptor alpha gene (ERα gene), 587
Etanercept therapy
  multiomics and machine learning accurately predict clinical response to, 728–730
Etheostoma olmste, 540
ETI. See Effector-triggered immunity (ETI)
Euchromatic gene expression, 227–228
Euchromatin, 89, 357–358, 361
Euchromatin-heterochromatin transition zone (Eu-Het transition zone), 220
Eukaryotic Chromosomal Organization, 384
Eukaryotic genomes, 27
Eukaryotic nuclei, 357
European Group for the Study of Insulin Resistance (EGIR), 611
European Molecular Quality Network (EMQN), 785–786
Eutherian mammals, 419
Event-free survival (EFS), 835
Evolution of epigenetic mechanisms in animal kingdom, 521–524
Evolutionary models, 552
EWAS. See Epigenetic-wide association studies (EWAS); Epigenome-wide association studies (EWAS)
Exaptation, 323–324
Excitatory postsynaptic potential (EPSP), 390
Exerting environment, 642–643
EXITS. See Escape from X-Inactivation Tumor-Suppressor (EXITS)
Exosomes, 305–306
Expanded Disability Status Scale (EDSS), 393–394
Expansion phase, genetic alterations, 703–704
Experience-dependent epigenetic inheritance, 471–472
Experimental allergic encephalomyelitis (EAE), 724
Exposome, 589
Expressed sequence tag (EST), 15
Extended plant homeodomain finger (ePHD), 503
Extensive epigenetic re-programming, 250
Extracellular matrix (ECM), 406, 725–726

Extracellular signal-regulated kinase pathway (ERK pathway), 721
Extracellular signal-regulated kinase/mitogen activated protein kinase (ERK/MAPK), 449
Extracellular vesicles (VE), 412
Extrinsic epigenetic age acceleration, 483
Ezh2 gene, 115–116, 720

F

F/Y-rich C-terminus (FYRC), 503
F/Y-rich N-terminus (FYRN), 503
FAANG. See Functional Annotation of Animal Genome Consortium (FAANG)
FABP4. See Fatty acid binding protein 4 (FABP4)
Facio-Scapulo-Humeral dystrophy (FSHD), 101
FAD. See Flavin adenosine dinucleotide (FAD)
FAIRE-seq. See Formaldehyde-assisted isolation of regulatory elements-seq (FAIRE-seq)
FAK. See Focal adhesion kinase (FAK)
FAM208A (factor for gene silencing), 92
FAP. See Fibroadipogenic progenitors (FAP)
Farm management, 592
Farydak. See Panobinostat
FAS. See Fetal alcohol syndrome (FAS)
Fasciola hepatica, 682
FASD. See Fetal alcohol spectrum disorders (FASD)
FASTKD1 gene, 145–147
Fatty acid binding protein 4 (FABP4), 386–387
Fatty acids, 360
FDA. See US Food and Drug Administration (FDA)
FDRA. See Friedreich ataxia (FDRA)
Fecal immunochemical testing (FIT), 307
Fecal occult-blood testing, 307
Feed-back amplifying mechanism, 331
Feeding practices, 592
Fertilization, 289–290
Fetal alcohol spectrum disorders (FASD), 644–645, 654–655
Fetal alcohol syndrome (FAS), 662
Fetal metabolic programming
  epigenetic modifications, 612t
  maternal diet and metabolic epigenome, 616–619
    B Vitamins, 616
    choline and betaine, 616–617
    dietary fiber and SCFA, 618–619
    high fat/high sugar diet, 617–618
    ketogenic diet, 618
    vitamin D, 617
  metabolic sensing by epigenetic mechanisms, 612–615
  prenatal nutrition impact on fetal reprogramming, 615
Fetal reprogramming, prenatal nutrition impact on, 615
FFARs. See Free fatty-acid receptors (FFARs)

FFAs. *See* Free fatty acids (FFAs)
Fgf1. *See* Fibrous growth factor 1 (Fgf1)
Fibroadipogenic progenitors (FAP), 412
Fibroblasts, 716–717
Fibrous growth factor 1 (Fgf1), 645–646
Field cancerization, 306–307
Firre locus, 426
First generation epigenetic clocks, 480–484
　Hannum's epigenetic clock, 483–484
　Horvath's epigenetic clock, 483
　initial studies, 480–483
　　DNA methylation clocks, 481t
　performance comparisons, 484
Fisher's exact test, 195
Fisher's method, 195
FIT. *See* Fecal immunochemical testing (FIT)
FK228. *See* Romidepsin
Flavin adenosine dinucleotide (FAD), 36, 362, 614, 616
Flavin mononucleotide (FMN), 616
"Flexibility loci", 81
Fli-1 proto-oncogene (FLI1), 725–726
FloChIP method, 144
Flowering Locus C (FLC), 267, 503
Flowering Wageningen (FWA), 270
5-fluorouracil (5-FU), 802
Fly nervous system, 216
Fly synapses, 216–217
FMN. *See* Flavin mononucleotide (FMN)
FMR1 gene, 96–97
Focal adhesion kinase (FAK), 406, 408
Folate, 363–364, 699
Folic acid, 640
Food animal industry, 569–570
Food fermentation, 677
Forkhead box (FOXO), 404–405
Forkhead box P1 (FOXP1), 692
Forkhead Box P3 (Foxp3), 719
Formaldehyde-assisted isolation of regulatory elements-seq (FAIRE-seq), 275
5-formylcytosine (5fC), 3–4, 12, 130, 290, 329, 455, 613–614, 738
Fos-related antigen 2-dependent manner (FRA2-dependent manner), 726
FosB expression, 631
4D chromatin structures in cell nuclei, 690–691
FRA2-dependent manner. *See* Fos-related antigen 2-dependent manner (FRA2-dependent manner)
Fragaria vesca. *See* Strawberry (*Fragaria vesca*)
Fragile X syndrome (FXS), 96–97, 313, 852
Fragile X tremor/ataxia syndrome (FXTAS), 313
Frailty, 488
Free fatty acids (FFAs), 618
Free fatty-acid receptors (FFARs), 676
Friedreich ataxia (FDRA), 96
Frontotemporal lobar degeneration (FTLD), 742
FSHD. *See* Facio-Scapulo-Humeral dystrophy (FSHD)
FTLD. *See* Frontotemporal lobar degeneration (FTLD)

Full-length RNA single-nucleus RNA-seq-based methods (FlsnRNA-seq-based methods), 277
Functional Annotation of Animal Genome Consortium (FAANG), 591
"Functional DMR", 640
Fusion RNA detection, algorithm for, 171–172, 172t
"Fuzzy boundaries", 90
FWA. *See* Flowering Wageningen (FWA)
FXN gene, 96–97
FXS. *See* Fragile X syndrome (FXS)
FXTAS. *See* Fragile X tremor/ataxia syndrome (FXTAS)

G

G9a (epigenetic factor), 20
G9a-related methyltransferase, 295
GA. *See* Gibberellic acid (GA)
GABA. *See* γ-aminobutyric acid (GABA)
GAD1. *See* Glutamate decarboxylase 1 (*GAD1*)
Gadd45b isoform, 455
Gadd45g isoform, 455
Gallus gallus, 525, 538
Gametes maturation, 289–290
Gametogenesis, 251
　epigenetic programming in, 250–251
Ganciclovir (GCV), 397
Gastric cancer (GC), 303, 306–307, 679
Gastric lymphoma, 679
Gastrointestinal tract (GI tract), 675
GATA binding protein 3 (Gata3), 719
Gata3. *See* GATA binding protein 3 (Gata3)
Gaussian distribution model, 190
Gaussian mathematical model, 187
GB
　MSC transplantations for, 396
　neural stem cell transplantation for, 397
GBLUP. *See* Genomically best linear unbiased prediction (GBLUP)
GbM. *See* Gene-body methylation (GbM)
GC. *See* Gastric cancer (GC); Granulosa cells (GC)
Gcn5-related acetyl transferases (NAT), 30
GCV. *See* Ganciclovir (GCV)
GDM. *See* Gestational diabetes mellitus (GDM)
gDMR. *See* Germline DMR (gDMR)
gDMRs. *See* Germline differentially methylated regions (gDMRs)
GDNF. *See* Glial cell line-derived neurotrophic factor (GDNF)
GEBV. *See* Genomic estimated breeding value (GEBV)
Gene
　activation, 226
　duplication, 501–502
　epigenetic regulation
　　of transposable elements and interactions with genes, 269–271
　　of transposable elements located within genes, 270–271
　gene-targeted approaches, 126–127
　in glioblastoma, 395–396

　hypermethylated status of MGMT Gene, 396
　hypermethylation status of IDH 1 and 2, 395–396
　promoter, 554
　repression, 223–226
　transposable elements impact on genetics and epigenetics of gene regulatory regions, 337–340
Gene editing approach, 570–571
Gene expression, 112–114, 187, 449–450, 465–466, 505, 738
　for cell, 187
　data, 187–189
　modulation, 822
　patterns, 227
　profiles, 189
　regulation of gene expression by transposable element-derived cis-elements, 337–340
Gene ontology (GO), 395, 575–576
Gene regulatory mechanisms, 111
Gene-body methylation (GbM), 264
Genetic biomarkers, 303–304
Genetic inheritance, 588
Genetic markers, 570
Genetic mechanisms of mutation, 699
Genetic risk factors in autoimmune diseases unraveling cell-type specific effects of, 717–719
Genetic variants alter histone modifications, 149–152
Genetics, 59–60
　transposable elements impact on genetics of gene regulatory regions, 337–340
Genome, 384, 571
Genome defense model, 329
Genome editing technologies in livestock breeding, application of, 595
Genome editing tools, 844
Genome imprinting, 266, 268
Genome topology, 92–93
Genome-wide analysis, 304, 315–316, 338, 386
　of DNA methylation, 676
Genome-wide approach to determine RNA stability, 172–173
Genome-wide assay, 308
Genome-wide association studies (GWAS), 149, 393, 490, 570–571, 632–633, 689, 708, 717
Genome-wide ChIP-on-CHIP data, 203
Genome-wide DNA methylation, 721, 723–725, 765–766
Genome-wide mapping of histone modifications, 145–149
　acylation, 145–148
　homocysteinylation, 148
　monoaminylation, 148
　O-GlcNAcylation, 148–149
Genome-wide screens, 256
Genome-wide sequencing method, 704
Genome-wide transcriptome, 163–164
Genomic estimated breeding value (GEBV), 570–571

Genomic imprinting, 60, 250, 253–259, 289–290, 292–293, 342, 779
  animal breeding, 589–590
  cataloging imprinted domains, 255
  discovery of, 253–254
  disorders, 308
  embryo manipulation techniques in mouse, 254f
  function of imprinted genes, 256–257
  mechanisms of, 255
  noncanonical imprinting, 257–258
  tissue and cell-specific imprinting, 256
  transient imprinting, 258–259
Genomic platforms, 700
Genomic selection, 570–571
Genomically best linear unbiased prediction (GBLUP), 570–571
Genotype Tissue Expression Project (GTEx Project), 338, 689
Germline differentially methylated domains, 255
  molecular mechanisms of maternal genomic imprinting, 256f
Germline differentially methylated regions (gDMRs), 252
Germline DMR (gDMR), 342, 779
Germline-mediated transgenerational inheritance, 470–471
Gestational diabetes mellitus (GDM), 647, 767–768
GFP. *See* Green fluorescent protein (GFP)
GH-CSH. *See* Growth hormone-chorionic somatomammotropin hormone (GH-CSH)
GI tract. *See* Gastrointestinal tract (GI tract)
Giardia intestinalis, 682
Gibberellic acid (GA), 506
Givinostat, 794–795
GlcNAcylation, 614
Glial cell line-derived neurotrophic factor (GDNF), 388, 655, 746–747
Glioblastoma
  epigenetics and, 394
  genes involved in, 395–396
  glioblastoma multiforme and stem cell therapy, 394
  stem cell transplantations in, 396–397
Glioma stem cells (GSCs), 394–395
  epigenetic changes in, 394–395
Global DNA methylation, 532, 542
  measurement of, 532
Global methylation analysis of B cell, 724–725
Global run-on sequencing (GRO-seq), 173, 231–232
Global transcriptome, 723
Global-based ceRNA prediction approaches, 190
β-globin gene, 340
Glucocorticoid receptor (GR), 647–648, 748
Glucocorticoids, 681
Glucose, 360
Glucose transporter 4 (*GLUT4*), 764
GLUT4. *See* Glucose transporter 4 (*GLUT4*)
Glutamate decarboxylase 1 (*GAD1*), 468–469

Glutamines (Q), 76, 364–367
Glutarylation, 145
Glycerol-phosphate acyltransferase 1 (Gpam), 615
GO. *See* Gene ontology (GO)
Gossypium raimondii. *See* Cotton (*Gossypium raimondii*)
Gpam. *See* Glycerol-phosphate acyltransferase 1 (Gpam)
GR. *See* Glucocorticoid receptor (GR)
Granulosa cells (GC), 410
Green fluorescent protein (GFP), 332
Green tea (*Camellia sinensis*), 827
GrimAge, 485
gRNA. *See* Guide RNA (gRNA)
GRO-seq. *See* Global run-on sequencing (GRO-seq)
Growth arrest and DNA damage gene (*Gadd45*), 645–646
Growth hormone-chorionic somatomammotropin hormone (GH-CSH), 647
GSCs. *See* Glioma stem cells (GSCs)
GTEx Project. *See* Genotype Tissue Expression Project (GTEx Project)
Guanidine hydrochlorid, 76
Guanines (G), 700
Guanosine to adenosine mutations (G-to-A mutations), 331–332
Guide RNA (gRNA), 595, 844–845
Gut microbiome, 650, 676
  early microbiome exposure and epigenetic influence, 675
  gut microbiota, inflammation, and colorectal carcinogenesis, 677–680
  human gut microflora, 675–677
  microbiome, epigenetics, and effect on metabolism, 677
  pathogenic infections and cancer, 680–683
  pathogenic infections and epigenetic modifications, 680–681
GWAS. *See* Genome wide association studies (GWAS)

H

Heligmosomoides polygyrus, 682–683
H. polygyrus TGF-β mimic (Hp-TGM), 682–683
H2A monoubiquitination (H2Aub), 41
H2A monoubiquitination at lysine 119 (H2AUb), 405
γH2A. X, 27–28
H2A/H4 phosphoserine 1 (H2A/H4S1ph), 298
H2AFY gene, 97–98
H2AK119ub1 (gene promoter), 114–115
H2B monoubiquitination (H2Bub), 41
H2BK120ub (gene promoter), 37–38
H3 at position 4 (H3K4), 64
H3 glutamine 5 (H3Q5), 631
H3 Ser-10 phosphorylation (H3S10p), 273
H3 with unmethylated arginine 2 (H3R2me0), 14–15
H3 with unmodified lysine 4 (H3K4me), 15

H3K27ac. *See* Histone mark histone 3 lysine 27 acetylation (H3K27ac)
H3K27ac ChIP-seq, 152
H3K36 tri-methylation (H3K36me3), 37–38, 59–60, 266–267
H3K4me3-K27me3, 38–39
H3K9 methylation, 448–449
  function in early embryo development, 295
H3K9 methyltransferase (HMTase), 90, 258
H3K9K14Ac, 273
H3K9me2/3, 38
H3K9me3, 59–60, 294–295, 340–341, 500
H4 lysine 20 (H4K20), 500
H4K8hib, 147–148
HA. *See* Hemagglutinin (HA); Histone acetylases (HA)
Haemophilus influenzae, 207
Hanks-type Ser/Thr kinases (STKs), 202
Hannum clocks, 484
Hannum's epigenetic clock, 483–484
HAS. *See* High-affinity sites (HAS)
HAT. *See* Histone acetyltransferase (HAT)
HAT inhibitor, 445–446
Hazardous environmental pollutants and chemicals, 650–658
  addictive substances, 654–658
    alcohol, 654–655
    cannabinoids, 657
    cocaine, 655–657
    nicotine, 657–658
    opioids, 657
  heavy metals and pollutants, 651–654
    air pollution, 653–654
    arsenic and cadmium, 651–652
    EDC, 652–653
    lead, 651
    methylmercury, 652
    pesticides, 653
  prescription medicine, 654
    γ-hydroxybutyrate, 654
    VPA, 654
Hazardous substances, 650–651
HBV. *See* Hepatitis B virus (HBV)
HCC. *See* Hepatocellular carcinoma (HCC)
HCMV. *See* Human cytomegalovirus (HCMV)
HCP. *See* High-CpG density promoters (HCP)
HCS. *See* Holocarboxylase synthetase (HCS)
HD. *See* Huntington disease (HD)
HDAC. *See* Histone deacetylases (HDAC)
HDAC inhibitor (HDACi), 412, 447, 727, 822, 829–831
HDACi. *See* HDAC inhibitor (HDACi); Histone deacetylase inhibitors (HDACi); Inhibitors against HDAC (HDACi)
HDL. *See* High-density lipoproteins (HDL)
HDL-C. *See* High-density lipoprotein cholesterol (HDL-C)
HDLs. *See* High density lipoproteins (HDLs)
HDR pathway. *See* DNA homology-dependent repair pathway (HDR pathway)

INDEX

Heading date 3a (*Hd3a*), 504–505, 508
Heart failure (HF), 311
Heat shock proteins, 272
Heat stress, 272
Heavy metals, 640, 651–654
    toxicants, 651, 653–654
Hebbian learning rule, 628
Hebbian plasticity, 628
HECT. *See* Homologous to E6AP-C terminus (HECT)
Helicobacter pylori, 206–207, 303, 678
Helicobacterioceae, 618
Heligmosomoides polygyrus, 682–683
Hemagglutinin (HA), 456
Hematological malignancies treatment, HDACi for, 795–801, 797t
Hematopoietic stem cells (HSCs), 114, 116–117, 383–384, 392
HEME OXYGENASE 1 (*HY1*), 506
Hepatic enzymes, 689
Hepatitis B virus (HBV), 625, 682
Hepatocellular carcinoma (HCC), 60, 117, 305–306
Hepatocyte growth factor (HGF), 406
Hepatocyte nuclear factor-4 alpha (HNF-4α), 647
Hereditary sensory, and autonomic neuropathy IE (HSAN IE), 13
Herpes simplex virus type 1 (HSV-1), 334–335, 846
Herpes viruses, 334–335
Herpesviridae, 334–335
HERV-W family members. *See* Human ERV-W family members (HERV-W family members)
hESCs. *See* Human embryonic stem cells (hESCs)
Heterochromatin, 27, 89–90, 220, 264, 266–267, 357–358, 444, 702
    within *Drosophila* genome, 220–221
    spreading, 340–341
Heterochromatin protein 1 (HP1), 220, 267, 331
Heterokaryon incompatibility, 77
Heterozygotic (HZ), 721
HF. *See* Heart failure (HF)
HFO. *See* High-fat overfeeding (HFO)
HGF. *See* Hepatocyte growth factor (HGF)
HHV6. *See* Human herpes virus type 6 (HHV6)
HHV8. *See* Human herpesvirus type 8 (HHV8)
Hi-C analysis of *Arabidopsis*, 267–268
HIF. *See* Hypoxiainducible factor (HIF)
HIF-1a. *See* Hypoxia inducible factor-1 alpha (HIF-1a)
HIF3A. *See* Hypoxia inducible factor 3 alpha (*HIF3A*)
High density lipoproteins (HDLs), 304
High fat/high sugar diet, 617–618
High throughput sequencing (HTS), 123
High throughput sequencing, 230–231
High-Affinity K1 Channel 1 (HKT1), 272–273
High-affinity sites (HAS), 230–231
High-CpG density promoters (HCP), 329
High-density lipoprotein cholesterol (HDL-C), 617–618
High-density lipoproteins (HDL), 490
High-fat diet-induced obesity mouse model, 764
High-fat overfeeding (HFO), 764
High-sensitivity Ms, 145
High-throughput assays for mapping histone modifications, 139–145
    histone modification profiling with low input and single cell, 140–144
        ACT-seq and scChIC-seq, 143
        ChIL-seq and scChIL-seq, 140
        CoBATCH and itChIP-seq, 143
        CUT&RUN and CUT&Tag, 140–143
        microfluidics-based ChIP-seq platforms, 143–144
    joint profiling of histone modification and transcriptome in single cell, 144–145
High-throughput sequencing technologies, 4–5, 220–221, 263, 268
Higher-order mammalian models, 653–654
HilD, 204
Hindlimb suspension (HLS), 412
Hippocampal histone acetylation, 445–446
Hippocampus, 628
Histidine phosphorylation, 382
Histone, 444f, 465–466, 499–500, 528
    epigenetic modification of histone proteins, 217–220
    metabolic and dietary control of, 359–360
    posttranslational modifications, 444–445
        epigenetic modifications and chromatin remodeling, 444f
    protein, 410
    PTMs, 746–747
    and structure, 384
Histone 2 (H2), 382
Histone 3 (H3), 382, 465–466
Histone 3 acetylation (H3Ac), 649
Histone 3 at lysine 9 (H3K9me3), 14–15, 294, 329–330, 500, 552
Histone 3 lysine 27 trimethylation (H3K27me3), 38, 59–60, 139, 167–168, 465–466, 720
    deposition, 225–226
    for mammalian cell lineage commitment, 296
    methyltransferase, 64
    regulates XCI and imprinting, 296
Histone 3 lysine 4 monomethylation (H3K4me1), 138, 720
Histone 3 lysine 4 trimethylation (H3K4me3), 37, 59–60, 273, 720
Histone 4 (H4), 574–575
Histone acetyl transferases (HATs), 90, 444–445, 739
Histone acetylases (HA), 642–643
Histone acetylation, 28–36, 218–219, 297, 367–373, 382, 393, 444–445, 614, 720, 821
    chromatin structure regulates transcriptional activity, 218f
    dynamic changes in early embryo development, 296–297
    and epigenetic modifiers, 829–833
    and hats, 445–447
    role in DNA repair, 35–36
    targeted therapeutics for neurodegenerative disorders, 235–237
    transcription regulation, role in, 34–35
    writers and erasers of, 28–34
        histone modifications, 31t
        histone N-terminal tails in humans and rodents, 29f
        N-terminal tails of core histone, 30f
Histone acetyltransferase (HAT), 28–30, 218–219, 233–234, 266–267, 290–291, 367, 370–371, 382, 514–515, 614, 700, 793, 829
    HAT-containing complexes, 335–336
    inhibitors, 370–371
Histone acylation, 614
Histone alterations, 305
Histone arginine methylation, 39
Histone BDNF, 746–747
Histone chaperones, 272
"Histone code", 28, 37, 357–358, 500, 702
Histone deacetylase 2C (HD2C), 272
Histone deacetylase 3 (HDAC3), 677
Histone deacetylase inhibitors (HDACi), 640
    clinical applications of HDACi for noncancer diseases, 807–809
    and inhibitors, 794t
    inhibitors for hematological malignancies treatment, 795–801
        belinostat, 800–801
        entinostat, 800
        others HDACi, 801
        panobinostat, 799
        romidepsin, 799–800
        valproic acid, 800
        vorinostat, 796–798
    inhibitors in solid tumors treatment, 801–807, 803t
        belinostat, 806–807
        entinostat, 805–806
        others HDACi, 807
        panobinostat, 804–805
        romidepsin, 805
        valproic acid, 806
        vorinostat, 802–804
Histone deacetylases (HDAC), 4, 7, 28–30, 218–219, 266–267, 290–291, 310, 330, 360, 371–372, 382, 444–445, 447–448, 514–515, 630–631, 642–643, 649, 676, 699–700, 710, 719, 739, 793
Histone deacetylation, 362–367, 371–373, 447–448
    and epigenetic modifiers, 828–833
Histone H2A N-terminal residues, 298
Histone H3 lysine 4 (H3K4me3), 500
Histone H3 lysine 9 (H3K9), 828, 848
Histone H3 lysine9 methylation (H3K9me), 264, 828
Histone H4 N-terminal residues, 298
Histone homocysteinylation, 614–615
Histone hyperacetylation, 297, 822

Histone lactylation, 614–615
Histone lysine 2-hydroxyisobutyrylation (Khib), 147–148
Histone lysine glutarylation (Kglu), 147
Histone lysine methylation, 36–39, 500
Histone lysine methyltransferases (HKMTs), 36, 500, 710
  phylogenetic relationships between histone lysine methyltransferases methylating H3K4, 502
  in plants, 500–502
  SET-domain genes in several plants, 501t
Histone lysine β-hydroxybutyrylation (Kbhb), 148
Histone mark, 28, 138, 500
Histone mark histone 3 lysine 27 acetylation (H3K27ac), 720
Histone methylases, 267
Histone methylation (HMT), 36–39, 137, 219, 360–362, 382, 448–449, 500, 526, 828
  epigenetic modifiers, 828–829
  H3K27me1/2/3, 266–267
  H3K36me2/3, 266–267
  H3K4me1/2/3, 266–267
  H3K79me1/2/3, 266–267
  H3K9me1/2/3, 266–267
  H4K20me1/2/3, 266–267
  inhibitors, 828–829
  role in DNA repair, 39
  role in transcription regulation, 36–39
    histone arginine methylation, 39
    histone lysine methylation, 36–39
  writers and erasers of, 36
Histone methyltransferases (HMT), 358–359, 405–406, 444–445, 642–643, 702, 739
Histone modification, 271–272, 569–570, 573, 614–615, 702, 848–850
  aberrant histone modifications role in disease, 45
  acetylation, 28–36
    role in DNA repair, 35–36
    transcription regulation, role in, 34–35
    writers and erasers of, 28–34
  in AD, 742–744
  ADP-ribosylation, 44
  analysis, 276
  catalog of, 149
  chip-seq data analysis workflow, 154–157
  and chromatin structure, 266–268
    histone variants, 267
    regulation of higher-order chromatin structures in plants, 267–268
  citrullination, 44–45
  in depression, 746–747
  dynamic changes and function in early embryo development, 293–298
  for epigenetic treatment, 828–833
  epigenetics and, 28
  genome-wide mapping of histone modifications, 145–149
  high-throughput assays for, 139–145
  main histone modifications, 28–45
  methylation acetylation, 36–39
    role in DNA repair, 39
    role in transcription regulation, 36–39
  writers and erasers of, 36
  and modifying enzymes, 266–267
  nucleosomes and chromatin organization, 27–28
  phosphorylation, 39–41
    role in DNA damage response, 40–41
    role in transcription regulation, 40
  proline isomerization, 45
  in regulatory regions, 138–139
  sumoylation, 43–44
  ubiquitination, 41–43
    role in DNA damage response, 43
    writers and erasers of, 41–42
  variation of, 149–153
    alteration of histone modifications in disease, 153
    genetic variants alter histone modifications, 149–152
Histone modification QTLs (hQTLs), 151
Histone modifications, 4, 111–112, 217–218, 272, 290–291, 309–310, 330, 333, 357–359, 449–450, 523, 528, 551–552, 630–631, 639–640, 681, 702, 843
  brain, 738–739
  non-coding RNA, 739
  manipulating histone modifications, 450–452
    cognitive epigenetic priming, 450f
    reported memory modulators, 451f
Histone modifications in early embryo development, 293–298
  H3K27 methylation dynamic changes in early embryo development, 295–296
  H3K27me3
    necessary for mammalian cell lineage commitment, 296
    regulates XCI and imprinting, 296
  H3K4 methylation
    dynamic changes in early embryo development, 293–294
    function in early embryo development, 294
  H3K9 methylation
    dynamic changes in early embryo development, 294–295
    function in early embryo development, 295
  histone acetylation
    dynamic changes in early embryo development, 296–297
    and embryo development, 297–298
    imprinting histone acetylation and XCI, 297
  histone phosphorylation
    dynamic changes in early embryo development, 298
    function in early embryo development, 298
Histone modifier, 698
  in AD, 742–744
  in depression, 746–747
Histone monoaminylation, 148
Histone phosphorylation, 39–41, 382
  dynamic changes in early embryo development, 298
  function in early embryo development, 298
  role in DNA damage response, 40–41
  role in transcription regulation, 40
Histone protein H3 at position 27 (H3K27), 64
  dimethylation, 746–747
  H3K27-specific HAT CBP, 226
  methylase MEDEA, 268
  methylation, 38–39
    dynamic changes in early embryo development, 295–296
Histone protein H3 at position 4 (H3K4), 64
  dimethylation, 828
  methylation, 36–37
    dynamic changes in early embryo development, 293–294
    function in early embryo development, 294
  plant histone lysine methyltransferases methylating H3K4, 502–508
Histone sumoylation, 43–44
Histone ubiquitination, 41–43, 382–383
  role in DNA damage response, 43
  writers and erasers of, 41–42
    nonexhaustive list of histone modifying enzymes and readers, 42t
Histone variants, 267, 272
  exchange, 456–457
  H2A. X, 28
  H2A. Z, 28
  H3. 3, 28
  linker histone, 147
  MacroH2A, 89
Histone-modifying complexes, 357–358
Histone-modifying enzymes for epigenetic treatment, 828–833
Histone-modifying systems, 500
Histones 3 (H3), 721
HIV. See Human immunodeficiency virus (HIV)
HKT1. See High-Affinity K1 Channel 1 (HKT1)
HLA. See Human leukocyte antigen (HLA)
HLA gene, 724–725
HLS. See Hindlimb suspension (HLS)
hMeDIP, 130
HML2. See Polymorphic HERV-K (HML2)
HMT. See Histone methylation (HMT); Histone methyltransferases (HMT)
HMT inhibitor (HMTi), 822
HMTase. See H3K9 methyltransferase (HMTase)
HMTi. See HMT inhibitor (HMTi)
HNF-4α. See Hepatocyte nuclear factor-4 alpha (HNF-4α)
hnRNPK protein, 425
Hodgkin lymphoma, 339, 800
Holocarboxylase synthetase (HCS), 329–330
Homeotic genes (Hox genes), 111–112, 216
  clusters organization in Drosophila, 224f
  epigenetic modifications maintain patterns of Hox gene expression, 223
Homo sapiens, 98–99
Homocysteine (Hcy), 148, 387–388

Homocysteinylation, 148, 360
Homologous recombination (HR), 39
Homologous to E6AP-C terminus (HECT), 41
Hordeum vulgare. See Barley (*Hordeum vulgare*)
Horvath's epigenetic clock, 483
Host epigenetics, influence of transposable elements on, 335–343
   genomic imprinting and epialleles, 342
   heterochromatin spreading, 340–341
   regulation of gene expression by TE regulatory RNA networks, 342–343
   transposable elements impact on genetics and epigenetics of gene regulatory regions, 337–340
      regulation of gene expression by TE-associated epigenetic marks, 339–340
      regulation of gene expression by transposable element-derived cis-elements, 337–340
   transposable elements influence on chromatin structure, 335–337
      centromere and telomere structures, 337
   TAD, 335–337
   XCI, 341–342
Hox antisense intergenic RNA (*HOTAIR*), 164, 166, 311
Hp-TGM. See *H. polygyrus* TGF-β mimic (Hp-TGM)
HPA. See Hypothalamic-pituitary-adrenal (HPA)
hPSCs. See Human pluripotent stem cells (hPSCs)
HPV. See Human papilloma viruses (HPV)
hQTLs. See Histone modification QTLs (hQTLs)
HR. See Homologous recombination (HR); Hypersensitive response (HR)
HSAN IE. See Hereditary sensory, and autonomic neuropathy IE (HSAN IE)
HSCs. See Hematopoietic stem cells (HSCs)
Hsp104 gene, 76
Hsp40 gene, 76
hsp60 mitochondrial chaperonin, self-driven assembly of, 77–78
hsp60ts mutant strains, 78
Hsp70 gene, 76
HSV-1. See Herpes simplex virus type 1 (HSV-1)
hTER. See Telomerase RNA template (hTER)
HTLV-1. See Human T-cell lymphotropic virus type 1 (HTLV-1)
HTS. See High throughput sequencing (HTS)
HTT gene. See *Huntingtin* gene (HTT gene)
Human androgen receptor (HUMARA), 429
Human *ATRN* gene, 339
Human brain, epigenetics and, 383–394
   DNA Methylation in MS, 392
   epigenetics and neurological disorders, 384–385
   epigenetics in HD, 387
   epigenetics in MS, 394
   epigenetics in PD, 389
   epigenetics in RTT, 390–392

Eukaryotic Chromosomal Organization, 384
   histone acetylation, 393
   histones and structure, 384
   Huntingtin Gene, 385–386
   impairment of one-carbon metabolism in PD, 387–388
   iPSC Models of RTT, 390
   MeCP2 Gene and Function, 389–390
   Micro-RNA, 393
   PARK Gene, 388
   SNCA Gene, 388
   stem cell therapy for HD, 386–387
   stem cell therapy for MS, 393–394
   stem cell therapy for PD, 388–389
   stem cells, 383–384
Human cancers, 306–307
Human carboxylesterase (CE-NSCs), 397
Human cells, 92
Human cytomegalovirus (HCMV), 681
Human diseases, 384
Human embryonic stem cells (hESCs), 116–117, 427
Human epigenome projects and population pharmacoepigenomics, 688–689
Human ERV-W family members (HERV-W family members), 327
Human gut microflora, 675–677
Human herpes virus type 6 (HHV6), 334–335
Human herpesvirus type 8 (HHV8), 678
Human immunodeficiency virus (HIV), 625
   HDAC inhibitor dependent reactivation of latent HIV, 809
Human leukocyte antigen (HLA), 717
Human neurodegenerative diseases, use of *Drosophila* disease models to study dysregulation of HAT activity in etiology of, 234–235
Human papilloma viruses (HPV), 678
Human pathologies, chromosomal position effect in, 93–98
Human pluripotent stem cells (hPSCs), 336–337
HUman Silencing Hub (HUSH), 333
Human T-cell lymphotropic virus type 1 (HTLV-1), 678
HUMARA. See Human androgen receptor (HUMARA)
Huntingtin (HTT), 742
Huntingtin gene (HTT gene), 235, 384–386
Huntington disease (HD), 7, 234, 382, 384–385
   epigenetics in, 387
   Parkinson disease and neural stem cells, 387
   stem cell therapy for, 386–387
HUSH. See HUman Silencing Hub (HUSH)
HY1. See *HEME OXYGENASE 1* (*HY1*)
Hyaloperonospora arabidopsidis, 274–275
5-hydorxy methyl cytosine (5hmC), 3–4, 290, 455, 650
Hydralazine, 721, 826
Hydrogen peroxide (H_2O_2), 362
Hydrolytic deamination of 5mC, 543

β-hydroxybutyrate, 372
γ-hydroxybutyrate, 654
β-hydroxybutyrylation, 145
6-hydroxydopamine (6-OHDA), 388
2-hydroxyglutarate (2-HG), 362–363
2-hydroxyisobutyrylation, 145
Hydroxymethylcytosine, 540
5-hydroxymethylcytosine (5-hmC), 12, 124, 130, 329, 389–390, 613–614, 721, 738, 848
   mapping methodologies, 130–131
5-hydroxytryptamine, 148
Hyperacetylated promoters, 725
Hypergeometric test-coexpression approach, 194
Hypergeometric text-based approach, 194
Hyperglycemia, 361
Hypermethylated status
   of IDH1, 395–396
   of IDH2, 395–396
   of MGMT Gene, 396
Hypermethylation, 304, 480, 680
Hypersensitive response (HR), 274
Hypertension, 769
Hypomethylating agents, 827–828
Hypomethylation, 306–307, 826
Hypothalamic-pituitary-adrenal (HPA), 647, 746–747
Hypoxia, 363
Hypoxia inducible factor 3 alpha (*HIF3A*), 770
Hypoxia inducible factor-1 alpha (HIF-1a), 395
Hypoxiainducible factor (HIF), 363, 578–580
Hysteresis of PrB activity, 79
HZ. See Heterozygotic (HZ)

I

IAP. See Intracisternal A particle (IAP)
IBD. See Inflammatory bowel disease (IBD)
IC1. See Imprinting center 1 (IC1)
ICARS. See International Cooperative Ataxia Rating Scale (ICARS)
iChIP. See Insertional ChIP (iChIP)
ICM. See Inner cell mass (ICM)
ICR. See Imprinting control region (ICR)
ICSS. See Intracranial self-stimulation (ICSS)
ICT. See Immune checkpoint therapy (ICT)
IDH. See Isocitrate dehydrogenase (IDH)
IDH1. See Isocitrate dehydrogenase 1 (IDH1)
IDH2. See Isocitrate dehydrogenase 2(IDH2)
IEG. See Immediate early genes (IEG)
IFI44L promoter methylation, 726–727
IFITM1. See Interferon induced transmembrane protein 1 and 3 genes (IFITM1)
IFITM3. See Interferon induced transmembrane protein 3 (IFITM3)
IFN-γ. See Interferon gamma (IFN-γ)
IFN-ϒ. See Interferon-ϒ (IFN-ϒ)
IG-DMR. See Intergenic differentially methylated region (IG-DMR)
IGF1. See Insulin-like growth factor 1 (IGF1)
IGF2. See Imprinted gene insulin-like growth factor 2 (*IGF2*); Insulin like growth factor gene 2 (IGF2)

IGF2R. *See* Insulin-like growth factor II receptor (IGF2R)
IGN. *See* Imprinted gene network (IGN)
IHEC. *See* International Human Epigenome Consortium (IHEC)
Ihh gene, 97
IHH gene, 97
IL-17A. *See* Interleukin-17A (IL-17A)
IL-2. *See* Interleukin-2 (IL-2)
IL-2RB gene, 724–725
IL-4. *See* Interleukin-4 (IL-4)
IL2RA. *See* Interleukin-2 receptor α gene (IL2RA)
IL33. *See* Interleukin-33 (IL33)
IL7RA. *See* Interleukin-7 receptor α gene (IL7RA)
Illumina Infinium array, 127
Illumina sequencing method, 534–536
Immediate early genes (IEG), 449–450
Immune cells, epigenetics of, 719–720
  B cells, 719–720
  dendritic cells, 720
  macrophages, 720
  T cells, 719
Immune checkpoint therapy (ICT), 710
"Immune memory" system, 330–331
Immune system, 725
Immunodeficiency, Centromeric instability, and Facial anomalies (ICF), 16–17
ImpDis. *See* Imprinting disorders (ImpDis)
Imprinted domains, 251
Imprinted gDMRs, 255
Imprinted gene insulin-like growth factor 2 (IGF2), 764
Imprinted gene network (IGN), 784
Imprinted genes, function of, 256–257
Imprinted silencing of paternal X chromosome, 258
Imprinting
  H3K27me3 regulates, 296
  histone acetylation and XCI, 297
Imprinting center 1 (IC1), 784
Imprinting control region (ICR), 11, 254, 342, 652
Imprinting disorders (ImpDis), 779–782
  Beckwith–Wiedemann syndrome, 781–782
  causes of disturbed imprinting, 782–783
  cis-acting factors, 783–784
  clinical findings in, 779–781
    ImpDis, 780t
  clinical management, 787
  differentially methylated region, 779
  genetic and reproductive counseling, 787
  maternal effect mutations and multilocus imprinting disturbance, 784–785
  molecular findings in, 781–782
    types of molecular alterations, 782f
  Prader–Willi syndromes, 782
  trans-acting factors, 784
  transient neonatal diabetes mellitus, 781–782
  translational use of new findings in ImpDis and new methodologies, 785–786
    molecular diagnostics, 785–786

IMSGC. *See* International Multiple Sclerosis Genetics Consortium (IMSGC)
In porcine intestinal epithelial (IPEC-J2), 577
In vitro enChIP, 860–862, 861f
Induced epigenetic changes, 552–554
  Baldwin effect, 554
  intergenerational effect, 553–554
  maternal effect, 553–554
  paternal effect, 553–554
  phenotypic plasticity, 553–554
  in plant, 552–553
  transgenerational effect, 553–554
Induced neural stem cells (iNSCs), 383–384, 397
Induced pluripotent stem cells (iPSCs), 383–384, 389, 427, 452–454, 784, 851
  models of RTT, 390
Induced resistance (IR), 274
Inducible nitrous oxide (iNOS), 653
Infinium methylation bead chips, 127
Inflammation, 362, 677–680
Inflammatory bowel disease (IBD), 677
Inflammatory diseases, 311
Inflammatory genes, 308
Inherited RNA molecules, 551–552
Inhibitors against HDAC (HDACi), 722
Inhibitory postsynaptic potential (IPSP), 390
Ink4α gene, 117–118
Innate immunity, 681
Inner cell mass (ICM), 116
iNOS. *See* Inducible nitrous oxide (iNOS)
Input cell, histone modification profiling with low, 140–144
INS gene, 724–725
iNSCs. *See* Induced neural stem cells (iNSCs)
Insertional ChIP (iChIP), 858–859
INSR gene, 770
Insulin (INS), 770–771
Insulin like growth factor gene 2 (IGF2), 308, 466–467
Insulin-like growth factor II receptor (IGF2R), 467–468
Insulin-like growth factor 1 (IGF1), 391, 763
INTACT method. *See* Isolation of Nuclei TAgged in specific Cell Types method (INTACT method)
Interferon gamma (IFN-γ), 392, 848
Interferon induced transmembrane protein 1 and 3 genes (IFITM1), 338
Interferon induced transmembrane protein 3 (IFITM3), 338
Interferon regulator factor (IRF), 719
Interferon regulatory factor 1 (IRF-1), 859
Interferon-ϒ (IFN-ϒ), 660
Intergenic differentially methylated region (IG-DMR), 61
Intergenic ncRNAs (lincRNAs), 58
Interleukin-17A (IL-17A), 618–619
Interleukin-2 (IL-2), 392
Interleukin-2 receptor α gene (IL2RA), 393
Interleukin-33 (IL33), 683
Interleukin-4 (IL-4), 392
Interleukin-6
  C-reactive protein, 490
Interleukin-7 receptor α gene (IL7RA), 393

International Cooperative Ataxia Rating Scale (ICARS), 391
International Human Epigenome Consortium (IHEC), 144
International Multiple Sclerosis Genetics Consortium (IMSGC), 393
Intestinal parasites, 682
Intracellular metabolic compounds, 359
Intracisternal A particle (IAP), 251–252, 331–332
IAPs, 333
Intracranial self-stimulation (ICSS), 448–449
Intratumoral heterogeneity and therapeutic resistance, epigenetics in, 704–705
"Intrinsic epigenetic age acceleration", 483
Invasive phase, genetic alterations, 704
IPC. *See* CpG-density promoters (IPC)
Ipomoea tricolor, 269
iPSCs. *See* Induced pluripotent stem cells (iPSCs)
IPSP. *See* Inhibitory postsynaptic potential (IPSP)
IR. *See* Induced resistance (IR)
IRF. *See* Interferon regulator factor (IRF)
IRF-1. *See* Interferon regulatory factor 1 (IRF-1)
ISG15 gene, 100
Isocitrate dehydrogenase (IDH), 362–363, 395, 827–828
Isocitrate dehydrogenase 1 (IDH1), 709–710
  hypermethylation status of, 395–396
Isocitrate dehydrogenase 2 (IDH2), 709–710
  hypermethylation status of, 395–396
Isolation of Nuclei TAgged in specific Cell Types method (INTACT method), 277
Isopentanol, 371
Istodax. *See* Romidepsin

J

Janus kinase 2 (JAK2), 40
Janus kinase-STAT signalling (JAK-STAT signalling), 722
Jasmonic acid, 274
JmjC-domain containing histone demethylases, 362–363
Johne disease (JD), 596–597
Jumeaux gene, 221
Jumonji AT-rich interactive domain 1 (JARID1), 36
Jumonji C domain–containing proteins (JMD2), 36
Jumonji domain containing 1 C (JMJD1c), 405–406
Jumonji domain-containing protein 3 (JMJD3), 726
Jumonji domain-containing proteins (JMJs), 267
JumonjiC (JmjC), 362
Jun N-terminal kinase (JNK), 617–618

K

Kagami–Ogata syndrome (KOS14), 782
K562 cell line, 831
KAP1. *See* KRAB-associated protein 1 (KAP1)

KAT inhibitors (KATi), 412
Kbhb. See Histone lysine β-hydroxybutyrylation (Kbhb)
Kcnh7 gene, 740
KCNMB2 gene, 576
Kcnq1ot1, 166
KDACs, 28–30
Kdm5c gene, 426, 433
KDM6A gene, 433
KDM6B, 363
KEGG. See Kyoto Encyclopedia of Genes and Genomes (KEGG)
Kenyon cells, 217
Kerb's cycle/tricarboxylic acid (TCA), 395
Ketogenic diet, 618
Kglu. See Histone lysine glutarylation (Kglu)
KLF14. See Krüppel-like factor 14 gene (*KLF14*)
Klf4 gene, 390
Klinefelter syndromes, 423–424
KRAB. See *Krüppel* associated box (KRAB)
KRAB-associated protein 1 (KAP1), 17, 331
KRAB-containing zinc finger proteins (KRAB-ZFPs), 330, 332
KRAB-ZFPs. See KRAB-containing zinc finger proteins (KRAB-ZFPs)
Krüppel associated box (KRAB), 17, 278, 331
Krüppel-like factor 14 gene (*KLF14*), 763
containing family of zinc finger proteins, 331
KRAB-ZFP/KAP1/SETDB1 axis, 334–335
Kryptonite (KYP), 270–271
Kyoto Encyclopedia of Genes and Genomes (KEGG), 575–576

L

L&M. See Learning and memory (L&M)
L1 retrotransposition, 327
L3MBTL1 gene, 766
L3MBTL1. See Lethal 3 mbt-like 1 (L3MBTL1)
Laboratory specimens, 534
Lactate, 373–374
Lactobacillus, 617–618, 682
Lactose operon, 79–80
Lactylation, 145, 360
LAD. See Lamin associated domains (LAD)
Lamarck's theory, 588
Lamin associated domains (LAD), 92, 294
Large offspring syndrome, 589–590
Late Elongated Hypocotyl (LHY), 266–267
Latent membrane protein 1 (LMP1), 681
Lateral hypothalamusmedial forebrain bundle (LH-MFB), 448–449
LBH589. See Panobinostat
LBW. See Low birth weight (LBW)
LCR. See Locus control region (LCR)
LDL. See Low-density lipoprotein (LDL)
LDL1. See Lysine-specific histone demethylase 1 (LDL1)
LDL2. See Lysine-specific histone demethylase 2 (LDL2)
Lead (Pb), 651
Learning, 232
Learning and addiction, 628–629
Learning and memory (L&M), 626–628
Learning in *Drosophila*, environmental enrichment improves, 237–238
Legionella, 680–681
Lenalidomide, 805
Leptin (*LEP*), 764
Leptospira species, 207–208
Lesch-Nyhan syndrome, 429
Lethal 3 mbt-like 1 (L3MBTL1), 38
Leucine rich repeat receptors (NLRs), 274
Leucine-rich repeat kinase 2 gene (LRRK2), 387
Leucovorin, 805
Leukemia, 822
LH-MFB. See Lateral hypothalamusmedial forebrain bundle (LH-MFB)
LHY. See Late Elongated Hypocotyl (LHY)
Licking/grooming (LG), 468–469
Lifestyle factors on epigenome, 762–766
Lifetime influence, 642
Light compound-mediated inducible systems, 855–856
Limited cutaneous SSc (lSSc), 725
LINC00152 expression level, 306
lincRNAs. See Intergenic ncRNAs (lincRNAs); Long intergenic RNAs (lincRNAs)
LINE-1. See Long interspersed nucleotide element-1 (LINE-1)
LINEs. See Long interspersed nuclear elements (LINEs)
Linker histone, 147
Liparis tunicatus, 523
genome size of, 523
methylation level of, 522–523
Lipid metabolism and livestock products, 583–587
adipogenesis and meat, 586–587
lipid metabolism, milk and constituents, 583–586
epigenetic alterations, 584*t*
wool and egg production, 587
Lipid Transfer Protein (*LTP*), 503
Lipopolysaccharide (LPS), 362, 576–577, 720
Lipoprotein complexes, 304
Listeria monocytogenes, 681
Litchi (*Litchi chinensis*), 500–501
Litchi chinensis. See Litchi (*Litchi chinensis*)
Liver cancer, 306–307, 310
Liver flukes, 678
Livestock breeding. See also Animal breeding
components of inclusive heritability in variance component calculations for livestock quantitative traits, 594*f*
epigenetic mechanisms role in, 591–595
genome and epigenome editing technologies in, 595
sources of epigenetic variation, 593*f*
Livestock epigenetics case study, 595–598
audience, 596
challenges and solutions, 598
expected results and deliverables, 597
learning and knowledge outcomes, 598
objective, 596
rationale, 596–597
results, 598
safety considerations, 597–598
scope, 596
Livestock health and production
epigenetic mechanisms impact lipid metabolism and, 583–587
epigenetic mechanisms regulate, 576–583
epigenetic mechanisms regulate livestock reproduction and growth, 573–575
phenotypic outcome, 574*f*
epigenetic source of phenotypic variation in, 571–587
parental nutritional impacts on livestock epigenome and production, 575–576
LMP1. See Latent membrane protein 1 (LMP1)
lncRNAs. See Long noncoding RNAs (lncRNAs)
Locus control region (LCR), 856
Locus-specific ChIP development, 858–859
Locus-specific epigenome editing, 845–858
advanced epigenome editing tools, 854–857
cancer biology, 850–851
characteristics of engineered DNA-binding molecules in, 857–858
locus-specific identification of epigenetic molecules interact with target genomic regions, 858–862
molecules fused to engineered DNA/RNA-binding molecules for, 847*t*
neurological diseases, 851
other applications, 852
other diseases, 851–852
potential problems with, 852–854
long-term effects, 853
off-target effects, 853
other factors affecting, 853–854
proof-of-concept research on, 846–850
Locus-specific HTS-based bisulfite sequencing, 126–127
Locusta migratoria, 169–170
LOF. See Loss-of-function (LOF)
Lomustine, 394
Long intergenic RNAs (lincRNAs), 163–164
Long interspersed nuclear elements (LINEs), 327, 333, 338, 420, 426
LINE-2 element, 343
Long interspersed nucleotide element-1 (LINE-1), 307, 327, 723
Long noncoding RNAs (lncRNAs), 55, 163–164, 185, 219, 269, 272, 304, 311, 419–420, 526, 573, 615, 703, 738, 854
algorithm for lncRNA detection, 169
biomarkers, 316
epigenetic regulation of, 64–65
X-chromosome, 65*f*
epitranscriptomics of, 65
regulation of epigenetics by, 65–66
Long terminal repeats (LTR), 324
retroelements, 338
Long-distance interactions, 92–93
Long-range intra-chromosomal interaction studies, 338
Long-term depression (LTD), 628

Long-term memory (LTM), 232–233, 445–446
　DNA methylation and, 452–455
　regulation, 444–445
Long-term potentiation (LTP), 445–446, 449–450, 628, 644
"Loop extrusion" model, 335–336
Loss-of-function (LOF), 221
Low birth weight (LBW), 766
Low-abundance RNA species, detection of, 167–169
　RNA-seq methods, 168f
Low-density lipoprotein (LDL), 651
Low-density lipoprotein receptor related protein (LRP-1), 744
LPS. *See* Lipopolysaccharide (LPS)
LRP-1. *See* Low-density lipoprotein receptor related protein (LRP-1)
LRRK2. *See* Leucine-rich repeat kinase 2 gene (LRRK2)
LSD. *See* Lysine specific demethylase (LSD); Lysine-specific demethylases (LSD)
LSD 1. *See* Lysine specific demethylase1 (LSD 1)
LSD1. *See* Lysine-specific histone demethylase 1 A (LSD1)
LSS. *See* Lysostaphin (LSS)
lSSc. *See* Limited cutaneous SSc (lSSc)
LTD. *See* Long-term depression (LTD)
LTM. *See* Long-term memory (LTM)
LTP. *See* Lipid Transfer Protein (*LTP*); Long-term potentiation (LTP)
LTR. *See* Long terminal repeats (LTR)
LTR retrotransposons, 324–327, 340–341
Lung neoplasia, 307
Luteolin, 827
Lymphoblastoid cell lines, 144
Lymphotoxin alpha gene (*Lta* gene), 647
Lysine (K), 36, 500, 738–739
　lactylation, 614–615
　residue, 207
Lysine 4 (K4), 4, 465–466
Lysine 9 (K9), 4
Lysine acetylation, 204, 614
　of PhoP, 203–204
Lysine acetyltransferase (KAT), 28–30, 445–446
Lysine demethylases, 36, 219
Lysine methylation, 36–39, 500
Lysine methyltransferases (KMT), 207, 219
Lysine specific demethylase (LSD), 267
Lysine specific demethylase1 (LSD 1), 448–449
Lysine-specific demethylases (LSD), 362, 614
Lysine-specific histone demethylase 1 (LDL1), 269
Lysine-specific histone demethylase 1 A (LSD1), 720
Lysine-specific histone demethylase 2 (LDL2), 269
Lysostaphin (LSS), 595

M

M-phase phosphoprotein (MPP8), 92
m5C. *See* Methylation of cytosine at C5 position (m5C)
m6A. *See* Methylated at adenosine sites (m6A)
m6A individual nucleoside resolution and crosslinking immunoprecipitation (miCLIP), 316
m7G. *See* 7-methylguanosine (m7G)
Machine learning, 727–730
Macrophages, 716–717, 720
MADs Affecting Flowering4 (*MAF4*), 503
MADS-box TFs, 277
MAF4. *See MADs Affecting Flowering*4 (*MAF4*)
Maintenance methylation, 3–4
Maize (*Zea mays*), 500–501
Major depressive disorder (MDD), 745–746
Major histocompatibility complex class II (MHC class-II), 392
Male-specific lethal 1 (msl1), 229
Male-specific lethal 2 (msl2), 229
Male-specific lethal 3 (msl3), 229
Maleless (mle), 229
Malonylation, 145
Mamestra brassicae, 523
Mammalian apparent LTR retrotransposons (MaLR), 339
Mammalian cell lineage commitment, H3K27me3 is necessary for, 296
Mammalian development, 466, 469
Mammalian early embryo development, DNA methylation during, 291–293
Mammalian embryonic genome, 249
Mammalian genomes, 571
Mammalian germ cells, 552
Mammalian prions, 77
Mammalian targets of rapamycin (mTOR), 740–742
MAMPs. *See* Microbial-associated molecular patterns (MAMPs)
MAP. *See Mycobacterium avium* subsp *paratuberculosis* (MAP)
MAP kinase kinase (MAPKK), 80–81
MAP kinase kinase kinase (MAPKKK), 80–81
Marbled crayfish, 528, 532, 540–541
Marchantia polymorpha, 500–501
Marek disease (MD), 577
Marker assisted selection (MAS), 570
MARTs. *See* Mono-ADP-ribosyl transferases (MARTs)
MAS. *See* Marker assisted selection (MAS)
Maspin, 850
Mass spectrometry (MS), 145, 844
Massively parallel sequencing technologies, 164
MAT. *See* Methionine adenosyl transferase (MAT)
Maternal cannabis, 657
Maternal diet, 616–619
Maternal effect mutations, 784–785
Maternal ketone body, 618
Maternal nutritional environment, 466–467
Maternal paternal, 552–553
Maternal stressors, 573–574
Maternal vitamin D deficiency, 617
Maternally expressed genes (MEGs), 342

Mathematical modeling, 190
Mathematical models for competitive endogenous RNAs network analysis, 187–190, 188t
　comparison of, 190
MBD. *See* Methyl CpG—binding domain (MBD); Methyl-CpG binding domain (MBD)
MBD protein-based affinity pulldown, 124–125
　MeDIP-seq, 125f
MBD proteins. *See* Methyl-CpG binding domain proteins (MBD proteins)
MBD1. *See* Methyl CpG-binding domain protein 1 (MBD1)
MBDs. *See* Methyl binding domain proteins (MBDs)
MBP. *See* Methyl-CpG-binding proteins (MBP); Myelin basic protein (MBP)
mcad. *See* Medium chain acyl coa decarboxylase (mcad)
McCune-Albright syndrome, 589–590
MCK. *See* Muscle creatine kinase gene (MCK)
MD. *See* Marek disease (MD)
MDD. *See* Major depressive disorder (MDD)
MDS. *See* Myelodysplastic syndrome (MDS)
MDSCs. *See* Myeloid-derived suppressor cells (MDSCs)
MeCP2 gene. *See* Methyl-CpG-binding protein 2 gene (*MeCP2* gene)
MeCP2-null mice, 740
MeCP2. *See* Methyl CpG binding protein 2 (MeCP2); Methylation-CpG binding-protein 2 (MeCP2)
Medauroidea extradentata, 525
Medial prefrontal cortex (mPFC), 445–446, 457, 626–628, 657
Medial preoptic area (MPOA), 471–472
Medical Genome Project, 688
MeDIP. *See* Methylated DNA immunoprecipitation (MeDIP)
Medium chain acyl coa decarboxylase (mcad), 412
MEF2. *See* Myocyte enhancer factor 2 (MEF2)
Mef2c gene, 115–116
MEFs. *See* Mouse embryonic fibroblasts (MEFs)
MEGs. *See* Maternally expressed genes (MEGs)
Melipona scutellaris, 528
Memory, 232
　and addiction, 628–629
　consolidation, 443
　environmental enrichment improves memory in *Drosophila*, 237–238
　formation, 443
　process, 443
Mendelian inheritance patterns, 387
Mental health, 491
Mental or physiological environment, 643–650
　activity and experience-dependent changes, 645–646
　　activity related neuronal plasticity, 645–646

experience and memory formation, 646
enriched environment, 643–645
exercise, 649
nutrition, 650
stress, 646–649
childhood adversity or early life stress, 647–649
prenatal stress, 646–647
mEPSCs. *See* Miniature excitatory postsynaptic currents (mEPSCs)
Mercury, 651
MeRIP-Seq. *See* Methylated RNA immunoprecipitation sequencing (MeRIP-Seq)
Merkel cells, 116
Mesenchymal stem cells (MSCs), 116, 383–385, 390–391, 396
TRAIL producing, 396–397
transplantations for GB, 396
Mesocorticolimbic dopaminergic pathway, 626–628
Mesoderm-specific transcript (MEST), 467–468
Messenger RNA (mRNA), 185, 323–324
MEST. *See* Mesoderm-specific transcript (MEST)
MEST gene, 768
MET1. *See* Methyltransferase1 (MET1)
Metabolic adaptation, 699
Metabolic diseases, 311, 573–574, 676
epigenetic features of, 769–771
metabolic syndrome, 769
obesity, 769–770
type 2 diabetes, 770–771
epigenetic memory, prenatal exposure, and risk of, 766–768
diabetes in pregnancy, 767–768
thrifty phenotype and birthweight, 766–767
epigenetics of, 761–762
age and epigenome, 762–763
impact of age and lifestyle factors on epigenome, 762–766
development, 762f
diet, metabolism, and epigenome, 763–764
metabolic syndrome, 761
obesity, 761
type 2 diabetes, 761
physical activity, inactivity, and the epigenome, 764–766
lifestyle factors and metabolic disease affects epigenetics, 765t
Metabolic epigenome, 616–619
Metabolic memory, 725
Metabolic regulation of JHDM, 363
Metabolic syndrome (MetS), 611, 761, 769
definition of, 769
diagnosis of, 611
epigenetic mechanism and, 612–615
DNA methylation, 612–614
histone methylation/demethylation, 611–612
histone modification, 614–615
miRNAs, 769

non-coding RNA modification, 615
epigenetic modifications and functions, 769
Metabolism, 359
microbiome, epigenetics, and effect on, 677
Metabolites, 677
Metaboloepigenetics, 359
Metal-dependent protein phosphatase, 201–202
Metallothionein 2 A (*Mt2a*), 652
Metaphase II (MII), 293–294
Metastability, 252
Metastable epialleles, 250–253
discovery of, 251–252
environmental modulation of epigenetic inheritance, 252–253
mechanisms of metastability, 252
Methanosarcana antivorans, 208
Methanosarcana mazei, 208
Methionine, 363–364
Methionine adenosyl transferase (MAT), 360–361
Methyl binding domain proteins (MBDs), 699–700
Methyl CpG binding protein 2 (MeCP2), 124–125, 389, 452–454, 465–466, 632, 725
Gene and function, 389–390
interactions with chromatin modifiers, 740
mutations, 740
Methyl CpG-binding domain protein 1 (MBD1), 747
Methyl CpG–binding domain (MBD), 738
Methyl donors, 465–466, 650, 764
Methyl groups (CH_3), 360–361, 716
1-methyl-4-phenyl-1,2,3,6-tetrahydropyridine (MPTP), 388
Methyl-BEAMing, 315
Methyl-CpG binding domain (MBD), 124–125
Methyl-CpG binding domain proteins (MBD proteins), 125, 452–454, 524–526, 612–613
MBD1 proteins, 124–125
MBD2 proteins, 124–125
MBD3 proteins, 124–125
MBD4 proteins, 124–125
Methyl-CpG-binding protein 2 gene (*MeCP2* gene), 739
Methyl-CpG-binding proteins (MBP), 330
Methyl-directed mismatch repair (MMR), 205
Methyl-Seq method, 127–128
"Methyl/phos switch", 228–229
Methylated at adenosine sites (m6A), 62
Methylated CpGs, 542
Methylated cytosines, 65
Methylated DNA fragments, 124
Methylated DNA immunoprecipitation (MeDIP), 124, 534–536
Methylated RNA immunoprecipitation sequencing (MeRIP-Seq), 316
Methylated-CpG island recovery assay (MIRA), 125
MIRAseq, 126f

Methylated-CPG island recovery assay, 125
Methylation, 11, 36–39, 112–114, 305, 384, 526
in animals, 525
biomarkers, 308
of histone H3, 361
of histone H4, 361
profiles, 394–395
status, 708
Methylation of cytosine at C5 position (m5C), 59–60
Methylation variable positions (MVPs), 724–725
Methylation-CpG binding-protein 2 (MeCP2), 646
Methylation-sensitive representational difference analysis (Ms-RDA), 124
Methylation-sensitive restriction enzyme digestion, 843
Methylation-sensitive restriction fingerprinting (MSRF), 124
Methylation-specific affinity purification, 843
Methylation-specific digital karyotyping (MSDK), 124
Methylation-specific multiplex ligation-dependent probe amplification (Ms-MLPA), 785–786
5-methylcytosine (5mC), 3–4, 11, 125–126, 276, 329, 389–390, 700, 721, 738
5mC-specific antibody, 124
7-methylguanosine (m7G), 62
MethyLight, 315
Methylmercury (MeHg), 652
Methylome wide association study (MWAS), 653
Methyltransferase-like (METTL), 615
Methyltransferase-like 3 (METTL3), 62, 425
Methyltransferase1 (MET1), 264
Methyltransferases (MTases), 845–846
MetS. *See* Metabolic syndrome (MetS)
MHC. *See* Myosin heavy chain (MHC)
MHC class-II. *See* Major histocompatibility complex class II (MHC class-II)
miCLIP. *See* m6A individual nucleoside resolution and crosslinking immunoprecipitation (miCLIP)
Microarray platforms, 123, 844
Microbial-associated molecular patterns (MAMPs), 681
Microbiome, 618, 673–674
and effect on metabolism, 677
Microbiota, 669–673
Microfluidics-based ChIP-seq platforms, 143–144
Micromonospora carbonacea, 208
MicroRNA (miRNA), 163–164, 185, 219, 304–306, 385–386, 388, 393, 524, 573, 591, 615, 631–632, 646–647, 688, 703, 738, 763, 766
biomarkers, 316
epigenetic regulation of miRNA expression, 60–62
epitranscriptomics of, 62
profiles, 639–640
regulation of epigenetics by, 62–64

MicroRNA (miRNA) (Continued)
  roles of miRNA types in MS, 393t
  targeting, 64
Microsatellite instability (MSI), 699–700, 705–708
MIEN1. See Migration and invasion enhancer 1 (MIEN1)
Migration and invasion enhancer 1 (MIEN1), 339–340
MII. See Metaphase II (MII)
Mineralocorticoid receptor (MR), 748
Mineralocorticoids, 748
Miniature excitatory postsynaptic currents (mEPSCs), 452–454
Miniature inverted repeat transposable elements (MITEs), 324
Minimal residual disease (MRD), 835
miR-124 promoter, 60–61
miR-127, 61
MiR-135 promoter, 747–748
miR-136, 61
miR-34, 61
miR-375, 60
MIRA. See Methylated-CpG island recovery assay (MIRA)
miRDeep, 169
miRNA. See MicroRNA (miRNA)
MiRNA response elements (MREs), 185
"Mismatch recognition protein", 205
MITEs. See Miniature inverted repeat transposable elements (MITEs)
Mitochondrial oxidative metabolism, 367
Mitogen-and stress-activated protein kinase 1 (MSK1), 449
Mitosis, 89
Mixed heredity, 78–79
Mixed-effect models, 592–595
Mixed-lineage leukemia 1 (MLL1), 502
Mixed-lineage leukemia gene (MLL gene), 18, 455
  MLL1, 405
  MLL2, 405
MLID. See Multilocus Imprinting Disturbance (MLID)
MLL gene. See Mixed-lineage leukemia gene (MLL gene)
MLL1. See Mixed-lineage leukemia 1 (MLL1)
MM. See Multiple myeloma (MM)
MMLV. See Moloney murine leukemia virus (MMLV)
MMR. See Methyl-directed mismatch repair (MMR)
Mobile sRNAs, 268–269
Mocetinostat, 794–795, 807
moDC. See Monocytes DC (moDC)
Modeling evolution, 552
Modern animal breeding, 570, 588
MODIFIER OF SNC1, 9 (MOS9), 506–507
Molecular diagnostics of ImpDis, 785–786
  SCMC and clinical outcome in case of mutations, 786t
Molecular genetic techniques, 570
Molecular graphics program, 210
Molecular modeling, protocols for, 210–211
"Molecular parasites", 323–324

Molecular titration mechanism, 188
Molecules, 360
Moloney murine leukemia virus (MMLV), 167
Mommes genes, 252
Mono-ADP-ribosyl transferases (MARTs), 44
Monoallelic silencing, 677
Monocytes DC (moDC), 720
Monozygotic (MZ), 717
Monozygotic twins, 677
MOS9. See Modifier Of Snc1, 9 (MOS9)
Mosaic methylation, 525
Mouse embryonic fibroblasts (MEFs), 13
Mouse epigenetic inheritance
  epigenetic programming in gametogenesis, 250–251
  epigenetic re-programming during embryogenesis, 251
  genomic imprinting, 253–259
  metastable epialleles, 251–253
Mouse model, 249
MOWChIP-seq method, 143–144
mPFC. See Medial prefrontal cortex (mPFC)
MPOA. See Medial preoptic area (MPOA)
MPP8. See M-phase phosphoprotein (MPP8)
MPTP. See 1-methyl-4-phenyl-1,2,3,6-tetrahydropyridine (MPTP)
MRD. See Minimal residual disease (MRD)
MRE. See MSL recognition element (MRE)
MREs. See MiRNA response elements (MREs)
MRF4 gene, 405
MRF4. See Muscle-specific regulatory factor 4 (MRF4)
MRFs. See Myogenic regulatory factors (MRFs)
mRNA, 55, 305–306
  m6A modification, 309
  methylation, 64
mRNA. See Messenger RNA (mRNA)
Ms. See Mass spectrometry (Ms)
MS. See Multiple sclerosis (MS)
Ms-275. See Entinostat
Ms-MLPA. See Methylation-specific multiplex ligation-dependent probe amplification (Ms-MLPA)
Ms-RDA. See Methylation-sensitive representational difference analysis (Ms-RDA)
MSCs. See Mesenchymal stem cells (MSCs)
MSDK. See Methylation-specific digital karyotyping (MSDK)
Msh homeobox 1 (MSX1), 407
MSI. See Microsatellite instability (MSI)
MSK1. See Mitogen-and stress-activated protein kinase 1 (MSK1)
MSL recognition element (MRE), 230–231
MSP. See Conventional methylation-specific PCR (MSP)
MSRF. See Methylation-sensitive restriction fingerprinting (MSRF)
MSX1. See Msh homeobox 1 (MSX1)
MTases. See Methyltransferases (MTases)
Mucor circinelloides, 554–555
Multicellular organisms, 276

Multilocus Imprinting Disturbance (MLID), 781–782, 784–785
Multimeric PcG complexes, 223
Multiomics, 728–730
Multiple biochemical technologies, 843
Multiple DNA methyltransferase isoforms, 290
Multiple myeloma (MM), 795–796
Multiple sclerosis (MS), 186, 334–335, 382, 392, 717, 723–724
  DNA Methylation in, 392
  epigenetics in, 394
  stem cell therapy for, 393–394
Multivariate Gaussian approximation, 189
Multivariate Gaussian distribution, 190
Mus musculus, 229
Muscle creatine kinase gene (MCK), 406
Muscle regeneration, 412
Muscle resident stem cells, 403
Muscle stem cells, 404
Muscle-specific genes, 406
Muscle-specific regulatory factor 4 (MRF4), 403–404
Muscle-wasting diseases, 404
Mushroom body (MB), 217, 237–238
Mutant β-synuclein, 235
Mutations, 304–305
  in Arabidopsis PICKLE gene, 272
Mutations, epigenetic
  distribution of mutational effects, 556
  stability of, 556
  supply of, 556
MutH (MutatorH), 205
MutL (MutatorL), 205
MutS (MutatorS), 205
MutS HOMOLOG1 (MSH1), 268–269
MVPs. See Methylation variable positions (MVPs)
MWAS. See Methylome wide association study (MWAS)
Mycale massa, 523
Mycobacterium avium subsp paratuberculosis (MAP), 596
Mycobacterium bovis, 576–577
Mycobacterium tuberculosis, 203
Myelin basic protein (MBP), 392, 723–724
Myelin-associated proteases, 392
Myelodysplastic syndrome (MDS), 709, 795–796
Myeloid-derived suppressor cells (MDSCs), 805–806
MYF5 gene, 405
Myf5. See Myogenic factor 5 (Myf5)
MYH4 gene, 409
Myoblast determination protein (MyoD), 403–406
  mRNA, 405
Myocyte enhancer factor 2 (MEF2), 403–404
MyoD. See Myoblast determination protein (MyoD)
Myofibers, 403
Myogenesis, 403–404
Myogenic factor 5 (Myf5), 403–404
Myogenic regulatory factors (MRFs), 405
Myogenin (MYOG), 405–406, 408–410

INDEX 891

Myosin heavy chain (MHC), 409
Myosin heavy chain 2b (MYH4), 406
MYST family, 218–219
MZ. *See* Monozygotic (MZ)

N

Nanog demethylation, 293
N terminal tails, 137
N-4-methylcytosine (4mC), 204
N-6 adenine-specific DNA methyltransferase 1 (N6AMT1), 382
N-acetyltransferase 10 (NAT10), 615
N-methyl-D-aspartate (NMDA), 646
N-terminal. *See* Amino-terminal (N-terminal)
N-terminal domain, 357
N-terminal RING-B box-coiled coil (RBCC), 331
N-terminal triple Tudor domain (TTD), 331
N6 methyladenosine modification (m⁶A modification), 615
N6-adenine methylation (6 Ma), 264
N^6-methyladenosine (m6A), 59–60, 67, 334, 382, 457
  modification, 308–309
N6AMT1. *See* N-6 adenine-specific DNA methyltransferase 1 (N6AMT1)
NAc. *See* Nucleus accumbens (NAc)
NAD. *See* Nicotinamide adenine dinucleotide (NAD); Nucleolar associated domains (NAD)
NAFLD. *See* Nonalcoholic fatty liver disease (NAFLD)
Naip. *See* Neuronal apoptosis inhibitory protein (Naip)
Named Ovation RNA-seq system (NuGEN), 166
Nanog gene, 116–117
Nanopore sequencing, 131–132
Nascent transcripts
  estimation of transcription rate by detection of, 173–174
NAT10. *See* N-acetyltransferase 10 (NAT10)
National Cholesterol Education Program—Third Adult Treatment Panel (NCEP ATP III), 611
National Institute on Alcohol Abuse and Alcoholism (NIAAA), 654–655
National Research Council (NRC), 575–576
National Survey on Drug Use and Health (NSDUH), 655–657
Native elongating transcript sequencing (NET-seq), 173
NATs. *See* Natural antisense transcripts (NATs)
Natural antisense transcripts (NATs), 269
NAWM. *See* Normal appearing white matter (NAWM)
NBW. *See* Normal birthweight (NBW)
NCED. *See* Nine-*cis*-epoxycarotenoid dioxygenase (*NCED*)
NCEP ATP III. *See* National Cholesterol Education Program—Third Adult Treatment Panel (NCEP ATP III)

NCoR. *See* Nuclear receptor corepressor complex (NCoR)
ncRNAs. *See* Non-coding RNAs (ncRNAs)
NEAT1. *See* Nuclear paraspeckle assembly transcript 1 (*NEAT1*)
Necrosis, 305
Nectria haematococca, 80–81
Neisseria meningitides, 850
Nematostella vectensis. *See* Sea anemone (*Nematostella vectensis*)
Neofunctionalization, 501–502
Neoplasm, 306
Neoplastic cells, epigenetic reprogramming of, 704
Neoplastic diseases, 677
Nerve growth factor (NGF), 644
NET-seq. *See* Native elongating transcript sequencing (NET-seq)
Neural networks, 232–233
Neural plasticity markers, 133
  BDNF, 133
  NGF, 133
Neural stem cells (NSCs), 116, 383–384, 387, 394
  transplantation for GB, 397
Neuro-2a cells, 452–454
Neuroblastoma, 153
Neurodegenerative disorders, 234, 310, 313, 383, 387–388
  epigenetic dysregulation in, 742–745
    DNA methylation in AD, 744–745
    histone acetylation targeted therapeutics for, 235–237
    histone modifications and histone modifier in AD, 742–744
Neurodevelopmental disorders, epigenetic dysregulation in, 739–742
  DNA methylation, 739–740
  MeCP2 interactions with chromatin modifiers, 740
  MeCP2 mutations, 740
  non-coding RNA in RS, 740–742
Neuroendocrine protein 1, 647
Neuroepigenetics, 443
Neurofibrillary tangles (NFTs), 742
Neurog2 genes, 383
Neurological diseases, 851
  epigenetics and, 384–385
    Huntington disease and mesenchymal stem cells, 384–385
Neuronal apoptosis inhibitory protein (Naip), 339
Neuronal gene, 383
Neuronal stimulation, 443
Neuropeptide-Y (Npy), 655
Next-generation sequencing (NGS), 315, 844
Nfil3 gene, 449–450
NFTs. *See* Neurofibrillary tangles (NFTs)
NFκB. *See* Nuclear factor kappa B cells (NFκB)
NGF. *See* Nerve growth factor (NGF)
NGS. *See* Next-generation sequencing (NGS)
NHEJ. *See* Nonhomologous end joining (NHEJ)

NIAAA. *See* National Institute on Alcohol Abuse and Alcoholism (NIAAA)
Nickel, 651
Nicotiana benthamiana, 278
Nicotinamide adenine dinucleotide (NAD), 358–359
Nicotine, 657–658
Nigrostriatal dopamine system, 388
Nine Cisepoxycarotenoid Dioxygenase 3 gene, 273
Nine-*cis*-epoxycarotenoid dioxygenase (*NCED*), 503
Nitric oxide (NO), 361, 406
Nitric oxide synthase (NOS), 406
NLS. *See* Nuclear localization signal (NLS)
NMDA. *See* N-methyl-D-aspartate (NMDA)
NO. *See* Nitric oxide (NO)
NOD-like receptors (NLRs), 681
Non-autonomous elements, 324
Non-CG methylation, 267
Non-coding RNAs (ncRNAs), 4, 55–59, 163–164, 185, 219, 269, 303, 342–343, 465–466, 523–524, 528, 573, 683, 698–700, 703, 738–739, 779, 854
  in AD, 745
  analysis of RNA-seq data for ncRNAs, 169–172
  cDNA library construction, 164–169
  circRNA, piRNA and, 59f
  CircRNAs and epigenetics, 67
  in depression, 747–748
  dyrec-seq protocol, 175–177
  in early-life stress, 749
  epigenetic regulation of lncRNAs, 64–65
  epigenetic regulation of miRNA expression, 60–62
  epigenetics, piRNAs and, 66
  estimating transcription and degradation rates of noncoding transcripts, 172–174
  genetics and epigenetics, 59–60
  lncRNA
    epitranscriptomics of, 65
    involved in epigenetic regulation, 58t
  miRNAs, 57t
    epitranscriptomics of, 62
    modification, 615
  ncRNAs associated with epigenetic phenomenon, 56f
  plausible interdependence between miRNA targeting and mRNA methylation, 64
  regulation, 569–570
    DNMT3a/b by miR-290 cluster, 63f
    of epigenetics by lncRNAs, 65–66
    of epigenetics by miRNAs, 62–64
  in RS, 740–742
  tethering of, 854
Non-histone protein, 410
Non-malignant mutations, 304
Non-metal pollutants, 651
Non-obese diabetic mice (NOD mice), 724
Non-random XCI, 429
Nonalcoholic fatty liver disease (NAFLD), 614, 690

"Nonamyloid prions", 78–79
Noncancer diseases
  clinical applications of HDACi for, 807–809
    phenylbutyrate, 808
    valproic acid, 807–808
  clinical evaluation of valproic acid and phenylbutyrate, 808
  HDACi dependent reactivation of latent HIV, 809
  others HDACi, 808–809
Noncanonical Imprinting, 257–258
Noncoding transcripts
  estimating transcription and degradation rates of, 172–174
    detecting unstable ncRNAs by inhibiting decay, 173
    estimation of transcription rate by detection of nascent transcripts, 173–174
    genome-wide approach to determine RNA stability, 172–173
    kinetic determination of RNA production and degradation, 174
Nongenetic inheritance, 588
Nonhistone proteins, 36, 207
Nonhomologous end joining (NHEJ), 35
Non–small cell lung cancer (NSCLC), 61, 305, 310, 691–692, 823
NOR. See Novel object recognition (NOR)
Normal appearing white matter (NAWM), 723–724
Normal birthweight (NBW), 767
NOS. See Nitric oxide synthase (NOS)
Not-so-random (NSR), 165–166
Novel object recognition (NOR), 445–446
Npy. See Neuropeptide-Y (Npy)
NR3C1 DNA methylation, 468–469
NR3C1. See Nuclear receptor subfamily 3 group C member 1 (NR3C1)
Nr4a1 gene, 449–450
Nr4a2 gene, 449–450
NRC. See National Research Council (NRC)
nrf. See Nuclear respiratory factor (nrf)
NRO-RNA. See Nuclear run-on RNA (NRO-RNA)
NSCLC. See Non–small cell lung cancer (NSCLC)
NSCs. See Neural stem cells (NSCs)
NSD1, 37–38
NSDUH. See National Survey on Drug Use and Health (NSDUH)
NSR. See Not-so-random (NSR)
[NU$^+$] prion, 76–77
Nuclear factor kappa B cells (NFκB), 617–618, 745
Nuclear hormone receptors, 44
Nuclear localization signal (NLS), 12–13
Nuclear paraspeckle assembly transcript 1 (NEAT1), 164–165, 407
Nuclear receptor (Nurr1), 388–389
Nuclear receptor corepressor complex (NCoR), 740
Nuclear receptor subfamily 3 group C member 1 (NR3C1), 467, 649, 768

Nuclear respiratory factor (nrf), 412
Nuclear run-on RNA (NRO-RNA), 173
Nucleation process, 341–342
Nucleolar associated domains (NAD), 92
Nucleosomes, 27–28, 89, 217–218, 357, 444
  nucleosome-remodeling ATPase brahma, 220
  reorganization, 227–228
Nucleotide-binding oligomerization domain proteins (NODs), 681
Nucleus accumbens (NAc), 455, 626–628
NuGEN. See Named Ovation RNA-seq system (NuGEN)
Numerous ChIP-seq alternatives, 140
NuRD complex, 341
Nurr1. See Nuclear receptor (Nurr1)
Nutrients, 360
Nutriepigenetics, 335
Nutritional epigenetics
  epigenetic modifications, 612t
  maternal diet and metabolic epigenome, 616–619
    B Vitamins, 616
    choline and betaine, 616–617
    dietary fiber and SCFA, 618–619
    high fat/high sugar diet, 617–618
    ketogenic diet, 618
    vitamin D, 617
  metabolic sensing by epigenetic mechanisms, 612–615
  prenatal nutrition impact on fetal reprogramming, 615

O

O-6-methylguanine DNA methyltransferase (Mgmt), 652
  MGMT enzyme, 396
O-GlcNAcylation, 148–149, 614
O-linked N-acetylglucosamine residue (GlcNAc), 614
O-linked β-d-N-acetylglucosamine transferase (OGT), 329, 331
OA. See Osteoarthritis (OA)
Obesity, 676, 761, 769–770
Object location memory (OLM), 447–448
Object recognition memory (ORM), 445–446
Oct3/4 gene, 390
Oct4 demethylation, 293
Oct4 gene, 116–117
Offspring development, paternal influence on, 469
OGT. See O-linked β-d-N-acetylglucosamine transferase (OGT)
OLM. See Object location memory (OLM)
OmpB. See Outer membrane protein precursor (OmpB)
Omps. See Outer membrane proteins (Omps)
Oncogene, 153
"Oncohistone", 702
Oncorhynchus keta, 523
Oncorhynchus mykiss, 523
One-carbon metabolism, 360–362
  pathway, 612–613
  in PD, impairment of, 387–388
One-carbon nutrients, 616

Oocytes, 17
Operant models of drug self-administration, 629
OPI. See Overproduction inhibition (OPI)
Opioid mu1 receptor, 657
Opioids, 657
OPN. See Osteopontin (OPN)
Oreochromis niloticus, 540
Organogenesis, 528
Organosulfur compound, 372–373
Organs, 654
ORM. See Object recognition memory (ORM)
Oryza sativa. See Rice (Oryza sativa)
OS. See Overall survival (OS)
Os_SDG711 gene, 514
Osteoarthritis (OA), 722
Osteopontin (OPN), 386–387
Ostreococcus tauri, 509
Outer membrane protein precursor (OmpB), 207
Outer membrane proteins (Omps), 207
Overall survival (OS), 835
Overproduction inhibition (OPI), 332
Oxaliplatin, 802
Oxidative bisulfite sequencing (OxBS-Seq), 131
Oxidative phosphorylation (OXPHOS), 832–833
  combined OXPHOS and epigenetic treatment impact on tumor growth suppression, 833f
Oxidative stress, 372
2-oxoglutarate (2OG), 18
Oxytocin, 748
Oxytocin receptor gene (OXTR), 468

P

P-cadherin, 306–307
P-element-Induced Wimpy testis (PIWI), 330–331
  PIWI-piRNA pathway, 330–333
p14ARF cell cycle, 116
p16 tumor suppressor gene, 851
p16INKA cell cycle, 116
P21 gene, 78
PA-m6A-Seq. See Photo-crosslinking-assisted m6A sequencing (PA-m6A-Seq)
pA-Tn5. A-Tn5 transposase (pA-Tn5);. See Tn5 transposase-protein A (pA-Tn5)
Paclitaxel, 830–831
PAD2. See Peptidyl argininedeiminase 2 (PAD2)
PADs. See Peptidylarginine deiminases (PADs)
pAG-MNase. See Protein, A-protein GMNase (pAG-MNase)
Pair-wise correlation, 190
Paired box 5 (PAX5), 719
Paired-Tag, 145
Pairedbox protein 7 (Pax7), 403–404
pan-HDACis, 450–451
Pancreas cancer, 60–61, 117–118, 310
Pancreatic ductal adenocarcinoma (PDAC), 310
  PSPC1 bind LTR retrotransposons, 333

Panobinostat, 794–795, 799, 804–805, 830–831
Paramecium
  P. aurelia, 78
  P. bursaria chlorella virus NMR, 208
  P. bursaria chlorella virus X-ray, 208
Paramutation, 268
Paraoxonase 1 gene (Pon1), 652
Parasitic infections, 683
Paraspeckle component 1 (PSPC1), 333
"Parental conflict hypothesis", 257
Parental gametes, 253
Parental nutritional impacts on livestock epigenome and production, 575–576
Parental predisposition, 641
PARK gene, 387–388
Parkinson disease (PD), 186, 234, 313, 382, 387, 651, 851
  epigenetics in, 389
    Rett Syndrome and Induced Pluripotent Stem Cells, 389
  impairment of one-carbon metabolism in, 387–388
  stem cell therapy for, 388–389
PARPs. *See* Poly ADP-ribose polymerases (PARPs)
PARs. *See* Pseudoautosomal regions (PARs)
"Party drug", 654
Passive DNA demethylation, 17–18
Paternal influence
  on offspring development, 469
  transgenerational effects of, 469–470
Paternally expressed 10 (PEG10), 342
Paternally expressed genes (PEGs), 342
Pathogen triggered immunity (PTI), 274
  and epigenetic modifications, 680–681
    bacterial infections, 681–682
    parasitic infections, 683
    viral infections, 682–683
Pattern recognition receptors (PRRs), 274, 681
Pax3 gene, 97
Pax6 gene, 385–386
PAX7 methylation, 405
PB. *See* Phenylbutyrate (PB)
PBD. *See* PCNA binding domain (PBD); Proliferating binding domain (PBD)
PBMC. *See* Peripheral blood mononuclear cells (PBMC)
PBS. *See* Primer binding site (PBS)
Pc system. *See* Polycomb system (Pc system)
PCa. *See* Prostate cancer (PCa)
PcG. *See* Polycomb group (PcG)
PcG response element/TrxG response element (PRE/TRE), 166
PCGF1. *See* Polycomb group ring finger 1 (PCGF1)
Pcgf3/5 knockout gene, 114–115
PCNA. *See* Proliferating cell nuclear antigen (PCNA)
PCNA binding domain (PBD), 12–13
PD. *See* Parkinson disease (PD)
PDAC. *See* Pancreatic ductal adenocarcinoma (PDAC)
PDB. *See* Protein Data Bank (PDB)
pDC. *See* Plasmacytoid DC (pDC)

PDMS. *See* Polydimethyl siloxane (PDMS)
PE. *See* Primitive endoderm (PE)
Pearson's correlation coefficient, 187–188, 190, 194–195
PEGs. *See* Paternally expressed genes (PEGs)
Pelizaeus–Merzbacher disease (PMD), 98
Penelope like element (PLE), 324
Peptic ulcer disease, 679
Peptidilprolyl isomerases (PPIases), 45
Peptidoglycan, 204
Peptidyl argininedeiminase 2 (*PAD2*), 392, 723–724
Peptidylarginine deiminases (PADs), 44
Peripheral blood mononuclear cells (PBMC), 334–335, 726–728, 796
Peripheral T-cell lymphoma (PTCL), 795–796, 823
Periphilin, 92
Permissive chromatin environment, 333
Peroxiredoxin1 gene promoter, 367–368
Peroxisome proliferator-activated receptors gamma 2 (PPAR-γ2), 386–387
Pesticides, 651, 653
Petunia (*Petunia hybrida*), 513
Petunia hybrida. *See* Petunia (*Petunia hybrida*)
PEV. *See* Position effect variegation (PEV)
PFC. *See* Prefrontal cortex (PFC)
PFS. *See* Progression free survival (PFS)
PGC-1α. *See* Proliferator-activated receptor gamma coactivator-1α (PGC-1α)
PGC7/Stella gene, 292–293
PGCs. *See* Primordial germ cells (PGCs)
Phalaenopsis equestris, 500–501
Pharmacodynamics, population pharmacoepigenomics in relation to, 689–690
  evidence for epigenetic basis for patient variables affecting pharmacodynamics, 690t
Pharmacoepigenetics, 687
Pharmacoepigenomics, 687
Pharmacokinetics, population pharmacoepigenomics in relation to, 689
  evidence for epigenetic basis for patient variables affecting pharmacokinetics, 689t
Phase-variable regulon, 207
Phasevarion, 207
PHD. *See* Plant homeodomain (PHD)
Phenethyl isothiocyanate, 372–373
PhenoAge, 484–485
Phenotypic aging, 484–485
Phenotypic lag, 79
Phenotypic plasticity, 531–534, 553–554
  marbled crayfish, 533f
Phenotypic variation
  epigenetic source of, 571–587
  MicroRNAs, as potential noninvasive biomarkers of fertility, 591
  cell–cell signaling, 591
  developmental abnormalities, 591
  diagnosis of cancer and fertility, 591
  embryonic development and cell communication, 591

in populations by epigenetic mechanisms, generation of, 530–534, 531f
phenotypic plasticity, 531–534
polyphenism in insects, 530–531
Phenotypically plastic, 553
Phenylbutyrate (PB), 808, 827
  clinical evaluation of, 808
Pheophytin Pheophorbide Hydrolase (PPH), 269–270
Phloem, 268–269
Pho repressive complex (PhoRC), 223–225
PhoP/PhoQ, 203–204
PhoRC. *See* Pho repressive complex (PhoRC)
Phosphate (PO_4), 39–40
Phosphorylation, 39–41, 112–114, 137, 371, 384
  phosphatase, 201–202
Phosphorylation of the Ser-10 of histone H3 (H3S10ph), 40, 298
Photo-crosslinking-assisted m6A sequencing (PA-m6A-Seq), 316
PHP1b. *See* Pseudohypoparathyroidism 1B (PHP1b)
Phthalates, 651–653
Physcomitrella patens, 505–506
Physical disability, 488
Physical environmental stressors, 699
Phytophora sojae, 554–555
PIC. *See* Pre-initiation complex (PIC)
PICKLE gene (*PKL*), 272
"Ping pong" amplification loop, 331
Pioglitazone, 615
Pipecolic acid, 274
piRISC. *See* piRNA-induced silencing complex (piRISC)
piRNA-induced silencing complex (piRISC), 330–331
piRNAs. *See* PIWI interacting RNAs (piRNAs)
Pistillata (*PI*), 503
PIWI. *See* P-element-Induced Wimpy testis (PIWI)
PIWI interacting RNAs (piRNAs), 163–164, 169–170, 219, 221–222, 330–331, 573, 749–750
  piRNA clusters, 58–59
  piRNAs and epigenetics, 66
PKC. *See* Protein kinase C (PKC)
PknB. *See* Protein kinase B (PknB)
Placental 11b-hydroxysteroid dehydrogenase type 2 (HSD11B2), 647
Plant histone lysine methyltransferases methylating H3K4, 502–508
  ATX1-like histone lysine methyltransferases in, 502–505
  ATX5-like histone lysine methyltransferases in, 505–506
  ATXR3-like histone lysine methyltransferases in, 507–508
  phylogenetic relationships between histone lysine methyltransferases methylating H3K4, 502
  SET1-like histone lysine methyltransferases in, 506–507
Plant homeodomain (PHD), 14–15, 30–34

Plant(s), 263
  abiotic stress, plant epigenetics and, 271–273
    cold stress, 272
    drought stress, 273
    heat stress, 272
    salt stress, 272–273
  based compounds, 370–371
  biotic stress, plant epigenetics and, 273–276
    chromatin modification and plant immune priming, 274–275
    DNA methylation and transgenerational-induced resistance, 275
    epigenetic memory and plant immune priming, 274
    epigenetic regulation of R-Genes, 275–276
    plant immune system, 273–274
  development, 512
  DNA methylation dynamics during plant life cycle, 264–266
  editing chromatin topology of, 278
  epigenetic mechanisms in
    evolution of plant histone lysine methyltransferases methylating H3K4, 502–508
    histone lysine methylation, 500
    histone lysine methyltransferases in plants, 500–502
  epigenetic regulation, 271–273
    biotic stress, 273–276
    in genic regions, epigenetic regulation of transposable elements, 269–270
    and interactions with genes, epigenetic regulation of transposable elements, 269–271
    located within genes, epigenetic regulation of transposable elements, 270–271
    of stress responses in, 271–276
    stress responses in plants, epigenetic regulation of, 271–276
  epigenome
    DNA methylation, 264–266
    epigenetic phenomena in plants, 268–269
    epigenome studies in plants, emerging technologies for, 276–278
    histone modifications and chromatin structure, 266–268
    mechanisms of plant epigenome regulation, 264–268
    regulation, basic mechanisms of, 264–268
  evolution, 509–510
  genomes, 264
  histone lysine methyltransferases in, 500–502
  immune priming, 274–275
  immune system, 273–274
  plant life cycle and epigenome dynamics, 264–266
  plant-specific RNA polymerase V, 264
  plant-specific short motifs, 278
  regulation of higher-order chromatin structures in, 267–268
  reproductive tissues, 265
  usable light-switch elements, 278
Plasma, 305
  plasma-proteins, 485
"Plasma DNA tissue mapping", 308
Plasmacytoid DC (pDC), 720
Plasmodesmata, 268–269
Plasmodiophora brassicae, 275
Plasticity
  epigenetics, plasticity, and evolving of inheritance, 472–473
  eukaryotic genome, 323–324
  phenotypic, 531–534
  stem cell, 383
  XCI, 434
PLE. *See* Penelope like element (PLE)
Plectosphaerella cucumerina, 275
PLPFC. *See* Pre-limbic prefrontal cortex (PLPFC)
Plug and play-ChIP-seq (PnP-ChIP-seq), 144
PMD. *See* Pelizaeus–Merzbacher disease (PMD)
PNMS. *See* Prenatal maternal stress-specific (PNMS)
Podospora anserina, 73–76
  crippled growth, self-sustained and mitotically inheritable signaling pathway in, 80–81
"Poised" genes, 38–39
Polistes dominula, 531
Pollutants, 640, 650–654
Poly ADP-ribose polymerases (PARPs), 44, 614
Polyadenylated (polyA), 164
Polycomb, 4, 223–225
Polycomb gene (Pc gene), 111–112
Polycomb group (PcG), 166, 223, 500
  PcG genes, 111–112
  PcG proteins, 111–112, 223–226
    binding, 114
    cancer cells, gene expression in, 117–118
    differentiation, 116
    Drosophila, PcG protein homologous sequences, 112
    embryonic stem cells, gene expression in, 116–117
    and functions, 113t
    histone modifying proteins, 112t
    mammalian PcG recruitment, 111
    mechanism of gene regulation by, 112–114
    polycomb repressive complexes, 112
    PRC1, 112
    PRC2, 112
    silencing and activation by PcG and TrxG protein complexes, 225f
    stem cells to maintain homeostasis, gene expression in, 114–116
Polycomb group protein Eed-Ezh2 complex, 296
Polycomb group response elements (PREs), 226
Polycomb group ring finger 1 (PCGF1), 296
Polycomb repressive complex where EZH2 subunit in proliferation (PRC2-EZH2), 409
Polycomb repressive complexes (PRCs), 112, 406, 424–425, 823–826
Polycomb repressive complexes 1(PRC1), 112, 221–225
Polycomb repressive complexes 2(PRC2), 112, 139, 167–168, 221–225, 257, 267, 296, 329, 702
Polycomb response elements (PREs), 111–112
Polycomb system (Pc system), 111–112
  PhoRC, 223–225
  PRC1, 223–225
  PRC2, 223–225
Polycomb-repressive complex 2 (PRC2), 65
Polycythemia vera (PV), 795–796
Polydimethyl siloxane (PDMS), 144
Polymorphic HERV-K (HML2), 333
Polyphenism in insects, 530–531
Polyploids, 540
  speciation, 540–542
Polytene chromosome, 220–221
Pomc. *See* Pro-opiomelanocortin (Pomc)
Poplar (*Populus trichocarpa*), 500–501
Population epigenetics, 687–688
Population pharmacoepigenomics, 687
  3D AND 4D chromatin structures in nuclei of cells in liver and body organs and, 690–691
  evidence for, 691t
  case study, 691–693
    audience, 692
    expected results and deliverables, 692
    learning and knowledge outcomes, 692
    objective, 691–692
    process, workflow and actions taken, 692
    rationale, 692
    results, 692
    safety considerations, 692
    scope, 692
    tools and materials used, 692
  epigenetic patterns, population variations of, 688
  general aspects of, 688
  human epigenome projects and, 688–689
  in relation to adverse drug reactions and drug interactions, 691
  in relation to pharmacodynamics, 689–690
  in relation to pharmacokinetics, 689
Population pharmacokinetics, 689
Populus trichocarpa, 273
Populus trichocarpa. *See* Poplar (*Populus trichocarpa*)
Position effect variegation (PEV), 90, 216, 220–223, 340
  Drosophila, epigenetic histone modifications regulate, 221–223
  PEV-mediated gene silencing in *Drosophila*, 221
Postimplantation, 257–258

Postmitotic neurons underlying cognitive function, epigenetic language in, 232–238
Postnatal influence, 642
Postnatal maternal LG, 471–472
Postnatal to adulthood, 642
Postprandial triglycerides, 490
Posttranslational histone modifications, 716
Posttranslational methylation of histones, 219
Posttranslational modifications (PTMs), 201, 217–218, 329–330, 357, 444, 499–500, 628, 738–739, 821
Posttraumatic stress disorder (PTSD), 449–451, 646
Potamopyrgus antipodarum, 534
Potassium chloride cotransporter 2 (Kcc2), 652–653
Potential DNA methylation biomarkers in cancer and diseases, 307–308
Potential histone modification biomarkers in cancer and diseases, 309–310
Potential lncRNA biomarkers in cancer and diseases, 311–313
  representative potential MiRNA biomarkers in cancer and diseases, 312t
Potential M6A methylation biomarkers in cancer and diseases, 308–309
Potential miRNA biomarkers in cancer and other diseases, 310–311
Potential non-coding RNA biomarkers in diseases, 310–313
PP1. *See* Protein phosphatase 1 (PP1)
PPAR. *See* Proliferator-activated receptor (PPAR)
PPAR-γ2. *See* Peroxisome proliferator-activated receptors gamma 2 (PPAR-γ2)
PPARGC1A, 764–767, 770–771
Ppargc1a gene, 768
PPARα gene, 467
PPH. *See* Pheophytin Pheophorbide Hydrolase (PPH)
PPIases. *See* Peptidilprolyl isomerases (PPIases)
Prader–Willi syndromes (PWS), 7, 308, 589–590, 782
PRC1. *See* Polycomb repressive complexes 1 (PRC1)
PRC2. *See* Polycomb repressive complexes 2 (PRC2); Polycomb-repressive complex 2 (PRC2)
PRCs. *See* Polycomb repressive complexes (PRCs)
Pre-initiation complex (PIC), 34–35
Pre-limbic prefrontal cortex (PLPFC), 455
pre-miRNA. *See* Precursor miRNA (pre-miRNA)
PRE/TRE. *See* PcG response element/TrxG response element (PRE/TRE)
Precision nuclear run-on sequencing (PRO-seq), 173
Precursor miRNA (pre-miRNA), 58
Prefrontal cortex (PFC), 144
Pregnancy, diabetes in, 767–768

Pregnancy-specific glycoprotein (PSG), 647
Prenatal alcohol, 655
Prenatal exposure, 641–642, 766–768
Prenatal maternal conditions, 588–589
Prenatal maternal exposures, epigenetic consequences of, 466–468
  parental targets of environmental exposure, 466f
  postnatal maternal regulation of epigenome, 468–469
Prenatal maternal stress-specific (PNMS), 647
Prenatal nutrition impact on fetal reprogramming, 615
Prenatal protein restriction, 467
Prenatal stress, 646–647
PREs. *See* Polycomb group response elements (PREs); Polycomb response elements (PREs)
PREs-TrxG-mediated gene expression, 226–227
Prescription medicine, 654
Prescriptive medicine, 640
Presenilin-1 (PS1), 742–744
Presenilin-2 (PS2), 742
Present-day advanced animal breeding, 570
Prevotella, 372–373
Primary Sjögren syndrome (pSS), 313
Primer binding site (PBS), 324–326
Primitive endoderm (PE), 289–290
Primordial germ cells (PGCs), 12, 17, 250, 332–333, 526
PRMTs. *See* Protein arginine methyltransferases (PRMTs)
Pro-inflammatory cytokine interferon (IFN-γ), 367–368
Pro-opiomelanocortin (Pomc), 340, 650
PRO-seq. *See* Precision nuclear run-on sequencing (PRO-seq)
Procainamide, 721, 826
Procambarus virginalis, 522, 526–528, 532
PROGENSA PCA3 test, 311
Prognostication, limitations to epigenetic classification and, 708
Programmed cell death, 274
Progression free survival (PFS), 796
Prokaryote, protein phosphorylation in, 201–204
Proliferating binding domain (PBD), 12–13
Proliferating cell nuclear antigen (PCNA), 12–13
Proliferative vitreoretinopathy (PVR), 313
Proliferator-activated receptor (PPAR), 614
Proliferator-activated receptor gamma coactivator-1α (PGC-1α), 614
Proline isomerization, 45
Proline-tryptophan-tryptophan-proline domain (PWWP domain), 15, 503
Promoter
  dual activity of, 139
  hypermethylation, 61
  shadow enhancer, 138–139
  super-enhancer, 138
Proof-of-concept research on locus-specific epigenome editing, 846–850
  DNA demethylation, 848

  DNA methylation, 846–848
  histone modification, 848–850
Propionate, 675
Propionylation, 145, 147
Proportion-based approach, 194
Prostate cancer (PCa), 310–311
  PCA3, 311
Prostate-specific antigen (PSA), 806
Protein, A-protein GMNase (pAG-MNase), 140
Protein acetylation in bacteria, 203–204
Protein acetyltransferases (KAT), 412
Protein arginine methyltransferases (PRMTs), 36, 207, 361, 614
  Prmt5, 361–362
Protein Data Bank (PDB), 210
Protein kinase B (PknB), 203
Protein kinase C (PKC), 370
Protein methylation, 207–209
Protein methyltransferases, 207
Protein phosphatase 1 (PP1), 449
Protein trans-splicing system (PTS system), 856–857
Proteins, 74, 112, 305–306, 455, 689
  phosphorylation in prokaryote, 201–204
  protein acetylation in bacteria, 203–204
  residues, 202t
  protein–protein interactions, 201
Protein-tyrosine phosphatase, 201–202
PRRs. *See* Pattern recognition receptors (PRRs)
PS1. *See* Presenilin-1 (PS1)
PS2. *See* Presenilin-2 (PS2)
PSA. *See* Prostate-specific antigen (PSA)
Psd95 gene, 452–454
Pseudoautosomal regions (PARs), 420–421
Pseudohypoparathyroidism 1B (PHP1b), 781–782
PSG. *See* Pregnancy-specific glycoprotein (PSG)
PSPC1. *See* Paraspeckle component 1 (PSPC1)
pSS. *See* Primary Sjögren syndrome (pSS)
Psychiatric disorders, epigenetic dysregulation in, 745–748
  DNA Methylation in depression, 747
  histone modifications and histone modifier in depression, 746–747
  non-coding RNA in depression, 747–748
PTCL. *See* Peripheral T-cell lymphoma (PTCL)
PTI. *See* Pathogen triggered immunity (PTI)
PTMs. *See* Posttranslational modifications (PTMs)
PTS system. *See* Protein trans-splicing system (PTS system)
PTSD. *See* Posttraumatic stress disorder (PTSD)
PV. *See* Polycythemia vera (PV)
PVR. *See* Proliferative vitreoretinopathy (PVR)
PWS. *See* Prader–Willi syndromes (PWS)
PWWP domain. *See* Proline-tryptophan-tryptophan-proline domain (PWWP domain)

PXD101. *See* Belinostat
Pyrimidine borane-based method, 129–130
Pyrosequencing, 315
Python script, 210

Q

Quantitative trait loci (QTL), 570
Quiescence, epigenetic control in maintenance of, 404–406
    myogenesis, 404f
Quisinostat, 794–795

R

R gene. *See* Resistance gene (R gene)
R/R AML. *See* Relapsed/refractory AML (R/R AML)
RA. *See* Rheumatoid arthritis (RA)
RAGE. *See* Receptor for advanced glycosylation end products (RAGE)
Rana temporaria, 523
Raphanus sativus L. cultivar, 269
RAR-related orphan receptor C (Rorγt), 719
RASSF1A gene, 766
Rate-equation based miRNA-ceRNA interaction model, 187
Ratio-based approach, 194
RBC. *See* Red blood cells (RBC)
RBL1. *See* Retinoblastoma-like protein 1 (RBL1)
RBL2. *See* Retinoblastoma-like protein 2 (RBL2)
RBM15. *See* RNA-binding motif protein 15 (RBM15)
RBM4. *See* RNA-binding motif protein 4 (RBM4)
RBPs. *See* RNAbinding proteins (RBPs)
RdDM. *See* RNA-directed DNA methylation (RdDM)
RdDM pathway, 272
RDRs polymerases. *See* RNA dependent RNA polymerases (RDRs polymerases)
RE. *See* Repetitive elements (RE)
Reactive nitrogen species (RNS), 361
Reactive oxygen species (ROS), 361–363
Really Interesting New Gene domain (RING domain), 14–15
Receptor for advanced glycosylation end products (RAGE), 745
Recognition of Peronospora Parasitica 7 (RPP7), 270
Recognition of Peronospora Parasitica4 (RPP4), 506–507
Red blood cells (RBC), 252
"Redox switch", 362–363
Reduced representation bisulfite sequencing (RRBS), 126, 293, 315
REF6. *See* Relative of ELF6 (REF6)
Regulatory elements, 90
Regulatory factor X1 (RFX1), 721–722
Regulatory inheritance, 79
    Candida albicans, white/opaque switch of, 81–82
    lactose operon and positive feedback loop, 79–80

Podospora anserina, crippled growth, self-sustained and mitotically inheritable signaling pathway in, 80–81
Regulatory regions, histone modifications in, 138–139
    promoter and enhancer, 138–139
    silencer, 139
Regulatory T cells (Treg), 716–717, 852
Relapsed/refractory AML (R/R AML), 827–828
Relative of ELF6 (REF6), 267
REML. *See* Residual or restricted maximum likelihood (REML)
Renal cancer, 306–307
REP. *See* Roadmap Epigenomics Project (REP)
RepA gene, 167–168
Repbase, 329
RepeatMasker, 329
Repetitive elements (RE), 94–96
Repetitive elements 1-silencing TF (REST), 35
Replication foci-targeting sequence (RFTS), 12–13
Representation bisulfite sequencing (RRBS), 124
Repressive chromatin marks, 652
Repressor element-1 silencing transcription factor/neuron-restrictive silencer factor (REST/NRSF), 386
Repressor Of Silencing 1 (ROS1), 265
Residual or restricted maximum likelihood (REML), 570
Resistance gene (R gene), 274
    epigenetic regulation of, 275–276
REST. *See* Repetitive elements 1-silencing TF (REST)
REST/NRSF. *See* Repressor element-1 silencing transcription factor/neuron-restrictive silencer factor (REST/NRSF)
Restriction landmark genomic scanning (RLGS), 124
Restriction-like endonuclease (RLE), 327
Resveratrol, 373
Retinoblastoma-associated protein (RB), 408
Retinoblastoma-like protein 1 (RBL1), 408
Retinoblastoma-like protein 2 (RBL2), 408
Retinoid acid receptor-related orphan receptor α (RORA), 391
Retrotransposon-like 1 (RLT1), 342
Retroviruses, 220–221
Rett syndrome (RS), 7, 383–384, 389, 434, 739
    epigenetics in, 390–392
        epigenetics in SCA, 391–392
        multiple sclerosis and hematopoietic stem cells, 392
        Spinocerebellar Ataxia and mesenchymal stem cells, 390–391
    iPSC Models of, 390
    non-coding RNA in, 740–742
Reverse transcriptase (RT), 324
    RT-qPCR, 170–171
Reversible protein phosphorylation, 201–202
Revolutionary approach, 315

Reward circuits, DA and, 626–628
RFT1. See Rice Flowering Time Locus T1 (RFT1)
RFTS. *See* Replication foci-targeting sequence (RFTS)
RFX1. *See* Regulatory factor X1 (RFX1)
Rheumatoid arthritis (RA), 717, 728
    adalimumab and etanercept therapy in, 728–730
        audience, 728
        challenges, 729
        expected results and deliverables, 728
        learning and knowledge outcomes, 729–730
        objective, 728
        process, workflow and actions taken, 729
        rationale, 728
        results, 729
        safety considerations, 729
        scope, 728
        tools and materials used, 729
    epigenetics of, 722–723
Ribo-zero kit, 165
Ribo-zero treatment, 165
Riboflavin, 616
Ribonucleoproteins (RNPs), 164–165, 327, 853, 861
Ribosomal RNA (rRNA), 55, 165
    rRNA depletion, 167
    rRNA-depleted RNA, 168
Ribosomal RNA from cDNA libraries, depletion of, 165–166
    cDNA libraries without rRNA, 165f
Rice (*Oryza sativa*), 269, 500–501
Rice Flowering Time Locus T1 (RFT1), 504–505
Rickettsia prowazekii, 207–208
Ricolinostat, 801
Ring chromosomes, 100–101
RING domain. *See* Really Interesting New Gene domain (RING domain)
"Ring syndrome", 100–101
Ring1b gene, 114
RIP. *See* RNA immunoprecipitation (RIP)
RISC. *See* RNA-induced silencing complex (RISC)
RLE. *See* Restriction-like endonuclease (RLE)
RLGS. *See* Restriction landmark genomic scanning (RLGS)
RLT1. *See* Retrotransposon-like 1 (RLT1)
RMTg. *See* Rostromedial tegmental nucleus (RMTg)
RNA dependent RNA polymerases (RDRs polymerases), 264
RNA epigenetics, 308–309, 457
RNA immunoprecipitation (RIP), 167–168
    RIP-seq, 167–168
RNA interference-based mechanism (RNAi-based mechanism), 264
RNA methylation, 59–60, 308–309, 569–570, 573, 577
RNA polymerase II (RNAPII), 139, 169
RNA polymerase IV, 264
RNA production and degradation, kinetic determination of, 174

RNA recognition motif (RRM), 506–507
RNA sequencing (RNA-seq), 169, 723–724
　data for ncRNAs, analysis of, 169–172
　　circular RNAs detection, algorithm for, 170–171
　　fusion RNA detection, algorithm for, 171–172
　　lncRNA detection, algorithm for, 169
　　small ncRNA detection, algorithm for, 169–170
　system, 166
　technology, 164
RNA stability, genome-wide approach to determining, 172–173
RNA tagging and recovery of associated proteins (RNA TRAP), 92
RNA TRAP. See RNA tagging and recovery of associated proteins (RNA TRAP)
RNA-binding motif protein 15 (RBM15), 425
RNA-binding motif protein 4 (RBM4), 331
RNA-directed DNA methylation (RdDM), 264
RNA-directed methylation pathway, 552–553
RNA-induced silencing complex (RISC), 631–632, 739
RNA-seq. See RNA sequencing (RNA-seq)
RNAbinding proteins (RBPs), 164–165, 304
RNAi-based mechanism. See RNA interference-based mechanism (RNAi-based mechanism)
RNAPII. See RNA polymerase II (RNAPII)
RNPs. See Ribonucleoproteins (RNPs)
RNS. See Reactive nitrogen species (RNS)
Roadmap Epigenomics Project (REP), 338
Rocilinostat, 794–795
Romidepsin, 799–800, 805, 809, 830
RORA. See Retinoid acid receptor-related orphan receptor α (RORA)
Rorγt. See RAR-related orphan receptor C (Rorγt)
ROS. See Reactive oxygen species (ROS)
ROS1. See Repressor Of Silencing 1 (ROS1)
Rostromedial tegmental nucleus (RMTg), 626–628
Rotavirus, 682
Roux-en-Y gastric bypass (RYGB), 770
roX gene, 231
RPP4. See Recognition of Peronospora Parasitica4 (RPP4)
RPP7. See Recognition of Peronospora Parasitica 7 (RPP7)
RRBS. See Reduced representation bisulfite sequencing (RRBS); Representation bisulfite sequencing (RRBS)
RRM. See RNA recognition motif (RRM)
rRNA. See Ribosomal RNA (rRNA)
rrp6Δ depltrf4 strain, 173
RS. See Rett syndrome (RS)
RSTS. See Rubinstein-Taybi syndrome (RSTS)
RT. See Reverse transcriptase (RT)
Rubinstein-Taybi syndrome (RSTS), 384, 445–446
Rybp, 116–117
RYGB. See Roux-en-Y gastric bypass (RYGB)

S

5S ribosomal RNA (rRNA), 327–328
S-5′-adnosyl-L-methionine (SAM), 207
S-adenosyl-L-methionine (AdoMet or SAM), 12, 204
S-adenosylhomocysteine (SAH), 66, 360–361, 387–388
S-adenosylhomocysteine hydrolase (SAHH), 66, 360–361
S2 *Drosophila*, 227
Saccharomyces cerevisiae, 74–76, 98–99, 173, 502
　prion and prion-like in different organisms, 75t
SAGA-complex associated factor 9 (SGF9), 34
SAHA. See Suberoylanilide hydroxamic acid (SAHA)
Salicylic acid (SA), 274
Salmonella, 203–204
Salpingoeca rosetta, 522–523
Salt Overly Sensitive pathway, 272–273
Salt stress, 272–273
SAR. See Systemic acquired resistance (SAR)
SARS-CoV-2, 432
SAT. See Subcutaneous adipose tissue (SAT)
Satellite cells (SCs), 403
　stemness, 404
"Satellite" DNA sequences, 220–221
sc-itChIP-seq methods, 143
SCA. See Spinocerebellar ataxia (SCA)
scATAC-seq. See Single-cell ATACseq (scATAC-seq); Single-cell version of ATAC-seq (scATAC-seq)
scBS-seq. See Single-cell bisulfite sequencing (scBS-seq)
scChIC-seq, 140, 143
scChIL-seq method, 140
scDNase-seq. See Single-cell DNase sequencing (scDNase-seq)
SCFAs. See Short-chain fatty acids (SCFAs)
Schistocerca gregaria, 525
Schistosoma mansoni, 555
Schistosomes, 678
Schizophrenia (SZ), 334–335, 383
Schizosaccharomyces pombe, 38
sci-ATAC-seq. See Split-pool method for combinatorial indexing (sci-ATAC-seq)
sci-MET. See Single-cell combinatorial indexing for methylation analysis (sci-MET)
SCID. See Severe immune deficient (SCID)
sciHi-C. See Single-cell combinational index Hi-C (sciHi-C)
SCMC. See Subcortical maternal membrane (SCMC)
SCNT. See Somatic cell nuclear transfer (SCNT)
Screening tests, 307
scRNA-Seq. See Single-cell RNA sequencing (scRNA-Seq)
scRRBS. See Single-cell reduced representation bisulfate sequencing (scRRBS)
SCs. See Satellite cells (SCs)
SCS. See Somatic cell score (SCS)
scTHS-seq. See Single-cell transposome hypersensitive site sequencing (scTHS-seq)
Scylla paramamosain, 525
SDG723/OsTrx1 Interaction Protein 1 (SIP 1), 504–505
SDGs. See SET-domain group (SDGs)
sDMRs. See Somatic imprinted DMRs (sDMRs)
SDPV. See Stochastic developmental phenotypic variation (SDPV)
Sea anemone (*Nematostella vectensis*), 525
Sea squirt (*Ciona intestinalis*), 525
Second generation epigenetic clocks, 484–485
　GrimAge, 485
　PhenoAge, 484–485
Secreted frizzled-related protein 1 (SFRP1), 725
Selective serotonin reuptake inhibitor (SSRI), 747
Self-tolerance, 716
Sensitivity-based correlation approach, 194–195
Sensor histidine kinases (SHK), 202
Sensory inputs, 645–646
Septin 9 (SEPT9), 307
Sequential progression models, 699
Ser-10 of histone H3 (H3S10ph), 290–291
Serine (S), 290–291, 738–739
Serotonylation, 631
Serotonylation of H3Q5 (H3Q5ser), 632
Serum lncRNA LIPCAR, 313
Sessile taxa, 536
SET and RING associated (SRA), 14–15
SET-domain group (SDGs), 267, 500–501
Severe immune deficient (SCID), 723
Sex bias and diseases, 432–434
Sex-lethal (SXL), 230
SF. See Synovial fibroblasts (SF)
SFN. See Sulforaphane (SFN)
SFRP1. See Secreted frizzled-related protein 1 (SFRP1)
SGF9. See SAGA-complex associated factor 9 (SGF9)
sgRNA. See Single chimeric gRNA (sgRNA); Single-stranded guide RNA (sgRNA)
sgRNAs, tethering molecules on, 854–855
SGZ. See Subgranular zone (SGZ)
Shadow enhancer, 138–139
Shank2 gene, 452–454
SHFM1. See Split-hand/foot malformation (SHFM1)
Shh. See Sonic hedgehog (Shh)
SHK. See Sensor histidine kinases (SHK)
Shoot apical meristem (SAM), 5–6, 264–265, 360
Short interspersed nuclear element (SINE), 269–270, 327–328, 333
Short tandem repeat (STR), 96
Short-chain fatty acids (SCFAs), 618–619, 675–676, 794–795
Short-interference RNAsm (siRNAs), 445–446

Short-term memory (STM), 443, 445–446
SHOX2. *See* Stature homeobox 2 (SHOX2)
Signal transducer and activator of transcription (STAT), 719
Signal transduction, 229
SILAC. *See* Stable isotope labeling using amino acids in cell culture (SILAC)
Silencer, 139
Silver-Russell syndrome (SRS), 7, 308, 781–782
Simulaid program, 210
Simultaneous indexing and tagmentation-based ChIP-seq (itChIP-seq), 143
Sindbis virus (SINV), 334
SINE. *See* Short interspersed nuclear element (SINE)
Sine oculis homeobox homolog 1 (SIX1), 409–410
Single cell
 epigenomic informatics, 276–278
 histone modification and transcriptome in, 144–145
  CoTECH, 144–145
  paired-tag, 145
 histone modification profiling with low, 140–144
 mapping genome structure of, 277–278
 nucleosome-occupancy profiling methods, 277
 transcriptome data, 293
 whole genome methylation analysis, 132–133, 132f
Single chimeric gRNA (sgRNA), 844–845
Single molecule real-time sequencing (SMRTseq), 131–132
Single nucleotide polymorphisms (SNP), 255, 326–328, 393, 422, 570, 709, 717
Single nucleotide variants (SNVs), 532–534, 781–782
Single-cell ATACseq (scATAC-seq), 139–140
Single-cell bisulfite sequencing (scBS-seq), 276
Single-cell combinational index Hi-C (sciHi-C), 277
Single-cell combinatorial indexing for methylation analysis (sci-MET), 276
Single-cell DNase sequencing (scDNase-seq), 277
Single-cell Hi-C (scHi-C), 277
Single-cell reduced representation bisulfate sequencing (scRRBS), 276
Single-cell RNA sequencing (scRNA-Seq), 424, 429
Single-cell transposome hypersensitive site sequencing (scTHS-seq), 277
Single-cell version of ATAC-seq (scATAC-seq), 277
Single-Molecule Real-Time (SMRT), 130
Single-nucleotide polymorphisms (SNPs), 689
Single-stranded guide RNA (sgRNA), 278
SINV. *See* Sindbis virus (SINV)
SIP 1. *See* SDG723/OsTrx1 Interaction Protein 1 (SIP 1)
siRISC. *See* SiRNA induced silencing complex (siRISC)

siRNA. *See* Small interfering RNA (siRNA)
SiRNA induced silencing complex (siRISC), 58–59, 331
siRNAs. *See* Short-interference RNAsm (siRNAs)
Sirtuin (SIRT), 614
 SIRT1, 722, 763–764
  SIRT1-activating compounds, 373
Site-specific lysine acetylation, 203
SIX1. *See* Sine oculis homeobox homolog 1 (SIX1)
Skeletal muscle regeneration (SkM regeneration), 403
 epigenetic control
  of activation and proliferation of SCS, 406–408
  in maintenance of quiescence, 404–406
  of SCS differentiation, 408–410
  small molecules as therapeutic alternative in, 410–412
 mass, 766
SkM regeneration. *See* Skeletal muscle regeneration (SkM regeneration)
SLE. *See* Systemic lupus erythematosus (SLE)
Slow-twitch fibers, 403
SMA. *See* Spinal muscular atrophy (SMA)
Small interfering RNA (siRNA), 219, 264, 330–331, 526, 573
Small molecules as therapeutic alternative in epigenetic control of regeneration, 410–412
 compiled modifiers, 411t
Small non-coding RNAs (sncRNAs), 55, 342–343
Small nuclear ribonucleoprotein polypeptide N (SNRPN), 308
Small nuclear RNAs (snRNAs), 170, 310
 detection, algorithm for, 169–170
Small nucleolar RNAs (snoRNAs), 58–59, 170, 573
Small RNAs, 272
 small RNA-based mechanisms, 332
Small-ubiquitin like modifier (SUMO), 43, 331, 382–383, 738–739
Small-vault RNAs (svRNAs), 55–57
SMART method. *See* Switching mechanism at 5′ end of RNA transcript (SMART method)
SMN1. *See* Survival motor neuron 1 (SMN1)
SMRT. *See* Single-Molecule Real-Time (SMRT)
SMRTseq. *See* Single molecule real-time sequencing (SMRTseq)
SNC1. *See* Suppressor Of NPR1–1, Constitutive1 (SNC1)
SNCA. *See* α-synuclein gene (SNCA)
sncRNAs. *See* Small non-coding RNAs (sncRNAs)
SND-275. *See* Entinostat
SNF. *See* Sucrose non-fermentable (SNF)
snoRNAs. *See* Small nucleolar RNAs (snoRNAs)
SNP. *See* Single nucleotide polymorphisms (SNP)

SNpc. *See* Substantia nigra, pars compacta (SNpc)
SNPs. *See* Single-nucleotide polymorphisms (SNPs)
snRNAs. *See* Small nuclear RNAs (snRNAs); Spliceosomal small nuclear RNAs (snRNAs)
SNRPN. *See* Small nuclear ribonucleoprotein polypeptide N (SNRPN)
SNVs. *See* Single nucleotide variants (SNVs)
SOCS3 gene. *See* Suppressor of cytokine signaling 3 gene (*SOCS3* gene)
Sodium bisulfite-based approaches, 128–129
 HTS-compatible library preparation methods, 128f
Solanum lycopersicum, 269–270
Solanum lycopersicum. *See* Tomato (*Solanum lycopersicum*)
Solitary LTRs (Solo-LTR), 326–327
Solo-LTR. *See* Solitary LTRs (Solo-LTR)
Somatic cell nuclear transfer (SCNT), 258
Somatic cell score (SCS), 588–589
 epigenetic control of activation and proliferation of SCS, 406–408
 epigenetic control of SCS differentiation, 408–410
Somatic imprinted DMRs (sDMRs), 257–258
Sonic hedgehog (Shh), 116
Sox2 gene, 116–117, 383, 385–386, 390, 850
Speciation, polyploid, 540–542
Species-barrier crossing, 77
Sperm genomes, 291
Spermatogenesis, 251
Spermatogonia, 251
Spinal muscular atrophy (SMA), 808
Spinocerebellar ataxia (SCA), 383–384, 390–391
 epigenetics in, 391–392
 SCA1, 391
 SCA7, 391
 SCA8, 391
Spliceosomal small nuclear RNAs (snRNAs), 167
Split ends homolog (SPEN), 425
Split-hand/foot malformation (SHFM1), 97
Split-pool method for combinatorial indexing (sci-ATAC-seq), 277
Sprouty 1 (SPRY1), 404–405
SPRY1. *See* Sprouty 1 (SPRY1)
SRA. *See* SET and RING associated (SRA)
SRC. *See* Steroid receptor coactivator (SRC)
SRDX motif, 278
SRS. *See* Silver-Russell syndrome (SRS)
SSc. *See* Systemic sclerosis (SSc)
SSRI. *See* Selective serotonin reuptake inhibitor (SSRI)
Stable isotope labeling using amino acids in cell culture (SILAC), 859
Stable position effect, 90
Staphylococcus aureus, 204, 577
STAT. *See* Signal transducer and activator of transcription (STAT)
Stature homeobox 2 (SHOX2), 307
Stegodyphus dumicola, 525
Steinert myotonic dystrophy (DM), 96

Stem cell therapy
  for HD, 386–387
  for PD, 388–389
  for SCA, 391
Stem cells, 6, 116, 383–384
  epigenetic mechanisms associated with neurodegenerative diseases and glioblastoma, 384t
  transplantations in glioblastoma, 396–397
    MSC transplantations for GB, 396
    TRAIL producing MSCs, 396–397
Steroid receptor coactivator (SRC), 445
  SRC1, 34
Stimuli, 766
STKs. See Hanks-type Ser/Thr kinases (STKs)
STM. See Short-term memory (STM)
Stochastic developmental phenotypic variation (SDPV), 531
Stool DNA testing, 307
STR. See Short tandem repeat (STR)
Strains, 77
Strand-specific RNA-seq, library preparation for, 166–167
  cDNA Library Construction Without rRNAs, 166t
Strawberry (*Fragaria vesca*), 500–501
Streptococcus
  S. pneumoniae, 681
  S. pyogenes, 844–845
  S. suis, 202–203
Stress, 646–649
Stress-tolerant poplar *Populus simonii*, 271–272
Structural heredity, 74
  cytotaxis of cilia and complex structures, 78
  Saccharomyces cerevisiae, *Podospora anserina* prions, 74–76
  self-driven assembly of hsp60 mitochondrial chaperonin, 77–78
Structural inheritance, 73–74
Structural variations (SVs), 90
Stylophora pistillata, 536
Su(var) genes, 221
Su(var)3–9
  enhancer of zeste, and trithorax (SET), 361
  domain, 500
  SET1-like histone lysine methyltransferases in plants, 506–507
  SET8, 38
  SETDB1, 333
Subcortical maternal membrane (SCMC), 784–785
Subcutaneous adipose tissue (SAT), 770
Suberoylanilide hydroxamic acid (SAHA), 823, 830
Suberoylanilide hydroxamic acid. See Vorinostat
Subgranular zone (SGZ), 388
Substance use disorders (SUDs), 625
  utilizing epigenetic targets for diagnosis and treatments to combat addiction and, 632–634
Substantia nigra, pars compacta (SNpc), 387

Subtelomeric regions, 100
Subventricular zone (SVZ), 388, 394
Succinylation, 145
Sucrose non-fermentable (SNF), 408
SUDs. See Substance use disorders (SUDs)
Sulfate-reducing bacteria, 677
Sulforaphane (SFN), 274–275, 372–373
Sulindac, 827
SUMO. See Small-ubiquitin like modifier (SUMO)
Sumoylation, 112–114, 382–383
Sup35p, 74–76
Super-enhancer, 138
Support vector machine (SVM), 169
Suppressor of cytokine signaling 3 gene (*SOCS3* gene), 769
Suppressor Of NPR1-1, Constitutive1 (*SNC1*), 506–507
Suppressor of variegation (Su(var)), 90
Survival motor neuron 1 (SMN1), 808
SVM. See Support vector machine (SVM)
svRNAs. See Small-vault RNAs (svRNAs)
SVs. See Structural variations (SVs)
SVZ. See Subventricular zone (SVZ)
SWI/SNF. See Switch/sucrose non-fermentable (SWI/SNF)
Switch/sucrose non-fermentable (SWI/SNF), 408
Switching mechanism at 5′ end of RNA transcript (SMART method), 167
  SMART-MSP, 315
SYISL. See SYNPO2 intron sense-overlapping lncRNA (SYISL)
Syn1 gene, 452–454
Synaptic plasticity, 217, 232–234, 447–448, 628–629
Synovial fibroblasts (SF), 717–719
Synovial tissue, 722
SYNPO2 intron sense-overlapping lncRNA (SYISL), 407
α-synuclein gene (SNCA), 387–388
α-synuclein protein, 388
Syp gene, 452–454
Systemic acquired resistance (SAR), 274
Systemic lupus erythematosus (SLE), 62, 717
  epigenetics of, 720–722
Systemic sclerosis (SSc), 717, 725–726
SZ. See Schizophrenia (SZ)

T

T cells, 719
T helper (Th), 719
T-associated maternal effect (Tme), 253–254
T-cell lymphomas (TCL), 831
  cycle, 367
T-helper cells, 392
T1D. See Type 1 diabetes mellitus (T1D)
T2D. See Type 2 diabetes mellitus (T2D)
T4 phage β-glucosyltransferase (T4-βGT), 129
T4-βGT. See T4 phage β-glucosyltransferase (T4-βGT)
TAB-seq. See Ten-eleven translocations-assisted Bisulfite Sequencing (TAB-seq)

TAD. See Topologically associated domain (TAD)
TAFs. See TATA binding protein-associated factors (TAFs)
TAL proteins. See Transcription activator-like effector (TAL proteins)
TALEN. See Transcription activator-like effector nucleases (TALEN)
Tandem tudor domain (TTD), 14–15
TAPS. See Ten-eleven translocations-assisted pyrimidine borane sequencing (TAPS)
TAR RNA-binding protein (TRBP), 739
  of 43 kDa, 742
Target site duplication (TSD), 324
Target-site primed reverse transcription (TPRT), 327
Targeted bisulfite sequencing, 125–127
Targeted therapy, 822–823
TAS. See Telomere associated sequences (TAS)
Tat-interactive protein 60 (TIP60), 218–219, 719
TATA binding protein (TBP), 34
TATA binding protein-associated factors (TAFs), 34
Tatton-Brown-Rahman syndrome, 16
TCA. See Kerb's cycle/tricarboxylic acid (TCA)
TCGA. See The Cancer Genome Atlas (TCGA)
TCL. See T-cell lymphomas (TCL)
TCS. See Two-component system (TCS)
TDG. See Thymine DNA glycosylase (TDG)
TE. See Transposable element (TE)
TE-encoded RT. See Ten-eleven-encoded reverse transcriptase (TE-encoded RT)
Technical analyses of epigenetic alterations, 316
TEI. See Transgenerational epigenetic inheritance (TEI)
Telomerase RNA template (hTER), 337
Telomere associated sequences (TAS), 99
Telomere induced foci (TIF), 337
Telomere structures, 337
Telomeric position effect (TPE), 90, 98–101
Telomeric position effect-over long distances (TPE-OLD), 100
Telomeric repeats RNA (TERRA), 99–100
Temozolomide (TMZ), 394
Template switching reaction, 167
Temple syndrome (TS14), 781–782
Ten-eleven regulatory RNA networks, regulation of gene expression by, 342–343
Ten-eleven silencing mechanisms, 269
Ten-eleven translocation (TET), 290, 329, 333, 454–455, 613–614, 645–646, 738, 822
  enzymes, 721
  proteins, 465–466
  TET proteins, 18
  TET proteins in development, 20
  TET-dependent demethylation pathways, 18–20
  TET-mediated 5mC oxidation, 20
  TET2 bind LTR retrotransposons, 333

900 INDEX

Ten-eleven translocation 2 (TET2), 129
Ten-eleven translocations inhibitor (TETi), 822
Ten-eleven translocations-assisted Bisulfite Sequencing (TAB-seq), 131
Ten-eleven translocations-assisted pyrimidine borane sequencing (TAPS), 130
Ten-eleven-associated epiallele, 269–270
Ten-eleven-associated epigenetic marks, regulation of gene expression by, 339–340
Ten-eleven-encoded reverse transcriptase (TE-encoded RT), 324
Ten-eleven-translocation enzymes (Tets), 523
Terminal inverted repeat (TIR), 328
TERRA. See Telomeric repeats RNA (TERRA)
TET. See Ten-eleven translocation (TET)
TET1 gene, 18
TET2 gene, 18
TET2. See Ten-eleven translocation 2 (TET2)
TET3 gene, 18
Tet3-deficient zygotes, 293
TETi. See Ten-eleven translocations inhibitor (TETi)
2,3,7,8-tetrachlorodibenzo-p-dioxin, 766–767
Tetrahydrofolate (THF), 360–361
Tetrahymena, 78
Tets. See Ten-eleven-translocation enzymes (Tets)
TFBSs. See Transcription factor binding sites (TFBSs)
TFIID. See Transcription factors polymerase II D (TFIID)
TFs. See Transcription factors (TFs)
The Cancer Genome Atlas (TCGA), 139, 153
Therapeutic resistance, epigenetics in intratumoral heterogeneity and, 704–705
6-thioguanosine (6sG), 174
Thiol(SH)-linked alkylation for metabolic sequencing of RNA (SLAM-seq), 174
Thiouridine-to-cytidine sequencing (TUC-seq), 174
    TUC-seq dual, 174
3D chromatin structures in nuclei of cells in liver and body organs and population pharmacoepigenomics, 690–691
Three proinflammatory cytokines (TNF), 576–577
Three-dimensional chromatin architecture (3D chromatin architecture), 335
Threonine (T), 290–291, 738–739
"Thrifty phenotype hypothesis", 359, 766
Thymidine kinase (TK), 397
Thymine DNA glycosylase (TDG), 18–20, 455, 738, 848
TIF. See Telomere induced foci (TIF)
Timing of CAB EXPRESSION 1 (TOC1), 266–267
Tip60, 35–36, 218–219, 234, 370–371
TIP60. See Tat-interactive protein 60 (TIP60)
TIR. See Terminal inverted repeat (TIR)
Tissue analysis, 397
Tissue-specific imprinting, 256

TK. See Thymidine kinase (TK)
TLR4. See Toll-like receptor 4 (TLR4)
TLRs. See Toll-like receptors (TLRs)
Tme. See T-associated maternal effect (Tme)
Tn5 transposase-protein A (pA-Tn5), 140
TNDM. See Transient neonatal diabetes mellitus (TNDM)
TNF. See Three proinflammatory cytokines (TNF); Tumor necrosis factor (TNF)
TNF-α. See Tumor necrosis factor-α (TNF-α)
TNM system. See Tumor-Node-Metastasis system (TNM system)
Toll-like receptor 4 (TLR4), 617–618
Toll-like receptors (TLRs), 681, 720
Tomato (*Solanum lycopersicum*), 500–501
Topologically associated domain (TAD), 139, 335–337, 422–423
    TAD-like structures, 267–268
Toxicodendron, 372–373
TPE. See Telomeric position effect (TPE)
TPE-OLD. See Telomeric position effect-over long distances (TPE-OLD)
TPRT. See Target-site primed reverse transcription (TPRT)
tracrRNA. See Trans-activating crRNA (tracrRNA)
Traditional RNA-seq, 164–165
Traffic-related air pollution (TRAP), 653
TRAIL. See Tumor necrosis-factor-α-related apoptosis-inducing ligand (TRAIL)
Trans-activating crRNA (tracrRNA), 844–845
Transcribed pseudogenic RNAs, 186
Transcription, 111
    of eukaryotic genes, 337–338
Transcription activator-like effector (TAL proteins), 843–844
Transcription activator-like effector nucleases (TALEN), 595
Transcription factor binding sites (TFBSs), 336–337
Transcription factors (TFs), 27–28, 138, 338
Transcription factors polymerase II D (TFIID), 34
Transcription start site (TSS), 28, 64, 154–155, 270, 327, 336–338, 456–457, 647, 709
Transcriptional activation, 500
    of active x-linked genes by dosage compensation complex, 231–232
Transcriptional dynamics, metabolic and dietary control of, 359–360
Transcriptional mechanisms, 362
Transcriptional regulators, implementation of, 278
Transcriptional repressor domain (TRD), 389, 739
Transcriptome analyses, 163–164, 386
Transfer RNAs (tRNAs), 55–57, 163–164, 170, 324–326, 615
Transformation/transcription domain associated protein (TRRAP), 34
TGF-beta pathway, 652
Transgenerational epigenetic inheritance (TEI), 227, 542
    animal breeding, 589–590

Transgenerational epigenetics
    epigenetic consequences of prenatal maternal exposures, 466–468
    epigenetics, plasticity, and evolving of inheritance, 472–473
    experience-dependent epigenetic inheritance, 471–472
    germline-mediated transgenerational inheritance, 470–471
    paternal influence on offspring development, 469
    postnatal maternal regulation of epigenome, 468–469
    transgenerational effects of parental influence, 469–470
Transgenerational inheritance of epigenetically determined phenotypes, 542–543
Transgenerational-induced resistance, 275
Transient imprinting, 258–259
Transient neonatal diabetes mellitus (TNDM), 7, 589–590, 781–782, 784
Transient transcriptome sequencing (TT-seq), 174
"Transitivity", 341
Transmissible spongiform encephalopathies (TSEs), 74
Transposable element (TE), 94–96, 220–222, 264, 323–324, 507, 540
    chromatin structure, influence on, 335–337
    classification and structure of, 324–328
        class I transposable elements, 324–328
        class II TEs, 328
    epigenetic control of, 329–334
    factors influencing transposable elements epigenetics, 334–335
        environmental stressors, 334
        viral infections, 334–335
    in genic regions, epigenetics regulation of, 269–270
    genomic annotation, 328–329
    impact on genetics and epigenetics of gene regulatory regions, 337–340
    influence on host epigenetic, 335–343
    and interactions with genes, epigenetic regulation of, 269–271
    located within genes, epigenetics regulation of, 270–271
    TE-derived cis-elements, regulation of gene expression by, 337–340
Transposase-accessible chromatin using sequencing (ATACseq), 717–719
Transposons, 525
TRAP. See Traffic-related air pollution (TRAP)
TRBP. See TAR RNA-binding protein (TRBP)
TRD. See Transcriptional repressor domain (TRD)
Treg specific demethylated region (TSDR), 682–683
Tregs, 682–683
TREs. See Trithorax group response elements (TREs)
tRFs. See tRNA-derived fragments (tRFs)
Tricarboxylic acid cycle (TCA cycle), 359, 613–614

Trichinella
  T. pseudospiralis, 538
  T. spiralis, 538
Trichostatin A (TSA), 794–795, 801, 827
Trim28. *See* Tripartite motif containing 28 (Trim28)
Trimethylamine N-oxide (TMAO), 618–619
Trimethylation
  of histone H3 lysine 4, 500
  lysine, 382
Tripartite motif containing 28 (Trim28), 90–91, 331
Triploid marbled crayfish, 542
Trithorax (TrxG), 226
  TrxG gene *brm* (*brahma*), 226
  TrxG proteins, 226
    complexes, 223
Trithorax gene (*Trx* gene), 167–168, 424, 500
  proteins, 221
Trithorax group (TrxG), 166, 223, 500
Trithorax group response elements (TREs), 226
Trithorax-related (Trr), 502
Triticum aestivum. *See* Wheat (*Triticum aestivum*)
tRNA-derived fragments (tRFs), 55–57
tRNAs. *See* Transfer RNAs (tRNAs)
TRRAP. *See* Transformation/transcription domain associated protein (TRRAP)
Trx gene. *See* *Trithorax* gene (*Trx* gene)
TrxG. *See* Trithorax group (TrxG)
Trypanosoma, 78
TS14. *See* Temple syndrome (TS14)
TSA. *See* Trichostatin A (TSA)
TSD. *See* Target site duplication (TSD)
TSDR. *See* Treg specific demethylated region (TSDR)
TSEs. *See* Transmissible spongiform encephalopathies (TSEs)
TSG. *See* Tumor suppressor gene (TSG)
TSS. *See* Transcription start site (TSS)
TT-seq. *See* Transient transcriptome sequencing (TT-seq)
TTD. *See* N-terminal triple Tudor domain (TTD); Tandem tudor domain (TTD)
TUC-seq. *See* Thiouridine-to-cytidine sequencing (TUC-seq)
Tumor microenvironment, 307
Tumor necrosis factor (TNF), 722
Tumor necrosis factor-α (TNF-α), 392, 490, 677
Tumor necrosis-factor-α-related apoptosis-inducing ligand (TRAIL), 396–397
  producing MSCs, 396–397
Tumor necrosis-factor-α-related apoptosis-inducing ligand (TRAIL), 396–397
Tumor suppressor gene (TSG), 307
Tumor tissue biopsies, 305
Tumor-Node-Metastasis system (TNM system), 705, 708
Tumors, 705
Turmeric (*Curcuma longa*), 412
Turner syndromes, 423–424, 433
Two-component system (TCS), 202

Type 1 diabetes mellitus (T1D), 592, 717, 724–725, 767–768
Type 2 diabetes mellitus (T2D), 359, 363, 479–480, 488, 761, 770–771
Tyr kinases, 202
Tyrosine (Y), 738–739
Tyrosine hydroxylase (TH), 388–389

U
4s U-seq, 174
uaRNAs. *See* Upstream antisense RNAs (uaRNAs)
Ubiquitin conjugating enzyme 9 (UBC9), 43
Ubiquitin-interacting motif (UIM), 12–13
Ubiquitin-like domain (UBL domain), 14–15
Ubiquitin-like with PHD and RING finger domains 1 (UHRF1), 12
Ubiquitin-specific-processing protease 8 (UBP8), 34
Ubiquitination, 137
Ubiquitinylation, 112–114
UBL domain. *See* Ubiquitin-like domain (UBL domain)
UBP8. *See* Ubiquitin-specific-processing protease 8 (UBP8)
UC. *See* Umbilical cord (UC)
UC-MSCs. *See* Umbilical cord-derived mesenchymal stem cells (UC-MSCs)
UCA1. *See* Urothelial carcinoma-associated 1 (UCA1)
ucp. *See* Uncoupling protein (ucp)
UCSC. *See* University of California Santa Cruz (UCSC)
UDP-N-acetyl-glucosamine. *See* Uridine diphosphate-N-acetyl-glucosamine (UDP-N-acetyl-glucosamine)
UHRF1 gene, 13–15
UHRF1. *See* Ubiquitin-like with PHD and RING finger domains 1 (UHRF1)
UIM. *See* Ubiquitin-interacting motif (UIM)
Ulcerative colitis, 306–307
Ultra-low input-CUT&RUN (uli-CUT&RUN), 277
Ultraviolet light (UV light), 717
Umbilical cord (UC), 386
Umbilical cord-derived mesenchymal stem cells (UC-MSCs), 383–384
Uncoupling protein (ucp), 412
UNG. *See* Uracil-N-glycosylase (UNG)
Uniparental disomies (UPDs), 253–254, 781–782
University of California Santa Cruz (UCSC), 689
UNR. *See* Upstream N-Ras (UNR)
Unstable ncRNAs by inhibiting decay, 173
5′ untranslated region (5′UTR), 327
UPDs. *See* Uniparental disomies (UPDs)
Upstream antisense RNAs (uaRNAs), 139
Upstream N-Ras (UNR), 230
Upstream stimulatory factor (USF), 339–340
URA3 gene, 98–99
Uracil, 129
Uracil-N-glycosylase (UNG), 167
Uridine diphosphate-N-acetyl-glucosamine (UDP-N-acetyl-glucosamine), 203

Urinary tract diseases, 311
Urolithins, 677
Urothelial carcinoma-associated 1 (UCA1), 311
US Food and Drug Administration (FDA), 311, 709, 795–796, 823
USF. *See* Upstream stimulatory factor (USF)
USP9X gene, 432
Utx gene, 426
UV light. *See* Ultraviolet light (UV light)

V
VAMP-associated protein, 235
Valproic acid (VPA), 654, 800, 806–808
  clinical evaluation of, 808
Variants of uncertain significance (VUS), 785–786
Vascular disease (VD), 651
Vascular endothelial growth factor (VEGF), 391
Vegetative incompatibility, 77
VEGF-A gene, 848–850
Ventral tegmental area (VTA), 626–628
Vernalization, 268
Vernicifluum, 372–373
Vicia faba, 500
Vinclozolin, 653
  exposure, 653
Viral infections, 334–335, 682–683
Virus-targeted mucosa, 682
Viruses, 669–673
Vitamin folate pathway, 616
Vitamins, 360, 699
  vitamin B2, 616
  vitamin B9, 616
  vitamin B12, 616
  vitamin C, 362–363
  vitamin D, 617
Vitis vinifera, 500–501
Volasertib, 826–827
Vorinostat, 794–798, 802–804, 830
  hydroxymate HDAC inhibitormm, 796
VP16 transactivation domain, 270–271
VPS39, 771

W
Warburg effect, 677
WD40 repeat-containing protein, 272
Wheat (*Triticum aestivum*), 269, 500–501
White gene (*wt* gene), 90, 221
Whole genome bisulfite sequencing (WGBS), 126–128, 293
Whole genome duplications (WGD), 501–502
Whole genome HTS, 123
Whole genome-wide DNA methylation, 587
Whole-genome methylation analysis, 676
Whole-genome shotgun bisulfite sequencing (WGBS-seq), 315
Wilms tumor 1 associated protein (WTAP), 425
Wnt pathway, 116
Wnt6 gene, 97
WOR1 transcription factor, 81
World Health Organization (WHO), 611

X

X active specific transcript (XACT), 424
X-autosome dosage compensation, 427–428
  global XCI regulation steps, 428f
  XCI differences between mice and humans, 428t
X-chromosome, 3–4, 6, 11, 65, 114–115, 123, 419–420, 526
  evolution and incomplete nature of XCI, 420–422
  regulation of dosage compensation complex targeting to, 230–231
    model of dosage compensation, 230f
  sex bias and diseases, 432–434
  Xa, 419–420
  Xi, 419–420
  Xi^P, 422
X-chromosome inactivation (XCI), 289–292, 297, 340–342, 419, 422, 571, 722
  differences between mice and humans, 427
  H3K27me3 regulates, 296
  historical perspective of, 420
  imprinted and random XCI, 422–423
  mechanisms for, 65f
  methods for exploring XCI status, 429
  physiological and pathogenic XCI skewing, 429–432
    human X-linked diseases, 431t
  plasticity, 434
  regulation and main epigenetic steps, 423–427
  skewing, 429
  X-autosome dosage compensation, 427–428
  X-chromosome, sex bias and diseases, 432–434
  X-chromosome evolution and incomplete nature of, 420–422
X-dampening, 424
X-encoded *roX* genes, 230
X-inactivation center (Xic), 422–423
X-inactive specific transcript, 422–423
X-linked diseases, 422
XCI. *See* X-chromosome inactivation (XCI)
Xenopus eggs, 80–81
Xic. *See* X-inactivation center (Xic)
XIST gene, 167–168, 291–292, 424
  interactome, 425
  RNA accumulation, 426
Xrn1-sensitive unstable transcripts (XUTs), 173
XUTs. *See* Xrn1-sensitive unstable transcripts (XUTs)

Y

Yeast metabolic cycle (YMC), 147
Yeast prion, 74, 77
Ying Yang protein (YYI protein), 114, 408
Yolk sac (YC), 117
Yoruban, 688
Yy1 gene, 114

Z

ZBT38, 408
Zdbf2/Gpr1 locus, 259
Zea mays. *See* Maize (*Zea mays*)
Zebrafish (*Danio rerio*), 542
Zebularin, 410–412
Zebularine, 826–827
ZFP57 gene, 17
Zif268 protein, 846
Zinc finger nucleases (ZFN), 595
Zinc finger protein ZFP57, 784
Zinc finger proteins (ZF proteins), 633, 843–844
Zip53 protein, 846
ZNF445 gene, 17
Zolinza, 830
Zygotic genome activation (ZGA), 289–290

Printed in the United States
by Baker & Taylor Publisher Services